Recommended Dietary Allowances (RDA) and Adequate Intakes (AI) for Vitamins

Age (yr)	Thiamin (mg/day) RDA	Riboflavin (mg/day) RDA	Niacin (mg/day) RDA[a]	Biotin (µg/day) AI	Pantothenic acid (mg/day) AI	Vitamin B6 (mg/day) RDA	Folate (µg/day) RDA[b]	Vitamin B12 (µg/day) RDA	Choline (mg/day) AI	Vitamin C (mg/day) RDA	Vitamin A (µg/day) RDA[c]	Vitamin D (µg/day)[d]	Vitamin E (mg/day)[e]	Vitamin K (µg/day)
Infants														
0–0.5	0.2	0.3	2	5	1.7	0.1	65	0.4	125	40	400	10	4	2.0
0.5–1	0.3	0.4	4	6	1.8	0.3	80	0.5	150	50	500	10	5	2.5
Children														
1–3	0.5	0.5	6	8	2	0.5	150	0.9	200	15	300	15	6	30
4–8	0.6	0.6	8	12	3	0.6	200	1.2	250	25	400	15	7	55
Males														
9–13	0.9	0.9	12	20	4	1.0	300	1.8	375	45	600	15	11	60
14–18	1.2	1.3	16	25	5	1.3	400	2.4	550	75	900	15	15	75
19–30	1.2	1.3	16	30	5	1.3	400	2.4	550	90	900	15	15	120
31–50	1.2	1.3	16	30	5	1.3	400	2.4	550	90	900	15	15	120
51–70	1.2	1.3	16	30	5	1.7	400	2.4	550	90	900	15	15	120
>70	1.2	1.3	16	30	5	1.7	400	2.4	550	90	900	20	15	120
Females														
9–13	0.9	0.9	12	20	4	1.0	300	1.8	375	45	600	15	11	60
14–18	1.0	1.0	14	25	5	1.2	400	2.4	400	65	700	15	15	75
19–30	1.1	1.1	14	30	5	1.3	400	2.4	425	75	700	15	15	90
31–50	1.1	1.1	14	30	5	1.3	400	2.4	425	75	700	15	15	90
51–70	1.1	1.1	14	30	5	1.5	400	2.4	425	75	700	15	15	90
>70	1.1	1.1	14	30	5	1.5	400	2.4	425	75	700	20	15	90
Pregnancy														
≤18	1.4	1.4	18	30	6	1.9	600	2.6	450	80	750	15	15	75
19–30	1.4	1.4	18	30	6	1.9	600	2.6	450	85	770	15	15	90
31–50	1.4	1.4	18	30	6	1.9	600	2.6	450	85	770	15	15	90
Lactation														
≤18	1.4	1.6	17	35	7	2.0	500	2.8	550	115	1200	15	19	75
19–30	1.4	1.6	17	35	7	2.0	500	2.8	550	120	1300	15	19	90
31–50	1.4	1.6	17	35	7	2.0	500	2.8	550	120	1300	15	19	90

NOTE: For all nutrients, values for infants are AI. The glossary on the inside back cover defines units of nutrient measure.

[a] Niacin recommendations are expressed as niacin equivalents (NE), except for recommendations for infants younger than 6 months, which are expressed as preformed niacin.

[b] Folate recommendations are expressed as dietary folate equivalents (DFE).

[c] Vitamin A recommendations are expressed as retinol activity equivalents (RAE).

[d] Vitamin D recommendations are expressed as cholecalciferol and assume an absence of adequate exposure to sunlight.

[e] Vitamin E recommendations are expressed as α-tocopherol.

Recommended Dietary Allowances (RDA) and Adequate Intakes (AI) for Minerals

Age (yr)	Sodium (mg/day) AI	Chloride (mg/day) AI	Potassium (mg/day) AI	Calcium (mg/day) RDA	Phosphorus (mg/day) RDA	Magnesium (mg/day) RDA	Iron (mg/day) RDA	Zinc (mg/day) RDA	Iodine (µg/day) RDA	Selenium (µg/day) RDA	Copper (µg/day) RDA	Manganese (mg/day) AI	Fluoride (mg/day) AI	Chromium (µg/day) AI	Molybdenum (µg/day) RDA
Infants															
0–0.5	120	180	400	200	100	30	0.27	2	110	15	200	0.003	0.01	0.2	2
0.5–1	370	570	700	260	275	75	11	3	130	20	220	0.6	0.5	5.5	3
Children															
1–3	1000	1500	3000	700	460	80	7	3	90	20	340	1.2	0.7	11	17
4–8	1200	1900	3800	1000	500	130	10	5	90	30	440	1.5	1.0	15	22
Males															
9–13	1500	2300	4500	1300	1250	240	8	8	120	40	700	1.9	2	25	34
14–18	1500	2300	4700	1300	1250	410	11	11	150	55	890	2.2	3	35	43
19–30	1500	2300	4700	1000	700	400	8	11	150	55	900	2.3	4	35	45
31–50	1500	2300	4700	1000	700	420	8	11	150	55	900	2.3	4	35	45
51–70	1300	2000	4700	1000	700	420	8	11	150	55	900	2.3	4	30	45
>70	1200	1800	4700	1200	700	420	8	11	150	55	900	2.3	4	30	45
Females															
9–13	1500	2300	4500	1300	1250	240	8	8	120	40	700	1.6	2	21	34
14–18	1500	2300	4700	1300	1250	360	15	9	150	55	890	1.6	3	24	43
19–30	1500	2300	4700	1000	700	310	18	8	150	55	900	1.8	3	25	45
31–50	1500	2300	4700	1000	700	320	18	8	150	55	900	1.8	3	25	45
51–70	1300	2000	4700	1200	700	320	8	8	150	55	900	1.8	3	20	45
>70	1200	1800	4700	1200	700	320	8	8	150	55	900	1.8	3	20	45
Pregnancy															
≤18	1500	2300	4700	1300	1250	400	27	12	220	60	1000	2.0	3	29	50
19–30	1500	2300	4700	1000	700	350	27	11	220	60	1000	2.0	3	30	50
31–50	1500	2300	4700	1000	700	360	27	11	220	60	1000	2.0	3	30	50
Lactation															
≤18	1500	2300	5100	1300	1250	360	10	13	290	70	1300	2.6	3	44	50
19–30	1500	2300	5100	1000	700	310	9	12	290	70	1300	2.6	3	45	50
31–50	1500	2300	5100	1000	700	320	9	12	290	70	1300	2.6	3	45	50

NOTE: For all nutrients, values for infants are AI. The glossary on the inside back cover defines units of nutrient measure.

B

Tolerable Upper Intake Levels (UL) for Vitamins

Age (yr)	Niacin (mg/day)[a]	Vitamin B₆ (mg/day)	Folate (µg/day)[a]	Choline (mg/day)	Vitamin C (mg/day)	Vitamin A (IU/day)[b]	Vitamin D (µg/day)	Vitamin E (mg/day)[c]
Infants								
0–0.5	—	—	—	—	—	600	25	—
0.5–1	—	—	—	—	—	600	38	—
Children								
1–3	10	30	300	1000	400	600	63	200
4–8	15	40	400	1000	650	900	75	300
9–13	20	60	600	2000	1200	1700	100	600
Adolescents								
14–18	30	80	800	3000	1800	2800	100	800
Adults								
19–70	35	100	1000	3500	2000	3000	100	1000
>70	35	100	1000	3500	2000	3000	100	1000
Pregnancy								
≤18	30	80	800	3000	1800	2800	100	800
19–50	35	100	1000	3500	2000	3000	100	1000
Lactation								
≤18	30	80	800	3000	1800	2800	100	800
19–50	35	100	1000	3500	2000	3000	100	1000

[a]The UL for niacin and folate apply to synthetic forms obtained from supplements, fortified foods, or a combination of the two.

[b]The UL for vitamin A applies to the preformed vitamin only.

[c]The UL for vitamin E applies to any form of supplemental α-tocopherol, fortified foods, or a combination of the two.

Tolerable Upper Intake Levels (UL) for Minerals

Age (yr)	Sodium (mg/day)	Chloride (mg/day)	Calcium (mg/day)	Phosphorus (mg/day)	Magnesium (mg/day)[d]	Iron (mg/day)	Zinc (mg/day)	Iodine (µg/day)	Selenium (µg/day)	Copper (µg/day)	Manganese (mg/day)	Fluoride (mg/day)	Molybdenum (µg/day)	Boron (mg/day)	Nickel (mg/day)	Vanadium (mg/day)
Infants																
0–0.5	—	—	1000	—	—	40	4	—	45	—	—	0.7	—	—	—	—
0.5–1	—	—	1500	—	—	40	5	—	60	—	—	0.9	—	—	—	—
Children																
1–3	1500	2300	2500	3000	65	40	7	200	90	1000	2	1.3	300	3	0.2	—
4–8	1900	2900	2500	3000	110	40	12	300	150	3000	3	2.2	600	6	0.3	—
9–13	2200	3400	3000	4000	350	40	23	600	280	5000	6	10	1100	11	0.6	—
Adolescents																
14–18	2300	3600	3000	4000	350	45	34	900	400	8000	9	10	1700	17	1.0	—
Adults																
19–50	2300	3600	2500	4000	350	45	40	1100	400	10,000	11	10	2000	20	1.0	1.8
51–70	2300	3600	2000	4000	350	45	40	1100	400	10,000	11	10	2000	20	1.0	1.8
>70	2300	3600	2000	3000	350	45	40	1100	400	10,000	11	10	2000	20	1.0	1.8
Pregnancy																
≤18	2300	3600	3000	3500	350	45	34	900	400	8000	9	10	1700	17	1.0	—
19–50	2300	3600	2500	3500	350	45	40	1100	400	10,000	11	10	2000	20	1.0	—
Lactation																
≤18	2300	3600	3000	4000	350	45	34	900	400	8000	9	10	1700	17	1.0	—
19–50	2300	3600	2500	4000	350	45	40	1100	400	10,000	11	10	2000	20	1.0	—

[d]The UL for magnesium applies to synthetic forms obtained from supplements or drugs only.

NOTE: An Upper Limit was not established for vitamins and minerals not listed and for those age groups listed with a dash (—) because of a lack of data, not because these nutrients are safe to consume at any level of intake. All nutrients can have adverse effects when intakes are excessive.

SOURCE: Adapted from the *Dietary Reference Intakes series,* National Academies Press. National Academies of Sciences.

Understanding Nutrition

Whitney | Rolfes

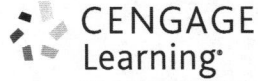
CENGAGE
Learning·

Australia • Brazil • Japan • Korea • Mexico • Singapore • Spain • United Kingdom • United States

Understanding Nutrition: 14th Edition

Understanding Nutrition, 14th Edition
Elli Whitney | Sharon Rady Rolfes

© 2016, 2013 Cengage Learning. All rights reserved.

For product information and technology assistance, contact us at
Cengage Learning Customer & Sales Support, 1-800-354-9706

For permission to use material from this text or product,
submit all requests online at **cengage.com/permissions**
Further permissions questions can be emailed to
permissionrequest@cengage.com

This book contains select works from existing Cengage Learning resources and was produced by Cengage Learning Custom Solutions for collegiate use. As such, those adopting and/or contributing to this work are responsible for editorial content accuracy, continuity and completeness.

Compilation © 2015 Cengage Learning

ISBN: 978-1-305-77136-9

WCN: 01-100-101

Cengage Learning
20 Channel Center Street
Boston, MA 02210
USA

Cengage Learning is a leading provider of customized learning solutions with office locations around the globe, including Singapore, the United Kingdom, Australia, Mexico, Brazil, and Japan. Locate your local office at:
www.international.cengage.com/region.

Cengage Learning products are represented in Canada by Nelson Education, Ltd.

For your lifelong learning solutions, visit **www.cengage.com/custom.**

Visit our corporate website at **www.cengage.com.**

Brief Contents

Preface

Nutrition is a science. The details of a nutrient's chemistry or a cell's biology can be overwhelming and confusing to some, but it needn't be. When the science is explained step by step and the facts are connected one by one, the details become clear and understandable. By telling stories about fat mice, using analogies of lamps, and applying guidelines to groceries, we make the science of nutrition meaningful and memorable. That has been our mission since the first edition—to reveal the fascination of science and share the excitement of nutrition with readers. We have learned from the thousands of professors and more than a million students who have used this book through the years that readers want an *understanding* of nutrition so they can make healthy choices in their daily lives. We hope that this book serves you well.

A Book Tour of This Edition

Understanding Nutrition presents the core information of an introductory nutrition course. The early chapters introduce the nutrients and their work in the body, and the later chapters apply that information to people's lives—describing the role of foods and nutrients in energy balance and weight control, in physical activity, in the life cycle, in disease prevention, in food safety, and in hunger.

The Chapters Chapter 1 begins by exploring why we eat the foods we do and continues with a brief overview of the nutrients, the science of nutrition, recommended nutrient intakes, assessment, and important relationships between diet and health. Chapter 2 describes the diet-planning principles and food guides used to create diets that support good health and includes instructions on how to read a food label. In Chapter 3 readers follow the journey of digestion and absorption as the body breaks down foods into nutrients. Chapters 4, 5, and 6 describe carbohydrates, fats, and proteins—their chemistry, roles in the body, and places in the diet. Then Chapter 7 shows how the body derives energy from these three nutrients. Chapters 8 and 9 continue the story with a look at energy balance, the factors associated with overweight and underweight, and the benefits and dangers of weight loss and weight gain. Chapters 10, 11, 12, and 13 complete the introductory lessons by describing the vitamins, the minerals, and water—their roles in the body, deficiency and toxicity symptoms, and sources.

The next seven chapters weave that basic information into practical applications, showing how nutrition influences people's lives. Chapter 14 describes how physical activity and nutrition work together to support fitness. Chapters 15, 16, and 17 present the special nutrient needs of people through the life cycle—pregnancy and lactation; infancy, childhood, and adolescence; and adulthood and the later years. Chapter 18 focuses on the dietary risk factors and recommendations associated with chronic diseases, and Chapter 19 addresses consumer concerns about the safety of the food and water supply. Chapter 20 closes the book by examining hunger and the global environment.

The Highlights Every chapter is followed by a highlight that provides readers with an in-depth look at a current, and often controversial, topic that relates to its companion chapter. For example, Highlight 4 examines the scientific evidence behind some of the current controversies surrounding carbohydrates and their role in weight gain and weight loss. New to this edition are Critical Thinking Questions designed to encourage readers to develop clear, rational, open-minded, and informed thoughts based on the evidence presented in the highlight.

Special Features The art and layout in this edition have been carefully designed to be inviting while enhancing student learning. For example, numbered steps have been added to several figures to clarify sequences and processes. In addition, special features help readers identify key concepts and apply nutrition knowledge. For example, when a new term is introduced, it is printed in bold type, and a **definition** is provided. These definitions often include pronunciations and derivations to facilitate understanding. The glossary at the end of the book includes all defined terms.

definition (DEF-eh-NISH-en): the meaning of a word.

- **de** = from
- **finis** = boundary

LEARNING GPS

The opening page of each chapter provides a Learning GPS that serves as an outline and directs readers to the main heads (and subheads) within the chapter. Each main head is followed by a Learn It—a learning objective for the content covered in that section. The Learn It also appears within the text at the start of each main section as well as at the start of each Review It. After reading the chapter, students will be able to demonstrate competency in the Learn It objectives.

Nutrition in Your Life

The opening paragraph of each chapter—called Nutrition in Your Life—introduces the chapter's content in a friendly and familiar way. This short paragraph closes with a preview of how readers might apply that content to their daily lives by inviting them to use the Nutrition Portfolio section at the end of the chapter.

Nutrition Portfolio

The Nutrition Portfolio section at the end of each chapter prompts readers to consider whether their personal choices are meeting the goals presented in the chapter. Most of these assignments include instructions that use the Diet & Wellness Plus program. Such tools help students assess their current choices and make informed decisions about healthy options.

> REVIEW IT Each major section within a chapter concludes with a Review It paragraph that summarizes key concepts. Similarly, Review It tables cue readers to important summaries.

Also featured in this edition are the 2010 *Dietary Guidelines for Americans*, which are introduced in Chapter 2 and presented throughout the text whenever their subjects are discussed. Look for the following design.

> DIETARY GUIDELINES FOR AMERICANS
These guidelines provide science-based advice to promote health and to reduce the risk of chronic disease through diet and physical activity.

>How To

Many of the chapters include "How To" features that guide readers through problem-solving tasks. For example, a "How To" in Chapter 1 presents the steps in calculating energy intake from the grams of carbohydrate, fat, and protein in a food.

> TRY IT Each "How To" feature ends with a "Try It" activity that gives readers an opportunity to practice these new lessons.

STUDY CARD > 1

> STUDY IT To review the key points of this chapter and take a practice quiz, go to the Study Cards at the end of the book.

Study cards appear at the back of the text—one for each chapter. The Study It side of each card presents a review of the chapter's core concepts, and perhaps a table or figure to remind readers of key points. The Test It side of the study card provides essay and multiple-choice questions to help prepare students for exams.

The Appendixes

The appendixes are valuable references for a number of purposes. Appendix A summarizes background information on the hormonal and nervous systems, complementing Appendixes B and C on basic chemistry, the chemical structures of nutrients, and major metabolic pathways. Appendix D describes measures of protein quality. Appendix E provides detailed coverage of nutrition assessment, and Appendix F presents the estimated energy requirements for men and women at various levels of physical activity. Appendix G presents the 2008 *Choose Your Foods: Exchange List for Diabetes.* Appendix H is a 4000-item food composition table. Appendix I lists nutrition recommendations from the World Health Organization (WHO), and Appendix J presents the 2020 Healthy People nutrition-related objectives. Appendix K features aids to calculations, a short tutorial on converting metric measures and handling basic math problems commonly found in the world of nutrition.

The Inside Covers

The inside covers put commonly used information at your fingertips. The inside front covers (pp. A–C) present the current nutrient recommendations; the inside back covers feature the Daily Values used on food labels and a glossary of nutrient measures (p. Y on the left) as well as suggested weight ranges for various heights (p. Z on the right).

Notable Changes in This Edition

Because nutrition is an active science, staying current is paramount. Just as nutrition research continuously adds to and revises the accepted body of knowledge, this edition builds on the science of previous editions with the latest in nutrition research. Much has changed in the world of nutrition and in our daily lives since the first edition. The number of foods has increased dramatically—even as we spend less time than ever in the kitchen preparing meals. The connections between diet and disease have become more apparent—and consumer interest in making smart health choices has followed. More people are living longer and healthier lives. The science of nutrition has grown rapidly, with new "facts" emerging daily. In this edition, as with all previous editions, every chapter has been revised to enhance learning by presenting current information accurately and attractively. For all chapters and highlights we have:

- Reviewed and updated content
- Created several new figures and tables and revised others to enhance learning
- Added Critical Thinking Questions to the highlights
- Revised figures to reflect new food and supplement labels and tables to include nutrients of concern (vitamin D, calcium, potassium, and iron)

Chapter 1

- Created table to summarize ways to describe six classes of nutrients
- Introduced *registered dietitian nutritionist (RDN)*, another term to describe an RD

Chapter 2

- Revised figure comparing nutrient density of two breakfasts to include potassium and vitamin D
- Introduced new food labels and revised figure to illustrate differences
- Introduced front-of-package labeling and added figure to illustrate

Chapter 3

- Introduced *microbiome* and revised section on gastrointestinal bacteria

Chapter 4

- Revised table showing nutrients in sugars and other foods to include potassium and vitamin D
- Created tables to define glucose for normal and diabetes; to show the glycemic index of a few common foods; to list the functions of sugars in foods; to present ways to prevent dental caries
- Included fructose metabolism in the highlight

Chapter 5

- Created tables to define blood lipids for heart health; to list fat choices among protein foods and among milk products; to show omega-3 fatty acid quantities in a variety of fish and seafood
- Created new figure on how to read fish oil supplement labels
- Added definitions for resistin and adiponectin

Chapter 6

- Expanded discussion on the association between dietary protein and body weight

Chapter 7

- Created new figure illustrating labels on beer, wine, and liquor

Chapter 8

- Discussed "3500 kcalorie rule" and its limitations
- Created new tables for estimating energy expended on basal metabolism and on thermic effect of foods and for percent body fat at various BMI
- Revised section on female athlete triad to include new expanded term—Relative Energy Deficiency in Sports (RED-S)—and created new table of its adverse consequences
- Added discussion of food addiction to section on binge eating disorder

Chapter 9

- Added discussion of brite adipocytes to section on brown adipocyte tissue and uncoupling proteins
- Updated table on FDA-approved weight loss drugs
- Revised figure on gastric surgery used to treat obesity
- Deleted discussion and figure on unrealistic expectations
- Created new table of national strategies to prevent obesity
- Updated table on popular weight loss diets

Chapter 11

- Added a paragraph on "golden rice," a genetically modified rice used in the worldwide fight against vitamin A deficiency
- Added details on vitamin D's non-bone-related roles
- Rewrote the introduction to vitamin E
- Rewrote the food sources of vitamin K paragraph to include the terms phylloquinone (vitamin K_1) and menaquinone (vitamin K_2)

Chapter 12

- Revised calcium balance figure

Chapter 13

- Created table of factors influencing iron absorption

Chapter 14

- Reorganized, shortened, and rewrote section on Developing Fitness

- Reorganized, clarified, and rewrote section entitled Energy Systems and Fuels to Support Activity
- Enhanced and reorganized table of primary fuels used for activities of different intensities and durations
- Created new table comparing symptoms of heat stroke and hypothermia
- Added discussion and definition of beta-alanine to highlight on ergogenic aids

Chapter 15

- Created several new tables: benefits of WIC, risk factors for gestational diabetes, signs and symptoms of pre-eclampsia, complications from smoking during pregnancy, tips to prevent listeriosis
- Reorganized sections on fetal programming and fetal development of chronic diseases

Chapter 16

- Created several new tables: protective factors in breast milk, tips for picky eaters, examples of foods and non-food items children can choke on, iron recommendations for adolescents
- Added information about fluoride and formula preparation
- Added brief discussion about new AAP guidelines for reduced, low-fat, and fat-free milk for toddlers
- Added discussion of new school meal initiatives

Chapter 17

- Created new figure comparing healthy lens with cataract lens

Chapter 18

- Created new tables for selected nutrient roles in immune function and criteria for diagnosis of diabetes
- Revised, reworked, and updated tables for recommendations and strategies to reduce the risk of CVD and recommendations and strategies to reduce the risk of cancer
- Emphasized role of obesity throughout the chapter as major risk for other chronic diseases
- Included information related to 2013 American College of Cardiology/American Heart Association guidelines for assessment of CVD risk and lifestyle modifications for reducing risk of heart disease
- Added discussion and definitions of fasting plasma glucose test and A1C test for diabetes

Chapter 19

- Added brief discussion in Nutritional Adequacy of Foods section on ultraprocessed foods

- Added a paragraph in the Environmental Contaminants section on arsenic
- Added more information throughout Pesticide section on children's health and consumers' perceptions
- Added brief discussion in Food Additives section comparing natural and artificial additives
- Added a paragraph in Food Additives section on edible packaging

Chapter 20

- Added discussion on controversy surrounding whether SNAP participants should be allowed to purchase soft drinks and other non-nutritious foods and beverages
- Added tips to reduce food waste to the How To feature on p. 667
- Added discussion on eco-friendly products and packaging to the highlight

Student and Instructor Resources

MindTap: A new approach to highly personalized online learning. Beyond an eBook, homework solution, digital supplement, or premium website, MindTap is a digital learning platform that works alongside your campus LMS to deliver course curriculum across the range of electronic devices in your life. MindTap is built on an "app" model allowing enhanced digital collaboration and delivery of engaging content accross a spectrum of Cengage and non-Cengage resources.

Instructor Companion Site: Everything you need for your course in one place! This collection of book-specific lecture and class tools is available online via www.cengage.com /login. Access and download PowerPoint presentations, images, instructor's manual, videos, and more.

Test Bank with Cognero: Cengage Learning Testing Powered by Cognero is a flexible, online system that allows you to:

- author, edit, and manage test bank content from multiple Cengage Learning solutions
- create multiple test versions in an instant
- deliver tests from your LMS, your classroom, or wherever you want

Diet & Wellness Plus: Diet & Wellness Plus helps you understand how nutrition relates to your personal health goals. Track your diet and activity, generate reports, and analyze the nutritional value of the food you eat. Diet & Wellness Plus includes over 55,000 foods as well as custom food and recipe features. The new Behavior Change Planner helps you identify risks in your life and guides you through the key steps to make positive changes.

Global Nutrition Watch: Bring currency to the classroom with Global Nutrition Watch from Cengage Learning. This user-friendly website provides convenient access to thousands of trusted sources, including academic journals, newspapers, videos, and podcasts, for you to use for research projects or classroom discussion. Global Nutrition Watch is updated daily to offer the most current news about topics related to nutrition.

Closing Comments

We have taken great care to provide accurate information and have included many references at the end of each chapter and highlight. To keep the number of references manageable over the decades, however, many statements that appeared in previous editions with references now appear without them. All statements reflect current nutrition knowledge, and the authors will supply references upon request. In addition to supporting text statements, the end-of-chapter references provide readers with resources for finding a good overview or more details on the subject. Nutrition is a fascinating subject, and we hope our enthusiasm for it comes through on every page.

Ellie Whitney
Sharon Rady Rolfes
January 2015

Acknowledgments

To produce a book requires the coordinated effort of a team of people—and, no doubt, each team member has another team of support people as well. We salute, with a big round of applause, everyone who has worked so diligently to ensure the quality of this book.

We thank our partners and friends, Linda DeBruyne and Fran Webb, for their valuable consultations and contributions; working together over the past 30 years has been a most wonderful experience. We especially appreciate Linda's research assistance on several chapters. Special thanks to our colleagues Kathy Pinna for her insightful comments and Sylvia Crews for her careful review of the math explanations in Appendix K. Thank you to Taylor Newman for clarifying content and providing a student's perspective, Chelsea Mackenzie for assisting in numerous office tasks, and Marni Jay Rolfes for offering behind-the-scenes editorial suggestions.

We also thank those who prepared content for instructor and student resources: Samantha Yunko, instructor resources, internet exercises, and case studies; Pat Colsher, case studies; and Jennifer Peterson, PowerPoint slides. Thanks also to Miriam Myers, and the folks at Axxya Systems for their assistance in creating the food composition appendix and developing the computerized Diet & Wellness Plus program that accompanies this book.

Our heartfelt thanks to our editorial team for their efforts in creating an outstanding nutrition textbook—Peggy Williams for gently pushing through resistance for changes that enhance this edition in amazing ways; Nedah Rose for her calming presence, delightful humor, and thoughtful suggestions; Carol Samet for her management of this project; Tom Ziolkowski for his energetic efforts in marketing; Miriam Myers for her dedication in developing online resources and study tools; John Sarantakis and Christine Myaskovsky for their assistance in obtaining permissions; and Casey Lozier and Kellie Petruzzelli for their competent editing of ancillaries.

We also thank Gary Hespenheide and John Walker for creatively designing these pages; Jill Traut for her diligent attention to the innumerable details involved in production; Priya Subbrayal for selecting photographs that deliver nutrition messages attractively; Debbie Stone for copyediting close to 1000 manuscript pages; Heather Mann for proofreading close to 1000 final text pages; and Cheryl Duksta for composing a thorough and useful index. To the hundreds of others involved in production and sales, we tip our hats in appreciation.

We are especially grateful to our friends and families for their continued encouragement and support. We also thank our many reviewers for their comments and contributions.

Reviewers of *Understanding Nutrition*

Becky Alejandre
American River College

Janet B. Anderson
Utah State University

Sandra D. Baker
University of Delaware

Angelina Boyce
Hillsborough Community College

Lynn S. Brann
Syracuse University

Shalon Bull
Palm Beach Community College

Dorothy A. Byrne
University of Texas, San Antonio

John R. Capeheart
University of Houston, Downtown

Leah Carter
Bakersfield College

James F. Collins
University of Florida

Diane Curis
Los Rios Community College District

Lisa K. Diewald
Montgomery County Community College

Kelly K. Eichmann
Fresno City College

Shannon Fenster
Bellevue College

Shawn Flanagan
University of Iowa

Mary Flynn
Brown University

Betty J. Forbes
West Virginia University

Sue Fredstrom
Minnesota State University, Mankato

Trish Froehlich
Palm Beach Community College

Stephen P. Gagnon
Hillsborough Community College

Leonard Gerber
University of Rhode Island

Jill Golden
Orange Coast College

Barbara J. Goldman
Palm Beach State College

Kathleen Gould
Towson University

Margaret Gunther
Palomar College

Charlene Hamilton
University of Delaware

D. J. Hennager
Kirkwood Community College

Catherine Hagen Howard
Texarkana College

Ernest B. Izevbigie
Jackson State University

Craig Kasper
Hillsborough Community College

Shawnee Kelly
Pennsylvania State University

Younghee Kim
Bowling Green State University

Rebecca A. Kleinschmidt
University of Alaska Southeast

Vicki Kloosterhouse
Oakland Community College

Donna M. Kopas
Pennsylvania State University

Susan M. Krueger
University of Wisconsin, Eau Claire

Barbara Lange
College of Central Florida

Melissa Langone
Pasco-Hernando Community College

Grace Lasker
Lake Washington Institute of Technology

Darlene M. Levinson
Oakland Community College, Orchard Ridge

Kimberly Lower
Collin County Community College

Melissa B. McGuire
Maple Woods Community College

Diane L. McKay
Tufts University

Anne Miller
De Anza College

Anahita M. Mistry
Eastern Michigan University

Lisa Morse
Arizona State University

Mithia Mukutmoni
Sierra College

Steven Nizielski
Grand Valley State University

Yvonne Ortega
Santa Monica College

Jane M. Osowski
University of Southern Mississippi

Sarah Panarello
Yakima Valley Community College

Ryan Paruch
Tulsa Community College

Jill Patterson
Pennsylvania State University

Gina Pazzaglia
West Chester University

Julie Priday
Centralia College

Kelsey Rami
South Dakota State University

Jennifer Rogers
University of Iowa

Kathy L. Sedlet
Collin County Community College

Vidya Sharma
San Antonio College

Melissa Shock
University of Central Arkansas

Krista M. Simonetti
Maricopa County College

LuAnn Soliah
Baylor University

Kenneth Strothkamp
Lewis & Clark College

Andrea Villarreal
Phoenix College

Terry Weideman
Oakland Community College, Highland Lake

H. Garrison Wilkes
University of Massachusetts, Boston

Lynne C. Zeman
Kirkwood Community College

Maureen Zimmerman
Mesa Community College

Understanding Nutrition

1

An Overview of Nutrition

Nutrition in Your Life

Believe it or not, you have probably eaten at least 20,000 meals in your life. Without any conscious effort on your part, your body uses the nutrients from those foods to make all its components, fuel all its activities, and defend itself against diseases. How successfully your body handles these tasks depends, in part, on your food choices. Nutritious food choices support healthy bodies. In the Nutrition Portfolio at the end of this chapter, you can see how your current food choices are influencing your health and risk of chronic diseases.

Nutrition has always played a significant role in your life. Every day, several times a day, you select **foods** that influence your body's health. Each day's food choices may benefit or harm health only a little, but over time, the consequences of these choices become major. That being the case, paying close attention to good eating habits now supports health benefits later. Conversely, carelessness about food choices can contribute to **chronic diseases.** Of course, some people will become ill or die young no matter what choices they make, and others will live long lives despite making poor choices. For most of us, however, the food choices we make will benefit or impair our health in proportion to the wisdom of those choices.

Although most people realize food habits affect health, they often choose foods for other reasons. After all, foods bring pleasures, traditions, and associations as well as nourishment. The challenge, then, is to combine favorite foods and fun times with a nutritionally balanced **diet.** Take a moment to review the definition and note that *diet* does *not* mean a restrictive food plan designed for weight loss. It simply refers to the foods and beverages a person consumes. Whether it's a vegetarian diet, a weight-loss diet, or any other kind of diet depends on the types of foods and beverages a person chooses.

nutrition: the science of the nutrients in foods and their actions within the body. A broader definition includes the study of human behaviors related to food and eating.

foods: products derived from plants or animals that can be taken into the body to yield energy and nutrients for the maintenance of life and the growth and repair of tissues.

chronic diseases: diseases characterized by slow progression and long duration. Examples include heart disease, diabetes, and some cancers.

• chronos = time

diet: the foods and beverages a person eats and drinks.

1.1 Food Choices

> LEARN IT Describe how various factors influence personal food choices.

People decide what to eat, when to eat, how much to eat, and even whether to eat in highly personal ways based on a complex interaction of genetic, behavioral, or social factors rather than on an awareness of nutrition's importance to health.[1] A variety of food choices can support good health, and an understanding of human nutrition helps you make sensible selections more often.

Preferences As you might expect, the number one reason most people choose certain foods is taste—they like the flavor. Two widely shared preferences are for the sweetness of sugar and the savoriness of salt.[2] High-fat foods also appear to be a universally common preference. Other preferences might be for the hot peppers common in Mexican cooking or the curry spices of Indian cuisine. Research suggests that genetics may influence taste perceptions and therefore food likes and dislikes.[3] Similarly, the hormones of pregnancy seem to influence food cravings and aversions (see Chapter 15).

Habit People sometimes select foods out of habit. They eat cereal every morning, for example, simply because they have always eaten cereal for breakfast. Eating a familiar food and not having to make any decisions can be comforting.

Ethnic Heritage and Regional Cuisines Among the strongest influences on food choices are ethnic heritage and regional cuisines. People tend to prefer the foods they grew up eating. Every country, and in fact every region of a country, has its own typical foods and ways of combining them into meals. These cuisines reflect a unique combination of local ingredients and cooking styles. Chowder in New England is made with clams, but in the Florida Keys conch is the featured ingredient. The Pacific Northwest is as famous for its marionberry pie as Georgia is for its peach cobbler. Philly has its cheesesteaks and New Orleans has its oyster po'boys. The "American diet" includes many ethnic foods and regional styles, all adding variety to the diet.

Enjoying traditional **ethnic foods** provides an opportunity to celebrate a person's heritage. People offering ethnic foods share a part of their culture with others, and those accepting the foods learn about another's way of life. Developing **cultural competence** honors individual preferences and is particularly important for professionals who help others plan healthy diets.

Social Interactions Most people enjoy companionship while eating. It's fun to go out with friends for a meal or share a snack when watching a movie together. Meals are often social events, and sharing food is part of hospitality. Social customs invite people to accept food or drink offered by a host or shared by a group—regardless of hunger signals. Chapter 9 describes how people tend to eat more food when socializing with others.

Availability, Convenience, and Economy People often eat foods that are accessible, quick and easy to prepare, and within their financial means. Consumers who value convenience frequently eat out, bring home ready-to-eat meals, or have food delivered. Even when they venture into the kitchen, they want to prepare a meal in 15 to 20 minutes, using less than a half dozen ingredients—and those "ingredients" are often semiprepared foods, such as canned soups and frozen foods.

Consumer emphasis on convenience limits food choices to the selections offered on menus and products designed for quick preparation. Whether decisions based on convenience meet a person's nutrition needs depends on the choices made. Eating a banana or a candy bar may be equally convenient, but the fruit provides more vitamins and minerals and less sugar and fat.

Rising food costs have shifted some consumers' priorities and changed their shopping habits.[4] They are less likely to buy higher-priced convenience foods and

An enjoyable way to learn about a culture is to taste the ethnic foods.

© Corbis Premium RF/Alamy

ethnic foods: foods associated with particular cultural groups.

cultural competence: having an awareness and acceptance of cultures and the ability to interact effectively with people of diverse cultures.

more likely to buy less-expensive store brand items and prepare home-cooked meals. In fact, more than 70 percent of meals are prepared in the home.[5] Those who frequently prepare their own meals report more positive emotions and healthier food choices.[6] They tend to eat fast food less often and are more likely to meet dietary guidelines for fat, calcium, fruits, vegetables, and whole grains. Not surprisingly, when eating out, consumers choose low-cost fast-food outlets over more expensive fine-dining restaurants. Foods eaten away from home, especially fast-food meals, tend to be high in kcalories, total fat, saturated fat, and *trans* fat—which can contribute to a variety of health problems.[7]

Unfortunately, healthful diets that include plenty of fruits and vegetables tend to be more costly than less healthful diets featuring foods containing solid fats and added sugars.[8] These low-cost foods are often high in kcalories and low in nutrients.[9] Consumers can improve the quality of their diets without increasing their spending by choosing more plant-based foods, such as nuts, legumes, and whole grains, and fewer refined grains, red and processed meats, and high-fat milk products.[10]

Positive and Negative Associations People tend to like particular foods associated with happy occasions—such as hot dogs at ball games or cake and ice cream at birthday parties. By the same token, people can develop aversions and dislike foods that they ate when they felt sick or that they were forced to eat in negative situations. Similarly, children learn to like and dislike certain foods when their parents use foods as rewards or punishments. Negative experiences can have long-lasting influences on food preferences. More than 50 years after World War II, veterans who had experienced intense combat in the Pacific dislike Chinese and Japanese food significantly more than their peers who were not engaged in battle or those who fought elsewhere.

Emotions Emotions guide food choices and eating behaviors.[11] Some people cannot eat when they are emotionally upset. Others may eat in response to a variety of emotional stimuli—for example, to relieve boredom or depression or to calm anxiety. A depressed person may choose to eat rather than to call a friend. A person who has returned home from an exciting evening out may unwind with a late-night snack. These people may find emotional comfort, in part, because foods can influence the brain's chemistry and the mind's response. Carbohydrates and alcohol, for example, tend to calm, whereas proteins and caffeine are more likely to stimulate. Eating in response to emotions and stress can easily lead to overeating and obesity, but it may be helpful at times. For example, sharing food at times of bereavement serves both the giver's need to provide comfort and the receiver's need to be cared for and to interact with others as well as to take nourishment.

© Wave Royalty Free/Alamy

Values Food choices may reflect people's religious beliefs, political views, or environmental concerns. For example, some Christians forgo meat on Fridays during Lent (the period prior to Easter), Jewish law includes an extensive set of dietary rules that govern the use of foods derived from animals, and Muslims fast between sunrise and sunset during Ramadan (the ninth month of the Islamic calendar). Some vegetarians select foods based on their concern for animal rights. A concerned consumer may boycott fruit picked by migrant workers who have been exploited. People may buy vegetables from local farmers to save the fuel and environmental costs of foods shipped from far away. They may also select foods packaged in containers that can be reused or recycled. Some consumers accept or reject foods that have been irradiated, grown organically, or genetically modified, depending on their approval of these processes (see Chapter and Highlight 19 for a complete discussion).

To enhance your health, keep nutrition in mind when selecting foods. To protect the environment, shop at local markets and reuse cloth shopping bags.

Body Weight and Image Sometimes people select certain foods and supplements that they believe will improve their physical appearance and avoid those they believe might be detrimental. Such decisions can be beneficial when based on sound nutrition and fitness knowledge, but decisions based on fads or carried to extremes undermine good health, as pointed out in later discussions of eating disorders (Highlight 8) and dietary supplements commonly used by athletes (Highlight 14).

Nutrition and Health Benefits Finally, of course, many consumers make food choices they believe will improve their health.[12] Food manufacturers and restaurant chefs have responded to scientific findings linking health with nutrition by offering an abundant selection of health-promoting foods and beverages.[13] Foods that provide health benefits beyond their nutrient contributions are called **functional foods.** Functional foods include whole foods as well as fortified, enriched, or enhanced foods.[14] Whole foods—as natural and familiar as oatmeal or tomatoes—are the simplest functional foods. In some cases, foods have been modified to provide health benefits, perhaps by lowering the *trans* fat contents. In other cases, manufacturers have fortified foods by adding nutrients or **phytochemicals** that provide health benefits (see Highlight 13). Examples of these functional foods include orange juice fortified with calcium to help build strong bones and margarine made with a plant sterol that lowers blood cholesterol.

Consumers typically welcome new foods into their diets, provided that these foods are reasonably priced, clearly labeled, easy to find in the grocery store, and convenient to prepare. These foods must also taste good—as good as the traditional choices. Of course, a person need not eat any "special" foods to enjoy a healthy diet; many "regular" foods provide numerous health benefits as well. In fact, "regular" foods such as whole grains; vegetables and legumes; fruits; seafood, meats, poultry, eggs, nuts, and seeds; and milk products are among the healthiest choices a person can make.

› REVIEW IT Describe how various factors influence personal food choices.
A person selects foods for a variety of reasons. Whatever those reasons may be, food choices influence health. Individual food selections neither make nor break a diet's healthfulness, but the balance of foods selected over time can make an important difference to health.[15] For this reason, people are wise to think "nutrition" when making their food choices.

1.2 The Nutrients

› LEARN IT Name the six major classes of nutrients and identify which are organic and which yield energy.

Biologically speaking, people eat to receive nourishment. Do you ever think of yourself as a biological being made of carefully arranged atoms, molecules, cells, tissues, and organs? Are you aware of the activity going on within your body even as you sit still? The atoms, molecules, and cells of your body continuously move and change, even though the structures of your tissues and organs and your external appearance remain relatively constant. The ongoing growth, maintenance, and repair of the body's tissues depend on the **energy** and the **nutrients** received from foods.

Nutrients in Foods and in the Body Amazingly, our bodies can derive all the energy, structural materials, and regulating agents we need from the foods we eat. This section introduces the nutrients that foods deliver and shows how they participate in the dynamic processes that keep people alive and well.

Nutrient Composition of Foods Chemical analysis of a food such as a tomato shows that it is composed primarily of water (95 percent). Most of the solid materials are carbohydrates, lipids (fats), and proteins. If you could remove these materials, you would find a tiny residue of vitamins, minerals, and other

Foods bring pleasure—and nutrients.

functional foods: foods that have a potentially beneficial effect on health when consumed as part of a varied diet on a regular basis at effective levels.

phytochemicals (FIE-toe-KEM-ih-cals): nonnutrient compounds found in plants. Some phytochemicals have biological activity in the body.

• **phyto** = plant

energy: the capacity to do work. The energy in food is chemical energy. The body can convert this chemical energy to mechanical, electrical, or heat energy.

nutrients: chemical substances obtained from food and used in the body to provide energy, structural materials, and regulating agents to support growth, maintenance, and repair of the body's tissues. Nutrients may also reduce the risks of some diseases.

> FIGURE 1-1   **Body Composition of Healthy-Weight Men and Women**

The human body is made of compounds similar to those found in foods—mostly water (60 percent) and some fat (18 to 21 percent for young men, 23 to 26 percent for young women), with carbohydrate, protein, vitamins, minerals, and other minor constituents making up the remainder. (Chapter 8 describes the health hazards of too little or too much body fat.)

Key:
- % Carbohydrate, protein, vitamins, minerals in the body
- % Fat in the body
- % Water in the body

© Cengage Learning

compounds. Water, carbohydrates, lipids, proteins, vitamins, and some of the minerals found in foods represent the six classes of nutrients—substances the body uses for the growth, maintenance, and repair of its tissues.

This book focuses mostly on the nutrients, but foods contain other compounds as well—fibers, phytochemicals, pigments, additives, alcohols, and others. Some are beneficial, some are neutral, and a few are harmful. Later sections of the book touch on these compounds and their significance.

Nutrient Composition of the Body A chemical analysis of your body would show that it is made of materials similar to those found in foods (see Figure 1-1). A healthy 150-pound body contains about 90 pounds of water and about 20 to 45 pounds of fat. The remaining pounds are mostly protein, carbohydrate, and the major minerals of the bones. Vitamins, other minerals, and incidental extras constitute a fraction of a pound.

Chemical Composition of Nutrients The simplest of the nutrients are the minerals. Each mineral is a chemical element; its atoms are all alike. As a result, its identity never changes. For example, iron may have different electrical charges, but the individual iron atoms remain the same when they are in a food, when a person eats the food, when the iron becomes part of a red blood cell, when the cell is broken down, and when the iron is lost from the body by excretion. The next simplest nutrient is water, a compound made of two elements—hydrogen and oxygen. Minerals and water are **inorganic** nutrients—which means they do not contain carbon.

The other four classes of nutrients (carbohydrates, lipids, proteins, and vitamins) are more complex. In addition to hydrogen and oxygen, they all contain carbon, an element found in all living things; they are therefore called **organic** compounds (meaning, literally, "alive").* This chemical definition of *organic* differs from the agricultural definition. As Chapter 19 explains, organic farming refers to growing crops and raising livestock according to standards set by the

inorganic: not containing carbon or pertaining to living organisms. The two classes of nutrients that are inorganic are minerals and water.

- **in** = not

organic: in chemistry, substances or molecules containing carbon-carbon bonds or carbon-hydrogen bonds that are characteristic of living organisms. The four classes of nutrients that are organic are carbohydrates, lipids (fats), proteins, and vitamins.

---

* Note that this definition of *organic* excludes coal, diamonds, and a few carbon-containing compounds that contain only a single carbon and no hydrogen, such as carbon dioxide (CO_2), calcium carbonate ($CaCO_3$), magnesium carbonate ($MgCO_3$), and sodium cyanide ($NaCN$).

TABLE 1-1 The Six Classes of Nutrients

Nutrient	Organic	Inorganic	Energy-yielding	Macronutrient	Micronutrient
Carbohydrates	✓		✓	✓	
Lipids (fats)	✓		✓	✓	
Proteins	✓		✓	✓	
Vitamins	✓				✓
Minerals		✓			✓
Water		✓			

© 2016 Cengage Learning

US Department of Agriculture (USDA). Protein and some vitamins also contain nitrogen and may contain other elements such as sulfur as well.

Essential Nutrients The body can make some nutrients, but it cannot make all of them. Also, it makes some in insufficient quantities to meet its needs and, therefore, must obtain these nutrients from foods. The nutrients that foods must supply are **essential nutrients.** When used to refer to nutrients, the word *essential* means more than just "necessary"; it means "needed from outside the body"— normally, from foods.

The Energy-Yielding Nutrients: Carbohydrate, Fat, and Protein

In the body, three of the organic nutrients can be used to provide energy: carbohydrate, fat, and protein. In contrast to these **energy-yielding nutrients,** vitamins, minerals, and water do not yield energy in the human body.

Carbohydrate, fat, and protein are sometimes called *macronutrients* because the body requires them in relatively large amounts (many grams daily). In contrast, vitamins and minerals are *micronutrients,* required only in small amounts (milligrams or micrograms daily). Table 1-1 summarizes some of the ways the six classes of nutrients can be described.

Energy Measured in kCalories The energy released from carbohydrates, fats, and proteins can be measured in **calories**—tiny units of energy so small that a single apple provides tens of thousands of them. To ease calculations, energy is expressed in 1000-calorie metric units known as kilocalories (shortened to kcalories, but commonly called "calories"). When you read in popular books or magazines that an apple provides "100 calories," it actually means 100 kcalories. This book uses the term *kcalorie* and its abbreviation *kcal* throughout, as do other scientific books and journals. The "How To" on p. 9 provides a few tips on "thinking metric."

Energy from Foods The amount of energy a food provides depends on how much carbohydrate, fat, and protein it contains. When completely broken down in the body, a gram of carbohydrate yields about 4 kcalories of energy; a gram of protein also yields 4 kcalories; and a gram of fat yields 9 kcalories (see Table 1-2).* The "How To" on p. 10 explains how to calculate the energy available from foods.

Because fat provides more energy per gram, it has a greater **energy density** than either carbohydrate or protein. Figure 1-2 on p. 10 compares the energy density of two breakfast options, and later chapters describe how foods with a high energy density help with weight *gain,* whereas those with a low energy density help with weight *loss.*

One other substance contributes energy—alcohol. Alcohol, however, is not considered a nutrient. Unlike the nutrients, alcohol does not sustain life. In fact, it interferes with the growth, maintenance, and repair of the body. Its only common characteristic with nutrients is that it yields energy (7 kcalories per gram) when metabolized in the body.

TABLE 1-2 kCalorie Values of Energy Nutrients

Nutrients	Energy
Carbohydrate	4 kcal/g
Fat	9 kcal/g
Protein	4 kcal/g

NOTE: Alcohol contributes 7 kcal/g that can be used for energy, but it is not considered a nutrient because it interferes with the body's growth, maintenance, and repair.

© Cengage Learning

essential nutrients: nutrients a person must obtain from food because the body cannot make them for itself in sufficient quantity to meet physiological needs; also called *indispensable nutrients.* About 40 nutrients are currently known to be essential for human beings.

energy-yielding nutrients: the nutrients that break down to yield energy the body can use:
- Carbohydrate
- Fat
- Protein

calories: a measure of *heat* energy. Energy provided by foods and beverages is measured in *kilocalories* (1000 calories equal 1 kilocalorie), abbreviated *kcalories* or *kcal.* One kcalorie is the amount of heat necessary to raise the temperature of 1 kilogram (kg) of water 1°C. The scientific use of the term *kcalorie* is the same as the popular use of the term *calorie.*

energy density: a measure of the energy a food provides relative to the weight of the food (kcalories per gram).

*For those using kilojoules: 1 g carbohydrate = 17 kJ; 1 g protein = 17 kJ; 1 g fat = 37 kJ; and 1 g alcohol = 29 kJ.

>How To

Think Metric

Like other scientists, nutrition scientists use metric units of measure. They measure food energy in kilocalories, people's height in centimeters, people's weight in kilograms, and the weights of foods and nutrients in grams, milligrams, or micrograms. For ease in using these measures, it helps to remember that the prefixes imply 1000. For example, a *kilo*gram is 1000 grams, a *milli*gram is 1/1000 of a gram, and a *micro*gram is 1/1000 of a milligram.

Most food labels and many recipes provide "dual measures," listing both household measures, such as cups, quarts, and teaspoons, and metric measures, such as milliliters, liters, and grams. This practice gives people an opportunity to gradually learn to "think metric."

A person might begin to "think metric" by simply observing the measure—by noticing the amount of soda in a 2-liter bottle, for example. Through such experiences, a person can become familiar with a measure without having to do any conversions.

The international unit for measuring food energy is the joule—the amount of energy expended when 1 kilogram is moved 1 meter by a force of 1 newton. The joule is thus a measure of work energy, whereas the kcalorie is a measure of heat energy. While many scientists and journals report their findings in kilojoules (kJ), many others, particularly those in the United States, use kcalories (kcal). To convert energy measures from kcalories to kilojoules, multiply by 4.2; to convert kilojoules to kcalories, multiply by 0.24. For example, a 50-kcalorie cookie provides 210 kilojoules:

$$50 \text{ kcal} \times 4.2 = 210 \text{ kJ}$$

Appendix K provides assistance and conversion factors for these and other units of measure.

Volume: Liters (L)

1 L = 1000 milliliters (mL)
0.95 L = 1 quart
1 mL = 0.03 fluid ounces
240 mL = 1 cup

A liter of liquid is approximately one US quart. (Four liters are only about 5 percent more than a gallon.)

One cup of liquid is about 240 milliliters; a half-cup of liquid is about 120 milliliters.

Weight: Grams (g)

1 g = 1000 milligrams (mg)
1 g = 0.04 ounce (oz)
1 oz = 28.35 g (or 30 g)
100 g = 3½ oz
1 kilogram (kg) = 1000 g
1 kg = 2.2 pounds (lb)
454 g = 1 lb

A kilogram is slightly more than 2 lb; conversely, a pound is about ½ kg.

A half-cup of vegetables weighs about 100 grams; one pea weighs about ½ gram.

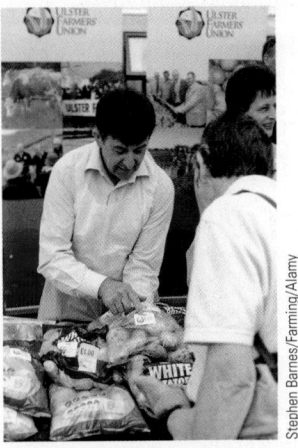

A 5-pound bag of potatoes weighs about 2 kilograms, and a 176-pound person weighs 80 kilograms.

> **TRY IT** Convert your body weight from pounds to kilograms and your height from inches to centimeters.

Most foods contain the energy-yielding nutrients, as well as vitamins, minerals, water, and other substances. For example, meat contains water, fat, vitamins, and minerals as well as protein. Bread contains water, a trace of fat, a little protein, and some vitamins and minerals in addition to its carbohydrate. Only a few foods are exceptions to this rule, the common ones being sugar (pure carbohydrate) and oil (essentially pure fat).

>How To

Calculate the Energy Available from Foods

To calculate the energy available from a food, multiply the number of grams of carbohydrate, protein, and fat by 4, 4, and 9, respectively. Then add the results together. For example, 1 slice of bread with 1 tablespoon of peanut butter on it contains 16 grams carbohydrate, 7 grams protein, and 9 grams fat:

$$16 \text{ g carbohydrate} \times 4 \text{ kcal/g} = 64 \text{ kcal}$$
$$7 \text{ g protein} \times 4 \text{ kcal/g} = 28 \text{ kcal}$$
$$9 \text{ g fat} \times 9 \text{ kcal/g} = 81 \text{ kcal}$$
$$\text{Total} = 173 \text{ kcal}$$

From this information, you can calculate the percentage of kcalories each of the energy nutrients contributes to the total. To determine the percentage of kcalories from fat, for example, divide the 81 fat kcalories by the total 173 kcalories:

$$81 \text{ fat kcal} \div 173 \text{ total kcal} = 0.468$$
$$(\text{rounded to } 0.47)$$

Then multiply by 100 to get the percentage:

$$0.47 \times 100 = 47\%$$

Dietary recommendations that urge people to limit fat intake to 20 to 35 percent of kcalories refer to the day's total energy intake, not to individual foods. Still, if the proportion of fat in each food choice throughout a day exceeds 35 percent of kcalories, then the day's total surely will, too. Knowing that this snack provides 47 percent of its kcalories from fat alerts a person to the need to make lower-fat selections at other times that day.

> **TRY IT** Calculate the energy available from a bean burrito with cheese (55 grams carbohydrate, 15 grams protein, and 12 grams fat). Determine the percentage of kcalories from each of the energy nutrients.

> FIGURE 1-2 **Energy Density of Two Breakfast Options Compared**

Gram for gram, ounce for ounce, and bite for bite, foods with a high energy density deliver more kcalories than foods with a low energy density. Both of these breakfast options provide 500 kcalories, but the cereal with milk, fruit salad, scrambled egg, turkey sausage, and toast with jam offers three times as much food as the doughnuts (based on weight); it has a lower energy density than the doughnuts. Selecting a variety of foods also helps to ensure nutrient adequacy.

LOWER ENERGY DENSITY
This 450-gram breakfast delivers 500 kcalories, for an energy density of 1.1
$$(500 \text{ kcal} \div 450 \text{ g} = 1.1 \text{ kcal/g}).$$

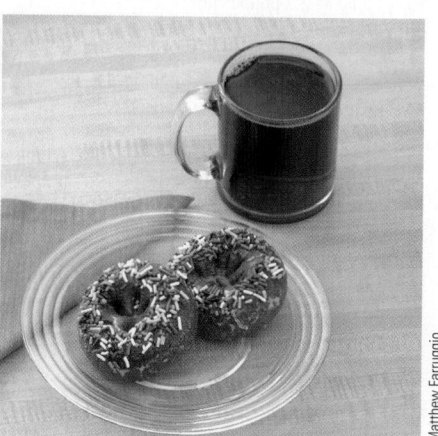

HIGHER ENERGY DENSITY
This 144-gram breakfast delivers 500 kcalories, for an energy density of 3.5
$$(500 \text{ kcal} \div 144 \text{ g} = 3.5 \text{ kcal/g}).$$

Energy in the Body When the body uses carbohydrate, fat, or protein to fuel its activities, the bonds between the nutrient's atoms break. As the bonds break, they release energy. Some of this energy is released as heat, but some is used to send electrical impulses through the brain and nerves, to synthesize body compounds, and to move muscles. Thus the energy from food supports every activity from quiet thought to vigorous sports.

If the body does not use these nutrients to fuel its current activities, it converts them into storage compounds (such as body fat), to be used between meals and overnight when fresh energy supplies run low. If more energy is consumed than expended, the result is an increase in energy stores and weight gain. Similarly, if less energy is consumed than expended, the result is a decrease in energy stores and weight loss.

When consumed in excess of energy needs, alcohol, too, can be converted to body fat and stored. When alcohol contributes a substantial portion of the energy in a person's diet, the harm it does far exceeds the problems of excess body fat. (Highlight 7 describes the effects of alcohol on health and nutrition.)

Other Roles of Energy-Yielding Nutrients In addition to providing energy, carbohydrates, fats, and proteins provide the raw materials for building the body's tissues and regulating its many activities. In fact, protein's role as a fuel source is relatively minor compared with both the other two energy-yielding nutrients and its other roles. Proteins are found in structures such as the muscles and skin and help to regulate activities such as digestion and energy metabolism. (Chapter 6 presents a full discussion on proteins.)

The Vitamins

The **vitamins** are also organic, but they do not provide energy. Instead, they facilitate the release of energy from carbohydrate, fat, and protein and participate in numerous other activities throughout the body.

Each of the 13 vitamins has its own special roles to play.* One vitamin enables the eyes to see in dim light, another helps protect the lungs from air pollution, and still another helps make the sex hormones—among other things. When you cut yourself, one vitamin helps stop the bleeding and another helps repair the skin. Vitamins busily help replace old red blood cells and the lining of the digestive tract. Almost every action in the body requires the assistance of vitamins.

Vitamins can function only if they are intact, but because they are complex organic molecules, they are vulnerable to destruction by heat, light, and chemical agents. This is why the body handles them carefully, and why nutrition-wise cooks do, too. The strategies of cooking vegetables at moderate temperatures for short times and using small amounts of water help to preserve the vitamins.

The Minerals

In the body, some **minerals** are put together in orderly arrays in such structures as bones and teeth. Minerals are also found in the fluids of the body, which influences fluid balance and distribution. Whatever their roles, minerals do not yield energy.

Only 16 minerals are known to be essential in human nutrition.** Others are being studied to determine whether they play significant roles in the human body. Still other minerals, such as lead, are environmental contaminants that displace the nutrient minerals from their workplaces in the body, disrupting body functions. The problems caused by contaminant minerals are described in Chapter 13.

---

*The water-soluble vitamins are vitamin C and the eight B vitamins: thiamin, riboflavin, niacin, vitamins B_6 and B_{12}, folate, biotin, and pantothenic acid. The fat-soluble vitamins are vitamins A, D, E, and K. The water-soluble vitamins are the subject of Chapter 10 and the fat-soluble vitamins, of Chapter 11.

**The major minerals are calcium, phosphorus, potassium, sodium, chloride, magnesium, and sulfate. The trace minerals are iron, iodine, zinc, chromium, selenium, fluoride, molybdenum, copper, and manganese. Chapters 12 and 13 are devoted to the major and trace minerals, respectively.

vitamins: organic, essential nutrients required in small amounts by the body for health.

minerals: inorganic elements. Some minerals are essential nutrients required in small amounts by the body for health.

Because minerals are inorganic, they are indestructible and need not be handled with the special care that vitamins require. Minerals can, however, be bound by substances that interfere with the body's ability to absorb them. They can also be lost during food-refining processes or during cooking when they leach into water that is discarded.

Water Water provides the environment in which nearly all the body's activities are conducted. It participates in many metabolic reactions and supplies the medium for transporting vital materials to cells and carrying waste products away from them. Water is discussed fully in Chapter 12, but it is mentioned in every chapter. If you watch for it, you cannot help but be impressed by water's participation in all life processes.

Water is an essential nutrient and naturally carries varying amounts of several minerals.

> **REVIEW IT** Name the six major classes of nutrients and identify which are organic and which yield energy.

Foods provide nutrients—substances that support the growth, maintenance, and repair of the body's tissues. The six classes of nutrients include:

- Carbohydrates
- Vitamins
- Lipids (fats)
- Minerals
- Proteins
- Water

Foods rich in the energy-yielding nutrients (carbohydrate, fat, and protein) provide the major materials for building the body's tissues and yield energy for the body's use or storage. Energy is measured in kcalories—a measure of heat energy. Vitamins, minerals, and water do not yield energy; instead they facilitate a variety of activities in the body.

Without exaggeration, nutrients provide the physical and metabolic basis for nearly all that we are and all that we do. The next section introduces the science of nutrition with emphasis on the research methods scientists have used in uncovering the wonders of nutrition.

1.3 The Science of Nutrition

> **LEARN IT** Explain the scientific method and how scientists use various types of research studies and methods to acquire nutrition information.

The science of nutrition is the study of the nutrients and other substances in foods and the body's handling of them. Its foundation depends on several other sciences, including biology, biochemistry, and physiology. As sciences go, nutrition is young, but as you can see from the size of this book, much has happened in nutrition's short life. And it is currently experiencing a tremendous growth spurt as scientists apply knowledge gained from sequencing the human **genome**. The integration of nutrition, genomics, and molecular biology has opened a whole new world of study called **nutritional genomics**—the science of how nutrients affect the activities of genes and how genes affect the interactions between diet and disease. Highlight 6 describes how nutritional genomics is shaping the science of nutrition, and examples of nutrient–gene interactions appear throughout later chapters of the book.

Conducting Research Consumers sometimes depend on personal experience or reports from friends to gather information on nutrition. Such a personal account of an experience or event is known as an **anecdote** and is not accepted as reliable scientific information (see the glossary on p. 13 for definitions of research terms). In contrast, researchers use the scientific method to guide their work (see Figure 1-3). As the figure shows, research always begins with a problem or a question. For example, "What foods or nutrients might protect against the common cold?" In search of an answer, scientists make an educated guess **(hypothesis)**, such as "foods rich in vitamin C reduce the number of common colds." Then they systematically conduct research studies to collect data that will

genome (GEE-nome): the complete set of genetic material (DNA) in an organism or a cell. The study of genomes is called *genomics*.

nutritional genomics: the science of how nutrients affect the activities of genes (*nutrigenomics*) and how genes affect the interactions between diet and disease (*nutrigenetics*).

> FIGURE 1-3 **The Scientific Method**

Research scientists follow the scientific method. Note that most research generates new questions, not final answers. Thus the sequence begins anew, and research continues in a somewhat cyclical way.

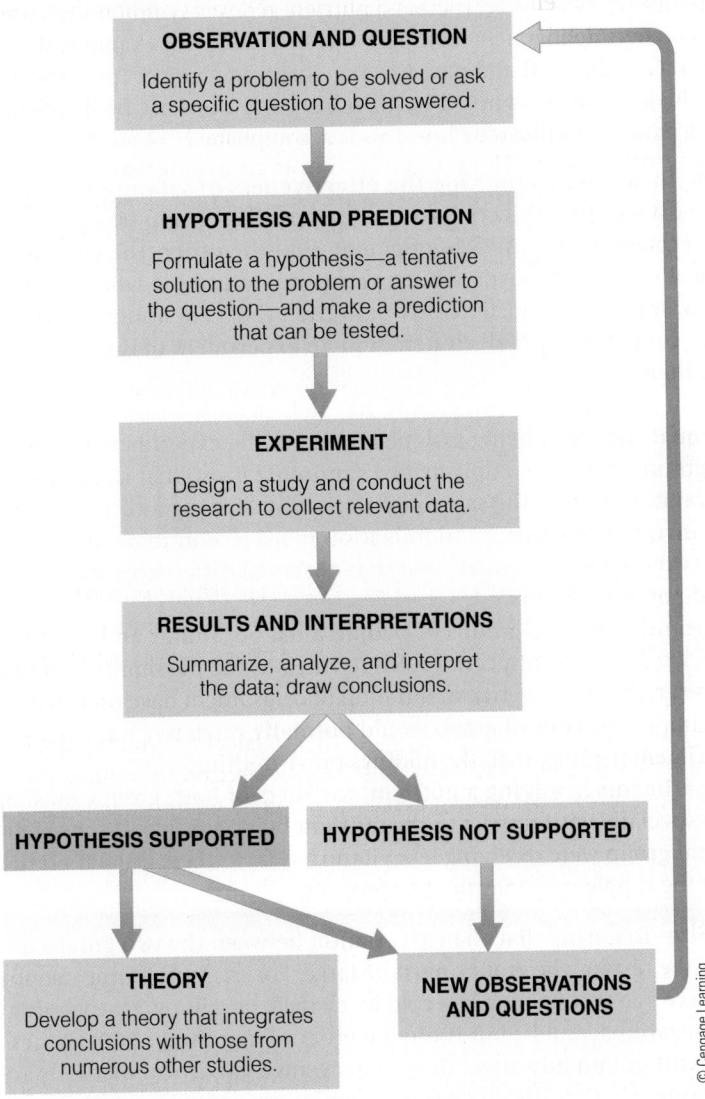

© Cengage Learning

anecdote: a personal account of an experience or event; not reliable scientific information.

blind experiment: an experiment in which the subjects do not know whether they are members of the experimental group or the control group.

control group: a group of individuals similar in all possible respects to the experimental group except for the treatment. Ideally, the control group receives a placebo while the experimental group receives a real treatment.

correlation (CORE-ee-LAY-shun): the simultaneous increase, decrease, or change in two variables. If A increases as B increases, or if A decreases as B decreases, the correlation is *positive*. (This does not mean that A causes B or vice versa.) If A increases as B decreases, or if A decreases as B increases, the correlation is *negative*. (This does not mean that A prevents B or vice versa.) Some third factor may account for both A and B.

double-blind experiment: an experiment in which neither the subjects nor the researchers know which subjects are members of the experimental group and which are serving as control subjects, until after the experiment is over.

experimental group: a group of individuals similar in all possible respects to the control group except for the treatment. The experimental group receives the real treatment.

hypothesis (hi-POTH-eh-sis): an unproven statement that tentatively explains the relationships between two or more variables.

peer review: a process in which a panel of scientists rigorously evaluates a research study to ensure that the scientific method was followed.

placebo (pla-SEE-bo): an inert, harmless medication given to provide comfort and hope; a sham treatment used in controlled research studies.

placebo effect: a change that occurs in response to expectations about the effectiveness of a treatment that actually has no pharmaceutical effects.

randomization (RAN-dom-ih-ZAY-shun): a process of choosing the members of the experimental and control groups without bias.

replication (REP-lih-KAY-shun): repeating an experiment and getting the same results.

subjects: the people or animals participating in a research project.

theory: a tentative explanation that integrates many and diverse findings to further the understanding of a defined topic.

validity (va-LID-ih-tee): having the quality of being founded on fact or evidence.

variables: factors that change. A variable may depend on another variable (for example, a child's height depends on his age), or it may be independent (for example, a child's height does not depend on the color of her eyes). Sometimes both variables correlate with a third variable (a child's height and eye color both depend on genetics).

test the hypothesis. Some examples of various types of research designs are presented in Figure 1-4. Because each type of study has strengths and weaknesses, some provide stronger evidence than others, as Figure 1-4 explains.

In attempting to discover whether a nutrient relieves symptoms or cures a disease, researchers deliberately manipulate one variable (for example, the amount of vitamin C in the diet) and measure any observed changes (perhaps the number of colds). As much as possible, all other conditions are held constant. The following paragraphs illustrate how this is accomplished.

Controls In studies examining the effectiveness of vitamin C, researchers typically divide the **subjects** into two groups. One group (the **experimental group**) receives a vitamin C supplement, and the other (the **control group**) does not. Researchers observe both groups to determine whether one group has fewer, milder, or shorter colds than the other. The following discussion describes some of the pitfalls inherent in an experiment of this kind and ways to avoid them.

In sorting subjects into two groups, researchers must ensure that each person has an equal chance of being assigned to either the experimental group or the control group. This is accomplished by **randomization**; that is, the subjects are chosen randomly from the same population by flipping a coin or some other method involving chance. Randomization helps to eliminate bias and ensure that the two groups are "equal" and that observed differences reflect the treatment and not other factors.[16]

Importantly, the two groups of people must be similar and must have the same track record with respect to colds to rule out the possibility that observed differences in the rate, severity, or duration of colds might have occurred anyway. If, for example, the control group would normally catch twice as many colds as the experimental group, then the findings prove nothing.

In experiments involving a nutrient, the diets of both groups must also be similar, especially with respect to the nutrient being studied. If those in the experimental group were receiving less vitamin C from their usual diet, then any effects of the supplement may not be apparent.

Sample Size To ensure that chance variation between the two groups does not influence the results, the groups must be large. For example, if one member of a group of five people catches a bad cold by chance, he will pull the whole group's average toward bad colds; but if one member of a group of 500 catches a bad cold, she will not unduly affect the group average. Statistical methods are used to determine whether differences between groups of various sizes support a hypothesis.

Placebos If people who take vitamin C for colds *believe* it will cure them, their chances of recovery may improve. Taking pills believed to be beneficial may shorten the duration and lessen the severity of illness regardless of whether the pills contain active ingredients.[17] This phenomenon, the result of expectations, is known as the **placebo effect**. In experiments designed to determine vitamin C's effect on colds, this mind-body effect must be rigorously controlled. Severity of symptoms is often a subjective measure, and people who believe they are receiving treatment may report less severe symptoms.

One way experimenters control for the placebo effect is to give pills to all participants. Those in the experimental group, for example, receive pills containing vitamin C, and those in the control group receive a **placebo**—pills of similar appearance and taste containing an inactive ingredient. This way, the expectations of both groups will be equal. It is not necessary to convince all subjects that they are receiving vitamin C, but the extent of belief or unbelief must be the same in both groups. A study conducted under these conditions is called a **blind experiment**—that is, the subjects do not know (are blind to) whether they are members of the experimental group (receiving treatment) or the control group (receiving the placebo).

> FIGURE 1-4 **Examples of Research Designs**

EPIDEMIOLOGICAL STUDIES research the incidence, distribution, and control of disease in a population. Epidemiological studies include cross-sectional, case-control, and cohort studies.

Strengths:
- Can narrow down the list of possible causes
- Can raise questions to pursue through other research

Weaknesses:
- Cannot control variables that may influence the development or the prevention of a disease
- Cannot prove cause and effect

CROSS-SECTIONAL STUDIES

Researchers observe how much and what kinds of foods a group of people eat and how healthy those people are. Their findings identify factors that might influence the incidence of a disease in various populations.

Example. Many people in the Mediterranean region drink more wine, eat more fat from olive oil, and yet have a lower incidence of heart disease than northern Europeans and North Americans.

CASE-CONTROL STUDIES

Researchers compare people who do and do not have a given condition such as a disease, closely matching them in age, gender, and other key variables so that differences in other factors will stand out. These differences may account for the condition in the group that has it.

Example. People with goiter lack iodine in their diets.

COHORT STUDIES

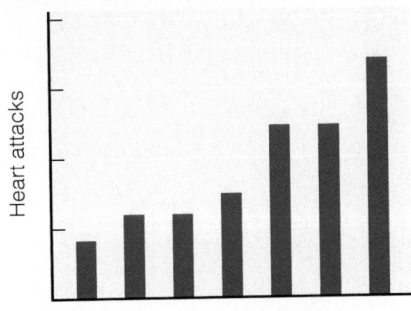

Researchers analyze data collected from a selected group of people (a cohort) at intervals over a certain period of time.

Example. Data collected periodically over the past several decades from more than 5000 people randomly selected from the town of Framingham, Massachusetts, in 1948 have revealed that the risk of heart attack increases as blood cholesterol increases.

EXPERIMENTAL STUDIES test cause-and-effect relationships between variables. Experimental studies include laboratory-based studies—on animals or in test tubes (in vitro)—and human intervention (or clinical) trials.

Strengths:
- Can control conditions (for the most part)
- Can determine effects of a variable
- Can apply some findings on human beings to some groups of human beings

Weaknesses:
- Cannot apply results from test tubes or animals to human beings
- Cannot generalize findings on human beings to all human beings
- Cannot use certain treatments for clinical or ethical reasons

LABORATORY-BASED
ANIMAL STUDIES

Researchers feed animals special diets that provide or omit specific nutrients and then observe any changes in health. Such studies test possible disease causes and treatments in a laboratory where all conditions can be controlled.

Example. Mice fed a high-fat diet eat less food than mice given a lower-fat diet, so they receive the same number of kcalories—but the mice eating the fat-rich diet become severely obese.

LABORATORY-BASED
IN VITRO STUDIES

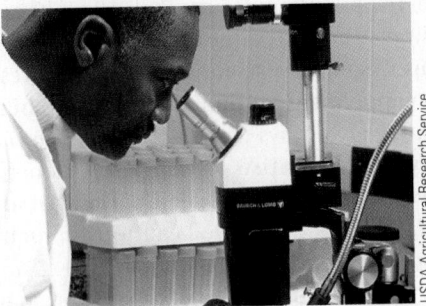

Researchers examine the effects of a specific variable on a tissue, cell, or molecule isolated from a living organism.

Example. Laboratory studies find that fish oils inhibit the growth and activity of the bacteria implicated in ulcer formation.

HUMAN INTERVENTION
(OR CLINICAL) TRIALS

Researchers ask people to adopt a new behavior (for example, eat a citrus fruit, take a vitamin C supplement, or exercise daily). These trials help determine the effectiveness of such interventions on the development or prevention of disease.

Example. Heart disease risk factors improve when men drink fresh-squeezed orange juice daily for 2 months compared with those on a diet low in vitamin C—even when both groups follow a diet high in saturated fat.

Double Blind When both the subjects and the researchers do not know which subjects are in which group, the study is called a **double-blind experiment.** Being fallible human beings and having an emotional and sometimes financial investment in a successful outcome, researchers might record and interpret results with a bias in the expected direction. To prevent such bias, the pills are coded by a third party, who does not reveal to the experimenters which subjects are in which group until all results have been recorded.

Analyzing Research Findings Research findings must be analyzed and interpreted with an awareness of each study's limitations. Scientists must be cautious about drawing any conclusions until they have accumulated a body of evidence from multiple studies that have used various types of research designs. As evidence accumulates, scientists begin to develop a **theory** that integrates the various findings and explains the complex relationships.

Correlations and Causes Researchers often examine the relationships between two or more **variables**—for example, daily vitamin C intake and the number of colds or the duration and severity of cold symptoms. Importantly, researchers must be able to observe, measure, or verify the variables selected. Findings sometimes suggest no **correlation** between variables (regardless of the amount of vitamin C consumed, the number of colds remains the same). Other times, studies find either a **positive correlation** (the more vitamin C, the more colds) or a **negative correlation** (the more vitamin C, the fewer colds). Notice that in a positive correlation, both variables change in the same direction, regardless of whether the direction is "more" or "less"—"the more vitamin C, the more colds" is a positive correlation, just as is "the less vitamin C, the fewer colds." In a negative correlation, the two variables change in opposite directions: "the less vitamin C, the more colds" or "the more vitamin C, the fewer colds." Also notice that a positive correlation does not necessarily reflect a desired outcome, nor does a negative correlation always reflect an unwanted outcome.

Correlational evidence proves only that variables are associated, not that one is the cause of the other. To actually prove that A causes B, scientists have to find evidence of the *mechanism*—that is, an explanation of how A might cause B.

Cautious Conclusions When researchers record and analyze the results of their experiments, they must exercise caution in their interpretation of the findings. For example, in an epidemiological study, scientists may use a specific segment of the population—say, men 18 to 30 years old. When the scientists draw conclusions, they are careful not to generalize the findings to men and women of all ages. Similarly, scientists performing research studies using animals are cautious in applying their findings to human beings. Conclusions from any one research study are always tentative and take into account findings from studies conducted by other scientists as well. As evidence accumulates, scientists gain confidence about making recommendations that affect people's health and lives. Still, their statements are worded cautiously, such as "A diet high in fruits and vegetables *may* protect against *some* cancers."

Quite often, as scientists approach an answer to one research question, they raise several more questions, so future research projects are never lacking. Further scientific investigation then seeks to answer questions, such as "What substance or substances within fruits and vegetables provide protection?" If those substances turn out to be the vitamins found so abundantly in fresh produce, then "How much is needed to offer protection?" "How do these vitamins protect against cancer?" "Is it their action as antioxidant nutrients?" "If not, might it be another action or even another substance that accounts for the protection fruits and vegetables provide against cancer?" (Highlight 11 explores the answers to these questions and reviews recent research on antioxidant nutrients and disease.)

TABLE 1-3 Parts of a Research Article

- *Abstract:* The abstract provides a brief overview of the article.
- *Introduction:* The introduction clearly states the purpose of the current study and provides a comprehensive review of the relevant literature.
- *Methodology:* The methodology section defines key terms and describes the study design, subjects, and procedures used in conducting the study.
- *Results:* The results report the findings and may include tables and figures that summarize the information.
- *Discussion*: The discussion draws tentative conclusions that are supported by the data and reflect the original purpose as stated in the introduction. Usually, it answers a few questions and raises several more.
- *References*: The references reflect the investigator's knowledge of the subject and should include an extensive list of relevant studies (including key studies several years old as well as current ones).

© Cengage Learning

Publishing Research The findings from a research study are submitted to a board of reviewers composed of other scientists who rigorously evaluate the study to ensure that the scientific method was followed—a process known as **peer review.** The reviewers critique the study's hypothesis, methodology, statistical significance, and conclusions. They also note the funding source, recognizing that financial support may bias scientific conclusions. If the reviewers consider the conclusions to be well supported by the evidence—that is, if the research has **validity**—they endorse the work for publication in a scientific journal where others can read it. This raises an important point regarding information found on the Internet: much gets published without the rigorous scrutiny of peer review. Consequently, readers must assume greater responsibility for examining the data and conclusions presented—often without the benefit of journal citations. Highlight 1 offers guidance in determining whether website information is reliable. Table 1-3 describes the parts of a typical research article.

Even when a new finding is published or released to the media, it is still only preliminary and not very meaningful by itself. Other scientists will need to confirm or disprove the findings through **replication.** To be accepted into the body of nutrition knowledge, a finding must stand up to rigorous, repeated testing in experiments performed by several different researchers. What we "know" in nutrition results from years of replicating study findings. Communicating the latest finding in its proper context without distorting or oversimplifying the message is a challenge for scientists and journalists alike. For a helpful scientific overview of current topics in nutrition, look for review articles in scholarly journals such as *Nutrition Reviews.* Similar to a review article, a meta-analysis study uses the power of a computer to combine and reanalyze the results of many previously published studies on a single topic. Keep in mind that a meta-analysis study is useful in providing an overview of the averages, but its results may not apply to all individuals or cases.[18]

With each report from scientists, the field of nutrition changes a little—each finding contributes another piece to the whole body of knowledge. People who know how science works understand that single findings, like single frames in a movie, are just small parts of a larger story. Over years, the picture of what is "true" in nutrition gradually changes, and dietary recommendations change to reflect the current understanding of scientific research. Highlight 5 provides a detailed look at how dietary fat recommendations have evolved over the past several decades as researchers have uncovered the relationships between the various kinds of fat and their roles in supporting or harming health.

> **REVIEW IT** Explain the scientific method and how scientists use various types of research studies and methods to acquire nutrition information.

Scientists learn about nutrition by conducting experiments that follow the protocol of scientific research. In designing their studies, researchers randomly assign control and experimental groups, seek large sample sizes, provide placebos, and remain blind to treatments. Their findings must be reviewed and replicated by other scientists before being accepted as valid.

© 2016 Cengage Learning

Knowledge about the nutrients and their effects on health comes from scientific studies.

The characteristics of well-designed research have enabled scientists to study the actions of nutrients in the body. Such research has laid the foundation for quantifying how much of each nutrient the body needs.

1.4 Dietary Reference Intakes

> **LEARN IT** Define the four categories of the DRI and explain their purposes.

Using the results of thousands of research studies, nutrition experts have produced a set of standards that define the amounts of energy, nutrients, other dietary components, and physical activity that best support health. These recommendations are called **Dietary Reference Intakes (DRI),** and they reflect the collaborative efforts of researchers in both the United States and Canada.*[19] The inside front cover of this book provides a handy reference for DRI values.

Establishing Nutrient Recommendations The DRI Committee consists of highly qualified scientists who base their estimates of nutrient needs on careful examination and interpretation of scientific evidence. These recommendations apply to healthy people and may not be appropriate for people with diseases that increase or decrease nutrient needs. The next several paragraphs introduce the four categories of the DRI, explain their purposes, and discuss specific aspects of how the committee goes about establishing these values:

- Estimated Average Requirements (EAR)
- Recommended Dietary Allowances (RDA)
- Adequate Intakes (AI)
- Tolerable Upper Intake Levels (UL)

Estimated Average Requirements (EAR) The committee reviews hundreds of research studies to determine the **requirement** for a nutrient—how much is needed in the diet. The committee selects a different criterion for each nutrient based on its roles in supporting various activities in the body and in reducing disease risks.

An examination of all the available data reveals that each person's body is unique and has its own set of requirements. Men differ from women, and needs change as people grow from infancy through old age. For this reason, the committee clusters its recommendations for people into groups based on gender and age. Even so, the exact requirements for people of the same gender and age are likely to be different. Person A might need 40 units of a particular nutrient each day; person B might need 35; and person C might need 57. Looking at enough people might reveal that their individual requirements fall into a symmetrical distribution, with most near the midpoint and only a few at the extremes (see the left side of Figure 1-5). Using this information, the committee determines an **Estimated Average Requirement (EAR)** for each nutrient—the average amount that appears sufficient for half of the population. In Figure 1-5, the EAR is shown as 45 units.

Recommended Dietary Allowances (RDA) Once a nutrient *requirement* is established, the committee must decide what intake to *recommend* for everybody—the **Recommended Dietary Allowance (RDA).** As you can see by the distribution in Figure 1-5, the EAR (shown in the figure as 45 units) is probably closest to everyone's need. If people consumed exactly the average requirement of a given nutrient each day, however, approximately half of the population would develop deficiencies of that nutrient—in Figure 1-5, for example, person C would be among them. Recommendations are therefore set greater than the EAR to meet the needs of most healthy people.

Small amounts greater than the daily requirement do no harm, whereas amounts less than the requirement may lead to health problems. When people's

Dietary Reference Intakes (DRI): a set of nutrient intake values for healthy people in the United States and Canada. These values are used for planning and assessing diets and include:

- Estimated Average Requirements (EAR)
- Recommended Dietary Allowances (RDA)
- Adequate Intakes (AI)
- Tolerable Upper Intake Levels (UL)

requirement: the lowest continuing intake of a nutrient that will maintain a specified criterion of adequacy.

Estimated Average Requirement (EAR): the average daily amount of a nutrient that will maintain a specific biochemical or physiological function in half the healthy people of a given age and gender group.

Recommended Dietary Allowance (RDA): the average daily amount of a nutrient considered adequate to meet the known nutrient needs of practically all healthy people; a goal for dietary intake by individuals.

---

*The DRI reports are produced by the Food and Nutrition Board, Institute of Medicine of the National Academies, with active involvement of scientists from the United States and Canada.

> FIGURE 1-5 **Estimated Average Requirements (EAR) and Recommended Dietary Allowances (RDA) Compared**

Each square in the graphs below represents a person with unique nutritional requirements. (The text discusses three of these people—A, B, and C.) Some people require only a small amount of nutrient X and some require a lot. Most people, however, fall somewhere in the middle.

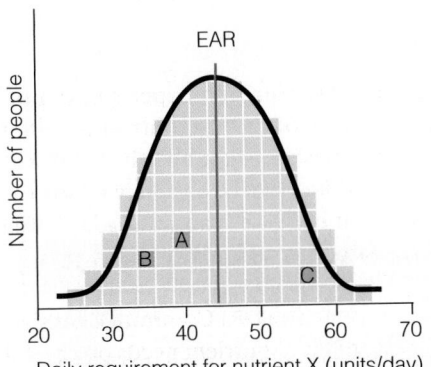

The Estimated Average Requirement (EAR) for a nutrient is the amount that meets the needs of about half of the population (shown here by the red line).

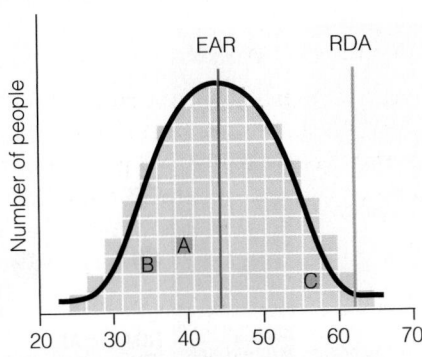

The Recommended Dietary Allowance (RDA) for a nutrient (shown here in green) is set well above the EAR, meeting the needs of about 98% of the population.

© Cengage Learning

nutrient intakes are consistently **deficient** (less than the requirement), their nutrient stores decline, and over time this decline leads to poor health and deficiency symptoms. Therefore, to ensure that the nutrient RDA meet the needs of as many people as possible, the RDA are set near the top end of the range of the population's estimated requirements.

In this example, a reasonable RDA might be 63 units a day (see the right side of Figure 1-5). Such a point can be calculated mathematically so that the needs of about 98 percent of a population are included. Almost everybody—including person C whose needs were more substantial than the average—would consume enough of the nutrient if they met this dietary goal. Relatively few people's requirements would exceed this recommendation, and even then, they wouldn't exceed by much.

Adequate Intakes (AI) For some nutrients, such as vitamin K, there is insufficient scientific evidence to determine an EAR (which is needed to set an RDA). In these cases, the committee establishes an **Adequate Intake (AI)** instead of an RDA. An AI reflects the average amount of a nutrient that a group of healthy people consumes. Like the RDA, the AI may be used as nutrient goals for individuals.

Although both the RDA and the AI serve as nutrient intake goals for individuals, their differences are noteworthy. An RDA for a given nutrient is based on enough scientific evidence to expect that the needs of almost all healthy people will be met. An AI, on the other hand, must rely more heavily on scientific judgments because sufficient evidence is lacking. For this reason, AI values are more tentative than RDA values. The table on the inside front cover identifies which nutrients have an RDA and which have an AI. Later chapters present the RDA and AI values for vitamins and minerals.

Tolerable Upper Intake Levels (UL) As mentioned earlier, the recommended intakes for nutrients are generous, yet they may not be sufficient for every individual for every nutrient. Nevertheless, it is probably best not to exceed these recommendations by very much or very often. Individual tolerances for high doses of nutrients vary, and somewhere beyond the recommended intake is a point beyond which a nutrient is likely to become toxic. This point is known as the **Tolerable Upper Intake Level (UL).** It is naïve—and inaccurate—to think of recommendations as minimum amounts. A more accurate view is to see a person's nutrient

deficient: inadequate; a nutrient amount that fails to meet the body's needs and eventually results in deficiency symptoms.

Adequate Intake (AI): the average daily amount of a nutrient that appears sufficient to maintain a specified criterion; a value used as a guide for nutrient intake when an RDA cannot be determined.

Tolerable Upper Intake Level (UL): the maximum daily amount of a nutrient that appears safe for most healthy people and beyond which there is an increased risk of adverse health effects.

> FIGURE 1-6 **Inaccurate versus Accurate View of Nutrient Intakes**

The RDA (or AI) for a given nutrient represents a point that lies within a range of appropriate and reasonable intakes between toxicity and deficiency. Both of these recommendations are high enough to provide reserves in times of short-term dietary inadequacies, but not so high as to approach toxicity. Nutrient intakes above or below this range may be equally harmful.

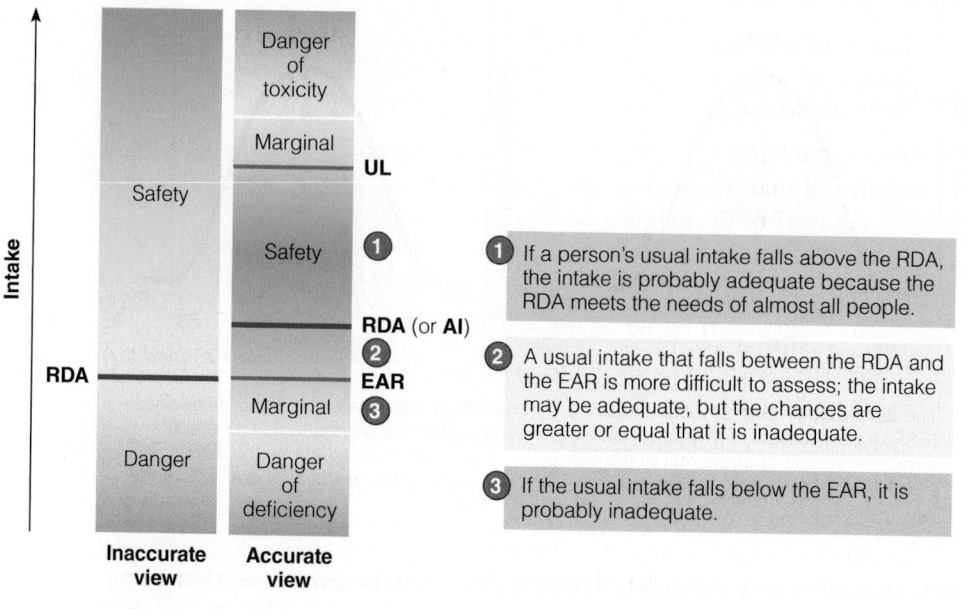

needs as falling within a range, with marginal and danger zones at each end for intakes that are either inadequate or excessive (see Figure 1-6).

Paying attention to upper levels is particularly useful in guarding against the overconsumption of nutrients, which may occur when people use large-dose dietary supplements and fortified foods regularly. Later chapters discuss the dangers associated with excessively high intakes of vitamins and minerals, and the inside front cover (p. C) presents tables of upper levels for selected nutrients.

Establishing Energy Recommendations

In contrast to the RDA and AI values for nutrients, the recommendation for energy is not generous. Excess energy cannot be readily excreted and is eventually stored as body fat. These reserves may be beneficial when food is scarce, but they can also lead to obesity and its associated health consequences.

Estimated Energy Requirement (EER) The energy recommendation—called the **Estimated Energy Requirement (EER)**—represents the average dietary energy intake (kcalories per day) that will maintain energy balance in a person who has a healthy body weight and level of physical activity. Balance is key to the energy recommendation. Enough food energy is needed to sustain a healthy and active life, but too much can lead to weight gain and obesity. Because *any* amount in excess of energy needs will result in weight gain, no upper level for energy has been determined.

Acceptable Macronutrient Distribution Ranges (AMDR) People don't eat energy directly; they derive energy from foods containing carbohydrates, fats, and proteins. Each of these three energy-yielding nutrients contributes to the total energy intake, and those contributions vary in relation to one another. The DRI committee has determined that the composition of a diet that provides adequate energy and nutrients and reduces the risk of chronic diseases is:

- 45 to 65 percent kcalories from carbohydrate
- 20 to 35 percent kcalories from fat
- 10 to 35 percent kcalories from protein

Estimated Energy Requirement (EER): the average dietary energy intake that maintains energy balance and good health in a person of a given age, gender, weight, height, and level of physical activity.

These values are known as **Acceptable Macronutrient Distribution Ranges (AMDR).**

Using Nutrient Recommendations

Although the intent of nutrient recommendations seems simple, they are the subject of much misunderstanding and controversy. Perhaps the following facts will help put them in perspective:

1. Estimates of adequate energy and nutrient intakes apply to *healthy* people. They need to be adjusted for malnourished people or those with medical problems who may require supplemented or restricted dietary intakes.

2. *Recommendations* are not minimum requirements, nor are they necessarily optimal intakes for all individuals. Recommendations can target only "most" of the people and cannot account for individual variations in nutrient needs—yet. Given the recent explosion of knowledge about genetics, the day may be fast approaching when nutrition scientists will be able to determine an individual's optimal nutrient needs. Until then, qualified health-care professionals can help determine if recommendations should be adjusted to meet individual needs. (Highlight 1 introduces the college-educated food and nutrition specialists who are qualified to evaluate people's nutritional health and needs.)

3. Most nutrient goals are intended to be met through diets composed of a variety of *foods* whenever possible. Because foods contain mixtures of nutrients and nonnutrients, they deliver more than just those nutrients covered by the recommendations. Excess intakes of vitamins and minerals are unlikely when they come from foods. Using dietary supplements to meet nutrient goals raises the risks of toxicity.

4. Recommendations apply to *average* daily intakes. Trying to meet the recommendations for every nutrient every day is difficult and unnecessary. The length of time over which a person's intake can deviate from the average without risk of deficiency or overdose varies for each nutrient, depending on how the body uses and stores the nutrient. For most nutrients (such as thiamin and vitamin C), deprivation would lead to rapid development of deficiency symptoms (within days or weeks); for others (such as vitamin A and vitamin B$_{12}$), deficiencies would develop more slowly (over months or years).

5. Each of the DRI categories serves a unique purpose. For example, the EAR are most appropriately used to develop and evaluate nutrition programs for *groups* such as schoolchildren or military personnel. The RDA (or AI if an RDA is not available) can be used to set goals for *individuals*. The UL serve as a reminder to keep nutrient intakes less than amounts that increase the risk of toxicity—not a common problem when nutrients derive from foods, but a real possibility for some nutrients if supplements are used regularly. With these understandings, professionals can use the DRI for a variety of purposes.[20]

Comparing Nutrient Recommendations

At least 40 different nations and international organizations have published nutrient standards similar to those used in the United States. Slight differences may be apparent, reflecting differences both in the interpretation of the data from which the standards were derived and in the food habits and physical activities of the populations they serve.

Many countries use the recommendations developed by two international groups: FAO (Food and Agriculture Organization) and WHO (World Health Organization). The FAO/WHO nutrient recommendations are considered sufficient to maintain health in nearly all healthy people worldwide and are provided in Appendix I.

© Photodisc/Getty Images

The DRI "alphabet soup" of nutrient intake standards makes sense when you learn their purposes.

Acceptable Macronutrient Distribution Ranges (AMDR): ranges of intakes for the energy nutrients that provide adequate energy and nutrients and reduce the risk of chronic diseases.

> **REVIEW IT** Define the four categories of the DRI and explain their purposes.
The Dietary Reference Intakes (DRI) are a set of nutrient intake values that can be used to plan and evaluate diets for healthy people. The Estimated Average Requirement (EAR) defines the amount of a nutrient that supports a specific function in the body for half of the population. The Recommended Dietary Allowance (RDA) is based on the Estimated Average Requirement and establishes a goal for dietary intake that will meet the needs of almost all healthy people. An Adequate Intake (AI) serves a similar purpose when an RDA cannot be determined. The Estimated Energy Requirement (EER) defines the average amount of energy intake needed to maintain energy balance, and the Acceptable Macronutrient Distribution Ranges (AMDR) define the proportions contributed by carbohydrate, fat, and protein to a healthy diet. The Tolerable Upper Intake Level (UL) establishes the highest amount that appears safe for regular consumption.

1.5 Nutrition Assessment

> **LEARN IT** Explain how the four assessment methods are used to detect energy and nutrient deficiencies and excesses.

What happens when a person doesn't consume enough or consumes too much of a specific nutrient or energy? If the deficiency or excess is significant over time, the person experiences symptoms of **malnutrition**. With a deficiency of energy, the person may display the symptoms of **undernutrition** by becoming extremely thin, losing muscle tissue, and becoming prone to infection and disease. With a deficiency of a nutrient, the person may experience skin rashes, depression, hair loss, bleeding gums, muscle spasms, night blindness, or other symptoms. Similarly, over time, regular intakes in excess of needs may also have adverse effects. With an excess of energy, the person may become obese and vulnerable to diseases associated with **overnutrition**, such as heart disease and diabetes. With a sudden nutrient overdose, the person may experience hot flashes, yellowing skin, a rapid heart rate, low blood pressure, or other symptoms.

Malnutrition symptoms—such as diarrhea, skin rashes, and fatigue—are easy to miss because they resemble the symptoms of other diseases. But a person who has learned how to use assessment techniques to detect malnutrition can identify when these conditions are caused by poor nutrition and can recommend steps to correct it. This discussion presents the basics of nutrition assessment; many more details are offered in later chapters and in Appendix E.

Nutrition Assessment of Individuals
To prepare a **nutrition assessment**, a trained health-care professional uses:

- Historical information
- Anthropometric measurements
- Physical examinations
- Laboratory tests

Each of these methods involves collecting data in various ways and interpreting each finding in relation to the others to create a total picture.

Historical Information
One step in evaluating nutrition status is to obtain information about a person's history with respect to health status, socioeconomic status, drug use, and diet. The health history reflects a person's medical record and may reveal a disease that interferes with the person's ability to eat or the body's use of nutrients. The person's family history of major diseases is also noteworthy, especially for conditions such as heart disease that have a genetic tendency to run in families. Economic circumstances may show a financial inability to buy foods or inadequate kitchen facilities in which to prepare them. Social factors such as marital status, ethnic background, and educational level also influence food choices and nutrition status. A drug history, including all prescribed and

malnutrition: any condition caused by excess or deficient food energy or nutrient intake or by an imbalance of nutrients.

- **mal** = bad

undernutrition: deficient energy or nutrients.

overnutrition: excess energy or nutrients.

nutrition assessment: a comprehensive analysis of a person's nutrition status that uses health, socioeconomic, drug, and diet histories; anthropometric measurements; physical examinations; and laboratory tests.

over-the-counter medications, may highlight possible interactions that lead to nutrient deficiencies (as described in Highlight 17). A diet history that examines a person's intake of foods, beverages, and dietary supplements may reveal either a surplus or inadequacy of nutrients or energy.

To take a diet history, the assessor collects data about the foods a person eats. The data may be collected by recording the foods the person has eaten over a period of 24 hours, 3 days, or a week or more or by asking what foods the person typically eats and how much of each. The days in the record must be fairly typical of the person's diet, and portion sizes must be recorded accurately. To determine the amounts of nutrients consumed, the assessor usually enters the foods and their portion sizes into a computer using a diet analysis program. This step can also be done manually by looking up each food in a table of food composition such as Appendix H in this book. The assessor then compares the calculated nutrient intakes with the DRI to determine the probability of adequacy. Alternatively, the diet history might be compared against standards such as the USDA Food Patterns or *Dietary Guidelines* (described in Chapter 2).

An estimate of energy and nutrient intakes from a diet history, when combined with other sources of information, can help confirm or rule out the *possibility* of suspected nutrition problems. A sufficient intake of a nutrient does not guarantee adequacy, and an insufficient intake does not always indicate a deficiency. Such findings, however, warn of possible problems.

Anthropometric Measurements A second technique that may help to reveal nutrition problems is taking **anthropometric** measures such as height and weight. The assessor compares a person's measurements with standards specific for gender and age or with previous measures on the same individual. (Chapter 8 presents information on body weight and its standards, and Appendix E includes growth charts for children.)

Measurements taken periodically and compared with previous measurements reveal patterns and indicate trends in a person's overall nutrition status, but they provide little information about specific nutrients. Instead, measurements out of line with expectations may reveal such problems as growth failure in children, wasting or swelling of body tissues in adults, and obesity—conditions that may reflect energy or nutrient deficiencies or excesses.

Physical Examinations A third nutrition assessment technique is a physical examination looking for clues to poor nutrition status. Visual inspection of the hair, eyes, skin, posture, tongue, and fingernails can provide such clues. In addition, information gathered from an interview can help identify symptoms. The examination requires skill because many physical signs and symptoms reflect more than one nutrient deficiency or toxicity—or even nonnutrition conditions. Like the other assessment techniques, a physical examination alone does not yield firm conclusions. Instead, physical examinations reveal possible imbalances that must be confirmed by other assessment techniques, or they confirm results from other assessment measures.

Laboratory Tests A fourth way to detect a developing deficiency, imbalance, or toxicity is to take samples of blood or urine, analyze them in the laboratory, and compare the results with normal values for a similar population. Laboratory tests are most useful in uncovering early signs of malnutrition before symptoms appear. In addition, they can confirm suspicions raised by other assessment methods.

Iron, for Example The mineral iron can be used to illustrate the stages in the development of a nutrient deficiency and the assessment techniques useful in detecting them. The **overt**, or outward, signs of an iron deficiency appear at the end of a long sequence of events. Figure 1-7 on p. 24 describes what happens in the body as a nutrient deficiency progresses and shows which assessment methods can reveal those changes.

anthropometric (AN-throw-poe-MET-rick): relating to measurement of the physical characteristics of the body, such as height and weight.
- **anthropos** = human
- **metric** = measuring

overt (oh-VERT): out in the open and easy to observe.
- **ouvrir** = to open

> FIGURE 1-7 Stages in the Development of a Nutrient Deficiency

Internal changes precede outward signs of deficiencies. Outward signs of sickness, however, need not appear before a person takes corrective measures. Laboratory tests can help determine nutrient status in the early stages.

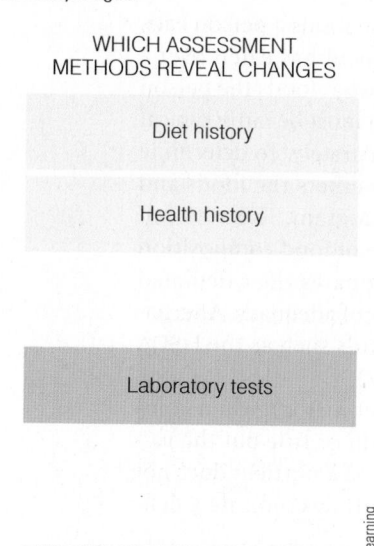

WHAT HAPPENS IN THE BODY

Primary deficiency caused by inadequate diet
or
Secondary deficiency caused by problem inside the body

Declining nutrient stores (subclinical)
and
Abnormal functions inside the body (covert)

Physical signs and symptoms (overt)

WHICH ASSESSMENT METHODS REVEAL CHANGES

Diet history

Health history

Laboratory tests

Physical examination and anthropometric measures

© Cengage Learning

First, the body has too little iron—either because iron is lacking in the person's diet (a **primary deficiency**) or because the person's body doesn't absorb enough, excretes too much, or uses iron inefficiently (a **secondary deficiency**). A diet history provides clues to primary deficiencies; a health history provides clues to secondary deficiencies.

Next, the body begins to use up its stores of iron. At this stage, the deficiency might be described as a **subclinical deficiency**. It exists as a **covert** condition, and although it might be detected by laboratory tests, outward signs are not yet apparent.

Finally, the body's iron stores are exhausted. Now, it cannot make enough iron-containing red blood cells to replace those that are aging and dying. Iron is needed in red blood cells to carry oxygen to all the body's tissues. When iron is lacking, fewer red blood cells are made, the new ones are pale and small, and every part of the body feels the effects of oxygen shortage. At this point in time, the overt symptoms of deficiency appear—weakness, fatigue, pallor, and headaches, reflecting the iron-deficient state of the blood. A physical examination and interview will reveal these symptoms.

Nutrition Assessment of Populations To assess a population's nutrition status, researchers conduct surveys using techniques similar to those used on individuals. The data collected are then used by various agencies for numerous purposes, including the development of national health goals.

National Nutrition Surveys National nutrition surveys gather information about the population's dietary, nutritional, and related health status. One

primary deficiency: a nutrient deficiency caused by inadequate dietary intake of a nutrient.

secondary deficiency: a nutrient deficiency caused by something other than an inadequate intake such as a disease condition or drug interaction that reduces absorption, accelerates use, hastens excretion, or destroys the nutrient.

subclinical deficiency: a deficiency in the early stages, before the outward signs have appeared.

covert (KOH-vert): hidden, as if under covers.

• **couvrir** = to cover

A peek inside the mouth provides clues to a person's nutrition status. An inflamed tongue may indicate a deficiency of one of the B vitamins, and mottled teeth may reveal fluoride toxicity, for example.

© Blend Images/Alamy

survey collects data on the kinds and amounts of foods people eat.* Another survey examines the people themselves, using anthropometric measurements, physical examinations, and laboratory tests.** The data provide valuable information on several nutrition-related conditions, such as growth retardation, heart disease, and nutrient deficiencies. National nutrition surveys often oversample high-risk groups (low-income families, pregnant women, adolescents, the elderly, African Americans, and Mexican Americans) to glean an accurate estimate of their health and nutrition status.

The resulting wealth of information from the national nutrition surveys is used for a variety of purposes. For example, Congress uses this information to establish public policy on nutrition education, food assistance programs, and the regulation of the food supply. Scientists use the information to establish research priorities. The food industry uses these data to guide decisions in public relations and product development. The Dietary Reference Intakes and other major reports that examine the relationships between diet and health depend on information collected from these nutrition surveys. These data also provide the basis for developing and monitoring national health goals.

National Health Goals The **Healthy People** program sets priorities and guides policies that "increase the quality and years of healthy life" and "eliminate health disparities." At the start of each decade, the program sets goals for improving the nation's health during the next ten years. Nutrition is one of many topic areas, each with numerous objectives. Table 1-4 lists the nutrition and weight status

National surveys provide valuable information about the kinds of foods people eat.

TABLE 1-4 Healthy People 2020 Nutrition and Weight Status Objectives

- Increase the proportion of adults who are at a healthy weight
- Reduce the proportion of adults who are obese
- Reduce iron deficiency among young children and females of childbearing age
- Reduce iron deficiency among pregnant females
- Reduce the proportion of children and adolescents who are overweight or obese
- Increase the contribution of fruits to the diets of the population aged 2 years and older
- Increase the variety and contribution of vegetables to the diets of the population aged 2 years and older
- Increase the contribution of whole grains to the diets of the population aged 2 years and older
- Reduce consumption of saturated fat in the population aged 2 years and older
- Reduce consumption of sodium in the population aged 2 years and older
- Increase consumption of calcium in the population aged 2 years and older
- Increase the proportion of worksites that offer nutrition or weight management classes or counseling
- Increase the proportion of physician office visits that include counseling or education related to nutrition or weight
- Eliminate very low food security among children in US households
- Prevent inappropriate weight gain in youth and adults
- Increase the proportion of primary care physicians who regularly measure the body mass index (BMI) of their patients
- Reduce consumption of kcalories from solid fats and added sugars in the population aged 2 years and older
- Increase the number of states that have state-level policies that incentivize food retail outlets to provide foods that are encouraged by the *Dietary Guidelines*
- Increase the number of states with nutrition standards for foods and beverages provided to preschool-aged children in childcare
- Increase the percentage of schools that offer nutritious foods and beverages outside of school meals

NOTE: Nutrition and Weight Status is one of 38 topic areas, each with numerous objectives. Several of the other topic areas have nutrition-related objectives, and these are presented in Appendix J.

SOURCE: www.healthypeople.gov

Healthy People: a national public health initiative under the jurisdiction of the US Department of Health and Human Services (DHHS) that identifies the most significant preventable threats to health and focuses efforts toward eliminating them.

---

*This survey is called *What We Eat in America*.
**This survey is known as the National Health and Nutrition Examination Survey (NHANES).

objectives for 2020, and Appendix J lists nutrition-related objectives from other topic areas.

Progress in meeting the 2010 goals was mixed. A few objectives were met, about half made some progress, and several showed no progress—or even moved in the wrong direction.[21] The objective to reduce average blood cholesterol levels was achieved, but objectives to eat more fruits, vegetables, and whole grains and to increase physical activity showed little or no improvement. Trends in overweight and obesity actually worsened. Clearly, "what we eat in America" must change if we hope to meet the Healthy People goals.

National Trends What do we eat in America and how has it changed over the past 45 years? The short answer to both questions is "a lot." We eat more meals away from home, particularly at fast-food restaurants. We eat larger portions. We drink more sweetened beverages and eat more energy-dense, nutrient-poor foods such as candy and chips. We snack frequently. As a result of these dietary habits, our energy intake has risen and, consequently, so has the incidence of overweight and obesity. Overweight and obesity, in turn, profoundly influence our health—as the next section explains.

> **REVIEW IT** Explain how the four assessment methods are used to detect energy and nutrient deficiencies and excesses.

People become malnourished when they get too little or too much energy or nutrients. Deficiencies, excesses, and imbalances of nutrients lead to malnutrition diseases. To detect malnutrition in individuals, health-care professionals use a combination of four nutrition assessment methods. Reviewing historical information on diet and health may suggest a possible nutrition problem. Laboratory tests may detect a possible nutrition problem in its earliest stages, whereas anthropometric measurements and physical examinations pick up on the problem only after it causes symptoms. National surveys use similar assessment methods to measure people's food consumption and to evaluate the nutrition status of populations.

1.6 Diet and Health

> **LEARN IT** Identify several risk factors and explain their relationships to chronic diseases.

Foods play a vital role in supporting health. Early nutrition research focused on identifying the nutrients in foods that would prevent such common diseases as rickets and scurvy, the vitamin D– and vitamin C–deficiency diseases. With this knowledge, developed countries have successfully defended against nutrient deficiency diseases. World hunger and nutrient deficiency diseases still pose a major health threat in developing countries, however, but not because of a lack of nutrition knowledge (as Chapter 20 explains). More recently, nutrition research has focused on chronic diseases associated with energy and nutrient excesses. Chronic diseases are responsible for 7 out of 10 deaths among US adults. Once thought to be "rich countries' problems," chronic diseases have now become epidemic in developing countries as well—contributing to three out of five deaths worldwide.[22]

Chronic Diseases Table 1-5 lists the ten leading causes of death in the United States. These "causes" are stated as if a single condition such as heart disease caused death, but most chronic diseases arise from multiple factors over many years. A person who died of heart disease may have been overweight, had high blood pressure, been a cigarette smoker, and spent years eating a diet high in saturated fat and getting too little exercise.

Of course, not all people who die of heart disease fit this description, nor do all people with these characteristics die of heart disease. People who are overweight might die from the complications of diabetes instead, or those who smoke

TABLE 1-5 Leading Causes of Death in the United States

	Percentage of Total Deaths
1. **Heart disease**	23.7
2. **Cancers**	22.9
3. Chronic lung diseases	5.7
4. **Strokes**	5.1
5. Accidents	4.9
6. Alzheimer's disease	3.4
7. **Diabetes mellitus**	2.9
8. Pneumonia and influenza	2.1
9. Kidney disease	1.8
10. Suicide	1.5

NOTE: The diseases highlighted in bold have relationships with diet.

SOURCE: Deaths: Preliminary data for 2011, *National Vital Statistics Reports*, October 10, 2012, Centers for Disease Control and Prevention, www.cdc.gov/nchs.

© Cengage Learning

might die of cancer. They might even die from something totally unrelated to any of these factors, such as an automobile accident. Still, statistical studies have shown that certain conditions and behaviors are linked to certain diseases.

Table 1-5 highlights four of the top seven causes of death as having a link with diet. Notice that these four diseases—heart disease, cancers, strokes, and diabetes—account for more than half of the deaths each year.

Risk Factors for Chronic Diseases Factors that increase or reduce the *risk* of developing chronic diseases can be identified by analyzing statistical data. A strong association between a **risk factor** and a disease means that when the factor is present, the *likelihood* of developing the disease increases. It does not mean that all people with the risk factor will develop the disease. Similarly, a lack of risk factors does not guarantee freedom from a given disease. On the average, though, the more risk factors in a person's life, the greater that person's chances of developing the disease. Conversely, the fewer risk factors in a person's life, the better the chances for good health.

Risk Factors Persist Risk factors tend to persist over time. Without intervention, a young adult with high blood pressure will most likely continue to have high blood pressure as an older adult, for example. Thus, to minimize the damage, early intervention is most effective.

Risk Factors Cluster Risk factors tend to cluster. For example, a person who is obese may be physically inactive, have high blood pressure, and have high blood cholesterol—all risk factors associated with heart disease. Multiple risk factors act synergistically to increase the risk of disease dramatically. Intervention that focuses on one risk factor often benefits the others as well. For example, physical activity can help reduce weight. Physical activity and weight loss will, in turn, help to lower blood pressure and blood cholesterol.

Risk Factors in Perspective The most prominent factor—contributing to one of every five deaths each year in the United States—is tobacco use, followed closely by diet and activity patterns, and then alcohol use (see Table 1-6). Risk factors such as smoking, poor dietary habits, physical inactivity, and alcohol consumption are personal behaviors that can be changed. Decisions to not smoke, to eat a well-balanced diet, to engage in regular physical activity, and to drink alcohol in moderation (if at all) improve the likelihood that a person will enjoy good health. Other risk factors, such as genetics, gender, and age, also play important roles in the development of chronic diseases, but they cannot be changed. Health recommendations acknowledge the influence of such factors on the development of disease, but they must focus on the factors that are changeable.

Health Behaviors in the United States Despite evidence linking certain behaviors with chronic diseases, many Americans continue to engage in unhealthy behaviors.[23] An estimated 20 percent of US adults consume five or more drinks in a single day at least once a year; 20 percent are cigarette smokers; 40 percent are physically inactive; 60 percent are either overweight or obese; and 30 percent average 6 hours or less of sleep per day.[24] For the two out of three Americans who do not smoke or drink alcohol excessively, the one choice that can influence long-term health prospects more than any other is diet.

> **REVIEW IT** Identify several risk factors and explain their relationships to chronic diseases.
Within the range set by genetics, a person's choice of diet influences long-term health. Diet has no influence on some diseases but is linked closely to others. Personal life choices, such as engaging in physical activity and using tobacco or alcohol, also affect health for the better or worse.

TABLE 1-6 **Factors Contributing to Deaths in the United States**

Factors	Percentage of Deaths
Tobacco	18
Poor diet/inactivity	15
Alcohol	4
Microbial agents	3
Toxic agents	2
Motor vehicles	2
Firearms	1
Sexual behavior	<1
Illicit drugs	<1

© Cengage Learning

SOURCE: A. H. Mokdad and coauthors, Actual causes of death in the United States, 2000, *Journal of the American Medical Association* 291 (2004): 1238–1245, with corrections from *Journal of the American Medical Association* 293 (2005): 298.

risk factor: a condition or behavior associated with an elevated frequency of a disease but not proved to be causal. Leading risk factors for chronic diseases include obesity, cigarette smoking, high blood pressure, high blood cholesterol, physical inactivity, and a diet high in saturated fats and low in vegetables, fruits, and whole grains.

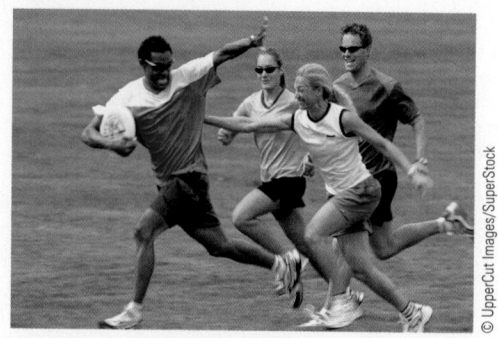

Physical activity can be both fun and beneficial.

The next several chapters provide many more details about nutrients and how they support health. Whenever appropriate, the discussion shows how diet influences each of today's major diseases. Dietary recommendations appear again and again, as each nutrient's relationships with health are explored. Most people who follow the recommendations will benefit and can enjoy good health into their later years.

Nutrition Portfolio

Each chapter in this book ends with simple Nutrition Portfolio activities that invite you to review key messages and consider whether your personal choices are meeting the dietary goals introduced in the text. By using the information you are recording in Diet & Wellness Plus (the dietary tracking software that accompanies this text) and keeping a journal of these Nutrition Portfolio assignments, you can examine how your knowledge and behaviors change as you progress in your study of nutrition.

Your food choices play a key role in keeping you healthy and reducing your risk of chronic diseases. After you have recorded at least one day's foods in Diet & Wellness Plus, look at that day's choices and record your answers to the following in your journal:

- Identify the factors that most influence your food choices for meals and snacks.

- List the chronic disease risk factors and conditions (listed in the definition of *risk factor*, p. 27) that you have.

- Describe lifestyle changes you can make to improve your chances of enjoying good health.

 To complete this exercise, go to your Diet and Wellness Plus at www.cengagebrain.com.

› STUDY IT To review the key points of this chapter and take a practice quiz, go to the study cards at the end of the book.

REFERENCES

1. E. R. Grimm and N. I. Steinle, Genetics of eating behavior: Established and emerging concepts, *Nutrition Reviews* 69 (2011): 52–60.
2. A. Drewnowski and coauthors, Sweetness and food preference, *Journal of Nutrition* 142 (2012): 1142S–1148S; J. E. Hayes, B. S. Sullivan, and V. B. Duffy, Explaining variability in sodium intake through oral sensory phenotype, salt sensation and liking, *Physiology and Behavior* 100 (2010): 369–380.
3. J. E. Hayes and R. S. Keast, Two decades of supertasting: Where do we stand? *Physiology and Behavior* 104 (2011): 1072–1074.
4. Food Marketing Institute, US grocery shopper trends 2012: Executive summary, www.fmi.org/research, 2012.
5. A. E. Sloan, Not too basic, *Food Technology* 65 (2011): 21.
6. J. Lu, C. Huet, and L. Dube, Emotional reinforcement ass a protective factor for healthy eating in home settings, *American Journal of Clinical Nutrition* 94 (2011): 254–261.
7. A. Jaworowska and coauthors, Nutritional challenges and health implications of takeaway and fast food, *Nutrition Reviews* 71 (2013): 310–318; J. E. Todd, L. Mancino, and B. Lin, The impact of food away from home on adult diet quality, *Economic Research Report* ERR-90, February 2010.
8. C. D. Rehm, P. Monsivais, and A. Drewnowski, The quality and monetary value of diets consumed by adults in the United States, *American Journal of Clinical Nutrition* 94 (2011):1333–1339; C. N. Mhurchu, Food costs and healthful diets: The need for solution-oriented research and policies, *American Journal of Clinical Nutrition* 92 (2010): 1007–1008.
9. A. Drewnowski, The cost of US foods as related to their nutritive value, *American Journal of Clinical Nutrition* 92 (2010): 1181–1188.
10. A. M. Bernstein and coauthors, Relation of food cost to healthfulness of diet among US women, *American Journal of Clinical Nutrition* 92 (2010): 1197–1203.
11. C. Jacquier and coauthors, Improving the effectiveness of nutritional information policies: Assessment of unconscious pleasure mechanisms involved in food-choice decisions, *Nutrition Reviews* 70 (2012): 118–131.
12. International Food Information Council Foundation, *2013 Food & Health Survey*, www.foodinsight.org.
13. J. P. Koplan and K. D. Brownell, Response of the food and beverage industry to the obesity threat, *Journal of the American Medical Association* 304 (2010): 1487–1488.
14. Position of the Academy of Nutrition and Dietetics: Functional foods, *Journal of the Academy of Nutrition and Dietetics* 113 (2013): 1096–1103.
15. Position of the Academy of Nutrition and Dietetics: Total diet approach to healthy eating, *Journal of the Academy of Nutrition and Dietetics* 113 (2013): 307–317.
16. Laura Mauri, Why we still need randomized trials to compare effectiveness, *New England Journal of Medicine* 366 (2012): 1538–1540.
17. B. Barrett and coauthors, Placebo effects and the common cold: A randomized controlled trial, *Annals of Family Medicine* 9 (2011): 312–322.
18. S. M. Chang, Should meta-analyses trump observational studies? *American Journal of Clinical Nutrition* 97 (2013): 237–238.

19. P. R. Trumbo and coauthors, Dietary Reference Intakes: Cases of appropriate and inappropriate uses, *Nutrition Reviews* 71 (2013): 657–664; S. A. Atkinson, Defining the process of Dietary Reference Intakes: Framework for the United States and Canada, *American Journal of Clinical Nutrition* 94 (2011): 655S–657S.

20. Practice paper of the American Dietetic Association: Using the Dietary Reference Intakes, *Journal of the American Dietetic Association* 111 (2011): 762–770.

21. E. J. Sondik and coauthors, Progress toward the Healthy People 2010 goals and objectives, *Annual Review of Public Health* 31 (2010): 271–281.

22. B. M. Popkin, L. S. Adair, and S. W. Ng, Global nutrition transition and the pandemic of obesity in developing countries, *Nutrition Reviews* 70 (2012): 3–21; M. H. Fernstrom and coauthors, Communication strategies to help reduce the prevalence of non-communicable diseases: Proceedings from the inaugural IFIC Foundation Global Diet and Physical Activity Communications Summit, *Nutrition Reviews* 70 (2012): 301–310; K. M. V. Narayan, M. K. Ali, and J. P. Koplan, Global noncommunicable diseases: Where worlds meet, *New England Journal of Medicine* 363 (2010): 1196–1198.

23. A. H. Mokdad and P. L. Remington, Measuring health behaviors in populations, *Preventing Chronic Diseases* 7 (2010): A75.

24. C. A. Schoenborn and P. F. Adams, Health behaviors of adults: United States, 2005–2007, National Center for Health Statistics, *Vital and Health Statistics*, 2010, www.cdc.gov/nchs/data/series/sr_10/sr10_245.pdf

HIGHLIGHT > 1
Nutrition Information and Misinformation

> **LEARN IT** Recognize misinformation and describe how to identify reliable nutrition information.

How can people distinguish valid nutrition information from misinformation? One excellent approach is to notice *who* is providing the information. The "who" behind the information is not always evident, though, especially in the world of electronic media. Keep in mind that *people* create websites on the Internet, just as people write books and report the news. In all cases, consumers need to determine whether the person is qualified to provide nutrition information.

This highlight begins by examining the unique potential as well as the problems of relying on the Internet and the media for nutrition information. It continues with a discussion of how to identify reliable nutrition information that applies to all resources, including the Internet and the news. (The accompanying glossary defines related terms.)

Nutrition on the Internet

Got a question? The **Internet** has an answer. An estimated two out of three US adults use the Internet to look up health information or learn about health topics in online chat groups.[1] The Internet offers endless opportunities to obtain high-quality information, but it also delivers an abundance of incomplete, misleading, and inaccurate information.[2] Simply put: anyone can publish anything.

Will & Deni McIntyre/Science Source

With hundreds of millions of **websites,** searching for nutrition information can be an overwhelming experience, with no guarantees of finding accurate information. When using the Internet, keep in mind that the quality of health-related information available covers a broad range. You must evaluate websites for their accuracy, just as you would any other source. The accompanying "How To" provides tips for determining whether a website is reliable.

One of the most trustworthy sites used by scientists and others is the US National Library of Medicine's PubMed, which provides free access to more than 23 million abstracts of research papers published in scientific journals around the world. Many abstracts provide links to the full articles. Figure H1-1 (p. 32) introduces this valuable resource.

Did you receive an e-mail warning of the health dangers associated with reusing or freezing plastic water bottles? If so, you've

GLOSSARY

Academy of Nutrition and Dietetics: the professional organization of dietitians in the United States; formerly the American Dietetic Association.

accredited: approved; in the case of medical centers or universities, certified by an agency recognized by the US Department of Education.

certified nutritionist or **certified nutritional consultant** or **certified nutrition therapist:** a person who has been granted a document declaring his or her authority as a nutrition professional.

dietetic technician: a person who has completed a minimum of an associate's degree from an accredited university or college and an approved dietetic

technician program that includes a supervised practice experience. See also *dietetic technician, registered.*

dietetic technician, registered (DTR): a dietetic technician who has passed a national examination and maintains registration through continuing professional education.

dietitian: a person trained in nutrition, food science, and diet planning. See also *registered dietitian nutritionist.*

diploma mills: entities without valid accreditation that provide worthless degrees.

DTR: see *dietetic technician, registered.*

fraudulent: the promotion, for financial gain, of devices, treatments, services, plans, or products (including diets and supplements) that alter or claim to alter a human condition without proof of safety or effectiveness.

Internet (the Net): a worldwide network of millions of computers linked together to share information.

license to practice: permission under state or federal law, granted on meeting specified criteria, to use a certain title (such as dietitian) and offer certain services. *Licensed dietitians* may use the initials *LD* after their names.

misinformation: false or misleading information.

public health dietitians: dietitians who specialize in providing nutrition services through organized community efforts.

RDN: see *registered dietitian nutritionist.*

registered dietitian nutritionist (RDN): a person who has completed a minimum of a bachelor's degree from an accredited university or college, has completed approved course work and a

supervised practice program, has passed a national examination, and maintains registration through continuing professional education; also called *registered dietitian (RD).*

registered dietitian (RD): an alternative term for an RDN.

registration: listing; with respect to health professionals, listing with a professional organization that requires specific course work, experience, and passing of an examination.

websites: Internet resources composed of text and graphic files, each with a unique URL (Uniform Resource Locator) that names the site (for example, www.usda.gov).

been a victim of urban scarelore. When nutrition information arrives in unsolicited e-mails, be suspicious if:

- The person sending it to you didn't write it and you cannot determine who did or if that person is a nutrition expert
- The phrase "Forward this to everyone you know" appears
- The phrase "This is not a hoax" appears because chances are good that it is
- The news is sensational and you've never heard about it from legitimate sources
- The language is emphatic and the text is sprinkled with capitalized words and exclamation marks
- No references are given or, if present, are of questionable validity when examined
- The message has been debunked on websites such as **www.quackwatch.org, www.snopes.com,** or **www.urbanlegends.about.com**

Nutrition in the News

Consumers get much of their nutrition information from Internet websites, television news, and magazine articles, which have heightened awareness of how diet influences the development of diseases. Consumers benefit from news coverage of nutrition when they learn to make lifestyle changes that will improve their health. Sometimes, however, popular reports mislead consumers and create confusion. They often tell a lopsided story quickly instead of presenting the integrated results of research studies or a balance of expert opinions.

Tight deadlines and limited understanding sometimes make it difficult to provide a thorough report. Hungry for the latest news, the media often report scientific findings quickly and prematurely—without benefit of careful interpretation, replication, or peer review. Usually, the reports present findings from a single, recently released study, making the news current and controversial. Consequently, the public receives diet and health news fast, but not always in perspective. Reporters may twist inconclusive findings into "meaningful discoveries" when pressured to write catchy headlines and sensational stories.

As a result "surprising new findings" sometimes seem to contradict one another, and consumers may feel frustrated and betrayed. Occasionally, the reports are downright false, but more often the apparent contradictions are simply the normal result of science at work. A single study contributes to the big picture, but when viewed alone, it can easily distort the image. To be meaningful the conclusions of any study must be presented cautiously within the context of other research findings.

>How To

Determine Whether a Website Is Reliable

To determine whether a website offers reliable nutrition information, ask the following questions:

- **Who?** Who is responsible for the site? Is it staffed by qualified professionals? Look for the authors' names and credentials. Have experts reviewed the content for accuracy?
- **When?** When was the site last updated? Because nutrition is an ever-changing science, sites need to be dated and updated frequently.
- **Where?** Where is the information coming from? The three letters following the dot in a Web address identify the site's affiliation. Addresses ending in "gov" (government), "edu" (educational institute), and "org" (organization) generally provide reliable information; "com" (commercial) sites represent

businesses and, depending on their qualifications and integrity, may or may not offer dependable information.

- **Why?** Why is the site giving you this information? Is the site providing a public service or selling a product? Many commercial sites provide accurate information, but some do not. When money is the prime motivation, be aware that the information may be biased.

If you are satisfied with the answers to all of the previous questions, then ask this final question:

- **What?** What is the message, and is it in line with other reliable sources? Information that contradicts common knowledge should be questioned. Many reliable sites provide links to other sites to facilitate your quest for knowledge, but this provision alone does not guarantee a reputable intention. Be aware that any site can link to any other site without permission.

> **TRY IT** Visit a nutrition website and answer the five "W" questions to determine whether it is a reliable resource.

Identifying Nutrition Experts

Regardless of whether the medium is electronic, print, or video, consumers need to ask whether the person behind the information is qualified to speak on nutrition. If the creator of an Internet website recommends eating three pineapples a day to lose weight, a trainer at the gym praises a high-protein diet, or a health-food store clerk suggests an herbal supplement, should you believe these people? Can you distinguish between accurate news reports and infomercials on television? Have you noticed that many televised nutrition messages are presented by celebrities, athletes, psychologists, food editors, and chefs—that is, almost anyone except a **dietitian**? When you are confused or need sound dietary advice, whom should you ask?

Physicians and Other Health-Care Professionals

Many people turn to physicians or other health-care professionals for dietary advice, expecting them to know about all health-related matters. But are they the best sources of accurate and current information on nutrition? Only about 30 percent of all medical schools in the

> **FIGURE H1-1**   **PubMed: Internet Resource for Scientific Nutrition References**

> **FIGURE H1-1**   **PubMed: Internet Resource for Scientific Nutrition References**

The US National Library of Medicine's PubMed website (www.pubmed.gov) offers tutorials to help teach beginners to use the search system effectively. Often, simply visiting the site, typing a query in the "Search for" box, and clicking "Go" will yield satisfactory results.

For example, to find research concerning calcium and bone health, typing "calcium bone" yields almost 50,000 results. Try setting limits on dates, types of articles, languages, and other criteria to obtain a more manageable number of abstracts to peruse.

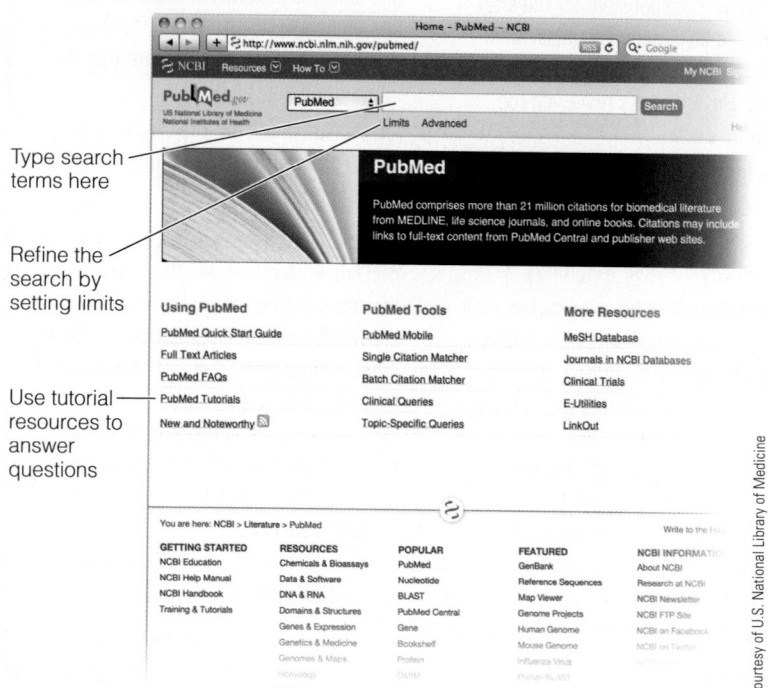

Type search terms here

Refine the search by setting limits

Use tutorial resources to answer questions

Courtesy of U.S. National Library of Medicine

nutrition are especially well qualified to speak on the subject. Few, however, have the time or experience to develop diet plans and provide detailed diet instructions for clients. Often they wisely refer clients to a qualified nutrition expert—a **registered dietitian nutritionist (RDN).**

Registered Dietitian Nutritionist (RDN)

To help consumers recognize credentialed dietitians and nutritionists, the Academy of Nutrition and Dietetics recently approved the optional use of both terms—registered dietitian nutritionist (RDN) and **registered dietitian (RD).** The meanings of RDN and RD are identical. A registered dietitian nutritionist (RDN) has the educational background necessary to deliver reliable nutrition advice and care.[3] To become an RDN, a person must earn an undergraduate degree requiring about 60 credit hours in nutrition, food science, and other related subjects; complete a year's clinical internship or the equivalent; pass a national examination administered by the Academy of Nutrition and Dietetics; and maintain up-to-date knowledge and **registration** by participating in required continuing education activities, such as attending seminars, taking courses, or conducting research.

Some states allow anyone to use the title dietitian or nutritionist, but others allow only an RDN or people with specified qualifications to call themselves dietitians. Many states provide a further guarantee: a state registration, certification, or **license to practice.** In this way, states identify people who have met minimal standards of education and experience. Still, these state standards may fall short of those defining an RDN. Similarly, some alternative educational programs qualify a graduate as a **certified nutritionist, certified nutritional consultant,** or **certified nutrition therapist**—terms that sound authoritative but lack the credentials of an RDN.

Dietitians perform a multitude of duties in many settings in most communities.[4] They work in the food industry, pharmaceutical companies, home health agencies, long-term care institutions, private practice, public health departments, research centers, education settings, fitness centers, and hospitals. Depending on their work settings, dietitians can assume a number of different job responsibilities and positions. In hospitals, administrative dietitians manage the foodservice system; clinical dietitians provide client care; and nutrition support team dietitians coordinate nutrition care with other health-care professionals. In the food industry, dietitians conduct research, develop products, and market services.

Public health dietitians who work in government-funded agencies such as health departments or clinics play a key role in delivering nutrition services to people in the community. Among their many roles, public health dietitians help plan, coordinate, and evaluate food assistance programs; act as consultants to other agencies; manage finances; and much more.

United States require students to take a separate nutrition course; less than half require the minimum 25 hours of nutrition instruction recommended by the National Academy of Sciences. By comparison, most students reading this text are taking a nutrition class that provides an average of 45 hours of instruction.

The **Academy of Nutrition and Dietetics** (formerly the American Dietetic Association) asserts that standardized nutrition education should be included in the curricula for all health-care professionals: physicians, nurses, physician's assistants, dental hygienists, physical and occupational therapists, social workers, psychologists, and all others who provide services directly to clients. When these professionals understand the relevance of nutrition in the treatment and prevention of diseases and have command of reliable nutrition information, then all the people they serve will also be better informed.

Most health-care professionals appreciate the connections between health and nutrition. Those who have specialized in clinical

Dietetic Technician, Registered (DTR)

In some facilities, a **dietetic technician** assists an RDN in both administrative and clinical responsibilities. A dietetic technician has been educated and trained to work under the guidance of an RDN; upon passing a national examination, the title changes to **dietetic technician, registered (DTR).**

Other Dietary Employees

In addition to the dietetic technician, other dietary employees may include clerks, aides, cooks, porters, and assistants. These dietary employees do not have extensive formal training in nutrition, and their ability to provide accurate information may be limited.

Identifying Fake Credentials

In contrast to an RDN, thousands of people obtain fake nutrition degrees and claim to be nutrition consultants or doctors of "nutrimedicine." These and other such titles may sound meaningful, but most of these people lack the established credentials and training of an RDN. If you look closely, you can see signs of their fake expertise.

Consider educational background, for example. The minimum standards of education for an RDN specify a bachelor of science (BS) degree in food science and human nutrition or related fields from an **accredited** college or university.* Such a degree generally requires 4 to 5 years of study. Similarly, minimum standards of education for a dietetic technician specify an associate degree that typically requires 2 years of study. In contrast, a fake nutritionist may display a degree from a 6-month course. Such a degree simply falls short. In some cases, businesses posing as schools offer even less—they sell certificates to anyone who pays the fees. To obtain these "degrees," a candidate need not attend any classes, read any books, or pass any examinations.

To safeguard educational quality, an accrediting agency recognized by the US Department of Education (DOE) certifies that certain schools meet criteria established to ensure that an institution provides complete and accurate schooling. Unfortunately, fake nutrition degrees are available from schools "accredited" by phony accrediting agencies. Acquiring false degrees and credentials is especially easy today, with **diploma mills** and **fraudulent** businesses operating via the Internet.[5]

Knowing the qualifications of someone who provides nutrition information can help you determine whether that person's advice might be harmful or helpful. Don't be afraid to ask for credentials. Table H1-1 lists credible sources of nutrition information.

TABLE H1-1 Credible Sources of Nutrition Information

Government agencies, volunteer associations, consumer groups, and professional organizations provide consumers with reliable health and nutrition information. Credible sources of nutrition information include:

- Nutrition and food science departments at a university or community college

- Local agencies such as the health department or County Cooperative Extension Service

- Government resources such as:

Centers for Disease Control and Prevention (CDC)	**www.cdc.gov**
Department of Agriculture (USDA)	**www.usda.gov**
Department of Health and Human Services (DHHS)	**www.hhs.gov**
Dietary Guidelines for Americans	**fnic.nal.usda.gov /dietary-guidance**
Food and Drug Administration (FDA)	**www.fda.gov**
Health Canada	**www.hc-sc.gc.ca/index-eng.php**
Healthy People	**www.healthypeople.gov**
Let's Move!	**www.letsmove.gov**
MyPlate	**www.choosemyplate.gov**
National Institutes of Health	**www.nih.gov**
Physical Activity Guidelines for Americans	**www.health.gov/paguidelines**

- Volunteer health agencies such as:

American Cancer Society	**www.cancer.org**
American Diabetes Association	**www.diabetes.org**
American Heart Association	**www.americanheart.org**

- Reputable consumer groups such as:

American Council on Science and Health	**www.acsh.org**
International Food Information Council	**www.foodinsight.org**

- Professional health organizations such as:

Academy of Nutrition and Dietetics	**www.eatright.org**
American Medical Association	**www.ama-assn.org**
Dietitians of Canada	**www.dietitians.ca**

- Journals such as:

American Journal of Clinical Nutrition	**www.ajcn.org**
Journal of the Academy of Nutrition and Dietetics	**www.andjrnl.org**
New England Journal of Medicine	**www.nejm.org**
Nutrition Reviews	**www.ilsi.org**

---

*To ensure the quality and continued improvement of nutrition and dietetics education programs, an agency of the Academy of Nutrition and Dietetics known as the Accreditation Council for Education in Nutrition and Dietetics (ACEND) establishes and enforces eligibility requirements and accreditation standards for programs preparing students for careers as registered dietitian nutritionists or dietetics technicians. Programs meeting those standards are accredited by ACEND.

> FIGURE H1-2   **Red Flags of Nutrition Quackery**

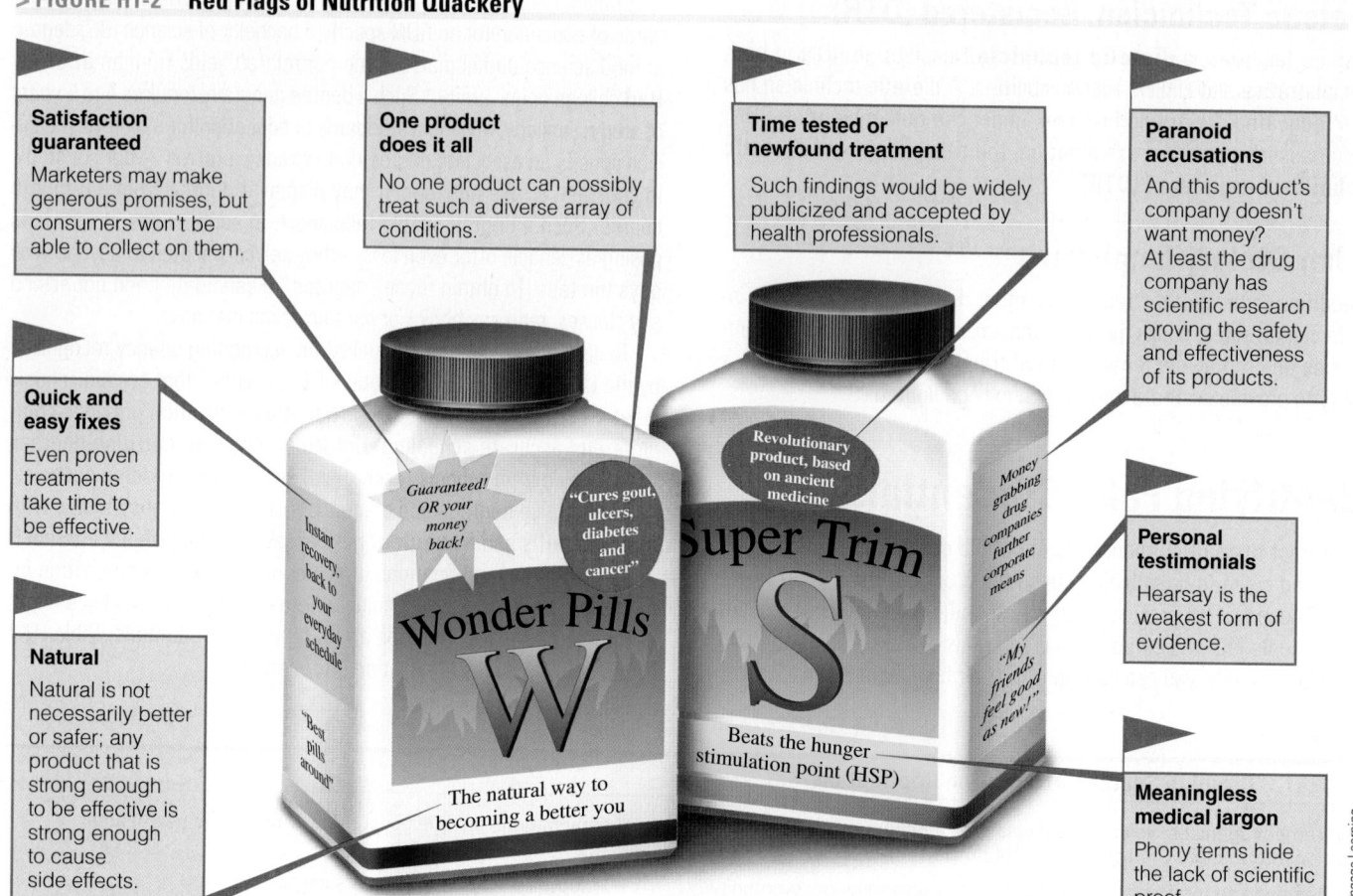

Red Flags of Nutrition Quackery

Figure H1-2 features eight red flags consumers can use to identify nutrition **misinformation.** Sales of unproven and dangerous products have always been a concern, but the Internet now provides merchants with an easy and inexpensive way to reach millions of customers around the world. Because of the difficulty in regulating the Internet, fraudulent and illegal sales of medical products have hit a bonanza. As is the case with the air, no one owns the Internet, and similarly, no one has control over the pollution. Countries have different laws regarding sales of drugs, dietary supplements, and other health products, but applying these laws to the Internet marketplace is almost impossible. Even if illegal activities could be defined and identified, finding the person responsible for a particular website is not always possible. Websites can appear and disappear in a blink of a cursor. Now, more than ever, consumers must heed the caution "Buyer beware."

In summary, when you hear nutrition news, consider its source. Ask yourself these two questions: Is the person providing the information qualified to speak on nutrition? Is the information based on valid scientific research? If not, find a better source. After all, your health depends on it.

CRITICAL THINKING QUESTIONS

A. How would you judge the accuracy or validity of nutrition information?
B. You have just received a forwarded e-mail from a friend warning that the artificial sweetener aspartame is a TOXIN that causes muscle spasms, leg numbness, stomach cramps, vertigo, dizziness, headaches, tinnitus, joint pain, depression, anxiety, slurred speech, blurred vision, and memory loss. It goes on to say that this DEADLY POISON causes blindness, multiple sclerosis, brain tumors, and cancer! The message alleges that aspartame remains on the market because of a conspiracy between the FDA and the manufacturer to keep these dangers hidden from the public. How can you determine whether these claims are legitimate warnings or an irresponsible hoax?

REFERENCES

1. R. A. Cohen and P. F. Adams, Use of the Internet for health information: United States, 2009, *NCHS Data Brief*, July 2011.
2. Practice paper of the Academy of Nutrition and Dietetics abstract: Communicating accurate food and nutrition information, *Journal of the Academy of Nutrition and Dietetics* 112 (2012): 759.
3. Position of the Academy of Nutrition and Dietetics: The role of nutrition in health promotion and chronic disease prevention, *Journal of the Academy of Nutrition and Dietetics* 113 (2013): 972–979.
4. Comprehensive scope of practice resources for the registered dietitian or registered dietitian nutritionist, *Journal of the Academy of Nutrition and Dietetics*, June 2013, Supplement 2.
5. E. B. Cohen and R. Winch, Diploma and accreditation mills: New trends in credential abuse, March 2011, www.accredibase.com.

2

Planning a Healthy Diet

Nutrition in Your Life

You make food choices—deciding what to eat and how much to eat—more than 1000 times every year. We eat so frequently that it's easy to choose a meal without giving any thought to its nutrient contributions or health consequences. Even when we want to make healthy choices, we may not know which foods to select or how much to consume. With a few tools and tips, you can learn to plan a healthy diet. In the Nutrition Portfolio at the end of this chapter, you can compare your current diet against a healthy eating plan.

Chapter 1 explained that the nutrients delivered by the foods people eat support the body's many activities. Food choices made over years and decades influence the body's health, and consistently poor choices increase the risks of developing chronic diseases. Stated positively, optimal nourishment supports a robust life of vigorous activity and good health. This chapter shows how a person can select from the tens of thousands of available foods to create a nutritionally balanced diet that meets the body's nutrient and energy needs. Fortunately, most foods provide several nutrients, so one trick for wise diet planning is to select a combination of foods that deliver a full array of nutrients.

This chapter begins by introducing the diet-planning principles and dietary guidelines that promote good health and reduce disease risks. It continues by showing how people can use diet-planning guides to create meal patterns that will deliver sufficient nutrients without excess energy (kcalories). Learning how to read food labels eases the task of making healthy selections at the market.

To ensure an adequate and balanced diet, eat a variety of foods daily, choosing different foods from each group.

2.1 Principles and Guidelines

> **LEARN IT** Explain how each of the diet-planning principles can be used to plan a healthy diet.

How well you nourish yourself does not depend on the selection of any one food. Instead, it depends on the overall **eating pattern**—the combination of many different foods and beverages at numerous meals over days, months, and years.[1] Diet-planning principles and dietary guidelines are key concepts to keep in mind whenever you are selecting foods—whether shopping at the grocery store, choosing from a restaurant menu, or preparing a home-cooked meal.

Diet-Planning Principles

Diet planners have developed several ways to select foods. Whatever plan or combination of plans they use, though, they keep in mind these basic diet-planning principles:

- Adequacy
- Balance
- kCalorie (energy) control
- Nutrient density
- Moderation
- Variety

Adequacy **Adequacy** reflects a diet that provides sufficient energy and enough of all the nutrients to meet the needs of healthy people. Take the essential nutrient iron, for example. Because the body loses some iron each day, people have to replace it by eating foods that contain iron. A person whose diet fails to provide enough iron-rich foods may develop the symptoms of iron-deficiency anemia: the person may feel weak, tired, and listless; have frequent headaches; and find that even the smallest amount of muscular work brings disabling fatigue. To prevent these deficiency symptoms, a person must include foods that supply adequate iron. The same is true for all the other essential nutrients introduced in Chapter 1.

Balance **Balance** in the diet helps to ensure adequacy. The art of balancing the diet involves consuming enough—but not too much—of different types of foods in proportion to one another. In a balanced diet, foods rich in some nutrients do not crowd out foods that are rich in other nutrients. The essential minerals calcium and iron, taken together, illustrate the importance of dietary **balance**. Meat is rich in iron but poor in calcium. Conversely, milk is rich in calcium but poor in iron. Use some meat for iron; use some milk for calcium; and save some space for other foods, too, because a diet consisting of milk and meat alone would not be adequate. For the other nutrients, people need to eat other protein foods, whole grains, vegetables, and fruits.

kCalorie (Energy) Control Designing an adequate diet within a reasonable kcalorie allowance requires careful planning. Once again, balance plays a key role. The amount of energy coming into the body from foods should balance with the amount of energy being used by the body to sustain its metabolic and physical activities. Upsetting this balance leads to gains or losses in body weight. The discussion of energy balance and weight control in Chapters 8 and 9 examines this issue in more detail, but one key to **kcalorie control** is to select foods of high nutrient density.

Nutrient Density **Nutrient density** promotes adequacy and kcalorie control. To eat well without overeating, select nutrient-dense foods—that is, foods that deliver the most nutrients for the least food energy.[2] Consider foods containing calcium, for example. You can get about 300 milligrams of calcium from either 1½ ounces of cheddar cheese or 1 cup of fat-free milk, but the cheese delivers about twice as much food energy (kcalories) as the milk. The fat-free milk, then, is twice as calcium dense as the cheddar cheese; it offers the same amount of calcium

eating pattern: customary intake of foods and beverages over time.

adequacy (dietary): providing all the essential nutrients, fiber, and energy in amounts sufficient to maintain health.

balance (dietary): providing foods in proportion to one another and in proportion to the body's needs.

kcalorie (energy) **control:** management of food energy intake.

nutrient density: a measure of the nutrients a food provides relative to the energy it provides. The more nutrients and the fewer kcalories, the higher the nutrient density.

>How To

Compare Foods Based on Nutrient Density

One way to evaluate foods is simply to notice their nutrient contribution *per serving:* 1 cup of milk provides about 300 milligrams of calcium, and ½ cup of fresh, cooked turnip greens provides about 100 milligrams. Thus a serving of milk offers three times as much calcium as a serving of turnip greens. To get 300 milligrams of calcium, a person could choose either 1 cup of milk or 1½ cups of turnip greens.

Another valuable way to evaluate foods is to consider their nutrient density—their nutrient contribution *per kcalorie.* Fat-free milk delivers about 85 kcalories with its 300 milligrams of calcium. To calculate the nutrient density, divide milligrams by kcalories:

$$\frac{300 \text{ mg calcium}}{85 \text{ kcal}} = 3.5 \text{ mg per kcal}$$

Do the same for the fresh turnip greens, which provide 15 kcalories with the 100 milligrams of calcium:

$$\frac{100 \text{ mg calcium}}{15 \text{ kcal}} = 6.7 \text{ mg per kcal}$$

The more milligrams per kcalorie, the greater the nutrient density. Turnip greens are more calcium dense than milk. They provide more calcium *per kcalorie* than milk, but milk offers more calcium *per serving.* Both approaches offer valuable information, especially when combined with a realistic appraisal. What matters most is which are you more likely to consume—1½ cups of turnip greens or 1 cup of milk? You can get 300 milligrams of calcium from either, but the greens will save you about 40 kcalories (the savings would be even greater if you usually use whole milk).

Keep in mind, too, that calcium is only one of the many nutrients that foods provide. Similar calculations for protein, for example, would show that fat-free milk provides more protein both *per kcalorie* and *per serving* than turnip greens—that is, milk is more protein dense. Combining variety with nutrient density helps to ensure the adequacy of all nutrients.

› TRY IT Compare the thiamin density of 3 ounces of lean T-bone steak (174 kcalories, 0.09 milligrams thiamin) with ½ cup of fresh cooked broccoli (27 kcalories, 0.05 milligrams thiamin).

---

for half the kcalories. Both foods are excellent choices for adequacy's sake alone, but to achieve adequacy while controlling kcalories, the fat-free milk is the better choice. (Alternatively, a person could select a low-fat cheddar cheese with its kcalories comparable to fat-free milk.) The accompanying "How To" describes how to compare foods based on nutrient density.

Just as a financially responsible person pays for rent, food, clothes, and tuition on a limited budget, healthy people obtain iron, calcium, and all the other essential nutrients on a limited energy (kcalorie) allowance. Success depends on getting many nutrients for each kcalorie "dollar." As Figure 2-1 illustrates on p. 40, a breakfast of cereal, fruit, egg, and sausage delivers many more nutrients than a couple of doughnuts—even though they both provide about the same number of kcalories. A person who makes nutrient-dense choices can meet daily nutrient needs on a lower energy budget. Such choices support good health.

Foods that are notably low in nutrient density—such as potato chips, candy, and colas—are called **empty-kcalorie foods.** The kcalories these foods provide are called "empty" because they deliver a lot of energy (from added sugars, solid fats, or both) but little, or no, protein, vitamins, or minerals.

The concept of nutrient density is relatively simple when examining the contributions of one nutrient to a food or diet. With respect to calcium, milk ranks high and meats rank low. With respect to iron, meats rank high and milk ranks low. But it is a more complex task to answer the question, which food is more

empty-kcalorie foods: a popular term used to denote foods that contribute energy but lack protein, vitamins, and minerals.

> **FIGURE 2-1**   **Nutrient Density of Two Breakfast Options Compared**

Chapter 1 presented these two breakfasts to illustrate energy density—that for the same number of kcalories, the breakfast on the left delivered less energy per gram of food, which benefits weight management. These two breakfasts also illustrate nutrient density—that for the same number of kcalories, the breakfast on the left delivers more nutrients per kcalorie.

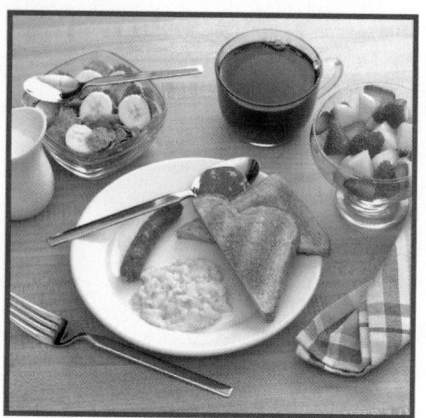

NUTRIENT-DENSE BREAKFAST

Key:
■ Nutrient-dense breakfast
□ Nutrient-poor breakfast

(bar chart: y-axis "Daily recommendations (%)" from 0 to 100; x-axis categories: Protein, Fiber, Vitamin D, Calcium, Iron, Potassium)

NUTRIENT-POOR BREAKFAST

Photos: © Matthew Farruggio; Art © Cengage Learning

nutritious? To answer that question, we need to consider several nutrients—including both nutrients that may harm health as well as those that may be beneficial. Ranking foods based on their overall nutrient composition is known as **nutrient profiling.** Researchers have yet to agree on an ideal way to rate foods based on the nutrient profile, but when they do, nutrient profiling will be quite useful in helping consumers identify nutritious foods and plan healthy diets.[3]

Moderation   **Moderation** contributes to adequacy, balance, and kcalorie control. Foods rich in fat and sugar often provide enjoyment and energy but relatively few nutrients; in addition, they promote weight gain when eaten in excess. A person practicing moderation eats such foods only on occasion and regularly selects foods low in **solid fats** and **added sugars**, a practice that automatically improves nutrient density. Returning to the example of cheddar cheese versus fat-free milk, the fat-free milk not only offers the same amount of calcium for less energy, but it also contains much less fat than the cheese.

Variety   **Variety** improves nutrient adequacy. A diet may have all of the virtues just described and still lack variety, if a person eats the same foods day after day. People should select foods from each of the food groups daily and vary their choices within each food group from day to day for several reasons. First, different foods within the same group contain different arrays of nutrients. Among the fruits, for example, strawberries are especially rich in vitamin C while apricots are rich in vitamin A. Second, no food is guaranteed entirely free of substances that, in excess, could be harmful. The strawberries might contain trace amounts of one contaminant, the apricots another. By alternating fruit choices, a person will ingest very little of either contaminant. (Contamination of foods is discussed in Chapter 19.) Third, as the adage goes, variety is the spice of life. A person who eats beans frequently can enjoy pinto beans in Mexican burritos today, garbanzo beans in a Greek salad tomorrow, and baked beans with barbecued chicken on the weekend. Eating nutritious meals need never be boring.

Dietary Guidelines for Americans   What should a person eat to stay healthy? The answers can be found in the *Dietary Guidelines for Americans.* These guidelines translate the *nutrient* recommendations of the DRI (presented in Chapter 1) into *food* recommendations.[4] The result is evidence-based advice designed to help people attain and maintain a healthy weight, reduce the risk of chronic diseases, and promote overall health through diet and physical activity.[5] In general, a healthy diet:

nutrient profiling: ranking foods based on their nutrient composition.

moderation (dietary): providing enough but not too much of a substance.

solid fats: fats that are not usually liquid at room temperature; commonly found in most foods derived from animals and vegetable oils that have been hydrogenated. Solid fats typically contain more saturated and *trans* fats than most oils (Chapter 5 provides more details).

added sugars: sugars and other kcaloric sweeteners that are added to foods during processing, preparation, or at the table. Added sugars do not include the naturally occurring sugars found in fruits and milk products.

variety (dietary): eating a wide selection of foods within and among the major food groups.

- Emphasizes a variety of fruits, vegetables, whole grains, and fat-free and low-fat milk products.
- Includes lean meats, poultry, seafood, legumes, eggs, seeds, and nuts.
- Is low in saturated and *trans* fats, cholesterol, salt (sodium), and added sugars.
- Stays within your daily energy needs for your recommended body weight.

Table 2-1 presents the key recommendations of the *Dietary Guidelines for Americans*, clustered into four major topic areas. The first area focuses on

---

TABLE 2-1 Key Recommendations of the *Dietary Guidelines for Americans*

Balancing kCalories to Manage Weight

- Prevent and/or reduce overweight and obesity through improved eating and physical activity behaviors (see Chapter 9).
- Control total kcalorie intake to manage body weight. For people who are overweight or obese, this will mean consuming fewer kcalories from foods and beverages (see Chapter 9).
- Increase physical activity and reduce time spent in sedentary behaviors (see Chapter 14).
- Maintain appropriate kcalorie balance during each stage of life—childhood, adolescence, adulthood, pregnancy and breastfeeding, and older age (see Chapters 15–17).

Foods and Food Components to Reduce

- Reduce daily sodium intake to less than 2300 milligrams and further reduce intake to 1500 milligrams among persons who are 51 and older and those of any age who are African American or have hypertension, diabetes, or chronic kidney disease (see Chapter 12).
- Consume less than 10 percent of kcalories from saturated fatty acids by replacing them with monounsaturated and polyunsaturated fatty acids (see Chapter 5).
- Consume less than 300 milligrams per day of dietary cholesterol (see Chapter 5).
- Keep *trans*-fatty acid consumption as low as possible by limiting foods that contain synthetic sources of *trans* fats, such as partially hydrogenated oils, and by limiting other solid fats (see Chapter 5).
- Reduce the intake of kcalories from solid fats and added sugars (see Chapters 4 and 5).
- Limit the consumption of foods that contain refined grains, especially refined grain foods that contain solid fats, added sugars, and sodium (see Chapters 4, 5, and 12).
- If alcohol is consumed it should be consumed in moderation—up to one drink per day for women and two drinks per day for men—and only by adults of legal drinking age (see Highlight 7).

Foods and Nutrients to Increase

- Increase vegetable and fruit intake.
- Eat a variety of vegetables, especially dark-green and red and orange vegetables and beans and peas.
- Consume at least half of all grains as whole grains. Increase whole-grain intake by replacing refined grains with whole grains.
- Increase intake of fat-free or low-fat milk and milk products, such as milk, yogurt, cheese, or fortified soy beverages.
- Choose a variety of protein foods, which include seafood, lean meat and poultry, eggs, beans and peas, soy products, and unsalted nuts and seeds.
- Increase the amount and variety of seafood consumed by choosing seafood in place of some meat and poultry.
- Replace protein foods that are higher in solid fats with choices that are lower in solid fats and kcalories and/or are sources of oils.
- Use oils to replace solid fats where possible (see Highlight 5).
- Choose foods that provide more potassium, dietary fiber, calcium, and vitamin D, which are nutrients of concern in American diets (see Chapters 4, 11, and 12). These foods include vegetables, fruits, whole grains, and milk and milk products.

Building Healthy Eating Patterns

- Select an eating pattern that meets nutrient needs over time at an appropriate kcalorie level.
- Account for all foods and beverages consumed and assess how they fit within a total healthy eating pattern.
- Follow food safety recommendations when preparing and eating foods to reduce the risk of foodborne illnesses (see Chapter 19).

© Cengage Learning

NOTE: These guidelines are intended for adults and healthy children ages 2 and older.
SOURCE: The *Dietary Guidelines for Americans*, available at www.healthierus.gov/dietaryguidelines.

The *Dietary Guidelines* encourage Americans to increase the energy (kcalories) they expend through physical activity.

balancing kcalories to manage a healthy body weight by improving eating habits and engaging in regular physical activity. The second area advises people to reduce their intakes of such foods and food components as sodium, solid fats (with their saturated fats, *trans* fats, and cholesterol), added sugars, refined grain products, and alcoholic beverages (for those who partake). The third area encourages people to consume a variety of fruits and vegetables, whole grains, and low-fat milk products and protein foods (including seafood). The fourth area helps consumers build healthy eating patterns that meet energy and nutrient needs while reducing the risk of foodborne illnesses. Together, the *Dietary Guidelines for Americans* point the way toward longer, healthier, and more active lives. These key recommendations, along with additional recommendations for specific population groups, appear throughout the text as their subjects are discussed.

By law, the *Dietary Guidelines for Americans* are reviewed and revised as needed every five years. Each edition shares some similarities with previous editions but also sets precedent in new ways.[6] Perhaps most noteworthy to the current edition is the overarching focus on curbing the obesity epidemic and improving the health of the American population.[7]

Some people might wonder why *dietary* guidelines include recommendations for physical activity. The simple answer is that most people who maintain a healthy body weight do more than eat right. They also exercise—the equivalent of 30 to 60 minutes or more of moderately intense physical activity on most days.[8] As you will see repeatedly throughout this text, food and physical activity choices are integral partners in supporting good health.

> **REVIEW IT** Explain how each of the diet-planning principles can be used to plan a healthy diet.

A well-planned diet delivers adequate nutrients, a balanced array of nutrients, and an appropriate amount of energy. It is based on nutrient-dense foods, moderate in substances that can be detrimental to health, and varied in its selections. The *Dietary Guidelines* apply these principles, offering practical advice on how to eat for good health.

2.2 Diet-Planning Guides

> **LEARN IT** Use the USDA Food Patterns to develop a meal plan within a specified energy allowance.

To plan a diet that achieves all of the dietary ideals just outlined, a person needs tools as well as knowledge. Among the most widely used tools for diet planning are **food group plans** that build a diet from clusters of foods that are similar in nutrient content. Thus each food group represents a set of nutrients that differs somewhat from the nutrients supplied by the other groups. Selecting foods from each of the groups eases the task of creating an adequate and balanced diet.

USDA Food Patterns The *Dietary Guidelines* encourage consumers to adopt a balanced eating pattern, using the USDA's Food Patterns. The USDA Food Patterns assign foods to five major groups—fruits, vegetables, grains, protein foods, and milk and milk products—and recommend daily amounts of foods from each group to meet nutrient needs. Figure 2-2 (pp. 44–45) presents the food groups, the most notable nutrients of each group, the serving equivalents, and the foods within each group. Chapter 16 provides a food guide for young children.

Recommended Amounts All food groups offer valuable nutrients, and people should make selections from each group daily. The amounts from each food group needed daily to create a healthful diet differ depending on a person's

food group plans: diet-planning tools that sort foods into groups based on nutrient content and then specify that people should eat certain amounts of foods from each group.

energy (kcalorie) needs. Table 2-2 presents estimated daily energy needs for sedentary adults. As Table 2-3 shows, a person needing 2000 kcalories a day, for example, would select 2 cups of fruit; 2½ cups of vegetables; 6 ounces of grain foods; 5½ ounces of protein foods; and 3 cups of milk or milk products.* Additionally, a small amount of unsaturated oil, such as vegetable oil, or the oils of nuts, olives, or fatty fish, is required to supply needed nutrients.

> DIETARY GUIDELINES FOR AMERICANS
Select an eating pattern that meets nutrient needs over time at an appropriate kcalorie level.

All vegetables provide an array of nutrients, but some vegetables are especially good sources of certain vitamins, minerals, and beneficial phytochemicals. For this reason, the vegetable group is sorted into five subgroups. The dark-green vegetables deliver the B vitamin folate; the red and orange vegetables provide vitamin A; legumes supply iron and protein; the starchy vegetables contribute carbohydrate energy; and the other vegetables fill in the gaps and add more of these same nutrients.

In a 2000-kcalorie diet, then, the recommended 2½ cups of daily vegetables should be varied among the subgroups over a week's time. In other words, consuming 2½ cups of potatoes or even nutrient-rich spinach every day for seven days does *not* meet the recommended amounts for vegetables. Potatoes and spinach make excellent choices when consumed in balance with vegetables from the other subgroups. One way to help ensure selections for all of the subgroups is to eat vegetables of various colors—for example, green broccoli, orange sweet potatoes, black beans, yellow corn, and white cauliflower. Intakes of vegetables are appropriately averaged over a week's time—it is not necessary to include every subgroup every day.

For similar reasons, the protein foods group is sorted into three subgroups. Perhaps most notably, each of these subgroups contributes a different assortment of fats. Table 2-4 on p. 46 presents the recommended *weekly* amounts for each of the subgroups for vegetables and protein foods.

Notable Nutrients As Figure 2-2 notes, each food group contributes key nutrients. This feature provides flexibility in diet planning because a person can select any food from a food group (or its subgroup) and receive similar nutrients. For example, a person can choose milk, cheese, or yogurt and receive the same key nutrients. Importantly, foods provide not only these key nutrients, but small amounts of other nutrients and phytochemicals as well.

Legumes contribute the same key nutrients—notably, protein, iron, and zinc—as meats, poultry, and seafood. They are also excellent sources of fiber, folate, and potassium, which are commonly found in vegetables. To encourage frequent consumption of these nutrient-rich foods, legumes are included as a subgroup of both the vegetable group and the protein foods group. Thus legumes can be counted in either the vegetable group or the protein foods group.[9]

*Milk and milk products also can be referred to as dairy products.

legumes (lay-GYOOMS or LEG-yooms): plants of the bean and pea family, with seeds that are rich in protein compared with other plant-derived foods.

TABLE 2-2 Estimated Energy Needs for Sedentary Adults

	Energy (kcal/day)
Women	
19–30 yr	2000
31–50 yr	1800
51+ yr	1600
Men	
19–30 yr	2400
31–50 yr	2200
51+ yr	2000

NOTE: Sedentary describes a lifestyle that includes the activities typical of daily living with less than 30 minutes of moderate activity on most days. Because physical activity increases energy needs, people who are more physically active need more kcalories per day. Chapter 8 and Appendix F provide more details.

© Cengage Learning

TABLE 2-3 USDA Food Patterns: Recommended Daily Amounts from Each Food Group

	1600 kcal	1800 kcal	2000 kcal	2200 kcal	2400 kcal	2600 kcal	2800 kcal	3000 kcal
Fruits	1½ c	1½ c	2 c	2 c	2 c	2 c	2½ c	2½ c
Vegetables	2 c	2½ c	2½ c	3 c	3 c	3½ c	3½ c	4 c
Grains	5 oz	6 oz	6 oz	7 oz	8 oz	9 oz	10 oz	10 oz
Protein foods	5 oz	5 oz	5½ oz	6 oz	6½ oz	6½ oz	7 oz	7 oz
Milk and milk products	3 c	3 c	3 c	3 c	3 c	3 c	3 c	3 c
Oils	5 tsp	5 tsp	6 tsp	6 tsp	7 tsp	8 tsp	8 tsp	10 tsp
Discretionary kcalories	121 kcal	161 kcal	258 kcal	266 kcal	330 kcal	362 kcal	395 kcal	459 kcal

© Cengage Learning

> FIGURE 2-2   USDA Food Patterns: Food Groups and Subgroups

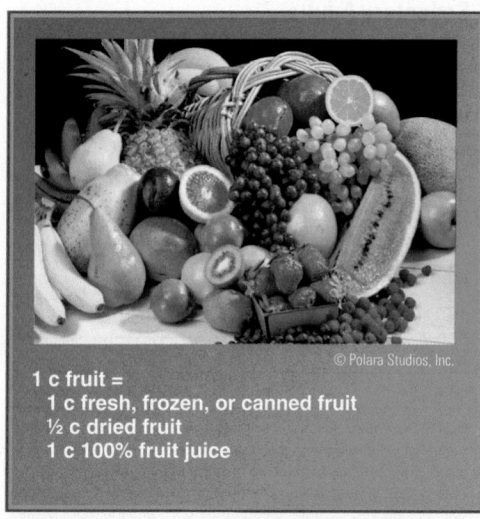

1 c fruit =
1 c fresh, frozen, or canned fruit
½ c dried fruit
1 c 100% fruit juice

© Polara Studios, Inc.

Fruits contribute folate, vitamin A, vitamin C, potassium, and fiber.

Consume a variety of fruits, and choose whole or cut-up fruits more often than fruit juice.

Apples, apricots, avocados, bananas, blueberries, cantaloupe, cherries, grapefruit, grapes, guava, honeydew, kiwi, mango, nectarines, oranges, papaya, peaches, pears, pineapples, plums, raspberries, strawberries, tangerines, watermelon; dried fruit (dates, figs, prunes, raisins); 100% fruit juices

Limit these fruits that contain solid fats and/or added sugars:
Canned or frozen fruit in syrup; juices, punches, ades, and fruit drinks with added sugars; fried plantains

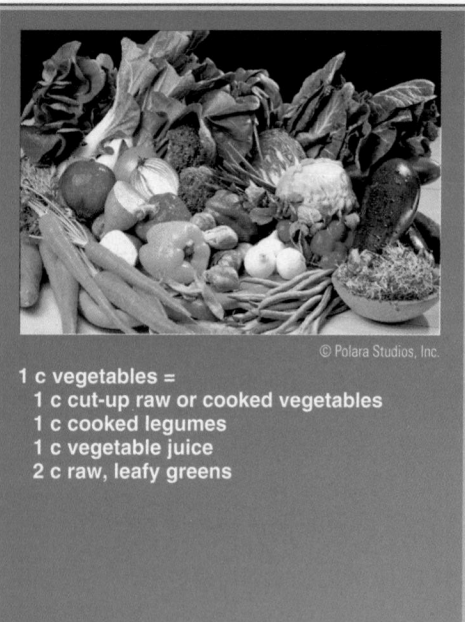

1 c vegetables =
1 c cut-up raw or cooked vegetables
1 c cooked legumes
1 c vegetable juice
2 c raw, leafy greens

© Polara Studios, Inc.

Vegetables contribute folate, vitamin A, vitamin C, vitamin K, vitamin E, magnesium, potassium, and fiber.

Consume a variety of vegetables each day, and choose from all five subgroups several times a week.

Dark-green vegetables: Broccoli and leafy greens such as arugula, beet greens, bok choy, collard greens, kale, mustard greens, romaine lettuce, spinach, turnip greens, watercress

Red and orange vegetables: Carrots, carrot juice, pumpkin, red bell peppers, sweet potatoes, tomatoes, tomato juice, vegetable juice, winter squash (acorn, butternut)

Legumes: Black beans, black-eyed peas, garbanzo beans (chickpeas), kidney beans, lentils, navy beans, pinto beans, soybeans and soy products such as tofu, split peas, white beans

Starchy vegetables: Cassava, corn, green peas, hominy, lima beans, potatoes

Other vegetables: Artichokes, asparagus, bamboo shoots, bean sprouts, beets, brussels sprouts, cabbages, cactus, cauliflower, celery, cucumbers, eggplant, green beans, green bell peppers, iceberg lettuce, mushrooms, okra, onions, seaweed, snow peas, zucchini

Limit these vegetables that contain solid fats and/or added sugars:
Baked beans, candied sweet potatoes, coleslaw, french fries, potato salad, refried beans, scalloped potatoes, tempura vegetables

1 oz grains =
1 slice bread
½ c cooked rice, pasta, or cereal
1 oz dry pasta or rice
1 c ready-to-eat cereal
3 c popped popcorn

© Polara Studios, Inc.

Grains contribute folate, niacin, riboflavin, thiamin, iron, magnesium, selenium, and fiber.

Make most (at least half) of the grain selections whole grains.

Whole grains: amaranth, barley, brown rice, buckwheat, bulgur, cornmeal, millet, oats, quinoa, rye, wheat, wild rice and whole-grain products such as breads, cereals, crackers, and pastas; popcorn

Enriched refined products: bagels, breads, cereals, pastas (couscous, macaroni, spaghetti), pretzels, white rice, rolls, tortillas

Limit these grains that contain solid fats and/or added sugars:
Biscuits, cakes, cookies, cornbread, crackers, croissants, doughnuts, fried rice, granola, muffins, pastries, pies, presweetened cereals, taco shells

Art © Cengage Learning

> FIGURE 2-2 USDA Food Patterns: Food Groups and Subgroups (*continued*)

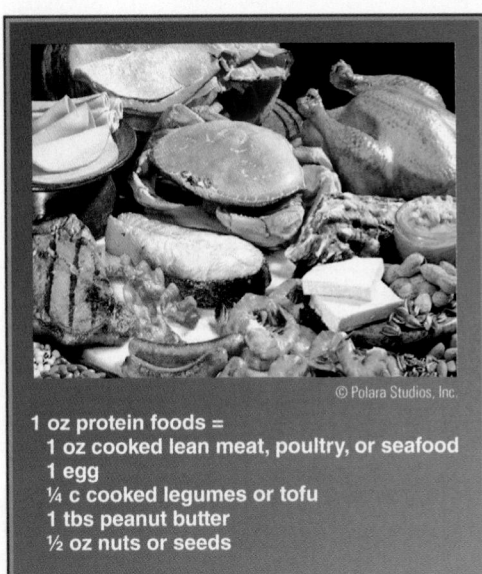

© Polara Studios, Inc.

1 oz protein foods =
1 oz cooked lean meat, poultry, or seafood
1 egg
¼ c cooked legumes or tofu
1 tbs peanut butter
½ oz nuts or seeds

Protein foods contribute protein, essential fatty acids, niacin, thiamin, vitamin B_6, vitamin B_12, iron, magnesium, potassium, and zinc.

Choose a variety of protein foods from the three subgroups, including seafood in place of meat or poultry twice a week.

Seafood: Fish (catfish, cod, flounder, haddock, halibut, herring, mackerel, pollock, salmon, sardines, sea bass, snapper, trout, tuna), shellfish (clams, crab, lobster, mussels, oysters, scallops, shrimp)

Meats, poultry, eggs: Lean or low-fat meats (fat-trimmed beef, game, ham, lamb, pork, veal), poultry (no skin), eggs

Nuts, seeds, soy products: Unsalted nuts (almonds, cashews, filberts, pecans, pistachios, walnuts), seeds (flaxseeds, pumpkin seeds, sesame seeds, sunflower seeds), legumes, soy products (textured vegetable protein, tofu, tempeh), peanut butter, peanuts

Limit these protein foods that contain solid fats and/or added sugars:
Bacon; baked beans; fried meat, seafood, poultry, eggs, or tofu; refried beans; ground beef; hot dogs; luncheon meats; marbled steaks; poultry with skin; sausages; spare ribs

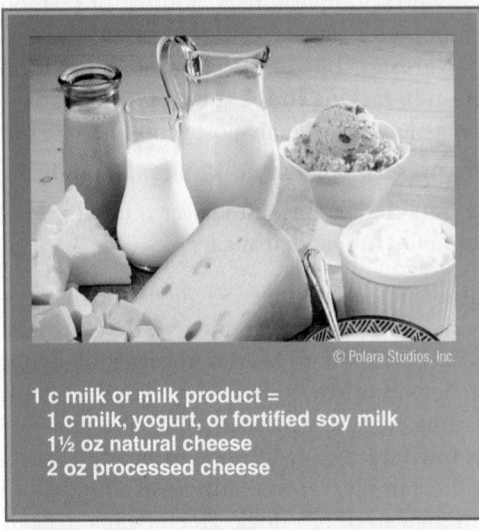

© Polara Studios, Inc.

1 c milk or milk product =
1 c milk, yogurt, or fortified soy milk
1½ oz natural cheese
2 oz processed cheese

Milk and milk products contribute protein, riboflavin, vitamin B_12, calcium, potassium, and, when fortified, vitamin A and vitamin D.

Make fat-free or low-fat choices. Choose other calcium-rich foods if you don't consume milk.

Fat-free or 1% low-fat milk and fat-free or 1% low-fat milk products such as buttermilk, cheeses, cottage cheese, yogurt; fat-free fortified soy milk

Limit these milk products that contain solid fats and/or added sugars:
2% reduced-fat milk and whole milk; 2% reduced-fat and whole-milk products such as cheeses, cottage cheese, and yogurt; flavored milk with added sugars such as chocolate milk, custard, frozen yogurt, ice cream, milk shakes, pudding, sherbet; fortified soy milk

© Matthew Farruggio

1 tsp oil =
1 tsp vegetable oil
1 tsp soft margarine
1 tbs low-fat mayonnaise
2 tbs light salad dressing

Oils are not a food group, but are featured here because they contribute vitamin E and essential fatty acids.

Use oils instead of solid fats, when possible.

Liquid vegetable oils such as canola, corn, flaxseed, nut, olive, peanut, safflower, sesame, soybean, sunflower oils; mayonnaise, oil-based salad dressing, soft *trans*-fat-free margarine; unsaturated oils that occur naturally in foods such as avocados, fatty fish, nuts, olives, seeds (flaxseeds, sesame seeds), shellfish

Limit these solid fats:
Butter, animal fats, stick margarine, shortening

Art © Cengage Learning

TABLE 2-4 USDA Food Patterns: Recommended Weekly Amounts from the Vegetable and Protein Foods Subgroups

Table 2-3 specifies the recommended amounts of total vegetables and protein foods per *day*. This table shows those amounts dispersed among five vegetable and three protein foods subgroups per *week*.

	1600 kcal	1800 kcal	2000 kcal	2200 kcal	2400 kcal	2600 kcal	2800 kcal	3000 kcal
Vegetable Subgroups								
Dark green	1½ c	1½ c	1½ c	2 c	2 c	2½ c	2½ c	2½ c
Red and orange	4 c	5½ c	5½ c	6 c	6 c	7 c	7 c	7½ c
Legumes	1 c	1½ c	1½ c	2 c	2 c	2½ c	2½ c	3 c
Starchy	4 c	5 c	5 c	6 c	6 c	7 c	7 c	8 c
Other	3½ c	4 c	4 c	5 c	5 c	5½ c	5½ c	7 c
Protein Foods Subgroups								
Seafood	8 oz	8 oz	8 oz	9 oz	10 oz	10 oz	11 oz	11 oz
Meats, poultry, eggs	24 oz	24 oz	26 oz	29 oz	31 oz	31 oz	34 oz	34 oz
Nuts, seeds, soy products	4 oz	4 oz	4 oz	4 oz	5 oz	5 oz	5 oz	5 oz

© Cengage Learning

In general, people who regularly eat meat, poultry, and seafood count legumes as a vegetable, and vegetarians and others who seldom eat meat, poultry, or seafood count legumes in the protein foods group.

The USDA Food Patterns encourage greater consumption from certain food groups to provide the nutrients most often lacking in the diets of Americans—dietary fiber, vitamin D, calcium, and potassium. In general, most people need to eat:

- *More* vegetables, fruits, whole grains, seafood, and milk and milk products.
- *Less* sodium, saturated fat, *trans* fat, cholesterol, and *fewer* refined grains and foods and beverages with solid fats and added sugars.

Nutrient-Dense Choices A healthy eating pattern emphasizes nutrient-dense options within each food group.[10] By consistently selecting nutrient-dense foods, a person can obtain all the nutrients needed and still keep kcalories under control. In contrast, eating foods that are low in nutrient density makes it difficult to get enough nutrients without exceeding energy needs and gaining weight. For this reason, consumers should select low-fat foods from each group and foods without solid fats or added sugars—for example, fat-free milk instead of whole milk, baked chicken without the skin instead of hot dogs, green beans instead of french fries, orange juice instead of fruit punch, and whole-wheat bread instead of biscuits. Notice that Figure 2-2 indicates which foods *within each group* contain solid fats and/or added sugars and therefore should be limited. Oil is a notable exception: even though oil is pure fat and therefore rich in kcalories, a small amount of oil from sources such as nuts, fish, or vegetable oils is necessary every day to provide nutrients lacking from other foods. Consequently, these high-fat foods are listed among the nutrient-dense foods (see Highlight 5 to learn why).

> DIETARY GUIDELINES FOR AMERICANS
Consume foods from all food groups in nutrient-dense forms and in recommended amounts. Reduce the intake of kcalories from solid fats and added sugars.

Discretionary kCalories People who consistently choose nutrient-dense foods may be able to meet most of their nutrient needs without consuming their full allowance of kcalories. The difference between the kcalories needed to supply nutrients and those needed to maintain weight might be considered **discretionary kcalories** (see Figure 2-3). Table 2-3 (p. 43) includes discretionary kcalories for each kcalorie level.

discretionary kcalories: the kcalories remaining in a person's energy allowance after consuming enough nutrient-dense foods to meet all nutrient needs for a day.

Discretionary kcalories allow a person to choose whether to:

- Eat additional nutrient-dense foods, such as an extra serving of skinless chicken or a second ear of corn.
- Select a few foods with fats or added sugars, such as reduced-fat milk or sweetened cereal.
- Add a little fat or sugar to foods, such as butter or jelly on toast.
- Consume some alcohol. (Highlight 7 explains why this may not be a good choice for some individuals.)

Alternatively, a person wanting to lose weight might choose to:

- *Not* use discretionary kcalories.

> FIGURE 2-3 **Discretionary kCalories in a 2000-kCalorie Diet**

> **DIETARY GUIDELINES FOR AMERICANS**
> For most people, no more than about 5 to 15 percent of kcalories from solid fats and added sugars (empty kcalories) can be reasonably accommodated in the USDA Food Patterns, which are designed to meet nutrient needs within kcalorie limits.

Serving Equivalents Recommended serving amounts for fruits, vegetables, and milk are measured in cups, and those for grains and protein foods, in ounces. Figure 2-2 (pp. 44–45) provides the **serving sizes** and equivalent measures for foods in each group specifying, for example, that 1 ounce of grains is equivalent to 1 slice of bread or ½ cup of cooked rice.

Consumers using the USDA Food Patterns can learn how standard serving sizes compare with their personal **portion sizes** by determining the answers to questions such as these: What portion of a cup is a small handful of raisins? Is a "helping" of mashed potatoes more or less than a half cup? How many ounces of cereal do you typically pour into the bowl? How many ounces is the steak at your favorite restaurant? How many cups of milk does your glass hold?

Ethnic Food Choices People can use the USDA Food Patterns and still enjoy a diverse array of culinary styles by sorting ethnic foods into their appropriate food groups. For example, a person eating Mexican foods would find tortillas in the grains group, jicama in the vegetable group, and guava in the fruit group. Table 2-5 (p. 48) features some ethnic food choices.

Vegetarian Food Guide Vegetarian diets are plant-based eating patterns that rely mainly on grains, vegetables, legumes, fruits, seeds, and nuts. Some vegetarian diets include eggs, milk products, or both. People who do not eat meats or milk products can still use the USDA Food Patterns to create an adequate diet.[11] The subgroups for protein foods have been reorganized to eliminate meats, poultry, and seafood (see Table H2-1 on p. 66). The other food groups and the recommended daily amounts for each food group remain the same. Highlight 2 defines vegetarian terms and provides details on planning healthy vegetarian diets.

Mixtures of Foods Some foods—such as casseroles, soups, and sandwiches—fall into two or more food groups. With a little practice, consumers can learn to see these mixtures of foods as items from various food groups. For example, from the USDA Food Patterns point of view, a taco represents four different food groups: the taco shell from the grains group; the onions, lettuce, and tomatoes from the vegetables group; the ground beef from the protein foods group; and the cheese from the milk group.

MyPlate The USDA created an educational tool called MyPlate to illustrate the five food groups. Figure 2-4 (p. 48) shows the MyPlate icon, which was designed to remind consumers to make healthy food choices.

Most bagels today weigh in at 4 ounces or more—meaning that a person eating one of these large bagels for breakfast is actually getting four or more grain servings, not one.

---

serving sizes: the standardized quantity of a food; such information allows comparisons when reading food labels and consistency when following the *Dietary Guidelines*.

portion sizes: the quantity of a food served or eaten at one meal or snack; *not* a standard amount.

TABLE 2-5 **Ethnic Food Choices**

	Grains	Vegetables	Fruits	Protein Foods	Milk and Milk Products
Asian	Rice, noodles, millet	Amaranth, baby corn, bamboo shoots, chayote, bok choy, mung bean sprouts, sugar peas, straw mushrooms, water chestnuts, kelp	Carambola, guava, kumquat, lychee, persimmon, melons, mandarin orange	Soybeans and soy products such as soy milk and tofu, squid, duck eggs, pork, poultry, fish and other seafood, peanuts, cashews	Usually excluded
Mediterranean	Pita pocket bread, pastas, rice, couscous, polenta, bulgur, focaccia, Italian bread	Eggplant, tomatoes, peppers, cucumbers, grape leaves	Olives, grapes, figs	Fish and other seafood, gyros, lamb, chicken, beef, pork, sausage, lentils, fava beans	Ricotta, provolone, parmesan, feta, mozzarella, and goat cheeses; yogurt
Mexican	Tortillas (corn or flour), taco shells, rice	Chayote, corn, jicama, tomato salsa, cactus, cassava, tomatoes, yams, chilies	Guava, mango, papaya, avocado, plantain, bananas, oranges	Refried beans, fish, chicken, chorizo, beef, eggs	Cheese, custard

© Josh Resnick/Shutterstock.com

© PhotoDisc/Getty Images

© PhotoDisc/Getty Images

© Cengage Learning

> FIGURE 2-4 **MyPlate**

source: USDA, www.choosemyplate.gov.

Healthy Eating Index: a measure that assesses how well a diet meets the recommendations of the *Dietary Guidelines for Americans.*

The MyPlate icon divides a plate into four sections, each representing a food group—fruits, vegetables, grains, and protein foods. The sections vary in size, indicating the relative proportion each food group contributes to a healthy diet. A circle next to the plate represents the milk group (dairy).

The MyPlate icon does not stand alone as an educational tool. A wealth of information can be found at the website (**www.choosemyplate.gov**). Consumers can choose the kinds and amounts of foods they need to eat each day based on their height, weight, age, gender, and activity level. Information is also available for children, pregnant and breastfeeding women, and college students. In addition to creating a personal plan, consumers can find daily tips to help them improve their diet and increase physical activity. A key message of the website is to enjoy food, but eat less by avoiding oversized portions.

Recommendations versus Actual Intakes The USDA Food Patterns and MyPlate were developed to help people choose a balanced and healthful diet. Are consumers actually eating according to these recommendations? The short answer is "not really." In general, consumers are not selecting the most nutrient-dense items from the food groups.[12] Instead, they are consuming too many foods high in solid fats and added sugars—soft drinks, desserts, whole milk products, and fatty meats. They are also not selecting the suggested quantities from each of the food groups, typically eating too few fruits, vegetables, whole grains, and milk products (see Figure 2-5).

An assessment tool, called the **Healthy Eating Index,** can be used to measure how well a diet meets the recommendations of the *Dietary Guidelines.*[13] Various components of the diet are given scores that reflect the quantities consumed. For most components, higher intakes result in higher scores. For example, selecting at least 3 ounces of whole grains (per 2000 kcalories) gives a score of 10 points,

> FIGURE 2-5   **Recommended and Actual Intakes Compared**

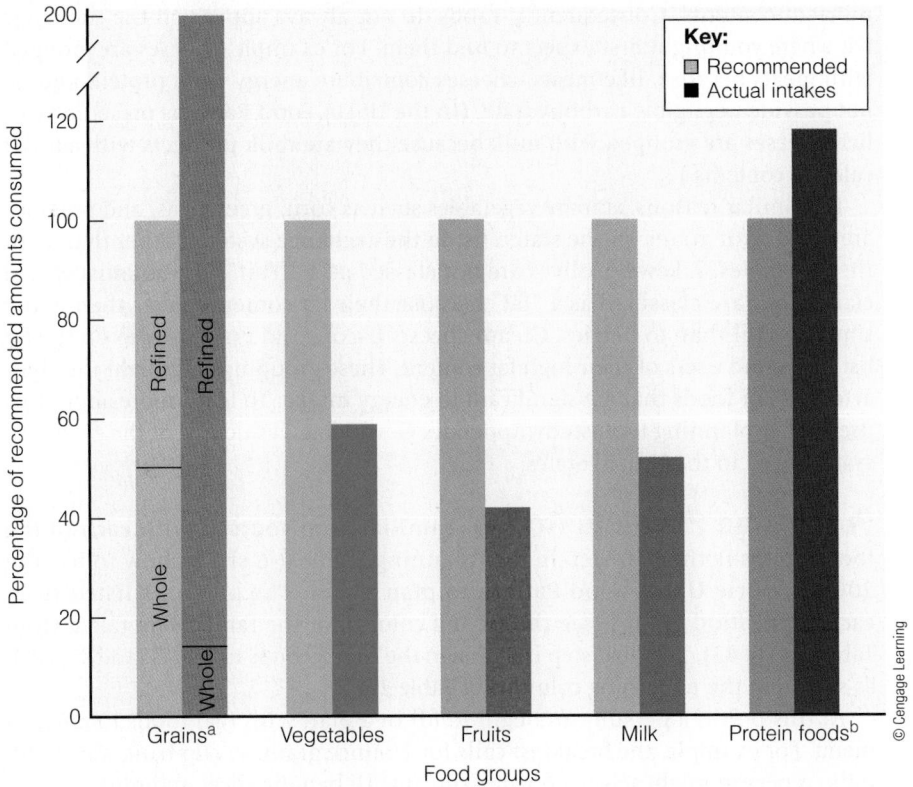

Percentage of recommended amounts consumed

Food groups

© Cengage Learning

[a] At least half of the grain selections should be whole grains.
[b] On average, actual intakes of all protein foods is close to recommended levels, but actual intakes of the seafood subgroup is only 44 percent of recommended levels.

whereas selecting no whole grains gives a score of 0 points. For a few components, lower intakes provide higher scores. For example, less than 2.2 grams of sodium (per 2000 kcalories) receives 10 points, but more than 4 grams gets 0 points. An assessment of recent nutrition surveys using the Healthy Eating Index reports that the American diet scores 54 out of a possible 100 points.[14] To improve this score, the American diet needs to decrease kcalories from solid fats and added sugars by about 60 percent; increase fruits by 100 percent and vegetables and milk products by 70 percent; maintain the quantity of grains but shift the quality to four times as many whole grains; and reduce salt by more than half.[15]

MyPlate Shortcomings MyPlate is not perfect and critics are quick to point out its flaws.[16] The first main criticism is that MyPlate fails to convey enough information to help consumers choose a healthy diet. MyPlate contains few words and depends on its website to provide key information—which is helpful for those who have Internet access and are willing to take the time to become familiar with its teachings. The second main criticism is that MyPlate fails to recognize that some foods within a food group are healthier choices than others. For example, MyPlate does not distinguish between fish sticks and salmon or between broccoli and french fries. Many of the upcoming chapters examine the links between diet and health, and Chapter 18 presents a complete summary, including a look at an alternative Healthy Eating Plate created by the faculty members in the Harvard School of Public Health.

Exchange Lists Food group plans are particularly well suited to help a person achieve dietary adequacy, balance, and variety. **Exchange lists** provide additional help in achieving kcalorie control and moderation. Originally developed as a meal planning guide for people with diabetes, exchange lists have proved useful for general diet planning as well.

exchange lists: diet-planning tools that organize foods by their proportions of carbohydrate, fat, and protein. Foods on any single list can be used interchangeably.

Unlike the USDA Food Patterns, which sort foods primarily by their vitamin and mineral contents, the exchange system sorts foods according to their energy-nutrient contents. Consequently, foods do not always appear on the exchange list where you might first expect to find them. For example, cheeses are grouped with meats because, like meats, cheeses contribute energy from protein and fat but provide negligible carbohydrate. (In the USDA Food Patterns presented earlier, cheeses are grouped with milk because they are milk products with similar calcium contents.)

For similar reasons, starchy vegetables such as corn, green peas, and potatoes are listed with grains on the starch list in the exchange system, rather than with the vegetables. Likewise, olives are not classed as a "fruit" as a botanist would claim; they are classified as a "fat" because their fat content makes them more similar to oil than to berries. Cream cheese, bacon, and nuts are also on the fat list to remind users of their high fat content. These groupings highlight the characteristics of foods that are significant to energy intake. To learn more about this useful diet-planning tool, study Appendix G, which gives details of the exchange system used in the United States.

Putting the Plan into Action   Familiarizing yourself with each of the food groups is the first step in diet planning. Table 2-6 shows how to use the 2000-kcalorie USDA Food Pattern to plan a diet. The amounts listed from each of the food groups (see the second column of the table) were taken from Table 2-3 (p. 43). The next step is to assign the food groups to meals (and snacks), as shown in the remaining columns of Table 2-6.

At this point, a person can begin to fill in a plan with real foods to create a menu. For example, the breakfast calls for 1 ounce grain, ½ cup fruit, and 1 cup milk. A person might select a bowl of cereal with banana slices and milk:

1 cup cereal = 1 ounce grain

½ large banana = ½ cup fruit

1 cup fat-free milk = 1 cup milk

Or ½ English muffin and a bowl of strawberries topped with yogurt:

½ English muffin = 1 ounce grain

½ cup strawberries = ½ cup fruit

1 cup fat-free plain yogurt = 1 cup milk

Then the person can continue to create a diet plan by creating menus for lunch, dinner, and snacks. The final menu might look like the one presented in Table 2-7.

As you can see, we all make countless food-related decisions daily—whether we have a plan or not. Following an eating pattern that incorporates health recommendations and diet-planning principles helps a person make wise nutrition decisions.

TABLE 2-6 **Diet Planning Using the 2000-kCalorie USDA Food Pattern**

This diet plan is one of many possibilities. It follows the amounts of foods suggested for a 2000-kcalorie diet as shown in Table 2-3 (p. 43), with a little less oil.

Food Group	Amounts	Breakfast	Lunch	Snack	Dinner	Snack
Fruits	2 c	½ c		½ c	1 c	
Vegetables	2½ c		1 c		1½ c	
Grains	6 oz	1 oz	2 oz	½ oz	2 oz	½ oz
Protein foods	5½ oz		2 oz		3½ oz	
Milk and milk products	3 c	1 c		1 c		1 c
Oils	6 tsp		1½ tsp		4 tsp	

TABLE 2-7 A Sample Menu

This sample menu provides about 1850 kcalories and meets the recommendations to provide 45 to 65 percent of kcalories from carbohydrate, 20 to 35 percent from fat, and 10 to 35 percent from protein.

Amounts	Sample Menu	Energy (kcal)
Breakfast		
1 oz whole grains	1 c whole-grain cereal	108
1 c milk	1 c fat-free milk	100
½ c fruit	1 medium banana (sliced)	105
Lunch		
2 oz meats, 2 oz whole grains	1 turkey sandwich on whole-wheat roll	272
1½ tsp oils	1½ tbs low-fat mayonnaise	71
1 c vegetables	1 c vegetable juice	50
Snack		
½ oz whole grains	4 whole-wheat reduced-fat crackers	86
1 c milk	1½ oz low-fat cheddar cheese	74
½ c fruit	1 medium apple	72
Dinner		
½ c vegetables	1 c raw spinach leaves	8
¼ c vegetables	¼ c shredded carrots	11
1 oz meats	¼ c garbanzo beans	71
2 tsp oils	2 tbs oil-based salad dressing and olives	76
¾ c vegetables, 2½ oz meat, 2 oz enriched grains	Spaghetti with meat and tomato sauce	425
½ c vegetables	½ c green beans	22
2 tsp oils	2 tsp soft margarine	67
1 c fruit	1 c strawberries	49
Snack		
½ oz whole grains	3 graham crackers	90
1 c milk	1 c fat-free milk	100

© Cengage Learning

From Guidelines to Groceries

Dietary recommendations emphasize nutrient-rich foods such as whole grains, fruits, vegetables, lean meats, poultry, seafood, and low-fat milk products. You can design such a diet for yourself, but how do you begin? Start with the foods you regularly enjoy eating and then try to make a few improvements.[17] For most people that will mean eating less red meat, cheeses, and salted snacks and more fruits, vegetables, whole grains, legumes, nuts, milk products, and seafood. Such small changes can dramatically improve the diet. When shopping, think of the food groups, and choose nutrient-dense foods within each group.

Be aware that many of the tens of thousands of food options available today are **processed foods** that have lost valuable nutrients and gained sugar, fat, and salt as they were transformed from farm-fresh foods to those found in the bags, boxes, and cans that line grocery-store shelves. Their value in the diet depends on the original food and how it was prepared or processed. By eating more fresh foods and fewer processed foods, consumers can reduce their intakes of added sugars, solid fats, and sodium for relatively little effort. Sometimes processed foods have been **fortified** to improve their nutrient contents, which can be helpful in increasing dietary intake of specific vitamins and minerals.

Grains When shopping for grain products, you will find them described as *refined*, *enriched*, or *whole grain*. These terms refer to the milling process and the making of grain products, and they have different nutrition implications (see Figure 2-6, p. 52). **Refined** grains have lost many nutrients during processing; **enriched** grains have had some nutrients added back; and **whole-grain** products have all the nutrients and fiber found in the original grain. As such, whole-grain

processed foods: foods that have been treated to change their physical, chemical, microbiological, or sensory properties.

fortified: the addition to a food of nutrients that were either not originally present or present in insignificant amounts. Fortification can be used to correct or prevent a widespread nutrient deficiency or to balance the total nutrient profile of a food.

refined: the process by which the coarse parts of a food are removed. When wheat is refined into flour, the bran, germ, and husk are removed, leaving only the endosperm.

enriched: the addition to a food of specific nutrients to replace losses that occur during processing so that the food will meet a specified standard.

whole grain: a grain that maintains the same relative proportions of starchy endosperm, germ, and bran as the original (all but the husk); not refined.

> FIGURE 2-6   A Wheat Plant

The protective coating of **bran** around the kernel of grain is rich in nutrients and fiber.

The **endosperm** contains starch and proteins.

The **germ** is the seed that grows into a wheat plant, so it is especially rich in vitamins and minerals to support new life.

The outer **husk** (or **chaff**) is the inedible part of a grain.

Whole-grain products contain much of the germ and bran, as well as the endosperm; that is why they are so nutritious. Refined grain products contain only the endosperm. Even with nutrients added back, they are not as nutritious as whole-grain products, as the next figure shows.

© Thomas Harm & Tom Peterson/Quest Photographic, Inc.

Common types of flour:

- **Refined flour:** finely ground endosperm that is usually enriched with nutrients and bleached for whiteness; sometimes called *white flour.*
- **Wheat flour:** any flour made from the endosperm of the wheat kernel.
- **Whole-wheat flour:** any flour made from the entire wheat kernel.

The difference between *white flour* and *white wheat* is noteworthy. Typically, *white flour* refers to refined flour (as defined above). Most flour—whether refined, white, or whole wheat—is made from red wheat. Whole-grain products made from red wheat are typically brown and full flavored.

To capture the health benefits of whole grains for consumers who prefer white bread, manufacturers use an albino variety of wheat called *white wheat.* Whole-grain products made from white wheat provide the nutrients and fiber of a whole grain with a light color and natural sweetness. Read labels carefully—white bread is a whole-grain product only if it is made from whole white wheat.

© Cengage Learning

products support good health and should account for at least half of the grains daily. Adding more whole grains to the diet can be as easy as eating oatmeal for breakfast and popcorn for a snack or substituting brown rice for white rice and whole-wheat bread for enriched white bread. To find whole-grain products, read food labels and select those that name a whole-grain first in the ingredient list. Examples of whole grains include:

- Amaranth
- Barley
- Buckwheat
- Bulgur
- Corn (and popcorn)
- Millet
- Oats (and oatmeal)
- Quinoa
- Rice (brown or wild)
- Whole rye
- Whole wheat

Products described as "multi-grain," "stone-ground," or "100% wheat" are usually *not* whole-grain products. Brown color is also not a useful hint, but fiber content often is.

When it became a common practice to refine the wheat flour used for bread by milling it and throwing away the bran and the germ, consumers suffered a tragic loss of many nutrients. As a consequence, in the early 1940s Congress passed legislation requiring that all grain products that cross state lines be enriched with iron, thiamin, riboflavin, and niacin. In 1996 this legislation was amended to include folate, a vitamin considered essential in the prevention of some birth defects. Most grain products that have been refined, such as rice, pastas such as macaroni and spaghetti, and cereals (both cooked and ready-to-eat types), have

© Roman Barnes Photo Research

When shopping for bread, look for the descriptive words *whole grain* or *whole wheat* and check the fiber content on the Nutrition Facts panel of the label—the more fiber, the more likely the bread is a whole-grain product.

subsequently been enriched. Food labels must specify that products have been enriched and include the enrichment nutrients in the ingredients list.

Enrichment doesn't make a slice of bread rich in these added nutrients, but people who eat several slices a day obtain significantly more of these nutrients than they would from unenriched bread. Even though the enrichment of flour helps to prevent deficiencies of these nutrients, it fails to compensate for losses of many other nutrients and fiber. As Figure 2-7 shows, whole-grain items deliver many more nutrients than the enriched ones. Only *whole-grain* flour contains all of the nutritive portions of the grain. Whole-grain products, such as brown rice and oatmeal, provide more nutrients and fiber and contain less salt, sugar, and fat than refined grain products. This helps to explain why diet quality tends to be better for consumers who eat more whole grains.[18]

> **DIETARY GUIDELINES FOR AMERICANS**
Increase whole-grain intake. Consume at least half of all grains as whole grains. Whenever possible, replace refined grains with whole grains.

Speaking of processed foods, ready-to-eat breakfast cereals are the most highly fortified foods on the market. Like an enriched food, a *fortified* food has had nutrients added during processing, but in a fortified food, the added nutrients may not have been present in the original product. (The terms *fortified* and *enriched* may be used interchangeably.[19]) Some breakfast cereals made from refined flour and fortified with high doses of vitamins and minerals are actually more like dietary supplements disguised as cereals than they are like whole grains. They may be nutritious—with respect to the nutrients added—but they still may fail to convey the full spectrum of nutrients that a whole-grain food or a mixture of such foods might provide. Still, fortified foods help people meet their vitamin and mineral needs.

> **FIGURE 2-7   Nutrients in Bread**
Whole-grain bread is more nutritious than other breads, even enriched bread. For iron, thiamin, riboflavin, niacin, and folate, enriched bread provides about the same quantities as whole-grain bread and significantly more than unenriched bread. For fiber and the other nutrients (those shown here as well as those not shown), enriched bread provides less than whole-grain bread.

Percentage of nutrients as compared with whole-grain bread

Key:
Whole-grain bread
Enriched bread
Unenriched bread

Consumers can remember to eat a variety of fruits and vegetables every day by selecting from each of five colors.

Vegetables Posters in the produce section of grocery stores encourage consumers to "think variety, think color." Such efforts are part of a national educational campaign to increase fruit and vegetable consumption. Easy ways to effectively increase vegetable consumption include serving a variety of vegetables at meals, increasing the portion sizes, and adding pureed vegetables to recipes such as muffins or casseroles.[20]

Choose fresh vegetables often, especially dark-green leafy and red and orange vegetables such as spinach, broccoli, tomatoes, and sweet potatoes. Cooked or raw, vegetables are good sources of vitamins, minerals, and fiber. Frozen and canned vegetables without added salt are acceptable alternatives to fresh. To control fat, energy, and sodium intakes, limit butter and salt on vegetables.

Choose often from the variety of legumes available:

- Adzuki beans
- Black beans
- Black-eyed peas
- Fava beans
- Garbanzo beans
- Great northern beans
- Kidney beans

- Lentils
- Lima beans
- Navy beans
- Peanuts
- Pinto beans
- Soybeans
- Split peas

Legumes are an economical, low-fat, nutrient- and fiber-rich food choice. Combining legumes with foods from other food groups creates delicious meals (see Figure 2-8).

 > DIETARY GUIDELINES FOR AMERICANS
Increase vegetable intake. Eat recommended amounts of vegetables and include a variety of vegetables, especially dark-green vegetables, red and orange vegetables, and legumes.

Fruit Choose fresh fruits often. Frozen, dried, and canned fruits without added sugar are acceptable alternatives to fresh. Fruits supply valuable vitamins, minerals, fibers, and phytochemicals. They add flavors, colors, and textures to meals, and their natural sweetness makes them enjoyable as snacks or desserts.

Fruit juices are healthy beverages but contain little dietary fiber compared with whole fruits. Whole fruits satisfy the appetite better than juices, thereby helping people to limit food energy intakes. For people who need extra food energy, though, 100 percent fruit juices are a good choice. Be aware that sweetened fruit "drinks" or "ades" contain mostly water, sugar, and a little juice for flavor. Some may have been fortified with vitamin C or calcium but lack any other significant nutritional value.

 > DIETARY GUIDELINES FOR AMERICANS
Increase fruit intake. Eat recommended amounts of fruits and choose a variety of fruits. Choose whole or cut-up fruits more often than fruit juice.

Protein Foods Protein foods include seafood, meats, poultry, eggs, legumes, soy products, nuts, and seeds. In addition to protein, these foods provide B vitamins,

> FIGURE 2-8 **Meals Featuring Legumes**

Add rice to red beans for a hearty meal.

Enjoy a Greek salad topped with garbanzo beans for a little ethnic diversity.

A bit of meat and lots of spices turn kidney beans into chili con carne.

vitamin E, iron, zinc, and magnesium. To buy and prepare these foods without adding excess energy, fat, and sodium takes a little knowledge and planning.

When shopping in the meat department, choose lean cuts of beef and pork named "round" or "loin" (as in top round or pork tenderloin). As a guide, "prime" and "choice" cuts generally have more fat than "select" cuts. Restaurants usually serve prime cuts. Ground beef, even "lean" ground beef, derives most of its food energy from fat. Have the butcher trim and grind a lean round steak instead. Alternatively, soy products such as **textured vegetable protein** can be used instead of ground beef in a casserole, spaghetti sauce, or chili, saving fat kcalories. Because nuts and seeds are energy dense, they need to be consumed in small quantities and in place of—not in addition to—other protein foods. To lower sodium intake, choose unsalted nuts and seeds.

Serving sizes for meats, poultry, and seafood reflect weight after cooking and without bones. In general, 4 ounces of raw meat is equal to about 3 ounces of cooked meat. Some examples of 3-ounce portions include 1 medium pork chop, ½ chicken breast, or 1 steak or fish filet about the size of a deck of cards. To keep fat intake moderate, bake, roast, broil, grill, or braise meats, poultry, and seafood (but do not fry them in fat); remove the skin from poultry after cooking; trim visible fat before cooking; and drain fat after cooking. Chapter 5 offers many additional strategies for moderating fat intake.

> DIETARY GUIDELINES FOR AMERICANS
Choose a variety of protein foods, which include seafood, lean meats and poultry, eggs, legumes, soy products, and unsalted nuts and seeds. Increase the amount and variety of seafood consumed by choosing seafood in place of some meat and poultry.

Milk and Milk Products Shoppers find a variety of fortified foods in the dairy case. Examples are milk, to which vitamins A and D have been added, and soy milk, to which calcium, vitamin D, and vitamin B_{12} have been added. Be aware that not all soy beverages have been fortified. Read labels carefully.

In addition, shoppers may find **imitation foods** (such as cheese products), **food substitutes** (such as egg substitutes), and functional foods (such as margarine with added plant sterols). As food technology advances, many such foods offer alternatives to traditional choices that may help people reduce their saturated fat and cholesterol intakes. Chapter 5 provides other examples.

Milk is often described by its fat contents:

- Fat-free milk (also called nonfat, skim, zero-fat, or no-fat)
- Low-fat milk (also called 1% milk)
- Reduced-fat milk (also called 2% milk)
- Whole milk

When shopping, choose fat-free or low-fat milk, yogurt, and cheeses. Such selections help consumers meet their vitamin and mineral needs within their energy and fat allowances. Milk products are important sources of calcium but can provide too much sodium and fat if not selected with care.

> DIETARY GUIDELINES FOR AMERICANS
Increase intake of fat-free or low-fat milk and milk products—such as milk, yogurt, cheese, or fortified soy milk—and replace whole milk products with fat-free or low-fat options.

› REVIEW IT Use the USDA Food Patterns to develop a meal plan within a specified energy allowance.
Food group plans such as the USDA Food Patterns help consumers select the types and amounts of foods to provide adequacy, balance, and variety in the diet. They make it easier to plan a diet that includes a balance of grains, vegetables, fruits, protein foods, and milk and milk products. In making any food choice, remember to view the food in the context of the total diet. The combination of many different foods provides the array of nutrients that is so essential to a healthy diet.

textured vegetable protein: processed soybean protein used in vegetarian products such as soy burgers.

2.3 Food Labels

› LEARN IT Compare the information on food labels to make selections that meet specific dietary and health goals.

Many consumers, especially those interested in preventing chronic diseases, read food labels to help them make healthy choices.[21] Food labels appear on virtually all packaged foods, and posters or brochures provide similar nutrition information for fresh fish, fruits, and vegetables. A few foods need not carry nutrition labels: those contributing few nutrients, such as plain coffee, tea, and spices; those produced by small businesses; and those prepared and sold in the same establishment. Markets selling nonpackaged items may voluntarily present nutrient information, either in brochures or on signs posted at the point of purchase.

Restaurants with 20 or more locations must provide menu listings of an item's kcalories, grams of saturated fat, and milligrams of sodium.[22] Other restaurants need not supply nutrition information for menu items unless claims such as "low-fat" or "heart healthy" have been made. When ordering such items, keep in mind that restaurants tend to serve extra-large portions—two to three times standard serving sizes.

The Ingredient List *All* packaged foods must list *all* ingredients—including additives used to preserve or enhance foods, such as vitamins and minerals added to enrich or fortify products. The ingredients are listed on the label in descending order of predominance by weight. Knowing that the first ingredient predominates by weight, consumers can glean much information. Compare these products, for example:

- A beverage powder that contains "sugar, citric acid, natural flavors . . ." versus a juice that contains "water, tomato concentrate, concentrated juices of carrots, celery . . ."
- A cereal that contains "puffed milled corn, sugar, corn syrup, molasses, salt . . ." versus one that contains "100 percent rolled oats"
- A canned fruit that contains "sugar, apples, water" versus one that contains simply "apples, water"

In each of these comparisons, consumers can see that the second product is more nutrient dense.

Nutrition Facts Panel The Nutrition Facts panel provides valuable nutrition information such as serving sizes, nutrient quantities, and Daily Values. Recent revisions to the nutrition facts panel reflect current nutrition science, updated serving sizes, and an improved design (see Figure 2-9).

Serving Sizes Because labels present nutrient information based on one serving, they must identify the size of the serving. The Food and Drug Administration (FDA) has established serving sizes for various foods and requires that all labels for a given product use the same serving size. For example, the new standard serving size for all ice creams is 1 cup. This facilitates comparison shopping. Consumers can see at a glance which brand has more or fewer kcalories or grams of added sugars, for example.

When examining the nutrition facts on a food label, consumers need to compare the serving size on the label with how much they actually eat and adjust their calculations accordingly. For example, if the serving size is four cookies and you eat only two, then you need to cut the nutrient and kcalorie values in half; similarly, if you eat eight cookies, then you need to double the values. Packages, such as a 15-ounce can of soup, that contain more than one but less than two servings and are commonly eaten in one sitting are labeled as one serving. For

imitation foods: foods that substitute for and resemble another food, but are nutritionally inferior to it with respect to vitamin, mineral, or protein content. If the substitute is not inferior to the food it resembles and if its name provides an accurate description of the product, it need not be labeled "imitation."

food substitutes: foods that are designed to replace other foods.

> **FIGURE 2-9** **Example of a Food Label**

Original Label

Nutrition Facts

Serving size and number of servings per container →

Serving Size 2/3 cup (55g)
Servings Per Container About 8

Amount Per Serving	
Calories 230	Calories from Fat 40

kCalories per serving and kcalories from fat →

	% Daily Value*
Total Fat 8g	**12%**
Saturated Fat 1g	**5%**
Trans Fat 0g	
Cholesterol 0mg	**0%**
Sodium 160mg	**7%**
Total Carbohydrate 37g	**12%**
Dietary Fiber 4g	**16%**
Sugars 1g	
Protein 3g	
Vitamin A	10%
Vitamin C	8%
Calcium	20%
Iron	45%

Nutrient quantities per serving listed in actual amounts and in % Daily Values based on 2000-kcalorie diet →

Nutrients required for Daily Values →

* Percent Daily Values are based on a 2,000 calorie diet. Your daily value may be higher or lower depending on your calorie needs.

	Calories:	2,000	2,500
Total Fat	Less than	65g	80g
Sat Fat	Less than	20g	25g
Cholesterol	Less than	300mg	300mg
Sodium	Less than	2,400mg	2,400mg
Total Carbohydrate		300mg	375mg
Dietary Fiber		25g	30g

Daily Values reminder for selected nutrients for a 2000- and a 2500-kcalorie diet →

Updated Label

Nutrition Facts

8 servings per container

Serving size	2/3 cup (55g)

Amount Per 2/3 cup

Calories   **230**

% DV*	
12%	**Total Fat** 8g
5%	Saturated Fat 1g
	Trans Fat 0g
0%	**Cholesterol** 0mg
7%	**Sodium** 160mg
12%	**Total Carbs** 37g
14%	Dietary Fiber 4g
	Sugars 1g
	Added Sugars 0g
	Protein 3g
10%	**Vitamin D** 2mcg
20%	**Calcium** 260mg
45%	**Iron** 8mg
5%	**Potassium** 235mg

* Footnote to help consumers understand the %DV will be inserted here.

← Servings per container in large, bold type; serving sizes revised to reflect actual portion sizes

← Serving size lists amount per quantity, not per serving; kcalories in large, bold type; kcalories from fat not listed

← Daily Values revised and reformatted to list % DV first

← Separate listing for added sugars

← Nutrients required for Daily Values revised to reflect nutrients of concern

← Footnote explains how to use Daily Values

© 2016 Cengage Learning

---

packages that contain two to four servings, food labels present two columns, listing information both "per serving" and "per package." Such dual listings are particularly helpful for people who may consume the entire package in a single sitting. Examples include pints of ice cream and 20-ounce sodas.

Nutrient Quantities In addition to the serving size and the servings per container, the FDA requires that the Nutrition Facts panel on food labels present nutrient information in two ways—in quantities (such as grams) and as percentages of standards called the **Daily Values.** The Nutrition Facts panel must provide the nutrient amount, **percent Daily Value,** or both for the following:

- Total food energy (kcalories)
- Total fat (grams and percent Daily Value)
- Saturated fat (grams and percent Daily Value)
- *Trans* fat (grams)
- Cholesterol (milligrams and percent Daily Value)
- Sodium (milligrams and percent Daily Value)
- Total carbohydrate, which includes starch, sugar, and fiber (grams and percent Daily Value)
- Dietary fiber (grams and percent Daily Value)

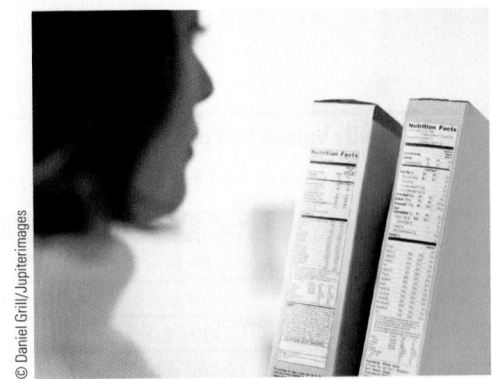

© Daniel Grill/Jupiterimages

Consumers read food labels to learn about the nutrient contents of a food or to compare similar foods.

Daily Values (DV): reference values developed by the FDA specifically for use on food labels.

percent Daily Value (%DV): the percentage of a Daily Value recommendation found in a specified serving of food for key nutrients based on a 2000-kcalorie diet.

- Sugars, which includes both those naturally present in and those added to the food (grams)
- Added sugars, which includes only those added to the food (grams)
- Protein (grams)

The labels must also present nutrient content information as a percent Daily Value for the following nutrients of concern:

- Vitamin D
- Calcium
- Iron
- Potassium

The Daily Values Table 2-8 presents the Daily Value standards for nutrients that are required to provide this information. Food labels list the amount of some nutrients in a product as a percentage of its Daily Value, which makes the numbers more meaningful to consumers. A person reading a food label might wonder, for example, whether 1 milligram of iron or calcium is a little or a lot. As Table 2-8 shows, the Daily Value for iron is 18 milligrams, so 1 milligram of iron is enough to notice—it is more than 5 percent, and that is what the food label will say. But because the Daily Value for calcium on food labels is 1300 milligrams, 1 milligram of calcium is insignificant, and the food label will read "0%."

The Daily Values reflect dietary recommendations for nutrients and dietary components that have important relationships with health. For example, for heart health, consumers are advised to limit saturated fat to 10 percent of energy intake. For a 2000-kcalorie diet, 10 percent is 200 kcalories, or 22 grams of fat. (Remember that fats deliver 9 kcalories per gram.) As Table 2-8 shows, the Daily Value for saturated fat has been rounded down to 20 grams.

The "% Daily Value" column on a label provides a ballpark estimate of how individual foods contribute to the total diet. It compares key nutrients in a serving of food with the goals of a person consuming 2000 kcalories per day. A 2000-kcalorie diet is considered about right for sedentary younger women, active older women, and sedentary older men. Young children and sedentary older women may need fewer kcalories. By comparison, a 2500-kcalorie diet is considered about right for many men, teenage boys, and active younger women. People who are exceptionally active may have still higher energy needs.

People who consume 2000 kcalories a day can simply add up all of the "% Daily Values" for a particular nutrient to see if their diet for the day fits recommendations. People who require more or less than 2000 kcalories daily must do some calculations to see how foods compare with their personal nutrition goals. Those interested can use the Calculation Factors column in Table 2-8 or the suggestions presented in the "How To" feature.

Daily Values help consumers readily see whether a food contributes "a little" or "a lot" of a nutrient. For example, the "% Daily Value" column on a package of frozen macaroni and cheese may say 20 percent for saturated fat. This tells the consumer that each serving of this food contains about 20 percent of the day's allotted 20 grams of saturated fat. Be aware that for some nutrients (such as saturated fat and sodium) you will want to select foods with a low "% Daily Value" and for others (such as calcium and fiber) you will want a high "% Daily Value." To determine whether a particular food is a wise choice, a consumer needs to consider its place in the diet among all the other foods eaten during the day.

TABLE 2-8 Daily Values for Food Labels

Food labels must present the "% Daily Value" for these nutrients.

Nutrient	Daily Value	Calculation Factors
Fat (total)	65 g	30% of kcalories
Saturated fat	20 g	10% of kcalories
Cholesterol	300 mg	—
Sodium	2300 mg	—
Carbohydrate (total)	300 g	60% of kcalories
Fiber	28 g	14 g per 1000 kcalories
Vitamin D	20 µg	—
Calcium	1300 mg	—
Iron	18 mg	—
Potassium	4700 mg	—

NOTE: Daily Values were established for adults and children aged 4 years and older and are based on an energy intake of 2000 kcalories a day.

© Cengage Learning

Daily Values also make it easy to compare foods. For example, a consumer might discover that frozen macaroni and cheese has a Daily Value for saturated fat of 20 percent, whereas macaroni and cheese prepared from a boxed mix has a Daily Value of 15 percent. By comparing labels, consumers who are concerned about their saturated fat intakes can make informed decisions.

Front-of-Package Labels Some consumers find the many numbers on Nutrition Facts panels overwhelming. They want an easier and quicker way to interpret information and select products. Food manufacturers responded by creating front-of-package labels that incorporate text, color, and icons to present key nutrient facts.[23] Without any regulations or oversight, however, different companies used a variety of different symbols to describe how healthful their products were. To calm the chaos and maintain the voluntary status of front-of-package labels, major food industry associations created a standardized presentation of nutrient information called Facts Up Front (see Figure 2-10). Whether consumers find this approach to be more helpful remains to be seen.[24] The FDA is currently evaluating the program and reviewing recommendations from the Institute of Medicine to determine the best way to present front-of-package information.[25]

> FIGURE 2-10   **Facts Up Front**

This example of front-of-package labeling (created by Grocery Manufacturers Association and the Food Marketing Institute) presents key nutrient facts.

Created by Grocery Manufacturers Association and the Food Marketing Institute

Claims on Labels In addition to the Nutrition Facts panel, consumers may find various claims on labels. These claims include nutrient claims, health claims, and structure-function claims.

Nutrient Claims Have you noticed phrases such as "good source of fiber" on a box of cereal or "rich in calcium" on a package of cheese? These and other **nutrient claims** may be used on labels so long as they meet FDA definitions, which include the conditions under which each term can be used. For example, in addition to having less than 2 milligrams of cholesterol, a "cholesterol-free" product may not contain more than 2 grams of saturated fat and *trans* fat combined per serving. The accompanying glossary defines nutrient terms on food labels, including criteria for foods described as "low," "reduced," and "free." When nutrients have been added to enriched or fortified products, they must appear in the ingredients list.

Some descriptions *imply* that a food contains, or does not contain, a nutrient. Implied claims are prohibited unless they meet specified criteria. For example, a claim that a product "contains no oil" *implies* that the food contains no fat. If the product is truly fat-free, then it may make the no-oil claim, but if it contains another source of fat, such as butter, it may not.

Health Claims **Health claims** describe a relationship between a food (or food component) and a disease or health-related condition. In some cases, the FDA authorizes health claims based on an extensive review of the scientific literature. For example, the health claim that "Diets low in sodium may reduce the risk of high blood pressure" is based on enough scientific evidence to establish a clear link between diet and health. In cases where there is emerging—but not established—evidence for a relationship between a food or food component and disease, the FDA allows the use of *qualified* health claims that must use specific language indicating that the evidence supporting the claim is limited. A qualified health claim might claim that "Very limited and preliminary research suggests that eating one-half to one cup of tomatoes and/or tomato sauce a week may reduce the risk of prostate cancer. The FDA concludes that there is little scientific evidence supporting the claim."

Structure-Function Claims Unlike health claims, which require food manufacturers to collect scientific evidence and petition the FDA, **structure-function claims** can be made without any FDA approval. Product labels can claim to "slow aging," "improve memory," and "build strong bones" without any proof. The only criterion for a structure-function claim is that it must not mention a disease or symptom. Unfortunately, structure-function claims can be deceptively similar to health claims, and most consumers do not distinguish between different types of claims.[26] Consider these statements:

- "May reduce the risk of heart disease"
- "Promotes a healthy heart"

The first is a health claim that requires FDA approval and the second is an unproven, but legal, structure-function claim. Figure 2-11 compares label claims.

Consumer Education Food labels are a primary source of information for consumers trying to make healthy diet choices, which is why FDA recently updated labels to place a bigger emphasis on total kcalories, added sugars, and nutrients of concern, such as vitamin D and potassium. In addition, the FDA has designed several programs to educate consumers. Consumers who understand how to read labels are best able to apply the information to achieve and maintain healthful dietary practices. Table 2-9 (p. 62) shows how the messages from the *Dietary Guidelines*, the USDA Food Patterns, and food labels coordinate with one another.

> REVIEW IT Compare the information on food labels to make selections that meet specific dietary and health goals.

Food labels provide consumers with information they need to select foods that will help them meet their nutrition and health goals. When labels contain relevant information presented in a standardized, easy-to-read format, consumers are well prepared to plan and create healthful diets.

nutrient claims: statements that characterize the quantity of a nutrient in a food.

health claims: statements that characterize the relationship between a nutrient or other substance in a food and a disease or health-related condition.

structure-function claims: statements that characterize the relationship between a nutrient or other substance in a food and its role in the body.

> FIGURE 2-11 Label Claims

Nutrient claims characterize the level of a nutrient in the food—for example, "fat free" or "less sodium."

Health claims characterize the relationship of a food or food component to a disease or health-related condition—for example, "soluble fiber from oatmeal daily in a diet low in saturated fat and cholesterol may reduce the risk of heart disease" or "a diet low in total fat may reduce the risk of some cancers."

Structure/function claims describe the effect that a substance has on the structure or function of the body and do not make reference to a disease—for example, "supports immunity and digestive health" or "calcium builds strong bones."

GLOSSARY
OF TERMS ON FOOD LABELS

GENERAL TERMS

free: "nutritionally trivial" and unlikely to have a physiological consequence; synonyms include *without, no,* and *zero.* A food that does not contain a nutrient naturally may make such a claim, but only as it applies to all similar foods (for example, "applesauce, a fat-free food").

gluten-free: a food that contains less that 20 parts per million of gluten from any source; synonyms include *no gluten, free of gluten,* or *without gluten.*

good source of: the product provides between 10 and 19 percent of the Daily Value for a given nutrient per serving.

healthy: a food that is low in fat, saturated fat, cholesterol, and sodium and that contains at least 10 percent of the Daily Values for vitamin D, potassium, iron, calcium, protein, or fiber.

high: 20 percent or more of the Daily Value for a given nutrient per serving; synonyms include *rich in* or *excellent source.*

less: at least 25 percent less of a given nutrient or kcalories than the comparison food (see individual nutrients); synonyms include *fewer* and *reduced.*

light or **lite:** one-third fewer kcalories than the comparison food; 50 percent or less of the fat or sodium than the comparison food; any use of the term other than as defined must specify what it is referring to (for example, "light in color" or "light in texture").

low: an amount that would allow frequent consumption of a food without exceeding the Daily Value for the nutrient. A food that is naturally low in a nutrient may make such a claim, but only as it applies to all similar foods (for example, "fresh cauliflower, a low-sodium food"); synonyms include *little, few,* and *low source of.*

more: at least 10 percent more of the Daily Value for a given nutrient than the comparison food; synonyms include *added* and *extra.*

organic: on food labels, that at least 95 percent of the product's ingredients have been grown and processed according to USDA regulations defining the use of fertilizers, herbicides, insecticides, fungicides, preservatives, and other chemical ingredients (see Chapter 19).

ENERGY

kcalorie-free: fewer than 5 kcalories per serving.

low kcalorie: 40 kcalories or less per serving.

reduced kcalorie: at least 25 percent fewer kcalories per serving than the comparison food.

FAT AND CHOLESTEROL[a]

percent fat-free: may be used only if the product meets the definition of *low fat* or *fat-free* and must reflect the amount of fat in 100 grams (for example, a food that contains 2.5 grams of fat per 50 grams can claim to be "95 percent fat-free").

fat-free: less than 0.5 gram of fat per serving (and no added fat or oil); synonyms include *zero-fat, no-fat,* and *nonfat.*

low fat: 3 grams or less of fat per serving.

less fat: 25 percent or less fat than the comparison food.

saturated fat-free: less than 0.5 gram of saturated fat and 0.5 gram of *trans* fat per serving.

low saturated fat: 1 gram or less of saturated fat and less than 0.5 gram of *trans* fat per serving.

less saturated fat: 25 percent or less of saturated fat and *trans* fat combined than the comparison food.

***trans* fat-free:** less than 0.5 gram of *trans* fat and less than 0.5 gram of saturated fat per serving.

cholesterol-free: less than 2 milligrams of cholesterol per serving and 2 grams or less of saturated fat and *trans* fat combined per serving.

low cholesterol: 20 milligrams or less of cholesterol per serving and 2 grams or less of saturated fat and *trans* fat combined per serving.

less cholesterol: 25 percent or less cholesterol than the comparison food (reflecting a reduction of at least 20 milligrams per serving), and 2 grams or less of saturated fat and *trans* fat combined per serving.

extra lean: less than 5 grams of fat, 2 grams of saturated fat and *trans* fat combined, and 95 milligrams of cholesterol per serving and per 100 grams of meat, poultry, and seafood.

lean: less than 10 grams of fat, 4.5 grams of saturated fat and *trans* fat combined, and 95 milligrams of cholesterol per serving and per 100 grams of meat, poultry, and seafood. For mixed dishes such as burritos and sandwiches, less than 8 grams of fat, 3.5 grams of saturated fat, and 80 milligrams of cholesterol per reference amount customarily consumed.

CARBOHYDRATES: FIBER AND SUGAR

high fiber: 5 grams or more of fiber per serving. A high-fiber claim made on a food that contains more than 3 grams of fat per serving and per 100 grams of food must also declare total fat.

sugar-free: less than 0.5 gram of sugar per serving.

SODIUM

sodium-free and **salt-free:** less than 5 milligrams of sodium per serving.

low sodium: 140 milligrams or less per serving.

very low sodium: 35 milligrams or less per serving.

[a]Foods containing more than 13 grams total fat per serving or per 50 grams of food must indicate those contents immediately after a cholesterol claim. As you can see, all cholesterol claims are prohibited when the food contains more than 2 grams saturated fat and *trans* fat combined per serving.

TABLE 2-9 From Guidelines to Groceries

Dietary Guidelines	USDA Food Patterns/MyPlate	Food Labels
Balancing kcalories to manage weight	Enjoy your food, but eat less. Select the recommended amounts from each food group at the energy level appropriate for your energy needs; meet, but do not exceed, energy needs. Limit foods and beverages with solid fats and added sugars. Use appropriate portion sizes; avoid oversized portions. Increase physical activity and reduce time spent in sedentary behaviors.	Read the Nutrition Facts to see how many kcalories are in a serving and the number of servings that are in a package. Look for foods that describe their kcalorie contents as *free, low, reduced, light,* or *less.*
Foods and food components to reduce	Choose foods within each group that are low in salt or sodium. Choose foods within each group that are lean, low fat, or fat free and have little solid fat (sources of saturated and *trans* fats); use unsaturated oils instead of solid fats whenever possible. Choose foods and beverages within each group that have little added sugars; drink water instead of sugary beverages. If alcohol is consumed by adults, use in moderation (no more than one drink a day for women and two drinks a day for men).	Read the Nutrition Facts to see how much sodium, saturated fat, *trans* fat, and cholesterol is in a serving of food. Look for foods that describe their salt and sodium contents as *free, low,* or *reduced;* foods that describe their saturated fat, *trans* fat, and cholesterol contents as *free, less, low, light, reduced, lean,* or *extra lean;* foods that describe their added sugar contents as *free* or *reduced.* Look for foods that provide no more than 5 percent of the Daily Value for sodium, saturated fat, and cholesterol. A food may be high in solid fats if its ingredients list begins with or contains several of the following: *beef fat (tallow, suet), butter, chicken fat, coconut oil, cream, hydrogenated oils, palm kernel oil, palm oil, partially hydrogenated oils, pork fat (lard), shortening,* or *stick margarine.* A food most likely contains *trans* fats if its ingredients list includes: *partially hydrogenated oils.* A food may be high in added sugars if its ingredients list begins with or contains several of the following: *brown sugar, confectioner's powdered sugar, corn syrup, dextrin, fructose, high-fructose corn syrup, honey, invert sugar, lactose, malt syrup, maltose, molasses, nectars, sucrose, sugar,* or *syrup.* Light beverages contain fewer kcalories and less alcohol than regular versions.
Foods and nutrients to increase	Make half your plate fruits and vegetables. Choose a variety of vegetables from all five subgroups (dark green, red and orange, legumes, starchy vegetables, and other vegetables) several times a week. Choose a variety of fruits; consume whole or cut-up fruits more often than fruit juice. Choose potassium-rich foods such as fruits and vegetables often. Choose fiber-rich fruits, vegetables, and whole grains often. Choose whole grains; make at least half of the grain selections whole grains by replacing refined grains with whole grains whenever possible. Choose fat-free or low-fat milk and milk products. Choose a variety of protein foods; increase the amount and variety of seafood by choosing seafood in place of some meat and poultry.	Look for foods that describe their fiber, calcium, potassium, iron, and vitamin D contents as *good, high,* or *excellent.* Look for foods that provide at least 10 percent of the Daily Value for fiber, calcium, potassium, iron, and vitamin D from a variety of sources. A food may be a good source of whole grains if its ingredients list begins with or contains several of the following: *barley, brown rice, buckwheat, bulgur, corn, millet, oatmeal, popcorn, quinoa, rolled oats, rye, sorghum, triticale, whole wheat,* or *wild rice.*
Building healthy eating patterns	Select nutrient-dense foods and beverages within and among the food groups. Keep foods safe.	Look for foods that describe their vitamin, mineral, or fiber contents as a *good source* or *high.* Follow the *safe handling instructions* on packages of meat and other safety instructions, such as *keep refrigerated,* on packages of perishable foods.

This chapter provides the links to go from dietary guidelines to buying groceries and offers helpful tips for selecting nutritious foods. For additional information on foods, including organic foods, irradiated foods, genetically modified foods, and more, turn to Chapter 19.

Nutrition Portfolio

The secret to making healthy food choices is learning to incorporate the *Dietary Guidelines for Americans* and the USDA Food Patterns into your decision-making process.

Go to Diet & Wellness Plus and choose one of the days on which you have tracked your diet for the entire day. Choose the MyPlate Report and, looking at it, record in your journal the answers to the following:

- How do the foods you consumed on the day you have chosen stack up with the daily goals (the percentages) in the MyPlate breakdown? Which food groups are over- or under-represented?

- Think about your choices within each food group for the day you recorded. Are they typical of the foods you choose from day to day? Are there simple and realistic ways to enhance the variety in your diet?

- Write yourself a letter describing the dietary changes you can make to improve your chances of enjoying good health.

DIET & WELLNESS PLUS+ To complete this exercise, go to your Diet & Wellness Plus at www.cengagebrain.com.

> **STUDY IT** To review the key points of this chapter and take a practice quiz, go to the study cards at the end of the book.

REFERENCES

1. Position of the Academy of Nutrition and Dietetics: Total diet approach to healthy eating, *Journal of the Academy of Nutrition and Dietetics* 113 (2013): 307–317.
2. Practice paper of the American Dietetic Association, Nutrient density: Meeting nutrient goals within calorie needs, *Journal of the American Dietetic Association* 107 (2007): 860–869.
3. The science behind current nutrition profiling systems to promote consumer intake of nutrient-dense foods, *American Journal of Clinical Nutrition* 91 (2010): entire supplement.
4. D. Mozaffarian and D. S. Ludwig, Dietary Guidelines in the 21st century: A time for food, *Journal of the American Medical Association* 304 (2010): 681–682.
5. J. M. Spahn and coauthors, The systematic review methodology used to support the 2010 Dietary Guidelines advisory committee, *Journal of the American Dietetic Association* 111 (2011): 520–523.
6. M. L. Watts and coauthors, The art of translating nutritional science into dietary guidance: History and evolution of the Dietary Guidelines for Americans, *Nutrition Reviews* 69 (2011): 404–412.
7. L. V. Horn, Development of the 2010 US Dietary Guidelines Advisory Committee Report: Perspectives from a registered dietitian, *Journal of the American Dietetic Association* 110 (2010): 1638–1645.
8. US Department of Health and Human Services, *2008 Physical Activity Guidelines for Americans*, www.health.gov/paguidelines; US Department of Agriculture and US Department of Health and Human Services, *Dietary Guidelines for Americans, 2010*, www.dietaryguidelines.gov.
9. www.choosemyplate.gov/food-groups/vegetables-beans-peas.html, accessed August 22, 2013.
10. P. Britten and coauthors, Updated US Department of Agriculture food patterns meet goals of the 2010 Dietary Guidelines, *Journal of the Academy of Nutrition and Dietetics* 112 (2012): 1648–1655.
11. Position of the American Dietetic Association: Vegetarian diets, *Journal of the American Dietetic Association* 109 (2009): 1266–1282.
12. P. Britten and coauthors, Impact of typical rather than nutrient-dense food choices in the US Department of Agriculture food patterns, *Journal of the Academy of Nutrition and Dietetics* 112 (2012): 1560–1569.
13. P. M. Guenther and coauthors, Update of the Healthy Eating Index: HEI-2010, *Journal of the Academy of Nutrition and Dietetics* 113 (2013): 569–580.
14. Center for Nutrition Policy and Promotion, Diet quality of Americans in 2001–02 and 2007–2008 as measured by the Healthy Eating Index—2010, www.cnpp.usda.gov, April 2013.
15. S. M. Krebs-Smith, J. Reedy, and C. Bosire, Healthfulness of the US food supply: Little improvement despite decades of dietary guidance, *American Journal of Preventive Medicine* 38 (2010): 472–477.
16. Harvard School of Public Health, *The Nutrition Source: Healthy Eating Plate vs. USDA's MyPlate*, www.hsph.harvard.edu, accessed October 4, 2011.
17. M. Maillot and coauthors, Individual diet modeling translates nutrient recommendations into realistic and individual-specific food choices, *American Journal of Clinical Nutrition* 91 (2010): 421–430.
18. C. E. O'Neil and coauthors, Whole-grain consumption is associated with diet quality and nutrient intake in adults: The National Health and Nutrition Examination Survey, 1999–2004, *Journal of the American Dietetic Association* 110 (2010): 1461–1468.
19. As cited in 21 Code of Federal Regulations—Food and Drugs, Section 104.20, 45 *Federal Register* 6323, January 25, 1980, as amended in 58 *Federal Register* 2228, January 6, 1993.
20. J. S. Meengs, L. S. Roe, and B. J. Rolls, Vegetable variety: An effective strategy to increase vegetable intake in adults, *Journal of the Academy of Nutrition and Dietetics* 112 (2012): 1211–1215; A. D. Blatt, L. S. Roes, and B. J. Rolls, Hidden vegetables: An effective strategy to reduce energy intake and increase vegetable intake in adults, *American Journal of Clinical Nutrition* 93 (2011): 756–763; B. J. Rolls, L. S. Roe, and J. S. Meengs, Portion size can be used strategically to increase vegetable consumption in adults, *American Journal of Clinical Nutrition* 91 (2010): 913–922.
21. N. J. Ollberding, R. L. Wolf, and I. Contento, Food label use and its relation to dietary intake among US adults, *Journal of the American Dietetic Association* 111 (2011): S47–S51.
22. L. Marr, National restaurant menu labeling legislation: Public nutrition education and professional opportunities, *Journal of the American Dietetic Association* 111 (2011): S7; K. Stein, A national approach to restaurant menu labeling: The Patient Protection and Affordable Health Care Act, section 4205, *Journal of the American Dietetic Association* 110 (2010): 1280–1286.
23. J. C. Hersey and coauthors, Effects of front-of-package and shelf nutrition labeling systems on consumers, *Nutrition Reviews* 71 (2013): 1–14.
24. K. D. Brownell and J. P. Koplan, Front-of-package nutrition labeling—An abuse of trust by the food industry? *New England Journal of Medicine* 364 (2011): 2373–2375.
25. E. A. Wartella and coauthors, *Front-of-Package Nutrition Rating Systems and Symbols: Promoting Healthier Choices* (Washington, D.C.: National Academies Press, 2011), available at www.iom.edu/Reports/2011/Front-of-Package-Nutrition-Rating-Systems-and-Symbols-Promoting-Healthier-Choices.aspx.
26. C. L. Wong and coauthors, Consumer attitudes and understanding of low-sodium claims on food: An analysis of healthy and hypertensive individuals, *American Journal of Clinical Nutrition* 97 (2013): 1288–1298.

HIGHLIGHT > 2
Vegetarian Diets

> **LEARN IT** Develop a well-balanced vegetarian meal plan.

The waiter presents this evening's specials: a fresh spinach salad topped with mandarin oranges, raisins, and sunflower seeds, served with a bowl of pasta smothered in a mushroom and tomato sauce and topped with grated parmesan cheese. Then this one: a salad made of chopped parsley, scallions, celery, and tomatoes mixed with bulgur wheat and dressed with olive oil and lemon juice, served with a spinach and feta cheese pie. Do these meals sound good to you? Or is something missing . . . a pork chop or chicken breast, perhaps?

Would vegetarian fare be acceptable to you some of the time? Most of the time? Ever? The health benefits of a primarily **vegetarian diet** seem to have encouraged many people to eat more plant-based meals. The popular press sometimes refers to individuals who eat small amounts of meat, seafood, or poultry from time to time as "flexitarians."

People who choose to exclude meat and other animal-derived foods from their diets today do so for many of the same reasons the Greek philosopher Pythagoras cited in the sixth century B.C.: physical health, ecological responsibility, and philosophical concerns. They might also cite world hunger issues, economic reasons, ethical concerns, or religious beliefs as motivating factors. Whatever their reasons—and even if they don't have a particular reason—people who exclude meat will be better prepared to plan well-balanced meals if they understand the nutrition and health implications of their choices.

Vegetarian diets generally are categorized, not by a person's motivations but by the foods that are excluded (see the accompanying glossary). Some diets exclude red meat only; some also exclude poultry or seafood; others also exclude eggs; and still others exclude milk and milk products as well. In contrast, **omnivorous** diets do not exclude foods, but include many foods derived from both animals and plants.

As you will see, though, the foods a person *excludes* are not nearly as important as the foods a person *includes* in the diet. **Plant-based diets** that include a variety of whole grains, vegetables, legumes, fruits, and nuts and seeds offer abundant complex carbohydrates and fibers,

an assortment of vitamins and minerals, a mixture of phytochemicals, and little saturated fat—characteristics that reflect current dietary recommendations aimed at maintaining good health and an appropriate body weight. Each of these foods—whole grains, vegetables, legumes, fruits, and nuts and seeds—independently reduces the risk for several chronic diseases. This highlight examines the health benefits and potential problems of vegetarian diets and shows how to plan a well-balanced vegetarian diet. Highlight 20 includes a discussion of the environmental benefits of a plant-based diet.

Health Benefits of Vegetarian Diets

Research findings suggest that well-planned vegetarian diets offer sound nutrition and health benefits to adults.[1] Eating patterns that include very little, if any, meat are associated with a lower rate of death from all causes.[2] Some researchers estimate that for the general population, the risk of dying could be lowered by 7 to 19 percent by eliminating just one serving of meat a day.[3]

GLOSSARY

lactovegetarian diet: an eating pattern that includes milk and milk products, but excludes meat, poultry, seafood, and eggs from the diet.

- **lacto** = milk

lacto-ovo-vegetarian diet: an eating pattern that includes milk, milk products, and eggs, but excludes meat, poultry, and seafood from the diet.

- **ovo** = egg

macrobiotic diet: a philosophical eating pattern based on mostly plant foods such as whole grains, legumes, and vegetables, with small amounts of fish, fruits, nuts, and seeds.

- **macro** = large, great
- **biotic** = life

meat replacements: products formulated to look and taste like meat, seafood, or poultry; usually made of textured vegetable protein.

omnivorous: an eating pattern that includes foods derived from both animals and plants.

- **omni** = all
- **vores** = to eat

plant-based diets: an eating pattern that derives most of its protein from plant products (although some animal products may be included).

tempeh (TEM-pay): a fermented-soybean food, rich in protein and fiber.

tofu (TOE-foo): a curd made from soybeans, rich in protein and often fortified with calcium; used in many Asian and vegetarian dishes in place of meat.

vegan (VEE-gan) **diet:** an eating pattern that excludes all animal-derived foods (including meat, poultry, fish, eggs, and dairy products); also called *pure vegetarian, strict vegetarian,* or *total vegetarian.*

vegetarian diet: a general term used to describe an eating pattern that excludes meat, poultry, fish, or other animal-derived foods from the diet.

Obesity

Vegetarians tend to maintain a lower and healthier body weight than nonvegetarians. In general, those who eat meat have higher energy intakes and body weights. Vegetarians' lower body weights correlate with their high intakes of fiber and low intakes of fat. In general, their diets tend to be nutrient-dense and consistent with the *Dietary Guidelines'* recommendations for weight management.[4] Because obesity impairs health in a number of ways, vegetarian diets offer several health advantages.

Diabetes

Obesity and weight gains are strong risk factors for diabetes, which partially explains why nonvegetarian diets are more often associated with diabetes than vegetarian diets. Even when body weight and lifestyle factors are taken into account, vegetarian eating patterns seem to protect against diabetes.[5]

Hypertension

Vegetarians tend to have lower blood pressure and lower rates of hypertension than nonvegetarians.[6] Appropriate body weight helps to maintain a healthy blood pressure, as does a diet low in saturated fat and cholesterol and high in fiber, fruits, vegetables, whole grains, low-fat milk products, and protein from plant sources.[7] Lifestyle factors also influence blood pressure: smoking and alcohol intake raise blood pressure, and physical activity lowers it.

Heart Disease

Meat is associated with an increased risk of heart disease and stroke.[8] The incidence of heart disease and related deaths and the concentrations of blood cholesterol are lower for vegetarians than for nonvegetarians, which can partly be explained by their avoidance of meat. The dietary factor most directly related to heart disease is saturated animal fat, and in general, vegetarian diets are lower in total fat, saturated fat, and cholesterol than typical meat-based diets. The fats common in plant-based diets—the monounsaturated fats of olives, seeds, and nuts and the polyunsaturated fats of vegetable oils—are associated with a decreased risk of heart disease. Furthermore, vegetarian diets are generally higher in dietary fiber, antioxidant vitamins, and phytochemicals—all factors that help control blood lipids and protect against heart disease.

Many vegetarians include soy products such as **tofu** in their diets. Soy products—with their polyunsaturated fats, fibers, vitamins, and minerals, and little saturated fat—may help to protect against heart disease.[9]

Cancer

Vegetarians have a lower overall cancer incidence than the general population.[10] Their low cancer rates may be due to their high intakes of fruits and vegetables (as Highlight 11 explains). In fact, the ratio of vegetables to meat may be the most relevant dietary factor responsible for cancer prevention.

Some scientific findings indicate that vegetarian diets are associated not only with lower cancer mortality in general, but also with lower incidence of cancer at specific sites as well, most notably, colon cancer. People with colon cancer seem to eat more meat. Some cancer experts recommend limiting consumption of red meat to no more than 11 ounces a week, with very little (if any) processed meat.

Other Diseases

In addition to obesity, diabetes, hypertension, heart disease, and some cancers, vegetarian diets may help prevent osteoporosis, diverticular disease, gallstones, cataracts, and rheumatoid arthritis. Health benefits of a vegetarian diet depend on wise diet planning.

Vegetarian Diet Planning

The vegetarian has the same meal-planning task as everyone else—using a variety of foods to deliver all the needed nutrients within an energy allowance that maintains a healthy body weight (as discussed in Chapter 2). Vegetarians who include milk, milk products, and eggs can meet recommendations for most nutrients about as easily as nonvegetarians. Such eating patterns may rely on some fortified foods, but generally provide enough energy, protein, and other nutrients to support the health of adults and the growth of children and adolescents. The USDA Food Patterns for vegetarians are flexible enough that a variety of people can use them: people who have adopted various vegetarian diets, those who want to make the transition to a vegetarian diet, and those who simply want to reduce their meat intake and include more plant-based meals in their diets.

Vegan diets exclude milk, milk products, and eggs and include protein foods such as legumes, nuts, and seeds as well as foods made from them, such as peanut butter, **tempeh,** and tofu. Vegans who do not use milk can use soy "milk"—a product made from soybeans that provides similar nutrients if fortified with calcium, vitamin D, and vitamin B_{12} (see Figure H2-1 on p. 66). Similarly, "milks" made from rice, almonds, and oats are reasonable alternatives, if adequately fortified. Vegan eating patterns must include fortified foods or supplements to provide adequate intakes of all essential nutrients.

MyPlate includes tips for planning vegetarian diets using an adaptation of the USDA Food Patterns. The recommended daily amounts from the food groups is the same for both vegetarians and nonvegetarians (see Table 2-3, p. 43). Selections from within the food groups may differ, of course. For example, the milk group features fortified soy milks for those who do not use milk, cheese, or yogurt. When selecting from the vegetable and fruit groups, vegetarians may want to emphasize particularly good sources of calcium and iron. Green leafy vegetables provide almost five times as much calcium per serving as other vegetables. Similarly, dried fruits deserve special notice in the fruit group because they deliver six times as much iron as other fruits. The protein foods group includes eggs (for those who use them),

> **FIGURE H2-1** **Low-Fat Milk and Soy Milk Compared**

A comparison of low-fat milk and enriched soy milk shows that they provide similar amounts of key nutrients.

Low-Fat Milk

Nutrition Facts		
8 servings per container		
Serving size		1 cup (240 mL)

Amount Per 1 cup	
Calories	**102**

% DV*		
4%	**Total Fat** 8g	
8%	Saturated Fat 1g	
	Trans Fat 0g	
4%	**Cholesterol** 12mg	
5%	**Sodium** 107mg	
4%	**Total Carbs** 12g	
0%	Dietary Fiber 0g	
	Sugars 1g	
	Added Sugars 0g	
	Protein 8g	
15%	**Vitamin D** 3µg	
23%	**Calcium** 305mg	
0%	**Iron** 0mg	
8%	**Potassium** 366mg	

Soy Milk

Nutrition Facts		
8 servings per container		
Serving size		1 cup (240 mL)

Amount Per 1 cup	
Calories	**100**

% DV*		
6%	**Total Fat** 8g	
3%	Saturated Fat 1g	
	Trans Fat 0g	
0%	**Cholesterol** 0mg	
5%	**Sodium** 120mg	
3%	**Total Carbs** 37g	
4%	Dietary Fiber 1g	
	Sugars 1g	
	Added Sugars 0g	
	Protein 7g	
15%	**Vitamin D** 2.9µg	
23%	**Calcium** 299mg	
6%	**Iron** 1mg	
6%	**Potassium** 299mg	
38%	**Riboflavin** 0.5mg	
125%	**Vitamin B12** 3µg	

© Cengage Learning

legumes, soy products, and nuts and seeds. Table H2-1 provides recommended *weekly* amounts of protein food subgroups for both vegetarians and vegans.

Most vegetarians easily obtain large quantities of the nutrients that are abundant in plant foods, including carbohydrate, fiber, thiamin, folate, vitamin B_6, vitamin C, vitamin A, and vitamin E. Well-planned vegetarian eating patterns help to ensure adequate intakes of the nutrients vegetarian diets might otherwise lack, including protein, iron, zinc, calcium, vitamin B_{12}, vitamin D, and omega-3 fatty acids. Table H2-2 presents good vegetarian sources of these key nutrients.

Protein

The protein RDA for vegetarians is the same as for others, although some have suggested that it should be higher because plant proteins are not digested as completely. **Lacto-ovo-vegetarian diets** that include animal-derived foods such as milk and eggs, deliver high-quality proteins and are likely to meet protein needs. Even vegetarians who adopt only plant-based diets are likely to meet protein needs provided that their energy intakes are adequate and the protein sources varied.[11] The proteins of whole grains, vegetables, legumes, and nuts and seeds can provide adequate amounts of all the amino acids. An advantage of many vegetarian sources of protein is that they are generally lower in saturated fat than meats and are often higher in fiber and richer in some vitamins and minerals.

Vegetarians sometimes use **meat replacements** made of textured vegetable protein (soy protein). These foods are formulated to look and taste like meat, seafood, or poultry. Many of these products are fortified to provide the vitamins and minerals found in animal sources of protein. Some may be high in salt, sugars, and saturated fats. A wise vegetarian learns to read labels and use a variety of whole, unrefined foods often and commercially prepared foods less frequently. Vegetarians may also use soy products such as tofu to bolster protein intake.

TABLE H2-1 **USDA Food Patterns: Recommended Weekly Amounts of Protein Foods for Vegetarians and Vegans**

The daily amounts for protein foods (listed in the top row) are the same for both vegetarians and nonvegetarians, but the subgroups and weekly amounts for protein foods differ (listed in the body of the table). The recommended daily amounts from each of the other food groups—fruits, vegetables, grains, and milk products—are the same (see Table 2-3, p. 43).

Protein Foods	1600 kcal	1800 kcal	2000 kcal	2200 kcal	2400 kcal	2600 kcal	2800 kcal	3000 kcal
Daily Amounts	5 oz	5 oz	5½ oz	6 oz	6½ oz	6½ oz	7 oz	7 oz
Vegetarian Subgroups								
Eggs	4 oz	4 oz	4 oz	4 oz	5 oz	5 oz	5 oz	5 oz
Legumes	9 oz	9 oz	10 oz	10 oz	11 oz	11 oz	12 oz	12 oz
Soy products	11 oz	11 oz	12 oz	13 oz	14 oz	14 oz	15 oz	15 oz
Nuts and seeds	12 oz	12 oz	13 oz	15 oz	16 oz	16 oz	17 oz	17 oz
Vegan Subgroups								
Legumes	12 oz	12 oz	13 oz	15 oz	16 oz	16 oz	17 oz	17 oz
Soy products	9 oz	9 oz	10 oz	11 oz	11 oz	11 oz	12 oz	12 oz
Nuts and seeds	14 oz	14 oz	15 oz	17 oz	18 oz	18 oz	20 oz	20 oz

© Cengage Learning

NOTE: Total recommended amounts for legumes include the sum of both the vegetables and protein foods. An ounce-equivalent of legumes in the protein foods group is ¼ cup. For a 2000-kcal vegan diet, that's 3¼ cups of legumes for protein foods plus 1½ cups of legumes for vegetables (see Table 2-4, p. 46), or about almost 5 cups of legumes weekly.

TABLE H2-2 **Good Vegetarian Sources of Key Nutrients**

Nutrients	Grains	Vegetables	Fruits	Protein Foods	Milk	Oils
Protein[a]	Whole grains			Legumes, seeds, nuts, soy products (tempeh, tofu, veggie burgers) Eggs (for ovo-vegetarians)	Milk, cheese, yogurt (for lactovegetarians)	
Iron	Fortified cereals, enriched and whole grains	Dark green leafy vegetables (spinach, turnip greens)	Dried fruits (apricots, prunes, raisins)	Legumes (black-eyed peas, kidney beans, lentils)		
Zinc	Fortified cereals, whole grains			Legumes (garbanzo beans, kidney beans, navy beans), nuts, seeds (pumpkin seeds)	Milk, cheese, yogurt (for lactovegetarians)	
Calcium	Fortified cereals	Dark-green leafy vegetables (bok choy, broccoli, collard greens, kale, mustard greens, turnip greens, watercress)	Fortified juices, figs	Fortified soy products, nuts (almonds), seeds (sesame seeds)	Milk, cheese, yogurt (for lactovegetarians) Fortified soy milk	
Vitamin B$_{12}$	Fortified cereals			Eggs (for ovo-vegetarians) Fortified soy products	Milk, cheese, yogurt (for lactovegetarians) Fortified soy milk	
Vitamin D					Milk, cheese, yogurt (for lactovegetarians) Fortified soy milk	
Omega-3 fatty acids				Flaxseed, walnuts, soybeans		Flaxseed oil, walnut oil, soybean oil

© Cengage Learning

[a]As Chapter 6 explains, many plant proteins do not contain all the essential amino acids in the amounts and proportions needed by human beings. To improve protein quality, vegetarians can eat grains and legumes together, for example, although it is not necessary if protein intake is varied and energy intake is sufficient.

Iron

Getting enough iron can be a problem even for meat eaters, and those who eat no meat must pay special attention to their iron intake. The iron in plant foods such as legumes, dark-green leafy vegetables, iron-fortified cereals, and whole-grain breads and cereals is poorly absorbed. Because iron absorption from a vegetarian diet is low, the iron RDA for vegetarians is higher than for others (see Chapter 13 for more details).

Fortunately, the body seems to adapt to a low-iron vegetarian diet by increasing iron absorption and decreasing iron losses. Furthermore, iron absorption is enhanced by vitamin C, and vegetarians typically eat many vitamin C–rich fruits and vegetables. Consequently, vegetarians are no more iron deficient than other people.

Zinc

Zinc is similar to iron in that meat is its richest food source, and zinc from plant sources is not well absorbed. In addition, phytates, fiber, and calcium, which are common in vegetarian diets, interfere with zinc absorption. Nevertheless, most vegetarian adults are not zinc deficient. Perhaps the best advice to vegetarians regarding zinc is to eat a variety of nutrient-dense foods; include whole grains, nuts, and legumes such as black-eyed peas, pinto beans, and kidney beans; and maintain an adequate energy intake. For those who include seafood in their diets, oysters, crabmeat, and shrimp are rich in zinc.

Calcium

The calcium intakes of those following a **lactovegetarian diet** are similar to those of the general population, but vegans who use no milk or milk products may risk inadequate intakes. To ensure adequate intakes, vegans can select calcium-rich foods, such as calcium-fortified juices, soy milk, and breakfast cereals, in ample quantities regularly. This advice is especially important for children and adolescents. Soy formulas for infants are fortified with calcium and can be used in cooking, even for adults. Other good calcium sources include figs, some legumes, some green vegetables such as broccoli and turnip greens, some nuts such as almonds, certain seeds such as sesame seeds, and calcium-set tofu.* The choices should be varied because calcium absorption from some plant foods may be limited (as Chapter 12 explains).

Vitamin B$_{12}$

The requirement for vitamin B$_{12}$ is small, but this vitamin is found only in animal-derived foods. Consequently, vegetarians, in general, and vegans who eat no foods of animal original, in particular, may not get enough vitamin B$_{12}$ in their diets.[12] Fermented soy products such as tempeh may contain some vitamin B$_{12}$ from the bacteria, but unfortunately, much of the vitamin B$_{12}$ found in these products may be an inactive form. Seaweeds such as nori and chlorella supply some vitamin B$_{12}$, but not much, and excessive intakes of these foods can lead to iodine toxicity. To defend against vitamin B$_{12}$ deficiency, vegans must rely on vitamin B$_{12}$–fortified sources (such as soy milk or breakfast cereals) or supplements. Without vitamin B$_{12}$, the nerves suffer damage, leading to such health consequences as loss of vision.

Vitamin D

The vitamin D status of vegetarians is similar to that of nonvegetarians. People who do not use vitamin D–fortified foods and do not receive enough exposure to sunlight to synthesize adequate vitamin D may need supplements to defend against bone loss. This is particularly important for infants, children, and older adults. In northern climates during winter months, young children on vegan diets can readily develop rickets, the vitamin D–deficiency disease.

Omega-3 Fatty Acids

Both Chapter 5 and Highlight 5 describe the health benefits of unsaturated fats, most notably the omega-3 fatty acids commonly found in fatty fish. A diet that includes some meat, fish, and eggs provides much more omega-3 fatty acids than a vegetarian diet, but the *blood* differences between those eating fish and others is relatively small.[13] Researchers speculate that the smaller-than-expected differences may reflect a more efficient conversion of plant-derived fats to omega-3 fats in non-fish eaters. Vegetarians can obtain sufficient amounts of the essential omega-3 fatty acids from plant sources such as flaxseed, walnuts, soy, and canola oil. Supplements derived from marine algae that contain omega-3 fatty acids may also be beneficial.[14]

Healthy Food Choices

Later chapters provide details on how vegetarian diets can meet nutrient needs for various stages of the life cycle, including pregnancy, lactation, infancy, childhood, and adolescence. In general, well-planned

vegetarian eating patterns may lower the risk of mortality and several chronic diseases, including obesity, diabetes, high blood pressure, heart disease, and some cancers.[15] But there is nothing mysterious or magical about a vegetarian eating pattern. A dietary pattern that includes small amounts of meat can be equally healthy.[16] The quality of the diet depends not on whether it includes meat but on whether the other food choices are nutritionally sound. A plant-based eating pattern that includes ample fruits, vegetables, whole grains, legumes, nuts, and seeds is higher in fiber, antioxidant vitamins, and phytochemicals and lower in saturated fats and cholesterol than meat-based diets. Variety is key to nutritional adequacy in a vegetarian diet. Restrictive plans that limit selections to a few grains and vegetables cannot possibly deliver a full array of nutrients.

Vegetarianism is not a religion like Buddhism or Hinduism, but merely an eating pattern that selects plant foods to deliver needed nutrients. That said, some vegetarians choose to follow a **macrobiotic diet.** Those following a macrobiotic diet select natural, organic foods and embrace a Zen-like spirituality. In other words, a macrobiotic diet represents a way of life, not just an eating pattern. Such a diet emphasizes whole grains, legumes, and vegetables, with small amounts of fish, fruits, nuts, and seeds. Practices include selecting locally grown foods, eating foods in their most natural state, and balancing cold, sweet, and passive foods with hot, salty, and aggressive ones. Some items, such as processed foods, alcohol, hot spices, and potatoes are excluded from the diet. Early versions of the macrobiotic diet followed a progression that ended with the "ultimate" diet of brown rice and water—a less than nutritionally balanced diet. Today's version reflects a modified vegetarian approach with an appreciation of how foods can enhance health. With careful planning, a macrobiotic diet can provide an array of nutrients that support good health.

If not properly balanced, any diet—vegetarian, macrobiotic, or otherwise—can lack nutrients. Poorly planned vegetarian diets typically lack iron, zinc, calcium, vitamin B$_{12}$, and vitamin D; without planning, meat-based diets may lack vitamin A, vitamin C, folate, and fiber, among others. Quite simply, the negative health aspects of any diet, including vegetarian diets, reflect poor diet planning. Careful attention to energy intake and specific nutrients of concern can ensure adequacy.

Keep in mind, too, that diet is only one factor influencing health. Whatever a diet consists of, its context is also important: no smoking, alcohol consumption in moderation (if at all), regular physical activity, adequate rest, and medical attention when needed all contribute to good health. Establishing these healthy habits early in life seems to be the most important step one can take to reduce the risks of chronic diseases later in life (as Highlight 16 explains).

---

*Calcium salts are often added during processing to coagulate the tofu.

CRITICAL THINKING QUESTIONS

A. What are the strengths and weaknesses of vegetarian diets?
B. Your interest in nutrition has been piqued by the concept of a vegetarian diet, and you wisely recognize that a well-planned diet involves more than simply replacing a turkey sandwich with peanut butter crackers. Design and follow a vegetarian meal plan for 3 days, including at least 1 vegan day. Outline the social, personal, and nutritional challenges you faced and describe how you might partially or fully integrate vegetarian meals into your current meal plan.

REFERENCES

1. Position of the American Dietetic Association: Vegetarian diets, *Journal of the American Dietetic Association* 109 (2009): 1266–1282.
2. M. J. Orlich and coauthors, Vegetarian dietary patterns and mortality in Adventist Health Study 2, *JAMA Internal Medicine* 173 (2013): 1230–1238.
3. A. Pan and coauthors, Red meat consumption and mortality, *Archives of Internal Medicine* 172 (2012): 555–563.
4. B. Farmer and coauthors, A vegetarian dietary pattern as a nutrient-dense approach to weight management: An analysis of the National health and Nutrition Examination Survey 1999–2004, *Journal of the American Dietetic Association* 111 (2011): 819–827.
5. S. Tonstad and coauthors, Vegetarian diets and incidence of diabetes in the Adventist Health Study-2, *Nutrition, Metabolism, and Cardiovascular Diseases* 23 (2013): 292–299.
6. B. J. Pettersen and coauthors, Vegetarian diets and blood pressure among white subjects: Results from the Adventist Health Study-2 (AHS-2), *Public Health Nutrition* 15 (2012): 1909–1916.
7. D. G. Hackam and coauthors, The 2010 Canadian Hypertension Education Program recommendations for the management of hypertension: Part 2, therapy, *Canadian Journal of Cardiology* 26 (2010): 249–258.
8. J. Kaluza, A. Wolk, and S. C. Larsson, Red meat consumption and risk of stroke: A meta-analysis of prospective studies, Stroke 43 (2012): 2556–2560; P. M. Clifton, Protein and coronary heart disease: The role of different protein sources, *Current Atherosclerosis Reports* 13 (2011): 493–498.
9. M. Messina, Insights gained from 20 years of soy research, *Journal of Nutrition* 140 (2010): 2289S–2295S.
10. T. Huang and coauthors, Cardiovascular disease mortality and cancer incidence in vegetarians: A meta-analysis and systematic review, *Annals of Nutrition and Metabolism* 60 (2012): 233–240.
11. Position of the American Dietetic Association, 2009.
12. R. Pawlak and coauthors, How prevalent is vitamin B_{12} deficiency among vegetarians? *Nutrition Reviews* 71 (2013): 110–117.
13. A. A. Welch and coauthors, Dietary intake and status of n-3 polyunsaturated fatty acids in a population of fish-eating and non-fish-eating meat-eaters, vegetarians, and vegans and the precursor product ratio of α-linolenic acid to long-chain n 3 polyunsaturated fatty acids: Results from the EPIC-Norfolk cohort, *American Journal of Clinical Nutrition* 92 (2010): 1040–1051.
14. A. M. Bernstein and coauthors, A meta-analysis shows that docosahexaenoic acid from algal oil reduces serum triglycerides and increases HDL-cholesterol and LDL-cholesterol in persons without coronary heart disease, *Journal of Nutrition* 142 (2012): 99–104.
15. W. J. Craig, Nutrition concerns and health effects of vegetarian diets, *Nutrition in Clinical Practice* 25 (2010): 613–620.
16. C. T. McEvoy, N. Temple, and J. V. Woodside, Vegetarian diets, low-meat diets and health: A review, *Public Health Nutrition* 15 (2012): 2287–2294.

Digestion, Absorption, and Transport

Nutrition in Your Life

Have you ever wondered what happens to the food you eat after you swallow it? Or how your body extracts nutrients from food? Have you ever marveled at how it all just seems to happen? Follow foods as they travel through the digestive system. Learn how a healthy digestive system takes whatever food you give it—whether sirloin steak and potatoes or tofu and brussels sprouts—and extracts the nutrients that will nourish the cells of your body. In the Nutrition Portfolio at the end of the chapter, you can determine whether your current eating habits are supporting a healthy digestive system.

Each cell in the body needs a continuous supply of many specific nutrients to maintain itself and carry out its work. These nutrients derive from the foods a person eats, but before the body's cells can use the nutrients, foods must first be broken down mechanically and chemically. This chapter follows the journey that breaks down foods into the nutrients featured in the later chapters. Then it follows the nutrients as they travel through the intestinal cells and into the body to do their work.

As you read about the complexities and intricacies of these processes, take a moment to appreciate the beauty and wisdom of the body. Recognize that the activities of the digestive system are finely coordinated and fully integrated with those of the circulatory, nervous, and hormonal systems. Then be thankful that your body can efficiently take care of its business without any direction from you, but know that it performs its best when you have given it optimal nourishment. This introduction presents a general overview of the processes common to all nutrients; later chapters discuss the specifics of digesting and absorbing individual nutrients.

The process of digestion breaks down all kinds of *foods* into *nutrients*.

3.1 Digestion

> **LEARN IT** Explain how foods move through the digestive system, describing the actions of the organs, muscles, and digestive secretions along the way.

Digestion is the body's ingenious way of breaking down foods into nutrients in preparation for **absorption.** In the process, the body overcomes many challenges without any conscious effort. Consider these challenges:

1. Human beings breathe, eat, and drink through their mouths. Air taken in through the mouth must go to the lungs; food and liquid must go to the stomach. The throat must be arranged so that swallowing and breathing don't interfere with each other.

2. Below the lungs lies the diaphragm, a dome of muscle that separates the upper half of the major body cavity from the lower half. The body needs a passageway that will allow food from the mouth to pass through the diaphragm to reach the stomach below.

3. The contents of the digestive tract should be kept moving forward, slowly but steadily, at a pace that permits all reactions to reach completion.

4. To move through the system, food must be lubricated with fluids. Too much would form a liquid that would flow too rapidly; too little would form a paste too dry and compact to move at all. The amount of fluids must be regulated to keep the intestinal contents at the right consistency to move along smoothly.

5. For digestive enzymes to work, foods must be broken down into small particles and suspended in enough liquid so that every particle is accessible. Once digestion is complete and nutrients have been absorbed from the GI tract into the body, the remaining waste must be excreted. Excreting all the water along with the solid residue, however, would be both wasteful and messy. Some water must be withdrawn, leaving a solid waste product that is easy to pass.

6. The digestive enzymes are designed to digest carbohydrate, fat, and protein. The cells of the GI tract are also made of carbohydrate, fat, and protein. These cells must be protected against the powerful digestive juices that they secrete.

7. Once waste matter has reached the end of the GI tract, it must be excreted, but it would be inconvenient and embarrassing if this function occurred continuously. Evacuation needs to occur periodically.

The following sections show how the body elegantly and efficiently handles these challenges. Each section follows the GI tract from one end to the other—first describing its anatomy, then its muscular actions, and finally its secretions.

Anatomy of the Digestive Tract The **gastrointestinal (GI) tract** is a flexible muscular tube that extends from the mouth, through the esophagus, stomach, small intestine, large intestine, and rectum to the anus. Figure 3-1 on p. 74 traces the path followed by food from one end to the other. In a sense, the human body surrounds the GI tract. The inner space within the GI tract, called the **lumen,** is continuous from one end to the other. (GI anatomy terms appear in boldface type and are defined in the accompanying glossary.) Only when a nutrient or other substance finally penetrates the GI tract's wall does it enter the body proper; many materials pass through the GI tract without being digested or absorbed.

Mouth The process of digestion begins in the **mouth.** During chewing, teeth crush large pieces of food into smaller ones, and fluids from foods, beverages, and salivary glands blend with these pieces to ease swallowing.* Fluids also help dissolve the food so that the tongue can taste it; only particles in solution can react with taste buds. When stimulated, the taste buds detect one, or a combination, of the five basic

digestion: the process by which food is broken down into absorbable units.

• **digest** = take apart

absorption: the uptake of nutrients by the cells of the small intestine for transport into either the blood or the lymph.

• **absorb** = suck in

gastrointestinal (GI) tract: the digestive tract. The principal organs are the stomach and intestines.

• **gastro** = stomach

• **intestinalis** = intestine

*The process of chewing is called *mastication* (mass-tih-KAY-shun).

anus (AY-nus): the terminal outlet of the GI tract.

appendix: a narrow blind sac extending from the beginning of the colon that contains bacteria and lymph cells.

duodenum (doo-oh-DEEN-um or doo-ODD-num): the top portion of the small intestine (about "12 fingers' breadth" long in ancient terminology).

- **duodecim** = twelve

epiglottis (epp-ih-GLOTT-iss): cartilage in the throat that guards the entrance to the trachea and prevents fluid or food from entering it when a person swallows.

- **epi** = upon (over)
- **glottis** = back of tongue

esophageal (ee -SOFF-ah-GEE-al) **sphincter:** a sphincter muscle at the upper or lower end of the esophagus. The *lower esophageal sphincter* is also called the *cardiac sphincter* because of its proximity to the heart.

esophagus (ee-SOFF-ah-gus): the food pipe; the conduit from the mouth to the stomach.

gallbladder: the organ that stores and concentrates bile. When it receives the signal that fat is present in the duodenum, the gallbladder contracts and squirts bile through the bile duct into the duodenum.

ileocecal (ill-ee-oh-SEEK-ul) **valve:** the sphincter separating the small and large intestines.

ileum (ILL-ee-um): the last segment of the small intestine.

jejunum (je-JOON-um): the first two-fifths of the small intestine beyond the duodenum.

large intestine or **colon** (COAL-un): the lower portion of intestine that completes the digestive process. Its segments are the *ascending colon*, the *transverse colon*, the *descending colon*, and the *sigmoid colon*.

- **sigmoid** = shaped like the letter S (sigma in Greek)

lumen (LOO-men): the space within a vessel such as the intestine.

mouth: the oral cavity containing the tongue and teeth.

pancreas: a gland that secretes digestive enzymes and juices into the duodenum. (The pancreas also secretes hormones into the blood that help to maintain glucose homeostasis.)

pharynx (FAIR-inks): the passageway leading from the nose and mouth to the larynx and esophagus, respectively.

pyloric (pie-LORE-ic) **sphincter:** the circular muscle that separates the stomach from the small intestine and regulates the flow of partially digested food into the small intestine; also called *pylorus* or *pyloric valve*.

- **pylorus** = gatekeeper

rectum: the muscular terminal part of the intestine, extending from the sigmoid colon to the anus.

small intestine: a 10-foot length of small-diameter intestine that is the major site of digestion of food and absorption of nutrients. Its segments are the *duodenum, jejunum,* and *ileum*.

sphincter (SFINK-ter): a circular muscle surrounding, and able to close, a body opening. Sphincters are found at specific points along the GI tract and regulate the flow of food particles.

- **sphincter** = band (binder)

stomach: a muscular, elastic, saclike portion of the digestive tract that grinds and churns swallowed food, mixing it with acid and enzymes to form chyme.

---

taste sensations: sweet, sour, bitter, salty and umami (oo-MOM-ee), a savory flavor commonly associated with monosodium glutamate.[1] In addition to these chemical triggers, aroma, appearance, texture, and temperature also affect a food's flavor.

The tongue provides taste sensations and moves food around the mouth, facilitating chewing and swallowing. When a mouthful of food is swallowed, it passes through the **pharynx,** a short tube that is shared by both the **digestive system** and the respiratory system. To bypass the entrance to the lungs, the **epiglottis** closes off the airway so that choking doesn't occur when swallowing, thus resolving the first challenge. (Choking is discussed on pp. 90–91.) After a mouthful of food has been chewed and swallowed, it is called a **bolus.**

Esophagus The **esophagus** has a **sphincter** muscle at each end. During a swallow, the upper **esophageal sphincter** opens. The bolus then slides down the esophagus, which passes through a hole in the diaphragm (challenge 2) to the stomach. The lower esophageal sphincter at the entrance to the stomach closes behind the bolus so that it proceeds forward and doesn't slip back into the esophagus (challenge 3).

Stomach The **stomach** retains the bolus for a while in its upper portion. Little by little, the stomach transfers the food to its lower portion, adds juices to it, and grinds it to a semiliquid mass called **chyme.** Then, bit by bit, the stomach releases the chyme through the **pyloric sphincter,** which opens into the **small intestine** and then closes behind the chyme.

Small Intestine At the beginning of the small intestine, the chyme bypasses the opening from the common bile duct, which is dripping fluids (challenge 4) into the small intestine from two organs outside the GI tract—the **gallbladder** and the **pancreas.** The chyme travels on down the small intestine through its three segments—the **duodenum,** the **jejunum,** and the **ileum**—almost 10 feet of tubing coiled within the abdomen.*

Large Intestine (Colon) Having traveled the length of the small intestine, the remaining contents arrive at another sphincter (challenge 3 again): the **ileocecal valve,** located at the beginning of the **large intestine (colon)** in the lower right side of the abdomen. Upon entering the colon, the contents pass another opening. Should

digestive system: all the organs and glands associated with the ingestion and digestion of food.

bolus (BOH-lus): a portion; with respect to food, the amount swallowed at one time.

- **bolos** = lump

chyme (KIME): the semiliquid mass of partly digested food expelled by the stomach into the duodenum.

- **chymos** = juice

---

*The small intestine is almost two and a half times shorter in living adults than it is at death, when muscles are relaxed and elongated.

> FIGURE 3-1   The Gastrointestinal Tract

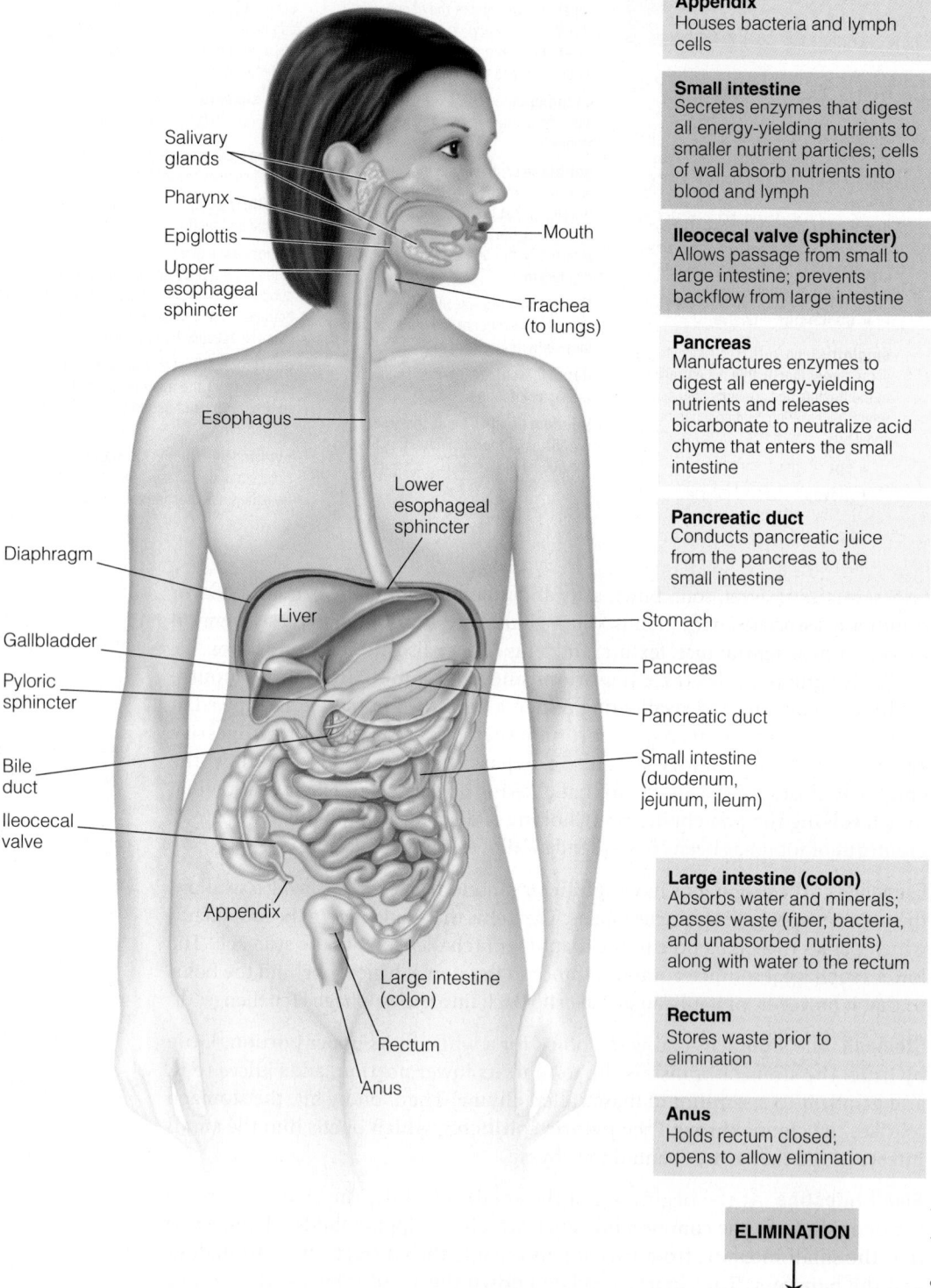

INGESTION

Mouth
Chews and mixes food with saliva

Pharynx
Directs food from mouth to esophagus

Salivary glands
Secrete saliva

Epiglottis
Protects airways during swallowing

Trachea
Allows air to pass to and from lungs

Esophagus
Passes food from the mouth to the stomach

Esophageal sphincters
Allow passage from mouth to esophagus and from esophagus to stomach; prevent backflow from stomach to esophagus and from esophagus to mouth

Diaphragm
Separates the abdomen from the thoracic cavity

Stomach
Churns, mixes, and grinds food to a liquid mass; adds acid, enzymes, and fluid

Pyloric sphincter
Allows passage from stomach to small intestine; prevents backflow from small intestine

Liver
Manufactures bile salts, detergent-like substances, to help digest fats

Gallbladder
Stores bile until needed

Bile duct
Conducts bile from the gallbladder to the small intestine

Appendix
Houses bacteria and lymph cells

Small intestine
Secretes enzymes that digest all energy-yielding nutrients to smaller nutrient particles; cells of wall absorb nutrients into blood and lymph

Ileocecal valve (sphincter)
Allows passage from small to large intestine; prevents backflow from large intestine

Pancreas
Manufactures enzymes to digest all energy-yielding nutrients and releases bicarbonate to neutralize acid chyme that enters the small intestine

Pancreatic duct
Conducts pancreatic juice from the pancreas to the small intestine

Large intestine (colon)
Absorbs water and minerals; passes waste (fiber, bacteria, and unabsorbed nutrients) along with water to the rectum

Rectum
Stores waste prior to elimination

Anus
Holds rectum closed; opens to allow elimination

ELIMINATION

Salivary glands
Pharynx
Epiglottis
Upper esophageal sphincter
Mouth
Trachea (to lungs)
Esophagus
Lower esophageal sphincter
Diaphragm
Liver
Gallbladder
Pyloric sphincter
Bile duct
Ileocecal valve
Appendix
Stomach
Pancreas
Pancreatic duct
Small intestine (duodenum, jejunum, ileum)
Large intestine (colon)
Rectum
Anus

any intestinal contents slip into this opening, it would end up in the **appendix,** a blind sac about the size of your little finger. Normally, the contents bypass this opening, however, and travel along the large intestine up the right side of the abdomen, across the front to the left side, down to the lower left side, and finally below the other folds of the intestines to the back of the body, above the **rectum** (see Figure 3-2).

As the intestinal contents pass to the rectum, the colon withdraws water, leaving semisolid waste (challenge 5). The strong muscles of the rectum and anal canal hold back this waste until it is time to defecate. Then the rectal muscles relax (challenge 7), and the two sphincters of the **anus** open to allow passage of the waste.

The Muscular Action of Digestion

In the mouth, chewing, the addition of saliva, and the action of the tongue transform food into a coarse mash that can be swallowed. After swallowing, all the activity that follows occurs without much conscience thought. As is the case with so much else that happens in the body, the muscles of the digestive tract meet internal needs without any concerted effort on your part. They keep things moving at just the right pace, slow enough to get the job done and fast enough to make progress.*

Peristalsis The entire GI tract is ringed with circular muscles. Surrounding these rings of muscle are longitudinal muscles. When the rings tighten and the long muscles relax, the tube is constricted. When the rings relax and the long muscles tighten, the tube bulges. This action—called **peristalsis**—occurs continuously and pushes the intestinal contents along (challenge 3 again). (If you have ever watched a lump of food pass along the body of a snake, you have a good picture of how these muscles work.)

The waves of contraction normally ripple along the GI tract at varying rates and intensities depending on the part of the GI tract and on whether food is present. Factors such as stress, medicines, and medical conditions may interfere with normal GI tract contractions.

Stomach Action The stomach has the thickest walls and strongest muscles of all the GI tract organs. In addition to the circular and longitudinal muscles, it has a third layer of diagonal muscles that also alternately contract and relax (see Figure 3-3). These three sets of muscles work to force the chyme downward, but the pyloric sphincter usually remains tightly closed, preventing the chyme from passing into the duodenum of the small intestine. As a result, the chyme is churned and forced down, hits the pyloric sphincter, and remains in the stomach. Meanwhile, the stomach wall releases gastric juices. When the chyme is completely liquefied with gastric juices, the pyloric sphincter opens briefly, about three times a minute, to allow small portions of chyme to pass through. At this point, the chyme no longer resembles food in the least.

Segmentation The circular muscles of the intestines rhythmically contract and squeeze their contents. These contractions, called **segmentation,** mix the chyme and promote close contact with the digestive juices and the absorbing cells of the intestinal walls before letting the contents move slowly along.

Sphincter Contractions Sphincter muscles periodically open and close, allowing the contents of the GI tract to move along at a controlled pace (challenge 3 again). At the top of the esophagus, the upper esophageal sphincter opens in response to swallowing. At the bottom of the esophagus, the lower esophageal sphincter (sometimes called the cardiac sphincter because of its proximity to the heart) prevents **reflux** of the stomach contents. At the bottom of the stomach, the pyloric sphincter, which stays closed most of the time, holds the chyme in the stomach long enough for it to be thoroughly mixed with gastric juice and liquefied. The pyloric sphincter also prevents the intestinal contents from backing up into the stomach. At the end of the small intestine, the ileocecal valve performs a similar function, allowing

_____
*The spontaneous movement of the GI tract muscles is called *motility* (moh-TIL-ih-tee).

> **FIGURE 3-2** **The Colon**

The colon begins with the ascending colon rising upward toward the liver. It becomes the transverse colon as it turns and crosses the body toward the spleen. The descending colon turns downward and becomes the sigmoid colon, which extends to the rectum. Along the way, the colon mixes the intestinal contents, absorbs water and salts, and forms stools.

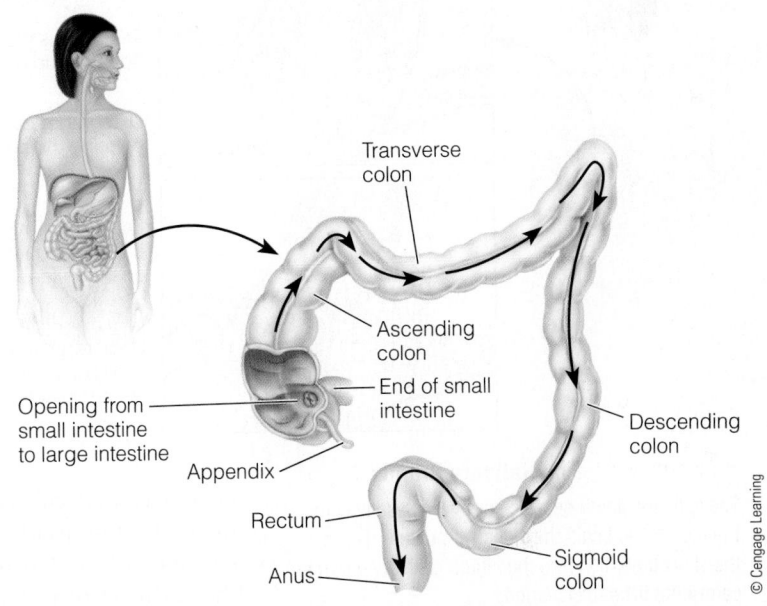

© Cengage Learning

> **FIGURE 3-3** **Stomach Muscles**

The stomach has three layers of muscles.

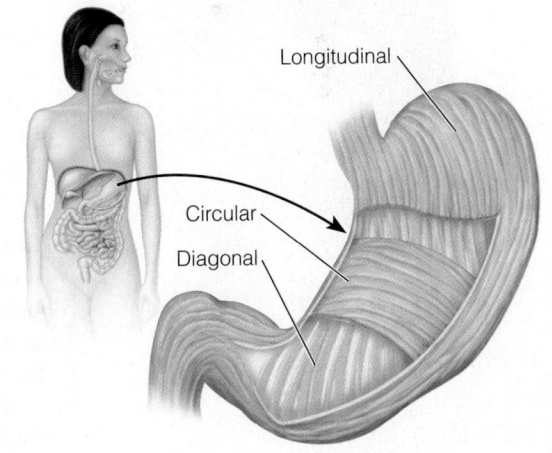

© Cengage Learning

peristalsis (per-ih-STALL-sis): wavelike muscular contractions of the GI tract that push its contents along.

• **peri** = around

• **stellein** = wrap

segmentation (SEG-men-TAY-shun): a periodic squeezing or partitioning of the intestine at intervals along its length by its circular muscles.

reflux: a backward flow.

• **re** = back

• **flux** = flow

> FIGURE 3-4   **An Example of a Sphincter Muscle**

When the circular muscles of a sphincter contract, the passage closes; when they relax, the passage opens.

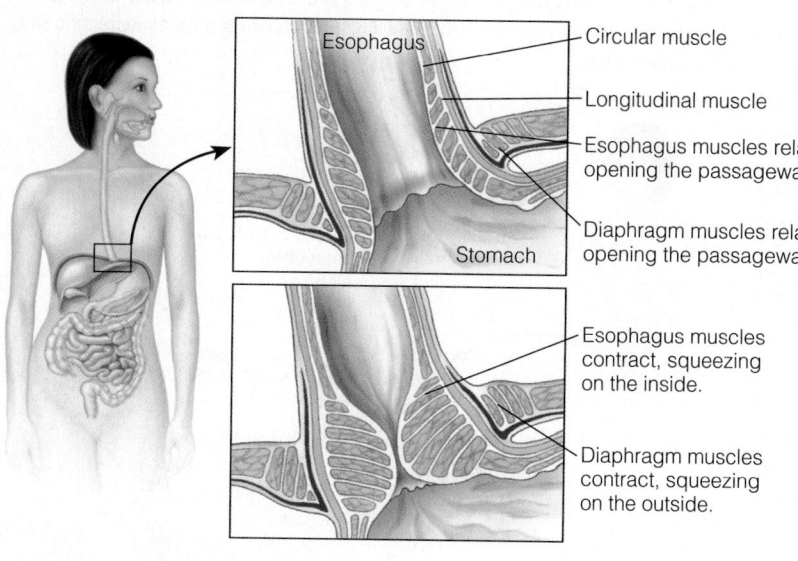

Esophagus

Circular muscle

Longitudinal muscle

Esophagus muscles relax, opening the passageway.

Diaphragm muscles relax, opening the passageway.

Stomach

Esophagus muscles contract, squeezing on the inside.

Diaphragm muscles contract, squeezing on the outside.

© Cengage Learning

> FIGURE 3-5   **The Salivary Glands**

The salivary glands secrete enzyme-rich saliva into the mouth and begin the digestive process. Given the short time food is in the mouth, salivary enzymes contribute little to digestion.

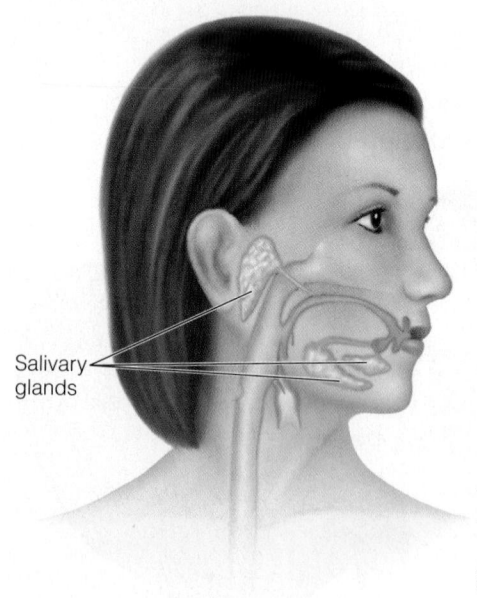

Salivary glands

© Cengage Learning

catalyst (CAT-uh-list): a compound that facilitates chemical reactions without itself being changed in the process.

the contents of the small intestine to empty into the large intestine. Finally, the tightness of the rectal muscle acts as a kind of safety device; together with the two sphincters of the anus, it prevents continuous elimination (challenge 7). Figure 3-4 illustrates how sphincter muscles contract and relax to close and open passageways.

The Secretions of Digestion The breakdown of food into nutrients requires secretions from five different organs: the salivary glands, the stomach, the pancreas, the liver (via the gallbladder), and the small intestine. These secretions enter the GI tract at various points along the way, bringing an abundance of water (challenge 4) and a variety of enzymes.

Enzymes are formally introduced in Chapter 6, but for now a simple definition will suffice. An enzyme is a protein that facilitates a chemical reaction—making a molecule, breaking a molecule apart, changing the arrangement of a molecule, or exchanging parts of molecules. As a **catalyst,** the enzyme itself remains unchanged. The enzymes involved in digestion facilitate a chemical reaction known as **hydrolysis**— the addition of water *(hydro)* to break *(lysis)* a molecule into smaller pieces. The glossary below describes how to identify some of the common **digestive enzymes** and related terms; later chapters introduce specific enzymes. When learning about enzymes, it helps to know that the word ending *-ase* denotes an enzyme. Enzymes are often identified by the organ they come from and the compounds they work on. *Gastric lipase,* for example, is a stomach enzyme that acts on lipids, whereas *pancreatic lipase* comes from the pancreas (and also works on lipids).

Saliva The **salivary glands,** shown in Figure 3-5, squirt just enough **saliva** to moisten each mouthful of food so that it can pass easily down the esophagus (challenge 4). (Digestive **glands** and their secretions are defined in the glossary on p. 77.) The saliva contains water, salts, mucus, and enzymes that initiate the digestion of carbohydrates. Saliva also protects the teeth and the linings of the mouth, esophagus, and stomach from substances that might cause damage.

Gastric Juice In the stomach, **gastric glands** secrete **gastric juice,** a mixture of water, enzymes, and **hydrochloric acid,** which acts primarily in protein digestion. The acid is so strong that it causes the sensation of heartburn if it happens to reflux into the esophagus. Highlight 3, following this chapter, discusses heartburn, ulcers, and other common digestive problems.

The strong acidity of the stomach prevents bacterial growth and kills most bacteria that enter the body with food. It would destroy the cells of the stomach as well, but for their natural defenses. To protect themselves from gastric juice, the cells of the stomach wall (in fact, of the entire gastrointestinal lining) secrete **mucus,** a

GLOSSARY
OF DIGESTIVE ENZYMES

-ase (ACE): suffix denoting an enzyme. The root of the word often identifies the compound the enzyme works on. Examples include:

- *carbohydrase* (KAR-boe-HIGH-drase), an enzyme that hydrolyzes carbohydrates.

- *lipase* (LYE-pase), an enzyme that hydrolyzes lipids (fats).

- *protease* (PRO-tee-ase), an enzyme that hydrolyzes proteins.

digestive enzymes: proteins found in digestive juices that act on food substances, causing them to break down into simpler compounds.

hydrolysis (high-DROL-ih-sis): a chemical reaction in which one molecule is split into two molecules, with hydrogen (H) added to one and a hydroxyl group (OH) added to the other (from water, H_2O). (The noun is *hydrolysis*; the verb is *hydrolyze*.)

- **hydro** = water

- **lysis** = breaking

thick, slippery, white substance that coats the cells, protecting them from the acid, enzymes, and disease-causing bacteria that might otherwise cause harm (challenge 6).

Figure 3-6 shows how the strength of acids is measured—in **pH** units. Note that the acidity of gastric juice registers below 2 on the pH scale—stronger than vinegar. The stomach enzymes work most efficiently in the stomach's strong acid, but the salivary enzymes, which are swallowed with food, do not work in acid this strong. Consequently, the salivary digestion of carbohydrates gradually ceases when the stomach acid penetrates each newly swallowed bolus of food. Once in the stomach, salivary enzymes simply become other proteins to be digested.

Pancreatic Juice and Intestinal Enzymes By the time food leaves the stomach, digestion of all three energy nutrients (carbohydrates, fats, and proteins) has begun, and the action gains momentum in the small intestine. There the pancreas contributes digestive juices by way of ducts leading into the duodenum. The **pancreatic juice** contains enzymes that act on all three energy nutrients, and the cells of the intestinal wall also possess digestive enzymes on their surfaces.

In addition to enzymes, the pancreatic juice contains sodium **bicarbonate,** which is basic or alkaline—the opposite of the stomach's acid (review Figure 3-6). The pancreatic juice thus neutralizes the acidic chyme arriving in the small intestine from the stomach. From this point on, the chyme remains at a neutral or slightly alkaline pH. The enzymes of both the intestine and the pancreas work best in this environment.

Bile **Bile** also flows into the duodenum. The **liver** continuously produces bile, which is then concentrated and stored in the gallbladder. The gallbladder squirts bile into the duodenum of the small intestine when fat arrives there. Bile is not an enzyme; it is an **emulsifier** that brings fats into suspension in water so that enzymes can break them down into their component parts. A summary of digestive secretions and their actions is presented in Table 3-1 on p. 78.

The Final Stage At this point, the three energy-yielding nutrients—carbohydrate, fat, and protein—have been digested and are ready to be absorbed. Some vitamins and minerals are altered slightly during digestion, but most are absorbed as they are. Undigested residues, such as some fibers, are not absorbed. Instead, they continue through the digestive tract, carrying some minerals, bile acids, additives, and contaminants out of the body. This semisolid mass helps exercise the GI muscles and keep them strong enough to perform peristalsis efficiently. Fiber also retains water, accounting for the consistency of **stools.**

By the time the contents of the GI tract reach the end of the small intestine, little remains but water, a few dissolved salts and body secretions, and undigested materials such as fiber (with some fat, cholesterol, and a few minerals bound to it). All of this remaining matter enters the large intestine (colon).

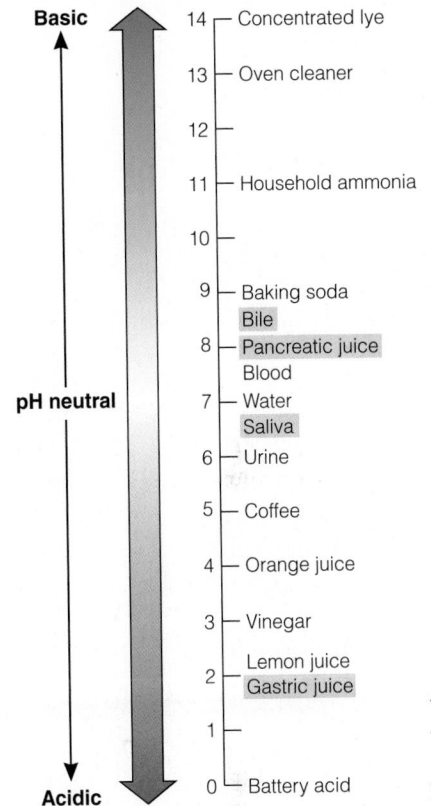

> **FIGURE 3-6** **The pH Scale**

A substance's acidity or alkalinity is measured in pH units. The pH is the negative logarithm of the hydrogen ion concentration. Each increment represents a tenfold increase in concentration of hydrogen ions, meaning, for example, that a pH of 2 is 1000 times stronger than a pH of 5.

pH of common substances:

Basic
14 — Concentrated lye
13 — Oven cleaner
12
11 — Household ammonia
10
9 — Baking soda
 Bile
8 — Pancreatic juice
 Blood
pH neutral
7 — Water
 Saliva
6 — Urine
5 — Coffee
4 — Orange juice
3 — Vinegar
2 — Lemon juice
 Gastric juice
1
Acidic
0 — Battery acid

© Cengage Learning

pH: the unit of measure expressing a substance's acidity or alkalinity. The lower the pH, the higher the H⁺ ion concentration and the stronger the acid. A pH above 7 is alkaline, or base (a solution in which OH⁻ ions predominate).

stools: waste matter discharged from the colon; also called *feces* (FEE-seez).

GLOSSARY
OF DIGESTIVE GLANDS AND THEIR SECRETIONS

bicarbonate: an alkaline compound with the formula HCO₃ that is secreted from the pancreas as part of the pancreatic juice. (Bicarbonate is also produced in all cell fluids from the dissociation of carbonic acid to help maintain the body's acid-base balance.)

bile: an emulsifier that prepares fats and oils for digestion; an exocrine secretion made by the liver, stored in the gallbladder, and released into the small intestine when needed.

emulsifier (ee-MUL-sih-fire): a substance with both water-soluble and fat-soluble portions that promotes the mixing of oils and fats in a watery solution.

gastric glands: exocrine glands in the stomach wall that secrete gastric juice into the stomach.

- **gastro** = stomach

gastric juice: the digestive secretion of the gastric glands of the stomach.

glands: cells or groups of cells that secrete materials for special uses in the body. Glands may be *exocrine* (EKS-oh-crin) *glands,* secreting their materials "out" (into the digestive tract or onto the surface of the skin), or *endocrine*

(EN-doe-crin) *glands,* secreting their materials "in" (into the blood).

- **exo** = outside
- **endo** = inside
- **krine** = to separate

hydrochloric acid: an acid composed of hydrogen and chloride atoms (HCl) that is normally produced by the gastric glands.

liver: the organ that manufactures bile, among many other functions (described in Chapter 7).

mucus (MYOO-kus): a slippery substance secreted by cells of the GI lining (and other body linings) that protects the cells from exposure to digestive juices (and other destructive

agents). The lining of the GI tract with its coat of mucus is a *mucous membrane.* (The noun is *mucus;* the adjective is *mucous.*)

pancreatic (pank-ree-AT-ic) **juice:** the exocrine secretion of the pancreas that contains both enzymes for the digestion of carbohydrate, fat, and protein as well as bicarbonate, a neutralizing agent. The juice flows from the pancreas into the small intestine through the pancreatic duct. (The pancreas also has an endocrine function, the secretion of insulin and other hormones.)

saliva: the secretion of the salivary glands. Its principal enzyme begins carbohydrate digestion.

salivary glands: exocrine glands that secrete saliva into the mouth.

TABLE 3-1 Summary of Digestive Secretions and Their Major Actions

Organ or Gland	Target Organ	Secretion	Action
Salivary glands	Mouth	Saliva	Fluid eases swallowing; salivary enzyme breaks down some **carbohydrate.***
Gastric glands	Stomach	Gastric juice	Fluid mixes with bolus; hydrochloric acid uncoils **proteins;** enzymes break down proteins; mucus protects stomach cells.*
Pancreas	Small intestine	Pancreatic juice	Bicarbonate neutralizes acidic gastric juices; pancreatic enzymes break down **carbohydrates, fats,** and **proteins.**
Liver	Gallbladder	Bile	Bile is stored until needed.
Gallbladder	Small intestine	Bile	Bile emulsifies **fat** so that enzymes can have access to break it down.
Intestinal glands	Small intestine	Intestinal juice	Intestinal enzymes break down **carbohydrate, fat,** and **protein** fragments; mucus protects the intestinal wall.

*Saliva and gastric juice also contain lipases, but most fat breakdown occurs in the small intestine.

© Cengage Learning

In the colon, intestinal bacteria ferment some fibers, producing water, gas, and small fragments of fat that provide energy for the cells of the colon. The colon itself retrieves all materials that the body can recycle—water and dissolved salts. The waste that is finally excreted has little or nothing of value left in it. The body has extracted all that it can use from the food. Figure 3-7 summarizes digestion by following a sandwich through the GI tract and into the body.

› REVIEW IT Explain how foods move through the digestive system, describing the actions of the organs, muscles, and digestive secretions along the way.
As Figure 3-1 shows, food enters the mouth and travels down the esophagus and through the upper and lower esophageal sphincters to the stomach, then through the pyloric sphincter to the small intestine, on through the ileocecal valve to the large intestine, past the appendix to the rectum, ending at the anus. The wavelike contractions of peristalsis and the periodic squeezing of segmentation keep things moving at a reasonable pace. Along the way, secretions from the salivary glands, stomach, pancreas, liver (via the gallbladder), and small intestine deliver fluids and digestive enzymes.

3.2 Absorption

› LEARN IT Describe the anatomical details of the intestinal cells that facilitate nutrient absorption.

Within three or four hours after a person has eaten a dinner of beans and rice (or spinach lasagna, or steak and potatoes) with vegetable, salad, beverage, and dessert, the body must find a way to absorb the molecules derived from carbohydrate, protein, and fat digestion—and the vitamin and mineral molecules as well. Most absorption takes place in the small intestine, one of the most elegantly designed organ systems in the body. Within its 10-foot length, which provides a surface area equivalent to a tennis court, the small intestine traps and absorbs the nutrient molecules. To remove the absorbed molecules rapidly and provide room for more to be absorbed, a rush of circulating blood continuously washes the underside of this surface, carrying the absorbed nutrients away to the liver and other parts of the body. Figure 3-8 on p. 80 describes how most nutrients are absorbed by simple diffusion, facilitated diffusion, or active transport. Later

amana images/Jupiter Images

Foods must first be digested and nutrients must be absorbed before the body can use them.

> FIGURE 3-7 **The Digestive Fate of a Sandwich**

To review the digestive processes, follow a peanut butter and banana sandwich on whole-wheat, sesame seed bread through the GI tract. As the graph on the right illustrates, digestion of the energy nutrients begins in different parts of the GI tract, but all are ready for absorption by the time they reach the end of the small intestine.

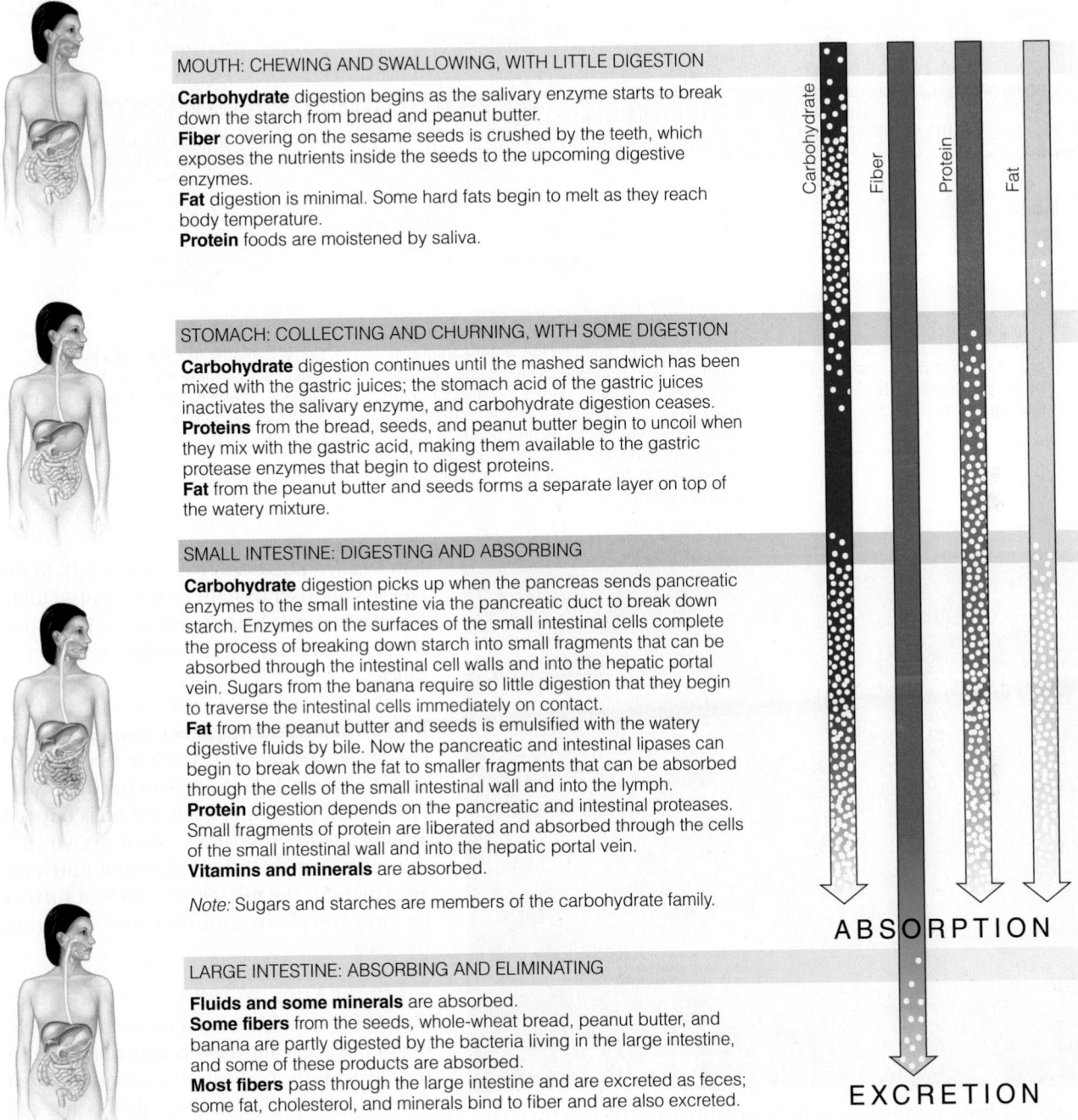

MOUTH: CHEWING AND SWALLOWING, WITH LITTLE DIGESTION

Carbohydrate digestion begins as the salivary enzyme starts to break down the starch from bread and peanut butter.
Fiber covering on the sesame seeds is crushed by the teeth, which exposes the nutrients inside the seeds to the upcoming digestive enzymes.
Fat digestion is minimal. Some hard fats begin to melt as they reach body temperature.
Protein foods are moistened by saliva.

STOMACH: COLLECTING AND CHURNING, WITH SOME DIGESTION

Carbohydrate digestion continues until the mashed sandwich has been mixed with the gastric juices; the stomach acid of the gastric juices inactivates the salivary enzyme, and carbohydrate digestion ceases.
Proteins from the bread, seeds, and peanut butter begin to uncoil when they mix with the gastric acid, making them available to the gastric protease enzymes that begin to digest proteins.
Fat from the peanut butter and seeds forms a separate layer on top of the watery mixture.

SMALL INTESTINE: DIGESTING AND ABSORBING

Carbohydrate digestion picks up when the pancreas sends pancreatic enzymes to the small intestine via the pancreatic duct to break down starch. Enzymes on the surfaces of the small intestinal cells complete the process of breaking down starch into small fragments that can be absorbed through the intestinal cell walls and into the hepatic portal vein. Sugars from the banana require so little digestion that they begin to traverse the intestinal cells immediately on contact.
Fat from the peanut butter and seeds is emulsified with the watery digestive fluids by bile. Now the pancreatic and intestinal lipases can begin to break down the fat to smaller fragments that can be absorbed through the cells of the small intestinal wall and into the lymph.
Protein digestion depends on the pancreatic and intestinal proteases. Small fragments of protein are liberated and absorbed through the cells of the small intestinal wall and into the hepatic portal vein.
Vitamins and minerals are absorbed.

Note: Sugars and starches are members of the carbohydrate family.

LARGE INTESTINE: ABSORBING AND ELIMINATING

Fluids and some minerals are absorbed.
Some fibers from the seeds, whole-wheat bread, peanut butter, and banana are partly digested by the bacteria living in the large intestine, and some of these products are absorbed.
Most fibers pass through the large intestine and are excreted as feces; some fat, cholesterol, and minerals bind to fiber and are also excreted.

ABSORPTION

EXCRETION

© Cengage Learning

chapters provide details on specific nutrients. Before following nutrients through the body, we must look more closely at the anatomy of the absorptive system.

Anatomy of the Absorptive System The inner surface of the small intestine looks smooth and slippery, but when viewed through a microscope, it turns out to be wrinkled into hundreds of folds. Each fold is contoured into thousands of fingerlike projections, as numerous as the hairs on velvet fabric. These small intestinal projections are called **villi.** A single villus, magnified still more, turns out to be composed of hundreds of cells, each covered with its own microscopic

villi (VILL-ee or **VILL-eye):** fingerlike projections from the folds of the small intestine; singular *villus.*

> FIGURE 3-8 **Absorption of Nutrients**

Absorption of nutrients into intestinal cells typically occurs by simple diffusion, facilitated diffusion, or active transport. Occasionally, a large molecule is absorbed by *endocytosis*—a process in which the cell membrane engulfs the molecule, forming a sac that separates from the membrane and moves into the cell.

Outside
cell

Cell
membrane

Inside
cell

SIMPLE
DIFFUSION

Carrier loads
nutrient on
outside of cell . . .

. . . and then
releases it on
inside of cell.

FACILITATED
DIFFUSION

Carrier loads
nutrient on
outside of cell . . .

Energy

. . . and then
releases it on
inside of cell.

ACTIVE
TRANSPORT

Some nutrients (such as water and small lipids) are absorbed by simple diffusion. They cross into intestinal cells freely.

Some nutrients (such as the water-soluble vitamins) are absorbed by facilitated diffusion. They need a specific carrier to transport them from one side of the cell membrane to the other. (Alternatively, facilitated diffusion may occur when the carrier changes the cell membrane in such a way that the nutrients can pass through.)

Some nutrients (such as glucose and amino acids) must be absorbed actively. These nutrients move against a concentration gradient, which requires energy.

© Cengage Learning

If you have ever watched a sea anemone with its fingerlike projections in constant motion, you have a good picture of how the intestinal villi move.

© Greg Amptman/Shutterstock.com

hairs, called **microvilli** (see Figure 3-9). In the crevices between the villi lie the **crypts**—tubular glands that secrete the intestinal juices into the small intestine. Nearby **goblet cells** secrete mucus.

The villi are in constant motion. Each villus is lined by a thin sheet of muscle, so it can wave, squirm, and wriggle like the tentacles of a sea anemone. Any nutrient molecule small enough to be absorbed is trapped among the microvilli and then drawn into the cells. Some partially digested nutrients are caught in the microvilli, digested further by enzymes there, and then absorbed into the cells.

A Closer Look at the Intestinal Cells

The cells of the villi are among the most amazing in the body, for they recognize and select the nutrients the body needs and regulate their absorption. As already described, each cell of a villus is coated with thousands of microvilli, which project from the cell's membrane (review Figure 3-9). In these microvilli, and in the membrane, lie hundreds of different kinds of enzymes and "pumps," which recognize and act on different nutrients. Descriptions of specific enzymes and pumps for each nutrient are presented in later chapters where appropriate; the point here is that the cells are equipped to handle all kinds and combinations of foods and their nutrients.

Specialized Cells A further refinement of the system is that the cells of successive portions of the intestinal tract are specialized to absorb different nutrients. The nutrients that are ready for absorption early are absorbed near the top of the GI

microvilli (MY-cro-VILL-ee or MY-cro-VILL-eye): tiny, hairlike projections on each cell of every villus that can trap nutrient particles and transport them into the cells; singular *microvillus*.

crypts (KRIPTS): tubular glands that lie between the intestinal villi and secrete intestinal juices into the small intestine.

goblet cells: cells of the GI tract (and lungs) that secrete mucus.

> FIGURE 3-9   **The Small Intestinal Villi**

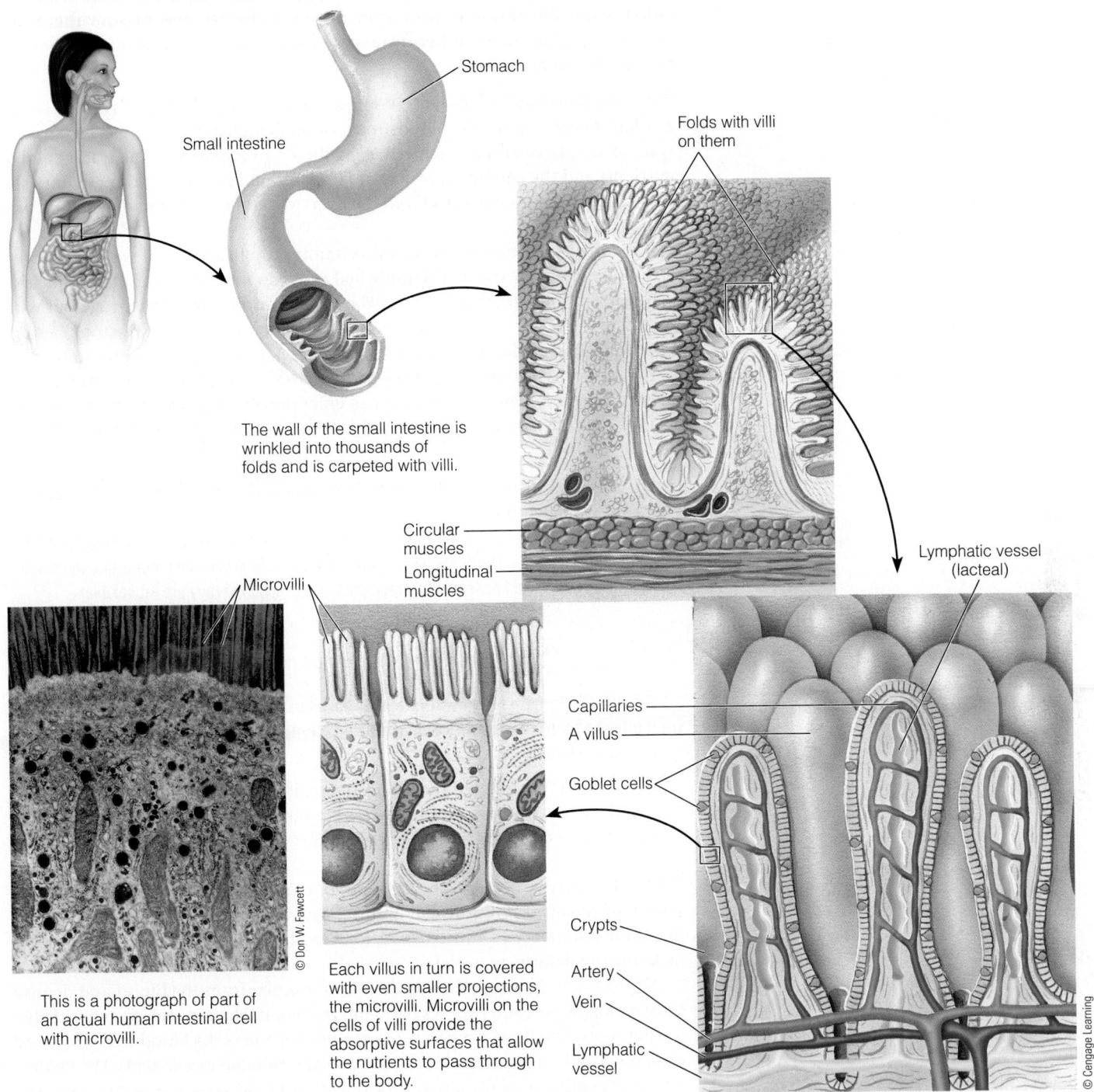

Stomach

Small intestine

Folds with villi on them

The wall of the small intestine is wrinkled into thousands of folds and is carpeted with villi.

Circular muscles

Longitudinal muscles

Lymphatic vessel (lacteal)

Microvilli

Capillaries

A villus

Goblet cells

Crypts

Artery

Vein

Lymphatic vessel

© Don W. Fawcett

© Cengage Learning

This is a photograph of part of an actual human intestinal cell with microvilli.

Each villus in turn is covered with even smaller projections, the microvilli. Microvilli on the cells of villi provide the absorptive surfaces that allow the nutrients to pass through to the body.

tract; those that take longer to be digested are absorbed farther down. Health-care professionals who treat digestive disorders learn the specialized absorptive functions of different parts of the GI tract so that if one part becomes dysfunctional, the diet can be adjusted accordingly.

Food Combining The idea that people should not eat certain food combinations (for example, fruit and meat) at the same meal, because the digestive system cannot handle more than one task at a time, is a myth. The art of "food combining"—which actually emphasizes "food separating"—is based on this myth, and it represents faulty logic and a gross underestimation of the body's capabilities. In fact, the contrary is often true; foods eaten together can enhance each other's use by the body. For example, vitamin C in a pineapple or other citrus fruit can enhance the absorption of iron from a meal of chicken and rice or other iron-containing foods. Many other instances of mutually beneficial interactions are presented in later chapters.

Preparing Nutrients for Transport When a nutrient molecule has crossed the cell of a villus, it enters either the bloodstream or the lymphatic system. Both transport systems supply vessels to each villus, as shown in Figure 3-9. The water-soluble nutrients and the smaller products of fat digestion are released directly into the bloodstream and guided directly to the liver, where their fate and destination will be determined.

The larger fats and the fat-soluble vitamins are insoluble in water, however, and blood is mostly water. The intestinal cells assemble many of the products of fat digestion into larger molecules. These larger molecules cluster together with special proteins, forming chylomicrons. Chylomicrons (kye-lo-MY-cronz) are defined and described in more detail in Chapter 5. For now, keep in mind that because chylomicrons carry fats, they are released into the lymphatic system. They move through the lymph until they can enter the bloodstream at a point near the heart. Consequently, chylomicrons bypass the liver at first. Details follow.

> **REVIEW IT** Describe the anatomical details of the intestinal cells that facilitate nutrient absorption.

The many folds and villi of the small intestine dramatically increase its surface area, facilitating nutrient absorption. Nutrients pass through the cells of the villi and enter either the blood (if they are water soluble or small fat fragments) or the lymph (if they are fat soluble).

3.3 The Circulatory Systems

> **LEARN IT** Explain how nutrients are routed in the circulatory systems from the GI tract into the body and identify which nutrients enter the blood directly and which must first enter the lymph.

Once a nutrient has entered the bloodstream, it may be transported to any of the cells in the body, from the tips of the toes to the roots of the hair. The circulatory systems deliver nutrients wherever they are needed.

The Vascular System The vascular, or blood circulatory, system is a closed system of vessels through which blood flows continuously, with the heart serving as the pump (see Figure 3-10). As the blood circulates through this system, it picks up and delivers materials as needed.

All the body tissues derive nutrients and oxygen from the blood and deposit carbon dioxide and other wastes back into the blood. The digestive system supplies the nutrients. The lungs exchange oxygen (which enters the blood to be delivered to all cells) and carbon dioxide (which leaves the blood to be exhaled). The kidneys filter wastes other than carbon dioxide out of the blood to be excreted in the urine.

Blood leaving the right side of the heart circulates through the lungs and then back to the left side of the heart. The left side of the heart then pumps the blood out of the **aorta** through **arteries** to all systems of the body. The blood circulates in the **capillaries,** where it exchanges material with the cells and then collects into **veins,** which return it again to the right side of the heart. In short, blood travels this simple route:

Heart to arteries to capillaries to veins to heart

The routing of the blood leaving the digestive system has a special feature. The blood is carried to the digestive system (as to all organs) by way of an artery, which (as in all organs) branches into capillaries to reach every cell. Blood leaving the digestive system, however, goes by way of a vein. The **hepatic portal vein** directs blood not back to the heart but to another organ, the liver. This

aorta (ay-OR-tuh): the large, primary artery that conducts blood from the heart to the body's smaller arteries.

arteries: vessels that carry blood from the heart to the tissues.

capillaries (CAP-ill-aries): small vessels that branch from an artery. Capillaries connect arteries to veins. Exchange of oxygen, nutrients, and waste materials takes place across capillary walls.

veins (VANES): vessels that carry blood to the heart.

hepatic portal vein: the vein that collects blood from the GI tract and conducts it to the liver.

• **portal** = gateway

> FIGURE 3-10 The Vascular System

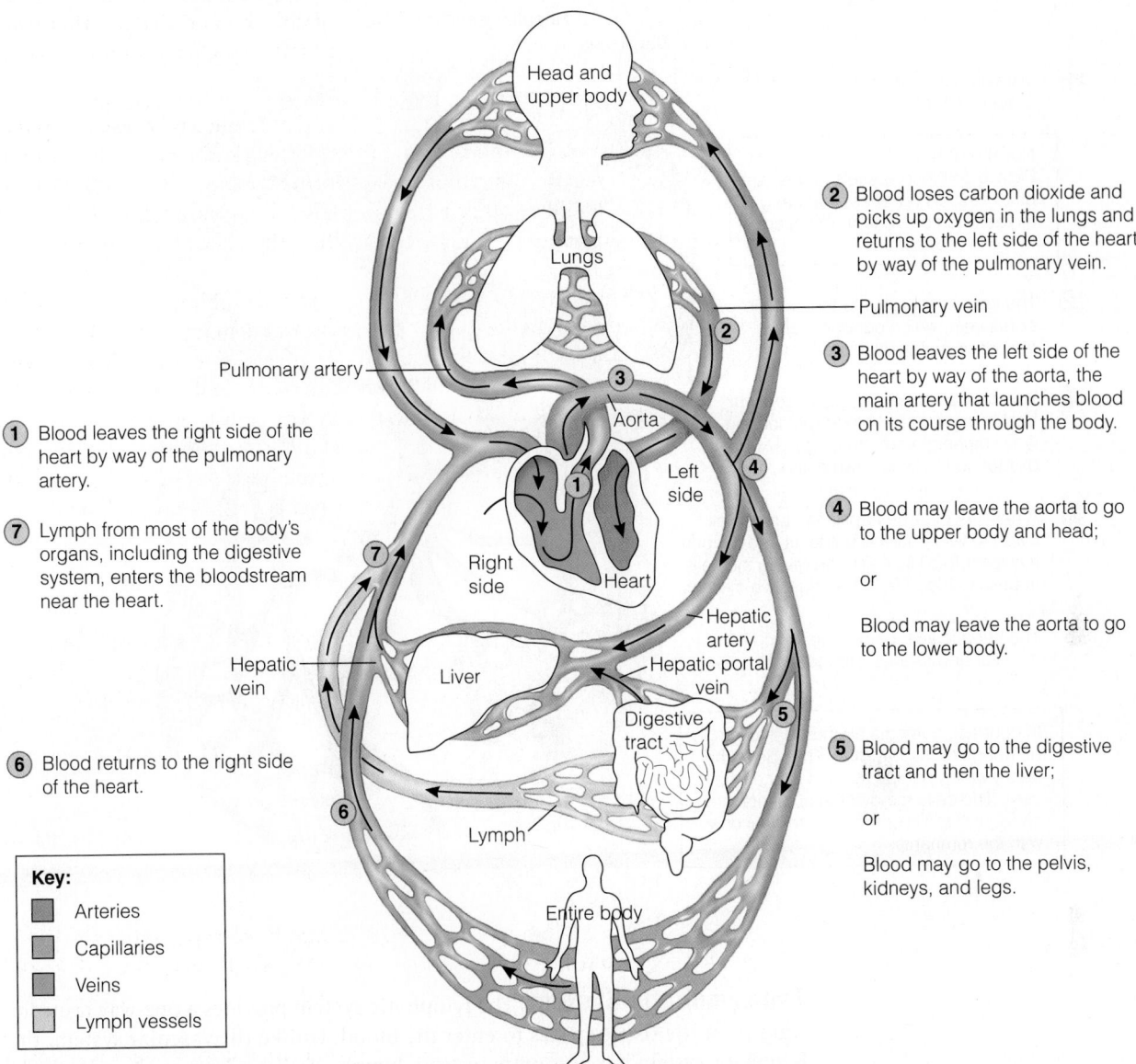

Head and upper body

Lungs

Pulmonary artery

(1) Blood leaves the right side of the heart by way of the pulmonary artery.

(7) Lymph from most of the body's organs, including the digestive system, enters the bloodstream near the heart.

Hepatic vein

(6) Blood returns to the right side of the heart.

Aorta

(3)

(1) **Right side**   **Left side**   **Heart**

Liver

Lymph

Entire body

Digestive tract

Hepatic artery

Hepatic portal vein

Key:
- Arteries
- Capillaries
- Veins
- Lymph vessels

(2) Blood loses carbon dioxide and picks up oxygen in the lungs and returns to the left side of the heart by way of the pulmonary vein.

— Pulmonary vein

(3) Blood leaves the left side of the heart by way of the aorta, the main artery that launches blood on its course through the body.

(4) Blood may leave the aorta to go to the upper body and head;

or

Blood may leave the aorta to go to the lower body.

(5) Blood may go to the digestive tract and then the liver;

or

Blood may go to the pelvis, kidneys, and legs.

© Cengage Learning

vein branches into a network of large capillaries so that every cell of the liver has access to the blood. Blood leaving the liver then collects into the **hepatic vein,** which returns blood to the heart. The route is:

Heart to arteries to capillaries (in intestines) to hepatic portal vein to capillaries (in liver) to hepatic vein to heart

Figure 3-11 on p. 84 shows the liver's key position in nutrient transport. An anatomist studying this system knows there must be a reason for this special arrangement. The liver's placement ensures that it will be first to receive the nutrients absorbed from the GI tract. In fact, the liver has many jobs to do in preparing the absorbed nutrients for use by the body. Of all the body's organs, the liver is the most metabolically active.

In addition, the liver defends the body by detoxifying substances that might cause harm and preparing waste products for excretion. This is why, when people ingest poisons that succeed in passing the first barrier (the intestinal cells), the liver quite often suffers the damage—from viruses such as hepatitis, from drugs such as barbiturates or alcohol, from toxins such as pesticide residues, and from contaminants such as mercury. Perhaps, in fact, you have been undervaluing your liver, not knowing what heroic tasks it quietly performs for you.

hepatic vein: the vein that collects blood from the liver and returns it to the heart.

• **hepatic** = liver

> FIGURE 3-11 The Liver

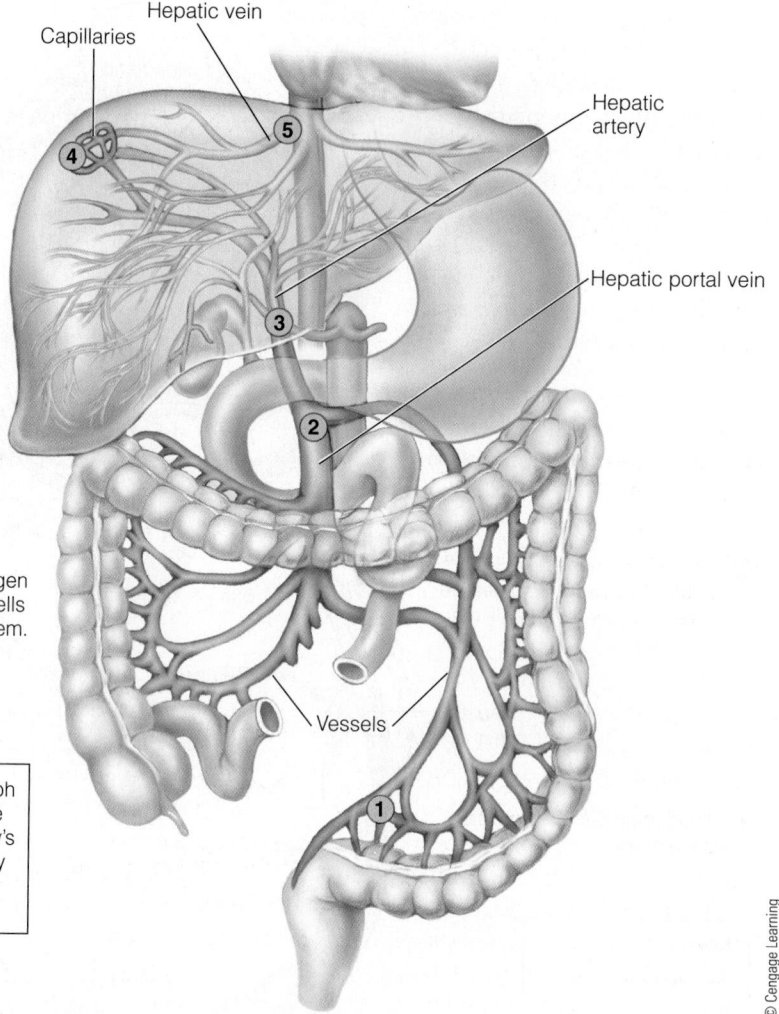

Hepatic vein

Capillaries

Hepatic artery

Hepatic portal vein

Vessels

(1) Vessels gather up nutrients from the digestive tract.

Not shown here:
Parallel to these vessels (veins) are other vessels (arteries) that carry oxygen-rich blood from the heart to the intestines.

(2) The vessels merge into the hepatic portal vein, which conducts all absorbed materials to the liver.

(3) The hepatic artery brings a supply of freshly oxygenated blood (not loaded with nutrients) from the lungs to supply oxygen to the liver's own cells.

(4) A network of large capillaries branch all over the liver, making nutrients and oxygen available to all its cells and giving the cells access to blood from the digestive system.

(5) The hepatic vein gathers up blood in the liver and returns it to the heart.

In contrast, nutrients absorbed into lymph do not go to the liver first. They go to the heart, which pumps them to all the body's cells. The cells remove the nutrients they need, and the liver then has to deal only with the remnants.

© Cengage Learning

The Lymphatic System The **lymphatic system** provides a one-way route for fluid from the tissue spaces to enter the blood. Unlike the vascular system, the lymphatic system has no pump; instead, **lymph** circulates *between* the cells of the body and collects into tiny vessels. The fluid moves from one portion of the body to another as muscles contract and create pressure here and there. Ultimately, much of the lymph collects in the **thoracic duct** behind the heart. The thoracic duct opens into the **subclavian vein,** where the lymph enters the bloodstream. Thus nutrients from the GI tract that enter lymphatic vessels (large fats and fat-soluble vitamins) ultimately enter the bloodstream, circulating through arteries, capillaries, and veins like the other nutrients, with a notable exception—they bypass the liver at first.*

Once inside the vascular system, the nutrients can travel all over the body, where they can be taken into cells and used as needed. What becomes of them is described in later chapters.

lymphatic (lim-FAT-ic) **system:** a loosely organized system of vessels and ducts that convey fluids toward the heart. The GI part of the lymphatic system carries the products of fat digestion into the bloodstream.

lymph (LIMF): a clear yellowish fluid that is similar to blood except that it contains no red blood cells or platelets. Lymph from the GI tract transports fat and fat-soluble vitamins to the bloodstream via lymphatic vessels.

thoracic (thor-ASS-ic) **duct:** the main lymphatic vessel that collects lymph and drains into the left subclavian vein.

subclavian (sub-KLAY-vee-an) **vein:** the vein that provides passageway from the lymphatic system to the vascular system.

> REVIEW IT Explain how nutrients are routed in the circulatory systems from the GI tract into the body and identify which nutrients enter the blood directly and which must first enter the lymph.

Nutrients leaving the digestive system via the blood are routed directly to the liver before being transported to the body's cells. Those leaving via the lymphatic system (large fats and fat-soluble vitamins) eventually enter the vascular system but bypass the liver at first.

*The lymphatic vessels of the intestine that take up nutrients and pass them to the lymph circulation are called *lacteals* (LACK-tee-als).

3.4 The Health and Regulation of the GI Tract

> **LEARN IT** Describe how bacteria, hormones, and nerves influence the health and activities of the GI tract.

This section describes the bacterial conditions and hormonal regulation of a healthy GI tract, but many factors can influence normal GI function. For example, peristalsis and sphincter action are poorly coordinated in newborns, so infants tend to "spit up" during the first several months of life. Older adults often experience constipation, in part because the intestinal wall loses strength and elasticity with age, which slows GI motility. Diseases can also interfere with digestion and absorption and often lead to malnutrition. Lack of nourishment, in general, and lack of certain dietary constituents such as fiber, in particular, alter the structure and function of GI cells. Quite simply, GI tract health depends on adequate nutrition.

Eaten regularly, yogurt can alleviate common digestive problems.

Gastrointestinal Microbiome A healthy GI tract is home to a vibrant community of some 100 trillion **microbes**—bacteria, viruses, fungi, protozoa, and other microorganisms, collectively known as the **human microbiome**. Weighing less than a pound in total, these microbial cells outnumber the body's cells tenfold. The bacteria alone represent more than 400 different species and subspecies. The prevalence of different microbes in various parts of the GI tract depends on such factors as pH, peristalsis, diet, and other microbes. Relatively few microbes can live in the low pH of the stomach with its somewhat rapid peristalsis, whereas the neutral pH and slow peristalsis of the lower small intestine and the large intestine permit the growth of a diverse and abundant population.

Recent research has revealed that a person's health reflects the relative stability, disturbance, and resilience of the microbiome.[2] Its composition and activity may contribute to dozens of common diseases, including inflammatory bowel disease and obesity.[3]

The microbiome population and environment change dramatically in response to diet—both in the short term (daily meals) and in the long term (habitual diet patterns).[4] Consider, for example, that fibers that cannot be digested by the human body provide a major source of energy for bacteria, fostering their growth. As GI bacteria digest and metabolize fibers and other nutrients, they produce compounds such as short fragments of fat, which can influence energy metabolism and immunity.[5] Fibers and some other food components are called **prebiotics** because they encourage the growth and activity of bacteria. Research suggests that prebiotics may reduce the risk of GI infections, inflammation, and disorders; increase the bioavailability of nutrients; and regulate appetite and satiety.[6]

Some foods contain **probiotics,** live microbes that change the conditions in the GI tract in ways that seem to benefit health.[7] For example, **yogurt** contains *Lactobacillus* and other living bacteria. The potential GI health benefits of probiotics or products of their metabolism include helping to alleviate diarrhea, constipation, inflammatory bowel disease, ulcers, allergies, lactose intolerance, and infant colic; enhance immune function; and protect against colon cancer.[8] Research studies continue to explore how diet influences GI bacteria and which foods—with their prebiotics and probiotics—affect GI health. In addition, research studies are beginning to reveal several health benefits beyond the GI tract—such as improving blood pressure and immune responses.

Bacteria in the GI tract also produce several vitamins, including biotin, folate, pantothenic acid, riboflavin, thiamin, vitamin B_6, vitamin B_{12}, and vitamin K. Because the amount produced is insufficient to meet the body's needs, these vitamins are considered essential nutrients that must be provided by the diet.

Gastrointestinal Hormones and Nerve Pathways The ability of the digestive tract to handle its ever-changing contents illustrates an important physiological principle that governs the way all living things function—the principle of **homeostasis.** Simply stated, survival depends on body conditions staying about the same; if they deviate too far from the norm, the body must

microbes (MY-krobes): microscopically small organisms including bacteria, viruses, fungi, and protozoa; also called *microorganisms.*

- **mikros** = small

human microbiome: the collection of microbes found in or on the human body.

prebiotics: food components (such as fibers) that are not digested by the human body but are used as food by the GI bacteria to promote their growth and activity.

probiotics: living microorganisms found in foods and dietary supplements that, when consumed in sufficient quantities, are beneficial to health.

- **pro** = for
- **bios** = life

yogurt: milk product that results from the fermentation of lactic acid in milk by *Lactobacillus bulgaricus* and *Streptococcus thermophilus.*

homeostasis (HOME-ee-oh-STAY-sis): the maintenance of constant internal conditions (such as blood chemistry, temperature, and blood pressure) by the body's control systems. A homeostatic system is constantly reacting to external forces to maintain limits set by the body's needs.

- **homeo** = like, similar
- **stasis** = staying

> FIGURE 3-12 **An Example of a Negative Feedback Loop**

ON Food in the stomach causes the cells of the stomach wall to start releasing gastrin.

OFF Acidity in the stomach causes the cells of the stomach wall to stop releasing gastrin.

Gastrin stimulates stomach glands to release the components of hydrochloric acid.

NEGATIVE FEEDBACK

Stomach pH reaches 1.5 acidity.

© Cengage Learning

"do something" to bring them back to normal. The body's regulation of digestion is one example of homeostatic regulation. The body also regulates its temperature, its blood pressure, and all other aspects of its blood chemistry in similar ways.

Two intricate and sensitive systems coordinate all the digestive and absorptive processes: the hormonal (or endocrine) system and the nervous system. Even before the first bite of food is taken, the mere thought, sight, or smell of food can trigger a response from these systems. Then, as food travels through the GI tract, it either stimulates or inhibits digestive secretions by way of messages that are carried from one section of the GI tract to another by both **hormones** and nerve pathways. (Appendix A presents a brief summary of the body's hormonal system and nervous system.)

Notice that the kinds of regulation described below are all examples of *feedback* mechanisms. A certain condition demands a response. The response changes that condition, and the change then cuts off the response. Thus the system is self-correcting.

- *The stomach normally maintains a pH between 1.5 and 1.7. How does it stay that way?* Food entering the stomach stimulates cells in the stomach wall to release the hormone **gastrin**. Gastrin, in turn, stimulates the stomach glands to secrete the components of hydrochloric acid. When pH 1.5 is reached, the acid itself turns off the gastrin-producing cells; they stop releasing gastrin, and the glands stop producing hydrochloric acid. Thus the system adjusts itself, as Figure 3-12 shows.

Nerve receptors in the stomach wall also respond to the presence of food and stimulate the gastric glands to secrete juices and the muscles to contract. As the stomach empties, the receptors are no longer stimulated, the flow of juices slows, and the stomach quiets down.

- *The pyloric sphincter opens to let out a little chyme, then closes again. How does it know when to open and close?* When the pyloric sphincter relaxes, acidic chyme slips through. The cells of the pyloric muscle on the intestinal side sense the acid, causing the pyloric sphincter to close tightly. Only after the chyme has been neutralized by pancreatic bicarbonate and the juices surrounding the pyloric sphincter have become alkaline can the muscle relax again. This process ensures that the chyme will be released slowly enough to be neutralized as it flows through the small intestine. This is important because the small intestine has less of a mucous coating than the stomach does and so is not as well protected from acid.

- *As the chyme enters the small intestine, the pancreas adds bicarbonate to it so that the intestinal contents always remain at a slightly alkaline pH. How does the pancreas know how much to add?* The presence of chyme stimulates the cells of the duodenal wall to release the hormone **secretin** into the blood. When secretin reaches the pancreas, it stimulates the pancreas to release its bicarbonate-rich juices. Thus, whenever the duodenum signals that acidic chyme is present, the pancreas responds by sending bicarbonate to neutralize it. When the need has been met, the cells of the duodenal wall are no longer stimulated to release secretin, the hormone no longer flows through the blood, and the pancreas no longer receives the message and stops sending pancreatic juice. Nerves also regulate pancreatic secretions.

- *Pancreatic secretions contain a mixture of enzymes to digest carbohydrate, fat, and protein. How does the pancreas know how much of each type of enzyme to provide?* This is one of the most interesting questions physiologists have asked. Clearly, the pancreas does know what its owner has been eating, and it secretes enzyme mixtures tailored to handle the food mixtures that have been arriving recently (over the past several days). Enzyme activity changes proportionately in response to the amounts of carbohydrate, fat, and protein in the diet. If a person has been eating mostly carbohydrates, the pancreas makes and secretes mostly carbohydrases; if the person's diet has been high in fat, the pancreas produces more lipases; and so forth. Hormones from the GI tract, secreted in response to meals, keep the pancreas informed as to its digestive tasks. The day or two lag between the time a person's diet changes dramatically

hormones: chemical messengers. Hormones are secreted by a variety of glands in response to altered conditions in the body. Each hormone travels to one or more specific target tissues or organs, where it elicits a specific response to maintain homeostasis.

gastrin: a hormone secreted by cells in the stomach wall. Target organ: the glands of the stomach. Response: secretion of gastric acid.

secretin (see-CREET-in): a hormone produced by cells in the duodenum wall. Target organ: the pancreas. Response: secretion of bicarbonate-rich pancreatic juice.

and the time digestion of the new diet becomes efficient explains why dietary changes can "upset digestion" and should be made gradually.

• *Why don't the digestive enzymes damage the pancreas?* The pancreas protects itself from harm by producing an inactive form of the enzymes.* It releases these proteins into the small intestine, where they are activated to become enzymes. In pancreatitis, the digestive enzymes become active within the infected pancreas, causing inflammation and damaging the delicate pancreatic tissues.

• *When fat is present in the intestine, the gallbladder contracts to squirt bile into the intestine to emulsify the fat. How does the gallbladder get the message that fat is present?* Fat in the intestine stimulates cells of the intestinal wall to release the hormone **cholecystokinin (CCK).** This hormone travels by way of the blood to the gallbladder and stimulates it to contract, which releases bile into the small intestine. Cholecystokinin also travels to the pancreas and stimulates it to secrete its juices, which releases bicarbonate and enzymes into the small intestine. Once the fat in the intestine is emulsified and enzymes have begun to work on it, the fat no longer provokes release of the hormone, and the message to contract is canceled. (By the way, fat emulsification can continue even after a diseased gallbladder has been surgically removed because the liver can deliver bile directly to the small intestine.)

• *Fat and protein take longer to digest than carbohydrate does. When fat or protein is present, intestinal motility slows to allow time for its digestion. How does the intestine know when to slow down?* Cholecystokinin is released in response to fat or protein in the small intestine. In addition to its role in fat emulsification and digestion, cholecystokinin slows GI tract motility. Slowing the digestive process helps to maintain a pace that allows all reactions to reach completion. Hormonal and nervous mechanisms like these account for much of the body's ability to adapt to changing conditions.

Table 3-2 summarizes the actions of these three GI hormones. Gastrin, secretin, and cholecystokinin are among the most studied GI hormones, but the GI tract releases more than 20 hormones. In addition to assisting with digestion and absorption, many of these hormones regulate food intake and influence satiation—the feeling of satisfaction and fullness that occurs during a meal and halts eating. Current research is focusing on the roles these hormones may play in the development of obesity and its treatments (more details provided in Chapter 8).

Discovering the answers to questions like these has led some people to devote their whole lives to the study of physiology. For now, however, these few examples illustrate how all the processes throughout the digestive system are precisely and automatically regulated without any conscious effort.

The System at Its Best This chapter describes the anatomy of the digestive tract on several levels: the sequence of digestive organs, the cells and structures of the villi, and the selective machinery of the cell membranes. The intricate architecture of the digestive system makes it sensitive and responsive to conditions in

cholecystokinin (COAL-ee-SIS-toe-KINE-in), or **CCK:** a hormone produced by cells of the intestinal wall. Target organ: the gallbladder. Response: release of bile and slowing of GI motility.

---

*The inactive precursor of an enzyme is called a *zymogen* (ZYE-mo-jen).

TABLE 3-2 The Primary Actions of Selected GI Hormones

Hormone	Responds to	Secreted from	Stimulates	Response
Gastrin	Food in the stomach	Stomach wall	Stomach glands	Hydrochloric acid secreted into the stomach to maintain an acidic pH
Secretin	Acidic chyme in the small intestine	Duodenal wall	Pancreas	Bicarbonate-rich juices secreted into the small intestine to maintain a slightly alkaline pH
Cholecystokinin	Fat or protein in the small intestine	Intestinal wall	Gallbladder	Bile secreted into the duodenum to emulsify fats
			Pancreas	Bicarbonate- and enzyme-rich juices secreted into the small intestine to maintain a slightly alkaline pH, digest fats and proteins, and slow GI tract motility

Nourishing foods and pleasant conversations support a healthy digestive system.

its environment. Several different kinds of GI tract cells confer specific immunity against intestinal diseases such as inflammatory bowel disease. In addition, secretions from the GI tract—saliva, mucus, gastric acid, and digestive enzymes—not only help with digestion, but also defend against foreign invaders. Together the GI's team of bacteria, cells, and secretions defend the body against many illnesses.

One indispensable condition is good health of the digestive system itself. Like all the other organs of the body, the GI tract depends on a healthy supply of blood. The cells of the GI tract become weak and inflamed when blood flow is diminished, as may occur in heart disease when arteries become clogged or blood clots form. Just as a diminished blood flow to the heart or brain can cause a heart attack or stroke, respectively, too little blood to the intestines can also be damaging—or even fatal. A diminished blood flow to the intestines—called **intestinal ischemia**—is characterized by abdominal pain, forceful bowel movements, and blood in the stool.

The health of the digestive system is also affected by such lifestyle factors as sleep, physical activity, and state of mind. Adequate sleep allows for repair and maintenance of tissue and removal of wastes that might impair efficient functioning. Activity promotes healthy muscle tone. Stress alters GI motility, secretions, permeability, blood flow, and bacteria.[9] For healthy digestion, mealtimes should be relaxed and tranquil. Pleasant conversations and peaceful environments during meals ease the digestive process.

Another factor in GI health is the kind of foods eaten. Among the characteristics of meals that promote optimal absorption of nutrients are those mentioned in Chapter 2: balance, moderation, variety, and adequacy. Balance and moderation require having neither too much nor too little of anything. For example, too much fat can be harmful, but some fat is beneficial in slowing down intestinal motility and providing time for absorption of some of the nutrients that are slow to be absorbed.

Variety is important for many reasons, but one is that some food constituents interfere with nutrient absorption. For example, some compounds common in high-fiber foods such as whole-grain cereals, certain leafy green vegetables, and legumes bind with minerals. To some extent, then, the minerals in those foods may become unavailable for absorption. These high-fiber foods are still valuable, but they need to be balanced with a variety of other foods that can provide the minerals.

As for adequacy—in a sense, this entire book is about dietary adequacy. A diet must provide all the essential nutrients, fiber, and energy in amounts sufficient to maintain health. But here, at the end of this chapter, is a good place

intestinal ischemia (is-KEY-me-ah): a diminished blood flow to the intestines that is characterized by abdominal pain, forceful bowel movements, and blood in the stool.

to emphasize the interdependence of the nutrients. It could almost be said that every nutrient depends on every other. All the nutrients work together, and all are present in the cells of a healthy digestive tract. To maintain health and promote the functions of the GI tract, make balance, moderation, variety, and adequacy features of every day's meals.

> **REVIEW IT** Describe how bacteria, hormones, and nerves influence the health and activities of the GI tract.

A diverse and abundant bacteria population supports GI health. The regulation of GI processes depends on the coordinated efforts of the hormonal system and the nervous system. Together, digestion and absorption break down foods into nutrients for the body's use. To function optimally, a healthy GI tract needs a balanced diet, adequate rest, and regular physical activity.

Nutrition Portfolio

A digestive system that is well cared for most of the time can adjust to handle almost any diet or combination of foods with ease on occasion. Go to Diet & Wellness Plus and choose one of the days on which you have tracked your diet for the entire day. Choose the day you thought you ate most poorly, and looking at it, record in your journal answers to the following:

- Describe the physical and emotional environment that typically surrounds your meals, including how it affects you and how it might be improved.

- Did you experience any GI discomfort on that day? Do you experience any GI discomfort regularly? If so, which of the foods that you ate might have contributed to your discomfort? What can you do to prevent or alleviate GI problems in the future? Use Table H3-1 (p. 95) as a guide.

- List any changes you can make in your eating habits to promote overall GI health.

DIET & WELLNESS **PLUS** To complete this exercise, go to your Diet & Wellness Plus at www.cengagebrain.com.

> **STUDY IT** To review the key points of this chapter and take a practice quiz, go to the study cards at the end of the book.

REFERENCES

1. X. Chen and coauthors, A gustotopic map of taste qualities in the mammalian brain, *Science* 333 (2011): 1262-1266; N. Chaudhari and S. D. Roper, The cell biology of taste, *Journal of Cell Biology* 190 (2010): 285–296.

2. D. A. Relman, The human microbiome: Ecosystem resilience and health, *Nutrition Reviews* 70 (2012): S2–S9.

3. E. LeChatelier and coauthors, Richness of human gut microbiome correlates with metabolic markers, *Nature* 500 (2013): 541–546; F. Bäckhed, Host responses to the human microbiome, *Nutrition Reviews* 70 (2012): S14–S17; W. M. deVos and E. A. J. deVos, Role of the intestinal microbiome in health and disease: From correlation to causation, *Nutrition Reviews* 70 (2012): S45–S56.

4. H. J. Flint, The impact of nutrition on the human microbiome, *Nutrition Reviews* 70 (2012): S10–S13; G. D. Wu and coauthors, Linking long-term dietary patterns with gut microbial enterotypes, *Science* 334 (2011): 105–108; R. Jumpertz and coauthors, Energy-balance studies reveal associations between gut microbes, caloric load, and nutrient absorption in humans, *American Journal of Clinical Nutrition* 94 (2011): 58–65.

5. N. M. Delzenne and P. D. Cani, Interaction between obesity and the gut microbiota: Relevance in nutrition, *Annual Review of Nutrition* 31 (2011): 15–31.

6. A. M. Brownawell and coauthors, Prebiotics and the health benefits of fiber: Current regulatory status, future research, and goals, *Journal of Nutrition* 142 (2012): 962–974.

7. T. C. Wallace and coauthors, Human gut microbiota and its relationship to health and disease, *Nutrition Reviews* 69 (2011): 392–403; S. C. Bischoff and M. Zeitz, Scientific evidence for the medical use of probiotics, *Annals of Nutrition and Metabolism* 57 (2010): S1–S5; N. T. Williams, Probiotics, *American Journal of Health System Pharmacy* 15 (2010): 449–458.

8. S. Hempel and coauthors, Probiotics for the prevention and treatment of antibiotic-associated diarrhea: A systematic review and meta-analysis, *Journal of the American Medical Association* 307 (2012): 1959–1969; B. C. Johnston and coauthors, Probiotics for the prevention of *Clostridium difficile*–associated diarrhea: A systematic review and meta-analysis, *Annals of Internal Medicine* 157 (2012): 878–888; M. Kumar and coauthors, Probiotic metabolites as epigenetic targets in the prevention of colon cancer, *Nutrition Reviews* 71 (2012): 23–34.

9. P. C. Konturek, T. Brzozowski, and S. J. Donturek, Stress and the gut: Pathophysiology, clinical consequences, diagnostic approach and treatment options, *Journal of Physiology and Pharmacology* 62 (2011): 591–599; C. Hughes and coauthors, Galactooligosaccharide supplementation reduces stress-induced gastrointestinal dysfunction and days of cold or flu: A randomized double-blind, controlled trial in healthy university students, *American Journal of Clinical Nutrition* 93 (2011): 1305–1311.

HIGHLIGHT > 3
Common Digestive Problems

© Corbis Super RF/Alamy

> LEARN IT Outline strategies to prevent or alleviate common GI problems.

The facts of anatomy and physiology presented in Chapter 3 permit easy understanding of some common problems that occasionally arise in the digestive tract. Food may slip into the airways instead of the esophagus, causing choking. Bowel movements may be loose and watery, as in diarrhea, or painful and hard, as in constipation. Some people complain about belching, while others are bothered by intestinal gas. Sometimes people develop medical problems such as ulcers. This highlight describes some of the symptoms of these common digestive problems and suggests strategies for preventing them (the accompanying glossary defines related terms).

Choking

Sometimes a sip of a beverage or a tiny bit of food "slips down the wrong pipe." The body's first response is to cough, and quite often coughing clears the passage. When someone is truly choking, however, food has slipped into the **trachea** and completely blocked the air passageways (see Figure H3-1). Thus the person cannot cough—or even breathe. Without oxygen, the person may suffer permanent brain damage within 5 minutes or may even die. For this reason, it is imperative that everyone learn to recognize the universal distress signal for choking (shown in Figure H3-2) and act promptly.

Because the **larynx** is in the trachea and makes sounds only when air is pushed across it, a person choking will be unable to speak. For this reason, to help a person who is choking, first ask "Can you speak?" If the person is coughing, breathing adequately, or able to speak, do not interfere. Whatever you do, do not hit him on the back as the particle may become lodged more firmly in his air passageway. If the person cannot speak or cough, shout for help and perform the **Heimlich maneuver** (described in Figure H3-2). Almost any food can cause choking, although some are cited more often than others: chunks of meat, hot dogs, nuts,

> FIGURE H3-1 **Normal Swallowing and Choking**

- Tongue
- Food
- Larynx rises
- Epiglottis closes over larynx
- Esophagus (to stomach)
- Trachea (to lungs)

Swallowing. The epiglottis closes over the larynx, blocking entrance to the lungs via the trachea. The red arrow shows that food is heading down the esophagus normally.

Choking. A choking person cannot speak or gasp because food lodged in the trachea blocks the passage of air. The red arrow points to where the food should have gone to prevent choking.

whole grapes, raw carrots, marshmallows, hard or sticky candies, gum, popcorn, and peanut butter. These foods are particularly difficult for young children (especially those 4 years of age and younger) to safely chew and swallow. Each year more than 10,000 children (14 years old or younger) in the United States choke; more than half choke on food. Every 5 days, a child in the United States chokes to death on food.[1] An adult should be present and alert to the dangers of choking whenever young children are eating. To prevent choking, cut food into small pieces, chew thoroughly before swallowing, don't talk or laugh with food in your mouth, and don't eat when breathing hard.

Vomiting

Vomiting can be a symptom of many different diseases or may arise in situations that upset the body's equilibrium, such as air or sea travel. For whatever reason, the contents of the stomach are propelled up through the esophagus to the mouth and expelled. Sometimes the muscular contractions will extend

> FIGURE H3-2 **First Aid for Choking**

First aid for choking relies on abdominal thrusts, sometimes called the Heimlich maneuver. If abdominal thrusts are not successful and the person loses consciousness, lower him to the floor, call 911, remove the object blocking the airway if possible, and begin CPR. Because there is no time for hesitation when called upon to perform this death-defying act, you would do well to take a life-saving course to learn these techniques.

The universal signal for choking alerts others to the need for assistance.

Stand behind the person with your arms wrapped around him. Make a fist with one hand and place the thumb side snugly against the body, slightly above the navel and below the breastbone.

Grasp the fist with your other hand and make a quick upward and inward thrust. Repeat thrusts until the object is dislodged.

To perform abdominal thrusts on yourself, make a fist and place the thumb below your breastbone and above your navel. Grasp your fist with your other hand and press inward with a quick upward thrust. Alternatively, quickly thrust your upper body against a table edge, chair, or railing.

beyond the stomach and carry the contents of the duodenum, with its green bile, into the stomach and then up the esophagus. Although certainly unpleasant and wearying for the nauseated person, vomiting is often not a cause for alarm. Vomiting is one of the body's adaptive mechanisms to rid itself of something irritating. The best advice is to rest and drink small amounts of liquids as tolerated until the nausea subsides.

A physician's care may be needed, however, if vomiting causes such large losses of fluid as to threaten dehydration. As fluid is lost from the GI tract, the body's other fluids redistribute themselves, taking fluid from every cell of the body. Fluid leaving the cells is accompanied by salts that are absolutely essential to the life of the cells. Replacing salts and fluid is difficult if the vomiting continues, and intravenous feedings of saline and glucose may be necessary. Vomiting and dehydration are especially serious in an infant, and a physician should be contacted without delay.

Self-induced vomiting, such as occurs in bulimia nervosa, also has serious consequences. In addition to fluid and salt imbalances, repeated vomiting can cause irritation and infection of the pharynx, esophagus, and salivary glands; erosion of the teeth and gums; and dental caries. The esophagus may rupture or tear, as may the stomach. Sometimes the eyes become red from pressure during vomiting. Bulimic behavior reflects underlying psychological problems that require intervention. (Bulimia nervosa is discussed fully in Highlight 8.)

Diarrhea

Diarrhea is characterized by frequent, loose, watery stools. Such stools indicate that the intestinal contents have moved too quickly through the intestines for fluid absorption to take place or that water has been drawn from the cells lining the intestinal tract and added to the food residue. Like vomiting, diarrhea can lead to considerable fluid and salt losses, but the composition of the fluids is different. Stomach fluids lost in vomiting are highly acidic, whereas intestinal fluids lost in diarrhea are nearly neutral. When fluid losses require medical attention, correct replacement is crucial.

Diarrhea is a symptom of various medical conditions and treatments. It may occur abruptly in a healthy person as a result of infections (such as foodborne illness) or as a side effect of medications. When used in large quantities, food ingredients such as the sugar alternative sorbitol and the fat alternative olestra may also cause diarrhea in some people. If a food is responsible, then that food must be omitted from the diet, at least temporarily. If medication is responsible, a different medicine, when possible, or a different form (injectable versus oral, for example) may alleviate the problem. Diarrhea may also occur as a result of disorders of the GI tract, such as irritable bowel syndrome or colitis.

Irritable Bowel Syndrome

Irritable bowel syndrome is one of the most common GI disorders and is characterized by frequent or severe abdominal discomfort and a disturbance in the motility of the GI tract.[2] In most cases, GI contractions are stronger and last longer than normal, forcing intestinal contents through quickly and causing gas, **bloating,** and diarrhea. In some cases, however, GI contractions are weaker than normal, slowing the passage of intestinal contents and causing constipation. The exact cause of irritable bowel syndrome is not known, but researchers are actively investigating the role of the nervous system.[3] The condition seems to worsen for some people when they eat certain foods or during stressful events. These triggers seem to aggravate symptoms but not cause them. Dietary treatment hinges on identifying and avoiding individual foods that aggravate symptoms; small meals may also be beneficial. Other treatments that may be effective include antispasmodic drugs and peppermint oil.

Colitis

People with **colitis,** an inflammation of the large intestine, may also suffer from severe diarrhea. They often benefit from complete bowel rest and medication. If treatment fails, surgery to remove the colon and rectum may be necessary.

Celiac Disease

Celiac disease is an autoimmune disease characterized by inflammation of the small intestine that occurs in response to foods that contain gluten, a protein commonly found in wheat, barley, rye, and possibly oats. The prevalence of celiac disease in the United States is estimated at 1 in 141.[4] In people with celiac disease, gluten triggers an immune system reaction in the small intestine that causes inflammation, which damages the villi and decreases nutrient absorption. Common symptoms include abdominal pains, bloating and gas, and diarrhea—making it commonly misdiagnosed as irritable bowel syndrome. Treatment focuses on a gluten-free diet.[5] Despite the growing popularity of gluten-free products, there is no evidence to suggest that a gluten-free diet is beneficial for the general population.[6]

Personal hygiene (such as regular hand washing with soap and water) and safe food preparation (as described in Chapter 19) are easy and effective steps to take in preventing diarrheal diseases.

Treatment

Treatment for diarrhea depends on cause and severity, but it always begins with rehydration. Mild diarrhea may subside with simple rest and extra liquids (such as clear juices and soups) to replace fluid losses. If diarrhea is bloody or if it worsens or persists—especially in an infant, young child, elderly person, or person with a compromised immune system—call a physician. Severe diarrhea can be life threatening.

Constipation

Like diarrhea, **constipation** describes a symptom, not a disease. Each person's GI tract has its own cycle of waste elimination, which depends on its owner's health, the type of food eaten, when it was eaten, and when the person takes time to **defecate.** What's normal for some people may not be normal for others. Some people have bowel movements three times a day; others may have them three times a week. The symptoms of constipation include straining during bowel movements, hard stools, and infrequent bowel movements (fewer than three per week). Abdominal discomfort, headaches, backaches, and the passing of gas sometimes accompany constipation.

Often a person's lifestyle may cause constipation. Being too busy to respond to the defecation signal is a common complaint. If a person receives the signal to defecate and ignores it, the signal may not return for several hours. In the meantime, fluids continue to be withdrawn from the fecal matter, so when the person does defecate, the stools are dry and hard. In such a case, a person's daily regimen may need to be revised to allow time to have a bowel movement when the body sends its signal.

Although constipation usually reflects lifestyle habits, in some cases it may be a side effect of medication or a medical problem such as bowel obstruction. If discomfort is associated with passing fecal matter, seek medical advice to rule out disease. Once this has been done, simple treatments, such as increased fiber, fluids, and exercise, are recommended before the use of medications.

One dietary measure that may be appropriate is to increase dietary fiber to 25 to 28 grams per day gradually over the course of a week or two. Fibers found in fruits, vegetables, and whole grains help to prevent constipation by increasing fecal mass. In the GI tract, fiber attracts water, creating soft, bulky stools that stimulate bowel contractions to push the contents along. These contractions strengthen the intestinal muscles. The improved muscle tone, together with the water content of the stools, eases elimination, reducing the pressure in the rectal veins and helping to prevent **hemorrhoids.** Chapter 4 provides more information on fiber's role in maintaining a healthy colon and reducing the risks of colon cancer and diverticulosis. **Diverticulosis** is a condition in which the intestinal walls develop bulges in weakened areas, most commonly in the colon (see Figure H3-3). These bulging pockets, known as **diverticula,** can worsen constipation, entrap feces, and become painfully infected and inflamed **(diverticulitis).** Treatment may require hospitalization, antibiotics, or surgery.

Drinking plenty of water in conjunction with eating high-fiber foods also helps to prevent constipation. The increased bulk physically

> FIGURE H3-3   **Diverticula in the Colon**

Diverticula may develop anywhere along the GI tract, but they are most common in the colon.

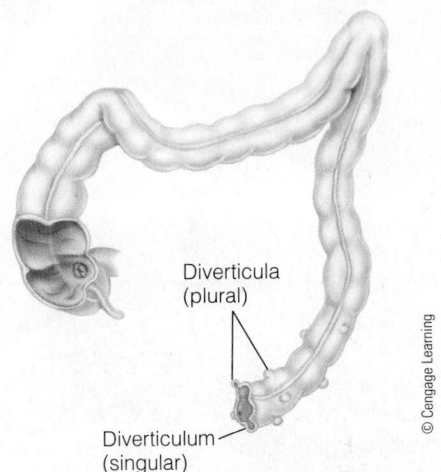

Diverticula (plural)

Diverticulum (singular)

© Cengage Learning

stimulates the upper GI tract, promoting peristalsis throughout. Similarly, physical activity improves the muscle tone and motility of the digestive tract. As little as 30 minutes of physical activity a day can help prevent or alleviate constipation.

Eating prunes—or "dried plums" as some have renamed them—can also be helpful. Prunes are high in fiber and also contain a laxative substance.* If a morning defecation is desired, a person can drink prune juice at bedtime; if the evening is preferred, the person can drink prune juice with breakfast.

If these suggested changes in lifestyle or diet do not correct constipation, then a physician might recommend the use of stool softeners, **laxatives,** or **mineral oil.** These products are best used for brief periods. If needed for extended times, they should be used under physician supervision. Frequent use of laxatives can lead to dependency and upset the body's fluid, salt, and mineral balances. Mineral oil interferes with the absorption of fat-soluble vitamins.

One potentially harmful but currently popular practice is **colonic irrigation**—the internal washing of the large intestine with a powerful **enema** machine. Such an extreme cleansing is not only unnecessary, but it can be hazardous, especially for those with a history of digestive diseases. Side effects may be relatively minor (cramping, abdominal pain, bloating, nausea, and vomiting) or quite severe (infections, kidney failure, pancreatitis, and heart failure), sometimes leading to death.[7] Common problems include equipment contamination, electrolyte abnormalities, and intestinal perforation. Less extreme practices can cause problems, too.

Belching and Gas

Many people complain of problems that they attribute to excessive gas. For some, belching is the complaint. Others blame intestinal gas for abdominal discomforts and embarrassment.

*This laxative substance is *dihydroxyphenyl isatin.*

People troubled by intestinal gas need to determine which foods bother them and then eat those foods in moderation.

Belching

Belching results from swallowing air. Everyone swallows a little bit of air with each mouthful of food, but people who eat too fast may swallow too much air. Ill-fitting dentures, carbonated beverages, and chewing gum can also contribute to the swallowing of air with resultant belching. The best advice for belching seems to be to eat slowly, chew thoroughly, and relax while eating.

Intestinal Gas

Although **flatulence** can be an embarrassing experience, it is quite normal. (People who experience painful bloating from malabsorption diseases, however, require medical treatment.) Healthy people expel several hundred milliliters of intestinal gas several times a day. Almost all (99 percent) of the gases expelled—nitrogen, oxygen, hydrogen, methane, and carbon dioxide—are odorless. The remaining "volatile" gases are the infamous ones.

Foods that produce gas usually must be determined individually. The most common offenders are foods rich in the carbohydrates—sugars, starches, and fibers. When partially digested carbohydrates reach the large intestine, bacteria digest them, giving off gas as a by-product. People can test foods suspected of forming gas by omitting them individually for a trial period to see if there is any improvement.

Gastroesophageal Reflux

Almost everyone has experienced heartburn at one time or another, usually soon after eating a meal. Medically known as **gastroesophageal reflux,** heartburn is the painful sensation a person feels behind the breastbone when the lower esophageal sphincter allows the stomach contents to reflux into the esophagus (see Figure H3-4). This may happen if a person eats or drinks too much (or both). Tight clothing and even changes of position (lying down, bending over) can cause it, too, as can some medications and smoking. Weight gain and overweight increase the frequency, severity, and duration of heartburn symptoms. A defect of the sphincter muscle itself is a possible, but less common, cause.

If heartburn is not caused by an anatomical defect, treatment is fairly simple. To avoid such misery in the future, the person needs to learn to eat less at a sitting, chew food more thoroughly, and eat more slowly. Additional strategies are presented in Table H3-1.

People who overeat or eat too quickly are likely to suffer from **indigestion.** The muscular reaction of the stomach to unchewed lumps or to being overfilled may be so intense that it upsets normal peristalsis. When this happens, overeaters may taste the stomach acid and feel pain. Over-the-counter **antacids** and **acid controllers**

> FIGURE H3-4   **Gastroesophageal Reflux**

- Esophagus
- Reflux
- Diaphragm
- Weakened lower esophageal sphincter
- Acidic stomach contents

Stomach

TABLE H3-1 Strategies to Prevent or Alleviate Common GI Problems

GI Problem	Strategies	GI Problem	Strategies
Choking	• Take small bites of food. • Chew thoroughly before swallowing. • Don't talk or laugh with food in your mouth. • Don't eat when breathing hard.	Heartburn	• Eat small meals. • Drink liquids between meals. • Sit up while eating; elevate your head when lying down. • Wait 3 hours after eating before lying down. • Wait 2 hours after eating before exercising. • Refrain from wearing tight-fitting clothing. • Avoid foods, beverages, and medications that aggravate your heartburn. Common irritants include foods that are fried or high in fat; chocolate and peppermint; coffee, alcoholic beverages, and carbonated beverages; mustard, ketchup, and tomato sauces; acidic substances such as vinegar, citrus juices, and citrus fruits. • Refrain from smoking cigarettes or using tobacco products. • Lose weight if overweight. • Take medicine as prescribed by your physician.
Diarrhea	• Avoid strenuous activity. • Rest. • Drink fluids to replace losses. • Call for medical help if diarrhea persists.		
Constipation	• Eat a high-fiber diet. • Drink plenty of fluids. • Exercise regularly. • Respond promptly to the urge to defecate.		
Belching	• Eat slowly. • Chew thoroughly. • Relax while eating.	Ulcer	• Avoid coffee and caffeine- and alcohol-containing beverages. • Avoid foods that aggravate your ulcer. • Minimize aspirin, ibuprofen, and naproxen use. • Refrain from smoking cigarettes.
Intestinal gas	• Eat bothersome foods in moderation.		

© Cengage Learning

may provide relief but should be used only infrequently for occasional heartburn; they may mask or cause problems if used regularly. If problems continue, people who suffer from frequent and regular bouts of heartburn and indigestion may need to see a physician, who can prescribe specific medication to control gastroesophageal reflux. Without treatment, the repeated splashes of acid can severely damage the cells of the esophagus, creating a condition known as Barrett's esophagus. At that stage, the risk of cancer in the esophagus increases dramatically.[8] To repeat, if symptoms persist, see a doctor—don't self-medicate.

Ulcers

Ulcers are another common digestive problem, affecting an estimated 1 out of every 12 adults in the United States. An **ulcer** is a lesion (a sore), and a **peptic ulcer** is a lesion in the lining of the stomach (gastric ulcers) or the duodenum of the small intestine (duodenal ulcers). The compromised lining is left unprotected and exposed to gastric juices, which can be painful. In some cases, ulcers can cause internal bleeding. If GI bleeding is excessive, iron deficiency may develop. Ulcers that perforate the GI lining can pose life-threatening complications.

Many people naively believe that an ulcer is caused by stress or spicy foods, but this is not the case. The stomach lining in a healthy person is well protected by its mucous coat. What, then, causes ulcers to form?

Three major causes of ulcers have been identified: bacterial infection with *Helicobacter pylori* (commonly abbreviated *H. pylori*); the use of certain anti-inflammatory drugs such as aspirin, ibuprofen, and naproxen; and disorders that cause excessive gastric acid secretion.

Most commonly, ulcers develop in response to *H. pylori* infection. The cause of the ulcer dictates the type of medication used in treatment. For example, people with ulcers caused by infection receive antibiotics, whereas those with ulcers caused by medicines discontinue their use. In addition, all treatment plans aim to relieve pain, heal the ulcer, and prevent recurrence.

The regimen for ulcer treatment is to treat for infection, eliminate any food that routinely causes indigestion or pain, and avoid coffee and caffeine- and alcohol-containing beverages. Both regular and decaffeinated coffee stimulate acid secretion and so aggravate *existing* ulcers.

Ulcers and their treatments highlight the importance of not self-medicating when symptoms persist. People with *H. pylori* infection often take over-the-counter acid controllers to relieve the pain of their ulcers when, instead, they need physician-prescribed antibiotics. Suppressing gastric acidity not only fails to heal the ulcer, but it also actually worsens inflammation during an *H. pylori* infection. Furthermore, *H. pylori* infection has been linked with stomach cancer, making prompt diagnosis and appropriate treatment essential.

Table H3-1 summarizes strategies to prevent or alleviate common GI problems. Many of these problems reflect hurried lifestyles. For this reason, many of their remedies require that people slow down and take the time to eat at a leisurely pace; chew food thoroughly to prevent choking, heartburn, and acid indigestion; rest until vomiting and diarrhea subside; and heed the urge to defecate. In addition, people must learn how to handle life's day-to-day problems and challenges without overreacting and becoming upset; learn how to relax, get enough sleep, and enjoy life. Remember, "what's eating you" may cause more GI distress than what you eat.

CRITICAL THINKING QUESTIONS

A. What strategies would be most helpful in preventing common digestive problems?

B. You've noticed the abundance of gluten-free foods on the grocery store shelves. The demand for gluten-free products has increased dramatically over the past decade as gluten-free diets have gained in popularity. Although a gluten-free diet is the best treatment for people with celiac disease, it has been adopted by millions of other people for a variety of other reasons. Compare the kcalories, fiber, added sugars, and saturated fat on the labels of two similar products—one whole grain and the other gluten free—and determine what benefits and risks might accompany a gluten-free diet for those with celiac disease and for others. Which product would you now be more likely to buy? Why?

REFERENCES

1. American Academy of Pediatrics, Policy statement: Prevention of choking among children, *Pediatrics* 125 (2010): 601–607.
2. D. Keszthelyi, F. J. Troost, and A. A. Masclee, Irritable bowel syndrome: Methods, mechanisms, and pathophysiology. Methods to assess visceral hypersensitivity in irritable bowel syndrome, *American Journal of Physiology: Gastrointestinal and Liver Physiology* 303 (2012): G141-G154; M. Camilleri, Peripheral mechanisms in irritable bowel syndrome, *New England Journal of Medicine* 367 (2012): 1626–1635.
3. C. M. Surawicz, Mechanisms of diarrhea, *Current Gastroenterology Reports* 12 (2010): 236–241.
4. A. Rubio-Tapia and coauthors, The prevalence of celiac disease in the United States, *American Journal of Gastroenterology* 107 (2012): 1538–1544.
5. P. Fric, D. Gabrovska, and J. Nevoral, Celiac disease, gluten-free diet, and oats, *Nutrition Reviews* 69 (2011): 107–115.
6. G. A. Gaesser and S. S. Angadi, Gluten-free diet: Imprudent dietary advice for the general population? *Journal of the Academy of Nutrition and Dietetics* 112 (2012): 1330–1333.
7. R. Mishori, A. Otubu, A. A. Jones, The dangers of colon cleansing, *Journal of Family Practice* 60 (2011): 454–457.
8. S. J. Spechler, Barrett esophagus and risk of esophageal cancer: A clinical review, *Journal of the American Medical Association* 310 (2013): 627–636; F. Hvid-Jensen and coauthors, Incidence of adenocarcinoma among patients with Barrett's esophagus, *New England Journal of Medicine* 365 (2011): 1375–1383.

The Carbohydrates: Sugars, Starches, and Fibers

Nutrition in Your Life

Whether you are studying for an exam or daydreaming about your next vacation, your brain needs carbohydrate to power its activities. Your muscles need carbohydrate to fuel their work, too, whether you are racing up the stairs to class or moving on the dance floor to your favorite music. Where can you get carbohydrate? Are some foods healthier choices than others? As you will learn from this chapter, whole grains, vegetables, legumes, and fruits naturally deliver ample carbohydrate and fiber with valuable vitamins and minerals and little or no fat. Milk products typically lack fiber, but they also provide carbohydrate along with an assortment of vitamins and minerals. In the Nutrition Portfolio at the end of this chapter, you can examine whether your current carbohydrate choices are meeting dietary goals.

A student, quietly studying, is seldom aware of the billions of glucose molecules in his brain cells that provide the energy to learn. Yet glucose fuels nearly all of the brain's activities. Similarly, a marathon runner, triumphantly crossing the finish line, seldom gives credit to the glycogen her muscles have used to fuel the race. Yet, together, these two **carbohydrates**—glucose and its storage form glycogen—provide about half of all the energy muscles and other body tissues use. The other half comes mostly from fat.

People don't eat glucose and glycogen. When they eat foods rich in carbohydrates, their bodies receive glucose for immediate energy and convert some glucose into glycogen for reserve energy. All plant foods—whole grains, vegetables, legumes, and fruits—provide carbohydrate. Milk also contains carbohydrate.

Some people mistakenly think of carbohydrates as "fattening" and avoid them when trying to lose weight. This strategy may help if the carbohydrates are the added sugars of soft drinks, candies, and cookies, but it is counterproductive if the carbohydrates are from whole grains, vegetables, and legumes. As the next section explains, not all carbohydrates are created equal.

carbohydrates: compounds composed of carbon, oxygen, and hydrogen arranged as monosaccharides or multiples of monosaccharides. Most, but not all, carbohydrates have a ratio of one carbon molecule to one water molecule: $(CH_2O)_n$.
- **carbo** = carbon (C)
- **hydrate** = with water (H_2O)

4.1 The Chemist's View of Carbohydrates

> **LEARN IT** Identify the monosaccharides, disaccharides, and polysaccharides common in nutrition by their chemical structures and major food sources.

The dietary carbohydrate family includes:

- Monosaccharides: single sugars
- Disaccharides: sugars composed of pairs of monosaccharides
- Polysaccharides: large molecules composed of chains of monosaccharides

Monosaccharides and disaccharides (the sugars) are sometimes called *simple carbohydrates*, and polysaccharides (starches and fibers) are sometimes called *complex carbohydrates*.

To understand the structure of carbohydrates, look at the atoms within them. Each atom can form a certain number of chemical bonds with other atoms:

- Hydrogen atoms, one
- Oxygen atoms, two
- Nitrogen atoms, three
- Carbon atoms, four

Chemists represent the bonds as lines between the chemical symbols (such as H, O, N, and C) that stand for the atoms (see Figure 4-1).

Atoms form molecules in ways that satisfy the bonding requirements of each atom. Figure 4-1 includes the structure of ethyl alcohol, the active ingredient of alcoholic beverages, as an example. The two carbons each have four bonds represented by lines; the oxygen has two; and each hydrogen has one bond connecting it to other atoms. Chemical structures always bond according to these rules.

The following list of the most important **sugars** in nutrition symbolizes them as hexagons and pentagons of different colors.* Three are monosaccharides:

- Glucose
- Fructose
- Galactose

Three are disaccharides:

- Maltose (glucose + glucose)
- Sucrose (glucose + fructose)
- Lactose (glucose + galactose)

Monosaccharides The three **monosaccharides** most important in nutrition all have the same numbers and kinds of atoms—each contains 6 carbon atoms, 12 hydrogens, and 6 oxygens (written in shorthand as $C_6H_{12}O_6$). The monosaccharides differ in their arrangements of the atoms. These chemical differences account for the differing sweetness of the monosaccharides. A pinch of purified glucose on the tongue gives only a mild sweet flavor, and galactose hardly tastes sweet at all. Fructose, however, is as intensely sweet as honey and, in fact, is the sugar primarily responsible for honey's sweetness.

Glucose Chemically, **glucose** is a larger and more complicated molecule than the ethyl alcohol shown in Figure 4-1, but it obeys the same rules of chemistry: each carbon atom has four bonds; each oxygen, two bonds; and each hydrogen, one bond. Figure 4-2 illustrates the chemical structure of a glucose molecule.

Commonly known as blood sugar, glucose serves as an essential energy source for all the body's activities. Its significance to nutrition is tremendous.

> **FIGURE 4-1**   **Atoms and Their Bonds**

The four main types of atoms found in nutrients are hydrogen (H), oxygen (O), nitrogen (N), and carbon (C).

$$H- \qquad -O- \qquad -N- \qquad -C-$$

1	2	3	4

Each atom has a characteristic number of bonds it can form with other atoms.

Notice that in this simple molecule of ethyl alcohol, each H has one bond, O has two, and each C has four.

© Cengage Learning

sugars: simple carbohydrates composed of monosaccharides, disaccharides, or both.

monosaccharides (mon-oh-SACK-uh-rides): carbohydrates of the general formula $C_nH_{2n}O_n$ that typically form a single ring. The monosaccharides important in nutrition are *hexoses*, sugars with six atoms of carbon and the formula $C_6H_{12}O_6$. See Appendix C for the chemical structures of the monosaccharides.

- **mono** = one
- **saccharide** = sugar
- **hex** = six

glucose (GLOO-kose): a monosaccharide; sometimes known as *blood sugar* in the body or *dextrose* in foods.

- **ose** = carbohydrate
- = glucose

*Fructose is shown as a pentagon, but like the other monosaccharides, it has six carbons (as you will see in Figure 4-3). The disaccharides are illustrated with a simple bond, but actual linkages differ (as shown in Appendix C).

> FIGURE 4-2 **Chemical Structure of Glucose**

The diagram of a glucose molecule on the left shows all the bonds between the 6 carbon (C), 12 hydrogen (H), and 6 oxygen (O) atoms. It proves simple on examination, but chemists have adopted shortcuts to depict chemical structures. The middle and right diagrams also present the chemical structure of glucose, but as simplified versions with fewer symbols and bonds showing.

On paper, the structure of glucose has to be drawn flat, but in nature the five carbons and oxygen are roughly in a plane. The atoms attached to the ring carbons extend above and below the plane.

The lines representing some of the bonds and the carbons at the corners are not shown.

Now the single hydrogens are not shown, but lines still extend upward or downward from the ring to show where they belong.

Later sections explain that glucose is one of the two sugars in every disaccharide and the unit from which the polysaccharides are made almost exclusively. One of these polysaccharides, starch, is the chief food source of energy for all the world's people; another, glycogen, is an important storage form of energy in the body. Glucose reappears frequently throughout this chapter and all those that follow.

Fructose Fructose is the sweetest of the sugars. Curiously, fructose has exactly the same chemical *formula* as glucose—$C_6H_{12}O_6$—but its *structure* differs (see Figure 4-3). The arrangement of the atoms in fructose stimulates the taste buds on the tongue to produce the sweet sensation. Fructose occurs naturally in fruits and honey; other sources include products such as soft drinks, ready-to-eat cereals, and desserts that have been sweetened with high-fructose corn syrup (defined on p. 114).

Galactose The monosaccharide **galactose** occurs naturally in foods as a single sugar only in very small amounts. Galactose has the same numbers and kinds of atoms as glucose and fructose in yet another arrangement. Figure 4-3 shows galactose beside a molecule of glucose for comparison.

Fruits package their sugars with fibers, vitamins, and minerals, making them a sweet and healthy snack.

> FIGURE 4-3 **The Monosaccharides**

Notice the similarities—all three monosaccharides have 6 carbons (those shown plus one in each corner), 12 hydrogens (those shown plus one at the end of each single line), and 6 oxygens (all shown). Also notice the differences compared with glucose—in fructose, the ring is five-sided and in galactose, the position of one OH group differs slightly.

Fructose

Glucose

Galactose

fructose (FRUK-tose or FROOK-tose): a monosaccharide; sometimes known as *fruit sugar* or *levulose*. Fructose is found abundantly in fruits, honey, and saps.

- **fruct** = fruit
- ⬠ = fructose

galactose (ga-LAK-tose): a monosaccharide; part of the disaccharide lactose.

- ⬡ = galactose

> FIGURE 4-4   Condensation of Two Monosaccharides to Form a Disaccharide

Glucose + glucose → Maltose

An OH group from one glucose and an H atom from another glucose combine to create a molecule of H_2O.

The two glucose molecules bond together with a single O atom to form the disaccharide maltose.

© Cengage Learning

Disaccharides

Disaccharides   The **disaccharides** are pairs of the three monosaccharides just described. Glucose occurs in all three; the second member of the pair is fructose, galactose, or another glucose. These carbohydrates—and all the other energy nutrients—are put together and taken apart by similar chemical reactions: condensation and hydrolysis.

Condensation   To make a disaccharide, a chemical reaction known as **condensation** links two monosaccharides together (see Figure 4-4). A hydroxyl (OH) group from one monosaccharide and a hydrogen atom (H) from the other combine to create a molecule of water (H_2O). The two originally separate monosaccharides link together with a single oxygen (O).

Hydrolysis   To break a disaccharide in two, a chemical reaction known as **hydrolysis** occurs (see Figure 4-5). A molecule of water (H_2O) splits to provide the H and OH needed to complete the resulting monosaccharides. Hydrolysis reactions commonly occur during digestion.

Maltose   The disaccharide **maltose** consists of two glucose units. Maltose is produced whenever starch breaks down—as happens in human beings during carbohydrate digestion. It also occurs during the fermentation process that yields alcohol. Maltose is only a minor constituent of a few foods, most notably barley.

Sucrose   Fructose and glucose together form the disaccharide **sucrose.** Sucrose is the sweetest of the disaccharides because it contains fructose, the sweetest of the monosaccharides. These sugars account for the natural sweetness of fruits, vegetables, and grains. To make table sugar, sucrose is refined from the juices of sugarcane and sugar beets, then granulated. Depending on the extent to which it

> FIGURE 4-5   Hydrolysis of a Disaccharide

Maltose → Glucose + glucose

The disaccharide maltose splits into two glucose molecules with H added to one and OH to the other (from the water molecule).

© Cengage Learning

disaccharides (dye-SACK-uh-rides): pairs of monosaccharides linked together. See Appendix C for the chemical structures of the disaccharides.
- **di** = two

condensation: a chemical reaction in which water is released as two molecules combine to form one larger product.

hydrolysis (high-DROL-ih-sis): a chemical reaction in which one molecule is split into two molecules, with hydrogen (H) added to one and a hydroxyl group (OH) to the other (from water, H_2O). (The noun is *hydrolysis*; the verb is *hydrolyze*.)
- **hydro** = water
- **lysis** = breaking

maltose (MAWL-tose): a disaccharide composed of two glucose units; sometimes known as *malt sugar.*
- ⬡⬡ = maltose

sucrose (SUE-krose): a disaccharide composed of glucose and fructose; commonly known as *table sugar, beet sugar,* or *cane sugar.* Sucrose also occurs in many fruits and some vegetables and grains.
- **sucro** = sugar
- ⬡⬠ = sucrose

is refined, the product becomes the familiar brown, white, and powdered sugars available at grocery stores.

Lactose The combination of galactose and glucose makes the disaccharide **lactose,** the principal carbohydrate of milk. Known as milk sugar, lactose contributes half of the energy (kcalories) provided by fat-free milk.

Polysaccharides
In contrast to the simple carbohydrates just mentioned—the monosaccharides glucose, fructose, and galactose and the disaccharides maltose, sucrose, and lactose—the **polysaccharides** are slightly more complex, containing many glucose units and, in some cases, a few other monosaccharides strung together. Three types of polysaccharides are important in nutrition: glycogen, starches, and fibers.

Glycogen is a storage form of energy in the body; starch is the storage form of energy in plants; and fibers provide structure in stems, trunks, roots, leaves, and skins of plants. Both glycogen and starch are built of glucose units; fibers are composed of a variety of monosaccharides and other carbohydrate derivatives.

Glycogen **Glycogen** is found to only a limited extent in meats and not at all in plants.* For this reason, food is not a significant source of glycogen. Glycogen performs an important role in the body, however: it stores glucose for future use. Glycogen is made of many glucose molecules linked together in highly branched chains (see the left side of Figure 4-6). When the hormonal message "release energy" arrives at a liver or muscle cell, enzymes respond by attacking the many branches of glycogen simultaneously, making a surge of glucose available.**

Starches The human body stores glucose as glycogen, but plant cells store glucose as **starches**—long, branched or unbranched chains of hundreds or thousands of glucose molecules linked together (see the middle and right side of Figure 4-6). These giant starch molecules are packed side by side in grains such as wheat or rice, in root crops and tubers such as yams and potatoes, and in legumes such as peas and beans. When you eat the plant, your body hydrolyzes the starch to glucose and uses the glucose for its own energy purposes.

All starchy foods come from plants. Grains are the richest food source of starch, providing much of the food energy for people all over the world—rice in Asia; wheat in Canada, the United States, and Europe; corn in much of Central

© Polara Studios, Inc.

Major sources of starch include grains (such as rice, wheat, millet, rye, barley, and oats), legumes (such as kidney beans, black-eyed peas, pinto beans, navy beans, and garbanzo beans), tubers (such as potatoes), and root crops (such as yams and cassava).

---

lactose (LAK-tose): a disaccharide composed of glucose and galactose; commonly known as *milk sugar*.

- **lac** = milk

 = lactose

polysaccharides: compounds composed of many monosaccharides linked together. An intermediate string of 3 to 10 monosaccharides is an *oligosaccharide*.

- **poly** = many
- **oligo** = few

glycogen (GLY-ko-jen): an animal polysaccharide composed of glucose; a storage form of glucose manufactured and stored in the liver and muscles. Glycogen is not a significant food source of carbohydrate and is not counted as a dietary carbohydrate in foods.

- **glyco** = glucose
- **gen** = gives rise to

starches: plant polysaccharides composed of many glucose molecules.

---

*Glycogen in animal muscles rapidly breaks down after slaughter.
**Normally, liver cells produce glucose from glycogen to be sent directly to the blood; muscle cells can also produce glucose from glycogen, but must use it themselves. Muscle cells can restore the blood glucose level indirectly, however, as Chapter 7 explains.

> FIGURE 4-6   **Glycogen and Starch Compared**

For details of the chemical structures, see Appendix C.

Glycogen

A glycogen molecule contains hundreds of glucose units in highly branched chains. Each new glycogen molecule needs a special protein (shown here in red) for the attachment of the first glucose.

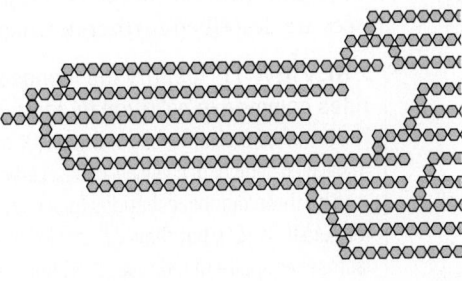

Starch (amylopectin)   Starch (amylose)

A starch molecule contains hundreds of glucose molecules in either occasionally branched chains (amylopectin) or unbranched chains (amylose).

© Cengage Learning

> FIGURE 4-7 **The Bonds of Starch and Cellulose Compared**

Human enzymes can digest starch but they cannot digest cellulose because the bonds that link the glucose molecules together are different. See Appendix C for chemical structures and descriptions of linkages.

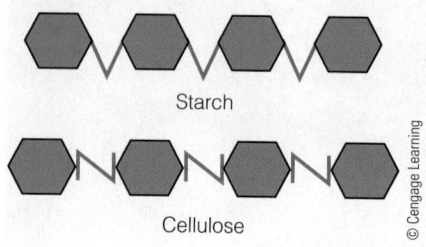

Starch

Cellulose

© Cengage Learning

and South America; and millet, rye, barley, and oats elsewhere. Legumes and tubers are also important sources of starch.

Fibers **Dietary fibers** are the structural parts of plants and thus are found in all plant-derived foods—vegetables, fruits, whole grains, and legumes. Most dietary fibers are polysaccharides. As mentioned earlier, starches are also polysaccharides, but dietary fibers differ from starches in that the bonds between their monosaccharides cannot be broken down by digestive enzymes in the body. For this reason, dietary fibers are often described as *nonstarch polysaccharides*.* Figure 4-7 illustrates the difference in the bonds that link glucose molecules together in starch with those found in the fiber cellulose. Because dietary fibers pass through the body undigested, they contribute no monosaccharides, and therefore little or no energy.

Even though most foods contain a variety of fibers, researchers often sort dietary fibers into two groups according to their solubility. Such distinctions help to explain their actions in the body.

Some dietary fibers dissolve in water (**soluble fibers**), form gels (**viscous**), and are easily digested by bacteria in the colon (**fermentable**).** Commonly found in oats, barley, legumes, and citrus fruits, soluble fibers are most often associated with protecting against heart disease and diabetes by lowering blood cholesterol and glucose levels, respectively.

Other fibers do not dissolve in water (**insoluble fibers**), do not form gels (nonviscous), and are less readily fermented. Found mostly in whole grains (bran) and vegetables, insoluble fibers promote bowel movements, alleviate constipation, and prevent diverticular disease.

As mentioned, *dietary fibers* occur naturally in plants. When these fibers have been extracted from plants or are manufactured and then added to foods or used in supplements, they are called *functional fibers*—if they have beneficial health effects. Cellulose in cereals, for example, is a dietary fiber, but when consumed as a supplement to alleviate constipation, cellulose is considered a functional fiber. *Total fiber* refers to the sum of dietary fibers and functional fibers.

A few starches are classified as dietary fibers. Known as **resistant starches**, these starches escape digestion and absorption in the small intestine. Starch may resist digestion for several reasons, including the body's digestive activities and the food's physical properties. Resistant starch is common in whole or partially milled grains, legumes, and just-ripened bananas. Cooked potatoes, pasta, and rice that have been chilled also contain resistant starch. Similar to insoluble fibers, resistant starch may support a healthy colon.

Phytic acid is not a dietary fiber, but it is often found in fiber-rich foods. Because of this close association, researchers have been unable to determine whether it is the dietary fiber, the phytic acid, or both, that binds with minerals, preventing their absorption. This binding presents a risk of mineral deficiencies, but the risk is minimal when total fiber intake is reasonable (less than 40 grams a day) and mineral intake is adequate. The nutrition consequences of mineral losses are described further in Chapters 12 and 13.

dietary fibers: in plant foods, the *nonstarch polysaccharides* that are not digested by human digestive enzymes, although some are digested by GI tract bacteria.

soluble fibers: nonstarch polysaccharides that dissolve in water to form a gel. An example is pectin from fruit, which is used to thicken jellies.

viscous: a gel-like consistency.

fermentable: the extent to which bacteria in the GI tract can break down fibers to fragments that the body can use.

insoluble fibers: nonstarch polysaccharides that do not dissolve in water. Examples include the tough, fibrous structures found in the strings of celery and the skins of corn kernels.

resistant starches: starches that escape digestion and absorption in the small intestine of healthy people.

phytic (FYE-tick) **acid:** a nonnutrient component of plant seeds; also called *phytate* (FYE-tate). Phytic acid occurs in the husks of grains, legumes, and seeds and is capable of binding minerals such as zinc, iron, calcium, magnesium, and copper in insoluble complexes in the intestine, which the body excretes unused.

> REVIEW IT Identify the monosaccharides, disaccharides, and polysaccharides common in nutrition by their chemical structures and major food sources.
The carbohydrates are made of carbon (C), oxygen (O), and hydrogen (H). Each of these atoms can form a specified number of chemical bonds: carbon forms four, oxygen forms two, and hydrogen forms one.

The three monosaccharides (glucose, fructose, and galactose) all have the same chemical formula ($C_6H_{12}O_6$), but their structures differ. The three disaccharides (maltose, sucrose, and lactose) are pairs of monosaccharides, each containing a glucose paired with one of the three monosaccharides. The sugars derive primarily from plants, except for lactose and its component galactose, which come from milk and milk products. Two monosaccharides can be linked together by a condensation reaction to form a disaccharide and water. A disaccharide, in turn, can be broken into its two monosaccharides by a hydrolysis reaction using water.

---

*The nonstarch polysaccharide fibers include cellulose, hemicelluloses, pectins, gums, and mucilages. Fibers also include some *nonpolysaccharides* such as lignins, cutins, and tannins.
**Dietary fibers are fermented by bacteria in the colon to short-chain fatty acids, which are absorbed and metabolized by cells in the GI tract and liver (Chapter 5 describes fatty acids).

The polysaccharides are chains of monosaccharides and include glycogen, starches, and dietary fibers. Both glycogen and starch are storage forms of glucose—glycogen in the body, and starch in plants—and both yield energy for human use. The dietary fibers also contain glucose (and other monosaccharides), but their bonds cannot be broken by human digestive enzymes, so they yield little, if any, energy. Table 4-1 summarizes the carbohydrate family of compounds.

4.2 Digestion and Absorption of Carbohydrates

> **LEARN IT** Summarize carbohydrate digestion and absorption.

The ultimate goal of digestion and absorption of sugars and starches is to break them into small molecules—chiefly glucose—that the body can absorb and use. The large starch molecules require extensive breakdown; the disaccharides need be broken only once and the monosaccharides not at all. The details follow.

Carbohydrate Digestion Figure 4-8 (p. 106) traces the digestion of carbohydrates through the GI tract. When a person eats foods containing starch, enzymes hydrolyze the long chains to shorter chains, the short chains to disaccharides, and, finally, the disaccharides to monosaccharides.* This process begins in the mouth.

In the Mouth In the mouth, thoroughly chewing high-fiber foods slows eating and stimulates the flow of saliva. The salivary enzyme **amylase** starts to work, hydrolyzing starch to shorter polysaccharides and to the disaccharide maltose. In fact, you can taste the change if you chew a piece of starchy food like a cracker and hold it in your mouth for a few minutes without swallowing it—the cracker begins tasting sweeter as the enzyme acts on it. Because food is in the mouth for a relatively short time, very little carbohydrate digestion takes place there; it begins again in the small intestine.

In the Stomach Carbohydrate digestion ceases in the stomach. The activity of salivary amylase diminishes as the stomach's acid and protein-digesting enzymes inactivate the enzyme. The stomach's digestive juices contain no enzymes to break down carbohydrates. Fibers are not digested, but because they linger in the stomach, they delay gastric emptying, thereby providing a feeling of fullness and **satiety.**

In the Small Intestine The small intestine performs most of the work of carbohydrate digestion. A major carbohydrate-digesting enzyme, pancreatic amylase, enters the intestine via the pancreatic duct and continues breaking down the polysaccharides to shorter glucose chains and maltose. The final step takes place on the outer membranes of the intestinal cells. There specific enzymes break down specific disaccharides:

- **Maltase** breaks maltose into two glucose molecules.
- **Sucrase** breaks sucrose into one glucose and one fructose molecule.
- **Lactase** breaks lactose into one glucose and one galactose molecule.

At this point, all polysaccharides and disaccharides have been broken down to monosaccharides—mostly glucose molecules, with some fructose and galactose molecules as well.

In the Large Intestine Within 1 to 4 hours after a meal, all the sugars and most of the starches have been digested. Only the fibers remain in the digestive tract. Fibers in the large intestine attract water, which softens the stools for passage without straining. Also, bacteria in the GI tract ferment some fibers. This process

---
*The short chains of glucose units that result from the breakdown of starch are known as *dextrins*. The word sometimes appears on food labels because dextrins can be used as thickening agents in processed foods.

TABLE 4-1 The Carbohydrate Family	
Monosaccharides	
Glucose	
Fructose	
Galactose	
Disaccharides	
Maltose (glucose + glucose)	
Sucrose (glucose + fructose)	
Lactose (glucose + galactose)	
Polysaccharides	
Glycogen[a]	
Starches (amylose and amylopectin)	
Fibers (soluble and insoluble)	

© Cengage Learning

[a]Glycogen is a polysaccharide, but not a common dietary source of carbohydrate.

© oliveromg/Shutterstock.com

When a person eats carbohydrate-rich foods, the body receives a valuable commodity—glucose.

amylase (AM-ih-lace): an enzyme that hydrolyzes amylose (a form of starch). Amylase is a *carbohydrase*, an enzyme that breaks down carbohydrates.

satiety (sah-TIE-eh-tee): the feeling of fullness and satisfaction that occurs after a meal and inhibits eating until the next meal. Satiety determines how much time passes between meals.

- **sate** = to fill

maltase: an enzyme that hydrolyzes maltose.

sucrase: an enzyme that hydrolyzes sucrose.

lactase: an enzyme that hydrolyzes lactose.

> FIGURE 4-8   Carbohydrate Digestion in the GI Tract

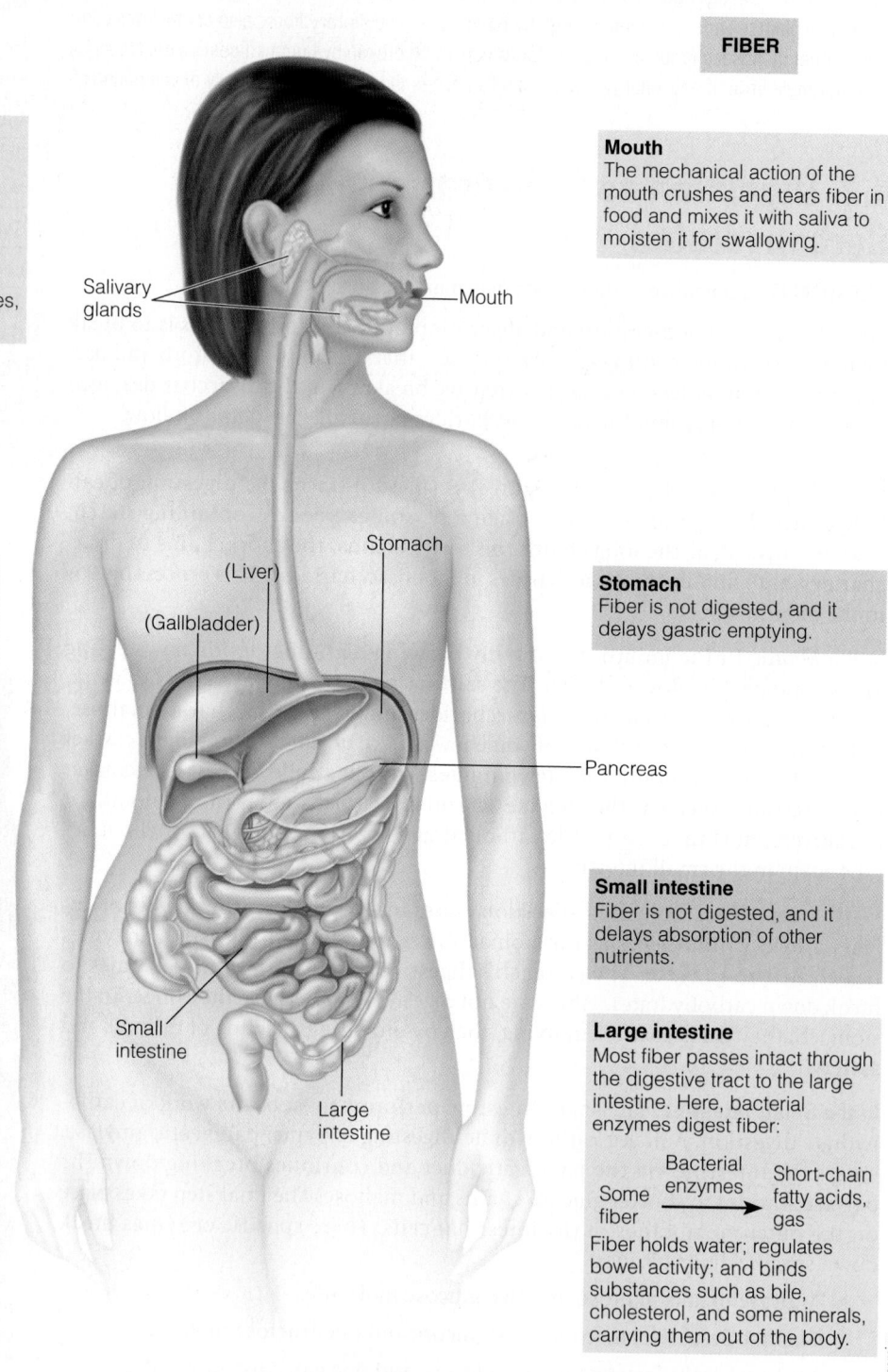

STARCH

Mouth and salivary glands
The salivary glands secrete saliva into the mouth to moisten the food. The salivary enzyme amylase begins digestion:

$$\text{Starch} \xrightarrow{\text{Amylase}} \begin{array}{l}\text{Small} \\ \text{polysaccharides,} \\ \text{maltose}\end{array}$$

Stomach
Stomach acid inactivates salivary enzymes, halting starch digestion.

Small intestine and pancreas
The pancreas produces an amylase that is released through the pancreatic duct into the small intestine:

$$\text{Starch} \xrightarrow{\begin{array}{c}\text{Pancreatic} \\ \text{amylase}\end{array}} \begin{array}{l}\text{Small} \\ \text{polysac-} \\ \text{charides,} \\ \text{maltose}\end{array}$$

Then disaccharidase enzymes on the surface of the small intestinal cells hydrolyze the disaccharides into mono-saccharides:

$$\text{Maltose} \xrightarrow{\text{Maltase}} \begin{array}{l}\text{Glucose} \\ + \\ \text{Glucose}\end{array}$$

$$\text{Sucrose} \xrightarrow{\text{Sucrase}} \begin{array}{l}\text{Fructose} \\ + \\ \text{Glucose}\end{array}$$

$$\text{Lactose} \xrightarrow{\text{Lactase}} \begin{array}{l}\text{Galactose} \\ + \\ \text{Glucose}\end{array}$$

Intestinal cells absorb these monosaccharides.

Salivary glands

Mouth

(Liver)

(Gallbladder)

Stomach

Pancreas

Small intestine

Large intestine

FIBER

Mouth
The mechanical action of the mouth crushes and tears fiber in food and mixes it with saliva to moisten it for swallowing.

Stomach
Fiber is not digested, and it delays gastric emptying.

Small intestine
Fiber is not digested, and it delays absorption of other nutrients.

Large intestine
Most fiber passes intact through the digestive tract to the large intestine. Here, bacterial enzymes digest fiber:

$$\begin{array}{l}\text{Some} \\ \text{fiber}\end{array} \xrightarrow{\begin{array}{c}\text{Bacterial} \\ \text{enzymes}\end{array}} \begin{array}{l}\text{Short-chain} \\ \text{fatty acids,} \\ \text{gas}\end{array}$$

Fiber holds water; regulates bowel activity; and binds substances such as bile, cholesterol, and some minerals, carrying them out of the body.

© Cengage Learning

generates water, gas, and short-chain fatty acids (described in Chapter 5).* The cells of the colon use these small fat molecules for energy. Metabolism of short-chain fatty acids also occurs in the cells of the liver. Fibers, therefore, can contribute some energy (1.5 to 2.5 kcalories per gram), depending on the extent to which they are broken down by bacteria and the fatty acids are absorbed. How much energy fiber contributes to a person's daily intake remains unclear.

*The short-chain fatty acids produced by GI bacteria are primarily acetic acid, propionic acid, and butyric acid.

Carbohydrate Absorption Glucose is unique in that it can be absorbed to some extent through the lining of the mouth, but for the most part, nutrient absorption takes place in the small intestine. Glucose and galactose enter the cells lining the small intestine by active transport; fructose is absorbed by facilitated diffusion.

As the blood from the small intestine circulates through the liver, cells there take up fructose and galactose and most often convert them to compounds within the same metabolic pathways as glucose. Figure 4-9 shows that fructose and galactose are mostly metabolized in the liver, whereas glucose is sent out to the body's cells for energy. In the end, all disaccharides provide at least one glucose molecule directly, and they can provide the equivalent of another one indirectly—through the metabolism of fructose and galactose in the liver.

Lactose Intolerance Normally, the intestinal cells produce enough of the enzyme lactase to ensure that the disaccharide lactose found in milk is both digested and absorbed efficiently. Lactase activity is highest immediately after birth, as befits an infant whose first and only food for a while will be breast milk or infant formula. In the great majority of the world's populations, lactase activity declines dramatically during childhood and adolescence to about 5 to 10 percent of the activity at birth. Only a relatively small percentage (about 30 percent) of the people in the world retain enough lactase to digest and absorb lactose efficiently throughout adult life.

Symptoms When more lactose is consumed than the available lactase can handle, lactose molecules remain in the intestine undigested, attracting water and causing bloating, abdominal discomfort, and diarrhea—the symptoms of **lactose intolerance.** The undigested lactose becomes food for intestinal bacteria, which multiply and produce irritating acid and gas, further contributing to the discomfort and diarrhea.

Causes As mentioned, lactase activity commonly declines with age. **Lactase deficiency** may also develop when the intestinal villi are damaged by disease, certain medicines, prolonged diarrhea, or malnutrition. Depending on the extent of the intestinal damage, lactose malabsorption may be temporary or permanent. In extremely rare cases, an infant is born with a lactase deficiency, making feeding a challenge.

Prevalence The prevalence of lactose intolerance varies widely among ethnic groups, indicating that the trait has a genetic component. The prevalence of

lactose intolerance: a condition that results from the inability to digest the milk sugar lactose; characterized by bloating, gas, abdominal discomfort, and diarrhea. Lactose intolerance differs from milk allergy, which is caused by an immune reaction to the protein in milk.

lactase deficiency: a lack of the enzyme required to digest the disaccharide lactose into its component monosaccharides (glucose and galactose).

> FIGURE 4-9　**Absorption of Monosaccharides**

① Monosaccharides, the end products of carbohydrate digestion, enter the capillaries of the intestinal villi.

④ Glucose is used by most cells in the body.

③ In the liver, galactose and fructose share metabolic pathways with glucose.

Small intestine

② Monosaccharides travel to the liver via the portal vein.

Key:
⬡ Glucose
⬠ Fructose
⬡ Galactose

© Cengage Learning

lactose intolerance is lowest among Scandinavians and other northern Europeans and highest among native North Americans and Southeast Asians. An estimated 30 million to 50 million people in the United States are lactose intolerant.

Dietary Changes Managing lactose intolerance requires some dietary changes, although total elimination of milk products usually is not necessary.[1] Excluding all milk products from the diet can lead to nutrient deficiencies because these foods are a major source of several nutrients, notably the mineral calcium, vitamin D, and the B vitamin riboflavin. Fortunately, many people with lactose intolerance can consume foods containing up to 6 grams of lactose (½ cup milk) without symptoms. The most successful strategies are to increase intake of milk products gradually, consume them with other foods in meals, and spread their intake throughout the day. In addition, yogurt containing live bacteria seems to improve lactose intolerance. A change in the type, number, and activity of GI bacteria—not the reappearance of the missing enzyme—accounts for the ability to adapt to milk products. Importantly, most lactose-intolerant individuals need to *manage* their dairy consumption rather than *restrict* it.

In many cases, lactose-intolerant people can tolerate fermented milk products such as yogurt and **kefir**. The bacteria in these products digest lactose for their own use, thus reducing the lactose content. Even when the lactose content is equivalent to milk's, yogurt produces fewer symptoms. Hard cheeses, such as cheddar, and cottage cheese are often well tolerated because most of the lactose is removed with the whey during manufacturing. Lactose continues to diminish as cheese ages.

Many lactose-intolerant people use commercially prepared milk products (such as Lactaid) that have been treated with an enzyme that breaks down the lactose. Alternatively, they take enzyme tablets with meals or add enzyme drops to their milk. The enzyme hydrolyzes much of the lactose in milk to glucose and galactose, which lactose-intolerant people can absorb without ill effects.

Because people's tolerance to lactose varies widely, lactose-restricted diets must be highly individualized. A completely lactose-free diet can be difficult because lactose appears not only in milk and milk products but also as an ingredient in many nondairy foods such as breads, cereals, breakfast drinks, salad dressings, and cake mixes (see Table 4-2). People on strict lactose-free diets need to read labels and avoid foods that include milk, milk solids, whey (milk liquid), and casein (milk protein, which may contain traces of lactose). They also need to check all medications with the pharmacist because 20 percent of prescription drugs and 5 percent of over-the-counter drugs contain lactose as a filler.

People who consume few milk products must take care to meet riboflavin, vitamin D, and calcium needs. Later chapters on the vitamins and minerals offer help with finding good nonmilk sources of these nutrients.

TABLE 4-2 Lactose in Selected Foods

Foods	Lactose (g)
Whole-wheat bread, 1 slice	0.5
Dinner roll, 1	0.5
Cheese, 1 oz	
Cheddar or American	0.5
Parmesan or cream	0.8
Doughnut (cake type), 1	1.2
Chocolate candy, 1 oz	2.3
Sherbet, 1 c	4.0
Cottage cheese (low-fat), 1 c	7.5
Ice cream, 1 c	9.0
Milk, 1 c	12.0
Yogurt (low-fat), 1 c	15.0

NOTE: Yogurt is often enriched with nonfat milk solids, which increase its lactose content to a level higher than milk's.

© Cengage Learning

> **REVIEW IT** Summarize carbohydrate digestion and absorption.

In the digestion and absorption of carbohydrates, the body breaks down starches into the disaccharide maltose. Maltose and the other disaccharides (lactose and sucrose) from foods are broken down into monosaccharides, which are absorbed. The fibers help to regulate the passage of food through the GI tract and slow the absorption of glucose, but they contribute little, if any, energy.

Lactose intolerance is a common condition that occurs when there is insufficient lactase to digest the disaccharide lactose found in milk and milk products. Symptoms are limited to GI distress. Because treatment requires limiting milk and milk products in the diet, other sources of riboflavin, vitamin D, and calcium must be included.

4.3 Glucose in the Body

> **LEARN IT** Explain how the body maintains its blood glucose concentration and what happens when blood glucose rises too high or falls too low.

The primary role of carbohydrates in the body is to supply the cells with glucose for energy. Scientists have long known that providing energy is glucose's primary role in the body, but they have recently uncovered additional roles that

kefir (keh-FUR): a fermented milk created by adding *Lactobacillus acidophilus* and other bacteria that break down lactose to glucose and galactose, producing a sweet, lactose-free product.

glucose and other sugars perform in the body.* When sugar molecules adhere to the body's protein and fat molecules, the consequences can be dramatic. Sugars attached to a protein change the protein's shape and function; when they bind to lipids in a cell's membranes, sugars alter the way cells recognize one another.**

A Preview of Carbohydrate Metabolism Glucose plays the central role in carbohydrate metabolism. This brief discussion provides just enough information about carbohydrate metabolism to illustrate that the body needs and uses glucose as a chief energy nutrient. Chapter 7 provides a full description of energy metabolism.

Storing Glucose as Glycogen After a meal, blood glucose rises, and liver cells link excess glucose molecules by condensation reactions into long, branching chains of glycogen (review Figure 4-6, p.103). When blood glucose falls, the liver cells break down glycogen by hydrolysis reactions into single molecules of glucose and release them into the bloodstream. Thus glucose becomes available to supply energy to the brain and other tissues regardless of whether the person has eaten recently.

The liver stores about one-third of the body's total glycogen and releases glucose into the bloodstream as needed. Muscle cells can also store glucose as glycogen (the other two-thirds), but muscles hoard most of their supply, using it just for themselves during exercise. The brain maintains a small amount of glycogen, which is thought to provide an emergency energy reserve during times of severe glucose deprivation.

Glycogen holds water and, therefore, is rather bulky. The body can store only enough glycogen to provide energy for relatively short periods of time—less than a day during rest and a few hours at most during exercise. For its long-term energy reserves, for use over days or weeks of food deprivation, the body uses its abundant, water-free fuel, fat, as Chapter 5 describes.

Using Glucose for Energy Glucose fuels the work of most of the body's cells and is the preferred energy source for brain cells, other nerve cells, and developing red blood cells. Inside a cell, a series of reactions can break glucose into smaller compounds that yield energy when broken down completely to carbon dioxide and water (see Chapter 7).

Making Glucose from Protein As mentioned, the liver's glycogen stores are limited, and the brain needs glucose to fuel its activities. To keep providing glucose to meet energy needs, a person has to eat carbohydrate-rich foods frequently. Yet people who do not always attend faithfully to their bodies' carbohydrate needs still survive. How do they manage without glucose from dietary carbohydrate? Do they simply draw energy from the other two energy-yielding nutrients, fat and protein? They do draw energy from them, but not simply.

Fat cannot make glucose to any significant extent. The amino acids of protein can be used to make glucose to some extent, but amino acids and proteins have jobs of their own that no other nutrient can perform. Still, when a person does not replenish glucose by eating carbohydrate, body proteins are broken down to make glucose to fuel the brain and other special cells. These body proteins derive primarily from the liver and skeletal muscles.

The conversion of protein to glucose is called **gluconeogenesis**—literally, the making of new glucose. Only adequate dietary carbohydrate can prevent this use of protein for energy, and this role of carbohydrate is known as its **protein-sparing action.**

Making Ketone Bodies from Fat Fragments An inadequate supply of carbohydrate can shift the body's energy metabolism in a precarious direction. With less carbohydrate providing glucose to meet the brain's energy needs, fat takes an alternative

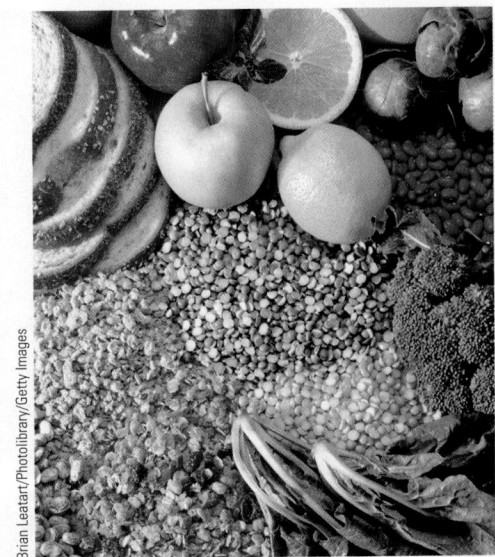

Brian Leatart/Photolibrary/Getty Images

The carbohydrates of grains, vegetables, fruits, and legumes supply most of the energy in a healthful diet.

gluconeogenesis (gloo-ko-nee-oh-JEN-ih-sis): the making of glucose from a noncarbohydrate source such as amino acids or glycerol (described in more detail in Chapter 7).

- **gluco** = glucose
- **neo** = new
- **genesis** = making

protein-sparing action: the action of carbohydrate (and fat) in providing energy that allows protein to be used for other purposes.

---

*The study of sugars and their derivatives is known as *glycobiology.*
**These combination molecules are known as *glycoproteins* and *glycolipids,* respectively.

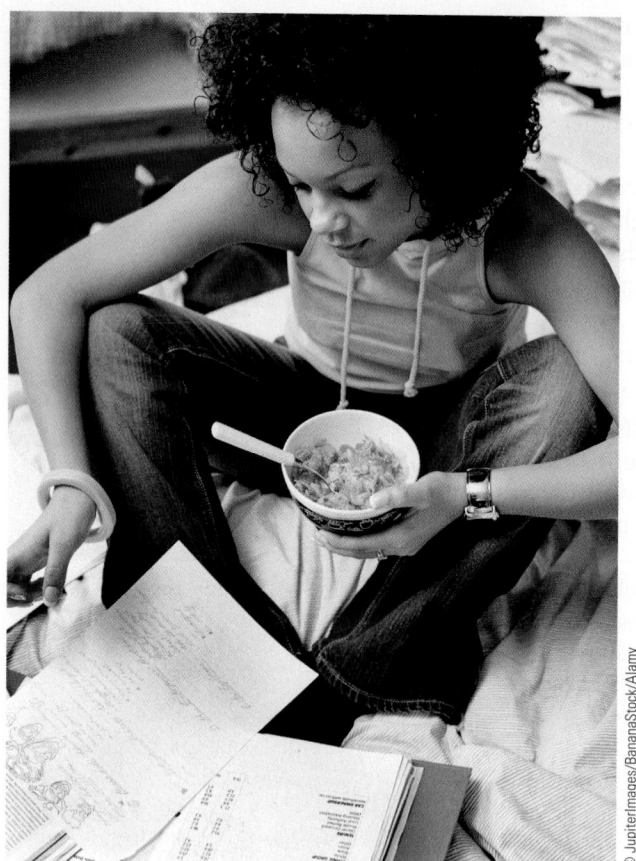
The brain uses glucose as its primary fuel for energy.

metabolic pathway; instead of entering the main energy pathway, fat fragments combine with one another, forming **ketone bodies.** Ketone bodies provide an alternative fuel source during starvation, but when their production exceeds their use, they accumulate in the blood, causing **ketosis.** Because most ketone bodies are acidic, ketosis disturbs the body's normal **acid-base balance.** (Chapter 7 explores ketosis and the metabolic consequences of low-carbohydrate diets further.)

To spare body protein and prevent ketosis, the body needs at least 50 to 100 grams of carbohydrate a day. Dietary recommendations urge people to select abundantly from carbohydrate-rich foods to provide for considerably more.

Using Glucose to Make Fat After meeting its immediate energy needs and filling its glycogen stores to capacity, the body must find a way to handle any extra glucose. When glucose is abundant, energy metabolism shifts to use more glucose instead of fat. If that isn't enough to restore glucose balance, the liver breaks glucose into smaller molecules and puts them together into the more permanent energy-storage compound—fat. Thus when carbohydrate is abundant, fat is either conserved (by using more carbohydrate in the fuel mix) or created (by using excess carbohydrate to make body fat). The fat then travels to the fatty tissues of the body for storage. Unlike the liver cells, which can store only enough glycogen to meet less than a day's energy needs, fat cells can store seemingly unlimited quantities of fat.

The Constancy of Blood Glucose
Every body cell depends on glucose for its fuel to some extent, and the cells of the brain and the rest of the nervous system depend almost exclusively on glucose for their energy. The activities of these cells never cease, and they have limited ability to store glucose. Day and night, they continually draw on the supply of glucose in the fluid surrounding them. To maintain the supply, a steady stream of blood moves past these cells bringing more glucose from either the small intestine (food) or the liver (via glycogen breakdown or gluconeogenesis).

Maintaining Glucose Homeostasis To function optimally, the body must maintain blood glucose within limits that permit the cells to nourish themselves. If blood glucose falls below normal, a person may become dizzy and weak; if it rises above normal, a person may become fatigued. Left untreated, fluctuations to the extremes—either high or low—can be fatal.

The Regulating Hormones Blood glucose homeostasis is regulated primarily by two hormones: *insulin,* which moves glucose from the blood into the cells, and *glucagon,* which brings glucose out of storage when necessary. Figure 4-10 depicts these hormonal regulators at work.

After a meal, as blood glucose rises, special cells of the pancreas respond by secreting **insulin** into the blood.* In general, the amount of insulin secreted corresponds with the rise in glucose. As the circulating insulin contacts the body's cells, receptors respond by ushering glucose from the blood into the cells. Most of the cells take only the glucose they can use for energy right away, but the liver and muscle cells can assemble the small glucose units into long, branching chains of glycogen for storage. The liver cells also convert extra glucose to fat.[2] Thus elevated blood glucose returns to normal levels as excess glucose is stored as glycogen and fat.

When blood glucose falls (as occurs between meals), other special cells of the pancreas respond by secreting **glucagon** into the blood.** Glucagon raises blood glucose by signaling the liver to break down its glycogen stores and release glucose into the blood for use by all the other body cells.

ketone (KEE-tone) **bodies:** acidic compounds produced by the liver during the breakdown of fat when carbohydrate is not available.

ketosis (kee-TOE-sis): an undesirably high concentration of ketone bodies in the blood and urine.

acid-base balance: the equilibrium in the body between acid and base concentrations (see Chapter 12).

insulin (IN-suh-lin): a hormone secreted by special cells in the pancreas in response to (among other things) elevated blood glucose concentration. Insulin controls the transport of glucose from the bloodstream into the muscle and fat cells.

glucagon (GLOO-ka-gon): a hormone secreted by special cells in the pancreas in response to low blood glucose concentration. Glucagon elicits release of glucose from liver glycogen stores.

---

*The *beta* (BAY-tuh) *cells,* one of several types of cells in the pancreas, secrete insulin in response to elevated blood glucose concentration.
**The *alpha cells* of the pancreas secrete glucagon in response to low blood glucose concentration.

> FIGURE 4-10 **Maintaining Blood Glucose Homeostasis**

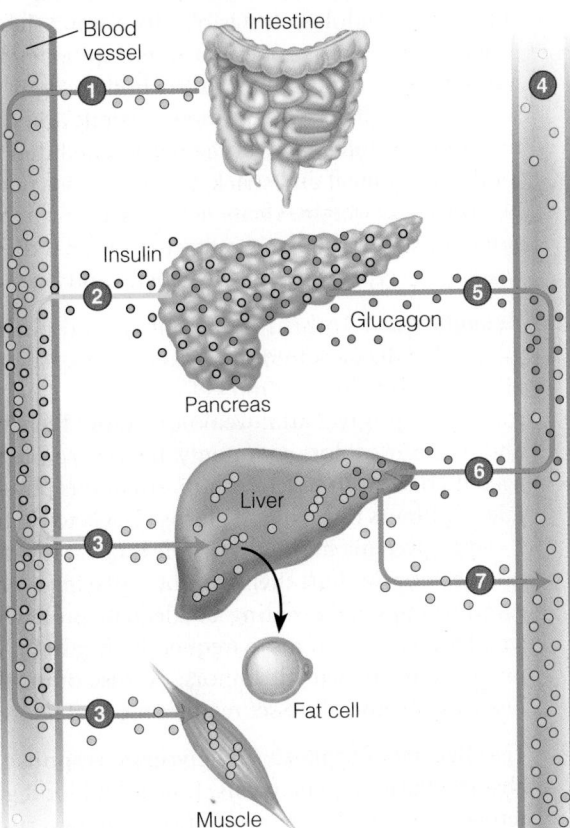

1 When a person eats, blood glucose rises.

2 High blood glucose stimulates the pancreas to release insulin into the bloodstream.

3 Insulin stimulates the uptake of glucose into cells and storage as glycogen in the liver and muscles. Insulin also stimulates the conversion of excess glucose into fat for storage.

4 As the body's cells use glucose, blood levels decline.

5 Low blood glucose stimulates the pancreas to release glucagon into the bloodstream.

6 Glucagon stimulates liver cells to break down glycogen and release glucose into the blood.[a]

7 Blood glucose begins to rise.

Labels: Intestine, Insulin, Glucagon, Pancreas, Liver, Fat cell, Muscle, Blood vessel

Key:
- Glucose
- Insulin
- Glucagon
- Glycogen

[a]The stress hormone epinephrine and other hormones also bring glucose out of storage.

Another hormone that signals the liver cells to release glucose is the "fight-or-flight" hormone, **epinephrine.** When a person experiences stress, epinephrine acts quickly to ensure that all the body cells have energy fuel in emergencies. Among its many roles in the body, epinephrine works to release glucose from liver glycogen to the blood.

Balancing within the Normal Range The maintenance of normal blood glucose depends on foods and hormones. When blood glucose falls below normal, food can replenish it, or in the absence of food, glucagon can signal the liver to break down glycogen stores. When blood glucose rises above normal, insulin can signal the cells to take in glucose for energy. Eating balanced meals that provide abundant carbohydrates, including fibers, and a little fat help to slow down the digestion and absorption of carbohydrate so that glucose enters the blood gradually. Eating at regular intervals also helps the body maintain a balance between the extremes.

Falling outside the Normal Range In some people, blood glucose regulation fails. When this happens, either of two conditions can result: diabetes or hypoglycemia. People with these conditions need to plan their diets and physical activities to help maintain their blood glucose within a normal range. Table 4-3 presents the blood glucose levels defining normal, prediabetes, and diabetes.

Diabetes In **diabetes,** blood glucose rises after a meal and remains above normal levels because insulin is either inadequate or ineffective. Elevated blood glucose is a characteristic of two main types of diabetes. In **type 1 diabetes,** the less common type, the pancreas fails to produce insulin. Although the exact cause is unclear, some research suggests that in genetically susceptible people, certain viruses activate the immune system to attack and destroy cells in the pancreas as if they were foreign cells. In **type 2 diabetes,** the more common type of diabetes, the cells fail to respond to insulin. This condition tends to occur as a consequence

TABLE 4-3 Fasting Blood Glucose

Normal	70–99 mg/dL
Prediabetes	100–125 mg/dL
Diabetes	≥126 mg/dL

epinephrine (EP-ih-NEFF-rin): a hormone of the adrenal gland that modulates the stress response; formerly called *adrenaline*. When administered by injection, epinephrine counteracts anaphylactic shock by opening the airways and maintaining heartbeat and blood pressure.

diabetes (DYE-ah-BEE-teez): metabolic disorder characterized by elevated blood glucose resulting from insufficient insulin, ineffective insulin, or both; the complete medical term is *diabetes mellitus* (meh-LIE-tus). When blood glucose levels are higher than normal, but below the diagnosis of diabetes, the condition is called *prediabetes*.

type 1 diabetes: the less common type of diabetes in which the pancreas produces little or no insulin. Type 1 diabetes usually results from autoimmune destruction of pancreatic beta cells.

type 2 diabetes: the more common type of diabetes in which the cells fail to respond to insulin. Type 2 diabetes usually accompanies obesity and results from insulin resistance coupled with insufficient insulin secretion.

of obesity. As the incidence of obesity in the United States has risen in recent decades, so too has the incidence of diabetes. This trend is most notable among children and adolescents as obesity among the nation's youth reaches epidemic proportions. Because obesity can precipitate type 2 diabetes, the best preventive measure is to maintain a healthy body weight. To manage diabetes and ensure stable blood glucose levels, food portions and choices must be balanced. It helps to eat meals and snacks at regularly scheduled times, to eat similar amounts of food at each meal and snack, and to choose nutritious foods that will support a healthy body weight. Chapter 15 describes the type of diabetes that develops in some women during pregnancy (gestational diabetes), and Chapter 18 gives full coverage to type 1 and type 2 diabetes and their associated problems.

Hypoglycemia In healthy people, blood glucose rises after eating and then gradually falls back into the normal range. The transition occurs without notice. Should blood glucose drop below normal, a person would experience the symptoms of **hypoglycemia**: weakness, rapid heartbeat, sweating, anxiety, hunger, and trembling. Most commonly, hypoglycemia is a consequence of poorly managed diabetes: too much insulin, strenuous physical activity, inadequate food intake, or illness cause blood glucose levels to plummet.

Hypoglycemia in healthy people is rare. Most people who experience hypoglycemia need only adjust their diets by replacing refined carbohydrates with fiber-rich carbohydrates and ensuring an adequate protein intake at each meal. In addition, smaller meals eaten more frequently may help. Hypoglycemia caused by certain medications, pancreatic tumors, overuse of insulin, alcohol abuse, uncontrolled diabetes, or other illnesses requires medical intervention.

The Glycemic Response The **glycemic response** refers to how quickly glucose is absorbed after a person eats, how high blood glucose rises, and how quickly it returns to normal. Slow absorption, a modest rise in blood glucose, and a smooth return to normal are desirable (a low glycemic response). Fast absorption, a surge in blood glucose, and an overreaction that plunges glucose below normal are less desirable (a high glycemic response). The glycemic response may be particularly important to people with diabetes, who may benefit from limiting foods that produce too great a rise, or too sudden a fall, in blood glucose.

Different foods elicit different glycemic responses; the **glycemic index** classifies foods accordingly (see Table 4-4). Some studies have shown that selecting

hypoglycemia (HIGH-po-gly-SEE-me-ah): an abnormally low blood glucose concentration.

glycemic (gly-SEEM-ic) **response**: the extent to which a food raises the blood glucose concentration and elicits an insulin response.

glycemic index: a method of classifying foods according to their potential for raising blood glucose.

TABLE 4-4 Glycemic Index of Selected Common Foods

Glycemic Index	Grains	Fruits	Vegetables	Milk Products	Protein Foods[a]	Other
Low	Barley, chapati, corn tortilla, rice noodles, rolled oats, udon noodles, spaghetti	Apple, apple juice, banana, dates, mango, orange, orange juice, peaches (canned), strawberry jam	Carrots, corn	Ice cream, milk, soy milk, yogurt	Legumes	Chocolate
Medium	Brown rice, couscous	Pineapple	Potatoes (french fries), sweet potatoes			Popcorn, potato chips, soft drinks
High	Breads, breakfast cereals, white rice	Watermelon	Potatoes (boiled)			Rice crackers

NOTE: Using the glucose reference scale, foods are classified as low (55 or less), medium (56 to 69), or high (70 or greater).

[a]Protein foods that contain little or no carbohydrate (such as meats, poultry, fish, and eggs) do not raise blood glucose, and therefore do not have a glycemic index.

SOURCE: Adapted from F. S. Atkinson, K. Foster-Powell, and J. C. Brand-Miller, International tables of glycemic index and glycemic load values: 2008, *Diabetes Care* 31 (2008): 2281–2283.

foods with a low glycemic index is a practical way to improve dietary adequacy and glucose control.[3] Lowering the glycemic index of the diet may improve blood lipids, reduce inflammation, and lower the risk of heart disease as well.[4] A low glycemic diet may also help with appetite regulation and weight management, although research findings are mixed.[5]

Researchers debate whether selecting foods based on the glycemic index is practical or offers any real health benefits. Those opposing the use of the glycemic index argue that it is not sufficiently supported by scientific research. The glycemic index has been determined for relatively few foods, and when the glycemic index has been established, it is based on an average of multiple tests with wide variations in their results. Values vary because of differences in the physical and chemical characteristics of foods, testing methods of laboratories, and digestive processes of individuals. Calculating the glycemic index for meals or diets based on individual foods greatly overestimates the values.[6]

Furthermore, the practical utility of the glycemic index is limited because this information is neither provided on food labels nor intuitively apparent. Indeed, a food's glycemic index is not always what one might expect. Ice cream, for example, is a high-sugar food but produces less of a glycemic response than baked potatoes, a high-starch food. Perhaps most relevant to real life, a food's glycemic effect differs depending on plant variety, food processing, cooking method, and whether it is eaten alone or with other foods. Most people eat a variety of foods, cooked and raw, that provide different amounts of carbohydrate, fat, and protein—all of which influence the glycemic index of a meal.

Paying attention to the glycemic index may be unnecessary because current guidelines already suggest many low and moderate glycemic index choices: whole grains, legumes, vegetables, fruits, and milk and milk products. In addition, eating frequent, small meals spreads glucose absorption across the day and thus offers similar metabolic advantages to eating foods with a low glycemic response. People wanting to follow a low glycemic diet should be careful not to adopt a low-carbohydrate diet as well. Highlight 4 explores the controversies surrounding low-carbohydrate diets.

> **REVIEW IT** Explain how the body maintains its blood glucose concentration and what happens when blood glucose rises too high or falls too low.

Dietary carbohydrates provide glucose that can be used by the cells for energy, stored by the liver and muscles as glycogen, or converted into fat if intakes exceed needs. All of the body's cells depend on glucose; those of the brain and central nervous system are especially dependent on it. Without glucose, the body is forced to break down its protein tissues to make glucose and to alter energy metabolism to make ketone bodies from fats. Blood glucose regulation depends primarily on two pancreatic hormones: insulin to move glucose from the blood into the cells when levels are high and glucagon to free glucose from glycogen stores and release it into the blood when levels are low.

4.4 Health Effects and Recommended Intakes of Sugars

> **LEARN IT** Describe how added sugars can contribute to health problems.

Almost everyone finds pleasure in sweet foods—after all, the taste preference for sweets is inborn. To a child, the sweeter the food, the better. In adults, this preference is somewhat diminished, but most adults still enjoy at least an occasional sweet food or beverage.

In the United States, the natural sugars of milk, fruits, vegetables, and grains account for about half of the sugar intake; the other half consists of concentrated sugars that have been refined and added to foods for a variety of purposes (see Table 4-5 on p.114). The use of added sugars has risen steadily over the past several decades, both in the United States and around the world, with soft drinks and sugared fruit drinks accounting for most of the increase. An estimated 75 percent of the packaged foods in the United States contain sweeteners, mostly added

TABLE 4-5 Functions of Sugar in Foods

- Acts as a bulking agent in ice cream and baked goods
- Adds texture and color to baked goods
- Balances the acidity of tomato- and vinegar-based products such as sauces, salad dressings, and other condiments
- Enhances flavor
- Imparts a creamy consistency in frozen desserts
- Inhibits microbial growth by binding with water in jams and jellies
- Maintains the natural color and texture of preserved fruits
- Provides fuel for yeast fermentation, causing bread to rise or producing alcohol

sugars.[7] On average, US adults consume almost 15 percent of their daily energy intake from added sugars.[8] These added sugars assume various names on food labels: sucrose, invert sugar, corn sugar, corn syrups and solids, high-fructose corn syrup, and honey. A food is likely to be high in added sugars if its ingredient list starts with any of the monosaccharides or disaccharides already defined or any of the sugars named in the accompanying glossary, or if it includes several of them.

Health Effects of Sugars In moderate amounts, sugars add pleasure to meals without harming health.[9] In excess, however, sugars can be detrimental, and the average American diet currently delivers excessive amounts. The *Dietary Guidelines* caution that added sugars may increase the risk of certain chronic diseases—even in the absence of overweight or obesity.[10]

Obesity and Chronic Diseases Over the past several decades, as obesity rates increased sharply, consumption of added sugars reached an all-time high—much of it because of the surge in high-fructose corn syrup use, especially in beverages.[11] High-fructose corn syrup is composed of fructose and glucose in a ratio of roughly 50:50. Compared with sucrose, high-fructose corn syrup is less expensive, easier to use, and more soluble. Manufacturers prefer high-fructose corn syrup because it retains moisture, resists drying out, controls crystallization, prevents microbial growth and blends easily with other sweeteners, acids, and flavorings. In addition to being used in beverages, high-fructose corn syrup sweetens candies, baked goods, and hundreds of other foods.

In general, the energy intake of people who drink soft drinks, fruit punches, and other sugary beverages is greater than those who choose differently. Adolescents, for example, who drink as much as 26 ounces or more (about two cans) of sugar-sweetened soft drinks daily, consume 400 more kcalories a day than teens who don't. Not too surprisingly, they also tend to weigh more. Overweight

© Polara Studios, Inc.

Almost half of the added sugars in our diet come from sugar-sweetened beverages, but baked goods, ice cream, candy, and breakfast cereals also make substantial contributions.

GLOSSARY
OF ADDED SUGARS

brown sugar: refined white sugar crystals to which manufacturers have added molasses syrup with natural flavor and color; 91 to 96 percent pure sucrose.

confectioners' sugar: finely powdered sucrose, 99.9 percent pure.

corn sweeteners: corn syrup and sugars derived from corn.

corn syrup: a syrup made from cornstarch that has been treated with acid, high temperatures, and enzymes to produce glucose, maltose, and dextrins. It may be dried and used as *corn syrup solids*. See also *high-fructose corn syrup (HFCS)*.

dextrose: the name food manufacturers use for the sugar that is chemically the same as glucose; *anhydrous dextrose* is similar, differing primarily in the temperature of crystallization.

high-fructose corn syrup (HFCS): a syrup made from cornstarch that has been treated with an enzyme that converts some of the glucose to the sweeter fructose; made especially for use in processed foods and beverages, where it is the predominant sweetener. With a chemical structure similar to sucrose, most HFCS has a fructose content of 42 or 55 percent, with glucose making up the remainder.

honey: sugar (mostly sucrose) formed from nectar gathered by bees. Composition and flavor vary, but honey always contains a mixture of sucrose, fructose, and glucose.

invert sugar: a mixture of glucose and fructose formed by the hydrolysis of sucrose in a chemical process; sold only in liquid form and sweeter than sucrose. Invert sugar is used as a food additive to help preserve freshness and prevent shrinkage.

levulose: an older name for fructose.

malt syrup: a sweetener made from sprouted barley and containing mostly maltose.

maple sugar: a sugar (mostly sucrose) purified from the concentrated sap of the sugar maple tree.

molasses: the thick brown syrup produced during sugar refining. Molasses retains residual sugar and other by-products and a few minerals; blackstrap molasses contains significant amounts of calcium and iron.

nectar: a sugary fluid secreted by plants to encourage pollination by insects.

raw sugar: the first crop of crystals harvested during sugar processing. Raw sugar cannot be sold in the United States because it contains too much filth (dirt, insect fragments, and the like). Sugar sold as "raw sugar" domestically has actually gone through more than half of the refining steps.

tagatose (TAG-ah-tose): poorly absorbed monosaccharide similar in structure to fructose; naturally occurring or derived from lactose.

turbinado (ter-bih-NOD-oh) **sugar:** sugar produced using the same refining process as white sugar, but without the bleaching and anticaking treatment. Traces of molasses give turbinado its sandy color.

white sugar: granulated sucrose or "table sugar," produced by dissolving, concentrating, and recrystallizing raw sugar.

children and adolescents consume more sweet desserts and soft drinks than their normal-weight peers. Research confirms that consuming sugary beverages correlates with increases in energy intake, body weight, and associated diseases.[12]

Some research suggests that added sugars in general, and fructose in particular, favor the fat-making pathways and impair the fat-clearing pathways in the liver.[13] The resulting blood lipid profile increases the risk of heart disease.[14] As the liver busily makes lipids, its handling of glucose becomes unbalanced and insulin resistance develops—an indicator of prediabetes. All in all, research is finding links between added sugars and the risk of diabetes, inflammation, hypertension, and heart disease.[15] Importantly, moderate intakes of sugars do not cause these health problems.[16] For this reason, researchers suggest replacing sugar-sweetened beverages with water, and the American Heart Association recommends limiting added sugars to no more than 100 kcalories per day for women and 150 kcalories per day for men (which is about 5 percent of a 2000- and 2500-kcalorie diet, respectively).[17]

Nutrient Deficiencies Foods such as whole grains, vegetables, legumes, and fruits that contain some natural sugars and lots of starches and fibers provide vitamins and minerals as well. By comparison, foods and beverages that contain lots of added sugars such as cakes, candies, and sodas provide the body with glucose and energy, but few, if any, other essential nutrients or fiber. The more added sugars (and solid fats) in the diet, the more difficult it is to meet recommendations for dietary fiber, vitamins, and minerals and still stay within kcalorie limits.

A person spending 200 kcalories of a day's energy allowance on a 16-ounce soda gets little of value for those kcalories. In contrast, a person using 200 kcalories on three slices of whole-wheat bread gets 9 grams of protein, 6 grams of fiber, plus several of the B vitamins with those kcalories. For the person who wants something sweet, a reasonable compromise might be two slices of bread with a teaspoon of jam on each. The amount of sugar a person can afford to eat depends on how many discretionary kcalories are available beyond those needed to deliver indispensable vitamins and minerals.

By following the USDA Food Pattern and making careful food selections, a typical adult can obtain all the needed nutrients within an allowance of about 1500 kcalories. An inactive older woman who is limited to fewer than 1500 kcalories a day can afford to eat only the most nutrient-dense foods—with few, or no, discretionary kcalories available. In contrast, an active teenage boy may need as many as 3000 kcalories a day. If he chooses wisely, then he may use discretionary kcalories for nutrient-dense foods that contain added sugars—or even an occasional empty kcalorie choice such as a cola beverage. Examples of nutrient-dense foods containing some added sugars include whole-grain breakfast cereals and vanilla yogurt.

Some people believe that because honey is a natural food, it is nutritious—or, at least, more nutritious than sugar.* A look at their chemical structures reveals the truth. Honey, like table sugar, contains glucose and fructose. The primary difference is that in table sugar the two monosaccharides are bonded together as the disaccharide sucrose, whereas in honey some of the monosaccharides are free. Whether a person eats monosaccharides individually, as in honey, or linked together, as in table sugar, they end up the same way in the body: as glucose and fructose.

Honey does contain a few vitamins and minerals, but not many. Honey is denser than crystalline sugar, too, so it provides more energy per spoonful. Table 4-6 on p. 116 shows that honey and white sugar are similar nutritionally—and both fall short of milk, legumes, fruits, grains, and vegetables.

Although the body cannot distinguish whether fructose and glucose derive from honey or table sugar, this is not to say that all sugar sources are alike. Some sugar sources are more nutritious than others. Consider a fruit, say, an orange. The fruit may give you the same amounts of fructose and glucose and the same number of kcalories as a spoonful of sugar or honey, but the packaging is more valuable nutritionally. The fruit's sugars arrive in the body diluted in a large

You receive about the same amount and kinds of sugars from an orange as from a tablespoon of honey, but the packaging makes a big nutrition difference.

---

*Honey should never be fed to infants because of the risk of botulism. Chapters 16 and 19 provide more details.

TABLE 4-6 Sample Nutrients in Sugar and Other Foods

The indicated portion of any of these foods provides approximately 100 kcalories. Notice that for a similar number of kcalories and grams of carbohydrate, foods such as milk, legumes, fruits, grains, and vegetables offer more of the other nutrients than do the sugars.

	Size of 100 kcal Portion	Carbohydrate (g)	Protein (g)	Calcium (mg)	Iron (mg)	Potassium (mg)	Vitamin D (µg)
Foods							
Milk, 1% low-fat	1 c	12	8	300	0.1	397	2
Kidney beans	½ c	20	7	45	1.6	303	0
Apricots	6	24	3	30	0.8	544	0
Bread, whole-wheat	1½ slices	20	4	77	1.2	122	0
Broccoli, cooked	2 c	20	7	125	2.1	914	0
Sugars							
Sugar, white	2 tbs	24	0	0	trace	0	0
Molasses	2 tbs	28	0	82	1.9	586	0
Cola beverage	1 c	26	0	6	trace	7	0
Honey	1½ tbs	26	trace	2	0.1	16	0

© Cengage Learning

> **FIGURE 4-11 Dental Caries**

Dental caries begins when acid dissolves the enamel that covers the tooth. If not repaired, the decay may penetrate the dentin and spread into the pulp of the tooth, causing inflammation, abscess, and possible loss of the tooth.

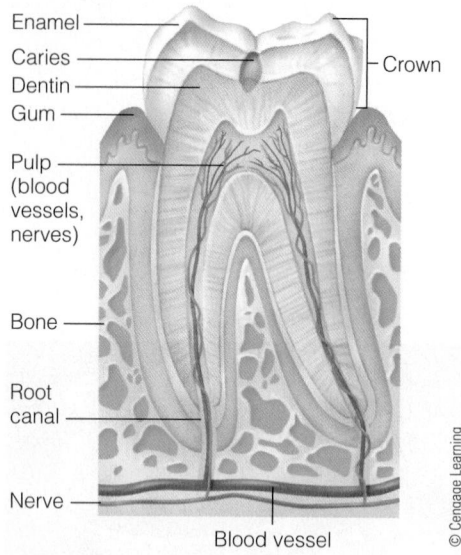

Enamel
Caries
Dentin
Gum
Crown
Pulp (blood vessels, nerves)
Bone
Root canal
Nerve
Blood vessel

© Cengage Learning

volume of water, packaged in fiber, and mixed with essential vitamins, minerals, and phytochemicals.

As these comparisons illustrate, the significant difference between sugar sources is not between "natural" honey and "purified" sugar but between concentrated added sugars and the dilute, naturally occurring sugars that sweeten foods. You can suspect an exaggerated nutrition claim when someone asserts that one product is more nutritious than another because it contains honey.

Added sugars contribute to nutrient deficiencies by displacing nutrients. For nutrition's sake, the appropriate attitude to take is not that sugar is "bad" and must be avoided, but that nutritious foods must come first. If nutritious foods crowd sugar out of the diet, that is fine—but not the other way around. As always, balance, variety, and moderation guide healthy food choices.

Dental Caries Both naturally occurring and added sugars from foods and from the breakdown of starches in the mouth can contribute to tooth decay. Bacteria in the mouth ferment the sugars and, in the process, produce an acid that erodes tooth enamel (see Figure 4-11), causing **dental caries,** or tooth decay. People can eat sugar without this happening, though. Much depends on how long foods stay in the mouth. Sticky foods stay on the teeth longer and continue to yield acid longer than foods that are readily cleared from the mouth. For that reason, sugar in a juice consumed quickly, for example, is less likely to cause dental caries than sugar in a pastry. By the same token, the sugar in sticky foods such as raisins can be more detrimental than the quantity alone would suggest.

Another concern is how often people eat sugar. Bacteria produce acid for 20 to 30 minutes after each exposure. If a person eats three pieces of candy at one time, the teeth will be exposed to approximately 30 minutes of acid destruction. But, if the person eats three pieces at half-hour intervals, the time of exposure increases to 90 minutes. Likewise, slowly sipping a sugary sports beverage may be more harmful than drinking quickly and clearing the mouth of sugar. Nonsugary foods can help remove sugar from tooth surfaces; hence, it is better to eat sugar with meals than between meals. Foods such as milk and cheese may be particularly helpful in protecting against dental caries by neutralizing acids, stimulating salivary flow, inhibiting bacterial activity, and promoting remineralization of damaged enamel.

Beverages such as soft drinks, orange juice, and sports drinks not only contain sugar but also have a low pH. These acidic drinks can erode tooth enamel and may explain why the prevalence of dental erosion is growing steadily.

dental caries: decay of teeth.
• caries = rottenness

The development of caries depends on several factors: the bacteria that reside in **dental plaque,** the saliva that cleanses the mouth, the minerals that form the teeth, and the foods that remain after swallowing. For most people, good oral hygiene will prevent dental caries (see Table 4-7). In fact, regular brushing (twice a day, with a fluoride toothpaste) and flossing may be more effective in preventing dental caries than restricting sugary foods. Still, nutrition is a key component of dental health.[18] *The Dietary Guidelines for Americans* recommend a combined approach to prevent dental caries—practicing good oral hygiene, drinking fluoridated water, and consuming sugar- and starch-containing foods and beverages less frequently.

Recommended Intakes of Sugars Estimates indicate that, on average, each person in the United States consumes about 30 teaspoons (about 120 grams) of sugars a day.[19] Most of the sugars in the average American diet are added to foods and beverages by manufacturers during processing; major sources of added sugars include sugar-sweetened beverages (sodas, energy drinks, sports drinks, fruit drinks), desserts, and candy. Some sugars are also added by consumers during food preparation and at the table. Because added sugars deliver kcalories, but few or no nutrients or fiber, the *Dietary Guidelines for Americans* urge consumers to "reduce the intake of kcalories from added sugars." These added sugar kcalories (and those from solid fats and alcohol) are considered discretionary kcalories—and most people need to limit their intake. By reducing the intake of foods and beverages with added sugars, consumers can lower the kcalorie content of the diet without compromising the nutrient content. The "How To" on p. 118 provides strategies for reducing the intake of added sugars.

> **DIETARY GUIDELINES FOR AMERICANS**
Reduce the intake of kcalories from added sugars.

To help consumers make healthier choices, food labels include the amounts of *added sugars* in a serving. Food labels also list the *total* grams of sugar a food provides, which reflects both added sugars and those occurring naturally in foods. To help estimate sugar and energy intakes accurately, keep in mind that 1 teaspoon of sugar—whether white sugar, brown sugar, corn syrup, honey, nectar, jam, jelly, maple syrup, or molasses—provides about 5 grams of carbohydrate and *about* 20 kcalories per teaspoon. Some are lower (16 kcalories for table sugar) and others are higher (22 kcalories for honey), but a 20-kcalorie average is an acceptable approximation. For a person who uses ketchup liberally, it may help to remember that 1 tablespoon of ketchup supplies about 1 teaspoon of sugar. And those who drink soft drinks regularly should keep in mind that a 12-ounce can of soda delivers about 10 teaspoons of sugar.

The Dietary Reference Intakes (DRI) committee did not publish a Tolerable Upper Intake Level (UL) for sugar, but as mentioned, excessive intakes can interfere with sound nutrition and good health. Few people can eat lots of sugary treats and still meet all of their nutrient needs without exceeding their kcalorie allowance. Specifically, the DRI suggests that added sugars should account for no more than 25 percent of the day's total energy intake.[20] One out of eight in the US population exceeds this maximum intake.[21] When added sugars occupy this much of a diet, intakes from the five food groups usually fall below recommendations. For a person consuming 2000 kcalories a day, 25 percent represents 500 kcalories (that is, 125 grams, or 31 teaspoons) from concentrated sugars—and that's a lot of sugar. Consider that 500 kcalories of sugar is 40 ounces of cola, 1/2 cup of honey, 125 jelly beans, 23 marshmallows, or 30 teaspoons of sugar. Perhaps an athlete in training whose energy needs are high can afford the added sugars from sports drinks without compromising nutrient intake, but most people do better by limiting their use of added sugars. Added sugars contribute an average of 16 percent of the total energy in the typical American diet. The World Health Organization (WHO) suggests limiting added sugars to less than 10 percent of energy intake,

© 2016 Cengage Learning

TABLE 4-7 Behaviors to Prevent Dental Caries

- Limit between meal sugar-sweetened beverages (such as carbonated beverages and fruit drinks), sticky foods (such as raisins and caramels), slow-dissolving candies (such as lollipops and jaw breakers), and snacks containing sugars and starches (such as cookies and cakes).
- Brush with a fluoride toothpaste and floss teeth regularly.
- If brushing and flossing are not possible, drink some milk, rinse with water, or chew sugarless gum immediately after a meal or snack.
- Get a dental checkup regularly.
- Drink fluoridated water.

Soft drinks deliver a startling amount of sugar.

dental plaque: a gummy mass of bacteria that grows on teeth and can lead to dental caries and gum disease.

Reduce the Intake of Added Sugars

- Use less table sugar when preparing meals and at the table.

- Use your sugar kcalories to sweeten nutrient-dense foods (such as oatmeal) instead of consuming empty kcalorie foods and beverages (such as candy and soda).

- Drink fewer regular sodas, sports drinks, energy drinks, and fruit drinks; choose water, fat-free milk, 100 percent fruit juice, or unsweetened tea or coffee instead. If you do drink sugar-sweetened beverages, have a small portion.

- Select fruit for dessert. Eat less cake, cookies, ice cream, other desserts, and candy. If you do eat these foods, have a small portion.

- Read the Nutrition Facts on a label to choose foods with fewer added sugars. Select the unsweetened version of a food (such as cornflakes) instead of the sweetened version (such as frosted cornflakes) to reduce the quantity of added sugars in the diet.

- The ingredients list also helps to identify foods with added sugars. A food is likely to be high in added sugars if its ingredient list starts with any of the sugars named in the glossary on p. 114, or if it includes several of them.

> **TRY IT** Compare the energy contents, added sugars, and ingredients lists of 1 cup of the following foods: fruit-flavored yogurt and plain yogurt, sugar-frosted cornflakes and plain cornflakes, orange soda and orange juice.

Consumers use artificial sweeteners to help them limit kcalories and minimize sugar intake.

© Matthew Farruggio

and notes that 5 percent or less provides additional benefits, which is in line with American Heart Association recommendations mentioned earlier.

Alternative Sweeteners To control weight gain, blood glucose, and dental caries, many consumers turn to alternative sweeteners to help them limit kcalories and minimize added sugars in the diet.[22] In doing so, they encounter three sets of alternative sweeteners: artificial sweeteners, an herbal sweetener, and sugar alcohols.

Artificial Sweeteners **Artificial sweeteners** are sometimes called **nonnutritive sweeteners** because they provide virtually no energy. Table 4-8 provides general details about each of the sweeteners approved for use in the United States, including their **Acceptable Daily Intakes (ADI)**. (The table does not include the nonnutritive sweeteners alitame, cyclamate, neohesperidine, and thaumatin, which are approved in other countries, but not in the United States.) Chapter 9 includes a discussion of their use in weight control and Chapter 19 focuses on some of the safety issues surrounding their use. Considering that all substances are toxic at some dose, it is little surprise that large doses of artificial sweeteners (or their components or metabolic by-products) may have adverse effects. The question to ask is whether their ingestion is safe for human beings in quantities people normally use (and potentially abuse).

Other High-Intensity Sweeteners Stevia leaves and monk fruit have long been used by the people of South America and China, respectively, to sweeten their foods and beverages. The FDA has granted these sweeteners the status of "generally recognized as safe," and they can be used as additives in a variety of foods and beverages.

Sugar Alcohols Some "sugar-free" or reduced-kcalorie products contain **sugar alcohols.** The sugar alcohols (or polyols) occur naturally in fruits and vegetables; manufacturers also use sugar alcohols in many processed foods to add bulk and texture, to provide a cooling effect or sweet taste, to inhibit browning from heat, and to retain moisture. These products may claim to be "sugar-free" on their labels, but in this case, "sugar-free" does not mean free of kcalories. Sugar alcohols do

artificial sweeteners: sugar substitutes that provide negligible, if any, energy; sometimes called *nonnutritive sweeteners.*

nonnutritive sweeteners: sweeteners that yield no energy (or insignificant energy in the case of aspartame).

Acceptable Daily Intake (ADI): the estimated amount of a sweetener that individuals can safely consume each day over the course of a lifetime without adverse effect.

sugar alcohols: sugarlike compounds that can be derived from fruits or commercially produced from dextrose; also called *polyols.* Examples include *erythritol, isomalt, lactitol, maltitol, mannitol, sorbitol,* and *xylitol.*

TABLE 4-8 **Nonnutritive Alternative Sweeteners**

Sweetener	Chemical Composition	Body's Response	Relative Sweetness[a]	Energy (kcal/g)	ADI and (Estimated Equivalent)[b]	Comments
Acesulfame potassium or Acesulfame K[c] (AY-sul-fame)	Potassium salt	Not digested or absorbed	200	0	15 mg/kg body weight[d] (30 cans diet soda)	Approved for general use (except in meats and poultry); combines well with other sweeteners; heat stable at baking temperatures
Advantame (ad-VAN-tame)	Aspartame derivative, similar to neotame	Rapidly, but poorly absorbed	20,000	0	32.8 mg/kg body weight (4000 packets of sweetener)	Approved for general use (except in meat and poultry); heat stable at baking temperatures
Aspartame[e] (ah-SPAR-tame or ASS-par-tame)	Amino acids (phenyl-alanine and aspartic acid) and a methyl group	Digested and absorbed	200	4[f]	50 mg/kg body weight[g] (18 cans diet soda)	Approved for general use; degrades at high temperatures
Luo han guo[h]	Cucurbitane glycosides extracts from *Siraitia grosvenorii* swingle fruit (also known as monk fruit)	Digested and absorbed	225	1	Not determined	GRAS;[i] general use as a tabletop sweetener and food ingredient
Neotame (NEE-oh-tame)	Aspartame with an additional side group attached	Not digested or absorbed	10,000	0	18 mg/day	Approved for general use (except in meats and poultry); used minimally in food processing
Saccharin[j] (SAK-ah-ren)	Benzoic sulfimide	Rapidly absorbed and excreted	300	0	5 mg/kg body weight (10 packets of sweetener)	Restricted use in beverages, in individual packages, and in processed foods
Stevia[k] (STEE-vee-ah)	Glycosides found in the leaves of the *Stevia rebaudiana* herb	Digested and absorbed	300	0	4 mg/kg body weight	GRAS
Sucralose[l] (SUE-kra-lose)	Sucrose with Cl atoms instead of OH groups	Not digested or absorbed	600	0	5 mg/kg body weight (6 cans diet soda)	Approved for general use; heat stable at baking temperatures

© Cengage Learning

[a]Relative sweetness is determined by comparing the approximate sweetness of a sugar substitute with the sweetness of pure sucrose, which has been defined as 1.0. Chemical structure, temperature, acidity, and other flavors of the foods in which the substance occurs all influence relative sweetness.

[b]The Acceptable Daily Intake (ADI) is the estimated amount of a sweetener that individuals can safely consume each day over the course of a lifetime without adverse effects. The Estimated Equivalent is based on a person weighing 70 kg (154 lb).

[c]Marketed under the trade names Sunett and Sweet One.

[d]Recommendations from the World Health Organization limit acesulfame K intake to 9 mg per kilogram of body weight per day.

[e]Marketed under the trade names NutraSweet, Equal, and Sugar Twin.

[f]Aspartame provides 4 kcal per gram, as does protein, but because so little is used, its energy contribution is negligible. In powdered form, it is sometimes mixed with lactose, however, so a 1-g packet may provide 4 kcal.

[g]Recommendations from the World Health Organization and in Europe and Canada limit aspartame intake to 40 mg per kilogram of body weight per day.

[h]Marketed under the trade name Fruit-Sweetness.

[i]GRAS = generally recognized as safe. The GRAS list is subject to revision as new facts become known.

[j]Marketed under the trade names Sweet'N Low, Sweet Twin, and Necta Sweet.

[k]Marketed under the trade names SweetLeaf, Purevia, Truvia, and Honey Leaf.

[l]Marketed under the trade name Splenda.

provide kcalories (0.2 to 2.6 kcalories per gram), but fewer than the sugars. Because sugar alcohols yield energy, they are sometimes referred to as **nutritive sweeteners.**

Sugar alcohols evoke a low glycemic response. The body partially absorbs some sugar alcohols and absorbs others slowly; consequently, they are slower to enter the bloodstream than other sugars. Unabsorbed sugar alcohols may be metabolized by bacteria in the GI tract, producing side effects such as intestinal gas, abdominal discomfort, and diarrhea. For this reason, regulations require food labels to state "Excess consumption may have a laxative effect" if reasonable consumption of that food could result in the daily ingestion of 50 grams of a sugar alcohol. For perspective, a low-carbohydrate energy bar or shake may contain 10 to 15 grams of a sugar alcohol.

The real benefit of using sugar alcohols is that they do not contribute to dental caries. Bacteria in the mouth cannot metabolize sugar alcohols as rapidly as sugar. Sugar alcohols are therefore valuable in chewing gums, breath mints, and other products that people keep in their mouths for a while. Figure 4-12 presents labeling information for products using sugar alternatives.

nutritive sweeteners: sweeteners that yield energy, including both sugars and sugar alcohols.

> **FIGURE 4-12** Sugar Alternatives on Food Labels

Products containing sugar replacers may claim to "not promote tooth decay" if they meet FDA criteria for dental plaque activity.

Products containing aspartame must carry a warning for people with phenylketonuria.

This ingredient list includes both sugar alcohols and artificial sweetenters.

INGREDIENTS: SORBITOL, MALTITOL, GUM BASE, MANNITOL, ARTIFICIAL AND NATURAL FLAVORING, ACACIA, SOFTENERS, TITANIUM DIOXIDE (COLOR), ASPARTAME, ACESULFAME POTASSIUM AND CANDELILLA WAX. **PHENYLKETONURICS: CONTAINS PHENYLALANINE.**

35% FEWER CALORIES THAN SUGARED GUM.

Products that claim to be "reduced kcalories" must provide at least 25% fewer kcalories per serving than the comparison item.

Nutrition Facts

6 servings per container

Serving size	2 pieces (3g)

Amount Per 2 pieces

Calories	**5**

% DV		
0%	**Total Fat** 0g	
0%	**Sodium** 0mg	
1%	**Total Carbs** 2g	
	Sugars 0g	
	Added Sugars 0g	
	Sugar Alcohol 2g	
	Protein 0g	

Not a significant source of other nutrients.

Products containing less than 0.5 g of sugar per serving can claim to be "sugarless" or "sugar-free."

© Craig M. Moore

© Cengage Learning

For consumers choosing to use alternative sweeteners, the Academy of Nutrition and Dietetics wisely advises that they be used in moderation and only as part of a well-balanced nutritious diet.[23] When used in moderation, these sweeteners will do no harm. In fact, they may even help, by providing an alternative to sugar for people with diabetes, by inhibiting caries-causing bacteria, and by limiting energy intake. People may find it appropriate to choose from among any of the sweeteners at times: artificial sweeteners, an herbal sweetener, sugar alcohols, and sugar itself.

> **REVIEW IT** Describe how added sugars can contribute to health problems.
Sugars increase the risk of dental caries; excessive intakes displace needed nutrients and fiber and contribute to obesity when energy intake exceeds needs. A person deciding to limit daily sugar intake should recognize that not all sugars need to be restricted, just concentrated sweets, which are relatively empty of other nutrients and high in kcalories. Sugars that occur naturally in fruits, vegetables, and milk are acceptable. Alternative sweeteners may help to limit kcalories and sugar intake.

4.5 Health Effects and Recommended Intakes of Starch and Fibers

> **LEARN IT** Identify the health benefits of, and recommendations for, starches and fibers.

Carbohydrates and fats are the two major sources of energy in the diet. When one is high, the other is usually low—and vice versa. A diet that provides abundant carbohydrate (45 to 65 percent of energy intake) and some fat (20 to 35 percent of energy intake) within a reasonable energy allowance best supports good health.

To increase carbohydrates in the diet, focus on whole grains, vegetables, legumes, and fruits—foods noted for their starch, fibers, and naturally occurring sugars.

Health Effects of Starch and Fibers In addition to starch, fibers, and natural sugars, remember that whole grains, vegetables, legumes, and fruits supply valuable vitamins and minerals and little or no fat. The following paragraphs describe some of the health benefits of diets that include a variety of these foods daily.

Heart Disease Unlike high-carbohydrate diets rich in added sugars that can alter blood lipids to favor heart disease, those rich in whole grains, legumes, vegetables, and fruits may protect against heart attack and stroke by lowering blood pressure, improving blood lipids, and reducing inflammation.[24] Such diets are low in animal fat and cholesterol and high in dietary fibers, vegetable proteins, and phytochemicals—all factors associated with a lower risk of heart disease. (The role of animal fat and cholesterol in heart disease is discussed in Chapter 5. The role of vegetable proteins in heart disease is presented in Chapter 6. The benefits of phytochemicals in disease prevention are featured in Highlight 13.)

Oatmeal was one of the first foods recognized for its ability to reduce cholesterol and the risk of heart disease. Foods rich in soluble fibers (such as oat bran, barley, and legumes) lower blood cholesterol by binding with bile acids in the GI tract and thereby increasing their excretion. Consequently, the liver must use its cholesterol to make new bile acids. In addition, the bacterial by-products of fiber fermentation in the colon also inhibit cholesterol synthesis in the liver. The net result is that soluble fibers such as those found in oats lower blood cholesterol.[25]

Several researchers have speculated that fiber may also exert its effect by displacing fats in the diet. Although this is certainly helpful, even when dietary fat is low, fibers exert a separate and significant cholesterol-lowering effect. In other words, a high-fiber diet helps to decrease the risk of heart disease independent of fat intake.

Diabetes High-fiber foods—especially whole grains—play a key role managing and preventing type 2 diabetes. When soluble fibers trap nutrients and delay their transit through the GI tract, glucose absorption is slowed, which helps to prevent glucose surge and rebound.

GI Health Dietary fibers also enhance the health of the large intestine. The healthier the intestinal walls, the better they can block absorption of unwanted constituents. Taken with ample fluids, insoluble fibers such as cellulose (as in cereal brans, fruits, and vegetables) increase stool weight, ease passage, and reduce transit time.

Large, soft stools ease elimination for the rectal muscles and reduce pressure in the lower bowel, preventing constipation and making it less likely that rectal veins will swell (hemorrhoids). Fiber prevents compaction of the intestinal contents, which could obstruct the appendix and permit bacteria to invade and infect it (appendicitis). In addition, fiber stimulates the GI tract muscles so that they retain their strength and resist bulging out into pouches known as diverticula (illustrated in Figure H3-3 on p. 93).[26] Recommendations typically suggest increasing fiber to protect against diverticular disease, although research findings are inconsistent.[27]

Cancer Research studies suggest that a high-fiber diet protects against colon cancer. When a large study of diet and cancer examined the diets of more than a half million people in ten countries for several years, the researchers found an inverse association between dietary fiber and colon cancer. People who ate the most dietary fiber (35 grams per day) reduced their risk of colon cancer by 40 percent compared with those who ate the least fiber (15 grams per day). Importantly, the study focused on dietary fiber, not fiber supplements or additives, which lack valuable nutrients and phytochemicals that also help protect against cancer. Plant foods—vegetables, fruits, and whole-grain products—reduce the risks of colon and rectal cancers.

Fibers may help prevent colon cancer by diluting, binding, and rapidly removing potential cancer-causing agents from the colon. In addition, soluble fibers stimulate bacterial fermentation of resistant starch and fiber in the colon, a process

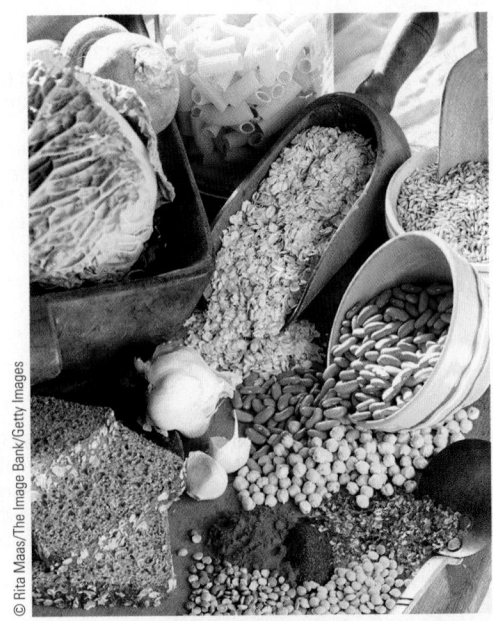

Foods rich in starch and fiber offer many health benefits.

that produces short-chain fatty acids that lower the pH. These small fat molecules activate cancer-killing enzymes and inhibit inflammation in the colon.[28]

Weight Management High-fiber and whole-grain foods may help a person to maintain a healthy body weight.[29] Foods rich in fiber tend to be low in solid fats and added sugars and therefore prevent weight gain and promote weight loss by delivering less energy per bite. In addition, as fibers absorb water from the digestive juices, they swell, creating feelings of fullness, lowering food intake, and delaying hunger.

Many weight-loss products on the market today contain bulk-inducing fibers such as methylcellulose, but buying pure fiber compounds like this is neither necessary nor advisable. Instead of fiber supplements, consumers should select whole grains, legumes, fruits, and vegetables. High-fiber foods not only add bulk to the diet but are economical and nutritious as well.

Dietary fiber provides numerous health benefits. Table 4-9 summarizes fiber characteristics, food sources, actions in the body, and their health benefits.

Harmful Effects of Excessive Fiber Intake Despite fibers' benefits to health, a diet excessively high in fiber also has a few drawbacks. A person who has a small capacity and eats mostly high-fiber foods may not be able to eat enough food to meet energy or nutrient needs. The malnourished, the elderly, and young children adhering to all-plant (vegan) diets are especially vulnerable to this problem.

Switching from a low-fiber diet to a high-fiber diet suddenly can cause temporary bouts of abdominal discomfort, gas, and diarrhea and, more seriously, can obstruct the GI tract. To prevent such complications, a person adopting a high-fiber diet can take the following precautions:

- Increase fiber intake gradually over several weeks to give the GI tract time to adapt.
- Drink plenty of liquids to soften the fiber as it moves through the GI tract.
- Select fiber-rich foods from a variety of sources—fruits, vegetables, legumes, and whole-grain breads and cereals.

TABLE 4-9 Characteristics, Sources, and Health Effects of Fibers

	Major Food Sources	Types of Fibers	Actions in the Body	Probable Health Benefits
	Viscous, Soluble, More Fermentable			
	Barley, oats, oat bran, rye, fruits (apples, citrus), legumes (especially young green peas and black-eyed peas), seaweeds, seeds and husks, many vegetables, fibers used as food additives	Gums Pectins Psyllium[a] Some hemicellulose	Lower blood cholesterol by binding bile	Lower risk of heart disease
			Slow glucose absorption	Lower risk of diabetes
			Slow transit of food through upper GI tract	Lower risk of colon and rectal cancer
			Hold moisture in stools, softening them	Increased satiety, and may help with weight management
			Yield small fat molecules after fermentation that the colon can use for energy	
			Increase satiety	
	Nonviscous, Insoluble, Less Fermentable			
	Brown rice, fruits, legumes, seeds, vegetables (cabbage, carrots, brussels sprouts), wheat bran, whole grains, extracted fibers used as food additives	Cellulose Lignins Resistant starch Hemicellulose	Increase fecal weight and speed fecal passage through colon	Alleviate constipation
			Provide bulk and feelings of fullness	May lower risk of diverticulosis, hemorrhoids, and appendicitis
				Lower risk of colon and rectal cancer

[a]Psyllium, a soluble fiber derived from seeds, is used as a laxative and food additive.

Brian Leatart/Photolibrary/Getty Images

© Stockbyte/Getty Images

© Cengage Learning

Some fibers can limit the absorption of nutrients by speeding the transit of foods through the GI tract and by binding to minerals. When mineral intake is adequate, however, a *reasonable* intake of high-fiber foods (less than 40 grams a day) does not compromise mineral balance.

Clearly, fiber is like all nutrients in that "more" is "better" only up to a point. Again, the key dietary goals are balance, moderation, and variety. Table 4-10 presents a list of fiber sources and tips to increase fiber intake.

Recommended Intakes of Starch and Fibers The DRI suggest that carbohydrates provide about half (45 to 65 percent) of the energy requirement. (The remainder comes from fat, with 20 to 35 percent, and protein, with 10 to 35 percent.) A person consuming 2000 kcalories a day should therefore have 900 to 1300 kcalories of carbohydrate, or about 225 to 325 grams. (Appendix K explains how to calculate such math problems.) This amount is more than adequate to meet the RDA for carbohydrate, which is set at 130 grams per day, based on the average minimum amount of glucose used by the brain.

On food labels, the Food and Drug Administration (FDA) uses the guideline of 60 percent of a 2000-kcalorie diet in setting the Daily Value for carbohydrate at 300 grams per day. To meet this goal, the *Dietary Guidelines* encourage people to choose a variety of whole grains, vegetables, fruits, and legumes daily.

> **DIETARY GUIDELINES FOR AMERICANS**
Choose foods that provide more dietary fiber, a nutrient of concern in American diets. Dietary fiber is found in plants, notably legumes, vegetables, fruits, whole grains, and nuts.

TABLE 4-10 Fiber in Selected Foods

Grains	Tips to Increase Fiber Intake
Whole-grain products provide about 1 to 2 g (or more) of fiber per serving: • 1 slice whole-wheat, pumpernickel, rye bread • 1 oz ready-to-eat cereal (100% bran cereals contain 10 g or more) • ½ c cooked barley, bulgur, grits, oatmeal 	Eat whole-grain breads that contain ≥3 g fiber per serving. Eat whole-grain cereals that contain ≥5 g fiber per serving.

Vegetables	Tips to Increase Fiber Intake
Most vegetables contain about 2 to 3 g of fiber per serving: • 1 c raw bean sprouts • ½ c cooked broccoli, brussels sprouts, cabbage, carrots, cauliflower, collards, corn, eggplant, green beans, green peas, kale, mushrooms, okra, parsnips, potatoes, pumpkin, spinach, sweet potatoes, swiss chard, winter squash • ½ c chopped raw carrots, peppers 	Eat raw vegetables. Eat vegetables (such as potatoes and zucchini) with their skins.

Fruit	Tips to Increase Fiber Intake
Fresh, frozen, and dried fruits have about 2 g of fiber per serving: • 1 medium apple, banana, kiwi, nectarine, orange, pear • ½ c applesauce, blackberries, blueberries, raspberries, strawberries • Fruit juices contain very little fiber 	Eat fresh and dried fruit for snacks. Eat fruits (such as apples and pears) with their skins.

Legumes	Tips to Increase Fiber Intake
Many legumes provide about 6 to 8 g of fiber per serving: • ½ c cooked baked beans, black beans, black-eyed peas, kidney beans, navy beans, pinto beans Some legumes provide about 5 g of fiber per serving: • ½ c cooked garbanzo beans, great northern beans, lentils, lima beans, split peas 	Add legumes to soups, salads, and casseroles.

NOTE: Appendix H provides fiber grams for more than 2000 foods.

Some food labels use a "whole-grain stamp" to help consumers identify whole-grain foods.

Recommendations for fiber suggest the same foods just mentioned: whole grains, vegetables, fruits, and legumes, which also provide minerals and vitamins. The FDA sets the Daily Value for fiber at 28 grams for a 2000-kcalorie intake. The DRI recommendation is in agreement at 14 grams per 1000-kcalorie intake. These recommendations are almost two times higher than the usual intake in the United States.[30] An effective way to add fiber while lowering fat is to substitute plant sources of proteins (legumes) for animal sources (meats).

Because high-fiber foods are so filling, they are not likely to be eaten in excess. Too much fiber can cause GI problems for some people, but it generally does not have adverse effects in most healthy people. For these reasons, the DRI committee did not set an Upper Level for fiber.

From Guidelines to Groceries A diet following the USDA Food Patterns, which include several servings of fruits, vegetables, and whole grains daily, can easily supply the recommended amount of carbohydrates and fiber. In selecting high-fiber foods, keep in mind the principle of variety. The fibers in oats lower cholesterol, whereas those in bran help promote GI tract health. (Review Table 4-9 to see the diverse health effects of various fibers.)

Grains An ounce-equivalent of most foods in the grain group (for example, one slice of bread) provides about 15 grams of carbohydrate, mostly as starch. Be aware that some foods in this group, especially snack crackers and baked goods such as biscuits, croissants, and muffins, contain added sugars, solid fats, and sodium. When selecting from the grain group, limit refined grains and be sure to include at least half as whole-grain products (see Figure 4-13). The "three are key" message may help consumers to remember to choose a whole-grain cereal for breakfast, a whole-grain bread for lunch, and a whole-grain pasta or rice for dinner. Because whole grains are typically high in fiber, nutrients, and antioxidants, consumers who eat more whole grains tend to have healthier diets and reduced risks for heart disease, diabetes, and certain cancers.[31]

> **DIETARY GUIDELINES FOR AMERICANS**
Limit the consumption of foods that contain refined grains, especially refined-grain foods that contain solid fats, added sugars, and sodium. Replace refined grains with whole grains.

Vegetables The amount of carbohydrate a serving of vegetables provides depends primarily on its starch content. Starchy vegetables—corn, peas, or potatoes—provide about 15 grams of carbohydrate per half-cup serving. A serving of most other *nonstarchy* vegetables—such as a half-cup of broccoli, green beans, or tomatoes—provides about 5 grams.

Fruits A typical fruit serving—a small banana, apple, or orange or a half-cup of most canned or fresh fruit—contains an average of about 15 grams of carbohydrate, mostly as sugars, including the fruit sugar fructose. Fruits vary greatly in their water and fiber contents and, therefore, in their sugar concentrations.

Milks and Milk Products A serving (a cup) of milk or yogurt provides about 12 grams of carbohydrate. Cottage cheese provides about 6 grams of carbohydrate per cup, but most other cheeses contain little, if any, carbohydrate.

Protein Foods With two exceptions, protein foods deliver almost no carbohydrate to the diet. The exceptions are nuts, which provide a little starch and fiber along with their abundant fat, and legumes, which provide an abundance of

> FIGURE 4-13 **Bread Labels Compared**

Although breads may appear similar, their ingredients vary widely. Breads made mostly from whole-grain flours provide more benefits to the body than breads made of enriched, refined, wheat flours.

Some "high-fiber" breads may contain purified cellulose or more nutritious whole grains. "Low-carbohydrate" breads may be regular white bread, thinly sliced to reduce carbohydrates per serving, or may contain soy flour, barley flour, or flaxseed to reduce starch content.

A trick for estimating a bread's content of a nutritious ingredient, such as whole-grain flour, is to read the ingredients list (ingredients are listed in order of predominance). Bread recipes generally include one teaspoon of salt per loaf. Therefore, when a bulky nutritious ingredient, such as whole grain, is listed after the salt, you'll know that less than a teaspoonful of the nutritious ingredient was added to the loaf—not enough to significantly improve the nutrient value of one slice of bread.

Whole Grain
WHOLE WHEAT

Nutrition Facts

18 servings per container
Serving size   1 slice (30g)

Amount per 1 slice

Calories   **90**

% DV	
2%	**Total Fat** 1.5g
	Trans Fat 0g
6%	**Sodium** 135mg
5%	**Total Carbohydrate** 15g
7%	Dietary Fiber 2g
	Sugars 2g
	Added Sugars 2g
	Protein 4g

MADE FROM: UNBROMATED STONE GROUND 100% WHOLE WHEAT FLOUR, WATER, CRUSHED WHEAT, HIGH-FRUCTOSE CORN SYRUP, PARTIALLY HYDROGENATED VEGETABLE SHORTENING (SOYBEAN AND COTTONSEED OILS), RAISIN JUICE CONCENTRATE, WHEAT GLUTEN, YEAST, WHOLE WHEAT FLAKES, UNSULPHURED MOLASSES, SALT, HONEY, VINEGAR, ENZYME MODIFIED SOY LECITHIN, CULTURED WHEY, UNBLEACHED WHEAT FLOUR AND SOY LECITHIN.

Natural
Wheat Bread

Nutrition Facts

15 servings per container
Serving size   1 slice (30g)

Amount per 1 slice

Calories   **90**

% DV	
2%	**Total Fat** 1.5g
	Trans Fat 0g
10%	**Sodium** 220mg
5%	**Total Carbohydrate** 15g
2%	Dietary Fiber Less than 1g
	Sugars 2g
	Added Sugars 2g
	Protein 4g

INGREDIENTS: UNBLEACHED ENRICHED WHEAT FLOUR [MALTED BARLEY FLOUR, NIACIN, REDUCED IRON, THIAMIN MONONITRATE (VITAMIN B1), RIBOFLAVIN (VITAMIN B2), FOLIC ACID], WATER, HIGH-FRUCTOSE CORN SYRUP, MOLASSES, PARTIALLY HYDROGENATED SOYBEAN OIL, YEAST, CORN FLOUR, SALT, GROUND CARAWAY, WHEAT GLUTEN, CALCIUM PROPIONATE (PRESERVATIVE), MONOGLYCERIDES, SOY LECITHIN.

Multi-fiber
Low carb

Nutrition Facts

21 servings per container
Serving size   1 slice (30g)

Amount per 1 slice

Calories   **60**

% DV	
2%	**Total Fat** 1.5g
	Trans Fat 0g
6%	**Sodium** 135mg
3%	**Total Carbohydrate** 9g
11%	Dietary Fiber 3g
	Sugars 0g
	Added Sugars 0g
	Protein 5g

INGREDIENTS: UNBLEACHED ENRICHED WHEAT FLOUR, WATER, WHEAT GLUTEN, CELLULOSE, YEAST, SOYBEAN OIL, CRACKED WHEAT, SALT, BARLEY, NATURAL FLAVOR PRESERVATIVES, MONOCALCIUM PHOSPHATE, MILLET, CORN, OATS, SOYBEANS, BROWN RICE, FLAXSEED, SUCRALOSE.

© Cengage Learning

both starch and fiber. Just a half-cup serving of legumes provides about 20 grams of carbohydrate, a third from fiber.

Read Food Labels Food labels list the amount, in grams, of *total* carbohydrate—including starch, fibers, and sugars—per serving (review Figure 4-13). Fiber grams are also listed separately, as are the grams of sugars and added sugars. With this information, you can calculate starch grams by subtracting the grams of fibers and sugars from the total carbohydrate. Using the first label in Figure 4-13 as an example, subtracting the 4 grams of fibers and sugars from the 15 grams of total carbohydrate leaves 11 grams of starch. Total carbohydrate and dietary fiber are also expressed as "% Daily Values" for a person consuming 2000 kcalories; there is no Daily Value for sugars or added sugars.

> **REVIEW IT** Identify the health benefits of, and recommendations for, starches and fibers.

Clearly, a diet rich in starches and fibers supports efforts to control body weight and prevent heart disease, some cancers, diabetes, and GI disorders. For these reasons, recommendations urge people to eat plenty of whole grains, vegetables, legumes, and fruits—enough to provide 45 to 65 percent of the daily energy intake from carbohydrate.

In today's world, there is one other reason why plant foods rich in complex carbohydrates and natural sugars are a better choice than animal foods or foods high in concentrated sugars: in general, less energy and fewer resources are required to grow and process plant foods than to produce sugar or foods derived from animals. Chapter 20 takes a closer look at the environmental impacts of food production and use.

Nutrition Portfolio

Foods that derive from plants—whole grains, vegetables, legumes, and fruits—naturally provide ample carbohydrates and fiber with little or no fat. Refined foods often contain added sugars and solid fats.

Go to Diet & Wellness Plus and choose one of the days on which you have tracked your diet for the entire day. Go to the Intake Spreadsheet report. Scroll down until you see: carb (g).

- Which of your foods for this day were highest in carbohydrate? Which of these foods also contain added sugars and solid fats? List better alternatives.

- List the types and amounts of grain products you ate on that day, making note of which are whole-grain or refined foods and how your choices could include more whole-grain options.

- List the types and amounts of fruits and vegetables you ate on that day, making note of how many are dark green, red and orange, or deep yellow, how many are starchy or legumes, and how your choices could include more of these options.

- Describe choices you can make in selecting and preparing foods and beverages to lower your intake of added sugars.

DIET & WELLNESS **PLUS** To complete this exercise, go to your Diet & Wellness Plus at **www.cengagebrain.com.**

> **STUDY IT** To review the key points of this chapter and take a practice quiz, go to the study cards at the end of the book.

REFERENCES

1. NIH Consensus Development Conference: Lactose intolerance and health, http://consensus.nih.gov/2010/lactosestatement.htm.
2. K. Sevastianova and coauthors, Effect of short-term carbohydrate overfeeding and long-term weight loss on liver fat in overweight humans, *American Journal of Clinical Nutrition* 96 (2012): 727–734.
3. G. Livesey and coauthors, Is there a dose-response relation of dietary glycemic load to risk of type 2 diabetes? Meta-analysis of prospective cohort studies, *American Journal of Clinical Nutrition* 97 (2013): 584–596; A. Pande, G. Krishnamoorthy, and N. D. Moulickm, Hypoglycaemic and hypolipidaemic effects of low GI and medium GL Indian diets in type 2 diabetics for a period of 4 weeks: A prospective study, *International Journal of Food Sciences and Nutrition* 63 (2012): 649–658; J. C. Y. Louie and coauthors, The link between dietary glycemic index and nutrient adequacy, *American Journal of Clinical Nutrition* 95 (2012): 694–702; A. N. Fabricatore and coauthors, Continuous glucose monitoring to assess the ecologic validity of dietary glycemic index and glycemic load, *American Journal of Clinical Nutrition* 94 (2011): 1519–1524.
4. L. M. Goff and coauthors, Low glycaemic index diets and blood lipids: A systematic review and meta-analysis of randomised controlled trials, *Nutrition, Metabolism, and Cardiovascular Diseases* 23 (2013): 1–10; Pande, Krishnamoorthy, and Moulickm, 2012; M. L. Neuhouser and coauthors, A low-glycemic load diet reduces serum C-reactive protein and modestly increases adiponectin in overweight and obese adults, *Journal of Nutrition* 142 (2012): 369–374; J. Brand-Miller and A. E. Buyken, The glycemic index issue, *Current Opinion in Lipidology* 23 (2012): 62–67; O. Gögebakan and coauthors, Effects of weight loss and long-term weight maintenance

with diets varying in protein and glycemic index on cardiovascular risk factors: The Diet, Obesity, and Genes (DiOGenes) study: A randomized, controlled trial, *Circulation* 124 (2011): 2829–2838; E. Denova-Gutiérrez and coauthors, Dietary glycemic index, dietary glycemic load, blood lipids, and coronary heart disease, *Journal of Nutrition and Metabolism* (2010): doi:10.1155/2010/170680. Epub February 28, 2010.
5. G. M. Turner-McGrievy and coauthors, Decreases in dietary glycemic index are related to weight loss among individuals following therapeutic diets for type 2 diabetes, *Journal of Nutrition* 141 (2011): 1469–1474.
6. H. Dodd and coauthors, Calculating meal glycemic index by using measured and published food values compared with directly measured meal glycemic index, *American Journal of Clinical Nutrition* 94 (2011): 992–996.
7. S. W. Ng, M. M. Slining, and B. M. Popkin, Use of caloric and noncaloric sweeteners in US consumer packaged foods, 2005-2009, *Journal of the Academy of Nutrition and Dietetics* 112 (2012): 1828–1834.
8. Position of the Academy of Nutrition and Dietetics: Use of nutritive and nonnutritive sweeteners, *Journal of the Academy of Nutrition and Dietetics* 112 (2012): 739–758.
9. S. W. Rizkalla, Health implications of fructose consumption: A review of recent data, *Nutrition and Metabolism* 4 (2010): 82–98.
10. US Department of Agriculture and US Department of Health and Human Services, *Dietary Guidelines for Americans, 2010*, www.dietaryguidelines .gov; R. E. Kavey, How sweet it is: Sugar sweetened beverage consumption, obesity, and cardiovascular risk in childhood, *Journal of the American Dietetic Association* 110 (2010): 1456–1460.

11. O. I. Bermudez and X. Gao, Greater consumption of sweetened beverages and added sugars is associated with obesity among US young adults, *Annals of Nutrition and Metabolism* 57 (2010): 211–218.

12. V. S. Malik and coauthors, Sugar-sweetened beverages and weight gain in children and adults: A systematic review and meta-analysis, *American Journal of Clinical Nutrition* 98 (2013): 1084-1102; C. A. Grimes and coauthors, Dietary salt intake, sugar-sweetened beverage consumption, and obesity risk, *Pediatrics* 131 (2013): 14–21; D. I. Jalal and coauthors, Increased fructose associates with elevated blood pressure, *Journal of the American Society of Nephrology* 21 (2010): 1543–1549; R. K. Johnson and B. A. Yon, Weighing in on added sugars and health, *Journal of the American Dietetic Association* 110 (2010): 1296–1299; V. S. Malik and coauthors, Sugar-sweetened beverages and risk of metabolic syndrome and type 2 diabetes: A meta-analysis, *Diabetes Care* 33 (2010): 2477–2483; L. Tappy and coauthors, Fructose and metabolic diseases: New findings, new questions, *Nutrition* 26 (2010): 1044–1049.

13. M. Maersk and coauthors, Sucrose-sweetener beverages increase fat storage in the liver, muscle, and visceral fat depot: A 6-mo randomized intervention study, *American Journal of Clinical Nutrition* 95 (2012): 283–289; K. Sevastianova and coauthors, Effect of short-term carbohydrate overfeeding and long-term weight loss on liver fat in overweight humans, *American Journal of Clinical Nutrition* 96 (2012): 727–734; M. J. Dekker and coauthors, Fructose: A highly lipogenic nutrient implicated in insulin resistance, hepatic steatosis, and the metabolic syndrome, *American Journal of Physiology, Endocrinology and Metabolism* 299 (2010): E685–E694; M. E. Bocarsly and coauthors, High-fructose corn syrup causes characteristics of obesity in rats: Increased body weight, body fat and triglyceride levels, *Pharmacology Biochemistry and Behavior* 97 (2010): 101–106.

14. K. L. Stanhope, Role of fructose-containing sugars in the epidemics of obesity and metabolic syndrome, *Annual Review of Medicine* 63 (2012): 19.1–19.15; K. L. Stanhope and coauthors, Consumption of fructose and high fructose corn syrup increase postprandial triglycerides, LDL-cholesterol, and apolipoprotein-B in young men and women, *Journal of Clinical Endocrinology and Metabolism* 96 (2011): 1596–1605; J. A. Welsh and coauthors, Consumption of added sugars and indicators of cardiovascular disease risk among US adolescents, *Circulation* 123 (2011): 249–257; J. A. Welsh and coauthors, Caloric sweetener consumption and dyslipidemia among US adults, *Journal of the American Medical Association* 303 (2010): 1490–1497; N. Wiernsperger, A. Geloen, and J. R. Rapin, Fructose and cardiometabolic disorders: The controversy will, and must, continue, *Clinics* 65 (2010): 729–738.

15. E. S. Eshak and coauthors, Soft drink intake in relation to incident ischemic heart disease, stroke, and stroke subtypes in Japanese men and women: The Japan Public Health Centre-based study cohort I, *American Journal of Clinical Nutrition* 96 (2012): 1390–1397; L. de Koning and coauthors, Sweetened beverage consumption, incident coronary heart disease and biomarkers of risk in men, *Circulation* 125 (2012): 1735–1741; L. de Koning and coauthors, Sugar-sweetened and artificially sweetened beverage consumption and risk of type 2 diabetes in men, *American Journal of Clinical Nutrition* 93 (2011): 1321–1327; L. Tappy and coauthors, Fructose and metabolic diseases: New findings, new questions, *Nutrition* 26 (2010): 1044–1049; K. J. Duffey and coauthors, Drinking caloric beverages increases the risk of adverse cardiometabolic outcomes in the Coronary Artery Risk Development in Young Adults (CARDIA) Study, *American Journal of Clinical Nutrition* 92 (2010): 954–959; F. B. Hu and V. S. Malik, Sugar-sweetened beverages and risk of obesity and type 2 diabetes: Epidemiologic evidence, *Physiology and Behavior* 100 (2010): 47–54.

16. L. C. Dolan, S. M. Potter, and G. A. Burdock, Evidence-based review on the effect of normal dietary consumption of fructose on blood lipids and body weight of overweight and obese individuals, *Critical Reviews in Food Science and Nutrition* 50 (2010): 889–918.

17. F. B. Hu and V. S. Malik, Sugar-sweetened beverages and risk of obesity and type 2 diabetes: Epidemiologic evidence, *Physiology and Behavior* 100 (2010): 47–54; R. K. Johnson and coauthors, Dietary sugars intake and cardiovascular health: A scientific statement from the American Heart Association, *Circulation* 120 (2009): 1011–1020.

18. Position of the Academy of Nutrition and Dietetics: Oral health and nutrition, *Journal of the Academy of Nutrition and Dietetics* 113 (2013): 693–701.

19. US Department of Agriculture, Agricultural Research Service, Beltsville Human Nutrition Research Center, Food Surveys Research Group (Beltsville, MD) and US Department of Health and Human Services, Centers for Disease Control and Prevention, National Center for Health Statistics (Hyattsville, MD), *What We Eat in America*, NHANES 2007–2008, www.ars.usda.gov/ba/bhnrc/fsrg, published 2010.

20. Committee on Dietary Reference Intakes, *Dietary Reference Intakes: Energy, Carbohydrate, Fiber, Fat, Fatty Acids, Cholesterol, Protein, and Amino Acids*, Washington, D.C.: National Academies Press, 2005.

21. B. P. Marriott and coauthors, Intake of added sugars and selected nutrients in the United States, National Health and Nutrition Examination Survey (NHANES) 2003–2006, *Critical Reviews in Food Science and Nutrition* 50 (2010): 228–258.

22. C. Gardner and coauthors, Nonnutritive sweeteners: Current use and health perspectives, A scientific statement from the American Heart Association and the American Diabetes Association, *Diabetes Care* 35 (2012): 1798–1808.

23. Position of the Academy of Nutrition and Dietetics: Use of nutritive and nonnutritive sweeteners, *Journal of the Academy of Nutrition and Dietetics* 112 (2012): 739–758.

24. M. U. Jakobsen and coauthors, Intake of carbohydrates compared with intake of saturated fatty acids and risk of myocardial infarction: Importance of the glycemic index, *American Journal of Clinical Nutrition* 91 (2010): 1764–1768.

25. R. A. Othman, M. H. Moghadasian, and P. J. H. Jones, Cholesterol lowering effects of oat β-glucan, *Nutrition Reviews* 69 (2011): 299–309.

26. S. Tarleton and J. K. Dibaise, Low-residue diet in diverticular disease: Putting an end to a myth, *Nutrition in Clinical Practice* 26 (2011): 137–142.

27. C. Ünlü and coauthors, A systematic review of high-fibre dietary therapy in diverticular disease, International Journal of Colorectal Disease 27 (2012): 419–427; L. L. Strate, Lifestyle factors and the course of diverticular disease, *Digestive Diseases* 30 (2012): 35–45; A. F. Peery and coauthors, A high-fiber diet does not protect against asymptomatic diverticulosis, *Gastroenterology* 142 (2012): 266–272; F. L. Crowe and coauthors, Diet and risk of diverticular disease in Oxford cohort of European Prospective Investigation into Cancer and Nutrition (EPIC): Prospective study of British vegetarians and non-vegetarians, *British Medical Journal* 343 (2011): d4131; J. E. Ravikoff and J. R. Korzenik, The role of fiber in diverticular disease, *Journal of Clinical Gastroenterology* 45 (2011): S7–S11.

28. M. H. Pan and coauthors, Molecular mechanisms for chemoprevention of colorectal cancer by natural dietary compounds, *Molecular Nutrition and Food Research* 55 (2011): 32–45.

29. H. Du and coauthors, Dietary fiber and subsequent changes in body weight and waist circumference in European men and women, *American Journal of Clinical Nutrition* 91 (2010): 329–336.

30. D. E. King, A. G. Mainous, and C. A. Lambourne, Trends in dietary fiber intake in the United States, 1999–2008, *Journal of the Academy of Nutrition and Dietetics* 112 (2012): 642–648.

31. T. Wirström and coauthors, Consumption of whole grain reduces risk of deteriorating glucose tolerance, including progression to prediabetes, *American Journal of Clinical Nutrition* 97 (2013): 179–187; E. Q. Ye and coauthors, Greater whole-grain intake is associated with lower risk of type 2 diabetes, cardiovascular disease, and weight gain, *Journal of Nutrition* 142 (2012): 1304–1313; M. Lefevre and S. Jonnalagadda, Effect of whole grains on markers of subclinical inflammation, *Nutrition Reviews* 70 (2012): 387–396; I. Y. Hur and M. Reicks, Relationship between whole-grain intake, chronic disease risk indicators, and weight status among adolescents in the National Health and Nutrition Examination Survey, 1999–2004, *Journal of the Academy of Nutrition and Dietetics* 112 (2012): 46–55; S. S. Jonnalagadda and coauthors, Putting the whole grain puzzle together: Health benefits associated with whole grains—Summary of American Society for Nutrition 2010 Satellite Symposium, *Journal of Nutrition* 141 (2011): 1011S–1022S; C. E. O'Neil and coauthors, Consumption of whole grains is associated with improved diet quality and nutrient intake in children and adolescents: The National Health and Nutrition Examination Survey 1999–2004, *Public Health Nutrition* 14 (2011): 347–355; C. E. O'Neil and coauthors, Whole-grain consumption is associated with diet quality and nutrient intake in adults: The National Health and Nutrition Examination Survey, 1999–2004, *Journal of the American Dietetic Association* 110 (2010): 1461–1468.

HIGHLIGHT > 4
Carbs, kCalories, and Controversies

> **LEARN IT** Summarize the key scientific evidence behind some of the current controversies surrounding carbohydrates and their kcalories.

Carbohydrate-rich foods are easy to like. Mashed potatoes, warm muffins, blueberry pancakes, freshly baked bread, and tasty rice and pasta dishes tempt most people's palates. In recent years, such homey foods have been blamed for causing weight gain and harming health. Popular writers have persuaded consumers that carbohydrates are "bad." In contrast, the *Dietary Guidelines* urge people to consume plenty of fruits, vegetables, legumes, and whole grains—all carbohydrate-rich foods.

Do carbohydrate-rich foods cause obesity and related health problems? Should people "cut carbs" to lose weight and protect their health? Many popular diet books espouse a carbohydrate-restricted or carbohydrate-modified diet. Some claim that all or some types of carbohydrates are bad. Some go so far as to equate carbohydrates with toxic poisons or addictive drugs. "Bad" carbohydrates—such as sugar, white flour, and potatoes—are considered evil because they are absorbed easily and raise blood glucose. The pancreas then responds by secreting insulin—and insulin is touted as the real villain responsible for our nation's obesity epidemic. Whether restricting overall carbohydrate intake or replacing certain "bad" carbohydrates with "good" carbohydrates, many of these popular diets tend to distort the facts. This highlight examines the scientific evidence behind some of the current controversies surrounding carbohydrates and their kcalories.

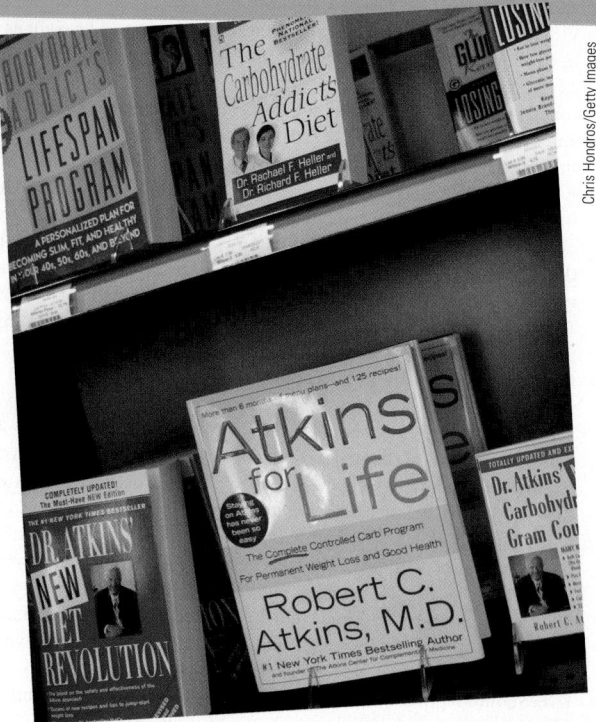

Carbohydrates' kCalorie Contributions

The incidence of obesity in the United States has risen dramatically over the past several decades.[1] Popular diet books often blame carbohydrates for this increase in obesity. One way researchers can explore whether the amount of carbohydrate in the diet contributes to increases in body weight over time is by reviewing national food intake survey records, such as NHANES (introduced in Chapter 1). Figure H4-1 presents a summary of energy nutrient data over the past three decades. Since the 1970s, kcalories from carbohydrates increased from 42 percent to 51 percent today.[2] At the same time, kcalories from fat dropped from 41 percent to 34 percent. The percentage of protein intake stayed about the same.

A closer look at the data reveals that, as the percentage of kcalories from the three energy nutrients shifted slightly, total daily energy intake increased significantly. In general, as food became more readily available in this nation, consumers began to eat more than they had in the past. Since the 1970s, total energy intakes have increased by about 200 to 300 kcalories a day (see Figure H4-2).[3] All of the increase in kcalories came from an increase in carbohydrate kcalories. At the same time, most people were not active enough to use up those extra

kcalories; in fact, activity levels declined.[4] Consequently, the average body weight for adults increased over these decades by about 25 to 30 pounds (see Figure H4-3).

Might too many carbohydrates in the diet be to blame for weight gains? Interestingly, epidemiological studies find an *inverse* relationship between carbohydrate intake and body weight. Those with the highest carbohydrate intake have the lowest body weight and vice versa. Dietary fiber, which favors a healthy body weight, explains some but not all of this relationship.

Might a low-carbohydrate diet support weight losses? For the most part, weight loss is similar for people following either a low-carbohydrate

> **FIGURE H4-1 Energy Nutrients over Time**

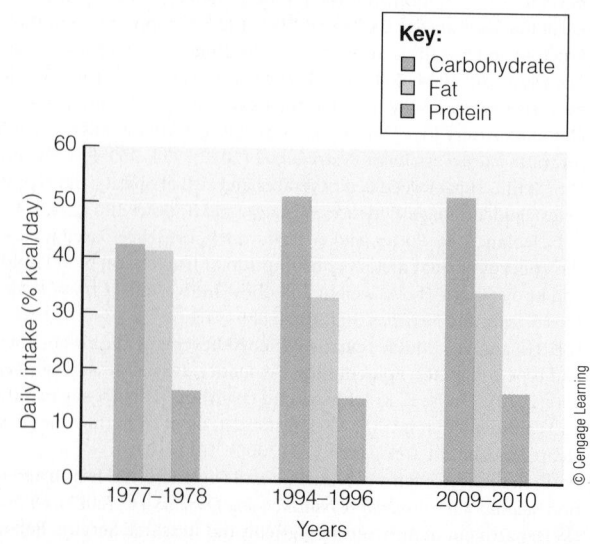

128

> **FIGURE H4-2**   **Daily Energy Intake over Time**

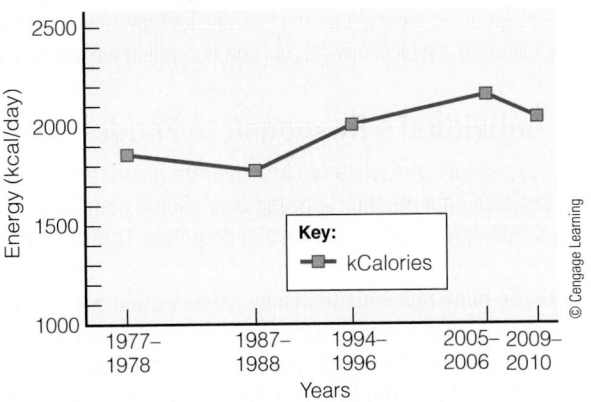

> **FIGURE H4-3**   **Increases in Adult Body Weight over Time**

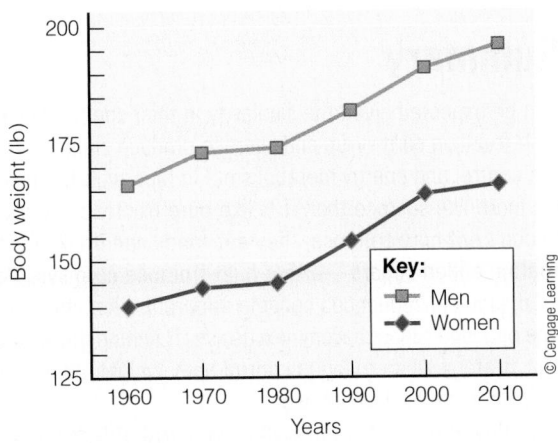

diet or a low-fat diet.[5] This is an important point. Weight losses reflect restricted kcalories—not the proportion of energy nutrients in the diet. Any diet can produce weight loss, at least temporarily, if energy intake is restricted. And most weight-loss diets also restrict sugars.

Sugars' Share in the Problem

As the chapter mentioned, the use of high-fructose corn syrup sweetener parallels unprecedented increases in the incidence of obesity, but does it mean that the increasing sugar intakes are responsible for the increase in body fat and its associated health problems?[6] Excess sugar in the diet is associated with more fat on the body. When eaten in excess of need, energy from added sugars contributes to body fat stores, just as excess energy from other sources does. Added sugars provide excess energy, raising the risk of weight gain. When total energy intake is controlled, however, *moderate* amounts of sugar do not *cause* obesity.[7] In other words, foods containing added sugars are no more likely to contribute to weight gain than any other foods.[8] Yet moderating sugar intake can be a challenge.

The liquid form of sugar in soft drinks makes it especially easy to overconsume kcalories. Swallowing liquid kcalories requires little effort. The sugar kcalories of sweet beverages also cost less than many other energy sources, and they are widely available. Also, beverages are energy-dense, providing more than 150 kcalories per 12-ounce can, and many people drink several cans a day. Drinking these beverages seems to correlate with a higher energy intake from foods as well, thus raising energy intake in two ways.[9] The convenience, economy, availability, and flavors of sugary foods and beverages make overconsumption especially likely.

Limiting selections of foods and beverages high in added sugars can be an effective weight-loss strategy, especially for people whose excess kcalories come primarily from added sugars. Replacing sodas with water every day, for example, can help a person lose a pound (or at least not gain a pound) in 1 month.[10] That may not sound like much, but it adds up to more than 10 pounds a year, for very little effort.

Cravings and Addictions

Some people describe themselves as having "carbohydrate cravings" or being "sugar addicts." One frequently noted theory is that people seek carbohydrates as a way to increase their levels of the brain neurotransmitter serotonin, which elevates mood. Interestingly, when those with self-described carbohydrate cravings indulge, they tend to eat more of everything; the percentage of energy from carbohydrates remains unchanged.

One reasonable explanation for the carbohydrate cravings that some people experience involves the self-imposed labeling of a food as both "good" and "bad"—that is, one that is desirable but should be eaten with restraint. Restricting intake heightens the desire further (a "craving"). Then "addiction" is used to explain why resisting the food is so difficult and, sometimes, even impossible. Carbohydrates, and sugars more specifically, are not addictive in the same ways that drugs are, but they share some of the same biological and psychological systems that are involved in rewards and self-control. Highlight 8 includes more details on the concept of food addictions.

Fructose Metabolism

Unlike glucose, which is metabolized by all the body's cells, fructose is metabolized primarily in the liver. When the diet delivers high intakes of added sugars (which is half fructose), the liver handles the excess by making fat. This fat is either retained in the liver or transported out, raising blood lipids and increasing fat stores—all risk factors for chronic diseases.

Appetite Control

Recall from Chapter 4 that glucose stimulates the release of insulin from the pancreas. Insulin, in turn, sets off a sequence of hormonal actions that suppress the appetite. Fructose, in contrast, does not stimulate the release of insulin, and therefore does not suppress appetite.

Whether the meal or snack is liquid or solid may also affect appetite. Even when kcaloric intake is the same, a fresh apple suppresses appetite more than apple juice. Consequently, beverages can influence weight gains both by providing energy and by not satisfying hunger.

Insulin's Response

Several popular diet books hold insulin responsible for the obesity problem and a low-glycemic diet as the weight-loss solution. Yet, among nutrition researchers, controversy continues to surround the questions of whether insulin promotes weight gain or a low-glycemic diet fosters weight loss.

Recall that just after a meal, blood glucose rises and insulin responds. How high insulin levels surge may influence whether the body stores or uses its glucose and fat supplies. What does insulin do? Among its roles, insulin facilitates the transport of glucose into the cells, the storage of fatty acids as fat, and the synthesis of protein. It is an anabolic hormone that builds and stores. True—but there's more to the story. Insulin is only one of many factors involved in the body's metabolism of nutrients and regulation of body weight.

Most importantly, insulin is critical to maintaining health, as any person with type 1 diabetes can attest. Insulin causes problems only when a person develops insulin resistance—that is, when the body's cells do not respond to the large quantities of insulin that the pancreas continues to pump out in an effort to get a response. Insulin resistance is a major health problem—but it is not caused by carbohydrate, or by protein, or by fat. It most often results from being obese. Importantly, when a person loses weight, insulin response usually improves.

The Glycemic Index and Body Weight

As Chapter 4's discussion of the glycemic index explained, the glycemic effect of a particular food varies. The glycemic effect of a food depends on how the food is ripened, processed, and cooked; the time of day the food is eaten; the other foods eaten with it; and the presence or absence of certain diseases such as type 2 diabetes in the person eating the food. All these factors influence a food's glycemic index, yet diet books often mislead people by claiming that each food has a specific glycemic index.

Even if a true glycemic index is known, what is the relationship between a diet's glycemic index and body weight? In general, studies find that diets with a high glycemic index are positively associated with body weight.

Might a low-glycemic diet foster weight loss? In general, research examining the use of low glycemic diets for weight loss finds inconsistent results.[11] Still, a low-glycemic diet may offer other advantages. A low-glycemic meal seems to curb appetite and limit energy intake of the next meal. Low-glycemic diets are also more likely to be rich in nutrients and fiber than high-glycemic diets.

Clearly, if kcalories are low, obese people on either a low-glycemic diet or a traditional low-fat diet can lose weight. Overweight people can lose as much or more weight by emphasizing low-glycemic foods as they can by following a typical low-fat, portion-controlled weight-loss diet.

The Individual's Response to Foods

The body's insulin response to carbohydrate depends not only on a food, but also on a person's metabolism.[12] Some people react to dietary carbohydrate with a low insulin response. Others have a high insulin response. How energy is stored after a meal depends in part on how the body responds to insulin. After eating a high-carbohydrate meal, normal-weight people who are insulin resistant tend to synthesize about half as much glycogen in muscles and make about twice as much fat in the liver as people who are insulin sensitive. Some research suggests that restricting carbohydrate intake may improve glucose control, insulin response, and blood lipids.

In Summary

As might be expected given the similarity in their chemical composition, high-fructose corn syrup and sucrose produce similar effects in appetite control and energy metabolism.[13] In fact, high-fructose corn syrup is more like sucrose than it is like pure fructose. Importantly, people don't eat pure fructose; they eat foods and drink beverages that contain added sugars—either high-fructose corn syrup or sucrose. Adverse consequences become apparent when intakes of either type of added sugars become excessive.[14] Limiting these sugars is a helpful strategy when trying to control body weight, but restricting all carbohydrates would be unwise.

The quality of the diet suffers when carbohydrates are restricted. Without fruits, vegetables, and whole grains, low-carbohydrate diets lack not only carbohydrate, but fiber, vitamins, minerals, and phytochemicals as well—all dietary factors protective against disease. The DRI recommends that carbohydrates contribute between 45 and 65 percent of daily energy intake. Intakes within this range can support healthy body weight and do not contribute to obesity—when added sugar intake is moderate and total energy intake is appropriate. Similarly, added sugars increase energy intake, but need not contribute to obesity—when added sugar intake is moderate and total energy intake is appropriate. When choosing carbohydrates, emphasize a variety of naturally occurring carbohydrates—such as whole grains, legumes, vegetables, and fruits—and limit foods and beverages with added sugars.

CRITICAL THINKING QUESTIONS

A. How are sugars, starches, and fibers related to weight gains and losses?
B. Your mom wants to lose 20 pounds before her high school reunion next month. She has done some research on the Internet, and has discovered an easy diet that seems to offer great success. The basis of the diet is that a person needs to eat "slow carb" foods and avoid sweets (including fruits) and starches. All white foods—such as potatoes, rice, pastas, tofu, breads, cereals, and milk—are banned and two to three workouts each week are encouraged. What evidence supports or contradicts a "slow carb" diet for weight loss?

REFERENCES

1. K. M. Flegal and coauthors, Prevalence and trends in obesity among US adults, 1999–2008, *Journal of the American Medical Association* 303 (2010): 235–241.
2. US Department of Agriculture, Agricultural Research Service, Beltsville Human Nutrition Research Center, Food Surveys Research Group (Beltsville, MD) and US Department of Health and Human Services, Centers for Disease Control and Prevention, National Center for Health Statistics (Hyattsville, MD), *What We Eat in America,* NHANES 2009–2010, www.ars.usda.gov/, published July 2012.
3. *What We Eat in America,* 2012.
4. US Department of Health and Human Services, *2008 Physical Activity Guidelines for Americans* Summary, www.health.gov/PAGuidelines /guidelines/summary.aspx.
5. G. D. Foster and coauthors, Weight and metabolic outcomes after 2 years on a low-carbohydrate versus low-fat diet: A randomized trial, *Annals of Internal Medicine* 153 (2010): 147–157.
6. R. D. Mattes and coauthors, Nutritively sweetened beverage consumption and body weight: A systematic review and meta-analysis of randomized experiments, *Obesity Reviews* 12 (2011): 346–365.
7. L. T. Morenga, S. Mallard, J. Mann, Dietary sugars and body weight: Systematic review and meta-analyses of randomised controlled trials and cohort studies, *British Medical Journal* 346 (2013): e7492; J. L. Sievenpiper and coauthors, Effect of fructose on body weight in controlled feeding trials: A systematic review and meta-analysis, *Annals of Internal Medicine* 156 (2012): 291–304.
8. R. D. Mattes and coauthors, Nutritively sweetened beverage consumption and body weight: A systematic review and meta-analysis of randomized experiments, *Obesity Reviews* 12 (2011): 346–365.
9. A. K. Kant, B. I. Graubard, and R. D. Mattes, Association of food form with self-reported 24-h energy intake and meal patterns in US adults: NHANES 2003–2008, *American Journal of Clinical Nutrition* 96 (2012): 1369–1378.
10. D. F. Tate and coauthors, Replacing caloric beverages with water or diet beverages for weight loss in adults: Main results of the Choose Healthy Options Consciously Everyday (CHOICE) randomized clinical trial, *American Journal of Clinical Nutrition* 95 (2012): 555–563.
11. A. Esfahani and coauthors, The application of the glycemic index and glycemic load in weight loss: A review of the clinical evidence, *International Union of Biochemistry and Molecular Biology* 63 (2011): 7–13.
12. W. J. Whelan and coauthors, The glycemic response is a personal attribute, *International Union of Biochemistry and Molecular Biology* 62 (2010): 637–641.
13. K. L. Stanhope and P. J. Havel, Fructose consumption: Recent results and their potential implications, *Annals of the New York Academy of Sciences* 1190 (2010): 15–24.
14. L. Tappy and K. Lê, Metabolic effects of fructose and the worldwide increase in obesity, *Physiology Reviews* 90 (2010): 23–46.

5

The Lipids: Triglycerides, Phospholipids, and Sterols

Nutrition in Your Life

Most likely, you know what you don't like about body fat, but do you appreciate how it insulates you against the cold or powers your hike around a lake? And what about food fat? You're right to credit fat for providing the delicious flavors and aromas of buttered popcorn and fried chicken—and to criticize it for contributing to the weight gain and heart disease so common today. The challenge is to strike a healthy balance of enjoying some fat, but not too much. Learning which kinds of fats are beneficial and which are most harmful will also help you make wise decisions. In the Nutrition Portfolio at the end of this chapter, you can examine whether your current fat choices are meeting dietary goals.

No doubt you have heard that fats can contribute to the development of several chronic diseases, but did you realize that some fats are also essential to good health? Most people are surprised to learn that fat has virtues and that a well-balanced diet needs at least a little fat. Getting enough fat is rarely a problem. At least traces of fat can be found in almost all foods. In our society of abundance, people are more likely to consume too much fat, or too much of some kinds of fat—with consequent health problems. Learning which kinds of fats are harmful or helpful is key to healthy diet planning.

Fat refers to the class of nutrients known as lipids. The lipid family includes triglycerides (fats and oils), phospholipids, and sterols. Triglycerides are most abundant, both in foods and in the body. The following sections describe the similarities and differences among the remarkably diverse members of the lipid family.[1]

5.1 The Chemist's View of Fatty Acids and Triglycerides

> **LEARN IT** Recognize the chemistry of fatty acids and triglycerides and differences between saturated and unsaturated fats.

Like carbohydrates, **lipids** are composed of carbon (C), hydrogen (H), and oxygen (O). Because lipids have many more carbons and hydrogens in proportion to their oxygens, they can supply more energy per gram than carbohydrates can (Chapter 7 provides details).

The many names and relationships in the lipid family can seem overwhelming—like meeting a friend's extended family for the first time. To ease the introductions, this chapter first presents each of the lipids from a chemist's point of view using both words and diagrams. Then the chapter follows the lipids through digestion and absorption and into the body to examine their roles in health and disease. For people who think more easily in words than in chemical symbols, this *preview* of the upcoming chemistry may be helpful:

1. Every triglyceride contains one molecule of glycerol and three fatty acids (basically, chains of carbon atoms).

2. Fatty acids may be 4 to 24 (even numbers of) carbons long, the 18-carbon ones being the most common in foods and especially noteworthy in nutrition.

3. Fatty acids may be saturated or unsaturated. Unsaturated fatty acids may have one or more points of unsaturation—that is, they may be *mono*unsaturated or *poly*unsaturated.

4. Of special importance in nutrition are the polyunsaturated fatty acids known as omega-3 fatty acids and omega-6 fatty acids.

5. The 18-carbon polyunsaturated fatty acids are linolenic acid (omega-3) and linoleic acid (omega-6). Both are essential fatty acids that the body cannot make. Each is the primary member of a family of longer-chain fatty acids that help to regulate blood pressure, blood clotting, and other body functions important to health.

The paragraphs, definitions, and diagrams that follow present this information again in much more detail.

Fatty Acids All **fatty acids** have the same basic structure—a chain of carbon and hydrogen atoms with an acid group (COOH) at one end and a methyl group (CH_3) at the other end. Fatty acids may differ from one another, however, in the length of their carbon chains and in the number and location of their double bonds, as the following paragraphs describe. (Fatty acids and related terms are defined in the accompanying glossary.)

The Length of the Carbon Chain Most naturally occurring fatty acids contain even numbers of carbons in their chains—up to 24 carbons in length. This discussion begins with the 18-carbon fatty acids, which are abundant in our food supply. Stearic acid is the simplest of the 18-carbon fatty acids; the bonds between its carbons are all alike:

Stearic acid, an 18-carbon saturated fatty acid

© Cengage Learning

As you can see, stearic acid is 18 carbons long, and each atom meets the rules of chemical bonding described in Figure 4-1 on p. 100. The following structure

lipids: a family of compounds that includes triglycerides, phospholipids, and sterols. Lipids are characterized by their insolubility in water. (Lipids also include the fat-soluble vitamins, described in Chapter 11.)

fatty acids: organic compounds composed of a carbon chain with hydrogens attached and an acid group (COOH) at one end and a methyl group (CH$_3$) at the other end.

monounsaturated fatty acid: a fatty acid that lacks two hydrogen atoms and has one double bond between carbons; abbreviated *MUFA*. Examples include palmitoleic acid and oleic acid. A *monounsaturated fat* is composed of triglycerides in which most of the fatty acids are monounsaturated.

- **mono** = one

point of unsaturation: the double bond of a fatty acid, where hydrogen atoms can easily be added to the structure.

polyunsaturated fatty acid: a fatty acid that lacks four or more hydrogen atoms and has two or more double bonds between carbons; abbreviated *PUFA*. Examples include linoleic acid (two double bonds) and linolenic acid (three double bonds). A *polyunsaturated fat* is composed of triglycerides in which most of the fatty acids are polyunsaturated.

- **poly** = many

saturated fatty acid: a fatty acid carrying the maximum possible number of hydrogen atoms—for example, stearic acid. A *saturated fat* is composed of triglycerides in which most of the fatty acids are saturated.

unsaturated fatty acid: a fatty acid that lacks hydrogen atoms and has at least one double bond between carbons (includes monounsaturated and polyunsaturated fatty acids). An *unsaturated fat* is composed of triglycerides in which most of the fatty acids are unsaturated.

also depicts stearic acid, but in a simpler way, with each "corner" on the zigzag line representing a carbon atom with two attached hydrogens:

Stearic acid (simplified structure)

© Cengage Learning

As mentioned, the carbon chains of fatty acids vary in length. The long-chain (12 to 24 carbons) fatty acids of meats, seafood, and vegetable oils are most common in the diet. Smaller amounts of medium-chain (6 to 10 carbons) and short-chain (fewer than 6 carbons) fatty acids also occur, primarily in dairy products. (Tables C-1 and C-2 in Appendix C provide the names, chain lengths, and sources of fatty acids commonly found in foods.)

The Number of Double Bonds Stearic acid (described and shown previously) is a **saturated fatty acid**. A saturated fatty acid is fully loaded with all its hydrogen atoms and contains only single bonds between its carbon atoms. If two hydrogens were missing from the middle of the carbon chain, the remaining structure might be:

An impossible chemical structure

© Cengage Learning

Notice that in the impossible chemical structure shown above, two of the carbons have only three bonds each. Such a compound cannot exist because every carbon must have four bonds. To satisfy this rule, the two carbons form a double bond:

Oleic acid, an 18-carbon monounsaturated fatty acid

© Cengage Learning

The same structure drawn more simply looks like this:*

Oleic acid (simplified structure)

© Cengage Learning

---

*Remember that each "corner" on the zigzag line represents a carbon atom with two attached hydrogens.

Although drawn straight here, the actual shape bends at the double bond. The double bond is a **point of unsaturation**. A fatty acid like this—with two hydrogens missing and a double bond—is an **unsaturated fatty acid**. This one is the 18-carbon **mono-unsaturated fatty acid** oleic acid, which is abundant in olive oil and canola oil.

A **polyunsaturated fatty acid** has two or more carbon-to-carbon double bonds. **Linoleic acid**, the 18-carbon fatty acid common in vegetable oils, lacks four hydrogens and has two double bonds:

Linoleic acid, an 18-carbon polyunsaturated fatty acid

Drawn more simply, linoleic acid looks like this (though the actual shape would bend at the double bonds):

Linoleic acid (simplified structure)

A fourth 18-carbon fatty acid is **linolenic acid,** which has three double bonds. Table 5-1 presents the 18-carbon fatty acids.

The Location of Double Bonds Fatty acids differ not only in the length of their chains and their degree of saturation, but also in the locations of their double bonds. Chemists identify polyunsaturated fatty acids by the position of the double bond closest to the methyl (CH_3) end of the carbon chain, which is described by an **omega** number. A polyunsaturated fatty acid with its closest double bond three carbons away from the methyl end is an **omega-3 fatty acid.** Similarly, an **omega-6 fatty acid** is a polyunsaturated fatty acid with its closest double bond six carbons away from the methyl end. Figure 5-1 compares two 18-carbon fatty acids— linolenic acid (an omega-3 fatty acid) and linoleic acid (an omega-6 fatty acid).

Monounsaturated fatty acids tend to belong to the omega-9 group, with their closest (and only) double bond nine carbons away from the methyl end. Oleic acid—the 18 carbon monounsaturated fatty acid common in olive oil mentioned earlier—is an omega-9 fatty acid. It is also the most predominant monounsaturated fatty acid in the diet.

Triglycerides Few fatty acids occur free in foods or in the body. Most often, they are incorporated into **triglycerides**—lipids composed of three fatty acids attached to a **glycerol**.* Figure 5-2 presents a glycerol molecule.

To make a triglyceride, a series of **condensation** reactions combine a hydrogen atom (H) from the glycerol and a hydroxyl (OH) group from a fatty acid, forming a molecule of water (H_2O) and leaving a bond between the two molecules (see the left side of Figure 5-3). Most triglycerides contain a mixture of more than one type of fatty acid (as shown on the right side of Figure 5-3).

Characteristics of Solid Fats and Oils The chemistry of a fatty acid— whether it is short or long, saturated or

linoleic (lin-oh-LAY-ick) **acid:** an essential fatty acid with 18 carbons and two double bonds.

linolenic (lin-oh-LEN-ick) **acid:** an essential fatty acid with 18 carbons and three double bonds.

omega: the last letter of the Greek alphabet (ω), used by chemists to refer to the position of the closest double bond to the methyl (CH_3) end of a fatty acid.

omega-3 fatty acid: a polyunsaturated fatty acid in which the closest double bond to the methyl (CH_3) end of the carbon chain is three carbons away.

omega-6 fatty acid: a polyunsaturated fatty acid in which the closest double bond to the methyl (CH_3) end of the carbon chain is six carbons away.

triglycerides (try-GLISS-er-rides): the chief form of fat in the diet and the major storage form of fat in the body; composed of a molecule of glycerol with three fatty acids attached; also called *triacylglycerols* (try-ay-seel-GLISS-er-ols).

• **tri** = three

• **glyceride** = of glycerol

glycerol (GLISS-er-ol): an alcohol composed of a three-carbon chain, which can serve as the backbone for a triglyceride.

condensation: a chemical reaction in which water is released as two molecules combine to form one larger product.

TABLE 5-1 18-Carbon Fatty Acids

Name	Number of Carbon Atoms	Number of Double Bonds	Saturation	Common Food Sources
Stearic acid	18	0	Saturated	Most animal fats
Oleic acid	18	1	Monounsaturated	Olive and canola oils
Linoleic acid	18	2	Polyunsaturated	Sunflower, safflower, corn, and soybean oils
Linolenic acid	18	3	Polyunsaturated	Soybean and canola oils, flaxseed, walnuts

NOTE: Chemists use a shorthand notation to describe fatty acids. The first number indicates the number of carbon atoms; the second, the number of the double bonds. For example, the notation for stearic acid is 18:0.

*Research scientists commonly use the term *triacylglycerols;* this book continues to use the more familiar term *triglycerides,* as do many other health and nutrition books and journals.

> FIGURE 5-1 Omega-3 and Omega-6 Fatty Acids Compared

The omega number indicates the position of the double bond closest to the methyl (CH$_3$) end. The fatty acids of an omega family may have different lengths and different numbers of double bonds, but the location of the double bond closest to the methyl end is the same in all of them. These structures are drawn linearly here to ease counting carbons and locating double bonds, but their shapes actually bend at the double bonds.

Linolenic acid, an 18-carbon, omega-3 fatty acid

Omega carbon

Methyl end

Acid end

Linoleic acid, an 18-carbon, omega-6 fatty acid

Omega carbon

Methyl end

Acid end

© Cengage Learning

unsaturated, with its closest double bond at carbon 3 or carbon 6—influences the characteristics of foods and the health of the body. A section later in this chapter explains how these features affect health; this section describes how the chemistry influences the **fats** and **oils** in foods.

Firmness The degree of unsaturation influences the firmness of fats at room temperature (see Figure 5-4 on p. 138). Generally speaking, most polyunsaturated vegetable oils are liquid at room temperature, and the more saturated animal fats are solid. Some oils—notably, cocoa butter, palm oil, palm kernel oil, and coconut oil—are saturated; they are firmer than most vegetable oils because of their saturation, but softer than most animal fats because of their shorter carbon chains (8 to 14 carbons long). Generally, the shorter the carbon chain, the softer the fat is at room temperature. Fatty acid compositions of selected fats and oils are shown in Figure 5-5 (p. 138), and Appendix H provides the fat and fatty acid contents of many other foods.

> FIGURE 5-2 Glycerol

When glycerol is free, an OH group is attached to each carbon. When glycerol is part of a triglyceride, each carbon is attached to a fatty acid (as shown in the next figure).

© Cengage Learning

fats: lipids that are solid at room temperature (77°F, or 25°C).

oils: lipids that are liquid at room temperature (77°F, or 25°C).

> FIGURE 5-3 Condensation of Glycerol and Fatty Acids to Form a Triglyceride

To make a triglyceride, three fatty acids attach to glycerol in condensation reactions.

Glycerol + three fatty acids ⟶ Triglyceride + three water molecules

© Cengage Learning

An H atom from glycerol and an OH group from a fatty acid combine to create water, leaving the O on the glycerol and the C at the acid end of each fatty acid to form a bond.

Three fatty acids attached to a glycerol form a triglyceride and yield water. In this example, the triglyceride includes (from top to bottom) a saturated fatty acid, a monounsaturated fatty acid, and a polyunsaturated fatty acid.

> FIGURE 5-4   **Diagram of Saturated and Unsaturated Fatty Acids Compared**

Saturated fatty acids tend to stack together. Consequently, saturated fats tend to be solid (or more firm) at room temperature.

Double bond

This mixture of saturated and unsaturated fatty acids does not stack neatly because unsaturated fatty acids bend at the double bond(s). Consequently, unsaturated fats tend to be liquid (or less firm) at room temperature.

© Cengage Learning

oxidation (OKS-ee-day-shun): the process of a substance combining with oxygen; oxidation reactions involve the loss of electrons.

Stability The degree of unsaturation also influences stability. All fats become spoiled when exposed to oxygen. The **oxidation** of fats produces a variety of compounds that smell and taste rancid. (Other types of spoilage can occur due to microbial growth.) Polyunsaturated fats spoil most readily because their double

> FIGURE 5-5   **Fatty Acid Composition of Common Food Fats**

Most fats are a mixture of saturated, monounsaturated, and polyunsaturated fatty acids.

Key:

- ■ Saturated fatty acids
- □ Monounsaturated fatty acids
- ■ Polyunsaturated, omega-6 fatty acids
- ■ Polyunsaturated, omega-3 fatty acids

Animal fats and the tropical oils of coconut and palm contain mostly saturated fatty acids.

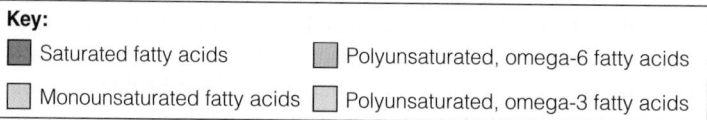

Coconut oil
Butter
Beef tallow (beef fat)
Palm oil
Lard (pork fat)
Chicken fat

Some vegetable oils, such as olive and canola, are rich in monounsaturated fatty acids.

Olive oil
Canola oil
Peanut oil

Many vegetable oils are rich in omega-6 polyunsaturated fatty acids.

Safflower oil[a]
Sunflower oil
Corn oil
Soybean oil
Walnut oil
Cottonseed oil

Only a few oils provide significant omega-3 polyunsaturated fatty acids.

Flaxseed oil
Fish oil (salmon)

[a]Salad or cooking type over 70% linoleic acid.

© Cengage Learning

At room temperature, saturated fats (such as those commonly found in butter and other animal fats) are solid, whereas unsaturated fats (such as those found in vegetable oils) are usually liquid.

© Polara Studios, Inc.

> FIGURE 5-6   **Hydrogenation**

Double bonds carry a slightly negative charge and readily accept positively charged hydrogen atoms, creating a saturated fatty acid. Most often, fat is *partially* hydrogenated, creating a *trans*-fatty acid (shown in the next figure).

Polyunsaturated fatty acid ⟶ Hydrogenated (saturated) fatty acid

© Cengage Learning 2013

bonds are unstable; monounsaturated fats are slightly less susceptible. Saturated fats are most resistant to oxidation and thus least likely to become rancid.

Manufacturers can protect fat-containing products against rancidity in three ways—none of which are perfect. First, products may be sealed in air-tight, non-metallic containers, protected from light, and refrigerated—an expensive and inconvenient storage system. Second, manufacturers may add **antioxidants** to compete for the oxygen and thus protect the oil (examples are the additives BHA and BHT and vitamin E).* The advantages and disadvantages of antioxidant additives in food processing are presented in Chapter 19. Third, products may undergo a process known as hydrogenation.

Hydrogenation During **hydrogenation,** some or all of the points of unsaturation are saturated by adding hydrogen molecules. Hydrogenation offers two advantages. First, it protects against oxidation (thereby prolonging shelf life) by making polyunsaturated fats more saturated. Second, it alters the texture of foods by making liquid vegetable oils more solid (as in margarine and shortening). Hydrogenated fats improve the texture of foods, making margarines spreadable, pie crusts flaky, and puddings creamy.

Figure 5-6 illustrates the *total* hydrogenation of a polyunsaturated fatty acid to a saturated fatty acid. Total hydrogenation rarely occurs during food processing. Most often, a fat is *partially* hydrogenated, and some of the double bonds that remain after processing change their configuration from *cis* to *trans.*

***Trans*-Fatty Acids** In nature, most double bonds are *cis*—meaning that the hydrogens next to the double bonds are on the same side of the carbon chain. Only a few fatty acids (notably a small percentage of those found in milk and meat products) naturally occur as ***trans*-fatty acids**—meaning that the hydrogens next to the double bonds are on opposite sides of the carbon chain (see Figure 5-7). In the body, *trans*-fatty acids behave more like saturated fats, increasing blood cholesterol and the risk of heart disease (as a later section describes).[2]

> FIGURE 5-7   ***Cis-* and *Trans*-Fatty Acids Compared**

This example compares the *cis* configuration for an 18-carbon monounsaturated fatty acid (oleic acid) with its corresponding *trans* configuration (elaidic acid).

cis-fatty acid

A *cis*-fatty acid has its hydrogens on the same side of the double bond; *cis* molecules bend into a U-like formation. Most naturally occuring unsaturated fatty acids in foods are *cis.*

trans-fatty acid

A *trans*-fatty acid has its hydrogens on the opposite sides of the double bond; *trans* molecules are more linear. The *trans* form typically occurs in partially hydrogenated foods when hydrogen atoms shift around some double bonds and change the configuration from *cis* to *trans.*

© Cengage Learning

antioxidants: as a food additive, preservatives that delay or prevent rancidity of fats in foods and other damage to food caused by oxygen.

hydrogenation (HIGH-dro-jen-AY-shun or high-DROJ-eh-NAY-shun): a chemical process by which hydrogens are added to monounsaturated or polyunsaturated fatty acids to reduce the number of double bonds, making the fats more saturated (solid) and more resistant to oxidation (protecting against rancidity). Hydrogenation produces *trans*-fatty acids.

cis: on the near side of; refers to a chemical configuration in which the hydrogen atoms are located on the same side of a double bond.

trans: on the other side of; refers to a chemical configuration in which the hydrogen atoms are located on opposite sides of a double bond.

***trans*-fatty acids:** fatty acids with hydrogens on opposite sides of the double bond.

---

*BHA is butylated hydroxyanisole; BHT is butylated hydroxytoluene.

Without help from emulsifiers, fats and water don't mix.

© Matthew Farruggio

conjugated linoleic acids: several fatty acids that have the same chemical formula as linoleic acid (18 carbons, two double bonds) but with different configurations (the double bonds occur on adjacent carbons).

phospholipid (FOS-foe-LIP-id): a compound similar to a triglyceride but having a phosphate and choline (or another nitrogen-containing compound) in place of one of the fatty acids.

lecithin (LESS-uh-thin): one of the phospholipids. Lecithin acts as an emulsifier to combine water-soluble and fat-soluble ingredients that do not ordinarily mix, such as water and oil.

choline (KOH-leen): a nitrogen-containing compound found in foods and made in the body from the amino acid methionine. Choline is part of the phospholipid lecithin and the neurotransmitter acetylcholine.

hydrophobic (high-dro-FOE-bick): water-fearing, or non-water-soluble, substances; also known as *lipophilic* (fat loving).

hydrophilic (high-dro-FIL-ick): water-loving, or water-soluble, substances.

emulsifier: a substance with both water-soluble and fat-soluble portions that promote the mixing of oils and fats in watery solutions.

sterols (STARE-ols or STEER-ols): compounds containing a four-ring carbon structure with side chains attached.

> **FIGURE 5-8   Lecithin**

Lecithin is similar to a triglyceride but contains only two fatty acids. The third position is occupied by a phosphate group and a molecule of choline. Other phospholipids have different fatty acids and different groups attached to phosphate.

From 2 fatty acids

The plus charge on the N is balanced by a negative ion—usually chloride.

From choline

From glycerol   From phosphate

© Cengage Learning

Some research suggests that both naturally occurring and commercially created *trans* fats change blood lipids similarly; other research suggests that the negative effects are specific to only the commercial *trans* fats.[3] In any case, the important distinction is that a relatively small amount of *trans*-fat in the diet comes from natural sources.* At current levels of consumption, natural *trans* fats have little, if any, effect on blood lipids. Some naturally occurring *trans*-fatty acids, known as **conjugated linoleic acids,** may even have health benefits.[4] Conjugated linoleic acids are not counted as *trans* fats on food labels.

> **REVIEW IT** Recognize the chemistry of fatty acids and triglycerides and differences between saturated and unsaturated fats.

The predominant lipids both in foods and in the body are triglycerides: a molecule of glycerol with three fatty acids attached. Fatty acids vary in the length of their carbon chains, their degrees of unsaturation (number of double bonds), and the location of their double bond(s). Those that are fully loaded with hydrogens are saturated; those that are missing hydrogens and therefore have double bonds are unsaturated (monounsaturated or polyunsaturated). The vast majority of triglycerides contain more than one type of fatty acid. Fatty acid saturation affects fats' physical characteristics and storage properties. Hydrogenation, which converts polyunsaturated fats to saturated fats, protects fats from oxidation and alters the texture by making liquid vegetable oils more solid. In the process, hydrogenation creates *trans*-fatty acids that damage health in ways similar to those of saturated fatty acids.

5.2 The Chemist's View of Phospholipids and Sterols

> **LEARN IT** Describe the chemistry, food sources, and roles of phospholipids and sterols.

The preceding pages have been devoted to one of the classes of lipids, the triglycerides, and their component parts, glycerol and the fatty acids. The other lipids, the phospholipids and sterols, make up only 5 percent of the lipids in the diet.

Phospholipids The best-known **phospholipid** is **lecithin** (see Figure 5-8). Notice that lecithin has one glycerol with two of its three attachment sites occupied by fatty acids like those in triglycerides. The third site is occupied by a phosphate group and a molecule of **choline.** The **hydrophobic** fatty acids make phospholipids soluble in fat; the **hydrophilic** phosphate group allows them to dissolve in water. Such versatility enables the food industry to use phospholipids as an **emulsifier** to mix fats with water in such products as mayonnaise, salad dressings, and candy bars.

Phospholipids in Foods In addition to the phospholipids used by the food industry as emulsifiers, phospholipids are also found naturally in foods. The richest food sources of lecithin are eggs, liver, soybeans, wheat germ, and peanuts.

Roles of Phospholipids Lecithin and other phospholipids are constituents of cell membranes (see Figure 5-9). Because phospholipids are soluble in both water and fat, they can help fat-soluble substances, including vitamins and hormones, to pass easily in and out of cells. Phospholipids also act as emulsifiers in the body, helping to keep fats suspended in the blood and body fluids.

Sterols In addition to triglycerides and phospholipids, the lipids include the **sterols,** compounds with a

---

*For example, most dairy products contain less than 0.5 gram of naturally occurring *trans* fat per serving.

multiple-ring structure.* The most well-known sterol is **cholesterol;** Figure 5-10 shows its chemical structure.

Sterols in Foods Foods derived from both plants and animals contain sterols, but only those from animals contain significant amounts of cholesterol—meats, eggs, seafood, poultry, and dairy products. Some people, confused about the distinction between dietary cholesterol and blood cholesterol, have asked which foods contain the "good" cholesterol. "Good" cholesterol is not a type of cholesterol found in foods, but it refers to the way the body transports cholesterol in the blood, as explained in a later section of this chapter.

Sterols other than cholesterol are naturally found in plants. Being structurally similar to cholesterol, plant sterols interfere with cholesterol absorption. By inhibiting cholesterol absorption, a diet rich in plant sterols lowers blood cholesterol levels. Food manufacturers have fortified foods such as margarine with plant sterols, creating a functional food that helps to reduce blood cholesterol.

Roles of Sterols Many vitally important body compounds are sterols. Among them are bile acids, the sex hormones (such as testosterone, androgen, and estrogen), the adrenal hormones (such as cortisol, cortisone, and aldosterone), and vitamin D, as well as cholesterol itself. Cholesterol in the body can serve as the starting material for the synthesis of these compounds or as a structural component of cell membranes; more than 90 percent of all the body's cholesterol is found in the cells. Despite common misconceptions, cholesterol is not a villain lurking in some evil foods—it is a compound the body makes and uses. The chemical structure is the same, but cholesterol that is made in the body is referred to as **endogenous,** whereas cholesterol from outside the body (from foods) is referred to as **exogenous.** Right now, as you read, your liver is manufacturing cholesterol from fragments of carbohydrate, protein, and fat. In fact, the liver makes about 800 to 1500 milligrams of cholesterol per day, thus contributing much more to the body's total than does the diet. For perspective, the Daily Value on food labels for cholesterol is 300 milligrams per day.

Cholesterol's harmful effects in the body occur when it accumulates in the artery walls and contributes to the formation of **plaque.** These plaque deposits lead to **atherosclerosis,** a disease that causes heart attacks and strokes. Chapter 18 provides many more details.

> **REVIEW IT** Describe the chemistry, food sources, and roles of phospholipids and sterols.

Phospholipids, including lecithin, have a unique chemical structure that allows them to be soluble in both water and fat. The food industry uses phospholipids as emulsifiers, and in the body, phospholipids are part of cell membranes. Sterols have a multiple-ring structure that differs from the structure of other lipids. In the body, sterols include cholesterol, bile, vitamin D, and some hormones. Animal-derived foods are rich sources of cholesterol. Table 5-2 summarizes the lipid family of compounds.

> **FIGURE 5-9** **Phospholipids of a Cell Membrane**

A cell membrane is made of phospholipids assembled into an orderly formation called a bilayer. The fatty acid "tails" orient themselves away from the watery fluid inside and outside of the cell. The glycerol and phosphate "heads" are attracted to the watery fluid.

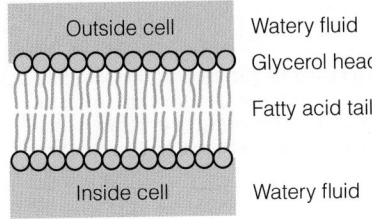

© Cengage Learning

> **FIGURE 5-10** **Cholesterol**

Notice how different cholesterol is from the triglycerides and phospholipids. The fat-soluble vitamin D is synthesized from cholesterol; notice the many structural similarities. The only difference is that cholesterol has a closed ring (highlighted in red), whereas vitamin D's is open, accounting for its vitamin activity.

Cholesterol

Vitamin D₃

© Cengage Learning

TABLE 5-2 **The Lipid Family**

Triglycerides

- 1 Glycerol (per triglyceride) and
- 3 Fatty acids (per triglyceride); depending on the number of double bonds, fatty acids may be:
  - *Saturated* (no double bonds)
  - *Monounsaturated* (one double bond)
  - *Polyunsaturated* (more than one double bond); depending on the location of the double bonds, polyunsaturated fatty acids may be:
    - *Omega-3* (double bond closest to methyl end is 3 carbons away)
    - *Omega-6* (double bond closest to methyl end is 6 carbons away)

Phospholipids (such as lecithin)

Sterols (such as cholesterol)

© Cengage Learning

cholesterol (koh-LESS-ter-ol): one of the sterols containing a four-ring carbon structure with a carbon side chain.

endogenous (en-DODGE-eh-nus): from within the body.
- **endo** = within

exogenous (eks-ODGE-eh-nus): from outside the body.
- **exo** = outside

plaque (PLACK): an accumulation of fatty deposits, smooth muscle cells, and fibrous connective tissue that develops in the artery walls in atherosclerosis; also known as *atheromatous* (ATH-er-OH-ma-tus) *plaque.*

atherosclerosis (ATH-er-oh-scler-OH-sis): a type of artery disease characterized by plaques (accumulations of lipid-containing material) on the inner walls of the arteries.

*The four-ring core structure identifies a steroid; sterols are alcohol derivatives with a steroid ring structure.

5.3 Digestion, Absorption, and Transport of Lipids

> **LEARN IT** Summarize fat digestion, absorption, and transport.

Each day, the GI tract receives, on average from the food we eat, 50 to 100 grams of triglycerides, 4 to 8 grams of phospholipids, and 200 to 350 milligrams of cholesterol. These lipids are hydrophobic, whereas the digestive enzymes are hydrophilic. As you read, notice how the body elegantly meets the challenges of keeping the lipids mixed in the watery fluids of the GI tract and facilitating the work of the **lipases**.

Lipid Digestion   Figure 5-11 traces the digestion of fat through the GI tract. The goal of fat digestion is to dismantle triglycerides into small molecules that

lipases (LYE-pasez): enzymes that hydrolyze lipids. *Lingual lipase* is a fat-digesting enzyme secreted from the salivary gland at the base of the tongue; *gastric lipase* is a fat-digesting enzyme secreted from the cells of the stomach.

> **FIGURE 5-11**   **Fat Digestion in the GI Tract**

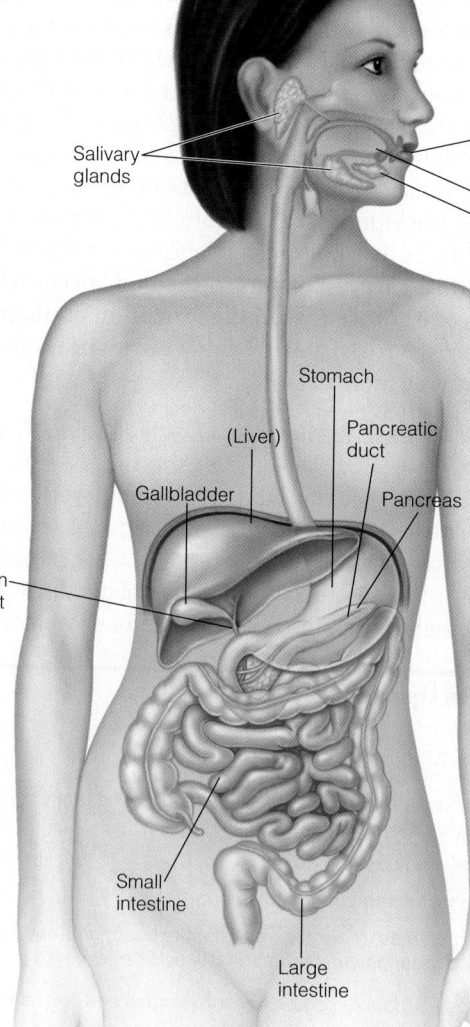

FAT

Mouth and salivary glands
Some hard fats begin to melt as they reach body temperature. The sublingual salivary gland in the base of the tongue secretes lingual lipase. The degree of hydrolysis by lingual lipase is slight for most fats but may be appreciable for milk fats.

Stomach
The stomach's churning action mixes fat with water and acid. A gastric lipase accesses and hydrolyzes (only a very small amount of) fat.

Small intestine
Cholecystokinin (CCK) signals the gallbladder to release bile (via the common bile duct):

Fat $\xrightarrow{\text{Bile}}$ Emulsified fat

Pancreatic lipase flows in from the pancreas (via the pancreatic duct):

Emulsified fat (triglycerides) $\xrightarrow{\text{Pancreatic (and intestinal) lipase}}$ Monoglycerides, glycerol, fatty acids (absorbed)

Large intestine
Some fat and cholesterol, trapped in fiber, exit in feces.

Labels on figure: Salivary glands, Mouth, Tongue, Sublingual salivary gland, Stomach, Pancreatic duct, (Liver), Pancreas, Gallbladder, Common bile duct, Small intestine, Large intestine

© Cengage Learning

the body can absorb and use—namely, **monoglycerides**, fatty acids, and glycerol. The following paragraphs provide the details.

In the Mouth Fat digestion starts off slowly in the mouth, with some hard fats beginning to melt when they reach body temperature. A salivary gland at the base of the tongue releases an enzyme (lingual lipase) that plays an active role in fat digestion in infants, but a relatively minor role in adults. In infants, this enzyme efficiently digests the short- and medium-chain fatty acids found in milk.

In the Stomach In a quiet stomach, fat would float as a layer above the watery components of swallowed food. But whenever food is present, the stomach becomes active. The strong muscle contractions of the stomach propel its contents toward the pyloric sphincter. Some chyme passes through the pyloric sphincter periodically, but the remaining partially digested food is propelled back into the body of the stomach. This churning grinds the solid pieces to finer particles, mixes the chyme, and disperses the fat into small droplets. These actions help to expose the fat for attack by the gastric lipase enzyme—an enzyme that performs best in the acidic environment of the stomach. Still, little fat digestion takes place in the stomach; most of the action occurs in the small intestine.

In the Small Intestine When fat enters the small intestine, it triggers the release of the hormone cholecystokinin (CCK), which signals the gallbladder to release its stores of bile. (Remember that the liver makes bile, and the gallbladder stores bile until it is needed.) Among bile's many ingredients are bile acids, which are made in the liver from cholesterol and have a similar structure. In addition, bile acids often pair up with an amino acid (a building block of protein). The amino acid end is hydrophilic, and the sterol end is hydrophobic. This structure enables bile to act as an emulsifier, drawing fat molecules into the surrounding watery fluids. There, the fats are fully digested as they encounter lipase enzymes from the pancreas and small intestine. The process of emulsification is diagrammed in Figure 5-12.

Most of the hydrolysis of triglycerides occurs in the small intestine. The major fat-digesting enzymes are pancreatic lipases; some intestinal lipases are also active. These enzymes remove each of a triglyceride's outer fatty acids one at a time, leaving a monoglyceride. Occasionally, enzymes remove all three fatty acids, leaving a free molecule of glycerol. Hydrolysis of a triglyceride is shown in Figure 5-13 (p. 144).

Phospholipids are digested similarly—that is, their fatty acids are removed by hydrolysis. The two fatty acids and the remaining glycerol and phosphate fragments are then absorbed. Most sterols can be absorbed as is; if any fatty acids are attached, they are first hydrolyzed off.

monoglycerides: molecules of glycerol with one fatty acid attached. A molecule of glycerol with two fatty acids attached is a *diglyceride*.

- **mono** = one
- **di** = two

> FIGURE 5-12 **Emulsification of Fat by Bile**

Like bile, detergents are emulsifiers and work the same way, which is why they are effective in removing grease spots from clothes. Molecule by molecule, the grease is dissolved out of the spot and suspended in the water, where it can be rinsed away.

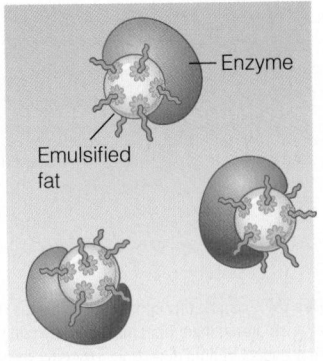

In the stomach, the fat and watery GI juices tend to separate. The enzymes in the GI juices can't get at the fat.

When fat enters the small intestine, the gallbladder secretes bile. Bile has an affinity for both fat and water, so it can bring the fat into the water.

Bile's emulsifying action converts large fat globules into small droplets that repel one another.

After emulsification, more fat is exposed to the enzymes, making fat digestion more efficient.

© Cengage Learning

The Lipids: Triglycerides, Phospholipids, and Sterols **143**

> FIGURE 5-13   **Digestion (Hydrolysis) of a Triglyceride**

Bonds break

Triglyceride

The triglyceride and two molecules of water are split. The H and OH from water complete the structures of two fatty acids and leave a monoglyceride.

Bonds break

Monoglyceride + two fatty acids

These products may pass into the intestinal cells, but sometimes the monoglyceride is split with another molecule of water to give a third fatty acid and glycerol. Fatty acids, monoglycerides, and glycerol are absorbed into intestinal cells.

© Cengage Learning

> FIGURE 5-14   **Enterohepatic Circulation**

Most of the bile released into the small intestine is reabsorbed and sent back to the liver to be reused. This cycle is called the *enterohepatic circulation* of bile. Some bile is excreted.

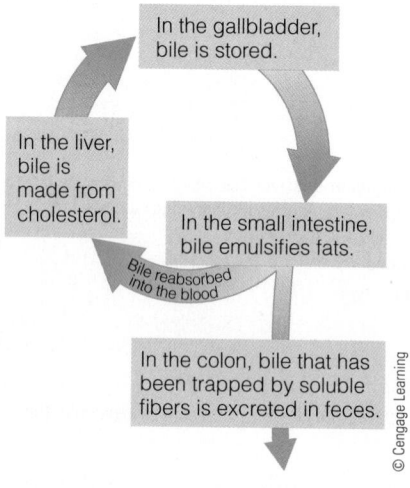

In the gallbladder, bile is stored.

In the liver, bile is made from cholesterol.

In the small intestine, bile emulsifies fats.

Bile reabsorbed into the blood

In the colon, bile that has been trapped by soluble fibers is excreted in feces.

© Cengage Learning

micelles (MY-cells): tiny spherical complexes of emulsified fat that arise during digestion; most contain bile salts and the products of lipid digestion, including fatty acids, monoglycerides, and cholesterol.

lipoproteins (LIP-oh-PRO-teenz): clusters of lipids associated with proteins that serve as transport vehicles for lipids in the lymph and blood.

Bile's Routes   After bile enters the small intestine and emulsifies fat, it has two possible destinations, illustrated in Figure 5-14. Most of the bile is reabsorbed from the small intestine and recycled. The other possibility is that some of the bile can be trapped by dietary fibers in the large intestine and excreted. Because cholesterol is needed to make bile, the excretion of bile effectively reduces blood cholesterol. As Chapter 4 explains, the dietary fibers most effective at lowering blood cholesterol this way are the soluble fibers commonly found in fruits, whole grains, and legumes.

Lipid Absorption   Figure 5-15 illustrates the absorption of lipids. Small molecules (glycerol and short- and medium-chain fatty acids) can diffuse easily into the intestinal cells; they are absorbed directly into the bloodstream. Larger molecules (monoglycerides and long-chain fatty acids) are emulsified by bile, forming spherical complexes known as **micelles.** The micelles diffuse into the intestinal cells, where the monoglycerides and long-chain fatty acids are reassembled into new triglycerides.

Within the intestinal cells, the newly made triglycerides and other lipids (cholesterol and phospholipids) are packed with protein into transport vehicles known as chylomicrons. The intestinal cells then release the chylomicrons into the lymphatic system. The chylomicrons glide through the lymph until they reach a point of entry into the bloodstream at the thoracic duct near the heart. (Recall from Chapter 3 that nutrients from the GI tract that enter the lymph system initially bypass the liver.) The blood carries these lipids to the rest of the body for immediate use or storage. A look at these lipids in the body reveals the kinds of fat the diet has been delivering. The blood, fat stores, and muscle cells of people who eat a diet rich in unsaturated fats, for example, contain more unsaturated fats than those of people who select a diet high in saturated fats.

Lipid Transport   The chylomicrons are one of several clusters of lipids and proteins that are used as transport vehicles for fats. As a group, these vehicles are known as **lipoproteins,** and they solve the body's challenge of transporting fat through the watery bloodstream. The body makes four main types of lipoproteins, distinguished by their size and density.* Each type contains different kinds and amounts of lipids and proteins. The more lipids, the less dense; the

---

*Chemists can identify the various lipoproteins by their density. They place a blood sample below a thick fluid in a test tube and spin the tube in a centrifuge. The most buoyant particles (highest in lipids) rise to the top and have the lowest density; the densest particles (highest in proteins) remain at the bottom and have the highest density. Others distribute themselves in between.

> **FIGURE 5-15** **Absorption of Fat**

The end products of fat digestion are mostly monoglycerides, some fatty acids, and very little glycerol. Their absorption differs depending on their size. (In reality, molecules of fatty acid are too small to see without a powerful microscope, whereas villi are visible to the naked eye.)

2 **Large lipids** such as monoglycerides and long-chain fatty acids combine with bile, forming micelles that are sufficiently water soluble to penetrate the watery solution that bathes the absorptive cells. There the lipid contents of the micelles diffuse into the cells.

1 **Glycerol and small lipids** such as short- and medium-chain fatty acids can move directly into the bloodstream.

© Cengage Learning

more proteins, the more dense. Figure 5-16 (p. 146) shows the relative compositions and sizes of the lipoproteins.

Chylomicrons The **chylomicrons** are the largest and least dense of the lipoproteins. They transport *diet*-derived lipids (mostly triglycerides) from the small intestine (via the lymph system) to the rest of the body. Cells all over the body remove triglycerides from the chylomicrons as they pass by, so the chylomicrons get smaller and smaller. Within 14 hours after absorption, most of the triglycerides have been depleted, and only a few remnants of protein, cholesterol, and phospholipid remain. Special protein receptors on the membranes of the liver cells recognize and remove these chylomicron remnants from the blood.

VLDL (Very-Low-Density Lipoproteins) Meanwhile, in the liver—the most active site of lipid synthesis—cells are making cholesterol, fatty acids, and other lipid compounds. Ultimately, the lipids made in the liver and those collected from chylomicron remnants are packaged with proteins as a **VLDL (very-low-density lipoprotein)** and shipped to other parts of the body.

As the VLDL travel through the body, cells remove triglycerides. As they lose triglycerides, the VLDL shrink and the proportion of lipids shifts. Cholesterol becomes the predominant lipid, and the lipoprotein becomes smaller and more dense. As this occurs, the VLDL becomes an **LDL (low-density lipoprotein)**, loaded with cholesterol, but containing relatively few triglycerides.*

---

*Before becoming LDL, the VLDL are first transformed into intermediate-density lipoproteins (IDL), sometimes called VLDL remnants. Some IDL may be picked up by the liver and rapidly broken down; those IDL that remain in circulation continue to deliver triglycerides to the cells and eventually become LDL. Researchers debate whether IDL are simply transitional particles or a separate class of lipoproteins; normally, IDL do not accumulate in the blood. Measures of blood lipids include IDL with LDL.

chylomicrons (kye-lo-MY-cronz): the class of lipoproteins that transport lipids from the intestinal cells to the rest of the body.

VLDL (very-low-density lipoprotein): the type of lipoprotein made primarily by liver cells to transport lipids to various tissues in the body; composed primarily of triglycerides.

LDL (low-density lipoprotein): the type of lipoprotein derived from very-low-density lipoproteins (VLDL) as triglycerides are removed and broken down; composed primarily of cholesterol.

> **FIGURE 5-16**   **Sizes and Compositions of the Lipoproteins**

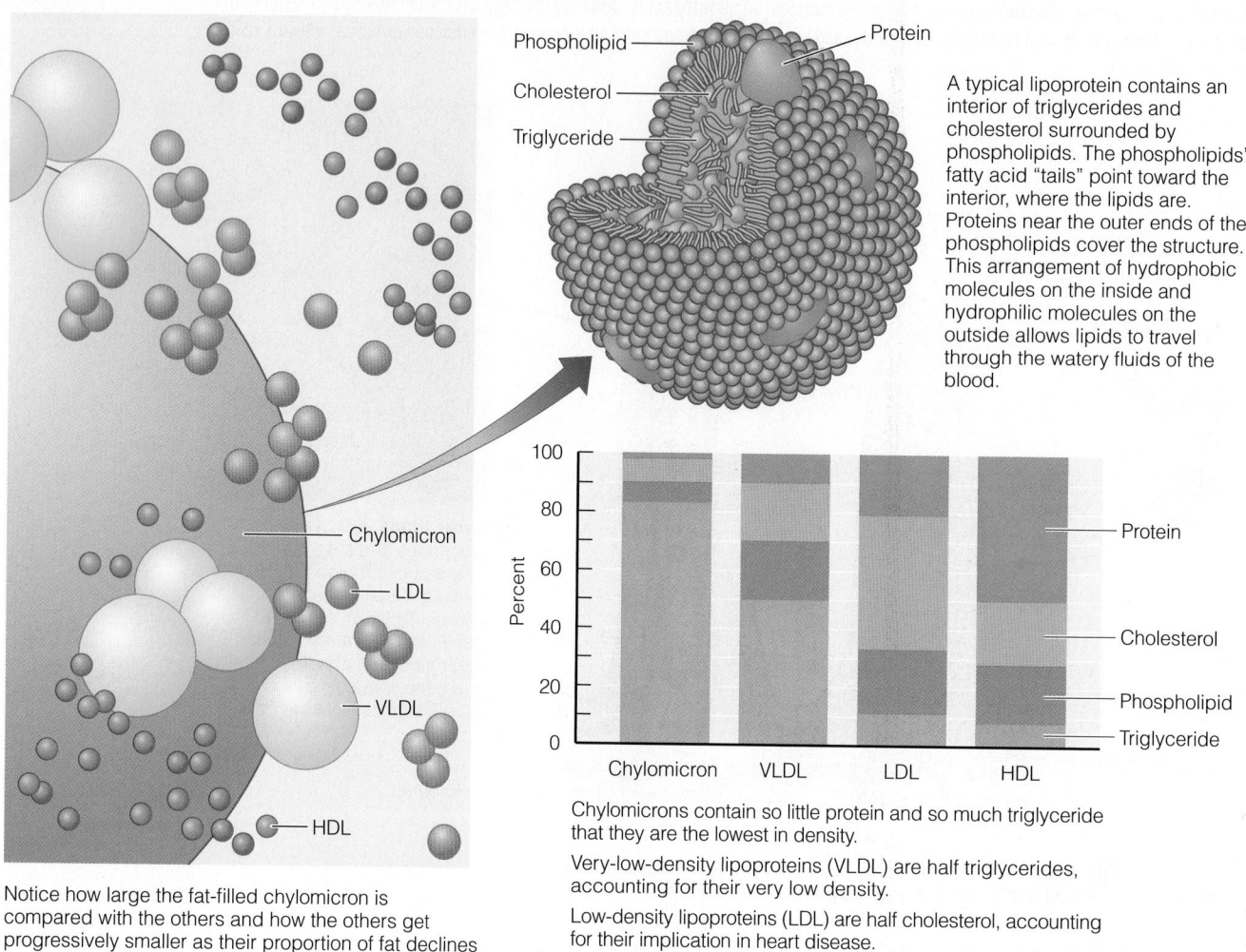

A typical lipoprotein contains an interior of triglycerides and cholesterol surrounded by phospholipids. The phospholipids' fatty acid "tails" point toward the interior, where the lipids are. Proteins near the outer ends of the phospholipids cover the structure. This arrangement of hydrophobic molecules on the inside and hydrophilic molecules on the outside allows lipids to travel through the watery fluids of the blood.

Notice how large the fat-filled chylomicron is compared with the others and how the others get progressively smaller as their proportion of fat declines and protein increases.

Chylomicrons contain so little protein and so much triglyceride that they are the lowest in density.

Very-low-density lipoproteins (VLDL) are half triglycerides, accounting for their very low density.

Low-density lipoproteins (LDL) are half cholesterol, accounting for their implication in heart disease.

High-density lipoproteins (HDL) are half protein, accounting for their high density.

© Cengage Learning

LDL (Low-Density Lipoproteins)   The LDL circulate throughout the body, making their contents available to the cells of all tissues—muscles (including the heart muscle), fat stores, the mammary glands, and others. The cells take triglycerides, cholesterol, and phospholipids to use for energy, make hormones or other compounds, or build new membranes. Special LDL receptors on the liver cells play a crucial role in the control of blood cholesterol concentrations by removing LDL from circulation.

HDL (High-Density Lipoproteins)   The liver makes **HDL (high-density lipoprotein)** to remove cholesterol from the cells and carry it back to the liver for recycling or disposal. By efficiently clearing cholesterol, HDL lowers the risk of heart disease.[5] In addition, HDL have anti-inflammatory properties that seem to keep artery-clogging plaque from breaking apart and causing heart attacks.[6] Figure 5-17 summarizes lipid transport via the lipoproteins.

Health Implications   The distinction between LDL and HDL has implications for the health of the heart and blood vessels. The blood lipid linked most directly to heart disease is LDL cholesterol. As mentioned, HDL also carry cholesterol, but elevated HDL represent cholesterol returning from the rest of the body to the liver for breakdown and excretion. The transport of cholesterol from the tissues back to the liver is sometimes called *reverse cholesterol transport* or the *scavenger pathway*.

High LDL and low HDL cholesterol are both associated with a high risk of heart disease. Having adequate HDL is beneficial, but having high HDL is not

HDL (high-density lipoprotein): the type of lipoprotein that transports cholesterol back to the liver from the cells; composed primarily of protein.

> **FIGURE 5-17** **Lipid Transport via Lipoproteins**

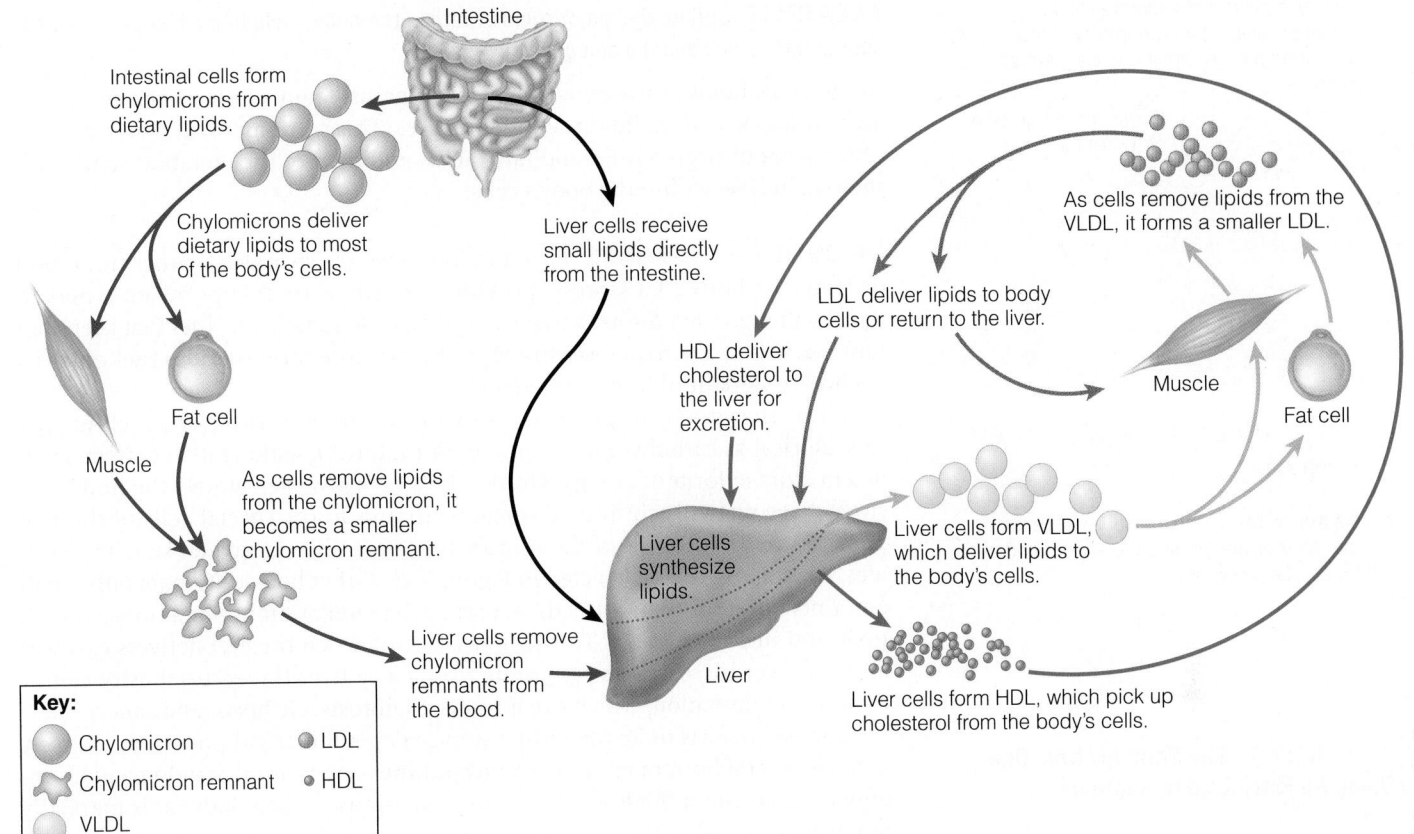

necessarily more beneficial. Some people think of HDL as healthy and LDL as lousy, or refer to LDL as "bad," and HDL as "good," cholesterol. Keep in mind that the cholesterol itself is the same and that the differences between LDL and HDL reflect the *proportions* and *types* of lipids and proteins within them—not the type of cholesterol. The following factors help to lower LDL and/or raise HDL:

- Weight control
- Monounsaturated or polyunsaturated, instead of saturated, fat in the diet
- Soluble dietary fibers
- Phytochemicals
- *Moderate* alcohol consumption
- Physical activity

Chapter 18 provides many more details.

Not too surprisingly, numerous genes influence how the body handles the synthesis, transport, and degradation of lipids and lipoproteins. Much current research is focused on how nutrient-gene interactions may direct the progression of heart disease.

> **REVIEW IT** Summarize fat digestion, absorption, and transport.

The body makes special arrangements to digest and absorb lipids. It provides the emulsifier bile to make them accessible to the fat-digesting lipases that dismantle triglycerides, mostly to monoglycerides and fatty acids, for absorption by the intestinal cells. Four types of lipoproteins transport all classes of lipids (triglycerides, phospholipids, and cholesterol), but the chylomicrons are the largest and contain mostly triglycerides from the diet; VLDL are smaller and are about half triglycerides; LDL are smaller still and contain mostly cholesterol; and HDL are the densest and are rich in protein. High LDL cholesterol indicates increased risk of heart disease, whereas high HDL cholesterol has a protective effect.

> FIGURE 5-18 **An Adipose Cell**

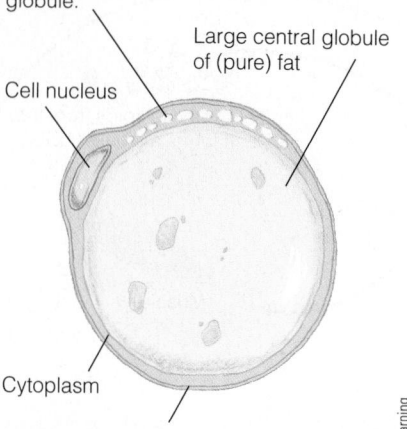

Newly imported triglycerides first form small droplets at the periphery of the cell, then merge with the large, central globule.

Large central globule of (pure) fat

Cell nucleus

Cytoplasm

As the central globule enlarges, the fat cell membrane expands to accommodate its swollen contents.

© Cengage Learning

> FIGURE 5-19 **The Pathway from One Omega-6 Fatty Acid to Another**

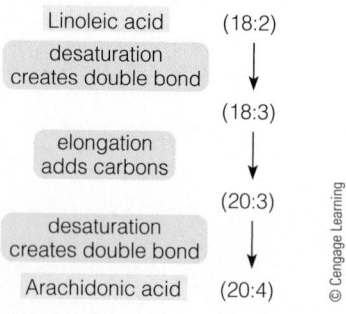

Linoleic acid   (18:2)

desaturation creates double bond

(18:3)

elongation adds carbons

(20:3)

desaturation creates double bond

Arachidonic acid   (20:4)

© Cengage Learning

The first number indicates the number of carbons and the second, the number of double bonds. Similar reactions occur when the body makes the omega-3 fatty acids EPA and DHA from linolenic acid.

adipose (ADD-ih-poce) **tissue:** the body's fat tissue; consists of masses of triglyceride-storing cells.

adipokines (ADD-ih-poe-kines): proteins synthesized and secreted by adipose cells.

resistin (ree-ZIS-tin): a protein produced by adipose cells that promotes inflammation and causes insulin resistance.

adiponectin: a protein produced by adipose cells that inhibits inflammation and protects against insulin resistance, type 2 diabetes, and cardiovascular disease.

essential fatty acids: fatty acids that the body requires but cannot make, and so must be obtained from the diet; both linoleic acid and linolenic acid are essential fatty acids.

5.4 Lipids in the Body

› LEARN IT Outline the major roles of fats in the body, including a discussion of essential fatty acids and the omega fatty acids.

In the body, lipids provide energy, insulate against temperature extremes, protect against shock, and maintain cell membranes. This section provides an overview of the roles of triglycerides and fatty acids and then of the metabolic pathways they can follow within the body's cells.

Roles of Triglycerides

First and foremost, triglycerides—either from food or from the body's fat stores—provide the cells with energy. When a person dances all night, her dinner's triglycerides provide some of the fuel that keeps her moving. When a person loses his appetite, his stored triglycerides fuel much of his body's work until he can eat again.

Recall that gram for gram, fat provides more than twice as much energy (9 kcalories) as carbohydrate or protein (4 kcalories), making it an extremely efficient storage form of energy. Unlike the liver's glycogen stores, the body's fat stores have virtually unlimited capacity, thanks to the special cells of the **adipose tissue.** The fat cells of the adipose tissue readily take up and store triglycerides. An adipose cell is depicted in Figure 5-18. Other body cells store only small amounts of fat for their immediate use; fat accumulation in nonadipose cells is toxic and impairs health.[7] This scenario occurs when the diet delivers excesses and the liver increases its fat production. Fatty liver linked to obesity causes chronic inflammation, which can advance to fibrosis, cirrhosis, and cancer.[8]

Adipose tissue is more than just a storage depot for fat. Adipose tissue actively secretes several hormones known as **adipokines**—proteins that help regulate energy balance and influence several body functions.[9] When body fat is markedly reduced or excessive, the type and quantity of adipokine secretions change, with consequences for the body's health. Researchers are currently exploring how adipokines influence the links between obesity and chronic diseases such as type 2 diabetes, hypertension, and heart disease.[10] Obesity, for example, increases the release of the adipokine **resistin** that promotes inflammation and insulin resistance—factors that predict heart disease and diabetes. Similarly, obesity decreases the release of the adipokine **adiponectin** that protects against inflammation, diabetes, and heart disease.

Fat serves other roles in the body as well. Because fat is a poor conductor of heat, the layer of fat beneath the skin insulates the body from temperature extremes. Fat pads also serve as natural shock absorbers, providing a cushion for the bones and vital organs. Fat provides the structural material for cell membranes and participates in cell signaling pathways.

Essential Fatty Acids

The human body needs fatty acids, and it can make all but two of them—linoleic acid (the 18-carbon omega-6 fatty acid) and linolenic acid (the 18-carbon omega-3 fatty acid). These two fatty acids must be supplied by the diet and are therefore **essential fatty acids.** The cells do not possess the enzymes to make any of the omega-6 or omega-3 fatty acids from scratch, nor can they convert an omega-6 fatty acid to an omega-3 fatty acid or vice versa. Cells *can*, however, use the 18-carbon member of an omega family from the diet to make the longer fatty acids of that family by forming double bonds (desaturation) and lengthening the chain two carbons at a time (elongation), as shown in Figure 5-19. This is a slow process because the omega-3 and omega-6 families compete for the same enzymes. Too much of a fatty acid from one family can create a deficiency of the other family's longer fatty acids, which becomes critical only when the diet fails to deliver adequate supplies. Therefore, the most effective way to maintain body supplies of all the omega-6 and omega-3 fatty acids is to obtain them directly from foods—most notably, from vegetable oils, seeds, nuts, fish, and other seafoods.

Linoleic Acid and the Omega-6 Family Linoleic acid is an essential fatty acid and the primary member of the omega-6 fatty acid family. When the body receives linoleic acid from the diet, it can make other members of the omega-6 family—such as the 20-carbon polyunsaturated fatty acid, **arachidonic acid** (as shown in Figure 5-19). Should a linoleic acid deficiency develop, arachidonic acid, and all other omega-6 fatty acids that derive from linoleic acid, would also become essential and have to be obtained from the diet. A nonessential nutrient (such as arachidonic acid) that must be supplied by the diet in special circumstances (as in a linoleic acid deficiency) is considered a **conditionally essential nutrient.** Normally, vegetable oils and meats supply enough omega-6 fatty acids to meet the body's needs.

Linolenic Acid and the Omega-3 Family Linolenic acid is an essential fatty acid and the primary member of the omega-3 fatty acid family.* Like linoleic acid, linolenic acid cannot be made in the body and must be supplied by foods. Given the 18-carbon linolenic acid, the body can make small amounts of the 20- and 22-carbon members of the omega-3 family, **eicosapentaenoic acid (EPA)** and **docosahexaenoic acid (DHA),** respectively. These omega-3 fatty acids play critical roles in the optimal structure and function of cells.[11] Found abundantly in the eyes and brain, the omega-3 fatty acids are essential for normal growth, visual acuity, and cognitive development.[12] They may also play an important role in the prevention and treatment of heart disease, as later sections explain.

Eicosanoids The body uses the longer omega-3 and omega-6 fatty acids to make substances known as **eicosanoids.** Eicosanoids are a diverse group of more than 100 compounds. Sometimes described as "hormonelike," eicosanoids differ from hormones in important ways. For one, hormones are secreted in one location and travel to affect cells all over the body, whereas eicosanoids appear to affect only the cells in which they are made or nearby cells in the same localized environment. For another, hormones elicit the same response from all their target cells, whereas eicosanoids often have different effects on different cells.

The actions of various eicosanoids sometimes oppose one another. For example, one causes muscles to relax and blood vessels to dilate, whereas another causes muscles to contract and blood vessels to constrict. Certain eicosanoids participate in the immune response to injury and infection, producing fever, inflammation, and pain. One of the ways aspirin relieves these symptoms is by slowing the synthesis of these eicosanoids.

Eicosanoids that derive from omega-3 fatty acids differ from those that derive from omega-6 fatty acids, with the omega-3 family providing greater health benefits. The omega-3 eicosanoids help lower blood pressure, prevent blood clot formation, protect against irregular heartbeats, and reduce inflammation, whereas the omega-6 eicosanoids tend to promote clot formation, inflammation, and blood vessel constriction.[13]

Omega-6 to Omega-3 Ratio Because omega-6 and omega-3 fatty acids compete for the same enzymes and their actions often oppose each other, researchers have studied whether there is an ideal ratio that best supports health.[14] Suggested ratios range from 4:1 to 10:1; while some researchers support such recommendations, others find the ratio of little value in improving health or predicting risk.[15] Increasing the amount of omega-3 fatty acids in the diet is clearly beneficial, but reducing the amount of omega-6 fatty acids in the diet to improve the ratio may not be helpful. Omega-6 fatty acids protect heart health by lowering LDL cholesterol and improving insulin resistance.[16]

Fatty Acid Deficiencies Most diets in the United States meet the minimum essential fatty acid requirement adequately. Historically, deficiencies have developed only in infants and young children who have been fed fat-free milk and low-fat diets or in hospital clients who have been mistakenly fed formulas

*This omega-3 linolenic acid is known as alpha-linolenic acid and is the fatty acid referred to in this chapter. Another fatty acid, also with 18 carbons and three double bonds, belongs to the omega-6 family and is known as gamma-linolenic acid.

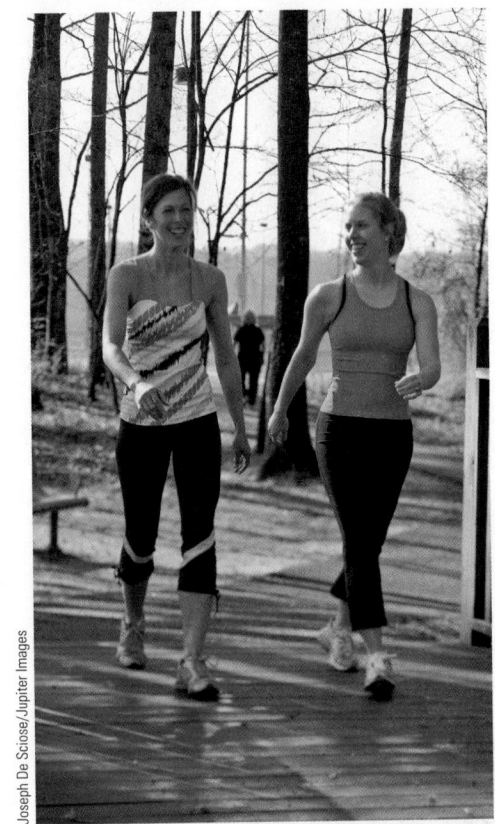

Joseph De Sciose/Jupiter Images

Double thanks: The body's fat stores provide energy for a walk, and fat pads on the heels provide cushion against the hard pavement.

arachidonic (a-RACK-ih-DON-ic) **acid:** an omega-6 polyunsaturated fatty acid with 20 carbons and four double bonds; present in small amounts in meat and other animal products and synthesized in the body from linoleic acid.

conditionally essential nutrient: a nutrient that is normally nonessential, but must be supplied by the diet in special circumstances when the need for it exceeds the body's ability to produce it.

eicosapentaenoic (EYE-cossa-PENTA-ee-NO-ick) **acid (EPA):** an omega-3 polyunsaturated fatty acid with 20 carbons and five double bonds; present in fatty fish and synthesized in limited amounts in the body from linolenic acid.

docosahexaenoic (DOE-cossa-HEXA-ee-NO-ick) **acid (DHA):** an omega-3 polyunsaturated fatty acid with 22 carbons and six double bonds; present in fatty fish and synthesized in limited amounts in the body from linolenic acid.

eicosanoids (eye-COSS-uh-noyds): derivatives of 20-carbon fatty acids; biologically active compounds that help to regulate blood pressure, blood clotting, and other body functions. They include *prostaglandins* (PROS-tah-GLAND-ins), *thromboxanes* (throm-BOX-ains), and *leukotrienes* (LOO-ko-TRY-eens).

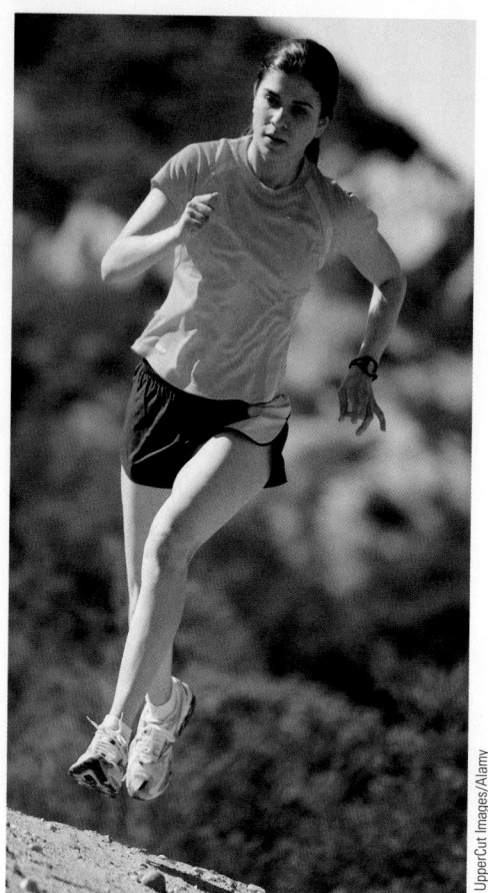

Fat supplies most of the energy during a long-distance run.

that provided no polyunsaturated fatty acids for long periods of time. Classic deficiency symptoms include growth retardation, reproductive failure, skin lesions, kidney and liver disorders, and subtle neurological and visual problems.

A Preview of Lipid Metabolism This preview of fat metabolism describes how the cells store and release energy from fat. Chapter 7 provides details.

Storing Fat as Fat When meals deliver more energy than the body needs, the excess is stored as fat in the adipose cells for later use. An enzyme—**lipoprotein lipase (LPL)**—hydrolyzes triglycerides from circulating lipoproteins, releasing fatty acids, diglycerides, and monoglycerides into the adipose cells. Enzymes inside the adipose cells reassemble these fatty acids, diglycerides, and monoglycerides into triglycerides again for storage. As Figure 5-18 (p. 148) shows, triglycerides fill the adipose cells, storing a lot of energy in a relatively small space.

Using Fat for Energy After meals, the blood delivers chylomicrons and VLDL loaded with triglycerides to the body's cells for energy. Fat supplies about 60 percent of the body's ongoing energy needs during rest. During prolonged light to moderately intense exercise or extended periods of food deprivation, fat may make a slightly greater contribution to energy needs.

During energy deprivation, several lipase enzymes (most notably **hormone-sensitive lipase**) inside the adipose cells respond by dismantling stored triglycerides and releasing the glycerol and fatty acids directly into the blood. Energy-hungry cells anywhere in the body can then capture these compounds and take them through a series of chemical reactions to yield energy, carbon dioxide, and water.

A person who fasts (drinking only water) will rapidly metabolize body fat. Even with abundant body fat, the person has to obtain some energy from lean protein tissue because the brain, nerves, and red blood cells need glucose—and without carbohydrate, only protein and the small glycerol molecule of a triglyceride can be converted to glucose; fatty acids cannot be. Still, in times of severe hunger and starvation, a fatter person can survive longer than a thinner person thanks to this energy reserve. But as Chapter 7 explains, fasting for too long will eventually cause death, even if the person still has ample body fat.

> **REVIEW IT** Outline the major roles of fats in the body, including a discussion of essential fatty acids and the omega fatty acids.

In the body, triglycerides provide energy, insulate against temperature extremes, protect against shock, provide structural material for cell membranes, and participate in cell signaling pathways. Linoleic acid (18 carbons, omega-6) and linolenic acid (18 carbons, omega-3) are essential fatty acids. They serve as structural parts of cell membranes and as precursors to the longer fatty acids that can make eicosanoids—powerful compounds that participate in blood pressure regulation, blood clot formation, and the immune response to injury and infection. Because essential fatty acids are common in the diet and stored in the body, deficiencies are unlikely. The body can easily store unlimited amounts of fat if given excesses, and this body fat is used for energy when needed.

lipoprotein lipase (LPL): an enzyme that hydrolyzes triglycerides passing by in the bloodstream and directs their parts into the cells, where they can be metabolized for energy or reassembled for storage.

hormone-sensitive lipase: an enzyme inside adipose cells that responds to the body's need for fuel by hydrolyzing triglycerides so that their parts (glycerol and fatty acids) escape into the general circulation and thus become available to other cells for fuel. The signals to which this enzyme responds include epinephrine and glucagon, which oppose insulin (see Chapter 4).

solid fats: fats that are not usually liquid at room temperature; commonly found in most foods derived from animals and vegetable oils that have been hydrogenated. Solid fats typically contain more saturated and *trans* fats than most oils.

5.5 Health Effects and Recommended Intakes of Saturated Fats, *Trans* Fats, and Cholesterol

> **LEARN IT** Explain the relationships among saturated fats, *trans* fat, and cholesterol and chronic diseases, noting recommendations.

Some fats in the diet are essential for good health, but others can be harmful. The current American diet delivers excessive amounts of **solid fats,** representing an average of almost one-fifth of the day's total kcalories. Major sources of solid fats in the American diet include desserts, pizza, cheese, and processed and fatty meats (sausages, hot dogs, bacon, ribs). Because foods made with solid fats provide abundant energy, but few if any essential nutrients, they contribute to weight gain and make it difficult to meet nutrient needs. Solid fats also provide abundant saturated fat, *trans* fat, and cholesterol. Even without overweight or obesity, high intakes of solid fats increase the risk of some chronic diseases. The easiest way to control saturated fat, *trans* fat, cholesterol, and kcalories is to limit solid fats in the diet.

Health Effects of Saturated Fats, *Trans* Fats, and Cholesterol

Hearing a physician say, "Your blood lipid profile looks fine," is reassuring (see Table 5-3). The **blood lipid profile** reveals the concentrations of various lipids in the blood, notably triglycerides and cholesterol, and their lipoprotein carriers (VLDL, LDL, and HDL). This information alerts people to possible disease risks and perhaps to a need for changing their physical activity and eating habits. Both the amounts and types of fat in the diet influence the risk for disease.

Heart Disease As mentioned earlier, elevated LDL cholesterol is a major risk factor for **cardiovascular disease (CVD)**.[17] As LDL cholesterol accumulates in the arteries, blood flow becomes restricted and blood pressure rises. The consequences are deadly; in fact, heart disease is the nation's number-one killer of adults. LDL cholesterol is often used to predict the likelihood of a person's suffering a heart attack or stroke; the higher the LDL, the earlier and more likely the tragedy. Much of the effort to prevent and treat heart disease focuses on lowering LDL cholesterol.[18]

Saturated fats are most often implicated in raising LDL cholesterol. In general, the more saturated fat in the diet, the more LDL cholesterol in the blood. Not all saturated fats have the same cholesterol-raising effect, however. Most notable among the saturated fatty acids that raise blood cholesterol are lauric, myristic, and palmitic acids (12, 14, and 16 carbons, respectively). In contrast, stearic acid (18 carbons) seems to have little or no effect on blood cholesterol.[19] Making such distinctions may be impractical in diet planning, however, because these saturated fatty acids typically appear together in the same foods. In addition to raising blood cholesterol, saturated fatty acids contribute to heart disease by promoting blood clotting. Fats from animal sources (meats, milk, and milk products) are the main sources of saturated fats in most people's diets. Selecting lean cuts of meat, skinless poultry, and fat-free milk products helps to lower saturated fat intake and the risk of heart disease.

Research also suggests an association between dietary *trans* fats and heart disease.[20] In the body, *trans* fats alter blood cholesterol the same way some saturated fats do: they raise LDL cholesterol and lower HDL cholesterol.[21] Limiting the intake of *trans* fats can improve blood cholesterol and lower the risk of heart disease. To that end, many restaurants and manufacturers have taken steps to eliminate or greatly reduce *trans* fats in foods.[22] The decrease in *trans* fatty acids in the food supply is apparent in a decrease in plasma concentrations of *trans* fatty acids in consumers.[23]

Unlike saturated fat and *trans* fat, dietary cholesterol raises blood cholesterol very little, if at all.[24] Less clear is its role in heart disease.[25]

Cancer The links between dietary fats and cancer are not as evident as they are for heart disease. Dietary fat does not seem to *initiate* cancer development but, instead, may *promote* cancer once it has arisen. Stronger risk factors for cancer include smoking, alcohol, and environmental contaminants. (Chapter 18 provides many more details about these risk factors and the development of cancer.)

The relationship between dietary fat and the risk of cancer differs for various types of cancers. In the case of breast cancer, evidence has been weak and inconclusive. Some studies indicate an association between dietary fat and breast cancer; more convincing evidence indicates that body fatness contributes to the risk. In the case of colon cancer, limited evidence suggests a harmful association with foods containing animal fats.

TABLE 5-3 Desirable Blood Lipid Profile

Total cholesterol	<200 mg/dL
LDL cholesterol	<100 mg/dL
HDL cholesterol	>60 mg/dL
Triglycerides	<100 mg/dL

© 2016 Cengage Learning

blood lipid profile: results of blood tests that reveal a person's total cholesterol, triglycerides, and various lipoproteins.

cardiovascular disease (CVD): diseases of the heart and blood vessels throughout the body. Atherosclerosis is the main cause of CVD. When the arteries that carry blood to the heart muscle become blocked, the heart suffers damage known as *coronary heart disease (CHD)*.

- **cardio** = heart
- **vascular** = blood vessels

The relationship between dietary fat and the risk of cancer differs for various types and combinations of fats as well. The increased risk in cancer from fat appears to be due primarily to saturated fats or dietary fat from meats (which is mostly saturated). Fat from milk or fish has not been implicated in cancer risk. Olive oil seems to have a protective effect.[26]

Obesity Remember that fat contributes more than twice as many kcalories per gram as either carbohydrate or protein. Consequently, people who eat high-fat diets regularly may exceed their energy needs and gain weight, especially if they are inactive. Because fat boosts energy intake, cutting fat from the diet can be an effective strategy in cutting kcalories. In some cases, though, choosing a fat-free food offers no kcalorie savings. Fat-free frozen desserts, for example, often have so much sugar added that the kcalorie count can be as high as in the regular-fat product. In this case, cutting fat and adding carbohydrate offers no kcalorie savings or weight-loss advantage. In fact, it may even raise energy intake and exacerbate weight problems. Later chapters revisit the role of dietary fat in the development of obesity.

Recommended Intakes of Saturated Fat, *Trans* Fat, and Cholesterol

Defining the exact amount of saturated fat, *trans* fat, or cholesterol that begins to harm health is difficult.[27] For this reason, no RDA or Upper Level has been set. Instead, the DRI and *Dietary Guidelines* suggest a diet that provides 20 to 35 percent of the daily energy intake from fat, less than 10 percent of daily energy intake from saturated fat, as little *trans* fat as possible, and less than 300 milligrams of cholesterol. These recommendations recognize that diets with up to 35 percent of kcalories from fat can be compatible with good health if energy intake is reasonable and saturated fat, *trans* fat, and cholesterol intakes are low. When total fat exceeds 35 percent, however, saturated fat usually rises to unhealthy levels. For a 2000-kcalorie diet, 20 to 35 percent represents 400 to 700 kcalories from fat (roughly 45 to 75 grams).

> DIETARY GUIDELINES FOR AMERICANS

Consume less than 10 percent of kcalories from saturated fat. Consume less than 300 milligrams per day of dietary cholesterol. Keep *trans*-fat consumption as low as possible by limiting foods that contain synthetic sources of *trans* fats, such as partially hydrogenated oils, and by limiting other solid fats.

According to surveys, diets in the United States provide about 34 percent of their total energy from fat, with saturated fat contributing about 11 percent of the total.[28] The average daily intake of *trans*-fatty acids in the United States is about 5 grams per day—mostly from products that have been hydrogenated. Cholesterol intakes in the United States average 224 milligrams a day for women and 333 for men.

Although it is very difficult to do, some people actually manage to eat too little fat—to their detriment. Among them are people with eating disorders, described in Highlight 8, and athletes. Athletes following a diet too low in fat (less than 20 percent of total kcalories) fall short on energy, vitamins, minerals, and essential fatty acids as well as on performance.[29] As a practical guideline, it is wise to include the equivalent of at least a teaspoon of fat in every meal—a little peanut butter on toast or mayonnaise in tuna salad, for example. Dietary recommendations that limit fat are designed for healthy people older than age 2; Chapter 16 discusses the fat needs of infants and young children.

Beware of fast-food meals delivering too much fat, especially saturated fat. This double bacon cheeseburger, fries, and milkshake provide more than 1600 kcalories, with almost 90 grams of fat and more than 30 grams of saturated fat—far exceeding dietary fat guidelines for the entire day.

› REVIEW IT Explain the relationships among saturated fats, *trans* fat, and cholesterol and chronic diseases, noting recommendations.

Although some fat in the diet is necessary, too much fat adds kcalories without nutrients, which leads to obesity and nutrient inadequacies. Too much saturated fat, *trans* fat, and cholesterol increases the risk of heart disease and possibly cancer. For these reasons, health authorities recommend a diet moderate in total fat and low in saturated fat, *trans* fat, and cholesterol.

5.6 Health Effects and Recommended Intakes of Monounsaturated and Polyunsaturated Fats

> **LEARN IT** Explain the relationships between monounsaturated and polyunsaturated fats and health, noting recommendations.

Whereas saturated fats, *trans* fats, and cholesterol are implicated in chronic diseases, monounsaturated and polyunsaturated fats seem to offer health benefits. For this reason, dietary recommendations suggest replacing sources of saturated fats, *trans* fats, and cholesterol with foods rich in monounsaturated and polyunsaturated fats—foods such as seafood, nuts, seeds, and vegetable oils. Table 5-4 lists major food sources of these various lipids.

Health Effects of Monounsaturated and Polyunsaturated Fats
Researchers examining eating patterns from around the world have noted that some diets support good health despite being high in fat. As Highlight 5 explains, the *type* of fat may be more important than the *amount* of fat.

Heart Disease Replacing saturated fats with unsaturated fats reduces LDL cholesterol and lowers the risk of heart disease.[30] To replace saturated fats with unsaturated fats, sauté foods in olive oil instead of butter, garnish salads with sunflower seeds instead of bacon, snack on mixed nuts instead of potato chips, use avocado instead of cheese on a sandwich, and eat salmon instead of steak. Table 5-5 (p. 154) shows how these simple substitutions can lower the saturated fat and raise the unsaturated fat in a meal. Highlight 5 provides more details about the benefits of healthy fats in the diet.

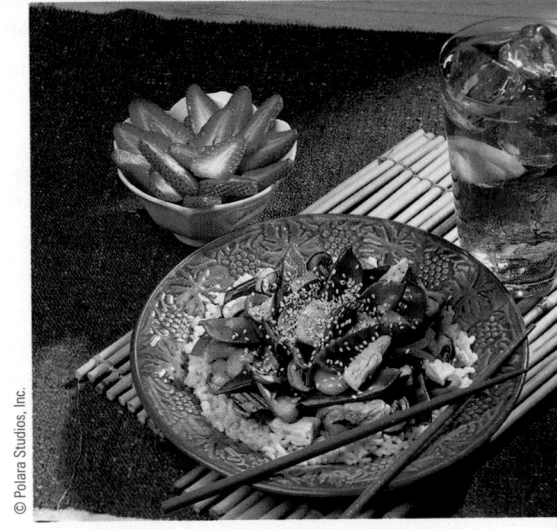

Well-balanced, healthy meals provide some fat with an emphasis on monounsaturated and polyunsaturated fats.

© Polara Studios, Inc.

TABLE 5-4 **Major Sources of Various Lipids**

Potentially Healthful Lipids Monounsaturated Fats	Omega-6 Polyunsaturated Fats	Omega-3 Polyunsaturated Fats
Avocado	Margarine (nonhydrogenated)	Fatty fish (herring, mackerel, salmon, sardines, tuna)
Nuts (almonds, cashews, filberts, hazelnuts, macadamia nuts, peanuts, pecans, pistachios)	Mayonnaise	Flaxseed, chia seed
	Nuts (pine nuts, walnuts)	Marine algae
Oils (canola, olive, peanut, sesame)	Oils (corn, cottonseed, safflower, soybean)	Nuts (walnuts)
Olives	Salad dressing	Oils (canola, flaxseed)
Peanut butter	Seeds (pumpkin, sunflower)	Yeast
Seeds (sesame)		

Potentially Harmful Lipids Saturated Fats	*Trans* Fats	Cholesterol
Bacon	Commercial baked goods (including doughnuts, cakes, cookies, pastries)	Eggs
Butter	Fried foods (hydrogenated shortening)	Meat, poultry, shellfish
Cheese		Milk and milk products
Chocolate	Many fast foods	
Coconut	Many snack foods (including microwave popcorn, chips, crackers)	
Cream cheese	Margarine (hydrogenated or partially hydrogenated)	
Cream, half-and-half	Nondairy creamers	
Ice cream	Shortening	
Lard		
Meats (fatty cuts of pork and beef)		
Milk and milk products (whole)		
Oils (coconut, palm, palm kernel) and products containing them (such as candies, cookies, doughnuts, pastries, pies)		
Shortening		
Sour cream		

© Cengage Learning

NOTE: Keep in mind that foods contain a mixture of fatty acids.

TABLE 5-5 Replacing Saturated Fat with Unsaturated Fat

Portion sizes have been adjusted so that each of these foods provides approximately 100 kcalories. Notice that for a similar number of kcalories and total grams of fat, the second choices offer less saturated fat and more unsaturated fat.

Replace these foods . . .

	Saturated Fat (g)	Unsaturated Fat (g)	Total Fat (g)
Butter (1 tbs)	7	4	11
Bacon (2 slices)	3	6	9
Potato chips (10 chips)	2	5	7
Cheese (1 slice)	4	4	8
Steak (1½ oz)	2	3	5
Totals	**18**	**22**	**40**

. . . with these foods.

	Saturated Fat (g)	Unsaturated Fat (g)	Total Fat (g)
Olive oil (1 tbs)	2	9	11
Sunflower seeds (2 tbs)	1	7	8
Mixed nuts (2 tbs)	1	8	9
Avocado (6 slices)	2	8	10
Salmon (2 oz)	1	3	4
Totals	**7**	**35**	**42**

© Cengage Learning

Research on the different types of fats has spotlighted the many beneficial effects of the omega-3 polyunsaturated fatty acids. Regular consumption of omega-3 fatty acids may help to prevent blood clots, protect against irregular heartbeats, improve blood lipids, and lower blood pressure, especially in people with hypertension or atherosclerosis.[31] In addition, omega-3 fatty acids support a healthy immune system and suppress inflammation.

Cancer The omega-3 fatty acids of fatty fish may protect against some cancers as well, perhaps by suppressing inflammation.[32] Even when omega-3 fats do not protect against cancer development, there seems to be a significant reduction in cancer-related deaths.[33] Thus dietary advice to reduce cancer risks parallels that given to reduce heart disease risks: reduce saturated fats and increase omega-3 fatty acids. Evidence does not support omega-3 supplementation.

Other Diseases Limited research suggests that the omega-3 fatty acids of fish may protect against bone loss, asthma, periodontal diseases, and eye diseases.[34] Omega-3 fats also appear to play a role in improving memory and cognition, but not depression.[35] Evidence on associations between fish consumption, omega-3 fatty acids, and diabetes is inconsistent.[36]

Omega-3 Supplements Omega-3 fatty acids are available in capsules of fish oil supplements, although routine supplementation is not recommended. High intakes of omega-3 polyunsaturated fatty acids may increase bleeding time, interfere with wound healing, raise LDL cholesterol, and suppress immune function. Such findings reinforce the concept that too much of a good thing can sometimes be harmful. People with heart disease, however, may benefit from doses greater than can be achieved through diet alone. Because high intakes of omega-3 fatty acids can cause excessive bleeding, supplements should be used only under close medical supervision.[37] For those who decide to use fish oil supplements, Figure 5-20 explains how to read the label.

Recommended Intakes of Monounsaturated and Polyunsaturated Fats
The 20 to 35 percent of kcalories from fat recommendation provides for the essential fatty acids—linoleic acid and linolenic acid—and Adequate Intakes (AI) have been established for these two fatty acids (see the inside front cover for details). The DRI suggest that linoleic acid provide 5 to 10 percent of the daily energy intake and linolenic acid 0.6 to 1.2 percent.

From Guidelines to Groceries
Fats accompany protein in foods derived from animals such as meat, seafood, poultry, and eggs, and fats accompany carbohydrate in foods derived from some plants such as avocados and coconuts. Fats carry with them the four fat-soluble vitamins—A, D, E, and K—together with many of the compounds that give foods their flavor, texture, and palatability. Fat is responsible for the delicious aromas associated with sizzling bacon,

> FIGURE 5-20 **How to Read a Fish Oil Supplement Label**

Supplement Facts

Serving Size: 1 Softgel
Servings Per Container: 120

	Amount Per Serving	% Daily Value
Calories	10	*
Total Fat	1	2%
Saturated Fat	0.5 g	3%
Trans Fat	0g	*
Polyunsaturated Fat	0.5g	*
Monounsaturated Fat	0 g	*
Cholesterol	5 mg	2%
Omega-3 Fatty Acids	300 mg	*
EPA (Eicosapentaenoic Acid)	180 mg	*
DHA (Docosahexaenoic Acid)	120 mg	*

* Daily Value not established.

Notice that this supplement offers 1000 mg of fish oil concentrate per capsule, but the oils offering the most health benefits are EPA and DHA. That information is in the Supplement Facts panel on the back.

Notice that one capsule of this supplement offers 180 mg of EPA and 120 mg of DHA, for a total of 300 mg of omega-3 oils—not 1000 mg. The recommended intake for omega-3 fatty acids is 500 mg per day. For heart health, consumers may need more, perhaps 2 to 4 grams (2000 to 4000 mg) per day.

hamburgers on the grill, onions being sautéed, and vegetables in a stir-fry. The essential oils of many spices are fat-soluble. Of course, these wonderful characteristics lure people into eating too much from time to time. With careful selections, a diet can support good health and still meet fat recommendations.

As the photos in Figure 5-21 (p. 156) show, fat accounts for much of the energy in foods, and removing the fat from foods cuts energy and saturated fat intakes dramatically. To reduce dietary fat, eliminate fat as a seasoning and in cooking; remove the fat from high-fat foods; replace high-fat foods with low-fat alternatives; and emphasize whole grains, fruits, and vegetables. The "How To" feature on p. 157 suggests additional heart-healthy choices by food group.

In general, except for seafood, animal fats tend to have a higher proportion of saturated fatty acids. Except for the tropical oils, plant foods tend to have a higher proportion of monounsaturated and polyunsaturated fatty acids. Consumers can find an abundant array of fresh, unprocessed foods that are naturally low in saturated fat, *trans* fat, cholesterol, and total fat. In addition, many familiar foods have been processed to provide less fat. For example, fat can be removed by skimming milk or trimming meats. Manufacturers can dilute fat by adding water or whipping in air. They can use fat-free milk in creamy desserts and lean meats in frozen entrées. Sometimes manufacturers simply prepare the products differently. For example, fat-free potato chips may be baked instead of fried. Such choices make healthy eating easy.

Protein Foods The fats in seafood, nuts, and seeds are considered oils, whereas the fats in meat, poultry, and eggs are considered solid fats because of their high fat, saturated fat, and cholesterol content. Because these meats provide high-quality protein and valuable vitamins and minerals, however, they can be included in a healthy diet if a person makes lean choices (see Table 5-6), prepares them using the suggestions

TABLE 5-6 Fat Options among the Protein Foods

Very lean options	Chicken (white meat, no skin)
	Cod, flounder, trout, tuna (canned in water)
	Legumes
Lean options	Beef or pork "round" or "loin" cuts
	Chicken (dark meat, no skin)
	Herring, salmon, tuna (canned in oil)
Medium-fat options	Ground beef
	Eggs
	Tofu
High-fat options	Bacon, hot dogs, luncheon meats, sausage
	Peanut butter
	Nuts

> FIGURE 5-21 **Cutting Fat Cuts kCalories—and Saturated Fat**

Pork chop with fat (340 kcal, 19 g fat, 7 g saturated fat)

Potato with 1 tbs butter and 1 tbs sour cream (350 kcal, 14 g fat, 10 g saturated fat)

Whole milk, 1 c (150 kcal, 8 g fat, 5 g saturated fat)

Pork chop with fat trimmed off (230 kcal, 9 g fat, 3 g saturated fat)

Savings:
110 kcal, 10 g fat, 4 g saturated fat

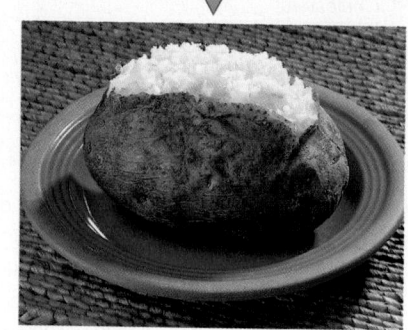

Plain potato (200 kcal, <1 g fat, 0 g saturated fat)

Savings:
150 kcal, 13 g fat, 10 g saturated fat

Fat-free milk, 1 c (90 kcal, <1 g fat, <1 g saturated fat)

Savings:
60 kcal, 7 g fat, 4 g saturated fat

outlined in the "How To" feature, and eats small portions. Selecting wild game or grass-fed cattle or bison instead of grain-fed livestock offers the nutrient advantages of being lower in fat and higher in omega-3 polyunsaturated fatty acids.[38] Another strategy to lower blood cholesterol is to prepare meals using soy protein instead of animal protein. When preparing meat, fish, or poultry, consider grilling, baking, or broiling, but not frying. Fried fish does not benefit heart disease.[39] Fried fish from fast-food restaurants and frozen fried fish products are often low in omega-3 fatty acids and high in *trans*- and saturated fatty acids.

Table 5-4 (p. 153) includes sources of omega-3 and omega-6 fatty acids, and Table 5-7 sorts fish and seafood by their quantity of omega 3 fatty acids. Fatty fish are among the best sources of omega-3 fatty acids, and Highlight 5 features their role in supporting heart health. The American Heart Association recommends eating at least two servings of fish a week, with an emphasis on fatty fish (salmon, herring, and mackerel, for example). Fish provides many minerals (except iron) and

TABLE 5-7 Omega 3 Fatty Acids in Fish and Seafood

3.5-oz serving	
500 mg	European sea bass (bronzini), herring (Atlantic and Pacific), mackerel, oysters (Pacific wild), salmon (wild and farmed), sardines, toothfish (includes Chilean sea bass), trout (wild and farmed)
150–500 mg	Black bass, catfish (wild and farmed), clams, cod (Atlantic), crab (Alaskan king), croakers, flounder, haddock, hake, halibut, oysters (eastern and farmed), perch, scallops, shrimp (mixed varieties), sole, swordfish, tilapia (farmed)
<150 mg	Cod (Pacific), grouper, lobster, mahi mahi, monkfish, red snapper, skate, triggerfish, tuna, wahoo

Make Heart-Healthy Choices—by Food Group

In General

- Select the most nutrient-dense foods from all food groups.
- Consume fewer and smaller portions of foods and beverages that contain solid fats.
- Check the Nutrition Facts label to choose foods with little or no saturated fat and no *trans* fat.

Grains

- Select breads, cereals, and crackers that are low in saturated and *trans* fat (for example, bagels instead of croissants).
- Prepare pasta with a tomato sauce instead of a cheese or cream sauce.
- Limit intake of cookies, doughnuts, pastries, and croissants.

Vegetables and Fruits

- Enjoy the natural flavor of steamed or roasted vegetables (without butter) for dinner and fruits for dessert.
- Eat at least two vegetables (in addition to a salad) with dinner.
- Snack on raw vegetables or fruits instead of high-fat items like potato chips.
- Buy frozen vegetables without sauce.

Milk and Milk Products

- Switch from whole milk to reduced-fat, from reduced-fat to low-fat, and from low-fat to fat-free (nonfat).
- Use fat-free and low-fat cheeses (such as part-skim ricotta and low-fat mozzarella) instead of regular cheeses.
- Use fat-free or low-fat yogurt or sour cream instead of regular sour cream.
- Use evaporated fat-free milk instead of cream.
- Enjoy fat-free frozen yogurt, sherbet, or ice milk instead of ice cream.

Protein Foods

- Fat adds up quickly, even with lean meat; limit intake to about 6 ounces (cooked weight) daily.
- Eat at least two servings of fish per week (particularly fish such as mackerel, lake trout, herring, sardines, and salmon).
- Choose fish, poultry, or lean cuts of pork or beef; look for unmarbled cuts named *round* or *loin* (eye of round, top round, bottom round, round tip, tenderloin, sirloin, center loin, and top loin).
- Choose processed meats such as lunch meats and hot dogs that are low in saturated fat and cholesterol.
- Trim the fat from pork and beef; remove the skin from poultry.
- Grill, roast, broil, bake, stir-fry, stew, or braise meats; don't fry. When possible, place food on a rack so that fat can drain.
- Use lean ground turkey or lean ground beef in recipes; brown ground meats without added fat, then drain off fat.
- Select tuna, sardines, and other canned meats packed in water; rinse oil-packed items with hot water to remove much of the fat.
- Fill kabob skewers with lots of vegetables and slivers of meat; create main dishes and casseroles by combining a little meat, fish, or poultry with a lot of pasta, rice, or vegetables.
- Use legumes often.
- Eat a meatless meal or two daily.
- Use egg substitutes in recipes instead of whole eggs or use two egg whites in place of each whole egg.

Fats and Oils

- Use small amounts of vegetable oils in place of solid fats.
- Use butter or stick margarine sparingly; select soft margarines instead of hard margarines.
- When selecting margarine, look for soft (liquid or tub) instead of hard (stick), ≤2 g saturated fat, liquid vegetable oil (not hydrogenated or partially hydrogenated) as the first ingredient, and "*trans*-fat free."
- Use fruit butters, reduced-kcalorie margarines, or butter replacers instead of butter.
- Use low-fat or fat-free mayonnaise and salad dressing instead of regular.
- Limit use of lard and meat fat.
- Limit use of products made with coconut oil, palm kernel oil, and palm oil (read labels on bakery goods, processed foods, popcorn oils, and nondairy creamers).
- Reduce use of hydrogenated shortenings and stick margarines and products that contain them (read labels on crackers, cookies, and other commercially prepared baked goods); use vegetable oils instead.

Miscellaneous

- Use a nonstick pan or coat the pan lightly with vegetable oil.
- Refrigerate soups and stews; when the fat solidifies, remove it before reheating.
- Use wine; lemon, orange, or tomato juice; herbs; spices; fruits; or broth instead of butter or margarine when cooking.
- Stir-fry in a small amount of oil; add moisture and flavor with broth, tomato juice, or wine.
- Use variety to enhance enjoyment of the meal: vary colors, textures, and temperatures—hot cooked versus cool raw foods—and use garnishes to complement food.
- Omit high-fat meat gravies and cheese sauces.
- Order pizzas with lots of vegetables, a little lean meat, and half the cheese.

> **TRY IT** Compare the total kcalories, grams of fat, and percent kcalories from fat for 1 cup of whole milk, reduced-fat milk, low-fat milk, and nonfat milk.

vitamins. Because fish is leaner than most other animal-protein sources it can help with weight-loss efforts. The combination of losing weight and eating fish improves blood lipids even more effectively than can be explained by either the weight loss or the omega-3 fats of the fish. Chapter 19 discusses the adverse consequences of mercury, an environmental contaminant common in some fish; in general, mercury is relatively high in tilefish (also called golden snapper or golden bass), swordfish, king mackerel, and shark and relatively low in cod, haddock, pollock, salmon, sole, tilapia, and most shellfish. Most healthy people who eat two servings of fish a week can maximize the health benefits while incurring minimal risks. Nonfish sources of omega-3 fatty acids such as flaxseed may have less benefit.[40]

Recall that cholesterol is found in all foods derived from animals. Consequently, eating fewer meats, eggs, and milk products helps lower dietary cholesterol intake (as well as total and saturated fat intakes). Most foods that are high in cholesterol are also high in saturated fat, but eggs are an exception. An egg contains only 1 gram of saturated fat but has a little more than 200 milligrams of cholesterol—roughly two-thirds of the recommended daily limit. Still, for the general population, eating eggs does not seem to increase the risk of heart disease.[41] For people with high blood cholesterol, however, limiting daily cholesterol intake to less than 200 milligrams may be beneficial. When eggs are included in the diet, other sources of cholesterol may need to be limited on that day. Eggs are a valuable part of the diet because they are inexpensive, useful in cooking, and a source of high-quality protein, other nutrients, and phytochemicals. To help consumers improve their omega-3 fatty acid intake, hens fed flaxseed, fish oil, or marine algae produce eggs rich in omega-3 fatty acids (up to 200 milligrams per egg).[42] Including even one enriched egg in the diet daily can significantly increase a person's intake of omega-3 fatty acids. Food manufacturers have produced several fat-free, cholesterol-free egg substitutes.

> **DIETARY GUIDELINES FOR AMERICANS**
Replace protein foods that are higher in solid fats (meat, poultry, and eggs) with choices that are lower in solid fats and higher in oils (seafood, nuts, and seeds).

Milk and Milk Products Like meats, milk and milk products should also be selected with an awareness of their fat, saturated fat, and cholesterol contents (see Table 5-8). Keep in mind that the fat in milk is a solid fat; it is apparent as butter, but less so when suspended in homogenized milk. Fat-free and low-fat milk products provide as much or more protein, calcium, and other nutrients as their whole-milk versions—but with little or no saturated fat. Selecting fermented milk products, such as yogurt, may also help to lower blood cholesterol. These foods increase the population and activity of bacteria in the colon that use cholesterol. Interestingly, cheese does not seem to raise LDL cholesterol as its saturated fat content might predict, perhaps because its calcium promotes fat excretion in the GI tract.[43] Such findings suggest that specific foods or nutrients within them influence the actions of their associated saturated fats.[44]

Vegetables, Fruits, and Grains Most vegetables and fruits naturally contain little or no fat. Although avocados and olives are exceptions, most of their fat is unsaturated, which is not harmful to heart health. Most grains contain only small amounts of fat. Consumers need to read food labels carefully, though, because many refined grain products such as fried taco shells, croissants, and biscuits are high in saturated fat, and pastries, crackers, and cookies may contain *trans* fats. Similarly, many people add butter, margarine, or cheese sauce to grains and vegetables, which raises the saturated- and *trans*-fat contents. Because fruits are often eaten without added fat, a diet that includes several servings of fruit daily can help a person meet the dietary recommendations for fat.

A diet rich in vegetables, fruits, whole grains, and legumes also offers abundant vitamin C, folate, vitamin A, vitamin E, and dietary fiber—all important

TABLE 5-8 Fat Options among Milk and Milk Products

Fat-free and low-fat options	Fat-free (skim) or 1% (low-fat) milk or yogurt (plain)
Reduced-fat options	2% milk or yogurt (plain)
High-fat options	Whole milk, yogurt Most cheeses

© 2016 Cengage Learning

in supporting health. Consequently, such a diet protects against disease by reducing saturated fat, cholesterol, and total fat as well as by increasing nutrients. It also provides valuable phytochemicals, which help defend against heart disease.

Solid Fats and Oils Solid fats include the fats in meat and poultry (as in poultry skin, luncheon meats, and sausage); the fats in whole milk, cheeses, and butter; shortening (as in fried foods and baked goods); and hard margarines. Because solid fats deliver an abundance of saturated fatty acids, they are considered discretionary kcalories. The fats of fish, nuts, and vegetable oils are *not* counted as discretionary kcalories because they provide valuable omega-3 fatty acids, essential fatty acids, and vitamin E. When discretionary kcalories are available, they may be used to add fats in cooking or at the table or to select higher fat items from the food groups.

> **DIETARY GUIDELINES FOR AMERICANS**
Reduce intake of solid fats (major sources of saturated and *trans* fats). Replace solid fats with oils (major sources of polyunsaturated and monounsaturated fats) when possible.

Some solid fats, such as butter and the fat trimmed from meat, are easy to see. Others—such as the fat that "marbles" a steak or is hidden in foods such as cheese—are less apparent and can be present in foods in surprisingly high amounts. Any *fried* food contains abundant solid fats—potato chips, french fries, fried wontons, and fried fish. Many *baked* goods, too, are high in solid fats—pie crusts, pastries, crackers, biscuits, cornbread, doughnuts, sweet rolls, cookies, and cakes.

Reports on *trans*-fatty acids raise the question whether margarine or butter is a better choice for heart health. The American Heart Association has stated that because butter is rich in both saturated fat and cholesterol whereas margarine is made from vegetable fat with no dietary cholesterol, margarine is still preferable to butter. Be aware that soft margarines (liquid or tub) are less hydrogenated and relatively lower in *trans*-fatty acids; consequently, they do not raise blood cholesterol as much as the saturated fats of butter or the *trans* fats of hard (stick) margarines do. Many manufacturers are now offering nonhydrogenated margarines that are "*trans*-fat free." In addition, manufacturers have developed margarines fortified with plant sterols that lower blood cholesterol. (Highlight 13 explores these and other functional foods designed to support health.) Whichever you decide to use, remember to use them sparingly.

Read Food Labels Labels list total fat, saturated fat, *trans* fat, and cholesterol contents of foods. Because each package provides information for a single serving and because serving sizes are standardized, consumers can easily compare similar products.

Total fat, saturated fat, and cholesterol are also expressed as "% Daily Values" for a person consuming 2000 kcalories. Using 30 percent of energy intake as the guideline for fat, the Daily Value is 65 grams of fat; using 10 percent for saturated fat, the Daily Value is 20 grams of saturated fat. The Daily Value for cholesterol is 300 milligrams regardless of energy intake. There is no Daily Value for *trans* fat, but consumers should try to keep intakes as low as possible and within the 10 percent allotted for saturated fat. People who consume more or less than 2000 kcalories daily can calculate their personal Daily Value for fat as described in the "How To" feature (p. 160).

Be aware that the "% Daily Value" for fat is not the same as "% kcalories from fat." This important distinction is explained in the "How To" feature on p. 161. Because recommendations apply to average daily intakes rather than individual food items, food labels do not provide "% kcalories from fat." Still, you can get an idea of whether a particular food is high or low in fat.

Calculate a Personal Daily Value for Fat

The % Daily Value for fat on food labels is based on 65 grams. To know how your intake compares with this recommendation, you can either count grams until you reach 65 or add the "% Daily Values" until you reach 100 percent—if your energy intake is 2000 kcalories a day. If your energy intake is more or less, you can calculate your personal daily fat allowance in grams. Suppose your energy intake is 1800 kcalories per day and your goal is 30 percent kcalories from fat. Multiply your total energy intake by 30 percent, then divide by 9:

1800 total kcal × 0.30 from fat = 540 fat kcal
540 fat kcal ÷ 9 kcal/g = 60 g fat

(In familiar measures, 60 grams of fat is about the same as ⅔ stick of butter or ¼ cup of oil.)

The accompanying table shows the numbers of grams of fat allowed per day for various energy intakes. With one of these numbers in mind, you can quickly evaluate the number of fat grams in foods you are considering eating.

Energy (kcal/day)	20% kCal from Fat	35% kCal from Fat	Fat (g/day)
1200	240	420	27–47
1400	280	490	31–54
1600	320	560	36–62
1800	360	630	40–70
2000	400	700	44–78
2200	440	770	49–86
2400	480	840	53–93
2600	520	910	58–101
2800	560	980	62–109
3000	600	1050	67–117

> **TRY IT** Calculate a personal daily fat allowance for a person with an energy intake of 2100 kcalories and a goal of 25 percent kcalories from fat.

Fat Replacers Some foods are made with **fat replacers**—ingredients that provide some of the taste and texture of fats, but with fewer kcalories. Because the body may digest and absorb some of these fat replacers, they may contribute energy, although significantly less energy than fat's 9 kcalories per gram.

Some fat replacers are derived from carbohydrate, protein, or fat. Carbohydrate-based fat replacers are used primarily as thickeners or stabilizers in foods such as soups and salad dressings. Protein-based fat replacers provide a creamy feeling in the mouth and are often used in foods such as ice creams and yogurts. Fat-based replacers act as emulsifiers and are heat stable, making them most versatile in shortenings used in cake mixes and cookies.

Fat replacers offering the sensory and cooking qualities of fats but none of the kcalories are called **artificial fats.** A familiar example of an artificial fat that has been approved for use in snack foods such as potato chips, crackers, and tortilla chips is **olestra.** Olestra's chemical structure is similar to that of a triglyceride but with important differences. A triglyceride is composed of a glycerol molecule with three fatty acids attached, whereas olestra is made of a sucrose molecule with six to eight fatty acids attached. Enzymes in the digestive tract cannot break the bonds of olestra, so unlike sucrose or fatty acids, olestra passes through the digestive system unabsorbed.

The FDA's evaluation of olestra's safety addressed two questions. First, is olestra toxic? Research on both animals and human beings supports the safety of olestra as a partial replacement for dietary fats and oils, with no reports of cancer or birth defects. Second, does olestra affect either nutrient absorption or the health of the digestive tract? When olestra passes through the digestive tract unabsorbed, it binds with some of the fat-soluble vitamins, A, D, E, and K, and carries them out of the body, robbing the person of these valuable nutrients. To compensate for these losses, the FDA requires manufacturers to fortify olestra

fat replacers: ingredients that replace some or all of the functions of fat and may or may not provide energy.

artificial fats: zero-energy fat replacers that are chemically synthesized to mimic the sensory and cooking qualities of naturally occurring fats but are totally or partially resistant to digestion.

olestra: a synthetic fat made from sucrose and fatty acids that provides 0 kcalories per gram; also known as *sucrose polyester.*

Understand "% Daily Value" and "% kCalories from Fat"

The "% Daily Value" that is used on food labels to describe the amount of fat in a food is not the same as the "% kcalories from fat" that is used in dietary recommendations to describe the amount of fat in the diet. They may appear similar, but their difference is worth understanding. Consider, for example, a piece of lemon meringue pie that provides 140 kcalories and 12 grams of fat. Because the Daily Value for fat is 65 grams for a 2000-kcalorie intake, 12 grams represent about 18 percent:

$$12 \text{ g} \div 65 \text{ g} = 0.18$$
$$0.18 \times 100 = 18\%$$

The pie's "% Daily Value" is 18 percent, or almost one-fifth, of the day's fat allowance.

Uninformed consumers may mistakenly believe that this food meets recommendations to limit fat to "20 to 35 percent kcalories," but it doesn't—for two reasons. First, the pie's 12 grams of fat contribute 108 of the 140 kcalories, for a total of 77 percent kcalories from fat:

$$12 \text{ g fat} \times 9 \text{ kcal/g} = 108 \text{ kcal}$$
$$108 \text{ kcal} \div 140 \text{ kcal} = 77\%$$

Second, the "percent kcalories from fat" guideline applies to a day's total intake, not to an individual food. Of course, if every selection throughout the day exceeds 35 percent kcalories from fat, you can be certain that the day's total intake will, too.

© PhotoDisc/Getty Images

Whether a person's energy and fat allowance can afford a piece of a lemon meringue pie depends on the other food and activity choices made that day.

> **TRY IT** Calculate the percent Daily Value and the percent kcalories from fat for ½ cup frozen yogurt that provides 115 kcalories and 4 grams of fat.

---

with vitamins A, D, E, and K. Saturating olestra with these vitamins does not make the product a good source of vitamins, but it does block olestra's ability to bind with the vitamins from other foods. An asterisk in the ingredients list informs consumers that these added vitamins are "dietarily insignificant."

Consumers need to keep in mind that low-fat and fat-free foods still deliver kcalories. Alternatives to fat can help to lower energy intake and support weight loss only when they actually *replace* fat and energy in the diet.[45]

> **REVIEW IT** Explain the relationships between monounsaturated and polyunsaturated fats and health, noting recommendations.
Some fat in the diet has health benefits, especially the monounsaturated and polyunsaturated fats that protect against heart disease and possibly cancer. For this reason, *Dietary Guidelines* recommend replacing saturated fats with monounsaturated and polyunsaturated fats, particularly omega-3 fatty acids from foods such as fatty fish, not from supplements. Many selection and preparation strategies can help bring these goals within reach, and food labels help to identify foods consistent with these guidelines.

Perhaps the best advice for consumers regarding fat in the diet would be to replace saturated fat with unsaturated fat. Sometimes these choices can be difficult, though, because fats make foods taste delicious. To maintain good health, must a person give up all high-fat foods forever—never again to eat marbled steak, hollandaise sauce, or gooey chocolate cake? Not at all. These foods bring pleasure to a meal and can be enjoyed as part of a healthy diet when eaten occasionally in small quantities; but they should not be everyday foods. The key dietary principle for fat is *moderation*, not *deprivation*. Appreciate the energy and enjoyment that fat provides, but take care not to exceed your needs.

Nutrition Portfolio

To maintain good health, eat enough, but not too much, fat and select the right kinds. Go to Diet & Wellness Plus and choose one of the days on which you have tracked your diet for the entire day. Go to the Intake Spreadsheet report. Scroll down until you see: fat (g), sat fat (g), mono fat (g), poly fat (g), and chol (g), which stand for grams of total fat, saturated fat, monounsaturated fat, polyunsaturated fat, and cholesterol, respectively. Use these columns to answer the following questions:

- List the types and amounts of fats and oils you ate on that day, making note of which are saturated, monounsaturated, or polyunsaturated and how your choices could include fewer saturated options.

- List the types and amounts of milk and milk products, meats, fish, and poultry you eat daily, noting how your choices could include more low-fat options.

- Describe choices you can make in selecting and preparing foods to lower your intake of solid fats.

DIET & WELLNESS PLUS To complete this exercise, go to your Diet & Wellness Plus at **www.cengagebrain.com**.

> **STUDY IT** To review the key points of this chapter and take a practice quiz, go to the study cards at the end of the book.

REFERENCES

1. O. Quehenberger and E. A. Dennis, The human plasma lipidome, *New England Journal of Medicine* 365 (2011): 1812–1823.

2. V. Remig and coauthors, *Trans* fats in America: A review of their use, consumption, health implications, and regulation, *Journal of the American Dietetic Association* 110 (2010): 585–592.

3. I. A. Brouwer, A. J. Wanders, and M. B. Katan, Effect of animal and industrial *trans* fatty acids on HDL and LDL cholesterol levels in humans—A quantitative review, *PLos One* 5 (2010): e9434.

4. S. W. Ing and M. A. Belury, Impact of conjugated linoleic acid on bone physiology: Proposed mechanism involving inhibition of adipogenesis, *Nutrition Reviews* 69 (2011): 123–131.

5. A. V. Khera and coauthors, Cholesterol efflux capacity, high-density lipoprotein function, and atherosclerosis, *New England Journal of Medicine* 364 (2011): 127–135.

6. X. Zhu and J. S. Parks, New roles of HDL in inflammation and hematopoiesis, *Annual Review of Nutrition* 32 (2012): 161–182.

7. D. M. Muoio, Metabolism and vascular fatty acid transport, *New England Journal of Medicine* 363 (2010): 291–293.

8. M. Krawczyk, L. Bonfrate, and P. Portincasa, Nonalcoholic fatty liver disease, *Best Practice and Research, Clinical Gastroenterology* 24 (2010): 695–708; G. Tarantino, S. Savastano, and A. Colao, Hepatic steatosis, low-grade chronic inflammation and hormone/growth factor/adipokine imbalance, *World Journal of Gastroenterology* 16 (2010): 4773–4783.

9. Y. Deng and P. E. Scherer, Adipokines as novel biomarkers and regulators of the metabolic syndrome, *Annals of the New York Academy of Sciences* 1212 (2010): E1–E19.

10. N. Ouchi and coauthors, Adipokines in inflammation and metabolic disease, *Nature Reviews, Immunology* 11 (2011): 85–97; C. Stryjecki and D. M. Mutch, Fatty acid-gene interactions, adipokines and obesity, *European Journal of Clinical Nutrition* 65 (2011): 285–297.

11. P. C. Calder, Mechanisms of action of (n-3) fatty acids, *Journal of Nutrition* 142 (2012): 592S–599S.

12. N. G. Bazan, M. F. Molina, and W. C. Gordon, Docosahexaenoic acid signalolipidomics in nutrition: Significance in aging, neuroinflammation, macular degeneration, Alzheimer's, and other neurodegenerative diseases, *Annual Review of Nutrition* 31 (2011): 321–351; E. E. Birch and coauthors, The DIAMOND (DHA Intake and Measurement of Neural Development) Study: A double-masked, randomized controlled clinical trial of the maturation of infant visual acuity as a function of the dietary level of docosahexaenoic acid, *American Journal of Clinical Nutrition* 91 (2010): 848–859; R. K. McNamara and coauthors, Docosahexaenoic acid supplementation increases prefrontal cortex activation during sustained attention in healthy boys: A placebo-controlled, dose-ranging, functional magnetic resonance imaging study, *American Journal of Clinical Nutrition* 91 (2010): 1060–1067.

13. R. Wall and coauthors, Fatty acids from fish: The anti-inflammatory potential of long-chain omega-3 fatty acids, *Nutrition Reviews* 68 (2010): 280–289.

14. A. P. Simopoulos, Evolutionary aspects of diet: The omega-6/omega-3 ratio and the brain, *Molecular Neurobiology* 44 (2011): 203–215.

15. K. Zelman, The great fat debate: A closer look at the controversy—Questioning the validity of age-old dietary guidance, *Journal of the American Dietetic Association* 111 (2011): 655–658.

16. P. Kris-Etherton, J. Fleming, and W. S. Harris, The debate about n-6 polyunsaturated fatty acid recommendations for cardiovascular health, *Journal of the American Dietetic Association* 110 (2010): 201–204.

17. L. H. Kuller, The great fat debate: Reducing cholesterol, *Journal of the American Dietetic Association* 111 (2011): 663–664.

18. Vital signs: Prevalence, treatment, and control of high levels of low-density lipoprotein cholesterol—United States, 1999–2002 and 2005–2008, *Morbidity and Mortality Weekly Report* 60 (2011): 109–114.

19. J. E. Hunter, J. Zhang, and P. M. Kris-Etherton, Cardiovascular disease risk of dietary stearic acid compared with *trans*, other saturated, and unsaturated fatty acids: A systematic review, *American Journal of Clinical Nutrition* 91 (2010): 46–63.

20. J. N. Kiage and coauthors, Intake of trans fat and all-cause mortality in the Reasons for Geographical and Racial Differences in Stroke (REGARDS) cohort, *American Journal of Clinical Nutrition* 97 (2013): 1121–1128; F. Imamura and coauthors, Novel circulating fatty acid patterns and risk of cardiovascular disease: The Cardiovascular Health Study, *American Journal of Clinical Nutrition* 96 (2012): 1252–1261.

21. I. A. Brouwer, A. J. Wanders, and M. B. Katan, Effect of animal and industrial trans fatty acids on HDL and LDL cholesterol levels in humans—A quantitative review, *PLos One* 5 (2010): e9434.

22. F. O. Otite and coauthors, Trends in trans fatty acids reformulations of US supermarket and brand-name foods from 2007 through 2011, *Preventing Chronic Disease* 10 (2013): 120198; W. H. Dietz and K. S. Scanlon, Eliminating the use of partially hydrogenated oil in food production and preparation, *Journal of the American Medical Association* 308 (2012): 143–144; I. Rahkovsky, S. Martinez, and F. Kuchler, *New Food Choices Free of Trans Fats Better Align US Diets with Health Recommendations*, EIB-95, US Department of Agriculture, Economic Research Service, April 2012.

23. A. Baylin, Secular trends in trans fatty acids: Decreased trans fatty acids in the food supply are reflected in decreased *trans* fatty acids in plasma, *American Journal of Clinical Nutrition* 97 (2013): 665–666.

24. J. M. Lecerf and M. deLorgeril, Dietary cholesterol: From physiology to cardiovascular risk, *British Journal of Nutrition* 106 (2011): 6–14.

25. A. M. Brownawell and M. C. Falk, Cholesterol: Where science and public health policy intersect, *Nutrition Reviews* 68 (2010): 355–364; M. L. Fernandez and M. Calle, Revisiting dietary cholesterol recommendations: Does the evidence support a limit of 300 mg/d? *Current Atherosclerosis Reports* 12 (2010): 377–383.

26. T. Psaltopoulou and coauthors, Olive oil intake is inversely related to cancer prevalence: A systematic review and a meta-analysis of 13800 patients and 23340 controls in 19 observational studies, *Lipids in Health and Disease* 10 (2011): 127.

27. P. R. Trumbo and T. Shimakawa, Tolerable upper intake levels for *trans* fat, saturated fat, and cholesterol, *Nutrition Reviews* 69 (2011): 270–278.

28. US Department of Agricultural Research Service, Nutrient intakes from food: Mean amounts consumed per individual, 2009–2010, www.ars.usda.gov/ba/bhnrc/fsrg, updated July 2012.

29. Position of the American Dietetic Association, Dietitians of Canada, and the American College of Sports Medicine: Nutrition and athletic performance, *Journal of the American Dietetic Association* 100 (2000): 1543–1556.

30. A. Astrup, The role of reducing intakes of saturated fats in the prevention of cardiovascular disease: Where does the evidence stand in 2010? *American Journal of Clinical Nutrition* 93 (2011): 684–688.

31. A. C. Skulas-Ray and coauthors, Dose-response effects of omega-3 fatty acids on triglycerides, inflammation, and endothelial function in healthy persons with moderate hypertriglyceridemia, *American Journal of Clinical Nutrition* 93 (2011): 243–252; K. Musa-Velosa and coauthors, Long-chain omega-3 fatty acids eicosapentaenoic acid and docosahexaenoic acid dose-dependently reduce fasting serum triglycerides, *Nutrition Reviews* 68 (2010): 155–167.

32. J. Zheng and coauthors, Intake of fish and marine n-3 polyunsaturated fatty acids and risk of breast cancer: Meta-analysis of data from 21 independent prospective cohort studies, *British Medical Journal* 346 (2013): f3706; K. He and coauthors, Types of fish consumed and fish preparation methods in relation to pancreatic cancer incidence: The VITAL Cohort Study, *American Journal of Epidemiology* 177 (2013): 152–160; S. Wu and coauthors, Fish consumption and colorectal cancer risk in humans: A systematic review and meta-analysis, *American Journal of Medicine* 125 (2012): 551–559; V. M. Heinze and A. B. Actis, Dietary conjugated linoleic acid and long-chain n-3 fatty acids in mammary and prostate cancer protection: A review, *International Journal of food Sciences and Nutrition* 63 (2012): 66–78; M. Touvier and coauthors, Modulation of the association between plasma intercellular adhesion molecule-1 and cancer risk by n-3 PUFA intake: A nested case-control study, *American Journal of Clinical Nutrition* 95 (2012): 944–950; T. M. Brasky and coauthors, Specialty supplements and breast cancer risk in the VITamins And Lifestyle (VITAL) Cohort, *Cancer Epidemiology, Biomarkers and Prevention* 19 (2010): 1696–1708.

33. K. M. Szymanski, D. C. Wheeler, and L. A. Mucci, Fish consumption and prostate cancer risk: A review and meta-analysis, *American Journal of Clinical Nutrition* 92 (2010): 1223–1233.

34. J. Li and coauthors, Intakes of long-chain omega-3 (n-3) PUFAs and fish in relation to incidence of asthma among American young adults: The CARDIA study, *American Journal of Clinical Nutrition* 97 (2013): 173–178; E. K. Farina and coauthors, Protective effects of fish intake and interactive effects of long-chain polyunsaturated fatty acid intakes on hip bone mineral density in older adults: The Framingham Osteoporosis Study, *American Journal of Clinical Nutrition* 93 (2011): 1142–1151; W. G. Christen and coauthors, Dietary v-3 fatty acid and fish intake and incident age-related macular degeneration in women, *Archives of Ophthalmology* 129 (2011): 921–929; E. Y. Chew, Fatty acids and retinopathy, *New England Journal of Medicine* 364 (2011): 1970–1971; A. Liu and coauthors, Long-chain and very long-chain polyunsaturated fatty acids in ocular aging and age-related macular degeneration, *Journal of Lipid Research* 51 (2010): 3217–3229; E. K. Kaye, n-3 Fatty acid intake and periodontal disease, *Journal of the American Dietetic Association* 110 (2010): 1650–1652.

35. W. Stonehouse, DHA supplementation improved both memory and reaction time in healthy young adults: A randomized controlled trial, *American Journal of Clinical Nutrition* 97 (2013): 1134–1143; L. J. Frensham, J. Bryan, and N. Parletta, Influences of micronutrient and omega-3 fatty acid supplementation on cognition, learning, and behavior: Methodical considerations and implications for children and adolescents in developed societies, *Nutrition Reviews* 70 (2012): 594–610; C. Chiu and coauthors, Associations between n-3 PUFA concentrations and cognitive function after recovery from late-life depression, *American Journal of Clinical Nutrition* 95 (2012): 420–427; M. Lucas and coauthors, Dietary intake of n-3 and n-6 fatty acids and the risk of clinical depression in women: A 10-y prospective follow-up study, *American Journal of Clinical Nutrition* 93 (2011): 1337–1343.

36. P. Xun and K. He, Meta-analysis of data from 438,000 individuals in 12 independent prospective cohorts with an average 11-year follow-up, *Diabetes Care* 35 (2012): 930–938; A. Wallin and coauthors, Fish consumption, dietary long-chain n-3 fatty acids, and risk of type 2 diabetes: Systematic review and meta-analysis of prospective studies, *Diabetes Care* 32 (2012): 918–929; A. Nanri and coauthors, Fish intake and type 2 diabetes in Japanese men and women: The Japan Public Health Center-based Prospective Study, *American Journal of Clinical Nutrition* 94 (2011): 884–891; R. Villegas and coauthors, Fish, shellfish, and long-chain n-3 fatty acid consumption and risk of incident type 2 diabetes in middle-aged Chinese men and women, *American Journal of Clinical Nutrition* 94 (2011): 543–551; D. P. Brostow and coauthors, Omega-3 fatty acids and incident type 2 diabetes: The Singapore Chinese Health Study, *American Journal of Clinical Nutrition* 94 (2011): 520–526; L. Djoussé and coauthors, Plasma omega-3 fatty acids and incident diabetes in older adults, *American Journal of Clinical Nutrition* 94 (2011): 527–533.

37. American Heart Association, Fish 101, www.heart.org, updated March 2013.

38. A. J. McAfee and coauthors, Red meat from animals offered a grass diet increases plasma and platelet n-3 PUFA in healthy consumers, *British Journal of Nutrition* 105 (2011): 80–89.

39. R. J. Belin and coauthors, Fish intake and risk of incident heart failure: The Women's Health Initiative, *Circulation: Heart Failure* 4 (2011): 404–413.

40. D. B. Jump, C. M. Depner, and S. Tripathy, Omega-3 fatty acid supplementation and cardiovascular disease, *Journal of Lipid Research* 53 (2012): 2525–2545.

41. J. Y. Shin and coauthors, Egg consumption in relation to risk of cardiovascular disease and diabetes: A systematic review and meta-analysis, *American Journal of Clinical Nutrition* 98 (2013): 146–159; Y. Rong and coauthors, Egg consumption and risk of coronary heart disease and stroke: Dose-response meta-analysis of prospective cohort studies, *British Medical Journal* 346 (2013): e8539.

42. I. Fraeye and coauthors, Dietary enrichment of eggs with omega-3 fatty acids: A review, *Food Research International* 48 (2012): 961–969.

43. J. Hjerpsted, E. Leedo, and T. Tholstrup, Cheese intake in large amounts lowers LDL-cholesterol concentrations compared with butter intake of equal fat content, *American Journal of Clinical Nutrition* 94 (2011): 1479–1484.

44. M. C. Otto and coauthors, Dietary intake of saturated fat by food source and incident cardiovascular disease: the Multi-Ethnic Study of Atherosclerosis, *American Journal of Clinical Nutrition* 96 (2012): 397–404.

45. Position of the American Dietetic Association: Fat replacers, *Journal of the American Dietetic Association* 105 (2005): 266–275.

HIGHLIGHT > 5
High-Fat Foods—Friend or Foe?

> **LEARN IT** Identify which fats support health and which impair it.

Eat less fat. Eat more fatty fish. Give up butter. Use margarine. Give up margarine. Use olive oil. Steer clear of saturated. Seek out omega-3. Stay away from *trans*. Stick with monounsaturated and polyunsaturated. Keep fat intake moderate. Today's fat messages seem to be forever multiplying and changing. No wonder some people feel confused about dietary fat. The confusion stems in part from the complexities of fat and in part from the nature of recommendations. As Chapter 5 explained, "dietary fat" refers to several kinds of fats. Some fats support health whereas others impair it, and foods typically provide a mixture of fats in varying proportions. Researchers have spent decades sorting through the relationships among the various kinds of fat and their roles in supporting or harming health. Translating these research findings into dietary recommendations is challenging. Too little information can mislead consumers, but too much detail can overwhelm them. As research findings accumulate, recommendations slowly evolve and become more refined. Fortunately, that's where we are with fat recommendations today—refining them from the general to the specific. Though they may seem to be "forever multiplying and changing," in fact, they are becoming more meaningful.

This highlight begins with a look at the dietary guidelines for fat intake. It continues by identifying which foods provide which fats and presenting the Mediterranean diet, an example of an eating pattern that embraces the heart-healthy fats. It closes with strategies to help consumers choose the right amounts of the right kinds of fats for a healthy diet.

Guidelines for Fat Intake

Dietary recommendations for fat have shifted emphasis from lowering total fat, in general, to limiting saturated and *trans* fat, specifically. Instead of urging people to cut back on all fats, recommendations suggest carefully replacing the "bad" saturated fats with the "good" unsaturated fats and enjoying them in moderation.[1] The goal is to create a diet moderate in kcalories that provides enough of the fats that support good health, but not too much of those that harm health. (Turn to pp. 150–154 for a review of the health consequences of each type of fat.)

With these findings and goals in mind, the Dietary Reference Intakes (DRI) committee suggests a healthy range of 20 to 35 percent of energy intake from fat. This range appears to be compatible with low rates of heart disease, diabetes, obesity, and cancer. Heart-healthy recommendations suggest that within this range, consumers should try to minimize their intakes of saturated fat, *trans* fat, and cholesterol and use monounsaturated and polyunsaturated fats instead.

Asking consumers to limit their total fat intake is less than perfect advice, but it is straightforward—find the fat and cut back. Asking consumers to keep their intakes of saturated fats, *trans* fats, and

cholesterol low and to use monounsaturated and polyunsaturated fats instead is more on target with heart health, but it also makes diet planning a bit more challenging. To make appropriate selections, consumers must first learn which foods contain which fats.

High-Fat Foods and Heart Health

Avocados, bacon, walnuts, potato chips, and mackerel are all high-fat foods, yet some of these foods have detrimental effects on heart health when consumed in excess, whereas others seem neutral or even beneficial. This section presents some of the accumulating evidence that help distinguish which high-fat foods belong in a healthy diet and which ones need to be kept to a minimum. As you will see, fat in the diet can be compatible with heart health, but only if most of it is unsaturated.

Cook with Olive Oil

As it turns out, the traditional diets of Greece and other countries in the Mediterranean region offer an excellent example of eating patterns that use "good" fats liberally. The primary fat in these diets is olive oil, which seems to play a key role in providing health benefits.[2] A classic study of the world's people, the Seven Countries Study, found that death rates from heart disease were strongly associated with diets high in saturated fats but only weakly linked with total fat.[3] In fact, the two countries with the highest fat intakes, Finland and the Greek island of Crete, had the highest (Finland) and lowest (Crete) rates of heart disease deaths. In both countries, the people consumed 40 percent or more of their kcalories from fat. Clearly, a high-fat diet is not the primary problem.[4] When researchers refocused their attention on the *type* of fat, they began to notice the benefits of olive oil.

A diet that uses olive oil instead of other fats, especially butter, stick margarine, and meat fats, offers numerous health benefits. Olive oil, canola oil, and other oils rich in monounsaturated fatty acids help to protect against heart disease and stroke by:[5]

- Lowering total and LDL cholesterol and not lowering HDL cholesterol or raising triglycerides
- Lowering LDL cholesterol susceptibility to oxidation
- Lowering blood-clotting factors
- Providing phytochemicals that act as antioxidants (see Highlight 11)
- Lowering blood pressure
- Interfering with the inflammatory response

When compared with other fats, olive oil seems to be a wise choice, but controlled clinical trials are too scarce to support population-wide recommendations to switch to a high-fat diet rich in olive oil. Importantly, olive oil is not a magic potion; drizzling it on foods does not make them healthier. Like other fats, olive oil delivers 9 kcalories per gram, which can contribute to weight gain in people who fail to balance their energy intake with physical activity. Its role in a healthy diet is to *replace* the saturated fats. Other vegetable oils, such as canola or safflower oil, are also generally low in saturated fats and high in unsaturated fats. For this reason, heart-healthy diets use these unsaturated vegetable oils to replace the more saturated fats of butter, hydrogenated stick margarine, lard, or shortening. (Remember that the tropical oils—coconut, palm, and palm kernel—are too saturated to be included with the heart-healthy vegetable oils.)

Nibble on Nuts

Tree nuts and peanuts are traditionally excluded from low-fat diets, and for good reasons. Nuts provide up to 80 percent of their kcalories from fat, and a quarter cup (about an ounce) of mixed nuts provides more than 200 kcalories. Frequent nut consumption, however, correlates with lower risk of mortality and chronic diseases, such as diabetes and heart disease.[6]

Olives and their oil may benefit heart health.

For heart health, snack on a few nuts instead of potato chips. Because nuts are energy dense (high in kcalories per ounce), it is especially important to keep portion size in mind when eating them.

Benefits are seen for a variety of nuts commonly eaten in the United States: almonds, Brazil nuts, cashews, hazelnuts, macadamia nuts, pecans, pistachios, walnuts, and even peanuts. On average, these nuts contain mostly monounsaturated fat (59 percent), some polyunsaturated fat (27 percent), and little saturated fat (14 percent). Nuts also provide valuable fiber, vegetable protein, vitamin E, minerals, and phytochemicals.

Including nuts may be a wise diet strategy against heart disease. Nuts may protect against heart disease by:[7]

- Lowering blood cholesterol
- Lowering blood pressure
- Limiting oxidative stress and inflammation

Some research suggests that a diet that includes nuts may benefit other diseases as well.

Before advising consumers to include nuts in their diets, however, a caution is in order. As mentioned, most of the energy nuts provide comes from fats. Consequently, they deliver many kcalories per bite. Incorporating nuts in the diet, however, does not necessarily lead to weight gains and may even help with weight control.[8] Consumers can enjoy nuts without increasing total kcalories by using nuts *instead of, not in addition to,* other foods (such as meats, potato chips, oils, margarine, and butter).

Feast on Fish

Research into the health benefits of the long-chain omega-3 polyunsaturated fatty acids began with a simple observation: the native peoples of Alaska, northern Canada, and Greenland, who eat a traditional diet rich in omega-3 fatty acids, notably EPA (eicosapentaenoic acid) and DHA (docosaheaenoic acid), have a remarkably low rate of heart disease even though their diets are relatively high in fat. These omega-3 fatty acids help to protect against heart disease by:[9]

- Reducing blood triglycerides
- Stabilizing plaque

- Lowering blood pressure and resting heart rate
- Reducing inflammation
- Serving as precursors to eicosanoids

For people with hypertension or atherosclerosis, these actions can be life saving.

Research studies have provided strong evidence that increasing omega-3 fatty acids in the diet supports heart health and lowers the rate of deaths from heart disease.[10] For this reason, the American Heart Association recommends including fish in a heart-healthy diet. People who eat some fish each week can lower their risks of heart attack and stroke.

Fish is the best source of EPA and DHA in the diet, but it is also a source of mercury, an environmental contaminant. Most fish contain at least trace amounts of mercury, but some have especially high levels. For this reason, the FDA advises pregnant and lactating women, women of childbearing age who may become pregnant, and young children to include fish in their diets, but to avoid:

- Tilefish (also called golden snapper or golden bass), swordfish, king mackerel, marlin, and shark

And to limit average weekly consumption of:

- A variety of ocean fish and shellfish to 12 ounces (cooked or canned)
- White (albacore) tuna to 6 ounces (cooked or canned)

Commonly eaten seafood relatively low in mercury include shrimp, catfish, pollock, salmon, and canned light tuna.

In addition to the direct toxic effects of mercury, some research suggests that mercury may diminish the health benefits of omega-3

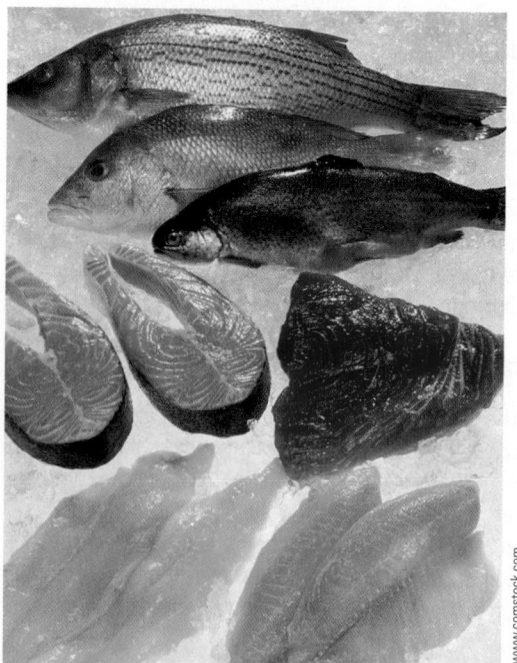

© www.comstock.com

Fish is a good source of the omega-3 fatty acids.

fatty acids. Such findings serve as a reminder that our health depends on the health of our planet. The protective effect of fish in the diet is available, provided that the fish and their surrounding waters are not heavily contaminated. (Chapter 19 discusses the adverse consequences of mercury, and Chapter 20 presents the relationships between diet and the environment in more detail.)

In an effort to limit exposure to pollutants, some consumers choose farm-raised fish. Compared with fish caught in the wild, farm-raised fish tend to be lower in mercury, but they are also lower in omega-3 fatty acids. When selecting fish, keep the diet strategies of variety and moderation in mind. Varying choices and eating moderate amounts helps to limit the intake of contaminants such as mercury.

High-Fat Foods and Heart Disease

The number-one dietary determinant of LDL cholesterol is saturated fat. Each 1 percent increase in energy from saturated fatty acids in the diet produces a 2 percent jump in heart disease risk by elevating LDL cholesterol. Conversely, reducing saturated fat intake by 1 percent can be expected to produce a 2 percent drop in heart disease risk by the same mechanism. Even a 2 percent drop in LDL represents a significant improvement for heart health. Like saturated fats, *trans* fats also raise heart disease risk by elevating LDL cholesterol. A heart-healthy diet limits foods rich in these two types of fat.

Limit Fatty Meats, Whole-Milk Products, and Tropical Oils

The major sources of saturated fats in the US diet are fatty meats, whole milk, tropical oils, and products made from any of these foods. To limit saturated fat intake, consumers must choose carefully among these high-fat foods. More than a third of the fat in most meats is saturated. Similarly, more than half of the fat is saturated in whole milk and other high-fat milk products, such as cheese, butter, cream, half-and-half, cream cheese, sour cream, and ice cream. The tropical oils of palm, palm kernel, and coconut, which are rarely used by consumers in the kitchen, are used heavily by food manufacturers, and are commonly found in many commercially prepared foods.

When choosing meats, milk products, and commercially prepared foods, look for those lowest in saturated fat. Labels provide a useful guide for comparing products in this regard, and Appendix H lists the saturated fat in several thousand foods.

Even with careful selections, a nutritionally adequate diet will provide some saturated fat. Zero saturated fat is not possible even when experts design menus with the mission to keep saturated fat as low as possible. Because most saturated fats come from animal foods, vegetarian diets can, and usually do, deliver fewer saturated fats than mixed diets.

Limit Hydrogenated Foods

Chapter 5 explained that solid shortening and margarine are made from vegetable oil that has been hardened through hydrogenation. This process both saturates some of the unsaturated fatty acids and introduces *trans*-fatty acids. Many convenience foods contain *trans* fats, including:

- Fried foods such as french fries, chicken, and other commercially fried foods
- Commercial baked goods such as cookies, doughnuts, pastries, breads, and crackers
- Snack foods such as chips
- Imitation cheeses

To keep *trans*-fat intake low, use these foods sparingly.

Table 5-4 (p. 153) summarizes which foods provide which fats. Substituting unsaturated fats for saturated fats at each meal and snack can help protect against heart disease. Figure H5-1 compares two meals and shows how such substitutions can lower saturated fat and raise unsaturated fat—even when total fat and kcalories remain unchanged.

The Mediterranean Diet

The links between good health and traditional Mediterranean eating patterns of the mid-1900s were introduced earlier with regard to olive oil. For people who eat these diets, the incidence of heart disease, some cancers, diabetes, and other chronic inflammatory diseases is low, and life expectancy is high.[11] Some research suggests that the health benefits of the Mediterranean eating pattern are partially due to its favorable effects on body weight.[12]

Although each of the many countries that border the Mediterranean Sea has its own culture, traditions, and dietary habits, their similarities are much greater than the use of olive oil alone. In fact, no one factor alone can be credited with reducing disease risks—the association holds true only when the overall eating pattern is present. Apparently, each of the foods contributes small benefits that harmonize to produce either a substantial cumulative or synergistic effect.

The Mediterranean eating pattern features fresh, whole foods. The people select crusty breads, whole grains, potatoes, and pastas; a variety of vegetables (including wild greens) and legumes; feta and mozzarella cheeses and yogurt; nuts; and fruits (especially grapes and figs). They eat some fish, other seafood, poultry, a few eggs, and little meat. Along with olives and olive oil, their principal sources of fat are nuts and fish; they rarely use butter or encounter hydrogenated fats. They commonly use herbs and spices instead of salt. Consequently, traditional Mediterranean diets are:

- Low in saturated fat
- Very low in *trans* fat

> FIGURE H5-1 **Two Meals Compared: Replacing Saturated Fat with Unsaturated Fat**

Examples of ways to replace saturated fats with unsaturated fats include sautéing vegetables in olive oil instead of butter, garnishing salads with avocado and sunflower seeds instead of bacon and blue cheese, and eating salmon instead of steak. Each of these meals provides roughly the same number of kcalories and grams of fat, but the one on the left has almost four times as much saturated fat and only half as many omega-3 fatty acids.

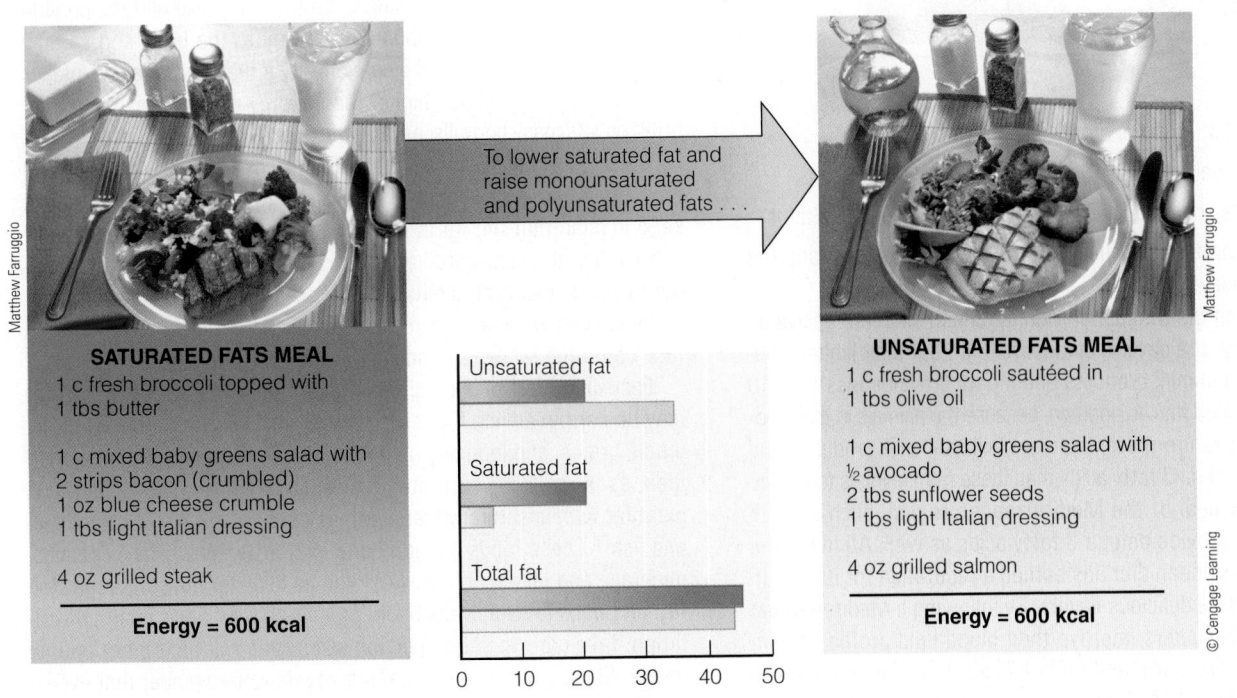

SATURATED FATS MEAL
1 c fresh broccoli topped with
1 tbs butter

1 c mixed baby greens salad with
2 strips bacon (crumbled)
1 oz blue cheese crumble
1 tbs light Italian dressing

4 oz grilled steak

Energy = 600 kcal

UNSATURATED FATS MEAL
1 c fresh broccoli sautéed in
1 tbs olive oil

1 c mixed baby greens salad with
½ avocado
2 tbs sunflower seeds
1 tbs light Italian dressing

4 oz grilled salmon

Energy = 600 kcal

To lower saturated fat and raise monounsaturated and polyunsaturated fats . . .

Unsaturated fat

Saturated fat

Total fat

0 10 20 30 40 50
GRAMS

Matthew Farruggio

© Cengage Learning

> **FIGURE H5-2** **Mediterranean Diet Pyramid**

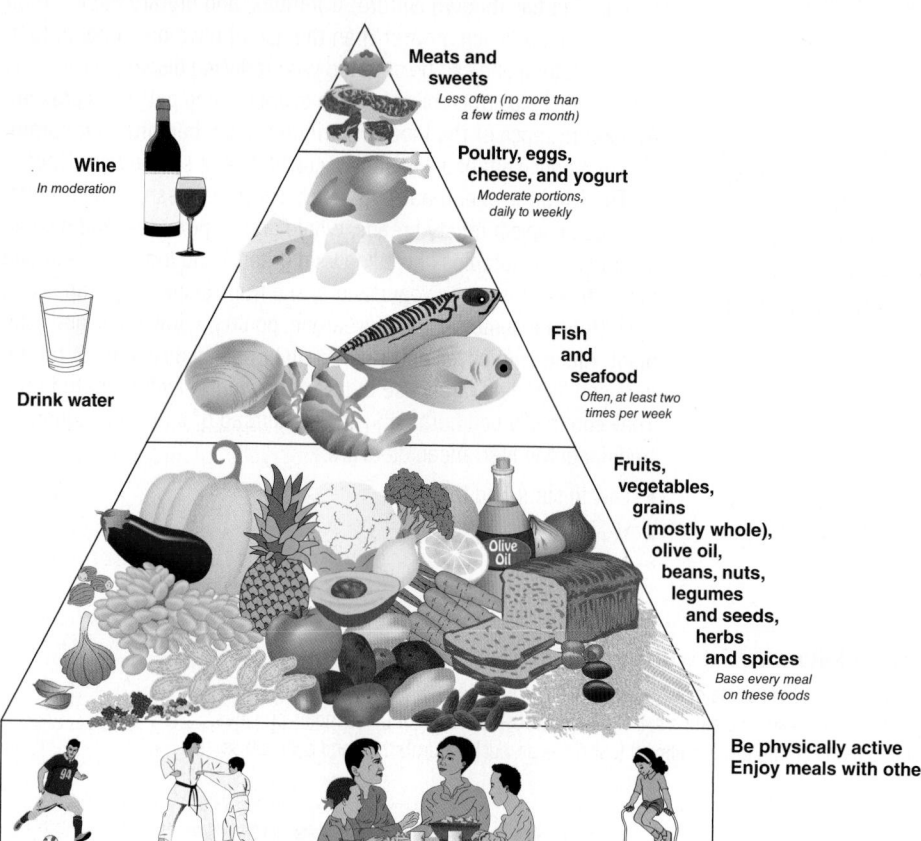

Mediterranean Diet Pyramid
A contemporary approach to delicious, healthy eating

Meats and sweets
Less often (no more than a few times a month)

Wine
In moderation

Poultry, eggs, cheese, and yogurt
Moderate portions, daily to weekly

Drink water

Fish and seafood
Often, at least two times per week

Fruits, vegetables, grains (mostly whole), olive oil, beans, nuts, legumes and seeds, herbs and spices
Base every meal on these foods

Olive Oil

Be physically active
Enjoy meals with others

© Cengage Learning

restaurants, however, has been adjusted to popular tastes. Quite often, these meals are much higher in saturated fats and meats—and much lower in the potentially beneficial constituents—than the traditional fare. Unfortunately, it appears that people in the Mediterranean region who are replacing some of their traditional dietary habits with those of the United States are losing the health benefits previously enjoyed. Figure H5-2 presents a Mediterranean Diet Pyramid. Notice the emphasis on abundant plant foods.

Conclusion

Are some fats "good," and others "bad" from the body's point of view? The saturated and *trans* fats indeed seem mostly bad for the health of the heart. Aside from providing energy, which unsaturated fats can do equally well, saturated and *trans* fats bring no indispensable benefits to the body. Furthermore, no harm can come from consuming diets low in them. Still, foods rich in these fats are often delicious, giving them an occasional place in the diet.

In contrast, the unsaturated fats are mostly good for heart health when consumed in moderation. To date, their one proven fault seems to be that they, like all fats, provide abundant energy to the body and so may promote obesity if they drive kcalorie intakes higher than energy needs. Obesity, in turn, often begets many body ills, as Chapters 8 and 9 describe.

Clearly, different fatty acids have different actions in the body and risks of chronic diseases.[14] When judging foods by their fatty acids, keep in mind that the fat in foods is a mixture of "good" and "bad," providing both unsaturated and saturated fatty acids. Even predominantly monounsaturated olive oil delivers some saturated fat. Consequently, even when a person chooses foods with mostly unsaturated fats, saturated fat can still add up if total fat is too high.

Focusing all efforts on simply lowering saturated fat in the diet may be narrow advice for heart health.[15] Including vegetables, fruits, whole grains, and legumes as part of a balanced daily diet is a good idea, as is *replacing* saturated fats such as butter, shortening, and meat fat with unsaturated fats such as olive oil and the oils from nuts and fish.[16] These foods provide beneficial fatty acids, fiber, vitamins, minerals, and phytochemicals as well as little (or no) salt, saturated fat, and *trans* fat—all valuable in protecting the body's health. In addition, take care to select portion sizes that will best meet energy needs. And enjoy some physical activity daily. Remember that even a healthy eating pattern can be detrimental if eaten in excess.[17]

- Rich in monounsaturated and polyunsaturated fat
- Rich in complex carbohydrate and fiber
- Rich in nutrients and phytochemicals that support good health

As a result, lipid profiles improve, inflammation diminishes, and the risk of heart disease declines.

People following the traditional Mediterranean diet can receive as much as 40 percent of a day's kcalories from fat, but their limited consumption of milk and milk products and meats provides less than 10 percent from saturated fats. In addition, because the animals in the Mediterranean region pasture-graze, the meat, milk and milk products, and eggs are richer in omega-3 fatty acids than those from animals fed grain.

Other foods typical of the Mediterranean region, such as wild plants and snails, provide omega-3 fatty acids as well. All in all, the traditional Mediterranean diet has earned a reputation for its health benefits as well as its delicious flavors. By following a Mediterranean eating pattern, consumers improve their blood lipid profile, insulin resistance, blood pressure, and body weight.[13] Consumers need to beware that the typical Mediterranean-style cuisine available in US

CRITICAL THINKING QUESTIONS

A. What are the features of a healthy high-fat diet?

B. Heart disease is rare among the Inuit people of Alaska who continue to eat their traditional diet of seal meat and blubber. A traditional Nordic diet of game meats, berries, root vegetables, and legumes helps to lower blood cholesterol and reduce heart disease risk. People following the traditional Mediterranean diet that emphasizes fruits, vegetables, whole grains, beans, nuts and seeds, and olive oil also enjoy good heart health. How are dietary fats related to heart health? How is it that such diverse diets can have such similar health outcomes?

REFERENCES

1. D. Mozaffarian, R. Micha, and S. Wallace, Effects on coronary heart disease of increasing polyunsaturated fat in place of saturated fat: A systematic review and meta-analysis of randomized controlled trials, *PLoS Medicine* 7 (2010): e10000252.

2. B. Bendinelli and coauthors, Fruit, vegetables, and olive oil and risk of coronary heart disease in Italian women: The EPICOR Study, *American Journal of Clinical Nutrition* 93 (2011): 275–283.

3. A. Keys, *Seven Countries: A Multivariate Analysis of Death and Coronary Heart Disease* (Cambridge: Harvard University Press, 1980).

4. W. C. Willet, The great fat debate: Total fat and health, *Journal of the American Dietetic Association* 111 (2011): 660–662.

5. L. Lin, Evidence of health benefits of canola oil, *Nutrition Reviews* 71 (2013): 370–385; L. Lucas, A. Russell, and R. Keast, Molecular mechanisms of inflammation: Anti-inflammatory benefits of virgin olive oil and the phenolic compound oleocanthal, *Current Pharmaceutical Design* 17 (2011): 754–768; C. Samieri and coauthors, Olive oil consumption, plasma oleic acid, and stroke incidence: The Three-City Study, *Neurology* 77 (2011): 1–8; D. Bester and coauthors, Cardiovascular effects of edible oils: A comparison between four popular edible oils, *Nutrition Research Reviews* 23 (2010): 334–348.

6. Y. Bao and coauthors, Association of nut consumption with total and cause-specific mortality, *New England Journal of Medicine* 369 (2013): 2001–2011; J. Sabaté and M. Wien, Nuts, blood lipids and cardiovascular disease, *Asia Pacific Journal of Clinical Nutrition* 19 (2010): 131–136.

7. C. E. Berryman and coauthors, Effects of almond consumption on the reduction of LDL-cholesterol: A discussion of potential mechanisms and future research directions, *Nutrition Reviews* 69 (2011): 171–185; E. Ros, L. C. Tapsell, and J. Sabaté, Nuts and berries for heart health, *Current Atherosclerosis Reports* 12 (2010): 397–406; J. Sabaté, K. Oda, and E. Ros, Nut consumption and blood lipid levels: A pooled analysis of 25 intervention trials, *Archives of Internal Medicine* 170 (2010): 821–827.

8. G. Flores-Mateo and coauthors, Nut intake and adiposity: Meta-analysis of clinical trials, *American Journal of Clinical Nutrition* 97 (2013): 1346–1355; V. Vadivel, C. N. Kunyanga, and H. K. Biesalski, Health benefits of nut consumption with special reference to body weight control, *Nutrition* 28 (2012): 1089–1097; M. Fogelhom and coauthors, Dietary macronutrients and food consumption as determinants of long-term weight change in adult populations: A systematic literature review, *Food and Nutrition Research* 56 (2012): doi 10.3402.

9. A. C. Skulas-Ray and coauthors, Dose-response effects of omega-3 fatty acids on triglycerides, inflammation, and endothelial function in healthy persons with moderate hypertriglyceridemia, *American Journal of Clinical Nutrition* 93 (2011): 243–252; P. C. Calder and P. Yaqoob, Omega-3 (n-3) fatty acids, cardiovascular disease and stability of atherosclerotic plaques, *Journal of Molecular Cell Biology* 56 (2010): 28–37; F. Dangardt and coauthors, Omega-3 fatty acid supplementation improves vascular function and reduces inflammation in obese adolescents, *Atherosclerosis* 212 (2010): 580–585; M. N. DiMinno and coauthors, Exploring newer cardioprotective strategies: ω-3 Fatty acids in perspective, *Journal of Thrombosis and Haemostasis* 104 (2010): 664–680.

10. D. Mozaffarian and J. H. Y. Wu, Omega-3 fatty acids and cardiovascular disease: Effects on risk factors, molecular pathways, and clinical events, *Journal of the American College of Cardiology* 58 (2011): 2047–2067.

11. K. Esposito and coauthors, Prevention and control of type 2 diabetes by Mediterranean diet: A systematic review, *Diabetes Research and Clinical Practice* 89 (2010): 97–102; P. P. McKeown and coauthors, Session 4: CVD, diabetes and cancer: Evidence for the use of the Mediterranean diet in patients with CHD, *Proceedings of the Nutrition Society* 69 (2010): 45–60; F. Sofi and coauthors, Accruing evidence on benefits of adherence to the Mediterranean diet on health: An updated systematic review and meta-analysis, *American Journal of Clinical Nutrition* 92 (2010): 1189–1196; L. Verberne and coauthors, Association between the Mediterranean diet and cancer risk: A review of observational studies, *Nutrition and Cancer* 62 (2010): 860–870.

12. J. J. Beunza and coauthors, Adherence to the Mediterranean diet, long-term weight change, and incident overweight or obesity: The Seguimiento Universidad de Navarra (SUN) cohort, *American Journal of Clinical Nutrition* 92 (2010): 1484–1493.

13. R. Estruch, Anti-inflammatory effects of the Mediterranean diet: The experience of the PREDIMED study, *Proceedings of the Nutrition Society* 69 (2010): 333–340.

14. S. J. Baum and coauthors, Fatty acids in cardiovascular health and disease: A comprehensive update, *Journal of Clinical Lipidology* 6 (2012): 216–234.

15. D. J. A. Jenkins, Effect of a dietary portfolio of cholesterol-lowering foods given at 2 levels of intensity of dietary advice on serum lipids in hyperlipidemia: A randomized controlled trial, *Journal of the American Medical Association* 306 (2011): 831–839.

16. Position of the Academy of Nutrition and Dietetics: Dietary fatty acids for healthy adults, *Journal of the Academy of Nutrition and Dietetics* 114 (2014): 136–153; D. Kromhout and coauthors, The confusion about dietary fatty acids recommendations for CHD prevention, *British Journal of Nutrition* 106 (2011): 627–632; D. Mozaffarian, The great fat debate: Taking the focus off of saturated fat, *Journal of the American Dietetic Association* 111 (2011): 665–666.

17. A. H. Lichtenstein, The great fat debate: The importance of message translation, *Journal of the American Dietetic Association* 111 (2011): 667–670.

6

Protein: Amino Acids

Nutrition in Your Life

The versatility of proteins in the body is impressive. They help your muscles to contract, your blood to clot, and your eyes to see. They keep you alive and well by facilitating chemical reactions and defending against infections. Without them, your bones, skin, and hair would have no structure. No wonder they were named *proteins*, meaning "of prime importance." Does that mean proteins deserve top billing in your diet as well? Are the best sources of protein beef, beans, or broccoli? Learn which foods will supply you with enough, but not too much, high-quality protein. In the Nutrition Portfolio at the end of this chapter, you can determine whether your diet is meeting your protein needs.

A few misconceptions surround the roles of protein in the body and the importance of protein in the diet. For example, people who associate meat with protein and protein with strength may eat steak to build muscles. Their thinking is only partly correct, however. Protein is a vital structural and working substance in all cells—not just muscle cells. To build strength, muscle cells need physical activity and all the nutrients—not just protein. Furthermore, protein is found in milk, eggs, legumes, and many grains and vegetables—not just meat. By overvaluing protein and overemphasizing meat in the diet, a person may mistakenly crowd out other, equally important nutrients and foods. As this chapter describes the various roles of protein in the body and food sources in the diet, keep in mind that protein is one of many nutrients needed to maintain good health.

6.1 The Chemist's View of Proteins

› **LEARN IT** Recognize the chemical structures of amino acids and proteins.

Chemically, **proteins** contain nitrogen (N) atoms in addition to the same atoms as carbohydrates and lipids—carbon (C), hydrogen (H), and oxygen (O). These nitrogen atoms give the name *amino* (nitrogen containing) to the amino acids that make the links in the chains of proteins.

Amino Acids All **amino acids** have the same basic structure—a central carbon (C) atom with a hydrogen atom (H), an amino group (NH_2), and an acid group (COOH) attached to it. Remember, however, that carbon atoms must have four bonds, so a fourth attachment is necessary. This fourth site distinguishes each amino acid from the others. Attached to the central carbon at the fourth bond is a distinct atom, or group of atoms, known as the *side group* or *side chain* (see Figure 6-1).

Unique Side Groups The side groups on the central carbon vary from one amino acid to the next, making proteins more complex than either carbohydrates or lipids. A polysaccharide (starch, for example) may be several thousand units long, but each unit is a glucose molecule just like all the others. A protein, on the other hand, is made up of about 20 different amino acids, each with a different side group. Table 6-1 lists the amino acids most common in proteins.*

The simplest amino acid, glycine, has a hydrogen atom as its side group. A slightly more complex amino acid, alanine, has an extra carbon with three hydrogen atoms. Other amino acids have more complex side groups (see Figure 6-2 for examples). Thus, although all amino acids share a common structure, they differ in size, shape, electrical charge, and other characteristics because of differences in these side groups.

Nonessential Amino Acids More than half of the amino acids are *nonessential*, meaning that the body can synthesize them for itself. Proteins in foods usually deliver these amino acids, but it is not essential that they do so. The body can make all **nonessential amino acids,** given nitrogen to form the amino group and fragments from carbohydrate or fat to form the rest of the structure.

> FIGURE 6-1 Amino Acid Structure

All amino acids have a central carbon with an amino group (NH_2), an acid group (COOH), a hydrogen (H), and a side group attached. The side group is a unique chemical structure that differentiates one amino acid from another.

© Cengage Learning

proteins: compounds composed of carbon, hydrogen, oxygen, and nitrogen atoms, arranged into amino acids linked in a chain. Some amino acids also contain sulfur atoms.

amino (a-MEEN-oh) **acids:** building blocks of proteins. Each contains an amino group, an acid group, a hydrogen atom, and a distinctive side group, all attached to a central carbon atom.

• **amino** = containing nitrogen

nonessential amino acids: amino acids that the body can make (see Table 6-1); also called *dispensable amino acids.*

TABLE 6-1 Amino Acids

Proteins are made up of about 20 common amino acids. The first column lists the *essential amino acids* for human beings (those the body cannot make—that must be provided in the diet). The second column lists the *nonessential amino acids*. In special cases, some nonessential amino acids may become *conditionally essential*. In a newborn, for example, only five amino acids are truly nonessential; the other nonessential amino acids are conditionally essential until the metabolic pathways are developed enough to make those amino acids in adequate amounts.

Essential Amino Acids		Nonessential Amino Acids	
Histidine	(HISS-tuh-deen)	Alanine	(AL-ah-neen)
Isoleucine	(eye-so-LOO-seen)	Arginine	(ARJ-ih-neen)
Leucine	(LOO-seen)	Asparagine	(ah-SPAR-ah-geen)
Lysine	(LYE-seen)	Aspartic acid	(ah-SPAR-tic acid)
Methionine	(meh-THIGH-oh-neen)	Cysteine	(SIS-teh-een)
Phenylalanine	(fen-il-AL-ah-neen)	Glutamic acid	(GLU-tam-ic acid)
Threonine	(THREE-oh-neen)	Glutamine	(GLU-tah-meen)
Tryptophan	(TRIP-toe-fan, TRIP-toe-fane)	Glycine	(GLY-seen)
Valine	(VAY-leen)	Proline	(PRO-leen)
		Serine	(SEER-een)
		Tyrosine	(TIE-roe-seen)

© Cengage Learning

---

*These 20 amino acids can all be commonly found in proteins. In addition, other amino acids do not occur in proteins but can be found individually (for example, taurine and ornithine). Some amino acids occur in related forms (for example, proline can acquire an OH group to become hydroxyproline).

> FIGURE 6-2 **Examples of Amino Acids**

Note that all amino acids have a common chemical structure but that each has a different side group. Appendix C presents the chemical structures of the 20 amino acids most common in proteins.

Glycine   Alanine   Aspartic acid   Phenylalanine

Essential Amino Acids There are nine amino acids that the human body either cannot make at all or cannot make in sufficient quantity to meet its needs. These nine amino acids must be supplied by the diet; they are *essential*. The first column in Table 6-1 presents the **essential amino acids**. Some researchers refer to essential amino acids as *indispensable* and to nonessential amino acids as *dispensable*.

Conditionally Essential Amino Acids Sometimes a nonessential amino acid becomes essential under special circumstances. For example, the body normally uses the essential amino acid phenylalanine to make tyrosine (a nonessential amino acid). But if the diet fails to supply enough phenylalanine, or if the body cannot make the conversion for some reason (as happens in the inherited disease phenylketonuria, described in Highlight 6), then tyrosine becomes a **conditionally essential amino acid**.

Proteins Cells link amino acids end-to-end in a variety of sequences to form thousands of different proteins. A **peptide bond** unites each amino acid to the next.

Amino Acid Chains Condensation reactions connect amino acids, just as they combine two monosaccharides to form a disaccharide and three fatty acids with a glycerol to form a triglyceride. Two amino acids bonded together form a **dipeptide** (see Figure 6-3). By another such reaction, a third amino acid can be added to the chain to form a **tripeptide**. As additional amino acids join the chain, a **polypeptide** is formed. Most proteins are a few dozen to several hundred amino acids long. Figure 6-4 (p. 174) illustrates the protein insulin.

Primary Structure—Amino Acid Sequence The primary structure of a protein is determined by the sequence of amino acids. If a person could walk along a carbohydrate molecule like starch, the first stepping stone would be a glucose.

> FIGURE 6-3 **Condensation of Two Amino Acids to Form a Dipeptide**

Amino acid   +   Amino acid   HOH Water   Dipeptide

An OH group from the acid end of one amino acid and an H atom from the amino group of another join to form a molecule of water.

A peptide bond (highlighted in red) forms between the two amino acids, creating a dipeptide.

essential amino acids: amino acids that the body requires but cannot make, and so must be obtained from the diet (see Table 6-1); also called *indispensable amino acids.*

conditionally essential amino acid: an amino acid that is normally nonessential, but must be supplied by the diet in special circumstances when the need for it exceeds the body's ability to make it.

peptide bond: a bond that connects the acid end of one amino acid with the amino end of another, forming a link in a protein chain.

dipeptide (dye-PEP-tide)**:** two amino acids bonded together.

- **di** = two
- **peptide** = amino acid

tripeptide: three amino acids bonded together.

- **tri** = three

polypeptide: many (10 or more) amino acids bonded together.

- **poly** = many

> FIGURE 6-4 Amino Acid Sequence of Human Insulin

Human insulin is a relatively small protein that consists of 51 amino acids in two short polypeptide chains. (For amino acid abbreviations, see Appendix C.) Two bridges link the two chains. A third bridge spans a section within the short chain. Known as *disulfide bridges,* these links form between the cysteine (Cys) amino acids, whose side group contains sulfur (S).

© Cengage Learning

The next stepping stone would also be a glucose, and it would be followed by a glucose, and yet another glucose. But if a person were to walk along a polypeptide chain, each stepping stone would be one of 20 different amino acids. The first stepping stone might be the amino acid methionine. The second might be an alanine. The third might be a glycine, the fourth tryptophan, and so on. Walking along another polypeptide path, a person might step on a phenylalanine, then a valine, then a glutamine. In other words, amino acid sequences within proteins vary.

The amino acids can act somewhat like the letters in an alphabet. If you had only the letter G, all you could write would be a string of Gs: G–G–G–G–G–G–G. But with 20 different letters available, you can create poems, songs, and novels. Similarly, the 20 amino acids can be linked together in a variety of sequences—even more than are possible for letters in a word or words in a sentence. Thus the variety of possible sequences for polypeptide chains is tremendous.

Secondary Structure—Polypeptide Shapes The secondary structure of proteins is determined not by chemical bonds as between the amino acids but by weak electrical attractions within the polypeptide chain. As positively charged hydrogens attract nearby negatively charged oxygens, sections of the polypeptide chain twist into a helix or fold into a pleated sheet, for example. These shapes give proteins strength and rigidity.

Tertiary Structure—Polypeptide Tangles The tertiary structure of proteins occurs as long polypeptide chains twist and fold into a variety of complex, tangled shapes. The unique side group of each amino acid gives it characteristics that attract it to, or repel it from, the surrounding fluids and other amino acids. Some amino acid side groups are attracted to water molecules; they are *hydrophilic.* Other side groups are repelled by water; they are *hydrophobic.* As amino acids are linked together to make a polypeptide, the chain folds so that its hydrophilic side groups are on the outer surface near water; the hydrophobic groups tuck themselves inside, away from water. Similarly, the disulfide bridges in insulin (see Figure 6-4) determine its tertiary structure. The extraordinary and unique shapes of proteins enable them to perform their various tasks in the body. Some form globular or spherical structures that can carry and store materials within them, and some, such as those of tendons, form linear structures that are more than 10 times as long as they are wide. The intricate shape a protein finally assumes gives it maximum stability.

Quaternary Structure—Multiple Polypeptide Interactions Some polypeptides are functioning proteins just as they are; others need to associate with other polypeptides to form larger working complexes. The quaternary structure of proteins involves the interactions between two or more polypeptides. One molecule of **hemoglobin**—the large, globular protein molecule that, by the billions, packs the red blood cells and carries oxygen—is made of four associated polypeptide chains, each holding the mineral iron (see Figure 6-5).

Protein Denaturation When proteins are subjected to heat, acid, or other conditions that disturb their stability, they undergo **denaturation**—that is, they uncoil and lose their shapes and, consequently, also lose their ability to function. Past a certain point, denaturation is irreversible. Familiar examples of denaturation include the hardening of an egg when it is cooked, the curdling of milk when acid is added, and the stiffening of egg whites when they are whipped. In the body, proteins are denatured when they are exposed to stomach acid.

© Matthew Farruggio

Cooking an egg denatures its proteins.

hemoglobin (HE-moh-GLO-bin): the globular protein of the red blood cells that transports oxygen from the lungs to tissues throughout the body; hemoglobin accounts for 80 percent of the body's iron.

• **hemo** = blood

• **globin** = globular protein

denaturation (dee-NAY-chur-AY-shun): the change in a protein's shape and consequent loss of its function brought about by heat, agitation, acid, base, alcohol, heavy metals, or other agents.

> **REVIEW IT** Recognize the chemical structures of amino acids and proteins. Chemically speaking, proteins are more complex than carbohydrates or lipids; they are made of some 20 different amino acids, 9 of which the body cannot make (the essential amino acids). Each amino acid contains an amino group, an acid group, a hydrogen atom, and a distinctive side group, all attached to a central carbon atom. Peptide bonds link amino acids together in a series of condensation reactions to create proteins. The distinctive sequence of amino acids in each protein determines its unique shape and function.

6.2 Digestion and Absorption of Proteins

> **LEARN IT** Summarize protein digestion and absorption.

Proteins in foods do not become body proteins directly. Instead, dietary proteins supply the amino acids from which the body makes its own proteins. When a person eats foods containing protein, enzymes break the long polypeptides into short polypeptides, the short polypeptides into tripeptides and dipeptides, and, finally, the tripeptides and dipeptides into individual amino acids.

Protein Digestion Figure 6-6 (p. 176) illustrates the digestion of protein through the GI tract and includes the names and actions of protein's digestive enzymes. Proteins are crushed and moistened in the mouth, but the real action begins in the stomach.

In the Stomach The major event in the stomach is the partial breakdown (hydrolysis) of proteins. Hydrochloric acid uncoils (denatures) each protein's tangled strands so that digestive enzymes can attack the peptide bonds. The hydrochloric acid also converts the inactive form of the enzyme pepsinogen to its active form, **pepsin.*** Pepsin cleaves proteins—large polypeptides—into smaller polypeptides and some amino acids.

In the Small Intestine When polypeptides enter the small intestine, several pancreatic and intestinal **proteases** hydrolyze them further into short peptide chains, tripeptides, dipeptides, and amino acids.** Then **peptidase** enzymes on the membrane surfaces of the intestinal cells split most of the dipeptides and tripeptides into single amino acids. Only a few peptides escape digestion and enter the blood intact.

Protein Absorption A number of specific carriers transport amino acids (and some dipeptides and tripeptides) into the intestinal cells. Once inside the intestinal cells, amino acids may be used for energy or to synthesize needed compounds. Amino acids that are not used by the intestinal cells are transported across the cell membrane into the surrounding fluid where they enter the capillaries on their way to the liver.

Consumers lacking nutrition knowledge may fail to realize that most proteins are broken down to amino acids before absorption. They may be misled by advertisements urging them to "Take this enzyme supplement to help you digest your food." Or "Don't eat this food that contains these enzymes that will digest cells in your body." In reality, enzymes in supplements and foods are proteins that are digested to amino acids, just as all proteins are. Even the digestive enzymes—which function optimally at their specific pH—are denatured and digested when the pH of their environment changes. The enzyme pepsin, for example, which works best in the low pH of the stomach becomes inactive and digested when it enters the higher pH of the small intestine.

Another misconception is that eating predigested proteins (amino acid supplements) saves the body from having to digest proteins and keeps the digestive system from "overworking." Such a belief grossly underestimates the body's abilities. As a matter of fact, the digestive system handles whole proteins *better* than predigested ones because it dismantles and absorbs the amino acids at rates that are optimal for the body's use. (The last section of this chapter discusses protein and amino acid supplements further.)

---

*The inactive form of an enzyme is called a *proenzyme* or a *zymogen* (ZYE-moh-jen).
**A short peptide chain of four to nine amino acids is called an *oligopeptide* (OL-ee-go-PEP-tide); *oligo* means few.

> **FIGURE 6-5** **The Structure of Hemoglobin**

The shape of each polypeptide chain is determined by an amino acid sequence (primary structure) that twists into a helix (secondary structure) and bends itself into a ball shape (tertiary structure). Together, the four polypeptide chains make the globular hemoglobin protein (quaternary structure).

Iron

Heme, the nonprotein portion of hemoglobin, holds iron.

© Cengage Learning

pepsin: a gastric enzyme that hydrolyzes protein. Pepsin is secreted in an inactive form, *pepsinogen*, which is activated by hydrochloric acid in the stomach.

proteases (PRO-tee-aces): enzymes that hydrolyze protein.

peptidase: a digestive enzyme that hydrolyzes peptide bonds. *Tripeptidases* cleave tripeptides; *dipeptidases* cleave dipeptides.

- **tri** = three
- **di** = two

> FIGURE 6-6   Protein Digestion in the GI Tract

Mouth and salivary glands

Chewing and crushing moisten protein-rich foods and mix them with saliva to be swallowed

HYDROCHLORIC ACID AND THE DIGESTIVE ENZYMES

Stomach

Hydrochloric acid (HCl) uncoils protein strands and activates stomach enzymes:

$$\text{Protein} \xrightarrow{\text{Pepsin, HCl}} \text{Smaller polypeptides}$$

In the stomach:

Hydrochloric acid (HCl)
- Denatures protein structure
- Activates pepsinogen to pepsin

Pepsin
- Cleaves proteins to smaller polypeptides and some free amino acids
- Inhibits pepsinogen synthesis

Small intestine and pancreas

Pancreatic and small intestinal enzymes split polypeptides further:

$$\text{Poly-peptides} \xrightarrow{\substack{\text{Pancreatic} \\ \text{and} \\ \text{intestinal} \\ \text{proteases}}} \substack{\text{Tripeptides,} \\ \text{dipeptides,} \\ \text{amino acids}}$$

Then enzymes on the surface of the small intestinal cells hydrolyze these peptides and the cells absorb them:

$$\text{Peptides} \xrightarrow{\substack{\text{Intestinal} \\ \text{tripeptidases} \\ \text{and} \\ \text{dipeptidases}}} \substack{\text{Amino acids} \\ \text{(absorbed)}}$$

In the small intestine:

Enteropeptidase
- Converts pancreatic trypsinogen to trypsin

Trypsin
- Inhibits trypsinogen synthesis
- Cleaves peptide bonds next to the amino acids lysine and arginine
- Converts pancreatic procarboxypeptidases to carboxypeptidases
- Converts pancreatic chymotrypsinogen to chymotrypsin

Chymotrypsin
- Cleaves peptide bonds next to the amino acids phenylalanine, tyrosine, tryptophan, methionine, asparagine, and histidine

Carboxypeptidases
- Cleave amino acids from the acid (carboxyl) ends of polypeptides

Elastase and collagenase
- Cleave polypeptides into smaller polypeptides and tripeptides

Intestinal tripeptidases
- Cleave tripeptides to dipeptides and amino acids

Intestinal dipeptidases
- Cleave dipeptides to amino acids

Intestinal aminopeptidases
- Cleave amino acids from the amino ends of small polypeptides (oligopeptides)

© Cengage Learning

> **REVIEW IT** Summarize protein digestion and absorption.

Digestion is facilitated mostly by the stomach's acid and enzymes, which first denature dietary proteins, then cleave them into smaller polypeptides and some amino acids. Pancreatic and intestinal enzymes split these short polypeptides further, to tripeptides and dipeptides, and then split most of these to single amino acids. Then carriers in the membranes of intestinal cells transport the amino acids into the cells, where they are released into the bloodstream.

6.3 Proteins in the Body

> LEARN IT Describe how the body makes proteins and uses them to perform various roles.

The human body has an estimated 20,000 to 25,000 genes that code for hundreds of thousands of proteins. Relatively few proteins have been studied in detail, although this number is growing rapidly with the surge in knowledge gained from sequencing the human genome. The relatively few proteins described in this chapter illustrate the versatility, uniqueness, and importance of proteins. As you will see, each protein has a specific function, and that function is determined during protein synthesis.

Protein Synthesis Each human being is unique because of small differences in the body's proteins. These differences are determined by the amino acid sequences of proteins, which, in turn, are determined by genes. The following paragraphs describe in words the ways cells synthesize proteins; Figure 6-7 (p. 178) provides a pictorial description. Protein synthesis depends on a diet that provides adequate protein and all the essential amino acids.

The instructions for making every protein in a person's body are transmitted by way of the genetic information received at conception. This body of knowledge, which is filed in the DNA (deoxyribonucleic acid) within the nucleus of every cell, never leaves the nucleus.

Delivering the Instructions Transforming the information in DNA into the appropriate sequence of amino acids needed to make a specific protein requires two major steps:

$$DNA \xrightarrow{\text{transcription}} RNA \xrightarrow{\text{translation}} protein.$$

In the first step, known as **transcription,** a stretch of DNA is used as a template to make messenger RNA. Messenger RNA then carries the code across the nuclear membrane into the body of the cell, where it seeks out and attaches itself to one of the ribosomes (a protein-making machine, which is itself composed of RNA and protein). There the second step, known as **translation,** takes place. Situated on a ribosome, messenger RNA specifies the sequence in which the amino acids line up for the synthesis of a protein.

Lining Up the Amino Acids Other forms of RNA, called transfer RNA, collect amino acids from the cell fluid and take them to messenger RNA. Each of the 20 amino acids has a specific transfer RNA. Thousands of transfer RNA, each carrying its amino acid, cluster around the ribosomes, awaiting their turn to unload. When the messenger RNA calls for a specific amino acid, the transfer RNA carrying that amino acid moves into position. Then the next loaded transfer RNA moves into place and then the next and the next. In this way, the amino acids line up in the sequence that is genetically determined, and enzymes bind them together. Finally, the completed protein strand is released, and the transfer RNA are freed to return for another load of amino acids.

Sequencing Errors The sequence of amino acids in each protein determines its shape, which supports a specific function. An error in the amino acid sequence results in an altered protein—sometimes with dramatic consequences. The protein hemoglobin offers one example of such a genetic variation. In a person with **sickle-cell anemia,** two of hemoglobin's four polypeptide chains (described earlier on pp. 174–175) have the normal sequence of amino acids, but the other two chains do not—they have the amino acid valine in a position that is normally occupied by glutamic acid (see Figure 6-8, p. 179). This single alteration in the amino acid sequence changes the characteristics and shape of hemoglobin so much that it loses its ability to carry oxygen effectively. The red blood cells filled with this abnormal hemoglobin stiffen into elongated sickle, or crescent, shapes instead of maintaining their normal pliable disc shape—hence the name, sickle-cell anemia. Sickle-cell anemia raises energy needs, causes many medical problems, and can be fatal.[1] Caring for people with sickle-cell anemia includes diligent attention to factors such as infection, stress, and dehydration, all of which can trigger a crisis.

transcription: the process of messenger RNA being made from a template of DNA.

translation: the process of messenger RNA directing the sequence of amino acids and synthesis of proteins.

sickle-cell anemia: a hereditary form of anemia characterized by abnormal sickle- or crescent-shaped red blood cells. Sickled cells interfere with oxygen transport and blood flow. Symptoms are precipitated by dehydration and insufficient oxygen (as may occur at high altitudes) and include hemolytic anemia (red blood cells burst), fever, and severe pain in the joints and abdomen.

> FIGURE 6-7 **Protein Synthesis**

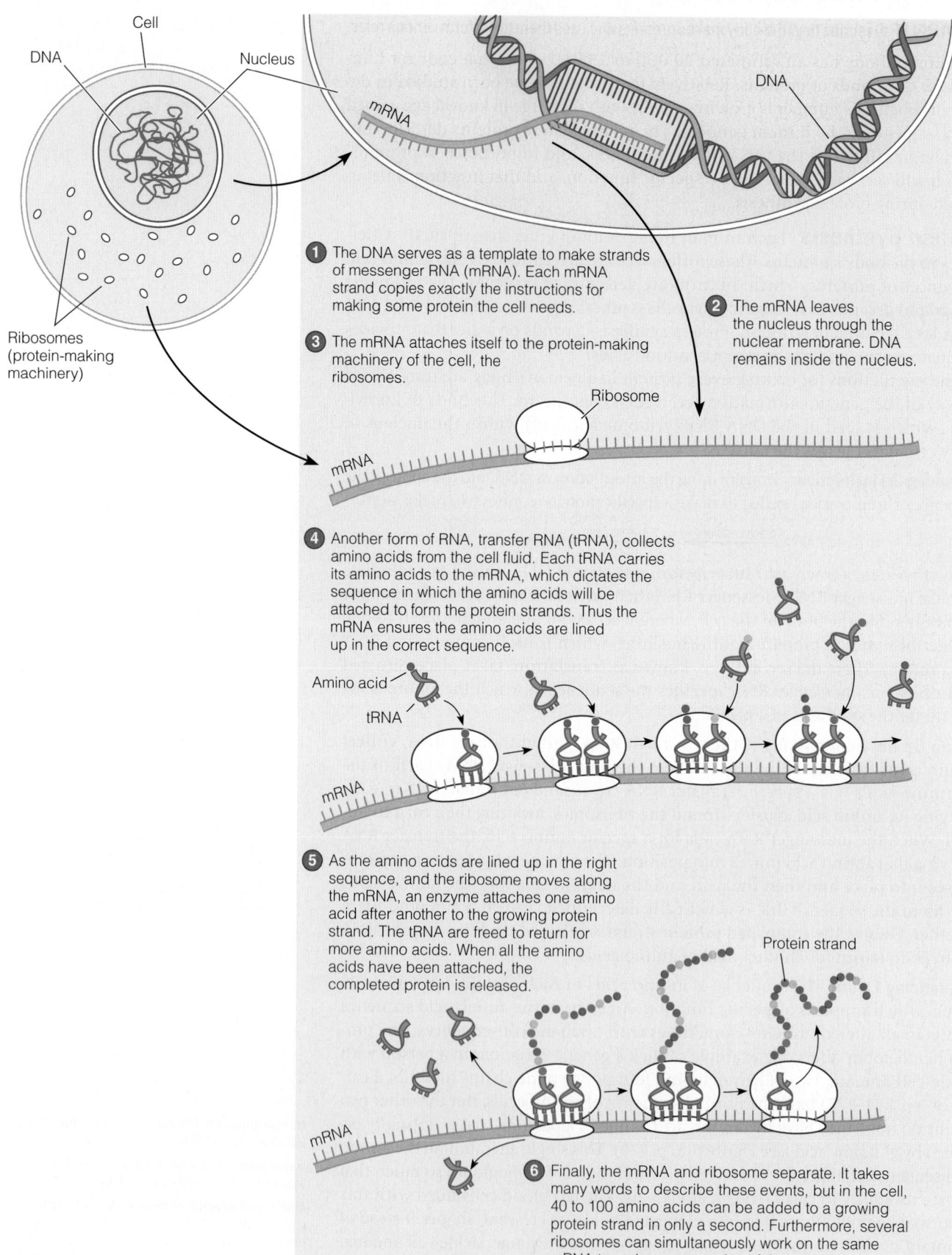

Cell

DNA

Nucleus

DNA

mRNA

Ribosomes (protein-making machinery)

1 The DNA serves as a template to make strands of messenger RNA (mRNA). Each mRNA strand copies exactly the instructions for making some protein the cell needs.

2 The mRNA leaves the nucleus through the nuclear membrane. DNA remains inside the nucleus.

3 The mRNA attaches itself to the protein-making machinery of the cell, the ribosomes.

Ribosome

mRNA

4 Another form of RNA, transfer RNA (tRNA), collects amino acids from the cell fluid. Each tRNA carries its amino acids to the mRNA, which dictates the sequence in which the amino acids will be attached to form the protein strands. Thus the mRNA ensures the amino acids are lined up in the correct sequence.

Amino acid

tRNA

mRNA

5 As the amino acids are lined up in the right sequence, and the ribosome moves along the mRNA, an enzyme attaches one amino acid after another to the growing protein strand. The tRNA are freed to return for more amino acids. When all the amino acids have been attached, the completed protein is released.

Protein strand

mRNA

6 Finally, the mRNA and ribosome separate. It takes many words to describe these events, but in the cell, 40 to 100 amino acids can be added to a growing protein strand in only a second. Furthermore, several ribosomes can simultaneously work on the same mRNA to make many copies of the protein.

Gene Expression When a cell makes a protein as described earlier, scientists say that the gene for that protein has been "expressed." Cells can regulate **gene expression** to make the type of protein, in the amounts and at the rate, they need. Nearly all of the body's cells possess the genes for making all human proteins, but each type of cell makes only the proteins it needs. For example, cells of the pancreas express the gene for insulin; in other cells, that gene is idle. Similarly, the cells of the pancreas do not make the protein hemoglobin, which is needed only by the red blood cells.

Recent research has unveiled some of the fascinating ways nutrients regulate gene expression and protein synthesis (see Highlight 6). Because diet plays an ongoing role in our lives from conception to death, it has a major influence on gene expression and disease development. The benefits of polyunsaturated fatty acids in defending against heart disease, for example, are partially explained by their role in influencing gene expression for lipid enzymes. Later chapters provide additional examples of relationships among nutrients, genes, and disease development.

Roles of Proteins
Whenever the body is growing, repairing, or replacing tissue, proteins are involved. Sometimes their role is to facilitate or to regulate; other times it is to become part of a structure. Versatility is a key feature of proteins.

As Structural Materials From the moment of conception, proteins form the building blocks of muscles, blood, and skin—in fact, protein is the major structural component of all the body's cells. To build a bone or a tooth, for example, cells first lay down a **matrix** of the protein **collagen** and then fill it with crystals of calcium, phosphorus, magnesium, fluoride, and other minerals.

Collagen also provides the material of ligaments and tendons and the strengthening "glue" between the cells of the artery walls that enables the arteries to withstand the pressure of the blood surging through them with each heartbeat. Also made of collagen are scars that knit the separated parts of torn tissues together.

Proteins are also needed for replacing dead or damaged cells. The average life span of a skin cell is only about 30 days. As old skin cells are shed, new cells made largely of protein grow from underneath to replace them. Cells in the deeper skin layers synthesize new proteins to form hair and fingernails. Muscle cells make new proteins to grow larger and stronger in response to exercise. Cells of the GI tract are replaced every few days. Both inside and outside, the body continuously uses protein to create new cells that replace those that have been lost.

As Enzymes Some proteins act as **enzymes.** Digestive enzymes have appeared in every chapter since Chapter 3, but digestion is only one of the many processes facilitated by enzymes. Enzymes not only break down substances, but they also build substances (such as bone) and transform one substance into another (amino acids into glucose, for example). Breaking down reactions are *catabolic*, whereas building up reactions are *anabolic*. (Chapter 7 provides more details.) Figure 6-9 diagrams a synthesis reaction.

An analogy may help to clarify the role of enzymes. Enzymes are comparable to the clergy and judges who make and dissolve marriages. When a minister marries two people, they become a couple, with a new bond between them. They are joined together—but the minister remains unchanged. The minister represents enzymes that synthesize large compounds from smaller ones. One minister can perform thousands of marriage ceremonies, just as one enzyme can expedite billions of reactions.

> **FIGURE 6-8** **Sickle Cell Compared with Normal Red Blood Cell**

Normally, red blood cells are disc-shaped, but in the inherited disorder sickle-cell anemia, red blood cells are sickle- or crescent-shaped. This alteration in shape occurs because valine replaces glutamic acid in the amino acid sequence of two of hemoglobin's polypeptide chains. As a result of this one alteration, the hemoglobin has a diminished capacity to carry oxygen.

© Cengage Learning

Normal red blood cell   Sickle-shaped blood cell

Amino acid sequence of normal hemoglobin:

Val —His —Leu — Thr — Pro — Glu —Glu

Amino acid sequence of sickle-cell hemoglobin:

Val —His —Leu — Thr — Pro — Val —Glu

gene expression: the process by which a cell converts the genetic code into RNA and protein.

matrix (MAY-tricks): the basic substance that gives form to a developing structure; in the body, the formative cells from which teeth and bones grow.

collagen (KOL-ah-jen): the structural protein from which connective tissues such as scars, tendons, ligaments, and the foundations of bones and teeth are made.

enzymes: proteins that facilitate chemical reactions without being changed in the process; protein catalysts.

> **FIGURE 6-9** **Enzyme Action**

Each enzyme facilitates a specific chemical reaction. In this diagram, an enzyme enables two compounds to make a more complex structure, but the enzyme itself remains unchanged.

  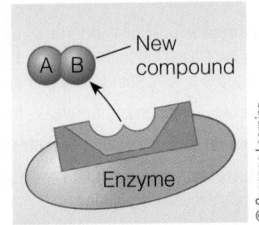

© Cengage Learning

The separate compounds, A and B, are attracted to the enzyme's active site, making a reaction likely.

The enzyme forms a complex with A and B.

The enzyme is unchanged, but A and B have formed a new compound, AB.

TABLE 6-2 Examples of Hormones and Their Actions

Hormones	Actions
Oxytocin and prolactin	Support lactation (see Chapter 15)
Growth hormone	Promotes growth
Insulin and glucagon	Regulate blood glucose (see Chapter 4)
Thyroxin	Regulates the body's metabolic rate (see Chapter 8)
Calcitonin and parathyroid hormone	Regulate blood calcium (see Chapter 12)
Angiotensin, renin, and antidiuretic hormone	Regulate fluid and electrolyte balance (see Chapter 12)

© Cengage Learning

Similarly, a judge who lets married couples separate may decree many divorces before retiring. The judge represents enzymes that hydrolyze larger compounds to smaller ones; for example, the digestive enzymes. The point is that, like the minister and the judge, enzymes themselves are not altered by the reactions they facilitate. They are catalysts, permitting reactions to occur more quickly and efficiently than if substances depended on chance encounters alone.

As Hormones The body's many hormones are messenger molecules, and *some* hormones are proteins. (Recall from Chapter 5 that some hormones, such as estrogen and testosterone, are made from the lipid cholesterol.) Various endocrine glands in the body release hormones in response to changes that challenge the body. The blood carries the hormones from these glands to their target tissues, where they elicit the appropriate responses to restore and maintain normal conditions.

The hormone insulin provides a familiar example. After a meal, when blood glucose rises, the pancreas releases insulin. Insulin stimulates the transport proteins of the muscles and adipose tissue to pump glucose into the cells faster than it can leak out. After acting on the message, the cells destroy the insulin. As blood glucose falls, the pancreas slows its release of insulin. Many other proteins act as hormones, regulating a variety of actions in the body (see Table 6-2 for examples).

As Regulators of Fluid Balance Proteins help to maintain the body's **fluid balance.** Normally, proteins are found primarily within the cells and in the plasma (essentially blood without its red blood cells). Being large, proteins do not normally cross the walls of the blood vessels. During times of critical illness or protein malnutrition, however, plasma proteins leak out of the blood vessels into the spaces between the cells. Because proteins attract water, fluid accumulates and causes swelling. Swelling due to an excess of fluid in the tissues is known as **edema.** The protein-related causes of edema include:

- Excessive protein losses caused by inflammation and critical illnesses
- Inadequate protein synthesis caused by liver disease
- Inadequate dietary intake of protein

Whatever the cause of edema, the result is the same: a diminished capacity to deliver nutrients and oxygen to the cells and to remove wastes from them. As a consequence, cells fail to function adequately.

As Acid-Base Regulators Proteins also help to maintain the balance between **acids** and **bases** within the body fluids. Normal body processes continuously produce acids and bases, which the blood carries to the kidneys and lungs for excretion. The challenge is to maintain acid-base balance as conditions continually change.

An acid solution contains an abundance of hydrogen ions (H+); the greater the concentration of hydrogen ions, the more acidic the solution and the lower the pH. Proteins, which have negative charges on their surfaces, attract hydrogen ions, which have positive charges. By accepting and releasing hydrogen ions, proteins act as **buffers,** maintaining the acid-base balance of the blood and body fluids.

The blood's acid-base balance is tightly controlled to maintain pH within the narrow range of between 7.35 and 7.45. Outside this range, either **acidosis** or **alkalosis** can lead to coma and death, largely by denaturing proteins. Denaturing a protein changes its shape and renders it useless. To give just one example, denatured hemoglobin loses its capacity to carry oxygen.

As Transporters Some proteins move about in the body fluids, carrying nutrients and other molecules. The protein hemoglobin carries oxygen from the lungs to the cells. The lipoproteins transport lipids around the body. Special transport proteins carry vitamins and minerals.

SPL/Science Source

In critical illness and protein malnutrition, blood vessels become "leaky" and allow plasma proteins to move into the tissues. Because proteins attract water, the tissues swell, causing edema.

fluid balance: maintenance of the proper types and amounts of fluid in each compartment of the body fluids (see also Chapter 12).

edema (eh-DEEM-uh): the swelling of body tissue caused by excessive amounts of fluid in the interstitial spaces; seen in protein deficiency (among other conditions).

acids: compounds that release hydrogen ions in a solution.

bases: compounds that accept hydrogen ions in a solution.

buffers: compounds that keep a solution's pH constant when acids or bases are added.

acidosis (assi-DOE-sis): higher-than-normal acidity in the blood and body fluids.

alkalosis (alka-LOE-sis): higher-than-normal alkalinity (base) in the blood and body fluids.

The transport of the mineral iron provides an especially good illustration of these proteins' specificity and precision. When iron is absorbed, it is captured in an intestinal cell by a protein. Before leaving the intestinal cell, iron is attached to another protein that carries it through the bloodstream to the cells. Once iron enters a cell, it is attached to a storage protein that will hold the iron until it is needed. When it is needed, iron is incorporated into proteins in the red blood cells and muscles that assist in oxygen transport and use. (Chapter 13 provides more details on how these protein carriers transport and store iron.)

Some transport proteins reside in cell membranes and act as "pumps," picking up compounds on one side of the membrane and releasing them on the other as needed. Each transport protein is specific for a certain compound or group of related compounds. Figure 6-10 illustrates how a membrane-bound transport protein helps to maintain the sodium and potassium concentrations in the fluids inside and outside cells. The balance of these two minerals is critical to nerve transmissions and muscle contractions; imbalances can cause irregular heartbeats, muscular weakness, kidney failure, and even death.

As Antibodies Proteins also defend the body against disease. A virus—whether it is one that causes flu, smallpox, measles, or the common cold—enters the cells and multiplies there. One virus may produce 100 replicas of itself within an hour or so. Each replica can then burst out and invade 100 different cells, soon yielding 10,000 viruses, which invade 10,000 cells. Left free to do their worst, they will soon overwhelm the body with disease.

Fortunately, when the body detects these invading **antigens,** it manufactures **antibodies,** giant protein molecules designed specifically to combat them. The antibodies work so swiftly and efficiently that in a healthy individual, most diseases never get started. Without sufficient protein, though, the body cannot maintain its army of antibodies to resist infectious diseases.

Each antibody is designed to destroy a specific antigen. Once the body has manufactured antibodies against a particular antigen (such as the measles virus), it "remembers" how to make them. Consequently, the next time the body encounters that same antigen, it produces antibodies even more quickly. In other words, the body develops a molecular memory, known as **immunity.** (Chapter 16 describes food allergies—the immune system's response to food antigens.)

As a Source of Energy and Glucose Without energy, cells die; without glucose, the brain and nervous system falter. Even though proteins are needed to do the

antigens: substances that elicit the formation of antibodies or an inflammation reaction from the immune system. A bacterium, a virus, a toxin, and a protein in food that causes allergy are all examples of antigens.

antibodies: large proteins of the blood and body fluids, produced by the immune system in response to the invasion of the body by foreign molecules (usually proteins called *antigens*). Antibodies combine with and inactivate the foreign invaders, thus protecting the body.

immunity: the body's ability to defend itself against diseases (see also Chapter 18).

> FIGURE 6-10   **An Example of a Transport Protein**

This transport protein resides within a cell membrane and acts as a two-door passageway. Molecules enter on one side of the membrane and exit on the other, but the protein doesn't leave the membrane. This example shows how the transport protein moves sodium and potassium in opposite directions across the membrane to maintain a high concentration of potassium and a low concentration of sodium within the cell. This active transport system requires energy.

Key:
- Sodium
- Potassium

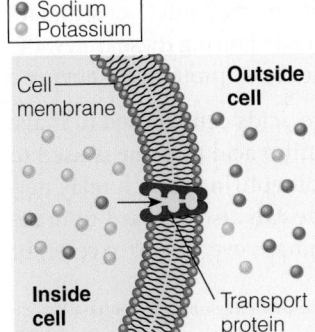
The transport protein picks up sodium from inside the cell.

The protein changes shape and releases sodium outside the cell.

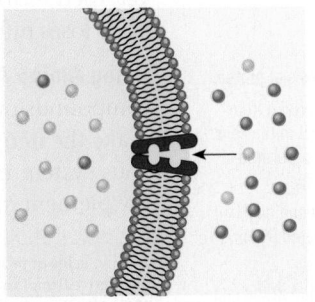
The transport protein picks up potassium from outside the cell.

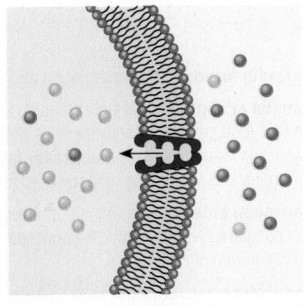
The protein changes shape and releases potassium inside the cell.

© Cengage Learning

Growing children end each day with more bone, blood, muscle, and skin cells than they had at the beginning of the day.

work that only they can perform, they will be sacrificed to provide energy and glucose during times of starvation or insufficient carbohydrate intake. The body will break down its tissue proteins to make amino acids available for energy or glucose production (a process known as *gluconeogenesis*). In this way, protein can maintain blood glucose levels, but at the expense of losing lean body tissue. Chapter 7 provides many more details on energy metabolism.

Other Roles As mentioned earlier, proteins form integral parts of most body structures such as skin, muscles, and bones. They also participate in some of the body's most amazing activities such as blood clotting and vision. When a tissue is injured, a rapid chain of events leads to the production of fibrin, a stringy, insoluble mass of protein fibers that forms a solid clot from liquid blood. Later, more slowly, the protein collagen forms a scar to replace the clot and permanently heal the wound. The light-sensitive pigments in the cells of the eye's retina are molecules of the protein opsin. Opsin responds to light by changing its shape, thus initiating the nerve impulses that convey the sense of sight to the brain.

The amino acids are as versatile as the proteins. In addition to serving as building blocks for proteins in the body, amino acids have multiple roles in regulating pathways that support growth, reproduction, metabolism, and immunity.

A Preview of Protein Metabolism This section previews protein metabolism; Chapter 7 provides a full description. Cells have several metabolic options, depending on their protein and energy needs.

Protein Turnover and the Amino Acid Pool Within each cell, proteins are continually being made and broken down, a process known as **protein turnover.** Protein breakdown releases amino acids. These amino acids mix with amino acids from dietary protein to form an **"amino acid pool"** within the cells and circulating blood.* The rate of protein degradation and the amount of protein intake may vary, but the pattern of amino acids within the pool remains fairly constant. Regardless of their source, any of these amino acids can be used to make body proteins or other nitrogen-containing compounds, or they can be stripped of their nitrogen and used for energy (either immediately or stored as fat for later use).

Nitrogen Balance Protein turnover and **nitrogen balance** go hand in hand. In healthy adults, protein synthesis balances with degradation, and protein intake from food balances with nitrogen excretion in the urine, feces, and sweat. When nitrogen intake equals nitrogen output, the person is in nitrogen equilibrium, or zero nitrogen balance. Researchers use nitrogen balance studies to estimate protein requirements.**

If the body synthesizes more than it degrades, then protein is added and nitrogen status becomes positive. Nitrogen status is positive in growing infants, children, adolescents, pregnant women, and people recovering from protein deficiency or illness; their nitrogen intake exceeds their nitrogen excretion. They are retaining protein in new tissues as they add blood, bone, skin, and muscle cells to their bodies.

If the body degrades more than it synthesizes, then protein is being lost and nitrogen status becomes negative. Nitrogen status is negative in people who are starving or suffering other severe stresses such as burns, injuries, infections, and fever; their nitrogen excretion exceeds their nitrogen intake. During these times, the body loses nitrogen as it breaks down muscle and other body proteins for energy.

Using Amino Acids to Make Other Compounds Amino acids can be used to make compounds other than proteins. For example, the amino acid tyrosine is used to make the **neurotransmitters** norepinephrine and epinephrine, which relay nervous system messages throughout the body. Tyrosine can also be used to make the pigment melanin, which is responsible for brown hair, eye, and skin color, or

protein turnover: the degradation and synthesis of protein.

amino acid pool: the supply of amino acids derived from either food proteins or body proteins that collect in the cells and circulating blood and stand ready to be incorporated in proteins and other compounds or used for energy.

nitrogen balance: the amount of nitrogen consumed (N in) as compared with the amount of nitrogen excreted (N out) in a given period of time.

neurotransmitters: chemicals that are released at the end of a nerve cell when a nerve impulse arrives there. They diffuse across the gap to the next cell and alter the membrane of that second cell to either inhibit or excite it.

*Amino acids or proteins that derive from within the body are *endogenous* (en-DODGE-eh-nus). In contrast, those that derive from foods are *exogenous* (eks-ODGE-eh-nus).
**The genetic materials DNA and RNA contain nitrogen, but the quantity is insignificant compared with the amount in protein. Protein is 16 percent nitrogen. Said another way, the average protein weighs about 6.25 times as much as the nitrogen it contains, so scientists can estimate the amount of protein in a sample of food, body tissue, or other material by multiplying the weight of the nitrogen in it by 6.25.

the hormone thyroxine, which helps to regulate the metabolic rate. For another example, the amino acid tryptophan serves as a precursor for the vitamin niacin and for **serotonin,** a neurotransmitter important in sleep regulation, appetite control, and sensory perception.

Using Amino Acids for Energy and Glucose As mentioned earlier, when glucose or fatty acids are limited, cells are forced to use amino acids for energy and glucose. The body does not have a specialized storage site for protein as it does for carbohydrate and fat. Recall that glucose is stored as glycogen in the liver and fat as triglycerides in adipose tissue, but protein is not stored as such. When the need arises, the body breaks down its working and structural proteins and uses the amino acids for energy or glucose. Thus, over time, energy deprivation (fasting or starvation) always causes wasting of lean body tissue as well as fat loss. An adequate supply of carbohydrates and fats spares amino acids from being used for energy and allows proteins to perform their unique roles.

Using Amino Acids to Make Fat Amino acids may be converted to fat when energy and protein intakes exceed needs and carbohydrate intake is adequate. In this way, protein-rich foods can contribute to weight gain.

Deaminating Amino Acids When amino acids are broken down (as occurs when they are used for energy or to make glucose or fat), they are first deaminated— stripped of their nitrogen-containing amino groups (see Figure 6-11). Two products result from **deamination:** one is **ammonia** (NH_3); the other product is the carbon structure without its amino group—often a **keto acid.** Keto acids may enter a number of metabolic pathways—for example, they may be used for energy or for the production of glucose, ketones, cholesterol, or fat.* They may also be used to make nonessential amino acids.

Using Amino Acids to Make Proteins and Nonessential Amino Acids As mentioned, cells can assemble amino acids into the proteins they need to do their work. If an essential amino acid is missing, the body may break down some of its own proteins to obtain it. If a particular nonessential amino acid is not readily available, cells can make it from a keto acid—if a nitrogen source is available. Cells can also make a nonessential amino acid by transferring an amino group from one amino acid to its corresponding keto acid, as shown in Figure 6-12. Through many such **transamination** reactions, involving many different keto acids, the liver cells can synthesize the nonessential amino acids.

Converting Ammonia to Urea As mentioned earlier, deamination produces ammonia. Ammonia is a toxic compound chemically identical to the strong-smelling ammonia in bottled cleaning solutions. Because ammonia is a base, excessive

> **FIGURE 6-11** **Deamination and Synthesis of a Nonessential Amino Acid**

The deamination of an amino acid produces ammonia (NH_3) and a keto acid.

Given a source of NH_3, the body can make nonessential amino acids from keto acids.

> **FIGURE 6-12** **Transamination and Synthesis of a Nonessential Amino Acid**

Keto acid A + Amino acid B ⟶ Amino acid A + Keto acid B

The body can transfer amino groups (NH_2) from an amino acid to a keto acid, forming a new *nonessential* amino acid and a new keto acid. Transamination reactions require the vitamin B_6 coenzyme.

serotonin (SER-oh-TONE-in): a neurotransmitter important in sleep regulation, appetite control, and sensory perception, among other roles. Serotonin is synthesized in the body from the amino acid tryptophan with the help of vitamin B_6.

deamination (dee-AM-ih-NAY-shun): removal of the amino (NH_2) group from a compound such as an amino acid.

ammonia: a compound with the chemical formula NH_3, produced during the deamination of amino acids.

keto (KEY-toe) **acid:** an organic acid that contains a carbonyl group (C=O).

transamination (TRANS-am-ih-NAY-shun): the transfer of an amino group from one amino acid to a keto acid, producing a new nonessential amino acid and a new keto acid.

*Chemists sometimes classify amino acids according to the destinations of their carbon fragments after deamination. If the fragment leads to the production of glucose, the amino acid is called *glucogenic*; if it leads to the formation of ketone bodies, fats, and sterols, the amino acid is called *ketogenic*. There is no sharp distinction between glucogenic and ketogenic amino acids, however. A few are both, most are considered glucogenic, only leucine and lysine are clearly ketogenic.

> FIGURE 6-13 **Urea Synthesis**

Ammonia is produced when amino acids are deaminated. The liver detoxifies ammonia by combining it with another waste product, carbon dioxide, to produce urea. See Appendix C for details.

> FIGURE 6-14 **Urea Excretion**

When amino acids are deaminated (stripped of their nitrogen), ammonia is released. The liver converts ammonia to urea, and the kidneys excrete urea. In this way the body disposes of excess nitrogen. (Figure 12-2 provides details of how the kidneys work.)

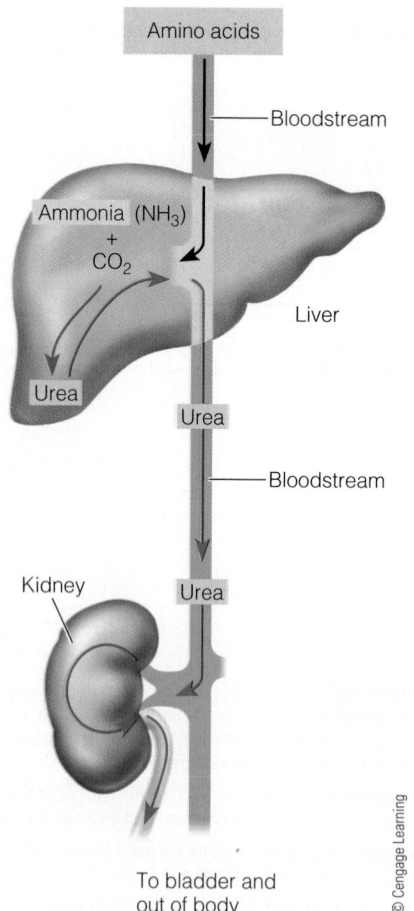

urea (you-REE-uh): the principal nitrogen-excretion product of protein metabolism. Two ammonia fragments are combined with carbon dioxide to form urea.

quantities upset the blood's critical acid-base balance. To prevent such a crisis, the liver combines ammonia with carbon dioxide to make **urea,** a much less toxic compound. Figure 6-13 provides a greatly oversimplified diagram of urea synthesis; details are shown in Appendix C. The production of urea increases as dietary protein increases, until production hits its maximum rate at intakes approaching 250 grams of protein per day. (For perspective, the average daily intake of protein in the United States is 80 grams.[2])

Excreting Urea Liver cells release urea into the blood, where it circulates until it passes through the kidneys (see Figure 6-14). The kidneys then filter urea out of the blood for excretion in the urine. Normally, the liver efficiently captures all the ammonia, makes urea from it, and releases the urea into the blood; then the kidneys clear all the urea from the blood. This division of labor allows easy diagnosis of diseases of both organs. In liver disease, blood ammonia is high; in kidney disease, blood urea is high.

Urea is the body's principal vehicle for excreting unused nitrogen, and the amount of urea produced increases with protein intake. To keep urea in solution, the body needs water. For this reason, a person who regularly consumes a high-protein diet (say, 100 grams a day or more) must drink plenty of water to dilute and excrete urea from the body. Without extra water, a person on a high-protein diet risks dehydration because the body uses its water to rid itself of urea. This explains some of the water loss that accompanies high-protein diets. Such losses may make high-protein diets *appear* to be effective, but water loss, of course, is of no value to the person who wants to lose body fat (as Highlight 9 explains).

> REVIEW IT Describe how the body makes proteins and uses them to perform various roles.

Cells synthesize proteins according to genetic information that dictates the sequence in which amino acids are linked together. Each protein plays a specific role. Table 6-3 summarizes some of the many roles proteins play and conveys a sense of the immense variety and importance of proteins in the body. Proteins are constantly being synthesized and broken down as needed. The body's assimilation of amino acids into proteins and its release of amino acids via protein breakdown and excretion can be tracked by measuring nitrogen balance, which should be positive during growth and steady in adulthood. An energy deficit or an inadequate protein intake may force the body to use amino acids as fuel, creating a negative nitrogen balance. Protein eaten in excess of need is broken down and stored as body fat.

TABLE 6-3 Protein Functions in the Body

Structural materials	Proteins form integral parts of most body tissues and provide strength and shape to skin, tendons, membranes, muscles, organs, and bones.
Enzymes	Proteins facilitate chemical reactions.
Hormones	Proteins regulate body processes. (Some, but not all, hormones are proteins.)
Fluid balance	Proteins help to maintain the volume and composition of body fluids.
Acid-base balance	Proteins help to maintain the acid-base balance of body fluids by acting as buffers.
Transportation	Proteins transport substances, such as lipids, vitamins, minerals, and oxygen, around the body.
Antibodies	Proteins inactivate foreign invaders, thus protecting the body against diseases.
Energy and glucose	Proteins provide some fuel, and glucose if needed, for the body's energy needs.
Other	The protein fibrin creates blood clots; the protein collagen forms scars; the protein opsin participates in vision.

6.4 Protein in Foods

> **LEARN IT** Explain the differences between high-quality and low-quality proteins, including notable food sources of each.

In the United States and other countries where nutritious foods are abundant, most people eat protein in such large quantities that they receive all the amino acids they need. In countries where food is scarce and the people eat only marginal amounts of protein-rich foods, however, the *quality* of the protein becomes crucial.

Protein Quality The protein quality of the diet determines, in large part, how well children grow and how well adults maintain their health. Put simply, **high-quality proteins** provide enough of all the essential amino acids needed to support the body's work, and low-quality proteins don't. Two factors influence protein quality—the protein's digestibility and its amino acid composition.

Digestibility As explained earlier, proteins must be digested before they can provide amino acids. **Protein digestibility** depends on such factors as the protein's source and the other foods eaten with it. The digestibility of most animal proteins is high (90 to 99 percent); plant proteins are less digestible (70 to 90 percent for most, but more than 90 percent for soy and other legumes).

Amino Acid Composition To make proteins, a cell must have all the needed amino acids available simultaneously. The liver can make any nonessential amino acid that may be in short supply so that the cells can continue linking amino acids into protein strands. If an essential amino acid is missing, though, a cell must dismantle its own proteins to obtain it. Therefore, to prevent protein breakdown in the body, dietary protein must supply at least the nine essential amino acids plus enough nitrogen-containing amino groups and energy for the synthesis of the nonessential ones. If the diet supplies too little of any essential amino acid, protein synthesis will be limited. The body makes whole proteins only; if one amino acid is missing, the others cannot form a "partial" protein. An essential amino acid supplied in less than the amount needed to support protein synthesis is called a **limiting amino acid.**

Reference Protein The quality of a food protein is determined by comparing its amino acid composition with the essential amino acid requirements of preschool-age children. Such a standard is called a **reference protein.** The rationale behind using the requirements of this age group is that if a protein will effectively support a young child's growth and development, then it will meet or exceed the requirements of older children and adults.

High-Quality Proteins As mentioned earlier, a high-quality protein contains all the essential amino acids in relatively the same amounts and proportions that human beings require; it may or may not contain all the nonessential amino acids. Proteins that are low in an essential amino acid cannot, by themselves, support protein synthesis. Generally, foods derived from animals (meat, seafood, poultry, eggs, and milk and milk products) provide high-quality proteins, although gelatin is an exception. Gelatin lacks tryptophan and cannot support growth and health as a diet's sole protein. Proteins from plants (vegetables, nuts, seeds, grains, and legumes) have more diverse amino acid patterns and tend to be limiting in one or more essential amino acids. Some plant proteins are notoriously low quality (for example, corn protein). A few others are high quality (for example, soy protein).

Researchers have developed several methods for evaluating the quality of food proteins and identifying high-quality proteins. Appendix D provides details.

Complementary Proteins In general, plant proteins are lower quality than animal proteins, and plants also offer less protein (per weight or measure of food). For this reason, many vegetarians improve the quality of proteins in their diets by combining plant-protein foods that have different but complementary amino acid

---

high-quality proteins: dietary proteins containing all the essential amino acids in relatively the same amounts that human beings require. They may also contain nonessential amino acids.

protein digestibility: a measure of the amount of amino acids absorbed from a given protein intake.

limiting amino acid: the essential amino acid found in the shortest supply relative to the amounts needed for protein synthesis in the body. Four amino acids are most likely to be limiting:

- Lysine
- Methionine
- Threonine
- Tryptophan

reference protein: a standard against which to measure the quality of other proteins.

Black beans and rice, a favorite Hispanic combination, together provide a balanced array of amino acids.

> FIGURE 6-15 **Complementary Proteins**

In general, legumes provide plenty of isoleucine (Ile) and lysine (Lys) but fall short in methionine (Met) and tryptophan (Trp). Grains have the opposite strengths and weaknesses, making them a perfect match for legumes.

	Ile	Lys	Met	Trp
Legumes	✓	✓		
Grains			✓	✓
Together	✓	✓	✓	✓

complementary proteins: two or more dietary proteins whose amino acid assortments complement each other in such a way that the essential amino acids missing from one are supplied by the other.

patterns. This strategy yields **complementary proteins** that together contain all the essential amino acids in quantities sufficient to support health. The protein quality of the combination is greater than either food alone (see Figure 6-15).

Some people have long believed that combining plant proteins at every meal is critical to protein nutrition. For most healthy vegetarians, though, it is *not* necessary to balance amino acids at each meal if protein intake is varied and energy intake is sufficient.[3] Vegetarians can receive all the amino acids they need over the course of a day by eating a variety of whole grains, legumes, seeds, nuts, and vegetables. Protein deficiency will develop, however, when fruits and certain vegetables make up the core of the diet, severely limiting both the *quantity* and *quality* of protein. Highlight 2 describes how to plan a nutritious vegetarian diet.

> **REVIEW IT** Explain the differences between high-quality and low-quality proteins, including notable food sources of each.

A diet that supplies all of the essential amino acids in adequate amounts ensures protein synthesis. The best guarantee of amino acid adequacy is to eat foods containing high-quality proteins or mixtures of foods containing complementary proteins that can each supply the amino acids missing in the other. In addition to its amino acid content, the quality of protein is measured by its digestibility and its ability to support growth. Such measures are of great importance in dealing with malnutrition worldwide, but in countries where protein deficiency is not common, the protein quality of individual foods deserves little emphasis.

6.5 Health Effects and Recommended Intakes of Protein

> **LEARN IT** Identify the health benefits of, and recommendations for, protein.

As you know by now, protein is indispensable to life. This section examines the health effects and recommended intakes of protein.

Health Effects of Protein It should come as no surprise that protein deficiency can have devastating effects on people's health. But like the other nutrients, protein in excess can also be harmful. High-protein diets have been implicated in several chronic diseases, including heart disease, cancer, osteoporosis, obesity, and kidney stones, but evidence is insufficient to establish an Upper Level (UL).[4]

Protein Deficiency Protein deficiency develops when the diet consistently supplies too little protein or lacks essential amino acids. When this occurs, the synthesis of body proteins decreases and degradation increases to provide cells with the amino acids they need. Without proteins to perform their critical roles, many of the body's activities come to a halt. The consequences of protein deficiency include slowed growth, impaired brain and kidney functions, poor immunity, and inadequate nutrient absorption.

The term *protein-energy malnutrition* has traditionally been used to describe the condition that develops when the diet delivers too little protein, too little energy, or both. The causes and consequences are complex, but clearly, such malnutrition reflects insufficient food intake. Importantly, not only are protein and energy inadequate, but so are many, if not all, of the vitamins and minerals. For this reason, severe malnutrition and its clinical forms—marasmus and kwashiorkor—are included in Chapter 20's discussion of world hunger.

Heart Disease In the United States and other developed countries, protein is so abundant that problems of excess are more common than deficiency. Depending on the food source, a high-protein diet may contribute to the progression of heart disease. As Chapter 5 mentions, foods rich in animal protein also tend to be rich in saturated fats. Consequently, it is not surprising to find a correlation between animal-protein intake (red meats and milk products) and heart disease.[5] On the other hand, substituting vegetable protein (legumes and nuts) for animal protein and using low-fat milk, poultry, and fish may improve blood pressure and blood lipids and decrease heart disease mortality.[6]

Many observational studies suggest that elevated levels of the amino acid homocysteine may be an independent risk factor for heart disease, heart attacks, and sudden death in patients with heart disease; findings from prospective studies, however, are far less conclusive.[7] Researchers do not yet fully understand the many factors—including a diet high in saturated fatty acids—that can raise homocysteine in the blood or whether elevated levels are a cause or an effect of heart disease.[8] Elevated homocysteine is associated with increased oxidative stress and inflammation.[9] Until researchers can determine the exact role homocysteine plays in heart disease, they are following several leads in pursuit of the answers. Coffee's role in heart disease has been controversial, but research suggests it is among the most influential factors in raising homocysteine, which may explain some of the adverse health effects of heavy consumption. Elevated homocysteine levels are among the many adverse health consequences of smoking cigarettes and drinking alcohol as well. Homocysteine is also elevated with inadequate intakes of B vitamins and can usually be lowered with fortified foods or supplements of vitamin B_{12}, vitamin B_6, and folate.[10] Lowering homocysteine, however, may not help in lowering risks or preventing heart attacks.[11] Supplements of the B vitamins do not always benefit those with heart disease and, in fact, may actually increase risks.[12]

In contrast to homocysteine, the amino acid arginine may help protect against heart disease by lowering blood pressure and homocysteine levels.[13] Additional research is needed to confirm the benefits of arginine. In the meantime, it is unwise for consumers to use supplements of arginine, or any other amino acid for that matter (as pp. 190–191 explain). Physicians, however, may consider the benefits of adding arginine supplements to their heart patients' treatment plan.

Cancer Protein does not seem to increase the risk of cancer, but some protein-rich foods do. For example, evidence suggests a correlation between high intakes of red meat and processed meats with cancer of the colon, pancreas, and ovaries.[14]* In contrast, protein-rich legumes, fish, and milk may lower the risk of some cancers. Chapter 18 discusses dietary links with cancer, and Chapter 19 presents food safety issues of processed meats and their additives.

Adult Bone Loss (Osteoporosis) Chapter 12 presents calcium metabolism, and Highlight 12 elaborates on the main factors that influence osteoporosis. This section briefly describes the relationships between protein intake and bone loss. When protein intake is high, calcium excretion increases. Whether excess protein depletes the bones of their chief mineral may depend upon the ratio of calcium intake to protein intake. After all, bones need both protein and calcium. An ideal ratio has not been determined, but a young woman whose intake meets recommendations for both nutrients has a calcium-to-protein ratio of more than 20 to 1 (milligrams to grams), which probably provides adequate protection for the bones. For most women in the United States, however, average calcium intakes are lower and protein intakes are higher, yielding a 13-to-1 ratio, which may produce calcium losses significant enough to compromise bone health. In other words, the problem may reflect too little calcium, not too much protein. In establishing recommendations, the DRI Committee considered protein's effect on calcium metabolism and bone health, but it did not find sufficient evidence to warrant an adjustment for calcium or a UL for protein.[15]

Importantly, adequate protein does not harm bones and may even improve bone mineral density, whereas *inadequate* intakes of protein may compromise bone health.[16] Osteoporosis is particularly common in elderly women and in adolescents with anorexia nervosa—groups who typically receive less protein than they need. For these people, a high-protein diet may be just what they need to protect their bones.[17]

---

*Processed meats include ham, bacon, pastrami, salami, sausage, bratwurst, and hot dogs that have been preserved by smoking, curing, salting, or adding preservatives.

Weight Control Research on the associations between protein intake and body weight has revealed some interesting, although often inconsistent, findings. One study suggests that inconsistent findings may reflect differences between animal and vegetable proteins, with animal proteins having a positive association with overweight and vegetable proteins having a negative one.[18] Another study examined people who were deliberately overfed by 1000 kcalories daily.[19] Not too surprisingly, they all gained weight, but those receiving a low-protein diet gained about half as much weight as those receiving an adequate- or high-protein diet. A look at their body composition revealed that the low-protein group stored almost all their excess kcalories as fat and lost a little lean body tissue. By comparison, the other protein groups stored about half the excess kcalories as fat and gained lean body tissue. Importantly, the excess kcalories increased total body fat similarly for all groups; the different amounts of dietary protein affected changes in lean body mass.[20] These findings highlight the importance of distinguishing between body weight and body fat—a point revisited in Chapters 8 and 9.

Fad weight-loss diets that encourage a high-protein, low-carbohydrate diet may be effective, but only because they are low-kcalorie diets. Diets that provide adequate protein (at least 65 to 70 grams a day), moderate fat, and sufficient energy from carbohydrates can better support weight loss and good health. Including protein at each meal may help with weight loss by providing satiety.[21] Selecting too many protein-rich foods may crowd out fruits, vegetables, and whole grains, making the diet inadequate in other nutrients.

Kidney Disease Excretion of the end products of protein metabolism depends, in part, on an adequate fluid intake and healthy kidneys. A high protein intake does not cause kidney disease, but it does increase the work of the kidneys. It may also accelerate kidney deterioration in people with chronic kidney disease. Restricting dietary protein may help to slow the progression of kidney disease in people who have this condition.

Recommended Intakes of Protein
As mentioned earlier, the body continuously breaks down and loses some protein and it cannot store proteins or amino acids. To replace protein, the body needs dietary protein for two reasons. First, dietary protein is the only source of the *essential* amino acids, and second, it is the only practical source of *nitrogen* with which to build the nonessential amino acids and other nitrogen-containing compounds the body needs.

Given recommendations that fat should contribute 20 to 35 percent of total food energy and carbohydrate should contribute 45 to 65 percent, that leaves 10 to 35 percent for protein. In a 2000-kcalorie diet, that represents 200 to 700 kcalories from protein, or 50 to 175 grams. The average intake in the United States is 80 grams per day.

Protein RDA The protein RDA for adults is 0.8 grams per kilogram of healthy body weight per day. For infants and children, the RDA is slightly higher. The table on the inside front cover lists the RDA for males and females at various ages in two ways—grams per day based on reference body weights and grams per kilogram of body weight per day.

The RDA covers the needs for replacing worn-out tissue, so it increases for larger people; it also covers the needs for building new tissue during growth, so it increases for infants, children, adolescents, and pregnant and lactating women. The accompanying "How To" feature explains how to calculate your RDA for protein.

The protein RDA is the same for athletes as for others, even though athletes may need more protein and many fitness authorities recommend a higher range of protein intakes for athletes pursuing different activities (see Table 14-5 in Chapter 14 for details). Most athletes in training typically don't need to actually increase their protein intakes, however, because the additional foods they eat to meet their high energy needs deliver protein as well. Importantly, these higher recommendations still fall within the 10 to 35 percent Acceptable Macronutrient Distribution Range (AMDR).

In setting the RDA, the DRI Committee assumes that people are healthy and do not have unusual metabolic needs for protein, that the protein eaten will be of

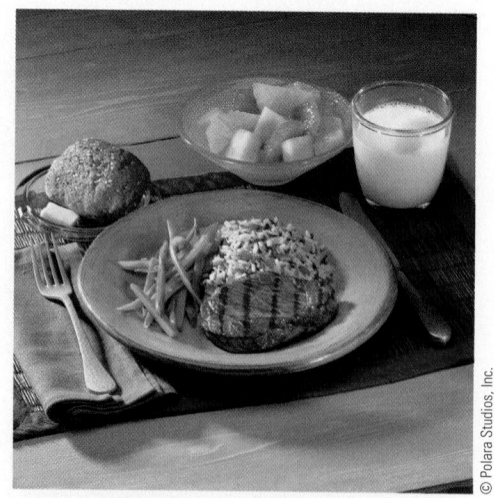

For many people, this 5-ounce steak provides almost all of the meat and much of the protein recommended for a day's intake.

mixed quality (from both high- and low-quality sources), and that the body will use the protein efficiently. In addition, the committee assumes that the protein is consumed along with sufficient carbohydrate and fat to provide adequate energy and that other nutrients in the diet are also adequate.

Adequate Energy Note the qualification "adequate energy" in the preceding statement, and consider what happens if energy intake falls short of needs. An intake of 50 grams of protein provides 200 kcalories, which represents 10 percent of the total energy from protein, if the person receives 2000 kcalories a day. But if the person cuts energy intake drastically—to, say, 800 kcalories a day—then an intake of 200 kcalories from protein is suddenly 25 percent of the total; yet it's still the same amount of protein (number of grams). The protein intake is reasonable, but the energy intake is not. The low energy intake forces the body to use the protein to meet energy needs rather than to replace lost body protein. Similarly, if the person's energy intake is high—say, 4000 kcalories—the 50-gram protein intake represents only 5 percent of the total; yet it *still* is a reasonable protein intake. Again, the energy intake is unreasonable for most people, but in this case, it permits the protein to be used to meet the body's needs.

Be careful when judging protein (or carbohydrate or fat) intake as a percentage of energy. Always consider the number of grams as well, and compare it with the RDA or another standard stated in grams. A recommendation stated as a percentage of energy intake is useful only if the energy intake is within reason.

From Guidelines to Groceries A diet following the USDA Food Patterns can easily supply the recommended amount of protein. In selecting foods for protein, keep in mind the principles of variety and moderation.

Protein Foods An ounce of most protein foods delivers about 7 grams of protein. The USDA Food Patterns encourage a variety by sorting protein foods into three subgroups (review Figure 2-2, p. 44 and Table 2-4, p. 46). Over a week's time, the total recommended intake of protein foods should be about 20 percent from seafood; almost 70 percent from meat, poultry, and eggs; and 10 percent from nuts, seeds, and legumes.

Either plant or animal sources of protein can support a healthy eating pattern, but some protein foods—notably those derived from animals—may be high in saturated fat. To minimize saturated fat intake, select lean meats and poultry. Trim fat from meats before cooking and drain fat from meat after cooking. Remove skin from poultry before eating. Include plant sources of protein as well. By selecting a variety of protein foods, consumers can improve their nutrient

Vegetarians obtain their protein from whole grains, legumes, nuts, vegetables, and, in some cases, eggs and milk products.

intake and incur health benefits. Highlight 5 describes how nuts and fish can reduce the risks of heart disease when consumed in place of other protein foods.

Milk and Milk Products The only other food group to provide significant amounts of protein per serving is the milk and milk products group. A serving (a cup) of milk or yogurt provides about 8 grams of protein.

Fruits, Vegetables, and Grains Fruits do not contain protein. A serving of vegetables or grains provides 2 to 3 grams of protein, respectively.

Read Food Labels
Food labels state the quantity of protein in grams. The "% Daily Value" for protein is not mandatory on all labels but is required whenever a food makes a protein claim or is intended for consumption by children younger than 4 years old. Whenever the Daily Value percentage is declared, researchers must determine the *quality* of the protein. Thus, when a % Daily Value is stated for protein, it reflects both quantity and quality.

To illustrate how easy it is to get enough protein, consider the amounts recommended by the USDA Food Pattern for a 2000-kcalorie diet. Six ounces of grains provide about 18 grams of protein; 2½ cups of vegetables deliver about 10 grams; 3 cups of milk offer 24 grams; and 5½ ounces of protein foods supply 38 grams. This totals 90 grams of protein—higher than the protein RDA for most people.

People in the United States typically get more protein than they need. If they have an adequate *food* intake, they have a more-than-adequate protein intake. The key diet-planning principle to emphasize for protein is moderation. Even though most people receive plenty of protein, some feel compelled to take supplements as well, as the next section describes.

Protein and Amino Acid Supplements
Websites, health-food stores, and popular magazine articles advertise a wide variety of protein supplements, and consumers spend billions of dollars taking these supplements for many different reasons. Athletes take protein powders to build muscle. Dieters take them to spare their bodies' protein while losing weight. Women take them to strengthen their fingernails. People take individual amino acids, too—to cure herpes, to make themselves sleep better, to lose weight, and to relieve pain and depression. Like many other magic solutions to health problems, protein and amino acid supplements don't work these miracles.

Protein Powders Because the body builds muscle protein from amino acids, many athletes take protein powders with the false hope of stimulating muscle growth. Muscle work builds muscle; protein supplements do not, and most athletes do not need them. Getting enough protein to support protein synthesis in the muscles is certainly important, but ingesting "more than enough" protein does not further enhance muscle growth or function. (Highlight 14 presents more information on other supplements athletes commonly use.) Protein powders can supply amino acids to the body, but nature's protein sources—lean meat, milk, eggs, and legumes—supply all these amino acids and more.

Whey protein appears to be particularly popular among athletes hoping to achieve greater muscle gains. A waste product of cheese manufacturing, whey protein is a common ingredient in many low-cost protein powders. When

whey protein: a by-product of cheese production; falsely promoted as increasing muscle mass. Whey is the watery part of milk that separates from the curds.

combined with strength training, whey supplements may increase protein synthesis slightly, but they do not seem to enhance athletic performance. To build stronger muscles, athletes need to eat food with adequate energy and protein to support the weight-training work that does increase muscle mass. Those who still think they need more whey can drink a glass of milk; one cup provides 1.5 grams of whey.

Amino Acid Supplements Single amino acids do not occur naturally in foods and offer no benefit to the body; in fact, they may be harmful. The body was not designed to handle the high concentrations and unusual combinations of amino acids found in supplements. Large doses of amino acids cause diarrhea. An excess of one amino acid can create such a demand for a carrier that it limits the absorption of another amino acid, presenting the possibility of a deficiency. Those amino acids winning the competition enter in excess, creating the possibility of toxicity. Anyone considering taking amino acid supplements should be cautious not to exceed levels normally found in foods.[22]

Most healthy athletes eating well-balanced diets do not need amino acid supplements. Advertisers point to research that identifies the **branched-chain amino acids** as the main ones used as fuel by exercising muscles. What the ads leave out is that compared to glucose and fatty acids, branched-chain amino acids provide very little fuel and that ordinary foods provide them in abundance anyway. Large doses of branched-chain amino acids can raise plasma ammonia concentrations, which can be toxic to the brain. Branched-chain amino acid supplements may be beneficial in conditions such as liver disease, but otherwise, they are not routinely recommended.

In two cases, recommendations for single amino acid supplements have led to widespread public use—lysine to prevent or relieve the infections that cause herpes cold sores on the mouth or genital organs, and tryptophan to relieve depression and insomnia. In both cases, enthusiastic popular reports preceded careful scientific experiments and health recommendations. Research has not determined that lysine suppresses herpes infections, but it appears safe (up to 3 grams per day) when taken in divided doses with meals.

Tryptophan may be effective with respect to inducing drowsiness, but caution is still advised. About 25 years ago, more than 1500 people who had taken tryptophan supplements developed a rare blood disorder known as eosinophilia-myalgia syndrome (EMS). EMS is characterized by severe muscle and joint pain, extremely high fever, and, in more than three dozen cases, death. Treatment for EMS usually involves physical therapy and low doses of corticosteroids to relieve symptoms temporarily. The Food and Drug Administration implicated impurities in the supplements and issued a recall of all products containing manufactured tryptophan. A recent review of the effects and side effects of tryptophan supplements currently on the market found only modest, short-lived side effects at doses typical of use (up to 5 grams per day).[23] People taking serotonin reuptake inhibitor drugs should consult with their physicians before taking tryptophan supplements.

> **› REVIEW IT** Identify the health benefits of, and recommendations for, protein.
> Protein deficiency impairs the body's ability to grow and function optimally. Excesses of protein offer no advantage; in fact, overconsumption of protein-rich foods may incur health problems as well. The optimal diet is adequate in energy from carbohydrate and fat and delivers 0.8 grams of protein per kilogram of healthy body weight each day. US diets are typically more than adequate in this respect. Normal, healthy people do not need protein or amino acid supplements.

As is true for the other nutrients as well, it is safest to obtain amino acids and protein from foods, eaten with abundant carbohydrate and some fat to facilitate their use in the body. With all that we know about science, it is hard to improve on nature.

branched-chain amino acids: the essential amino acids leucine, isoleucine, and valine, which are present in large amounts in skeletal muscle tissue; falsely promoted as fuel for exercising muscles.

Nutrition Portfolio

Foods that derive from animals—meats, fish, poultry, eggs, and milk products—provide plenty of protein but are often accompanied by fat. Those that derive from plants—whole grains, vegetables, and legumes—may provide less protein but also less fat.

Go to Diet & Wellness Plus and choose one of the days on which you have tracked your diet for the entire day. Go to the Intake Spreadsheet report. Scroll down until you see: protein (g).

- Which of your food choices provided you with the most protein on that day? Does that food also have a lot of fat? Refer to the fat (g) column for this information.

- Describe your dietary sources of proteins and whether you use mostly plant-based or animal-based protein foods in your diet.

Now take a look at the Intake vs. Goals report.

- How do your protein needs compare with your protein intake? Consider whether you receive enough, but not too much, protein daily. Remember, 100 percent means your intake is meeting your needs based on your intake and profile information.

- If your protein intake exceeds 100 percent, consider the possible negative consequences of a high protein intake over many years.

- Debate the risks and benefits of taking protein or amino acid supplements.

 DIET & WELLNESS **PLUS** To complete this exercise, go to your Diet & Wellness Plus at **www.cengagebrain.com.**

> **STUDY IT** To review the key points of this chapter and take a practice quiz, go to the study cards at the end of the book.

REFERENCES

1. M. Brown, Managing the acutely ill adult with sickle cell disease, *British Journal of Nursing* 21 (2012): 90–96.
2. US Department of Agriculture, Agricultural Research Service, Nutrient Intakes from Food, *What We Eat in America*, NHANES 2009–2010, www. ars.usda.gov/ba/bhnrc/fsrg, published 2012.
3. Position of the American Dietetic Association and Dietitians of Canada: Vegetarian diets, *Journal of the American Dietetic Association* 109 (2009): 1266–1282.
4. Committee on Dietary Reference Intakes, *Dietary Reference Intakes: Energy, Carbohydrate, Fiber, Fat, Fatty Acids, Cholesterol, Protein, and Amino Acids* (Washington, D.C.: National Academies Press, 2005), p. 694.
5. S. R. Preis and coauthors, Dietary protein and risk of ischemic heart disease in middle-aged men, *American Journal of Clinical Nutrition* 92 (2010): 1265–1272.
6. P. M. Clifton, Protein and coronary heart disease: The role of different protein sources, *Current Atherosclerosis Reports* 13 (2011): 493–498; D. G. Hackam and coauthors, The 2010 Canadian Hypertension Education Program recommendations for the management of hypertension: Part 2—Therapy, *Canadian Journal of Cardiology* 26 (2010): 249–258.
7. J. B. J. vanMeurs and coauthors, Common genetic loci influencing plasma homocysteine concentrations and their effect on risk of coronary artery disease, *American Journal of Clinical Nutrition* 98 (2013): 668–676; Q. Yang and coauthors, Prospective study of methylenetetrahydrofolate reductase (*MTHFR*) variant C677T and risk of all-cause and cardiovascular disease mortality among US adults, *American Journal of Clinical Nutrition* 95 (2012): 1245–1253.
8. R. J. Glynn, Complex relations of genetic polymorphisms with nutritionally influenced biomarkers, *American Journal of Clinical Nutrition* 95 (2012): 1001–1002; R. Clarke and coauthors, Homocysteine and vascular disease: Review of published results of the homocysteine-lowering trials, *Journal of Inherited Metabolic Disease* 34 (2011): 83–91.
9. M. Hoffman, Hypothesis: Hyperhomocysteinemia is an indicator of oxidant stress, *Medical Hypotheses* 77 (2011): 1088–1093.
10. P. Tighe and coauthors, A dose-finding trial of the effect of long-term folic acid intervention: Implications for food fortification policy, *American Journal of Clinical Nutrition* 93 (2011): 11–18.
11. J. D. Spence and M. J. Stampfer, Understanding the complexity of homocysteine lowering with vitamins: The potential role of subgroup analyses, *Journal of the American Medical Association* 306 (2011): 2610–2611; S. Eilat-Adar and U. Goldbourt, Nutritional recommendations for preventing coronary heart disease in women: Evidence concerning whole foods and supplements, *Nutrition, Metabolism, and Cardiovascular Disease* 20 (2010): 459–466.
12. L. Chao-Qiang, *MAT1A* variants are associated with hypertension, stroke, and markers of DNA damage and are modulated by plasma vitamin B-6 and folate, *American Journal of Clinical Nutrition* 91 (2010): 1377–1386; J. M. Armitage and coauthors, Effects of homocysteine-lowering with folic acid plus vitamin B$_{12}$ vs placebo on mortality and major morbidity in myocardial infarction survivors: A randomized trial, *Journal of the American Medical Association* 303 (2010): 2486–2494.
13. U. N. Das and coauthors, L-arginine, NO and asymmetrical dimethylarginine in hypertension and type 2 diabetes, *Frontiers in Bioscience* 16 (2011): 13–20; D. Tousoulis and coauthors, Novel therapeutic strategies targeting vascular endothelium in essential hypertension, *Expert Opinion on Investigational Drugs* 19 (2010): 1395–1412.

14. B. Magalhães, B. Peleteiro, and N. Lunet, Dietary patterns and colorectal cancer: Systematic review and meta-analysis, *European Journal of Cancer Prevention* 21 (2012): 15–23; S. C. Larsson and A. Wolk, Red and processed meat consumption and risk of pancreatic cancer: Meta-analysis of prospective studies, *British Journal of Cancer* 106 (2012): 603–607; R. Takachi and coauthors, Red meat intake may increase the risk of colon cancer in Japanese, a population with relatively low red meat consumption, *Asia Pacific Journal of Clinical Nutrition* 20 (2011): 603–612; A. T. Chan and E. L. Giovannucci, Primary prevention of colorectal cancer, *Gastroenterology* 138 (2010): 2029–2043.

15. Committee on Dietary Reference Intakes, 2005, p. 841; Committee on Dietary Reference Intakes, *Dietary Reference Intakes for Calcium and Vitamin D* (Washington, D.C.: National Academies Press, 2011).

16. J. Calvez and coauthors, Protein intake, calcium balance and health consequences, *European Journal of Clinical Nutrition* 66 (2012): 281–295; M. P. Thorpe and E. M. Evans, Dietry protein and bone health: Harmonizing conflicting theories, *Nutrition Reviews* 9 (2011): 215–230; J. J. Cao, L. K. Johnson, and J. R. Hunt, A diet high in meat protein and potential renal acid load increases fractional calcium absorption and urinary calcium excretion without affecting markers of bone resorption or formation in postmenopausal women, *Journal of Nutrition* 141 (2011): 391–397; J. M. Beasley and coauthors, Is protein intake associated with bone mineral density in young women? *American Journal of Clinical Nutrition* 91 (2010): 1311–1316.

17. J. J. Cao and F. H. Nielsen, Acid diet (high-meat protein) effects on calcium metabolism and bone health, *Current Opinion in Clinical Nutrition and Metabolic Care* 13 (2010): 698–702.

18. D. Bujnowski and coauthors, Longitudinal association between animal and vegetable protein intake and obesity among men in the United States: The Chicago Western Electric Study, *Journal of the American Dietetic Association* 111 (2011): 1150–1155.

19. G. A. Bray and coauthors, Effect of dietary protein content on weight gain, energy expenditure, and body composition during overeating: A randomized controlled trial, *Journal of the American Medical Association* 307 (2012): 47–55.

20. Z. Li and D. Heber, Overeating and overweight: Extra calories increase fat mass while protein increases lean mass, *Journal of the American Medical Association* 307 (2012): 86–87.

21. A. Belza and coauthors, Contribution of gastroenteropancreatic appetite hormones to protein-induced satiety, *American Journal of Clinical Nutrition* 97 (2013): 980–989.

22. Committee on Dietary Reference Intakes, *Dietary Reference Intakes: The Essential Guide to Nutrient Requirements* (Washington, D.C.: National Academies Press, 2006), p. 152.

23. J. D. Fernstrom, Effects and side effects associated with the non-nutritional use of tryptophan by humans, *Journal of Nutrition* 142 (2012): 2236S–2244S.

HIGHLIGHT > 6
Nutritional Genomics

> **LEARN IT** Explain how nutrients influence gene activity (nutrigenomics) and how genes influence the activities of nutrients (nutrigenetics).

Imagine this scenario: A physician scrapes a sample of cells from inside your cheek and submits it to a **genomics** lab. The lab returns a report based on your genetic profile that reveals which diseases you are most likely to develop, and your physician recommends specific lifestyle changes and medical treatments that can help you maintain good health. You may also be given a prescription for an individualized diet and dietary supplements that will best meet your personal nutrient requirements. This scenario may one day become a common reality as scientists uncover the relationships among **genetics**, diet, and disease.[1] Such genetic testing holds great promise, but consumers need to know that current genetic test kits commonly available to the public are unproven and may create more problems than they resolve; the American Academy of Pediatrics strongly discourages direct-to-consumer testing and advises against testing children for diseases that typically develop in adulthood.[2]

Figure H6-1 introduces **nutritional genomics,** a new field of study that examines how nutrients influence gene activity (nutrigenomics) and how **genes** influence the activities of nutrients (nutrigenetics). The accompanying glossary defines related terms.

The recent surge in genomics research grew from the Human Genome Project, an international effort by industry and government scientists to identify and describe all of the genes in the **human genome**—that is, all the genetic information contained within a person's cells. Completed in 2003, this project developed many of the research technologies needed to study genes and genetic variation. Scientists are now working on the human **proteome** and hope to identify each of the proteins made by the genes, the genes associated with aging and diseases, and the dietary and lifestyle choices that most influence the expression of those genes. Such information

will have major implications for society in general, and for health care in particular.[3]

A Genomics Primer

Figure H6-2 (p. 196) shows the relationships among the materials that comprise the genome. As Chapter 6's discussion of protein synthesis points out, genetic information is encoded in DNA molecules within the nucleus of cells. The **DNA (deoxyribonucleic acid)** molecules and associated proteins are packed within 46 **chromosomes.** The genes are segments of a DNA strand that can eventually be translated into one or more proteins. The sequence of **nucleotide bases** within each gene determines the amino acid sequence of a particular protein. Scientists currently estimate that there are between 20,000 and 25,000 protein-coding genes in the human genome.

GLOSSARY

chromosomes: structures within the nucleus of a cell made of DNA and associated proteins. Human beings have 46 chromosomes in 23 pairs. Each chromosome has many genes.

DNA (deoxyribonucleic acid): the double helix molecules of which genes are made.

epigenetics: the study of heritable changes in gene function that occur without a change in the DNA sequence.

gene expression: the process by which a cell converts the genetic code into RNA and protein.

genes: sections of chromosomes that contain the instructions needed to make one or more proteins.

genetics: the study of genes and inheritance.

genomics: the study of all the genes in an organism and their interactions with environmental factors.

human genome (GEE-nome): the complete set of genetic material (DNA) in a human being.

methylation: the addition of a methyl group (CH_3).

microarray technology: research tools that analyze the expression of thousands of genes simultaneously and search for particular gene changes

associated with a disease. DNA microarrays are also called *DNA chips.*

mutations: permanent changes in the DNA that can be inherited.

nucleotide bases: the nitrogen-containing building blocks of DNA and RNA—cytosine (C), thymine (T), uracil (U), guanine (G), and adenine (A). In DNA, the base pairs are A–T and C–G and in RNA, the base pairs are A–U and C–G.

nucleotides: the subunits of DNA and RNA molecules, composed of a phosphate group, a 5-carbon sugar (deoxyribose for DNA and ribose for RNA), and a nitrogen-containing base.

nutritional genomics: the science of how nutrients affect the activities of genes

(nutrigenomics) and how genes affect the activities of nutrients *(nutrigenetics).*

phenylketonuria (FEN-il-KEY-toe-NEW-ree-ah) or **PKU:** an inherited disorder characterized by failure to metabolize the amino acid phenylalanine to tyrosine.

proteome: all proteins in a cell. The study of all proteins produced by a species is called *proteomics.*

RNA (ribonucleic acid): a compound similar to DNA, but RNA is a single strand with a ribose sugar instead of a deoxyribose sugar and uracil instead of thymine as one of its bases.

> **FIGURE H6-1**   **Nutritional Genomics**

Genes   Food and nutrients

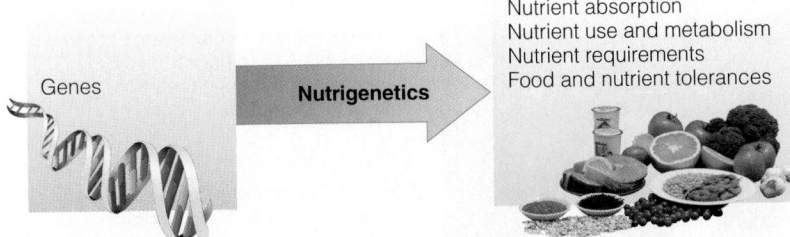

Nutritional genomics

Nutritional genomics examines the interactions of genes and nutrients. These interactions include both nutrigenetics and nutrigenomics.

Genes   **Nutrigenetics**   Nutrient absorption
Nutrient use and metabolism
Nutrient requirements
Food and nutrient tolerances

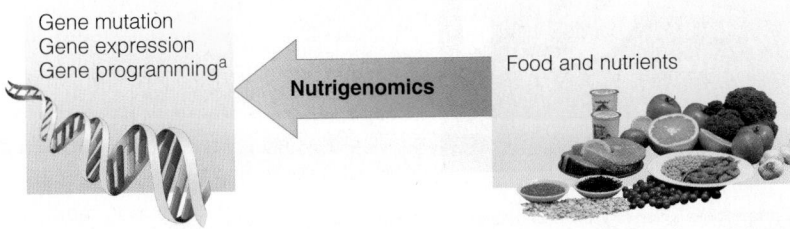

Nutrigenetics (or nutritional genetics) examines how genes influence the activities of nutrients.

Gene mutation
Gene expression
Gene programming[a]   **Nutrigenomics**   Food and nutrients

Nutrigenomics, which includes epigenetics, examines how nutrients influence the activities of genes.

[a]Chapter 15 introduces fetal programming and describes how a mother's nutrition can permanently change gene expression in the fetus with consequences for future generations.

As Figure 6-7 (p. 178) explains, when cells make proteins, a DNA sequence is used to make messenger **RNA (ribonucleic acid).** The **nucleotide** sequence in messenger RNA then determines the amino acid sequence to make a protein. This process—from genetic information to protein synthesis—is known as **gene expression.** Gene expression can be determined by measuring the amounts of messenger RNA in a tissue sample. **Microarray technology** (see photo on the previous page) allows researchers to detect messenger RNA and analyze the expression of thousands of genes simultaneously. These patterns of gene expression help to explain the development of diseases and relationships between diet and diseases.[4]

Simply having a certain gene does not determine that its associated trait will be expressed; the gene has to be activated. (Similarly, owning lamps does not ensure you will have light in your home unless you turn them on.) Nutrients are among many environmental factors that play key roles in either activating or silencing genes. Switching genes on and off does not change the DNA itself, but it can have dramatic consequences for a person's health.

The area of study that examines how environmental factors influence gene expression without changing the DNA is known as **epigenetics.**[5] To turn genes on, enzymes attach proteins near the beginning of a gene. If enzymes attach a methyl group (CH_3) instead, the protein is blocked from binding to the gene and the gene remains switched off. Other factors influence gene expression as well, but methyl groups are currently the most well understood.[6] They also are known to have dietary connections.[7]

The accompanying photo of two mice illustrates epigenetics and how diet can influence genetic traits such as hair color and body weight. Both mice have a gene that tends to produce fat, yellow pups, but their mothers were given different diets during pregnancy. The mother of the mouse on the right was given a dietary supplement containing the B vitamins folate and vitamin B_{12}. These nutrients silenced the gene for "yellow and fat," resulting in brown pups with normal appetites. As Chapter 10 explains, one of the main roles of these B vitamins is to transfer methyl groups. In the case of the supplemented mice, methyl groups migrated onto DNA and silenced several genes, thus producing brown coats and protecting against the development of obesity (and consequently, some related diseases). Keep in mind that these changes occurred epigenetically. In other words, the DNA sequence within the genes of the mice remained the same. Nutrition and other environmental factors can influence genes in a way that creates inheritable changes in the body's metabolism and susceptibility to disease. In this way, the dietary habits of parents, and even grandparents, can influence future generations.

Both of these mice have the gene that tends to produce fat, yellow pups, but their mothers had different diets. The mother of the mouse on the right received a dietary supplement, which silenced the gene, resulting in brown pups with normal appetites.

> FIGURE H6-2   **The Human Genome**

1. The human genome is a complete set of genetic material organized into 46 chromosomes, located within the nucleus of a cell.

2. A chromosome is made of DNA and associated proteins.

3. The double helical structure of a DNA molecule is made up of two long chains of nucleotides. Each nucleotide is composed of a phosphate group, a 5-carbon sugar, and a base.

4. The sequence of nucleotide bases (C, G, A, T) determines the amino acid sequence of proteins. These bases are connected by hydrogen bonding to form base pairs—adenine (A) with thymine (T) and guanine (G) with cytosine (C).

5. A gene is a segment of DNA that includes the information needed to synthesize one or more proteins.

© Cengage Learning

SOURCE: Adapted from "A Primer: From DNA to Life," Human Genome Project, US Department of Energy Office of Science, accessed at www.orn.gov/sci/techresources/human_genome/primer-pic.shtml.

Many nutrients and phytochemicals regulate gene expression and influence health through their involvement in DNA **methylation.** Some, such as folate, silence genes and protect against some cancers by increasing methylation.[8] Others, such as a phytochemical found in green tea, activate genes and protect against some cancers by inhibiting methylation activity. Whether silencing or activating a gene is beneficial or harmful depends on what the gene does. Silencing a gene that stimulates cancer growth, for example, would be beneficial, but silencing a gene that suppresses cancer growth would be harmful. Similarly, activating a gene that defends against obesity would be beneficial, but activating a gene that promotes obesity would be harmful. Figure H6-3 illustrates how nutrient regulation of gene expression can influence a person's health. Much research is under way to determine which nutrients activate or silence which genes. Such knowledge is expected to help researchers reverse the epigenetic changes that lead to cancer.[9] Similarly, researchers exploring how kcalorie-restricted diets influence DNA methylation are gaining new insights on the regulation of appetite and the metabolism of weight loss.[10]

Genetic Variation and Disease

Except for identical twins, no two persons are genetically identical. Even then, a particular gene may become active in one twin and silenced in the other because of epigenetic changes.

The variation in the genomes of any two persons is only about 0.1 percent, a difference of only one nucleotide base in every 1000. Yet it is this incredibly small difference that makes each of us unique and explains why, given the same environmental influences, some of us develop certain diseases and others do not. Similarly, genetic variation explains why some of us respond to interventions such as diet and others do not. For example, following a diet low in saturated fats will significantly lower LDL cholesterol for most people, but the degree of change varies dramatically among individuals, with some people having only a small decrease or even a slight increase. In other words, dietary factors may be more helpful or more harmful depending on a person's particular genetic variations. Such findings help to explain some of the conflicting results from research studies. One of the goals of nutritional genomics is to custom design *specific* recommendations that fit the needs of *each* individual. Such personalized recommendations are expected to provide more effective disease prevention and treatment solutions.

Diseases characterized by a single-gene disorder are genetically predetermined, usually exert their effects early in life, and greatly affect those touched by them; such diseases are relatively rare. The cause and effect of single-gene disorders is clear—those with the genetic defect get the disease and those without it don't. In contrast, the more common diseases, such as heart disease and cancer, are influenced by many genes and typically develop over several decades. These chronic diseases have multiple genetic components that *predispose* the prevention or development of a disease, depending on a variety of environmental factors (such as smoking, diet, and physical activity). Both types of diseases are of interest to researchers studying nutritional genomics.

Single-Gene Disorders

Some disorders are caused by **mutations** in single genes that are inherited at birth. The consequences of a missing or malfunctioning protein can seriously disrupt metabolism and may require significant dietary or medical intervention. A classic example of a diet-related, single-gene disorder is **phenylketonuria,** or **PKU.**

Approximately one in every 15,000 infants in the United States is born with PKU. PKU arises from mutations in the gene that codes for the enzyme that converts the essential amino acid phenylalanine to the amino acid tyrosine. Without this enzyme, phenylalanine and its metabolites accumulate and damage the nervous system, resulting in

mental retardation, seizures, and behavior abnormalities. At the same time, the body cannot make tyrosine or compounds made from it (such as the neurotransmitter epinephrine). Consequently, tyrosine becomes a conditionally essential amino acid: because the body cannot make it, the diet must supply it.

Although the most debilitating effect is on brain development, other symptoms of PKU become evident if the condition is left untreated. Infants with PKU may have poor appetites and grow slowly. They may be irritable or have tremors or seizures. Their bodies and urine may have a musty odor. Their skin coloring may be unusually pale, and they may develop skin rashes.

The effect of nutrition intervention in PKU is remarkable. In fact, the only current treatment for PKU is a diet that restricts phenylalanine and supplies tyrosine to maintain blood levels of these amino acids within safe ranges. Because all foods containing protein provide phenylalanine, the diet must depend on a special formula to supply a phenylalanine-free source of energy, amino acids, vitamins, and minerals. If the restricted diet is conscientiously followed, the symptoms can be prevented. Because phenylalanine is an essential amino acid, the diet cannot exclude it completely. Children with PKU need phenylalanine to grow, but they cannot handle excesses without detrimental effects. Therefore, their diets must provide enough phenylalanine to support normal growth and health but not enough to cause harm. The diet must also provide tyrosine. To ensure that blood concentrations of phenylalanine and tyrosine are close to normal, children and adults who have PKU must have blood tests periodically and adjust their diets as necessary.

Multigene Disorders

In multigene disorders, several genes can influence the progression of a disease, but no single gene causes the disease on its own. For this reason, genomics researchers must study the expression and interactions of *multiple* genes. Because multigene disorders are often sensitive to interactions with environmental influences, they are not as straightforward as single-gene disorders.[11]

Heart disease provides an example of a chronic disease with multiple gene and environmental influences.[12] Consider that major risk factors for heart disease include elevated blood cholesterol levels, obesity, diabetes, and hypertension. Each of these risk factors has multiple underlying genetic and environmental causes, many of which are not completely understood. Research in nutritional genomics involves coordinating multiple findings on each of these risk factors and explaining the interactions among several genes, biological pathways, and nutrients in relatively little time. Studies have been quite successful in examining the genome and identifying multiple pathways in the development of complex diseases.[13] This information could then guide physicians and dietitians to prescribe the most appropriate medical and dietary interventions from among many possible solutions. Finding the best option for each person is a challenge given the many possible interactions between genes and environmental factors and the millions of possible gene variations in the human genome that make each individual unique.

The results of genomic research are helping to explain findings from previous nutrition research. Consider dietary fat and heart disease, for example. As Highlight 5 explains, epidemiological and clinical studies have found that a diet high in omega-3 polyunsaturated fatty acids benefits heart health. Now genetic studies offer an underlying explanation of this relationship: diets rich in omega-3 polyunsaturated fatty acids alter gene expression of immune cells to suppress inflammation and inhibit plaque build-up. Both actions support a healthy heart.

To learn more about how individuals respond to diet, researchers examine the genetic differences among people. The most common genetic differences involve a change in a single nucleotide base located in a particular region of a DNA strand—thymine replacing cytosine, for example. Such variations are called single nucleotide polymorphisms (SNPs), and they commonly occur throughout the genome. Many SNPs (commonly pronounced "snips") have no effect on cell activity. In fact, SNPs are significant only if they affect the amino acid sequence of a protein in a way that alters its function *and* if that function is critical to the body's well-being. In these cases, SNPs may reveal fascinating answers to previously unexplained findings. Consider that research on a gene that plays a key role in lipid metabolism reveals differences in a person's response to diet depending on whether the gene has a common SNP. People with the SNP have lower LDL when eating a diet rich in polyunsaturated fatty acids—and higher LDL with a low intake—than those without the SNP. These findings clearly show how diet (in this case, polyunsaturated fat) interacts with a gene (in this case, a fat metabolism gene with a SNP) to influence the development of a disease (changing blood lipids implicated in heart disease).[14]

Clinical Concerns

Because multigene, chronic diseases are common, an understanding of the human genome will have widespread ramifications for health care.[15] This new understanding of the human genome is expected to change health care by:

- Providing knowledge of an individual's genetic predisposition to specific diseases[16]
- Allowing physicians to develop "designer" therapies—prescribing the most effective schedule of screening, behavior changes (including diet), and medical interventions based on each individual's genetic profile

- Enabling manufacturers to create new medications for each genetic variation so that physicians can prescribe the best medicine in the exact dose and frequency to enhance effectiveness and minimize the risks of side effects
- Providing a better understanding of how nutrition influences the biological pathways of diseases

Enthusiasm surrounding genomic research needs to be put into perspective, however, given the scope of its promises and the reality of its limitations.[17] Critics have questioned whether genetic markers for disease would be more useful than simple family history and clinical measurements, which reflect both genetic *and* environmental influences. In other words, knowing that a person is genetically predisposed to diabetes is not necessarily more useful than knowing the person's actual risk factors.[18] Furthermore, if a disease has many genetic risk factors, each gene that contributes to susceptibility may have little influence on its own, so the benefits of identifying an individual genetic marker might be small. The long-range possibility is that many genetic markers will eventually be identified, and the hope is that the combined information will be a useful and accurate predictor of disease. Of course, the flood of information may also be overwhelming, offer no benefit, and create anxiety.

Having the knowledge to prevent disease and actually taking action do not always coincide. Despite the abundance of current dietary recommendations, many people are unwilling to make behavior changes known to improve their health—especially when they can simply blame their genes.[19] For example, it has been estimated that heart disease and type 2 diabetes are 90 percent preventable when people adopt an appropriate diet, maintain a healthy body weight, and exercise regularly. Yet these two diseases remain among the leading causes of death. Given the difficulty that many people have with current recommendations, it may be unrealistic to expect that they will enthusiastically adopt an even more detailed list of lifestyle modifications. Then again, compliance may be better when it is supported by information based on a person's own genetic profile and the knowledge that the epigenetic profile can be changed.

The debate over nature versus nurture—whether genes or the environment are more influential—has quieted. The focus has shifted. Scientists acknowledge the important roles of each and understand the real answers lie within the myriad interactions. Current research is sorting through how nutrients and other dietary factors interact with genes to confer health benefits or risks. Answers from genomic research may not become apparent for years to come, but the opportunities and rewards may prove well worth the efforts.

CRITICAL THINKING QUESTIONS

A. How might nutritional genomics influence health care in the future?
B. You may have heard about the diet that is based on a person's blood type and claims to restore the body's natural genetic rhythms and improve health. Research may one day reveal exactly which foods might best turn on and off specific genes to defend against specific chronic diseases. No doubt marketers will rush to fill grocery shelves with foods manufactured to match genetic profiles. Why do you think these genetic approaches to diet and health might be more or less appealing than eating patterns that include a variety of fruits, vegetables, whole grains, milk products, and meats?

REFERENCES

1. W. G. Feero, A. E. Guttmacher, and F. S. Collins, Genomic medicine: An updated primer, *New England Journal of Medicine* 362 (2010): 2001–2011.

2. American Academy of Pediatrics, Policy statement: Ethical and policy issues in genetic testing and screening of children, *Pediatrics* 131 (2013): 620–622; Government Accountability Office, *Direct-to-consumer genetic tests: Misleading test results are further complicated by deceptive marketing and other questionable practices*, GAO-10-847T (Washington D.C.: July 22, 2010); J. P. Annes, M. A. Giovanni, and M. F. Murray, Risks of presymptomatic direct-to-consumer genetic testing, *New England Journal of Medicine* 363 (2010): 1100–1101; L. Esserman and V. Kaklamani, Lessons learned from genetic testing, *Journal of the American Medical Association* 304 (2010): 1011–1012; J. P. Evans, D. C. Dale, and C. Fomous, Preparing for a consumer-driven genomic age, *New England Journal of Medicine* 363 (2010): 1099–1103.

3. Position of the Academy of Nutrition and Dietetics: Nutritional genomics, *Journal of the Academy of Nutrition and Dietetics* 114 (2014): 299–312; G. S. Ginsburg, Realizing the opportunities of genomics in health care, *Journal of the American Medical Association* 309 (2013): 1463–1464; K. L. Hudson, Genomics, health care, and society, *New England Journal of Medicine* 365 (2011): 1033–1041; H. Varmus, Ten years on: The human genome and medicine, *New England Journal of Medicine* 362 (2010): 2028–2029.

4. J. C. Jiménez-Chillarón and coauthors, The role of nutrition on epigenetic modifications and their implications on health, *Biochimie* 94 (2012): 2242–2263; M. P. Keller and A. D. Attie, Physiological insights gained from gene expression analysis in obesity and diabetes, *Annual Review of Nutrition* 30 (2010): 341–364; M. I. McCarthy, Genomics, type 2 diabetes, and obesity, *New England Journal of Medicine* 363 (2010): 2339–2350.

5. S. W. Choi and S. Friso, Epigenetics: A new bridge between nutrition and health, *Advances in Nutrition* 1 (2010): 8–16.

6. O. S. Anderson, K. E. Sant, D. C. Dolinoy, Nutrition and epigenetics: An interplay of dietary methyl donors, one-carbon metabolism and DNA methylation, *Journal of Nutritional Biochemistry* 23 (2012): 853–859.

7. L. K. Park, S. Friso, and S. W. Choi, Vitamins, infectious and chronic disease during adulthood and aging: Nutritional influences on epigenetics and age-related disease, *Proceedings of the Nutrition Society* 71 (2012): 75–83.

8. R. A. Stein, Epigenetics—The link between infectious diseases and cancer, *Journal of the American Medical Association* 305 (2011): 1484–1485.

9. M. A. Dawson, T. Kouzarides, and B. J. P. Huntly, Targeting epigenetic readers in cancer, *New England Journal of Medicine* 367 (2012): 647–657; S. Sharma, T. K. Kelly, and P. A. Jones, Epigenetics in cancer, *Carcinogenesis* 31 (2010): 27–36.

10. L. Bouchard and coauthors, Differential epigenomic and transcriptomic responses in subcutaneous adipose tissue between low and high responders to caloric restriction, *American Journal of Clinical Nutrition* 91 (2010): 309–320.

11. H. G. Brunner, The variability of genetic disease, *New England Journal of Medicine* 367 (2012): 1350–1352.

12. W. G. Feero and A. E. Guttmacher, Genomics of cardiovascular disease, *New England Journal of Medicine* 365 (2011): 2098–2109.

13. T. A. Manolio, Genomewide association studies and assessment of the risk of disease, *New England Journal of Medicine* 363 (2010): 166–176.

14. L. A. Afman and M. Müller, Human nutrigenomics of gene regulation by dietary fatty acids, *Progress in Lipid Research* 51 (2012): 63–70; R. Do and coauthors, The effect of chromosome 9p21 variants on cardiovascular disease may be modified by dietary intake: Evidence from a case/control and a prospective study, *PLoS Medicine* 9 (2011): e1001106.

15. W. G. Feero and E. D. Green, Genomics education for health care professionals in the 21st century, *Journal of the American Medical Association* 306 (2011): 989–990.

16. R. P. Lifton, Individual genomes on the horizon, *New England Journal of Medicine* 362 (2010): 1235–1236.

17. C. Klein, K. Lohmann, and A. Ziegler, The promise and limitations of genome-wide association studies, *Journal of the American Medical Association* 308 (2012): 1867–1868.

18. N. P. Paynter and coauthors, Association between a literature-based genetic risk score and cardiovascular events in women, *Journal of the American Medical Association* 303 (2010): 631–637.

19. S. C. O'Neill and coauthors, Preferences for genetic and behavioral health information: The impact of risk factors and disease attributions, *Annals of Behavioral Medicine* 40 (2010): 127–173.

Net Wt. 25 Kg.

Batch No.

Energy Metabolism

Nutrition in Your Life

You eat breakfast and hustle off to class. After lunch, you study for tomorrow's exam. Dinner is followed by an evening of dancing. Do you ever think about how the food you eat powers the activities of your life? What happens when you don't eat—or when you eat too much? Learn how the cells of your body transform carbohydrates, fats, and proteins into energy—and what happens when you give your cells too much or too little of any of these nutrients. Discover the metabolic pathways that lead to body fat and those that support physical activity. It's really quite fascinating. In the Nutrition Portfolio at the end of this chapter, you can determine whether your diet provides a healthy balance of the energy nutrients.

Energy makes it possible for people to breathe, ride bicycles, compose music, and do everything else they do. As Chapter 1 explains, *energy* is the capacity to do work. Although every aspect of our lives depends on energy, the concept of energy can be difficult to grasp because it cannot be seen or touched, and it manifests in various forms, including heat, mechanical, electrical, and chemical energy. In the body, heat energy maintains a constant body temperature, mechanical energy moves muscles, and electrical energy sends nerve impulses. Energy is stored in foods and in the body as chemical energy. This chemical energy powers the myriad activities of all cells.

By studying metabolism, you will understand how the body uses foods to meet its needs and why some foods meet those needs better than others. Readers who are interested in weight control will discover which foods contribute most to body fat and which to select when trying to gain or lose weight safely. Readers who are physically active will discover which foods best support endurance activities and which to select when trying to build lean body mass.

7.1 Chemical Reactions in the Body

> **LEARN IT** Identify the nutrients involved in energy metabolism and the high-energy compound that captures the energy released during their breakdown.

Earlier chapters introduced some of the body's chemical reactions: the making and breaking of the bonds in carbohydrates, fats, and proteins. Metabolism is the sum of these and all the other chemical reactions that go on in living cells; *energy metabolism* includes all the ways the body obtains and uses energy from food.

All the energy that sustains human life initially comes from the sun—the ultimate source of energy. During **photosynthesis,** plants make simple sugars from carbon dioxide and capture the sun's light energy in the chemical bonds of those sugars. Then human beings eat either the plants or animals that have eaten the plants. These foods provide energy, but how does the body obtain that energy from foods? This chapter answers that question by following the nutrients that provide the body with **fuel** through a series of reactions that release energy from their chemical bonds. As the bonds break, they release energy in a controlled version of the same process by which wood burns in a fire. Both wood and food have the potential to provide energy. When wood burns in the presence of oxygen, it generates heat and light (energy), steam (water), and some carbon dioxide and ash (waste). Similarly, during **metabolism,** the body releases energy, water, and carbon dioxide (and other waste products).

The Site of Metabolic Reactions—Cells The human body is made up of trillions of cells, and each cell busily conducts its metabolic work all the time. (Appendix A presents a brief summary of the structure and function of the cell.) Figure 7-1 depicts a typical cell and shows where the major reactions of energy metabolism take place. The type and extent of metabolic activities vary depending on the type of cell, but of all the body's cells, the liver cells are the most versatile and metabolically active. Table 7-1 offers insights into the liver's work.

The Building Reactions—Anabolism Earlier chapters describe how condensation reactions combine molecules to build body compounds. Glucose molecules may be joined together to make glycogen chains. Glycerol and fatty acids may be assembled into triglycerides. Amino acids may be linked together to make proteins. Each of these reactions starts with small, simple compounds and uses them as building

photosynthesis: the process by which green plants use the sun's energy to make carbohydrates from carbon dioxide and water.
- **photo** = light
- **synthesis** = put together (making)

fuel: compounds that cells can use for energy. The major fuels include glucose, fatty acids, and amino acids; other fuels include ketone bodies, lactate, glycerol, and alcohol.

metabolism: the sum total of all the chemical reactions that go on in living cells. *Energy metabolism* includes all the reactions by which the body obtains and expends the energy from food.
- **metaballein** = change

> FIGURE 7-1 **A Typical Cell (Simplified Diagram)**

Inside the cell membrane lies the cytoplasm, a lattice-type structure that supports and controls the movement of the cell's structures. A protein-rich jelly-like fluid called cytosol fills the spaces within the lattice. The cytosol contains the enzymes involved in glucose breakdown (glycolysis).

A separate inner membrane encloses the cell's nucleus.

Inside the nucleus are the chromosomes, which contain the genetic material DNA.

Known as the "powerhouses" of the cells, the mitochondria are intricately folded membranes that house all the enzymes involved in the conversion of pyruvate to acetyl CoA, fatty acid oxidation, the TCA cycle, and the electron transport chain (described later in the chapter).

This network of membranes is known as smooth endoplasmic reticulum—the site of lipid synthesis.

A membrane encloses each cell's contents and regulates the passage of molecules in and out of the cell.

Rough endoplasmic reticulum is dotted with ribosomes—the site of protein synthesis (described in Chapter 6).

© Cengage Learning

TABLE 7-1 Metabolic Work of the Liver

The liver is the most active processing center in the body. When nutrients enter the body from the digestive tract, the liver receives them first; then it metabolizes, packages, stores, or ships them out for use by other tissues. When alcohol, drugs, or poisons enter the body, they are also sent directly to the liver; here they are detoxified and their by-products shipped out for excretion. An enthusiastic anatomy and physiology professor once remarked that given the many vital activities of the liver, we should express our feelings for others by saying, "I love you with all my liver" instead of "with all my heart." Granted, this declaration lacks romance, but it makes a valid point. Here are just some of the many jobs performed by the liver. To renew your appreciation for this remarkable organ, review figure 3-11 (p. 84).

Carbohydrates

- Metabolizes fructose, galactose, and glucose
- Makes and stores glycogen
- Breaks down glycogen and releases glucose
- Breaks down glucose for energy when needed
- Makes glucose from some amino acids and glycerol when needed
- Converts excess glucose and fructose to fatty acids

Lipids

- Builds and breaks down triglycerides, phospholipids, and cholesterol as needed
- Breaks down fatty acids for energy when needed
- Packages lipids in lipoproteins for transport to other body tissues
- Manufactures bile to send to the gallbladder for use in fat digestion
- Makes ketone bodies when necessary

Proteins

- Manufactures nonessential amino acids that are in short supply
- Removes from circulation amino acids that are present in excess of need and converts them to other amino acids or deaminates them and converts them to glucose or fatty acids
- Removes ammonia from the blood and converts it to urea to be sent to the kidneys for excretion
- Makes other nitrogen-containing compounds the body needs (such as bases used in DNA and RNA)
- Makes many proteins

Other

- Detoxifies alcohol, other drugs, and poisons; prepares waste products for excretion
- Helps dismantle old red blood cells and captures the iron for recycling
- Stores most vitamins and many minerals
- Activates vitamin D

blocks to form larger, more complex structures. Because such reactions involve doing work, they require energy. The building up of body compounds is known as **anabolism**. Anabolic reactions are represented in this book, wherever possible, with "up" arrows in chemical diagrams (such as those shown at the top of Figure 7-2).

anabolism (an-AB-o-lism): reactions in which small molecules are put together to build larger ones. Anabolic reactions require energy.

- **ana** = (build) up

> FIGURE 7-2 Anabolic and Catabolic Reactions Compared

ANABOLIC REACTIONS

Anabolic reactions include the making of glycogen, triglycerides, and protein; these reactions require differing amounts of energy.

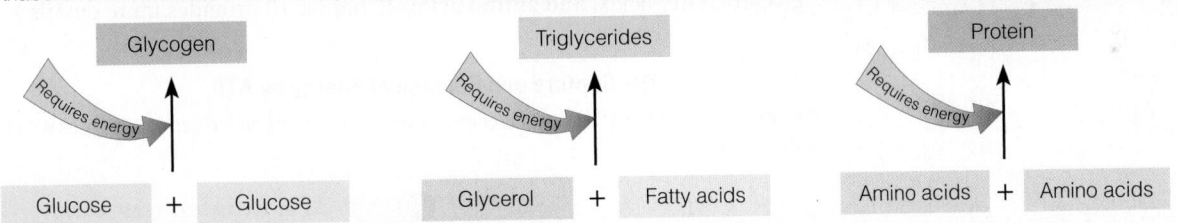

CATABOLIC REACTIONS

Catabolic reactions include the breakdown of glycogen, triglycerides, and protein; the further catabolism of glucose, glycerol, fatty acids, and amino acids releases differing amounts of energy. Much of the energy released is captured in the bonds of adenosine triphosphate (ATP).

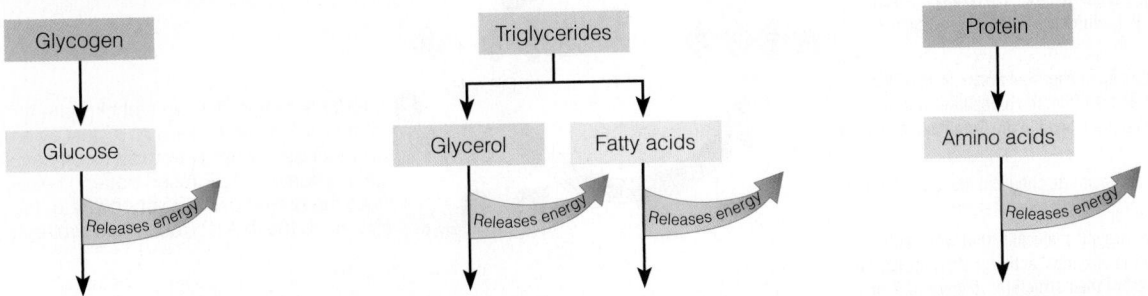

NOTE: You need not memorize a color code to understand the figures in this chapter, but you may find it helpful to know that blue is used for carbohydrates, yellow for fats, and red for proteins.

> FIGURE 7-3   ATP (Adenosine Triphosphate)

Notice that the bonds connecting the three phosphate groups have been drawn as wavy lines, indicating a high-energy bond. When these bonds are broken, energy is released.

Adenosine   +   3 phosphate groups

© Cengage Learning

The Breakdown Reactions—Catabolism

The breaking down of body compounds is known as **catabolism;** catabolic reactions release energy and are represented, wherever possible, by "down" arrows in chemical diagrams (as in the bottom of Figure 7-2, p. 203). Earlier chapters describe how hydrolysis reactions break down glycogen to glucose, triglycerides to fatty acids and glycerol, and proteins to amino acids. When the body needs energy, it breaks down these molecules further.

The Transfer of Energy in Reactions—ATP

Some of the energy released during the breakdown of glucose, glycerol, fatty acids, and amino acids is captured in the high-energy compound **ATP (adenosine triphosphate).** ATP, as its name indicates, contains three phosphate groups (see Figure 7-3). The negative charges on the phosphate groups make ATP vulnerable to hydrolysis. When the bonds between the phosphate groups are hydrolyzed, they readily break, splitting off one or two phosphate groups and releasing energy. In this way, ATP provides the energy that powers all the activities of living cells. Figure 7-4 describes how the body captures and releases energy in the bonds of ATP.

Quite often, the hydrolysis of ATP occurs simultaneously with reactions that will use that energy—a metabolic duet known as **coupled reactions.** In essence, the body uses ATP to transfer the energy released during catabolic reactions to power anabolic reactions that require energy. The body converts the chemical energy of food to the chemical energy of ATP with about 50 percent efficiency, radiating the rest as heat. Some energy is lost as heat again when the body uses the chemical energy of ATP to do its work—moving muscles, synthesizing compounds, or transporting nutrients, for example.

The Helpers in Metabolic Reactions—Enzymes and Coenzymes

Metabolic reactions almost always require **enzymes** to facilitate their action. In many cases, the enzymes need assistants to help them. Enzyme helpers are called **coenzymes.***

Coenzymes are complex organic molecules that associate closely with enzymes but are not proteins themselves. The relationships between various coenzymes and their respective enzymes may differ in detail, but one thing is true of all: without its coenzyme, an enzyme cannot function. Some of the B vitamins serve as coenzymes that participate in the energy metabolism of glucose, glycerol, fatty acids, and amino acids. (Chapter 10 provides more details.)

> FIGURE 7-4   **The Capture and Release of Energy by ATP**

It may help to think of ATP as a rechargeable battery—capturing and releasing energy as it does the body's work.

catabolism (ca-TAB-o-lism): reactions in which large molecules are broken down to smaller ones. Catabolic reactions release energy.

• **kata** = (break) down

ATP, or **adenosine** (ah-DEN-oh-seen) **triphosphate** (try-FOS-fate): a common high-energy compound composed of a purine (adenine), a sugar (ribose), and three phosphate groups. ATP = A–P~P~P, with each ~ denoting a "high-energy" bond.

coupled reactions: pairs of chemical reactions in which some of the energy released from the breakdown of one compound is used to create a bond in the formation of another compound.

enzymes: proteins that facilitate chemical reactions without being changed in the process; protein catalysts.

coenzymes: complex organic molecules that work with enzymes to facilitate the enzymes' activity. Many coenzymes have B vitamins as part of their structures. (Figure 10-2 on p. 305 illustrates coenzyme action.)

• **co** = with

1 Energy is released when a high-energy phosphate bond in ATP is broken. Just as a battery can be used to provide energy for a variety of uses, the energy from ATP can be used to do most of the body's work—contract muscles, transport compounds, make new molecules, and more. With the loss of a phosphate group, high-energy ATP (charged battery) becomes low-energy ADP (used battery).

2 Energy is required when a phosphate group is attached to ADP, making ATP. Just as a used battery needs energy from an electrical outlet to get recharged, ADP (used battery) needs energy from the breakdown of carbohydrate, fat, and protein to make ATP (recharged battery).

© Cengage Learning

ATP breakdown   Releases energy

ⒶⓅⓅⓅ   ⒶⓅⓅ + Ⓟ
ATP   ADP + P

ATP synthesis   Requires energy

*The general term for substances that facilitate enzyme action is *cofactors;* they include both organic coenzymes made from vitamins and inorganic substances such as minerals.

> **REVIEW IT** Identify the nutrients involved in energy metabolism and the high-energy compound that captures the energy released during their breakdown. During digestion the energy-yielding nutrients—carbohydrates, fats, and proteins—are broken down to glucose (and other monosaccharides), glycerol, fatty acids, and amino acids. With the help of enzymes and coenzymes, the cells use these molecules to build more complex compounds (anabolism) or break them down further to release energy (catabolism). High-energy compounds such as ATP may capture the energy released during catabolism and provide the energy needed for anabolism.

7.2 Breaking Down Nutrients for Energy

> **LEARN IT** Summarize the main steps in the energy metabolism of glucose, glycerol, fatty acids, and amino acids.

Chapters 4, 5, and 6 provide previews of metabolism; a brief review may be helpful. During digestion, the body breaks down the three energy-yielding nutrients—carbohydrates, fats, and proteins—into smaller molecules that can be absorbed:

- From carbohydrates—glucose (and other monosaccharides)
- From fats (triglycerides)—glycerol and fatty acids
- From proteins—amino acids

Each molecule of glucose, glycerol, fatty acids, and amino acids is composed of atoms—carbons, nitrogens, oxygens, and hydrogens. During catabolism, the bonds between these atoms break, releasing energy. To follow this action, recall how many carbons are in each of these molecules:

- Glucose has 6 carbons:

- Glycerol has 3 carbons:

- A fatty acid usually has an even number of carbons, commonly 16 or 18 carbons:*

- An amino acid has 2, 3, or more carbons with a nitrogen attached:**

Full chemical structures and reactions appear both in the earlier chapters and in Appendix C. This chapter diagrams the reactions using just the compounds' carbon and nitrogen atoms.

As you will see, each of these molecules—glucose, glycerol, fatty acids, and amino acids—starts down a different path, but they all can end up in the same place. (Similarly, three people entering an interstate highway at three different locations can all travel to the same destination.) Along the way, two new names appear—**pyruvate** (a 3-carbon structure) and **acetyl CoA** (a 2-carbon structure with a coenzyme, **CoA,** attached)—and the rest of the story falls into place around them.*** Two major points to notice in the following discussion:

- Pyruvate can be used to make glucose.
- Acetyl CoA cannot be used to make glucose.

---

*The figures in this chapter show 16- or 18-carbon fatty acids. Fatty acids may have 4 to 20 or more carbons, with chain lengths of 16 and 18 carbons most prevalent.
**The figures in this chapter usually show amino acids as compounds of 2, 3, or 5 carbons arranged in a straight line, but in reality amino acids may contain other numbers of carbons and assume other structural shapes (see Appendix C).
***The term *pyruvate* means a salt of *pyruvic acid.* (Throughout this book, the ending *-ate* is used interchangeably with *-ic acid*; for our purposes they mean the same thing.)

pyruvate (PIE-roo-vate): a 3-carbon compound that plays a key role in energy metabolism.

acetyl CoA (ASS-eh-teel or ah-SEET-il, coh-AY): a 2-carbon compound (acetate or acetic acid) to which a molecule of CoA is attached.

CoA (coh-AY): coenzyme A; the coenzyme derived from the B vitamin pantothenic acid and central to energy metabolism.

All the energy used to keep the heart beating, the brain thinking, and the body moving comes from the carbohydrates, fats, and proteins in foods.

Learning which fuels can be converted to glucose and which cannot is a major key to understanding energy metabolism. Amino acids and glycerol can be converted to pyruvate and therefore *can* provide glucose for the body. Fatty acids are converted to acetyl CoA and therefore *cannot* make glucose. Acetyl CoA can readily make fat. Whereas most of the body's cells can use glucose, fat, or both for energy, the body *must* have glucose to fuel the activities of the central nervous system and red blood cells. Without glucose from food, the body will break down its own lean (protein-containing) tissue to get the amino acids needed to make glucose. To protect this protein tissue, the body needs foods that provide glucose—primarily carbohydrate. Eating only fat provides abundant acetyl CoA, but forces the body to break down protein tissue to make glucose. Eating only protein requires the body to convert protein to glucose. Clearly, the best diet provides ample carbohydrate (45 to 65 percent of kcalories), adequate protein (10 to 35 percent of kcalories), and some fat (20 to 35 percent of kcalories).

Figure 7-5 provides a simplified overview of the energy-yielding pathways. Upcoming sections of the chapter describe how each of the energy-yielding nutrients follows its pathway as it is broken down to acetyl CoA. Their paths merge at acetyl CoA, where the real action begins. Acetyl CoA enters the **TCA cycle,** and energy is harnessed through the **electron transport chain.** The TCA cycle and electron transport chain have central roles in energy metabolism and receive full attention later in the chapter—after following each of the energy nutrient pathways to acetyl CoA.

Glucose   What happens to glucose, glycerol, fatty acids, and amino acids during energy metabolism can best be understood by starting with glucose. This discussion features glucose because of its central role in all cells' metabolism and because liver cells can convert the monosaccharides fructose and galactose to compounds that can enter the same energy pathways.

Glucose-to-Pyruvate   The first pathway glucose takes on its way to yield energy is called **glycolysis** (glucose splitting).* Figure 7-6 (p. 208) shows a simplified drawing of glycolysis. (This pathway actually involves several more steps and several enzymes, which are detailed in Appendix C.) In a series of reactions, the 6-carbon glucose is converted to similar 6-carbon compounds before being split in half, forming two 3-carbon compounds. These 3-carbon compounds continue along the pathway until they are converted to pyruvate. Thus the net yield of one glucose molecule is two pyruvate molecules.

The net yield of energy at this point is small; to start glycolysis, the cell needs a little energy and then releases only a little more than it invested initially.** In addition, as glucose breaks down to pyruvate, hydrogen atoms with their electrons are released and carried to the electron transport chain by coenzymes made from the B vitamin niacin. A later section of the chapter explains how oxygen accepts the electrons and combines with the hydrogens to form water and how the process captures energy in the bonds of ATP.

This discussion focuses primarily on the breakdown of glucose for energy, but if needed, cells in the liver (and to some extent, the kidneys) can make glucose again from pyruvate in a process similar to the reversal of glycolysis. For this reason, the

TCA cycle or **tricarboxylic** (try-car-box-ILL-ick) **acid cycle:** a series of metabolic reactions that break down molecules of acetyl CoA to carbon dioxide and hydrogen atoms; also called the *citric acid cycle* or the *Krebs cycle* after the biochemist who elucidated its reactions.

electron transport chain: the final pathway in energy metabolism that transports electrons from hydrogen to oxygen and captures the energy released in the bonds of ATP; also called the *respiratory chain.*

glycolysis (gly-COLL-ih-sis): the metabolic breakdown of glucose to pyruvate. Glycolysis does not require oxygen (anaerobic).

• **glyco** = glucose
• **lysis** = breakdown

---

*Glycolysis takes place in the cytosol of the cell (see Figure 7-1, p. 202).
**The cell uses two ATP to begin the breakdown of glucose to pyruvate, but it then gains four ATP, for a net gain of two ATP.

arrows between glucose and pyruvate could point up as well as down. Making glucose requires energy, however, and different enzymes. Still, depending on the cell's needs, glucose may go "down" to make pyruvate, or pyruvate may go "up" to make glucose.

Pyruvate's Options—Anaerobic or Aerobic Whenever carbohydrates, fats, or proteins are broken down to provide energy, oxygen is always ultimately involved in the process. The role of oxygen in metabolism is worth noticing, for it helps our understanding of physiology and metabolic reactions. Chapter 14 describes the body's use of the energy nutrients to fuel physical activity, but the facts presented here offer a sneak preview.

When the body needs energy quickly—as occurs when you run a quarter mile as fast as you can—pyruvate is converted to lactate. The breakdown of glucose-to-pyruvate-to-lactate proceeds without oxygen—it is **anaerobic.** This anaerobic pathway yields energy quickly, but it cannot be sustained for long—a couple of minutes at most.

When energy expenditure proceeds at a slower pace—as occurs when you jog around the track for an hour—pyruvate breaks down to acetyl CoA in an **aerobic** pathway. Aerobic pathways produce energy more slowly, but because they can be sustained for a long time, their total energy yield is greater. The following paragraphs provide more details.

Pyruvate-to-Lactate (Anaerobic) As mentioned earlier, coenzymes carry the hydrogens from glucose breakdown to the electron transport chain. If the electron transport chain is unable to accept these hydrogens, as may occur when cells lack sufficient **mitochondria** (review Figure 7-1, p. 202) or in the absence of sufficient oxygen, pyruvate can accept the hydrogens. By accepting the hydrogens, pyruvate becomes **lactate,** and the coenzymes are freed to return to glycolysis to pick up more hydrogens (see the left side of Figure 7-7 on p. 209). In this way, glucose can continue providing energy anaerobically for a while.

The production of lactate occurs to a limited extent even at rest. During high-intensity exercise, however, the muscles rely heavily on anaerobic glycolysis to produce ATP quickly, and the concentration of lactate increases dramatically. The rapid rate of glycolysis produces abundant pyruvate and releases hydrogen-carrying coenzymes more rapidly than the mitochondria can handle. To enable exercise to continue at this intensity, pyruvate is converted to lactate and coenzymes are released, which allows glycolysis to continue. The accumulation of lactate in the muscles coincides with—but does not seem to be the cause of—the subsequent drop in blood pH, burning pain, and fatigue that are commonly associated with intense exercise. In fact, making lactate from pyruvate removes two hydrogen ions, which actually diminishes acidity and improves the performance of tired muscles. A person performing the same exercise following endurance training actually experiences less discomfort—in part because the number of mitochondria in the muscle cells has increased. This adaptation

> FIGURE 7-5 **Simplified Overview of Energy-Yielding Pathways**

① All of the energy-yielding nutrients—protein, carbohydrate, and fat—can be broken down to acetyl CoA.

② Acetyl CoA can enter the TCA cycle.

③ Most of the reactions above release hydrogen atoms with their electrons, which are carried by coenzymes to the electron transport chain.

④ ATP is synthesized.

⑤ Hydrogen atoms react with oxygen to produce water.

© Cengage Learning

anaerobic (AN-air-ROE-bic): not requiring oxygen.

• **an** = not

aerobic (air-ROE-bic): requiring oxygen.

mitochondria (my-toh-KON-dree-uh): the cellular organelles responsible for producing ATP aerobically; made of membranes with enzymes mounted on them. (The singular is *mitochondrion.*)

• **mitos** = thread (referring to their slender shape)

• **chondros** = cartilage (referring to their external appearance)

lactate: a 3-carbon compound produced from pyruvate during anaerobic metabolism.

> FIGURE 7-6 Glycolysis: Glucose-to-Pyruvate

This simplified overview of glycolysis illustrates the steps in the process of converting glucose to pyruvate. (Appendix C provides more details.) Notice that these arrows point down, indicating the breakdown of glucose to pyruvate during energy metabolism. (Alternatively, the arrows could point up, indicating the making of glucose from pyruvate, but that is not the focus of this discussion.)

Glucose

1 A little energy (ATP) is required to start glycolysis. Galactose and fructose enter glycolysis at different places, but all continue on the same pathway.

Requires energy

2 In a series of reactions, the 6-carbon glucose is converted to other 6-carbon compounds, which eventually split into two interchangeable 3-carbon compounds.

Requires energy

3 Coenzymes carry the hydrogens and their electrons to the electron transport chain.

Coenzymes

Coenzymes H^+ e^- → To electron transport chain

4 A little energy (ATP) is released.

Releases energy

5 The 3-carbon compounds go through a series of conversions, producing other 3-carbon compounds, each slightly different.

6 Eventually, the 3-carbon compounds are converted to pyruvate. Glycolysis of one molecule of glucose produces two molecules of pyruvate.

Releases energy

2 Pyruvate

© Cengage Learning

NOTE: The cell uses a little energy (–2 ATP) to begin the breakdown of glucose to pyruvate, but then it gains a little more energy (+4 ATP), for a small net gain (of 2 ATP).

improves the mitochondria's ability to keep pace with the muscles' demand for energy.

One possible fate of lactate is to be transported from the muscles to the liver. The liver can convert the lactate produced in muscles to glucose, which can then be returned to the muscles. (Muscle cells cannot convert lactate to glucose because they lack the necessary enzyme.) This recycling process is called the **Cori cycle** (see the right side of Figure 7-7).

Pyruvate-to-Acetyl CoA (Aerobic) If a cell needs energy and oxygen is available, pyruvate molecules enter the mitochondria of the cell. There a carbon group (COOH) from the 3-carbon pyruvate is removed to produce a 2-carbon compound that bonds with a molecule of CoA, becoming acetyl CoA. The carbon group from pyruvate becomes carbon dioxide (CO_2), which is released into the blood, circulated to the lungs, and breathed out. Figure 7-8 (p. 210) diagrams the pyruvate-to-acetyl CoA reaction.

Figure 7-9 (p. 210) shows that many of the body's metabolic pathways are reversible, but the step from pyruvate to acetyl CoA is not one of them. A cell cannot retrieve the carbons from carbon dioxide to remake pyruvate and then glucose. It is one way only.

The story of acetyl CoA continues on p. 211 after a discussion of how fat and protein arrive at the same crossroads. For now, know that when acetyl CoA continues on its energy-yielding pathway, much more ATP is produced than during glycolysis.

> REVIEW IT

The glucose-to-energy pathway begins with glycolysis—the breakdown of glucose to pyruvate. Pyruvate may be converted to lactate anaerobically or to acetyl CoA aerobically. The pathway from pyruvate to acetyl CoA is irreversible. Once the commitment to acetyl CoA is made, glucose is not retrievable; acetyl CoA cannot go back to glucose. Glucose can be synthesized only from pyruvate or compounds earlier in the pathway. Figure 7-10 (p. 210) summarizes the metabolism of glucose for energy.

Glycerol and Fatty Acids Recall that triglycerides can break down to glycerol and fatty acids. They enter energy metabolism via different pathways.

Glycerol-to-Pyruvate Glycerol is a 3-carbon compound like pyruvate but with a different arrangement of H and OH on the C. As such, glycerol can easily be converted to

Cori cycle: the pathway in which glucose is metabolized to lactate (by anaerobic glycolysis) in the muscle, lactate is converted back to glucose in the liver, and then glucose is returned to the muscle; named after the scientist who elucidated this pathway.

> FIGURE 7-7   **Pyruvate-to-Lactate and Lactate-to-Glucose (the Cori Cycle)**

Because muscle cells lack the enzyme to convert lactate to glucose, lactate must first travel to the liver. The process of converting lactate from the muscles to glucose in the liver that can be returned to the muscles is known as the Cori cycle.

1 Working muscles break down most of their glucose molecules to pyruvate, releasing energy (ATP).

Glucose returns to the muscles

In the liver:

Glucose

Glycolysis in the muscle:

Glucose

Coenzymes

Releases energy

Coenzymes   H⁺ e⁻

Coenzymes

1

Pyruvate

2

Lactate

Lactate travels to the liver

Requires energy

3

Lactate

2 If the cells lack sufficient mitochondria or in the absence of sufficient oxygen, pyruvate can accept the hydrogens from glucose breakdown and become lactate. This conversion frees the coenzymes so that glycolysis can continue.

3 Liver enzymes can convert lactate to glucose, but this reaction requires energy (ATP).

© Cengage Learning

another 3-carbon compound that can go either "up" to glucose or "down" to pyruvate and then to acetyl CoA (see Figure 7-9, p. 210).

Fatty Acids-to-Acetyl CoA Fatty acids are taken apart two carbons at a time in a series of reactions known as **fatty acid oxidation.*** Figure 7-11 (p. 211) illustrates fatty acid oxidation and shows that in the process, each 2-carbon fragment splits off and combines with a molecule of CoA to make acetyl CoA. As each 2-carbon fragment breaks off, hydrogens and their electrons are released and carried to the electron transport chain by coenzymes made from the B vitamins riboflavin and niacin.

Fatty Acids Cannot Make Glucose As mentioned earlier, red blood cells and the brain and nervous system depend primarily on glucose as fuel. When carbohydrate is unavailable, liver cells can make glucose from pyruvate and other 3-carbon compounds, such as glycerol. Importantly, cells cannot make glucose from the 2-carbon fragments of fatty acids.

Remember that almost all dietary fats are triglycerides and that triglycerides contain only one small molecule of glycerol with three fatty acids. The glycerol can yield glucose, but that represents only 3 of the 50 or so carbons in a triglyceride—about 5 percent of its weight. The other 95 percent cannot be used to make glucose.

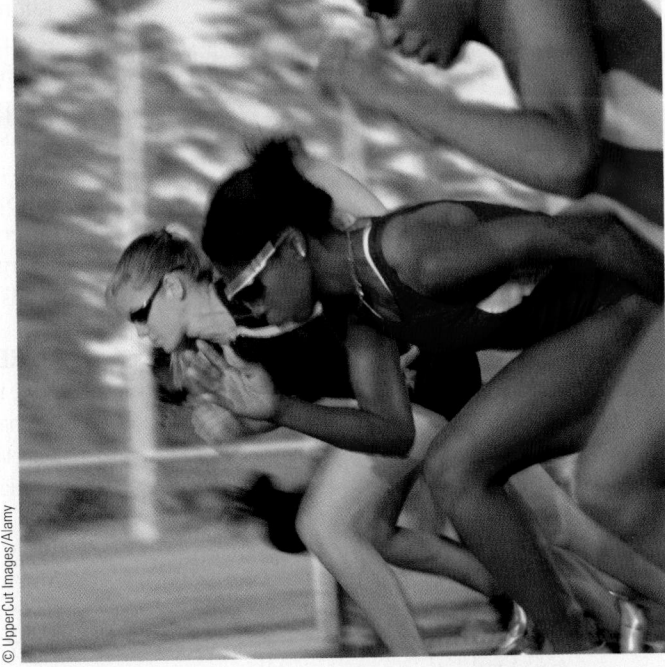

© UpperCut Images/Alamy

The anaerobic breakdown of glucose-to-pyruvate-to-lactate is the major source of energy for short, intense exercise.

❭ REVIEW IT

The body can convert the small glycerol portion of a triglyceride to either pyruvate (and then glucose) or acetyl CoA. The fatty acids of a triglyceride, on the other hand, cannot make glucose, but they can provide abundant acetyl CoA. Acetyl CoA may then enter the TCA cycle to release energy or combine with other molecules of acetyl CoA to make body fat. Figure 7-12 (p. 212) summarizes the metabolism of fats for energy.

fatty acid oxidation: the metabolic breakdown of fatty acids to acetyl CoA; also called *beta oxidation*.

---

*Oxidation of fatty acids occurs in the mitochondria of the cells (review Figure 7-1, p. 202).

> FIGURE 7-8 Pyruvate-to-Acetyl CoA

The pyruvate-to-acetyl CoA reaction is not reversible.

1 Pyruvate picks up a molecule of CoA.

2 A carbon is lost as carbon dioxide.

To electron transport chain

Acetyl CoA

To TCA cycle

© Cengage Learning

> FIGURE 7-10 Glucose Enters the Energy Pathway

Glucose

Coenzymes

To electron transport chain

2 Pyruvate

2 CoA

Coenzymes

To electron transport chain

2 Carbon dioxide

2 Acetyl CoA

To TCA cycle

© Cengage Learning

> REVIEW IT

1 glucose yields 2 pyruvate, which yield 2 acetyl CoA.

> FIGURE 7-9 The Paths of Pyruvate and Acetyl CoA

Pyruvate may follow several reversible paths, but the path from pyruvate to acetyl CoA is irreversible. Notice that fatty acids cannot be used to make glucose.

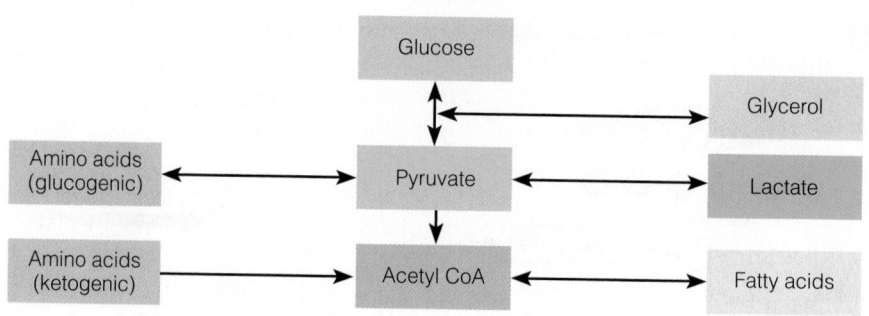

Glucose

Amino acids (glucogenic)

Pyruvate

Glycerol

Lactate

Amino acids (ketogenic)

Acetyl CoA

Fatty acids

NOTE: Amino acids that can be used to make glucose are called *glucogenic*; amino acids that are converted to acetyl CoA are called *ketogenic*.

© Cengage Learning

Amino Acids The preceding two sections have described how the breakdown of carbohydrate and fat produces acetyl CoA. One energy-yielding nutrient remains: protein or, rather, the amino acids of protein.

Amino Acid Deamination Before entering the metabolic pathways, amino acids are deaminated (that is, they lose their nitrogen-containing amino group). Chapter 6 describes how deamination produces ammonia (NH_3), which provides the nitrogen needed to make nonessential amino acids and other nitrogen-containing compounds. Any remaining ammonia is cleared from the body via urea synthesis in the liver and excretion in the kidneys.

Amino Acid Pathways Amino acids can enter the energy pathways in several ways. Some amino acids can be converted to pyruvate, others are converted to acetyl CoA, and still others enter the TCA cycle directly as compounds other than acetyl CoA.

As you might expect, amino acids that are used to make pyruvate can provide glucose, whereas those used to make acetyl CoA can provide additional energy or make body fat but cannot make glucose. Amino acids entering the TCA cycle directly can continue in the cycle and generate energy; alternatively, they can generate glucose.* Thus protein, unlike fat, is a fairly good source of glucose when carbohydrate is not available.

> REVIEW IT

The body can use some amino acids to make glucose, whereas others can be used either to provide energy or to make fat. Before an amino acid enters any of these metabolic pathways, its nitrogen-containing amino group must be removed through deamination. Figure 7-13 (p. 212) summarizes the metabolism of amino acids for energy.

Table 7-2 (p. 213) reviews the ways the body can use the energy-yielding nutrients. To obtain energy, the body uses glucose and fatty acids as its primary fuels and amino acids to a lesser extent. To make glucose, the body can use all carbohydrates, most amino acids, and the glycerol portion of a triglyceride. Fatty acids cannot make glucose. To make proteins, the body needs amino acids. It can use glucose and glycerol to make some nonessential amino acids when nitrogen is available; it cannot use fatty acids to make body proteins. Finally, when energy intake exceeds the body's needs, all three energy-yielding nutrients can contribute to body fat stores.

---

*Amino acids that can make glucose via either pyruvate or TCA cycle intermediates are *glucogenic*; amino acids that are degraded to acetyl CoA are *ketogenic*.

> FIGURE 7-11 **Fatty Acid-to-Acetyl CoA**

Fatty acids are broken apart into 2-carbon fragments that combine with CoA to make acetyl CoA.

16-carbon fatty acid

1 The fatty acid is first activated by coenzyme A, which requires energy.

2 As each carbon-carbon bond is broken, hydrogens and their electrons are released, and coenzymes pick them up.

3 Another CoA joins the chain, and the bond at the second carbon (the beta-carbon) weakens. Acetyl CoA splits off, leaving a fatty acid that is two carbons shorter.

4 The shorter fatty acid enters the pathway and the cycle repeats, releasing more hydrogens with their electrons to coenzymes and producing more acetyl CoA. The molecules of acetyl CoA enter the TCA cycle, and the coenzymes carry the hydrogens and their electrons to the electron transport chain.

Coenzymes

Coenzymes → To electron transport chain

To TCA cycle

Net result from a 16-carbon fatty acid:	14-carbon fatty acid CoA	+	1 acetyl CoA
4 Cycle repeats, leaving:	12-carbon fatty acid CoA	+	2 acetyl CoA
Cycle repeats, leaving:	10-carbon fatty acid CoA	+	3 acetyl CoA
Cycle repeats, leaving:	8-carbon fatty acid CoA	+	4 acetyl CoA
Cycle repeats, leaving:	6-carbon fatty acid CoA	+	5 acetyl CoA
Cycle repeats, leaving:	4-carbon fatty acid CoA	+	6 acetyl CoA
Cycle repeats, leaving:	2-carbon fatty acid CoA*	+	7 acetyl CoA

*Notice that 2-carbon fatty acid CoA = acetyl CoA, so that the final yield from a 16-carbon fatty acid is 8 acetyl CoA.

The Final Steps of Energy Metabolism Thus far the discussion has followed each of the energy-yielding nutrients down three different pathways, all arriving at acetyl CoA. Acetyl CoA has two main options—it may be used to synthesize fats or to generate the high-energy compound ATP. When ATP is abundant, acetyl CoA makes fat, the most efficient way to store energy for later use when energy may be needed. Thus any molecule that can make acetyl CoA—including glucose, glycerol, fatty acids, and amino acids—can make fat. In reviewing Figure 7-9 notice that acetyl CoA can be used as a building block for fatty acids, but it cannot be used to make glucose or amino acids. When ATP is low and the cells need energy, acetyl CoA may proceed through the TCA cycle, releasing hydrogens with their electrons to the electron transport chain.

The TCA Cycle The TCA cycle is the final common metabolic pathway for carbohydrates, fats, and amino acids.[1] Its reactions take place in the inner compartment of the mitochondria. Examine the structure of the mitochondria shown in Figure 7-14 (p. 213). The significance of its structure will become evident as details unfold.

When cells need energy, acetyl CoA enters the TCA cycle, a busy metabolic traffic center. The TCA cycle is a circular path, but that doesn't mean it regenerates acetyl

> FIGURE 7-12   **Fats Enter the Energy Pathway**

① Glycerol enters the glycolysis pathway about midway between glucose and pyruvate.

② Fatty acids are broken down into 2-carbon fragments that combine with CoA to form acetyl CoA (shown in Figure 7-11).

REVIEW IT   16-carbon fatty acid yields 8 acetyl CoA.

CoA. Acetyl CoA goes one way only—down to two carbon dioxide molecules and a coenzyme (CoA). The TCA cycle is a circular path because a 4-carbon compound known as **oxaloacetate** is needed in the first step and it is synthesized in the last step.

Oxaloacetate's role in replenishing the TCA cycle is critical. When oxaloacetate is insufficient, the TCA cycle slows down, and the cells face an energy crisis. Oxaloacetate is made primarily from pyruvate, although it can also be made from certain

oxaloacetate (OKS-ah-low-AS-eh-tate): a carbohydrate intermediate of the TCA cycle.

> FIGURE 7-13   **Amino Acids Enter the Energy Pathway**

① Most amino acids can be converted to pyruvate, which can be used to make glucose; they are glucogenic.

② Some amino acids are converted directly to acetyl CoA; they are ketogenic.

③ Some amino acids can enter the TCA cycle directly; they are glucogenic.

NOTE: Deamination and the synthesis of urea are discussed and illustrated in Chapter 6, Figure 6-11 and Figure 6-13 (pp.183–184). The arrows from pyruvate and the TCA cycle to amino acids are possible only for *nonessential* amino acids; remember, the body cannot make essential amino acids.

amino acids. Importantly, oxaloacetate cannot be made from fat. That oxaloacetate must be available for acetyl CoA to enter the TCA cycle underscores the importance of carbohydrates in the diet. A diet that provides ample carbohydrate ensures an adequate supply of oxaloacetate—because glucose produces pyruvate during glycolysis. (This chapter closes with a discussion of the consequences of low-carbohydrate diets.)

As Figure 7-15 (p. 214) shows, oxaloacetate is the first 4-carbon compound to enter the TCA cycle. Oxaloacetate picks up acetyl CoA (a 2-carbon compound), drops off one carbon (as carbon dioxide), then another carbon (as carbon dioxide), and returns to pick up another acetyl CoA. As for the acetyl CoA, its carbons go only one way—to carbon dioxide (see Appendix C for additional details).*

As compounds in the TCA cycle lose a carbon to carbon dioxide, hydrogen atoms with their electrons are carried off by coenzymes made from the B vitamins niacin and riboflavin to the electron transport chain—much like a taxicab that picks up passengers in one location and drops them off in another. Each turn of the TCA cycle releases a total of eight electrons.

The Electron Transport Chain The electron transport chain captures energy in the high-energy bonds of ATP. The electron transport chain consists of a series of proteins that serve as electron "carriers." These carriers are mounted in sequence on the inner membrane of the mitochondria (shown in Figure 7-14). As the coenzymes deliver their electrons from the TCA cycle, glycolysis, and fatty acid oxidation to the electron transport chain, each carrier receives the electrons and passes them on to the next carrier. These electron carriers continue passing the electrons down until they reach oxygen. Oxygen (O) accepts the electrons and combines with hydrogen atoms (H) to form water (H_2O). That oxygen must be available for energy metabolism explains why it is essential to life.

As electrons are passed from carrier to carrier, hydrogen ions are pumped across the membrane to the outer compartment of the mitochondria. The rush of hydrogen ions back into the inner compartment powers the synthesis of ATP. In this way, energy is captured in the bonds of ATP. The ATP leaves the mitochondria and enters the cytoplasm, where it can be used for energy. Figure 7-16 (p. 215) provides a simple diagram of the electron transport chain (see Appendix C for details).

The kCalories-per-Gram Secret Revealed Of the three energy-yielding nutrients, fat provides the most energy per gram. The reason may be apparent in Figure 7-17, (p. 215) which compares a fatty acid with

TABLE 7-2 Review of Energy-Yielding Nutrient End Points

Nutrient	Yields energy?	Yields glucose?	Yields amino acids and body proteins?	Yields fat stores?
Carbohydrates (glucose)	Yes	Yes	Yes—when nitrogen is available, can yield *nonessential* amino acids	Yes
Lipids (fatty acids)	Yes	No	No	Yes
Lipids (glycerol)	Yes	Yes—when carbohydrate is unavailable	Yes—when nitrogen is available, can yield *nonessential* amino acids	Yes
Proteins (amino acids)	Yes	Yes—when carbohydrate is unavailable	Yes	Yes

© Cengage Learning

> **FIGURE 7-14   A Mitochondrion**

Outer compartment

Outer membrane (site of fatty acid activation)

Cytosol (site of glycolysis)

A typical cell

A mitochondrion

Inner membrane (site of electron transport chain)

Inner compartment (site of pyruvate-to-acetyl CoA, fatty acid oxidation, and TCA cycle)

© Cengage Learning

---

*Actually, the carbons that enter the cycle in acetyl CoA may not be the exact ones that are given off as carbon dioxide. In one of the steps of the cycle, a 6-carbon compound of the cycle becomes symmetrical, both ends being identical. Thereafter it loses carbons to carbon dioxide at one end or the other. Thus only half of the carbons from acetyl CoA are given off as carbon dioxide in any one turn of the cycle; the other half become part of the compound that returns to pick up another acetyl CoA. It is true to say, though, that for each acetyl CoA that enters the TCA cycle, two carbons are given off as carbon dioxide. It is also true that with each turn of the cycle, the energy equivalent of one acetyl CoA is released.

> FIGURE 7-15   **The TCA Cycle**

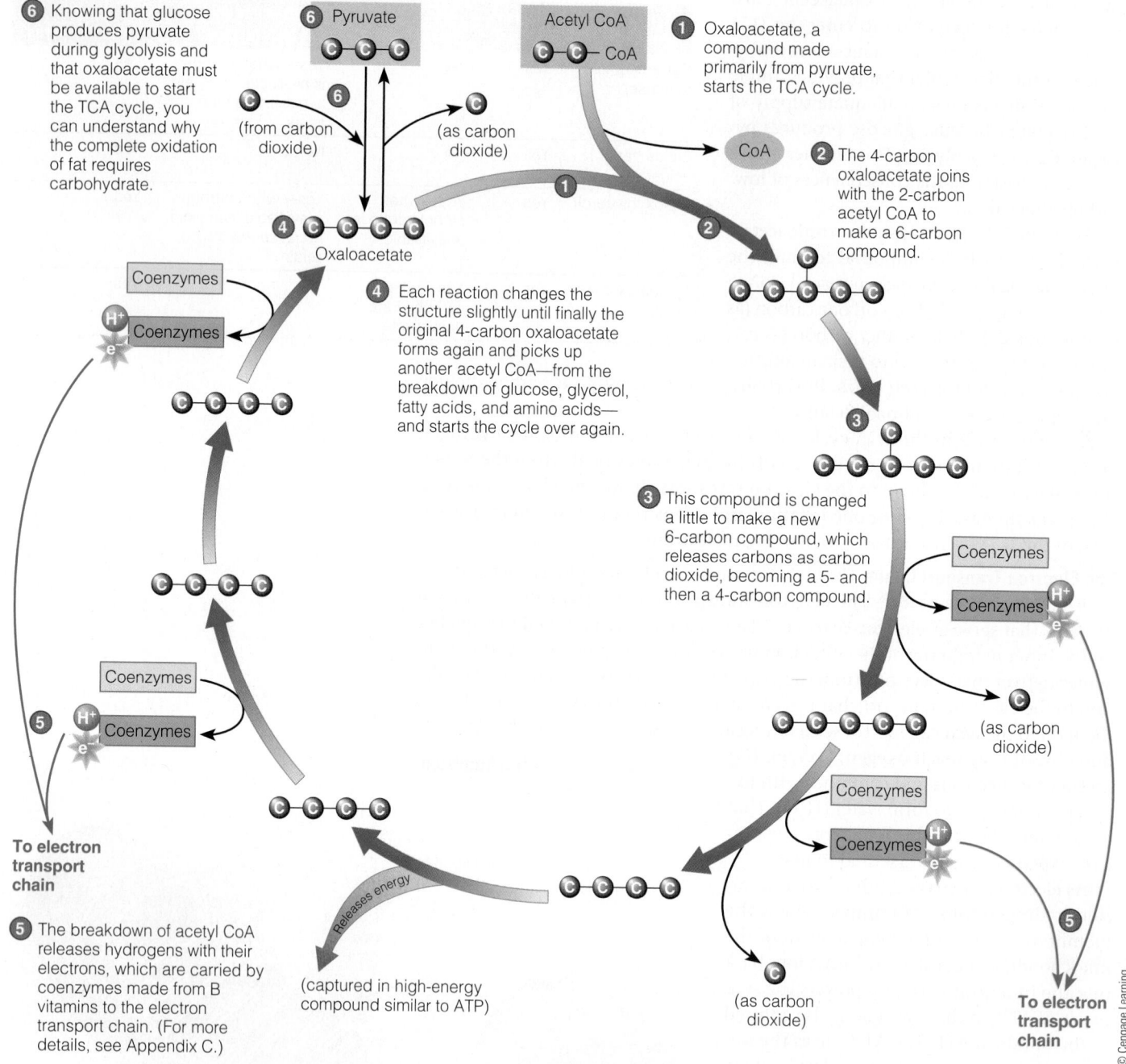

6 Knowing that glucose produces pyruvate during glycolysis and that oxaloacetate must be available to start the TCA cycle, you can understand why the complete oxidation of fat requires carbohydrate.

6 Pyruvate

C — C — C

C (from carbon dioxide)

6

C (as carbon dioxide)

Acetyl CoA

C — C — CoA

1 Oxaloacetate, a compound made primarily from pyruvate, starts the TCA cycle.

CoA

2 The 4-carbon oxaloacetate joins with the 2-carbon acetyl CoA to make a 6-carbon compound.

4 C — C — C — C
Oxaloacetate

Coenzymes

H⁺ e⁻ Coenzymes

4 Each reaction changes the structure slightly until finally the original 4-carbon oxaloacetate forms again and picks up another acetyl CoA—from the breakdown of glucose, glycerol, fatty acids, and amino acids—and starts the cycle over again.

C — C — C — C — C

C
C — C — C — C — C

3 This compound is changed a little to make a new 6-carbon compound, which releases carbons as carbon dioxide, becoming a 5- and then a 4-carbon compound.

Coenzymes

Coenzymes H⁺ e⁻

C (as carbon dioxide)

Coenzymes

Coenzymes H⁺ e⁻

C — C — C — C — C

C — C — C — C

5 H⁺ e⁻ Coenzymes

Coenzymes

To electron transport chain

5 The breakdown of acetyl CoA releases hydrogens with their electrons, which are carried by coenzymes made from B vitamins to the electron transport chain. (For more details, see Appendix C.)

C — C — C — C

Releases energy

(captured in high-energy compound similar to ATP)

C (as carbon dioxide)

5

To electron transport chain

© Cengage Learning

a glucose molecule. Notice that nearly all the bonds in the fatty acid are between carbons and hydrogens. Oxygen can be added to all of them—forming carbon dioxide (CO_2) with the carbons and water (H_2O) with the hydrogens. As this happens, hydrogens are released to coenzymes heading for the electron transport chain. In glucose, on the other hand, an oxygen is already bonded to each carbon. Thus there is less potential for oxidation, and fewer hydrogens are released when the remaining bonds are broken.

Because fat contains many carbon-hydrogen bonds that can be readily oxidized, it sends numerous coenzymes with their hydrogens and electrons to the electron transport chain, where that energy can be captured in the bonds of ATP. This explains why fat yields more kcalories per gram than carbohydrate or protein. (Remember that each ATP holds energy and that kcalories measure energy; thus the more ATP, the more kcalories.) For example, one glucose

> **FIGURE 7-16** **Electron Transport Chain and ATP Synthesis**

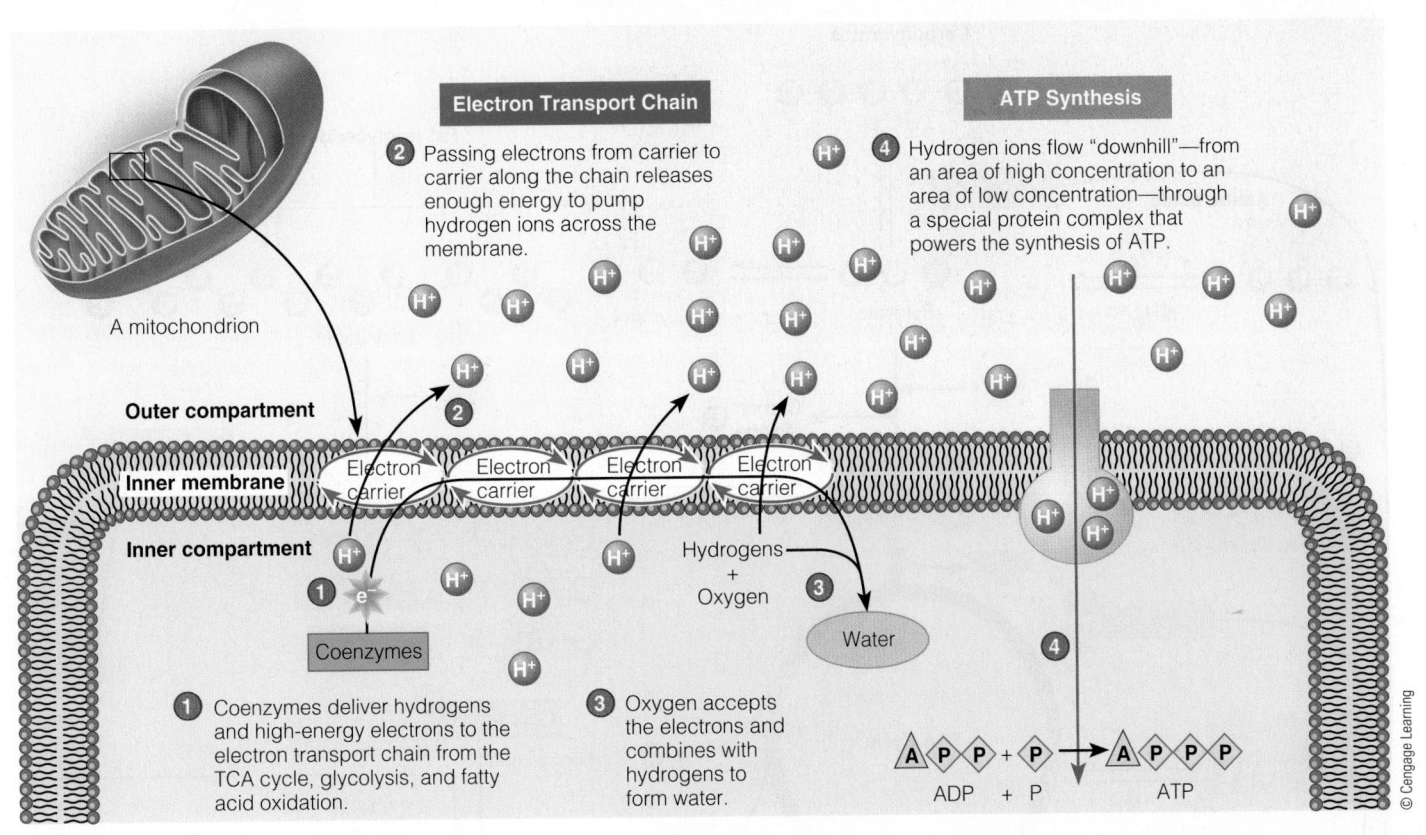

A mitochondrion

Electron Transport Chain

2 Passing electrons from carrier to carrier along the chain releases enough energy to pump hydrogen ions across the membrane.

ATP Synthesis

4 Hydrogen ions flow "downhill"—from an area of high concentration to an area of low concentration—through a special protein complex that powers the synthesis of ATP.

Outer compartment

Inner membrane

Inner compartment

Electron carrier

Coenzymes

Hydrogens + Oxygen

Water

1 Coenzymes deliver hydrogens and high-energy electrons to the electron transport chain from the TCA cycle, glycolysis, and fatty acid oxidation.

3 Oxygen accepts the electrons and combines with hydrogens to form water.

ADP + P → ATP

© Cengage Learning

> **FIGURE 7-17** **Chemical Structures of a Fatty Acid and Glucose Compared**

To ease comparison, the structure shown here for glucose is not the ring structure shown in Chapter 4, but an alternative way of drawing its chemical structure.

Fatty acid

Glucose

© Cengage Learning

molecule will yield 30 to 32 ATP when completely oxidized. In comparison, one 16-carbon fatty acid molecule will yield 129 ATP when completely oxidized. Fat is a more efficient fuel source. Gram for gram, fat can provide much more energy than either of the other two energy-yielding nutrients, making it the body's preferred form of energy storage. (Similarly, you might prefer to fill your car with a fuel that provides 130 miles per gallon versus one that provides 30 miles per gallon.)

> **REVIEW IT** Summarize the main steps in the energy metabolism of glucose, glycerol, fatty acids, and amino acids.

Carbohydrate, fat, and protein take different paths to acetyl CoA, but once there, the final pathways—the TCA cycle and electron transport chain—are shared. All of the pathways, which are shown as a simplified overview in Figure 7-5 (p. 207), are shown again in more detail in Figure 7-18 (p. 216). Instead of dismissing this figure as "too busy," take a few moments to appreciate the busyness of it all. Consider that this figure is merely an overview of energy metabolism, and then imagine how busy a living cell really is during the metabolism of hundreds of compounds, each of which may be involved in several reactions, each requiring specific enzymes.

> FIGURE 7-18   The Central Pathways of Energy Metabolism

REVIEW IT
- All of the energy-yielding nutrients—protein, carbohydrates, and fat—can be broken down to acetyl CoA.
- Acetyl CoA can enter the TCA cycle or it can make fat.
- Many of these reactions release hydrogen atoms with their electrons, which are carried by coenzymes to the electron transport chain.
- In the end, oxygen is consumed, water and carbon dioxide are produced, and energy is captured in ATP.
- Some amino acids, pyruvate, and glycerol can be used to make glucose.
- Fatty acids cannot be used to make glucose.

© Cengage Learning

7.3 Feasting and Fasting

> **LEARN IT** Explain how an excess of any of the three energy-yielding nutrients contributes to body fat and how an inadequate intake of any of them shifts metabolism.

Every day, a healthy diet delivers more than a thousand kcalories of energy—from carbohydrate, fat, and protein—to fuel the physical activity and metabolic work of the body. The details of energy metabolism have already been described; this discussion examines what happens when energy intake is excessive or inadequate and how metabolism shifts when the three energy-yielding nutrients are out of balance.

Feasting—Excess Energy When a person eats too much, metabolism favors fat formation. Fat cells enlarge regardless of whether the excess in kcalories derives from protein, carbohydrate, or fat. The pathway from dietary fat to body fat, however, is the most direct (requiring only a few metabolic steps) and the most efficient (costing only a few kcalories). To convert a dietary triglyceride to a triglyceride in adipose tissue, the body removes two of the fatty acids from the glycerol, absorbs the parts, and puts them (and others) together again. By comparison, to convert a molecule of sucrose, the body has to split glucose from fructose, absorb them, dismantle them to pyruvate and acetyl CoA, assemble many acetyl CoA molecules into fatty acid chains, and finally attach fatty acids to a glycerol molecule to make a triglyceride for storage in adipose tissue. Quite simply, the body uses much less energy to convert dietary fat to body fat than it does to convert dietary carbohydrate to body fat. On average, storing excess energy from dietary fat as body fat uses only 5 percent of the ingested energy intake, but storing excess energy from dietary carbohydrate as body fat requires 25 percent of the ingested energy intake.

The pathways from excess protein and excess carbohydrate to body fat are not only indirect and inefficient, but they are also less preferred by the body (having other priorities for using these nutrients). Before entering fat storage, protein must first tend to its many roles in the body's lean tissues, and carbohydrate must fill the glycogen stores. Simply put, using these two nutrients to make fat is a low priority for the body. Still, if eaten in abundance, any of the energy-yielding nutrients will be converted to fat for storage.

This chapter has described each of the energy-yielding nutrients individually, but cells use a mixture of these fuels. How much of which nutrient is in the fuel mix depends, in part, on its availability from the diet. (The proportion of each fuel also depends on physical activity, as Chapter 14 explains.) Usually, protein's contribution to the fuel mix is relatively minor and fairly constant, but protein oxidation does increase when protein is eaten in excess. Similarly, carbohydrate eaten in excess significantly enhances carbohydrate oxidation. In contrast, fat oxidation does *not* respond to dietary fat intake. The more protein or carbohydrate in the fuel mix, the less fat contributes to the fuel mix. Instead of being oxidized, fat accumulates in storage. Details follow.

Excess Protein Recall from Chapter 6 that the body cannot store excess amino acids as such; it has to convert them to other compounds. Contrary to popular opinion, a person cannot grow muscle simply by overeating protein. Lean tissue such as muscle develops in response to a stimulus such as hormones or physical activity. When a person overeats protein, the body uses the surplus first by replacing normal daily losses and then by increasing protein oxidation. An increase in protein oxidation uses some excess protein, but it displaces fat in the fuel mix. If excess protein is still available, the amino acids are deaminated and the remaining carbons are used to make fatty acids, which are stored as triglycerides in adipose tissue. Thus a person can grow fat by eating too much protein.

People who eat huge portions of meat and other protein-rich foods may wonder why they have weight problems. Not only does the fat in those foods lead to body fat, but the protein can, too, when energy intake exceeds energy needs. Many fad weight-loss diets encourage high protein intakes based on the false assumption that protein builds only muscle, not fat.

Excess Carbohydrate Compared with protein, the proportion of carbohydrate in the fuel mix changes more dramatically when a person overeats. The body handles abundant carbohydrate by first storing it as glycogen, but glycogen storage areas are limited and fill quickly. Because maintaining glucose balance is critical, the body uses glucose frugally when the diet provides only small amounts and freely when supplies are abundant. In other words, glucose oxidation rapidly adjusts to the dietary intake of carbohydrate.

Like protein, excess glucose can also be converted to fat directly. This pathway is relatively minor, however. As mentioned earlier, converting glucose to fat is energetically expensive and does not occur until after glycogen stores have been filled. Still, new body fat is made whenever carbohydrate intake is excessive.

People can enjoy bountiful meals such as this without storing body fat, provided they expend as much energy as they take in.

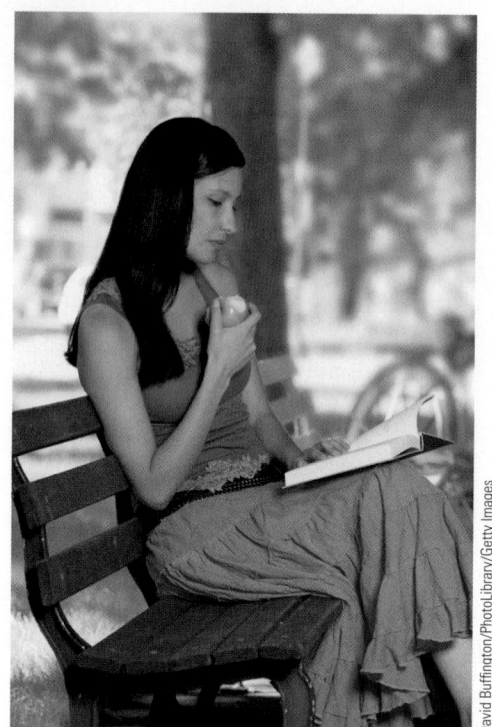

The brain and nerve cells depend on glucose—either directly from carbohydrates or indirectly from proteins (through gluconeogenesis). Importantly, fatty acids cannot provide glucose.

Excess dietary carbohydrate can also displace fat in the fuel mix. When this occurs, carbohydrate spares both dietary fat and body fat from oxidation—an effect that may be more pronounced in overweight people than in lean people. The net result: excess carbohydrate contributes to obesity or at least to the maintenance of an overweight body.

Excess Fat Unlike excess protein and carbohydrate, which both increase oxidation, eating too much fat does not promote fat oxidation. Instead, excess dietary fat moves efficiently into the body's fat stores; almost all of the excess is stored.

The Transition from Feasting to Fasting Figure 7-19 shows the metabolic pathways operating in the body as it shifts from feasting (part A) to fasting (parts B and C). After a meal, glucose, glycerol, and fatty acids from foods are used as needed and then stored. Later, as the body shifts from a fed state to a fasting one, it begins drawing on these stores. Glycogen and fat are released from storage to provide more glucose, glycerol, and fatty acids for energy.

Energy is needed all the time. Even when a person is asleep and totally relaxed, the cells are hard at work. In fact, this work—the cells' work that maintains all life processes without any conscious effort—represents about two-thirds of the total energy a person expends in a day.* The relatively small remainder is the work that a person's muscles perform voluntarily during waking hours.

The body's top priority is to meet the cells' needs for energy, and it normally does this by periodic refueling—that is, by eating several times a day. When food is not available, the body turns to its own tissues for fuel. If people choose not to eat, we say they are fasting; if they have no choice, we say they are starving. The body makes no such distinction. In either case, the body must draw on its reserves of carbohydrate and fat and, within a day or so, on its vital protein tissues as well.

Fasting—Inadequate Energy During fasting, carbohydrate, fat, and protein are all eventually used for energy—fuel must be delivered to every cell. As the fast begins, glucose from the liver's stored glycogen and fatty acids from the adipose tissue's stored fat travel to the cells. As described earlier, these molecules are broken down to acetyl CoA, which enters the energy pathways that power the cells' work. Several hours later, however, liver glycogen is depleted and blood glucose begins to fall. The body must adjust its normal metabolism to survive without food. Starvation demands cells to degrade their components for fuel.

Adaptation: Making Glucose At this point, most cells are using fatty acids for their fuel. But, as mentioned earlier, red blood cells and the cells of the nervous system need glucose. Glucose is their primary energy fuel. Normally, the brain and nerve cells—which weigh only about three pounds—consume about half of the total *glucose* used each day (about 500 kcalories' worth). About one-fourth of the *energy* the adult body uses when it is at rest is spent by the brain.

During a fast, the need for glucose poses a major problem. The body can use its stores of fat, which may be quite generous, to furnish most of its cells with energy, but the red blood cells are completely dependent on glucose, and the brain and nerves prefer energy in the form of glucose.** Amino acids that yield pyruvate can be used for **gluconeogenesis**—the making of glucose from noncarbohydrate sources. The liver is the major site of gluconeogenesis, but the kidneys become increasingly involved under certain circumstances, such as starvation.

The glycerol portion of a triglyceride and most amino acids can be used to make glucose (review Figure 7-9, p. 210). To obtain the amino acids, body proteins must be broken down.*** For this reason, protein tissues such as muscle and liver always break down to some extent during fasting. The amino acids that cannot be used to make glucose are used as an energy source for other body cells.

The breakdown of body protein is an expensive way to obtain glucose. In the first few days of a fast, body protein provides about 90 percent of the needed glucose;

gluconeogenesis (gloo-ko-nee-oh-JEN-ih-sis): the making of glucose from a noncarbohydrate source such as amino acids or glycerol.

- **gluco** = glucose
- **neo** = new
- **genesis** = making

---

*The cells' work that maintains all life processes refers to the body's *basal metabolism*, which is described in Chapter 8.
**Red blood cells contain no mitochondria. Review Figure 7-1 (p. 202) to fully appreciate why red blood cells must depend on glucose for energy.
***An estimated 1 gram of protein can make ½ gram of glucose.

> **FIGURE 7-19** **Feasting and Fasting**

A Feasting: When a person eats in excess of energy needs, the body stores a small amount of glycogen and much larger quantities of fat.

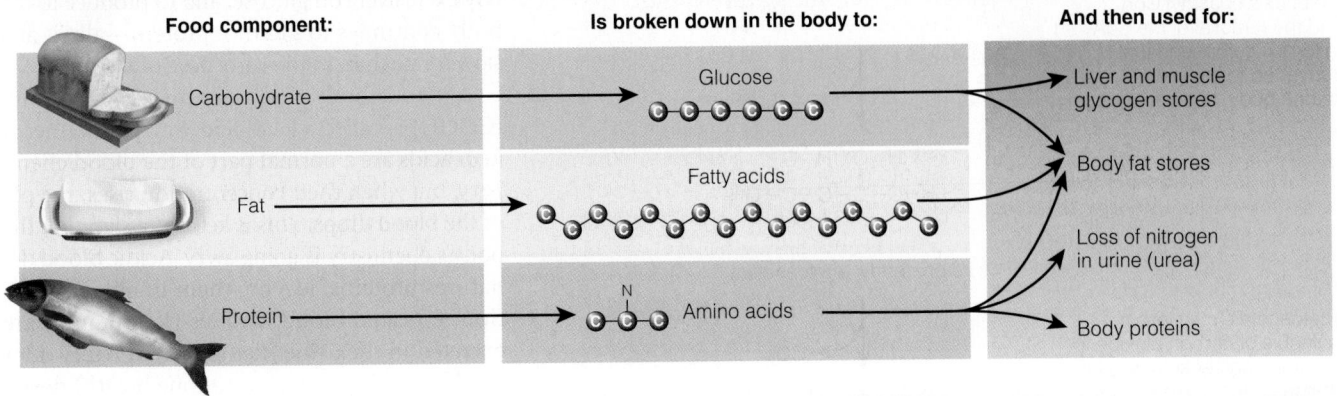

B Fasting: When nutrients from a meal are no longer available to provide energy (about 2 to 3 hours after a meal), the body draws on its glycogen and fat stores for energy.

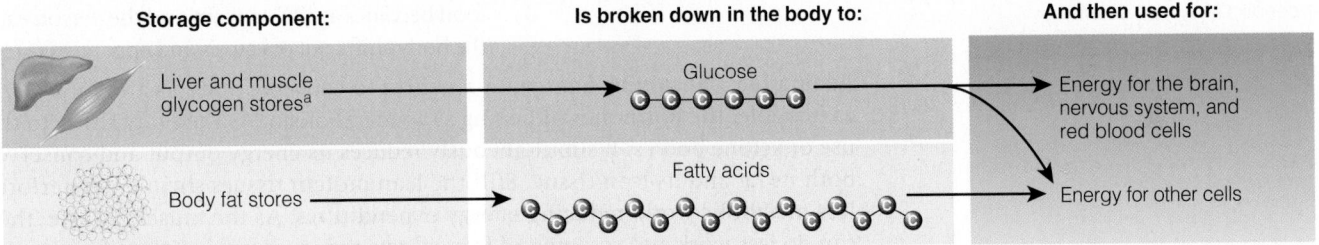

C Fasting beyond glycogen depletion: As glycogen stores dwindle (after about 24 hours of starvation), the body begins to break down its protein (muscle and lean tissue) to amino acids to synthesize glucose needed for brain and nervous system energy. In addition, the liver converts fats to ketone bodies, which serve as an alternative energy source for the brain, thus slowing the breakdown of body protein.

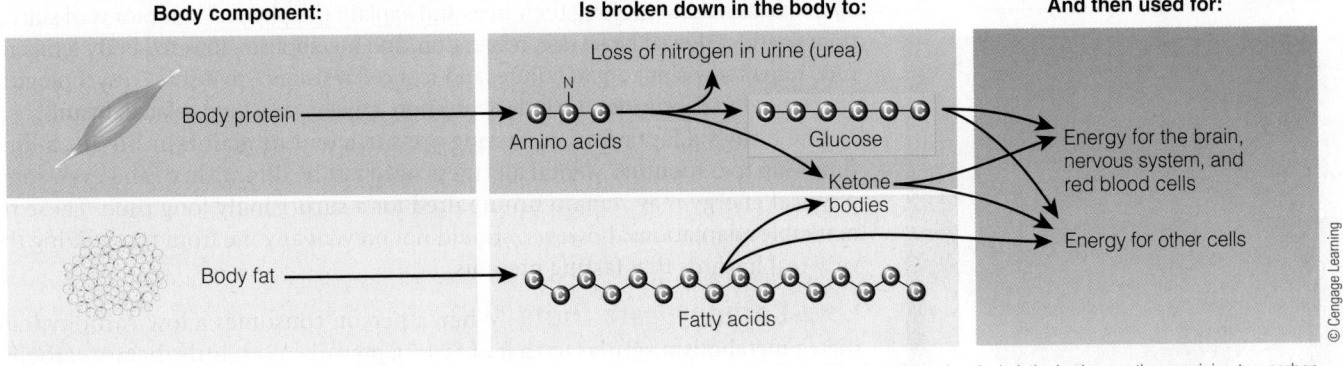

NOTE: Alcohol is not included because it is a toxin and not a nutrient, but it does contribute energy to the body. After detoxifying the alcohol, the body uses the remaining two carbon fragments to build fatty acids and stores them as fat.

aThe muscles' stored glycogen provides glucose only for the muscle in which the glycogen is stored.

© Cengage Learning

glycerol, about 10 percent. If body protein losses were to continue at this rate, death would follow within three weeks, regardless of the quantity of fat a person had stored. Fortunately, fat breakdown also increases with fasting—in fact, fat breakdown almost doubles, providing energy for other body cells and glycerol for glucose production.

Adaptation: Creating an Alternative Fuel As the fast continues, the body finds a way to use its fat to fuel the brain. It adapts by combining acetyl CoA fragments derived from fatty acids to produce an alternative energy source, **ketone bodies** (see Figure 7-20, p. 220). Normally produced and used only in small quantities, ketone bodies can

ketone (KEE-tone) **bodies:** acidic compounds produced by the liver during the incomplete breakdown of fat when carbohydrate is not available.

> FIGURE 7-20 **Ketone Body Formation**

① The first step in the formation of ketone bodies is the condensation of two molecules of acetyl CoA and the removal of the CoA to form a compound that is converted to the first ketone body, acetoacetate.

H–C–C–CoA + H–C–C–CoA + H₂O

Acetyl CoA   Acetyl CoA

→ 2 CoA

H–C–C–C–C–OH

A ketone body, acetoacetate

② Acetoacetate may lose a molecule of carbon dioxide to become another ketone body, acetone.

→ CO₂

③ Or, acetoacetate may add two hydrogens, becoming another ketone body (beta-hydroxybutyrate). See Appendix C for more details.

H–C–C–C–H

A ketone body, acetone

© Cengage Learning

efficiently provide fuel for brain cells. Ketone body production rises until, after about 10 days of fasting, it is meeting much of the nervous system's energy needs. Still, many areas of the brain rely exclusively on glucose, and to produce it, the body continues to sacrifice protein—albeit at a slower rate than in the early days of fasting.

A ketone body that contains an acid group (COOH) is called a **keto acid.** Small amounts of keto acids are a normal part of the blood chemistry, but when their concentration rises, the pH of the blood drops. This is ketosis, a sign that the body's chemistry is going awry. Acidic blood denatures proteins, leaving them unable to function. Elevated blood ketones (ketonemia) are excreted in the urine (ketonuria). A fruity odor on the breath (known as acetone breath) develops, reflecting the presence of the ketone acetone.

Ketosis induces a loss of appetite. As starvation continues, this loss of appetite becomes an advantage to a person without access to food. When food becomes available again and the person eats, the body shifts out of ketosis and appetite returns.

Adaptation: Conserving Energy In an effort to conserve body tissues for as long as possible, the hormones of fasting slow metabolism. As the body shifts to the use of ketone bodies, it simultaneously reduces its energy output and conserves both its fat and its lean tissue. Still the lean protein tissues shrink and perform less metabolic work, reducing energy expenditures. As the muscles waste, they can do less work and so demand less energy, reducing expenditures further. Although fasting may promote dramatic *weight* loss, a low-kcalorie diet and physical activity better support *fat* loss while retaining lean tissue.

These adaptations of fasting—slowing of energy output and reduction in fat loss—occur in the starving child, the hungry homeless adult, the fasting religious person, the adolescent with anorexia nervosa, and the malnourished hospital patient. Such adaptations help to prolong their lives and explain the physical symptoms of starvation: wasting; slowed heart rate, respiration, and metabolism; lowered body temperature; impaired vision; organ failure; and reduced resistance to disease. Psychological effects of food deprivation include depression, anxiety, and food-related dreams.

The body's adaptations to fasting are sufficient to maintain life for a long time—up to 2 months. Mental alertness need not be diminished, and even some physical energy may remain unimpaired for a surprisingly long time. These remarkable adaptations, however, should not prevent anyone from recognizing the very real hazards that fasting presents.

Low-Carbohydrate Diets When a person consumes a low-carbohydrate diet, a metabolism similar to that of fasting prevails. With little dietary carbohydrate coming in, the body uses its glycogen stores to provide glucose for the cells of the brain, nerves, and blood. Once the body depletes its glycogen reserves, it begins making glucose from the amino acids of protein (gluconeogenesis). A low-carbohydrate diet may provide abundant protein from food, but the body still uses some protein from body tissues.

Dieters can know glycogen depletion has occurred and gluconeogenesis has begun by monitoring their urine. Whenever glycogen or protein is broken down, water is released and urine production increases. Low-carbohydrate diets also induce ketosis, and ketones can be detected in the urine. Ketones form whenever glucose is lacking and fat breakdown is incomplete.

Many fad diets regard ketosis as the key to losing weight, but studies comparing weight-loss diets find no relation between ketosis and weight loss. People in ketosis

Low-carbohydrate meals overemphasize meat, fish, poultry, eggs, and cheeses, and shun breads, pastas, fruits, and starchy vegetables.

© Matthew Farruggio

keto (KEY-toe) **acid:** an organic acid that contains a carbonyl group (C=O).

may experience a loss of appetite and a dramatic weight loss within the first few days. They should know that much of this weight loss reflects the loss of glycogen and protein together with large quantities of body fluids and important minerals. They need to appreciate the difference between loss of *fat* and loss of *weight*. Fat losses on ketogenic diets are no greater than on other diets providing the same number of kcalories. Once the dieter returns to well-balanced meals that provide adequate energy, carbohydrate, fat, protein, vitamins, and minerals, the body avidly retains these needed nutrients. The weight will return, quite often to a level higher than the starting point. In addition to weight loss, ketogenic diets are often used in the treatment of several diseases, most notably epilepsy.[2] Table 7-3 lists some of the consequences of a ketogenic diet.

TABLE 7-3 Adverse Side Effects of Low-Carbohydrate, Ketogenic Diets

- Nausea
- Fatigue (especially if physically active)
- Constipation
- Low blood pressure
- Elevated uric acid (which may exacerbate kidney disease and cause inflammation of the joints in those predisposed to gout)
- Stale, foul taste in the mouth (bad breath)
- In pregnant women, fetal harm and stillbirth

© Cengage Learning

> **REVIEW IT** Explain how an excess of any of the three energy-yielding nutrients contributes to body fat and how an inadequate intake of any of them shifts metabolism.

When energy intake exceeds energy needs, the body makes fat—regardless of whether the excess intake is from protein, carbohydrate, or fat. The only difference is that the body is much more efficient at storing energy when the excess derives from dietary fat.

When fasting, the body makes a number of adaptations: increasing the breakdown of fat to provide energy for most of the cells, using glycerol and amino acids to make glucose for the red blood cells and central nervous system, producing ketones to fuel the brain, suppressing the appetite, and slowing metabolism. All of these measures conserve energy and minimize losses. Low-carbohydrate diets incur similar changes in metabolism.

This chapter has probed the intricate details of metabolism at the level of the cells. Several upcoming chapters and highlights build on this information. The highlight that follows this chapter focuses on how alcohol disrupts metabolism. Chapter 8 describes how a person's intake and expenditure of energy are reflected in body weight and body composition. Chapter 9 examines the consequences of unbalanced energy budgets—overweight and underweight. Chapter 10 shows the vital roles the B vitamins play as coenzymes assisting in all the metabolic pathways described here. And Chapter 14 revisits metabolism to show how it supports the work of physically active people and how athletes can best apply that information in their choices of foods to eat.

Nutrition Portfolio

All day, every day, your cells dismantle carbohydrates, fats, and proteins, with the help of vitamins, minerals, and water, releasing energy to meet your body's immediate needs or storing it as fat for later use. Go to Diet & Wellness Plus and choose one of the days on which you have tracked your diet for the entire day. Go to the Intake vs. Goals report and answer the following questions. Keep in mind that in this report 100 percent means you are meeting your needs perfectly.

- How close were you to 100 percent for: carbohydrates, fats, proteins, vitamins, minerals, and water? In general, which category was lowest? Which category was highest?
- Describe what types of foods best support aerobic and anaerobic activities.
- Consider whether you eat more protein, carbohydrate, or fat than your body needs.
- Explain how a low-carbohydrate diet forces your body into ketosis.

> **STUDY IT** To review the key points of this chapter and take a practice quiz, go to the study cards at the end of the book.

REFERENCES

1. M. Akram, Citric acid cycle and role of its intermediates in metabolism, *Cell Biochemistry and Biophysics* 68 (2014): 475–478.

2. A. Paoli and coauthors, Beyond weight loss: A review of the therapeutic uses of very-low-carbohydrate (ketogenic) diets, *European Journal of Clinical Nutrition* 67 (2013): 789–796.

HIGHLIGHT > 7
Alcohol in the Body

> LEARN IT Describe how alcohol disrupts metabolism and impairs health.

With the understanding of metabolism gained from Chapter 7, you are in a position to understand how the body handles alcohol, how alcohol interferes with metabolism, and how alcohol impairs health and nutrition. Before examining alcohol's damaging effects, it may be appropriate to mention that drinking alcohol in *moderation* may have some health benefits, including reduced risks of heart disease, diabetes, and osteoporosis.[1] Moderate alcohol consumption may lower mortality from all causes, but only in adults aged 35 and older. Although health benefits may be apparent in younger adults, the protective effects are negligible compared with the increased risks of traffic accidents and cancer associated with alcohol consumption at that age.[2] Similarly, health benefits begin to disappear in older age, as metabolism changes and organs become more sensitive to toxic substances.[3] Importantly, protective effects disappear when moderate alcohol consumption is mixed with irregular heavy drinking occasions.[4] Furthermore, any benefits of moderate alcohol use must be weighed against the many harmful effects of excessive alcohol use described in this highlight, as well as the possibility of alcohol abuse.[5]

Alcohol in Beverages

To the chemist, **alcohol** refers to a class of organic compounds containing hydroxyl (OH) groups (the accompanying glossary defines *alcohol* and related terms). The glycerol to which fatty acids are attached

GLOSSARY

acetaldehyde (ass-et-AL-duh-hide): an intermediate in alcohol metabolism.

alcohol: a class of organic compounds containing hydroxyl (OH) groups.

- **ol** = alcohol

alcohol abuse: a pattern of drinking that includes failure to fulfill work, school, or home responsibilities; drinking in situations that are physically dangerous (as in driving while intoxicated); recurring alcohol-related legal problems (as in aggravated assault charges); or continued drinking despite ongoing social problems that are caused by or worsened by alcohol.

alcohol dehydrogenase (dee-high-DROJ-eh-nayz): an enzyme active in the stomach and the liver that converts ethanol to acetaldehyde.

alcoholism: a pattern of drinking that includes a strong craving for alcohol, a loss of control and an inability to stop drinking once begun, withdrawal symptoms (nausea, sweating, shakiness, and anxiety) after heavy drinking, and the need for increasing amounts of alcohol to feel "high."

antidiuretic hormone (ADH): a hormone produced by the pituitary gland in response to dehydration (or a high sodium concentration in the blood) that stimulates the kidneys to reabsorb more water and therefore to excrete less. In addition to its antidiuretic effect, ADH elevates blood pressure and so is also called *vasopressin* (VAS-oh-PRES-in).

beer: an alcoholic beverage traditionally brewed by fermenting malted barley and adding hops for flavor.

binge drinking: pattern of drinking that raises blood alcohol concentration to 0.08 percent or higher; usually corresponds to four or more drinks for women and five or more drinks for men on a single occasion, generally within a couple of hours.

cirrhosis (seer-OH-sis): advanced liver disease in which liver cells turn orange, die, and harden, permanently losing their function; often associated with alcoholism.

- **cirrhos** = an orange

drink: a dose of any alcoholic beverage that delivers ½ ounce of pure ethanol:

- 5 ounces of wine
- 10 ounces of wine cooler
- 12 ounces of beer
- 1½ ounces of liquor (80 proof whiskey, scotch, rum, or vodka)

drug: a substance that can modify one or more of the body's functions.

ethanol: a particular type of alcohol found in beer, wine, and liquor; also called *ethyl alcohol* (see Figure H7-1).

excessive drinking: heavy drinking, binge drinking, or both.

fatty liver: an early stage of liver deterioration seen in several diseases, including obesity and alcoholic liver disease. Fatty liver is characterized by an accumulation of fat in the liver cells.

fibrosis (fye-BROH-sis): an intermediate stage of liver deterioration seen in several diseases, including viral hepatitis and alcoholic liver disease. In fibrosis, the liver cells lose their function and assume the characteristics of connective tissue cells (fibers).

heavy drinking: more than three drinks on any day for women and more than four drinks on any day for men.

liquor or **distilled spirits:** an alcoholic beverage traditionally made by fermenting and distilling a carbohydrate source such as molasses, potatoes, rye, beets, barley, or corn.

MEOS or **microsomal** (my-krow-SO-mal) **ethanol-oxidizing system:** a system of enzymes in the liver that oxidize not only alcohol but also several classes of drugs.

moderation (alcohol): up to one drink per day for women and up to two drinks per day for men.

narcotic (nar-KOT-ic): a drug that dulls the senses, induces sleep, and becomes addictive with prolonged use.

proof: a way of stating the percentage of alcohol in distilled liquor. Liquor that is 100 proof is 50 percent alcohol; 90 proof is 45 percent, and so forth.

Wernicke-Korsakoff (VER-nee-key KORE-sah-kof) **syndrome:** a neurological disorder typically associated with chronic alcoholism and caused by a deficiency of the B vitamin thiamin; also called *alcohol-related dementia*.

wine: an alcoholic beverage traditionally made by fermenting a sugar source such as grape juice.

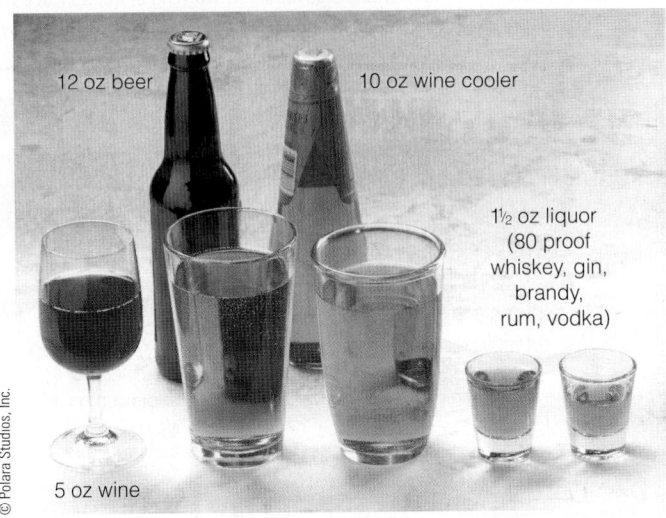

> **FIGURE H7-1** **Two Alcohols: Glycerol and Ethanol**

Glycerol is the alcohol used to make triglycerides.

Ethanol is the alcohol in beer, wine, and liquor.

© Cengage Learning

> **FIGURE H7-2**

Each of these servings equals one drink. Moderation is up to one drink per day for women and two drinks per day for men.

12 oz beer

10 oz wine cooler

1½ oz liquor (80 proof whiskey, gin, brandy, rum, vodka)

5 oz wine

© Polara Studios, Inc.

in triglycerides is an example of an alcohol to a chemist. To most people, though, *alcohol* refers to the intoxicating ingredient in **beer, wine,** and **liquor (distilled spirits).** The chemist's name for this particular alcohol is *ethyl alcohol,* or **ethanol.** Glycerol has three carbons with three hydroxyl groups attached; ethanol has only two carbons and one hydroxyl group (see Figure H7-1). The remainder of this highlight talks about the particular alcohol ethanol but refers to it simply as *alcohol.*

Alcohols affect living things profoundly, partly because they act as lipid solvents. Their ability to dissolve lipids out of cell membranes allows alcohols to penetrate rapidly into cells, destroying cell structures and thereby killing the cells. For this reason, most alcohols are toxic in relatively small amounts; by the same token, because they kill microbial cells, they are useful as skin disinfectants.

Ethanol is less toxic than the other alcohols. Sufficiently diluted and taken in small enough doses, its action in the brain produces an effect that people seek—not with zero risk, but with a low enough risk (if the doses are low enough) to be tolerable. Used in this way, alcohol is a **drug**—that is, a substance that modifies body functions. Like all drugs, alcohol both offers benefits and poses hazards. The *Dietary Guidelines for Americans* advise "if alcohol is consumed, it should be consumed in moderation."

> **DIETARY GUIDELINES FOR AMERICANS**

- If alcohol is consumed, it should be consumed in moderation—up to one drink per day for women and two drinks per day for men—and only by adults of legal drinking age.
- Alcoholic beverages should not be consumed by some individuals, including those who cannot restrict their alcohol intake to moderate levels, women of childbearing age who may become pregnant, pregnant and lactating women, children and adolescents, individuals taking medications that can interact with alcohol, and those with specific medical conditions.
- Alcoholic beverages should not be consumed by individuals engaging in activities that require attention, skill, or coordination, such as driving or swimming.

The term **moderation** is important when describing alcohol use. How many drinks constitute moderate use, and how much is "a drink"?

First, a **drink** is any alcoholic beverage that delivers ½ ounce of *pure ethanol* (see Figure H7-2):

- 5 ounces of wine
- 10 ounces of wine cooler
- 12 ounces of beer
- 1½ ounces of liquor (80 proof whiskey, scotch, rum, or vodka)

As a practical tip, prevent overpouring by measuring liquids and using tall, narrow glasses.

Beer, wine, and liquor deliver different amounts of alcohol. The amount of alcohol in liquor is stated as **proof:** 100 proof liquor is 50 percent alcohol, 80 proof is 40 percent alcohol, and so forth. Wine and beer have less alcohol than liquor, although some fortified wines and beers have more alcohol than the regular varieties. Figure H7-3 (p. 224) presents examples of labels for beer, wine, and liquor.

Second, because people have different tolerances for alcohol, it is impossible to name an exact daily amount of alcohol that is appropriate for everyone. Authorities have attempted to identify amounts that are acceptable for most healthy people. An accepted definition of *moderation* is up to two drinks per day for men and up to one drink per day for women. (Pregnant women are advised to abstain from alcohol, as Highlight 15 explains.) Notice that this advice is stated as a maximum, not as an average; seven drinks one night a week would not be considered moderate, even though one a day would be. Doubtless, some people could consume slightly more; others could not handle nearly so much without risk. The amount a person can drink safely is highly individual, depending on genetics, health, gender, body composition, age, and family history.[6] A recent US survey reports that most men and women do not exceed the limits suggested by the *Dietary Guidelines for Americans;* an estimated 8 percent of men and 3 percent of women drink excessively (defined as more than four drinks for men and three drinks for women).[7]

HIGHLIGHT > 7

> **FIGURE H7-3** **Example of Alcohol Beverage Labels**

All alcohol beverage labels provide the brand name, which identifies and markets the product; the name and address, which identifies the bottler or importer; and a health warning, which provides a government warning of health issues associated with alcohol beverages containing >0.5 percent alcohol by volume. Additional information is provided depending on whether the beverage is beer, wine, or liquor.

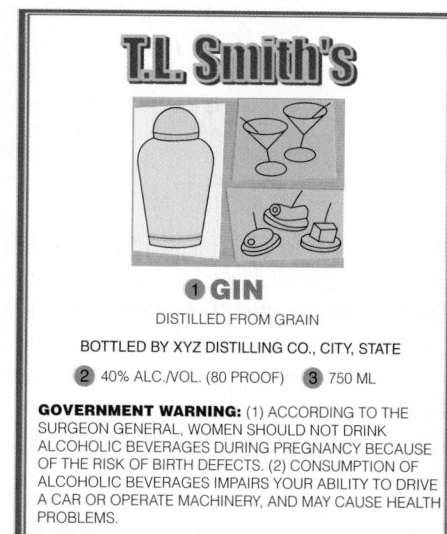

① DRAFT/DRAUGHT
Means the product has not been pasteurized and another method of controlling bacteria has been used; also used to describe beer packaged in large containers and drawn off through a tap

② CLASS DESIGNATION
Identifies the product based on the ingredients and processes used (for example, ales, stouts, and porters are fermented at relatively high temperatures and wheat beer is made from a fermentable base that is ≥ 25 percent malted wheat)

③ ALCOHOL CONTENT
Beers typically contain <5 percent alcohol by volume and malt liquors 5 to 8 percent; regulations vary, with some states requiring alcohol content on beer labels and others prohibiting such statements

④ NET CONTENTS
States the bottle quantity in English units (for example, pints or fluid ounces)

① VINTAGE DATE
Indicates year of grape harvest

② APPELLATION OF ORIGIN
Locates where most of the grapes were grown

③ VARIETAL DESIGNATION
Names the dominant grapes in the wine (for example, chardonnay, merlot, or zinfandel)

④ ALCOHOL CONTENT
Wines contain 7 to 24 percent alcohol by volume; wines with <14 percent may simply state "table wine" or "light wine"

⑤ NET CONTENTS
States the bottle quantity in metric units

⑥ DECLARATION OF SULFITES
Contains ≥10 ppm of sulfur dioxide

① CLASS/TYPE DESIGNATION
Identifies the product based on the ingredients and processes used (for example, gin's flavor derives from juniper berries, rum derives from fermented juice of sugar cane products, and tequila derives from the agave plant)

② ALCOHOL CONTENT
Most common liquors contain about 40 percent alcohol (80 proof), but this varies; alcohol content is stated as a percent by volume and by proof

③ NET CONTENTS
States the bottle quantity in metric units

© Cengage Learning

Alcohol's Influence

From the moment an alcoholic beverage enters the body, alcohol is treated as if it has special privileges. Its influence is most apparent in the GI tract, the liver, and the brain.

In the GI Tract

Unlike foods, which require time for digestion, alcohol needs no digestion and is quickly absorbed across the walls of an empty stomach, reaching the brain within a few minutes. Consequently, a person can immediately feel euphoric when drinking, especially on an empty stomach.

When the stomach is full of food, alcohol has less chance of touching the walls and diffusing through, so its influence on the brain is slightly delayed. This information leads to another practical tip: eat snacks when drinking alcoholic beverages. Carbohydrate snacks slow alcohol absorption and high-fat snacks slow peristalsis, keeping the alcohol in the stomach longer. Salty snacks make a person thirsty; to quench thirst, drink water instead of more alcohol.

The stomach begins to break down alcohol with its **alcohol dehydrogenase** enzyme. Women produce less of this stomach enzyme than men; consequently, more alcohol reaches the intestine for absorption into the bloodstream. As a result, women absorb more alcohol than men of the same size who drink the same amount of alcohol. Consequently, they are more likely to become more intoxicated on less alcohol than men. Such differences between men and women help explain why women have a lower alcohol tolerance and a lower guideline for moderate intake.

In the small intestine, alcohol is rapidly absorbed. From this point on, alcohol receives priority treatment: it gets absorbed and metabolized before most nutrients. Alcohol's priority status helps to ensure a speedy disposal and reflects two facts: alcohol cannot be stored in the body, and it is potentially toxic.

In the Liver

As Chapter 3 explains, the capillaries of the digestive tract merge into veins that carry blood first to the liver. These veins branch and rebranch into a capillary network that touches every liver cell. Consequently, liver cells are the first to receive alcohol-laden blood. Liver cells are also the only other cells in the body that can make enough of the alcohol dehydrogenase enzyme to oxidize alcohol at an appreciable rate. The routing of blood through the liver cells gives them the chance to dispose of some alcohol before it moves on.

Alcohol affects every organ of the body, but the most dramatic evidence of its disruptive behavior appears in the liver. If liver cells could talk, they would describe alcohol as demanding, egocentric, and disruptive of the liver's efficient way of running its business. For example, liver cells normally prefer fatty acids as their fuel, and they like to package excess fatty acids into triglycerides and ship them out to other tissues. When alcohol is present, however, the liver cells metabolize alcohol first and let the fatty acids accumulate, sometimes in huge stockpiles. Alcohol metabolism can also permanently change liver cell structure, impairing the liver's ability to metabolize fats. As a result, heavy drinkers develop fatty livers.

The liver is the primary site of alcohol metabolism. It can process about ½ ounce of *ethanol* per hour (the amount defined as a drink), depending on the person's body size, previous drinking experience, food intake, and general health. This maximum rate of alcohol breakdown is determined by the amount of alcohol dehydrogenase available. If more alcohol arrives at the liver than the enzymes can handle, the extra alcohol travels around the body, circulating again and again until liver enzymes are finally available to process it. Another practical tip derives from this information: drink slowly enough to allow the liver to keep up—no more than one drink per hour.

The amount of alcohol dehydrogenase enzyme present in the liver varies with individuals, depending on the genes they have inherited and on how recently they have eaten. Fasting for as little as a day prompts the body to degrade its proteins, including the alcohol-processing enzymes, and this can slow the rate of alcohol metabolism by half. Drinking after not eating all day thus causes the drinker to feel the effects more promptly for two reasons: rapid absorption and slowed breakdown.

Figure H7-4 provides a simplified diagram of alcohol metabolism; Appendix C provides the chemical details. The alcohol dehydrogenase enzyme breaks down alcohol by removing hydrogens in two steps. In the first step, alcohol dehydrogenase oxidizes alcohol to **acetaldehyde**—a highly reactive and toxic compound. High concentrations of acetaldehyde in the brain and other tissues are responsible for many of the damaging effects of **alcohol abuse.**

In the second step, a related enzyme, acetaldehyde dehydrogenase, converts acetaldehyde to acetate, which is then converted to either carbon dioxide (CO_2) or acetyl CoA—the compound that plays such a central role in energy metabolism, as described in Chapter 7. The reactions from alcohol to acetaldehyde to acetate produce hydrogens (H^+) and electrons. The B vitamin niacin, in its role as a coenzyme, helpfully picks up these hydrogens and electrons and escorts them to the electron transport chain. (Chapter 10 presents information on the coenzyme roles of the B vitamins.)

During alcohol metabolism, the multitude of other metabolic processes for which the niacin coenzyme is required, including glycolysis, the TCA cycle, and the electron transport chain, falter. Its presence is sorely missed in these energy pathways because it is the chief carrier of the hydrogens that travel with their electrons along the electron transport chain. Without adequate coenzymes, these energy pathways cannot function. Traffic either backs up or an alternate route is taken.

Such changes in the normal flow of energy pathways have striking metabolic consequences. For one, the accumulation of

> FIGURE H7-4 **Alcohol Metabolism**

The conversion of alcohol to acetyl CoA requires the B vitamin niacin in its role as a coenzyme. When the enzymes oxidize alcohol, they remove H atoms and attach them to the niacin coenzyme.

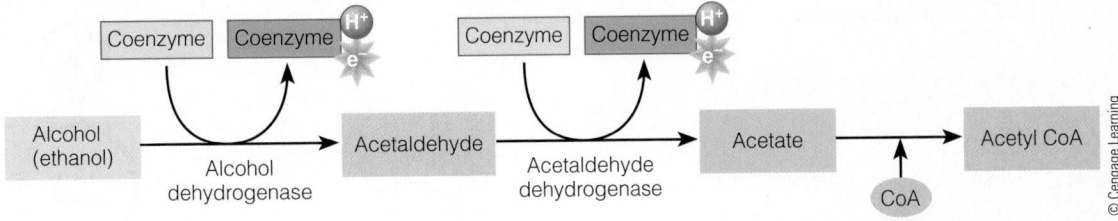

hydrogen ions during alcohol metabolism shifts the body's acid-base balance toward acid. For another, alcohol's interference with energy metabolism promotes the making of lactate from pyruvate. The conversion of pyruvate to lactate uses some of the excess hydrogens, but a lactate build-up has serious consequences of its own—it adds still further to the body's acid burden and interferes with the excretion of another acid, uric acid, causing inflammation of the joints.

Alcohol alters both amino acid and protein metabolism. Synthesis of proteins important in the immune system slows down, weakening the body's defenses against infections. Evidence of protein deficiency becomes apparent, both from a diminished synthesis of proteins and from a poor diet. Normally, the cells would at least use the amino acids from the protein foods a person eats, but the drinker's liver deaminates the amino acids and uses the carbon fragments primarily to make fat or ketone bodies. Eating well does not protect the drinker from protein depletion; a person has to stop drinking alcohol.

The accumulation of coenzymes with their hydrogens and electrons slows the TCA cycle, so pyruvate and acetyl CoA build up. Excess acetyl CoA then takes the pathway to fatty acid synthesis (as Figure H7-5 illustrates), and fat clogs the liver. As you might expect, a liver overburdened with fat cannot function properly. Liver cells become less efficient at performing a number of tasks. Much of this inefficiency impairs a person's nutritional health in ways that cannot be corrected by diet alone. For example, the liver has difficulty activating vitamin D, as well as producing and releasing bile. The fatty liver has difficulty making glucose from protein. Without gluconeogenesis, blood glucose can plummet, leading to irreversible damage to the central nervous system. The lack of glucose together with the overabundance of acetyl CoA sets the stage for ketosis. The body uses excess acetyl CoA to make ketone bodies; their acidity pushes the acid-base balance further toward acid and suppresses nervous

system activity. To overcome such problems, a person needs to stop drinking alcohol.

The synthesis of fatty acids accelerates with exposure to alcohol. Fat accumulation can be seen in the liver after a single night of heavy drinking. **Fatty liver,** the first stage of liver deterioration seen in heavy drinkers, interferes with the distribution of nutrients and oxygen to the liver cells. Fatty liver is reversible with abstinence from alcohol. If fatty liver lasts long enough, however, the liver cells will die and form fibrous scar tissue. This second stage of liver deterioration is called **fibrosis.** Some liver cells can regenerate with good nutrition and abstinence from alcohol, but in the most advanced stage, **cirrhosis,** damage is the least reversible.

The liver's priority treatment of alcohol affects its handling of drugs as well as nutrients. In addition to the dehydrogenase enzymes already described, the liver possesses an enzyme system that metabolizes *both* alcohol and several other types of drugs. Called the **MEOS (microsomal ethanol-oxidizing system),** this system handles about one-fifth of the total alcohol a person consumes. At high blood concentrations or with repeated exposures, alcohol stimulates the synthesis of enzymes in the MEOS. The result is a more efficient metabolism of alcohol and tolerance to its effects.

As a person's blood alcohol rises, alcohol competes with—and wins out over—other drugs whose metabolism also relies on the MEOS. If a person drinks and uses another drug at the same time, the MEOS will dispose of alcohol first and metabolize the drug more slowly. While the drug waits to be handled later, the dose may build up so that its effects are greatly amplified—sometimes to the point of being fatal. Many drug labels provide warnings to avoid alcohol while taking the drug.

In contrast, once a heavy drinker stops drinking and alcohol is no longer competing with other drugs, the enhanced MEOS metabolizes

> FIGURE H7-5   **Alternate Route for Acetyl CoA: To Fat**

Acetyl CoA molecules are blocked from getting into the TCA cycle by the low level of coenzymes. Instead of being used for energy, the acetyl CoA molecules become building blocks for fatty acids.

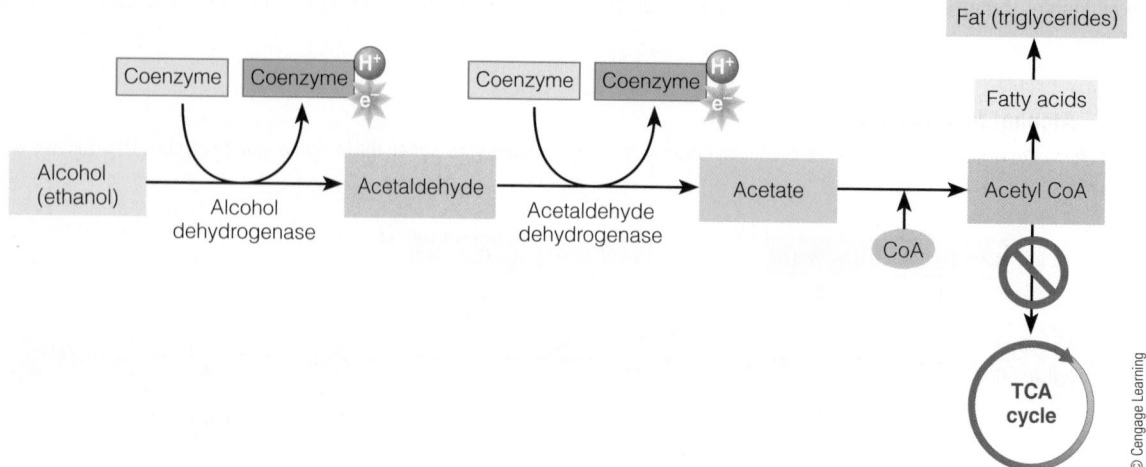

drugs much faster than before. As a result, determining the correct dosages of medications can be challenging.

This discussion has emphasized the major way that the blood is cleared of alcohol—metabolism by the liver—but there is another way. About 10 percent of the alcohol leaves the body through the breath and in the urine. This is the basis for the breath and urine tests for drunkenness. The amounts of alcohol in the breath and in the urine are in proportion to the amount still in the bloodstream and brain. In all states, legal drunkenness is set at 0.08 percent or less, reflecting the relationship between alcohol use and traffic and other accidents.

In the Brain

Figure H7-6 describes alcohol's effects on the brain. Alcohol is a **narcotic.** People used it for centuries as an anesthetic because it can deaden pain. But alcohol was a poor anesthetic because one could never be sure how much a person would need and how much would be a fatal dose. Today's anesthetics provide a more predictable response. Alcohol continues to be used socially to help people relax or to relieve anxiety. People think that alcohol is a stimulant because it seems to relieve inhibitions. Actually, though, it accomplishes this by sedating *inhibitory* nerves, which are more numerous than excitatory nerves. Ultimately, alcohol acts as a depressant and affects all the nerve cells.

It is lucky that the brain centers respond to a rising blood alcohol concentration in the order described in Figure H7-6 because a person usually passes out before managing to drink a lethal dose. It is possible, though, to drink so fast that the effects of alcohol continue to accelerate after the person has passed out. Occasionally, a person drinks so much as to stop breathing and die. Table H7-1 (p. 228) shows the blood alcohol levels that correspond to progressively greater intoxication, and Table H7-2 (p. 228) shows the brain responses that occur at these blood levels.

Like liver cells, brain cells die with excessive exposure to alcohol. Liver cells may be replaced, but not all brain cells can regenerate. Thus some heavy drinkers suffer permanent brain damage. Whether alcohol impairs cognition in moderate drinkers is unclear.

Alcohol's Damage

As alcohol busily disrupts cellular activities, the physical consequences become apparent. People become dehydrated and malnourished; their alcohol use brings both short- and long-term effects.

Dehydration

People who drink alcoholic beverages may notice that they urinate more, but they may be unaware of the vicious cycle that results.

> FIGURE H7-6 **Alcohol's Effects on the Brain**

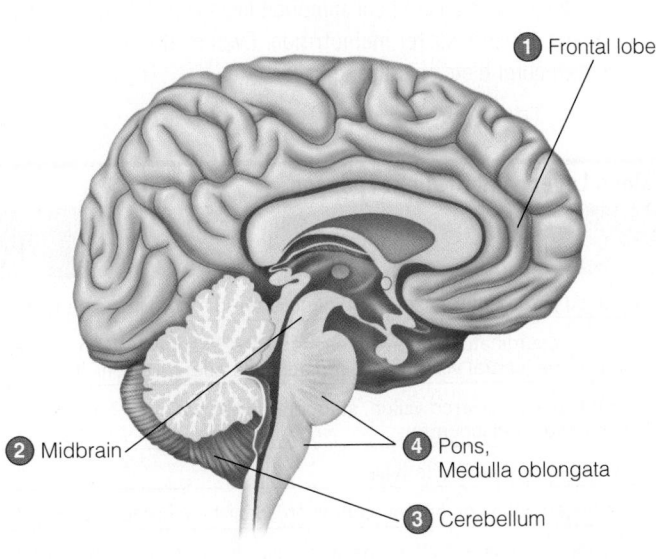

1 Frontal lobe

2 Midbrain

4 Pons, Medulla oblongata

3 Cerebellum

1 Judgment and reasoning centers are most sensitive to alcohol. When alcohol flows to the brain, it first sedates the frontal lobe, the center of all conscious activity. As alcohol diffuses into the cells of these lobes, it interferes with reasoning and judgment.

2 Speech and vision centers in the midbrain are affected next. If the drinker drinks faster than the rate at which the liver can oxidize the alcohol, blood alcohol concentrations rise: the speech becomes challenging and vision becomes blurry.

3 Voluntary muscular control is then affected. At still higher concentrations, the cells in the cerebellum responsible for coordination of voluntary muscles are affected, including those used in speech, eye-hand coordination, and limb movements. At this point people under the influence stagger or weave when they try to walk, or they may slur their speech.

4 Respiration and heart action are the last to be affected. Finally, the conscious brain is completely subdued, and the person passes out. Now the person can drink no more; this is fortunate because higher doses would anesthetize the deepest brain centers that control breathing and heartbeat, causing death.

© Cengage Learning

TABLE H7-1 Alcohol Doses and Approximate Blood Level Percentages for Men and Women

Drinks[a]	Body Weight in Pounds—Men								
	100	120	140	160	180	200	220	240	
	00	00	00	00	00	00	00	00	ONLY SAFE DRIVING LIMIT
1	.04	.03	.03	.02	.02	.02	.02	.02	IMPAIRMENT BEGINS
2	.08	.06	.05	.05	.04	.04	.03	.03	
3	.11	.09	.08	.07	.06	.06	.05	.05	DRIVING SKILLS SIGNIFICANTLY AFFECTED
4	.15	.12	.11	.09	.08	.08	.07	.06	
5	.19	.16	.13	.12	.11	.09	.09	.08	
6	.23	.19	.16	.14	.13	.11	.10	.09	
7	.26	.22	.19	.16	.15	.13	.12	.11	LEGALLY INTOXICATED
8	.30	.25	.21	.19	.17	.15	.14	.13	
9	.34	.28	.24	.21	.19	.17	.15	.14	
10	.38	.31	.27	.23	.21	.19	.17	.16	

Drinks[a]	Body Weight in Pounds—Women									
	90	100	120	140	160	180	200	220	240	
	00	00	00	00	00	00	00	00	00	ONLY SAFE DRIVING LIMIT
1	.05	.05	.04	.03	.03	.03	.02	.02	.02	IMPAIRMENT BEGINS
2	.10	.09	.08	.07	.06	.05	.05	.04	.04	
3	.15	.14	.11	.10	.09	.08	.07	.06	.06	DRIVING SKILLS SIGNIFICANTLY AFFECTED
4	.20	.18	.15	.13	.11	.10	.09	.08	.08	
5	.25	.23	.19	.16	.14	.13	.11	.10	.09	
6	.30	.27	.23	.19	.17	.15	.14	.12	.11	
7	.35	.32	.27	.23	.20	.18	.16	.14	.13	LEGALLY INTOXICATED
8	.40	.36	.30	.26	.23	.20	.18	.17	.15	
9	.45	.41	.34	.29	.26	.23	.20	.19	.17	
10	.51	.45	.38	.32	.28	.25	.23	.21	.19	

© Cengage Learning

NOTE: Driving under the influence is proved when an adult's blood contains 0.08 percent alcohol. Many states have adopted a "zero-tolerance" policy for drivers under age 21, using 0.00 to 0.02 percent as the limit.

[a]Taken within an hour or so; each drink equivalent to ½ ounce pure ethanol.

SOURCE: National Clearinghouse for Alcohol and Drug Information.

Alcohol depresses production of **antidiuretic hormone (ADH),** a hormone produced by the pituitary gland that retains water—consequently, with less ADH, more water is lost. Loss of body water leads to thirst, and thirst leads to more drinking. Water will relieve dehydration, but the thirsty drinker may drink alcohol instead, which only worsens the problem. Such information provides another practical tip: drink water when thirsty and before each alcoholic drink. Drink an extra glass or two before going to bed. This strategy will help lessen the effects of a hangover.

Water loss is accompanied by the loss of important minerals. As Chapters 12 and 13 explain, these minerals are vital to the body's fluid balance and to many chemical reactions in the cells, including muscle action. Detoxification treatment includes restoration of mineral balance as quickly as possible.

Malnutrition

For some light-to-moderate drinkers, alcohol may suppress food intake and prevent weight gains.[8] For others, however, alcohol may actually stimulate appetite. Moderate drinkers usually consume alcohol as *added* energy—on top of their normal food intake. In addition, alcohol in moderate doses is efficiently metabolized. Consequently, alcohol can contribute to body fat and weight gain—either by inhibiting oxidation or by being converted to fat. Metabolically, alcohol is almost as efficient as fat in promoting obesity; each ounce of alcohol represents about a half-ounce of fat. Alcohol's contribution to body fat is most evident in the abdominal obesity that commonly accompanies alcohol consumption, popularly known as the "beer belly."[9] Although not required, some beer, wine, and liquor labels may now provide information on the serving size, servings per container, kcalories, carbohydrates, protein, and fat per serving.

Alcohol in heavy doses, though, is not efficiently metabolized, generating more heat than fat. Heavy drinkers usually consume alcohol as *substituted* energy—instead of their normal food intake. Diet quality declines as alcohol consumption increases.[10] Consequently, many heavy drinkers suffer malnutrition. Even moderate drinkers tend to have poorer diets on drinking days.[11]

TABLE H7-2 Alcohol Blood Levels and Brain Responses

Blood Alcohol Concentration	Effect on Brain
0.05	Impaired judgment, relaxed inhibitions, altered mood, increased heart rate
0.10	Impaired coordination, delayed reaction time, exaggerated emotions, impaired peripheral vision, impaired ability to operate a vehicle
0.15	Slurred speech, blurred vision, staggered walk, seriously impaired coordination and judgment
0.20	Double vision, inability to walk
0.30	Uninhibited behavior, stupor, confusion, inability to comprehend
0.40 to 0.60	Unconsciousness, shock, coma, death (cardiac or respiratory failure)

© Cengage Learning

NOTE: Blood alcohol concentration depends on a number of factors, including alcohol in the beverage, the rate of consumption, the person's gender, and body weight. For example, a 100-pound female can become legally drunk (≥0.08 concentration) by drinking two beers in an hour, whereas a 220-pound male consuming that amount at the same rate would have a 0.03 blood alcohol concentration.

On average, adults in the United States consume almost 100 kcalories from alcohol daily.[12] Alcohol is rich in energy (7 kcalories per gram), but as with pure sugar or fat, the kcalories are empty of nutrients. The more alcohol people drink, the less likely that they will eat enough food to obtain adequate nutrients. The more kcalories used for alcohol, the fewer kcalories available to use from nutritious foods. Table H7-3 shows the kcalorie amounts of typical alcoholic beverages.

Chronic alcohol abuse not only displaces nutrients from the diet, but it also interferes with the body's metabolism of nutrients. Most dramatic is alcohol's effect on the B vitamin folate. The liver loses its ability to retain folate, and the kidneys increase their excretion of it. Alcohol abuse creates a folate deficiency that devastates digestive system function. The small intestine normally releases and retrieves folate continuously, but it becomes damaged by folate deficiency and alcohol toxicity, so it fails to retrieve its own folate and misses any that may trickle in from food as well. Alcohol also interferes with the action of folate in converting the amino acid homocysteine to methionine. The result is an excess of homocysteine, which has been linked to heart disease, and an inadequate supply of methionine, which slows the production of new cells, especially the rapidly dividing cells of the intestine and the blood. The combination of poor folate status and alcohol consumption has also been implicated in promoting colorectal cancer.

The inadequate food intake and impaired nutrient absorption that accompany chronic alcohol abuse frequently lead to a deficiency of another B vitamin—thiamin. In fact, the cluster of thiamin-deficiency symptoms commonly seen in chronic **alcoholism** has its own name— **Wernicke-Korsakoff syndrome.** This syndrome is characterized by paralysis of the eye muscles, poor muscle coordination, impaired memory, and damaged nerves; it and other alcohol-related memory problems may respond to thiamin supplements.

Acetaldehyde, an intermediate in alcohol metabolism (review Figure H7-4), interferes with nutrient use, too. For example, acetaldehyde dislodges vitamin B_6 from its protective binding protein so that it is destroyed, causing a vitamin B_6 deficiency and, thereby, lowered production of red blood cells.

Malnutrition occurs not only because of lack of intake and altered metabolism but because of direct toxic effects as well. Alcohol causes stomach cells to oversecrete both gastric acid and histamine, an immune system agent that produces inflammation. Beer in particular stimulates gastric acid secretion, irritating the linings of the stomach and esophagus and making them vulnerable to ulcer formation.

Overall, nutrient deficiencies are virtually inevitable in alcohol abuse, not only because alcohol displaces food but also because alcohol directly interferes with the body's use of nutrients, making them ineffective even if they are present. Intestinal cells fail to absorb B vitamins, notably, thiamin, folate, and vitamin B_{12}. Liver cells lose efficiency in activating vitamin D. Cells in the retina of the eye, which normally process the alcohol form of vitamin A (retinol) to the aldehyde form needed in vision (retinal), find themselves processing ethanol to acetaldehyde instead. Likewise, the liver

cannot convert the aldehyde form of vitamin A to its acid form (retinoic acid), which is needed to support the growth of its (and all) cells. Regardless of dietary intake, excessive drinking over a lifetime creates deficits of all the nutrients mentioned in this discussion and more. No diet can compensate for the damage caused by heavy alcohol consumption.

TABLE H7-3 kCalories in Alcoholic Beverages and Mixers

Beverage	Amount (oz)	Energy (kcal)	Alcohol (g)
Beer			
Regular	12	153	14
Light	12	103	11
Nonalcoholic	12	32	0
Cocktails			
Daiquiri, canned	6.8	259	20
Daiquiri, from recipe	4.5	223	28
Piña colada, canned	6.8	526	20
Piña colada, from recipe	4.5	245	14
Tequila sunrise, canned	6.8	232	20
Whiskey sour, canned	6.8	249	20
Liquor (gin, rum, vodka, whiskey)			
80 proof	1.5	97	14
86 proof	1.5	105	15
90 proof	1.5	110	16
94 proof	1.5	116	17
100 proof	1.5	124	18
Sake	1.5	58	7
Liqueurs			
Coffee and cream liqueur, 34 proof	1.5	154	7
Coffee liqueur, 53 proof	1.5	170	11
Coffee liqueur, 63 proof	1.5	160	14
Crème de menthe, 72 proof	1.5	186	15
Mixers			
Club soda	12	0	0
Cola	12	136	0
Cranberry juice cocktail	4	72	0
Ginger ale or tonic water	12	124	0
Grapefruit juice	4	48	0
Orange juice	4	56	0
Tomato or vegetable juice	4	21	0
Wine			
Champagne	5	105	13
Cooking	5	72	5
Dessert, dry	5	224	23
Dessert, sweet	5	236	23
Red or rosé	5	125	16
White	5	121	15
Wine cooler	10	150	11

Short-Term Effects

The effects of abusing alcohol may be apparent immediately, or they may not become evident for years to come. Among the immediate consequences, all of the following involve alcohol use:

- 20 percent of all boating fatalities
- 33 percent of all suicides
- 31 percent of all traffic fatalities
- 40 percent of all residential fire fatalities
- 47 percent of all homicides and purposeful injuries
- 55 percent of all domestic violence incidents

These statistics are sobering. The consequences of heavy drinking touch all races and all segments of society—men and women, young and old, rich and poor. One group particularly hard hit by heavy drinking is college students—not because they are prone to alcoholism, but because they live in an environment and are in a developmental stage of life in which risk-taking behaviors are common and heavy drinking is acceptable.

Excessive drinking—including both **heavy drinking** and **binge drinking**—is widespread on college campuses and poses serious health and social consequences to drinkers and nondrinkers alike. In fact, binge drinking can kill: the respiratory center of the brain becomes anesthetized, and breathing stops. It can also cause coronary artery spasms, leading to a heart attack and death.

Binge drinking is especially common among college students, especially males. Compared with nondrinkers or moderate drinkers, people who frequently binge drink (at least three times within two weeks) are more likely to engage in unprotected sex, have multiple sex partners, damage property, and assault others. On average, *every day* alcohol is involved in the:

- Death of 5 college students
- Sexual assault of 266 college students
- Injury of 1641 college students
- Assault of 1890 college students

Binge drinkers skew the statistics on college students' alcohol use. The median number of drinks consumed by college students is 1.5 per week, but for binge drinkers, it is 14.5. Nationally, only 20 percent of all students are frequent binge drinkers; yet they account for two-thirds of all the alcohol students report consuming and most of the alcohol-related problems.

The dangers of binge drinking have been amplified by the use of beverages that contain caffeine as an additive. The caffeine seems to mask the sensory cues that an individual normally relies on to determine intoxication. Consequently, individuals drinking these beverages typically consume more alcohol and become more intoxicated than they realize. The Food and Drug Administration (FDA) has warned manufacturers of packaged caffeinated alcoholic beverages to stop sales. The combination of alcohol and added caffeine has not been approved because these products are associated with risky behaviors that may lead to hazardous and life-threatening situations.[13]

For the same reasons, individuals should not mix alcohol with high-energy drinks.

Another phenomenon closely related to binge drinking is "drunkorexia," a popular term used to describe the restrictive eating, excessive exercising, and abusive drinking patterns of some college students.[14] In an effort to compensate for the kcalories of an evening's drinking binge and to amplify the high, these students skip meals and exercise vigorously during the day; they might also drink excessively in an effort to vomit any food previously eaten. These students may perceive themselves to be "weight conscious" drinkers, but in reality, their restrictive eating, excessive exercising, and abusive drinking patterns are not supportive of healthy weight loss.

Binge drinking is not limited to college campuses, of course, but it is most common among 18 to 34 year olds.[15] That age group and campus environment seem most accepting of such behavior despite its problems. Social acceptance may make it difficult for binge drinkers to recognize themselves as problem drinkers. For this reason, interventions must focus both on educating individuals and on changing the campus social environment. The damage alcohol causes becomes worse if the pattern is not broken. Alcohol abuse sets in much more quickly in young people than in adults. Those who start drinking at an early age more often suffer from alcoholism than others. Table H7-4 lists the key signs of alcoholism.

Long-Term Effects

The most devastating long-term effect of alcohol is the damage done to a child whose mother drinks alcohol during pregnancy. The effects of alcohol on the unborn and the message that pregnant women

---

TABLE H7-4 Signs of Alcoholism

- Tolerance: the person needs higher and higher intakes of alcohol to achieve intoxication.
- Withdrawal: the person who stops drinking experiences anxiety, agitation, increased blood pressure, or seizures, or seeks alcohol to relieve these symptoms.
- Impaired control: the person intends to have 1 or 2 drinks, but has many more instead, or the person tries to control or quit drinking, but efforts are unsuccessful.
- Disinterest: the person neglects important social, family, job, or school activities because of drinking.
- Time: the person spends a great deal of time obtaining and drinking alcohol or recovering from excessive drinking.
- Cravings: the person has strong urges to use alcohol.
- Impaired ability: the person's intoxication or withdrawal symptoms interfere with work, school, or home.
- Problems: the person continues drinking despite physical hazards or medical, legal, psychological, family, employment, or school problems caused or exacerbated by alcohol.

These conditions suggest that a person may have an alcohol problem and might benefit from an abstinence program or professional help.

SOURCE: Adapted from *Diagnostic and Statistical Manual of Mental Disorders*, 4th ed. (Washington, D.C.: American psychiatric Association, 1994).

© Cengage Learning

should not drink alcohol are presented in Highlight 15. Quite simply, there is no safe amount of drinking during pregnancy.[16]

For nonpregnant adults, a drink or two sets in motion many destructive processes in the body, but the next day's abstinence reverses them. As long as the doses are moderate, the time between them is ample, and nutrition is adequate, recovery is probably complete.

If the doses of alcohol are heavy and the time between them short, complete recovery cannot take place. Repeated onslaughts of alcohol gradually take a toll on all parts of the body and increase the risks of several chronic diseases (see Table H7-5). Compared with nondrinkers and moderate drinkers, heavy drinkers have significantly greater risks of dying from all causes. Excessive alcohol consumption is the third leading preventable cause of death in the United States. Worldwide, alcohol contributes to 1 out of 25 deaths.[17]

Personal Strategies

One obvious option available to people attending social gatherings is to enjoy the conversation, eat the food, and drink nonalcoholic beverages. Several nonalcoholic beverages are available that mimic the look and taste of their alcoholic counterparts. For those who enjoy champagne or beer, sparkling ciders and beers without alcohol are available. Instead of drinking a cocktail, a person can sip tomato juice with a slice of lime and a stalk of celery or just a plain cola beverage. The person who chooses to drink alcohol should sip each drink slowly accompanied by food and water.

If you want to help sober up a friend who has had too much to drink, don't bother walking arm in arm around the block. Walking muscles have to work harder, but muscle cells don't have the enzymes to metabolize alcohol; only liver cells can clear alcohol from the blood. Remember that each person has a limited amount of the alcohol dehydrogenase enzyme, which clears the blood at a steady rate. Time alone will do the job. Nor will it help to give your friend a cup of coffee. Caffeine is a stimulant, but it won't speed up alcohol metabolism. Table H7-6 (p. 232) presents other alcohol myths.

People who have passed out from drinking need 24 hours to sober up completely. Let them sleep, but watch over them. Encourage them to lie on their sides, instead of their backs. That way, if they vomit, they won't choke.

Don't drive after drinking. Every day, an estimated 30 people in the United States die in traffic accidents that involve an alcohol-impaired driver; said another way, an alcohol-related auto accident kills someone every 48 minutes. The lack of glucose for the brain to function and the length of time to clear the blood of alcohol make alcohol's adverse effects linger long after its blood concentration has fallen. Driving coordination is still impaired the morning *after* a night of drinking, even if the drinking was moderate. Responsible aircraft pilots know that they must allow 24 hours for their bodies to clear alcohol completely, and they do not fly any sooner. The Federal Aviation Administration and major airlines enforce this rule.

Look again at the drawing of the brain in Figure H7-6 (p. 227), and note that when someone drinks, judgment fails first. Judgment might tell a person to limit alcohol consumption to two drinks at a party, but if the first drink takes judgment away, many more drinks may follow. The failure to stop drinking as planned, on repeated occasions, is a warning sign that the person may have an alcohol abuse problem.

Ethanol interferes with a multitude of metabolic reactions in the body—many more than have been enumerated here. With heavy alcohol consumption, the potential for harm is great. If you drink alcoholic beverages, do so with care, and in moderation.

TABLE H7-5 Health Effects of Heavy Alcohol Consumption

Health Problem	Effects of Alcohol
Arthritis	Increases the risk of inflamed joints.
Bone loss	Decreases bone mass and strength.
Cancer	Increases the risk of cancer of the liver, breast, mouth, pharynx, larynx, esophagus, colon, and rectum.
Fetal alcohol syndrome	Causes physical and behavioral abnormalities in the fetus (see Highlight 15).
Heart disease	In heavy drinkers, raises blood pressure, blood lipids, and the risk of stroke and heart disease; when compared with those who abstain, heart disease risk is generally lower in light-to-moderate drinkers.
Hyperglycemia	Raises blood glucose.
Hypoglycemia	Lowers blood glucose, especially in people with diabetes.
Infertility	Increases the risks of menstrual disorders and spontaneous abortions (in women); suppresses luteinizing hormone (in women) and testosterone (in men).
Kidney disease	Enlarges the kidneys, alters hormone functions, and increases the risk of kidney failure.
Liver disease	Causes fatty liver, alcoholic hepatitis, and cirrhosis.
Malnutrition	Increases the risk of malnutrition; low intakes of protein, calcium, iron, vitamin A, vitamin C, thiamin, vitamin B_6, and riboflavin; and impaired absorption of calcium, phosphorus, vitamin D, and zinc.
Nerve disorders	Causes neuropathy and dementia; impairs balance and memory.
Obesity	Increases energy intake, but is not a primary cause of obesity.
Psychological disturbances	Causes depression, anxiety, and insomnia.

NOTE: This list is by no means all-inclusive. Alcohol has direct toxic effects on all body systems.

© Cengage Learning

TABLE H7-6 Myths and Truths Concerning Alcohol

Myth:	Liquors such as rum, vodka, and tequila are more harmful than wine and beer.
Truth:	The damage caused by alcohol depends largely on the *amount* consumed. Compared with liquor, beer and wine have relatively low percentages of alcohol, but they are often consumed in larger quantities.
Myth:	Consuming alcohol with raw seafood diminishes the likelihood of getting hepatitis.
Truth:	People have eaten contaminated oysters while drinking alcoholic beverages and not gotten as sick as those who were not drinking. But do not be misled: hepatitis is too serious an illness for anyone to depend on alcohol for protection.
Myth:	Alcohol stimulates the appetite.
Truth:	For some people, alcohol may stimulate appetite, but it seems to have the opposite effect in heavy drinkers. Heavy drinkers tend to eat poorly and suffer malnutrition.
Myth:	Drinking alcohol is healthy.
Truth:	Moderate alcohol consumption is associated with a lower risk for heart disease. Higher intakes, however, raise the risks for high blood pressure, stroke, heart disease, some cancers, accidents, violence, suicide, birth defects, and deaths in general. Furthermore, excessive alcohol consumption damages the liver, pancreas, brain, and heart. No authority recommends that nondrinkers begin drinking alcoholic beverages to obtain health benefits.
Myth:	Wine increases the body's absorption of minerals.
Truth:	Wine may increase the body's absorption of potassium, calcium, phosphorus, magnesium, and zinc, but the alcohol in wine also promotes the body's excretion of these minerals, so no benefit is gained.
Myth:	Alcohol is legal and, therefore, not a drug.
Truth:	Alcohol is legal for adults 21 years old and older, but it is also a drug—a substance that alters one or more of the body's functions.
Myth:	A shot of alcohol warms you up.
Truth:	Alcohol diverts blood flow to the skin making you *feel* warmer, but it actually cools the body.
Myth:	Wine and beer are mild; they do not lead to alcoholism.
Truth:	Alcoholism is not related to the kind of beverage, but rather to the quantity and frequency of consumption.
Myth:	Mixing different types of drinks gives you a hangover.
Truth:	Too much alcohol in any form produces a hangover.
Myth:	Alcohol is a stimulant.
Truth:	People think alcohol is a stimulant because it seems to relieve inhibitions, but it does so by depressing the activity of the brain. Alcohol is medically defined as a depressant drug.
Myth:	Beer is a great source of carbohydrate, vitamins, minerals, and fluids.
Truth:	Beer does provide some carbohydrate, but most of its kcalories come from alcohol. The few vitamins and minerals in beer cannot compete with rich food sources. And the diuretic effect of alcohol causes the body to lose more fluid in urine than is provided by the beer.

© Cengage Learning

CRITICAL THINKING QUESTIONS

A. If body organs could talk, what might some of them say to alcohol about its role in metabolism and disease development?

B. Some people choose to abstain from drinking alcoholic beverages. Others overindulge, to the detriment of their health and safety.

Moderation lies somewhere between the two ends of the spectrum. How would you plan a social gathering that would ensure that guests of legal age could enjoy an evening that includes alcoholic beverages in a safe and responsible way?

REFERENCES

1. D. J. McLernon and coauthors, Do lifestyle choices explain the effect of alcohol on bone mineral density in women around menopause? *American Journal of Clinical Nutrition* 95 (2012): 1261–1269; P. E. Ronksley and coauthors, Association of alcohol consumption with selected cardiovascular disease outcomes: A systematic review and meta-analysis, *British Medical Journal* 342 (2011): doi101136/bmj.d671; Q. Sun and coauthors, Alcohol consumption at midlife and successful ageing in women: A prospective cohort analysis in the Nurses' Health Study, *PLoS Medicine* 8 (2011): e1001090; U. Benedetto and coauthors, Alcohol intake and outcomes following coronary artery bypass grafting, *Circulation* 122 (2010): A14440; D. A. Boggs and coauthors, Coffee, tea, and alcohol intake in relation to risk of type 2 diabetes in African American women, *American Journal of Clinical Nutrition* 92 (2010): 960–966; M. M. Joosten and coauthors, Combined effect of alcohol consumption and lifestyle behaviors on risk of type 2 diabetes, *American Journal of Clinical Nutrition* 91 (2010): 1777–1783.

2. U. A. Hvidtfeldt and coauthors, Alcohol intake and risk of coronary heart disease in younger, middle-aged, and older adults, *Circulation* 121 (2010): 1589–1597.

3. A. J. Barnes and coauthors, Prevalence and correlates of at-risk drinking among older adults: The project SHARE study, *Journal of General Internal Medicine* 25 (2010): 840–846.

4. M. Roerecke and J. Rehm, Irregular heavy drinking occasions and risk of ischemic heart disease: A systematic review and meta-analysis, *American Journal of Epidemiology* 171 (2010): 633–644.

5. K. J. Mukamal, A 42-year-old man considering whether to drink alcohol for his health, *Journal of the American Medical Association* 303 (2010): 2065–2073.

6. X. Yang, Common variants at 12q24 are associated with drinking behavior in Han Chinese, *American Journal of Clinical Nutrition* 97 (2013): 545–551; A. Agrawal and coauthors, Measuring alcohol consumption for genomic meta-analyses of alcohol intake: Opportunities and challenges, *American Journal of Clinical Nutrition* 95 (2012): 539–547; I. Baik and coauthors, Genome-wide association studies identify genetic loci related to alcohol consumption in Korean men, *American Journal of Clinical Nutrition* 93 (2011): 809–816.

7. P. M. Guenther, E. L. Ding, and E. B. Rimm, Alcoholic beverage consumption by adults compared to dietary guidelines: Results of the National Health and Nutrition Examination Survey, 2009–2010, *Journal of the Academy of Nutrition and Dietetics* 113 (2013): 546–550.

8. C. Sayon-Orea, M. A. Martinez-Gonzalez, and M. Bes-Rastrollo, Alcohol consumption and body weight: A systematic review, *Nutrition Reviews* 69 (2011): 419–431; L. Wang and coauthors, Alcohol consumption, weight gain, and risk of becoming overweight in middle-aged and older women, *Archives of Internal Medicine* 170 (2010): 453–461.

9. N. T. Bendsen and coauthors, Is beer consumption related to measures of abdominal and general obesity? A systematic review and meta-analysis, *Nutrition Reviews* 71 (2013): 67–87.

10. R. A. Breslow and coauthors, Alcoholic beverage consumption, nutrient intakes, and diet quality in the US adult population, 1999–2006, *Journal of the American Dietetic Association* 110 (2010): 551–562.

11. R. A. Breslow and coauthors, Diets of drinkers on drinking and nondrinking days: NHANES 2003–2008, *American Journal of Clinical Nutrition* 97 (2013): 1068–1075.

12. S. J. Nielsen and coauthors, Calories consumed from alcoholic beverages by US adults, 2007–2010, www.cdc.gov/nchs/data/databriefs/db110.htm, published November 2012.

13. J. Howland and D. J. Rohsenow, Risks of energy drinks mixed with alcohol, *Journal of the American Medical Association* 309 (2013): 245–246; A. M. Arria and M. C. O'Brien, The "high" risk of energy drinks, *Journal of the American Medical Association* 305 (2011): 600–601; *Questions and Answers: Caffeinated Alcoholic Beverages,* www.fda.gov/Food/FoodIngredientsPackaging/ucm233726.htm, accessed November 17, 2010.

14. A. E. Barry and A. K. Piazza-Gardner, Drunkorexia: Understanding the co-occurrence of alcohol consumption and eating/exercise weight management behaviors, *Journal of American College Health* 60 (2012): 236–243.

15. Centers for Disease Control and Prevention, Vital signs: Binge drinking prevalence, frequency, and intensity among adults—United States, 2010, *Morbidity and Mortality Weekly Report* 61 (2012): 14–19.

16. S. Feldman, Prenatal alcohol exposure patterns and alcohol-related birth defects and growth deficiencies: A prospective study, *Alcoholism: Clinical and Experimental Research* 36 (2012): 670–676.

17. World Health Organization, *Global Status Report on Alcohol and Health*, 2011.

Energy Balance and Body Composition

Nutrition in Your Life

It's simple: energy balance occurs when energy in = energy out. The reality, of course, is much more complex. One day you may devour a dozen doughnuts at midnight and sleep through your morning workout—tipping the scales toward weight gain. Another day you may snack on veggies and train for this weekend's 10K race—shifting the balance toward weight loss. Your body weight—especially as it relates to your body fat—and your level of fitness have consequences for your health. So, how are you doing? In the Nutrition Portfolio at the end of this chapter, you can see how your "energy in" and "energy out" balance and whether your body weight and fat measures are consistent with good health.

As Chapter 7 explains, the body's remarkable metabolism can cope with variations in the diet. When the diet delivers too little energy, carbohydrate, or protein, the body uses its fat to meet energy needs and degrades its lean tissue to meet glucose and protein needs. When the diet delivers too much energy—whether from excess carbohydrate, excess protein, or excess fat—the body stores fat.

Both excessive and deficient body fat result from an energy imbalance. The simple picture is as follows. People who consume more food energy than they expend store the surplus as body fat. To reduce body fat, they need to expend more energy than they take in from food. In contrast, people who consume too little food energy to support their bodies' activities must rely on their bodies' fat stores and possibly some of their lean tissues as well. To gain weight, these people need to take in more food energy than they expend. As you will see, though, the details of energy balance and weight regulation are quite complex.[1] This chapter describes energy balance and body composition and examines the health problems associated with having too much or too little body fat. The next chapter presents strategies toward resolving these problems.

ENERGY

IN   OUT

© Cengage Learning

When energy in balances with energy out, a person's body weight is stable.

> FIGURE 8-1   **Bomb Calorimeter**

When food is burned, energy is released in the form of heat. Heat energy is measured in kcalories.

Insulated container keeps heat from escaping

Thermometer measures temperature changes

Motorized stirrer

Reaction chamber (bomb)

Food is burned

Heating element

Water in which temperature increase from burning food is measured

© Cengage Learning

energy balance: the energy (kcalories) consumed from foods and beverages compared with the energy expended through metabolic processes and physical activities.

bomb calorimeter (KAL-oh-RIM-eh-ter): an instrument that measures the heat energy released when foods are burned, thus providing an estimate of the potential energy of the foods.

• **calor** = heat
• **metron** = measure

8.1 Energy Balance

> **LEARN IT** Describe energy balance and the consequences of not being in balance.

People expend energy continuously and eat periodically to refuel. Ideally, their energy intakes cover their energy expenditures with little, or no, excess. Excess energy is stored as fat, and stored fat is used for energy between meals. The fat stores of even a healthy-weight adult represent an ample reserve of energy—50,000 to 200,000 kcalories.

The amount of body fat a person deposits in, or withdraws from, storage on any given day depends on the **energy balance** for that day—the amount consumed (energy in) versus the amount expended (energy out). When a person is maintaining weight, energy in equals energy out. When the balance shifts, weight changes.

A classic rule states that for each 3500 kcalories eaten in excess, a pound of body fat is stored; similarly, a pound of fat is lost for each 3500 kcalories expended beyond those consumed.* To that end, many diet plans recommend lowering energy intake by 500 kcalories a day to incur a weight loss of 1 pound per week. This "3500 kcalorie rule" has been used for more than 50 years, but it has several limitations.[2] For one, as a person loses weight, the deficit in energy needed to continue losing weight shifts; in general, the kcalorie deficit is relatively low and weight loss is relatively rapid in the early phase but then it is followed by a markedly slower weight loss that plateaus as the kcalorie deficit needed to continue losing weight gradually increases. For another, body composition differs dramatically for men and women and for obese and lean people; in general, the kcalorie deficit needed for weight loss is relatively larger for women than for men and for obese than for lean people. Understanding the dynamic nature of weight loss may help people adopt more realistic expectations than a fixed 3500-kcalorie rule provides.

Quick changes in body weight are not simple changes in fat stores. Weight gained or lost rapidly includes some fat, large amounts of fluid, and some lean tissues such as muscle proteins and bone minerals. Because water constitutes about 60 percent of an adult's body weight, retention or loss of water can greatly influence body weight. Even over the long term, the composition of weight gained or lost is normally about 75 percent fat and 25 percent lean. During starvation, losses of fat and lean are about equal. (Recall from Chapter 7 that without adequate carbohydrate, protein-rich lean tissues break down to provide glucose.) Invariably, though, *fat* gains and losses are gradual. The next two sections examine the two sides of the energy-balance equation—energy in and energy out. As you read, keep in mind that this simple equation falls short of fully explaining the many metabolic changes that cause obesity.[3]

> **REVIEW IT** Describe energy balance and the consequences of not being in balance.

When energy consumed equals energy expended, a person is in energy balance and body weight is stable. If more energy is taken in than is expended, a person gains weight. If more energy is expended than is taken in, a person loses weight.

8.2 Energy In: The kCalories Foods Provide

> **LEARN IT** Discuss some of the physical, emotional, and environmental influences on food intake.

Foods and beverages provide the "energy in" part of the energy-balance equation. How much energy a person receives depends on the composition of the foods and beverages and on the amount the person eats and drinks.

Food Composition   To find out how many kcalories a food provides, a scientist can burn the food in a **bomb calorimeter** (see Figure 8-1). When the

*Body fat, or adipose tissue, is composed of a mixture of mostly fat, some protein, and water. A pound of body fat (454 g) is approximately 87 percent fat, or (454 × 0.87) 395 g, and 395 g × 9 kcal/g = 3555 kcal.

food burns, energy is released in the form of heat. The amount of heat given off provides a *direct* measure of the food's energy value (remember that kcalories are units of heat energy).* In addition to releasing heat, these reactions generate carbon dioxide and water—just as the body's cells do when they metabolize the energy-yielding nutrients from foods. Details of the chemical reactions in a calorimeter and in the body differ, but the overall process is similar: when the food burns and the chemical bonds break, the carbons (C) and hydrogens (H) combine with oxygens (O) to form carbon dioxide (CO_2) and water (H_2O). The amount of oxygen consumed gives an *indirect* measure of the amount of energy released.

A bomb calorimeter measures the available energy in foods but overstates the **physiological fuel value**—the amount of energy that the human body derives from foods. The body is less efficient than a calorimeter and cannot metabolize all of the energy-yielding nutrients in a food completely. Researchers can correct for this discrepancy mathematically to create useful tables of the energy values of foods (such as Appendix H). These values provide reasonable estimates, but they do not reflect the *precise* amount of energy a person will derive from the foods consumed.

The energy values of foods can also be computed from the amounts of carbohydrate, fat, and protein (and alcohol, if present) in the foods.** For example, a food containing 12 grams of carbohydrate, 5 grams of fat, and 8 grams of protein will provide 48 carbohydrate kcalories, 45 fat kcalories, and 32 protein kcalories, for a total of 125 kcalories. (To review how to calculate the energy foods provide, turn to the "How To" feature on p. 10.)

Food Intake To achieve energy balance, the body must meet its needs without taking in too much or too little energy. **Appetite** prompts a person to eat—or not to eat. Somehow the body decides how much and how often to eat—when to start eating and when to stop. As you will see, many signals—from both the environment and genetics—initiate or delay eating.[4]

Hunger People eat for a variety of reasons, most obviously (although not necessarily most commonly) because they are hungry. Most people recognize **hunger** as an irritating feeling that prompts thoughts of food and motivates them to start eating. In the body, hunger is the physiological response to a need for food triggered by nerve signals and chemical messengers originating and acting in the brain, primarily in the **hypothalamus**. Hunger can be influenced by the presence or absence of nutrients in the bloodstream, the size and composition of the preceding meal, customary eating patterns, climate (heat reduces food intake; cold increases it), physical activity, hormones, and illnesses. Hunger determines what to eat, when to eat, and how much to eat.

The stomach is ideally designed to handle periodic batches of food, and people typically eat meals at roughly 4-hour intervals. Four hours after a meal, most, if not all, of the food has left the stomach. Most people do not feel like eating again until the stomach is either empty or almost so. Even then, a person may not feel hungry for quite a while.

Satiation During the course of a meal, as food enters the GI tract and hunger diminishes, **satiation** occurs. As receptors in the stomach stretch and hormones such as cholecystokinin become active, the person begins to feel full. The response: satiation, which prompts the person to stop eating.

Satiety After a meal, the feeling of **satiety** continues to suppress hunger and allows a person to not eat again for a while. Whereas *satiation* tells us to "stop eating," *satiety* reminds us to "not start eating again." Figure 8-2 (p. 238) summarizes the relationships among hunger, satiation, and satiety. Of course, people can override these signals, especially when presented with stressful situations or favorite foods.

physiological fuel value: the number of kcalories that the body derives from a food, in contrast to the number of kcalories determined by calorimetry.

appetite: the integrated response to the sight, smell, thought, or taste of food that initiates or delays eating.

hunger: the painful sensation caused by a lack of food that initiates food-seeking behavior.

hypothalamus (high-po-THAL-ah-mus): a brain center that controls activities such as maintenance of water balance, regulation of body temperature, and control of appetite.

satiation (say-she-AY-shun): the feeling of satisfaction and fullness that occurs during a meal and halts eating. Satiation determines how much food is consumed during a meal.

satiety (sah-TIE-eh-tee): the feeling of fullness and satisfaction that occurs after a meal and inhibits eating until the next meal. Satiety determines how much time passes between meals.

---

*As Chapter 1 mentions, many scientists measure food energy in *kilojoules* (a measure of work energy). Conversion factors for these and other measures can be found in Appendix K.

**Some of the food energy values in the table of food composition in Appendix H were derived by bomb calorimetry, and many were calculated from their energy-yielding nutrient contents.

> FIGURE 8-2   **Hunger, Satiation, and Satiety**

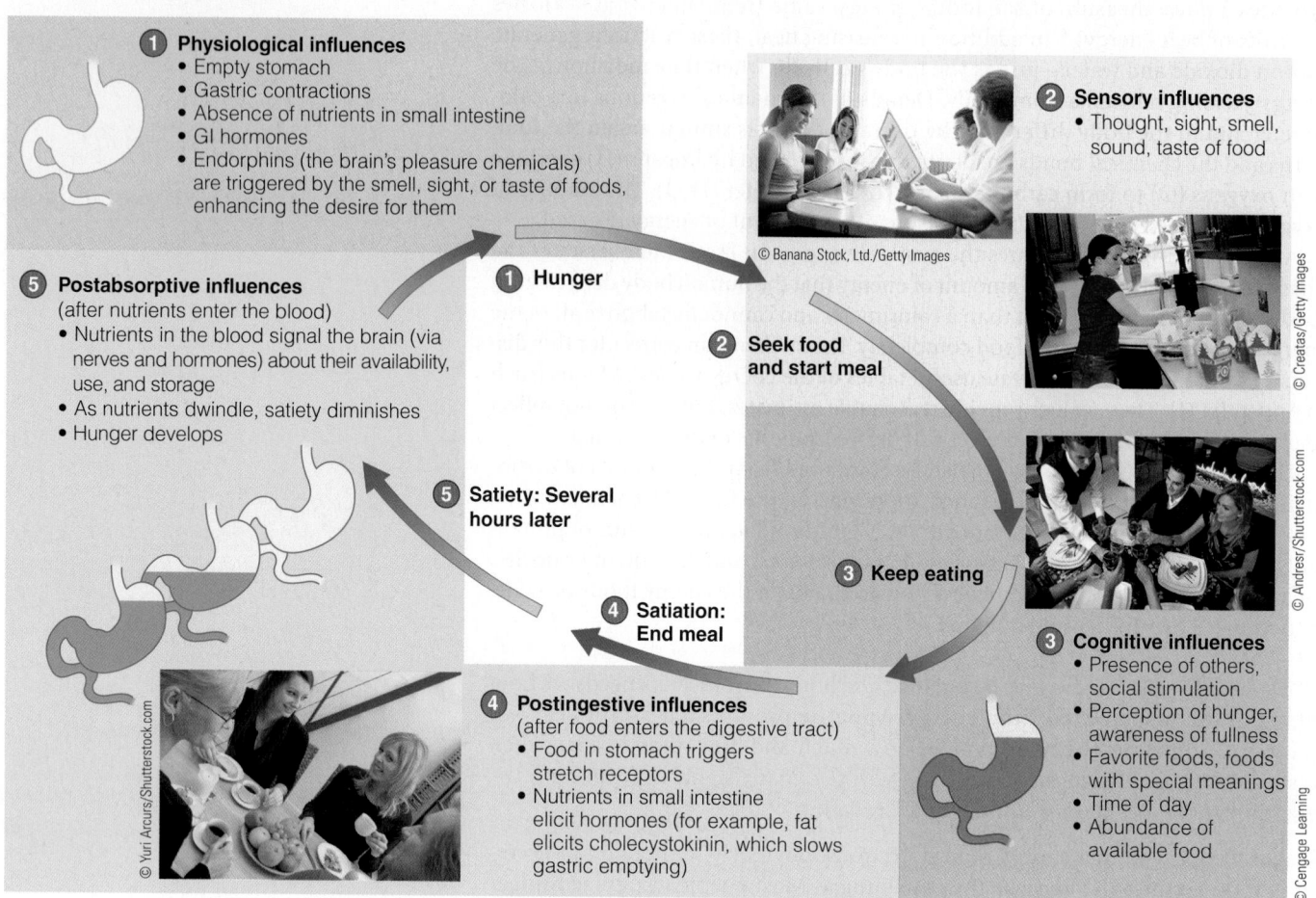

① **Physiological influences**
• Empty stomach
• Gastric contractions
• Absence of nutrients in small intestine
• GI hormones
• Endorphins (the brain's pleasure chemicals)
  are triggered by the smell, sight, or taste of foods,
  enhancing the desire for them

② **Sensory influences**
• Thought, sight, smell,
  sound, taste of food

⑤ **Postabsorptive influences**
(after nutrients enter the blood)
• Nutrients in the blood signal the brain (via
  nerves and hormones) about their availability,
  use, and storage
• As nutrients dwindle, satiety diminishes
• Hunger develops

① **Hunger**

② **Seek food
and start meal**

⑤ **Satiety: Several
hours later**

③ **Keep eating**

④ **Satiation:
End meal**

③ **Cognitive influences**
• Presence of others,
  social stimulation
• Perception of hunger,
  awareness of fullness
• Favorite foods, foods
  with special meanings
• Time of day
• Abundance of
  available food

④ **Postingestive influences**
(after food enters the digestive tract)
• Food in stomach triggers
  stretch receptors
• Nutrients in small intestine
  elicit hormones (for example, fat
  elicits cholecystokinin, which slows
  gastric emptying)

© Banana Stock, Ltd./Getty Images
© Creatas/Getty Images
© Andresr/Shutterstock.com
© Yuri Arcurs/Shutterstock.com
© Cengage Learning

Overriding Hunger and Satiety Not surprisingly, eating can be triggered by signals other than hunger, even when the body does not need food. Some people experience food cravings when they are bored or anxious. In fact, they may eat in response to any kind of stress, negative or positive. ("What do I do when I'm grieving? Eat. What do I do when I'm celebrating? Eat!") Not too surprisingly, repeatedly eating to relieve chronic stress can lead to overeating and weight gain.

Many people respond to external cues such as the time of day ("It's time to eat") or the availability, sight, and taste of food ("I'd love a piece of chocolate even though I'm full"). Environmental influences such as large portion sizes, favorite foods, or an abundance or variety of foods stimulate eating and increase energy intake. Cognitive influences—such as perceptions, memories, intellect, and social interactions—can easily lead to weight gain. Those who are overweight or obese may be especially susceptible to external cues that trigger hunger and the desire to eat.[5]

Eating can also be suppressed by signals other than satiety, even when a person is hungry. People with the eating disorder anorexia nervosa, for example, use tremendous discipline to ignore the pangs of hunger. Some people simply cannot eat during times of stress, negative or positive. ("I'm too sad to eat." "I'm too excited to eat!") Why some people overeat in response to stress and others cannot eat at all remains a bit of a mystery, although researchers are beginning to understand the connections between stress hormones, brain activity, and "comfort foods." Factors that appear to be involved include how the person perceives the stress and whether usual eating behaviors are restrained. (Highlight 8 features anorexia nervosa and other eating disorders.)

Sustaining Satiation and Satiety The extent to which foods produce satiation and sustain satiety depends in part on the nutrient composition of a meal. Of the

Regardless of hunger, people typically overeat when offered the abundance and variety of a buffet. To limit unhealthy weight gains, listen to hunger and satiety signals.

three energy-yielding nutrients, protein is considered the most **satiating.** In fact, too little protein in the diet can leave a person feeling hungry. Including some protein—such as drinking milk—provides satiety and decreases energy intake at the next meal.[6] In contrast, fructose in a sugary fruit drink seems to stimulate appetite and increase food intake.

Chapter 1 explains that energy density is a measure of the energy a food provides relative to the amount of food (kcalories per gram). Foods with a high energy density provide more kcalories, and those with low energy density provide fewer kcalories, for the same amount of food. Foods low in energy density are also more satiating. High-fiber foods effectively provide satiation by filling the stomach and delaying the absorption of nutrients. For this reason, eating a large salad as a first course helps a person eat less during the meal. In contrast, fat has a weak effect on satiation; consequently, eating high-fat foods may lead to passive overconsumption. High-fat foods are flavorful, which stimulates the appetite and entices people to eat more. High-fat foods are also energy dense; consequently, they deliver more kcalories per bite. (Chapter 9 describes how considering a food's energy density can help with weight management.) Although fat provides little satiation during a meal, it produces strong satiety signals once it enters the intestine. Fat in the intestine triggers the release of cholecystokinin— a hormone that signals satiety and inhibits food intake.

Eating high-fat foods while trying to limit energy intake requires small portion sizes, which can leave a person feeling unsatisfied. Portion size correlates directly with a food's satiety. Instead of eating small portions of high-fat foods and feeling deprived, a person can feel satisfied by eating large portions of low-fat, high-fiber, and low-energy-density foods. Figure 8-3 (p. 240) illustrates how fat influences portion size.

Message Central—The Hypothalamus As you can see, eating is a complex behavior controlled by a variety of genetic, psychological, social, metabolic, and physiological factors.[7] The hypothalamus appears to be the control center, integrating messages about energy intake, expenditure, and storage from other parts of the brain and from the mouth, GI tract, and liver.[8] Some of these messages influence satiation, which helps control the size of a meal; others influence satiety, which helps determine the frequency of meals.*

Dozens of gastrointestinal hormones influence appetite control and energy balance.[9] By understanding the action of these hormones, researchers may one day be able to develop anti-obesity treatments. An added challenge is to sort out

---

*Gastrointestinal hormones that regulate food intake include amylin, cholecystokinin (CCK), enterostatin, ghrelin, glucagon-like peptide-1 (GLP-1), oxyntomodulin, pancreatic polypeptide (PP), and peptide YY (PYY).

satiating: having the power to suppress hunger and inhibit eating.

> FIGURE 8-3   How Fat Influences Portion Sizes

For the same size portion, peanuts deliver more than 15 times the kcalories and 20 times the fat of popcorn.

For the same number of kcalories, a person can have a few high-fat peanuts or almost 2 cups of high-fiber popcorn. (This comparison used oil-based popcorn; using air-popped popcorn would double the amount of popcorn in this example.)

neuropeptide Y: a chemical produced in the brain that stimulates appetite, diminishes energy expenditure, and increases fat storage.

thermogenesis: the generation of heat; used in physiology and nutrition studies as an index of how much energy the body is expending.

basal metabolism: the energy needed to maintain life when a body is at complete digestive, physical, and emotional rest.

> FIGURE 8-4   **Components of Energy Expenditure**

The amount of energy expended in voluntary physical activities has the greatest variability, depending on a person's activity patterns. For a sedentary person, physical activities may account for less than half as much energy as basal metabolism, whereas an extremely active person may expend as much on physical activity as for basal metabolism.

The amount of energy expended in a day differs for each individual, but in general, basal metabolism is the largest component of energy expenditure and thermic effect of food is the smallest.

the many actions of related brain chemicals. For example, one brain chemical, **neuropeptide Y,** causes carbohydrate cravings, initiates eating, decreases energy expenditure, and increases fat storage—all factors favoring a positive energy balance and weight gain.

> **REVIEW IT** Discuss some of the physical, emotional, and environmental influences on food intake.

A mixture of signals governs a person's eating behaviors. Hunger and appetite initiate eating, whereas satiation and satiety stop and delay eating, respectively. Each responds to messages from the nervous and hormonal systems. Superimposed on these signals are complex factors involving emotions, habits, and other aspects of human behavior.

8.3 Energy Out: The kCalories the Body Expends

> **LEARN IT** List the components of energy expenditure and factors that might influence each.

Chapter 7 explains that heat is released whenever the body breaks down carbohydrate, fat, or protein for energy and again when that energy is used to do work. The generation of heat, known as **thermogenesis,** can be measured to determine the amount of energy expended. The total energy a body expends reflects three main categories of thermogenesis:

- Energy expended for basal metabolism
- Energy expended for physical activity
- Energy expended for food consumption

A fourth category is sometimes involved:

- Energy expended for adaptation

Components of Energy Expenditure   People expend energy when they are physically active, of course, but they also expend energy when they are resting quietly. In fact, quiet metabolic activities account for the largest share of most people's energy expenditures, as Figure 8-4 shows.

Basal Metabolism   About two-thirds of the energy the average person expends in a day supports the body's **basal metabolism.** Metabolic activities include the lungs inhaling and exhaling air, the bone marrow making new red blood cells,

TABLE 8-1 Estimating Energy Expended on Basal Metabolism

	BMR Estimates	BMR Equations
Men	Slightly >1 kcal/min (1.1 to 1.3 kcal/min) or 24 kcal/kg/day	$(10 \times wt) + (6.25 \times ht) - (5 \times age) + 5$
Women	Slightly <1 kcal/min (0.8 to 1.0 kcal/min) or 23 kcal/kg/day	$(10 \times wt) + (6.25 \times ht) - (5 \times age) - 161$
Note	For perspective, a burning candle or a 75-watt light bulb releases about 1 kcal/min	Use actual weight in kilograms, height in centimeters, and age in years

© Cengage Learning

the heart beating 100,000 times a day, and the kidneys filtering wastes—in short, they support all the basic processes of life.

The **basal metabolic rate (BMR)** is the rate at which the body expends energy for these life-sustaining activities. The rate may vary from person to person and may vary for the same individual with a change in circumstance or physical condition. The rate is slowest when a person is sleeping undisturbed, but it is usually measured in a room with a comfortable temperature when the person is awake, but lying still, after a restful sleep and an overnight (12 to 14 hour) fast. A similar measure of energy output—called the **resting metabolic rate (RMR)**—is slightly higher than the BMR because its criteria for recent food intake and physical activity are not as strict. When energy needs cannot be measured, equations can provide reasonably accurate estimates (see Table 8-1).

In general, the more a person weighs, the more *total* energy is expended on basal metabolism, but the amount of energy *per pound* of body weight may be lower. For example, an adult's BMR might be 1500 kcalories per day and an infant's only 500, but compared to body weight, the infant's BMR is more than twice as fast. Similarly, a normal-weight adult may have a metabolic rate one and a half times that of an obese adult when compared to body weight because lean tissue is metabolically more active than body fat.

Table 8-2 (p. 242) summarizes the factors that raise and lower the BMR. For the most part, the BMR is highest in people who are growing (children, adolescents, and pregnant women) and in those with considerable **lean body mass** (physically fit people and males). One way to increase the BMR, then, is to participate in endurance and strength-training activities regularly to maximize lean body mass. The BMR is also high in people with fever or under stress and in people with highly active thyroid glands. The BMR slows down with a loss of lean body mass and during fasting and malnutrition.

Physical Activity The second component of a person's energy output is physical activity: voluntary movement of the skeletal muscles and support systems. Physical activity is the most variable—and the most changeable—component of energy expenditure. Consequently, its influence on both weight gain and weight loss can be significant.

During physical activity, the muscles need extra energy to move, and the heart and lungs need extra energy to deliver nutrients and oxygen and dispose of wastes. The amount of energy needed for any activity, whether playing tennis or studying for an exam, depends on three factors: muscle mass, body weight, and activity. The larger the muscle mass and the heavier the weight of the body part being moved, the more energy is expended. Table 8-3 (p. 242) gives average energy expenditures for various activities. The activity's duration, frequency, and intensity also influence energy expenditure: the longer, the more frequent, and the more intense the activity, the more kcalories expended. (Chapter 14 describes how an activity's duration, frequency, and intensity also influence the body's use of the energy-yielding nutrients.)

basal metabolic rate (BMR): the rate of energy use for metabolism under specified conditions: after a 12-hour fast and restful sleep, without any physical activity or emotional excitement, and in a comfortable setting. It is usually expressed as kcalories per kilogram body weight per hour.

resting metabolic rate (RMR): similar to the basal metabolic rate (BMR), a measure of energy use for a person at rest in a comfortable setting, but with less stringent criteria for recent food intake and physical activity. Consequently, the RMR is slightly higher than the BMR.

lean body mass: the body minus its fat.

TABLE 8-2 Factors that Affect the BMR

Factor	Effect on BMR
Age	Lean body mass diminishes with age, slowing the BMR.[a]
Height	In tall, thin people, the BMR is higher.[b]
Growth	In children, adolescents, and pregnant women, the BMR is higher.
Body composition (gender)	The more lean tissue, the higher the BMR (which is why males usually have a higher BMR than females). The more fat tissue, the lower the BMR.
Fever	Fever raises the BMR.[c]
Stresses	Stresses (including many diseases and certain drugs) raise the BMR.
Environmental temperature	Both heat and cold raise the BMR.
Fasting/starvation	Fasting/starvation lowers the BMR.[d]
Malnutrition	Malnutrition lowers the BMR.
Hormones	The thyroid hormone thyroxine, for example, can speed up or slow down the BMR.[e] Premenstrual hormones slightly raise the BMR.
Smoking	Nicotine increases energy expenditure.
Caffeine	Caffeine increases energy expenditure.
Sleep	BMR is lowest when sleeping.

[a]The BMR begins to decrease in early adulthood (after growth and development cease) at a rate of about 2 percent/ decade. A reduction in voluntary activity as well brings the total decline in energy expenditure to about 5 percent/decade.
[b]If two people weigh the same, the taller, thinner person will have the faster metabolic rate, reflecting the greater skin surface, through which heat is lost by radiation, in proportion to the body's volume (see Figure 8-5, p. 244).
[c]Fever raises the BMR by 7 percent for each degree Fahrenheit.
[d]Prolonged starvation reduces the total amount of metabolically active lean tissue in the body, although the decline occurs sooner and to a greater extent than body losses alone can explain. More likely, the neural and hormonal changes that accompany fasting are responsible for changes in the BMR.
[e]The thyroid gland releases hormones that travel to the cells and influence cellular metabolism. Thyroid hormone activity can speed up or slow down the rate of metabolism by as much as 50 percent.

© Cengage Learning

TABLE 8-3 Estimating Energy Expended on Physical Activities

The values listed in this table reflect both the energy expended in physical activity and the amount used for BMR. To calculate kcalories spent per minute of activity for your own body weight, multiply kcal/lb/min (or kcal/kg/min) by your exact weight and then multiply that number by the number of minutes spent in the activity. For example, if you weigh 142 pounds, and you want to know how many kcalories you spent doing 30 minutes of vigorous aerobic dance: 0.062 × 142 = 8.8 kcalories per minute; 8.8 × 30 minutes = 264 total kcalories expended.

Activity	kCal/lb min	kCal/kg min	Activity	kCal/lb min	kCal/kg min	Activity	kCal/lb min	kCal/kg min
Aerobic dance (vigorous)	.062	.136	Handball	.078	.172	Table tennis (skilled)	.045	.099
Basketball (vigorous, full court)	.097	.213	Horseback riding (trot)	.052	.114	Tennis (beginner)	.032	.070
Bicycling			Rowing (vigorous)	.097	.213	Vacuuming and other household tasks	.030	.066
13 mph	.045	.099	Running			Walking		
15 mph	.049	.108	5 mph	.061	.134	3.5 mph	.035	.077
17 mph	.057	.125	6 mph	.074	.163	4.5 mph	.048	.106
19 mph	.076	.167	7.5 mph	.094	.207	Weight lifting		
21 mph	.090	.198	9 mph	.103	.227	light-to-moderate	.024	.053
23 mph	.109	.240	10 mph	.114	.251	vigorous	.048	.106
25 mph	.139	.306	11 mph	.131	.288	Wheelchair basketball	.084	.185
Canoeing, flat water, moderate pace	.045	.099	Soccer (vigorous)	.097	.213	Wheeling self in wheelchair	.030	.066
Cross-country skiing 8 mph	.104	.229	Studying	.011	.024	Wii games		
			Swimming			bowling	.021	.046
Gardening	.045	.099	20 yd/min	.032	.070	boxing	.021	.047
Golf (carrying clubs)	.045	.099	45 yd/min	.058	.128	tennis	.022	.048
			50 yd/min	.070	.154			

© Cengage Learning

It feels like work and it may make you tired, but studying requires only one or two kcalories per minute.

Thermic Effect of Food When a person eats, the GI tract muscles speed up their rhythmic contractions, the cells that manufacture and secrete digestive juices become active, and some nutrients require energy to be absorbed. This acceleration of activity requires energy and produces heat; it is known as the **thermic effect of food (TEF).**

The thermic effect of food is proportional to the food energy taken in and is usually estimated at 10 percent of energy intake. Thus a person who ingests 2000 kcalories probably expends about 200 kcalories on the thermic effect of food. The proportions vary for different foods, however, and are also influenced by factors such as meal size and frequency. In general, the thermic effect of food is greater for high-protein foods than for high-fat foods (see Table 8-4) and for a meal eaten all at once rather than spread out over a couple of hours. For most purposes, however, the thermic effect of food can be ignored when estimating energy expenditure because its contribution to total energy output is smaller than the probable errors involved in estimating overall energy intake and output.

Adaptive Thermogenesis Additional energy is expended when circumstances in the body are dramatically changed. A body challenged to physical conditioning, extreme cold, overfeeding, starvation, trauma, or other types of stress must adapt; it has extra work to do and uses extra energy to build the tissues and produce the enzymes and hormones necessary to cope with the demand. This energy is known as **adaptive thermogenesis,** and in some circumstances (for example, in burn victims), it makes a considerable difference in the total energy expended. Because this component of energy expenditure is so variable and

TABLE 8-4 Estimating Energy Expended on Thermic Effect of Foods

Food Component	Energy Expended
Carbohydrate	5–10%
Fat	0–5%
Protein	20–30%
Alcohol	15–20%

NOTE: Percentages are calculated by dividing the energy expended during digestion and absorption (above basal) by the energy content of the food.

thermic effect of food (TEF): an estimation of the energy required to process food (digest, absorb, transport, metabolize, and store ingested nutrients); also called the *specific dynamic effect (SDE)* of food or the *specific dynamic activity (SDA)* of food. The sum of the TEF and any increase in the metabolic rate due to overeating is known as *diet-induced thermogenesis (DIT).*

adaptive thermogenesis: adjustments in energy expenditure related to changes in environment such as extreme cold and to physiological events such as overfeeding, trauma, and changes in hormone status.

> FIGURE 8-5 **How Body Size Influences BMR**

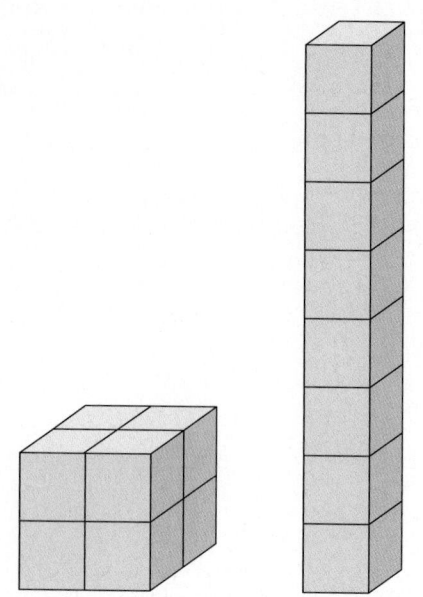

Each of these structures is made of eight blocks. They weigh the same, but they are arranged differently. The short, wide structure has 24 sides exposed and the tall, thin one has 34. Because the tall, thin structure has a greater surface area, it will lose more heat (expend more energy) than the short, wide one. Similarly, two people of different heights might weigh the same, but the taller, thin one will have a higher BMR (expending more energy) because of the greater skin surface.

© Cengage Learning

specific to individuals, it is not included when estimating energy requirements for most healthy people.

Estimating Energy Requirements In estimating energy requirements, the DRI Committee developed equations based on research measuring total daily energy expenditure. These equations consider how the following factors influence BMR and consequently energy expenditure:

- *Gender.* In general, women have a lower BMR than men, in large part because men typically have more lean body mass. Two sets of energy equations—one for men and one for women—were developed to accommodate the influence of gender on energy expenditure (provided on the next page).

- *Growth.* The BMR is high in people who are growing. For this reason, pregnant and lactating women, infants, children, and adolescents have their own sets of energy equations.

- *Age.* The BMR declines during adulthood as lean body mass diminishes. This change in body composition occurs, in part, because some hormones that influence appetite, body weight, and metabolism become more, or less, active with age. Physical activities tend to decline as well, bringing the average reduction in energy expenditure to about 5 percent per decade. The decline in BMR that occurs when a person becomes less active reflects the loss of lean body mass and may be minimized with ongoing physical activity. Because age influences energy expenditure, it is also factored into the energy equations.

- *Physical activity.* Using individual values for various physical activities (as in Table 8-3 on p. 242) is time-consuming and impractical for estimating the energy needs of a population. Instead, various activities are clustered according to the typical intensity of a day's efforts. Energy equations include a physical activity factor for various levels of intensity for each gender.

- *Body composition and body size.* The BMR is high in people who are tall and so have a large surface area, as illustrated in Figure 8-5. Similarly, the more a person weighs, the more energy is expended on basal metabolism. For these reasons, energy equations include a factor for both height and weight.

As just explained, energy needs vary between individuals depending on such factors as gender, growth, age, physical activity, and body size and composition. Even when two people are similarly matched, however, their energy needs still differ because of genetic differences. Perhaps one day genetic research will reveal how to estimate requirements for each individual. For now, the "How To" feature provides instructions on estimating energy requirements using the DRI equations and physical activity factors. Appendix F presents a table that provides a shortcut to estimating daily energy needs by age, gender, and activity level, based on the DRI equations using reference heights and weights.

> **REVIEW IT** List the components of energy expenditure and factors that might influence each.

A person in energy balance takes in energy from food and expends much of it on basal metabolism, some of it on physical activities, and a little on the thermic effect of food. Energy requirements vary from person to person, depending on such factors as gender, age, weight, and height as well as the intensity and duration of physical activity. All of these factors must be considered when estimating energy requirements.

>How To

Estimate Energy Requirements

To determine your estimated energy requirement (EER), use the appropriate equation, inserting your age in years, weight (wt) in kilograms, height (ht) in meters, and physical activity (PA) factor from the accompanying table. (To convert pounds to kilograms, divide by 2.2; to convert inches to meters, divide by 39.37.)

- For men 19 years and older:
  $$EER = [662 - (9.53 \times age)] + PA \times [(15.91 \times wt) + (539.6 \times ht)]$$
- For women 19 years and older:
  $$EER = [354 - (6.91 \times age)] + PA \times [(9.36 \times wt) + (726 \times ht)]$$

For example, consider an active 30-year-old male who is 5 feet 11 inches tall and weighs 178 pounds. First, he converts his weight from pounds to kilograms and his height from inches to meters, if necessary:

$$178\ lb \div 2.2 = 80.9\ kg$$

$$71\ in \div 39.37 = 1.8\ m$$

Next, he considers his level of daily physical activity and selects the appropriate PA factor from the accompanying table. (In this example, 1.25 for an active male.) Then, he inserts his age, PA factor, weight, and height into the appropriate equation:

$$EER = [662 - (9.53 \times 30)] + 1.25 \times [(15.91 \times 80.9) + (539.6 \times 1.8)]$$

(A reminder: Do calculations within the parentheses first.) He calculates:

$$EER = [662 - 286] + 1.25 \times [1287 + 971]$$

(Another reminder: Do calculations within the brackets next.)

$$EER = 376 + 1.25 \times 2258$$

(One more reminder: Do multiplication before addition.)

$$EER = 376 + 2823$$

$$EER = 3199$$

The estimated energy requirement for an active 30-year-old male who is 5 feet 11 inches tall and weighs 178 pounds is about 3200 kcalories/day. His actual requirement probably falls within a range of 200 kcalories above and below this estimate.

NOTE: Appendix F provides estimates of energy needs based on EER equations, using reference heights and weights for each age-gender group.

Physical Activity (PA) Factors for EER Equations			
	Men	Women	Physical Activity
Sedentary	1.0	1.0	Typical daily living activities
Low active	1.11	1.12	plus 30–60 min moderate activity
Active	1.25	1.27	plus ≥ 60 min moderate activity
Very active	1.48	1.45	plus ≥ 60 min moderate activity and 60 min vigorous or 120 min moderate activity

NOTE: Moderate activity is equivalent to walking at 3 to 4½ mph.

> TRY IT Estimate your energy requirement based on your current age, weight, height, and activity level.

8.4 Body Weight and Body Composition

> LEARN IT Distinguish between body weight and body composition, including methods to assess each.

A person 5 feet 10 inches tall who weighs 150 pounds may carry only about 30 of those pounds as fat.* The rest is mostly water and lean tissues—muscles, organs such as the heart and liver, and the bones of the skeleton. Direct measures of **body composition** are impossible in living human beings; instead, researchers assess body composition indirectly based on the following assumption:

Body weight = fat + lean tissue (including water)

Weight gains and losses tell us nothing about how the body's composition may have changed, yet weight is the measure most people use to judge their "fatness." For many people, overweight is overfat, but this is not always the case. Athletes with dense bones and well-developed muscles may be overweight by some standards but have little body fat. Conversely, inactive people may seem to have acceptable weights, when, in fact, they may have too much body fat.

*In metric terms, a person 1.78 meters tall who weights 68 kilograms may carry only about 14 of those kilograms as fat.

body composition: the proportions of muscle, bone, fat, and other tissue that make up a person's total body weight.

Defining Healthy Body Weight How much should a person weigh? How can a person know if her weight is appropriate for her height? How can a person know if his weight is jeopardizing his health? Such questions seem so simple, yet the answers can be complex—and quite different depending on whom you ask.

The Criterion of Fashion In asking what is ideal, people often mistakenly turn to friends and fashion for the answer and judge body weight by appearances. No doubt our society sets unrealistic ideals for body weight, especially for women. Magazines, movies, and television all convey the message that to be thin is to be beautiful and happy. As a result, the media have a great influence on the weight concerns and dieting patterns of people of all ages, but most tragically on young, impressionable children and adolescents.

Importantly, perceived body image may have little to do with actual body weight or size. People of all shapes, sizes, and ages—including extremely thin fashion models with anorexia nervosa and fitness instructors with ideal body composition—have learned to be unhappy with their "overweight" bodies. Such dissatisfaction can lead to damaging behaviors, such as starvation diets, diet pill abuse, and health-care avoidance. The first step toward making healthy changes may be self-acceptance. Keep in mind that fashion is fickle; the body shapes valued by our society change with time. Furthermore, body shapes valued by one society differ from those of other societies. The standards defining "ideal" are subjective and may have little in common with health. Table 8-5 offers some tips for adopting health as an ideal.

The Criterion of Health Even if our society were to accept fat as beautiful, obesity is still a major risk factor for several life-threatening diseases, including heart disease, type 2 diabetes, and some cancers. For this reason, the most important criterion for determining how much a person should weigh and how much body fat a person needs is not appearance but good health and longevity. Ideally, a person has enough fat to meet basic needs but not so much as to incur health risks. This range of healthy body weights has been identified using a common measure of weight and height—the body mass index.

Body Mass Index The **body mass index (BMI)** describes relative weight for height:

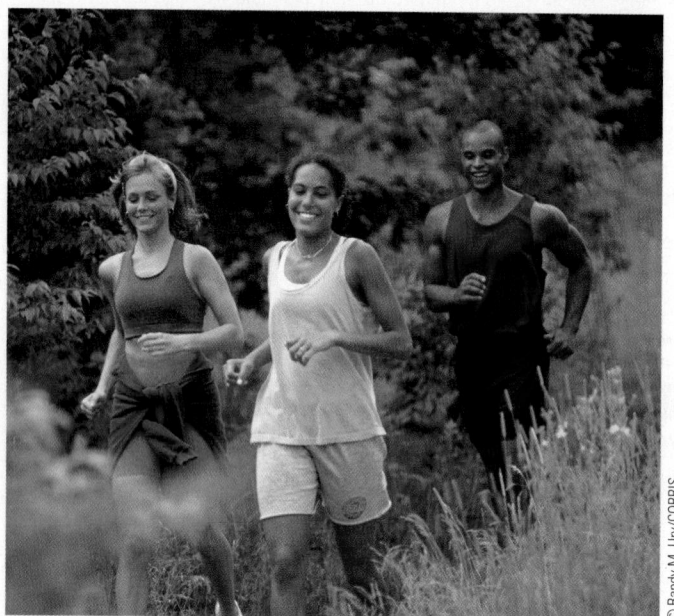

A healthy body contains enough lean tissue to support health and the right amount of fat to meet body needs.

© Randy M. Ury/CORBIS

body mass index (BMI): a measure of a person's weight relative to height; determined by dividing the weight (in kilograms) by the square of the height (in meters).

underweight: body weight lower than the weight range that is considered healthy; BMI less than 18.5.

overweight: body weight greater than the weight range that is considered healthy; BMI 25 to 29.9.

obese: too much body fat with adverse health effects; BMI 30 or more.

$$\text{BMI} = \frac{\text{weight (kg)}}{\text{height (m)}^2} \text{ or } \frac{\text{weight (lb)}}{\text{height (in)}^2} \times 703$$

Weight classifications based on BMI are presented in Table 8-6. Notice that healthy weight falls between a BMI of 18.5 and 24.9, with **underweight** below 18.5, **overweight** above 25, and **obese** above 30. Figure 8-6 shows examples of body shapes with different BMI. More than two-thirds of adults in the United States have a BMI greater than 25, as Figure 8-7 shows.[10]

TABLE 8-5 Tips for Accepting a Healthy Body Weight

- Value yourself and others for human attributes other than body weight. Realize that prejudging people by weight is as harmful as prejudging them by race, religion, or gender.
- Use positive, nonjudgmental descriptions of your body.
- Accept positive comments from others.
- Focus on your whole self, including your intelligence, social grace, and professional and scholastic achievements.
- Accept that no magic diet exists.
- Stop dieting to lose weight. Adopt a lifestyle of healthy eating and physical activity permanently.

- Follow the USDA Food Patterns. Never restrict food intake below the minimum levels that meet nutrient needs.
- Become physically active, not because it will help you get thin but because it will make you feel good and improve your health.
- Seek support from loved ones. Tell them of your plan for a healthy life in the body you have been given.
- Seek professional counseling from someone who can help you make gains in self-esteem without weight as the primary focus.
- Appreciate body weight for its influence on health, not appearance.

© Cengage Learning

TABLE 8-6 Body Mass Index (BMI)

Height	Under-weight (<18.5)	Healthy Weight (18.5–24.9)						Overweight (25–29.9)					Obese (≥30)										
	18	19	20	21	22	23	24	25	26	27	28	29	30	31	32	33	34	35	36	37	38	39	40
											Body weight (pounds)												
4'10"	86	91	96	100	105	110	115	119	124	129	134	138	143	148	153	158	162	167	172	177	181	186	191
4'11"	89	94	99	104	109	114	119	124	128	133	138	143	148	153	158	163	168	173	178	183	188	193	198
5'0"	92	97	102	107	112	118	123	128	133	138	143	148	153	158	163	168	174	179	184	189	194	199	204
5'1"	95	100	106	111	116	122	127	132	137	143	148	153	158	164	169	174	180	185	190	195	201	206	211
5'2"	98	104	109	115	120	126	131	136	142	147	153	158	164	169	175	180	186	191	196	202	207	213	218
5'3"	102	107	113	118	124	130	135	141	146	152	158	163	169	175	180	186	191	197	203	208	214	220	225
5'4"	105	110	116	122	128	134	140	145	151	157	163	169	174	180	186	192	197	204	209	215	221	227	232
5'5"	108	114	120	126	132	138	144	150	156	162	168	174	180	186	192	198	204	210	216	222	228	234	240
5'6"	112	118	124	130	136	142	148	155	161	167	173	179	186	192	198	204	210	216	223	229	235	241	247
5'7"	115	121	127	134	140	146	153	159	166	172	178	185	191	198	204	211	217	223	230	236	242	249	255
5'8"	118	125	131	138	144	151	158	164	171	177	184	190	197	203	210	216	223	230	236	243	249	256	262
5'9"	122	128	135	142	149	155	162	169	176	182	189	196	203	209	216	223	230	236	243	250	257	263	270
5'10"	126	132	139	146	153	160	167	174	181	188	195	202	209	216	222	229	236	243	250	257	264	271	278
5'11"	129	136	143	150	157	165	172	179	186	193	200	208	215	222	229	236	243	250	257	265	272	279	286
6'0"	132	140	147	154	162	169	177	184	191	199	206	213	221	228	235	242	250	258	265	272	279	287	294
6'1"	136	144	151	159	166	174	182	189	197	204	212	219	227	235	242	250	257	265	272	280	288	295	302
6'2"	141	148	155	163	171	179	186	194	202	210	218	225	233	241	249	256	264	272	280	287	295	303	311
6'3"	144	152	160	168	176	184	192	200	208	216	224	232	240	248	256	264	272	279	287	295	303	311	319
6'4"	148	156	164	172	180	189	197	205	213	221	230	238	246	254	263	271	279	287	295	304	312	320	328
6'5"	151	160	168	176	185	193	202	210	218	227	235	244	252	261	269	277	286	294	303	311	319	328	336
6'6"	155	164	172	181	190	198	207	216	224	233	241	250	259	267	276	284	293	302	310	319	328	336	345

© Cengage Learning

Obesity-related diseases become evident beyond a BMI of 25. For this reason, a BMI of 25 for adults represents a healthy goal for overweight people and an upper limit for others. The lower end of the healthy range may be a reasonable target for severely underweight people. BMI values slightly below the healthy range may

> FIGURE 8-6 **BMI and Body Shapes**

Standard silhouette figures such as those shown below are commonly used in research studies (without the BMI numbers) to determine how accurately people perceive their body size.

20   21   22   24   26   28   32   35   42

18   19   21   23   26   30   34   39   45

Source: A. J. Stunkard, T. Sorensen, and F. Schulsinger, Use of the Danish Adoption Register for the study of obesity and thinness, Research Publications: Association for Research in Nervous and Mental Disorders 60 (1983): 115–120.

© Cengage Learning

> FIGURE 8-7 **Distribution of Body Weights in US Adults**

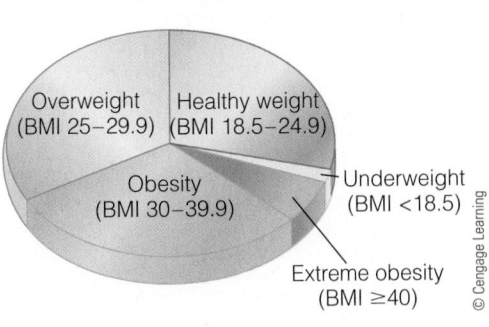

Overweight (BMI 25–29.9)
Healthy weight (BMI 18.5–24.9)
Obesity (BMI 30–39.9)
Underweight (BMI <18.5)
Extreme obesity (BMI ≥40)

© Cengage Learning

TABLE 8-7 Percent Body Fat at Various BMI

	BMI 18.5	BMI 25	BMI 30	BMI 35	BMI 40
Men	12–19%	23–28%	27–32%	31–35%	34–38%
Women	25–32%	35–40%	40–44%	43–47%	46–49%

NOTE: In general, women have roughly 12% more body fat than men at the same BMI.

SOURCE: Adapted from M. Heo and coauthors, Percentage of body fat cutoffs by sex, age, and race-ethnicity in the US adult population from NHANES 1999–2004, *American Journal of Clinical Nutrition* 95 (2012): 594–602.

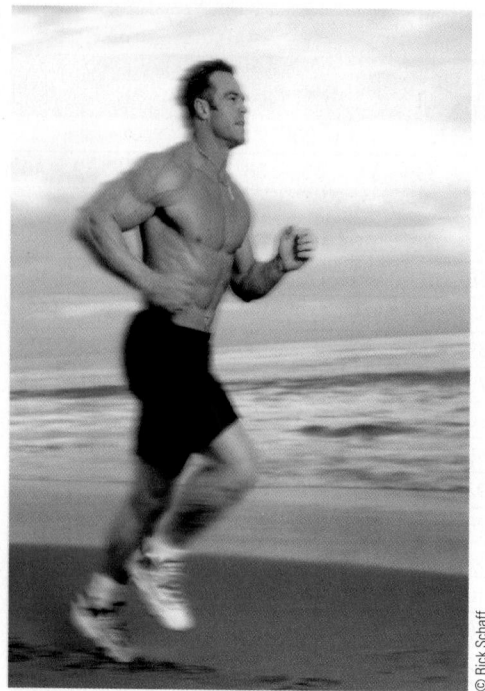

At 6 feet 4 inches tall and 250 pounds (1.93 meters and 113 kilograms), this runner would be considered over*weight* by most standards. Yet he is clearly not over*fat*.

be compatible with good health if food intake is adequate, but signs of illness, reduced work capacity, and poor reproductive function become apparent when BMI is below 17. The "How To" feature describes how to determine your BMI and how to find a goal weight based on a desired BMI.

Keep in mind that BMI reflects height and weight measures and not body composition. Consequently, muscular athletes may be classified as over*weight* by BMI standards and not be over*fat*. At the peak of his bodybuilding career, Arnold Schwarzenegger won the Mr. Olympia competition with a BMI of 31, the same BMI as the man running in the accompanying photo. Yet neither would be considered obese. Striking differences in body composition are also apparent among people of different ages and various ethnic and racial groups, making standard BMI guidelines inappropriate for some populations. For example, blacks tend to have a greater bone density and protein content than whites; consequently, using BMI as the standard may overestimate the prevalence of overweight and obesity among blacks.

Body Fat and Its Distribution Although weight measures are inexpensive, easy to take, and highly accurate, they fail to reveal two valuable pieces of information in assessing disease risk: how much of the weight is fat and where the fat is located.[11] The ideal amount of body fat depends partly on the person. Table 8-7 shows the percent body fat in the US population at various BMI and Table 8-8 compares percent body fat values of healthy weight, average fitness individuals with averages from national surveys.

Some People Need Less Body Fat For many athletes, a lower percentage of body fat may be ideal—just enough fat to provide fuel, insulate and protect the body, assist in nerve impulse transmissions, and support normal hormone activity, but not so much as to burden the body with excess bulk. Percent body fat for athletes, then, might be 7 to 16 percent for young men and 15 to 22 percent for young women. (Review the runner's photo to appreciate what 8 percent body fat looks like—even with a BMI greater than 30.)

Some People Need More Body Fat For an Alaska fisherman, a higher percentage of body fat is probably beneficial because fat provides an insulating blanket to prevent excessive loss of body heat in cold climates. A woman starting a

TABLE 8-8 Percent Body Fat: Ideal vs Actual

Age (yr)	Ideal (Healthy weight, average fitness)	Actual (US average)
Male		
20–39	18–21%	26%
40–59	22–25%	29%
60+	24–27%	31%
Female		
20–39	23–26%	38%
40–59	28–32%	41%
60+	31–34%	42%

SOURCE: L. G. Borrud and coauthors, Body composition data for individuals 8 years of age and older: US population, 1999–2004, *Vital and Health Statistics* 11 (2010): 1–87; *ACSM's Health-Related Physical Fitness Assessment Manual*, 2nd ed. (Baltimore, MD: Lippincott Williams & Wilkins, 2008), p. 59.

pregnancy needs sufficient body fat to support conception and fetal growth. Below a certain threshold for body fat, hormone synthesis falters, and individuals may become infertile, develop depression, experience abnormal hunger regulation, or become unable to keep warm. These thresholds differ for each function and for each individual; much remains to be learned about them.

Fat Distribution The location of fat on the body may influence health as much, or more than, total fat alone. **Visceral fat** that is stored around the organs of the abdomen is referred to as **central obesity** or upper-body fat (see Figure 8-8 on p. 250). Much research supports the widely held belief that central obesity—significantly and independently of BMI—contributes to heart disease, cancers, diabetes, and related deaths.[12]

Visceral fat is most common in men and to a lesser extent in women past menopause. Even when total body fat is similar, men have more visceral fat than women. **Subcutaneous fat** around the hips and thighs, sometimes referred to as lower-body fat, is most common in women during their reproductive years, and is associated with lower heart disease risks.[13] Figure 8-9 (p. 250) compares the body shapes of people with upper-body fat and lower-body fat.

Waist Circumference A person's **waist circumference** is a good indicator of central obesity and its associated health risks.[14] In general, women with a waist circumference of greater than 35 inches (88 centimeters) and men with a waist circumference of greater than 40 inches (102 centimeters) have a high risk of central obesity–related health problems. To simplify the message, waist circumference should be less than half of a person's height; the waist-to-height ratio is also a useful measure of disease risks.[15] As waist circumference increases, disease risks increase. Appendix E includes instructions for measuring waist circumference and assessing abdominal fat.

Some researchers use the waist-to-hip ratio as an indicator of disease risks. The ratio requires another step or two (measuring the hips and comparing that

visceral fat: fat stored within the abdominal cavity in association with the internal abdominal organs; also called *intra-abdominal fat.*

central obesity: excess fat around the trunk of the body; also called *abdominal fat* or *upper-body fat.*

subcutaneous fat: fat stored directly under the skin.
- **sub** = beneath
- **cutaneous** = skin

waist circumference: an anthropometric measurement used to assess a person's abdominal fat.

> FIGURE 8-8   **Central Obesity**

In healthy-weight people, some fat is stored around the organs of the abdomen.

In overweight people, excess abdominal fat increases the risks of diseases.

© Cengage Learning

> FIGURE 8-9   **"Apple" and "Pear" Body Shapes Compared**

Popular articles sometimes call bodies with upper-body fat "apples" and those with lower-body fat, "pears." Researchers sometimes refer to upper-body fat as "android" (manlike) obesity and to lower-body fat as "gynoid" (womanlike) obesity.

Upper-body fat is more common in men than in women and may be more closely associated with chronic diseases.

Lower-body fat is more common in women than in men and is not usually associated with chronic diseases.

© Cengage Learning

SOURCE: R.E.C. Wildman and D. M. Medeiros, *Advanced Human Nutrition* (Boca Raton, FL: CRC Press, 2000), pp. 321–323. Copyright © 2000 Taylor and Francis Books LLC. Reprinted with permission.

measurement to the waist measurement), but it does not provide any additional information. Therefore, waist circumference alone is the preferred method for assessing abdominal fat in a clinical setting.

Other Measures of Body Composition   Health-care professionals commonly use BMI and waist circumference measurements because they are relatively easy and inexpensive. Together, these two measurements prove most valuable in assessing a person's health risks and monitoring changes over time.[16] Researchers needing more precise measures of body composition may choose any of several other techniques to estimate body fat and its distribution (see Figure 8-10). Mastering these techniques requires proper instruction and practice to ensure reliability. In addition to the methods shown in Figure 8-10, researchers sometimes estimate body composition using these methods: total body water, radioactive potassium count, near-infrared spectrophotometry, ultrasound, computed tomography, and magnetic resonance imaging. Each method has advantages and disadvantages with respect to cost, technical difficulty, and precision of estimating body fat. Appendix E provides additional details and includes many of the tables and charts routinely used in assessment procedures.

> **REVIEW IT** Distinguish between body weight and body composition, including methods to assess each.

The body mass index (BMI) is based on weight relative to height and serves as a reliable indicator of chronic disease risks, but it says little about body composition. The ideal amount of body fat varies from person to person, but researchers have found that body fat in excess of 22 percent for young men and 27 percent for young women (the levels rise slightly with age) poses health risks. Central obesity is measured by waist circumference and indicates excess abdominal fat distributed around the trunk of the body. Central obesity contributes to chronic diseases.

> **FIGURE 8-10** **Common Methods Used to Assess Body Fat**

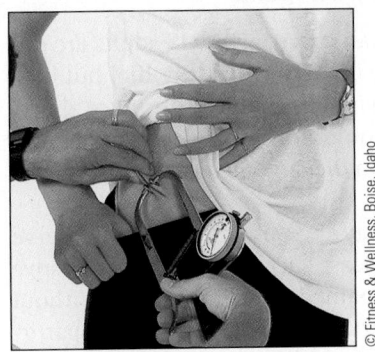

Skinfold measures estimate body fat by using a caliper to gauge the thickness of a fold of skin on the back of the arm (over the triceps), below the shoulder blade (subscapular), and in other places (including lower-body sites), and then comparing these measurements with standards.

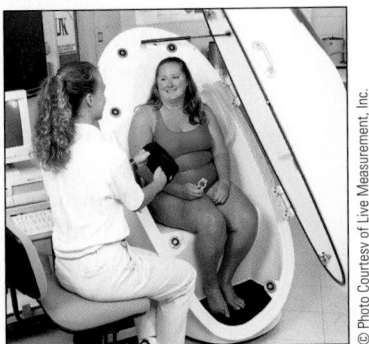

Air displacement plethysmography (commonly called the *bod pod*) estimates body composition by having a person sit inside a chamber while computerized sensors determine the amount of air displaced by the person's body.

Hydrodensitometry measures body density by weighing the person first on land and then again while submerged in water. The difference between the person's actual weight and underwater weight provides a measure of the body's volume. A mathematical equation using the two measurements (volume and actual weight) determines body density, from which the percentage of body fat can be estimated.

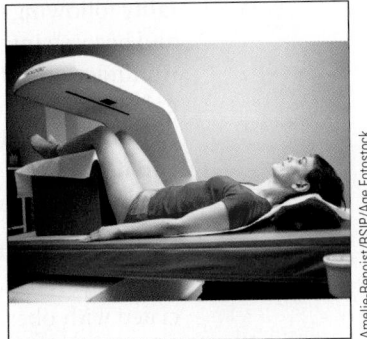

Dual energy X-ray absorptiometry (DEXA) uses two low-dose X-rays that differentiate among fat-free soft tissue (lean body mass), fat tissue, and bone tissue, providing a precise measurement of total fat and its distribution in all but extremely obese subjects.

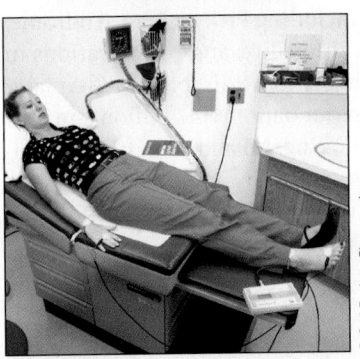

Bioelectrical impedance measures body fat by using a low-intensity electrical current. Because electrolyte-containing fluids, which readily conduct an electrical current, are found primarily in lean body tissues, the leaner the person, the less resistance to the current. The measurement of electrical resistance is then used in a mathematical equation to estimate the percentage of body fat.

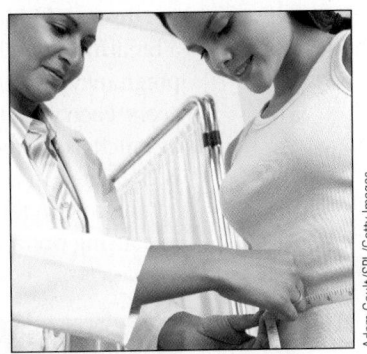

Waist circumference measures central obesity by placing a nonstretchable measuring tape around the waist just above the bony crest of the hip. The tape runs parallel to the floor and is snug, but does not compress the skin.

8.5 Health Risks Associated with Body Weight and Body Fat

> **LEARN IT** Identify relationships between body weight and chronic diseases.

Body weight and body fat correlate with disease risks and life expectancy. The correlation suggests a greater *likelihood* of developing chronic diseases and shortening life expectancy for those with a higher BMI and waist circumference.[17] Not all overweight and underweight people will get sick and die before their time nor will all normal-weight people live long healthy lives. *Correlations* are not *causes*. For the most part though, people with a BMI between 18.5 and 24.9 have relatively few weight-related health risks; risks increase as BMI falls below or rises above this range, indicating that both too little and too much body fat impair health. Epidemiological data show a J- or U-shaped relationship between body weights and mortality (see Figure 8-11).[18] People who are extremely underweight or extremely obese carry higher risks of early deaths than those whose weights fall within the healthy or even the slightly overweight range.[19] These mortality risks decline with age.

Independently of BMI, factors such as smoking habits raise health risks, and physical fitness lowers them. A man with a BMI of 22 who smokes two packs of

> **FIGURE 8-11** **BMI and Mortality**

This J-shaped curve describes the relationship between body mass index (BMI) and mortality and shows that both underweight and overweight present risks of a premature death.

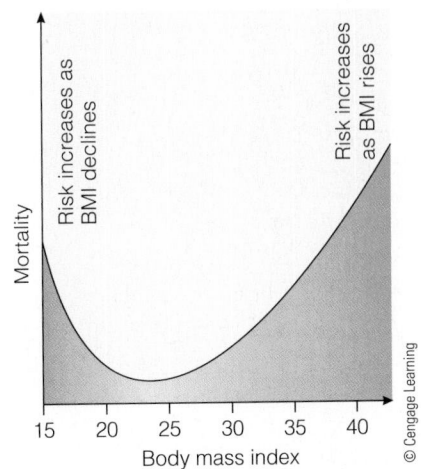

cigarettes a day is jeopardizing his health, whereas a woman with a BMI of 32 who walks briskly for an hour a day is improving her health.

Health Risks of Underweight   Fewer than 2 percent of US adults are underweight.[20] Some underweight people enjoy an active, healthy life, but others are underweight because of malnutrition, smoking habits, substance abuse, or illnesses. Weight and fat measures alone would not reveal these underlying causes, but a complete assessment that includes a diet and medical history, physical examination, and laboratory tests would.

An underweight person, especially an older adult, may be unable to preserve lean tissue during the fight against a wasting disease such as cancer or a digestive disorder, especially when the disease is accompanied by malnutrition. Without adequate nutrient and energy reserves, an underweight person will have a particularly tough battle against such medical stresses and face increased risks of mortality following surgeries.[21] Underweight women develop menstrual irregularities and become infertile. Those who do conceive may give birth to unhealthy infants. An underweight woman can improve her chances of having a healthy infant by gaining weight prior to conception, during pregnancy, or both. Underweight and significant weight loss are also associated with osteoporosis and bone fractures. For all these reasons, underweight people may benefit from enough of a weight gain to provide an energy reserve and protective amounts of all the nutrients.

Health Risks of Overweight   As for excessive body fat, the health risks are so many that it has been designated a disease—obesity. Among the health risks associated with obesity are diabetes, hypertension, cardiovascular disease, sleep apnea (abnormal ceasing of breathing during sleep), osteoarthritis, some cancers, gallbladder disease, kidney stones, respiratory problems (including Pickwickian syndrome, a breathing blockage linked with sudden death), infertility, and complications in pregnancy and surgery. Obese people are more likely to be disabled in their later years. Each year, these obesity-related illnesses cost our nation $147 billion—in fact, as much as, or more than, the medical costs of smoking. An additional $73 billion is estimated in a loss of productivity at work due to mortality and disability.[22]

The cost in terms of lives is also great. In fact, obesity is second only to tobacco in causing premature deaths.[23]

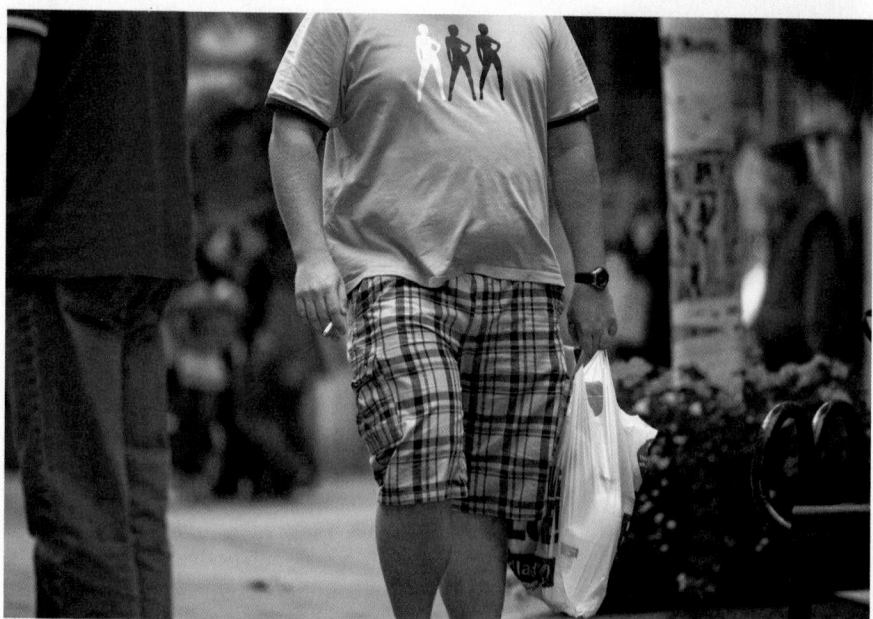

© Craig Stephen/Alamy

Smoking is the leading cause of preventable illnesses and early deaths. Obesity is a close second. A BMI of 40 or greater is equivalent to a lifetime of smoking, representing 10 years' loss of life.

Cardiovascular Disease The relationship between obesity and cardiovascular disease risk is strong, with links to both elevated blood cholesterol and hypertension.[24] Central obesity may raise the risk of heart attack and stroke as much as the three leading risk factors (high LDL cholesterol, hypertension, and smoking) do. In addition to body fat, weight gain also increases the risk of cardiovascular disease. Weight loss, on the other hand, can effectively reverse atherosclerosis and lower both blood cholesterol and blood pressure in overweight and obese people.[25] Of course, lean and normal-weight people may also have high blood cholesterol and blood pressure, and these factors are just as dangerous in lean people as in obese people. Obese people who do not have high blood cholesterol, high blood pressure, or other indicators of heart disease tend to have more a favorable fat distribution and may be described as "metabolically healthy" with lower risks for heart disease.[26]

Type 2 Diabetes The incidence of diabetes has risen dramatically in recent decades, as the nation's population has grown more overweight. Most adults with type 2 diabetes are overweight or obese.[27] Type 2 diabetes is three times more likely to develop in an obese person than in a nonobese person. Furthermore, the person with type 2 diabetes often has central obesity. Central-body fat cells appear to be larger and more insulin-resistant than lower-body fat cells. The association between **insulin resistance** and obesity is strong, and both are major risk factors for the development of type 2 diabetes.

Diabetes appears to be influenced by weight gains as well as by body weight. A weight gain of more than 10 pounds (4.5 kilograms) after the age of 18 doubles the risk of developing diabetes, even in adults of average weight. In contrast, weight loss is effective in improving glucose tolerance and insulin resistance.[28]

Inflammation and the Metabolic Syndrome Chronic **inflammation** accompanies obesity, and inflammation contributes to chronic diseases.[29] As a person grows fatter, lipids first fill the adipose tissue and then migrate into other tissues such as the muscles and liver. Fatty liver is a major contributor to the many diseases associated with obesity.[30] This accumulation of fat, especially in the abdominal region, changes the body's metabolism, resulting in insulin resistance (and high blood glucose), low HDL cholesterol, high triglycerides, and high blood pressure.[31] This cluster of symptoms—collectively known as the metabolic syndrome—increases the risks for diabetes, hypertension, and atherosclerosis. Fat accumulation, especially in the abdominal region, activates genes that code for proteins (adipokines) involved in inflammation.[32] Furthermore, although relatively few immune cells are commonly found in adipose tissue, weight gain significantly increases their number and their role in inflammation. Elevated blood lipids—whether due to obesity or to a high-fat diet—also promote inflammation. Together, these factors help to explain why chronic inflammation accompanies obesity and how obesity contributes to the metabolic syndrome and the progression of chronic diseases.[33] Even in healthy youngsters, body fat correlates positively with chronic inflammation. As might be expected, weight loss improves insulin resistance, reduces the number of immune cells in adipose tissue, and changes gene expression to reduce inflammation.

Cancer The risk of some cancers increases with both body weight and weight gain, but researchers do not fully understand the relationships.[34] One possible explanation may be that obese people have elevated levels of hormones that could influence cancer development. For example, adipose tissue is the major site of estrogen synthesis in women, obese women have elevated levels of estrogen, and estrogen has been implicated in the development of cancers of the female reproductive system—cancers that account for half of all cancers in women. Another possible explanation may be that the chronic inflammation that accompanies obesity is a risk factor for several cancers.[35]

Fit and Fat versus Sedentary and Slim
Importantly, BMI and weight gains and losses do not tell the whole story. Cardiorespiratory and muscular fitness play major roles in health and longevity, independently of body weight. Normal-weight people who are fit have a lower risk of mortality than normal-weight

insulin resistance: the condition in which a normal amount of insulin produces a subnormal effect in muscle, adipose, and liver cells, resulting in an elevated fasting glucose; a metabolic consequence of obesity that precedes type 2 diabetes.

inflammation: an immunological response to cellular injury characterized by an increase in white blood cells.

people who are unfit. Furthermore, overweight but fit people have lower risks than normal-weight, unfit ones.[36] Fit people are also likely to gain less weight over the years. Clearly, a healthy body weight is good, but it may not be good enough. Fitness, in and of itself, offers many health benefits, as Chapter 14 confirms.

> **REVIEW IT** Identify relationships between body weight and chronic diseases.
The weight appropriate for an individual depends largely on factors specific to that individual, including body fat distribution, family health history, and current health status. At the extremes, both overweight and underweight carry clear risks to health.

This chapter has described energy balance and body composition with a focus on the health problems associated with too much or too little body weight and body fat. Highlight 8 examines the health problems that arise when efforts to control body weight become eating disorders. The next chapter continues the discussion with a look at weight management and the benefits of choosing nutritious foods and exercising regularly.

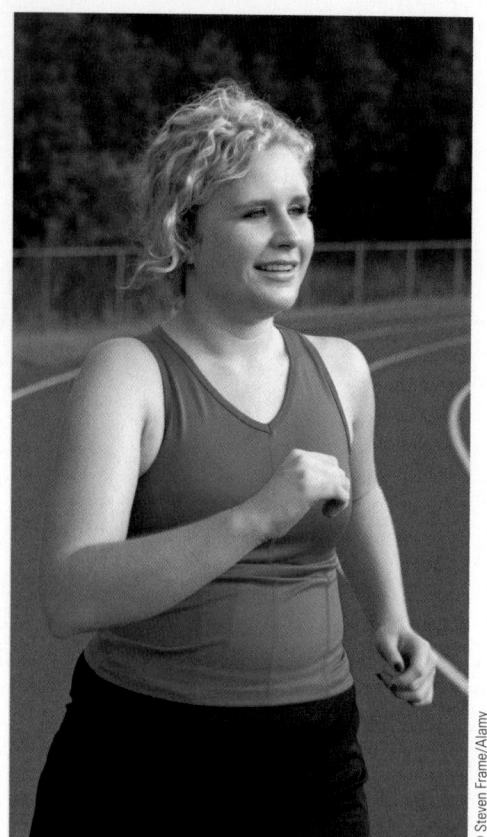

Being active—even if overweight—is healthier than being sedentary.

Nutrition Portfolio

When combined with fitness, a healthy body weight will help you to defend against chronic diseases. Go to Diet & Wellness Plus and choose one of the days on which you have tracked your diet for the entire day. Go to the Energy Balance report; use this report to help you answer the following questions:

- Describe how your daily food intake and physical activity balance with each other.
- What did the diet analysis program estimate as your daily energy requirement? What information was this based on?
- Describe any health risks that may be of concern for a person who continuously has inadequate or excessive energy intakes for many years.

DIET & WELLNESS PLUS To complete this exercise, go to your Diet & Wellness Plus at **www.cengagebrain.com**.

> **STUDY IT** To review the key points of this chapter and take a practice quiz, go to the study cards at the end of the book.

REFERENCES

1. M. B. Katan and D. S. Ludwig, Extra calories cause weight gain: But how much? *Journal of the American Medical Association* 303 (2010): 65–66.
2. S. B. Heymsfield and coauthors, Energy content of weight loss: Kinetic features during voluntary caloric restriction, *Metabolism* 61 (2012): 937–943; K. D. Hall and coauthors, Energy balance and its components: Implications for body weight regulation, *American Journal of Clinical Nutrition* 95 (2012): 989–994.
3. K. D. Hall, Modeling metabolic adaptations and energy regulation in humans, *Annual Review of Nutrition* 32 (2012): 35–54; J. C. K. Wells and M. Siervo, Obesity and energy balance: Is the tail wagging the dog? *European Journal of Clinical Nutrition* 65 (2011): 1173–1189.
4. E. R. Grimm and N. I. Steinle, Genetics of eating behavior: Established and emerging concepts, *Nutrition Reviews* 69 (2011): 52–60.
5. D. Ferriday and J. M. Brunstrom, "I just can't help myself": Effects of food-cue exposure in overweight and lean individuals, *International Journal of Obesity* 35 (2011): 142–149; L. B. Shomaker and coauthors, Eating in the absence of hunger in adolescents: Intake after a large-array meal compared with that after a standardized meal, *American Journal of Clinical Nutrition* 92 (2010): 697–703.
6. J. A. Gilbert and coauthors, Milk supplementation facilitates appetite control in obese women during weight loss: A randomized, single-blind, placebo-controlled trial, *British Journal of Nutrition* 105 (2011): 133–143.
7. E. Egecioglu and coauthors, Hedonic and incentive signals for body weight control, *Reviews in Endocrine and Metabolic Disorders* 12 (2011): 141–151.
8. J. A. Parker and S. R. Bloom, Hypothalamic neuropeptides and the regulation of appetite, *Neuropharmacology* 63 (2012): 18–30.
9. H. Schloegl and coauthors, Peptide hormones regulating appetite: Focus on neuroimaging studies in humans, *Diabetes/Metabolism Research and Reviews* 27 (2011): 104–112; K. A. Simpson and S. R. Bloom, Appetite and hedonism: Gut hormones and the brain, *Endocrinology and Metabolism Clinics of North America* 39 (2010): 729–743; S. Zac-Varghese, T. Tan, and S. R. Bloom, Hormonal interactions between gut and brain, *Discovery Medicine* 10 (2010): 543–552.

10. C. D. Fryar, M. D. Carroll, and C. L. Ogden, Prevalence of overweight, obesity, and extreme obesity among adults: United States, trends 1960–1962 through 2009–2010, *NCHS Health E-Stats*, September 2012.

11. A. G. Dulloo and coauthors, Body composition phenotypes in pathways to obesity and the metabolic syndrome, *International Journal of Obesity* 3 (2010): S4–S17.

12. K. A. Britton and coauthors, Body fat distribution, incident cardiovascular disease, cancer, and all-cause mortality, *Journal of the American College of Cardiology* 62 (2013): 921–925; A. M. Sironi and coauthors, Impact of increased visceral and cardiac fat on cardiometabolic risk and disease, *Diabetic Medicine* 29 (2012): 622–627; The InterAct Consortium, Long-term risk of incident type 2 diabetes and measures of overall and regional obesity: The EPIC-InterAct case-cohort study, *PLoS Medicine* 9 (2012): e1001230; T. Coutinho and coauthors, Central obesity and survival in subjects with coronary artery disease: A systematic review of the literature and collaborative analysis with individual subject data, *Journal of the American College of Cardiology* 57 (2011): 1877–1886; D. Sluik and coauthors, Associations between general and abdominal adiposity and mortality in individuals with diabetes mellitus, *American Journal of Epidemiology* 174 (2011): 22–34; B. J. Arsenault and coauthors, Physical inactivity, abdominal obesity and risk of coronary heart disease in apparently healthy men and women, *International Journal of Obesity* 34 (2010): 340–347; E. J. Jacobs and coauthors, Waist circumference and all-cause mortality in a large US cohort, *Archives of Internal Medicine* 170 (2010): 1293–1301.

13. K. Karastergiou and coauthors, Sex differences in human adipose tissues—the biology of pear shape, *Biology of Sex Differences* 3 (2013): 13.

14. P. T. Katzmarzyk, S. B. Heymsfield, and C. Bouchard, Clinical utility of visceral adipose tissue for the identification of cardiometabolic risk in white and African American adults, *American Journal of Clinical Nutrition* 97 (2013): 480–486.

15. M. Ashwell, P. Gunn, and S. Gibson, Waist-to-height ratio is a better screening tool than waist circumference and BMI for adult cardiometabolic risk factors: Systematic review and meta-analysis, *Obesity Reviews* 13 (2012): 275–286.

16. Position of the American Dietetic Association: Weight management, *Journal of the American Dietetic Association* 109 (2009): 330–346.

17. A. E. Staiano and coauthors, Body mass index versus waist circumference as predictors of mortality in Canadian adults, *International Journal of Obesity* 36 (2012): 1450–1454; N. Y. Krakauer and J. C. Krakauer, A new body shape index predicts mortality hazard independently of body mass index, *PLoS One* 7 (2012): e39504.

18. A. Berrington de Gonzalez and coauthors, Body-mass index and mortality among 1.46 million white adults, *New England Journal of Medicine* 363 (2010): 2211–2219.

19. K. M. Flegal and coauthors, Association of all-cause mortality with overweight and obesity using standard body mass index categories: A systematic review and meta-analysis, *Journal of the American Medical Association* 309 (2013): 71–82.

20. C. D. Fryar and C. L. Ogden, Prevalence of underweight among adults aged 20 years and over: United States, 1960–1962 through 2007–2010, *NCHS Health E-Stat*, September 2012.

21. F. E. Turrentine and coauthors, The relationship between body mass index and 30-day mortality risk, by principal surgical procedure, *Archives of Surgery* 147 (2012): 236–242; R. Gupta and coauthors, The effect of low body mass index on outcome in critically ill surgical patients, *Nutrition in Clinical Practice* 26 (2011): 593–597.

22. E. A. Finkelstein and coauthors, The costs of obesity in the workplace, *Journal of Occupational and Environmental Medicine* 52 (2010): 971–976; Society of Actuaries, Obesity and its relation to mortality and morbidity costs, www.soa.org/files/research/projects/research-2011-obesity-relation-mortality.pdf, December 2010.

23. H. Jia and E. I. Lubetkin, Trends in quality-adjusted life-years lost contributed by smoking and obesity, *American Journal of Preventive Medicine* 38 (2010): 138–144.

24. C. W. Mende, Obesity and hypertension: A common coexistence, *Journal of Clinical Hypertension* 14 (2012): 137–138.

25. C. DeCiuceis and coauthors, Effects of weight loss on structural and functional alterations of subcutaneous small arteries in obese patients, *Hypertension* 58 (2011): 29–36; I Shai and coauthors, Dietary intervention to reverse carotid atherosclerosis, *Circulation* 121 (2010): 1200–1208.

26. M. Hamer and E. Stamatakis, Metabolically healthy obesity and risk of all-cause and cardiovascular disease mortality, *Journal of Clinical Endocrinology and Metabolism* 97 (2012): 2482–2488.

27. M. L. Biggs and coauthors, Association between adiposity in midlife and older age and risk of diabetes in older adults, *Journal of the American Medical Association* 303 (2010): 2504–2512.

28. B. Kowall and coauthors, Impact of weight and weight change on normalization of prediabetes and on persistence of normal glucose tolerance in an older population: The KORA S4/F4 Study, *International Journal of Obesity* 36 (2012): 826–833.

29. F. P. deHeredia, S. C. Gómez-Martínezm, and A. Marcos, Chronic and degenerative diseases: Obesity, inflammation, and the immune system, *Proceedings of the Nutrition Society* 71 (2012): 332–338; N. Lumeng and A. R. Saltiel, Inflammatory links between obesity and metabolic disease, *Journal of Clinical Investigation* 121 (2011): 2111–2117; B. B. Aggarwal, Targeting inflammation-induced obesity and metabolic diseases by curcumin and other nutraceuticals, *Annual Review of Nutrition* 30 (2010): 173–199.

30. E. M. McCarthy and M. E. Rinella, The role of diet and nutrient composition in nonalcoholic fatty liver disease, *Journal of the Academy of Nutrition and Dietetics* 112 (2012): 401–409; K. L. Kopec and D. Burns, Nonalcoholic fatty liver disease: A review of the spectrum of disease, diagnosis, and therapy, *Nutrition in Clinical Practice* 26 (2011): 565–576; N. N. Kumashiro and coauthors, Cellular mechanism of insulin resistance in nonalcoholic fatty liver disease, *Proceedings of the National Academy of Sciences* 108 (2011): 16381–16385; J. C. Cohen, J. D. Horton, and H. H. Hobbs, Human fatty liver disease: Old questions and new insights, *Science* 332 (2011): 1519–1523.

31. E. J. Gallagher, D. Leroith, and E. Karnieli, Insulin resistance in obesity as the underlying cause for the metabolic syndrome, *Mt. Sinai Journal of Medicine* 77 (2010): 511–523; E. W. Demerath, Causes and consequences of human variation in visceral adiposity, *American Journal of Clinical Nutrition* 91 (2010): 1–2.

32. J. M. Northcott and coauthors, Adipokines and the cardiovascular system: Mechanisms mediating health and disease, *Canadian Journal of Physiology and Pharmacology* 90 (2012): 1029–1059; R. Stienstra and coauthors, The inflammasome puts obesity in the danger zone, *Cell Metabolism* 15 (2012): 10–18; E. Dalmas and coauthors, Variations in circulating inflammatory factors are related to changes in calorie and carbohydrate intakes early in the course of surgery-induced weight reduction, *American Journal of Clinical Nutrition* 94 (2011): 450–458.

33. S. Sun and coauthors, Mechanisms of inflammatory responses in obese adipose tissue, *Annual Review of Nutrition* 32 (2012): 261–286; R. Lorenzet and coauthors, Thrombosis and obesity: Cellular bases, *Thrombosis Research* 129 (2012): 285–289; A. Das and S. Mukhopadhyay, The evil axis of obesity, inflammation and type-2 diabetes, *Endocrine, Metabolic and Immune Disorders Drug Targets* 11 (2011): 23–31; A. L. Marsland and coauthors, Systemic inflammation and the metabolic syndrome among middle-aged community volunteers, *Metabolism: Clinical and Experimental* 59 (2010): 1801–1808.

34. Institute of Medicine, *The role of obesity in cancer survival and recurrence: Workshop summary*, Washington, DC: National Academies Press, 2012; C. Eheman and coauthors, Annual report to the nation on the status of cancer, 1975–2008, featuring cancers associated with excess weight and lack of sufficient physical activity, *Cancer* 118 (2012): 2338–2366; N. Parekh, U. Chandran, and E. V, Bandera, Obesity in cancer survival, *Annual Review of Nutrition* 32 (2012): 311–342.

35. S. Pendyala and coauthors, Diet-induced weight loss reduces colorectal inflammation: Implications for colorectal carcinogenesis, *American Journal of Clinical Nutrition* 93 (2011): 234–242.

36. D. E. Larson-Meyer and coauthors, Caloric restriction with or without exercise: The fitness versus fatness debate, *Medicine and Science in Sports and Exercise* 42 (2010): 152–159.

HIGHLIGHT > 8
Eating Disorders

> **LEARN IT** Compare the diagnoses, characteristics, and treatments of the different eating disorders.

For some people, the struggle with body weight manifests itself as an **eating disorder**. (The accompanying glossary defines this and related terms.) Three eating disorders—anorexia nervosa, bulimia nervosa, and binge eating disorder—are relatively uncommon, but present real concerns because of their health consequences.[1] Findings from large national surveys suggest that 0.9 percent of women and 0.3 percent of men suffer from anorexia nervosa at some time in their lives. Prevalence of bulimia nervosa is slightly higher, with 1.5 percent of women and 0.5 percent of men. Binge eating disorder is higher still, with 3.5 percent of women and 2 percent of men. Many more suffer from other unspecified eating disorders that do not meet the strict diagnostic criteria but still imperil a person's well-being.[2]

Why do so many people in our society suffer from eating disorders? Most experts agree that the causes include multiple factors: sociocultural, psychological, and perhaps neurochemical. Excessive pressure to be thin is at least partly to blame. Family attitudes concerning body shape and eating habits can have profound effects. Young people may have learned to identify discomforts such as anger, jealousy, or disappointment with "feeling fat." They often have other psychological issues such as depression, anxiety, or substance abuse. As weight issues become more of a focus, psychological problems worsen, and the likelihood of developing eating disorders intensifies. Unfortunately, few seek health care for eating disorders. Athletes and dancers are among those most likely to develop eating disorders.

© BananaStock/SuperStock

Disordered Eating in Athletes

At age 14, Suzanne was a top contender for a spot on the state gymnastics team. Each day her coach reminded team members that they must weigh no more than their assigned weights to qualify for competition. The coach chastised gymnasts who gained weight, and

Suzanne was terrified of being singled out. Convinced that the less she weighed the better she would perform, Suzanne weighed herself several times a day to confirm that she had not exceeded her 80-pound limit. Driven to excel in her sport, Suzanne kept her weight down by eating very little and training very hard. Unlike many of her friends, Suzanne never began to menstruate. A few months before her fifteenth birthday, Suzanne's coach dropped her back to the second-level team. Suzanne blamed her poor performance on a slow-healing stress fracture. Mentally stressed and physically exhausted, she quit gymnastics and began overeating between periods of self-starvation. Suzanne had developed the dangerous combination of problems originally known as the **female athlete triad,** which focused

GLOSSARY

amenorrhea (ay-MEN-oh-REE-ah): the absence of or cessation of menstruation. *Primary amenorrhea* is menarche delayed beyond 15 years of age. *Secondary amenorrhea* is the absence of three consecutive menstrual cycles.

anorexia (an-oh-RECK-see-ah) **nervosa:** an eating disorder characterized by a refusal to maintain a minimally normal body weight and a distortion in perception of body shape and weight.

- **an** = without
- **orex** = mouth
- **nervos** = of nervous origin

binge-eating disorder: an eating disorder characterized by recurring episodes of eating a significant amount of food in a short period of time with marked feelings of lack of control.

bulimia (byoo-LEEM-ee-ah) **nervosa:** an eating disorder characterized by repeated episodes of binge eating usually followed by self-induced vomiting, misuse of laxatives or diuretics, fasting, or excessive exercise.

cathartic (ka-THAR-tik): a strong laxative.

disordered eating: eating behaviors that are neither normal nor healthy, including restrained eating, fasting, binge eating, and purging.

eating disorder: any of several psychological disorders characterized by serious disturbances in eating behavior that jeopardize a person's physical or psychological health.

emetic (em-ETT-ic): an agent that causes vomiting.

female athlete triad: a potentially fatal combination of three medical problems—disordered eating, amenorrhea, and osteoporosis.

muscle dysmorphia (dis-MORE-fee-ah): a psychiatric disorder characterized by a preoccupation with building body mass.

relative energy deficiency in sport (RED-S): a syndrome of impaired physiological functions caused by relative energy deficiency (too little energy intake for the energy expended).

stress fractures: bone damage or breaks caused by stress on bone surfaces during exercise.

on disordered eating, amenorrhea, and osteoporosis.[3] Because the problems reach beyond these three components and male athletes are also affected, a more comprehensive term is now being used: **Relative Energy Deficiency in Sport (RED-S).**

Relative Energy Deficiency

Central to RED-S is an energy deficiency—the athlete's diet is providing too little energy given the amount of energy expended to support health, activities of daily living, growth, and sports. Sometimes energy deficiencies develop as the result of mismanaged athletic programs to quickly reduce body weight. Consider David, for example. Each week throughout the season, David drastically restricts his food and fluid intake before a wrestling match in an effort to "make weight." He believes that competing in a lower weight class will give him a competitive advantage over smaller opponents. To that end, David intensifies his exercise, skips meals, restricts fluids, practices in plastic suits, and trains in heated rooms to lose 4 to 7 pounds rapidly.[4] He hopes to replenish the lost fluids, glycogen, and lean tissue during the hours between weigh-in and competition, but the body needs days to correct this metabolic mayhem. Reestablishing fluid and electrolyte balances may take 1 to 2 days, replenishing glycogen stores may take 2 to 3 days, and replacing lean tissue may take even longer.

Ironically, the combination of food deprivation and dehydration impairs physical performance by reducing muscle strength, decreasing anaerobic power, and reducing endurance capacity. For optimal performance, athletes need to first achieve their competitive weight during the off-season and then eat well-balanced meals and drink plenty of fluids during the competitive season.

Energy deficiencies sometimes occur when athletes participate in unsupervised weight loss regimens or fail to eat enough during times of extreme exercise. Most often, however, **disordered eating** underlies energy deficiencies in athletes.

Disordered Eating

One reason many athletes engage in disordered eating is that they and their coaches have embraced unsuitable weight standards. An athlete's body must be heavier for a given height than a nonathlete's body because the athlete's body is dense, containing more healthy bone and muscle and less fat. When athletes rely only on the scales, they may mistakenly believe they are too fat because weight standards, such as the BMI, do not provide adequate information about body composition.

Many young athletes severely restrict energy intakes to improve performance, enhance appearance, or meet the weight guidelines of a specific sport. They fail to realize that the loss of lean tissue that accompanies energy restriction actually impairs their physical performance. Risk factors for eating disorders among athletes include:

- Young age (adolescence)
- Pressure to excel at a chosen sport
- Focus on achieving or maintaining an "ideal" body weight or body fat percentage
- Participation in sports or competitions that emphasize a lean appearance or judge performance on aesthetic appeal such as gymnastics, wrestling, figure skating, or dance
- Weight-loss dieting at an early age
- Unsupervised dieting

Disordered eating among athletes usually involves energy deficits and weight loss, but some athletes, usually males, go to extreme measures to bulk up and *gain* weight. People afflicted with **muscle dysmorphia** eat high-protein diets, take dietary supplements, weight train for hours at a time, and often abuse steroids in an attempt to increase muscle mass. Their bodies are large and muscular, yet they see themselves as puny 90-pound weaklings. They are preoccupied with the idea that their bodies are too small or inadequately muscular. Like others with distorted body images, people with muscle dysmorphia weigh themselves frequently and center their lives on diet and exercise. Paying attention to diet and pumping iron for fitness is admirable, but obsessing over it can cause serious social, occupational, and physical problems.

Adverse Consequences

A prolonged, inadequate energy intake has numerous adverse consequences, as outlined in Table H8-1. It leads to nutrient deficiencies (including anemia), chronic fatigue, and increased risk of infections and illnesses. Protein synthesis decreases and blood lipids increase, favoring heart disease. All of these consequences harm health and impair performance.

As mentioned earlier, females commonly develop **amenorrhea.** The prevalence of amenorrhea among premenopausal women in the United States is about 2 to 5 percent overall, but among female athletes, it may be as high as 65 to 70 percent. Body fat stores and hormone levels are too inadequate to support normal menstruation. Amenorrhea is often accompanied by bone mineral losses.

TABLE H8-1 Consequences of Relative Energy Deficiency in Sport (RED-S)

Physiological functions	Psychological problems	Physical performance
Altered hormone activities	Decreased concentration	Decreased coordination
Anemia	Depression	Decreased endurance
Bone loss	Impaired judgment	Decreased muscle strength
Decreased glycogen stores	Irritability	Decreased training response
Decreased protein synthesis		Increased injuries
Impaired metabolism		
Menstrual dysfunction		
Poor growth		

© Cengage Learning

SOURCE: Adapted from M. Mountjoy and coauthors, The IOC statement: Beyond the Female Athlete Triad—Relative Energy Deficiency in Sport (RED-S), *British Journal of Sports Medicine* 48 (2014): 491–497.

In general, weight-bearing physical activity, dietary calcium, and the hormone estrogen protect against the bone loss of osteoporosis, but in women with disordered eating and amenorrhea, strenuous activity can increase bone turnover, impair bone health, and increase the risks of **stress fractures.**[5] To grow strong bones, athletes should be encouraged to consume 1300 milligrams of calcium each day, to eat nutrient-dense foods, and to obtain enough energy to support both a healthy body weight and the energy expended in physical activity. Nutrition is critical to bone recovery.

Preventing Eating Disorders in Athletes

To prevent eating disorders in athletes and dancers, the performers, their coaches, and their parents must learn about inappropriate body weight ideals, improper weight-loss techniques, eating disorder development, proper nutrition, and safe weight-management strategies. Young people naturally search for identity and will often follow the advice of a person in authority without question. Therefore, coaches and dance instructors should never encourage unhealthy weight loss to qualify for competition or to conform to distorted artistic ideals. Athletes who need to lose weight for health's sake should try to do so during the off-season and under the supervision of a health-care professional.

Table H8-2 includes suggestions to help athletes and dancers protect themselves against developing eating disorders. The remaining sections describe eating disorders that anyone, athlete or nonathlete, may experience.

Anorexia Nervosa

Julie, 18 years old, is a superachiever in school. She watches her diet with great care, and she exercises daily, maintaining a rigorous schedule of self-discipline. She is thin, but she is determined to lose more weight.

TABLE H8-2 Tips for Combating Eating Disorders

General Guidelines

- Never restrict food amounts to below those suggested for adequacy by the USDA Food Patterns (see Table 2-3 on p. 43).
- Eat frequently. Include healthy snacks between meals. The person who eats frequently never gets so hungry as to allow hunger to dictate food choices.
- If not at a healthy weight, establish a reasonable weight goal based on a healthy body composition.
- Allow a reasonable time to achieve the goal. A reasonable loss of excess fat can be achieved at the rate of about 10 percent of body weight in 6 months.
- Establish a weight-maintenance support group with people who share interests.

Specific Guidelines for Athletes and Dancers

- Adopt realistic and health-promoting goals related to weight and body composition instead of weight restrictive guidelines.
- Disregard critical comments about weight and body composition.
- Recognize that eating disorders impair health and physical performance. Seek professional treatment if needed.
- Emphasize nutrition as an important key to optimal performance.

© Cengage Learning

She is 5 feet 6 inches tall and weighs 104 pounds (roughly 1.68 meters and 47 kilograms). Her BMI is less than 17. She has **anorexia nervosa.**

Characteristics of Anorexia Nervosa

Julie is unaware that she is undernourished, and she sees no need to obtain treatment. She developed amenorrhea several months ago and has become moody and chronically depressed. She views normal healthy body weight as too fat and insists that she needs to lose weight, although her eyes are sunk in deep hollows in her face. Julie denies that she is ever tired, although she is close to physical exhaustion and no longer sleeps easily. Her family is concerned, and though reluctant to push her, they have finally insisted that she see a psychiatrist. Julie's psychiatrist has diagnosed anorexia nervosa using specific criteria that describe such characteristics as a significantly low body weight caused by persistent restriction of energy intake; an intense fear of gaining weight or becoming fat, or persistent behaviors that interfere with weight gains; and a disturbance in self-perceived weight or shape.[6] She is prescribed group therapy as a start and if she does not begin to gain weight soon, she may need to enter a residential program or be hospitalized.

Central to the diagnosis of anorexia nervosa is a distorted body image that overestimates personal body fatness. When Julie looks at herself in the mirror, she sees a "fat" 104-pound body. The more Julie overestimates her body size, the more resistant she is to treatment, and the more unwilling she is to examine her faulty values and misconceptions. In fact, she finds value in her condition. Malnutrition and weight loss affect brain functioning and judgment in this way, causing lethargy, confusion, and delirium and influencing mood, anxiety, and emotions.

Anorexia nervosa cannot be self-diagnosed. Many people in our society are engaged in the pursuit of thinness, and denial runs high among people with anorexia nervosa. Some women have all the attitudes and behaviors associated with the condition, but without the dramatic weight loss.

How can a person as thin as Julie continue to starve herself? Julie uses tremendous discipline against her hunger to strictly limit her portions of low-fat, high-fiber, low-kcalorie foods. She will deny her hunger, and having adapted to eating so little food, she feels full after nibbling on a few carrot sticks. She knows the kcalorie intake of various foods and the kcalorie expenditure of different physical activities. If she feels that she has gained an ounce of weight, she runs or jumps rope until she is sure she has exercised it off. If she fears that the food she has eaten outweighs her physical activity, she may take laxatives to hasten the passage of food from her system. She drinks water incessantly to fill her stomach, risking dangerous mineral imbalances. She is desperately hungry. In fact, she is starving, but she doesn't eat because her need for self-control dominates.

Many people, on learning of this disorder, say they wish they had "a touch" of it to get thin. They mistakenly think that people with anorexia nervosa feel no hunger. They also fail to recognize the pain of the associated psychological and physical trauma.

The starvation of anorexia nervosa damages the body just as the starvation of war and poverty does. In fact, most people with anorexia enzymes and absorptive surfaces for handling any food that is eaten. The person may suffer from diarrhea, further worsening malnutrition.

Other effects of starvation include altered blood lipids, high blood vitamin A and vitamin E, low blood proteins, dry thin skin, abnormal nerve functioning, reduced bone density, low body temperature, low blood pressure, and the development of fine body hair (the body's attempt to keep warm). The electrical activity of the brain becomes abnormal, and insomnia is common. Both women and men lose their sex drives. In short, the metabolic mayhem of anorexia nervosa results in numerous physical complications, many of them life-threatening.[7]

Women with anorexia nervosa commonly develop amenorrhea. In young girls, the onset of menstruation is delayed. Menstrual periods typically resume with recovery, although some women never restart even after they have gained weight. Should an underweight woman with anorexia nervosa become pregnant, she is likely to give birth to an underweight baby—and low-birthweight babies face many health problems (as Chapter 15 explains). Mothers with anorexia nervosa may underfeed their children, who then fail to grow and may also suffer the other consequences of starvation.

Treatment of Anorexia Nervosa

Treatment of eating disorders requires a multidisciplinary approach.[8] Teams of physicians, nurses, psychiatrists, family therapists, and dietitians work together to resolve two sets of issues and behaviors: those relating to food and weight and those involving relationships with oneself and others.

The first dietary objective is to stop weight loss while establishing regular eating patterns. Appropriate diet is crucial to recovery and must be tailored to each individual's needs. Because body weight is low and fear of weight gain is high, initial food intake may be small— perhaps only 1200 kcalories per day. A variety of foods and foods with a higher energy density help to ensure greater success.[9] As eating becomes more comfortable, clients should gradually increase energy intake. Initially, clients may be unwilling to eat for themselves. Those who do eat will have a good chance of recovering without additional interventions. Even after recovery, however, energy intakes and eating behaviors may not fully return to normal. Furthermore, weight gains may be slow because energy needs may be slightly elevated due to anxiety, abdominal pain, and cigarette smoking.

Because anorexia nervosa is like starvation physically, health-care professionals classify clients based on indicators of malnutrition. Low-risk clients need dietary counseling. Intermediate-risk clients may need supplements such as high-kcalorie, high-protein formulas in addition to regular meals. High-risk clients may require hospitalization and may need to be fed by tube at first to prevent death. Residential programs that provide intensive behavioral treatment may be most appropriate for those who do not respond to less intensive approaches.

Denial runs high among those with anorexia nervosa. Few seek treatment on their own. About half of the women who are treated can maintain their body weight at 85 percent or more of a healthy weight, and at that weight, many of them may begin menstruating again. The other half have poor to fair treatment outcomes,

People with anorexia nervosa see themselves as fat, even when they are dangerously underweight.

Sniegirova Mariia/Shutterstock.com

nervosa are malnourished. Their bodies have been depleted of both body fat and protein. Victims are dying to be thin—quite literally. In young people, growth ceases and normal development falters. They lose so much lean tissue that their basal metabolic rate slows. In addition, the heart pumps inefficiently and irregularly, the heart muscle becomes weak and thin, the chambers diminish in size, and the blood pressure falls. Minerals that help to regulate heartbeat become unbalanced. Many deaths occur because of multiple organ system failure when the heart, kidneys, and liver cease to function.

Starvation brings other physical consequences as well, such as loss of brain tissue, impaired immune response, anemia, and a loss of digestive functions that worsen malnutrition. Peristalsis becomes sluggish, the stomach empties slowly, and the lining of the intestinal tract atrophies. The pancreas slows its production of digestive enzymes. The deteriorated GI tract fails to provide sufficient digestive relapse into abnormal eating behaviors, or die. Anorexia nervosa has one of the highest mortality rates among psychiatric disorders—most commonly from cardiac complications or by suicide.[10] Much like treatment for drug addictions, treatment for eating disorders engages family members. Therapists help family members to understand how their past interactions have enabled the client to continue destructive behaviors and how new ways of interacting can support change.

Before drawing conclusions about someone who is extremely thin or who eats very little, remember that diagnosis requires professional assessment. Several national organizations offer information for people who are seeking help with anorexia nervosa, either for themselves or for others.

Bulimia Nervosa

Kelly is a charming, intelligent, 30-year-old flight attendant of normal weight who thinks constantly about food. She alternates between starving herself and secretly bingeing, and when she has eaten too much, she makes herself vomit. Most readers recognize these symptoms as those of **bulimia nervosa.**

Characteristics of Bulimia Nervosa

Bulimia nervosa is distinct from anorexia nervosa and is more prevalent, although the true incidence is difficult to establish because bulimia nervosa is not as physically apparent. More men suffer from bulimia nervosa than from anorexia nervosa, but bulimia nervosa is still more common in women than in men. The secretive nature of bulimic behaviors makes recognition of the problem difficult, but once it is recognized, diagnosis is based on such criteria as number and frequency of binge eating episodes, inappropriate compensatory behaviors to prevent weight gain (such as self-induced vomiting or misuse of laxatives), and self-evaluation unduly influenced by body shape and weight.[11]

Like the typical person with bulimia nervosa, Kelly is single, female, and white. She is well educated and close to her ideal body weight, although her weight fluctuates over a range of 10 pounds or so every few weeks. She prefers to weigh less than the weight that her body maintains naturally.

Kelly seldom lets her eating disorder interfere with work or other activities, although a third of all bulimics do. From early childhood, she has been a high achiever and emotionally dependent on her parents. As a young teen, Kelly frequently followed severely restricted diets but could never maintain the weight loss. Kelly feels anxious at social events and cannot easily establish close personal relationships. She is usually depressed, is often impulsive, and has low self-esteem. When crisis hits, Kelly responds by replaying events, worrying excessively, and blaming herself but never asking for help—behaviors that interfere with effective coping.

Like the person with anorexia nervosa, the person with bulimia nervosa spends much time thinking about body image and food. The preoccupation with food manifests itself in secret binge-eating episodes, which usually progress through several emotional stages: anticipation and planning, anxiety, urgency to begin, rapid and uncontrollable consumption of food, relief and relaxation, disappointment, and finally shame or disgust.

A bulimic binge is characterized by a sense of no control over eating. During a binge, the person consumes food for its emotional comfort and cannot stop eating or control what or how much is eaten. A typical binge occurs periodically, in secret, usually at night, and lasts an hour or more. Because a binge frequently follows a period of restrictive dieting, eating is accelerated by intense hunger. Energy restriction followed by bingeing can set in motion a pattern of weight cycling, which may make weight loss and maintenance more difficult over time.

During a binge, Kelly consumes thousands of kcalories of easy-to-eat, low-fiber, high-fat, and, especially, high-carbohydrate foods. Typically, she chooses cookies, cakes, and ice cream—and she eats the entire bag of cookies, the whole cake, and every last spoonful in a carton of ice cream. After the binge, Kelly pays the price with swollen hands and feet, bloating, fatigue, headache, nausea, and pain.

To purge the food from her body, Kelly may use a **cathartic**—a strong laxative that can injure the lower intestinal tract. Or she may induce vomiting, with or without the use of an **emetic**—a drug intended as first aid for poisoning. These purging behaviors are often accompanied by feelings of shame or guilt. Hence a vicious cycle develops: negative self-perceptions followed by dieting, bingeing, and purging, which in turn lead to negative self-perceptions (see Figure H8-1).

On first glance, purging seems to offer a quick and easy solution to the problems of unwanted kcalories and body weight. Many people perceive such behavior as neutral or even positive, when, in fact, binge eating and purging have serious physical consequences. Signs of subclinical malnutrition are evident in a compromised immune system. Fluid and mineral imbalances caused by vomiting or diarrhea can lead to abnormal heart rhythms and injury to the kidneys. Urinary tract infections can lead to kidney failure. Vomiting causes irritation and infection of the pharynx, esophagus, and salivary glands; painful sores in the mouth; erosion of the teeth; and dental caries. The esophagus

> FIGURE H8-1   **The Vicious Cycle of Restrictive Dieting and Binge Eating**

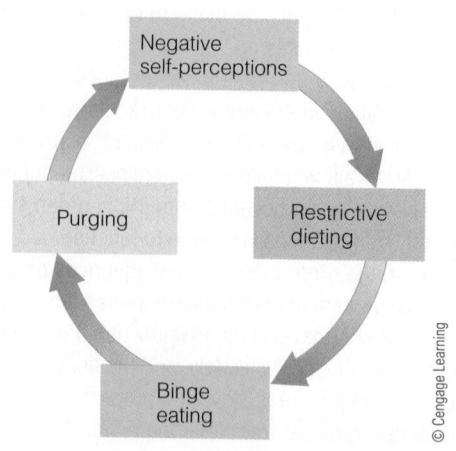

© Cengage Learning

may rupture or tear, as may the stomach. Sometimes the eyes become red from pressure during vomiting. The hands may be calloused or cut by the teeth while inducing vomiting. Overuse of emetics depletes potassium concentrations and can lead to death by heart failure.

Unlike Julie, Kelly is aware that her behavior is abnormal, and she is deeply ashamed of it. She wants to recover, and this makes recovery more likely for her than for Julie, who clings to denial. Feeling inadequate ("I can't even control my eating"), Kelly tends to be passive and to look to others for confirmation of her sense of worth. When she experiences rejection, either in reality or in her imagination, her bulimia nervosa becomes worse. If Kelly's depression deepens, she may seek solace in drug or alcohol abuse or in other addictive behaviors. Clinical depression is common in people with bulimia nervosa, and the rates of substance abuse are high.

Treatment of Bulimia Nervosa

Kelly needs to establish regular eating patterns. She may also benefit from a regular exercise program. Weight maintenance, rather than cyclic weight gains and losses, is the treatment goal. Major steps toward recovery include discontinuing purging and restrictive dieting habits and learning to eat three meals a day plus snacks. Initially, energy intake should provide enough food to satisfy hunger and maintain body weight. Table H8-3 offers diet strategies to correct the eating problems of bulimia nervosa. Most women diagnosed with bulimia nervosa recover within 5 to 10 years, with or without treatment, but treatment probably speeds the recovery process. Cognitive behavioral therapy may be more effective than other types of treatment.[12] A mental health professional should be on the treatment team to help clients with their depression and addictive behaviors.

Anorexia nervosa and bulimia nervosa are distinct eating disorders, yet they sometimes overlap in important ways. Anorexia victims may purge, and victims of both disorders are overly concerned with body image and have a tendency to drastically undereat. Many perceive foods as "forbidden" and "give in" to an eating binge. The two disorders can also appear in the same person, or one can lead to the other. Treatment is challenging and relapses are common. Another common eating disorder is **binge-eating disorder.**

Binge-Eating Disorder

Charlie is a 40-year-old schoolteacher who has been overweight all his life. His friends and family are forever encouraging him to lose weight, and he has come to believe that if he only had more willpower, dieting would work. He periodically gives dieting his best shot—restricting energy intake for a day or two only to succumb to uncontrollable cravings, especially for high-fat foods. Like Charlie, up to half of the obese people who try to lose weight periodically binge; unlike people with bulimia nervosa, however, they typically do not purge. Binge-eating disorder has its own specific diagnostic criteria based on recurring episodes of binge eating, with a marked sense of lack of control. It can occur in people of normal weight as well as those who are severely overweight. Obesity alone is not an eating disorder.

TABLE H8-3 **Diet Strategies for Combating Bulimia Nervosa**

Planning Principles

- Plan meals and snacks; record plans in a food diary prior to eating.
- Plan meals and snacks that require eating at the table and using utensils.
- Refrain from finger foods.
- Refrain from "dieting" or skipping meals.

Nutrition Principles

- Eat a well-balanced diet and regularly timed meals consisting of a variety of foods.
- Include raw vegetables, salad, or raw fruit at meals to prolong eating times.
- Choose whole-grain, high-fiber breads, pasta, rice, and cereals to increase bulk.
- Consume adequate fluid, particularly water.

Other Tips

- Choose foods that provide protein and fat for satiety and bulky, fiber-rich carbohydrates for immediate feelings of fullness.
- Try including soups and other water-rich foods for satiety.
- Choose portions that meet the definition of "a serving" according to the USDA Food Patterns (pp. 42–43).
- For convenience (and to reduce temptation) select foods that naturally divide into portions. Select one potato, rather than rice or pasta that can be overloaded onto the plate; purchase yogurt and cottage cheese in individual containers; look for small packages of precut steak or chicken; choose frozen dinners with measured portions.
- Include 30 minutes of physical activity every day—exercise may be an important tool in defeating bulimia.

© Cengage Learning

Clinicians note differences between people with bulimia nervosa and those with binge-eating disorder. People with binge-eating disorder typically consume less during a binge, rarely purge, and exert less restraint during times of dieting. Similarities also exist, including feeling out of control, disgusted, depressed, embarrassed, guilty, or distressed because of their self-perceived gluttony.

There are also differences between obese binge eaters and obese people who do not binge. Those with binge-eating disorder report higher rates of self-loathing, disgust about body size, depression, and anxiety. Their eating habits differ as well. Obese binge eaters tend to consume more kcalories and more dessert and snack-type foods during regular meals and binges than obese people who do not binge. Binge eating may incur health risks greater than those of obesity alone.[13]

Some of the characteristics seen in people with binge-eating disorder are similar to those seen in people with substance-use disorders: strong cravings, poor self-control, a diminished sensitivity to pleasure, and patterns of compulsive use.[14] These resemblances have given rise to the concept of *food addictions* and may reflect the same biological and psychological systems that are involved in rewards and self-control.[15] When the reward of delicious foods tempts a person, the ability to resist depends on self-control.[16] Neural images show that certain foods (especially those with added sugars and solid fats) have effects on the brain similar to those seen with addictive drugs. Dopamine activity—which helps to regulate emotional and motivational behavior—is also similarly altered in both drug addicts and those with compulsive eating behaviors.[17]

Binge eating can be resolved with treatment. Reducing binge eating makes participation in weight-control programs easier. It also improves physical health, mental health, and the chances of success in breaking the cycle of rapid weight losses and gains.

Eating Disorders in Society

Society plays a central role in eating disorders. Consider that the average US woman is 5 feet 4 inches tall and weighs 140 pounds, whereas the average US model is 7 inches taller and weighs 23 pounds less. Adolescent girls and women of all ages who obsess over weight loss and envy beautiful models are likely to engage in unhealthy eating habits. Further proof of society's influence is found in the demographic distribution of eating disorders—they are known only in developed nations, and they become more prevalent as wealth increases and food becomes plentiful. Some people point to the vomitoriums of ancient times and claim that bulimia nervosa is not new, but the two are actually distinct. Ancient people were eating for pleasure, without guilt, and in the company of others; they vomited so that they could rejoin the feast. Bulimia nervosa is a disorder of isolation and is often accompanied by low self-esteem.

Chapter 8 describes how our society sets unrealistic ideals for body weight, especially in women, and devalues those who do not conform to them. Anorexia nervosa and bulimia nervosa are not a form of rebellion against these unreasonable expectations, but rather an exaggerated acceptance of them. In fact, some people fail to recognize the health dangers and endorse eating disorders as a lifestyle choice. Some 200 websites encourage, support, and motivate users to continue their lives with anorexia and bulimia.[18]

The incidence and prevalence of eating disorders in young people has increased steadily since the 1950s.[19] Most alarming is the rising prevalence at progressively younger ages. Restrained eating, fasting, binge eating, purging, fear of fatness, and distortion of body image are extraordinarily common among children and adolescents. Most are "on diets," and many are poorly nourished. Some eat too little food to support normal growth, thus they miss out on their adolescent growth spurts and may never catch up. Many eat so little that hunger propels them into binge-purge cycles. Disordered eating behaviors set a pattern that likely continues into young adulthood.[20]

Perhaps a person's best defense against these disorders is to learn to appreciate his or her own uniqueness. When people discover and honor their body's real physical needs, they become unwilling to sacrifice health for conformity. To respect and value oneself may be lifesaving.

CRITICAL THINKING QUESTIONS

A. How do eating disorders affect health?
B. You overheard someone saying that eating disorders aren't really diseases and that if people who have anorexia or bulimia would just eat normally, they'd be cured. What is your opinion of this position? How can you make a distinction between someone with an eating disorder and others who are concerned with managing their body weight?

REFERENCES

1. F. R. E. Smink, D. vanHocken, and H. W. Hock, Epidemiology of eating disorders: Incidence, prevalence and mortality rates, *Current Psychiatry Reports* 14 (2012): 406–414.
2. R. D. Grave, Eating disorders: Progress and challenges, *European Journal of Internal Medicine* 22 (2011): 153–160.
3. J. C. Gibbs, N. I. Williams, and M. J. deSouza, Prevalence of individual and combined components of the female athlete triad, *Medicine and Science in Sports and Exercise* 45 (2013): 985–996; T. G. Nazem and K. E. Ackerman, The female athlete triad, *Sports Health* 4 (2012): 302–311.
4. G. G. Artioli and coauthors, Prevalence, magnitude, and methods of rapid weight loss among judo competitors, *Medicine and Science in Sports and Exercise* 42 (2010): 436–442.
5. C. A. Hincapié and J. D. Cassidy, Disordered eating, menstrual disturbances, and low bone mineral density in dancers: A systematic review, *Archives of Physical Medicine and Rehabilitation* 91 (2010): 1777–1789; E. Waugh and coauthors, Effects of exercise on bone mass in young women with anorexia nervosa, *Medicine and Science in Sports and Exercise* 43 (2011): 755–763; M. T. Barrack and coauthors, Physiologic and behavioral indicators of energy deficiency in female adolescent runners with elevated bone turnover, *American Journal of Clinical Nutrition* 92 (2010): 652–659.
6. American Psychiatric Association, *Diagnostic and Statistical Manual of Mental Disorders,* (Washington, D.C.: American Psychiatric Publishing, 2013).
7. A. P. Winston, The clinical biochemistry of anorexia nervosa, *Annals of Clinical Biochemistry* 49 (2012): 132–143.
8. Position of the American Dietetic Association: Nutrition intervention in the treatment of eating disorders, *Journal of the American Dietetic Association* 111 (2011): 1236–1241.
9. J. E. Schebendach and coauthors, Food choice and diet variety in weight-restored patients with anorexia nervosa, *Journal of the American Dietetic Association* 111 (2011): 732–736.
10. M. J. Krantz and coauthors, Factors influencing QT prolongation in patients hospitalized with severe anorexia nervosa, *General Hospital Psychiatry* 34 (2012): 173–177.
11. American Psychiatric Association, *Diagnostic and Statistical Manual of Mental Disorders* (Washington, D.C.: American Psychiatric Publishing, 2013).
12. ECRI Institute, *Bulimia Nervosa: Comparative Efficacy of Available Psychological and Pharmacological Treatments,* as cited in M. Mitka, Reports weighs options for bulimia nervosa treatment, *Journal of the American Medical Association* 305 (2011): 875.
13. J. I. Hudson and coauthors, Longitudinal study of the diagnosis of components of the metabolic syndrome in individuals with binge-eating disorder, *American Journal of Clinical Nutrition* 91 (2010): 1568–1573.
14. American Psychiatric Association, *Diagnostic and Statistical Manual of Mental Disorders* (Washington, D.C.: American Psychiatric Publishing), 2013.

15. D. G. Smith and T. W. Robbins, The neurobiological underpinnings of obesity and binge eating: A rationale for adopting the food addiction model, *Biological Psychiatry* 73 (2013): 804–810; N. D. Volkow and coauthors, Obesity and addiction: Neurobiological overlaps, *Obesity Reviews* 14 (2013): 2–18; J. L. Fortuna, The obesity epidemic and food addiction: Clinical similarities to drug dependence, *Journal of Psychoactive Drugs* 44 (2012): 56–63; C. Moreno and R. Randon, Should overeating and obesity be classified as an addictive disorder in DSM-5? *Current Pharmaceutical Design* 17 (2011): 1128–1131.

16. N. D. Volkow and coauthors, The addictive dimensionality of obesity, *Biological Psychiatry* 73 (2013): 811–818.

17. J. H. Baik, Dopamine signaling in food addiction: Role of Dopamine D2 receptors, *BMB Reports* (2013): pii: 2509.

18. D. L. G. Borzekowski and coauthors, e-Ana and e-Mia: A content analysis of pro-eating disorder web sites, *American Journal of Public Health* 100 (2010): 1526–1534.

19. D. S. Rosen and the Committee on Adolescence, Clinical report: Identification and management of eating disorders in children and adolescents, *Pediatrics* 126 (2010): 1240–1253.

20. D. Neumark-Sztainer and coauthors, Dieting and disordered eating behaviors from adolescence to young adulthood: Findings from a 10-year longitudinal study, *Journal of the American Dietetic Association* 111 (2011): 1004–1011.

9

Weight Management: Overweight, Obesity, and Underweight

Nutrition in Your Life

Are you pleased with your body weight? If so, you are a rare individual. Most people in our society think they should weigh more or less (mostly less) than they do. Usually, their primary concern is appearance, but they often understand that physical health is also somehow related to body weight. One does not necessarily cause the other—that is, an ideal body weight does not ensure good health. Instead, both depend on diet and physical activity. A well-balanced diet and active lifestyle support good health—and help maintain body weight within a reasonable range. In the Nutrition Portfolio at the end of this chapter, you can consider whether your eating habits and physical activities are supporting good health and a reasonable body weight.

The previous chapter described how body weight is stable when energy in equals energy out. Weight gains occur when energy intake exceeds energy expended, and conversely, weight losses occur when energy expended exceeds energy intake. At the extremes, both overweight and underweight present health risks. **Weight management** is a key component of good health. To that end, this chapter offers strategies to help achieve and maintain a healthy body weight. It also explores overweight and obesity by examining some of the causes, consequences, and treatments.

This chapter emphasizes overweight (BMI 25 to 29.9) and obesity (BMI ≥30), partly because they have been more intensively studied and partly because they represent a major health problem in the United States and a growing concern worldwide. Underweight (BMI <18.5) is a far less prevalent problem. Information on underweight is presented at the end of the chapter. The highlight that follows this chapter examines fad diets.

weight management: maintaining body weight in a healthy range by preventing gradual weight gains over time and losing weight if overweight, and by preventing weight losses and gaining weight if underweight.

> FIGURE 9-1 **Increasing Prevalence of Obesity (BMI ≥30) among US Adults**

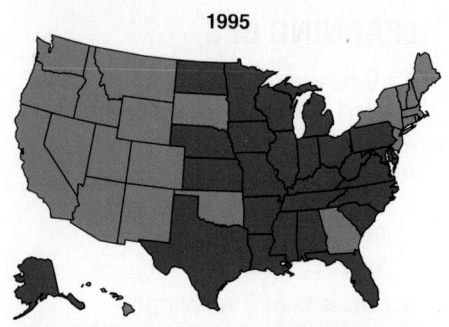

1995

10%–14%	15%–19%	20%–24%	25%–29%	≥30%
23 states	27 states	0 states	0 states	0 states

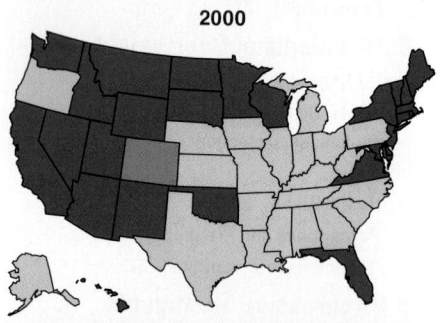

2000

10%–14%	15%–19%	20%–24%	25%–29%	≥30%
1 state	27 states	22 states	0 states	0 states

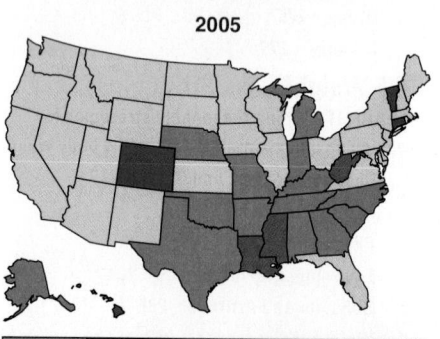

2005

10%–14%	15%–19%	20%–24%	25%–29%	≥30%
0 states	4 states	29 states	14 states	3 states

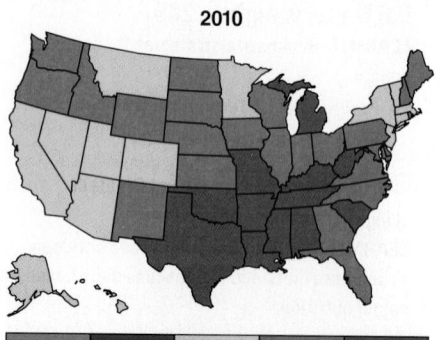

2010

10%–14%	15%–19%	20%–24%	25%–29%	≥30%
0 states	0 states	14 states	24 states	12 states

SOURCE: www.cdc.gov/obesity/data/adult.html

9.1 Overweight and Obesity

> **LEARN IT** Describe how body fat develops and why it can be difficult to maintain weight gains and losses.

Despite our preoccupation with body image and weight loss, the prevalence of overweight and obesity in the United States continues to be high.[1] In the past four decades, obesity increased in every state, in both genders, and across all ages, races, and educational levels (see Figure 9-1).* An estimated 69 percent of the adults in the United States are now considered overweight or obese, as defined by a BMI of 25 to 29.9, or 30 and greater, respectively.[2] The prevalence of overweight is especially high among women, the poor, blacks, and Mexican Americans.

The prevalence of overweight among children in the United States has also risen at an alarming rate. An estimated 32 percent of children and adolescents aged 2 to 19 years are either overweight or obese.[3] Chapter and Highlight 16 present information on overweight during childhood and adolescence.

Obesity in the United States is widespread. Prevalence increased rapidly over the past four decades, but seems to have leveled out in recent years.[4] This **epidemic** of obesity has spread worldwide, affecting 1.4 billion adults and 40 million children younger than age 5.[5] Increasing rates of obesity in countries around the world reflects a global food system that delivers an abundance of energy-dense, processed, affordable, and effectively marketed products.[6] Before examining the suspected causes of obesity and the various strategies used to treat it, it is helpful to understand the development and metabolism of body fat.

Fat Cell Development When "energy in" exceeds "energy out," much of the excess energy is stored in the fat cells of adipose tissue. The amount of fat in adipose tissue reflects both the number and the size of the fat cells.** The number of fat cells increases most rapidly during the growing years of late childhood and early puberty. After growth ceases, fat cell numbers may continue to increase whenever energy balance is positive.[7] Obese people have more fat cells than healthy-weight people; their fat cells are also larger.

As fat cells accumulate triglycerides, they expand in size (review Figure 5-18 on p. 148). When the cells enlarge, they stimulate cell proliferation so that their numbers increase again. Thus obesity develops when a person's fat cells increase in number, in size, or quite often both. Figure 9-2 illustrates fat cell development.

When "energy out" exceeds "energy in," the size of fat cells dwindles, but not their number. People with extra fat cells tend to regain lost weight rapidly; with weight gain, their many fat cells readily fill. In contrast, people with an average number of enlarged fat cells may be more successful in maintaining weight losses; when their cells shrink, both cell size and number are normal. Prevention of obesity is most critical, then, during the growing years of childhood and adolescence when fat cells increase in number. Researchers are exploring ways to induce fat cell death—which would decrease the number.[8]***

As mentioned, excess fat first fills the body's natural storage site—adipose tissue. If fat is still abundant, the excess is deposited in organs such as the heart and liver and plays a key role in the development of diseases such as heart failure and fatty liver, respectively.[9]**** As adipose tissue produces adipokines, metabolic changes that indicate disease risk—such as insulin resistance—become apparent and chronic inflammation develops.[10] The adipokine profile begins to improve with as little as a 5 percent weight loss and a decrease in fat cell size, suggesting that other metabolic changes might also occur at that time to improve disease risks.

---

*Figure 9.1 maps offer a glimpse at the dramatic changes in obesity prevalence over the past several decades. Because of changes in methodology, however, estimates of obesity prevalence from 2011 forward cannot be compared to estimates from previous years.
**Obesity due to an increase in the *number* of fat cells is *hyperplastic obesity*. Obesity due to an increase in the *size* of fat cells is *hypertrophic obesity*.
***Cell death is known as *apoptosis*.
****The adverse effect of fat in nonadipose tissue is known as *lipotoxicity*.

> **FIGURE 9-2**   **Fat Cell Development**

Fat cells are capable of increasing their size by 20-fold and their number by several thousandfold.

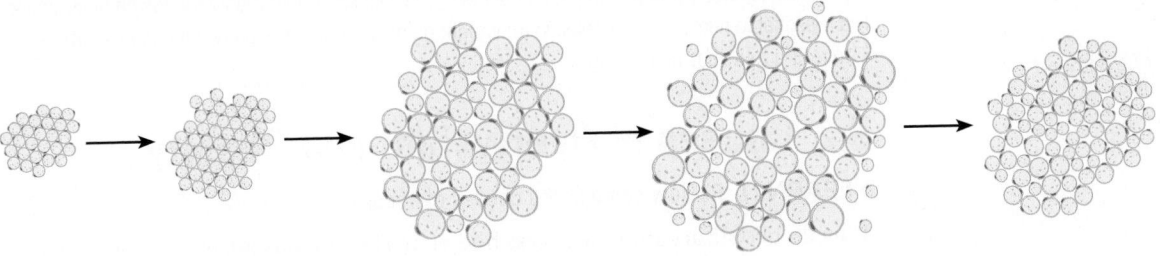

During growth, fat cells increase in number.

When energy intake exceeds expenditure, fat cells increase in size.

When fat cells have enlarged and energy intake continues to exceed energy expenditure, fat cells may increase in number again.

With fat loss, the size of the fat cells shrinks but not the number.

© Cengage Learning

Fat Cell Metabolism   The enzyme **lipoprotein lipase (LPL)** plays a major role in the metabolism and transport of lipids, and consequently is a participant in the development of obesity.[11] One of its roles is to remove triglycerides from the blood for storage in both adipose tissue and muscle cells. Obese people generally have much more LPL activity in their adipose cells than lean people do (their muscle cell LPL activity is similar, though). This high LPL activity makes fat storage especially efficient. Consequently, even modest excesses in energy intake have a more dramatic impact on obese people than on lean people. When obese people eat less in an effort to lose weight, their LPL activity diminishes.

The activity of LPL in different regions of the body is partially influenced by gender. In women, fat cells in the breasts, hips, and thighs produce abundant LPL, storing fat in those body sites; in men, fat cells in the abdomen produce abundant LPL. This enzyme activity explains why men tend to develop central obesity around the abdomen (apple-shaped) whereas women more readily develop lower-body fat around the hips and thighs (pear-shaped).

Gender differences are also apparent in the activity of the lipase enzymes controlling the release and breakdown of fat in various parts of the body. The release of lower-body fat is less active in women than in men, whereas the release of upper-body fat is similar. Furthermore, the rate of fat breakdown is lower in women than in men. Consequently, women may have a more difficult time losing fat in general, and from the hips and thighs in particular.

Enzyme activity may also explain why some people who lose weight regain it so easily. After weight loss and weight stabilization, adipose tissue LPL is increased and its response to meals is heightened. It's as if the LPL enzyme gene is saying "Make more fat-storing enzymes." People easily regain weight after having lost it because they are battling against enzymes that want to store fat. Fat storage is efficient, and fat oxidation is not. Dietary fat oxidation correlates negatively with body fatness: obese people have the least activity. The activities of these and other proteins provide an explanation for the observation that some biological mechanism seems to set a person's body weight or composition at a fixed point; the body will make adjustments to restore that **set point** if the person tries to change it.

Set-Point Theory   Many physiological variables, such as blood glucose, blood pH, and body temperature, remain fairly stable under a variety of conditions. The hypothalamus and other regulatory centers constantly monitor and delicately adjust conditions to maintain homeostasis. The stability of such complex systems may depend on set-point regulators that maintain variables within specified limits.

Researchers have confirmed that after weight losses, the body adjusts its metabolism. The decrease in the metabolic rate after weight loss is greater than would be expected based on body composition alone.[12] This adaptation helps to explain why it can be difficult for an overweight person to maintain weight losses. While set point answers some questions regarding the biology of energy balance, it fails to explain the many other influences contributing to the population's obesity epidemic.[13]

epidemic (ep-ih-DEM-ick): the appearance of a disease (usually infectious) or condition that attacks many people at the same time in the same region.

● **epi** = upon

● **demos** = people

lipoprotein lipase (LPL): an enzyme that hydrolyzes triglycerides passing by in the bloodstream and directs their parts into the cells, where they can be metabolized or reassembled for storage.

set point: the point at which controls are set (for example, on a thermostat). The set-point theory that relates to body weight proposes that the body tends to maintain a certain weight by means of its own internal controls.

> **REVIEW IT** Describe how body fat develops and why it can be difficult to maintain weight gains and losses.

Fat cells develop by increasing in number and size. Obesity prevention depends on maintaining a reasonable number of fat cells. With weight gains or losses, the body adjusts in an attempt to return to its set-point weight.

9.2 Causes of Overweight and Obesity

> **LEARN IT** Review some of the causes of obesity.

Why do people accumulate excess body fat? The obvious answer is that they take in more energy from foods and beverages than they expend in physical activity and metabolic processes. But that answer falls short of explaining why they do this. Is it genetic? Environmental? Cultural? Behavioral? Socioeconomic? Psychological? Metabolic? All of these? Most likely the latter. Many factors contribute to the development of obesity and most are interrelated. This section reviews the two major contributing and interacting factors—genetics and the environment.[14]

Genetics and Epigenetics Genetics plays a true causative role in relatively few cases of obesity, for example, in Prader-Willi syndrome—a genetic disorder characterized by excessive appetite, massive obesity, short stature, and often mental retardation. Most cases of obesity, however, do not stem from a single gene, yet multiple genetic influences do seem to be involved. Highlight 6 describes epigenetics—the influence of environmental factors, such as diet and physical activity, on gene expression. Obesity provides a classic example of epigenetic regulation.[15]

Researchers have found that adopted children tend to be more similar in weight to their biological parents than to their adoptive parents.[16] Studies of twins yield similar findings: compared with fraternal twins, identical twins are far more likely to weigh the same.[17] These findings suggest an important role for genetics in determining a person's *susceptibility* to obesity.[18] In other words, genes interact with the diet and activity patterns that lead to obesity and the metabolic pathways that influence satiety and energy balance. Even identical twins with identical genes become different over the years as epigenetic changes accumulate. This raises an important point: you cannot change the genome you inherit, but you can influence the epigenome. Physical activity, for example, can minimize the genetic influences on BMI.[19] Likewise, high-fat diets, sugar-sweetened beverages, and low physical activity can accentuate the genetic influences on obesity.[20]

Clearly, something genetic makes a person more or less likely to gain or lose weight when overeating or undereating. Some people gain more weight than others on comparable energy intakes. Given an extra 1000 kcalories a day for 100 days, some pairs of identical twins gain less than 10 pounds while others gain up to 30 pounds. Within each pair, the amounts of weight gained, percentages of body fat, and locations of fat deposits are similar. Similarly, some people lose more weight than others following comparable exercise routines.

Researchers have been examining the human genome in search of genetic and epigenetic answers to obesity questions. As the section on protein synthesis in Chapter 6 describes, each cell expresses only the genes for the proteins it needs, and each protein performs a unique function. The following paragraphs describe only a couple of the proteins that help explain appetite control, energy regulation, and obesity development.

Leptin Researchers have identified an obesity gene, called *ob*, that is expressed primarily in the adipose tissue and codes for the protein **leptin**. Leptin acts as a hormone, primarily in the hypothalamus. Leptin maintains homeostasis by regulating food intake and energy expenditure in response to adipose tissue. When body fat increases, leptin increases—which suppresses appetite. When body fat decreases, leptin decreases—which stimulates appetite and suppresses energy expenditure.

Mice with a defective *ob* gene do not produce leptin and can weigh up to three times as much as normal mice and have five times as much body fat (see Figure 9-3).

leptin: a protein produced by fat cells under direction of the *ob* gene that decreases appetite and increases energy expenditure.

• **leptos** = thin

When injected with a synthetic form of leptin, the mice rapidly lose body fat. (Because leptin is a protein, it would be destroyed during digestion if given orally; consequently, it must be given by injection.) The fat cells not only lose fat, but they self-destruct (reducing cell number), which may explain why weight gains are delayed when the mice are fed again.

Although extremely rare, a genetic deficiency of leptin or genetic mutation of its receptor has been identified in human beings as well. Extremely obese children with barely detectable blood levels of leptin have little appetite control; they are constantly hungry and eat considerably more than their siblings or peers. Given daily injections of leptin, these children lose a substantial amount of weight, confirming leptin's role in regulating appetite and body weight.

Very few obese people have a leptin deficiency, however. In fact, leptin levels increase as BMI increases. Leptin rises but fails to suppress appetite or enhance energy expenditure—a condition researchers describe as leptin resistance.[21] With weight loss, leptin levels decline, which reduces satiation and challenges weight loss maintenance; leptin injections effectively increase satiation after weight loss.[22]

Ghrelin Another protein, known as **ghrelin**, also acts as a hormone primarily in the hypothalamus. In contrast to leptin, ghrelin is secreted primarily by the stomach cells and promotes eating and weight gain by increasing smell sensitivity, stimulating appetite, and promoting efficient energy storage.[23]

Ghrelin triggers the desire to eat. Blood levels of ghrelin typically rise before and fall after a meal—reflecting the hunger and satiety that precede and follow eating. On average, ghrelin levels are high whenever the body is in negative energy balance, as occurs during low-kcalorie diets, for example. This response may help explain why weight loss is so difficult to maintain. Weight loss is more successful with exercise and after gastric bypass surgery, in part because ghrelin levels are relatively low. Ghrelin levels decline again whenever the body is in positive energy balance, as occurs with weight gains.

Findings from an interesting research study suggest that a person's mindset also influences ghrelin's response to a meal.[24] Young adults were given beverages on two separate occasions—either a beverage labeled as a high-fat, 620-kcalorie "indulgent" milkshake or one labeled as a low-fat, 140-kcalorie "sensible" milkshake. Ghrelin's rise in anticipation of drinking the indulgent milkshake and its decline afterward was much steeper than for the sensible milkshake. In reality, though, the two milkshakes were identical (380 kcalories). Not only was the ghrelin response different, but the participants' satiety differed, reflecting their perceptions of the products. Drinking the sensible milkshake was not as satisfying, despite having the exact same nutrient contents as the indulgent milkshake. The sensible mindset with its relatively flat ghrelin response leaves a person with an increased appetite. These researchers suggest there may be a physiological benefit to adopting a psychological mindset of indulgence when eating low-energy-density, healthy foods.

Some research indicates that ghrelin also promotes sleep. Interestingly, a lack of sleep increases the hunger hormone ghrelin and decreases the satiety hormone leptin—which may help to explain the association between inadequate sleep and overweight.[25]

These two proteins—leptin and ghrelin—illustrate some of the complex factors involved in the regulation of food intake and energy homeostasis. Scientists have identified numerous proteins expressed by dozens of genes linked to obesity and several others associated with fat distribution in the body. Each of these genes has slight variations that differ among individuals.[26] Furthermore, these genes interact with one another and with the environment. The complexity of it all creates a multitude of possible genetic explanations.[27]

> **FIGURE 9-3** **Mice with and without Leptin Compared**

Both of these mice have a defective *ob* gene. Consequently, they do not produce leptin. They both became obese, but the one on the right received daily injections of leptin, which suppressed food intake and increased energy expenditure, resulting in weight loss.

Without leptin, this mouse weighs almost three times as much as a normal mouse.

With leptin treatment, this mouse lost a significant amount of weight but still weighs almost one and a half times as much as a normal mouse.

ghrelin (GRELL-in): a protein produced by the stomach cells that enhances appetite and decreases energy expenditure.

• **ghre** = growth

The food industry spends billions of dollars a year on advertising. The message? "Eat more."

Uncoupling Proteins Genes also code for proteins involved in energy metabolism. These proteins may influence the storing or expending of energy with different efficiencies or in different types of fat. The body has two main types of fat: white and **brown adipose tissue.** White adipose tissue stores fat for other cells to use for energy; brown adipose tissue releases stored energy as heat, thus defending against cold and preventing obesity. Recall from Chapter 7 that when fat is oxidized, some of the energy is released in heat and some is captured in ATP. In brown adipose tissue, oxidation is uncoupled from ATP formation, producing heat only.* By radiating energy away as heat, the body expends, rather than stores, energy. In contrast, efficient coupling facilitates synthesis reactions, including the making of fat for storage. In other words, weight gains or losses may depend on whether the body dissipates the energy from an ice cream sundae as heat or stores it in body fat.

Brown fat and heat production is particularly important in newborns and in animals exposed to cold weather, especially those that hibernate. They have plenty of brown adipose tissue. In contrast, human adults have little brown fat, stored primarily around the neck and clavicles.[28] Brown fat is most metabolically active during exposure to cold.[29] Importantly, brown fat activity declines with age and with obesity; overweight and obese individuals have less brown fat activity than others.[30] The role of brown fat in body weight regulation is not yet fully understood, but such an understanding may prove most useful in developing obesity treatments.[31]

Recent research has revealed that some white fat cells can undergo a process known as browning as they take on characteristics of brown fat, most notably the activity of uncoupled proteins; these fat cells have been named **brite adipocytes.**[32] Brite fat cells are far more abundant than brown fat cells in adults.[33] By learning how browning is regulated, researchers hope to tilt energy balance from storage to expenditure in the effort to fight obesity. Interestingly, among the factors that trigger browning is physical activity. During exercise, muscle cells release a protein (the myokine irisin) that triggers the transformation of white fat cells into brite fat cells.[34] Such findings help to explain one of the many ways physical activity expends energy and supports weight management.

Environment
With obesity rates rising and the **gene pool** remaining relatively unchanged, environment must also play a role in obesity. Obesity reflects the interactions between genes and the environment. An **obesogenic environment** includes all of the circumstances that we encounter daily that push us toward fatness. Over the past 4 decades, the demand for physical activity has decreased as the abundance of food has increased.

Keep in mind that genetic and environmental factors are not mutually exclusive; in fact, their *interactions* create the epigenetics that provide a greater understanding of obesity and related diseases. Genes can influence eating behaviors, for example, and food and activity behaviors influence the genes that regulate body weight. Interestingly, even social relationships can influence the development of obesity.[35] The likelihood that a person will become obese increases when a friend, sibling, or spouse becomes obese.

Overeating
One explanation for obesity is that overweight people overeat, although diet histories may not always reflect high intakes. Diet histories are not always accurate records of actual intakes; both normal-weight and obese people commonly misreport their dietary intakes. Most importantly, current dietary intakes may not reflect the eating habits that led to obesity. Obese people who had a positive energy balance for years and accumulated excess body fat may not currently have a positive energy balance. This reality highlights an important point: the energy-balance equation must consider time. Both present *and* past eating and activity patterns influence current body weight.

brown adipose tissue: masses of specialized fat cells packed with pigmented mitochondria that produce heat instead of ATP.

brite adipocytes: white fat cells with brown fat cell characteristics; also called *beige adipocytes*.

gene pool: all the genetic information of a population at a given time.

obesogenic (oh-BES-oh-JEN-ick) **environment:** all the factors surrounding a person that promote weight gain, such as increased food intake, especially of unhealthy choices, and decreased physical activity.

*In *coupled reactions,* the energy released from the breakdown of one compound is used to create a bond in the formation of another compound. In *uncoupled reactions,* the energy is released as heat.

We live in an environment that exposes us to an abundance of high-kcalorie, high-fat foods that are readily available, relatively inexpensive, heavily advertised, and reasonably tasty. Food is available everywhere, all the time—thanks largely to fast food. Our highways are lined with fast-food restaurants. Convenience stores and service stations offer fast food and snacks as well. Fast food is available in our schools, malls, and airports. The mere proximity of fast food increases the risk of obesity. It's convenient and it's available morning, noon, and night—and all times in between. Consequently, we are eating more meals more frequently than in decades past—and energy intake has risen accordingly.[36]

Most alarming are the extraordinarily large portions and ready-to-go combomeals. Eating large portion sizes multiple times a day accounts for much of the weight increase seen over the decades.[37] People buy the large portions and combinations, perceiving them to be a good value, but then they eat more than they need—a bad deal. In fact, one research study calculated that for the 67 cents extra to upsize a meal, consumers receive an extra 400 kcalories, an extra 36 grams of body fat, and an extra $1 to $7 in health-care costs.[38]

Simply put, large portion sizes deliver more kcalories. And portion sizes of virtually all foods and beverages have increased markedly in the past several decades, most notably at fast-food restaurants. Not only have portion sizes increased over time, but they are now two to eight times larger than standard serving sizes. The trend toward large portion sizes parallels the increasing prevalence of overweight and obesity in the United States, beginning in the 1970s, increasing sharply in the 1980s, and continuing today.

Restaurant food, especially fast food, contributes significantly to the development of obesity.[39] Fast food is often energy-dense food, which increases energy intake, BMI, and body fatness. The combination of large portions and energy-dense foods is a double whammy. Reducing portion sizes is somewhat helpful, but the real kcalorie savings come from lowering the energy density. Low-energy-density foods such as fruits and vegetables can help with weight loss.

"Want fries with that?" A supersize portion delivers more than 600 kcalories.

Physical Inactivity Our environment fosters physical inactivity as well.[40] Life requires little exertion—escalators carry us up stairs, automobiles take us across town, and remote controls change television channels from a distance. Modern technology has replaced physical activity at home, at work, and in transportation.[41] Inactivity contributes to weight gain and poor health. Most physical inactivity occurs when watching television, playing video games, and using the computer. The more time people spend in these sedentary activities, the more likely they are to be overweight—and to incur the metabolic risk factors of heart disease (high blood lipids, high blood pressure, and high blood glucose).[42]

Sedentary activities contribute to weight gain in several ways. First, they require little energy beyond the resting metabolic rate. Second, they replace time spent in more vigorous activities. Third, watching television influences food purchases and correlates with between-meal snacking on the high-kcalorie, solid fat and added sugars foods and beverages most heavily advertised.

Some obese people are so extraordinarily inactive that even when they eat less than lean people, they still have an energy surplus. Reducing their food intake further would incur nutrient deficiencies and jeopardize health. Physical activity is a necessary component of nutritional health. People must be physically active if they are to eat enough food to deliver all the nutrients they need without unhealthy weight gain. In fact, *to prevent weight gain*, the DRI suggests an accumulation of 60 minutes of moderately intense physical activities every day in addition to the less intense activities of daily living. Recommendations *to lose weight* encourage even greater duration, intensity, or frequency of physical activity (as a later section of the chapter discusses).

People may be obese, therefore, not because they eat too much, but because they move too little—both in purposeful exercise and in the activities of daily life. Studies report that the differences in the time obese and lean people spend lying, sitting, standing, and moving accounts for about 350 kcalories a day. In general, lean people tend to be more spontaneously active in their occupations and their leisure time. The energy expended in these everyday spontaneous activities—called *nonexercise activity thermogenesis (NEAT)*—plays a pivotal role in energy balance and weight management.

> **› REVIEW IT** Review some of the causes of obesity.
>
> Obesity has many causes and most interact, creating a complex scenario. Environmental factors, such as overeating and physical inactivity, may influence a person's genetic susceptibility to obesity.

Lack of physical activity fosters obesity.

© Istockphoto.com/Spauln

9.3 Problems of Overweight and Obesity

› LEARN IT Discuss the physical, social, and psychological consequences of overweight and obesity.

Millions of US adults are trying to lose weight on any given day. Some of these people may not even need to lose weight. Others may benefit from weight loss, but they will not be successful. Relatively few people succeed in losing weight, and even fewer succeed permanently. For many, improving diet and activity habits to simply prevent further weight gains may be sufficient. Whether a person will benefit from weight loss is a question of health.

Health Risks Chapter 8 describes some of the health problems that commonly accompany obesity. In evaluating the risks to health from obesity, health-care professionals use three indicators:

- Body mass index (25 to 29.9 for overweight and ≥30 for obese)
- Waist circumference (>40 inches for men and >35 inches for women)
- Disease risk profile

Importantly, the disease risk profile takes into account family history, life-threatening diseases, and common risk factors for chronic diseases (such as blood lipid profile). The higher the BMI, the greater the waist circumference, and the more risk factors—the greater the urgency to treat obesity.

People can best decide whether weight loss might be beneficial by considering their health status. People who are overweight by BMI standards, but otherwise in good health, might not benefit from losing weight; they might focus on preventing further weight gains instead. In contrast, those who are obese and suffering from a life-threatening disease such as diabetes might improve their health substantially by adopting a diet and activity plan that supports weight loss.

Overweight in Good Health Often a person's motivations for weight loss have nothing to do with health. A healthy young woman with a BMI of 26 might want to lose a few pounds for spring break, but doing so might not improve her health. In fact, if she opts for a starvation diet or diet pills, she would be healthier *not* trying to lose weight. In any case, she should try to avoid additional weight gains.

Obese or Overweight with Risk Factors Weight loss is recommended for people who are obese and those who are overweight with one or more of the following risk factors for chronic diseases:

- Hypertension
- Cigarette smoking
- Abnormal blood lipids
- Diabetes or prediabetes
- Family history of heart disease
- Men 45 or older and women 55 or older

A 50-year-old man with a BMI of 28 who has high blood pressure and a family history of heart disease can improve his health by adjusting his diet and engaging in a regular exercise plan.

Obese or Overweight with Life-Threatening Condition Weight loss is also recommended for a person who is either obese or overweight and suffering from a life-threatening condition such as heart disease, type 2 diabetes, or sleep apnea. The health benefits of weight loss are clear. For example, a 30-year-old man with a BMI of 40 might be able to prevent or control diabetes by losing 75 pounds. Although the effort required to do so may be great, it may be no greater than the effort and consequences of living with diabetes.

Perceptions and Prejudices
Many people assume that every obese person can achieve slenderness and should pursue that goal. First consider that most obese people do not—for whatever reason—successfully lose weight and maintain their losses. Then consider the prejudice involved in that assumption. People come with varying weight tendencies, just as they come with varying potentials for height and physical talents, yet we do not expect tall people to shrink or fast runners to slow down in an effort to become "normal."

Social Consequences Large segments of our society place such enormous value on thinness that obese people face prejudice and discrimination on the job, at school, and in social situations: they are judged on their appearance more than on their character. Socially, obese people are negatively stereotyped as lazy and lacking in self-control. Such a critical view of overweight is not prevalent in many other cultures, including segments of our own society. Instead, overweight is simply accepted or even embraced as a sign of robust health and beauty. To free society of its obsession with body weight and prejudice against obesity, people must first learn to judge others—and themselves—for who they are and not for what they weigh.

Psychological Problems Psychologically, obese people may suffer embarrassment when others treat them with hostility and contempt, and many have come to view their own bodies as flawed. Feelings of rejection, shame, and depression are common among obese people. Anxiety and depression, in turn, may contribute to the development of obesity, which perpetuates the problem.[43]

Most weight-loss programs assume that the problem can be solved simply by applying willpower and hard work. If determination were the only factor involved, though, the success rate would be far greater than it is. Overweight people may readily assume blame for failure to lose weight and maintain the losses when, in fact, it is the programs that have failed. Ineffective treatment and its associated sense of failure add to a person's psychological burden. Figure 9-4 illustrates how the devastating psychological effects of obesity and dieting perpetuate themselves.

Dangerous Interventions
Some people attach so many dreams of happiness to weight loss that they willingly risk huge sums of money for the slightest chance of success. As a result, weight-loss schemes flourish. Of the tens of thousands of claims, treatments, and theories for losing weight, few are effective—and many are downright dangerous. The negative consequences must be carefully considered before embarking on any weight-loss program. Some interventions entail greater dangers than the risk of being overweight. Physical, metabolic, and psychosocial problems may arise from fad diets and "yo-yo" dieting.[44] Wise consumers scrutinize fad diets, magic potions, and wonder gizmos with a healthy dose of skepticism.

Some of the nation's most popular diet books and weight-loss programs have misled consumers with unsubstantiated claims and deceptive testimonials. Furthermore, they fail to provide an assessment of the short- and long-term results of their treatment plans, even though such evaluations are possible and would permit consumers to make informed decisions. Of course, some weight-loss programs are better than others in terms of cost, approach, and customer satisfaction. Reputable weight-loss programs will explain the risks associated with their plans and provide honest predictions of success.

Fad Diets **Fad diets** often sound good, but they typically fall short of delivering on their promises. They espouse exaggerated or false theories of weight loss and advise consumers to follow inadequate diets. Some fad diets are hazardous to

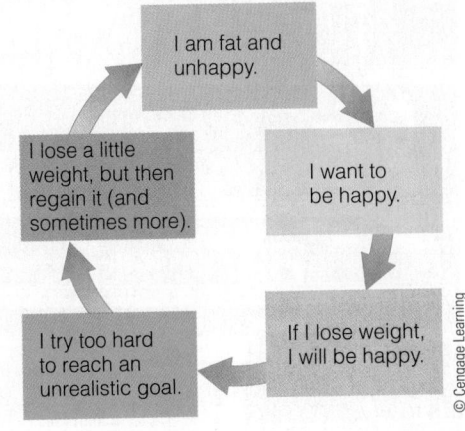

> FIGURE 9-4 **The Psychology of Weight Cycling**

I am fat and unhappy.

I want to be happy.

If I lose weight, I will be happy.

I try too hard to reach an unrealistic goal.

I lose a little weight, but then regain it (and sometimes more).

© Cengage Learning

fad diets: popular eating plans that promise quick weight loss. Most fad diets severely limit certain foods or overemphasize others (for example, never eat potatoes or pasta, or eat cabbage soup daily).

So many promises, so little success.

health as Highlight 9 explains. Adverse reactions can be as minor as headaches, nausea, and dizziness or as serious as death. The "How To" on p. 299 offers guidelines for identifying unsound weight-loss schemes and fad diets.

Weight-Loss Products Millions of people in the United States use over-the-counter weight-loss products. Most users are women, especially young overweight women, but almost 10 percent are of normal weight.

In their search for weight-loss magic, some consumers turn to "natural" herbal products and dietary supplements, even though few have proved to be effective and many have proved to be harmful.[45] For example, in addition to the many cautions that accompany the use of all herbal remedies, consumers should be aware that St. John's wort is often prepared in combination with the herbal stimulant ephedrine. Ephedrine-containing supplements promote modest short-term weight loss (about 2 pounds a month), but with great risks. These supplements have been implicated in numerous heart attacks and seizures, resulting in about 100 deaths. For this reason, the FDA has banned the sale of ephedrine-containing supplements, but they are still readily available on the Internet.* Similarly, the FDA has issued warnings for another herbal weight loss supplement called Que She, which contains not only ephedrine but two weight-loss drugs that have been withdrawn from the market and another drug used to treat heart conditions.

Highlight 18 explores the possible benefits and potential dangers of herbal products and other alternative therapies. As it explains, dietary supplements do not need to be approved by the FDA, and manufacturers do not need to test the safety or effectiveness of any product. In other words, consumers cannot assume that an herbal product or dietary supplement is safe or effective just because it is available on the market. In fact, the FDA has identified more than 75 products that contain undeclared, active pharmaceutical ingredients that can have serious consequences such as seizures and heart attacks.[46] These ingredients are not listed on the labels, and consumers have no way of knowing what the products actually contain. Anyone considering whether to use dietary supplements for weight loss should consult with a physician and research the product with the FDA (www.fda.gov).

Other Gimmicks Other gimmicks don't help with weight loss either. Hot baths do not speed up metabolism so that pounds can be lost in hours. Steam and sauna baths do not melt the fat off the body, although they may dehydrate people so that they lose water weight. Brushes, sponges, wraps, creams, and massages intended to move, burn, or break up fat do nothing of the kind.

> **REVIEW IT** Discuss the physical, social, and psychological consequences of overweight and obesity.

The question of whether a person should lose weight depends on many factors: among them are the extent of overweight, age, health, and genetic makeup. Not all obesity will cause disease or shorten life expectancy. Just as there are unhealthy, normal-weight people, there are healthy, overweight people. Some people may risk more in the process of losing weight than in remaining overweight. Fad diets and weight-loss supplements can be as physically and psychologically damaging as excess body weight.

9.4 Aggressive Treatments for Obesity

> **LEARN IT** Explain the risks and benefits, if any, of aggressive ways to treat obesity.

The appropriate strategies for weight loss depend on the degree of obesity and the risk of disease. An overweight person in good health may need only to improve eating habits and increase physical activity, but someone with **clinically severe obesity** may need more aggressive treatment options—drugs or surgery. Drugs appear

clinically severe obesity: a BMI of 40 or greater or a BMI of 35 or greater with additional medical problems. A less preferred term used to describe the same condition is *morbid obesity*.

---

*Ephedrine is an amphetamine-like substance extracted from the Chinese ephedra herb *ma huang*. The FDA has banned the sale of *ma huang* in the United States.

Bill Aron / PhotoEdit

to be modestly effective and safe, at least in the short term; surgery appears to be dramatically effective but can have severe complications, at least for some people.

Drugs Based on new understandings of obesity's genetic basis and its classification as a chronic disease, much research effort has focused on drug treatments for obesity. Experts reason that if obesity is a chronic disease, it should be treated as such—and the treatment of most chronic diseases includes drugs. The challenge, then, is to develop an effective drug—or more likely, a combination of drugs—that can be used over time without adverse side effects or the potential for abuse. Weight-loss drugs should be prescribed only to those with medical risks—not for cosmetic reasons—and in tandem with a healthy diet and activity program.

Several drugs for weight loss have been tried over the years, with varying degrees of effectiveness and safety.[47] When used as part of a long-term, comprehensive weight-loss program, drugs can help with modest weight loss. Because weight regain commonly occurs with the discontinuation of drug therapy, treatment must be long term. Yet the long-term use of drugs poses risks. We don't yet know whether a person would be harmed more from maintaining a 100-pound excess or from taking a drug for a decade to keep the 100 pounds off. Physicians must prescribe drugs appropriately, inform consumers of the potential risks, and monitor side effects carefully. Table 9-1 presents the drugs to treat obesity that meet the FDA mandate that "benefits must exceed risks."[48]

Some physicians prescribe drugs that have not been approved for weight loss, a practice known as "off-label" use. These drugs have been approved for other conditions (such as seizures) and incidentally cause modest weight loss. Physicians using off-label drugs must be well-informed of the drugs' use and effects and monitor their patients' responses closely.

Surgery The US prevalence of clinically severe obesity (BMI >40) is estimated at 6 percent.[49] At this level of obesity, lifestyle changes and modest weight losses can improve disease risks a little, but the most effective treatment is surgery.[50] Surgery may be an option for people with all of the following conditions:

- Unable to achieve adequate weight loss with diet and exercise
- BMI ≥40 or BMI ≥35 with weight-related health problems (such as diabetes or hypertension)
- No medical or psychological contraindications
- Understanding of risks and strong motivation to comply with post-surgery treatment plan

More than 100,000 such surgeries are performed in the United States annually.[51] As Figure 9-5 (p. 276) shows, the two most common surgical procedures effectively

TABLE 9-1 FDA-Approved Drugs for Weight Loss

Drug	Action	Side Effects
Orlistat (OR-leh-stat), trade names: Alli, Xenical	Inhibits pancreatic lipase activity in the GI tract, thus blocking digestion and absorption of dietary fat and limiting energy intake	GI cramping, diarrhea, gas, frequent bowel movements, reduced absorption of fat-soluble vitamins; rare cases of liver injury
Phentermine (FEN-ter-mean), diethylpropion (DYE-eth-ill-PRO-pee-on), phendimetrazine (FEN-dye-MEH-tra-zeen)	Enhances the release of the neurotransmitter norepinephrine, which suppresses appetite	Increased blood pressure and heart rate, insomnia, nervousness, dizziness, headache
Lorcaserin hydrochloride, trade name: Belviq (BELL-veek)	Interacts with brain serotonin receptors to increase satiety and reduce food intake	Headache, dizziness, fatigue, nausea, dry mouth, and constipation; low blood glucose in people with diabetes; serotonin syndrome, including agitation, confusion, fever, loss of coordination, rapid or irregular heart rate, shivering, seizures, and unconsciousness; cannot be safely used by pregnant or lactating women or people with heart-valve problems; high doses cause hallucinations
Phentermine (an appetite suppressant) and topiramate (a seizure/migraine medication) combination, trade name: Qsymia (kyoo-sim-EE-uh)	Enhances the release of the neurotransmitter norepinephrine, which suppresses appetite, and increases the feeling of being full, making foods taste less appealing	Increased heart rate; can cause birth defects if taken in the first weeks or months of pregnancy; increased heart rate; suicidal thoughts; may worsen glaucoma and other eye problems

NOTE: Weight-loss drugs are most effective when taken as directed and used in combination with reduced-kcalorie diet and increased physical activity.

© Cengage Learning

> FIGURE 9-5 **Gastric Surgery Used in the Treatment of Clinically Severe Obesity**
Both of these surgical procedures limit the amount of food that can be comfortably eaten.

In gastric bypass, the surgeon constructs a small stomach pouch and creates an outlet directly to the small intestine, bypassing most of the stomach, the entire duodenum, and some of the jejunum. (Dark areas highlight the flow of food through the GI tract; pale areas indicate bypassed sections.)

Advantages:
• No foreign object in abdomen or need for adjustments
• More durable, reliable, and effective

In gastric banding, the surgeon uses a gastric band to create a small stomach pouch. The size of the opening can be adjusted by inflating or deflating the band by way of a port placed in the abdomen just beneath the skin.

Advantages:
• No malabsorption
• More flexible, less invasive, safer

© Cengage Learning

limit food intake by reducing the capacity of the stomach. In addition, gastric bypass suppresses hunger by changing production of gastrointestinal hormones.[52] Changes in food preferences and GI microbes may also influence weight losses.[53] The results are significant: depending on the type of surgery, nearly 50 percent of the excess weight remains lost after 15 years.[54] Importantly, most people experience dramatic and lasting improvements in their diabetes, blood lipids, and blood pressure—even before significant weight loss.[55] Improvements in depression and anxiety are not as likely. Whether surgery is a reasonable option for obese teens is the subject of much debate among pediatricians and **bariatric** surgeons (see Chapter 16).[56]

Because the long-term safety and effectiveness of surgery depend, in large part, on compliance with dietary instructions, nutrition care plays an important role in follow-up treatment.[57] Vitamin and mineral deficiencies are common, and dietary supplements are routinely prescribed.[58] Weight regain may occur and psychological problems—such as disordered eating behaviors—may also develop.[59] Lifelong medical supervision is necessary, but the possible health benefits of weight loss—improved blood lipid profile, blood pressure, and insulin sensitivity—may outweigh the risks. Overall risk of death and heart disease is lower for obese people after successful surgery than for obese people who do not undergo surgery.[60]

Another surgical procedure removes some fat deposits by liposuction. This cosmetic procedure has little effect on body weight (less than 10 pounds), but can alter body shape slightly in specific areas. Liposuction is a popular procedure in part because of its perceived safety, but immediate and delayed complications can arise.[61] Furthermore, removing adipose tissue by way of liposuction does not provide the health benefits that typically accompany weight loss. In other words, liposuction does not improve blood pressure, inflammation, blood lipid profile, or insulin sensitivity. Perhaps most surprisingly, a year after liposuction, body fat returns and redistributes itself from the thighs to the abdomen.[62]

> REVIEW IT Explain the risks and benefits, if any, of aggressive ways to treat obesity.

Overweight and obese people may benefit most from improving eating and activity habits. Those with clinically severe obesity and high risks of medical problems may need more aggressive treatment, including drugs or surgery. Such treatments may offer benefits, but also incur some risks.

9.5 Weight-Loss Strategies

> LEARN IT Outline reasonable strategies for achieving and maintaining a healthy body weight.

From the bustling activity of a cell making fat to the inactivity of a person watching television, the factors contributing to obesity are numerous and complex. Each

bariatric: pertaining to the field of medicine that specializes in treating obesity.

interacts with many others. Efforts to combat obesity must integrate healthy eating patterns, physical activities, supportive environments, and psychosocial support.[63]

Changes, Losses, and Goals Successful weight-loss strategies embrace changes, celebrate losses, and set goals. A comprehensive lifestyle approach that includes low-kcalorie, nutrient-dense foods and regular physical activity supports both weight loss and health benefits. In keeping with this philosophy, the *Dietary Guidelines for Americans* advise those who need to lose weight to "consume fewer kcalories from foods and beverages, increase physical activity, and reduce time in sedentary behaviors."

> **DIETARY GUIDELINES FOR AMERICANS**
Strive to achieve and maintain a healthy body weight through improved eating and physical activity behaviors.

Even modest weight loss brings health benefits. Modest weight loss, even when a person is still overweight, can improve blood glucose and reduce the risks of heart disease by lowering blood pressure and blood cholesterol, especially for those with central obesity. Improvements in physical capabilities and quality of life become evident with even a 5 percent weight loss.[64] For these reasons, parameters such as blood pressure, blood cholesterol, or even vitality are more useful than body weight in marking success. People less concerned with disease risks may prefer to set goals for personal fitness, such as being able to play with children or climb stairs without becoming short of breath. Importantly, they can focus on healthy eating and activity habits instead of weight loss.

Depending on initial body weight, a reasonable rate of loss for overweight adults is ½ to 2 pounds a week, or 5 to 10 percent of body weight over 6 months. For a person weighing 250 pounds, a 10 percent loss is 25 pounds, or about 1 pound a week for 6 months. Such gradual weight losses are more likely to be maintained than rapid losses. Keep in mind that pursuing good health is a lifelong journey. Most adults are keenly aware of their body weights and shapes and realize that what they eat and what they do can make a difference to some extent. Those who are most successful at weight management seem to have fully incorporated healthful eating and physical activity into their daily lives.

Eating Patterns Contrary to the claims of fad diets, no single food plan is magical, and no specific food must be included or avoided in a weight-management program. In designing an eating pattern, people need only consider foods that they like or can learn to like, that are available, and that are within their means. Creating a healthful eating pattern is the first step. The important next step is following it for the rest of one's life. Achieving and maintaining a healthy weight requires permanent lifestyle changes.

Be Realistic about Energy Intake The main characteristic of a weight-loss diet is that it provides less energy than the person needs to maintain present body weight. If food energy is restricted too severely, dieters may not receive sufficient nutrients. Rapid weight loss usually means excessive loss of lean tissue, a lower BMR, and rapid weight regains to follow. The composition of regained weight is more fat and less lean than the composition of the originally lost weight.[65] In addition, restrictive eating may create stress or foster unhealthy behaviors of eating disorders as described in Highlight 8.[66]

Energy intake should provide nutritional adequacy without excess—that is, somewhere between deprivation and complete freedom to eat whatever, whenever. A reasonable suggestion for overweight and obese adults is to increase activity and reduce food intake enough to create a deficit of 500 to 750 kcalories per day. Such a deficit produces a weight loss of 1 to 2 pounds per week—a rate that supports the loss of fat efficiently while retaining lean tissue.[67] In general, weight-loss diets need to provide about 1200 to 1500 kcalories per day for women and 1500 to 1800 kcalories a day for men.

Some people skip meals, typically breakfast, in an effort to reduce energy intake and lose weight. Research does not support such a causal relationship between breakfast and body weight, but it does suggest some interesting associations.[68] Breakfast frequency is inversely associated with obesity and its associated risk factors—that is, people who frequently eat breakfast have a lower BMI, blood pressure, and blood cholesterol than those who tend to skip breakfast.[69] Furthermore, eating breakfast, especially a protein-rich breakfast, improves satiety and diet quality—two factors that support healthy body weight.[70] One study found that even when total kcalories were the same on two weight-loss diets, the "breakfast diet" (big breakfast, medium lunch, and small dinner) had better results than the "dinner diet" (small breakfast, medium lunch, and big dinner).[71] After three months, the women on the breakfast diet lost 10 more pounds than the others; had better triglycerides, HDL, blood glucose, and waist circumference measures; and were less hungry.

> DIETARY GUIDELINES FOR AMERICANS
Control total kcalorie intake to manage body weight. For people who are overweight or obese, this will mean consuming fewer kcalories from foods and beverages.

Emphasize Nutritional Adequacy Healthy diet plans make nutritional adequacy a priority. Nutritional adequacy is difficult to achieve on fewer than 1200 kcalories a day, and most healthy adults need never consume any less. A plan that provides an adequate intake supports a healthier and more successful weight loss than a restrictive plan that creates feelings of starvation and deprivation, which can lead to an irresistible urge to binge.

Table 9-2 specifies the amounts of foods from each food group for diets providing 1200 to 1800 kcalories. Such an intake would allow most people to lose weight and still meet their nutrient needs with careful, low-kcalorie, nutrient-dense food selections. Keep in mind, too, that well-balanced diets that emphasize fruits, vegetables, whole grains, lean protein foods, and low-fat milk products offer many health rewards even when they don't result in weight loss. A dietary supplement providing vitamins and minerals—especially iron and calcium for women—at or below 100 percent of the Daily Values can help people following low-kcalorie diets to achieve nutrient adequacy.

Eat Small Portions As mentioned earlier, portion sizes at markets, at restaurants, and even at home have increased dramatically over the years, contributing significantly to energy intake and weight gains.[72] We have come to expect large portions, and we have learned to clean our plates. Many of us pay more attention to these external cues defining how much to eat than to our internal cues of hunger and satiety. For health's sake, we may need to learn to eat less food at each meal—one piece of chicken for dinner instead of two, a teaspoon of butter on vegetables instead of a tablespoon, and one cookie for dessert instead of six. The goal is to eat enough food for adequate energy, abundant vitamins and minerals, and some pleasure, but not more. This amount should leave a person feeling satisfied—not stuffed. A saying credited to Confucius captures this concept—hara hachi bu—which translates to "eat until you are 80 percent full."

TABLE 9-2 Daily Amounts from Each Food Group for 1200- to 1800-kCalorie Diets

Food Group	1200 kCalories	1400 kCalories	1600 kCalories	1800 kCalories
Fruit	1 c	1½ c	1½ c	1½ c
Vegetables	1½ c	1½ c	2 c	2½ c
Grains	4 oz	5 oz	5 oz	6 oz
Protein foods	3 oz	4 oz	5 oz	5 oz
Milk and milk products	2½ c	2½ c	3 c	3 c
Oils	4 tsp	4 tsp	5 tsp	5 tsp

© Cengage Learning

Keep in mind that even fat-free and low-fat foods can deliver a lot of kcalories when a person eats large quantities. A low-fat cookie or two can be a sweet treat even on a weight-loss diet, but larger portions defeat the savings.

People who have difficulty making low-kcalorie selections or controlling portion sizes may find it easier to use prepared meal plans. Prepared meals that provide low-kcalorie, nutritious meals or snacks can support weight loss while easing the task of diet planning.[73] Ideally, those using a prepared meal plan will also receive counsel from a registered dietitian nutritionist to learn how to select appropriately from conventional food choices as well.

Slow Down Eating can be a pleasurable experience, and taking the time to savor the flavors can help with weight management. Eating slowly, taking small bites, and chewing thoroughly all help to decrease food intake.[74] A person who slows down and savors each bite eats less before hormones signal satiety and the end of a meal.[75] Consequently, energy intake is lower when meals are eaten slowly. Savoring each bite also activates the pleasure centers of the brain. Some research suggests that people may overeat when the brain doesn't sense enough gratification from food. Faster eating correlates with higher BMI.[76]

Lower Energy Density Most people take their cues about how much to eat based on portion sizes, and the larger the portion size, the more they eat. To lower energy intake, a person can either reduce the portion size or reduce the energy density. Reducing energy density while maintaining or even increasing food quantity, especially by reducing fat and including fruits and vegetables, seems to be a successful strategy to control hunger and manage weight.[77] This concept of using large quantities of low-energy-density foods is sometimes referred to as *volumetrics*. Figure 9-6 illustrates how water, fiber, and fat influence energy density, and the accompanying "How To" feature (p. 280) compares foods based on their energy density. Foods containing water, those rich in fiber, and those low in fat help to lower energy density, providing more satiety for fewer kcalories. Because a low-energy-density diet is a low-fat, high-fiber diet rich in many vitamins and minerals, it supports good health in addition to weight loss.

Remember Water In addition to lowering the energy density of foods, water seems to help those who are trying to lose or maintain weight.[78] For one, foods with high water content (such as broth-based soups) increase fullness, reduce hunger, and consequently reduce energy intake. For another, drinking a large glass of water before a meal eases hunger, fills the stomach, and consequently reduces energy intake.[79] Importantly, water adds no kcalories. The average US diet delivers an estimated 75 to 150 kcalories a day from sweetened beverages. Simply replacing nutrient-poor, energy-dense beverages with water can help a person achieve a 5 percent weight loss at 6 months.[80] Water also helps the GI tract adapt to a high-fiber diet.

> FIGURE 9-6 **Energy Density**

Decreasing the energy density (kcal/g) of foods allows a person to eat satisfying portions while still reducing energy intake. To lower energy density, select foods high in water or fiber and low in fat.

100 grams delivers

299 kcal vs. 67 kcal

Selecting grapes with their high water content instead of raisins increases the volume and cuts the energy intake.

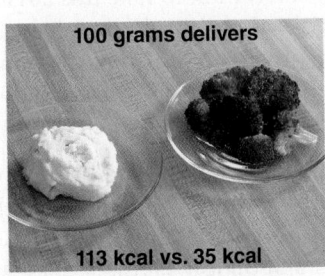

100 grams delivers

113 kcal vs. 35 kcal

Even at the same weight and similar serving sizes, the fiber-rich broccoli delivers twice the fiber for about one-third the energy.

100 grams delivers

198 kcal vs. 116 kcal

By selecting the water-packed tuna (on the right) instead of the oil-packed tuna (on the left), a person can enjoy the same amount for fewer kcalories.

Compare Foods Based on Energy Density

Chapter 2 describes how to evaluate foods based on their nutrient density—their nutrient contribution per kcalorie. Another way to evaluate foods is to consider their energy density—their energy contribution per gram. This example compares carrot sticks with french fries. The conclusion is no surprise, but understanding the mathematics may offer valuable insight into the concept of energy density. A carrot weighing 72 grams delivers 31 kcalories. To calculate the energy density, divide kcalories by grams:

$$\frac{31 \text{ kcal}}{72 \text{ g}} = 0.43 \text{ kcal/g}$$

Do the same for french fries weighing 50 grams and contributing 167 kcalories:

$$\frac{167 \text{ kcal}}{50 \text{ g}} = 3.34 \text{ kcal/g}$$

The more kcalories per gram, the greater the energy density. French fries are more energy dense than carrots. They provide more energy per gram—and per bite. Considering a food's energy density is especially useful in planning diets for weight management. Foods with a high energy density help with weight gain, whereas foods with a low energy density help with weight loss.

© Matthew Farruggio

> **TRY IT** Compare the energy density of a hard-boiled egg (50 grams and 78 kcalories) with light tuna canned in water (57 grams and 66 kcalories).

Focus on Fiber High-fiber foods such as fresh fruits, vegetables, legumes, and whole grains may help with weight management. By offering abundant vitamins, minerals, and fiber but little fat, these foods tend to be relatively low in energy and high in nutrients. Eating high-fiber foods also takes time, which eases hunger and promotes satiety.

Choose Fats Sensibly One way to lower energy intake is to lower fat intake. Lowering the fat content of a food lowers its energy density—for example, selecting fat-free milk instead of whole milk. That way, a person can consume the usual amount (say, a cup of milk) at a lower energy intake (85 instead of 150 kcalories).

Fat has a weak satiating effect, and satiation plays a key role in determining food intake during a meal. Consequently, a person eating a high-fat meal increases energy intake in two ways—more food and more fat kcalories. For these reasons, measure fat with extra caution. (Review p. 157 for strategies to lower fat in the diet.) Be careful not to take this advice to extremes, however; too little fat incurs health risks as well, as Chapter 5 explains.

Lowering the amount of fat in the diet can lead to weight loss, but an important point to notice in any discussion on weight-loss diets is total energy intake.[81] A low-fat diet supports weight loss only when energy intake is less than energy expenditure.

Select Carbohydrates Carefully Another popular way to lower energy intake is to lower carbohydrate intake. Highlight 4's discussion of carbohydrate-restricted and carbohydrate-modified diets reaches the same conclusion as the previous paragraph on low-fat diets: they work only when energy intake is less than energy expenditure.

Chapter 4 describes how foods with added sugars increase energy intake and contribute to weight gain. Limiting consumption of foods with added sugars can help with weight management. One way people try to control weight is to use

© Corbis

If you want to lose weight, steer clear of the empty kcalories in fancy coffee drinks. A 16-ounce caffè mocha delivers 400 kcalories—half of them from fat and the other half from sugar.

foods and beverages sweetened with artificial sweeteners. Using artificial sweeteners instead of sugars can lower energy intake and may support modest weight loss, or at least prevent weight gain, although evidence is inconsistent; in fact, some research indicates that artificial sweeteners may stimulate appetite and lead to weight gain.[82] One study offers a possible explanation. People who regularly drink diet sodas have decreased activity in the brain center that signals reward and controls food intake.[83] Such an alteration makes it more likely that these people would eat more later in the day.

To what extent artificial sweeteners can help someone lose weight depends in part on the person's motivations and actions. For example, one person might drink an artificially sweetened beverage now so as to be able to eat a high-kcalorie food later. This person's energy intake might stay the same or increase. A person trying to control energy intake might drink an artificially sweetened beverage now and choose a low-kcalorie food later. This plan would help reduce the person's total energy intake. Using artificial sweeteners will not automatically lower energy intake. To control energy intake successfully, a person needs to make informed diet and activity decisions throughout the day.

Watch for Other Empty kCalories A person trying to achieve or maintain a healthy weight needs to pay attention not only to fat and sugar, but to alcohol too. Not only does alcohol add kcalories, but accompanying mixers can also add both kcalories and fat, especially in creamy drinks such as piña coladas (review Table H7-3 on p. 229). Furthermore, drinking alcohol reduces a person's inhibitions, which can lead to excessive eating.[84]

A person who adopts a lifelong "eating plan for good health" rather than a "diet for weight loss" will be more likely to keep the lost weight off. Table 9-3 provides several tips for successful weight management.

Physical Activity Whether trying to minimize weight gains or support weight losses, the best approach includes physical activity.[85] To prevent weight gains and support weight losses, current recommendations advise 200 to 300 minutes of moderately intense physical activity a week in addition to activities of daily life.[86] People who combine diet and exercise typically lose more fat, retain more muscle, and regain less weight than those who only follow a weight-loss diet. Even when they do not lose more weight, they seem to follow their diet plans more closely and maintain their losses better than those who do not exercise. Consequently, they benefit from taking in a little less energy from the diet

TABLE 9-3 Weight-Loss Strategies

Food	Activities
• To maintain weight, consume foods and drinks to meet, not exceed, kcalorie needs. To lose weight, energy out should exceed energy in by about 500 kcalories/day.	• Limit screen time.
• Emphasize foods with a low energy density and a high nutrient density; make legumes, whole grains, vegetables, and fruits central to your diet plan.	• Choose moderate- or vigorous-intensity physical activities.
• Eat slowly.	• Avoid inactivity. Some physical activity is better than none.
• Drink water before you eat and while you eat; drink plenty of water throughout the day.	• Gradually increase the frequency, intensity, and duration of physical activities.
• Track food and kcalorie intake.	
• Plan ahead to make better food choices.	
• Limit kcalorie intake from solid fats and added sugars.	
• Reduce portions, especially of high-kcalorie foods.	
• Cook and eat more meals at home, instead of eating out. When eating out, think about choosing healthy options.	

as well as from expending a little more energy in physical activity. Importantly, those who exercise reap important health benefits—reduced abdominal obesity and improved blood pressure, insulin resistance, and cardiorespiratory fitness—regardless of weight loss.[87] Fitness benefits—such as strength and balance—also improve when exercise is part of a weight-loss program.[88] Chapter 14 presents the many benefits of physical activity; the focus here is on its role in weight management.

> DIETARY GUIDELINES FOR AMERICANS
Increase physical activity and reduce time spent in sedentary behaviors.

Activity and Energy Expenditure Table 8-3 (p. 242) shows how much energy each of several activities uses. The number of kcalories spent in an activity depends on body weight, intensity, and duration. For example, a person who weighs 150 pounds and walks 3½ miles in 60 minutes expends about 315 kcalories. That same person running 3 miles in 30 minutes uses a similar amount. By comparison, a 200-pound person running 3 miles in 30 minutes expends an additional 100 kcalories or so. The goal is to expend as much energy as your time allows. The greater the energy deficit created by exercise, the greater the fat loss. And be careful not to compensate for the energy expended in exercise by eating more food.[89] Otherwise, energy balance won't shift, and fat loss will be less significant.

Activity and Discretionary kCalories Chapter 2 introduced the concept of discretionary kcalories as the difference between the kcalories needed to supply nutrients and those needed to maintain energy balance. Because exercise expends energy, the energy allowance to maintain weight increases with increased physical activity—yet the energy needed to deliver needed nutrients remains about the same. In this way, physical activity increases discretionary kcalories (see Figure 9-7). Having more discretionary kcalories puts a little wiggle room in a weight-loss diet for such options as second helpings, sweet treats, or alcoholic beverages on occasion. Of course, selecting nutrient-dense foods and *not* using discretionary kcalories will maximize weight loss.

Activity and Metabolism Activity also contributes to energy expenditure in an indirect way—by speeding up metabolism. It does this both immediately and over the long term. On any given day, metabolism remains elevated for several hours after vigorous and prolonged exercise.[90] This postexercise effect may raise the energy expenditure of exercise up to 15 percent. Over the long term, a person who engages in daily vigorous activity gradually develops more lean tissue. Metabolic rate rises accordingly, and this supports continued weight loss or maintenance.

> FIGURE 9-7 **Influence of Physical Activity on Discretionary kCalories**

Activity and Body Composition Physically active people have less body fat than sedentary people do—even if they have the same BMI. Physical activity, even without weight loss, changes body composition: body fat decreases and lean body mass increases; high-intensity intermittent exercises may be even more effective at reducing body fat than other types of exercise.[91] Furthermore, physical activity reduces abdominal fat even without weight loss.[92]

Activity and Appetite Control Some people think that being active will increase hunger, but research does not show that exercise causes overeating; in fact, when sedentary people participate in an ongoing activity program, they reduce their energy intake.[93] Active people do have healthy appetites, but appetite is suppressed after an aerobic workout.[94] The body has released fuels from storage to support the exercise, so glucose and fatty acids are abundant in the blood. At the same time, the body has suppressed its digestive functions. Hard physical work and eating are not compatible. A person must calm down, put energy fuels back in storage, and relax before eating. At that time, a physically active person may eat more than a sedentary person, but not so much as to fully compensate for the energy expended in exercise.

Exercise may also help curb the inappropriate appetite that accompanies boredom, anxiety, or depression. Weight-management programs encourage people who feel the urge to eat when not particularly hungry to exercise instead. The activity passes time, relieves anxiety, and prevents inappropriate eating.

Activity and Psychological Benefits Activity also helps reduce stress, which is especially helpful for people who respond to stress with inappropriate eating. In addition, physical activity helps to improve body image and separate the connections between body weight and self-worth.[95] A physically active person begins to look and feel healthy and, as a result, gains self-esteem. High self-esteem motivates a person to continue seeking good health and fitness, which keeps the beneficial cycle going. The benefits of physical activity in a weight-management program include:

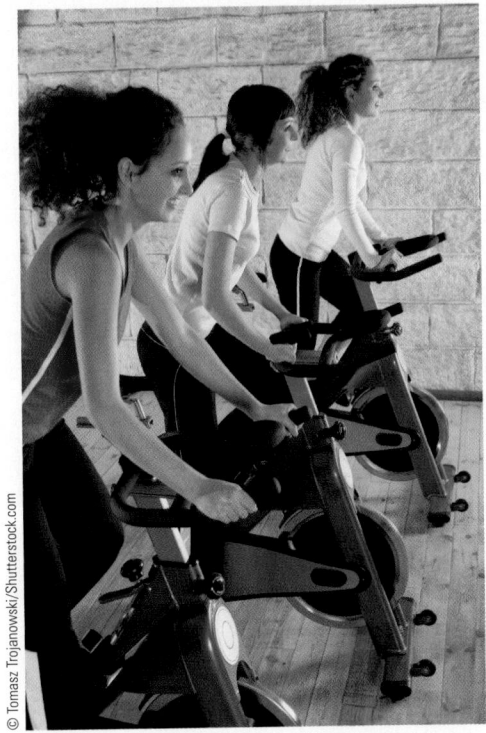

- Short-term increase in energy expenditure (from exercise and from a slight rise in metabolism)
- Long-term increase in BMR (from an increase in lean tissue)
- Improved body composition
- Appetite control
- Stress reduction and control of stress eating
- Physical, and therefore psychological, well-being
- Improved self-esteem

Chapter 14 presents additional benefits of physical activity.

The key to good health is to combine sensible eating with regular exercise.

Choosing Activities Clearly, physical activity is a plus in a weight-management program. What kind of physical activity is best? People should choose activities that they enjoy and are willing to do regularly. What schedule of physical activity is best? It doesn't matter; a person can benefit from either several short bouts of exercise or one continuous workout. Any activity is better than being sedentary. For an active life, limit sedentary activities, engage in strength and flexibility activities, enjoy leisure activities often, engage in vigorous activities regularly, and be as active as possible every day.

Health-care professionals frequently advise people to engage in activities of low-to-moderate intensity for a long duration, such as an hour-long, fast-paced walk. The reasoning behind such advice is that walking offers the health benefits of aerobic physical activity with low risk of injury. It can be done almost anywhere at any time. A person who stays with an activity routine long enough to enjoy the rewards will be less inclined to give it up and will, over the long term, reap many health benefits. A regular walking program can prevent or slow the weight gain that commonly occurs in most adults. An average of 60 minutes a day of moderate-intensity activity or an expenditure of at least 2000 kcalories per week is especially helpful for weight management.[96] Higher levels of duration, frequency, or intensity produce greater losses.

In addition to exercise, a person can incorporate hundreds of energy-expending activities into daily routines: take the stairs instead of the elevator, walk to the neighbor's apartment instead of making a phone call, and rake the leaves instead of using a blower. Remember that sitting uses more kcalories than lying down, standing uses more kcalories than sitting, and moving uses more kcalories than standing. A 175-pound person who replaces a 30-minute television program with a 2-mile walk a day can expend enough energy to lose (or at least not gain) 18 pounds in a year. Even walking in place during the commercials of a one-hour program can increase activity time by 25 minutes, steps taken by 2100, and kcalories expended by 150.[97] Meeting an activity goal of 10,000 steps a day is an excellent way to support a healthy BMI. By wearing a pedometer, a person can easily increase physical activity, lose weight, and lower blood pressure without measuring miles or watching the clock. The point is to be active. Walk. Run. Swim. Dance. Cycle. Climb. Skip. Do whatever you enjoy doing—and do it often.

Spot Reducing People sometimes ask about "spot reducing." Unfortunately, muscles do not "own" the fat that surrounds them. Fat cells all over the body release fat in response to the demand of physical activity for use by whatever muscles are active. Specific exercises—whether moderate or intense—do not influence the site of adipose tissue loss.

Exercise can help with trouble spots in another way, though. The "trouble spot" for most men is the abdomen, their primary site of fat storage. During aerobic exercise, abdominal fat readily releases its stores, providing fuel to the physically active body. With regular exercise and weight loss, men will deplete these abdominal fat stores before those in the lower body. Women may also deplete abdominal fat with exercise, but their "trouble spots" are more likely to be their hips and thighs.

In addition to aerobic activity, strength training can help to improve the tone of muscles in a trouble area, and stretching to gain flexibility can help with associated posture problems. A combination of aerobic, strength, and flexibility workouts best improves fitness and physical appearance.

Environmental Influences

Chapter 8 describes how hormones regulate hunger, satiety, and satiation, but people don't always pay close attention to such internal signals. Instead, their eating behaviors are often dictated by environmental factors—those surrounding the eating experience as well as those pertaining to the food itself. Changing any of these factors can influence how much a person eats.

Atmosphere The environment surrounding a meal or snack influences its duration. When the lighting, décor, aromas, and sounds of an environment are pleasant and comfortable, people tend to spend more time eating and thus eat more. A person needn't eat under neon lights with offensive music to eat less, of course. Instead, after completing a meal, remove food from the table and enjoy the ambience—without the presence of visual cues to stimulate additional eating.

Accessibility Among the strongest influences on how much we eat are the accessibility, ease, and convenience of obtaining food. In general, the less effort needed to obtain food, the more likely food will be eaten. Think about it. Are you more likely to eat if half a leftover pizza is in your refrigerator or if you have to drive to the grocery store, buy a frozen pizza, and bake it for 45 minutes? Similarly you are more likely to reach for a second helping of potatoes or another piece of chicken if they are on the dining table in front of you than if the leftovers have already been wrapped and refrigerated. Having food nearby and visible encourages eating—regardless of hunger. The message is clear. For people wanting to eat fewer empty-kcalorie or high-kcalorie foods, keep them out of sight in an inconvenient place, or better yet, don't even bring them home. In contrast, a bowl of fruit on the counter and vegetables in the refrigerator promote healthy eating options.

Socializing People tend to eat more when socializing with others. Pleasant conversations extend the duration of a meal, allowing a person more time to eat more, and the longer the meal, the greater the consumption. In addition, by taking a visual cue from companions, a person might eat more when others at the table eat large portions or go to the buffet line for seconds.[98] One way to eat less is to pace yourself with the person who seems to be eating the least and slowest.

Social interactions also distract a person from paying attention to how much has been eaten. In some cases, socializing with friends during a meal may provide comfort and lower a person's motivation to limit consumption. In other cases, socializing with unfamiliar people during a meal—during a job interview or blind date, for example—may create stress and reduce food consumption. To eat less while socializing, pay attention to portion size.

Distractions Distractions influence food intake by initiating eating, interfering with internal controls to stop eating, and extending the duration of eating. Some people start eating dinner when a favorite television program comes on, regardless of hunger. Other people continue eating breakfast until they finish reading the newspaper. Such mindless eating can easily become overeating. Distractions interfere with a person's ability to perceive and regulate how much is consumed. Not only do people tend to eat more and feel less full after eating a meal while distracted, they tend to eat more at the next meal.[99] If distractions are a part of the eating experience, extra care is needed to control portion sizes.

Multiple Choices When offered a large assortment of foods, or several flavors of the same food, people tend to eat more. To limit intake, then, focus on a limited number of foods per meal; eating the same meal everyday tends to lower energy intake.[100] Be careful not to misunderstand and abandon variety in diet planning. Eating a variety of nutrient-dense foods from each of the food groups is still a healthy plan.

Package and Portion Sizes As noted earlier, the sizes of packages in grocery stores as well as portion sizes at restaurants and at home have increased dramatically in recent decades, contributing to the increase in obesity in the United States. Put simply, we tend to clean our plates and finish the package. The larger the bag of potato chips, the greater the intake. To keep from overeating, repackage snacks into smaller containers or eat a measured portion from a plate, not directly from the package.

Serving Containers We often use plates, utensils, and glasses as visual cues to guide our decisions on how much to eat and drink. If you plan to eat a bowl of ice cream, it matters whether the bowl you select holds 8 ounces or 24 ounces. Large dinner plates and wide glasses create illusions and misperceptions about quantities consumed. A scoop of mashed potatoes on a small plate looks larger than the same-size scoop on a large plate, leading a person to underestimate the amount of food eaten. To control portion sizes, use small bowls and plates, small serving spoons, and tall, narrow glasses. Of course, using a small plate will not result in less food eaten if multiple servings are taken.

Behavior and Attitude Changes in behavior and attitude can be very effective in supporting efforts to achieve and maintain appropriate body weight and composition. **Behavior modification** focuses on how to change behaviors to increase energy expenditure and decrease energy intake. A person must commit to taking action. Adopting a positive, matter-of-fact attitude helps to ensure success. Healthy eating and activity choices are an essential part of healthy living and should simply be incorporated into the day—much like brushing one's teeth or wearing a safety belt.

Become Aware of Behaviors To solve a problem, a person must first identify all the behaviors that created the problem. Keeping a record will help to identify eating and exercise behaviors that may need changing (see Figure 9-8, p. 286). Such self-monitoring raises awareness, establishes a baseline against which to measure future progress, and improves compliance.[101]

In this era of technology, many companies have developed weight-loss applications for smartphones to help users manage their daily food and physical activity behaviors.* Applications include diet analysis tools that can track eating habits, scanning devices that can quickly enter food data, customized activity and meal

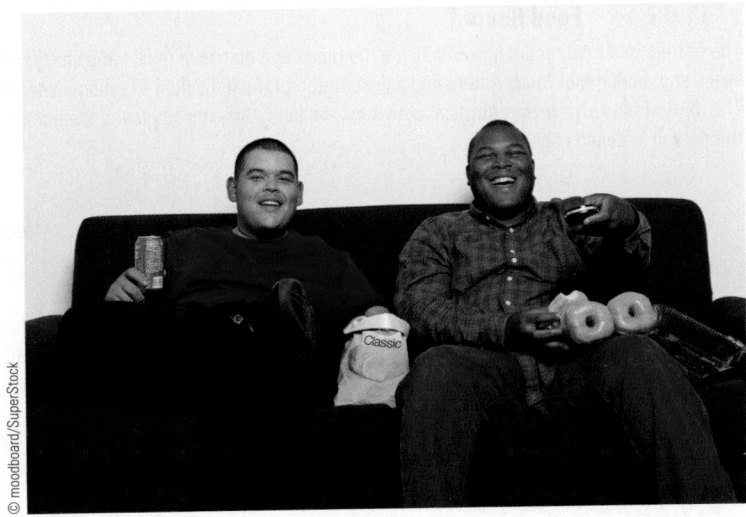

Eating from the package while distracted by television is a weight-gaining combination.

behavior modification: the changing of behavior by the manipulation of antecedents (cues or environmental factors that trigger behavior), the behavior itself, and consequences (the penalties or rewards attached to behavior).

---

*Reliable reviews of food and nutrition apps are available at www.eatright.org/appreviews.

> FIGURE 9-8 Food Record

The entries in a food record should include the times and places of meals and snacks, the types and amounts of foods eaten, and a description of the individual's feelings when eating. The diary should also record physical activities: the kind, the intensity level, the duration, and the person's feelings about them.

Time	Place	Activity or food eaten	People present	Mood
10:30–10:40	School vending machine	6 peanut butter crackers and 12 oz. cola	by myself	starved
12:15–12:30	Restaurant	Sub sandwich and 12 oz. cola	friends	relaxed & friendly
3:00–3:45	Gym	Weight training	work out partner	tired
4:00–4:10	Snack bar	Small frozen yogurt	by myself	OK

© Cengage Learning

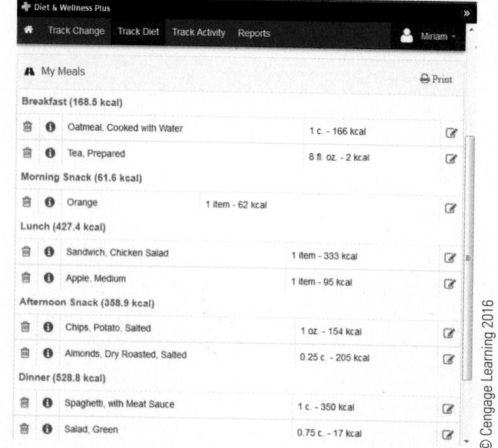

Diet analysis programs help people identify high-kcalorie foods and monitor their eating habits.

© Cengage Learning 2016

plans that can be sent to users, and support programs that deliver encouraging messages and helpful tips. Social media sites allow users to upload progress reports and receive texts. Using these applications can help a person become more aware of behaviors that lead to weight gains and losses.[102]

Change Behaviors Behavior modification strategies focus on learning desired eating and activity behaviors and eliminating unwanted behaviors. Examples include not grocery shopping when hungry and exercising when watching television. With so many possible behavior changes, a person can feel overwhelmed. Start with small time-specific goals for each behavior—for example, "I'm going to take a 30-minute walk after dinner every evening" instead of "I'm going to run in a marathon someday." Practice desired behaviors until they become routine. Addressing multiple behaviors that focus on a common goal simultaneously may better support changes than taking on one at a time. Using a reward system also seems to effectively support weight-loss efforts.

Cognitive Skills Successful behavior changes depend in part on two cognitive skills—problem solving and cognitive restructuring. Problem-solving skills enable a person to identify the problem, generate potential solutions, list the pros and cons of each, implement the most feasible solution, and evaluate whether behaviors should be continued or abandoned. Cognitive restructuring requires a person to replace negative thoughts that derail success with positive thoughts that support behavior change. In general, people who believe they can complete tasks and reach goals are more likely to follow a diet plan and achieve success than those lacking that confidence.[103] Cognitive behavioral treatment for weight loss can be most effective in helping families work together to reduce energy intake and increase physical activity.[104]

The effectiveness of cognitive behavioral treatment in weight-loss extends to other health behaviors as well. Overweight smokers who participate in a cognitive program for weight management lose weight, make healthy food choices, increase their confidence in managing their eating and smoking habits, decrease the number of cigarettes smoked, and increase their readiness to quit smoking. Such findings highlight the need to include dietary strategies in smoking cessation programs. Smoking a cigarette overrides feelings of hunger. When smokers receive a hunger signal, they can quiet it with cigarettes instead of food. Such behavior ignores body signals and postpones energy and nutrient intake. Indeed, smokers tend to weigh less than nonsmokers and to gain weight when they stop smoking. People contemplating giving up cigarettes should know that the average weight gain is about 10 pounds in the first year. Smokers wanting to quit should prepare for the possibility of weight gain and adjust their diet and activity habits so as to maintain weight during and after quitting.

Personal Attitude For many people, overeating and being overweight have become an integral part of their identity. Those who fully understand their personal relationships with food are best prepared to make healthful changes in eating and activity behaviors.

Sometimes habitual behaviors that are hazardous to health, such as smoking or drinking alcohol, contribute positively by helping people adapt to stressful situations. Similarly, many people overeat to cope with the stresses of life. Weight gains, in turn, contribute to psychosocial stress, thus creating an unhealthy cycle.

To break out of that pattern, they must first identify the particular stressors that trigger the urge to overeat. Then, when faced with these situations, they must learn and practice problem-solving skills that will help them to respond appropriately. Learning to reduce episodes of emotional eating can help lead to weight loss.

All this is not to imply that psychotherapy holds the magic answer to a weight problem. Still, efforts to improve one's general well-being may result in healthy eating and activity habits even when weight loss is not the primary goal. When the problems that trigger the urge to overeat are resolved in alternative ways, people may find they eat less. They may begin to respond appropriately to internal cues of hunger rather than inappropriately to external cues of stress. Sound emotional health supports a person's ability to take care of physical health in all ways—including nutrition, weight management, and fitness.

Support Groups Group support can prove helpful when making life changes. Some people find it useful to join a group such as Take Off Pounds Sensibly (TOPS), Weight Watchers (WW), Overeaters Anonymous (OA), or others. Some dieters prefer to form their own self-help groups or find support online. The Internet offers numerous opportunities for weight-loss education and counseling that may be effective alternatives to face-to-face or telephone counseling programs.[105] As always, consumers need to choose wisely and avoid rip-offs.

Weight Maintenance

The prevalence of **successful weight-loss maintenance** is difficult to determine, in part because researchers have used different criteria. Some look at success after 1 year and others after 5 years; some quantify success as 10 or more pounds lost and others as 5 or 10 percent of initial body weight lost. Furthermore, most research studies examine the success of one episode of weight loss in a structured program, but this scenario does not necessarily reflect the experiences of the general population. In reality, most people have lost weight several times in their lifetimes and did so on their own, not in a formal program. An estimated one out of every six overweight adults in the United States has successfully maintained at least a 10 percent loss for at least a year.[106]

Those who are successful in maintaining their weight loss have established regular exercise regimens and careful eating patterns, taking in less energy than the national average. Because formerly overweight people are more efficient at

© Blend Images/Dan Bannister/Jupiterimages

Maintaining a healthy body weight requires maintaining the vigorous physical activities and careful eating habits that supported weight loss.

successful weight-loss maintenance: achieving a weight loss of at least 5 to 10 percent of initial body weight and maintaining the loss for at least 1 year.

storing fat, they do not have the same flexibility in their food and activity habits as their friends who have never been overweight. With weight loss, hormones involved in appetite regulation shift in a way that encourages weight gain, and metabolism shifts downward so that formerly overweight people require less energy than might be expected given their current body weight and body composition.[107] These hormonal and metabolic changes persist over time.[108] Consequently, to keep weight off, they must either eat less or exercise more than people the same size who have never been obese. Put simply, it takes more effort to prevent weight regain than to prevent weight gain.

Physical activity plays a key role in preventing weight gains and maintaining weight losses.[109] Those who consistently exercise are far more successful than those who are inactive. Weight maintenance may require a person to expend at least 2500 kcalories in physical activity per week. To accomplish this, a person might exercise either moderately (such as brisk walking at 4 miles per hour) for 60 minutes a day or vigorously (such as fast bicycling at 18 miles per hour) for 30 minutes a day, for example. Being active during both work hours and leisure time also helps a person expend more energy and maintain weight loss.[110]

In addition to limiting energy intake and exercising regularly, one other strategy helps with weight maintenance: frequent self-monitoring. People who weigh themselves periodically and monitor their eating and exercise habits regularly can detect weight gains in the early stages and promptly initiate changes to prevent relapse.

Losing weight and maintaining the loss may not be easy, but it is possible. The National Weight Control Registry tracks over 10,000 individuals who have maintained a significant weight loss over time. Strategies of those who have been successful may differ in the details, but in general, most do the following:[111]

- Eat a low-kcalorie diet (usually small portions four to five times a day).
- Follow a diet that is high in nutrient density and low in energy density.
- Eat breakfast (curbs hunger).
- Engage in physical activity regularly (at least 60 minutes of moderate activity daily).
- Monitor weight frequently (at least weekly) and take prompt action with small gains.
- Use productive problem-solving skills and positive self-talk.
- Limit television time (less than 10 hours a week).
- Consult a registered dietitian nutritionist, physician, or other support person (or group).

Importantly, people who are successful losing weight find that it gets easier with time—the changes in diet and activity patterns become permanent.

Prevention Given the information presented up to this point in the chapter, the adage "An ounce of prevention is worth a pound of cure" seems particularly apropos. Obesity is a major risk factor for numerous diseases, and losing weight is challenging and often temporary. Many of the strategies for preventing weight gain are very similar to those for losing weight, with one exception: they begin early. Over the years, these strategies become an integral part of a person's life:

- Eat regular meals and limit snacking
- Drink water instead of high-kcalorie beverages
- Select sensible portion sizes and limit daily energy intake to no more than energy expended
- Become physically active and limit sedentary activities

It is much easier for a person to resist doughnuts for breakfast if he rarely eats them. Similarly, a person will have little trouble walking each morning if she has always been active.

TABLE 9-4 National Strategies to Prevent Obesity

- Provide a variety of opportunities to help make physical activity an integral and routine part of life.
- Create environments that ensure healthy foods and beverages are visible, attractive, and easy-to-obtain.
- Encourage media messages that promote physically active lifestyles and nutritionally healthy diets.
- Support health care providers in offering information on weight management and employers in offering wellness programs.
- Make schools centers for health and wellness.

© Cengage Learning

SOURCE: *Accelerating Progress in Obesity Prevention: Solving the Weight of the Nation*, (Washington, DC: Institute of Medicine of the National Academies), 2012.

> **DIETARY GUIDELINES FOR AMERICANS**
Maintain appropriate kcalorie balance during each stage of life—childhood, adolescence, adulthood, pregnancy and lactation, and older age.

Community Programs Reversing the US obesity epidemic is a challenge in an environment of abundant food and physical inactivity. Success may depend on community actions to promote healthy lifestyle choices. Table 9-4 lists health strategies to speed the progress in obesity prevention in the United States.[112] Whether changes in public policy—such as providing pedestrian-friendly streets or taxing sugar-sweetened beverages and high-fat snacks—will influence activity or diet habits remains to be seen.[113] Clearly, effective strategies will need to reach beyond individuals to address social networks, community institutions, and government policies.

> **REVIEW IT** Outline reasonable strategies for achieving and maintaining a healthy body weight.
A surefire remedy for obesity has yet to be found, although many people find a combination of approaches to be most effective. Diet and exercise shift energy balance so that more energy is expended than is taken in. Behavior modification and cognitive restructuring retrain habits to support a healthy eating and activity plan. Such a plan requires time, individualization, and sometimes the assistance of a registered dietitian nutritionist or support group.

9.6 Underweight

> **LEARN IT** Summarize strategies for gaining weight.

Underweight is a far less prevalent problem than overweight, affecting no more than 2 percent of US adults (review Figure 8-7 on p. 247). Whether an underweight person needs to gain weight is a question of health and, like weight loss, a highly individual matter. There are no compelling reasons for people who are healthy at their present weight to try to gain weight. Those who are thin because of malnourishment or illness, however, might benefit from a diet that supports weight gain. Medical advice can help make the distinction.

Thin people may find gaining weight difficult. Unlike the genes expressed in obesity, the genes in lean people protect against energy excesses. Those who wish to gain weight for appearance's sake or to improve their athletic performance need to be aware that healthful weight gains can be achieved only by physical conditioning combined with high energy intakes. On a high-kcalorie diet alone, a person may gain weight, but it will be mostly fat. Even if the gain improves appearance, it can be detrimental to health and might impair athletic performance. Therefore, in weight gain, as in weight loss, physical activity and energy intake are essential components of a sound plan.

Problems of Underweight The causes of underweight may be as diverse as those of overweight—genetic tendencies; hunger, appetite, and satiety

underweight: body weight lower than the weight range that is considered healthy; BMI less than 18.5.

irregularities; psychological traits; and metabolic factors. Habits learned early in childhood, especially food aversions, may perpetuate themselves.

The high demand for energy to support physical activity and growth may contribute to underweight. An active, growing boy may need more than 4000 kcalories a day to maintain his weight and may be too busy to take time to eat adequately. In addition, underweight people may find it hard to gain weight because they are expending energy in adaptive thermogenesis. So much energy may be expended adapting to a higher food intake that at first as many as 750 to 800 extra kcalories a day may be needed to gain a pound a week. Like those who want to lose weight, people who want to gain must learn new habits and learn to like new foods. They are also similarly vulnerable to potentially harmful schemes.

As described in Highlight 8, the underweight condition anorexia nervosa sometimes develops in people who employ self-denial to control their weight. They go to such extremes that they become severely undernourished, achieving final body weights of 70 pounds or even less. One difference between a person with anorexia nervosa and other underweight people is that starvation is intentional. (See Highlight 8 for a review of anorexia nervosa and other eating disorders.)

Weight-Gain Strategies Adequacy and balance are the key diet-planning strategies for weight gain. Meals focus on energy-dense foods to provide many kcalories in a small volume and exercise to build muscle. By using the USDA Food Pattern recommendations for the higher kcalorie levels (see Table 2-3 on p. 43), a person can gain weight while meeting nutrient needs.

Energy-Dense Foods Energy-dense foods (the very ones eliminated from a successful weight-loss diet) hold the key to weight gain. Pick the highest-kcalorie items from each food group—that is, milk shakes instead of fat-free milk, salmon instead of snapper, avocados instead of cucumbers, a cup of grape juice instead of a small apple, and whole-wheat muffins instead of whole-wheat bread. Because fat provides more than twice as many kcalories per teaspoon as sugar does, fat adds kcalories without adding much bulk.

Although eating high-kcalorie, high-fat foods is not healthy for most people, it may be essential for an underweight individual who needs to gain weight. An underweight person who is physically active and eating a nutritionally adequate diet can afford a few extra kcalories from fat. For health's sake, it is wise to select foods with monounsaturated and polyunsaturated fats instead of those with saturated or *trans* fats: for example, sautéing vegetables in olive oil instead of butter or hydrogenated margarine.

Regular Meals Daily People who are underweight need to make meals a priority and take the time to plan, prepare, and eat each meal. They should eat at least three healthy meals every day. Another suggestion is to eat meaty appetizers or the main course first and leave the soup or salad until later.

Large Portions Underweight people need to learn to eat more food at each meal. For example, they can add extra slices of ham and cheese on a sandwich for lunch, drink milk from a larger glass, and eat cereal from a larger bowl.

The person should expect to feel full. Most underweight individuals are accustomed to small quantities of food. When they begin eating significantly more, they feel uncomfortable. This is normal and passes over time.

Extra Snacks Because a substantially higher energy intake is needed each day, in addition to eating more food at each meal, it is necessary to eat more frequently. Between-meal snacks can readily lead to weight gains. For example, a student might make three sandwiches in the morning and eat them between classes in addition to the day's three regular meals. Snacking on dried fruit, nuts, and seeds is also an easy way to add kcalories.

Juice and Milk Beverages provide an easy way to increase energy intake. Consider that 6 cups of cranberry juice add almost 1000 kcalories to the day's intake.

TABLE 9-5 **Weight-Gain Strategies**

- Energy in should exceed energy out by at least 500 kcalories/day. Eat enough to store more energy than you expend in exercise. Exercise and eat to build muscles.
- Expect weight gain to take time (1 pound per month would be reasonable).
- Emphasize energy-dense foods.
- Eat at least three meals a day.
- Eat large portions of foods and expect to feel full.
- Eat snacks between meals.
- Drink plenty of juice and milk.

© Cengage Learning

kCalories can be added to milk by mixing in powdered milk or packets of instant breakfast.

For people who are underweight because of illness, liquid dietary supplements are often recommended because a weak person can swallow them easily. Used in addition to regular meals, these high-protein, high-kcalorie formulas can help an underweight person maintain or gain weight easily.

Exercising to Build Muscles To gain weight, use strength training primarily, and increase energy intake to support that exercise. Eating extra food to provide an additional 500 to 1000 kcalories a day above normal energy needs can support the exercise as well as build muscle.

> **REVIEW IT** Summarize strategies for gaining weight.

Both the incidence of underweight and the health problems associated with it are less prevalent than overweight and its associated problems. To gain weight, a person must train physically and increase energy intake by selecting energy-dense foods, eating regular meals, taking larger portions, and consuming extra snacks and beverages. Table 9-5 includes a summary of weight-gain strategies.

Achieving and maintaining a healthy weight requires vigilant attention to diet and physical activity. Taking care of oneself is a lifelong responsibility.

Nutrition Portfolio

To enjoy good health and maintain a reasonable body weight, combine sensible eating habits and regular physical activity. Go to Diet & Wellness Plus and choose one of the days on which you have tracked your diet for the entire day. Go to the Energy Balance and Intake vs. Goals reports.

- Calculate your BMI and consider whether you need to maintain, lose, or gain weight for the sake of good health. If you do need to gain or lose weight, do the Diet Analysis reports give you insight into why you may be overweight or underweight?
- Reflect on your weight over the past year or so and explain any weight gains or losses. Using the Intake vs. Goals report, can you identify areas in which you need to adjust your food intake, perhaps eating more or less?
- Describe the potential risks and possible benefits of fad diets and over-the-counter weight-loss drugs or herbal supplements.

 DIET & WELLNESS **PLUS**   To complete this exercise, go to your Diet & Wellness Plus at **www.cengagebrain.com.**

> **STUDY IT** To review the key points of this chapter and take a practice quiz, go to the study cards at the end of the book.

REFERENCES

1. K. M. Flegal and coauthors, Prevalence of obesity and trends in the distribution of body mass index among US adults, 1999–2010, *Journal of the American Medical Association* 307 (2012): 491–497.

2. Centers for Disease Control and Prevention, Obesity and overweight, www.cdc.gov/nchs/fastats/overwt.htm, accessed November 2013.

3. C. L. Ogden and coauthors, Prevalence of obesity and trends in body mass index among US children and adolescents, 1999–2010, *Journal of the American Medical Association* 307 (2012): 483–490.

4. C. L. Ogden and coauthors, Prevalence of obesity in the United States, 2009–2010, NCHS data brief, no 82 (Hyattsville, MD: National Center for Health Statistics), 2012.

5. World Health Organization, www.who.int/features/factfiles/obesity, accessed November 2013.

6. B. M. Popkin, L. S. Adair, and S. W. Ng, Global nutrition transition and the pandemic of obesity in developing countries, *Nutrition Reviews* 70 (2012): 3–21; B. A. Swinburn and coauthors, The global obesity pandemic: Shaped by global drivers and local environments, *Lancet* 378 (2011): 804–814; R. W. Kimokoti and B. E. Millen, Diet, the global obesity epidemic, and prevention, *Journal of the American Dietetic Association* 111 (2011): 1137–1140.

7. Y. D. Tchoukalova and coauthors, Regional differences in cellular mechanisms of adipose tissue gain with overfeeding, *Proceedings of the National Academic of Sciences of the United States of America* 107 (2010): 18226–18231.

8. C. A. Baile and coauthors, Effect of resveratrol on fat mobilization, *Annals of the New York Academy of Sciences* 1215 (2011): 40–47.

9. M. Krawczyk, L. Bonfrate, and P. Portincasa, Nonalcoholic fatty liver disease, *Best Practice and Research, Clinical Gastroenterology* 24 (2010): 695–708; D. M. Muoio, Metabolism and vascular fatty acid transport, *New England Journal of Medicine* 363 (2010): 291–293; G. Tarantino, S. Savastano, and A. Colao, Hepatic steatosis, low-grade chronic inflammation and hormone/growth factor/adipokine imbalance, *World Journal of Gastroenterology* 16 (2010): 4773–4783.

10. P. Trayhurn, C. A. Drevon, and J. Eckel, Secreted proteins from adipose tissue and skeletal muscle: Adipokines, myokines and adipose/muscle cross-talk, *Archives of Physiology and Biochemistry* 117 (2011): 47–56; N. Ouchi and coauthors, Adipokines in inflammation and metabolic disease, *Nature Reviews Immunology* 11 (2011): 85–97; C. Stryjecki and D. M. Mutch, Fatty acid-gene interactions, adipokines and obesity, *European Journal of Clinical Nutrition* 65 (2011): 285–297.

11. H. Wong and R. H. Eckel, Lipoprotein lipase in the brain and nervous system, *Annual Review of Nutrition* 32 (2012): 147–160.

12. S. Camps, S. Verhoef, and K. R. Westerterp, Weight loss, weight maintenance, and adaptive thermogenesis, *American Journal of Clinical Nutrition* 97 (2013): 990–994.

13. J. R. Speakman and coauthors, Set points, settling points and some alternative models: Theoretical options to understand how genes and environments combine to regulate body adiposity, *Disease Models and Mechanisms* 4 (2011): 733–745.

14. J. R. Speakman, Evolutionary perspectives on the obesity epidemic: Adaptive, maladaptive, and neutral viewpoints, *Annual Review of Nutrition* 33 (2013): 289–317; L. Dubois and coauthors, Genetic and environmental contributions to weight, height, and BMI from birth to 19 years of age: An international study of over 12,000 twin pairs, *PLoS One* 7 (2012): e30153; B. Levin, Developmental gene x environment interactions affecting systems regulating energy homeostasis and obesity, *Frontiers in Neuroendocinology* 31 (2010): 270–283.

15. C. Lavebratt, M. Almgren, and T. J. Ekström, Epigenetic regulation in obesity, *International Journal of Obesity* 36 (2012): 757–765.

16. K. Silventoinen and coauthors, The genetic and environmental influences on childhood obesity: A systematic review of twin and adoption studies, *International Journal of Obesity* 34 (2010): 29–40.

17. J. Naukkarinen and coauthors, Causes and consequences of obesity: The contribution of recent twin studies, *International Journal of Obesity* 36 (2012): 1017–1024.

18. J. Cecil and coauthors, Obesity and eating behaviour in children and adolescents: Contribution of common gene polymorphisms, *International Review of Psychiatry* 24 (2012): 200–210; M. Manco and B. Dallapiccola, Genetics of pediatric obesity, *Pediatrics* 130 (2012): 123–133.

19. P. T. Williams, Attenuated inheritance of body weight by running in monozygotic twins, *Medicine and Science in Sports and Exercise* 44 (2012): 98–103; S. Li and coauthors, Physical activity attenuates the genetic predisposition to obesity in 20,000 men and women from EPIC-Norfolk prospective population study, *PLoS Medicine* 7 (2010): e1000331.

20. J. Mattei and coauthors, *TCF7L2* genetic variants modulate the effect of dietary fat intake on changes in body composition during a weight-loss intervention, *American Journal of Clinical Nutrition* 96 (2012): 1129–1136; Q. Qibin and coauthors, Sugar-sweetened beverages and genetic risk of obesity, *New England Journal of Medicine* 367 (2012): 1387–1396.

21. Y. Xu and Q. Tong, Expanding neurotransmitters in the hypothalamic neurocircuitry for energy balance, *Protein Cell* 2 (2011): 800–813; L. Gautron and J. K. Elmquist, Sixteen years and counting: An update on leptin in energy balance, *Journal of Clinical Investigation* 121 (2011): 2087–2093.

22. H. R. Kissileff and coauthors, Leptin reverses declines in satiation in weight-reduced obese humans, *American Journal of Clinical Nutrition* 95 (2012): 309–317.

23. J. Tong and coauthors, Ghrelin enhances olfactory sensitivity and exploratory sniffing in rodents and humans, *Journal of Neuroscience* 31 (2011): 5841–5846; T. R. Castañeda and coauthors, Ghrelin in the regulation of body weight and metabolism, *Frontiers in Neuroendocrinology* 31 (2010): 44–60.

24. A. J. Crum and coauthors, Mind over milkshakes: Mindsets, not just nutrients, determine ghrelin response, *Health Psychology* 30 (2011): 424–429.

25. M. St. Onge and coauthors, Sleep restriction leads to increased activation of brain regions sensitive to food stimuli, *American Journal of Clinical Nutrition* 95 (2012): 818–824; H. K. J. Gonnissen and coauthors, Effect of a phase advance and phase delay of the 24-h cycle on energy metabolism, appetite, and related hormones, *American Journal of Clinical Nutrition* 96 (2012): 689–697; C. Benedict and coauthors, Acute sleep deprivation reduces energy expenditure in healthy men, *American Journal of Clinical Nutrition* 93 (2011): 1229–1236; R. Hursel and coauthors, Effects of sleep fragmentation in healthy men on energy expenditure, substrate oxidation, physical activity, and exhaustion measured over 48 h in a respiratory chamber, *American Journal of Clinical Nutrition* 94 (2011): 804–808; P. Lyytikäinen and coauthors, Sleep problems and major weight gain: A follow-up study, *International Journal of Obesity* 35 (2011): 109–114; L. Brondel and coauthors, Acute partial sleep deprivation increases food intake in healthy men, *American Journal of Clinical Nutrition* 91 (2010): 1550–1559.

26. S. Li and coauthors, Cumulative effects and predictive value of common obesity-susceptibility variants identified by genome-wide association studies, *American Journal of Clinical Nutrition* 91 (2010): 184–190.

27. M. M. Hetherington and J. E. Cecil, Gene-environment interactions in obesity, *Forum of Nutrition* 63 (2010): 195–203; C. Bouchard, Defining the genetic architecture of the predisposition to obesity: A challenging but not insurmountable task, *American Journal of Clinical Nutrition* 91 (2010): 5–6.

28. M. E. Lidell and coauthors, Evidence for two types of brown adipose tissue in humans, *Nature Medicine* 19 (2013): 631–634; B. Cannon and J. Nedergaard, Yes, even human brown fat is on fire! *Journal of Clinical Investigation* 122 (2012): 486–489.

29. K. A. Virtanen and P. N. Muutila, Brown adipose tissue in humans, *Current Opinion in Lipidology* 22 (2011): 49–54.

30. A. Bartelt and J. Heeren, The holy grail of metabolic disease: Brown adipose tissue, *Current Opinion in Lipidology* 23 (2012): 190–195.

31. T. J. Schulz and Y. H. Tseng, Brown adipose tissue: Development, metabolism and beyond, *Biochemical Journal* 453 (2013): 167–178; E. Ravussin and J. E. Galgani, The implication of brown adipose tissue for humans, *Annual Review of Nutrition* 31 (2011): 33–47.

32. K. A. Lo and L. Sun, Turning WAT into BAT: A review on regulators controlling the browning of white adipocytes, *Bioscience Reports* 33 (2013): 711–719; G. E. Beranger, In vitro brown and "brite"/"beige" adipogenesis: Human cellular models and molecular aspects, *Biochimica et Biophysica Acta* 1831 (2013): 905–914; M. L. Bonet, P. Oliver, and A. Palou, Pharmacological and nutritional agents promoting browning of white adipose tissue, *Biochimica et Biophysica Acta* 1831 (2013): 969–985; J. Wu and coauthors, Beige adipocytes are a distinct type of thermogenic fat cell in mouse and human, *Cell* 150 (2012): 366–376.

33. L. Z. Sharp and coauthors, Human BAT possesses molecular signatures that resemble beige/brite cells, *PLoS One* 7 (2012): doi 10.1371.

34. B. K. Pedersen, A muscular twist on the fate of fat, *New England Journal of Medicine* 366 (2012): 1544–1545.

35. J. M. McCaffery and coauthors, Effects of social contact and zygosity on 21-y weight change in male twins, *American Journal of Clinical Nutrition* 94 (2011): 404–409.

36. B. M. Popkin and K. J. Duffey, Does hunger and satiety drive eating anymore? Increasing eating occasions and decreasing time between eating occasions in the United States, *American Journal of Clinical Nutrition* 91 (2010): 1342–1347.

37. K. J. Duffey and B. M. Popkin, Energy density, portion size, and eating occasions: Contributions to increased energy intake in the United States, 1977–2006, *PLoS Medicine* 8 (2011): e1001050.

38. R. N. Close and D. A. Schoeller, The financial reality of overeating, *Journal of the American College of Nutrition* 25 (2006): 203–209.

39. I. N. Bezerra, C. Curioni, and R. Sichieri, Association between eating out of home and body weight, *Nutrition Reviews* 70 (2012): 65–79.

40. J. F. Sallis and coauthors, Role of built environments in physical activity, obesity, and cardiovascular disease, *Circulation* 125 (2012): 729–737.

41. T. S. Church and coauthors, Trends over 5 decades in US occupation-related physical activity and their association with obesity, *PLoS ONE* 6 (2011): e19657.

42. K. Wijndaele and coauthors, Increased cardiometabolic risk is associated with increased TV viewing time, *Medicine and Science in Sports and Exercise* 42 (2010): 1511–1518.

43. H. Konttinen and coauthors, Emotional eating and physical activity self-efficacy as pathways in the association between depressive symptoms and adiposity indicators, *American Journal of Clinical Nutrition* 92 (2010): 1031–1039.

44. D. P. Beavers and coauthors, Cardiometabolic risk after weight loss and subsequent weight regain in overweight and obese postmenopausal women, *Journals of Gerontology Series A: Biological Sciences and medical Sciences* 68 (2013): 691–698; K. Stohacker and B. K. McFarlin, Influence of obesity, physical inactivity, and weight cycling on chronic inflammation, *Frontiers in Bioscience (Elite Edition)* 2 (2010): 98–104.

45. K. Poddar and coauthors, Nutraceutical supplements for weight loss: A systematic review, *Nutrition in Clinical Practice* 26 (2011): 539–552.

46. FDA, Tainted weight loss products, http://www.fda.gov/drugs/resources foryou/consumers/buyingusingmedicinesafely/medicationhealthfraud /ucm234592.htm, accessed November 2013; M. H. Tang and coauthors, Case series on a diversity of illicit weight-reducing agents: From the well known to the unexpected, *British Journal of Clinical Pharmacology* 71 (2011): 250–253.

47. A. Astrup, Drug management of obesity: Efficacy versus safety, *New England Journal of Medicine* 363 (2010): 288–290.

48. E. H. Morrato and D. B. Allison, FDA approval of obesity drugs: A difference in risk-benefit perceptions, *Journal of the American Medical Association* 308 (2012): 1097–1098; E. Colman and coauthors, The FDA's assessment of two drugs for chronic weight management, *New England Journal of Medicine* 367 (2012): 1577–1579.

49. C. D. Fryar, M. D. Carroll, and C. L. Ogden, Prevalence of overweight, obesity, and extreme obesity among adults: United States, trends 1960–1962 through 2009–2010, *NCHS Health E-Stats*, September 2012.

50. R. Padwal and coauthors, Bariatric surgery: A systematic review of the clinical and economic evidence, *Journal of Internal Medicine* 26 (2011): 1183–1194; A. Nagle, Bariatric surgery: A surgeon's perspective, *Journal of the American Dietetic Association* 110 (2010): 520–523; G. L. Blackburn, S. Wollner, and S. B. Heymsfield, Lifestyle interventions for the treatment of class III obesity: A primary target for nutrition medicine in the obesity epidemic, *American Journal of Clinical Nutrition* 91 (2010): 289S–292S.

51. E. H. Livingston, The incidence of bariatric surgery has plateaued in the U.S., *American Journal of Surgery* 200 (2010): 378–385.

52. B. Laferrère, Diabetes remission after bariatric surgery: Is it just the incretins? *International Journal of Obesity* 35 (2011): S22–S25; L. M. Beckman, T. R. Beckman, and C. P. Earthman, Changes in gastrointestinal hormones and leptin after Roux-en-Y gastric bypass procedure: A review, *Journal of the American Dietetic Association* 110 (2010): 571–584.

53. L. Kong and coauthors, Gut microbiota after gastric bypass in human obesity: Increased richness and associations of bacterial genera with adipose tissue genes, *American Journal of Clinical Nutrition* 98 (2013): 16–24; A. D. Miras and coauthors, Gastric bypass surgery for obesity decreases the reward value of a sweet-fat stimulus as assessed in a progressive ratio task, *American Journal of Clinical Nutrition* 96 (2012): 467–473.

54. P. E. O'Brien and coauthors, Long-term outcomes after bariatric surgery: Fifteen-year follow-up of adjustable gastric banding and a systematic review of the bariatric surgical literature, *Annals of Surgery* 257 (2013): 87–94.

55. T. D. Adams, Health benefits of gastric bypass surgery after 6 years, *Journal of the American Medical Association* 308 (2012): 1122–1131; G. Mingrone and coauthors, Bariatric surgery versus conventional medical therapy for type 2 diabetes, *New England Journal of Medicine* 366 (2012): 1577–1585; D. Sandoval, Bariatric surgeries: Beyond restriction and malabsorption, *International Journal of Obesity* 35 (2011): S45–S49.

56. E. H. Livingston, Surgical treatment of obesity in adolescence, *Journal of the American Medical Association* 303 (2010): 559–560.

57. L. Beckman and C. Earthman, Nutritional implications of bariatric surgery and the role of registered dietitians, *Journal of the Academy of Nutrition and Dietetics* 113 (2013): 398–399; Y. Chen, Acute bariatric surgery complications: Managing parenteral nutrition in the morbidly obese, *Journal of the American Dietetic Association* 110 (2010): 1734–1737; D. Kulick, L. Hark, and D. Deen, The bariatric surgery patient: A growing role for registered dietitians, *Journal of the American Dietetic Association* 110 (2010): 593–599; G. Snyder-Marlow, D. Taylor, and J. Lenhard, Nutrition care for patients undergoing laparoscopic sleeve gastrectomy for weight loss, *Journal of the American Dietetic Association* 110 (2010): 600–607.

58. E. Saltzman and J. P. Karl, Nutrient deficiencies after gastric bypass surgery, *Annual Review of Nutrition* 33 (2013): 183–203; S. P. Donadelli and coauthors, Daily vitamin supplementation and hypovitaminosis after obesity surgery, *Nutrition* 28 (2012): 391–396; R. Welbourn and D. Pournaras, Bariatric surgery: A cost-effective intervention for morbid obesity; functional and nutritional outcomes, *Proceedings of the Nutrition Society* 69 (2010): 528–535.

59. M. Kruseman and coauthors, Dietary, weight, and psychological changes among patients with obesity, 8 years after gastric bypass, *Journal of the American Dietetic Association* 110 (2010): 527–534.

60. L. Sjöström and coauthors, Bariatric surgery and long-term cardiovascular events, *Journal of the American Medical Association* 301 (2012): 56–65.

61. P. J. Stephan and J. M. Kenkel, Updates and advances in liposuction, *Aesthetic Surgery Journal* 30 (2010) 83–97.

62. T. L. Hernandez and coauthors, Fat redistribution following suction lipectomy: Defense of body fat and patterns of restoration, *Obesity* 19 (2011): 1388–1395.

63. D. Laddu and coauthors, A review of evidence-based strategies to treat obesity in adults, *Nutrition in Clinical Practice* 26 (2011): 512–525; D. Heber, An integrative view of obesity, *American Journal of Clinical Nutrition* 91 (2010): 280S–283S.

64. M. D. Jensen and coauthors, *2013 AHA/ACC/TOS Guideline for the Management of Overweight and Obesity in Adults*, 2013.

65. K. M. Beavers and coauthors, Is lost lean mass from intentional weight loss recovered during weight regain in postmenopausal women? *American Journal of Clinical Nutrition* 94 (2011): 767–774.

66. A. J. Tomiyama and coauthors, Low calorie dieting increases cortisol, *Psychosomatic Medicine* 72 (2010): 357–364.

67. H. M. Seagle and coauthors, Position of the American Dietetic Association: Weight management, *Journal of the American Dietetic Association* 109 (2009): 330–346.

68. A. W. Brown, M. M. B. Brown, and D. B. Allison, Belief beyond the evidence: Using the proposed effect of breakfast on obesity to show 2 practices that distort scientific evidence, *American Journal of Clinical Nutrition* 98 (2013): 1298–1308.

69. P. Deshmukh-Taskar and coauthors, The relationship of breakfast skipping and type of breakfast consumed with overweight/obesity, abdominal obesity, other cardiometabolic risk factors and the metabolic syndrome in young adults. The National Health and Nutrition Examination Survey (NHANES): 1999–2006, *Public Health Nutrition* 16 (2013): 2073–2082; S. P. P. Tin and coauthors, Breakfast skipping and change in body mass index in young children, *International Journal of Obesity* 35 (2011): 899–906; K. J. Smith and coauthors, Skipping breakfast: Longitudinal associations with cardiometabolic risk factors in the Childhood Determinants of Adult Health Study, *American Journal of Clinical Nutrition* 92 (2010): 1316–1325.

70. H. J. Leidy and coauthors, Beneficial effects of a higher-protein breakfast on the appetitive, hormonal, and neural signals controlling energy intake regulation in overweight/obese, "breakfast skipping," late-adolescent girls, *American Journal of Clinical Nutrition* 97 (2013): 677–688; P. R. Deshmukh-Taskar and coauthors, The relationship of breakfast skipping and type of breakfast consumption with nutrient intake and weight status in children and adolescents: The National Health and Nutrition Examination Survey 1999–2006, *Journal of the American Dietetic Association* 110 (2010): 869–878.

71. D. Jakubowicz and coauthors, High caloric intake at breakfast vs. dinner differentially influences weight loss of overweight and obese women, *Obesity* (2013): doi:10.1002/oby.20460.

72. K. J. Duffey and B. M. Popkin, Energy density, portion size, and eating occasions: Contributions to increased energy intake in the United States, 1977–2006, *PLoS Medicine* 8 (2011): e1001050.

73. C. L. Rock and coauthors, Effect of a free prepared meal and incentivized weight loss program on weight loss and weight loss maintenance in obese and overweight women: A randomized controlled trial, *Journal of the American Medical Association* 304 (2010): 1803–1810; Position of the American Dietetic Association: Weight management, *Journal of the American Dietetic Association* 109 (2009): 330–346.

74. J. Li and coauthors, Improvement in chewing activity reduces energy intake in one meal and modulates plasma gut hormone concentrations in obese and lean young Chinese men, *American Journal of Clinical Nutrition* 94 (2011): 709–716.

75. A. Kokkinos and coauthors, Eating slowly increases the postprandial response of the anorexigenic gut hormones, peptide YY and glucagon-like peptide-1, *Journal of Clinical Endocrinology and Metabolism* 95 (2010): 333–337.

76. S. L. Leong and coauthors, Faster self-reported speed of eating is related to higher body mass index in a nationwide survey of middle-aged women, *Journal of the American Dietetic Association* 111 (2011): 1192–1197.

77. R. A. Williams, L. S. Roe, and B. J. Rolls, comparison of three methods to reduce energy density: Effects on daily energy intake, *Appetite* 66 (2013): 75–83; R. Pérez-Escamilla and coauthors, Dietary energy density and body weight in adults and children: A systematic review, *Journal of the Academy of Nutrition and Dietetics* 112 (2012): 671–684; H. A. Raynor and coauthors, The effects of an energy density prescription on diet quality and weight loss: A pilot randomized controlled trial, *Journal of the Academy of Nutrition and Dietetics* 112 (2012): 1397–1402.

78. R. Muckelbauer and coauthors, Association between water consumption and body weight outcomes: A systematic review, *American Journal of Clinical Nutrition* 98 (2013): 282–299.

79. M. C. Daniels and B. M. Popkin, Impact of water intake on energy intake and weight status: A systematic review, *Nutrition Reviews* 68 (2010): 505–521.

80. D. F. Tate and coauthors, Replacing caloric beverages with water or diet beverages for weight loss in adults: Main results of the Choose Healthy Options Consciously Everyday (CHOICE) randomized clinical trial, *American Journal of Clinical Nutrition* 95 (2012): 555–563.

81. L. Hooper and coauthors, Effect of reducing total fat intake on body weight: Systematic review and meta-analysis of randomised controlled trials and cohort studies, *British Medical Journal* 345 (2012): e766.

82. M. A. Pereira, Diet beverages and the risk of obesity, diabetes, and cardiovascular disease: A review of the evidence, *Nutrition Reviews* 71 (2013): 433–440.

83. E. Green and C. Murphy, Altered processing of sweet taste in the brain of diet soda drinkers, *Physiology and Behavior* 107 (2012): 560–567.

84. C. D. Chapman and coauthors, Lifestyle determinants of the drive to eat: A meta-analysis, *American Journal of Clinical Nutrition* 96 (2012): 492–497.

85. B. H. Goodpaster and coauthors, Effects of diet and physical activity interventions on weight loss and cardiometabolic risk factors in severely obese adults: A randomized study, *Journal of the American Medical Association* 304 (2010): 1795–1802; A. L. Hankinson and coauthors, Maintaining a high physical activity level over 20 years and weight gain, *Journal of the American Medical Association* 304 (2010): 2603–2610.

86. J. E. Donnelly and coauthors, American College of Sports Medicine Position Stand: Appropriate physical activity intervention strategies for weight loss and prevention of weight regain for adults, *Medicine and Science in Sports and Exercise* 41 (2009): 459–471; Committee on Dietary Reference Intakes, *Dietary Reference Intakes for Energy, Carbohydrate, Fiber, Fat, Fatty Acids, Cholesterol, Protein, and Amino Acids* (Washington, D.C.: National Academies Press, 2005).

87. Goodpaster and coauthors, 2010; M. Hamer and G. O'Donovan, Cardiorespiratory fitness and metabolic risk factors in obesity, *Current Opinion in Lipidology* 21 (2010): 1–7; D. E. Larson-Meyer and coauthors, Caloric restriction with or without exercise: The fitness versus fatness debate, *Medicine and Science in Sports and Exercise* 42 (2010): 152–159.

88. D. T. Villareal and coauthors, Weight loss, exercise, or both and physical function in obese older adults, *New England Journal of Medicine* 364 (2011): 1218–1229.

89. J. E. Turner and coauthors, Nonprescribed physical activity energy expenditure is maintained with structured exercise and implicates a compensatory increase in energy intake, *American Journal of Clinical Nutrition* 92 (2010): 1009–1016.

90. A. M. Knab and coauthors, A 45-minute vigorous exercise bout increases metabolic rate for 14 hours, *Medicine and Science in Sports and Exercise* 43 (2011): 1643–1648.

91. S. H. Boutcher, High-intensity intermittent exercise and fat loss, *Journal of Obesity* 2011 (2011): doi 10.1155.2011.868305.

92. U. Ekelund and coauthors, Physical activity and gain in abdominal adiposity and body weight: Prospective cohort study in 288,498 men and women, *American Journal of Clinical Nutrition* 93 (2011): 826–835; C. A. Slentz and coauthors, The effects of aerobic versus resistance training on visceral and liver fat stores, liver enzymes and HOMA from STRRIDE AT/RT: A randomized trial, *American Journal of Physiology: Endocrinology and Metaolism* 301 (2011): E1033-E1039; J. W. Bea and coauthors, Resistance training predicts 6-yr body composition change in postmenopausal women, *Medicine and Science in Sports and Exercise* 42 (2010): 1286–1295.

93. C. W. Bales and coauthors, Aerobic and resistance training effects on energy intake: The STRRIDE-AT/RT Study, *Medicine and Science in Sports and Exercise* 44 (2012): 2033–2039; D. Stensel, Exercise, Appetite and appetite-regulating hormones: Implications for food intake and weight control, *Annals of Nutrition and Metabolism* 57 (2010): 36–42; J. A. King and coauthors, Influence of brisk walking on appetite, energy intake, and plasma acylated ghrelin, *Medicine and Science in Sports and Exercise* 42 (2010): 485–492.

94. K. J. Guelfi, C. E. Conges, R. Duffield, Beneficial effects of 12 weeks of aerobic compared with resistance exercise training on perceived appetite in previously sedentary overweight and obese men, *Metabolism: Clinical and Experimental* 62 (2013): 235–243; K. Deighton, J. C. Zahra, and D. J. Stensel, Appetite, energy intake and resting metabolic responses to 60 min treadmill running performed in a fasted versus a postprandial state, *Appetite* 58 (2012): 946–954.

95. E. V. Carraca and coauthors, Physical activity predicts changes in body image during obesity treatment in women, *Medicine and Science in Sports and Exercise* 44 (2012): 1604–1612.

96. I. M. Lee and coauthors, Physical activity and weight gain prevention, *Journal of the American Medical Association* 303 (2010): 1173–1179.

97. J. A. Steeves, D. L. Thompson, and D. R. Bassett Jr., Energy cost of stepping in place while watching television commercials, *Medicine and Science in Sports and Exercise* 44 (2012): 330–335.

98. B. McFerran and coauthors, I'll have what she's having: Effects of social influence and body type on the food choices of others, *Journal of Consumer Research* 36 (2010): 915–929.

99. E. Robinson and coauthors, Eating attentively: A systematic review and meta-analysis of the effect of food intake memory and awareness on eating, *American Journal of Clinical Nutrition* 97 (2013): 728–742; R. E. Oldham-Cooper and coauthors, Playing a computer game during lunch affects fullness, memory for lunch, and later snack intake, *American Journal of Clinical Nutrition* 93 (2011): 308–313.

100. L. H. Epstein and coauthors, Long-term habituation to food in obese and nonobese women, *American Journal of Clinical Nutrition* 94 (2011): 371–376.

101. A. Kong and coauthors, Self-monitoring and eating-related behaviors are associated with 12-month weight loss in postmenopausal overweight-to-obese women, *Journal of the Academy of Nutrition and Dietetics* 112 (2012): 1428–1435; M. B. Conroy and coauthors, Physical activity self-monitoring and weight loss: 6-month results of the SMART Trial, *Medicine and Science in Sports and Exercise* 43 (2011): 1568–1574; L. E. Burke, J. Wang, and M. A. Sevick, Self-monitoring in weight loss: A systematic review of the literature, *Journal of the American Dietetic Association* 111 (2011): 92–102.

102. S. D. Acharya and coauthors, Using a personal digital assistant for self-monitoring influences diet quality in comparison to a standard paper record among overweight/obese adults, *Journal of the American Dietetic Association* 111 (2011): 583–588.

103. H. Shin and coauthors, Self-efficacy improves weight loss in overweight/obese postmenopausal women during a 6-month weight loss intervention, *Nutrition Research* 31 (2011): 822–828.

104. R. Rossini and coauthors, Effects of cognitive-behavioral treatment for weight loss in family members, *Journal of the American Dietetic Association* 111 (2011): 1712–1719.

105. S. Kodama and coauthors, Effect of web-based lifestyle modification on weight control: A meta-analysis, *International Journal of Obesity* 36 (2012): 675–685; L. J. Appel and coauthors, Comparative effectiveness of weight-loss interventions in clinical practice, *New England Journal of Medicine* 365 (2011): 1959–1968.

106. J. L. Kraschnewski and coauthors, Long-term weight loss maintenance in the United States, *International Journal of Obesity* 34 (2010): 1644–1654.

107. C. B. Ebbeling and coauthors, Effects of dietary composition on energy expenditure during weight-loss maintenance, *Journal of the American Medical Association* 307 (2012): 2627–2634.

108. P. Sumithran and coauthors, Long-term persistence of hormonal adaptations to weight loss, *New England Journal of Medicine* 365 (2011): 1597–1604.

109. J. L. Unick, J. M. Jakicic, and B. H. Marcus, Contribution of behavior intervention components to 24-month weight loss, *Medicine and Science in Sports and Exercise* 42 (2010): 745–753.

110. E. Manthou and coauthors, Behavioral compensatory adjustments to exercise training in overweight women, *Medicine and Science in Sports and Medicine* 42 (2010): 1221–1228.

111. National Weight Loss Registry, www.nwlr.ws, accessed November 2013; S. F. L. Kirk and coauthors, Effective weight management practice: A review of the lifestyle intervention evidence, *International Journal of Obesity* 36 (2012): 178–185; N. R. Reyes and coauthors, Similarities and differences between weight loss maintainers and regainers: A qualitative analysis, *Journal of the Academy of Nutrition and Dietetics* 112 (2012): 499–505; J. P. Moreno and C. A. Johnston, Successful habits of weight loss, *American Journal of Lifestyle Medicine* 6 (2012): 113–115; J. L. Bachman and coauthors, Eating frequency is higher in weight loss maintainers and normal-weight individuals than in overweight individuals, *Journal of the American Dietetic Association* 111 (2011): 1730–1734; S. N. Grief and R. L. Miranda, Weight loss maintenance, *American Family Physician* 82 (2010): 630–634.

112. *Accelerating Progress in Obesity Prevention: Solving the Weight of the Nation*, (Washington, DC: Institute of Medicine of the National Academies), 2012.

113. J. C. Giesen and coauthors, Exploring how calorie information and taxes on high-calorie foods influence lunch decisions, *American Journal of Clinical Nutrition* 93 (2011): 689–694; E. A. Finkelstein and coauthors, Impact of targeted beverage taxes on higher- and lower-income households, *Archives of Internal Medicine* 170 (2010): 2028–2034.

HIGHLIGHT > 9
The Latest and Greatest Weight-Loss Diet—Again

> **LEARN IT** Contrast the differences between popular fad diets and weight-loss diets based on sound nutrition.

© Jerry Arcieri/Corbis

To paraphrase William Shakespeare, "A fad diet by any other name would still be a fad diet." Year after year, "new and improved" diets appear on bookstore shelves and circulate among friends.* People of all sizes eagerly try the best diet ever on the market, hoping that this one will really work. Sometimes these diets seem to work for a while, but more often than not, their success is short-lived. Then another diet takes the spotlight. Here's how Dr. K. Brownell, an obesity researcher and dean at Duke University's Sanford School of Public Policy, describes this phenomenon: "When I get calls about the latest diet fad, I imagine a trick birthday cake candle that keeps lighting up and we have to keep blowing it out."

Realizing that fad diets do not offer a safe and effective long-term plan for weight loss, health professionals speak out, but they never get the candle blown out permanently. New fad diets can keep making outrageous claims because no one requires their advocates to prove what they say. Fad diet gurus do not have to conduct credible research on the benefits or dangers of their diets. They can simply make recommendations and then later, if questioned, search for bits and pieces of research that support the conclusions they have already reached. That's backward. Diet and health recommendations should *follow* years of sound scientific research *before* being offered to the public.

Because anyone can publish anything—in books or on the Internet—peddlers of fad diets can make unsubstantiated statements that fall far short of the truth but sound impressive to the uninformed. They often offer distorted bits of legitimate research. They may start with one or more actual facts but then leap from one erroneous conclusion to the next. Anyone who wants to believe these claims has to wonder how the thousands of scientists working on obesity research over the past century could possibly have missed such obvious connections.

Fad diets come in almost as many shapes and sizes as the people who search them out. Some restrict fats or carbohydrates, some limit portion sizes, some focus on food combinations, and some claim that a person's genetic type or blood type determines the foods best suited to manage weight and prevent disease. A lack of scientific evidence just doesn't seem to stop diets from making claims.[1] Table H9-1 compares some of today's popular diets.

Fad Diets' Appeal

With more than half of our nation's adults overweight and many more concerned about their weight, the market for a weight-loss book, product, or program is huge (no pun intended). Americans spend an estimated $33 billion a year on weight-loss books and products. Even a plan that offers only minimal weight-loss success easily attracts a following.

Perhaps the greatest appeal of fad diets is that they tend to ignore dietary recommendations. Foods such as meats and milk products that need to be selected carefully to limit saturated fat can be eaten with abandon. Whole grains, legumes, vegetables, and fruits that should be eaten in abundance can now be bypassed. For some people, this is a dream come true: steaks without the potatoes, ribs without the coleslaw, and meatballs without the pasta. Who can resist the promise of weight loss while eating freely from a list of favorite foods?

Dieters are also lured into fad diets by sophisticated—yet often erroneous—explanations of the metabolic consequences of eating certain foods. Terms such as *eicosanoids* and *de novo lipogenesis* are scattered about, often intimidating readers into believing that the authors must be right given their brilliance in understanding the body.

If fad diets were as successful as some people claim, then consumers who tried them would lose weight, and their obesity problems would be solved. But this is not the case. Similarly, if fad diets were as worthless as others claim, then consumers would eventually stop pursuing them. Clearly, this is not happening either. Most fad diets have enough going for them that they work for some people at least for a short time, but they fail to produce long-lasting results for most people.

Don't Count kCalories

Who wants to count kcalories? Even experienced dieters find counting kcalories burdensome, not to mention timeworn. They want a new, easy way to lose weight, and fad diet plans seem to offer this boon. But, though fad diets often claim to disregard kcalories, their design typically ensures a low energy intake. Most of the sample menu plans,

---

*The Academy of Nutrition and Dietetics offers evaluations of popular diets for your review. Look for reviews of popular diets at their website, **www.eatright.org/dietreviews**.

TABLE H9-1 Popular Diets Compared

Diet	Claim(s)	Strong Point(s)	Weak Point(s)
The 4-Hour Body	• Small, simple changes produce big, long-lasting results. • The Slow-Carb diet supports a 20-pound weight loss in 30 days without exercise.	• Encourages lean proteins, legumes, and vegetables. • Organized format provides simple plan.	• Excludes fruit, whole grains, and milk (except cottage cheese), which may lead to nutrient deficiencies. • Lacks variety.
The 17 Day Diet	• Changing the way you eat every few days creates "body confusion," which prevents metabolism from settling into homeostasis. • You can boost metabolism by "eating clean," which means no sugar, no processed food, and no fried foods.	• Prevents boredom by alternating between cycles. • Fairly well-balanced diet promotes healthy eating.	• Lacks scientific evidence that changing the diet creates "body confusion." • Does not provide individualized kcalorie goals. • Promotes its own processed foods.
The 100	• Restricting sugar consumption to 100 kcalories a day will reduce insulin levels (the weight-gaining hormone), speed weight loss, improve health, and boost metabolism. • The real problem causing weight gains is hidden sugars in healthy foods such as yogurt and fruit.	• Limits processed foods. • Encourages consumption of vegetables and fiber-rich foods.	• Focuses on very basic food choices and lacks variety. • Limits important food groups such as low-fat dairy, legumes, and fruits, which may lead to nutrient deficiencies.
Biggest Loser Diet	• Lose weight, gain health, feel young, and take control of your life using portion control, progressively lowering energy intake, and a following a customized food pyramid.	• Provides motivation and promotes selecting low-fat foods and drinking water. • Stresses the importance of exercise.	• Recommends energy intakes below the recommended minimum of 1200 kcalories a day, which may lead to nutrient deficiencies.
Cinch!	• A nutrient-dense diet composed mainly of plant-based foods will help you lose weight and lower the risk of disease.	• Recommends a plant-based, nutrient-dense diet. • Stresses the importance of exercise.	• A little confusing and dense with facts.
The Dukan Diet	• A high-protein, low-kcalorie diet promotes rapid and permanent weight loss.	• Encourages daily exercise, moderate salt intake, and lifelong weight management. • Provides a highly structured plan.	• Restricts carbohydrates to induce ketosis, which can cause nausea, light-headedness, and fatigue and can worsen medical problems such as kidney disease. • Not suited for vegetarians and others who prefer not to emphasize animal proteins.
The Fast Diet	• Lose weight by eating "normally" for five days while choosing two non-consecutive days to "fast"—limiting kcalories to 500 for women and 600 for men.	• Focuses only on two days of the week. • Promotes lean protein and low glycemic foods on "fast" days.	• Little research supporting the health benefits of intermittent fasting diets. • Fasting can cause irritability, sleeplessness or sleepiness, and dehydration.
New Sonoma Diet	• Enjoying portion-controlled Coastal California style foods supports weight loss and promotes good health.	• Emphasizes nutrient-dense foods. • Limits processed foods.	• No individualized kcalorie plan.
Wheat Belly	• Lose weight and reverse health problems by eliminating all forms of wheat.	• Creates a low-kcalorie diet.	• Restrictive diet would likely be low in B vitamins, calcium, and vitamin D.
The Zen Diet Revolution	• Combine spiritual wisdom and dietary adjustments to reduce fat cells without counting kcalories. • Visualization, meditation, and mindfulness can change mental, dietary, lifestyle, and activity habits.	• Offers a basic nutrition approach that does no harm. • Reminds you to make food meaningful and treat your body with respect.	• Recommendations to use fat-burning herbs, supplements, and green tea to decrease fat cells are unsubstantiated and expensive.

SOURCE: Adapted from Academy of Nutrition and Dietetics, *Consumer Diet and Lifestyle Book Reviews*, www.eatright.org/dietreviews, accessed November 2013.

© Cengage Learning

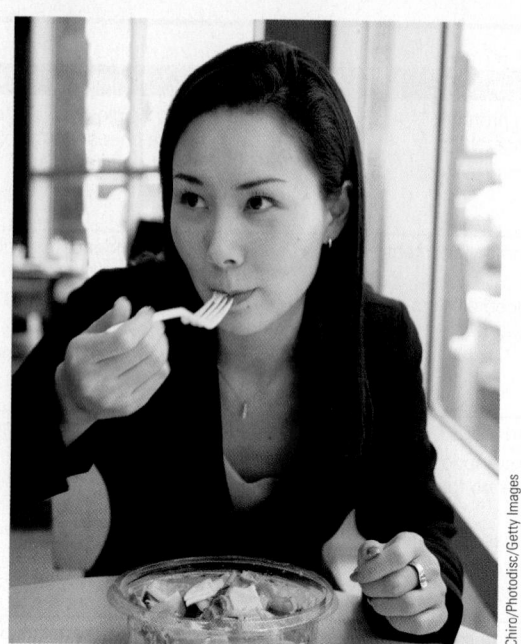

The wise consumer seeks a diet that supports not only weight loss, but also health gains.

When food choices are limited, nutrient intakes may be inadequate. To help shore up some of these inadequacies, fad diets often recommend a dietary supplement. Conveniently, many of the companies selling fad diets also peddle these supplements. But as Highlights 10 and 11 explain, foods offer many more health benefits than any supplement can provide. Quite simply, if the diet is inadequate, it needs to be improved.

Follow a Plan

Most people need specific instructions and examples to make dietary changes. Popular diets offer dieters a plan. The user doesn't have to decide what foods to eat, how to prepare them, or how much to eat. Unfortunately, these instructions serve only short-term weight-loss needs. They do not provide for long-term changes in lifestyle that will support weight maintenance or health goals.

The success of any weight-loss diet depends on the person adopting the plan and sticking with it. People who prefer a high-protein, low-carbohydrate diet over a high-carbohydrate, low-fat diet, for example, may have more success at sticking with it, perhaps because of protein's role in providing satiety.[3] Keep in mind, though, that weight loss occurs because of the duration of a low-kcalorie plan—not the proportion of energy nutrients.[4]

especially in the early stages, are designed to deliver an average of 1200 kcalories a day.

Even when counting kcalories is truly not necessary, total kcalories tend to be low simply because food intake is so limited. Diets that omit hundreds of foods and several food groups limit a person's options and lack variety. Chapter 2 praises variety as a valuable way to ensure an adequate intake of nutrients, but variety also entices people to eat more food and gain more weight. Without variety, some people lose interest in eating, which further reduces energy intake. Even if the allowed foods are favorites, eating the same foods week after week can become monotonous.

Without its refried beans, tortilla wrapping, and chopped vegetables, a burrito is reduced to a pile of ground beef. Without the baked potato, there's no need for butter and sour cream. Weight loss occurs because of the low energy intake. This is an important point. Any diet can produce weight loss, at least temporarily, if intake is restricted. The real value of a diet is determined by its ability to maintain weight loss and support good health over the long term. The goal is not simply weight loss, but health gains—and most fad diets cannot support optimal health over time. In fact, some weight-loss diets can create or exacerbate health problems.[2]

The Real Deal

Fad diets attribute magical powers to their weight-loss plans, but in reality, the magic is in tipping the energy balance so that metabolic and physical activities expend more kcalories than foods bring in. Because new diets emerge in the market regularly, it can be challenging to sort the fad diets from the healthy options. Furthermore, it can be difficult determining how a diet's overall quality rates and how it compares with others.

Keep in mind that healthy weight loss requires long-term lifestyle changes in eating and activity habits—not quick, short-term fixes. A healthy plan may not be quick, but it allows for flexibility and a variety of foods, including some favorite treats on occasion.

Some currently popular diet plans offer a sensible approach to weight loss and healthy eating. The challenge is sorting through "the good, the bad, and the ugly." The accompanying "How To" feature offers tips for identifying fad diets and other weight-loss scams. Fad diets may not harm healthy people if used for only a little while, but they cannot support optimal health for long. Chapter 9 includes reasonable approaches to weight management and concludes that the ideal diet is one you can live with for the rest of your life. Keep that criterion in mind when you evaluate the next "latest and greatest weight-loss diet" that comes along.

>How To

Identify a Fad Diet or Weight-Loss Scam

It may be a fad diet or weight-loss scam if it:

- Sounds too good to be true.
- Recommends using a single food consistently as the key to the program's success.
- Promises quick and easy weight loss with no effort. "Lose weight while you sleep!"
- Eliminates an entire food group such as grains or milk and milk products.
- Guarantees an unrealistic outcome in an unreasonable time period. "Lose 10 pounds in 2 days!"
- Bases evidence for its effectiveness solely on anecdotal stories.
- Requires you to buy special products that are not readily available in the marketplace at affordable prices.
- Specifies a proportion for the energy nutrients that falls outside the recommended ranges—carbohydrate (45 to 65 percent), fat (20 to 35 percent), and protein (10 to 35 percent).
- Claims to alter your genetic code or reset your metabolism.
- Fails to mention potential risks or additional costs.
- Promotes products or procedures that have not been proven safe and effective.
- Neglects plans for weight maintenance following weight loss.

> **TRY IT** Review an advertisement for a popular weight-loss plan and explain why you think it might—or might not—be a fad diet.

CRITICAL THINKING QUESTIONS

A. What patterns are evident in fad weight-loss diets?

B. A fad diet website says it all: 30 ways to lose 5 pounds in a week. Millions of people have tried hundreds of crash diets in search of short-term fixes to their weight problems. It's amazing to think about what people are willing to do to avoid eating a well-balanced diet and exercising regularly—wire their jaws closed, swallow a parasite, drink nothing but salt water for a week. Why do you think consumers continue to pursue fad diets? How would you design a weight loss plan that would be appealing to most people *and* produce long-lasting results?

REFERENCES

1. L. Cusack and coauthors, Blood type diets lack supporting evidence: A systematic review, *American Journal of Clinical Nutrition* 98 (2013): 99–104.
2. P. Sjögren and coauthors, Mediterranean and carbohydrate-restricted diets and mortality among elderly men: A cohort study in Sweden, *American Journal of Clinical Nutrition* 92 (2010): 967–974.
3. M. S. Westerterp-Plantenga, S. G. Lemmens, and K. R. Westerterp, Dietary protein: Its role in satiety, energetics, weight loss and health, *British Journal of Nutrition* 108 (2012): S105–S112; T. M. Larsen and coauthors, Diets with high or low protein content and glycemic index for weight-loss maintenance, *New England Journal of Medicine* 363 (2010): 2102–2113.
4. W. S. Yancy and coauthors, A randomized trial of a low-carbohydrate diet vs orlistat plus a low-fat diet for weight loss, *Archives of Internal Medicine* 170 (2010): 121–123.

10

The Water-Soluble Vitamins: B Vitamins and Vitamin C

Nutrition in Your Life

If you were playing a word game and your partner said "vitamins," how would you respond? If "pills" and "supplements" immediately come to mind, you may be missing the main message of the vitamin story—that hundreds of foods deliver more than a dozen vitamins that participate in thousands of activities throughout your body. Quite simply, foods supply vitamins to support all that you are and all that you do—and supplements of any one of them, or even a combination of them, can't compete with foods in keeping you healthy. In the Nutrition Portfolio at the end of this chapter, you can determine whether the foods you are eating are meeting your water-soluble vitamin needs.

Earlier chapters focused on the energy-yielding nutrients—carbohydrates, fats, and proteins. This chapter begins with an overview of the **vitamins** and then examines each of the water-soluble vitamins; the next chapter features the fat-soluble vitamins. Researchers first recognized in the early 1900s that foods contain substances that are "vital to life."[1] Since then, the world of vitamins has opened up dramatically.

Vitamins are powerful, as their *absence* attests. Vitamin A deficiency can cause blindness; a lack of the B vitamin niacin can cause dementia; and without vitamin D, bones fail to grow. The *presence* of vitamins also attests to their power. The B vitamin folate helps to prevent birth defects, and vitamin K creates blood clots. Every year, people spend billions of dollars on supplements, hoping to cure their ailments (see Highlight 10). Vitamins do support good health, but they do not cure all ills nor do supplements provide all of the many disease-preventing benefits of vitamin-rich foods such as vegetables, fruits, and whole grains (as Highlight 11 explains).

vitamins: organic, essential nutrients required in small amounts by the body for health. Vitamins regulate body processes that support growth and maintain life.

- **vita** = life
- **amine** = containing nitrogen (the first vitamins discovered contained nitrogen)

10.1 The Vitamins—An Overview

> LEARN IT Describe how vitamins differ from the energy nutrients and how fat-soluble vitamins differ from water-soluble vitamins.

The vitamins differ from carbohydrates, fats, and proteins in the following ways:

- *Structure.* Vitamins are individual units; they are not linked together (as are molecules of glucose or amino acids). Appendix C presents the chemical structure for each of the vitamins.
- *Function.* Vitamins do not yield energy when metabolized; many of them do, however, assist the enzymes that participate in the release of energy from carbohydrates, fats, and proteins.
- *Food contents.* The amounts of vitamins people ingest from foods and the amounts they require daily are measured in *micrograms* (µg) or *milligrams* (mg), rather than grams (g).*

The vitamins are similar to the energy-yielding nutrients, though, in that they are essential, organic, and available from foods.

Bioavailability Some water-soluble vitamins are synthesized by GI tract bacteria and absorbed by the large intestine, but not in quantities great enough to meet the body's needs; foods must supply these essential nutrients.[2] The amount of vitamins available from foods depends not only on the quantity provided by a food but also on the amount absorbed and used by the body—referred to as the vitamins' **bioavailability.** The quantity of vitamins in a food can be determined relatively easily. Researchers analyze foods to determine the vitamin contents and publish the results in tables of food composition such as Appendix H. Determining the bioavailability of a vitamin is a more complex task because it depends on many factors, including:

- Efficiency of digestion and time of transit through the GI tract
- Previous nutrient intake and nutrition status
- Method of food preparation (raw, cooked, or processed)
- Source of the nutrient (synthetic, fortified, or naturally occurring)
- Other foods consumed at the same time

Chapters 10 through 13 describe factors that inhibit or enhance the absorption of individual vitamins and minerals. Experts consider these factors when estimating recommended intakes.

Precursors Some of the vitamins are available from foods in inactive forms known as **precursors,** or provitamins. Once inside the body, the precursor is converted to an active form of the vitamin. For example, beta-carotene, a red-orange pigment found in fruits and vegetables, is a precursor to vitamin A. Thus, in measuring a person's vitamin intake, it is important to count both the amount of the active vitamin and the potential amount available from its precursors. The discussions and summary tables throughout this chapter and the next indicate which vitamins have precursors.

Organic Nature Fresh foods naturally contain vitamins, but because these vitamins are organic, they can be readily destroyed during processing. Therefore, processed foods should be used sparingly, and fresh foods should be handled with care during storage and in cooking. Prolonged heating may destroy much of the thiamin in food. Because riboflavin can be destroyed by the ultraviolet rays of the sun or by fluorescent light, foods stored in transparent glass containers are most likely to lose riboflavin. Oxygen destroys vitamin C, so losses occur when foods are cut, processed, and stored; these losses may be enough to reduce its action in the body. Table 10-1 summarizes ways to minimize nutrient losses in the kitchen, and Chapter 19 provides more details.

Solubility As you may recall, carbohydrates and proteins are hydrophilic and lipids are hydrophobic. The vitamins divide along the same lines—the hydrophilic,

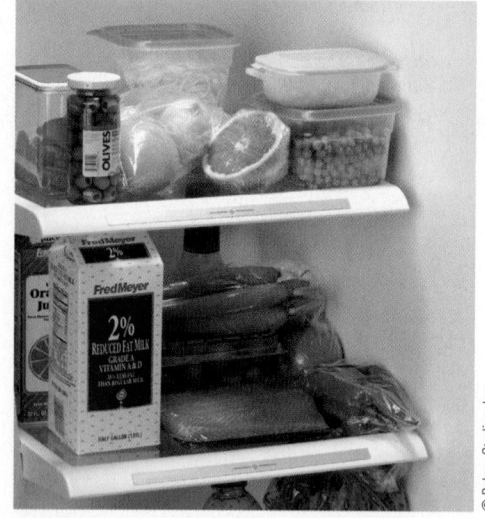

To minimize vitamin losses, wrap cut fruits and vegetables or store them in airtight containers.

© Polara Studios, Inc.

bioavailability: the rate at and the extent to which a nutrient is absorbed and used.

precursors: substances that precede others; with regard to vitamins, compounds that can be converted into active vitamins; also known as *provitamins.*

*For perspective, a dollar bill weighs about 1 g; 1 g = 1000 mg, and 1 mg = 1000 µg. Appendix K explains how to convert a measurement from one unit of measure to another.

TABLE 10-1 Minimizing Nutrient Losses

- To slow the degradation of vitamins, refrigerate (most) fruits and vegetables.
- To minimize the oxidation of vitamins, store fruits and vegetables that have been cut in airtight wrappers, and store juices that have been opened in closed containers (and refrigerate them).
- To prevent vitamin losses during washing, rinse fruits and vegetables before cutting (not after).
- To minimize vitamin losses during cooking, use a microwave oven or steam vegetables in a small amount of water. Add vegetables after water has come to a boil. Use the cooking water in mixed dishes such as casseroles and soups. Avoid high temperatures and long cooking times.

© Cengage Learning

water-soluble ones are the B vitamins (thiamin, riboflavin, niacin, biotin, pantothenic acid, vitamin B_6, folate, and vitamin B_{12}) and vitamin C; the hydrophobic, fat-soluble ones are vitamins A, D, E, and K. As each vitamin was discovered, it was given a name and sometimes a letter and number as well. Many of the vitamins have multiple names, which has led to some confusion. The summary tables throughout this chapter and the next provide both the standard and the common alternative names.

Solubility is apparent in the food sources of the different vitamins, and it affects their absorption, transport, storage, and excretion by the body. The water-soluble vitamins are found in the watery compartments of foods; the fat-soluble vitamins usually occur together in the fats and oils of foods. On being absorbed, the water-soluble vitamins move directly into the blood. Like fats, the fat-soluble vitamins must first enter the lymph, then the blood. Once in the blood, many of the water-soluble vitamins travel freely, whereas many of the fat-soluble vitamins require transport proteins. Upon reaching the cells, water-soluble vitamins freely circulate in the water-filled compartments whereas fat-soluble vitamins are held in fatty tissues and the liver until needed. The kidneys, monitoring the blood that flows through them, detect and remove small excesses of water-soluble vitamins; large excesses, however, may overwhelm the system, creating adverse effects. Fat-soluble vitamins tend to remain in fat-storage sites in the body rather than being excreted, and so are more likely to reach toxic levels when consumed in excess.

Because the body stores fat-soluble vitamins, they can be eaten in large amounts once in a while and still meet the body's needs over time. Water-soluble vitamins are retained for varying lengths of time in the body. The water-soluble vitamins must be eaten more regularly than the fat-soluble vitamins, although a single day's omission from the diet does not create a deficiency.

Toxicity Knowledge about some of the amazing roles of vitamins has prompted many people to take vitamin supplements, assuming that "more is better." Just as an inadequate intake can cause harm, so can an excessive intake. Even some of the water-soluble vitamins have adverse effects when taken in large doses.

That a vitamin can be both essential and harmful may seem surprising, but the same is true of most nutrients. The effects of every substance depend on its dose, and this is one reason consumers should not self-prescribe supplements. Figure 10-1 (p. 304) shows three possible relationships between dose levels and effects. The third diagram in Figure 10-1 represents the situation with nutrients—more is better up to a point, but beyond that point, still more can be harmful.

The Committee on Dietary Reference Intakes (DRI) addresses the possibility of adverse effects from high doses of nutrients by establishing Tolerable Upper Intake Levels (UL). The UL defines the highest amount of a nutrient that is likely not to cause harm for most healthy people when consumed daily. The risk of harm increases as intakes rise above the UL. Of the nutrients discussed in this chapter, niacin, vitamin B_6, folate, choline, and vitamin C have UL, and these values are presented in their respective summary tables. Data are lacking to establish UL for the remaining B vitamins, but this does not mean that excessively high intakes would be without risk. (The inside front cover pages present UL for the vitamins and minerals.)

> FIGURE 10-1 **Dose Levels and Effects**

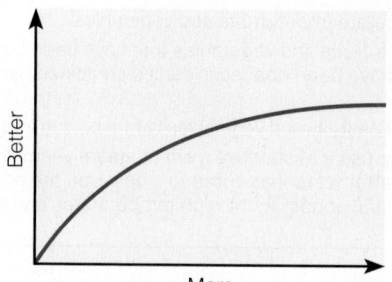

As you progress in the direction of more, the effect gets better and better, with no end in sight (real life is seldom, if ever, like this).

As you progress in the direction of more, the effect reaches a maximum and then a plateau, becoming no better with higher doses.

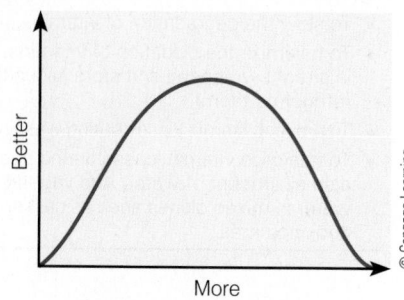

As you progress in the direction of more, the effect reaches an optimum at some intermediate dose and then declines, showing that more is better up to a point and then harmful. That too much can be as harmful as too little represents the situation with most nutrients.

© Cengage Learning

TABLE 10-2 **Water-Soluble and Fat-Soluble Vitamins Compared**

	Water-Soluble Vitamins: B Vitamins and Vitamin C	Fat-Soluble Vitamins: Vitamins A, D, E, and K
Absorption	Directly into the blood	First into the lymph, then the blood
Transport	Travel freely	Many require transport proteins
Storage	Circulate freely in water-filled parts of the body	Stored in the cells associated with fat
Excretion	Kidneys detect and remove excess in urine	Less readily excreted; tend to remain in fat-storage sites
Toxicity	Possible to reach toxic levels when consumed from supplements	Likely to reach toxic levels when consumed from supplements
Requirements	Needed in frequent doses (perhaps 1 to 3 days)	Needed in periodic doses (perhaps weeks or even months)

© Cengage Learning

NOTE: Exceptions occur, but these differences between the water-soluble and fat-solule vitamins are valid generalizations.

> **REVIEW IT** Describe how vitamins differ from the energy nutrients and how fat-soluble vitamins differ from water-soluble vitamins.

The vitamins are essential nutrients needed in tiny amounts in the diet both to prevent deficiency diseases and to support optimal health. The water-soluble vitamins are the B vitamins and vitamin C; the fat-soluble vitamins are vitamins A, D, E, and K. Table 10-2 summarizes the differences between the water-soluble and fat-soluble vitamins.

The discussion of B vitamins that follows begins with a brief description of each of them, then offers a look at the ways they work together. Thus, a preview of the individual vitamins is followed by a discussion of their interactions.

10.2 The B Vitamins

> **LEARN IT** Identify the main roles, deficiency symptoms, and food sources for each of the B vitamins.

Despite supplement advertisements that claim otherwise, the vitamins do not provide the body with fuel for energy. It is true, though, that without B vitamins the body would lack energy. The energy-yielding nutrients—carbohydrate, fat, and protein—are used for fuel; the B vitamins help the body to use that fuel. Several of the B vitamins—thiamin, riboflavin, niacin, pantothenic acid, and biotin—form part of the **coenzymes** that assist enzymes in the release of energy from carbohydrate, fat, and protein. Other B vitamins play other indispensable

coenzymes: complex organic molecules that work with enzymes to facilitate the enzymes' activity. Many coenzymes have B vitamins as part of their structures.

> FIGURE 10-2   **Coenzyme Action**

Some vitamins form part of the coenzymes that enable enzymes either to synthesize compounds (as illustrated by the lower enzymes in this figure) or to dismantle compounds (as illustrated by the upper enzymes).

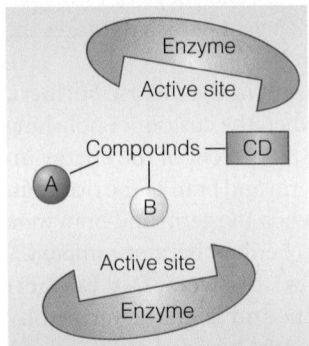

Without coenzymes, compounds A, B, and CD don't respond to their enzymes.

With the coenzymes in place, compounds are attracted to their sites on the enzymes . . .

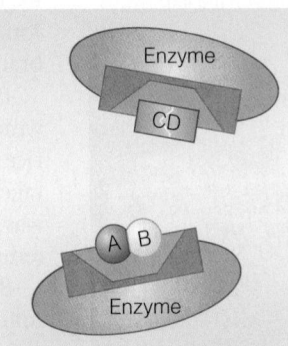

. . . and the reactions proceed instantaneously. The coenzymes often donate or accept electrons, atoms, or groups of atoms.

The reactions are completed with either the formation of a new product, AB, or the breaking apart of a compound into two new products, C and D, and the release of energy.

© Cengage Learning

roles in metabolism. Vitamin B_6 assists enzymes that metabolize amino acids. Folate and vitamin B_{12} help cells to multiply. Among these cells are the red blood cells and the cells lining the GI tract—cells that deliver energy to all the others.

The vitamin portion of a coenzyme allows a chemical reaction to occur; the remaining portion of the coenzyme binds to the enzyme. Without its coenzyme, an enzyme cannot function. Thus symptoms of B vitamin deficiencies directly reflect the disturbances of metabolism caused by a lack of coenzymes. Figure 10-2 illustrates coenzyme action.

The following sections describe the roles of individual B vitamins and note many coenzymes and metabolic pathways. Keep in mind that a later discussion assembles these pieces of information into a whole picture. The following sections also present the recommendations, deficiency and toxicity symptoms, and food sources for each vitamin. For thiamin, riboflavin, niacin, vitamin B_6, folate, vitamin B_{12}, and vitamin C, sufficient data were available to establish an RDA; for biotin, pantothenic acid, and choline, an Adequate Intake (AI) was set; only niacin, vitamin B_6, folate, choline, and vitamin C have UL. These values appear in the summary tables and figures that follow and on the pages of the inside front cover.

Thiamin   Thiamin is the vitamin part of the coenzyme TPP (thiamin pyrophosphate) that assists in energy metabolism. The TPP coenzyme participates in the conversion of pyruvate to acetyl CoA (described in Chapter 7). Recall how important this step is in allowing carbohydrate fuel to enter the TCA cycle and produce much more ATP than during glycolysis. The reaction removes 1 carbon from the 3-carbon pyruvate to make the 2-carbon acetyl CoA and carbon dioxide (CO_2). In a similar step in the TCA cycle, TPP helps convert a 5-carbon compound to a 4-carbon compound. Besides playing these pivotal roles in energy metabolism, thiamin occupies a special site on the membranes of nerve cells. Consequently, nerve activity and muscle activity in response to nerves depend heavily on thiamin.

Thiamin Recommendations   Dietary recommendations are based primarily on thiamin's role in enzyme activity. Generally, thiamin needs will be met if a person eats enough food to meet energy needs—if that energy comes from nutritious foods. The average thiamin intake in the United States meets or exceeds recommendations.

Thiamin Deficiency and Toxicity   People who fail to eat enough food to meet energy needs risk nutrient deficiencies, including thiamin deficiency. Inadequate thiamin intakes have been reported among the nation's malnourished and homeless people. Similarly, people who derive most of their energy from empty-kcalorie foods and beverages risk thiamin deficiency. Alcohol provides a good example of how empty

thiamin (THIGH-ah-min): a B vitamin. The coenzyme form is *TPP* (*thiamin pyrophosphate*).

> **FIGURE 10-3** **Thiamin-Deficiency Symptom—The Edema of Beriberi**

Physical examination confirms that this person has wet beriberi. Notice how the impression of the physician's thumb remains on the leg.

© NMSB/Custom Medical Stock Photo

beriberi: the thiamin-deficiency disease characterized by muscle weakness, edema, or both.

- **beri** = weakness
- **beriberi** = "I can't, I can't"

kcalories can lead to thiamin deficiency. Alcohol contributes energy but provides few, if any, nutrients and often displaces food. In addition, alcohol impairs thiamin absorption and enhances thiamin excretion in the urine, doubling the risk of deficiency. An estimated four out of five alcoholics are thiamin deficient, which damages the brain's structure and impairs its function.[3]*

Prolonged thiamin deficiency can result in the disease **beriberi,** which was first observed in Indonesia when the custom of polishing rice became widespread. Rice provided 80 percent of the energy intake of the people of that area, and the germ and bran of the rice grain was their principal source of thiamin. When the germ and bran were removed in the preparation of white rice, beriberi became rampant.

Beriberi is often described as "dry" or "wet." Dry beriberi reflects damage to the nervous system and is characterized by muscle weakness in the arms and legs. Wet beriberi reflects damage to the cardiovascular system and is characterized by dilated blood vessels, which cause the heart to work harder and the kidneys to retain salt and water, resulting in edema. Typically, both types of beriberi appear together, with one set of symptoms predominating. Figure 10-3 presents the edema of beriberi. No adverse effects have been associated with excesses of thiamin, and no UL has been determined.

Thiamin Food Sources Before examining Figure 10-4, you may want to read the accompanying "How To" feature, which describes the content in this and similar figures found in this chapter and the next three chapters. When you look at Figure 10-4, notice that thiamin occurs in small quantities in many nutritious foods. The long red bar near the bottom of the graph shows that meats in the

---

*Severe thiamin deficiency in alcohol abusers is called the *Wernicke-Korsakoff* (VER-nee-key KORE-sah-kof) *syndrome.* Symptoms include disorientation, loss of short-term memory, jerky eye movements, and staggering gait.

>How To

Evaluate Foods for Their Nutrient Contributions

Figure 10-4 is the first of a series of figures in this and the next three chapters that present the vitamins and minerals in foods. Each figure presents the same 24 foods, which were selected to ensure a variety of choices representative of each of the food groups as suggested by the USDA Food Patterns. For example, a bread, a cereal, and a pasta were chosen from the grain group. The suggestion to include a variety of vegetables was also considered: dark green vegetables (broccoli); orange and red vegetables (carrots); starchy vegetables (potatoes); legumes (pinto beans) and other vegetables (tomato juice). The selection of fruits followed suggestions to use whole fruits (bananas); citrus fruits (oranges); melons (watermelon); and berries (strawberries). Items were selected from the

milk group and protein foods in a similar way. In addition to the 24 foods that appear in all of the figures, three different foods were selected for each of the nutrients to add variety and often reflect excellent, and sometimes unusual, sources.

Notice that the figures list the food, the serving size, and the food energy (kcalories) on the left. The amount of the nutrient per serving is presented in the graph on the right along with the RDA (or AI) for adults, so you can see how many servings would be needed to meet recommendations.

The colored bars show at a glance which food groups best provide a nutrient: yellow for grains; green for vegetables; purple for fruits; white for milk and milk products; brown for legumes; and red for protein foods. Because the USDA Food Patterns include legumes with both the protein foods group and the vegetable group and because

legumes are especially rich in many vitamins and minerals, they have been given their own color to highlight their nutrient contributions.

Notice how the bar graphs shift in the various figures. Careful study of all of the figures taken together will confirm that variety is the key to nutrient adequacy.

Another way to evaluate foods for their nutrient contributions is to consider their nutrient density (their thiamin *per 100 kcalories*, for example). Quite often, vegetables rank higher on a nutrient-per-kcalorie list than they do on a nutrient-per-serving list (see p. 39 to review how to evaluate foods based on nutrient density). The left column in the figure highlights about five foods that offer the best nutrient density. Notice how many of them are vegetables.

Realistically, people cannot eat for single nutrients. Fortunately, most foods deliver more than one nutrient, allowing people to combine foods into nourishing meals.

> **TRY IT** Calculate which food provides more riboflavin per 1-ounce serving—a pork chop (3 oz, 291 kcal, 0.25 mg riboflavin) or cheddar cheese (1½ oz, 165 kcal, 0.11 mg riboflavin). Which food is more nutrient dense with respect to riboflavin?

> FIGURE 10-4   **Thiamin in Selected Foods**

Milligrams

Food	Serving size (kcalories)	Thiamin (mg)
Bread, whole wheat	1-oz slice (70 kcal)	
Cornflakes, fortified	1 oz (110 kcal)	
Spaghetti pasta	½ c cooked (99 kcal)	
Tortilla, flour	1 10" round (234 kcal)	
Broccoli	½ c cooked (22 kcal)	
Carrots	½ c shredded raw (24 kcal)	
Potato	1 medium baked w/skin (133 kcal)	
Tomato juice	¾ c (31 kcal)	
Banana	1 medium raw (109 kcal)	
Orange	1 medium raw (62 kcal)	
Strawberries	½ c fresh (22 kcal)	
Watermelon	1 slice (92 kcal)	
Milk	1 c reduced-fat 2% (121 kcal)	
Yogurt, plain	1 c low-fat (155 kcal)	
Cheddar cheese	1½ oz (171 kcal)	
Cottage cheese	½ c low-fat 2% (101 kcal)	
Pinto beans	½ c cooked (117 kcal)	
Peanut butter	2 tbs (188 kcal)	
Sunflower seeds	1 oz dry (165 kcal)	
Tofu (soybean curd)	½ c (76 kcal)	
Ground beef, lean	3 oz broiled (244 kcal)	
Chicken breast	3 oz roasted (140 kcal)	
Tuna, canned in water	3 oz (99 kcal)	
Egg	1 hard cooked (78 kcal)	
Excellent, and sometimes unusual, sources:		
Pork chop, lean	3 oz broiled (169 kcal)	
Soy milk	1 c (81 kcal)	
Squash, acorn	½ c baked (69 kcal)	

RDA for men

RDA for women

THIAMIN
Many different foods contribute some thiamin, but few are rich sources. Together, several servings of a variety of nutritious foods will help meet thiamin needs. Grain selections should be either whole grain or enriched.

Key:
- Grains
- Vegetables
- Fruits
- Milk and milk products
- Legumes, nuts, seeds
- Meats, poultry, seafood

Best sources per kcalorie

© Cengage Learning

pork family are exceptionally rich in thiamin. Yellow bars confirm that grains—whole grains or enriched—are a reliable source of thiamin.

As mentioned earlier, prolonged cooking can destroy thiamin. Also, like other water-soluble vitamins, thiamin leaches into water when foods are boiled or blanched. Cooking methods that require little or no water such as steaming and microwave heating conserve thiamin and other water-soluble vitamins. The accompanying table provides a summary of thiamin.

> REVIEW IT Thiamin

Other Names

Vitamin B₁

RDA

Men: 1.2 mg/day

Women: 1.1 mg/day

Chief Functions in the Body

Part of coenzyme TPP (thiamin pyrophosphate) used in energy metabolism

Significant Sources

Whole-grain, fortified, or enriched grain products; moderate amounts in all nutritious food; pork

Easily destroyed by heat

Deficiency Disease

Beriberi (wet, with edema; dry, with muscle wasting)

Deficiency Symptoms[a]

Enlarged heart, cardiac failure; muscular weakness; apathy, poor short-term memory, confusion, irritability; anorexia, weight loss

Toxicity Symptoms

None reported

© Polara Studios, Inc.

Pork is the richest source of thiamin, but enriched or whole-grain products typically make the greatest contribution to a day's intake because of the quantities eaten. Legumes such as split peas are also valuable sources of thiamin.

[a]Severe thiamin deficiency is often related to heavy alcohol consumption with limited food consumption (Wernicke-Korsakoff syndrome).

> FIGURE 10-5 **Riboflavin Coenzyme, Accepting and Donating Hydrogens**

This figure shows the chemical structure of the riboflavin portion of the coenzyme only; the remainder of the coenzyme structure is represented by dotted lines (see Appendix C for the complete chemical structures of FAD and FMN). The reactive sites that accept and donate hydrogens are highlighted in white.

FAD

During the TCA cycle, compounds release hydrogens, and the riboflavin coenzyme FAD picks up two of them. As it accepts two hydrogens, FAD becomes $FADH_2$.

$FADH_2$

$FADH_2$ carries the hydrogens to the electron transport chain. At the end of the electron transport chain, the hydrogens are accepted by oxygen, creating water, and $FADH_2$ becomes FAD again. For every $FADH_2$ that passes through the electron transport chain, two ATP are generated.

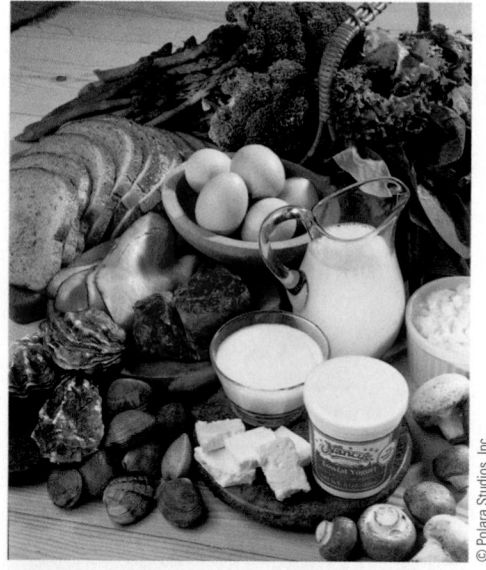

All of these foods are rich in riboflavin, but milk and milk products provide much of the riboflavin in the diets of most people.

Riboflavin Like thiamin, **riboflavin** serves as a coenzyme in many reactions, most notably in energy metabolism. The coenzyme forms of riboflavin are FMN (flavin mononucleotide) and FAD (flavin adenine dinucleotide); both can accept and then donate two hydrogens (see Figure 10-5). During energy metabolism, FAD picks up two hydrogens (with their electrons) from the TCA cycle and delivers them to the electron transport chain (described in Chapter 7).

Riboflavin Recommendations Like thiamin's RDA, riboflavin's RDA is based primarily on its role in enzyme activity. Most people in the United States meet or exceed riboflavin recommendations.

Riboflavin Deficiency and Toxicity Riboflavin deficiency most often accompanies other nutrient deficiencies.* Lack of the vitamin causes inflammation of the membranes of the mouth, skin, eyes, and GI tract. Excesses of riboflavin appear to cause no harm, and no UL has been established.

Riboflavin Food Sources The greatest contributions of riboflavin come from milk and milk products (see Figure 10-6). Whole-grain or enriched grains are also valuable sources because of the quantities people typically consume. When riboflavin sources are ranked by nutrient density (per kcalorie), many dark green, leafy vegetables (such as broccoli, turnip greens, asparagus, and spinach) appear high on the list. Vegans and others who don't use milk must rely on ample servings of dark greens and enriched grains for riboflavin. Nutritional yeast is another good source.

Ultraviolet light and irradiation destroy riboflavin. For these reasons, milk is sold in cardboard or opaque plastic containers, instead of clear glass bottles. In contrast, riboflavin is stable to heat, so cooking does not destroy it. The accompanying table provides a summary of riboflavin.

riboflavin (RYE-boh-flay-vin): a B vitamin. The coenzyme forms are *FMN (flavin mononucleotide)* and *FAD (flavin adenine dinucleotide)*.

_____

*Riboflavin deficiency is called *ariboflavinosis* (ay-RYE-boh-FLAY-vin-oh-sis).

> **FIGURE 10-6** **Riboflavin in Selected Foods**

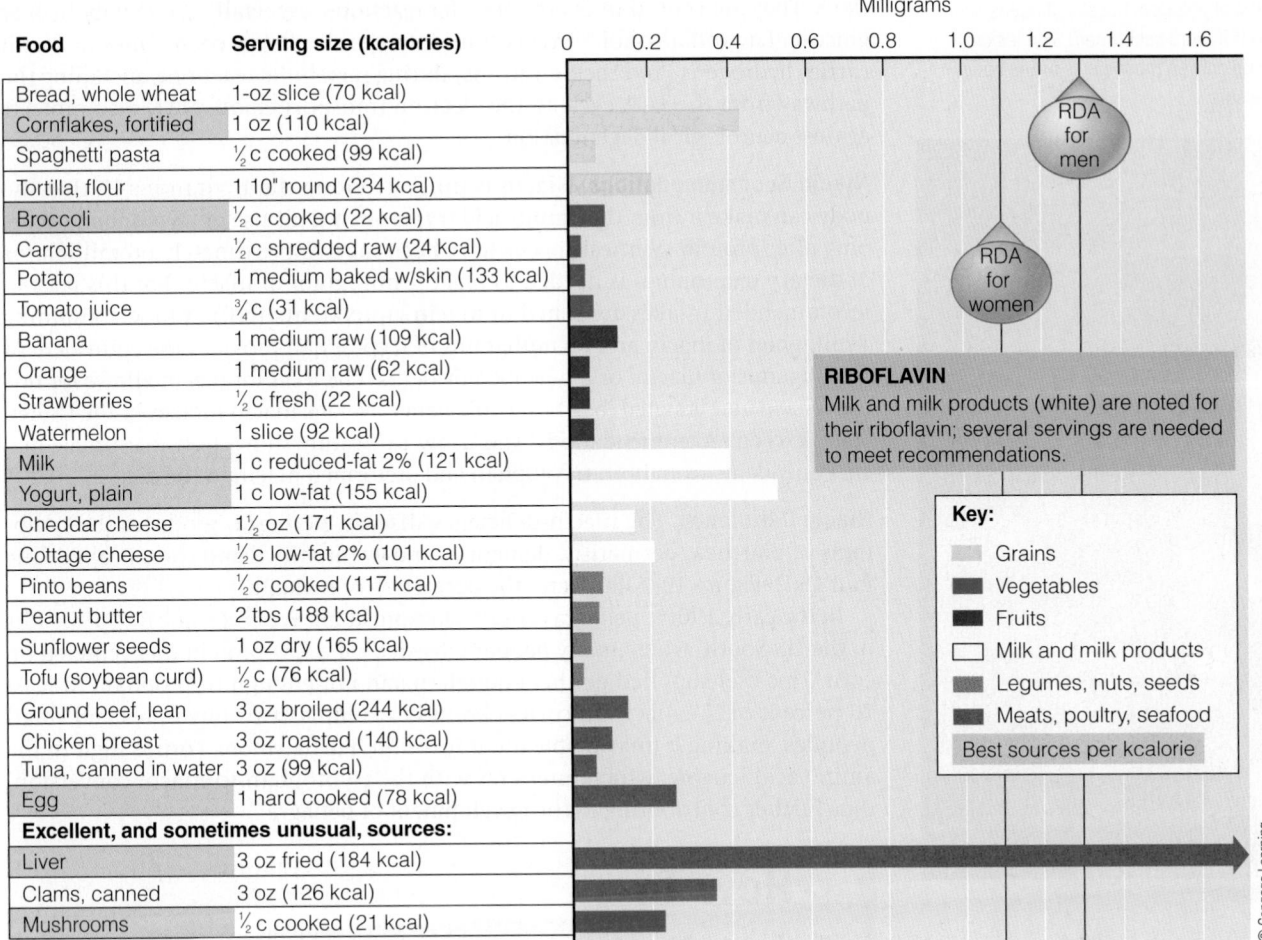

Food	Serving size (kcalories)
Bread, whole wheat	1-oz slice (70 kcal)
Cornflakes, fortified	1 oz (110 kcal)
Spaghetti pasta	½ c cooked (99 kcal)
Tortilla, flour	1 10" round (234 kcal)
Broccoli	½ c cooked (22 kcal)
Carrots	½ c shredded raw (24 kcal)
Potato	1 medium baked w/skin (133 kcal)
Tomato juice	¾ c (31 kcal)
Banana	1 medium raw (109 kcal)
Orange	1 medium raw (62 kcal)
Strawberries	½ c fresh (22 kcal)
Watermelon	1 slice (92 kcal)
Milk	1 c reduced-fat 2% (121 kcal)
Yogurt, plain	1 c low-fat (155 kcal)
Cheddar cheese	1½ oz (171 kcal)
Cottage cheese	½ c low-fat 2% (101 kcal)
Pinto beans	½ c cooked (117 kcal)
Peanut butter	2 tbs (188 kcal)
Sunflower seeds	1 oz dry (165 kcal)
Tofu (soybean curd)	½ c (76 kcal)
Ground beef, lean	3 oz broiled (244 kcal)
Chicken breast	3 oz roasted (140 kcal)
Tuna, canned in water	3 oz (99 kcal)
Egg	1 hard cooked (78 kcal)
Excellent, and sometimes unusual, sources:	
Liver	3 oz fried (184 kcal)
Clams, canned	3 oz (126 kcal)
Mushrooms	½ c cooked (21 kcal)

RIBOFLAVIN
Milk and milk products (white) are noted for their riboflavin; several servings are needed to meet recommendations.

Key:
- Grains
- Vegetables
- Fruits
- Milk and milk products
- Legumes, nuts, seeds
- Meats, poultry, seafood
- Best sources per kcalorie

© Cengage Learning

› REVIEW IT Riboflavin

Other Names

Vitamin B_2

RDA

Men: 1.3 mg/day

Women: 1.1 mg/day

Chief Functions in the Body

Part of coenzymes FMN (flavin mono-nucleotide) and FAD (flavin adenine dinucleotide) used in energy metabolism

Significant Sources

Milk products (yogurt, cheese); whole-grain, fortified, or enriched grain products; liver

Easily destroyed by ultraviolet light and irradiation

Deficiency Disease

Ariboflavinosis (ay-RYE-boh-FLAY-vin-oh-sis)

Deficiency Symptoms

Sore throat; cracks and redness at corners of mouth;[a] painful, smooth, purplish red tongue;[b] inflammation characterized by skin lesions covered with greasy scales

Toxicity Symptoms

None reported

[a]Cracks at the corners of the mouth are called *angular stomatitis* or *cheilosis* (kye-LOH-sis or kee-LOH-sis).

[b]Smoothness of the tongue is caused by loss of its surface structures and is termed *glossitis* (gloss-EYE-tis).

Niacin Niacin refers to two chemical structures: nicotinic acid and nicotinamide (also known as niacinamide). The body can easily convert nicotinic acid to nicotinamide, which is the major form of niacin in the blood.

niacin (NIGH-a-sin): a B vitamin. The coenzyme forms are NAD (*nicotinamide adenine dinucleotide*) and NADP (*the phosphate form of NAD*). Niacin can be eaten preformed or made in the body from its precursor, tryptophan, an essential amino acid.

> FIGURE 10-7 **Niacin-Deficiency Symptom—The Dermatitis of Pellagra**

In the dermatitis of pellagra, the skin darkens and flakes away as if it were sunburned. Skin lesions typically develop only on those parts of the body exposed to the sun.

Dr. M.A. Ansary/Science Source

The two coenzyme forms of niacin, NAD (nicotinamide adenine dinucleotide) and NADP (the phosphate form), participate in numerous metabolic reactions. They are central in energy-transfer reactions, especially the metabolism of glucose, fat, and alcohol. NAD is similar to the riboflavin coenzymes in that it carries hydrogens (and their electrons) during metabolic reactions, including the pathway from the TCA cycle to the electron transport chain. NAD also protects against neurological degeneration.

Niacin Recommendations Niacin is unique among the B vitamins in that the body can make it from the amino acid tryptophan. This use of tryptophan occurs only after protein synthesis needs have been met. Approximately 60 milligrams of dietary tryptophan is needed to make 1 milligram of niacin. For this reason, recommended intakes are stated in **niacin equivalents (NE).** A food containing 1 milligram of niacin and 60 milligrams of tryptophan provides the equivalent of 2 milligrams of niacin, or 2 niacin equivalents. The RDA for niacin allows for this conversion and is stated in niacin equivalents; average niacin intakes in the United States exceed recommendations. The "How To" feature shows how to estimate niacin equivalents from both tryptophan and preformed niacin in the diet.

Niacin Deficiency The niacin-deficiency disease, **pellagra,** produces the symptoms of diarrhea, dermatitis, dementia, and eventually death (often called "the four Ds"). Figure 10-7 illustrates the dermatitis of pellagra.

In the early 1900s, pellagra caused widespread misery and some 87,000 deaths in the US South, where many people subsisted on a low-protein diet centered on corn. This diet supplied neither enough niacin nor enough tryptophan. At least 70 percent of the niacin in corn is bound to complex carbohydrates and small peptides, making it unavailable for absorption. Furthermore, corn is high in the amino acid leucine, which interferes with the tryptophan-to-niacin conversion, thus further contributing to the development of pellagra.

>How To

Estimate Niacin Equivalents

Niacin recommendations are expressed as niacin equivalents (NE), but diet analysis programs and food composition tables report only preformed niacin. To estimate niacin equivalents from the tryptophan in dietary protein:

- Assume that most dietary proteins contain about 1 percent tryptophan. To determine the amount of tryptophan in protein, divide grams of protein by 100.

- Multiply by 1000 to convert grams of tryptophan to milligrams.

- Because it takes 60 milligrams of tryptophan to make 1 milligram of niacin, divide milligrams of tryptophan by 60 to get niacin equivalents.

- Add the amount of preformed niacin obtained in the diet.

Consider, for example, a person who consumes 80 grams of protein and 5 milligrams of preformed niacin.

- Estimate the amount of tryptophan in 80 grams of protein and convert to milligrams:

  80 g protein ÷ 100 = 0.8 g tryptophan

  0.8 g tryptophan × 1000 = 800 mg tryptophan

- Convert milligrams of tryptophan to niacin equivalents:

  800 mg tryptophan ÷ 60 = 13 mg NE

To determine the total amount of niacin available from the diet, add the amount available from tryptophan to the amount preformed in the diet.

  13 mg NE + 5 mg preformed niacin = 18 mg NE

> **TRY IT** Calculate how many niacin equivalents a person receives from a diet that delivers 60 grams of protein and 6 milligrams of niacin.

niacin equivalents (NE): the amount of niacin present in food, including the niacin that can theoretically be made from its precursor, tryptophan, present in the food.

- 1 NE = 1 mg niacin or 60 mg tryptophan

pellagra (pell-AY-gra): the niacin-deficiency disease, characterized by diarrhea, dermatitis, dementia, and eventually death.

- **pellis** = skin

- **agra** = rough

Pellagra was originally believed to be caused by an infection. Medical researchers spent many years and much effort searching for infectious microbes until they realized that the problem was not what was *present* in the food but what was *absent* from it. That a disease such as pellagra could be caused by diet inadequacies—and not by pathogens—was a groundbreaking discovery. It contradicted commonly held medical opinions that diseases were caused only by infectious agents. By carefully following the scientific method (as described in Chapter 1), researchers advanced the science of nutrition dramatically.

Niacin Toxicity When a normal dose of a nutrient (levels commonly found in foods) provides a normal blood concentration, the nutrient is having a *physiological* effect. When a large dose (levels commonly available only from supplements) overwhelms the body and raises blood concentrations to abnormally high levels, the nutrient is acting like a drug and having a *pharmacological* effect. Naturally occurring niacin from foods has a physiological effect that causes no harm. Large doses of nicotinic acid from supplements or drugs, however, produce a variety of pharmacological effects, most notably **"niacin flush."** Niacin flush occurs when nicotinic acid is taken in doses only three to four times the RDA. It dilates the capillaries and causes a tingling sensation that can be painful. The nicotinamide form does not produce this effect.

Large doses of nicotinic acid can effectively lower LDL cholesterol and triglycerides and raise HDL cholesterol—all factors that help to protect against heart disease.[4] As effective as niacin therapy is in improving blood lipids, however, it may not benefit patients with heart disease whose blood lipids are already being controlled with statin drugs.[5] The use of niacin as a drug may benefit other patients, but its use must be closely monitored. People with the following conditions may be particularly susceptible to the toxic effects of niacin: liver disease, diabetes, peptic ulcers, gout, irregular heartbeats, inflammatory bowel disease, migraine headaches, and alcoholism. The nicotinamide form does not improve blood cholesterol levels.[6]

Niacin Food Sources Tables of food composition typically list preformed niacin only, but as mentioned, niacin can also be made in the body from the amino acid tryptophan. Dietary tryptophan could meet about half the daily niacin need for most people, but the average diet easily supplies enough preformed niacin.

Figure 10-8 (p. 312) presents niacin in selected foods. Meat, poultry, fish, legumes, and enriched and whole grains contribute about half the niacin people consume. Mushrooms, potatoes, and tomatoes are among the richest vegetable sources, and they can provide abundant niacin when eaten in generous amounts.

Niacin is less vulnerable to losses during food preparation and storage than other water-soluble vitamins. Being fairly heat resistant, niacin can withstand reasonable cooking times, but like other water-soluble vitamins, it will leach into cooking water. The accompanying table provides a summary of niacin.

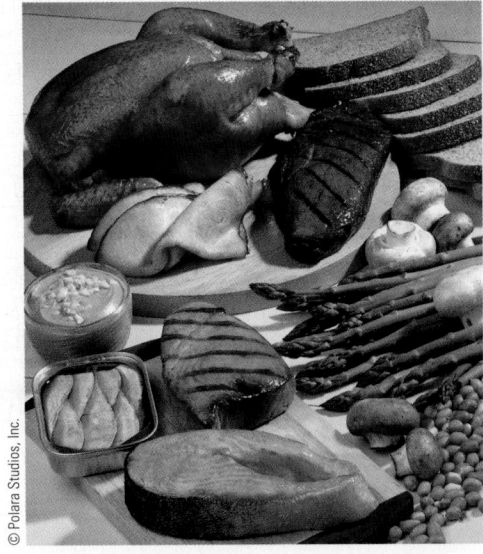

© Polara Studios, Inc.

Protein-rich foods such as meat, fish, poultry, and peanut butter contribute much of the niacin in people's diets. Enriched breads and cereals and a few vegetables are also rich in niacin.

› REVIEW IT Niacin

Other Names

Nicotinic acid, nicotinamide, niacinamide, vitamin B₃; precursor is dietary tryptophan (an amino acid)

RDA

Men: 16 mg NE/day

Women: 14 mg NE/day

UL^a

Adults: 35 mg/day

Chief Functions in the Body

Part of coenzymes NAD (nicotinamide adenine dinucleotide) and NADP (its phosphate form) used in energy metabolism

Significant Sources

Milk, eggs, meat, poultry, fish; whole-grain, fortified, and enriched grain products; nuts and all protein-containing foods

Deficiency Disease

Pellagra

Deficiency Symptoms

Diarrhea, abdominal pain, vomiting; inflamed, swollen, smooth, bright red tongue;^b depression, apathy, fatigue, loss of memory, headache; bilateral symmetrical rash on areas exposed to sunlight

Toxicity Symptoms

Painful flush, hives, and rash ("niacin flush"); nausea and vomiting; liver damage; impaired glucose tolerance

niacin flush: a temporary burning, tingling, and itching sensation that occurs when a person takes a large dose of nicotinic acid; often accompanied by a headache and reddened face, arms, and chest.

^aThe UL applies to synthetic forms obtained from supplements, fortified foods, or a combination.
^bSmoothness of the tongue is caused by loss of its surface structures and is termed *glossitis* (gloss-EYE-tis).

> FIGURE 10-8   Niacin in Selected Foods

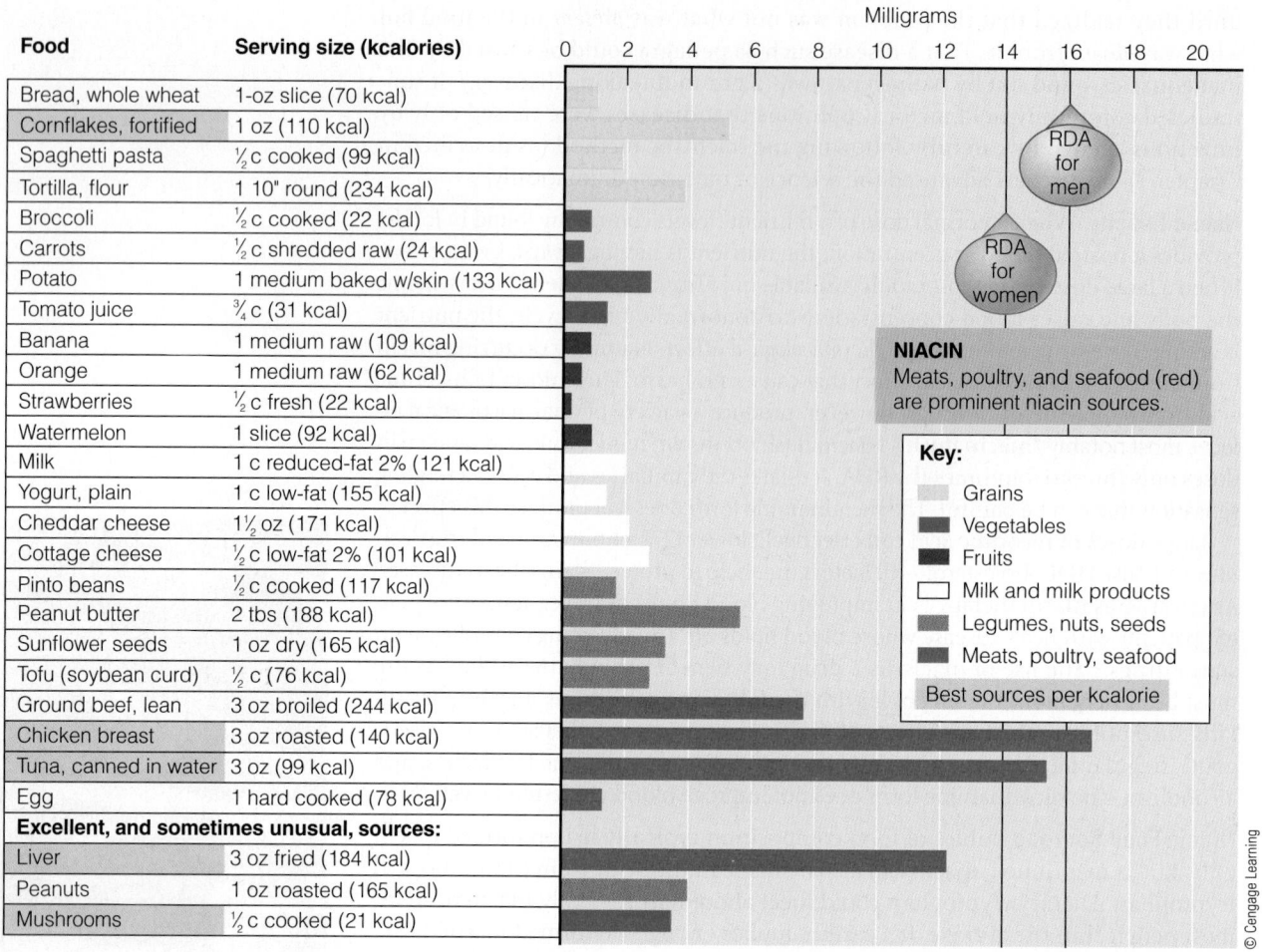

Food	Serving size (kcalories)
Bread, whole wheat	1-oz slice (70 kcal)
Cornflakes, fortified	1 oz (110 kcal)
Spaghetti pasta	½ c cooked (99 kcal)
Tortilla, flour	1 10" round (234 kcal)
Broccoli	½ c cooked (22 kcal)
Carrots	½ c shredded raw (24 kcal)
Potato	1 medium baked w/skin (133 kcal)
Tomato juice	¾ c (31 kcal)
Banana	1 medium raw (109 kcal)
Orange	1 medium raw (62 kcal)
Strawberries	½ c fresh (22 kcal)
Watermelon	1 slice (92 kcal)
Milk	1 c reduced-fat 2% (121 kcal)
Yogurt, plain	1 c low-fat (155 kcal)
Cheddar cheese	1½ oz (171 kcal)
Cottage cheese	½ c low-fat 2% (101 kcal)
Pinto beans	½ c cooked (117 kcal)
Peanut butter	2 tbs (188 kcal)
Sunflower seeds	1 oz dry (165 kcal)
Tofu (soybean curd)	½ c (76 kcal)
Ground beef, lean	3 oz broiled (244 kcal)
Chicken breast	3 oz roasted (140 kcal)
Tuna, canned in water	3 oz (99 kcal)
Egg	1 hard cooked (78 kcal)
Excellent, and sometimes unusual, sources:	
Liver	3 oz fried (184 kcal)
Peanuts	1 oz roasted (165 kcal)
Mushrooms	½ c cooked (21 kcal)

RDA for men

RDA for women

NIACIN Meats, poultry, and seafood (red) are prominent niacin sources.

Key:
- Grains
- Vegetables
- Fruits
- Milk and milk products
- Legumes, nuts, seeds
- Meats, poultry, seafood
- Best sources per kcalorie

© Cengage Learning

Biotin Biotin plays an important role in metabolism as a coenzyme that carries activated carbon dioxide. This role is critical in the TCA cycle: biotin delivers a carbon to 3-carbon pyruvate, thus replenishing oxaloacetate, the 4-carbon compound needed to combine with acetyl CoA to keep the TCA cycle turning (review Figure 7-15 on p. 214). The biotin coenzyme also participates in gluconeogenesis, fatty acid synthesis, and the breakdown of certain fatty acids and amino acids.

Biotin Recommendations Biotin is needed in very small amounts. Because there is insufficient research on biotin requirements, an Adequate Intake (AI) has been determined, instead of an RDA.

Biotin Deficiency and Toxicity Biotin deficiencies rarely occur. Researchers can induce a biotin deficiency in animals or human beings by feeding them raw egg whites, which contain a protein that binds biotin and thus prevents its absorption.* Biotin-deficiency symptoms include skin rash, hair loss, and neurological impairment. More than two dozen raw egg whites must be consumed daily for several months to produce these effects; cooking eggs denatures the binding protein. Because no adverse effects have been reported from high biotin intakes, a UL has not been set.

Biotin Food Sources Biotin is widespread in foods (including egg yolks), so eating a variety of foods protects against deficiencies. Some biotin is also synthesized by GI tract bacteria, but this amount does not contribute much to the biotin absorbed. The accompanying table provides a summary of biotin.

biotin (BY-oh-tin): a B vitamin that functions as a coenzyme in metabolism.

*The protein *avidin* (AV-eh-din) in egg whites binds biotin.

AI	Deficiency Symptoms
Adults: 30 µg/day	Depression, lethargy, hallucinations, numb or tingling sensation in the arms and legs; red, scaly rash around the eyes, nose, and mouth; hair loss
Chief Functions in the Body	
Part of a coenzyme used in energy metabolism, fat synthesis, amino acid metabolism, and glycogen synthesis	**Toxicity Symptoms**
Significant Sources	None reported
Widespread in foods; liver, egg yolks, soybeans, fish, whole grains; also produced by GI bacteria	

Pantothenic Acid **Pantothenic acid** is part of the chemical structure of coenzyme A—the same CoA that forms acetyl CoA, a key compound in several metabolic pathways featured in Chapter 7, including the TCA cycle. (Appendix C presents the chemical structures of these two molecules and shows that coenzyme A is made up in part of pantothenic acid.) As such, it is involved in more than 100 different steps in the synthesis of lipids, neurotransmitters, steroid hormones, and hemoglobin.

Pantothenic Acid Recommendations An Adequate Intake (AI) for pantothenic acid has been set. It reflects the amount needed to replace daily losses.

Pantothenic Acid Deficiency and Toxicity Pantothenic acid deficiency is rare. Its symptoms involve a general failure of all the body's systems and include fatigue, GI distress, and neurological disturbances. The "burning feet" syndrome that affected prisoners of war in Asia during World War II is thought to have been caused by pantothenic acid deficiency. No toxic effects have been reported, and no UL has been established.

Pantothenic Acid Food Sources Pantothenic acid is widespread in foods, and typical diets seem to provide adequate intakes. Beef, poultry, whole grains, potatoes, tomatoes, and broccoli are particularly good sources. Losses of pantothenic acid during food production can be substantial because it is readily destroyed by the freezing, canning, and refining processes. The accompanying table provides a summary of pantothenic acid.

AI	Deficiency Symptoms
Adults: 5 mg/day	Vomiting, nausea, stomach cramps; insomnia, fatigue, depression, irritability, restlessness, apathy; hypoglycemia, increased sensitivity to insulin; numbness, muscle cramps, inability to walk
Chief Functions in the Body	
Part of coenzyme A, used in energy metabolism	
Significant Sources	**Toxicity Symptoms**
Widespread in foods; chicken, beef, potatoes, oats, tomatoes, liver, egg yolk, broccoli, whole grains	None reported
Easily destroyed by food processing	

Vitamin B₆ Vitamin B_6 occurs in three forms—pyridoxal, pyridoxine, and pyridoxamine. All three can be converted to the coenzyme PLP (pyridoxal phosphate), which is active in more than 100 reactions, including carbohydrate, fatty acid, and amino acid metabolism.[7] Because PLP can transfer amino groups (NH_2) from an amino acid to a keto acid, the body can make nonessential amino acids (review Figure 6-12 on p. 183). The ability to add and remove amino groups makes PLP valuable in protein and urea metabolism as well. The conversions of the amino acid tryptophan to niacin or to the neurotransmitter serotonin also depend on PLP.

pantothenic (PAN-toe-THEN-ick) **acid:** a B vitamin. The principal active form is part of coenzyme A, called "CoA" throughout Chapter 7.

• **pantos** = everywhere

vitamin B₆: a family of compounds—pyridoxal, pyridoxine, and pyridoxamine. The primary active coenzyme form is *PLP* (*pyridoxal phosphate*).

In addition, PLP participates in the synthesis of heme (the nonprotein portion of hemoglobin), nucleic acids (such as DNA and RNA), and lecithin (a phospholipid).

Vitamin B$_6$ Recommendations The RDA for vitamin B$_6$ is based on the amounts needed to maintain adequate levels of its coenzymes. Unlike other water-soluble vitamins, vitamin B$_6$ is stored extensively in muscle tissue. Research does not support claims, however, that large doses of vitamin B$_6$ enhance muscle strength or physical endurance.

Vitamin B$_6$ Deficiency Without adequate vitamin B$_6$, synthesis of key neurotransmitters diminishes, and abnormal compounds produced during tryptophan metabolism accumulate in the brain. Early symptoms of vitamin B$_6$ deficiency include depression and confusion; advanced symptoms include abnormal brain wave patterns and convulsions. Low levels of vitamin B$_6$ are associated with increased risks of some cancers and cardiovascular disease.[8]

Alcohol contributes to the destruction and loss of vitamin B$_6$ from the body. As Highlight 7 describes, when the body breaks down alcohol, it produces acetaldehyde. If allowed to accumulate, acetaldehyde dislodges the PLP coenzyme from its enzymes; once loose, PLP breaks down and is excreted.

Another drug that acts as a vitamin B$_6$ **antagonist** is isoniazid, a medication that inhibits the growth of the tuberculosis bacterium.* This drug has saved countless lives, but because isoniazid binds and inactivates vitamin B$_6$, it can induce a deficiency. Whenever isoniazid is used to treat tuberculosis, vitamin B$_6$ supplements must be given to protect against deficiency. Oral contraceptives also seem to decrease the body's vitamin B$_6$ reserves.[9]

Vitamin B$_6$ Toxicity The first major report of vitamin B$_6$ toxicity appeared in the early 1980s. Until that time, most researchers and dietitians believed that, like the other water-soluble vitamins, vitamin B$_6$ could not reach toxic concentrations in the body. The report described neurological damage in people who had been taking more than 2 *grams* of vitamin B$_6$ daily (20 times the current UL of 100 *milligrams* per day) for 2 months or more.

Vitamin B$_6$ Food Sources As you can see from the colors in Figure 10-9, meats, fish, and poultry (red bars), potatoes and a few other vegetables (green bars), and fruits (purple bars) offer vitamin B$_6$. As is true of most of the other vitamins, fruits and vegetables rank considerably higher when foods are judged by nutrient density (vitamin B$_6$ per kcalorie). Several servings of vitamin B$_6$–rich foods are needed to meet recommended intakes.

Foods lose vitamin B$_6$ when heated. Information is limited, but vitamin B$_6$ bioavailability from plant-derived foods seems to be lower than from animal-derived foods. Fiber does not appear to interfere with vitamin B$_6$ absorption. The accompanying table provides a summary of vitamin B$_6$.

Most protein-rich foods such as meat, fish, and poultry provide ample vitamin B$_6$; some vegetables and fruits are good sources too.

> REVIEW IT Vitamin B$_6$

Other Names

Pyridoxine, pyridoxal, pyridoxamine

RDA

Adults (19–50 yr): 1.3 mg/day

UL

Adults: 100 mg/day

Chief Functions in the Body

Part of coenzymes PLP (pyridoxal phosphate) and PMP (pyridoxamine phosphate) used in amino acid and fatty acid metabolism; helps to convert tryptophan to niacin and to serotonin; helps to make red blood cells

Significant Sources

Meats, fish, poultry, potatoes and other starchy vegetables, legumes, noncitrus fruits, fortified cereals, liver, soy products

Easily destroyed by heat

Deficiency Symptoms

Scaly dermatitis; anemia (small-cell type);[a] depression, confusion, convulsions

Toxicity Symptoms

Depression, fatigue, irritability, headaches, nerve damage causing numbness and muscle weakness leading to an inability to walk and convulsions; skin lesions

[a]Small-cell-type anemia is called *microcytic anemia*.

*Isoniazid (eye-so-NYE-uh-zid) is also known as INH (isonicotinic acid hydrazide).

antagonist: a competing factor that counteracts the action of another factor. When a drug displaces a vitamin from its site of action, the drug renders the vitamin ineffective and thus acts as a vitamin antagonist.

> **FIGURE 10-9** **Vitamin B_6 in Selected Foods**

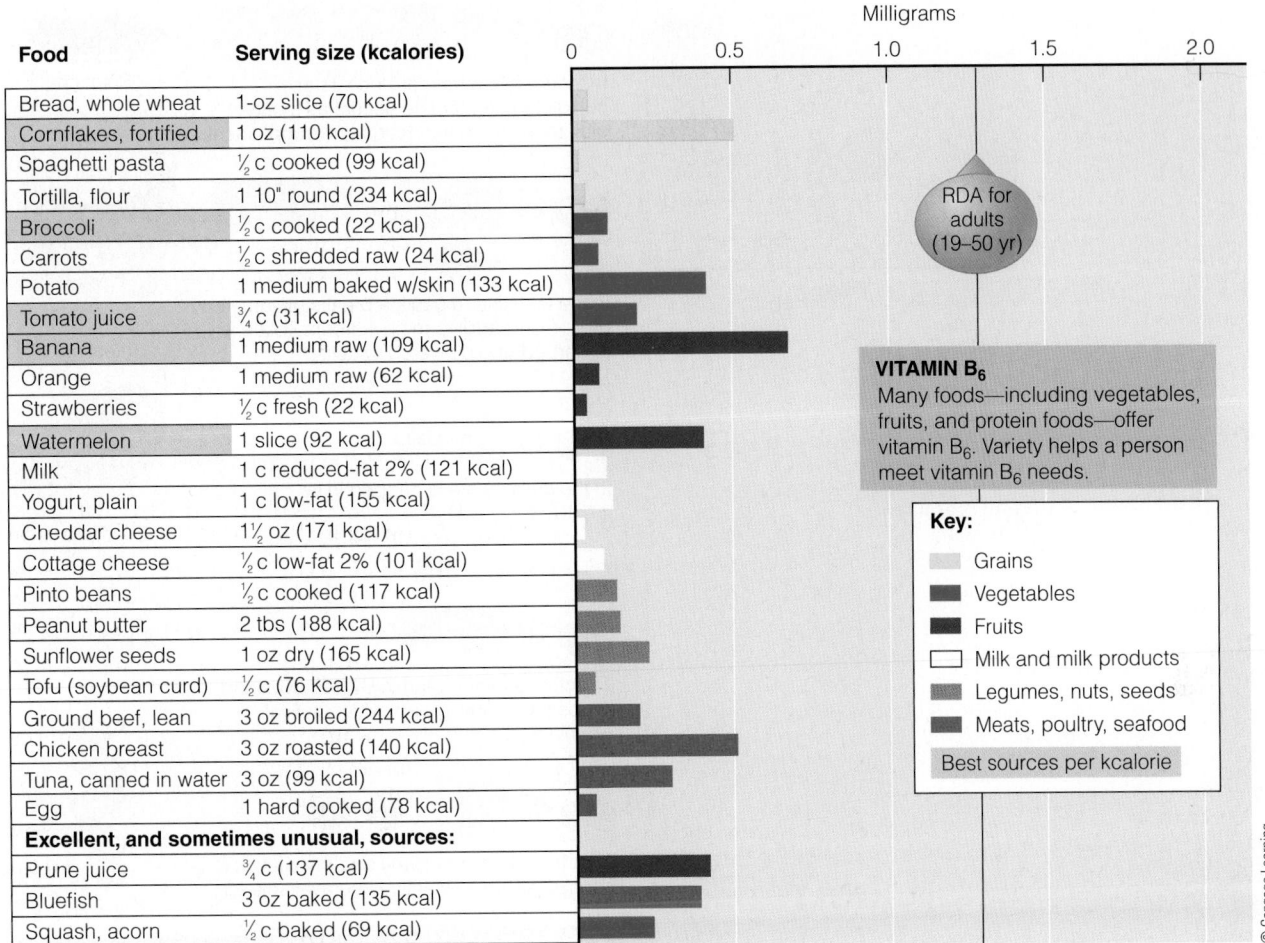

Food	Serving size (kcalories)
Bread, whole wheat	1-oz slice (70 kcal)
Cornflakes, fortified	1 oz (110 kcal)
Spaghetti pasta	½ c cooked (99 kcal)
Tortilla, flour	1 10" round (234 kcal)
Broccoli	½ c cooked (22 kcal)
Carrots	½ c shredded raw (24 kcal)
Potato	1 medium baked w/skin (133 kcal)
Tomato juice	¾ c (31 kcal)
Banana	1 medium raw (109 kcal)
Orange	1 medium raw (62 kcal)
Strawberries	½ c fresh (22 kcal)
Watermelon	1 slice (92 kcal)
Milk	1 c reduced-fat 2% (121 kcal)
Yogurt, plain	1 c low-fat (155 kcal)
Cheddar cheese	1½ oz (171 kcal)
Cottage cheese	½ c low-fat 2% (101 kcal)
Pinto beans	½ c cooked (117 kcal)
Peanut butter	2 tbs (188 kcal)
Sunflower seeds	1 oz dry (165 kcal)
Tofu (soybean curd)	½ c (76 kcal)
Ground beef, lean	3 oz broiled (244 kcal)
Chicken breast	3 oz roasted (140 kcal)
Tuna, canned in water	3 oz (99 kcal)
Egg	1 hard cooked (78 kcal)
Excellent, and sometimes unusual, sources:	
Prune juice	¾ c (137 kcal)
Bluefish	3 oz baked (135 kcal)
Squash, acorn	½ c baked (69 kcal)

RDA for adults (19–50 yr)

VITAMIN B_6
Many foods—including vegetables, fruits, and protein foods—offer vitamin B_6. Variety helps a person meet vitamin B_6 needs.

Key:
- Grains
- Vegetables
- Fruits
- Milk and milk products
- Legumes, nuts, seeds
- Meats, poultry, seafood

Best sources per kcalorie

© Cengage Learning

Folate

Folate, also known as folacin or folic acid, has a chemical name that would fit a flying dinosaur: pteroylglutamic acid (PGA for short). Its primary coenzyme form, THF (tetrahydrofolate), serves as part of an enzyme complex that transfers 1-carbon compounds that arise during metabolism.[10] This action converts vitamin B_{12} to one of its coenzyme forms, synthesizes the DNA required for all rapidly growing cells, and regenerates the amino acid methionine from homocysteine.

Figure 10-10 (p. 316) summarizes folate's absorption, activation, and relationship with vitamin B_{12}. It explains that foods deliver folate mostly in the "bound" form—that is, combined with a string of amino acids (all glutamate), known as polyglutamate. (See Appendix C for the chemical structure.) Enzymes on the intestinal cell surfaces hydrolyze the polyglutamate to monoglutamate— folate with only one glutamate attached—and several single glutamates. The monoglutamate is then attached to a methyl group (CH_3) and delivered to the liver and other body cells. To activate folate, the methyl group must be removed by an enzyme that requires the help of vitamin B_{12}. Without that help, folate becomes trapped inside cells in its methyl form, unavailable to support DNA synthesis and cell growth.

The liver incorporates excess folate into bile that is then sent to the gallbladder and GI tract. Thus folate travels in the same enterohepatic circulation as bile (review Figure 5-14 on p. 144).

This complicated system for handling folate is vulnerable to GI tract injuries. Because folate is actively secreted back into the GI tract with bile, it can be

folate (FOLE-ate): a B vitamin; also known as folic acid, folacin, or pteroylglutamic (tare-o-EEL-glue-TAM-ick) acid (PGA). The coenzyme forms are *DHF (dihydrofolate)* and *THF (tetrahydrofolate)*.

> FIGURE 10-10 **Folate's Absorption and Activation**

Spinach

In foods, folate naturally occurs as polyglutamate. (Folate occurs as mono-glutamate in fortified foods and supplements.)

+ CH₃

In the intestine, digestion breaks glutamates off . . . and adds a methyl group. Folate is absorbed and delivered to cells.

Intestine

—CH₃

In the cells, folate is trapped in its inactive form.

Cell

B₁₂

—CH₃

To activate folate, vitamin B₁₂ removes and keeps the methyl group, which activates vitamin B₁₂.

B₁₂—CH₃

Both the folate coenzyme and the vitamin B₁₂ coenzyme are now active and available for DNA synthesis.

DNA

© Cengage Learning

In the folate ring structure diagram:

COOH
|
CH₂
|
CH₂
|
O   CH
‖ |
C — NH — CH
     |
     COOH

Ring structure + Glutamate

Folate

reabsorbed repeatedly. If the GI tract cells are damaged, then folate is lost. Such is the case in alcohol abuse; folate deficiency rapidly develops and, ironically, further damages the GI tract. Remember, folate is active in cell multiplication—and the cells lining the GI tract are among the most rapidly replaced cells in the body. When unable to make new cells, the GI tract deteriorates and not only loses folate, but fails to absorb other nutrients as well.

Folate Recommendations The bioavailability of folate ranges from 50 percent for foods to 100 percent for supplements taken on an empty stomach. These differences in bioavailability must be considered when establishing folate recommendations.[11] The DRI committee gives naturally occurring folate from foods full credit. Synthetic folate from fortified foods and supplements is given extra credit because, on average, it is 1.7 times more available than naturally occurring food folate. Thus a person consuming 100 micrograms of folate from foods and 100 micrograms from a supplement (multiplied by 1.7) receives 270 **dietary folate equivalents (DFE).** The "How To" feature (p. 317) describes how to estimate dietary folate equivalents. The need for folate rises considerably during pregnancy and whenever cells are multiplying, so the recommendations for pregnant women are considerably higher than for other adults.

Folate and Neural Tube Defects The brain and spinal cord develop from the **neural tube,** and defects in its orderly formation during the early weeks of pregnancy may result in various central nervous system disorders and death. (Figure 15-5 in Chapter 15 includes an illustration of spina bifida, a neural tube defect.)

dietary folate equivalents (DFE): the amount of folate available to the body from naturally occurring sources, fortified foods, and supplements, accounting for differences in the bioavailability from each source.

• DFE = μg food folate + (1.7 × μg synthetic folate)

neural tube: the embryonic tissue that forms the brain and spinal cord.

Estimate Dietary Folate Equivalents

Folate is expressed in terms of DFE (dietary folate equivalents) because synthetic folate from supplements and fortified foods is absorbed at almost twice (1.7 times) the rate of naturally occurring folate from other foods. Use the following equation to calculate:

$$DFE = \mu g \text{ food folate} + (1.7 \times \mu g \text{ synthetic folate})$$

Consider, for example, a pregnant woman who takes a supplement and eats a bowl of fortified cornflakes, 2 slices of fortified bread, and a cup of fortified pasta. From the supplement and fortified foods, she obtains synthetic folate:

Supplement	100 μg folate
Fortified cornflakes	100 μg folate
Fortified bread	40 μg folate
Fortified pasta	60 μg folate
	300 μg folate

To calculate the DFE, multiply the amount of synthetic folate by 1.7:

$$300 \text{ μg} \times 1.7 = 510 \text{ μg DFE}$$

Now add the naturally occurring folate from the other foods in her diet—in this example, another 90 μg of folate.

$$510 \text{ μg DFE} + 90 \text{ μg} = 600 \text{ μg DFE}$$

Notice that if we had not converted synthetic folate from supplements and fortified foods to DFE, then this woman's intake would appear to fall short of the 600 μg recommendation for pregnancy (300 μg + 90 μg = 390 μg). But as our example shows, her intake does meet the recommendation. Recent revisions to food labels now list folate in μg DFE, making such calculations unnecessary.

> **TRY IT** Calculate how many dietary folate equivalents a person receives from 200 μg of folate from a supplement, 75 μg of folate from fortified cereal, and 120 μg of folate from other foods.

Folate supplements taken 1 month before conception and continued throughout the first trimester of pregnancy can help prevent **neural tube defects.** For this reason, all women of childbearing age who are capable of becoming pregnant should consume 0.4 milligram (400 micrograms) of folate daily—easily accomplished by eating folate-rich foods, folate-fortified foods, or a multivitamin supplement daily. Because half of the pregnancies each year are unplanned and because neural tube defects occur early in development before most women realize they are pregnant, the Food and Drug Administration (FDA) has mandated that grain products be fortified to deliver folate to the US population.* Labels on fortified products may claim that "adequate intake of folate has been shown to reduce the risk of neural tube defects." Fortification has improved folate status in women of childbearing age and lowered the prevalence rate of neural tube defects, as Figure 10-11 (p. 318) shows.

© Wellcome Trust Library/Custom Medical Stock Photo

Folate helps to protect against spina bifida, a neural tube defect characterized by the incomplete closure of the spinal cord and its bony encasement.

Some research suggests that folate taken before and during pregnancy may also prevent congenital birth defects, such as cleft lip and cleft palate, and neurodevelopmental disorders, such as autism.[12] Such findings strengthen recommendations for women to pay attention to their folate needs.

Folate fortification raises safety concerns as well. Because high intakes of folate can mask a vitamin B_{12} deficiency, folate consumption should not exceed 1 milligram daily without close medical supervision. The risks and benefits of folate fortification continue to be a topic of current debate, especially given that 5 percent of the US population exceed the UL for folate.[13]

neural tube defects: malformations of the brain, spinal cord, or both during embryonic development that often result in lifelong disability or death. The two main types of neural tube defects are *spina bifida* (literally "split spine") and *anencephaly* ("no brain").

---

*Bread products, flour, corn grits, cornmeal, farina, rice, macaroni, and noodles must be fortified with 140 micrograms of folate per 100 grams of grain. For perspective, 100 grams is roughly 3 slices of bread; 1 cup of flour; ½ cup of corn grits, cornmeal, farina, or rice; or ¾ cup of macaroni or noodles.

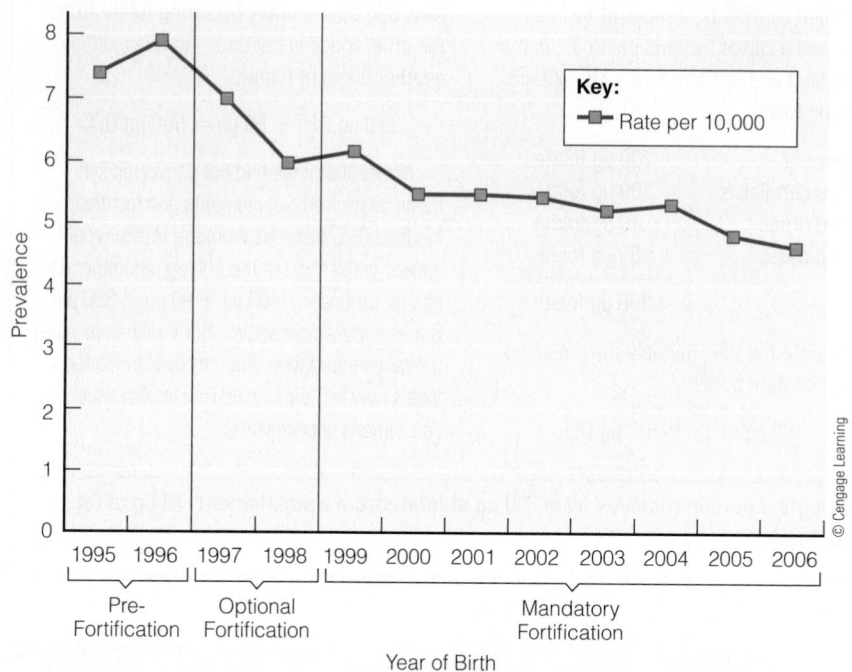

SOURCE: National Center for Health Statistics, Centers for Disease Control and Prevention, www.cdc.gov, updated January 2010.

Folate and Heart Disease The FDA's decision to fortify grain products with folate was strengthened by research suggesting a role for folate in protecting against heart disease.[14] One of folate's key roles in the body is to break down the amino acid homocysteine. Without folate, homocysteine accumulates, which seems to enhance formation of blood clots and atherosclerotic lesions. Fortified foods and folate supplements raise blood folate and reduce blood homocysteine, but do not seem to reduce the risk of heart attacks, strokes, or death from cardiovascular causes.[15]

Folate and Cancer Because the synthesis of DNA and the transfer of methyl groups depend on folate, its relationships with cancer are complex, depending on the type of cancer and the timing of folate supplementation. Some research suggests that sufficient folate may protect against the initiation of cancer, whereas other studies report that high intakes may enhance progression once cancer has begun.[16] In general, foods containing folate probably reduce the risk of pancreatic cancer.[17] Limited evidence suggests that folate may also reduce the risk of esophageal and colorectal cancer.[18]

Folate Deficiency Folate deficiency impairs cell division and protein synthesis—processes critical to growing tissues. In a folate deficiency, the replacement of red blood cells and GI tract cells falters. Not surprisingly, then, two of the first symptoms of a folate deficiency are **anemia** and GI tract deterioration.

The anemia of folate deficiency is known as *macrocytic* or *megaloblastic anemia* and is characterized by large, immature red blood cells (see Figure 10-12). Without folate, DNA damage destroys many of the red blood cells as they attempt to divide and mature. The result is fewer, but larger, red blood cells that cannot carry oxygen or travel through the capillaries as efficiently as normal red blood cells. Since the implementation of folate fortification in the United States, the prevalence of macrocytic anemia has decreased dramatically.[19]

Primary folate deficiencies may develop from inadequate intake and have been reported in infants who were fed goat's milk, which is notoriously low in folate. Secondary folate deficiencies may result from impaired absorption or an unusual metabolic need for the vitamin. Metabolic needs increase in situations where cell multiplication must speed up, such as pregnancies involving twins and triplets; cancer; skin-destroying diseases such as chicken pox and measles; and burns, blood loss, GI tract damage, and the like.

Of all the vitamins, folate appears to be most vulnerable to interactions with drugs, which can also lead to a secondary deficiency. Some medications, notably anticancer drugs, have a chemical structure similar to folate's structure and can displace the vitamin from enzymes and interfere with normal metabolism. Like all cells, cancer cells need the real vitamin to multiply—without it, they die. Unfortunately, anticancer drugs affect both cancerous cells and healthy cells, creating a folate deficiency for all cells. (Highlight 17 discusses nutrient-drug interactions, and Figure H17-1 illustrates the similarities between the vitamin folate and the anticancer drug methotrexate.)

anemia (ah-NEE-me-ah): literally, "too little blood." Anemia is any condition in which too few red blood cells are present, or the red blood cells are immature (and therefore large) or too small or contain too little hemoglobin to carry the normal amount of oxygen to the tissues. Anemia is not a disease itself but can be a consequence of many different disease conditions, including many nutrient deficiencies, bleeding, excessive red blood cell destruction, and defective red blood cell formation.

• **an** = without

• **emia** = blood

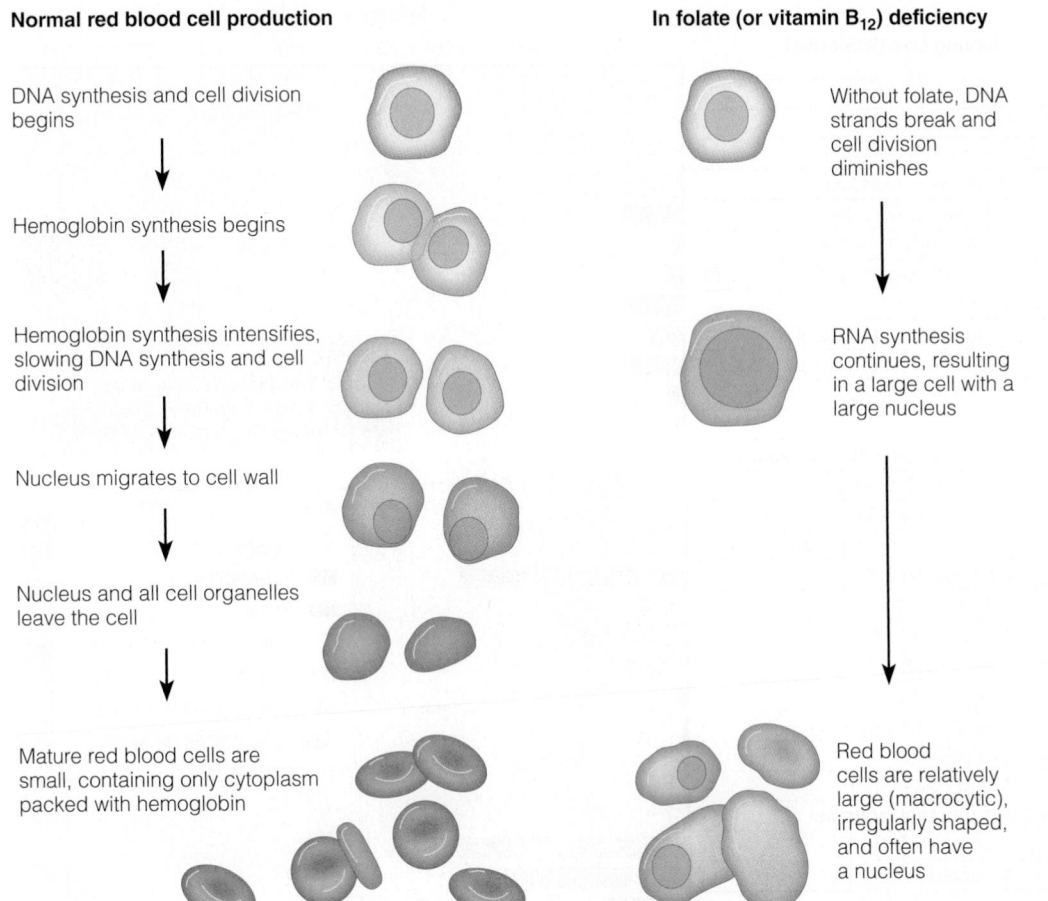

Normal red blood cell production

DNA synthesis and cell division begins

Hemoglobin synthesis begins

Hemoglobin synthesis intensifies, slowing DNA synthesis and cell division

Nucleus migrates to cell wall

Nucleus and all cell organelles leave the cell

Mature red blood cells are small, containing only cytoplasm packed with hemoglobin

In folate (or vitamin B₁₂) deficiency

Without folate, DNA strands break and cell division diminishes

RNA synthesis continues, resulting in a large cell with a large nucleus

Red blood cells are relatively large (macrocytic), irregularly shaped, and often have a nucleus

© Cengage Learning

Aspirin and antacids also interfere with the body's folate status: aspirin inhibits the action of folate-requiring enzymes, and antacids limit the absorption of folate. Healthy adults who use these drugs to relieve an occasional headache or upset stomach need not be concerned, but people who rely heavily on aspirin or antacids should be aware of the nutrition consequences.

Folate Toxicity A UL has been established for folate from fortified foods or supplements (see the inside front cover). Commonly consumed amounts of folate from both natural sources and fortified foods appear to cause no harm. The small percentage of adults who also take high-dose folate supplements, however, can reach levels that are high enough to obscure a vitamin B₁₂ deficiency and delay diagnosis of neurological damage.[20]

Folate Food Sources Figure 10-13 (p. 320) shows that folate is especially abundant in legumes, fruits, and vegetables. The vitamin's name suggests the word *foliage*, and indeed, dark green, leafy vegetables are outstanding sources. With fortification, grain products also contribute folate. The small red and white bars in Figure 10-13 indicate that meats and milk products are poor folate sources. Heat and oxidation during cooking and storage can destroy as much as half of the folate in foods. The table on p. 320 provides a summary of folate.

© Polara Studios, Inc.

Dark green and leafy vegetables (such as spinach and broccoli), legumes (such as black beans, kidney beans, and black-eyed peas), liver, and some fruits (notably citrus fruits and juices) are naturally rich in folate.

> **FIGURE 10-13** **Folate in Selected Foods**

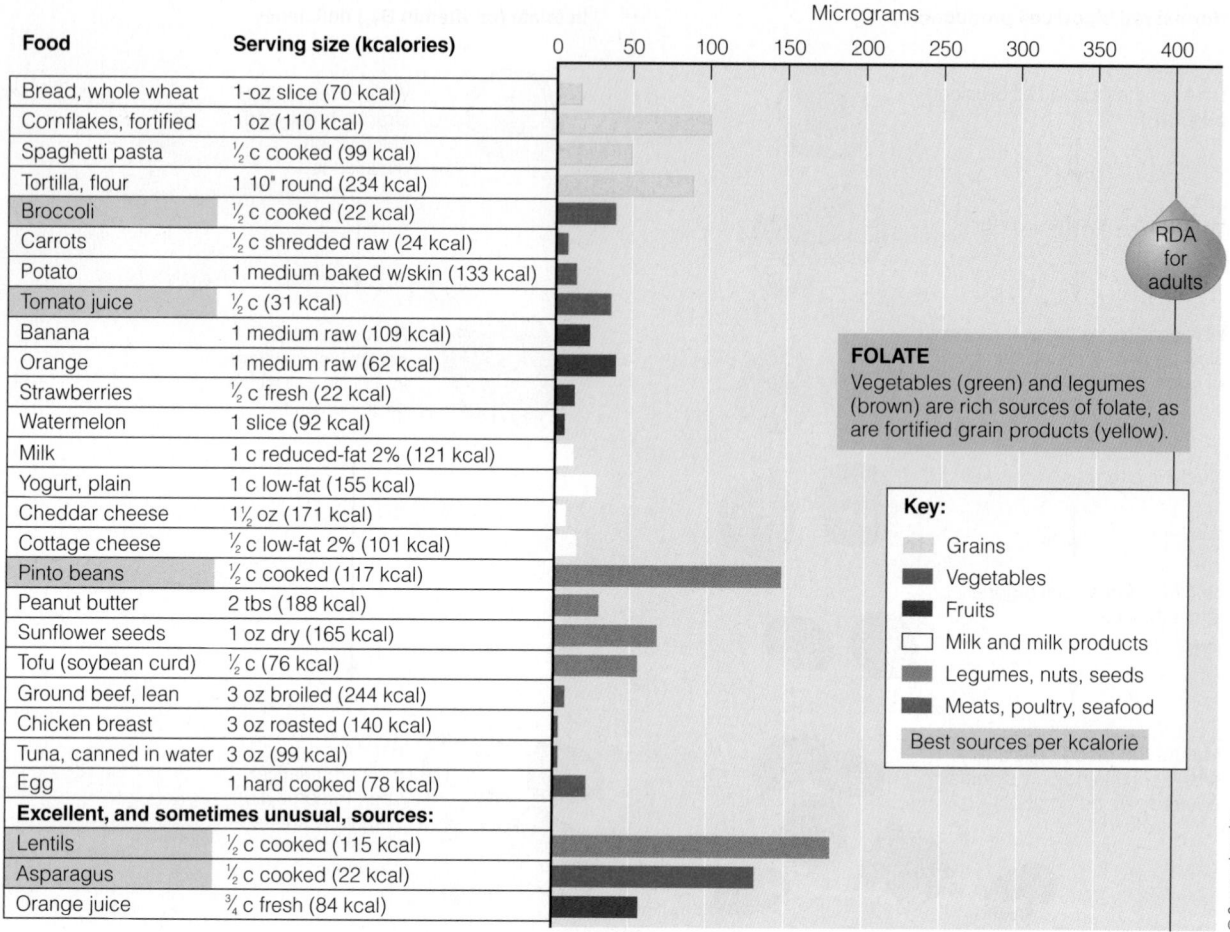

Micrograms

Food	Serving size (kcalories)
Bread, whole wheat	1-oz slice (70 kcal)
Cornflakes, fortified	1 oz (110 kcal)
Spaghetti pasta	½ c cooked (99 kcal)
Tortilla, flour	1 10" round (234 kcal)
Broccoli	½ c cooked (22 kcal)
Carrots	½ c shredded raw (24 kcal)
Potato	1 medium baked w/skin (133 kcal)
Tomato juice	½ c (31 kcal)
Banana	1 medium raw (109 kcal)
Orange	1 medium raw (62 kcal)
Strawberries	½ c fresh (22 kcal)
Watermelon	1 slice (92 kcal)
Milk	1 c reduced-fat 2% (121 kcal)
Yogurt, plain	1 c low-fat (155 kcal)
Cheddar cheese	1½ oz (171 kcal)
Cottage cheese	½ c low-fat 2% (101 kcal)
Pinto beans	½ c cooked (117 kcal)
Peanut butter	2 tbs (188 kcal)
Sunflower seeds	1 oz dry (165 kcal)
Tofu (soybean curd)	½ c (76 kcal)
Ground beef, lean	3 oz broiled (244 kcal)
Chicken breast	3 oz roasted (140 kcal)
Tuna, canned in water	3 oz (99 kcal)
Egg	1 hard cooked (78 kcal)
Excellent, and sometimes unusual, sources:	
Lentils	½ c cooked (115 kcal)
Asparagus	½ c cooked (22 kcal)
Orange juice	¾ c fresh (84 kcal)

RDA for adults

FOLATE
Vegetables (green) and legumes (brown) are rich sources of folate, as are fortified grain products (yellow).

Key:
Grains
Vegetables
Fruits
Milk and milk products
Legumes, nuts, seeds
Meats, poultry, seafood
Best sources per kcalorie

© Cengage Learning

> **REVIEW IT** Folate

Other Names

Folic acid, folacin, pteroylglutamic acid (PGA)

RDA

Adults: 400 µg/day

UL[a]

Adults: 1000 µg/day

Chief Functions in the Body

Part of coenzymes THF (tetrahydrofolate) and DHF (dihydrofolate) used in DNA synthesis and therefore important in new cell formation

Significant Sources

Fortified grains, leafy green vegetables, legumes, seeds, liver

Easily destroyed by heat and oxygen

Deficiency Symptoms

Anemia (large-cell type);[b] smooth, red tongue;[c] mental confusion, weakness, fatigue, irritability, headache; shortness of breath; elevated homocysteine

Toxicity Symptoms

Masks vitamin B$_{12}$–deficiency symptoms

[a]The UL applies to synthetic forms obtained from supplements, fortified foods, or a combination.
[b]Large-cell-type anemia is known as either *macrocytic* or *megaloblastic anemia*.
[c]Smoothness of the tongue is caused by loss of its surface structures and is termed *glossitis* (gloss-EYE-tis).

Vitamin B$_{12}$ Vitamin B$_{12}$ and folate are closely related: each depends on the other for activation. Recall that vitamin B$_{12}$ removes a methyl group to activate the folate coenzyme. When folate gives up its methyl group, the vitamin B$_{12}$ coenzyme becomes activated (review Figure 10-10 on p. 316).

The regeneration of the amino acid methionine and the synthesis of DNA and RNA depend on both folate and vitamin B$_{12}$.* In addition, without any help from folate, vitamin B$_{12}$ maintains the sheath that surrounds and protects nerve fibers

vitamin B$_{12}$: a B vitamin characterized by the presence of cobalt (see Figure 13-2). The active forms of coenzyme B$_{12}$ are *methylcobalamin* and *deoxyadenosylcobalamin*.

*In the body, methionine serves as a methyl (CH_3) donor. In doing so, methionine can be converted to other amino acids. Some of these amino acids can regenerate methionine, but methionine is still considered an essential amino acid that is needed in the diet.

and promotes their normal growth. Bone cell activity and metabolism also depend on vitamin B_{12}.

The digestion and absorption of vitamin B_{12} depends on several steps. In the stomach, hydrochloric acid and the digestive enzyme pepsin release vitamin B_{12} from the proteins to which it is attached in foods. Then as vitamin B_{12} passes from the stomach to the small intestine, it binds with a stomach secretion called **intrinsic factor.** Bound together, intrinsic factor and vitamin B_{12} travel to the end of the small intestine, where receptors recognize the complex. Importantly, the receptors do not recognize vitamin B_{12} without intrinsic factor. The vitamin is gradually absorbed into the bloodstream as the intrinsic factor is degraded. Transport of vitamin B_{12} in the blood depends on specific binding proteins.

Like folate, vitamin B_{12} enters the enterohepatic circulation—continuously being secreted into bile and delivered to the intestine, where it is reabsorbed. Because most vitamin B_{12} is reabsorbed, healthy people rarely develop a deficiency even when their intake is minimal.

Vitamin B_{12} Recommendations The RDA for adults is only 2.4 micrograms of vitamin B_{12} a day—just over two-millionths of a gram. The ink in the period at the end of this sentence may weigh about that much. As tiny as this amount appears to the human eye, it contains billions of molecules of vitamin B_{12}, enough to provide coenzymes for all the enzymes that need its help.

Vitamin B_{12} Deficiency and Toxicity Most vitamin B_{12} deficiencies reflect inadequate absorption, not poor intake. Inadequate absorption typically occurs for one of two reasons: a lack of hydrochloric acid or a lack of intrinsic factor. Without hydrochloric acid, the vitamin is not released from the dietary proteins and so is not available for binding with the intrinsic factor. Without the intrinsic factor, the vitamin cannot be absorbed.

Vitamin B_{12} deficiency is common among the elderly. Many older adults develop **atrophic gastritis,** a condition that damages the cells of the stomach. Atrophic gastritis may also develop in response to iron deficiency or infection with *Helicobacter pylori*, the bacterium implicated in ulcer formation. Without healthy stomach cells, production of hydrochloric acid and intrinsic factor diminishes. Even with an adequate intake from foods, vitamin B_{12} status suffers. The vitamin B_{12} deficiency caused by atrophic gastritis and a lack of intrinsic factor is known as **pernicious anemia.**

Some people inherit a defective gene for the intrinsic factor. In such cases, or when the stomach has been injured and cannot produce enough of the intrinsic factor, vitamin B_{12} must be given by injection to bypass the need for intestinal absorption. Alternatively, the vitamin may be delivered by nasal spray; absorption is rapid, high, and well tolerated.

Because vitamin B_{12} is found primarily in foods derived from animals, people who follow a vegetarian diet may develop a vitamin B_{12} deficiency.[21] It may take several years for people who stop eating animal-derived foods to develop deficiency symptoms because the body recycles much of its vitamin B_{12}, reabsorbing it over and over again. Even when the body fails to absorb vitamin B_{12}, deficiency may take up to 3 years to develop because the body conserves its supply. Neurological degeneration, a sign of vitamin B_{12} deficiency, appears more rapidly in infants born to mothers with unsupplemented vegan diets or untreated pernicious anemia.

Because vitamin B_{12} is required to convert folate to its active form, one of the most obvious vitamin B_{12}–deficiency symptoms is the anemia commonly seen in folate deficiency. This anemia is characterized by large, immature red blood cells, which indicate slow DNA synthesis and an inability to divide (see Figure 10-12, p. 319). When folate is trapped in its inactive (methyl folate) form because of vitamin B_{12} deficiency or is unavailable because of folate deficiency itself, DNA synthesis slows.

First to be affected in a vitamin B_{12} or folate deficiency are the rapidly growing blood cells. Either vitamin B_{12} or folate will clear up the anemia, but if folate is given when vitamin B_{12} is needed, the result is disastrous: devastating neurological symptoms. Remember that vitamin B_{12}, but not folate, maintains the sheath that

intrinsic factor: a glycoprotein (a protein with short polysaccharide chains attached) secreted by the stomach cells that binds with vitamin B_{12} in the small intestine to aid in the absorption of vitamin B_{12}.

- **intrinsic** = on the inside

atrophic (a-TRO-fik) **gastritis** (gas-TRY-tis): chronic inflammation of the stomach accompanied by a diminished size and functioning of the mucous membranes and glands. This condition is also characterized by inadequate hydrochloric acid and intrinsic factor—two substances needed for vitamin B_{12} absorption.

- **atrophy** = wasting
- **gastro** = stomach
- **itis** = inflammation

pernicious (per-NISH-us) **anemia:** a blood disorder that reflects a vitamin B_{12} deficiency caused by lack of intrinsic factor and characterized by abnormally large and immature red blood cells. Other symptoms include muscle weakness and irreversible neurological damage.

- **pernicious** = destructive

surrounds and protects nerve fibers and promotes their normal growth. Folate "cures" the *blood* symptoms of a vitamin B_{12} deficiency, but cannot stop the *nerve* symptoms from progressing. By doing so, folate "masks" a vitamin B_{12} deficiency.

Marginal vitamin B_{12} deficiency impairs cognition.[22] Advanced neurological symptoms include a creeping paralysis that begins at the extremities and works inward and up the spine. Early detection and correction are necessary to prevent permanent nerve damage and paralysis. With sufficient folate in the diet, the neurological symptoms of vitamin B_{12} deficiency can develop without evidence of anemia and the cognitive decline is especially rapid. Such interactions between folate and vitamin B_{12} highlight some of the safety issues surrounding the use of supplements and the fortification of foods. No adverse effects have been reported for excess vitamin B_{12}, and no UL has been set.

Vitamin B_{12} Food Sources Vitamin B_{12} is unique among the vitamins in being found almost exclusively in foods derived from animals. Its bioavailability is greatest from milk and fish. Anyone who eats reasonable amounts of animal-derived foods is most likely to have an adequate intake, including vegetarians who use milk products or eggs. Vegans, who restrict all foods derived from animals, need a reliable source, such as vitamin B_{12}–fortified soy milk or vitamin B_{12} supplements. Yeast grown on a vitamin B_{12}–enriched medium and mixed with that medium provides some vitamin B_{12}, but yeast itself does not contain active vitamin B_{12}. Similarly, neither fermented soy products such as miso (a soybean paste) nor sea algae such as spirulina provide active vitamin B_{12}. Extensive research shows that the amounts listed on the labels of these plant products are inaccurate and misleading because the vitamin B_{12} is in an inactive, unavailable form.

As mentioned earlier, the water-soluble vitamins are particularly vulnerable to losses in cooking. For most of these nutrients, microwave heating minimizes losses as well as, or better than, traditional cooking methods. Such is not the case for vitamin B_{12}, however. Microwave heating inactivates vitamin B_{12}. To preserve this vitamin, use the oven or stovetop instead of a microwave to cook meats and milk products (major sources of vitamin B_{12}). The accompanying table provides a summary of vitamin B_{12}.

> **REVIEW IT** Vitamin B_{12}

Other Names	**Significant Sources**
Cobalamin (and related forms)	Foods of animal origin (meat, fish, poultry, shellfish, milk, cheese, eggs), fortified cereals
RDA	
Adults: 2.4 µg/day	Easily destroyed by microwave cooking
Chief Functions in the Body	**Deficiency Disease**
Part of coenzymes methylcobalamin and deoxyadenosylcobalamin used in new cell synthesis; helps to maintain nerve cells; reforms folate coenzyme; helps to break down some fatty acids and amino acids	Pernicious anemia[a]
	Deficiency Symptoms
	Anemia (large-cell type);[b] fatigue, degeneration of peripheral nerves progressing to paralysis; sore tongue, loss of appetite, constipation
	Toxicity Symptoms
	None reported

[a]The name *pernicious anemia* refers to the vitamin B_{12} deficiency caused by atrophic gastritis and a lack of intrinsic factor, but not to that caused by inadequate dietary intake.
[b]Large-cell-type anemia is known as either *macrocytic* or *megaloblastic anemia*.

Choline Although not defined as a vitamin, choline is an essential nutrient that is commonly grouped with the B vitamins. The body uses choline to make the neurotransmitter acetylcholine and the phospholipid lecithin. During fetal development, choline supports the structure and function of the brain and spinal cord, by supporting neural tube closure and enhancing learning performance.

Choline Recommendations The body can make choline from the amino acid methionine, but without dietary choline, synthesis alone appears to be insufficient to meet the body's needs. For this reason, the DRI Committee established an Adequate Intake (AI) for choline.

Choline Deficiency and Toxicity Average choline intakes fall below the AI, but the impact of deficiencies are not fully understood. The UL for choline is based on its life-threatening effect in lowering blood pressure.

Choline Food Sources Choline is found in a variety of common foods such as milk, eggs, and peanuts and as part of lecithin, a food additive commonly used as an emulsifying agent (review Figure 5-8 on p. 140). The accompanying table provides a summary of choline.

> **REVIEW IT** Choline

AI	Deficiency Symptoms
Men: 550 mg/day	Liver damage
Women: 425 mg/day	**Toxicity Symptoms**
UL	Body odor, sweating, salivation, reduced growth rate, low blood pressure, liver damage
Adults: 3500 mg/day	**Significant Sources**
Chief Functions in the Body	Milk, liver, eggs, peanuts
Needed for the synthesis of the neurotransmitter acetylcholine and the phospholipid lecithin	

Nonvitamins Some substances have been mistaken for vitamins, but they are not essential nutrients. Among them are the compounds **inositol** and **carnitine**, which can be made by the body. Inositol is a part of cell membrane structures, and carnitine transports long-chain fatty acids from the cytosol to the mitochondria for oxidation. Other nonvitamins include PABA (para-aminobenzoic acid, a component of folate's chemical structure), the bioflavonoids (vitamin P or hesperidin), pyrroloquinoline quinone (methoxatin), orotic acid, lipoic acid, and ubiquinone (coenzyme Q_{10}). Other names erroneously associated with vitamins are "vitamin O" (oxygenated saltwater), "vitamin B_5" (another name for pantothenic acid), "vitamin B_{15}" (also called "pangamic acid," a hoax), and "vitamin B_{17}" (laetrile, an alleged "cancer cure" and not a vitamin or a cure by any stretch of the imagination—in fact, laetrile is a potentially dangerous substance).

Interactions among the B Vitamins This chapter has described some of the impressive ways that vitamins work individually, as if their many actions in the body could easily be disentangled. In fact, it is often difficult to tell which vitamin is truly responsible for a given effect because the nutrients are interdependent; the presence or absence of one affects another's absorption, metabolism, and excretion. You have already seen this interdependence with folate and vitamin B_{12}.

Riboflavin and vitamin B_6 provide another example. One of the riboflavin coenzymes, FMN, assists the enzyme that converts vitamin B_6 to its coenzyme form PLP. Consequently, a severe riboflavin deficiency can impair vitamin B_6 activity. Thus a deficiency of one nutrient may alter the action of another. Furthermore, a deficiency of one nutrient may create a deficiency of another. For example, both riboflavin and vitamin B_6 (as well as iron) are required for the conversion of tryptophan to niacin. Consequently, an inadequate intake of either riboflavin or vitamin B_6 can diminish the body's niacin supply. These interdependent relationships are evident in many of the roles B vitamins play in the body.

B Vitamin Roles Figure 10-14 (p. 324) summarizes the metabolic pathways introduced in Chapter 7 and conveys an *impression* of the many ways B vitamins assist in

inositol (in-OSS-ih-tall): a nonessential nutrient that can be made in the body from glucose. Inositol is a part of cell membrane structures.

carnitine (CAR-neh-teen): a nonessential, nonprotein amino acid made in the body from lysine that helps transport fatty acids across the mitochondrial membrane.

> FIGURE 10-14 **Metabolic Pathways Involving B Vitamins**

These metabolic pathways are introduced in Chapter 7 and are presented here to highlight the many coenzymes that facilitate the reactions. These coenzymes depend on the following vitamins:

- NAD and NADP: niacin
- TPP: thiamin
- CoA: pantothenic acid
- B_{12}: vitamin B_{12}
- FMN and FAD: riboflavin
- THF: folate
- PLP: vitamin B_6
- Biotin

Pathways leading toward acetyl CoA and the TCA cycle are catabolic, and those leading toward amino acids, glycogen, and fat are anabolic. For further details, see Appendix C.

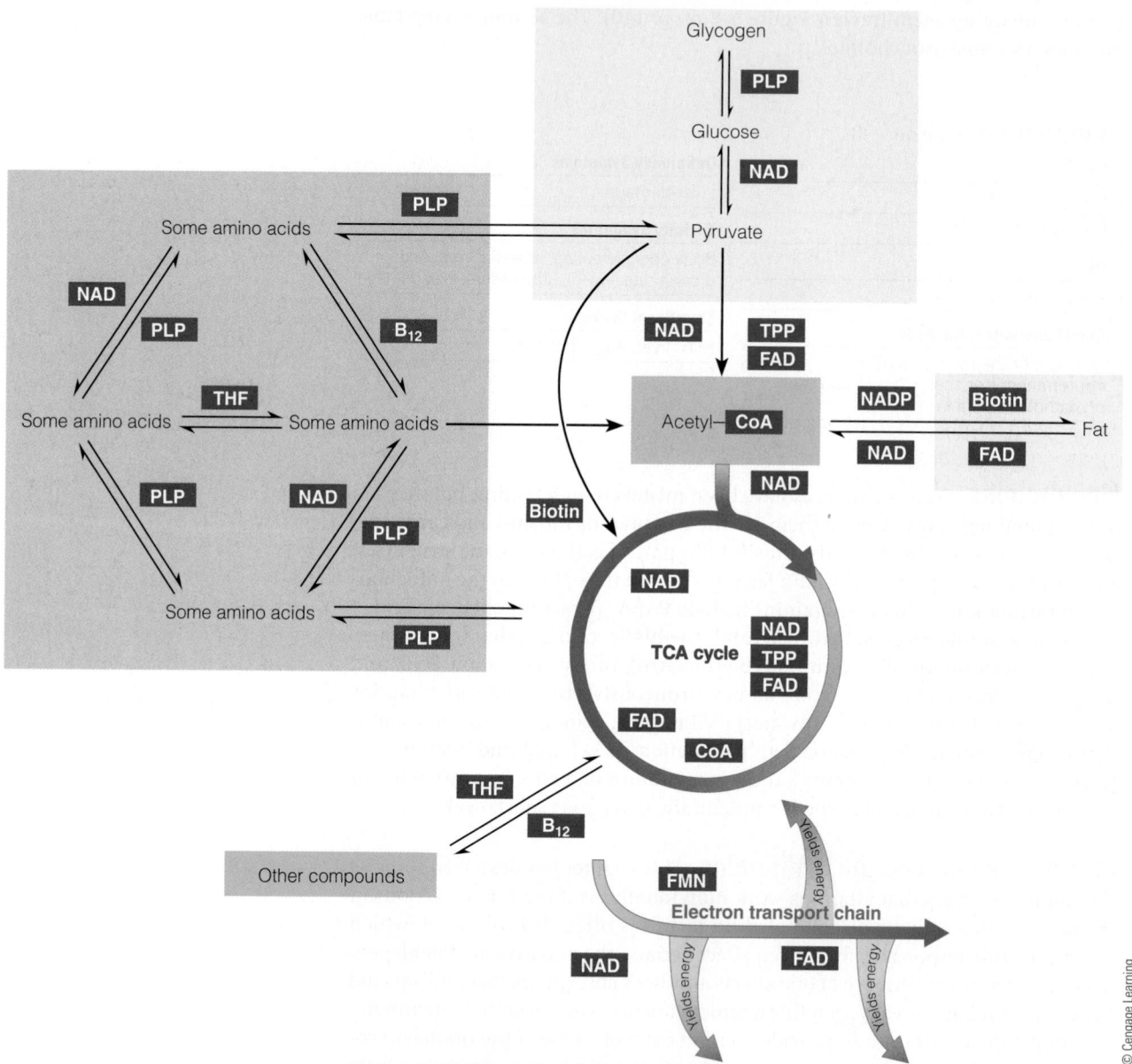

metabolic pathways. Metabolism is the body's work, and the B vitamin coenzymes are indispensable to every step. In scanning the pathways of metabolism depicted in the figure, note the many abbreviations for the coenzymes that keep the processes going.

Look at the now-familiar pathway of glucose breakdown. To break down glucose to pyruvate, the cells must have certain enzymes. For the enzymes to work, they must have the niacin coenzyme NAD. Cells can make NAD, but only if they have enough niacin (or enough of the amino acid tryptophan to make niacin).

The next step is the breakdown of pyruvate to acetyl CoA. The enzymes involved in this step require both NAD and the thiamin and riboflavin coenzymes

TPP and FAD, respectively. The cells can manufacture the enzymes they need from the vitamins, if the vitamins are in the diet.

Another coenzyme needed for this step is CoA. Predictably, the cells can make CoA except for an essential part that must be obtained in the diet—pantothenic acid. Another coenzyme requiring biotin serves the enzyme complex involved in converting pyruvate to oxaloacetate, the compound that combines with acetyl CoA to start the TCA cycle.

These and other coenzymes participate throughout all the metabolic pathways. Vitamin B_6 is an indispensable part of PLP—a coenzyme required for many amino acid conversions, for a crucial step in the making of the iron-containing portion of hemoglobin for red blood cells, and for many other reactions. Folate becomes THF—the coenzyme required for the synthesis of new genetic material and therefore new cells. The vitamin B_{12} coenzyme, in turn, regenerates THF to its active form; thus vitamin B_{12} is also necessary for the formation of new cells.

Thus each of the B vitamin coenzymes is involved, directly or indirectly, in energy metabolism. Some facilitate the energy-releasing reactions themselves; others help build new cells to deliver the oxygen and nutrients that allow the energy reactions to occur.

B Vitamin Deficiencies Now suppose the body's cells lack one of these B vitamins—niacin, for example. Without niacin, the cells cannot make NAD. Without NAD, the enzymes involved in every step of the glucose-to-energy pathway cannot function. Then, because all the body's activities require energy, literally everything begins to grind to a halt. This is no exaggeration. The deadly disease pellagra, caused by niacin deficiency, produces the "devastating four Ds": dermatitis, which reflects a failure of the skin; dementia, a failure of the nervous system; diarrhea, a failure of digestion and absorption; and eventually, as would be the case for any severe nutrient deficiency, death. These symptoms are the obvious ones, but a niacin deficiency affects all other organs, too, because all are dependent on the energy pathways.

All the vitamins are as essential as niacin. With any B vitamin deficiency, many body systems become deranged, and similar symptoms may appear. A lack of any of them can have disastrous and far-reaching effects.

Deficiencies of single B vitamins seldom show up in isolation, however. After all, people do not eat nutrients singly; they eat foods, which contain mixtures of nutrients. Only in two cases described earlier—beriberi and pellagra—have dietary deficiencies associated with single B vitamins been observed on a large scale in human populations. Even in these cases, several vitamins were lacking even though one vitamin stood out above the rest. When foods containing the vitamin known to be needed were provided, the other vitamins that were in short supply came as part of the package.

Major deficiency diseases of epidemic proportions such as pellagra and beriberi are no longer seen in the United States, but lesser deficiencies of nutrients, including the B vitamins, sometimes occur in people whose food choices are poor because of poverty, ignorance, illness, or poor health habits like alcohol abuse. (Review Highlight 7 to fully appreciate how alcohol induces vitamin deficiencies and interferes with energy metabolism.) Remember from Chapter 1 that deficiencies can arise not only from deficient intakes (primary causes), but also for other (secondary) reasons.

In identifying nutrient deficiencies, it is important to realize that a particular sign or symptom may not always have the same cause. The skin and the tongue (shown in Figure 10-15 on p. 326) appear to be especially sensitive to B vitamin deficiencies, but focusing on these body parts gives them undue emphasis. Both the skin and the tongue are readily visible in a physical examination.* The physician sees and reports the deficiency's outward signs, but the full impact of a vitamin deficiency occurs inside the cells of the body. If the skin develops a rash or lesions, other tissues beneath it may be degenerating too. Similarly, the mouth

---

*The two common signs of B vitamin deficiencies are *glossitis* (gloss-EYE-tis), an inflammation of the tongue, and *cheilosis* (kye-LOH-sis or kee-LOH-sis), a condition of reddened lips with cracks at the corners of the mouth.

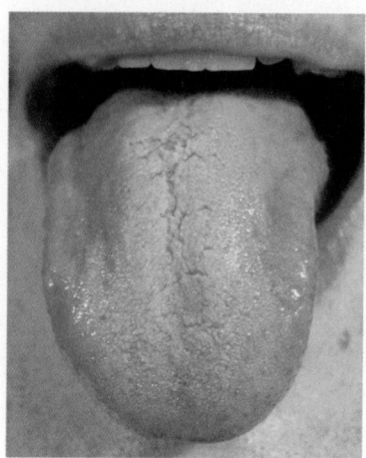

A healthy tongue has a rough and somewhat bumpy surface.

In a B vitamin deficiency, the tongue becomes smooth and swollen due to atrophy of the tissue (glossitis).

In a B vitamin deficiency, the corners of the mouth become irritated and inflamed (cheilosis).

and tongue are the visible part of the digestive system; if they are abnormal, most likely the rest of the GI tract is as well.

Keep in mind that the cause of a sign or symptom is not always apparent. The summary tables in this chapter show that deficiencies of riboflavin, niacin, biotin, and vitamin B_6 can all cause skin rashes. So can a deficiency of protein, linoleic acid, or vitamin A. Because skin is on the outside and easy to see, it is a useful indicator of "things going wrong inside cells." By itself, a skin condition says nothing about its possible cause.

The same is true of anemia. Anemia is often caused by iron deficiency, but it can also be caused by a folate or vitamin B_{12} deficiency; by digestive tract failure to absorb any of these nutrients; or by such nonnutritional causes as infections, parasites, cancer, or loss of blood. No single nutrient will always cure a given symptom.

A person who feels chronically tired may be tempted to self-diagnose iron-deficiency anemia and self-prescribe an iron supplement. But this will relieve tiredness only if the cause is indeed iron-deficiency anemia. If the cause is a folate deficiency, taking iron will only prolong the fatigue. A person who is better informed may decide to take a vitamin supplement with iron, covering the possibility of a vitamin deficiency. But the symptom may have a nonnutritional cause. If the cause of the tiredness is actually hidden blood loss due to cancer, the postponement of a diagnosis may be life-threatening. When fatigue is caused by a lack of sleep, of course, no nutrient or combination of nutrients can replace a good night's rest. A person who is chronically tired should see a physician rather than self-prescribe. If the condition is nutrition-related, a registered dietitian nutritionist should be consulted as well.

B Vitamin Toxicities Toxicities of the B vitamins from foods alone are unknown, but they can occur when people overuse dietary supplements. With supplements, the quantities can quickly overwhelm the cells. Consider that one small capsule can easily deliver 2 milligrams of vitamin B_6, but it would take more than 3000 bananas, 6600 cups of rice, or 3600 chicken breasts to supply an equivalent amount. When the cells become oversaturated with a vitamin, they must work to eliminate the excess. The cells dispatch water-soluble vitamins to the urine for excretion, but sometimes they cannot keep pace with the onslaught. Homeostasis becomes disturbed and symptoms of toxicity develop.

B Vitamin Food Sources Significantly, deficiency diseases, such as beriberi and pellagra, were eliminated by providing foods. Dietary supplements advertise that vitamins are indispensable to life, but human beings obtained their nourishment from foods for centuries before supplements existed. If the diet lacks a vitamin, the first solution is to adjust food intake to obtain that vitamin.

The bar graphs of selected foods in this chapter, taken together, sing the praises of a balanced diet. The grains deliver thiamin, riboflavin, niacin, and folate. The fruit and vegetable groups excel in folate. Protein foods serve thiamin, niacin, vitamin B_6, and vitamin B_{12} well. The milk group stands out for riboflavin and vitamin B_{12}. A diet that offers a variety of foods from each group, prepared with reasonable care, serves up ample B vitamins.

> **REVIEW IT** Identify the main roles, deficiency symptoms, and food sources for each of the B vitamins.

The B vitamins serve as coenzymes that facilitate the work of every cell. They are active in carbohydrate, fat, and protein metabolism and in the making of DNA and thus new cells. Historically famous B vitamin–deficiency diseases are beriberi (thiamin), pellagra (niacin), and pernicious anemia (vitamin B_{12}). Pellagra can be prevented by an adequate protein intake because the amino acid tryptophan can be converted to niacin in the body. A high intake of folate can mask the blood symptoms of a vitamin B_{12} deficiency, but it will not prevent the associated nerve damage. Vitamin B_6 participates in amino acid metabolism and can be harmful in excess. Biotin and pantothenic acid serve important roles in energy metabolism and are common in a variety of foods. Many substances that people claim as B vitamins are not. Fortunately, a variety of foods from each of the food groups provides an adequate supply of all of the B vitamins.

10.3 Vitamin C

> **LEARN IT** Identify the main roles, deficiency symptoms, and food sources for vitamin C.

For many centuries, any man who joined the crew of a seagoing ship knew he had at best a 50–50 chance of returning alive—not because he might be slain by pirates or die in a storm, but because he might contract **scurvy.** As many as two-thirds of a ship's crew could die of scurvy during a long voyage. Only men on short voyages, especially around the Mediterranean Sea, were free of scurvy. No one knew the reason: that on long ocean voyages, the ship's cook used up the fresh fruits and vegetables early and then served only cereals and meats until the return to port.

In the mid-1700s, James Lind, a British physician serving in the navy, devised an experiment to find a cure for scurvy. He divided 12 sailors with scurvy into 6 pairs. Each pair received a different supplemental ration: cider, vinegar, sulfuric acid, seawater, oranges and lemons, or a strong laxative. Those receiving the citrus fruits quickly recovered, but sadly, it was almost 50 years before the British navy required all vessels to provide every sailor with lemon or lime juice daily. The tradition of providing British sailors with citrus juice daily to prevent scurvy gave them the nickname "limeys."

The antiscurvy "something" in citrus and other foods was dubbed the **antiscorbutic factor.** Centuries later, the factor was isolated and found to be a 6-carbon compound similar to glucose; it was named **ascorbic acid.**

Vitamin C Roles Vitamin C parts company with the B vitamins in its mode of action. In some settings, vitamin C serves as a **cofactor** helping a specific enzyme perform its job, but in others, it acts as an antioxidant participating in more general ways.

As an Antioxidant Vitamin C loses electrons easily, a characteristic that allows it to perform as an antioxidant. In the body, **antioxidants** defend against free radicals. Free radicals are discussed fully in Highlight 11, but for now, a simple definition will suffice. A **free radical** is a molecule with one or more unpaired electrons, which makes it unstable and highly reactive. Antioxidants can neutralize free radicals by donating an electron or two. In doing so, antioxidants protect other substances from free radical damage. Figure 10-16 (p. 328) illustrates how vitamin C can give up electrons and then accept them again to become reactivated. This recycling of vitamin C

scurvy: the vitamin C–deficiency disease.

antiscorbutic (AN-tee-skor-BUE-tik) **factor:** the original name for vitamin C.

- **anti** = against
- **scorbutic** = causing scurvy

ascorbic acid: one of the two active forms of vitamin C (see Figure 10-16). Many people refer to vitamin C by this name.

- **a** = without
- **scorbic** = having scurvy

cofactor: a small, inorganic or organic substance that facilitates the action of an enzyme.

antioxidants: in the body, substances that significantly decrease the adverse effects of free radicals on normal physiological functions.

free radical: an unstable molecule with one or more unpaired electrons.

> FIGURE 10-16   **Active Forms of Vitamin C**

The two hydrogens highlighted in yellow give vitamin C its acidity and its ability to act as an antioxidant.

© Cengage Learning

Ascorbic acid protects against oxidative damage by donating its two hydrogens with their electrons to free radicals (molecules with unpaired electrons). In doing so, ascorbic acid becomes dehydroascorbic acid.

Dehydroascorbic acid can readily accept hydrogens to become ascorbic acid. The reversibility of this reaction is key to vitamin C's role as an antioxidant.

is key to limiting losses and maintaining a reserve of antioxidants in the body. Other key antioxidant nutrients include vitamin E, beta-carotene, and selenium.

Vitamin C is like a bodyguard for water-soluble substances; it stands ready to sacrifice its own life to save theirs. In the cells and body fluids, vitamin C protects tissues from the **oxidative stress** of free radicals and thus may play an important role in preventing diseases.[23] In the intestines, vitamin C enhances iron absorption by protecting iron from oxidation. (Chapter 13 provides more details about the relationship between vitamin C and iron.)

As a Cofactor in Collagen Formation Vitamin C helps to form the fibrous structural protein of connective tissues known as **collagen.** Collagen serves as the matrix on which bones and teeth are formed. When a person is wounded, collagen glues the separated tissues together, forming scars. Cells are held together largely by collagen; this is especially important in the walls of the blood vessels, which must withstand the pressure of blood surging with each beat of the heart.

Chapter 6 describes how the body makes proteins by stringing together chains of amino acids. During the synthesis of collagen, each time a proline or lysine is added to the growing protein chain, an enzyme hydroxylates it (adds an OH group), making the amino acid hydroxyproline or hydroxylysine, respectively. These two special amino acids facilitate the binding together of collagen fibers to make strong, ropelike structures. The conversion of proline to hydroxyproline requires both vitamin C and iron. Iron works as a cofactor in the reaction, and vitamin C protects iron from oxidation, thereby allowing iron to perform its duty. Without vitamin C and iron, the hydroxylation step does not occur.

As a Cofactor in Other Reactions Vitamin C also serves as a cofactor in the synthesis of several other compounds. As in collagen formation, vitamin C helps in the hydroxylation of carnitine, a compound that transports fatty acids, especially long-chain fatty acids, across the inner membrane of mitochondria in cells. It also participates in the conversions of the amino acids tryptophan and tyrosine to the neurotransmitters serotonin and norepinephrine, respectively. Vitamin C also assists in the making of hormones, including thyroxine, which regulates the metabolic rate; when metabolism speeds up in times of extreme physical stress, the body's use of vitamin C increases.

In Stress Among the stresses known to increase vitamin C needs are infections; burns; extremely high or low temperatures; intakes of toxic heavy metals such as lead, mercury, and cadmium; the chronic use of certain medications, including aspirin, barbiturates, and oral contraceptives; and cigarette smoking. During stress, the adrenal glands—which contain more vitamin C than any other organ in the body—release vitamin C and hormones into the blood.*

oxidative stress: a condition in which the production of oxidants and free radicals exceeds the body's ability to handle them and prevent damage.

collagen: the structural protein from which connective tissues such as scars, tendons, ligaments, and the foundations of bones and teeth are made.

---

*High amounts of vitamin C are also found in the pituitary glands; medium amounts in the liver, spleen, heart, kidneys, lungs, pancreas, and white blood cells; and small amounts in the muscles and red blood cells.

When immune system cells are called into action, they use a great deal of oxygen and produce free radicals. In this case, free radicals are helpful. They act as ammunition in an "oxidative burst" that demolishes the offending viruses and bacteria and destroys the damaged cells. Vitamin C steps in as an antioxidant to control this oxidative activity.

In the Prevention and Treatment of the Common Cold Vitamin C has been a popular option for the prevention and treatment of the common cold for decades, but research supporting such claims has been conflicting and controversial. Some studies find no relationship between vitamin C and the occurrence of the common cold, whereas others report modest benefits—fewer colds, fewer days, and shorter duration of severe symptoms, especially for those exposed to physical and environmental stresses. A review of the research on vitamin C in the treatment and prevention of the common cold reveals a slight, but consistent reduction in the duration of the common cold in favor of those taking a daily dose of at least 200 milligrams of vitamin C. The question for consumers to consider is, "Is this enough to warrant routine daily supplementation?"

Discoveries about how vitamin C works in the body provide possible links between the vitamin and the common cold. Anyone who has ever had a cold knows the discomfort of a runny or stuffed-up nose. Nasal congestion develops in response to elevated blood **histamine,** and people commonly take antihistamines for relief. Like an antihistamine, vitamin C comes to the rescue and deactivates histamine.

In Disease Prevention Whether vitamin C may help in preventing or treating cancer, heart disease, cataract, and other diseases is still being studied, and findings are presented in Highlight 11's discussion on antioxidants. Conducting research in the United States can be difficult, however, because diets typically contribute enough vitamin C to provide optimal health benefits.

Vitamin C Recommendations
For decades, vitamin C ranked at the top of dietary supplement sales. How much vitamin C does a person need? As is true of all the vitamins, recommendations are set generously above the minimum requirement to prevent deficiency disease and well below the toxicity level (see Figure 10-17).

The requirement—the amount needed to prevent the overt symptoms of scurvy—is only 10 milligrams daily. Consuming 10 milligrams a day does not saturate all the body tissues, however; higher intakes will increase the body's total vitamin C. At about 100 milligrams per day, 95 percent of the population reaches tissue saturation. (For perspective, 1 cup of orange juice provides more than 100 milligrams of vitamin C.) Recommendations are slightly lower, based on the amounts needed to provide antioxidant protection. At about 200 milligrams, absorption reaches a maximum, and there is little, if any, increase in blood concentrations at higher doses. Excess vitamin C is readily excreted.

As mentioned earlier, cigarette smoking increases the need for vitamin C. Cigarette smoke contains oxidants, which greedily deplete this potent antioxidant. Exposure to cigarette smoke, especially when accompanied by low dietary intakes of vitamin C, depletes the body's vitamin C in both active and passive smokers. People who chew tobacco also have low levels of vitamin C. Because people who smoke cigarettes regularly suffer significant oxidative stress, their requirement for vitamin C is increased an additional 35 milligrams; non-smokers regularly exposed to cigarette smoke should also be sure to meet their RDA for vitamin C. Smokers are among those most likely to suffer vitamin C deficiency.

Vitamin C Deficiency
Early signs of nutrient deficiencies can be difficult to recognize. Two of the most notable signs of a vitamin C deficiency reflect its role in maintaining the integrity of blood vessels. The gums bleed easily around the teeth, and capillaries under the skin break spontaneously, producing pinpoint hemorrhages (see Figure 10-18 on p. 330).

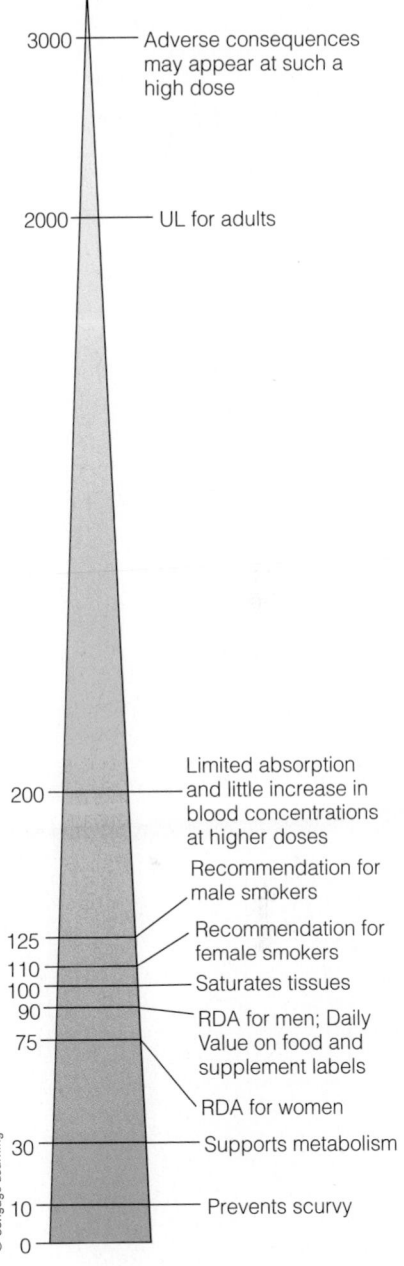

> FIGURE 10-17 Vitamin C Intake (mg/day)

Recommendations for vitamin C are set generously above the minimum requirement and well below the toxicity level.

3000 — Adverse consequences may appear at such a high dose

2000 — UL for adults

200 — Limited absorption and little increase in blood concentrations at higher doses

— Recommendation for male smokers

125 — Recommendation for female smokers

110 — Saturates tissues

100 — RDA for men; Daily Value on food and supplement labels

90

75 — RDA for women

30 — Supports metabolism

10 — Prevents scurvy

0

© Cengage Learning

histamine (HISS-tah-mean or HISS-tah-men): a substance produced by cells of the immune system as part of a local immune reaction to an antigen.

> FIGURE 10-18 **Vitamin C–Deficiency Symptoms—Scorbutic Gums and Pinpoint Hemorrhages**

Scorbutic gums. Unlike other lesions of the mouth, scurvy presents a symmetrical appearance without infection.

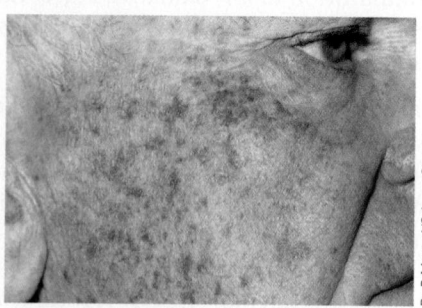

Pinpoint hemorrhages. Small red spots appear in the skin, indicating spontaneous bleeding internally.

When vitamin C concentrations fall to about a fifth of optimal levels (this may take more than a month on a diet lacking vitamin C), scurvy symptoms begin to appear. Inadequate collagen synthesis causes further hemorrhaging. Muscles, including the heart muscle, degenerate. The skin becomes rough, brown, scaly, and dry. Wounds fail to heal because scar tissue will not form. Bone rebuilding falters; the ends of the long bones become softened, malformed, and painful, and fractures develop. The teeth become loose as the cartilage around them weakens. Anemia and infections are common. There are also characteristic psychological signs, including hysteria and depression. Sudden death is likely, caused by massive internal bleeding.

Once diagnosed, scurvy is readily resolved by increasing vitamin C intake. Moderate doses in the neighborhood of 100 milligrams per day are sufficient, curing the scurvy within about 5 days. Such an intake is easily achieved by including vitamin C–rich foods in the diet.

Vitamin C Toxicity The availability of vitamin C supplements and the publication of books recommending vitamin C to prevent colds and cancer have led many people to take large doses of vitamin C. Not surprisingly, side effects of vitamin C supplementation such as gastrointestinal distress and diarrhea have been reported. The UL for vitamin C was established based on these symptoms.

Several instances of interference with medical regimens are also known. Large amounts of vitamin C excreted in the urine obscure the results of tests used to detect glucose or ketones in the diagnosis of diabetes. In some instances, excess vitamin C gives a **false positive** result; in others, a **false negative.** People taking anticlotting medications may unwittingly counteract the effect if they also take massive doses of vitamin C. Those with kidney disease, a tendency toward gout, or a genetic abnormality that alters vitamin C's breakdown to its excretion products are prone to forming kidney stones if they take large doses of vitamin C.* Vitamin C supplements may adversely affect people with iron overload. As Chapter 13 explains, vitamin C enhances iron absorption and releases iron from body stores; too much free iron causes the kind of cellular damage typical of free radicals. These adverse consequences illustrate how vitamin C can act as a *prooxidant* when quantities exceed the body's needs.

Vitamin C Food Sources Fruits and vegetables can easily provide a generous amount of vitamin C. A cup of orange juice at breakfast, a salad for lunch, and a stalk of broccoli and a potato for dinner alone provide more than 300 milligrams. (For perspective, review Figure 10-17, p. 329.) Clearly, a person making such food choices does not need vitamin C supplements.

Figure 10-19 shows the amounts of vitamin C in various common foods. The overwhelming abundance of purple and green bars reveals not only that the

false positive: a test result indicating that a condition is present (positive) when in fact it is not present (therefore false).

false negative: a test result indicating that a condition is not present (negative) when in fact it is present (therefore false).

*Vitamin C is inactivated and degraded by several routes, and sometimes oxalate, which can form kidney stones, is produced along the way. People may also develop oxalate crystals in their kidneys regardless of vitamin C status.

> FIGURE 10-19 **Vitamin C in Selected Foods**

Milligrams

Food	Serving size (kcalories)	Vitamin C (mg)
Bread, whole wheat	1-oz slice (70 kcal)	
Cornflakes, fortified	1 oz (110 kcal)	
Spaghetti pasta	½ c cooked (99 kcal)	
Tortilla, flour	1 10" round (234 kcal)	
Broccoli	½ c cooked (22 kcal)	~58
Carrots	½ c shredded raw (24 kcal)	~5
Potato	1 medium baked w/skin (133 kcal)	~17
Tomato juice	¾ c (31 kcal)	~33
Banana	1 medium raw (109 kcal)	~11
Orange	1 medium raw (62 kcal)	~70
Strawberries	½ c fresh (22 kcal)	~42
Watermelon	1 slice (92 kcal)	~27
Milk	1 c reduced-fat 2% (121 kcal)	
Yogurt, plain	1 c low-fat (155 kcal)	
Cheddar cheese	1½ oz (171 kcal)	
Cottage cheese	½ c low-fat 2% (101 kcal)	
Pinto beans	½ c cooked (117 kcal)	
Peanut butter	2 tbs (188 kcal)	
Sunflower seeds	1 oz dry (165 kcal)	
Tofu (soybean curd)	½ c (76 kcal)	
Ground beef, lean	3 oz broiled (244 kcal)	
Chicken breast	3 oz roasted (140 kcal)	
Tuna, canned in water	3 oz (99 kcal)	
Egg	1 hard cooked (78 kcal)	
Excellent, and sometimes unusual, sources:		
Red bell pepper	½ c raw chopped (20 kcal)	(off chart)
Kiwi	1 (46 kcal)	~75
Brussels sprouts	½ c cooked (30 kcal)	~48

RDA for men

RDA for women

VITAMIN C
Meeting vitamin C needs without fruits (purple) and vegetables (green) is almost impossible. Many of them provide the entire RDA in one serving, and others provide at least half. Most meats, legumes, breads, and milk products are poor sources.

Key:
- Grains
- Vegetables
- Fruits
- ☐ Milk and milk products
- Legumes, nuts, seeds
- Meats, poultry, seafood
- Best sources per kcalorie

© Cengage Learning

citrus fruits are justly famous for being rich in vitamin C, but that other fruits and vegetables are in the same league. A half cup of broccoli, bell pepper, or strawberries provides more than 50 milligrams of the vitamin (and an array of other nutrients). Because vitamin C is vulnerable to heat, raw fruits and vegetables usually have a higher nutrient density than their cooked counterparts. Similarly, because vitamin C is readily destroyed by oxygen, foods and juices should be stored properly and consumed within a week of opening.

When dietitians say "vitamin C," people think "citrus fruits" . . .

© Photodisc/Getty Images

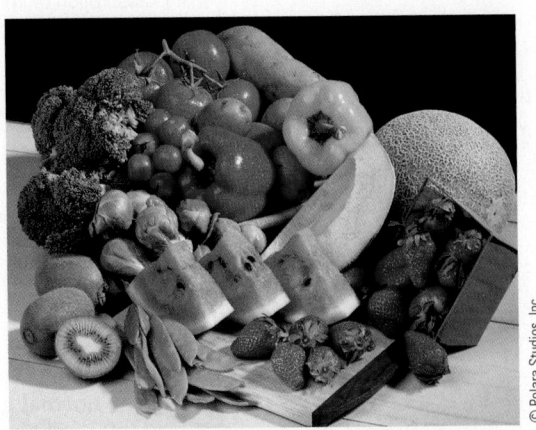

. . . but these foods are also rich in vitamin C.

© Polara Studios, Inc.

The potato is an important source of vitamin C, not because one potato by itself meets the daily need, but because potatoes are such a common staple that they make significant contributions. In fact, scurvy was unknown in Ireland until the potato blight of the mid-1840s, when some 2 million people died of malnutrition and infection.

The lack of yellow, white, brown, and red bars in Figure 10-19 (p. 331) confirms that grains, milk and milk products (except breast milk), and most protein foods are notoriously poor sources of vitamin C. Organ meats (liver, kidneys, and others) and raw meats contain some vitamin C, but most people don't eat large quantities of these foods. Raw meats and fish contribute enough vitamin C to be significant sources in parts of Alaska, Canada, and Japan, but elsewhere fruits and vegetables are necessary to supply sufficient vitamin C.

Because of vitamin C's antioxidant property, food manufacturers sometimes add a variation of vitamin C to some beverages and most cured meats, such as luncheon meats, to prevent oxidation and spoilage. This compound safely preserves these foods, but it does not have vitamin C activity in the body. Simply put, "Ham and bacon cannot replace fruits and vegetables."

> **REVIEW IT** Identify the main roles, deficiency symptoms, and food sources for vitamin C.

Vitamin C acts primarily as an antioxidant and a cofactor. Recommendations are set well above the amount needed to prevent the deficiency disease scurvy. A variety of fruits and vegetables—most notably citrus fruits—provide generous amounts of vitamin C. The accompanying table provides a summary of vitamin C.

Vitamin C

Other Names

Ascorbic acid

RDA

Men: 90 mg/day

Women: 75 mg/day

Smokers: +35 mg/day

UL

Adults: 2000 mg/day

Chief Functions in the Body

Collagen synthesis (strengthens blood vessel walls, forms scar tissue, provides matrix for bone growth), antioxidant, thyroxine synthesis, amino acid metabolism, strengthens resistance to infection, helps in absorption of iron

Significant Sources

Citrus fruits, cabbage-type vegetables (such as brussels sprouts and cauliflower), dark green vegetables (such as bell peppers and broccoli), cantaloupe, strawberries, lettuce, tomatoes, potatoes, papayas, mangoes

Easily destroyed by heat and oxygen

Deficiency Disease

Scurvy

Deficiency Symptoms

Anemia (small-cell type),[a] atherosclerotic plaques, pinpoint hemorrhages; bone fragility, joint pain; poor wound healing, frequent infections; bleeding gums, loosened teeth; muscle degeneration, pain, hysteria, depression; rough skin, blotchy bruises

Toxicity Symptoms

Nausea, abdominal cramps, diarrhea; headache, fatigue, insomnia; hot flashes; rashes; interference with medical tests, aggravation of gout symptoms, urinary tract problems, kidney stones[b]

[a]Small-cell-type anemia is *microcytic anemia.*

[b]People with kidney disease, a tendency toward gout, or a genetic abnormality that alters the breakdown of vitamin C are prone to forming kidney stones. Vitamin C is inactivated and degraded by several routes, sometimes producing oxalate, which can form stones in the kidneys.

Vita means life. After this discourse on the vitamins, who could dispute that they deserve their name? Their regulation of metabolic processes makes them vital to the normal growth, development, and maintenance of the body. The accompanying table condenses the information provided in this chapter for a quick review. The remarkable roles of the vitamins continue in the next chapter.

Vitamin and Chief Functions	Deficiency Symptoms	Toxicity Symptoms	Food Sources
Thiamin Part of coenzyme TPP in energy metabolism	Beriberi (edema or muscle wasting), anorexia, weight loss, neurological disturbances, muscular weakness, heart enlargement and failure	None reported	Enriched, fortified, or whole-grain products; pork
Riboflavin Part of coenzymes FAD and FMN in energy metabolism	Inflammation of the mouth, skin, and eyelids	None reported	Milk products; enriched, fortified, or whole-grain products; liver
Niacin Part of coenzymes NAD and NADP in energy metabolism	Pellagra (diarrhea, dermatitis, and dementia)	Niacin flush, liver damage, impaired glucose tolerance	Protein-rich foods
Biotin Part of coenzyme in energy metabolism	Skin rash, hair loss, neurological disturbances	None reported	Widespread in foods; GI bacteria synthesis
Pantothenic acid Part of coenzyme A in energy metabolism	Digestive and neurological disturbances	None reported	Widespread in foods
Vitamin B$_6$ Part of coenzymes used in amino acid and fatty acid metabolism	Scaly dermatitis, depression, confusion, convulsions, anemia	Nerve degeneration, skin lesions	Protein-rich foods
Folate Activates vitamin B$_{12}$; helps synthesize DNA for new cell growth	Anemia, glossitis, neurological disturbances, elevated homocysteine	Masks vitamin B$_{12}$ deficiency	Legumes, vegetables, fortified grain products
Vitamin B$_{12}$ Activates folate; helps synthesize DNA for new cell growth; protects nerve cells	Anemia; nerve damage and paralysis	None reported	Foods derived from animals
Vitamin C Synthesis of collagen, carnitine, hormones, neurotransmitters; antioxidant	Scurvy (bleeding gums, pinpoint hemorrhages, abnormal bone growth, and joint pain)	Diarrhea, GI distress	Fruits and vegetables

Nutrition Portfolio

To obtain all the vitamins you need each day, be sure to select from a variety of foods from all the food groups. Go to Diet & Wellness Plus and choose one of the days on which you have tracked your diet for the entire day. Go to the Intake vs. Goals report. Near the bottom of this report, you will see all of the vitamins grouped together; using this section of the report for reference, answer the following questions:

- How was your vitamin intake overall?
- Did you consume too much or too little of any vitamin?
- Which vitamins concerned you most?

Next go to the Intake Spreadsheet report, and looking at each of the vitamins, answer the following questions:

- Which of your foods provided high intakes of vitamins?
- Which of your foods provided few or no vitamins?
- How do your daily choices of whole or enriched grains, dark green vegetables, citrus fruits, and legumes contribute to your vitamin intakes?

- If you are a woman of childbearing age, how many dietary folate equivalents did you receive from folate-rich foods, fortified foods, and supplements? How does this compare to your RDA?
- How do your vitamin intakes from supplements compare with their UL?

DIET & WELLNESS
PLUS To complete this exercise, go to your Diet & Wellness Plus at **www.cengagebrain.com.**

› **STUDY IT** To review the key points of this chapter and take a practice quiz, go to the study cards at the end of the book.

REFERENCES

1. A. Piro and coauthors, Casimir Funk: His discovery of the vitamins and their deficiency disorders, *Annals of Nutrition and Metabolism* 57 (2010): 85–88.

2. H. M. Said, Intestinal absorption of water-soluble vitamins in health and disease, *Biochemical Journal* 437 (2011): 357–372.

3. S. S. Jhala and A. S. Hazell, Modeling neurodegenerative disease pathophysiology in thiamine deficiency: Consequences of impaired oxidative metabolism, *Neurochemistry International* 58 (2011): 248–260.

4. J. E. Digby, N. Ruparelia, and R. P. Choudhury, Niacin in cardiovascular disease: Recent preclinical and clinical developments, *Arteriosclerosis, Thrombosis, and Vascular Biology* 32 (2012): 582–588; J. C. Creider, R. A. Hegele, and T. R. Joy, Niacin: Another look at an underutilized lipid-lowering medication, *Nature Reviews: Endocrinology* 8 (2012): 517–528; J. M. Backes, R. J. Padley, and P. M. Moriarty, Important considerations for treatment with dietary supplement versus prescription niacin products, *Postgraduate Medicine* 123 (2011): 70–83.

5. K. M. Ali and coauthors, Cardiovascular risk and HDL cholesterol, *British Journal of Pharmacology* 167 (2012): 1177–1194; M. R. Kolber, N. Ivers, and G. M. Allan, Niacin added to statins for cardiovascular disease, *Canadian Family Physician* 58 (2012): 842; The AIM-HIGH Investigators, Niacin in patients with low HDL cholesterol levels receiving intensive statin therapy, *New England Journal of Medicine* 365 (2011): 2255–2267.

6. D. MacKay, J. Hathcock, and E. Guarneri, Niacin: Chemical forms, bioavailability, and health effects, *Nutrition Reviews* 70 (2012): 357–366.

7. F. G. Bowling, Pyridoxine supply in human development, *Seminars in Cell and Developmental Biology* 22 (2011): 611–618.

8. S. C. Larsson, N. Orsini, and A. Wolk, Vitamin B$_6$ and risk of colorectal cancer: A meta-analysis of prospective studies, *Journal of the American Medical Association* 303 (2010): 1077–1083; J. Shen and coauthors, Association of vitamin B-6 status with inflammation, oxidative stress, and chronic inflammatory conditions: The Boston Puerto Rican Health Study, *American Journal of Clinical Nutrition* 91 (2010): 337–342.

9. S. M. C. Wilson and coauthors, Oral contraceptive use: Impact on folate, vitamin B$_6$, and vitamin B$_{12}$ status, *Nutrition Reviews* 69 (2011): 572–583.

10. A. S. Tibbetts and D. R. Appling, Compartmentalization of mammalian folate-mediated one-carbon metabolism, *Annual Review of Nutrition* 30 (2010): 57–81.

11. M. A. Caudill, Folate bioavailability: Implications for establishing dietary recommendations and optimizing status, *American Journal of Clinical Nutrition* 91 (2010): 1455S–1460S.

12. P. Surén and coauthors, Association between maternal use of folic acid supplements and risk of autism spectrum disorders in children, *Journal of the American Medical Association* 309 (2013): 570–577; R. J. Berry, K. S. Crider, and M. Yeargin-Allsopp, Periconceptional folic acid and risk of autism spectrum disorders, *Journal of American Medical Association* 309 (2013): 611–612; R. J. Schmidt and coauthors, Maternal periconceptional folic acid intake and risk of autism spectrum disorders and developmental delay in the CHARGE (CHildhood Autism Risks from Genetics and Environment) case-control study, *American Journal of Clinical Nutrition* 96 (2012): 80–89; S. H. Blanton and coauthors, Folate pathway and nonsyndromic cleft lip and palate, *Birth Defects Research, Part A, Clinical and Molecular Teratology* 91 (2011): 50–60.

13. P. Verhoef, New insights on the lowest dose for mandatory folic acid fortification? *American Journal of Clinical Nutrition* 93 (2011): 1–2; R. L. Bailey and coauthors, Total folate and folic acid intake from foods

and dietary supplements in the United States: 2003–2006, *American Journal of Clinical Nutrition* 91 (2010): 231–237.

14. R. Cui and coauthors, Dietary folate and vitamin B6 and B12 intake in relation to mortality from cardiovascular diseases: Japan collaborative cohort study, *Stroke* 41 (2010): 1285–1289; A. Imamura and coauthors, Low folate levels may be an atherogenic factor regardless of homocysteine levels in young healthy nonsmokers, *Metabolism* 59 (2010): 728–733.

15. J. M. Artmitage and the Study of Effectiveness of Additional Reductions in Cholesterol and Homocysteine (SEARCH) Collaborative Group, Effects of homocysteine-lowering with folic acid plus vitamin B$_{12}$ vs placebo on mortality and major morbidity in myocardial infarction survivors: A randomized trial, *Journal of the American Medical Association* 303 (2010): 2486–2494; P. Tighe and coauthors, A dose-finding trial of the effect of the long-term folic acid intervention: Implications for food fortification policy, *American Journal of Clinical Nutrition* 93 (2011): 11–18.

16. J. B. Mason, Folate consumption and cancer risk: A confirmation and some reassurance, but we're not out of the woods yet, *American Journal of Clinical Nutrition* 94 (2011): 965–966; V. L. Stevens and coauthors, Folate and other one-carbon metabolism-related nutrients and risk of postmenopausal breast cancer in the Cancer Prevention Study II Nutrition Cohort, *American Journal of Clinical Nutrition* 91 (2010): 1708–1715.

17. B. M. Oaks and coauthors, Folate intake, post-folic acid grain fortification, and pancreatic cancer risk in the Prostate, Lung, Colorectal, and Ovarian Cancer Screening Trial, *American Journal of Clinical Nutrition* 91 (2010): 449–455.

18. T. M. Gibson and coauthors, Pre- and postfortification intake of folate and risk of colorectal cancer in a large prospective cohort study in the United States, *American Journal of Clinical Nutrition* 94 (2011): 1053–1062; J. E. Lee and coauthors, Folate intake and risk of colorectal cancer and adenoma: Modification by time, *American Journal of Clinical Nutrition* 93 (2011): 817–825.

19. O. A. Odewole and coauthors, Near-elimination of folate-deficiency anemia by mandatory folic acid fortification in older US adults: Reasons for Geographic and Racial Differences in Stroke study 2003–2007, *American Journal of Clinical Nutrition* 98 (2013): 1042–1047; Centers for Disease Control and Prevention, *Second National Report on Biochemical Indicators of Diet and Nutrition in the U.S. Population, 2012: Executive Summary*, www.cdc.gov/nutritionreport.

20. R. L. Bailey and coauthors, Total folate and folic acid intake from foods and dietary supplements in the United States: 2003–2006, *American Journal of Clinical Nutrition* 91 (2010): 231–237; Q. Yang and coauthors, Folic acid source, usual intake, and folate and vitamin B-12 status in US adults: National Health and Nutrition Examination Survey (NHANES) 2003–2006, *American Journal of Clinical Nutrition* 91 (2010): 64–72.

21. R. Pawlak and coauthors, How prevalent is vitamin B$_{12}$ deficiency among vegetarians? *Nutrition Reviews* 71 (2013): 110–117.

22. M. S. Morris, J. Selhub, and P. F. Jacques, Vitamin B-12 and folate status in relation to decline in scores on the Mini-Mental State Examination in the Framingham Heart Study, *Journal of the American Geriatrics Society* 60 (2012): 1457–1464.

23. M. G. Traber and J. F. Stevens, Vitamins C and E: Beneficial effects from a mechanistic perspective, *Free Radical Biology and Medicine* 51 (2011): 1000–1013.

HIGHLIGHT > 10
Vitamin and Mineral Supplements

> **LEARN IT** Present arguments for and against the use of dietary supplements.

An estimated 75,000 supplements are currently on the US market. More than half of the adults in the United States take a **dietary supplement** regularly, spending almost $24 billion each year.[1] Many people take supplements as dietary insurance—in case they are not meeting their nutrient needs from foods alone. Others take supplements as health insurance—to protect against certain diseases.

An estimated 40 percent of US adults take multivitamin-mineral supplements regularly. Others take large doses of single nutrients, most commonly, vitamin D and calcium. In many cases, taking supplements is a costly but harmless practice; sometimes, it is both costly and harmful to health.[2]

For the most part, people self-prescribe supplements, taking them on the advice of friends, advertisements, websites, or books that may or may not be reliable. Sometimes, they take supplements on the recommendation of a physician. When such advice follows a valid nutrition assessment, supplementation may be warranted, but even then the preferred course of action is to improve food choices and eating habits.[3] Without an assessment, the advice to take supplements may be inappropriate. A registered dietitian nutritionist can help with the decision.

When people think of dietary supplements, they often think of vitamins, but a diet that lacks vitamins probably lacks several minerals as well. This highlight asks several questions related to vitamin-mineral supplements. (The accompanying glossary defines dietary supplements and related terms.) What are the arguments *for* taking supplements? What are the arguments *against* taking them? Finally, if people do take supplements, how can they choose the appropriate ones? (Amino acid supplements and herbal supplements are discussed in Chapter 6 and Highlight 18, respectively.)

Arguments for Supplements

Vitamin-mineral supplements may be appropriate in some circumstances. In some cases, they can prevent or correct deficiencies; in others, they can reduce the risk of diseases. Consumers should discuss supplement use with their health-care providers, who can help monitor for adverse effects or nutrient-drug interactions.

Tanya Constantine/Brand X Pictures/Getty Images

Correct Overt Deficiencies

In the United States, adults rarely suffer nutrient deficiency diseases such as scurvy, pellagra, and beriberi, but nutrient deficiencies do still occur. To correct an overt deficiency disease, a physician may prescribe therapeutic doses two to ten times the RDA (or AI) of a nutrient. At such high doses, the supplement is having a pharmacological effect and acting as a drug.

Support Increased Nutrient Needs

As Chapters 15 through 17 explain, nutrient needs increase during certain stages of life, making it difficult to meet some of those needs without supplementation. For example, women who lose a lot of blood and therefore a lot of iron during menstruation each month may need an iron supplement. Women of childbearing age need folate supplements to reduce the risks of neural tube defects. Similarly, pregnant women and women who are breastfeeding their infants have exceptionally high nutrient needs and so usually need special supplements. Newborns routinely receive a single dose of vitamin K at birth to prevent abnormal bleeding. Infants may need other supplements as well, depending on whether they are breastfed or receiving formula, and on whether the water they drink contains fluoride.

Improve Nutrition Status

In contrast to the classical deficiencies, which present a multitude of symptoms and are relatively easy to recognize, subclinical deficiencies are subtle and easy to overlook—and they are also more likely to occur. Without fortification or supplementation, many adults in the

United States fall short of recommended intakes for several vitamins and minerals.[4] People who do not eat enough food to deliver the needed amounts of nutrients, such as habitual dieters and the elderly, risk developing subclinical deficiencies. Similarly, vegetarians who restrict their use of entire food groups without appropriate substitutions may fail to fully meet their nutrient needs. If there is no way for these people to eat enough nutritious foods to meet their needs, then vitamin-mineral supplements may be appropriate to help prevent nutrient deficiencies.

Improve the Body's Defenses

Health-care professionals may provide special supplementation to people being treated for addictions to alcohol or other drugs and to people with prolonged illnesses, extensive injuries, or other severe stresses such as surgery. Illnesses that interfere with appetite, eating, or nutrient absorption impair nutrition status. For example, the stomach condition atrophic gastritis often creates a vitamin B_{12} deficiency. In addition, nutrient needs are often heightened by diseases or medications. In all these cases, supplements are appropriate.

Reduce Disease Risks

Few people consume the optimal amounts of all the vitamins and minerals by diet alone. Inadequate intakes have been linked to chronic diseases such as heart disease, some cancers, and osteoporosis. For this reason, some physicians recommend that all adults take vitamin-mineral supplements. Such regular supplementation would provide an optimum intake to enhance metabolic harmony and prevent disease at relatively little cost. Others recognize the lack of conclusive evidence and the potential harm of supplementation and advise against such a recommendation.[5] A statement from the National Institutes of Health acknowledges that evidence is insufficient to recommend either for or against the use of supplements to prevent chronic diseases.

Highlight 11 reviews the relationships between supplement use and disease prevention. It describes some of the accumulating evidence suggesting that intakes of certain nutrients at levels much higher than can be attained from foods alone may be beneficial in reducing some disease risks. It also presents research confirming the associated risks. Clearly, consumers must be cautious in taking supplements to prevent disease.

Who Needs Supplements?

In summary, the following list acknowledges that in these specific conditions, these people may need to take supplements:

- People with specific nutrient deficiencies may need specific nutrient supplements.
- People whose energy intakes are particularly low (fewer than 1600 kcalories per day) may need multivitamin-mineral supplements.
- Vegetarians who eat all-plant diets (vegans) and older adults with atrophic gastritis may need vitamin B_{12}.
- People who have lactose intolerance or milk allergies or who otherwise do not consume enough milk products to forestall extensive bone loss may need calcium.

- People in certain stages of the life cycle who have increased nutrient requirements may need specific nutrient supplements. For example, infants may need vitamin D, iron, and fluoride; women of childbearing age and pregnant women may need folate and iron; and the elderly may need vitamin B_{12} and vitamin D.
- People who have inadequate intakes of milk or milk products, limited sun exposure, or heavily pigmented skin may need vitamin D.
- People who have diseases, infections, or injuries or who have undergone surgery that interferes with the intake, absorption, metabolism, or excretion of nutrients may need specific nutrient supplements.
- People taking medications that interfere with the body's use of specific nutrients may need specific nutrient supplements.

Except for people in these circumstances, most adults can get all the nutrients they need by eating a variety of nutrient-dense foods. Even athletes can meet their nutrient needs without the help of supplements, as Chapter 14 explains.

Arguments against Supplements

Foods rarely cause nutrient imbalances or toxicities, but supplements can. The higher the dose, the greater the risk of harm. People's tolerances for high doses of nutrients vary, just as their risks of deficiencies do. Amounts that some can tolerate may be harmful for others, and no one knows who falls where along the spectrum. It is difficult to determine just how much of a nutrient is enough—or too much. The Tolerable Upper Intake Levels (UL) of the DRI answer the question "How much is too much?" by defining the highest amount that appears safe for most healthy people. Table H10-1 presents UL and Daily Values for selected vitamins and minerals.

Who Should Not Take Supplements?

The following list recognizes that in certain circumstances, these people may need to avoid specific supplements:

- Men and postmenopausal women should not take iron supplements given that excess iron is harmful and generally more likely than inadequacies.
- Smokers should not take beta-carotene supplements given that high doses have been associated with increased lung cancer and mortality.
- Postmenopausal women should not take vitamin A supplements given that excess retinol has been associated with increased risk of hip fractures and reduced bone density.
- Surgery patients should not take vitamin E supplements during the week before surgery because vitamin E acts as a blood thinner.

Toxicity

Supplement users are more likely to have excessive intakes of certain nutrients—notably folate, vitamin A, vitamin B_6, vitamin C, calcium, magnesium, iron, and zinc.[6] The extent and severity of supplement toxicity remain unclear. Only a few alert health-care professionals can

TABLE H10-1 Vitamin and Mineral Intakes for Adults

Nutrient	Tolerable Upper Intake Levels[a]	Daily Values
Vitamins		
Vitamin A	3000 µg[b]	900 µg
Vitamin D (as cholecalciferol)	100 µg	20 µg
Vitamin E (as alpha-tocopherol)	1000 mg[b]	15 mg
Vitamin K	—[c]	120 µg
Thiamin	—[c]	1.2 mg
Riboflavin	—[c]	1.3 mg
Niacin (as niacinamide)	35 mg[b]	16 mg
Vitamin B$_6$ (as pyridoxine)	100 mg	1.7 mg
Folate	1000 µg[b]	400 µg
Vitamin B$_{12}$ (as cyanocobalamin)	—[c]	2.4 µg
Pantothenic acid	—[c]	5 mg
Biotin	—[c]	30 µg
Vitamin C (as ascorbic acid)	2000 mg	90 mg
Choline	3500 mg	550 mg
Minerals		
Chloride	3600 mg	2300 mg
Potasssium	—[c]	4700 mg
Calcium	2500 mg	1300 mg
Phosphorus	4000 mg	1250 mg
Magnesium	350 mg[b]	420 mg
Iron	45 mg	18 mg
Zinc	40 mg	11 mg
Iodine	1100 µg	150 µg
Selenium	400 µg	55 µg
Fluoride	10 mg	—
Copper	10 mg	0.9 mg
Manganese	11 mg	2.3 mg
Chromium	—[c]	35 µg
Molybdenum	2000 µg	45 µg

© Cengage Learning

[a] Unless otherwise noted, Upper Levels represent total intakes from food, water, and supplements.
[b] Upper Levels for vitamin A are for preformed vitamin A only; for vitamin E, niacin, and folate, the UL represent intakes from supplements, fortified foods, or both; for magnesium, the UL represent intakes from supplements only and do not include intakes from food and water.
[c] These nutrients have been evaluated by the DRI Committee for Tolerable Upper Intake Levels, but none were established because of insufficient data. No adverse effects have been reported with intakes of these nutrients at levels typical of supplements, but caution is still advised, given the potential for harm that accompanies excessive intakes.

recognize toxicity, even when it is acute. When it is chronic, with the effects developing subtly and progressing slowly, it often goes unrecognized and unreported.[7] In view of the potential hazards, some authorities believe supplements should bear warning labels, advising consumers that large doses may be toxic.

At a minimum, manufacturers should be held to the same standards required of the drug industry, which may help to prevent toxicities. Consider that more than 200 people reported symptoms of diarrhea, fatigue, hair loss, and joint pain when the selenium supplement they had taken delivered 200 times the selenium concentration listed on the label.[8]

Toxic overdoses of vitamins and minerals in children are more readily recognized and, unfortunately, fairly common. Fruit-flavored, chewable vitamins shaped like cartoon characters entice young children to eat them like candy in amounts that can cause poisoning. Iron supplements (30 milligrams of iron or more per tablet) are especially toxic and are the leading cause of accidental ingestion fatalities among children. Even mild overdoses cause GI distress, nausea, and black diarrhea, which reflects gastric bleeding. Severe overdoses result in bloody diarrhea, shock, liver damage, coma, and death.

Life-Threatening Misinformation

Another problem arises when people who are ill come to believe that high doses of vitamins or minerals can be therapeutic. Not only can high doses be toxic, but the person may take them instead of seeking medical help. Furthermore, there are no guarantees that the supplements will be effective. Taking vitamin supplements instead of medication may sound appealing, but they do not protect against the progression of heart disease or cancers.[9] In some cases, supplements may even be harmful.[10] Supplements of beta-carotene and vitamin A increase the risk of lung cancer and mortality, especially among smokers. Similarly, supplements of vitamin E increase the risk of prostate cancer among healthy men.[11]

Marketing materials for supplements often make health statements that are required to be "truthful and not misleading," but they often fall far short of both. Highlight 18 revisits this topic and includes a discussion of herbal preparations and other alternative therapies.

Unknown Needs

Another argument against the use of supplements is that there are no standards and no one knows exactly how to formulate the "ideal" supplement. What nutrients should be included? Which, if any, of the phytochemicals should be included? How much of each? On whose needs should the choices be based? Surveys have repeatedly shown little relationship between the supplements people take and the nutrients they actually need.

False Sense of Security

Another argument against supplement use is that it may lull people into a false sense of security. A person might eat irresponsibly, thinking, "My supplement will ensure my needs are met." Or, experiencing a warning symptom of a disease, a person might postpone seeking a diagnosis, thinking, "I probably just need a supplement to make this go away." Such self-diagnosis is potentially dangerous.

Other Invalid Reasons

Other invalid reasons people might use for taking supplements include:

- The belief that the food supply or soil contains inadequate nutrients
- The belief that supplements can provide energy

337

- The belief that supplements can enhance athletic performance or build lean body tissues without physical work or faster than work alone (see Highlight 14)
- The belief that supplements will help a person cope with stress
- The belief that supplements can prevent, treat, or cure conditions ranging from the common cold to cancer

Ironically, people with health problems are more likely to take supplements than other people, yet today's health problems are more likely to be due to overnutrition and poor lifestyle choices than to nutrient deficiencies. The truth—that most people would benefit from improving their eating and activity patterns—is harder to swallow than a supplement pill.

Bioavailability and Antagonistic Actions

In general, the body absorbs nutrients best from foods in which the nutrients are diluted and dispersed among other substances that may facilitate their absorption. Taken in pure, concentrated form, nutrients are likely to interfere with one another's absorption or with the absorption of nutrients in foods eaten at the same time. Documentation of these effects is particularly extensive for minerals: zinc hinders copper and calcium absorption, iron hinders zinc absorption, calcium hinders magnesium and iron absorption, and magnesium hinders the absorption of calcium and iron. Similarly, binding agents in supplements limit mineral absorption.

Although minerals provide the most-familiar and best-documented examples, interference among vitamins is now being seen as supplement use increases. The vitamin A precursor beta-carotene, long thought to be nontoxic, interferes with vitamin E metabolism when taken over the long term as a dietary supplement. Vitamin E, on the other hand, antagonizes vitamin K activity and so should not be used by people being treated for blood-clotting disorders. Consumers who want the benefits of optimal absorption of nutrients should eat foods selected for nutrient density and variety.

Whenever the diet is inadequate, the person should first attempt to improve it so as to obtain the needed nutrients from foods. If that is truly impossible, then the person needs a multivitamin-mineral supplement that supplies between 50 and 150 percent of the Daily Value for each of the nutrients. These amounts reflect the ranges commonly found in foods and therefore are compatible with the body's normal handling of nutrients (its physiologic tolerance). The next section provides some pointers to assist in the selection of an appropriate supplement.

Selection of Supplements

Whenever a physician or registered dietitian nutritionist recommends a supplement, follow the directions carefully. When selecting a supplement yourself, look for a single, balanced vitamin-mineral supplement. Supplements with a USP verification logo have been tested by the US Pharmacopeia (USP) to ensure that the supplement:

- Contains the declared ingredients and amounts listed on the label
- Does not contain harmful levels of contaminants
- Will disintegrate and release ingredients in the body
- Was made under safe and sanitary conditions

If you decide to take a vitamin-mineral supplement, ignore the eye-catching art and meaningless claims. Pay attention to the form the supplements are in, the list of ingredients, and the price. Here's where the truth lies, and from it you can make a rational decision based on facts. You have two basic questions to answer.

Form

The first question: What form do you want—chewable, liquid, or pills? If you'd rather drink your supplements than chew them, fine. If you choose a chewable form, though, be aware that chewable vitamin C can dissolve tooth enamel. If you choose pills, look for statements about the disintegration time. The USP suggests that supplements should completely disintegrate within 30 to 45 minutes. Obviously, supplements that don't dissolve have little chance of entering the bloodstream, so look for a brand that claims to meet USP disintegration standards.

Contents

The second question: What vitamins and minerals do you need? Generally, an appropriate supplement provides vitamins and minerals in amounts that do not exceed recommended intakes. Avoid supplements that, in a daily dose, provide more than the UL for *any* nutrient. Avoid preparations with more than 10 milligrams of iron per dose, except as prescribed by a physician. Iron is hard to get rid of once it's in the body, and an excess of iron can cause problems, just as a deficiency can (see Chapter 13).

Misleading Claims

Manufacturers of *organic* or natural vitamins boast that their pills are purified from real foods rather than synthesized in a laboratory. These supplements are no more effective than others and often cost more. The word *synthetic* may sound like "fake," but to synthesize just means to put together. Think back on the course of human evolution; it is not natural to take any kind of pill. In reality, the finest, most natural vitamin "supplements" available are whole grains, vegetables, fruits, meat, fish, poultry, eggs, legumes, nuts, and milk and milk products.

Avoid products that make **"high potency"** claims. More is not better (review Figure 10-1 on p. 304). Remember that foods are also providing these nutrients. Nutrients can build up and cause unexpected problems. For example, a man who takes vitamins and begins to lose his hair may think his hair loss means he needs more vitamins, when in fact it may be the early sign of a vitamin A overdose. (Of course, it may be completely unrelated to nutrition as well.)

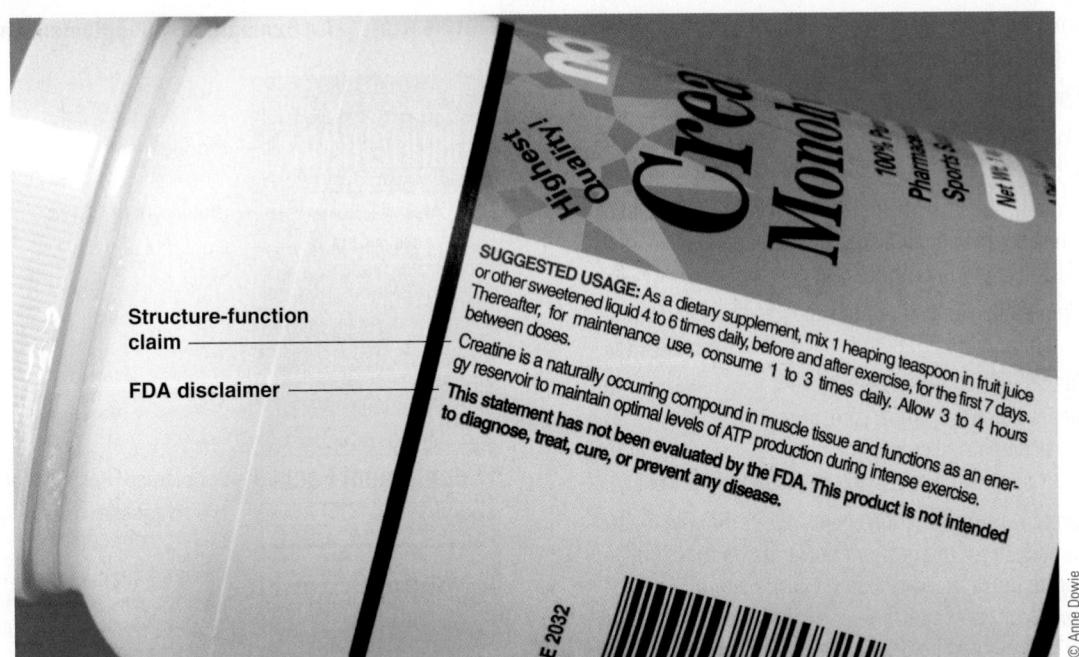

Structure-function claim

FDA disclaimer

SUGGESTED USAGE: As a dietary supplement, mix 1 heaping teaspoon in fruit juice or other sweetened liquid 4 to 6 times daily, before and after exercise, for the first 7 days. Thereafter, for maintenance use, consume 1 to 3 times daily. Allow 3 to 4 hours between doses.

Creatine is a naturally occurring compound in muscle tissue and functions as an energy reservoir to maintain optimal levels of ATP production during intense exercise.

This statement has not been evaluated by the FDA. This product is not intended to diagnose, treat, cure, or prevent any disease.

Structure-function claims do not need FDA authorization, but they must be accompanied by a disclaimer.

Be aware that fake vitamins and preparations that contain items not needed in human nutrition, such as carnitine and inositol, reflect a marketing strategy aimed at your pocket, not at your health. The manufacturer wants you to believe that its pills contain the latest "new" nutrient that other brands omit, but in reality, these substances are not known to be needed by human beings.

Realize that the claim that supplements "relieve stress" is another marketing ploy. If you give even passing thought to what people mean by "stress," you'll realize manufacturers could never design a supplement to meet everyone's needs. Is it stressful to take an exam? Well, yes. Is it stressful to survive a major car wreck with third-degree burns and multiple bone fractures? Definitely, yes. The body's responses to these stresses are different. The body does use vitamins and minerals in mounting a stress response, but a body fed a well-balanced diet can meet the needs of most minor stresses. For the major ones, medical intervention is needed. In any case, taking a dietary supplement won't make life any less stressful.

Other marketing tricks to sidestep are "green" pills that contain dehydrated, crushed parsley, alfalfa, and other fruit and vegetable extracts. The nutrients and phytochemicals advertised can be obtained from a serving of vegetables more easily and for less money. Such pills may also provide enzymes, but enzymes are inactivated in the stomach during protein digestion.

Recognize the latest nutrition buzzwords. Manufacturers were marketing "antioxidant" supplements before the print had time to dry on the first scientific reports of antioxidant vitamins' action in the body. Remember, too, that high doses can alter a nutrient's action in the body. An antioxidant in physiological quantities may be beneficial,

but in pharmacological quantities, it may act as a prooxidant and cause harm. Highlight 11 explores antioxidants and supplement use in more detail.

Similarly, manufacturers began making dietary supplements using **nanotechnology** before the FDA had created guidelines defining their use in consumer products. These **nanoceuticals** promise enhanced nutrient absorption and activity. Such claims may sound good, but again, more does not always mean better.

Finally, be aware that advertising on the Internet is cheap and not closely regulated. Promotional e-mails can be sent to millions of people in an instant. Internet messages can easily cite references and provide links to other sites, implying an endorsement when in fact none has been given. Be cautious when examining unsolicited information and search for a balanced perspective.

Cost

When shopping for supplements, remember that local or store brands may be just as good as nationally advertised brands. If they are less expensive, it may be because the price does not have to cover the cost of national advertising.

Regulation of Supplements

Dietary supplements are regulated by the **FDA (Food and Drug Administration)** as foods. Details of supplement regulation are defined in the Dietary Supplement Health and Education Act of 1994, which was intended to enable consumers to make informed choices about dietary supplements. The act subjects supplements to the same general labeling requirements that apply to foods. Specifically:

- Nutrition labeling for dietary supplements is required.

- Labels may make nutrient claims (as "high" or "low") according to specific criteria (for example, "an excellent source of vitamin C").

- Labels may claim that the lack of a nutrient can cause a deficiency disease, but if they do, they must also include the prevalence of that deficiency disease in the United States.

- Labels may make health claims that are supported by significant scientific agreement and are not brand specific (for example, "folate protects against neural tube defects").

- Labels may claim to diagnose, treat, cure, or relieve common complaints such as menstrual cramps or memory loss, but may *not* make claims about specific diseases (except as noted previously).

- Labels may make structure-function claims about the role a nutrient plays in the body, how the nutrient performs its function, and how consuming the nutrient is associated with general well-being. The manufacturer is responsible for ensuring that the claims are truthful and not misleading. Claims must be accompanied by an FDA disclaimer statement: "This statement has not been evaluated by the Food and Drug Administration. This product is not intended to diagnose, treat, cure, or prevent any disease." Figure H10-1 provides an example of a supplement label that complies with the requirements.

The multibillion-dollar-a-year supplement industry spends much money and effort influencing these regulations. The net effect of the Dietary Supplement Health and Education Act was a deregulation of the supplement industry. Unlike food additives or drugs, supplements do not need to be proved safe and effective, nor do they need the FDA's approval before being marketed. Furthermore, there are no standards for potency or dosage and no requirements for providing warnings of potential side effects. The FDA can only require good manufacturing practices: that dietary supplements be produced and packaged in a quality manner, do not contain contaminants or impurities, and are accurately labeled to reflect the actual contents.

Should a problem arise, the burden falls to the FDA to prove that the supplement poses a "significant or unreasonable risk of illness or injury." Only then would it be removed from the market. When asked, most Americans express support for greater regulation of dietary supplements. Health professionals agree.[12] To learn more about dietary supplements currently on the US market as well as those that have been recalled, consumers can visit the National Institutes of Health website (www.dsld.nlm.nih.gov/dsld/index.jsp).

If all the nutrients we need can come from food, why not just eat food? Foods have so much more to offer than supplements do. Nutrients in foods come in an infinite variety of combinations with a multitude of different carriers and absorption enhancers. They

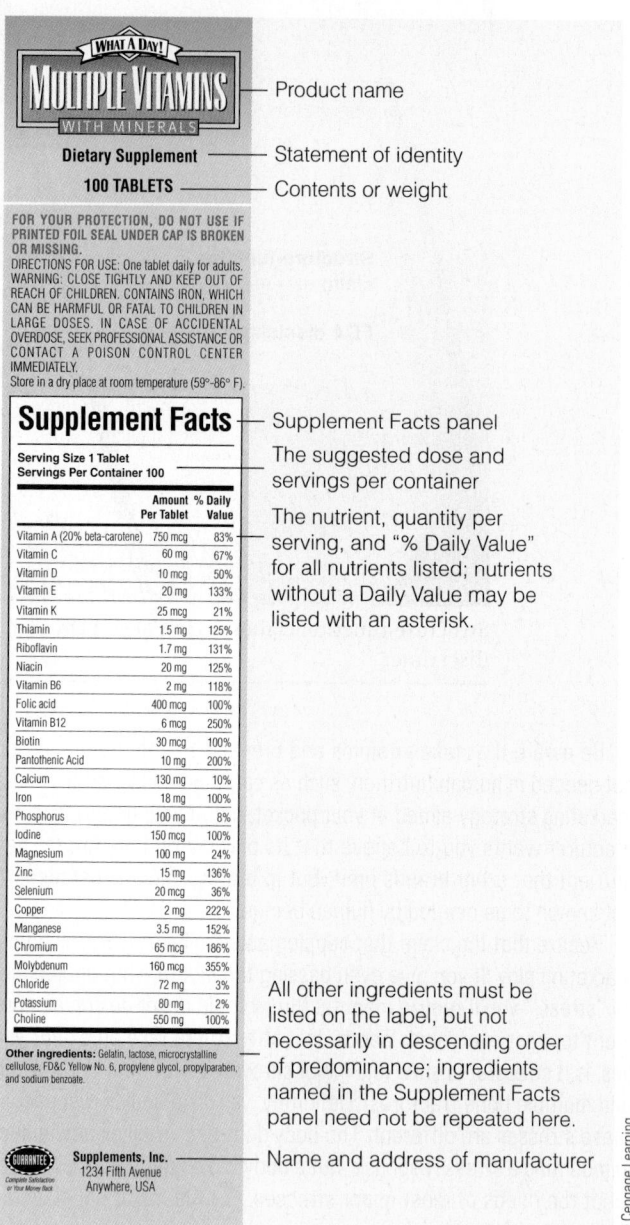

> FIGURE H10-1   **An Example of a Supplement Label**

come with water, fiber, and an array of beneficial phytochemicals. Foods stimulate the GI tract to keep it healthy. They provide energy, and as long as you need energy each day, why not have nutritious foods deliver it? Foods offer pleasure, satiety, and opportunities for socializing while eating. Quite simply, foods meet human health needs far better than dietary supplements. For further proof, read Highlight 11.

CRITICAL THINKING QUESTIONS

A. What are the arguments for and against the use of dietary supplements?

B. According to the Health and Education Act of 1994, dietary supplements with familiar ingredients may be marketed without any evidence of effectiveness or safety. Supplements with new ingredients are supposed to provide the FDA with evidence of safety, but this part of the law is rarely enforced. Both the industry and the FDA acknowledge that most supplements are currently on the market without any assessment of safety. What is your position on this situation and what changes to the law, if any, would you propose to support your position?

REFERENCES

1. J. Gahche and coauthors, Dietary supplement use among US adults has increased since NHANES III (1988–1994), *National Center for Health Statistics: Data Brief* 61 (2011): 1–8.

2. D. B. McCormick, Vitamin/mineral supplements: Of questionable benefit for the general population, *Nutrition Reviews* 68 (2010): 207–213.

3. Position of the American Dietetic Association: Nutrient supplementation, *Journal of the American Dietetic Association* 109 (2009): 2073–2085.

4. V. L. Fulgoni III and coauthors, Foods, fortificants, and supplements: Where do Americans get their nutrients? *Journal of Nutrition* 141 (2011): 1847–1854.

5. M. E. Martinez and coauthors, Dietary supplements and cancer prevention: Balancing potential benefits against proven harms, *Journal of the National Cancer Institute* 104 (2012): 732–739.

6. R. L. Bailey and coauthors, Examination of vitamin intakes among US adults by dietary supplement use, *Journal of the Academy of Nutrition and Dietetics* 112 (2012): 657–663; R. L. Bailey and coauthors, Dietary supplement use is associated with higher intakes of minerals from food sources, *American Journal of Clinical Nutrition* 94 (2011): 1376–1381.

7. M. Cellini and coauthors, Dietary supplements: Physician knowledge and adverse event reporting, *Medicine and Science in Sports and Exercise* 45 (2013): 23–28.

8. J. K. MacFarquhar and coauthors, Acute selenium toxicity associated with a dietary supplement, *Archives of Internal Medicine* 170 (2010): 256–261.

9. M. G. O'Doherty and coauthors, Effect of supplementation with B vitamins and antioxidants on levels of asymmetric dimethylarginine (ADMA) and C-reactive protein (CRP): A double-blind, randomized, factorial design, placebo-controlled trial, *European Journal of Nutrition* 49 (2010): 483–492; G. J. Hankey and VITATOPS Trial Study Group, B vitamins in patients with recent transient ischaemic attack or stroke in the VITAmins TO Prevent Stroke (VITATOPS) trial: A randomized, double-blind, parallel, placebo-controlled trial, *The Lancet Neurology* 9 (2010): 855–865.

10. G. Bjelakovic and C. Gluud, Vitamin and mineral supplement use in relation to all-cause mortality in the Iowa Women's Health Study, *Archives of Internal Medicine* 171 (2011): 1633–1634.

11. E. A. Klein and coauthors, Vitamin E and the risk of prostate cancer: The Selenium and Vitamin E Cancer Prevention Trial (SELECT), *Journal of the American Medical Association* 306 (2011): 1549–1556.

12. P. A. Cohen, Hazards of hindsight—Monitoring the safety of nutritional supplements, *New England Journal of Medicine* 370 (2014): 1277–1280.

11

The Fat-Soluble Vitamins: A, D, E, and K

Nutrition in Your Life

Realizing that vitamin A from vegetables participates in vision, a mom encourages her children to "eat your carrots" because "they're good for your eyes." A dad takes his children outside to "enjoy the fresh air and sunshine" because they need the vitamin D that is made with the help of the sun. A physician recommends that a patient use vitamin E to slow the progression of heart disease. Another physician gives a newborn a dose of vitamin K to protect against life-threatening blood loss. These common daily occurrences highlight some of the heroic work of the fat-soluble vitamins. In the Nutrition Portfolio at the end of this chapter, you can determine whether the foods you are eating are meeting your fat-soluble vitamin needs.

The fat-soluble vitamins A, D, E, and K differ from the water-soluble vitamins in several significant ways (review Table 10-2 on p. 304). Being insoluble in the watery juices of the GI tract, the fat-soluble vitamins require bile for their digestion and absorption. Upon absorption, fat-soluble vitamins travel through the lymphatic system within chylomicrons before entering the bloodstream, where many of them require protein carriers for transport. The fat-soluble vitamins participate in numerous activities throughout the body, but excesses are stored primarily in the liver and adipose tissue. The body maintains blood concentrations by retrieving these vitamins from storage as needed; thus people can eat less than their daily need for days, weeks, or even months or years without ill effects. They need only ensure that, over time, *average* daily intakes approximate recommendations. By the same token, because fat-soluble vitamins are not readily excreted, the risk of toxicity is greater than it is for the water-soluble vitamins.

11.1 Vitamin A and Beta-Carotene

> **LEARN IT** Identify the main roles, deficiency symptoms, and food sources for vitamin A.

Vitamin A was the first fat-soluble vitamin to be recognized. More than a century later, vitamin A and its precursor, **beta-carotene,** continue to intrigue researchers with their diverse roles and profound effects on health.

Three different forms of vitamin A are active in the body: **retinol, retinal,** and **retinoic acid.** Collectively known as **retinoids,** these compounds are commonly found in foods derived from animals. Foods derived from plants provide **carotenoids,** some of which can be converted to vitamin A.[1] * The most studied of the carotenoids with **vitamin A activity** is beta-carotene, which can be split to form retinol in the intestine and liver.[2] Figure 11-1 illustrates the structural similarities and differences of these vitamin A compounds and the cleavage of beta-carotene.

The cells can convert retinol and retinal to the other active forms of vitamin A as needed. The conversion of retinol to retinal is reversible, but the further conversion of retinal to retinoic acid is irreversible (see Figure 11-2). This irreversibility is significant because each form of vitamin A performs a specific function that the others cannot.

Several proteins participate in the digestion and absorption of vitamin A. After absorption via the lymph system, vitamin A eventually arrives at the liver, where it is stored. There, a special transport protein, **retinol-binding protein (RBP),** picks up vitamin A from the liver and carries it in the blood. Cells that use vitamin A have special protein receptors for it, and its action within each cell may differ depending on the receptor. For example, retinoic acid can *stimulate* cell growth in the skin and *inhibit* cell growth in tumors.

Roles in the Body
Vitamin A is a versatile vitamin, known to regulate the expression of several hundred genes. Its major roles include:

- Promoting vision
- Participating in protein synthesis and cell differentiation, thereby maintaining the health of epithelial tissues and skin
- Supporting reproduction and regulating growth

As mentioned, each form of vitamin A performs specific tasks. Retinol supports reproduction and is the major transport and storage form of the vitamin.

vitamin A: all naturally occurring compounds with the biological activity of *retinol,* the alcohol form of vitamin A.

beta-carotene (BAY-tah KARE-oh-teen): one of the carotenoids; an orange pigment and vitamin A precursor found in plants.

retinol (RET-ih-nol): the alcohol form of vitamin A.

retinal (RET-ih-nal): the aldehyde form of vitamin A.

retinoic (RET-ih-NO-ick) **acid:** the acid form of vitamin A.

retinoids (RET-ih-noyds): chemically related compounds with biological activity similar to that of retinol; metabolites of retinol.

carotenoids (kah-ROT-eh-noyds): pigments commonly found in plants and animals, some of which have vitamin A activity. The carotenoid with the greatest vitamin A activity is beta-carotene.

vitamin A activity: a term referring to both the active forms of vitamin A and the precursor forms in foods without distinguishing between them.

retinol-binding protein (RBP): the specific protein responsible for transporting retinol.

*Carotenoids with vitamin A activity include alpha-carotene, beta-carotene, and beta-cryptoxanthin; carotenoids with no vitamin A activity include the phytochemicals lycopene, lutein, and zeaxanthin.

> **FIGURE 11-1** **Forms of Vitamin A**

In this diagram, corners represent carbon atoms, as in all previous diagrams in this book. A further simplification here is that methyl groups (CH_3) are understood to be at the ends of the lines extending from corners. (See Appendix C for complete structures.)

Retinol, the alcohol form

Retinal, the aldehyde form

Retinoic acid, the acid form

Cleavage at this point can yield two molecules of vitamin A*

Beta-carotene, a precursor

*Sometimes cleavage occurs at other points as well, so that one molecule of beta-carotene may yield only one molecule of vitamin A. Furthermore, not all beta-carotene is converted to vitamin A, and absorption of beta-carotene is not as efficient as that of vitamin A. For these reasons, 12 μg of beta-carotene are equivalent to 1 μg of vitamin A. Conversion of other carotenoids to vitamin A is even less efficient.

© Cengage Learning

> FIGURE 11-2   Conversion of Vitamin A Compounds

Notice that the conversion from retinol to retinal is reversible, whereas the pathway from retinal to retinoic acid is not.

Retinal is active in vision and is also an intermediate in the conversion of retinol to retinoic acid (review Figure 11-2). Retinoic acid acts like a hormone, regulating cell differentiation, growth, and embryonic development. Animals raised on retinoic acid as their only source of vitamin A can grow normally, but they become blind because retinoic acid cannot be converted to retinal (review Figure 11-2).

Vitamin A in Vision   Vitamin A plays two indispensable roles in the eye: it helps maintain a crystal-clear outer window, the **cornea,** and it participates in the conversion of light energy into nerve impulses at the **retina** (see Figure 11-3 for details). Some of the photosensitive cells of the retina contain **pigment** molecules called **rhodopsin.** Each rhodopsin molecule is composed of a protein called **opsin** bonded to a molecule of retinal, which plays a central role in vision.[3] When light passes through the cornea of the eye and strikes the retina, rhodopsin responds. As it does, opsin is released and retinal shifts from a *cis* to a *trans* configuration, just as fatty acids do during hydrogenation (see p. 139). These changes generate an electrical impulse that conveys the message to the brain. Much of the retinal is then converted back to its active *cis* form and combined with the opsin protein to regenerate rhodopsin. Some retinal, however, may be oxidized to retinoic acid, a biochemical dead end for the visual process. Visual activity leads to repeated small losses of retinal, necessitating its constant replenishment either directly from foods or indirectly from retinol stores.

Vitamin A in Protein Synthesis and Cell Differentiation   Despite its important role in vision, only one-thousandth of the body's vitamin A is in the retina. Much more is in the cells lining the body's surfaces. There, the vitamin participates in protein synthesis and **cell differentiation,** a process by which each type of cell develops to perform a specific function.

cornea (KOR-nee-uh): the transparent membrane covering the outside of the eye.

retina (RET-in-uh): the innermost membrane of the eye, composed of several layers, including one that contains the rods and cones.

pigment: a molecule capable of absorbing certain wavelengths of light so that it reflects only those that we perceive as a certain color.

rhodopsin (ro-DOP-sin): a light-sensitive pigment of the retina that contains the retinal form of vitamin A and the protein opsin.

• **rhod** = red (pigment)
• **opsin** = visual protein

opsin (OP-sin): the protein portion of visual pigment molecules.

cell differentiation (DIF-er-EN-she-AY-shun): the process by which immature cells develop specific functions different from those of the original that are characteristic of their mature cell type.

> FIGURE 11-3   **Vitamin A's Role in Vision**

More than 100 million photosensitive cells reside in the retina, and each contains about 30 million molecules of vitamin A–containing visual pigments. The rods contain the rhodopsin pigment and respond to faint light; the cones contain the iodopsin pigment and function in color vision.

As light enters the eye, the cells of the retina convert images into electrical impulses.

The cells of the retina contain rhodopsin, a molecule composed of opsin (a protein) and *cis*-retinal (vitamin A).

cis-Retinal   *trans*-Retinal

As rhodopsin absorbs light, retinal changes from *cis* to *trans*, which triggers an electrical impulse that carries visual information to the brain through the optic nerve.

> FIGURE 11-4 **Mucous Membrane Integrity**

Vitamin A maintains healthy cells in the mucous membranes.

Without vitamin A, the normal structure and function of the cells in the mucous membranes are impaired.

© Cengage Learning

Mucus   Goblet cells

All body surfaces, both inside and out, are covered by layers of cells known as **epithelial cells.** The **epithelial tissue** on the outside of the body is, of course, the skin—and vitamin A and beta-carotene help to protect against skin damage from sunlight.[4] The epithelial tissues that line the inside of the body are the **mucous membranes:** the linings of the mouth, stomach, and intestines; the linings of the lungs and the passages leading to them; the linings of the urinary bladder and urethra; the linings of the uterus and vagina; and the linings of the eyelids and sinus passageways. Within the body, the mucous membranes of the GI tract alone line an area larger than a quarter of a football field, and vitamin A helps to maintain their integrity (see Figure 11-4).

Vitamin A promotes differentiation of epithelial cells and goblet cells, one-celled glands that synthesize and secrete mucus. Mucus coats and protects the epithelial cells from invasive microorganisms and other potentially damaging substances, such as gastric juices.

Vitamin A in Reproduction and Growth As mentioned, vitamin A also supports reproduction and regulates growth.[5] In men, retinol participates in sperm development, and in women, vitamin A supports normal fetal development during pregnancy. Children lacking vitamin A fail to grow; given vitamin A supplements, these children gain weight and grow taller.

The growth of bones illustrates that growth is a complex phenomenon of **remodeling.** To convert a small bone into a large bone, some bone cells must "undo" parts of the bone before other cells can build new bone, and vitamin A participates in the dismantling.* The cells that break down bone contain acid and enzymes that dissolve the minerals and digest the matrix.** With the help of vitamin A, these bone-dismantling cells destroy selected sites in the bone, removing the parts that are not needed. After completing their work, the bone-dismantling cells die, leaving their excavation site to be rebuilt by the bone-building cells.

Beta-Carotene as a Precursor and an Antioxidant Beta-carotene plays two primary roles in the body.[6] First, it serves as a vitamin A precursor. Second, some beta-carotene acts as an antioxidant capable of protecting the body against disease. (Highlight 11 provides details.)

Vitamin A Deficiency
Vitamin A status depends mostly on the adequacy of vitamin A stores, 90 percent of which are in the liver. Vitamin A status also depends on a person's protein status because retinol-binding protein serves as the vitamin's transport carrier inside the body.

If a person were to stop eating vitamin A–containing foods, deficiency symptoms would not begin to appear until after stores were depleted—1 to 2 years for a healthy adult but much sooner for a growing child. Then the consequences would

epithelial (ep-i-THEE-lee-ul) **cells:** cells on the surface of the skin and mucous membranes.

epithelial tissue: the layer of the body that serves as a selective barrier between the body's interior and the environment. Examples are the cornea of the eyes, the skin, the respiratory lining of the lungs, and the lining of the digestive tract.

mucous (MYOO-kus) **membranes:** the membranes, composed of mucus-secreting cells, that line the surfaces of body tissues.

remodeling: the dismantling and re-formation of a structure.

---

*The cells that dismantle bone during growth are *osteoclasts;* those that build bone are *osteoblasts.*
**The degradative enzymes are contained within *lysosomes* (LYE-so-zomes).

be profound and severe. Vitamin A deficiency is uncommon in the United States, but it is a major nutrition problem in many developing countries, responsible for a million or more unnecessary deaths and cases of blindness each year.[7] Routine vitamin A supplementation and food fortification can be a life-saving intervention.[8]

Infectious Diseases Vitamin A supports immune function and inhibits replication of the measles virus.[9] In developing countries around the world, measles is a devastating infectious disease, killing 430 children each day.[10] The severity of the illness often correlates with the degree of vitamin A deficiency; deaths are usually due to related infections such as pneumonia and severe diarrhea. Providing measles vaccinations and large doses of vitamin A reduces the risk of dying from these infections by more than half.[11]

The World Health Organization (WHO) and UNICEF (the United Nations International Children's Emergency Fund) have made the control of vitamin A deficiency a major goal in their quest to improve child health and survival throughout the developing world. They recommend two doses of vitamin A supplements, given 24 hours apart, for all children with measles. In the United States, the American Academy of Pediatrics recommends vitamin A supplements for certain groups of measles-infected infants and children. Vitamin A supplements also protect against blindness and the complications of other life-threatening infections, including malaria, lung diseases, and HIV (human immunodeficiency virus, the virus that causes AIDS).

Night Blindness **Night blindness** is one of the first detectable signs of vitamin A deficiency and permits early diagnosis. In night blindness, the person loses the ability to recover promptly from the temporary blinding that follows a flash of bright light at night or to see after dark. In many parts of the world, after the sun goes down, vitamin A–deficient people become night-blind. They often cling to others or sit still, afraid that they may trip and fall or lose their way if they try to walk alone.

Blindness (Xerophthalmia) Beyond night blindness is total blindness—failure to see at all. Night blindness is caused by a lack of vitamin A at the back of the eye, the retina; total blindness is caused by a lack of vitamin A at the front of the eye, the cornea. Severe vitamin A deficiency is the leading cause of preventable blindness in the world, causing as many as half a million preschool children to lose their sight each year.

Blindness due to vitamin A deficiency, known as **xerophthalmia**, develops in stages. At first, the cornea becomes dry and hard because of inadequate mucous production—a condition known as **xerosis**. Then xerosis quickly progresses to **keratomalacia**, the softening of the cornea that leads to irreversible blindness.

Keratinization Elsewhere in the body, vitamin A deficiency affects other surfaces. On the body's outer surface, the epithelial cells change shape and begin to secrete the protein **keratin**—the hard, inflexible protein of hair and nails. As Figure 11-5 shows, the skin becomes dry, rough, and scaly as lumps of keratin accumulate (**keratinization**). Without vitamin A, the goblet cells in the GI tract diminish in number and activity, limiting the secretion of mucus. With less mucus, normal digestion and absorption of nutrients falter, and this, in turn, worsens malnutrition by limiting the absorption of whatever nutrients the diet may deliver. Similar changes in the cells of other epithelial tissues weaken defenses, making infections of the respiratory tract, the GI tract, the urinary tract, the vagina, and inner ear likely.

Vitamin A Toxicity
Just as a deficiency of vitamin A affects all body systems, so does a toxicity. Symptoms of toxicity begin to develop when all the binding proteins are loaded, and vitamin A is free to damage cells. Such effects are unlikely when a person depends on a balanced diet for nutrients, but toxicity is a real possibility when concentrated amounts of **preformed vitamin A** in foods derived from animals, fortified foods, or supplements are consumed. Children

© H. Sanstead / U. of Texas/Galveston

In vitamin A deficiency, the epithelial cells secrete the protein keratin in a process known as *keratinization*. (Keratinization doesn't occur in the GI tract, but mucus-producing cells dwindle and mucus production declines.) The extreme of this condition is *hyperkeratinization* or *hyperkeratosis*. When keratin accumulates around hair follicles, the condition is known as *follicular hyperkeratosis*.

night blindness: slow recovery of vision after flashes of bright light at night or an inability to see in dim light; an early symptom of vitamin A deficiency.

xerophthalmia (zer-off-THAL-mee-uh): progressive blindness caused by inadequate mucus production due to severe vitamin A deficiency.

- **xero** = dry
- **ophthalm** = eye

xerosis (zee-ROW-sis): abnormal drying of the skin and mucous membranes; a sign of vitamin A deficiency.

keratomalacia (KARE-ah-toe-ma-LAY-shuh): softening of the cornea that leads to irreversible blindness; a sign of severe vitamin A deficiency.

keratin (KARE-uh-tin): a water-insoluble protein; the normal protein of hair and nails.

keratinization: accumulation of keratin in a tissue; a sign of vitamin A deficiency.

preformed vitamin A: dietary vitamin A in its active form.

> FIGURE 11-6 **Symptom of Beta-Carotene Excess—Discoloration of the Skin**

James Stevenson/Science Source

The hand on the right shows the skin yellowing that occurs when blood levels of beta-carotene rise in response to a diet that features carrots, pumpkins, and orange juice. (The hand on the left belongs to someone else and is shown here for comparison.)

are most vulnerable to toxicity because they need less vitamin A and are more sensitive to overdoses. An Upper Level (UL) has been set for preformed vitamin A (see inside front cover). Even multivitamin supplements typically provide 1500 micrograms—much more vitamin A than most people need. (For perspective, the RDA for vitamin A is 700 micrograms for women and 900 micrograms for men.)

Beta-carotene, which is found in a wide variety of fruits and vegetables, is not converted efficiently enough in the body to cause vitamin A toxicity; instead, it is stored in the fat just under the skin. Although overconsumption of beta-carotene from foods may turn the skin yellow, this is not harmful (see Figure 11-6). In contrast, overconsumption of beta-carotene from supplements may be quite harmful. In excess, this antioxidant may act as a prooxidant (as Highlight 11 explains). Adverse effects of beta-carotene supplements are most evident in people who drink alcohol and smoke cigarettes.

Bone Defects Excessive intakes of vitamin A over the years may weaken the bones and contribute to fractures and osteoporosis.[12] Vitamin A suppresses bone-building activity, stimulates bone-dismantling activity, and interferes with vitamin D's ability to maintain normal blood calcium.

Birth Defects Excessive vitamin A during pregnancy leads to abnormal cell death in the spinal cord, which increases the risk of birth defects such as spina bifida and cleft palate.[13] In such cases, vitamin A is considered a **teratogen**. High intakes (daily supplemental intakes of vitamin A equivalent to roughly four times the RDA for women) before the seventh week of pregnancy appear to be the most damaging. For this reason, vitamin A is not given as a supplement in the first trimester of pregnancy without specific evidence of deficiency, which is rare.

Not for Acne Adolescents need to know that massive doses of vitamin A have no beneficial effect on **acne**. The prescription medicine Accutane is made from vitamin A but is chemically different.* Taken orally, Accutane is effective against the deep lesions of cystic acne. It is highly toxic, however, especially during growth, and has caused birth defects in infants when women have taken it during their pregnancies. For this reason, women taking Accutane must agree to pregnancy testing and to using two forms of contraception from at least 1 month before taking the drug through at least 1 month after discontinuing its use. Should they become pregnant, they need to stop taking Accutane immediately and notify their physician.

Another vitamin A relative, Retin-A, fights acne, the wrinkles of aging, and other skin disorders.** Applied topically, this ointment smooths and softens skin; it also lightens skin that has become darkly pigmented after inflammation. During treatment, the skin becomes red and tender and peels.

Vitamin A Recommendations Because the body can derive vitamin A from both retinoids and carotenoids, its content in foods and its recommendations are expressed as **retinol activity equivalents (RAE)**. One microgram of retinol counts as 1 RAE, as does 12 micrograms of dietary beta-carotene.*** This difference recognizes that beta-carotene's absorption and conversion are significantly less efficient than those of the retinoids. Until recently, food and supplement labels reported vitamin A contents using International Units (IU), a measure of vitamin activity used before direct chemical analysis was possible. The glossary on the inside back page provides factors that can be used to convert IU to a weight measurement.

---

teratogen (ter-AT-oh-jen): a substance that causes abnormal fetal development and birth defects.

acne: a chronic inflammation of the skin's follicles and oil-producing glands, which leads to an accumulation of oils inside the ducts that surround hairs; usually associated with the maturation of young adults.

retinol activity equivalents (RAE): a measure of vitamin A activity; the amount of retinol that the body will derive from a food containing preformed retinol or its precursor, beta-carotene.

---

*The generic name for Accutane is *isotretinoin*.
**The generic name for Retin-A is *tretinoin topical*.
***For beta-carotene from supplements, 2 micrograms equal 1 μg RAE and for other vitamin A precursor carotenoids, 24 micrograms equal 1 μg RAE.

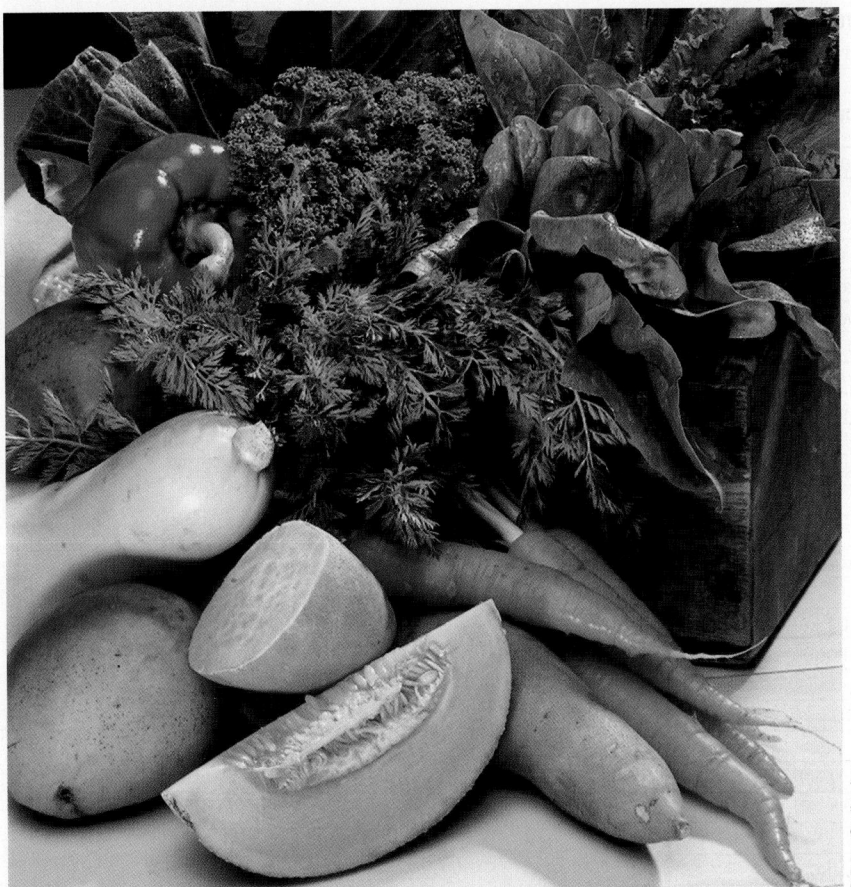

The carotenoids in foods bring colors to meals; the retinoids in our eyes allow us to see them.

Vitamin A in Foods The richest sources of the retinoids are foods derived from animals—liver, fish liver oils, milk and milk products, butter, and eggs. Because vitamin A is fat soluble, it is lost when milk is skimmed. To compensate, reduced-fat, low-fat, and fat-free milks are fortified so as to provide the amount found in whole milk. Margarine is usually fortified to provide the same amount of vitamin A as butter.

Plants contain no retinoids, but many vegetables and some fruits contain vitamin A precursors—the carotenoids. Only a few carotenoids have vitamin A activity; the carotenoid with the greatest vitamin A activity is beta-carotene. Beta-carotene is a rich, deep yellow, almost orange, compound. The beta-carotene in dark green, leafy vegetables is abundant, but masked by large amounts of the green pigment **chlorophyll**. Attractive meals that include colorful fruits and vegetables rich in beta-carotene are likely to provide vitamin A.[14]

The Colors of Vitamin A Foods Dark leafy greens (like broccoli and spinach—not celery or cabbage) and rich yellow or deep orange vegetables and fruits (such as cantaloupe, carrots, and sweet potatoes—not corn or bananas) help people meet their vitamin A needs (see Figure 11-7 on p. 350). A diet including several servings of such carotene-rich sources helps to ensure a sufficient intake.

Bright color is not always a sign of vitamin A activity, however. Beets and corn, for example, derive their colors from the red and yellow **xanthophylls**, which have no vitamin A activity. As for white plant foods such as potatoes, cauliflower, pasta, and rice, they also offer little or no vitamin A. Similarly, fast foods often lack vitamin A. Anyone who dines frequently on hamburgers, french fries, and colas is wise to emphasize colorful vegetables and fruits at other meals.

chlorophyll (KLO-row-fil): the green pigment of plants, which absorbs light and transfers the energy to other molecules, thereby initiating photosynthesis.

xanthophylls (ZAN-tho-fills): pigments found in plants responsible for the color changes seen in autumn leaves.

> FIGURE 11-7   Vitamin A in Selected Foods

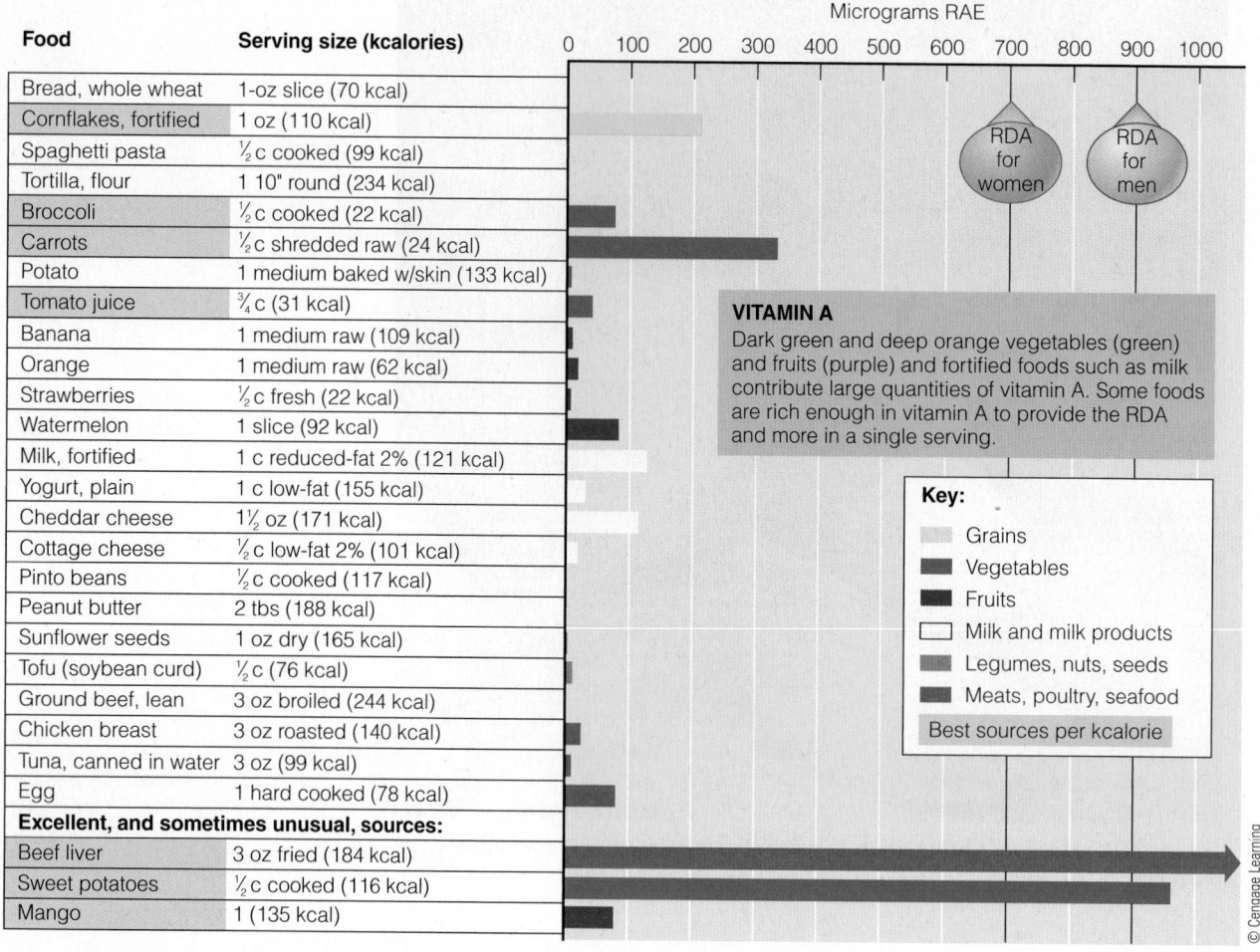

Vitamin A–Rich Liver People sometimes wonder if eating liver too frequently can cause vitamin A toxicity. Liver is a rich source because vitamin A is stored in the livers of animals, just as in humans.* Arctic explorers who have eaten large quantities of polar bear liver have become ill with symptoms suggesting vitamin A toxicity. Liver offers many nutrients, and eating it periodically may improve a person's nutrition status, but caution is warranted not to eat too much too often, especially for pregnant women. With 1 ounce of beef liver providing more than three times the RDA for vitamin A, intakes can rise quickly.

Golden Rice As mentioned earlier, vitamin A deficiency is a major problem in developing countries, impairing growth, causing blindness, and suppressing the immune system. In these developing regions of the world, fruits and vegetables are a scarcity, and rice, which contains no beta-carotene or vitamin A, is the staple food. Through biotechnology, scientists have been able to genetically modify rice to be a significant source of beta-carotene. Commonly called *golden rice* because of its yellowish tinge, this rice offers a promising solution to world malnutrition, but it also raises questions about the potential risks to the environment. More details are provided in Highlight 19's review of food biotechnology and Chapter 20's presentation of world hunger and possible solutions.

---

*The liver is not the only organ that stores vitamin A. The kidneys, adrenal glands, and other organs do, too, but the liver stores the most and is the most commonly eaten organ meat.

> **REVIEW IT** Identify the main roles, deficiency symptoms, and food sources for vitamin A.

Vitamin A is found in the body in three forms: retinol, retinal, and retinoic acid. Together, they are essential to vision, healthy epithelial tissues, and growth. Vitamin A deficiency is a major health problem worldwide, leading to infections, blindness, and keratinization. Toxicity can also cause problems and is most often associated with supplement abuse. Animal-derived foods such as liver and whole or fortified milk provide retinoids, whereas brightly colored plant-derived foods such as spinach, carrots, and pumpkins provide beta-carotene and other carotenoids. In addition to serving as a precursor for vitamin A, beta-carotene acts as an antioxidant in the body. The accompanying table provides a summary of vitamin A.

Vitamin A

Other Names	**Deficiency Disease**
Retinol, retinal, retinoic acid; precursors are carotenoids such as beta-carotene	Hypovitaminosis A
RDA	**Deficiency Symptoms**
Men: 900 µg RAE/day Women: 700 µg RAE/day	Night blindness, corneal drying (xerosis), triangular gray spots on eye (Bitot's spots), softening of the cornea (keratomalacia), and corneal degeneration and blindness (xerophthalmia); impaired immunity (infectious diseases); plugging of hair follicles with keratin, forming white lumps (hyperkeratosis)
UL	
Adults: 3000 µg/day	
Chief Functions in the Body	**Toxicity Disease**
Vision; maintenance of cornea, epithelial cells, mucous membranes, skin; bone and tooth growth; reproduction; immunity	Hypervitaminosis A[a]
	Chronic Toxicity Symptoms
Significant Sources	Increased activity of osteoclasts[b] causing reduced bone density; liver abnormalities; birth defects
Retinol: fortified milk, cheese, cream, butter, fortified margarine, eggs, liver	**Acute Toxicity Symptoms**
Beta-carotene: spinach and other dark green, leafy vegetables, broccoli, deep orange fruits (apricots, cantaloupe) and vegetables (squash, carrots, sweet potatoes, pumpkin)	Blurred vision, nausea, vomiting, vertigo; increase of pressure inside skull, mimicking brain tumor; headaches; muscle incoordination

[a]A related condition, *hypercarotenemia*, is caused by the accumulation of too much of the vitamin A precursor beta-carotene in the blood, which turns the skin noticeably yellow. Hypercarotenemia is not, strictly speaking, a toxicity symptom.
[b]*Osteoclasts* are the cells that destroy bone during its growth. Those that build bone are *osteoblasts*.

11.2 Vitamin D

> **LEARN IT** Identify the main roles, deficiency symptoms, and sources for vitamin D.

Vitamin D differs from the other nutrients in that the body can synthesize it, with the help of sunlight, from a precursor that the body makes from cholesterol. Therefore, vitamin D is not an essential nutrient; given enough time in the sun, people need no vitamin D from foods.

Also known as **calciferol**, vitamin D comes in two major forms.[15] **Vitamin D$_2$** derives primarily from plant foods in the diet. **Vitamin D$_3$** derives from animal foods in the diet and from synthesis in the skin. These two forms of vitamin D are similar and both must be activated before they can fully function.

Figure 11-8 (p. 352) diagrams the pathway for making and activating vitamin D in the body. To make vitamin D, ultraviolet rays from the sun hit a precursor in the skin and convert it to previtamin D$_3$, which is converted to vitamin D$_3$ with the help of the body's heat. To activate vitamin D—whether made in the body or consumed from the diet—two hydroxylation reactions must occur. First, the liver adds an OH group, and then the kidneys add another OH group to produce the active vitamin. As you might expect, diseases affecting either the liver or the kidneys can interfere with the activation of vitamin D and produce symptoms of deficiency.

Roles in the Body Though called a vitamin, the active form of vitamin D is actually a hormone—a compound manufactured by one part of the body that

calciferol (kal-SIF-er-ol): vitamin D.

vitamin D$_2$: vitamin D derived from plants in the diet; also called *ergocalciferol* (ER-go-kal-SIF-er-ol).

vitamin D$_3$: vitamin D derived from animals in the diet or made in the skin from 7-dehydrocholesterol, a precursor of cholesterol, with the help of sunlight; also called *cholecalciferol* (KO-lee-kal-SIF-er-ol) or *calciol*. After hydroxylation in the liver, calciol becomes *calcidiol* and after hydroxylation in the kidneys, calcidiol becomes *calcitriol*.

> **FIGURE 11-8** **Vitamin D Synthesis and Activation**

The final activation step in the kidneys is tightly regulated by hormones.

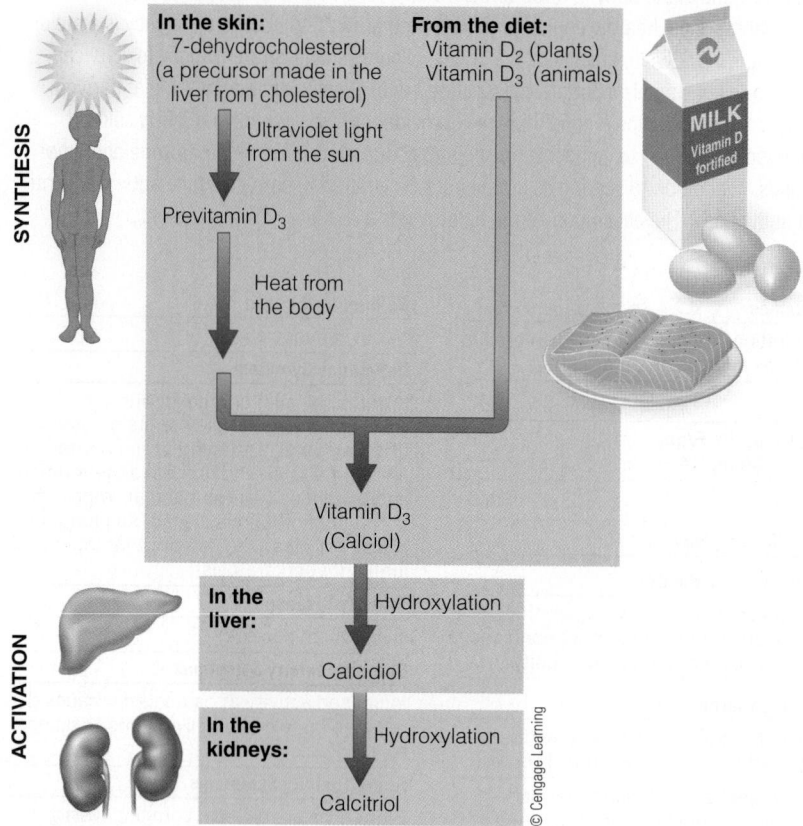

travels through the blood and causes another body part to respond. Like vitamin A, vitamin D has a binding protein that carries it to the target organs—most notably, the intestines, the kidneys, and the bones. All respond to vitamin D by making the minerals needed for bone growth and maintenance available.

Vitamin D in Bone Growth Vitamin D is a member of a large and cooperative bone-making and maintenance team composed of nutrients and other compounds, including vitamins A and K; the hormones parathyroid hormone and calcitonin; the protein collagen; and the minerals calcium, phosphorus, magnesium, and fluoride. Vitamin D's special role in bone health is to assist in the absorption of calcium and phosphorus, thus helping to maintain blood concentrations of these minerals. The bones grow denser and stronger as they absorb and deposit these minerals. Details of calcium balance and mineral deposition appear in Chapter 12, but here's a sneak preview: adequate nutrition and regular exercise are essential to achieving peak bone mass before age 30.

Vitamin D raises blood concentrations of bone minerals in three ways. When the diet is sufficient, vitamin D enhances mineral absorption from the GI tract. When the diet is insufficient, vitamin D provides the needed minerals from other sources: reabsorption by the kidneys and mobilization from the bones into the blood. The vitamin may work alone, as it does in the GI tract, or in combination with parathyroid hormone, as it does in the bones and kidneys.

Vitamin D in Other Roles Scientists have discovered many other tissues that respond to vitamin D, as the following examples describe. In the brain and nerve cells, vitamin D protects against cognitive decline and slows the progression of Parkinson disease.[16] Vitamin D in muscle cells encourages growth in children and preserves strength in adults.[17] Vitamin D signals cells of the immune system to defend against infectious diseases.[18] Vitamin D may also regulate the cells of the adipose tissue in ways that might influence the development of obesity.[19]

In many cases, vitamin D enhances or suppresses the activity of genes that regulate cell growth. As such, it may be valuable in treating a number of diseases. Recent research suggests that vitamin D may protect against metabolic syndrome, type 2 diabetes, tuberculosis, inflammation, multiple sclerosis, macular degeneration, hypertension, and some cancers.[20] Even so, evidence does not support vitamin D supplementation to improve health beyond correcting deficiencies.[21] In fact, some evidence suggests certain cancers are associated with both too little and too much vitamin D, making routine supplementation potentially harmful.[22]

Vitamin D Deficiency Overt signs of vitamin D deficiency are relatively rare, but vitamin D insufficiency is remarkably common.[23] Almost 10 percent of the US population is deficient and another 25 percent are marginal.[24] Factors that contribute to vitamin D deficiency include dark skin, breastfeeding without supplementation, lack of sunlight, and not using fortified milk. In vitamin D deficiency, production of **calbindin,** a protein that binds calcium in the intestinal cells, slows. Thus, even when calcium in the diet is adequate, it passes through the GI tract unabsorbed, leaving the bones undersupplied. Consequently, a vitamin D deficiency creates a calcium deficiency and increases the risks of several chronic diseases and osteoporosis. Vitamin D–deficient adolescents do not reach their peak bone mass.

Rickets Worldwide, the prevalence of the vitamin D–deficiency disease **rickets** is extremely high, affecting more than half of the children in countries such as China and Mongolia, and regions such as sub-Saharan Africa, the Middle East, and Latin America.[25] In the United States, rickets is not common, but when it occurs, black children and adolescents—especially females and overweight teens—are the ones most likely to be affected. To prevent rickets, the American Academy of Pediatrics recommends a supplement for all infants, children, and adolescents who do not receive enough vitamin D.

In rickets, the bones fail to calcify normally, causing growth retardation and skeletal abnormalities. The bones become so weak that they bend when they have to support the body's weight (see Figure 11-9). A child with rickets who is

> **FIGURE 11-9 Vitamin D–Deficiency Symptoms—Bowed Legs and Beaded Ribs of Rickets**

Bowed legs. In rickets, the poorly formed long bones of the legs bend outward as weight-bearing activities such as walking begin.

Beaded ribs. In rickets, a series of "beads" develop where the cartilages and bones attach.

calbindin: a calcium-binding transport protein that requires vitamin D for its synthesis.

rickets: the vitamin D–deficiency disease in children characterized by inadequate mineralization of bone (manifested in bowed legs or knock-knees, outward-bowed chest, and "beads" on ribs). A rare type of rickets, not caused by vitamin D deficiency, is known as *vitamin D–refractory rickets.*

old enough to walk characteristically develops bowed legs, often the most obvious sign of the disease. Another sign is the beaded ribs that result from the poorly formed attachments of the bones to the cartilage.*

Osteomalacia In adults, the poor mineralization of bone results in the painful bone disease **osteomalacia.** The bones become increasingly soft, flexible, brittle, and deformed.

Osteoporosis Any failure to synthesize adequate vitamin D or obtain enough from foods sets the stage for a loss of calcium from the bones, which can result in fractures. Highlight 12 describes the many factors that lead to osteoporosis, a condition of reduced bone density.

The Elderly Vitamin D deficiency is especially likely in older adults for several reasons. For one, the skin, liver, and kidneys lose their capacity to make and activate vitamin D with advancing age. For another, older adults typically drink little or no milk—the main dietary source of vitamin D. And finally, older adults typically spend much of the day indoors, and when they do venture outside, many of them cautiously wear protective clothing or apply sunscreen to all sun-exposed areas of their skin. Dark-skinned adults living in northern regions are particularly vulnerable. All of these factors increase the likelihood of vitamin D deficiency and its consequences: bone losses, osteoporotic fractures, and muscle weakness.[26] Vitamin D supplementation helps to raise blood levels, reduce bone loss, improve muscle performance, and lower the risks of falls and fractures in elderly persons.[27]

Vitamin D Toxicity
Vitamin D clearly illustrates how nutrients in optimal amounts support health, but both inadequacies and excesses create harm. Vitamin D is among the most likely of the vitamins to have toxic effects when consumed in excessive amounts. The amounts of vitamin D made by the skin and found in foods are well within the safe limits set by the UL, but supplements containing the vitamin in concentrated form should be kept out of the reach of children and used cautiously by adults.

Excess vitamin D raises the concentration of blood calcium.** Excess blood calcium tends to precipitate in the soft tissue, forming stones, especially in the kidneys where calcium is concentrated in an effort to excrete it. Calcification may also harden the blood vessels and is especially dangerous in the major arteries of the brain, heart, and lungs, where it can cause death.

Vitamin D Recommendations and Sources
Only a few foods contain vitamin D naturally. Fortunately, the body can make vitamin D with the help of a little sunshine. In setting dietary recommendations, however, the DRI Committee assumed that no vitamin D was available from skin synthesis. In order to reach sufficient levels of vitamin D in the blood without contributions from the sun, dietary recommendations were recently increased.[28] Some research suggests that vitamin D recommendations should be higher still.[29]

Vitamin D in Foods The *Dietary Guidelines* advise consumers to drink at least 2 cups of vitamin D–fortified milk a day. The fortification of milk and other foods with vitamin D is the best guarantee that people will meet their needs.[30]*** Consumers using alternatives such as soy milk or almond milk need to read labels carefully to ensure they are getting vitamin D-fortified products. Despite vitamin D fortification, the average intake in the United States falls short of recommendations. Egg yolks and oily fish such as salmon, mackerel, and sardines are the best natural sources of vitamin D.

Meeting vitamin D needs is difficult without adequate sunshine, fortification, or supplementation. Importantly, feeding infants and young children

A cold glass of milk refreshes as it replenishes vitamin D and other bone-building nutrients.

© Westend61 GmbH/Alamy

---

osteomalacia (OS-tee-oh-ma-LAY-shuh): a bone disease characterized by softening of the bones. Symptoms include bending of the spine and bowing of the legs. The disease occurs most in adult women.

- **osteo** = bone
- **malacia** = softening

---

*Because the poorly formed rib attachments resemble rosary beads, this symptom is commonly known as *rachitic* (ra-KIT-ik) *rosary* ("the rosary of rickets").

**High blood calcium is known as *hypercalcemia* and may develop from a variety of disorders, including vitamin D toxicity. It does *not* develop from too much calcium in the diet.

***Vitamin D fortification of milk in the United States is 10 micrograms per quart.

> FIGURE 11-10   **Vitamin D Synthesis and Latitude**

Above 40° north latitude (and below 40° south latitude in the southern hemisphere), vitamin D synthesis essentially ceases for the 4 months of winter. Synthesis increases as spring approaches, peaks in summer, and declines again in the fall. People living in regions of extreme northern (or extreme southern) latitudes may miss as much as 6 months of vitamin D production.

nonfortified "health beverages" instead of milk or infant formula can create severe nutrient deficiencies, including rickets.

Vitamin D from the Sun   Most of the world's population relies on natural exposure to sunlight to maintain adequate vitamin D nutrition. The sun imposes no risk of vitamin D toxicity; prolonged exposure to sunlight degrades the vitamin D precursor in the skin, preventing its conversion to the active vitamin.

Prolonged exposure to sunlight can, however, prematurely wrinkle the skin and cause skin cancer. Sunscreens help reduce these risks, but sunscreens with a sun protection factor (SPF) of 8 and higher can also reduce vitamin D synthesis. Still, even with an SPF 15 to 30 sunscreen, sufficient vitamin D synthesis can be obtained in 10 to 20 minutes of sun exposure. Alternatively, a person could apply sunscreen after enough time has elapsed to provide sufficient vitamin D synthesis. For most people, exposing hands, face, and arms on a clear summer day for 5 to 10 minutes two or three times a week should be sufficient to maintain vitamin D nutrition.

The pigments of dark skin provide some protection from the sun's damage, but they also reduce vitamin D synthesis. Dark-skinned people require more sunlight exposure than light-skinned people—perhaps as much as 4 to 6 times longer.[31] Latitude, season, and time of day also have dramatic effects on vitamin D synthesis and status (see Figure 11-10). Heavy cloud cover, smoke, or smog block the ultraviolet (UV) rays of the sun that promote vitamin D synthesis. People who stay in the shade and wear long-sleeved clothing are twice as likely to develop vitamin D deficiency as those who rarely do so.[32] Vitamin D deficiency is especially prevalent in the winter and in the Arctic and Antarctic regions of the world.[33] To ensure an adequate vitamin D status, supplements may be needed. The body's vitamin D supplies from summer synthesis alone are insufficient to meet winter needs.[34]

> **DIETARY GUIDELINES FOR AMERICANS**

Choose foods that provide more vitamin D, a nutrient of concern in American diets. Most dietary vitamin D derives from fortified foods such as milk, yogurt, and breakfast cereals; natural dietary sources of vitamin D include oily fish and egg yolks.

Depending on the radiation used, the UV rays from tanning lamps and tanning beds may also stimulate vitamin D synthesis and increase bone density. The potential

The sunshine vitamin—vitamin D.

The Fat-Soluble Vitamins: A, D, E, and K   355

hazards of skin damage, however, may outweigh any possible benefits.* The Food and Drug Administration (FDA) warns that if the lamps are not properly filtered, people using tanning booths risk burns, damage to the eyes and blood vessels, and skin cancer.

Vitamin D from Supplements As mentioned, some people may benefit from taking vitamin D supplements. Vitamin D can be found in multivitamin-mineral supplements as well as a high-dose single supplement. As a single supplement, vitamin D_3 is less expensive, more commonly available, and more effective than vitamin D_2.[35] Taking vitamin D supplements with the largest meal of the day improves absorption, resulting in a 50 percent increase in blood levels.[36]

> **> REVIEW IT** Identify the main roles, deficiency symptoms, and sources for vitamin D. Vitamin D can be synthesized in the body with the help of sunlight or obtained from some foods, most notably fortified milk. Vitamin D sends signals to three primary target sites: the GI tract to absorb more calcium and phosphorus, the bones to release more, and the kidneys to retain more. These actions maintain blood calcium concentrations and support bone formation. A deficiency causes rickets in childhood and osteomalacia in later life. The accompanying table provides a summary of vitamin D.

Vitamin D

Other Names

calciferol (vitamin D)

ergocalciferol (vitamin D₂): vitamin D derived from plants in the diet and made from the yeast and plant sterol ergosterol.

cholecalciferol (vitamin D₃ or calciol): vitamin D derived from animals in the diet or made in the skin from 7-dehydrocholesterol, a precursor of cholesterol, with the help of sunlight.

calcidiol (25-hydroxyvitamin D): vitamin D found in the blood that is made from the hydroxylation of calciol in the liver.

calcitriol (1,25-dihydroxyvitamin D): vitamin D that is made from the hydroxylation of calcidiol in the kidneys; the biologically active hormone, sometimes called *active vitamin D.*

RDA

Adults: 15 µg/day or 600 IU/day (19–70 yr)
　　　　20 µg/day or 800 IU/day (>70 yr)

UL

Adults: 100 µg/day or 4000 IU/day

Chief Functions in the Body

Mineralization of bones (raises blood calcium and phosphorus by increasing absorption from digestive tract, withdrawing calcium from bones, stimulating retention by kidneys)

Significant Sources

Synthesized in the body with the help of sunlight; fortified milk, margarine, butter, juices, cereals, and chocolate mixes; veal, beef, egg yolks, liver, fatty fish (herring, salmon, sardines) and their oils

Deficiency Diseases

Rickets, osteomalacia

Deficiency Symptoms

Rickets in children:

Inadequate calcification, resulting in misshapen bones (bowing of legs); enlargement of ends of long bones (knees, wrists); deformities of ribs (bowed, with beads or knobs);[a] delayed closing of fontanel, resulting in rapid enlargement of head (see figure below); lax muscles resulting in protrusion of abdomen; muscle spasms

Osteomalacia or osteoporosis in adults:

Loss of calcium, resulting in soft, flexible, brittle, and deformed bones; progressive weakness; pain in pelvis, lower back, and legs

Toxicity Disease

Hypervitaminosis D

Toxicity Symptoms

Elevated blood calcium; calcification of soft tissues (blood vessels, kidneys, heart, lungs, tissues around joints)

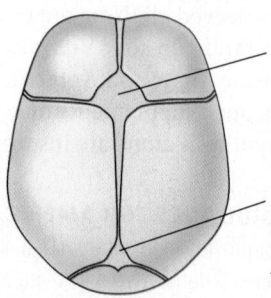

Fontanel
A fontanel is an open space in the top of a baby's skull before the bones have grown together. In rickets, closing of the fontanel is delayed.

Anterior fontanel normally closes by the end of the second year.

Posterior fontanel normally closes by the end of the first year.

© Cengage Learning

[a]Bowing of the ribs causes the symptoms known as *pigeon breast*. The beads that form on the ribs resemble rosary beads; thus this symptom is known as *rachitic* (ra-KIT-ik) *rosary* ("the rosary of rickets").

---

*The best wavelengths for vitamin D synthesis are UV-B rays between 290 and 310 nanometers. Some tanning parlors advertise "UV-A rays only, for a tan without the burn," but UV-A rays can damage the skin.

11.3 Vitamin E

> LEARN IT Identify the main roles, deficiency symptoms, and food sources for vitamin E.

The vitamin E family consists of two subgroups—the **tocopherols** and the **tocotrienols**—each containing four members designated by letters of the Greek alphabet (alpha, beta, gamma, and delta). All consist of a complex ring structure with a long saturated (in tocopherols) or unsaturated (in tocotrienols) side chain. The positions of methyl groups (CH_3) on the side chain and their chemical rotations distinguish the four members within each subgroup. (Appendix C provides the chemical structures.)

Of all the members of the vitamin E family, only **alpha-tocopherol** is maintained in the body and can meet the body's needs for the vitamin. The others are not converted to alpha-tocopherol in the body, nor are they recognized by its transport protein. For these reasons, the RDA is based only on alpha-tocopherol.

Most vitamin E research has focused on alpha-tocopherol, but recent studies suggest that the other tocopherols and tocotrienols might also be beneficial. For example, gamma-tocopherol and possibly delta-tocopherol, appear to be most effective in inhibiting inflammation and cancer growth.[37] In addition to preventing cancer, tocotrienols may also protect against osteoporosis, diabetes, heart disease, and neurological disorders.[38]

Vitamin E as an Antioxidant

Vitamin E is a fat-soluble antioxidant and one of the body's primary defenders against the adverse effects of free radicals. Its main action is to stop the chain reaction of free radicals from producing more free radicals (see Highlight 11). In doing so, vitamin E protects the vulnerable components of the cells and their membranes from destruction. Most notably, vitamin E prevents the oxidation of the polyunsaturated fatty acids, but it protects other lipids and related compounds (for example, vitamin A) as well.

Accumulating evidence suggests that vitamin E may reduce the risk of heart disease by protecting low-density lipoproteins (LDL) against oxidation and reducing inflammation. The oxidation of LDL and inflammation have been implicated as key factors in the development of heart disease. Highlight 11 explains how vitamin E and other antioxidants might protect against chronic diseases, such as heart disease and cancer, and explores whether foods or supplements might be most helpful—or harmful.

Vitamin E Deficiency

A primary deficiency of vitamin E (from poor dietary intake) is rare; deficiency is usually associated with diseases of fat malabsorption such as cystic fibrosis. Without vitamin E, the red blood cells break and spill their contents, probably because of oxidation of the polyunsaturated fatty acids in their membranes. This classic sign of vitamin E deficiency, known as **erythrocyte hemolysis**, is seen in premature infants born before the transfer of vitamin E from the mother to the infant that takes place in the last weeks of pregnancy. Vitamin E treatment corrects **hemolytic anemia.**

Prolonged vitamin E deficiency, as can occur with some genetic disorders, also causes neuromuscular dysfunction.[39] Common symptoms include loss of muscle coordination and reflexes and impaired vision and speech. Vitamin E treatment helps to correct these neurological symptoms of vitamin E deficiency.

Two other conditions seem to respond to vitamin E treatment, although results are inconsistent. One is **fibrocystic breast disease,** a nonmalignant breast disease. The other is **intermittent claudication,** an abnormality of blood flow that causes cramping in the legs.

Vitamin E Toxicity

Vitamin E supplement use has risen in recent years as its protective actions against chronic diseases have been recognized. Fortunately, the liver carefully regulates vitamin E concentrations. Toxicity is rare, and vitamin E appears safe across a broad range of intakes. The UL for vitamin E (1000 milligrams)

tocopherols (tuh-KOFF-uh-rawls): members of the vitamin E family having the chemical structure of a complex ring structure with a long saturated side chain. (See Appendix C for chemical structures.)

tocotrienols (TOE-koh-try-EE-nawls): members of the vitamin E family having the chemical structure of a complex ring structure with a long unsaturated side chain. (See Appendix C for chemical structures.)

alpha-tocopherol: the active vitamin E compound.

erythrocyte (eh-RITH-ro-cite) **hemolysis** (he-MOLL-uh-sis): the breaking open of red blood cells (erythrocytes); a symptom of vitamin E–deficiency disease in human beings.

- **erythro** = red
- **cyte** = cell
- **hemo** = blood
- **lysis** = breaking

hemolytic (HE-moh-LIT-ick) **anemia:** the condition of having too few red blood cells as a result of erythrocyte hemolysis.

fibrocystic (FYE-bro-SIS-tik) **breast disease:** a harmless condition in which the breasts develop lumps, sometimes associated with caffeine consumption. In some, it responds to abstinence from caffeine; in others, it can be treated with vitamin E.

- **fibro** = fibrous tissue
- **cyst** = closed sac

intermittent claudication (klaw-dih-KAY-shun): severe calf pain caused by inadequate blood supply. It occurs when walking and subsides during rest.

- **intermittent** = at intervals
- **claudicare** = to limp

Fat-soluble vitamin E is found predominantly in vegetable oils, seeds, and nuts.

is more than 65 times greater than the recommended intake for adults (15 milligrams). Extremely high doses of vitamin E may interfere with the blood-clotting action of vitamin K and enhance the effects of drugs used to oppose blood clotting, causing hemorrhage.

Vitamin E Recommendations The RDA for vitamin E is based on the alpha-tocopherol form only. As mentioned earlier, the other tocopherols and tocotrienols cannot be converted to alpha-tocopherol, nor do they perform the same metabolic roles in the body. A person who consumes large quantities of polyunsaturated fatty acids needs more vitamin E. Fortunately, vitamin E and polyunsaturated fatty acids tend to occur together in the same foods.

Vitamin E in Foods Vitamin E is widespread in foods. Much of the vitamin E in the diet comes from vegetable oils and products made from them, such as margarine and salad dressings. Wheat germ oil is especially rich in vitamin E.

Because vitamin E is readily destroyed by heat and oxidation, fresh foods are preferable sources. Most processed and convenience foods do not contribute enough vitamin E to ensure an adequate intake.

> **REVIEW IT** Identify the main roles, deficiency symptoms, and foods sources for vitamin E.

Vitamin E acts as an antioxidant, defending lipids and other components of the cells against oxidative damage. Deficiencies are rare, but they do occur in premature infants, the primary symptom being erythrocyte hemolysis. Vitamin E is found predominantly in vegetable oils and appears to be one of the least toxic of the fat-soluble vitamins. The accompanying table provides a summary of vitamin E.

Vitamin E

Other Names	Significant Sources
Alpha-tocopherol	Polyunsaturated plant oils (margarine, salad dressings), dark green, leafy vegetables (spinach, turnip greens, collard greens, broccoli), wheat germ, whole grains, liver, egg yolks, nuts, seeds, fatty meats
RDA	
Adults: 15 mg/day	Easily destroyed by heat and oxygen
UL	**Deficiency Symptoms**
Adults: 1000 mg/day	Red blood cell breakage,[a] nerve damage
Chief Functions in the Body	**Toxicity Symptoms**
Antioxidant (stabilization of cell membranes, regulation of oxidation reactions, protection of polyunsaturated fatty acids [PUFA] and vitamin A)	Augments the effects of anticlotting medication

[a]The breaking of red blood cells is called *erythrocyte hemolysis*.

11.4 Vitamin K

> **LEARN IT** Identify the main roles, deficiency symptoms, and sources for vitamin K.

Vitamin K appropriately gets its name from the Danish word *koagulation* ("coagulation" or "clotting"). Its primary action is blood clotting, where its presence can make the difference between life and death. Blood has a remarkable ability to remain liquid, but it can clot within seconds when the integrity of that system is disturbed.

Roles in the Body More than a dozen different proteins and the mineral calcium are involved in making a blood clot. Vitamin K is essential for the activation of several of these proteins, among them prothrombin, made by the

> FIGURE 11-11 **Blood-Clotting Process**

Vitamin K is essential for the synthesis of prothrombin and several other clotting factors. Blood clots are formed by a cascade of reactions, with each step creating a compound that activates the next step.

liver as a precursor of the protein thrombin (see Figure 11-11). When any of the blood-clotting factors is lacking, **hemorrhagic disease** results. If an artery or vein is cut or broken, bleeding goes unchecked. Of course, this is not to say that hemorrhaging is always caused by vitamin K deficiency. Another cause is the genetic disorder **hemophilia,** which is neither caused nor cured by vitamin K.

Vitamin K also participates in the metabolism of bone proteins, most notably **osteocalcin.** Without vitamin K, osteocalcin cannot bind to the minerals that normally form bones, resulting in low bone density.* An adequate intake of vitamin K helps to decrease bone turnover and protect against fractures. The effectiveness of vitamin K supplements on bone health is inconclusive.[40]

Vitamin K is historically known for its role in blood clotting, and more recently for its participation in bone building, but researchers continue to discover proteins needing vitamin K's assistance. These proteins have been identified in the plaques of atherosclerosis, the kidneys, and the nervous system.

Vitamin K Deficiency Chapter 1 explains that a *primary deficiency* develops in response to an inadequate dietary intake whereas a *secondary deficiency* occurs for other reasons. A primary deficiency of vitamin K is rare, but a secondary deficiency may occur in two circumstances. First, whenever fat absorption falters, as occurs when bile production fails, vitamin K absorption diminishes. Second, some drugs disrupt vitamin K's synthesis and action in the body: antibiotics kill the vitamin K–producing bacteria in the intestine, and anticoagulant drugs interfere with vitamin K metabolism and activity. Excessive bleeding due to a vitamin K deficiency can be fatal.

Newborn infants present a unique case of vitamin K nutrition because they are born with a **sterile** intestinal tract, and the vitamin K–producing bacteria take weeks to establish themselves. Furthermore, vitamin K is minimally transported across the placenta and its concentration in breast milk is low. At the same time, plasma prothrombin concentrations are low, which reduces the likelihood of fatal blood clotting during the stress of birth. To prevent hemorrhagic disease in the newborn, a single dose of vitamin K is given at birth by intramuscular injection.[41] Concerns that vitamin K given at birth raises the risks of childhood cancer are unfounded.

Vitamin K Toxicity Toxicity is not common, and no adverse effects have been reported with high intakes of vitamin K. Therefore, a UL has not been established. High doses of vitamin K can, however, reduce the effectiveness of

Soon after birth, newborn infants receive a dose of vitamin K to prevent hemorrhagic disease.

hemorrhagic (hem-oh-RAJ-ik) **disease:** a disease characterized by excessive bleeding.

hemophilia (HE-moh-FEEL-ee-ah): a hereditary disease in which the blood is unable to clot because it lacks the ability to synthesize certain clotting factors.

osteocalcin (os-teo-KAL-sen): a calcium-binding protein in bones, essential for normal mineralization.

sterile: free of microorganisms, such as bacteria.

*Vitamin K is a cofactor for a carboxylase enzyme. When vitamin K is inadequate, osteocalcin is undercarboxylated and therefore less effective in binding calcium.

Notable food sources of vitamin K include green vegetables such as collards, spinach, bib lettuce, brussels sprouts, and cabbage and vegetable oils such as soybean oil and canola oil.

anticoagulant drugs used to prevent blood clotting. People taking these drugs can continue eating their usual diets. Their blood clotting times should be monitored closely and drug dosages adjusted accordingly.

Vitamin K Recommendations and Sources

Like vitamin D, vitamin K can be obtained both from foods and from a nonfood source. Bacteria in the GI tract synthesize vitamin K, although the amount is insufficient to meet the body's needs and its bioavailability is limited. Therefore the diet must also supply vitamin K, which is found primarily in leafy green vegetables such as spinach and kale, fruits such as avocado and kiwi, and some vegetable oils such as soybean oil. Naturally occurring vitamin K in foods is **phylloquinone** (sometimes called vitamin K_1), whereas vitamin K produced by GI bacteria is **menaquinone** (sometimes called vitamin K_2).

> **REVIEW IT** Identify the main roles, deficiency symptoms, and sources for vitamin K. Vitamin K helps with blood clotting, and its deficiency causes hemorrhagic disease (uncontrolled bleeding). Bacteria in the GI tract can make the vitamin; people typically receive about half of their requirements from bacterial synthesis and half from foods such as green vegetables and vegetable oils. Because people depend on bacterial synthesis for vitamin K, deficiency is most likely in newborn infants and in people taking antibiotics. The accompanying table provides a summary of vitamin K.

Vitamin K

Other Names	Significant Sources
Phylloquinone (vitamin K_1), menaquinone (vitamin K_2), menadione (in supplements)	Bacterial synthesis in the digestive tract;[a] liver; dark green, leafy vegetables, cabbage-type vegetables; milk

AI	Deficiency Symptoms
Men: 120 µg/day	Hemorrhaging
Women: 90 µg/day	

Chief Functions in the Body	Toxicity Symptoms
Synthesis of blood-clotting proteins and bone proteins	None known

[a]Vitamin K needs cannot be met from bacterial synthesis alone.

The four fat-soluble vitamins play many specific roles in the growth and maintenance of the body. Their presence affects the health and function of the eyes, skin, GI tract, lungs, bones, teeth, nervous system, and blood; their deficiencies become apparent in these same areas. Toxicities of the fat-soluble vitamins are possible, especially when people use supplements, because the body stores excesses.

As with the water-soluble vitamins, the function of one fat-soluble vitamin often depends on the presence of another. Recall that vitamin E protects vitamin A from oxidation. In vitamin E deficiency, vitamin A absorption and storage are impaired. Three of the four fat-soluble vitamins—A, D, and K—play important roles in bone growth and remodeling. As mentioned, vitamin K helps synthesize a specific bone protein, and vitamin D regulates that synthesis. Vitamin A, in turn, may control which bone-building genes respond to vitamin D. Vitamin E and vitamin K share some metabolic pathways, which can create problems, especially in blood clotting.

Fat-soluble vitamins also interact with minerals. Vitamin D and calcium cooperate in bone formation, and zinc is required for the synthesis of vitamin A's transport protein, retinol-binding protein. Zinc also assists the enzyme that regenerates retinal from retinol in the eye. Vitamin A deficiency and iron deficiency often occur together and each seems to interfere with the other's metabolism.

phylloquinone (fill-oh-KWYN-own): the plant form of vitamin K; also called *vitamin K_1*.

menaquinone (men-ah-KWYN-own): the bacteria-produced form of vitamin K; also called *vitamin K_2*.

© Matthew Farruggio

The roles of the fat-soluble vitamins differ from those of the water-soluble vitamins, and they appear in different foods—yet they are just as essential to life. The need for them underlines the importance of eating a wide variety of nourishing foods daily. The accompanying table provides a summary of the fat-soluble vitamins.

› REVIEW IT The Fat-Soluble Vitamins

Vitamin and Chief Functions	Deficiency Symptoms	Toxicity Symptoms	Significant Sources
Vitamin A Vision; maintenance of cornea, epithelial cells, mucous membranes, skin; bone and tooth growth; reproduction; immunity	Infectious diseases, night blindness, blindness (xerophthalmia), keratinization	Reduced bone mineral density, liver abnormalities, birth defects	Retinol: milk and milk products Beta-carotene: dark green, leafy and deep yellow/orange vegetables
Vitamin D Mineralization of bones (raises blood calcium and phosphorus by increasing absorption from digestive tract, withdrawing calcium from bones, stimulating retention by kidneys)	Rickets, osteomalacia	Calcium imbalance (calcification of soft tissues and formation of stones)	Synthesized in the body with the help of sunshine; fortified milk
Vitamin E Antioxidant (stabilization of cell membranes, regulation of oxidation reactions, protection of polyunsaturated fatty acids [PUFA] and vitamin A)	Erythrocyte hemolysis, nerve damage	Hemorrhagic effects	Vegetable oils
Vitamin K Synthesis of blood-clotting proteins and bone proteins	Hemorrhage	None known	Synthesized in the body by GI bacteria; dark green, leafy vegetables

Nutrition Portfolio

For the fat-soluble vitamins, select colorful fruits and vegetables, fortified milk or soy products, and vegetable oils; use supplements with caution, if at all. Go to Diet & Wellness Plus and choose one of the days on which you tracked your diet for an entire day. Select the MyPlate report and then consider the following questions:

- How was your overall intake in the vegetable group? Do you need improvement in this area? If so, what are some changes you could make?

Now look at the report titled Intake Spreadsheet to answer the following questions:

- Examine your weekly choices of vegetables and evaluate whether you meet the recommendations for dark green or orange and deep yellow vegetables.
- Consider whether you drink enough vitamin D–fortified milk or go outside in the sunshine regularly.
- Describe the vegetable oils you use when you cook and their vitamin contributions.

DIET & WELLNESS PLUS+ To complete this exercise, go to your Diet & Wellness Plus at www.cengagebrain.com.

› **STUDY IT** To review the key points of this chapter and take a practice quiz, go to the study cards at the end of the book.

REFERENCES

1. G. Tang, Bioconversion of dietary provitamin A carotenoids to vitamin A in humans, *American Journal of Clinical Nutrition* 91 (2010): 1468S–1473S.

2. J. von Lintig, Colors with functions: Elucidating the biochemical and molecular basis of carotenoid metabolism, *Annual Review of Nutrition* 30 (2010): 35–56.

3. J. C. Saari, Vitamin A metabolism in rod and cone visual cycles, *Annual Review of Nutrition* 32 (2012): 125–145; J. von Lintig, Metabolism of carotenoids and retinoids related to vision, *Journal of Biological Chemistry* 287 (2012): 1627–1634.

4. W. Stahl and H. Sies, β-Carotene and other carotenoids in protection from sunlight, *American Journal of Clinical Nutrition* 96 (2012): 1179S–1184S.

5. M. Clagett-Dame and D. Knutson, Vitamin A in reproduction and development, *Nutrients* 3 (2011): 385–428; N. Noy, Between death and survival: Retinoic acid in regulation of apoptosis, *Annual Review of Nutrition* 30 (2010): 201–217.

6. J. von Lintig, Provitamin A metabolism and functions in mammalian biology, *American Journal of Clinical Nutrition* 96 (2012): 1234S–1244S.

7. A. Sommer and K. S. Vyas, A global clinical view on vitamin A and carotenoids, *American Journal of Clinical Nutrition* 96 (2012): 1204S–1206S.

8. J. C. Sherwin and coauthors, Epidemiology of vitamin A deficiency and xerophthalmia in at-risk populations, *Transactions of the Royal Society of Tropical Medicine and Hygiene* 106 (2012): 205–214; E. Mayo-Wilson and coauthors, Vitamin A supplements for preventing mortality, illness, and blindness in children aged under 5: Systematic review and meta-analysis, *British Medical Journal* 343 (2011): d5094.

9. A. C. Ross, Vitamin A and retinoic acid in T cell-related immunity, *American Journal of Clinical Nutrition* 96 (2012): 1166S–1172S.

10. World Health Organization, Measles fact sheet, February 2013, www.who.int/mediacentre/factsheets/fs286/en/index.html.

11. C. R. Sudfeld, A. M. Navar, and M. A. Halsey, Effectiveness of measles vaccination and vitamin A treatment, *International Journal of Epidemiology* 39 (2010): i48–i55.

12. S. A. Tanumihardjo, Vitamin A and bone health: The balancing act, *Journal of Clinical Densitometry* 16 (2013): 414–419; H. Ahmadieh and A. Arabi, Vitamins and bone health: Beyond calcium and vitamin D, *Nutrition Reviews* 69 (2011): 584–598.

13. M. M. G. Ackermans and coauthors, Vitamin A and clefting: Putative biological mechanisms, *Nutrition Reviews* 69 (2011): 613–624.

14. M. J. Haskell, The challenge to reach nutritional adequacy for vitamin A: β-carotene bioavailability and conversion—evidence in humans, *American Journal of Clinical Nutrition* 96 (2012): 1193S–1203S.

15. G. Jones, Extrarenal vitamin D activation and interactions between vitamin D_2, vitamin D_3, and vitamin D analogs, *Annual Review of Nutrition* 33 (2013): 23–44.

16. X. Cui and coauthors, Low vitamin D concentration exacerbates adult brain dysfunction, *American Journal of Clinical Nutrition* 97 (2013): 907–908; D. J. Llewellyn and coauthors, Vitamin D and risk of cognitive decline in elderly persons, *Archives of Internal Medicine* 170 (2010): 1135–1141.

17. A. S. Grimaldi and coauthors, 25(OH) Vitamin D is associated with greater muscle strength in healthy men and women, *Medicine and Science in Sports and Exercise* 45 (2013): 157–162; T. J. Hazell, J. R. DeGuire, and H. A. Weiler, Vitamin D: An overview of its role in skeletal muscle physiology in children and adolescents, *Nutrition Reviews* 70 (2012): 520–533.

18. I. Laaksi, Vitamins, infectious and chronic disease during adulthood and aging: Vitamin D and respiratory infection in adults, *Proceedings of the Nutrition Society* 71 (2012): 90–97; M. E. Sundaram and L. A. Coleman, Vitamin D and influenza, *Advances in Nutrition: An International Review Journal* 3 (2012): 517–525; M. Hewison, Vitamin D and innate and adaptive immunity, *Vitamins and Hormones* 86 (2011): 23–62; F. Baeke and coauthors, Human T lymphocytes are direct targets of 1,25-dihydroxyvitamin D3 in the immune system, *Journal of Steroid Biochemistry and Molecular Biology* 121 (2010): 221–227.

19. C. P. Earthman and coauthors, The link between obesity and low circulating 25-hydroxyvitamin D concentrations: Considerations and implications, *International Journal of Obesity* 36 (2012): 387–396; C. Ding and coauthors, Vitamin D signalling in adipose tissue, *British Journal of Nutrition* 108 (2012): 1915–1923.

20. S. Lim and coauthors, Association of vitamin D deficiency with incidence of type 2 diabetes in high-risk Asian subjects, *American Journal of Clinical Nutrition* 97 (2013): 524–530; T. D. Cheng and coauthors, Vitamin D intake and lung cancer risk in the Women's Health Initiative, *American Journal of Clinical Nutrition* 98 (2013): 1002–1011; F. M. Yousef and coauthors, Vitamin D status and breast cancer in Saudi Arabian women: Case-control study, *American Journal of Clinical Nutrition* 98 (2013): 105–110; R. E. Stubbins, A. Hakeem, and N. P. Núñez, Using components of the vitamin D pathway to prevent and treat colon cancer, *Nutrition Reviews* 70 (2012): 721–729; G. J. Fung and coauthors, Vitamin D intake is inversely related to risk of developing metabolic syndrome in African American and white men and women over 20 y: The Coronary Artery Risk Development in Young Adults study, *American Journal of Clinical Nutrition* 96 (2012): 24–29; V. Ganji and coauthors, Serum 25-hydroxyvitamin D concentrations are associated with prevalence of metabolic syndrome and various cardiometabolic risk factors in US children and adolescents based on assay-adjusted serum 25-hydroxyvitamin D data from NHANES 2001–2006, *American Journal of Clinical Nutrition* 94 (2011): 225–233; S. A. Chacko and coauthors, Serum 25-hydroxyvitamin D concentrations in relation to cardiometabolic risk factors and metabolic syndrome in postmenopausal women, *American Journal of Clinical Nutrition* 94 (2011): 209–217; J. Mitri, M. D. Muraru, and A. G. Pittas, Vitamin D and type 2 diabetes: A systematic review, *European Journal of Clinical Nutrition* 65 (2011): 1005–1015; E. M. Mowry, Vitamin D: Evidence for its role as a prognostic factor in multiple sclerosis, *Journal of Neurological Science* 311 (2011): 19–22; M. H. Hopkins and coauthors, Effects of supplemental vitamin D and calcium on biomarkers of inflammation in colorectal adenoma patients: A randomized, controlled clinical trial, *Cancer Prevention Research* 4 (2011): 1645–1654; K. Luong and L. T. Nguyen, Impact of vitamin D in the treatment of tuberculosis, *American Journal of Medical Sciences* 341 (2011): 493–498; A. E. Millen and coauthors, Vitamin D status and early age-related macular degeneration in postmenopausal women, *Archives of Ophthalmology* 129 (2011): 481–489; Y. Ma and coauthors, Association between vitamin D and risk of colorectal cancer: A systematic review of prospective studies, *Journal of Clinical Oncology* 29 (2011): 3775–3782; L. N. Anderson and coauthors, Vitamin D and calcium intakes and breast cancer risk in pre- and postmenopausal women, *American Journal of Clinical Nutrition* 91 (2010): 1699–1707; N. Parekh, Protective role of vitamin D against age-related macular degeneration: A hypothesis, *Topics in Clinical Nutrition* 25 (2010): 290–301; A. G. Pittas and coauthors, Systematic review: Vitamin D and cardiometabolic outcomes, *Annals of Internal Medicine* 152 (2010): 307–314; C. D. Toner, C. D. Davis, and J. A. Milner, The vitamin D and cancer conundrum: Aiming at a moving target, *Journal of the American Dietetic Association* 110 (2010): 1492–1500.

21. R. Jorde and G. Grimmes, Vitamin D and metabolic health with special reference to the effect of vitamin D on serum lipids, *Progress in Lipid Research* 50 (2011): 303–312; A. Grey and M. Bolland, Vitamin D: A place in the sun? *Archives of Internal Medicine* 170 (2010): 1099–1100.

22. C. D. Davis and J. A. Milner, Nutrigenomics, vitamin D and cancer prevention, *Journal of Nutrigenetics and Nutrigenomics* 4 (2011): 1–11; K. Michaëlsson and coauthors, Plasma vitamin D and mortality in older men: A community-based prospective cohort study, *American Journal of Clinical Nutrition* 92 (2010): 841–848.

23. C. J. Rosen, Vitamin D insufficiency, *New England Journal of Medicine* 364 (2011): 248–254; A. A. Ginde , M. C. Liu, and C. A. Camargo, Demographic differences and trends of vitamin D insufficiency in the US population, 1988–2004, *Archives of Internal Medicine* 169 (2009): 626–632; S. A. Bowden and coauthors, Prevalence of vitamin D deficiency and insufficiency in children with osteopenia or osteoporosis referred to a pediatric metabolic bone clinic, *Pediatrics* 121 (2008):

e1585–e1590; M. L. Neuhouser and coauthors, Vitamin D insufficiency in a multiethnic cohort of breast cancer survivors, *American Journal of Clinical Nutrition* 88 (2008): 133–139.

24. A. C. Looker and coauthors, Vitamin D status: United States, 2001–2006, *NCHS Data Brief* 59 (2011): 1–8.

25. A. Arabi, R. El Rassi, and G. El-Hajj Fuleihan, Hypovitaminosis D in developing countries–prevalence, risk factors and outcomes, *Nature Reviews: Endocrinology* 6 (2010): 550–561.

26. H. A. Bischoff-Ferrari and coauthors, A pooled analysis of vitamin D dose requirements for fracture prevention, *New England Journal of Medicine* 367 (2012): 40–49; S. R. Mastaglia and coauthors, Effect of vitamin D nutritional status on muscle function and strength in healthy women aged over sixty-five years, *Journal of Nutrition, Health and Aging* 15 (2011): 349–354.

27. P. Lips and coauthors, Once-weekly dose of 8400 IU vitamin D_3 compared with placebo: Effects on neuromuscular function and tolerability in older adults with vitamin D insufficiency, *American Journal of Clinical Nutrition* 91 (2010): 985–991.

28. Committee on Dietary Reference Intakes, *Dietary Reference Intakes for Calcium and Vitamin D*, (Washington, D.C.: National Academies Press, 2011), pp. 75–124.

29. L. M. Hall and coauthors, Vitamin D intake needed to maintain target serum 25-hydroxyvitamin D concentrations in participants with low sun exposure and dark skin pigmentation is substantially higher than current recommendations, *Journal of Nutrition* 140 (2010): 542–550.

30. K. H. Madsern and coauthors, Randomized controlled trial of the effects of vitamin D-fortified milk and bread on serum 25-hydroxyvitamin D concentrations in families in Denmark during winter: The VitmaD study, *American Journal of Clinical Nutrition* 98 (2013): 374–382; R. M. Biancuzzo and coauthors, Fortification of orange juice with vitamin D_2 or vitamin D_3 is as effective as an oral supplement in maintaining vitamin D status in adults, *American Journal of Clinical Nutrition* 91 (2010): 1621–1626.

31. M. D. Farrar and coauthors, Recommended summer sunlight exposure amounts fail to produce sufficient vitamin D status in UK adults of South Asian origin, *American Journal of Clinical Nutrition* 94 (2011): 1219–1224.

32. E. Linos and coauthors, Sun protective behaviors and vitamin D levels in the US population: NHANES 2003–2006, *Cancer Causes and Control* 23 (2012): 133–140.

33. S. Sharma and coauthors, Vitamin D deficiency and disease risk among aboriginal Arctic populations, *Nutrition Reviews* 69 (2011): 468–478.

34. L. A. Houghton and coauthors, Predictors of vitamin D status and its association with parathyroid hormone in young New Zealand children, *American Journal of Clinical Nutrition* 92 (2010): 69–76.

35. U. Lehmann and coauthors, Bioavailability of vitamin D_2 and D_3 in healthy volunteers, a randomised placebo-controlled trial, *Journal of Clinical Endocrinology and Metabolism* 98 (2013): 4339–4345; L. Tripkovic and coauthors, comparison of vitamin D_2 and vitamin D_3 supplementation in raising serum 25-hydroxyvitamin D status: A systematic review and meta-analysis, *American Journal of Clinical Nutrition* 95 (2012): 1357–1364; R. P. Heaney and coauthors, Vitamin D(3) is more potent than vitamin D(2) in humans, *Journal of Clinical Endocrinology and Metabolism* 96 (2011): E447–E452.

36. G. B. Mulligan and A. Licata, Taking vitamin D with the largest meal improves absorption and results in higher serum levels of 25-hydroxyvitamin D, *Journal of Bone and Mineral Research* 25 (2010): 928–930.

37. A. K. Smolarek and N. Suh, chemopreventive activity of vitamin E breast cancer: A focus on γ- and δ-tocopherol, *Nutrients* 3 (2011): 962–986; C. S. Yang and coauthors, Inhibition of inflammation and carcinogenesis in the lung and colon by tocopherols, *Annals of the New York Academy of Sciences* 1203 (2010): 29–34; J. Ju and coauthors, Cancer-preventive activities of tocopherols and tocotrienols, *Carcinogenesis* 31 (2010): 533–542.

38. R. S. Y. Wong and A. K. Radhakrishnan, Tocotrienol research: Past into present, *Nutrition Reviews* 70 (2012): 483–490; P. W. Sylvester and coauthors, The value of tocotrienols in the prevention and treatment of cancer, *Journal of the American College of Nutrition* 29 (2010): 324S–333S; B. B. Aggarwal and coauthors, Tocotrienols, the vitamin E of the 21st century: Its potential against cancer and other chronic disease, *Biochemical Pharmacology* 80 (2010): 1613–1631.

39. R. F. Pfeiffer, Neurologic manifestations of malabsorption syndromes, *Handbook of Clinical Neurology* 120 (2014): 621–632; D. Bromley, P. C. Anderson, and V. Daggett, Sturctural consequences of mutations to the α-tocopherol transfer protein associated with the neurodegenerative disease ataxia with vitamin E deficiency, *Biochemistry* 52 (2013): 4264–4273.

40. M. S. Hamidi, O. Gajic-Velijanoski, and A. M. Cheung, Vitamin K and bone health, *Journal of Clinical Densitometry* 16 (2013): 409–413.

41. M. J. Shearer, X. Fu, and S. L. Booth, Vitamin K nutrition, metabolism, and requirements: Current concepts and future research, *Advances in Nutrition: An International Review Journal* 3 (2012): 182–195; H. J. Ipema, Use of oral vitamin K for prevention of late vitamin K deficiency bleeding in neonates when injectable vitamin K is not available, *Annals of Pharmacotherapy* 46 (2012): 879–883; G. Lippi and M. Franchini, Vitamin K in neonates, Facts and myths, *Blood Transfusion* 9 (2011): 4–9.

HIGHLIGHT > 11
Antioxidant Nutrients in Disease Prevention

> **LEARN IT** Describe how antioxidants defend against free radicals that contribute to diseases.

Count on supplement manufacturers to exploit the day's hot topics in nutrition. The moment bits of research news surface, new supplements appear—and terms such as *antioxidants* and *lycopene* become household words. Friendly faces in TV commercials try to persuade us that these supplements hold magic in the fight against aging and disease. New supplements hit the market and sales soar.

In the meantime, scientists and medical experts around the world continue their work to clarify and confirm the roles of antioxidants in preventing chronic diseases. This highlight summarizes some of the accumulating evidence. It also revisits the advantages of foods over supplements. But first it is important to introduce the troublemaker—an unstable molecule known as a **free radical.** (The accompanying glossary defines free radical and related terms.)

Free Radicals and Disease

Chapter 7 describes how the body's cells use oxygen in metabolic reactions. In the process, oxygen reacts with body compounds and produces highly unstable molecules known as free radicals. In addition to normal body processes, environmental factors such as ultraviolet radiation, air pollution, and tobacco smoke generate free radicals.

A free radical is a molecule with one or more unpaired electrons.* An electron without a partner is unstable and highly reactive. To regain its stability, the free radical quickly finds a stable but vulnerable compound from which to steal an electron.

With the loss of an electron, the formerly stable molecule becomes a free radical itself and steals an electron from another nearby molecule. Thus an electron-snatching chain reaction is under way with free radicals producing more free radicals. **Antioxidants** neutralize free radicals by donating one of their own electrons, thus ending the

© Istockphoto.com/Nicole S. Young

chain reaction. When they lose electrons, antioxidants do not become free radicals because they are stable in either form. (Review Figure 10-16 on p. 328 to see how ascorbic acid can give up two hydrogens with their electrons and become dehydroascorbic acid.)

Free radicals attack. Occasionally, these free-radical attacks are helpful. For example, cells of the immune system use free radicals as ammunition in an "oxidative burst" that demolishes disease-causing viruses and bacteria. Most often, however, free-radical attacks cause widespread damage. They commonly damage the polyunsaturated fatty acids in lipoproteins and in cell membranes, disrupting the transport of substances into and out of cells. Free radicals also alter DNA, RNA, and proteins, creating excesses and deficiencies of specific proteins, impairing cell functions, and eliciting an inflammatory response. All of these actions contribute to cell damage, disease progression, and aging (see Figure H11-1).

The body's natural defenses and repair systems try to control the destruction caused by free radicals, but these systems are not 100 percent effective. In fact, they become less effective with age, and the unrepaired damage accumulates. To some extent, dietary antioxidants defend the body against **oxidative stress,** but if antioxidants are unavailable or if free-radical production becomes excessive, health problems may develop.[1] Oxygen-derived free radicals may cause diseases, not only by indiscriminately destroying the valuable components

---

*Many free radicals exist, but oxygen-derived free radicals are most common in the human body. Examples of oxygen-derived free radicals include superoxide radical (O_2^{-}), hydroxyl radical (OH·), and nitric oxide (NO·). (The dots in the symbols represent the unpaired electrons.) Technically, hydrogen peroxide (H_2O_2) and singlet oxygen are not free radicals because they contain paired electrons, but the unstable conformation of their electrons makes radical-producing reactions likely. Scientists sometimes use the term *reactive oxygen species (ROS)* to describe all of these compounds.

Free radicals are highly reactive. They might attack the polyunsaturated fatty acids in a cell membrane, which generates lipid radicals that damage cells and accelerate disease progression. Free radicals might also attack and damage DNA, RNA, and proteins, which interferes with the body's ability to maintain normal cell function, causing disease and premature aging.

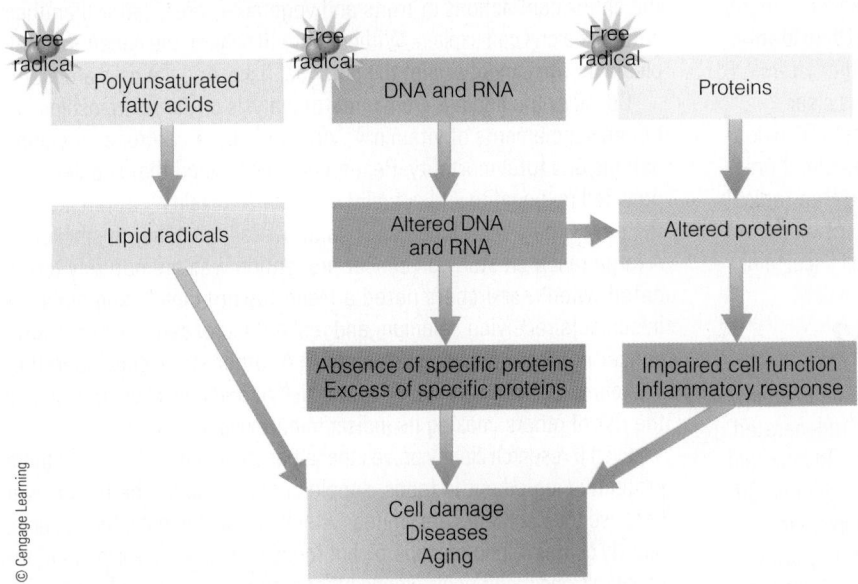

© Cengage Learning

of cells, but also by serving as signals for specific activities within the cells. Scientists have identified oxidative stress as a causative factor and antioxidants as a protective factor in cognitive performance and the aging process as well as in the development of diseases such as cancer, arthritis, cataracts, diabetes, hypertension, and heart disease.[2]

Defending against Free Radicals

The body maintains a couple lines of defense against free-radical damage. A system of enzymes disarms the most harmful **oxidants.*** The action of these enzymes depends on the minerals selenium, copper, manganese, and zinc. If the diet fails to provide adequate supplies of these minerals, this line of defense weakens. The body also uses the antioxidant vitamins—vitamin E, beta-carotene, and vitamin C. Vitamin E defends the body's lipids (cell membranes, nervous tissues, and lipoproteins, for example) by efficiently stopping the free-radical chain reaction.[3] Beta-carotene also acts as an antioxidant in lipid membranes. Vitamin C protects other tissues, such as the skin and fluid of the blood, against free-radical attacks. Vitamin C seems especially adept at neutralizing free radicals from polluted air and cigarette smoke; it also restores oxidized vitamin E to its active state.

Dietary antioxidants also include some of the **phytochemicals** (featured in Highlight 13). Together, nutrients and phytochemicals with antioxidant activity minimize damage and prevent disease in the following ways:[4]

- Limiting free-radical formation
- Destroying free radicals or their precursors

*These enzymes include *glutathione peroxidase, thioredoxin reductase, superoxide dismutase,* and *catalase.*

- Stimulating antioxidant enzyme activity
- Repairing oxidative damage
- Stimulating repair enzyme activity
- Supporting a healthy immune system

These actions play key roles in defending the body against chronic diseases such as cancer and heart disease.

Defending against Cancer

Cancers arise when cellular DNA is damaged—sometimes by free-radical attacks. Antioxidants may reduce cancer risks by protecting DNA from this damage. Many researchers have reported low rates of cancer in people whose diets include abundant vegetables and fruits, rich in antioxidants. Preliminary reports suggest an inverse relationship between DNA damage and vegetable intake and a positive relationship with beef and pork intake.[5]

Foods rich in vitamin C seem to protect against certain types of cancers, especially those of the esophagus. Such a correlation may reflect the benefits of a diet rich in fruits and vegetables and low in fat; evidence that vitamin C supplements reduce the risk of cancer is lacking.

Researchers hypothesize that vitamin E might inhibit cancer formation by attacking free radicals that damage DNA. Evidence that vitamin E supplements help guard against cancer, however, is lacking.

Several studies report a cancer-preventing benefit of vegetables and fruits rich in beta-carotene and the other carotenoids as well. Carotenoids may protect against oxidative damage to DNA. Some research suggests that high concentrations of beta-carotene and the other carotenoids are associated with lower rates of some cancers.[6] Studies do not, however, find a reduction in cancer risk with beta-carotene supplementation. Benefits most likely reflect a healthy diet abundant in fruits and vegetables. In fact, a major review of several large research studies concluded that none produced evidence to justify the use of antioxidant supplements for cancer prevention.[7]

Defending against Heart Disease

Decades of research have contributed to our understanding of how oxidative stress contributes to atherosclerosis and how antioxidants might protect against heart disease, yet questions remain.[8] High blood cholesterol carried in LDL (low-density lipoproteins) is a major risk factor for cardiovascular disease, but how do LDL exert their damage? One scenario is that free radicals within the arterial walls oxidize LDL, changing their structure and function. The oxidized LDL then accelerate the formation of artery-clogging plaques. These free radicals also oxidize the polyunsaturated fatty acids of the cell membranes, sparking additional changes in the arterial walls, which impede the flow of blood. Susceptibility to such oxidative damage within the arterial walls is heightened by a diet high in saturated fat and by cigarette smoke. In contrast, diets that include plenty of fruits and vegetables, especially when saturated fat is low, strengthen antioxidant defenses against LDL oxidation.

Antioxidants, especially vitamin E, may protect against hypertension and cardiovascular disease.[9] Epidemiological studies suggest that people who eat foods rich in vitamin E have relatively few atherosclerotic plaques and low rates of death from heart disease. Among its many protective roles, vitamin E defends against LDL oxidation, inflammation, arterial injuries, and blood clotting. Whether vitamin E supplements slow the progression of heart disease is less clear.

Some studies suggest that vitamin C protects against LDL oxidation, raises HDL, lowers total cholesterol, and improves blood pressure. Vitamin C may also minimize inflammation and the free-radical action within the arterial wall. Like vitamin E, the role of vitamin C supplements in reducing the risk of heart disease remains uncertain.

Foods, Supplements, or Both?

In the process of scavenging and quenching free radicals, antioxidants themselves become oxidized. To some extent, they can be regenerated, but losses still occur and free radicals attack continuously. To maintain defenses, a person must replenish dietary antioxidants regularly. But should antioxidants be replenished from foods or from supplements?

Foods—especially fruits and vegetables—offer not only antioxidants, but an array of other valuable vitamins and minerals as well. Importantly, deficiencies of these nutrients can damage DNA as readily as free radicals can. Eating fruits and vegetables in abundance protects against both deficiencies and diseases—and may protect against inflammation and DNA damage.[10] A major review of the evidence gathered from metabolic studies, epidemiologic studies, and dietary intervention trials identified three dietary strategies most effective in preventing heart disease:

- Use unsaturated fats instead of saturated or *trans* fats (see Highlight 5).
- Select foods rich in omega-3 fatty acids (see Chapter 5).
- Consume a diet high in fruits, vegetables, nuts, and whole grains and low in refined grain products.

Such a diet combined with exercise, weight control, and not smoking serves as the best prescription for health. Notably, taking supplements is not among these disease-prevention recommendations.

Diets that deliver sufficient quantities of antioxidant vitamins may protect against cancer and heart disease—but only a small fraction of the US population consumes recommended amounts. Some research suggests a protective effect from as little as a daily glass of orange juice or carrot juice (rich sources of vitamin C and beta-carotene, respectively). Other intervention studies, however, have used levels of nutrients that far exceed current recommendations and can be achieved only by taking supplements. In making their recommendations for the antioxidant nutrients, members of the DRI Committee considered whether these studies support substantially higher intakes to help protect against chronic diseases. They did raise the recommendations for vitamins C and E, but they do not support taking supplements over eating a healthy diet.

Though fruits and vegetables containing many antioxidant nutrients and phytochemicals have been associated with a diminished risk of many chronic diseases, supplements have not always proved beneficial.[11] In fact, sometimes the benefits are more apparent when the vitamins come from foods rather than from supplements. In other words, the antioxidant actions of fruits and vegetables are greater than their nutrients alone can explain. Without data to confirm the benefits of supplements, we cannot accept the potential risks. And the risks are real.

Consider the findings from a meta-analysis of the relationships between supplements of vitamin A, vitamin E, beta-carotene, or combinations and total mortality. Researchers concluded that supplements provided no benefits and actually *increased* mortality.[12] Beta-carotene *increases* the risk of lung cancer and overall mortality in smokers.[13] A large research study on cancer prevention was prematurely terminated when researchers noted a trend toward developing diabetes in subjects receiving selenium and a slight increased risk of prostate cancer in those receiving vitamin E.[14] Another study concluded that vitamin E supplements increase the risk of some strokes, but reduce the risk of others, making its indiscriminate use unwise.[15]

Even if research clearly proves that a particular nutrient is the ultimate protective ingredient in foods, supplements would not be the answer because their contents are limited. Vitamin E supplements, for example, usually contain alpha-tocopherol, but foods provide an assortment of tocopherols and tocotrienols among other nutrients, many of which provide valuable protection against free-radical damage. In addition to a full array of nutrients, foods provide phytochemicals that also fight against many

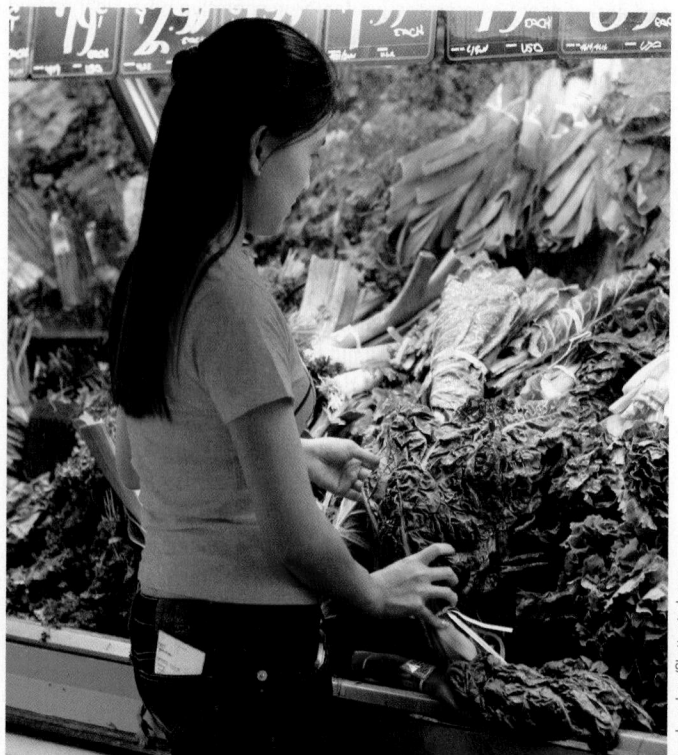

Many cancer-fighting products are available now at your local produce counter.

TABLE H11-1 Antioxidants and Chronic Disease Risk

Antioxidant	Disease	Risk from Foods	Risk from Supplements
Vitamin C	Coronary heart disease	Inconsistent results	Inconsistent results
	Breast cancer	Inconsistent results	—
	Colorectal cancer	Inconsistent results	—
	Gastrointestinal cancer	—	Not known
	Lung cancer	No effect	Not known
Vitamin E	Coronary heart disease	Inconsistent results	No effect or possible increased risk
	Breast cancer	—	No effect
	Colorectal cancer	Inconsistent results	—
	Gastrointestinal cancer	—	No effect
	Lung cancer	No effect	No effect
	Prostate cancer	Decreased risk	Decreased risk in smokers
Beta-carotene	Coronary heart disease	Decreased risk	No effect in nonsmokers, increased risk in smokers
	Lung cancer	Inconsistent results	No effect in nonsmokers, increased risk in smokers
	Colorectal cancer	Decreased risk	—
	Gastrointestinal cancer	—	No effect
	Prostate cancer	No effect	—
Other carotenoids	Lung cancer	Decreased risk for beta-cryptoxanthin	—
	Colorectal cancer	Decreased risk	—
	Prostate cancer	Decreased risk for lycopene	—
Fruits and vegetables	Coronary heart disease	Decreased risk	
	Breast cancer	No effect	
	Colorectal cancer	Inconsistent results	
	Gastric and esophageal cancer	Decreased risk	
	Lung cancer	Decreased risk for fruits, no effect for vegetables	
	Prostate cancer	No effect	
Supplement containing a combination of antioxidants	Coronary heart disease		Possibly increased risk
	Gastrointestinal cancer		Possibly increased risk
	Lung cancer		No effect in nonsmokers, increased risk in smokers

SOURCE: Adapted from H. Verhagen and coauthors, The state of antioxidant affairs, *Nutrition Today* 41 (2006): 244–249.

© Cengage Learning

diseases. Supplements shortchange users. Furthermore, supplements should be used only as an adjunct to other measures such as smoking cessation, weight control, physical activity, and medication as needed.

Clearly, much more research is needed to define optimal and harmful levels of intake. This much we know: antioxidants behave differently under various conditions. At physiological levels typical of a healthy diet, they act as antioxidants, but at pharmacological doses typical of supplements, they may act as **prooxidants,** stimulating the production of free radicals and altering metabolism in a way that may promote disease. A high intake of vitamin C from supplements, for example, may *increase* the risk of heart disease in women with diabetes. Until the optimum intake of antioxidant nutrients can be determined, the risks of supplement use remain unclear. Table H11-1 presents a summary of the relationships between antioxidants and chronic diseases—sorted by foods or supplements. As you can see, many studies report either no effect or inconsistent results. Any decrease in risk is attributed to

foods 9 out of 10 times. Any increase in risk is always from supplements, and often in smokers. Clearly, the best way to add antioxidants to the diet is to eat generous servings of fruits and vegetables daily.

It should be clear by now that we cannot know the identity and action of every chemical in every food. Even if we did, why create a supplement to replicate a food? Why not eat foods and enjoy the pleasure, nourishment, and health benefits they provide? The beneficial constituents in foods are widespread among plants. Among the fruits, pomegranates, berries, and citrus rank high in antioxidants; top antioxidant vegetables include kale, spinach, and brussels sprouts; millet and oats contain the most antioxidants among the grains; pinto beans and soybeans are outstanding legumes; and walnuts outshine the other nuts. But don't try to single out one particular food for its "magical" nutrient, antioxidant, or phytochemical. Instead, eat a wide variety of fruits, vegetables, grains, legumes, and nuts every day—and get *all* the benefits these foods have to offer.

CRITICAL THINKING SKILLS

A. What are the arguments for obtaining antioxidants from foods, supplements, or both?

B. The American Heart Association and other health organizations have concluded that consumers should get their antioxidants from foods rather than supplements. They add that taking supplements may even be harmful. Supplement manufacturers claim that such statements are unfair and that

their natural botanical extracts provide numerous health benefits. In fact, some suggest that beneficial effects can only be achieved by taking high-dose supplements. Given that there are currently no DRI defining the kinds of antioxidants or the daily quantities needed, how might you ensure a healthy intake of antioxidants? If you decided to take an antioxidant supplement, how might you research the product to determine its safety and effectiveness?

REFERENCES

1. B. Halliwell, Free radicals and antioxidants: Updating a personal view, *Nutrition Reviews* 70 (2012): 257–265.

2. A. Whaley-Connell, P. A. McCullough, and J. R. Sowers, The role of oxidative stress in the metabolic syndrome, *Reviews in Cardiovascular Medicine* 12 (2011): 21–29.

3. M. G. Traber and J. F. Stevens, Vitamins C and E: Beneficial effects from a mechanistic perspective, *Free Radical Biology and Medicine* 51 (2011): 1000–1013.

4. H. Yao and coauthors, Dietary flavonoids as cancer prevention agents, *Environmental Carcinogenesis and Ecotoxicology Reviews* 29 (2011): 1–31.

5. P. Riso and coauthors, DNA damage and repair activity after broccoli intake in young healthy smokers, *Mutagenesis* 25 (2010): 595–602; A. Brevik and coauthors, Polymorphisms in base excision repair genes as colorectal cancer risk factors and modifiers of the effects of diets high in red meat, *Cancer Epidemiology, Biomarkers and Prevention* 19 (2010): 3167–3173.

6. A. H. Eliassen and coauthors, Circulating carotenoids and risk of breast cancer: Pooled analysis of eight prospective studies, *Journal of the National Cancer Institute* 104 (2012): 1905–1916.

7. M. Goodman and coauthors, Clinical trials of antioxidants as cancer prevention agents: Past, present, and future, *Free Radical Biology and Medicine* 51 (2011): 1068–1084.

8. G. Riccioni and coauthors, Carotenoids and vitamins C and E in the prevention of cardiovascular disease, *International Journal for Vitamin and Nutrition Research* 82 (2012): 15–26; D. Farbstein, A. Kozak-Blickstein, and A. P. Levy, Antioxidant vitamins and their use in preventing cardiovascular disease, *Molecules* 15 (2010): 8098–8110.

9. E. L. Schiffrin, Antioxidants in hypertension and cardiovascular disease, *Molecular Interventions* 10 (2010): 354–362.

10. S. N. Bhupathiraju and K. L. Tucker, Greater variety in fruit and vegetable intake is associated with lower inflammation in Puerto Rican adults, *American Journal of Clinical Nutrition* 93 (2011): 37–46; M. K. Shanmugam, R. Kannaiyan, and G. Sethi, Targeting cell signaling and apoptotic pathways by dietary agents: Role in the prevention and treatment of cancer, *Nutrition and Cancer* 63 (2011): 161–173.

11. B. Halliwell, Free radicals and antioxidants: Quo vadis? *Trends in Pharmacological Sciences* 32 (2011): 125–130.

12. G. Bjelakovic, D. Nikolova, and C. Gluud, Antioxidant supplements to prevent mortality, *Journal of the American Medical Association* 310 (2013): 1178–1179.

13. A. M. Mondul and coauthors, Metabolomic profile of response to supplementation with β-carotene in the Alpha-Tocopherol, Beta-Carotene Cancer Prevention Study, *American Journal of Clinical Nutrition* 98 (2013): 488–493.

14. M. C. Ledesma and coauthors, Selenium and vitamin E for prostate cancer: Post-SELECT (Selenium and Vitamin E Cancer Prevention Trial) status, *Molecular Medicine* 17 (2011): 134–143.

15. M. Schürks and coauthors, Effects of vitamin E on stroke subtypes: Meta-analysis of randomised controlled trials, *British Medical Journal* 341 (2010): c5702.

Water and the Major Minerals

Nutrition in Your Life

What's your beverage of choice? If you said water, then congratulate yourself for recognizing its importance in maintaining your body's fluid balance. If you answered milk, then pat yourself on the back for taking good care of your bones. Without water, you would realize within days how vital it is to your survival. The consequences of a lack of milk (or other calcium-rich foods) are also dramatic, but may not become apparent for decades. Water, calcium, and all the other major minerals support fluid balance and bone health. Before getting too comfortable reading this chapter, pour yourself a glass of water or milk. Your body will thank you. In the Nutrition Portfolio at the end of this chapter, you can determine whether the foods you are eating are meeting your water and major mineral needs.

Water is an essential nutrient, more important to life than any of the others. The body needs more water each day than any other nutrient. Furthermore, you can survive only a few days without water, whereas a deficiency of the other nutrients may take weeks, months, or even years to develop.

This chapter begins with a look at water and the body's fluids. The body maintains an appropriate balance and distribution of fluids with the help of another class of nutrients—the minerals. In addition to introducing the minerals that help regulate body fluids, this chapter describes many of the other important functions minerals perform in the body. Chapter 19 revisits water as a beverage and addresses consumer concerns about its safety; Chapter 20 looks at water usage in agriculture and food production and examines the impending water shortages we face.

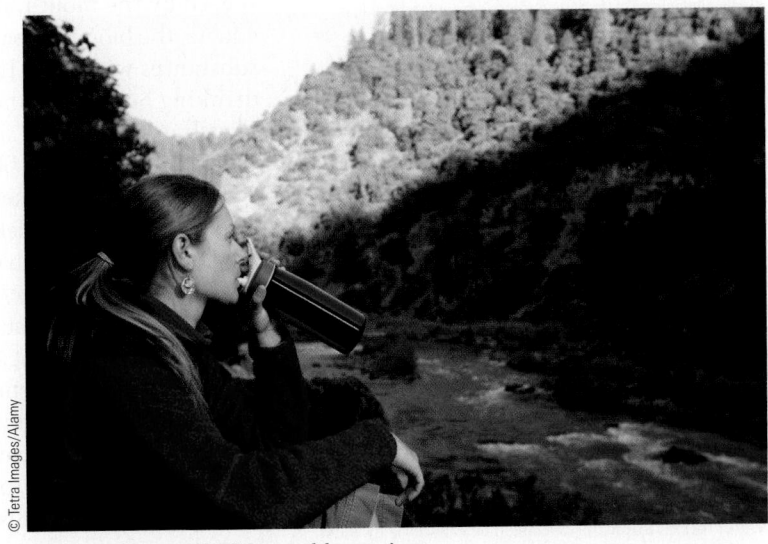

© Tetra Images/Alamy

Water is the most indispensable nutrient.

> FIGURE 12-1 **One Cell and Its Associated Fluids**

Extracellular fluid between the cells (intercellular or interstitial)

Cell membrane

Nucleus

Intracellular fluid within the cell

Extracellular fluid (plasma) within the blood vessels (intravascular)

Blood vessel

© Cengage Learning

water balance: the balance between water intake and output (losses).

intracellular fluid: fluid inside the cells, usually high in potassium and phosphate. Intracellular fluid accounts for approximately two-thirds of the body's water.

• **intra** = within

extracellular fluid: fluid outside the cells. Extracellular fluid includes two main components—the interstitial fluid between cells and the intravascular fluid inside blood vessels. Extracellular fluid accounts for approximately one-third of the body's water.

• **extra** = outside

interstitial (IN-ter-STISH-al) **fluid:** fluid between the cells (intercellular), usually high in sodium and chloride. Interstitial fluid is a large component of extracellular fluid.

• **inter** = in the midst, between

intravascular fluid: fluid within blood vessels.

• **intra** = within

thirst: a conscious desire to drink.

hypothalamus (high-po-THAL-ah-mus): a brain center that controls activities such as maintenance of water balance, regulation of body temperature, and control of appetite.

dehydration: the condition in which body water output exceeds water input. Symptoms include thirst, dry skin and mucous membranes, rapid heartbeat, low blood pressure, and weakness.

water intoxication: the rare condition in which body water contents are too high in all body fluid compartments.

hyponatremia (HIGH-po-na-TREE-me-ah): a decreased concentration of sodium in the blood.

12.1 Water and the Body Fluids

> **LEARN IT** Explain how the body regulates fluid balance.

Water constitutes about 60 percent of an adult's body weight and a higher percentage of a child's (see Figure 1-1, p. 7). Because water makes up about 75 percent of the weight of lean tissue and less than 25 percent of the weight of fat, a person's body composition influences how much of the body's weight is water. The proportion of water is generally smaller in females, obese people, and the elderly because of their smaller proportion of lean tissue.

In the body, water is the fluid in which all life processes occur. The water in the body fluids:

• Carries nutrients and waste products throughout the body

• Maintains the structure of large molecules such as proteins and glycogen

• Participates in metabolic reactions

• Serves as the solvent for minerals, vitamins, amino acids, glucose, and many other small molecules so that they can participate in metabolic activities

• Acts as a lubricant and cushion around joints and inside the eyes, the spinal cord, and, in pregnancy, the amniotic sac surrounding the fetus in the womb

• Aids in the regulation of normal body temperature, as the evaporation of sweat from the skin removes excess heat from the body

• Maintains blood volume

To support these and other vital functions, the body actively maintains an appropriate **water balance** between intake and output.

Water Balance and Recommended Intakes Every cell contains fluid of the exact composition that is best for that cell. Fluid inside cells is called **intracellular fluid,** whereas fluid outside cells is called **extracellular fluid.** The extracellular fluid that surrounds each cell is called **interstitial fluid,** whereas the extracellular fluid in the blood vessels is called **intravascular fluid.** Figure 12-1 illustrates a cell and its associated fluids. The compositions of intercellular and extracellular fluids differ from each other. They continuously lose and replace their components, yet the composition in each compartment remains remarkably constant under normal conditions. Because imbalances can be devastating, the body quickly responds by adjusting both water intake and excretion as needed. Consequently, the entire system of cells and fluids remains in a delicate, but controlled, state of homeostasis.

Water Intake Thirst and satiety influence water intake in response to changes sensed by the mouth, **hypothalamus,** and nerves. When water intake is inadequate, the blood becomes concentrated (having lost water but not the dissolved substances within it), the mouth becomes dry, and the hypothalamus initiates drinking behavior. When water intake is excessive, the stomach expands and stretch receptors send signals to stop drinking. Similar signals are sent from receptors in the heart as blood volume increases.

When too much water is lost from the body and not replaced, **dehydration** develops. A first sign of dehydration is thirst, the signal that the body has lost some fluid. If a person is unable to obtain water or, as in many elderly people, fails to perceive the thirst message, the symptoms of dehydration may progress rapidly from thirst to weakness, exhaustion, and delirium—and end in death if not corrected (see Table 12-1). Notice that an early sign of dehydration is fatigue; keep that in mind when considering caffeinated beverages for an afternoon "pick-me-up" and choose water instead. Dehydration develops with either inadequate water intake or excessive water losses. (Chapter 14 revisits dehydration and the fluid needs of athletes.)

Water intoxication, on the other hand, is rare but can occur with excessive water intake and kidney disorders that reduce urine production. The symptoms may include confusion, convulsions, and even death in extreme cases. Excessive water ingestion (10 to 20 liters) within a few hours dilutes the sodium concentration of the blood and contributes to a dangerous condition known as **hyponatremia.**

TABLE 12-1 Signs of Dehydration

Body Weight Lost (%)	Symptoms
1–2	Thirst, fatigue, weakness, vague discomfort, loss of appetite
3–4	Impaired physical performance, dry mouth, reduction in urine, flushed skin, impatience, apathy
5–6	Difficulty concentrating, headache, irritability, sleepiness, impaired temperature regulation, increased respiratory rate
7–10	Dizziness, spastic muscles, loss of balance, delirium, exhaustion, collapse

© Cengage Learning

NOTE: The onset and severity of symptoms at various percentages of body weight lost depend on the activity, fitness level, degree of acclimation, temperature, and humidity. If not corrected, dehydration can lead to death.

For this reason, guidelines suggest limiting fluid intake during times of heavy sweating to between 1 and 1.5 liters per hour. (Chapter 14 revisits hyponatremia as sometimes seen in endurance athletes.)

Water Sources The obvious dietary source of water is water itself, which provides about one-third of the total water intake in the United States.[1] In addition, other beverages and nearly all foods also contain water. Most fruits and vegetables contain up to 90 percent water, and many meats and cheeses contain at least 50 percent. See Table 12-2 for selected foods and Appendix H for many more. Also, **metabolic water** is generated as an end product during condensation reactions and the oxidation of energy-yielding nutrients. Recall from Chapter 7 that when the energy-yielding nutrients break down, their carbons and hydrogens combine with oxygen to yield carbon dioxide (CO_2) and water (H_2O). As Table 12-3 shows, the water derived daily from these three sources—beverages, foods, and metabolism—averages about 2500 milliliters (roughly 2.5 quarts or 10.5 cups).

Water Losses At the very least, the body must excrete enough water to carry away the waste products generated by a day's metabolic activities. This **obligatory water excretion** is a minimum of about 500 milliliters (about 2 cups) of water each day. Above this amount, excretion adjusts to balance intake. If a person drinks more water, the kidneys excrete more urine, and the urine becomes more dilute. In addition to urine, water is lost from the lungs as vapor and from the skin as sweat; some is also lost in feces.* The amount of fluid lost from each source varies, depending on the environment (such as heat or humidity) and the body's physical condition (such as exercise or fever). On average, daily losses total about 2500 milliliters. Table 12-3 shows how daily water losses and intakes balance; maintaining this balance requires healthy kidneys and an adequate intake of fluids. An adequate intake of fluids, in turn, helps to maintain healthy kidneys and prevent kidney stone formation.[2]

Water Recommendations Because water needs vary depending on diet, activity, environmental temperature, and humidity, a general water requirement is difficult to establish. Recommendations are sometimes expressed in proportion to the amount of energy expended under average environmental conditions; for

TABLE 12-2 Percentage of Water in Selected Foods

100%	Water
90–99%	Fat-free milk, strawberries, watermelon, lettuce, cabbage, celery, spinach, broccoli
80–89%	Fruit juice, yogurt, apples, grapes, oranges, carrots
70–79%	Shrimp, bananas, corn, potatoes, avocados, cottage cheese, ricotta cheese
60–69%	Pasta, legumes, salmon, ice cream, chicken breast
50–59%	Ground beef, hot dogs, feta cheese
40–49%	Pizza
30–39%	Cheddar cheese, bagels, bread
20–29%	Pepperoni sausage, cake, biscuits
10–19%	Butter, margarine, raisins
1–9%	Crackers, cereals, pretzels, taco shells, peanut butter, nuts
0%	Oils, sugars

© Cengage Learning

TABLE 12-3 Water Balance

Water Sources	Amount (mL)	Water Losses	Amount (mL)
Beverages	550 to 1500	Kidneys (urine)	500 to 1400
Foods	700 to 1000	Skin (sweat)	450 to 900
Metabolism	200 to 300	Lungs (breath)	350
		GI tract (feces)	150
Total	1450 to 2800	Total	1450 to 2800

© Cengage Learning

NOTE: For perspective, 100 milliliters is a little less than ½ cup and 1000 milliliters is a little more than 1 quart (1 mL = 0.03 oz).

*Water lost from the lungs and skin accounts for almost half of the daily losses even when a person is not visibly perspiring; these losses are commonly referred to as *insensible water losses*.

metabolic water: water generated during metabolism.

obligatory (ah-BLIG-ah-TORE-ee) **water excretion:** the minimum amount of water the body has to excrete each day to dispose of its wastes—about 500 milliliters (about 2 cups, or 1 pint).

adults, for example, 1.0 to 1.5 milliliters per kcalorie expended (roughly one-half cup per 100 kcalories). The recommended water intake for a person who expends 2000 kcalories a day, then, is 2 to 3 liters of water (about 8 to 12 cups). This recommendation is in line with the Adequate Intake (AI) for *total* water set by the DRI Committee. Total water includes not only drinking water, but water in other beverages and in foods as well. Only one in five adults in the United States report drinking at least 8 cups of water a day.[3]

Because a wide range of water intakes will prevent dehydration and its harmful consequences, the AI is based on average intakes. People who are physically active or who live in hot environments may need more.

Which beverages are best? Any beverage can readily meet the body's fluid needs, but those with few or no kcalories do so without contributing to weight gain. Given that obesity is a major health problem and that beverages currently represent more than 20 percent of the total energy intake in the United States, water is the best choice for most people. Other choices include tea, coffee, nonfat and low-fat milk and soymilk, artificially sweetened beverages, fruit and vegetable juices, sports drinks, and lastly, sweetened nutrient-poor beverages.

Some research indicates that people who drink caffeinated beverages lose a little more fluid than when drinking water because caffeine acts as a diuretic. The DRI Committee considered such findings in their recommendations for water intake and concluded that caffeinated beverages contribute to the daily total water intake similar to that contributed by non-caffeinated beverages. In other words, it doesn't seem to matter whether people rely on caffeine-containing beverages or other beverages to meet their fluid needs.

As Highlight 7 explains, alcohol acts as a diuretic and can impair a person's health. Alcohol should not be used to meet fluid needs.

Health Effects of Water Water supports good health.[4] Physical and mental performances depend on it, as does the optimal functioning of the GI tract, kidneys, heart, and other body systems.

The kind of water a person drinks may also make a difference to health. Water is usually either hard or soft. **Hard water** has high concentrations of calcium and magnesium; the principal mineral of **soft water** is sodium or potassium. (See the accompanying glossary for other common terms used to describe water.) In practical terms, soft water makes more bubbles with less soap; hard water leaves a ring on the tub, a crust of rocklike crystals in the teakettle, and a gray residue in the laundry.

Soft water may seem more desirable around the house, and some homeowners purchase water softeners that replace magnesium and calcium with sodium. In the body, however, soft water with sodium may aggravate hypertension and heart disease. In contrast, the minerals in hard water may benefit these conditions.

Soft water also more easily dissolves certain contaminant minerals, such as cadmium and lead, from old plumbing pipes. As Chapter 13 explains, these contaminant minerals harm the body by displacing the nutrient minerals from their normal sites of action. People who live in buildings with old plumbing should run the cold water tap a minute or two to flush out harmful minerals whenever the water faucet has been off for more than 6 hours.[5]

Many people select **bottled water,** believing it to be safer than tap water and therefore worth its substantial cost. Chapter 19 offers a discussion of bottled water safety and regulations.

Blood Volume and Blood Pressure Fluids maintain the blood volume, which in turn influences blood pressure. The kidneys are central to the regulation of blood volume and blood pressure. All day, every day, the kidneys reabsorb needed substances and water and excrete wastes with some water in the urine (see Figure 12-2). The kidneys meticulously adjust the volume and the concentration of the urine to accommodate changes in the body, including variations in the day's food and beverage intakes. Instructions on whether to retain or release substances or water come from ADH, renin, angiotensin, and aldosterone.

hard water: water with a high calcium and magnesium content.

soft water: water with a high sodium or potassium content.

bottled water: drinking water sold in bottles.

> **FIGURE 12-2** **A Nephron, One of the Kidney's Many Functioning Units**

A nephron (a working unit of the kidney). Each kidney contains more than 1 million nephrons.

Blood vessel — Glomerulus

1 Blood flows into the glomerulus, and some of its fluid, with dissolved substances, is absorbed into the tubule.

Capillaries of glomerulus

Tubule

2 Then the fluid and substances needed by the body are returned to the blood in vessels alongside the tubule.

To the body

3 The tubule passes waste materials on to the bladder.

To the bladder

Kidney
Ureter
Pelvis
Bladder

Kidney, sectioned to show location of nephrons

Renal artery
Renal vein

The cleansing of blood in the nephron is roughly analogous to the way you might clean your car. First 1 you remove all your possessions and trash so that the car can be vacuumed. Then 2 you put back in the car what you want to keep and 3 throw away the trash.

© Cengage Learning

ADH Whenever blood volume or blood pressure falls too low, or whenever the extracellular fluid becomes too concentrated, the hypothalamus signals the pituitary gland to release **antidiuretic hormone (ADH)**. ADH is a water-conserving hormone that stimulates the kidneys to reabsorb water. Consequently, the more water you need, the less your kidneys excrete. These events also trigger thirst. Drinking water and retaining fluids raise the blood volume and dilute the concentrated fluids, thus helping to restore homeostasis. (Recall from Highlight 7 that alcohol depresses ADH activity, thus promoting fluid losses and dehydration.)

Renin Cells in the kidneys respond to low blood pressure by releasing an enzyme called **renin**. Through a complex series of events, renin causes the kidneys to reabsorb sodium. Sodium reabsorption, in turn, is always accompanied by water retention, which helps to raise blood volume and blood pressure.

antidiuretic hormone (ADH): a hormone produced by the pituitary gland in response to dehydration (or a high sodium concentration in the blood) that stimulates the kidneys to reabsorb more water and therefore to excrete less. In addition to its antidiuretic effect, ADH elevates blood pressure and so is also called *vasopressin* (VAS-oh-PRES-in).

- **vaso** = vessel
- **press** = pressure

renin (REN-in): an enzyme from the kidneys that hydrolyzes the protein angiotensinogen to angiotensin I, which results in the kidneys reabsorbing sodium.

GLOSSARY OF WATER TERMS

artesian water: water drawn from a well that taps a confined aquifer in which the water is under pressure.

carbonated water: water that contains carbon dioxide gas, either naturally occurring or added, that causes bubbles to form in it; also called *bubbling* or *sparkling water*. The FDA defines seltzer, soda, and tonic waters

as soft drinks; they are not regulated as water.

distilled water: water that has been vaporized and recondensed, leaving it free of dissolved minerals.

filtered water: water treated by filtration, usually through *activated carbon filters* that reduce the lead in tap water, or by *reverse osmosis* units that force pressurized water across a membrane removing lead, arsenic, and some microorganisms from tap water.

mineral water: water from a spring or well that naturally contains at least

250 parts per million (ppm) of minerals. Minerals give water a distinctive flavor. Many mineral waters are high in sodium.

natural water: water obtained from a spring or well that is certified to be safe and sanitary. The mineral content may not be changed, but the water may be treated in other ways such as with ozone or by filtration.

public water: water from a municipal or county water system that has been treated and disinfected.

purified water: water that has been treated by distillation or other physical

or chemical processes that remove dissolved solids. Because purified water contains no minerals or contaminants, it is useful for medical and research purposes.

spring water: water originating from an underground spring or well. It may be bubbly (carbonated), or "flat" or "still," meaning not carbonated. Brand names such as "Spring Pure" do not necessarily mean that the water comes from a spring.

well water: water drawn from groundwater by tapping into an aquifer.

Angiotensin In addition to its role in sodium retention, renin hydrolyzes a protein from the liver called **angiotensinogen** to **angiotensin I**. Angiotensin I is inactive until another enzyme converts it to its active form—**angiotensin II**. Angiotensin II is a powerful **vasoconstrictor** that narrows the diameters of blood vessels, thereby raising the blood pressure.

Aldosterone In addition to acting as a vasoconstrictor, angiotensin II stimulates the release of the hormone **aldosterone** from the **adrenal glands.** Aldosterone signals the kidneys to excrete potassium and to retain more sodium, and therefore water, because when sodium moves, water follows. Again, the effect is that when more water is needed, less is excreted.

All of these actions are presented in Figure 12-3 and help to explain why high-sodium diets aggravate conditions such as hypertension and edema. Too much sodium causes water retention and an accompanying rise in blood pressure or swelling in the interstitial spaces. Chapter 18 discusses hypertension in detail.

Fluid and Electrolyte Balance Maintaining a balance of about two-thirds of the body fluids inside the cells and one-third outside is vital to the life of the cells. If too much water were to enter the cells, they might rupture; if too much water were to leave, they would collapse. To control the movement of water, the

> FIGURE 12-3 **How the Body Regulates Blood Volume and Blood Pressure**

The renin-angiotensin-aldosterone system helps regulate blood volume and therefore blood pressure.

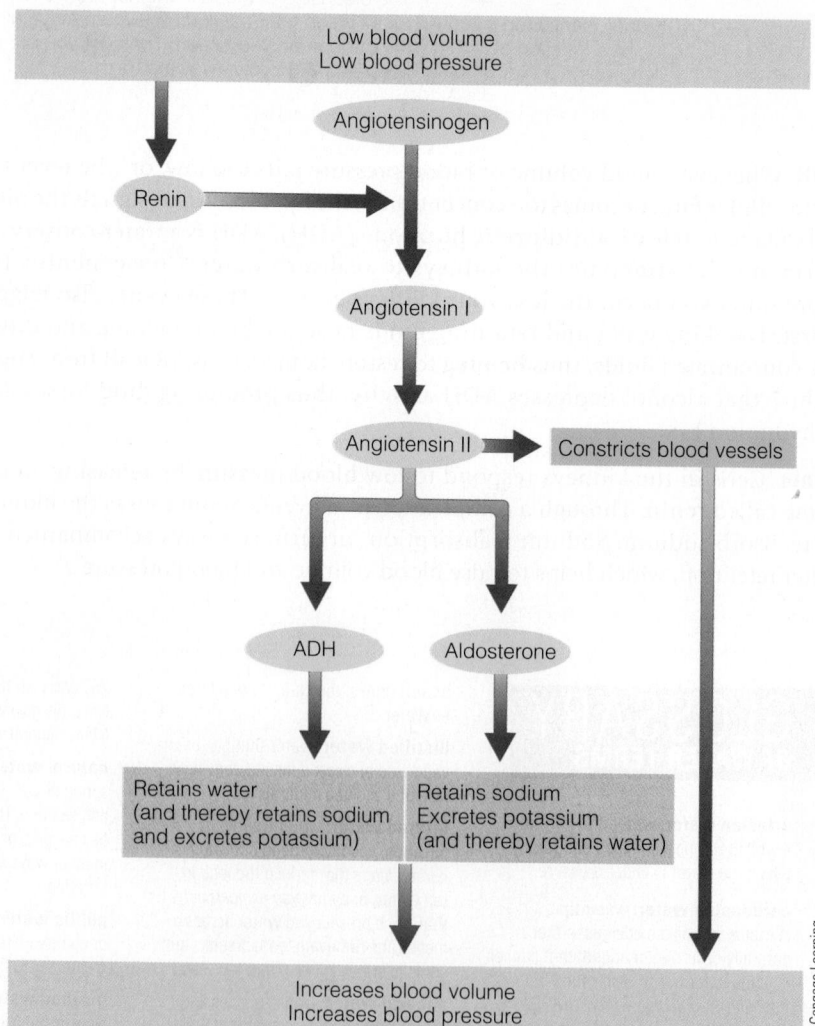

angiotensinogen: a precursor protein that is hydrolyzed to angiotensin I by renin.

angiotensin I (AN-gee-oh-TEN-sin): an inactive precursor that is converted by an enzyme to yield active angiotensin II.

angiotensin II: a hormone involved in blood pressure regulation.

vasoconstrictor (VAS-oh-kon-STRIK-tor): a substance that constricts or narrows the blood vessels.

aldosterone (al-DOS-ter-own): a hormone secreted by the adrenal glands that regulates blood pressure by increasing the reabsorption of sodium by the kidneys. Aldosterone also regulates chloride and potassium concentrations.

adrenal glands: glands adjacent to, and just above, each kidney.

cells direct the movement of the major minerals—sodium, chloride, potassium, calcium, phosphorus, magnesium, and sulfur.

Dissociation of Salt in Water When a mineral **salt** such as sodium chloride (NaCl) dissolves in water, it separates (**dissociates**) into **ions**—positively and negatively charged particles (Na^+ and Cl^-). The positive ions are **cations**; the negative ones are **anions**. (To remember the difference between cations and anions, think of the "t" in cations as a "plus" sign and the "n" in anions as a "negative.") Unlike pure water, which conducts electricity poorly, ions dissolved in water carry electrical current. For this reason, salts that dissociate into ions are called **electrolytes**, and fluids that contain them are **electrolyte solutions**.

In all electrolyte solutions, anion and cation concentrations are balanced (the number of negative and positive charges are equal). If a fluid contains 1000 negative charges, it must contain 1000 positive charges too. If an anion enters the fluid, a cation must accompany it or another anion must leave so that electrical neutrality will be maintained. Thus, whenever sodium (Na^+) ions leave a cell, potassium (K^+) ions enter, for example. In fact, it's a good bet that whenever Na^+ and K^+ ions are moving, they are going in opposite directions.

Table 12-4 shows that, indeed, the positive and negative charges inside and outside cells are perfectly balanced even though the numbers of each kind of ion differ over a wide range. Inside the cells, the positive charges total 202 and the negative charges balance these perfectly. Outside the cells, the amounts and proportions of the ions differ from those inside, but again the positive and negative charges balance. Scientists count these charges in **milliequivalents per liter (mEq/L)**.

Electrolytes Attract Water Electrolytes attract water. Each water molecule has a net charge of zero, but the oxygen side of the molecule has a slight negative charge, and the hydrogens have a slight positive charge. Figure 12-4 (p. 378) shows the result in an electrolyte solution: both positive and negative ions attract clusters of water molecules around them. This attraction dissolves salts in water and enables the body to move fluids into appropriate compartments.

Water Follows Electrolytes As Figure 12-5 (p. 378) shows, some electrolytes reside primarily outside the cells (notably, sodium, chloride, and calcium), whereas others reside predominantly inside the cells (notably, potassium, magnesium,

TABLE 12-4 Important Body Electrolytes

Electrolytes	Intracellular (inside cells) Concentration (mEq/L)	Extracellular (outside cells) Concentration (mEq/L)
Cations (positively charged ions)		
Sodium (Na^+)	10	142
Potassium (K^+)	150	5
Calcium (Ca^{++})	2	5
Magnesium (Mg^{++})	40	3
	202	155
Anions (negatively charged ions)		
Chloride (Cl^-)	2	103
Bicarbonate (HCO_3^-)	10	27
Phosphate ($HPO_4^=$)	103	2
Sulfate ($SO_4^=$)	20	1
Organic acids (lactate, pyruvate)	10	6
Proteins	57	16
	202	155

© Cengage Learning

NOTE: The numbers of positive and negative charges in a given fluid are the same. For example, in extracellular fluid, the cations and anions both equal 155 milliequivalents per liter (mEq/L). Of the cations, sodium ions make up 142 mEq/L; and potassium, calcium, and magnesium ions make up the remainder. Of the anions, chloride ions number 103 mEq/L; bicarbonate ions number 27; and the rest are provided by phosphate ions, sulfate ions, organic acids, and protein.

salt: a compound composed of a positive ion other than H^+ and a negative ion other than OH^-. An example is sodium chloride (Na^+Cl^-).

- **Na** = sodium
- **Cl** = chloride

dissociates (dis-SO-see-aites): physically separates.

ions (EYE-uns): atoms or molecules that have gained or lost electrons and therefore have electrical charges. Examples include the positively charged sodium ion (Na^+) and the negatively charged chloride ion (Cl^-). For a closer look at ions, see Appendix B.

cations (CAT-eye-uns): positively charged ions.

anions (AN-eye-uns): negatively charged ions.

electrolytes: salts that dissolve in water and dissociate into charged particles called ions.

electrolyte solutions: solutions that can conduct electricity.

milliequivalents per liter (mEq/L): the concentration of electrolytes in a volume of solution. Milliequivalents reveal characteristics about the solution that are not evident when the concentration is expressed in terms of weight.

> FIGURE 12-4 Water Dissolves Salts and Follows Electrolytes

The structural arrangement of the two hydrogen atoms and one oxygen atom enables water to dissolve salts. Water's role as a solvent is one of its most valuable characteristics.

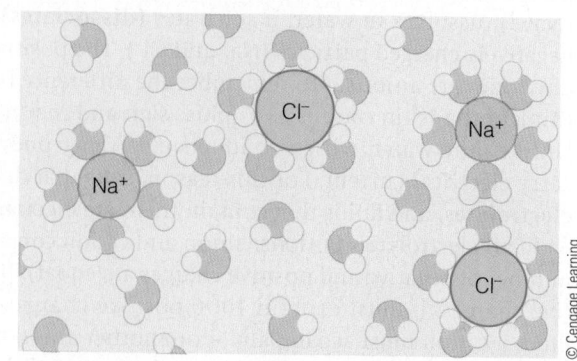

The negatively charged electrons that bond the hydrogens to the oxygen spend most of their time near the oxygen atom. As a result, the oxygen is slightly negative, and the hydrogens are slightly positive (see Appendix B).

In an electrolyte solution, water molecules are attracted to both anions and cations. Notice that the negative oxygen atoms of the water molecules are drawn to the sodium cation (Na⁺), whereas the positive hydrogen atoms of the water molecules are drawn to the chloride ions (Cl⁻).

> FIGURE 12-5 A Cell and Its Electrolytes

All of these electrolytes are found both inside and outside the cells, but each can be found mostly on one side or the other of the cell membrane.

Chemical symbols:
Ca = calcium
Cl = chloride
K = potassium
Mg = magnesium
Na = sodium
P = phosphorus
S = sulfate

Key:
○ Cations
○ Anions

solutes (SOLL-yutes): the substances that are dissolved in a solution. The number of molecules in a given volume of fluid is the *solute concentration*.

osmosis: the movement of water across a membrane *toward* the side where the solutes are more concentrated.

osmotic pressure: the amount of pressure needed to prevent the movement of water across a membrane.

phosphate, and sulfate). Cell membranes are *selectively permeable*, meaning that they allow the passage of some molecules, but not others. Whenever electrolytes move across the membrane, water follows.

The movement of water across a membrane toward the more concentrated **solutes** is called **osmosis**. The amount of pressure needed to prevent the movement of water across a membrane is called the **osmotic pressure**. Figure 12-6 presents osmosis, and the photos of salted eggplant and rehydrated raisins provide familiar examples.

Proteins Regulate Flow of Fluids and Ions Chapter 6 describes how proteins attract water and help to regulate fluid movement. It explains that when proteins leak out of the blood vessels into the spaces between the cells, fluids follow and cause the swelling of edema. In addition, transport proteins in the cell membranes regulate the passage of positive ions and other substances from one side of the membrane to the other. Negative ions follow positive ions, and water flows toward the more concentrated solution.

An example of a protein that regulates the flow of fluids and ions in and out of cells is the sodium-potassium pump. The pump actively exchanges sodium for potassium across the cell membrane, using ATP as an energy source. Figure 6-10 on p. 181 illustrates this action.

Regulation of Fluid and Electrolyte Balance The amounts of various minerals in the body must remain nearly constant. Regulation occurs chiefly at two sites: the GI tract and the kidneys.

Minerals in foods enter the body by way of the GI tract. In addition, the digestive juices of the GI tract contain minerals. These minerals and those from foods are absorbed in the large intestine or excreted as needed. Each day, 8 liters of fluids and associated minerals are recycled this way, providing ample opportunity for the regulation of electrolyte balance.

The kidneys' control of the body's *water* content by way of the hormone ADH has already been described (see p. 375). The kidneys regulate the *electrolyte* contents by responding to the hormone aldosterone (also explained on p. 376). If the body's sodium is low, aldosterone stimulates sodium reabsorption from the kidneys. As sodium is reabsorbed, potassium (another positive ion) is excreted in accordance with the rule that total positive charges must remain in balance with total negative charges.

Fluid and Electrolyte Imbalance

Normally, the body defends itself successfully against fluid and electrolyte imbalances. Certain situations and some medications, however, may overwhelm the body's ability to compensate. Severe, prolonged vomiting and diarrhea as well as heavy sweating, burns, and traumatic wounds may incur such great fluid and electrolyte losses as to precipitate a medical emergency.

Different Solutes Lost by Different Routes

Different solutes are lost depending on why fluid is lost. If fluid is lost by vomiting or diarrhea, sodium is lost indiscriminately. If the adrenal glands oversecrete aldosterone, as may occur when they develop a tumor, the kidneys may excrete too much potassium. A person with uncontrolled diabetes may lose glucose, a solute not normally excreted, and large amounts of fluid with it. Each situation results in dehydration, but drinking water alone will not restore electrolyte balance. Medical intervention is required.

Replacing Lost Fluids and Electrolytes

In many cases, people can replace the fluids and minerals lost in sweat or in a temporary bout of diarrhea by drinking plain cool water and eating regular foods. Some cases, however, demand rapid replacement of fluids and electrolytes—for example, when diarrhea threatens the life of a malnourished child. Caregivers around the world have learned to use **oral rehydration therapy (ORT)**—a simple solution of sugar, salt, and water, taken by mouth—to treat dehydration caused by diarrhea. These lifesaving formulas do not require hospitalization and can be prepared from ingredients available locally. Caregivers need only learn to measure ingredients carefully and use sanitary water. Once rehydrated, a person can begin eating foods. (Chapter 14 presents a discussion of sport drinks.)

Acid-Base Balance

The body uses its ions not only to help maintain fluid and electrolyte balance, but also to regulate the acidity (pH) of its fluids. The pH scale introduced in Chapter 3 is repeated here, in Figure 12-7 (p. 380), with the

> **FIGURE 12-6   Osmosis**

Water flows in the direction of the more highly concentrated solution.

  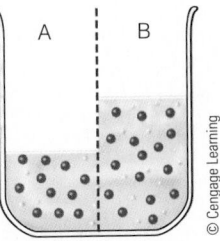

1. With equal numbers of solute particles on both sides of the semipermeable membrane, the concentrations are equal, and the tendency of water to move in either direction is about the same.

2. Now additional solute is added to side B. Solute cannot flow across the divider (in the case of a cell, its membrane).

3. Water can flow both ways across the divider, but has a greater tendency to move from side A to side B, where there is a greater concentration of solute. The volume of water becomes greater on side B, and the concentrations on side A and B become equal.

© Cengage Learning

Physically active people must remember to replace their body fluids.

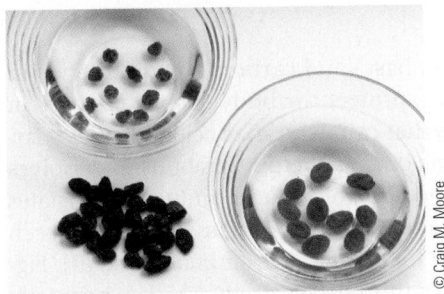

When immersed in water, raisins become plump because water moves toward the higher concentration of sugar inside the raisins.

When sprinkled with salt, eggplant and other vegetables "sweat" because water moves toward the higher concentration of salt outside the vegetable.

oral rehydration therapy (ORT): the administration of a simple solution of sugar, salt, and water, taken by mouth, to treat dehydration caused by diarrhea. A simple ORT recipe (cool before giving):

- ½ L boiling water
- A small handful of sugar (4 tsp)
- 3 pinches of salt (½ tsp)

> FIGURE 12-7   The pH Scale

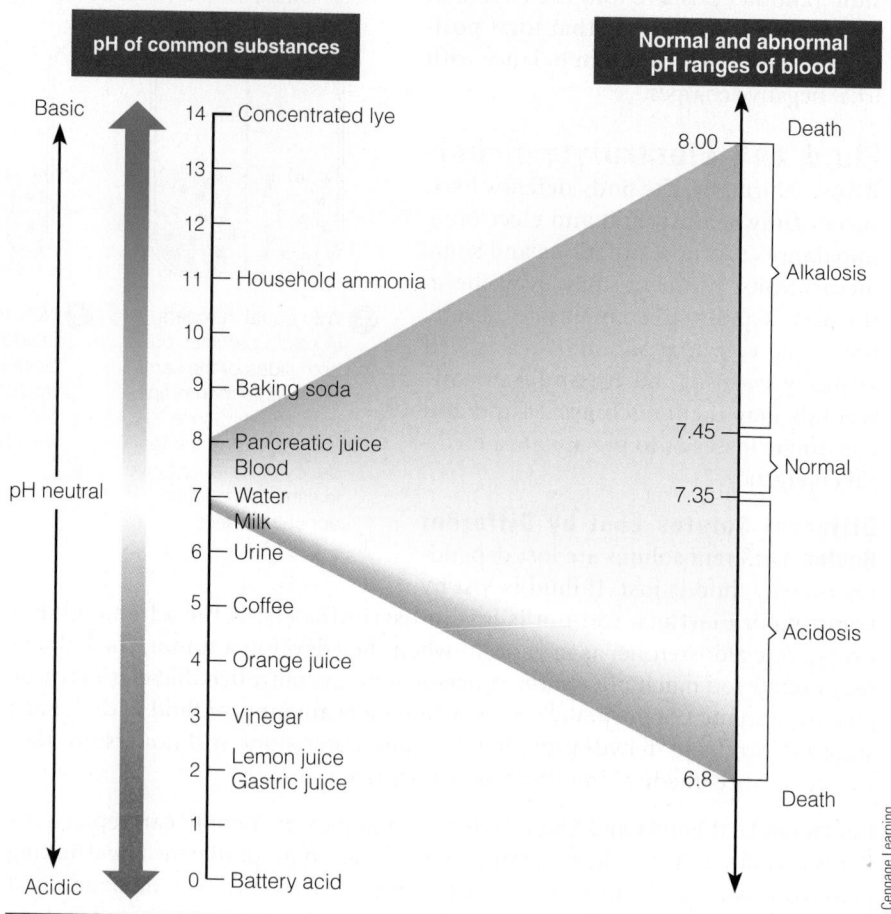

NOTE: Each step is 10 times as concentrated in base (¹⁄₁₀ as much as acid, or H⁺) as the one below it.

normal and abnormal pH ranges of the blood added. As you can see, the body must maintain the pH within a narrow range to avoid life-threatening consequences. Slight deviations in either direction can denature proteins, rendering them useless. Enzymes couldn't catalyze reactions and hemoglobin couldn't carry oxygen—to name just two examples.

The acidity of the body's fluids is determined by the concentration of hydrogen ions (H⁺).* A high concentration of hydrogen ions is acidic. Normal energy metabolism generates hydrogen ions, as well as many other acids, that must be neutralized. Three systems defend the body against fluctuations in pH—buffers in the blood, respiration in the lungs, and excretion in the kidneys.

Regulation by the Buffers **Bicarbonate** (a base) and **carbonic acid** (an acid) in the body fluids, as well as some proteins, protect the body against changes in acidity by acting as **buffers**—substances that can neutralize acids or bases. Carbon dioxide, which is formed all the time during energy metabolism, dissolves in water to form carbonic acid in the blood. Carbonic acid, in turn, dissociates to form hydrogen ions and bicarbonate ions. The appropriate balance between carbonic acid and bicarbonate is essential to maintaining optimal blood pH. Figure 12-8 presents the chemical reactions of this buffer system, which is primarily under the control of the lungs and kidneys.

bicarbonate: an alkaline compound with the formula HCO_3 that is produced in all cell fluids from the dissociation of carbonic acid to help maintain the body's acid-base balance. Bicarbonate is also secreted from the pancreas as part of the pancreatic juice.

carbonic acid: a compound with the formula H_2CO_3 that results from the combination of carbon dioxide (CO_2) and water (H_2O); of particular importance in maintaining the body's acid-base balance.

buffers: compounds that keep a solution's pH constant when acids or bases are added.

*The lower the pH, the higher the H⁺ ion concentration and the stronger the acid. A pH above 7 is alkaline, or base—a solution in which OH⁻ ions predominate.

Respiration in the Lungs The lungs control the concentration of carbonic acid by raising or slowing the respiration rate, depending on whether the pH needs to be increased or decreased. If too much carbonic acid builds up, the respiration rate speeds up; this hyperventilation increases the amount of carbon dioxide exhaled, thereby lowering the carbonic acid concentration and restoring homeostasis. Conversely, if bicarbonate builds up, the respiration rate slows; carbon dioxide is retained and forms more carbonic acid. Again, homeostasis is restored.

Excretion in the Kidneys The kidneys control the concentration of bicarbonate by either reabsorbing or excreting it, depending on whether the pH needs to be increased or decreased, respectively. Their work is complex, but the net effect is easy to sum up. The *body's* total acid burden remains nearly constant; the acidity of the *urine* fluctuates to accommodate that balance.

> **REVIEW IT** Explain how the body regulates fluid balance.

Water makes up about 60 percent of the adult body's weight. It assists with the transport of nutrients and waste products throughout the body, participates in chemical reactions, acts as a solvent, serves as a shock absorber, and regulates body temperature. To maintain water balance, intake from liquids, foods, and metabolism must equal losses from the kidneys, skin, lungs, and GI tract. Whenever the body experiences low blood volume, low blood pressure, or highly concentrated body fluids, the actions of ADH, renin, angiotensin, and aldosterone restore homeostasis. Electrolytes (charged minerals) in the fluids help distribute the fluids inside and outside the cells, thus ensuring the appropriate water balance and acid-base balance to support all life processes. Excessive losses of fluids and electrolytes upset these balances, and the kidneys play a key role in restoring homeostasis.

> FIGURE 12-8 **Bicarbonate–Carbonic Acid Buffer System**

The reversible reactions of the bicarbonate–carbonic acid buffer system help to regulate the body's pH and maintain homeostasis. Recall from Chapter 7 that carbon dioxide and water are formed during energy metabolism.

Carbon dioxide (CO_2) is a volatile gas that quickly dissolves in water (H_2O), forming carbonic acid (H_2CO_3), which lowers the body's pH:

$$CO_2 \quad + \quad H_2O \quad \longleftrightarrow \quad H_2CO_3$$

carbon dioxide   +   water   ⟷   carbonic acid

Carbonic acid readily dissociates to a hydrogen ion (H^+) and a bicarbonate ion (HCO_3^-), which raises the body's pH:

$$H_2CO_3 \quad \longleftrightarrow \quad H^+ \quad + \quad HCO_3^-$$

carbonic acid   ⟷   hydrogen ion   +   bicarbonate ion

© Cengage Learning

12.2 The Minerals—An Overview

> **LEARN IT** List some of the ways minerals differ from vitamins and other nutrients.

Figure 12-9 shows the amounts of the **major minerals** found in the body and, for comparison, some of the **trace minerals**. The distinction between the major and trace minerals does not mean that one group is more important than the other—all minerals are vital. The major minerals are so named because they are present, and needed, in larger amounts in the body. They are shown at the top of the figure and are discussed in this chapter. The trace minerals, shown at the bottom of the figure, are discussed in Chapter 13. A few generalizations pertain to all of the minerals and distinguish them from the vitamins. Especially notable is their chemical nature.

major minerals: essential mineral nutrients the human body requires in relatively large amounts (greater than 100 milligrams per day); sometimes called *macrominerals.*

trace minerals: essential mineral nutrients the human body requires in relatively small amounts (less than 100 milligrams per day); sometimes called *microminerals.*

Inorganic Elements Unlike the organic vitamins, which are easily destroyed, minerals are inorganic elements that always retain their chemical identity. Once minerals enter the body, they remain there until excreted; they cannot be changed into anything else. Iron, for example, may temporarily combine with other charged elements in salts, but it is always iron. Neither can minerals be destroyed by heat, air, acid, or mixing.

> FIGURE 12-9 **Minerals in a 60-kilogram (132-pound) Human Body**

Not only are the major minerals needed by the body in larger amounts, but they are also present in the body in larger amounts than the trace minerals.

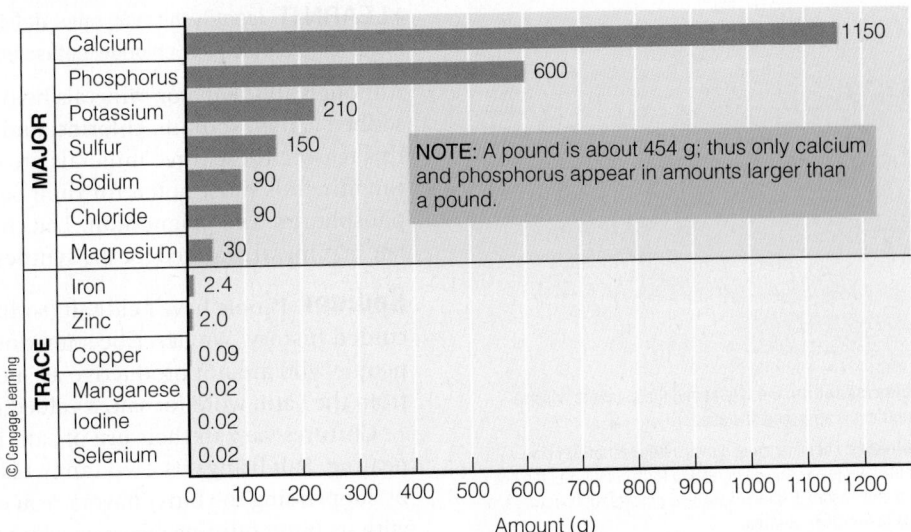

© Cengage Learning

NOTE: A pound is about 454 g; thus only calcium and phosphorus appear in amounts larger than a pound.

Consequently, little care is needed to preserve minerals during food preparation. In fact, the ash that remains when a food is burned contains all the minerals that were in the food originally. Minerals can be lost from food only when they leach into cooking water that is then poured down the drain.

The Body's Handling of Minerals The minerals also differ from the vitamins in the amounts the body can absorb and in the extent to which they must be specially handled. Some minerals, such as potassium, are easily absorbed into the blood, transported freely, and readily excreted by the kidneys, much like the water-soluble vitamins. Other minerals, such as calcium, are more like fat-soluble vitamins in that they must have carriers to be absorbed and transported. And, like some of the fat-soluble vitamins, minerals consumed in excess can be toxic.

Variable Bioavailability The **bioavailability** of minerals varies. Some foods contain **binders** that combine chemically with minerals, preventing their absorption and carrying them out of the body with other wastes. Examples of binders include phytates, which are found primarily in legumes, seeds, nuts, and grains, and oxalates, which are present in rhubarb, beet greens, sweet potatoes, and spinach, among other vegetables. These foods contain more minerals than the body actually receives for use.

Nutrient Interactions Chapter 10 describes how the presence or absence of one vitamin can affect another's absorption, metabolism, and excretion. The same is true of the minerals. The interactions between sodium and calcium, for example, cause both to be excreted when sodium intakes are high. Phosphorus binds with magnesium in the GI tract, so magnesium absorption is limited when phosphorus intakes are high. These are just two examples of the interactions involving minerals featured in this chapter. Discussions in both this chapter and the next point out additional problems that arise from such interactions. Notice how often they reflect an excess of one mineral creating an inadequacy of another and how supplements—not foods—are most often to blame.

> **REVIEW IT** List some of the ways minerals differ from vitamins and other nutrients.
Compared with the trace minerals, major minerals are found, and needed, in larger quantities in the body. Unlike vitamins and the energy-yielding nutrients, minerals are inorganic elements that retain their chemical identities. Minerals usually receive special handling and regulation in the body, and they may bind with other substances or interact with other minerals, thus limiting their absorption.

12.3 The Major Minerals

> **LEARN IT** Identify the main roles, deficiency symptoms, and food sources for each of the major minerals (sodium, chloride, potassium, calcium, phosphorus, magnesium, and sulfate).

Although all the major minerals help to maintain the body's fluid balance as described earlier, sodium, chloride, and potassium are most noted for that role. For this reason, these three minerals are discussed first here. Later sections describe the minerals most noted for their roles in bone growth and health—calcium, phosphorus, and magnesium. The chapter closes with a brief discussion on sulfate, a mineral required for the synthesis of several sulfur-containing compounds.

Sodium People have held salt (sodium chloride) in high regard throughout recorded history. We describe someone we admire as "the salt of the earth" and people who are not productive as "not worth their salt." The word *salary* comes from the Latin word for salt, a valued commodity.

Cultures vary in their use of salt, but most people find its taste innately appealing. Salt brings its own tangy taste and enhances other flavors, most likely by suppressing the bitter flavors. You can taste this effect for yourself: tonic water with its bitter quinine tastes sweeter with a little salt added.

bioavailability: the rate at and the extent to which a nutrient is absorbed and used.

binders: chemical compounds in foods that combine with nutrients (especially minerals) to form complexes the body cannot absorb. Examples include *phytates* (FYE-tates) and *oxalates* (OCK-sa-lates).

Sodium Roles in the Body **Sodium** is the principal cation of the extracellular fluid and the primary regulator of its volume. Sodium also helps maintain acid-base balance and is essential to nerve impulse transmission and muscle contraction.*

Sodium is readily absorbed by the intestinal tract and travels freely in the blood until it reaches the kidneys, which filter all the sodium out of the blood. Then, with great precision, the kidneys return to the blood the exact amount of sodium the body needs. Normally, the amount excreted is approximately equal to the amount ingested on a given day. When blood sodium rises, as when a person eats salted foods, thirst signals the person to drink until the appropriate sodium-to-water concentration is restored. Then the kidneys excrete both the excess water and the excess sodium together. Both too much and too little sodium in the diet increase the risk of heart disease.[6] The key to good health, then, is finding the balance that meets the relatively small need for this essential nutrient but does not exceed the amount that leads to hypertension and heart disease.[7]

Sodium Recommendations Diets rarely lack sodium, and even when intakes are low, the body adapts by reducing sodium losses in urine and sweat, thus making deficiencies unlikely. Sodium recommendations are set low enough to protect against high blood pressure, but high enough to allow an adequate intake of other nutrients with a typical diet. Because high sodium intakes correlate with high blood pressure, the Upper Level (UL) for adults is set at 2300 milligrams per day, as is the Daily Value used on food labels. The average sodium intake in the United States is 3400 milligrams, which exceeds recommendations—and most adults will develop hypertension at some point in their lives.[8]

Sodium and Hypertension For years, a high *sodium* intake was considered the primary factor responsible for high blood pressure. Then research pointed to *salt* (sodium chloride) as the dietary culprit. Salt has a greater effect on blood pressure than either sodium or chloride alone or in combination with other ions. The response to a high salt meal may be immediate, reducing blood flow through arteries; this condition is reversible if such meals are not habitual.[9] The elevation of blood pressure in response to a high-salt diet over years is progressive, and the damage caused to blood vessels is irreversible.

Blood pressure increases in response to excesses in salt intake—most notably for those with hypertension, African Americans, and people older than 40 years of age. For them, a high salt intake correlates strongly with heart disease, and salt restriction (to no more than 1500 milligrams of sodium per day) helps to lower blood pressure.

A salt-restricted diet lowers blood pressure and improves heart disease risk in people without hypertension as well.[10] Because reducing salt intake causes no harm and diminishes the risk of hypertension and heart disease, the *Dietary Guidelines for Americans* advise limiting daily *salt* intake to about 1 teaspoon (the equivalent of about 2.3 grams or 2300 milligrams of *sodium*).[11] The American Heart Association goal is to lower blood pressure by reducing sodium intake to less than 1500 milligrams a day.[12] The "How To" feature on p. 384 offers strategies for cutting salt (and therefore sodium) intake.

> **DIETARY GUIDELINES FOR AMERICANS**
Choose foods low in sodium and prepare foods with little salt. Reduce daily sodium intake to less than 2300 milligrams and further reduce intake to 1500 milligrams among persons who are 51 and older and those of any age who are African American or have hypertension, diabetes, or chronic kidney disease.

Given the current US food supply and typical eating habits, creating a nutritionally balanced diet that meets sodium recommendations can be quite a challenge.[13] One eating pattern, known as the DASH (Dietary Approaches to Stop Hypertension) Eating Plan, is especially effective in lowering blood pressure.[14]

---

*One of the ways the kidneys regulate acid-base balance is by excreting hydrogen ions (H⁺) in exchange for sodium ions (Na⁺).

sodium: the principal cation in the extracellular fluids of the body; critical to the maintenance of fluid balance, nerve impulse transmissions, and muscle contractions.

>How To

Cut Salt (and Sodium) Intake

Salt (sodium chloride) is about 40% sodium and 60% chloride.

- 1 g salt contributes about 400 mg sodium and 600 mg chloride
- 6 g salt = 1 tsp
- 1 tsp salt contributes about 2300 mg sodium and 3700 mg chloride

Most people eat more salt (and therefore sodium) than they need. Some people can lower their blood pressure by avoiding highly salted foods and removing the salt shaker from the table. Foods eaten without salt may seem less tasty at first, but with repetition, people can learn to enjoy the natural flavors of many unsalted foods. Strategies to cut salt intake include:

- Select fresh or frozen vegetables. If buying canned vegetables, drain and rinse in water to remove some of the sodium or select those labeled low-sodium or no-salt-added.
- Cook with little or no added salt.
- Prepare foods with sodium-free herbs and spices such as basil, bay leaves, curry, garlic, ginger, mint, oregano, pepper, rosemary, and thyme; lemon juice; vinegar; or wine.
- Add little or no salt at the table; taste foods before adding salt.
- Read labels with an eye open for sodium. (See the glossary on p. 61 for terms used to describe the sodium contents of foods on labels.)
- Select low-salt or salt-free products when available.

Use these foods sparingly:

- Foods prepared in brine, such as pickles, olives, and sauerkraut
- Salty or smoked meats, such as bologna, corned or chipped beef, bacon, frankfurters, ham, lunchmeats, salt pork, sausage, and smoked tongue
- Salty or smoked fish, such as anchovies, caviar, salted and dried cod, herring, sardines, and smoked salmon
- Snack items such as potato chips, pretzels, salted popcorn, salted nuts, and crackers
- Condiments such as bouillon cubes; seasoned salts; MSG; soy, teriyaki, Worcestershire, and barbeque sauces; prepared horseradish, ketchup, and mustard
- Cheeses, especially processed types
- Canned and instant soups
- Packaged instant or flavored rice, pasta, and cereal mixes

> **TRY IT** Compare the sodium contents of 1 ounce of the following foods: a plain bagel, potato chips, and animal crackers.

Like other USDA Food Patterns, the DASH Eating Plan reflects the *Dietary Guidelines* and allows people to stay within their energy allowance, meet nutrient needs, and reduce chronic disease risk. The DASH approach emphasizes potassium-rich fruits, vegetables, and low-fat milk products; includes whole grains, nuts, poultry, and fish; and calls for reduced intakes of sodium, red and processed meats, sweets, and sugar-containing beverages. In combination with a reduced sodium intake, DASH is even more effective at lowering blood pressure than either strategy alone. In addition, DASH lowers the risk of some cancers, heart disease, and stroke.[15] Chapter 18 offers a complete discussion of hypertension and the dietary recommendations for its prevention and treatment.

Sodium and Bone Loss (Osteoporosis) A high salt intake is also associated with increased calcium excretion, but its influence on bone loss is less clear. In addition, potassium may prevent the calcium excretion caused by a high-salt diet. For these reasons, dietary advice to prevent bone loss parallels that suggested for hypertension—a DASH eating pattern that is low in sodium and abundant in potassium-rich fruits and vegetables and calcium-rich low-fat milk.

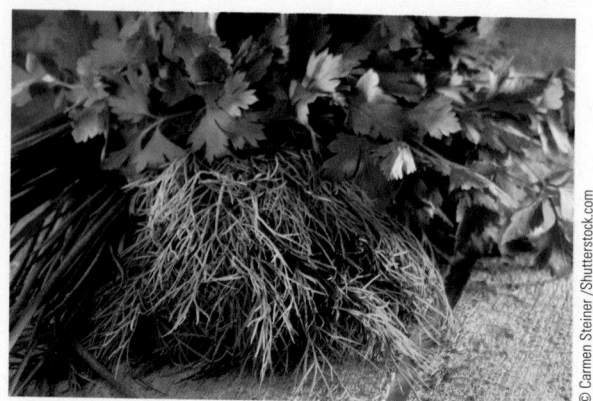

Fresh herbs add flavor to a recipe without adding salt.

© Carmen Steiner /Shutterstock.com

Sodium in Foods In general, processed foods have the most sodium, whereas unprocessed foods such as fresh fruits and vegetables have the least. In fact, as much as 75 percent of the sodium in people's diets comes from salt added to foods by manufacturers; about 15 percent comes from salt added during cooking and at the table; and only 10 percent comes from the natural content in foods. Among foods with the highest sodium density (milligrams of sodium per kcalorie) are those from fast food and pizza restaurants.[16] Because sodium intake tends to increase as kcalories increase, making food choices based on low sodium density is a practical and effective way to lower sodium intake.[17]

To help consumers limit their intake, public health organizations and policymakers worldwide are calling for manufacturers and restaurants to reduce sodium in the food supply.[18] In addition to reducing the sodium content of foods, food scientists are designing products to enhance salty perceptions with less salt.[19] Reducing the sodium content in processed foods could prevent an estimated 100,000 deaths and save up to $24 billion in health care costs in the United States annually.[20]

Because processed foods may contain sodium without chloride, as in additives such as sodium bicarbonate or sodium saccharin, they do not always taste salty. Most people are surprised to learn that 1 ounce of some cereals contains more sodium than 1 ounce of salted peanuts—and that ½ cup of instant chocolate pudding contains still more. The peanuts taste saltier because the salt is all on the surface, where the tongue's taste receptors immediately pick it up.

Figure 12-10 shows that processed foods not only contain more sodium than their less-processed counterparts but also have less potassium. Low potassium may be as significant as high sodium when it comes to blood pressure regulation, so processed foods have (at least) two strikes against them.

Sodium Deficiency Sodium deficiency does not develop from an inadequate diet. The body needs so little and diets provide enough. Blood sodium may drop with vomiting, diarrhea, or heavy sweating, and in these cases, both sodium and water must be replenished. Under normal conditions of sweating due to physical activity, salt losses can easily be replaced later in the day with ordinary foods. Salt tablets are not recommended because too much salt, especially if taken with too little water, can induce dehydration. During intense activities, such as ultra-endurance events, athletes can lose so much sodium and drink so much water that they develop hyponatremia—the dangerous condition of having too little sodium in the blood. Symptoms of hyponatremia include headache, confusion, stupor, seizures, and coma. Importantly, hyponatremia is caused by excessive sodium losses, not from inadequate sodium intake. (Chapter 14 offers details about hyponatremia and guidelines for ultra-endurance athletes.)

Sodium Toxicity and Excessive Intakes The immediate symptoms of acute sodium toxicity are edema and high blood pressure. Prolonged excessive sodium intake may contribute to hypertension in some people, as explained earlier.

> FIGURE 12-10   **What Processing Does to the Sodium and Potassium Contents of Foods**

People who eat foods high in salt often happen to be eating fewer potassium-containing foods at the same time. Notice how potassium is lost and sodium is gained as foods become more processed, causing the potassium-to-sodium ratio to fall dramatically. Even when potassium isn't lost, the addition of sodium still lowers the potassium-to-sodium ratio. Selecting fresh, unprocessed foods lowers blood pressure in two ways, then—by lowering sodium intakes and by raising potassium intakes.

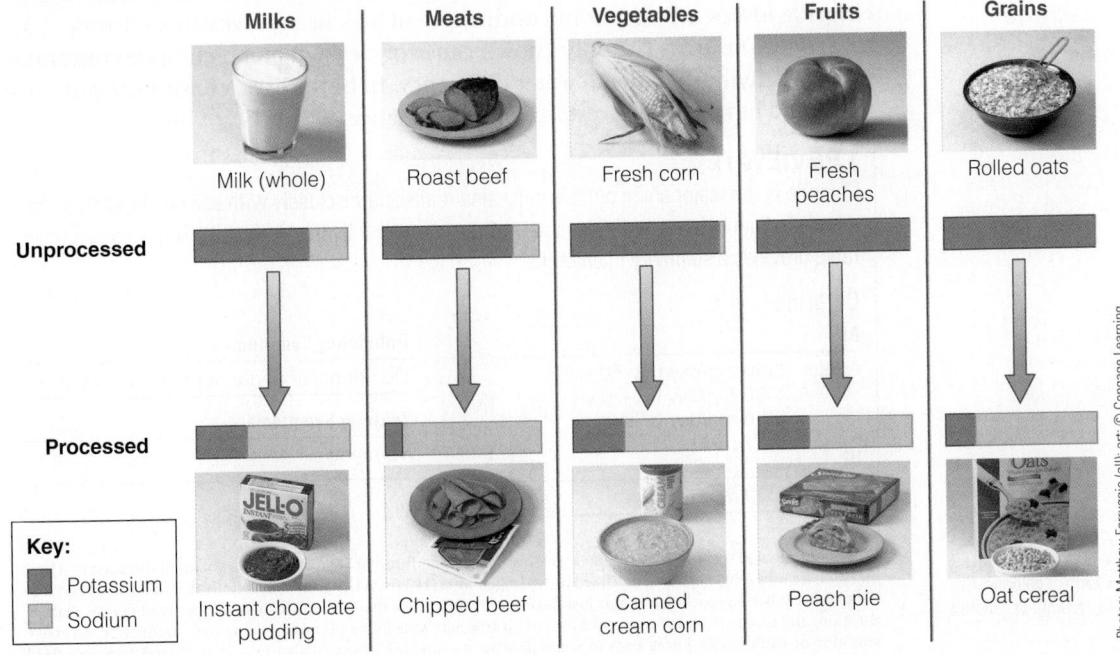

Sodium is the main cation outside cells and one of the primary electrolytes responsible for maintaining fluid balance. Dietary deficiency is unlikely, and excesses raise blood pressure in many people. For this reason, health professionals advise a diet moderate in salt and sodium. The accompanying table provides a summary of sodium.

Sodium

AI	Deficiency Symptoms
Adults: 1500 mg/day (19–50 yr) 1300 mg/day (51–70 yr) 1200 mg/day (>70 yr)	Not from inadequate intakes Hyponatremia from excessive losses
UL	**Toxicity Symptoms**
Adults: 2300 mg/day	Edema, acute hypertension
Chief Functions in the Body	**Significant Sources**
Maintains normal fluid and electrolyte balance; assists in nerve impulse transmission and muscle contraction	Table salt, soy sauce; moderate amounts in meats, milks, breads, and vegetables; large amounts in processed foods

Chloride

The element *chlorine* (Cl_2) is a poisonous gas. When chlorine reacts with sodium or hydrogen, however, it forms the negative chloride ion (Cl^-). *Chloride*, an essential nutrient, is required in the diet.

Chloride Roles in the Body Chloride is the major anion of the extracellular fluids (outside the cells), where it occurs mostly in association with sodium. Chloride moves passively across membranes through channels and so also associates with potassium inside cells. Like sodium and potassium, chloride maintains fluid and electrolyte balance.

In the stomach, the chloride ion is part of hydrochloric acid, which maintains the strong acidity of gastric juice. One of the most serious consequences of vomiting is the loss of this acid from the stomach, which upsets the acid-base balance.* Such imbalances are commonly seen in bulimia nervosa, as described in Highlight 8.

Chloride Recommendations and Intakes Chloride is abundant in foods (especially processed foods) as part of sodium chloride and other salts. Chloride recommendations are slightly higher than, but still equivalent to, those of sodium. In other words, ¾ teaspoon of salt will deliver some sodium, more chloride, and still meet the AI for both.

Chloride Deficiency and Toxicity Diets rarely lack chloride. Like sodium losses, chloride losses may occur in conditions such as heavy sweating, chronic diarrhea, and vomiting. The only known cause of elevated blood chloride concentrations is dehydration due to water deficiency. In both cases, consuming ordinary foods and beverages can restore chloride balance.

Chloride is the major anion outside cells, and it associates closely with sodium. In addition to its role in fluid balance, chloride is part of the stomach's hydrochloric acid. The accompanying table provides a summary of chloride.

Chloride

AI	Deficiency Symptoms
Adults: 2300 mg/day (19–50 yr) 2000 mg/day (51–70 yr) 1800 mg/day (>70 yr)	Do not occur under normal circumstances
UL	**Toxicity Symptoms**
Adults: 3600 mg/day	Vomiting

chloride (KLO-ride): the major anion in the extracellular fluids of the body. Chloride is the ionic form of chlorine, Cl^-. See Appendix B for a description of the chlorine-to-chloride conversion.

---

*Hydrochloric acid secretion into the stomach involves the addition of bicarbonate ions (base) to the plasma. These bicarbonate ions (HCO_3^-) are neutralized by hydrogen ions (H^+) from the gastric secretions that are reabsorbed into the plasma. When hydrochloric acid is lost during vomiting, these hydrogen ions are no longer available for reabsorption, and so, in effect, the concentrations of bicarbonate ions in the plasma are increased. In this way, excessive vomiting of acidic gastric juices leads to *metabolic alkalosis*—an above-normal alkalinity in the blood and body fluids.

Chief Functions in the Body	Significant Sources
Maintains normal fluid and electrolyte balance; part of hydrochloric acid found in the stomach, necessary for proper digestion	Table salt, soy sauce; moderate amounts in meats, milks, eggs; large amounts in processed foods

Potassium Like sodium, **potassium** is a positively charged ion. In contrast to sodium, potassium is the body's principal intracellular cation, *inside* the body cells.

Potassium Roles in the Body Potassium plays a major role in maintaining fluid and electrolyte balance and cell integrity. During nerve transmissions and muscle contractions, potassium and sodium briefly trade places across the cell membrane. The cell then quickly pumps them back into place. Controlling potassium distribution is a high priority for the body because it affects many aspects of homeostasis, including a steady heartbeat.

Potassium Recommendations and Intakes Potassium is abundant in all living cells. Because cells remain intact unless foods are processed, the richest sources of potassium are *fresh* foods—as Figure 12-11 shows. In contrast, most processed foods such as canned vegetables, ready-to-eat cereals, and luncheon meats contain less potassium—and more sodium (recall Figure 12-10, p. 385). To meet the AI for potassium, most people need to increase their daily intake of fruits and vegetables.

> DIETARY GUIDELINES FOR AMERICANS

Choose foods that provide more potassium, a nutrient of concern in American diets. Potassium is found in all food groups, notably vegetables, fruits, and milk and milk products.

potassium: the principal cation within the body's cells; critical to the maintenance of fluid balance, nerve impulse transmissions, and muscle contractions.

> FIGURE 12-11 **Potassium in Selected Foods**

Food	Serving size (kcalories)
Bread, whole wheat	1-oz slice (70 kcal)
Cornflakes, fortified	1 oz (110 kcal)
Spaghetti pasta	½ c cooked (99 kcal)
Tortilla, flour	1 10" round (234 kcal)
Broccoli	½ c cooked (22 kcal)
Carrots	½ c shredded raw (24 kcal)
Potato	1 medium baked w/skin (133 kcal)
Tomato juice	¾ c (31 kcal)
Banana	1 medium raw (109 kcal)
Orange	1 medium raw (62 kcal)
Strawberries	½ c fresh (22 kcal)
Watermelon	1 slice (92 kcal)
Milk	1 c reduced-fat 2% (121 kcal)
Yogurt, plain	1 c low-fat (155 kcal)
Cheddar cheese	1½ oz (171 kcal)
Cottage cheese	½ c low-fat 2% (101 kcal)
Pinto beans	½ c cooked (117 kcal)
Peanut butter	2 tbs (188 kcal)
Sunflower seeds	1 oz dry (165 kcal)
Tofu (soybean curd)	½ c (76 kcal)
Ground beef, lean	3 oz broiled (244 kcal)
Chicken breast	3 oz roasted (140 kcal)
Tuna, canned in water	3 oz (99 kcal)
Egg	1 hard cooked (78 kcal)
Excellent, and sometimes unusual, sources:	
Squash, acorn	½ c baked (69 kcal)
Soybeans	½ c cooked (149 kcal)
Artichoke	1 (60 kcal)

Milligrams: 0, 200, 400, 600, 800

The AI for potassium is 4700 mg per day.

POTASSIUM
Fresh fruits (purple); vegetables (green); legumes, nuts, and seeds (brown); and meats, poultry, and seafood (red) contribute potassium to the diet.

Key:
- Grains
- Vegetables
- Fruits
- Milk and milk products
- Legumes, nuts, seeds
- Meats, poultry, seafood
- Best sources per kcalorie

© Cengage Learning

Fresh foods, especially fruits and vegetables, provide potassium in abundance.

© Polara Studios, Inc.

Potassium and Hypertension Diets low in potassium, especially when combined with high sodium intakes, raise blood pressure and increase the risk of death from heart disease.[21] In contrast, high potassium intakes, especially when combined with low sodium intakes, appear to both prevent and correct hypertension. Unfortunately, most US adults consume too much sodium and too little potassium.[22] Recall that the DASH eating pattern described earlier is used to lower blood pressure and emphasizes potassium-rich foods such as fruits and vegetables. Potassium-rich fruits and vegetables also appear to reduce the risk of strokes and heart attacks—more so than can be explained by the reduction in blood pressure alone.[23]

Potassium Deficiency Potassium deficiency is characterized by an increase in blood pressure, kidney stones, and bone turnover. As deficiency progresses, symptoms include irregular heartbeats, muscle weakness, and glucose intolerance.

Potassium Toxicity Potassium toxicity does not result from overeating foods high in potassium; therefore a UL has not been set. It can result from overconsumption of potassium salts or supplements (including some "energy fitness shakes") and from certain diseases or treatments. Given more potassium than the body needs, the kidneys accelerate excretion. If potassium is injected directly into a vein, however, it can stop the heart.

> ### › REVIEW IT

Potassium, like sodium and chloride, is an electrolyte that plays an important role in maintaining fluid balance. Potassium is the primary cation inside cells; fresh foods, notably fruits and vegetables, are its best sources. The accompanying table provides a summary of potassium.

Potassium

AI	Toxicity Symptoms
Adults: 4700 mg/day	Muscular weakness; vomiting; if given into a vein, can stop the heart
Chief Functions in the Body	
Maintains normal fluid and electrolyte balance; facilitates many reactions; supports cell integrity; assists in nerve impulse transmission and muscle contractions	**Significant Sources**
	All whole foods: meats, milks, fruits, vegetables, grains, legumes
Deficiency Symptoms[a]	
Irregular heatbeat, muscular weakness, glucose intolerance	

[a]Deficiency accompanies dehydration.

Calcium

Calcium is the most abundant mineral in the body. It receives much emphasis in this chapter and in the highlight that follows because an adequate intake helps grow a healthy skeleton in early life and minimize bone loss in later life.

Calcium Roles in the Body Only 1 percent of the body's calcium is in the body fluids. The remaining 99 percent of the body's calcium is in the bones (and teeth), where it plays two roles. First, it is an integral part of bone structure, providing a rigid frame that holds the body upright and serves as attachment points for muscles, making motion possible. Second, it serves as a calcium bank, offering a readily available source of calcium to the body fluids should a drop in blood calcium occur.

As bones begin to form, calcium salts form crystals, called **hydroxyapatite,** on a matrix of the protein collagen. During **mineralization,** as the crystals become denser, they give strength and rigidity to the maturing bones. As a result, the long leg bones of children can support their weight by the time they have learned to walk.

Many people have the idea that once a bone is built, it is inert like a rock. Actually, the bones are gaining and losing minerals continuously in an ongoing process of remodeling. Growing children gain more bone than they lose, and healthy adults maintain a reasonable balance. When withdrawals substantially exceed deposits, problems such as osteoporosis develop (as described in Highlight 12).

The formation of teeth follows a pattern similar to that of bones. The turnover of minerals in teeth is not as rapid as in bone, however; fluoride hardens and stabilizes the crystals of teeth, opposing the withdrawal of minerals from them.

calcium: the most abundant mineral in the body; found primarily in the body's bones and teeth.

hydroxyapatite (high-drox-ee-APP-ah-tite): crystals made of calcium and phosphorus.

mineralization: the process in which calcium, phosphorus, and other minerals crystallize on the collagen matrix of a growing bone, hardening the bone.

Although only 1 percent of the body's calcium circulates in the extracellular and intracellular fluids, its presence there is vital to life. Cells throughout the body can detect calcium in the extracellular fluids and respond accordingly. Many of calcium's actions help to maintain normal blood pressure, perhaps by stabilizing the smooth muscle cells of the blood vessels or by releasing relaxing factors from the blood vessel cell walls. Extracellular calcium also participates in blood clotting.

The calcium in intracellular fluids binds to proteins within the cells and activates them. For example, when the protein **calmodulin** binds with calcium, it activates the enzymes involved in breaking down glycogen, which releases energy for muscle contractions. Many such proteins participate in the regulation of muscle contractions, the transmission of nerve impulses, the secretion of hormones, and the activation of some enzyme reactions.

Calcium in Disease Prevention Calcium may protect against some chronic diseases, including hypertension.[24] Considering the success of DASH in lowering blood pressure, restricting sodium to treat hypertension may be narrow advice. The DASH eating pattern is rich in calcium, as well as in magnesium and potassium—all of which help lower blood pressure.

Calcium-rich foods may play a role in reducing body fat, protecting lean tissue, and maintaining a healthy body weight.[25] Some epidemiological studies suggest an inverse relationship between calcium intake and body weight: the higher the calcium intake, the lower the prevalence of overweight. Clinical studies, however, report such small losses (1 to 2 pounds) as to be statistically insignificant.[26] Some would argue that the real-life benefits are significant in that weight gains are diminished and body composition is improved.[27] In addition, calcium-rich foods suppress the inflammation commonly associated with overweight, even without weight loss.[28] Importantly, calcium-rich foods help with weight loss only when used within an energy-restricted diet.[29]

Calcium Balance Calcium homeostasis involves a system of hormones and vitamin D. Whenever blood calcium falls too low or rises too high, three organ systems respond: the intestines, bones, and kidneys. Figure 12-12 illustrates how vitamin D and two hormones—**parathyroid hormone** and **calcitonin**—return blood calcium to normal.

The calcium in bones provides a nearly inexhaustible bank of calcium for the blood. The blood borrows and returns calcium as needed so that even with an inadequate diet, *blood* calcium remains normal—even as *bone* calcium diminishes (see Figure 12-13 on p. 390). Blood calcium changes only in response to abnormal regulatory control, not to diet. A person can have an inadequate calcium intake for years and have no noticeable symptoms. Only later in life does it become apparent that bone integrity has been compromised.

calmodulin (cal-MOD-you-lin): a calcium-binding protein that regulates such cell activities as muscle contractions.

parathyroid hormone: a hormone from the parathyroid glands that regulates blood calcium by raising it when levels fall too low; also known as *parathormone* (PAIR-ah-THOR-moan).

calcitonin (KAL-seh-TOE-nin): a hormone secreted by the thyroid gland that regulates blood calcium by lowering it when levels rise too high.

> **FIGURE 12-12 Calcium Balance**

Low blood calcium		High blood calcium
Signals the parathyroid glands to secrete parathyroid hormone into the blood	Thyroid gland with parathyroid glands embedded	Signals the thyroid gland to secrete calcitonin

Vitamin D	Parathyroid hormone		Calcitonin
	Stimulates the activation of vitamin D		Inhibits the activation of vitamin D
Stimulates calcium reabsorption from the kidneys into the blood	Stimulates calcium reabsorption from the kidneys into the blood	Kidneys	Prevents calcium reabsorption in the kidneys
Enhances calcium absorption in the intestines			Limits calcium absorption in the intestines
		Intestines	
Stimulates osteoclast cells to break down bone, releasing calcium into the blood	Stimulates osteoclast cells to break down bone, releasing calcium into the blood	Bones	Inhibits osteoclast cells from breaking down bone, preventing the release of calcium

End results		End results
Raised blood calcium	Raised blood calcium	Lower blood calcium
	Parathyroid hormone secretion inhibited	Calcitonin secretion inhibited

© Cengage Learning

NOTE: Calcitonin plays a major role in defending infants and young children against the dangers of rising blood calcium that can occur when regular feedings of milk deliver large quantities of calcium to a small body. In contrast, calcitonin plays a relatively minor role in adults because their absorption of calcium is less efficient and their bodies are larger, making elevated blood calcium unlikely.

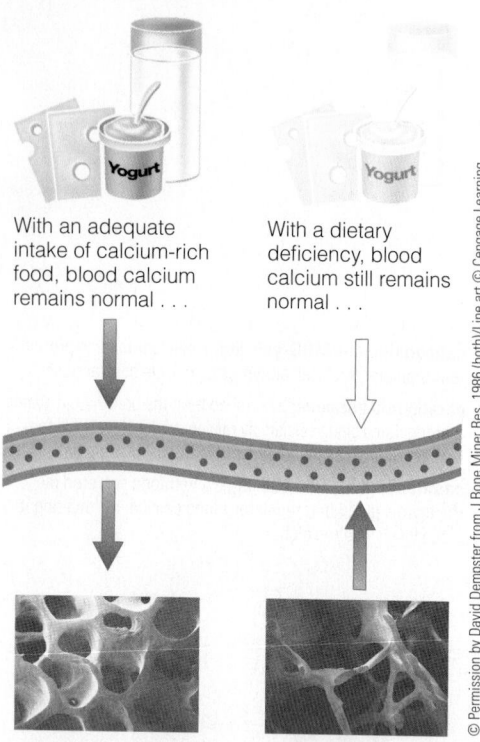

With an adequate intake of calcium-rich food, blood calcium remains normal . . .

With a dietary deficiency, blood calcium still remains normal . . .

. . . and bones deposit calcium. The result is strong, dense bones.

. . . because bones give up calcium to the blood. The result is weak, osteoporotic bones.

© Permission by David Dempster from J Bone Miner Res, 1986 (both)/Line art © Cengage Learning

Blood calcium above normal results in **calcium rigor:** the muscles contract and cannot relax. Similarly, blood calcium below normal causes **calcium tetany**—also characterized by uncontrolled muscle contraction. These conditions do *not* reflect a *dietary* excess or lack of calcium; they are caused by a lack of vitamin D or by abnormal secretion of the regulatory hormones. A chronic *dietary* deficiency of calcium, or a chronic deficiency due to poor absorption over the years, depletes the bones. Again: the *bones*, not the blood, are robbed by a calcium deficiency.

Calcium Absorption Because many factors affect calcium absorption, the most effective way to ensure adequacy is to increase calcium intake. On average, adults absorb about 30 percent of the calcium they ingest. The stomach's acidity helps to keep calcium soluble, and vitamin D helps to make the **calcium-binding protein** needed for absorption. This relationship explains why calcium-rich milk is a good choice for vitamin D fortification.

Whenever calcium is needed, the body increases its calcium absorption. The result is obvious in the case of a newborn infant, whose calcium absorption is 55 to 60 percent. Similarly, a pregnant woman doubles her absorption of calcium. Growing children and teens absorb up to 50 percent of the calcium they consume. Then, when bone growth slows or stops, absorption falls to the adult level of about 30 percent. In addition, absorption becomes more efficient during times of inadequate intakes.

Many of the conditions that enhance calcium absorption limit its absorption when they are absent. For example, sufficient vitamin D supports absorption, and a deficiency impairs it. In addition, fiber in general, and the binders phytate and oxalate in particular, interfere with calcium absorption, but their effects are relatively minor in typical US diets. Vegetables with oxalates and whole grains with phytates are nutritious foods, of course, but they are not useful calcium sources.

Calcium Recommendations Calcium is unlike most other nutrients in that hormones maintain its *blood* concentration regardless of dietary intake. As Figure 12-13 shows, when calcium intake is high, the *bones* benefit; when intake is low, the *bones* suffer. Calcium recommendations are therefore based on the amount needed to retain the most calcium in bones. By retaining the most calcium possible, the bones can develop to their fullest potential in size and density—their **peak bone mass**—within genetic limits.

Calcium recommendations have been set high enough to accommodate a 30 percent absorption rate. Because obtaining enough calcium during growth helps to ensure that the skeleton will be strong and dense, the recommendation for adolescents to the age of 18 years is 1300 milligrams daily. Between the ages of 19 and 50, recommendations are lowered to 1000 milligrams a day; for women older than 50 and all adults older than 70, recommendations are raised again to 1200 milligrams a day to minimize the bone loss that tends to occur later in life. Some authorities advocate as much as 1500 milligrams a day for women older than 50. Most people in the United States have calcium intakes below current recommendations. Those meeting recommendations for calcium are likely to be using calcium supplements.[30] High intakes of calcium from supplements may have adverse effects such as kidney stone formation. For this reason, a UL has been established.

A high-protein diet increases urinary calcium losses, but does not seem to impair bone health.[31] In fact, protein may even improve calcium absorption and bone strength. The DRI Committee considered these nutrient interactions in establishing the RDA for calcium and did not adjust dietary recommendations based on this information.[32]

Calcium Food Sources Figure 12-14 shows that calcium is found most abundantly in a single food group—milk and milk products. The person who doesn't like to drink milk may prefer to eat cheese or yogurt. Alternatively, milk and milk products can be concealed in foods. Powdered fat-free milk can be added to casseroles, soups, and other mixed dishes during preparation; 5 heaping tablespoons offer the equivalent of 1 cup of milk. This simple step is an excellent way for older women to obtain not only extra calcium, but more protein, vitamins, and minerals as well.

calcium rigor: hardness or stiffness of the muscles caused by high blood calcium concentrations.

calcium tetany (TET-ah-nee): intermittent spasm of the extremities due to nervous and muscular excitability caused by low blood calcium concentrations.

calcium-binding protein: a protein in the intestinal cells, made with the help of vitamin D, that facilitates calcium absorption.

peak bone mass: the highest attainable bone density for an individual, developed during the first three decades of life.

> FIGURE 12-14 **Calcium in Selected Foods**

Milligrams

Food	Serving size (kcalories)
Bread, whole wheat	1-oz slice (70 kcal)
Cornflakes, fortified	1 oz (110 kcal)
Spaghetti pasta	½ c cooked (99 kcal)
Tortilla, flour	1 10" round (234 kcal)
Broccoli	½ c cooked (22 kcal)
Carrots	½ c shredded raw (24 kcal)
Potato	1 medium baked w/skin (133 kcal)
Tomato juice	¾ c (31 kcal)
Banana	1 medium raw (109 kcal)
Orange	1 medium raw (62 kcal)
Strawberries	½ c fresh (22 kcal)
Watermelon	1 slice (92 kcal)
Milk	1 c reduced-fat 2% (121 kcal)
Yogurt, plain	1 c low-fat (155 kcal)
Cheddar cheese	1½ oz (171 kcal)
Cottage cheese	½ c low-fat 2% (101 kcal)
Pinto beans	½ c cooked (117 kcal)
Peanut butter	2 tbs (188 kcal)
Sunflower seeds	1 oz dry (165 kcal)
Tofu (soybean curd)[a]	½ c (76 kcal)
Ground beef, lean	3 oz broiled (244 kcal)
Chicken breast	3 oz roasted (140 kcal)
Tuna, canned in water	3 oz (99 kcal)
Egg	1 hard cooked (78 kcal)
Excellent, and sometimes unusual, sources:	
Sardines, with bones[b]	3 oz canned (176 kcal)
Bok choy (Chinese cabbage)	½ c cooked (10 kcal)
Almonds	1 oz (167 kcal)

CALCIUM

As in the riboflavin figure, milk and milk products (white) dominate the calcium figure. Most people need at least three selections from the milk group to meet recommendations.

[a]Values based on products containing added calcium salts; the calcium in ½ c soybeans is about two-thirds as much as in ½ c tofu.
[b]If bones are discarded, calcium declines dramatically.

RDA for adults 19–50
RDA for women 51+
RDA for men 51–70
RDA for men 71+

Key:
- Grains
- Vegetables
- Fruits
- Milk and milk products
- Legumes, nuts, seeds
- Meats, poultry, seafood

Best sources per kcalorie

© Cengage Learning

It is especially difficult for children who don't drink milk to meet their calcium needs. The consequences of drinking too little milk during childhood and adolescence persist into adulthood. Women who seldom drank milk as children have lower bone density and greater risk of fractures than those who drank milk regularly. It is possible for people who do not drink milk to obtain adequate calcium, but only if they carefully select other calcium-rich foods.

> **DIETARY GUIDELINES FOR AMERICANS**
Choose foods that provide more calcium, a nutrient of concern in American diets. The best sources of calcium are milk and milk products.

Many people, for a variety of reasons, cannot or do not drink milk. Some cultures do not use milk in their cuisines; some vegetarians exclude milk as well as meat; and some people are allergic to milk protein or are lactose intolerant. Others simply do not enjoy the taste of milk. These people need to find other foods to help meet their calcium needs. Some brands of tofu, corn tortillas, some nuts (such as almonds), and some seeds (such as sesame seeds) can supply calcium for the person who doesn't use milk products. A slice of most breads contains only about 5 to 10 percent of the calcium found in milk, but it can be a major source for people who eat many slices because the calcium is well absorbed. Oysters are also a rich source of calcium, as are small fish eaten with their bones, such as canned sardines.

Among the vegetables, mustard and turnip greens, bok choy, kale, parsley, watercress, and broccoli are good sources of available calcium. So are some seaweeds such as the nori popular in Japanese cooking. Some dark green, leafy

Milk and milk products are well known for their calcium, but calcium-set tofu, bok choy, kale, calcium-fortified orange juice, and broccoli are also rich in calcium.

© Matthew Farruggio

> FIGURE 12-15 **Bioavailability of Calcium from Selected Foods**

≥50% absorbed	Cauliflower, watercress, cabbage, brussels sprouts, rutabaga, kale, mustard greens, bok choy, broccoli, turnip greens
≈30% absorbed	Milk, calcium-fortified soy milk, calcium-set tofu, cheese, yogurt, calcium-fortified foods and beverages
≈20% absorbed	Almonds, sesame seeds, pinto beans, sweet potatoes
≤5% absorbed	Spinach, rhubarb, Swiss chard

© Cengage Learning

vegetables—notably spinach and Swiss chard—appear to be calcium-rich but actually provide little, if any, calcium because they contain binders that limit absorption. It would take 8 cups of spinach—containing six times as much calcium as 1 cup of milk—to deliver the equivalent in *absorbable* calcium.

With the exception of foods such as spinach that contain calcium binders, the calcium content of foods is usually more important than bioavailability. Consequently, recognizing that people eat a variety of foods containing calcium, the DRI Committee did not adjust for calcium bioavailability when setting recommendations. Figure 12-15 ranks selected foods according to their calcium bioavailability.

Some mineral waters provide as much as 500 milligrams of calcium per liter, offering a convenient way to meet both calcium and water needs. Similarly, calcium-fortified orange juice and other fruit and vegetable juices allow a person to obtain both calcium and vitamins easily. Other examples of calcium-fortified foods include high-calcium milk (milk with extra calcium added) and calcium-fortified cereals. Fortified juices and foods help consumers increase calcium intakes, but depending on the calcium sources, the bioavailability may be significantly less than quantities listed on food labels. The accompanying "How To" feature describes a quick way to estimate calcium intake. Highlight 12 discusses calcium supplements.

>How To

Estimate Your Calcium Intake

Most dietitians have developed useful shortcuts to help them estimate nutrient intakes and "see" inadequacies in the diet. They can tell at a glance whether a day's meals fall short of calcium recommendations, for example.

To estimate calcium intakes, keep two bits of information in mind:

- A cup of milk provides about 300 milligrams of calcium.
- Adults need between 1000 and 1200 milligrams of calcium per day, which represents 3 to 4 cups of milk—or the equivalent:

$$1000 \text{ mg} \div 300 \text{ mg/c} = 3\tfrac{1}{3} \text{ c}$$
$$1200 \text{ mg} \div 300 \text{ mg/c} = 4 \text{ c}$$

If a person drinks 3 to 4 cups of milk a day, it's easy to see that calcium needs are being met. If not, it takes some detective work to identify the other sources and estimate total calcium intake.

To estimate a person's daily calcium intake, use this shortcut, which compares the calcium in calcium-rich foods to the calcium content of milk. The calcium in a cup of milk is assigned 1 point,

and the goal is to attain 3 to 4 points per day. Foods are given points as follows:

- 1 c milk, yogurt, or fortified soy milk or 1½ oz cheese = 1 point
- 4 oz canned fish with bones (sardines) = 1 point
- 1 c ice cream, cottage cheese, or calcium-rich vegetable (see the text) = ½ point

Then, because other foods also contribute small amounts of calcium, together they are given a point.

- Well-balanced diet containing a variety of foods = 1 point

Now consider a day's meals with calcium in mind. Cereal with 1 cup of milk for breakfast (1 point for milk), a ham and cheese sub sandwich for lunch (1 point for cheese), and a cup of broccoli and lasagna for dinner (½ point for calcium-rich vegetable and 1 point for cheese in lasagna)—plus 1 point for all other foods eaten that day—adds up to 4½ points. This shortcut estimate indicates that calcium recommendations have been met, and a diet analysis of these few foods reveals a calcium intake of more than 1000 milligrams. By knowing the best sources of each nutrient, you can learn to scan the day's meals and quickly see if you are meeting your daily goals.

> **TRY IT** Compare the calcium contents of ½ cup of the following foods: almonds, broccoli, and yogurt.

> **FIGURE 12-16   Phases of Bone Development throughout Life**

The active growth phase occurs from birth to approximately age 20. The phase of peak bone mass development occurs between the ages of 12 and 30. The final phase, when bone resorption exceeds formation, begins between the ages of 30 and 40 and continues through the remainder of life.

A generalization that has been gaining strength throughout this book is supported by the information given here about calcium. A balanced diet that supplies a variety of foods is the best plan to ensure adequacy for all essential nutrients. All food groups should be included, and none should be overemphasized. In our culture, calcium intake is usually inadequate wherever milk is lacking in the diet. By contrast, iron is usually lacking whenever milk is overemphasized, as Chapter 13 explains.

Calcium Deficiency A low calcium intake during the growing years limits the bones' ability to reach their peak bone mass. Most people achieve a peak bone mass by their late 20s, and dense bones best protect against age-related bone loss and fractures (see Figure 12-16). All adults lose bone as they grow older, beginning between the ages of 30 and 40. When bone losses reach the point of causing fractures under common, everyday stresses, the condition is known as **osteoporosis**. Osteoporosis and low bone mass (osteopenia) affect an estimated 52 million people in the United States, mostly older women.[33]

Unlike many diseases that make themselves known through symptoms such as pain, shortness of breath, skin lesions, tiredness, and the like, osteoporosis is silent. The body sends no signals saying bones are losing their calcium and, as a result, their integrity. Blood samples offer no clues because blood calcium remains normal regardless of bone content, and measures of bone density are not routinely taken until later in life. Highlight 12 suggests strategies to protect against bone loss, of which eating calcium-rich foods is only one.

> **REVIEW IT**

Most of the body's calcium is in the bones, where it provides a rigid structure and a reservoir of calcium for the blood. Blood calcium participates in muscle contraction, blood clotting, and nerve impulses, and it is closely regulated by a system of hormones and vitamin D. Calcium is found predominantly in milk and milk products. Even when calcium intake is inadequate, blood calcium remains normal, but at the expense of bone loss, which can lead to osteoporosis. The accompanying table provides a summary of calcium.

Calcium

RDA	**Deficiency Symptoms**
Adults: 1000 mg/day (adults, 19–50 yr) 1000 mg/day (men, 51–70 yr) 1200 mg/day (men, ≥71 yr) 1200 mg/day (women, ≥51 yr)	Stunted growth in children; bone loss (osteoporosis) in adults
	Toxicity Symptoms
UL	Constipation; increased risk of urinary stone formation and kidney dysfunction; interference with absorption of other minerals
Adults: 2500 mg/day (adults, 19–50 yr) 2000 mg/day (adults, ≥51 yr)	
Chief Functions in the Body	**Significant Sources**
Mineralization of bones and teeth; also involved in muscle contraction and relaxation, nerve functioning, blood clotting, blood pressure	Milk and milk products, small fish (with bones), calcium-set tofu (bean curd), greens (bok choy, broccoli, chard, kale), legumes

osteoporosis (OS-tee-oh-pore-OH-sis): a disease in which the bones become porous and fragile due to a loss of minerals; also called *adult bone loss*.

- **osteo** = bone
- **porosis** = porous

Phosphorus

Phosphorus is the second most abundant mineral in the body. About 85 percent of it is found combined with calcium in the hydroxyapatite crystals of bones and teeth.

Phosphorus Roles in the Body Phosphorus is found not only in bones and teeth, but also in all body cells as part of a major buffer system. Phosphorus is also part of DNA and RNA and is therefore necessary for all growth.

Phosphorus assists in energy metabolism. The high-energy compound ATP uses three phosphate groups to do its work. Many enzymes and the B vitamins become active only when a phosphate group is attached.

Phospholipids provide stability to the lipoprotein vehicles that help to transport lipids in the blood. Phospholipids are also the major structural components of cell membranes, where they control the transport of nutrients into and out of the cells. Some proteins, such as the casein in milk, contain phosphorus as part of their structures (phosphoproteins).

Phosphorus Recommendations and Intakes Because phosphorus is commonly found in almost all foods, dietary deficiencies are unlikely. As Figure 12-17 shows, foods rich in proteins—such as meat, poultry, fish, milk, and cheese—are the best sources of phosphorus. Many processed foods and soft drinks contain phosphate-based additives, and phosphorus intakes in the United States have increased as consumption of these processed foods and beverages has increased.

Phosphate toxicity is rare and usually reflects a significant problem such as kidney failure.[34] Still, phosphorus intakes can be excessive, creating disruptions in normal hormonal functions that contribute to kidney failure, heart disease,

phosphorus: a major mineral found mostly in the body's bones and teeth.

> **FIGURE 12-17** **Phosphorus in Selected Foods**

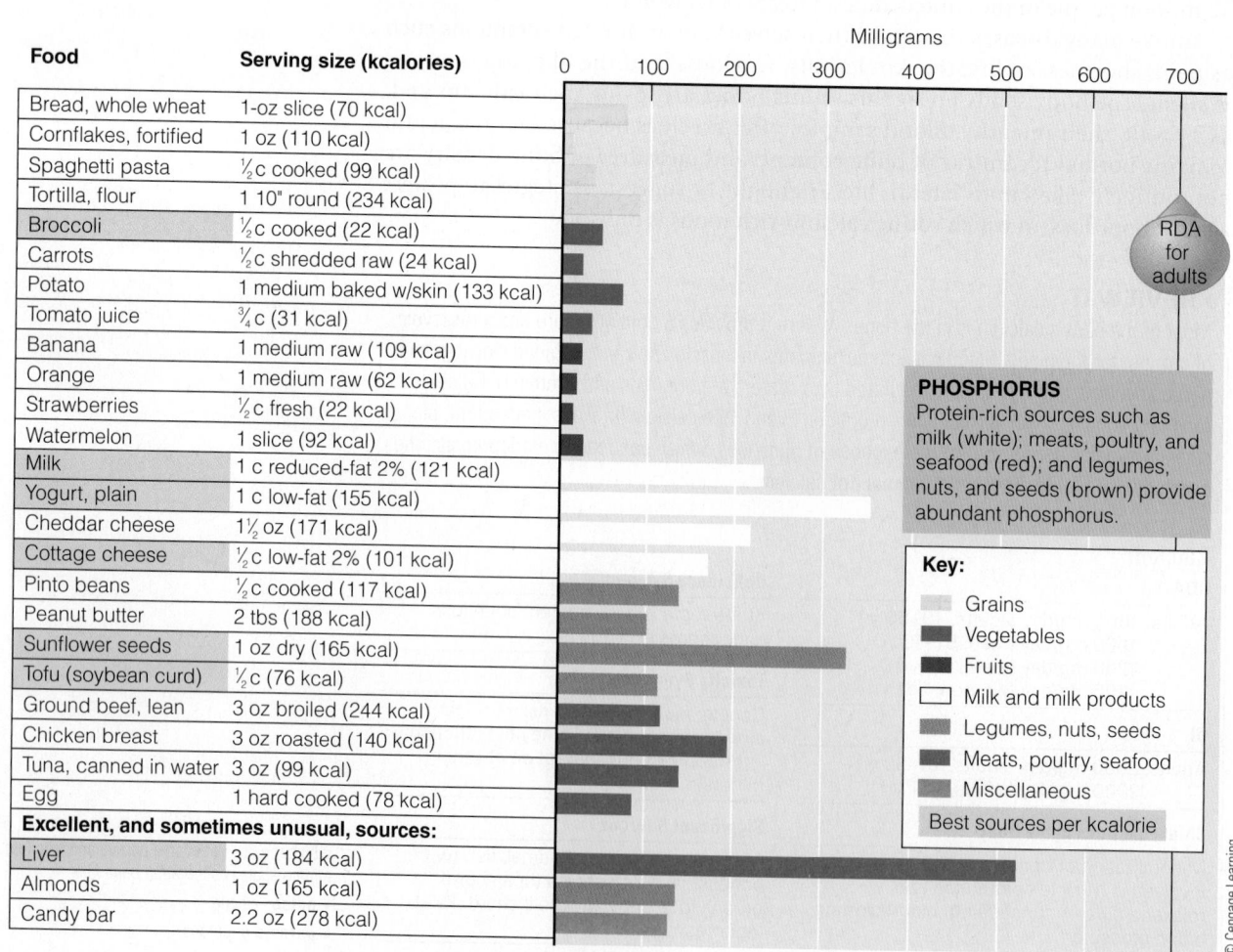

PHOSPHORUS
Protein-rich sources such as milk (white); meats, poultry, and seafood (red); and legumes, nuts, and seeds (brown) provide abundant phosphorus.

Food	Serving size (kcalories)
Bread, whole wheat	1-oz slice (70 kcal)
Cornflakes, fortified	1 oz (110 kcal)
Spaghetti pasta	½ c cooked (99 kcal)
Tortilla, flour	1 10" round (234 kcal)
Broccoli	½ c cooked (22 kcal)
Carrots	½ c shredded raw (24 kcal)
Potato	1 medium baked w/skin (133 kcal)
Tomato juice	¾ c (31 kcal)
Banana	1 medium raw (109 kcal)
Orange	1 medium raw (62 kcal)
Strawberries	½ c fresh (22 kcal)
Watermelon	1 slice (92 kcal)
Milk	1 c reduced-fat 2% (121 kcal)
Yogurt, plain	1 c low-fat (155 kcal)
Cheddar cheese	1½ oz (171 kcal)
Cottage cheese	½ c low-fat 2% (101 kcal)
Pinto beans	½ c cooked (117 kcal)
Peanut butter	2 tbs (188 kcal)
Sunflower seeds	1 oz dry (165 kcal)
Tofu (soybean curd)	½ c (76 kcal)
Ground beef, lean	3 oz broiled (244 kcal)
Chicken breast	3 oz roasted (140 kcal)
Tuna, canned in water	3 oz (99 kcal)
Egg	1 hard cooked (78 kcal)
Excellent, and sometimes unusual, sources:	
Liver	3 oz (184 kcal)
Almonds	1 oz (165 kcal)
Candy bar	2.2 oz (278 kcal)

Key:
- Grains
- Vegetables
- Fruits
- Milk and milk products
- Legumes, nuts, seeds
- Meats, poultry, seafood
- Miscellaneous
- Best sources per kcalorie

© Cengage Learning

and bone loss.[35] High intakes of phosphorus are not common when diets are based mostly on fresh foods, but can become excessive when processed foods take center stage. A UL of 4000 milligrams has been established.

› REVIEW IT

Phosphorus accompanies calcium both in the crystals of bone and in many foods such as milk. Phosphorus is also important in energy metabolism as part of ATP, in lipid structures as part of phospholipids, and in genetic materials as part of DNA and RNA. The accompanying table provides a summary of phosphorus.

Phosphorus

RDA	**Deficiency Symptoms**
Adults: 700 mg/day	Muscular weakness, bone pain[a]
UL	**Toxicity Symptoms**
Adults (19–70 yr): 4000 mg/day	Calcification of nonskeletal tissues, particularly the kidneys
Chief Functions in the Body	
Mineralization of bones and teeth; part of every cell; important in genetic material, part of phospholipids, used in energy transfer and in buffer systems that maintain acid-base balance	**Significant Sources**
	Foods derived from animals (meat, fish, poultry, eggs, milk)

[a]Dietary deficiency rarely occurs, but some drugs can bind with phosphorus making it unavailable and resulting in bone loss that is characterized by weakness and pain.

Magnesium Only about 1 ounce of **magnesium** is present in the body of a 132-pound person (review Figure 12-9, p. 381). More than half of the body's magnesium is in the bones. Much of the rest is in the muscles and soft tissues, with only 1 percent in the extracellular fluid. As with calcium, bone magnesium may serve as a reservoir to ensure normal blood concentrations.

Magnesium Roles in the Body In addition to maintaining bone health, magnesium acts in all the cells of the soft tissues, where it forms part of the protein-making machinery and is necessary for energy metabolism. It participates in hundreds of enzyme systems. A major role of magnesium is as a catalyst in the reaction that adds the last phosphate to the high-energy compound ATP, making it essential to the body's use of glucose; the synthesis of protein, fat, and nucleic acids; and the cells' membrane transport systems. Together with calcium, magnesium is involved in muscle contraction and blood clotting: calcium promotes the processes, whereas magnesium inhibits them. This dynamic interaction between the two minerals helps regulate blood pressure and lung function. Like many other nutrients, magnesium supports the normal functioning of the immune system.

Magnesium Intakes The brown bars in Figure 12-18 (p. 396) indicate that legumes, nuts, and seeds make significant magnesium contributions. Magnesium is part of the chlorophyll molecule, so dark green, leafy vegetables are also good sources. In areas with hard water, the water contributes both calcium and magnesium to daily intakes. Mineral waters noted earlier for their calcium content may also be magnesium-rich and can be important sources of this mineral for those who drink them. Bioavailability of magnesium from mineral water is about 50 percent, but it improves when the water is consumed with a meal.

Magnesium Deficiency Average magnesium intakes typically fall below recommendations, which may exacerbate inflammation and contribute to chronic diseases such as heart disease, stroke, hypertension, diabetes, and cancer.[36] A severe magnesium deficiency causes a tetany similar to the calcium tetany described earlier. Magnesium deficiencies also impair central nervous system activity and may be responsible for the hallucinations experienced during alcohol withdrawal.

Magnesium and Hypertension Magnesium is critical to heart function and seems to protect against hypertension and heart disease.[37] Interestingly, people living in

magnesium: a cation within the body's cells, active in many enzyme systems.

> **FIGURE 12-18** **Magnesium in Selected Foods**

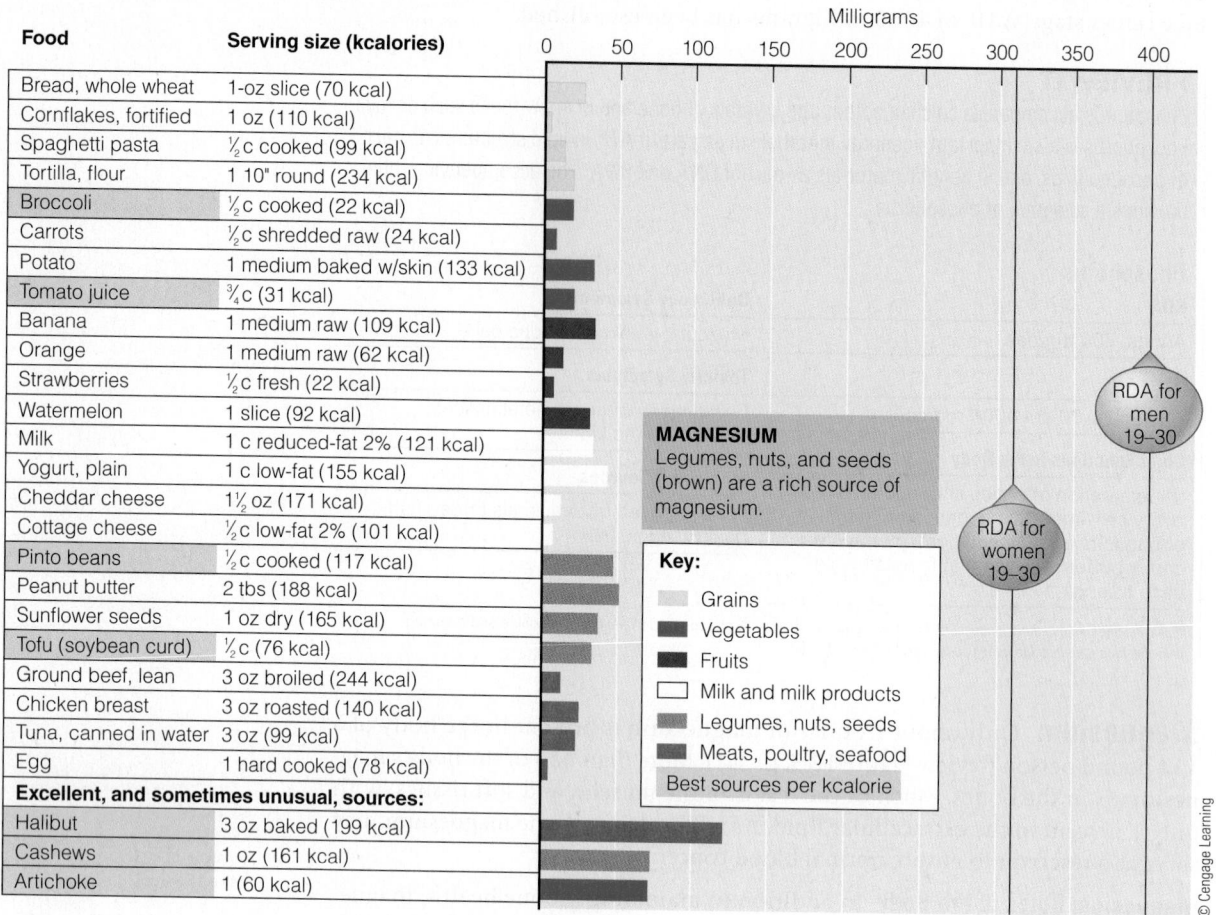

areas of the country with hard water, which contains high concentrations of calcium and magnesium, tend to have low rates of heart disease. With magnesium deficiency, the walls of the arteries and capillaries tend to constrict—a possible explanation for the hypertensive effect.

Magnesium Toxicity Magnesium toxicity is rare, but it can be fatal. The UL for magnesium applies only to nonfood sources such as supplements or magnesium salts.

› REVIEW IT

Like calcium and phosphorus, magnesium supports bone mineralization. Magnesium is also involved in numerous enzyme systems and in heart function. It is found abundantly in legumes and dark green, leafy vegetables and, in some areas, in water. The accompanying table offers a summary of magnesium.

Magnesium

RDA	**Deficiency Symptoms**
Men (19–30 yr): 400 mg/day	Weakness; confusion; if extreme, convulsions, bizarre muscle movements (especially of eye and face muscles), hallucinations, and difficulty in swallowing; in children, growth failure[a]
Women (19–30 yr): 310 mg/day	
UL	
Adults: 350 mg nonfood magnesium/day	
Chief Functions in the Body	**Toxicity Symptoms**
Bone mineralization, building of protein, enzyme action, normal muscle contraction, nerve impulse transmission, maintenance of teeth, and functioning of immune system	From nonfood sources only; diarrhea, alkalosis, dehydration
	Significant Sources
	Nuts, legumes, whole grains, dark green vegetables, seafood, chocolate, cocoa

[a]A still more severe deficiency causes tetany, an extreme, prolonged contraction of the muscles similar to that caused by low blood calcium.

Sulfate **Sulfate** is the oxidized form of the mineral **sulfur**, as it exists in foods and water. The body's need for sulfate is easily met by a variety of foods and beverages. In addition, the body receives sulfate from the amino acids methionine and cysteine, which are found in dietary proteins. These sulfur-containing amino acids help determine the contour of protein molecules. The sulfur-containing side chains in cysteine molecules can link to each other via disulfide bridges, which stabilize the protein structure. (See the drawing of insulin with its disulfide bridges in Figure 6-4 on p. 174.) Skin, hair, and nails contain some of the body's more rigid proteins, which have a high sulfur content.

Because the body's sulfate needs are easily met with normal protein intakes, there is no recommended intake for sulfate. Deficiencies do not occur when diets contain protein. Only when people lack protein to the point of severe deficiency will they lack the sulfur-containing amino acids.

> **REVIEW IT** Identify the main roles, deficiency symptoms, and food sources for each of the major minerals (sodium, chloride, potassium, calcium, phosphorus, magnesium, and sulfate).

Like the other nutrients, minerals' actions are coordinated to get the body's work done. The major minerals, especially sodium, chloride, and potassium, influence the body's fluid balance; whenever an anion moves, a cation moves—always maintaining homeostasis. Sodium, chloride, potassium, calcium, and magnesium are key members of the team of nutrients that direct nerve impulse transmission and muscle contraction. They are also the primary nutrients involved in regulating blood pressure. Phosphorus and magnesium participate in many reactions involving glucose, fatty acids, amino acids, and the vitamins. Calcium, phosphorus, and magnesium combine to form the structure of the bones and teeth. Each major mineral also plays other specific roles in the body. The table provides a summary of the major minerals.

sulfate: a salt produced from the oxidation of sulfur.

sulfur: a mineral present in the body as part of some proteins.

> **REVIEW IT** The Major Minerals

Chief Functions	Deficiency Symptoms	Toxicity Symptoms	Significant Sources
Sodium Maintains normal fluid and electrolyte balance; assists in nerve impulse transmission and muscle contraction	Muscle cramps, mental apathy, loss of appetite	Edema, acute hypertension	Table salt, soy sauce; moderate amounts in meats, milks, breads, and vegetables; large amounts in processed foods
Chloride Maintains normal fluid and electrolyte balance; part of hydrochloric acid found in the stomach, necessary for proper digestion	Do not occur under normal circumstances	Vomiting	Table salt, soy sauce; moderate amounts in meats, milks, eggs; large amounts in processed foods
Potassium Maintains normal fluid and electrolyte balance; facilitates many reactions; supports cell integrity; assists in nerve impulse transmission and muscle contractions	Irregular heartbeat, muscular weakness, glucose intolerance	Muscular weakness; vomiting; if injected into a vein, can stop the heart	All whole foods; meats, milks, fruits, vegetables, grains, legumes
Calcium Mineralization of bones and teeth; also involved in muscle contraction and relaxation, nerve functioning, blood clotting, and blood pressure	Stunted growth in children; bone loss (osteoporosis) in adults	Constipation; increased risk of urinary stone formation and kidney dysfunction; interference with absorption of other minerals	Milk and milk products, small fish (with bones), tofu, greens (bok choy, broccoli, chard), legumes
Phosphorus Mineralization of bones and teeth; part of every cell; important in genetic material, part of phospholipids, used in energy transfer and in buffer systems that maintain acid-base balance	Muscular weakness, bone pain[a]	Calcification of nonskeletal tissues, particularly the kidneys	All animal tissues (meat, fish, poultry, eggs, milk)
Magnesium Bone mineralization, building of protein, enzyme action, normal muscle contraction, nerve impulse transmission, maintenance of teeth, and functioning of immune system	Weakness; confusion; if extreme, convulsions, bizarre muscle movements (especially of eye and face muscles), hallucinations, and difficulty in swallowing; in children, growth failure[b]	From nonfood sources only; diarrhea, alkalosis, dehydration	Nuts, legumes, whole grains, dark green vegetables, seafood, chocolate, cocoa
Sulfate As part of proteins, stabilizes their shape by forming disulfide bridges; part of the vitamins biotin and thiamin and the hormone insulin	None known; protein deficiency would occur first	Toxicity would occur only if sulfur-containing amino acids were eaten in excess; this (in animals) suppresses growth	All protein-containing foods (meats, fish, poultry, eggs, milk, legumes, nuts)

[a]Dietary deficiency rarely occurs, but some drugs can bind with phosphorus, making it unavailable and resulting in bone loss that is characterized by weakness and pain.
[b]A still more severe deficiency causes tetany, an extreme, prolonged contraction of the muscles similar to that caused by low blood calcium.

With all of the tasks these minerals perform, they are of great importance to life. Consuming enough of each of them every day is easy, given a variety of foods from each of the food groups. Whole-grain breads supply magnesium; fruits, vegetables, and legumes provide magnesium and potassium too; milk products offer calcium and phosphorus; meats, poultry, and seafood offer phosphorus and sulfate as well; all foods provide sodium and chloride, with excesses being more problematic than inadequacies. The message is quite simple and has been repeated throughout this text: for an adequate intake of all the nutrients, including the major minerals, choose a variety of foods from each of the five food groups. And drink plenty of water.

Nutrition Portfolio

Many people may miss the mark when it comes to drinking enough water to keep their bodies well hydrated or obtaining enough calcium to promote strong bones; in contrast, sodium intakes often exceed those recommended for health. Go to Diet & Wellness Plus and choose one of the days on which you tracked your diet for an entire day. Select the Intake vs. Goals report and then consider the following questions.

- Did you exceed, fail to meet, or meet your goal for water intake? Was that a typical day for you? Describe your strategy for ensuring that you drink plenty of water—about eight glasses—every day.

- Take a look at your sodium intake in this report. Most people in the United States exceed the UL. Did you? Explain the importance of selecting and preparing foods with less salt.

- How was your intake of calcium for that day? If you are not getting enough calcium, consult Chapter 12 for ideas to help you get more, then list at least three foods or beverages you would be willing to eat or drink that would improve your intake.

DIET & WELLNESS PLUS To complete this exercise, go to your Diet & Wellness Plus at www.cengagebrain.com.

› **STUDY IT** To review the key points of this chapter and take a practice quiz, go to the study cards at the end of the book.

REFERENCES

1. R. S. Sebastian, C. W. Enns, and J. D. Goldman, Drinking water intake in the U.S., *What We Eat in America, NHANES 2005–2008*, September 2011.
2. W. F. Clark and coauthors, "Drink at least 8 glasses of water a day to be healthy???", *Nutrition Today* 48 (2013): S18–S21; G. F. M. Strippoli, Fluids, water, and nutrients and the risk of renal diseases, *Nutrition Today* 47 (2012): S17–S21; M. D. Sorensen and coauthors, Impact of nutritional factors on incident kidney stone formation: A report from the WHI OS, *Journal of Urology* 187 (2012): 1645–1649.
3. A. B. Goodman and coauthors, Behaviors and attitudes associated with low drinking water intake among US adults, food attitudes and behaviors survey, 2007, *Preventing Chronic Disease* 10 (2013): 120248.
4. B. M. Popkin, K. E. D'Anci, and I. H. Rosenberg, Water, hydration, and health, *Nutrition Reviews* 68 (2010): 439–458.
5. Centers for Disease Control and Prevention, http://www.cdc.gov/nceh/lead/tips/water.htm, updated October 15, 2013.
6. M. J. O'Donnell and coauthors, Urinary sodium and potassium excretion and risk of cardiovascular events, *Journal of the American Medical Association* 306 (2011): 2229–2238.
7. T. A. Kotchen, A. W. Cowley, and E. D. Frohlich, Salt in health and disease—A delicate balance, *New England Journal of Medicine* 368 (2013): 1229–1237.
8. A. Carriquiry and coauthors, Trends in the prevalence of excess dietary sodium intake—United States, 2003–2010, *Morbidity and Mortality Weekly Report* 62 (2013): 1021–1025.
9. K. M. Dickinson, P. M. Clifton, and J. B. Keogh, Endothelial function is impaired after a high-salt meal in healthy subjects, *American Journal of Clinical Nutrition* 93 (2011): 500–505.
10. S. C. Eufinger and coauthors, Habitual dietary sodium intake is inversely associated with coronary flow reserve in middle-aged male twins, *American Journal of Clinical Nutrition* 95 (2012): 572–579; J. P. Forman and coauthors, Association between sodium intake and change in uric acid, urine albumin excretion, and the risk of developing hypertension, *Circulation* 125 (2012): 3108–3116.
11. US Department of Agriculture and US Department of Health and Human Services, *Dietary Guidelines for Americans, 2010*, www.dietaryguidelines.gov.

12. P. K. Whelton and coauthors, Sodium, blood pressure, and cardiovascular disease: Further evidence supporting the American Heart Association sodium reduction recommendations, *Circulation* 126 (2012): 2880–2889; L. J. Appel and coauthors, The importance of population-wide sodium reduction as a means to prevent cardiovascular disease and stroke: A call to action from the American Heart Association, *Circulation* 123 (2011): 1138–1143.

13. M. Maillot and A. Drewnowski, A conflict between nutritionally adequate diets and meeting the 2010 Dietary Guidelines for sodium, *American Journal of Preventive Medicine* 42 (2012): 174–179.

14. D. E. Epstein and coauthors, Determinants and consequences of adherence to the Dietary Approaches to Stop Hypertension Diet in African-American and white adults with high blood pressure: Results from the ENCORE Trial, *Journal of the Academy of Nutrition and Dietetics* 112 (2012): 1763–1773.

15. A. Sherzai and coauthors, Stroke, food groups, and dietary patterns: A systematic review, *Nutrition Reviews* 70 (2012): 423–435; T. T. Fung and coauthors, The Mediterranean and Dietary Approaches to Stop Hypertension (DASH) diets and colorectal cancer, *American Journal of Clinical Nutrition* 92 (2010): 1429–1435; S. T. Chen, N. M. Maruthur, and L. J. Appel, The effect of dietary patterns on estimated coronary heart risk: Results from the Dietary Approaches to Stop Hypertension (DASH) Trial, *Circulation: Cardiovascular Quality and Outcomes* 3 (2010): 484–489.

16. A. J. Moshfegh and coauthors, Vital signs: Food categories contributing the most to sodium consumption—United States, 2007–2008, *Morbidity and Mortality Weekly Report* 61 (2012): 92–98.

17. P. M Guenther, J. M. G. Lyon, and L. J. Appel, Modeling dietary patterns to assess sodium recommendations for nutrient adequacy, *American Journal of Clinical Nutrition* 97 (2013): 842–847.

18. J. P. Gunn and coauthors, CDC grand rounds: Dietary sodium reduction—Time for choice, *Morbidity and Mortality Weekly Report* 61 (2012): 89–91; Usual sodium intakes compared with current dietary guidelines—United States, 2005–2008, *Morbidity and Mortality Weekly Report* 60 (2011): 1413–1417; C. N. Mhurchu and coauthors, Sodium content of processed foods in the United Kingdom: Analysis of 44,000 foods purchased by 21,000 households, *American Journal of Clinical Nutrition* 93 (2011): 594–600; C.A.M. Anderson and coauthors, Dietary sources of sodium in China, Japan, the United Kingdom, and the United States, women and men aged 40 to 59 years: The INTERMAP Study, *Journal of the American Dietetic Association* 110 (2010): 736–745; J. L. Webster, E. K. Dunford, and B. C. Neal, *American Journal of Clinical Nutrition* 91 (2010): 413–420.

19. J. L. H. C. Busch, F. Y. S. Yong, and S. J. Goh, Sodium reduction: Optimizing product composition and structure towards increasing saltiness perception, *Trends in Food Science and Technology* 29 (2013): 21–34.

20. K. Bibbins-Domingo and coauthors, Projected effect of dietary salt reductions on future cardiovascular disease, *New England Journal of Medicine* 362 (2010): 590–599; Institute of Medicine (US) Committee on Strategies to Reduce Sodium Intake, *Strategies to Reduce Sodium Intake in the United States* (Washington, D.C.: National Academies Press, 2010).

21. Q. Yang and coauthors, Sodium and potassium intake and mortality among US adults: Prospective data from the third National Health and Nutrition Examination Survey, *Archives of Internal Medicine* 171 (2011): 1183–1191.

22. M. E. Cogswell and coauthors, Sodium and potassium intakes among US adults: NHANES 2003–2008, *American Journal of Clinical Nutrition* 96 (2012): 647–657.

23. M. J. O'Donnell and coauthors, Urinary sodium and potassium excretion and risk of cardiovascular events, *Journal of the American Medical Association* 306 (2011): 2229–2238; M. C. Houston, The importance of potassium in managing hypertension, *Current Hypertension Reports* 13 (2011): 309–317.

24. J. Kaluza and coauthors, Dietary calcium and magnesium intake and mortality: A prospective study of men, *American Journal of Epidemiology* 171 (2010): 801–807; I. R. Reid and coauthors, Effects of calcium supplementation on lipids, blood pressure, and body composition in healthy older men: A randomized controlled trial, *American Journal of Clinical Nutrition* 91 (2010): 131–139.

25. J. L. Rosenblum and coauthors, Calcium and vitamin D supplementation is associated with decreased abdominal visceral adipose tissue in overweight and obese adults, *American Journal of Clinical Nutrition* 95 (2012): 101–108; D. R. Shahar and coauthors, Dairy calcium intake, serum vitamin D, and successful weight loss, *American Journal of Clinical Nutrition* 92 (2010): 1017–1022.

26. I. P. Onakpoya and coauthors, Efficacy of calcium supplementation for management of overweight and obesity: Systematic review of randomized clinical trials, *Nutrition Reviews* 69 (2011): 335–343; M. J. Soares, W. C. S. Ping-Delfos, and M. H. Ghanbari, Calcium and vitamin D for obesity: A review of randomized clinical trials, *European Journal of Clinical Nutrition* 65 (2011): 994–1004.

27. R. P. Heaney, Calcium and obesity: Effect size and clinical relevance, *Nutrition Reviews* 69 (2011): 333–334.

28. M. B. Zemel and coauthors, Effects of dairy compared with soy on oxidative and inflammatory stress in overweight and obese subjects, *American Journal of Clinical Nutrition* 91 (2010): 16–22.

29. M. Chen and coauthors, Effects of dairy intake on body weight and fat: A meta-analysis of randomized controlled trials, *American Journal of Clinical Nutrition* 96 (2012): 735–747.

30. R. L. Bailey and coauthors, Estimation of total usual calcium and vitamin D intakes in the United States, *Journal of Nutrition* 140 (2010): 817–822.

31. J. Calvez and coauthors, Protein intake, calcium balance and health consequences, *European Journal of Clinical Nutrition* 66 (2012): 281–295.

32. S. A. Abrams, Setting Dietary Reference Intakes with the use of bioavailability data: Calcium, *American Journal of Clinical Nutrition* 91 (2010): 1474S–1477S.

33. National Osteoporosis Foundation, www.nof.org, accessed January 2014.

34. M. S. Razzaque, Phosphate toxicity: New insights into an old problem, *Clinical Science* 120 (2011): 91–97.

35. M. S. Calvo and J. Uribarri, Public health impact of dietary phosphorus excess on bone and cardiovascular health in the general population, *American Journal of Clinical Nutrition* 98 (2013): 6–15; E. Takeda and coauthors, Dietary phosphorus in bone health and quality of life, *Nutrition Reviews* 70 (2012): 311–321.

36. M. M. Joosten and coauthors, Urinary and plasma magnesium and risk of ischemic heart disease, *American Journal of Clinical Nutrition* 97 (2013): 1299–1306; L. C. Del Gobbo and coauthors, Circulating and dietary magnesium and risk of cardiovascular disease: A systematic review and meta-analysis of prospective studies, *American Journal of Clinical Nutrition* 98 (2013): 160–173; W. B. Weglicki, Hypomagnesemia and inflammation: Clinical and basic aspects, *Annual Review of Nutrition* 32 (2012) 55–71; J. Sugimoto and coauthors, Magnesium decreases inflammatory cytokine production: A novel innate immunomodulatory mechanism, *Journal of Immunology* 188 (2012): 6338–6346; A. Rosanoff, C. M. Weaver, and R. K. Rude, Suboptimal magnesium status in the United States: Are the health consequences underestimated? *Nutrition Reviews* 70 (2012): 153–164; S. C. Larsson, N. Orsini, and A. Wolk, Dietary magnesium intake and risk of stroke: A meta-analysis of prospective studies, *American Journal of Clinical Nutrition* 95 (2012): 362–366; S. E. Chiuve and coauthors, Plasma and dietary magnesium and risk of sudden cardiac death in women, *American Journal of Clinical Nutrition* 93 (2011): 253–260; M. Shechter, Magnesium and cardiovascular system, *Magnesium Research* 23 (2010): 60–72; F. H. Nielsen, Magnesium, inflammation, and obesity in chronic disease, *Nutrition Reviews* 68 (2010): 333–340.

37. L. Kass, J. Weekes, and L. Carpenter, Effect of magnesium supplementation on blood pressure: A meta-analysis, *European Journal of Clinical Nutrition* 66 (2012): 411–418.

HIGHLIGHT > 12
Osteoporosis and Calcium

> **LEARN IT** Describe factors that contribute to the development of osteoporosis and strategies to prevent it.

Osteoporosis becomes apparent during the later years, but it develops much earlier—and without warning. Few people are aware that their bones are being robbed of their strength. The problem often first becomes evident when someone's hip suddenly gives way. People say, "She fell and broke her hip," but in fact the hip may have been so fragile that it broke *before* she fell. Even bumping into a table may be enough to shatter a porous bone into fragments so numerous and scattered that they cannot be reassembled. Removing them and replacing them with an artificial joint requires major surgery. An estimated 258,000 people in the United States are hospitalized each year because of hip fractures related to osteoporosis. About one in five die of complications within a year; one in three will never walk or live independently again.[1] Their quality of life slips downward.

This highlight examines low bone density and osteoporosis, one of the most prevalent diseases of aging, affecting an estimated 52 million people in the United States—most of them women older than 50.[2] It reviews the many factors that contribute to the 2 million fractures in the bones of the hips, vertebrae, wrists, arms, and ankles each year. And it presents strategies to reduce the risks, paying special attention to the role of dietary calcium.

© ONOKY-Photononstop/Alamy

Bone Development and Disintegration

Bone has two compartments: the outer, hard shell of **cortical bone** and the inner, lacy matrix of **trabecular bone.** (The glossary defines these and other bone-related terms.) Both can lose minerals, but in different ways and at different rates. The first photograph in Figure H12-1 shows a human leg bone sliced lengthwise, exposing the lacy, calcium-containing crystals of trabecular bone. These crystals give up calcium to the blood when the diet runs short, and they take up calcium again when the supply is plentiful (review Figure 12-13 on p. 390). For people who have eaten calcium-rich foods throughout the bone-forming years of their youth, these deposits make bones dense and provide a rich reservoir of calcium.

Surrounding and protecting the trabecular bone is a dense, ivorylike exterior shell—the cortical bone. Cortical bone composes the shafts of the long bones, and a thin cortical shell caps the ends of the bones too. Both compartments confer strength on bone: cortical bone provides the sturdy outer wall, and trabecular bone provides support along the lines of stress.

The two types of bone handle calcium in different ways. Supplied with blood vessels and metabolically active, trabecular bone is sensitive to hormones that govern day-to-day deposits and withdrawals of calcium. It readily gives up minerals whenever blood calcium needs replenishing. Losses of trabecular bone start becoming significant for men and women in their 30s, although losses can occur whenever calcium withdrawals exceed deposits. Cortical bone also gives up calcium, but slowly and at a steady pace. Cortical bone losses typically begin at about age 40 and continue slowly but surely thereafter.

As bone loss continues, **bone density** declines, and osteoporosis becomes apparent (see Figure H12-1). Bones become so fragile that even the body's own weight can overburden the spine—vertebrae may suddenly disintegrate and crush down, painfully pinching major nerves. Or the vertebrae may compress into wedge shapes, forming what is

GLOSSARY

antacids: medications used to relieve indigestion by neutralizing acid in the stomach. Calcium-containing preparations (such as Tums) contain available calcium. Antacids with aluminum or magnesium hydroxides (such as Rolaids) can accelerate calcium losses.

bone density: a measure of bone strength. When minerals fill the bone matrix (making it dense), they give it strength.

bone meal or **powdered bone:** crushed or ground bone preparations intended to supply calcium to the diet. Calcium from bone is not well absorbed and is often contaminated with toxic minerals such as arsenic, mercury, lead, and cadmium.

cortical bone: the very dense bone tissue that forms the outer shell surrounding trabecular bone and comprises the shaft of a long bone.

dolomite: a compound of minerals (calcium magnesium carbonate) found in limestone and marble. Dolomite is powdered and is sold as a calcium-magnesium supplement. However, it may be contaminated with toxic minerals, is not well absorbed, and interferes with absorption of other essential minerals.

osteoporosis (OS-tee-oh-pore-OH-sis)**:** a disease in which the bones become porous and fragile due to loss of minerals; also called *adult bone loss.*

oyster shell: a product made from the powdered shells of oysters that is sold as a calcium supplement, but it is not well absorbed by the digestive system.

trabecular (tra-BECK-you-lar) **bone:** the lacy inner structure of calcium crystals that supports the bone's structure and provides a calcium storage bank.

> **FIGURE H12-1**   **Healthy and Osteoporotic Trabecular Bones**

Trabecular bone is the lacy network of calcium-containing crystals that fills the interior. Cortical bone is the dense, ivorylike bone that forms the exterior shell.

Electron micrograph of healthy trabecular bone.

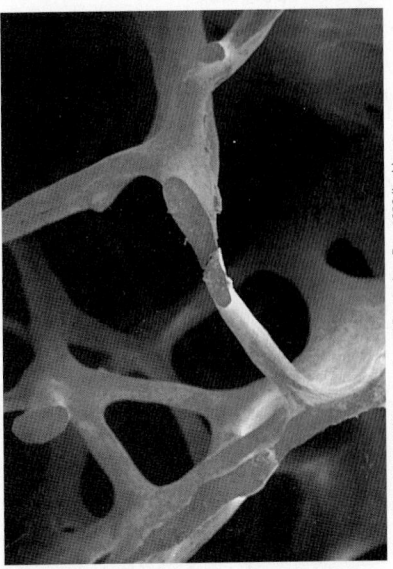

Electron micrograph of trabecular bone affected by osteoporosis.

often called a "dowager's hump," the posture many older people assume as they "grow shorter." Figure H12-2 (p. 402) shows the effect of compressed spinal bone on a woman's height and posture. Because both the cortical shell and the trabecular interior weaken, breaks most often occur in the hip, as mentioned in the introductory paragraph.

Physicians can determine bone loss and diagnose osteoporosis by measuring bone density using dual-energy X-ray absorptiometry (DEXA scan).[3] They also consider risk factors for osteoporosis, including age, personal and family history of fractures, and physical inactivity. Table H12-1 summarizes the major risk factors for

osteoporosis. The more risk factors that apply to a person, the greater the chances of bone loss. Notice that several risk factors that are influential in the development of osteoporosis—such as age, gender, and genetics—cannot be changed. Other risk factors—such as diet, physical activity, body weight, smoking, and alcohol use—are personal behaviors that can be changed. By eating a calcium-rich, well-balanced diet; being physically active; abstaining from smoking; and drinking alcohol in moderation (if at all), people can defend themselves against osteoporosis. These decisions are particularly important for those with other risk factors that cannot be changed.

Whether a person develops osteoporosis seems to depend on the interactions of several factors, including nutrition. The strongest predictor of bone density is age.

Using a DEXA (dual-energy X-ray absorpiometry) test to measure bone mineral density identifies osteoporosis, determines risks for fractures, and tracks responses to treatment.

TABLE H12-1   **Risk Factors for Osteoporosis**

Nonmodifiable	Modifiable
• Female gender	• Sedentary lifestyle
• Older age (>50 yr)	• Diet inadequate in calcium and vitamin D
• Small frame	• Diet excessive in protein, sodium, caffeine
• Caucasian, Asian, or Hispanic/Latino	
• Family history of osteoporosis or fractures	• Cigarette smoking
• Personal history of fractures	• Alcohol abuse
	• Low body weight
• Estrogen deficiency in women (amenorrhea or menopause, especially early or surgically induced); testosterone deficiency in men	• Certain medications, such as glucocorticoids, aluminum-containing antacids, and antiseizure drugs

401

HIGHLIGHT > 12

> **FIGURE H12-2** **Loss of Height in a Woman Caused by Osteoporosis**

The woman on the left is about 50 years old. On the right, she is 80 years old. Her legs have not grown shorter. Instead, her back has lost length due to collapse of her spinal bones (vertebrae). Collapsed vertebrae cannot protect the spinal nerves from pressure that causes excruciating pain.

50 years old   80 years old

Age and Bone Calcium

Two major stages of life are critical in the development of osteoporosis. The first is the bone-acquiring stage of childhood and adolescence. The second is the bone-losing decades of late adulthood, especially in women after menopause. The bones gain strength and density all through the growing years and into young adulthood. As people age, the cells that build bone gradually become less active, but those that dismantle bone continue working. The result is that bone loss exceeds bone formation. Some bone loss is inevitable, but losses can be curtailed by maximizing bone mass.

Maximizing Bone Mass

To maximize bone mass, the diet must deliver an adequate supply of calcium during the first three decades of life. Children and teens who

> **FIGURE H12-3** **Bone Losses over Time Compared**

Peak bone mass is achieved by age 30. Women gradually lose bone mass until menopause, when losses accelerate dramatically and then gradually taper off.

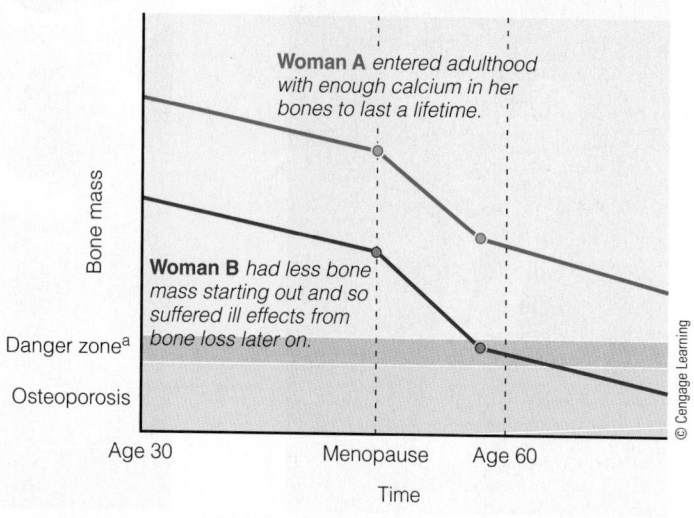

Woman A *entered adulthood with enough calcium in her bones to last a lifetime.*

Woman B *had less bone mass starting out and so suffered ill effects from bone loss later on.*

ªPeople with a moderate degree of bone mass loss are said to have *osteopenia* and are at increased risk of fractures.

consume milk products and get enough calcium have denser bones than those with inadequate intakes. With little or no calcium from the diet, the body must depend on bone to supply calcium to the blood—bone mass diminishes, and bones lose their density and strength. When people reach the bone-losing years of middle age, those who formed dense bones during their youth have the advantage. They simply have more bone starting out and can lose more before suffering ill effects. Figure H12-3 demonstrates this effect.

Minimizing Bone Loss

Not only does dietary calcium build strong bones in youth, but it remains important in protecting against losses in the later years. Unfortunately, calcium intakes of older adults are typically low, and calcium absorption declines after menopause. The kidneys do not activate vitamin D as well as they did earlier (recall that vitamin D enhances calcium absorption). Also, sunlight is needed to form vitamin D, and many older people spend little or no time outdoors in the sunshine. For these reasons, and because intakes of vitamin D are typically low anyway, blood levels of vitamin D decline.

Some of the hormones that regulate bone and calcium metabolism—parathyroid hormone, calcitonin, and estrogen—also change with age and accelerate bone loss. Together, these age-related factors contribute to bone loss: inefficient bone remodeling, reduced calcium intakes, impaired calcium absorption, poor vitamin D status, and hormonal changes that favor bone mineral withdrawal.

Gender and Hormones

After age, gender is the next strongest predictor of osteoporosis. The sex hormones play a major role in regulating the rate of bone turnover.[4] Men have greater bone density than women at maturity, and women have greater losses than men in later life. Consequently, men develop bone problems about 10 years later than women, and women account for two out of three cases of osteoporosis.

Menopause imperils women's bones. Bone dwindles rapidly when the hormone estrogen diminishes and menstruation ceases. The lack of estrogen contributes to the release of cytokines that produce inflammation and accelerate bone loss.[5] Women may lose up to 20 percent of their bone mass during the 6 to 8 years following menopause. Eventually, losses taper off so that women again lose bone at the same rate as men their age. Losses of bone minerals continue throughout the remainder of a woman's lifetime, but not at the free-fall pace of the menopause years (review Figure H12-3).

Rapid bone losses also occur when *young* women's ovaries fail to produce enough estrogen, causing menstruation to cease. In some cases, diseased ovaries are to blame and must be removed; in others, the ovaries fail to produce sufficient estrogen because the women suffer from anorexia nervosa and have unreasonably restricted their body weight (see Highlight 8). The amenorrhea and low body weights explain much of the bone loss seen in these young women, even years after diagnosis and treatment.

Estrogen therapy may help some women prevent further bone loss and reduce the incidence of fractures. Because estrogen therapy may increase the risks for breast cancer, women must carefully weigh any potential benefits against the possible dangers. A combination of drugs or of hormone replacement and a drug may be most beneficial.

Several drug therapies have been developed to inhibit bone loss and enhance bone formation.[6] The FDA has approved the following drugs to prevent or treat osteoporosis: biophosphates, calcitonin, estrogens, estrogen antagonists, and parathyroid hormone.[7]

Some women who choose not to use estrogen therapy turn to soy as an alternative treatment. Interestingly, the phytochemicals commonly found in soy mimic the actions of estrogen in the body. Research results have been mixed and controversial, but overall seem to indicate a lack of benefit for soy and its phytochemicals in helping to prevent the rapid bone losses of the menopause years.[8] As is true of all herbal products, there may be risks associated with their use, and in the case of soy, evidence is lacking that the benefits clearly outweigh the potential risks.[9] Because the risks and benefits vary depending on each person's medical history, women should discuss soy options with their physicians.

As in women, sex hormones appear to play a key role in men's bone loss as well.[10] Other common causes of osteoporosis in men include corticosteroid use and alcohol abuse.

Genetics

Risks of osteoporosis appear to run along racial lines and reflect genetic differences in bone development. African Americans, for example, seem to use and retain calcium more efficiently than Caucasians. Consequently, even though their calcium intakes are typically lower, black people have denser bones than white people do. Greater bone density expresses itself in less bone loss, fewer fractures, and a lower rate of osteoporosis among blacks.

The exact role of genetics is unclear.[11] Most likely, genes influence both the peak bone mass achieved during growth and the bone loss incurred during the later years. The extent to which a given genetic potential is realized, however, depends on many outside factors. Diet and physical activity, for example, can maximize peak bone density during growth, whereas alcohol and tobacco abuse can accelerate bone losses later in life. Importantly, these factors are within a person's control.

Physical Activity and Body Weight

Physical activity may be the single most important factor supporting bone growth during adolescence.[12] Muscle strength and bone strength go together. When muscles work, they pull on the bones, stimulating them to grow denser. The hormones that promote new muscle growth also favor the building of bone. As a result, active bones are denser and stronger than sedentary bones.

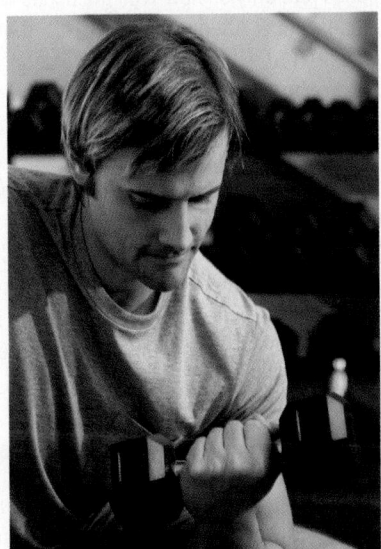

Strength training helps to build strong bones.

Both the muscle contraction and the gravitational pull of the body's weight create a load that benefits bone metabolism. To keep bones healthy, a person should engage in weight training or weight-bearing endurance activities (such as tennis and jogging or sprint cycling) regularly.[13] Regular physical activity combined with an adequate calcium intake helps to maximize bone density in adolescence. Adults can also maximize and maintain bone density with a regular program of weight training. Even past menopause, when most women are losing bone, weight training improves bone density.

Heavier body weights and weight gains place a similar stress on the bones and promote their density. In contrast, weight losses reduce bone density and increase the risk of fractures—in part because energy restriction diminishes calcium absorption and compromises calcium balance. As mentioned in Highlight 8, the relative energy deficiency that results from a combination of restricted energy intake and extreme daily exercise reliably predicts bone loss.

Smoking and Alcohol

Add bone damage to the list of ill consequences associated with smoking. The bones of smokers are less dense than those of nonsmokers—even after controlling for differences in age, body weight, and physical activity habits. Fortunately, the damaging effects can be reversed with smoking cessation. Blood indicators of beneficial bone activity are apparent 6 weeks after a person stops smoking. In time, bone density is similar for former smokers and nonsmokers.

People who abuse alcohol often suffer from osteoporosis and experience more bone breaks than others. Several factors appear to be involved. Alcohol enhances fluid excretion, leading to excessive calcium losses in the urine; upsets the hormonal balance required for healthy bones; slows bone formation, leading to lower bone density; stimulates bone breakdown; and increases the risk of falling.

Dietary Calcium

Diets that are habitually low in calcium increase the risk of fractures and osteoporosis.[14] For older adults, an adequate calcium intake alone cannot protect against bone fractures. Bone strength later in life depends primarily on how well the bones were built during childhood and adolescence. Adequate calcium nutrition during the growing years is essential to achieving optimal peak bone mass. Simply put, growing children who do not get enough calcium do not develop strong bones. For this reason, the DRI Committee recommends 1300 milligrams of calcium per day for everyone 9 through 18 years of age. Unfortunately, few girls meet the recommendations for calcium during these bone-forming years. (Boys generally obtain intakes close to those recommended because they eat more food.) Consequently, most girls start their adult years with less-than-optimal bone density. As adults, women rarely meet their recommended intakes of 1000 to 1200 milligrams from food. Some authorities suggest 1500 milligrams of calcium for postmenopausal women who are not receiving estrogen.

Other Nutrients

Much research has focused on calcium, but other nutrients support bone health too. Adequate protein protects bones and reduces the likelihood of hip fractures. As mentioned earlier, vitamin D is needed to maintain calcium metabolism and optimal bone health. Vitamin K decreases bone turnover and protects against hip fractures. Vitamin C may slow bone losses. The minerals magnesium and potassium also help to maintain bone mineral density. Vitamin A is needed in the bone-remodeling process, but too much vitamin A may be associated with osteoporosis. Carotenoids may inhibit bone loss. Omega-3 fatty acids may help preserve bone integrity. Additional research points to the bone benefits not of a specific nutrient, but of a diet rich in fruits, vegetables, and whole grains.[15] In contrast, diets containing too much salt are associated with bone losses. Similarly, diets containing too many colas or commercially baked snack and fried foods are associated with low bone mineral density. Clearly, a well-balanced diet that depends on all the food groups to supply a full array of nutrients is central to bone health.

A Perspective on Calcium Supplements

Bone health depends, in part, on calcium. People who do not consume milk products or other calcium-rich foods in amounts that provide even half the recommendation should consider consulting a registered dietitian nutritionist who can assess the diet and suggest food choices to correct any inadequacies. Calcium from foods may support bone health better than calcium from supplements. For those who are unable to consume enough calcium-rich foods, however, taking calcium supplements—especially in combination with vitamin D—may help to enhance bone density and protect against bone loss and fractures.[16] Because some research suggests that calcium supplements may increase the risk of heart attacks and strokes, women should consult their physicians when making this decision.[17]

An estimated 60 percent of women aged 60 and over take calcium supplements.[18] Selecting a calcium supplement requires a little investigative work to sort through the many options. Before examining calcium supplements, recognize that multivitamin-mineral pills contain little or no calcium. The label may list a few milligrams of calcium, but remember that the recommended intake is a gram (1000 milligrams) or more for adults.

Calcium supplements are typically sold as compounds of calcium carbonate (common in **antacids** and fortified chocolate candies), citrate, gluconate, lactate, malate, or phosphate. These supplements often include magnesium, vitamin D, or both. In addition, some calcium supplements are made from **bone meal, oyster shell,** or **dolomite** (limestone). Many calcium supplements, especially those derived from these natural products, contain lead—which impairs health in numerous ways, as Chapter 13 points out. Fortunately, calcium interferes with the absorption and action of lead in the body.

The first question to ask is how much calcium the supplement provides. Most calcium supplements provide between 250 and 1000 milligrams of calcium. To be safe, total calcium intake from both foods and supplements should not exceed the UL. Read the label to find out how much a dose supplies. Unless the label states otherwise, supplements of calcium carbonate are 40 percent calcium; those of calcium citrate are 21 percent; lactate, 13 percent; and gluconate, 9 percent. Select a low-dose supplement and take it several times a day rather than taking a large-dose supplement all at once. Taking supplements in doses of 500 milligrams or less improves absorption. Small doses also help ease the GI distress (constipation, intestinal bloating, and excessive gas) that sometimes accompanies calcium supplement use.

The next question to ask is how well the body absorbs and uses the calcium from various supplements. Most healthy people absorb calcium equally well from milk and any of these supplements: calcium carbonate, citrate, or phosphate. More important than supplement solubility is tablet disintegration. When manufacturers compress large quantities of calcium into small pills, the stomach acid has difficulty penetrating the pill. To test a supplement's ability to dissolve, drop it into a 6-ounce cup of vinegar, and stir occasionally. A high-quality formulation will dissolve within a half-hour.

Finally, people who choose supplements must take them regularly. Furthermore, consideration should be given to the best time to take the supplements. To circumvent adverse nutrient interactions, take calcium supplements between, not with, meals. (Importantly, do not take calcium supplements with iron supplements or iron-rich meals; calcium inhibits iron absorption.) To enhance calcium absorption, take supplements with meals. If such contradictory advice drives you crazy, reconsider the benefits of food sources of calcium. Most experts agree that foods are the best source of most nutrients.

Some Closing Thoughts

Unfortunately, many of the strongest risk factors for osteoporosis are beyond people's control: age, gender, and genetics. But several strategies are effective for prevention. First, ensure an optimal peak bone mass during childhood and adolescence by eating a balanced diet rich in calcium and vitamin D and by engaging in regular physical activity. Then, maintain that bone mass in early adulthood by continuing those healthy diet and activity habits, abstaining from cigarette smoking and using alcohol moderately, if at all. Finally, minimize bone loss in later life by maintaining an adequate nutrition and exercise regimen, and, especially for older women, consult a physician about bone density tests, calcium supplements, or other drug therapies that may be effective both in preventing bone loss and in restoring lost bone.[19] The reward is the best possible chance of preserving bone health throughout life.

CRITICAL THINKING QUESTIONS

A. What behaviors would be most helpful in preventing osteoporosis?

B. Osteoporosis typically develops in old age, yet the time to optimize bone density is during childhood and adolescence—decades away from the realities of hip fractures and spinal collapses. What plan of action might you develop to encourage teens to adopt strategies that will enhance bone development? Be sure to address potential obstacles and reluctances typical of that age.

REFERENCES

1. Centers for Disease Control and Prevention, Hip fractures among older adults, http://www.cdc.gov/homeandrecreationalsafety/falls/adulthipfx.html, September 30, 2013; A. Leboime and coauthors, Osteoporosis and mortality, *Joint Bone Spine* 77 (2010): S107–S112.

2. National Osteoporosis Foundation, www.nof.org, accessed January 2014; R. Nuti and coauthors, Bone fragility in men: Where are we? *Journal of Endocrinological Investigation* 33 (2010): 33–38.

3. R. Lorente-Ramos and coauthors, Dual-energy x-ray absorptiometry in the diagnosis of osteoporosis: A practical guide, *American Journal of Roentgenology* 196 (2011): 897–904.

4. B. Frenkel and coauthors, Regulation of adult bone turnover by sex steroids, *Journal of Cellular Physiology* 224 (2010): 305–310.

5. Y. Imai and coauthors, Minireview: Osteoprotective action of estrogens is mediated by osteoclastic estrogen receptor-alpha, *Molecular Endocrinology* 24 (2010): 877–885.

6. T. D. Rachner, S. Khosla, and L. C. Hofbauer, Osteoporosis: Now and the future, *Lancet* 377 (2011): 1276–1287.

7. National Osteoporosis Foundation, *Clinician's Guide to Prevention and Treatment of Osteoporosis* (Washington, D.C.: National Osteoporosis Foundation, 2010), pp. 21–24.

8. V. S. Lagari and S. Levis, Phytoestrogens in the prevention of postmenopausal bone loss, *Journal of Clinical Densitometry* 16 (2013): 445–449; A. Bitto and coauthors, Genistein aglycone: A dual mode of action anti-osteoporotic soy isoflavone rebalancing bone turnover towards bone formation, *Current Medicinal Chemistry* 17 (2010): 3007–3018.

9. E. Poluzzi and coauthors, Phytoestrogens in postmenopause: The state of the art from a chemical pharmacological and regulatory perspective, *Current Medicinal Chemistry* 21 (2014): 417–436; J. Pitkin, Alternative and complementary therapies for the menopause, *Menopause International* 18 (2012): 20–27.

10. E. Gielen and coauthors, Osteoporosis in men, *Best Practice and Research: Clinical Endocrinology and Metabolism* 25 (2011): 321–335; N. Ducharme, Male osteoporosis, *Clinics in Geriatric Medicine* 26 (2010): 301–309; S. Khosla, Update in male osteoporosis, *Journal of Clinical Endocrinology and Metabolism* 95 (2010): 3–10.

11. B. D. Mitchell and L. M. Yerges-Armstrong, The genetics of bone loss: Challenges and prospects, *Journal of Clinical Endocrinology and Metabolism* 96 (2011): 1258–1268.

12. K. F. Janz and coauthors, Early physical activity provides sustained bone health benefits later in childhood, *Medicine and Science in Sports and Exercise* 42 (2010): 1072–1078; A. Guadalupe-Grau and coauthors, Exercise and bone mass in adults, *Sports Medicine* 39 (2009): 439–468.

13. E. A. Marques and coauthors, Response of bone mineral density, inflammatory cytokines, and biochemical bone markers to a 32-week combined loading exercise programme in older men and women, *Archives of Gerontology and Geriatrics* 57 (2013): 226–233; M. T. Korhonen and coauthors, Bone density, structure and strength, and their determinants in aging sprint athletes, *Medicine and Science in Sports and Exercise* 44 (2012): 2340–2349; T. E. Howe and coauthors, Exercise for preventing and treating osteoporosis in postmenopausal women, *Cochrane Database of Systematic Reviews* 7 (2011): CD000333.

14. E. Warensjö and coauthors, Dietary calcium intake and risk of fracture and osteoporosis prospective longitudinal cohort study, *British Medical Journal* 342 (2011): d1473.

15. L. Langsetmo and coauthors, Dietary patterns and incident low-trauma fractures in postmenopausal women and men aged ≥50 y: A population-based cohort study, *American Journal of Clinical Nutrition* 93 (2011): 192–199.

16. R. L. Prentice and coauthors, Health risks and benefits from calcium and vitamin D supplementation Women's Health Initiative clinical trial and cohort study, *Osteoporosis International* 24 (2013): 567–580.

17. D. C. Bauer, Calcium supplements and fracture prevention, *New England Journal of Medicine* 369 (2013): 1537–1543; K. Li and coauthors, Associations of dietary calcium intake and calcium supplementation with myocardial infarction and stroke risk and overall cardiovascular mortality in the Heidelberg cohort of the European Prospective Investigation into Cancer and Nutrition study (EPIC-Heidelberg), *Heart* 98 (2012): 920–925; M. J. Bolland and coauthors, Calcium supplements with or without vitamin D and risk of cardiovascular events: Reanalysis of the Women's Health Initiative limited access dataset and meta-analysis, *British Medical Journal* 342 (2011): d2040.

18. J. Gahche and coauthors, Dietary supplement use among US adults has increased since NHANES III (1988–1994), *NCHS Data Brief* 61 (2011): 1–8.

19. R. C. Hamdy and coauthors, Algorithm for the management of osteoporosis, *Southern Medical Journal* 103 (2010): 1009–1015.

13

The Trace Minerals

Nutrition in Your Life

Trace—barely a perceptible amount. But the trace minerals tackle big jobs. Your blood can't carry oxygen without iron, and insulin can't deliver glucose without chromium. Teeth become decayed without fluoride, and thyroid glands develop goiter without iodine. Together, the trace minerals keep you healthy and strong. Where can you get these amazing minerals? A variety of foods, especially those from the protein foods group, sprinkled with a little iodized salt and complemented by a glass of fluoridated water will do the trick. It's remarkable what your body can do with only a few milligrams—or even micrograms—of the trace minerals. In the Nutrition Portfolio at the end of this chapter, you can determine whether the foods you are eating are meeting your trace mineral needs.

This chapter features the essential **trace minerals**—iron, zinc, iodine, selenium, copper, manganese, fluoride, chromium, and molybdenum. Figure 12-9 in Chapter 12 (p. 381) showed the tiny quantities of trace minerals in the human body. The trace minerals are so named because they are present, and needed, in relatively small amounts in the body. All together, they would hardly fill a teaspoon. Yet they are no less important than the major minerals or any of the other nutrients. Each of the trace minerals performs a vital role. A deficiency of any of them may be fatal, and excesses are equally deadly. Remarkably, a well-balanced diet supplies enough of these minerals to maintain health.

This chapter also mentions other trace minerals—such as arsenic, boron, nickel, bromine, and vanadium—that are not considered nutrients. These minerals may have beneficial roles in the body, but research on them is insufficient to determine essentiality. Also mentioned in this chapter are contaminant minerals that disrupt body processes and impair nutrition status. The highlight that follows examines phytochemicals—compounds that also are not essential nutrients, but that have biological activity in the body. Again, a well-balanced diet—especially one abundant in fruits and vegetables—supplies a full array of phytochemicals to support good health.

trace minerals: essential mineral nutrients the human body requires in relatively small amounts (less than 100 milligrams per day); sometimes called *microminerals*.

13.1 The Trace Minerals—An Overview

> **LEARN IT** Summarize key factors unique to the trace minerals.

The body requires the trace minerals in minuscule quantities. They participate in diverse tasks all over the body, each having special duties that only it can perform.

Food Sources The trace mineral contents of foods depend on soil and water composition and on how foods are processed. Furthermore, many factors in the diet and within the body affect the minerals' **bioavailability.** Still, outstanding food sources for each of the trace minerals, just like those for the other nutrients, include a wide variety of foods.

Deficiencies Assessing trace mineral status is challenging. Severe deficiencies of the better-known minerals are relatively easy to recognize. Deficiencies of the others may be harder to diagnose, and for all minerals, mild deficiencies are easy to overlook. Because the minerals are active in many body systems—digestive, cardiovascular, circulatory, muscular, skeletal, and nervous—deficiencies can have wide-reaching effects and can affect people of all ages. The most common result of a deficiency in children is failure to grow and thrive.

Toxicities Most of the trace minerals are toxic at intakes only two and a half to eleven times above current recommendations (see Figure 13-1). Thus it is important not to habitually exceed the Upper Level (UL) of recommended

> **FIGURE 13-1** **RDA (or AI) and UL Compared for Selected Trace Minerals**

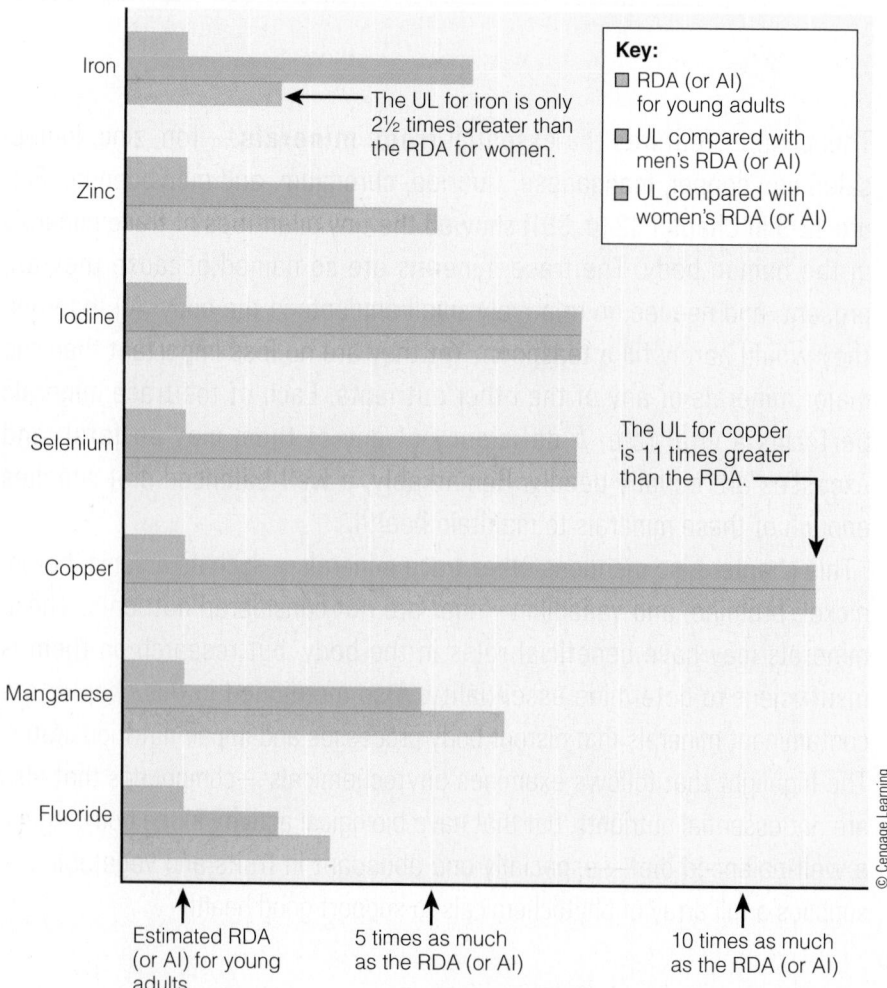

Key:
- RDA (or AI) for young adults
- UL compared with men's RDA (or AI)
- UL compared with women's RDA (or AI)

The UL for iron is only 2½ times greater than the RDA for women.

The UL for copper is 11 times greater than the RDA.

Estimated RDA (or AI) for young adults

5 times as much as the RDA (or AI)

10 times as much as the RDA (or AI)

© Cengage Learning

bioavailability: the rate at and the extent to which a nutrient is absorbed and used.

intakes (see inside front pages). Many dietary supplements contain trace minerals, making it easy for users to exceed their needs. Highlight 10 discusses supplement use and some of the regulations included in the Dietary Supplement Health and Education Act. As that discussion notes, consumers have demanded the freedom to choose their own doses of nutrients. By law, the Food and Drug Administration (FDA) has no authority to limit the amounts of trace minerals in supplements. Individuals who take supplements must therefore be aware of the possible dangers and select supplements that contain no more than 100 percent of the Daily Value. It is easier and safer to meet nutrient needs by selecting a variety of foods than by combining an assortment of supplements.

Interactions Interactions among the trace minerals are common and often well coordinated to meet the body's needs. For example, several of the trace minerals support insulin's work, influencing its synthesis, storage, release, and action.

At other times, interactions lead to nutrient imbalances. An excess of one may cause a deficiency of another. (A slight manganese overload, for example, may aggravate an iron deficiency.) A deficiency of one may interfere with the work of another. (A selenium deficiency halts the activation of the iodine-containing thyroid hormones.) A deficiency of a trace mineral may even open the way for a contaminant mineral to cause a toxic reaction. (Iron deficiency, for example, makes the body more vulnerable to lead poisoning.) These examples of nutrient interactions highlight one of the many reasons why people should use supplements conservatively, if at all: supplementation can easily create imbalances.

A good food source of one nutrient may be a poor food source of another, and factors that enhance the action of some trace minerals may interfere with others. Meats, for example, are a good source of iron but a poor source of calcium; vitamin C enhances the absorption of iron but hinders that of copper.

Nonessential Trace Minerals The essential trace minerals featured in this chapter have been well studied; researchers understand the primary roles in the body and the consequences of deficiencies and toxicities. Enough information is available to determine DRI. In contrast, research to determine whether other trace minerals are essential is challenging because quantities in the body are so small and also because human deficiencies are unknown. Identifying their functions in the body can be particularly problematic. Much of the available knowledge comes from research using animals.

Research is currently insufficient to determine the DRI for nickel, bromine, vanadium, cobalt, and boron, even though they may play beneficial roles in the human body. Nickel may serve as a cofactor for certain enzymes. Bromine is involved in the formation of collagen. Vanadium is necessary for growth and bone development and for normal reproduction. Cobalt is a key mineral in the large vitamin B_{12} molecule (see Figure 13-2). Boron may play a key role in bone health, brain activities, and immune response.

In the future, we may discover that these and other trace minerals are essential to growth and health. Even arsenic—famous as a poison used by murderers and known to be a carcinogen—may turn out to be essential for human beings in tiny quantities. It has already proved useful in the treatment of some types of leukemia.[1] Research on all the trace minerals is active, suggesting that we have much more to learn about them.

> **REVIEW IT** Summarize key factors unique to the trace minerals.
Although the body uses only tiny amounts of the trace minerals, they are vital to health. Because so little is required, the trace minerals can be toxic at levels not far above estimated requirements—a consideration for supplement users. Like the other nutrients, the trace minerals are best obtained by eating a variety of foods.

> **FIGURE 13-2   Cobalt in Vitamin B_{12}**

The intricate vitamin B_{12} molecule contains one atom of the mineral cobalt. The alternative name for vitamin B_{12}, cobalamin, reflects the presence of cobalt in its structure.

© Cengage Learning

13.2 The Trace Minerals

> **› LEARN IT** Identify the main roles, deficiency symptoms, and food sources for each of the essential trace minerals (iron, zinc, iodine, selenium, copper, manganese, fluoride, chromium, and molybdenum).

Iron Iron is an essential nutrient, vital to many of the cells' activities, but it poses a problem for millions of people. Some people simply don't eat enough iron-containing foods to support their health optimally, whereas others absorb so much iron that it threatens their health. Iron exemplifies the principle that both too little and too much of a nutrient in the body can be harmful. In its wisdom, the body has several ways to maintain iron balance, protecting against both deficiency and toxicity.

Iron Roles in the Body Iron has the knack of switching back and forth between two ionic states. In the reduced state, iron has lost two electrons and therefore has a net positive charge of two; it is known as *ferrous iron* (Fe^{++}). In the oxidized state, iron has lost a third electron, has a net positive charge of three, and is known as *ferric iron* (Fe^{+++}). Ferrous iron can be oxidized to ferric iron, and ferric iron can be reduced to ferrous iron. By doing so, iron can serve as a **cofactor** to enzymes involved in the numerous oxidation-reduction reactions that commonly occur in all cells. Enzymes involved in making amino acids, collagen, hormones, and neurotransmitters all require iron. (For details about ions, oxidation, and reduction, see Appendix B.)

Iron forms a part of the electron carriers that participate in the electron transport chain (discussed in Chapter 7).* These carriers transfer hydrogens and electrons to oxygen, forming water, and in the process, make ATP for the cells' energy use.

Most of the body's iron is found in two proteins: **hemoglobin** in the red blood cells and **myoglobin** in the muscle cells. In both, iron helps accept, carry, and then release oxygen.

Iron Absorption The body conserves iron. Because it is difficult to excrete iron once it is in the body, balance is maintained primarily through absorption. More iron is absorbed when stores are empty and less is absorbed when stores are full. Special proteins help the body absorb iron from food (see Figure 13-3).[2] The iron-storage protein **ferritin** captures iron from food and stores it in the cells of the small intestine. When the body needs iron, ferritin releases some iron to an iron transport protein called **transferrin.** If the body does not need iron, it is carried out when the intestinal cells are shed and excreted in the feces; intestinal cells are replaced about every 3 to 5 days. By holding iron temporarily, these cells control iron absorption by either delivering iron when the day's intake falls short or disposing of it when intakes exceed needs.

Iron absorption depends in part on its dietary source. Iron occurs in two forms in foods: as **heme iron,** which is found only in foods derived from the flesh of animals, such as meats, poultry, and fish and as **nonheme iron,** which is found in both plant-derived and animal-derived foods (see Figure 13-4). On average, heme iron represents about 10 percent of the iron a person consumes in a day. Even though heme iron accounts for only a small proportion of the intake, it is so well absorbed that it contributes significant iron. About 25 percent of heme iron and 17 percent of nonheme iron is absorbed, depending on dietary factors and the body's iron stores.[3] In iron deficiency, absorption increases. In iron overload, absorption declines.

Heme iron has a high bioavailability and is not influenced by dietary factors. In contrast, several dietary factors influence nonheme iron absorption (see Table 13-1).[4] Meat, fish, and poultry contain not only the well-absorbed heme iron, but also a peptide (sometimes called the **MFP factor**) that promotes the

---

TABLE 13-1 Factors That Influence Nonheme Iron Absorption

Enhancing Factors	Inhibiting Factors
• MFP factor	• Phytates (legumes, grains, nuts, seeds)
• Vitamin C (ascorbic acid)	• Vegetable proteins (soybeans, legumes, nuts)
• Acids (citric and lactic)	• Calcium (milk)
• Sugars (fructose)	• Tannic acid (and other polyphenols in tea and coffee)

© 2016 Cengage Learning

---

iron: an essential trace mineral that is needed for the transport of oxygen and the metabolism of energy nutrients.

cofactor: a small, inorganic or organic substance that facilitates the action of an enzyme.

hemoglobin (HE-moh-GLO-bin): the globular protein of the red blood cells that transports oxygen from the lungs to tissues throughout the body; hemoglobin accounts for 80 percent of the body's iron.

myoglobin: the oxygen-holding protein of the muscle cells.

• **myo** = muscle

ferritin (FAIR-ih-tin): the iron storage protein.

transferrin (trans-FAIR-in): the iron transport protein.

heme (HEEM) **iron:** the iron in foods that is bound to the hemoglobin and myoglobin proteins; found only in meat, fish, and poultry.

nonheme iron: the iron in foods that is not bound to proteins; found in both plant-derived and animal-derived foods.

MFP factor: a peptide released during the digestion of meat, fish, and poultry that enhances nonheme iron absorption.

---

*The iron-containing electron carriers of the electron transport chain are known as *cytochromes*. See Appendix C for details on the electron transport chain.

> FIGURE 13-3 **Iron Absorption**

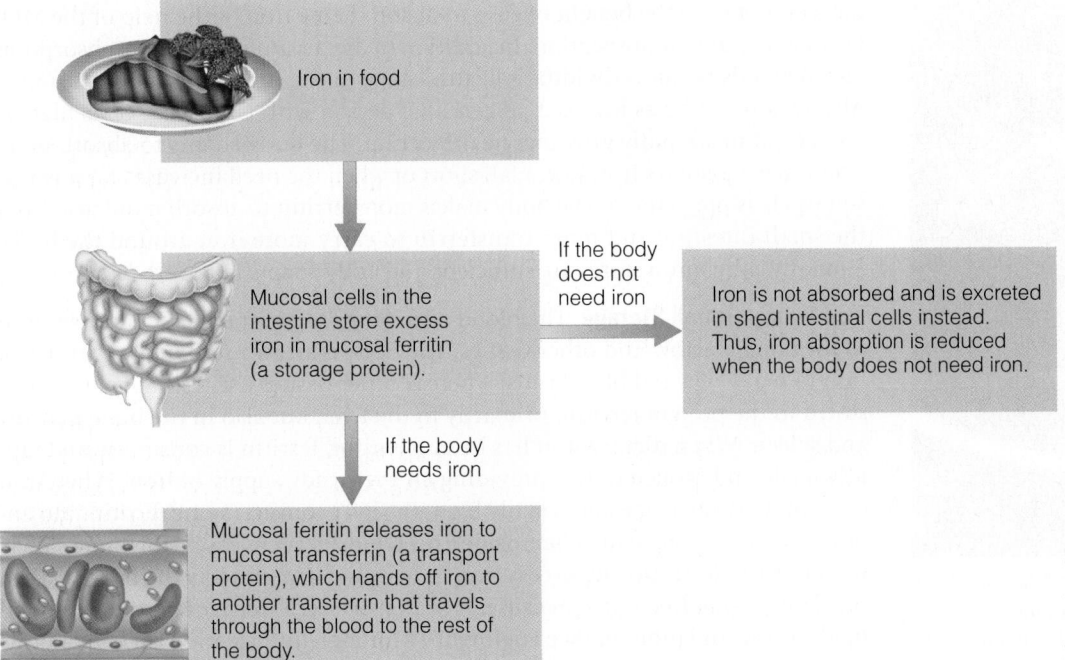

Iron in food

Mucosal cells in the intestine store excess iron in mucosal ferritin (a storage protein).

If the body does not need iron

Iron is not absorbed and is excreted in shed intestinal cells instead. Thus, iron absorption is reduced when the body does not need iron.

If the body needs iron

Mucosal ferritin releases iron to mucosal transferrin (a transport protein), which hands off iron to another transferrin that travels through the blood to the rest of the body.

© Cengage Learning

absorption of nonheme iron from other foods eaten at the same meal. Vitamin C (ascorbic acid) also enhances nonheme iron absorption from foods eaten at the same meal by capturing the iron and keeping it in the reduced ferrous form, ready for absorption. Some acids (such as citric acid) and sugars (such as fructose) also enhance nonheme iron absorption.

Some dietary factors bind with nonheme iron, inhibiting absorption. These factors include the phytates in legumes, whole grains, and rice; the vegetable proteins in soybeans, other legumes, and nuts; the calcium in milk; and the polyphenols (such as tannic acid) in tea, coffee, grain products, oregano, and red wine.

The many dietary enhancers, inhibitors, and their combined effects make it difficult to estimate iron absorption. Most of these factors exert a strong influence individually, but not when combined with the others in a meal. Furthermore, the impact of the combined effects diminishes when a diet is evaluated over several days. When multiple meals are analyzed together, three factors appear to be most relevant: MFP factor and vitamin C as enhancers and phytates as inhibitors.

> FIGURE 13-4 **Heme and Nonheme Iron in Foods**

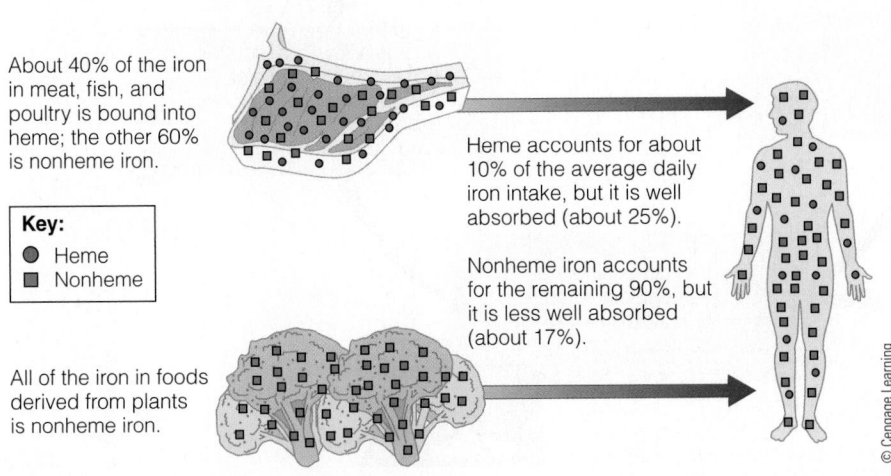

About 40% of the iron in meat, fish, and poultry is bound into heme; the other 60% is nonheme iron.

Key:
- ● Heme
- ■ Nonheme

All of the iron in foods derived from plants is nonheme iron.

Heme accounts for about 10% of the average daily iron intake, but it is well absorbed (about 25%).

Nonheme iron accounts for the remaining 90%, but it is less well absorbed (about 17%).

© Cengage Learning

This chili dinner provides several factors that may enhance iron absorption: heme and nonheme iron and the MFP factor from meat, nonheme iron from legumes, and vitamin C from tomatoes.

hemosiderin (heem-oh-SID-er-in): an iron-storage protein primarily made in times of iron overload.

Overall, about 18 percent of dietary iron is absorbed from mixed diets and only about 10 percent from vegetarian diets. As you might expect, vegetarian diets do not have the benefit of easy-to-absorb heme iron or the help of the MFP factor in enhancing absorption. In addition to dietary influences, iron absorption also depends on an individual's health, stage in the life cycle, and iron status. Absorption can be as low as 2 percent in a person with GI disease or as high as 35 percent in a rapidly growing, healthy child. The body adapts to absorb more iron when a person's iron stores fall short or when the need increases for any reason (such as pregnancy). The body makes more ferritin to absorb more iron from the small intestine and more transferrin to carry more iron around the body. Similarly, when iron stores are sufficient, the body adapts to absorb less iron.

Iron Transport and Storage The blood transport protein transferrin delivers iron to the bone marrow and other tissues. The bone marrow uses large quantities of iron to make new red blood cells, whereas other tissues use less. Surplus iron is stored in the protein ferritin, primarily in the liver, but also in the bone marrow and spleen. When dietary iron has been plentiful, ferritin is constantly and rapidly made and broken down, providing an ever-ready supply of iron. When iron concentrations become abnormally high, the liver converts some ferritin into another storage protein called **hemosiderin.** Hemosiderin releases iron more slowly than ferritin does. Storing excess iron in hemosiderin protects the body against the damage that free iron can cause. Free iron acts as a free radical, attacking cell lipids, DNA, and protein. (See Highlight 11 for more information on free radicals and the damage they can cause.)

The average red blood cell lives about 4 months; then the spleen and liver cells remove it from the blood, take it apart, and prepare the degradation products for excretion or recycling. The iron is salvaged: the liver attaches it to transferrin, which transports it back to the bone marrow to be reused in making new red blood cells. Thus, although red blood cells live for only about 4 months, the iron recycles through each new generation of cells (see Figure 13-5). The body loses some iron daily via the GI tract and, if bleeding occurs, in blood. Only

> FIGURE 13-5 **Iron Recycled in the Body**

Once iron enters the body, most of it is recycled. Some is lost with body tissues and must be replaced by eating iron-containing foods.

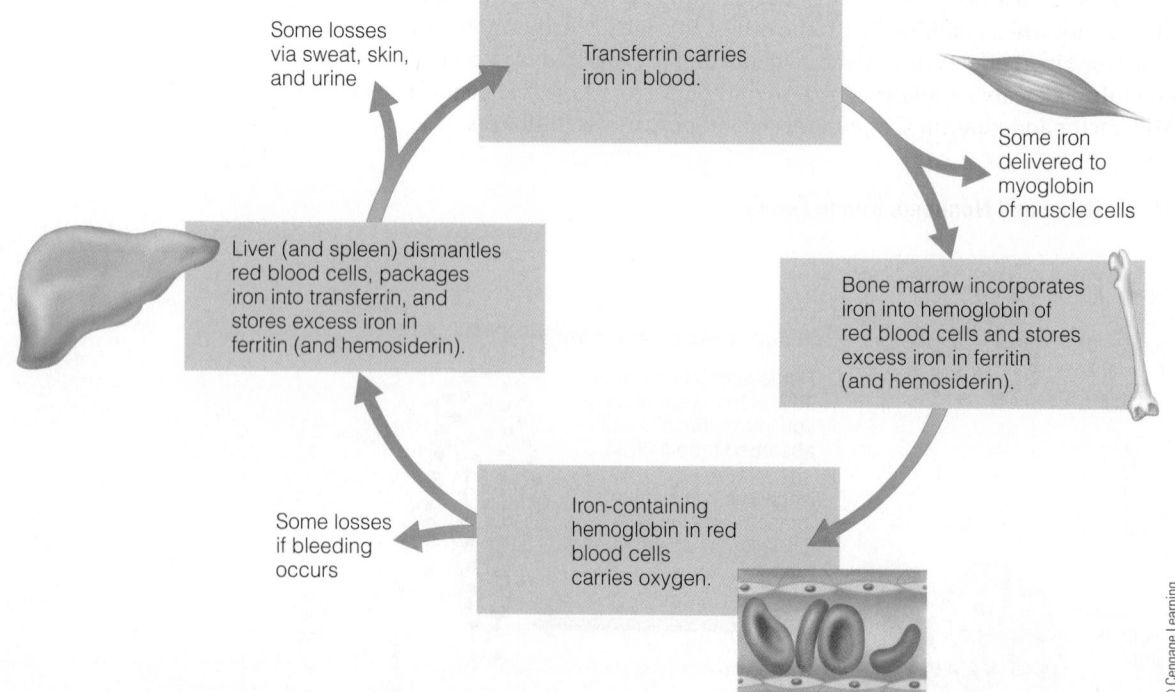

tiny amounts of iron are lost in urine, sweat, and shed skin. Iron excretion differs for men and women. On average, men and women lose about 1.0 milligram of iron per day, with women losing additional iron in menses; menstrual losses vary considerably, but over a month, they average about 0.5 milligram per day.

Maintaining iron balance depends on the careful regulation of iron absorption, transport, storage, recycling, and losses. Central to the regulation of iron balance is the hormone **hepcidin**.[5] Produced by the liver, hepcidin helps to maintain blood iron within the normal range by limiting absorption from the small intestine and controlling release from the liver, spleen, and bone marrow. Hepcidin production increases in iron overload and decreases in iron deficiency.[6]

Iron Deficiency Worldwide, **iron deficiency** is the most common nutrient deficiency, with **iron-deficiency anemia** affecting 1.5 to 2.0 billion people—mostly preschool children and pregnant women.[7] In the United States, iron deficiency is less prevalent, but it still affects about 10 percent of toddlers, adolescent girls, and women of childbearing age. Iron deficiency is also relatively common among those who are overweight. The association between iron deficiency and obesity has yet to be explained, but researchers are currently examining the relationships between the inflammation that develops with excess body fat and reduced iron absorption.[8] The increased production of hepcidin in obesity may also help to explain the relationship between obesity and iron deficiency.[9] Preventing and correcting iron deficiency are high priorities.

Some stages of life demand more iron but provide less, making deficiency likely.[10] Women in their reproductive years are especially prone to iron deficiency because of repeated blood losses during menstruation. Pregnancy demands additional iron to support the added blood volume, growth of the fetus, and blood loss during childbirth. Infants and young children receive little iron from their high-milk diets, yet need extra iron to support their rapid growth and brain development.* Iron deficiency among toddlers in the United States is common. The rapid growth of adolescence, especially for males, and the menstrual losses of females also demand extra iron that a typical teen diet may not provide. An adequate iron intake is especially important during these stages of life.

Bleeding from any site incurs iron losses.** In some cases, such as an active ulcer, the bleeding may not be obvious, but even small chronic blood losses significantly deplete iron reserves. In developing countries, blood loss is often brought on by malaria and parasitic infections of the GI tract. People who donate blood regularly also incur losses and may benefit from iron supplements. As mentioned, menstrual losses can be considerable as they tap women's iron stores regularly.

Assessment of Iron Deficiency Iron deficiency develops in stages. This section provides a brief overview of how to detect these stages, and Appendix E provides more details. In the first stage of iron deficiency, iron stores diminish. Measures of serum ferritin (in the blood) reflect iron stores and are most valuable in assessing iron status at this earliest stage. Unfortunately, serum ferritin increases with infections, which interferes with an accurate diagnosis and estimates of prevalence.[11]

The second stage of iron deficiency is characterized by a decrease in transport iron: serum iron falls, and the iron-carrying protein transferrin *increases* (an adaptation that enhances iron absorption). Together, measurements of serum iron and transferrin can determine the severity of the deficiency—the more transferrin and the less iron in the blood, the more advanced the deficiency is. Transferrin saturation—the percentage of transferrin that is saturated with iron—decreases as iron stores decline.

The third stage of iron deficiency occurs when the lack of iron limits hemoglobin production. Now the hemoglobin precursor, **erythrocyte protoporphyrin,** begins to accumulate as hemoglobin and **hematocrit** values decline.

hepcidin: a hormone produced by the liver that regulates iron balance.

iron deficiency: the state of having depleted iron stores.

iron-deficiency anemia: severe depletion of iron stores that results in low hemoglobin and small, pale red blood cells. Iron-deficiency anemia is a *microcytic* (my-cro-SIT-ic) *hypochromic* (high-po-KROME-ic) *anemia.*

- **micro** = small
- **cytic** = cell
- **hypo** = too little
- **chrom** = color

erythrocyte protoporphyrin (PRO-toe-PORE-fe-rin): a precursor to hemoglobin.

hematocrit (hee-MAT-oh-krit): the percentage of total blood volume that consists of red blood cells.

---

*The condition of developing iron-deficiency anemia because iron-poor milk displaces iron-rich foods in the diet is sometimes called *milk anemia.*

**The iron content of blood is about 0.5 miligram/100 milliliters of blood. A person donating a pint of blood (approximately 500 milliliters) loses about 2.5 milligrams of iron.

Hemoglobin and hematocrit tests are easy, quick, and inexpensive, so they are the tests most commonly used in evaluating iron status. Their usefulness in detecting iron deficiency is limited, however, because they are late indicators. Furthermore, other nutrient deficiencies and medical conditions can influence their values.

Iron Deficiency and Anemia Notice that iron deficiency and iron-deficiency anemia are not the same: people may be iron deficient without being anemic. The term *iron deficiency* refers to depleted body iron stores without regard to the degree of depletion or to the presence of anemia. The term *iron-deficiency anemia* refers to the severe depletion of iron stores that results in a low hemoglobin concentration. In iron-deficiency anemia, hemoglobin synthesis decreases, resulting in red blood cells that are pale (hypochromic) and small (microcytic), as shown in Figure 13-6. Without adequate iron, these cells can't carry enough oxygen from the lungs to the tissues. Energy metabolism in the cells falters. The result is fatigue, weakness, headaches, apathy, pallor, and poor resistance to cold temperatures. Because hemoglobin is the bright red pigment of the blood, the skin of a fair person who is anemic may become noticeably pale. In a dark-skinned person, the tongue and eye lining, normally pink, is very pale.

The fatigue that accompanies iron-deficiency anemia differs from the tiredness a person experiences from a simple lack of sleep. People with anemia feel fatigue only when they exert themselves. Consequently, their work productivity, voluntary activities, and athletic performance decline.[12] Iron supplementation can relieve the fatigue and improve the body's response to physical activity.[13] (The iron needs of physically active people and the special iron deficiency known as *sports anemia* are discussed in Chapter 14.)

> **FIGURE 13-6** **Normal Blood Cells and Blood Cells in Iron Deficiency Anemia Compared**

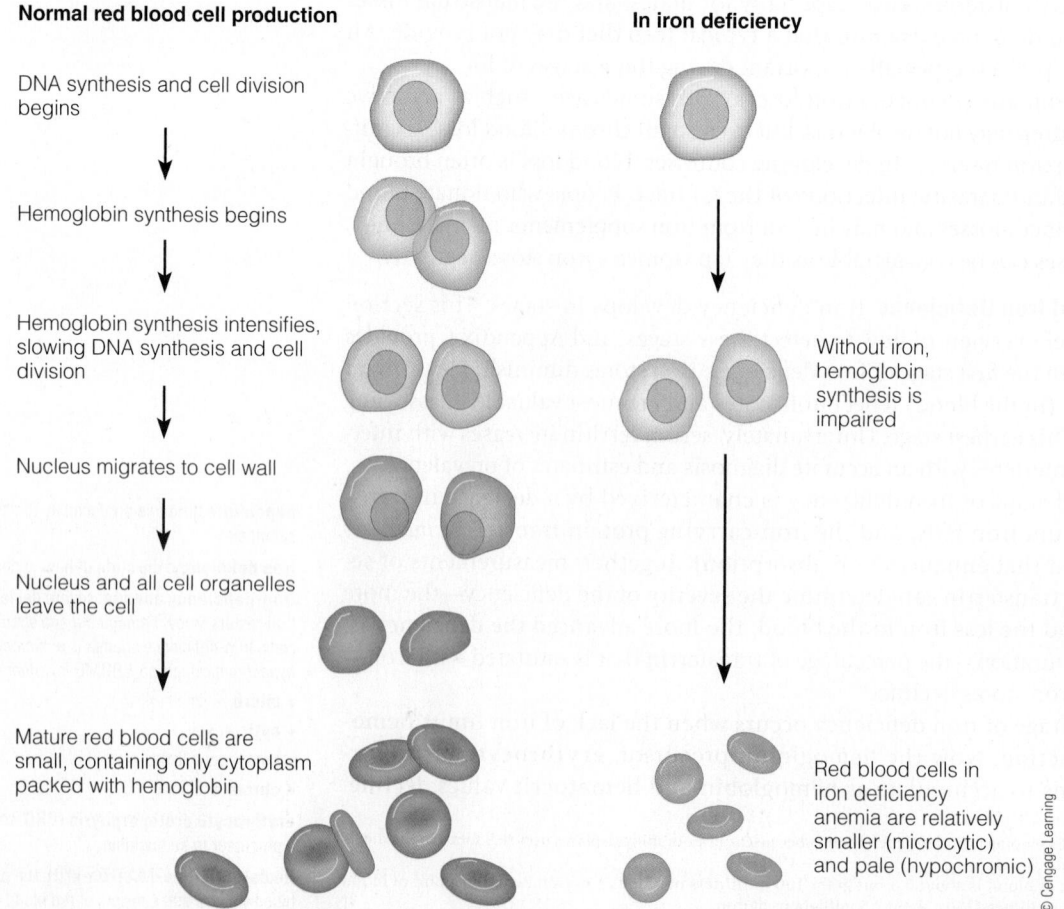

Normal red blood cell production

In iron deficiency

DNA synthesis and cell division begins

Hemoglobin synthesis begins

Hemoglobin synthesis intensifies, slowing DNA synthesis and cell division

Nucleus migrates to cell wall

Nucleus and all cell organelles leave the cell

Mature red blood cells are small, containing only cytoplasm packed with hemoglobin

Without iron, hemoglobin synthesis is impaired

Red blood cells in iron deficiency anemia are relatively smaller (microcytic) and pale (hypochromic)

© Cengage Learning

Iron Deficiency and Behavior Long before the red blood cells are affected and anemia is diagnosed, a developing iron deficiency affects behavior.[14] Even at slightly lowered iron levels, energy metabolism is impaired and neurotransmitter synthesis is altered, reducing physical work capacity and mental productivity.[15] Without the physical energy and mental alertness to work, plan, think, play, sing, or learn, people simply do less. They have no obvious deficiency symptoms; they just appear unmotivated and apathetic.

Many of the symptoms associated with iron deficiency are easily mistaken for behavioral or motivational problems. A restless child who fails to pay attention in class might be thought contrary. An apathetic homemaker who has let housework pile up might be thought lazy. No responsible dietitian would ever claim that all behavioral problems are caused by nutrient deficiencies, but poor nutrition is always a possible contributor to problems like these. When investigating a behavioral problem, check the adequacy of the diet and seek a routine physical examination before undertaking more expensive, and possibly more harmful, treatment options. If iron deficiency is the problem, then treatment with iron supplements may improve mood, cognitive skills, and physical performance. The effects of iron deficiency on children's behavior are discussed further in Chapter 16.

Iron Deficiency and Pica A curious behavior seen in some iron-deficient people, especially in women and children of low-income groups, is **pica**—the craving and consumption of ice, chalk, starch, and other nonfood substances. These substances contain no iron and cannot remedy a deficiency; in fact, clay actually inhibits iron absorption, which may explain the iron deficiency that accompanies such behavior. Pica is poorly understood. Its cause is unknown, but researchers hypothesize that it may be motivated by hunger, nutrient deficiencies, or an attempt to protect against toxins or microbes.[16] The consequence of pica is anemia.

Iron Overload As mentioned earlier, because too much iron can be toxic, its levels in the body are closely regulated and absorption normally decreases when iron stores are full.[17] Even a diet that includes fortified foods usually poses no risk for most people, but some individuals are vulnerable to excess iron. Once considered rare, **iron overload** has emerged as an important disorder of iron metabolism and regulation.

The iron overload disorder known as **hemochromatosis** is caused by a genetic failure to prevent unneeded iron in the diet from being absorbed.[18] Research suggests that just as insulin supports normal glucose homeostasis and its absence or ineffectiveness causes diabetes, the hormone hepcidin supports iron homeostasis and its deficiency or (rarely) resistance causes hemochromatosis.[19] Other causes of iron overload include repeated blood transfusions (which bypass the intestinal defense), massive doses of supplementary iron (which overwhelm the intestinal defense), and other rare metabolic disorders.

Some of the signs and symptoms of iron overload are similar to those of iron deficiency: apathy, lethargy, and fatigue. Therefore, taking iron supplements before assessing iron status is clearly unwise; hemoglobin tests alone would fail to make the distinction because excess iron accumulates in storage. Iron overload assessment tests measure transferrin saturation and serum ferritin.

Iron overload is characterized by a toxic accumulation of iron in the liver, heart, joints, and other tissues. Excess iron in these tissues causes free-radical damage.[20] Infections are likely because viruses and bacteria thrive on iron-rich blood. Symptoms are most severe in alcohol abusers because alcohol damages the small intestine, further impairing its defenses against absorbing excess iron. Untreated iron overload increases the risks of diabetes, liver cancer, heart disease, and arthritis.[21] Currently, treatment involves **phlebotomy**, which removes blood from the body, and chelation therapy, which uses a **chelate** to form a complex with iron and promote its excretion.[22] Research targeting the activity of hepcidin is active and promising.[23]

Iron overload is much more common in men than in women and is twice as prevalent among men as iron deficiency. The widespread fortification of foods with iron makes it difficult for people with hemochromatosis to follow a low-iron

pica (PIE-ka): a craving for and consumption of nonfood substances. Pica is known as *geophagia* (gee-oh-FAY-gee-uh) when referring to eating clay, baby powder, chalk, ash, ceramics, paper, paint chips, or charcoal; *pagophagia* (pag-oh-FAY-gee-uh) when referring to eating large quantities of ice; and *amylophagia* (AM-ee-low-FAY-gee-ah) when referring to eating uncooked starch (flour, laundry starch, or raw rice).

iron overload: toxicity from excess iron.

hemochromatosis (HE-moh-KRO-ma-toe-sis): a genetically determined failure to prevent absorption of unneeded dietary iron that is characterized by iron overload and tissue damage.

phlebotomy: the withdrawal of blood from the body.

chelate (KEY-late): a substance that can grasp the positive ions of a mineral.

• **chele** = claw

diet, and greater dangers lie in the indiscriminate use of iron and vitamin C supplements. Vitamin C not only enhances iron absorption, but also releases iron from ferritin, allowing free iron to wreak the damage typical of free radicals. Thus vitamin C acts as a *proox*idant when taken in high doses. (See Highlight 11 for a discussion of free radicals and their effects on disease development.)

Iron and Chronic Diseases Some research suggests a link between heart disease and excess iron.[24] Limited evidence suggests an association between iron and some cancers. Explanations for how iron might be involved in contributing to these chronic diseases focus on its free-radical activity. One of the benefits of a high-fiber diet may be that the accompanying phytates bind iron, making it less available for such reactions.

Iron Poisoning Large doses of iron supplements cause GI distress, including constipation, nausea, vomiting, and diarrhea. These effects may not be as serious as other consequences of iron toxicity, but they are consistent enough to establish a UL of 45 milligrams per day for adults.

Ingestion of iron-containing supplements is a common cause of accidental poisoning in young children.[25] Symptoms of toxicity include nausea, vomiting, diarrhea, a rapid heartbeat, a weak pulse, dizziness, shock, and confusion. As few as five iron tablets containing as little as 200 milligrams of iron have caused death in young children. The exact cause of death is uncertain, but excessive free-radical damage is thought to play a role in heart failure and respiratory distress. Autopsy reports reveal iron deposits and cell death in the stomach, small intestine, liver, and blood vessels (which can cause internal bleeding). As with medicines and other potentially toxic substances, keep iron-containing tablets out of the reach of children. If you suspect iron poisoning, call the nearest poison control center or a physician immediately.

Iron Recommendations The usual diet in the United States provides about 6 to 7 milligrams of iron for every 1000 kcalories. The recommended daily intake for men is 8 milligrams, and because most men eat more than 2000 kcalories a day, they can meet their iron needs with little effort. Women in their reproductive years, however, need 18 milligrams a day. The accompanying "How To" feature explains how the recommended intake was calculated.

>How To

Estimate the Recommended Daily Intake for Iron

To calculate the recommended daily iron intake, the DRI Committee considers a number of factors. For example, for a woman of childbearing age (19 to 50):

- Losses from feces, urine, sweat, and shed skin: 1.0 milligram
- Losses through menstruation: 0.5 milligram (about 14 milligrams total averaged over 28 days)

These losses reflect an average daily need (total) of 1.5 milligrams of *absorbed iron*.

An estimated average requirement is determined based on the daily need and the assumption that an average of 18 percent of ingested iron is absorbed:

$$\text{1.5 mg iron (needed)}$$
$$\div \text{ 0.18 (percent iron absorbed)}$$
$$= \text{8 mg iron (estimated average requirement)}$$

Then, a margin of safety is added to cover the needs of essentially all women of childbearing age, and the RDA is set at 18 milligrams.

Because the iron bioavailability of typical vegetarian diets is low, the recommendation for iron is 1.8 times higher for vegetarians. To calculate the RDA for vegetarians, multiply by 1.8:

- 8 mg × 1.8 = 14 mg/day (vegetarian men and women >50 yr)
- 18 mg × 1.8 = 32 mg/day (vegetarian women, 19 to 50 yr)

> **TRY IT** Calculate how many slices of whole-wheat bread, cups of broccoli, ounces of hamburger meat, and cups of milk it takes to provide 18 milligrams of iron.

Because women have higher iron needs and lower energy needs, they sometimes have trouble obtaining enough iron. On average, women receive only 12 to 13 milligrams of iron per day, which is not enough iron for women until after menopause. To meet their iron needs from foods, premenopausal women need to select iron-rich foods at every meal.

> **DIETARY GUIDELINES FOR AMERICANS**
Women capable of becoming pregnant should choose foods that supply heme iron, which is more readily absorbed by the body, additional iron sources, and enhancers of iron absorption such as vitamin C. If pregnant, women should take an iron supplement, as recommended by a health-care provider.

Vegetarians need 1.8 times as much iron to make up for the low bioavailability typical of their diets. Good vegetarian sources of iron include soy foods (such as soybeans and tofu), legumes (such as lentils and kidney beans), nuts (such as cashews and almonds), seeds (such as pumpkin seeds and sunflower seeds), cereals (such as cream of wheat and oatmeal), dried fruit (such as apricots and raisins), vegetables (such as mushrooms and potatoes), and blackstrap molasses.

Iron Food Sources To obtain enough iron, people must first select iron-rich foods—both naturally occurring and enriched or fortified—and then take advantage of factors that maximize iron absorption. This discussion begins by identifying iron-rich foods and then reviews the factors affecting absorption. Figure 13-7 (p. 418) shows the amounts of iron in selected foods. Meats, fish, and poultry contribute the most iron per serving; other protein-rich foods such as legumes and eggs are also good sources. Although an indispensable part of the diet, foods in the milk group are notoriously poor in iron. Grain products vary, with whole-grain, enriched, and fortified breads and cereals contributing significantly to iron intakes. Finally, dark greens (such as broccoli) and dried fruits (such as raisins) contribute some iron.

The FDA does not mandate iron enrichment, but most states require manufacturers to enrich flour and grain products with iron.* One serving of enriched bread or cereal provides only a little iron, but because people eat many servings of these foods, the contribution can be significant. Iron added to foods is nonheme iron, which is not absorbed as well as heme iron, but when eaten with absorption-enhancing foods, enrichment iron can increase iron stores and reduce iron deficiency. In cases of iron overload, enrichment may exacerbate the problem.

In general, the bioavailability of iron is high in meats, fish, and poultry, intermediate in grains and legumes, and low in most vegetables, especially those containing oxalates such as spinach. As mentioned earlier, the amount of iron ultimately absorbed from a meal depends on the combined effects of several enhancing and inhibiting factors. For maximum absorption of nonheme iron, eat meat for the MFP factor and fruits or vegetables for vitamin C. The iron of baked beans, for example, will be enhanced by the MFP factor in a piece of pork served with them. The iron of bread will be enhanced by the vitamin C in a slice of tomato on a sandwich.

Iron Contamination In addition to the iron from foods, **contamination iron** from nonfood sources of inorganic iron salts can contribute to the day's intakes. Foods cooked in iron cookware take up iron salts. The more acidic the food and the longer it is cooked in iron cookware, the higher the iron content. The iron content of eggs can triple in the time it takes to scramble them in an iron pan. Admittedly,

When the label on a grain product says "enriched," it means iron and several B vitamins have been added to meet FDA standards.

© Craig M. Moore

*Each pound of enriched flour contains at least 20 milligrams of iron.

contamination iron: iron found in foods as the result of contamination by inorganic iron salts from iron cookware, iron-containing soils, and the like.

> FIGURE 13-7 **Iron in Selected Foods**

Milligrams

Food	Serving size (kcalories)
Bread, whole wheat	1-oz slice (70 kcal)
Cornflakes, fortified	1 oz (110 kcal)
Spaghetti pasta	½ c cooked (99 kcal)
Tortilla, flour	1 10" round (234 kcal)
Broccoli	½ c cooked (22 kcal)
Carrots	½ c shredded raw (24 kcal)
Potato	1 medium baked w/skin (133 kcal)
Tomato juice	½ c (31 kcal)
Banana	1 medium raw (109 kcal)
Orange	1 medium raw (62 kcal)
Strawberries	½ c fresh (22 kcal)
Watermelon	1 slice (92 kcal)
Milk	1 c reduced-fat 2% (121 kcal)
Yogurt, plain	1 c low-fat (155 kcal)
Cheddar cheese	1½ oz (171 kcal)
Cottage cheese	½ c low-fat 2% (101 kcal)
Pinto beans	½ c cooked (117 kcal)
Peanut butter	2 tbs (188 kcal)
Sunflower seeds	1 oz dry (165 kcal)
Tofu (soybean curd)	½ c (76 kcal)
Ground beef, lean	3 oz broiled (244 kcal)
Chicken breast	3 oz roasted (140 kcal)
Tuna, canned in water	3 oz (99 kcal)
Egg	1 hard cooked (78 kcal)
Excellent, and sometimes unusual, sources:	
Clams, canned	3 oz (126 kcal)
Beef liver	3 oz fried (184 kcal)
Parsley	1 c raw (22 kcal)

RDA for women 51+

RDA for women 19–50

RDA for men

IRON
Protein foods (red and brown), and some vegetables (green) make the greatest contributions of iron to the diet.

Key:
- Grains
- Vegetables
- Fruits
- Milk and milk products
- Legumes, nuts, seeds
- Meats, poultry, seafood
- Best sources per kcalorie

© Cengage Learning

© Polara Sutdios, Inc.

An old-fashioned iron skillet adds iron to foods. Increase in iron content (mg) for selected foods (3 oz) after cooking in iron skillet:

Beef stew	0.66→3.40
Chili	0.96→6.27
Cornbread	0.67→0.86
Hamburger	1.49→2.29
Pancake	0.63→1.31
Rice	0.67→1.97
Scrambled egg	1.49→4.76
Spaghetti sauce	0.61→5.77

the absorption of this iron may be poor (perhaps only 1 to 2 percent), but every little bit helps a person who is trying to increase iron intake.

Iron Supplementation People who are iron deficient may need supplements as well as an iron-rich, absorption-enhancing diet. Many physicians routinely recommend iron supplements to pregnant women, infants, and young children. Iron from supplements is less well absorbed than that from food, so the doses must be high. The absorption of iron taken as ferrous sulfate is better than that from other iron supplements. Absorption also improves when supplements are taken between meals, at bedtime on an empty stomach, and with liquids (other than milk, tea, or coffee, which inhibit absorption). Taking iron supplements in a single dose instead of several doses per day is equally effective and may improve a person's willingness to take it regularly.

There is no benefit to taking iron supplements with orange juice because vitamin C does not enhance absorption from supplements as it does from foods. Vitamin C enhances iron absorption by converting insoluble ferric iron in foods to the more soluble ferrous iron, and supplemental iron is already in the ferrous form. Constipation is a common side effect of iron supplementation; drinking plenty of water may help to relieve this problem. The best strategy to ensure compliance is to individualize the dose, formulation, and schedule. Most importantly, iron supplements should be taken only when prescribed by a physician who has assessed an iron deficiency.

Most of the body's iron is in hemoglobin and myoglobin, where it carries oxygen for use in energy metabolism; some iron is also required for enzymes involved in a variety of reactions. Special proteins assist with iron absorption, transport, and storage—all helping to maintain an appropriate balance—because both too little and too much iron can be damaging. Iron deficiency is most common among infants and young children, teenagers, women of childbearing age, and pregnant women. Symptoms include fatigue and anemia. Iron overload is most common in men. Heme iron, which is found only in meat, fish, and poultry, is better absorbed than nonheme iron, which occurs in most foods. Nonheme iron absorption is improved by eating iron-containing foods with foods containing the MFP factor and vitamin C; absorption is limited by phytates and oxalates. The accompanying table provides a summary of iron.

Iron

RDA	Significant Sources
Men: 8 mg/day	Red meats, fish, poultry, shellfish, eggs, legumes, dried fruits
Women: 18 mg/day (19–50 yr) 8 mg/day (51+)	**Deficiency Symptoms**
UL	Anemia: weakness, fatigue, headaches; impaired work performance and cognitive function; impaired immunity; pale skin, nail beds, mucous membranes, and palm creases; concave nails; inability to regulate body temperature; pica
Adults: 45 mg/day	
Chief Functions in the Body	**Toxicity Symptoms**
Part of the protein hemoglobin, which carries oxygen in the blood; part of the protein myoglobin in muscles, which makes oxygen available for muscle contraction; necessary for the utilization of energy as part of the cells' metabolic machinery	GI distress
	Iron overload: infections, fatigue, joint pain, skin pigmentation, organ damage

Zinc
Zinc is an essential trace element required for numerous metabolic reactions.[26] Virtually all cells contain zinc, but the highest concentrations are found in muscle and bone.

Zinc Roles in the Body Zinc supports the work of hundreds of proteins in the body, such as the **metalloenzymes**, which participate in a variety of metabolic processes, and **transcription factors**, which regulate gene expression.* In addition, zinc stabilizes cell membranes and DNA, helping to strengthen antioxidant defenses against free-radical attacks. Zinc also assists in immune function and in growth and development. Zinc participates in the synthesis, storage, and release of the hormone insulin in the pancreas, although it does not appear to play a direct role in insulin's action. Zinc interacts with platelets in blood clotting, affects thyroid hormone function, and influences behavior and learning performance. It is needed to produce the active form of vitamin A (retinal) in visual pigments and the retinol-binding protein that transports vitamin A. It is essential to normal taste perception, wound healing, sperm production, and fetal development. A zinc deficiency impairs all these and other functions, underlining the vast importance of zinc in supporting the body's proteins.

Zinc Absorption The body's handling of zinc resembles that of iron in some ways and differs in others. A key difference is the circular passage of zinc from the small intestine to the body and back again.

The rate of zinc absorption varies from about 15 to 40 percent, depending on the amount of zinc consumed—as zinc intake increases, the rate of absorption decreases, and as zinc intake decreases, the rate of absorption increases.[27] Like iron, dietary factors such as phytates influence absorption, limiting its bioavailability.[28]

Upon absorption into an intestinal cell, zinc has two options. Zinc may participate in the metabolic functions of the intestinal cell itself, or it may be retained within the intestinal cells by **metallothionein** until the body needs zinc. Metallothionein plays a key role in storing and distributing zinc throughout the body.

zinc: an essential trace mineral that is part of many enzymes and a constituent of insulin.

metalloenzymes (meh-TAL-oh-EN-zimes): enzymes that contain one or more minerals as part of their structures.

transcription factors: proteins that bind to specific sites in DNA and alter gene expression.

metallothionein (meh-TAL-oh-THIGH-oh-neen): a sulfur-rich protein that avidly binds with and transports metals such as zinc.

- **metallo** = containing a metal
- **thio** = containing sulfur
- **ein** = a protein

---

*Among the metalloenzymes requiring zinc are carbonic anhydrase, deoxythymidine kinase, DNA and RNA polymerase, and alkaline phosphatase.

> FIGURE 13-8 **Enteropancreatic Circulation of Zinc**

Some zinc from food is absorbed by the small intestine and sent to the pancreas to be incorporated into digestive enzymes that return to the small intestine. This cycle is called the *enteropancreatic circulation* of zinc.

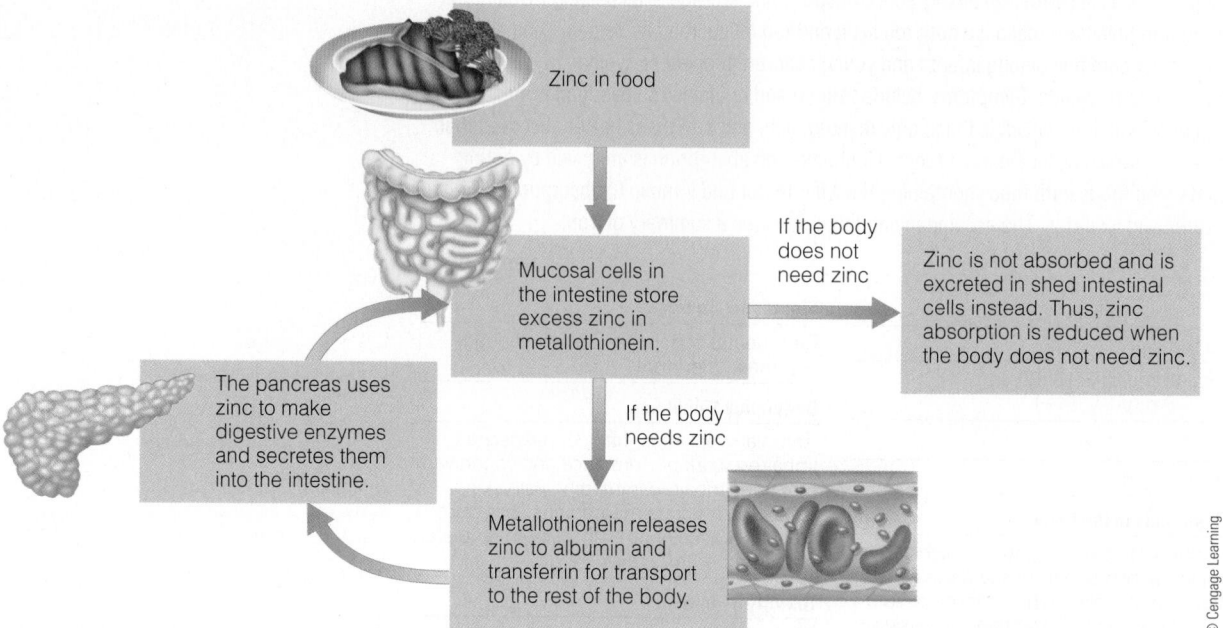

enteropancreatic (EN-ter-oh-PAN-kree-AT-ik)
circulation: the circulatory route from the pancreas to the small intestine and back to the pancreas.

> FIGURE 13-9 **Zinc-Deficiency Symptom— The Stunted Growth of Dwarfism**

The growth retardation, known as dwarfism, is rightly ascribed to zinc deficiency because it is partially reversible when zinc is restored to the diet.

The Egyptian man on the right is an adult of average height. The Egyptian boy on the left is 17 years old but is only 4 feet tall, like a 7-year-old in the United States. His genitalia are like those of a 6-year-old.

Zinc Transport After being absorbed, some zinc eventually reaches the pancreas, where it is incorporated into many of the digestive enzymes that the pancreas releases into the small intestine at mealtimes. The small intestine thus receives two doses of zinc with each meal—one from foods and the other from the zinc-rich pancreatic juices. The recycling of zinc in the body from the pancreas to the small intestine and back to the pancreas is referred to as the **enteropancreatic circulation** of zinc. Each time zinc circulates through the small intestine, it may be excreted in shed intestinal cells or reabsorbed into the body (see Figure 13-8). The body loses zinc primarily in feces. Smaller losses occur in urine, shed skin, hair, sweat, menstrual fluids, and semen.

Numerous proteins participate in zinc transport. Zinc's main transport vehicle in the blood is the protein albumin. Some zinc also binds to transferrin—the same transferrin that carries iron in the blood.

Zinc Deficiency Severe zinc deficiency is not widespread in developed countries, but in the developing world, nearly 2 billion people are zinc deficient.[29] Human zinc deficiency was first reported in the 1960s in children and adolescent boys in Egypt, Iran, and Turkey. Children have especially high zinc needs because they are growing rapidly and synthesizing many zinc-containing proteins, and the native diets among those populations were not meeting these needs. Middle Eastern diets are traditionally low in the richest zinc source, meats. Furthermore, the staple foods in these diets are legumes, unleavened breads, and other whole-grain foods—all high in fiber and phytates, which inhibit zinc absorption.*

Figure 13-9 shows the severe growth retardation and mentions the immature sexual development characteristic of zinc deficiency. In addition, zinc deficiency hinders digestion and absorption, causing diarrhea, which worsens malnutrition not only for zinc, but for other nutrients as well. It also impairs the immune response, making infections likely—among them, pneumonia and GI tract infections, which worsen malnutrition, including zinc malnutrition (a classic downward spiral of events).[30] Chronic zinc deficiency damages the central

---

*Unleavened bread contains no yeast, which normally breaks down phytates during fermentation.

> **FIGURE 13-10** **Zinc in Selected Foods**

Milligrams

Food	Serving size (kcalories)	0	2	4	6	8	10	12
Bread, whole wheat	1-oz slice (70 kcal)							
Cornflakes, fortified	1 oz (110 kcal)							
Spaghetti pasta	½ c cooked (99 kcal)							
Tortilla, flour	1 10" round (234 kcal)							
Broccoli	½ c cooked (22 kcal)							
Carrots	½ c shredded raw (24 kcal)							
Potato	1 medium baked w/skin (133 kcal)							
Tomato juice	¾ c (31 kcal)							
Banana	1 medium raw (109 kcal)							
Orange	1 medium raw (62 kcal)							
Strawberries	½ c fresh (22 kcal)							
Watermelon	1 slice (92 kcal)							
Milk	1 c reduced-fat 2% (121 kcal)							
Yogurt, plain	1 c low-fat (155 kcal)							
Cheddar cheese	1½ oz (171 kcal)							
Cottage cheese	½ c low-fat 2% (101 kcal)							
Pinto beans	½ c cooked (117 kcal)							
Peanut butter	2 tbs (188 kcal)							
Sunflower seeds	1 oz dry (165 kcal)							
Tofu (soybean curd)	½ c (76 kcal)							
Ground beef, lean	3 oz broiled (244 kcal)							
Chicken breast	3 oz roasted (140 kcal)							
Tuna, canned in water	3 oz (99 kcal)							
Egg	1 hard cooked (78 kcal)							
Excellent, and sometimes unusual, sources:								
Oysters	3 oz cooked (139 kcal)							
Sirloin steak, lean	3 oz broiled (172 kcal)							
Crab	3 oz cooked (94 kcal)							

RDA for men

RDA for women

ZINC
Meat, seafood, and poultry (red) are concentrated sources of zinc. Milk (white) and legumes, nuts, and seeds (brown) contain some zinc.

Key:
- Grains
- Vegetables
- Fruits
- Milk and milk products
- Legumes, nuts, seeds
- Meats, poultry, seafood
- Best sources per kcalorie

© Cengage Learning

nervous system and brain and may lead to poor motor development and cognitive performance. Because zinc deficiency directly impairs vitamin A metabolism, vitamin A–deficiency symptoms often appear. Zinc deficiency also disturbs thyroid function and the metabolic rate. It alters taste, causes loss of appetite, and slows wound healing—in fact, its symptoms are so pervasive that generalized malnutrition and sickness are more likely to be the diagnosis than simple zinc deficiency.

Zinc Toxicity High doses (more than 50 milligrams) of zinc may cause vomiting, diarrhea, headaches, exhaustion, and other symptoms. The UL for adults was set at 40 milligrams based on zinc's interference in copper metabolism—an effect that, in animals, leads to degeneration of the heart muscle.

Zinc Recommendations and Sources Figure 13-10 shows zinc amounts in selected foods per serving. Zinc is highest in protein-rich foods such as shellfish (especially oysters), meats, poultry, milk, and cheese. Legumes and whole-grain products are good sources of zinc if eaten in large quantities; in typical US diets, the phytate content of grains is not high enough to impair zinc absorption. Vegetables vary in zinc content depending on the soil in which they are grown. Average zinc intakes in the United States are slightly higher than recommendations.

Zinc Supplementation In developed countries, most people obtain enough zinc from the diet without resorting to supplements. In developing countries, zinc supplementation plays a major role in effectively reducing the incidence of disease and death associated with diarrhea and pneumonia.[31]

Zinc lozenges may shorten the duration, but not the severity, of common cold symptoms.[32] Lozenges of zinc acetate or zinc gluconate are most effective,

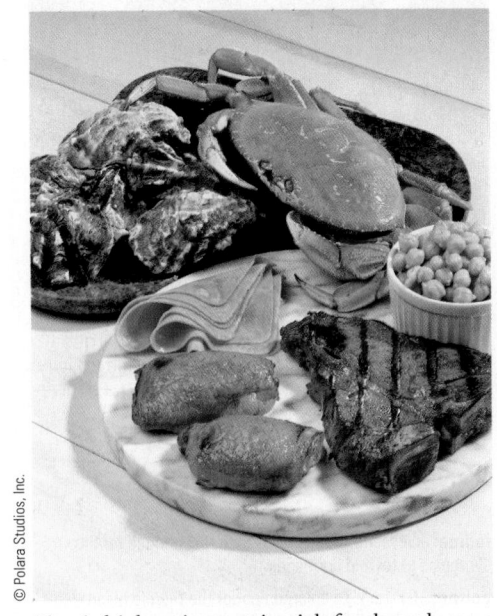

© Polara Studios, Inc.

Zinc is highest in protein-rich foods such as oysters, beef, poultry, legumes, and nuts.

whereas other zinc compounds, including those with flavor enhancers, are much less effective.[33] In addition to selecting the appropriate zinc formulation, consumers need to take relatively high doses (75 milligrams) of the lozenges within 24 hours of the onset of symptoms and continue daily throughout the duration of the cold.[34] Common side effects of zinc lozenges include nausea and bad taste reactions.

› REVIEW IT

Zinc-requiring enzymes participate in a multitude of reactions affecting growth, vitamin A activity, and pancreatic digestive enzyme synthesis, among others. After a meal, both dietary zinc and zinc-rich pancreatic secretions (via enteropancreatic circulation) are absorbed. Absorption is regulated by a special binding protein (metallothionein) in the small intestine. Protein-rich foods derived from animals are the best sources of bioavailable zinc. Fiber and phytates in cereals bind zinc, limiting absorption. Growth retardation and sexual immaturity are hallmark symptoms of zinc deficiency. The accompanying table provides a summary of zinc.

Zinc

RDA	Significant Sources
Men: 11 mg/day	Protein-containing foods: red meats, shellfish, whole grains; some fortified cereals
Women: 8 mg/day	
UL	**Deficiency Symptoms**[a]
Adults: 40 mg/day	Growth retardation, delayed sexual maturation, impaired immune function, hair loss, eye and skin lesions, loss of appetite
Chief Functions in the Body	
Part of many enzymes; associated with the hormone insulin; involved in making genetic material and proteins, immune reactions, transport of vitamin A, taste perception, wound healing, the making of sperm, and the normal development of the fetus	**Toxicity Symptoms**
	Loss of appetite, impaired immunity, low HDL, copper and iron deficiencies

[a]A rare inherited disease of zinc malabsorption, *acrodermatitis* (AK-roh-der-ma-TIE-tis) *enteropathica* (EN-teroh- PATH-ick-ah), causes additional and more severe symptoms.

> FIGURE 13-11   **Iodine-Deficiency Symptom—The Enlarged Thyroid of Goiter**

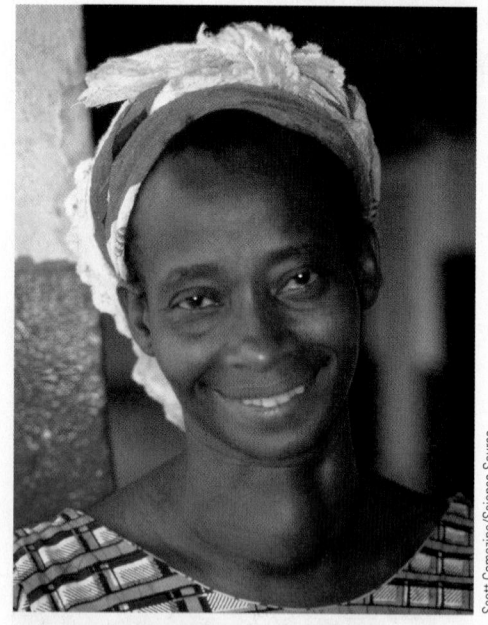

In iodine deficiency, the thyroid gland enlarges—a condition known as simple goiter. Iodine toxicity also enlarges the thyroid gland, creating a similar-looking goiter.

Scott Camazine/Science Source

Iodine   Traces of **iodine** are indispensable to life. In the GI tract, iodine from foods becomes iodide, which is readily absorbed.

Iodide Roles in the Body   Iodide is an integral part of the thyroid hormones that regulate body temperature, metabolic rate, reproduction, growth, blood cell production, nerve and muscle function, and more.* By controlling the rate at which the cells use oxygen, these hormones influence the amount of energy expended during basal metabolism.

Iodine Deficiency   The hypothalamus regulates thyroid hormone production by controlling the release of the pituitary's thyroid-stimulating hormone (TSH).** With iodine deficiency, thyroid hormone production declines, and the body responds by secreting more TSH in a futile attempt to accelerate iodide uptake by the thyroid gland. If a deficiency persists, the cells of the thyroid gland enlarge to trap as much iodide as possible. Sometimes the gland enlarges until it makes a visible lump in the neck, a **goiter** (shown in Figure 13-11).

Goiter afflicts about 200 million people the world over, many of them in South America, Asia, and Africa. In all but 4 percent of these cases, the cause is iodine deficiency. As for the 4 percent (8 million), most have goiter because they regularly eat excessive amounts of foods that contain an antithyroid substance

iodine: an essential trace mineral that is needed for the synthesis of thyroid hormones.

goiter (GOY-ter): an enlargement of the thyroid gland due to an iodine deficiency, malfunction of the gland, or overconsumption of a goitrogen. Goiter caused by iodine deficiency is sometimes called *simple goiter*.

*The thyroid gland releases tetraiodothyronine (T_4), commonly known as *thyroxine* (thigh-ROCKS-in), to its target tissues. Upon reaching the cells, T_4 loses one iodine, becoming triiodothyronine (T_3), which is the active form of the hormone.
**Thyroid-stimulating hormone is also called *thyrotropin*.

(goitrogen) whose effect is not counteracted by dietary iodine. Goitrogen-containing foods include vegetables such as cabbage, spinach, radishes, and rutabagas; legumes such as soybeans and peanuts; and fruits such as peaches and strawberries. The goitrogens present in plants remind us that even natural components of foods can cause harm when eaten in excess.

Goiter may be the earliest and most obvious sign of iodine deficiency, but the most tragic and prevalent damage occurs in the brain. Iodine deficiency is the most common cause of *preventable* mental retardation and brain damage in the world. Nearly one-third of the world's school-age children have iodine deficiency.[35] Children with even a mild iodine deficiency typically have goiters and perform poorly in school. With sustained treatment, however, mental performance in the classroom as well as thyroid function improves.

Even in the United States, pregnant women may not get as much iodine as they need.[36] A severe iodine deficiency during pregnancy causes the extreme and irreversible mental and physical retardation known as **cretinism.*** Cretinism affects approximately 6 million people worldwide and can be averted by the early diagnosis and treatment of maternal iodine deficiency. A worldwide effort to provide iodized salt to people living in iodine-deficient areas has been dramatically successful. An estimated 70 percent of all households in developing countries have access to iodized salt.[37] Because iron deficiency is common among people with iodine deficiency and because iron deficiency reduces the effectiveness of iodized salt, dual fortification with both iron and iodine may be most beneficial.

Iodine Toxicity Excessive intakes of iodine can interfere with thyroid function and enlarge the gland, just as deficiency can. During pregnancy, exposure to excessive iodine from foods, prenatal supplements, or medications is especially damaging to the developing infant. An infant exposed to toxic amounts of iodine during gestation may develop a goiter so severe as to block the airways and cause suffocation. The UL is 1100 micrograms per day for an adult—several times higher than average or recommended intakes (review Figure 13-1 on p. 408). For perspective, most foods provide 3 to 75 micrograms of iodine per serving.

Iodine Recommendations and Sources The ocean is the world's major source of iodine. In coastal areas, kelp, seafood, water, and even iodine-containing sea mist are dependable iodine sources. Further inland, the amount of iodine in foods is variable and generally reflects the amount present in the soil in which plants are grown or on which animals graze. Landmasses that were once under the ocean have soils rich in iodine; those in flood-prone areas where water leaches iodine from the soil are poor in iodine. In the United States, the iodization of salt provides about 60 micrograms of iodine per gram of salt. This tiny amount eliminated the widespread misery caused by iodine deficiency during the 1930s, but iodized salt is not available in many parts of the world. Some countries add iodine to bread, fish paste, or drinking water instead.

Although the average consumption of iodine in the United States exceeds recommendations, it falls below toxic levels. Some of the excess iodine in the US diet stems from fast foods, which use iodized salt liberally. Some iodine comes from bakery products and from milk. The baking industry uses iodates (iodine salts) as dough conditioners, and most dairies feed cows iodine-containing medications and use iodine to disinfect milking equipment. Processed foods in the United States use regular salt, not iodized salt.

The recommended intake of iodine for adults is a minuscule amount. The need for iodine is easily met by consuming seafood, vegetables grown in iodine-rich soil, and iodized salt. Just one-half teaspoon of iodized salt provides the RDA for iodine. In the United States, labels indicate whether salt is iodized.

goitrogen (GOY-troh-jen): a substance that enlarges the thyroid gland and causes *toxic goiter.* Goitrogens occur naturally in such foods as cabbage, kale, brussels sprouts, cauliflower, broccoli, and kohlrabi.

cretinism (CREE-tin-ism): a congenital disease characterized by mental and physical retardation and commonly caused by maternal iodine deficiency during pregnancy.

---

*The underactivity of the thyroid gland is known as *hypothyroidism* and may be caused by iodine deficiency or any number of other causes. Without treatment, an infant with *congenital hypothyroidism* will develop the physical and mental retardation of *cretinism.*

Only "iodized salt" has had iodine added.

© Craig M. Moore

Iodide, the ion of the mineral iodine, is an essential component of the thyroid hormones. An iodine deficiency can lead to simple goiter (enlargement of the thyroid gland) and can impair fetal development, causing cretinism. Iodization of salt has largely eliminated iodine deficiency in the United States. The accompanying table provides a summary of iodine.

Iodine

RDA	**Deficiency Disease**
Adults: 150 µg/day	Simple goiter, cretinism
UL	**Deficiency Symptoms**
1100 µg/day	Underactive thyroid gland, goiter, mental and physical retardation in infants (cretinism)
Chief Functions in the Body	
A component of two thyroid hormones that help to regulate growth, development, and metabolic rate	**Toxicity Symptoms**
	Underactive thyroid gland, elevated TSH, goiter
Significant Sources	
Iodized salt, seafood, bread, dairy products, plants grown in iodine-rich soil and animals fed those plants	

Selenium The essential mineral **selenium** shares some of the chemical characteristics of the mineral sulfur. This similarity allows selenium to substitute for sulfur in the amino acids methionine, cysteine, and cystine.

Selenium Roles in the Body Selenium is one of the body's antioxidant nutrients, working primarily as a part of proteins—most notably, the glutathione peroxidase enzymes.[38] Glutathione peroxidase and vitamin E work in tandem. Glutathione peroxidase prevents free-radical formation, thus blocking the chain reaction before it begins; if free radicals do form and a chain reaction starts, vitamin E stops it. (Highlight 11 describes free-radical formation, chain reactions, and antioxidant action in detail.) Other selenium-containing enzymes selectively activate or inactivate the thyroid hormones.

Selenium Deficiency Selenium deficiency is associated with **Keshan disease**—a heart disease that is prevalent in regions of China where the soil and foods lack selenium.[39] Although the primary cause of this heart disease is probably a virus or toxin, selenium deficiency appears to predispose people to it, and adequate selenium seems to prevent it.[40] Symptoms of selenium deficiency include impaired cognition and poor immunity.[41]

Selenium and Cancer Limited research suggests that the antioxidant action of selenium may protect against some types of cancers.[42] Selenium supplements, however, have not proved effective in preventing cancer and may in fact damage DNA and cause harm.[43]

Selenium Recommendations and Sources Selenium is found in the soil, and therefore in the crops grown for consumption. People living in regions with selenium-poor soil may still get enough selenium, partly because they eat vegetables and grains transported from other regions and partly because they eat meats, milk, and eggs, which are reliable sources of selenium. Eating as few as two Brazil nuts a day effectively improves selenium status. Average intakes in the United States exceed the RDA, which is based on the amount needed to maximize glutathione peroxidase activity.

Selenium Toxicity Because high doses of selenium are toxic, a UL has been set. Selenium toxicity causes loss and brittleness of hair and nails, garlic breath odor, and nervous system abnormalities.

selenium (se-LEEN-ee-um): an essential trace mineral that is part of an antioxidant enzyme.

Keshan (KESH-an or ka-SHAWN) **disease:** the heart disease associated with selenium deficiency; named for one of the provinces of China where it was first studied. Keshan disease is characterized by heart enlargement and insufficiency; fibrous tissue replaces the muscle tissue that normally composes the middle layer of the walls of the heart.

Selenium is an antioxidant nutrient that works closely with the glutathione peroxidase enzyme and vitamin E. Selenium is found in association with protein in foods. Deficiencies are associated with a predisposition to a type of heart abnormality known as Keshan disease. The accompanying table provides a summary of selenium.

Selenium

RDA	Deficiency Symptoms
Adults: 55 µg/day	Predisposition to heart disease characterized by cardiac tissue becoming fibrous (Keshan disease)
UL	
Adults: 400 µg/day	**Toxicity Symptoms**
Chief Functions in the Body	Loss and brittleness of hair and nails; skin rash, fatigue, irritability, and nervous system disorders; garlic breath odor
Defends against oxidation; regulates thyroid hormone	
Significant Sources	
Seafood, meat, whole grains, fruits, and vegetables (depending on soil content)	

Copper The body contains about 100 milligrams of **copper** in a variety of cells and tissues. Copper balance and transport depend on a system of proteins.

Copper Roles in the Body Copper serves as a constituent of several enzymes. The copper-containing enzymes have diverse metabolic roles with one common characteristic: all involve reactions that consume oxygen or oxygen radicals. For example, copper-containing enzymes catalyze the oxidation of ferrous iron to ferric iron, which allows iron to bind to transferrin. Copper's role in iron metabolism makes it a key factor in hemoglobin synthesis. Copper- and zinc-containing enzymes participate in the body's natural defenses against the oxidative damage of free radicals. Still other copper enzymes help to manufacture collagen, inactivate histamine, and degrade serotonin. Copper, like iron, is needed in many of the reactions involved in energy metabolism.

Copper Deficiency and Toxicity Typical US diets provide adequate amounts of copper, and deficiency is rare. In animals, copper deficiency raises blood cholesterol and damages blood vessels, raising questions about whether low dietary copper might contribute to cardiovascular disease in humans.

Some genetic disorders create a copper toxicity, but excessive intakes from foods are unlikely. Excessive intakes from supplements may cause liver damage, and therefore a UL has been set.

Two rare genetic disorders affect copper status in opposite directions.[44] In **Menkes disease,** the intestinal cells absorb copper, but cannot release it into circulation, causing a life-threatening deficiency. Treatment involves giving copper intravenously. In **Wilson's disease,** copper accumulates in the liver and brain, creating a life-threatening toxicity. Wilson's disease can be controlled by reducing copper intake, using chelating agents such as penicillamine, and taking zinc supplements, which interfere with copper absorption.

Copper Recommendations and Sources The richest food sources of copper are legumes, whole grains, nuts, shellfish, and seeds. More than half of the copper from foods is absorbed, and the major route of elimination appears to be bile. Water may also provide copper, depending on the type of plumbing pipe and the hardness of the water.

› **REVIEW IT**

Copper is a component of several enzymes, all of which are involved in some way with oxygen or oxidation. Some act as antioxidants; others are essential to iron metabolism. Legumes,

copper: an essential trace mineral that is part of many enzymes.

Menkes disease: a genetic disorder of copper transport that creates a copper deficiency and results in mental retardation, poor muscle tone, seizures, brittle kinky hair, and failure to thrive.

Wilson's disease: a genetic disorder of copper metabolism that creates a copper toxicity and results in neurologic symptoms such as tremors, impaired speech, inappropriate behaviors, and personality changes.

> REVIEW IT Copper (*continued*)

whole grains, and shellfish are good sources of copper. The accompanying table provides a summary of copper.

Copper

RDA	Significant Sources
Adults: 900 µg/day	Seafood, nuts, whole grains, seeds, legumes
UL	**Deficiency Symptoms**
Adults: 10,000 µg/day (10 mg/day)	Anemia, bone abnormalities
Chief Functions in the Body	**Toxicity Symptoms**
Necessary for the absorption and use of iron in the formation of hemoglobin; part of several enzymes	Liver damage

Manganese The human body contains a mere 20 milligrams of **manganese**. Most of it can be found in the bones and metabolically active organs such as the liver, kidneys, and pancreas.

Manganese Roles in the Body Manganese acts as a cofactor for many enzymes that facilitate the metabolism of carbohydrate, lipids, and amino acids. In addition, manganese-containing metalloenzymes assist in bone formation and the conversion of pyruvate to a TCA cycle compound.

Manganese Deficiency and Toxicity Manganese requirements are low, and many plant foods contain significant amounts of this trace mineral, so deficiencies are rare. As is true of other trace minerals, however, dietary factors such as phytates inhibit its absorption. In addition, high intakes of iron and calcium limit manganese absorption, so people who use supplements of those minerals regularly may impair their manganese status.

Manganese toxicity is more likely to occur from a contaminated environment than from an excessive dietary intake. Miners who inhale large quantities of manganese dust on the job over prolonged periods show symptoms of a brain disease, along with abnormalities in appearance and behavior. A UL has been established based on intakes from food, water, and supplements.

Manganese Recommendations and Sources Grain products make the greatest contribution of manganese to the diet. With insufficient information to establish an RDA, an AI was set based on average intakes.

> REVIEW IT

Manganese-dependent enzymes are involved in bone formation and various metabolic processes. Because manganese is widespread in plant foods, deficiencies are rare, although regular use of calcium and iron supplements may limit manganese absorption. The accompanying table provides a summary of manganese.

Manganese

AI	Significant Sources
Men: 2.3 mg/day	Nuts, whole grains, leafy vegetables, tea
Women: 1.8 mg/day	**Deficiency Symptoms**
UL	Rare
Adults: 11 mg/day	**Toxicity Symptoms**
Chief Functions in the Body	Nervous system disorders
Cofactor for several enzymes; bone formation	

manganese: an essential trace mineral that acts as a cofactor for many enzymes.

fluoride: an essential trace mineral that makes teeth stronger and more resistant to decay.

Fluoride Fluoride is present in virtually all soils, water supplies, plants, and animals. The body contains only a trace of fluoride, but with this amount, the crystalline deposits in teeth are larger and more perfectly formed.

Fluoride Roles in the Body As Chapter 12 explains, during the mineralization of bones and teeth, calcium and phosphorus form crystals called hydroxyapatite. Then fluoride replaces the hydroxyl (OH) portions of the hydroxyapatite crystal, forming **fluorapatite,** which makes the teeth stronger and more resistant to decay.

Dental caries ranks as the nation's most widespread public health problem: an estimated 95 percent of the population have decayed, missing, or filled teeth. These dental problems can quickly lead to a multitude of nutrition problems by interfering with a person's ability to chew and eat a wide variety of foods. Where fluoride is lacking, dental decay is common.

Drinking water is usually the best source of fluoride, and 75 percent of the US population served by community water systems receives optimal levels of fluoride (see Figure 13-12).[45] Fluoridation of drinking water (to raise the concentration to 0.7 milligram per liter of water) protects against dental caries and supports oral health.[46] By fluoridating the drinking water, a community offers its residents, particularly the children, a safe, economical, practical, and effective way to defend against dental caries. Most bottled waters lack fluoride.

Fluoride Toxicity Too much fluoride can damage the teeth, causing **fluorosis.**[47] For this reason, a UL has been established. In mild cases, the teeth develop small white flecks; in severe cases, the enamel becomes pitted and permanently stained (as shown in Figure 13-13). Fluorosis occurs only during tooth development and cannot be reversed, making its prevention during the first 3 years of life a high priority.[48] To limit fluoride ingestion, take care not to swallow fluoride-containing dental products such as toothpaste and mouthwash and use fluoride supplements only as prescribed by a physician.

Fluoride Recommendations and Sources As mentioned earlier, much of the US population has access to water with an optimal fluoride concentration, which typically delivers about 1 milligram per person per day. Fish and most teas contain appreciable amounts of natural fluoride.

> FIGURE 13-12 **US Population with Access to Fluoridated Water through Public Water Systems**

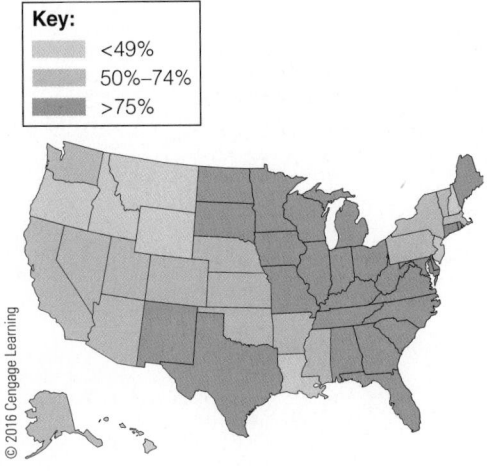

Key:
<49%
50%–74%
>75%

> FIGURE 13-13 **Fluoride-Toxicity Symptom—The Mottled Teeth of Fluorosis**

› REVIEW IT

Fluoride makes teeth stronger and more resistant to decay. Fluoridation of public water supplies can significantly reduce the incidence of dental caries, but excess fluoride during tooth development can cause fluorosis—discolored and pitted tooth enamel. The accompanying table provides a summary of fluoride.

Fluoride

AI	**Significant Sources**
Men: 4 mg/day	Drinking water (if fluoride containing or fluoridated), tea, seafood
Women: 3 mg/day	
UL	**Deficiency Symptoms**
Adults: 10 mg/day	Susceptibility to tooth decay
Chief Functions in the Body	**Toxicity Symptoms**
Strengthens teeth; helps to make teeth resistant to decay	Fluorosis (pitting and discoloration of teeth)

Chromium
Chromium is an essential mineral that participates in carbohydrate and lipid metabolism. Like iron, chromium assumes different charges. In chromium, the Cr^{+++} ion is the most stable and most commonly found in foods.

Chromium Roles in the Body Chromium helps maintain glucose homeostasis by enhancing the activity of the hormone insulin.* When chromium is lacking, a diabetes-like condition may develop, with elevated blood glucose and impaired

fluorapatite (floor-APP-uh-tite): the stabilized form of tooth crystal, in which fluoride has replaced the hydroxyl groups of hydroxyapatite.

fluorosis (floor-OH-sis): discoloration and pitting of tooth enamel caused by excess fluoride during tooth development.

chromium (KRO-mee-um): an essential trace mineral that enhances the activity of insulin.

---

*Small organic compounds that enhance insulin's actions are called *glucose tolerance factors (GTF)*. Some glucose tolerance factors contain chromium.

glucose tolerance, insulin response, and glucagon response. Some research suggests that chromium supplements lower blood glucose or improve insulin responses in type 2 diabetes, but findings have not been consistent.[49]

Chromium Recommendations and Sources Chromium is present in a variety of foods. The best sources are unrefined foods, particularly liver, brewer's yeast, and whole grains. The more refined foods people eat, the less chromium they ingest.

Chromium Supplements Supplement advertisements have succeeded in convincing consumers that they can lose fat and build muscle by taking chromium picolinate. Whether chromium supplements (either picolinate or plain) reduce body fat or improve muscle strength remains controversial. (Highlight 14 discusses chromium picolinate and other supplements athletes use in the hopes of improving their performance.)

> **REVIEW IT**

Chromium enhances insulin's action. A deficiency can impair glucose homeostasis. Chromium is widely available in unrefined foods including brewer's yeast, whole grains, and liver. The accompanying table provides a summary of chromium.

Chromium

AI	Significant Sources
Men: 35 µg/day	Meats (especially liver), whole grains, brewer's yeast
Women: 25 µg/day	
Chief Functions in the Body	**Deficiency Symptoms**
Enhances insulin action and may improve glucose tolerance	Diabetes-like condition
	Toxicity Symptoms
	None reported

Molybdenum Molybdenum acts as a working part of several metalloenzymes. Dietary deficiencies of molybdenum are unknown because the amounts needed are minuscule—as little as 0.1 part per million parts of body tissue. Legumes, breads and other grain products, leafy green vegetables, milk, and liver are molybdenum-rich foods. Average daily intakes fall within the suggested range of intakes.

Molybdenum toxicity in people is rare. It has been reported in animal studies, and a UL has been established. Characteristics of molybdenum toxicity include kidney damage and reproductive abnormalities.

> **REVIEW IT**

Molybdenum is found in a variety of foods and participates in several metabolic reactions. The accompanying table provides a summary of molybdenum.

Molybdenum

RDA	Significant Sources
Adult: 45 µg/day	Legumes, cereals, nuts
UL	**Deficiency Symptoms**
Adults: 2 mg/day	Unknown
Chief Functions in the Body	**Toxicity Symptoms**
Cofactor for several enzymes	None reported; reproductive effects in animals

molybdenum (mo-LIB-duh-num): an essential trace mineral that acts as a cofactor for many enzymes.

13.3 Contaminant Minerals

> **LEARN IT** Describe how contaminant minerals disrupt body processes and impair nutrition status.

Chapter 12 and this chapter explain the many ways minerals serve the body—maintaining fluid and electrolyte balance, providing structural support to the bones, transporting oxygen, and assisting enzymes. In contrast to the essential minerals that the body requires, contaminant minerals impair the body's growth, work capacity, and general health. Contaminant minerals include the **heavy metals** lead, mercury, and cadmium, which enter the food supply by way of soil, water, and air pollution. This section focuses on lead poisoning because it is a serious environmental threat to young children and because reducing blood lead levels in children is a goal of the Healthy People initiative. Much of the information on lead applies to the other contaminant minerals as well—they all disrupt body processes and impair nutrition status similarly.

Like other minerals, lead is indestructible; the body cannot change its chemistry. Chemically similar to nutrient minerals such as iron, calcium, and zinc (cations with two positive charges), lead displaces them from some of the metabolic sites they normally occupy so they are then unable to perform their roles. For example, lead competes with iron in heme, but it cannot carry oxygen. Similarly, lead competes with calcium in the brain, but it cannot signal messages from nerve cells. Excess lead in the blood also deranges the structure of red blood cell membranes, making them leaky and fragile. Lead interacts with white blood cells, too, impairing their ability to fight infection, and it binds to antibodies, thwarting their effort to resist disease.

Children with iron deficiency are particularly vulnerable to lead toxicity. Chapter 16 examines the damaging effects of iron deficiency and lead toxicity on a child's growth and development.[50]

> **REVIEW IT** Describe how contaminant minerals disrupt body processes and impair nutrition status.
> Lead typifies the ways all heavy metals behave in the body: they interfere with nutrients that are trying to do their jobs. The "good guy" nutrients are shoved aside by the "bad guy" contaminants. Then, when the contaminants cannot perform the roles of the nutrients, health diminishes. To safeguard our health, we must defend ourselves against contamination by eating nutrient-rich foods and preserving a clean environment.

This chapter completes the introductory lessons on the nutrients. Each nutrient from the amino acids to zinc has been described rather thoroughly—its chemistry, roles in the body, sources in the diet, symptoms of deficiency and toxicity, and influences on health and disease. Such a detailed examination is informative, but it can also be misleading. It is important to step back from the detailed study of the individual nutrients to look at them as a whole. After all, people eat foods, not nutrients, and most foods deliver dozens of nutrients. Furthermore, nutrients work cooperatively with one another in the body; their actions are most often *inter*actions. This chapter alone mentioned how iron depends on vitamin C to keep it in its active form and copper to incorporate it into hemoglobin, how zinc is needed to activate and transport vitamin A, and how both iodine and selenium are needed for the synthesis of thyroid hormones. The table on p. 430 provides a summary of the trace minerals for your review. Highlight 13 explores the benefits of phytochemicals.

heavy metals: mineral ions such as mercury and lead, so called because they are of relatively high atomic weight. Many heavy metals are poisonous.

Mineral and Chief Functions	Deficiency Symptoms	Toxicity Symptoms[a]	Significant Sources
Iron			
Part of the protein hemoglobin, which carries oxygen in the blood; part of the protein myoglobin in muscles, which makes oxygen available for muscle contraction; necessary for energy metabolism	Anemia: weakness, fatigue, headaches; impaired work performance; impaired immunity; pale skin, nail beds, mucous membranes, and palm creases; concave nails; inability to regulate body temperature; pica	GI distress; iron overload: infections, fatigue, joint pain, skin pigmentation, organ damage	Red meats, fish, poultry, shellfish, eggs, legumes, dried fruits
Zinc			
Part of insulin and many enzymes; involved in making genetic material and proteins, immune reactions, transport of vitamin A, taste perception, wound healing, the making of sperm, and normal fetal development	Growth retardation, delayed sexual maturation, impaired immune function, hair loss, eye and skin lesions, loss of appetite	Loss of appetite, impaired immunity, low HDL, copper and iron deficiencies	Protein-containing foods: red meats, fish, shellfish, poultry, whole grains; fortified cereals
Iodine			
A component of the thyroid hormones that help to regulate growth, development, and metabolic rate	Underactive thyroid gland, goiter, mental and physical retardation (cretinism)	Underactive thyroid gland, elevated TSH, goiter	Iodized salt; seafood; plants grown in iodine-rich soil and animals fed those plants
Selenium			
Part of an enzyme that defends against oxidation; regulates thyroid hormone	Associated with Keshan disease	Nail and hair brittleness and loss; fatigue, irritability, and nervous system disorders, skin rash, garlic breath odor	Seafoods, organ meats; other meats, whole grains, fruits, and vegetables (depending on soil content)
Copper			
Helps form hemoglobin; part of several enzymes	Anemia, bone abnormalities	Liver damage	Seafood, nuts, legumes, whole grains, seeds
Manganese			
Cofactor for several enzymes; bone formation	Rare	Nervous symptom disorders	Nuts, whole grains, leafy vegetables, tea
Fluoride			
Maintains health of bones and teeth; confers decay resistance on teeth	Susceptibility to tooth decay	Fluorosis (pitting and discoloration) of teeth	Drinking water (if fluoridated), tea, seafood
Chromium			
Enhances insulin action, may improve glucose intolerance	Diabetes-like condition	None reported	Meats (liver), whole grains, brewer's yeast
Molybdenum			
Cofactor for several enzymes	Unknown	None reported	Legumes, cereals, nuts

[a]Acute toxicities of many minerals cause abdominal pain, nausea, vomiting, and diarrhea.

Nutrition Portfolio

Trace minerals from a variety of foods, especially those in the protein foods group, support many of your body's activities. Go to Diet & Wellness Plus and choose one of the days on which you tracked your diet for an entire day. Select the Intake vs. Goals report and then consider the following questions. Remember that scoring 100 percent on this report means you met your goal.

- Your Intake vs. Goals report may only display your intake for two of the trace minerals: iron and zinc. How was your intake for these two trace minerals?

Now look at the Intake Spreadsheet report and consider the following questions:

- How often do you include meats, seafood, poultry, legumes, and enriched or fortified grain products weekly? These foods often contain trace minerals.

- What are the advantages of using iodized salt?
- Does your community provide fluoridated water?

DIET & WELLNESS PLUS+ To complete this exercise, go to your Diet & Wellness Plus at www.cengagebrain.com.

> **STUDY IT** To review the key points of this chapter and take a practice quiz, go to the study cards at the end of the book.

REFERENCES

1. A. Kritharis, T. P. Bradley, and D. R. Budman, The evolving use of arsenic in pharmacotherapy of malignant disease, *Annals of Hematology* 92 (2013): 719–730.
2. M. Muñoz, J. A. García-Erce, and A. F. Remacha, Disorders of iron metabolism. Part 1: Molecular basis of iron homeostasis, *Journal of Clinical Pathology* 64 (2011): 281–286; M. D. Knutson, Iron-sensing proteins that regulate hepcidin and enteric iron absorption, *Annual Review of Nutrition* 30 (2010): 149–171.
3. P. A. Sharp, Intestinal iron absorption: Regulation by dietary & systematic factors, *International Journal for Vitamin and Nutrition Research* 80 (2010): 231–242.
4. R. Collings and coauthors, The absorption of iron from whole diets: A systematic review, *American Journal of Clinical Nutrition* 98 (2013): 65–81.
5. K. E. Finberg, Unraveling mechanisms regulating systemic iron homeostasis, *American Society of Hematology Education Book* 2011 (2011): 532–537; T. Ganz and E. Nemeth, Hepcidin and disorders of iron metabolism, *Annual Review of Medicine* 62 (2011): 347–360; M. Wessling-Resnick, Iron homeostasis and the inflammatory response, *Annual Review of Nutrition* 30 (2010): 105–122.
6. T. Ganz, Hepcidin and iron regulation: 10 years later, *Blood* 117 (2011): 4425–4433; J. Kaplan, D. M. Ward, and I. De Domenico, The molecular basis of iron overload disorders and iron-linked anemias, *International Journal of Hematology* 93 (2011): 14–20; A. Pietrangelo, Hereditary hemochromatosis: Pathogenesis, diagnosis, and treatment, *Gastroenterology* 139 (2010): 393–408.
7. S. R. Lynch, Why nutritional iron deficiency persists as a worldwide problem, *Journal of Nutrition* 141 (2011): 763S–768S.
8. A. C. Cepeda-Lopez and coauthors, Sharply higher rates of iron deficiency in obese Mexican women and children are predicted by obesity-related inflammation rather than by differences in dietary iron intake, *American Journal of Clinical Nutrition* 93 (2011): 975–983.
9. L. Tussing-Humphreys and coauthors, Rethinking iron regulation and assessment in iron deficiency, anemia of chronic disease, and obesity: Introducing hepcidin, *Journal of the Academy of Nutrition and Dietetics* 112 (2012): 391–400.
10. N. Milman, Anemia: Still a major health problem in many parts of the world, *Annals of Hematology* 90 (2011): 369–377.
11. D. I. Thurnham and coauthors, Adjusting plasma ferritin concentrations to remove the effects of subclinical inflammation in the assessment of iron deficiency: A meta-analysis, *American Journal of Clinical Nutrition* 92 (2010): 546–555; M. A. Ayoya and coauthors, α_1-Acid glycoprotein, hepcidin, C-reactive protein, and serum ferritin are correlated in anemic schoolchildren with *Schistosoma haematobium*, *American Journal of Clinical Nutrition* 91 (2010): 1784–1790.
12. J. P. McClung and L. E. Murray-Kolb, Iron nutrition and premenopausl women: Effects of poor iron status on physical and neuropsychological performance, *Annual Review of Nutrition* 33 (2013): 271–288.
13. P. Vaucher and coauthors, Effect of iron supplementation on fatigue in nonanemic menstruating women with low ferritin: A randomized controlled trial, *Canadian Medical Association Journal* 184 (2012): 1247–1254.
14. L. E. Murray-Kolb, Iron status and neuropsychological consequences in women of reproductive age: What do we know and where are we headed? *Journal of Nutrition* 141 (2011): 747S–755S; K. Kordas, Iron, lead, and children's behavior and cognition, *Annual Review of Nutrition* 30 (2010): 123–148.
15. B. Lozoff, Early iron deficiency has brain and behavior effects consistent with dopaminergic dysfunction, *Journal of Nutrition* 141 (2011): 740S–746S.
16. S. L. Young, Pica in pregnancy: New ideas about an old condition, *Annual Review of Nutrition* 30 (2010): 403–422.
17. G. J. Anderson and F. Wang, Essential but toxic: Controlling the flux of iron in the body, *Clinical and Experimental Pharmacology and Physiology* 39 (2012): 719–724.
18. P. Brissot and coauthors, Molecular diagnosis of genetic iron-overload disorders, *Expert Review of Molecular Diagnostics* 10 (2010): 755–763.
19. N. C. Andrews, Closing the iron gate, *New England Journal of Medicine* 366 (2012): 376–377; C. Camaschella and E. Poggiali, Inherited disorders of iron metabolism, *Current Opinion in Pediatrics* 23 (2011): 14–20.
20. G. A. Ramm and R. G. Ruddell, Iron homeostasis, hepatocellular injury, and fibrogenesis in hemochromatosis: The role of inflammation in a noninflammatory liver disease, *Seminars in Liver Disease* 30 (2010): 271–287.
21. X. Zheng and coauthors, Hepatic iron stores are increased as assessed by magnetic resonance imaging in a Chinese population with altered glucose homeostasis, *American Journal of Clinical Nutrition* 94 (2011): 1012–1019.
22. R. E. Fleming and P. Ponka, Iron overload in human disease, *New England Journal of Medicine* 366 (2012): 348–359; G. M. Brittenham, Iron-chelating therapy for transfusional iron overload, *New England Journal of Medicine* 364 (2011): 146–156.
23. C. Camaschella, Treating iron overload, *New England Journal of Medicine* 368 (2013): 2325–2327.
24. N. Ahluwalia and coauthors, Iron status is associated with carotid atherosclerotic plaques in middle-aged adults, *Journal of Nutrition* 140 (2010): 812–816.
25. A. C. Bronstein and coauthors, 2009 Annual Report of the American Association of Poison Control Centers' National Poison Data System (NPDS): 27th Annual Report, *Clinical Toxicology* 28 (2010): 979–1178.
26. J. C. King, Zinc: An essential but elusive nutrient, *American Journal of Clinical Nutrition* 94 (2011): 679S–684S.
27. J. R. Hunt, Algorithms for iron and zinc bioavailability: Are they accurate? *International Journal of Vitamin and Nutrition Research* 80 (2010): 257–262; J. C. King, Does zinc absorption reflect zinc status? *International Journal of Vitamin and Nutrition Research* 80 (2010): 300–306.
28. K. M. Hambidge and coauthors, Zinc bioavailability and homeostasis, *American Journal of Clinical Nutrition* 91 (2010): 1478S–1483S.
29. A. S. Prasad, Discovery of human zinc deficiency: Its impact on human health and disease, *Advances in Nutrition* 4 (2013): 176–190.
30. J. B. Barnett, D. H. Hamer, and S. N. Meydani, Low zinc status: A new risk factor for pneumonia in the elderly? *Nutrition Reviews* 68 (2010): 30–37.
31. M. E. Penny, Zinc supplementation in public health, *Annals of Nutrition and Metabolism* 62 (2013): 31–42.

32. M. Science and coauthors, Zinc for the treatment of the common cold: A systematic review and meta-analysis of randomized controlled trials, *Canadian Medical Association Journal* 184 (2012): E551–E561.

33. G. A. Eby, Zinc lozenges as cure for the common cold—A review and hypothesis, *Medical Hypotheses* 74 (2010): 482–492.

34. M. Singh and R. R. Das, Zinc for the common cold, *Cochrane Database of Systemic Reviews* 6 (2013): CD001364.

35. M. Andersson, V. Karumbunathan, and M. B. Zimmerman, Global iodine status in 2011 and trends over the past decade, *Journal of Nutrition* 142 (2012): 744–750.

36. A. Stagnaro-Green, S. Sullivan, and E. N. Pearce, Iodine supplementation during pregnancy and lactation, *Journal of the American Medical Association* 308 (2012): 2463–2464; C. G. Perrine and coauthors, Some subgroups of reproductive age women in the United States may be at risk for iodine deficiency, *Journal of Nutrition* 140 (2010): 1489–1494.

37. GAIN-UNICEF Universal Salt Iodization Partnership Program, www .gainhealth.org/programs/USI, accessed January 2014.

38. S. J. Fairweather-Tait and coauthors, Selenium in human health and disease, *Antioxidants and Redox Signaling* 4 (2011): 1337–1383.

39. C. Lei and coauthors, Is selenium deficiency really the cause of Keshan disease? *Environmental Geochemistry and Health* 33 (2011): 183–188; J. Yang and coauthors, Selenium level surveillance for the year 2007 of Keshan disease in endemic areas and analysis on surveillance results between 2003 and 2007, *Biological Trace Element Research* 138 (2010): 53–59.

40. S. Sun, Chronic exposure to cereal mycotoxin likely citreoviridin may be a trigger for Keshan disease mainly through oxidative stress mechanism, *Medical Hypotheses* 74 (2010): 841–842.

41. M. P. Rayman, Selenium and human health, *Lancet* 379 (2012): 1256–1268.

42. C. D. Davis, P. A. Tsuji, and J. A. Milner, Selenoproteins and cancer prevention, *Annual Review of Nutrition* 32 (2012): 73–95; R. Hurst and coauthors, Selenium and prostate cancer: Systematic review and meta-analysis, *American Journal of Clinical Nutrition* 96 (2012): 111–122; G. Dennert and coauthors, Selenium for preventing cancer, *Cochrane Database of Systematic Reviews* 5 (2011): CD005195; S. J. Fairweather-Tait and coauthors, Selenium in human health and disease, *Antioxidants and Redox Signaling* 14 (2011): 1337–1383.

43. B. K. Dunn and coauthors, A nutrient approach to prostate cancer prevention: The Selenium and Vitamin E Cancer Prevention Trial (SE-LECT), *Nutrition and Cancer* 62 (2010): 896–918; J. Brozmanová and coauthors, Selenium: A double-edged sword for defense and offence in cancer, *Archives of Toxicology* 84 (2010): 919–938.

44. D. L. de Romaña and coauthors, Risks and benefits of copper in light of new insights of copper homeostasis, *Journal of Trace Elements in Medicine and Biology* 25 (2011): 3–13.

45. 2012 Water Fluoridation Statistics, www.cdc.gov/fluoridation/statistics/2012stats.htm, updated November 22, 2013.

46. Position of the Academy of Nutrition and Dietetics: The impact of fluoride on health, *Journal of the Academy of Nutrition and Dietetics* 112 (2012): 1443–1453.

47. E. D. Beltrán-Aguilar, L. Barker, and B. A. Dye, Prevalence and severity of dental fluorosis in the United States, *NCHS Data Brief* 53 (2010): 1–8.

48. M. A. Buzalaf and S. M. Levy, Fluoride intake of children: Considerations for dental caries and dental fluorosis, *Monographs in Oral Science* 22 (2011): 1–19.

49. Y. Hua and coauthors, Molecular mechanisms of chromium in alleviating insulin resistance, *Journal of Nutritional Biochemistry* 23 (2012): 313–319; Z. Q. Wang and W. T. Cefalu, Current concepts about chromium supplementation in type 2 diabetes and insulin resistance, *Current Diabetes Reports* 10 (2010): 145–151.

50. C. Warniment, K. Tsang, and S. S. Galazka, Lead poisoning in children, *American Family Physician* 81 (2010): 751–757.

HIGHLIGHT > 13
Phytochemicals and Functional Foods

> **LEARN IT** Define *phytochemicals* and explain how they might defend against chronic diseases.

Chapter 13 completes the introductory lessons on the six classes of nutrients—carbohydrates, lipids, proteins, vitamins, minerals, and water. In addition to these nutrients, foods contain thousands of other compounds, including the **phytochemicals.** Chapter 1 introduces the phytochemicals as compounds found in plant-derived foods that have biological activity in the body. Research on phytochemicals is unfolding daily, adding to our knowledge of their roles in human health, but there are still many questions and only tentative answers. Just a few of the tens of thousands of phytochemicals have been researched at all, and only a sampling are mentioned in this highlight—enough to illustrate their wide variety of food sources and roles in supporting health.

The concept that foods provide health benefits beyond those of the nutrients emerged from numerous epidemiological studies showing the protective effects of plant-based diets on cancer and heart disease. People have been using foods to maintain health and prevent disease for years, but now these foods have been given a name—they are called **functional foods.**[1] (The accompanying glossary defines this and other terms.) As Chapter 1 explains, functional foods include all foods (whole, fortified, enriched, or enhanced foods) that have a potentially beneficial effect on health.[2] Much of this text touts the benefits of nature's functional foods—whole grains rich in dietary fibers, oily fish rich in omega-3 fatty acids, and fresh fruits rich in phytochemicals, for example. This highlight begins with a look at some of these familiar functional foods, the phytochemicals they contain, and their roles in disease prevention. Then the discussion turns to examine the most controversial of functional foods—novel foods to which phytochemicals have been added to promote health. How these foods fit into a healthy diet is still unclear.

© HSNphotography/Shutterstock.com

The Phytochemicals

In foods, phytochemicals impart tastes, aromas, colors, and other characteristics. They give hot peppers their burning sensation, garlic its pungent flavor, and tomatoes their red color. In the body, phytochemicals can have profound physiological effects—acting as antioxidants, mimicking hormones, stimulating enzymes, interfering with DNA replication, suppressing inflammation, destroying bacteria, and binding to cell walls. Any of these actions may prevent the development of chronic diseases, depending in part on how genetic factors interact with the phytochemicals. Phytochemicals might also have adverse effects when consumed in excess. Table H13-1 (p. 434) presents the names, possible effects, and food sources of some of the better-known phytochemicals.

Defending against Cancer

A variety of phytochemicals from a variety of foods appear to protect against DNA damage and defend the body against cancer. A few examples follow.

TABLE H13-1 Phytochemicals—Their Food Sources and Actions

Name	Possible Effects	Food Sources
Alkylresorcinols (phenolic lipids)	May contribute to the protective effect of grains in reducing the risks of diabetes, heart disease, and some cancers.	Whole-grain wheat and rye
Allicin (organosulfur compound)	Antimicrobial that may reduce ulcers; may lower blood cholesterol.	Chives, garlic, leeks, onions, scallions
Capsaicin	Modulates blood clotting, possibly reducing the risk of fatal clots in heart and artery disease.	Hot peppers
Carotenoids (including beta-carotene, lycopene, lutein, zeaxanthin, and hundreds of related compounds)	Act as antioxidants, possibly reducing risks of cancer and other diseases.	Deeply pigmented fruits and vegetables (apricots, broccoli, cantaloupe, carrots, pink grapefruit, pumpkin, spinach, sweet potatoes, tomatoes, red peppers, watermelon)
Curcumin (polyphenol)	Acts as an antioxidant and anti-inflammatory agent; may reduce blood clot formation; may inhibit enzymes that activate carcinogens.	Turmeric, a yellow-colored spice common in curry powder
Flavonoids (including anthocyanins, flavones, flavonols, isoflavones, catechins, and others)	Act as antioxidants; scavenge carcinogens; bind to nitrates in the stomach, preventing conversion to nitrosamines; inhibit cell proliferation.	Berries, black tea, celery, citrus fruits, green tea, olives, onions, oregano, purple grapes, purple grape juice, soybeans and soy products, vegetables, whole wheat, wine
Genistein and daidzein (isoflavonoids)	Phytoestrogens that inhibit cell replication in GI tract; may reduce risk of breast, colon, ovarian, prostate, and other estrogen-sensitive cancers; may reduce cancer cell survival; may reduce risk of osteoporosis.	Soybeans, soy flour, soy milk, tofu, textured vegetable protein, other legume products
Indoles (organosulfur compound)	May trigger production of enzymes that block DNA damage from carcinogens; may inhibit estrogen action.	Cruciferous vegetables such as bok choy, broccoli, brussels sprouts, cabbage, cauliflower, collard greens, mustard greens, kale, swiss chard, watercress
Isothiocyanates (organosulfur compounds, including sulforaphane)	Act as antioxidants; inhibit enzymes that activate carcinogens; activate enzymes that detoxify carcinogens; may reduce risk of breast cancer, prostate cancer, and colorectal cancer.	Cruciferous vegetables such as bok choy, broccoli, broccoli sprouts, brussels sprouts, cabbage, cauliflower, collard greens, mustard greens, kale, swiss chard, watercress
Lignans (polyphenol)	Phytoestrogens that block estrogen activity in cells possibly reducing the risk of cancer of the breast, colon, ovaries, and prostate.	Flaxseed, whole grains
Monoterpenes (including limonene)	May trigger enzyme production to detoxify carcinogens; inhibit cancer promotion and cell proliferation.	Citrus fruits, cherries
Phenolic acids (including ellagic acid)	May trigger enzyme production to make carcinogens water soluble, facilitating excretion.	Coffee beans, fruits (apples, blueberries, cherries, grapes, oranges, pears, prunes), oats, potatoes, soybeans
Phytic acid (phenolic acid)	Binds to minerals, preventing free-radical formation, possibly reducing cancer risk.	Whole grains
Resveratrol (flavonoid)	Acts as an antioxidant; may inhibit cancer growth; reduce inflammation, LDL oxidation, and blood clot formation.	Red wine, peanuts, grapes, raspberries
Saponins (glucosides)	May interfere with DNA replication, preventing cancer cells from multiplying; stimulate immune response.	Alfalfa sprouts, other sprouts, green vegetables, potatoes, tomatoes
Tannins (flavonoid)	Act as antioxidants; may inhibit carcinogen activation and cancer promotion.	Black-eyed peas, grapes, lentils, red and white wine, tea

Soy may protect against breast and prostate cancers.[3] Soybeans—as well as other legumes, **flaxseeds,** whole grains, fruits, and vegetables—are a rich source of an array of phytochemicals, among them the **phytoestrogens.** Because the chemical structure of phytoestrogens is similar to the hormone estrogen, they can weakly mimic or modulate the effects of estrogen in the body. They also have antioxidant activity that appears to slow the growth of some cancers.

Soy foods appear to be most effective when consumed in moderation early and throughout life.[4] Importantly, soy extracts and phytoestrogen supplements are ill-advised—especially for women with breast cancer and those with high risk factors—as phytoestrogens may promote the growth of estrogen-dependent tumors (such as breast cancer).[5] The American Cancer Society recommends that women with breast cancer should consume only *moderate* amounts of soy as part of a healthy plant-based diet and should not intentionally ingest high levels of soy or supplements of phytoestrogens.

Limited evidence suggests that tomatoes may offer protection against some cancers.[6] Among the phytochemicals thought to be responsible for this effect is **lycopene,** one of the many **carotenoids.**[7] Lycopene is the pigment that gives apricots, guava, papaya, pink grapefruits, and watermelon their red color—and it is especially abundant in tomatoes. Because food processing and cooking can improve carotenoid absorption, cooked tomato products, such as spaghetti sauce, provide even more lycopene. Lycopene is a powerful antioxidant that seems to inhibit the growth of cancer cells. Importantly, the benefits of lycopene have been seen when people have eaten *foods* containing lycopene; lycopene supplements may interfere with cancer treatments.[8]

Soybeans and tomatoes are only two of the many fruits and vegetables credited with providing anticancer activity. Strong and convincing evidence shows that the risk of many cancers, and perhaps of cancer in general, decreases when diets include an abundance of fruits and vegetables.[9] To that end, current recommendations urge consumers to eat five to nine servings of fruits and vegetables a day.

Defending against Heart Disease

Diets based primarily on unprocessed foods appear to support heart health better than those founded on highly refined foods—perhaps because of the abundance of nutrients, fiber, or phytochemicals such as the **flavonoids.** Flavonoids, a large group of phytochemicals known for their health-promoting qualities, are found in whole grains, legumes, soy, vegetables, fruits, herbs, spices, teas, chocolate, nuts, olive oil, and red wines.[10] Flavonoids are powerful antioxidants that may help to protect LDL cholesterol against oxidation, minimize inflammation, and reduce blood platelet stickiness, thereby slowing the progression of atherosclerosis and making blood clots less likely. Whereas an abundance of flavonoid-containing *foods* in the diet may lower the risks of chronic diseases, no claims can be made for flavonoids themselves as the protective factor, particularly when they are extracted from foods and sold as supplements. In fact, purified flavonoids may even be harmful.[11]

In addition to flavonoids, fruits and vegetables are rich in carotenoids such as beta-carotene and **lutein.** Studies suggest that a diet rich in carotenoids may lower the risk of heart disease by decreasing inflammation and oxidative stress.[12]

The **plant sterols** of soy and the **lignans** of flaxseed may also protect against heart disease. These cholesterol-like molecules are naturally found in all plants and inhibit cholesterol absorption in the body. As a result, blood cholesterol levels decline.[13] These phytochemicals also seem to protect against heart disease by reducing inflammation and lowering blood pressure.[14]

Defending against Other Diseases

Most research on phytochemicals has focused on cancer and heart disease, but phytochemicals defend against other diseases as well. The orange-yellow pigment curcumin, commonly found in curry powder, may help reverse insulin resistance, inflammation, and other symptoms associated with obesity.[15] The carotenoids lutein and zeaxanthin may protect the eyes and skin from ultraviolet light damage and the bones from mineral loss.[16]

The Phytochemicals in Perspective

Because foods deliver thousands of phytochemicals in addition to dozens of nutrients, researchers must be careful in giving credit for particular health benefits to any one compound. Diets rich in whole grains, legumes, vegetables, fruits, and nuts seem to protect against heart disease and cancer, but identifying *the* specific foods or components of foods that are responsible is difficult. Each food possesses a unique array of phytochemicals—citrus fruits provide monoterpenes; grapes, resveratrol; and flaxseed, lignans. (Review Table H13-1, p. 434, for the possible effects and other food sources of these phytochemicals.) Broccoli may contain as many as 10,000 different phytochemicals—each with the potential to influence some action in the body. Beverages such as wine, spices such as oregano, and oils such as olive oil (especially virgin olive oil) contain many phytochemicals that may explain, in part, why people who eat a traditional Mediterranean diet have reduced risks of heart disease and cancer. Phytochemicals might also explain why the DASH diet is so effective in lowering blood pressure and blood lipids. Even identifying all of the phytochemicals and their effects doesn't answer all the questions because the actions of phytochemicals may be complementary or overlapping—which reinforces the principle of variety in diet planning. For an appreciation of the array of phytochemicals offered by a variety of foods, see Figure H13-1 (p. 436).

Functional Foods

Because foods naturally contain thousands of phytochemicals that are biologically active in the body, virtually all of them have some value in supporting health.[17] In other words, even simple, whole foods, in reality, are functional foods. Cranberries may help prevent urinary tract infections; garlic may lower blood cholesterol; grapes may reduce inflammation; and green tea may protect against nonalcoholic fatty liver disease, just to name a few examples.[18] Functional foods rich in phytochemicals are easy to find in the produce section of grocery stores. Just look for the colorful fruits and vegetables (see Table H13-2, p. 437). But food manufacturers continue to create new functional foods as well.

Many processed foods become functional foods when they are fortified with nutrients or enhanced with phytochemicals or herbs (calcium-fortified orange juice, for example). Less frequently, an entirely new food is created, as in the case of a meat substitute made of mycoprotein—a protein derived from a fungus.* This functional food not only provides dietary fiber, polyunsaturated fats, and high-quality protein, but it also lowers LDL cholesterol, raises HDL cholesterol, improves glucose response, and prolongs satiety after a meal. Such a novel functional food raises the question—is it a food or a drug?

---

*This mycoprotein product is marketed under the trade name Quorn (pronounced KWORN).

> **FIGURE H13-1**   **An Array of Phytochemicals in a Variety of Foods and Beverages**

Broccoli and broccoli sprouts (and brussels sprouts, bok choy, cabbage, cauliflower, kale, collard greens, swiss chard, turnips, and watercress) contain an abundance of the cancer-fighting phyto-chemicals sulforaphane and indoles.

An apple a day—rich in phenolic acids—may protect against heart disease.

The phytochemical resveratrol found in grapes (and nuts) protects against cancer by inhibiting cell growth and against heart disease by limiting clot formation and inflammation.

The ellagic acid of strawberries (and blackberries, blueberries, raspberries, and grapes) may inhibit certain types of cancer and decrease cholesterol levels.

The flavonoids in cocoa and chocolate defend against oxidation and reduce the tendency of blood to clot.

Tomatoes (and pink grapefruit, red peppers, and watermelons), with their abundant lycopene, may defend against cancer and heart disease by protecting DNA from oxidative damage.

Spinach (and collard greens, corn, swiss chard, and winter squash) contains the carotenoids lutein and zeaxanthin, which help protect the eyes against macular degeneration.

Quercetins—commonly found in kale (and onions, pears, and grapes)—reduce inflammation from allergies, inhibit tumor growth, and protect the lungs.

The phytoestrogens of soybeans seem to starve cancer cells and inhibit tumor growth; the plant sterols may lower blood cholesterol and protect cardiac arteries.

The monoterpenes of citrus fruits (and cherries) may protect the lungs.

Colorful foods such as apricots (and cantaloupes, carrots, kale, kiwifruit, mangoes, papaya, pumpkins, spinach, sweet potatoes, and winter squash) contain beta-carotene, which may help slow aging, protect against some cancers, improve lung function, and reduce complications of diabetes.

Garlic (and chives, leeks, onions, and scallions), with its abundant organosulfur compounds, may lower blood cholesterol and blood pres-sure and protect against stomach cancer.

The flavonoids in black tea may protect against heart disease, whereas those in green tea may defend against cancer.

Blueberries (and cherries, plums, and strawberries), a rich source of anthocyanins, may protect against the effects of aging.

Flaxseed, the richest source of lignans, may prevent the spread of cancer.

Red	White-Brown	Orange-Yellow	Blue-Purple	Green
Anthocyanins	Allicin	Beta-carotene	Anthocyanins	Beta-carotene
Lycopene	Allyl sulfides	Limonene	Ellagic acid	Lutein
			Phenolics	Indoles
Beets	Bananas	Apricots	Black currants	Artichokes
Cherries	Brown pears	Cantaloupe	Blackberries	Arugula
Cranberries	Cauliflower	Carrots	Blueberries	Asparagus
Pink grapefruit	Chives	Lemons	Dried plums	Avocados
Pomegranates	Dates	Mangoes	Eggplant	Broccoli
Radicchio	Garlic	Nectarines	Elderberries	Brussels sprouts
Radishes	Ginger	Oranges	Plums	Cabbage
Raspberries	Leeks	Papayas	Purple figs	Celery
Red apples	Mushrooms	Peaches	Purple peppers	Cucumbers
Red peppers	Onions	Persimmons	Raisins	Endive
Red potatoes	Parsnips	Pineapple	Purple cabbage	Green apples
Rhubarb	Shallots	Pumpkin	Purple grapes	Green beans
Strawberries	Turnips	Rutabagas		Green grapes
Tomatoes		Squash		Green onions
Watermelon		Sweet potatoes		Green pears
		Tangerines		Green peppers
		Yellow peppers		Honeydew melon
				Kiwifruit
				Leafy greens
				Limes
				Okra
				Peas
				Snow peas
				Spinach
				Sugar snap peas
				Watercress
				Zucchini

Foods as Pharmacy

Hippocrates is credited with saying, "Let food be thy medicine and medicine be thy food." This simple message, uttered thousands of years ago, recognizes how good food supports good health.

Not too long ago, most of us could agree on what was a food and what was a drug. Today, functional foods blur the distinctions. They have characteristics similar to both foods and drugs, but do not fit neatly into either category. Consider margarine, for example.

Eating nonhydrogenated margarine sparingly instead of butter generously may lower blood cholesterol slightly over several months and clearly falls into the food category. Taking a statin drug, on the other hand, lowers blood cholesterol significantly within weeks and clearly falls into the drug category. But margarine enhanced with a plant sterol that lowers blood cholesterol is in a gray area between the two. The margarine looks and tastes like a food, but it acts like a drug.

The use of functional foods as drugs creates a whole new set of diet-planning challenges. Not only must foods provide an adequate intake of all the nutrients to support good health, but they must also deliver druglike ingredients to protect against disease. Like drugs used to treat chronic diseases, functional foods may need to be eaten several times a day for several months or even years to have a beneficial effect. Sporadic users may be disappointed in the results. Margarine enriched with 2 to 3 grams of plant sterols may reduce cholesterol by up to 15 percent, much more than regular margarine does, but not nearly as much as the more than 30 percent reduction seen with cholesterol-lowering drugs. For this reason, functional foods may be

Nature offers a variety of functional foods that provide us with many health benefits.

Functional foods currently on the market promise to "enhance mood," "promote relaxation and good karma," "increase alertness," and "improve memory," among other claims.

more useful for prevention and mild cases of disease than for intervention and more severe cases. In any case, because prescription medicines are so much more effective and because people respond to plant sterols so differently, consumers should always make treatment decisions in consultation with their health care providers.[19]

Foods and drugs differ dramatically in cost as well. Functional foods such as fruits and vegetables incur no added costs, but foods that have been manufactured with added phytochemicals can be expensive, costing up to six times as much as their conventional counterparts. The price of functional foods typically falls between that of traditional foods and medicines.

Unanswered Questions

To achieve a desired health effect, which is the better choice: to eat a novel functional food created to affect a specific body function or to adjust the diet? Does it make more sense to use a margarine enhanced with a plant sterol that lowers blood cholesterol or simply to limit the amount of butter eaten?* Is it smarter to eat eggs enriched with omega-3 fatty acids or to restrict egg consumption? Might functional foods offer a sensible solution for improving our nation's health—if done correctly? Perhaps so, but the problem is that the food industry moves faster than either scientists or the Food and Drug Administration. Consumers were able to buy soup with St. John's wort that claimed to enhance mood and fruit juice with echinacea that was supposed to fight colds while scientists were still conducting their studies on these ingredients. Research to determine the safety and effectiveness of these substances is still in progress. Until this work is complete, consumers are on their own in finding answers to the following questions:

- *Does it work?* Research is generally lacking and findings are often inconclusive.
- *How much does it contain?* Food labels are not required to list the quantities of added phytochemicals. Even if they were, consumers have no standard for comparison and cannot deduce whether the

amounts listed are a little or a lot. Most importantly, until research is complete, food manufacturers do not know what amounts (if any) are most effective—or most toxic.

- *Is it safe?* Functional foods can act like drugs. They contain ingredients that can alter body functions and cause allergies, drug interactions, drowsiness, and other side effects. Yet, unlike drug labels, food labels do not provide instructions for the dosage, frequency, or duration of treatment.
- *Is it healthy?* Adding phytochemicals to a food does not magically make it a healthy choice. A candy bar fortified with phytochemicals is still made mostly of sugar and fat.

Critics suggest that the designation "functional foods" may be nothing more than a marketing tool. After all, even the most experienced researchers cannot yet identify the perfect combination of nutrients and phytochemicals to support optimal health. Yet manufacturers are freely experimenting with various concoctions as if they possessed that knowledge. Is it okay for them to sprinkle phytochemicals on fried snack foods or caramel candies and label them "functional," thus implying health benefits?

Future Foods

Nature has elegantly designed foods to provide us with a complex array of dozens of nutrients and thousands of additional compounds that may benefit health—most of which we have yet to identify or understand. Over the years, we have taken those foods, deconstructed them, and then reconstructed them in an effort to "improve" them. With new scientific understandings of how nutrients—and the myriad other compounds in foods—interact with genes, we may someday be able to design *specific* eating patterns to meet the *exact* health needs of *each* individual. Indeed, our knowledge of the human genome and of human nutrition may well merge to allow specific recommendations for individuals based on their predisposition to diet-related diseases.

---

*Margarine products that lower blood cholesterol contain either sterol esters from vegetable oils, soybeans, and corn or stanol esters from wood pulp.

If the present trend continues, someday physicians may be able to prescribe the perfect foods to enhance your health, and farmers will be able to grow them. As Highlight 19 explains, scientists have already developed gene technology to alter the composition of food crops. They can grow rice enriched with vitamin A and tomatoes containing a hepatitis vaccine, for example. It seems quite likely that foods can be created to meet every possible human need. But then, in a sense, that was largely true 100 years ago when we relied on the bounty of nature.

CRITICAL THINKING QUESTIONS

A. Which is the better choice—to eat processed foods that have been enhanced with phytochemicals or to eat natural foods that are rich in phytochemicals?

B. Some research suggests that cranberries may help prevent urinary tract infections, but what about cranberry supplements? Limited research suggests that *foods* rich in phytoestrogens may benefit heart, bone, breast, and menopausal health, but *phytoestrogen supplements* or *phytoestrogen-enhanced functional foods* are not recommended, particularly for women at high risk of breast cancer. How can you determine whether a phytochemical and/or functional food offers a safe and sensible solution to improving your health?

REFERENCES

1. W. R. Kapsak and coauthors, Functional foods: Consumer attitudes, perceptions, and behaviors in a growing market, *Journal of the American Dietetic Association* 111 (2011): 804–810.

2. Position of the Academy of Nutrition and Dietetics: Functional foods, *Journal of the Academy of Nutrition and Dietetics* 113 (2013): 1096–1103.

3. D. C. Vitale and coauthors, Isoflavones: Estrogenic activity, biological effect and bioavailability, *European Journal of Metabolism and Pharmacokinetics* 38 (2013): 15–25; M. Adjakly and coauthors, Genistein and daidzein: Different molecular effects on prostate cancer, *Anticancer Research* 33 (2013): 39–44; P. L. de Souza and coauthors, Clinical pharmacology of isoflavones and its relevance for potential prevention of prostate cancer, *Nutrition Reviews* 68 (2010): 542–555.

4. L. Hilakivi-Clarke, J. E. Andrade, and W. Helferich, Is soy consumption good or bad for the breast? *Journal of Nutrition* 140 (2010): 2326S–2334S.

5. S. Andres and coauthors, Risks and benefits of dietary isoflavones for cancer, *Critical Reviews in Toxicology* 41 (2011): 463–506.

6. N. P. Gullet and coauthors, Cancer prevention with natural compounds, *Seminars in Oncology* 37 (2010): 258–281.

7. J. Talvas and coauthors, Differential effects of lycopene consumed in tomato paste and lycopene in the form of a purified extract on target genes of cancer prostatic cells, *American Journal of Clinical Nutrition* 91 (2010): 1716–1724.

8. B. Cassileth, Lycopene, *Oncology* 24 (2010): 296.

9. T. J. Key, Fruit and vegetables and cancer risk, *British Journal of Cancer* 104 (2011): 6–11; J. M. Matés and coauthors, Anticancer antioxidant regulatory functions of phytochemicals, *Current Medicinal Chemistry* 18 (2011): 2315–2338.

10. O. K. Chun and coauthors, Estimation of antioxidant intakes from diet and supplements in US adults, *Journal of Nutrition* 140 (2010): 317–324.

11. S. Egert and G. Rimbach, Which sources of flavonoids: Complex diets or dietary supplements? *Advances in Nutrition* 2 (2011): 8–14.

12. G. Riccioni and coauthors, Novel phytonutrient contributors to antioxidant protection against cardiovascular disease, *Nutrition* 28 (2012): 605–610; P. Giordano and coauthors, Carotenoids and cardiovascular risk, *Current Pharmaceutical Design* 18 (2012): 5577–5589.

13. M. A. Shaghaghi, S. S. Abumweis, and P. J. H. Jones, Cholesterol-lowering efficacy of plant sterols/stanols provided in capsule and tablet formats: Results of a systematic review and meta-analysis, *Journal of the Academy of Nutrition and Dietetics* 113 (2013): 1494–1503; S. R. Eussen and coauthors, Dose-dependent cholesterol-lowering effects of phytosterol/phytostanol-enriched margarine in statin users and statin non-users under free-living conditions, *Public Health Nutrition* 14 (2011): 1823–1832; R. P. Mensink and coauthors, Plant stanols deose-dependently decrease LDL-cholesterol concentrations, but not cholesterol-standardized fat-soluble antioxidant concentrations, at intakes up to 9 g/d, *American Journal of Clinical Nutrition* 92 (2010): 24–33; S. B. Racette and coauthors, Dose effects of dietary phytosterols on cholesterol metabolism: A controlled feeding study, *American Journal of Clinical Nutrition* 91 (2010): 32–38.

14. R. A. Othman and M. H. Moghadasian, Beyond cholesterol-lowering effects of plant sterols: Clinical and experimental evidence of anti-inflammatory properties, *Nutrition Reviews* 69 (2011): 371–382.

15. B. B. Aggarwal, Targeting inflammation-induced obesity and metabolic diseases by curcumin and other nutraceuticals, *Annual Review of Nutrition* 30 (2010): 173–199; L. Alappat and A. B. Awad, Curcumin and obesity: Evidence and mechanisms, *Nutrition Reviews* 68 (2010): 729–738.

16. S. M. Abdel-Aal and coauthors, Dietary sources of lutein and zeaxanthin carotenoids and their role in eye health, *Nutrients* 5 (2013): 1169–1185.

17. A. S. Chang, B. Y. Yeong, and W. P. Koh, Symposium on plant polyphenols: Nutrition, health and innovations, June 2009, *Nutrition Reviews* 68 (2010): 246–252.

18. K. Ried, C. Toben, and P. Fakler, Effect of garlic on serum lipids: An updated meta-analysis, *Nutrition Reviews* 71 (2013): 282–299; C. Masterjohn and R. S. Bruno, Therapeutic potential of green tea in nonalcoholic fatty liver disease, *Nutrition Reviews* 70 (2012): 41–56; C. Wang and coauthors, Cranberry-containing products for prevention of urinary tract infections in susceptible populations, *Archives of Internal Medicine* 172 (2012): 988–996; C. Chuang and M. K. McIntosh, Potential mechanisms by which polyphenol-rich grapes prevent obesity-mediated inflammation and metabolic diseases, *Annual Review of Nutrition* 31 (2011): 155–176.

19. S. A. Doggrell, Lowering LDL cholesterol with margarine containing plant stanol/sterol esters: Is it still relevant in 2011? *Complementary Therapies in Medicine* 19 (2011): 37–46; T. C. Rideout and coauthors, High basal fractional cholesterol synthesis is associated with nonresponse of plasma LDL cholesterol to plant sterol therapy, *American Journal of Clinical Nutrition* 92 (2010): 41–46.

Fitness: Physical Activity, Nutrients, and Body Adaptations

Nutrition in Your Life

Every day, you choose whether to be physically active or not, and your choices over time can influence how well you feel and how long you live. Today's world makes it easy to be inactive—too easy in fact—but the many health rewards of being physically active make it well worth the effort. You may even discover how much fun it is to be active, and with a little perseverance, you may become physically fit as well. The choice is yours. In the Nutrition Portfolio at the end of this chapter, you can determine whether your physical activities meet current recommendations and whether your daily food, fluid, and carbohydrate intakes are appropriate to support those activities.

Are you physically fit? If so, your joints are flexible, your muscles are strong, and your body is lean with enough, but not too much, fat. You have the endurance to engage in daily physical activities with enough reserve energy to handle added challenges. Carrying heavy suitcases, opening a stuck window, or climbing four flights of stairs, which might strain an unfit person, is easy for you. What's more, you are prepared to meet mental and emotional challenges too. All these characteristics of **fitness** describe the same wonderful condition of a healthy body.

Or perhaps you are **sedentary**. Today's world encourages inactivity, and people who exert minimal physical effort become weak and unhealthy. In fact, a sedentary lifestyle fosters the development of several chronic diseases.[1]

Regardless of your level of fitness, this chapter is written for "you," whoever you are and whatever your goals—whether you want to improve your health, lose weight, hone your athletic skills, or simply adopt an active lifestyle. This chapter begins by discussing fitness and its benefits, and then goes on to explain how the body uses energy nutrients to fuel physical activity. Finally, it describes diets to support fitness.

fitness: the characteristics that enable the body to perform physical activity; more broadly, the ability to meet routine physical demands with enough reserve energy to rise to a physical challenge; or the body's ability to withstand stress of all kinds.

sedentary: physically inactive (literally, "sitting down a lot").

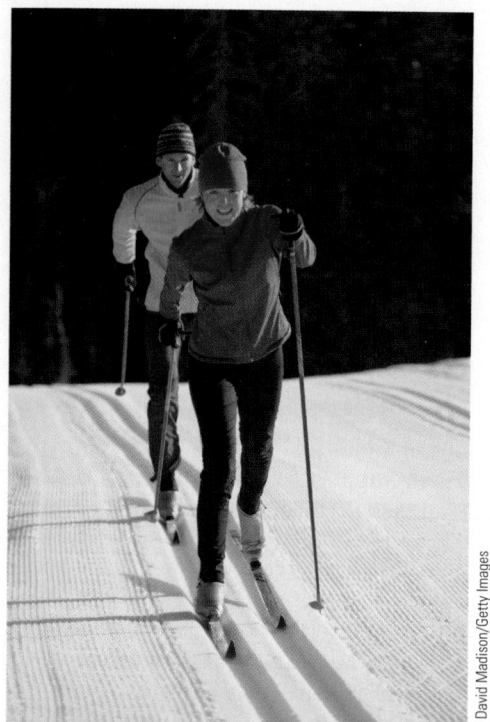

Physical activity, or lack of it, exerts a significant and pervasive influence on everyone's nutrition and overall health.

14.1 Fitness

> **LEARN IT** Describe the health benefits of being physically fit and explain how to develop the components of fitness.

Fitness depends on a certain minimum amount of **physical activity** or **exercise.** Both physical activity and exercise involve body movement, muscle contraction, and enhanced energy expenditure, but "exercise" is often used to describe structured, planned physical activity. This chapter focuses on how the active body uses energy nutrients—whether that body is pedaling a bike across campus or pedaling a stationary bike in a gym. Thus, for our purposes, the terms *physical activity* and *exercise* are used interchangeably.

Benefits of Fitness The *Dietary Guidelines for Americans* emphasize the benefits of increasing physical activities and reducing sedentary activities.[2] Extensive evidence confirms that regular physical activity promotes health and reduces the risk of developing a number of diseases.[3] Yet, despite an increasing awareness of the health benefits that physical activity confers, only about 20 percent of adults in the United States meet physical activity guidelines.[4] Like smoking and obesity, physical inactivity is linked to the major degenerative diseases—heart disease, cancer, stroke, diabetes, and hypertension—the primary killers of adults in developed countries.[5]

> **DIETARY GUIDELINES FOR AMERICANS**
Increase physical activity and reduce time spent in sedentary behaviors.

As a person becomes physically fit, the health of the entire body improves. In general, physically fit people enjoy:[6]

- *Restful sleep.* Rest and sleep occur naturally after periods of physical activity.
- *Nutritional health.* Physical activity expends energy and thus allows people to eat more food. If they choose wisely, active people will consume more nutrients and be less likely to develop nutrient deficiencies.
- *Improved body composition.* A balanced program of physical activity limits body fat and increases or maintains lean tissue. Thus physically active people have relatively less body fat than sedentary people at the same body weight.[7]
- *Improved bone density.* Weight-bearing physical activity builds bone strength and protects against osteoporosis.[8]
- *Resistance to colds and other infectious diseases.* Fitness enhances immunity.[9]*
- *Low risks of some cancers.* Lifelong physical activity may help to protect against colon cancer, breast cancer, and some other cancers.[10]
- *Strong circulation and lung function.* Physical activity that challenges the heart and lungs strengthens the circulatory system.
- *Low risk of cardiovascular disease.* Physical activity lowers blood pressure, slows resting pulse rate, and lowers blood cholesterol, thus reducing the risks of heart attacks and strokes.[11] Some research suggests that physical activity may reduce the risk of cardiovascular disease in another way as well—by reducing visceral fat stores.[12]
- *Low risk of type 2 diabetes.* Physical activity normalizes glucose tolerance.[13] Regular physical activity reduces the risk of developing type 2 diabetes and improves glucose control.
- *Reduced risk of gallbladder disease.* Regular physical activity reduces the risk of gallbladder disease—perhaps by facilitating weight control and lowering blood lipid levels.[14]

physical activity: bodily movement produced by muscle contractions that substantially increase energy expenditure.

exercise: planned, structured, and repetitive body movements that promote or maintain physical fitness.

---

*Moderate physical activity can stimulate immune function. Intense, vigorous, prolonged activity such as marathon running, however, may compromise immune function.

TABLE 14-1 Levels of Physical Activity Intensity Compared

Level of Intensity	Breathing and/or Heart Rate	Perceived Exertion (on a Scale of 0 to 10)	Talk Test	Energy Expenditure	Walking Pace
Light	Little to no increase	<5	Able to sing	<3.5 kcal/min	<3 mph
Moderate	Some increase	5 or 6	Able to have a conversation	3.5 to 7 kcal/min	3 to 4.5 mph
Vigorous	Large increase	7 or 8	Conversation is difficult or "broken"	>7 kcal/min	>4.5 mph

SOURCE: Centers for Disease Control and Prevention, www.cdc.gov/physicalactivity/everyone; updated March 30, 2011.

- *Low incidence and severity of anxiety and depression.* Physical activity may improve mood and enhance the quality of life by reducing depression and anxiety.[15]
- *Strong self-image.* The sense of achievement that comes from meeting physical challenges promotes self-confidence.
- *Long life and high quality of life in the later years.* Active people live longer, healthier lives than sedentary people do.[16] Even as little as 15 minutes a day of moderate-intensity activity can add years to a person's life.[17] In addition to extending longevity, physical activity supports independence and mobility in later life by reducing the risk of falls and minimizing the risk of injury should a fall occur.[18]

What does a person have to do to reap the health rewards of physical activity? To gain substantial *health* benefits, most guidelines recommend a minimum amount of time performing **aerobic physical activity.**[19] The minimum amount of time depends on whether the activity is **moderate-intensity physical activity** or **vigorous-intensity physical activity.** Table 14-1 compares intensity levels. Whether moderate or vigorous, a minimum length of 10 minutes for short bouts of aerobic physical activity is recommended.[20] Of course, more time and greater intensity bring even greater health benefits—maintaining a healthy body weight (BMI of 18.5 to 24.9) and further reducing the risk of chronic diseases.

> **DIETARY GUIDELINES FOR AMERICANS**
Choose moderate- or vigorous-intensity physical activities. Slowly build up the amount of physical activity you choose.

In addition to providing health benefits, physical activity helps to develop and maintain *fitness.* Table 14-2 (p. 444) presents the American College of Sports Medicine (ACSM) guidelines for physical activity.[21] Following these guidelines will help adults improve their cardiorespiratory endurance, body composition, strength, and flexibility. At this level of fitness, a person can reap still greater health benefits (substantially lower risk of premature death compared with those who are inactive, improved cardiorespiratory fitness, and improved body composition, for example).[22] Fitness and health both depend on maintaining an active lifestyle every day.

The bottom line is that any physical activity, even moderate activity, *beyond* activities of daily living, provides some health benefits, and these benefits follow a dose-response relationship. In other words, some activity is better than none, and more activity is better still—up to a point. Pursued in excess, intense physical activity, especially when combined with poor eating habits, can undermine health, as Highlight 8, Eating Disorders, explains (pp. 256–263).

Developing Fitness A person who practices a physical activity *adapts* by becoming better able to perform that activity after each session. People shape their bodies by what they choose to do (and not do). Muscle cells and tissues respond to a physical activity **overload** by building, within genetic limits, the structures needed to perform it.[23]

Muscles are continually under renovation. Every day, particularly during the fasting periods between meals, a healthy body degrades a small portion of its muscle protein and then rebuilds new muscle with available amino acids during

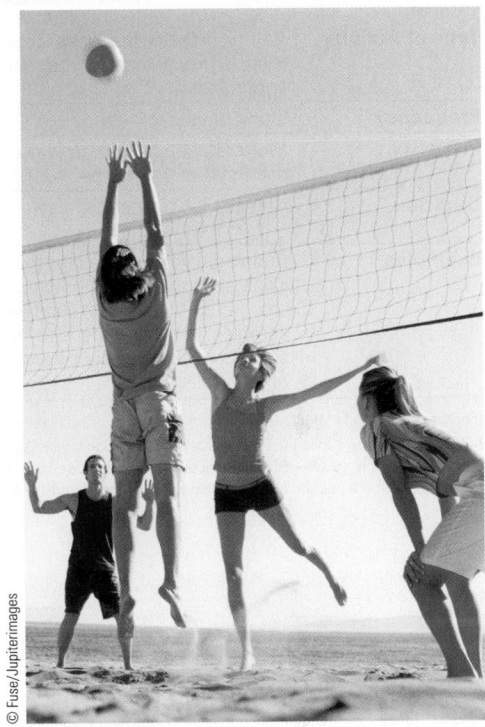

Physical activity helps you look good, feel good, and have fun, and it brings many long-term health benefits as well.

aerobic physical activity: activity in which the body's large muscles move in a rhythmic manner for a sustained period of time. Aerobic activity, also called *endurance activity,* improves cardiorespiratory fitness. Brisk walking, running, swimming, and bicycling are examples.

moderate-intensity physical activity: physical activity that requires some increase in breathing and/or heart rate and expends 3.5 to 7 kcalories per minute. Walking at a speed of 3 to 4.5 miles per hour (about 15 to 20 minutes to walk 1 mile) is an example.

vigorous-intensity physical activity: physical activity that requires a large increase in breathing and/or heart rate and expends more than 7 kcalories per minute. Walking at a very brisk pace (>4.5 miles per hour) or running at a pace of at least 5 miles per hour are examples.

overload: an extra physical demand placed on the body; an increase in the frequency, duration, or intensity of an activity. A principle of training that states for a body system to improve, it must be worked at frequencies, durations, or intensities that increase by increments.

TABLE 14-2 **ACSM Guidelines for Physical Fitness**

	Cardiorespiratory	Strength	Flexibility
	Steve Cole/Photodisc/Getty Images	© David Hanover Photography	© David Hanover Photography
Type of Activity	Aerobic activity that uses large-muscle groups and can be maintained continuously	Resistance activity that is performed at a controlled speed and through a full range of motion	Stretching activity that uses the major muscle groups
Frequency	5 to 7 days per week	2 to 3 nonconsecutive days per week	2 to 7 days per week
Intensity	Moderate (equivalent to walking at a pace of 3 to 4 miles per hour)[a]	Enough to enhance muscle strength and improve body composition	Enough to feel tightness or slight discomfort
Duration	At least 30 minutes per day	2 to 4 sets of 8 to 12 repetitions involving each major muscle group	2 to 4 repetitions of 15 to 30 seconds per muscle group
Examples	Running, cycling, dancing, swimming, inline skating, rowing, power walking, cross-country skiing, kickboxing, water aerobics, jumping rope; sports activities such as basketball, soccer, racquetball, tennis, volleyball	Pull-ups, push-ups, sit-ups, weightlifting, pilates	Yoga

[a]For those who prefer vigorous-intensity aerobic activity such as walking at a very brisk pace (>4.5 mph) or running (≥5 mph), a minimum of 20 minutes per day, 3 days per week is recommended.
SOURCE: American College of Sports Medicine position stand: Quantity and quality of exercise for developing and maintaining cardiorespiratory, musculoskeletal, and neuromotor fitness in apparently healthy adults—Guidance for prescribing exercise, *Medicine and Science in Sports and Exercise* 43 (2011): 1334–1359; W. L. Haskell and coauthors, Physical activity and public health: Updated recommendation for adults from the American College of Sports Medicine and the American Heart Association, *Medicine & Science in Sports & Exercise* 39 (2007): 1423–1434.

© Cengage Learning

feeding periods.[24] This balance between protein degradation and synthesis maintains the body's muscle tissue. To *gain* muscle tissue, protein synthesis must be greater than degradation, a condition called **hypertrophy**. When protein degradation is greater than synthesis, the result is **atrophy**. Physical activity tips the balance toward muscle hypertrophy. The opposite is also true: unused muscles diminish in size and weaken over time, tipping the balance toward muscle atrophy.

The muscles adapt and build only the proteins they need to cope with the work performed.[25] Muscles engaged in activities that require strength develop more of the proteins needed for greater muscle mass; body builders have large, well-developed muscles. By comparison, those engaged in endurance activities develop more of the proteins needed to combat muscle fatigue; distance cyclists can pedal for many hours before fatigue sets in.

Strategies to Build Fitness and Prevent Injuries A variety of physical activities produces the best overall fitness, and to this end, people need to work different muscle groups from day to day. This strategy provides a day or two of rest for different muscle groups, allowing time to replenish nutrients and to repair any minor damage incurred by the activity.

Other tips for building fitness and minimizing the risk of overuse injuries are:

- Be active all week, not just on the weekends.
- Use proper equipment and wear proper attire.
- Perform exercises using proper form.
- Include **warm-up** and **cool-down** activities in each session. Warming up helps to prepare muscles, ligaments, and tendons for the upcoming activity and mobilizes fuels to support strength and endurance activities. Cooling down reduces muscle cramping and allows the heart rate to slow gradually.
- Train hard enough to challenge your strength or endurance a few times each week rather than every time you work out. Between challenges, do moderate workouts and include at least one day of rest each week.

hypertrophy (high-PER-tro-fee): growing larger; with regard to muscles, an increase in size (and strength) in response to use.

atrophy (AT-ro-fee): becoming smaller; with regard to muscles, a decrease in size (and strength) because of disuse, undernutrition, or wasting diseases.

warm-up: 5 to 10 minutes of light activity, such as easy jogging or cycling, prior to a workout to prepare the body for more vigorous activity.

cool-down: 5 to 10 minutes of light activity, such as walking or stretching, following a vigorous workout to gradually return the body's core to near-normal temperature.

- Pay attention to body signals. Symptoms such as abnormal heartbeat, dizziness, light-headedness, cold sweat, confusion, or pain or pressure in the middle of the chest, teeth, jaw, neck, or arm demand immediate medical attention.
- Work out wisely. Do not start with activities so demanding that pain stops you within a day or two. Learn to enjoy small steps toward improvement. Fitness builds slowly.

Cautions on Starting a Fitness Program Before beginning a fitness program, make sure it is safe for you to do so. Most apparently healthy people can begin a moderate exercise program such as walking or increasing daily activities without a medical examination, but people with any of the following coronary risk factors may need medical advice:[26]

- Age (men ≥45 yr or women ≥55 yr)
- Family history of heart disease
- Cigarette smoking
- Hypertension
- Serum cholesterol ≥200 mg/dL, LDL ≥130 mg/dL or HDL ≤40 mg/dL, or taking lipid-lowering medication
- Prediabetes or diabetes (fasting plasma glucose ≥100 mg/dL)
- Sedentary lifestyle
- Obesity (BMI ≥30)

Mike Powell/Stockbyte/Getty Images

Choose an active lifestyle. Use the stairs; walk or bike to work, class, or shops; wash and wax the car; mow the grass; rake the leaves; shovel snow; walk the dog; play with children. Be active and have fun.

The Components of Fitness To be physically fit, a person must develop enough flexibility, muscle strength and endurance, and cardiorespiratory endurance to meet the everyday demands of life with some to spare, and achieve a healthy body composition. **Flexibility** allows the joints to move freely, reducing the risk of injury. **Cardiorespiratory endurance** supports the ongoing activity of the heart and lungs. **Muscle strength** and **muscle endurance** enable muscles to

flexibility: the capacity of the joints to move through a full range of motion; the ability to bend and recover without injury.

cardiorespiratory endurance: the ability to perform large-muscle, dynamic exercise of moderate to high intensity for prolonged periods.

muscle strength: the ability of muscles to work against resistance.

muscle endurance: the ability of a muscle to contract repeatedly without becoming exhausted.

© Kzenon/Shutterstock.com

© istockphoto.com/Steve Froebe

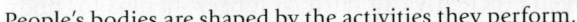

People's bodies are shaped by the activities they perform.

work harder and longer without fatigue. **Body composition** improves as physical activity supports lean body tissues and reduces excess body fat.

Although nutrition influences each component of fitness to some extent, its role in flexibility is not as apparent as for the others. For this reason, cardiorespiratory endurance and muscle strength and endurance are emphasized in the following sections. As the heart and lungs improve their capacity to sustain physical activity and the muscles become stronger and less readily fatigued, body composition improves: body fat decreases and lean body mass increases. Chapter 8 discusses body composition and the health risks of too much body fat.

Cardiorespiratory Endurance
The length of time a person can remain active with an elevated heart rate—that is, the ability of the heart, lungs, and blood to sustain a given demand—defines a person's cardiorespiratory endurance. Cardiorespiratory endurance training improves a person's ability to sustain vigorous activities such as running, brisk walking, or swimming. Such training enhances the capacity of the heart, lungs, and blood to deliver oxygen to, and remove waste from, the body's cells.[27] Cardiorespiratory endurance training, therefore, is *aerobic*, meaning oxygen requiring. As the cardiorespiratory system gradually adapts to the demands of aerobic activity, the body delivers oxygen more efficiently. In fact, the accepted measure of a person's cardiorespiratory fitness is maximal oxygen uptake (VO_{2max}). The benefits of cardiorespiratory training are not just physical, though, because all of the body's cells, including the brain cells, require oxygen to function. When the cells receive more oxygen more readily, both the body and the mind benefit.

Cardiorespiratory Conditioning
Cardiorespiratory conditioning occurs as aerobic workouts improve heart and lung function. **Cardiac output** increases, thus enhancing oxygen delivery. The heart becomes stronger, and each beat pumps more blood. Because the heart pumps more blood with each beat, fewer beats are necessary, and the resting heart rate slows down. The average resting pulse rate for adults is around 70 beats per minute, but people who achieve cardiorespiratory conditioning may have resting pulse rates of 50 or even lower. The muscles that work the lungs become stronger, too, so breathing becomes more efficient. Circulation through the arteries and veins improves. Blood moves easily, and blood pressure falls.[28] In short, cardiorespiratory conditioning:

- Increases cardiac output and oxygen delivery
- Increases blood volume per heart beat (stroke volume)
- Slows resting pulse rate
- Increases breathing efficiency
- Improves circulation
- Reduces blood pressure

Cardiorespiratory endurance reflects the health of the heart and circulatory system, on which all other body systems depend. To improve your cardiorespiratory endurance, activities must be sustained for 20 minutes or longer and use most of the large-muscle groups of the body (legs, buttocks, and abdomen). The level of training must be intense enough to elevate your heart rate.

A person's own perceived effort is usually a reliable indicator of the intensity of an activity. In general, workouts should be at an intensity that raises your heart rate but still leaves you able to talk comfortably. For those who are more competitive and want to work to their limits on some days, a treadmill test can reveal the maximum heart rate. Workouts are safe at up to 85 percent of that rate. Table 14-2 (p. 444) includes the ACSM guidelines for developing and maintaining cardiorespiratory fitness.

Muscle Conditioning
One of the benefits of cardiorespiratory training is that fit muscles use oxygen efficiently, reducing the heart's workload. An added bonus is that muscles that use oxygen efficiently can burn fat longer—a plus for body composition and weight control.

body composition: the proportions of muscle, bone, fat, and other tissues that make up a person's total body weight.

VO_{2max}: the maximum rate of oxygen consumption by an individual at sea level.

cardiorespiratory conditioning: improvements in heart and lung function and increased blood volume, brought about by aerobic training.

cardiac output: the volume of blood discharged by the heart each minute; determined by multiplying the stroke volume by the heart rate. The stroke volume is the amount of oxygenated blood the heart ejects toward the tissues at each beat. Cardiac output (volume/minute) = stroke volume (volume/beat) × heart rate (beats/minute)

The key to regular physical activity is finding an activity you enjoy.

Muscle Strength and Endurance **Resistance training** has long been recognized as a means to build and maintain muscle mass, strength, power, and endurance. Additional benefits of resistance training, however, have also emerged. Progressive resistance training helps prevent and manage several chronic diseases, including cardiovascular disease, and enhances psychological well-being.[29] Resistance training can also help to maximize and maintain bone mass.[30] Even in women past menopause (when most women are losing bone), resistance training can improve bone density, especially in combination with adequate dietary calcium and vitamin D intake.[31]

By promoting strong muscles in the back and abdomen, resistance training can improve posture and reduce the risk of back injury. Resistance training can also help prevent the decline in physical mobility that often accompanies aging.[32] Older adults, even those in their 80s, who participate in resistance training programs not only gain muscle strength but also improve their muscle endurance, which enables them to walk longer before exhaustion. Leg strength and walking endurance are powerful indicators of an older adult's physical abilities.

Resistance training builds muscle strength, **muscle power,** and muscle endurance. To emphasize muscle strength, combine high resistance (heavy weight) with a low number of repetitions (8 to 12).[33] To emphasize muscle power, combine moderate resistance (light to medium weight) with high velocity (as fast as safely possible). To emphasize muscle endurance, combine less resistance (lighter weight) with more repetitions (15 to 20). Resistance training enhances performance in other sports too. Swimmers can develop a more efficient stroke and tennis players a more powerful serve when they train with weights, for example.

A Balanced Fitness Program The intensity and type of physical activities that are best for one person may not be good for another. A person who has been sedentary will initially perform at a dramatically different level of intensity than a fit person.

The type of physical activity that is best for you depends, too, on what you want to achieve and what you enjoy doing. Some people love walking, whereas others prefer to dance or ride a bike. Those who want to be stronger and firmer, lift weights. Keep in mind that muscle is more metabolically active than body fat, so the more muscle you have, the more energy you'll expend.

In a balanced fitness program, aerobic activity improves cardiorespiratory fitness, stretching enhances flexibility, and resistance training develops muscle strength, muscle power, and muscle endurance. Table 14-3 provides an example of a balanced fitness program.

TABLE 14-3 A Sample Balanced Fitness Program

Monday, Tuesday, Wednesday, Thursday, Friday:
- 5 minutes of warm-up activity
- 30–60 minutes of aerobic activity
- 10 minutes of cool-down activity and stretching

Tuesday, Thursday, Saturday:
- 5 minutes of warm-up activity
- 30 minutes of resistance training
- 10 minutes of cool-down activity and stretching

Saturday and/or Sunday:
- Sports, walking, hiking, biking, or swimming

© Cengage Learning

---

resistance training: the use of free weights or weight machines to provide resistance for developing muscle strength, power, and endurance; also called *weight training*. A person's own body weight may also be used to provide resistance such as when a person does push-ups, pull-ups, or abdominal crunches.

muscle power: the product of force generation (strength) and movement velocity (speed); the speed at which a given amount of exertion is completed.

Physical activity brings good health and long life. To develop fitness—whose components are flexibility, cardiorespiratory endurance, muscle strength and endurance, and body composition—a person must condition the body, through training, to adapt to the activity performed.

14.2 Energy Systems and Fuels to Support Activity

> LEARN IT Identify the factors that influence fuel use during physical activity and the types of activities that depend more on glucose or fat, respectively.

Nutrition and physical activity go hand in hand. Activity demands carbohydrate and fat as fuel, protein to build and maintain lean tissues, vitamins and minerals to support both energy metabolism and tissue building, and water to help distribute the fuels and to dissipate the resulting heat and wastes. This section describes how nutrition supports a person who decides to get up and go.

The Energy Systems of Physical Activity As Chapter 7 describes, the high-energy compound ATP provides the energy that powers all the activities of living cells. ATP is present in small amounts in all body tissues all the time, and it can deliver energy instantly. Muscle cells may need additional ATP to power their many activities. When an ATP molecule is split in the muscle cells, the energy released contracts muscles and generates heat. Thus physical exertion consumes ATP. Three major energy systems enable muscle cells to regenerate ATP during different types of physical activity:

- the phosphagen system (also called the creatine phosphate system)
- the lactic acid system (anaerobic glycolysis)
- the aerobic system (aerobic glycolysis, fatty acid oxidation, and TCA cycle)

All three energy systems function at all times, but depending on the intensity of the activity and the conditioning of the athlete, one system will predominate at any given time. The following sections describe each of these systems.

The Phosphagen System Immediately after the onset of a demand, before muscle ATP pools dwindle, a muscle enzyme begins to break down another high-energy compound that is stored in the muscle, **creatine phosphate (CP).** CP is made from creatine, a compound commonly found in muscles, with a phosphate group attached. CP can split anaerobically (not requiring oxygen) to release phosphate, which can be used to replenish ATP. Supplies of CP in a muscle last for only about 10 seconds, producing enough quick energy, without oxygen, for a 100-meter dash.

When activity ceases and the muscles are resting, ATP gives up one of its phosphate groups to creatine. Thus CP is produced during rest by reversing the process that occurs during muscular activity. (Highlight 14 includes creatine supplements in its discussion of substances commonly used in the pursuit of fitness.)

The Lactic Acid System After the first 10 seconds or so of intense activity, energy from the phosphagen system diminishes, so muscle cells call upon the lactic acid system to produce more ATP. As Chapter 7 describes, the lactic acid system involves the anaerobic breakdown of glucose to pyruvate (and then of pyruvate to lactate). The primary source of glucose is muscle glycogen. This system can generate a small amount of ATP quickly for high-intensity activity lasting up to 3 minutes. A later section describes the consequences of lactate accumulation.

The Aerobic System To meet the more prolonged demands of sustained activity, the muscles rely on the aerobic system to provide ATP for muscle contraction. As Chapter 7 describes, carbohydrate, fat, and some amino acids are continuously oxidized to ensure an uninterrupted supply of ATP.

creatine phosphate (CP): a high-energy compound in muscle cells that acts as a reservoir of energy that can maintain a steady supply of ATP. CP provides the energy for short bursts of activity; also called *phosphocreatine*.

- During rest: ATP + creatine → CP
- During activity: CP → ATP + creatine

Sustained muscular efforts as in a long-distance rowing event or a cross-country run involve *aerobic* work.

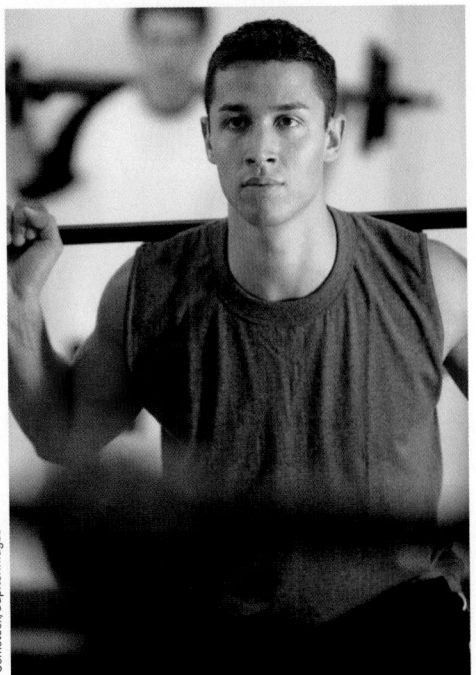

Split-second surges of power as in the heave of a barbell or jump of a basketball player involve *anaerobic* work.

During rest, the body derives more than half of its ATP from the oxidation of fatty acids, most of the rest from the complete oxidation of glucose, and a small percentage from the oxidation of amino acids. During physical activity, the body adjusts its mixture of fuels. Muscles always use a mixture of fuels—never just one. How much of which fuel the muscles use during physical activity depends on an interplay among the fuels available from the diet, the intensity and duration of the activity, and the degree to which the body is conditioned to perform that activity. The next sections explain these relationships by examining each of the energy-yielding nutrients individually, but keep in mind that although one fuel may predominate at a given time, the other two will still be involved. Table 14-4 shows how fuel use changes according to the intensity and duration of the activity.

Anaerobic versus Aerobic Oxygen—or its lack—is central to the energy systems just described. The phosphagen and lactic acid systems are anaerobic—meaning without oxygen. As its name suggests, the aerobic system uses oxygen. As you read the following pages, notice how the fuel mixture shifts depending on whether the activity is anaerobic or aerobic.

Anaerobic activities are associated with strength, agility, and split-second surges of power. The jump of a slam dunk, the power of a tennis serve, and the heave of a

TABLE 14-4 Primary Fuels Used for Activities of Different Intensities and Durations

Activity Intensity	Activity Duration	Energy System	Preferred Fuel Source	Oxygen Needed?	Activity Example
Extreme	5 to 10 sec	Phosphagen system	ATP-CP (immediate availability)	No	100-meter sprint, shot put, golf or baseball bat swing, tennis or volleyball serve
Very high	20 sec to 2 min	Lactic acid system	ATP from carbohydrate (anaerobic glycolysis)	No	400-meter run, 100-meter swim, gymnastic routines
High	2 min to 20 min	Aerobic system	ATP from carbohydrate (glycolysis and TCA cycle)	Yes	Cycling, swimming, running
Moderate	> 20 min	Aerobic system	ATP from fat (fatty acid-oxidation and TCA cycle)	Yes	Hiking

NOTE: All energy systems function at all times, but depending on the intensity of the activity and the conditioning of the athlete, one system will predominate at any given time.

bench press all involve anaerobic work. Such high-intensity, short-duration activities depend mostly on glucose as the chief energy fuel in the lactic acid system.

Endurance activities of low to moderate intensity and long duration depend more on fat to provide energy through the aerobic system. The ability to continue swimming to the shore, to keep on hiking to the top of the mountain, or to continue pedaling all the way home reflects aerobic capacity. As mentioned earlier, aerobic capacity is also crucial to maintaining a healthy heart and circulatory system. Because various physical activities use different energy systems and quantities of nutrients to various degrees, a person's food choices can greatly influence performance.

Glucose Use during Physical Activity
Glucose, stored in the liver and muscles as glycogen, is vital to physical activity. During exertion, the liver breaks down its glycogen and releases the glucose into the bloodstream. The muscles use this glucose as well as their own private glycogen stores to fuel their work. Glycogen supplies can easily support everyday activities but are limited to less than 2000 kcalories of energy, enough for about 20 miles of running. The more athletes can fill muscle glycogen stores, the longer the glycogen will last during physical activity, which in turn influences performance. When glycogen is depleted, the muscles become fatigued.

Diet Affects Glycogen Storage and Use
How much carbohydrate a person eats influences how much glycogen is stored. A classic study compared fuel use during activity among three groups of runners on different diets.[34] For several days before testing, one group consumed a normal mixed diet, a second group consumed a high-carbohydrate diet, and the third group consumed a no-carbohydrate diet (fat and protein diet). As Figure 14-1 shows, the high-carbohydrate diet allowed the runners to keep going longer before exhaustion. This study and many others that followed have confirmed that high-carbohydrate diets enhance endurance by ensuring ample glycogen stores. Thus, to fill glycogen stores, eat plenty of carbohydrate-rich foods.

Intensity of Activity Affects Glycogen Use
How long an exercising person's glycogen will last depends not only on diet, but also on the intensity of the activity. Moderate activities such as jogging use glycogen slowly. The lungs and circulatory system have no trouble keeping up with the muscles' need for oxygen. The individual breathes easily, and the heart beats steadily—the activity is aerobic. The muscles derive their energy from both glucose and fatty acids. By depending partly on fatty acids, moderate aerobic activity conserves glycogen.

Intense activities—the kind that make it difficult "to catch your breath," such as a quarter-mile race—use glycogen quickly. In such activities, the muscles rely on the lactic acid system to break down glucose to pyruvate anaerobically, producing ATP quickly.

Lactate When the rate of glycolysis exceeds the capacity of the mitochondria to accept hydrogens with their electrons for the electron transport chain, the accumulating pyruvate molecules are converted to lactate. Lactate is the product of anaerobic glycolysis. At low intensities, lactate is readily cleared from the blood, but at higher intensities, lactate accumulates. When the rate of lactate production exceeds the rate of clearance, intense activity can be maintained for only 1 to 3 minutes (as in a 400- to 800-meter race or a boxing match). Lactate has long been blamed for muscle fatigue, but research disputes this idea. Working muscles may produce lactate and experience fatigue, but the lactate does not cause the fatigue.[35]

> **FIGURE 14-1   The Effect of Diet on Physical Endurance**

A high-carbohydrate diet can increase an athlete's endurance. In this study, the fat and protein diet provided 94 percent of kcalories from fat and 6 percent from protein; the normal mixed diet provided 55 percent of kcalories from carbohydrate; and the high-carbohydrate diet provided 83 percent of kcalories from carbohydrate.

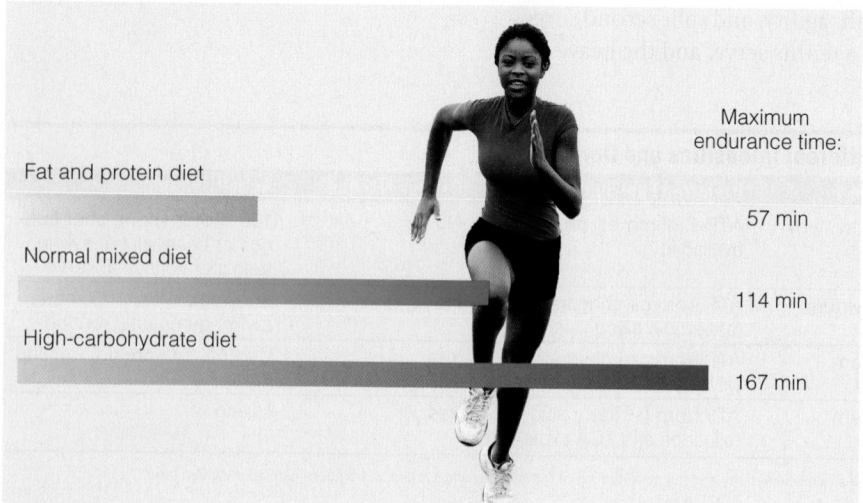

Fat and protein diet

Normal mixed diet

High-carbohydrate diet

Maximum endurance time:

57 min

114 min

167 min

Photo: Photos.com (RF)/Getty Images/Jupiterimages; Line art: © Cengage Learning

Lactate quickly leaves the muscles and travels in the blood to the liver. There, liver enzymes convert the lactate back into glucose. Glucose can then return to the muscles to fuel additional activity. (The recycling process that regenerates glucose from lactate is known as the *Cori cycle*, as shown in Figure 7-7 on p. 209.)

Duration of Activity Affects Glycogen Use Glycogen use depends not only on the intensity of an activity, but also on its duration. Initially, a person uses mostly glycogen for fuel. As the muscles devour their own glycogen, they become ravenous for more glucose. Glucagon signals the liver to empty out its glycogen stores to provide the muscles with more glucose.

A person who continues exercising moderately (mostly aerobically) begins to use less and less glucose and more and more fat for fuel (review Table 14-4 on p. 449). Still, glucose use continues, and if the activity lasts long enough and is intense enough, blood glucose declines and muscle and liver glycogen stores become depleted. Physical activity can continue for a short time thereafter only because glucagon signals the liver to produce, from lactate and certain amino acids, the minimum amount of glucose needed to briefly forestall total depletion.

Moderate- to high-intensity aerobic exercises that can be sustained for only a short time (less than 20 minutes) use some fat, but more glucose for fuel.

Glucose Depletion Gluconeogenesis could support an activity of low intensity, such as walking, but cannot support continued strenuous activity. After a couple of hours of strenuous activity, glucose stores are depleted. When depletion occurs, it brings nervous system function to a near halt, making continued exertion at the same intensity almost impossible. Marathon runners refer to this point of glucose exhaustion as "hitting the wall."

To avoid such debilitation, endurance athletes try to maintain their blood glucose for as long as they can. The following guidelines will help endurance athletes maximize glucose supply:

- Eat a high-carbohydrate diet (approximately 8 grams of carbohydrate per kilogram of body weight or about 70 percent of energy intake) regularly.*
- Consume glucose (usually in sports drinks) periodically during activities that last for 1 hour or more.
- Eat carbohydrate-rich foods (approximately 60 grams of carbohydrate) immediately following activity; for example, a 16-ounce sports drink and a small bagel, 16 ounces of milk and four oatmeal cookies, or 8 ounces of pineapple juice and a granola bar.

Another strategy, **carbohydrate loading**, involves training the muscles to store as much glycogen as they can, while supplying the dietary glucose to enable them to do so. This strategy benefits endurance athletes who exercise at high intensity for longer than 90 minutes and who cannot meet their carbohydrate needs during competition. Although carbohydrate-loading can increase an individual's glycogen stores, performance may or may not improve. Those who exercise for shorter times or at a slower pace are better served by a regular high-carbohydrate diet. The last section of this chapter, "Diets for Physically Active People," discusses how to design a high-carbohydrate diet for performance.

Glucose during Activity Muscles can obtain the glucose they need not only from glycogen stores, but also from foods and beverages consumed during activity. Consuming carbohydrate is especially useful during exhausting endurance activities (lasting more than 1 hour) and during games such as soccer or hockey, which last for hours and demand repeated bursts of intense activity.[36]

---

*Percentage of energy intake is meaningful only when total energy intake is known. Consider that at high energy intakes (say, 5000 kcalories/day), even a moderate-carbohydrate diet (50 percent of energy intake) supplies 625 grams of carbohydrate—enough for a 165-pound (75 kilogram) athlete in heavy training. By comparison, at a moderate energy intake (2000 kcalories/day), a high-carbohydrate intake (70 percent of energy intake) supplies 350 grams—plenty of carbohydrate for most people, but not enough for athletes in heavy training.

carbohydrate loading: a regimen of moderate exercise followed by the consumption of a high-carbohydrate diet that enables muscles to store glycogen beyond their normal capacities; also called *glycogen loading* or *glycogen super compensation*.

Endurance athletes often run short of glucose by the end of competitive events. To ensure optimal carbohydrate intake, sports nutrition experts recommend 30 to 60 grams of carbohydrate per hour during prolonged events.[37] Carbohydrate-based sports drinks offer a convenient way to meet this recommendation and also help replace water and electrolyte losses, as a later section explains. Thus, to ensure optimal hydration and carbohydrate intake, endurance athletes are advised to drink one-half to one liter of a 4 to 8 percent carbohydrate-based sports drink per hour, in small, frequent doses during activity.[38] During the last stages of an endurance competition, when glycogen is running low, glucose consumed during the event can slowly make its way from the digestive tract to the muscles and augment the body's supply of glucose enough to forestall exhaustion.

Some researchers have questioned whether adding protein to carbohydrate-containing sports beverages would offer a performance advantage to endurance athletes.[39] Evidence so far suggests that when carbohydrate intake is optimal, protein provides no additional performance benefit.

Glucose after Activity Muscles depleted of glycogen have greater insulin sensitivity, which enhances glucose uptake and promotes glycogen synthesis. Thus, eating high-carbohydrate foods *after* physical activity also enlarges glycogen stores. To accelerate the rate of glycogen storage, train normally; then, within 2 hours after physical activity, consume a high-carbohydrate meal, such as a glass of orange juice and some graham crackers, toast, or cereal.[40] After 2 hours, the rate of glycogen storage declines by almost half. This strategy is particularly important to athletes who train hard more than once a day.

Chapter 4 introduces the glycemic response and discusses the possible health benefits of eating a *low*-glycemic diet. Such a diet may also benefit endurance performance. Some research indicates that foods with a low glycemic response enhance fatty acid availability and use during subsequent activity, thereby reducing reliance on the muscles' own lipid and glycogen stores.[41] More research is needed to confirm these findings.

Training Affects Glycogen Use Training, too, affects how much glycogen muscles will store. Muscle cells that repeatedly deplete their glycogen through hard work adapt to store greater amounts of glycogen to support that work.

Conditioned muscles also rely less on glycogen and more on fat for energy, so glycogen breakdown and glucose use occur more slowly in trained than in untrained individuals at a given work intensity. A person attempting an activity for the first time uses much more glucose than an athlete who is trained to perform the same activity. Oxygen delivery to the muscles by the heart and lungs plays a role, but equally importantly, trained muscles are better equipped to use the oxygen because their cells contain more mitochondria, the structures within a cell responsible for producing ATP (see Figure 7-14 on p. 213). Untrained muscles depend more heavily on anaerobic glucose breakdown, even when physical activity is just moderate.

Fat Use during Physical Activity
Sports nutrition experts recommend that athletes consume 20 to 35 percent of their energy from fat to meet nutrient and energy needs—the same recommendation as for others.[42] Athletes who restrict fat below 20 percent of total energy intake may fail to consume adequate energy and nutrients. Recommendations to include vegetable oils, nuts, olives, fatty fish, and other sources of health-promoting fats in the diet apply to athletes as well as to everyone else.

Diets high in saturated fat carry risks of heart disease. Physical activity offers some protection against cardiovascular disease, but athletes may still suffer heart attacks and strokes. Limiting saturated fat intake to less than 10 percent of total energy intake is another way athletes can protect themselves from heart disease.

In contrast to *dietary* fat, *body* fat stores are extremely important during physical activity, as long as the activity is not too intense. Unlike glycogen stores, the body's fat stores can usually provide more than 70,000 kcalories and fuel hours of activity without running out.

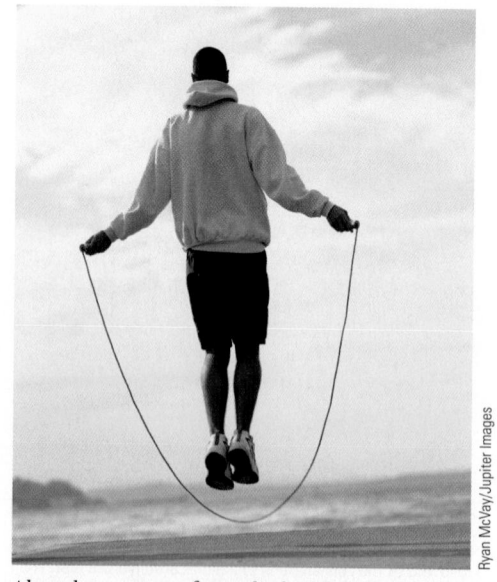

Abundant energy from the breakdown of fat comes from aerobic metabolism.

The fat used in physical activity is liberated as fatty acids from the internal fat stores and from the fat under the skin. Areas that have the most fat to spare donate the greatest amounts to the blood (although they may not be the areas that appear fattiest). Thus "spot reducing" doesn't work because muscles do not "own" the fat that surrounds them. Fat cells release fatty acids into the blood, not into the underlying muscles. Then the blood gives to each muscle the amount of fat that it needs. Proof of this is found in a tennis player's arms—the skinfold measures of fat are the same in both arms, even though the muscles of one arm work much harder and may be larger than those of the other. A balanced fitness program that includes strength training will tighten muscles underneath the fat, improving the overall appearance. Keep in mind that some body fat is essential to good health. (Chapter 8 discusses the health risks of too little body fat.)

Duration of Activity Affects Fat Use Early in an activity, as the muscles draw on fatty acids, blood levels fall. If the activity continues for more than a few minutes, the hormone epinephrine signals the fat cells to begin breaking down their stored triglycerides and liberating fatty acids into the blood. As physical activity continues, the blood fatty acid concentration surpasses the normal resting concentration. Thereafter, sustained, moderate activity uses body fat stores as its major fuel.

Intensity of Activity Affects Fat Use The intensity of physical activity also affects fat use. As the intensity of activity increases, fat makes less and less of a contribution to the fuel mixture. Remember that fat can be broken down for energy only by aerobic metabolism. For fat to fuel activity, then, oxygen must be abundantly available. If a person is breathing easily during activity, the muscles are getting all the oxygen they need and are able to use more fat in the fuel mixture.

Training Affects Fat Use Training—repeated aerobic activity—produces the adaptations that permit the body to draw more heavily on fat for fuel. Training stimulates the muscle cells to manufacture more and larger mitochondria, the "powerhouse" structures of the cells that produce ATP for energy. Another adaptation: the heart and lungs become stronger and better able to deliver oxygen to muscles at high activity intensities. Still another: hormones in the body of a trained person slow glucose release from the liver and speed up the use of fat instead. These adaptations reward not only trained athletes but all active people; a person who trains in aerobic activities such as distance running or cycling becomes well suited to the activity.

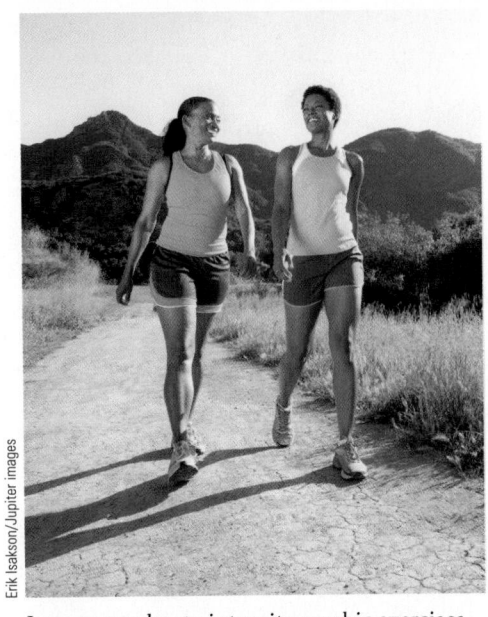

Low- to moderate-intensity aerobic exercises that can be sustained for a long time (more than 20 minutes) use some glucose, but more fat, for fuel.

Protein Use during Physical Activity—and between Times

Table 14-4 on p. 449 summarizes the fuel uses discussed so far, but does not include the third energy-yielding nutrient, protein, because protein is not a major fuel for physical activity. Nevertheless, physically active people use protein just as other people do—to build muscle and other lean tissues and, to some extent, to fuel activity. The body does, however, handle protein differently during activity than during rest.

Protein Used in Muscle Building Synthesis of body proteins is suppressed during activity. In the hours of recovery following activity, though, protein synthesis accelerates beyond normal resting levels. As noted earlier, eating high-carbohydrate foods immediately after exercise accelerates muscle glycogen storage. Similarly, research shows that eating high-quality protein, either by itself or together with carbohydrate, enhances muscle protein synthesis.[43] With food intake, of course, insulin is secreted. Insulin not only stimulates muscle protein synthesis, but also greatly inhibits muscle protein breakdown.[44] Remember that the body adapts and builds the molecules, cells, and tissues it needs for the next period of activity. Whenever the body remodels a part of itself, it tears down old structures to make way for new ones. Repeated activity, with just a slight overload, triggers the protein-dismantling and protein-synthesizing equipment of each muscle cell to make needed changes—that is, to adapt.

The physical work of each muscle cell acts as a signal to its DNA and RNA to begin producing the kinds of proteins that will best support that work. Take running, for example. In the first difficult sessions, the body is not yet equipped to perform aerobic work easily, but with each session, the cells' genetic material gets the message that an overhaul is needed. In the hours that follow the session, the genes send molecular messages to the protein-building equipment that tell it what old structures to break down and what new structures to build. Within the limits of its genetic potential, the body responds. Running (or any aerobic activity) stimulates synthesis of mitochondria to facilitate efficient aerobic metabolism. Over time, the body adapts and running becomes easier.

The body of a weightlifter responds to training as well, but the response differs from that of aerobic training. Weightlifting stimulates synthesis of muscle fiber protein to enhance muscle mass and strength—with little change in mitochondria. An athlete may add between ¼ ounce and 1 ounce (between 7 and 28 grams) of protein to muscle mass each day during active muscle-building phases of training.

Protein Used as Fuel Not only do athletes retain more protein in their muscles, but they also use more protein as fuel. Muscles speed up their use of amino acids for energy during physical activity, just as they speed up their use of fat and carbohydrate. Still, protein contributes only about 10 percent of the total fuel used, both during activity and during rest. The most active people of all—endurance athletes—use large amounts of all energy fuels, including protein, during performance, but they also eat more food and therefore usually consume enough protein.

Diet Affects Protein Use during Activity Diet influences how much protein is used during physical activity. People who consume diets adequate in energy and rich in *carbohydrate* use less protein than those who eat protein- and fat-rich diets. Recall that carbohydrates spare proteins from being broken down to make glucose when needed. Because physical activity requires glucose, a diet lacking in carbohydrate necessitates the conversion of amino acids to glucose. The same is true for a diet high in fat because fatty acids can never provide glucose. In short, to conserve protein, eat a diet adequate in energy and rich in carbohydrate.

Intensity and Duration of Activity Affect Protein Use during Activity The intensity and duration of activity also modifies protein use. Endurance athletes who train for more than an hour a day, engaging in aerobic activity of moderate intensity and long duration, may deplete their glycogen stores by the end of their workouts and become somewhat more dependent on body protein for energy.

In contrast, anaerobic strength training does not use more protein for energy, but it does demand more protein to build muscle. Thus the protein needs of both endurance and strength athletes are slightly higher than those of sedentary people, but certainly not as high as the protein intakes many athletes consume.

Protein Recommendations for Active People As mentioned, all active people, and especially athletes in training, probably need more protein than sedentary people do. Endurance athletes, such as long-distance runners and cyclists, use more protein for fuel than strength or power athletes do, and they retain some, especially in the muscles used for their sport. Strength athletes, such as weightlifters, and power athletes, such as football players, use less protein for fuel, but they still use some and retain much more. Therefore, *all* athletes in training should attend to protein needs, but they should first meet their energy needs with adequate carbohydrate intakes. Without adequate carbohydrate intake, athletes will burn off as fuel the very protein they wish to retain in muscle.

How much protein, then, should an active person consume? The DRI Committee does not recommend greater than normal protein intakes for athletes, but other authorities do.[45] Recommendations specify different protein intakes for athletes pursuing different activities (see Table 14-5).[46] Even the highest protein recommendations can be met without protein supplements, or even excessive servings of meat. Chapter 6 reviews the potential dangers of using protein and amino acid supplements (see pp. 190–191).

TABLE 14-5 Recommended Protein Intakes for Athletes

	Recommendations (g/kg/day)	Protein Intakes (g/day)	
		Males	*Females*
RDA for adults	0.8	56	44
Recommended intake for power (strength or speed) athletes	1.2–1.7	84–119	66–94
Recommended intake for endurance athletes	1.2–1.4	84–98	66–77
US average intake		102	70

© Cengage Learning

NOTE: Daily protein intakes are based on a 70-kilogram (154-pound) man and 55-kilogram (121-pound) woman.
SOURCES: Position of the American Dietetic Association, Dietitians of Canada, and the American College of Sports Medicine: Nutrition and athletic performance, *Journal of the American Dietetic Association* 109 (2009): 509–527; US Department of Agriculture, Agricultural Research Service, 2008, Nutrient Intakes from Food: Mean Amounts Consumed per Individual, One Day, 2005–2006, www.ars.usda.gov/ba/bhnrc/fsrg; Committee on Dietary Reference Intakes, *Dietary Reference Intakes for Energy, Carbohydrate, Fiber, Fat, Fatty Acids, Cholesterol, Protein and Amino Acids* (Washington, D.C.: National Academies Press, 2005), pp. 660–661.

> **REVIEW IT** Identify the factors that influence fuel use during physical activity and the types of activities that depend more on glucose or fat, respectively.

The mixture of fuels the muscles use during physical activity depends on diet, the intensity and duration of the activity, and training. During intense activity, the fuel mix contains mostly glucose, whereas during less intense, moderate activity, fat makes a greater contribution. With endurance training, muscle cells adapt to store more glycogen and to rely less on glucose and more on fat for energy. Athletes in training may need more protein than sedentary people do, but they typically eat more food as well and therefore obtain enough protein without supplements.

14.3 Vitamins and Minerals to Support Activity

> **LEARN IT** List which vitamin and mineral supplements, if any, athletes may need and why.

Many of the vitamins and minerals assist in releasing energy from fuels and in transporting oxygen. This knowledge has led many people to believe, mistakenly, that vitamin and mineral *supplements* offer physically active people both health benefits and athletic advantages. (Review Highlight 10 for a discussion of vitamin and mineral supplements, and see Highlight 14, which explores supplements and other products people use in the hope of enhancing athletic performance.)

Dietary Supplements Nutrient supplements do not enhance the performance of well-nourished people. Deficiencies of vitamins and minerals, however, do impede performance. Regular, strenuous, physical activity increases the demand for energy, and athletes and active people tend to eat more food. If they select nutrient-dense foods to meet those increased energy needs, they will, in most cases, also meet their vitamin and mineral needs. Remember, however, that nutrient recommendations are based on the needs of healthy people in different life stages and do not reflect the potentially greater needs that some athletes may incur.

As Highlight 8 mentions, some athletes who struggle to meet low body-weight requirements may eat so little food that they fail to obtain all the nutrients they need.[47] The practice of "making weight" is opposed by many health and fitness organizations, but for athletes who choose this course of action, a single daily multivitamin-mineral supplement that provides no more than the DRI recommendations for nutrients may be beneficial. In addition, some athletes simply do not eat enough food to maintain a healthy body weight during times of intense training or competition. For these athletes as well, a daily multivitamin-mineral supplement can be helpful.

© IM_photo/Shutterstock.com

For perfect functioning, every nutrient is needed.

Some athletes believe that taking vitamin or mineral supplements directly before competition will enhance performance. These beliefs are contrary to scientific reality. Most vitamins and minerals function as small parts of larger working units. After absorption into the blood, these nutrients travel to the cells for assembly with their appropriate other parts before they can do their work. This takes time—hours or days. Vitamins or minerals taken right before an event are useless for improving performance, even if the person is actually suffering deficiencies of them.

Nutrients of Concern In general, then, active people who eat well-balanced meals do not need vitamin or mineral supplements. Two nutrients, vitamin E and iron, do merit special mention here, however, each for a different reason. Vitamin E is discussed because so many athletes take vitamin E supplements. Iron is discussed because some athletes may be unaware that they need iron supplements.

Vitamin E During prolonged, high-intensity physical activity, the muscles' consumption of oxygen increases tenfold or more, which increases the production of free radicals in the body.[48] As Highlight 11 explains, vitamin E is a potent antioxidant that vigorously defends cell membranes against the oxidative damage of free radicals.

Does vitamin E supplementation protect against exercise-induced oxidative stress? Some studies find that it does; others show no effect, and still others report enhanced oxidative stress.[49] Recent research may offer some insight into these inconsistencies. Although free radicals are usually damaging, during repeated episodes of endurance activities, they may actually be beneficial. Free radicals activate powerful antioxidant enzymes, which may enhance the athlete's tolerance to oxidative stresses.[50] Researchers speculate that antioxidant supplements such as vitamin E interfere with this protective adaptation. This may explain why, in some studies, athletes taking vitamin E show signs of increased oxidative stress. Clearly, more research is needed on supplements, but in the meantime, active people can benefit by using vitamin E–rich vegetable oils and eating generous servings of antioxidant-rich fruits and vegetables regularly. Beta-carotene and vitamin C in fruits and vegetables, as well as many of the phytochemicals in these foods, are also potent antioxidants.

Iron Deficiency Physically active young women, especially those who engage in endurance activities such as distance running, are prone to iron deficiency.[51] Habitually low intakes of iron-rich foods, high iron losses through menstruation, and the high demands of muscles for the iron-containing electron carriers of the mitochondria and the muscle protein myoglobin can contribute to iron deficiency in physically active young women. In addition, endurance activities temporarily increase the release of hepcidin, which limits iron absorption (see Chapter 13).[52]

Adolescent female athletes who eat vegetarian diets may be particularly vulnerable to iron deficiency. As Chapter 13 explains, the bioavailability of iron is often poor in vegetarian diets.[53] To protect against iron deficiency, vegetarian athletes need to select good dietary sources of iron (fortified cereals, legumes, nuts, and seeds) and include vitamin C–rich foods with each meal. So long as vegetarian athletes, like all athletes, consume enough nutrient-dense foods, they can perform as well as anyone.

Iron-Deficiency Anemia Iron-deficiency anemia impairs physical performance because iron is an essential component of hemoglobin. Without adequate iron, hemoglobin in red blood cells cannot deliver oxygen to the cells for energy metabolism. Without adequate oxygen, an active person cannot perform aerobic activities and tires easily. Iron deficiency without clinical signs of anemia may also impair physical performance.[54]

Sports Anemia Early in training, athletes may develop low blood hemoglobin for a while. This condition, sometimes called **sports anemia**, is not a true iron-deficiency condition. Strenuous aerobic activity promotes destruction of the more fragile, older red blood cells, and the resulting cleanup work reduces the blood's iron content temporarily. Strenuous activity also expands the blood's plasma volume, thereby reducing the red blood cell count per unit of blood. This low hematocrit looks like iron-deficiency anemia, but it is not the same.

sports anemia: a transient condition of low hemoglobin in the blood, associated with the early stages of sports training or other strenuous activity.

In sports anemia, the red blood cells do not diminish in size or number as in anemia, so the oxygen-carrying capacity is not hindered. Most researchers view sports anemia as an *adaptive,* temporary response to endurance training. Iron-deficiency anemia requires iron supplementation, but sports anemia does not.

Iron Recommendations for Athletes The best strategy for maintaining adequate iron nutrition depends on the individual. Menstruating women may border on iron deficiency even without the iron losses incurred by physical activity. Active teens of both genders have high iron needs because they are growing. Especially for women and teens, then, prescribed supplements may be needed to correct iron deficiencies. Physicians use the results of blood tests to determine whether such supplementation is needed. (Review Chapter 13 for many more details about iron, and see Appendix E for a description of the tests used in assessing anemia.)

> **REVIEW IT** List which vitamin and mineral supplements, if any, athletes may need and why.

With the possible exception of iron, well-nourished active people and athletes do not need nutrient supplements. Women and teens may need to pay special attention to their iron needs.

14.4 Fluids and Electrolytes to Support Activity

> **LEARN IT** Identify the factors that influence an athlete's fluid needs and describe the differences between water and sports drinks.

The need for water far surpasses the need for any other nutrient. The body relies on watery fluids as the medium for all of its life-supporting activities, and if it loses too much water, its well-being will be compromised.

Obviously, the body loses water via sweat. Breathing uses water, too, exhaled as vapor. During physical activity, water losses from both routes are significant, and dehydration becomes a threat. Dehydration's first symptom is fatigue: a water loss of greater than 2 percent of body weight can reduce a person's capacity to do muscular work.[55] With a water loss of about 7 percent, a person is likely to collapse. For perspective, a water loss of 2 percent in a person weighing 150 pounds is equal to 3 pounds; a water loss of 7 percent is equal to 10.5 pounds.

Temperature Regulation As Chapter 7 discusses, working muscles produce heat as a by-product of energy metabolism. During intense activity, muscle heat production can be 15 to 20 times greater than at rest. The body cools itself by sweating. Each liter of sweat dissipates almost 600 kcalories of heat, preventing a rise in body temperature of almost 10 degrees on the Celsius scale.* The body routes its blood supply through the capillaries just under the skin, and the skin secretes sweat to evaporate and cool the skin and the underlying blood. The blood then flows back to cool the deeper body chambers.

Hyperthermia In hot, humid weather, sweat doesn't evaporate well because the surrounding air is already laden with water. In **hyperthermia,** body heat builds up and triggers maximum sweating, but without sweat evaporation, little cooling takes place. In such conditions, active people must take precautions to prevent **heat stroke.** To reduce the risk of heat stroke, drink enough fluid before and during the activity, rest in the shade when tired, and wear lightweight clothing that allows sweat to evaporate. (Hence the danger of rubber or heavy suits that supposedly promote weight loss during physical activity—they promote profuse sweating, prevent sweat evaporation, and invite heat stroke.) If you ever experience any of the symptoms of heat stroke listed in Table 14-6 (p. 458) stop your activity, ask for help, and cool down as quickly as possible. Heat stroke can be fatal, people often die of it, and these symptoms demand attention.

Hypothermia In cold weather, **hypothermia,** or low body temperature, can be as serious as heat stroke is in hot weather. Inexperienced, slow runners participating in long races on cold or wet, chilly days are especially vulnerable to

To prevent dehydration and the fatigue that accompanies it, drink liquids before, during, and after physical activity.

hyperthermia: an above-normal body temperature.

heat stroke: a dangerous accumulation of body heat with accompanying loss of body fluid.

hypothermia: a below-normal body temperature.

---

*10 degrees on the Celcius scale is about 18 degrees on the Fahrenheit scale.

TABLE 14-6 Symptoms of Heat Stroke and Hypothermia Compared

Heat Stroke	Hypothermia
• Clumsiness, stumbling	• Clumsiness, loss of coordination
• Confusion, dizziness, other mental changes, loss of consciousness	• Confusion, disorientation, other mental changes, loss of consciousness
• Flushed skin (skin may turn red as body temperature rises)	• Drowsiness
• Muscle cramping (early symptom)	• Shivering (early symptom)
• Nausea and vomiting	• Cessation of shivering (late symptom)
• Rapid breathing	• Slurred speech
• Rapid heart rate	• Slow breathing
• Sudden cessation of sweating (hot, dry, skin)	• Slow heart rate
• Throbbing headache	

SOURCE: Adapted from Centers for Disease Control and Prevention, www.cdc.gov.

hypothermia. Slow runners who produce little heat can become too cold if clothing is inadequate. Early symptoms of hypothermia include feeling cold, shivering, apathy, and social withdrawal.[56] As body temperature continues to fall, shivering may stop, and disorientation, slurred speech, and change in behavior or appearance set in (see Table 14-6). People with these symptoms soon become helpless to protect themselves from further body heat losses. Even in cold weather, however, the active body still sweats and still needs fluids. The fluids should be warm or at room temperature to help protect against hypothermia.

Fluid Replacement via Hydration Endurance athletes can easily lose 1.5 liters or more of fluid during *each hour* of activity. Table 14-7 presents a suggested schedule of hydration for physical activity. To prepare for fluid losses, a person must hydrate before activity; to replace fluid losses, the person must rehydrate during and after activity. Even then, in hot weather, the GI tract may not be able to absorb enough water fast enough to keep up with sweat losses, and some degree of dehydration may be inevitable. Athletes who know their body's **hourly sweat rate** can strive to replace the total amount of fluid lost during activity to prevent dehydration.

Athletes who are preparing for competition are often advised to drink extra fluids in the *days* immediately before the event, especially if they are still training. The extra water is not stored in the body, but drinking extra water ensures maximum hydration at the start of the event. Full hydration is imperative for every athlete both in training and in competition. The athlete who arrives at an event even slightly dehydrated begins with a disadvantage.

TABLE 14-7 Suggested Hydration Schedule for Physical Activity

Timing	Recommended Intakes (based on body weight)	Males	Females
≥4 hours before activity	1 oz/10 lb	16 oz	12 oz
2 hours before activity, if heavy sweating is expected	Add 0.6 oz/10 lb	+9 oz	+7 oz
Every 15 min during activity	Drink enough to minimize loss of body weight, but don't drink too much	8 oz	4 oz
After activity	≥2 cups for each pound of body weight lost[a]		

NOTE: Intakes are based on a 70-kilogram (154-pound) man and a 55-kilogram (121-pound) woman.

[a]Drinking 2 cups of fluid every 20 to 30 minutes after exercise until the total amount required is consumed is more effective for rehydration than drinking the needed amount all at once. Rapid fluid replacement after exercise stimulates urine production and results in less body water retention.

SOURCE: C. A. Rosenbloom and E. J. Coleman, eds., *Sports Nutrition: A Practice Manual for Professionals* (Chicago: Academy of Nutrition and Dietetics, 2012), p. 115; Position of the American Dietetic Association, Dietitians of Canada, and the American College of Sports Medicine: Nutrition and athletic performance, *Journal of the American Dietetic Association* 109 (2009): 509–527; American College of Sports Medicine, Position stand; Exercise and fluid replacement, *Medicine and Science in Sports and Exercise* 39 (2007): 377–390.

hourly sweat rate: the amount of weight lost plus fluid consumed during exercise per hour. One pound equals roughly 2 cups (500 milliliters) of fluid.

What is the best fluid for an exercising body? For noncompetitive, everyday active people, plain, cool water is recommended, especially in warm weather, for two reasons: (1) water rapidly leaves the digestive tract to enter the tissues where it is needed, and (2) it cools the body from the inside out. For endurance athletes, carbohydrate-containing beverages may be appropriate. Fluid ingestion during the event has the dual purposes of replenishing water lost through sweating and providing a source of carbohydrate to supplement the body's limited glycogen stores. Many carbohydrate-containing sports drinks are marketed for active people; a later section compares them with water.

Water is the best fluid for most physically active people, but some consumers prefer the flavors of sports drinks.

Electrolyte Losses and Replacement When a person sweats, small amounts of electrolytes—the electrically charged minerals sodium, potassium, chloride, and magnesium—are lost from the body along with water. Losses are greatest in beginners; training improves electrolyte retention.

To replenish lost electrolytes, a person ordinarily needs only to eat a regular diet that meets energy and nutrient needs. In events lasting more than 1 hour, sports drinks may be needed to replace fluids and electrolytes. Salt tablets can worsen dehydration and impair performance; they increase potassium losses, irritate the stomach, and cause vomiting. Endurance athletes considering the use of salt tablets to replace sodium are advised to consult with a health care professional before doing so.

Hyponatremia When athletes compete in endurance sports lasting longer than 3 hours, replenishing electrolytes is crucial. If athletes sweat profusely over a long period of time and do not replace lost sodium, a dangerous condition known as **hyponatremia** may result. Research shows that some athletes who sweat profusely may also lose more sodium in their sweat than others—and are prone to debilitating heat cramps.[57] These athletes lose twice as much sodium in sweat as athletes who don't cramp. Depending on individual variation, exercise intensity, and changes in ambient temperature and humidity, sweat rates for these athletes can exceed 2 liters per hour.

Hyponatremia may also occur when endurance athletes drink such large amounts of water over the course of a long event that they overhydrate, diluting the body's fluids to such an extent that the sodium concentration becomes extremely low. During long competitions, when athletes lose sodium through heavy sweating *and* consume excessive amounts of liquids, especially water, hyponatremia becomes likely.

Some athletes may still be vulnerable to hyponatremia even when they consume sports drinks during an event. Sports drinks do contain sodium, but as a later section points out, the sodium content of sports drinks is low and, in some cases, too low to replace sweat losses. Still, sports drinks do offer more sodium than plain water.

To prevent hyponatremia, athletes need to replace sodium during prolonged events. Sports drinks, salty pretzels, and other sodium sources can provide sodium during long competitions. Some athletes may need beverages with higher sodium concentrations than commercial sports drinks. In the days before the event, especially an event in the heat, athletes should not restrict salt in their diets. The symptoms of hyponatremia are similar to, but not the same as, those of dehydration (see Table 14-8).

Sports Drinks Hydration is critical to optimal performance. As stated earlier, water best meets the fluid needs of most people, yet manufacturers market many good-tasting sports drinks that deliver both fluid and carbohydrate for active people. The term *sports drink* generally refers to beverages that contain carbohydrates and electrolytes in specific concentrations, and they are the focus of this discussion.

TABLE 14-8 Symptoms of Hyponatremia

- Bloating, puffiness from water retention (shoes tight, rings tight)
- Confusion
- Seizure
- Severe headache
- Vomiting

© 2016 Cengage Learning

hyponatremia (HIGH-poe-na-TREE-mee-ah): a decreased concentration of sodium in the blood.
- **hypo** = below
- **natrium** = sodium (Na)
- **emia** = blood

>How To

Calculate the Carbohydrate Concentration of Sports Drinks

Sports drinks with a carbohydrate concentration of 6 to 8 percent best support athletes participating in events lasting 1 hour or longer. To calculate the carbohydrate concentration of a sports drink, divide the grams of carbohydrates per serving by the serving size (in milliliters) and multiply by 100.

Consider a sports drink that contains 22 grams of carbohydrates per serving and lists the serving size as 360 milliliters (12 ounces):

$$\frac{22 \text{ grams carbohydrate}}{360 \text{ milliliters}} = 0.06$$

$$0.06 \times 100 = 6\% \text{ carbohydrates}$$

> **TRY IT** Calculate the carbohydrate concentration of a sports drink that contains 14 grams of carbohydrate per serving (240 milliliters).

Many sports drinks compete for their share of the multi-billion-dollar market. What do sports drinks have to offer?

Fluid Sports drinks offer fluids to help offset the loss of fluids during physical activity, but plain water can do this too. Alternatively, diluted fruit juices or flavored water can be used if preferred to plain water.

Glucose Sports drinks offer simple sugars or **glucose polymers** that help maintain hydration and blood glucose and enhance performance as effectively as, or in some circumstances, even better than, water. Such measures are especially beneficial for strenuous endurance activities lasting longer than 1 hour, during intense activities, or during prolonged competitive games that demand repeated intermittent activity.[58] Sports drinks may also be used for events lasting less than 1 hour, although plain water is appropriate as well.

Fluid transport to the tissues from beverages containing up to 8 percent glucose is rapid. Most sports drinks contain about 7 percent carbohydrate (about half the sugar of ordinary soft drinks, or about 5 teaspoons in each 12 ounces). Less than 6 percent carbohydrate may not enhance performance, and more than 8 percent may cause abdominal cramps, nausea, and diarrhea. The accompanying "How To" feature describes how to calculate the carbohydrate concentration of sports drinks.

Although glucose can enhance endurance and performance in strenuous competitive events, for the moderate exerciser, it can be counterproductive if weight loss is the goal. Glucose is sugar, and like candy, it provides only empty kcalories—no vitamins or minerals. Most sports drinks provide between 50 and 100 kcalories per 8-ounce cup.

Sodium and Other Electrolytes Sports drinks offer sodium and other electrolytes to help replace those lost during physical activity. Sodium in sports drinks also helps to increase the rate of fluid absorption from the GI tract and maintain plasma volume during activity and recovery. Most physically active people do not need to replace the minerals lost in sweat immediately; a meal eaten within hours of competition replaces these minerals soon enough. Most sports drinks are relatively low in sodium, however, so those who choose to use these beverages run little risk of excessive intake.

Good Taste Manufacturers reason that if a drink tastes good, people will drink more, thereby ensuring adequate hydration. For athletes who prefer the flavors of sports drinks over water, it may be worth paying for good taste to replace lost fluids.

For athletes who exercise for 1 hour or more, sports drinks provide an advantage over water. Sports drinks may also be beneficial for athletes who:

- Exercise on an empty stomach
- Don't eat enough carbohydrate

glucose polymers: compounds that supply glucose, not as single molecules, but linked in chains somewhat like starch. The objective is to attract less water from the body into the digestive tract (osmotic attraction depends on the number, not the size, of particles).

- Want to load carbohydrates
- Want to gain weight
- Train at altitude or in extreme weather
- Had diarrhea (or vomiting) recently
- Refuse to drink adequate amounts of water

For most physically active people, though, water is the best fluid to replenish the body's lost fluids. The most important thing to do is drink—even if you don't feel thirsty.

Enhanced Water Another beverage often marketed to athletes and active people is **enhanced water.** Enhanced waters are lightly flavored waters with added sugars, vitamins, minerals, and in some cases, protein. They are neither natural nor healthy, as marketing ads claim. In fact, most enhanced waters contain small amounts of only a few minerals, some of the B vitamins, and sometimes vitamin C or vitamin E. In the context of daily needs, the vitamins and minerals in these drinks do not add up to much. For example, it takes a quart of most of these beverages to provide only 10 percent of the RDA for iron or calcium. Quite simply, enhanced waters are not a substitute for eating nutrient-rich fruits and vegetables. Enhanced waters may not be harmful (except maybe to the wallet), but most people do not need them. Plain water can meet fluid needs. If the flavor of enhanced waters encourages greater fluid intake, then they may offer some advantage. Serious endurance athletes need the carbohydrate-electrolyte sports drinks discussed earlier.

Poor Beverage Choices: Caffeine and Alcohol
Athletes, like others, sometimes drink beverages that contain caffeine or alcohol. Each of these substances can influence physical performance.

Caffeine Caffeine is a stimulant, and athletes sometimes use it to enhance performance, as Highlight 14 explains. Carbonated soft drinks, with or without caffeine, may not be a wise choice for athletes: bubbles make a person feel full quickly and so limit fluid intake. Some of the increasingly popular beverages, called energy drinks, contain amounts of caffeine equivalent to a cup or two of coffee. When used in excess or in combination with stimulants or other unregulated substances, energy drinks can hinder performance and are potentially dangerous.[59] Another reason energy drinks should not be used for fluid replacement during athletic events is that the carbohydrate concentrations are too high for optimal fluid absorption. The caffeine contents of selected energy drinks are listed in the table at the beginning of Appendix H.

Alcohol Some athletes mistakenly believe that they can replace fluids and load up on carbohydrates by drinking beer. Beer is not carbohydrate-rich. A 12-ounce beer provides 13 grams of carbohydrate—one-third the amount of carbohydrate in a glass of orange juice the same size. In addition to carbohydrate, beer also contains alcohol, of course. Energy from alcohol breakdown generates heat, but it does not fuel muscle work because alcohol is metabolized in the liver.

Beer is not rich in vitamins or minerals. Beer contains traces of some B vitamins, but it cannot compete with food sources. As for minerals, beer contains a few minerals, but to replace the minerals lost in sweat, athletes need good sources such as fruit juices or sports drinks.

Alcohol's diuretic effect impairs the body's fluid balance, making dehydration likely. After physical activity, a person needs to replace fluids, not lose them by drinking beer. Alcohol also impairs the body's ability to regulate its temperature, increasing the likelihood of hypothermia or heat stroke.

It is difficult to overstate alcohol's detrimental effects on physical activity. Alcohol alters perceptions; slows reaction time; reduces strength, power, and endurance; and hinders accuracy, balance, eye-hand coordination, and coordination in general—all opposing optimal athletic performance. In addition, it deprives people of their judgment, thereby compromising safety in sports. Many sports-related fatalities and injuries involve alcohol or other drugs.

enhanced water: water that is fortified with ingredients such as vitamins, minerals, protein, oxygen, or herbs. Enhanced water is marketed as *vitamin water, sports water, oxygenated water,* and *protein water.*

Clearly, alcohol impairs performance. For those who do drink, do not drink alcohol before exercising and drink plenty of water after exercising before drinking alcohol.

> **REVIEW IT** Identify the factors that influence an athlete's fluid needs and describe the differences between water and sports drinks.

Active people need to drink plenty of water; endurance athletes need to drink both water and carbohydrate-containing beverages, especially during training and competition. During events lasting longer than 3 hours, athletes need to pay special attention to replace sodium losses to prevent hyponatremia.

14.5 Diets for Physically Active People

> **LEARN IT** Discuss an appropriate daily eating pattern for athletes and list one example of a recommended pregame and recovery meal.

No one diet best supports physical performance. Active people who choose foods within the framework of the diet-planning principles presented in Chapter 2 can design many excellent diets.

Choosing a Diet to Support Fitness Above all, keep in mind that water is depleted more rapidly than any other nutrient. A diet to support fitness must provide water, energy, and all the other nutrients.[60]

Water Even casual exercisers must attend conscientiously to their fluid needs. Physical activity blunts the thirst mechanism, especially in cold weather. During activity, thirst signals come too late, so don't wait to feel thirsty before drinking. To find out how much water is needed to replenish activity losses, weigh yourself before and after the activity—the difference is almost all water. One pound equals roughly 2 cups (500 milliliters) of fluid.

Nutrient Density A healthful diet is based on nutrient-dense foods—foods that supply adequate vitamins and minerals for the energy they provide. Active people need to eat both for nutrient adequacy and for energy—and energy needs can be extremely high. For example, during training, meals for some Olympic athletes provide as much as 12,000 kcalories a day. Still, a nutrient-rich diet remains central for adequacy's sake. Though vital, energy alone is not enough to support performance. Table 14-9 offers convenient, nutrient-dense snack ideas for athletes and active people.[61]

Carbohydrate A diet that is high in carbohydrate (60 to 70 percent of total kcalories), moderate in fat (20 to 35 percent), and adequate in protein (10 to 20 percent) ensures full glycogen and other nutrient stores. On two occasions, however, the active person's regular high-carbohydrate, fiber-rich diet may require temporary adjustment. Both of these exceptions involve training for competition rather than

---

TABLE 14-9 Nutrient-Dense Snacks for Athletes and Active People

One ounce of almonds provides protein, fiber, calcium, vitamin E, and healthy unsaturated fats. Similar choices include other nuts or trail mix consisting of dried fruit, nuts, and seeds.	Low-fat Greek yogurt contains more protein per serving than regular yogurt, but a little less calcium. A small amount of fresh fruit adds fiber and vitamins. A similar choice is low-fat cheese paired with fresh fruit.	Low-fat milk or chocolate milk together with fig bars or oatmeal-raisin cookies offer protein and fiber. A similar choice is whole-grain cereal with low-fat milk.	Popcorn offers fiber and a fruit smoothie quenches thirst and provides valuable vitamins. A similar choice is pretzels and fruit juice.

for fitness in general. One special occasion is the pregame meal, when fiber-rich, bulky foods are best avoided. The pregame meal is discussed in a later section.

The other occasion is during intensive training, when energy needs may be so high as to outstrip the person's capacity to eat enough food to meet them. In this case, the athlete can add concentrated carbohydrate foods, such as dried fruits, sweet potatoes, and nectars, and even high-fat foods, such as avocados and nuts. Some athletes use commercial high-carbohydrate liquid supplements to obtain the carbohydrate and energy needed for intense training and top performance. These supplements do not *replace* regular food; they are meant to be used in *addition* to it. Unlike the sports drinks discussed earlier, these high-carbohydrate supplement beverages are too concentrated in carbohydrate to be used for fluid replacement.

Protein In addition to carbohydrate and some fat (and the energy they provide), physically active people need protein. Meats and milk products are rich protein sources, but recommending that active people emphasize these foods is narrow advice. As mentioned repeatedly, active people need diets rich in carbohydrate, and of course, meats have none to offer. Legumes, whole grains, and vegetables provide some protein with abundant carbohydrate.

Meals before and after Competition
No single food improves speed, strength, or skill in competitive events, although some *kinds* of foods do support performance better than others, as already explained. Still, a competitor may eat a particular food before or after an event for psychological reasons. One eats a steak the night before wrestling. Another eats a spoonful of honey within 5 minutes of diving. So long as these practices remain harmless, they should be respected.

Pregame Meals Science indicates that the pregame meal or snack should include plenty of fluids and be light and easy to digest. It should provide between 300 and 800 kcalories, primarily from carbohydrate-rich foods that are familiar and well tolerated by the athlete. The meal should end 3 to 4 hours before competition to allow time for the stomach to empty before exertion.

Breads, potatoes, pasta, and fruit juices—that is, carbohydrate-rich foods low in fat and fiber—form the basis of the best pregame meal (see Figure 14-2 on p. 464 for some examples). Bulky, fiber-rich foods such as raw vegetables or high-bran cereals, although usually desirable, are best avoided just before competition. Fiber in the digestive tract attracts water and can cause stomach discomfort during performance.

© Polara Studios Inc.

A variety of foods is the best source of nutrients for athletes.

Liquid meals are easy to digest, and many such meals are commercially available. Alternatively, athletes can mix fat-free milk or fruit juices, frozen fruits, and flavorings in a blender. For example, mixing fat-free milk or apple juice with a frozen banana and cinnamon or vanilla makes a tasty, high-carbohydrate, liquid pregame meal.

Recovery Meals Athletes who perform intense practice sessions several times a day or compete for hours on consecutive days need to quickly replenish glycogen to be ready for the next activity. As mentioned earlier, eating high-carbohydrate foods *after* physical activity enhances glycogen storage. Because people are usually not hungry immediately following physical activity, carbohydrate-containing beverages such as sports drinks or fruit juices may be preferred. If an active person does feel hungry after an event, then foods high in carbohydrate, moderate in protein, and low in fat and fiber are the ones to choose—similar to those recommended prior to competition.

> **REVIEW IT** Discuss an appropriate daily eating pattern for athletes and list one example of a recommended pregame and recovery meal.

The person who wants to excel physically will apply accurate nutrition knowledge along with dedication to rigorous training. A diet that provides ample fluid and includes a variety of nutrient-dense foods in quantities to meet energy needs will enhance not only athletic performance, but overall health as well. Carbohydrate-rich foods that are light and easy to digest are recommended for both the pregame and recovery meal.

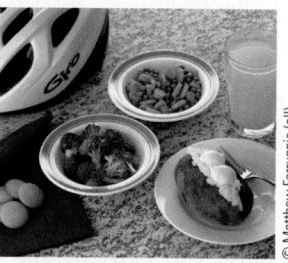

> FIGURE 14-2 **Examples of High-Carbohydrate Pregame Meals**

Pregame meals should be eaten 3 to 4 hours before the event and provide 300 to 800 kcalories, primarily from carbohydrate-rich foods. Each of these sample meals provides at least 65 percent of total kcalories from carbohydrate.

300-kcalorie meal
1 large apple
4 saltine crackers
1½ tbs reduced-fat peanut butter

500-kcalorie meal
1 large whole-wheat bagel
2 tbs jelly
1½ c low-fat milk

750-kcalorie meal
1 large baked potato
2 tsp margarine
1 c steamed broccoli
1 c mixed carrots and green peas
5 vanilla wafers
1½ c apple or pineapple juice

© Matthew Farruggio (all)

Training and genetics being equal, who will win a competition? The athlete who habitually consumes inadequate amounts of needed nutrients will lag behind the competitor who arrives at the event with a long history of full nutrient stores and well-met metabolic needs. Some athletes learn that nutrition can support physical performance and turn to pills and powders instead of foods. In case you need further convincing that a healthful diet surpasses such potions, the following highlight addresses this issue.

Nutrition Portfolio

The foods and beverages you eat and drink provide fuel and other nutrients to support your physical activity. Go to Diet & Wellness Plus and choose one of the days on which you tracked your diet and activity for an entire day. Select the Activities Spreadsheet report to help you answer the following.

- Describe your daily physical activities and how they compare with recommendations to be physically active for at least 30 minutes, and preferably 60 minutes, a day on most or all days of the week.

Now click on the Intake vs. Goals report and consider the following:

- Estimate your daily fluid intake, making note of whether you drink fluids, especially water, before, during, and after physical activity.

- Evaluate the carbohydrate contents of your diet and consider whether it would meet the needs of a physically active person.

To complete this exercise, go to your Diet & Wellness Plus at **www.cengagebrain.com**.

> **STUDY IT** To review the key points of this chapter and take a practice quiz, go to the study cards at the end of the book.

REFERENCES

1. P. T. Katzmarzyk and I. M. Lee, Sedentary behaviour and life expectancy in the USA: A cause-deleted life table analysis, *BMJ Open* 2 (2012): doi: 10.1136; A. Grøntved and F. B. Hu, Television viewing and risk of type 2 diabetes, cardiovascular disease, and all-cause mortality, *Journal of the American Medical Association* 305 (2011) 2448–2455; T. Y. Warren and coauthors, Sedentary behaviors increase risk of cardiovascular disease mortality in men, *Medicine and Science in Sports and Exercise* 42 (2010): 879–885; A. V. Patel and coauthors, Leisure time spent sitting in relation to total mortality in a prospective cohort of adults, *American Journal of Epidemiology* 172 (2010): 419–429.

2. Centers for Disease Control and Prevention, www.cdc.gov/physicalactivity/everyone/health; updated February 16, 2011; US Department of Agriculture and US Department of Health and Human Services, *Dietary Guidelines for Americans*, 2010, www.dietaryguidelines.gov.

3. E. H. Bell and coauthors, Physical activity and cardiovascular disease in African Americans in Atherosclerosis Risk in Communities, *Medicine and Science in Sports and Exercise* 45 (2013): 901–907; C. R. Mikus and coauthors, Lowering physical activity impairs glycemic control in healthy volunteers, *Medicine and Science in Sports and Exercise* 44 (2012): 225–231; J. Sattelmair and coauthors, Dose response between physical activity and risk of coronary heart disease: A meta-analysis, *Circulation* 124 (2011): 789–795; Centers for Disease Control and Prevention, 2011; M. Hamer and coauthors, Physical activity and cardiovascular mortality risk: Possible protective mechanisms, *Medicine and Science in Sports and Exercise* 44 (2012): 84–88.

4. C. D. Harris and coauthors, Adult participation in aerobic and muscle-strengthening physical activities—United States, 2011, *Morbidity and Mortality Weekly Report* 62 (2013): 326–330.

5. Katzmarzyk and Lee, 2012; Hamer and coauthors, 2012; H. M. Ahmed and coauthors, Effects of physical activity on cardiovascular disease, *American Journal of Cardiology* 109 (2012): 288–295; T. Boyle and coauthors, Timing and intensity of recreational physical activity and the risk of subsite-specific colorectal cancer, *Cancer Causes and Control* 22 (2011): 1647–1658; Warren and coauthors, 2010; Patel and coauthors, 2010.

6. Centers for Disease Control and Prevention, 2011.

7. American College of Sports Medicine, Position paper: Appropriate physical activity intervention strategies for weight loss and prevention of weight regain for adults, *Medicine and Science in Sports and Exercise* 41 (2009): 459–471.

8. M. T. Korhonen and coauthors, Bone density, structure and strength, and their determinants in aging sprint athletes, *Medicine and Science in Sports and Exercise* 44 (2012): 2340–2349.

9. J. Romeo and coauthors, Physical activity, immunity and infection, *Proceedings of the Nutrition Society* 69 (2010): 390–399.

10. C. Eheman and coauthors, Annual Report to the Nation on the status of cancer, 1975–2008, featuring cancers associated with excess weight and lack of sufficient physical activity, *Cancer* 118 (2012): 2338–2366; X. Sui and coauthors, Influence of cardiorespiratory fitness on lung cancer mortality, *Medicine and Science in Sports and Exercise* 42 (2010): 872–878; J. B. Peel and coauthors, A prospective study of cardiorespiratory fitness and breast cancer mortality, *Medicine and Science in Sports and Exercise* 41 (2009): 742–748; S. Y. Pan and M. DesMeules, Energy intake, physical activity energy balance, and cancer: Epidemiologic evidence, *Methods in Molecular Biology* 472 (2009): 191–215; World Cancer Research Fund/American Institute for Cancer Research, *Food, Nutrition, Physical Activity, and the Prevention of Cancer: A Global Perspective* (Washington, D.C.: AICR, 2007), pp. 244–321.

11. Hamer and coauthors, 2012; N. T. Artinian and coauthors, Interventions to promote physical activity and dietary lifestyle changes for cardiovascular risk factor reduction in adults: A scientific statement from the American Heart Association, *Circulation* 122 (2010): 406–441.

12. U. Ekelund and coauthors, Physical activity and gain in abdominal adiposity and body weight: Prospective cohort study in 288,498 men and women, *American Journal of Clinical Nutrition* 93 (2011): 826–835; B. H. Goodpaster and coauthors, Effects of diet and physical activity interventions on weight loss and cardiometabolic risk factors in severely obese adults, *Journal of the American Medical Association* 304 (2010): 1795–1802.

13. C. K. Roberts, J. P. Little, and J. P. Thyfault, Modification of insulin sensitivity and glycemic control by activity and exercise, *Medicine and Science in Sports and Exercise* 45 (2013): 1868–1877; M. C. Audelin and coauthors, Change of energy expenditure from physical activity is the most powerful determinant of improved insulin sensitivity in overweight patients with coronary artery disease participating in an intensive lifestyle modification program, *Metabolism* 61 (2012): 672–679; C. R. Mikus and coauthors, Lowering physical activity impairs glycemic control in healthy volunteers, *Medicine and Science in Sports and Exercise* 44 (2012): 225–231; J. Ralph and coauthors, Low-intensity exercise reduces the prevalence of hyperglycemia in type 2 diabetes, *Medicine and Science in Sports and Exercise* 42 (2010): 219–225.

14. P. J. Banim and coauthors, Physical activity reduces the risk of symptomatic gallstones: A prospective cohort study, *European Journal of Gastroenterology and Hepatology* 22 (2010): 983–988.

15. J. C. Sieverdes and coauthors, Association between leisure-time physical activity and depressive symptoms in men, *Medicine and Science in Sports and Exercise* 44 (2012): 260–265; M. O. Melancon, D. Lorrain, and I. J. Dionne, Exercise increases tryptophan availability to the brain in older men age 57–70 years, *Medicine and Science in Sports and Exercise* 44 (2012): 881–887.

16. L. M. León-Muñoz and coauthors, Continued sedentariness, change in sitting time and mortality in older adults, *Medicine and Science in Sports and Exercise* 45 (2013): 1501–1507; S. C. Moore and coauthors, Leisure time physical activity of moderate to vigorous intensity and mortality: A large pooled cohort analysis, *PLOS Medicine* 9 (2012): e1001335; B. L. Willis and coauthors, Midlife fitness and the development of chronic conditions in later life, *Archives of Internal Medicine* 172 (2012): 1333–1340; R. Liu and coauthors, Cardiorespiratory fitness as a predictor of dementia mortality in men and women, *Medicine and Science in Sports and Exercise* 44 (2012): 253–259.

17. C. P. Wen and coauthors, Minimum amount of physical activity for reduced mortality and extended life expectancy: A prospective cohort study, *The Lancet* 378 (2011): 1244–1253.

18. M. D. Graminer and coauthors, Task-specific training reduces trip-related fall risk in women, *Medicine and Science in Sports and Exercise* 44 (2012): 2410–2414; American College of Sports Medicine position stand: Quantity and quality of exercise for developing and maintaining cardiorespiratory, musculoskeletal, and neuromotor fitness in apparently healthy adults—Guidelines for prescribing exercise, *Medicine and Science in Sports and Exercise* 43 (2011): 1334–1359.

19. Centers for Disease Control and Prevention, 2011.

20. American College of Sports Medicine position stand, Quantity and quality of exercise for developing and maintaining cardiorespiratory, musculoskeletal, and neuromotor fitness in apparently healthy adults: Guidance for prescribing exercise, *Medicine and Science in Sports and Exercise* 43 (2011): 1334–1359.

21. American College of Sports Medicine position stand, 2011.

22. Centers for Disease Control and Prevention, 2011; American College of Sports Medicine position stand, 2011.

23. P. J. Atherton and K. Smith, Muscle protein synthesis in response to nutrition and exercise, *Journal of Physiology* 590.5 (2012): 1049–1047.

24. B. E. Phillips, D. S. Hill, and P. J. Atherton, Regulation of muscle protein synthesis in humans, *Current Opinion in Clinical Nutrition and Metabolic Care* 15 (2012): 58–63.

25. Atherton and Smith, 2012.

26. American College of Sports Medicine, *ACSM's Guidelines for Exercise Testing and Prescription*, 8th ed. (Philadelphia, Pa.: Lippincott, Williams, & Wilkins, 2010), pp. 18–39.

27. L. A. Bateman and coauthors, Comparison of aerobic versus resistance exercise training effects on metabolic syndrome (from the studies of a targeted risk reduction intervention through defined exercise—STRRIDE-AT/RT), *American Journal of Cardiology* 108 (2011): 838–844; S. Marwood and coauthors, Faster pulmonary oxygen uptake kinetics in trained versus untrained male adolescents, *Medicine and Science in Sports and Exercise* 42 (2010): 127–134.

28. D. M. Bhammar, S. S. Angadi, and G. A. Gaesser, Effects of fractionized and continuous exercise on 24-h ambulatory blood pressure, *Medicine and Science in Sports and Exercise* 44 (2012): 2270–2276; S. Liu and coauthors, Blood pressure responses to acute and chronic exercise are related in prehypertension, *Medicine and Science in Sports and Exercise* 44 (2012): 1644–1652; D. I. Musa and C. A. Williams, Cardiorespiratory fitness, fatness, and blood pressure associations in Nigerian youth, *Medicine and Science in Sports and Exercise* 44 (2012): 1978–1985; E. G. Ciolac and coauthors, Effects of high-intensity aerobic interval training vs. moderate exercise on hemodynamic, metabolic and neuro-humoral abnormalities of young normotensive women at high familial risk for hypertension, *Hypertension Research* 33 (2010): 836–843.

29. American College of Sports Medicine position stand, 2011; American College of Sports Medicine position stand: Progression models in resistance training for healthy adults, *Medicine and Science in Sports and Exercise* 41 (2009): 687–708.

30. American College of Sports Medicine position stand, 2011.

31. E. A. Marques and coauthors, Response of bone mineral density, inflammatory cytokines, and biochemical bone markers to a 32-week combined loading exercise programme in older men and women, *Archives of Gerontology and Geriatrics* 57 (2013): 226–233; T. E. Howe and coauthors, Exercise for preventing and treating osteoporosis in postmenopausal women, *Cochrane Database of Systematic Reviews* 7 (2011): CD000333.

32. L. Krist, F. Dimeo, and T. Keil, Can progressive resistance training twice a week improve mobility, muscle strength, and quality of life in very elderly nursing-home residents with impaired mobility? A pilot study, *Clinical Interventions in Aging* 8 (2013): 443–448; L. Clemson and coauthors, Integration of balance and strength training into daily life activity to reduce rate of falls in older people (the LiFE study): Randomized parallel trial, *British Medical Journal* 7 (2012): e4547.

33. American College of Sports Medicine position stand, Progression models in resistance training for healthy adults, *Medicine and Science in Sports and Exercise* 41 (2009): 687–708.

34. J. Bergstrom and coauthors, Diet, muscle glycogen and physical performance, *Acta Physiologica Scandanavica* 71 (1967): 140–150.

35. J. F. Moxnes and Ø. Sandbakk, The kinetics of lactate production and removal during whole-body exercise, *Theoretical Biology and Medical Modeling* 9 (2012): 7.

36. J. Temesi and coauthors, Carbohydrate ingestion during endurance exercise improves performance in adults, *Journal of Nutrition* 141 (2011): 890–897; L. M. Burke, Fueling strategies to optimize performance: Training high or training low? *Scandinavian Journal of Medicine and Science in Sports* 20 (2010): 48–58.

37. C. Rosenbloom, Food and fluid guidelines before, during, and after exercise, *Nutrition Today* 47 (2012): 63–69; Position of the American Dietetic Association, Dietitians of Canada, and the American College of Sports Medicine: Nutrition and Athletic Performance, *Journal of the American Dietetic Association* 109 (2009): 509–527.

38. Rosenbloom, 2012.

39. L. Breen, K. D. Tipton, and A. E. Jeukendrup, No effect of carbohydrate-protein on cycling performance and indices of recovery, *Medicine and Science in Sports and Exercise* 42 (2010): 1140–1148.

40. Position of the American Dietetic Association, Dietitians of Canada, and the American College of Sports Medicine: Nutrition and athletic performance, *Journal of the American Dietetic Association* 109 (2009): 509–527.

41. C. M. Perry, T. L. Perry, and M.C. Rose, Glycemic index and endurance performance, *International Journal of Sport Nutrition and Exercise Metabolism* 20 (2010): 154–165.

42. Position of the American Dietetic Association, Dietitians of Canada, and the American College of Sports Medicine, 2009.

43. W. R. Lunn and coauthors, Chocolate milk and endurance exercise recovery: Protein balance, glycogen, and performance, *Medicine and Science in Sports and Exercise* 44 (2012): 682–691; A. R. Josse and coauthors, Body composition and strength changes in women with milk and resistance exercise, *Medicine and Science in Sports and Exercise* 42 (2010): 1122–1130.

44. Atherton and Smith, 2012.

45. Position of the American Dietetic Association, Dietitians of Canada, and the American College of Sports Medicine, 2009; Committee on Dietary Reference Intakes, *Dietary Reference Intakes for Energy, Carbohydrate, Fiber, Fat, Fatty Acids, Cholesterol, Protein, and Amino Acids* (Washington, D.C.: National Academies Press, 2002), pp. 660–661.

46. Position of the American Dietetic Association, Dietitians of Canada, and the American College of Sports Medicine, 2009.

47. Position of the American Dietetic Association: Nutrition intervention in the treatment of eating disorders, *Journal of the American Dietetic Association* 111 (2011): 1236–1241; Position of the American Dietetic Association, Dietitians of Canada, and the American College of Sports Medicine, 2009.

48. J. E. Turner and coauthors, Prolonged depletion of antioxidant capacity after ultraendurance exercise, *Medicine and Science in Sports and Exercise* 43 (2011): 1770–1776; S. K. Powers, E. E. Talbert, and P. J. Adhihetty, Reactive oxygen and nitrogen species as intracellular signals in skeletal muscle, *Journal of Physiology* 589 (2011): 2129–2138.

49. T. T. Peternelj and J. S. Coombes, Antioxidant supplementation during exercise training: Beneficial or detrimental? *Sports Medicine* 41 (2011): 1043–1069; N. A. Strobel and coauthors, Antioxidant supplementation reduces skeletal muscle mitochondrial biogenesis, *Medicine and Science in Sports and Exercise* 43 (2011): 1017–1024.

50. T. M. Farney and coauthors, Absence of blood oxidative stress in trained men after strenuous exercise, *Medicine and Science in Sports and Exercise* 44 (2012): 1855–1863; Peternelj and Coombes, 2011; Strobel and coauthors, 2011.

51. D. M. Dellavalle and J. D. Haas, Iron status is associated with endurance performance and training in female rowers, *Medicine and Science in Sports and Exercise* 44 (2012): 1552–1559.

52. M. K. Newlin and coauthors, The effects of acute exercise bouts on hepcidin in women, *International Journal of Sport Nutrition and Exercise Metabolism* 22 (2012): 79–88.

53. Position of the American Dietetic Association, Dietitians of Canada, and the American College of Sports Medicine, 2009.

54. Dellavalle and Haas, 2012; D. M. DellaValle and J. D. Haas, Impact of iron depletion without anemia on performance in trained endurance athletes at the beginning of a training session: A study of female collegiate rowers, *International Journal of Sport Nutrition and Exercise Metabolism* 21 (2011): 501–506.

55. Position of the American Dietetic Association, Dietitians of Canada, and the American College of Sports Medicine, 2009; American College of Sports Medicine position stand: Exercise and fluid replacement, 2007; Committee on Dietary Reference Intakes, *Dietary Reference Intakes for Water, Potassium, Sodium, Chloride, and Sulfate* (Washington, D.C.: National Academies Press, 2005), pp. 108–110.

56. American College of Sports Medicine position stand: Prevention of cold injuries during exercise, *Medicine and Science in Sports and Exercise* 38 (2006): 2012–2029.

57. American College of Sports Medicine position stand: Exercise and fluid replacement, 2007.

58. American College of Sports Medicine position stand: Exercise and fluid replacement, 2007.

59. Position of the American Dietetic Association, Dietitians of Canada, and the American College of Sports Medicine, 2009.

60. S. A. Herring and coauthors, Selected issues for nutrition and the athlete: A Team Physician Consensus Statement, *Medicine and Science in Sports and Exercise* 45 (2013): 2378–2386.

61. C. Rosenbloom, Snack attack: Evaluating and rating snacks for athletes, *Nutrition Today* 46 (2011): 106–115.

HIGHLIGHT > 14
Supplements as Ergogenic Aids

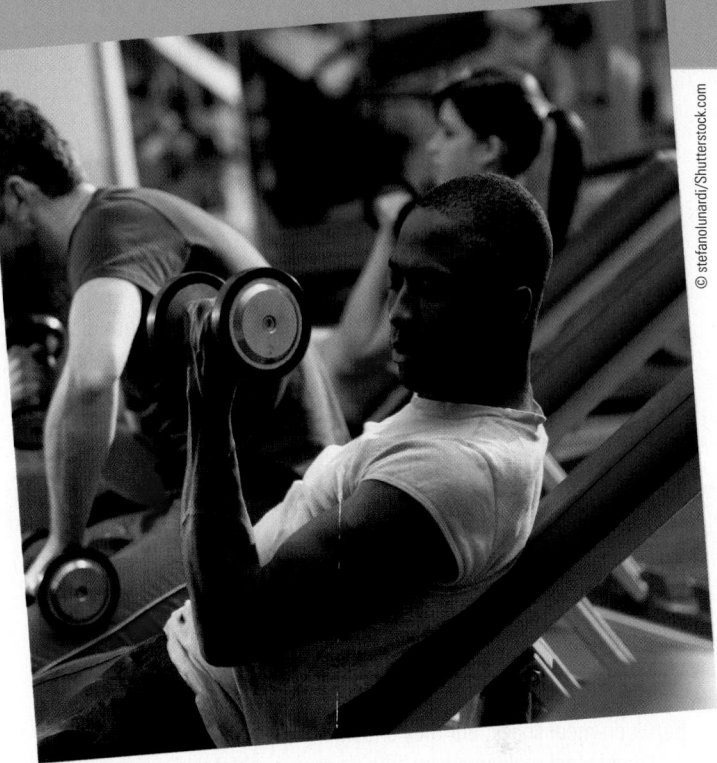

> **LEARN IT** Present arguments for and against the use of ergogenic aids.

Athletes gravitate to promises that they can enhance their performance by taking pills, powders, or potions. Unfortunately, they often hear such promises from their coaches and peers, who advise them to use dietary supplements, take drugs, or follow procedures that claim to deliver results with little effort.[1] When such performance-enhancing aids are harmless, they are only a waste of money; when they impair performance or harm health, they waste athletic potential and cost lives. This highlight looks at scientific evidence for and against dietary supplements and hormonal preparations available to athletes.

Many substances or treatments claim to be *ergogenic*, meaning work enhancing. In connection with athletic performance, **ergogenic aids** are substances or treatments that purportedly improve athletic performance above and beyond what is possible through training. For practical purposes, most ergogenic aids can be categorized as follows:[2]

- Those that perform as claimed
- Those that may perform as claimed but for which there is insufficient evidence at this time
- Those that do not perform as claimed
- Those that are dangerous, banned, or illegal, and therefore should not be used

The accompanying glossary defines several of the commonly used ergogenic aids discussed in this highlight.

For the vast majority of ergogenic aids, research findings do not support the claims made.[3] Athletes who hear that a product is ergogenic should ask who is making the claim and who will profit from the sale. Chapter 6 includes a discussion on protein powders and amino acid supplements (pp. 190–191).

Athletes who supplement their diets with products promoted to improve athletic performance should be aware that some supplements are contaminated with illegal substances such as steroids or stimulants that are not listed on the label.[4] Supplements contaminated with illegal substances pose health risks to those who use them as well as the risk of positive drug testing for athletes subject to such tests.

GLOSSARY

anabolic steroids: drugs related to the male sex hormone, testosterone, that stimulate the development of lean body mass.

- **anabolic** = promoting growth
- **sterols** = compounds chemically related to cholesterol

androstenedione: *See DHEA.*

beta-alanine: a nonessential amino acid that is the rate-limiting precursor for the synthesis of the dipeptide carnosine. Carnosine acts primarily as a buffer in skeletal muscle. Beta-alanine supplements raise carnosine concentrations, which enhance the muscles' buffering capacity.

beta-hydroxymethylbutyrate (HMB): a metabolite of the amino acid leucine promoted to increase muscle mass and strength.

caffeine: a natural stimulant found in many common foods and beverages, including coffee, tea, and chocolate; may enhance endurance by stimulating fatty acid release. High doses cause headaches, trembling, rapid heart rate, and other undesirable side effects.

carnitine: a nonessential nonprotein amino acid made in the body from lysine that helps transport fatty acids across the mitochondrial membrane. As a supplement, carnitine supposedly "burns" fat and spares glycogen during endurance events, but in reality it does neither.

chromium picolinate (CROW-mee-um pick-oh-LYN-ate): a trace mineral supplement; falsely promoted as building muscle, enhancing energy, and burning fat. *Picolinate* is a derivative of the amino acid tryptophan that seems to enhance chromium absorption.

convenient dietary supplements: liquid meal replacers, energy drinks, energy bars, and energy gels that athletes and active people use to replenish energy and nutrients when time is limited.

creatine (KREE-ah-tin): a nitrogen-containing compound that combines with phosphate to form the high-energy compound creatine phosphate (or phosphocreatine) in muscles.

DHEA (dehydroepiandrosterone) and **androstenedione:** hormones made in the adrenal glands that serve as precursors to the male sex hormone, testosterone; falsely promoted as burning fat, building muscle, and slowing aging.

ergogenic (ER-go-JEN-ick) **aids:** substances or techniques used in an attempt to enhance physical performance.

- **ergo** = work
- **genic** = gives rise to

hGH (human growth hormone): a hormone produced by the brain's pituitary gland that regulates normal growth and development; also called *somatotropin*.

ribose: a naturally occurring 5-carbon sugar needed for the synthesis of ATP.

sodium bicarbonate (baking soda): a white crystalline powder that is used to buffer acid that accumulates in the muscles and blood during high-intensity exercise.

Dietary Supplements that Perform as Claimed

Among the extensive array of dietary supplements and other ergogenic aids that athletes use, a few seem to live up to the claims made for them, based on research so far. Convenient dietary supplements, caffeine, creatine, sodium bicarbonate, and beta-alanine are the examples discussed here.

Convenient Dietary Supplements

Ready-to-drink supplements such as liquid meal replacers and energy drinks, energy bars, and energy gels are **convenient dietary supplements** for athletes and active people, especially when time is limited. Many such products appeal to athletes by claiming to provide "complete" nutrition. These supplements usually taste good and provide extra carbohydrate and food energy, but they fall short of providing "complete" nutrition. They can be useful as a pregame meal or a between-meal snack, but they should not replace regular meals.

Liquid meal replacers may help a nervous athlete who cannot tolerate solid food on the day of an event. A liquid meal 2 to 3 hours before competition can supply some of the fluid and carbohydrate needed in a pregame meal, but a shake of fat-free milk or juice (such as apple or papaya) and frozen fat-free yogurt or frozen fruit (such as strawberries or bananas) can do the same thing less expensively.

Caffeine

Some research supports the use of **caffeine** to enhance endurance and, to some extent, to enhance short-term, high-intensity exercise performance.[5] Caffeine may stimulate fatty acid release during endurance activity, but in contrast to what was previously thought, caffeine does not slow muscle glycogen use.[6] Light activity before a workout also stimulates fat release, and unlike caffeine, also warms the muscles and connective tissues, making them flexible and resistant to injury.

Caffeine is a stimulant that elicits a number of physiological and psychological effects in the body. Caffeine may enhance alertness and reduce the perception of fatigue.[7] The possible benefits of caffeine use must be weighed against its adverse effects—stomach upset, nervousness, irritability, headaches, and diarrhea. Caffeine-containing beverages should be used in moderation, if at all, and *in addition* to other fluids, not as a substitute for them.

Caffeine is a *restricted* substance by the National Collegiate Athletic Association, which allows urine concentrations of 15 milligrams per liter or less (equivalent to about 5 cups of coffee consumed within a few hours before testing). Urine tests that detect more caffeine than this disqualify athletes from competition. (The table at the start of Appendix H provides a list of common caffeine-containing items and the doses they deliver.)

Creatine

Interest in—and use of—**creatine** supplements to enhance performance during intense activity has grown dramatically in the last few years. Power athletes such as weightlifters use creatine supplements to enhance stores of the high-energy compound creatine phosphate (CP) in muscles. Theoretically, the more creatine phosphate in muscles, the higher the intensity at which an athlete can train. High-intensity training stimulates the muscles to adapt, which in turn, improves performance.

Research suggests that creatine supplementation does enhance performance of short-term, repetitive, high-intensity activity such as weightlifting or sprinting.[8] Creatine may improve performance by increasing muscle strength and size, cell hydration, or glycogen loading capacity. In contrast, creatine supplementation has not been shown to benefit endurance activity.

The question of whether short-term use (up to a year) of creatine supplements (up to 5 grams per day) is safe continues to be studied, but so far, the supplements are considered safe for healthy adults.[9] More research is needed, however, to confirm the safety of larger doses and long-term use. Creatine supplementation may pose risks to athletes with kidney disease or other conditions.[10] One side effect of creatine supplementation that no one disputes is weight gain. For some athletes, weight gain, especially muscle gain, is beneficial, but for others, it is not.

Some medical and fitness experts voice concern that, like many performance enhancement supplements before it, creatine is being taken in huge doses (up to 30 *grams* per day) before evidence of its value has been ascertained. Even people who eat red meat, which is a creatine-rich food, do not consume nearly the amount supplements provide. (Creatine content varies, but on average, pork, chicken, and beef provide 65 to 180 *milligrams* per ounce.) Despite the uncertainties, creatine supplements are not illegal in international competition. The American Academy of Pediatrics strongly discourages the use of creatine supplements, as well as the use of any performance-enhancing substance in adolescents younger than 18 years old.

Sodium Bicarbonate

During short-term, high-intensity activity, acid and carbon dioxide accumulate in the blood and muscles. **Sodium bicarbonate** (0.3 gram per kilogram of body weight) ingested prior to high-intensity sports performance buffers the acid and neutralizes the carbon dioxide, thereby maintaining muscle pH levels closer to normal and enhancing exercise capacity.[11] Sodium bicarbonate supplementation may cause unpleasant side effects such as diarrhea in some athletes.

Beta-Alanine

Beta-alanine is a nonessential amino acid that has recently received much attention from exercise researchers.[12] Beta-alanine is the rate-limiting precursor for the synthesis of the dipeptide carnosine. Carnosine occurs in high concentrations in skeletal muscle and is one of the primary buffering substances available in muscles. Supplementation with beta-alanine has been shown to raise the concentration of muscle carnosine, which enhances the muscles' buffering capacity.[13] This enhanced buffering is especially beneficial to high-intensity exercise performance such as sprinting. The only known adverse side effect of beta-alanine supplementation is a "pins and needles" sensation that occurs with high (more than 800 milligrams) single doses. Additional research is needed to confirm the safety of beta-alanine supplementation.[14]

Dietary Supplements that May Perform as Claimed

For some dietary supplements, it is just too early to tell whether they deliver on the promises made for them because research thus far is inconclusive. Examples of supplements that may perform as claimed, but for which there is insufficient evidence of efficacy, include beta-hydroxymethylbutyrate and ribose.

Beta-hydroxymethylbutyrate

Beta-hydroxymethylbutyrate (HMB) is a metabolite of the amino acid leucine. Supplementing the diet with HMB during training has been shown to increase muscle mass and strength, especially among untrained individuals.[15] Additional research is needed, however, to determine whether HMB supplementation in trained athletes enhances training adaptations.

Ribose

Manufacturers of ribose supplements claim that **ribose** increases ATP and improves performance. Ribose, a naturally occurring sugar, does help resynthesize muscle ATP. Based on research thus far, however, ribose supplements do not improve athletic performance.[16]

Dietary Supplements that Do Not Perform as Claimed

Most of the dietary supplements promoted as ergogenic aids fall into the category of "those that do not perform as claimed." Carnitine and chromium picolinate are two examples of ineffective supplements discussed here, but others include bee pollen, boron, coenzyme Q, ginseng, pyruvate, and vanadium.[17]

Carnitine

In the body, **carnitine** facilitates the transfer of fatty acids across the mitochondrial membrane. Supplement manufacturers suggest that with more carnitine available, fat oxidation will be enhanced, but this does not seem to be the case. Carnitine supplements neither raise muscle carnitine concentrations nor enhance exercise performance. Milk and meat products are good sources of carnitine, and supplements are not needed.

Chromium Picolinate

Chapter 13 introduces chromium as an essential trace mineral involved in carbohydrate and lipid metabolism. Claims that **chromium picolinate,** which is more easily absorbed than chromium alone, builds muscle, enhances energy, and burns fat derive from one or two early studies reporting that men who strength-trained while taking chromium picolinate supplements increased lean body mass and reduced body fat. Most subsequent studies, however, show no effects of chromium picolinate on strength, lean body mass, or body fat. Chromium-rich foods include whole grains, liver, and nuts.

Dangerous, Banned, or Illegal Supplements

The dietary supplements discussed thus far may or may not help athletic performance, but in the doses commonly taken, they seem to cause little harm. The remaining discussion features hormonal preparations that are clearly damaging: anabolic steroids, DHEA (dehydroepiandrosterone), androstenedione, and hGH (human growth hormone). All of these ergogenic aids are dangerous to use and are banned by most professional sports leagues and the World Anti-Doping Agency (WADA) established by the International Olympic Committee. The American Academy of Pediatrics and the American College of Sports Medicine also condemn athletes' use of these substances.

Anabolic Steroids

Among the most dangerous and illegal ergogenic practices is the taking of **anabolic steroids.** These drugs are derived from the male sex hormone testosterone, which promotes the development of male characteristics and lean body mass. Athletes who take steroids do so to stimulate muscle bulking.

The known toxic side effects of steroids include, but are not limited to, extreme aggression and hostility, heart disease, and liver damage. Taking these drugs is a form of cheating. Other athletes are put in the difficult position of either conceding an unfair advantage to competitors who use steroids or taking steroids and accepting the risk of harmful side effects. Athletes, especially young athletes, should not be forced to make such a choice.

The price for the potential competitive edge that steroids confer is high—sometimes it is life itself. Steroids are not simple pills that build bigger muscles. They are complex chemicals to which the body reacts in many ways, particularly when bodybuilders and other athletes take large amounts. The safest, most effective way to build muscle has always been through hard training and a sound diet, and—despite popular misconceptions—it still is.

Some manufacturers peddle specific herbs as legal substitutes for steroid drugs. They falsely claim that these herbs contain hormones, enhance the body's hormonal activity, or both. In some cases, an herb may contain plant sterols, such as gamma-oryzanol, but these compounds are poorly absorbed. Even if absorption occurs, the body cannot convert herbal compounds to anabolic steroids. None of these products has any proven anabolic steroid activity, none enhances muscle strength, and some contain natural toxins. In short, "natural" does not mean "harmless."

DHEA and Androstenedione

Some athletes use **DHEA** and **androstenedione** as alternatives to anabolic steroids. DHEA (dehydroepiandrosterone) and androstenedione are hormones made in the adrenal glands that serve as precursors to the male hormone testosterone. Advertisements claim the hormones "burn fat," "build muscle," and "slow aging," but evidence to support such claims is lacking.

Short-term side effects of DHEA and androstenedione may include oily skin, acne, body hair growth, liver enlargement, testicular shrinkage, and aggressive behavior. Long-term effects such as serious liver damage may take years to become evident. The potential for harm from DHEA and androstenedione supplements is great, and athletes, as well as others, should avoid them.

Human Growth Hormone

A wide range of athletes, including weightlifters, baseball players, cyclists, and track and field participants use **hGH (human growth hormone)** to build lean tissue and improve athletic performance. The athletes use hGH, believing the injectable hormone will provide the benefits of anabolic steroids without the dangerous side effects.

Taken in large quantities, hGH causes the disease acromegaly, in which the body becomes huge and the organs and bones overenlarge. Other effects include diabetes, thyroid disorder, heart disease, menstrual irregularities, diminished sexual desire, and shortened life span.

Sometimes it is difficult to distinguish valid claims from bogus ones. Fitness magazines and Internet websites are particularly troublesome

because many of them present both valid and invalid nutrition information along with slick advertisements for nutrition products. Advertisements often feature colorful anatomical figures, graphs, and tables that appear scientific. Some ads even include references, citing or linking to such credible sources as the *American Journal of Clinical Nutrition* and the *Journal of the American Medical Association.* These ads create the illusion of endorsement and credibility to gain readers' trust. Keep in mind, however, that the ads are created not to teach, but to sell. A careful reading of the cited research might reveal that the ads have presented the research findings out of context. For example, an ad might use a research article to conclude that its human growth hormone supplement "increases lean body mass and bone mineral," when in fact, the researchers would conclude that "its general use now or in the immediate future is not justified." Scientific facts are often exaggerated and twisted to promote sales. Highlight 1 describes ways to recognize misinformation and quackery.

The search for a single food, nutrient, drug, or technique that will safely and effectively enhance athletic performance will no doubt continue as long as people strive to achieve excellence in sports. When athletic performance does improve after use of an ergogenic aid, the improvement can often be attributed to the placebo effect, which is strongly at work in athletes. Even if a reliable source reports a performance boost from a newly tried product, give the effect time to fade away. Chances are excellent that it simply reflects the power of the mind over the body.

The overwhelming majority of performance-enhancing aids sold for athletes are frauds. Wishful thinking will not substitute for talent, hard training, adequate diet, and mental preparedness in competition. But don't discount the power of mind over body for a minute—it is formidable, and sports psychologists dedicate their work to harnessing it. You can use it by imagining yourself a winner and visualizing yourself excelling in your sport. You don't have to buy magic to obtain a winning edge; you already possess it—your physically fit mind and body.

CRITICAL THINKING QUESTIONS

A. What are the moral implications of using ergogenic aids?
B. Because of serious doping violations, Lance Armstrong was stripped of his seven Tour de France titles and banned from cycling for life. For him, these actions are "equivalent to a death sentence." What is your opinion of Armstrong's doping violations and the imposed consequences? If you were a member of the International Cycling Union, how would you propose determining which, if any, substances to allow and/or ban?

REFERENCES

1. J. R. Hoffman and coauthors, Position stand on androgen and human growth hormone use, *Journal of Strength and Conditioning Research* 23 (2009): S1–S59.
2. Position of the American Dietetic Association, Dietetians of Canada, and the American College of Sports Medicine: Nutrition and athletic performance, *Journal of the American Dietetic Association* 109 (2009): 509–527.
3. R. B. Kreider and coauthors, ISSN exercise and sport nutrition review: Research and recommendations, *Journal of the International Society of Sports Nutrition* 7 (2010): www.jissn.com/content/7/1/7.
4. M. Dunford and E. J. Coleman, Ergogenic aids, dietary supplements, and exercise, in *Sports Nutrition: A Practice Manual for Professionals*, 5th ed., eds. C. A. Rosenbloom and E. J. Coleman (Chicago, IL: Academy of Nutrition and Dietetics, 2012): pp. 128–161; Position of the American Dietetic Association, Dietitians of Canada, and the American College of Sports Medicine, 2009.
5. T. A. Astorino and coauthors, Increases in cycling performance in response to caffeine ingestion are repeatable, *Nutrition Research* 32 (2012): 78–84; Kreider and coauthors, 2010; E. R. Goldstein and coauthors, International Society of Sports Nutrition position stand: Caffeine and performance, *Journal of the International Society of Sports Nutrition* 7 (2010): www.jissn.com/content/7/1/5; G. L. Warren and coauthors, Effect of caffeine ingestion on muscular strength and endurance: A meta-analysis, *Medicine and Science in Sports and Exercise* 42 (2010): 1375–1387.

6. Dunford and Coleman, 2012.

7. S. H. Backhouse and coauthors, Caffeine ingestion, affect and perceived exertion during prolonged cycling, *Appetite* 57 (2011): 247–252; Goldstein and coauthors, 2010.

8. R. Jager and coauthors, Analysis of the efficacy, safety, and regulatory status of novel forms of creatine, *Amino Acids* 40 (2011): 1369–1383; M. A. Tarnopolsky, Caffeine and creatine use in sport, *Annals of Nutrition and Metabolism* 57 (2010): 1–8 (Supplement 2); D. H. Fukuda and coauthors, The effects of creatine loading and gender on anaerobic running capacity, *Journal of Strength and Conditioning Research* 24 (2010): 1826–1833.

9. Kreider and coauthors, 2010; Position of the American Dietetic Association, Dietetians of Canada, and the American College of Sports Medicine, 2009.

10. Position of the American Dietetic Association, Dietitians of Canada, and the American College of Sports Medicine, 2009.

11. J. C. Siegler and D. O. Gleadall-Siddall, Sodium bicarbonate ingestion and repeated swim sprint performance, *Journal of Strength and Conditioning Research* 24 (2010): 3105–3111; S. L. Cameron and coauthors, Increased blood pH but not performance in elite rugby union players, *International Journal of Sport Nutrition and Exercise Metabolism* 20 (2010): 307–321; J. C. Siegler and K. Hirscher, Sodium bicarbonate ingestion and boxing performance, *Journal of Strength and Conditioning Research* 24 (2010): 103–108.

12. R. C. Harris and T. Stellingwerff, Effect of β-alanine supplementation on high-intensity exercise performance, *Nestle Nutrition Institute Workshop Series* 76 (2013): 61–71; C. Sale and coauthors, Carnosine: From exercise performance to health, *Amino Acids* 44 (2013): 1477–1491; W. Derave, Use of β-alanine as an ergogenic aid, *Nestle Nutrition Institute Workshop Series* 75 (2013): 99–108.

13. Derave, 2013; G. G. Artioli and coauthors, Role of β-alanine supplementation on muscle carnosine and exercise performance, *Medicine and Science in Sports and Exercise* 42 (2010): 1162–1173.

14. J. J. Quesnele and coauthors, The effects of beta alanine supplementation on performance: A systematic review of the literature, *International Journal of Sport Nutrition and Exercise Metabolism* 24 (2014): 14–27.

15. J. M. Wilson and coauthors, β-hydroxy-β-methylbutyrate free acid reduces markers of exercise-induced muscle damage and improves recovery in resistance-trained men, *British Journal of Nutrition* 110 (2013): 538–544; Kreider and coauthors, 2010.

16. Kreider and coauthors, 2010; D. Bishop, Dietary supplements and team-sport performance, *Sports Medicine* 40 (2010): 995–1017.

17. Kreider and coauthors, 2010; Position of the American Dietetic Association, Dietetians of Canada, and the American College of Sports Medicine, 2009.

15

Life Cycle Nutrition: Pregnancy and Lactation

Nutrition in Your Life

Food choices have consequences. Sometimes they are immediate, such as when you get heartburn after eating a pepperoni pizza. Other times they sneak up on you, such as when you gain weight after repeatedly overindulging in hot fudge sundaes. Quite often, they are temporary and easily resolved, such as when hunger pangs strike after you skip lunch. During pregnancy, however, the consequences of a woman's food choices are dramatic. They affect not only her health, but also the growth and development of another human being—and not just for today, but for years to come. Making smart food choices is a huge responsibility, but fortunately, it's fairly simple. In the Nutrition Portfolio at the end of this chapter, you can determine how well your current diet might support the needs of a pregnant woman.

Each person enters this world with a unique genetic map that determines the primary ways that person's physical and mental characteristics will develop throughout life. Some of those characteristics cannot be changed, but others can be influenced within genetically defined limits. One of several ways to ensure the optimal growth, maintenance, and health of the body is through proper nutrition. Ideally, a person's diet supplies sufficient amounts of all the nutrients to meet the needs incurred by the physiological demands of pregnancy, lactation, growth, and aging.

All people—pregnant and lactating women, infants, children, adolescents, and adults—need the same nutrients, but the amounts they need vary depending on their stage of life. This chapter focuses on nutrition in preparation for, and support of, pregnancy and lactation. The next two chapters address the needs of infants, children, adolescents, and older adults.

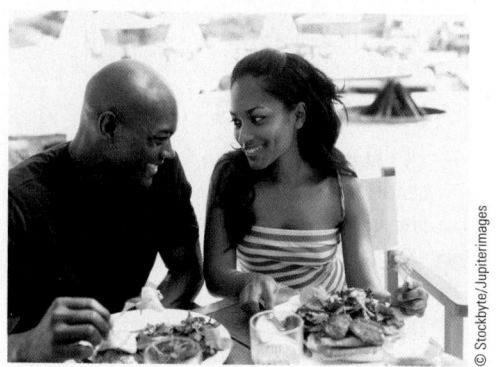

Young adults can prepare for a healthy pregnancy by taking care of themselves today.

15.1 Nutrition prior to Pregnancy

› LEARN IT List the ways men and women can prepare for a healthy pregnancy.

Both a man's and a woman's nutrition may affect **fertility** and possibly the genetic contributions they make to their children, but it is the woman's nutrition that has the most direct influence on the developing fetus.[1] Her body provides the environment for the growth and development of a new human being. Prior to pregnancy, however, both men and women have a unique opportunity to prepare physically, mentally, and emotionally for the many changes to come.[2] In preparation for a healthy pregnancy, they can establish the following habits:

- *Achieve and maintain a healthy body weight.* Both underweight and overweight are associated with infertility. Overweight and obese men have low sperm counts and hormonal changes that reduce fertility.[3] Excess body fat in women disrupts menstrual regularity and ovarian hormone production.[4] Should a pregnancy occur, mothers, both underweight and overweight, and their newborns, face increased risks of complications.

- *Choose an adequate and balanced diet.* Malnutrition reduces fertility and impairs the early development of an infant should a woman become pregnant. In contrast, a healthy diet that includes a full array of vitamins and minerals can favorably influence fertility.[5] Men with diets rich in antioxidant nutrients and low in saturated fats have higher sperm numbers and motility.[6]

- *Be physically active.* A woman who wants to be physically active *when* she is pregnant needs to become physically active *beforehand*.

- *Receive regular medical care.* Regular health-care visits help ensure a healthy start to pregnancy.

- *Manage chronic conditions.* Conditions such as diabetes, hypertension, HIV/AIDS, phenylketonuria (PKU), and sexually transmitted diseases can adversely affect a pregnancy and need close medical attention to help ensure a healthy outcome.

- *Avoid harmful influences.* Both maternal and paternal ingestion of, or exposure to, harmful substances (such as cigarettes, alcohol, drugs, or environmental contaminants) can cause miscarriage or abnormalities, alter genes or their expression, and may interfere with fertility.[7]

Young adults who nourish and protect their bodies do so not only for their own sakes, but also for future generations.

› DIETARY GUIDELINES FOR AMERICANS

Women who are capable of becoming pregnant should:

- Achieve and maintain a healthy weight before becoming pregnant.
- Choose foods that supply heme iron, which is more readily absorbed by the body, additional iron sources, and enhancers of iron absorption such as vitamin C–rich foods.
- Consume 400 micrograms per day of synthetic folate from fortified foods and/or supplements in addition to folate from a varied diet.

› REVIEW IT List the ways men and women can prepare for a healthy pregnancy.

Prior to pregnancy, the health and behaviors of both men and women can influence fertility and fetal development. In preparation, they can achieve and maintain a healthy body weight, choose an adequate and balanced diet, be physically active, receive regular medical care, manage chronic diseases, and avoid harmful influences.

fertility: the capacity of a woman to produce a normal ovum periodically and of a man to produce normal sperm; the ability to reproduce.

15.2 Growth and Development during Pregnancy

> **LEARN IT** Describe fetal development from conception to birth and explain how maternal malnutrition can affect critical periods.

A whole new life begins at **conception.** Organ systems develop rapidly, and nutrition plays many supportive roles. This section describes placental development and fetal growth, paying close attention to times of intense developmental activity.

Placental Development In the early days of pregnancy, a spongy structure known as the **placenta** develops in the **uterus.** Two associated structures also form (see Figure 15-1). One is the **amniotic sac,** a fluid-filled balloonlike structure that houses the developing fetus. The other is the **umbilical cord,** a ropelike structure containing fetal blood vessels that extends through the fetus's "belly button" (the umbilicus) to the placenta. These three structures play crucial roles during pregnancy, and then are expelled from the uterus during childbirth.

The placenta develops as an interweaving of fetal and maternal blood vessels embedded in the uterine wall. The maternal blood transfers oxygen and nutrients to the fetus's blood and picks up fetal waste products. By exchanging oxygen, nutrients, and waste products, the placenta performs the respiratory, absorptive, and excretory functions that the fetus's lungs, digestive system, and kidneys will provide after birth.

The placenta is a versatile, metabolically active organ. Like all body tissues, the placenta uses energy and nutrients to support its work. It produces an array of hormones that maintain pregnancy and prepare the mother's breasts for lactation (making milk). A healthy placenta is essential for the developing fetus to attain its full potential.[8]

Fetal Growth and Development Fetal development begins with the fertilization of an **ovum** by a **sperm.** Three stages follow: the zygote, the embryo, and the fetus (see Figure 15-2, p. 476).

conception: the union of the male sperm and the female ovum; fertilization.

placenta (plah-SEN-tuh): the organ that develops inside the uterus early in pregnancy, through which the fetus receives nutrients and oxygen and returns carbon dioxide and other waste products to be excreted.

uterus (YOU-ter-us): the muscular organ within which the infant develops before birth.

amniotic (am-nee-OTT-ic) **sac:** the "bag of waters" in the uterus, in which the fetus floats.

umbilical (um-BILL-ih-cul) **cord:** the ropelike structure through which the fetus's veins and arteries reach the placenta; the route of nourishment and oxygen to the fetus and the route of waste disposal from the fetus. The scar in the middle of the abdomen that marks the former attachment of the umbilical cord is the *umbilicus* (um-BILL-ih-cus), commonly known as the "belly button."

ovum (OH-vum): the female reproductive cell, capable of developing into a new organism upon fertilization; commonly referred to as an egg.

sperm: the male reproductive cell, capable of fertilizing an ovum.

> **FIGURE 15-1** **The Placenta and Associated Structures**

To understand how placental villi absorb nutrients without maternal and fetal blood interacting directly, think of how the intestinal villi work. The GI side of the intestinal villi is bathed in a nutrient-rich fluid (chyme). The intestinal villi absorb the nutrient molecules and release them into the body via capillaries. Similarly, the maternal side of the placental villi is bathed in nutrient-rich maternal blood. The placental villi absorb the nutrient molecules and release them to the fetus via fetal capillaries.

The arrows indicate the direction of blood flow.

Pool of mother's blood

Fetal vein   Fetal artery

Umbilical cord
Uterine wall
Placenta

Umbilical cord
Umbilical vein
Umbilical arteries

Amniotic sac

In the placenta, maternal blood vessels lie side by side with fetal blood vessels that reach the fetus through the umbilical cord.

Fingerlike projections (called placental villi) contain fetal blood vessels and extend into the pool of mother's blood. No actual mingling of fetal and maternal blood occurs, but substances pass back and forth.

Thus, oxygen and nutrients from the mother's blood enter fetal vessels, and waste products are removed.

Mother's veins carry fetal wastes away.

Mother's arteries bring fresh blood with oxygen and nutrients to the fetus.

Fetal portion of placenta   Maternal portion of placenta

© Cengage Learning

> FIGURE 15-2 **Stages of Embryonic and Fetal Development**

1 A newly fertilized ovum is called a **zygote** and is about the size of the period at the end of this sentence. Less than 1 week after fertilization, these cells have rapidly divided multiple times to become a blastocyst ready for implantation.

3 A **fetus** after 11 weeks of development is just over an inch long. Notice the umbilical cord and blood vessels connecting the fetus with the placenta.

2 After implantation, the placenta develops and begins to provide nourishment to the developing embryo. An **embryo** 5 weeks after fertilization is about ½ inch long.

4 A **newborn infant** after 9 months of development measures close to 20 inches in length. From 8 weeks to term, this infant grew 20 times longer and 50 times heavier.

zygote (ZY-goat): the initial product of the union of ovum and sperm; a fertilized ovum.

blastocyst (BLASS-toe-sist): the developmental stage of the zygote when it is about 5 days old and ready for implantation.

implantation (IM-plan-TAY-shun): the embedding of the blastocyst in the inner lining of the uterus.

embryo (EM-bree-oh): the developing infant from 2 to 8 weeks after conception.

fetus (FEET-us): the developing infant from 8 weeks after conception until term.

full term: births occurring at 39 through 40 weeks of gestation.

critical periods: finite periods during development in which certain events occur that will have irreversible effects on later developmental stages; usually a period of rapid cell division.

gestation (jes-TAY-shun): the period from conception to birth. For human beings, the average length of a healthy gestation is 40 weeks. Pregnancy is often divided into 3-month periods, called *trimesters*.

The Zygote The newly fertilized ovum is called a **zygote**. It begins as a single cell and rapidly divides to become a **blastocyst**. During that first week, the blastocyst floats down into the uterus, where it will embed itself in the inner uterine wall—a process known as **implantation**. Cell division continues at an amazing rate as each set of cells divides into many other cells.

The Embryo At first, the number of cells in the **embryo** doubles approximately every 24 hours; later the rate slows, and only one doubling occurs during the final 10 weeks of pregnancy. At 8 weeks, the 1¼-inch embryo has a complete central nervous system, a beating heart, a digestive system, well-defined fingers and toes, and the beginnings of facial features.

The Fetus The **fetus** continues to grow during the next 7 months. Each organ grows to maturity according to its own schedule, with greater intensity at some times than at others. As Figure 15-2 shows, fetal growth is phenomenal: weight increases from less than an ounce to about 7½ pounds (3500 grams). Most successful pregnancies are **full term**—defined as births occurring at 39 through 40 weeks—and produce a healthy infant weighing 6½ to 8 pounds.[9]

Critical Periods Times of intense development and rapid cell division are called **critical periods**—critical in the sense that those cellular activities can occur only at those times. If cell division and number are limited during a critical period, full recovery is not possible (see Figure 15-3). Damage during these critical times of pregnancy has permanent consequences for the life and health of the fetus.[10]

The development of each organ and tissue is most vulnerable to adverse influences (such as nutrient deficiencies or toxins) during its own critical period (see Figure 15-4). The neural tube, for example, is the structure that eventually becomes the brain and the spinal cord, and its critical period of development is from 17 to 30 days of **gestation**. Consequently, neural tube development is most vulnerable to nutrient deficiencies, nutrient excesses, or toxins during this critical time—when most women do not yet even realize they are pregnant. Any abnormal development of the neural tube or its failure to close completely can cause a major defect in the central nervous system.

Petit Format/Nestle/Science Source (photos 1, 2, 3)

© Cindy Vannelli

> **FIGURE 15-3** **The Concept of Critical Periods in Fetal Development**

Critical periods occur early in fetal development. An adverse influence felt early in pregnancy can have a much more severe and prolonged impact than one felt later on.

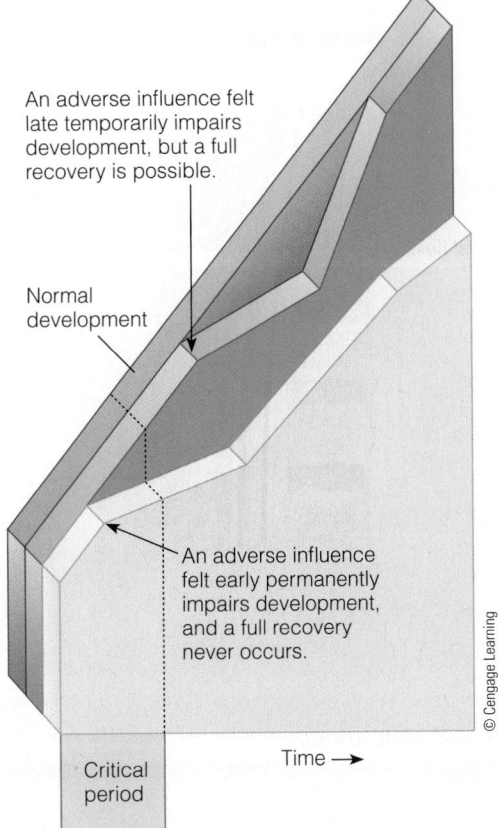

An adverse influence felt late temporarily impairs development, but a full recovery is possible.

Normal development

An adverse influence felt early permanently impairs development, and a full recovery never occurs.

Critical period

Time →

© Cengage Learning

> **FIGURE 15-4** **Critical Periods of Development**

During embryonic development (from 2 to 8 weeks), many of the tissues are in their critical periods (purple area of the bars); events occur that will have irreversible effects on the development of those tissues. In the later stages of development (green area of the bars), the tissues continue to grow and change, but the events are less critical in that they are relatively minor or reversible.

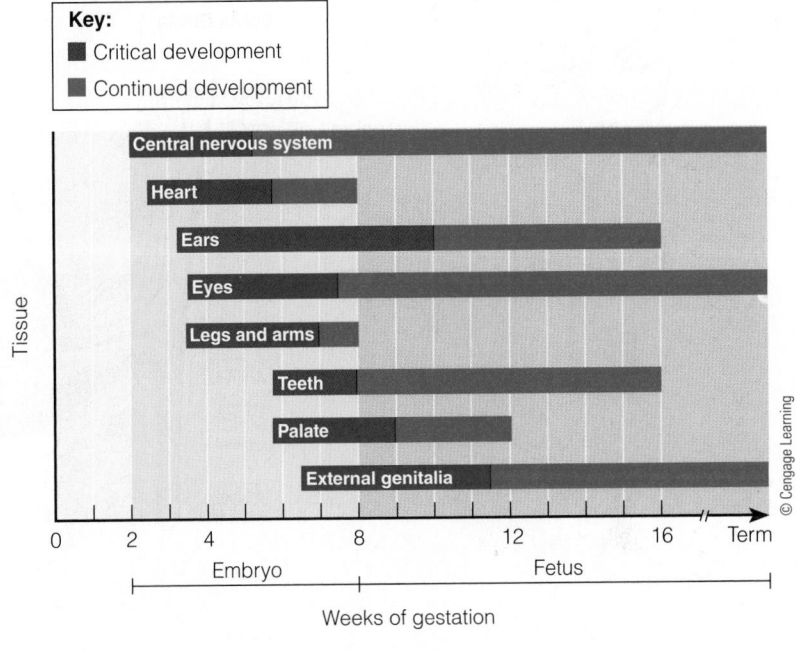

Key:
■ Critical development
■ Continued development

Central nervous system
Heart
Ears
Eyes
Legs and arms
Teeth
Palate
External genitalia

Tissue

0   2   4   8   12   16   Term
Embryo   Fetus

Weeks of gestation

© Cengage Learning

SOURCE: Adapted from *Before We Are Born: Essentials of Embryology and Birth Defects* by K. L. Moore and T.V.N. Persaud (W. B. Saunders, 2003).

Neural Tube Defects Each year in the United States, approximately 3000 pregnancies are affected by a **neural tube defect**—a malformation of the brain, spinal cord, or both during embryonic development.* The two most common types of neural tube defects are anencephaly (no brain) and spina bifida (split brain). In **anencephaly,** the upper end of the neural tube fails to close. Consequently, the brain is either missing or fails to develop. Pregnancies affected by anencephaly often end in miscarriage; infants born with anencephaly die shortly after birth.

Spina bifida is characterized by incomplete closure of the spinal cord and its bony encasement (see Figure 15-5, p. 478). The meninges membranes covering the spinal cord often protrude as a sac, which may rupture and lead to meningitis, a life-threatening infection. Spina bifida is accompanied by varying degrees of paralysis, depending on the extent of the spinal cord damage. Mild cases may not even be noticed, but severe cases lead to death. Common problems include clubfoot, dislocated hip, kidney disorders, curvature of the spine, muscle weakness, mental impairments, and motor and sensory losses.

The cause of neural tube defects is unknown, but researchers are examining several gene-gene, gene-nutrient, and gene-environment interactions. A pregnancy affected by a neural tube defect can occur in any woman, but these factors make it more likely:[11]

- A personal or family history of a neural tube defect
- Maternal diabetes or gestational diabetes
- Maternal Hispanic ethnicity

neural tube defect: malformations of the brain, spinal cord, or both during embryonic development that often results in lifelong disability or death.

anencephaly (AN-en-SEF-a-lee): an uncommon and always fatal type of neural tube defect, characterized by the absence of a brain.

- **an** = not (without)
- **encephalus** = brain

spina (SPY-nah) **bifida** (BIFF-ih-dah): one of the most common types of neural tube defects, characterized by the incomplete closure of the spinal cord and its bony encasement.

- **spina** = spine
- **bifida** = split

*Worldwide, some 300,000 pregnancies are affected by neural tube defects each year.

> **FIGURE 15-5** **Spina Bifida**

Spina bifida, a common neural tube defect, occurs when the vertebrae of the spine fail to close around the spinal cord, leaving it unprotected. The B vitamin folate—consumed prior to and during pregnancy—helps prevent spina bifida and other neural tube defects.

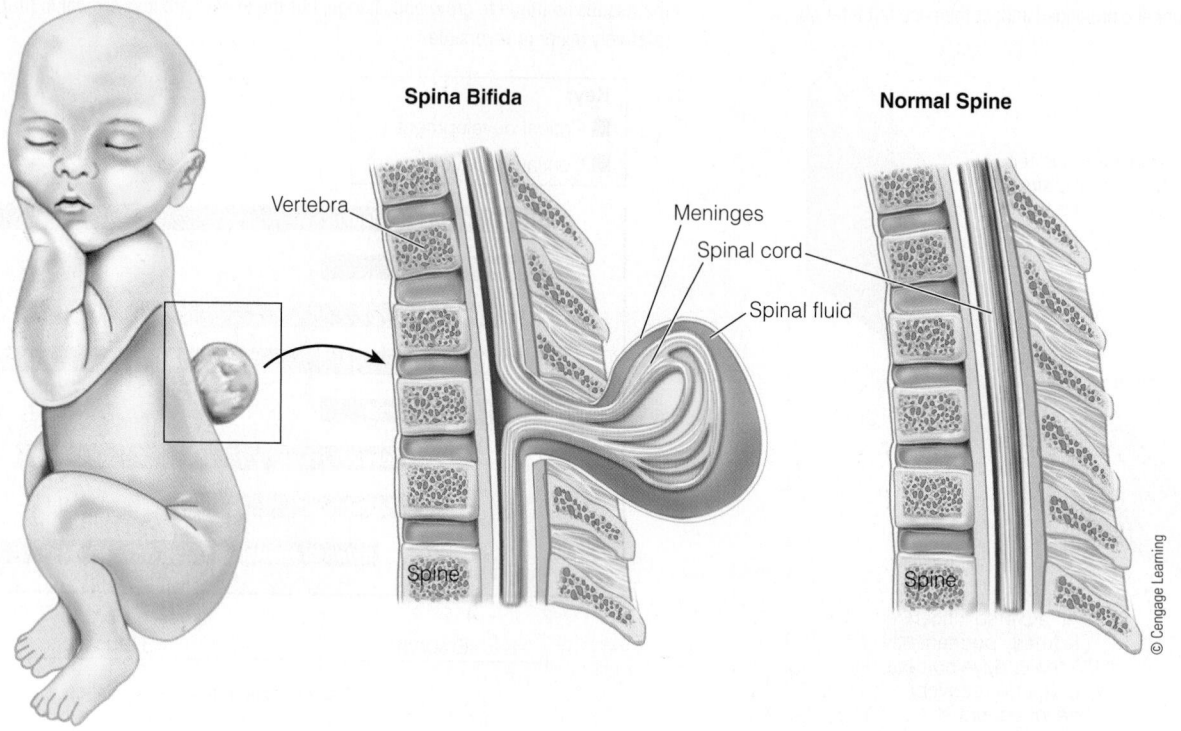

- Maternal use of certain antiseizure medications
- Inadequate folate
- Maternal obesity

Not all cases of neural tube defects can be prevented, but folate supplementation greatly reduces the incidence and severity.[12]

Folate Supplementation Chapter 10 describes how folate supplements taken 1 month before conception and continued throughout the first trimester can help support a healthy pregnancy, prevent neural tube defects, and reduce the severity of defects that do occur. For this reason, all women of childbearing age who are capable of becoming pregnant should consume 400 micrograms (0.4 milligrams) of folate daily. A woman who has previously had an infant with a neural tube defect may be advised by her physician to take folate supplements in doses ten times larger— 4 milligrams daily. Because high doses of folate can mask the symptoms of pernicious anemia associated with a vitamin B_{12} deficiency, quantities of 1 milligram or more require a prescription. Most over-the-counter multivitamin-mineral supplements contain 400 micrograms of folate; prenatal supplements usually contain 800 micrograms.

Because half of the pregnancies each year are unplanned and because neural tube defects occur early in development before most women realize they are pregnant, grain products in the United States are fortified with folate to help ensure an adequate intake. Labels on fortified products may claim that an "adequate intake of folate has been shown to reduce the risk of neural tube defects." Fortification has improved folate status in women of childbearing age and lowered the number of neural tube defects that occur each year, as shown in Figure 10-11 (p. 318).

> **DIETARY GUIDELINES FOR AMERICANS**
Women who are pregnant are advised to consume 600 micrograms of dietary folate equivalents from all sources.

Fetal Programming Clearly, substances such as nutrients influence the growth and development of an infant, but recent genetic research is beginning to explain how they might influence the infant's development of obesity and diseases in *adulthood*. This process is commonly known as **fetal programming**, although "developmental origins of disease" may more appropriately describe the ever-changing interactions involved in disease development.[13] In the case of pregnancy, the mother's nutrition can change gene expression in the fetus.[14] Such epigenetic changes during pregnancy can affect the infant's development of obesity and related adult diseases.[15] Some research suggests that these epigenetic changes during pregnancy may even influence succeeding generations. (See Highlight 6 for further discussion of epigenetics.)

Chronic Diseases Much research suggests that dietary influences at critical times during fetal development program the infant's future development, metabolism, and health.[16] Maternal diet during pregnancy may alter the infant's bodily functions by permanently changing an organ's structure and resulting secretions, altering gene expression through epigenetics, or influencing the regulation of cellular aging.[17] For example, undernutrition may limit liver growth and program lipid metabolism in such a way that the infant will develop risk factors for cardiovascular disease as an adult.[18] Similarly, overnutrition and maternal obesity may program the fetus to develop chronic diseases later in life.[19] On a positive note, a maternal diet rich in nutrients such as folate can have epigenetic effects that protect the developing fetus against some cancers into adulthood.[20] (See Highlight 6 for a detailed discussion of epigenetics and folate's role in methylation reactions.)

Malnutrition during the critical period of pancreatic cell growth provides an example of how type 2 diabetes may develop in adulthood. The pancreatic cells responsible for producing insulin (the beta cells) normally increase more than 130-fold between 12 weeks of gestation and 5 months after birth. Nutrition is a primary determinant of beta cell growth, and infants who have suffered prenatal malnutrition have significantly fewer beta cells than well-nourished infants. They are also more likely to be low-birthweight infants—and low birthweight correlates with insulin resistance later in life.[21] One hypothesis suggests that diabetes may develop from the interaction of inadequate nutrition early in life (low birthweight) with abundant nutrition later in life (overweight adult): the small mass of beta cells developed in times of undernutrition during fetal development may be insufficient in times of overnutrition during adulthood when the body needs more insulin.

Hypertension may develop from a similar scenario of inadequate growth during placental and gestational development followed by accelerated growth during early childhood: the small mass of kidney cells developed during malnutrition may be insufficient to handle the excessive demands of later life. As adults, low-birthweight infants may be particularly sensitive to the blood-pressure raising effects of salt.[22]

> **REVIEW IT** Describe fetal development from conception to birth and explain how maternal malnutrition can affect critical periods.

Maternal nutrition before and during pregnancy affects both the mother's health and the infant's growth. As the infant develops through its three stages—the zygote, embryo, and fetus—its organs and tissues grow, each on its own schedule. Times of intense development are critical periods that depend on nutrients to proceed smoothly. Without folate, for example, the neural tube fails to develop completely during the first month of pregnancy, prompting recommendations that all women of childbearing age take folate daily.

Because critical periods occur throughout pregnancy, a woman should continuously take good care of her health. That care should include achieving and maintaining a healthy body weight prior to pregnancy and gaining sufficient weight during pregnancy to support a healthy infant.

fetal programming: the influence of substances during fetal growth on the development of diseases in later life.

Fetal growth and maternal health depend on a sufficient weight gain during pregnancy.

15.3 Maternal Weight

> **LEARN IT** Explain how both underweight and overweight can interfere with a healthy pregnancy and how weight gain and physical activity can support maternal health and infant growth.

Birthweight is the most reliable indicator of an infant's health. As a later section of this chapter explains, compared with a normal-weight infant, an underweight infant is more likely to have physical and mental abnormalities, suffer illnesses, and die. In general, higher birthweights present fewer risks for infants. Two characteristics of the mother's weight influence an infant's birthweight: her weight *prior* to conception and her weight gain *during* pregnancy.

Weight prior to Conception A woman's weight prior to conception influences fetal growth. Even with the same weight gain during pregnancy, underweight women tend to have smaller babies than heavier women. Ideally, before a woman becomes pregnant, she will have established diet and activity habits to support an adequate, and not excessive, weight gain during pregnancy.[23]

Underweight An underweight woman (BMI <18.5) has a high risk of having a low-birthweight infant, especially if she is malnourished or unable to gain sufficient weight during pregnancy. In addition, the rates of **preterm** births and infant deaths are higher for underweight women. An underweight woman improves her chances of having a healthy infant by gaining sufficient weight prior to conception or by gaining extra pounds during pregnancy. To gain weight and ensure nutrient adequacy, an underweight woman can follow the dietary recommendations for pregnant women (described on pp. 484–487).

Overweight and Obesity An estimated one-third of all pregnant women in the United States are obese (BMI ≥30), which can create problems related to pregnancy, infancy, and childbirth.[24] Obese women have an especially high risk of medical complications such as gestational hypertension, gestational diabetes, and postpartum infections. Compared with other women, obese women are also more likely to have other complications of labor and delivery. Complications in women after gastric bypass surgery are lower than in obese women, but their infants are more likely to be small, perhaps because of limited nutrient absorption; careful monitoring during pregnancy is advised.[25]

Infants of obese women are more likely to be large for gestational age, weighing more than 9 pounds.[26] Problems associated with **macrosomia** include increases in the likelihood of a difficult labor and delivery, birth trauma, and **cesarean delivery,** which presents a high risk in obese women.[27] Consequently, these infants have a greater risk of poor health and death than infants of normal weight.

Of greater concern than infant birthweight is the poor development of infants born to obese mothers.[28] Obesity may double the risk for neural tube defects. Folate's role has been examined, but a more likely explanation seems to be poor glycemic control. Undiagnosed diabetes might also explain why obese women have a greater risk of giving birth to infants with heart defects and other abnormalities.

Even moderate overweight increases the risks for some complications, such as gestational hypertension, gestational diabetes, preterm births, and cesarean deliveries; risks tend to increase as BMI increases.[29] Health-care providers have traditionally advised against weight-loss dieting during pregnancy. Limited research, however, suggests that following a well-balanced, kcalorie-restricted diet and regular exercise program can support a healthy pregnancy with little or no weight gain. Ideally, overweight and obese women will achieve a healthier body weight before becoming pregnant and avoid excessive weight gain during pregnancy.[30]

Weight Gain during Pregnancy Fetal growth and maternal health depend on a sufficient weight gain during pregnancy. Maternal weight gain during pregnancy correlates closely with infant birthweight, which is a strong predictor of the health and subsequent development of the infant.[31]

preterm (premature): births occurring before 37 weeks of gestation; births occurring at 37 to 38 weeks of gestation are designated *early term.*

macrosomia (mak-roh-SO-me-ah): abnormally large body size. In the case of infants, a birthweight at the 90th percentile or higher for gestational age (roughly 9 lb—or 4000 g—or more); macrosomia results from prepregnancy obesity, excessive weight gain during pregnancy, or uncontrolled gestational diabetes.

- **macro** = large
- **soma** = body

cesarean (si-ZAIR-ee-un) **delivery:** a surgically assisted birth involving removal of the fetus by an incision into the uterus, usually by way of the abdominal wall.

TABLE 15-1 Recommended Weight Gains Based on Prepregnancy Weight

Prepregnancy Weight	Recommended Weight Gain	
	For single birth	For twin birth
Underweight (BMI <18.5)	28 to 40 lb (12.5 to 18.0 kg)	Insufficient data to make recommendation
Healthy weight (BMI 18.5 to 24.9)	25 to 35 lb (11.5 to 16.0 kg)	37 to 54 lb (17.0 to 25.0 kg)
Overweight (BMI 25.0 to 29.9)	15 to 25 lb (7.0 to 11.5 kg)	31 to 50 lb (14.0 to 23.0 kg)
Obese (BMI ≥30)	11 to 20 lb (5.0 to 9.0 kg)	25 to 42 lb (11.0 to 19.0 kg)

SOURCE: Institute of Medicine, *Weight Gain during Pregnancy: Reexamining the Guidelines* (Washington, D.C.: National Academies Press, 2009).

> **DIETARY GUIDELINES FOR AMERICANS**

Maintain appropriate kcalorie balance during pregnancy. Pregnant women are encouraged to gain weight within the gestational weight gain guidelines (see Table 15-1).

Recommended Weight Gains Table 15-1 presents recommended weight gains for various prepregnancy weights. The recommended gain for a woman who begins pregnancy at a healthy weight and is carrying a single fetus is 25 to 35 pounds. An underweight woman needs to gain between 28 and 40 pounds; and an overweight woman, between 15 and 25 pounds. About one-third of US women gain weight within these recommended ranges; about half gain more than recommended.[32] Appropriate weight gains reduce complications, help women limit weight retention and gains after pregnancy, and help their infants prevent obesity during childhood.[33] To limit excessive weight gains, pregnant women can select foods with a high nutrient density (nutrient per kcalorie) but a low energy density (kcalorie per gram). Physical activity also plays a key role in preventing excessive weight gains during pregnancy and minimizing weight retention after the birth.[34]

Weight-Gain Patterns For the normal-weight woman, weight gain ideally follows a pattern of 3½ pounds during the first trimester and 1 pound per week thereafter. Health-care professionals monitor weight gain using a prenatal weight-gain grid (see Figure 15-6). Identifying inadequate or excessive weight gains by the second trimester allows sufficient time for adjustments in diet and activity.[35]

> **FIGURE 15-6 Recommended Prenatal Weight Gain Based on Prepregnancy Weight**

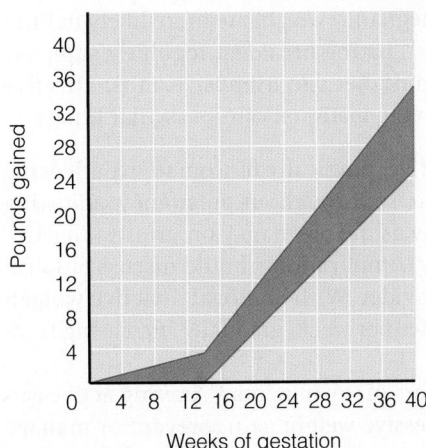

Normal-weight women should gain about 3½ pounds in the first trimester and just under 1 pound/week thereafter, achieving a total gain of 25 to 35 pounds by term.

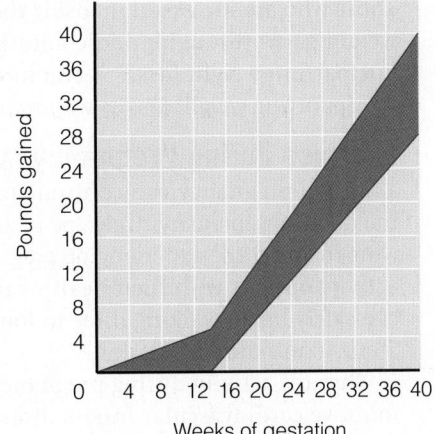

Underweight women should gain about 5 pounds in the first trimester and just over 1 pound/week thereafter, achieving a total gain of 28 to 40 pounds by term.

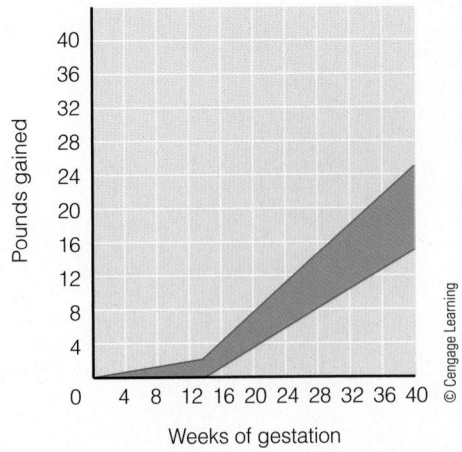

Overweight women should gain about 2 pounds in the first trimester and ⅔ pound/week thereafter, achieving a total gain of 15 to 25 pounds.

© Cengage Learning

> **FIGURE 15-7** **Components of Weight Gain during Pregnancy**

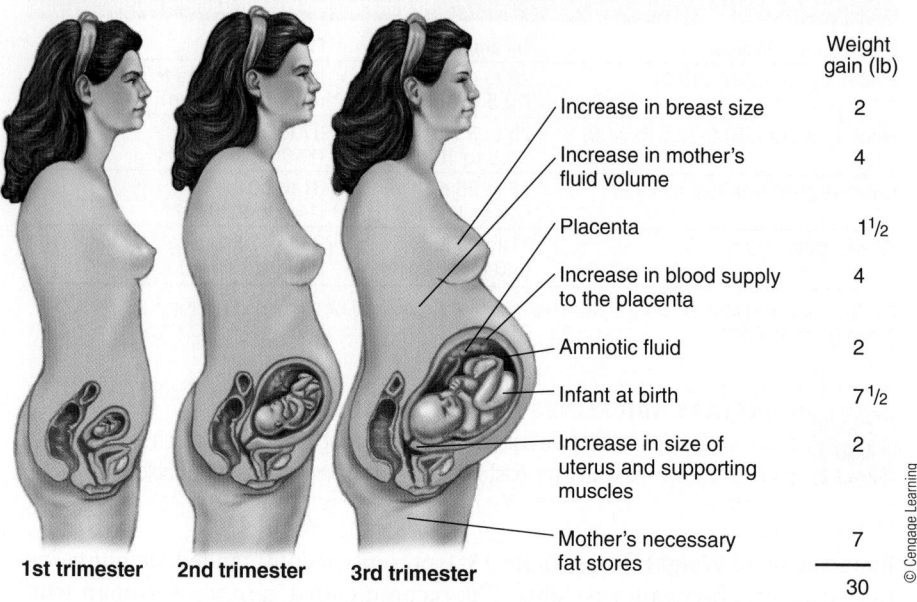

	Weight gain (lb)
Increase in breast size	2
Increase in mother's fluid volume	4
Placenta	1½
Increase in blood supply to the placenta	4
Amniotic fluid	2
Infant at birth	7½
Increase in size of uterus and supporting muscles	2
Mother's necessary fat stores	7
	30

1st trimester   2nd trimester   3rd trimester

© Cengage Learning

Components of Weight Gain Women often express concern about the weight gain that accompanies a healthy pregnancy. They may find comfort by remembering that most of the gain supports the growth and development of the placenta, uterus, blood, and breasts, the increase in blood supply and fluid volume, as well as a healthy 7½-pound infant. A small amount goes into maternal fat stores, and even that fat has a special purpose—to provide energy for growth, labor, and lactation. Figure 15-7 shows the components of a healthy 30-pound weight gain.

Weight Loss after Pregnancy The pregnant woman loses some weight at delivery. In the following weeks, she loses more as her blood volume returns to normal and she sheds accumulated fluids. Quite likely, her goal is to return to her pre-pregnancy weight, but that may depend in part on whether she stayed within the pregnancy weight gain recommendations. In general, the more weight a woman gains beyond the needs of pregnancy, the more she retains and the more likely she will continue to gain over the next several years.[36] Even with an average weight gain during pregnancy, most women tend to retain a couple of pounds with each pregnancy.[37] When those couple of pounds become several more, complications such as diabetes and hypertension in future pregnancies as well as chronic diseases in later life become more likely—even for women who are not overweight. Those who are successful in losing their pregnancy weight are more likely to limit weight gains through middle adulthood.[38] Eating breakfast regularly supports postpartum weight loss.[39] A combination of diet and exercise is most effective in supporting weight loss as well as improving maternal cardiovascular fitness.[40]

Exercise during Pregnancy An active, physically fit woman experiencing a normal pregnancy can continue to exercise throughout pregnancy, adjusting the duration, intensity, and type of activity as the pregnancy progresses. Inactive women and those experiencing pregnancy complications should discuss physical activity options with their health-care provider. With approval, inactive women can safely begin walking three to four times per week, gradually increasing from 25 to 40 minutes per session.[41]

Physical activity during pregnancy offers many benefits.[42] Staying active can improve cardiovascular fitness, limit excessive weight gain, prevent or manage gestational diabetes and gestational hypertension, and reduce stress.[43] Women who exercise during pregnancy report fewer discomforts throughout their pregnancies. Regular exercise develops the strength and endurance a woman needs to carry the extra weight through pregnancy and to labor through an intense

> FIGURE 15-8   Exercise Guidelines during Pregnancy

DO

Do begin to exercise gradually.

Do exercise regularly (most, if not all, days of the week).

Do warm up with 5 to 10 minutes of light activity.

Do 30 minutes or more of moderate physical activity.

Do cool down with 5 to 10 minutes of slow activity and gentle stretching.

Do drink water before, after, and during exercise.

Do eat enough to support the needs of pregnancy plus exercise.

Do rest adequately.

Pregnant women can enjoy the benefits of exercise.

DON'T

Don't exercise vigorously after long periods of inactivity.

Don't exercise in hot, humid weather.

Don't exercise when sick with fever.

Don't exercise while lying on your back after the 1st trimester of pregnancy or stand motionless for prolonged periods.

Don't exercise if you experience any pain, discomfort, or fatigue.

Don't participate in activities that may harm the abdomen or involve jerky, bouncy movements.

Don't scuba dive.

delivery. It also maintains the habits that help a woman lose excess weight and get back into shape after the birth.

A pregnant woman should participate in low-impact activities and avoid sports in which she might fall or be hit by other people or objects. For example, playing singles tennis with one person on each side of the net is safer than a fast-moving game of racquetball in which the two competitors can collide. Swimming and water aerobics are particularly beneficial because they allow the body to remain cool and move freely with the water's support, thus reducing back pain. Figure 15-8 provides some guidelines for exercise during pregnancy. Several of the guidelines are aimed at preventing excessively high internal body temperature and dehydration, both of which can harm fetal development. To this end, pregnant women should also stay out of saunas, steam rooms, and hot tubs or hot whirlpool baths.

> **REVIEW IT** Explain how both underweight and overweight can interfere with a healthy pregnancy and how weight gain and physical activity can support maternal health and infant growth.
A healthy pregnancy depends on a sufficient weight gain. Women who begin their pregnancies at a healthy weight need to gain about 30 pounds, which covers the growth and development of the placenta, uterus, blood, breasts, and infant. By remaining active throughout pregnancy, a woman can develop the strength she needs to carry the extra weight and maintain habits that will help her lose weight after the birth.

15.4 Nutrition during Pregnancy

> **LEARN IT** Summarize the nutrient needs of women during pregnancy.

A woman's body changes dramatically during pregnancy. Her uterus and its supporting muscles increase in size and strength; her blood volume increases by half to carry the additional nutrients and other materials; her joints become more flexible in preparation for childbirth; her feet swell in response to high concentrations of the hormone estrogen, which promotes water retention and helps to ready the uterus for delivery; and her breasts enlarge in preparation for lactation. The hormones that mediate all these changes may influence her mood. She can best prepare to handle these changes given a nutritious diet, regular physical activity, plenty of rest, and caring companions. This section highlights the role of nutrition.

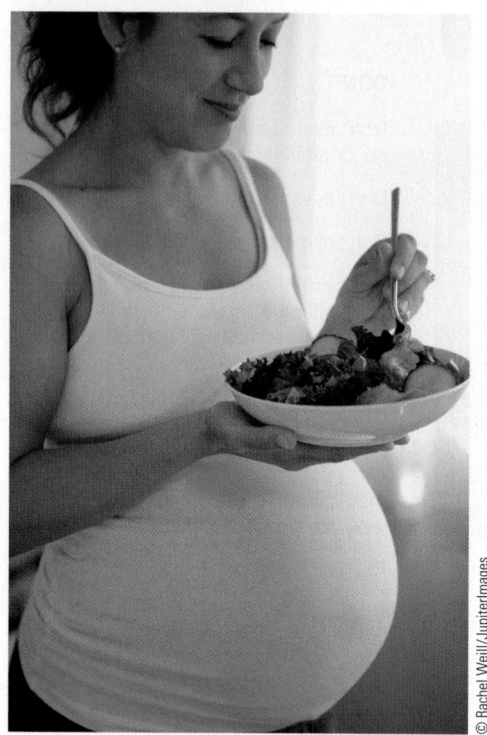

A pregnant woman's food choices support both her health and her infant's growth and development.

In general, the following guidelines will allow most women to enjoy a healthy pregnancy:

- Strive for good nutrition and health prior to pregnancy and get prenatal care during pregnancy.
- Gain a healthy amount of weight.
- Eat a balanced diet, safely prepared, and engage in physical activity regularly.
- Take prenatal vitamin and mineral supplements as prescribed.
- Refrain from cigarettes, alcohol, and drugs (including herbal remedies, unless prescribed by a physician).

An adequate diet may also help a woman manage the challenges and possible depression that can arise *after* the infant arrives. Details follow.

Energy and Nutrient Needs during Pregnancy From conception to birth, all parts of the infant—bones, muscles, blood cells, skin, and all other tissues—are made from nutrients in the foods the mother eats. For most women, nutrient needs during pregnancy and lactation are higher than at any other time (see Figure 15-9). Yet intakes do not consistently meet recommendations for energy and key nutrients.[44] To meet the high nutrient demands of pregnancy, a woman will need to make careful food choices, but her body will also help by maximizing absorption and minimizing losses. The Dietary Reference Intakes (DRI) table on the inside front cover provides separate listings for women during pregnancy and lactation, reflecting their heightened nutrient needs.

Energy The enhanced work of pregnancy raises the woman's basal metabolic rate dramatically and demands extra energy. After the first trimester, energy needs of pregnant women are greater than those of nonpregnant women—an additional 340 kcalories per day during the second trimester and an extra 450 kcalories per day during the third trimester. A woman can easily get these added kcalories with nutrient-dense selections from the five food groups. See Table 2-3 (p. 43) for

> FIGURE 15-9 **Comparison of Nutrient Recommendations for Nonpregnant, Pregnant, and Lactating Women**

For actual values, turn to the table on the inside front cover. For vitamins and minerals not shown here, the values do not change for pregnant and lactating women.

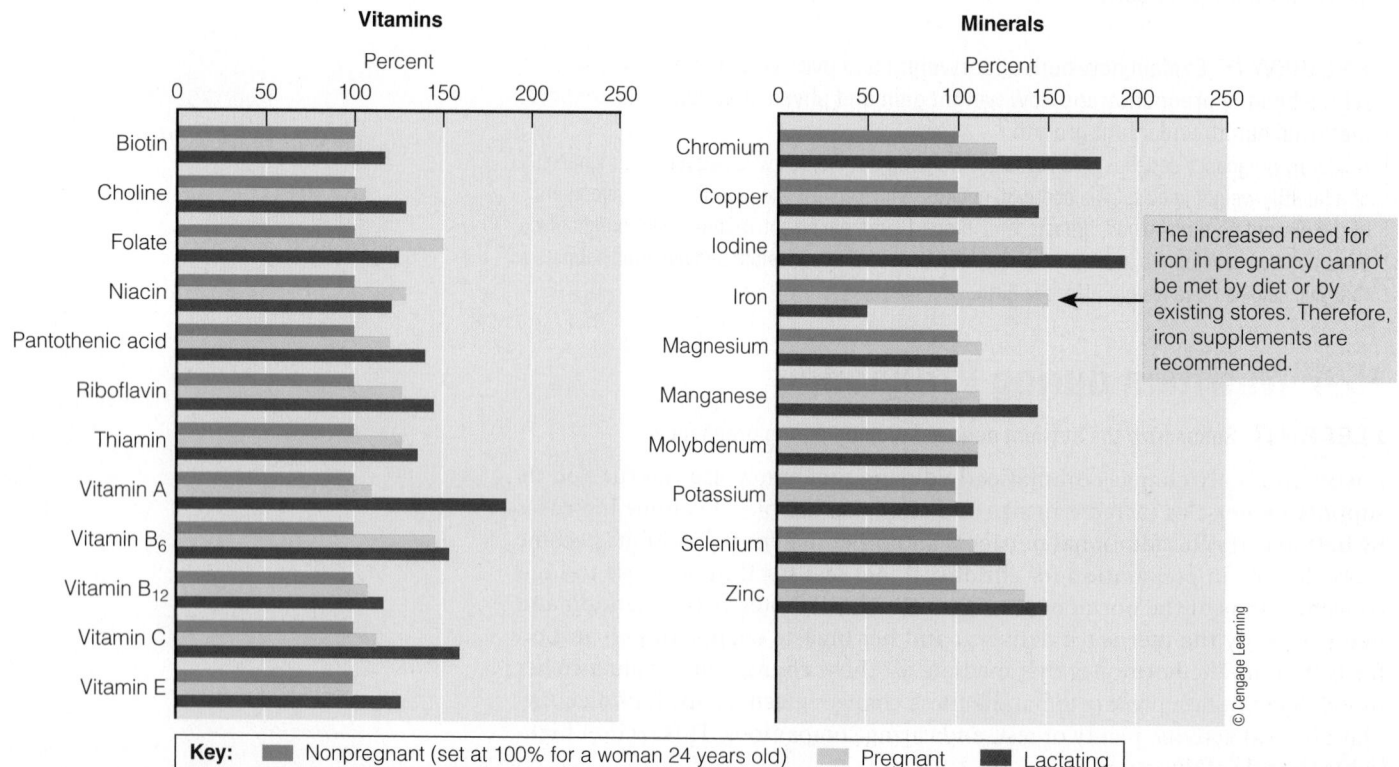

> FIGURE 15-10 **Daily Food Choices for Pregnant and Lactating Women**

Food Group	Amount	SAMPLE MENU	
Fruits	2 c	**Breakfast**	**Dinner**
		1 whole-wheat English muffin	Chicken cacciatore
		2 tbs peanut butter	3 oz chicken
Vegetables	2½–3 c	1 c low-fat vanilla yogurt	½ c stewed tomatoes
		½ c fresh strawberries	1 c rice
		1 c orange juice	½ c summer squash
Grains	6–8 oz	**Midmorning snack**	1½ c salad (spinach, mushrooms, carrots)
		½ c cranberry juice	1 tbs salad dressing
		1 oz pretzels	1 slice Italian bread
Protein foods	5½–6½ oz	**Lunch**	2 tsp soft margarine
		Sandwich (tuna salad on whole-wheat bread)	1 c low-fat milk
		½ carrot (sticks)	
Milk	3 c	1 c low-fat milk	

© Cengage Learning

NOTE: The range of recommended amounts reflects the differences of the first trimester versus the second and third trimesters. This sample meal plan provides about 2500 kcalories (55% from carbohydrate, 20% from protein, and 25% from fat) and meets most of the vitamin and mineral needs of pregnant and lactating women.

suggested dietary patterns for several kcalorie levels. A sample menu for pregnant and lactating women is presented in Figure 15-10.

For a 2000-kcalorie daily intake, these added kcalories represent about 15 to 20 percent more food energy than before pregnancy. The increase in nutrient needs is often greater than this, so nutrient-dense foods should be chosen to supply the extra kcalories: foods such as whole-grain breads and cereals, legumes, dark green vegetables, citrus fruits, low-fat milk and milk products, and lean meats, fish, poultry, and eggs.

Carbohydrate Ample carbohydrate (ideally, 175 grams or more per day and certainly no less than 135 grams) is necessary to fuel the fetal brain. Sufficient carbohydrate also ensures that the protein needed for growth will not be broken down and used to make glucose.

Protein The protein RDA for pregnancy is an additional 25 grams per day higher than for nonpregnant women. Pregnant women can easily meet their protein needs by selecting meats, milk products, and protein-containing plant foods such as legumes, whole grains, nuts, and seeds. Because use of high-protein supplements during pregnancy may be harmful to the infant's development, it is discouraged unless medically prescribed and carefully monitored to treat fetal growth problems.[45]

Essential Fatty Acids The high nutrient requirements of pregnancy leave little room in the diet for excess fat, but the essential long-chain polyunsaturated fatty acids are particularly important to the growth and development of the fetus.[46] The brain is largely made of lipid material, and it depends heavily on the long-chain omega-3 and omega-6 fatty acids for its growth, function, and structure.[47] (See Table 5-4 on p. 153 for a list of good food sources of the omega fatty acids.)

Nutrients for Blood Production and Cell Growth New cells are laid down at a tremendous pace as the fetus grows and develops. At the same time, the mother's red blood cell mass expands. All nutrients are important in these processes, but for folate, vitamin B_{12}, iron, and zinc, the needs are especially great because of their key roles in the synthesis of DNA and new cells.

The requirement for folate increases dramatically during pregnancy (from 400 micrograms to 600 micrograms daily). It is best to obtain sufficient folate from a combination of supplements, fortified foods, and a diet that includes fruits, juices, green vegetables, and whole grains. The "How To" feature in Chapter 10 on p. 317 describes how folate from each of these sources contributes to a day's intake.

The pregnant woman also has a slightly greater need for the B vitamin that activates the folate enzyme—vitamin B_{12}. Generally, even modest amounts of meat, fish, eggs, or milk products together with body stores easily meet the need for vitamin B_{12}. Vegans who exclude all foods of animal origin, however, need daily supplements of vitamin B_{12} or vitamin B_{12}–fortified foods to prevent the neurological complications of a deficiency.

Pregnant women need iron to support their increased blood volume and to provide for placental and fetal needs.[48] The developing fetus draws on maternal iron stores to create sufficient stores of its own to last through the first 4 to 6 months after birth. Ideally, a woman enters pregnancy with adequate iron stores and maintains sufficient iron nutrition throughout the pregnancy.[49] The transfer of significant amounts of iron to the fetus is regulated by the placenta, which gives the iron needs of the fetus priority over those of the mother.[50] Women with inadequate iron stores are left with too little iron to meet their own health needs. In addition, blood losses are inevitable at birth and can further drain the mother's iron supply.

During pregnancy, the body makes several adaptations to help meet the exceptionally high need for iron. Menstruation, the major route of iron loss in women, ceases, and iron absorption improves thanks to an increase in transferrin, the body's iron-absorbing and iron-carrying protein. Without sufficient intake, though, iron stores quickly dwindle. Women with iron-deficiency anemia are likely to give birth to low-birthweight infants.[51]

Few women enter pregnancy with adequate iron stores, so a daily iron supplement is recommended early in pregnancy, if not before.[52] To enhance iron absorption, the supplement should be taken between meals or at bedtime and with liquids other than milk, coffee, or tea, which inhibit iron absorption. Drinking orange juice does not enhance iron absorption from supplements as it does from foods; vitamin C enhances iron absorption by converting iron from ferric to ferrous, but supplemental iron is already in the ferrous form. Vitamin C is helpful, however, in preventing the premature rupture of amniotic membranes.

> **DIETARY GUIDELINES FOR AMERICANS**
Women who are pregnant should take an iron supplement as recommended by an obstetrician or other health-care provider.

Zinc is required for DNA and RNA synthesis and thus for protein synthesis and cell development. Typical zinc intakes for pregnant women are lower than recommendations, but fortunately, zinc absorption increases when intakes are low.

Nutrients for Bone Development Vitamin D and the bone-building minerals calcium, phosphorus, magnesium, and fluoride are in great demand during pregnancy. All are needed to produce healthy fetal bones and teeth.[53]

Vitamin D plays a central role in calcium absorption and utilization. Consequently, severe maternal vitamin D deficiency interferes with normal calcium metabolism, resulting in rickets in the infant and osteomalacia in the mother.[54] Regular exposure to sunlight and consumption of vitamin D–fortified milk are usually sufficient to provide the recommended amount of vitamin D during pregnancy, which is the same as for nonpregnant women.[55] Pregnant women who do not receive sufficient dietary vitamin D or enough exposure to sunlight may need a supplement.

Calcium absorption and retention increases dramatically in pregnancy, helping the mother to meet the calcium needs of pregnancy. During the last trimester, as the fetal bones begin to calcify, up to 350 milligrams a day are transferred to the fetus. If the diet is inadequate in calcium, the mother's bones give up their calcium to meet fetal needs and become less dense.[56] Recommendations to ensure an adequate calcium intake during pregnancy help to conserve maternal bones while meeting fetal needs.

Calcium intakes for pregnant women typically fall below recommendations. Because bones are still actively depositing minerals until about age 30, adequate calcium is especially important for young women. Pregnant women younger than age 25 who receive less than 600 milligrams of dietary calcium daily need to increase their intake of milk, cheese, yogurt, and other calcium-rich foods. The USDA Food Patterns suggest consuming 3 cups per day of fat-free or low-fat milk or the equivalent in milk products. Alternatively, and less preferably, they may need a daily supplement of 600 milligrams of calcium, taken with meals.

Other Nutrients The nutrients mentioned here are those most intensely involved in blood production, cell growth, and bone development. Of course, other vitamins and minerals are also needed during pregnancy to support the growth and health of both fetus and mother.[57] Even with adequate nutrition, repeated pregnancies within a short time span can deplete nutrient reserves. Short intervals between pregnancies compromise the growth of the fetus and health of the mother.[58] The optimal interval between pregnancies is 18 to 23 months.

Nutrient Supplements A healthy pregnancy and optimal infant development depend on the mother's diet. Pregnant women who make wise food choices can meet most of their nutrient needs, with the possible exception of iron. Even so, physicians routinely recommend daily multivitamin-mineral supplements for pregnant women. Prenatal supplements typically contain greater amounts of folate, iron, and calcium than regular multivitamin-mineral supplements. These supplements are particularly beneficial for women who do not eat adequately and for those in high-risk groups: women carrying multiple fetuses, cigarette smokers, and alcohol and drug abusers. The use of prenatal supplements may help reduce the risks of preterm delivery, low infant birthweights, and birth defects.[59] Prenatal supplement use seems to support the infant's physical growth and cognitive development for the first couple of years as well.[60]

Vegetarian Diets during Pregnancy and Lactation
In general, a well-planned vegetarian diet can support a healthy pregnancy and successful lactation if it provides adequate energy and contains a wide variety of legumes, whole grains, nuts, seeds, fruits, and vegetables.[61] Many vegetarian women are well nourished, with nutrient intakes from diet alone exceeding the RDA for most vitamins and minerals except iron, which is low for most women. In contrast, vegan women who restrict themselves to an exclusively plant-based diet generally have low energy intakes and are underweight. For pregnant women, this can be a problem. Women with low prepregnancy weights and insufficient weight gains during pregnancy jeopardize a healthy pregnancy.

Vegan diets may require supplementation with vitamin B_{12}, calcium, and vitamin D, or the addition of foods fortified with these nutrients. Infants may suffer spinal cord damage and develop severe psychomotor retardation due to a lack of vitamin B_{12} in the mother's diet during pregnancy. Breastfed infants of vegan mothers have been reported to develop vitamin B_{12} deficiency and severe movement disorders. Giving infants vitamin B_{12} supplements corrects the blood and neurological symptoms of the deficiency, as well as the structural abnormalities, but cognitive and language development delays may persist. A pregnant woman needs a regular source of vitamin B_{12}-fortified foods or a supplement that provides 2.6 micrograms daily.

Common Nutrition-Related Concerns of Pregnancy
Nausea, constipation, heartburn, and food sensitivities are common nutrition-related concerns during pregnancy. A few simple strategies can help alleviate maternal discomforts (see Table 15-2).

TABLE 15-2 Strategies to Alleviate Maternal Discomforts

To Alleviate the Nausea of Pregnancy	To Prevent or Alleviate Constipation	To Prevent or Relieve Heartburn
• On waking, arise slowly. • Eat dry toast or crackers. • Chew gum or suck hard candies. • Eat small, frequent meals. • Avoid foods with offensive odors. • When nauseated, drink carbonated beverages instead of citrus juice, water, milk, coffee, or tea.	• Eat foods high in fiber (fruits, vegetables, and whole grains). • Exercise regularly. • Drink at least eight glasses of liquids a day. • Respond promptly to the urge to defecate. • Use laxatives only as prescribed by a physician; do not use mineral oil, because it interferes with absorption of fat-soluble vitamins.	• Relax and eat slowly. • Chew food thoroughly. • Eat small, frequent meals. • Drink liquids between meals. • Avoid spicy or greasy foods. • Sit up while eating; elevate the head while sleeping. • Wait 3 hours after eating before lying down. • Wait 2 hours after eating before exercising.

© Cengage Learning

Nausea and Vomiting Not all women have queasy stomachs in the early months of pregnancy, but many do. The nausea of "morning sickness" may actually occur anytime and ranges from mild queasiness to debilitating nausea and vomiting. Severe and continued vomiting may require hospitalization if it results in acidosis, dehydration, or excessive weight loss. The hormonal changes of early pregnancy seem to be responsible for a woman's sensitivities to the appearance, texture, or smell of foods. The problem typically peaks at 9 weeks of gestation and resolves within a month or two.[62] Traditional strategies for quelling nausea are listed in Table 15-2 (p. 487), but there is little evidence to support such advice.[63] Many women benefit most from simply resting when nauseous and eating the foods they want when they feel like eating. They may also find comfort in a clean, quiet, and temperate environment.

Constipation and Hemorrhoids As the hormones of pregnancy alter muscle tone and the growing fetus crowds intestinal organs, an expectant mother may experience constipation. She may also develop hemorrhoids (swollen veins of the rectum). Hemorrhoids can be painful, and straining during bowel movements may cause bleeding. She can gain relief by following the strategies listed in Table 15-2.

Heartburn Heartburn is another common complaint during pregnancy. The hormones of pregnancy relax the digestive muscles, and the growing fetus puts increasing pressure on the mother's stomach. This combination causes gastroesophageal reflux, the painful sensation a person feels behind the breastbone when stomach acid splashes back up into the lower esophagus (see Highlight 3). Tips to help relieve heartburn are included in Table 15-2.

Food Cravings and Aversions Some women develop cravings for, or aversions to, particular foods and beverages during pregnancy. **Food cravings** and **food aversions** are fairly common, but they do not seem to reflect real physiological needs. In other words, a woman who craves pickles does not necessarily need salt. Similarly, cravings for ice cream are common in pregnancy but do not signify a calcium deficiency. Cravings and aversions that arise during pregnancy are most likely due to hormone-induced changes in sensitivity to taste and smell.

Nonfood Cravings Some pregnant women develop cravings for nonfood items such as freezer frost, laundry starch, clay, soil, or ice—a practice known as **pica**.[64] Pica is a cultural phenomenon that reflects a society's folklore; it is especially common among African American women. Pica is often associated with iron-deficiency anemia, but whether iron deficiency leads to pica or pica leads to iron deficiency is unclear. Eating clay or soil may interfere with iron absorption and displace iron-rich foods from the diet.

> **REVIEW IT** Summarize the nutrient needs of women during pregnancy.
Energy and nutrient needs are high during pregnancy. A balanced diet that includes an extra serving from each of the five food groups can usually meet these needs, with the possible exception of iron and folate (supplements are recommended). The nausea, constipation, and heartburn that sometimes accompany pregnancy can usually be alleviated with a few simple strategies. Food cravings do not typically reflect physiological needs.

15.5 High-Risk Pregnancies

> **LEARN IT** Identify factors predicting low-risk and high-risk pregnancies and describe ways to manage them.

Some pregnancies jeopardize the life and health of the mother and infant. Table 15-3 identifies several risk factors for a **high-risk pregnancy**. A woman with none of these risk factors is said to have a **low-risk pregnancy**. The more factors that apply, the higher the risk. All pregnant women, especially those in high-risk categories, need prenatal medical care, including the following nutrition advice:

- Eat well-balanced meals.
- Gain enough weight to support fetal growth.

---

food cravings: strong desires to eat particular foods.

food aversions: strong desires to avoid particular foods.

pica (PIE-ka): a craving for and consumption of nonfood substances. Pica is known as *geophagia* (gee-oh-FAY-gee-uh) when referring to eating clay, baby powder, chalk, ash, ceramics, paper, paint chips, or charcoal; *pagophagia* (pag-oh-FAY-gee-uh) when referring to eating large quantities of ice; and *amylophagia* (AM-ee-low-FAY-gee-ah) when referring to eating uncooked starch (flour, laundry starch, or raw rice).

high-risk pregnancy: a pregnancy characterized by risk factors that make it likely the birth will be surrounded by problems such as premature delivery, difficult birth, restricted growth, birth defects, and early infant death.

low-risk pregnancy: a pregnancy characterized by factors that make it likely the birth will be normal and the infant healthy.

TABLE 15-3 High-Risk Pregnancy Factors

Factor	Condition that Raises Risk
Maternal weight	
• Prior to pregnancy	Prepregnancy BMI either <18.5 or ≥25
• During pregnancy	Insufficient or excessive pregnancy weight gain (see Table 15-1, p. 481)
Maternal nutrition	Nutrient deficiencies or toxicities; eating disorders
Socioeconomic status	Poverty, lack of family support, low level of education, limited food availability
Lifestyle habits	Smoking, alcohol or other drug use
Age	Teens, especially 15 years or younger; women 35 years or older
Previous pregnancies	
• Number	Many previous pregnancies (3 or more to mothers younger than age 20; 4 or more to mothers age 20 or older)
• Interval	Short or long intervals between pregnancies (<18 months or >59 months)
• Outcomes	Previous history of problems
• Multiple births	Twins or triplets
• Birthweight	Low- or high-birthweight infants
Maternal health	
• High blood pressure	Development of gestational hypertension
• Diabetes	Development of gestational diabetes
• Chronic diseases	Diabetes; heart, respiratory, and kidney disease; certain genetic disorders; special diets and medications

© Cengage Learning

- Take prenatal supplements as prescribed.
- Stop drinking alcohol.

The Infant's Birthweight

A high-risk pregnancy is likely to produce an infant with **low birthweight (LBW)**. Low-birthweight infants, defined as infants who weigh 5½ pounds or less, are classified according to their gestational age. Preterm infants are born before they are fully developed; they are often underweight and have trouble breathing because their lungs are immature. Preterm infants may be small, but if their size and weight are appropriate for their gestational age, they can catch up in growth given adequate nutrition support. In contrast, small-for-gestational-age infants have suffered growth failure in the uterus and do not catch up as well. For the most part, survival improves with increased gestational age and birthweight.

Low-birthweight infants are more likely to experience complications during delivery than normal-weight babies. They also have a statistically greater chance of having physical and mental birth defects, becoming ill, and dying early in life. Of infants who die before their first birthdays, about two-thirds were low-birthweight newborns. Very-low-birthweight infants (3½ pounds or less) struggle not only for their immediate physical health and survival, but for their future cognitive development and abilities as well.

A strong association is seen between socioeconomic disadvantage and low birthweight. Low socioeconomic status impairs fetal development by causing stress and by limiting access to medical care and nutritious foods. Low socioeconomic status often accompanies teen pregnancies, smoking, and alcohol and drug abuse—all predictors of low birthweight.

Malnutrition and Pregnancy

Good nutrition clearly supports a healthy pregnancy. In contrast, malnutrition interferes with the ability to conceive, the likelihood of implantation, and the subsequent development of a fetus should conception and implantation occur.

Malnutrition and Fertility

The nutrition habits and lifestyle choices people make can influence the course of a pregnancy they are not even planning at the time. Inadequate nutrition and food deprivation can reduce fertility because women

© Photodisc/Jupiterimages

Low-birthweight babies need special care and nourishment.

low birthweight (LBW): a birthweight of 5½ pounds (2500 grams) or less; indicates probable poor health in the newborn and poor nutrition status in the mother during pregnancy, before pregnancy, or both. Optimal birthweight for a full-term baby is about 6½ to 8 pounds.

The nutrition education and nutritious foods WIC provides to infants, children to age 5, and pregnant and breastfeeding women improves health and saves lives.

may develop amenorrhea—the temporary or permanent absence of menstrual periods.* Men who are poorly nourished may be unable to produce viable sperm. Furthermore, both men and women lose sexual interest during times of starvation. Starvation arises predictably during famines, wars, and droughts, but it can also occur amid peace and plenty. Many young women who diet excessively are starving and suffering from malnutrition (see Highlight 8).

Malnutrition and Early Pregnancy If a poorly nourished woman does become pregnant, she faces the challenge of supporting both the growth of a baby and her own health with inadequate nutrient stores. Inadequate nutrition prior to and around conception prevents the placenta from developing fully. A poorly developed placenta cannot deliver optimal nourishment to the fetus, and the infant will be born small and possibly with physical and cognitive abnormalities. If this small infant is a female, she may develop poorly and have an elevated risk of developing a chronic condition that could impair her ability to give birth to a healthy infant. Thus a woman's poor nutrition status can adversely affect not only her children but also her *grandchildren*.

Malnutrition and Fetal Development Without adequate nutrition during pregnancy, fetal growth and infant health are compromised.[65] In general, consequences of inadequate nutrition during pregnancy include fetal growth restriction, congenital malformations (birth defects), spontaneous abortion and stillbirth, preterm birth, and low infant birthweight. Nutrient deficiencies, coupled with low birthweight, are responsible for more than half of all deaths of children younger than 4 years of age worldwide.

Food Assistance Programs
Women in high-risk pregnancies can find assistance from the WIC program—a high-quality, cost-effective health-care and nutrition services program for women, infants, and children in the United States. Formally known as the Special Supplemental Nutrition Program for Women, Infants, and Children, WIC provides nutrition education and nutritious foods to infants, children to age 5, and pregnant and breastfeeding women who qualify financially and have a high risk of medical or nutritional problems. The program is both remedial and preventive: services include health-care referrals, nutrition education, and food packages or vouchers for specific foods. These foods supply nutrients known to be lacking in the diets of the target population—most notably, protein, calcium, iron, vitamin A, and vitamin C. WIC-sponsored foods include tuna, tofu, fruits, vegetables, eggs, milk, iron-fortified cereal, whole-grain breads, vitamin C–rich juices, cheeses, legumes, peanut butter, and iron-fortified infant formula and cereal.

More than 9 million people—most of them young children—receive WIC benefits each month. Prenatal WIC participation can effectively reduce iron deficiency, infant mortality, low birthweight, and maternal and newborn medical costs. In 2012, Congress appropriated more than $6.6 billion for WIC. For every dollar spent on WIC, an estimated $3 in medical costs are saved in the first 2 months after birth. Table 15-4 presents some of the many benefits of WIC.

Maternal Health
Medical disorders can threaten the life and health of both mother and fetus. If diagnosed and treated early, many diseases can be managed to ensure a healthy outcome—another strong argument for early prenatal care. Furthermore, the changes in pregnancy can reveal disease risks, making screening important and early intervention possible.

Preexisting Diabetes The risks of diabetes depend on how well it is managed before and during pregnancy. Without proper management of maternal diabetes, women face high infertility rates, and those who do conceive may experience episodes of severe hypoglycemia or hyperglycemia, preterm labor, and pregnancy-related hypertension. Infants may be large, suffer physical and mental

TABLE 15-4 Benefits of WIC

- Earlier prenatal care
- Better diet during pregnancy
- Better weight gain during pregnancy
- Longer duration of pregnancy
- Fewer fetal and infant deaths
- Fewer low-birthweight infants
- Better growth in infants and children
- Less iron-deficiency anemia in children
- Better diet for children
- Better medical care for children
- Better preparation for school
- Improved intellectual development

*Amenorrhea is normal before puberty, after menopause, during pregnancy, and during lactation; otherwise it is abnormal.

abnormalities, and experience other complications such as severe hypoglycemia or respiratory distress, both of which can be fatal. Signs of fetal health problems are apparent even when maternal glucose is above normal but still below the diagnosis of diabetes. To minimize complications, a woman needs to achieve glucose control before conception and continued glucose control throughout pregnancy.[66]

Gestational Diabetes An estimated 2 to 10 percent of pregnancies in the United States are complicated by a condition known as **gestational diabetes.**[67] Gestational diabetes usually develops during the second half of pregnancy, with subsequent return to normal after childbirth. Some women with gestational diabetes, however, develop diabetes (usually type 2) after pregnancy, especially if they are overweight. For this reason, health-care professionals strongly advise against excessive weight gain during—and after—pregnancy. Weight gains after pregnancy increase the risk of gestational diabetes in the next pregnancy.[68]

The most common consequences of gestational diabetes are complications during labor and delivery and a high infant birthweight. Birth defects associated with gestational diabetes include heart damage, limb deformities, and neural tube defects. To ensure that the problems of gestational diabetes are dealt with promptly, physicians screen for the risk factors listed in Table 15-5 and test high-risk women for glucose intolerance immediately and average-risk women between 24 and 28 weeks gestation.[69]

Dietary recommendations should meet the needs of pregnancy and control maternal blood glucose.[70] Diet and moderate exercise may control gestational diabetes, but if blood glucose fails to normalize, insulin or other drugs may be required. Importantly, treatment reduces preeclampsia, birth complications, large newborns, and infant deaths.[71]

Chronic Hypertension Hypertension complicates pregnancy and affects its outcome in different ways, depending on when the hypertension first develops and on how severe it becomes.[72] In addition to the threats hypertension always carries (such as heart attack and stroke), high blood pressure increases the risks of fetal growth restriction, preterm birth, and separation of the placenta from the wall of the uterus before the birth, resulting in stillbirth.[73] To minimize complications, blood pressure needs to be under control before a woman with hypertension becomes pregnant.

Gestational Hypertension Women with chronic hypertension have a greater likelihood of developing **gestational hypertension**—high blood pressure during the second half of pregnancy.* For some women with gestational hypertension, the rise in blood pressure is mild and does not affect the pregnancy adversely. Blood pressure usually returns to normal during the first few weeks after childbirth. For others, gestational hypertension increases the risks of subsequent hypertension and type 2 diabetes. Gestational hypertension is also an early sign of the most serious maternal complication of pregnancy—preeclampsia.

Preeclampsia **Preeclampsia** is a condition characterized not only by gestational hypertension but also by protein in the urine. Table 15-6 presents the signs and symptoms of preeclampsia. The cause of preeclampsia remains unclear, but it usually occurs with first pregnancies and most often after 20 weeks of gestation. Obesity may increase the risk.[74] Symptoms typically regress within 2 days of delivery. Both men and women who were born of pregnancies complicated by preeclampsia are more likely to have a child born of a pregnancy complicated by preeclampsia, suggesting a genetic predisposition. They also tend to have a higher BMI and increased blood pressure during childhood and adolescence, indicating a greater risk for heart disease.[75] Black women have a much greater risk of preeclampsia than white women.

Preeclampsia affects almost all of the mother's organs—the circulatory system, liver, kidneys, and brain. Blood flow through the vessels that supply oxygen and nutrients to the placenta diminishes. For this reason, preeclampsia often restricts fetal growth. It also seems to increase the risk of epilepsy for the infant. In some cases, the placenta separates from the uterus, resulting in preterm birth or stillbirth.

---

*Blood pressure of 140/90 millimeters of mercury or greater during the second half of pregnancy in a woman who has not previously exhibited hypertension indicates high blood pressure.

TABLE 15-5 Risk Factors for Gestational Diabetes

- Age 25 or older
- BMI ≥25 or excessive weight gain
- Complications in previous pregnancies, including gestational diabetes or high-birthweight infant (>9 lb)
- Prediabetes or symptoms of diabetes
- Family history of type 2 diabetes
- Hispanic, African American, Native American, Asian, Pacific Islander

© 2016 Cengage Learning

TABLE 15-6 Signs and Symptoms of Preeclampsia

- Hypertension
- Protein in the urine
- Upper abdominal pain
- Severe headaches
- Swelling of hands, feet, and face
- Vomiting
- Blurred vision
- Sudden weight gain (1 lb/day)
- Fetal growth restriction

© 2016 Cengage Learning

gestational diabetes: glucose intolerance with onset or first recognition during pregnancy.

gestational hypertension: high blood pressure that develops in the second half of pregnancy and resolves after childbirth, usually without affecting the outcome of the pregnancy.

preeclampsia (PRE-ee-KLAMP-see-ah): a condition characterized by high blood pressure and some protein in the urine.

Pregnancy during the teen years has major health, emotional, social, and financial consequences for young mothers and their children.

Preeclampsia can progress rapidly to **eclampsia**—a condition characterized by seizures and coma. Maternal death during pregnancy and childbirth is rare in developed countries, but when it does occur, eclampsia is a common cause.[76] The rate of death for black women with eclampsia is more than four times the rate for white women.

Preeclampsia demands prompt medical attention. Treatment focuses on controlling blood pressure and preventing seizures. If preeclampsia develops early and is severe, induced labor or cesarean delivery may be necessary, regardless of gestational age. The infant will be preterm, with all of the associated problems, including poor lung development and special care needs. Several dietary factors have been studied, but none have proved beneficial in preventing preeclampsia.

The Mother's Age
Maternal age also influences the course of a pregnancy. Compared with women of the physically ideal childbearing age of 20 to 25, both younger and older women face more complications of pregnancy.

Pregnancy in Adolescents Almost 40 percent of adolescents are sexually active before age 19.[77] More than 350,000 adolescent girls give birth each year in the United States; almost 20 percent of them already have at least one child.[78] Nourishing a growing fetus adds to a teenage girl's nutrition burden, especially if her growth is still incomplete. Simply being young and physically immature increases the risks of pregnancy complications. Pregnant teens are less likely to receive early prenatal care and are more likely to smoke during pregnancy—two factors that predict low birthweight and infant death.

The typical energy-dense, but nutrient-poor diet of pregnant adolescents increases the risk of low-birthweight infants. Common complications among adolescent mothers include iron-deficiency anemia (which may reflect poor diet and inadequate prenatal care) and prolonged labor (which reflects the mother's physical immaturity). On a positive note, maternal death is lowest for mothers younger than age 20.

The rates of stillbirths, preterm births, and low-birthweight infants are high for teenagers—both for teen moms and for teen dads. Many of these infants suffer physical problems, require intensive care, and die within the first year. The care of teen mothers and their infants costs our society billions of dollars annually. Because teenagers have few financial resources, they cannot pay these costs. Furthermore, their low economic status contributes significantly to the complications surrounding their pregnancies. At a time when prenatal care is most important, it is less accessible. And the pattern of teenage pregnancies continues from generation to generation, with daughters of teenage mothers more likely to become teenage mothers themselves. Clearly, teenage pregnancy is a major public health problem.

To support the needs of both mother and fetus, young teenagers (13 to 16 years old) are encouraged to strive for the highest weight gains recommended for pregnancy. For a teen who enters pregnancy at a healthy body weight, a weight gain of approximately 35 pounds is recommended; this amount minimizes the risk of delivering a low-birthweight infant. Pregnant and lactating teenagers can use the food patterns presented in Table 2-3 (p. 43), making sure to select a high enough kcalorie level to support adequate weight gain.

Without the appropriate economic, social, and physical support, a young mother will not be able to care for herself during her pregnancy and for her child after the birth. To improve her chances for a successful pregnancy and a healthy infant, she must seek prenatal care. WIC provides health-care referrals and helps pregnant teenagers obtain adequate food for themselves and their infants. (WIC is introduced on p. 490.)

Pregnancy in Older Women In the past several decades, many women have delayed childbearing while they pursue education and careers. As a result, the number of first births to women 35 and older has increased dramatically. Most of these women, even those older than age 50, have healthy pregnancies.

eclampsia (eh-KLAMP-see-ah): a condition characterized by extremely high blood pressure, elevated protein in the urine, seizures, and possibly coma.

The few complications associated with later childbearing often reflect chronic conditions such as hypertension and diabetes, which can complicate an otherwise healthy pregnancy. These complications may result in a cesarean delivery, which is twice as common in women older than 35 as among younger women. For all these reasons, maternal death rates are higher in women older than 35 than in younger women.

The babies of older mothers face problems of their own, including higher rates of preterm births and low birthweight. Their rates of birth defects are also high. Because 1 out of 50 pregnancies in older women produces an infant with genetic abnormalities, obstetricians routinely screen women older than 35. For a 40-year-old mother, the risk of having a child with **Down syndrome,** for example, is about 1 in 100 compared with 1 in 300 for a 35-year-old and 1 in 10,000 for a 20-year-old. In addition, fetal death is twice as high for women 35 years and older than for younger women. Why this is so remains unclear. One possibility is that the uterine blood vessels of older women may not fully adapt to the increased demands of pregnancy.

Practices Incompatible with Pregnancy

Besides malnutrition, a variety of lifestyle factors can have adverse effects on pregnancy, and some may be **teratogenic,** causing abnormal fetal development and birth defects. By practicing healthy behaviors, people who are planning to have children can reduce the risks.

Alcohol One out of 13 pregnant women drinks alcohol at some time during her pregnancy; 1 out of 75 reports binge drinking.[79] Alcohol consumption during pregnancy can cause the irreversible mental and physical retardation of the fetus known as fetal alcohol syndrome (FAS). Of the leading causes of mental retardation, FAS is the only one that is totally *preventable.* To that end, the surgeon general urges all pregnant women to refrain from drinking alcohol. Fetal alcohol syndrome is the topic of Highlight 15, which includes mention of how alcohol consumption by men may also affect fertility and fetal development. In addition to fetal alcohol syndrome, maternal alcohol use has been associated with SIDS and leukemia.[80]

> **DIETARY GUIDELINES FOR AMERICANS**
> Women who are pregnant should not drink alcohol.

Medicinal Drugs Drugs other than alcohol can also cause complications during pregnancy, problems in labor, and serious birth defects. For these reasons, pregnant women should not take any medicines without consulting their physicians, who must weigh the benefits against the risks.

Herbal Supplements Similarly, pregnant women should seek a physician's advice before using herbal supplements. Women sometimes seek herbal preparations during their pregnancies to quell nausea, induce labor, aid digestion, promote water loss, support restful sleep, and fight depression. As Highlight 18 explains, some herbs may be safe, but others may be harmful.

Illicit Drugs The recommendation to avoid drugs during pregnancy also includes illicit drugs, of course. Unfortunately, use of illicit drugs, such as cocaine, methamphetamine, and marijuana, is common among some pregnant women.[81]

Drugs of abuse, such as cocaine, easily cross the placenta and impair fetal growth and development. Furthermore, they are responsible for preterm births, low-birthweight infants, **perinatal** deaths, and sudden infant deaths. If these newborns survive, central nervous system damage is evident: their cries, sleep, and behaviors early in life are abnormal, and their cognitive development later in life is impaired.[82] They may be hypersensitive or underaroused; many suffer the symptoms of withdrawal. Their growth and development throughout childhood and adolescence continues to be delayed.[83]

Smoking and Chewing Tobacco Unfortunately, an estimated 12 percent of pregnant women in the United States smoke.[84] Smoking cigarettes and chewing tobacco at any time exert harmful effects, and pregnancy dramatically magnifies the hazards of these practices. Smoking restricts the blood supply to the growing fetus

Down syndrome: a genetic abnormality that causes mental retardation, short stature, and flattened facial features.

teratogenic (ter-AT-oh-jen-ik): causing abnormal fetal development and birth defects.

perinatal: referring to the time between the twenty-eighth week of gestation and 1 month after birth.

and thus limits oxygen and nutrient delivery and waste removal. A mother who smokes is more likely to have a complicated birth and a low-birthweight infant. Indeed, of all preventable causes of low birthweight in the United States, smoking is at the top of the list. Although most infants born to cigarette smokers are low birthweight, some are not, suggesting that the effect of smoking on birthweight also depends, in part, on genes involved in the metabolism of smoking toxins.

In addition to contributing to low birthweight, smoking interferes with heart and lung growth and arterial structure and function; consequently, it increases the risks of heart defects, poor lung function, respiratory infections, and childhood asthma.[85] It can also cause death in an otherwise healthy fetus or newborn. A positive relationship exists between **sudden infant death syndrome (SIDS)** and both cigarette smoking during pregnancy and postnatal exposure to passive smoke. Smoking during pregnancy may reduce brain size and impair the intellectual and behavioral development of the child later in life. Table 15-7 lists complications of smoking during pregnancy.

Alternatives to smoking—such as using snuff or chewing tobacco—are not safe during pregnancy.[86] The safety and effectiveness of nicotine-replacement patches has not been determined, and therefore this therapy cannot be recommended at this time.[87] Any woman who uses nicotine in any form and is considering pregnancy or who is already pregnant needs to quit. Avoiding secondhand smoke is also advised. Pregnant women exposed to secondhand smoke during pregnancy are more likely to experience complications such as stillbirth and the birth of an infant with congenital malformations.[88]

Environmental Contaminants Proving that environmental contaminants cause reproductive damage is difficult, but evidence in wildlife is established and seems likely for human beings. Infants and young children of pregnant women exposed to environmental contaminants such as lead show signs of delayed mental and psychomotor development. During pregnancy, lead readily crosses into the placenta, inflicting severe damage on the developing fetal nervous system. In addition, infants exposed to even low levels of lead during gestation weigh less at birth and consequently struggle to survive. For these reasons, it is particularly important that pregnant women receive foods and beverages grown and prepared in environments free of contamination.

Mercury is another contaminant of concern. As Chapter 5 mentions, fatty fish are a good source of omega-3 fatty acids, but some fish contain large amounts of the pollutant mercury, which can impair fetal growth and harm the developing brain and nervous system. Because the benefits of seafood consumption seem to outweigh the risks, pregnant (and lactating) women need reliable information on which fish are safe to eat.[89] In general, they need to do the following:

- Avoid shark, swordfish, king mackerel, and tilefish (also called golden snapper or golden bass).
- Limit average weekly consumption to 12 ounces (cooked or canned) of seafood *or* to 6 ounces (cooked or canned) of white (albacore) tuna.

Ideally, pregnant (and lactating) women will select fish that are both high in omega-3 fatty acids and low in mercury.[90] Supplements of fish oil are not recommended because they may contain concentrated toxins and because their effects on pregnancy remain unknown.

TABLE 15-7 Complications Associated with Smoking during Pregnancy

- Fetal growth restriction
- Preterm birth
- Low birthweight
- Premature separation of the placenta
- Miscarriage
- Stillbirth
- Sudden infant death syndrome (SIDS)
- Congenital malformations

© 2016 Cengage Learning

> **DIETARY GUIDELINES FOR AMERICANS**
>
> Women who are pregnant or breastfeeding should consume 8 to 12 ounces of seafood per week from a variety of seafood types. Fish relatively high in omega-3 fatty acids and low in mercury include anchovies, bonito, eel, herring, mackerel, pollock, salmon, sardines, smelt, tilapia, and trout. Because of the high methyl mercury content of some types of fish, pregnant or breastfeeding women should avoid tilefish, shark, swordfish, and king mackerel and should limit white (albacore) tuna to 6 ounces per week.

sudden infant death syndrome (SIDS): the unexpected and unexplained death of an apparently well infant; the most common cause of death of infants between the second week and the end of the first year of life; also called *crib death*.

Foodborne Illness As Chapter 19 explains, foodborne illnesses arise when people eat foods that contain infectious microbes or microbes that produce toxins. At best, the vomiting and diarrhea associated with these illnesses can leave a pregnant woman exhausted and dehydrated; at worst, foodborne illnesses can cause meningitis, pneumonia, or even fetal death. Pregnant women are about 20 times more likely than other healthy adults to get the foodborne illness **listeriosis.** Table 15-8 presents tips to prevent listeriosis; Chapter 19 includes precautions to minimize the risks of other common foodborne illnesses.

> DIETARY GUIDELINES FOR AMERICANS

Women who are pregnant should:

- Eat foods containing seafood, meat, poultry, or eggs only if cooked to recommended safe minimum internal temperatures.

- Take special precautions not to consume unpasteurized juice or milk products.

- Reheat deli and luncheon meats and hot dogs to steaming hot and not eat raw sprouts.

Vitamin-Mineral Megadoses Pregnant women who are trying to eat well may mistakenly assume that more is better when it comes to multivitamin-mineral supplements. This is simply not true; many vitamins and minerals are toxic when taken in excess. Excessive vitamin A is particularly infamous for its role in fetal malformations.[91] Intakes before the seventh week appear to be the most damaging. (Review Figure 15-4 on p. 477 to see how many tissues are in their critical periods prior to the seventh week.) For this reason, vitamin A supplements are not given during pregnancy unless there is specific evidence of deficiency, which is rare. A pregnant woman can obtain all the vitamin A and most of the other vitamins and minerals she needs by making wise food choices. She should take supplements only on the advice of a registered dietitian nutritionist or physician.

Caffeine Caffeine crosses the placenta, and the developing fetus has a limited ability to metabolize it. Research studies have not proved that caffeine (even in high doses) causes birth defects or preterm births in human infants (as it does in animals), but limited evidence suggests that heavy use increases the risk of hypertension, miscarriage, and stillbirth.[92] (In these studies, heavy caffeine use is defined as the equivalent of more than 3 cups of coffee a day.) Depending on the quantities consumed and the mother's metabolism, caffeine may also interfere with fetal growth.[93] Lower doses of caffeine—say, at levels equivalent to 3 cups of coffee or less—appear to be compatible with healthy pregnancies.[94] All things considered, it may be most sensible to limit caffeine consumption to the equivalent of a cup or two of coffee a day. (The caffeine contents of selected beverages, foods, and drugs are listed at the beginning of Appendix H.)

Restrictive Dieting Restrictive dieting, even for short periods, can be hazardous during pregnancy. Low-carbohydrate diets or fasts that cause ketosis deprive the fetal brain of needed glucose and may impair cognitive development. Such diets are also likely to lack other nutrients vital to fetal growth. Regardless of prepregnancy weight, pregnant women need an adequate diet to support healthy fetal development.

Sugar Substitutes Artificial sweeteners have been approved by the FDA and are generally considered safe to use during pregnancy.[95] Recent studies, however, have reported an association between a high intake of artificially sweetened soft drinks and an increased risk of preterm births.[96] Pregnant women might be wise to use sweeteners in moderation, if at all, and within an otherwise nutritious and well-balanced diet. Women with the inherited disease phenylketonuria (PKU) should not use the artificial sweetener aspartame. Aspartame contains the amino acid phenylalanine, and people with PKU are unable to dispose of any excess phenylalanine. The accumulation of phenylalanine and its by-products is toxic to the developing nervous system, causing irreversible brain damage.

TABLE 15-8 Tips to Prevent Listeriosis

- Use only pasteurized juices and dairy products; do not eat soft cheeses such as feta, brie, Camembert, Panela, "queso blanco," "queso fresco," and blue-veined cheeses such as Roquefort; do not drink raw (unpasteurized) milk or eat foods that contain it.
- Thoroughly cook meat, poultry, eggs, and seafood.
- Do not eat hot dogs or luncheon meats unless heated until steaming hot.
- Wash all fruits and vegetables.
- Do not eat refrigerated pâtés or meat spreads.
- Do not eat refrigerated smoked seafood such as salmon or trout, or any fish labeled "nova," "lox," or "kippered," unless prepared in a cooked dish.

© 2016 Cengage Learning

Prenatal dietary supplements may be beneficial, but megadoses can be toxic.

© nata-lunata/Shutterstock.com

listeriosis (lis-TEAR-ee-OH-sis): an infection caused by eating food contaminated with the bacterium *Listeria monocytogenes,* which can be killed by pasteurization and cooking but can survive at refrigerated temperatures; certain ready-to-eat foods, such as hot dogs and deli meats, may become contaminated after cooking or processing, but before packaging.

> **REVIEW IT** Identify factors predicting low-risk and high-risk pregnancies and describe ways to manage them.

High-risk pregnancies, especially for teenagers, threaten the life and health of both mother and infant. Proper nutrition and abstinence from smoking, alcohol, and other drugs improve the outcome. In addition, prenatal care includes monitoring pregnant women for gestational diabetes, gestational hypertension, and preeclampsia.

15.6 Nutrition during Lactation

> **LEARN IT** Summarize the nutrient needs of women during lactation.

Childbirth marks the end of pregnancy and the beginning of a new set of parental responsibilities—including feeding the newborn. Before the end of her pregnancy, a woman needs to consider whether to feed her infant breast milk, infant formula, or both. These options are the only recommended foods for an infant during the first 4 to 6 months of life. The current rate of breastfeeding met the Healthy People goal of 75 percent at birth, but it falls short of goals at 3 months, 6 months, and 1 year.[97] This section focuses on how the mother's nutrition supports the making of breast milk, and the next chapter describes how the infant benefits from drinking breast milk.

In many countries around the world, a woman breastfeeds her newborn without considering the alternatives or making a conscious decision. In other parts of the world, a woman feeds her newborn formula simply because she knows so little about breastfeeding. She may have misconceptions or feel uncomfortable about a process she has never seen or experienced. Breastfeeding offers many health benefits to both mother and infant, and every pregnant woman should seriously consider it (see Table 15-9).[98] Even so, women's choices are often influenced by factors other than health and science—factors such as culture, politics, religion, and marketing. In any case, keep in mind that mothers may have valid reasons for not breastfeeding and that formula-fed infants grow and develop into healthy children.

lactation: production and secretion of breast milk for the purpose of nourishing an infant.

mammary glands: glands of the female breast that secrete milk.

prolactin (pro-LAK-tin): a hormone secreted from the anterior pituitary gland that acts on the mammary glands to promote the production of milk. The release of prolactin is mediated by *prolactin-inhibiting hormone (PIH)*.

- **pro** = promote
- **lacto** = milk

oxytocin (OCK-see-TOH-sin): a hormone that stimulates the mammary glands to eject milk during lactation and the uterus to contract during childbirth.

let-down reflex: the reflex that forces milk to the front of the breast when the infant begins to nurse.

TABLE 15-9 Benefits of Breastfeeding

For Infants

- Provides the appropriate composition and balance of nutrients with high bioavailability
- Provides hormones that promote physiological development
- Improves cognitive development
- Protects against a variety of infections and illnesses, including diarrhea, ear infections, and pneumonia
- May protect against some chronic diseases—such as diabetes (both types), obesity, atherosclerosis, asthma, some cancers, and hypertension—later in life
- Protects against food allergies
- Reduces the risk of SIDS
- Supports healthy weight

For Mothers

- Contracts the uterus
- Delays the return of regular ovulation, thus lengthening birth intervals (this is not, however, a dependable method of contraception)
- Conserves iron stores (by prolonging amenorrhea)
- May protect against breast and ovarian cancer and reduce the risk of diabetes (type 2)
- Increases energy expenditure, which may contribute to weight loss

Other

- Cost and time savings from not needing medical treatment for childhood illnesses or leaving work to care for sick infants
- Cost and time savings from not needing to purchase and prepare formula (even after adjusting for added foods in the diet of a lactating mother)[a]
- Environmental savings to society from not needing to manufacture, package, and ship formula and dispose of the packaging
- Convenience of not having to shop for and prepare formula

[a]Estimated savings of $1200–$1500 in the first year.

© Cengage Learning

Lactation: A Physiological Process

Lactation naturally follows pregnancy, as the mother's body continues to nourish the infant. The **mammary glands** secrete milk for this purpose. The mammary glands develop during puberty but remain fairly inactive until pregnancy. During pregnancy, hormones promote the growth and branching of a duct system in the breasts and the development of the milk-producing cells.

The hormones **prolactin** and **oxytocin** finely coordinate lactation. The infant's demand for milk stimulates the release of these hormones, which signal the mammary glands to supply milk. Prolactin is responsible for milk production. As long as the infant is nursing, prolactin concentrations remain high, and milk production continues.

The hormone oxytocin causes the mammary glands to eject milk into the ducts, a response known as the **let-down reflex.** The mother feels this reflex as a contraction of the breast, followed by the flow of milk and the release of pressure. By relaxing and eating well, the nursing mother promotes easy let-down of milk and greatly enhances her chances of successful lactation.

Breastfeeding: A Learned Behavior Lactation is an automatic physiological process that virtually all mothers are capable of doing. Breastfeeding, on the other hand, is a learned behavior that not all mothers decide to do. Of women who do breastfeed, those who receive early and repeated information and support breastfeed their infants longer than others. Mothers who are confident and committed are most successful in breastfeeding, especially when challenged by obstacles such as a lack of support from friends and family. Health-care professionals play an important role in providing encouragement and accurate information on breastfeeding. Especially helpful are **certified lactation consultants,** who specialize in helping new mothers establish a healthy breastfeeding relationship with their newborn. These consultants are often registered nurses with specialized training in breast and infant anatomy and physiology. Women who have been successful breastfeeding can offer advice and dispel misperceptions about lifestyle issues. Table 15-10 lists tips to promote successful breastfeeding among new mothers.

The mother's partner also plays an important role in encouraging breastfeeding. When partners support the decision, mothers are more likely to start and continue breastfeeding. Clearly, educating those closest to the mother could change attitudes and promote breastfeeding.

Most healthy women who want to breastfeed can do so with a little preparation. Physical obstacles to breastfeeding are rare, although most nursing mothers quit within a few months because of perceived difficulties. Obese mothers seem to have a particularly difficult time because of both biological and sociocultural factors.[99] Successful breastfeeding requires adequate nutrition and rest. This, plus the support of all who care, will help to enhance the well-being of mother and infant.

Maternal Energy and Nutrient Needs during Lactation
Ideally, the mother who chooses to breastfeed her infant will continue to eat nutrient-dense foods throughout lactation. An adequate diet is needed to support the stamina, patience, and self-confidence that nursing an infant demands.

Energy Intake and Exercise A nursing mother produces about 25 ounces of milk per day, with considerable variation from woman to woman and in the same woman from time to time, depending primarily on the infant's demand for milk. To produce an adequate supply of milk, a woman needs extra energy—almost 500 kcalories a day above her regular need during the first 6 months of lactation. To meet this energy need, she can eat an extra 330 kcalories of food each day during the first 6 months and an extra 400 kcalories each day during the second 6 months and the fat reserves she accumulated during pregnancy can provide the rest. Most women need at least 1800 kcalories a day to receive all the nutrients required for successful lactation. Severe energy restriction may hinder milk production.

After the birth of the infant, many women actively try to lose the extra weight and body fat they accumulated during pregnancy. How much weight a woman retains after pregnancy depends on her gestational weight gain and the duration and intensity of breastfeeding.[100] Many women who follow recommendations for gestational weight gain and breastfeeding can readily return to prepregnancy weight by 6 months. Neither the quality nor the quantity of breast milk is adversely affected by moderate weight loss, and infants grow normally.

A woman who decides to breastfeed provides her infant with a full array of nutrients and protective factors to support optimal health and development.

TABLE 15-10 Tips for Successful Breastfeeding

- Learn about the benefits of breastfeeding
- Initiate breastfeeding within 1 hour after birth
- Ask a health-care professional to explain how to breastfeed and how to maintain lactation
- Give newborn infants no food or drink other than breast milk, unless medically indicated
- Breastfeed on demand
- Give no artificial nipples or pacifiers to breastfeeding infants[a]
- Find breastfeeding support groups, books, or websites to help troubleshoot breastfeeding problems

[a]Compared with nonusers, infants who use pacifiers breastfeed less frequently and stop breastfeeding at a younger age.

© Cengage Learning

certified lactation consultants: health-care providers who specialize in helping new mothers establish a healthy breastfeeding relationship with their newborn. These consultants are often registered nurses with specialized training in breast and infant anatomy and physiology.

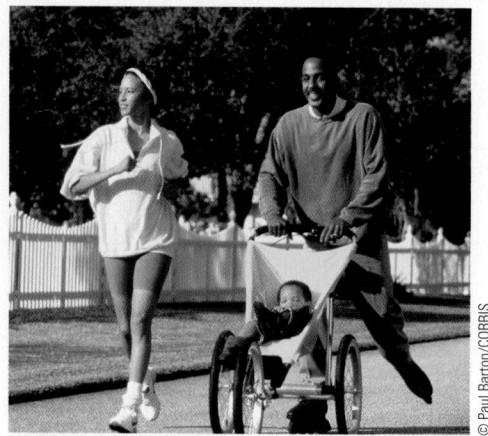

A jog through the park provides an opportunity for physical activity and fresh air.

Women often exercise to lose weight and improve fitness, and this is compatible with breastfeeding and infant growth.[101] Because intense physical activity can raise the lactate concentration of breast milk and influence the milk's taste, some infants may prefer milk produced prior to exercise. In these cases, mothers can either breastfeed before exercise or express their milk before exercise for use afterward.

> DIETARY GUIDELINES FOR AMERICANS
Maintain appropriate kcalorie balance during breastfeeding.

Energy Nutrients Recommendations for protein and fatty acids remain about the same during lactation as during pregnancy, but they increase for carbohydrates. Nursing mothers need additional carbohydrate to replace the glucose used to make the lactose in breast milk. The fiber recommendation is 1 gram higher simply because it is based on kcalorie intake, which increases during lactation.

Vitamins and Minerals A question often raised is whether a mother's milk may lack a nutrient if she fails to get enough in her diet. The answer differs from one nutrient to the next, but in general, nutrient inadequacies reduce the *quantity*, not the *quality*, of breast milk. Women can produce milk with adequate protein, carbohydrate, fat, and most minerals, even when their own supplies are limited. For these nutrients and for the vitamin folate as well, milk quality is maintained at the expense of maternal stores. This is most evident in the case of calcium: dietary calcium has no effect on the calcium concentration of breast milk, but maternal bones lose some density during lactation if calcium intakes are inadequate. Exercise may help protect against bone loss during lactation. The nutrients in breast milk that are most likely to decline in response to prolonged inadequate intakes are the vitamins—especially vitamins B_6, B_{12}, A, and D. Review Figure 15-9 (p. 484) to compare a lactating woman's nutrient needs with those of pregnant and nonpregnant women.

Water Despite misconceptions, a mother who drinks more fluid does not produce more breast milk. To protect herself from dehydration, however, a lactating woman needs to drink plenty of fluids. The recommendation for *total* water (including drinking water, other beverages, and foods) during lactation is 3.8 liters per day. Because foods provide about 20 percent of total water intake, beverages—including drinking water—should provide about 3.1 liters per day (roughly 13 cups). A sensible guideline is to drink a glass of milk, juice, or water at each meal and each time the infant nurses.

Nutrient Supplements Most lactating women can obtain all the nutrients they need from a well-balanced diet without taking multivitamin-mineral supplements. Nevertheless, some may need iron supplements, not to enhance the iron in breast milk, but to refill depleted iron stores. The mother's iron stores dwindle during pregnancy as she supplies the developing fetus with enough iron to last through the first 4 to 6 months of the infant's life. In addition, childbirth may have caused blood losses. Thus a woman may need iron supplements during lactation even though, until menstruation resumes, her iron requirement is about half that of other nonpregnant women her age.

Food Assistance Programs In general, women most likely to participate in the food assistance program WIC—those who are poor and have little education—are less likely to breastfeed.[102] Furthermore, WIC provides infant formula at no cost. Because WIC recognizes the many benefits of breastfeeding, efforts are made to overcome this dilemma.[103] In addition to nutrition education and encouragement, breastfeeding mothers receive the following WIC incentives:

- Higher priority in certification into WIC
- Greater quantity and variety of foods
- Longer eligibility to participate in WIC

- Support from peers and experts
- Breast pumps and other support materials

Together, these efforts help to provide nutrition support and encourage WIC mothers to breastfeed.

Particular Foods Foods with strong or spicy flavors (such as garlic) may alter the flavor of breast milk. A sudden change in the taste of the milk may annoy some infants. Familiar flavors may enhance enjoyment. Flavors in breast milk from the mother's diet can influence the infant's later food preferences.[104]

Current evidence does not support a major role for maternal dietary restrictions during lactation to prevent or delay the onset of food allergy in infants. Infants who develop symptoms of food allergy, however, may be more comfortable if the mother's diet excludes the most common offenders—cow's milk, eggs, fish, peanuts, and tree nuts. Generally, infants with a strong family history of food allergies benefit from breastfeeding.

Nutritious foods support successful lactation.

A nursing mother can usually eat whatever nutritious foods she chooses. If she suspects a particular food is causing the infant discomfort, her physician may recommend a dietary challenge: eliminate the food from the diet to see if the infant's reactions subside, then return the food to the diet and again monitor the infant's reactions. If a food must be eliminated for an extended time, appropriate substitutions must be made to ensure nutrient adequacy.

Maternal Health
If a woman has an ordinary cold, she can continue nursing without worry. If susceptible, the infant will catch it from her anyway. Thanks to the immunological protection of breast milk, the baby may be less susceptible than a formula-fed baby would be. With appropriate treatment, a woman who has an infectious disease such as tuberculosis or hepatitis can breastfeed; transmission is rare. Women with HIV (human immunodeficiency virus) infections, however, should consider other options.

HIV Infection and AIDS Mothers with HIV infections can transmit the virus (which causes AIDS) to their infants through breast milk, especially during the early months of breastfeeding. In developed countries such as the United States, where safe alternatives are available, HIV-positive women should *not* breastfeed their infants. In developing countries, where the feeding of inappropriate or contaminated formulas causes more than 1 million infant deaths each year, breastfeeding can be critical to infant survival. Thus, in making the decision of whether to breastfeed, HIV-infected women in developing countries must weigh the potential risks and benefits. The World Health Organization (WHO) recommends exclusive breastfeeding for infants of HIV-infected women for the first six months of life unless formula feeding is acceptable, feasible, affordable, sustainable, and safe before that time. Alternatively, HIV-exposed infants may be protected by receiving antiretroviral treatment while being breastfed.

Diabetes Women with diabetes (type 1) may need careful monitoring and counseling to ensure successful lactation. These women need to adjust their energy intakes and insulin doses to meet the heightened needs of lactation. Maintaining good glucose control helps to initiate lactation and support milk production.

Postpartum Amenorrhea Women who breastfeed experience prolonged **postpartum amenorrhea.** Absent menstrual periods, however, do not protect a woman from pregnancy. To prevent pregnancy, a couple must use some form of contraception. Breastfeeding women who use oral contraceptives should use progestin-only agents for at least the first 6 months. Estrogen-containing oral contraceptives reduce the volume and the protein content of breast milk.

Breast Health Some women fear that breastfeeding will cause their breasts to sag. The breasts do swell and become heavy and large immediately after the birth, but even

postpartum amenorrhea (ay-MEN-oh-REE-ah): the normal temporary absence of menstrual periods immediately following childbirth.

when they produce enough milk to nourish a thriving infant, they eventually shrink back to their prepregnancy size. Given proper support, diet, and exercise, breasts often return to their former shape and size when lactation ends. Breasts change their shape as the body ages, but breastfeeding does not accelerate this process.

Whether the physical and hormonal events of pregnancy and lactation protect women from later breast cancer is an area of active research. Some research suggests a protective effect between breastfeeding and breast cancer.[105]

Practices Incompatible with Lactation Some substances impair milk production or enter breast milk and interfere with infant development. This section discusses practices that a breastfeeding mother should avoid.

Alcohol Alcohol easily enters breast milk, and its concentration peaks within an hour after ingestion. Infants drink less breast milk when their mothers have consumed even small amounts of alcohol (equivalent to a can of beer). Three possible reasons, acting separately or together, may explain why. For one, the alcohol may have altered the flavor of the breast milk and thereby the infants' acceptance of it. For another, because infants metabolize alcohol inefficiently, even low doses may be potent enough to suppress their feeding and cause sleepiness. Third, the alcohol may have interfered with lactation by inhibiting the hormone oxytocin.

In the past, alcohol has been recommended to mothers to facilitate lactation despite a lack of scientific evidence that it does so. The research summarized here suggests that alcohol actually hinders breastfeeding. An occasional alcoholic beverage may be within safe limits, but breastfeeding should be delayed for at least 2 hours afterward.

Medicinal Drugs Most medicines are compatible with breastfeeding, but some are contraindicated, either because they suppress lactation or because they are secreted into breast milk and can harm the infant.[106] As a precaution, a nursing mother should consult with her physician prior to taking any drug, including herbal supplements.

Illicit Drugs Illicit drugs, of course, are harmful to the physical and emotional health of both the mother and the nursing infant. Breast milk can deliver such high doses of illicit drugs as to cause irritability, tremors, hallucinations, and even death in infants. Women whose infants have overdosed on illicit drugs contained in breast milk have been convicted of murder. Women who use methadone to control withdrawal symptoms for opiate addiction can safely breastfeed their infants.

Smoking Many women who quit smoking during pregnancy relapse after delivery. Smoking following childbirth may help them with weight loss at that time, but it harms the infant as well as the mother.[107] Because cigarette smoking lowers the concentrations of lipids and proteins and reduces milk volume, the breast milk of smokers may not meet their infants' energy needs.[108] The milk they do produce contains nicotine, which alters its smell and flavor. Furthermore, smoking increases the risk for SIDS.[109] Breastfeeding helps to protect against SIDS, but infant exposure to passive smoke negates this protective effect.[110]

Environmental Contaminants Chapter 19 discusses environmental contaminants in the food supply. Some of these environmental contaminants, such as DDT, PCBs, and dioxin, can find their way into breast milk. Inuit mothers living in Arctic Québec who eat seal and beluga whale blubber have high concentrations of DDT and PCBs in their breast milk, but the impact on infant development is unclear. Preliminary studies indicate that the children of these Inuit mothers are developing normally. Researchers speculate that the abundant omega-3 fatty acids of the Inuit diet may protect against damage to the central nervous system. Breast milk tainted with dioxin interferes with tooth development during early infancy, producing soft, mottled teeth that are vulnerable to dental caries. To limit mercury intake, lactating women should heed the fish restrictions for pregnant women that were mentioned previously (see p. 494).

Caffeine Caffeine enters breast milk and may make an infant irritable and wakeful. As during pregnancy, caffeine consumption should be moderate—the equivalent of 1 to 2 cups of coffee a day. Larger doses of caffeine may interfere with the bioavailability of iron from breast milk and impair the infant's iron status.

> **REVIEW IT** Summarize the nutrient needs of women during lactation.
The lactating woman needs extra fluid and enough energy and nutrients to produce about 25 ounces of milk a day. Breastfeeding is contraindicated for those with HIV/AIDS. Alcohol, other drugs, smoking, and contaminants may reduce milk production or enter breast milk and impair infant development.

This chapter has focused on the nutrition needs of the mother during pregnancy and lactation. The next chapter explores the dietary needs of infants, children, and adolescents.

Nutrition Portfolio

The choices a woman makes in preparation for, and in support of, pregnancy and lactation can influence both her health and her infant's development—today and for decades to come. Go to Diet & Wellness Plus and choose one of the days on which you tracked your diet and activity for an entire day. Select the Intake vs. Goals report to help you answer the following questions:

- For women of childbearing age, determine whether you consume at least 400 micrograms of dietary folate equivalents daily.

- For women who are pregnant, evaluate whether you are meeting your nutrition needs and gaining the amount of weight recommended.

- For women who are about to give birth, carefully consider all the advantages of breastfeeding your infant and obtain the needed advice to support you.

DIET & WELLNESS **PLUS**➕ To complete this exercise, go to your Diet & Wellness Plus at **www.cengagebrain.com.**

> **STUDY IT** To review the key points of this chapter and take a practice quiz, go to the study cards at the end of the book.

REFERENCES

1. R. Sharma and coauthors, Lifestyle factors and reproductive health: Taking control of your fertility, *Reproductive Biology and Endocrinology* 11 (2013): 66; N. Mmbaga and J. Luk, The impact of preconceptual diet on the outcome of reproductive treatments, *Current Opinion in Obstetrics and Gynecology* 24 (2012): 127–131; C. Dupont and coauthors, Maternal environment and the reproductive function of the offspring, *Theriogenology* 78 (2012): 1405–1414.

2. V. Berghella and coauthors, Preconception care, *Obstetric and Gynecological Survey* 65 (2010): 119–131.

3. U. Paasch and coauthors, Obesity and age affect male fertility potential, *Fertility and Sterility* 94 (2010): 2898–2901.

4. C. J. Brewer and A. H. Balen, The adverse effects of obesity on conception and implantation, *Reproduction* 140 (2010): 347–364.

5. E. Toledo and coauthors, Dietary patterns and difficulty conceiving: A nested case-control study, *Fertility and Sterility* 96 (2011): 1149–1153; I. Cetin, C. Berti, and S. Calabrese, Role of micronutrients in the periconceptional period, *Human Reproduction Update* 16 (2010): 80–95.

6. P. Zareba and coauthors, Semen quality in relation to antioxidant intake in a healthy male population, *Fertility and Sterility* 100 (2013): 1572–1579; T. K. Jensen and coauthors, High dietary intake of saturated fat is associated with reduced semen quality among 701 young Danish men from the general population, *American Journal of Clinical Nutrition* 97 (2013): 411–418; J. Mendiola and coauthors, A low intake of antioxidant nutrients is associated with poor semen quality in patients attending fertility clinics, *Fertility and Sterility* 93 (2010): 1128–1133.

7. J. C. Sadeu and coauthors, Alcohol, drugs, caffeine, tobacco, and environmental contaminant exposure: Reproductive health consequences and clinical implications, *Critical Reviews in Toxicology* 40 (2010): 633–652.

8. M. Desforges and C. P. Sibley, Placental nutrient supply and fetal growth, *The International Journal of Developmental Biology* 54 (2010): 377–390.

9. C. Y. Spong, Defining "term" pregnancy: Recommendations from the Defining "Term" Pregnancy Workgroup, *Journal of the American Medical Association* 309 (2013): 2445–2446.

10. D. O. Mook-Kanamori and coauthors, Risk factors and outcomes associated with first-trimester fetal growth restriction, *Journal of the American Medical Association* 303 (2010): 527–534.

11. A. J. Agopian and coauthors, Proportion of neural tube defects attributable to known risk factors, *Birth Defects Research. Part A, Clinical and Molecular Teratology* 97 (2013): 42–46.

12. G. C. Burdge and K. A. Lillycrop, Folic acid supplementation in pregnancy: Are there devils in the detail? *British Journal of Nutrition* 108 (2012): 1924–1930; H. Heseker, Folic acid and other potential measures in the prevention of neural tube defects, *Annals of Nutrition Metabolism* 59 (2011): 41–45; D. Taruscio and coauthors, Folic acid and primary prevention of birth defects, *Biofactors* 37 (2011): 280–284.

13. M. L. de Gusmão Correia and coauthors, Developmental origins of health and disease: Experimental and human evidence of fetal programming for metabolic syndrome, *Journal of Human Hypertension* 26 (2012): 405–419.

14. G. C. Burdge and K. A. Lillycrop, Nutrition, epigenetics, and developmental plasticity: Implications for understanding human disease, *Annual Review of Nutrition* 30 (2010): 315–339.

15. S. M. Ruchat, M. F. Hivert, and L. Bouchard, Epigenetic programming of obesity and diabetes by in utero exposure to gestational diabetes mellitus, *Nutrition Reviews* 71 (2013): S88–S94; J. E. Wiedmeier and coauthors, Early postnatal nutrition and programming of the preterm neonate, *Nutrition Reviews* 69 (2011): 76–82.

16. B. Koletzko and coauthors, The Early Nutrition Programming Project (EARNEST): 5 y of successful multidisciplinary collaborative research, *American Journal of Clinical Nutrition* 94 (2011): 1749S–1753S.

17. S. M. Ruchat, M. F. Hivert, and L. Bouchard, Epigenetic programming of obesity and diabetes by in utero exposure to gestational diabetes mellitus, *Nutrition Reviews* 71 (2013): S88–S94; J. L. Tarry-Adkins and S. E. Ozanne, Mechanisms of early life programming: Current knowledge and future directions, *American Journal of Clinical Nutrition* 94 (2011): 1765S–1771S; A. Gabory, L. Attig, and C. Junien, Developmental programming and epigenetics, *American Journal of Clinical Nutrition* 94 (2011):1943S–1952S; S. Sebert and coauthors, The early programming of metabolic health: Is epigenetic setting the missing link? *American Journal of Clinical Nutrition* 94 (2011): 1953S–1958S.

18. J. G. Eriksson, Early growth and coronary heart disease and type 2 diabetes: Findings from the Helsinki Birth Cohort Study (HBCS), *American Journal of Clinical Nutrition* 94 (2011): 1799S–1802S.

19. P. W. Nathanielsz and coauthors, Interventions to prevent adverse fetal programming due to maternal obesity during pregnancy, *Nutrition Reviews* 71 (2013): S78–S87.

20. P. Kaur and coauthors, The epigenome as a potential mediator of cancer and disease prevention in prenatal development, *Nutrition Reviews* 71 (2013): 441–457; E. D. Ciappio, J. B. Mason, and J. W. Crott, Maternal one-carbon nutrient intake and cancer risk in offspring, *Nutrition Reviews* 69 (2011): 561–571.

21. B. Reusens and coauthors, Maternal malnutrition programs the endocrine pancreas in progeny, *American Journal of Clinical Nutrition* 94 (2011): 1824S–1829S.

22. S. Wesseling, M. P. Koeners, and J. A. Joles, Salt sensitivity of blood pressure: Developmental and sex-related effects, *American Journal of Clinical Nutrition* 94 (2011): 1928S–1932S; M. M. Perälä and coauthors, The association between salt intake and adult systolic blood pressure is modified by birth weight, *American Journal of Clinical Nutrition* 93 (2011): 422–426.

23. Position of the American Dietetic Association and American Society for Nutrition: Obesity, reproduction, and pregnancy outcomes, *Journal of the American Dietetic Association* 109 (2009): 918–927.

24. J. F. Mission, N. E. Marshall, and A. B. Caughey, Obesity in pregnancy: A big problem and getting bigger, *Obstetrical and Gynocological Survey* 68 (2013): 389–399; P. M. Nodine and M. Hastings-Tolsma, Maternal obesity: Improving pregnancy outcomes, *American Journal of Maternal Child Nursing* 37 (2012): 110–115.

25. K. Willis and E. Sheiner, Bariatric surgery and pregnancy: The magical solution? *Journal of Perinatal Medicine* 41 (2013): 133–143; Bariatric surgery: Impact on pregnancy outcomes, *Current Diabetes Reports* 13 (2013): 19–26; R. Magdaleno, Jr. and coauthors, Pregnancy after bariatric surgery: A current view of maternal, obstetrical and perinatal challenges,

Archives of Gynecology and Obstetrics 285 (2012): 559–566; M. A. Kominiarek, Pregnancy after bariatric surgery, *Obstetrics & Gynecology Clinics of North America* 37 (2010): 305–320.

26. J. M. Walsh and F. M. McAuliffe, Prediction and prevention of the macrosomic fetus, *European Journal of Obstetrics, Gynecology, and Reproductive Biology* 162 (2012): 125–130.

27. B. P. Wispelwey and E. Sheiner, Cesarean delivery in obese women: A comprehensive review, *Journal of Maternal and Fetal Neonatal Medicine* 26 (2013): 547–551.

28. J. L. Mills and coauthors, Maternal obesity and congenital heart defects: A population-based study, *American Journal of Clinical Nutrition* 91 (2010): 1543–1549.

29. S. Cnattingius and coauthors, Maternal obesity and risk of preterm delivery, *Journal of the American Medical Association* 309 (2013): 2362–2370; S. D. McDonald and coauthors, Overweight and obesity in mothers and risk of preterm birth and low birth weight infants: Systematic review and meta-analyses, *British Medical Journal* 341 (2010): c3428.

30. G. A. L. Davies and coauthors, SOGC Clinical Practice Guideline: Obesity in pregnancy, *Journal of Obstetrics and Gynaecology Canada* 110 (2010): 165–173.

31. T. A. Simas and coauthors, Prepregnancy weight, gestational weight gain, and risk of growth affected neonates, *Journal of Women's Health* 21 (2012): 410–417.

32. C. K. McClure and coauthors, Associations between gestational weight gain and BMI, abdominal adiposity, and traditional measures of cardiometabolic risk in mothers 8 y postpartum, *American Journal of Clinical Nutrition* 98 (2013): 1218–1225.

33. D. S. Ludwig, H. L. Rouse, and J. Currie, Pregnancy weight gain and childhood body weight comparison, *PLoS Medicine* 10 (2013): e1001521; K. Schellong and coauthors, Birth weight and long-term overweight risk: Systematic review and a meta-analysis including 643,902 persons from 66 studies and 26 countries globally, *PLos One* 7 (2012): e47776; A. A. Mamun and coauthors, Associations of maternal pre-pregnancy obesity and excess pregnancy weight gains with adverse pregnancy outcomes and length of hospital stay, *BMC Pregnancy and Childbirth* 11 (2011): 62; J. Josefson, The impact of pregnancy nutrition on offspring obesity, *Journal of the American Dietetic Association* 111 (2011): 50–52; S. R. Crozier and coauthors, Weight gain in pregnancy and childhood body composition: Findings from the Southhampton Women's Survey, *American Journal of Clinical Nutrition* 91 (2010): 1745–1751; M. F. Mottola and coauthors, Nutrition and exercise prevent excess weight gain in overweight pregnant women, *Medicine and Science in Sports and Exercise* 42 (2010): 265–272.

34. M. F. Mottola and coauthors, Nutrition and exercise prevent excess weight gain in overweight pregnant women, *Medicine and Science in Sports and Exercise* 42 (2010): 265–272; I. Streuling, A. Beyerlein, and R. von Kries, Can gestational weight gain be modified by increasing physical activity and diet counseling? A meta-analysis of interventional trials, *American Journal of Clinical Nutrition* 92 (2010): 678–687.

35. A. Chmitorz and coauthors, Do trimester-specific cutoffs predict whether women ultimately stay within the Institute of Medicine/National Research Council guidelines for gestational weight gain? Findings of a retrospective cohort study, *American Journal of Clinical Nutrition* 95 (2012): 1432–1437.

36. I. Nehring and coauthors, Gestational weight gain and long-term postpartum weight retention: A meta-analysis, *American Journal of Clinical Nutrition* 94 (2011): 1225–1231; A. A. Mamun and coauthors, Associations of excess weight gain during pregnancy with long-term maternal overweight and obesity: Evidence from 21 y postpartum follow-up, *American Journal of Clinical Nutrition* 91 (2010): 1336–1341.

37. M. Mannan, S. A. R. Doi, and A. A. Mamun, Association between weight gain during pregnancy and postpartum weight retention and obesity: A bias-adjusted meta-analysis, *Nutrition Reviews* 71 (2013): 343–352.

38. P. van der Pligt and coauthors, Systematic review of lifestyle interventions to limit postpartum weight retention: Implications for future opportunities to prevent maternal overweight and obesity following childbirth, *Obesity Reviews* 14 (2013): 792–805.

39. D. Haire-Joshu and coauthors, Postpartum teens' breakfast consumption is associated with snack and beverage intake and body mass index, *Journal of the American Dietetic Association* 111 (2011): 124–130.

40. S. L. Nascimento and coauthors, The effect of physical exercise strategies on weight loss in postpartum women: A systematic review and meta-analysis, *International Journal of Obesity* (2013): doi:10.1038; A. R. A. Adegboye and Y. M. Linne, Diet or exercise, or both, for weight reduction in women after childbirth, *Cochrane Database of Systematic Reviews* 7 (2013): CD005627.

41. M. F. Mottola, Physical activity and maternal obesity: Cardiovascular adaptations, exercise recommendations, and pregnancy outcomes, *Nutrition Reviews* 71 (2013): S31–S36.

42. L. M. Mudd and coauthors, Health benefits of physical activity during pregnancy: An international perspective, *Medicine and Science in Sports and Exercise* 45 (2013): 268–277.

43. S. M. Ruchat and coauthors, Nutrition and exercise reduce excessive weight gain in normal-weight pregnant women, *Medicine and Science in Sports and Exercise* 44 (2012): 1419–1426; K. Melzer and coauthors, Physical activity and pregnancy: Cardiovascular adaptations, recommendations and pregnancy outcomes, *Sports Medicine* 40 (2010): 493–507.

44. M. L. Blumfield and coauthors, Systematic review and meta-analysis of energy and macronutrient intakes during pregnancy in developed countries, *Nutrition Reviews* 70 (2012): 322–336.

45. L. D. Brown and coauthors, Maternal amino acid supplementation for intrauterine growth restriction, *Frontiers in Bioscience* 3 (2011): 428–444.

46. J. F. W. Cohen and coauthors, Maternal *trans* fatty acid intake and fetal growth, *American Journal of Clinical Nutrition* 94 (2011): 1241–1247; P. Haggarty, Fatty acid supply to the human fetus, *Annual Review of Nutrition* 30 (2010): 237–255.

47. A. S. de Souza, F. S. Fernandes, and M. das Graças Tavares do Carmo, Effects of maternal malnutrition and postnatal nutritional rehabilitation on brain fatty acids, learning, and memory, *Nutrition Reviews* 69 (2011): 132–144.

48. C. Cao and K. O. O'Brien, Pregnancy and iron homeostasis: An update, *Nutrition Reviews* 71 (2013): 35–51.

49. F. E. Viteri, Iron endowment at birth: Maternal iron status and other influences, *Nutrition Reviews* 69 (2011): S3–S16.

50. H. J. McArdle and coauthors, Role of the placenta in regulation of fetal iron status, *Nutrition Reviews* 69 (2011): S17–S22.

51. N. Kozuki and coauthors, Moderate to severe, but not mild, maternal anemia is associated with increased risk of small-for-gestational-age outcomes, *Journal of Nutrition* 142 (2012): 358–362.

52. T. O. Scholl, Maternal iron status: Relation to fetal growth, length of gestation, and iron endowment of the neonate, *Nutrition Reviews* 69 (2011): S23–S29; J. Berger and coauthors, Strategies to prevent iron deficiency and improve reproductive health, *Nutrition Reviews* 69 (2011): S78–S86.

53. B. E. Young and coauthors, Maternal vitamin D status and calcium intake interact to affect fetal skeletal growth in utero in pregnant adolescents, *American Journal of Clinical Nutrition* 95 (2012): 1103–1112.

54. P. M. Brannon and M. F. Picciano, Vitamin D in pregnancy and lactation in humans, *Annual Review of Nutrition* 31 (2011): 89–115; D. K. Dror and L. H. Allen, Vitamin D inadequacy in pregnancy: Biology, outcomes, and interventions, *Nutrition Reviews* 68 (2010): 465–477.

55. Committee on Dietary Reference Intakes, *Dietary Reference Intakes for Calcium and Vitamin D* (Washington, D.C.: National Academies Press, 2011).

56. A. N. Hacker, E. B. Fung, and J. C. King, Role of calcium during pregnancy: Maternal and fetal needs, *Nutrition Reviews* 70 (2012): 397–409.

57. P. Christian, Micronutrients, birth weight, and survival, *Annual Review of Nutrition* 30 (2010): 83–104.

58. A. Wendt and coauthors, Impact of increasing inter-pregnancy interval on maternal and infant health, *Paediatric and Perinatal Epidemiology* 26 (1012): 239–258; B. Z. Shachar and D. J. Lyell, Interpregnancy interval and obstetrical complications, *Obstetrical and Gynecological Survey* 67 (2012): 584–596.

59. J. M. Catov and coauthors, Periconceptional multivitamin use and risk of preterm or small-for-gestational-age births in the Danich National Birth Cohort, *American Journal of Clinical Nutrition* 94 (2011): 906–912.

60. D. Roberfroid and coauthors, Impact of prenatal multiple micronutrients on survival and growth during infancy: A randomized controlled trial, *American Journal of Clinical Nutrition* 95 (2012): 916–924; E. L. Prado and coauthors, Maternal multiple micronutrient supplements and child cognition: A randomized trial in Indonesia, *Pediatrics* 130 (2012): e536–e546.

61. Position of the American Dietetic Association: Vegetarian diets, *Journal of the American Dietetic Association* 109 (2009): 1266–1282.

62. J. R. Niebyl, Nausea and vomiting in pregnancy, *New England Journal of Medicine* 363 (2010): 1544–1550.

63. A. Matthews and coauthors, Interventions for nausea and vomiting in early pregnancy, *Cochrane Database of Systemic Reviews* 8 (2010): CD007575.

64. S. L. Young, Pica in pregnancy: New ideas about an old condition, *Annual Review of Nutrition* 30 (2010): 403–422.

65. F. H. Bloomfield, How is maternal nutrition related to preterm birth? *Annual Review of Nutrition* 31 (2011): 235–261.

66. J. L. Kitzmiller and coauthors, Preconception care for women with diabetes and prevention of major congenital malformations, *Birth Defects Research Part A: Clinical and Molecular Teratology* 88 (2010): 791–803.

67. Centers for Disease Control and Prevention, Diabetes Public Health Resource, http://www.cdc.gov/diabetes/news/docs/gestational.htm, September 2013.

68. S. F. Ehrlich and coauthors, Change in body mass index between pregnancies and the risk of gestational diabetes in a second pregnancy, *Obstetrics and Gynecology* 117 (2011): 1323–1330.

69. C. Zhang and Y. Ning, Effect of dietary and lifestyle factors on the risk of gestational diabetes: Review of epidemiologic evidence, *American Journal of Clinical Nutrition* 94 (2011): 1975S–1979S.

70. S. H. Ley and coauthors, Effect of macronutrient intake during the second trimester on glucose metabolism later in pregnancy, *American Journal of Clinical Nutrition* 94 (2011): 1232–1240.

71. L. Hartling and coauthors, Benefits and harms of treating gestational diabetes mellitus: A systematic review and meta-analysis for the U. S. Preventive Services Task Force and the national Institutes of health Office of Medical Applications of Research, *Annals of Internal Medicine* 159 (2013): 123–129; T. Sathyapalan, D. Mellor, and S. L. Atkin, Obesity and gestational diabetes, *Seminars in Fetal and Neonatal Medicine* 15 (2010): 89–93.

72. R. Mustafa and coauthors, A comprehensive review of hypertension in pregnancy, *Journal of Pregnancy* 2012 (2012): 105918.

73. E. W. Seely and J. Ecker, Chronic hypertension in pregnancy, *New England Journal of Medicine* 365 (2011): 439–446.

74. A. Jeyabalan, Epidemiology of preeclampsia: Impact of obesity, *Nutrition Reviews* 71 (2013): S18–S25.

75. E. F. Davis and coauthors, Cardiovascular risk factors in children and young adults born to preeclamptic pregnancies: A systematic review, *Pediatrics* 129 (2012): e1552–e1561.

76. A. Paxton and T. Wardlaw, Are we making progress in maternal mortality? *New England Journal of Medicine* 364 (2011): 1990–1993.

77. G. M. Martinez, C. E. Copen, and J. C. Abma, Teenagers in the United States: Sexual activity, contraceptive use, and childbearing, 2006–2010, *Morbidity and Mortality Weekly Report* 60 (2011): 1460.

78. One in 5 teens giving birth already has a child, *Journal of the American Medical Association* 309 (2013): 1987; US Department of Health and Human Services, Trends in teen pregnancy and childbearing, www.hhs.gov/ash/oah/adolescent-health-topics/reproductive-health/teen-pregnancy/trends.html, December 20, 2013.

79. Alcohol use and binge drinking among women of childbearing age: United States, 2006–2010, *Mortality and Morbidity Weekly Report* 61 (2012): 534.

80. C. M. O'Leary and coauthors, Maternal alcohol-use and sudden infant death syndrome and infant mortality excluding SIDS, *Pediatrics* 131 (2013): e770–e778; P. Latino-Martel and coauthors, Maternal alcohol consumption during pregnancy and risk of childhood leukemia: Systematic review and meta-analysis, *Cancer Epidemiology, Biomarkers and Prevention* 19 (2010): 1238–1260.

81. M. Behnke, V. C. Smith, Committee on Substance Abuse and Committee on Fetus and Newborn, Prenatal substance abuse: Short- and long-term effects on the exposed fetus, *Pediatrics* 131 (2013): e1009–e1024.

82. L. L. LaGasse and coauthors, Prenatal methamphetamine exposure and childhood behavior at 3 and 5 years of age, *Pediatrics* 129 (2012): 681–688; J. P. Ackerman, T. Riggins, and M. M. Black, A review of the effects of prenatal cocaine exposure among school-aged children, *Pediatrics* 125 (2010): 554–565.

83. S. Buckingham-Howes and coauthors, Systematic review of prenatal cocaine exposure and adolescent development, *Pediatrics* 131 (2013): e1917–e1936.

84. V. T. Tong and coauthors, Trends in smoking before, during, and after pregnancy—Pregnancy Risk Assessment Monitoring System, United States, 40 sites, 2000–2010, *Morbidity and Mortality Weekly Report* 62 (2013): 1–19.

85. C. C. Geerts and coauthors, Parental smoking and vascular damage in their 5-year-old children, *Pediatrics* 129 (2012): 45–54; H. Burke and coauthors, Prenatal and passive smoke exposure and incidence of asthma and wheeze: Systematic review and meta-analysis, *Pediatrics* 129 (2012): 735–744; A. Bjerg and coauthors, A strong synergism of low birth weight and prenatal smoking on asthma in schoolchildren, *Pediatrics* 127 (2011): e905–e912; C. J. Alverson and coauthors, Maternal smoking and congenital heart defects in the Baltimore-Washington infant study, *Pediatrics* 127 (2011): e647–e653.

86. A. Gunnerbeck and coauthors, Relationship of maternal snuff use and cigarette smoking with neonatal apnea, *Pediatrics* 128 (2011): 503–509.

87. T. Coleman and coauthors, A randomized trial of nicotine-replacement therapy patches in pregnancy, *New England Journal of Medicine* 366 (2012): 808–818; C. Oncken, Nicotine replacement for smoking cessation during pregnancy, *New England Journal of Medicine* 366 (2012): 846–847; I. Milidou and coauthors, Nicotine replacement therapy during pregnancy and infantile colic in the offspring, *Pediatrics* 129 (2012): e652–e658.

88. J. Leonardi-Bee, J. Britton, and A. Venn, Secondhand smoke and adverse fetal outcomes in nonsmoking pregnant women: A meta-analysis, *Pediatrics* 127 (2011): 734–741.

89. A. Bloomingdale and coauthors, A qualitative study of fish consumption during pregnancy, *American Journal of Clinical Nutrition* 92 (2010): 1234–1240.

90. K. R. Mahaffey and coauthors, Balancing the benefits of n-3 polyunsaturated fatty acids and the risks of methylmercury exposure from fish consumption, *Nutrition Reviews* 69 (2011): 493–508.

91. M. M. G. Ackermans and coauthors, Vitamin A and clefting: Putative biological mechanisms, *Nutrition Reviews* 69 (2011): 613–624.

92. R. Bakker and coauthors, Maternal caffeine intake, blood pressure, and the risk of hypertensive complications during pregnancy: The Generation R Study, *American Journal of Hypertension* 24 (2011): 421–428; E. Maslova and coauthors, Caffeine consumption during pregnancy and risk of preterm birth: A meta-analysis, *American Journal of Clinical Nutrition* 92 (2010): 1120–1132; D. C. Greenwood and coauthors, Caffeine intake during pregnancy, late miscarriage and stillbirth, *European Journal of Epidemiology* 25 (2010): 275–280.

93. R. Bakker and coauthors, Maternal caffeine intake from coffee and tea, fetal growth, and the risks of adverse birth outcomes: The Generation R Study, *American Journal of Clinical Nutrition* 91 (2010): 1691–1698.

94. M. Jarosz, R. Wierzejska, and M. Siuba, Maternal caffeine intake and its effect on pregnancy outcomes, *European Journal of obstetrics and Gynecology and Reproductive Biology* 160 (2012): 156–160.

95. Position of the Academy of Nutrition and Dietetics: Use of nutritive and nonnutritive sweeteners, *Journal of the Academy of Nutrition and Dietetics* 112 (2012): 739–758.

96. L. Englund-Ögge and coauthors, Association between intake of artificially sweetened and sugar-sweetened beverages and preterm delivery: A large prospective cohort study, *American Journal of Clinical Nutrition* 96 (2012): 552–559; T. I. Halldorsson and coauthors, Intake of artificially sweetened soft drinks and risk of preterm delivery: A prospective cohort study in 59,334 Danish pregnant women, *American Journal of Clinical Nutrition* 92 (2010): 626–633.

97. Centers for Disease Control and Prevention, Breastfeeding report card: United States, 2013, www.cdc.gov/breastfeeding/data/reportcard.htm, July 2013.

98. US Department of Health and Human Services, *The Surgeon General's Call to Action to Support Breastfeeding,* (Washington, D.C.: U.S. Department of Health and Human Services, Office of the Surgeon General, 2011); Position of the American Dietetic Association: Promoting and supporting breastfeeding, *Journal of the American Dietetic Association* 109 (2009): 1926–1942; Breastfeeding, *Pediatric Nutrition*, 7th ed., ed. R. E. Kleinman (Elk Grove Village, IL: American Academy of Pediatrics, 2014), pp. 41–59.

99. L. A. Nommsen-Rivers and coauthors, Delayed onset of lactogenesis among first-time mothers is related to maternal obesity and factors associated with ineffective breastfeeding, *American Journal of Clinical Nutrition* 92 (2010): 574–584.

100. R. Sámano and coauthors, Effects of breastfeeding on weight loss and recovery of pregestational weight in adolescent and adult mothers, *Food and Nutrition Bulletin* 34 (2013): 123–130.

101. A. J. Daley and coauthors, Maternal exercise and growth in breastfed infants: A meta-analysis of randomized controlled trials, *Pediatrics* 130 (2012): 108–114.

102. K. M. Ziol-Guest and D. C. Hernandez, First- and second-trimester WIC participation is associated with lower rates of breastfeeding and early introduction of cow's milk during infancy, *Journal of the American Dietetic Association* 110 (2010): 702–709.

103. M. Murimi and coauthors, Factors that influence breastfeeding decisions among Special Supplemental Nutrition Program for Women, Infants, and Children participants from Central Louisiana, *Journal of the American Dietetic Association* 110 (2010): 624–627.

104. G. K. Beauchamp and J. A. Mennella, Flavor perception in human infants: Development and functional significance, *Digestion* 83 (2011): 1–6.

105. K. N. Anderson, R. B. Schwab, and M. E. Martinez, Reproductive risk factors and breast cancer subtypes: A review of the literature, *Breast Cancer Research and Treatment* (2014): PMID24477977; H. Pan and coauthors, Reproductive factors and breast cancer risk among BRCA1 or BRCA2 mutation carriers: Results from ten studies, *Cancer Epidemiology* (2013): doi:10.1016.

106. H. C. Sachs and Committee on Drugs, The transfer of drugs and therapeutics into human breast milk: An update on selected topics, *Pediatrics* 132 (2013): e796–e809.

107. M. D. Levine and coauthors, Relapse to smoking and postpartum weight retention among women who quit smoking during pregnancy, *Obesity* 20 (2012): 457–459.

108. P. Bachour and coauthors, Effects of smoking, mother's age, body mass index, and parity number on lipid, protein, and secretory immunoglobulin A concentrations of human milk, *Breastfeeding Medicine* 7 (2012): 179–188.

109. G. Liebrechts-Akkerman and coauthors, Postnatal parental smoking: An important risk factor for SIDS, *European Journal of Pediatrics* 170 (2011): 1281–1291; B. M. Ostfeld and coauthors, Concurrent risks in sudden infant death syndrome, *Pediatrics* 125 (2010): 447–453.

110. F. R. Hauck and coauthors, Breastfeeding and reduced risk of sudden infant death syndrome: A meta-analysis, *Pediatrics* 128 (2011): 103–110.

HIGHLIGHT > 15
Fetal Alcohol Syndrome

> **LEARN IT** Explain how drinking alcohol endangers the fetus and how women can prevent fetal alcohol syndrome.

As Chapter 15 mentions, drinking alcohol during pregnancy endangers the fetus.[1] Alcohol crosses the placenta freely and deprives the developing fetus of both nutrients and oxygen. The damaging effects of alcohol on the developing fetus cover a range of abnormalities referred to as **fetal alcohol spectrum disorder** (see the accompanying glossary).[2] Those at the most severe end of the spectrum are described as having **fetal alcohol syndrome (FAS),** a cluster of physical, mental, and neurobehavioral symptoms that includes:[3]

- Prenatal and postnatal growth restriction
- Abnormalities of the brain and central nervous system, with consequent impairment in cognition and behavior
- Physical abnormalities of the face and skull that alter normal patterns of symmetry (see Figure H15-1, p. 506)[4]
- Increased frequency of major birth defects such as cleft palate, heart defects, and defects in ears, eyes, genitals, and urinary system

Those with more severe physical abnormalities have more cognitive limitations. Tragically, the damage evident at birth persists: children with FAS never fully recover.

Each year, an estimated 6000 infants are born in the United States with FAS because their mothers drank too much alcohol during pregnancy.[5] In addition, at least three times as many infants have had enough **prenatal alcohol exposure** to result in some symptoms of fetal alcohol spectrum disorder. The cluster of *mental* problems associated with prenatal alcohol exposure is known as **alcohol-related neurodevelopmental disorder (ARND),** and the *physical* malformations are referred to as **alcohol-related birth defects (ARBD).** Some children with ARBD and ARND have no outward signs; others may be short or have only minor facial abnormalities. Diagnosis is often overlooked until after 1 year of age.[6] Children commonly go undiagnosed even when they develop learning difficulties in the early school years. Mood disorders and problem behaviors, such as aggression, are common. These children typically need support and guidance to function and participate in daily activities.

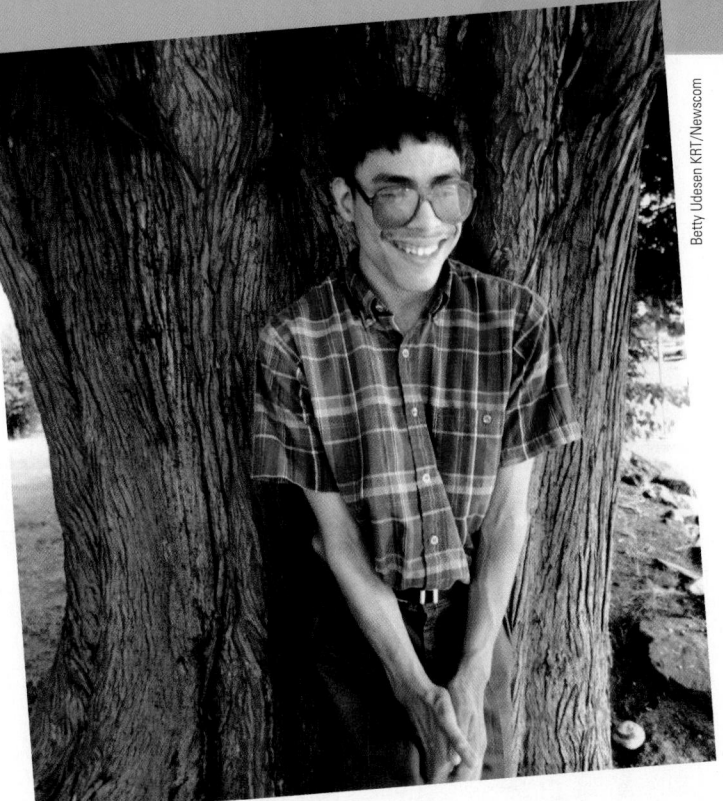

Betty Udesen KRT/Newscom

The surgeon general states that pregnant women should abstain from alcohol. Abstinence from alcohol is the best policy for pregnant women both because alcohol consumption during pregnancy has such severe consequences and because FAS can only be prevented—it cannot be treated. Further, because the most severe damage occurs around the time of conception—*before a woman may even realize that she is pregnant*—the warning to abstain includes women who may become pregnant.

> **DIETARY GUIDELINES FOR AMERICANS**
Women who are pregnant should not drink alcohol.

Drinking during Pregnancy

As mentioned in Chapter 15, 1 out of 13 pregnant women drinks alcohol at some time during her pregnancy; 1 out of 75 admits to binge drinking.[7] When a woman drinks during pregnancy, she causes damage in two ways: directly, by intoxication, and indirectly, by malnutrition.

GLOSSARY

alcohol-related birth defects (ARBD): malformations in the skeletal and organ systems (heart, kidneys, eyes, ears) associated with prenatal alcohol exposure.

alcohol-related neurodevelopmental disorder (ARND): abnormalities in the central nervous system and cognitive development associated with prenatal alcohol exposure.

fetal alcohol spectrum disorder: a range of physical, behavioral, and cognitive abnormalities caused by prenatal alcohol exposure.

fetal alcohol syndrome (FAS): a cluster of physical, behavioral, and cognitive abnormalities associated with prenatal alcohol exposure, including facial malformations, growth retardation, and central nervous disorders.

prenatal alcohol exposure: subjecting a fetus to a pattern of excessive alcohol intake characterized by substantial regular use or heavy episodic drinking.

Note: See Highlight 7 for other alcohol-related terms and information.

> **FIGURE H15-1 Typical Facial Characteristics of FAS**

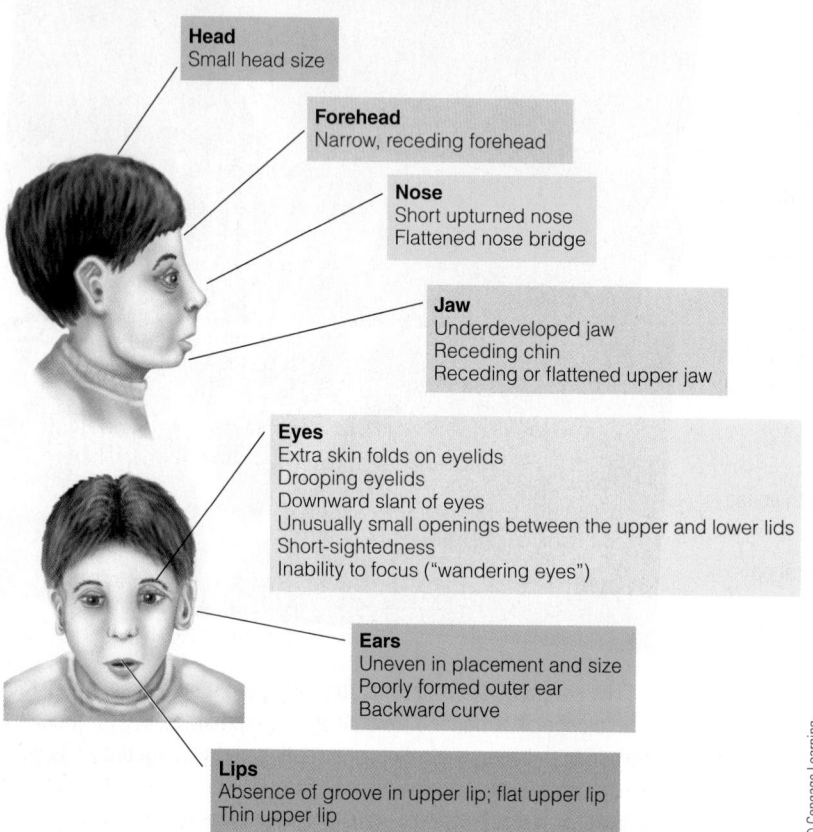

Head
Small head size

Forehead
Narrow, receding forehead

Nose
Short upturned nose
Flattened nose bridge

Jaw
Underdeveloped jaw
Receding chin
Receding or flattened upper jaw

Eyes
Extra skin folds on eyelids
Drooping eyelids
Downward slant of eyes
Unusually small openings between the upper and lower lids
Short-sightedness
Inability to focus ("wandering eyes")

Ears
Uneven in placement and size
Poorly formed outer ear
Backward curve

Lips
Absence of groove in upper lip; flat upper lip
Thin upper lip

© Cengage Learning

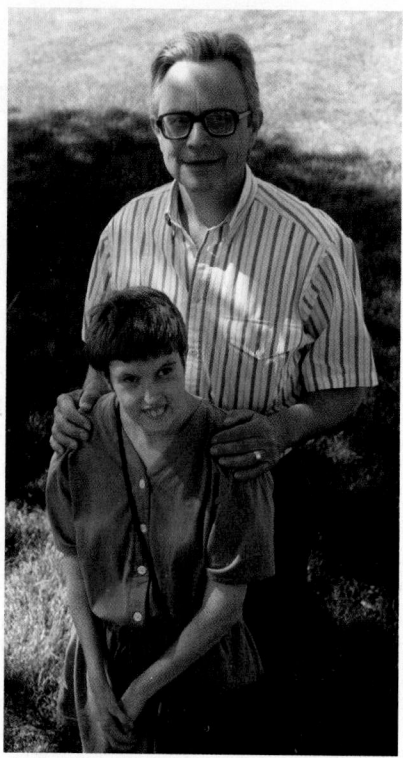

Stuart Wong Krt/Newscom

As can be seen from this 25-year-old woman with FAS, the characteristic facial features may diminish with time, but physical impairments continue, resulting in short height.

Prior to the complete formation of the placenta (approximately 12 weeks), alcohol diffuses directly into the tissues of the developing embryo, causing incredible damage. (Review Figure 15-4 on p. 477 and note that the critical periods for most tissues occur during this time of embryonic development.) Alcohol interferes with the orderly development of tissues during their critical periods, reducing the number of cells and damaging those that are produced. The damage of alcohol toxicity during brain development is apparent in its reduced size and impaired function.

When alcohol crosses the placenta, fetal blood alcohol rises until it reaches equilibrium with maternal blood alcohol. The mother may not even appear drunk, but the fetus may be poisoned. The fetus's body is small, its detoxification system is immature, and alcohol remains in fetal blood long after it has been cleared from maternal blood.

A pregnant woman harms her unborn child not only by consuming alcohol but also by not consuming food. This combination enhances the likelihood of malnutrition and a poorly developed infant. It is important to realize, however, that malnutrition is not the cause of FAS. It is true that mothers of FAS children often have unbalanced diets and nutrient deficiencies. It is also true that nutrient deficiencies may exacerbate the clinical signs seen in these children, but it is the *alcohol* that causes the damage. An adequate diet alone will not prevent FAS if alcohol use continues.

How Much Is Too Much?

A pregnant woman need not have an alcohol-abuse problem to give birth to a baby with FAS. She need only drink in excess of her liver's capacity to detoxify alcohol. The damaging effects on the developing fetus are dose dependent, becoming greater as the dose increases.[8] Even one drink a day threatens neurological development and behaviors. Four drinks a day dramatically increase the risk of having an infant with physical malformations.

In addition to total alcohol intake, drinking patterns play an important role. Most FAS studies report their findings in terms of average intake per day, but people often drink more heavily on some days than on others. For example, a woman who drinks an *average* of 1 ounce of alcohol (2 drinks) a day may not drink at all during the week, but then have 10 drinks on Saturday night, exposing the fetus to extremely toxic quantities of alcohol. Whether various drinking patterns incur damage depends on the frequency of consumption, the quantity consumed, and the stage of fetal development at the time of each drinking episode.

An occasional drink may be innocuous, but researchers are unable to say how much alcohol is safe to consume during pregnancy. For this reason, health-care professionals urge women to stop drinking alcohol as soon as they realize they are pregnant, or better, as soon as they *plan* to become pregnant. Why take any risk? The only sure way to protect an infant from alcohol damage is for the mother to abstain.

Children born with FAS must live with the long-term consequences of alcohol damage. This 20-year-old man was born 3 months preterm with a 0.237 blood alcohol level, which took his 2-pound body 4 days to clear from his system.

All containers of beer, wine, and liquor warn women not to drink alcoholic beverages during pregnancy because of the risk of birth defects.

When Is the Damage Done?

The type of abnormality observed in an FAS infant depends on the developmental events occurring at the times of alcohol exposure. During the first trimester, developing organs such as the brain, heart, and kidneys may be malformed. During the second trimester, the risk of spontaneous abortion increases. During the third trimester, body and brain growth may be retarded. Quite simply, the risk begins with any use.[9]

The father's alcohol ingestion may also affect fertility and fetal development. Animal studies have found smaller litter sizes, lower birthweights, reduced survival rates, and impaired learning ability in the offspring of males consuming alcohol prior to conception. An association between paternal alcohol intake 1 month prior to conception and low infant birthweight has been reported in human beings. Alcohol use creates epigenetic changes in sperm DNA that may alter gene expression and result in features of fetal alcohol spectrum disorders.

In view of the damage caused by FAS, prevention efforts focus on educating women not to drink during pregnancy. Public service announcements and alcohol beverage warning labels help to raise awareness. Everyone should hear the message loud and clear: don't drink alcohol during pregnancy.

CRITICAL THINKING QUESTIONS

A. What are the moral implications of using alcohol during pregnancy?
B. Clearly, drinking alcohol during pregnancy endangers the developing fetus, yet controversy surrounds the questions as to how much alcohol causes damage and when the damage is done. What difficulties might researchers encounter in trying to determine the exact amount of alcohol and the timing during pregnancy that might be reasonably safe for fetal development?

REFERENCES

1. F. Foltran and coauthors, Effect of alcohol consumption in prenatal life, childhood, and adolescence on child development, *Nutrition Reviews* 69 (2011): 642–659.
2. S. N. Mattson, N. Crocker, and T. T. Nguyen, Fetal alcohol spectrum disorders: Neuropsychological and behavioral features, *Neuropsychology Review* 21 (2011): 81–101.
3. E. P. Riley, M. A. Infante, and K. R. Warren, Fetal alcohol spectrum disorders: An overview, *Neuropsychological Review* 21 (2011): 73–80; K. L. Jones and coauthors, Fetal alcohol spectrum disorders: Extending the range of structural defects, *American Journal of Medical Genetics* 152A (2010): 2731–2735.
4. M. Suttie and coauthors, Facial dysmorphism across the fetal alcohol spectrum, *Pediatrics* 131 (2013): e779–e788; C. P. Klingenberg and coauthors, Prenatal alcohol exposure alters the patterns of facial asymmetry, *Alcohol* 44 (2010): 649–657; P. A. May and coauthors, Population differences in dysmorphic features among children with fetal alcohol spectrum disorders, *Journal of Developmental and Behavioral Pediatrics* 31 (2010): 304–316.
5. Centers for Disease Control and Prevention, *Fetal Alcohol Spectrum Disorders*, www.cdc.gov/ncbddd/fasd/data.html, updated October 6, 2010.
6. C. Bower and coauthors, Age at diagnosis of birth defects, *Birth Defects Research Part A: Clinical and Molecular Teratology* 88 (2010): 251–255.
7. Alcohol use and binge drinking among women of childbearing age: United States, 2006–2010, *Mortality and Morbidity Weekly Report* 61 (2012): 534–538.
8. A. Ornoy and Z. Ergaz, Alcohol abuse in pregnant women: Effects on the fetus and newborn, mode of action and maternal treatment, *International Journal of Environmental Research and Public Health* 7 (2010): 364–379.
9. H. S. Feldman and coauthors, Prenatal alcohol exposure patterns and alcohol-related birth defects and growth deficiencies: A prospective study, *Alcoholism: Clinical and Experimental Research* 36 (2012): 670–676.

16

Life Cycle Nutrition: Infancy, Childhood, and Adolescence

Nutrition in Your Life

Much of this book has focused on you—your food choices and how they might affect your health. This chapter shifts the focus from you the recipient to you the caregiver. One day (if not already), children may depend on you to feed them well and teach them wisely. The responsibility of nourishing children can seem overwhelming at times, but the job is fairly simple. Offer children a variety of nutritious foods to support their growth and teach them how to make healthy food and activity choices. Presenting foods in a relaxed and supportive environment nourishes both physical and emotional well-being. In the Nutrition Portfolio at the end of this chapter, you can plan a day's menu for a child 4 to 8 years of age and determine whether it meets nutrient requirements to support healthy growth but not so much as to promote obesity.

Nutrient needs change throughout the growing years and vary from individual to individual. The first year of life (infancy) is a time of phenomenal growth and development. The infant's high nutrient needs and developing maturity determine which foods are most appropriate to meet the needs for each stage of the first year. After the first year, a child continues to grow and change, but more slowly. Still, the cumulative effects over the next decade are remarkable. Then as the child enters the teen years, the pace toward adulthood accelerates dramatically. The physical changes of adolescents make their nutrient needs high, and their emotional, intellectual, and social changes make meeting those needs a challenge. Sound nutrition throughout infancy, childhood, and adolescence promotes normal growth and development, facilitates academic and physical performance, and may help prevent the development of obesity, diabetes, heart disease, some cancers, and other chronic diseases in adulthood. This chapter examines the special nutrient needs of infants, children, and adolescents.

> FIGURE 16-1   **Weight Gain of Infants in Their First Five Years of Life**

In the first year, an infant's birthweight may triple, but over the following several years, the rate of weight gain gradually diminishes.

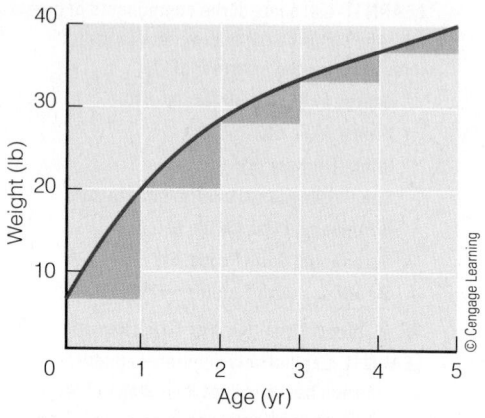

TABLE 16-1   **Infant and Adult Heart Rate, Respiration Rate, and Energy Needs Compared**

	Infants	Adults
Heart rate (beats/minute)	120 to 140	70 to 80
Respiration rate (breaths/minute)	20 to 40	15 to 20
Energy needs (kcal/body weight)	45/lb (100/kg)	<18/lb (<40/kg)

After 6 months, energy saved by slower growth is spent in increased activity.

16.1 Nutrition during Infancy

> **LEARN IT** List some of the components of breast milk and describe the appropriate foods for infants during the first year of life.

Initially, the infant drinks only breast milk or formula but later begins to eat some foods, as appropriate. Common sense in the selection of infant foods—along with a nurturing, relaxed environment—supports an infant's health and well-being.

Energy and Nutrient Needs   An infant grows fast during the first year, as Figure 16-1 shows. Growth directly reflects nutrient intake and is an important factor in assessing the nutrition status of infants and children. Health-care professionals measure the height and weight of an infant or child at intervals and compare the measurements with standard growth charts for gender and age and with previous measures of that infant or child (see the accompanying "How To" feature).

Energy Intake and Activity   A healthy infant's birthweight doubles by about 5 months of age and triples by 1 year, typically reaching 20 to 25 pounds. The infant's length changes more slowly than weight, increasing about 10 inches from birth to 1 year. By the end of the first year, infant growth slows considerably; during the second year, an infant typically gains less than 10 pounds and grows about 5 inches in length. At the age of 2, healthy children have attained approximately half of their adult height.

Not only do infants grow rapidly, but their energy requirement is remarkably high—about twice that of an adult, based on body weight. A newborn baby requires about 450 kcalories per day, whereas most adults require about 2000 kcalories per day. In terms of body weight, the difference is remarkable. Infants require about 100 kcalories per kilogram of body weight per day, whereas most adults need fewer than 40 (see Table 16-1). If an infant's energy needs were applied to an adult, a 170-pound adult would require more than 7000 kcalories a day. After 6 months, the infant's energy needs decline as the growth rate slows, but some of the energy saved by slower growth is spent in increased activity.

Energy Nutrients   Recommendations for the energy nutrients—carbohydrate, fat, and protein—during the first 6 months of life are based on the average intakes of healthy, full-term infants fed breast milk.[1] During the second 6 months of life, recommendations reflect typical intakes from solid foods as well as breast milk.

As Chapter 4 discusses, carbohydrates provide energy to all the cells of the body, but those in the brain depend primarily on glucose to fuel activities. Relative to the size of the body, the size of an infant's brain is greater than that of an adult's. An infant's brain weight is about 12 percent of its body weight, whereas an adult's brain weight is about 2 percent. Thus, an infant's brain uses *relatively* more glucose—about 60 percent of the day's total energy intake.

Fat provides most of the energy in breast milk and standard infant formula. Its high energy density supports the rapid growth of early infancy. Fat also provides the essential fatty acids needed for normal growth and development. As Chapter 5 mentions, DHA (docosahexaenoic acid) is the most abundant fatty acid in the brain and is also present in the retina of the eye, contributing to neural and visual development. DHA accumulation in the brain is greatest during fetal development and early infancy.[2]

No single nutrient is more essential to growth than protein; it is the basic building material of the body's tissues. All of the body's cells and most of its fluids contain protein. Consequently, inadequate protein intake has widespread effects, limiting brain function, weakening immune defenses, and disrupting digestion and absorption. The term *failure to thrive* is used to describe the many problems associated with infants and children suffering from protein deficiency. Excess dietary protein can cause problems, too, especially in a small infant. Too much protein stresses the liver and kidneys, which have to metabolize and excrete the excess nitrogen. Signs of protein overload include acidosis, dehydration, diarrhea, elevated blood ammonia, elevated blood urea, and fever. Such

>How To

Plot Measures on a Growth Chart

You can assess the growth of infants and children by plotting their measurements on a percentile graph. Percentile graphs divide the measures of a population into 100 equal divisions so that half of the population falls at or above the 50th percentile and half falls below. Using percentiles allows for comparisons among people of the same age and gender.

To plot measures on a growth chart, follow these steps:

- Select the appropriate chart based on age and gender. For this example, use the accompanying chart, which gives percentiles for weight for girls from birth to 36 months. (Appendix E provides other growth charts for both boys and girls of various ages.)

- Locate the infant's age along the horizontal axis at the bottom of the chart (in this example, 6 months).

- Locate the infant's weight in pounds or kilograms along the vertical axis of the chart (in this example, 17 pounds or 7.7 kilograms).

- Mark the chart where the age and weight lines intersect (shown here with a red dot), and follow the curved line to find the percentile.

This 6-month-old infant is at the 75th percentile. Her pediatrician will weigh her again over the next few months and expect the growth curve to follow the same percentile throughout the first year. In general, dramatic changes or measures much above the 80th percentile or much below the 10th percentile may be cause for concern.

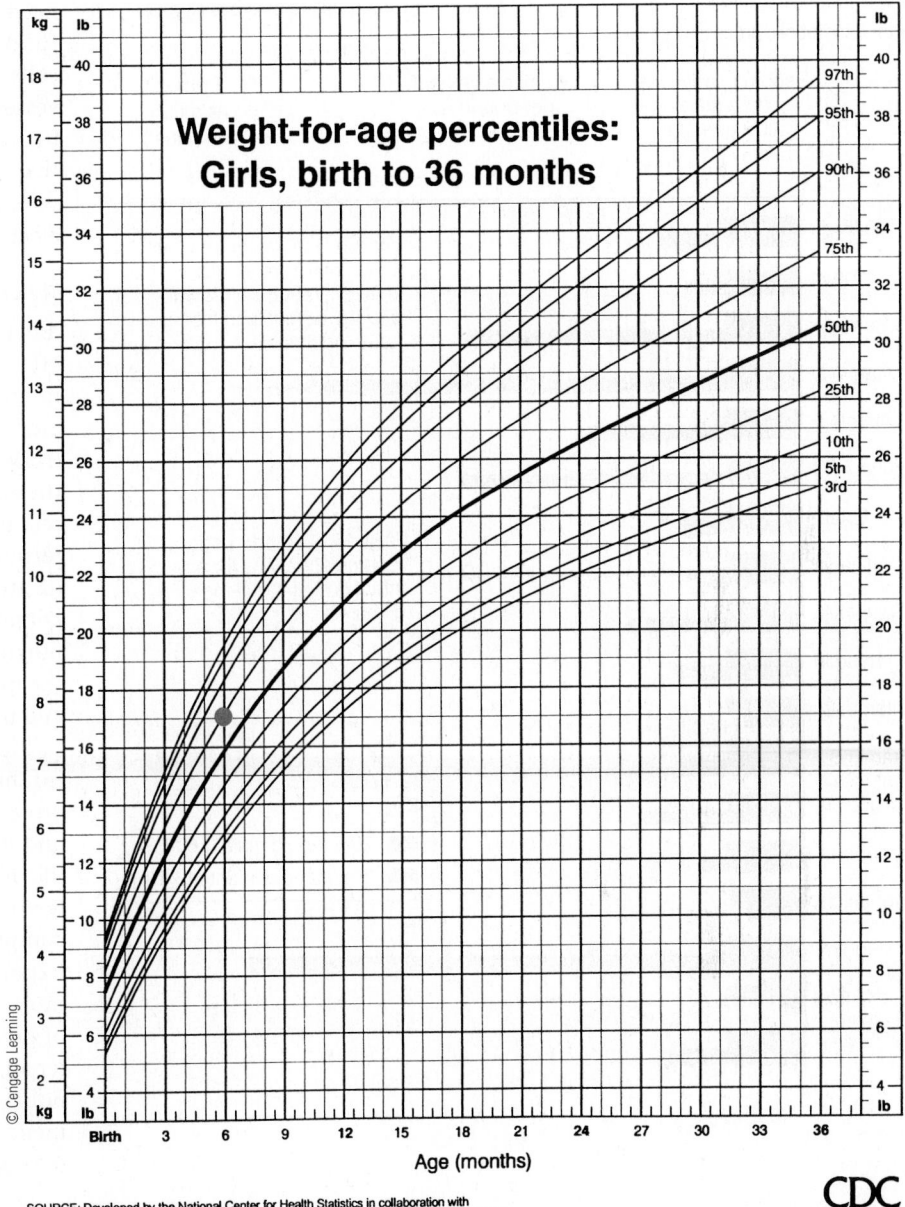

Weight-for-age percentiles: Girls, birth to 36 months

Age (months)

© Cengage Learning

SOURCE: Developed by the National Center for Health Statistics in collaboration with the National Center for Chronic Disease Prevention and Health Promotion (2000).

CDC

> **TRY IT** Determine the percentile for a 12-month-old girl who weighs 21 pounds.

problems are not common, but they have been observed in infants fed inappropriate foods, such as fat-free milk or concentrated formula.

Vitamins and Minerals An infant's needs for most nutrients, in proportion to body weight, are more than double those of an adult. Figure 16-2 (p. 512) illustrates this by comparing a 5-month-old infant's needs per unit of body weight with those of an adult man. Some of the differences are extraordinary. Infant recommendations are based on the average amount of nutrients consumed by thriving infants breastfed by well-nourished mothers.

Water One of the most essential nutrients for infants, as for everyone, is water. The younger the infant, the greater the percentage of body weight is water. During

> **FIGURE 16-2** **Recommended Intakes of an Infant and an Adult Compared on the Basis of Body Weight**

Because infants are small, they need smaller total amounts of the nutrients than adults do, but when comparisons are based on body weight, infants need more than twice as much of many nutrients. Infants use large amounts of energy and nutrients, in proportion to their body size, to keep all their metabolic processes going.

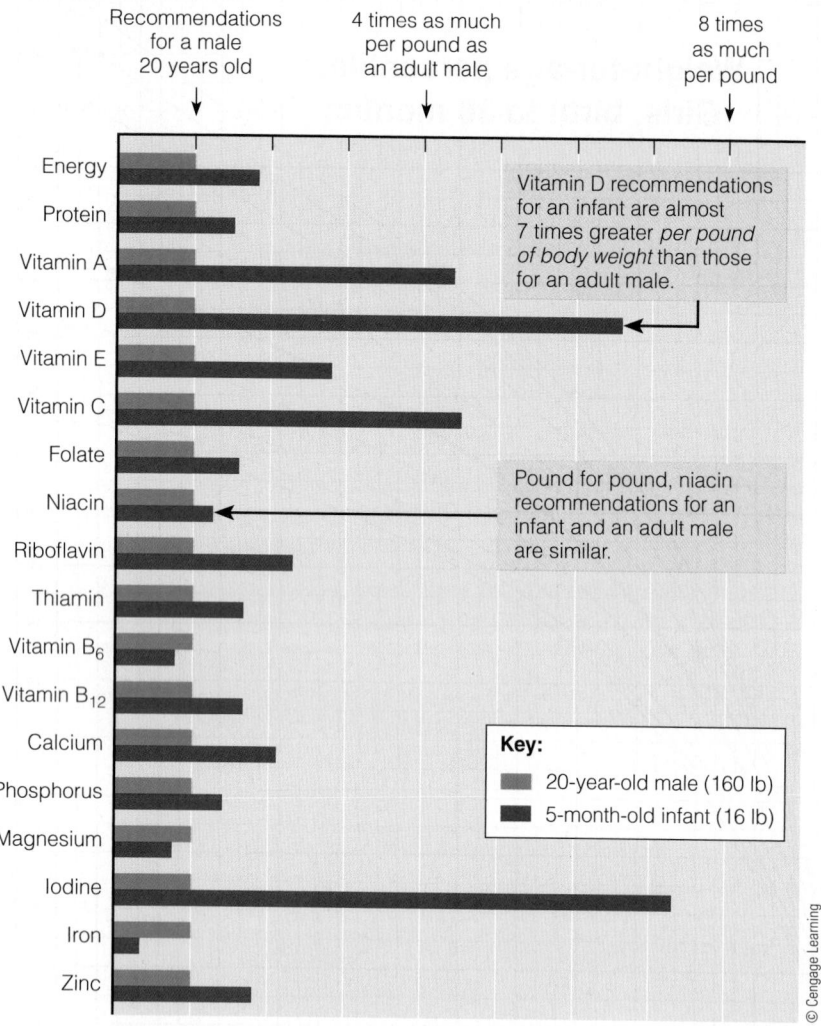

Recommendations for a male 20 years old

4 times as much per pound as an adult male

8 times as much per pound

Vitamin D recommendations for an infant are almost 7 times greater *per pound of body weight* than those for an adult male.

Pound for pound, niacin recommendations for an infant and an adult male are similar.

Energy
Protein
Vitamin A
Vitamin D
Vitamin E
Vitamin C
Folate
Niacin
Riboflavin
Thiamin
Vitamin B$_6$
Vitamin B$_{12}$
Calcium
Phosphorus
Magnesium
Iodine
Iron
Zinc

Key:
- 20-year-old male (160 lb)
- 5-month-old infant (16 lb)

© Cengage Learning

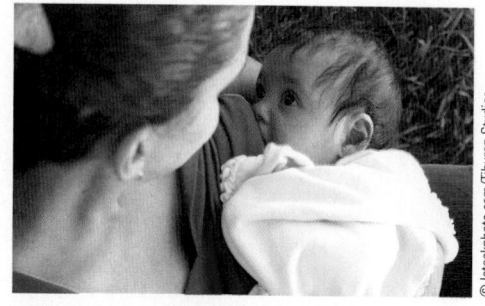

© Istockphoto.com/Tiburon Studios

Women are encouraged to breastfeed whenever possible because breast milk provides infants with many nutrient and health advantages.

early infancy, breast milk or infant formula normally provides enough water to replace fluid losses in a healthy infant. If the environmental temperature is extremely high, however, infants need supplemental water.[3] Because much of the fluid in an infant's body is located *outside* the cells—between the cells and within the blood vessels—rapid fluid losses and the resulting dehydration can be life-threatening. Conditions that cause rapid fluid loss, such as diarrhea or vomiting, require prompt treatment with an electrolyte solution designed for infants.

Breast Milk In the United States, the two dietary practices that have the most significant effect on an infant's nutrition are the milk the infant receives and the age at which solid foods are introduced. A later section discusses the introduction of solid foods, but as to the milk, the American Academy of Pediatrics strongly recommends breastfeeding for healthy full-term infants, except where specific contraindications exist. The Academy of Nutrition and Dietetics also advocates breastfeeding for the nutritional health of the infant as well as for the many other benefits it provides both infant and mother (review Table 15-9 on p. 496).[4]

Breast milk excels as a source of nutrients for infants. Its unique nutrient composition and protective factors promote optimal infant health and development throughout the first year of life. Ideally, infants will receive exclusive breastfeeding for 6 months, and breastfeeding with complementary foods for at least 12 months.[5] Experts add, though, that iron-fortified formula, which imitates the nutrient composition of breast milk, is an acceptable alternative. After all, the primary goal is to provide the infant nourishment in a relaxed and loving environment. Chapter 15 discusses maternal nutrition to support successful breastfeeding.

Frequency and Duration of Breastfeeding Breast milk is more easily and completely digested than formula, so breastfed infants usually need to eat more frequently than formula-fed infants do. During the first few weeks, approximately 8 to 12 feedings a day, on demand, as soon as the infant shows early signs of hunger such as increased alertness, activity, or suckling motions. Such a schedule promotes optimal milk production and infant growth. Crying is a late indicator of hunger. An infant who nurses every 2 to 3 hours and sleeps contentedly between feedings is adequately nourished. As the infant gets older, stomach capacity enlarges and the mother's milk production increases, allowing for longer intervals between feedings.

Even though the infant obtains about half the milk from the breast during the first 2 to 3 minutes of suckling, the infant should be encouraged to breastfeed on the first breast for as long as he or she is actively suckling, before being offered the second breast. The next feeding begins on the breast offered last. The infant's suckling, as well as the complete removal of milk from the breast, stimulates milk production.

Energy Nutrients The energy-nutrient composition of breast milk differs dramatically from that recommended for adult diets (see Figure 16-3). Yet for infants, breast milk is nature's most nearly perfect food, providing the clear lesson that people at different stages of life have different nutrient needs.

The main carbohydrate in breast milk (and standard infant formula) is the disaccharide lactose. In addition to being easily digested, lactose enhances calcium absorption. The carbohydrate component of breast milk also contains abundant oligosaccharides, which are present only in trace amounts in cow's milk and infant formula made from cow's milk.[6] Breast milk oligosaccharides help protect the infant from infection by preventing the binding of pathogens to the infant's intestinal cells.[7]

The amount of protein in breast milk is less than in cow's milk, but this quantity is actually beneficial because it places less stress on the infant's immature kidneys to excrete urea, the major end product of protein metabolism. Much of the protein in breast milk is **alpha-lactalbumin,** which is efficiently digested and absorbed.

As for the lipids, breast milk contains a generous proportion of the essential fatty acids linoleic acid and linolenic acid, as well as their longer-chain derivatives arachidonic acid and DHA. DHA accumulation in the brain is higher in breastfed infants than in formula-fed infants.[8] Research has focused on the mental and visual development of breastfed infants and infants fed standard formula with and without DHA added.[9] Results of studies for visual acuity are mixed, perhaps because of factors such as the amount of DHA provided, the source of the DHA, and the sensitivity of different measures for visual acuity.[10] Some evidence from studies examining the effects of DHA status during fetal and infant development on cognitive function suggests that DHA supplementation can make a positive difference.[11] Adding DHA to standard infant formulas has no adverse effects; most standard formulas are currently fortified with both DHA and arachidonic acid.

Vitamins With the exception of vitamin D, the vitamins in breast milk are ample to support infant growth. The vitamin D in breast milk is low, however, and vitamin D deficiency impairs bone mineralization.[12] Vitamin D deficiency is most likely in infants who are not exposed to sunlight daily, have darkly pigmented skin, and receive breast milk without vitamin D supplementation. Reports of infants in the United States developing the vitamin D–deficiency disease rickets and recommendations to keep infants younger than 6 months of age out of direct sunlight prompted revisions in vitamin D guidelines. The American Academy of Pediatrics currently recommends a vitamin D supplement for all infants who are breastfed exclusively and for all infants who do not receive at least 1 liter (1000 milliliters, roughly 1 quart or 32 ounces) of vitamin D–fortified formula daily.[13]

Minerals The calcium content of breast milk is ideal for infant bone growth, and the calcium is well absorbed. Breast milk contains relatively small amounts of iron, but the iron has a high bioavailability. Zinc also has a high bioavailability, thanks to the presence of a zinc-binding protein. Breast milk is low in sodium, another benefit for the infant's immature kidneys. Fluoride promotes the development of strong teeth, but breast milk is not a good source.

Supplements Many pediatricians routinely prescribe liquid supplements containing vitamin D, iron, and fluoride as outlined in Table 16-2. In addition, infants receive a single dose of vitamin K at birth to protect them from bleeding to death. (See Chapter 11 for a description of vitamin K's role in blood clotting.)

Immunological Protection In addition to its nutritional benefits, breast milk offers immunological protection. Not only is breast milk sterile, but it also actively fights disease and protects infants from illnesses.[14] Such protection is most valuable

> **FIGURE 16-3** **Percentages of Energy-Yielding Nutrients in Breast Milk and in Recommended Adult Diets**

The proportions of energy-yielding nutrients in human breast milk differ from those recommended for adults.

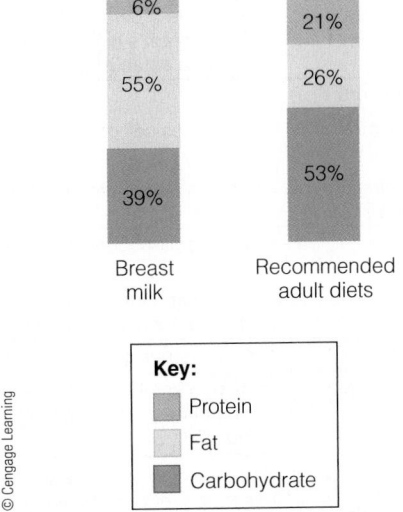

Key:
- Protein
- Fat
- Carbohydrate

NOTE: The values listed for adults represent approximate midpoints of the acceptable ranges for protein (10 to 35 percent), fat (20 to 35 percent), and carbohydrate (45 to 65 percent).

alpha-lactalbumin (lact-AL-byoo-min): a major protein in human breast milk, as opposed to *casein* (CAY-seen), a major protein in cow's milk.

TABLE 16-2 Supplements for Full-Term Infants

	Vitamin D[a]	Iron[b]	Fluoride[c]
Breastfed infants			
Birth to 6 months of age	✓		
6 months to 1 year	✓	✓	✓
Formula-fed infants			
Birth to 6 months of age			
6 months to 1 year		✓	✓

[a]Vitamin D supplements are recommended for all infants who are exclusively breastfed and for any infants who do not receive at least 1 liter (1000 milliliters, roughly 1 quart or 32 ounces) of vitamin D–fortified formula per day.

[b]All infants 6 months of age need additional iron, preferably in the form of iron-fortified infant cereal and/or infant meats. Formula-fed infants need iron-fortified infant formula.

[c]At 6 months of age, breastfed infants and formula-fed infants who receive ready-to-use formulas (these are prepared with water low in fluoride) or formula mixed with water that contains little or no fluoride (less than 0.3 ppm) need supplements.

SOURCE: Adapted from Committee on Nutrition, American Academy of Pediatrics, *Pediatric Nutrition, 7th* ed., ed. R. E. Kleinman (Elk Grove Village, Ill.: American Academy of Pediatrics, 2014).

TABLE 16-3 Protective Factors in Breast Milk

Factor(s)	Action(s)
Antibodies	Offer protection in the upper respiratory tract and gastrointestinal tract by inhibiting pathogen attachment to the mucosa and protecting against invasive infections; may stimulate the infant's immune system
Bifidus factors	Favor the growth of the "friendly" bacterium *Lactobacillus bifidus* in the infant's digestive tract so that other, harmful bacteria cannot become established
Growth factors:	
Epidermal growth factor	Regulates cell growth, proliferation, and differentiation
Transforming growth factor-beta (TGF-β)	Inhibits inflammatory bowel diseases; supports a healthy epithelial barrier
Lactadherin	Inhibits pathogen attachment to the intestinal mucosa
Lactoferrin	Prevents bacteria from getting the iron needed to grow; helps absorb iron into the infant's bloodstream; kills some bacteria directly; inhibits viral activity
Lysozyme	Kills bacteria by assisting lactoferrin
Oligosaccharides	Help to establish and maintain growth of desired bacteria in gastrointestinal tract; inhibit pathogen attachment to the intestinal mucosa

SOURCE: Adapted from B. Lonnerdal, Bioactive proteins in breast milk, *Journal of Paediatrics and Child Health* (Supplement S1) 49 (2013): 1–7; D. E. W. Chatterton and coauthors, Anti-inflammatory mechanisms of bioactive milk proteins in the intestine of newborns, *International Journal of Biochemistry and Cell Biology* 45 (2013): 1730–1747; P. V. Jeurink and coauthors, Mechanisms underlying immune effects of dietary oligosaccharides, *American Journal of Clinical Nutrition* 98 (2013): 572S–577S; A. Walker, Breast milk as the gold standard for protective nutrients, *Journal of Pediatrics* 156 (2010): 53–57.

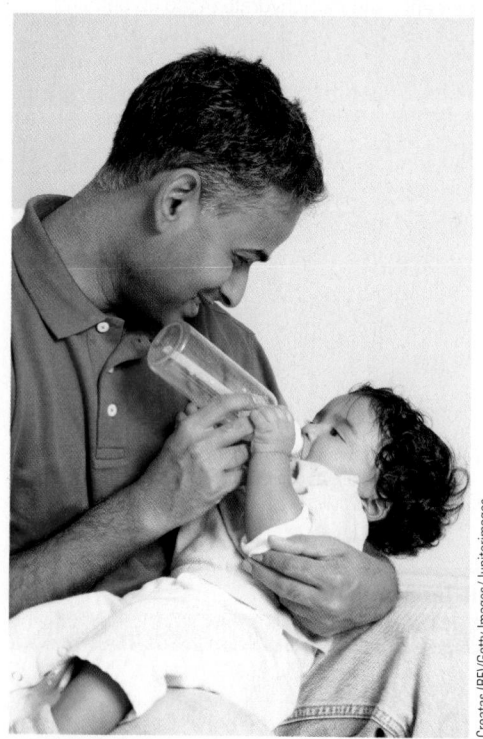

Creatas (RF)/Getty Images/Jupiterimages

The infant thrives on infant formula offered with affection.

colostrum (ko-LAHS-trum): a milklike secretion from the breast, present during the first few days after delivery before milk appears; rich in protective factors.

breast milk bank: a service that collects, screens, processes, and distributes donated human milk.

during the first year, when the infant's immune system is not fully prepared to mount a response against infections.

During the first 2 to 3 days after delivery, the breasts produce **colostrum**, a pre-milk substance containing mostly serum with antibodies and white blood cells. Colostrum (like breast milk) helps protect the newborn from infections against which the mother has developed immunity. The maternal antibodies in the breast milk inactivate disease-causing bacteria within the infant's digestive tract before they can start infections.[15] This explains, in part, why breastfed infants have fewer intestinal infections than formula-fed infants. In addition to antibodies, colostrum and breast milk provide other powerful agents that help to fight against bacterial and viral infection (see Table 16-3).

Allergy and Disease Protection In addition to protection against infections, breast milk may offer protection against the development of allergies.[16] Compared with formula-fed infants, breastfed infants have a lower incidence of allergic reactions, such as recurrent wheezing and skin rashes. This protection is especially noticeable among infants with a family history of allergies.[17] Breastfeeding may also reduce the risk of sudden infant death syndrome (SIDS).[18] The protective effect against SIDS is stronger when infants are exclusively breastfed, but breastfeeding to any extent for any duration offers some protection. Similarly, breast milk may offer protection against the development of cardiovascular disease in adulthood, but additional research, using consistent and precise definitions of breastfeeding, is needed to confirm this effect.[19]

Other Potential Benefits Breastfeeding may offer some protection against excessive weight gain later, although findings are inconsistent.[20] For example, some research suggests that the longer the duration of breastfeeding, the lower the risk of overweight in childhood, while other research conflicts with such findings.[21] Researchers note that many other factors—socioeconomic status, other infant and child feeding practices, and especially the mother's weight—strongly predict a child's body weight.[22]

Many studies suggest a beneficial effect of breastfeeding on intelligence, but when subjected to strict standards of methodology (for example, large sample size and appropriate intelligence testing), the evidence is less than convincing.[23] Nevertheless, the possibility that breastfeeding may positively affect later intelligence is intriguing. It may be that some specific component of breast milk, such as DHA, stimulates brain development or that certain factors associated with the feeding process itself promote intellect.[24] Most likely, a combination of factors is involved. More large, well-controlled studies are needed to confirm the effects, if any, of breastfeeding on brain development and intelligence.

Breast Milk Banks Similar to blood banks that collect blood from individuals to give to others in need, **breast milk banks** receive milk from lactating women who have an abundant supply to give to infants whose own mothers' milk is unavailable or insufficient.[25] The women who donate breast milk are carefully screened to exclude those who smoke cigarettes, use illicit drugs, take medications (including high doses of dietary supplements), drink more than two alcoholic beverages a day, or have communicable diseases. The breast milk from several donors is pooled to ensure an even distribution of all components, pasteurized to destroy bacteria, checked for contamination, and frozen before being shipped overnight to hospitals, where it is dispensed by physician prescription. In the absence of a mother's own breast milk, donor milk may be the life-saving solution for fragile infants, most notably those with very low birthweight or unusual medical conditions.

Infant Formula

A woman who breastfeeds for a year can **wean** her infant to cow's milk, bypassing the need for infant formula. A woman who decides to feed her infant formula from birth, to wean to formula after less than a year of breastfeeding, or to substitute formula for breastfeeding on occasion must select an appropriate infant formula and learn to prepare it. Cow's milk is inappropriate during the first year of life.

Infant Formula Composition Formula manufacturers attempt to copy the nutrient composition of breast milk as closely as possible. Figure 16-4 illustrates the energy-nutrient balance of breast milk, infant formula, and cow's milk. All formula-fed infants should be given iron-fortified infant formulas.[26] The increasing use of iron-fortified formulas over the past few decades is responsible for the decline in iron-deficiency anemia among infants in the United States.

Risks of Formula Feeding Infant formulas contain no protective antibodies for infants, but in general, vaccinations, purified water, and clean environments in developed countries help protect infants from infections. Formulas can be prepared safely by following the rules of proper food handling and by using water that is free of contamination. Of particular concern is lead-contaminated water, a major source of lead poisoning in infants. Because the first water drawn from the tap each day is highest in lead, a person living in a house with old, lead-soldered plumbing should let the water run a few minutes before drinking or using it to prepare formula or food.

Water supplies contain variable concentrations of minerals, including fluoride. As mentioned in Chapter 13, optimal levels of fluoride protect against dental caries, but too much fluoride during tooth development can cause defects in the teeth known as fluorosis. Inadequate fluoride after 6 months of age may also be a concern. Thus, health experts urge caregivers to have their well water analyzed for mineral contents and to check with the local health department to determine the fluoride content of the community water supply. Caregivers should reconstitute powdered or concentrated liquid formulas with optimally fluoridated water (0.7 to 1.2 parts per million).[27] If fluoride levels are excessive, ready-to-feed formulas or formulas prepared with fluoride-free or low-fluoride water can be used. Such waters are labeled "purified," "demineralized," "deionized," or "distilled." If fluoride levels are inadequate, formulas may need to be prepared with fluoridated bottled water once the infant is 6 months of age.

In developing countries and in poor areas of the United States, formula may be unavailable, prepared with contaminated water, or overdiluted in an attempt to save money. Contaminated formulas may cause infections, leading to diarrhea, dehydration, and malabsorption. Without sterilization and refrigeration, formula is an ideal breeding ground for bacteria. Whenever such risks are present, breastfeeding can be a life-saving option: breast milk is sterile, and its antibodies enhance an infant's resistance to infections.

Infant Formula Standards National and international standards have been set for the nutrient contents of infant formulas. In the United States, the standard developed by the American Academy of Pediatrics reflects "human milk taken from well-nourished mothers during the first or second month of lactation, when the infant's growth rate is high." The Food and Drug Administration (FDA) mandates the safety and nutritional quality of infant formulas. Formulas meeting these standards have similar nutrient compositions. Small differences among formulas are sometimes confusing, but they are usually unimportant.

Special Formulas Standard formulas are inappropriate for some infants. Special formulas have been designed to meet the dietary needs of infants with specific conditions such as prematurity or inherited diseases. Most infants allergic to milk protein can drink formulas based on soy protein.[28] Soy formulas also

> **FIGURE 16-4** **Percentages of Energy-Yielding Nutrients in Breast Milk, Infant Formula, and Cow's Milk**

The average proportions of energy-yielding nutrients in human breast milk and formula differ slightly. In contrast, cow's milk provides too much protein and too little carbohydrate.

Key:
- Protein
- Fat
- Carbohydrate

wean: to gradually replace breast milk with infant formula or other foods appropriate to an infant's diet.

> **FIGURE 16-5** **Nursing Bottle Tooth Decay**

This child was frequently put to bed sucking on a bottle filled with apple juice, so the teeth were bathed in carbohydrate for long periods of time—a perfect medium for bacterial growth. The upper teeth show signs of decay.

© E. H. Gill/Custom Medical Stock Photo

use cornstarch and sucrose instead of lactose and so are recommended for infants with lactose intolerance as well. They are also useful as an alternative to milk-based formulas for vegan families. Despite these limited uses, soy formulas account for about 15 percent of the infant formulas sold today. Although soy formulas support the normal growth and development of infants, for infants who don't need them, they offer no advantage over milk formulas.

Some infants who are allergic to cow's milk protein may also be allergic to soy protein.[29] For these infants, special formulas based on hydrolyzed protein are available. The protein in these formulas is a mixture of free amino acids, dipeptides, tripeptides, and short-chain peptides that do not elicit an allergic reaction in most infants.

Inappropriate Formulas Caregivers must use only products designed for infants; soy *beverages*, for example, are nutritionally incomplete and inappropriate for infants. Goat's milk is also inappropriate for infants in part because of its low folate content.[30] An infant receiving goat's milk is likely to develop "goat's milk anemia," an anemia characteristic of folate deficiency.

Nursing Bottle Tooth Decay An infant cannot be allowed to sleep with a bottle because of the potential damage to developing teeth. Salivary flow, which normally cleanses the mouth, diminishes as the infant falls asleep. Prolonged sucking on a bottle of formula, milk, or juice bathes the upper teeth in a carbohydrate-rich fluid that nourishes decay-producing bacteria. (The tongue covers and protects most of the lower teeth, but they, too, may be affected.) The result is extensive and rapid tooth decay (see Figure 16-5). To prevent **nursing bottle tooth decay,** no infant should be put to bed with a bottle of nourishing fluid.

Special Needs of Preterm Infants
An estimated one out of eight pregnancies in the United States results in a preterm birth.[31] The terms *preterm* and *premature* imply incomplete fetal development, or immaturity, of many body systems. As might be expected, preterm birth is a leading cause of infant deaths. Preterm infants face physical independence from their mothers before some of their organs and body tissues are ready. The rate of weight gain in the fetus is greater during the last trimester of gestation than at any other time. Therefore, a preterm infant is most often a low-birthweight infant as well. A premature birth deprives the infant of the nutritional support of the placenta during a time of maximal growth.

The last trimester of gestation is also a time of building nutrient stores. Being born with limited nutrient stores intensifies the already precarious situation for the infant. The physical and metabolic immaturity of preterm infants further compromises their nutrition status. Nutrient absorption, especially of fat and calcium, from an immature GI tract is limited. Consequently, preterm, low-birthweight infants are candidates for nutrient imbalances. Deficiencies of the fat-soluble vitamins, calcium, iron, and zinc are common.

Preterm breast milk is well suited to meet a preterm infant's needs. During early lactation, preterm breast milk contains higher concentrations of protein and is lower in volume than term breast milk. The low milk volume is advantageous because preterm infants consume small quantities of milk per feeding, and the higher protein concentration allows for better growth. In many instances, supplements of nutrients specifically designed for preterm infants are added to the mother's expressed breast milk and fed to the infant from a bottle. When fortified with a preterm supplement, preterm breast milk supports growth at a rate that approximates the growth rate that would have occurred within the uterus.[32]

Introducing Cow's Milk
The age at which cow's milk should be introduced to the infant's diet has long been a source of controversy. The American Academy of Pediatrics advises that cow's milk is not appropriate during the first year.[33] For some infants, particularly those younger than 6 months of age, cow's milk may cause intestinal bleeding, which can lead to iron deficiency.[34] Cow's milk is also a poor source of iron. Consequently, it both causes iron loss and fails to replace iron. Furthermore,

nursing bottle tooth decay: extensive tooth decay due to prolonged tooth contact with formula, milk, fruit juice, or other carbohydrate-rich liquid offered to an infant in a bottle.

the bioavailability of iron from infant cereal and other foods is reduced when cow's milk replaces breast milk or iron-fortified formula during the first year. Compared with breast milk or iron-fortified formula, cow's milk is higher in calcium and lower in vitamin C, characteristics that further reduce iron absorption. In addition, the higher protein concentration of cow's milk can stress the infant's kidneys. In short, infants need breast milk or iron-fortified infant formula, *not* cow's milk.

> **DIETARY GUIDELINES FOR AMERICANS**
Children 2 to 3 years of age should consume 2 cups of milk or milk products per day, and children 4 to 8 years of age should consume 2½ cups per day.

Ideally, by 1 year, the infant is obtaining at least two thirds of the total daily food energy from a balanced mixture of cereals, vegetables, fruits, and other foods. At this time, reduced-fat or low-fat cow's milk is an acceptable and recommended beverage to accompany a diet that supplies 30 percent of kcalories from fat.[35] After the age of 2, a transition to fat-free milk can take place, but care should be taken to avoid excessive restriction of dietary fat.

Until recently, the American Academy of Pediatrics recommended the use of whole cow's milk for children between the ages of 1 and 2 years. That guideline was dropped in light of accumulating evidence that atherosclerosis begins in childhood and that risk factors (such as the saturated fat of whole milk) should be limited as early as possible. Among the new guidelines is the recommendation that children 1 to 2 years of age consume reduced-fat (2 percent fat) or low-fat (1 percent fat) milk and that children 2 years of age and older consume fat-free milk to limit saturated fat in the diet.

Introducing Solid Foods
The high nutrient needs of infancy are met first by breast milk or formula only and then by the limited addition of selected foods over time. Infants gradually develop the ability to chew, swallow, and digest the wide variety of foods available to adults. The caregiver's selection of appropriate foods at the appropriate stages of development is prerequisite to the infant's optimal growth and health.

When to Begin In addition to breast milk or formula, an infant can begin eating solid foods between 4 and 6 months.[36] The American Academy of Pediatrics supports exclusive breastfeeding for 6 months but recognizes that infants are often developmentally ready to accept complementary foods between 4 and 6 months of age. The main purpose of introducing solid foods is to provide needed nutrients that are no longer supplied adequately by breast milk or formula alone.[37] The foods chosen must be those that the infant is developmentally capable of handling both physically and metabolically. As digestive secretions gradually increase throughout the first year of life, the digestion of solid foods becomes more efficient. The exact timing depends on the individual infant's needs and developmental readiness (see Table 16-4, p. 518), which vary from infant to infant because of differences in growth rates, activities, and environmental conditions. In addition to the infant's nutrient needs and physical readiness to handle different forms of foods, the need to detect and control allergic reactions should also be considered when introducing solid foods.

Food Allergies To prevent allergy and to facilitate its prompt identification should it occur, experts recommend introducing single-ingredient foods, one at a time, in small portions, and waiting 3 to 5 days before introducing the next new food.[38] For example, rice cereal is usually the first cereal introduced because it is the least allergenic. When it is clear that rice cereal is not causing an allergy, another grain, perhaps barley or oat is introduced. Wheat cereal is offered last because it is the most common offender. If a cereal causes an allergic reaction such as a skin rash, digestive upset, or respiratory discomfort, it should be discontinued before introducing the next food. A later section in this chapter offers more information about food allergies.

TABLE 16-4 Infant Development and Recommended Foods

Because each stage of development builds on the previous stage, the foods from an earlier stage continue to be included in all later stages.

Age (mo)	Feeding Skill	Appropriate Foods Added to the Diet
0–4	Turns head toward any object that brushes cheek. Initially swallows using back of tongue; gradually begins to swallow using front of tongue as well. Strong reflex to push food out (extrusion) during first 2 to 3 months.	Feed breast milk or infant formula.
4–6	Extrusion reflex diminishes, and the ability to swallow nonliquid foods develops. Indicates desire for food by opening mouth and leaning forward. Indicates satiety or disinterest by turning away and leaning back. Sits erect with support at 6 months. Begins chewing action. Brings hand to mouth. Grasps objects with palm of hand.	Begin iron-fortified cereal mixed with breast milk, formula, or water. Begin pureed meats, legumes, vegetables, and fruits.
6–8	Able to self-feed finger foods. Develops pincer (finger to thumb) grasp. Begins to drink from cup.	Begin textured vegetables and fruits. Begin unsweetened, diluted fruit juices from cup.
8–10	Begins to hold own bottle. Reaches for and grabs food and spoon. Sits unsupported.	Begin breads and cereals from table. Begin yogurt. Begin pieces of soft, cooked vegetables and fruit from table. Gradually begin finely cut meats, fish, casseroles, cheese, eggs, and mashed legumes.
10–12	Begins to master spoon, but still spills some.	Add variety. Gradually increase portion sizes.[a]

© Cengage Learning

[a]Portion sizes for infants and young children are smaller than those for an adult. For example, a grain serving might be ½ slice of bread instead of 1 slice, or ¼ cup rice instead of ½ cup.
SOURCE: Adapted in part from Committee on Nutrition, American Academy of Pediatrics, *Pediatric Nutrition*, 7th ed., ed. R. E. Kleinman (Elk Grove Village, Ill.: American Academy of Pediatrics, 2014), pp. 113–142.

Choice of Infant Foods Infant foods should be selected to provide variety, balance, and moderation. Commercial baby foods offer a wide variety of palatable, nutritious foods in a safe and convenient form. Parents or caregivers should not feed directly from the jar; instead, spoon the needed portion into a dish and feed from there, leaving the leftovers in the jar uncontaminated by a used spoon. Homemade infant foods can be as nutritious as commercially prepared ones, so long as the cook minimizes nutrient losses during preparation. Ingredients for homemade foods should be fresh, whole foods without added salt, sugar, or seasonings. Pureed food can be frozen in ice cube trays, providing convenient-size blocks of food that can be thawed, warmed, and fed to the infant. To guard against foodborne illnesses, hands and equipment must be kept clean.

Foods to Provide Iron Rapid growth demands iron. At about 4 to 6 months of age, the infant begins to need more iron than body stores plus breast milk or iron-fortified formula can provide. In addition to breast milk or iron-fortified formula, infants can receive iron from iron-fortified cereals and, once they readily accept solid foods, from meats or legumes. Iron-fortified cereals contribute a significant amount of iron to an infant's diet, but the iron's bioavailability is poor.[39] Caregivers can enhance iron absorption from iron-fortified cereals by serving vitamin C–rich foods with meals.

Foods to Provide Vitamin C The best sources of vitamin C are fruits and vegetables (see pp. 330–332 in Chapter 10). It has been suggested that infants who are introduced to fruits before vegetables may develop a preference for sweets and find the vegetables less palatable, but there is no evidence to support offering these foods in a particular order.[40]

Fruit juice is a good source of vitamin C, but excessive juice intake can lead to diarrhea in infants and young children.[41] Furthermore, too much fruit juice contributes excessive kcalories and displaces other nutrient-rich foods. The American Academy of Pediatrics recommends limiting juice consumption for infants and children (2 to 10 years of age) to 4 ounces per day.[42] Fruit juices should be diluted and served in a cup, not a bottle, once the infant is 6 months of age or older.

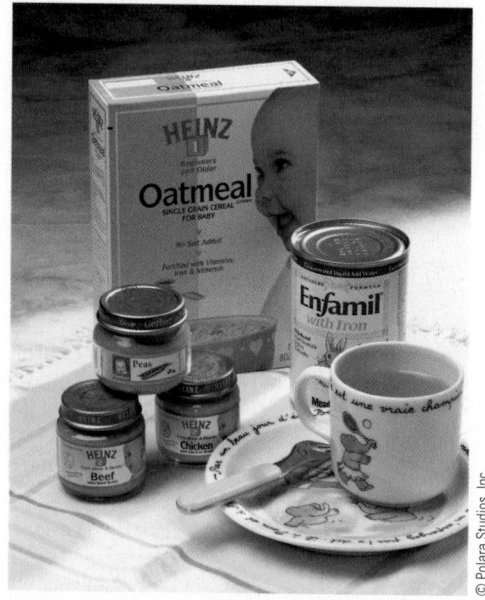

© Polara Studios, Inc.

Foods such as iron-fortified cereals and formulas, mashed legumes, and strained meats provide iron.

TABLE 16-5 **Examples of Common Choking Items**

Monitor intake of 100 percent fruit juice for children, especially those who are overweight or obese.

Foods to Omit Concentrated sweets, including baby food "desserts," have no place in an infant's diet. They convey no nutrients to support growth, and the extra food energy can promote obesity. Products containing sugar alcohols such as sorbitol should also be limited because they may cause diarrhea. Canned vegetables are also inappropriate for infants because they often contain too much sodium. Honey and corn syrup should never be fed to infants because of the risk of **botulism.*** Infants and young children are vulnerable to foodborne illnesses, and the *Dietary Guidelines for Americans* address this risk.

Infants and young children should not eat or drink unpasteurized milk, milk products, or juices; raw or undercooked eggs, meat, poultry, fish, or shellfish; or raw sprouts.

Infants and even young children cannot safely chew and swallow any of the foods listed in Table 16-5; they can easily choke on these foods, a risk not worth taking. Nonfood items may present even greater choking hazards to infants and young children. Parents and caregivers must pay careful attention to eliminate choking hazards in children's environments.

Vegetarian Diets during Infancy The newborn infant is a lacto-vegetarian. As long as the infant has access to sufficient quantities of either iron-fortified infant formula or breast milk (plus a vitamin D supplement) from a mother who eats an adequate diet, the infant will thrive during the early months. "Health-food beverages," such as rice milk, are inappropriate choices because they lack the protein, vitamins, and minerals infants and toddlers need; in fact, their use can lead to nutrient deficiencies.

Infants older than about 6 months of age present a greater challenge in terms of meeting nutrient needs by way of vegetarian and, especially, vegan diets. Continued breastfeeding or formula feeding is recommended, but supplementary feedings are necessary to ensure adequate energy and iron intakes. Infants and young children in vegetarian families should be given iron-fortified infant cereals well into the second year. Mashed or pureed legumes, tofu, and cooked eggs can be added to their diets in place of meats.[43]

Infants who receive a well-balanced vegetarian diet that includes milk products and a variety of other foods can easily meet their nutritional requirements for growth. This is not always true for vegan infants; the growth of vegan infants slows significantly when weaning from breast milk to solid foods. Deficiencies of protein, vitamin D, vitamin B_{12}, iron, and calcium have been reported in infants fed vegan diets. Vegan diets that are high in fiber, other complex carbohydrates, and water will fill infants' stomachs before meeting energy needs. This problem can be partially alleviated by providing more energy-dense foods, such as mashed legumes, tofu, and avocado. Using soy formulas (or milk) fortified with calcium, vitamin B_{12}, and vitamin D and including vitamin C–containing foods at meals to enhance iron absorption will help prevent some nutrient deficiencies in vegan diets. Parents or caregivers who choose to feed their infants vegan diets should consult with their pediatrician and a registered dietitian nutritionist frequently to ensure a nutritionally adequate diet that will support growth.

Foods at 1 Year At 1 year of age, reduced-fat or low-fat cow's milk can become a primary source of most of the nutrients an infant needs; 2 to 3 cups a day meets those needs sufficiently. Ingesting more milk than this can displace iron-rich foods, which can lead to **milk anemia.** If powdered milk is used, it should contain some fat.

Foods	
• Gum	• Raw carrots
• Hard or gel-type candies	• Raw celery
• Hot dog slices	• Sausage sticks or slices
• Large raw apple slices	• Whole beans
• Marshmallows	• Whole cherries
• Nuts	• Whole grapes
• Peanut butter	
• Popcorn	

Nonfood items	
• Balloons	• Small balls and marbles
• Coins	• Other items of similar size
• Pen tops	

Ideally, a 1-year-old eats many of the same foods as the rest of the family.

botulism (BOT-chew-lism): an often fatal foodborne illness caused by the ingestion of foods containing a toxin produced by bacteria that grow without oxygen. (See Chapter 19 for details.)

milk anemia: iron-deficiency anemia that develops when an excessive milk intake displaces iron-rich foods from the diet.

*In infants, but not in older individuals, ingestion of *Clostridium botulinum* spores can cause illness when the spores germinate in the intestine and produce a toxin, which is absorbed. Symptoms include poor feeding, constipation, loss of tension in the arteries and muscles, weakness, and respiratory compromise. Infant botulism has been implicated in 5 percent of cases of sudden infant death syndrome (SIDS).

> FIGURE 16-6   **Sample Meal Plan for a 1-Year-Old**

SAMPLE MENU	
Breakfast	1 scrambled egg 1 slice whole-wheat toast ½ c reduced-fat milk
Morning snack	½ c yogurt ¼ c fruitᵃ
Lunch	½ grilled cheese sandwich: 1 slice whole-wheat bread with ½ slice cheese ½ c vegetablesᵇ (steamed carrots) ¼ c 100% fruit juice (diluted)
Afternoon snack	½ c fruitᵃ ½ c toasted oat cereal
Dinner	1 oz chopped meat or ¼ c well-cooked mashed legumes ½ c rice or pasta ½ c vegetablesᵇ (chopped broccoli) ½ c reduced-fat milk

NOTE: This sample menu provides about 1000 kcalories.
ᵃInclude citrus fruits, melons, and berries.
ᵇInclude dark-green, leafy, and deep-yellow vegetables.

© Cengage Learning

George Doyle/Stockbyte/Getty Images

Let toddlers explore and enjoy their food.

Other foods—meats, other protein foods, iron-fortified cereals, enriched or whole-grain breads, fruits, and vegetables—should be supplied in variety and in amounts sufficient to round out total energy needs. Ideally, a 1-year-old will sit at the table, eat many of the same foods everyone else eats, and drink liquids from a cup, not a bottle. Figure 16-6 shows a meal plan that meets a 1-year-old's requirements.

Mealtimes with Toddlers The nurturing of a young child involves more than nutrition. Those who care for young children are responsible not only for providing nutritious foods, milk, and water, but also a safe, loving, secure environment in which the children may grow and develop. In light of toddlers' developmental and nutrient needs and their often contrary and willful behavior, a few feeding guidelines may be helpful:

- *Discourage unacceptable behavior, such as standing at the table or throwing food.* Be consistent and firm, not punitive. For example, instead of saying "You make me mad when you don't sit down," say "The fruit salad tastes good; please sit down and eat some with me." The child will soon learn to sit and eat.

- *Let toddlers explore and enjoy food, even if this means eating with fingers for a while.* Learning to use a spoon will come in time. Children who are allowed to touch, mash, and smell their food while exploring it are more likely to accept it.

- *Don't force food on children.* Rejecting new foods is normal, and acceptance is more likely as children become familiar with new foods through repeated opportunities to taste them. Instead of saying "You cannot go outside to play until you taste your carrots," say "You can try the carrots again another time."

- *Provide nutritious foods and let children choose which ones, and how much, they will eat.* Gradually, they will acquire a taste for different foods.

- *Limit sweets.* Infants and young children have little room for empty-kcalorie foods in their daily energy allowance. Do not use sweets as a reward for eating meals.

- *Don't turn the dining table into a battleground.* Make mealtimes enjoyable. Teach healthy food choices and eating habits in a pleasant environment. Mealtimes are not the time to fight, argue, or scold.

> **REVIEW IT** List some of the components of breast milk and describe the appropriate foods for infants during the first year of life.
The primary food for infants during the first 12 months is either breast milk or iron-fortified formula. In addition to nutrients, breast milk also offers immunological protection. At about 4 to 6 months of age, infants should gradually begin eating solid foods. By 1 year, they are drinking from a cup and eating many of the same foods as the rest of the family.

16.2 Nutrition during Childhood

> **LEARN IT** Explain how children's appetites and nutrient needs reflect their stage of growth and why iron deficiency and obesity are often concerns during childhood.

Each year from age 1 to adolescence, a child typically grows taller by 2 to 3 inches and heavier by 5 to 6 pounds. Growth charts provide valuable clues to a child's health. Weight gains out of proportion to height gains may reflect overeating and inactivity, whereas measures significantly less than the standard suggest inadequate nutrition.

Increases in height and weight are only two of the many changes growing children experience (see Figure 16-7). At age 1, children can stand alone and are beginning to toddle; by 2, they can walk and are learning to run; and by 3, they can jump and climb with confidence. Bones and muscles increase in mass and density to make these accomplishments possible. Thereafter, lengthening of the long bones and increases in musculature proceed unevenly and more slowly until adolescence.

Energy and Nutrient Needs Children's appetites begin to diminish around 1 year, consistent with the slowing growth. Thereafter, children spontaneously vary their food intakes to coincide with their growth patterns; they demand more food during periods of rapid growth than during slow growth. Sometimes they seem insatiable, and other times they seem to live on air and water.

Children's energy intakes also vary widely from meal to meal. Even so, their total daily intakes remain remarkably constant. If children eat less at one meal, they typically eat more at the next, and vice versa. Overweight children do not always adjust their energy intakes appropriately, however, and may eat in response to external cues such as television commercials, disregarding hunger and satiety signals.[44]

Energy Intake and Activity As mentioned, children's energy needs vary widely, depending on their growth and physical activity. A 1-year-old child needs about 800 kcalories a day; an active 6-year-old child needs twice as many kcalories a day. By age 10, an active child needs about 2000 kcalories a day. Total energy needs increase slightly with age, but energy needs per kilogram of body weight actually decline gradually.

Physically active children of any age need more energy because they expend more, and inactive children can become obese even when they eat less food than the average. Unfortunately, our nation's children are becoming less and less active; child care programs and schools would serve our children well by offering more activities to promote physical fitness.[45] Children who learn to enjoy physical play and exercise, both at home and at school, are best prepared to maintain active lifestyles as adults.

> FIGURE 16-7 **Body Shape of 1-Year-Old and 2-Year-Old Compared**
The body shape of a 1-year-old (left) changes dramatically by age 2 (right). The 2-year-old has lost much of the baby fat; the muscles (especially in the back, buttocks, and legs) have firmed and strengthened; and the leg bones have lengthened.

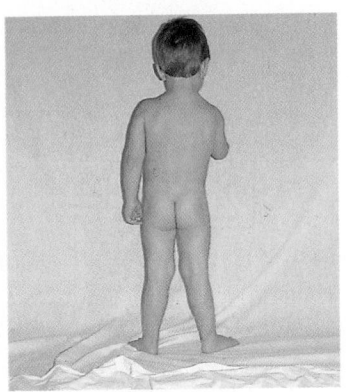

© Anthony M. Vannelli (all)

 > **DIETARY GUIDELINES FOR AMERICANS**
Children should do 60 minutes or more of physical activity daily. Children are encouraged to spend no more than 1 to 2 hours each day watching television, playing electronic games, or using the computer (other than for homework).

Some children, notably those adhering to a vegan diet, may have difficulty meeting their energy needs. Grains, vegetables, and fruits provide plenty of fiber, adding bulk, but may provide too little energy to support growth. Soy products, other legumes, and nut or seed butters offer more concentrated sources of energy to support optimal growth and development.[46]

Carbohydrate and Fiber Carbohydrate recommendations are based on the brain's glucose needs. After 1 year of age, the brain's use of glucose remains fairly constant and is within the adult range. Carbohydrate recommendations for children older than 1 year are therefore the same as for adults (see inside front cover).

Fiber recommendations derive from adult intakes shown to reduce the risk of heart disease and are based on energy intakes. Consequently, fiber recommendations for younger children with low energy intakes are less than those for older ones with high energy intakes (see inside front cover).

> **DIETARY GUIDELINES FOR AMERICANS**
Children and adolescents should consume whole-grain products often, and at least half of the grains should be whole grains.

Fat and Fatty Acids No RDA for total fat has been established, but the DRI Committee recommends a fat intake of 30 to 40 percent of energy for children 1 to 3 years of age and 25 to 35 percent for children 4 to 18 years of age. As long as children's energy intakes are adequate, however, fat intakes less than 30 percent of total energy do not impair growth.[47] Children who eat low-fat diets, however, tend to have low intakes of some vitamins and minerals. Recommended intakes of the essential fatty acids are based on average intakes (see inside front cover).

Protein Like energy needs, total protein needs increase slightly with age, but when the child's body weight is considered, the protein requirement actually declines slightly (see inside front cover). Protein recommendations must consider the requirements for maintaining nitrogen balance, the quality of protein consumed, and the added needs of growth.

Vitamins and Minerals The vitamin and mineral needs of children increase with age (see inside front cover). A balanced diet of nutritious foods can meet children's needs for these nutrients, with the notable exception of iron, and possibly vitamin D. Iron-deficiency anemia is a major problem worldwide and is prevalent among US children, especially toddlers 1 to 3 years of age.[48] During the second year of life, toddlers progress from a diet of iron-rich infant foods such as breast milk, iron-fortified formula, and iron-fortified infant cereal to a diet of adult foods and iron-poor cow's milk. In addition, their appetites often fluctuate—some become finicky about the foods they eat, and others prefer milk and juice to solid foods. These situations can interfere with children eating iron-rich foods at a critical time for brain growth and development.

To prevent iron deficiency, children's foods must deliver 7 to 10 milligrams of iron per day. To achieve this goal, snacks and meals should include iron-rich foods, and milk intake should be reasonable so that it will not displace lean meats, fish, poultry, eggs, legumes, and whole-grain or enriched products. (Chapter 13 describes iron-rich foods and ways to maximize iron absorption.)

The DRI committee recently revised their recommendations for vitamin D intakes for healthy Americans.[49] Children typically obtain most of their vitamin D from fortified milk (2.5 micrograms per 1 cup serving) and dry cereals (1 microgram per ½ cup serving). Children who do not meet their RDA from these sources should receive a vitamin D supplement.[50] Remember that sunlight is also a source of vitamin D, especially in tropical climates and warm seasons.

Supplements With the exception of specific recommendations for fluoride, iron, and vitamin D during infancy and childhood, the American Academy of Pediatrics and other professional groups agree that well-nourished children do not need vitamin and mineral supplements. Despite this, many children and adolescents take supplements.[51] Researchers are still studying the safety of supplement use by children. The Federal Trade Commission has warned parents about giving supplements advertised to prevent or cure childhood illnesses such as colds, ear infections, or asthma. Dietary supplements on the market today include many herbal products that have not been tested for safety and effectiveness in children.

Planning Children's Meals To provide all the needed nutrients, children's meals should include a variety of foods from each food group—in amounts suited to their appetites and needs. Table 16-6 provides USDA Food Patterns for several kcalorie levels. Estimated daily kcalorie needs for active and sedentary children of various ages are shown in Table 16-7. MyPlate online resources for preschoolers (2 to 5 years) translate the eating patterns into messages that can help parents ensure that the foods they provide meet their child's needs. For older children (6 to 11 years), the site provides an interactive

TABLE 16-6 USDA Food Patterns: Recommended Daily Amounts from Each Food Group (1000 to 1800 kCalories)

Food Group	1000 kcal	1200 kcal	1400 kcal	1600 kcal	1800 kcal
Fruits	1 c	1 c	1½ c	1½ c	1½ c
Vegetables	1 c	1½ c	1½ c	2 c	2½ c
Grains	3 oz	4 oz	5 oz	5 oz	6 oz
Protein foods	2 oz	3 oz	4 oz	5 oz	5 oz
Milk	2 c	2½ c	2½ c	3 c	3 c

© Cengage Learning

"Blast Off" nutrition teaching game and other resources for teachers, parents, and children themselves (Figure 16-8). These guidelines and resources also stress the importance of balancing kcalorie intake with kcalorie expenditure through adequate physical activity to promote growth without increasing the risks of developing obesity. Childhood obesity is the topic of a later section in this chapter.

Children whose diets follow the patterns presented in Table 16-6 can meet their nutrient needs fully, but few children eat according to these recommendations.[52] One analysis of the quality of children's diets found that most (up to 88 percent) children between 2 and 9 years of age have diets that need substantial improvement.[53] A comprehensive survey, called the Feeding Infants and Toddlers Study (FITS), assessed the food and nutrient intakes of more than 3000 infants and toddlers.[54] The survey found that fruit and vegetable intakes of infants and toddlers are limited, and in fact, about 25 percent of infants and toddlers older than 9 months did not eat a single serving of fruits or vegetables in a day.[55] Not only are nutritious fruits and vegetables lacking, but more than 80 percent of young preschoolers (2 to 3 years of age) consumed nutrient-poor, energy-dense beverages, desserts, and snack foods each day. The most popular vegetable among this age group is french fries. Parents and caregivers of infants and toddlers thus need to offer a much greater variety of nutrient-dense vegetables and fruits at meals and snacks to help ensure adequate nutrition. Among other nutrition concerns for US children are inadequate intakes of vitamin E, potassium, and fiber, and excessive intakes of sodium.[56]

Hunger and Malnutrition in Children
Most children in the United States have access to regular meals, but hunger and malnutrition are not uncommon, especially among children in very low-income families. More than 16 million US children are hungry at least some of the time and are living in poverty.[57] Chapter 20 examines the causes and consequences of hunger in the United States and around the world.

Hunger and Behavior Both short-term and long-term hunger exert negative effects on behavior and health. Short-term hunger, such as when a child misses a meal, impairs the child's ability to pay attention and be productive. Hungry

TABLE 16-7 Estimated Daily Energy Needs for Children

Children	Sedentary[a]	Active[b]
2 to 3 yr	1000 kcal/day	1400 kcal/day
Females		
4 to 8 yr	1200 kcal/day	1800 kcal/day
9 to 13 yr	1400 kcal/day	2200 kcal/day
Males		
4 to 8 yr	1200 kcal/day	2000 kcal/day
9 to 13 yr	1600 kcal/day	2600 kcal/day

[a] *Sedentary* describes a lifestyle that includes only the activities typical of day-to-day life.

[b] *Active* describes a lifestyle that includes at least 60 minutes per day of moderate physical activity (equivalent to walking more than 3 miles per day at 3 to 4 miles per hour) in addition to the activities of day-to-day life.

© 2016 Cengage Learning

> FIGURE 16-8   **MyPlate Resources for Children**

www.choosemyplate.gov

NOTE: Abundant MyPlate resources for preschool children and older children can be found at www.choosemyplate.gov.

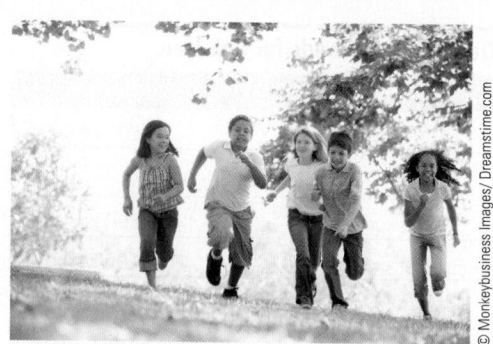

Healthy, well-nourished children are alert in the classroom and energetic at play.

children are irritable, apathetic, and uninterested in their environment. Long-term hunger impairs growth and immune defenses. Food assistance programs such as the WIC program (discussed in Chapter 15) and the School Breakfast and National School Lunch Programs (discussed later in this chapter) are designed to protect against hunger and improve children's health.[58]

Children who eat no breakfast are more likely to be overweight, perform poorly in tasks requiring concentration, have shorter attention spans, and achieve lower test scores than their well-fed peers.[59] A nutritious breakfast is a central feature of a diet that meets the needs of children and supports their healthy growth and development.[60] Children who skip breakfast typically do not make up the deficits at later meals—they simply have lower intakes of energy, vitamins, and minerals than those who eat breakfast. Poorly nourished children are particularly vulnerable. Common sense dictates that it is unreasonable to expect anyone to learn and perform without fuel. For the child who hasn't had breakfast, the morning's lessons may be lost altogether. Even if a child has eaten breakfast, discomfort from hunger may become distracting by late morning. Teachers aware of the late-morning slump in their classrooms wisely request that midmorning snacks be provided; snacks improve classroom performance all the way to lunchtime.[61]

> **DIETARY GUIDELINES FOR AMERICANS**
Eat a nutrient-dense breakfast. Not eating breakfast has been associated with excess body weight, especially among children and adolescents.

Iron Deficiency and Behavior Iron deficiency has well-known and widespread effects on children's behavior and intellectual performance.[62] In addition to carrying oxygen in the blood, iron transports oxygen within cells, which use it during energy metabolism. Iron is also used to make neurotransmitters—most notably, those that regulate the ability to pay attention, which is crucial to learning. Consequently, iron deficiency not only causes an energy crisis, but also directly impairs attention span and learning ability.

Iron deficiency is often diagnosed by a quick, easy, inexpensive hemoglobin or hematocrit test that detects a deficit of iron in the *blood*. A child's *brain*, however, is sensitive to low iron concentrations long before the blood effects appear. Iron deficiency lowers the motivation to persist in intellectually challenging tasks and impairs overall intellectual performance. Anemic children perform poorly on tests and are disruptive in the classroom; iron supplementation improves learning and memory. When combined with other nutrient deficiencies, iron-deficiency anemia has synergistic effects that are especially detrimental to learning. Furthermore, children who had iron-deficiency anemia *as infants* continue to perform poorly as they grow older, even if their iron status improves.[63] The long-term damaging effects on mental development make prevention and treatment of iron deficiency during infancy and early childhood a high priority.

Other Nutrient Deficiencies and Behavior A child with any of several nutrient deficiencies may be irritable, aggressive, and disagreeable, or sad and withdrawn. Such a child may be labeled "hyperactive," "depressed," or "unlikable," when in fact these traits may be due to simple, even marginal, malnutrition. Parents and medical practitioners often overlook the possibility that malnutrition may account for abnormalities of appearance and behavior. Any departure from normal healthy appearance and behavior is a sign of possible poor nutrition (see Table 16-8). In any such case, inspection of the child's diet by a registered dietitian nutritionist or other qualified health-care professional is in order. Any suspicion of dietary inadequacies, no matter what other causes may be implicated, should prompt steps to correct those inadequacies immediately.

The Malnutrition-Lead Connection Children who are malnourished are vulnerable to lead poisoning. They absorb more lead if their stomachs are empty;

TABLE 16-8 Physical Signs of Malnutrition in Children

	Well-Nourished	Malnourished	Possible Nutrient Deficiencies
Hair	Shiny, firm in the scalp	Dull, brittle, dry, loose; falls out	Protein
Eyes	Bright, clear pink membranes; adjust easily to light	Pale membranes; spots; redness; adjust slowly to darkness	Vitamin A, the B vitamins, zinc, and iron
Teeth and gums	No pain or caries, gums firm, teeth bright	Missing, discolored, decayed teeth; gums bleed easily and are swollen and spongy	Minerals and vitamin C
Face	Clear complexion without dryness or scaliness	Off-color, scaly, flaky, cracked skin	Protein, vitamin A, and iron
Glands	No lumps	Swollen at front of neck, cheeks	Protein and iodine
Tongue	Red, bumpy, rough	Sore, smooth, purplish, swollen	B vitamins
Skin	Smooth, firm, good color	Dry, rough, spotty; "sandpaper" feel or sores; lack of fat under skin	Protein, essential fatty acids, vitamin A, B vitamins, and vitamin C
Nails	Firm, pink	Spoon-shaped, brittle, ridged	Iron
Internal systems	Regular heart rhythm, heart rate, and blood pressure; no impairment of digestive function, reflexes, or mental status	Abnormal heart rate, heart rhythm, or blood pressure; enlarged liver, spleen; abnormal digestion; burning, tingling of hands, feet; loss of balance, coordination; mental confusion, irritability, fatigue	Protein and minerals
Muscles and bones	Muscle tone; posture, long bone development appropriate for age	"Wasted" appearance of muscles; swollen bumps on skull or ends of bones; small bumps on ribs; bowed legs or knock-knees	Protein, minerals, and vitamin D

© Cengage Learning

if they have low intakes of calcium, zinc, vitamin C, or vitamin D; and, of greatest concern because it is so common, if they have an iron deficiency. Iron deficiency weakens the body's defenses against lead absorption, and lead poisoning can cause iron deficiency.[64] Common to both iron deficiency and lead poisoning are a low socioeconomic background and a lack of immunizations against infectious diseases. Another common factor is pica—a craving for nonfood items. Many children with lead poisoning eat dirt or chips of old paint, two common sources of lead.

The anemia brought on by lead poisoning may be mistaken for a simple iron deficiency and therefore may be incorrectly treated. Like iron deficiency, mild lead toxicity has nonspecific symptoms, including diarrhea, irritability, and fatigue. Adding iron to the diet does not reverse the symptoms; exposure to lead must stop and treatment for lead poisoning must begin. With further exposure, the symptoms become more pronounced, and children develop learning disabilities and behavioral problems. Still more severe lead toxicity can cause irreversible nerve damage, paralysis, mental retardation, and death.

Approximately half a million children between the ages of 1 and 5 in the United States have blood lead concentrations above 5 micrograms per deciliter, the level at which the Centers for Disease Control and Prevention recommend public health actions be initiated.[65] Lead toxicity in young children comes from their own behaviors and activities—putting their hands in their mouths, playing in dirt and dust, and chewing on nonfood items. Unfortunately, the body readily absorbs lead during times of rapid growth and hoards it possessively thereafter. Lead is not easily excreted and accumulates mainly in the bones, but also in the brain, teeth, and kidneys. Tragically, a child's neuromuscular system is also maturing during these first few years of life. No wonder children with elevated lead levels experience impairment of balance, motor development, and the relaying of nerve messages to and from the brain. Deficits in intellectual development are only partially reversed when blood lead levels decline.

Federal laws mandating reductions in leaded gasoline, lead-based solder, and other products over the past four decades have helped to reduce the amounts of lead in food and in the environment in the United States. As a consequence, the prevalence of lead toxicity in children has declined dramatically for most of the United States, but lead exposure is still a threat in certain communities. The "How To" feature on p. 526 presents strategies for defending children against lead toxicity.

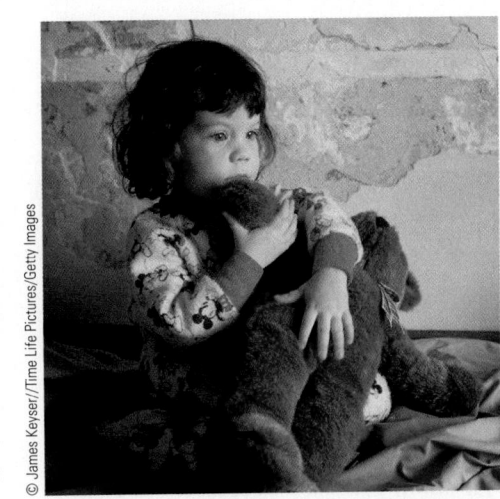

© James Keyser//Time Life Pictures/Getty Images

Old, lead-based paint threatens the health of an exploring child.

Protect against Lead Toxicity

Strategies to protect children from lead poisoning:

- Ask a pediatrician whether your child should be tested for lead poisoning.

- Prevent small children from putting dirty or old painted objects in their mouths, and make sure children wash their hands before eating. Similarly, keep small children from eating any nonfood items. Lead poisoning has been reported in young children who have eaten crayons or pool cue chalk.

- Wash floors, window sills, and other surfaces regularly. Use a mop or sponge with warm water and a general all-purpose cleaner.

- Wash children's bottles, pacifiers, and toys often.

- Feed children balanced, timely, meals with ample iron and calcium.

- Be aware that other countries do not have the same regulations protecting consumers against lead. Children have been poisoned by eating crayons made in China and drinking fruit juice canned in Mexico.

- Do not use lead-contaminated water to make infant formula.

- Have the water in your home tested by a competent laboratory.

- Use only cold water for drinking, cooking, and making formula (cold water absorbs less lead).

- When water has been standing in pipes for more than two hours, flush the coldwater pipes by running water through them for 30 seconds before using it for drinking, cooking, or mixing formulas.

By taking these steps, parents can protect their children from this preventable danger.

> **TRY IT** Visit the lead section of the website for the Environmental Protection Agency (www.epa.gov/lead) and identify the most common sources of lead poisoning.

Hyperactivity and "Hyper" Behavior All children are naturally active, and many of them become overly active on occasion—for example, in anticipation of a birthday party. Such behavior is markedly different from true **hyperactivity**.

Hyperactivity Hyperactive children have trouble sleeping, cannot sit still for more than a few minutes at a time, act impulsively, and have difficulty paying attention. These behaviors interfere with social development and academic progress. The cause of hyperactivity remains unknown, but it affects about 11 percent of young school-age children.[66] To resolve the problems surrounding hyperactivity, physicians often recommend specific behavioral strategies, special educational programs, and psychological counseling. If these interventions are ineffective, they may prescribe medication.[67]

Research on hyperactivity has focused on several nutritional factors as possible causes or treatments.[68] Parents often blame sugar. They mistakenly believe that simply eliminating candy and other sweet treats will solve the problem. This dietary change will not solve the problem, however, and studies have consistently found no convincing evidence that sugar causes hyperactivity or worsens behavior. Such speculation has been based on personal stories. No scientific evidence supports a relationship between sugar and hyperactivity or other misbehaviors.

Food additives have also been blamed for hyperactivity and other behavior problems in children, but scientific evidence to substantiate the connection has been elusive.[69] Limited research suggests that food additives such as artificial colors or sodium benzoate preservative (or both) may exacerbate hyperactive symptoms such as inattention and impulsivity in some children.[70] Additional studies are needed to confirm the findings and to determine which additives might be responsible for specific negative behaviors. A Food and Drug Administration (FDA) review determined that evidence linking color additives to hyperactivity is lacking.[71] The FDA did not rule out the possibility that some food additives, including food colorings, may aggravate hyperactivity and other behavioral problems in some susceptible children.

Misbehaving Even a child who is not truly hyperactive can be difficult to manage at times. Michael may act unruly out of a desire for attention, Jessica may be cranky because of a lack of sleep, Christopher may react violently after watching too much television, and Ashley may be unable to sit still in class because of a lack of exercise. All of these children may benefit from more consistent care—regular hours of sleep, regular mealtimes, and regular outdoor activity.

hyperactivity: inattentive and impulsive behavior that is more frequent and severe than is typical of others a similar age; professionally called *attention-deficit/hyperactivity disorder (ADHD)*.

Food Allergy and Intolerance **Food allergy** is frequently blamed for physical and behavioral abnormalities in children, but only 4 to 8 percent of children younger than 4 years of age are diagnosed with true food allergies.[72] Food allergies diminish with age, until in adulthood they affect less than 4 percent of the population. The prevalence of food allergy, especially peanut allergy, is on the rise, however.[73] Reasons for an increase in peanut allergy are not yet clear, but possible contributing factors include genetics, food preparation methods (roasting peanuts at very high temperatures makes them more allergenic), and exposure to medicinal skin creams containing peanut oil.

A true food allergy occurs when fractions of a food protein or other large molecule are absorbed into the blood and elicit an immunologic response. (Recall that proteins are normally dismantled in the digestive tract to amino acids that are absorbed without such a reaction.) The body's immune system reacts to these large food molecules as it does to other antigens—by producing antibodies, histamines, and other defensive agents.

Detecting Food Allergy Allergies may have one or two components. They always involve antibodies, but they may or may not involve symptoms.* Therefore, allergies cannot be diagnosed from symptoms alone. The National Institute of Allergy and Infectious Diseases (NIAID) has developed clinical guidelines for the diagnosis and management of food allergy.[74] Even symptoms exactly like those of an allergy may not be caused by an allergy. The NIAID recommends that food allergy should be considered when an individual experiences symptoms such as skin rash, respiratory difficulties, vomiting, diarrhea, or anaphylactic shock (described later) within minutes to hours of eating food, especially in young children.

Diagnosis of food allergy requires medical testing and food challenges. Once a food allergy has been diagnosed, the required treatment is strict elimination of the offending food. Children with allergies, like all children, need all their nutrients, so it is important to include other foods that offer the same nutrients as the omitted foods.[75] Nutritional counseling and growth monitoring are recommended for all children with food allergies.[76]

Allergic reactions to food may be immediate or delayed. In either case, the antigen interacts immediately with the immune system, but the timing of symptoms varies from minutes to 24 hours after consumption of the antigen. Identifying the food that causes an immediate allergic reaction is fairly easy because the symptoms appear shortly after the food is eaten. Identifying the food that causes a delayed reaction is more difficult because the symptoms may not appear until much later. By this time, many other foods may have been eaten, complicating the picture.

Anaphylactic Shock The life-threatening food allergy reaction of **anaphylactic shock** is most often caused by peanuts, tree nuts, milk, eggs, wheat, soy, fish, or shellfish. Among these foods, eggs, milk, soy, and peanuts most often cause problems in children.[77] Children are more likely to outgrow allergies to eggs, milk, and soy than allergies to peanuts. Peanuts cause more life-threatening reactions than do all other food allergies combined. Research is currently under way to help people with peanut allergies tolerate small doses, thus saving lives and minimizing reactions.[78] Families of children with a life-threatening food allergy and the school personnel who supervise those children must guard them against any exposure to the allergen. The child must learn to identify which foods pose a problem and then learn and use refusal skills for all foods that may contain the allergen.

Parents of children with allergies can pack safe lunches and snacks and ask school officials to strictly enforce a "no swapping" policy in the lunchroom. The child and caregivers must be able to recognize symptoms of impending anaphylactic shock:

- Tingling sensation in mouth
- Swelling of the tongue and throat
- Irritated, reddened eyes

These normally wholesome foods—peanuts, tree nuts, milk, eggs, wheat, soy, fish, and shellfish—may cause life-threatening symptoms in people with allergies.

food allergy: an adverse reaction to food that involves an immune response; also called *food-hypersensitivity reaction.*

anaphylactic (ana-fill-LAC-tic) **shock:** a life-threatening, whole-body allergic reaction to an offending substance.

---

*A person who produces antibodies *without* having any symptoms has an *asymptomatic allergy;* a person who produces antibodies *and* has symptoms has a *symptomatic allergy.*

- Difficulty breathing, asthma
- Hives, swelling, rashes
- Vomiting, abdominal cramps, diarrhea
- Drop in blood pressure
- Loss of consciousness

Any person with food allergies severe enough to cause anaphylactic shock should wear a medical alert bracelet or necklace. Finally, the responsible child and the school staff should be prepared with injections of epinephrine, which prevents anaphylactic shock after exposure to the allergen.* Many preventable deaths occur each year when people with food allergies accidentally ingest the allergen but have no epinephrine available.

Food Labeling Food labels must list the presence of common allergens in plain language, using the names of the eight most common allergy-causing foods. For example, a food containing "textured vegetable protein" must say "soy" on its label. Similarly, "casein" must be identified as "milk," and so forth. Food producers must also prevent cross-contamination during production and clearly label foods in which it is likely to occur. For example, equipment used for making peanut butter cookies must be scrupulously clean before being used to make oatmeal cookies; even then, the oatmeal cookie label warns consumers that this product "may contain peanuts" or "was made in a facility that uses peanuts."

Technology may soon offer new solutions. New drugs are being developed that may interfere with the immune response that causes allergic reactions. Also, through genetic engineering, scientists may one day create allergen-free peanuts, soybeans, and other foods to make them safer.

Food Intolerances Not all **adverse reactions** to foods are food allergies, although even physicians may describe them as such. Signs of adverse reactions to foods include stomachaches, headaches, rapid pulse rate, nausea, wheezing, hives, bronchial irritation, coughs, and other such discomforts. Among the causes may be reactions to chemicals in foods, such as the flavor enhancer monosodium glutamate (MSG), the natural laxative in prunes, or the mineral sulfur; digestive diseases, obstructions, or injuries; enzyme deficiencies, such as lactose intolerance; and even psychological aversions. These reactions involve symptoms but no antibody production. Therefore, they are **food intolerances,** not allergies.

Pesticides on produce may also cause adverse reactions. Pesticides that were applied in the fields may linger on foods. Health risks from pesticide exposure may be low for healthy adults, but children are vulnerable. Therefore, government agencies have set a **tolerance level** for each pesticide by first identifying foods that children commonly eat in large amounts and then considering the effects of pesticide exposure during each stage of development. Chapter 19 revisits the issues surrounding the use of pesticides on food crops.

Hunger, lead poisoning, hyperactivity, and allergic reactions can all adversely affect a child's nutrition status and health. Fortunately, each of these problems has solutions. They may not be easy solutions, but programs are in place and improvements are evident. As for the most pervasive health problem for children in the United States today—obesity—health experts and researchers understand much about the causes of, the consequences of, and even the solutions, but putting this knowledge into action that meets with success is an ongoing challenge.

Childhood Obesity The number of overweight children has increased dramatically over the past four decades (see Figure 16-9).[79] Like their parents, children in the United States are becoming fatter. An estimated 32 percent of US children and adolescents 2 to 19 years of age are overweight and 17 percent are obese.[80] Based on

adverse reactions: unusual responses to food (including intolerances and allergies).

food intolerances: adverse reactions to foods that do not involve the immune system.

tolerance level: the maximum amount of residue permitted in a food when a pesticide is used according to the label directions.

*Epinephrine is a hormone of the adrenal gland that modulates the stress response; formerly called *adrenaline*. When administered by injection, epinephrine counteracts anaphylactic shock by opening the airways and maintaining heartbeat and blood pressure.

data from the BMI-for-age growth charts, children and adolescents are categorized as *overweight* above the 85th percentile and as *obese* at the 95th percentile and above.[81] There are exceptions to the use of the 85th and 95th percentile cutoff points. For older adolescents, a BMI at the 95th percentile is higher than a BMI of 30, the adult obesity cutoff point. Therefore, obesity is defined as a BMI at the 95th percentile or a BMI of 30 or greater, whichever is lower. For children younger than 2 years of age, BMI values are not available. For this age group, weight-for-height values above the 95th percentile are classified as overweight. Figure 16-10 presents the BMI for children and adolescents, indicating cutoff points for obesity and overweight.

> FIGURE 16-9   **Trends in Childhood Obesity**

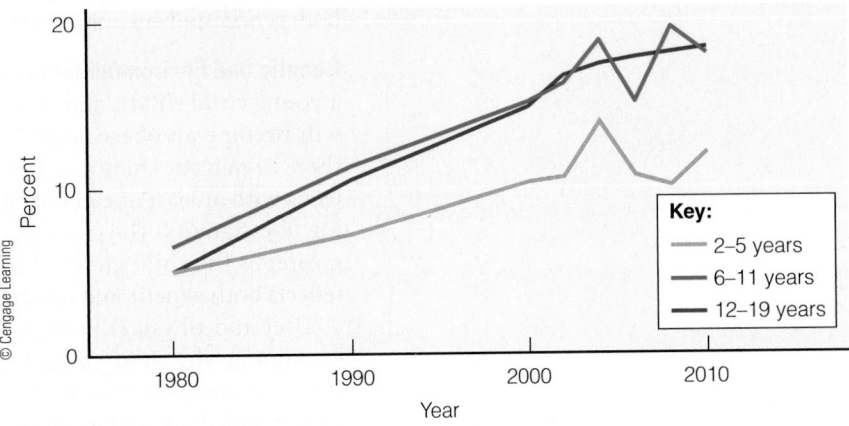

The Expert Committee of the American Medical Association recommends a third cutoff point (99th percentile) to define severe obesity in childhood. Unfortunately, severe obesity in children is becoming more prevalent. Many of these children have multiple risk factors for cardiovascular disease and a high risk of severe obesity in adulthood.[82] The special risks and treatment needs of severely obese children need to be recognized.

The problem of obesity in children is especially troubling because overweight children have the potential of becoming obese adults with all the social, economic, and medical ramifications that often accompany obesity. They have additional problems,

> FIGURE 16-10   **Body Mass Index-for-Age Percentiles: Boys and Girls, Ages 2 to 20**

Body mass index-for-age percentiles: Boys, 2 to 20 years

Body mass index-for-age percentiles: Girls, 2 to 20 years

Key:

Obese ≥95th percentile	Normal 10th to 85th percentile
Overweight >85th percentile	Underweight <10th percentile

too, arising from differences in their growth, physical health, and psychological development. In trying to explain the rise in childhood obesity, researchers point to both genetic and environmental factors.

Genetic and Environmental Factors Parental obesity predicts an early increase in a young child's BMI, and it more than doubles the chances that a young child will become an obese adult.[83] Children with neither parent obese have a less than 10 percent chance of becoming obese in adulthood, whereas overweight teens with at least one obese parent have a greater than 80 percent chance of being obese adults. The chances of an obese child becoming an obese adult grow greater as the child grows older.[84] The link between parental and child obesity reflects both genetic and environmental factors (as described in Chapter 9).

Diet and physical inactivity are the two strongest environmental factors explaining why children are heavier today than they were 40 or so years ago. In that time, the prevalence of childhood obesity throughout the United States more than doubled for young children and more than tripled for children 6 to 11 years of age and adolescents. As a society, our eating and activity patterns changed considerably. In many families, both parents work outside the home and work longer hours; more emphasis is placed on convenience foods and foods eaten away from home; meal choices at school are more diverse and often less nutritious; sedentary activities such as watching television and playing video or computer games occupy much of children's free time; and opportunities for physical activity and outdoor play both during and after school have declined.[85] All of these factors—and many others—influence children's eating and activity patterns.

Children learn food behaviors from their families, and research confirms the significant roles parents play in teaching their children about healthy food choices, providing nutrient-dense foods, and serving as role models.[86] When parents eat fruits and vegetables frequently, their children do too.[87] The more fruits and vegetables children eat, the more vitamins, minerals, and fiber, and the less saturated fat, in their diets.[88]

In children 2 to 18 years of age, about 40 percent of total energy intake comes from solid fats and added sugars—in other words—empty kcalories.[89] About half of these empty kcalories are contributed by six specific foods: soda, fruit drinks, dairy desserts (ice cream, frozen yogurt, sorbet, sherbet, pudding, and custard), grain desserts (cakes, cookies, pies, cobblers, donuts, and granola bars), pizza, and whole milk. Not surprisingly, when researchers ask "Are today's children eating more kcalories than those of 40 years ago?" the answer is, "Yes."

As Highlight 4 discusses, as the prevalence of obesity among both children and adults has surged over the past four decades, so has the consumption of added sugars and, especially, high-fructose corn syrup—the easily consumed, energy-dense liquid sugar added to soft drinks.[90] Each 12-ounce can of soft drink provides the equivalent of about 10 teaspoons of sugar and 150 kcalories. More than half of school-age children consume at least one soft drink each day at school; adolescent males consume the most—two or more cans daily. Research shows that soft drink consumption is associated with increased energy intake and body weight.[91]

No doubt, the tremendous increase in soft drink consumption plays a role, but much of the obesity epidemic can be explained by lack of physical activity. Children have become more sedentary, and sedentary children are more often overweight.[92] Television watching may contribute most to physical inactivity. Children 8 to 18 years of age spend an average of 4.5 hours per day watching television.[93] Longer television time is linked with overweight in children.[94] Television fosters overweight and obesity because it:

- Requires no energy beyond basal metabolism
- Replaces vigorous activities
- Encourages snacking
- Promotes a sedentary lifestyle

A child who spends more than an hour or two each day in front of a television, computer monitor, or other media can become overweight even while eating fewer kcalories than a more active child. Too much screen time and not enough activity time also contributes to a child's psychological distress.

Children who have television sets in their bedrooms spend more time watching TV, less time being physically active, and are more likely to be overweight than children who do not have televisions in their rooms.[95] Watching television influences food intake as well as physical activity.[96] Children who watch a great deal of television are most likely to be overweight and least likely to eat family meals or fruits and vegetables.[97] They often snack on the nutrient-poor, energy-dense foods that are advertised.[98]

The average child sees approximately 12 television food ads each day—most peddling foods high in sugar, saturated fat, and salt such as sugar-coated breakfast cereals, candy bars, chips, fast foods, and carbonated beverages.[99] More than half of all food advertisements are aimed specifically at children and market their products as fun and exciting, oftentimes using sports heroes. Not surprisingly, the more time children spend watching television, the more they request these advertised foods and beverages—and they get their requests about half of the time. The most popular foods and beverages are marketed to children and adolescents on the Internet as well, using "advergaming" (advertised product as part of a game), cartoon characters or "spokes-characters," and designated children's areas.[100] Food marketing to children, including TV ads and Internet ads, as well as marketing to children in their local communities by way of store giveaways, restaurant promotions, school activities, and sporting events has a profound effect on children's nutrition and health.[101] Despite initiatives by the food industry to answer public health concerns about child-targeted advertising, much remains to be done to reduce the marketing of unhealthy foods to children.

The physically inactive time spent watching television is second only to time spent sleeping. Children also spend more time playing computer and video games. In most cases, these activities use no more energy than resting, displace participation in more vigorous activities, and foster snacking on high-fat foods.[102] Compared to sedentary screen-time activities, playing active video games does expend a little more energy, but not enough to count toward the daily 60 minutes of moderate-to-vigorous physical activity recommended for children.[103] Simply reducing the amount of time spent watching television (and playing sedentary video games) can improve a child's BMI. The American Academy of Pediatrics recommends no television viewing before 2 years of age and thereafter limiting television and video time to 2 hours per day as a strategy to help prevent childhood obesity.[104]

Growth Overweight children develop a characteristic set of physical traits. They typically begin puberty earlier and so grow taller than their peers at first, but then they stop growing at a shorter height. They develop greater bone and muscle mass in response to the demand of having to carry more weight—both fat and lean weight. Consequently, they appear "stocky" even when they lose their excess fat.

Physical Health Like overweight adults, overweight and obese children display a blood lipid profile indicating that atherosclerosis is beginning to develop—high levels of total cholesterol, triglycerides, and LDL cholesterol. Overweight and obese children also tend to have high blood pressure; in fact, obesity is a leading cause of pediatric hypertension.[105] Their risks for developing type 2 diabetes and respiratory diseases (such as asthma) are also exceptionally high.[106] These relationships between childhood obesity and chronic diseases are discussed fully in Highlight 16.

Psychological Development In addition to the physical consequences, childhood obesity brings a host of emotional and social problems.[107] Because people frequently judge others on appearance more than on character, overweight and obese children are often victims of prejudice and bullying. Many suffer discrimination by adults and rejection by their peers. They may have poor self-images, a sense of failure, and a passive approach to life. Television shows, which are a major influence in children's lives, often portray the fat person as the bumbling misfit. Overweight children may come to accept this negative stereotype in themselves and in others,

Excessive television watching promotes physical inactivity and poor snacking habits.

TABLE 16-9 Recommended Eating and Physical Activity Behaviors to Prevent Obesity

The Expert Committee of the American Medical Association recommends the following healthy habits for children 2 to 18 years of age to help prevent childhood obesity:

- Limit consumption of sugar-sweetened beverages, such as soft drinks and fruit-flavored punches.
- Eat the recommended amounts of fruits and vegetables every day (2 to 4.5 cups per day based on age).
- Learn to eat age-appropriate portions of foods.
- Eat foods low in energy density such as those high in fiber and/or water and modest in fat.
- Eat a nutritious breakfast every day.
- Eat a diet rich in calcium.
- Eat a diet balanced in recommended proportions for carbohydrate, fat, and protein.
- Eat a diet high in fiber.
- Eat together as a family as often as possible.
- Limit the frequency of restaurant meals.
- Limit television watching or other screen time to no more than 2 hours per day and do not have televisions or computers in bedrooms.
- Engage in at least 60 minutes of moderate to vigorous physical activity every day.

SOURCE: S. E. Barlow, Expert Committee recommendations regarding the prevention, assessment, and treatment of child and adolescent overweight and obesity: Summary report, *Pediatrics* 120 (2007): S164–S192.

© Cengage Learning

which can lead to additional emotional and social problems. Researchers investigating children's reactions to various body types find that both normal-weight and underweight children respond unfavorably to overweight bodies.

Prevention and Treatment of Obesity Medical science has worked wonders in preventing or curing many of even the most serious childhood diseases, but obesity remains a challenge. Once excess fat has been stored, it is difficult to lose. In light of all this, parents are encouraged to make major efforts to prevent childhood obesity, starting at birth, or to begin treatment early—before adolescence. The Expert Committee of the American Medical Association recommends specific eating and physical activity behaviors to prevent obesity, for all children (see Table 16-9).

The main goal of obesity treatment is to improve long-term physical health through permanent changes in lifestyle habits. The most successful approach integrates diet, physical activity, psychological support, and behavioral changes.[108] As a first step, the Expert Committee recommends that overweight and obese children and their families adopt the same healthy eating and activity behaviors presented in Table 16-9 for obesity prevention. The goal for overweight and obese children is to improve BMI. If the child's BMI does not improve after several months, the Expert Committee recommends increasing the intensity of the treatment. The level of intensity depends on treatment response, age, degree of obesity, health risks, and the family's readiness to change.[109] Advanced treatment involves close follow-up monitoring by a health-care provider and greater support and structure for the child.

Diet The initial goal for overweight children is to reduce the rate of weight gain; that is, to maintain weight as the child grows taller. Continued growth will then accomplish the desired change in BMI. Weight loss is usually not recommended because diet restriction can interfere with growth and development. Intervention for some overweight children with accompanying medical conditions may warrant weight loss, but this treatment requires an individualized approach based on the degree of overweight and severity of the medical conditions.[110] Dietary strategies begin with those listed in Table 16-9 and progress to more structured family meal plans when necessary. For example, the child or the parent may be instructed to keep detailed records of dietary intake and physical activity.

> DIETARY GUIDELINES FOR AMERICANS

Children are encouraged to maintain kcalorie balance to support normal growth and development without promoting excess weight gain. Children who are overweight or obese should change their eating and physical activity behaviors so that their BMI-for-age percentile does not increase over time.

Physical Activity The many benefits of physical activity are well known but often are not enough to motivate overweight people, especially children. Yet regular vigorous activity can improve a child's weight, body composition, and physical fitness.[111] Ideally, parents will limit sedentary activities and encourage at least 1 hour of daily physical activity to promote strong skeletal, muscular, and cardiovascular development and instill in their children the desire to be physically active throughout life. Opportunities to be physically active can include team, individual, and recreational activities (see Table 16-10). Most importantly, parents need to set a good example. Physical activity is a natural and lifelong

TABLE 16-10 Examples of Aerobic, Muscle-Strengthening, and Bone-Strengthening Physical Activities for Children and Adolescents

Moderate-to-Vigorous Aerobic Activities	Muscle-Strengthening Activities	Bone-Strengthening Activities

  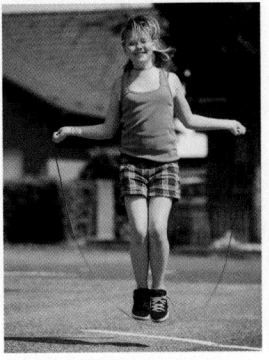

Moderate Active recreation such as hiking, skateboarding, rollerblading Bicycle riding[a] Brisk walking **Vigorous** Active games involving running and chasing, such as tag Bicycle riding[a] Cross-country skiing Jumping rope Martial arts Running Sports such as soccer, ice or field hockey, basketball, swimming, tennis	Games such as tug-of-war Modified push-ups (with knees on the floor) Resistance exercises using body weight, free weights, or resistance bands Rope or tree climbing Sit-ups (curl-ups or crunches) Swinging on playground equipment/bars	Games such as hopscotch Hopping, skipping, jumping Jumping rope Running Sports such as gymnastics, basketball, volleyball, tennis

© 2016 Cengage Learning

[a]Some activities, such as bicycling, can be moderate or vigorous, depending on level of effort.

behavior of healthy living. It can be as simple as riding a bike, playing tag, jumping rope, or doing chores. The American Academy of Pediatrics supports the efforts of schools to include more physical activity in the curriculum and encourages parents to support their children's participation.

> **PHYSICAL ACTIVITY GUIDELINES FOR AMERICANS**

- Children and adolescents should do 60 minutes (1 hour) or more of physical activity daily.
- *Aerobic.* Most of the 60 or more minutes a day should be either moderate- or vigorous-intensity aerobic physical activity and should include vigorous-intensity physical activity at least 3 days a week.
- *Muscle-strengthening.* As part of their 60 or more minutes of daily physical activity, children and adolescents should include muscle-strengthening physical activity on at least 3 days of the week.
- *Bone-strengthening.* As part of their 60 or more minutes of daily physical activity, children and adolescents should include bone-strengthening physical activity on at least 3 days of the week.

Psychological Support Weight-loss programs that involve parents and other caregivers in treatment report greater success than those without parental involvement. Because obesity in parents and their children tends to be positively correlated, both benefit when parents participate in a weight-loss program. Parental attitudes about food greatly influence children's eating behavior, so it is important that the influence be positive. Otherwise, eating problems may become exacerbated.

Behavioral Changes In contrast to traditional weight-loss programs that focus on *what* to eat, behavioral programs focus on *how* to eat. These techniques involve learning new habits that lead a child to make healthy choices.

Physical activity is fun—play games in the park, build a sandcastle or a snowman, row a boat, toss a Frisbee, run with the dog, or plant a garden.

Drugs The use of weight-loss drugs to treat obesity in children merits special concern because the long-term effects of these drugs on growth and development have not been studied. The drugs may be used in addition to structured lifestyle changes for carefully selected children or adolescents who are at high risk for severe obesity in adulthood. Orlistat (see Chapter 9) is the only prescription weight-loss medication that has been approved for use in adolescents 12 years of age and older.[112] Alli, the over-the-counter version of orlistat, should not be given to anyone younger than age 18.

Surgery The use of surgery to treat severe obesity in adults (see Chapter 9) has created interest in its use for adolescents. Limited research shows that after surgery extremely obese adolescents lose significant weight and experience improvements in type 2 diabetes and cardiovascular risk factors.[113] The selection criteria for surgery to treat obesity in adolescents are based on recommendations of a panel of pediatricians and surgeons and include the following:

- Physically mature
- BMI ≥40 or BMI >35 with significant weight-related health problems
- Failure in a formal, 6-month weight-loss program
- Capable of adhering to the long-term lifestyle changes required after surgery

Obesity is prevalent in our society. Because treatment of obesity is frequently unsuccessful, it is most important to prevent its onset. Above all, be sensible in teaching children how to maintain appropriate body weight. Children can easily get the impression that their worth is tied to their body weight. Parents and the media are most influential in shaping self-concept, weight concerns, and dieting practices.[114] Some parents fail to realize that society's ideal of slimness can be perilously close to starvation and that a child encouraged to "diet" cannot obtain the energy and nutrients required for normal growth and development. Weight loss in obese children can be managed without compromising growth, but it should be overseen by a health-care professional.

Mealtimes at Home Traditionally, parents served as **gatekeepers,** determining what foods and activities were available in their children's lives. Then the children made their own selections. Gatekeepers who wanted to promote nutritious choices and healthful habits provided access to nutrient-dense foods and opportunities for active play at home.

In today's consumer-oriented society, children have greater influence over family decisions concerning food—the fast-food restaurant the family chooses when eating out, the snacks the family eats at home, and the specific brands the family purchases at the grocery store. Parental guidance in food choices is still necessary, but teaching children consumer skills to help them make informed choices is equally important.

Honoring Children's Preferences Researchers attempting to explain children's food preferences encounter contradictions. Children say they like colorful foods, yet they most often reject green and yellow vegetables in favor of brown peanut butter and white potatoes, apple wedges, and bread. They seem to like raw vegetables better than cooked ones, so it is wise to offer vegetables that are raw or slightly undercooked, served separately, and easy to eat. Foods should be warm, not hot, because a child's mouth is much more sensitive than an adult's. The flavor should be mild because a child has more taste buds, and smooth foods such as mashed potatoes or split-pea soup should contain no lumps (a child wonders, with some disgust, what the lumps might be).

Make mealtimes fun for children. Young children like to eat at little tables and to be served small portions of food. They like sandwiches cut in different geometric shapes and common foods called silly names. They also like to eat with other children, and they tend to eat more when in the company of their friends. Children are also more likely to give up their prejudices against foods when they see their peers eating them.[115]

Eating is more fun for children when friends are there.

©istockphoto.com/monkeybusinessimages

gatekeepers: with respect to nutrition, key people who control other people's access to foods and thereby exert profound impacts on their nutrition. Examples are the spouse who buys and cooks the food, the parent who feeds the children, and the caregiver in a day-care center.

Learning through Participation Allowing children to help plan and prepare the family's meals provides enjoyable learning experiences and encourages children to eat the foods they have prepared. Vegetables are attractive, especially when fresh, and provide opportunities for children to learn about color, seeds, growing vegetables, and shapes and textures—all of which are fascinating to young children. Measuring, stirring, washing, and arranging foods are skills that even a young child can practice with enjoyment and pride (see Table 16-11).

Avoiding Power Struggles Problems over food often arise during the second or third year, when children begin asserting their independence. For example, very young children may suddenly refuse foods they once liked, but as they get a little older, these same children will begin to request favorite foods and make simple choices when given the opportunity. Many food choice problems stem from the conflict between children's developmental stages and capabilities and parents who, in attempting to do what they think is best for their children, try to control every aspect of eating. Such conflicts can disrupt children's abilities to regulate their own food intakes and to determine their own likes and dislikes. For example, many people share the misconception that children must be persuaded or coerced to try new foods. In fact, the opposite is true. When children are forced to try new foods, even by way of rewards, they are less likely to try those foods again than are children who are left to decide for themselves. Similarly, when children are restricted from eating their favorite foods, they are more likely to want those foods. Wise parents provide healthful foods and allow their child to determine *how much* and even *whether* to eat.

When introducing new foods, offer them one at a time and only in small amounts such as a small bite at first. The more often a food is presented to a young child, the more likely the child will accept that food.[116] Offer the new food at the beginning of the meal, when the child is hungry, and allow the child to make the decision to accept or reject it. Never make an issue of food acceptance. Table 16-12 (p. 536) offers tips for feeding picky eaters.

Choking Prevention Parents must always be alert to the dangers of choking. A choking child is silent, so an adult should be present whenever a child is eating. Make sure the child sits when eating; choking is more likely when a child is running or falling. See Table 16-5 on p. 519 for foods and nonfood items most likely to cause choking.

Playing First Children may be more relaxed and attentive during meals if outdoor play or other fun activities are scheduled before, rather than immediately after, mealtimes. Otherwise children "hurry up and eat" so that they can go play.

Snacking Parents may find that when their children snack, they aren't hungry at mealtimes. Instead of teaching children *not* to snack, parents are wise to teach them *how* to snack. Provide snacks that are as nutritious as the foods served at mealtime. Snacks can even be mealtime foods served individually over time, instead of all at once on one plate. When providing snacks to children, think of the five food groups and offer such snacks as pieces of cheese, tangerine slices, and egg salad on whole-wheat crackers (see Table 16-13, p. 537). Replacing nutrient-poor, high-kcalorie snacks such as potato chips with nutrient-rich, low-kcalorie snacks such as vegetables not only reduces children's energy intakes, but improves their nutrient intakes as well.[117] Snacks that are easy to prepare should be readily available to children, especially if they arrive home from school before their parents.

To ensure that children have healthy appetites and plenty of room for nutritious foods when they are hungry, parents and teachers must limit access to candy, soft drinks, and other concentrated sweets. Limiting access includes limiting the amount of pocket money children have to buy such foods themselves. If these foods are permitted in large quantities, the only possible outcomes are nutrient deficiencies, obesity, or both. The preference for sweets is innate; most children do not naturally select nutritious foods on the basis of taste. When children are allowed to create meals freely from a variety of foods, they typically select

TABLE 16-11 Food Skills of Preschool Children

Age 2 years, when large muscles develop:

- Uses a spoon
- Helps feed self
- Lifts and drinks from a cup
- Helps scrub fruits and vegetables, tear lettuce or greens, snap green beans, or dip foods
- Wipes table
- Places items in recycle bin or trash

Age 3 years, when medium hand muscles develop:

- Spears food with a fork
- Feeds self independently
- Adds ingredients to pancake batters, cookie recipes, salads or other mixed dishes
- Helps wrap, pour, mix, shake, stir, or spread foods
- Helps crack nuts with supervision
- Follows simple instructions

Age 4 years, when small finger muscles develop:

- Uses all utensils and napkin
- Helps roll, juice, or mash foods
- Helps measure dry ingredients
- Cracks egg shells
- Helps make sandwiches and toss salads
- Peels foods such as hard-boiled eggs and bananas
- Learns table manners

Age 5 years, when fine coordination of fingers and hands develops:

- Measures liquids
- Helps grind, grate, and cut (soft foods with dull knife)
- Uses hand mixer with supervision

© 2016 Cengage Learning

NOTE: These ages are approximate. Healthy children develop at their own pace.

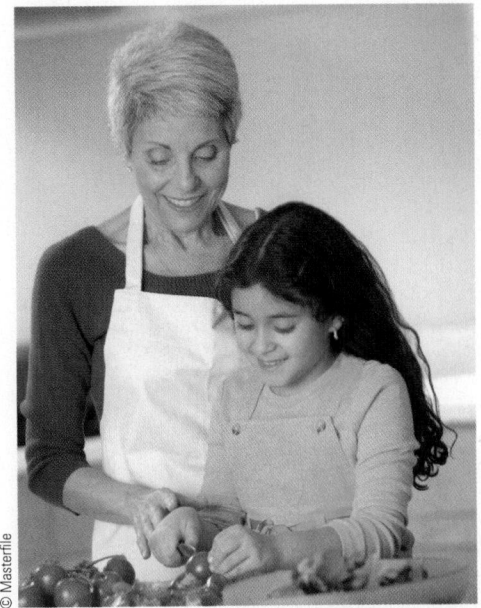

© Masterfile

Children enjoy eating the foods they help to prepare.

TABLE 16-12 Tips for Feeding Picky Eaters

Get Them Involved

- Encourage children to help with meal planning.
- Take children grocery shopping.
- Ask children to help with cooking.
- Help children garden and harvest foods they will eat.

Be Creative

- Try serving vegetables as finger foods with dips or spreads.
- Use cookie cutters to cut breads, fruits, and vegetables into fun shapes.
- Put healthy snacks in ice cube trays or muffin pans where children can easily reach them and graze as they play.
- Serve traditional meals out of order (for example, breakfast for dinner).
- Use healthy foods such as vegetables and whole grains in craft projects to help kids become familiar with them and encourage their interest in and enthusiasm for these foods.

Enhance Favorite Recipes

- Include sliced or shredded vegetables in sauces, casseroles, pancakes, and muffins.
- Serve fruit over cereal, yogurt, or ice cream.
- Bake brownies with black beans or cookies with lentils as an ingredient.

Model and Share

- Be a role model to children by eating healthy foods alongside them and offer to share your healthy snack with them.
- Make healthy options readily available and don't give up on repeatedly offering foods in which your child might not seem interested.
- Encourage your child to taste at least one bite of each food served at a meal.

Respect and Relax

- Remember that it is not uncommon for children to eat sporadically. They have smaller stomachs, and therefore are likely to feel full faster and become hungry again not long after a snack or meal.
- Focus on your child's overall weekly intake of foods and nutrients rather than daily consumptions.
- Discuss concerns with your child's doctor. It might be helpful to maintain a 3-day food record to review at the next appointment.

SOURCE: Adapted from Mayo Clinic Staff, Children's nutrition: 10 tips for picky eaters, 2011, www.mayoclinic.com/health /childrens-health/HQ01107.

foods that provide a lot of sugar. When their parents are watching, or even when they only think their parents are watching, children improve their selections.

Sweets need not be banned altogether. Children who are exceptionally active can enjoy high-kcalorie foods such as ice cream or pudding from the milk group or pancakes from the bread group. Sedentary children need to become more active so they can also enjoy some of these foods without unhealthy weight gain.

Preventing Dental Caries Children frequently snack on sticky, sugary foods that stay on the teeth and provide an ideal environment for the growth of bacteria that cause dental caries. Teach children to brush and floss after meals, to brush or rinse after eating snacks, to avoid sticky foods, and to select crisp or fibrous foods frequently.

Serving as Role Models In an effort to practice these many tips, parents may overlook perhaps the single most important influence on their children's food habits—themselves. Parents who don't eat carrots shouldn't be surprised when their children refuse to eat carrots. Likewise, parents who comment negatively on the smell of brussels sprouts may not be able to persuade children to try them. Children learn much through imitation. It is not surprising that children prefer the foods other family members enjoy and dislike foods that are never offered to them. Parents, older siblings, and other caregivers set an irresistible example by sitting with younger children, eating the same foods, and having pleasant conversations during mealtimes.

While serving and enjoying food, caregivers can promote both physical and emotional growth at every stage of a child's life. They can help their children develop both a positive self-concept and a positive attitude toward food. With good beginnings, children will grow without the conflicts and confusions about food that can lead to nutrition and health problems.

Nutrition at School While parents are doing what they can to establish good eating habits in their children at home, others are preparing and serving foods to their children at day-care centers and schools. In addition, children begin to learn about food and nutrition in the classroom. Meeting the nutrition and education needs of children is critical to supporting their healthy growth and development.[118]

The Academy of Nutrition and Dietetics has set nutrition standards for child-care programs. Among them, meal plans should:

- Be nutritionally adequate and consistent with the *Dietary Guidelines for Americans.*

- Emphasize fresh fruit, fresh and frozen vegetables, whole grains, and low-fat milk and milk products.

- Limit foods and beverages high in energy, added sugars, solid fats, and sodium, and low in vitamins and minerals.

- Provide foods and beverages in quantities and meal patterns appropriate to ensure optimal growth and development.

- Involve parents in planning.
- Provide furniture and eating utensils that are age appropriate and developmentally suitable to encourage children to accept and enjoy mealtime.

In addition, child-care providers can encourage active play for children by creating opportunities for children to engage in both structured and unstructured activity throughout the day.

Meals at School The US government assists schools financially so that every student can receive nutritious meals at school. Both the School Breakfast Program and the National School Lunch Program provide meals free or at reduced cost to children from low-income families. In addition, schools can obtain food commodities. Nationally, the US Department of Agriculture (USDA) administers the programs; on the state level, state departments of education operate them. The educational rewards of school meal programs are great. Several studies have reported that children who participate in school food programs perform better in the classroom.[119] Unfortunately, many school districts are finding it more and more difficult to operate the programs because the cost to produce lunches and breakfasts exceeds the reimbursement rate. Furthermore, the number of children qualifying for free and reduced-price school meals continues to rise.[120]

More than 31 million children receive lunches through the National School Lunch Program—more than half of them free or at a reduced price.[121] School lunches offer a variety of food choices and help children meet at least one third of their recommended intakes for energy, protein, vitamin A, vitamin C, iron, and calcium. Table 16-14 (p. 538) shows school lunch patterns for children of different ages and specifies the numbers of servings of milk, protein-rich foods (meat, poultry, fish, cheese, eggs, legumes, or peanut butter), vegetables, fruits, and whole grains. In an effort to help reduce disease risk, all government-funded meals served at schools must follow the *Dietary Guidelines for Americans*.

The USDA Food and Nutrition Service recently updated the meal patterns and nutrition standards for school meals.[122] Changes to meals include greater availability of fruits, vegetables, whole grains, and fat-free and low-fat milk; decreased levels of sodium, saturated fat, and *trans* fat; and guidelines for meeting nutrient needs within specified kcalorie ranges based on age/grade groups for school children.

Parents often rely on school lunches to meet a significant part of their children's nutrient needs on school days. Indeed, students who regularly eat school lunches have higher intakes of many nutrients and fiber than students who do not. Children don't always like what they are served, however, and school lunch programs must strike a balance between what children want to eat and what will nourish them and guard their health.

The School Breakfast Program is available in more than 80 percent of the nation's schools that offer school lunch, and close to 12 million children participate in

TABLE 16-13 Healthful Snack Ideas—Think Food Groups, Alone and in Combination

Selecting two or more foods from different food groups adds variety and nutrient balance to snacks. The combinations are endless, so be creative. Whenever possible, choose whole grains, low-fat or reduced-fat milk products, and lean meats.

Grains

Grain products are filling snacks, especially when combined with other foods:

- Cereal with fruit and milk
- Crackers and cheese
- Whole-grain toast with peanut butter
- Popcorn with grated cheese
- Oatmeal raisin cookies with milk

Vegetables

Cut-up, fresh, raw vegetables make great snacks alone or in combination with foods from other food groups:

- Celery with peanut butter
- Broccoli, cauliflower, and carrot sticks with a flavored cottage cheese dip

Fruits

Fruits are delicious snacks and can be eaten alone—fresh, dried, or juiced—or combined with other foods:

- Apples and cheese
- Bananas and peanut butter
- Peaches with yogurt
- Raisins mixed with sunflower seeds or nuts

Protein Foods

Seafood, meat, poultry, eggs, legumes, nuts, seeds, and soy products add protein to snacks:

- Refried beans with nachos and cheese
- Tuna on crackers
- Luncheon meat on whole-grain bread

Milk and Milk Products

Milk can be used as a beverage with any snack, and many other milk products, such as yogurt and cheese, can be eaten alone or with other foods as listed above.

© Cengage Learning

© Monkey Business Images/Dreamstime.com

School lunches provide children with nourishment at little or no charge.

TABLE 16-14 School Lunch Patterns

Food Group	Grades K–5	Grades 6–8	Grades 9–12
	Amount per week (minimum per day)		
Fruits[a] (cups)	2½ (½)	2½ (½)	5 (1)
Vegetables[a] (cups)	3¾ (¾)	3¾ (¾)	5 (1)
Dark green	≥½	≥½	≥½
Red and orange	≥¾	≥¾	≥1¼
Legumes	≥½	≥½	≥½
Starchy	≥½	≥½	≥½
Other	≥½	≥½	≥¾
Any additional vegetables to meet total requirement	1	1	1½
Grains (oz equivalents)	8–9 (1)	8–10 (1)	10–12 (2)
Protein foods (oz equivalents)	8–10 (1)	9–10 (1)	10–12 (2)
Fluid milk[b] (cups)	5 (1)	5 (1)	5 (1)
Other			
kCalories	550–650	600–700	750–850
Saturated fat (% of total kcalories)	<10	<10	<10
Sodium (mg)	≤640	≤710	≤740
Trans fat (g per serving)	0	0	0

[a]No more than half of the fruit or vegetable servings may be in the form of juice. All juice must be 100% full strength.

[b]Fluid milk must be low-fat (unflavored) or fat-free (flavored or unflavored).

SOURCE: US Department of Agriculture, *Nutrition Standards in the National School Lunch and School Breakfast Programs*, January 25, 2012.

© Cengage Learning

it.[123] At a minimum, the school breakfast must contain:

- One serving of fluid milk (either unflavored low-fat or flavored or unflavored fat-free)

- One serving of fruit or vegetable (no more than half of the servings may be 100% full-strength juice)

- One to two servings of whole grains; or one serving of whole grains and one serving of meat or meat alternatives

Unfortunately, for many children who need it, the School Breakfast Program is either unavailable or the children do not participate in it.[124] The majority of children who eat school breakfasts are from low-income families. As research results continue to emphasize the positive impact breakfast has on school performance and health, vigorous campaigns to expand and improve school breakfast programs are under way.[125]

Another federal program, the Child and Adult Care Food Program (CACFP), operates similarly and provides funds to organized child-care programs. All eligible children, centers, and family day-care homes may participate. Sponsors are reimbursed for most meal costs and may also receive USDA commodity foods.

Competing Influences at School Serving healthful lunches is only half the battle; students need to eat them too. Short lunch periods and long waiting lines prevent some students from eating a school lunch and leave others with too little time to finish their meals.[126] Nutrition efforts at schools are also undermined when students can buy what the USDA labels "competitive" or "nonreimbursable" foods—meals from fast-food restaurants or a la carte foods such as pizza or snack foods and carbonated beverages from snack bars, school stores, and vending machines.[127] When students have access to competitive foods, participation in the school lunch program decreases, nutrient intake from lunch declines, and more food is discarded.[128]

Increasingly, school-based nutrition issues are being addressed by legislation. Some states restrict the sale of competitive foods and have higher rates of participation in school meal programs than the national average. Federal legislation mandates that all school districts that participate in the USDA's National School Lunch Program develop and implement a local wellness policy.[129] By law, wellness policies must:

- Set goals for nutrition education, physical activity, and other school-based activities.

- Establish nutrition guidelines for all foods available on school campuses during the school day.

- Develop a plan to measure policy implementation.

School districts across the nation have made progress toward meeting these goals, but implementation is inconsistent, and because wellness policies are established locally, a great deal of variety exists among them.[130] Some are well defined and detailed, while others are vague and less detailed.[131] To enhance local wellness policies, standards for competitive foods and beverages served in schools define limits for fat, saturated fat, kcalories, sugar, and sodium.

Establishing and implementing these nutrition standards for competitive foods and beverages help to ensure that all foods served in schools are consistent and comply with the *Dietary Guidelines for Americans*.

> **REVIEW IT** Explain how children's appetites and nutrient needs reflect their stage of growth and why iron deficiency and obesity are often concerns during childhood.

Children's appetites and nutrient needs reflect their stage of growth. Those who are chronically hungry and malnourished suffer growth retardation; when hunger is temporary and nutrient deficiencies are mild, the problems are usually more subtle—such as poor academic performance. Iron deficiency is widespread and has many physical and behavioral consequences. Hyperactivity is not caused by poor nutrition; misbehavior may be due to lack of sleep, too little physical activity, or too much television, among other factors. Childhood obesity has become a major health problem. Adults at home and at school need to provide children with nutrient-dense foods and teach them how to make healthful diet and activity choices

16.3 Nutrition during Adolescence

> **LEARN IT** Describe some of the challenges in meeting the nutrient needs of adolescents.

Teenagers make many more choices for themselves than they did as children. They are not fed, they eat; they are not sent out to play, they choose to go. At the same time, social pressures thrust choices at them, such as whether to drink alcoholic beverages and whether to develop their bodies to meet extreme ideals of slimness or athletic prowess. Their interest in nutrition—both valid information and misinformation—derives from personal, immediate experiences. They are concerned with how diet can improve their lives now—they engage in fad dieting in order to fit into a new bathing suit, avoid greasy foods in an effort to clear acne, or eat a plate of spaghetti to prepare for a big sporting event. In presenting information on the nutrition and health of adolescents, this section includes many topics of interest to teens.

Growth and Development With the onset of **adolescence,** the steady growth of childhood speeds up abruptly and dramatically, and the growth patterns of females and males become distinct. Hormones direct the intensity of the adolescent growth spurt, profoundly affecting every organ of the body, including the brain. After 2 to 3 years of intense growth and a few more at a slower pace, physically mature adults emerge.

In general, the adolescent growth spurt begins at age 10 to 11 for females and at 12 to 13 for males. It lasts about 2½ years. Before **puberty,** male and female body compositions differ only slightly, but during the adolescent spurt, differences between the genders become apparent in the skeletal system, lean body mass, and fat stores. In females, fat assumes a larger percentage of total body weight, and in males, the lean body mass—principally muscle and bone—increases much more than in females (review Figure 1-1 on p. 7). During adolescence, males grow an average of 8 inches taller, and females, 6 inches taller. Males gain approximately 45 pounds, and females, about 35 pounds.

Energy and Nutrient Needs Energy and nutrient needs are greater during adolescence than at any other time of life, except pregnancy and lactation. In general, nutrient needs rise throughout childhood, peak in adolescence, and then level off or even diminish as the teen becomes an adult.

Energy Intake and Activity The energy needs of adolescents vary greatly, depending on the current rate of growth, gender, body composition, and physical activity.[132] Boys' energy needs may be especially high; they typically grow faster than girls and, as mentioned, develop a greater proportion of lean body mass. An exceptionally active boy of 15 may need 3500 kcalories or more a day just

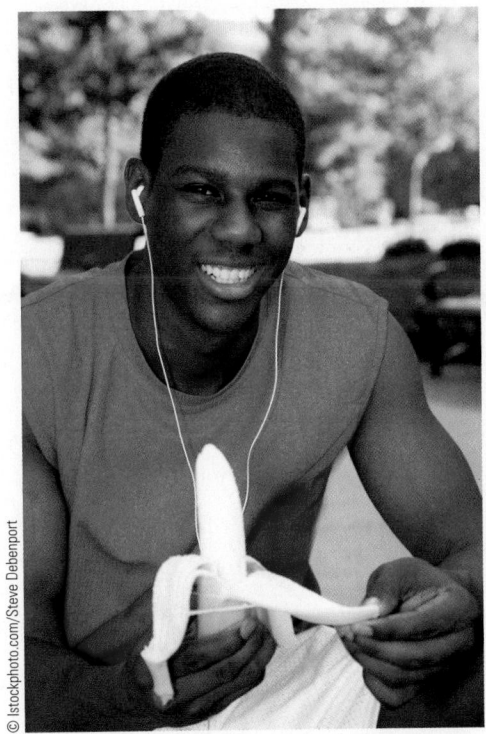

Nutritious snacks contribute valuable nutrients and energy to an active teen's diet.

adolescence: the period from the beginning of puberty until maturity.

puberty: the period in life in which a person becomes physically capable of reproduction.

to maintain his weight. Girls start growing earlier than boys and attain shorter heights and lower weights, so their energy needs peak sooner and decline earlier than those of their male peers. A sedentary girl of 15 whose growth is nearly at a standstill may need fewer than 1800 kcalories a day if she is to avoid excessive weight gain. Thus adolescent girls need to pay special attention to being physically active and selecting foods of high nutrient density so as to meet their nutrient needs without exceeding their energy needs.

The problems of obesity become ever more apparent in adolescence, especially for African American females and Hispanic teens of both genders. Without intervention, overweight adolescents face numerous physical and emotional consequences. The consequences of obesity are so dramatic and our society's attitude toward thin people is so positive that even teens of normal or below-normal weight may perceive a need to lose weight. When taken to extremes, restrictive diets bring dramatic physical consequences of their own, as Highlight 8 explains.

> **DIETARY GUIDELINES FOR AMERICANS**

- Adolescents are encouraged to maintain kcalorie balance to support normal growth and development without promoting excess weight gain. Adolescents who are overweight or obese should change their eating and physical activity behaviors so that their BMI-for-age percentile does not increase over time.

- Adolescents should do 60 minutes or more of physical activity daily. Adolescents are encouraged to spend no more than 1 to 2 hours each day watching television, playing electronic games, or using the computer (other than for homework).

Vitamins The RDA (or AI) for most vitamins increases during the adolescent years (see the inside front cover). Several of the vitamin recommendations for adolescents are similar to those for adults, including the recently revised recommendations for vitamin D.[133] Vitamin D is essential for bone growth and development. Recent studies of vitamin D status in adolescents show that many adolescents are vitamin D deficient; blacks, females, and overweight adolescents are most at risk.[134] Adolescents who do not receive enough vitamin D from fortified foods such as milk and cereals, or from sun exposure each day, may need a vitamin D supplement.[135]

Iron The need for iron increases during adolescence for both females and males, but for different reasons. Iron needs increase for females as they start to lose blood through menstruation and for males as their lean body mass develops. Hence the RDA increases at age 14 for both males and females. For females, the RDA remains high into late adulthood. For males, the RDA returns to preadolescent values in early adulthood.

In addition, iron needs increase when the adolescent growth spurt begins, whether that occurs before or after age 14. Therefore, boys in a growth spurt need an additional 2.9 milligrams of iron per day above the RDA for their age; girls need an additional 1.1 milligrams per day.

Furthermore, iron recommendations for girls before age 14 do not reflect the iron losses of menstruation. The average age of menarche (first menstruation) in the United States is 12.5 years. Therefore, for girls younger than the age of 14 who have started to menstruate, an additional 2.5 milligrams of iron per day is recommended. Thus the RDA for iron depends not only on age and gender but also on whether the individual is in a growth spurt or has begun to menstruate, as listed in Table 16-15.

Iron intakes often fail to keep pace with increasing needs, especially for females, who typically consume fewer iron-rich foods such as meat and fewer total kcalories than males. Not surprisingly, iron deficiency is most prevalent among adolescent girls. Iron-deficient children and teens score lower on standardized tests than those who are not iron deficient.

TABLE 16-15 Iron Recommendations for Adolescents

Males	Females
9–13 yr: 8 mg/day	9–13 yr: 8 mg/day
9–13 yr in growth spurt: 10.9 mg/day	9–13 yr in menarche: 10.5 mg/day
	9–13 yr in menarche and growth spurt: 11.6 mg/day
14–18 yr: 11 mg/day	14–18 yr: 15 mg/day
14–18 yr in growth spurt: 13.9 mg/day	14–18 yr in growth spurt: 16.1 mg/day

© 2016 Cengage Learning

> **DIETARY GUIDELINES FOR AMERICANS**
Adolescent girls should choose foods that supply heme iron, which is more readily absorbed by the body, additional iron sources, and enhancers of iron absorption such as vitamin C–rich foods.

Calcium Adolescence is a crucial time for bone development, and the requirement for calcium reaches its peak during these years.[136] Unfortunately, many adolescents, especially females, have calcium intakes below recommendations.[137] Low calcium intakes during times of active growth, especially if paired with physical inactivity, can compromise the development of peak bone mass, which is considered the best protection against adolescent fractures and adult osteoporosis. Increasing milk and milk products in the diet to meet calcium recommendations greatly increases bone density.[138] Once again, however, teenage girls are most vulnerable because their milk—and therefore their calcium—intakes begin to decline at the time when their calcium needs are greatest.[139] Furthermore, women have much greater bone losses than men in later life. In addition to dietary calcium, physical activity causes bones to grow stronger. Because some high schools do not require students to participate in physical education classes, however, many adolescents are not as physically active as healthy bones demand.

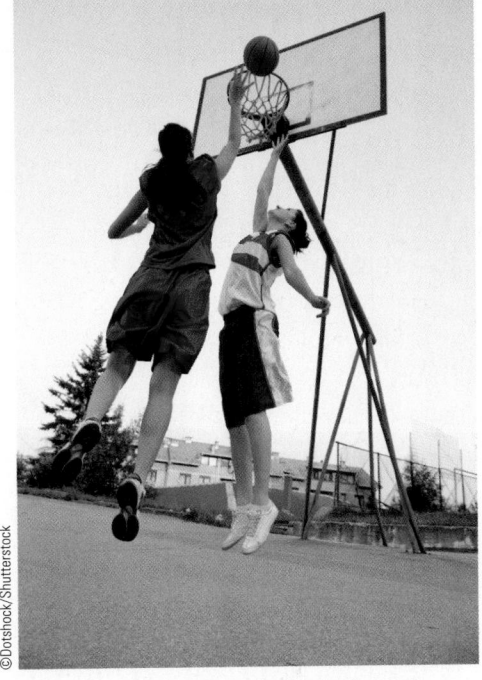

Bones grow stronger with physical activity.

> **DIETARY GUIDELINES FOR AMERICANS**
Children 9 years of age and older should consume 3 cups per day of fat-free or low-fat milk or equivalent milk products.

Food Choices and Health Habits Teenagers like the freedom to come and go as they choose. They eat what they want if it is convenient and if they have the time. With a multitude of after school, social, and job activities, they almost inevitably fall into irregular eating habits. At any given time on any given day, a teenager may be skipping a meal, eating a snack, preparing a meal, or consuming food prepared by a parent or restaurant.[140] Adolescents who frequently eat meals with their families, however, eat more fruits, vegetables, grains, and calcium-rich foods, and drink fewer soft drinks, than those who seldom eat with their families.[141] Some research shows that the more often teenagers eat dinner with their families, the less likely they are to smoke, drink, or use drugs; other research supports these findings only in teenage girls.[142] Many adolescents also begin to skip breakfast on a regular basis, missing out on important nutrients that are not made up at later meals during the day. Compared with those who skip breakfast, teenagers who do eat breakfast have higher intakes of vitamin A, vitamin C, and riboflavin, as well as calcium, iron, and zinc.[143] Teenagers who eat breakfast are therefore more likely to meet their nutrient recommendations.

Breakfast skipping may also lead to weight gain in adolescents. Eating breakfast each day, especially a breakfast rich in fiber and protein, improves satiety and reduces hunger and the desire to eat throughout the day.[144] As adolescents make the transition to adulthood, not only do they skip breakfast more often, but they also eat fast food more often. Both skipping breakfast and eating fast foods lead to weight gain.[145]

Ideally, parents continue to play the role of gatekeepers, providing nutritious, easy-to-grab foods in the refrigerator (meats for sandwiches; low-fat cheeses; fresh, raw vegetables and fruits; fruit juices; and milk) and more in the cabinets (whole-grain breads and crackers, peanut butter, nuts, popcorn, and cereal). In many households today, adults work outside the home and teenagers help with some of the gatekeepers' tasks, such as shopping for groceries or choosing fast or prepared foods.

Snacks Snacks typically provide at least 25 percent of the average teenager's daily food energy intake. If chosen carefully, snacks can contribute some needed nutrients (see Table 16-13 on p. 537). A survey of more than 5000 adolescents, found that those who ate snacks more often were less likely to be overweight or obese compared with those who ate snacks less often.[146]

Beverages Most frequently, adolescents drink soft drinks instead of fruit juice or milk with lunch, supper, and snacks. About the only time they select fruit juices is at breakfast. When teens drink milk, they are more likely to consume it with a meal (especially breakfast) than as a snack. Because of their greater food intakes, boys are more likely than girls to drink enough milk to meet their calcium needs.[147]

Soft drinks, when chosen as the primary beverage, may affect bone density, partly because they displace milk from the diet. Over the past three decades, teens (especially girls) have been drinking more soft drinks and less milk. Adolescents who drink soft drinks regularly have a higher energy intake and a lower calcium intake than those who do not.[148]

Soft drinks containing caffeine present another problem if caffeine intake becomes excessive. Many adolescents consume energy drinks on a regular basis and these beverages contain much more caffeine than soft drinks.[149]* Caffeine seems to be relatively harmless when used in *moderate* doses (less than 100 milligrams per day, roughly the equivalent of fewer than three 12-ounce cola beverages a day). In greater amounts, however, caffeine can cause symptoms associated with anxiety, such as sweating, tenseness, and inability to concentrate. Teens with certain diseases such as diabetes, heart abnormalities, and mood disorders and those taking certain medications may encounter more severe consequences such as seizures, heart failure, and death.

> **DIETARY GUIDELINES FOR AMERICANS**
Reduce intake of sugar-sweetened beverages. Children and adolescents who consume more sugar-sweetened beverages have higher body weight compared to those who drink less.

Eating Away from Home Adolescents eat about one-third of their meals away from home, and their nutritional welfare is enhanced or hindered by the choices they make.[150] A lunch consisting of a hamburger, a chocolate shake, and french fries supplies substantial quantities of many nutrients at a kcalorie cost of about 800, an energy intake some adolescents can afford. When they eat this sort of lunch, teens can adjust their breakfast and dinner choices to include fruits and vegetables for vitamin A, vitamin C, folate, and fiber and lean meats and legumes for iron and zinc. (See Appendix H for the nutrient contents of fast foods.) Fortunately, many fast-food restaurants are offering more nutritious choices than the standard hamburger meal.

Peer Influence Physical maturity and growing independence present adolescents with new choices. The consequences of those choices will influence their health and nutrition status both today and throughout life. Many of the food and health choices adolescents make reflect the opinions and actions of their peers. When others perceive milk as "babyish," a teen may choose soft drinks instead; when others skip lunch and hang out in the parking lot, a teen may join in for the camaraderie, regardless of hunger. Some teenagers begin using drugs, alcohol, and tobacco; others wisely refrain. Adults can set up the environment so that nutritious foods are available and can stand by with reliable information and advice about health and nutrition, but the rest is up to the adolescents. Ultimately, they make the choices. (Highlight 8 examines the influence of social pressures on the development of eating disorders.)

Because their lunches rarely include fruits, vegetables, or milk, many teens fail to get all the vitamins and minerals they need each day.

*Caffeine-containing soft drinks typically deliver about 30 mg of caffeine per 12-ounce can; energy drinks typically deliver about 100 mg of caffeine for equivalent amounts. A pharmacologically active dose of caffeine is defined as 200 mg. Appendix H starts with a table listing the caffeine contents of selected foods, beverages, and drugs.

> **REVIEW IT** Describe some of the challenges in meeting the nutrient needs
of adolescents.
Nutrient needs rise dramatically as children enter the rapid growth phase of adolescence.
Teenagers' busy lifestyles add to the challenge of meeting their nutrient needs, especially for
iron and calcium.

The nutrition and lifestyle choices people make as children and adolescents have long-term, as well as immediate, effects on their health. Highlight 16 describes how sound choices and good habits during childhood and adolescence can help prevent chronic diseases later in life.

Nutrition Portfolio

Encouraging children to eat nutritious foods today helps them learn how to make healthy food choices tomorrow.

- If there are children in your life, think about the food they eat and consider whether they receive enough food for healthy growth, but not so much as to lead to obesity.

- Describe the advantages of physical activity to children's health and well-being.

- Plan a day's menu for a child 4 to 8 years of age, making sure to include foods that provide enough calcium and iron.

- Now, go to Diet & Wellness Plus and create a profile for a child 4 to 8 years of age. Enter the day's menu you suggested in the previous exercise and see if you met the basic requirements for that child.

DIET & WELLNESS **PLUS** ✚ To complete this exercise, go to your Diet & Wellness Plus at
www.cengagebrain.com.

> **STUDY IT** To review the key points of this chapter and take a practice quiz, go
to the study cards at the end of the book.

REFERENCES

1. Committee on Dietary Reference Intakes, *Dietary Reference Intakes for Energy, Carbohydrate, Fiber, Fat, Fatty Acids, Cholesterol, Protein, and Amino Acids* (Washington, D.C.: National Academies Press, 2005).

2. R. S. Kuipers and coauthors, Fetal intrauterine whole body linoleic, arachidonic and docosahexaenoic acid contents and accretion rates, *Prostaglandins, Leukotrienes, and Essential Fatty Acids* 86 (2012): 13–20.

3. Formula feeding of term infants, in *Pediatric Nutrition*, 7th ed., ed. R. E. Kleinman (Elk Grove Village, Ill.: American Academy of Pediatrics, 2014), pp. 61–81.

4. Position of the American Dietetic Association: Promoting and supporting breastfeeding, *Journal of the American Dietetic Association* 109 (2009): 1926–1942.

5. American Academy of Pediatrics, Policy statement: Breastfeeding and the use of human milk, *Pediatrics* 129 (2012): e827–e841, available at www.pediatrics.org/content/129/3/e827.full; Breastfeeding, in *Pediatric Nutrition*, 7th ed., ed. R. E. Kleinman (Elk Grove Village, Ill.: American Academy of Pediatrics, 2014), pp. 41–59.

6. P. V. Jeurink and coauthors, Mechanisms underlying immune effects of dietary oligosaccharides, *American Journal of Clinical Nutrition* 98 (2013): 572S–577S.

7. E. Castanys-Munoz, M. J. Martin, and P. A. Prieto, 2'-Fucosyllactose: An abundant, genetically determined soluble glycan present in human milk, *Nutrition Reviews* 71 (2013): 773–789; Jeurink and coauthors, 2013.

8. P. Willatts and coauthors, Effects of long-chain PUFA supplementation in infant formula on cognitive function in later childhood, *American Journal of Clinical Nutrition* 98 (2013): 536S–542S.

9. Willatts and coauthors, 2013; S. J. Meldrum and coauthors, Achieving definitive results in long-chain polyunsaturated fatty acid supplementation trials of term infants: Factors for consideration, *Nutrition Reviews* 69 (2011): 205–214; E. E. Birch and coauthors, The DIAMOND (DHA Intake and Measurement of Neural Development) Study: A double-masked, randomized controlled clinical trial of the maturation of infant visual acuity as a function the dietary level of docosahexaenoic acid, *American Journal of Clinical Nutrition* 91 (2010): 848–859.

10. Meldrum and coauthors, 2011.

11. Willatts and coauthors, 2013.

12. S. A. Abrams, What are the risks and benefits to increasing dietary bone minerals and vitamin D intake in infants and small children? *Annual Review of Nutrition* 31 (2011): 285–297.

13. S. Gallo and coauthors, Effect of different dosages of oral vitamin D supplementation on vitamin D status in healthy, breastfed infants, *Journal of the American Medical Association* 309 (2013): 1785–1792; American Academy of Pediatrics, 2012; C. L. Wagner, F. R. Greer, and the Section on Breastfeeding and Committee on Nutrition, Prevention of rickets and vitamin D deficiency in infants, children, and adolescents, *Pediatrics* 122 (2008): 1142–1152.

14. D. E. W. Chatterton and coauthors Anti-inflammatory mechanisms of bioactive milk proteins in the intestine of newborns, *International Journal of Biochemistry and Cell Biology* 45 (2013): 1730–1747; B. Lonnerdal, Bioactive proteins in breast milk, *Journal of Paediatrics and Child Health* (Supplement S1) 49 (2013): 1–7; American Academy of Pediatrics, 2012; A. Walker, Breast milk as the gold standard for

protective nutrients, *Journal of Pediatrics* 156 (2010): S3–S7; Position of the American Dietetic Association, 2009.

15. Walker, 2010.
16. American Academy of Pediatrics, 2012; F. R. Greer, S. H. Sicherer, and A. W. Burks, and the Committee on Nutrition and Section on Allergy and Immunology, Effects of early nutritional interventions on the development of atopic disease in infants and children: The role of maternal dietary restriction, breastfeeding, timing of introduction of complementary foods, and hydrolyzed formulas, *Pediatrics* 121 (2008): 183–191.
17. American Academy of Pediatrics, 2012.
18. F. R. Hauck and coauthors, Breastfeeding and reduced risk of sudden infant death syndrome: A meta-anaaylsis, *Pediatrics* 128 (2011): 103–110.
19. I. Labayen and coauthors, Association of exclusive breastfeeding duration and fibrinogen levels in childhood and adolescence: The European Youth Heart Study, *Archives of Pediatric and Adolescent Medicine* 166 (2012): 56–61; C. G. Owen, P. H. Whincup, and D. G. Cook, Breast-feeding and cardiovascular risk factors and outcomes in later life: Evidence from epidemiological studies, *Proceedings of the Nutrition Society* 70 (2011): 478–484; M. S. Fewtrell, Breastfeeding and later risk of CVD and obesity: Evidence from randomized trials, *Proceedings of the Nutrition Society* 70 (2011): 472–477
20. K. Casazza, J. R. Fernandez, and D. B. Allison, Modest protective effects of breastfeeding on obesity, *Nutrition Today* 47 (2012): 33–38; L. Shields and coauthors, Breastfeeding and obesity at 21 years: A cohort study, *Journal of Clinical Nursing* 19 (2010): 1612–1617; L. Schack-Nielsen and coauthors, Late introduction of complementary feeding, rather than duration breastfeeding, may protect against adult overweight, *American Journal of Clinical Nutrition* 91 (2010): 619–627; L. Twells and L. A. Newhook, Can exclusive breastfeeding reduce the likelihood of childhood obesity in some regions of Canada? *Canadian Journal of Public Health* 101 (2010): 36–39; P. Chivers and coauthors, Body mass index, adiposity rebound and early feeding in a longitudinal cohort (Raine Study), *International Journal of Obesity* 34 (2010): 1169–1176.
21. Chivers and coauthors, 2010; Shields and coauthors, 2010.
22. Casazza, Fernandez, and Allison, 2012; K. L. Whitaker and coauthors, Comparing maternal and paternal intergenerational transmission of obesity risk in a large population-based sample, *American Journal of Clinical Nutrition* 91 (2010): 1560–1567; R. Li, S. B. Fein, and L. M. Grummer-Strawn, Do infants fed from bottles lack self-regulation of milk intake compared with directly breastfed infants? *Pediatrics* 125 (2010): e1386–e1393.
23. W. Jedrychowski and coauthors, Effect of exclusive breastfeeding on the development of children's cognitive function in the Krakow prospective birth cohort study, *European Journal of Pediatrics* 171 (2012): 151–158; M. A. Quigley and coauthors, Breastfeeding is associated with improved child cognitive development: A population-based cohort study, *Journal of Pediatrics* 160 (2012): 25–32; C. McCrory and R. Layte, The effect of breast-feeding on children's educational test scores at nine years of age: Results of an Irish cohort study, *Social Science and Medicine* 72 (2011): 1515–1521.
24. M. Guxens and coauthors, Breastfeeding, long-chain polyunsaturated fatty acids in colostrum, and infant mental development, *Pediatrics* 128 (2011): e880–e889.
25. D. J. Miracle and coauthors, Contemporary ethical issues in human milk-banking in the United States, *Pediatrics* 128 (2011): 1186–1191.
26. Formula feeding of term infants, in *Pediatric Nutrition*, 7th ed., ed. R. E. Kleinman (Elk Grove Village, Ill.: American Academy of Pediatrics, 2014), pp. 61–81.
27. Position of the Academy of Nutrition and Dietetics: The impact of fluoride on health, *Journal of the Academy of Nutrition and Dietetics* 112 (2012): 1443–1453.
28. Formula feeding of term infants, 2014.
29. Formula feeding of term infants, 2014.
30. S. Basnet and coauthors, Fresh goat's milk for infants: Myths and realities: A review, *Pediatrics* 125 (2010): e973–e977.
31. B. E. Hamilton and coauthors, Annual summary of vital statistics; 2010–2011, *Pediatrics* 131 (2013): 548–558.
32. J. Miller and coauthors, Effect of increasing protein content of human milk fortifier on growth in preterm infnats born at <31 wk gestation: A randomized controlled trial, *American Journal of Clinical Nutrition* 95

(2012): 648–655; Nutritional needs of the preterm infant, in *Pediatric Nutrition*, 7th ed., ed. R. E. Kleinman (Elk Grove Village, Ill.: American Academy of Pediatrics, 2014), pp. 83–121.
33. Formula feeding of term infants, 2014.
34. E. E. Ziegler, Consumption of cow's milk as a cause of iron deficiency in infants and toddlers, *Nutrition Reviews* 69 (2011): S37–S42.
35. Expert Panel on Integrated Guidelines for Cardiovascular Health and Risk Reduction in Children and Adolescents: Summary report, *Pediatrics* 128 (2011): S213–S256.
36. Complementary feeding, in *Pediatric Nutrition*, 7th ed., ed. R. E. Kleinman (Elk Grove Village, Ill.: American Academy of Pediatrics, 2014), pp. 123–139.
37. H. Przyrembel, Timing of introduction of complementary food: Short- and long-term health consequences, *Annals of Nutrition and Metabolism* (supplement) 60 (2012): 8–20.
38. Complementary feeding, in *Pediatric Nutrition*, 7th ed., ed. R. E. Kleinman (Elk Grove Village, Ill.: American Academy of Pediatrics, 2014), pp. 123–139.
39. E. E. Ziegler, S. E. Nelson, and J. M. Jeter, Iron supplementation of breastfed infants, *Nutrition Reviews* 69 (2011): S71–S77; Complementary feeding, in *Pediatric Nutrition*, 7th ed., ed. R. E. Kleinman (Elk Grove Village, Ill.: American Academy of Pediatrics, 2014), pp. 123–139.
40. Complementary feeding, 2014.
41. Feeding the child, in *Pediatric Nutrition*, 7th ed., ed. R. E. Kleinman (Elk Grove Village, Ill.: American Academy of Pediatrics, 2014), pp. 143–173.
42. Expert Panel, 2011.
43. Position of the American Dietetic Association: Vegetarian diets, *Journal of the American Dietetic Association* 109 (2009): 1266–1282.
44. D. M. Hoelscher and coauthors, Position of the Academy of Nutrition and Dietetics: Interventions for the prevention and treatment of pediatric overweight and obesity, *Journal of the Academy of Nutrition and Dietetics* 113 (2013): 1375–1394; American Academy of Pediatrics, Policy Statement—Children, adolescents, obesity, and the media, *Pediatrics* 128 (2011): 201–208; V. C. Strasburger, A. B. Jordan, and E. Donnerstein, Health effects of media on children and adolescents, *Pediatrics* 125 (2010): 756–767; Position of the American Dietetic Association: Nutrition guidance for healthy children ages 2 to 11 years, *Journal of the American Dietetic Association* 108 (2008): 1038–1047.
45. J. L. Foltz and coauthors, Population-level intervention strategies and examples for obesity prevention in children, *Annual Review of Nutrition* 32 (2012): 391–415; Position of the American Dietetic Association: Benchmarks for nutrition in child care, *Journal of the American Dietetic Association* 111 (2011): 607–615; N. Larson and coauthors, What role can child-care settings play in obesity prevention? A review of the evidence and call for research efforts, *Journal of the American Dietetic Association* 111 (2011): 1343–1362; J. E. Fulton and coauthors, Physical activity levels of high school students: United States, 2010, *Morbidity and Mortality Weekly Report* 60 (2011): 773–777.
46. Nutritional aspects of vegetarian diets, in *Pediatric Nutrition*, 7th ed., ed. R. E. Kleinman (Elk Grove Village, Ill.: American Academy of Pediatrics, 2014), pp. 241–264.
47. Committee on Dietary Reference Intakes, 2005, Chapter 8.
48. R. D. Baker, F. R. Greer, and the Committee on Nutrition, Clinical Report: Diagnosis and prevention of iron deficiency and iron-deficiency anemia in infants and young children (0–3 years of age), *Pediatrics* 126 (2010): 1040–1050.
49. Committee on Dietary Reference Intakes, *Dietary Reference Intakes for Calcium and Vitamin D* (Washington, D.C.: National Academies Press, 2011), pp. 345–402.
50. Wagner, Greer, and the Section on Breastfeeding and Committee on Nutrition, 2008.
51. J. Dwyer and coauthors, Prevalence and predictors of children's dietary supplement use: The 2007 National Health Interview Survey, *American Journal of Clinical Nutrition* 97 (2013): 1331–1337; R. L. Bailey and coauthors, Do dietary supplements improve micronutrient sufficiency in children and adolescents? *Journal of Pediatrics* 161 (2012): 837–842.
52. C. N. Ford, M. M. Slining, and B. M. Popkin, Trends in dietary intake among US 2- to 6-year-old children, 1989–2008, *Journal of the Academy of Nutrition and Dietetics* 113 (2013): 35–42.

53. Position of the American Dietetic Association, 2008.

54. R. R. Briefel and coauthors, The Feeding Infants and Toddlers Study 2008: Study design and methods, *Journal of the American Dietetic Association* 110 (2010): S16–S26.

55. M. K. Fox and coauthors, Food consumption patterns of young preschoolers: Are they starting off on the right path? *Journal of the American Dietetic Association* 110 (2010): S52–S59.

56. N. Tian and coauthors, Sodium and potassium intakes among US infants and preschool children, 2003–2010, *American Journal of Clinical Nutrition* 98 (2013): 1113–1122; Dwyer and coauthors, 2010.

57. Bread for the World, Hunger facts: Domestic, www.bread.org/learn /hunger-basics/hunger-facts-domestic.html, accessed January 25, 2014.

58. Position of the American Dietetic Association: Child and adolescent nutrition assistance programs, *Journal of the American Dietetic Association* 110 (2010): 791–799.

59. C. E. Basch, Breakfast and the achievement gap among urban minority youth, *Journal of School Health* 81 (2011): 635–640; S. B. Cooper, S. Bandelow, and M. E. Nevill, Breakfast consumption and cognitive function in adolescent schoolchildren, *Physiology & Behavior* 103 (2011): 431–439; S. P. P. Tin and coauthors, Breakfast skipping and change in body mass index in young children, *International Journal of Obesity* 35 (2011): 899–906; P. R. Deshmukh-Taskar and coauthors, The relationship of breakfast skipping and type of breakfast consumption with nutrient intake and weight status in children and adolescents: The National Health and Nutrition Examination Survey 1999–2006, *Journal of the American Dietetic Association* 110 (2010): 869–878.

60. T. Coppinger and coauthors, Body mass, frequency of eating and breakfast consumption in 9–13-year-olds, *Journal of Human Nutrition and Dietetics* 25 (2012): 43–49; T.V.E. Kral and coauthors, Effects of eating breakfast compared with skipping breakfast on ratings of appetite and intake at subsequent meals in 8- to 10-y-old children, *American Journal of Clinical Nutrition* 93 (2011): 284–291; K. J. Smith and coauthors, Skipping breakfast: Longitudinal associations with cardiometabolic risk factors in the Childhood Determinants of Adult Health Study, *American Journal of Clinical Nutrition* 92 (2010): 1316–1325.

61. F. Koohdani and coauthors, Midmorning snack programs have a beneficial effect on cognitive performance of students from high socioeconomic background, *Nutrition Today* doi: 10.1097/ NT.0b013e318298e7dd.

62. K. Kordas, Iron, lead, and children's behavior and cognition, *Annual Review of Nutrition* 30 (2010): 123–148.

63. M. M. Black and coauthors, Iron deficiency and iron-deficiency anemia in the first two years of life: Strategies to prevent loss of developmental potential, *Nutrition Reviews* 69 (2011): S64–S70; M. K. Georgieff, Long-term brain and behavioral consequences of early iron deficiency, *Nutrition Reviews* 69 (2011): S43–S48; C. S. Wang and coauthors, Iron-deficiency anemia in infancy and social emotional development in preschool-aged Chinese children, *Pediatrics* 127 (2011): e927–e933; F. Corapci and coauthors, Longitudinal evaluation of externalizing and internalizing behavior problems following iron deficiency in infancy, *Journal of Pediatric Psychology* 35 (2010): 296–305.

64. Kordas, 2010.

65. Centers for Disease Control and Prevention, Lead, www.cdc.gov/nceh /lead/.htm, updated on December 18, 2013.

66. Centers for Disease Control and Prevention, Attention-Deficit/ Hyperactivty Disorder, Data and Statistics, www.cdc.gov/ncbddd/ adhd/data.html, updated November 13, 2013.

67. Subcommittee on Attention-Deficit/Hyperactivity Disorder, Steering Committee on Quality Improvement and Management, ADHD: Clinical practice guideline for the diagnosis, evaluation, and treatment of attention-deficit/hyperactivity disorder in children and adolescents, *Pediatrics* 128 (2011): 1007–1022.

68. J. G. Millichap and M. M. Yee, The diet factor in attention-deficit/ hyperactivity disorder, *Pediatrics* 129 (2012): 330–337; L. F. Martí, Effectiveness of nutritional interventions on the functioning of children with ADHD and/or ASD: An updated review of research evidence, *Bulletin of the Porto Rico Medical Association* 102 (2010): 31–42.

69. L. J. Stevens and coauthors, Mechanisms of behavioral, atopic, and other reactions to artificial food colors in children, *Nutrition Reviews*

71 (2013): 268–281; R. B. Kanarek, Artificial food dyes and attention deficit hyperactivity disorder, *Nutrition Reviews* 69 (2011): 385–391; R. E. Kleinman and coauthors, A research model for investigating the effects of artificial food colorings on children with ADHD, *Pediatrics* 127 (2011): e1575–e1584; A. Connolly and coauthors, Pattern of intake of food additives associated with hyperactivity in Irish children and teenagers, *Food Additives and Contaminants: Part A, Chemistry, Analysis, Control, Exposure and Risk Assessment* 27 (2010): 447–456.

70. Kanarek, 2011.

71. Food and Drug Administration, Food ingredients and colors, www.fda. gov. revised April 2010.

72. R. S. Gupta and coauthors, The prevalence, severity, and distribution of childhood food allergy in the United States, *Pediatrics* 128 (2011): e9–e17; Food and Drug Administration, Food allergies: Reducing the risks, www .fda.gov/consumer/updates/foodallergies012209.html, January 23, 2009.

73. Food and Drug Administration, Food allergies; Reducing the risks, www.fda.gov/consumer/updates/foodallergies012209.html, January 23, 2009.

74. J. A. Boyce and coauthors, Guidelines for the diagnosis and management of food allergy in the United States: Summary of the NIAID-Sponsored Expert Panel Report, *Journal of the American Dietetic Association* 111 (2011): 17–27.

75. H. N. Cho and coauthors, Nutritional status according to sensitized food allergens in children with atopic dermatitis, *Allergy, Asthma, and Immunology Research* 3 (2010): 53–57.

76. Boyce and coauthors, 2011.

77. Gupta and coauthors, 2011.

78. M. Pansare and D. Kamat, Peanut allergy, *Current Opinions in Pediatrics* 22 (2010): 642–646; A. T. Clark and coauthors, Successful oral tolerance induction in severe peanut allergy, *Allergy* 64 (2009): 1218–1220.

79. C. D. Fryar, M. D. Carroll, and C. L. Ogden, Prevalence of obesity among children and adolescents: United States, trends 1963–1965 through 2009–2010, *NCHS Health E-Stat*, www.cdc.gov/nchs/data /hestat/obesity_child_09_10/obesity_child_09_10.pdf.

80. C. L. Ogden, M. D. Carroll, and K. M. Flegal, Prevalence of childhood and adult obesity in the United States, 2011–2012, *Journal of the American Medical Association* 311 (2014): 806–814; S. A. Cunningham, M. R. Kramer, and K. M. Venkat Narayan, Incidence of childhood obesity in the United States, *New England Journal of Medicine* 370 (2014): 403–411.

81. Expert Panel on Integrated Guidelines for Cardiovascular Health and Risk Reduction in Children and Adolescents, Summary Report, *Pediatrics* 128 (2011): S213–S256; J. J. Reilly, Assessment of obesity in children and adolescents: Synthesis of recent systematic reviews and clinical guidelines, *Journal of Human Nutrition and Dietetics* 23 (2010): 205–211.

82. A. S. Kelly and coauthors, Severe obesity in children and adolescents: Identification, associated health risks, and treatment approaches: A Scientific Statement from the American Heart Association, *Circulation* 128 (2013): 1689–1712; A. K. Gulati, D. W. Kaplan, and S. R. Daniels, Clinical tracking of severely obese children: A new growth chart, *Pediatrics* 130 (2012): 1136–1140; N. S. The and coauthors, Association of adolescent obesity with risk of severe obesity in adulthood, *Journal of the American Medical Association* 304 (2010): 2042–2057; F. M. Biro and M. Wien, Childhood obesity and adult morbidities, *American Journal of Clinical Nutrition* 91 (2010): 1499S–1505S.

83. A. L. Thompson, Intergenerational impact of maternal obesity and postnatal feeding practices on pediatric obesity, *Nutrition Reviews* 71(2013): S55–S61; A. Jaaskelainen and coauthors, Intergenerational transmission of overweight among Finnish adolescents and their parents: A 16-year follow-up study, *International Journal of Obesity* 35 (2011): 1289–1294; R. Cooper and coauthors, Associations between parental and offspring adiposity up to midlife: The contribution of adult lifestyle factors in the 1958 British Birth Cohort Study, *American Journal of Clinical Nutrition* 92 (2010): 946–953; Whitaker and coauthors, 2010.

84. Expert Panel, 2011; Biro and Wien, 2010.

85. Centers for Disease Control and Prevention, Overweight and obesity: A growing problem, www.cdc.gov/obesity/childhood/problem.html, updated April 27, 2012; Foltz and coauthors, 2012.

86. L. McGowan and coauthors, Healthy feeding habits: Efficacy results from a cluster-randomized, controlled exploratory trial of a novel, habit-based intervention with parents, *American Journal of Clinical*

Nutrition 98 (2013): 769–777; S. J. Salvy and coauthors, Influence of parents and friends on children's and adolescents' food intake and food selection, *American Journal of Clinical Nutrition* 93 (2011): 87–92; L. Hall and coauthors, Children's intake of fruit and selected energy-dense nutrient-poor foods is associated with fathers' intake, *Journal of the American Dietetic Association* 111 (2011): 1039–1044; H. A. Raynor and coauthors, The relationship between child and parent food hedonics and parent and child food group intake in children with overweight/obesity, *Journal of the American Dietetic Association* 111 (2011): 425–430; C. Sweetman and coauthors, Characteristics of family mealtimes affecting children's vegetable consumption and liking, *Journal of the American Dietetic Association* 111 (2011): 269–273.

87. L. R. Jones and coauthors, Influences on child fruit and vegetable intake: Sociodemographic, parental and child factors in a longitudinal cohort study, *Public Health Nutrition* 13 (2010): 1122–1130.

88. C. R. McGill and coauthors, Improved diet quality and increased nutrient intakes associated with grape product consumption by U.S. children and adults: National Health and Nutrition Examination Survey 2003–2008, *Journal of Food Science* (Suppl 1) 78 (2013): A1–A4; J. S. Savage and coauthors, Serving smaller age-appropriate entrée portions to children aged 3–5 y increases fruit and vegetable intake and reduces energy density and energy intake at lunch, *American Journal of Clinical Nutrition* 95 (2012): 335–341.

89. J. Reedy and S. M. Krebs-Smith, Dietary sources of energy, solid fats, and added sugars among children and adolescents in the United States, *Journal of the American Dietetic Association* 110 (2010): 1477–1484.

90. R. B. Ervin and coauthors, Consumption of added sugar among U.S. children and adolescents, 2005–2008, *NCHS Data Brief* 87 (2012), available at www.cdc.gov/nchs/data/databriefs/db87.htm; J. A. Welsh and coauthors, Consumption of added sugars is decreasing in the United States, *American Journal of Clinical Nutrition* 94 (2011): 726–734.

91. R. R. Briefel and coauthors, Reducing calories and added sugars by improving children's beverage choices, *Journal of the Academy of Nutrition and Dietetics* 113 (2013): 269–275; M. D. BeBoer, R. J. Scharf, and R. T. Demmer, Sugar-sweetened beverages and weight gain in 2- to 5-year old children, *Pediatrics* 132 (2013): 413–420; G. L. Ambrosini and coauthors, Prospective associations between sugar-sweetened beverage intakes and cardiometabolic risk factors in adolescents, *American Journal of Clinical Nutrition* 98 (2013): 327–334; V. S. Malik and coauthors, Sugar-sweetened beverages and weight gain in children and adults: A systematic review and meta-analysis, *American Journal of Clinical Nutrition* 98 (2013): 1084–1092; D. F. Tate and coauthors, Replacing caloric beverages with water or diet beverages for weight loss in adults: Main results of the Choose Healthy Options Consciously Everyday (CHOICE) randomized clinical trial, *American Journal of Clinical Nutrition* 95 (2012): 555–563; V. S. Malik and coauthors, sugar-sweetened beverages, obesity, type 2 diabetes mellitus, and cardiovascular disease risk, *Circulation* 121 (2010): 1356–1364; F. B. Hu and V. S. Malik, Sugar-sweetened beverages and risk of obesity and type 2 diabetes: epidemiologic evidence, *Physiology and Behavior* 100 (2010): 47–54; O. L. Bermudez and X. Gao, Greater consumption of sweetened beverages and added sugars is associated with obesity among US young adults, *Annals of Nutrition and Metabolism* 57 (2010): 211–218.

92. U. Ekelund and coauthors, Moderate to vigorous physical activity and sedentary time and caridometabolic risk factors in children and adolescents, *Journal of the American Medical Association* 307 (2012): 704–712.

93. Centers for Disease Control and Prevention, Overweight and obesity: A growing problem, www.cdc.gov/obesity/childhood/problem.html, updated April 27, 2012.

94. M. Govindan and coauthors, Gender differences in physiologic markers and health behaviors associated with childhood obesity, *Pediatrics* 132 (2013): 468–474; J. Falbe and coauthors, Adiposity and different types of screen time, *Pediatrics* 132 (2013): e1497–e1505; American Academy of Pediatrics, Council on Communications and Media, Policy statement—Children, adolescents, obesity, and the media, *Pediatrics* 128 (2012): 201–208.

95. A. E. Staiano and coauthors, Television, adiposity, and cardiometabolic risk in children and adolescents, *American Journal of Preventive Medicine* 44 (2013): 40–47; American Academy of Pediatrics, Council on Communications and Media, 2011; S. B. Sisson and coauthors, TVs in the bedrooms of children: Does it impact health and behavior? *Preventive Medicine* 52 (2011): 104–108.

96. S. B. Sisson and coauthors, Television-viewing time and dietary quality among U.S. children and adults, *American Journal of Preventive Medicine* 43 (2012): 196–200; American Academy of Pediatrics, Council on Communications and Media, 2011; M. Skatrud-Mickelson, A. M. Adachi-Mejia, and L. A. Sutherland, Tween sex differences in snacking preferences during television viewing, *Journal of the American Dietetic Association* 111 (2011): 1385–1390.

97. Sisson and coauthors, 2012.

98. N. Pearson and S. J. Biddle, Sedentary behavior and dietary intake in children, adolescents, and adults: A systematic review, *American Journal of Preventive Medicine* 41 (2011): 178–188.

99. L. M. Powell, G. Szczypka, and F. J. Chaloupka, Trends in exposure to television food advertisements among children and adolescents in the United States, *Archives of Pediatric and Adolescent Medicine* 164 (2010): 794–802.

100. F. Folkvord and coauthors, The effect of playing advergames that promote energy-dense snacks or fruit on actual food intake among children, *American Journal of Clinical Nutrition* 97 (2013): 239–245.

101. J. L. Harris and S. K. Graff, Protecting children from harmful food marketing: Options for local government to make a difference, Centers for Disease Control and Prevention, Preventing Chronic Disease, 8 (September 2011), www.cdc.gov/pcd/issues/2011/sep/10_0272.htm.

102. J. P. Chaput and coauthors, Video game playing increases food intake in adolescents: A randomized crossover study, *American Journal of Clinical Nutrition* 93 (2011): 1196–1203.

103. K. White, G. Schofield, and A. E. Kilding, Energy expended by boys playing active video games, *Journal of Science and Medicine in Sport* 14 (2011): 130–134.

104. American Academy of Pediatrics, Council on Communications and Media, 2011; Policy statement: Media use by children younger than 2 years, *Pediatrics* 128 (2011): 1040–1045.

105. W. Tu and coauthors, Intensified effect of adiposity on blood pressure in overweight and obese children, *Hypertension* 58 (2011): 818–824; M. Salvadori and coauthors, Elevated blood pressure in relation to overweight and obesity among children in a rural Canadian community, *Pediatrics* 122 (2008): e821–e827.

106. A. L. May, E. V. Kuklina, and P. W. Yoon, Prevalence of cardiovascular disease risk factors among US adolescents, 1999–2008, *Pediatrics* 129 (2012): 1035–1041; J. Ohman and coauthors, Early childhood overweight and asthma and allergic sensitization at 8 years of age, *Pediatrics* 129 (2012): 70–76; A. Tirosh and coauthors, Adolescent BMI trajectory and risk of diabetes versus coronary disease, *New England Journal of Medicine* 364 (2011): 1315–1325; J. C. Han, D. A. Lawlor, and S. Y. S. Kimm, Childhood obesity, *Lancet* 375 (2010): 1737–1748; F. M. Biro and M. Wien, Childhood obesity and adult morbidities, *American Journal of Clinical Nutrition* 91 (2010): 1499S–1505S; P. W. Franks and coauthors, Childhood obesity, other cardiovascular risk factors, and premature death, *New England Journal of Medicine* 362 (2010): 458–493.

107. S. A. Taylor and coauthors, A qualitative study of the day-to-day lives of obese Mexican-American adolescent females, *Pediatrics* 131 (2013): 1132–1138.

108. D. M. Hoelscher and coauthors, Position of the Academy of Nutrition and Dietetics: Interventions for the prevention and treatment of pediatric overweight and obesity, *Journal of the Academy of Nutrition and Dietetics* 113 (2013): 1375–1394.

109. Hoelscher and coauthors, 2013.

110. Hoelscher and coauthors, 2013.

111. P. J. Colllings and coauthors, Physical activity intensity, sedentary time, and body composition in preschoolers, *American Journal of Clinical Nutrition* 97 (2013): 1020–1028; C. L. Davis and coauthors, Exercise dose and diabetes risk in overweight and obese children, *Journal of American Medical Association* 308 (2012): 1103–1112

112. Hoelscher and coauthors, 2013.

113. S. Bondada, H. C. Jen, and D. A. Deugarte, Outcomes of bariatric surgery in adolescents, *Current Opinion in Pediatrics* 23 (2011): 552–556; P. E. O'Brien and coauthors, Laparoscopic adjustable gastric banding in severely obese adolescents, *Journal of the American Medical Association* 303 (2010): 519–526.

114. K. W. Bauer, J. M. Berge, and D. Neumark-Sztainer, The importance of families to adolescents' physical activity and dietary intake, *Adolescent Medicine: State of the Art Reviews* 22 (2011): 601–613; S. L. Anzman and coauthors, Parental influence on children's early eating environments and obesity risk: Implications for prevention, *International Journal of Obesity* 34 (2010): 1116–1124.

115. M. L. O'Connell and coauthors, Repeated exposure in a natural setting: A preschool intervention to increase vegetable consumption, *Journal of the Academy of Nutrition and Dietetics* 112 (2012): 230–234.

116. Position of the American Dietetic Association, 2008.

117. B. Wansink, M. Shimizu, and A. Brumberg, Association of nutrient-dense snack combinations with calories and vegetable intake, *Pediatrics* 131 (2013): 22–29.

118. Position of the American Dietetic Association: Benchmarks for nutrition in child care, *Journal of the American Dietetic Association* 111 (2011): 607–615; Position of the American Dietetic Association, School Nutrition Association, and Society of Nutrition Education: Comprehensive school nutrition services, *Journal of the American Dietetic Association* 110 (2010): 1738–1749.

119. Position of the American Dietetic Association, School Nutrition Association, and Society of Nutrition Education, 2010.

120. Position of the American Dietetic Association, School Nutrition Association, and Society of Nutrition Education, 2010.

121. National School Lunch Program, www.fns.usda.gov/cnd/lunch /AboutLunce/NSLPFactSheet.pdf, updated August, 2012; Position of the American Dietetic Association, School Nutrition Association, and Society for Nutrition Education, 2010; Position of the American Dietetic Association: Local support for nutrition integrity in schools, *Journal of the American Dietetic Association* 110 (2010): 1244–1254.

122. Food and Nutrition Services (FNS) USDA, Nutrition standards in the National School Lunch and School Breakfast Programs. Final rule, *Federal Register* 77 (2012): 4088–4167.

123. The School Breakfast Program, www.fns.usda.gov/cnd/breakfast.htm, updated August 5, 2012; Position of the American Dietetic Association: Local support for nutrition integrity in schools, *Journal of the American Dietetic Association* 110 (2010): 1244–1254.

124. Position of the American Dietetic Association, 2010.

125. Centers for Disease Control and Prevention, School health guidelines to promote health eating and physical activity, *Morbidity and Mortality Weekly Report* 60 (Supplement), September 16, 2011, pp. 1–76.

126. Position of the American Dietetic Association, 2010.

127. Position of the American Dietetic Association, 2010.

128. M. Kakarala, D. R. Keast, and S. Hoerr, Schoolchildren's consumption of competitive foods and beverages, excluding á la carté, *Journal of School Health* 80 (2010): 429–435; Position of the American Dietetic Association: Local support for nutrition integrity in schools, *Journal of the American Dietetic Association* 110 (2010): 1244–1254.

129. Position of the American Dietetic Association, School Nutrition Association, and Society for Nutrition Education: Comprehensive school nutrition services, 2010; Position of the American Dietetic Association, 2010.

130. R. Turner and F. J. Chaloupka, Student access to competitive foods in elementary schools, trends over time and regional differences, *Archives of Pediatrics and Adolescent Medicine* 166 (2012): 164–169.

131. Position of the American Dietetic Association, School Nutrition Association, and Society for Nutrition Education, 2010.

132. L. B. Shomaker and coauthors, Puberty and observed energy intake: Boy, can they eat! *American Journal of Clinical Nutrition* 92 (2010): 123–129; Committee on Dietary Reference Intakes, 2005, Chapter 5.

133. Committee on Dietary Reference Intakes, 2011.

134. The Society for Adolescent Health and Medicine, Recommended vitamin D intake and management of low vitamin D status in adolescents: A position statement of the Society for Adolescent Health and Medicine, *Journal of Adolescent Health* 52 (2013): 801–803; L. E. Au and coauthors, Associations of vitamin D intake with 25-hydroxyvitamin D in overweight and racially/ethnically diverse US children, *Journal of the Academy of Nutrition and Dietetics* 113 (2013): 1511–1516.

135. The Society for Adolescent Health and Medicine, 2013.

136. Committee on Dietary Reference Intakes, 2011, pp. 35–74.

137. M. Mesias, I. Seiquer, and M. P. Navarro, Calcium nutrition in adolescence, *Critical Reviews in Food Science and Nutrition* 51 (2011): 195–209; Committee on Dietary Reference Intakes, 2011, pp. 458–460; R. L. Bailey and coauthors, Estimation of total usual calcium and vitamin D intakes in the United States, *Journal of Nutrition* 140 (2010): 817–822.

138. D. K. Dror and L. H. Allen, Dairy product intake in children and adolescents in developed countries: Trends, nutritional contribution, and a review of association with health outcomes, *Nutrition Reviews* 72 (2014): 68–81; Mesias, Seiquer, and Navarro, 2011.

139. L. M. Fiorito and coauthors, Girls' early sweetened carbonated beverage intake predicts different patterns of beverage and nutrient intake across childhood and adolescence, *Journal of the American Dietetic Association* 110 (2010): 543–550.

140. A. Moag-Stahlberg, *Family Nutrition and Physical Activity Survey*, January 2011, available at www.eatright.org; G. J. Cutler and coauthors, Multiple sociodemographic and socioenvironmental characteristics are correlated with major patterns of dietary intake in adolescents, *Journal of the American Dietetic Association* 111 (2011): 230–240.

141. J. M. Berge and coauthors, Structural and interpersonal characteristics of family meals: Associations with adolescent body mass index and dietary patterns, *Journal of the Academy of Nutrition and Dietetics* 113 (2013): 816–822; A. J. Hammons and B. H. Fiese, Is frequency of shared family meals related to the nutritional health of children and adolescents? *Pediatrics* 127 (2011): e1565–e1574; D. Neumark-Sztainer and coauthors, Family meals and adolescents: What have we learned from Project EAT (Eating Among Teens)? *Public Health Nutrition* 13 (2010): 1113–1121.

142. Neumark-Sztainer and coauthors, 2010; J. White and E. Halliwell, Alcohol and tobacco use during adolescence: The importance of the family mealtime environment, *Journal of Health Psychology* 15 (2010): 526–532; B. Sen, The relationship between frequency of family dinner and adolescent problem behaviors after adjusting for other family characteristics, *Journal of Adolescence* 33 (2010): 187–196.

143. P. R. Deshmukh-Taskar and coauthors, The relationship of breakfast skipping and type of breakfast consumption with nutrient intake and weight status in children and adolescents: The National Health and Nutrition examination Survey 1999–2006, *Journal of the American Dietetic Association* 110 (2010): 869–878; K. J. Smith and coauthors, Skipping breakfast: Longitudinal associations with cardiometabolic risk factors in the Childhood Determinants of Adult Health Study, *American Journal of Clinical Nutrition* 92 (2010): 1316–1325.

144. H. J. Leidy and coauthors, Beneficial effects of a higher-protein breakfast on the appetitive, hormonal, and neural signals controlling energy intake regulation in overweight/obese, "breakfast-skipping" late-adolescent girls, *American Journal of Clinical Nutrition* 97 (2013): 677–688; L. W. Jackson, The most important meal of the day: Why children skip breakfast and what can be done about it, *Pediatric Annals* 42 (2013): 184–187.

145. J. M. Poti, K. J. Duffey, and B. M. Popkin, The association of fast food consumption with poor dietary outcomes and obesity among children: Is it the fast food or the remainder of the diet? *American Journal of Clinical Nutrition* 99 (2014): 162–171; Smith and coauthors, 2010.

146. D. R. Keast, T. A. Nicklas, and C. E. O'Neil, Snacking is associated with reduced risk of overweight and reduced abdominal obesity in adolescents: National Health and Nutrition Examination survey (NHANES) 1999–2004, *American Journal of Clinical Nutrition* 92 (2010): 428–435.

147. N. D. Brener and coauthors, Beverage consumption among high school students: United States, 2010, *Morbidity and Mortality Weekly Report* 60 (2011): 778–780.

148. E. Han and L. M. Powell, Consumption patterns of sugar-sweetened beverages in the United States, *Journal of the Academy of Nutrition and Dietetics* 113 (2013): 43–53; D. T. Levy, K. B. Friend, and Y. C. Wang, A review of the literature on policies directed at the youth consumption of sugar sweetened beverages, *Advances in Nutrition* 2 (2011): 182S–200S.

149. S. M. Seifert and coauthors, Health effects of energy drinks on children, adolescents, and young adults, *Pediatrics* 127 (2011): 511–528.

150. Poti, Duffey, and Popkin, 2014.

HIGHLIGHT > 16
Childhood Obesity and the Early Development of Chronic Diseases

> **LEARN IT** Describe the lifestyle factors that can help prevent childhood obesity and the development of type 2 diabetes and heart disease.

When people think about the health problems of children and adolescents, they typically think of ear infections, colds, and acne—not heart disease, diabetes, or hypertension. Today, however, unprecedented numbers of US children are being diagnosed with obesity and serious "adult diseases," such as **type 2 diabetes,** that accompany overweight.[1] When type 2 diabetes develops before the age of 20, the incidence of diabetic kidney disease and death in middle age increases dramatically, largely because of the long duration of the disease. For children born in the United States in the year 2000, the risk of developing type 2 diabetes sometime in their lives is estimated to be 30 percent for boys and 40 percent for girls. US children are not alone—rapidly rising rates of obesity threaten the health of an alarming number of children around the globe.[2] Without immediate intervention, millions of children are destined to develop type 2 diabetes and hypertension in childhood followed by **cardiovascular disease (CVD)** in early adulthood.[3] (See the accompanying glossary for this and related terms.)

This highlight focuses on efforts to prevent childhood obesity and the development of heart disease and type 2 diabetes, but the benefits extend to other obesity-related diseases as well. The years of childhood (ages 2 to 18) are emphasized here because the earlier in life health-promoting habits become established, the better they will stick. Chapter 18 fills in the rest of the story of nutrition's role in reducing chronic disease risk.

Invariably, questions arise as to what extent genetics is involved in disease development. For heart disease and type 2 diabetes, genetics does not appear to play a *determining* role; that is, a person is not simply destined at birth to develop these diseases. Instead, genetics appears to play a *permissive* role—the potential is inherited and will develop if given a push by poor health choices such as excessive weight gain, poor diet, sedentary lifestyle, and cigarette smoking.[4]

Many experts agree that preventing or treating obesity in childhood will reduce the rate of chronic diseases in adulthood. Without

© Zhudifeng/Dreamstime.com

intervention, most obese children become obese adolescents who become obese adults, and being obese exacerbates every chronic disease that adults face.[5] Fatty liver, a condition that correlates directly with BMI, was not even recognized in pediatric research until recently. Today, fatty liver disease has a high prevalence in obese children.[6]

Early Development of Type 2 Diabetes

In recent years, type 2 diabetes, a chronic disease closely linked with obesity, has been on the rise among children and adolescents as the prevalence of obesity in US youth has increased.[7] Obesity is the most important risk factor for type 2 diabetes; it is most likely to occur in those who are obese and sedentary and have a family history of

GLOSSARY

atherosclerosis (ATH-er-OH-scler-OH-sis): a type of artery disease characterized by plaques (accumulations of lipid-containing material) on the inner walls of the arteries (see Chapter 18).

- **athero** = porridge or soft
- **scleros** = hard
- **osis** = condition

cardiovascular disease (CVD): diseases of the heart and blood vessels throughout the body. Atherosclerosis is the main cause of CVD. When the arteries that carry blood to the heart muscle become blocked, the heart suffers damage known as *coronary heart disease (CHD).*

- **cardio** = heart
- **vascular** = blood vessels

fatty streaks: accumulations of cholesterol and other lipids along the walls of the arteries.

plaque (PLACK): an accumulation of fatty deposits, smooth muscle cells, and fibrous connective tissue that develops in the artery walls in atherosclerosis. Plaque associated with atherosclerosis is known as *atheromatous* (ATH-er-OH-ma-tus) *plaque.*

type 2 diabetes: the more common type of diabetes in which the cells fail to respond to insulin. Type 2 diabetes usually accompanies obesity and results from insulin resistance coupled with insufficient insulin secretion.

diabetes. Diagnoses typically are made during puberty, but as younger children become more obese and less active, the trend is shifting to younger ages.

In type 2 diabetes, the cells become insulin-resistant—that is, the cells become less sensitive to insulin, reducing the amount of glucose entering the cells from the blood. The combination of obesity and insulin resistance produces a cluster of symptoms, including high blood cholesterol and high blood pressure, which, in turn, promotes the development of atherosclerosis and the early development of heart disease.[8] Other common problems evident by early adulthood include kidney disease, blindness, and miscarriages. The complications of diabetes, especially when encountered at a young age, can shorten life expectancy.

Prevention and treatment of type 2 diabetes depend on weight management, which can be particularly difficult in a child's world of food advertising, video games, and pocket money for candy bars. The activity and dietary suggestions to help defend against heart disease later in this highlight apply to type 2 diabetes as well.

Early Development of Heart Disease

Most people consider heart disease to be an adult disease because its incidence rises with advancing age, and symptoms rarely appear before age 30. The disease process actually begins much earlier.

Atherosclerosis

Most cardiovascular disease involves **atherosclerosis**. Atherosclerosis develops when regions of an artery's walls become progressively thickened with **plaque**—an accumulation of fatty deposits, smooth muscle cells, and fibrous connective tissue. If it progresses, atherosclerosis may eventually block the flow of blood to the heart and cause a heart attack or cut off blood flow to the brain and cause a stroke. Infants are born with healthy, smooth, clear arteries, but within the first decade of life, **fatty streaks** may begin to appear (see Figure H16-1). During adolescence, these fatty streaks may begin to accumulate fibrous connective tissue. By early adulthood, the fibrous plaques may begin to calcify and become raised lesions, especially in boys and young men. As the lesions grow more numerous and enlarge, the heart disease rate begins to rise, most dramatically at about age 45 in men and 55 in women. From this point on, arterial damage and blockage progress rapidly, and heart attacks and strokes threaten life. In short, the consequences of atherosclerosis, which become apparent only in adulthood, have their beginnings in the first decades of life.[9]

Atherosclerosis is not inevitable; people can grow old with relatively clear arteries. Early lesions may either progress or regress, depending on several factors, many of which reflect lifestyle behaviors. Smoking, for example, is strongly associated with the prevalence of fatty streaks and raised lesions, even in young adults.

Blood Cholesterol

As blood cholesterol rises, atherosclerosis worsens. Cholesterol values at birth are similar in all populations; differences emerge in early childhood. Standard values for cholesterol in children and adolescents (ages 2 to 18 years) are listed in Table H16-1 on p. 550. Cholesterol concentrations change with age in children and adolescents, however, and are especially variable during puberty.[10] Thus, use of a single cutoff point for all pediatric age groups has limitations.

In general, blood cholesterol tends to rise as dietary saturated fat intakes increase. Blood cholesterol also correlates with childhood obesity, especially abdominal obesity; LDL cholesterol rises and HDL declines. These relationships are apparent throughout childhood, and their magnitude increases with age.

> FIGURE H16-1   **The Formation of Plaques in Atherosclerosis**

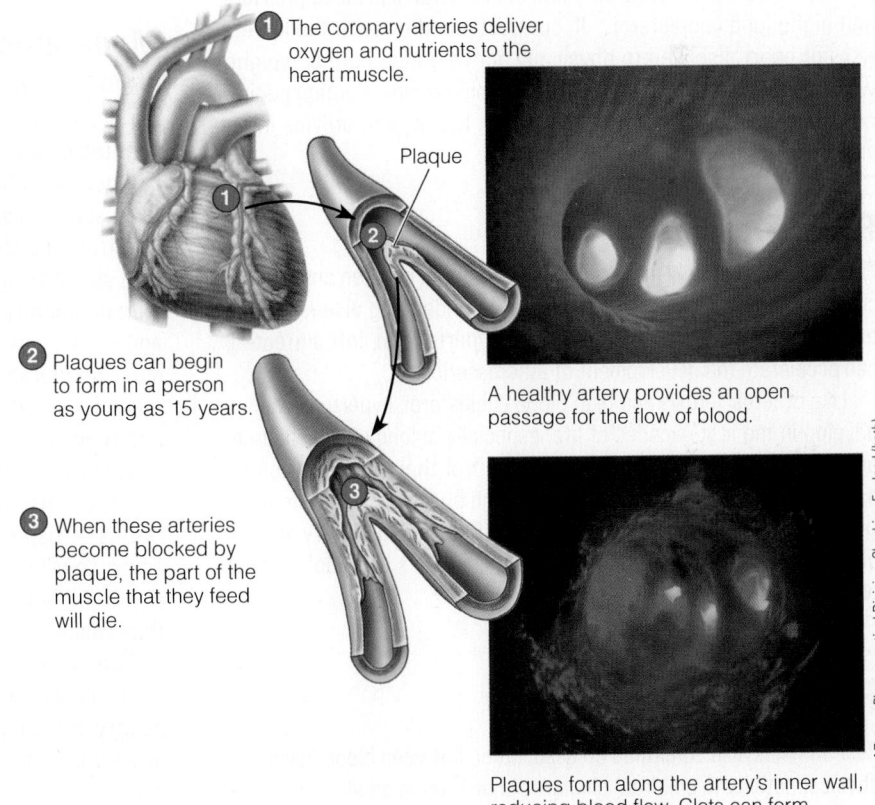

1 The coronary arteries deliver oxygen and nutrients to the heart muscle.

Plaque

2 Plaques can begin to form in a person as young as 15 years.

3 When these arteries become blocked by plaque, the part of the muscle that they feed will die.

A healthy artery provides an open passage for the flow of blood.

Plaques form along the artery's inner wall, reducing blood flow. Clots can form, aggravating the problem.

© Courtesy of Zeneca Pharmaceutical Division, Cheshire, England (both)

TABLE H16-1 Cholesterol Values for Children and Adolescents

Disease Risk	Total Cholesterol (mg/dL)	LDL Cholesterol (mg/dL)
Acceptable	<170	<110
Borderline	170–199	110–129
High	≥200	≥130

NOTE: Adult values appear in Chapter 18.

© Cengage Learning

Children who are both overweight or obese and have high blood cholesterol are likely to have parents who develop heart disease early. For this reason, selective screening is recommended for children and adolescents of any age who are overweight or obese; those whose parents (or grandparents) have premature heart disease (≤55 years of age for men and ≤65 years of age for women); those whose parents have elevated blood cholesterol; those who have other risk factors for heart disease such as hypertension, cigarette smoking, or diabetes; and those whose family history is unavailable. Because blood cholesterol in children is a good predictor of adult values, some experts recommend universal screening for all children aged 9 to 11.[11]

Early—but not advanced—atherosclerotic lesions are reversible, making screening and education a high priority. Both those with family histories of heart disease and those with multiple risk factors need intervention. Children with the highest risks of developing heart disease are sedentary and obese, with high blood pressure and high blood cholesterol.[12] In contrast, children with the lowest risks of heart disease are physically active and of normal weight, with low blood pressure and favorable lipid profiles. Routine pediatric care should identify these known risk factors and provide intervention when needed.

Blood Pressure

Pediatricians routinely monitor blood pressure in children and adolescents. High blood pressure may signal an underlying disease or the early onset of hypertension. Childhood hypertension, left untreated, can accelerate the development of atheroscerlosis.

Like atherosclerosis and high blood cholesterol, hypertension may develop in the first decades of life, especially among obese children, and worsen with time. Children can control their hypertension by participating in regular aerobic activity and by losing weight or maintaining their weight as they grow taller. Restricting dietary sodium also causes an immediate drop in most children's and adolescents' blood pressure.[13]

Physical Activity

Research has also confirmed an association between blood lipids and physical activity in children, similar to that seen in adults. Physically active children have a better lipid profile and lower blood pressure than physically inactive children, and these positive findings often persist into adulthood.[14] The *Physical Activity Guidelines for Americans* recommendations for children and adolescents are listed in Table 16-10 on p. 533.

Just as blood cholesterol and obesity track over the years, so does a child's level of physical activity. Those who are inactive now are likely to still be inactive years later. Similarly, those who are physically active now tend to remain so. Compared with inactive teens, those who are physically active weigh less, smoke less, eat a diet lower in saturated fats, and have better blood lipid profiles. Both obesity and blood cholesterol correlate with the inactive pastime of watching television. The message is clear: physical activity offers numerous health benefits, and children who are active today are most likely to be active for years to come.

Dietary Recommendations for Children

Regardless of family history, experts agree that all children older than age 2 should eat a variety of foods and maintain desirable weight (see Table H16-2). For heart health, children (2 to 18 years of age) should receive at least 25 percent and no more than 30 percent of total energy from fat, less than 10 percent from saturated fat, and less than 300 milligrams of cholesterol per day.[15]

Moderation, Not Deprivation

Healthy children older than age 2 can begin the transition to eating according to recommendations by selecting more fruits and vegetables and fewer foods high in saturated fat. Healthy meals can occasionally include moderate amounts of a child's favorite food, such as ice cream, even if it is high in saturated fat. A steady diet from the children's menus in some restaurants—which feature chicken nuggets, hot dogs, and french fries—easily exceeds a prudent intake of saturated fat, *trans* fat, and kcalories, however, and invites both nutrient shortages and weight gains.[16] Fortunately, most restaurant chains are changing children's menus to include steamed vegetables, fruit cups, and broiled or grilled chicken—additions welcomed by busy parents who often dine out or purchase take-out foods.

Other fatty foods, such as nuts, vegetable oils, and some varieties of fish such as tuna or salmon, contribute essential fatty acids. Low-fat milk and milk products also deserve special attention in a child's diet for the needed calcium and other nutrients they supply.

Parents and caregivers play a key role in helping children establish healthy eating habits. Balanced meals need to provide lean meat, poultry, fish, and legumes; fruits and vegetables; whole grains; and low-fat milk products. Such meals can provide enough energy and nutrients to support growth and maintain blood cholesterol within a healthy range.

TABLE H16-2 American Heart Association Dietary Guidelines and Strategies for Children

- Balance dietary kcalories with physical activity to maintain normal growth.
- Every day, engage in 60 minutes of moderate to vigorous play or physical activity.
- Eat vegetables and fruits daily. Serve fresh, frozen, or canned vegetables and fruits at every meal; limit those with added fats, salt, and sugar.
- Limit juice intake (4 ounces per day for children 2 to 10 years of age, 4 to 6 ounces for children 11 to 18 years of age).
- Use vegetable oils (canola, soybean, olive, safflower, or other unsaturated oils) and soft margarines low in saturated fat and *trans*-fatty acids instead of butter or most other animal fats in the diet.
- Choose whole-grain breads and cereals rather than refined products; read labels and make sure that "whole grain" is the first ingredient.
- Limit or avoid the intake of sugar-sweetened beverages; encourage water.
- Consume low-fat and non-fat milk and milk products daily.
- Include two servings of fish per week, especially fatty fish such as broiled or baked salmon.
- Choose legumes and tofu in place of meat for some meals.
- Choose only lean cuts of meat and reduced-fat meat products; remove the skin from poultry.
- Use less salt, including salt from processed foods. Breads, breakfast cereals, and soups may be high in salt and/or sugar, so read food labels and choose high-fiber, low-salt, low-sugar alternatives.
- Limit the intake of high-kcalorie add-ons such as gravy, Alfredo sauce, cream sauce, cheese sauce, and hollandaise sauce.
- Serve age-appropriate portion sizes on appropriately sized plates and bowls.

© Cengage Learning

NOTE: These guidelines are for children 2 years of age and older.
SOURCE: Adapted from Expert Panel on Integrated Guidelines for Cardiovascular Health and Risk Reduction in Children and Adolescents, Summary report, *Pediatrics* 128 (2011): S213–S256; American Heart Association, Samuel S. Gidding, and coauthors, Dietary recommendations for children and adolescents: A guide for practitioners, *Pediatrics* 117 (2006): 544–559.

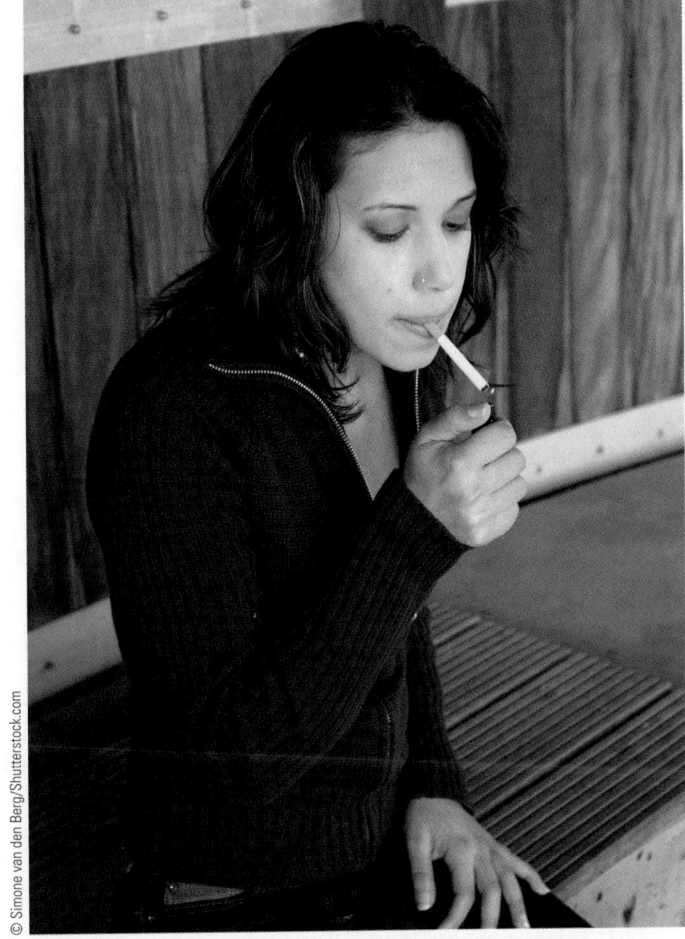

© Simone van den Berg/Shutterstock.com

Cigarette smoking is the number one preventable cause of deaths.

Pediatricians warn parents to avoid extremes. Although intentions may be good, excessive food restriction may create nutrient deficiencies and impair growth. Furthermore, parental control over eating may instigate battles and foster attitudes about foods that can lead to inappropriate eating behaviors.

Diet First, Drugs Later

Experts agree that children with high blood cholesterol should first be treated with diet. If high blood cholesterol persists despite dietary intervention in children 10 years of age and older, then drugs may be necessary to lower blood cholesterol. Drugs can effectively lower blood cholesterol without interfering with adolescent growth or development.

Smoking

Even though the focus of this text is nutrition, another risk factor for heart disease that starts in childhood and carries over into adulthood must also be addressed—cigarette smoking. Each day nearly 4000 young people between the ages of 12 and 17 light up for the first time, and an estimated 1000 become daily cigarette smokers.[17] Among high school students, one in seven smokes regularly.[18] Approximately 90 percent of all adult smokers began smoking before the age of 18.

Of those teenagers who continue smoking, half will eventually die of smoking-related causes. Efforts to teach children about the dangers of smoking need to be aggressive. Children are not likely to consider the long-term health consequences of tobacco use. They are more likely to be struck by the immediate health consequences, such as shortness of breath when playing sports, or social consequences, such as having bad breath. Whatever the context, the message to all children and teens should be clear: don't start smoking. If you've already started, quit.

In conclusion, *adult* heart disease is a major *pediatric* problem. Without intervention, some 60 million children are destined to suffer its consequences within the next 30 years. Optimal prevention efforts focus on children, especially on those who are obese.

Just as young children receive vaccinations against infectious diseases, they need screening for, and education about, chronic diseases. Many health education programs have been implemented in schools around the country. These programs are most effective when they include education in the classroom, heart-healthy meals in the lunch room, fitness activities on the playground, and parental involvement at home.[19]

CRITICAL THINKING QUESTIONS

A. How does childhood obesity influence a person's health and a country's health care system?
B. Child abuse is defined as either action or neglect that damages a child or puts the child at risk of injury. If found guilty of child abuse, parents can lose their parental rights and face criminal charges. Recently, the mother of a 14-year-old was arrested and charged with criminal neglect because her son weighed more than 500 pounds. Is severe childhood obesity a life-threatening form of abuse that justifies removing a child from his or her parents?

REFERENCES

1. C. L. Ogden, M. D. Carroll, and K. M. Flegal, Prevalence of childhood and adult obesity in the United States, 2011–2012, *Journal of the American Medical Association* 311 (2014): 806–814; Position of the Academy of Nutrition and Dietetics: The role of nutrition in health promotion and chronic disease prevention, *Journal of the Academy of Nutrition and Dietetics* 113 (2013): 972–979S; Saydah and coauthors, Cardiometabolic risk factors among US adolescents and young adults and risk of early mortality, *Pediatrics* 131 (2013): e679–e686; J. P. Reiss and coauthors, Association between duration of overall and abdominal obesity beginning in young adulthood and coronary artery calcification in middle age, *Journal of the American Medical Association* 310 (2013): 280–288; A. Tirosh and coauthors, Adolescent BMI trajectory and risk of diabetes versus coronary disease, *New England Journal of Medicine* 364 (2011): 1315–1325; J. C. Han, D. A. Lawlor, and S.Y.S. Kimm, Childhood obesity, *Lancet* 375 (2010): 1737–1748; F. M. Biro and M. Wien, Childhood obesity and adult morbidities, *American Journal of Clinical Nutrition* 91 (2010): 1499S–1505S; P. W. Franks and coauthors, Childhood obesity, other cardiovascular risk factors, and premature death, *New England Journal of Medicine* 362 (2010): 485–493.
2. M. de Onis and coauthors, Global prevalence and trends of overweight and obesity among preschool children, *American Journal of Clinical Nutrition* 92 (2010): 1257–1264; B. M. Popkin, Recent dynamics suggest selected countries catching up to US obesity, *American Journal of Clinical Nutrition* 91 (2010): 284S–288S.
3. T. H. Inge and coauthors, the effect of obesity in adolescence on adult health status, *Pediatrics* 132 (2013): 1098–1104; A. L. May, E. V. Kuklina, and P. W. Yoon, Prevalence of cardiovascular disease risk factors among US adolescents, 1999–2008, *Pediatrics* 129 (2012): 1035–1041; C. Friedemann and coauthors, Cardiovascular disease risk in healthy children and its association with body mass index: Systematic review and meta-analysis, *British Medical Journal* 345 (2012): e4759; Expert Panel on Integrated Guidelines for Cardiovascular Health and Risk Reduction in Children and Adolescents, Summary report, *Pediatrics* 128 (2011): S213–S256; Tirosh and coauthors, 2011; P. W. Franks and coauthors, Childhood obesity, other cardiovascular risk factors, and premature death, *New England Journal of Medicine* 362 (2010): 485–493.
4. R. N. H. Touwslager and coauthors, Genetic and environmental factors in associations between infant growth and adult cardiometabolic risk profile in twins, *American Journal of Clinical Nutrition* 98 (2013): 994–1001;

L. Brazionis and coauthors, Diet spanning infancy and toddlerhood is associated with child blood pressure at age 7.5 y, *American Journal of Clinical Nutrition* 97 (2013): 1375–1386; D. K. Arnett and S. A. Claas, Preventing and controlling hypertension in the era of genomic innovation and environmental transformation, *Journal of the American Medical Association* 308 (2012): 1745–1746; M. Manco and B. Dallapiccola, Genetics of pediatric obesity, *Pediatrics* 130 (2012): 123–133; C. H. Llewellyn and coauthors, Inherited behavioral susceptibility to adiposity in infancy: A multivariate genetic analysis of appetite and weight in the Gemini birth cohort, *American Journal of Clinical Nutrition* 95 (2012): 633–639.
5. J. P. Reiss and coauthors, Association between duration of overall and abdominal obesity beginning in young adulthood and coronary artery calcification in middle age, *Journal of the American Medical Association* 310 (2013): 280–288; M. Juonaloa and coauthors, Childhood adiposity, adult adiposity, and cardiovascular risk factors, *New England Journal of Medicine* 365 (2011): 1876–1885; P. W. Franks and coauthors, Childhood obesity, other cardiovascular risk factors, and premature death, *New England Journal of Medicine* 362 (2010): 485–493.
6. B. G. Koot and coauthors, Lifestyle intervention for non-alcoholic fatty liver disease: Prospective cohort study of its efficacy and factors related to improvement, *Archives of Disease in Childhood* 96 (2011): 669–674.
7. M. Juonala and coauthors, Childhood adiposity, adult adiposity, and cardiovascular risk factors, *New England Journal of Medicine* 365 (2011): 1876–1885; Biro and Wien, 2010.
8. Biro and Wien, 2010.
9. Friedemann and coauthors, 2012; Expert Panel on Integrated Guidelines for Cardiovascular Health and Risk Reduction in Children and Adolescents, 2011; G. Raghuveer, Lifetime cardiovascular risk of childhood obesity, *American Journal of Clinical Nutrition* 91 (2010): 1514S–1519S; A. C. Skinner and coauthors, Multiple markers of inflammation and weight status: Cross-sectional analyses throughout childhood, *Pediatrics* 125 (2010): e801–e809.
10. Expert Panel on Integrated Guidelines for Cardiovascular Health and Risk Reduction in Children and Adolescents, 2011.
11. Expert Panel on Integrated Guidelines for Cardiovascular Health and Risk Reduction in Children and Adolescents, 2011.
12. Centers for Disease Control and Prevention, Prevalence of abnormal lipid levels among youths: United States, 1999–2006, *Morbidity and Mortality Weekly Report* 59 (2010): 29–33.

13. Q. Yang and coauthors, Sodium intake and blood pressure among US children and adolescents, *Pediatrics* 130 (2012): 611–619S; Stabouli and coauthors, The role of obesity, salt and exercise on blood pressure in children and adolescents, *Expert Review of Cardiovascular Therapy* 9 (2011): 753–761; J. Feber and M. Ahmed, Hypertension in children: New trends and challenges, *Clinical Science* 119 (2010): 151–161.

14. U. Ekelund and coauthors, Moderate to vigorous physical activity and sedentary time and cardiometabolic risk factors in children and adolescents, *Journal of the American Medical Association* 15 (2012): 704–712.

15. Expert Panel on Integrated Guidelines for Cardiovascular Health and Risk Reduction in Children and Adolescents, 2011.

16. K. N. Boutelle and coauthors, Nutritional quality of lunch meal purchased for children at a fast-food restaurant, *Childhood Obesity* 7 (2011): 316–322.

17. Centers for Disease Control and Prevention, Youth and tobacco use, www.cdc.gov/tobacco/data_statistics/fact_sheets/youth_data/tobacco_use/index.htm, June 10, 2013.

18. R. A. Arrazola, S. R. Dube; and B. A. King, Tobacco product use among middle and high school students—United States, 2011 and 2012, *Morbidity and Mortality Weekly Report* 62 (2013): 893–897.

19. D. M. Hoelscher and coauthors, Position of the Academy of Nutrition and Dietetics: Interventions for the prevention and treatment of pediatric overweight and obesity, *Journal of the Academy of Nutrition and Dietetics* 113 (2013): 1375–1394.

Life Cycle Nutrition: Adulthood and the Later Years

Nutrition in Your Life

Take a moment to envision yourself at age 60, 75, or even 90. Are you physically fit and healthy? Can you see yourself walking on the beach with friends or tossing a ball with children? Are you able to climb stairs and carry your own groceries? Importantly, are you enjoying life? If you're lucky, you will enjoy old age in good health. Making nutritious foods and physical activities a priority in your life can help bring rewards of continued health and enjoyment throughout life. In the Nutrition Portfolio at the end of this chapter, you can examine the nutritional health and concerns of an older adult.

Much of this text has focused on nutrition to support health, and Chapter 18 features prevention of chronic diseases such as cancer and heart disease. This chapter focuses on aging and the nutrition needs of older adults. As you will see, the same diet and behaviors that reduce disease risks also slow aging.

Our society uses the arbitrary age of 65 years to define the transition point between middle age and old age, but growing "old" happens day by day, with changes occurring gradually over time. Since 1950 the population of those older than 65 has almost tripled. Remarkably, the fastest-growing age group has been people older than 85 years; since 1950 their numbers have increased sevenfold. The number of people in the United States age 100 or older doubled in the last decade. Similar trends are occurring in populations worldwide. Interestingly, there are places around the world where people live long and, importantly, healthy and active lives even after the age of 100 years. These places have been called *Blue Zones,* and include Sardinia, Italy; Okinawa, Japan; Loma Linda, California; Nicoya, Costa Rica; and Ikaria Island, Greece. The "secrets" of these peoples' **longevity** are presented on p. 557.

longevity: long duration of life.

17.1 Nutrition and Longevity

> **LEARN IT** Describe the role nutrition plays in longevity.

Figure 17-1 shows how the US population is growing older. The majority is now middle-aged, and the ratio of old people to young is increasing. In 1900, only 1 out of 25 people was 65 or older. In 2000, 1 out of 8 had reached age 65. Projections for 2030 are 1 out of 5. **Life expectancy** in the United States is 79 years: 81 years for white women and 78 years for black women, 77 years for white men and 72 years for black men.[1] All of these record highs are much higher than the average life expectancy of 47 years in 1900. Women who live to 75 can expect to survive an additional 13 years, on average; men, an additional 11 years.[2] Advances in medical science—antibiotics and other treatments—are largely responsible for almost doubling the life expectancy in the 20th century. Improved nutrition and an abundant food supply have also contributed to lengthening life expectancy. Ironically, an abundant food supply has also jeopardized the chances of lengthening life expectancy as obesity rates increase.

The **life span** has not lengthened as dramatically; human longevity appears to have an upper limit. The maximum potential human life span is currently about 130 years. The verifiably oldest person died in 1997 at age 122. With recent advances in medical technology and genetic knowledge, researchers may one day be able to extend the life span even further by slowing, or perhaps preventing, aging and its accompanying diseases.

Research in the field of aging is active—and difficult. Researchers are challenged by the diversity of older adults. When older adults experience health problems, it is hard to know whether to attribute these problems to genetics, aging, or environmental factors such as nutrition. The idea that nutrition can influence the

> **FIGURE 17-1** **The Aging of the US Population**

In general, the percentage of older people in the population has increased over the decades whereas the percentage of younger people has decreased.

Key:
- ≥65 years
- 45–64 years
- 25–44 years
- 15–24 years
- <15 years

	1900	1910	1920	1930	1940	1950	1960	1970	1980	1990	2000	2010
≥65	4.1	4.3	4.7	5.4	6.8	8.1	9.2	9.9	11.3	12.6	12.4	13
45–64	13.7	14.6	16.1	17.5	19.8	20.3	20.1	20.6	19.6	18.6	22.0	26.4
25–44	28.1	29.2	29.6	29.5	30.1	30.0	26.2	23.6	27.7	32.5	30.2	26.6
15–24	19.6	19.7	17.7	18.3	18.2	14.7	13.4	17.4	18.8	14.8	13.9	9.9
<15	34.5	32.1	31.8	29.4	25.0	26.9	31.1	28.5	22.6	21.5	21.4	24.0

NOTE: Data for 2010 split age groups slightly differently. Blue represents 18–24 years and purple represents <18 years.
SOURCE: US Census Bureau.

© Cengage Learning

life expectancy: the average number of years lived by people in a given society.

life span: the maximum number of years of life attainable by a member of a species.

aging process is particularly appealing because people can control and change their eating habits. The questions being asked include:

- To what extent is aging inevitable, and can it be slowed through changes in lifestyle and environment?
- What role does nutrition play in the aging process, and what role can it play in slowing aging?

With respect to the first question, it seems that aging is an inevitable, natural process, programmed into the genes at conception. People can, however, slow the process within genetic limits by adopting healthy lifestyle habits such as eating nutritious foods and engaging in physical activities. In fact, a person's life expectancy depends on both individual health-related behaviors and genes.[3]

With respect to the second question, good nutrition helps to maintain a healthy body and can therefore ease the aging process in many significant ways. Clearly, nutrition can improve the **quality of life** in the later years.

Observation of Older Adults   The strategies adults use to meet two goals that motivate them to make changes—promoting health and slowing aging—are actually very much the same. What to eat, how physically active to be, and other lifestyle choices greatly influence both physical health and the aging process.

Healthy Habits   A person's **physiological age** reflects his or her health status and may or may not reflect the person's **chronological age**. Quite simply, some people seem younger, and others older, than their years. Some behaviors seem to have the greatest influence on people's health and therefore on their physiological age:[4]

- Following a healthy plant-based eating pattern, such as the Mediterranean diet (rich in fruits, vegetables, whole grains, poultry, fish, and low fat milk products)[5]
- Engaging in moderate physical activity daily
- Not smoking
- Not using alcohol, or using it in moderation
- Maintaining a healthy body weight
- Sleeping regularly and adequately
- Having a sense of purpose
- Relieving stress (through meditation, prayer, naps, or other calming activity)
- Belonging to a community of loving family and friends (home, church, or other social networks)

Over the years, the effects of these lifestyle choices accumulate—that is, people who follow most of these practices live longer and have a better quality of life as they age.[6] They are in better health, even when older in chronological age, than people who do not adopt these behaviors. Even though people cannot change their birth dates, they may be able to add years to, and enhance the quality of, their lives. Physical activity seems to be most influential in preventing or slowing the many changes that define a stereotypical "old" person. After all, many of the physical limitations that accompany aging occur because people become inactive, not because they become older.

Physical Activity   The many remarkable benefits of regular physical activity outlined in Chapter 14 are not limited to the young. Compared with those who are inactive, older adults who are active weigh less; have greater flexibility, more endurance, better balance, and better health; and they live longer. Perhaps most importantly, they enjoy better overall health.[7] They reap additional benefits from various activities as well: aerobic activities improve cardiorespiratory endurance, blood pressure, and blood lipid concentrations; moderate-endurance activities improve the quality of sleep; and strength training improves posture and mobility. In

quality of life: a person's perceived physical and mental well-being.

physiological age: a person's age as estimated from her or his body's health and probable life expectancy.

chronological age: a person's age in years from his or her date of birth.

fact, regular physical activity is the most powerful predictor of a person's mobility in the later years. Mobility, in turn, is closely associated with longevity.[8] Physical activity also increases blood flow to the brain, thereby preserving mental ability, alleviating depression, supporting independence, and improving quality of life.[9]

Muscle mass and muscle strength tend to decline with aging, making older people vulnerable to falls and immobility. Falls are a major cause of fear, injury, disability, and even death among older adults. Many lose their independence as a result of falls. Regular physical activity tones, firms, and strengthens muscles, helping to improve balance, restore confidence, reduce the risk of falling, and lessen the risk of injury should a fall occur.

Even without a fall, older adults may become so weak that they can no longer perform life's daily tasks, such as climbing stairs, carrying packages, and opening jars. Resistance training helps older adults to maintain independence by improving

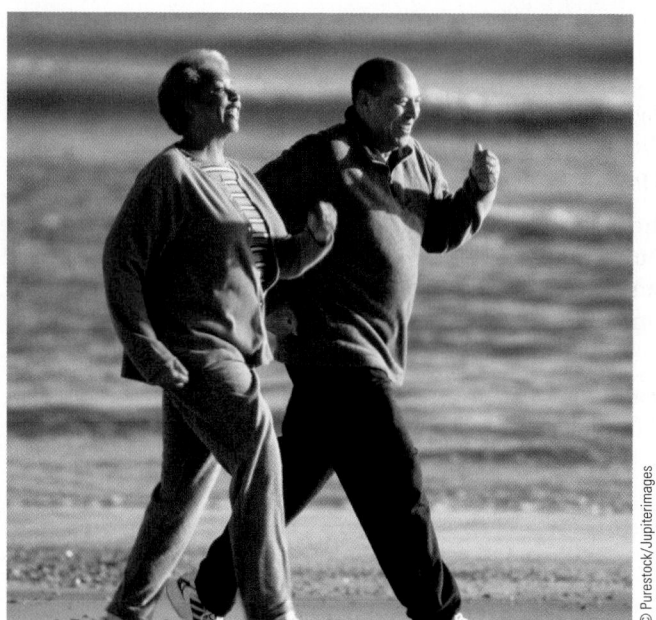

Regular physical activity promotes a healthy, independent lifestyle.

© Purestock/Jupiterimages

mobility and muscle strength to perform these tasks.[10] Even in frail, elderly people older than 85 years of age, strength training not only improves balance, muscle strength, and mobility, but it also increases energy expenditure and energy intake, thereby enhancing nutrient intakes. This finding highlights another reason to be physically active: a person expending energy can afford to eat more food and thus receives more nutrients. People who are committed to an ongoing fitness program can benefit from higher energy and nutrient intakes and still maintain healthy body weights.

Ideally, physical activity should be part of each day's schedule and should be intense enough to prevent muscle atrophy and to speed the heartbeat and respiration rate. Although aging reduces both speed and endurance to some degree, older adults can still train and achieve exceptional performances. Some older adults may enjoy pumping iron and running marathons, but such activities are not essential to good health. Daily activities can be as simple as growing gardens and doing household chores.

Healthy older adults who have not been active can ease into a suitable routine, becoming as physically active as their abilities allow. They can start by walking short distances until they are walking at least 10 minutes continuously, and then gradually increase their distance to a 30- to 60-minute workout at least 5 days a week. Table 17-1 provides exercise goals and guidelines for older adults. Relatively few older adults meet these goals. People with medical conditions should check with a physician before beginning an exercise routine, as should sedentary men older than 40 and sedentary women older than 50 who want to participate in a vigorous program.

> **DIETARY GUIDELINES FOR AMERICANS**
Older adults should be as physically active as their abilities and conditions will allow.

Manipulation of Diet In their efforts to understand longevity, researchers have not only observed people, but they have also manipulated influencing factors, such as diet, in animals. This research has given rise to some interesting and suggestive findings.

Energy Restriction in Animals Decades of research has revealed that animals live longer and have fewer age-related diseases when their energy intakes are restricted.[11] These life-prolonging benefits become evident when the diet provides enough food to prevent malnutrition and an energy intake of about 70 percent of normal; benefits decline as the age of starting the energy restriction is delayed. Exactly how energy restriction prolongs life remains unexplained, although gene activity appears to play a key role. The genetic activity of old mice differs from that

TABLE 17-1 Exercise Guidelines for Older Adults

	Aerobic	Strength	Balance	Flexibility
Examples				
Start easy and progress gradually	Be active 5 minutes on most or all days	Using 0- to 2-pound weights, do 1 set of 8–12 repetitions twice a week	Hold onto table or chair with one hand, then with one finger	Hold stretch for 10 seconds; do each stretch 3 times
Frequency	At least 5 days per week of moderate activity or at least 4 days per week of vigorous activity	At least 2 (nonconsecutive) days per week	2 to 3 days each week	At least 2 days per week; preferably on all days that aerobic or strength activities are performed
Intensity[a]	Moderate, vigorous, or combination	Moderate to high; 10 to 15 repetitions per exercise; gradually increase weights		Moderate
Duration	At least 30 minutes of moderate activity in bouts of at least 10 minutes each or at least 20 minutes of continuous vigorous activity	8 to 10 exercises involving the major muscle groups	At least 20 to 30 minutes	Stretch major muscle groups for 10–30 seconds, repeating each stretch 3 to 4 times
Cautions and comments	Stop if you are breathing so hard you can't talk or if you feel dizziness or chest pain	Breathe out as you contract and in as you relax (do not hold breath); use smooth, steady movements	Incorporate balance techniques with strength exercises as you progress	Stretch after strength and endurance exercises for 20 minutes, 3 times a week; use slow, steady movements; bend joints slightly

[a]On a 10-point scale, where sitting = 0 and maximum effort = 10, moderate intensity = 5 to 6 and vigorous intensity = 7 to 8.

NOTE: Activity recommendations are in addition to routine activities of daily living (such as getting dressed, cooking, grocery shopping) and moderate activities lasting less than 10 minutes.

SOURCE: Centers for Disease Control and Prevention, Division of Nutrition Physical Activity and Obesity, National Center for Chronic Disease prevention and Health Promotion, 2011, www.cdc.gov/physicalactivity/everyone/guidelines/olderadults.html; C. E. Garber and coauthors, Quantity and quality of exercise for developing and maintaining cardiorespiratory, musculoskeletal, and neuromotor fitness in apparently healthy adults: Guidance for prescribing exercise, *Medicine & Science in Sports & Exercise* 43 (2011): 1334–1359.

of young mice, with some genes becoming more active with age and others less active. With an energy-restricted diet, many of the genetic activities of older mice revert to those of younger mice. These "slow-aging" genetic changes are apparent in as little as 1 month on an energy-restricted, but still nutritionally adequate, diet.

The consequences of energy restriction in animals include a delay in the onset, or prevention, of chronic diseases such as cancer and atherosclerosis and age-related conditions such as neuron degeneration; prolonged growth and development; and improved blood glucose, insulin sensitivity, and blood lipids. In addition, energy metabolism slows and body temperature drops—indications of a reduced rate of oxygen consumption. As Highlight 11 explains, the use of oxygen during energy metabolism produces free radicals, which have been implicated in the aging process. Restricting energy intake in animals not only produces fewer free radicals, but also increases antioxidant activity and enhances DNA repair. Reducing oxidative stress may at least partially explain how restricting energy intake lengthens life expectancy.

Interestingly, longevity appears to depend on restricting energy intake and not on energy balance or body composition.[12] Genetically obese rats live longer when given a restricted diet even though their body fat is similar to that of other rats allowed to eat freely.

Energy Restriction in Human Beings Research on a variety of species—including mice, rats, monkeys, spiders, and fish—confirms the relationship between energy restriction and longevity.[13] Applying the results of animal studies to human beings is problematic, however, and conducting studies on human beings raises numerous questions—beginning with how to define *energy restriction*.[14] Does it mean eating less or just weighing less? Is it less than you want or less than the average? Does eating less have to result in weight loss? Does it matter whether weight loss results from more exercise or from less food? Or whether weight loss is intentional or unintentional? Answers await additional research.

Extreme starvation to extend life, like any extreme, is rarely, if ever, worth the price. Hunger is persistent when energy is restricted by 30 percent. Furthermore, using animal data to extrapolate to humans, researchers estimate that it would take 30 years of such energy-restricted dieting to increase life expectancy by less than 3 years.

Moderation, on the other hand, may be valuable. Many of the physiological responses to energy restriction seen in animals also occur in people whose intakes are *moderately* restricted.[15] When people cut back on their usual energy intake by 10 to 20 percent, body weight, body fat, inflammatory proteins, growth factors, and blood pressure drop, and blood lipids and insulin response improve—favorable changes for preventing chronic diseases such as some cancers, type 2 diabetes, hypertension, and heart disease.[16] (For perspective, a person with a usual energy intake of 2000 kcalories might cut back to 1600 to 1800 kcalories.) Some research suggests that fasting on alternative days may provide similar benefits.[17]

The reduction in oxidative damage that occurs with energy restriction in animals also occurs in people whose diets include antioxidant nutrients and phytochemicals. Diets, such as the Mediterranean diet, which include an abundance of fruits, vegetables, olive oil, and moderate amounts of red wine—with their array of phytochemicals that have antioxidant activity—support good health and long life.[18] Clearly, nutritional adequacy is essential to living a long and healthy life.

> **REVIEW IT** Describe the role nutrition plays in longevity.

Life expectancy in the United States increased dramatically in the 20th century. Factors that enhance longevity include well-balanced meals, regular physical activity, abstinence from smoking, limited or no alcohol use, healthy body weight, adequate sleep, and strong social relationships. Energy restriction in animals seems to lengthen their lives. Whether such dietary intervention in human beings is beneficial remains unknown. At the very least, nutrition—especially when combined with regular physical activity—can influence aging and longevity in human beings by supporting good health and preventing disease.

17.2 The Aging Process

> **LEARN IT** Summarize how nutrition interacts with the physical, psychological, economic, and social changes involved in aging.

As people get older, each person becomes less and less like anyone else. The older people are, the more time has elapsed for such factors as nutrition, genetics, physical activity, and everyday **stress** to influence physical and psychological aging.

Stress contributes to a variety of age-related conditions. Both physical **stressors** (such as alcohol abuse, other drug abuse, smoking, pain, and illness) and psychological stressors (such as exams, divorce, moving, and the death of a loved one) elicit the body's **stress response.** The body responds to such stressors with an elaborate series of physiological steps, as the nervous and hormonal systems bring about defensive readiness in every body part. These effects favor physical action—the classic fight-or-flight response. Prolonged or severe stress can drain the body of its reserves and leave it weakened, aged, and vulnerable to illness, especially if physical action is not taken. As people age, they lose their ability to adapt to both external and internal disturbances. When disease strikes, the reduced ability to adapt makes the aging individual more vulnerable to death than a younger person. Strategies to preserve health forestall disease, disability, and death.

Because the stress response is mediated by hormones, it differs between men and women. The fight-or-flight response may be more typical of men than of women. Women's reactions to stress more typically follow a pattern of "tend-and-befriend." Women *tend* by nurturing and protecting themselves, their children,

stress: any threat to a person's well-being; a demand placed on the body to adapt.

stressors: environmental elements, physical or psychological, that cause stress.

stress response: the body's response to stress, mediated by both nerves and hormones.

and other loved ones. These actions promote safety and reduce stress. Women *befriend* by creating and maintaining a social group that can help in the process.

Highlight 11 describes the oxidative stresses and cellular damage that occur when free radicals exceed the body's ability to defend itself. Increased free-radical activity and decreased antioxidant protection are common features of aging—and foods rich in antioxidants may help slow the aging process and improve cognition.[19] Such findings seem to suggest that the fountain of youth may actually be a cornucopia of fruits and vegetables rich in antioxidants. (Return to Highlight 11 for more details on the antioxidant action of fruits and vegetables in defending against oxidative stress.)

Physiological Changes As aging progresses, inevitable changes in each of the body's organ systems contribute to the body's declining function. These physiological changes influence nutrition status, just as growth and development do in the earlier stages of the life cycle.[20]

Body Weight An estimated 35 percent of older adults in the United States are now considered obese. Chapter 8 presents the many health problems that accompany obesity and the BMI guidelines for a healthy body weight (18.5 to 24.9). These guidelines apply to all adults, regardless of age, but they may be too restrictive for older adults. The importance of body weight in defending against chronic diseases differs for older adults.[21] Being *moderately overweight* may not be harmful. For adults older than 65, the lowest mortality correlates with a higher BMI (23.5 to 27.5).[22] Older adults who are *obese*, however, may still face serious medical complications, at least until age 85.[23]

For some older adults, a low body weight may be more detrimental than a high one. Low body weight often reflects malnutrition and the trauma associated with a fall. Many older adults experience unintentional weight loss, in large part because of inadequate food intake. Without adequate body fat and nutrient reserves, an underweight person may be unprepared to fight against diseases. For underweight people, even a slight weight loss (5 percent) increases the likelihood of disease and premature death, making every meal a life-saving event. Drinking liquid nutritional supplements and snacking between meals can help older adults obtain needed nutrients and energy.

> DIETARY GUIDELINES FOR AMERICANS
Maintain appropriate kcalorie balance during older age. Adults aged 65 and older who are overweight are encouraged to not gain additional weight.

Body Composition In general, older people tend to lose bone and muscle and gain body fat. Many of these changes occur because some hormones that regulate appetite and metabolism become less active with age, whereas others become more active.

Loss of muscle, known as **sarcopenia,** can be significant in the later years, and its consequences can be quite dramatic (see Figure 17-2 on p. 562).[24] As muscles diminish and weaken, people lose the ability to move and maintain balance—making falls likely. The limitations that accompany the loss of muscle mass and strength play a key role in the diminishing health that often accompanies aging. Optimal nutrition with sufficient protein at each meal along with regular strength-building physical activity can help maintain muscle mass and strength and minimize the changes in body composition associated with aging.[25] Omega-3 fatty acids may also stimulate muscle protein synthesis in older adults and help prevent sarcopenia.[26]

Immunity and Inflammation As people age, the immune system loses function. As they become ill, the immune system becomes overstimulated. The combination of an inefficient and overactive response in aging—known as "inflammaging"—results in a chronic inflammation that accompanies frailty, illness, and death.

sarcopenia (SAR-koh-PEE-nee-ah): loss of skeletal muscle mass, strength, and quality.
- **sarco** = flesh
- **penia** = loss or lack

> FIGURE 17-2   **Sarcopenia**

These cross sections of two women's thighs may appear to be about the same size from the outside, but the 20-year-old woman's thigh (left) is dense with muscle tissue. The 64-year-old woman's thigh (right) has lost muscle and gained fat, changes that may be largely preventable with good nutrition and strength-building physical activities.

© Courtesy of Dr. William Evans (both)

Most diseases common in older adults—such as atherosclerosis, Alzheimer's disease, obesity, and rheumatoid arthritis—are different in obvious ways, but they all reflect an underlying inflammatory process. Because of this association with diseases, inflammation is often perceived as a harmful process, yet it is critical in supporting health as the immune system destroys invading organisms and repairs damaged tissues. Thus inflammation presents a challenge to identify factors that will both protect the beneficial effects and limit the harmful consequences.

In addition to aging and diseases, the immune system is compromised by nutrient deficiencies. Thus the combination of age, illness, and poor nutrition makes older people particularly vulnerable to infectious diseases. Adding insult to injury, antibiotics often are not effective against infections in people with compromised immune systems. Consequently, infectious diseases are a major cause of death in older adults. Older adults may improve their immune system responses with diets rich in fruits and vegetables and regular physical activity.[27]

GI Tract In the GI tract, numerous changes take place, contributing to poor appetite, early satiety, and malnutrition.[28] The intestinal wall loses strength and elasticity with age, and GI hormone secretions change. All of these actions slow motility. Constipation is much more common in the elderly than in the young. Changes in GI hormone secretions also diminish appetite, leading to decreased energy intake and unintentional weight loss.

Atrophic gastritis, a condition that affects almost one-third of those older than 60, is characterized by an inflamed stomach, bacterial overgrowth, and a lack of hydrochloric acid and intrinsic factor. All of these factors can impair the digestion and absorption of nutrients, most notably, vitamin B_{12}, but also biotin, folate, calcium, iron, and zinc.

Difficulty swallowing, medically known as **dysphagia,** occurs in all age groups, but especially in the elderly. Being unable to swallow a mouthful of food can be scary, painful, and dangerous. Even swallowing liquids can be a problem for some people. Consequently, the person may eat less food and drink fewer beverages, resulting in weight loss, malnutrition, and dehydration. Dietary intervention for dysphagia is highly individualized based on the person's abilities and tolerances. The diet typically provides moist, soft-textured, tender-cooked, or pureed foods and thickened liquids.

Tooth Loss Regular dental care over a lifetime protects against tooth loss and gum disease, which are common in old age. These conditions make chewing difficult or painful. Dentures, even when they fit properly, are less effective than natural teeth, and inefficient chewing can cause choking. Chewing crushes foods into smaller pieces in preparation for digestion. Inefficient chewing leaves larger pieces of food moving from the stomach into the small intestine, thus limiting enzyme accessibility. Simple changes in food texture might improve chewing efficiency and benefit digestion. For example, research suggests that, for older adults, chopped beef is more rapidly digested and absorbed than beef steak, resulting in increased amino acid availability.[29]

People with tooth loss, gum disease, and ill-fitting dentures tend to limit their food selections to soft foods.* If foods such as corn on the cob and apples are replaced by creamed corn and applesauce, then nutrition status may not be greatly affected. However, when food groups are eliminated and variety is limited, poor nutrition follows. People without teeth typically eat fewer fruits and vegetables and have less variety in their diets. Consequently, they have low intakes of fiber and vitamins, which exacerbates their dental and overall health problems.

dysphagia (dis-FAY-jah): difficulty swallowing.

*The medical term for lack of teeth is *edentulous* (ee-DENT-you-lus).

The following conditions help to determine whether a visit to the dentist is needed:

- Dry mouth
- Eating difficulty
- No dental care within 2 years
- Tooth or mouth pain
- Altered food selections
- Lesions, sores, or lumps in mouth

Sensory Losses and Other Physical Problems Sensory losses and other physical problems can also interfere with an older person's ability to obtain adequate nourishment. Failing eyesight, for example, can make driving to the grocery store impossible and shopping for food a frustrating experience. It may become so difficult to read food labels and count money that the person doesn't buy needed foods. Carrying bags of groceries may be an unmanageable task. Similarly, a person with limited mobility may find cooking and cleaning up too hard to do. Not too surprisingly, the prevalence of undernutrition is high among those who are home-bound.

Sensory losses can also interfere with a person's ability or willingness to eat. Taste and smell sensitivities tend to diminish with age and may make eating less enjoyable. If a person eats less, then weight loss and nutrient deficiencies may follow. Loss of vision and hearing may contribute to social isolation, and eating alone may lead to poor intake.

Other Changes

In addition to the physiological changes that accompany aging, adults change in many other ways that influence their nutrition status.[30] Psychological, economic, and social factors play major roles in a person's ability and willingness to eat.

Psychological Changes Late-life depression is associated with an increased risk of mortality.[31] Although not an inevitable component of aging, depression is common among older adults, especially among those in poor health and those living in long-term nursing homes.[32] Relatively few receive adequate treatment from either antidepressant medication or mental health counseling.[33]

Depressed people, even those without physical disabilities, lose their ability to perform simple physical tasks. They frequently lose their appetite and the motivation to cook or even to eat. An overwhelming sense of grief and sadness at the death of a spouse, friend, or family member may leave a person, especially an elderly person, feeling powerless to overcome depression. When a person is suffering the heartache and loneliness of bereavement, cooking meals may not seem worthwhile. The support and companionship of family and friends, especially at mealtimes, can help overcome depression and enhance appetite.

Several nutrient interventions to relieve depression have been studied, but evidence of effectiveness is inconclusive. A balanced, healthy diet may be the best nutritional approach to reducing symptoms of depression and improving quality of life.[34]

Economic Changes Overall, older adults today have higher incomes than their cohorts of previous generations. Still, 9 percent of the people older than age 65 live in poverty.[35] Factors such as living arrangements and income make significant differences in the food choices, eating habits, and nutrition status of older adults, especially those older

Shared meals can brighten the day and enhance the appetite.

than age 80. People of low socioeconomic means are likely to have inadequate food and nutrient intakes. Only about one-third of eligible seniors participate in the Supplemental Nutrition Assistance Program (SNAP).

Social Changes Malnutrition is most likely to occur among those living alone, especially men; those with the least education; those living in federally funded housing (an indicator of low income); and those who have recently experienced a change in lifestyle. Adults who live alone do not necessarily make poor food choices, but they often consume too little food. Loneliness is directly related to nutritional inadequacies, especially of energy intake. Feeling lonely is also associated with a decline in activities of daily living and mobility as well as an increased risk of death.[36]

> **REVIEW IT** Summarize how nutrition interacts with the physical, psychological, economic, and social changes involved in aging.
Many changes that accompany aging can impair nutrition status. Among physiological changes, hormone activity alters body composition, immune system changes raise the risk of infections, atrophic gastritis interferes with digestion and absorption, and tooth loss limits food choices. Psychological changes such as depression, economic changes such as loss of income, and social changes such as loneliness contribute to poor food intake.

17.3 Energy and Nutrient Needs of Older Adults

> **LEARN IT** Explain why the needs for some nutrients increase or decrease during aging.

Knowledge about the nutrient needs and nutrition status of older adults has grown considerably in recent years. The Dietary Reference Intakes (DRI) cluster people older than 50 into two age categories—one group of 51 to 70 years and one of 71 and older.

Setting standards for older people is difficult because individual differences become more pronounced as people grow older. People start out with different genetic predispositions and ways of handling nutrients, and the effects of these differences become magnified with years of unique dietary habits. For example, one person may tend to omit fruits and vegetables from his diet, and by the time he is old, he may have a set of nutrition problems associated with a lack of fiber and antioxidants. Another person may have omitted milk and milk products all her life—her nutrition problems may be related to a lack of calcium. Also, as people age, they suffer different chronic diseases and take various medicines—both of which will affect nutrient needs. For all of these reasons, researchers have difficulty even defining *healthy aging,* a prerequisite to developing recommendations to meet the "needs of practically all healthy persons." The following discussion gives special attention to the nutrients of greatest concern.

Water Despite real fluid needs, many older people do not seem to feel thirsty or notice mouth dryness. Many nursing home employees say it is hard to persuade their elderly clients to drink enough water and fruit juices. Older adults may find it difficult and bothersome to get a drink or to get to a bathroom. Those who have lost bladder control may be afraid to drink too much water.

Dehydration is a risk for older adults. Total body water decreases as people age, so even mild stresses such as fever or hot weather can precipitate rapid dehydration in older adults. Dehydrated older adults seem to be more susceptible to urinary tract infections, pneumonia, **pressure ulcers,** and confusion and disorientation. To prevent dehydration, older adults need to drink *at least* six glasses of water or other beverages every day. Emphasizing foods with high-water content, such as melons and soups, can also be helpful.

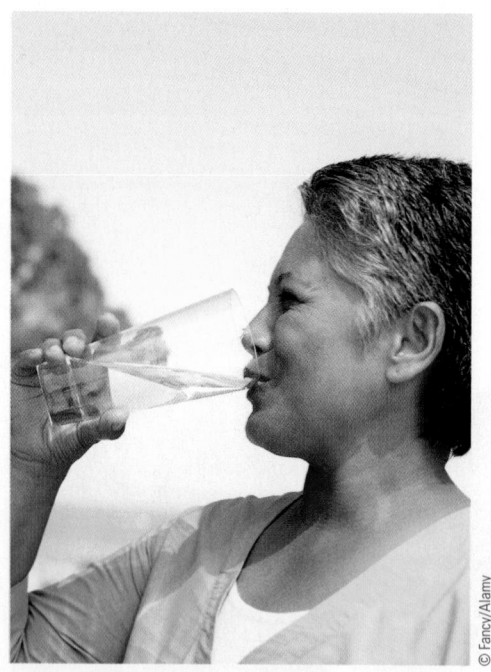
© Fancy/Alamy

To ensure adequate hydration, keep a glass of water next to you at home, drink from water fountains whenever you walk by, and put a bottle of water in your car.

pressure ulcers: damage to the skin and underlying tissues as a result of compression and poor circulation; commonly seen in people who are bedridden or chair-bound.

Energy and Energy Nutrients On average, energy needs decline an estimated 5 percent per decade. One reason is that people usually reduce their physical activity as they age, although they need not do so. Another reason is that basal metabolic rate declines 1 to 2 percent per decade in part because lean body mass and thyroid hormones diminish.

The lower energy expenditure of older adults means that they need to eat less food to maintain their weight. Accordingly, the estimated energy requirements for adults decrease steadily after age 19.

Older adults need fewer kcalories as they age, but their nutrient needs remain high. For this reason, it is most important that they select mostly nutrient-dense foods. There is little leeway for added sugars, solid fats, or alcohol; such nutrient-poor selections can easily lead to weight gain and malnutrition. The USDA Food Patterns (Table 2-3 on p. 43) offer a dietary framework for adults of all ages.

Growing old can be enjoyable for people who take care of their health and live each day fully.

Protein Because energy needs decrease, protein must be obtained from low-kcalorie sources of high-quality protein, such as lean meats, poultry, fish, and eggs; fat-free and low-fat milk products; and legumes. Protein is especially important for the elderly to support a healthy immune system, prevent muscle wasting, and optimize bone mass. Maintaining muscles helps to support protein metabolism and immune function.

Underweight or undernourished older adults need protein- and energy-dense snacks such as hard-boiled eggs, tuna salad, peanut butter on wheat toast, and hearty soups. Drinking liquid nutritional supplements between meals can also boost energy and nutrient intakes. Importantly, the diet should provide enjoyment as well as nutrients.[37]

Carbohydrate and Fiber As always, abundant carbohydrate is needed to protect protein from being used as an energy source. Carbohydrate-rich foods such as legumes, vegetables, whole grains, and fruits are also rich in fiber and essential vitamins and minerals. Average fiber intakes among older adults are lower than current recommendations (14 grams per 1000 kcalories). Eating high-fiber foods and drinking water can alleviate constipation—a condition common among older adults, especially nursing home residents. (Physical inactivity and medications also contribute to the high incidence of constipation.)

Fat As is true for people of all ages, fat intake needs to be moderate in the diets of most older adults—enough to enhance flavors and provide valuable nutrients, but not so much as to raise the risks of atherosclerosis and other degenerative diseases. This recommendation should not be taken too far; limiting fat too severely may lead to nutrient deficiencies and weight loss—two problems that carry greater health risks in the elderly than being overweight.

Vitamins and Minerals Most people can achieve adequate vitamin and mineral intakes simply by including foods from all food groups in their diets, but older adults often omit fruits and vegetables. Similarly, few older adults consume the recommended amounts of milk or milk products.

Vitamin B₁₂ An estimated 10 to 30 percent of adults older than 50 have **atrophic gastritis;** as Chapter 10 explains, people with atrophic gastritis are particularly vulnerable to vitamin B_{12} deficiency. The bacterial overgrowth that accompanies this condition uses up the vitamin, and without hydrochloric acid and intrinsic factor, digestion and absorption of vitamin B_{12} are inefficient. Given the poor cognition, anemia, and devastating neurological effects associated with a vitamin B_{12} deficiency, an adequate intake is imperative. The RDA for older adults is the same as for younger adults, but with the added suggestion to obtain most of a day's intake from vitamin B_{12}–fortified foods and supplements. The bioavailability of vitamin B_{12} from these sources is better than from foods.

atrophic (a-TRO-fik) **gastritis** (gas-TRY-tis): chronic inflammation of the stomach accompanied by a diminished size and functioning of the mucous membranes and glands. This condition is also characterized by inadequate hydrochloric acid and intrinsic factor— two substances needed for vitamin B_{12} absorption.

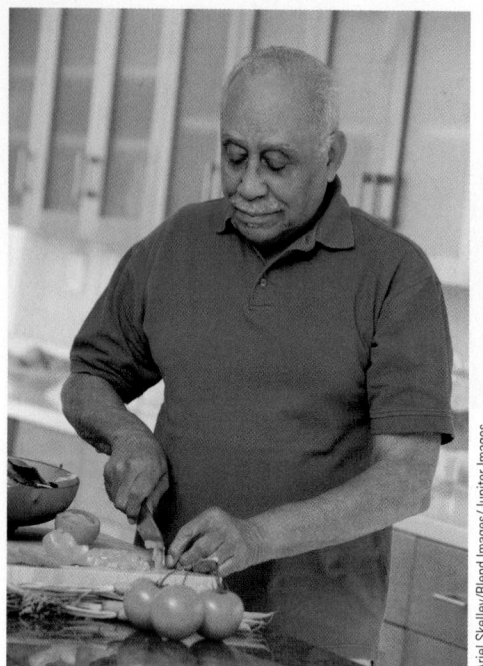

Taking time to nourish your body well is a gift you give yourself.

> **DIETARY GUIDELINES FOR AMERICANS**
Individuals aged 50 years and older should consume foods fortified with vitamin B_{12}, such as fortified cereals, or take dietary supplements.

Vitamin D Vitamin D deficiency is a problem for many older adults. Vitamin D–fortified milk is the most reliable source of vitamin D, but many older adults drink little or no milk. Further compromising the vitamin D status of many older people, especially those in nursing homes, is their limited exposure to sunlight. Finally, aging reduces the skin's capacity to make vitamin D and the kidneys' ability to convert it to its active form. Not only are older adults not getting enough vitamin D, but they may actually need more to improve both muscle and bone strength. To prevent bone loss and to maintain vitamin D status, especially in those who engage in minimal outdoor activity, adults 51 to 70 years old need 15 micrograms daily, and those 71 and older need 20 micrograms. Supplements may be needed to achieve adequate levels of vitamin D.[38]

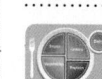
> **DIETARY GUIDELINES FOR AMERICANS**
Choose fortified foods and dietary supplements to provide more vitamin D, a nutrient of concern.

Folate As is true of vitamin B_{12}, folate intakes of older adults typically fall short of recommendations. The elderly are also more likely to have medical conditions or to take medications that can compromise folate status (see Highlight 17).

Calcium Both Chapter 12 and Highlight 12 emphasize the importance of abundant dietary calcium throughout life, especially for women after menopause, to protect against osteoporosis. The DRI Committee recommends 1200 milligrams of calcium daily for women older than 50 and men older than 70, but the calcium intakes of older people in the United States are well below recommendations. Some older adults avoid milk and milk products because they dislike these foods or associate them with stomach discomfort. Simple solutions include using calcium-fortified juices, adding powdered milk to recipes, and taking supplements. Chapter 12 offers many other strategies for including nonmilk sources of calcium for those who do not drink milk.

Iron The iron needs of men remain unchanged throughout adulthood. For women, iron needs decrease substantially at menopause when blood loss through menstruation ceases. Consequently, iron-deficiency anemia is less common in older adults than in younger people. In fact, elevated iron stores are more likely than deficiency in older people, especially for those who take iron supplements, eat red meat regularly, and include vitamin C–rich fruits in their daily diet.

Nevertheless, iron deficiency may develop in older adults, especially when their food energy intakes are low. Aside from diet, two other factors may lead to iron deficiency in older people: chronic blood loss from diseases and medicines and poor iron absorption due to reduced stomach acid secretion and antacid use. Iron deficiency impairs immunity and leaves older adults vulnerable to infectious diseases. Anyone concerned with older people's nutrition should keep these possibilities in mind.

Zinc Zinc intake is commonly low in older people. Zinc deficiency can depress the appetite and blunt the sense of taste, thereby reducing food intake and worsening zinc status. Many medications that older adults commonly use can impair zinc absorption or enhance its excretion and thus lead to deficiency. Low zinc status impairs immune function and increases the risk of pneumonia and death.[39]

Dietary Supplements People judge for themselves how to manage their nutrition, and more than half of older adults turn to dietary supplements.[40] When

TABLE 17-2 Nutrient Concerns of Aging

Nutrient	Effect of Aging	Comments
Water	Lack of thirst and decreased total body water make dehydration likely.	Mild dehydration is a common cause of confusion. Difficulty obtaining water or getting to the bathroom may compound the problem.
Energy	Need decreases as muscle mass decreases (sarcopenia).	Physical activity moderates the decline.
Fiber	Likelihood of constipation increases with low intakes and changes in the GI tract.	Inadequate water intakes and lack of physical activity, along with some medications, compound the problem.
Protein	Needs may stay the same or increase slightly.	Low-fat, high-fiber legumes and grains meet both protein and other nutrient needs.
Vitamin B$_{12}$	Atrophic gastritis is common.	Deficiency causes neurological damages; supplements may be needed.
Vitamin D	Increased likelihood of inadequate intake; skin synthesis declines.	Daily sunlight exposure in moderation or supplements may be beneficial.
Calcium	Intakes may be low; osteoporosis is common.	Stomach discomfort commonly limits milk intake; calcium substitutes or supplements may be needed.
Iron	In women, status improves after menopause; deficiencies are linked to chronic blood losses and low stomach acid output.	Adequate stomach acid is required for absorption; antacid or other medicine use may aggravate iron deficiency; vitamin C and meat increase absorption.
Zinc	Intakes are often inadequate and absorption may be poor, but needs may also increase.	Medications interfere with absorption; deficiency may depress appetite and sense of taste.

© Cengage Learning

recommended by a physician or registered dietitian nutritionist, vitamin D and calcium supplements for osteoporosis or vitamin B$_{12}$ for pernicious anemia may be beneficial. Many health-care professionals recommend a daily multivitamin-mineral supplement that provides 100 percent or less of the Daily Value for the listed nutrients. They reason that such a supplement is more likely to be beneficial than to cause harm. Supplement use may help older adults obtain enough of some nutrients, but it may also lead to excessive intakes of others.[41]

People with small energy allowances would do well to become more active so they can afford to eat more food. Food is the best source of nutrients for everybody. Supplements are just that—supplements to foods, not substitutes for them. For anyone who is motivated to obtain the best possible health, it is never too late to learn to eat well, drink water, exercise regularly, and adopt other lifestyle habits such as quitting smoking and moderating alcohol use.

> **REVIEW IT** Explain why the needs for some nutrients increase or decrease during aging.
Table 17-2 provides a summary of the nutrient concerns of aging. Although some nutrients need special attention in the diet, supplements are not routinely recommended. The ever-growing number of older people creates an urgent need to learn more about how their nutrient requirements differ from those of others and how such knowledge can enhance their health.

17.4 Nutrition-Related Concerns of Older Adults

> **LEARN IT** Identify how nutrition might contribute to, or prevent, the development of age-related problems associated with vision, arthritis, the brain, and alcohol use.

Nutrition may play a greater role than has been realized in preventing many changes once thought to be inevitable consequences of growing older. The following discussions of vision, arthritis, the aging brain, and alcohol use show how nutrition interacts with these conditions.

Vision One key aspect of healthy aging is maintaining good vision. Age-related eye diseases that impair vision, such as cataracts and macular degeneration, correlate with poor survival that cannot be explained by other risk factors.

> FIGURE 17-3   Healthy Lens and Cataract Lens Compared

Healthy lens

Clouded lens

Cataract lens

© 2016 Cengage Learning

The healthy lens (on the left) focuses light, producing clear, sharp images on the retina. A lens affected by cataracts (on the right) scatters the light, resulting in blurred vision.

© Panu Ruangjan/Shutterstock.com

Vision through a healthy lens (on the left) is crisp and clear, whereas vision through cataracts (on the right) is cloudy.

Cataracts Cataracts are age-related cloudy areas in the lenses of the eyes that impair vision (see Figure 17-3). If not surgically removed, they ultimately lead to blindness. Cataracts may develop as a result of ultraviolet light exposure, oxidative stress, injury, viral infections, toxic substances, and genetic disorders. Most cataracts, however, are vaguely called senile cataracts—meaning "caused by aging." In the United States, more than half of all adults 65 and older have a cataract.

Oxidative stress appears to play a significant role in the development of cataracts, but supplements of the antioxidant nutrients (vitamin C, vitamin E, and carotenoids) do not seem to prevent or slow the progression.[42] By comparison, a healthy diet that includes plenty of fruits and vegetables rich in these antioxidant nutrients does seem to slow the progression or reduce the risk of developing cataracts.[43] A word of caution: vitamin C supplements in high doses (1000 milligrams) and long duration (several years) may *increase* the risk of cataracts.[44]

One other diet-related factor may play a role in the development of cataracts—obesity. Obesity appears to be associated with cataracts, but its role has not been identified. Risk factors that typically accompany obesity, such as inactivity, diabetes, or hypertension, do not explain the association.

Macular Degeneration The leading cause of visual loss among older people is age-related **macular degeneration,** a deterioration of the macular region of the retina. As with cataracts, risk factors for age-related macular degeneration include oxidative stress from sunlight. Preventive factors may include supplements of the omega-3 fatty acids, some B vitamins (folate, vitamin B_6, and vitamin B_{12}), antioxidants (vitamin C, vitamin E, and beta-carotene), zinc, and the carotenoids lutein and zeaxanthin.[45]

Arthritis More than 50 million people in the United States have some form of **arthritis.**[46] As the population ages, it is expected that the prevalence will increase to 70 million by 2030. Arthritis pain and fear of further damage limit physical activity.

Osteoarthritis The most common type of arthritis that disables older people is **osteoarthritis,** a painful deterioration of the cartilage in the joints. During movement, the ends of bones are normally protected from wear by cartilage and by small sacs of fluid that act as a lubricant. With age, cartilage begins to deteriorate, and the joints may become malformed and painful to move.

Obesity is common among adults with arthritis.[47] Weight loss may relieve some of the pain for overweight persons with osteoarthritis, partly because the joints affected are often weight-bearing joints that are stressed and irritated by having to carry excess pounds. Interestingly, though, weight loss often relieves

cataracts (KAT-ah-rakts): clouding of the eye lenses that impairs vision and can lead to blindness.

macular (MACK-you-lar) **degeneration:** deterioration of the macular area of the eye that can lead to loss of central vision and eventual blindness. The *macula* is a small, oval, yellowish region in the center of the retina that provides the sharp, straight-ahead vision so critical to reading and driving.

arthritis: inflammation of a joint, usually accompanied by pain, swelling, and structural changes.

osteoarthritis: a painful, degenerative disease of the joints that occurs when the cartilage in a joint deteriorates; joint structure is damaged, with loss of function; also called *degenerative arthritis.*

much of the pain of arthritis in the hands as well, even though they are not weight-bearing joints. Importantly, walking and other weight-bearing exercises do not worsen arthritis. In fact, low-impact aerobic activity and resistance strength training offer improvements in physical performance and pain relief, especially when accompanied by even modest weight loss.[48]

Rheumatoid Arthritis Another type of arthritis known as **rheumatoid arthritis** has possible links to diet through the immune system. In rheumatoid arthritis, the immune system mistakenly destroys bone and cartilage as if they were made of foreign tissue.[49]

The omega-3 fatty acids commonly found in fatty fish reduce joint tenderness and improve mobility in some people with rheumatoid arthritis. The same diet recommended for heart health—a Mediterranean-type diet low in saturated fat from meats and milk products and high in vegetables, olive oil, and omega-3 fats from fish—helps prevent or reduce the inflammation in the joints that makes arthritis so painful.

Another possible link between nutrition and rheumatoid arthritis involves the oxidative damage to the membranes within joints that causes inflammation and swelling. The antioxidant vitamins C and E and the carotenoids defend against oxidation, and increased intakes of these nutrients may help prevent or relieve the pain of rheumatoid arthritis.

Gout Another form of arthritis, which most commonly affects men, is **gout,** a condition characterized by deposits of uric acid crystals in the joints. Uric acid derives from the breakdown of **purines,** primarily from those made by the body but also from those found in foods. Recommendations to lower uric acid levels and the risk of gout include limiting alcohol and excessive amounts of meat, seafood, and sugar-sweetened beverages.[50] For most people, however, such strategies are insufficient, and drugs are needed to control symptoms.[51]

Treatment Treatment for arthritis—dietary or otherwise—may help relieve discomfort and improve mobility, but it does not cure the condition. Traditional medical intervention for arthritis includes medication and surgery. Alternative therapies to treat arthritis abound, but none have proved safe and effective in scientific studies. Popular supplements—glucosamine, chondroitin, or a combination—may relieve pain and improve mobility as well as over-the-counter pain relievers, but mixed reports from studies emphasize the need for additional research.[52]

The Aging Brain
Dementia affects an estimated 15 percent of adults older than 70 years of age in the United States and represents a financial burden of between $157 billion and $215 billion.[53] The brain, like all of the body's organs, ages in response to both genetic and environmental factors—such as physical activities, intellectual challenges, social interactions, and nutritious diets—that enhance or diminish its amazing capacities.[54] One of the challenges researchers face when studying the human brain is to distinguish among normal age-related physiological changes, changes caused by diseases, and changes that result from cumulative, environmental factors such as diet.

The brain normally changes in some characteristic ways as it ages. For one thing, its blood supply decreases. For another, the number of **neurons,** the brain cells that specialize in transmitting information, diminishes as people age. When the number of neurons in one part of the cerebral cortex diminishes, hearing and speech are affected. Losses of neurons in other parts of the cortex can impair memory and cognitive function. When the number of neurons in the cerebellum diminishes, balance and posture are affected. Losses of neurons in other parts of the brain affect still other functions. Some of the cognitive loss and forgetfulness generally attributed to aging may be due in part to environmental, and therefore controllable, factors—including nutrient deficiencies.

Nutrient Deficiencies and Brain Function Nutrients influence the development and activities of the brain. The ability of neurons to synthesize specific neurotransmitters depends in part on the availability of precursor nutrients that are

Both foods and mental challenges nourish the brain.

rheumatoid (ROO-ma-toyd) **arthritis:** a disease of the immune system involving painful inflammation of the joints and related structures.

gout (GOWT): a common form of arthritis characterized by deposits of uric acid crystals in the joints.

purines: compounds of nitrogen-containing bases such as adenine, guanine, and caffeine. Purines that originate from the body are *endogenous* and those that derive from foods are *exogenous*.

neurons: nerve cells; the structural and functional units of the nervous system. Neurons initiate and conduct nerve impulse transmissions.

obtained from the diet. The neurotransmitter serotonin, for example, derives from the amino acid tryptophan. To function properly, the enzymes involved in neurotransmitter synthesis require vitamins such as vitamin C and pantothenic acid. The B vitamins folate, vitamin B_6, and vitamin B_{12} slow brain atrophy and improve cognition and memory.[55] The essential fatty acid DHA counteracts the cognitive decline commonly seen in elderly adults.[56] Thus nutrient deficiencies may contribute to the loss of memory and cognition that some older adults experience. Such losses may be preventable or at least diminished or delayed through diet and exercise. Healthy eating patterns, such as the Mediterranean diet, seem to slow cognitive decline and lower the risk of dementia.[57]

In some instances, the degree of cognitive loss is extensive. Some **senile dementia** may be attributable to a specific disorder such as a brain tumor or Alzheimer's disease.

Alzheimer's Disease Much attention has focused on the *abnormal* deterioration of the brain called **Alzheimer's disease,** which affects one out of eight US adults older than age 65 and almost half of adults aged 85 and older.[58] Nerve cells in the brain die, and communication between the cells breaks down. Diagnosis of Alzheimer's disease depends on its characteristic symptoms: the victim gradually loses memory and reasoning, the ability to communicate, physical capabilities, and eventually life itself.[59] Table 17-3 compares the signs of Alzheimer's disease with typical age-related changes.

The primary risk factor for Alzheimer's disease is age, but the exact cause remains uknown.[60] Clearly, genetic factors are involved. Free radicals and oxidative stress also seem to be involved. Nerve cells in the brains of people with Alzheimer's disease show evidence of free-radical attack—damage to DNA, cell membranes, and proteins. They also show evidence of the minerals that trigger free-radical attacks—iron, copper, zinc, and aluminum. Increasing evidence also suggests that obesity in middle age is associated with cognitive decline and dementia in general, and with Alzheimer's disease in particular.[61]

In Alzheimer's disease, the brain develops **senile plaques** and **neurofibrillary tangles** (see Figure 17-4).[62] Senile plaques are clumps of a protein fragment called beta-amyloid, whereas neurofibrillary tangles are snarls of the fibers that extend from the nerve cells. Biochemical markers of beta-amyloid deposits and impaired brain metabolism are apparent decades before the onset of symptoms.[63] Oxidative stress seems to be a contributing factor, but antioxidant supplements do not seem to be effective in preventing the progression of the disease.[64] The accumulation of beta-amyloid seems to be a problem of impaired clearance more than excessive production.[65] Much treatment research focuses on lowering beta-amyloid levels.[66] Interestingly, the fat cell hormone leptin that decreases appetite and increases energy expenditure, also promotes beta-amyloid clearance in animals and lowers the risk of Alzheimer's disease.

Late in the course of the disease there is a decline in the activity of the enzyme that assists in the production of the neurotransmitter acetylcholine from choline and acetyl CoA. Acetylcholine is essential to memory, but supplements of choline (or of lecithin, which contains choline) have

senile dementia: the loss of brain function beyond the normal loss of physical adeptness and memory that occurs with aging.

Alzheimer's (AHLZ-high-merz) **disease:** a degenerative disease of the brain involving memory loss and major structural changes in neuron networks; also known as *senile dementia of the Alzheimer's type (SDAT), primary degenerative dementia of senile onset,* or *chronic brain syndrome.*

senile plaques: clumps of the protein fragment beta-amyloid on the nerve cells, commonly found in the brains of people with Alzheimer's dementia.

neurofibrillary tangles: snarls of the threadlike strands that extend from the nerve cells, commonly found in the brains of people with Alzheimer's dementia.

TABLE 17-3 Signs of Alzheimer's and Typical Age-Related Changes Compared

Signs of Alzheimer's	Typical Age-Related Changes
Memory loss that disrupts daily life such as asking for the same information repeatedly or asking others to handle tasks of daily living	Forgetting a name or missing an appointment
Challenges in planning or solving problems such as following a recipe or paying monthly bills	Missing a monthly payment or making an error when balancing the checkbook
Difficulty completing familiar tasks at home such as using the microwave, at work such as preparing a report, or at leisure such as playing a game	Needing help recording a television program
Confusion with time or place including current season and location	Not knowing today's date
Trouble understanding visual images and spatial relationships such as judging distances and recognizing self in a mirror	Experiencing visual changes due to cataracts
New problems with words in speaking or writing such as knowing the name of a common object	Being unable to find the right word to use
Misplacing things and losing the ability to retrace steps such as putting the milk in the closet and having no idea when or where the milk was last seen	Misplacing a pair of glasses or the car keys
Decreased or poor judgment such as giving large sums of money to strangers	Making a bad decision on occasion
Withdrawal from work projects or social activities	Feeling too tired to participate in work, family, or social activities
Changes in mood and personality such as confusion, suspicion, depression, and anxiety especially when in unfamiliar places or with unfamiliar people	Becoming irritable when routines are disrupted

© Cengage Learning

SOURCE: Adapted from Alzheimer's Association, www.alz.org/alzheimers_disease_10_signs_of_alzheimers.asp.

> **FIGURE 17.4** **Alzheimer's and Healthy Brains Compared**

Healthy brain   Alzheimer's brain

Plaques—clumps of beta-amyloid protein pieces—block cell-to-cell synapse signals. Tangles—twisted strands of protein—destroy the cell transport system. As plaques and tangles block essential nutrients from reaching the nerve cells, they eventually die.

As nerve cells die, the brain shrinks and loses its ability to think, plan, remember, and form new memories. The fluid-filled spaces within the brain grow larger.

SOURCE: Alzheimer's Association, www.alz.org/research/science/alzheimers_brain_tour.asp.

no effect on memory or on the progression of the disease. Drugs that inhibit the breakdown of acetylcholine, on the other hand, have proved beneficial.

Research suggests that cardiovascular disease risk factors such as high blood pressure, diabetes, obesity, smoking, and physical inactivity may be related to the development of dementia and Alzheimer's disease.[67] Heart healthy diets that include the omega-3 fatty acid DHA and vitamin E may benefit brain health as well, although supplements may not be beneficial in treating Alzheimer's, given that they are often initiated "too late" in the disease process.[68] Similarly, physical activity supports heart health and slows cognitive decline.[69] Just as saturated fatty acids contribute to heart disease, they also worsen cognition and memory and contribute to Alzheimer's disease.[70]

Treatment for Alzheimer's disease includes providing care to clients and support to their families. Drugs may be used to improve or at least to slow the loss of short-term memory and cognition, but they do not cure the disease. Other drugs may be used to control depression, anxiety, and behavior problems.

Maintaining appropriate body weight may be the most important nutrition concern for the person with Alzheimer's disease. Depression and forgetfulness can lead to changes in eating behaviors and poor food intake. Perhaps the best that a caregiver can do nutritionally for a person with Alzheimer's disease is to supervise food planning and mealtimes. Providing well-liked and well-balanced meals and snacks in a cheerful atmosphere encourages food consumption. To minimize confusion, offer a few ready-to-eat foods, in bite-size pieces, with seasonings and sauces. To avoid mealtime disruptions, control distractions such as music, television, children, and the telephone. One study reported that simply having an aquarium in the dining area increased food intake and helped to maintain body weight, perhaps because it provided a naturally attractive and calming environment.[71]

Alcohol Highlight 7 presented information on alcohol metabolism and some of the health consequences of excessive use. Among the consequences of chronic alcohol use are impaired memory and cognition, which can complicate the diagnosis and treatment of age-related dementia.[72]

A variety of tools can be used to diagnose alcohol abuse, but simply asking a question or two can identify hazardous drinking behaviors and potential problems in the elderly.[73] "In the past year, how often did you drink four (for women, and five for men) or more drinks? What is the maximum number of drinks you consumed on any given day?" Such questions help to identify regular heavy use of alcohol and binge drinking.

Although the age group with the most binge drinkers is adults younger than 35, the age group that binge drinks most often is adults 65 and older.[74] Excessive alcohol use among elderly adults is associated with other risk factors as well, including illicit drug use, tobacco use, and misuse of prescription medications—all factors exacerbating overall health, independence, and health-care costs. Fortunately, alcohol-dependent elderly adults seeking treatment seem to experience less intense alcohol cravings during withdrawal therapy than younger adults.[75]

> **REVIEW IT** Identify how nutrition might contribute to, or prevent, the development of age-related problems associated with vision, arthritis, the brain, and alcohol use.
Senile dementia and other losses of brain function, including the impaired memory and cognition of alcohol use, afflict millions of older adults, and others face loss of vision due to cataracts or macular degeneration or cope with the pain of arthritis. As the number of people older than age 65 continues to grow, the need for solutions to these problems becomes urgent. Some problems may be inevitable, but others are, preventable, and good nutrition may play a key role.

17.5 Food Choices and Eating Habits of Older Adults

> **LEARN IT** Instruct an adult on how to shop for groceries and prepare healthy meals for one person on a tight budget.

Older people are an incredibly diverse group, and for the most part, they are independent, socially sophisticated, mentally lucid, fully participating members of society who report themselves to be happy and healthy. In fact, the quality of life among the elderly has improved, and their chronic disabilities have declined dramatically in recent years. By practicing stress-management skills, maintaining physical fitness, participating in activities of interest, and cultivating spiritual health, as well as obtaining adequate nourishment, people can support a high quality of life into old age (see Table 17-4 for some strategies).

Compared with other age groups, older people spend more money per person on foods to eat at home and less money on foods away from home. Manufacturers would be wise to cater to the preferences of older adults by providing good-tasting, nutritious foods in easy-to-open, single-serving packages with labels that are easy to read. Such services enable older adults to maintain their independence

---

TABLE 17-4 Strategies for Growing Old Healthfully

- Choose nutrient-dense foods.
- Be physically active. Walk, run, dance, swim, bike, or row for aerobic activity. Lift weights, do calisthenics, or pursue some other activity to tone, firm, and strengthen muscles. Practice balancing on one foot or doing simple movements with your eyes closed. Modify activities to suit changing abilities and preferences.
- Maintain appropriate body weight.
- Reduce stress—cultivate self-esteem, maintain a positive attitude, manage time wisely, know your limits, practice assertiveness, release tension, and take action.
- For women, discuss with a physician the risks and benefits of estrogen replacement therapy.
- For people who smoke, discuss with a physician strategies and programs to help you quit.
- Expect to enjoy sex, and learn new ways of enhancing it.
- Use alcohol only moderately, if at all; use drugs only as prescribed.
- Take care to prevent accidents.
- Expect good vision and hearing throughout life; obtain glasses and hearing aids if necessary.
- Take care of your teeth; obtain dentures if necessary.

- Be alert to confusion as a disease symptom, and seek diagnosis.
- Take medications as prescribed; see a physician before self-prescribing medicines or herbal remedies and a registered dietitian nutritionist before self-prescribing supplements.
- Control depression through activities and friendships; seek professional help if necessary.
- Drink six to eight glasses of water every day.
- Practice mental skills. Keep on solving math problems and crossword puzzles, playing cards or other games, reading, writing, imagining, and creating.
- Make financial plans early to ensure security.
- Accept change. Work at recovering from losses; make new friends.
- Cultivate spiritual health. Cherish personal values. Make life meaningful.
- Go outside for sunshine and fresh air as often as possible.
- Be socially active—play bridge, join an exercise or dance group, take a class, teach a class, eat with friends, volunteer time to help others.
- Stay interested in life—pursue a hobby, spend time with grandchildren, take a trip, read, grow a garden, or go to the movies.
- Enjoy life.

and to feel a sense of control and involvement in their own lives. Another way older adults can take care of themselves is by remaining or becoming physically active. As mentioned earlier, physical activity helps preserve one's ability to perform daily tasks and so promotes independence.

Familiarity, taste, and health beliefs are most influential on older people's food choices. Eating foods that are familiar, especially ethnic foods that recall family meals and pleasant times, can be comforting. Older adults are less likely to diet to lose weight than younger people are, but they are more likely to diet in pursuit of medical goals such as controlling blood glucose and cholesterol.

Malnutrition As mentioned, most older adults are adequately nourished, but an estimated one out of six is malnourished. Chronic illnesses, medications, depression, and social isolation can all contribute to malnutrition. Malnutrition limits a person's ability to function and diminishes quality of life by:[76]

- Impairing muscle function
- Decreasing bone mass
- Limiting immune defenses
- Reducing cognitive abilities
- Delaying wound healing
- Slowing recovery from surgery
- Increasing hospitalizations

Healthy snacks or liquid nutrition supplements between meals enhance energy and nutrient intakes, which improves body weight and body composition as well as physical and cognitive functioning.[77]

The Nutrition Screening Initiative is part of a national effort to identify and treat nutrition problems in older adults; it uses a screening checklist to *determine* the risk of malnutrition (see Table 17-5). Providing access to safe, adequate food and nutrition programs and services can help ensure healthful aging.[78]

Food Assistance Programs An integral component of the Older Americans Act (OAA) is the OAA Nutrition Program. Its services are designed to improve older people's nutrition status and enable them to avoid medical problems, continue living in communities of their own choice, and stay out of institutions. Its specific goals are to provide low-cost, nutritious meals; opportunities for social interaction; homemaker education and shopping assistance; counseling and referral to social services; and transportation. The program's mission has always been to provide "more than a meal."

The OAA Nutrition Program provides for **congregate meals** at group settings such as community centers. Administrators try to select sites for congregate meals

---

TABLE 17-5 Risk Factors for Malnutrition in Older Adults

	These questions help *determine* the risk of malnutrition in older adults:
Disease	• Do you have an illness or condition that changes the types or amounts of foods you eat?
Eating poorly	• Do you eat fewer than two meals a day? Do you eat fruits, vegetables, and milk products daily?
Tooth loss or mouth pain	• Is it difficult or painful to eat?
Economic hardship	• Do you have enough money to buy the food you need?
Reduced social contact	• Do you eat alone most of the time?
Multiple medications	• Do you take three or more different prescribed or over-the-counter medications daily?
Involuntary weight loss or gain	• Have you lost or gained 10 pounds or more in the last 6 months?
Needs assistance	• Are you physically able to shop, cook, and feed yourself?
Elderly person	• Are you older than 80?

© Cengage Learning

congregate meals: nutrition programs that provide food for the elderly in conveniently located settings such as community centers.

Social interactions at a congregate meal site can be as nourishing as the foods served.

where as many eligible people as possible can participate. Volunteers may also deliver meals to those who are home-bound either permanently or temporarily; these home-delivered meals are known as **Meals on Wheels.** Home-delivered meals help older adults live independently in the community.[79] Although the home-delivery program ensures nutrition, its recipients miss out on the social benefits of the congregate meals. Therefore, every effort is made to persuade older people to come to the shared meals, if they can. All persons aged 60 years and older and their spouses are eligible to receive meals from these programs, regardless of their income. Priority is given to those who are economically and socially needy. An estimated 3 million of our nation's older adults benefit from these meals.

These programs provide at least one meal a day that meets one-third of the RDA for this age group, and they operate five or more days a week. Many programs voluntarily offer additional services designed to appeal to older adults: provisions for special diets (to meet medical needs or religious preferences), food pantries, ethnic meals, and delivery of meals to the homeless. Adding breakfast to the service increases energy and nutrient intakes, which helps to relieve hunger and depression.

Older adults can also take advantage of the Senior Farmers Market Nutrition Program, which provides low-income older adults with coupons that can be exchanged for fresh fruits, vegetables, and herbs at community-supported farmers' markets and roadside stands. This program increases fresh fruit and vegetable consumption, provides nutrition information, and even reaches the home-bound elderly, a group of people who normally do not have access to farmers' markets.

Older adults can learn about the available programs in their communities by contacting the Eldercare section of the Department of Health and Human Services.* In addition, the local senior center and hospital can usually direct people to programs that provide nutrition and other health-related services.

In addition to programs designed specifically for older adults, the Supplemental Nutrition Assistance Program (SNAP) offers services to eligible people of all ages. As mentioned earlier, though, the participation rate for eligible seniors is only about 30 percent.

Meals for Singles Many older adults live alone, and singles of all ages face challenges in purchasing, storing, and preparing food. Large packages of meat and vegetables are often intended for families of four or more, and even a head of lettuce can spoil before one person can use it all. Many singles live in small dwellings and have little storage space for foods. A limited income presents additional obstacles. This section offers suggestions that can help to solve some of the problems singles face, beginning with a special note about the dangers of foodborne illnesses.

Foodborne Illnesses The risk of older adults getting a foodborne illness is greater than for other adults. The consequences of an upset stomach, diarrhea, fever, vomiting, abdominal cramps, and dehydration are oftentimes more severe, sometimes leading to paralysis, meningitis, or even death. For these reasons, older adults need to carefully follow the food safety suggestions presented in Chapter 19.

> **DIETARY GUIDELINES FOR AMERICANS**

- Older adults should not eat or drink unpasteurized milk, milk products, or juices; raw or undercooked eggs, meat, poultry, fish, or shellfish; or raw sprouts.

- Older adults should eat deli meats and frankfurters only if they have been reheated to steaming hot.

Meals on Wheels: a nutrition program that delivers food for the elderly to their homes.

*Call Eldercare Locator at (800) 677–1116 or search www.eldercare.gov.

Spend Wisely People who have the means to shop and cook for themselves can cut their food bills simply by being wise shoppers. Large supermarkets are usually less expensive than convenience stores. Shopping when not hungry and having a grocery list helps reduce impulse buying, and specials and coupons can save money when the items featured are those that the shopper needs and uses.

Buying the right amount so as not to waste any food is a challenge for people eating alone. They can buy fresh milk in the size best suited for personal needs. Boxes of milk that have been exposed to temperatures above those of pasteurization just long enough to sterilize the milk—a process called *ultrahigh temperature (UHT)*—are available and can be stored unopened on a shelf for as long as 3 months without refrigeration.

Foods in bulk are usually less expensive than packaged items. Staples such as rice, pastas, oatmeal, dry powdered milk, and dried legumes can be purchased in bulk and stored for months at room temperature.

A person who has ample freezer space can buy large packages of meat or whole chickens when they are on sale. Then the meat or chicken can be portioned and immediately wrapped into individual servings for the freezer. All the individual servings can be put in a bag marked appropriately with the contents and the date.

Frozen vegetables are more economical in large bags than in small boxes. After the amount needed is taken out, the bag can be closed tightly with a twist tie or rubber band. If the package is returned quickly to the freezer each time, the vegetables will stay fresh for a long time.

Finally, breads and cereals usually must be purchased in larger quantities. Again the amount needed for a few days can be taken out and the rest stored in the freezer. Consider buying day-old bread and baked goods for added savings.

Grocers will break open a package of wrapped meat and rewrap the portion needed. Similarly, eggs can be purchased by the half-dozen. Eggs do keep for long periods, though, if stored properly in the refrigerator.

Fresh fruits and vegetables generally cost less when they are in season. A person can buy individual pieces of fresh fruit at various stages of ripeness: a ripe one to eat right away, a semiripe one to eat soon after, and a green one to ripen on the windowsill. If vegetables are packaged in large quantities, the grocer can divide the package so that a smaller amount can be purchased. Small cans of fruits and vegetables, even though they are more expensive per unit, are a reasonable alternative, considering that it is expensive to buy a regular-size can and let the unused portion spoil.

Be Creative Creative chefs think of various ways to use foods when only large amounts are available. For example, a head of cauliflower can be divided into thirds. Then one-third is cooked and eaten hot. Another third is put into a vinegar and oil marinade for use in a salad. And the last third can be used in a casserole or stew.

A variety of vegetables and meats can be enjoyed stir-fried; inexpensive vegetables such as cabbage, celery, and onion are delicious when sautéed in a little oil with herbs or lemon added. Interesting frozen vegetable mixtures are available in larger grocery stores. Cooked, leftover vegetables can be dropped in at the last minute. A bonus of a stir-fried meal is that there is only one pan to wash. Similarly, a microwave oven allows a chef to use fewer pots and pans. Meals and leftovers can also be frozen or refrigerated in microwavable containers to reheat as needed.

Many frozen dinners offer nutritious options. Adding a fresh salad, a whole-wheat roll, and a glass of milk can make a nutritionally balanced meal.

Finally, single people might want to invite someone to share meals with them whenever there is enough food. It's likely that the person will return the invitation, and both parties will get to enjoy companionship and a meal prepared by others.

Noel Hendrickson/Blend Images/Getty Images

Buy only what you will use.

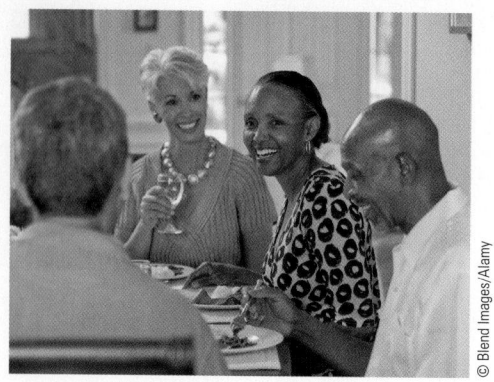
Invite guests to share a meal.

> **REVIEW IT** Instruct an adult on how to shop for groceries and prepare healthy meals for one person on a tight budget.

Older people can benefit from both the nutrients provided and the social interaction available at congregate meals. Other government programs deliver meals to those who are home-bound. With creativity and careful shopping, those living alone can prepare nutritious, inexpensive meals.

Healthy meal patterns throughout adulthood support good health and long life. Physical activity, mental challenges, stress management, and social activities can also help people grow old comfortably. The next chapter describes how similar lifestyle choices help prevent chronic diseases as well.

Nutrition Portfolio

By eating a balanced diet, maintaining a healthy body weight, and engaging in a variety of physical, social, and mental activities, you can enjoy good health in later life. Visit older adults in your community and do the following:

- Consider whether they have the financial means, physical ability, and social support they need to eat adequately.

- Note whether they have experienced an unintentional loss of weight recently.

- Discuss how they occupy their time physically, socially, and mentally.

- Offer to analyze the diet of an older adult who you care about using Diet & Wellness Plus. Have the person write down 1, 2, or even 3 days of their food and beverage intake, and then enter it into the program and print out a 3-day average report of the results. It will be fun and educational to go over it together. Remind the person that you are not a doctor and that this is a learning tool for an introductory nutrition course.

 To complete this exercise, go to your Diet & Wellness Plus at www.cengagebrain.com.

> **STUDY IT** To review the key points of this chapter and take a practice quiz, go to the study cards at the end of the book.

REFERENCES

1. D. L. Hoyert and J. Xu, Deaths: Preliminary data for 2011, *National Vital Statistics Reports* 61 (2012):1–51.
2. National Center for Health Statistics, *Health, United States, 2012: With Special Feature on Emergency Care*, 2013, pp. 76–77.
3. N. R. Eaton and coauthors, Genes, environments, personality, and successful aging: Toward a comprehensive developmental model in later life, *Journal of Gerontology* 67 (2012): 480–488.
4. D. B. Panagiotakos and coauthors, Sociodemographic and lifestyle statistics of oldest old people (>80 years) living in Ikaria island: The Ikaria Study, *Cardiology Research and Practice* 2011 (2011): 679187; E. S. Ford and coauthors, Low-risk lifestyle behaviors and all-cause mortality: Findings from the National Health and Nutrition Examination Survey III Mortality Study, *American Journal of Public Health* 101 (2011): 1922–1929.
5. S. A. McNaughton, C. J. Bates, and G. D. Mishra, Diet quality is associated with all-cause mortality in adults aged 65 years and older, *Journal of Nutrition* 142 (2012): 320–325; A. L. Anderson and coauthors, Dietary patterns and survival of older adults, *Journal of the American Dietetic Association* 111 (2011): 84–91.
6. S. Sabia and coauthors, Influence of individual and combined healthy behaviours on successful aging, *Journal of the Canadian Medical Association* 184 (2012): 1985–1992.
7. S. Q. Townsend and coauthors, Physical activity at midlife in relation to successful survival in women at age 70 years or older, *Archives of Internal Medicine* 170 (2010): 194–201.
8. S. Studenski and coauthors, Gait speed and survival in older adults, *Journal of the American Medical Association* 305 (2011): 50–58.
9. L. D. Baker and coauthors, Effects of aerobic exercise on mild cognitive impairment, *Archives of Neurology* 67 (2010): 71–79.
10. L. Krist, F. Dimeo, and T. Keil, Can progressive resistance training twice a week improve mobility, muscle strength, and quality of life in a very elderly nursing-home residents with impaired mobility? A pilot study, *Clinical Interventions in Aging* 8 (2013): 443–448.
11. H. W. Park, Longevity, aging, and caloric restriction: Clive Maine McCay and the construction of a multidisciplinary research program, *Historical Studies in the Natural Sciences* 40 (2010): 79–124.
12. D. M. Huffman, Exercise as a calorie restriction mimetic: Implications for improving healthy aging and longevity, *Interdisciplinary Topics in Gerontology* 37 (2010): 157–174.
13. R. J. Colman and R. M. Anderson, Nonhuman primate calorie restriction, *Antioxidants and Redox Signaling* 14 (2011): 229–239.
14. R. K. Minor and coauthors, Dietary interventions to extend life span and health span based on calorie restriction, *Journal of Gerontology: Biological Sciences* 65 (2010): 695–703.

15. J. D. Kark and coauthors, Energy intake and leukocyte telomere length in young adults, *American Journal of Clinical Nutrition* 95 (2012): 479–487; Y. Kagawa, From clock genes to telomeres in the regulation of the healthspan, *Nutrition Reviews* 70 (2012): 459–471; A. Soare and coauthors, Long-term calorie restriction, but not endurance exercise lowers core body temperature in humans, *Aging* 3 (2011): 374–379.

16. R. Pallavi, M. Giorgio, and P. G. Pellicci, Insights into the beneficial effect of caloric/dietary restriction fro a healthy and prolonged life, *Frontiers in Physiology* 3 (2012): 318; L. M. Redman and E. Ravussin, Caloric restriction in humans: Impact on physiological, psychological, and behavioral outcomes, *Antioxidants and Redox Signaling* 14 (2011): 275–287.

17. O. Froy and R. Miskin, Effect of feeding regimens on circadian rhythms: Implications for aging and longevity, *Aging* 11 (2010): 7–27; Minor and coauthors, 2010.

18. C. Samieri and coauthors, The association between dietary patterns at midlife and health in aging: An observational study, *Annals of Internal Medicine* 159 (2013): 584–591; F. Sofi and coauthors, Accruing evidence on benefits of adherence to the Mediterranean diet on health: An updated systematic review and meta-analysis, *American Journal of Clinical Nutrition* 92 (2010): 1189–1196.

19. M. E. Obrenovich and coauthors, Antioxidants in health, disease and aging, *CNS and Neurological Disorders Drug Targets* 10 (2011): 192–207; K. Chong-Han, Dietary lipophilic antioxidants: Implications and significance in the aging process, *Critical Reviews in Food Science and Nutrition* 50 (2010): 931–937.

20. T. Ahmed and N. Haboubi, Assessment and management of nutrition in older people and its importance to health, *Clinical Interventions* 5 (2010): 207–216.

21. D. K. Childers and D. B. Allison, The "obesity paradox": A parsimonious explanation for relations among obesity, mortality rate and aging? *International Journal of Obesity* 34 (2010): 1231–1238.

22. L. M. Donini and coauthors, A systematic review of the literature concerning the relationship between obesity and mortality in the elderly, *Journal of Nutrition, Health and Aging* 16 (2012): 89–98.

23. J. Cohen-Mansfield and R. Perach, Is there a reversal in the effect of obesity on mortality in old age? *Journal of Aging Research* 2011 (2011): 765071.

24. P. Szulc and coauthors, Rapid loss of appendicular skeletal muscle mass is associated with higher all-cause mortality in older men: The prospective MINOS study, *American Journal of Clinical Nutrition* 91 (2010): 1227–1236.

25. Z. Li and D. Heber, Sarcopenic obesity in the elderly and strategies for weight management, *Nutrition Reviews* 70 (2012): 57–64; L. A. Burton and D. Sumukadas, Optimal management of sarcopenia, *Clinical Interventions in Aging* 2010 (2010): 217–228; G. A. Power and coauthors, Motor unit number estimates in masters runners: Use it or lose it? *Medicine and Science in Sports and Exercise* 42 (2010): 1644–1650.

26. G. I. Smith and coauthors, Dietary omega-3 fatty acid supplementation increases the rate of muscle protein synthesis in older adults: A randomized controlled trial, *American Journal of Clinical Nutrition* 93 (2011): 402–412.

27. A. Gibson and coauthors, Effect of fruit and vegetable consumption on immune function in older people: A randomized controlled trial, *American Journal of Clinical Nutrition* 96 (2012): 1429–1436.

28. E. Britton and J. T. Mclaughlin, Ageing and the gut, *Proceedings of the Nutrition Society* 72 (2012): 173–177.

29. B. Pennings and coauthors, Minced beef is more rapidly digested and absorbed than beef steak, resulting in greater postprandial protein retention in older men, *American Journal of Clinical Nutrition* 98 (2013): 121–128.

30. Position of the Academy of Nutrition and Dietetics: Food and Nutrition for older adults—Promoting health and wellness, *Journal of the Academy of Nutrition and Dietetics* 112 (2012): 1255–1277.

31. M. Hamer, C. J. Bates, and G. D. Mishra, Depression, physical function, and risk of mortality: National Diet and Nutrition Survey in adults older than 65 years, *American Journal of Geriatric Psychiatry* 19 (2011): 72–78.

32. C. Caffrey, Ten most common chronic conditions among persons living in residential care facilities—National Survey of Residential Care Facilities, United States, 2010, *Morbidity and Mortality Weekly Report* 31 (2012): 603; H. Chang-Quan and coauthors, Health status and risk for depression among the elderly: A meta-analysis of published literature, *Age and Ageing* 39 (2010): 23–30; V. Colasanti and coauthors, Tests for the evaluation of depression in the elderly: A systematic review, *Archives of Gerontology and Geriatrics* 50 (2010): 227–230; D. R. Hoover and coauthors, Depression in the first year of stay for elderly long-term nursing home residents in the USA, *International Psychogeriatric Association* 22 (2010): 1161–1167.

33. R. S. Shim and coauthors, Prevalence, treatment, and control of depressive symptoms in the United States: Results from the National Health and Nutrition Examination Survey (NHANES), 2005–2008, *Journal of the American Board of Family Medicine* 24 (2011): 33–38.

34. M. F. Kuczmarski and coauthors, Higher Health Eating Index-2005 scores associated with reduced symptoms of depression in an urban population: Findings from the Healthy Aging in Neighborhoods of Diversity Across the Life Span (HANDLS) study, *Journal of the American Dietetic Association* 110 (2010): 383–389.

35. C. Denavas-Walt, B. D. Proctor, and J. C. Smith, US Census Bureau, *Income, Poverty, and Health Insurance Coverage in the United States: 2010* (Washington, D.C.: US Government Printing Office, 2011).

36. C. M. Perissinotto, I. S. Cenzer, and K. E. Covinsky, Loneliness in older persons: A predictor of functional decline and death, *Archives of Internal Medicine* 172 (2012): 1078–1084.

37. Position of the American Dietetic Association: Individualized nutrition approaches for older adults in health care communities, *Journal of the American Dietetic Association* 110 (2010): 1549–1553.

38. R. Baraké and coauthors, Vitamin D supplement consumption is required to achieve a minimal target 25-hydroxyvitamin D concentration of >75 nmol/L in older people, *Journal of Nutrition* 140 (2010): 551–556.

39. J. B. Barnett, D. H. Hamer, and S. N. Meydani, Low zinc status: A new risk factor for pneumonia in the elderly? *Nutrition Reviews* 68 (2010): 30–37.

40. R. L. Bailey and coauthors, Dietary supplement use in the United States, 2003–2006, *Journal of Nutrition* 141 (2011): 261–266.

41. A. Weeden and coauthors, Vitamin and mineral supplements have a nutritionally significant impact on micronutrient intakes of older adults attending senior centers, *Journal of Nutrition for the Elderly* 29 (2010): 241–254.

42. M. C. Mathew and coauthors, Antioxidant vitamin supplementation for preventing and slowing the progression of age-related cataract, *Cochrane Database of Systematic Reviews* 6 (2012): CD004567.

43. K. A. Weikel and coauthors, Nutritional modulation of cataract, *Nutrition Reviews* 72 (2014): 30–47; Y. Cui, C. Jing, and H. Pan, Association of blood antioxidants and vitamin with risk of age-related cataract: A meta-analysis of observational studies, *American Journal of Clinical Nutrition* 98 (2013): 778–786.

44. S. Rautiainen and coauthors, Vitamin C supplements and the risk of age-related cataract: A population-based prospective cohort study in women, *American Journal of Clinical Nutrition* 91 (2010): 487–493.

45. B. Gopinath and coauthors, Homocysteine, folate, vitamin B-12, and 10-y incidence of age-related macular degeneration, *American Journal of Clinical Nutrition* 98 (2013): 129–135; The Age-Related Eye Disease Study 2 (AREDS2) Research Group, Lutein + zeaxanthin and omega-3 fatty acids for Age-Related Eye Disease Study 2 (AREDS2) randomized clinical trial, *Journal of the American Medical Association* 309 (2013): 2005–2015; J. R. Evans and J. G. Lawrenson, Antioxidant vitamin and mineral supplements for slowing the progression of age-related macular degeneration, *Cochrane Database of Systematic Reviews* 11 (2012): CD000254; J. P. SanGiovanni and M. Neuringer, The putative role of lutein and zeaxanthin as protective agents against age-related macular degeneration: Promise of molecular genetics for guiding mechanistic and translational research in the field, *American Journal of Clinical Nutrition* 96 (2012): 1223S–1233S; L. Ho and coauthors, Reducing the genetic risk of age-related macular degeneration with dietary antioxidants, zinc, and ω-3 fatty acids: The Rotterdam study, *Epidemiology* 129 (2011): 758–766; E. J. Johnson, Age-related macular degeneration and antioxidant vitamins: Recent findings, *Current Opinion in Clinical Nutrition and Metabolic Care* 13 (2010): 28–33.

46. K. E. Barbour and coauthors, Prevalence of doctor-diagnosed arthritis and arthritis-attributable activity limitation: United States, 2010–2012, *Morbidity and Mortality Weekly Report* 62 (2013): 869–873.

47. Prevalence of obesity among adults with arthritis—United States, 2003–2009, *Morbidity and Mortality Weekly Report* 60 (2011): 509–513.

48. E. M. Roos and C. B. Juhl, Osteoarthritis 2012 year in review: Rehabilitation and outcomes, *Osteoarthritis and Cartilage* 20 (2012): 1477–1483; K. R. Vincent and H. K. Vincent, Resistance exercise for knee osteoarthritis, *Journal of Injury, Function, and Rehabilitation* 4 (2012): S45–S52; K. L. Bennell and R. S. Hinman, A review of the clinical evidence for exercise in osteoarthritis of the hip and knee, *Journal of Science and Medicine in Sport* 14 (2011): 4–9.

49. I. B. McInnes and G. Schett, The pathogenesis of rheumatoid arthritis, *New England Journal of Medicine* 365 (2011): 2205–2219.

50. T. Neogi, Gout, *New England Journal of Medicine* 364 (2011): 443–452.

51. D. Khanna and coauthors, 2012 American College of Rheumatology guidelines for management of gout, *Arthritis Care Research* 64 (2012): 1431–1446.

52. R. L. Ragle and A. D. Sawitzke, Nutraceuticals in the management of osteoarthritis: A critical review, *Nutrition Reviews* 29 (2012): 717–731.

53. M. D. Hurd and coauthors, Monetary costs of dementia in the United States, *New England Journal of Medicine* 368 (2013): 1326–1334.

54. O. P. Ottersen, How hardwired is the brain? Technological advances provide new insight into brain malleability and neurotransmission, *Nutrition Reviews* 68 (2010): S60–S64.

55. J. G. Walker and coauthors, Oral folic acid and vitamin B-12 supplementation to prevent cognitive decline in community-dwelling older adults with depressive symptoms—The Beyond Ageing Project: A randomized controlled trial, *American Journal of Clinical Nutrition* 95 (2012): 194–203; A. D. Smith and coauthors, Homocysteine–lowering by B vitamins slows the rate of accelerated brain atrophy in mild cognitive impairment: A randomized controlled trial, *PLoS One* 5 (2010): e12244; J. Selhub, A. Troen, and I. H. Rosenberg, B vitamins and the aging brain, *Nutrition Reviews* 68 (2010): S112–S118.

56. N. G. Bazan, M. F. Molina, and W. C. Gordon, Docosahexaenoic acid signalolipidomics in nutrition: Significance in aging, neuroinflammation, macular degeneration, Alzheimer's, and other neurodegenerative diseases, *Annual Review of Nutrition* 31 (2011): 321–351.

57. I. Lourida and coauthors, Mediterranean diet, cognitive function, and dementia: A systematic review, *Epidemiology* 24 (2013): 479–489; M. Ozawa and coauthors, Dietary patterns and risk of dementia in an elderly Japanese population: The Hisayama Study, *American Journal of Clinical Nutrition* 97 (2013): 1076–1082; E. Kesse-Guyot and coauthors, Adherence to nutritional recommendations and subsequent cognitive performance: Findings from the prospective Supplementation with Antioxidant Vitamins and Minerals 2 (SU.VI.MAX 2) study, *American Journal of Clinical Nutrition* 93 (2011): 200–210.

58. Alzheimer's Association, 2011 Alzheimer's Disease Facts and Figures, *Alzheimer's and Dementia*, 2011.

59. R. Mayeux, Early Alzheimer's disease, *New England Journal of Medicine* 362 (2010): 2194–2201.

60. A. Burns and S. Iliffe, Clinical review: Alzheimer's disease, *British Medical Journal* 338 (2009): b158; H. W. Querfurth and F. M. LaFerla, Mechanisms of disease: Alzheimer's disease, *New England Journal of Medicine* 362 (2010): 329–344.

61. A. Singh-Manoux and coauthors, obesity phenotypes in midlife and cognition in early old age: The Whitehall II Cohort Study, *Neurology* 79 (2012): 755–762.

62. D. P. Perl, Neuropathology of Alzheimer's disease, *Mt. Sinai Journal of Medicine* 77 (2010): 32–42.

63. R. J. Bateman and coauthors, Clinical and biomarker changes in dominantly inherited Alzheimer's disease, *New England Journal of Medicine* 367 (2012): 795–804.

64. M. Ramamoorthy and coauthors, Sporadic Alzheimer disease fibroblasts display an oxidative stress phenotype, *Free Radical Biology and Medicine* 53 (2012): 1371–1380; J. Viña and coauthors, Antioxidant pathways in Alzheimer's disease: Possibilities of intervention, *Current Pharmaceutical Design* 17 (2011): 3861–3864.

65. K. G. Mawuenyega and coauthors, Decreased clearance of CNS beta-amyloid in Alzheimer's disease, *Science* 330 (2010): 1774.

66. L. Lannfelt, F. E. Pettersson, and L. N. G. Nilsson, Translating research on brain aging into public health: A new type of immunotherapy for Alzheimer's disease, *Nutrition Reviews* 68 (2010): S128–S134.

67. P. K. Crane and coauthors, Glucose levels and risk of dementia, *New England Journal of Medicine* 369 (2013): 540–548; D. E. Barnes and K. Yaffe, The projected effect of risk factor reduction on Alzheimer's disease prevalence, *Lancet Neurology* 10 (2011): 819–828; J. A. Luchsinger and coauthors, Improved diabetes control in the elderly delays global cognitive decline, *Journal of Nutrition, Health and Aging* 15 (2011): 445–449.

68. M. W. Dysken and coauthors, Effect of vitamin E and memantine on functional decline in Alzheimer disease, *Journal of the American Medical Association* 311 (2014): 33–44; Z. S. Tan and coauthors, Red blood cell ω-3 fatty acid levels and markers of accelerated brain aging, *Neurology* 78 (2012): 658–664; J. F. Quinn and coauthors, Docosahexaenoic acid supplementation and cognitive decline in Alzheimer disease: A randomized trial, *Journal of the American Medical Association* 304 (2010): 1903–1911; K. Yaffe, Treatment of Alzheimer disease and prognosis of dementia: Time to translate research to results, *Journal of the American Medical Association* 304 (2010): 1952–1953; F. Mangialasche and coauthors, High plasma levels of vitamin E forms and reduced Alzheimer's disease risk in advanced age, *Journal of Alzheimer's Disease* 20 (2010): 1029–1037; G. M. Cole and S. A. Frautschy, DHA may prevent age-related dementia, *Journal of Nutrition* 140 (2010): 869–874.

69. L. E. Middleton and coauthors, Activity energy expenditure and incident cognitive impairment in older adults, *Archives of Internal Medicine* 171 (2011): 1251–1257.

70. A. J. Hanson and coauthors, Effect of apolipoprotein E genotype and diet on apolipoprotein E lipidation and amyloid peptides, *JAMA Neurology* 70 (2013): 972–980; O. I. Okereke and coauthors, Dietary fat types and 4-year cognitive change in community-dwelling older women, *Annals of Neurology* 72 (2012): 124–134.

71. N. E. Edwards and A. M. Beck, The influence of aquariums on weight in individuals with dementia, *Alzheimer Disease and Associated Disorders* 27 (2013): 379–383.

72. E. Sinforiani and coauthors, The effects of alcohol on cognition in the elderly: From protection to neurodegeneration, *Functional Neurology* 26 (2011): 103–106.

73. D. A. Dawson, A. J. Pulay, and B. F. Grant, A comparison of two single-item screeners for hazardous drinking and alcohol use disorder, *Alcoholism: Clinical and Experimental Research* 34 (2010): 364–374.

74. Centers for Disease Control and Prevention, Binge drinking, http://www.cdc.gov/vitalsigns/BingeDrinking/index.html, October 10, 2013.

75. A. K. Hintzen and coauthors, Does alcohol craving decrease with increasing age? Results from a cross-sectional study, *Journal of Studies on Alcohol and Drugs* 72 (2011): 158–162; A. J. Barnes and coauthors, Prevalence and correlates of at-risk drinking among older adults: The Project SHARE Study, *Journal of General Internal Medicine* 25 (2010): 840–846.

76. T. Ahmed and N. Haboubi, Assessment and management of nutrition in older people and its importance to health, *Clinical Interventions in Aging* 5 (2010): 207–216.

77. C. A. Zizza, D. D. Arsiwalla, and K. J. Ellison, Contribution of snacking to older adults' vitamin, carotenoid, and mineral intakes, *Journal of the American Dietetic Association* 110 (2010): 768–772.

78. Position of the American Dietetic Association, American Society for Nutrition, and Society for Nutrition Education: Food and nutrition programs for community-residing older adults, *Journal of the American Dietetic Association* 110 (2010): 463–472.

79. K. S. Thomas and V. Mor, Providing more home-delivered meals is one way to keep older adults with low care needs out of nursing homes, *Health Affairs* 32 (2013): 1796–1802.

HIGHLIGHT > 17
Nutrient-Drug Interactions

> **LEARN IT** Explain why certain nutrients and medications interact.

As people age, their likelihood of taking medicines increases. People older than the age of 65 take about 35 percent of all the over-the-counter drugs and 40 percent of the prescription drugs sold in the United States. They receive an average of 14 to 18 prescriptions a year and may take as many as 6 drugs at a time.[1] Older adults taking five or more medications have high risks of frailty, disability, falls, and mortality.[2] They take a variety of non-vitamin-mineral supplements, such as glucosamine, as well.

Most often, older adults take multiple drugs and supplements for heart disease, but also to treat arthritis, respiratory problems, and gastrointestinal disorders. They often go to different doctors for each condition and receive different prescriptions from each. Furthermore, physiologic changes associated with aging may alter drug metabolism and excretion, which may in turn, diminish drug effectiveness or create potential toxicities. For all these reasons, physicians need to "start low and go slow" when prescribing for older adults.

An estimated 35 percent of all reported adverse drug reactions involve older adults. To avoid harmful drug interactions, consumers need to inform all of their physicians and pharmacists of all the medicines being taken. These medicines enable people of all ages to enjoy better health, but they also bring side effects and risks.

This highlight focuses on some of the nutrition-related consequences of medical drugs, both prescription drugs and over-the-counter drugs. Highlight 7 describes the relationships between nutrition and the drug alcohol, and Highlight 18 presents information on herbal supplements and other alternative therapies.

The Actions of Drugs

Most people think of drugs either as medicines that help them recover from illnesses or as illegal substances that lead to bodily harm and addiction. Actually, both uses of the term *drug* are correct because any substance that modifies one or more of the body's functions is, technically, a drug. Even medical drugs have both desirable and undesirable consequences within the body.

Consider aspirin. One action of aspirin is to limit the production of certain prostaglandins (biologically active compounds derived from the omega-3 fatty acid EPA). Some prostaglandins help to produce fevers, some sensitize pain receptors, some cause contractions of the uterus, some stimulate digestive tract motility, some control nerve impulses, some regulate blood pressure, some promote blood clotting, and some cause inflammation. By interfering with prostaglandin actions, aspirin reduces fever and inflammation, relieves pain, and slows blood clotting, among other actions.

A person cannot use aspirin to produce one of its effects without producing all of its other effects. Someone who is prone to strokes and heart attacks might take aspirin to prevent blood clotting, but it will also ease that person's sense of pain. Another person who

takes aspirin only for pain will also experience slow blood clotting. The anticlotting effect might be dangerous if it causes abnormal bleeding. A single two-tablet dose of aspirin doubles the bleeding time of wounds, an effect that lasts from 4 to 7 hours. For this reason, physicians instruct clients to refrain from taking aspirin before surgery.

The Interactions between Drugs and Nutrients

Hundreds of drugs and nutrients interact, and these interactions can lead to nutrient imbalances or interfere with drug effectiveness. Adverse nutrient-drug interactions are most likely if drugs are taken over long periods, if several drugs are taken, or if nutrition status is poor or deteriorating. Understandably, then, elderly people with chronic diseases are most vulnerable.

Nutrients and medications may interact in many ways:[3]

- Drugs can alter food intake and the absorption, metabolism, and excretion of nutrients.

- Foods and nutrients can alter the absorption, metabolism, and excretion of drugs.

- Combinations can be toxic.

The following paragraphs describe these interactions, and Table H17-1 (p. 580) summarizes this information and provides specific examples.[4]

Drugs Alter Food Intake

Some medications can make eating difficult or unpleasant. They may suppress appetite, alter taste sensations, induce nausea or vomiting, cause mouth dryness, or create inflammation or lesions in the mouth, stomach, or intestinal lining. Side effects, such as abdominal discomfort, constipation, and diarrhea may worsen when food is eaten. Medications that cause drowsiness may make a person too tired to eat. All

TABLE H17-1 Examples of Diet-Drug Interactions

Drugs May Alter Food Intake by

- Altering the appetite (Amphetamines suppress appetite; corticosteroids increase appetite.)
- Interfering with taste or smell (Amphetamines change taste perceptions.)
- Inducing nausea or vomiting (Digitalis may do both.)
- Interfering with oral function (Some antidepressants may cause dry mouth.)
- Causing sores or inflammation in the mouth (Methotrexate may cause painful mouth ulcers.)

Drugs May Alter Nutrient Absorption by

- Changing the acidity of the digestive tract (Antacids may interfere with iron and folate absorption.)
- Damaging mucosal cells (Cancer chemotherapy may damage mucosal cells.)
- Binding to nutrients (Bile acid binders bind to fat-soluble vitamins.)

Foods and Nutrients May Alter Drug Absorption by

- Stimulating secretion of gastric acid (The antifungal agent ketoconazole is absorbed better with meals due to increased acid secretion.)
- Altering rate of gastric emptying (Intestinal absorption of drugs may be delayed when they are taken with food.)
- Binding to drugs (Calcium binds to tetracycline, reducing both drug and calcium absorption.)
- Competing for absorption sites in the small intestine (Dietary amino acids interfere with levodopa absorption.)

Drugs and Nutrients May Interact and Alter Metabolism by

- Acting as structural analogs (Warfarin and vitamin K are structural analogs.)
- Using similar enzyme systems (Phenobarbital induces liver enzymes that increase metabolism of folate, vitamin D, and vitamin K.)
- Competing for transport on serum proteins (Fatty acids and drugs may compete for the same sites on the serum protein albumin.)

Drugs May Alter Nutrient Excretion by

- Altering nutrient reabsorption in the kidneys (Some diuretics increase the excretion of sodium and potassium.)
- Causing diarrhea or vomiting (Diarrhea and vomiting may cause electrolyte losses.)

Foods May Alter Medication Excretion by

- Inducing activities of liver enzymes that metabolize drugs, increasing drug excretion (Components of charcoal-broiled meats increase metabolism of warfarin, theophylline, and acetaminophen.)

Diet and Drug Interactions May Cause Toxicity

- Increasing side effects of the drug (Caffeine in beverages can increase adverse effects of stimulants.)
- Increasing drug action to excessive levels (Grapefruit components inhibit the enzymes that degrade certain drugs, increasing drug concentrations in the body.)

© Cengage Learning

of these complications limit food intake and can lead to weight loss and malnutrition if not resolved.

Some medications stimulate appetite and cause weight gain. Unintentional weight gain may result from the use of some antipsychotics, antidepressants, and corticosteroids (for example, prednisone).[5] People using these drugs do not feel satiated and sometimes gain 40 to 60 pounds in just a few months. For many adults, fear of weight gain is a common reason for avoiding treatment. For patients with diseases that cause wasting, such as cancer or AIDS, weight gain may be desirable. They may be prescribed appetite enhancers, such as megestrol acetate (Megace), a progesterone analog, or dronabinol (Marinol), which is derived from the active ingredient in marijuana.

Drugs Alter Nutrient Absorption

Nutrient malabsorption is most likely to occur with medications that upset GI function or damage the intestinal mucosa. Antineoplastic and antiretroviral drugs are especially detrimental, although nonsteroidal anti-inflammatory drugs (NSAIDs) and some antibiotics can have similar, though milder, effects.

Some medications bind to nutrients in the GI tract, preventing their absorption. For example, bile acid binders (such as cholestyramine, or Questran), which are used to reduce cholesterol levels, also bind to the fat-soluble vitamins A, D, E, and K. Some antibiotics, notably tetracycline and ciprofloxacin (Cipro), bind to the calcium in foods and supplements, reducing the absorption of both the drug and the calcium. Other minerals, such as iron, magnesium, and zinc, may also bind to antibiotics. For this reason, pharmacists advise consumers to use dairy products and all mineral supplements at least 2 hours apart from these medications.

Medications that reduce stomach acidity may interfere with the absorption of vitamin B_{12}, folate, and iron. Examples include antacids, which neutralize stomach acid by acting as weak bases, and antiulcer drugs (such as proton pump inhibitors and H2 blockers), which interfere with acid secretion.

Several drugs impede nutrient absorption by interfering with the intestinal metabolism or transport into mucosal cells. For example, the antibiotics trimethoprim (Proloprim) and pyrimethamine (Daraprim) compete with folate for absorption into intestinal cells. The anti-inflammatory medication colchicine, a treatment for gout, inhibits vitamin B_{12} absorption.

Diets Alter Drug Absorption

Major influences on drug absorption include the stomach-emptying rate, level of acidity in the stomach, and direct interactions with dietary components. The drug's formulation may also influence its absorption, and pharmacists often provide instructions advising whether food should be eaten or avoided when using a medication.

Drugs reach the small intestine more quickly when the stomach is empty. Therefore, taking a medication with meals may delay its absorption, even though the total amount absorbed may not be lower. As an example, aspirin works faster when taken on an empty stomach, but taking it with food is often encouraged to reduce stomach irritation. Slow stomach emptying can sometimes enhance drug absorption because the drug's absorption sites in the small intestine are less likely to become saturated. A slow drug absorption rate due to slow stomach emptying can be a problem if high drug concentrations are needed for effectiveness, as when a hypnotic is given to induce sleep.

Some drugs are better absorbed in an acidic environment, whereas others are better absorbed in alkaline conditions. For example, reduced stomach acidity (due to disorders or antacid medications) may reduce the absorption of ketoconazole (Nizoral, an antifungal medication) and atazanavir (an antiretroviral medication), but increase the absorption of digoxin (Lanoxin, which treats heart failure) and alendronate (Fosamax, which treats osteoporosis). Some drugs can be damaged by acid and are available in coated forms that resist the action of the stomach's acidity.

Both nutrients and nonnutrients may bind to drugs and inhibit their absorption. For example, high-fiber meals may decrease the absorption of some tricyclic antidepressants. Conversely, the absorption of many lipophilic drugs is improved when the drugs are taken with a fat-contaning meal.[6] As mentioned earlier, minerals may bind to some antibiotics, reducing absorption of both the minerals and the drugs.

Drugs Alter Nutrient Metabolism

Drugs and nutrients share similar enzyme systems in the small intestine and the liver. Consequently, some drugs may enhance or inhibit the activities of enzymes that are needed for nutrient metabolism. For example, the anticonvulsants phenobarbital and phenytoin increase levels of the liver enzymes that metabolize folate, vitamin D, and vitamin K; therefore, persons using these drugs may require supplements of these vitamins.

The drug methotrexate, used to treat cancer and some inflammatory conditions, acts by interfering with folate metabolism and thus depriving rapidly dividing cancer cells of the folate they need to multiply. Methotrexate resembles folate in structure (see Figure H17-1) and competes with folate for the enzyme that converts folate to its active form. The adverse effects of using methotrexate therefore include symptoms of folate deficiency. These adverse effects can be reduced by using a pre-activated form of folate (called leucovorin), which is often prescribed along with methotrexate to ensure that the body's rapidly dividing cells (such as cells of the digestive tract, skin cells, and red blood cells) receive adequate folate. Isoniazid, an antibacterial agent used to treat and prevent tuberculosis, inhibits the conversion of vitamin B_6 to its coenzyme form (pyridoxal phosphate), which is involved in neurotransmitter synthesis. To prevent the resulting peripheral neuropathy, supplements of vitamin B_6 are routinely given during the course of isoniazid treatment.[7]

Diet Alters Drug Metabolism

Some foods affect the activities of enzymes that metabolize drugs or may counteract the drugs' effects in other ways. For example, compounds in grapefruit and grapefruit juice interfere with enzymes that metabolize a number of drugs. As a result of reduced enzyme action, blood concentrations of the drugs increase, leading to stronger physiological effects. The effect of the grapefruit juice lasts for a substantial period, possibly as long as several days after the juice is

> FIGURE H17-1 **Folate and Methotrexate**

By competing for the enzyme that activates folate, methotrexate prevents cancer cells from obtaining the folate they need to multiply. This interference with folate metabolism creates a secondary deficiency of folate that deprives normal cells of the folate they need as well. Notice the similarities in their chemical structures.

Folate

Methotrexate

© Cengage Learning

TABLE H17-2 **Grapefruit Juice–Drug Interactions—Selected Examples**

Drug Category	Drugs Affected by Grapefruit Juice	Drugs Unaffected by Grapefruit Juice
Anticoagulants	—	Acenocoumarol Warfarin
Anti-diabetic drugs	Repaglinide Saxagliptin	Glyburide Metformin
Anti-infective drugs	Erythromycin Saquinavir	Clarithromycin Indinivir
Cardiovascular drugs	Amiodarone Felodipine Nicardipine	Amlodipine Digoxin Diltiazem
Central nervous system drugs	Buspirone Carbamazepine Diazepam	Haloperidol Lorazapam Risperidone
Cholesterol-lowering drugs	Atorvastatin Lovastatin Simvastatin	Fluvastatin Pravastatin Rosuvastatin
Immunosuppressants	Cyclosporine Tacrolimus	Prednisone

© Cengage Learning

consumed; thus, the interaction cannot be avoided by simply separating grapefruit juice consumption from drug usage.[8] Table H17-2 provides examples of drugs that interact with grapefruit juice, as well as some drugs that are not affected.

A number of dietary factors affect the activity of the anticoagulant drug warfarin (Coumadin). One important interaction is with vitamin K, which is structurally similar to warfarin. Warfarin acts by blocking the enzyme that activates vitamin K, thereby preventing the synthesis of blood-clotting factors. The amount of warfarin prescribed is dependent, in part, on how much vitamin K is in the diet. If vitamin K consumption from foods or supplements changes substantially, it can alter the effect of the drug. Individuals using warfarin are advised to consume similar amounts of vitamin K daily to keep warfarin activity stable. The dietary sources highest in vitamin K are dark-green, leafy vegetables.

Drugs Alter Nutrient Excretion

Drugs that increase urine production may reduce nutrient reabsorption in the kidneys, resulting in greater urinary losses of the nutrients. For example, some diuretics accelerate the losses of calcium, potassium, magnesium, and thiamin; thus, dietary supplements may be needed to avoid deficiency. Risk of mineral depletion is highest if multiple drugs with the same effect are used, if kidney function is impaired, or if medications are used for a long time. Note that some diuretics may cause certain minerals to be retained, rather than excreted.

Corticosteroids, which are used as anti-inflammatory agents and immunosuppressants, promote sodium and water retention and increase urinary potassium excretion.[9] Long-term use of corticosteroids can have multiple adverse effects, which include muscle wasting, bone loss, weight gain, and hyperglycemia, with eventual development of osteoporosis and diabetes.

Diets Alter Drug Excretion

Inadequate excretion of medications can cause toxicity, whereas excessive losses may reduce the amount available for therapeutic effect. Some food components influence drug excretion by altering the amount of drug reabsorbed by the kidneys. For example, the amount of lithium (a mood stabilizer) reabsorbed by the kidneys correlates with the amount of sodium reabsorbed. Consequently, both dehydration and sodium depletion, which increase sodium reabsorption, may result in lithium retention. Similarly, a person with a high sodium intake will excrete more sodium in the urine, and therefore more lithium. Individuals using lithium are advised to maintain a consistent sodium intake from day to day in order to maintain stable blood concentrations of lithium.

Urine acidity can also affect drug excretion due to the effects of pH on a compound's chemical structure. The medication quinidine, used to treat arrhythmias, is excreted more readily in acidic urine. Foods or drugs that cause urine to become more alkaline may reduce quinidine excretion and raise its blood levels.

Diet-Drug Toxicities

Some interactions between foods and drugs can cause toxicity or exacerbate a drug's side effects. The combination of tyramine, a compound in some foods, and monoamine oxidase (MAO) inhibitors, which treat depression and Parkinson's disease, can be fatal. MAO inhibitors block an enzyme that normally inactivates tyramine, as well as the hormones epinephrine and norepinephrine. When people who take MAO inhibitors consume excessive tyramine, the increased tyramine in the blood can cause a sudden release of accumulated norepinephrine. This surge in norepinephrine results in severe headaches, rapid heartbeat, and a dangerous increase in blood pressure. For this reason, people taking MAO inhibitors are advised to restrict their intakes of foods rich in tyramine (see Table H17-3). Tyramine occurs naturally in foods and is also formed when bacteria degrade the protein in

TABLE H17-3 **Examples of Foods with a High Tyramine Content**	
• Aged cheeses (cheddar, Gruyère)	• Fish or shrimp sauce
• Aged or cured meats (sausage, salami)	• Prepared soy foods (miso, tempeh, tofu)
• Beer	• Soy sauce
• Fermented vegetables (sauerkraut, kimchi)	• Yeast extract (Marmite, Vegemite)

NOTE: The tyramine content of foods depends on storage conditions and processing; thus the amounts in similar products can vary substantially.

© Cengage Learning

foods. Thus, the tyramine content of a food usually increases when a food ages or spoils. Individuals at risk of tyramine toxicity are advised to buy mainly fresh foods and consume them promptly.

The Inactive Ingredients in Drugs

Besides the active ingredients, medicines may contain other substances such as sugar, sorbitol, lactose, and sodium. For most people who use medicines on occasion and in small amounts, such ingredients pose no problem. When medicines are taken regularly or in large doses, however, people on special diets may need to be aware of these additional ingredients and their effects.

Sugar, Sorbitol, and Lactose

Many liquid preparations contain sugar or sorbitol to make them taste better. For people who must regulate their intakes of carbohydrates, such as people with diabetes, the amount of sugar in these medicines may need to be considered. Large doses of liquids containing sorbitol may cause diarrhea. The lactose added as filler to some medications may cause problems for people who are lactose intolerant.

Sodium

Antibiotics and antacids often contain sodium. People who take Alka Seltzer, for example, may not realize that a single two-tablet dose may exceed their recommended sodium intake for a whole day. In addition, antacids neutralize stomach acid, and many nutrients depend on acid for their digestion. Taking any antacid regularly will reduce the absorption of many nutrients.

Nutrient interactions and risks are not unique to prescription drugs. People who buy over-the-counter drugs also need to protect themselves. The increasing availability of over-the-counter drugs allows people to treat themselves for many ailments from arthritis to yeast infections. Consumers need to ask their physicians about potential interactions and check with their pharmacists for instructions on taking drugs with foods. If problems arise, they should seek professional care without delay.

CRITICAL THINKING QUESTIONS

A. What could be done to minimize the risks associated with nutrient-drug interactions?

B. Drug packaging commonly includes an oversized sheet of paper with small writing that includes a description of the many possible side effects, including possible nutrient-drug interactions. Unfortunately, few people take the time to carefully read and follow these instructions. How might you improve patient education and compliance with these instructions?

REFERENCES

1. American Society of Consultant Pharmacists, ASCP Fact Sheet, https://www.ascp.com/articles/about-ascp/ascp-fact-sheet, accessed February 2014.

2. D. Gnjidic and coauthors, Polypharmacy cutoff and outcomes: Five or more medicines were used to identify community-dwelling older men at risk of different adverse outcomes, *Journal of Clinical Epidemiology* 65 (2012): 989–995.

3. J. I. Boullata and L. M. Hudson, Drug-nutrient interactions: A broad view with implications for practice, *Journal of the Academy of Nutrition and Dietetics* 112 (2012): 506–517.

4. L. N. Chan, Drug-nutrient interactions, in A. C. Ross and coeditors, eds., *Modern Nutrition in Health and Disease* (Baltimore: Lippincott Williams & Wilkins, 2014), pp. 1440–1452; J. I. Boullata and L. M. Hudson, Drug-nutrient interactions: A broad view with implications for practice, *Journal of the Academy of Nutrition and Dietetics* 112 (2012): 506–517.

5. J. M. Gervasio, Drug-induced changes to nutritional status, in J. I. Boullata and V. T. Armenti, eds., *Handbook of Drug-Nutrient Interactions* (New York: Humana Press, 2010), pp. 427–445.

6. D. Fleisher and coauthors, Drug absorption with food, in J. I. Boullata and V. T. Armenti, eds., *Handbook of Drug-Nutrient Interactions* (New York: Humana Press, 2010), pp. 209–241.

7. I. F. Btaiche, B. V. Sweet, and M. D. Kraft, Positive drug-nutrient interactions, in J. I. Boullata and V. T. Armenti, eds., *Handbook of Drug-Nutrient Interactions* (New York: Humana Press, 2010), pp. 303–339.

8. D. G. Bailey, Grapefruit and other fruit juices interactions with medications, in J. I. Boullata and V. T. Armenti, eds., *Handbook of Drug-Nutrient Interactions* (New York: Humana Press, 2010), pp. 267–302.

9. G. W. Cannon, Immunosuppressing drugs including corticosteroids, in L. Goldman and A. I. Schafer, eds., *Goldman's Cecil Medicine* (Philadelphia: Saunders, 2012), pp. 159–165.

18

Diet and Health

Nutrition in Your Life

No doubt, you're familiar with the recommendations. Eat more veggies. Eat more fiber. Eat more fish. Put down the saltshaker. Limit saturated fats. Be active. Don't smoke. And don't drink too much alcohol. What's the deal? If you follow this advice, will it really make a difference in how well or how long you live? In a word, yes. You can bet your life on it. If you could grow old in good health without having a heart attack or stroke, or getting diabetes, hypertension, or cancer, wouldn't you be willing to do just about anything—including improving your diet and activity habits? Of course, you would. And you can start today. In the Nutrition Portfolio at the end of this chapter, you can determine your risk factors for chronic diseases and learn how to minimize those risks.

Much of this text has described how good nutrition supports good health. This chapter examines some of the relationships between nutrition and disease—exploring how poor nutrition may promote the progression of diseases and how good nutrition may guard against the development of diseases. The bulk of this chapter focuses on the chronic diseases that pose the greatest threat to the lives of most people in developed countries, but it begins with a description of the immune system and its inflammatory response. As you will see, inflammation underlies the development of many chronic diseases. Chronic diseases develop over a lifetime as a result of metabolic abnormalities induced by such factors as genetics, age, gender, and lifestyle. As you have learned, diet is among the many lifestyle factors that influence the development of chronic diseases.[1]

In their efforts to prevent and treat chronic diseases, many people in the United States turn to complementary and alternative medicine—which includes products such as herbs and practices such as acupuncture. Because complementary and alternative approaches range from useful folk medicine to dangerous frauds, consumers need to become informed about the products and practices they are considering and discuss their decisions with their health care providers. Highlight 18 explores the usefulness and safety of complementary and alternative medicine and its role in improving health care.

18.1 Nutrition and Infectious Diseases

> **› LEARN IT** Identify factors that protect people from the spread of infectious diseases and describe the role of nutrition in immunity.

Infectious diseases such as smallpox once claimed the lives of many children and limited the average life expectancy of adults. Thanks to medical science's ability to identify disease-causing microorganisms and develop preventive strategies, most people now live well into their later years, and the average life expectancy far exceeds that of our ancestors. In developed nations, purification of water and safe handling of foods help prevent the spread of infection. Antibiotics and immunizations provide additional protection for individuals.[2]

Despite these advances, some infectious diseases still endanger many lives today. Disease strains such as tuberculosis and some foodborne infections, for example, have become resistant to antibiotics, while other diseases such as measles and pertussis, despite vaccine availability, have resurged periodically.[3] Public health strategies help the entire country defend against the spread of infection, and each individual's immune system provides a personal line of defense. A strong immune system depends on adequate nutrition. Poor nutrition weakens the immune system, which increases susceptibility to infections.[4]

The Immune System The **immune system** defends the body so diligently that people do not even notice the thousands of enemy attacks mounted against them every day (the accompanying glossary defines immune system terms). If the immune system fails, though, the body suddenly becomes vulnerable to every wayward disease-causing agent that comes its way. Infectious disease invariably follows.

The body's first lines of defense—the skin, mucous membranes, and GI tract—normally deter foreign substances. If these barriers fail, then the organs (spleen, lymph nodes, and thymus) and cells of the immune system race into action. Foreign substances that gain entry into the body and elicit such a response—the **immune response**—are called **antigens**. Examples include bacteria, viruses, toxins, and food proteins that cause allergies.

Of the 100 trillion cells that make up the human body, one in every hundred is a white blood cell. Two types of white blood cells, the phagocytes and lymphocytes, defend the body against infectious diseases.

Phagocytes: Neutrophils and Macrophages **Phagocytes,** the scavengers of the immune system, are the first to arrive at the scene if an invader, such as a microorganism, gains entry. Upon recognizing the foreign invader, the phagocyte engulfs and digests it, if possible, in a process called **phagocytosis.** Two types of immune system cells ingest and destroy foreign antigens by phagocytosis: **neutrophils** and **macrophages.** Neutrophils are the most common type of white blood cell and are responsible for much of the body's protection against

infectious diseases: diseases caused by bacteria, viruses, parasites, or other microorganisms that can be transmitted from one person to another through air, water, or food; by contact; or through vector organisms such as mosquitoes.

GLOSSARY
OF IMMUNE SYSTEM TERMS

antibodies: large proteins of the blood and body fluids, produced by the immune system in response to the invasion of the body by foreign molecules (usually proteins called *antigens*). Antibodies combine with and inactivate the foreign invaders, thus protecting the body.

antigens: substances that elicit the formation of antibodies or an inflammation reaction from the immune system. A bacterium, a virus, a toxin, and a protein in food that causes allergy are all examples of antigens.

B-cells: lymphocytes that produce antibodies. *B* stands for *bone marrow*, where the B-cells develop and mature.

cytokines (SIGH-toe-kines): special proteins that direct immune and inflammatory responses.

immune response: the body's reaction to foreign antigens, which neutralizes or eliminates them, thus preventing damage.

immune system: the body's natural defense against foreign materials that have penetrated the skin or mucous membranes.

immunoglobulins (IM-you-noh-GLOB-you-linz): proteins capable of acting as antibodies.

lymphocytes (LIM-foh-sites): white blood cells that participate in acquired immunity; B-cells and T-cells.

macrophages (MAK-roe-fay-jez): large phagocytic cells that serve as scavengers of the blood, clearing it of old or abnormal cells, cellular debris, and antigens.

neutrophils (NEW-tro-fills): the most common type of white blood cell. Neutrophils destroy antigens by phagocytosis.

phagocytes (FAG-oh-sites): white blood cells (neutrophils and macrophages) that have the ability to ingest and destroy foreign substances.

- **phagein** = to eat

phagocytosis (FAG-oh-sigh-TOH-sis): the process by which phagocytes engulf and destroy foreign materials.

T-cells: lymphocytes that attack antigens. *T* stands for the *thymus gland*, where the T-cells mature.

infection. Macrophages move and kill bacteria more slowly than neutrophils, but they are larger and can engulf larger targets, including the body's dead and damaged cells. Phagocytes also secrete special proteins called **cytokines** that activate the metabolic and immune responses to infection.

Lymphocytes: B-cells There are two distinct types of **lymphocytes:** B-cells and T-cells. **B-cells** respond to infection by rapidly dividing and producing large proteins known as **antibodies.** Antibodies travel in the bloodstream to the site of infection. There they stick to the surfaces of antigens and kill or otherwise inactivate them, making them easy for phagocytes to ingest.

The antibodies are members of a class of proteins known as **immunoglobulins**—literally, large globular proteins that provide immunity. Antibodies react selectively to a specific foreign organism, and the B-cells retain a memory of how to make them. Consequently, the immune system can respond with greater speed the next time it encounters the same foreign organism. By doing so, B-cells play a major role in resistance to infection.

Lymphocytes: T-cells The **T-cells** travel directly to the invasion site to battle the invaders. T-cells recognize the antigens displayed on the surfaces of phagocyte cells and multiply in response. Then they release powerful cytokines to destroy all the foreign particles that have this antigen on their surfaces. As the T-cells begin to win the battle against infection, they release signals to slow down the immune response.

Unlike the phagocytes, which are capable of inactivating many different types of invaders, T-cells are highly specific. Each T-cell can attack only one type of antigen. This specificity is remarkable, for nature creates millions of antigens. After destroying a particular antigen, some T-cells retain the necessary information to serve as memory cells so that the immune system can rapidly produce the same type of T-cells again if the identical infection recurs.

T-cells actively defend the body against fungi, viruses, parasites, and a few types of bacteria; they can also destroy cancer cells. In organ transplant patients, T-cells participate in the rejection of newly transplanted tissues, which is why physicians prescribe immunosuppressive drugs following such surgery.

Each of the immune system cells plays a key role in fighting infectious disease. The differentiation and multiplication of these cells depends on a full array of nutrients.

Nutrition and Immunity The immune system is sensitive to subtle changes in nutrition status. Malnutrition—whether undernutrition or overnutrition—compromises both immune system tissues and immune responses.[5] From day to day, the immune system requires adequate amounts of protein, fat, carbohydrate, vitamins, and minerals to maintain its tissues and mount a response. Immune cells in tissues, lymph, and blood constantly survey their surroundings to detect foreign invaders and destroy those invaders before any symptoms of illness set in. If some of these foreign invaders escape destruction, then the immune system initiates a response. The immune response demands greater amounts of nutrients for the synthesis of antibodies and cytokines, for cell multiplication, for free radical generation, and for the active process of ending the response.[6] Exactly which nutrients are needed in greater quantities, and how much, is not yet fully known, but Table 18-1 (p. 588) lists several nutrients known to play key roles in immunity.

Impaired immunity opens the way for infectious diseases, which typically raise nutrient needs and reduce food intake. Consequently, nutrition status suffers further. Thus disease and malnutrition create a **synergistic** downward spiral that must be broken for recovery to occur.

Quite simply, optimal immunity depends on optimal nutrition—enough, but not too much, of each of the nutrients. People with weakened immune systems, such as the elderly, benefit from a nutritious diet and supplements of selected nutrients.

synergistic (SIN-er-JIS-tick): multiple factors operating together in such a way that their combined effects are greater than the sum of their individual effects.

TABLE 18-1 Selected Nutrients in Immune Function

The immune system requires all nutrients for optimal functioning. The nutrients listed here have well-known, specific roles in immunity.

Nutrient	Key Role(s) in Immune Function
Protein	Maintains healthy skin and other epithelial tissues (barriers to infection); participates in the synthesis and function of the organs and cells of the immune system and antibody production
Omega-3 fatty acids	Help to resolve inflammation when it is no longer needed for the immune response through production of lipid mediators known as resolvins and protectins, which also play a role in enhancing host defense
Vitamin A	Maintains healthy skin and other epithelial tissues (barriers to infection); participates in the cell replication and specialization that supports immune-cell and antibody production and the anti-inflammatory response
Vitamin D	Regulates immune cell (T-cell) responses; participates in antibody production
Vitamins C and E	Protect against oxidative damage
Vitamin B_6	Helps maintain an effective immune response; participates in antibody production
Vitamin B_{12} and folate	Assist in cell replication and specialization that support immune cell and antibody production
Selenium	Protects against oxidative damage
Zinc	Helps maintain an effective immune response; participates in antibody production

© 2016 Cengage Learning

HIV and AIDS

Perhaps the most well-known infectious disease today is **AIDS (acquired immune deficiency syndrome)**. AIDS develops from infection by **HIV (human immunodeficiency virus)**, which is transmitted by direct contact with contaminated body fluids, including semen, vaginal secretions, and blood (but not saliva), or by passage of the infection from a mother to her infant during pregnancy, birth, or breastfeeding. HIV attacks the immune system and disables the body's defenses against other diseases. Then these diseases, which would produce only mild, if any, illness in people with healthy immune systems, destroy health and life.

Even though remarkable progress has been made in understanding and treating HIV infection, the disease still has no cure. Without a cure, the best course is prevention. Unlike the chronic diseases featured in the remainder of this chapter, AIDS prevention does not in any way depend on good nutrition. Although good nutrition cannot prevent or cure AIDS, an adequate diet may improve responses to drugs, shorten hospital stays, promote independence, and improve the quality of life.[7] In addition, because common food bacteria can easily overwhelm a compromised immune system, attention to food safety is critical. (Chapter 19 provides food safety strategies.)

Inflammation and Chronic Diseases

The immune system's response to infection or injury results in inflammation. The blood supply to the infected area increases and the blood vessels become permeable, which allows the white blood cells to rush to the site. As phagocytes engulf the offending microbes, they release oxidative molecules, such as hydrogen peroxide, that kill the microbes. In this acute phase, inflammation fights off the infection or injury, removes damaged tissue, heals wounds, and promotes recovery from external stressors.[8] In this way, acute inflammation and oxidative stress are beneficial.

When the inflammatory process persists, however, chronic inflammation is harmful. Cells of chronically inflamed tissues produce cytokines, oxidative molecules, blood clotting factors, and other bioactive chemicals that sustain the inflammatory response.[9] Chronic inflammation is apparent in obesity, and helps to explain why obesity underlies so many of the chronic diseases discussed in the remainder of this chapter.

> **REVIEW IT** Identify factors that protect people from the spread of infectious diseases and describe the role of nutrition in immunity.

Public health measures such as purification of water and safe handling of food help prevent the spread of infection in developed nations, and immunizations and antibiotics protect individuals. Nevertheless, some infectious diseases still endanger people today. Nutrition cannot prevent or cure infectious diseases, but adequate intakes of all the nutrients can help support the immune system as the body defends against disease-causing agents. If the immune system is impaired because of malnutrition or diseases such as AIDS, a person becomes vulnerable to infectious disease. Inflammation underlies obesity and many chronic diseases.

AIDS (acquired immune deficiency syndrome): the late stage of HIV infection, in which severe complications of opportunistic infections and cancers develop.

HIV (human immunodeficiency virus): the virus that destroys lymphocytes and impairs immunity, eventually causing AIDS.

18.2 Nutrition and Chronic Diseases

> **LEARN IT** List the leading nutrition-related causes of death in the United States.

Figure 18-1 shows the 10 leading causes of death in the United States.[10] Many of these deaths reflect chronic diseases that developed in response to lifestyle factors such as diet, physical inactivity, overweight, tobacco use, and alcohol and drug abuse. Note that four of these causes of death, including the top two, have some relationship with diet. Taken together, these four conditions account for 60 percent of the nation's more than 2 million deaths each year. Worldwide, statistics are similar, with developing nations sharing many of the same chronic diseases as developed nations.[11]

This chapter explains how the major chronic diseases develop and summarizes their most important links with nutrition. Earlier chapters that describe the connections between individual nutrients and diseases may have left the mistaken impression of "one disease–one nutrient" relationships. Indeed, valid links do exist between saturated fat and heart disease, calcium and osteoporosis, and antioxidant nutrients and cancer, but focusing only on these links oversimplifies the story. In reality, each nutrient may have connections with several diseases because its role in the body is not specific to a disease but to a body function. Furthermore, each of the chronic diseases develops in response to multiple risk factors, including many nondietary factors such as genetics, physical inactivity, and smoking. This chapter presents an integrated and balanced approach to disease prevention, paying careful attention to all of the factors involved. Table 18-2 (p. 590) presents some of the relationships between risk factors and chronic diseases. Figure 18-2 (p. 590) shows how many of the diseases themselves are risk factors for other chronic diseases. Notice that obesity is a gateway disease that contributes to several other diseases.

All of the diseases listed in Table 18-2 have a genetic component. A family history of a certain disease is a powerful indicator of a person's tendency to contract that disease. Still, lifestyle factors are often pivotal in determining whether that tendency will be expressed.[12] Genetics and lifestyle often work synergistically; for instance, cigarette smoking is especially likely to bring on heart disease in people who

© 2001 PhotoDisc, Inc.

Vegetables rich in fiber, phytochemicals, and the antioxidant nutrients (beta-carotene, vitamin C, and vitamin E) help to protect against chronic diseases.

> **FIGURE 18-1** **The 10 Leading Causes of Death in the United States**

Many deaths have multiple causes, but diet influences the development of several chronic diseases—notably, heart disease, some cancers, stroke, and diabetes.

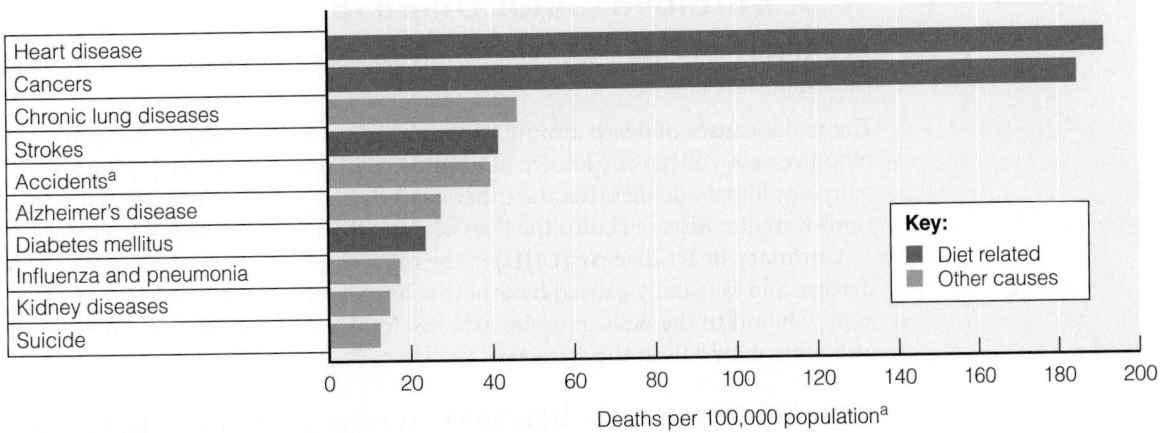

NOTE: Rates are age adjusted to allow relative comparisons of mortality among groups and over time.
[a]Motor vehicle and other accidents are the leading cause of death among people aged 15–24, followed by homicide, suicide, cancer, and heart disease. Alcohol contributes to about half of all accident fatalities.
SOURCE: Data from D. L. Hoyert, J. Xu, and the Division of Vital Statistics, Deaths: Preliminary data for 2011, *National Vital Statistics Reports*, October 10, 2012.

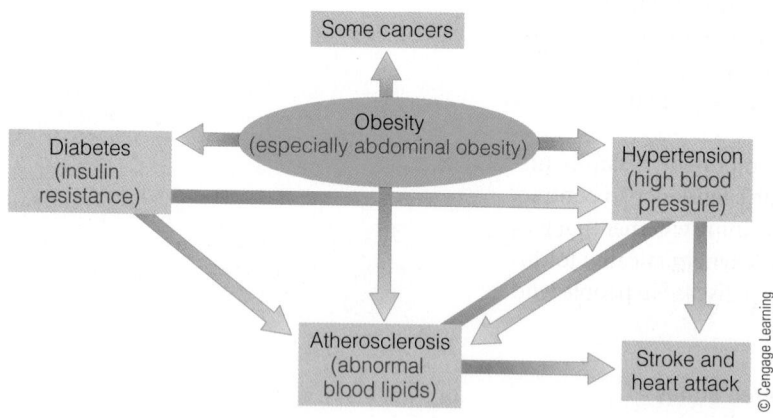

TABLE 18-2 Risk Factors and Chronic Diseases

	Cancers	Hypertension	Diabetes (type 2)	Atherosclerosis	Obesity	Stroke
Dietary Risk Factors						
Diets high in added sugars (beverages)					✓	
Diets high in salty or pickled foods	✓	✓				
Diets high in saturated and/or *trans* fat	✓	✓	✓	✓	✓	✓
Diets low in fruits, vegetables, and other foods rich in fiber and phytochemicals	✓		✓	✓	✓	✓
Diets low in vitamins and/or minerals	✓	✓		✓		
Excessive alcohol intake	✓	✓		✓	✓	✓
Other Risk Factors						
Age	✓	✓	✓	✓		✓
Environmental contaminants	✓					
Genetics	✓	✓	✓	✓	✓	✓
Sedentary lifestyle	✓	✓	✓	✓	✓	✓
Smoking and tobacco use	✓	✓		✓		✓
Stress		✓		✓		✓

© Cengage Learning

> FIGURE 18-2 **Interrelationships among Chronic Diseases**

Notice that many chronic diseases are themselves risk factors for other chronic diseases and that all of them are linked to obesity. The risk factors highlighted in blue define the metabolic syndrome.

are genetically predisposed to develop it. Not smoking would benefit everyone's health, of course, regardless of genetic predisposition, but some recommendations to prevent chronic diseases best meet an individual's needs when family history is considered. For example, women with a family history of breast cancer might reduce their risks if they abstain from alcohol, whereas those with a family history of heart disease might benefit from one or two glasses of wine a week.

> **REVIEW IT** List the leading nutrition-related causes of death in the United States.

Heart disease and cancers are the two leading causes of death in the United States, and strokes and diabetes also rank among the top 10. All four of these chronic diseases have significant links with nutrition. Other lifestyle risk factors and genetics are also important.

18.3 Cardiovascular Disease

> **LEARN IT** Describe how atherosclerosis develops and strategies to lower blood cholesterol levels.

The major causes of death around the world today are diseases of the heart and blood vessels, collectively known as **cardiovascular disease (CVD).** (The accompanying glossary defines this and other heart disease terms.) In the United States, cardiovascular disease claims the lives of more than 800,000 people each year.[13]

Coronary heart disease (CHD) is the most common form of cardiovascular disease and is usually caused by **atherosclerosis** in the **coronary arteries** that supply blood to the heart muscle. Atherosclerosis is the accumulation of lipids and other materials in the arteries.

How Atherosclerosis Develops
As Highlight 16 points out, no one is free of the fatty streaks that may one day become the **plaques** of atherosclerosis.*

---

*Plaque associated with atherosclerosis is known as *atheromatous* (ATH-er-OH-ma-tus) *plaque.*

For most adults, the question is not whether you have plaques, but how advanced they are and what you can do to slow or reverse their progression.

Atherosclerosis or "hardening of the arteries" usually begins with the accumulation of soft fatty streaks along the inner arterial walls, especially at branch points (review Figure H16-1 on p. 549). These fatty streaks gradually enlarge and harden as they fill with cholesterol, other lipids, and calcium, and they become encased in fibrous connective tissue, forming plaques. Plaques stiffen the arteries and narrow the passages through them. Most people have well-developed plaques by the age of 30. As Chapter 5 points out, a diet high in saturated and *trans* fats contributes to the development of plaques and the progression of atherosclerosis.[14] But atherosclerosis is much more than the simple accumulation of lipids within the artery wall—it is a complex inflammatory response to tissue damage. Indeed, extensive evidence confirms that inflammation is centrally involved in all stages of atherosclerosis.[15]

Inflammation The cells lining the blood vessels may incur damage from high LDL cholesterol, hypertension, toxins from cigarette smoking, obesity, elevated homocysteine, or some viral and bacterial infections.[16] Such damage increases the permeability of the blood vessel walls and elicits an inflammatory response. The immune system sends in macrophages (the large, phagocytic cells of the immune system mentioned earlier), and the smooth muscle cells of the artery wall try to repair the damage. Particles of LDL cholesterol become trapped in the blood vessel walls. Free radicals produced during inflammatory responses oxidize the LDL cholesterol, and the macrophages engulf it. The macrophages swell with large quantities of oxidized LDL cholesterol and eventually become the cells of plaque. Arterial damage and the inflammatory response also favor the formation of blood clots and allow minerals to harden plaque and form the fibrous connective tissue that encapsulates it.

The inflammatory response of atherosclerosis weakens the walls of the arteries and may cause an **aneurysm**—the abnormal bulging of a blood vessel wall. Aneurysms can rupture and lead to massive bleeding and death, particularly when a large blood vessel such as the aorta is affected. The central role of the inflammatory response in atherosclerosis has led researchers to look for signs or markers of inflammation in the blood vessel walls. One of these markers is a protein known as **C-reactive protein (CRP)**. High levels of CRP have proved to more accurately predict future heart attack than high LDL cholesterol, which has a strong relationship with atherosclerosis, as a later section explains.[17] Another important inflammatory marker, **lipoprotein-associated phospholipase A(2)** or

C-reactive protein (CRP): a protein released during the acute phase of infection or inflammation that enhances immunity by promoting phagocytosis and activating platelets. Its presence may be used to assess a person's risk of an impending heart attack or stroke.

lipoprotein-associated phospholipase A(2) or **(Lp-PLA(2):** a lipoprotein-bound enzyme that generates potent proinflammatory and proatherogenic products such as oxidized free fatty acids and lysophosphatidylcholine. LpPLA(2) is a specific marker of plaque inflammation.

GLOSSARY
OF HEART DISEASE TERMS

aneurysm (AN-you-rizm): an abnormal enlargement or bulging of a blood vessel (usually an artery) caused by damage to or weakness in the blood vessel wall.

angina (an-JYE-nah or AN-ji-nah): a painful feeling of tightness or pressure in and around the heart, often radiating to the back, neck, and arms; caused by a lack of oxygen to an area of heart muscle.

atherosclerosis (ATH-er-oh-scler-OH-sis): a type of artery disease characterized by plaques along the inner walls of the arteries.

cardiovascular disease (CVD): diseases of the heart and blood vessels throughout the body.

CHD risk equivalents: disorders that raise the risk of heart attacks, strokes, and other complications associated with cardiovascular disease to the same degree as existing CHD. These disorders include symptomatic carotid artery disease, peripheral arterial disease, abdominal aortic aneurysm, and diabetes mellitus.

coronary arteries: blood vessels that supply blood to the heart.

coronary heart disease (CHD): the damage that occurs when the blood vessels carrying blood to the heart (the *coronary arteries*) become narrow and occluded.

embolism (EM-boh-lizm): the obstruction of a blood vessel by an *embolus* (EM-boh-luss), or traveling clot, causing sudden tissue death.

- **embol** = to insert, plug

heart attack: sudden tissue death caused by blockages of vessels that feed the heart muscle; also called *myocardial* (my-oh-KAR-dee-al)

infarction (in-FARK-shun) or *cardiac arrest*.

- **myo** = muscle
- **cardial** = heart
- **infarct** = tissue death

hypertension: consistently higher-than-normal blood pressure. Hypertension that develops without an identifiable cause is known as *essential* or *primary hypertension*; hypertension that is caused by a specific disorder such as kidney disease is known as *secondary hypertension*.

plaques (PLACKS): the accumulation of fatty deposits, smooth muscle cells, and fibrous connective tissue that develops in the artery walls in atherosclerosis.

prehypertension: slightly higher than normal blood pressure, but not as high as hypertension (see Table 18-4, p. 593).

stroke: an event in which the blood flow to a part of the brain is cut off; also called *cerebrovascular accident (CVA)*.

- **cerebro** = brain
- **vascular** = blood vessels

thrombosis (throm-BOH-sis): the formation of a *thrombus* (THROM-bus), or a blood clot, that may obstruct a blood vessel, causing gradual tissue death.

- **thrombo** = clot

transient ischemic (is-KEY-mik) **attack (TIA):** a temporary reduction in blood flow to the brain, which causes temporary symptoms that vary depending on the part of the brain affected. Common symptoms include light-headedness, visual disturbances, paralysis, staggering, numbness, and inability to swallow.

Lp-PLA(2), appears to be a highly specific marker for plaque inflammation and the formation of plaques that are most susceptible to rupture.[18]

Plaques Once plaque has formed, a sudden spasm or surge in blood pressure in an artery can tear away part of its fibrous coat, causing it to rupture. Some plaques are more unstable than others and are therefore more vulnerable to rupture.[19]* Such plaques have a thin fibrous cap, a large lipid core, and an abundance of macrophages—characteristics that undermine plaque stability. Researchers now know that the *composition* of plaques—rather than the *size* of plaques—is a key predictor of plaque rupture and subsequent clot formation. When plaques rupture, the immune system responds to the damage as it would to other tissue injuries.

Blood Clots Blood clots form when **platelets** and other factors cover an injured or damaged area. Abnormal blood clotting can trigger life-threatening events. For example, a blood clot may gradually grow large enough to restrict or close off a blood vessel **(thrombosis).** A *coronary thrombosis* blocks blood flow through an artery that feeds the heart muscle. A *cerebral thrombosis* blocks blood flow through an artery that feeds the brain. A clot may also break free from an artery wall and travel through the circulatory system until it lodges in a small artery and suddenly shuts off blood flow to the tissues **(embolism).**

The action of platelets is under the control of certain eicosanoids, known as prostaglandins and thromboxanes, which are made from the 20-carbon omega-6 and omega-3 fatty acids (introduced in Chapter 5). Each eicosanoid plays a specific role in helping to regulate many of the body's activities, including the regulation of blood pressure, blood clot formation, blood vessel contractions, immune response, and nerve impulse transmissions. Sometimes their actions oppose each other.[20] For example, one eicosanoid prevents clot formation, and another promotes it. Similarly, one dilates the blood vessels, and another constricts them. When omega-3 fatty acids are abundant in the diet, they make more of the kinds of eicosanoids that favor heart health.[21] (Table 5-4 on p. 153 includes good sources of omega-3 fatty acids.)

Blood Pressure The stress of blood flow along artery walls can cause physical damage to arteries.[22] High blood pressure intensifies the stress of blood flow on arterial tissue, provoking a low-grade inflammatory state that may stimulate plaque formation and progression.[23]

The Result: Heart Attacks and Strokes When atherosclerosis in the coronary arteries becomes severe enough to restrict blood flow and deprive the heart muscle of oxygen, CHD develops. The person with CHD often experiences pain and pressure in the area around the heart **(angina).** A **heart attack** occurs when blood flow to the heart is cut off and that area of the heart muscle dies. Restricted blood flow to the brain causes a **transient ischemic attack (TIA)** or **stroke.** Coronary heart disease and strokes are the first and fourth leading causes of death, respectively, for adults in the United States.

Risk Factors for Coronary Heart Disease
Although atherosclerosis can develop in any blood vessel, the coronary arteries are most often affected, leading to CHD. Table 18-3 lists the major risk factors for CHD. Notice that some risk factors, such as diet and physical activity, are *modifiable*, meaning that they can be changed; others, such as age, gender, and family history are not modifiable. The criteria for defining blood lipids, blood pressure, and obesity in relation to CHD risk are shown in Table 18-4; Table H16-1 on p. 550 presents cholesterol standards for children and adolescents.

By age 20, half of the adults in the United States have at least one major risk factor for CHD, and many have more than one.[24] Public health officials in the

---

TABLE 18-3 Risk Factors for CHD

Major Risk Factors for CHD (not modifiable)

- Increasing age
- Male gender
- Family history of premature heart disease

Major Risk Factors for CHD (modifiable)

- High blood LDL cholesterol
- Low blood HDL cholesterol
- High blood triglyceride (VLDL) levels
- High blood pressure (hypertension)
- Diabetes
- Obesity (especially abdominal obesity)
- Physical inactivity
- Cigarette smoking
- An "atherogenic" diet (high in saturated fats and low in vegetables, fruits, and whole grains)

NOTE: Risk factors highlighted in yellow have relationships with diet.
SOURCE: A. S. Go and coauthors, Heart disease and stroke statistics—2014 update: A report from the American Heart Association, *Circulation* 129 (2014): e28-e292; Expert Panel on Detection, Evaluation, and Treatment of High Blood Cholesterol in Adults (Adult Treatment Panel III), *Third Report of the National Cholesterol Education Program (NCEP)*, NIH publication no. 02-5215 (Bethesda, Md.: National Heart, Lung, and Blood Institute, 2002), pp. II-15–II-20.

© Cengage Learning

---

platelets: tiny, disc-shaped bodies in the blood, important in blood clot formation.

*Plaque that is susceptible to rupture because it has only a thin fibrous barrier between its lipid-rich core and the artery lining is called *vulnerable plaque.*

TABLE 18-4 Standards for CHD Risk Factors

Risk Factors	Desirable	Borderline	High Risk
Total blood cholesterol (mg/dL)	<200	200–239	≥240
LDL cholesterol (mg/dL)	<100[a]	130–159	160–189[b]
HDL cholesterol (mg/dL)	≥60	59–40	<40
Triglycerides, fasting (mg/dL)	<150	150–199	200–499[c]
Body mass index (BMI)[d]	18.5–24.9	25–29.9	≥30
Blood pressure (systolic and/or diastolic pressure)[e]	<120/<80	120–139/80–89[f]	≥140/≥90[g]

© Cengage Learning

[a]100–129 mg/dL LDL indicates a near or above optimal level.
[b]≥190 mg/dL LDL indicates a very high risk.
[c]≥500 md/dL triglycerides indicates a very high risk.
[d]Body mass index (BMI) is defined in Chapter 8; BMI standards are found on the inside back cover.
[e]Blood pressure is measured in millimeters of mercury (mm Hg). Blood pressure is measured both when the heart muscle contracts (systolic blood pressure) and when it relaxes (diastolic blood pressure).
[f]These values indicate prehypertension.
[g]These values indicate stage one hypertension; ≥160/≥100 indicates stage two hypertension. Physicians use these classifications to determine medical treatment.

United States recommend screening to identify risk factors in individuals and offer preventive advice for the population.[25] Regular screening and early detection have proved successful: since 1960, both blood cholesterol levels and deaths from cardiovascular disease among US adults have shown a continuous and substantial downward trend.[26] These trends also reflect behavior changes in individuals. As adults grow older, many of them stop smoking, limit alcohol consumption, and become mindful that their food choices can improve their cardiovascular health.

Age, Gender, and Family History A review of Table 18-3 shows that three of the major risk factors for CHD cannot be modified by diet or otherwise: age, gender, and family history. As men and women grow older, the risk of CHD rises. The increasing risk of CHD with advancing age reflects the steady progression of atherosclerosis.[27] On average, older people have more atherosclerosis than younger people do.

In men, aging becomes a significant risk factor at age 45 and older. CHD occurs about 10 to 15 years later in women than in men. Women younger than 45 tend to have lower LDL cholesterol than men of the same age, but women's blood cholesterol typically begins to rise between ages 45 and 55. Thus aging becomes a significant risk factor for women who are 55 and older. The gender difference has been attributed to a protective effect of estrogen in women. After menopause, women's risks increase, in part, because low estrogen levels have unfavorable effects on lipoprotein levels and arterial function.[28] As in men, however, heart disease rates increase linearly with age. The speed at which atherosclerosis progresses, however, depends more on the presence or absence of risk factors such as hypertension, high blood cholesterol, diabetes, and smoking than on age alone.[29] Ultimately, CHD kills as many women as men—and kills more women in the United States than any other disease.

Nonetheless, at every age, men have a greater risk of CHD than women do. The reasons for this gender difference are not completely understood, but they can be partly explained by the earlier onset of risk factors such as elevated LDL cholesterol and blood pressure in men. Levels of the amino acid homocysteine, which damages artery walls and increases oxidative stress, rise with age and are generally higher in men. Researchers have not determined whether the damage is caused by homocysteine itself or by an associated factor.[30]

A history of early CHD in immediate family members is an independent risk factor even when other risk factors are considered. The more family members affected and the earlier the age of onset, the greater the risk.[31]

High LDL and Low HDL Cholesterol In population studies, the relationship between total blood cholesterol and atherosclerosis is strong—and most of the total cholesterol is made up of LDL cholesterol. The higher the LDL cholesterol, the greater the risk of CHD. In contrast, the lower the LDL cholesterol and blood pressure, the slower the progression of atherosclerosis.[32]

The LDL are clearly the most **atherogenic** lipoproteins. As Chapter 5 explains, HDL also carry cholesterol, but HDL are helpfully returning the cholesterol from the cells to the liver, where it will be used to make bile and excreted in the GI tract. Because HDL remove cholesterol from circulation and contain proteins that inhibit inflammation, LDL oxidation, and plaque accumulation, *low* HDL levels can contribute to the development of atherosclerosis as well.[33] Having adequate levels of HDL is beneficial, but high levels of HDL do not necessarily confer additional benefit.[34]

Any LDL cholesterol that remains in the blood after the body's cells take up the amount they need becomes vulnerable to oxidation. High blood levels of LDL cholesterol, especially oxidized LDL, trigger a series of events that promote plaque formation and contribute to plaque instability.[35] Oxidized LDL cholesterol stimulates production of atherogenic signaling molecules. These signaling molecules attract immune cells and allow them to adhere to, and penetrate, the innermost layer of the arterial wall. Within the arterial wall, these immune cells become macrophages that form the lipid-rich foam cells characteristic of fatty streaks. Macrophages within the arterial wall perpetuate chronic inflammation by releasing inflammatory cytokines that contribute to plaque rupture.[36] When plaques rupture, a heart attack or stroke may occur. In the early stages of atherosclerosis, the goal of treatment is to slow plaque development. In the later stages, the goal of treatment is to stabilize plaques.

LDL vary in size and density, and these LDL subtypes have differing effects on heart disease risk. The smallest, most dense LDL can slip into the artery walls easily and are therefore more atherogenic than the larger, less dense LDL.[37]

High Blood Triglycerides High blood triglycerides (found in VLDL) also promote atherosclerosis, either by influencing the production of other atherogenic lipoproteins or by causing molecular changes in endothelial cells and macrophages that promote inflammation and plaque development.[38] About one-third of adults in the United States have high blood triglyceride levels. High blood triglyceride levels are associated with a sedentary lifestyle, overweight and obesity (especially abdominal obesity), and type 2 diabetes.

High Blood Pressure (Hypertension) Atherosclerosis is frequently accompanied by chronic high blood pressure (**hypertension**). The higher blood pressure is above normal, the greater the risk of heart disease. However, even values only slightly higher than desirable—classified as **prehypertension** in Table 18-4—increase the risk of heart attack and stroke.[39] This relationship between hypertension and heart disease risk holds true for men and women, young and old. High blood pressure injures the artery walls and accelerates plaque formation, thus initiating or worsening the progression of atherosclerosis. Then the plaques and reduced blood flow raise blood pressure further, and hypertension and atherosclerosis become mutually aggravating conditions.

Diabetes Diabetes is a major independent risk factor for cardiovascular disease and substantially increases the risk of death from CHD.[40] In diabetes, blood vessels often become blocked and circulation diminishes. Atherosclerosis progresses rapidly. For many people with diabetes, the risk of heart attack is similar to that for people with CHD. In fact, physicians describe diabetes and other disorders that have risks similar to CHD as **CHD risk equivalents.** Treatment to lower LDL cholesterol in diabetes follows the same recommendations as in CHD.

Obesity and Physical Inactivity Obesity, especially abdominal obesity, and physical inactivity significantly increase risk factors for CHD, contributing to high

atherogenic: able to initiate or promote atherosclerosis.

LDL cholesterol, low HDL cholesterol, high triglycerides, hypertension, and diabetes.[41] Conversely, weight loss and physical activity protect against CHD by lowering LDL, raising HDL, lowering triglycerides, improving insulin sensitivity, and lowering blood pressure.[42]

Cigarette Smoking Cigarette smoking is a powerful risk factor for CHD and other forms of cardiovascular disease.[43] The risk increases the more a person smokes and is the same for men and women. Smoking damages the heart directly by increasing blood pressure and the heart's workload. It deprives the heart of oxygen and damages platelets, making blood clot formation likely. Toxins in cigarette smoke damage blood vessels, setting the stage for atherosclerosis. When people quit smoking, their risk of CHD begins to decline within a few months.

Atherogenic Diet Diet also influences the risk of CHD. An "atherogenic" diet—high in saturated fats, *trans* fats, and cholesterol and low in vegetables, fruits, and whole grains—elevates LDL cholesterol. Conversely, eating patterns rich in fruits, vegetables, and whole grains seem to lower the risk of CHD even more than might be expected based on risk factors such as LDL cholesterol alone.[44] A number of beneficial factors share the credit, including vitamins, minerals, fibers, phytochemicals, and omega-3 fatty acids. Research shows that a diet rich in raw vegetables and fruits may protect against CHD even in people who are genetically susceptible to the disease.[45] Dietary and other lifestyle strategies to reduce the risk of CHD are presented in a later section.

Metabolic Syndrome As Table 18-3 (p. 592) shows, most of the modifiable risk factors for CHD are directly related to diet. Several of these diet-related risk factors—low HDL, high blood triglycrides, high blood pressure, elevated fasting blood glucose, and abdominal obesity—comprise a cluster of health risks known as **metabolic syndrome.** As Figure 18-2 (p. 590) shows, the risks that define metabolic syndrome underlie several chronic diseases and increase the risks of CHD and type 2 diabetes.[46] Abdominal obesity and **insulin resistance** are both considered primary factors contributing to the metabolic syndrome.[47] Metabolic syndrome, like the chronic diseases associated with it, also includes markers of inflammation and thrombosis.[48] Overeating and physical inactivity play major roles in the development of metabolic syndrome. The prevalence of metabolic syndrome among US adults is high, and treatment to reduce these risk factors for heart disease and diabetes should begin early and focus on changes in lifestyle.

Recommendations for Reducing Cardiovascular Disease Risk

Recommendations to reduce cardiovascular disease (CVD) risk, including CHD, comprise both screening and intervention. The American Heart Association, together with the American College of Cardiology, developed guidelines for the assessment of CVD risk, the treatment of elevated blood cholesterol, and the management of lifestyle factors that reduce risk.[49]

Screening for Cardiovascular Disease Risk The American Heart Association has developed an online calculator and downloadable spreadsheet to estimate risk for CVD.* The risk factors included in the assessment are age, sex, race, total cholesterol, HDL cholesterol, blood pressure, blood pressure medication use, diabetes, and smoking. Once risks have been assessed, the guidelines define when physicians should prescribe the use of cholesterol-lowering medications (statins),

metabolic syndrome: a combination of risk factors—elevated fasting blood glucose, hypertension, abnormal blood lipids, and abdominal obesity—that greatly increase a person's risk of developing coronary heart disease; also called *Syndrome X, insulin resistance syndrome,* or *dysmetabolic syndrome.*

insulin resistance: the condition in which a normal amount of insulin produces a subnormal effect in muscle, adipose, and liver cells, resulting in an elevated fasting glucose; a metabolic consequence of obesity that precedes type 2 diabetes.

---

*To calculate your risk for CVD, visit http://my.americanheart.org/cvriskcalculator and http://www.cardiosource.org/science-and-quality/practice-guidelines-and-quality-standards/2013-prevention-guideline-tools.aspx.

TABLE 18-5 Recommendations and Strategies for Reducing CVD Risk

Recommendation	Strategy
Dietary patterns: Choose a dietary pattern that emphasizes vegetables, fruits, and whole grains; includes low-fat dairy products, poultry, fish, legumes, nontropical vegetable oils, and nuts; and limits intake of sweets, sugar-sweetened beverages, and red meats.	Adopt a heart-healthy eating pattern such as DASH, the USDA Food Pattern, Mediterranean diet, or other well-planned plant-based dietary pattern.
Energy: Balance energy intake and physical activity to prevent weight gain and to achieve or maintain a healthy body weight.	Engage in at least 30 minutes of moderate-intensity endurance activity or 15 minutes of vigorous-intensity endurance activity on most days of the week.
Saturated fat, *trans* fat, and cholesterol: Limit saturated fat to less than 10 percent of total kcalories, *trans* fat to less than 1 percent of total kcalories, and cholesterol to less than 300 milligrams a day. For those who have been advised to lower LDL cholesterol or those who have diabetes, limit saturated fat to 5 to 6 percent of total kcalories and cholesterol to less than 200 milligrams a day.	Replace solid fats (saturated and *trans* fats) with food or oils containing unsaturated fatty acids such as omega-3 fatty fish, avocado, nuts, olive oil, canola oil, safflower oil, corn oil, and soybean oil.
Sodium: Limit sodium intake to 1500 milligrams per day.	Choose and prepare foods with little or no salt.
Alcohol: If alcohol is consumed, practice moderation.	Limit alcohol intake to no more than one drink daily for women and two drinks daily for men.
Tobacco: Avoid exposure to any form of tobacco or tobacco smoke.	Do not smoke or use tobacco in any form and avoid exposure to second-hand smoke.

SOURCES: R. H. Eckel and coauthors, 2013 AHA/ACC Guidelines on lifestyle management to reduce cardiovascular risk: A report of the American College of Cardiology/American Heart Association Task Force on Practice Guidelines, *Circulation*, http://circ.ahajournals.org; M. R. Flock and P. M. Kris-Etherton, Dietary Guidelines for Americans 2010: Implications for cardiovascular disease, *Current Atherosclerosis Reports* 13 (2011): 499-507; D. Mozaffarian, L.J. Appel, and L. Van Horn, Recent advances in preventive cardiology and lifestyle medicine, *Circulation* 123 (2011): 2870-2891.

© Cengage Learning

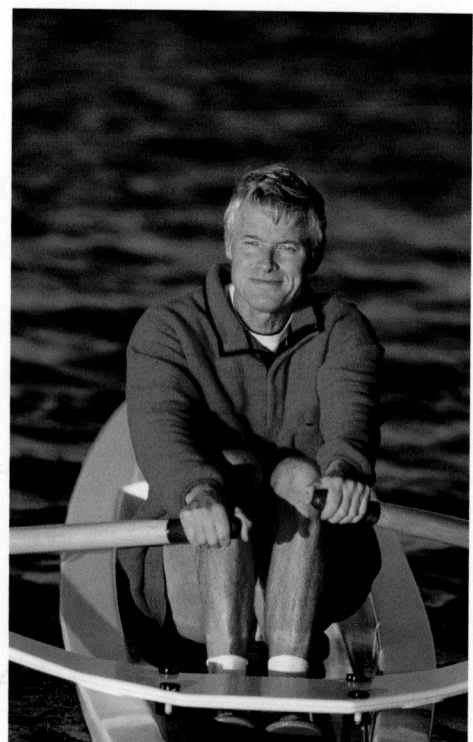

Regular aerobic exercise can help to defend against heart disease by strengthening the heart muscle, promoting weight loss, and improving blood lipids and blood pressure.

with cautions to use clinical judgments with each individual.[50] Thus, health care providers are encouraged to discuss options with their patients and to consider personal preferences, possible harms, costs, and inconveniences of taking medications. In this way, the assessment guidelines focus on patient-centered care. In addition to medications, treatment plans include major lifestyle changes in diet, physical activity, and smoking cessation.

Lifestyle Interventions To reduce the risk of CVD, people are encouraged to increase physical activity, lose weight (if necessary), implement dietary changes, and reduce exposure to tobacco smoke either by quitting smoking or by avoiding secondhand smoke. Altering one's lifestyle is challenging, and instruction and guidance are critical for success. Health professionals can explain the reasons for change, set obtainable goals, and offer practical suggestions. If lifestyle changes fail to lower LDL or blood pressure to acceptable levels, then medications are prescribed. Table 18-5 summarizes strategies to reduce the risk of heart disease.[51] The "How To" feature on p. 597 offers suggestions for implementing a heart-healthy diet.

> **DIETARY GUIDELINES FOR AMERICANS**
Consume less than 300 milligrams per day of dietary cholesterol. Individuals at high risk of cardiovascular disease should consume less than 200 milligrams of cholesterol daily.

> **REVIEW IT** Describe how atherosclerosis develops and strategies to lower blood cholesterol levels.

Atherosclerosis is characterized by plaque build-up in artery walls. Plaques rupturing or blood clotting can cause heart attacks and strokes. Dietary recommendations to lower the risks of cardiovascular disease are summarized in Table 18-5. Quitting smoking and engaging in regular physical activity also improve heart health.

> How To

Implement a Heart-Healthy Diet

Following a heart-healthy diet can require major changes in dietary choices. It helps to make a few changes at a time and to focus on positive choices (what to eat) first, rather than negative ones (what not to eat).

Grains

- Choose whole-grain breads and cereals that list "whole wheat" as the first ingredient on labels and provide at least 1.1 grams of dietary fiber per 10 grams of carbohydrate.
- Limit foods that list any *trans* fat in the Nutrition Facts panel or "hydrogenated oil" in the ingredients list.
- Limit products that contain tropical oils (coconut, palm, and palm kernel oil), which are high in saturated fat.

Fruits and Vegetables

- Consume fruits and vegetables frequently. Keep the refrigerator stocked with a variety of colorful fruits and vegetables (carrots, grapes, blueberries, melon).
- Incorporate at least one or two servings of fruits and vegetables into each meal.
- Choose canned products carefully. Canned vegetables (especially tomato-based products) may be high in sodium. Canned fruits may be high in added sugars.
- Limit high-sodium foods such as pickles, olives, sauerkraut, and kimchee.
- Limit french fries from fast-food restaurants, which are often loaded with *trans* fats.

Lunch and Dinner Entrées

- Limit meat, fish, and poultry servings to 5 ounces per day.
- Include at least two 3-ounce servings of fish rich in omega-3 fatty acids (salmon, mackerel, herring, trout) each week.

- Select lean cuts of beef, such as sirloin tip, round steak, and arm roast, and lean cuts of pork, such as center-cut ham, loin chops, and tenderloin. Trim visible fat before cooking.
- Select extra-lean ground meat and drain well after cooking. Use lean ground turkey, without skin added, in place of ground beef.
- Limit cholesterol-rich organ meats (liver, brain, sweetbreads).
- Limit egg yolks to no more than three per week; replace whole eggs in recipes with egg whites or commercial egg substitutes.
- Include more vegetarian entrées or legume dishes to reduce meat intake and boost vegetable, soluble fiber, and soy protein intakes.
- Restrict these high-sodium foods:
  - Cured or smoked meats such as beef jerky, bologna, corned or chipped beef, frankfurters, ham, luncheon meats, salt pork, and sausage
  - Salty or smoked fish, such as anchovies, caviar, salted or dried cod, herring, sardines, and smoked salmon
  - Packaged, canned, or frozen soups, sauces, and entrées

Milk Products

- To obtain two to three servings of milk daily, include a portion of fat-free or low-fat milk, yogurt, or cottage cheese in each meal.
- Use yogurt or fat-free sour cream to make dips or salad dressings. Substitute evaporated fat-free milk for heavy cream.
- Limit foods high in saturated fat or sodium, such as cheese, processed cheeses, ice cream, and other milk-based desserts.

Fats and Oils

- Add nuts (not salted) and avocados to meals to increase monounsaturated fat intakes and make meals more appetizing.

- Include unsaturated vegetable oils such as canola, corn, olive, peanut, safflower, sesame, soybean, and sunflower oils, in salad dressings and recipes.
- Use margarines with added plant sterols or stanols regularly.
- Select soft margarines in tubs or liquid form; limit stick margarines and solid vegetable shortenings.

Spices and Seasonings

- Use salt only at the end of cooking, and you will need to add much less. Use salt substitutes at the table.
- Spices and herbs improve the flavor of foods without adding sodium. Try using more garlic, ginger, basil, curry or chili powder, cumin, pepper, lemon, mint, oregano, rosemary, and thyme.
- Check the sodium content on labels. Flavorings and sauces that are usually high in sodium include bouillon cubes, soy sauce, steak and barbecue sauces, relishes, mustard, and catsup.

Snacks and Desserts

- Select low-sodium and low-saturated fat choices such as unsalted pretzels, nuts, popcorn, chips, and crackers.
- Choose canned or dried fruits and some raw vegetables to boost fruit and vegetable intake.
- Enjoy angel food cake, which is made without egg yolks and added fat.
- Select low-fat frozen desserts such as sherbet, sorbet, fruit bars, and some low-fat ice creams.

> **TRY IT** Plan heart-healthy meals for a day and analyze them using your personal profile. Discuss whether these meals meet the strategies listed in Table 18-5 (p. 596) and how to improve any shortcomings.

18.4 Hypertension

> **LEARN IT** Present strategies to lower blood pressure.

Anyone concerned about atherosclerosis and the risk it presents must also be concerned about hypertension. Together, the two are a life-threatening combination. The higher the blood pressure is above normal, the greater the risk. Low blood pressure is generally a sign of long life expectancy and low heart disease risk.[52]

Hypertension affects about one out of three adults in the United States.[53] It contributes to an estimated 1 million heart attacks and almost 800,000 strokes each year. People usually do not feel the physical effects of high blood pressure, but it can impair life's quality and end life prematurely.

How Hypertension Develops The underlying causes of most cases of hypertension are not fully understood, but much is known about the physiological factors that affect blood pressure. Blood pressure arises from contractions in the heart muscle that pump blood away from the heart (**cardiac output**) and the resistance blood encounters in the arterioles (**peripheral resistance**). When either cardiac output or peripheral resistance increases, blood pressure rises. Cardiac output is raised when heart rate or blood volume increases; peripheral resistance is affected mostly by the diameters of the arterioles. Blood pressure is therefore influenced by the nervous system, which regulates heart muscle contractions and the arteriole's diameters, and hormonal signals, which may cause fluid retention or blood vessel constriction. The kidneys play a key role in regulating blood pressure by controlling the secretion of the hormones involved in vasoconstriction and retention of sodium and water (review Figure 12-3 on p. 376).

Risk Factors for Hypertension Several major risk factors predicting the development of hypertension have been identified, including:

- *Aging.* Hypertension risk increases with age. An estimated two-thirds of persons 65 and older have hypertension.[54] Individuals who have normal blood pressure at age 55 still have a 90 percent risk of developing high blood pressure during their lifetimes.

- *Genetics.* Hypertension risk is similar among family members. It is also more prevalent and severe in certain ethnic groups: for African Americans in the United States, the prevalence of high blood pressure is among the highest in the world.[55] Compared with others, African Americans typically develop high blood pressure earlier in life, and their average blood pressure is much higher.

- *Obesity.* Most people with hypertension—an estimated 70 percent—are overweight or obese.[56] Obesity is recognized as a major cause of high blood pressure, and the combination of obesity and hypertension greatly increases the risk for CVD. Obesity raises blood pressure by altering kidney function, increasing blood volume, and promoting blood vessel damage through insulin resistance.[57]

- *Salt intake.* Excess intake of salt plays a major role in hypertension.[58] Reducing salt intake lowers blood pressure in people with or without hypertension, and also diminishes the age-related rise in blood pressure.

- *Alcohol.* Alcohol consumption, especially if consumed regularly in amounts greater than two drinks per day, is strongly associated with hypertension.[59] Alcohol is also associated with strokes independently of hypertension and its use may interfere with medications.

Treatment of Hypertension The single most effective step people can take against hypertension is to find out whether they have it. At checkup time, a health-care professional can provide an accurate resting blood pressure reading. Under normal conditions, blood pressure fluctuates continuously in response to a variety of factors, including stress, and such actions as talking or shifting position. Some people react emotionally to the procedure, which raises the blood

cardiac output: the volume of blood discharged by the heart each minute; determined by multiplying the stroke volume by the heart rate. The stroke volume is the amount of oxygenated blood the heart ejects toward the tissues at each beat. Cardiac output (volume/minute) = stroke volume (volume/beat) × heart rate (beats/minute).

peripheral resistance: the resistance to pumped blood in the small arterial branches (arterioles) that carry blood to the tissues.

pressure reading. For these reasons, if the resting blood pressure is above normal, the reading should be repeated before confirming the diagnosis of hypertension (see Table 18-4 on p. 593 for blood pressure standards). Thereafter, blood pressure should be checked regularly. Both lifestyle modifications and medications are used to treat hypertension. Recently, new guidelines for initiating medication in the treatment of hypertension in adults have been issued.[60] Table 18-6 describes the lifestyle changes that reduce blood pressure and the expected reduction in systolic blood pressure for each change.

Weight Control Efforts to lower blood pressure focus on weight control. Weight loss alone is one of the most effective nondrug treatments for hypertension.[61] Those who are using drugs to control their blood pressure can often reduce or discontinue the drugs when they lose weight. Even a modest weight loss of 5 to 10 percent of body weight can lower blood pressure significantly.

The DASH Eating Pattern Results of the Dietary Approaches to Stop Hypertension (DASH) trials show that an eating pattern rich in fruits, vegetables, low-fat milk products, whole grains, and nuts, and low in total fat and saturated fat can significantly lower blood pressure. In addition to lowering blood pressure, DASH lowers total cholesterol and LDL cholesterol, and reduces inflammation.[62] Compared to the typical American diet, DASH provides more fiber, potassium, magnesium, and calcium and less red meat, sweets, and sugar-containing beverages. Table 18-7 (p. 600) shows how DASH compares to the USDA Food Pattern (introduced in Chapter 2). Both eating patterns meet the goals specified in the *Dietary Guidelines for Americans*.[63]

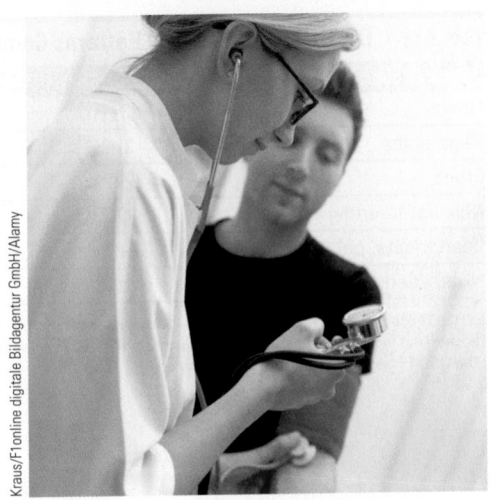
To guard against hypertension, have your blood pressure checked regularly.

> **DIETARY GUIDELINES FOR AMERICANS**
African Americans and individuals with hypertension should increase their intakes of potassium.

Salt/Sodium Intake Strong evidence supports the important role restricting sodium and/or salt plays in preventing and reducing hypertension. Lowering

TABLE 18-6 Lifestyle Modifications to Reduce Blood Pressure

Modification	Recommendation	Expected Reduction in Systolic Blood Pressure
Weight reduction	Maintain healthy body weight (BMI 18.5–24.9).	10 mm Hg/10 kg lost
DASH pattern	Adopt a diet rich in fruits, vegetables, and low-fat milk products to reduce saturated fat intake.	5–6 mm Hg
Sodium restriction	Reduce dietary sodium intake to less than 2300 mg sodium (less than 6 g salt) per day, and further reduce intake to 1500 mg among persons who are 51 and older and those of any age who are African American or have hypertension, diabetes, or chronic kidney disease.	2–7 mm Hg
Physical activity	Perform aerobic physical activity for at least 40 minutes per day, most days of the week.	2–5 mm Hg
Moderate alcohol consumption	Men: Limit to two drinks per day. Women and lighter-weight men: Limit to one drink per day.	2–4 mm Hg

SOURCE: Adapted from R. H. Eckel and coauthors, 2013 AHA/ACC Guidelines on lifestyle management to reduce cardiovascular risk: A report of the American College of Cardiology/American Heart Association Task Force on Practice Guidelines, *Circulation*, http:circ.ahajournals.org; L. Landberg and coauthors, Obesity-related hypertension: Pathogenesis, cardiovascular risk, and treatment, A position paper of The Obesity Society and the American Society of Hypertension, *Journal of Clinical Hypertension* 15 (2013): 14-33; US Department of Agriculture, US Department of Health and Human Services, *Dietary Guidelines for Americans*, 2010, www.dietaryguidelines.gov.

© Cengage Learning

TABLE 18-7 DASH and USDA Food Patterns Compared

Food Group	DASH	USDA
Grains	6–8 oz	6 oz
Vegetables	2–2½ c	2½ c
Fruits	2–2½ c	2 c
Milk (fat-free/low-fat)	2–3 c	3 c
Lean meats, poultry, fish	6 oz or less	5½ oz
Nuts, seeds, legumes	4–5 oz per week	—a

NOTE: These diet plans are based on 2000 kcalories per day. Both DASH and USDA Food Patterns recommend that fats and sugars be used sparingly and with discretion.
aThe USDA Food Patterns combine nuts, seeds, and legumes with meat, poultry, and fish.

© Cengage Learning

Increasing potassium in the diet helps to lower blood pressure and defend against losses caused by some hypertension medicines. The richest sources of potassium are fresh fruits and vegetables of all kinds.

© stocker1970/Shutterstock.com

sodium intake reduces blood pressure regardless of gender or race, presence or absence of hypertension, or whether people follow the DASH eating pattern or a typical American diet. The combination of DASH with a limited intake of sodium, however, improves blood pressure better than either strategy alone.[64] Furthermore, the lower the sodium intake, the greater the drop in blood pressure. (See the "How To" feature in Chapter 12 on p. 384 for suggestions to limit sodium intake.) Research shows that salt restriction provides additional protection against heart disease, beyond lowering blood pressure.[65]

> **DIETARY GUIDELINES FOR AMERICANS**
Reduce daily sodium intake to less than 2300 milligrams and further reduce intake to 1500 milligrams among persons who are 51 and older and those of any age who are African American or have hypertension, diabetes, or chronic kidney disease.

Physical Activity The higher the blood pressure and the less active a person is to begin with, the greater the effect physical activity has in reducing blood pressure. Physical activity helps with weight control, of course, but moderate aerobic activity, such as 30 to 60 minutes of brisk walking most days, also helps to lower blood pressure directly.[66] Those who engage in regular aerobic activity may be able to control mild hypertension without medication.

Medications When diet and physical activity fail to reduce blood pressure, diuretics and antihypertensive drugs may be prescribed. Diuretics lower blood pressure by increasing fluid loss and lowering blood volume.

Some diuretics can lead to a potassium deficiency. People taking these diuretics need to include rich food sources of potassium or supplements daily and watch for signs of potassium imbalances such as weakness (particularly of the legs), unexplained numbness or tingling sensation, cramps, irregular heartbeats, and excessive thirst and urination. Blood potassium should be monitored regularly. Although some diuretics can lead to a potassium deficiency, others spare potassium. A combination of these two types of diuretics may be prescribed to prevent potassium deficiency.

Most people with hypertension use two or more medications to meet their blood pressure goals. Using a combination of drugs with different modes of action can reduce the doses needed and minimize side effects.

> **REVIEW IT** Present strategies to lower blood pressure.
The most effective dietary strategy for preventing hypertension is weight control. Also beneficial are diets rich in fruits, vegetables, nuts, and low-fat milk products and low in fat, saturated fat, and sodium.

18.5 Diabetes Mellitus

> **LEARN IT** Compare the dietary strategies to manage type 1 diabetes with those to prevent and treat type 2 diabetes.

The incidence of diabetes among children and adults has risen dramatically in recent years (see Figure 18-3). An estimated 21 million people in the United States have been diagnosed with diabetes.[67] As many as 86 million adults have **prediabetes**—their blood glucose is elevated but not to such an extent as to be classified as diabetes.[68] People with prediabetes have a high risk of developing diabetes. The glossary on p. 602 defines diabetes terms.

Diabetes ranks seventh among the leading causes of death (review Figure 18-1 on p. 589). In addition, diabetes underlies, or contributes to, several other major diseases, including heart disease, stroke, hypertension, blindness, and kidney failure. Cardiovascular disease is the leading cause of diabetes-related deaths.[69]

How Diabetes Develops **Diabetes mellitus** includes several metabolic disorders characterized by high blood glucose concentrations and disturbed insulin metabolism. People with diabetes may have insufficient insulin, ineffective insulin, or a combination of the two. The result is **hyperglycemia,** a marked elevation in blood glucose that can ultimately damage blood vessels, nerves, and tissues.

To appreciate the problems presented by an absolute or relative lack of insulin, consider insulin's normal action. After a meal, insulin signals the body's cells to receive the energy nutrients from the blood—amino acids, glucose, and fatty acids. Insulin helps to maintain blood glucose within normal limits and stimulates protein synthesis, glycogen synthesis in liver and muscle, and fat synthesis. Without insulin, glucose regulation falters, and metabolism of the energy-yielding nutrients changes.

Diagnosis of Diabetes Diabetes can be diagnosed by one of several tests, including a **fasting plasma glucose** or a nonfasting **A1C**.[70] A fasting plasma glucose test measures plasma glucose after a person has fasted for at least 8 hours to determine whether current glucose levels are within the normal range (see Table 18-8 on p. 602). The A1C test measures the percentage of a person's hemoglobin that has glucose attached to it. The higher the percentage, the higher a person's blood glucose levels have been over time. Because red blood cells typically live for about 3 months, the A1C test provides a long-term evaluation of glucose control. Table 18-8 includes A1C values.

Types of Diabetes Mellitus Table 18-9 (p. 602) shows the distinguishing features of the two main forms of diabetes, type 1 diabetes and type 2 diabetes. As the following pages explain, the development of type 1 and type 2 diabetes differs, but some of their complications are similar.

Type 1 Diabetes In **type 1 diabetes,** the less common type of diabetes (about 5 to 10 percent of all diagnosed cases), the pancreas loses its ability to synthesize the hormone insulin. Type 1 diabetes is an **autoimmune disorder.** In most cases, the individual inherits a defect in which immune cells mistakenly attack and destroy the insulin-producing beta cells of the pancreas. The rate of beta cell destruction in type 1 diabetes varies. In some people (mainly infants and children), destruction is rapid; in others (mainly adults), it is slow. Type 1 diabetes commonly occurs in childhood and adolescence, but it can occur at any age, even late in life.

Without insulin, the body's energy metabolism changes, with such severe consequences as to threaten survival. The cells must have insulin to take up the needed fuels from the blood. People with type 1 diabetes must receive insulin either by injection or external pumps; insulin cannot be taken orally because it is a protein, and the enzymes of the GI tract would digest it.

Type 2 Diabetes **Type 2 diabetes** is the most prevalent form of diabetes, accounting for 90 to 95 percent of cases.[71] The primary defect in type 2 diabetes is insulin resistance, a reduced sensitivity to insulin. Consequently, muscle and adipose cells cannot remove glucose from the blood, and liver cells continue to make glucose. To compensate, the pancreas secretes larger amounts of insulin, and plasma insulin concentrations can rise to abnormally high levels (hyperinsulinemia). Over time, the pancreas becomes less able to compensate for the cells' reduced sensitivity to insulin, and hyperglycemia worsens. The high demand for insulin can eventually exhaust the beta cells of the pancreas and lead to impaired insulin secretion

> FIGURE 18-3 **Prevalence of Diabetes among Adults in the United States**

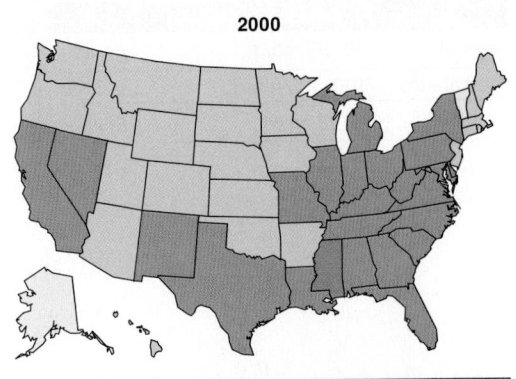

2000

<4.5%	4.5%–5.9%	6.0%–7.4%	7.5%–8.9%	≥9.0%
2 states	24 states	23 states	1 state	0 states

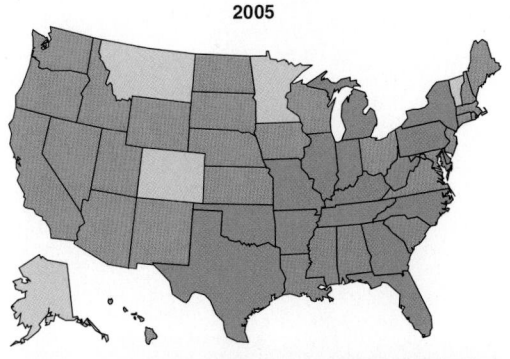

2005

<4.5%	4.5%–5.9%	6.0%–7.4%	7.5%–8.9%	≥9.0%
0 states	5 states	25 states	15 states	5 states

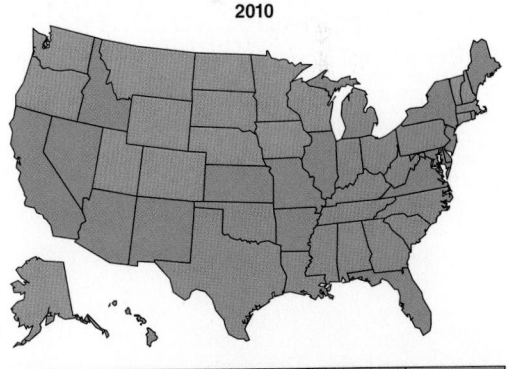

2010

<4.5%	4.5%–5.9%	6.0%–7.4%	7.5%–8.9%	≥9.0%
0 states	0 states	18 states	17 states	15 states

© Cengage Learning

hyperglycemia: elevated blood glucose concentrations.

fasting plasma glucose: a test that measures plasma glucose after a person has fasted (no food or kcaloric beverages) for at least 8 hours.

A1C: a test that measures the percentage of hemoglobin that has glucose attached, which helps to diagnose diabetes and evaluate long-term glycemic control.

autoimmune disorder: a condition in which the body develops antibodies to its own proteins and then proceeds to destroy cells containing these proteins. In type 1 diabetes, the body develops antibodies to its insulin and destroys the pancreatic cells that produce the insulin, creating an insulin deficiency.

TABLE 18-8 Criteria for the Diagnosis of Diabetes

Diagnosis	Fasting Plasma Glucose	A1C
Normal	70–99 mg/dL	<5.7%
Prediabetes	100–125 mg/dL	5.7–6.4%
Diabetes	≥126 mg/dL	≥6.5%

© 2016 Cengage Learning

TABLE 18-9 Features of Type 1 and Type 2 Diabetes

	Type 1	Type 2
Prevalence in diabetic population	5–10% of cases	90–95% of cases
Age of onset	<30 years	>40 years[a]
Associated conditions	Autoimmune diseases, viral infections, inherited factors	Obesity, aging, inherited factors
Major defect	Destruction of pancreatic beta cells; insulin deficiency	Insulin resistance; insulin deficiency (relative to needs)
Insulin secretion	Little or none	Varies; may be normal, increased, or decreased
Requirement for insulin therapy	Always	Sometimes
Older names	Juvenile-onset diabetes Insulin-dependent diabetes mellitus (IDDM)	Adult-onset diabetes Noninsulin-dependent diabetes mellitus (NIDDM)

[a]Incidence of type 2 diabetes is increasing in children and adolescence; in more than 90 percent of these cases, it is associated with overweight or obesity and a family history of type 2 diabetes.

© Cengage Learning

and reduced plasma insulin concentrations. Type 2 diabetes is therefore associated both with insulin resistance and with relative insulin deficiency; that is, the amount of insulin is insufficient to compensate for its diminished action in cells.

Although the actual causes of type 2 diabetes are unknown, the risk is substantially increased by obesity (especially abdominal obesity), poor dietary habits, smoking, excessive alcohol consumption, aging, and physical inactivity. When people take action to control their lifestyle choices, prevention of type 2 diabetes is not only possible, but likely.[72]

Most people with type 2 diabetes are obese, and obesity itself can directly cause some degree of insulin resistance.[73] As discussed in Highlight 16, obesity has led to a dramatic rise in the incidence of type 2 diabetes among children and adolescents during the past two decades.[74] Inherited factors also strongly influence risk, and type 2 diabetes is more common in certain ethnic populations, including Native Americans, Hispanic Americans, Mexican Americans, African Americans, Asian Americans, and Pacific Islanders.

Inflammation contributes to insulin resistance, and its many links with obesity, metabolic syndrome, and CVD make inflammation central to the development of diabetes as well.[75] Chronic inflammation correlates with increases in blood glucose and decreases in insulin effectiveness.

Complications of Diabetes In both types of diabetes, glucose fails to gain entry into the cells and consequently accumulates in the blood. These two problems lead to both acute and chronic complications. Figure 18-4 summarizes the metabolic changes and acute complications that can arise in uncontrolled diabetes. Notice that when some glucose enters the cells, as in type 2 diabetes, many of the symptoms of type 1 diabetes do not occur.

Over the long term, the person with diabetes suffers not only from the acute complications shown in Figure 18-4, but also from its chronic effects. Chronically elevated blood glucose alters glucose metabolism in virtually every cell of the body.

GLOSSARY
OF DIABETES TERMS

diabetes (DYE-uh-BEE-teez): metabolic disorder characterized by elevated blood glucose resulting from insufficient insulin, ineffective insulin, or both;

the complete medical term is *diabetes mellitus* (meh-LIE-tus).

- **mellitus** = honey-sweet (sugar in urine)

prediabetes: condition in which blood glucose levels are higher than normal but below the diagnosis of diabetes; formerly called *impaired glucose*

tolerance. Prediabetes is considered a major risk factor for future diabetes and cardiovascular diseases.

type 1 diabetes: the less common type of diabetes in which the pancreas produces little or no insulin. Type 1 diabetes usually results from autoimmune destruction of pancreatic beta cells.

type 2 diabetes: the more common type of diabetes in which the cells fail to respond to insulin. Type 2 diabetes usually accompanies obesity and results from insulin resistance coupled with insufficient insulin secretion.

> FIGURE 18-4   **Metabolic Consequences of Untreated Diabetes**

The metabolic consequences of type 1 diabetes differ from those of type 2. In type 1, no insulin is available to allow any glucose to enter the cells. When glucose cannot enter the cells, a cascade of metabolic changes quickly follows. In type 2 diabetes, some glucose enters the cells. Because the cells are not "starved" for glucose, the body does not shift into the metabolism of fasting (losing weight and producing ketones).

aHyperosmolar hyperglycemic state usually develops in the absence of ketosis and is most often associated with type 2 diabetes.

© Cengage Learning

Some cells begin to convert excess glucose to sugar alcohols, for example, causing toxicity and cell distention—distended cells in the lenses of the eyes, for example, cause blurry vision. Some cells produce glycoproteins by attaching excess glucose to an amino acid in a protein; the altered proteins cannot function normally, which leads to a host of other problems. (Attachment of glucose to the hemoglobin protein is the basis of the A1C test for diabetes diagnosis mentioned earlier.) The structures of the blood vessels and nerves become damaged, leading to loss of circulation and nerve function. Poor circulation coupled with glucose-rich blood and urine readily lead to infections. People with diabetes must pay special attention to hygiene and keep alert for early signs of infection. Early, aggressive treatment to control blood glucose significantly reduces the risk of long-term diabetes-related complications.

Diseases of the Large Blood Vessels As mentioned, atherosclerosis tends to develop early, progress rapidly, and be more severe in people with diabetes. The interrelationships among insulin resistance, obesity, hypertension, and atherosclerosis help explain why about 75 percent of people with diabetes die as a consequence of cardiovascular diseases, especially heart attacks. Intensive diabetes treatment that keeps blood glucose levels tightly controlled can reduce the risk of cardiovascular disease among some people with diabetes.[76]* Although intensive therapy is associated

---

*Intensive treatment may be inappropriate for some individuals with diabetes; examples include individuals with limited life expectancy or a history of hypoglycemia and middle-aged and older adults with previous heart disease or multiple heart disease risk factors.

with some risks, including an increased risk of abnormally low blood glucose (hypoglycemia), benefits usually outweigh these disadvantages.

Diseases of the Small Blood Vessels For people with diabetes, disorders of the small blood vessels (capillaries)—called **microangiopathies**—may also develop and lead to loss of kidney function and retinal degeneration with accompanying loss of vision. Diabetes is the leading cause of both kidney failure and blindness in adults in the United States.[77]

Diseases of the Nerves Nerve tissues may also deteriorate with diabetes, expressed at first as a painful prickling sensation, often in the arms and legs. Later, the person may lose sensation in the hands and feet. Injuries to these areas may go unnoticed, and infections can progress rapidly. With loss of both circulation and nerve function, undetected injury and infection may lead to death of tissue (gangrene), necessitating amputation of the limbs (most often the legs or feet). People with diabetes are advised to take conscientious care of their feet and visit a podiatrist regularly.

Recommendations for Diabetes
Diet is an important component of diabetes treatment. The American Diabetes Association recognizes that a variety of eating patterns—for example, DASH, the Mediterranean diet, and a plant-based diet—are all acceptable for the management of diabetes and can be designed based on personal preferences and metabolic goals.[78] Similarly, the percentage of kcalories from carbohydrate, protein, and fat is flexible and based on an individual's current eating patterns, metabolic goals, and personal preferences. Diabetes is best managed with the help of a registered dietitian nutritionist, who can provide education and support.

Carbohydrate Intake To maintain near-normal blood glucose levels, the diet is designed to deliver the same amount of carbohydrate each day, spaced evenly throughout the day. Several approaches can be used to plan such diets, but many people with diabetes learn to count carbohydrates using the exchange system that is presented in Appendix G.

Providing a consistent carbohydrate intake spaced throughout the day helps to maintain appropriate blood glucose levels and maximize the effectiveness of drug therapy. Eating too much carbohydrate at one time can raise blood glucose too high, stressing the already-compromised insulin-producing cells. Eating too little carbohydrate can lead to hypoglycemia. As is true for the general US population, carbohydrate intake from vegetables, fruits, whole grains, legumes, and low-fat or nonfat dairy products is recommended over intake from other carbohydrate sources, especially those that have added fats, sugars, or sodium.[79]

Carbohydrate Sources Different carbohydrate-containing foods have varying effects on blood glucose levels; for example, consuming a portion of white rice may cause blood glucose to rise higher and quicker than would a similar portion of barley. As Chapter 4 describes, this *glycemic effect* of foods is influenced by a food's fiber content, the preparation method, the other foods included in a meal, and individual tolerances. For individuals with diabetes, using the glycemic index may provide some additional benefit for achieving glucose control as compared with that obtained by considering only the amount of carbohydrate consumed. In addition, high-fiber, minimally processed foods—which typically have more moderate effects on blood glucose than do highly processed, starchy foods—are among foods frequently recommended for people with diabetes.

Sugar recommendations for people with diabetes are similar to those for the general population, which suggests limiting foods and beverages with added sugars. Of course, sugars and sugary foods must be counted as part of the daily carbohydrate allowance.

Dietary Fat People with diabetes are advised to follow the *Dietary Guidelines for Americans* regarding saturated fat, dietary cholesterol, and *trans* fat intake.[80] These recommendations include reducing saturated fat intake to less than 10 percent of kcalories, limiting daily cholesterol intake to less than 300 milligrams, and limiting *trans* fat as much as possible. As is true for the general population, foods rich

microangiopathies: disorders of the small blood vessels.
- **micro** = small
- **angeion** = vessel
- **pathos** = disease

in omega-3 fatty acids are recommended for those with diabetes because such foods have beneficial effects on lipoproteins and the prevention of heart disease.

Protein An ideal protein intake to control blood glucose or to improve CVD risk factors in diabetes has not been determined.[81] Protein intake goals should therefore be individualized. Because dietary protein increases the insulin response, but does not increase blood glucose concentrations, carbohydrates rich in protein should not be used to treat or prevent hypoglycemia.

Alcohol Adults with diabetes can drink alcohol in moderation. Guidelines are similar to those for the general population, which advise a daily limit of one drink for women and two drinks for men.

Recommendations for Type 1 Diabetes Normally, the body secretes a constant baseline amount of insulin at all times and secretes more as blood glucose rises following meals. People with type 1 diabetes, however, produce little or no insulin. They must learn to adjust the amount and schedule of their insulin doses to accommodate meals, physical activity, and health status. To maintain blood glucose within a fairly normal range requires a lifelong commitment to a carefully coordinated program of diet, physical activity, and insulin.

Nutrition therapy for type 1 diabetes focuses on maintaining optimal nutrition status, controlling blood glucose, achieving a desirable blood lipid profile, controlling blood pressure, and preventing and treating the complications of diabetes. In addition to meeting basic nutrient requirements, the diet must provide a fairly consistent carbohydrate intake from day to day and at each meal and snack to help minimize fluctuations in blood glucose. Further alterations in diet may be necessary for a person with chronic complications such as cardiovascular or kidney disease.

Participation in all levels of physical activity is possible for people with type 1 diabetes who have good blood glucose control and no complications, but they should check with their physician first. One potential problem is hypoglycemia, which can occur during, immediately after, or many hours after physical activity. To avoid hypoglycemia, the person must monitor blood glucose before and after activity to identify when adjustments in insulin or food intake are needed. Carbohydrate-rich foods should be readily available during and after activity to prevent hypoglycemia.

Recommendations for Type 2 Diabetes In overweight people with type 2 diabetes, even moderate weight loss (5 to 10 percent of body weight) can help improve insulin resistance, blood lipids, and blood pressure. Together with diet, a regular routine of moderate physical activity not only supports weight loss, but also improves blood glucose control, blood lipid profiles, and blood pressure. Thus the benefits of regular, long-term physical activity for the treatment and prevention of type 2 diabetes are substantial.[82]

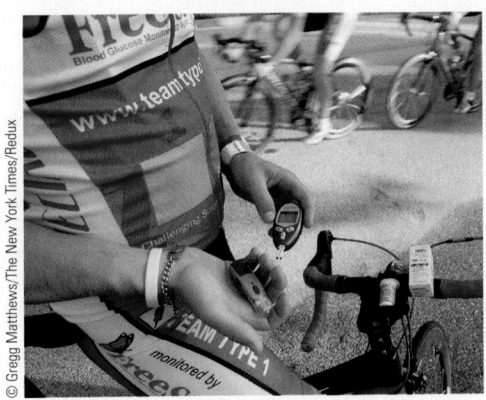

Phil Southerland has raced his bicycle across America and checks his blood glucose regularly. With appropriate diet, regular physical activity, frequent glucose checks, and treatment as needed, people with diabetes can achieve their dreams (www.teamtype1.org).

> **REVIEW IT** Compare the dietary strategies to manage type 1 diabetes with those to prevent and treat type 2 diabetes.
Diabetes is characterized by high blood glucose and insufficient insulin, ineffective insulin, or a combination of the two. People with type 1 diabetes coordinate diet, insulin, and physical activity to help control their blood glucose. Those with type 2 diabetes benefit most from a diet and physical activity program that controls glucose fluctuations and promotes weight loss.

18.6 Cancer

> **LEARN IT** Differentiate among cancer initiators, promoters, and antipromoters and describe how nutrients or foods might play a role in each category.

Cancer, the growth of **malignant** tissue, ranks just below cardiovascular disease as a cause of death in the United States. (See the glossary, p. 606, of cancer terms.) As with cardiovascular disease, the prognosis for cancer today is far brighter than in the past. Identification of risk factors, new detection techniques, and innovative therapies offer hope and encouragement.

antipromoters: factors that oppose the development of cancer.

cancers: malignant growths or tumors that result from abnormal and uncontrolled cell division.

carcinogenesis (CAR-sin-oh-JEN-eh-sis): the process of cancer development.

carcinogen (CAR-sin-oh-jen or car-SIN-oh-jen): a substance that can cause cancer; the adjective is *carcinogenic.*

- **carcin** = cancer
- **gen** = gives rise to

initiators: factors that cause mutations that give rise to cancer, such as radiation and carcinogens.

malignant (ma-LIG-nant): describes a cancerous cell or tumor, which can injure healthy tissue and spread cancer to other regions of the body.

metastasize (me-TAS-tah-size): the spread of cancer from one part of the body to another.

promoters: factors that favor the development of cancers once they have begun.

tumor: an abnormal tissue mass with no physiological function; also called a *neoplasm* (NEE-oh-plazm).

Cancers are classified by the tissues or cells from which they develop:

Adenomas (ADD-eh-NOH-mahz) arise from glandular tissues.

Carcinomas (KAR-see-NOH-mahz) arise from epithelial tissues.

Gliomas (gly-OH-mahz) arise from glial cells of the central nervous system.

Leukemias (loo-KEE-mee-ahz) arise from white blood cell precursors.

Lymphomas (lim-FOH-mahz) arise from lymph tissue.

Melanomas (MEL-ah-NOH-mahz) arise from pigmented skin cells.

Sarcomas (sar-KOH-mahz) arise from connective tissues, such as muscle or bone.

Cancer is not a single disorder. There are many **cancers;** that is, many different kinds of malignant growths. They have different characteristics, occur in different locations in the body, take different courses, and require different treatments.

How Cancer Develops The development of cancer, called **carcinogenesis,** begins as a single cell that loses control of its normal growth and replication processes. A cancer arises from mutations in the genes that control cell division in a single cell. These mutations may promote cellular replication, interfere with growth restraint, or prevent cellular death.[83] The affected cell thereby loses its built-in capacity for halting cell division, and it produces daughter cells with the same genetic defects. As the abnormal mass of cells, called a **tumor,** grows, a network of blood vessels develops to supply the tumor with the nutrients it needs to support its growth.* The tumor can disrupt the functioning of the normal tissue around it, and some tumor cells may **metastasize** to other regions of the body. Figure 18-5 illustrates cancer development. In leukemia (cancer affecting the white blood cells) the cells do not form a tumor, but rather accumulate in blood and other tissues.

The reasons cancers develop are numerous and varied. Vulnerability to cancer is sometimes inherited, such as when a person is born with a genetic defect that alters DNA structure, function, or repair. A few cancers are linked with microbial infections.[84]** Certain metabolic processes may initiate carcinogenesis, as when phagocytes of the immune system produce oxidants that cause DNA damage or when chronic inflammation enhances the rate of cell division and the risk of a damaging mutation. More often, cancers are caused by interactions between a person's genes and the environment. Exposure to a cancer-causing substance, or **carcinogen,** may either induce genetic mutations that lead to cancer or promote proliferation of cancerous cells.

Environmental Factors Among environmental factors, exposure to radiation and sun, water and air pollution, and smoking are known to cause cancer. Lack of physical activity may also play a role in the development of some cancers.[85] Men and women whose lifestyles include regular, vigorous physical activity have the lowest risk of colon cancer, for example.[86] Physical activity may also protect against breast cancer by reducing body weight and by other mechanisms, such as changes in hormone levels and immune functions not related to body weight.[87]

Obesity itself is clearly a risk factor for certain cancers (such as colon, breast in postmenopausal women, endometrial, pancreas, kidney, and esophageal) and possibly for other types (such as gallbladder) as well.[88] Because different cancers have various causes, obesity's influence on cancer development depends on the site as well as other factors such as hormonal interactions. In the case of breast

People with cancer take comfort from the support of others and from the knowledge that medical science is waging an unrelenting battle in their defense.

AP images/Michael Dwyer

---

*An abnormal mass of cells that is noncancerous is called a *benign* tumor.
**Examples include viral hepatitis and liver cancer, human papillomavirus and cervical cancer, and *Helicobacter pylori* bacterium (the ulcer bacterium) and stomach cancer.

> FIGURE 18-5 **Cancer Development**

Normal cells

Initiation →

Mutagens alter the DNA in a cell and induce abnormal cell division.

Promotion →

Promoters enhance the development of abnormal cells, resulting in formation of a tumor.

Further tumor development →

Normal cells
Malignant cells

The cancerous tumor releases cells into the bloodstream or lymphatic system (metastasis).

© Cengage Learning

cancer in postmenopausal women, for example, the hormone estrogen is implicated. Obese postmenopausal women have much higher levels of estrogen than lean women do because fat tissue produces estrogen. Researchers believe that the extended exposure to estrogen in obese women is linked to the increased risk of breast cancer after menopause.[89] The relationships between excessive body weight and certain cancers provide yet another reason to adopt a lifestyle that embraces physical activity and good nutrition.

As Table 18-10 shows, specific dietary constituents are associated with an increased risk of certain cancers. Some dietary factors may initiate cancer development (**initiators**), others may promote cancer development once it has started (**promoters**), and still others may protect against the development of cancer (**antipromoters**).

TABLE 18-10 **Factors Associated with Cancer at Specific Sites**

Cancer Sites	Risk Factors	Protective Factors
Breast (postmenopause)	Alcoholic drinks, body fatness, adult attained height,[a] abdominal fatness, adult weight gain	Lactation, physical activity
Breast (premenopause)	Alcoholic drinks, adult attained height,[a] greater birth weight	Lactation, body fatness
Colon and rectum	Red meat, processed meat, alcoholic drinks, body fatness, abdominal fatness, adult attained height[a]	Physical activity, foods containing dietary fiber, garlic, milk, calcium
Endometrium	Body fatness, abdominal fatness	Physical activity
Esophagus	Alcoholic drinks, body fatness, maté[b]	Nonstarchy vegetables, fruits, foods containing beta-carotene, foods containing vitamin C
Gallbladder	Body fatness	
Kidney	Body fatness	
Liver	Aflatoxins,[c] alcoholic drinks	
Lung	Arsenic in drinking water, beta-carotene supplements[d]	Fruits, foods containing carotenoids
Mouth, pharynx, and larynx	Alcoholic drinks	Nonstarchy vegetables, fruits, foods containing carotenoids
Nasopharynx	Cantonese-style salted fish	
Ovary	Adult attained height[a]	
Pancreas	Body fatness, abdominal fatness, adult attained height[a]	Foods containing folate
Prostate	Diets high in calcium	Foods containing lycopene, foods containing selenium, selenium[e]
Skin	Arsenic in drinking water	
Stomach	Salt, salty and salted foods	Nonstarchy vegetables, allium vegetables,[f] fruits

© Cengage Learning

[a]Adult attained height is unlikely to directly modify the risk of cancer. It is a marker for genetic, environmental, hormonal, and also nutritional factors affecting growth during the period from preconception to completion of linear growth.
[b]As drunk traditionally in parts of South America, scalding hot through a metal straw. Any increased risk of cancer is judged to be caused by epithelial damage resulting from the heat, and not by the herb itself.
[c]Aflatoxins are toxins produced by molds or fungi. The main foods that may be contaminated are all types of grains (wheat, rye, rice, corn, barley, oats) and legumes, notably peanuts.
[d]The evidence is derived from studies using high-dose supplements (20 mg/day for beta-carotene; 25,000 IU/day for retinol) in smokers.
[e]The evidence is derived from studies using supplements at a dose of 200 μg/day. Selenium is toxic at higher doses.
[f]This includes vegetables such as garlic, onions, leeks, and shallots.
NOTE: Strength of evidence for all these factors is either "convincing" or "probable."
SOURCE: World Cancer Research Fund/American Institute for Cancer Research, *Food, Nutrition, Physical Activity and the Prevention of Cancer: A Global Perspective* (Washington, DC: AICR, 2007).

Dietary Factors—Cancer Initiators Some experts estimate that diet may be linked to as many as one-third of all cancer cases.[90] Consequently, many people think that certain foods are carcinogenic, especially those that contain additives or pesticides. As Chapter 19 explains, our food supply is one of the safest in the world. Additives that have been approved for use in foods are not carcinogens. Some pesticides are carcinogenic at high doses, but not at the concentrations commonly found on fruits and vegetables. The benefits of eating fruits and vegetables are far greater than the current risks.

Cancers of the head and neck correlate strongly with the combination of alcohol and tobacco use and with low intakes of fruits and vegetables. Alcohol intake alone is associated with cancers of the mouth, throat, and breast, and alcoholism often damages the liver and precedes the development of liver cancer.[91] These findings illustrate clearly why any potential benefit of moderate alcohol consumption on cardiovascular disease must be weighed against the potential dangers.

Cooking meats at high temperatures (frying, broiling) causes amino acids and creatine in the meats to react together and form carcinogens.* Grilling meat, fish, or other foods over a direct flame causes fat and added oils to splash on the fire, creating carcinogens that vaporize onto the food.** Eating grilled food introduces these carcinogens to the digestive system, where they may damage the stomach and intestinal lining. Once these compounds are absorbed into the blood, however, they are detoxified by the liver. To minimize carcinogen formation when grilling, line the grill with foil or wrap the food in foil, take care not to burn foods, and marinate meats beforehand.

Evidence from population studies spanning the globe for more than 30 years strongly suggests that diets high in red meat and processed meat (meat preserved by smoking, curing, or salting, or by the addition of preservatives) are a cause of colon cancer.[92] Based on such evidence, replacing most servings of red meat with poultry, fish, or legumes and choosing only occasional servings of grilled, fried, blackened, or smoked foods is in the best interest of health.

Another reason to moderate consumption of fried foods such as french fries and potato chips is the presence of acrylamide, a potential carcinogen. Acrylamide is produced when certain starches such as potatoes are fried or baked at high temperatures. In the body, some acrylamide is metabolized to a substance that may mutate or damage genetic materal. As such, acrylamide is classified as "reasonably anticipated to be a human carcinogen."[93] Chapter 19 also discusses acrylamide in foods.

Dietary Factors—Cancer Promoters Unlike carcinogens, which initiate cancers, some dietary components promote cancers. That is, once the initiating step has taken place, these components may accelerate cancer development.

Although studies of animals suggest that high-fat diets may promote cancer, studies of human beings have not proved that the effects of fat are independent of the effects of energy intake and physical inactivity. Overall, the evidence associating fats and oils with cancer risk is limited.[94]

The type of fat in the diet, however, may influence cancer promotion or prevention. Studies of colon cancer implicate animal fats but not vegetable fat, and although a number of studies suggest that omega-3 fatty acids from fish may protect against some cancers, other studies do not support such findings.[95] Researchers note that factors such as total fat intake, the ratio of dietary omega-3 fatty acids to omega-6 fatty acids, an individual's genetic risk of cancer, body fatness, and gender, as well as the specific cancer studied, may all influence the relationship between omega-3 fatty acids and cancer. Such findings underscore the importance of consulting with a health care provider before taking fish oil or other supplements to prevent disease.

Dietary Factors—Antipromoters Some foods may contain antipromoters—dietary compounds that defend against cancer. Table 18-10 (p. 607) includes these protective dietary factors. Research on dietary patterns of populations has led to recommendations aimed at reducing cancer risks.

---

*These carcinogens are *heterocyclic amines*.
**These carcinogens are *polycyclic aromatic hydrocarbons*.

Recommendations for Reducing Cancer Risks A diet rich in fruits and vegetables may provide protection against the development of some cancers.[96] Fruits and vegetables contain both nutrients and phytochemicals with antioxidant activity, and antioxidants may prevent or reduce the oxidative reactions in cells that damage DNA. Phytochemicals may also enhance immune functions that protect against cancer development and promote enzyme reactions that inactivate carcinogens. The nutrients and phytochemicals in fruits and vegetables may also reduce the risk of some cancers by way of epigenetic actions such as DNA methylation and gene silencing.[97] For example, the **cruciferous vegetables**—cabbage, cauliflower, broccoli, and brussels sprouts—contain a variety of phytochemicals that defend against certain cancers. Further knowledge and understanding of the roles and interactions among environmental, genetic, and epigenetic factors will help to explain cancer development, predict a person's susceptibility, and provide specific treatment recommendations.

Fruits and vegetables, as well as legumes and whole grains, are rich in fiber. As Chapter 4 explains, fiber may protect against cancer by binding, diluting, and rapidly removing potential carcinogens from the GI tract. High-fiber and whole-grain foods also help a person to maintain a healthy body weight—another preventive measure against cancer. Physical activity also helps maintain a healthy body weight and reduce the risk of some cancers. Table 18-11 summarizes dietary and lifestyle recommendations for reducing cancer risk.

> **REVIEW IT** Differentiate among cancer initiators, promoters, and antipromoters and describe how nutrients or foods might play a role in each category.
Some dietary factors, such as alcohol and heavily smoked foods, may initiate cancer development; others, such as animal fats, may promote cancer once it has gotten started; and still others, such as fiber, antioxidant nutrients, and phytochemicals, may act as antipromoters that protect against the development of cancer. By eating many fruits, vegetables, legumes, and whole grains and reducing saturated fat intake, people obtain the best possible nutrition at the lowest possible risk. Minimizing weight gain through regular physical activity and a healthy diet is also beneficial.

© Polara Studios, Inc.

Cruciferous vegetables, such as cauliflower, broccoli, and brussels sprouts, contain nutrients and phytochemicals that may inhibit cancer development.

cruciferous vegetables: vegetables of the cabbage family, including cauliflower, broccoli, and brussels sprouts.

TABLE 18-11 Recommendations and Strategies for Reducing Cancer Risk

Recommendation	Strategy
Body fatness: Achieve and maintain a healthy body weight throughout life.	Follow the USDA food pattern for your appropriate energy level.
	Engage in regular physical activity.
	Limit consumption of energy-dense foods and avoid beverages with added sugars.
	Consume "fast foods" sparingly, if at all.
Physical activity: Adopt a physically active lifestyle.	Engage in at least 150 minutes of moderate-intensity physical activity or 75 minutes of vigorous-intensity activity or an equivalent combination throughout the week.
	Limit sedentary behaviors such as sitting, lying down, watching television, or other forms of screen-based recreation.
Plant foods: Consume a healthy diet with an emphasis on plant foods.	Eat at least the daily amounts of vegetables and fruits recommended by the USDA food patterns.
	Choose whole grains instead of refined grain products.
	Limit intake of red meat.
	Limit refined starchy foods.
Alcoholic drinks: If you drink alcoholic beverages, limit consumption.	Drink no more than two drinks a day for men and one drink a day for women.
Preservation, processing, preparation: Limit consumption of salt-cured foods and processed meats.	Avoid salt-preserved, salted, or salty foods.
	Limit consumption of processed foods with added salt to ensure an intake of less than 6 grams of salt (2.4 grams of sodium) a day.
	Avoid processed meats.
Dietary supplements: Aim to meet nutritional needs through diet.	Dietary supplements are not recommended for cancer prevention.

SOURCE: L. H. Kushi and coauthors, American Cancer Society Guidelines on Nutrition and Physical Activity for Cancer Prevention, CA: *Cancer Journal for Clinicians* 62 (2012): 30–67; World Cancer Research Fund/American Institute for Cancer Research, *Food, nutrition, physical activity and the prevention of cancer: A global perspective* (Washington, D.C.: AICR, 2007), pp. 373–390.

Dietary Guidelines	Heart Disease	Hypertension	Diabetes	Cancer
Prevent and/or reduce overweight and obesity.	✓	✓	✓	✓
Increase physical activity and reduce time spent in sedentary behaviors.	✓	✓	✓	✓
Keep total fat 20 to 35 percent of kcalories.	✓			
Consume less than 10 percent of kcalories from saturated fatty acids and keep *trans*-fatty acid consumption as low as possible.	✓		✓	✓
Increase vegetable and fruit intake. Consume at least half of all grains as whole grains.	✓	✓	✓	✓
Reduce the intake of kcalories from added sugars.	✓		✓	
Reduce daily sodium intake to less than 2300 milligrams.[a]	✓	✓	✓	✓
If alcohol is consumed, it should be consumed in moderation.	✓	✓	✓	✓

© Cengage Learning

[a]Further reduce daily sodium intake to 1500 milligrams among persons who are 51 and older and those of any age who are African American or have hypertension, diabetes, or chronic kidney disease.

18.7 Recommendations for Chronic Diseases

› LEARN IT Summarize dietary recommendations to prevent chronic diseases.

This chapter's discussion of chronic diseases described cardiovascular diseases, hypertension, diabetes, and cancer—four different conditions with distinct sets of causes. Yet dietary excesses, particularly excess food energy and saturated fat intakes, increase the likelihood of all of them.[98] Similarly, all are responsive to diet, and in most cases, the beneficial foods are similar.

Not all diet recommendations apply equally to all of the diseases or to all people with a particular disease, but fortunately for the consumer, dietary recommendations do not contradict one another. In fact, they support one another. Most people can gain some disease-prevention benefits by making dietary changes. To that end, the recommendations presented earlier for reducing the risks of heart disease (Table 18-5, p. 596), hypertension (Table 18-6, p. 599), and cancer (Table 18-11, p. 609) describe the kinds of foods people should include or limit. Table 18-12 compares the *Dietary Guidelines for Americans* with all these recommendations for chronic diseases. (A summary of the *Diet, Nutrition, and Prevention of Chronic Diseases* report from the World Health Organization is presented in Appendix I.)

Several recommendations are aimed at weight control. Obesity is common in the United States, and it is linked with most of the chronic diseases that threaten life (review Figure 18-2 on p. 590).[99] The problems of obesity multiply when medical conditions develop. For example, obese people readily develop diabetes, which is often accompanied by high blood pressure and high blood cholesterol.

Physical activity and a moderate weight loss of even 10 to 20 pounds can help improve blood glucose, blood lipids, and blood pressure.

Ryan McVay/Photodisc/Getty Images

> **FIGURE 18-6**   **The Healthy Eating Plate—An Alternative to USDA MyPlate**

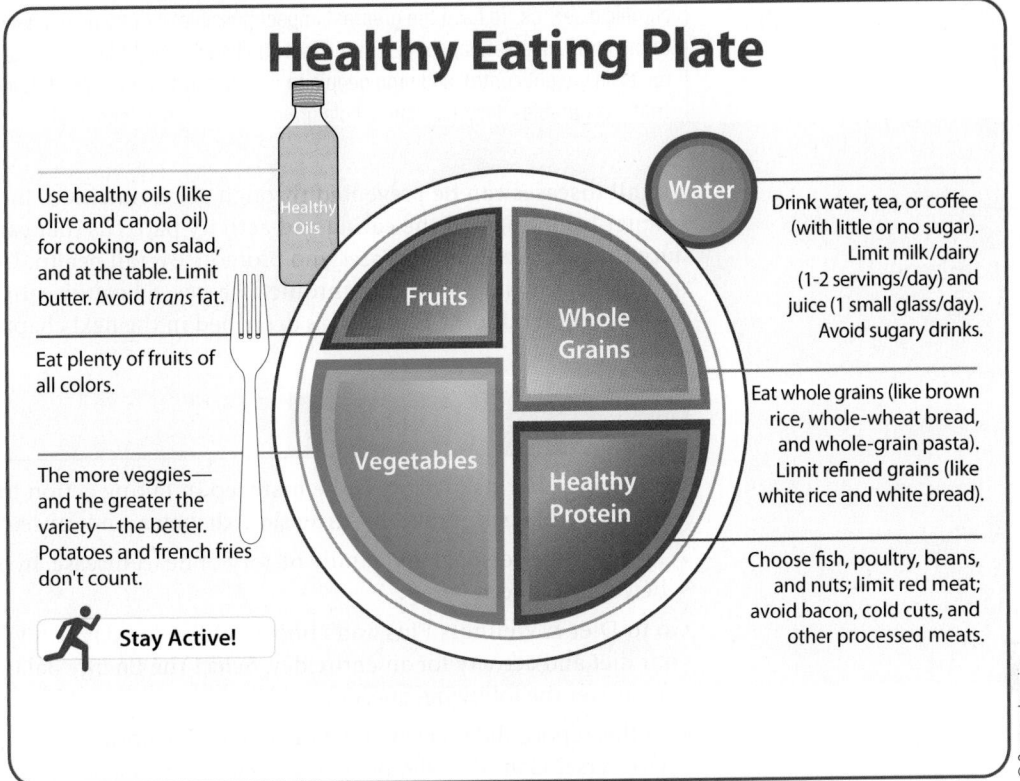

Healthy Eating Plate

Use healthy oils (like olive and canola oil) for cooking, on salad, and at the table. Limit butter. Avoid *trans* fat.

Eat plenty of fruits of all colors.

The more veggies— and the greater the variety—the better. Potatoes and french fries don't count.

Stay Active!

Healthy Oils

Fruits

Whole Grains

Vegetables

Healthy Protein

Water

Drink water, tea, or coffee (with little or no sugar). Limit milk/dairy (1-2 servings/day) and juice (1 small glass/day). Avoid sugary drinks.

Eat whole grains (like brown rice, whole-wheat bread, and whole-grain pasta). Limit refined grains (like white rice and white bread).

Choose fish, poultry, beans, and nuts; limit red meat; avoid bacon, cold cuts, and other processed meats.

© Cengage Learning

Such a combination of problems may require only one treatment: adopting a healthful diet and regular exercise program.

> **DIETARY GUIDELINES FOR AMERICANS**
Increase physical activity and reduce time spent in sedentary behaviors.

The USDA Food Patterns and MyPlate offer guidance on selecting a healthy diet, but as Chapter 2 discusses, critics have noted several shortcomings. An alternative plate was created by the nutrition faculty from the Harvard School of Public Health.[100] The *Healthy Eating Plate* (Figure 18-6) is based on their critical review of links between diet and health.[101]

Recommendations for the Population   Recommendations to prevent chronic diseases take a population approach and urge all people to make dietary changes believed to forestall or prevent diseases. Such a strategy is similar to national efforts to vaccinate to prevent measles, fluoridate water to prevent dental caries, and fortify grains with folate to prevent neural tube defects.

Recommendations for Individuals   Unlike nutrient-deficiency diseases, which develop when nutrients are lacking and disappear when the nutrients are provided, chronic diseases are neither caused nor prevented by diet alone. Many people have followed dietary advice and developed heart disease or cancer anyway; others have ignored all advice and lived long and healthy lives. For many people, though, diet does influence the time of onset and course of some chronic diseases.

To determine whether dietary recommendations are important to you personally, look at your family history to see which diseases are common to your relatives. In addition, examine your personal history, taking note of your body weight, blood pressure, blood lipid profile, and lifestyle habits such as smoking and physical activity.

> REVIEW IT Summarize dietary recommendations to prevent chronic diseases. Clearly, optimal nutrition plays a key role in keeping people healthy and reducing the risk of chronic diseases. To have the greatest impact possible, dietary recommendations are aimed at the entire population, not just at the individuals who might benefit most. Recommendations focus on weight control and urge people to limit saturated and *trans* fat; increase fiber-rich fruits, vegetables, legumes, and whole grains; and balance food intake with physical activity.

Not all diseases can be prevented through good nutrition, but many can. This chapter has presented the eating and activity patterns that can help people reduce the risk of chronic diseases and promote overall health. In addition to promoting health and preventing chronic diseases, a healthy eating pattern should also prevent foodborne illnesses, as described in the next chapter.

Nutrition Portfolio

Identifying your risk factors is the first step in taking action to defend yourself against heart attack, stroke, hypertension, diabetes, and cancer.

- Review your personal and family history of heart disease, hypertension, diabetes, and cancer.

Go to Diet & Wellness Plus and choose one of the days on which you tracked your diet and activity for an entire day. Select the Energy Balance report to help you answer the following questions:

- In this report, did you expend more energy (kcalories) than you consumed or vice versa? Consider whether you are sedentary or overweight and how you might become more physically active and achieve a healthy body weight. If you smoke cigarettes, develop a reasonable plan for quitting.
- Learn whether you have high blood cholesterol or high blood pressure.

Go to Diet & Wellness Plus and choose the Intake vs. Goals report to help you answer the following questions:

- In this report many dietary factors related to heart disease are assessed: sodium intake, fat intake, cholesterol intake, vitamin and mineral intake, and more. Describe how your Intake vs. Goals report relates to your risk factors. Is your diet helping to protect you or could it be putting you at higher risk?

Finally, go to Diet & Wellness Plus and choose the Fat Breakdown report.

- We know from Chapter 5 that some types of fats appear to be protective against heart disease (monounsaturated and polyunsaturated) and some appear to put us more at risk (saturated fats and cholesterol). How was your intake for each type of fat?

 To complete this exercise, go to your Diet & Wellness Plus at **www.cengagebrain.com**.

> STUDY IT To review the key points of this chapter and take a practice quiz, go to the study cards at the end of the book.

REFERENCES

1. Position of the Academy of Nutrition and Dietetics: The role of nutrition in health promotion and chronic disease prevention, *Journal of the Academy of Nutrition and Dietetics* 113 (2013): 972–979; M. Ezzati and E. Riboli, Behavioral and dietary risk factors for noncommunicable diseases, *New England Journal of Medicine* 369 (2013): 954–964; A. Fardet and Y. Boirie, Associations between diet-related diseases and impaired physiological mechanisms: A holistic approach based on meta-analyses to identify targets for preventive nutrition, *Nutrition Reviews* 71 (2013): 643–656; A. C. Vergnaud and coauthors, Adherence to the World Cancer Research Fund/American Institute for Cancer Research guidelines and risk of death in Europe: Results form the European Prospective Investigation into Nutrition and Cancer cohort study, *American Journal*

of *Clinical Nutrition* 97 (2013): 1107–1120; L. H. Kushi and coauthors, American Cancer Society Guidelines on Nutrition and Physical Activity for Cancer Prevention, *CA: Cancer Journal for Clinicians* 62 (2012): 30–67; A. Sherzai and coauthors, Stroke, food groups, and dietary patterns: A systematic review, *Nutrition Reviews* 70 (2012): 423–435; A. Oliveira and coauthors, Major habitual dietary patterns are associated with acute myocardial infarction and cardiovascular risk markers in a Southern European population, *Journal of the American Dietetic Association* 111 (2011): 241–250; S. C. Larsson, J. Virtamo, and A. Wolk, Red meat consumption and risk of stroke in Swedish men, *American Journal of Clinical Nutrition* 94 (2011): 417–421; D. J. A. Jenkins and coauthors, Effect of a dietary portfolio of cholesterol-lowering foods given at 2 levels of intensity of dietary advice on serum lipids in hyperlipidemia, *Journal of the American Medical Association* 306 (2011): 831–839.

2. W. G. van Panhuis and coauthors, Contagious diseases in the United States from 1888 to the present, *New England Journal of Medicine* 369 (2013): 2152–2158; Centers for Disease Control and Prevention, Ten great public health achievements: United States, 2001–2010, *Morbidity and Mortality Weekly Report*, 60 (2011): 619–623.

3. D. F. Warner and V. Mizrahi, Complex genetics of drug resistance in Mycobacterium tuberculosis, *Nature Genetics* 45 (2013): 1107–1108; van Panhuis and coauthors, 2013.

4. N. J. Afacan, C. D. Fjell, and R. E. W. Hancock, A systems biology approach to nutritional immunology—Focus on innate immunity, *Molecular Aspects of Medicine* 33 (2012): 14–25; I. Laaksi, Vitamin D and respiratory infection in adults, *Proceedings of the Nutrition Society* 71 (2012): 90–97; J. M. Monk, T. Y. Hou, and R. S. Chapkin, Recent advances in the field of nutritional immunology, *Expert Reviews of Clinical Immunology* 7 (2011): 747–749.

5. Afacan, Fjell, and Hancock, 2012; S. S. Percival, Nutrition and immunity: Balancing diet and immune function, *Nutrition Today* 46 (2011): 12–17.

6. D. E. Berryman and C. A. Taylor, Disturbances in energy balance, in *Biochemical, Physiological, and Molecular Aspects of Human Nutrition*, 3rd ed., eds. M. H. Stipanuk and M. A. Caudill (St. Louis, Mo.: Elsevier Saunders, 2013), pp. 519–535; B. H. Maskrey and coauthors, Emerging importance of omega-3 fatty acids in the innate immune response: Molecular mechanisms and lipidomic strategies for their analysis, *Molecular and Food Research* 57 (2013): 1390–1400; M. Spite, Deciphering the role of n-3 polyunsaturated fatty acid-derived lipid mediators in health and disease, *Proceedings of the Nutrition Society* 72 (2013): 441–450; L. Paul, P. M. Ueland, and J. Selhub, Mechanistic perspective on the relationship between pyridoxal 5'-phosphate and inflammation, *Nutrition Reviews* 71 (2013): 239–244; Afacan, Fjell, and Hancock, 2012; O. P. Garcia, Effect of vitamin A deficiency on the immune response in obesity, *Proceedings of the Nutrition Society*, 71 (2012): 290–297; Laaksi, 2012; M. Hewison, Vitamin D and innate and adaptive immunity, *Vitamins and Hormones* 86 (2011): 23–62; Percival, 2011.

7. Position of the American Dietetic Association: Nutrition intervention and human immunodeficiency virus infection, *Journal of the American Dietetic Association* 110 (2010): 1105–1119.

8. P. J. Murray and T. A. Wynn, Protective and pathogenic functions of macrophage subsets, *Nature Reviews Immunology* 11 (2011): 723–737.

9. B. K. Surmi and A. H. Hasty, The role of chemokines in recruitment of immune cells to the artery wall and adipose tissue, *Vascular Pharmacology* 52 (2010): 27–36.

10. D. L. Hoyert, and J. Xu, Deaths: Preliminary data for 2011, *National Vital Statistics Reports*, October 10, 2012.

11. World Health Organization, Chronic diseases and health promotion, www.who.int/chp/en, accessed February 19, 2013.

12. E. Incalcaterra and coauthors, Pro-inflammatory genetic markers of atherosclerosis, *Current Atherosclerosis Reports* 15 (2013): 329–335; F. C. McGillicuddy and H. M. Roche, Nutritional status, genetic susceptibility, and insulin-resistance—important precedents to atherosclerosis, *Molecular Nutrition and Food Research* 56 (2012): 1173–1184; P. Perez-Martinez and coauthors, Calpain-10 interacts with plasma saturated fatty acid concentrations to influence insulin resistance in individuals with the metabolic syndrome, *American Journal of Clinical Nutrition* 93 (2011): 1136–1141.

13. A. S. Go and coauthors, Heart disease and stroke statistics—2014 update: A report from the American Heart Association, *Circulation* 129 (2014): e28–e292.

14. S. E. Chiuve and coauthors, Dietary fat quality and risk of sudden cardiac death in women, *American Journal of Clinical Nutrition* 96 (2012): 498–507; D. Mozaffarian, L. J. Appel, and L. Van Horn, Recent advances in preventive cardiology and lifestyle medicine, *Circulation* 123 (2011): 2870–2891; M. R. Flock and P. M. Kris-Etherton, Dietary Guidelines for Americans 2010: Implications for cardiovascular disease, *Current Atherosclerosis Reports* 13 (2011): 499–507; A. Astrup and coauthors, The role of reducing intakes of saturated fat in the prevention of cardiovascular disease: Where does the evidence stand in 2010? *American Journal of Clinical Nutrition* 93 (2011): 684–688; R. J. Berlin and coauthors, Diet quality and the risk for cardiovascular disease: The Women's Health Initiative (WHI), *American Journal of Clinical Nutrition* 94 (2011): 49–57.

15. S. Lim and S. Park, Role of vascular smooth muscle cell in the inflammation of atherosclerosis, *BMB Reports* 47 (2014): 1–7; N. Alexopooulos, D. Katritsis, and P. Raggi, Visceral adipose tissue as a source of inflammation and promoter of atherosclerosis, *Atherosclerosis* 233 (2014): 104–112; A. Tuttolmondo and coauthors, Atherosclerosis as an inflammatory disease, *Current Pharmaceutical Design* 18 (2012): 4266–4288; M. Drechsler and coauthors, Neutrophilic granulocytes—promiscuous accelerators of atherosclerosis, *Thrombosis and Haemostasis* 106 (2011): 839–848; G. K. Hansson and A. Hermansson, The immune system in atherosclerosis, *Nature Immunology* 12 (2011): 204–212; P. J. Murray and T. A. Wynn, Protective and pathogenic functions of macrophage subsets, *Nature Reviews. Immunology* 11 (2011): 723–737.

16. S. M. Afonso and coauthors, The impact of dietary fatty acids on macrophage cholesterol homeostasis, *Journal of Nutritional Biochemistry* 25 (2014): 95–103; B. Messner and D. Bernhard, Smoking and cardiovascular disease: Mechanisms of endothelial dysfunction and early atherogenesis, *Arteriosclerosis, Thrombosis and Vascular Biology* 34 (2014): 509–515.

17. J. Yeboah and coauthors, Comparison of novel risk markers for improvement in cardiovascular risk assessment in intermediate-risk individuals, *Journal of the American Medical Association* 308 (2012): 788–795; W. Bekwelem and coauthors, White blood cell count, C-reactive protein, and incident heart failure in the Atherosclerosis Risk in Communities (ARIC) Study, *Annals of Epidemiology* 21 (2011): 739–748.

18. M. Katan and coauthors, Lipoprotein-associated phospholipase A2 is associated with atherosclerotic stroke risk: The northern Manhattan study, *PLoS One* 9 (2014): e83393; P. Burchardt and coauthors, Low-density lipoprotein, its susceptibility to oxidation and the role lipoprotein-associated phospholipase A2 and carboxyl ester lipase lipases in atherosclerotic plaque formation, *Archives of Medical Science* 9 (2013): 151–158; I. Ikonomidis and coauthors, The role of lipoprotein-associated phospholipase A2 (Lp-PLA$_2$) in cardiovascular disease, *Reviews on Recent Clinical Trials* 6 (2011): 108–113.

19. G. Niccoli and coauthors, Advances in mechanisms, imaging and management of the unstable plaque, *Atherosclerosis* 233 (2014): 467–477.

20. D. Mozaffarian and J. H. Y. Wu, Omega-3 fatty acids and cardiovascular disease, *Journal of the American College of Cardiology* 58 (2011): 2047–2067; W. S. Harris and coauthors, Omega-6 fatty acids and risk for cardiovascular disease, A Science Advisory from the American Heart Association Nutrition Subcommittee of the Council on Nutrition, Physical Activity, and Metabolism; Council on Cardiovascular Nursing; and Council on Epidemiology and Prevention, *Circulation* 119 (2009): 902–907.

21. A. C. Skulas-Ray and coauthors, Dose-response effects of omega-3 fatty acids on triglycerides, inflammation, and endothelial function in healthy persons with moderate hypertriglyceridemia, *American Journal of Clinical Nutrition* 93 (2011): 243–252; F. Dangardt and coauthors, Omega-3 fatty acid supplementation improves vascular function and reduces inflammation in obese adolescents, *Atherosclerosis* 12 (2010): 580–585; M. N. Di Minno and coauthors, Exploring newer cardioprotective strategies: ω-3 fatty acids in perspective, *Thrombosis and Haemostasis* 104 (2010): 664–680.

22. J. Ando and K. Yamamoto, Effects of shear stress and stretch on endothelial function, *Antioxidants and Redox Signaling* 15 (2011): 1389–1403.

23. E. L. Schiffrin, Immune mechanisms in hypertension and vascular injury, *Clinical Science* 126 (2014): 267–274; M. van Rooy and E. Pretorius, Obesity, hypertension, and hypercholesterolemia as risk factors for atherosclerosis leading to ischemic events, *Current Medicinal Chemistry* 21 (2014): 2121–2129.

24. A. L. Valderrama and coauthors, Million Hearts: Strategies to reduce the prevalence of leading cardiovascular disease risk factors—United States, 2011, *Morbidity and Mortality Weekly Report* 60 (2011): 1248–1251; N. T. Artinian and coauthors, Interventions to promote physical activity and dietary lifestyle changes for cardiovascular risk factor reduction in adults: A Scientific Statement from the American Heart Association, *Circulation* 122 (2010): 406–441.

25. D. J. Hunter and K. S. Reddy, Noncommunicable diseases, *New England Journal of Medicine* 369 (2013): 1136–1143; Institue of Medicine, *Living Well with Chronic Illness: A Call for Public Health Action*, January 31, 2012, www.iom.edu/Reports/2012/Living-Well-with-Chronic-Ilness /Report-Brief.aspx.

26. J. Fang and coauthors, Prevalence of Coronary Heart Disease: United States, 2006–2010, *Morbidity and Mortality Weekly Report* 60 (2011): 1377–1381.

27. Expert Panel on Detection, Evaluation, and Treatment of High Blood Cholesterol in Adults (Adult Treatment Panel III), *Third Report of the National Cholesterol Education Program (NCEP)*, NIH publication no. 02-5215 (Bethesda, Md.: National Heart, Lung, and Blood Institute, 2002), p. II-18.

28. O. M. Reslan and R. A. Khalil, Vascular effects of estrogenic menopausal hormone therapy, *Reviews on Recent Clinical Trials* 7 (2012): 47–70.

29. J. D. Berry and coauthors, Lifetime risks of cardiovascular disease, *New England Journal of Medicine* 366 (2012): 321–329.

30. J. Durga and coauthors, Effect of 3 y of folic acid supplementation on the progression of carotid intima-media thickness and carotid arterial stiffness in older adults, *American Journal of Clinical Nutrition* 93 (2011): 941–949; M. Hoffman, Hypothesis: Hyperhomocysteinemia is an indicator of oxidant stress, *Medical Hypotheses* 77 (2011): 1088–1093.

31. Go and coauthors, 2014.

32. R. C. Hooqeveen and coauthors, Small dense low-density lipoprotein cholesterol concentrations predict risk for coronary heart disease: The Atherosclerosis Risk in Communities (ARIC) Study, *Arteriosclerosis, Thrombosis, and Vascular Biology* 34 (2014): 1069–1077; N. B. Allen and coauthors, Blood pressure trajectories in early adulthood and subclinical atherosclerosis in middle age, *Journal of the American Medical Association* 311 (2014): 490–497.

33. K. Mahdy Ali and coauthors, Cardiovascular disease risk reduction by raising HDL cholesterol—Current therapies and future opportunities, *British Journal of Pharmacology* 167 (2012):1177–1194; D. Kothapalli and coauthors, Cardiovascular protection by apoE and apoE-HDL linked to suppression of ECM gene expression and arterial stiffening *Cell Reports* 2 (2012): 1–13.

34. B. F. Voight and coauthors, Plasma HDL cholesterol and risk of myocardial infarction: A mendelian randomisation study, *Lancet* 380 (2012): 572–580.

35. P. Burchardt and coauthors, Low-density lipoprotein, its susceptibility to oxidation and the role of lipoprotein-associated phospholipase A2 and carboxyl ester lipase lipases in atherosclerotic plaque formation, *Archives of Medical Science* 9 (2013): 151–158; A. A. Kalanuria, P. Nyquist, and G. Ling, The prevention and regression of atherosclerotic plaques: Emerging treatments, *Vascular Health and Risk Management* 8 (2012): 549–561.

36. Murray and Wynn, 2011; Surmi and Hasty, 2010.

37. Hoogeveen and coauthors, 2014.

38. M. Miller and coauthors, Triglycerides and cardiovascular disease: A scientific statement from the American Heart Associtaion, *Circulation* 123 (2011): 2292–2333.

39. Go and coauthors, 2014; P. A. James and coauthors, 2014 Evidence-based guidelines for the management of high blood pressure in adults: Report from the panel members appointed to the Eighth Joint National Committee (JNC8), *Journal of the American Medical Association* 311 (2014): 507–520.

40. Go and coauthors, 2014; E. S. Ford, Trends in the risk for coronary heart disease among adults with diagnosed diabetes in the US: Findings from the National Health and Nutrition Examination Survey, 1999–2008, *Diabetes Care* 34 (2011): 1337–1343; R. A. DeFronzo and M. Abdul-Ghani, Assessment and treatment of cardiovascular risk in prediabetes: Impaired glucose tolerance and impaired fasting glucose, *American Journal of Cardiology* 108 (2011): 3B–24B.

41. Go and coauthors, 2014; A. Fardet and Y. Boirie, Associations between diet-related diseases and impaired physiological mechanisms: A holistic approach based on meta-analyses to identify targets for preventive nutrition, *Nutrition Reviews* 71 (2013): 643–656; T. Coutinho and coauthors, Central obesity and survival in subjects with coronary artery disease: A systematic review of the literature and collaborative analysis with individual subject data, *Journal of the American College of Cardiology* 57 (2011): 1877–1886; F. Schouten and coauthors, Increases in central fat mass and decreases in peripheral fat mass are associated with accelerated arterial stiffening in healthy adults: The Amsterdam Growth and Health Longitudinal Study, *American Journal of Clinical Nutrition* 94 (2011): 40–48; T. Y. Warren and coauthors, Sedentary behaviors increase risk of cardiovascular disease mortality in men, *Medicine and Science in Sports and Exercise* 42 (2010): 879–885; J. M. Sacheck, J. F. Kuder, and C. D. Economos, Physical fitness, adiposity, and metabolic risk factors in young college students, *Medicine and Science in Sports and Exercise* 42 (2010): 1039–1044.

42. M. Hamer and coauthors, Physical activity and cardiovascular mortality risk: Possible protective mechanisms? *Medicine and Science in Sports and Exercise* 44 (2012): 84–88; J. Sattelmair and coauthors, Dose response between physical activity and risk of coronary heart disease: A meta-analysis, *Circulation* 124 (2011): 288–295; H. M. Ahmed and coauthors, Effects of physical activity on cardiovascular disease, *American Journal of Cardiology* 109 (2012): 288–295.

43. Go and coauthors, 2014.

44. D. Mozaffarian, L.J. Appel, and L. Van Horn, Recent advances in preventive cardiology and lifestyle medicine, *Circulation* 123 (2011): 2870–2891.

45. R. Do and coauthors, The effect of chromosome 9p21 variants on cardiovascular disease may be modified by dietary intake: Evidence from a case/control and a prospective study, *PLoS Medicine* 8 (2011): e1001106.

46. Go and coauthors, 2014; A. R. Aroor and coauthors, Maladaptive immune and inflammatory pathways lead to cardiovascular insulin resistance, *Metabolism* (2013): 62 (2013): 1543–1552; E. J. Gallagher, D. Leroith, and E. Karnieli, Insulin resistance in obesity as the underlying cause for the metabolic syndrome, *Mount Sinai Journal of Medicine* 77 (2010): 511–523.

47. M. Santaniemi and coauthors, Metabolic syndrome in the prediction of cardiovascular events: The potential additive role of hsCRP and adiponectin, *European Journal of Preventive Cardiology*, June 20, 2013, doi: 10.1177/2047487313494028; F. R. Jornayvaz, V. T. Samuel, and G. I. Shulman, The role of muscle insulin resistance in the pathogenesis of atherogenic dyslipidemia and nonalcoholic fatty liver disease associated with the metabolic syndrome, *Annual Review of Nutrition* 30 (2010): 273–290.

48. J. S. Chae and coauthors, Association of Lp-PLA(2) activity and LDL size with interleukin-6, an inflammatory cytokine and oxidized LDL, a marker of oxidative stress, in women with metabolic syndrome, *Atherosclerosis* 218 (2011): 499–506.

49. D. C. Goff and coauthors, 2013 ACC/AHA guideline on the assessment of cardiovascular risk: A report of the American College of Cardiology/ American Heart Association Task Force on Practice Guidelines, http:// circ.ahajournals.org/content/early/2013/11/11/01.cir.000048606.98. citation; R. H. Eckel and coauthors, 2013 AHA/ACC Guideline on lifestyle management to reduce cardiovascular risk: A report of the American College of Cardiology/American Heart Association Task Force on Practice Guidelines, http://circ.ahajournals.org/content /early/2013/11/11/01.cir.0000437740.48606.dl.citation; N. J. Stone and coauthors, ACC/AHA guideline on the treatment of blood cholesterol to reduce atherosclerotic cardiovascular risk in adults: A report of the American College of Cardiology/American Heart Association

Task Force on Practice Guidelines, *Circulation* doi:10.1161/01. cir.0000437738.638553.7a.

50. V. M. Montori, J. P. Brito, and H. H. Ting, Patient-centered and practical application of new high cholesterol guidelines to prevent cardiovascular disease, *Journal of the American Medical Association* 311 (2014): 464–466; J. F. Keaney, G.D. Curfman, and J. A. Jarcho, A pragmatic view of the new cholesterol treatment guidelines, *New England Journal of Medicine* 370 (2014): 275–278.

51. Eckel and coauthors, 2013.

52. Go and coauthors, 2014.

53. Centers for Disease Control and Prevention, High blood pressure facts, www.cdc.gov/bloodpressure/facts.htm, updated March 20, 2013.

54. Centers for Disease Control and Prevention, high blood pressure facts, www.cdc.gov/bloodpressure/facts.htm, updated March 20, 2013.

55. Go and coauthors, 2014.

56. L. Landsberg and coauthors, Obesity-related hypertension: Pathogenesis, cardiovascular risk, and treatment: A position paper of The Obesity Society and the American Society of Hypertension, *Journal of Clinical Hypertension* 15 (2013): 14–33.

57. Landsberg and coauthors, 2013.

58. T. A. Kotchen, A. W. Cowley, and E. D. Frohlich, Salt in health and disease—A delicate balance, *New England Journal of Medicine* 368 (2013): 1229–1237; P. K. Whelton and coauthors, AHA Presidential Advisory, Sodium, blood pressure, and cardiovascular disease, *Circulation* 126 (2012): 2880–2889; L. J. Appel and coauthors, The importance of population-wide sodium reduction as a means to prevent cardiovascular disease and stroke: A call to action from the American Heart Association, *Circulation* 123 (2011): 1138–1143.

59. C. Koliaki and N. Katsilambros, Dietary sodium, potassium, and alcohol: Key players in the pathophysiology, prevention, and treatment of human hypertension, *Nutrition Reviews* 71 (2013): 402–411; V. Savica, G. Bellinghieri, and J. D. Kopple, The effect of nutrition on blood pressure, *Annual Review of Nutrition* 30 (2010): 365–401.

60. P. A. James and coauthors, 2014 Evidence-based guidelines for the management of high blood pressure in adults, Report from the panel members appointed to the Eighth Joint National Committee (JNC 8), *Journal of the American Medical Association* 311 (2014): 507–520.

61. Landsberg and coauthors, 2013; L. J. Appel on behalf of the American Society of Hypertension Writing Group, ASH Position paper: Dietary approaches to lower blood pressure, *Journal of Clinical Hypertension* 11 (2009): 358–368.

62. Eckel and coauthors, 2013; D. E. Epstein and coauthors, Determinants and consequences of adherence to the Dietary Approaches to Stop Hypertension Diet in African-American and white adults with high blood pressure: Results from the ENCORE Trial, *Journal of the Academy of Nutrition and Dietetics* 112 (2012): 1763–1773; L. Azadbakht and coauthors, The Dietary Approaches to Stop Hypertension Eating Plan affects C-reactive protein, coagulation abnormalities, and hepatic function tests among type 2 diabetic patients, *Journal of Nutrition* 141 (2011): 1083–1088.

63. US Department of Agriculture, US Department of Health and Human Services, *Dietary Guidelines for Americans 2010*, www.dietaryguidelines.gov.

64. Eckel and coauthors, 2013.

65. Whelton and coauthors, 2012; Appel and coauthors, 2011.

66. Eckel and coauthors, 2013; S. Liu and coauthors, Blood pressure responses to acute and chronic exercise are related in prehypertension, *Medicine and Science in Sports and Exercise* 44 (2012): 1644–1652; E. G. Ciolac and coauthors, Effects of high-intensity aerobic interval training vs. moderate exercise on hemodynamic, metabolic and neuro-humoral abnormalities of young normotensive women at high familial risk for hypertension, *Hypertension Research* 33 (2010): 836–843.

67. Centers for Disease Control and Prevention, *National Diabetes Statistics Report: Estimates of Diabetes and Its Burden in the United States*, 2014, www.cdc.gov/diabetes/pubs/statsreport14/national-diabetes-report-web.pdf.

68. Centers for Disease Control, *National Diabetes Statistics Report: Estimates of Diabetes and Its Burden in the United States*, 2014, www.cdc.gov/diabetes/pubs/statsreport14/national-diabetes-report-web.pdf.

69. American Diabetes Association, Position statement: Standards of medical care in diabetes—2014, *Diabetes Care* 37 (2014): S14–S80.

70. American Diabetes Association, Position statement: Standards of medical care in diabetes—2014, *Diabetes Care* 37 (2014): S14–S80.

71. American Diabetes Association, Position statement: Diagnosis and classification of diabetes mellitus, *Diabetes Care* 37 (2014): S81–S90.

72. J. E. Fradkin, B. T. Roberts, and G. P. Rodgers, What's preventing us from preventing type 2 diabetes? *New England Journal of Medicine* 367 (2012): 1177–1179; J. P. Reis and coauthors, Lifestyle factors and risk for new-onset diabetes, *Annals of Internal Medicine* 155 (2011): 292–299.

73. American Diabetes Association, Position statement: Diagnosis and classification of diabetes mellitus, *Diabetes Care* 37 (2014): S81–S90.

74. B. Bennett and coauthors, Impaired insulin sensitivity and elevated ectopic fat in healthy obese vs. nonobese prepubertal children, *Obesity (Silver Spring)* 20 (2012): 371–375.

75. H. Johnston-Cox and coauthors, The A2b adenosine receptor modulates glucose homeostasis and obesity, *PLoS One* 7 (2012): e40584; A. Das and S. Mukhopadhyay, The evil axis of obesity, inflammation and type-2 diabetes, *Endocrine Metabolic and Immune Disorders Drug Targets* 11 (2011): 23–31; J. M. Olefsky and C. K. Glass, Macrophages, inflammation, and insulin resistance, *Annual Review of Physiology* 17 (2010): 219–246.

76. American Diabetes Association, Position statement: Standards of medical care in diabetes—2014, *Diabetes Care* 37 (2014): S14–S80; A. Brown, L. R. Reynolds, and D. Bruemmer, Intensive glycemic control and cardiovascular disease: An update, *Nature Reviews. Cardiology* 7 (2010): 369–375; D. Hill and M. Fisher, The effect of intensive glycaemic control on cardiovascular outcomes, *Diabetes, Obesity and Metabolism* 12 (2010): 641–647.

77. American Diabetes Association, Position statement: Standards of medical care in diabetes—2014, *Diabetes Care* 37 (2014): S14–S80; Department of Health and Human Services, Centers for Disease Control and Prevention, National Diabetes Fact Sheet, 2011, www.cdc.gov/diabetes/pubs/pdf/ndfs_2011.pdf, updated October 24, 2013.

78. A. B. Evert and coauthors, Nutrition therapy recommendations for the management of adults with diabetes: A position statement from the American Diabetes Association, *Diabetes Care* 36 (2013): 3821–3842.

79. Evert and coauthors, 2013.

80. Evert and coauthors, 2013.

81. Evert and coauthors, 2013.

82. American Diabetes Association, Position statement: Standards of medical care in diabetes—2014, *Diabetes Care* 37 (2014): S14–S80; C. K. Roberts , J. P. Little, and J. P. Thyfault, Modification of insulin sensitivity and glycemic control by activity and exercise, *Medicine and Science in Sports and Exercise* 45 (2013): 1868–1877; American College of Sports Medicine, American Diabetes Association, Joint position statement: Exercise and type 2 diabetes, *Medicine and Science in Sports and Exercise* 42 (2010): 2282–2303.

83. K. Brinkmann and H. Kashkar, Targeting the mitochondrial apoptotic pathway: A preferred approach in hematologic malignancies? *Cell Death and Disease* 5 (2014): e1098; D. T. Fisher, M. M. Appenheimer, and S. S. Evans, The two faces of IL-6 in the tumor microenvironment, *Seminars in Immunology* 26 (2014): 38–47; D. Hanahan and R. A. Weinberg, Hallmarks of cancer: The next generation, *Cell* 144 (2011): 646–674.

84. R. A. Stein, Epigenetics—The link between infectious diseases and cancer, *Journal of the American Medical Association* 305 (2011): 1484–1485.

85. L. H. Kushi and coauthors, American Cancer Society Guidelines on Nutrition and Physical Activity for Cancer Prevention, *CA: Cancer Journal for Clinicians* 62 (2012): 30–67; M. M. Heinen and coauthors, Physical activity, energy restriction, and the risk of pancreatic cancer: A prospective study in the Netherlands, *American Journal of Clinical Nutrition* 94 (2011): 1314–1323; X. Sui and coauthors, Influence of cardiorespiratory fitness on lung cancer mortality, *Medicine and Science in Sports and Exercise* 42 (2010): 872–878; World Cancer Research Fund/American Institute for Cancer Research, *Policy and Action for Cancer Prevention. Food, Nutrition, and Physical Activity: A Global Perspective* (Washington, DC: AICR, 2009), pp. 12–28.

86. T. Boyle and coauthors, Physical activity and risks of proximal and distal colon cancers: A systematic review and meta-analysis, *Journal of the National Cancer Institute* 104 (2012): 1548–1561; K. Y. Wolin, Y. Yan, and G. A. Colditz, Physical activity and risk of colon adenoma: A meta-analysis, *British Journal of Cancer* 104 (2011): 882–885.

87. H.-K. Na, and S. Oliynyk, Effects of physical activity on cancer prevention, *Annals of the New York Academy of Sciences* 1229 (2011): 176–183.

88. Y. Chen, C. Yu, and Y. Li, Physical activity and risks of esophageal and gastric cancers: A meta-analysis, *PLoS One* 9 (2014): e88082; Kushi and coauthors, 2012; C. Eheman and coauthors, Annual Report to the Nation on the Status of Cancer, 1975–2008, featuring cancers associated with excess weight and lack of sufficient physical activity, *Cancer* 118 (2012): 2338–2366; S. Pendyala and coauthors, Diet-induced weight loss reduces colorectal inflammation: Implications for colorectal carcinogenesis, *American Journal of Clinical Nutrition* 93 (2011): 234–242; A. Vrieling and E. Kampman, The role of body mass index, physical activity, and diet in colorectal cancer recurrence and survival: A review of the literature, *American Journal of Clinical Nutrition* 92 (2010): 471–490.

89. S. Ghosh and coauthors, Association of obesity and circulating adipose stromal cells among breast cancer survivors, *Molecular Biology Reports* 41 (2014): 2907–2916.

90. Kushi and coauthors, 2012.

91. Kushi and coauthors, 2012; W. Y. Chen and coauthors, Moderate alcohol consumption during adult life, drinking patterns, and breast cancer risk, *Journal of the American Medical Association* 306 (2011): 1884–1890.

92. D. Aune and coauthors, Red and processed meat intake and risk of colorectal adenomas: A systematic review and meta-analysis of epidemiological studies, *Cancer Causes and Control: CCC* 24 (2013): 611–627; M. Di Maso and coauthors, Red meat and cancer risk in a network of case-control studies focusing on cooking practices, *Annals of Oncology* 24 (2013): 3107–3112; D. S. Chan and coauthors, Red and processed meat and colorectal cancer incidence: Meta-analysis of prospective studies, *PLoS One* 6 (2011): e20456.

93. Centers for Disease Control and Prevention, National Center for Environmental Health, *Second National Report on Biochemical Indicators of Diet and Nutrition in the U.S. Population 2012*, Executive Summary, www.cdc.gov/nutritionreport; A. Brevik and coauthors, Polymorphisms in base excision repair genes as colorectal cancer risk factors and modifiers of the effect of diets high in red meat, *Cancer Epidemiology, Biomarkers, and Prevention* 19 (2010): 3167–3173.

94. Kushi and coauthors, 2012.

95. T. M. Brasky and coauthors, Associations of long-chain ω-3 fatty acids and fish intake with endometrial cancer risk in the VITamins And Lifestyle cohort, *American Journal of Clinical Nutrition* 99 (2014): 599–608; N. M. Iyengar, C. A. Hudis, and A. Gucalp, Omega-3 fatty acids for the prevention of breast cancer: An update and state of the science, *Current Breast Cancer Reports* 5 (2013): 247–254; N. Makarem and coauthors, Dietary fat in breast cancer survival, *Annual Review of Nutrition* 33 (2013): 319–348; E. D. Kantor and coauthors, Long-chain omega-3 polyunsaturated fatty acid intake and risk of colorectal cancer, *Nutrition and Cancer* 66 (2014): 716–727; T. M. Brasky and coauthors, Specialty supplements and breast cancer risk in the VITamins And Lifestyle (VITAL) Cohort, *Cancer Epidemiology, Biomarkers, and Prevention* 19 (2010): 1696–1708.

96. T. Norat and coauthors, Fruits and vegetables: Updating the epidemiologic evidence for the WCRF/AICR lifestyle recommendations for cancer prevention, *Cancer Treatment and Research* 159 (2014): 35–50; L. B. Link and coauthors, Dietary patterns and breast cancer risk in the California Teachers Study cohort, *American Journal of Clinical Nutrition* 98 (2013): 1524–1532; N. Annema and coauthors, Fruit and vegetable consumption and the risk of proximal colon, distal colon, and rectal cancers in a case-control study in Western Australia, *Journal of the American Dietetic Association* 111 (2011): 1479–1490; T. J. Key, Fruit and vegetables and cancer risk, *British Journal of Cancer* 104 (2010): 6–11.

97. G. Supic, M. Jagodic, and Z. Magic, Epigenetics: A new link between nutrition and cancer, *Nutrition and Cancer* 65 (2013): 781–792.

98. Appel and coauthors, 2013; Evert and coauthors, 2013; Kushi and coauthors, 2012; Mozzaffarian, Appel, and Van Horn, 2011.

99. A. Fardet and Y. Boirie, Associations between diet-related diseases and impaired physiological mechanisms: A holistic approach based on meta-analyses to identify targets for preventive nutrition, *Nutrition Reviews* 71 (2013): 643–656.

100. W. C. Willett and D. S. Ludwig, The 2010 Dietary Guidelines: The best recipe for health? *New England Journal of Medicine* 365 (2011): 1563–1565.

101. Harvard School of Public Health, The Nutrition Source, Healthy Eating Plate, www.hsph.harvard.edu/nutritionsource, accessed September 14, 2011.

HIGHLIGHT > **18**
Complementary and Alternative Medicine

> **LEARN IT** Present arguments for and against the use of complementary and alternative medicine.

If you suffered from migraine headaches or severe joint pain, where would you turn for relief? Would you visit a physician? Or are you more likely to go to an herbalist or an acupuncturist? Most physicians diagnose and treat medical conditions in ways that are accepted by the established medical community; herbalists and acupuncturists, among others, offer alternatives to standard medical practice. Instead of taking two aspirin, for example, you might be advised to chew two fresh leaves of the herb feverfew or to swallow a tincture of white willow bark. Or you might receive a massage and several acupuncture needles.

Complementary and alternative medicine (CAM) has become increasingly popular in recent decades (see the accompanying glossary for this and related terms).[1] People use these therapies for a variety of reasons. Some want to take more responsibility for both maintaining their own health and finding cures for their own diseases, especially when traditional medical therapies prove ineffective. Others have become distrustful of, and feel overwhelmed by, the high-tech diagnostic tests and costly treatments that **conventional medicine** offers. This highlight explores alternative therapies in search of their possible benefits and with an awareness of their potential harms.

John Block/The Image Bank/Getty Images

Defining Complementary and Alternative Medicine

By definition, complementary and alternative medicine is not conventional medicine because there is insufficient scientific evidence that it is safe and effective. It includes a variety of approaches, philosophies, and treatments, some of which are defined in the glossary of alternative therapies on p. 618. When these therapies are used instead of conventional medicine, they are called *alternative;* when used together with conventional medicine, they are called *complementary.* **Integrative medicine** combines conventional medicine and CAM treatments for which there is some high-quality scientific evidence of safety and effectiveness.[2]

A growing number of health-care professionals are learning about alternative therapies; more than half of US medical schools now offer elective courses in alternative medicine, and even more include discussions of these therapies in their required courses. More than ever before, health-care professionals are incorporating some of the beneficial alternative therapies into their practices, thus the gap between conventional medicine and CAM is narrowing.[3]

For some alternative therapies, preliminary and limited scientific evidence suggests some effectiveness; but for most, well-designed scientific studies have yet to determine safety and effectiveness. If proved safe and effective, an alternative therapy may be adopted by conventional medicine. Cancer radiation therapy, for example, was once considered an unconventional therapy, but it proved its clinical value and became part of accepted medical practice. In some cases, a therapy that is accepted by conventional medicine for a specific ailment is used for a different purpose in an alternative therapy. For example, chelation therapy, the preferred medical treatment for lead poisoning, is a common alternative therapy for cardiovascular disease.[4]

Sound Research, Loud Controversy

Much information on alternative therapies comes from folklore, tradition, and testimonial accounts. Relatively few clinical trials have been conducted. Consequently, scientific evidence proving the safety and effectiveness of many alternative therapies is lacking. Some say that alternative

GLOSSARY

complementary and alternative medicine (CAM): diverse medical and health-care systems, practices, and products that are not currently considered part of conventional medicine; also called *adjunctive, unconventional,* or *unorthodox therapies.*

conventional medicine: diagnosis and treatment of diseases as practiced by medical doctors (M.D.), doctors of osteopathy (D.O.), and allied health professionals such as physical therapists and registered nurses; also called *allopathy; Western, mainstream, orthodox,* or *regular medicine;* and *biomedicine.*

integrative medicine: care that combines conventional and complementary therapies for which there is some high-quality scientific evidence of safety and effectiveness. Integrative medicine emphasizes the importance of the relationship between the practitioner and the patient and focuses on wellness, healing, and the whole person.

therapies simply do not work; others suggest that these therapies have not been given a fair trial. In an effort to "explore complementary and alternative healing practices through vigorous science," the National Center for Complementary and Alternative Medicine supports clinical trials of these therapies. Articles reporting the results of these clinical trials are available at their website **(nccam.nih.gov/research).**

Sound research would answer two important questions. First, does the treatment offer better results than either doing nothing or giving a placebo? Second, do the benefits clearly outweigh the risks? Each of these points is worthy of elaboration.

Placebo Effect

Stories abound that credit alternative therapies with miraculous cures. Without scientific research to determine effectiveness, however, one is left to wonder whether it is the therapies or the placebo effect that produces the cure. Recall from Chapter 1 that giving a placebo often brings about a healing effect in people who believe they are receiving the treatment.[5] Conventional medicine tends to neglect this powerful remedy, whereas many alternative therapies embrace it.[6]

Risks versus Benefits

Ideally, a therapy provides benefits with little or no risk. Figure H18-1 presents several examples of herbal remedies that appear to be generally safe and possibly effective in treating various conditions.[7] Such findings, if replicated, hold promise that these alternative therapies may one day be integrated into conventional medicine.

Some alternative therapies are innocuous, providing little or no benefit for little or no risk. Sipping a cup of warm tea with a pleasant aroma, for example, won't cure heart disease, but it may improve one's mood and help relieve tension. Given no physical hazard and little financial risk, such therapies are acceptable.

In contrast, other products and procedures are downright dangerous, posing great risks while providing no benefits. One example is the folk practice of geophagia (eating earth or clay), which can cause GI impaction and impair iron absorption. Another is the taking of laetrile to treat cancer, which can cause cyanide poisoning. Clearly, such therapies are too harmful to be used.

Perhaps most controversial are alternative therapies that may provide benefits, but also carry significant, unknown, or debatable risks. Smoking marijuana is an example of such an alternative therapy.[8] The compounds in marijuana seem to provide relief from symptoms such as nausea, vomiting, and pain that commonly accompany cancer, AIDS, and other diseases, but marijuana use also poses risks that some people consider acceptable whereas others deem them intolerable. Figure H18-2 summarizes the relationships between risks and benefits.

GLOSSARY
OF ALTERNATIVE THERAPIES

acupuncture (AK-you-PUNK-cher): a technique that involves piercing the skin with long thin needles at specific anatomical points to relieve pain or illness. Acupuncture sometimes uses heat, pressure, friction, suction, or electromagnetic energy to stimulate the points.

aroma therapy: a technique that uses oil extracts from plants and flowers (usually applied by massage or baths) to enhance physical, psychological, and spiritual health.

ayurveda (AH-your-VAY-dah): a traditional Hindu system of improving health by using herbs, diet, meditation, massage, and yoga to stimulate the body, mind, and spirit to prevent and treat disease.

bioelectromagnetic medical applications: the use of electrical energy, magnetic energy, or both to stimulate bone repair, wound healing, and tissue regeneration.

biofeedback: the use of special devices to convey information about heart rate, blood pressure, skin temperature, muscle relaxation, and the like to enable a person to learn how to consciously control these medically important functions.

biofield therapeutics: a manual healing method that directs a healing force from an outside source (commonly God or another supernatural being) through the practitioner and into the client's body; commonly known as "laying on of hands."

cartilage therapy: the use of cleaned and powdered connective tissue, such as collagen, to improve health.

chelation (kee-LAY-shun) **therapy:** the use of ethylene diamine tetraacetic acid (EDTA) to bind with metallic ions, thus healing the body by removing toxic metals.

chiropractic (KYE-roh-PRAK-tik): a manual healing method of manipulating the spine to restore health.

faith healing: healing by invoking divine intervention without the use of medical, surgical, or other traditional therapy.

herbal (ERB-al) **medicine:** the use of plants to treat disease or improve health; also known as *botanical medicine* or *phytotherapy*.

homeopathy (hoh-me-OP-ah-thee): a practice based on the theory that "like cures like," that is, that substances that cause symptoms in healthy people can cure those symptoms when given in very dilute amounts.

- **homeo** = like
- **pathos** = suffering

hydrotherapy: the use of water (in whirlpools, as douches, or packed as ice, for example) to promote relaxation and healing.

hypnotherapy: a technique that uses hypnosis and the power of suggestion to improve health behaviors, relieve pain, and heal.

imagery: a technique that guides clients to achieve a desired physical, emotional, or spiritual state by visualizing themselves in that state.

iridology: the study of changes in the iris of the eye and their relationships to disease.

macrobiotic diet: a philosophical eating pattern based on mostly plant foods such as whole grains, legumes, and vegetables, with small amounts of fish, fruits, nuts, and seeds.

- **macro** = large, great
- **biotic** = life

massage therapy: a healing method in which the therapist manually kneads muscles to reduce tension, increase blood circulation, improve joint mobility, and promote healing of injuries.

meditation: a self-directed technique of relaxing the body and calming the mind.

naturopathic (nay-chur-oh-PATH-ick) **medicine:** a system that taps the natural healing forces within the body by integrating several practices, including traditional medicine, herbal medicine, clinical nutrition, homeopathy, acupuncture, East Asian medicine, hydrotherapy, and manipulative therapy.

orthomolecular medicine: the use of large doses of vitamins to treat chronic disease.

ozone therapy: the use of ozone gas to enhance the body's immune system.

qi gong (chée GUNG): a Chinese system that combines movement, meditation, and breathing techniques to enhance the flow of *qi* (vital energy) in the body.

> FIGURE H18-1   **Examples of Herbal Remedies**

Ginger may relieve nausea and vomiting due to motion sickness or pregnancy.

Ginkgo may slow the loss of cognitive function associated with aging.

St. John's wort may be effective in treating mild depression.

American ginseng may improve glucose control in people with type 2 diabetes.

Saw palmetto may improve the symptoms associated with an enlarged prostate.

The gel of an aloe vera plant soothes a minor burn.

> FIGURE H18-2   **Risk-Benefit Relationships**

	No (or little) **RISK** Much
BENEFIT Much	**Ideal situation**	**Cautionary situation**
	Benefits with little or no risk	Possible benefits with great or unknown risks
	(Accept)	(Consider carefully)
No (or little)	**Neutral situation**	**Dangerous situation**
	Little or no benefit with little or no risk	No benefits with great risks
	(Accept or reject as preferred)	(Reject)

© Cengage Learning

Nutrition-Related Alternative Therapies

Most alternative therapies fall outside the field of nutrition, but nutrition itself can be an alternative therapy. Furthermore, many alternative therapies prescribe specific dietary regimens even though most practitioners are not registered dietitian nutritionists (see Highlight 1). Nutrition-related alternative therapies include the use of foods, vitamin and mineral supplements, and herbs to prevent and treat illnesses.

Foods

The many dietary recommendations presented throughout this text are based on scientific evidence and do *not* fall into the alternative therapies category; strategies that are still experimental, however, do. For example, alternative therapists may recommend macrobiotic diets to help prevent chronic diseases, whereas most registered dietitian nutritionists would advise people to eat a balanced diet that includes servings from each of the five food groups. Similarly, enough scientific evidence is available to recommend including soy foods in the diet to protect against heart disease—but not to determine whether the phytoestrogens of soy supplements are safe or beneficial in managing the symptoms of menopause.[9]

Highlight 13 explores the potential health benefits of soy and many other functional foods and concludes that no one food is magical. As part of a balanced diet, these foods can support good health and protect against disease. Importantly, the benefits derive from a variety of *foods*. More research is needed to determine the safety and effectiveness of taking supplements of the nutrients and phytochemicals found in these foods.

Vitamin and Mineral Supplements

Like foods, vitamin and mineral supplements may fall into either the conventional or the alternative realm of medicine. For example, conventional advice recommends consuming 400 micrograms of folate to prevent neural tube defects, but not the taking of 1000 milligrams of vitamin C to prevent the common cold. Highlight 10 examines the appropriate use of supplements and potential dangers of excessive intakes.

As research on nutrition and chronic diseases has revealed many of the roles played by the vitamins and minerals in supporting health, conventional medicine has warmed up to the possibility that vitamin and mineral supplements might be an appropriate preventive therapy.[10] Some vitamin and mineral supplements appear to be in transition from alternative medicine to conventional medicine; that is, they have begun to prove their safety and effectiveness. Herbal remedies, however, still remain clearly in the realm of complementary and alternative medicine.

Herbal Remedies

From earliest times, people have used myriad herbs and other plants to cure aches and ills with varying degrees of success (review Figure H18-1). In fact, today's pharmaceutical industry originated from

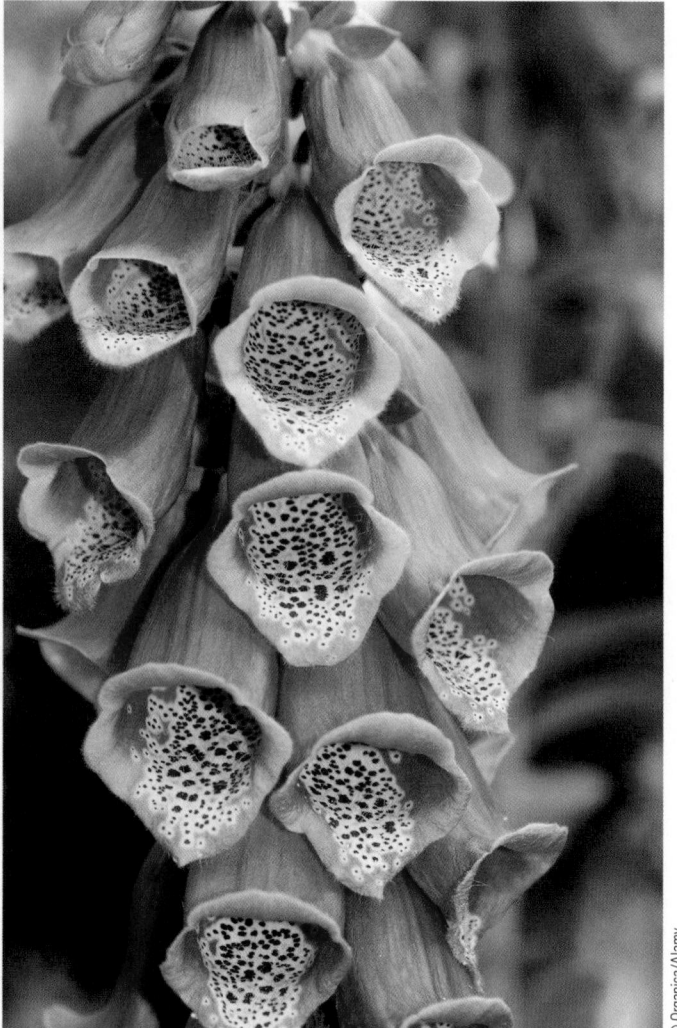

Digoxin, the most commonly prescribed heart medication, derives from the leaves of the foxglove plant *(Digitalis purpurea).*

the use of plant-derived products for human health.[11] Upon scientific study, dozens of these folk remedies reveal their secrets. For example, myrrh, a plant resin used as a painkiller in ancient times, does indeed have an analgesic effect. The herb valerian, which has long been used as a tranquilizer, contains oils that have a sedative effect. Senna leaves, brewed as a laxative tea, produce compounds that act as a potent cathartic drug. Green tea, brewed from the dried leaves of *Camellia sinensis*, contains phytochemicals that induce cancer cells to self-destruct. Naturally occurring salicylates provide the same protective effects as low doses of aspirin. Salicylates are found in spices such as curry, paprika, and thyme; fruits; vegetables; teas; and candies flavored with wintergreen (methylsalicylate).

Beneficial compounds from wild species contribute to about half of our modern medicines. By analyzing these compounds, pharmaceutical labs can synthesize pure forms of the drugs. Unlike herbs and wild species, which vary from batch to batch, synthesized medicines deliver exact dosages. By synthesizing drugs, we are also able to conserve endangered species. Consider that it took all of the bark from one 40-foot-tall, 100-year-old Pacific yew tree to produce one 300 milligram dose of the anticancer drug paclitaxel (Taxol), until scientists learned how to synthesize it. Many yet undiscovered cures may be forever lost as wild species are destroyed, long before their secrets are revealed to medicine.

Herbal Precautions

Plants are "natural," but that does not mean all plants are beneficial or even safe.[12] Nothing could be more natural—and deadly—than the poisonous herb hemlock. Several herbal remedies have toxic effects. The popular Chinese herbal potion jin bu huan, which is used as a pain and insomnia remedy, has been linked with several cases of acute hepatitis. Germanium, a nonessential mineral commonly found in many herbal products, has been associated with chronic kidney failure. Paraguay tea produces symptoms of agitation, confusion, flushed skin, and fever. Kombucha tea, commonly used in the hopes of preventing cancer, relieving arthritis, curing insomnia, and stimulating hair regrowth, can cause severe metabolic acidosis. Table H18-1 (p. 622) provides an overview of selected herbs.[13]

Although some people use herbs to treat or prevent disease, herbs are not regulated as drugs; they are considered dietary supplements. The Food and Drug Administration (FDA) does not evaluate dietary supplements for safety or effectiveness, nor does it monitor their contents. Under the Dietary Supplement Health and Education Act, rather than the herb manufacturers having to prove the safety of their products, the FDA has the burden of proving that a product is not safe. Consequently, consumers may lack information about or find discrepancies regarding:

- *True identification of herbs.* Most mint teas are safe, for instance, but some varieties contain the highly toxic pennyroyal oil. Mistakenly used to soothe a colicky baby, mint tea laden with pennyroyal has been blamed for the liver and neurological injuries of at least two infants, one of whom died.

- *Purity of herbal preparations.* A young child diagnosed with lead poisoning had been given an herbal vitamin that contained large quantities of lead and mercury for 4 years. Twelve cases of lead poisoning among adults using ayurvedic remedies were reported to the Centers for Disease Control and Prevention in recent years. In New York, investigators found six cases of lead poisoning in pregnant women who had taken ayurvedic medications made in India.[14] Some of the medications also contained mercury or arsenic.

- *Appropriate uses and contraindications of herbs.* Herbal remedies alone may be appropriate for minor ailments—a cup of chamomile tea to ease gastric discomfort or the gel of an aloe vera plant to soothe a sunburn, for example—but not for major health problems such as cancer or AIDS.

- *Effectiveness of herbs.* Herbal remedies may claim to work wonders without having to prove effectiveness. Research studies often report conflicting findings, with some suggesting a benefit and others indicating no effectiveness.

- *Variability of herbs.* Not all species are created equal. The various species of coneflower provide an example. *Echinacea purpurea,* for example, may help in the early treatment of colds, but *Echinacea augustifolia* may not. Similarly, not all parts of a plant provide the same compounds. Leaves, roots, and oils contain different compounds and extracts, and the temperatures used during manufacturing may affect their potency. Consumers are not always aware of such differences, and manufacturers do not always make such distinctions when preparing and labeling supplements.

- *Accuracy of labels.* Supplements may contain none of an herb or mixed species, and labels are often inaccurate. More often than not, supplements do not contain the species or the quantities of active ingredients stated on their labels.[15] In several cases, supplements do not even contain herbs, but drugs that are known to interact with prescription medicines and lower blood pressure to dangerous levels. Such discrepancies in the contents of supplements can be dangerous to the consumer, interfere with scientific research, and make it difficult to interpret the findings. Consumers may want to shop for supplements bearing a logo from either US Pharmacopeia or Consumer Lab indicating that the contents have been analyzed and found to contain the ingredients and quantities listed on the label.

- *Safe dosages of herbs.* Herbs may contain active ingredients—compounds that affect the body. Each of these active ingredients has a different potency, time of onset, duration of activity, and consequent effects, making the plant itself too unpredictable to be useful. Foxglove leaves, for example, contain dozens of compounds that have an effect on the heart; digoxin, a drug derived from foxglove, offers a standard dosage that allows for a more predictable cardiac response. Even when herbs are manufactured into capsules or liquids, their concentrations of active ingredients differ dramatically from batch to batch and from the quantities stated on the labels.

- *Interactions of herbs with medicines and other herbs.* Like drugs, herbs may interfere with, or potentiate, the effects of other herbs and drugs (see Table H18-2, p. 623).[16] A person taking both cardiac medication and the herb foxglove may be headed for disaster from the combined effect on the heart. Similarly, taking St. John's wort with medicines used to treat heart disease, depression, seizures, and certain cancers might diminish or exaggerate the intended effects.[17] Because *Ginkgo biloba* impairs blood clotting, it can cause bleeding problems for people taking aspirin or other blood-thinning medicines regularly.[18]

- *Adverse reactions and toxicity levels of herbs.* Herbs may produce undesirable reactions. The herbal root kava, commonly used to treat anxiety and insomnia, can cause liver abnormalities and may have such a sedating effect as to impair driving. Chinese herbal treatments containing *Aristolochia fangchi* are known to cause kidney damage and cancers.[19] Table H18-1 (p. 622) includes risks associated with commonly used herbs. To ensure the safety and standardization of herbal remedies, Congress may need to establish new regulations.

TABLE H18-1 Overview of Selected Herbs

Common and Scientific Names	Claims	Research Findings	Risks[a]
Aloe (gel) *Aloe vera*	Promote wound healing	May help heal minor burns and abrasions; may cause infections in severe wounds	Generally considered safe
Black cohosh (stems and roots) *Actaea racemosa, Cimifuga racemosa*	Ease menopause symptoms	Conflicting evidence	May cause headaches, stomach discomfort, liver damage
Chamomile (flowers) *Matricaria recutita, Chamomilla recutita*	Relieve indigestion	Little evidence available	Generally considered safe
Chaparral (leaves and twigs) *Larrea tridentata*	Slow aging, "cleanse" blood, heal wounds, cure cancer, treat acne	No evidence available	Acute, toxic hepatitis; liver damage
Cinnamon (bark) *Cinnamomum zeylanicum, C. cassia*	Relieve indigestion, lower blood glucose and blood lipids	May lower blood glucose in type 2 diabetes	May have a blood-thinning effect; not safe for pregnant women or those taking diabetes medication
Comfrey (leafy plant) *Symphytum officinale, S. asperum, S. x uplandicum*	Soothe nerves	No evidence available	Liver damage
Echinacea (roots) *Enchinacea angustifolia, E. pallida, E. purpurea*	Alleviate symptoms of colds, flus, and infections; promote wound healing; boost immunity	Ineffective in preventing colds or other infections	Generally considered safe; may cause headache, dizziness, nausea
Ephedra (stems) *Ephedra sinica*	Promote weight loss	Little evidence available; FDA has banned the sale of ephedra-containing products	Rapid heart rate, tremors, seizures, insomnia, headaches, hypertension
Feverfew (leaves) *Tanacetum parthenium*	Prevent migraine headaches	May prevent migraine headaches	Generally considered safe; may cause mouth irritation, swelling, ulcers, and GI distress
Garlic (bulbs) *Allium sativum*	Lower blood lipids and blood pressure	May lower blood cholesterol slightly; conflicting evidence on blood pressure	Generally considered safe; may cause garlic breath, body odor, gas, and GI distress; inhibits blood clotting
Ginger (roots) *Zingiber officinale*	Prevent motion sickness, nausea	May relieve pregnancy-induced nausea; conflicting evidence on nausea caused by motion, chemotherapy, or surgery	Generally considered safe
Ginkgo (tree leaves) *Ginkgo biloba*	Improve memory, relieve vertigo	Little evidence available	Generally considered safe; may cause headache, GI distress, dizziness; may inhibit blood clotting
Ginseng (roots) *Panax ginseng* (Asian), *P. quinquefolius* (American)	Boost immunity, increase endurance	Little evidence available	Generally considered safe; may cause insomnia, headaches, and high blood pressure
Goldenseal (roots) *Hydrastis canadensis*	Relieve indigestion, treat urinary infections	Little evidence available	Generally considered safe; not safe for people with hypertension or heart disease
Kava (roots) *Piper methysticum*	Relieves anxiety, promotes relaxation	Little evidence available	Liver failure
Saw palmetto (ripe fruits) *Serenoa repens*	Relieve symptoms of enlarged prostate; diuretic; enhance sexual vigor	Little evidence available	Generally considered safe; may cause nausea, vomiting, diarrhea
St. John's wort (leaves and tops) *Hypericum perforatum*	Relieve depression and anxiety	May relieve mild depression	Generally considered safe; may cause fatigue, increased sensitivity to sunlight, and GI distress
Turmeric (roots) *Curcuma longa*	Reduces inflammation; relieves heartburn; prevents or treats cancer	No evidence available	Generally considered safe; may cause indigestion; not safe for people with gallbladder disease
Valerian (roots) *Valeriana officinalis*	Calm nerves, improve sleep	Little evidence available	Long-term use associated with liver damage
Yohimbe (tree bark) *Pausinystalia yohimbe*	Enhance "male performance"	No evidence available	Kidney failure, seizures

[a]Allergies are always a possible risk; see Table H18-2 (p. 623) for drug interactions. Pregnant women should not use herbal supplements.

© Cengage Learning

TABLE H18-2 Herb and Drug Interactions

Herb	Drug	Interaction
American ginseng	Estrogens, corticosteroids	Enhances hormonal response
American ginseng	Breast cancer therapeutic agent	Synergistically inhibits cancer cell growth
American ginseng, karela	Blood glucose regulators	Affect blood glucose levels
Echinacea (possible immunostimulant)	Cyclosporine and corticosteroids (immunosuppressants)	May reduce drug effectiveness
Evening primrose oil, borage	Anticonvulsants	Lower seizure threshold
Feverfew	Aspirin, ibuprofen, and other nonsteroidal anti-inflammatory drugs	Negates the effect of the herb in treating migraine headaches
Feverfew, garlic, ginkgo, ginger, and Asian ginseng	Warfarin, coumarin (anticlotting drugs, "blood thinners")	Prolong bleeding time; increase likelihood of hemorrhage
Garlic	Protease inhibitor (HIV drug)	May reduce drug effectiveness
Kava, valerian	Anesthetics	May enhance drug action
Kelp (iodine source)	Synthroid or other thyroid hormone replacers	Interferes with drug action
Kyushin, licorice, plantain, uzara root, hawthorn, Asian ginseng	Digoxin (cardiac antiarrhythmic drug derived from the herb foxglove)	Interfere with drug action and monitoring
St. John's wort, saw palmetto, black tea	Iron supplements	Tannins in herbs inhibit iron absorption
St. John's wort	Protease inhibitors (HIV drugs), warfarin (anticlotting drug), digoxin (cardiac antiarrhythmic drug), oral contraceptives, tamoxifen (breast cancer drug)	May enhance or reduce drug effectiveness
Valerian	Barbiturates	Causes excessive sedation

© Cengage Learning

Because herbal medicines are sold as dietary supplements, their labels cannot claim to cure a disease, but they can make various other claims. Not surprisingly, when a label claims that an herbal product may strengthen immunity, improve memory, support eyesight, or maintain heart health, consumers believe that taking the product will provide those benefits. Beware. Manufacturers need not prove effectiveness; they need only state on the product label that this claim "Has not been evaluated by the FDA." Consumers who decide to use herbs need to become informed of the possible risks.

Internet Precautions

As Highlight 1 points out, just because something appears on the Internet, "it ain't necessarily so." Keep in mind that the thousands of websites touting the benefits of herbal medicines and other dietary supplements are marketing their products. Most product advertisements claim to prevent or treat specific diseases, but few include the FDA disclaimer statement. Many of the websites promote products by quoting researchers or physicians. Such quotations lend an air of authority to advertisements, but be aware that these sources may not even exist—and if they do, their comments may have been taken out of context. When asked, these experts may not agree at all with the claims attributed to them by the manufacturer.

Other deceits and dangers lurk in cyberspace as well. Potentially toxic substances, illegal and unavailable in many countries, are now easy to obtain via the Internet. Electronic access to products such as absinthe and oil of wormwood could be deadly. When the FDA discovers websites selling unapproved drugs, such as laetrile, it can order the business to shut down. But consumers need to remain vigilant because other similar businesses pop up quickly.

The Consumer's Perspective

Some health-care professionals may dismiss alternative therapies as ineffective and perhaps even dangerous, but many consumers think otherwise. About 4 out of 10 adults have used at least one alternative therapy for a variety of medical symptoms ranging from back pain and anxiety to heart disease and cancer. Interestingly, those who seek alternative therapies seem to do so not so much because they are dissatisfied with conventional medicine as because they find these alternatives more in line with their beliefs about health and life.

Most often, people use alternative therapies in addition to, rather than in place of, conventional therapies. Few consult an alternative therapist without also seeing a physician. In fact, most people seek alternative therapies for nonserious medical conditions or for health promotion. They simply want to feel better and access is easy. Sometimes their symptoms are chronic and subjective, such as pain and fatigue, and difficult to treat. In these cases, the chances of finding relief are often as good with a placebo, standard medical intervention, or even nonintervention.

Consumers spend billions of dollars on alternative health services and related products such as herbs, crystals, and aromas. As Highlight 1 points out, selecting a reliable practitioner depends on finding out about training, qualifications, and licenses. (To review how a person can identify health fraud and quackery, turn to p. 31. For a list of credible sources of nutrition information, see p. 33.)

In addition, consumers should inform their physicians about the use of any alternative therapies so that a comprehensive treatment plan can be developed and potential problems can be averted. Remember to include supplements, teas, and garden plants. Sometimes herbal products may need to be discontinued, especially before surgery when interactions with anesthesia or normal blood clotting can be life-threatening.

Alternative therapies come in a variety of shapes and sizes. Both their benefits and their risks may be small, none, or great. Wise consumers and health-care professionals accept the beneficial, or even neutral, practices with an open mind and reject those practices known to cause harm. Making healthful choices requires understanding all the choices.

CRITICAL THINKING QUESTIONS

A. How would you judge the accuracy or validity of complementary and alternative medicine practices?

B. A friend of yours was recently diagnosed with cancer. His physician discussed surgery, radiation, and possibly chemotherapy as recommended treatment options. His sister suggested that in *addition* to the physician's treatment, your friend might also benefit from a few acupuncture sessions and yoga classes. His roommate disagreed, suggesting that *instead* of following the physician's recommendations, your friend should start taking high doses of dietary supplements and doing colonic enemas daily. At this point, your friend is confused. What advice might you offer to help him make an informed decision?

REFERENCES

1. R. E. Wells and coauthors, Complementary and alternative medicine use among adults with migraines/severe headaches, *Headache* 51 (2011): 1087–1097.

2. R. Y. Teets, S. Dahmer, and E. Scott, Integrative medicine approach to chronic pain, *Primary Care* 37 (2010): 407–421; National Institutes of Health, National Center for Complementary and Alternative Medicine, Acupuncture for pain, http://nccam.nih.gov/health/acupuncture/acupuncture-for-pain.htm, updated August 2010.

3. A. Nerurkar and coauthors, When conventional medical providers recommend unconventional medicine: Results of a national study, *Archives of Internal Medicine* 171 (2011): 862–864.

4. G. A. Lamas and coauthors, Effect of disodium EDTA chelation regimen on cardiovascular events in patients with previous myocardial infarction: The TACT randomized trial, *Journal of the American Medical Association* 309 (2013): 1241–1250.

5. A. J. Vickers and coauthors, Acupuncture for chronic pain: Individual patient data meta-analysis, *Archives of Internal Medicine* 10 (2012): 1–10; M. E. Wechsler and coauthors, Active albuterol or placebo, sham acupuncture, or no intervention in asthma, *New England Journal of Medicine* 365 (2011): 119–126.

6. H. Brody and F. G. Miller, Lessons from recent research about the placebo effect—from art to science, *Journal of the American Medical Association* 306 (2011): 2612–2613.

7. N. A. Qureshi and A. M. Al-Bedah, Mood disorders and complementary and alternative medicine: A literature review, *Neuropsychiatric Disease and Treatment* 9 (2013): 639–658; C. M. Luberto and coauthors, Integrative medicine for treating depression: A update on the latest evidence, *Current Psychiatric Reports* 15 (2013): 391–392; M. J. Berry and coauthors, Effect of increasing doses of saw palmetto extract on lower urinary tract symptoms, *Journal of the American Medical Association* 306 (2011): 1344–1351; M. S. Baliga and coauthors, Update on the chemopreventive effects of ginger and its phytochemicals, *Critical Reviews in Food Science and Nutrition* 51 (2011): 499–523; R. Nahas, and O. Sheikh, Complementary and alternative medicine for the treatment of major depressive disorder, *Canadian Family Physician* 57 (2011): 659–663; M. P. Freeman and coauthors, Complementary and alternative medicine in major depressive disorder: The American Psychiatric Association Task Force report, *Journal of Clinical Psychiatry* 71 (2010): 669–681.

8. L. Leung, Cannabis and its derivatives: Review of medical use, *Journal of the American Board of Family Medicine* 24 (2011): 452–462.

9. E. Poluzzi and coauthors, Phytoestrogens in postmenopause: The state of the art from a chemical, pharmacological and regulatory perspective, *Current Medicinal Chemistry* 21 (2014): 417–436; J. M. W. Wong and coauthors, Equol status and blood lipid profile in hyperlipidemia after consumption of diets containing soy foods, *American Journal of Clinical Nutrition* 95 (2012): 564–571.

10. J. M. Gaziano and coauthors, Multivitamins in the prevention of cancer in men, *Journal of the American Medical Association* 308 (2012): 1871–1889; M. Goodman and coauthors, Clinical trials of antioxidants as cancer prevention agents: Past, present, and future, *Free Radical Biology and Medicine* 51 (2011): 1068–1084.

11. D. Singh, R. Gupta, and S. A. Saraf, Herbs—are they safe enough? An overview, *Critical Reviews in Food Science and Nutrition* 52 (2012): 876–898.

12. Singh, Gupta, and Saraf, 2012.

13. National Institutes of Health, National Center for Complementary and Alternative Medicine, Herbs at a glance, http://nccam.nih.gov/health/herbsataglance.htm, updated January 20, 2014.

14. P. Hore and coauthors, Lead poisoning in pregnant women who used Ayurvedic medications from India—New York City, 2011–2012, Centers for Disease Control and Prevention, *Morbidity and Mortality Weekly Report* 61 (2012): 641–646.

15. S. G. Newmaster and coauthors, DNA barcoding detects contamination and substitution in North American herbal products, *BMC Medicine* 11 (2013): 222.

16. Singh, Gupta, and Saraf, 2012; M. J. Stranger and coauthors, Anticoagulant activity of select dietary supplements, *Nutrition Reviews* 70 (2012): 107–117.

17. National Institutes of Health, National Center for Complementary and Alternative Medicine, Herbs at a glance, St. John's wort, http://nccam.nih .gov/health/stjohnswort/ataglance.htm, updated April 2012.

18. National Institutes of Health, National Center for Complementary and Alternative Medicine, Herbs at a glance: Ginkgo, http://nccam.nih .gov/health/ginkgo/ataglance.htm, updated June 2013; Stranger and coauthors, 2012.

19. S. Chen and A. Vieira, A meta-analysis of medicinal plants to assess the evidence for toxicity, *Interdisciplinary Toxicology* 3 (2010): 82–85.

Appendixes

© Kim D. French/Shutterstock.com

A-1

Appendix A Cells, Hormones, and Nerves

This appendix offers an understanding of how the body coordinates its activities. It presents a brief summary of the structure and function of the body's basic working unit (the cell) and of the body's two major regulatory systems (the hormonal system and the nervous system).

Cells

The body's organs are made up of millions of cells and of materials produced by them. Each **cell** is specialized to perform its organ's functions, but all cells have common structures (see the accompanying glossary and Figure A-l). Every cell is contained within a **cell membrane.** The cell membrane assists in moving materials into and out of the cell, and some of its special proteins act as "pumps" (described in Chapter 6). Some features of cell membranes, such as microvilli (described in Chapter 3), permit cells to interact with other cells and with their environments in highly specific ways.

Inside the membrane lies the **cytoplasm,** which is filled with **cytosol,** a jelly-like fluid. The cytoplasm contains much more than just cytosol, though. It is a highly organized system of fibers, tubes, membranes, particles, and subcellular **organelles** as complex as a city. These parts intercommunicate, manufacture and exchange materials, package and prepare materials for export, and maintain and repair themselves.

Within each cell is another membrane-enclosed body, the **nucleus.** Inside the nucleus are the **chromosomes,** which contain the genetic material, DNA. The DNA encodes all the instructions for carrying out the cell's activities. The role of DNA in coding for the synthesis of cell proteins is summarized in Figure 6-7 on p. 178. Chapter 6 also describes the variety of proteins produced by cells and some of the ways they perform the body's work.

Among the organelles within a cell are ribosomes, mitochondria, and lysosomes. Figure 6-7 briefly refers to the **ribosomes;** they assemble amino acids into proteins, following directions conveyed to them by RNA.

GLOSSARY OF CELL STRUCTURES

cell: the basic structural unit of all living things.

cell membrane: the thin layer of tissue that surrounds the cell and encloses its contents, made primarily of lipid and protein.

chromosomes: structures within the nucleus of a cell made of DNA and associated proteins. Human beings have 46 chromosomes in 23 pairs. Each chromosome has many genes.

cytoplasm (SIGH-toh-plazm): the cell contents, except for the nucleus.

cytosol: the fluid of cytoplasm that contains water, ions, nutrients, and enzymes.

endoplasmic reticulum (en-doh-PLAZ-mic reh-TIC-you-lum): a complex network of intracellular membranes. The *rough endoplasmic reticulum* is dotted with ribosomes, where protein synthesis takes place. The *smooth endoplasmic reticulum* bears no ribosomes.

Golgi (GOAL-gee) **apparatus:** a set of membranes within the cell where secretory materials are packaged for export.

lysosomes (LYE-so-zomes): cellular organelles; membrane-enclosed sacs of degradative enzymes.

mitochondria (my-toh-KON-dree-uh): the cellular organelles responsible for producing ATP aerobically; made of membranes with enzymes mounted on them. (The singular is *mitochondrion.*)

nucleus: a major membrane-enclosed body within cells, which contains the cell's genetic material (DNA) embedded in chromosomes.

organelles: subcellular structures such as ribosomes, mitochondria, and lysosomes.

ribosomes (RYE-boh-zomes): protein-making organelles in cells that are composed of RNA and protein.

The cell shown might be one in a gland (such as the pancreas) that produces secretory products (enzymes) for export (to the intestine). The rough endoplasmic reticulum with its ribosomes produces the enzymes; the smooth reticulum conducts them to the Golgi region; the Golgi membranes merge with the cell membrane, where the enzymes can be released into the extracellular fluid.

Cytoplasm

Golgi apparatus

Smooth endoplasmic reticulum

Lysosome

Cell membrane

Nucleus

Chromosomes

Rough endoplasmic reticulum

Ribosomes

Mitochondrion

The **mitochondria** are made of intricately folded membranes that bear thousands of highly organized sets of enzymes on their inner and outer surfaces. Mitochondria are crucial to energy metabolism (described in Chapter 7) and muscles conditioned to work aerobically are packed with them. Their presence is implied whenever the TCA cycle and electron transport chain are mentioned because the mitochondria house the needed enzymes.*

The **lysosomes** are membrane-enclosed sacs of degradative enzymes. When a cell needs to self-destruct or to digest materials in its surroundings, its lysosomes release their enzymes. Lysosomes are active when tissue repair or remodeling is taking place—for example, in cleaning up infections, healing wounds, shaping embryonic organs, and remodeling bones.

In addition to these and other cellular organelles, the cell's cytoplasm contains a highly organized system of membranes, the **endoplasmic reticulum.** The ribosomes, mentioned earlier, may either float freely in the cytoplasm or be mounted on the endoplasmic reticulum. A surface dotted with ribosomes looks speckled under the microscope and is called "rough" endoplasmic reticulum; such a surface without ribosomes is called "smooth." Some intracellular membranes are organized into tubules that collect cellular materials, merge with the cell membrane, and discharge their contents to the outside of the cell; these membrane systems are named the **Golgi apparatus,** after the scientist who first described them. The rough and smooth endoplasmic reticula and the Golgi apparatus are continuous with one another, so secretions produced deep in the interior of the cell can be efficiently transported and released to the outside. These and other cell structures enable cells to perform a multitude of specialized functions.

The actions of cells are coordinated by both hormones and nerves. Among the types of cellular organelles are receptors for the hormones delivering instructions that originate elsewhere in the body. Some hormones penetrate the cell and its nucleus and attach to receptors on chromosomes, where they activate certain genes to initiate, stop, speed up, or slow down synthesis of certain proteins as needed. Other hormones attach to receptors on the cell surface and transmit their messages from there. The hormones are described in the next section; the nerves, in the one following.

*For the reactions of glycolysis, the TCA cycle, and the electron transport chain, see Chapter 7 and Appendix C. The reactions of glycolysis take place in the cytoplasm; the conversion of pyruvate to acetyl CoA takes place in the mitochondria, as do the TCA cycle and electron transport chain reactions. The mitochondria then release carbon dioxide, water, and ATP as their end products.

Hormones

Hormones are chemical messengers secreted by a variety of glands in response to altered conditions in the body. Each hormone travels in the blood to all parts of the body, but only its specific target cells possess receptors to accept it. Only then can the hormone elicit a response to restore homeostasis.

The hormones, the glands they originate in, their target cells, and their effects are described in Table A-1. Figure A-2 identifies the glands that produce the hormones.

A hormone typically has one or more signals that turn it on and another (or others) that turns it off. Hormones are often turned off by their own effects; they are said to be regulated by *negative feedback* (see Figure 3-12 on p. 86). Consider, for example, the hormone prolactin, which promotes milk production. High prolactin levels ensure that milk is made; they also trigger the release of prolactin-inhibiting hormone (PIH), which ensures that prolactin levels don't get too high. But when the infant is suckling—and creating a demand for milk—PIH is not allowed to work (suckling turns off PIH). The consequence is that prolactin remains high, and milk production continues. Demand from the infant thus directly adjusts the supply of milk. The need is met through the interaction of the nerves and hormones.

Every body part is affected by hormones. Each different hormone has unique effects, and hormones that oppose each other are produced in carefully regulated amounts, so each can respond to the exact degree that is appropriate to the condition.

As Table A-1 summarizes, hormones have an enormous impact on body processes. The body's other overall regulating agency is the nervous system.

> FIGURE A-2 **The Endocrine System**

These organs and glands release hormones that regulate body processes. An *endocrine gland* secretes its product directly into *(endo)* the blood; for example, the pancreas cells that secrete insulin into the blood. An *exocrine gland* secretes its product(s) out *(exo)* to an epithelial surface either directly or through a duct; the sweat glands of the skin and the pancreas cells that secrete digestive enzymes into the gastrointestinal tract are both examples. The pancreas is therefore both an endocrine and an exocrine gland.

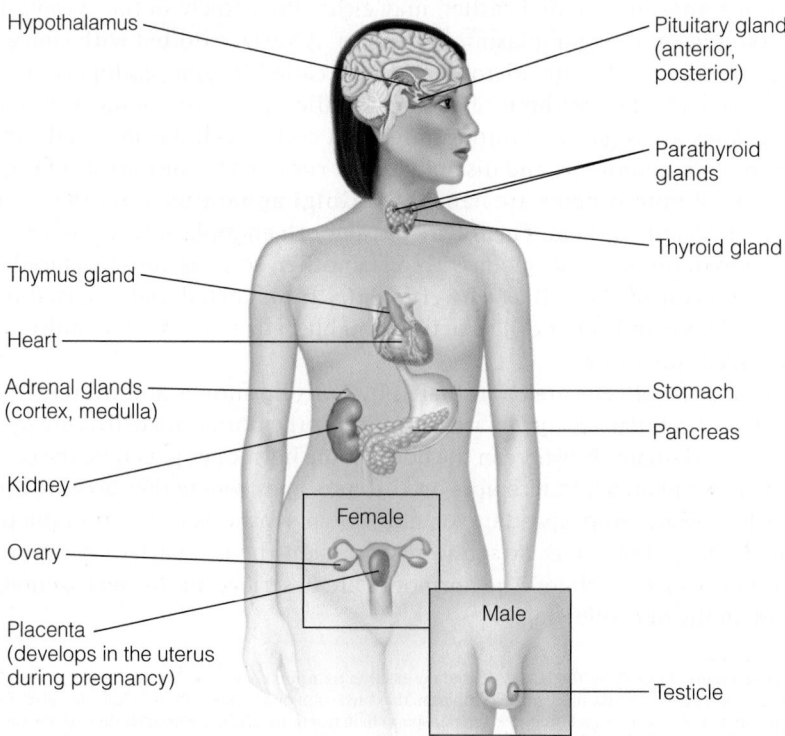

hormones: chemical messengers. Hormones are secreted by a variety of endocrine glands in response to altered conditions in the body. Each hormone travels to one or more specific target tissues or organs, where it elicits a specific response to maintain homeostasis. The study of hormones and their actions is called *endocrinology.*

TABLE A-1 Summary of Major Hormones

Gland	Hormone	Target Cells	Action
Anterior pituitary	Adrenocorticotropin (ACTH)	Adrenal cortex	Stimulates secretion of glucocorticoids and androgens
Adrenal cortex	Aldosterone	Kidneys	Stimulates sodium reabsorption, thereby regulating acid-base balance, blood volume, and blood pressure
Posterior pituitary	Antidiuretic hormone (ADH); also called vasopressin	Arteries	Causes vasoconstriction
		Kidneys	Promotes water retention
Thyroid gland	Calcitonin	Bones	Lowers blood calcium by moving calcium from the bloodstream into the bones whenever blood calcium rises above normal
		Kidneys	Increases excretion of calcium and phosphorus
Duodenum	Cholecystokinin	Gallbladder	Releases bile into the intestine
		Pancreas	Releases pancreatic juices into the intestine
Hypothalamus	Corticotropin-releasing hormone (CRH)	Anterior pituitary	Controls release of adrenocorticotropin (ACTH)
Kidneys	Erythropoietin	Bone marrow	Stimulates red blood cell production
Ovaries	Estrogens	Female sexual tissues	Promotes growth, development, and health of all tissues involved in female sexuality
Anterior pituitary	Follicle-stimulating hormone (FSH)	Ovaries (female)	Stimulates follicular development and ovulation
		Testicles (male)	Stimulates sperm production
Stomach, duodenum	Gastrin	Stomach	Stimulates gastric acid secretion; slows motility
Hypothalamus	Growth hormone releasing hormone (GHRH) and growth hormone inhibiting hormone (GHIH or somatostatin)	Anterior pituitary	Controls release of growth hormone (GH)
Pancreas (alpha cells)	Glucagon	Liver	Promotes the breakdown of glycogen to glucose
		Most cells	Increases use of fat and amino acids for energy
Adrenal cortex	Glucocorticoids	Most cells	Protects against stress; raises blood glucose
Hypothalamus	Gonadotropin-releasing hormone (GnRH)	Anterior pituitary	Controls release of follicle-stimulating hormone (FSH) and luteinizing hormone (LH)
Anterior pituitary	Growth hormone (GH); also called somatotropin	All tissues	Stimulates growth; regulates metabolism
Pancreas (beta cells)	Insulin	Most cells	Stimulates nutrient uptake into cells
Anterior pituitary	Luteinizing hormone (LH)	Ovaries (female)	Stimulates follicular development and ovulation
		Testicles (male)	Stimulates testosterone production
Adrenal medulla	Norepinephrine and epinephrine; formerly called noradrenalin and adrenalin, respectively	Many cells	Facilitates the body's readiness for fight or flight: maintains blood pressure, increases cardiac output, constricts blood vessels, keeps airways open, raises blood glucose levels
Posterior pituitary	Oxytocin	Uterus (female in late pregnancy)	Induces muscle contractions during childbirth
		Mammary glands (female in lactation)	Causes milk ejection during lactation
Parathyroid gland	Parathyroid hormone (PTH); also called parathormone	Bones	Releases stored calcium into the blood
		Kidneys	Slows calcium excretion
		Intestines	Increases calcium absorption
Corpus luteum, placenta	Progesterone	Uterus	Facilitates implantation at the start of pregnancy
		Mammary glands	Stimulates mammary gland development for lactation
Anterior pituitary	Prolactin	Mammary glands (female in lactation)	Stimulates milk production
Hypothalamus	Prolactin-inhibiting hormone (PIH)	Anterior pituitary	Controls release of prolactin
Duodenum	Secretin	Pancreas	Stimulates bicarbonate secretion into the intestine; slows stomach motility
Testicles	Testosterone	Male sexual tissues	Promotes growth, development, and health of all tissues involved in male sexuality
Anterior pituitary	Thyroid-stimulating hormone (TSH)	Thyroid gland	Stimulates synthesis and release of thyroid hormones (thyroxine and triiodothyronine)
Thyroid gland	Thyroxine	Many cells	Regulates metabolic rate, growth, and heat production
Hypothalamus	TSH-releasing hormone (TRH)	Anterior pituitary	Controls release of thyroid-stimulating hormone (TSH)
Skin	Vitamin D	Intestines	Increases calcium absorption

Nerves

The nervous system has a central control system that can evaluate information about conditions within and outside the body, and an expansive communication system that receives information and sends instructions. The control system is the brain and spinal cord, called the **central nervous system;** and the communication system between the center and the parts is the **peripheral nervous system.** The smooth functioning that results from the systems' adjustments to changing conditions is homeostasis.

The nervous system is best understood as two systems that use the same or similar pathways to receive and transmit their messages. The **somatic nervous system** controls the voluntary muscles; the **autonomic nervous system** controls the involuntary, internal muscles and organs.

When scientists were first studying the autonomic nervous system, they noticed that when something hurt one organ of the body, some of the other organs reacted as if in sympathy for the afflicted one. They therefore named the nerve network they were studying the **sympathetic nervous system.** The term is still used today to refer to the branch of the autonomic nervous system that responds to stressful conditions. The other branch that supports normal conditions is called the **parasympathetic nervous system.** (Think of the sympathetic branch as the emergency responder when homeostasis needs prompt restoring and the parasympathetic branch as the steady commander during normal times.) Both systems transmit their messages through the brain and spinal cord. Nerves of the two branches travel side by side along the same pathways to transmit their messages, but they oppose each other's actions (see Figure A-3).

An example will show how the sympathetic and parasympathetic nervous systems work to maintain homeostasis. When you go outside in cold weather, your skin's temperature receptors send "cold" messages to the spinal cord and brain. Your conscious mind may intervene at this point to tell you to zip your jacket, but let's say you have no jacket. Your sympathetic nervous system reacts to the external stressor, the cold. It signals your skin-surface capillaries to shut down so that your blood will circulate deeper in your tissues, where it will conserve heat. Your sympathetic nervous system also signals involuntary contractions of the small muscles just under the skin surface. The product of these muscle contractions is heat, and the visible result is goose bumps. If these measures do not raise your body temperature enough, then the sympathetic nerves signal your large muscle groups to shiver; the contractions of these large muscles produce still more heat. All of this activity helps to maintain your homeostasis (with respect to temperature) under conditions of external extremes (cold) that would throw it off balance. The cold was a stressor; the body's response was resistance.

Now let's say you come in and sit by a fire and drink hot cocoa. You are warm and no longer need all that sympathetic activity. At this point, your parasympathetic nerves take over; they signal your skin-surface capillaries to dilate again, your goose bumps to subside, and your muscles to relax. Your body is back to normal. This is recovery.

GLOSSARY
OF NERVOUS SYSTEM

autonomic nervous system: the division of the nervous system that controls the body's automatic responses. Its two branches are the *sympathetic* branch, which helps the body respond to stressors from the outside environment, and the *parasympathetic* branch, which regulates normal body activities between stressful times.

central nervous system: the central part of the nervous system; the brain and spinal cord.

parasympathetic nervous system: the part of the autonomic nervous system that dominates during nonstressful conditions and includes such effects as normal heart rate, pupil dilation, and peristalsis.

peripheral (puh-RIFF-er-ul) **nervous system:** the peripheral (outermost) part of the nervous system; the vast complex of wiring that extends from the central nervous system to the body's outermost areas. It contains both *somatic* and *autonomic* components.

somatic (so-MAT-ick) **nervous system:** the division of the nervous system that controls the voluntary muscles, as distinguished from the autonomic nervous system, which controls involuntary functions.

sympathetic nervous system: the part of the autonomic nervous system that dominates during stressful conditions and includes such effects as increased heart rate, dilated pupils, slowed peristalsis, and secretion of epinephrine and norepinephrine.

The brain and spinal cord evaluate information about conditions within and outside the body, and the peripheral nerves receive information and send instructions.

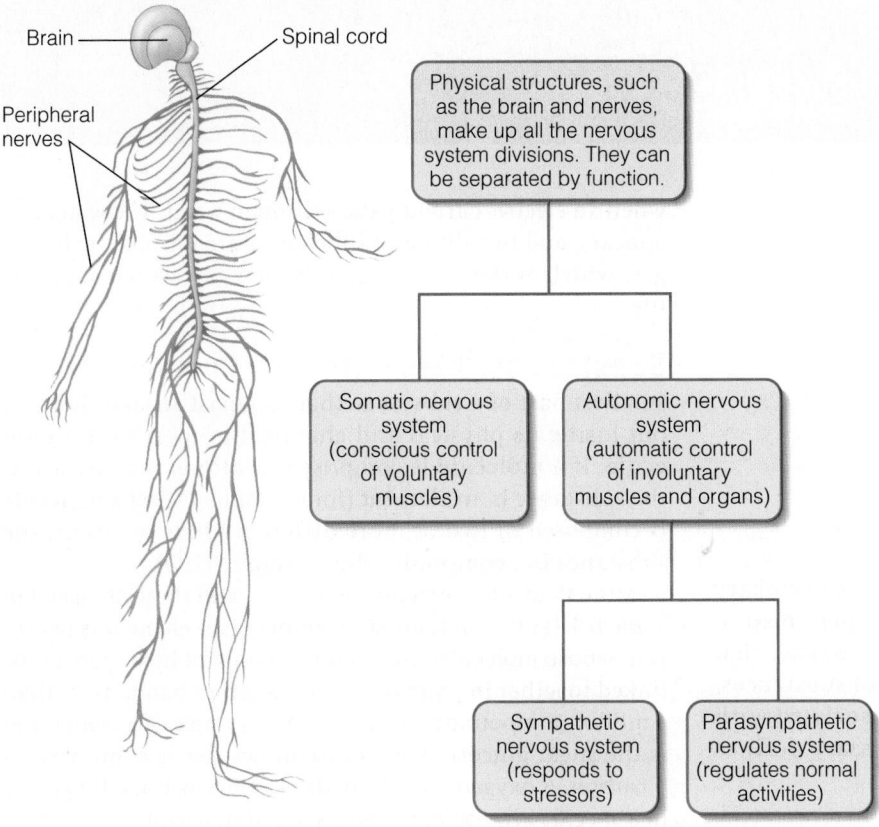

Brain

Spinal cord

Peripheral nerves

Physical structures, such as the brain and nerves, make up all the nervous system divisions. They can be separated by function.

Somatic nervous system (conscious control of voluntary muscles)

Autonomic nervous system (automatic control of involuntary muscles and organs)

Sympathetic nervous system (responds to stressors)

Parasympathetic nervous system (regulates normal activities)

Putting It Together

The hormonal and nervous systems coordinate body functions by transmitting and receiving messages. The point-to-point messages of the nervous system travel through the spinal cord and brain, whereas the messages of the hormonal system are sent through the bloodstream, and any organ with the appropriate receptors can pick them up. Nerve impulses travel faster than hormonal messages do—although both are remarkably swift. Whereas your brain's command to wiggle your toes reaches the toes within a fraction of a second and stops as quickly, a gland's message to alter a body condition may take several seconds or minutes to get started and may fade away equally slowly.

Together, the two systems possess every characteristic a superb communication network needs: varied speeds of transmission, along with private communication lines or public broadcasting systems, depending on the needs of the moment. The hormonal system, together with the nervous system, integrates the whole body's functioning so that all parts act smoothly together.

Appendix B   Basic Chemistry Concepts

This appendix provides the background in basic chemistry needed to understand the nutrition concepts presented in this book. Chemistry is the branch of natural science that deals with the composition and properties of substances, how substances interact, and the **energy** associated with these interactions. The accompanying glossary defines related terms.

The Properties of Atoms

Every substance has physical and chemical properties that distinguish it from all other substances and thus give it a unique identity. The physical properties include such characteristics as color, taste, texture, and odor, as well as the temperatures at which a substance changes its state (from a solid to a liquid or from a liquid to a gas) and the weight of a unit volume (its density). The chemical properties of a substance have to do with how it reacts with other substances or responds to a change in its environment.

A physical change does not change a substance's chemical composition. The three physical states—ice, water, and steam—all consist of two hydrogen atoms and one oxygen atom bound together. In contrast, a chemical change occurs when an electric current passes through water. The water disappears, and two different substances are formed: hydrogen gas, which is flammable, and oxygen gas, which supports life.

Substances: Elements and Compounds The smallest part of a substance that can exist separately without losing its physical and chemical properties is a **molecule**. If a molecule is composed of **atoms** that are alike, the substance is an **element** (for example, O_2). If a molecule is composed of two or more different kinds of atoms, the substance is a **compound** (for example, H_2O).

More than 100 elements are known, and these are listed in Table B-1 (p. B-1). A familiar example of an element is hydrogen, whose molecules are composed only of hydrogen atoms linked together in pairs (H_2). On the other hand, more than a million compounds are known. An example of a compound is the sugar glucose. Each of its molecules is composed of 6 carbon, 6 oxygen, and 12 hydrogen atoms linked together in a specific arrangement (as described in Chapter 4).

The Nature of Atoms Atoms themselves are made of smaller particles. Within an atom's nucleus are protons (positively charged particles), and surrounding the nucleus are an equal number of electrons (negatively charged particles). The number of protons in the nucleus of an atom determines the atomic number. The positive charge on a proton is equal to the negative charge on an electron, so the charges cancel each other out and leave the atom neutral to its surroundings.

The nucleus may also include neutrons, subatomic particles that have no charge. Protons and neutrons are of equal mass, and together they give an atom its atomic mass. Electrons bond atoms together to make molecules, and they are involved in chemical reactions.

GLOSSARY

anions (AN-eye-uns): negatively changed ions.

atoms: the smallest components of an element that have all of the properties of the element.

cations (CAT-eye-uns): positively changed ions.

compound: a substance composed of two or more different atoms—for example, water (H_2O).

covalent bonds: strong chemical bonds formed between atoms by sharing electrons.

element: a substance composed of atoms that are alike—for example, iron (Fe).

energy: the capacity to do work. The energy in food is chemical energy. The body can convert this chemical energy to mechanical, electrical, or heat energy.

ions (EYE-uns): atoms or molecules that have gained or lost one or more electrons and therefore have electrical charges. Examples include the positively charged sodium ion (Na^+) and the negatively charged chloride ion (Cl^-).

molecule: two or more atoms of the same or different elements joined by chemical bonds. Examples are molecules of the element oxygen, composed of two oxygen atoms (O_2), and molecules of the compound water, composed of two hydrogen atoms and one oxygen atom (H_2O).

TABLE B-1 The Elements

Number of Protons (Atomic Number)	Element	Number of Electrons in Outer Shell	Number of Protons (Atomic Number)	Element	Number of Electrons in Outer Shell	Number of Protons (Atomic Number)	Element	Number of Electrons in Outer Shell
1	Hydrogen (H)	1	38	Strontium (Sr)	2	75	Rhenium (Re)	2
2	Helium (He)	2	39	Yttrium (Y)	2	76	Osmium (Os)	2
3	Lithium (Li)	1	40	Zirconium (Zr)	2	77	Iridium (Ir)	2
4	Beryllium (Be)	2	41	Niobium (Nb)	1	78	Platinum (Pt)	1
5	Boron (B)	3	42	Molybdenum (Mo)	1	79	Gold (Au)	1
6	Carbon (C)	4	43	Technetium (Tc)	1	80	Mercury (Hg)	2
7	Nitrogen (N)	5	44	Ruthenium (Ru)	1	81	Thallium (Tl)	3
8	Oxygen (O)	6	45	Rhodium (Rh)	1	82	Lead (Pb)	4
9	Fluorine (F)	7	46	Palladium (Pd)	—	83	Bismuth (Bi)	5
10	Neon (Ne)	8	47	Silver (Ag)	1	84	Polonium (Po)	6
11	Sodium (Na)	1	48	Cadmium (Cd)	2	85	Astatine (At)	7
12	Magnesium (Mg)	2	49	Indium (In)	3	86	Radon (Rn)	8
13	Aluminum (Al)	3	50	Tin (Sn)	4	87	Francium (Fr)	1
14	Silicon (Si)	4	51	Antimony (Sb)	5	88	Radium (Ra)	2
15	Phosphorus (P)	5	52	Tellurium (Te)	6	89	Actinium (Ac)	2
16	Sulfur (S)	6	53	Iodine (I)	7	90	Thorium (Th)	2
17	Chlorine (Cl)	7	54	Xenon (Xe)	8	91	Protactinium (Pa)	2
18	Argon (Ar)	8	55	Cesium (Cs)	1	92	Uranium (U)	2
19	Potassium (K)	1	56	Barium (Ba)	2	93	Neptunium (Np)	2
20	Calcium (Ca)	2	57	Lanthanum (La)	2	94	Plutonium (Pu)	2
21	Scandium (Sc)	2	58	Cerium (Ce)	2	95	Americium (Am)	2
22	Titanium (Ti)	2	59	Praseodymium (Pr)	2	96	Curium (Cm)	2
23	Vanadium (V)	2	60	Neodymium (Nd)	2	97	Berkelium (Bk)	2
24	Chromium (Cr)	1	61	Promethium (Pm)	2	98	Californium (Cf)	2
25	Manganese (Mn)	2	62	Samarium (Sm)	2	99	Einsteinium (Es)	2
26	Iron (Fe)	2	63	Europium (Eu)	2	100	Fermium (Fm)	2
27	Cobalt (Co)	2	64	Gadolinium (Gd)	2	101	Mendelevium (Md)	2
28	Nickel (Ni)	2	65	Terbium (Tb)	2	102	Nobelium (No)	2
29	Copper (Cu)	1	66	Dysprosium (Dy)	2	103	Lawrencium (Lr)	2
30	Zinc (Zn)	2	67	Holmium (Ho)	2	104	Rutherfordium (Rf)	2
31	Gallium (Ga)	3	68	Erbium (Er)	2	105	Dubnium (Db)	2
32	Germanium (Ge)	4	69	Thulium (Tm)	2	106	Seaborgium (Sg)	2
33	Arsenic (As)	5	70	Ytterbium (Yb)	2	107	Bohrium (Bh)	2
34	Selenium (Se)	6	71	Lutetium (Lu)	2	108	Hassium (Hs)	2
35	Bromine (Br)	7	72	Hafnium (Hf)	2	109	Meitnerium (Mt)	2
36	Krypton (Kr)	8	73	Tantalum (Ta)	2	110	Darmstadtium (Ds)	2
37	Rubidium (Rb)	1	74	Tungsten (W)	2			

Key
Elements found in energy-yielding nutrients, vitamins, and water
Major minerals
Trace minerals

Each element has a characteristic number of protons in its atom's nucleus. For example, the hydrogen atom (the simplest of all) possesses a single proton, with a single electron associated with it:

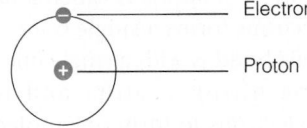

Electron

Proton

Hydrogen atom (H), atomic number 1

Just as hydrogen always has one proton, helium always has two, lithium three, and so on. The atomic number of each element is the number of protons in the nucleus of that atom, and this never changes in a chemical reaction; it gives the atom its identity. The atomic numbers for the known elements are listed in Table B-1.

In addition to hydrogen, the atoms most common in living things are carbon (C), nitrogen (N), and oxygen (O), whose atomic numbers are 6, 7, and 8, respectively. Their structures are more complicated than that of hydrogen, but each of them possesses the same number of electrons as there are protons in the nucleus. These electrons are found in orbits, or shells (shown below).

Carbon atom (C), atomic number 6   Nitrogen atom (N), atomic number 7   Oxygen atom (O), atomic number 8

In these and all diagrams of atoms that follow, only the protons and electrons are shown. The neutrons, which contribute only to atomic weight, not to charge, are omitted.

The most important structural feature of an atom for determining its chemical behavior is the number of electrons in its outermost shell. The first, or innermost, shell is full when it is occupied by two electrons; so an atom with two or more electrons has a filled first shell. When the first shell is full, electrons begin to fill the second shell.

The second shell is completely full when it has eight electrons. A substance that has a full outer shell tends not to enter into chemical reactions. Atomic number 10, neon, is a chemically inert substance because its outer shell is complete. Fluorine, atomic number 9, has a great tendency to attract an electron from other substances to complete its outer shell, and thus it is highly reactive. Carbon has a half-full outer shell, which helps explain its great versatility; it can combine with many other elements in a variety of ways to form a large number of compounds.

Atoms seek to reach a state of maximum stability or of lowest energy in the same way that a ball will roll down a hill until it reaches the lowest place. An atom achieves a state of maximum stability:

- By gaining or losing electrons to either fill or empty its outer shell.
- By sharing its electrons with other atoms and thereby completing its outer shell.

The number of electrons determines how the atom will chemically react with other atoms.

Chemical Bonding

Atoms often complete their outer shells by sharing electrons with other atoms. In order to complete its outer shell, a carbon atom requires four electrons. A hydrogen atom requires one. Thus, when a carbon atom shares electrons with four hydrogen atoms, each completes its outer shell (as shown in the next drawing). Electron sharing binds the atoms together and satisfies the conditions of maximum stability for the molecule. The outer shell of each atom is complete because hydrogen effectively has the required 2 electrons in its first (outer) shell, and carbon has 8 electrons in its second (outer)

shell; and the molecule is electrically neutral, with a total of 10 protons and 10 electrons.

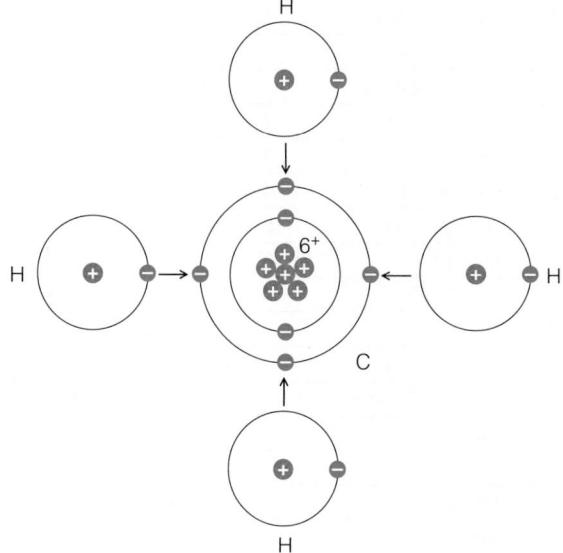

When a carbon atom shares electrons with four hydrogen atoms, a methane molecule is made.

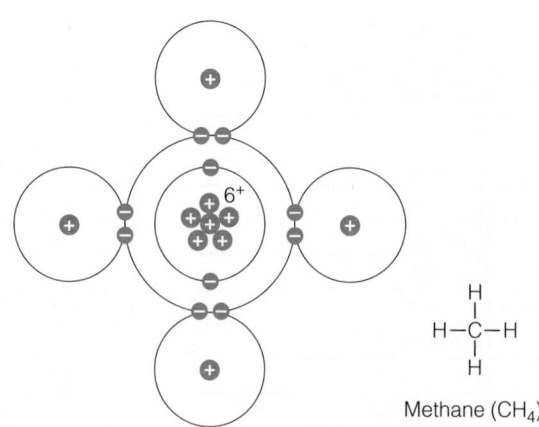

Methane (CH₄)

The chemical formula for methane is CH_4. Note that by sharing electrons, every atom achieves a filled outer shell.

Bonds that involve the sharing of electrons, like the bonds in methane between the one carbon and the four hydrogens, are the most stable kind of association that atoms can form with one another. These bonds are called **covalent bonds,** and the resulting combination of atoms is called a molecule. A single pair of shared electrons forms a single bond. A simplified way to represent a single bond is with a single line.

Similarly, one nitrogen atom and three hydrogen atoms can share electrons to form one molecule of ammonia (NH_3):

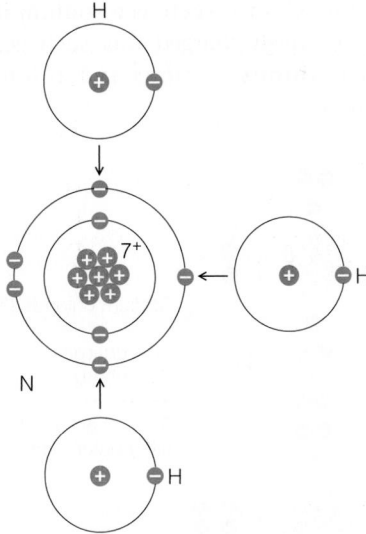

When a nitrogen atom shares electrons with three hydrogen atoms, an ammonia molecule is made.

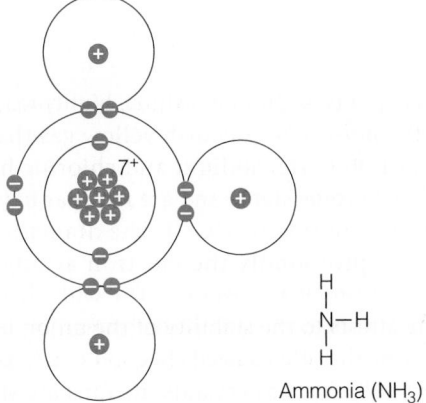

$$\begin{array}{c} H \\ | \\ H-N-H \end{array}$$

Ammonia (NH_3)

The chemical formula for ammonia is NH_3. Count the electrons in each atom's outer shell to confirm that it is filled.

One oxygen atom may be bonded to two hydrogen atoms to form one molecule of water (H_2O):

$$\begin{array}{c} H \\ | \\ H-O \end{array}$$

Water molecule (H_2O)

When two oxygen atoms form a molecule of oxygen, they must share two pairs of electrons. This double bond may be represented as two lines:

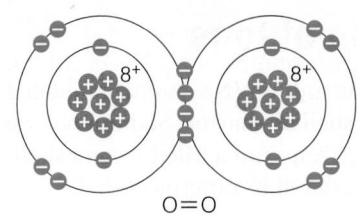

O=O

Oxygen molecule (O_2)

Small atoms form the tightest, most stable bonds. H, O, N, and C are the smallest atoms capable of forming one, two, three, and four electron-pair bonds, respectively. This is the basis for the statement in Chapter 4 that in drawings of compounds containing these atoms, hydrogen must always have one, oxygen two, nitrogen three, and carbon four bonds radiating to other atoms:

$$H- \qquad -O- \qquad -\overset{|}{N}- \qquad -\overset{|}{\underset{|}{C}}-$$

The stability of the associations between these small atoms (hydrogen, carbon, nitrogen, and oxygen) and the versatility with which they can combine make them very common in living things. Interestingly all cells—whether they come from animals, plants, or bacteria—contain the same elements in very nearly the same proportions. The elements commonly found in living things are shown in Table B-2.

TABLE B-2 Elemental Composition of the Human Body

Element	Chemical Symbol	By Weight (%)
Oxygen	O	65.0
Carbon	C	18.0
Hydrogen	H	10.0
Nitrogen	N	3.0
Calcium	Ca	1.5
Phosphorus	P	1.0
Potassium	K	0.4
Sulfur	S	0.3
Sodium	Na	0.2
Chloride	Cl	0.1
Magnesium	Mg	0.1
Total		99.6[a]

[a]The remaining 0.4 percent by weight is contributed by the trace elements: chromium (Cr), copper (Cu), zinc (Zn), selenium (Se), molybdenum (Mo), fluorine (F), iodine (I), manganese (Mn), and iron (Fe). Cells may also contain variable traces of some of the following: boron (B), cobalt (Co), Lithium (Li), strontium (Sr), aluminum (AL), silicon (Si), Lead (Pb), vanadium (V), arsenic (As), bromine (Br), and others.

Formation of Ions

An atom such as sodium (Na, atomic number 11) cannot easily fill its outer shell by sharing. Sodium possesses a filled first shell of two electrons and a filled second shell of eight; there is only one electron in its outermost shell:

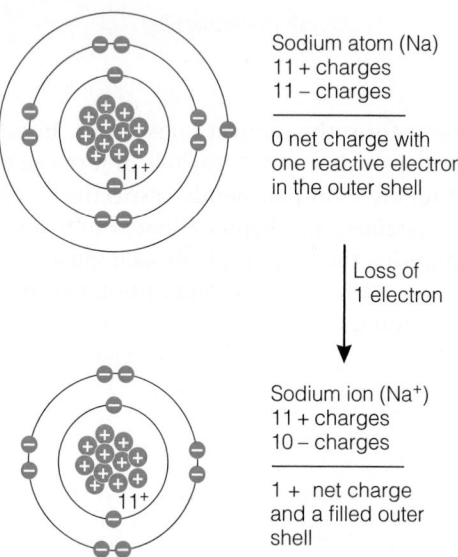

Sodium atom (Na)
11 + charges
11 − charges
___
0 net charge with one reactive electron in the outer shell

Loss of 1 electron

Sodium ion (Na⁺)
11 + charges
10 − charges
___
1 + net charge and a filled outer shell

If sodium loses this electron, it satisfies one condition for stability: a filled outer shell (now its second shell counts as the outer shell). However, it is not electrically neutral. It has 11 protons (positive) and only 10 electrons (negative). It therefore has a net positive charge. An atom or molecule that has lost or gained one or more electrons and so is electrically charged is called an **ion**.

An atom such as chlorine (Cl, atomic number 17), with seven electrons in its outermost shell, can share electrons to fill its outer shell, or it can gain one electron to complete its outer shell and thus give it a negative charge:

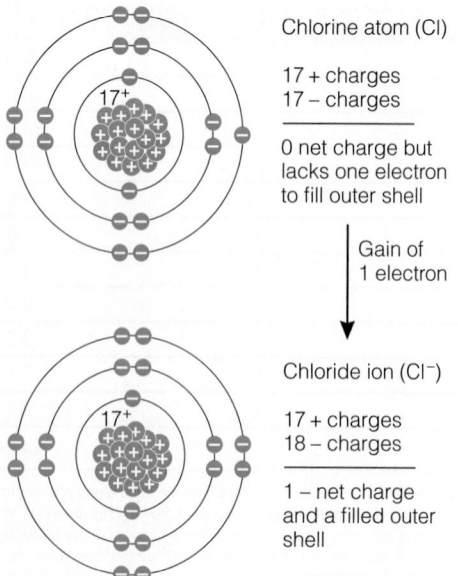

Chlorine atom (Cl)
17 + charges
17 − charges
___
0 net charge but lacks one electron to fill outer shell

Gain of 1 electron

Chloride ion (Cl⁻)
17 + charges
18 − charges
___
1 − net charge and a filled outer shell

Positively charged ions, such as a sodium ion (Na^+), are called **cations**; negatively charged ions, such as a chloride ion (Cl^-), are called **anions**. Cations and anions attract one another to form salts:

Sodium chloride (Na^+Cl^-)
28 + charges
28 − charges
___
0 net charge and filled outer shells

With all its electrons, sodium is a shiny, highly reactive metal; chlorine is the poisonous greenish yellow gas that was used in World War I. But after sodium and chlorine have shared electrons, they become stable and are familiar to you as table salt, or sodium chloride (Na^+Cl^-). The dramatic difference illustrates how profoundly the electron arrangement can influence the nature of a substance. The wide distribution of salt in nature attests to the stability of the union between the ions. Each meets the other's needs (a good marriage).

When dry, salt exists as crystals; its ions are stacked very regularly into a lattice, with positive and negative ions alternating in a three-dimensional checkerboard structure. In water, however, the salt quickly dissolves, and its ions separate from one another, forming an electrolyte solution in which they move about freely. Covalently bonded molecules rarely dissociate like this in a water solution. The most common exception is when they behave like acids and release H^+ ions, as discussed in the next section.

An ion can also be a group of atoms bound together in such a way that the group has a net charge and enters into reactions as a single unit. Many such groups are active in the fluids of the body. The bicarbonate ion is composed of five atoms—one H, one C, and three Os—and has a net charge of −1 (HCO_3^-). Another example is the phosphate ion with one H, one P, and four O, that has a net charge of −2 (HPO_4^{-2}).

Whereas many elements have only one configuration in the outer shell and thus only one way to bond with other elements, some elements have the possibility of varied configurations. Iron is such an element. Under some conditions iron loses two electrons, and under other circumstances it loses three. If iron loses two electrons, it then has a net charge of +2 and is called ferrous iron (Fe^{++}). If it loses three electrons, it becomes the +3 ion called ferric iron (Fe^{+++}).

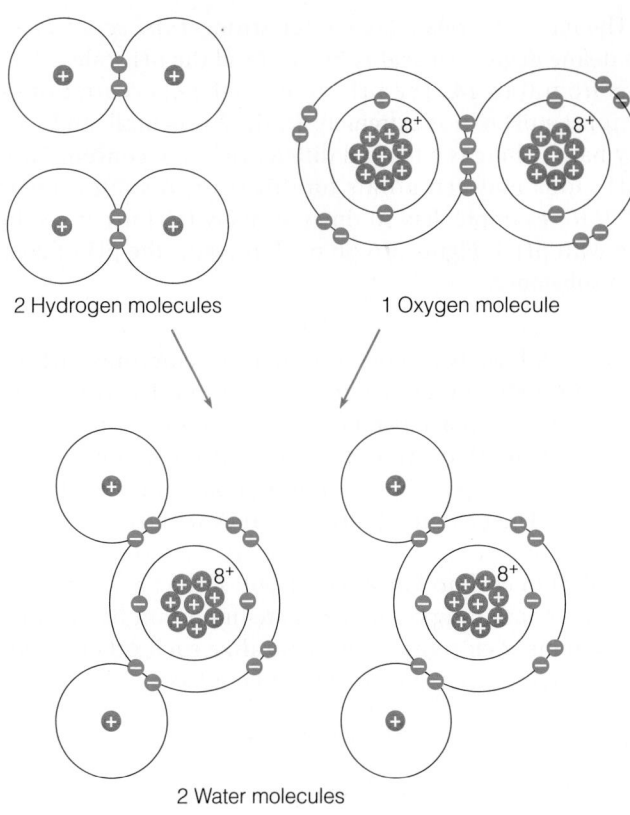

2 Hydrogen molecules   1 Oxygen molecule

2 Water molecules

Structures:

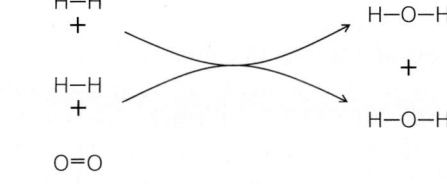

Formulas:

$$2H_2 + O_2 \longrightarrow 2H_2O$$

Hydrogen and oxygen react to form water.

Reactions in which the end products contain more energy than the reacting compounds started with are called endergonic, or "uphill," reactions and do not occur until an energy source is provided. An example of such an energy source is the sunlight used in photosynthesis reactions that combine carbon dioxide and water (low-energy compounds) to form glucose (a higher-energy compound). Conversely, the oxidation of glucose to carbon dioxide and water is an exergonic, or "downhill," reaction because the end products have less energy than the starting product. Oftentimes, but not always, reduction reactions are endergonic, resulting in an increase in the energy of the end products. Oxidation reactions often, but not always, are exergonic.

Chemical reactions tend to occur spontaneously if the end products are in a lower energy state and therefore are more stable than the reacting compounds. These reactions often give off energy in the form of heat as they occur. The generation of heat by wood burning in a fireplace and the maintenance of warmth in the human body both depend on energy-yielding chemical reactions. These downhill reactions occur easily although they may require some activation energy to get them started, just as a ball requires a push to start rolling.

Energy change as reaction occurs

Start of reaction	End of reaction
Reactants	Products
$2H_2 + O_2$	$2H_2O$

Formation of Free Radicals

Normally, when a chemical reaction takes place, bonds break and re-form to create new, stable compounds. Occasionally, bonds break in such a way as to create a free radical—a molecule with one or more unpaired electrons. When they do, free radicals are formed. Free radicals are highly unstable and quickly react with other compounds, forming more free radicals in a chain reaction. A cascade may ensue in which many highly reactive radicals are generated, resulting in damage that contributes to the development of many chronic diseases (see Highlight 11 for more details). Free radicals are of special interest in nutrition because the antioxidant properties of vitamins C and E as well as those of beta-carotene and the mineral selenium protect against the destructive effects of these free radicals.

Ferrous iron (Fe^{++})
(lost 2 outer-shell electrons)
26 + charges
24 − charges
_____
  2 + net charge

Ferric iron (Fe^{+++})
(lost 3 outer-shell electrons)
26 + charges
23 − charges
_____
  3 + net charge

Remember that a positive charge on an ion means that negative charges—electrons—have been lost and not that positive charges have been added to the nucleus.

Water, Acids, and Bases

Water The water molecule is electrically neutral, having equal numbers of protons and electrons. When a hydrogen atom shares its electron with oxygen, however, that electron will spend most of its time closer to the positively charged oxygen nucleus. This leaves the positive proton (nucleus of the hydrogen atom) exposed on the outer part of the water molecule. We know, too, that the two hydrogens both bond toward the same side of the oxygen. These two facts explain why water molecules are polar: they have regions that are more positively and more negatively charged.

Polar molecules like water are attracted to one another by the forces between the positive areas of one and the negative areas of another. These attractive forces, sometimes known as polar bonds or hydrogen bonds, occur among many molecules and also within different parts of the same molecule. Although very weak in comparison with covalent bonds, polar bonds may occur in such abundance that they become exceedingly important in determining the structure of large molecules such as proteins and DNA.

This diagram of a polar water molecule shows the negative area near the O and the positive area near the H atoms.

Water molecules have a slight tendency to ionize, separating into positive (H$^+$) and negative (OH$^-$) ions. In pure water, a small but constant number of these ions is present, and the number of positive ions exactly equals the number of negative ions.

Acids An acid is a substance that releases H$^+$ ions (protons) in a water solution. Hydrochloric acid (HCl$^-$) is such a substance because it dissociates in a water solution into H$^+$ and Cl$^-$ ions. Acetic acid is also an acid because it dissociates in water to acetate ions and free H$^+$:

$$\text{H-}\overset{\overset{\text{H}}{|}}{\underset{\underset{\text{H}}{|}}{\text{C}}}\text{-}\overset{\overset{\text{O}}{\|}}{\text{C}}\text{-O-H} \longrightarrow \text{H-}\overset{\overset{\text{H}}{|}}{\underset{\underset{\text{H}}{|}}{\text{C}}}\text{-}\overset{\overset{\text{O}}{\|}}{\text{C}}\text{-O}^- + \text{H}^+$$

Acetic acid dissociates into an acetate ion and a hydrogen ion.

The more H$^+$ ions released, the stronger the acid. Chemists define degrees of acidity by means of the pH scale, which runs from 0 to 14. The pH expresses the concentration of H$^+$ ions: a pH of 1 is extremely acidic, 7 is neutral, and 13 is very basic. There is a tenfold difference in the concentration of H$^+$ ions between points on this scale. A solution with pH 3, for example, has 10 times as many H$^+$ ions as a solution with pH 4. Figure 3-6 on p. 77 presents the pH of common substances.

Bases A base is a substance that can combine with H$^+$ ions, thus reducing the acidity of a solution. The compound ammonia is such a substance. The ammonia molecule has two electrons that are not shared with any other atom; a hydrogen ion (H$^+$) is a proton without an outer shell of electrons. The proton readily combines with the ammonia molecule to form an ion; thus a free H$^+$ is withdrawn from the solution and no longer contributes to its acidity. Many nitrogen-containing compounds are important bases in living systems. Acids and bases neutralize each other to produce substances that are neither acid nor base.

Chemical Reactions

A chemical reaction results in the breakdown or formation of substances. Almost all such reactions involve a change in the bonding of atoms. Old bonds are broken, and new ones are formed. The nuclei of atoms are never involved in chemical reactions—only the outer-shell electrons participate. At the end of a chemical reaction, the number of atoms of each type is always the same as at the beginning. For example, two hydrogen molecules (2H$_2$) can react with one oxygen molecule (O$_2$) to form two water molecules (2H$_2$O). In this reaction two substances (hydrogen and oxygen) disappear, and a new one (water) is formed, but at the end of the reaction there are still four H atoms and two O atoms, just as there were at the beginning. Because the atoms are now linked in a different way, their characteristics and properties have changed.

In many instances chemical reactions involve the exchange of electrons or protons between molecules. In such reactions the molecule that gains one or more electrons (or loses one or more protons) is said to be reduced; the molecule that loses electrons (or gains protons) is oxidized. A hydrogen ion is equivalent to a proton.

Oxidation and reduction reactions take place simultaneously because an electron or proton that is lost by one molecule is accepted by another. The addition of an atom of oxygen is also oxidation because oxygen (with six electrons in the outer shell) accepts two electrons in becoming bonded. Oxidation occurs, then, with the loss of electrons, the gain of protons, or the addition of oxygen (with six electrons); reduction occurs with the opposite—a gain of electrons, a loss of protons, or a loss of oxygen. The addition of hydrogen atoms to oxygen to form water can thus be described as the reduction of oxygen *or* the oxidation of hydrogen.

Appendix C   Biochemical Structures and Pathways

This appendix describes the biochemical structures and pathways most important to the study of nutrition. It begins by presenting diagrams of nutrients commonly found in the human diet. Following the diagrams of nutrients are sections on the major metabolic pathways mentioned in Chapter 7—glycolysis, fatty acid oxidation, amino acid degradation, the TCA cycle, and the electron transport chain—and a description of how alcohol interferes with these pathways. Discussions of the urea cycle and the formation of ketone bodies complete the appendix.

CONTENTS

APPENDIX C

Carbohydrates
Monosaccharides

Glucose (alpha form). The ring would be at right angles to the plane of the paper. The bonds directed upward are above the plane; those directed downward are below the plane. This molecule is considered an alpha form because the OH on carbon 1 points downward.

Glucose (beta form). The OH on carbon 1 points upward.
Fructose, galactose: see Chapter 4.

Glucose (alpha form) shorthand notation. This notation, in which the carbons in the ring and single hydrogens have been eliminated, will be used throughout this appendix.

Disaccharides

Maltose.

Lactose (alpha form).

Sucrose.

Polysaccharides: Starches As described in Chapter 4, starch, glycogen, and cellulose are all long chains of glucose molecules covalently linked together. (Appendix B discusses covalent bonding.)

Amylose (unbranched starch)

Starch. Two kinds of covalent bonds link the glucose molecules in starch, giving rise to two kinds of chains. Amylose is composed of straight chains, with carbon 1 of one glucose linked to carbon 4 of the next (α-1,4 linkage). Amylopectin is made up of straight chains like amylose, but has occasional branches where the carbon 6 of a glucose is also linked to the carbon 1 of another glucose (α-1,6 linkage).

Glycogen. The structure of glycogen is like amylopectin but with many more branches.

Cellulose. Like starch and glycogen, cellulose is also made of chains of glucose units, but there is an important difference: in cellulose, the OH on carbon 1 is in the beta position (see p. C-1). When carbon 1 of one glucose is linked to carbon 4 of the next, it forms a β-1,4 linkage, which cannot be broken by digestive enzymes in the human GI tract.

Amylopectin (branched starch)

Polysaccharides: Fibers Fibers, such as hemicelluloses, consist of long chains of various monosaccharides.

Monosaccharides common in the backbone chain of hemicelluloses:*

Xylose　　　Mannose　　　Galactose

---

*These structures are shown in the alpha form with the H on the carbon pointing upward and the OH pointing downward, but they may also appear in the beta form with the H pointing downward and the OH upward.

Monosaccharides common in the side chains of hemicelluloses:

Arabinose   Glucuronic acid

Galactose

Hemicelluloses. The most common hemicelluloses are composed of a backbone chain of xylose, mannose, and galactose, with branching side chains of arabinose, glucuronic acid, and galactose.

Lipids

TABLE C-1 Saturated Fatty Acids Found in Natural Fats

Saturated Fatty Acids	Chemical Formulas	Number of Carbons	Major Food Sources
Butyric	C_3H_7COOH	4	Butterfat
Caproic	$C_5H_{11}COOH$	6	Butterfat
Caprylic	$C_7H_{15}COOH$	8	Coconut oil
Capric	$C_9H_{19}COOH$	10	Palm oil
Lauric	$C_{11}H_{23}COOH$	12	Coconut oil, palm oil
Myristic	$C_{13}H_{27}COOH$	14	Coconut oil, palm oil
Palmitic	$C_{15}H_{31}COOH$	16	Palm oil
Stearic	$C_{17}H_{35}COOH$	18	Most animal fats
Arachidic	$C_{19}H_{39}COOH$	20	Peanut oil
Behenic	$C_{21}H_{43}COOH$	22	Seeds
Lignoceric	$C_{23}H_{47}COOH$	24	Peanut oil

NOTE: The most common fatty acids are myristic, palmitic, and stearic.

TABLE C-2 Unsaturated Fatty Acids Found in Natural Fats

Unsaturated Fatty Acids	Chemical Formulas	Number of Carbons	Number of Double Bonds	Standard Notation[a]	Omega Notation[b]	Major Food Sources
Palmitoleic	$C_{15}H_{29}COOH$	16	1	16:1;9	16:1ω7	Seafood, beef
Oleic	$C_{17}H_{33}COOH$	18	1	18:1;9	18:1ω9	Olive oil, canola oil
Linoleic	$C_{17}H_{31}COOH$	18	2	18:2;9,12	18:2ω6	Sunflower oil, safflower oil
Linolenic	$C_{17}H_{29}COOH$	18	3	18:3;9,12,15	18:3ω3	Soybean oil, canola oil
Arachidonic	$C_{19}H_{31}COOH$	20	4	20:4;5,8,11,14	20:4ω6	Eggs, most animal fats
Eicosapentaenoic	$C_{19}H_{29}COOH$	20	5	20:5;5,8,11,14,17	20:5ω3	Seafood
Docosahexaenoic	$C_{21}H_{31}COOH$	20	6	22:6;4,7,10,13,16,19	22:6ω3	Seafood

NOTE: A fatty acid has two ends; designated the methyl (CH_3) end and the carboxyl, or acid (COOH), end.
[a]Standard chemistry notation begins counting carbons at the acid end. The number of carbons the fatty acid contains comes first, followed by a colon and another number that indicates the number of double bonds; next comes a semicolon followed by a number or numbers indicating the positions of the double bonds. Thus the notation for linoleic acid, an 18-carbon fatty acid with two double bonds between carbons 9 and 10 and between carbons 12 and 13, is 18:2;9,12.
[b]Because fatty acid chains are lengthened by adding carbons at the acid end of the chain, chemists use the omega system of notation to ease the task of identifying them. The omega system begins counting carbons at the methyl end. The number of carbons the fatty acid contains comes first, followed by a colon and the number of double bonds; next come the omega symbol (ω) and a number indicating the position of the double bond nearest the methyl end. Thus linoleic acid with its first double bond at the sixth carbon from the methyl end would be noted 18:2ω6 in the omega system.

Protein: Amino Acids

The common amino acids may be classified into the seven groups listed below based on their structural similarities. Amino acids marked with an asterisk (*) are essential.

1. Amino acids with aliphatic side chains, which consist of hydrogen and carbon atoms (hydrocarbons):

Glycine (Gly)

Alanine (Ala)

Valine* (Val)

Leucine* (Leu)

Isoleucine* (Ile)

2. Amino acids with hydroxyl (OH) side chains:

Serine (Ser)

Threonine* (Thr)

3. Amino acids with side chains containing acidic groups or their amides, which contain the group NH_2:

Aspartic acid (Asp)

Glutamic acid (Glu)

Asparagine (Asn)

Glutamine (Gln)

4. Amino acids with basic side chains:

Lysine* (Lys)

Arginine (Arg)

Histidine* (His)

5. Amino acids with aromatic side chains, which are characterized by the presence of at least one ring structure:

Phenylalanine* (Phe)

Tyrosine (Tyr)

Tryptophan* (Trp)

6. Amino acids with side chains containing sulfur atoms:

Cysteine (Cys)

Methionine* (Met)

7. Imino acid:

Proline (Pro)

Proline has the same chemical structure as the other amino acids, but its amino group has given up a hydrogen to form a ring.

Vitamins and Coenzymes

Vitamin A: retinol. Retinol is the alcohol form of vitamin A.

Vitamin A: retinal. Retinal is the aldehyde form of vitamin A.

Vitamin A: retinoic acid. Retinoic acid is the acid form of vitamin A.

Vitamin A precursor: beta-carotene. Beta-carotene is the carotenoid with the most vitamin A activity.

Thiamin. Thiamin is part of the coenzyme thiamin pyrophosphate (TPP).

Thiamin pyrophosphate (TPP). TPP is a coenzyme that includes the thiamin molecule as part of its structure.

Riboflavin. Riboflavin is a part of two coenzymes—flavin mononucleotide (FMN) and flavin adenine dinucleotide (FAD).

Flavin mononucleotide (FMN). FMN is a coenzyme that includes the riboflavin molecule as part of its structure.

FAD can pick up hydrogens and carry them to the electron transport chain.

FAD (oxidized form)

FADH₂ (reduced form)

Flavin adenine dinucleotide (FAD). FAD is a coenzyme that includes the riboflavin molecule as part of its structure.

Nicotinic acid

Nicotinamide

Niacin (nicotinic acid and nicotinamide). Niacin is a part of two coenzymes—nicotinamide adenine dinucleotide (NAD$^+$) and nicotinamide adenine dinucleotide phosphate (NADP$^+$).

Nicotinamide

Adenine

D-ribose

D-ribose

Pyrophosphate

Nicotinamide adenine dinucleotide (NAD$^+$). NAD is a coenzyme that includes niacin as part of its structure. NADP has the same structure as NAD but with a phosphate group attached to the O instead of the (H).

NAD$^+$

NADH

Reduced NAD$^+$ (NADH). When NAD$^+$ is reduced by the addition of H$^+$ and two electrons, it becomes the coenzyme NADH. (The dots on the H entering this reaction represent electrons—see Appendix B.)

Pyridoxine

Pyridoxal

Pyridoxamine

Vitamin B$_6$. Vitamin B$_6$ is a general name for three compounds—pyridoxine, pyridoxal, and pyridoxamine, which become a part of two coenzymes—pyridoxal phosphate and pyridoxamine phosphate.

Pyridoxal phosphate   Pyridoxamine phosphate

Pyridoxal phosphate (PLP) and pyridoxamine phosphate. These coenzymes include vitamin B_6 as part of their structures.

Folate (folacin or folic acid). Folate consists of a double ring combined with a single ring and at least one glutamate (a nonessential amino acid highlighted in color). Folate is a part of the coenzyme tetrahydrofolate (THF).

Vitamin B_{12} (cyanocobalamin). The arrows in this diagram indicate that the spare electron pairs on the nitrogens attract them to the cobalt.

Tetrahydrofolate (THF). THF is the active coenzyme form of folate and has four added hydrogens. An intermediate form, dihydrofolate, has two added hydrogens.

Pantothenic acid. Pantothenic acid is part of coenzyme A (CoA).

Coenzyme A (CoA). Coenzyme A is a coenzyme that includes pantothenic acid as part of its structure.

Biotin.

Ascorbic acid
(reduced form)

Dehydroascorbic acid
(oxidized form)

Vitamin C. Two hydrogen atoms with their electrons are lost when ascorbic acid is oxidized and gained when it is reduced again.

7-dehydrocholesterol

Carbon #7

Ultraviolet light
on the skin

Vitamin D₃
(also called
cholecalciferol
or calciol)

Hydroxylation in
the liver

25-hydroxy-vitamin D₃
(also called calcidiol)

Carbon #25

Hydroxylation in
the kidneys

1,25-dihydroxy-vitamin D₃
(also called calcitriol)

Carbon #1

Vitamin D. The synthesis of active vitamin D begins with 7-dehydrocholesterol, a precursor made in the liver from cholesterol. (The carbon atoms at which changes occur are numbered.)

APPENDIX C

Tocotrienols contain double bonds here.

Vitamin E (alpha-tocopherol). Vitamin E consists of two subgroups—the tocopherols and the tocotrienols. All are made up of a complex ring structure with a long saturated (in tocopherols) or unsaturated (in tocotrienols) side chain. The number and positions of methyl groups (CH_3) distinguish the members within each subgroup.

Vitamin K (phylloquinone). Naturally occurring compounds with vitamin K activity include phylloquinones (from plants) and menaquinones (from bacteria). The chemical structure of menaquinones differs only slightly from that of phylloquinones.

Vitamin K (menadione). Menadione is a synthetic compound that has the same activity as natural vitamin K.

Triphosphate

Adenine

Ribose

Cleavage

Cleavage

$+ H-O-H$
(Water)

Phosphate
+

ADP

Adenosine diphosphate (ADP).

Adenosine triphosphate (ATP). The high-energy compound ATP releases energy when one or two phosphate groups split off. The cleavage point marks the bond that is broken when ATP splits to become ADP + P.

Glycolysis

Figure C-1 depicts glycolysis. The following text describes key steps as numbered on the figure.

> **FIGURE C-1** **Glycolysis**

Notice that galactose and fructose enter at different places but continue on the same pathway.

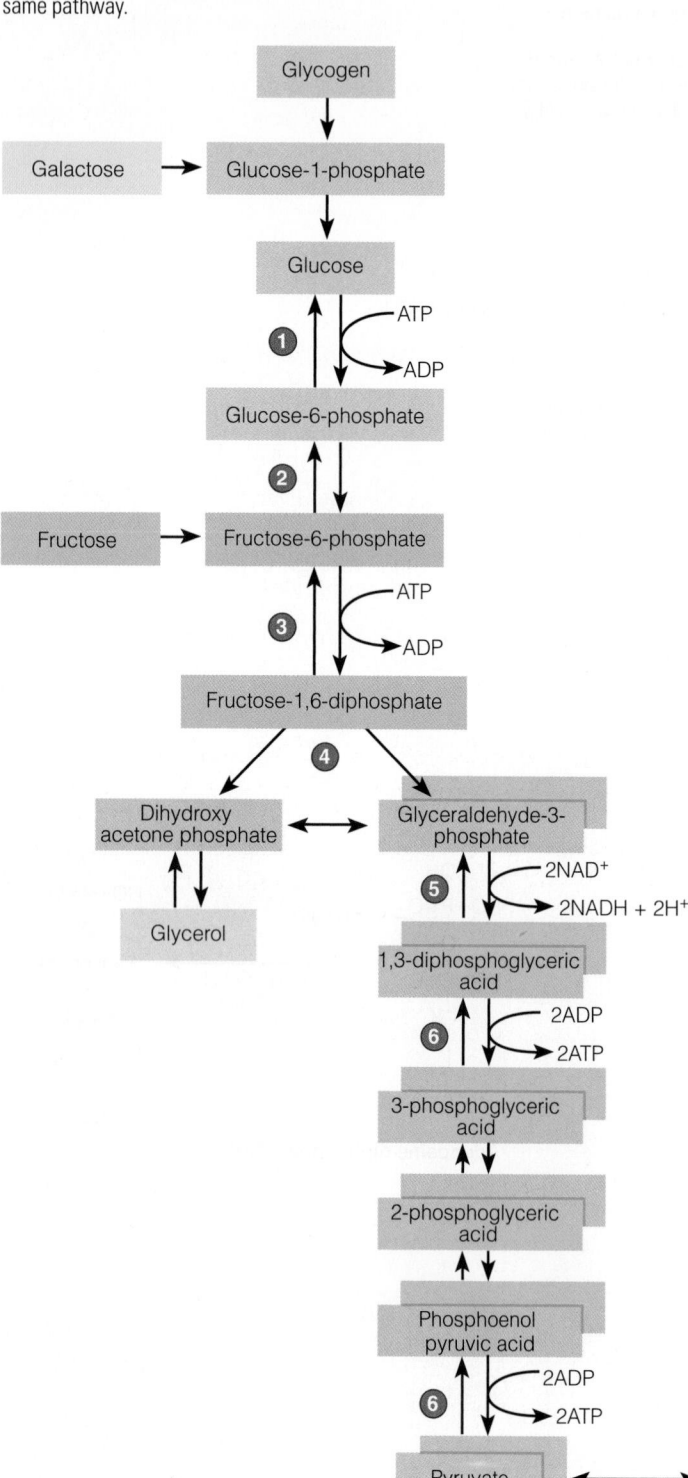

1. A phosphate is attached to glucose at the carbon that chemists call number 6 (review the first diagram of glucose on p. C-1 to see how chemists number the carbons in a glucose molecule). The product is called, logically enough, glucose-6-phosphate. One ATP molecule is used to accomplish this.

2. Glucose-6-phosphate is rearranged by an enzyme.

3. A phosphate is added in another reaction that uses another molecule of ATP. The resulting product is fructose-1,6-diphosphate. At this point the 6-carbon sugar has a phosphate group on its first and sixth carbons and is ready to break apart.

4. When fructose-1,6-diphosphate breaks in half, the two 3-carbon compounds are not identical. Each has a phosphate group attached, but only glyceraldehyde-3-phosphate converts directly to pyruvate. The other compound, however, converts easily to glyceraldehyde-3-phosphate.

5. In the next step, NAD^+ is reduced to $NADH + H^+$.

6. In two of the following steps ATP is regenerated.

Remember that in effect two molecules of glyceraldehyde-3-phosphate are produced from glucose; therefore, four ATP molecules are generated from each glucose molecule. Two ATP were needed to get the sequence started, so the net gain at this point is two ATP and two molecules of $NADH + H^+$. As you will see later, each $NADH + H^+$ moves to the electron transport chain to unload its hydrogens, producing more ATP.

Fatty Acid Oxidation

Figure C-2 presents fatty acid oxidation. The sequence is as follows.

1. The fatty acid is activated by combining with coenzyme A (CoA). In this reaction, ATP loses two phosphorus atoms and becomes AMP (adenosine monophosphate)—the equivalent of a loss of two ATP.

2. In the next reaction, two H with their electrons are removed and transferred to FAD, forming $FADH_2$.

3. In a later reaction, two H are removed and go to NAD^+ (forming $NADH + H^+$).

4. The fatty acid is cleaved at the "beta" carbon, the second carbon from the carboxyl (COOH) end. This break results in a fatty acid that is two carbons shorter than the previous one and a 2-carbon molecule of

> FIGURE C-2   Fatty Acid Oxidation

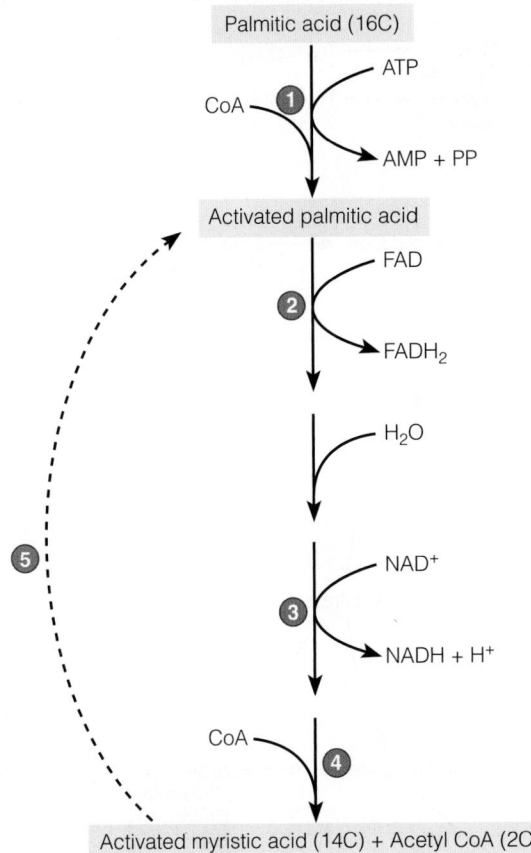

Palmitic acid (16C)

① CoA + ATP → AMP + PP

Activated palmitic acid

② FAD → FADH$_2$

H_2O

③ NAD$^+$ → NADH + H$^+$

④ CoA

Activated myristic acid (14C) + Acetyl CoA (2C)

⑤

acetyl CoA. At the same time, another CoA is attached to the fatty acid, thus activating it for its turn through the series of reactions.

5. The sequence is repeated with each cycle producing an acetyl CoA and a shorter fatty acid until only a 2-carbon fatty acid remains—acetyl CoA.

In the example shown in Figure C-2, palmitic acid (a 16-carbon fatty acid) will go through this series of reactions seven times, using the equivalent of two ATP for the initial activation and generating seven FADH$_2$, seven NADH + H$^+$, and eight acetyl CoA. As you will see later, each of the seven FADH$_2$ will enter the electron transport chain, yielding two ATP (for a total of 14). Similarly, each NADH + H$^+$ will enter the electron transport chain, yielding three ATP (for a total of 21). Thus the oxidation of a 16-carbon fatty acid uses 2 ATP and generates 35 ATP. When the eight acetyl CoA enter the TCA cycle, even more ATP will be generated, as a later section describes.

Amino Acid Degradation

The first step in amino acid degradation is the removal of the nitrogen-containing amino group through either deamination (Figure 6-11 on p. 183) or transamination (Figure 6-12 on

p. 183) reactions. Then the remaining carbon skeletons may enter the metabolic pathways at different places, as shown in Figure C-3 (p. C-12).

The TCA Cycle

The tricarboxylic acid, or TCA, cycle is the set of reactions that break down acetyl CoA to carbon dioxide and hydrogen atoms. Pyruvate derived from glycolysis does not enter the TCA cycle directly; instead pyruvate enters the mitochondrion, loses a carbon group, and bonds with a molecule of CoA to become acetyl CoA. The TCA cycle uses any substance that can be converted to acetyl CoA directly or indirectly through pyruvate.

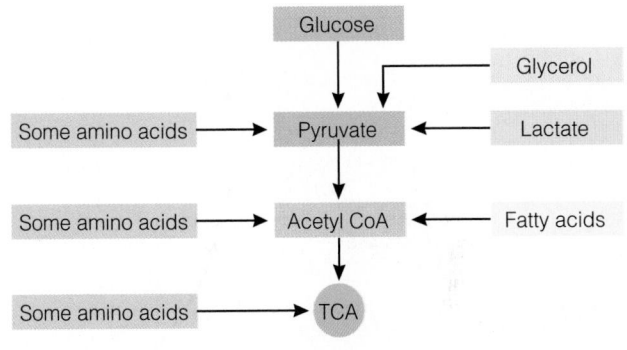

Any substance that can be converted to acetyl CoA directly, or indirectly through pyruvate, may enter the TCA cycle.

The pathway from pyruvate to acetyl CoA is complex. We have included only those steps that will help you understand the transfer of energy from the nutrients. Pyruvate loses a carbon to carbon dioxide and is attached to a molecule of CoA. In the process, NAD$^+$ picks up two hydrogens with their associated electrons, becoming NADH + H$^+$.

The step from pyruvate to acetyl CoA. (TPP and NAD are coenzymes containing the B vitamins thiamin and niacin, respectively.)

> FIGURE C-3 **Amino Acids Enter the Metabolic Pathways**

After losing their amino groups, carbon skeletons can be converted to one of seven molecules that can enter the TCA cycle (presented in Figure C-4).

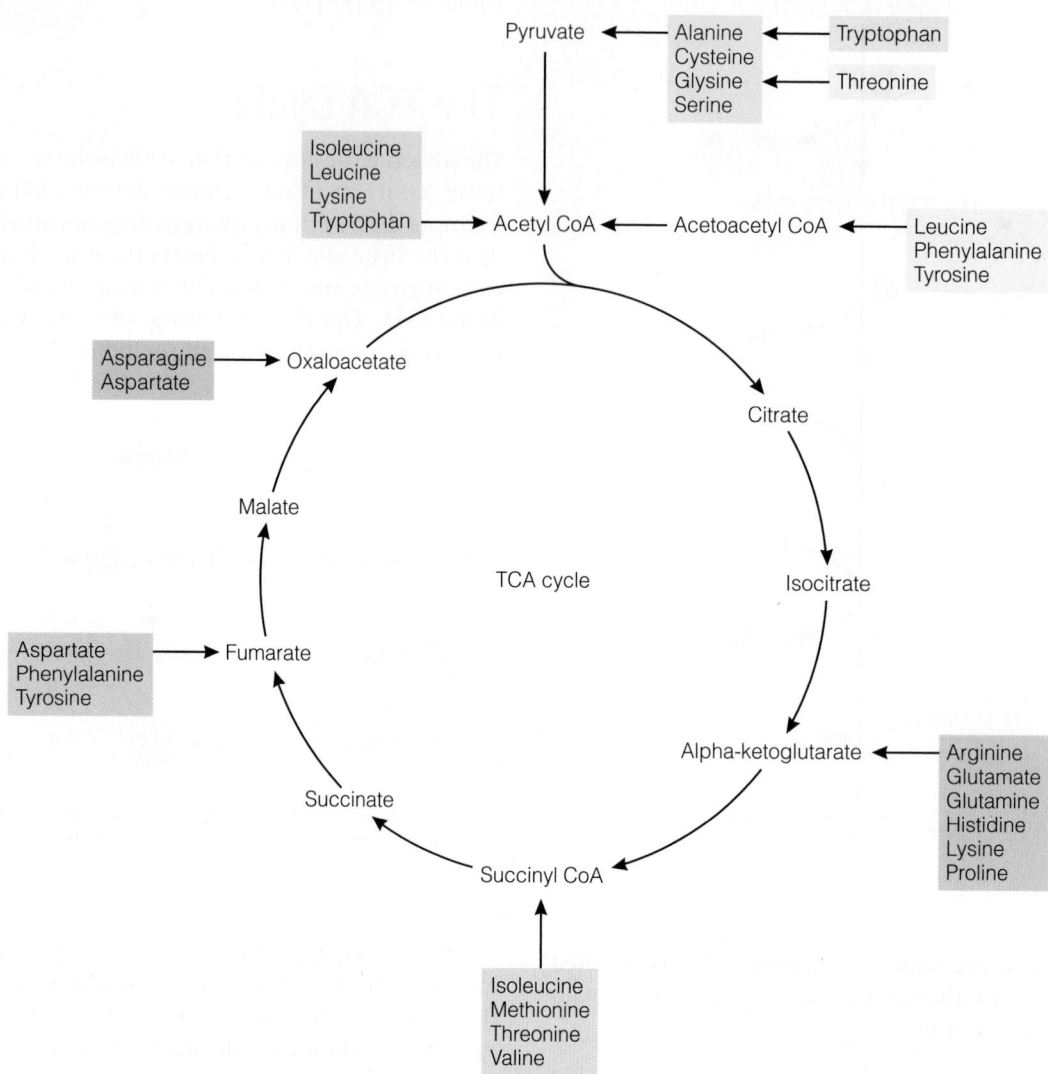

Let's follow the steps of the TCA cycle (see the corresponding numbers in Figure C-4 on the next page).

1. The 2-carbon acetyl CoA combines with a 4-carbon compound, oxaloacetate. The CoA comes off, and the product is a 6-carbon compound, citrate.

2. The atoms of citrate are rearranged to form isocitrate.

3. Now two H (with their two electrons) are removed from the isocitrate. NAD^+ accepts the hydrogens with their electrons and becomes $NADH + H^+$. (Remember this $NADH + H^+$, but let's follow the carbons first.) A carbon is combined with two oxygens, forming carbon dioxide (which diffuses away into the blood and is exhaled). What is left is the 5-carbon compound alpha-ketoglutarate.

4. Now two compounds interact with alpha-ketoglutarate—a molecule of CoA and a molecule of NAD^+. In this complex reaction, a carbon and two oxygens are removed (forming carbon dioxide); two hydrogens are removed and go to NAD^+ (forming $NADH + H^+$); and the remaining 4-carbon compound is attached to the CoA, forming succinyl CoA. (Remember this $NADH + H^+$ also. You will see later what happens to it.)

5. Now two molecules react with succinyl CoA—a molecule called GDP and one of phosphate (P). The CoA comes off, the GDP and P combine to form the high-energy compound GTP (similar to ATP), and succinate remains. (Remember this GTP.)

6. In the next reaction, two H with their electrons are removed from succinate and are transferred to a molecule of FAD (a coenzyme like NAD^+) to form $FADH_2$. The product that remains is fumarate. (Remember this $FADH_2$.)

7. Next a molecule of water is added to fumarate, forming malate.

> FIGURE C-4 **The TCA Cycle**

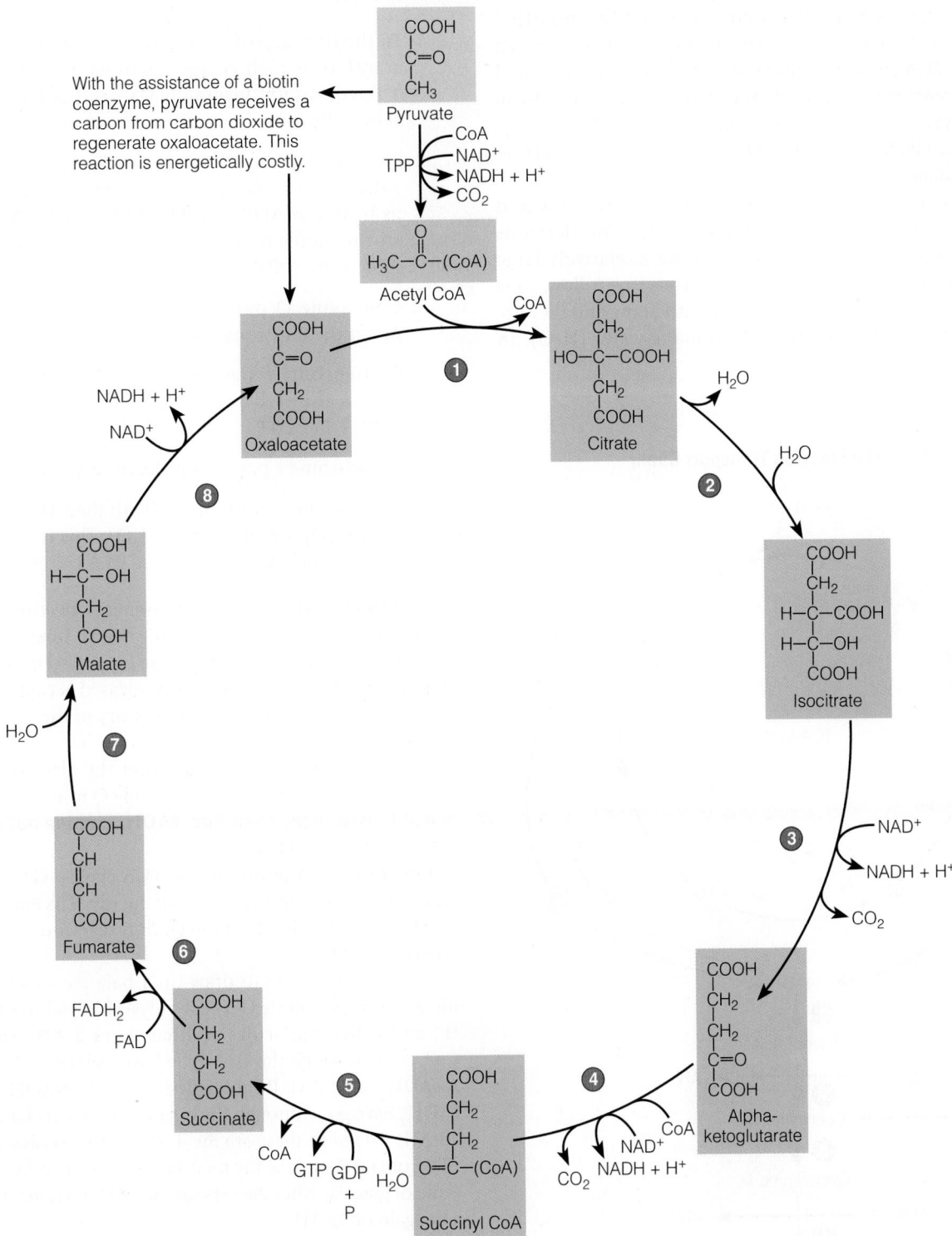

With the assistance of a biotin coenzyme, pyruvate receives a carbon from carbon dioxide to regenerate oxaloacetate. This reaction is energetically costly.

8. A molecule of NAD^+ accepts two H with their associated electrons and forms $NADH + H^+$. The product that remains is the 4-carbon compound oxaloacetate. (Remember this $NADH + H^+$.)

The cycle is complete and we are back where we started. The oxaloacetate can combine with another molecule of acetyl CoA (step 1), and the cycle can begin again.

So far, we have seen two carbons brought in with acetyl CoA and two carbons ending up in carbon dioxide. But where are the energy and the ATP that were promised?

A review of the eight steps of the TCA cycle shows that the compounds $NADH + H^+$ (three molecules), $FADH_2$, and GTP captured energy along the way. To see how this energy ends up in ATP, we must follow the electrons further—into the electron transport chain.

The Electron Transport Chain

The six reactions described here are a highly simplified overview of the electron transport chain, which is shown below the TCA cycle in Figure C-5. Since oxygen is required for these reactions, and ADP and P are combined to form ATP in several of them (ADP is phosphorylated), the reactions of the electron transport chain are also called *oxidative phosphorylation*.

An important concept to remember at this point is that an electron is not a fixed amount of energy. The electrons that bond the H to NAD^+ in NADH have a relatively large amount of energy. In the series of reactions that follow, they release this energy in small amounts, until at the end they are attached (with H) to oxygen (O) to make water (H_2O). In some of the steps, the energy they release is captured into ATP in coupled reactions.

1. In the first step of the electron transport chain, NADH transfers its high-energy electrons to a molecule called a flavoprotein, leaving NAD^+ and reduced flavoprotein. A little energy is released as heat in this reaction.

2. The flavoprotein passes on the electrons to a molecule called coenzyme Q. Again a little energy is released as heat, but ADP and P bond together and form ATP, storing much of the energy. This is a coupled reaction: $ADP + P \rightarrow ATP$.

3. Coenzyme Q passes the electrons to cytochrome *b*. Again the electrons release energy.

4. Cytochrome *b* passes the electrons to cytochrome *c* in a coupled reaction in which ATP is formed: $ADP + P \rightarrow ATP$.

5. Cytochrome *c* passes the electrons to cytochrome *a*.

6. Cytochrome *a* passes them (with their H) to an atom of oxygen (O), forming water (H_2O). This is a coupled reaction in which ATP is formed: $ADP + P \rightarrow ATP$.

At the end of the chain, the low-energy electrons are passed to oxygen, which combines with the free H^+ ions to form water. As Figure C-5 shows, each time the three NADH are oxidized (losing their electrons), the energy released is captured in three ATP molecules. This completes the story of the electrons from NADH.

As for $FADH_2$, its electrons enter the electron transport chain at coenzyme Q. From coenzyme Q to water, ATP is generated in two steps. Therefore, $FADH_2$ coming out of the TCA cycle yields two ATP molecules.

One other compound of the TCA cycle—GTP—does not enter the electron transport chain but gives its energy directly to ADP in a simple phosphorylation reaction. This reaction yields one ATP.

It is now possible to draw up a balance sheet of glucose metabolism (see Table C-3). Glycolysis has yielded 4 NADH + H^+ and 4 ATP molecules and has spent 2 ATP. The 2 acetyl CoA going through the TCA cycle have yielded 6 NADH + H^+, 2 $FADH_2$, and 2 GTP molecules. After the NADH + H^+ and $FADH_2$ have gone through the electron transport chain, there are 28 ATP. Added to these are the 4 ATP from glycolysis and the 2 ATP from GTP, making the total 34 ATP generated from one molecule of glucose. After the expense of 2 ATP is subtracted, there is a net gain of 32 ATP.*

A similar balance sheet from the complete breakdown of one 16-carbon fatty acid would show a net gain of 129 ATP.

> FIGURE C-5   **The Electron Transport Chain**

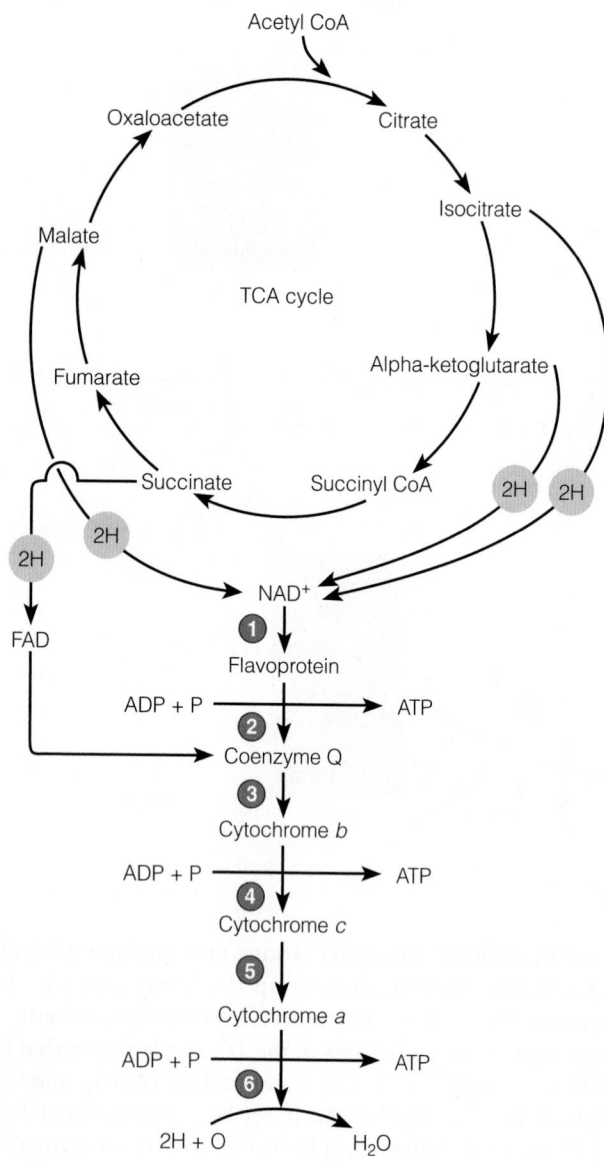

---

*The total may sometimes be 30 ATP. The NADH + H^+ generated in the cytoplasm during glycolysis pass their electrons on to shuttle molecules, which move them into the mitochondria. One shuttle, malate, contributes its electrons to the electron transport chain before the first site of ATP synthesis, yielding 5 ATP. Another, glycerol phosphate, adds its electrons into the chain beyond that first site, yielding 3 ATP. Thus sometimes 5, and sometimes 3, ATP result from the NADH + H^+ that arise from glycolysis. The amount depends on the cell.

TABLE C-3 Balance Sheet for Glucose Metabolism

		ATP
Glycolysis:	4 ATP − 2 ATP	2
1 glucose to 2 pyruvate	2 NADH + H$^+$	3–5[a]
2 pyruvate to 2 acetyl CoA	2 NADH + H$^+$	5
TCA cycle and electron transport chain:		
2 isocitrate	2 NADH + H$^+$	5
2 alpha-ketoglutarate	2 NADH + H$^+$	5
2 succinyl CoA	2 GTP	2
2 succinate	2 FADH$_2$	3
2 malate	2 NADH + H$^+$	5
Total ATP collected from one molecule glucose:		30–32

[a]Each NADH + H$^+$ from glycolysis can yield 1.5 or 2.5 ATP. See the accompanying text.

other compounds as well. During alcohol metabolism, however, NAD$^+$ becomes unavailable for the multitude of reactions for which it is required.

As the previous sections just explained, for glucose to be completely metabolized, the TCA cycle must be operating, and NAD$^+$ must be present. If these conditions are not met (and when alcohol is present, they may not be), the pathway will be blocked, and traffic will back up—or an alternate route will be taken. Think about this as you follow the pathway shown in Figure C-6.

In each step of alcohol metabolism in which NAD$^+$ is converted to NADH + H$^+$, hydrogen ions accumulate, resulting in a dangerous shift of the acid-base balance toward acid (Chapter 12 explains acid-base balance). The accumulation of NADH + H$^+$ slows TCA cycle activity, so pyruvate and acetyl CoA build up. This condition favors the conversion of

As mentioned earlier, 35 ATP were generated from the 7 FADH$_2$ and 7 NADH + H$^+$ produced during fatty acid oxidation. The 8 acetyl CoA produced will each generate 12 ATP as they go through the TCA cycle and the electron transport chain, for a total of 96 more ATP. After subtracting the 2 ATP needed to activate the fatty acid initially, the net yield from one 16-carbon fatty acid: 35 + 96 − 2 = 129 ATP.

These calculations help explain why fat yields more energy (measured as kcalories) per gram than carbohydrate or protein. The more hydrogen atoms a fuel contains, the more ATP will be generated during oxidation. The 16-carbon fatty acid molecule, with its 32 hydrogen atoms, generates 129 ATP, whereas glucose, with its 12 hydrogen atoms, yields only 32 ATP.

The TCA cycle and the electron transport chain are the body's major means of capturing the energy from nutrients in ATP molecules. Other means, such as anaerobic glycolysis, contribute energy quickly, but the aerobic processes are the most efficient.

Alcohol's Interference with Energy Metabolism

Highlight 7 provides an overview of how alcohol interferes with energy metabolism. With an understanding of the TCA cycle, a few more details may be appreciated. During alcohol metabolism, the enzyme alcohol dehydrogenase oxidizes alcohol to acetaldehyde while it simultaneously reduces a molecule of NAD$^+$ to NADH + H$^+$. The related enzyme acetaldehyde dehydrogenase reduces another NAD$^+$ to NADH + H$^+$ while it oxidizes acetaldehyde to acetyl CoA, the compound that enters the TCA cycle to generate energy. Thus, whenever alcohol is being metabolized in the body, NAD$^+$ diminishes, and NADH + H$^+$ accumulates, thus altering the body's "redox state." NAD$^+$ can oxidize, and NADH + H$^+$ can reduce, many

> FIGURE C-6 **Ethanol Enters the Metabolic Pathways**

This is a simplified version of the glucose-to-energy pathway showing the entry of ethanol. The coenzyme NAD (which is the active form of the B vitamin niacin) is the only one shown here; however, many others are involved.

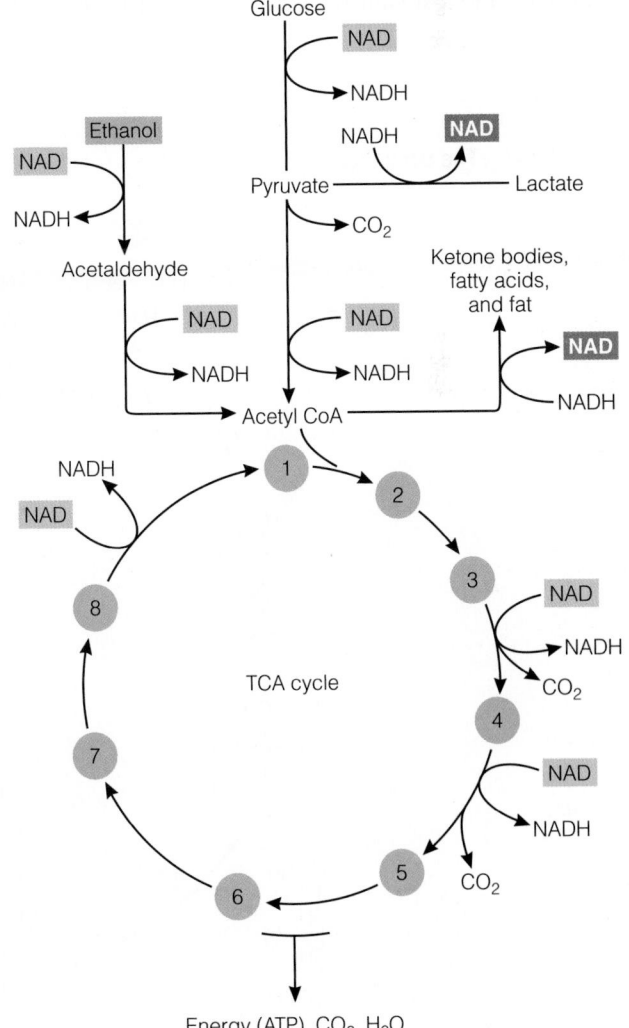

pyruvate to lactate, which serves as a temporary storage place for hydrogens from NADH + H⁺. The conversion of pyruvate to lactate restores some NAD⁺, but a lactate buildup has serious consequences of its own. It adds to the body's acid burden and interferes with the excretion of uric acid, causing goutlike symptoms. Molecules of acetyl CoA become building blocks for fatty acids or ketone bodies. The making of ketone bodies consumes acetyl CoA and generates NAD⁺; but some ketone bodies are acids, so they push the acid-base balance further toward acid.

Thus alcohol cascades through the metabolic pathways, wreaking havoc along the way. These consequences have physical effects, as Highlight 7 describes.

The Urea Cycle

Chapter 6 sums up the process by which waste nitrogen is eliminated from the body by stating that ammonia molecules combine with carbon dioxide to produce urea. This is true, but it is not the whole story. Urea is produced in a multistep process within the cells of the liver.

Ammonia, freed from an amino acid or other compound during metabolism anywhere in the body, arrives at the liver by way of the bloodstream and is taken into a liver cell. There,

it is first combined with carbon dioxide and a phosphate group from ATP to form carbamyl phosphate:

$$CO_2 + NH_3 \xrightarrow[2\ ATP]{\quad} 2\ ADP + P \quad H_2N-C-O-P-O^-$$

Carbon dioxide   Ammonia   Carbamyl phosphate

Figure C-7 shows the cycle of four reactions that follow.

1. Carbamyl phosphate combines with the amino acid ornithine, losing its phosphate group. The compound formed is citrulline.

2. Citrulline combines with the amino acid aspartic acid, to form argininosuccinate. The reaction requires energy from ATP. (In this reaction, ATP loses two phosphorus atoms, and becomes adenosine monophosphate, AMP.)

3. Argininosuccinate is split, forming another acid, fuma-rate, and the amino acid arginine.

4. Arginine loses its terminal carbon with two attached amino groups and picks up an oxygen from water. The

> FIGURE C-7   **The Urea Cycle**

end product is urea, which the kidneys excrete in the urine. The compound that remains is ornithine, identical to the ornithine with which this series of reactions began, ready to react with another molecule of carbamyl phosphate and turn the cycle again.

Formation of Ketone Bodies

Normally, fatty acid oxidation proceeds all the way to carbon dioxide and water. In ketosis, however, an intermediate is formed from the condensation of two molecules of acetyl CoA: acetoacetyl CoA. Figure C-8 shows the formation of ketone bodies from that intermediate.

1. Acetoacetyl CoA condenses with acetyl CoA to form a 6-carbon intermediate, beta-hydroxy-beta-methylglutaryl CoA.

2. This intermediate is cleaved to acetyl CoA and acetoacetate.

3. Acetoactate can be metabolized either to beta-hydroxybutyrate acid (step 3a) or to acetone (3b).

Acetoacetate, beta-hydroxybutyrate, and acetone are the ketone bodies of ketosis. Two are real ketones (they have a C=O group between two carbons); the other is an alcohol that has been produced during ketone formation—hence the term *ketone bodies*, rather than ketones, to describe the three of them. There are many other ketones in nature; these three are characteristic of ketosis in the body.

> FIGURE C-8   The Formation of Ketone Bodies

Appendix D   Measures of Protein Quality

CONTENTS

In a world where food is scarce and many people's diets contain marginal or inadequate amounts of protein, it is important to know which foods contain the highest-quality protein. Chapter 6 describes protein quality, and this appendix presents different measures researchers use to assess the quality of a food protein. Measures of protein quality aim to determine how well a food protein supports the body's normal metabolism and growth. The accompanying glossary defines related terms.

Amino Acid Score

The **amino acid score** predicts protein quality based on the pattern of essential amino acids. It compares the essential amino acid composition of a food protein with that of a reference protein (egg). The score of each amino acid in the food protein is described as a percentage of the amino acid in the reference protein. The amino acid with the lowest percentage score is the most limiting amino acid. For example, results might find that compared with the reference amino acids, leucine gets a 93, lysine gets an 80, and all the other essential amino acids get higher scores. Lysine is the limiting amino acid (the one that falls shortest compared with egg). If the protein's limiting amino acid is 80 percent of the amount found in the reference protein, it receives an amino acid score of 80. The advantages of amino acid scoring are that it is simple and inexpensive, it easily identifies the limiting amino acid, and it can be used to score mixtures of different proportions of multiple proteins mathematically without having to create a mixture to test. Its chief weakness is that it fails to estimate the digestibility of a protein, which may strongly affect the protein's quality.

PDCAAS

The **protein digestibility-corrected amino acid score,** or PDCAAS, is a widely used measurement of protein quality.[1] The PDCAAS compares the amino acid composition of a food protein with human amino acid requirements and corrects for digestibility. First the protein's essential amino acid composition is

GLOSSARY

amino acid score: a measure of protein quality assessed by comparing a protein's amino acid pattern with that of a reference protein; also called the *chemical score.*

biological value (BV): a measure of protein quality assessed by measuring the amount of protein nitrogen that is retained from a given amount of protein nitrogen absorbed.

DIAAS (digestible indispensable amino acid score): a measure of protein quality similar to PDCAAS, except it determines protein digestibility at the end of the small intestine, which more accurately reflects the extent of amino acid absorption.

net protein utilization (NPU): a measure of protein quality assessed by measuring the amount of protein nitrogen that is retained from a given amount of protein nitrogen eaten.

PDCAAS (protein digestibility-corrected amino acid score): a measure of protein quality assessed by comparing the amino acid score of a food protein with the amino acid requirements of preschool-age children and then correcting for the true digestibility of the protein.

protein efficiency ratio (PER): a measure of protein quality assessed by determining how well a given protein supports weight gain in growing rats; used to establish the protein quality for infant formulas and baby foods.

determined, and then it is compared against the amino acid requirements of preschool-aged children.[2] This comparison reveals the most limiting amino acid—the one that falls shortest compared with the reference. If a food protein's limiting amino acid is 70 percent of the amount found in the reference protein, it receives a score of 70. The amino acid score is multiplied by the food's protein digestibility percentage to determine the PDCAAS. The accompanying "How To" provides an example of how to calculate the PDCAAS, and Table D-1 lists the PDCAAS values of selected foods.

DIAAS

Recently, the Food and Agriculture Organization of the United Nations recommended that a new method, known as Digestible Indispensable Amino Acid Score (DIAAS), should replace PDCAAS as the preferred method to determine protein quality.[3] The DIAAS overcomes some of the weaknesses of the current PDCAAS method. Whereas PDCAAS estimates protein digestibility over the entire intestine and uses a single digestibility score, the DIAAS determines amino acid digestibility at the end of the small intestine, which more accurately reflects the extent of amino acids absorption, and considers the digestibility of individual amino acids, as opposed to the protein as a whole.

TABLE D-1 PDCAAS Values of Selected Foods

Casein (milk protein)	1.00
Egg white	1.00
Soybean (isolate)	.99
Beef	.92
Pea flour	.69
Kidney beans (canned)	.68
Chickpeas (canned)	.66
Pinto beans (canned)	.66
Rolled oats	.57
Lentils (canned)	.52
Peanut meal	.52
Whole wheat	.40

NOTE: 1.0 is the maximum PDCAAS a food protein can receive.

>How To

Measure Protein Quality Using PDCAAS

To calculate the PDCAAS (protein digestibility-corrected amino acid score), researchers first determine the amino acid profile of the test protein (in this example, pinto beans). The second column of the table below presents the essential amino acid profile for pinto beans. The third column presents the amino acid reference pattern.

To determine how well the food protein meets human needs, researchers calculate the ratio by dividing the second column by the third column (for example, $30 \div 18 = 1.67$). The amino acid with the lowest percentage is the most limiting amino acid—in this case, methionine. Its percentage is the amino acid score for the protein—in this case, 0.84.

The amino acid score alone, however, does not account for digestibility. Protein digestibility, as determined by animal studies, yields a value of 79 percent for pinto beans. Together, the amino acid score and the digestibility value determine the PDCAAS:

$$\text{PDCAAS} = \text{protein digestibility} \times \text{amino acid score}$$

$$\text{PDCAAS for pinto beans} = 0.79 \times 0.84 = 0.66$$

Thus the PDCAAS for pinto beans is 0.66 as Table D-1 shows.

Essential Amino Acids	Amino Acid Profile of Pinto Beans (mg/g protein)	Amino Acid Reference Pattern (mg/g protein)	Amino Acid Score
Histidine	30.0	18	1.67
Isoleucine	42.5	25	1.70
Leucine	80.4	55	1.46
Lysine	69.0	51	1.35
Methionine (+ cystine)	21.1	25	0.84
Phenylalanine (+ tyrosine)	90.5	47	1.93
Threonine	43.7	27	1.62
Tryptophan	8.8	7	1.26
Valine	50.1	32	1.57

Biological Value

The **biological value (BV)** of a food protein measures its efficiency in supporting the body's growth and maintenance. In a test of biological value, two nitrogen balance studies are done. In the first, no protein is fed, and nitrogen (N) excretions in the urine and feces are measured. It is assumed that under these conditions, N lost in the urine is the amount the body loses each day, regardless of food protein; this endogenous N is "urinary N on a zero-protein diet." The N lost in the feces is the amount the body loses each day, regardless of food protein; this metabolic N is "fecal N on a zero-protein diet."

In the second study, a diet containing the test protein in an amount equal to the requirement is fed. Intakes and losses are measured; then the BV is derived using this formula:

$$BV = \frac{N \text{ retained}}{N \text{ absorbed}} \times 100$$

The more nitrogen retained, the higher the protein quality. (Recall that when an essential amino acid is missing, protein synthesis stops, and the remaining amino acids are deaminated and the nitrogen excreted.)

Egg protein has a BV of 100, indicating that 100 percent of the nitrogen absorbed is retained. Supplied in adequate quantity, a protein with a BV of 70 or greater can support human growth as long as energy intake is adequate. Table D-2 presents the BV for selected foods.

TABLE D-2 Biological Values (BV) of Selected Foods

Egg	100
Milk	93
Beef	75
Fish	75
Corn	72

NOTE: 100 is the maximum BV a food protein can receive.

Net Protein Utilization

Like BV, **net protein utilization (NPU)** measures how efficiently a protein is used by the body and involves two balance studies. The difference is that NPU measures retention of food nitrogen consumed rather than food nitrogen absorbed (as in BV). The formula for NPU is:

$$NPU = \frac{N \text{ retained}}{N \text{ intake}} \times 100$$

The numerator is the same as for BV, but the denominator represents food N intake only—not N absorbed.

Protein Efficiency Ratio

The **protein efficiency ratio (PER)** measures the weight gain of a growing animal and compares it to the animal's protein intake. To determine the PER of a food protein, young, growing animals are given a standard diet containing about 10 percent (by weight) of the test protein. After a specified period of time, weight gain is measured and compared to the amount of test protein consumed. The PER is expressed as:

$$PER = \frac{\text{weight gain (g)}}{\text{protein intake (g)}}$$

Table D-3 presents PER values for selected foods.

TABLE D-3 Protein Efficiency Ratio (PER) Values of Selected Proteins

Casein (milk)	2.8
Soy	2.4
Glutein (wheat)	0.4

© Cengage Learning

REFERENCES

1. G. Schaafsma, Advantages and limitations of the protein digestibility-corrected amino acid score (PDCAAS) as a method for evaluating protein quality in human diets, *British Journal of Nutrition* 108 (2012): S333–S336.
2. D. J. Millward, Amino acid scoring patterns for protein quality assessment, *British Journal of Nutrition* 108 (2012): S31–S43.
3. S. Leser, The 2013 FAO report on dietary protein quality evaluation in human nutrition: Recommendations and implications, *Nutrition Bulletin* 38 (2013): 421–428.

Appendix E   Nutrition Assessment

Nutrition assessment evaluates a person's health from a nutrition perspective. Many factors influence or reflect nutrition status. Consequently, the assessor, usually a registered dietitian assisted by other qualified health-care professionals, gathers information from many sources, including:

- Historical information.
- Anthropometric measurements.
- Physical examinations.
- Biochemical analyses (laboratory tests).

Each of these methods involves collecting data in a variety of ways and interpreting each finding in relation to the others to create a total picture.

The accurate gathering of this information and its careful interpretation are the basis for a meaningful evaluation. The more information gathered about a person, the more accurate the assessment will be. Gathering information is a time-consuming process, however, and time is often a rare commodity in the health care setting. Nutrition care is only one part of total care. It may not be practical or essential to collect detailed information on each person.

A strategic compromise is to screen patients by collecting preliminary data. Data such as height-weight and hematocrit are easy to obtain and can alert health-care workers to potential problems. **Nutrition screening** identifies patients who will require additional nutrition assessment. This appendix provides a sample of the procedures, standards, and charts commonly used in nutrition assessment.

Historical Information

Clues about a person's nutrition status and nutrient requirements become evident with a careful review of a person's historical information (see Table E-1). A thorough history identifies risk factors associated with poor nutrition status (see Table E-2) and reveals personal preferences that need consideration when developing a nutrition care plan. Historical information is obtained from the medical record and interviews with the patient or caregiver.

CONTENTS

TABLE E-1 **Historical Information Used in Nutrition Assessments**

Type of History	What It Identifies
Medical history	Current and previous mental and physical health problems and surgeries, chronic disease risks, and family medical history that affect nutrient needs, nutrition status, or the need for intervention to prevent or alleviate health problems
Personal and social history	Personal, cultural, financial, and environmental influences on food intake, nutrient needs, and diet therapy options
Medication and supplement history	Medications (prescription and over-the-counter), illicit drugs, dietary supplements, and alternative therapies that affect nutrition status
Food and nutrition history	Nutrient intake excesses or deficiencies and reasons for imbalances

nutrition screening: the use of preliminary nutrition assessment techniques to identify people who are malnourished or are at risk for malnutrition.

TABLE E-2 Risk Factors for Malnutrition

Medical History

- Acquired immune deficiency syndrome (AIDS)
- Alcoholism
- Anorexia (lack of appetite)
- Anorexia nervosa
- Bulimia nervosa
- Burns (extensive or severe)
- Cancer
- Celiac disease
- Chewing or swallowing difficulties (including poorly fitted dentures, dental caries, missing teeth, and mouth ulcers)
- Chronic obstructive pulmonary disease
- Circulatory problems
- Constipation
- Crohn's disease
- Cystic fibrosis
- Decubitus ulcers (pressure sores)
- Dementia
- Depleted blood proteins
- Depression

- Diabetes mellitus
- Diarrhea, prolonged or severe
- Drug addiction
- Dysphagia
- Failure to thrive
- Feeding disabilities
- Fever
- GI tract disorders or surgery
- Heart disease
- HIV infection
- Hormonal imbalance
- Hyperlipidemia
- Hypertension
- Infections
- Inflammatory bowel diseases
- Kidney disease
- Liver disease
- Lung disease
- Malabsorption
- Mental illness

- Mental retardation
- Multiple pregnancies
- Nausea
- Neurologic disorders
- Organ failure
- Overweight
- Pancreatic insufficiency
- Paralysis
- Physical disability
- Pneumonia
- Pregnancy
- Radiation therapy
- Surgery (recent or major)
- Tobacco use
- Trauma
- Ulcerative colitis
- Ulcers
- Underweight
- Vomiting (prolonged or severe)

Personal and Social History

- Access to groceries
- Activities
- Age
- Cognitive abilities

- Economic status
- Education
- Employment
- Ethnic/cultural identity

- Home and family situation
- Kitchen facilities
- Religious beliefs

Medication and Supplement History

- Dietary and herbal supplements
- Illicit drug use

- Over-the-counter drugs
- Prescription medicines

Food and Nutrition History

- Deficient or excessive food intakes
- Food allergies or intolerances
- Food insecurity
- Frequently eating out

- Intravenous fluids (other than total parenteral nutrition) for 7 or more days
- Monotonous diet (lacking variety)
- Nutrition knowledge

- Poor appetite
- Restricted or fad diets
- Unbalanced diet (omitting any food group)
- Weight gains or losses (recent)

An adept history taker uses the interview both to gather facts and to establish a rapport with the patient. This section briefly reviews the major areas of nutrition concern in a person's history.

Medical History The assessor can obtain a **medical history** from records completed by the attending physician, nurse, or other health-care professional. In addition, conversations with the patient can uncover valuable information previously overlooked because no one thought to ask or because the patient was not thinking clearly when asked.

An accurate, complete medical history includes the family medical history as well and can reveal conditions that increase a patient's risk for malnutrition (review Table E-2). Diseases and their therapies can have either immediate or long-term effects on nutrition status by interfering with ingestion, digestion, absorption, metabolism, or excretion of nutrients.

Personal and Social History A **personal and social history** reveals factors that influence food choices and a person's ability to manage health

medical history: an account of a patient's current and past health status and disease risks.

personal and social history: a record of a person's social and economic background, including such factors as education, income, and ethnic identity.

and nutrition problems. The ethnic background and religious beliefs of both the patient and the other members of the household influence food choices. Financial concerns influence access to healthy food choices. In general, the quality of the diet declines as income falls. At some point, the ability to purchase the foods required to meet nutrient needs is lost; an inadequate income puts an adequate diet out of reach. Agencies use poverty indexes to identify people at risk for poor nutrition and to qualify people for government food assistance programs.

Low income affects not only the power to purchase foods but also the ability to shop for, store, and cook them; some individuals depend on others to procure and prepare meals. A skilled assessor will note whether a person has transportation to a grocery store that sells a sufficient variety of low-cost foods, and whether the person has access to a refrigerator and stove.

Medication and Supplement History

The many interactions of foods and drugs require that healthcare professionals take a **medication and supplement history** and pay special attention to any patient who takes drugs routinely. If a person is taking any drug, the assessor records the name of the drug; the dose, frequency, and duration of intake; the reason for taking the drug; and signs of any adverse effects.

The interactions of drugs and nutrients may take many forms:

- Drugs can alter food intake and the absorption, metabolism, and excretion of nutrients.
- Foods and nutrients can alter the absorption, metabolism, and excretion of drugs.

Highlight 17 discusses nutrient-drug interactions in more detail, and Table H17-1 (p. 580) summarizes the mechanisms by which these interactions occur and provides specific examples.

Food and Nutrition History

A **food and nutrition history** (often called a *diet history*) provides a record of a person's eating habits and food intake and can help identify possible nutrient imbalances. Food choices are an important part of lifestyle. The assessor who asks nonjudgmental questions about eating habits and food intake encourages trust and enhances the likelihood of obtaining accurate information.

Besides identifying possible nutrient imbalances, food and nutrition histories provide valuable clues about how a person will accept diet changes should they be necessary. Information about what and how a person eats provides the background for realistic and attainable nutrition goals. The following section describes the most common methods of gathering food intake information.

Food Intake Data

Assessors evaluate food intake using various tools such as the 24-hour recall, the food record, the food frequency questionnaire, and direct observation. Food models or photos and measuring devices can help patients identify the types of foods and quantities consumed. The assessor also needs to know how the foods are prepared and when they are eaten. In addition to asking about foods, assessors will ask about beverage consumption, including beverages containing alcohol or caffeine.

24-Hour Dietary Recall

The **24-hour dietary recall** provides data for one day only and is commonly used in nutrition surveys to obtain estimates of the typical food intakes for a population. The assessor asks an individual to recount all the foods and beverages consumed in the past 24 hours or for the previous day. The interview includes questions about the times of meals and snacks, amounts consumed, and methods of food preparation.

medication and supplement history: a record of all the drugs, over-the-counter and prescribed, as well as dietary and herbal supplements that a person takes routinely.

food and nutrition history: a record of eating behaviors and the foods a person eats.

24-hour dietary recall: a record of foods eaten by a person for one 24-hour period.

The multiple-pass method is the most effective approach for obtaining an accurate list of foods consumed. In this procedure, the interview includes four or five separate passes through the 24-hour period of interest. In the first pass, the person provides a "quick list" of foods consumed without being prompted by the interviewer. The second pass helps the person remember foods that are often forgotten, such as beverages, bread, additions to foods (such as butter on toast), savory snacks, and sweets. Third and fourth passes elicit additional details about the foods consumed, such as the amounts eaten, preparation methods, and places where foods were obtained or consumed. A final pass provides an opportunity to recall foods and to probe for additional details. The entire multiple-pass interview can be conducted in about 30 to 45 minutes.

An advantage of the 24-hour recall is that it is relatively easy to obtain. It does not, however, provide enough information to allow accurate generalizations about an individual's usual food intake. The previous day's intake may not be typical, for example, or the person may be unable to report portion sizes accurately or may conceal or forget information about foods eaten. This limitation is partially overcome when 24-hour dietary recalls are collected on several nonconsecutive days.

Food Record Another tool for history taking is the **food record,** in which the person records food and beverages consumed, including the quantity and method of preparation. Chapter 9 (Figure 9-8, p. 286) provides an example. A food record can help both the assessor and the patient to determine factors associated with eating that may affect dietary balance and adequacy.

Food records work especially well with cooperative people but require considerable time and effort on their part. A prime advantage is that the record keeper assumes an active role and may for the first time become aware of personal food habits and assume responsibility for them. It also provides the assessor with an accurate picture of the person's lifestyle and factors that affect food intake. For these reasons, a food record can be particularly useful in outpatient counseling for such nutrition problems as overweight, underweight, or food allergy. The major disadvantages stem from poor compliance in recording the data and conscious or unconscious changes in eating habits that may occur while the person is keeping the record.

Food Frequency Questionnaire An assessor uses a **food frequency questionnaire** to survey the foods and beverages regularly consumed during a specific time period. Some questionnaires are qualitative only: food lists contain common foods, organized by food group, with check boxes to indicate frequency of consumption. Other types of questionnaires can collect semiquantitative information by including portion sizes as well. This information helps pinpoint foods and food groups, and therefore nutrients, that may be excessive or deficient in the diet. That a person ate no vegetables yesterday may not seem particularly significant, but never eating vegetables is a warning of possible nutrient deficiencies. When used with the usual intake or 24-hour recall approach, the food frequency questionnaire enables the assessor to double-check the accuracy of the information obtained.

Direct Observation In facilities that serve meals, food intakes can be directly observed and analyzed. This method can also reveal a person's food preferences, changes in appetite, and any problems with a prescribed diet. Health practitioners use direct observation to estimate a patient's intake of food energy (and often, protein) during a single day or on several consecutive days. To estimate food and nutrient intakes, the clinician records the dietary items that a patient is given at meals and subtracts the amounts remaining after meals are completed; this procedure allows an estimate of the amounts of foods and beverages actually consumed. Although a useful means of discerning patients' intakes, direct observation requires regular and careful documentation and can be labor-intensive and costly.

Analysis of Food Intake Data After collecting food intake data, the assessor estimates nutrient intakes, either informally by using food guides or formally by

food record: an extensive, accurate log of all foods eaten over a period of several days or weeks. A food record that includes associated information such as when, where, and with whom each food is eaten is sometimes called a *food diary*.

food frequency questionnaire: a checklist of foods on which a person can record the frequency with which he or she eats each food.

using diet analysis programs. The assessor compares intakes with standards, usually nutrient recommendations or dietary guidelines, to determine how closely the person's diet meets the standards. Are the types and amounts of proteins, carbohydrates (including fiber), and fats (including cholesterol) appropriate? Are all food groups included in appropriate amounts? Is caffeine or alcohol consumption excessive? Are intakes of any vitamins or minerals (including sodium and iron) excessive or deficient? An informal evaluation is possible only if the assessor has enough prior experience with formal calculations to "see" nutrient amounts in reported food intakes without calculations. Even then, such an informal analysis is best followed by a spot check for key nutrients by actual calculation.

Formal calculations can be performed either manually (by looking up each food in a table of food composition, recording its nutrients, and adding them up) or by using a diet analysis program. The assessor then compares the intakes with standards such as the RDA.

Limitations of Food Intake Analysis Diet histories can be most informative, but the skillful assessor also keeps their limitations in mind. For example, a diet analysis program tends to imply greater accuracy than is possible to obtain from data as uncertain as the starting information. Nutrient contents of foods listed in tables of food composition or stored in computer databases are averages and, for some nutrients, incomplete. In addition, the available data on nutrient contents of foods do not reflect the amounts of nutrients a person actually absorbs. Iron is a case in point: its availability from a given meal may vary depending on the person's iron status; the relative amounts of heme iron, nonheme iron, vitamin C, meat, fish, and poultry eaten at the meal; and the presence of inhibitors of iron absorption such as tea, coffee, and nuts. (Chapter 13 describes the many factors that influence iron absorption from a meal.)

Furthermore, reported portion sizes may not be correct. The person who reports eating "a serving" of greens may not distinguish between ¼ cup and 2 cups; only individuals who have practice measuring food quantities can accurately report serving sizes. Children tend to remember the serving sizes of foods they like as being larger than serving sizes of foods they dislike.

An estimate of nutrient intakes from a food and nutrition history, combined with other sources of information, allows the assessor to confirm or eliminate the possibility of suspected nutrition problems. The assessor must constantly remember that nutrient intakes in adequate amounts do not guarantee adequate nutrient status for an individual. Likewise, insufficient intakes do not always indicate deficiencies, but instead alert the assessor to possible problems. Each person digests, absorbs, metabolizes, and excretes nutrients in a unique way; individual needs vary. Intakes of nutrients identified by diet histories are only pieces of a puzzle that must be put together with other indicators of nutrition status in order to extract meaning.

Anthropometric Measurements

Anthropometrics are physical measurements that reflect body composition and development (see Table E-3). They serve three main purposes: first, to evaluate the progress of growth in pregnant women, infants, children, and adolescents; second, to detect undernutrition and overnutrition in all age groups; and third, to measure changes in body composition over time.

Health-care professionals compare anthropometric measurements taken on an individual with population standards specific for gender and age or with previous measures of the individual. Measurements taken periodically and compared with previous measurements reveal changes in an individual's status.

Mastering the techniques for taking anthropometric measurements requires proper instruction and practice to ensure reliability. Once the correct techniques are learned, taking measurements is easy and requires minimal equipment.

anthropometrics (AN-throw-poe-MET-ricks): measurements of the physical characteristics of the body, such as height and weight.
- **anthropos** = human
- **metric** = measuring

> FIGURE E-1   **Length Measurement of an Infant**

An infant is measured lying down on a measuring board with a fixed headboard and a movable footboard. Note that two people are needed to measure the infant's length.

> FIGURE E-2   **Height Measurement of an Older Child or Adult**

Height is measured most accurately when the person stands against a flat wall to which a measuring tape has been affixed.

TABLE E-3 **Anthropometric Measurements Used in Nutrition Assessments**

Type of Measurement	What It Reflects
Abdominal girth measurement	Abdominal fluid retention and abdominal organ size
Height-weight	Overnutrition and undernutrition; growth in children
Head circumference	Brain growth and development in infants and children under age 2
Skinfold	Subcutaneous and total body fat
Waist circumference	Body fat distribution

© Cengage Learning

Height and weight are well-recognized anthropometrics; other anthropometrics include skinfold measurements and various measures of lean tissue. Other measures are useful in specific situations. For example, a head circumference measurement may help to assess brain development in an infant, and an abdominal girth measurement supplies information about abdominal fluid retention in individuals with liver disease.

Measures of Growth and Development   Height and weight are among the most common and useful anthropometric measurements. Length measurements for infants and children up to age 3 and height measurements for children over 3 are particularly valuable in assessing growth and therefore nutrition status. For adults, height measurements alone are not critical, but help to estimate healthy weight and to interpret other assessment data. Once adult height has been reached, changes in body weight provide useful information in assessing overnutrition and undernutrition.

Height   For infants and children younger than 3, health-care professionals may use special equipment to measure length. The assessor lays the barefoot infant on a measuring board that has a fixed headboard and movable footboard attached at right angles to the surface (see Figure E-1). Often two people are needed to obtain an accurate measurement: one to gently hold the infant's head against the headboard, and the other to straighten the infant's legs and move the footboard to the bottom of the infant's feet.

The procedure for measuring a child who can stand erect and cooperate is the same as for an adult. The best way to measure standing height is with the person's back against a flat wall to which a nonstretchable measuring tape or stick has been fixed (see Figure E-2). The person stands erect, without shoes, with heels together. The person's line of sight should be horizontal, with the heels, buttocks, shoulders, and head touching the wall. The assessor places a ruler or other inflexible object on top of the head at a right angle to the wall; carefully checks the height measurement; and records it immediately in either inches or centimeters so that the correct measurement will not be forgotten.

The measuring rod of a scale is commonly used, but is less accurate because it bends easily. The assessor follows the same general procedure, asking the person to face away from the scale and to take extra care to stand erect.

Unfortunately, many health-care professionals merely ask patients how tall they are rather than measuring their height. Self-reported height is often inaccurate and should be used only as a last resort when measurement is impractical (in the case of an uncooperative patient, an emergency admission, or the like).

Weight   Valid weight measurements require scales that have been carefully maintained, calibrated, and checked for accuracy at regular intervals. Beam balance and electronic scales are the most accurate. To measure infants' weight, assessors use special scales that allow infants to lie or sit (see Figure E-3). Weighing infants naked, without diapers, is standard procedure. Children who can stand are weighed in the same way as adults (see Figure E-4). To make repeated measures useful, standardized conditions are necessary. Each weighing should take

APPENDIX E

place at the same time of day (preferably before breakfast), in the same amount of clothing (without shoes), after the person has voided, and on the same scale. Special scales and hospital beds with built-in scales are available for weighing people who are bedridden. Bathroom scales are inaccurate and inappropriate in a professional setting. As with all measurements, the assessor records the observed weight immediately in either pounds or kilograms.

Head Circumference Assessors may also measure head circumference to confirm that infant growth is proceeding normally or to help detect malnutrition and evaluate the extent of its impact on brain size. To measure head circumference, the assessor places a nonstretchable tape so that it encircles the largest part of the infant's or child's head: just above the eyebrow ridges, just above the point where the ears attach, and around the occipital prominence at the back of the head. To ensure accurate recording, the assessor immediately notes the measure in either inches or centimeters.

Analysis of Measures in Infants and Children Growth retardation is a sign of poor nutrition status. Obesity is also a sign that dietary intervention may be needed.

Health professionals generally evaluate physical development by monitoring the growth rate of a child and comparing this rate with standard charts. Standard charts compare weight to age, height to age, and weight to height; ideally, height and weight are in roughly the same percentile. Although individual growth patterns may vary, a child's growth curve will generally stay at about the same percentile throughout childhood. In children whose growth has been retarded, nutrition rehabilitation will ideally induce height and weight to increase to higher percentiles. In overweight children, the goal is for weight to remain stable as height increases, until weight becomes appropriate for height.

To evaluate growth in infants, an assessor uses charts such as those in Figures E-5 (A and B) through E-10 (A and B). The assessor follows these steps to plot a weight measurement on a percentile graph:

- Select the appropriate chart based on age and gender.
- Locate the child's age along the horizontal axis on the bottom of the chart.
- Locate the child's weight in pounds or kilograms along the vertical axis.
- Mark the chart where the age and weight lines intersect, and read off the percentile.

To assess length, height, or head circumference, the assessor follows the same procedure, using the appropriate chart. (When length is measured, use the chart for birth to 36 months; when height is measured, use the chart for 2 to 20 years.) Head circumference percentile should be similar to the child's height and weight percentiles. With height, weight, and head circumference measures plotted on growth percentile charts, a skilled clinician can begin to interpret the data.

Percentile charts divide the measures of a population into 100 equal divisions. Thus half of the population falls above the 50th percentile, and half falls below. The use of percentile measures allows for comparisons among people of the same age and gender. For example, a 6-month-old female infant whose weight is at the 75 percentile weighs more than 75 percent of the female infants her age.

Head circumference is generally measured in children under 2 years of age. Because the brain grows rapidly before birth and during early infancy, extreme and chronic malnutrition during these times can impair brain development, curtailing the number of brain cells and the size of head circumference. Nonnutritional factors, such as certain disorders and genetic variation, can also influence head circumference.

Analysis of Measures in Adults For adults, health-care professionals typically compare weights with weight-for-height standards. One such standard is the body mass index (BMI), described in Chapter 8, which is useful for estimating the risk to health associated with overnutrition. The inside back cover shows BMI for various heights and weights.

> **FIGURE E-3 Weight Measurement of an Infant**

Infants sit or lie down on scales that are designed to hold them while they are being weighed.

> **FIGURE E-4 Weight Measurement of an Older Child or Adult**

Whenever possible, children and adults are measured on beam balance or electronic scales to ensure accuracy.

> **FIGURE E-5A** Weight-for-Age Percentiles: Boys, Birth to 36 Months

> **FIGURE E-5B** Weight-for-Age Percentiles: Girls, Birth to 36 Months

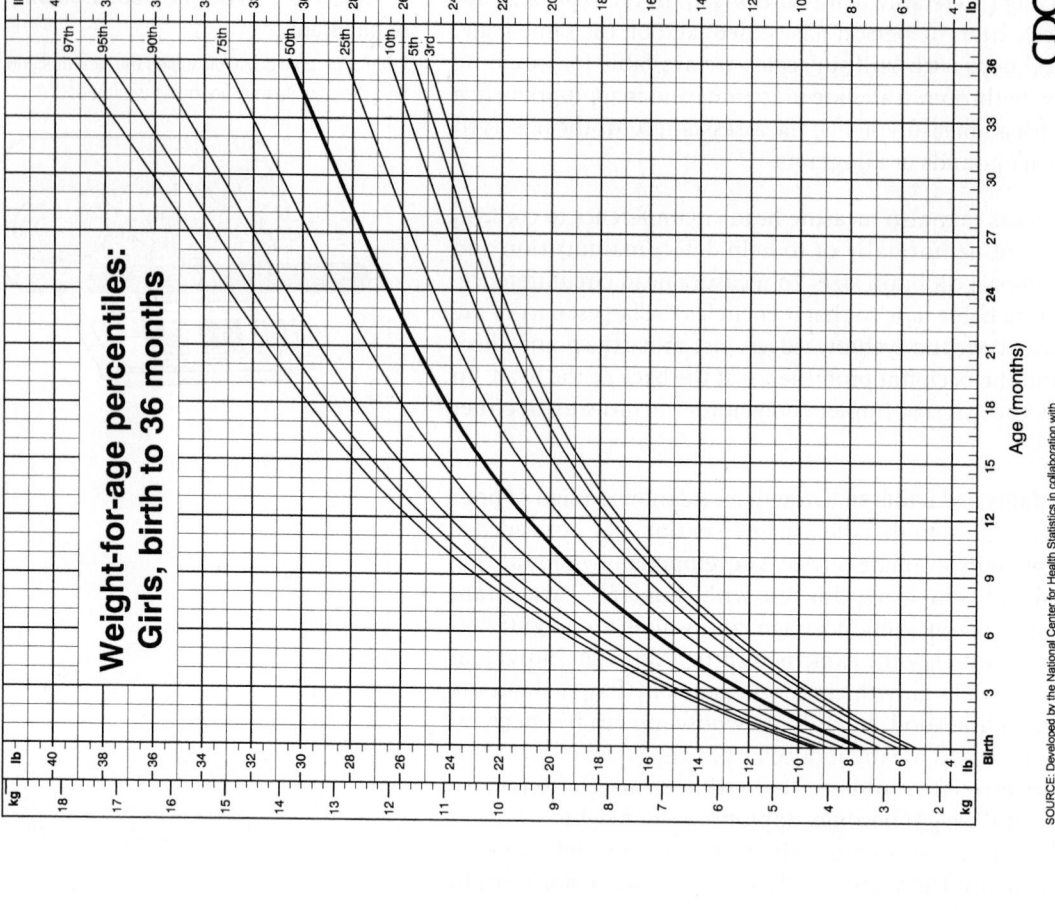

Weight-for-age percentiles:
Boys, birth to 36 months

Weight-for-age percentiles:
Girls, birth to 36 months

Age (months)

SOURCE: Developed by the National Center for Health Statistics in collaboration with
the National Center for Chronic Disease Prevention and Health Promotion (2000).

Figure 1. Weight-for-age percentiles, boys, birth to 36 months, CDC growth charts: United States

SOURCE: Developed by the National Center for Health Statistics in collaboration with
the National Center for Chronic Disease Prevention and Health Promotion (2000).

Figure 2. Weight-for-age percentiles, girls, birth to 36 months, CDC growth charts: United States

Length-for-age percentiles:
Girls, birth to 36 months

SOURCE: Developed by the National Center for Health Statistics in collaboration with
the National Center for Chronic Disease Prevention and Health Promotion (2000).

Figure 4. Length-for-age percentiles, girls, birth to 36 months, CDC growth charts: United States

Length-for-age percentiles:
Boys, birth to 36 months

SOURCE: Developed by the National Center for Health Statistics in collaboration with
the National Center for Chronic Disease Prevention and Health Promotion (2000).

Figure 3. Length-for-age percentiles, boys, birth to 36 months, CDC growth charts: United States

> FIGURE E-7A   Weight-for-Length Percentiles: Boys, Birth to 36 Months

> FIGURE E-7B   Weight-for-Length Percentiles: Girls, Birth to 36 Months

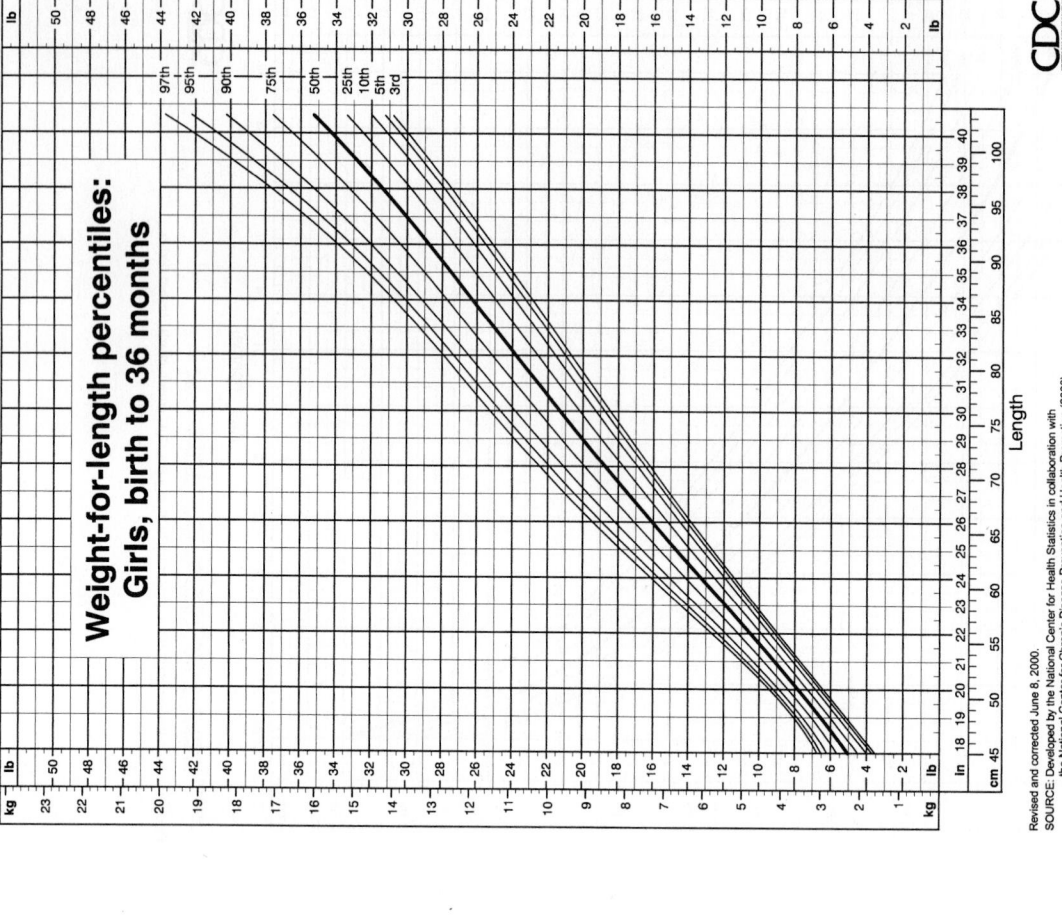

Weight-for-length percentiles: Boys, birth to 36 months

Revised and corrected June 8, 2000.
SOURCE: Developed by the National Center for Health Statistics in collaboration with
the National Center for Chronic Disease Prevention and Health Promotion (2000).

Figure 5. Weight-for-length percentiles, boys, birth to 36 months, CDC growth charts: United States

Weight-for-length percentiles: Girls, birth to 36 months

Revised and corrected June 8, 2000.
SOURCE: Developed by the National Center for Health Statistics in collaboration with
the National Center for Chronic Disease Prevention and Health Promotion (2000).

Figure 6. Weight-for-length percentiles, girls, birth to 36 months, CDC growth charts: United States

Weight-for-age percentiles:
Girls, 2 to 20 years

SOURCE: Developed by the National Center for Health Statistics in collaboration with
the National Center for Chronic Disease Prevention and Health Promotion (2000).

Figure 10. Weight-for-age percentiles, girls, 2 to 20 years, CDC growth charts: United States

Weight-for-age percentiles:
Boys, 2 to 20 years

SOURCE: Developed by the National Center for Health Statistics in collaboration with
the National Center for Chronic Disease Prevention and Health Promotion (2000).

Figure 9. Weight-for-age percentiles, boys, 2 to 20 years, CDC growth charts: United States

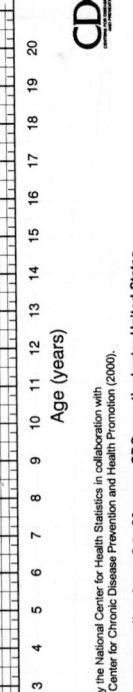

> FIGURE E-9A   Stature-for-Age Percentiles: Boys, 2 to 20 Years

Stature-for-age percentiles:
Boys, 2 to 20 years

Age (years)

SOURCE: Developed by the National Center for Health Statistics in collaboration with
the National Center for Chronic Disease Prevention and Health Promotion (2000).

Figure 11. Stature-for-age percentiles, boys, 2 to 20 years, CDC growth charts: United States

> FIGURE E-9B   Stature-for-Age Percentiles: Girls, 2 to 20 Years

Stature-for-age percentiles:
Girls, 2 to 20 years

Age (years)

SOURCE: Developed by the National Center for Health Statistics in collaboration with
the National Center for Chronic Disease Prevention and Health Promotion (2000).

Figure 12. Stature-for-age percentiles, girls, 2 to 20 years, CDC growth charts: United States

Weight-for-stature percentiles: Girls

SOURCE: Developed by the National Center for Health Statistics in collaboration with the National Center for Chronic Disease Prevention and Health Promotion (2000).

Figure 14. Weight-for-stature percentiles, girls, CDC growth charts: United States

Weight-for-stature percentiles: Boys

SOURCE: Developed by the National Center for Health Statistics in collaboration with the National Center for Chronic Disease Prevention and Health Promotion (2000).

Figure 13. Weight-for-stature percentiles, boys, CDC growth charts: United States

Measures of Body Fat and Lean Tissue Significant weight changes in both children and adults can reflect overnutrition and undernutrition. To estimate the degree to which fat stores or lean tissues are affected by overnutrition or malnutrition, several anthropometric measurements are useful (review Table E-3 on p. E-6).

Skinfold Measures Skinfold measures provide a good estimate of total body fat and a fair assessment of the fat's location. Approximately half the fat in the body lies directly beneath the skin, and the thickness of this subcutaneous fat reflects total body fat. In some parts of the body, such as the back and the back of the arm over the triceps muscle, this fat is loosely attached; a person can pull it up between the thumb and forefinger to obtain a measure of skinfold thickness. (Other common sites for skinfold measures include the biceps, subscapular area below the shoulder blade, suprailiac area above the hip bone, abdomen, and upper thigh.) To measure skinfold, a skilled assessor follows a standard procedure using reliable calipers (illustrated in Figure E-11) and then compares the measurement with standards.

Skinfold measurements correlate directly with the risk of heart disease. They assess central obesity and its associated risks better than do weight measures alone. If a person gains body fat, the skinfold increases proportionately; if the person loses fat, it decreases. Measurements taken from central-body sites (around the abdomen) better reflect changes in fatness than those taken from upper sites (arm and back). A major limitation of the skinfold test is that fat may be thicker under the skin in one area than in another. A pinch at the side of the waistline may not yield the same measurement as a pinch on the back of the arm. This limitation can be overcome by taking skinfold measurements at several

> FIGURE E-11 **How to Measure the Triceps Skinfold**

Clavicle
Acromion process
Midpoint
Olecranon process

A. Find the midpoint of the arm:
1. Ask the subject to bend his or her arm at the elbow and lay the hand across the stomach. (If he or she is right-handed, measure the left arm, and vice versa.)
2. Feel the shoulder to locate the acromion process. It helps to slide your fingers along the clavicle to find the acromion process. The olecranon process is the tip of the elbow.
3. Place a measuring tape from the acromion process to the tip of the elbow. Divide this measurement by 2, and mark the midpoint of the arm with a pen.

B. Measure the skinfold:
1. Ask the subject to let his or her arm hang loosely to the side.
2. Grasp a fold of skin and subcutaneous fat between the thumb and forefinger slightly above the midpoint mark. Gently pull the skin away from the underlying muscle. (This step takes a lot of practice. To be sure you don't have muscle as well as fat, ask the subject to contract and relax the muscle. You should be able to feel if you are pinching muscle.)
3. Place the calipers over the skinfold at the midpoint mark, and read the measurement to the near-

est 1.0 millimeter in two to three seconds. (If using plastic calipers, align pressure lines, and read the measurement to the nearest 1.0 millimeter in two to three seconds.)
4. Repeat steps 2 and 3 twice more. Add the three readings, and then divide by 3 to find the average.

(often three) different places on the body (including upper-, central-, and lower-body sites) and comparing each measurement with standards for that site. Multiple measures are not always practical in clinical settings, however, and most often, the triceps skinfold measurement alone is used because it is easily accessible. Skinfold measures are not useful in assessing changes in body fat over time.

Waist Circumference Chapter 8 describes how fat distribution correlates with health risks and mentioned that the waist circumference is a valuable indicator of fat distribution. To measure waist circumference, the assessor places a non-stretchable tape around the person's body, crossing just above the upper hip bones and making sure that the tape remains on a level horizontal plane on all sides (see Figure E-12). The tape is tightened slightly, but without compressing the skin.

Waist-to-Hip Ratio Alternatively, some clinicians measure both the waist and the hips. To calculate the waist-to-hip ratio, divide the waistline measurement by the hip measurement. For example, a woman with a 28-inch waist and 38-inch hips would have a ratio of $28 \div 38 = 0.74$. Like the waist circumference, the waist-to-hip ratio also assesses abdominal obesity, but provides no more information than using the waist circumference alone. In general, women with a waist-to-hip ratio of 0.80 or greater and men with a waist-to-hip ratio of 0.90 or greater have a high risk of health problems.

Hydrodensitometry To estimate body density using hydrodensitometry, the person is weighed twice—first on land and then again when submerged under water. Underwater weighing usually generates a good estimate of body fat and is useful in research, although the technique has drawbacks: it requires bulky, expensive, and

> FIGURE E-12 **How to Measure Waist Circumference**

Place the measuring tape around the waist just above the bony crest of the hip. The tape runs parallel to the floor and is snug (but does not compress the skin). The measurement is taken at the end of normal expiration.

SOURCE: National Institutes of Health Obesity Education Initiative, *Clinical Guidelines on the Identification, Evaluation, and Treatment of Overweight and Obesity in Adults* (Washington, D.C.: U.S. Department of Health and Human Services, 1998), p. 59.

TABLE E-4 Methods of Estimating Body Fat and Its Distribution

Method	Cost	Ease of Use	Accuracy	Measures Fat Distribution
Height and weight	Low	Easy	High	No
Skinfolds	Low	Easy	Low	Yes
Circumferences	Low	Easy	Moderate	Yes
Ultrasound	Moderate	Moderate	Moderate	Yes
Hydrodensitometry	Low	Moderate	High	No
Heavy water tritiated	Moderate	Moderate	High	No
Deuterium oxide, or heavy oxygen	High	Moderate	High	No
Potassium isotope (^{40}K)	Very high	Difficult	High	No
Total body electrical conductivity (TOBEC)	High	Moderate	High	No
Bioelectric impedance (BIA)	Moderate	Easy	High	No
Dual-energy X-ray absorptiometry (DEXA)	High	Easy	High	No
Computed tomography (CT)	Very high	Difficult	High	Yes
Magnetic resonance imaging (MRI)	Very high	Difficult	High	Yes

SOURCE: Adapted from G. A. Bray, a handout presented at the North American Association for the Study of Obesity and Emory University School of Medicine Conference on Obesity. Update: Pathophysiology, Clinical Consequences, and Therapeutic Options, Atlanta, Georgia, August 31–September 2, 1992.

nonportable equipment. Furthermore, submerging some people (especially those who are very young, very old, ill, or fearful) under water is not always practical.

Bioelectric Impedance To measure body fat using the bioelectric impedance technique, a very-low-intensity electrical current is briefly sent through the body by way of electrodes placed on the wrist and ankle. As is true of other anthropometric techniques, bioelectrical impedance requires standardized procedures and calibrated instruments to provide reliable results. Recent food intake and hydration status, for example, influence results. Bioelectrical impedance is most accurate for people within a normal fat range; it tends to overestimate fat in lean people and underestimate fat in obese people.

Chapter 8 (Figure 8-10, p. 251) illustrates several anthropometric measures of body fat and lean tissue. Clinicians use other methods to estimate body fat and its distribution as well. Each has its advantages and disadvantages as Table E-4 summarizes.

Physical Examinations

An assessor can use a physical examination to search for signs of nutrient deficiency or toxicity. Like the other assessment methods, such an examination requires knowledge and skill. Many physical signs are nonspecific; they can reflect any of several nutrient deficiencies as well as conditions not related to nutrition (see Table E-5). For example, cracked lips may be caused by sunburn, windburn, dehydration, or any of several B vitamin deficiencies, to name just a few possible causes. For this reason, physical findings are most valuable in revealing problems for other assessment techniques to confirm or for confirming other assessment measures.

With this limitation understood, physical symptoms can be most informative about nutrition health. Many tissues and organs can reflect signs of malnutrition. The signs appear most rapidly in parts of the body where cell replacement occurs at a high rate, such as in the hair, skin, and digestive tract (including the mouth and tongue). The summary tables in Chapters 10, 11, 12, and 13 list additional physical signs of vitamin and mineral malnutrition.

TABLE E-5 Physical Findings Used in Nutrition Assessments

Body System	Signs of Good Health	Signs of Malnutrition	Other Possible Causes
Hair	Shiny, firm in the scalp	Dull, brittle, dry, loose; falls out (severe acute malnutrition); corkscrew hair (vitamin C)	Excessive hair bleaching; hair loss from aging, chemotherapy, or radiation therapy
Eyes	Bright, clear, pink, moist membranes; adjust easily to light	Pale membranes (iron); spots, dryness, night blindness (vitamin A); redness at corners of eyes (B vitamins)	Anemia unrelated to nutrition; eye disorders; allergies; aging
Mouth	No sores, swelling, or bleeding; red tongue; normal sense of taste; no caries; ability to chew and swallow; smooth lips	Bleeding gums (vitamin C); smooth or magenta tongue (B vitamins); poor taste sensation (zinc); dry, cracked, or sores in the corners of the lips (B vitamins)	Sunburn, windburn, excessive salivation from ill-fitting dentures or various disorders, medications, periodontal diseases, poor dental hygiene
Skin	Smooth, firm, good color	Poor wound healing (severe acute malnutrition, vitamin C, zinc); dry, rough, lack of fat under skin (essential fatty acids, vitamin A, B vitamins); bruising or bleeding under skin (vitamin C and vitamin K); pale (iron)	Poor skin hygiene, diabetes mellitus, aging, medications
Nails	Firm, pink, smooth	Spoon-shaped, brittle, pale (iron); rigid (protein malnutrition)	

Biochemical Analyses

All of the approaches to nutrition assessment discussed so far are external approaches. Biochemical analyses or laboratory tests help to determine what is happening to the body internally. Common tests are based on analysis of blood and urine samples, which contain nutrients, enzymes, and metabolites that reflect nutrition status. Other tests, such as blood glucose, help pinpoint disease-related problems with nutrition implications. Tests that define fluid and electrolyte balance, acid-base balance, and organ function also have nutrition implications. Table E-6 lists biochemical tests most useful for assessing vitamin and mineral status.

The interpretation of biochemical data requires skill. Long metabolic sequences lead to the production of the end-products and metabolites seen in blood and urine. No single test can reveal nutrition status because many factors influence test results. The low blood concentration of a nutrient may reflect a primary deficiency of that nutrient, but it may also be secondary to the deficiency of one or several other nutrients or to a disease. Taken together with other assessment data, however, laboratory test results help to create a picture that becomes clear with careful interpretation. They are especially useful in helping to detect subclinical malnutrition by uncovering early signs of malnutrition before the clinical signs of a classic deficiency disease appear.

Laboratory tests used to assess vitamin and mineral status (review Table E-6) are particularly useful when combined with diet histories and physical findings. Vitamin and mineral levels present in the blood and urine sometimes reflect recent rather than long-term intakes. This makes detecting subclinical deficiencies difficult. Furthermore, many nutrients interact; therefore, the amounts of other nutrients in the body can affect a lab value for a particular nutrient. It is also important to remember that nonnutrient conditions such as diseases influence biochemical measures.

It is beyond the scope of this text to describe all lab tests and their relations to nutrition status. Instead, the emphasis is on lab tests used to detect protein malnutrition and nutritional anemias.

Protein Malnutrition No single biochemical analysis can adequately evaluate protein malnutrition. This discussion focuses on the measures commonly used today—transthyretin, retinol-binding protein, serum transferrin, and IGF-1 (insulin-like growth factor 1). Table E-7 provides standards for these indicators.

TABLE E-6 Biochemical Tests Useful for Assessing Vitamin and Mineral Status

Nutrient	Assessment Tests
Vitamins	
Vitamin A	Serum retinol, retinol-binding protein
Thiamin[a]	Erythrocyte (red blood cell) transketolase activity, erythrocyte thiamin pyrophosphate
Riboflavin[a]	Erythrocyte glutathione reductase activity
Vitamin B_6[a]	Urinary xanthurenic acid excretion after tryptophan load test, erythrocyte transaminase activity, plasma pyridoxal 5'-phosphate (PLP)
Niacin	Plasma or urinary metabolites NMN (N-methyl nicotinamide) or 2-pyridone, or preferably both expressed as a ratio
Folate[b]	Serum folate, erythrocyte folate (reflects liver stores)
Vitamin B_{12}[b]	Serum vitamin B_{12}, serum and urinary methylmalonic acid, Schilling test
Biotin	Urinary biotin, urinary 3-hydroxyisovaleric acid
Vitamin C	Plasma vitamin C[c], leukocyte vitamin C
Vitamin D	Serum vitamin D
Vitamin E	Serum α-tocopherol, erythrocyte hemolysis
Vitamin K	Serum vitamin K, plasma prothrombin; blood-clotting time (prothrombin time) is not an adequate indicator
Minerals	
Phosphorus	Serum phosphate
Sodium	Serum sodium
Chloride	Serum chloride
Potassium	Serum potassium
Magnesium	Serum magnesium, urinary magnesium
Iron	Hemoglobin, hematocrit, serum ferritin, total iron-binding capacity (TIBC), erythrocyte protoporphyrin, serum iron, transferrin saturation
Iodine	Serum thyroxine or thyroid-stimulating hormone (TSH), urinary iodine
Zinc	Plasma zinc, hair zinc
Copper	Erythrocyte superoxide dismutase, serum copper, serum ceruloplasmin
Selenium	Erythrocyte selenium, glutathione peroxidase activity

[a]Urinary measurements for these vitamins are common, but may be of limited use. Urinary measurements reflect recent dietary intakes and may not provide reliable information concerning the severity of a deficiency.
[b]Folate assessments should always be conducted in conjunction with vitamin B_{12} assessments (and vice versa) to help distinguish the cause of common deficiency symptoms.
[c]Vitamin C shifts between the plasma and the white blood cells known as leukocytes; thus a plasma determination may not accurately reflect the body's pool. A measurement of leukocyte vitamin C can provide information about the body's stores of vitamin C. A combination of both tests may be more reliable than either one alone.
SOURCE: Adapted from H. E. Sauberlich, *Laboratory Tests for the Assessment of Nutritional Status* (Boca Raton, FIA.: CRC Press, 1999).

Although serum albumin is easily and routinely measured, it lacks the sensitivity to assess protein malnutrition because of its long turnover rate.*

Transthyretin and Retinol-Binding Protein Transthyretin and retinol-binding protein occur as a complex in the plasma.** They have a rapid turnover and thus respond quickly to dietary protein inadequacy and therapy.*** Conditions other than malnutrition that lower transthyretin include metabolic stress, hemodialysis, and hypothyroidism; those that raise transthyretin include kidney disease and corticosteroid use. Conditions other than protein malnutrition that lower retinol-binding protein include vitamin A deficiency, metabolic stress, hyperthyroidism, liver disease, and cystic fibrosis; kidney disease raises retinol-binding protein levels.

TABLE E-7 Normal Values for Serum Proteins

Test	Normal Values
Albumin (g/dL)	3.5–5.0
Transferrin (mg/dL)	200–400
Transthyretin (mg/dL)	16–40
Retinol-binding protein (mg/dL)	3–7
IGF-1 (μg/L)	300

NOTE: Levels less than normal suggest compromised protein status.

*The half-life of albumin is 14 to 20 days, an indication of a slow degradation rate.
**Transthyretin is also known as *prealbumin* or *thyroxine-binding prealbumin*.
***The half-lives of transthyretin and retinol-binding protein are 2 days and 12 hours, respectively.

Serum Transferrin Serum transferrin transports iron; consequently, its concentrations reflect both protein and iron status. Using transferrin as an indicator of protein status is complicated when an iron deficiency is present. Transferrin rises as iron deficiency grows worse and falls as iron status improves. Markedly reduced transferrin levels indicate severe protein malnutrition; in mild-to-moderate protein malnutrition, transferrin levels may vary, limiting their usefulness. Conditions other than protein malnutrition that lower transferrin include liver disease, kidney disease, and metabolic stress; those that raise transferrin include pregnancy, iron deficiency, hepatitis, blood loss, and oral contraceptive use. Although transferrin breaks down in the body more quickly than albumin, it is still relatively slow to respond to changes in protein intake and is not a sensitive indicator of the response to therapy.*

IGF-1 (Insulin-like Growth Factor 1) IGF-1 (insulin-like growth factor 1) declines in protein malnutrition. IGF-1 has a relatively short half-life and responds specifically to dietary protein rather than energy.** For these reasons, it is a sensitive indicator of protein status and response to therapy. Conditions that decrease IGF-1 include anorexia nervosa, inflammatory bowel disease, celiac disease, HIV infection, and fasting.

Nutritional Anemias
Anemia, a symptom of a wide variety of nutrition- and nonnutrition-related disorders, is characterized by a reduced number of red blood cells. Iron, folate, and vitamin B_{12} deficiencies caused by inadequate intake, poor absorption, or abnormal metabolism of these nutrients are the most common nutritional anemias. The remainder of this appendix describes laboratory tests that distinguish among the various nutrition-related anemias. Some nonnutrition-related causes of anemia include massive blood loss, infections, hereditary blood disorders such as sickle-cell anemia, and chronic liver or kidney disease.

Assessment of Iron-Deficiency Anemia
Iron deficiency, a common mineral deficiency, develops in stages. First, iron stores diminish, then transport iron decreases, and finally hemoglobin production falls. Chapter 13 describes iron deficiency in detail. This section describes tests used to uncover iron deficiency as it progresses. Table E-8 (p. E-20) provides values used for assessing iron status. Although other tests are more specific in detecting early deficiencies, hemoglobin and hematocrit are the most commonly available tests.

Hemoglobin Iron forms an integral part of the hemoglobin molecule that transports oxygen to the cells. In iron deficiency, the body cannot synthesize hemoglobin. Low hemoglobin values signal depleted iron stores. Table E-8 provides normal hemoglobin values. Hemoglobin's usefulness in evaluating iron status is limited, however, because hemoglobin concentrations drop fairly late in the development of iron deficiency, and other nutrient deficiencies and medical conditions can also alter hemoglobin concentrations.

Hematocrit Hematocrit is commonly used to diagnose iron-deficiency anemia, even though it is an inconclusive measure of iron status. The hematocrit is the percentage of red blood cells in the total blood volume. Table E-8 on p. E-20 includes normal values for hematocrit. Low values indicate microcytic (abnormally small-celled), hypochromic (abnormally lacking in color) red blood cells.

Low hemoglobin and hematocrit values alert the assessor to the possibility of iron deficiency. However, many nutrients and other conditions can affect hemoglobin and hematocrit. The following tests of iron status help pinpoint true iron deficiency.

Serum Ferritin In the first stage of iron deficiency, iron stores diminish. Measures of serum ferritin provide an estimate of iron stores. Serum ferritin is

---

*The half-life of transferrin is 8 to 10 days.
**The half-life of IGF-1 is 12 to 15 hours.

Test	Normal Values
Hemoglobin (g/dL)	12–16 (female)
	13–17 (male)
Hematocrit (%)	35–45 (female)
	39–49 (male)
Serum ferritin (ng/mL)	30–300
Total iron-binding capacity (µg/dL)	250–450
Serum iron (mg/dL)	50–180
Mean corpuscualr volume (fL)	80–100
Transferrin (mg/dL)	200–360
Transferrin saturation (%)	20–50
Erythrocyte protoporphyrin (µg/dL RBC)	<70

not a reliable indicator of iron deficiency, however, because its concentrations are increased by infection, inflammation, alcohol consumption, and liver disease.

The second stage of iron deficiency is characterized by a decrease in transport iron. This is revealed by an increase in the iron-binding capacity of the protein transferrin and a decrease in serum iron. These changes are reflected by the transferrin saturation, which is calculated from the ratio of the other two values as described in the following paragraphs.

Total Iron-Binding Capacity (TIBC) Iron travels through the blood bound to the protein transferrin. TIBC is a measure of the total amount of iron that transferrin can carry. Lab technicians measure iron-binding capacity directly. Table E-8 includes the normal values for TIBC.

Serum Iron Lab technicians can also measure serum iron directly. Elevated values indicate iron overload; reduced values indicate iron deficiency. Table E-8 shows the normal values for serum iron.

Transferrin Saturation The percentage of transferrin that is saturated with iron is an indirect measure that is derived from the serum iron and total iron-binding capacity measures as follows:

$$\% \text{ Transferrin} = \frac{\text{serum iron}}{\text{total iron-binding capacity}} \times 100$$

Table E-8 shows normal values for transferrin saturation. During iron deficiency, transferrin saturation decreases. The transferrin saturation value is a useful indicator of iron status because it includes information about both the iron and transferrin content of the blood.

The third stage of iron deficiency occurs when the supply of transport iron diminishes to the point that it limits hemoglobin production. It is characterized by increases in erythrocyte protoporphyrin, a decrease in mean corpuscular volume, and decreased hemoglobin and hematocrit.

Erythrocyte Protoporphyrin The iron-containing portion of the hemoglobin molecule is heme. Heme is a combination of iron and protoporphyrin. Protoporphyrin accumulates in the blood when iron supplies are inadequate for the formation of heme. Protoporphyrin may also increase when hemoglobin synthesis is impaired for other reasons, such as lead poisoning or inflammation. Lab technicians can measure erythrocyte protoporphyrin directly in a blood sample. The normal values of erythrocyte protoporphyrin are shown in Table E-8.

Mean Corpuscular Volume (MCV) The hematocrit value divided by the red blood cell count provides a measure of the average size of a red blood cell, referred to as the mean corpuscular volume (MCV). Such a measure helps to classify the type of nutrient anemia. In iron deficiency, the red blood cells are smaller than average.

Assessment of Folate and Vitamin B_{12} Anemias

Folate deficiency and vitamin B_{12} deficiency present a similar clinical picture—an anemia characterized by abnormally large red blood cell precursors (megaloblasts) in the bone marrow and abnormally large, mature red blood cells (macrocytic cells) in the blood. Distinguishing between these two deficiencies is essential because their treatments differ. Giving folate to a person with vitamin B_{12} deficiency improves many of the lab test results indicative of vitamin B_{12} deficiency, but this is a dangerous error because vitamin B_{12} deficiency causes nerve damage that folate cannot correct. Thus inappropriate folate administration masks vitamin B_{12}–deficiency anemia, and nerve damage worsens. For this reason, it is critical to determine whether the anemia results from a folate deficiency or from a vitamin B_{12} deficiency. The following biochemical assessment measures help to make this distinction.

Mean Corpuscular Volume (MCV) As previously mentioned, the MCV is a measure of red blood cell size. In folate and vitamin B_{12} deficiencies, the red blood cells are larger than average (macrocytic). Additional tests must be performed to differentiate folate deficiency from vitamin B_{12} deficiency. Macrocytic cells may also result from a high alcohol intake, liver disease, and various medications.

Serum Folate and Vitamin B_{12} Levels Analyses of serum folate and vitamin B_{12} levels are usually among the first tests conducted to determine the cause of macrocytic red blood cells. Low serum levels of either nutrient is consistent with a deficiency, whereas adequate levels can help rule out deficiency. Folate levels are not a specific measure of folate status, however; they increase with folate consumption and decrease with alcohol consumption, pregnancy, or use of anticonvulsant medications. Folate depletion is characterized by a fall in the folate concentrations of red blood cells (erythrocytes). As erythrocyte folate levels diminish, folate-deficiency anemia develops. Because low erythrocyte folate concentrations also occur with vitamin B_{12} deficiency, serum vitamin B_{12} concentrations must also be measured. Table E-9 shows standards for folate and vitamin B_{12} assessment.

Methylmalonic Acid and Homocysteine Levels To determine whether a nutrient is deficient, measures are taken of substances that accumulate when the functions of the nutrient are impaired. For example, the amino acid homocysteine

TABLE E-9 **Normal Values for Folate and Vitamin B_{12}**

Test	Normal Values
Serum folate (ng/mL)[a]	3–16
Serum homocysteine (μmol/L)	5–14
Erythrocyte folate (ng/mL)[a]	140–628
Serum vitamin B_{12} (pg/mL)	200–835
Serum methylmalonic acid (nmol/L)	70–270

NOTE: A nanogram (ng) is one-billionth of a gram; a picogram (pg) is one-trillionth of a gram.
[a]To convert folate values (ng/mL) to international standard units (nmol/L), multiply by 2.266.

usually increase in both folate and vitamin B_{12} deficiencies because both nutrients are needed for its metabolism. Methylmalonic acid is a breakdown product of several amino acids and requires vitamin B_{12} for its metabolism; its concentrations are elevated in vitamin B_{12} deficiency, but not in folate deficiency. Thus this measure is useful in distinguishing between the two.

Schilling Test Vitamin B_{12} deficiency usually arises from malabsorption, not from poor intake. To determine whether malabsorption is the cause, a small oral dose of radioactive vitamin B_{12} is given, and urinary excretion is measured. This test is rarely performed, but it measures vitamin B_{12} absorption and is called a Schilling test.

Antibodies to Intrinsic Factor Serum antibodies for intrinsic factor can help confirm a diagnosis of pernicious anemia, an autoimmune disease characterized by destruction of the cells that produce intrinsic factor. Intrinsic factor is a protein required for vitamin B_{12} absorption, as Chapter 10 explains.

Cautions about Nutrition Assessment

The tests outlined here yield information that becomes meaningful only when conducted and interpreted by a skilled clinician. Potential sources of error may be introduced at any step, from the collection of samples and data to their reporting and analyses. Equipment must be regularly calibrated to ensure accuracy of measurements. In addition, the assessor must keep in mind that each assessment method and measure is useful to confirm or eliminate the possibility of suspected nutrition problems. For example, the assessor must constantly remember that a sufficient intake of a nutrient does not guarantee adequate nutrient status for an individual. Conversely, the apparent inadequate intake of a nutrient does not, by itself, establish that a deficiency exists.

Appendix F   Estimated Energy Needs

Chapter 8 described how to calculate estimated energy requirements (EER) by using an equation that accounts for gender, age, weight, height, and physical activity level. This appendix presents tables that provide a shortcut to estimating daily energy requirements, as developed by the *Dietary Guidelines for Americans*, and based on the EER equations of the Dietary Reference Intakes.

Table F-1 describes three activity levels: sedentary, moderately active, and active. Table F-2 presents estimated daily energy needs by age, gender, and these three levels of physical activity. Keep in mind that these values are estimates that have been rounded to the nearest 200 kcalories; an individual's energy needs may be higher or lower than these average estimates. EER equations for this table use reference heights and weights. For children and adolescents, reference heights and weights vary. For adults, the reference man is 5 feet 10 inches tall and weighs 154 pounds and the reference woman is 5 feet 4 inches tall and weighs 126 pounds. Estimates for women do not include women who are pregnant or breastfeeding.

TABLE F-1 Sedentary, Moderately Active, and Active People

Sedentary	A lifestyle that includes only the light physical activity associated with typical day-to-day life.
Moderately active	A lifestyle that includes physical activity equivalent to walking about 1.5 to 3 miles per day at 3 to 4 miles per hour in addition to the light physical activity associated with typical day-to-day life.
Active	A lifestyle that includes physical activity equivalent to walking more than 3 miles per day at 3 to 4 miles per hour in addition to the light physical activity associated with typical day-to-day life.

© Cengage Learning

TABLE F-2 Estimated Daily kCalorie Needs by Age, Gender, and Physical Activity Level

Age (years)	Gender/Activity Level					
	Male/Sedentary	Male/Moderately Active	Male/Active	Female/Sedentary	Female/Moderately Active	Female/Active
2	1000	1000	1000	1000	1000	1000
3	1200	1400	1400	1000	1200	1400
4	1200	1400	1600	1200	1400	1400
5	1200	1400	1600	1200	1400	1600
6	1400	1600	1800	1200	1400	1600
7	1400	1600	1800	1200	1600	1800
8	1400	1600	2000	1400	1600	1800
9	1600	1800	2000	1400	1600	1800
10	1600	1800	2200	1400	1800	2000

© Cengage Learning

Continued

TABLE F-2 Estimated Daily kCalorie Needs by Age, Gender, and Physical Activity Level *(continued)*

Age (years)	Male/Sedentary	Male/Moderately Active	Male/Active	Female/Sedentary	Female/Moderately Active	Female/Active
11	1800	2000	2200	1600	1800	2000
12	1800	2200	2400	1600	2000	2200
13	2000	2200	2600	1600	2000	2200
14	2000	2400	2800	1800	2000	2400
15	2200	2600	3000	1800	2000	2400
16–18	2400	2800	3200	1800	2000	2400
19–20	2600	2800	3000	2000	2200	2400
21–25	2400	2800	3000	2000	2200	2400
26–30	2400	2600	3000	1800	2000	2400
31–35	2400	2600	3000	1800	2000	2200
36–40	2400	2600	2800	1800	2000	2200
41–45	2200	2600	2800	1800	2000	2200
46–50	2200	2400	2800	1800	2000	2200
51–55	2200	2400	2800	1600	1800	2200
56–60	2200	2400	2600	1600	1800	2200
61–65	2000	2400	2600	1600	1800	2000
66–75	2000	2200	2600	1600	1800	2000
76+	2000	2200	2400	1600	1800	2000

SOURCE: U.S. Department of Agriculture and U.S. Department of Health and Human Services, *Dietary Guidelines for Americans 2010*, www.dietaryguidelines.gov.

Appendix G   Exchange Lists for Diabetes

Chapter 2 introduces the exchange system, and this appendix provides details from the *2008 Choose Your Foods: Exchange Lists for Diabetes.* Exchange lists can help people with diabetes to manage their blood glucose levels by controlling the amount and kinds of carbohydrates they consume. These lists can also help in planning diets for weight management by controlling kcalorie and fat intake.*

CONTENTS

The Exchange System
Serving Sizes
The Foods on the Lists
Controlling Energy, Fat, and Sodium
Planning a Healthy Diet

The Exchange System

The exchange system sorts foods into groups by their proportions of carbohydrate, fat, and protein (Table G-1). These groups may be organized into several exchange lists of foods (Tables G-2 through G-12 on pp. G-3–G-16). For example, the carbohydrate group includes these exchange lists:

- Starch
- Fruits
- Milk (fat-free, reduced-fat, and whole)
- Sweets, Desserts, and Other Carbohydrates
- Nonstarchy Vegetables

TABLE G-1 The Food Lists

Lists	Typical Item/Portion Size	Carbohydrate (g)	Protein (g)	Fat (g)	Energy[a] (kcal)
Carbohydrates					
Starch[b]	1 slice bread	15	0–3	0–1	80
Fruits	1 small apple	15	—	—	60
Milk					
Fat-free, low-fat, 1%	1 c fat-free milk	12	8	0–3	100
Reduced-fat, 2%	1 c reduced-fat milk	12	8	5	120
Whole	1 c whole milk	12	8	8	160
Sweets, desserts, and other carbohydrates[c]	2 small cookies	15	varies	varies	varies
Nonstarchy vegetables	½ c cooked carrots	5	2	—	25
Meat and Meat Substitutes					
Lean	1 oz chicken (no skin)	—	7	0–3	45
Medium-fat	1 oz ground beef	—	7	4–7	75
High-fat	1 oz pork sausage	—	7	8+	100
Plant-based proteins	½ c tofu	varies	7	varies	varies
Fats	1 tsp butter	—	—	5	45
Alcohol	12 oz beer	varies	—	—	100

[a]The energy value for each exchange list represents an approximate average for the group and does not reflect the precise number of grams of carbohydrate, protein, and fat. For example, a slice of bread contains 15 grams of carbohydrate (60 kcalories), 3 grams protein (12 kcalories), and a little fat—rounded to 80 kcalories for ease in calculating. A ½ cup of vegetables (not including starchy vegetables) contains 5 grams carbohydrate (20 kcalories) and 2 grams protein (8 more), which has been rounded down to 25 kcalories.
[b]The Starch list includes cereals, grains, breads, crackers, snacks, starchy vegetables (such as corn, peas, and potatoes), and legumes (dried beans, peas, and lentils).
[c]The Sweets, Desserts, and Other Carbohydrates list includes foods that contain added sugars and fats such as sodas, candy, cakes, cookies, doughnuts, ice cream, pudding, syrup, and frozen yogurt.

*The Exchange Lists are the basis of a meal planning system designed by a committee of the American Diabetes Association and The Academy of Nutrition and Dietetics. While designed primarily for people with diabetes and others who must follow special diets, the Exchange Lists are based on principles of good nutrition that apply to everyone. © 2008 by the American Diabetes Association and The American Dietetic Association.

Then any food on a list can be "exchanged" for any other on that same list. Another group for alcohol has been included as a reminder that these beverages often deliver substantial carbohydrate and kcalories, and therefore warrant their own list.

Serving Sizes

The serving sizes have been carefully adjusted and defined so that a serving of any food on a given list provides roughly the same amount of carbohydrate, fat, and protein, and, therefore, total energy. Any food on a list can thus be exchanged, or traded, for any other food on the same list without significantly affecting the diet's energy-nutrient balance or total kcalories. For example, a person may select 17 small grapes or ½ large grapefruit as one fruit exchange, and either choice would provide roughly 15 grams of carbohydrate and 60 kcalories. A whole grapefruit, however, would count as 2 fruit exchanges.

To apply the system successfully, users must become familiar with the specified serving sizes. A convenient way to remember the serving sizes and energy values is to keep in mind a typical item from each list (review Table G-1).

The Foods on the Lists

Foods do not always appear on the exchange list where you might first expect to find them. They are grouped according to their energy-nutrient contents rather than by their source (such as milks), their outward appearance, or their vitamin and mineral contents. For example, cheeses are grouped with meats (not milk) because, like meats, cheeses contribute energy from protein and fat but provide negligible carbohydrate.

For similar reasons, starchy vegetables such as corn, green peas, and potatoes are found on the Starch list with breads and cereals, not with the vegetables. Likewise, bacon is grouped with the fats and oils, not with the meats.

Diet planners learn to view mixtures of foods, such as casseroles and soups, as combinations of foods from different exchange lists. They also learn to interpret food labels with the exchange system in mind.

Controlling Energy, Fat, and Sodium

The exchange lists help people control their energy intakes by paying close attention to serving sizes. People wanting to lose weight can limit foods from the Sweets, Desserts, and Other Carbohydrates and Fats lists, and they might choose to avoid the Alcohol list altogether. The Free Foods list provide low-kcalorie choices.

By assigning items like bacon to the Fats list, the exchange lists alert consumers to foods that are unexpectedly high in fat. Even the Starch list specifies which grain products contain added fat (such as biscuits, cornbread, and waffles) by marking them with a symbol to indicate added fat (the symbols are explained in the table keys). In addition, the exchange lists encourage users to think of fat-free milk as milk and of whole milk as milk with added fat, and to think of lean meats as meats and of medium-fat and high-fat meats as meats with added fat. To that end, foods on the milk and meat lists are separated into categories based on their fat contents (review Table G-1). The Milk list is subdivided for fat-free, reduced fat, and whole; the meat list is subdivided for lean, medium fat, and high fat. The meat list also includes plant-based proteins, which tend to be rich in fiber. Notice that many of these foods (p. G-11) bear the symbol for "high fiber."

People wanting to control the sodium in their diets can begin by eliminating any foods bearing the "high sodium" symbol. In most cases, the symbol identifies foods that, in one serving, provide 480 milligrams or more of sodium. Foods on the Combination Foods or Fast Foods lists that bear the symbol provide more than 600 milligrams of sodium. Other foods may also contribute substantially to sodium (consult Chapter 12 for details).

Planning a Healthy Diet

To obtain a daily variety of foods that provide healthful amounts of carbohydrate, protein, and fat, as well as vitamins, minerals, and fiber, the meal plan for adults and teenagers should include at least:

- Two to three servings of nonstarchy vegetables
- Two servings of fruits
- Six servings of grains (at least three of whole grains), beans, and starchy vegetables
- Two servings of low-fat or fat-free milk
- About 6 ounces of meat or meat substitutes
- *Small* amounts of fat and sugar

The actual amounts are determined by age, gender, activity levels, and other factors that influence energy needs.

TABLE G-2 **Starch**

The Starch list includes bread, cereals and grains, starchy vegetables, crackers and snacks, and legumes (dried beans, peas, and lentils). 1 starch choice = 15 grams carbohydrate, 0–3 grams protein, 0–1 grams fat, and 80 kcalories.

NOTE: In general, one starch exchange is ½ cup cooked cereal, grain, or starchy vegetable; ⅓ cup cooked rice or pasta; 1 ounce of bread product; ¾ ounce to 1 ounce of most snack foods.

Bread		Cereals and Grains	
Food	**Serving Size**	**Food**	**Serving Size**
Bagel, large (about 4 oz)	¼ (1 oz)	Barley, cooked	⅓ cup
▽ Biscuit, 2½ inches across	1	Bran, dry	
Bread		☺ oat	¼ cup
☺ reduced-kcalorie	2 slices (1½ oz)	☺ wheat	½ cup
white, whole-grain, pumpernickel, rye, unfrosted raisin	1 slice (1 oz)	☺ Bulgur (cooked)	½ cup
Chapatti, small, 6 inches across	1	Cereals	
▽ Cornbread, 1¾ inch cube	1 (1½ oz)	☺ bran	½ cup
English muffin	½	cooked (oats, oatmeal)	½ cup
Hot dog bun or hamburger bun	½ (1 oz)	puffed	1½ cups
Naan, 8 inches by 2 inches	¼	shredded wheat, plain	½ cup
Pancake, 4 inches across, ¼ inch thick	1	sugar-coated	½ cup
Pita, 6 inches across	½	unsweetened, ready-to-eat	¾ cup
Roll, plain, small	1 (1 oz)	Couscous	⅓ cup
▽ Stuffing, bread	⅓ cup	Granola	
▽ Taco shell, 5 inches across	2	low-fat	¼ cup
Tortilla, corn, 6 inches across	1	▽ regular	¼ cup
Tortilla, flour, 6 inches across	1	Grits, cooked	½ cup
Tortilla, flour, 10 inches across	⅓	Kasha	½ cup
▽ Waffle, 4-inch square or 4 inches across	1	Millet, cooked	⅓ cup

(continued)

KEY

☺ = More than 3 grams of dietary fiber per serving.

▽ = Extra fat, or prepared with added fat. (Count as 1 starch + 1 fat.)

🧂 = 480 milligrams or more of sodium per serving.

Cereals and Grains—continued

Food	Serving Size
Muesli	¼ cup
Pasta, cooked	⅓ cup
Polenta, cooked	⅓ cup
Quinoa, cooked	⅓ cup
Rice, white or brown, cooked	⅓ cup
Tabbouleh (tabouli), prepared	½ cup
Wheat germ, dry	3 Tbsp
Wild rice, cooked	½ cup

Starchy Vegetables

Food	Serving Size
Cassava	⅓ cup
Corn	½ cup
on cob, large	½ cob (5 oz)
☺ Hominy, canned	¾ cup
☺ Mixed vegetables with corn, peas, or pasta	1 cup
☺ Parsnips	½ cup
☺ Peas, green	½ cup
Plantain, ripe	⅓ cup
Potato	
baked with skin	¼ large (3 oz)
boiled, all kinds	½ cup or ½ medium (3 oz)
▽ mashed, with milk and fat	½ cup
french fried (oven-baked)[a]	1 cup (2 oz)
☺ Pumpkin, canned, no sugar added	1 cup
Spaghetti/pasta sauce	½ cup
☺ Squash, winter (acorn, butternut)	1 cup
☺ Succotash	½ cup
Yam, sweet potato, plain	½ cup

Crackers and Snacks[b]

Food	Serving Size
Animal crackers	8
Crackers	
▽ round-butter type	6
saltine-type	6
▽ sandwich-style, cheese or peanut butter filling	3
▽ whole-wheat regular	2–5 (¾ oz)
☺ whole-wheat lower fat or crispbreads	2–5 (¾ oz)
Graham cracker, 2½-inch square	3
Matzoh	¾ oz
Melba toast, about 2-inch by 4-inch piece	4
Oyster crackers	20
Popcorn	3 cups
▽ ☺ with butter	3 cups
☺ no fat added	3 cups
☺ lower fat	3 cups
Pretzels	¾ oz
Rice cakes, 4 inches across	2
Snack chips	
fat-free or baked (tortilla, potato), baked pita chips	15–20 (¾ oz)
▽ regular (tortilla, potato)	9–13 (¾ oz)

Beans, Peas, and Lentils[c]

The choices on this list count as 1 starch + 1 lean meat.

Food	Serving Size
☺ Baked beans	⅓ cup
☺ Beans, cooked (black, garbanzo, kidney, lima, navy, pinto, white)	½ cup
☺ Lentils, cooked (brown, green, yellow)	½ cup
☺ Peas, cooked (black-eyed, split)	½ cup
▯ ☺ Refried beans, canned	½ cup

KEY

☺ = More than 3 grams of dietary fiber per serving.

▽ = Extra fat, or prepared with added fat. (Count as 1 starch + 1 fat.)

▯ = 480 milligrams or more of sodium per serving.

[a]Restaurant-style french fries are on the Fast Foods list.
[b]For other snacks, see the Sweets, Desserts, and Other Carbohydrates list. For a quick estimate of serving size, an open handful is equal to about 1 cup or 1 to 2 ounces of snack food.
[c]Beans, peas, and lentils are also found on the Meat and Meat Substitutes list.

Fruit[a]

The Fruits list includes fresh, frozen, canned, and dried fruits and fruit juices. 1 fruit choice = 15 grams carbohydrate, 0 grams protein, 0 grams fat, and 60 kcalories.

NOTE: In general, one fruit exchange is ½ cup canned or fresh fruit or unsweetened fruit juice; 1 small fresh fruit (4 ounces); 2 tablespoons dried fruit.

Food	Serving Size
Apple, unpeeled, small	1 (4 oz)
Apples, dried	4 rings
Applesauce, unsweetened	½ cup
Apricots	
canned	½ cup
dried	8 halves
☻ fresh	4 whole (5½ oz)
Banana, extra small	1 (4 oz)
☻ Blackberries	¾ cup
Blueberries	¾ cup
Cantaloupe, small	⅓ melon or 1 cup cubed (11 oz)
Cherries	
sweet, canned	½ cup
sweet fresh	12 (3 oz)
Dates	3
Dried fruits (blueberries, cherries, cranberries, mixed fruit, raisins)	2 Tbsp
Figs	
dried	1½
☻ fresh	1½ large or 2 medium (3½ oz)
Fruit cocktail	½ cup
Grapefruit	
large	½ (11 oz)
sections, canned	¾ cup
Grapes, small	17 (3 oz)
Honeydew melon	1 slice or 1 cup cubed (10 oz)
☻ Kiwi	1 (3½ oz)
Mandarin oranges, canned	¾ cup
Mango, small	½ (5½ oz) or ½ cup

Food	Serving Size
Nectarine, small	1 (5 oz)
☻ Orange, small	1 (6½ oz)
Papaya	½ or 1 cup cubed (8 oz)
Peaches	
canned	½ cup
fresh, medium	1 (6 oz)
Pears	
canned	½ cup
fresh, large	½ (4 oz)
Pineapple	
canned	½ cup
fresh	¾ cup
Plums	
canned	½ cup
dried (prunes)	3
small	2 (5 oz)
☻ Raspberries	1 cup
☻ Strawberries	1¼ cup whole berries
☻ Tangerines, small	2 (8 oz)
Watermelon	1 slice or 1¼ cups cubes (13½ oz)

Fruit Juice

Food	Serving Size
Apple juice/cider	½ cup
Fruit juice blends, 100% juice	⅓ cup
Grape juice	⅓ cup
Grapefruit juice	½ cup
Orange juice	½ cup
Pineapple juice	½ cup
Prune juice	⅓ cup

KEY

☻ = More than 3 grams of dietary fiber per serving.

▽ = Extra fat, or prepared with added fat. (Count as 1 starch + 1 fat.)

▤ = 480 milligrams or more of sodium per serving.

[a]The weight listed includes skin, core, seeds, and rind.

TABLE G-4 Milk

The Milk list groups milks and yogurts based on the amount of fat they have (fat-free/low fat, reduced fat, and whole). Cheeses are found on the Meat and Meat Substitutes list and cream and other dairy fats are found on the Fats list.

NOTE: In general, one milk choice is 1 cup (8 fluid ounces or ½ pint) milk or yogurt.

Milk and Yogurts

Food	Serving Size
Fat-free or low-fat (1%)	
1 fat-free/low-fat milk choice = 12 g carbohydrate, 8 g protein, 0–3 g fat, and 100 kcal.	
Milk, buttermilk, acidophilus milk, Lactaid	1 cup
Evaporated milk	½ cup
Yogurt, plain or flavored with an artificial sweetener	⅔ cup (6 oz)
Reduced-fat (2%)	
1 reduced-fat milk choice = 12 g carbohydrate, 8 g protein, 5 g fat, and 120 kcal.	
Milk, acidophilus milk, kefir, Lactaid	1 cup
Yogurt, plain	⅔ cup (6 oz)
Whole	
1 whole milk choice = 12 g carbohydrate, 8 g protein, 8 g fat, and 160 kcal.	
Milk, buttermilk, goat's milk	1 cup
Evaporated milk	½ cup
Yogurt, plain	8 oz

Dairy-Like Foods

Food	Serving Size	Count as
Chocolate milk		
fat-free	1 cup	1 fat-free milk + 1 carbohydrate
whole	1 cup	1 whole milk + 1 carbohydrate
Eggnog, whole milk	½ cup	1 carbohydrate + 2 fats
Rice drink		
flavored, low fat	1 cup	2 carbohydrates
plain, fat-free	1 cup	1 carbohydrate
Smoothies, flavored, regular	10 oz	1 fat-free milk + 2½ carbohydrates
Soy milk		
light	1 cup	1 carbohydrate + ½ fat
regular, plain	1 cup	1 carbohydrate + 1 fat
Yogurt		
and juice blends	1 cup	1 fat-free milk + 1 carbohydrate
low carbohydrate (less than 6 grams carbohydrate per choice)	⅔ cup (6 oz)	½ fat-free milk
with fruit, low-fat	⅔ cup (6 oz)	1 fat-free milk + 1 carbohydrate

TABLE G-5 Sweets, Desserts, and Other Carbohydrates

1 other carbohydrate choice = 15 grams carbohydrate, variable grams protein, variable grams fat, and variable kcalories.

NOTE: In general, one choice from this list can substitute for foods on the Starch, Fruits, or Milk lists.

Beverages, Soda, and Energy/Sports Drinks

Food	Serving Size	Count as
Cranberry juice cocktail	½ cup	1 carbohydrate
Energy drink	1 can (8.3 oz)	2 carbohydrates
Fruit drink or lemonade	1 cup (8 oz)	2 carbohydrates

Beverages, Soda, and Energy/Sports Drinks—continued		
Food	**Serving Size**	**Count as**
Hot chocolate		
regular	1 envelope added to 8 oz water	1 carbohydrate + 1 fat
sugar-free or light	1 envelope added to 8 oz water	1 carbohydrate
Soft drink (soda), regular	1 can (12 oz)	2½ carbohydrates
Sports drink	1 cup (8 oz)	1 carbohydrate

Brownies, Cake, Cookies, Gelatin, Pie, and Pudding		
Food	**Serving Size**	**Count as**
Brownie, small, unfrosted	1¼-inch square, ⅞ inch high (about 1 oz)	1 carbohydrate + 1 fat
Cake		
angel food, unfrosted	1⁄12 of cake (about 2 oz)	2 carbohydrates
frosted	2-inch square (about 2 oz)	2 carbohydrates + 1 fat
unfrosted	2-inch square (about 2 oz)	1 carbohydrate + 1 fat
Cookies		
chocolate chip	2 cookies (2¼ inches across)	1 carbohydrate + 2 fats
gingersnap	3 cookies	1 carbohydrate
sandwich, with crème filling	2 small (about ⅔ oz)	1 carbohydrate + 1 fat
sugar-free	3 small or 1 large (¾-1 oz)	1 carbohydrate + 1–2 fats
vanilla wafer	5 cookies	1 carbohydrate + 1 fat
Cupcake, frosted	1 small (about 1¾ oz)	2 carbohydrates + 1–1½ fats
Fruit cobbler	½ cup (3½ oz)	3 carbohydrates + 1 fat
Gelatin, regular	½ cup	1 carbohydrate
Pie		
commercially prepared fruit, 2 crusts	⅙ of 8-inch pie	3 carbohydrates + 2 fats
pumpkin or custard	⅛ of 8-inch pie	1½ carbohydrates + 1½ fats
Pudding		
regular (made with reduced-fat milk)	½ cup	2 carbohydrates
sugar-free or sugar- and fat-free (made with fat-free milk)	½ cup	1 carbohydrate

Candy, Spreads, Sweets, Sweeteners, Syrups, and Toppings		
Food	**Serving Size**	**Count as**
Candy bar, chocolate/peanut	2 "fun size" bars (1 oz)	1½ carbohydrates + 1½ fats
Candy, hard	3 pieces	1 carbohydrate
Chocolate "kisses"	5 pieces	1 carbohydrate + 1 fat
Coffee creamer		
dry, flavored	4 tsp	½ carbohydrate + ½ fat
liquid, flavored	2 Tbsp	1 carbohydrate
Fruit snacks, chewy (pureed fruit concentrate)	1 roll (¾ oz)	1 carbohydrate
Fruit spreads, 100% fruit	1½ Tbsp	1 carbohydrate
Honey	1 Tbsp	1 carbohydrate
Jam or jelly, regular	1 Tbsp	1 carbohydrate
Sugar	1 Tbsp	1 carbohydrate
Syrup		
chocolate	2 Tbsp	2 carbohydrates
light (pancake type)	2 Tbsp	1 carbohydrate
regular (pancake type)	1 Tbsp	1 carbohydrate

(continued)

Condiments and Sauces[a]		
Food	**Serving Size**	**Count as**
Barbeque sauce	3 Tbsp	1 carbohydrate
Cranberry sauce, jellied	¼ cup	1½ carbohydrates
🧂 Gravy, canned or bottled	½ cup	½ carbohydrate + ½ fat
Salad dressing, fat-free, low-fat, cream-based	3 Tbsp	1 carbohydrate
Sweet and sour sauce	3 Tbsp	1 carbohydrate

Doughnuts, Muffins, Pastries, and Sweet Breads		
Food	**Serving Size**	**Count as**
Banana nut bread	1-inch slice (1 oz)	2 carbohydrates + 1 fat
Doughnut		
cake, plain	1 medium (1½ oz)	1½ carbohydrates + 2 fats
yeast type, glazed	3¾ inches across (2 oz)	2 carbohydrates + 2 fats
Muffin (4 oz)	¼ muffin (1 oz)	1 carbohydrate + ½ fat
Sweet roll or Danish	1 (2½ oz)	2½ carbohydrates + 2 fats

Frozen Bars, Frozen Desserts, Frozen Yogurt, and Ice Cream		
Food	**Serving Size**	**Count as**
Frozen pops	1	½ carbohydrate
Fruit juice bars, frozen, 100% juice	1 bar (3 oz)	1 carbohydrate
Ice cream		
fat-free	½ cup	1½ carbohydrates
light	½ cup	1 carbohydrate + 1 fat
no sugar added	½ cup	1 carbohydrate + 1 fat
regular	½ cup	1 carbohydrate + 2 fats
Sherbet, sorbet	½ cup	2 carbohydrates
Yogurt, frozen		
fat-free	⅓ cup	1 carbohydrate
regular	½ cup	1 carbohydrate + 0–1 fat

Granola Bars, Meal Replacement Bars/Shakes, and Trail Mix		
Food	**Serving Size**	**Count as**
Granola or snack bar, regular or low-fat	1 bar (1 oz)	1½ carbohydrates
Meal replacement bar	1 bar (1⅓ oz)	1½ carbohydrates + 0–1 fat
Meal replacement bar	1 bar (2 oz)	2 carbohydrates + 1 fat
Meal replacement shake, reduced kcalorie	1 can (10–11 oz)	1½ carbohydrates + 0–1 fat
Trail mix		
candy/nut-based	1 oz	1 carbohydrate + 2 fats
dried fruit-based	1 oz	1 carbohydrate + 1 fat

KEY

🧂 = 480 milligrams or more of sodium per serving.

[a]You can also check the Fats list and Free Foods list for other condiments.

TABLE G-6 Nonstarchy Vegetables

The Nonstarchy Vegetables list includes vegetables that have few grams of carbohydrates or kcalories; starchy vegetables are found on the Starch list. 1 nonstarchy vegetable choice = 5 grams carbohydrate, 2 grams protein, 0 grams fat, and 25 kcalories.

NOTE: In general, one nonstarchy vegetable choice is ½ cup cooked vegetables or vegetable juice or 1 cup raw vegetables. Count 3 cups of raw vegetables or 1½ cups of cooked vegetables as one carbohydrate choice.

Nonstarchy Vegetables[a]	
Amaranth or Chinese spinach	Kohlrabi
Artichoke	Leeks
Artichoke hearts	Mixed vegetables (without corn, peas, or pasta)
Asparagus	Mung bean sprouts
Baby corn	Mushrooms, all kinds, fresh
Bamboo shoots	Okra
Beans (green, wax, Italian)	Onions
Bean sprouts	Oriental radish or daikon
Beets	Pea pods
☻ Borscht	☻ Peppers (all varieties)
Broccoli	Radishes
☻ Brussels sprouts	Rutabaga
Cabbage (green, bok choy, Chinese)	☻ Sauerkraut
☻ Carrots	Soybean sprouts
Cauliflower	Spinach
Celery	Squash (summer, crookneck, zucchini)
☻ Chayote	Sugar pea snaps
Coleslaw, packaged, no dressing	☻ Swiss chard
Cucumber	Tomato
Eggplant	Tomatoes, canned
Gourds (bitter, bottle, luffa, bitter melon)	☻ Tomato sauce
Green onions or scallions	☻ Tomato/vegetable juice
Greens (collard, kale, mustard, turnip)	Turnips
Hearts of palm	Water chestnuts
Jicama	Yard-long beans

KEY

☻ = More than 3 grams of dietary fiber per serving.

🔋 = 480 milligrams or more of sodium per serving.

[a]Salad greens (like chicory, endive, escarole, lettuce, romaine, spinach, arugula, radicchio, watercress) are on the Free Foods list.

TABLE G-7 Meat and Meat Substitutes

The Meat and Meat Substitutes list groups foods based on the amount of fat they have (lean meat, medium-fat meat, high-fat meat, and plant-based proteins).

Lean Meats and Meat Substitutes

1 lean meat choice = 0 grams carbohydrate, 7 grams protein, 0–3 grams fat, and 45 kcalories.

Food	Amount
Beef: Select or Choice grades trimmed of fat: ground round, roast (chuck, rib, rump), round, sirloin, steak (cubed, flank, porterhouse, T-bone), tenderloin	1 oz
▤ Beef jerky	1 oz
Cheeses with 3 grams of fat or less per oz	1 oz
Cottage cheese	¼ cup
Egg substitutes, plain	¼ cup
Egg whites	2
Fish, fresh or frozen, plain: catfish, cod, flounder, haddock, halibut, orange roughy, salmon, tilapia, trout, tuna	1 oz
▤ Fish, smoked: herring or salmon (lox)	1 oz
Game: buffalo, ostrich, rabbit, venison	1 oz
▤ Hot dog with 3 grams of fat or less per oz (8 dogs per 14 oz package) *Note: May be high in carbohydrate.*	1
Lamb: chop, leg, or roast	1 oz
Organ meats: heart, kidney, liver *Note: May be high in cholesterol.*	1 oz
Oysters, fresh or frozen	6 medium
Pork, lean	
▤ Canadian bacon	1 oz
rib or loin chop/roast, ham, tenderloin	1 oz
Poultry, without skin: Cornish hen, chicken, domestic duck or goose (well-drained of fat), turkey	1 oz
Processed sandwich meats with 3 grams of fat or less per oz: chipped beef, deli thin-sliced meats, turkey ham, turkey kielbasa, turkey pastrami	1 oz
Salmon, canned	1 oz
Sardines, canned	2 medium
▤ Sausage with 3 grams of fat or less per oz	1 oz
Shellfish: clams, crab, imitation shellfish, lobster, scallops, shrimp	1 oz
Tuna, canned in water or oil, drained	1 oz
Veal, lean chop, roast	1 oz

Medium-Fat Meat and Meat Substitutes

1 medium-fat meat choice = 0 grams carbohydrate, 7 grams protein, 4–7 grams fat, and 75 kcalories.

Food	Amount
Beef: corned beef, ground beef, meatloaf, Prime grades trimmed of fat (prime rib), short ribs, tongue	1 oz

Medium-Fat Meat and Meat Substitutes—continued

Food	Amount
Cheeses with 4–7 grams of fat per oz: feta, mozzarella, pasteurized processed cheese spread, reduced-fat cheeses, string	1 oz
Egg *Note: High in cholesterol, so limit to 3 per week.*	1
Fish, any fried product	1 oz
Lamb: ground, rib roast	1 oz
Pork: cutlet, shoulder roast	1 oz
Poultry: chicken with skin; dove, pheasant, wild duck, or goose; fried chicken; ground turkey	1 oz
Ricotta cheese	2 oz or ¼ cup
▤ Sausage with 4–7 grams of fat per oz	1 oz
Veal, cutlet (no breading)	1 oz

High-Fat Meat and Meat Substitutes

1 high-fat meat choice = 0 grams carbohydrate, 7 grams protein, 8+ grams fat, and 100 kcalories. These foods are high in saturated fat, cholesterol, and kcalories and may raise blood cholesterol levels if eaten on a regular basis. Try to eat 3 or fewer servings from this group per week.

Food	Amount
Bacon	
▤ pork	2 slices (16 slices per lb or 1 oz each, before cooking)
▤ turkey	3 slices (½ oz each before cooking)
Cheese, regular: American, bleu, brie, cheddar, hard goat, Monterey jack, queso, and Swiss	1 oz
▽▤ Hot dog: beef, pork, or combination (10 per lb-sized package)	1
▤ Hot dog: turkey or chicken (10 per lb-sized package)	1
Pork: ground, sausage, spareribs	1 oz
Processed sandwich meats with 8 grams of fat or more per oz: bologna, pastrami, hard salami	1 oz
▤ Sausage with 8 grams fat or more per oz: bratwurst, chorizo, Italian, knockwurst, Polish, smoked, summer	1 oz

(continued)

KEY

☻ = More than 3 grams of dietary fiber per serving.

▽ = Extra fat, or prepared with added fat. (Count as 1 starch + 1 fat.)

▤ = 480 milligrams or more of sodium per serving.

Plant-Based Proteins[a]

1 plant-based protein choice = variable grams carbohydrate, 7 grams protein, variable grams fat, and variable kcalories.
Because carbohydrate content varies among plant-based proteins, you should read the food label.

Food	Serving Size	Count as
"Bacon" strips, soy-based	3 strips	1 medium-fat meat
😊 Baked beans	⅓ cup	1 starch + 1 lean meat
😊 Beans, cooked: black, garbanzo, kidney, lima, navy, pinto, white[a]	½ cup	1 starch + 1 lean meat
😊 "Beef" or "sausage" crumbles, soy-based	2 oz	½ carbohydrate + 1 lean meat
"Chicken" nuggets, soy-based	2 nuggets (1½ oz)	½ carbohydrate + 1 medium-fat meat
😊 Edamame	½ cup	½ carbohydrate + 1 lean meat
Falafel (spiced chickpea and wheat patties)	3 patties (about 2 inches across)	1 carbohydrate + 1 high-fat meat
Hot dog, soy-based	1 (1½ oz)	½ carbohydrate + 1 lean meat
😊 Hummus	⅓ cup	1 carbohydrate + 1 high-fat meat
😊 Lentils, brown, green, or yellow	½ cup	1 carbohydrate + 1 lean meat
😊 Meatless burger, soy-based	3 oz	½ carbohydrate + 2 lean meats
😊 Meatless burger, vegetable- and starch-based	1 patty (about 2½ oz)	1 carbohydrate + 2 lean meats
Nut spreads: almond butter, cashew butter, peanut butter, soy nut butter	1 Tbsp	1 high-fat meat
😊 Peas, cooked: black-eyed and split peas	½ cup	1 starch + 1 lean meat
🧂 😊 Refried beans, canned	½ cup	1 starch + 1 lean meat
"Sausage" patties, soy-based	1 (1½ oz)	1 medium-fat meat
Soy nuts, unsalted	¾ oz	½ carbohydrate + 1 medium-fat meat
Tempeh	¼ cup	1 medium-fat meat
Tofu	4 oz (½ cup)	1 medium-fat meat
Tofu, light	4 oz (½ cup)	1 lean meat

KEY

😊 = More than 3 grams of dietary fiber per serving.

▽ = Extra fat, or prepared with added fat. (Add an additional fat choice to this food.)

🧂 = 480 milligrams or more of sodium per serving (based on the sodium content of a typical 3-oz serving of meat, unless 1 or 2 oz is the normal serving size).

[a]Beans, peas, and lentils are also found on the Starch list; nut butters in smaller amounts are found in the Fats list.

TABLE G-8 Fats

Fats and oils have mixtures of unsaturated (polyunsaturated and monounsaturated) and saturated fats. Foods on the Fats list are grouped together based on the major type of fat they contain. 1 fat choice = 0 grams carbohydrate, 0 grams protein, 5 grams fat, and 45 kcalories.

NOTE: In general, one fat exchange is 1 teaspoon of regular margarine, vegetable oil, or butter; 1 tablespoon of regular salad dressing.

When used in large amounts, bacon and peanut butter are counted as high-fat meat choices (see Meat and Meat Substitutes list). Fat-free salad dressings are found on the Sweets, Desserts, and Other Carbohydrates list. Fat-free products such as margarines, salad dressings, mayonnaise, sour cream, and cream cheese are found on the Free Foods list.

Monounsaturated Fats

Food	Serving Size
Avocado, medium	2 Tbsp (1 oz)
Nut butters (trans fat-free): almond butter, cashew butter, peanut butter (smooth or crunchy)	1½ tsp
Nuts	
almonds	6 nuts
Brazil	2 nuts
cashews	6 nuts
filberts (hazelnuts)	5 nuts
macadamia	3 nuts
mixed (50% peanuts)	6 nuts
peanuts	10 nuts
pecans	4 halves
pistachios	16 nuts
Oil: canola, olive, peanut	1 tsp
Olives	
black (ripe)	8 large
green, stuffed	10 large

Polyunsaturated Fats

Food	Serving Size
Margarine: lower-fat spread (30%–50% vegetable oil, trans fat-free)	1 Tbsp
Margarine: stick, tub (trans fat-free) or squeeze (trans fat-free)	1 tsp
Mayonnaise	
reduced-fat	1 Tbsp
regular	1 tsp
Mayonnaise-style salad dressing	
reduced-fat	1 Tbsp
regular	2 tsp
Nuts	
Pignolia (pine nuts)	1 Tbsp
walnuts, English	4 halves
Oil: corn, cottonseed, flaxseed, grape seed, safflower, soybean, sunflower	1 tsp
Oil: made from soybean and canola oil—Enova	1 tsp
Plant stanol esters	
light	1 Tbsp
regular	2 tsp

Polyunsaturated Fats—continued

Food	Serving Size
Salad dressing	
🥫 reduced-fat Note: May be high in carbohydrate.	2 Tbsp
🥫 regular	1 Tbsp
Seeds	
flaxseed, whole	1 Tbsp
pumpkin, sunflower	1 Tbsp
sesame seeds	1 Tbsp
Tahini or sesame paste	2 tsp

Saturated Fats

Food	Serving Size
Bacon, cooked, regular or turkey	1 slice
Butter	
reduced-fat	1 Tbsp
stick	1 tsp
whipped	2 tsp
Butter blends made with oil	
reduced-fat or light	1 Tbsp
regular	1½ tsp
Chitterlings, boiled	2 Tbsp (½ oz)
Coconut, sweetened, shredded	2 Tbsp
Coconut milk	
light	⅓ cup
regular	1½ Tbsp
Cream	
half and half	2 Tbsp
heavy	1 Tbsp
light	1½ Tbsp
whipped	2 Tbsp
whipped, pressurized	¼ cup
Cream cheese	
reduced-fat	1½ Tbsp (¾ oz)
regular	1 Tbsp (½ oz)
Lard	1 tsp
Oil: coconut, palm, palm kernel	1 tsp
Salt pork	¼ oz
Shortening, solid	1 tsp
Sour cream	
reduced-fat or light	3 Tbsp
regular	2 Tbsp

KEY

🥫 = 480 milligrams or more of sodium per serving.

TABLE G-9 Free Foods

A "free" food is any food or drink choice that has less than 20 kcalories and 5 grams or less of carbohydrate per serving.

- Most foods on this list should be limited to 3 servings (as listed here) per day. Spread out the servings throughout the day. If you eat all 3 servings at once, it could raise your blood glucose level.
- Food and drink choices listed here without a serving size can be eaten whenever you like.

Low Carbohydrate Foods

Food	Serving Size
Cabbage, raw	½ cup
Candy, hard (regular or sugar-free)	1 piece
Carrots, cauliflower, or green beans, cooked	¼ cup
Cranberries, sweetened with sugar substitute	½ cup
Cucumber, sliced	½ cup
Gelatin	
dessert, sugar-free	
unflavored	
Gum	
Jam or jelly, light or no sugar added	2 tsp
Rhubarb, sweetened with sugar substitute	½ cup
Salad greens	
Sugar substitutes (artificial sweeteners)	
Syrup, sugar-free	2 Tbsp

Modified Fat Foods with Carbohydrate

Food	Serving Size
Cream cheese, fat-free	1 Tbsp (½ oz)
Creamers	
nondairy, liquid	1 Tbsp
nondairy, powdered	2 tsp
Margarine spread	
fat-free	1 Tbsp
reduced-fat	1 tsp
Mayonnaise	
fat-free	1 Tbsp
reduced-fat	1 tsp
Mayonnaise-style salad dressing	
fat-free	1 Tbsp
reduced-fat	1 tsp
Salad dressing	
fat-free or low-fat	1 Tbsp
fat-free, Italian	2 Tbsp
Sour cream, fat-free or reduced-fat	1 Tbsp
Whipped topping	
light or fat-free	2 Tbsp
regular	1 Tbsp

Condiments

Food	Serving Size
Barbecue sauce	2 tsp
Catsup (ketchup)	1 Tbsp
Honey mustard	1 Tbsp

KEY

🧂 = 480 milligrams or more of sodium per serving.

Condiments—continued

Food	Serving Size
Horseradish	
Lemon juice	
Miso	1½ tsp
Mustard	
Parmesan cheese, freshly grated	1 Tbsp
Pickle relish	1 Tbsp
Pickles	
🧂 dill	1½ medium
sweet, bread and butter	2 slices
sweet, gherkin	¾ oz
Salsa	¼ cup
🧂 Soy sauce, light or regular	1 Tbsp
Sweet and sour sauce	2 tsp
Sweet chili sauce	2 tsp
Taco sauce	1 Tbsp
Vinegar	
Yogurt, any type	2 Tbsp

Drinks/Mixes

Any food on the list—without a serving size listed—can be consumed in any moderate amount.

- 🧂 Bouillon, broth, consommé
- Bouillon or broth, low-sodium
- Carbonated or mineral water
- Club soda
- Cocoa powder, unsweetened (1 Tbsp)
- Coffee, unsweetened or with sugar substitute
- Diet soft drinks, sugar-free
- Drink mixes, sugar-free
- Tea, unsweetened or with sugar substitute
- Tonic water, diet
- Water
- Water, flavored, carbohydrate free

Seasonings

Any food on this list can be consumed in any moderate amount.

- Flavoring extracts (for example, vanilla, almond, peppermint)
- Garlic
- Herbs, fresh or dried
- Nonstick cooking spray
- Pimento
- Spices
- Hot pepper sauce
- Wine, used in cooking
- Worcestershire sauce

Many foods are eaten in various combinations, such as casseroles. Because "combination" foods do not fit into any one choice list, this list of choices provides some typical combination foods.

Entrees

Food	Serving Size	Count as
🔋 Casserole type (tuna noodle, lasagna, spaghetti with meatballs, chili with beans, macaroni and cheese)	1 cup (8 oz)	2 carbohydrates + 2 medium-fat meats
🔋 Stews (beef/other meats and vegetables)	1 cup (8 oz)	1 carbohydrate + 1 medium-fat meat + 0–3 fats
Tuna salad or chicken salad	½ cup (3½ oz)	½ carbohydrate + 2 lean meats + 1 fat

Frozen Meals/Entrees

Food	Serving Size	Count as
🔋 ☺ Burrito (beef and bean)	1 (5 oz)	3 carbohydrates + 1 lean meat + 2 fats
🔋 Dinner-type meal	generally 14–17 oz	3 carbohydrates + 3 medium-fat meats + 3 fats
🔋 Entrée or meal with less than 340 kcalories	about 8–11 oz	2–3 carbohydrates + 1–2 lean meats
Pizza		
🔋 cheese/vegetarian, thin crust	¼ of a 12 inch (4½–5 oz)	2 carbohydrates + 2 medium-fat meats
🔋 meat topping, thin crust	¼ of a 12 inch (5 oz)	2 carbohydrates + 2 medium-fat meats + 1½ fats
🔋 Pocket sandwich	1 (4½ oz)	3 carbohydrates + 1 lean meat + 1–2 fats
🔋 Pot pie	1 (7 oz)	2½ carbohydrates + 1 medium-fat meat + 3 fats

Salads (Deli-Style)

Food	Serving Size	Count as
Coleslaw	½ cup	1 carbohydrate + 1½ fats
Macaroni/pasta salad	½ cup	2 carbohydrates + 3 fats
🔋 Potato salad	½ cup	1½–2 carbohydrates + 1–2 fats

Soups

Food	Serving Size	Count as
🔋 Bean, lentil, or split pea	1 cup	1 carbohydrate + 1 lean meat
🔋 Chowder (made with milk)	1 cup (8 oz)	1 carbohydrate + 1 lean meat + 1½ fats
🔋 Cream (made with water)	1 cup (8 oz)	1 carbohydrate + 1 fat
🔋 Instant	6 oz prepared	1 carbohydrate
🔋 with beans or lentils	8 oz prepared	2½ carbohydrates + 1 lean meat
🔋 Miso soup	1 cup	½ carbohydrate + 1 fat
🔋 Oriental noodle	1 cup	2 carbohydrates + 2 fats
Rice (congee)	1 cup	1 carbohydrate
🔋 Tomato (made with water)	1 cup (8 oz)	1 carbohydrate
🔋 Vegetable beef, chicken noodle, or other broth-type	1 cup (8 oz)	1 carbohydrate

KEY

☺ = More than 3 grams of dietary fiber per serving.

▽ = Extra fat, or prepared with added fat.

🔋 = 600 milligrams or more of sodium per serving (for combination food main dishes/meals).

TABLE G-11 Fast Foods

The choices on the Fast Foods list are not specific fast-food meals or items, but are estimates based on popular foods. Ask the restaurant or check its website for nutrition information about your favorite fast foods.

Breakfast Sandwiches

Food	Serving Size	Count as
▤ Egg, cheese, meat, English muffin	1 sandwich	2 carbohydrates + 2 medium-fat meats
▤ Sausage biscuit sandwich	1 sandwich	2 carbohydrates + 2 high-fat meats + 3½ fats

Main Dishes/Entrees

Food	Serving Size	Count as
▤ ☺ Burrito (beef and beans)	1 (about 8 oz)	3 carbohydrates + 3 medium-fat meats + 3 fats
▤ Chicken breast, breaded and fried	1 (about 5 oz)	1 carbohydrate + 4 medium-fat meats
Chicken drumstick, breaded and fried	1 (about 2 oz)	2 medium-fat meats
▤ Chicken nuggets	6 (about 3½ oz)	1 carbohydrate + 2 medium-fat meats + 1 fat
▤ Chicken thigh, breaded and fried	1 (about 4 oz)	½ carbohydrate + 3 medium-fat meats + 1½ fats
▤ Chicken wings, hot	6 (5 oz)	5 medium-fat meats + 1½ fats

Oriental

Food	Serving Size	Count as
▤ Beef/chicken/shrimp with vegetables in sauce	1 cup (about 5 oz)	1 carbohydrate + 1 lean meat + 1 fat
▤ Egg roll, meat	1 (about 3 oz)	1 carbohydrate + 1 lean meat + 1 fat
Fried rice, meatless	½ cup	1½ carbohydrates + 1½ fats
▤ Meat and sweet sauce (orange chicken)	1 cup	3 carbohydrates + 3 medium-fat meats + 2 fats
▤ ☺ Noodles and vegetables in sauce (chow mein, lo mein)	1 cup	2 carbohydrates + 1 fat

Pizza

Food	Serving Size	Count as
Pizza		
▤ cheese, pepperoni, regular crust	⅛ of a 14 inch (about 4 oz)	2½ carbohydrates + 1 medium-fat meat + 1½ fats
▤ cheese/vegetarian, thin crust	¼ of a 12 inch (about 6 oz)	2½ carbohydrates + 2 medium-fat meats + 1½ fats

Sandwiches

Food	Serving Size	Count as
▤ Chicken sandwich, grilled	1	3 carbohydrates + 4 lean meats
▤ Chicken sandwich, crispy	1	3½ carbohydrates + 3 medium-fat meats + 1 fat
Fish sandwich with tartar sauce	1	2½ carbohydrates + 2 medium-fat meats + 2 fats
Hamburger		
▤ large with cheese	1	2½ carbohydrates + 4 medium-fat meats + 1 fat
regular	1	2 carbohydrates + 1 medium-fat meat + 1 fat
▤ Hot dog with bun	1	1 carbohydrate + 1 high-fat meat + 1 fat
Submarine sandwich		
▤ less than 6 grams fat	6-inch sub	3 carbohydrates + 2 lean meats
▤ regular	6-inch sub	3½ carbohydrates + 2 medium-fat meats + 1 fat
Taco, hard or soft shell (meat and cheese)	1 small	1 carbohydrate + 1 medium-fat meat + 1½ fats

(continued)

KEY

☺ = More than 3 grams of dietary fiber per serving.

▽ = Extra fat, or prepared with added fat.

▤ = 600 milligrams or more of sodium per serving (for fast-food main dishes/meals).

TABLE G-11 Fast Foods (*continued*)

Salads

Food	Serving Size	Count as
🧂 😊 Salad, main dish (grilled chicken type, no dressing or croutons)		1 carbohydrate + 4 lean meats
Salad, side, no dressing or cheese	Small (about 5 oz)	1 vegetable

Sides/Appetizers

Food	Serving Size	Count as
▽ French fries, restaurant style	small	3 carbohydrates + 3 fats
	medium	4 carbohydrates + 4 fats
	large	5 carbohydrates + 6 fats
🧂 Nachos with cheese	small (about 4½ oz)	2½ carbohydrates + 4 fats
🧂 Onion rings	1 serving (about 3 oz)	2½ carbohydrates + 3 fats

Desserts

Food	Serving Size	Count as
Milkshake, any flavor	12 oz	6 carbohydrates + 2 fats
Soft-serve ice cream cone	1 small	2½ carbohydrates + 1 fat

KEY

😊 = More than 3 grams of dietary fiber per serving.

▽ = Extra fat, or prepared with added fat.

🧂 = 600 milligrams or more of sodium per serving (for fast-food main dishes/meals).

TABLE G-12 Alcohol

1 alcohol equivalent = variable grams carbohydrate, 0 grams protein, 0 grams fat, and 100 kcalories.

NOTE: In general, one alcohol choice (½ ounce absolute alcohol) has about 100 kcalories. For those who choose to drink alcohol, guidelines suggest limiting alcohol intake to 1 drink or less per day for women, and 2 drinks or less per day for men. To reduce your risk of low blood glucose (hypoglycemia), especially if you take insulin or a diabetes pill that increases insulin, always drink alcohol with food. While alcohol, by itself, does not directly affect blood glucose, be aware of the carbohydrate (for example, in mixed drinks, beer, and wine) that may raise your blood glucose.

Alcoholic Beverage	Serving Size	Count as
Beer		
light (4.2%)	12 fl oz	1 alcohol equivalent + ½ carbohydrate
regular (4.9%)	12 fl oz	1 alcohol equivalent + 1 carbohydrate
Distilled spirits: vodka, rum, gin, whiskey, 80 or 86 proof	1½ fl oz	1 alcohol equivalent
Liqueur, coffee (53 proof)	1 fl oz	1 alcohol equivalent + 1 carbohydrate
Sake	1 fl oz	½ alcohol equivalent
Wine		
dessert (sherry)	3½ fl oz	1 alcohol equivalent + 1 carbohydrate
dry, red or white (10%)	5 fl oz	1 alcohol equivalent

Appendix H   Table of Food Composition

This edition of the table of food composition includes a wide variety of foods. It is updated with each edition to reflect current nutrient data for foods, to remove outdated foods, and to add foods that are new to the marketplace.* The nutrient database for this appendix is compiled from a variety of sources, including the USDA Nutrient Database and manufacturers' data. The USDA database provides data for a wider variety of foods and nutrients than other sources. Because laboratory analysis for each nutrient can be quite costly, manufacturers tend to provide data only for those nutrients mandated on food labels. Consequently, data for their foods are often incomplete; any missing information on this table is designated as a dash. Keep in mind that a dash means only that the information is unknown and should not be interpreted as a zero. A zero means that the nutrient is not present in the food.

Whenever using nutrient data, remember that many factors influence the nutrient contents of foods. These factors include the mineral content of the soil, the diet fed to the animal or the fertilizer used on the plant, the season of harvest, the method of processing, the length and method of storage, the method of cooking, the method of analysis, and the moisture content of the sample analyzed. With so many influencing factors, users should view nutrient data as a close approximation of the actual amount.

Diet & Wellness Plus, the dietary analysis software that accompanies this text, contains a database of 55,000 foods. The following comments will help in using that program and this appendix.

- *Fats*   Total fats, as well as the breakdown of total fats to saturated, monounsaturated, and polyunsaturated are listed in the table. The fatty acids seldom add up to the total in part due to rounding but also because values may include some non-fatty acids, such as glycerol, phosphate, or sterols. *Trans*-fatty acids are not listed separately in this edition because newer hydrogenated fats generally add less than 0.5 grams *trans* fat to a serving of food, an amount often reported as 0.

- *Vitamin A, Vitamin E, and Folate*   In keeping with the RDA for vitamin A, this appendix presents data for vitamin A in micrograms (µg) RAE. Similarly because the RDA for vitamin E is based only on the alpha-tocopherol form of vitamin E, this appendix reports vitamin E data in milligrams alpha-tocopherol, listed on the table as Vit E (mg α). Folate values are listed in µg DFE, a measure that adjusts for lower bioavailability of naturally occurring folate from foods compared to that from fortified foods or supplements.

- *Bioavailability*   Keep in mind that the availability of nutrients from foods depends not only on the quantity provided by a food, but also on the amount absorbed and used by the body—the bioavailability. Chapters 10–13 provide conversion factors and additional details.

---

*This food composition table has been prepared by Cengage Learning. The nutritional data are supplied by Axxya Systems.

- *Using the Table*   The foods and beverages in this table are organized into several categories, which are listed at the head of each right-hand page. Page numbers are provided, and each group is color-coded to make it easier to find individual foods.

- *Caffeine Sources*   Caffeine occurs in several plants, including the familiar coffee bean, the tea leaf, and the cocoa bean from which chocolate is made. Most human societies use caffeine regularly, most often in beverages, for its stimulant effect and flavor. Caffeine contents of beverages vary depending on the plants they are made from, the climates and soils where the plants are grown, the grind or cut size, the method and duration of brewing, and the amounts served. The accompanying table shows that, in general, a cup of coffee contains the most caffeine; a cup of tea, less than half as much; and cocoa or chocolate, less still. As for cola beverages, they are made from kola nuts, which contain caffeine, but most of their caffeine is added, using the purified compound obtained from decaffeinated coffee beans. The FDA lists caffeine as a multipurpose GRAS (generally recognized as safe) substance that may be added to foods and beverages. Drug manufacturers also use caffeine in many products.

TABLE   **Caffeine Content of Selected Beverages, Foods, and Medications**

Beverages and Foods	Serving Size	Average (mg)
Coffee		
Brewed	8 oz	95
Decaffeinated	8 oz	2
Expresso	1 oz	64
Instant	8 oz	64
Tea		
Brewed, green	8 oz	30
Brewed, herbal	8 oz	0
Brewed, leaf or bag	8 oz	47
Instant	8 oz	26
Lipton, Nestea, bottled iced tea	12 oz	10
Snapple iced tea (all flavors)	16 oz	42

Beverages and Foods	Serving Size	Average (mg)
Soft Drinks		
A&W Creme Soda	12 oz	29
Barq's Root Beer	12 oz	18
Colas, Dr. Pepper, Mr. Pibb, Sunkist Orange	12 oz	30–40
A&W Root Beer, club soda, Fresca, ginger ale, 7-Up, Sierra Mist, Sprite, Squirt, tonic water, caffeine-free soft drinks	12 oz	0
Mello Yello, Mountain Dew	12 oz	45–55

TABLE **Caffeine Content of Selected Beverages, Foods, and Medications (*continued*)**

Beverages and Foods	Serving Size	Average (mg)
Energy Drinks		
Amp	16 oz	145
Full Throttle	16 oz	200
Monster	16 oz	160
NOS	16 oz	160
Red Bull	8.3 oz	75
Rockstar	16 oz	160
Xyience Xenergy	16 oz	185
Other Beverages		
Chocolate milk or hot cocoa	8 oz	5
Coffee liqueur	1 oz	50
Starbucks Frappuccino Mocha	9.5 oz	72
Starbucks Frappuccino Vanilla	9.5 oz	64
Yoohoo chocolate drink	9 oz	3
Candies		
Baker's chocolate	1 oz	26
Dark chocolate covered coffee beans	1 oz	235
Dark chocolate	1 oz	20
Gum, caffeinated	1 piece	95
Java pops	1 pop	60
Milk chocolate	1 oz	6
Milk chocolate covered coffee beans	1 oz	224
White chocolate	1 oz	0

Beverages and Foods	Serving Size	Average (mg)
Foods		
Frozen yogurt, Ben & Jerry's coffee fudge	1 cup	85
Frozen yogurt, Häagen-Dazs coffee	1 cup	40
Frozen yogurt, chocolate	1 cup	5
Ice cream, Starbucks coffee	1 cup	50
Ice cream, Starbucks Frappuccino bar	1 bar	15
Puddings, chocolate	1 cup	5
Yogurt, Dannon coffee flavored	1 cup	45

Drugs[a]	Serving Size	Average (mg)
Cold Remedies		
Coryban-D, Dristan	1 tablet	30
Diuretics		
Aqua-Ban	1 tablet	100
Pre-Mens Forte	1 tablet	100
Pain Relievers		
Anacin, BC Fast Pain Reliever	1 tablet	32
Excedrin, Midol, Midol Max Strength	1 tablet	65
Stimulants		
Awake, NoDoz	1 tablet	100
Awake Maximum Strength, Caffedrine, NoDoz Maximum Strength, Stay Awake, Vivarin	1 tablet	200
Weight-Control Aids		
Dexatrim	1 tablet	200

[a]A pharmacologically active dose of caffeine is defined as 200 milligrams.

NOTE: The FDA suggests a maximum of 65 milligrams per 12-ounce cola beverage but does not regulate the caffeine contents of other beverages. Because products change, contact the manufacturer for an update on products you use regularly.

SOURCE: Adapted from USDA database Release 18 (www.nal.usda.gov/fnic/foodcomp/Data/), Caffeine content of foods and drugs, Center for Science and the Public Interest (www.cspinet.org/new/cafchart.htm), and R. R. McCusker, B. A. Goldberger, and E. J. Cone, Caffeine content of energy drinks, carbonated sodas, and other beverages, *Journal of Analytical Toxicology* 30 (2006): 112–114.

DA+ Code	Food Description	Quantity	Measure	Wt (g)	H₂O (g)	Ener (kcal)	Prot (g)	Carb (g)	Fiber (g)	Fat (g)	Fat Breakdown (g) Sat	Mono	Poly
Breads, Baked Goods, Cakes, Cookies, Crackers, Chips, Pies												
	Bagels												
8534	Cinnamon and raisin	1	item(s)	71	22.7	195	7.0	39.2	1.6	1.2	0.2	0.1	0.5
8538	Oat bran	1	item(s)	71	23.4	181	7.6	37.8	2.6	0.9	0.1	0.2	0.3
4910	Plain, enriched	1	item(s)	71	25.8	182	7.1	35.9	1.6	1.2	0.3	0.4	0.5
4911	Plain, enriched, toasted	1	item(s)	66	18.7	190	7.4	37.7	1.7	1.1	0.2	0.3	0.6
72275	Whole wheat	1	item(s)	94	—	210	10.0	41.0	6.0	1.5	0	0	0.5
	Biscuits												
25008	Biscuits	1	item(s)	41	15.8	121	2.6	16.4	0.5	4.9	1.4	1.4	1.8
16729	Scone	1	item(s)	42	11.5	149	3.8	19.0	0.6	6.3	2.0	2.5	1.4
25166	Wheat biscuits	1	item(s)	55	21.0	162	3.6	21.9	1.4	6.7	1.9	1.9	2.5
	Bread												
325	Boston brown, canned	1	slice(s)	45	21.2	88	2.3	19.5	2.1	0.7	0.1	0.1	0.3
8716	Bread sticks, plain	4	item(s)	24	1.5	99	2.9	16.4	0.7	2.3	0.3	0.9	0.9
25176	Cornbread	2	piece(s)	55	25.9	141	4.6	18.4	0.6	5.4	2.1	1.5	1.4
327	Cracked wheat	1	slice(s)	25	8.9	65	2.2	12.4	1.4	1.0	0.2	0.5	0.2
9079	Croutons, plain	¼	cup(s)	8	0.4	31	0.9	5.5	0.4	0.5	0.1	0.2	0.1
8582	Egg	1	slice(s)	40	13.9	115	3.8	19.1	0.9	2.4	0.6	0.9	0.4
8585	Egg, toasted	1	slice(s)	37	10.5	117	3.9	19.5	0.9	2.4	0.6	1.1	0.4
329	French	1	slice(s)	32	8.9	92	3.8	18.1	0.8	0.6	0.2	0.1	0.3
8591	French, toasted	1	slice(s)	23	4.7	73	3.0	14.2	0.7	0.5	0.1	0.1	0.2
42096	Indian fry, made with lard (Navajo)	3	ounce(s)	85	26.9	281	5.7	41.0	—	10.4	3.9	3.8	0.9
332	Italian	1	slice(s)	30	10.7	81	2.6	15.0	0.8	1.0	0.3	0.2	0.4
1393	Mixed grain	1	slice(s)	26	9.6	69	3.5	11.3	1.9	1.1	0.2	0.2	0.5
8604	Mixed grain, toasted	1	slice(s)	24	7.6	69	3.5	11.3	1.9	1.1	0.2	0.2	0.5
8605	Oat bran	1	slice(s)	30	13.2	71	3.1	11.9	1.4	1.3	0.2	0.5	0.5
8608	Oat bran, toasted	1	slice(s)	27	10.4	70	3.1	11.8	1.3	1.3	0.2	0.5	0.5
8609	Oatmeal	1	slice(s)	27	9.9	73	2.3	13.1	1.1	1.2	0.2	0.4	0.5
8613	Oatmeal, toasted	1	slice(s)	25	7.8	73	2.3	13.2	1.1	1.2	0.2	0.4	0.5
1409	Pita	1	item(s)	60	19.3	165	5.5	33.4	1.3	0.7	0.1	0.1	0.3
7905	Pita, whole wheat	1	item(s)	64	19.6	170	6.3	35.2	4.7	1.7	0.3	0.2	0.7
338	Pumpernickel	1	slice(s)	32	12.1	80	2.8	15.2	2.1	1.0	0.1	0.3	0.4
334	Raisin, enriched	1	slice(s)	26	8.7	71	2.1	13.6	1.1	1.1	0.3	0.6	0.2
8625	Raisin, toasted	1	slice(s)	25	6.9	74	2.1	14.2	1.2	1.2	0.3	0.6	0.2
10168	Rice, white, gluten free, wheat free	1	slice(s)	38	—	100	1.0	17.0	1.0	3.5	0	—	—
8653	Rye	1	slice(s)	32	11.9	83	2.7	15.5	1.9	1.1	0.2	0.4	0.3
74338	Rye, light	1	slice(s)	43	—	100	3.0	20.0	1.0	0.5	0	—	—
8654	Rye, toasted	1	slice(s)	29	9.0	82	2.7	15.4	1.9	1.0	0.2	0.4	0.3
8588	Sourdough	1	slice(s)	25	7.0	72	2.9	14.1	0.6	0.5	0.1	0.1	0.2
8592	Sourdough, toasted	1	slice(s)	23	4.7	73	3.0	14.2	0.7	0.5	0.1	0.1	0.2
8596	Vienna, toasted	1	slice(s)	23	4.7	73	3.0	14.2	0.7	0.5	0.1	0.1	0.2
8670	Wheat	1	slice(s)	25	8.6	68	2.6	12.4	1.0	0.9	0.2	0.2	0.4
8671	Wheat, toasted	1	slice(s)	23	5.6	72	3.0	12.8	1.1	1.0	0.2	0.2	0.4
340	White	1	slice(s)	25	9.1	66	2.3	12.3	0.7	0.8	0.2	0.1	0.4
1395	Whole wheat	1	slice(s)	46	15.0	128	3.9	23.6	2.8	2.5	0.4	0.5	1.4
	Cakes												
386	Angel food, prepared from mix	1	piece(s)	50	16.5	129	3.0	29.4	0.1	0.2	0	0	0.1
8772	Butter pound, ready to eat, commercially prepared	1	slice(s)	75	18.5	291	4.1	36.6	0.4	14.9	8.7	4.4	0.8
28517	Carrot	1	slice(s)	131	56.4	340	4.8	56.8	1.8	11.1	1.0	6.1	3.6
4931	Chocolate with chocolate icing, commercially prepared	1	slice(s)	64	14.7	235	2.6	34.9	1.8	10.5	3.1	5.6	1.2
8756	Chocolate, prepared from mix	1	slice(s)	95	23.2	352	5.0	50.7	1.5	14.3	5.2	5.7	2.6
8757	Fruitcake, ready to eat, commercially prepared	1	piece(s)	43	10.9	139	1.2	26.5	1.6	3.9	0.5	1.8	1.4
14284	Lemon, prepared from mix	1	slice(s)	80	—	250	3.0	36.0	0.5	11.1	2.6	—	—
1397	Pineapple upside down, prepared from mix	1	slice(s)	115	37.1	367	4.0	58.1	0.9	13.9	3.4	6.0	3.8
411	Sponge, prepared from mix	1	slice(s)	63	18.5	187	4.6	36.4	0.3	2.7	0.8	1.0	0.4
8817	White with coconut frosting, prepared from mix	1	slice(s)	112	23.2	399	4.9	70.8	1.1	11.5	4.4	4.1	2.4
8819	Yellow with chocolate frosting, ready to eat, commercially prepared	1	slice(s)	64	14.3	243	2.0	35.4	1.0	11.4	3.7	4.6	3.0
8822	Yellow with vanilla frosting, ready to eat, commercially prepared	1	slice(s)	64	14.1	239	2.2	37.6	0.2	9.3	1.5	3.9	3.3

Chol (mg)	Calc (mg)	Iron (mg)	Magn (mg)	Pota (mg)	Sodi (mg)	Zinc (mg)	Vit A (µg)	Thia (mg)	Vit E (mg α)	Ribo (mg)	Niac (mg)	Vit B6 (mg)	Fola (µg)	Vit C (mg)	Vit B12 (µg)	Sele (µg)
0	13	2.70	19.9	105.1	244.2	0.80	14.9	0.27	0.22	0.20	2.19	0.04	123.5	0.5	0	22.0
0	9	2.19	22.0	81.7	418.9	0.64	0.7	0.24	0.23	0.24	2.10	0.03	95.1	0.1	0	24.3
0	63	4.30	15.6	53.3	338.7	1.35	0	0.43	0.07	0.18	2.82	0.05	160.5	0.7	0	16.2
0	65	2.97	15.8	56.1	354.4	0.86	0	0.40	0.08	0.18	2.89	0.05	134.0	0	0	16.6
0	40	2.70	—	—	420.0	—	0	—	—	—	—	—	—	0	—	—
0	37	0.94	6.0	45.9	204.5	0.20	0	0.16	0.05	0.13	1.20	0.01	59.3	0	0.1	7.3
43	79	1.34	7.1	49.1	281.8	0.31	66.4	0.15	0.55	0.16	1.20	0.03	49.1	0.1	0.1	10.8
0	57	1.22	16.1	81.2	321.2	0.42	0	0.20	0.12	0.15	1.65	0.04	63.2	0.1	0.1	12.2
0	32	0.94	28.4	143.1	284.0	0.22	11.3	0.01	0.14	0.05	0.50	0.04	6.3	0	0	9.9
0	5	1.03	7.7	29.8	171.1	0.21	0	0.14	0.24	0.13	1.27	0.02	61.2	0	0	9.0
21	94	0.94	10.1	69.2	208.8	0.49	0	0.14	0.40	0.16	1.06	0.04	61.7	1.5	0.2	6.8
0	11	0.70	13.0	44.3	134.5	0.31	0	0.09	—	0.06	0.92	0.08	19.0	0	0	6.3
0	6	0.31	2.3	9.3	52.3	0.07	0	0.05	—	0.02	0.41	0	15.7	0	0	2.8
20	37	1.22	7.6	46.0	152.0	0.32	25.2	0.18	0.10	0.17	1.94	0.03	52.0	0	0	12.0
21	38	1.24	7.8	46.6	154.3	0.32	25.5	0.14	0.11	0.16	1.77	0.02	47.4	0	0	12.2
0	14	1.16	9.0	41.0	164.2	0.30	0	0.14	0.06	0.09	1.52	0.03	73.6	0.1	0	8.7
0	11	0.89	7.1	32.2	165.6	0.24	0	0.10	0.04	0.09	1.24	0.02	49.9	0	0	6.8
6	48	3.44	15.3	65.5	279.8	0.30	—	0.37	0	0.18	3.91	0.03	166.7	—	0	15.8
0	23	0.88	8.1	33.0	183.9	0.26	0	0.14	0.09	0.09	1.31	0.01	91.2	0	0	8.2
0	27	0.65	20.3	59.8	99.1	0.44	0	0.07	0.10	0.03	1.05	0.07	19.5	0	0	8.6
0	27	0.65	20.4	60.0	99.4	0.44	0	0.06	0.10	0.03	1.05	0.07	16.8	0	0	8.6
0	20	0.94	10.5	44.1	105.9	0.27	0.6	0.15	0.13	0.10	1.45	0.02	36.0	0	0	9.0
0	19	0.93	9.2	33.2	104.5	0.28	0.5	0.12	0.13	0.09	1.29	0.01	28.1	0	0	8.9
0	18	0.73	10.0	38.3	120.7	0.28	1.4	0.11	0.13	0.06	0.85	0.02	23.5	0	0	6.6
0	18	0.74	10.3	38.5	121.5	0.28	1.3	0.09	0.13	0.06	0.77	0.02	18.8	0.1	0	6.7
0	52	1.57	15.6	72.0	321.6	0.50	0	0.36	0.18	0.20	2.78	0.02	99.0	0	0	16.3
0	10	1.96	44.2	108.8	284.2	0.97	0	0.22	0.39	0.05	1.82	0.17	22.4	0	0	28.2
0	22	0.92	17.3	66.6	190.7	0.47	0	0.10	0.13	0.10	0.99	0.04	42.9	0	0	7.8
0	17	0.75	6.8	59.0	90.2	0.19	0	0.09	0.07	0.10	0.90	0.02	40.6	0	0	5.2
0	18	0.79	7.0	61.5	94.3	0.19	0	0.07	0.08	0.10	0.85	0.02	37.0	0.1	0	5.4
0	20	0.72	—	—	120.0	—	0	0.15	—	0.07	1.20	—	—	0	—	—
0	23	0.91	12.8	53.1	193.0	0.36	0	0.14	0.11	0.11	1.22	0.02	48.3	0.1	0	9.9
0	0	1.08	—	—	220.0	—	0	—	—	—	—	—	—	1.2	—	—
0	23	0.90	12.5	53.1	192.6	0.36	0	0.11	0.11	0.10	1.09	0.02	42.9	0.1	0	9.9
0	11	0.91	7.0	32.0	128.3	0.23	0	0.11	0.05	0.07	1.19	0.03	57.5	0.1	0	6.8
0	11	0.89	7.1	32.2	165.6	0.24	0	0.10	0.04	0.09	1.24	0.02	49.9	0	0	6.8
0	11	0.89	7.1	32.2	165.6	0.24	0	0.10	0.04	0.09	1.24	0.02	49.9	0	0	6.8
0	35	0.88	11.5	45.5	129.8	0.29	0	0.12	0.05	0.07	1.48	0.03	24.8	0.1	0	7.2
0	38	0.94	13.6	51.3	138.2	0.34	0	0.10	0.06	0.09	1.44	0.04	23.0	0	0	7.7
0	65	0.90	6.3	28.8	122.8	0.21	0	0.13	0.05	0.06	1.20	0.02	42.8	0	0	5.5
0	15	1.43	37.3	144.4	159.2	0.69	0	0.14	0.35	0.10	1.83	0.09	35.9	0	0	17.8
0	42	0.12	4.0	67.5	255.5	0.06	0	0.05	0	0.10	0.09	0	14.5	0	0	7.7
166	26	1.03	8.3	89.3	298.5	0.34	111.8	0.10	—	0.17	0.98	0.03	46.5	0	0.2	6.6
0	64	1.87	19.2	250.5	370.2	0.44	0	0.24	1.96	0.21	1.74	0.11	77.7	4.8	0	14.6
27	28	1.41	21.8	128.0	213.8	0.44	16.6	0.02	0.63	0.09	0.37	0.03	14.9	0.1	0.1	2.1
55	57	1.53	30.4	133.0	299.3	0.66	38.0	0.13	—	0.20	1.08	0.04	37.0	0.2	0.2	11.3
2	14	0.89	6.9	65.8	43.4	0.12	3.0	0.02	0.39	0.04	0.34	0.02	13.8	0.2	0	0.9
54	40	0.72	—	—	312.0	—	0	0.12	—	0.14	0.80		—	0	—	—
25	138	1.70	14.9	128.8	366.9	0.36	71.3	0.18	—	0.18	1.37	0.04	44.8	1.4	0.1	10.8
107	26	1.00	5.7	88.8	143.6	0.37	48.5	0.10	—	0.19	0.76	0.04	33.4	0	0.2	11.7
1	101	1.30	13.4	110.9	318.1	0.37	13.4	0.14	0.13	0.21	1.19	0.03	57.1	0.1	0.1	12.0
10	20	1.30	12.8	119.7	198.4	0.27	21.1	0.03	2.88	0.06	0.51	0	20.5	0	0	3.8
35	40	0.68	3.8	33.9	220.2	0.16	12.2	0.06	—	0.04	0.32	0.02	25.6	0	0.1	3.5

APPENDIX H

DA+ Code	Food Description	Quantity	Measure	Wt (g)	H₂O (g)	Ener (kcal)	Prot (g)	Carb (g)	Fiber (g)	Fat (g)	Fat Breakdown (g)		
											Sat	Mono	Poly
Breads, Baked Goods, Cakes, Cookies, Crackers, Chips, Pies—*continued*													
Snack cakes													
8791	Chocolate snack cake, creme filled, with frosting	1	item(s)	50	9.3	200	1.8	30.2	1.6	8.0	2.4	4.3	0.9
25010	Cinnamon coffee cake	1	piece(s)	72	22.6	231	3.6	35.9	0.7	8.3	2.2	2.6	3.0
16777	Funnel cake	1	item(s)	90	37.7	275	7.3	29.2	0.9	14.3	2.7	5.5	5.4
8794	Sponge snack cake, creme filled	1	item(s)	43	8.3	159	1.5	27.2	0.4	4.9	1.8	2.1	0.8
Snacks, chips, pretzels													
57179	Bagel chips, plain	1	ounce(s)	28	—	132	4.1	19.2	1.0	4.6	0.5	3.5	0.5
57180	Bagel chips, toasted garlic	1	ounce(s)	28	—	132	4.1	19.2	1.0	4.6	0.5	3.5	0.5
38192	Chex traditional snack mix	1	cup(s)	52	—	220	4.0	38.0	1.0	6.0	1.0	—	—
654	Potato chips, salted	1	ounce(s)	28	0.6	154	1.9	14.4	1.2	10.3	1.1	4.5	4.5
8816	Potato chips, unsalted	1	ounce(s)	28	0.5	152	2.0	15.0	1.4	9.8	3.1	2.8	3.5
5096	Pretzels, plain, hard, twists	5	item(s)	30	1.0	114	3.1	23.9	0.9	0.8	0.2	0.4	0.3
4632	Pretzels, whole wheat	1	ounce(s)	28	1.1	103	3.1	23.0	2.2	0.7	0.2	0.3	0.2
4641	Tortilla chips, plain	6	item(s)	11	0.2	53	0.8	7.1	0.6	2.5	0.3	0.7	1.2
Cookies													
8859	Animal crackers	12	item(s)	30	1.2	134	2.1	22.2	0.3	4.1	1.0	2.3	0.6
8876	Brownie, prepared from mix	1	item(s)	24	3.0	112	1.5	12.0	0.5	7.0	1.8	2.6	2.3
25207	Chocolate chip cookies	1	item(s)	30	3.7	140	2.0	16.2	0.6	7.9	2.1	3.3	2.1
8915	Chocolate sandwich cookie with extra creme filling	1	item(s)	13	0.2	65	0.6	8.9	0.4	3.2	0.7	2.1	0.3
14145	Fig Newtons cookies	1	item(s)	16	—	55	0.5	11.0	0.5	1.0	0	—	—
8920	Fortune cookie	1	item(s)	8	0.6	30	0.3	6.7	0.1	0.2	0.1	0.1	0.0
25208	Oatmeal cookies	1	item(s)	69	12.5	234	5.4	45.3	3.2	4.2	0.7	1.3	1.8
25213	Peanut butter cookies	1	item(s)	35	4.1	162	4.2	16.9	0.9	9.2	1.7	4.7	2.4
33095	Sugar cookies	1	item(s)	16	4.1	61	1.1	7.4	0.1	3.0	0.6	1.4	0.8
9002	Vanilla sandwich cookie with creme filling	1	item(s)	10	0.2	48	0.5	7.2	0.1	2.0	0.3	0.8	0.8
Crackers													
9012	Cheese cracker sandwich with peanut butter	4	item(s)	28	0.9	139	3.5	15.9	1.0	7.0	1.2	3.6	1.4
9008	Cheese crackers (mini)	30	item(s)	30	1.1	147	3.3	17.8	0.7	6.8	1.6	1.7	3.0
33362	Cheese crackers, low sodium	1	serving(s)	30	0.9	151	3.0	17.5	0.7	7.6	2.9	3.6	0.7
8928	Honey graham crackers	4	item(s)	28	1.2	118	1.9	21.5	0.8	2.8	0.4	1.1	1.1
9016	Matzo crackers, plain	1	item(s)	28	1.2	112	2.8	23.8	0.9	0.4	0.1	0	0.2
9024	Melba toast	3	item(s)	15	0.8	59	1.8	11.5	0.9	0.5	0.1	0.1	0.2
9028	Melba toast, rye	3	item(s)	15	0.7	58	1.7	11.6	1.2	0.5	0.1	0.1	0.2
14189	Ritz crackers	5	item(s)	16	0.5	80	1.0	10.0	0	4.5	1.0	1.0	1.7
9014	Rye crispbread crackers	1	item(s)	10	0.6	37	0.8	8.2	1.6	0.1	0	0	<0.1
9040	Rye wafer	1	item(s)	11	0.6	37	1.1	8.8	2.5	0.1	0	0	0
432	Saltine crackers	5	item(s)	15	0.6	63	1.4	11.2	0.4	1.3	0.3	0.3	0.6
9046	Saltine crackers, low salt	5	item(s)	15	0.6	63	1.4	11.2	0.4	1.3	0.3	0.3	0.6
9052	Snack cracker sandwich with cheese filling	4	item(s)	28	1.1	134	2.6	17.3	0.5	5.9	1.7	3.2	0.7
9054	Snack cracker sandwich with peanut butter filling	4	item(s)	28	0.8	138	3.2	16.3	0.6	6.9	1.4	3.9	1.3
9048	Snack crackers, round	10	item(s)	30	1.0	151	2.0	18.4	0.6	7.7	1.6	2.0	3.8
9050	Snack crackers, round, low salt	10	item(s)	30	1.0	151	2.2	18.3	0.5	7.6	1.1	3.2	2.9
9044	Soda crackers	5	item(s)	15	0.6	63	1.4	11.2	0.4	1.3	0.3	0.3	0.6
9059	Wheat cracker sandwich with cheese filling	4	item(s)	28	0.9	139	2.7	16.3	0.9	7.0	1.2	2.9	2.6
9061	Wheat cracker sandwich with peanut butter filling	4	item(s)	28	1.0	139	3.8	15.1	1.2	7.5	1.3	3.3	2.5
9055	Wheat crackers	10	item(s)	30	1.0	137	2.7	20.2	1.1	5.2	0.8	1.3	2.8
9057	Wheat crackers, low salt	10	item(s)	30	0.9	142	2.6	19.5	1.4	6.2	1.6	3.4	0.8
9022	Whole wheat crackers	7	item(s)	28	0.8	120	3.0	19.5	2.9	4.0	0.6	0.9	1.9
Pastry													
16754	Apple fritter	1	item(s)	17	6.4	61	1.0	5.5	0.2	3.9	1.0	1.6	1.1
41565	Cinnamon rolls with icing, refrigerated dough	1	serving(s)	44	12.3	145	1.9	23.5	0.6	5.0	1.4	—	—
4945	Croissant, butter	1	item(s)	57	13.2	231	4.7	26.1	1.5	12.0	6.6	3.1	0.6
9096	Danish, nut	1	item(s)	65	13.3	280	4.6	29.7	1.3	16.4	3.8	8.9	2.8
9115	Doughnut with creme filling	1	item(s)	85	32.5	307	5.4	25.5	0.7	20.8	4.6	10.3	2.6
9117	Doughnut with jelly filling	1	item(s)	85	30.3	289	5.0	33.2	0.8	15.9	4.1	8.7	2.0
4947	Doughnut, cake	1	item(s)	47	10.8	196	2.8	21.4	0.8	11.1	3.3	6.0	1.2
9105	Doughnut, cake, chocolate glazed	1	item(s)	42	6.8	175	1.9	24.1	0.9	8.4	2.2	4.7	1.0

Chol (mg)	Calc (mg)	Iron (mg)	Magn (mg)	Pota (mg)	Sodi (mg)	Zinc (mg)	Vit A (µg)	Thia (mg)	Vit E (mg α)	Ribo (mg)	Niac (mg)	Vit B6 (mg)	Fola (µg)	Vit C (mg)	Vit B12 (µg)	Sele (µg)
0	58	1.80	18.0	88.0	166.0	0.52	0.5	0.02	0.55	0.04	0.46	0.07	17.5	0.9	0	1.7
26	54	1.20	8.3	72.6	275.8	0.30	0	0.18	0.35	0.17	1.30	0.03	66.1	0	0.1	9.8
55	127	1.90	16.2	147.6	271.8	0.69	56.7	0.24	1.58	0.33	1.86	0.05	75.6	0	0.3	17.5
17	10	0.58	3.4	30.2	199.8	0.25	2.1	0.08	0.26	0.07	0.66	0	23.0	0	0.1	1.5
0	0	1.09	—	—	313.9	—	0	—	—	—	—	—	—	0	—	—
0	0	1.09	—	—	303.8	—	0	—	—	—	—	—	—	0	—	—
0	0	0.72	—	—	480.0	—	0	—	—	—	—	—	—	0	—	—
0	7	0.46	19.8	465.5	136.1	0.68	0	0.02	1.91	0.07	1.19	0.20	21.3	5.3	0	2.3
0	7	0.46	19.0	361.5	2.3	0.31	0	0.05	2.58	0.06	1.08	0.19	12.8	8.8	0	2.3
0	5	1.56	8.7	40.8	407.1	0.43	0	0.15	0.10	0.10	1.54	0.01	85.6	0	0	1.8
0	8	0.76	8.5	121.9	57.6	0.18	0	0.12	—	0.08	1.85	0.08	15.3	0.3	0	
0	19	0.25	15.8	23.2	45.5	0.27	0	0	0.46	0.01	0.14	0.02	2.2	0	0	0.7
0	13	0.82	5.4	30.0	122.1	0.19	0	0.10	0.04	0.10	1.04	0.01	49.5	0	0	2.1
18	14	0.44	12.7	42.2	82.3	0.23	42.2	0.03	—	0.05	0.24	0.02	9.4	0.1	0	2.8
13	11	0.64	11.4	51.3	108.6	0.24	0	0.08	0.60	0.06	0.87	0.02	29.8	0	0	4.2
0	2	1.01	4.7	17.8	45.6	0.10	0	0.01	0.25	0.02	0.25	0	9.0	0	0	1.1
0	10	0.36	—	—	62.5	—	—	—	—	—	—	—	—	0	—	—
0	1	0.12	0.6	3.3	2.5	0.01	0.1	0.01	0	0.01	0.15	0	8.4	0	0	0.2
0	26	1.83	45.5	157.5	310.3	1.49	0	0.23	1.02	0.13	1.40	0.09	46.4	0.3	0	16.7
13	27	0.65	20.3	107.2	153.7	0.46	0	0.09	1.45	0.09	1.86	0.06	34.7	0.1	0.1	4.7
18	5	0.32	1.7	12.8	50.0	0.08	0	0.04	0.32	0.05	0.31	0.01	17.1	0	0	3.1
0	3	0.22	1.4	9.1	34.9	0.04	0	0.03	0.16	0.02	0.27	0	8.2	0	0	0.3
0	14	0.76	15.7	61.0	232.1	0.29	0.3	0.15	0.66	0.08	1.63	0.04	39.8	0	0.1	2.3
1	41	1.46	7.5	46.8	291.9	0.36	5.1	0.17	0.66	0.10	1.83	0.05	72.3	0	0.1	4.1
4	45	1.43	10.8	31.8	137.4	0.34	5.1	0.17	0.09	0.13	1.40	0.17	40.2	0	0.1	2.6
0	7	1.04	8.4	37.8	133.6	0.23	0	0.06	0.09	0.09	1.15	0.02	18.5	0	0	2.9
0	4	0.90	7.1	31.8	0	0.19	0	0.11	0.02	0.08	1.11	0.03	4.8	0	0	10.5
0	14	0.56	8.9	30.3	89.7	0.30	0	0.06	0.06	0.04	0.62	0.01	29.0	0	0	5.2
0	12	0.55	5.8	29.0	134.9	0.20	0	0.07	—	0.04	0.71	0.01	19.4	0	0	5.8
0	20	0.36	3.0	19.0	135.0	0.10	—	0.07	0.56	0.04	0.78	0.01	17.8	0	—	0.7
0	3	0.24	7.8	31.9	41.0	0.24	0	0.02	0.08	0.01	0.10	0.02	6.5	0	0	3.7
0	4	0.65	13.3	54.5	61.3	0.31	0	0.05	0.09	0.03	0.17	0.03	4.9	0	0	2.6
0	3	0.77	3.8	23.9	153.1	0.12	0	0.09	0.17	0.05	0.79	0.01	33.2	0	0	1.0
0	18	0.77	3.8	108.6	29.7	0.12	0	0.09	0.17	0.05	0.79	0.01	33.2	0	0	1.0
1	72	0.67	10.1	120.1	273.8	0.17	4.8	0.12	0.06	0.19	1.05	0.01	44.8	0	0	6.0
0	23	0.78	15.4	60.2	224.3	0.32	0.3	0.14	0.58	0.08	1.71	0.04	34.2	0	0	3.0
0	29	1.14	5.4	33.3	224.4	0.17	0	0.13	1.07	0.07	1.25	0.02	55.8	0	0	1.1
0	36	1.08	8.1	106.5	64.8	0.20	0	0.12	0.61	0.10	1.21	0.02	42.6	0	0	2.0
0	3	0.77	3.8	23.9	153.1	0.12	0	0.09	0.17	0.05	0.79	0.01	33.2	0	0	1.0
2	57	0.73	15.1	85.7	234.9	0.24	5.9	0.10	—	0.12	0.89	0.07	26.9	0.4	0	6.8
0	48	0.75	10.6	83.2	226.0	0.23	0	0.11	—	0.08	1.65	0.04	26.0	0	0	6.1
0	24	1.41	13.8	62.1	236.1	0.52	0	0.18	0.33	0.09	1.49	0.04	56.1	0	0.1	3.8
0	15	1.32	18.6	60.9	57.0	0.48	0	0.15	0.15	0.10	1.49	0.04	21.6	0	0	10.1
0	10	0.94	30.8	96.6	197.1	0.74	0	0.05	0.39	0.01	1.30	0.05	7.8	0	0	2.8
13	9	0.26	2.2	22.3	7.3	0.10	8.2	0.03	0.25	0.04	0.23	0.01	9.2	0.2	0	2.6
0	12	0.85	—	—	343.2	—	0	—	—	—	—	—	—	0	—	—
38	21	1.16	9.1	67.3	424.1	0.43	117.4	0.22	0.48	0.14	1.25	0.03	74.1	0.1	0.1	12.9
30	61	1.17	20.8	61.8	193.7	0.57	5.8	0.14	0.53	0.16	1.50	0.07	79.3	1.1	0.1	9.2
20	21	1.56	17.0	68.0	262.8	0.68	9.4	0.29	0.25	0.13	1.91	0.06	92.7	0	0.1	9.2
22	21	1.50	17.0	67.2	387.0	0.64	14.5	0.27	0.37	0.12	1.82	0.09	88.5	0	0.2	10.6
4	12	1.41	7.5	53.1	261.8	0.32	1.4	0.11	0.89	0.07	0.93	0.01	54.0	0.6	0	4.8
24	89	0.95	14.3	44.5	90.3	0.24	5.0	0.02	0.09	0.03	0.20	0.01	26.9	0	0	1.7

DA+ Code	Food Description	Quantity	Measure	Wt (g)	H₂O (g)	Ener (kcal)	Prot (g)	Carb (g)	Fiber (g)	Fat (g)	Fat Breakdown (g)		
											Sat	Mono	Poly
Breads, Baked Goods, Cakes, Cookies, Crackers, Chips, Pies—*continued*													
437	Doughnut, glazed	1	item(s)	60	15.2	242	3.8	26.6	0.7	13.7	3.5	7.7	1.7
10617	Toaster pastry, brown sugar cinnamon	1	item(s)	50	5.3	210	2.0	34.0	0.5	8.0	2.5	—	—
30928	Toaster pastry, cream cheese	1	item(s)	54	—	200	3.0	24.0	0.5	11.0	4.0	—	—
	Muffins												
25015	Blueberry	1	item(s)	63	29.8	160	3.4	23.0	0.8	6.0	0.9	1.6	3.0
9189	Corn, ready to eat	1	item(s)	57	18.6	174	3.4	29.0	1.9	4.8	0.8	1.2	1.8
9121	English muffin, plain, enriched	1	item(s)	57	24.0	134	4.4	26.2	1.5	1.0	0.1	0.2	0.5
29582	English muffin, toasted	1	item(s)	50	16.5	135	5.2	26.3	1.4	1.0	0.3	0.2	0.5
9145	English muffin, wheat	1	item(s)	57	24.1	127	5.0	25.5	2.6	1.1	0.2	0.2	0.5
8894	Oat bran	1	item(s)	57	20.0	154	4.0	27.5	2.6	4.2	0.6	1.0	2.4
	Granola bars												
34783	Kudos milk chocolate with M&Ms granola bar	1	item(s)	24	—	100	1.0	17.0	1.0	2.5	1.5	—	—
38187	Nature Valley fruit 'n' nut trail mix bar	1	item(s)	35	—	140	3.0	25.0	2.0	4.0	0.5	—	—
3436	Nature Valley oats 'n honey crunchy granola bars	2	item(s)	42	—	180	4.0	29.0	2.0	6.0	0.5	—	—
1383	Plain, hard	1	item(s)	25	1.0	115	2.5	15.8	1.3	4.9	0.6	1.1	3.0
4606	Plain, soft	1	item(s)	28	1.8	126	2.1	19.1	1.3	4.9	2.1	1.1	1.5
	Pies												
454	Apple pie, prepared from home recipe	1	slice(s)	155	73.3	411	3.7	57.5	2.3	19.4	4.7	8.4	5.2
470	Pecan pie, prepared from home recipe	1	slice(s)	122	23.8	503	6.0	63.7	—	27.1	4.9	13.6	7.0
33356	Pie crust mix, prepared, baked	1	slice(s)	20	2.1	100	1.3	10.1	0.4	6.1	1.5	3.5	0.8
9007	Pie crust, ready to bake, frozen, enriched, baked	1	slice(s)	16	1.1	81	1.0	9.0	0.5	4.6	1.5	2.2	0.6
472	Pumpkin pie, prepared from home recipe	1	slice(s)	155	90.7	316	7.0	40.9	—	14.4	4.9	5.7	2.8
	Rolls												
8555	Crescent dinner roll	1	item(s)	28	9.6	78	2.8	14.0	0.6	1.1	0.2	0.3	0.4
489	Hamburger roll or bun, plain	1	item(s)	43	14.7	120	4.2	21.6	0.9	1.6	0.4	0.4	0.7
490	Hard roll	1	item(s)	57	17.7	167	5.6	30.0	1.3	2.5	0.3	0.6	1.0
5127	Kaiser roll	1	item(s)	57	17.7	167	5.6	30.0	1.3	2.5	0.3	0.6	1.0
5130	Whole wheat roll or bun	1	item(s)	28	9.4	75	2.5	14.5	2.1	1.3	0.2	0.3	0.6
	Sport bars												
37026	Balance original chocolate bar	1	item(s)	50	—	200	14.0	21.0	2.0	7.0	4.0	—	—
37024	Balance original peanut butter bar	1	item(s)	50	—	200	15.0	21.0	0.5	7.0	3.0	—	—
36580	Clif Bar chocolate brownie energy bar	1	item(s)	68	—	240	10.0	44.0	5.0	5.0	1.5	—	—
36583	Clif Bar crunchy peanut butter energy bar	1	item(s)	68	—	250	11.0	42.0	5.0	6.0	1.0	—	—
36589	Clif Luna Nutz over Chocolate energy bar	1	item(s)	48	—	180	9.0	25.0	4.0	6.0	2.0	—	—
12005	PowerBar apple cinnamon	1	item(s)	65	—	230	8.0	45.0	2.0	3.5	0.5	—	—
16078	PowerBar banana	1	item(s)	65	—	240	8.0	46.0	1.0	3.5	0.5	—	—
16080	PowerBar chocolate	1	item(s)	65	—	240	8.0	45.0	3.0	3.0	1.0	—	—
29092	PowerBar peanut butter	1	item(s)	65	—	240	9.0	44.0	1.0	4.0	1.0	—	—
	Tortillas												
1391	Corn tortillas, soft	1	item(s)	24	11.0	52	1.4	10.7	1.5	0.7	0.1	0.2	0.3
1669	Flour tortilla	1	item(s)	30	9.5	90	2.4	15.4	0.7	2.0	0.5	1.1	0.4
1390	Taco shells, hard	1	item(s)	13	0.9	63	0.9	8.5	0.9	2.8	0.8	0.8	0.9
	Pancakes, waffles												
8926	Pancakes, blueberry, prepared from recipe	3	item(s)	114	60.6	253	7.0	33.1	0.8	10.5	2.3	2.6	4.7
13402	Pancakes, frozen	3	item(s)	105	—	240	5.0	41.0	1.0	6.0	1.0	—	—
9221	Waffle plain, frozen, toasted	1	item(s)	33	10.1	103	2.4	16.3	0.8	3.2	0.5	1.6	0.7
30311	Waffle, 100% whole grain	1	item(s)	182	85.2	426	16.1	60.7	4.2	14.2	3.5	5.5	4.1
500	Waffle, plain, prepared from recipe	1	item(s)	75	31.5	218	5.9	24.7	1.7	10.6	2.1	2.6	5.1
Cereal, Flour, Grain, Pasta, Noodles, Popcorn													
	Grain												
2861	Amaranth, dry	½	cup(s)	98	11.0	362	13.2	63.6	6.5	6.8	1.4	1.6	2.7
1953	Barley, pearled, cooked	½	cup(s)	79	54.0	97	1.8	22.2	3.0	0.3	0.1	0	0.2

Chol (mg)	Calc (mg)	Iron (mg)	Magn (mg)	Pota (mg)	Sodi (mg)	Zinc (mg)	Vit A (µg)	Thia (mg)	Vit E (mg α)	Ribo (mg)	Niac (mg)	Vit B$_6$ (mg)	Fola (µg)	Vit C (mg)	Vit B$_{12}$ (µg)	Sele (µg)
4	26	0.36	13.2	64.8	205.2	0.46	2.4	0.53	—	0.04	0.39	0.03	13.2	0.1	0.1	5.0
0	0	1.80	8.0	67.5	190.0	0.61	150.1	0.15	—	0.17	2.00	0.20	—	0	0	—
10	0	0.72	—	—	210.0	—	0	—	—	—	—	—	—	0	—	—
20	56	1.04	7.9	68.6	287.8	0.29	0	0.17	0.85	0.16	1.26	0.03	62.5	0.2	0.2	9.0
15	42	1.60	18.2	39.3	118.6	0.31	29.6	0.16	0.46	0.19	1.16	0.05	63.8	0	0.1	8.7
0	30	1.42	12.0	74.7	264.5	0.40	0	0.25	—	0.16	2.21	0.02	57.0	0	0	—
0	99	2.33	14.0	64.5	238.5	0.70	0	0.27	0.18	0.14	2.49	0.03	90.0	0.9	0	13.1
0	101	1.64	21.1	106.0	201.2	0.61	0	0.25	0.26	0.17	1.91	0.05	46.2	0	0	16.6
0	36	2.39	89.5	289.0	224.0	1.05	0	0.15	0.38	0.05	0.24	0.09	79.2	0	0	6.3
0	300	0.36	—	—	105.0	—	0	—	—	—	—	—	—	0	—	—
0	0	0.36	—	—	100.0	—	0	—	—	—	—	—	—	0	—	—
0	0	1.08	—	95.0	160.0	—	0	—	—	—	—	—	—	0	—	—
0	15	0.72	23.8	82.3	72.0	0.50	0.5	0.06	0.51	0.03	0.39	0.02	5.6	0.2	0	4.0
0	30	0.73	21.0	92.3	79.0	0.43	0	0.08	—	0.05	0.15	0.03	6.8	0	0.1	4.6
0	11	1.74	10.9	122.4	327.0	0.29	17.0	0.23	—	0.17	1.91	0.05	58.9	2.6	0	12.1
106	39	1.81	31.7	162.3	319.6	1.24	100.0	0.23	—	0.22	1.03	0.07	41.5	0.2	0.2	14.6
0	12	0.43	3.0	12.4	145.8	0.08	0	0.06	—	0.04	0.47	0.01	22.2	0	0	4.4
0	3	0.45	2.7	18.2	74.7	0.08	0	0.05	0.09	0.02	0.55	0.01	16.3	0	0	1.1
65	146	1.97	29.5	288.3	348.8	0.71	660.3	0.14	—	0.31	1.21	0.07	43.4	2.6	0.1	11.0
0	48	0.96	6.7	35.6	140.0	0.24	0	0.18	0.02	0.06	1.33	0.02	47.9	0.4	0.1	6.4
0	74	1.47	10.3	54.6	215.0	0.37	0	0.28	0.03	0.09	2.05	0.03	73.5	0.6	0.1	9.8
0	54	1.87	15.4	61.6	310.1	0.54	0	0.27	0.24	0.19	2.42	0.02	86.1	0	0	22.3
0	54	1.87	15.4	61.6	310.1	0.54	0	0.27	0.24	0.19	2.42	0.02	86.1	0	0	22.3
0	30	0.69	24.1	77.1	135.5	0.57	0	0.07	0.26	0.04	1.04	0.06	8.5	0	0	14.0
0	150	2.70	—	260.0	160.0	2.25	—	0.22	—	0.25	3.00	0.30	102.0	60.0	0.9	10.5
0	150	2.70	—	150.0	170.0	2.25	—	0.22	—	0.25	3.00	0.30	102.0	60.0	0.9	10.5
0	250	4.50	100.0	340.0	150.0	3.00	—	0.38	—	0.25	3.00	0.40	—	60.0	0.9	14.0
0	250	4.50	100.0	270.0	230.0	3.00	—	0.38	—	0.25	3.00	0.40	—	60.0	0.9	14.0
0	350	5.40	40.0	160.0	190.0	5.25	—	0.15	—	0.68	8.00	1.50	—	12.0	4.5	24.5
0	250	4.50	—	105.0	200.0	—	0	0.22	—	0.17	—	0.50	—	42.0	—	—
0	150	2.70	—	105.0	200.0	—	0	0.22	—	0.17	—	0.50	—	42.0	—	—
0	150	2.70	—	260.0	200.0	—	0	0.22	—	0.17	—	0.50	—	42.0	—	—
0	20	0.72	—	105.0	200.0	—	0	—	—	—	—	—	—	—	—	—
0	19	0.30	17.3	44.6	10.8	0.31	0	0.02	0.07	0.02	0.36	0.05	1.2	0	0	1.5
0	32	1.00	6.0	45.9	205.8	0.16	0	0.15	0.06	0.04	1.09	0.01	60.3	0	0	7.2
0	13	0.24	11.4	31.0	32.3	0.21	0.1	0.03	0.09	0.01	0.25	0.03	11.3	0	0	0.6
64	235	1.96	18.2	157.3	469.7	0.62	57.0	0.22	—	0.31	1.74	0.06	60.4	2.5	0.2	16.0
35	40	1.80	—	—	500.0	—	0	0.30	—	0.17	2.00	—	—	0	—	—
5	101	2.28	7.9	47.5	240.9	0.17	131.3	0.17	0.32	0.23	2.93	0.34	39.3	0	1.0	4.2
4	460	3.15	69.2	424.1	910.0	1.60	3.6	0.38	1.04	0.60	3.20	0.17	80.1	0	0.6	41.3
52	191	1.73	14.3	119.3	383.3	0.51	48.8	0.20	—	0.26	1.55	0.04	51.0	0.3	0.2	34.7
0	155	7.42	241.8	495.3	3.9	2.80	0	0.11	1.16	0.19	0.90	0.58	79.9	4.1	0	18.2
0	9	1.04	17.3	73.0	2.4	0.64	0	0.07	0.01	0.05	1.62	0.09	12.6	0	0	6.8

DA+ Code	Food Description	Quantity	Measure	Wt (g)	H$_2$O (g)	Ener (kcal)	Prot (g)	Carb (g)	Fiber (g)	Fat (g)	Fat Breakdown (g)		
											Sat	Mono	Poly
Cereal, Flour, Grain, Pasta, Noodles, Popcorn—*continued*													
1956	Buckwheat groats, cooked, roasted	½	cup(s)	84	63.5	77	2.8	16.7	2.3	0.5	0.1	0.2	0.2
1957	Bulgur, cooked	½	cup(s)	91	70.8	76	2.8	16.9	4.1	0.2	0	0	0.1
1963	Couscous, cooked	½	cup(s)	79	57.0	88	3.0	18.2	1.1	0.1	0	0	0.1
1967	Millet, cooked	½	cup(s)	120	85.7	143	4.2	28.4	1.6	1.2	0.2	0.2	0.6
1969	Oat bran, dry	½	cup(s)	47	3.1	116	8.1	31.1	7.2	3.3	0.6	1.1	1.3
1972	Quinoa, dry	½	cup(s)	85	11.3	313	12.0	54.6	6.0	5.2	0.6	1.4	2.8
	Rice												
129	Brown, long grain, cooked	½	cup(s)	98	71.3	108	2.5	22.4	1.8	0.9	0.2	0.3	0.3
2863	Brown, medium grain, cooked	½	cup(s)	98	71.1	109	2.3	22.9	1.8	0.8	0.2	0.3	0.3
37488	Jasmine, saffroned, cooked	½	cup(s)	79	—	85	2.0	19.5	0	0	0	0	0
30280	Pilaf, cooked	½	cup(s)	103	73.4	135	2.1	22.2	0.6	4.1	0.6	1.6	1.7
28066	Spanish, cooked	½	cup(s)	113	85.2	114	2.9	23.5	1.5	1.0	0.2	0.3	0.3
2867	White glutinous, cooked	½	cup(s)	87	66.7	84	1.8	18.3	0.9	0.2	0	0.1	0.1
484	White, long grain, boiled	½	cup(s)	79	54.1	103	2.1	22.3	0.3	0.2	0.1	0.1	0.1
482	White, long grain, enriched, instant, boiled	½	cup(s)	83	59.4	97	1.8	20.7	0.5	0.4	0	0.1	0
486	White, long grain, enriched, parboiled, cooked	½	cup(s)	79	55.6	97	2.3	20.6	0.7	0.3	0.1	0.1	0.1
1994	Wild brown, cooked	½	cup(s)	82	60.6	83	3.3	17.5	1.5	0.3	0	0	0.2
	Flour & grain fractions												
505	All-purpose flour, self rising, enriched	½	cup(s)	63	6.6	221	6.2	46.4	1.7	0.6	0.1	0.1	0.3
503	All-purpose flour, white, bleached, enriched	½	cup(s)	63	7.4	228	6.5	47.7	1.7	0.6	0.1	0.1	0.3
53271	Barley flour	½	cup(s)	74	9.0	255	7.8	55.1	7.5	1.2	0.2	0.2	0.6
383	Buckwheat flour, whole groat	½	cup(s)	60	6.7	201	7.6	42.4	6.0	1.9	0.4	0.6	0.6
504	Cake wheat flour, enriched	½	cup(s)	69	8.6	248	5.6	53.5	1.2	0.6	0.1	0.1	0.3
426	Cornmeal, degermed, enriched	½	cup(s)	69	7.7	255	4.9	54.8	2.7	1.2	0.1	0.2	0.4
424	Cornmeal, yellow whole grain	½	cup(s)	61	6.3	221	5.0	46.9	4.5	2.2	0.3	0.6	1.0
1978	Dark rye flour	½	cup(s)	64	6.9	208	10.2	43.9	15.2	1.4	0.2	0.2	0.7
1644	Masa corn flour, enriched	½	cup(s)	57	5.2	208	5.3	43.5	3.6	2.2	0.3	0.6	1.0
1976	Rice flour, brown	½	cup(s)	79	9.5	287	5.7	60.4	3.6	2.2	0.4	0.8	0.8
1645	Rice flour, white	½	cup(s)	79	9.4	289	4.7	63.3	1.9	1.1	0.3	0.3	0.3
1980	Semolina, enriched	½	cup(s)	84	10.6	301	10.6	60.8	3.3	0.9	0.1	0.1	0.4
2827	Soy flour, raw	½	cup(s)	42	2.2	185	14.7	14.9	4.1	8.8	1.3	1.9	4.9
1990	Wheat germ, crude	2	tablespoon(s)	14	1.6	52	3.3	7.4	1.9	1.4	0.2	0.2	0.9
506	Whole-wheat flour	½	cup(s)	60	6.4	204	7.9	43.2	6.4	1.5	0.3	0.2	0.7
	Breakfast bars												
39230	Atkins Day Break apple crisp bar	1	item(s)	35	—	130	10.0	17.0	7.0	5.0	2.0	—	—
10571	Nutri-Grain apple cinnamon cereal bar	1	item(s)	37	—	130	2.0	24.0	2.0	3.0	0.5	—	—
10647	Nutri-Grain blueberry cereal bar	1	item(s)	37	—	130	2.0	24.0	2.0	3.0	0.5	—	—
10648	Nutri-Grain raspberry cereal bar	1	item(s)	37	—	130	2.0	24.0	2.0	3.0	0.5	—	—
10649	Nutri-Grain strawberry cereal bar	1	item(s)	37	—	130	2.0	24.0	2.0	3.0	0.5	—	—
	Breakfast cereals, hot												
41046	Cream of Wheat, instant, prepared	½	cup(s)	121	101.8	75	2.2	15.8	0.7	0.3	0	0	0.2
365	Farina, enriched, cooked with water and salt	½	cup(s)	117	100.8	62	2.1	12.7	0.9	0.4	0.1	0	0.1
363	Grits, white corn, regular and quick, enriched, cooked with water and salt	½	cup(s)	121	100.3	86	2.1	17.9	1.0	0.6	0.1	0.1	0.2
8636	Grits, yellow corn, regular and quick, enriched, cooked with salt	½	cup(s)	121	102.1	79	1.5	16.8	0.8	0.5	0.1	0.1	0.2
8657	Oatmeal, cooked with water	½	cup(s)	117	97.8	83	3.0	14.0	2.0	1.8	0.4	0.5	0.7
5500	Oatmeal, maple and brown sugar, instant, prepared	1	item(s)	198	147.3	208	4.8	41.9	3.7	2.5	0.4	0.7	0.9
5510	Oatmeal, ready to serve, packet, prepared	1	item(s)	186	152.2	143	4.8	25.2	3.8	2.5	0.4	0.8	0.9
	Breakfast cereals, ready to eat												
1197	All-Bran	1	cup(s)	62	1.3	160	8.0	46.0	20.0	2.0	0	0.4	—
1200	All-Bran Buds	1	cup(s)	91	2.7	242	9.1	72.7	39.4	3.0	0	0.5	1.1

Chol (mg)	Calc (mg)	Iron (mg)	Magn (mg)	Pota (mg)	Sodi (mg)	Zinc (mg)	Vit A (µg)	Thia (mg)	Vit E (mg α)	Ribo (mg)	Niac (mg)	Vit B$_6$ (mg)	Fola (µg)	Vit C (mg)	Vit B$_{12}$ (µg)	Sele (µg)
0	6	0.67	42.8	73.9	3.4	0.51	0	0.03	0.08	0.03	0.79	0.06	11.8	0	0	1.8
0	9	0.87	29.1	61.9	4.6	0.52	0	0.05	0.01	0.03	0.91	0.08	16.4	0	0	0.5
0	6	0.30	6.3	45.5	3.9	0.20	0	0.05	0.10	0.02	0.77	0.04	11.8	0	0	21.6
0	4	0.76	52.8	74.4	2.4	1.09	0	0.13	0.02	0.10	1.60	0.13	22.8	0	0	1.1
0	27	2.54	110.4	266.0	1.9	1.46	0	0.55	0.47	0.10	0.44	0.08	24.4	0	0	21.2
0	40	3.89	167.5	478.8	4.3	2.64	0.9	0.31	2.08	0.27	1.29	0.41	156.5	—	0	7.2
0	10	0.41	41.9	41.9	4.9	0.61	0	0.09	0.03	0.02	1.49	0.14	3.9	0	0	9.6
0	10	0.52	42.9	77.0	1.0	0.60	0	0.10	—	0.01	1.30	0.15	3.9	0	0	38.0
0	—	0.54	—	195.0	—	—	—	—	—	—	—	—	—	—	—	—
0	12	1.15	9.3	53.6	366.7	0.37	0	0.13	0.50	0.02	1.24	0.06	73.1	0.4	0	4.3
0	17	0.73	44.1	157.6	14.7	0.65	126.7	0.13	0.28	0.03	1.72	0.18	8.8	6.3	0	6.8
0	2	0.12	4.3	8.7	4.3	0.36	0	0.02	0.03	0.01	0.25	0.02	0.9	0	0	4.9
0	8	0.95	9.5	27.6	0.8	0.39	0	0.13	0.03	0.01	1.17	0.07	76.6	0	0	5.9
0	7	1.46	4.1	7.4	3.3	0.40	0	0.06	0.01	0.01	1.43	0.04	97.3	0	0	4.0
0	15	1.43	7.1	44.2	1.6	0.29	0	0.17	0.01	0.02	1.82	0.12	107.4	0	0	7.3
0	2	0.49	26.2	82.8	2.5	1.10	0	0.04	0.20	0.07	1.06	0.11	21.3	0	0	0.7
0	211	2.92	11.9	77.5	745.6	0.39	0	0.42	0.03	0.26	3.64	0.03	191.9	0	0	21.5
0	9	2.90	13.8	66.9	1.3	0.44	0	0.49	0.04	0.31	3.69	0.03	181.9	0	0	21.2
0	24	1.98	71.0	228.6	3.0	1.48	0	0.27	0.42	0.08	4.64	0.29	5.9	0	0	27.9
0	25	2.44	150.6	346.2	6.6	1.87	0	0.25	0.19	0.11	3.69	0.35	32.4	0	0	3.4
0	10	5.01	11.0	71.9	1.4	0.42	0	0.61	0.01	0.29	4.65	0.02	193.2	0	0	3.4
0	2	3.01	22.1	98.0	4.8	0.46	7.6	0.38	0.08	0.26	3.43	0.13	231.1	0	0	7.2
0	4	2.10	77.5	175.1	21.4	1.11	6.7	0.23	0.26	0.12	2.22	0.19	15.3	0	0	9.5
0	24	3.18	102.4	458.9	1.3	3.23	0.6	0.20	1.75	0.16	2.73	0.28	21.1	0	0	11.5
0	78	4.27	53.0	149.9	2.8	1.03	0	0.84	0.07	0.46	5.66	0.27	190.9	0	0	8.0
0	9	1.56	88.5	228.3	6.3	1.94	0	0.35	0.95	0.06	5.01	0.58	12.6	0	0	—
0	8	0.28	27.6	60.0	0.0	0.63	0	0.11	0.09	0.02	2.05	0.34	3.2	0	0	11.9
0	14	3.64	39.2	155.3	0.8	0.88	0	0.68	0.22	0.48	5.00	0.09	217.9	0	0	74.6
0	87	2.70	182.0	1066.9	5.5	1.66	2.5	0.25	0.83	0.49	1.83	0.20	146.3	0	0	3.2
0	6	0.90	34.4	128.3	1.7	1.77	0	0.27	—	0.07	0.98	0.19	40.4	0	0	11.4
0	20	2.16	82.2	217.8	1.2	1.56	0	0.30	0.43	0.10	2.97	0.24	26.4	0	0	37.1
0	200	1.44	—	35.0	140.0	—	—	0.15	—	0.17	2.00	—	68.0	12.0	—	—
0	200	1.80	—	—	105.0	1.50	225.0	0.22	—	0.43	5.00	0.50	—	0	—	—
0	200	1.80	—	—	105.0	1.50	225.0	0.22	—	0.43	5.00	0.50	—	0	—	—
0	200	1.80	—	—	100.0	1.50	225.0	0.22	—	0.43	5.00	0.50	—	0	—	—
0	200	1.80	—	—	120.0	1.50	225.0	0.22	—	0.43	5.00	0.50	—	0	—	—
0	77	5.99	7.2	24.1	123.1	0.21	280.0	0.28	0.02	0.25	3.73	0.37	121.9	0	0	4.2
0	113	6.21	8.2	26.8	146.8	0.27	0	0.15	0.05	0.08	1.74	0.11	138.6	0	0	3.5
0	1	0.69	8.5	32.7	269.8	0.22	0	0.10	0.04	0.07	0.97	0.06	46.0	0	0	3.8
0	1	0.69	6.1	26.6	269.8	0.17	0	0.12	0.02	0.07	0.92	0.04	44.8	0	0	3.3
0	11	1.05	31.6	81.9	4.7	1.17	0	0.09	0.09	0.02	0.26	0.01	7.0	0	0	6.3
0	99	7.40	48.1	136.7	272.2	1.09	0	1.12	—	0.05	2.55	0.33	56.5	0	0	11.1
0	109	4.79	48.9	138.5	242.7	1.11	0	0.78	—	0.05	1.78	0.22	40.2	0	0	3.8
0	200	9.00	200.0	700.0	160.0	3.00	300.5	0.75	0.77	0.85	10.00	4.00	1362.8	12.4	12.0	5.8
0	0	13.64	186.4	—	636.4	4.55	455.0	1.14	0.65	1.27	15.15	6.06	2052.7	18.2	18.2	26.3

APPENDIX H

DA+ Code	Food Description	Quantity	Measure	Wt (g)	H₂O (g)	Ener (kcal)	Prot (g)	Carb (g)	Fiber (g)	Fat (g)	Fat Breakdown (g) Sat	Mono	Poly
	Cereal, Flour, Grain, Pasta, Noodles, Popcorn—*continued*												
1199	Apple Jacks	1	cup(s)	28	0.8	100	1.0	25.0	3.0	0.5	0	0.1	0.2
1204	Cap'n Crunch	1	cup(s)	36	0.9	147	1.3	30.7	1.3	2.0	1.3	0	0
1205	Cap'n Crunch Crunchberries	1	cup(s)	35	0.9	133	1.3	29.3	1.3	2.0	1.3	0.2	0.3
1206	Cheerios	1	cup(s)	28	1.1	100	3.0	20.0	3.0	2.0	0	0.5	0.5
3415	Cocoa Puffs	1	cup(s)	36	0.6	147	1.3	30.7	1.3	2.0	0	0.7	0.7
1207	Cocoa Rice Krispies	1	cup(s)	41	1.2	160	1.3	36.0	0.7	1.3	0.7	0.2	0.1
5522	Complete wheat bran flakes	1	cup(s)	39	1.1	120	4.0	30.7	6.7	0.7	0	0	0
1211	Corn Flakes	1	cup(s)	28	1.1	100	2.0	24.0	1.0	0	0	0	0
1247	Corn Pops	1	cup(s)	32	1.0	120	1.0	29.0	3.0	0	0	0	0
1937	Cracklin' Oat Bran	1	cup(s)	65	1.8	267	5.3	46.7	8.0	9.3	4.0	3.1	2.0
58208	Fiber One Honey Clusters	1	cup(s)	52	—	160	5.0	42.0	13.0	1.5	0	0	0.5
1220	Froot Loops	1	cup(s)	29	0.9	110	1.0	25.0	3.0	1.0	0.5	0.1	0.2
38214	Frosted Cheerios	1	cup(s)	37	0.8	147	2.7	30.7	2.7	1.3	0	0	0.7
372	Frosted Flakes	1	cup(s)	40	1.3	147	1.3	36.0	1.3	0	0	0	0
1221	Frosted Mini-Wheats	1	cup(s)	47	2.6	158	4.7	37.9	4.7	0.8	0	0	0.4
1223	Granola, prepared	½	cup(s)	61	3.3	298	9.1	32.5	5.5	14.7	2.5	5.8	5.6
2415	Honey Bunches of Oats honey roasted	1	cup(s)	40	1.2	160	2.7	33.3	2.7	2.0	0	1.3	0
1227	Honey Nut Cheerios	1	cup(s)	37	0.8	147	2.7	29.3	2.7	2.0	0	0.7	0.7
1248	Honey Smacks	1	cup(s)	36	0.6	133	2.7	32.0	1.3	0.7	0	0.2	0.3
2424	Honeycomb	1	cup(s)	21	0.3	87	1.3	18.7	0.7	0.7	0	0	0
10286	Kashi Whole Grain Puffs	1	cup(s)	19	0.6	70	2.0	15.0	1.0	0.5	0	0.2	0.2
41142	Kellogg's Mueslix	1	cup(s)	82	7.2	294	7.5	60.2	6.8	4.5	0.5	2.1	1.1
1231	Kix	1	cup(s)	24	0.7	88	1.6	20.0	2.4	0.8	0	0	0
30569	Life	1	cup(s)	43	1.8	160	4.0	33.3	2.7	2.0	0	0.7	0.7
1233	Lucky Charms	1	cup(s)	36	1.1	147	2.7	29.3	1.3	1.3	0	0	0
38220	Multi Grain Cheerios	1	cup(s)	29	0.7	110	2.0	24.0	3.0	1.0	0	0	0.5
1201	Multi-Bran Chex	1	cup(s)	63	1.6	213	4.0	52.0	8.0	2.0	0	0	0.7
13633	Post Bran Flakes	1	cup(s)	40	1.1	133	4.0	32.0	6.7	0.7	0	0	0
1241	Product 19	1	cup(s)	30	0.9	100	2.0	25.0	1.0	0	0	0	0
32432	Puffed rice, fortified	1	cup(s)	14	0.4	56	0.9	12.6	0.2	0.1	0	—	—
32433	Puffed wheat, fortified	1	cup(s)	12	0.4	44	1.8	9.6	0.5	0.1	0	—	—
5584	Quaker Oatmeal Squares, brown sugar	1	cup(s)	55	1.1	210	6.0	44.0	5.0	2.5	0.5	1.0	1.0
41095	Quaker Oatmeal Squares, cinnamon	1	cup(s)	60	1.6	227	6.7	46.8	5.2	2.9	0.5	1.0	0.9
2420	Raisin Bran	1	cup(s)	59	4.8	190	4.0	46.0	8.0	1.0	0	0	0.5
1244	Rice Chex	1	cup(s)	27	0.7	100	2.0	23.0	0.0	0.5	0	0	0
1245	Rice Krispies	1	cup(s)	26	0.9	104	1.6	23.2	0.4	0	0	0	0
13648	Shredded Wheat	1	cup(s)	54	0.9	184	5.7	46.0	6.9	1.1	0	0	0.6
1246	Special K	1	cup(s)	31	0.9	120	6.0	23.0	0.5	0.5	0	0.1	0.2
41119	Total Raisin Bran	1	cup(s)	55	4.6	172	3.1	42.2	5.0	0.9	0.2	0.2	0.4
1253	Total whole grain	1	cup(s)	40	1.1	133	2.7	30.7	4.0	0.7	0	0	0
1254	Trix	1	cup(s)	32	0.6	120	1.0	28.0	1.0	1.5	0	0.5	0.5
41030	Wheat Chex	1	cup(s)	30	0.8	104	2.9	24.7	3.7	0.5	0.1	0.1	0.3
382	Wheat germ, toasted	2	tablespoon(s)	14	0.8	54	4.1	7.0	2.1	1.5	0.3	0.2	0.9
1257	Wheaties	1	cup(s)	36	0.9	133	4.0	29.3	4.0	0.7	0	0	0
	Pasta, noodles												
449	Chinese chow mein noodles, cooked	½	cup(s)	23	0.3	103	2.4	15.1	0.6	4.0	0.4	1.2	2.4
1995	Corn pasta, cooked	½	cup(s)	70	47.8	88	1.8	19.5	3.4	0.5	0.1	0.1	0.2
448	Egg noodles, enriched, cooked	½	cup(s)	80	54.2	110	3.6	20.1	1.0	1.7	0.3	0.5	0.4
1563	Egg noodles, spinach, enriched, cooked	½	cup(s)	80	54.8	106	4.0	19.4	1.8	1.3	0.3	0.4	0.3
440	Macaroni, enriched, cooked	½	cup(s)	70	43.5	111	4.1	21.6	1.3	0.7	0.1	0.1	0.2
2000	Macaroni, tricolor vegetable, enriched, cooked	½	cup(s)	67	45.8	86	3.0	17.8	2.9	0.1	0	0	0
1996	Plain pasta, fresh-refrigerated, cooked	½	cup(s)	64	43.9	84	3.3	16.0		0.7	0.1	0.1	0.3
1725	Ramen noodles, cooked	½	cup(s)	114	90.9	104	3.0	15.4	1.0	4.3	0.2	0.2	0.2
2878	Soba noodles, cooked	½	cup(s)	95	69.4	94	4.8	20.4	—	0.1	0	0	0
2879	Somen noodles, cooked	½	cup(s)	88	59.8	115	3.5	24.2	—	0.2	0	0	0.1
2881	Spaghetti, enriched, cooked	½	cup(s)	70	43.5	111	4.1	21.6	1.3	0.7	0.1	0.1	0.2
2884	Spaghetti, whole wheat, cooked	½	cup(s)	70	47.0	87	3.7	18.6	3.2	0.4	0.1	0.1	0.1

Chol (mg)	Calc (mg)	Iron (mg)	Magn (mg)	Pota (mg)	Sodi (mg)	Zinc (mg)	Vit A (µg)	Thia (mg)	Vit E (mg α)	Ribo (mg)	Niac (mg)	Vit B$_6$ (mg)	Fola (µg)	Vit C (mg)	Vit B$_{12}$ (µg)	Sele (µg)
0	0	4.50	5.9	35.0	130.0	1.50	146.8	0.38	0.04	0.43	5.00	0.50	184.5	15.0	1.5	1.6
0	0	6.00	19.4	66.7	266.7	5.00	0	0.51	0.26	0.57	6.68	0.67	906.7	0	0	2.3
0	0	6.00	18.7	66.7	253.4	5.00	0	0.51	0.22	0.57	6.68	0.67	906.8	0	0	0
0	100	8.10	40.0	170.0	190.0	3.75	150.1	0.38	0.19	0.43	5.00	0.50	460.3	6.0	1.5	8.0
0	133	6.00	10.7	80.0	200.0	5.00	23.8	0.50	0.28	0.57	6.67	0.67	199.1	8.0	2.0	2.3
0	53	6.00	15.7	80.0	200.0	1.98	203.3	0.50	0.11	0.57	6.61	0.67	446.4	20.0	2.0	6.7
0	0	24.00	53.3	226.7	280.0	20.00	300.5	2.00	17.92	2.27	26.67	2.67	901.4	80.0	8.0	4.1
0	0	8.12	2.5	25.0	200.0	0.05	127.3	0.38	0.04	0.43	5.00	0.50	221.8	6.0	1.5	2.3
0	0	1.80	2.2	45.0	125.0	1.57	143.1	0.38	0.02	0.43	5.00	0.50	174.7	6.0	1.5	2.1
0	27	2.40	80.0	293.3	200.0	2.03	300.5	0.50	0.37	0.57	6.66	0.65	218.2	20.0	2.0	7.0
0	100	4.50	32.0	320.0	280.0	3.75	0	0.38	—	0.43	5.00	0.50		0	1.5	—
0	0	4.50	6.7	35.0	135.0	1.50	144.7	0.38	0.03	0.43	5.00	0.50	150.5	15.0	1.5	1.7
0	133	6.00	19.9	73.3	226.6	5.00	200.1	0.50	0.16	0.57	6.67	0.67	448.3	8.0	2.0	5.8
0	0	6.00	2.8	26.7	186.7	0.06	200.2	0.50	0.04	0.57	6.67	0.67	254.0	8.0	2.0	1.8
0	0	12.80	47.4	158.1	4.0	1.19	0	0.30	0	0.34	3.95	0.40	149.2	0	1.2	1.9
0	48	2.58	106.8	329.4	15.3	2.46	0.6	0.45	6.78	0.18	1.31	0.18	50.0	0.7	0	17.0
0	0	14.40	21.3	73.3	200.0	0.40	300.3	0.50	0.59	0.57	6.67	0.67	549.6	0	2.0	3.3
0	133	6.00	42.7	153.3	253.3	5.00	200.1	0.50	0.36	0.57	6.67	0.67	448.3	8.0	2.0	8.8
0	0	0.48	21.2	53.3	66.7	0.47	200.2	0.50	0.17	0.58	6.66	0.67	224.6	8.0	2.0	17.5
0	0	1.80	10.7	33.3	120.0	1.00	150.1	0.25	0.04	0.28	3.33	0.33	122.2	0	1.0	4.3
0	0	0.36	29.3	60.0	0	0.58	0	0.02	0.17	0.03	0.61	0.06	6.1	0	0	6.8
0	48	6.76	73.4	256.5	207.8	5.61	135.2	0.66	5.94	0.66	8.25	3.05	1023.3	0.3	9.1	14.3
0	120	6.48	12.2	36.0	168.0	3.00	116.3	0.30	0.06	0.34	4.00	0.40	317.8	4.8	1.2	2.0
0	133	10.80	40.5	120.0	213.4	5.00	0	0.50	0.26	0.57	6.67	0.67	544.0	0	0	1.4
0	133	6.00	10.7	60.0	253.3	5.00	201.2	0.50	0.10	0.57	6.67	0.67	451.1	8.0	2.0	5.9
0	97	18.00	16.0	85.0	200.0	15.00	150.2	1.50	13.50	1.70	20.00	2.00	676.3	15.0	6.0	4.9
0	133	21.60	80.0	253.3	413.4	5.00	184.2	0.50	0.19	0.57	6.67	0.67	888.7	8.0	2.0	5.0
0	0	10.80	80.0	240.0	293.3	2.00	300.3	0.50	0.93	0.57	6.67	0.67	453.3	0	2.0	3.2
0	0	18.00	15.9	50.0	210.0	15.00	225.0	1.50	20.10	1.70	20.01	2.00	675.9	60.0	6.0	3.6
0	1	4.44	3.5	15.8	0.4	0.14	0	0.36	—	0.25	4.94	0.01	2.7	0	0	1.5
0	3	3.80	17.4	41.8	0.5	0.28	0	0.31	—	0.22	4.24	0.02	3.8	0	0	14.8
0	100	16.20	60.0	210.0	250.0	3.75	150.4	0.38	1.35	0.43	5.00	0.50	680.0	6.0		3.8
0	127	17.71	69.6	218.4	204.6	4.51	180.6	0.45	1.85	0.51	6.01	0.60	732.6	7.2	0	3.8
0	20	10.80	100.0	310.0	250.0	2.25	225.2	0.38	0.48	0.43	5.00	0.50	333.9	0	1.5	3.5
0	100	9.00	24.0	45.0	250.0	3.75	150.1	0.38	0.18	0.43	5.00	0.50	339.1	6.0	1.5	1.1
0	0	7.20	6.9	24.0	176.0	0.33	300.2	0.30	0.03	0.34	4.00	0.40	241.0	12.0	1.2	4.8
0	23	1.24	68.9	218.3	0	1.72	0	0.14	0	0.07	3.45	0.62	23.2	0	0	1.5
0	0	8.10	19.2	50.0	220.0	0.60	225.2	0.53	4.74	0.59	7.00	2.00	675.8	21.0	6.0	7.0
0	1038	18.70	33.0	277.2	186.4	15.56	154.6	1.54	14.01	1.76	20.74	2.08	698.5	0	6.2	3.8
0	1333	24.00	42.7	120.0	253.3	20.00	200.4	2.00	18.00	2.27	26.67	2.67	901.2	80.0	8.0	1.6
0	100	4.50	8.0	40.0	180.0	3.75	150.1	0.38	0.38	0.43	5.00	0.50	165.8	6.0	1.5	2.1
0	64	9.18	25.5	110.7	171.0	3.36	95.1	0.24	0.20	0.27	3.18	0.32	429.9	3.8	1.0	1.5
0	6	1.28	45.2	133.7	0.6	2.35	0.7	0.24	2.26	0.12	0.79	0.14	49.7	0.8	0	9.2
0	27	10.80	32.0	126.7	253.3	10.00	200.4	1.00	0.43	1.13	13.33	1.33	403.6	8.0	4.0	1.7
0	5	1.06	11.7	27.0	190.6	0.31	0	0.13	0.53	0.09	1.34	0.03	38.0	0	0	9.7
0	1	0.17	25.2	21.7	0	0.44	2.1	0.04	—	0.02	0.39	0.04	4.2	0	0	2.0
23	10	1.18	16.8	30.4	4.0	0.52	4.8	0.23	0.14	0.11	1.66	0.04	110.4	0	0.1	19.1
26	15	0.87	19.2	29.6	9.6	0.50	8.0	0.20	0.44	0.10	1.18	0.09	75.2	0	0.1	17.4
0	5	0.90	12.6	30.8	0.7	0.36	0	0.19	0.04	0.10	1.18	0.03	83.3	0	0	18.5
0	7	0.33	12.7	20.8	4.0	0.29	3.3	0.08	0.14	0.04	0.72	0.02	71.0	0	0	13.3
21	4	0.73	11.5	15.4	3.8	0.36	3.8	0.13	—	0.10	0.63	0.02	66.6	0	0.1	—
18	9	0.89	8.5	34.5	414.5	0.31	—	0.08	0.11	0.05	0.71	0.03	—	0.1	0	
0	4	0.46	8.5	33.3	57.0	0.11	0	0.09	—	0.02	0.48	0.04	6.7	0	0	
0	7	0.46	1.8	25.5	141.7	0.19	0	0.02	—	0.03	0.09	0.01	1.8	0	0	
0	5	0.90	12.6	30.8	0.7	0.36	0	0.19	0.04	0.10	1.18	0.03	83.3	0	0	18.5
0	11	0.74	21.0	30.8	2.1	0.57	0	0.08	0.21	0.03	0.49	0.06	3.5	0	0	18.1

APPENDIX H

DA+ Code	Food Description	Quantity	Measure	Wt (g)	H₂O (g)	Ener (kcal)	Prot (g)	Carb (g)	Fiber (g)	Fat (g)	Fat Breakdown (g)		
											Sat	Mono	Poly
Cereal, Flour, Grain, Pasta, Noodles, Popcorn—*continued*													
Popcorn													
476	Air popped	1	cup(s)	8	0.3	31	1.0	6.2	1.2	0.4	0	0.1	0.2
4619	Caramel	1	cup(s)	35	1.0	152	1.3	27.8	1.8	4.5	1.3	1.0	1.6
4620	Cheese flavored	1	cup(s)	35	0.9	185	3.3	18.2	3.5	11.7	2.3	3.4	5.4
477	Popped in oil	1	cup(s)	11	0.1	64	0.8	5.0	0.9	4.8	0.8	1.1	2.6
Fruit and Fruit Juices													
Apples													
226	Applesauce, sweetened, canned	½	cup(s)	123	100.8	84	0.2	21.5	1.5	0.2	0	0	0.1
227	Applesauce, unsweetened, canned	½	cup(s)	122	107.6	51	0.2	13.7	1.3	0.1	0	0	0
38492	Crabapples	1	item(s)	35	27.6	27	0.1	7.0	0.9	0.1	0	0	0
948	Dried, sulfured	¼	cup(s)	22	6.8	52	0.2	14.2	1.9	0.1	0	0	0
952	Juice, prepared from frozen concentrate	½	cup(s)	120	105.0	56	0.2	13.8	0.1	0.1	0	0	0
225	Juice, unsweetened, canned	½	cup(s)	124	109.4	57	0.1	14.0	0.2	0.2	0	0	0
223	Raw medium, with peel	1	item(s)	182	155.7	95	0.5	25.1	4.4	0.3	0.1	0	0.1
224	Slices	½	cup(s)	55	46.6	28	0.1	7.5	1.3	0.1	0	0	0
946	Slices without skin, boiled	½	cup(s)	86	73.1	45	0.2	11.7	2.1	0.3	0	0	0.1
Apricot													
228	Fresh without pits	4	item(s)	140	120.9	67	2.0	15.6	2.8	0.5	0	0.2	0.1
229	Halves with skin, canned in heavy syrup	½	cup(s)	129	100.1	107	0.7	27.7	2.1	0.1	0	0	0
230	Halves, dried, sulfured	¼	cup(s)	33	10.1	79	1.1	20.6	2.4	0.2	0	0	0
Avocado													
233	California, whole, without skin or pit	½	cup(s)	115	83.2	192	2.3	9.9	7.8	17.7	2.4	11.3	2.1
234	Florida, whole, without skin or pit	½	cup(s)	115	90.6	138	2.6	9.0	6.4	11.6	2.3	6.3	1.9
2998	Pureed	½	cup(s)	115	84.2	184	2.3	9.8	7.7	16.9	2.4	11.3	2.1
Banana													
4580	Dried chips	¼	cup(s)	18	0.8	93	0.4	10.5	1.4	6.0	5.2	0.4	0.1
235	Fresh whole, without peel	1	item(s)	118	88.4	105	1.3	27.0	3.1	0.4	0.1	0	0.1
Blackberries													
958	Frozen, unsweetened	½	cup(s)	76	62.1	48	0.9	11.8	3.8	0.3	0	0	0.2
237	Raw	½	cup(s)	72	63.5	31	1.0	6.9	3.8	0.4	0	0	0.2
Blueberries													
959	Canned in heavy syrup	½	cup(s)	128	98.3	113	0.8	28.2	2.0	0.4	0	0.1	0.2
960	Frozen, unsweetened	½	cup(s)	78	67.1	40	0.3	9.4	2.1	0.5	0	0.1	0.2
238	Raw	½	cup(s)	74	62.3	42	0.5	10.7	1.8	0.2	0	0	0.1
Boysenberries													
961	Canned in heavy syrup	½	cup(s)	128	97.6	113	1.3	28.6	3.3	0.2	0	0	0.1
962	Frozen, unsweetened	½	cup(s)	66	56.7	33	0.7	8.0	3.5	0.2	0	0	0.1
35576	Breadfruit	1	item(s)	384	271.3	396	4.1	104.1	18.8	0.9	0.2	0.1	0.3
Cherries													
967	Sour red, canned in water	½	cup(s)	122	109.7	44	0.9	10.9	1.3	0.1	0	0	0
3000	Sour red, raw	½	cup(s)	78	66.8	39	0.8	9.4	1.2	0.2	0.1	0.1	0.1
3004	Sweet, canned in heavy syrup	½	cup(s)	127	98.2	105	0.8	26.9	1.8	0.2	0	0.1	0.1
969	Sweet, canned in water	½	cup(s)	124	107.9	57	1.0	14.6	1.9	0.2	0	0	0
240	Sweet, raw	½	cup(s)	77	63.3	49	0.8	12.3	1.6	0.2	0	0	0
Cranberries													
3007	Chopped, raw	½	cup(s)	55	47.9	25	0.2	6.7	2.5	0.1	0	0	0
1717	Cranberry apple juice drink	½	cup(s)	123	102.6	77	0	19.4	0	0.1	0	0	0.1
1638	Cranberry juice cocktail	½	cup(s)	127	109.0	68	0	17.1	0	0.1	0	0	0.1
241	Cranberry juice cocktail, low calorie, with saccharin	½	cup(s)	119	112.8	23	0	5.5	0	0	0	0	0
242	Cranberry sauce, sweetened, canned	¼	cup(s)	69	42.0	105	0.1	26.9	0.7	0.1	0	0	0
Dates													
244	Domestic, chopped	¼	cup(s)	45	9.1	125	1.1	33.4	3.6	0.2	0	0	0
243	Domestic, whole	¼	cup(s)	45	9.1	125	1.1	33.4	3.6	0.2	0	0	0
Figs													
975	Canned in heavy syrup	½	cup(s)	130	98.8	114	0.5	29.7	2.8	0.1	0	0	0.1
974	Canned in water	½	cup(s)	124	105.7	66	0.5	17.3	2.7	0.1	0	0	0.1
973	Raw, medium	2	item(s)	100	79.1	74	0.8	19.2	2.9	0.3	0.1	0.1	0.1

Chol (mg)	Calc (mg)	Iron (mg)	Magn (mg)	Pota (mg)	Sodi (mg)	Zinc (mg)	Vit A (µg)	Thia (mg)	Vit E (mg α)	Ribo (mg)	Niac (mg)	Vit B$_6$ (mg)	Fola (µg)	Vit C (mg)	Vit B$_{12}$ (µg)	Sele (µg)
0	1	0.26	11.5	26.3	0.6	0.25	0.8	0.01	0.02	0.01	0.18	0.01	2.5	0	0	0
2	15	0.61	12.3	38.4	72.5	0.20	0.7	0.02	0.42	0.02	0.77	0.01	1.8	0	0	1.3
4	40	0.79	32.0	91.9	312.9	0.71	13.4	0.04	—	0.08	0.51	0.08	3.9	0.2	0.2	4.2
0	0	0.22	8.7	20.0	116.4	0.34	0.9	0.01	0.27	0.01	0.13	0.01	2.8	0	0	0.2
0	4	0.15	3.7	92.3	2.5	0.04	0	0.02	0.22	0.03	0.09	0.03	1.2	2.1	0	0.4
0	5	0.28	3.7	90.3	2.4	0.04	1.2	0.03	0.20	0.04	0.10	0.03	3.7	1.2	0	0.4
0	6	0.13	2.5	67.9	0.3	—	0.7	0.01	0.21	0.01	0.04	—	2.0	2.8	0	—
0	3	0.30	3.4	96.8	18.7	0.04	0	0	0.11	0.03	0.20	0.03	0	0.8	0	0.3
0	7	0.31	6.0	150.6	8.4	0.05	0	0	0.01	0.02	0.05	0.04	0	0.7	0	0.1
0	10	0.15	6.2	125.2	5.0	0.02	0	0.03	0.01	0.02	0.09	0.02	0	1.1	0	0.1
0	11	0.22	9.1	194.7	1.8	0.07	5.5	0.03	0.33	0.05	0.17	0.07	5.5	8.4	0	0
0	3	0.07	2.7	58.3	0.5	0.02	1.6	0.01	0.10	0.01	0.05	0.02	1.6	2.5	0	0
0	4	0.16	2.6	75.2	0.9	0.03	1.7	0.01	0.04	0.01	0.08	0.04	0.9	0.2	0	0.3
0	18	0.55	14.0	362.6	1.4	0.28	134.4	0.04	1.25	0.06	0.84	0.08	12.6	14.0	0	0.1
0	12	0.39	9.0	180.6	5.2	0.14	80.0	0.03	0.77	0.03	0.49	0.07	2.6	4.0	0	0.1
0	18	0.87	10.5	381.5	3.3	0.13	59.1	0	1.42	0.02	0.85	0.05	3.3	0.3	0	0.7
0	15	0.70	33.4	583.1	9.2	0.78	8.1	0.09	2.27	0.16	2.20	0.33	102.4	10.1	0	0.5
0	12	0.20	27.6	403.7	2.3	0.46	8.1	0.02	3.06	0.06	0.77	0.09	40.2	20.0	0	—
0	14	0.63	33.4	557.8	8.1	0.74	8.0	0.08	2.38	0.15	2.00	0.30	93.2	11.5	0	0.5
0	3	0.22	13.7	96.4	1.1	0.13	0.7	0.02	0.04	0	0.13	0.05	2.5	1.1	0	0.3
0	6	0.31	31.9	422.4	1.2	0.18	3.5	0.04	0.12	0.09	0.78	0.43	23.6	10.3	0	1.2
0	22	0.60	16.6	105.7	0.8	0.19	4.5	0.02	0.88	0.03	0.91	0.05	25.7	2.3	0	0.3
0	21	0.45	14.4	116.6	0.7	0.38	7.9	0.01	0.84	0.02	0.47	0.02	18.0	15.1	0	0.3
0	6	0.42	5.1	51.2	3.8	0.09	2.6	0.04	0.49	0.07	0.14	0.05	2.6	1.4	0	0.1
0	6	0.14	3.9	41.8	0.8	0.05	1.5	0.02	0.37	0.03	0.40	0.05	5.4	1.9	0	0.1
0	4	0.21	4.4	57.0	0.7	0.12	2.2	0.03	0.42	0.03	0.31	0.04	4.4	7.2	0	0.1
0	23	0.55	14.1	115.2	3.8	0.24	2.6	0.03	—	0.04	0.29	0.05	43.5	7.9	0	0.5
0	18	0.56	10.6	91.7	0.7	0.15	2.0	0.03	0.57	0.02	0.51	0.04	41.6	2.0	0	0.1
0	65	2.07	96.0	1881.6	7.7	0.46	7.7	0.42	0.38	0.12	3.46	0.38	53.8	111.4	0	2.3
0	13	1.67	7.3	119.6	8.5	0.09	46.4	0.02	0.28	0.05	0.22	0.05	9.8	2.6	0	0
0	12	0.25	7.0	134.1	2.3	0.08	49.6	0.02	0.05	0.03	0.31	0.03	6.2	7.8	0	0
0	11	0.44	11.4	183.4	3.8	0.13	10.1	0.03	0.22	0.05	0.50	0.03	5.1	4.6	0	0
0	14	0.45	11.2	162.4	1.2	0.10	9.9	0.03	0.29	0.05	0.51	0.04	5.0	2.7	0	0
0	10	0.28	8.5	170.9	0	0.05	2.3	0.02	0.05	0.03	0.12	0.04	3.1	5.4	0	0
0	4	0.14	3.3	46.8	1.1	0.05	1.6	0.01	0.66	0.01	0.06	0.03	0.6	7.3	0	0.1
0	4	0.09	1.2	20.8	2.5	0.02	0	0	0.15	0	0	0	0	48.4	0	0
0	4	0.13	1.3	17.7	2.5	0.04	0	0	0.28	0	0.05	0	0	53.5	0	0.3
0	11	0.05	2.4	29.6	3.6	0.02	0	0	0.06	0	0	0	0	38.2	0	0
0	3	0.15	2.1	18.0	20.1	0.03	1.4	0.01	0.57	0.01	0.07	0.01	0.7	1.4	0	0.2
0	17	0.45	19.1	291.9	0.9	0.13	0	0.02	0.02	0.03	0.57	0.07	8.5	0.2	0	1.3
0	17	0.45	19.1	291.9	0.9	0.13	0	0.02	0.02	0.03	0.57	0.07	8.5	0.2	0	1.3
0	35	0.36	12.9	128.2	1.3	0.14	2.6	0.03	0.16	0.05	0.55	0.09	2.6	1.3	0	0.3
0	35	0.36	12.4	127.7	1.2	0.15	2.5	0.03	0.10	0.05	0.55	0.09	2.5	1.2	0	0.1
0	35	0.37	17.0	232.0	1.0	0.15	7.0	0.06	0.11	0.05	0.40	0.11	6.0	2.0	0	0.2

APPENDIX H

Table of Food Composition H-15

DA+ Code	Food Description	Quantity	Measure	Wt (g)	H₂O (g)	Ener (kcal)	Prot (g)	Carb (g)	Fiber (g)	Fat (g)	Fat Breakdown (g) Sat	Mono	Poly
	Fruit and Fruit Juices—*continued*												
	Fruit cocktail & salad												
245	Fruit cocktail, canned in heavy syrup	½	cup(s)	124	99.7	91	0.5	23.4	1.2	0.1	0	0	0
978	Fruit cocktail, canned in juice	½	cup(s)	119	103.6	55	0.5	14.1	1.2	0	0	0	0
977	Fruit cocktail, canned in water	½	cup(s)	119	107.6	38	0.5	10.1	1.2	0.1	0	0	0
979	Fruit salad, canned in water	½	cup(s)	123	112.1	37	0.4	9.6	1.2	0.1	0	0	0
	Gooseberries												
982	Canned in light syrup	½	cup(s)	126	100.9	92	0.8	23.6	3.0	0.3	0	0	0.1
981	Raw	½	cup(s)	75	65.9	33	0.7	7.6	3.2	0.4	0	0	0.2
	Grapefruit												
251	Juice, pink, sweetened, canned	½	cup(s)	125	109.1	57	0.7	13.9	0.1	0.1	0	0	0
249	Juice, white	½	cup(s)	124	111.2	48	0.6	11.4	0.1	0.1	0	0	0
3022	Pink or red, raw	½	cup(s)	114	100.8	48	0.9	12.2	1.8	0.2	0	0	0
247	Raw, white	½	cup(s)	115	104.1	38	0.8	9.7	1.3	0.1	0	0	0
248	Sections, canned in light syrup	½	cup(s)	127	106.2	76	0.7	19.6	0.5	0.1	0	0	0
983	Sections, canned in water	½	cup(s)	122	109.6	44	0.7	11.2	0.5	0.1	0	0	0
	Grapes												
255	American, slip skin	½	cup(s)	46	37.4	31	0.3	7.9	0.4	0.2	0.1	0	0
256	European, red or green, adherent skin	½	cup(s)	76	60.8	52	0.5	13.7	0.7	0.1	0	0	0
3159	Juice drink, canned	½	cup(s)	125	106.6	71	0	18.2	0.1	0	0	0	0
259	Juice, sweetened, with added vitamin C, prepared from frozen concentrate	½	cup(s)	125	108.6	64	0.2	15.9	0.1	0.1	0	0	0
3060	Raisins, seeded, packed	¼	cup(s)	41	6.8	122	1.0	32.4	2.8	0.2	0.1	0	0.1
987	**Guava, raw**	1	item(s)	55	44.4	37	1.4	7.9	3.0	0.5	0.1	0	0.2
35593	**Guava, strawberry**	1	item(s)	6	4.8	4	0	1.0	0.3	0	0	0	0
3027	**Jackfruit**	½	cup(s)	83	60.6	78	1.4	19.2	1.2	0.5	0.2	0.1	0.1
990	**Kiwi fruit or Chinese gooseberries**	1	item(s)	76	63.1	46	0.9	11.1	2.3	0.4	0	0	0.2
	Lemon												
262	Juice	1	tablespoon(s)	15	14.1	3	0.1	1.1	0	0	0	0	0
993	Peel	1	teaspoon(s)	2	1.6	1	0	0.3	0.2	0	0	0	0
992	Raw	1	item(s)	108	94.4	22	1.3	11.6	5.1	0.3	0	0	0.1
	Lime												
269	Juice	1	tablespoon(s)	15	14.0	4	0.1	1.3	0.1	0	0	0	0
994	Raw	1	item(s)	67	59.1	20	0.5	7.1	1.9	0.1	0	0	0
995	**Loganberries, frozen**	½	cup(s)	74	62.2	40	1.1	9.6	3.9	0.2	0	0	0.1
	Mandarin orange												
1038	Canned in juice	½	cup(s)	125	111.4	46	0.8	11.9	0.9	0	0	0	0
1039	Canned in light syrup	½	cup(s)	126	104.7	77	0.6	20.4	0.9	0.1	0	0	0
999	**Mango**	½	cup(s)	83	68.9	49	0.7	12.4	1.3	0.3	0.1	0.1	0.1
1005	**Nectarine, raw, sliced**	½	cup(s)	69	60.4	30	0.7	7.3	1.2	0.2	0	0.1	0.1
	Melons												
271	Cantaloupe	½	cup(s)	80	72.1	27	0.7	6.5	0.7	0.2	0	0	0.1
1000	Casaba melon	½	cup(s)	85	78.1	24	0.9	5.6	0.8	0.1	0	0	0
272	Honeydew	½	cup(s)	89	79.5	32	0.5	8.0	0.7	0.1	0	0	0.1
318	Watermelon	½	cup(s)	76	69.5	23	0.5	5.7	0.3	0.1	0	0	0
	Orange												
14412	Juice with calcium and vitamin D	½	cup(s)	120	105.5	55	1.0	13.0	0	0	0	0	0
29630	Juice, fresh squeezed	½	cup(s)	124	109.5	56	0.9	12.9	0.2	0.2	0	0	0
14411	Juice, not from concentrate	½	cup(s)	120	105.5	55	1.0	13.0	0	0	0	0	0
278	Juice, unsweetened, prepared from frozen concentrate	½	cup(s)	125	109.7	56	0.8	13.4	0.2	0.1	0	0	0
3040	Peel	1	teaspoon(s)	2	1.5	2	0	0.5	0.2	0	0	0	0
273	Raw	1	item(s)	131	113.6	62	1.2	15.4	3.1	0.2	0	0	0
274	Sections	½	cup(s)	90	78.1	42	0.8	10.6	2.2	0.1	0	0	0
	Papaya												
16830	Dried, strips	2	item(s)	46	9.6	131	1.4	33.0	5.2	0.8	0.2	0.2	0.2
282	Raw	½	cup(s)	70	61.6	30	0.3	7.6	1.2	0.2	0.1	0.1	0
35640	**Passion fruit, purple**	1	item(s)	18	13.1	17	0.4	4.2	1.9	0.1	0	0	0.1
	Peach												
285	Halves, canned in heavy syrup	½	cup(s)	131	103.9	97	0.6	26.1	1.7	0.1	0	0	0.1

Chol (mg)	Calc (mg)	Iron (mg)	Magn (mg)	Pota (mg)	Sodi (mg)	Zinc (mg)	Vit A (µg)	Thia (mg)	Vit E (mg α)	Ribo (mg)	Niac (mg)	Vit B6 (mg)	Fola (µg)	Vit C (mg)	Vit B12 (µg)	Sele (µg)
0	7	0.36	6.2	109.1	7.4	0.10	12.4	0.02	0.50	0.02	0.46	0.06	3.7	2.4	0	0.6
0	9	0.25	8.3	112.6	4.7	0.11	17.8	0.01	0.47	0.02	0.48	0.06	3.6	3.2	0	0.6
0	6	0.30	8.3	111.4	4.7	0.11	15.4	0.02	0.47	0.01	0.43	0.06	3.6	2.5	0	0.6
0	9	0.37	6.1	95.6	3.7	0.10	27.0	0.02	—	0.03	0.46	0.04	3.7	2.3	0	1.0
0	20	0.42	7.6	97.0	2.5	0.14	8.8	0.03	—	0.07	0.19	0.02	3.8	12.6	0	0.5
0	19	0.23	7.5	148.5	0.8	0.09	11.3	0.03	0.28	0.02	0.22	0.06	4.5	20.8	0	0.4
0	10	0.45	12.5	202.2	2.5	0.07	0	0.05	0.05	0.03	0.40	0.02	12.5	33.6	0	0.1
0	11	0.25	14.8	200.1	1.2	0.06	1.2	0.05	0.27	0.02	0.25	0.05	12.4	46.9	0	0.1
0	25	0.09	10.3	154.5	0	0.08	66.4	0.05	0.15	0.04	0.23	0.06	14.9	35.7	0	0.1
0	14	0.07	10.3	170.2	0	0.08	2.3	0.04	0.15	0.02	0.31	0.05	11.5	38.3	0	1.6
0	18	0.51	12.7	163.8	2.5	0.10	0	0.05	0.11	0.03	0.31	0.03	11.4	27.1	0	1.1
0	18	0.50	12.2	161.0	2.4	0.11	0	0.05	0.11	0.03	0.30	0.02	11.0	26.6	0	1.1
0	6	0.13	2.3	87.9	0.9	0.02	2.3	0.04	0.09	0.03	0.14	0.05	1.8	1.8	0	0
0	8	0.27	5.3	144.2	1.5	0.05	2.3	0.05	0.14	0.05	0.14	0.06	1.5	2.4	0	0.1
0	9	0.16	7.5	41.3	11.3	0.04	0	0.28	0	0.44	0.18	0.04	1.3	33.1	0	0.1
0	5	0.13	5.0	26.3	2.5	0.05	0	0.02	0	0.03	0.16	0.05	1.3	29.9	0	0.1
0	12	1.07	12.4	340.3	11.6	0.07	0	0.05	—	0.08	0.46	0.08	1.2	2.2	0	0.2
0	10	0.14	12.1	229.4	1.1	0.13	17.0	0.04	0.40	0.02	0.60	0.06	27.0	125.6	0	0.3
0	1	0.01	1.0	17.5	2.2	—	0.3	0	0	0	0.04	0.00	—	2.2	0	
0	20	0.19	23.9	369.6	1.6	0.11	4.1	0.09	0.28	0.05	0.76	0.27	19.8	11.3	0	0.5
0	26	0.24	12.9	237.1	2.3	0.11	3.0	0.02	1.11	0.02	0.26	0.05	19.0	70.5	0	0.2
0	1	0.01	0.9	15.7	0.2	0.01	0	0	0.02	0	0.01	0.01	3.0	5.9	0	0
0	3	0.02	0.3	3.2	0.1	0	0.1	0	0	0	0.01	0.0	0.3	2.6	0	0
0	66	0.76	13.0	156.6	3.2	0.11	2.2	0.05	—	0.04	0.22	0.12	—	83.2	0	1.0
0	2	0.01	1.2	18.0	0.3	0.01	0.3	0	0.03	0	0.02	0.01	1.5	4.6	0	0
0	22	0.40	4.0	68.3	1.3	0.07	1.3	0.02	0.15	0.01	0.13	0.03	5.4	19.5	0	0.3
0	19	0.47	15.4	106.6	0.7	0.25	1.5	0.04	0.64	0.02	0.62	0.05	19.1	11.2	0	0.1
0	14	0.34	13.7	165.6	6.2	0.63	53.5	0.10	0.12	0.04	0.55	0.05	6.2	42.6	0	0.5
0	9	0.47	10.1	98.3	7.6	0.30	52.9	0.07	0.13	0.06	0.56	0.05	6.3	24.9	0	0.5
0	9	0.13	8.3	138.6	0.8	0.07	44.5	0.02	0.74	0.03	0.55	0.10	35.5	30.0	0	0.5
0	4	0.19	6.2	138.7	0	0.12	11.7	0.02	0.53	0.02	0.78	0.02	3.5	3.7	0	0
0	7	0.17	9.6	213.6	12.8	0.14	135.2	0.03	0.04	0.02	0.59	0.06	16.8	29.4	0	0.3
0	9	0.29	9.4	154.8	7.7	0.06	0.0	0.01	0.04	0.03	0.20	0.14	6.8	18.5	0	0.3
0	5	0.15	8.8	201.8	15.9	0.08	2.7	0.03	0.02	0.01	0.37	0.08	16.8	15.9	0	0.6
0	5	0.18	7.6	85.1	0.8	0.08	21.3	0.03	0.04	0.02	0.14	0.03	2.3	6.2	0	0.3
0	175	0	12.0	225.0	0	0.06	0	0.08	0.24	0.03	0.40	0.06	—	36.0	0	0.1
0	14	0.25	13.6	248.0	1.2	0.06	12.4	0.11	0.05	0.04	0.50	0.05	37.2	62.0	0	0.1
0	10	0	12.4	225.0	0	0.06	0	0.08	0.24	0.03	0.40	0.06	—	36.0	0	0.1
0	11	0.12	12.4	236.6	1.2	0.06	6.2	0.10	0.25	0.02	0.25	0.05	54.8	48.4	0	0.1
0	3	0.02	0.4	4.2	0.1	0	0.4	0	0	0	0.02	0.00	0.6	2.7	0	0
0	52	0.13	13.1	237.1	0	0.09	14.4	0.11	0.24	0.05	0.37	0.08	39.3	69.7	0	0.7
0	36	0.09	9.0	162.9	0	0.06	9.9	0.08	0.16	0.04	0.25	0.05	27.0	47.9	0	0.5
0	61	0.76	63.9	554.3	24.4	0.24	71.8	0.05	0.92	0.07	0.98	0.10	56.6	37.1	0	1.8
0	14	0.17	14.7	127.4	5.6	0.06	32.9	0.02	0.21	0.02	0.25	0.03	25.9	42.6	0	0.4
0	2	0.29	5.2	62.6	5.0	0.02	11.5	0	0	0.02	0.27	0.02	2.5	5.4	0	0.1
0	4	0.35	6.6	120.5	7.9	0.12	22.3	0.01	0.64	0.03	0.80	0.02	3.9	3.7	0	0.4

DA+ Code	Food Description	Quantity	Measure	Wt (g)	H₂O (g)	Ener (kcal)	Prot (g)	Carb (g)	Fiber (g)	Fat (g)	Fat Breakdown (g)		
											Sat	Mono	Poly
Fruit and Fruit Juices—*continued*													
286	Halves, canned in water	½	cup(s)	122	113.6	29	0.5	7.5	1.6	0.1	0	0	0
283	Raw, medium	1	item(s)	150	133.3	59	1.4	14.3	2.3	0.4	0	0.1	0.1
290	Slices, sweetened, frozen	½	cup(s)	125	93.4	118	0.8	30.0	2.3	0.2	0	0.1	0.1
	Pear												
8672	Asian	1	item(s)	122	107.7	51	0.6	13.0	4.4	0.3	0	0.1	0.1
80684	Bartlett	1	item(s)	177	148.9	112	0.7	26.6	5.5	0.3	—	—	—
294	Halves, canned in heavy syrup	½	cup(s)	133	106.9	98	0.3	25.5	2.1	0.2	0	0	0
1012	Halves, canned in juice	½	cup(s)	124	107.2	62	0.4	16.0	2.0	0.1	0	0	0
291	Raw	1	item(s)	166	139.4	95	0.6	25.3	5.1	0.2	0	0.1	0.2
1017	**Persimmon**	1	item(s)	25	16.1	32	0.2	8.4	—	0.1	0	0	0
	Pineapple												
3053	Canned in extra heavy syrup	½	cup(s)	130	101.0	108	0.4	28.0	1.0	0.1	0	0	0
1019	Canned in juice	½	cup(s)	125	104.0	75	0.5	19.5	1.0	0.1	0	0	0
296	Canned in light syrup	½	cup(s)	126	108.0	66	0.5	16.9	1.0	0.2	0	0	0.1
1018	Canned in water	½	cup(s)	123	111.7	39	0.5	10.2	1.0	0.1	0	0	0
299	Juice, unsweetened, canned	½	cup(s)	125	108.0	66	0.4	16.1	0.3	0.2	0	0	0.1
295	Raw, diced	½	cup(s)	78	66.7	39	0.4	10.2	1.1	0.1	0	0	0
1024	**Plantain, cooked**	½	cup(s)	77	51.8	89	0.6	24.0	1.8	0.1	0.1	0	0
300	**Plum, raw, large**	1	item(s)	66	57.6	30	0.5	7.5	0.9	0.2	0	0.1	0
1027	**Pomegranate**	1	item(s)	282	219.8	234	4.7	52.7	11.3	3.3	0.3	0.3	0.2
	Prunes												
5644	Dried	2	item(s)	17	5.2	40	0.4	10.7	1.2	0.1	0	0	0
305	Dried, stewed	½	cup(s)	124	86.5	133	1.2	34.8	3.8	0.2	0	0.1	0
306	Juice, canned	1	cup(s)	256	208.0	182	1.6	44.7	2.6	0.1	0	0.1	0
	Raspberries												
309	Raw	½	cup(s)	62	52.7	32	0.7	7.3	4.0	0.4	0	0	0.2
310	Red, sweetened, frozen	½	cup(s)	125	90.9	129	0.9	32.7	5.5	0.2	0	0	0.1
39793	**Rhubarb, frozen, cooked with sugar**	½	cup(s)	120	81.3	139	0.5	37.4	2.4	0.1	0	0	0
	Strawberries												
313	Raw	½	cup(s)	72	65.5	23	0.5	5.5	1.4	0.2	0	0	0.1
315	Sweetened, frozen, thawed	½	cup(s)	128	99.5	99	0.7	26.8	2.4	0.2	0	0	0.1
16828	**Tangelo**	1	item(s)	95	82.4	45	0.9	11.2	2.3	0.1	0	0	0
	Tangerine												
1040	Juice	½	cup(s)	124	109.8	53	0.6	12.5	0.2	0.2	0	0	0
316	Raw	1	item(s)	88	74.9	47	0.7	11.7	1.6	0.3	0	0.1	0.1
Vegetables, Legumes													
	Amaranth												
1043	Leaves, boiled, drained	½	cup(s)	66	60.4	14	1.4	2.7	—	0.1	0	0	0.1
1042	Leaves, raw	1	cup(s)	28	25.7	6	0.7	1.1	—	0.1	0	0	0
8683	**Arugula leaves, raw**	1	cup(s)	20	18.3	5	0.5	0.7	0.3	0.1	0	0	0.1
	Artichoke												
1044	Boiled, drained	1	item(s)	120	100.9	64	3.5	14.3	10.3	0.4	0.1	0	0.2
2885	Hearts, boiled, drained	½	cup(s)	84	70.6	45	2.4	10.0	7.2	0.3	0.1	0.0	0.1
	Asparagus												
566	Boiled, drained	½	cup(s)	90	83.4	20	2.2	3.7	1.8	0.2	0	0	0.1
568	Canned, drained	½	cup(s)	121	113.7	23	2.6	3.0	1.9	0.8	0.2	0	0.3
40162	Frozen, boiled, drained	½	cup(s)	90	84.7	16	2.7	1.7	1.4	0.4	0.1	0	0.2
	Bamboo shoots												
1048	Boiled, drained	½	cup(s)	60	57.6	7	0.9	1.2	0.6	0.1	0	0	0.1
1049	Canned, drained	½	cup(s)	66	61.8	12	1.1	2.1	0.9	0.3	0.1	0	0.1
	Beans												
1801	Adzuki beans, boiled	½	cup(s)	87	57.7	111	6.5	21.5	6.4	0.1	0	0	0
511	Baked beans with franks, canned	½	cup(s)	130	89.8	184	8.7	19.9	8.9	8.5	3.0	3.7	1.1
513	Baked beans with pork in sweet sauce, canned	½	cup(s)	127	89.5	140	6.6	26.7	5.4	1.7	0.4	0.4	0.7
512	Baked beans with pork in tomato sauce, canned	½	cup(s)	127	93.0	119	6.5	23.6	5.1	1.2	0.4	0.4	0.2
1805	Black beans, boiled	½	cup(s)	86	56.5	114	7.6	20.4	7.5	0.5	0.1	0	0.2
14597	Chickpeas, garbanzo beans or bengal gram, boiled	½	cup(s)	82	49.4	134	7.3	22.5	6.2	2.1	0.2	0.5	0.9
569	Fordhook lima beans, frozen, boiled, drained	½	cup(s)	85	62.1	88	5.2	16.4	4.9	0.3	0.1	0	0.1
1806	French beans, boiled	½	cup(s)	89	58.9	114	6.2	21.3	8.3	0.7	0.1	0	0.4
2773	Great northern beans, boiled	½	cup(s)	89	61.1	104	7.4	18.7	6.2	0.4	0.1	0	0.2

Chol (mg)	Calc (mg)	Iron (mg)	Magn (mg)	Pota (mg)	Sodi (mg)	Zinc (mg)	Vit A (µg)	Thia (mg)	Vit E (mg α)	Ribo (mg)	Niac (mg)	Vit B$_6$ (mg)	Fola (µg)	Vit C (mg)	Vit B$_{12}$ (µg)	Sele (µg)
0	2	0.39	6.1	120.8	3.7	0.11	32.9	0.01	0.60	0.02	0.64	0.02	3.7	3.5	0	0.4
0	9	0.38	13.5	285.0	0	0.25	24.0	0.04	1.10	0.05	1.21	0.04	6.0	9.9	0	0.2
0	4	0.46	6.3	162.5	7.5	0.06	17.5	0.02	0.77	0.04	0.82	0.02	3.8	117.8	0	0.5
0	5	0	9.8	147.6	0	0.02	0	0.01	0.15	0.01	0.27	0.03	9.8	4.6	0	0.1
—	16	0.34	10.6	178.8	1.8	0.14	1.8	0.02	0.21	0.05	0.29	0.05	—	7.8	—	0.2
0	7	0.29	5.3	86.4	6.7	0.11	0	0.01	0.11	0.03	0.32	0.02	1.3	1.5	0	0
0	11	0.36	8.7	119.0	5.0	0.11	0	0.01	0.10	0.01	0.25	0.02	1.2	2.0	0	0
0	15	0.30	11.6	192.6	1.7	0.17	1.7	0.02	0.20	0.04	0.27	0.05	11.6	7.1	0	0.2
0	7	0.63	—	77.5	0.3	—	—	—	—	—	—	—	—	16.5	0	—
0	18	0.49	19.5	132.6	1.3	0.14	1.3	0.12	—	0.03	0.37	0.10	6.5	9.5	0	—
0	17	0.35	17.4	151.9	1.2	0.12	2.5	0.12	0.01	0.02	0.35	0.09	6.2	11.8	0	0.5
0	18	0.49	20.2	132.3	1.3	0.15	2.5	0.11	0.01	0.03	0.37	0.09	6.3	9.4	0	0.5
0	18	0.49	22.1	156.2	1.2	0.15	2.5	0.11	0.01	0.03	0.37	0.09	6.2	9.5	0	0.5
0	16	0.39	15.0	162.5	2.5	0.14	0	0.07	0.03	0.03	0.25	0.13	22.5	12.5	0	0.1
0	10	0.22	9.3	84.5	0.8	0.09	2.3	0.06	0.02	0.02	0.39	0.09	13.9	37.0	0	0.1
0	2	0.45	24.6	358.0	3.8	0.10	34.7	0.04	0.10	0.04	0.58	0.18	20.0	8.4	0	1.1
0	4	0.11	4.6	103.6	0	0.07	11.2	0.02	0.17	0.02	0.28	0.02	3.3	6.3	0	0
0	28	0.85	33.8	665.5	8.5	0.99	0	0.19	1.69	0.15	0.83	0.21	107.2	28.8	0	1.4
0	7	0.16	6.9	123.0	0.3	0.07	6.6	0.01	0.07	0.03	0.32	0.03	0.7	0.1	0	0.1
0	24	0.51	22.3	398.0	1.2	0.24	21.1	0.03	0.24	0.12	0.90	0.27	0	3.6	0	0.1
0	31	3.02	35.8	706.6	10.2	0.54	0	0.04	0.31	0.18	2.01	0.56	0	10.5	0	1.5
0	15	0.42	13.5	92.9	0.6	0.26	1.2	0.02	0.54	0.02	0.37	0.03	12.9	16.1	0	0.1
0	19	0.81	16.3	142.5	1.3	0.22	3.8	0.02	0.90	0.06	0.29	0.04	32.5	20.6	0	0.4
0	174	0.25	14.4	115.2	1.2	0.10	4.8	0.02	0.23	0.03	0.24	0.02	6.0	4.0	0	1.1
0	12	0.30	9.4	110.2	0.7	0.10	0.7	0.02	0.21	0.02	0.28	0.03	17.3	42.3	0	0.3
0	14	0.60	7.7	124.9	1.3	0.06	1.3	0.02	0.31	0.10	0.37	0.04	5.1	50.4	0	0.9
0	38	0.09	9.5	171.9	0	0.07	10.4	0.08	0.17	0.04	0.27	0.06	28.5	50.5	0	0.5
0	22	0.25	9.9	219.8	1.2	0.04	16.1	0.07	0.16	0.02	0.12	0.05	6.2	38.3	0	0.1
0	33	0.13	10.6	146.1	1.8	0.06	29.9	0.05	0.18	0.03	0.33	0.07	14.1	23.5	0	0.1
0	138	1.49	36.3	423.1	13.9	0.58	91.7	0.01	—	0.09	0.37	0.12	37.6	27.1	0	0.6
0	60	0.65	15.4	171.1	5.6	0.25	40.9	0.01	—	0.04	0.18	0.05	23.8	12.1	0	0.3
0	32	0.29	9.4	73.8	5.4	0.09	23.8	0.01	0.09	0.02	0.06	0.01	19.4	3.0	0	0.1
0	25	0.73	50.4	343.2	72.0	0.48	1.2	0.06	0.23	0.11	1.33	0.10	106.8	8.9	0	0.2
0	18	0.51	35.3	240.2	50.4	0.34	0.8	0.04	0.16	0.07	0.93	0.07	74.8	6.2	0	0.2
0	21	0.82	12.6	201.6	12.6	0.54	45.0	0.15	1.35	0.13	0.98	0.07	134.1	6.9	0	5.5
0	19	2.21	12.1	208.1	347.3	0.48	49.6	0.07	1.48	0.12	1.15	0.13	116.2	22.3	0	2.1
0	16	0.50	9.0	154.8	2.7	0.37	36.0	0.06	1.08	0.09	0.93	0.02	121.5	22.0	0	3.5
0	7	0.14	1.8	319.8	2.4	0.28	0	0.01	—	0.03	0.18	0.06	1.2	0	0	0.2
0	5	0.21	2.6	52.4	4.6	0.43	0.7	0.02	0.41	0.02	0.09	0.09	2.0	0.7	0	0.3
0	24	1.74	45.2	462.8	7.0	1.54	0	0.10	—	0.06	0.62	0.08	105.3	0	0	1.0
8	62	2.24	36.3	304.3	556.8	2.42	5.2	0.08	0.21	0.07	1.17	0.06	38.8	3.0	0.4	8.4
1	72	2.09	40.5	322.6	426.3	1.02	1.3	0.06	0.06	0.07	0.44	0.07	10.1	3.4	0	6.3
9	71	4.10	43.0	373.2	552.8	6.93	5.1	0.07	0.13	0.06	0.62	0.08	19.0	3.8	0	5.9
0	23	1.81	60.2	305.3	0.9	0.96	0	0.21	—	0.05	0.43	0.06	128.1	0	0	1.0
0	40	2.37	39.4	238.6	5.7	1.25	0.8	0.10	0.29	0.05	0.43	0.11	141.0	1.1	0	3.0
0	26	1.55	35.7	258.6	58.7	0.63	8.5	0.06	0.25	0.05	0.91	0.10	17.9	10.9	0	0.5
0	56	0.96	49.6	327.5	5.3	0.57	0	0.12	—	0.05	0.48	0.09	66.4	1.1	0	1.1
0	60	1.89	44.3	346.0	1.8	0.78	0	0.14	—	0.05	0.60	0.10	90.3	1.2	0	3.6

DA+ Code	Food Description	Quantity	Measure	Wt (g)	H_2O (g)	Ener (kcal)	Prot (g)	Carb (g)	Fiber (g)	Fat (g)	Sat	Mono	Poly
Vegetables, Legumes—*continued*													
2736	Hyacinth beans, boiled, drained	½	cup(s)	44	37.8	22	1.3	4.0	—	0.1	0.1	0.1	0
570	Lima beans, baby, frozen, boiled, drained	½	cup(s)	90	65.1	95	6.0	17.5	5.4	0.3	0.1	0	0.1
515	Lima beans, boiled, drained	½	cup(s)	85	57.1	105	5.8	20.1	4.5	0.3	0.1	0	0.1
579	Mung beans, sprouted, boiled, drained	½	cup(s)	62	57.9	13	1.3	2.6	0.5	0.1	0	0	0
510	Navy beans, boiled	½	cup(s)	91	58.1	127	7.5	23.7	9.6	0.6	0.1	0.1	0.4
32816	Pinto beans, boiled, drained, no salt added	½	cup(s)	63	58.8	14	1.2	2.6	—	0.2	0	0	0.1
1052	Pinto beans, frozen, boiled, drained	½	cup(s)	47	27.3	76	4.4	14.5	4.0	0.2	0	0	0.1
514	Red kidney beans, canned	½	cup(s)	128	99.8	104	6.7	19.0	6.8	0.5	0.1	0.1	0.3
71879	Red kidney beans, canned, reduced sodium	½	cup(s)	130	—	105	7.0	22.0	8.0	0.0	0	—	—
1810	Refried beans, canned	½	cup(s)	119	90.6	108	6.4	18.2	6.1	1.4	0.5	0.5	0.4
1053	Shell beans, canned	½	cup(s)	123	111.1	37	2.2	7.6	4.2	0.2	0	0	0.1
1670	Soybeans, boiled	½	cup(s)	86	53.8	149	14.3	8.5	5.2	7.7	1.1	1.7	4.4
1108	Soybeans, green, boiled, drained	½	cup(s)	90	61.7	127	11.1	9.9	3.8	5.8	0.7	1.1	2.7
1807	White beans, small, boiled	½	cup(s)	90	56.6	127	8.0	23.1	9.3	0.6	0.1	0.1	0.2
575	Yellow snap, string or wax beans, boiled, drained	½	cup(s)	63	55.8	22	1.2	4.9	2.1	0.2	0	0	0.1
576	Yellow snap, string or wax beans, frozen, boiled, drained	½	cup(s)	68	61.7	19	1.0	4.4	2.0	0.1	0	0	0.1
	Beets												
584	Beet greens, boiled, drained	½	cup(s)	72	64.2	19	1.9	3.9	2.1	0.1	0	0	0.1
2730	Pickled, canned with liquid	½	cup(s)	114	92.9	74	0.9	18.5	3.0	0.1	0	0	0
581	Sliced, boiled, drained	½	cup(s)	85	74.0	37	1.4	8.5	1.7	0.2	0	0	0.1
583	Sliced, canned, drained	½	cup(s)	85	77.4	26	0.8	6.1	1.5	0.1	0	0	0
580	Whole, boiled, drained	2	item(s)	100	87.1	44	1.7	10.0	2.0	0.2	0	0	0.1
585	**Cowpeas or black-eyed peas, boiled, drained**	½	cup(s)	83	62.3	80	2.6	16.8	4.1	0.3	0.1	0	0.1
	Broccoli												
588	Chopped, boiled, drained	½	cup(s)	78	69.6	27	1.9	5.6	2.6	0.3	0.1	0	0.1
590	Frozen, chopped, boiled, drained	½	cup(s)	92	83.5	26	2.9	4.9	2.8	0.1	0	0	0.1
587	Raw, chopped	½	cup(s)	46	40.6	15	1.3	3.0	1.2	0.2	0	0	0
16848	**Broccoflower, raw, chopped**	½	cup(s)	32	28.7	10	0.9	1.9	1.0	0.1	0	0	0
	Brussels sprouts												
591	Boiled, drained	½	cup(s)	78	69.3	28	2.0	5.5	2.0	0.4	0.1	0	0.2
592	Frozen, boiled, drained	½	cup(s)	78	67.2	33	2.8	6.4	3.2	0.3	0.1	0	0.2
	Cabbage												
595	Boiled, drained, no salt added	1	cup(s)	150	138.9	35	1.9	8.3	2.8	0.1	0	0	0
35611	Chinese (pak choi or bok choy), boiled with salt, drained	1	cup(s)	170	162.4	20	2.7	3.0	1.7	0.3	0	0	0.1
16869	Kim chee	1	cup(s)	150	137.6	32	2.4	5.9	1.8	0.3	0	0	0.2
594	Raw, shredded	1	cup(s)	70	64.5	18	0.9	4.1	1.8	0.1	0	0	0
596	Red, shredded, raw	1	cup(s)	70	63.3	22	1.0	5.2	1.5	0.1	0	0	0.1
597	Savoy, shredded, raw	1	cup(s)	70	63.7	19	1.4	4.3	2.2	0.1	0	0	0.1
35417	**Capers**	1	teaspoon(s)	4	—	2	0	0	0	0	0	0	0.0
	Carrots												
8691	Baby, raw	8	item(s)	80	72.3	28	0.5	6.6	2.3	0.1	0	0	0.1
601	Grated	½	cup(s)	55	48.6	23	0.5	5.3	1.5	0.1	0	0	0.1
1055	Juice, canned	½	cup(s)	118	104.9	47	1.1	11.0	0.9	0.2	0	0	0.1
600	Raw	½	cup(s)	61	53.9	25	0.6	5.8	1.7	0.1	0	0	0.1
602	Sliced, boiled, drained	½	cup(s)	78	70.3	27	0.6	6.4	2.3	0.1	0	0	0.1
32725	**Cassava or manioc**	½	cup(s)	103	61.5	165	1.4	39.2	1.9	0.3	0.1	0.1	0
	Cauliflower												
606	Boiled, drained	½	cup(s)	62	57.7	14	1.1	2.5	1.4	0.3	0	0	0.1
607	Frozen, boiled, drained	½	cup(s)	90	84.6	17	1.4	3.4	2.4	0.2	0	0	0.1
605	Raw, chopped	½	cup(s)	50	46.0	13	1.0	2.5	1.0	0.1	0	0	0
	Celery												
609	Diced	½	cup(s)	51	48.2	8	0.3	1.5	0.8	0.1	0	0	0
608	Stalk	2	item(s)	80	76.3	13	0.6	2.4	1.3	0.1	0	0	0.1
	Chard												
1057	Swiss chard, boiled, drained	½	cup(s)	88	81.1	18	1.6	3.6	1.8	0.1	0	0	0
1056	Swiss chard, raw	1	cup(s)	36	33.4	7	0.6	1.3	0.6	0.1	0	0	0

Chol (mg)	Calc (mg)	Iron (mg)	Magn (mg)	Pota (mg)	Sodi (mg)	Zinc (mg)	Vit A (µg)	Thia (mg)	Vit E (mg α)	Ribo (mg)	Niac (mg)	Vit B$_6$ (mg)	Fola (µg)	Vit C (mg)	Vit B$_{12}$ (µg)	Sele (µg)
0	18	0.33	18.3	114.0	0.9	0.17	3.0	0.02	—	0.04	0.21	0.01	20.4	2.2	0	0.7
0	25	1.76	50.4	369.9	26.1	0.50	7.2	0.06	0.58	0.05	0.69	0.10	14.4	5.2	0	1.5
0	27	2.08	62.9	484.8	14.5	0.67	12.8	0.12	0.12	0.08	0.88	0.16	22.1	8.6	0	1.7
0	7	0.40	8.7	62.6	6.2	0.29	0.6	0.03	0.04	0.06	0.51	0.03	18.0	7.1	0	0.4
	63	2.15	48.2	354.0	0	0.94	0	0.22	0.01	0.06	0.59	0.13	127.4	0.8	0	2.6
0	9	0.42	11.3	61.7	32.1	0.11	0	0.04	—	0.04	0.46	0.03	18.3	3.8	0	0.4
0	24	1.27	25.4	303.6	39.0	0.32	0	0.13	—	0.05	0.30	0.09	16.0	0.3	0	0.7
0	37	1.60	38.4	332.8	327.7	0.79	0	0.14	0.03	0.09	0.63	0.10	33.3	1.0	0	1.4
0	80	1.80	—	400.0	130.0	—	0	—	—	—	—	—	—	0	0	—
0	39	1.99	45.2	399.8	534.3	0.77	0	0.04	0.06	0.02	0.50	0.13	13.1	7.1	0	7.7
0	36	1.21	18.4	133.5	409.1	0.33	13.5	0.04	0.04	0.07	0.25	0.06	22.0	3.8	0	2.6
0	88	4.42	74.0	442.9	0.9	0.99	0	0.13	0.30	0.25	0.34	0.20	46.4	1.5	0	6.3
0	131	2.25	54.0	485.1	12.6	0.82	7.2	0.23	—	0.14	1.13	0.05	99.9	15.3	0	1.3
0	65	2.54	60.9	414.4	1.8	0.98	0	0.21	—	0.05	0.24	0.11	122.6	0	0	1.2
0	29	0.80	15.6	186.9	1.9	0.22	2.5	0.05	0.28	0.06	0.38	0.04	20.6	6.1	0	0.3
0	33	0.59	16.2	85.1	6.1	0.32	4.1	0.02	0.03	0.06	0.26	0.04	15.5	2.8	0	0.3
0	82	1.37	49.0	654.5	173.5	0.36	275.8	0.08	1.30	0.21	0.36	0.10	10.1	17.9	0	0.6
0	12	0.47	17.0	168.0	299.6	0.30	2.3	0.01	0.07	0.05	0.28	0.06	30.6	2.6	0	1.1
0	14	0.67	19.6	259.4	65.5	0.30	1.7	0.02	0.03	0.03	0.28	0.06	68.0	3.1	0	0.6
0	13	1.55	14.5	125.9	165.0	0.18	0.9	0.01	0.03	0.03	0.13	0.05	25.5	3.5	0	0.4
0	16	0.79	23.0	305.0	77.0	0.35	2.0	0.03	0.04	0.04	0.33	0.07	80.0	3.6	0	0.7
0	106	0.92	42.9	344.9	3.3	0.85	33.0	0.08	0.18	0.12	1.16	0.05	104.8	1.8	0	2.1
0	31	0.52	16.4	228.5	32.0	0.35	60.1	0.05	1.13	0.10	0.43	0.16	84.2	50.6	0	1.2
0	30	0.56	12.0	130.6	10.1	0.26	46.9	0.05	1.21	0.07	0.42	0.12	51.5	36.9	0	0.6
0	21	0.33	9.6	143.8	15.0	0.19	14.1	0.03	0.35	0.05	0.29	0.08	28.7	40.6	0	1.1
0	11	0.23	6.4	96.0	7.4	0.20	2.6	0.03	0.01	0.03	0.23	0.07	18.2	28.2	0	0.2
0	28	0.94	15.6	247.3	16.4	0.26	30.4	0.08	0.34	0.06	0.47	0.14	46.8	48.4	0	1.2
0	20	0.37	13.9	224.8	11.6	0.19	35.7	0.08	0.40	0.09	0.42	0.22	78.3	35.4	0	0.5
0	72	0.25	22.5	294.0	12.0	0.30	6.0	0.09	0.21	0.06	0.37	0.17	45.0	56.3	0	0.9
0	158	1.77	18.7	630.7	459.0	0.29	360.4	0.05	0.15	0.11	0.73	0.28	69.7	44.2	0	0.7
0	137	1.23	27.0	367.5	936.0	0.36	274.5	0.07	0.41	0.10	0.78	0.33	85.5	68.1	0	1.5
0	28	0.33	8.4	119.0	12.6	0.13	3.5	0.04	0.10	0.03	0.16	0.09	30.1	25.6	0	0.2
0	32	0.56	11.2	170.1	18.9	0.15	39.2	0.04	0.08	0.05	0.29	0.15	12.6	39.9	0	0.4
0	25	0.28	19.6	161.0	19.6	0.19	35.0	0.05	0.12	0.02	0.21	0.13	56.0	21.7	0	0.6
0	0	0	—	140.0	—	—	0	—	—	—	—	—	—	0	—	—
0	26	0.71	8.0	189.6	62.4	0.14	552.0	0.02	—	0.03	0.44	0.08	21.6	2.1	0	0.7
0	18	0.17	6.6	176.0	37.9	0.13	459.3	0.04	0.36	0.03	0.54	0.08	10.5	3.2	0	0.1
0	28	0.54	16.5	344.6	77.9	0.21	1128.1	0.11	1.37	0.06	0.46	0.26	4.7	10.0	0	0.7
0	20	0.18	7.3	195.2	42.1	0.15	509.4	0.04	0.40	0.04	0.60	0.08	11.6	3.6	0	0.1
0	23	0.27	7.8	183.3	45.2	0.16	664.6	0.05	0.80	0.03	0.50	0.12	10.9	2.8	0	0.5
0	16	0.28	21.6	279.1	14.4	0.35	1.0	0.09	0.20	0.05	0.88	0.09	27.8	21.2	0	0.7
0	10	0.20	5.6	88.0	9.3	0.11	0.6	0.03	0.04	0.03	0.25	0.11	27.3	27.5	0	0.4
0	15	0.37	8.1	125.1	16.2	0.12	0	0.03	0.05	0.05	0.28	0.08	36.9	28.2	0	0.5
0	11	0.21	7.5	149.5	15.0	0.14	0	0.03	0.04	0.03	0.25	0.09	28.5	24.1	0	0.3
0	20	0.10	5.6	131.3	40.4	0.07	11.1	0.01	0.14	0.03	0.16	0.04	18.2	1.6	0	0.2
0	32	0.16	8.8	208.0	64.0	0.10	17.6	0.02	0.22	0.05	0.26	0.06	28.8	2.5	0	0.3
0	51	1.98	75.3	480.4	156.6	0.29	267.8	0.03	1.65	0.08	0.31	0.07	7.9	15.8	0	0.8
0	18	0.65	29.2	136.4	76.7	0.13	110.2	0.01	0.68	0.03	0.14	0.04	5.0	10.8	0	0.3

APPENDIX H

DA+ Code	Food Description	Quantity	Measure	Wt (g)	H₂O (g)	Ener (kcal)	Prot (g)	Carb (g)	Fiber (g)	Fat (g)	Fat Breakdown (g) Sat	Mono	Poly
Vegetables, Legumes—*continued*													
	Collard greens												
610	Boiled, drained	½	cup(s)	95	85.7	31	2.6	5.4	3.8	0.7	0	0	0.2
611	Frozen, chopped, boiled, drained	½	cup(s)	85	75.2	31	2.5	6.0	2.4	0.3	0.1	0	0.2
	Corn												
29614	Yellow corn, fresh, cooked	1	item(s)	100	73.0	95	3.4	20.9	2.4	1.5	0.2	0.4	0.6
615	Yellow creamed sweet corn, canned	½	cup(s)	128	100.8	92	2.2	23.2	1.5	0.5	0.1	0.2	0.3
612	Yellow sweet corn, boiled, drained	½	cup(s)	82	60.2	79	2.8	17.2	2.0	1.2	0.2	0.3	0.5
614	Yellow sweet corn, frozen, boiled, drained	½	cup(s)	82	63.2	66	2.1	15.8	2.0	0.5	0.1	0.2	0.3
	Cucumber												
703	Pickled, dill	¼	cup(s)	39	36.6	5	0.2	1.0	0.4	0.1	0	0	0
2755	Pickled, dill, low sodium	¼	cup(s)	39	35.5	7	0.2	1.6	0.5	0.1	0	0	0
618	Raw	¼	item(s)	75	71.7	11	0.5	2.7	0.4	0.1	0	0	0
	Dandelion greens												
620	Chopped, boiled, drained	½	cup(s)	53	47.1	17	1.0	3.4	1.5	0.3	0.1	0	0.1
2734	Raw	1	cup(s)	55	47.1	25	1.5	5.1	1.9	0.4	0.1	0	0.2
1066	**Eggplant, boiled, drained**	½	cup(s)	50	44.4	17	0.4	4.3	1.2	0.1	0	0	0
621	**Endive or escarole, chopped, raw**	1	cup(s)	50	46.9	9	0.6	1.7	1.6	0.1	0	0	0
8784	**Jicama or yambean**	½	cup(s)	67	60.1	25	0.5	5.9	3.3	0.1	0	0	0.0
	Kale												
623	Frozen, chopped, boiled, drained	½	cup(s)	65	58.8	20	1.8	3.4	1.3	0.3	0	0	0.2
29313	Raw	1	cup(s)	67	56.3	33	2.9	5.9	1.3	0.6	0.1	0	0.2
	Kohlrabi												
1072	Boiled, drained	½	cup(s)	83	74.5	24	1.5	5.5	0.9	0.1	0	0	0
1071	Raw	1	cup(s)	135	122.8	36	2.3	8.4	4.9	0.1	0	0	0.1
	Leeks												
1074	Boiled, drained	½	cup(s)	52	47.2	16	0.4	4.0	0.5	0.1	0	0	0.1
1073	Raw	1	cup(s)	89	73.9	54	1.3	12.6	1.6	0.3	0	0	0.1
	Lentils												
522	Boiled	¼	cup(s)	50	34.5	57	4.5	10.0	3.9	0.2	0	0	0.1
1075	Sprouted	1	cup(s)	77	51.9	82	6.9	17.0	—	0.4	0	0.1	0.2
	Lettuce												
625	Butterhead leaves	11	piece(s)	83	78.9	11	1.1	1.8	0.9	0.2	0	0	0.1
624	Butterhead, Boston or Bibb	1	cup(s)	55	52.6	7	0.7	1.2	0.6	0.1	0	0	0.1
626	Iceberg	1	cup(s)	72	68.9	10	0.6	2.1	0.9	0.1	0	0	0.1
628	Iceberg, chopped	1	cup(s)	55	52.6	8	0.5	1.6	0.7	0.1	0	0	0
629	Looseleaf	1	cup(s)	36	34.2	5	0.5	1.0	0.5	0.1	0	0	0
1665	Romaine, shredded	1	cup(s)	47	44.5	8	0.6	1.5	1.0	0.1	0	0	0.1
	Mushrooms												
15585	Crimini (about 6)	3	ounce(s)	85	—	28	3.7	2.8	1.9	0	0	0	0
8700	Enoki	30	item(s)	90	79.5	33	2.4	7.0	2.4	0.3	0	0	0.1
1079	Mushrooms, boiled, drained	½	cup(s)	78	71.0	22	1.7	4.1	1.7	0.4	0	0	0.1
1080	Mushrooms, canned, drained	½	cup(s)	78	71.0	20	1.5	4.0	1.9	0.2	0	0	0.1
630	Mushrooms, raw	½	cup(s)	48	44.4	11	1.5	1.6	0.5	0.2	0	0	0.1
35465	Portabella, raw	1	item(s)	84	78.0	18	1.8	3.3	1.1	0.3	0.1	0	0.1
2743	Shiitake, cooked	½	cup(s)	73	60.5	41	1.1	10.4	1.5	0.2	0	0.1	0
	Mustard greens												
2744	Frozen, boiled, drained	½	cup(s)	75	70.3	14	1.7	2.3	2.1	0.2	0	0.1	0
29319	Raw	1	cup(s)	56	50.8	15	1.6	2.6	1.8	0.2	0	0.1	0
	Okra												
16866	Batter coated, fried	11	piece(s)	83	50.3	159	4.0	20.1	2.3	7.1	1.4	2.8	2.4
32742	Frozen, boiled, drained, no salt added	½	cup(s)	92	83.8	27	1.5	5.9	1.9	0.2	0.1	0	0.1
632	Sliced, boiled, drained	½	cup(s)	80	74.1	18	1.5	3.6	2.0	0.2	0	0	0
	Onions												
635	Chopped, boiled, drained	½	cup(s)	105	92.3	46	1.4	10.7	1.5	0.2	0	0	0.1
2748	Frozen, boiled, drained	½	cup(s)	106	97.8	30	0.8	7.0	1.9	0.1	0	0	0
1081	Onion rings, breaded and pan fried, frozen, heated	10	piece(s)	71	20.2	289	3.8	27.1	0.9	19.0	6.1	7.7	3.6
633	Raw, chopped	½	cup(s)	80	71.3	32	0.9	7.5	1.4	0.1	0	0	0
16850	Red onions, sliced, raw	½	cup(s)	58	51.2	23	0.6	5.4	1.0	0.1	0	0	0
636	Scallions, green or spring onions	2	item(s)	30	26.9	10	0.5	2.2	0.8	0.1	0	0	0

Chol (mg)	Calc (mg)	Iron (mg)	Magn (mg)	Pota (mg)	Sodi (mg)	Zinc (mg)	Vit A (µg)	Thia (mg)	Vit E (mg α)	Ribo (mg)	Niac (mg)	Vit B6 (mg)	Fola (µg)	Vit C (mg)	Vit B12 (µg)	Sele (µg)
0	134	1.07	20.0	111.2	14.3	0.22	361.0	0.04	0.84	0.10	0.55	0.12	15.2	17.3	0	0.5
0	179	0.95	25.5	213.5	42.5	0.23	489.0	0.04	1.06	0.10	0.54	0.10	64.6	22.5	0	1.3
0	3	0.45	26.0	217.0	227.0	0.62	13.0	0.09	0.09	0.06	1.67	0.14	23.0	5.5	0	0.2
0	4	0.49	21.8	171.5	364.8	0.68	5.1	0.03	0.09	0.07	1.23	0.08	57.6	5.9	0	0.5
0	2	0.37	21.3	178.8	0.8	0.51	10.7	0.08	0.07	0.05	1.38	0.11	18.9	4.5	0	0.2
0	2	0.39	23.0	191.1	0.8	0.52	8.2	0.02	0.06	0.05	1.08	0.08	28.7	2.9	0	0.6
0	16	0.14	2.7	35.7	339.1	0.04	3.5	0.01	0.03	0.01	0.04	0.01	0.4	0.3	0	0
0	3	0.21	4.3	45.0	7.0	0.05	5.0	0.01	—	0.01	0.02	0.01	0.4	0.7	0	0
0	12	0.21	9.8	110.6	1.5	0.15	3.8	0.02	0.02	0.02	0.07	0.03	5.3	2.1	0	0.2
0	74	0.94	12.6	121.8	23.1	0.15	179.6	0.07	1.28	0.09	0.27	0.08	6.8	9.4	0	0.2
0	103	1.70	19.8	218.3	41.8	0.23	279.4	0.10	1.89	0.14	0.44	0.14	14.8	19.3	0	0.3
0	3	0.12	5.4	60.9	0.5	0.06	1.0	0.04	0.20	0.01	0.30	0.04	6.9	0.6	0	0
0	26	0.41	7.5	157.0	11.0	0.40	54.0	0.04	0.22	0.04	0.20	0.01	71.0	3.3	0	0.1
0	8	0.40	8.0	100.0	2.7	0.11	0.7	0.01	0.31	0.02	0.13	0.03	8.0	13.5	0	0.5
0	90	0.61	11.7	208.6	9.8	0.12	477.8	0.03	0.60	0.07	0.44	0.06	9.1	16.4	0	0.6
0	101	0.98	31.5	329.0	25.5	0.38	335.0	0.07	—	0.09	0.67	0.18	20.8	80.4	0	0.6
0	21	0.33	15.7	280.5	17.3	0.26	1.6	0.03	0.43	0.02	0.32	0.13	9.9	44.5	0	0.7
0	32	0.54	25.6	472.5	27.0	0.04	2.7	0.07	0.65	0.03	0.54	0.20	21.6	83.7	0	0.9
0	16	0.57	7.3	45.2	5.2	0.03	21.3	0.01	0.26	0.01	0.10	0.06	12.5	2.2	0	0.3
0	53	1.87	24.9	160.2	17.8	0.11	73.9	0.05	0.82	0.03	0.36	0.21	57.0	10.7	0	0.9
0	9	1.65	17.8	182.7	1.0	0.63	0	0.08	0.05	0.04	0.52	0.09	89.6	0.7	0	1.4
0	19	2.47	28.5	247.9	8.5	1.16	1.5	0.18	—	0.10	0.87	0.15	77.0	12.7	0	0.5
0	29	1.02	10.7	196.4	4.1	0.17	136.9	0.05	0.15	0.05	0.29	0.07	60.2	3.1	0	0.5
0	19	0.68	7.2	130.9	2.8	0.11	91.3	0.03	0.10	0.03	0.20	0.05	40.1	2.0	0	0.3
0	13	0.30	5.0	101.5	7.2	0.11	18.0	0.03	0.13	0.02	0.09	0.03	20.9	2.0	0	0.1
0	10	0.23	3.9	77.5	5.5	0.08	13.8	0.02	0.10	0.01	0.07	0.02	15.9	1.5	0	0.1
0	13	0.31	4.7	69.8	10.1	0.06	133.2	0.03	0.08	0.03	0.14	0.03	13.7	3.3	0	0.2
0	16	0.46	6.6	116.1	3.8	0.11	204.9	0.03	0.06	0.03	0.15	0.03	63.9	1.9	0	0.2
0	0	0.67	—	—	32.6	—	0	—	—	—	—	—	—	0	0	—
0	0	1.03	14.4	323.1	2.7	0.58	—	0.20	0.01	0.18	6.33	0.09	43.2	0	0	2.0
0	5	1.36	9.4	277.7	1.6	0.68	0	0.06	0.01	0.23	3.48	0.07	14.0	3.1	0	9.3
0	9	0.62	11.7	100.6	331.5	0.56	0	0.07	0.01	0.02	1.24	0.05	9.4	0	0	3.2
0	1	0.24	4.3	152.6	2.4	0.25	0	0.04	0	0.19	1.73	0.05	8.2	1.0	0	4.5
0	3	0.26	9.2	305.8	7.6	0.45	0	0.05	0.02	0.11	3.77	0.12	23.5	0	0	15.6
0	2	0.32	10.1	84.8	2.9	0.96	0	0.03	0	0.12	1.09	0.12	15.2	0.2	0	18.0
0	76	0.84	9.8	104.3	18.8	0.15	265.5	0.03	1.01	0.04	0.19	0.08	52.5	10.4	0	0.4
0	64	0.92	17.9	215.0	11.2	0.14	84.6	0.04	1.13	0.06	0.45	0.10	6.7	39.2	0	0.5
40	47	1.54	34.7	190.6	142.7	0.57	28.9	0.21	0.69	0.16	1.52	0.14	95.7	8.6	0.1	7.3
0	68	0.48	36.8	169.3	2.8	0.45	13.8	0.07	0.29	0.09	0.57	0.03	92.0	8.8	0	0.6
0	62	0.22	28.8	108.0	4.8	0.34	11.2	0.11	0.22	0.04	0.70	0.15	36.8	13.0	0	0.3
0	23	0.25	11.5	174.3	3.1	0.22	0	0.04	0.02	0.02	0.17	0.14	15.8	5.5	0	0.6
0	17	0.32	6.4	114.5	12.7	0.07	0	0.02	0.01	0.03	0.15	0.07	13.8	2.8	0	0.4
0	22	1.20	13.5	91.6	266.3	0.30	7.8	0.20	—	0.10	2.56	0.05	73.1	1.0	0	2.5
0	18	0.17	8.0	116.8	3.2	0.14	0	0.04	0.02	0.02	0.09	0.10	15.2	5.9	0	0.4
0	13	0.12	5.8	83.9	2.3	0.10	0	0.03	0.01	0.02	0.07	0.07	10.9	4.3	0	0.3
0	22	0.44	6.0	82.8	4.8	0.12	15.0	0.02	0.17	0.02	0.16	0.02	19.2	5.6	0	0.2

APPENDIX H

DA+ Code	Food Description	Quantity	Measure	Wt (g)	H$_2$O (g)	Ener (kcal)	Prot (g)	Carb (g)	Fiber (g)	Fat (g)	Fat Breakdown (g)		
											Sat	Mono	Poly
Vegetables, Legumes—*continued*													
16860	**Palm hearts, cooked**	½	cup(s)	73	50.7	84	2.0	18.7	1.1	0.1	0	0	0.1
637	**Parsley, chopped**	1	tablespoon(s)	4	3.3	1	0.1	0.2	0.1	0	0	0	0
638	**Parsnips, sliced, boiled, drained**	½	cup(s)	78	62.6	55	1.0	13.3	2.8	0.2	0	0.1	0
	Peas												
639	Green peas, canned, drained	½	cup(s)	88	72.0	60	3.9	9.9	4.3	0.7	0.1	0.1	0.3
641	Green peas, frozen, boiled, drained	½	cup(s)	80	63.6	62	4.1	11.4	4.4	0.2	0	0	0.1
35694	Pea pods, boiled with salt, drained	½	cup(s)	80	71.1	32	2.6	5.2	2.2	0.2	0	0	0.1
1082	Peas and carrots, canned with liquid	½	cup(s)	128	112.4	48	2.8	10.8	2.5	0.3	0.1	0	0.2
1083	Peas and carrots, frozen, boiled, drained	½	cup(s)	80	68.6	38	2.5	8.1	2.5	0.3	0.1	0	0.2
2750	Snow or sugar peas, frozen, boiled, drained	½	cup(s)	80	69.3	42	2.8	7.2	2.5	0.3	0.1	0	0.1
640	Snow or sugar peas, raw	½	cup(s)	32	28.0	13	0.9	2.4	0.8	0.1	0	0	0
29324	Split peas, sprouted	½	cup(s)	60	37.4	74	5.3	16.3	—	0.4	0.1	0	0.2
	Peppers												
644	Green bell or sweet, boiled, drained	½	cup(s)	68	62.5	19	0.6	4.6	0.8	0.1	0	0	0.1
643	Green bell or sweet, raw	½	cup(s)	75	69.9	15	0.6	3.5	1.3	0.1	0	0	0
1664	Green hot chili	1	item(s)	45	39.5	18	0.9	4.3	0.7	0.1	0	0	0
1663	Green hot chili, canned with liquid	½	cup(s)	68	62.9	14	0.6	3.5	0.9	0.1	0	0	0
1086	Jalapeno, canned with liquid	½	cup(s)	68	60.4	18	0.6	3.2	1.8	0.6	0.1	0	0.3
8703	Yellow bell or sweet	1	item(s)	186	171.2	50	1.9	11.8	1.7	0.4	0.1	0	0.2
1087	**Poi**	½	cup(s)	120	86.0	134	0.5	32.7	0.5	0.2	0	0	0.1
	Potatoes												
1090	Au gratin mix, prepared with water, whole milk and butter	½	cup(s)	122	96.3	113	2.8	15.7	1.1	5.0	3.2	1.4	0.2
1089	Au gratin, prepared with butter	½	cup(s)	123	90.7	162	6.2	13.8	2.2	9.3	5.8	2.6	0.3
5791	Baked, flesh and skin	1	item(s)	202	151.3	188	5.1	42.7	4.4	0.3	0.1	0	0.1
645	Baked, flesh only	½	cup(s)	61	46.0	57	1.2	13.1	0.9	0.1	0	0	0
1088	Baked, skin only	1	item(s)	58	27.4	115	2.5	26.7	4.6	0.1	0	0	0
5795	Boiled in skin, flesh only, drained	1	item(s)	136	104.7	118	2.5	27.4	2.1	0.1	0	0	0.1
5794	Boiled, drained, skin and flesh	1	item(s)	150	115.9	129	2.9	29.8	2.5	0.2	0	0	0.1
647	Boiled, flesh only	½	cup(s)	78	60.4	67	1.3	15.6	1.4	0.1	0	0	0
648	French fried, deep fried, prepared from raw	14	item(s)	113	47.9	335	4.6	39.5	5.0	18.3	2.4	7.3	7.9
649	French fried, frozen, heated	14	item(s)	70	43.7	115	1.9	19.4	2.0	3.7	0.7	2.3	0.2
1091	Hashed brown	½	cup(s)	78	36.9	207	2.3	27.4	2.5	9.8	1.5	4.1	3.7
652	Mashed with margarine and whole milk	½	cup(s)	105	79.0	119	2.1	17.8	1.6	4.4	1.0	2.0	1.2
653	Mashed, prepared from dehydrated granules with milk, water, and margarine	½	cup(s)	105	79.8	122	2.2	16.9	1.4	5.0	1.2	2.2	1.4
2759	Microwaved	1	item(s)	202	145.5	212	4.9	49.0	4.6	0.2	0.1	0	0.1
2760	Microwaved in skin, flesh only	½	cup(s)	78	57.1	78	1.6	18.1	1.2	0.1	0	0	0
5804	Microwaved, skin only	1	item(s)	58	36.8	77	2.5	17.2	4.2	0.1	0	0	0
1097	Potato puffs, frozen, heated	½	cup(s)	64	38.2	122	1.3	17.8	1.6	5.5	1.2	3.9	0.3
1094	Scalloped mix, prepared with water, whole milk and butter	½	cup(s)	124	98.6	116	2.6	15.9	1.4	5.4	3.3	1.5	0.2
1093	Scalloped, prepared with butter	½	cup(s)	123	99.2	108	3.5	13.2	2.3	4.5	2.8	1.3	0.2
	Pumpkin												
656	Canned	½	cup(s)	123	110.2	42	1.3	9.9	3.6	0.3	0.2	0	0
1773	Boiled, drained	½	cup(s)	123	114.8	25	0.9	6.0	1.3	0.1	0	0	0
	Radicchio												
8731	Leaves, raw	1	cup(s)	40	37.3	9	0.6	1.8	0.4	0.1	0	0	0
2498	Raw	1	cup(s)	40	37.3	9	0.6	1.8	0.4	0.1	0	0	0
657	**Radishes**	6	item(s)	27	25.7	4	0.2	0.9	0.4	0	0	0	0
1099	**Rutabaga, boiled, drained**	½	cup(s)	85	77.9	26	0.8	5.8	1.5	0.2	0	0	0.1
658	**Sauerkraut, canned**	½	cup(s)	118	109.2	22	1.1	5.1	3.4	0.2	0	0	0.1
	Seaweed												
1102	Kelp	½	cup(s)	40	32.6	17	0.7	3.8	0.5	0.2	0.1	0	0
1104	Spirulina, dried	½	cup(s)	56	2.6	162	32.2	13.4	2.0	4.3	1.5	0.4	1.2
1106	**Shallots**	3	tablespoon(s)	30	23.9	22	0.8	5.0	1.0	0	0	0	0

Chol (mg)	Calc (mg)	Iron (mg)	Magn (mg)	Pota (mg)	Sodi (mg)	Zinc (mg)	Vit A (µg)	Thia (mg)	Vit E (mg α)	Ribo (mg)	Niac (mg)	Vit B_6 (mg)	Fola (µg)	Vit C (mg)	Vit B_{12} (µg)	Sele (µg)
0	13	1.23	7.3	1318.4	10.2	2.72	2.2	0.03	0.37	0.13	0.62	0.53	14.6	5.0	0	0.5
0	5	0.24	1.9	21.1	2.1	0.04	16.0	0	0.03	0	0.05	0	5.8	5.1	0	0
0	29	0.45	22.6	286.3	7.8	0.20	0	0.06	0.78	0.04	0.56	0.07	45.2	10.1	0	1.3
0	20	1.03	15.8	92.8	238.9	0.58	37.6	0.09	0.33	0.05	0.86	0.05	27.1	3.7	0	1.5
0	19	1.22	17.6	88.0	57.6	0.54	84.0	0.23	0.02	0.08	1.18	0.09	47.2	7.9	0	0.8
0	34	1.58	20.8	192.0	192.0	0.30	41.6	0.10	0.31	0.06	0.43	0.12	23.2	38.3	0	0.6
0	29	0.96	17.9	127.5	331.5	0.74	368.5	0.09	—	0.07	0.74	0.11	23.0	8.4	0	1.1
0	18	0.75	12.8	126.4	54.4	0.36	380.8	0.18	0.42	0.05	0.92	0.07	20.8	6.5	0	0.9
0	47	1.92	22.4	173.6	4.0	0.39	52.8	0.05	0.38	0.10	0.45	0.14	28.0	17.6	0	0.6
0	14	0.66	7.6	63.0	1.3	0.09	17.0	0.05	0.12	0.03	0.19	0.05	13.2	18.9	0	0.2
0	22	1.36	33.6	228.6	12.0	0.63	4.8	0.14	—	0.09	1.85	0.16	86.4	6.2	0	0.4
0	6	0.31	6.8	112.9	1.4	0.08	15.6	0.04	0.34	0.02	0.32	0.16	10.9	50.6	0	0.2
0	7	0.25	7.4	130.4	2.2	0.10	13.4	0.04	0.28	0.02	0.36	0.17	7.4	59.9	0	0
0	8	0.54	11.3	153.0	3.2	0.14	26.5	0.04	0.31	0.04	0.43	0.13	10.4	109.1	0	0.2
0	5	0.34	9.5	127.2	797.6	0.12	24.5	0.01	0.47	0.03	0.54	0.10	6.8	46.2	0	0.2
0	16	1.28	10.2	131.2	1136.3	0.23	57.8	0.03	0.47	0.03	0.27	0.13	9.5	6.8	0	0.3
0	20	0.86	22.3	394.3	3.7	0.32	18.6	0.05	—	0.05	1.66	0.31	48.4	341.3	0	0.6
0	19	1.06	28.8	219.6	14.4	0.26	3.6	0.16	2.76	0.05	1.32	0.33	25.2	4.8	0	0.8
18	101	0.39	18.3	267.2	535.5	0.29	63.4	0.02	—	0.10	1.15	0.05	8.5	3.8	0	3.3
28	146	0.78	24.5	485.1	530.4	0.85	78.4	0.08	—	0.14	1.22	0.21	15.9	12.1	0	3.3
0	30	2.18	56.6	1080.7	20.2	0.73	2.0	0.13	0.08	0.10	2.85	0.63	56.6	19.4	0	0.8
0	3	0.21	15.3	238.5	3.0	0.18	0	0.06	0.02	0.01	0.85	0.18	5.5	7.8	0	0.2
0	20	4.08	24.9	332.3	12.2	0.28	0.6	0.07	0.02	0.06	1.78	0.36	12.8	7.8	0	0.4
0	7	0.42	29.9	515.4	5.4	0.41	0	0.14	0.01	0.03	1.96	0.41	13.6	17.7	0	0.4
0	13	1.27	34.1	572.0	7.4	0.47	0	0.15	0.01	0.03	2.13	0.44	15.0	18.4	0	—
0	6	0.24	15.6	255.8	3.9	0.21	0	0.08	0.01	0.01	1.02	0.21	7.0	5.8	0	0.2
0	27	1.78	52.2	952.6	276.7	0.66	0	0.15	2.25	0.07	2.27	0.63	27.2	35.6	0	0.7
0	8	0.52	18.2	315.7	271.6	0.27	0	0.09	0.08	0.02	1.55	0.13	19.6	9.3	0	0.1
0	11	0.43	27.3	449.3	266.8	0.37	0	0.13	0.01	0.03	1.80	0.37	12.5	10.1	0	0.4
1	22	0.27	20.0	342.3	349.6	0.31	46.2	0.10	0.44	0.04	1.23	0.26	9.5	11.0	0.1	0.8
2	37	0.21	21.0	162.7	180.6	0.25	53.5	0.09	0.54	0.09	0.90	0.17	8.4	6.8	0.1	5.9
0	22	2.50	54.5	902.9	16.2	0.73	0	0.24	—	0.06	3.46	0.69	24.2	30.5	0	0.8
0	4	0.32	19.4	319.0	5.4	0.26	0	0.10	—	0.02	1.26	0.25	9.3	11.7	0	0.3
0	27	3.45	21.5	377.0	9.3	0.30	0	0.04	0.01	0.04	1.29	0.29	9.9	8.9	0	0.3
0	9	0.41	10.9	199.7	295.7	0.20	0	0.08	0.15	0.02	0.97	0.08	9.0	4.0	0	0.4
14	45	0.47	17.4	252.7	424.5	0.31	43.6	0.02	—	0.07	1.28	0.05	12.4	4.1	0	2.0
15	70	0.70	23.3	463.0	410.4	0.49	—	0.08	—	0.11	1.29	0.22	15.9	13.0	0	2.0
0	32	1.70	28.2	252.4	6.1	0.21	953.0	0.03	1.30	0.07	0.45	0.07	14.7	5.1	0	0.5
0	18	0.70	11.0	281.8	1.2	0.28	352.8	0.04	0.98	0.10	0.51	0.05	11.0	5.8	0	0.2
0	8	0.23	5.2	120.8	8.8	0.25	0.4	0.01	0.90	0.01	0.10	0.02	24.0	3.2	0	0.4
0	8	0.23	5.2	120.8	8.8	0.25	0.4	0.01	0.90	0.01	0.10	0.02	24.0	3.2	0	0.4
0	7	0.09	2.7	62.9	10.5	0.08	0	0	0.00	0.01	0.07	0.02	6.8	4.0	0	0.2
0	15	0.15	8.5	183.7	4.3	0.10	0	0.07	0.20	0.03	0.61	0.09	12.8	16.0	0	0.6
0	35	1.73	15.3	200.6	780.0	0.22	1.2	0.02	0.17	0.03	0.17	0.15	28.3	17.3	0	0.7
0	67	1.14	48.4	35.6	93.2	0.49	2.4	0.02	0.35	0.06	0.19	0	72.0	1.2	0	0.3
0	67	15.96	109.2	763.3	586.9	1.12	16.2	1.33	2.80	2.06	7.18	0.20	52.6	5.7	0	4.0
0	11	0.36	6.3	100.2	3.6	0.12	0	0.02	0.01	0.01	0.06	0.10	10.2	2.4	0	0.4

APPENDIX H

DA+ Code	Food Description	Quantity	Measure	Wt (g)	H₂O (g)	Ener (kcal)	Prot (g)	Carb (g)	Fiber (g)	Fat (g)	Fat Breakdown (g)		
											Sat	Mono	Poly
	Vegetables, Legumes—*continued*												
	Soybeans												
1670	Boiled	½	cup(s)	86	53.8	149	14.3	8.5	5.2	7.7	1.1	1.7	4.4
2825	Dry roasted	½	cup(s)	86	0.7	388	34.0	28.1	7.0	18.6	2.7	4.1	10.5
2824	Roasted, salted	½	cup(s)	86	1.7	405	30.3	28.9	15.2	21.8	3.2	4.8	12.3
8739	Sprouted, stir fried	½	cup(s)	62	41.7	77	8.1	5.8	0.5	4.4	0.6	1.0	2.5
	Soy products												
1813	Soy milk	1	cup(s)	240	211.3	130	7.8	15.1	1.4	4.2	0.5	1.0	2.3
2838	Tofu, dried, frozen (koyadofu)	3	ounce(s)	85	4.9	408	40.8	12.4	6.1	25.8	3.7	5.7	14.6
13844	Tofu, extra firm	3	ounce(s)	85	68.8	86	8.6	2.2	1.1	4.3	0.5	1.1	2.7
13843	Tofu, firm	3	ounce(s)	85	70.9	75	7.5	2.2	0.5	3.2	0	1.1	2.2
1816	Tofu, firm, with calcium sulfate and magnesium chloride (nigari)	3	ounce(s)	85	72.2	60	7.0	1.4	0.8	3.5	0.7	1.0	1.5
1817	Tofu, fried	3	ounce(s)	85	43.0	230	14.6	8.9	3.3	17.2	2.5	3.8	9.7
13841	Tofu, silken	3	ounce(s)	85	77.0	42	3.7	0.9	0	1.9	0	0.5	0.9
13842	Tofu, soft	3	ounce(s)	85	73.0	65	6.5	1.1	0.5	3.2	0	0.5	2.2
1671	Tofu, soft, with calcium sulfate and magnesium chloride (nigari)	3	ounce(s)	85	74.2	52	5.6	1.5	0.2	3.1	0.5	0.7	1.8
	Spinach												
663	Canned, drained	½	cup(s)	107	98.2	25	3.0	3.6	2.6	0.5	0.1	0	0.2
660	Chopped, boiled, drained	½	cup(s)	90	82.1	21	2.7	3.4	2.2	0.2	0	0	0.1
661	Chopped, frozen, boiled, drained	½	cup(s)	95	84.5	32	3.8	4.6	3.5	0.8	0.1	0	0.4
662	Leaf, frozen, boiled, drained	½	cup(s)	95	84.5	32	3.8	4.6	3.5	0.8	0.1	0	0.4
659	Raw, chopped	1	cup(s)	30	27.4	7	0.9	1.1	0.7	0.1	0	0	0
	Squash												
1662	Acorn winter, baked	½	cup(s)	102	87.0	58	1.2	15.3	4.6	0.1	0	0	0.1
29702	Acorn winter, boiled, mashed	½	cup(s)	170	152.5	58	1.1	14.9	4.4	0.1	0	0	0.1
1661	Butternut winter, baked	½	cup(s)	103	90.0	41	0.9	10.8	3.3	0.1	0	0	0
32773	Butternut winter, frozen, boiled, mashed, no salt added	½	cup(s)	121	106.4	47	1.5	12.2	—	0.1	0	0	0
29451	Butternut, frozen, boiled	½	cup(s)	120	105.4	47	1.5	12.1	1.8	0.1	0	0	0
29700	Crookneck and straightneck summer, boiled, drained	½	cup(s)	65	61.0	12	0.7	2.5	0.6	0.2	0.1	0	0.1
29703	Hubbard winter, baked	½	cup(s)	102	86.8	51	2.5	11.0	5.0	0.6	0.1	0	0.3
1660	Hubbard winter, boiled, mashed	½	cup(s)	118	107.5	35	1.7	7.6	3.4	0.4	0.1	0	0.2
29704	Spaghetti winter, boiled, drained, or baked	½	cup(s)	78	71.5	21	0.5	5.0	1.1	0.2	0	0	0.1
664	Summer, all varieties, sliced, boiled, drained	½	cup(s)	90	84.3	18	0.8	3.9	1.3	0.3	0.1	0	0.1
665	Winter, all varieties, baked, mashed	½	cup(s)	103	91.4	38	0.9	9.1	2.9	0.4	0.1	0	0.2
1112	Zucchini summer, boiled, drained	½	cup(s)	90	85.7	14	1.0	2.4	0.9	0.3	0.1	0	0.1
1113	Zucchini summer, frozen, boiled, drained	½	cup(s)	112	105.6	19	1.3	4.0	1.4	0.1	0	0	0.1
	Sweet potatoes												
666	Baked, peeled	½	cup(s)	100	75.8	90	2.0	20.7	3.3	0.2	0	0	0.1
667	Boiled, mashed	½	cup(s)	164	131.4	125	2.2	29.1	4.1	0.2	0.1	0	0.1
668	Candied, home recipe	½	cup(s)	91	57.3	150	0.8	29.3	1.9	3.2	2.0	0.8	0.2
670	Canned, vacuum pack	½	cup(s)	100	76.0	91	1.6	21.1	1.8	0.2	0	0	0.1
2765	Frozen, baked	½	cup(s)	88	64.5	88	1.5	20.5	1.6	0.1	0	0	0
1136	Yams, baked or boiled, drained	½	cup(s)	68	47.7	79	1.0	18.7	2.7	0.1	0	0	0
32785	**Taro shoots, cooked, no salt added**	½	cup(s)	70	66.7	10	0.5	2.2	—	0.1	0	0	0
	Tomatillo												
8774	Raw	2	item(s)	68	62.3	22	0.7	4.0	1.3	0.7	0.1	0.1	0.3
8777	Raw, chopped	½	cup(s)	66	60.5	21	0.6	3.9	1.3	0.7	0.1	0.1	0.3
	Tomato												
16846	Cherry, fresh	5	item(s)	85	80.4	15	0.7	3.3	1.0	0.2	0	0	0.1
671	Fresh, ripe, red	1	item(s)	123	116.3	22	1.1	4.8	1.5	0.2	0	0	0.1
675	Juice, canned	½	cup(s)	122	114.1	21	0.9	5.2	0.5	0.1	0	0	0
75	Juice, no salt added	½	cup(s)	122	114.1	21	0.9	5.2	0.5	0.1	0	0	0
1699	Paste, canned	2	tablespoon(s)	33	24.1	27	1.4	6.2	1.3	0.2	0	0	0.1
1123	Paste, canned, no salt added	2	tablespoon(s)	32	23.5	26	1.4	6.1	1.3	0.2	0	0	0.1
1700	Puree, canned	¼	cup(s)	63	54.9	24	1.0	5.6	1.2	0.1	0	0	0.1
1124	Puree, canned, no salt added	¼	cup(s)	63	54.9	24	1.0	5.6	1.2	0.1	0	0	0.1
1118	Red, boiled	½	cup(s)	120	113.2	22	1.1	4.8	0.8	0.1	0	0	0

Chol (mg)	Calc (mg)	Iron (mg)	Magn (mg)	Pota (mg)	Sodi (mg)	Zinc (mg)	Vit A (µg)	Thia (mg)	Vit E (mg α)	Ribo (mg)	Niac (mg)	Vit B$_6$ (mg)	Fola (µg)	Vit C (mg)	Vit B$_{12}$ (µg)	Sele (µg)
0	88	4.42	74.0	442.9	0.9	0.99	0	0.13	0.30	0.25	0.34	0.20	46.4	1.5	0	6.3
0	120	3.40	196.1	1173.0	1.7	4.10	0	0.37	—	0.65	0.91	0.19	176.3	4.0	0	16.6
0	119	3.35	124.7	1264.2	140.2	2.70	8.6	0.09	0.78	0.12	1.21	0.18	181.5	1.9	0	16.4
0	51	0.25	59.5	351.5	8.7	1.30	0.6	0.26	—	0.12	0.68	0.10	78.7	7.4	0	0.4
0	60	1.54	60.0	283.2	122.4	0.29	0	0.14	0.26	0.17	1.23	0.18	43.2	0	0	11.5
0	310	8.27	50.2	17.0	5.1	4.17	22.1	0.42	—	0.27	1.01	0.24	78.2	0.6	0	46.2
0	65	1.55	—	—	0	—	0	—	—	—	—	—	—	0	—	—
0	108	1.16	—	—	0	—	0	—	—	—	—	—	—	0	—	—
0	171	1.37	31.5	125.9	10.2	0.71	0	0.05	0.01	0.05	0.09	0.06	16.2	0.2	0	8.4
0	316	4.14	51.0	124.2	13.6	1.69	0.9	0.14	0.03	0.04	0.09	0.08	23.0	0	0	24.2
0	56	0.67	—	—	0	—	0	—	—	—	—	—	—	0	—	—
0	108	1.16	—	—	0	—	0	—	—	—	—	—	—	0	—	—
0	94	0.94	23.0	102.0	6.8	0.54	0	0.04	0.01	0.03	0.45	0.04	37.4	0.2	0	7.6
0	136	2.46	81.3	370.2	344.5	0.49	524.3	0.02	2.08	0.15	0.42	0.11	104.9	15.3	0	1.5
0	122	3.21	78.3	419.4	63.0	0.68	471.6	0.09	1.87	0.21	0.44	0.22	131.4	8.8	0	1.4
0	145	1.86	77.9	286.9	92.2	0.47	572.8	0.07	3.36	0.17	0.42	0.13	114.9	2.1	0	5.2
0	145	1.86	77.9	286.9	92.2	0.47	572.8	0.07	3.36	0.17	0.42	0.13	114.9	2.1	0	5.2
0	30	0.81	23.7	167.4	23.7	0.16	140.7	0.02	0.61	0.06	0.22	0.06	58.2	8.4	0	0.3
0	46	0.98	45.2	459.0	4.2	0.18	22.1	0.18	—	0.01	0.93	0.20	20.0	11.4	0	0.7
0	44	0.95	44.2	447.1	5.1	0.19	69.7	0.17	—	0.01	0.90	0.20	18.7	11.1	0	0.7
0	42	0.62	29.7	291.1	4.1	0.13	572.0	0.07	1.32	0.02	0.99	0.13	19.5	15.5	0	0.5
0	23	0.70	10.9	161.2	2.4	0.15	202.4	0.06	—	0.05	0.56	0.08	19.4	4.2	0	0.6
0	23	0.70	10.8	159.6	2.4	0.14	200.4	0.06	0.14	0.05	0.56	0.08	19.2	4.2	0	0.6
0	14	0.28	12.9	143.6	1.3	0.19	5.2	0.03	0.08	0.03	0.29	0.07	12.3	12.5	0	0.1
0	17	0.48	22.4	365.1	8.2	0.15	341.7	0.08	0.20	0.05	0.57	0.18	16.3	9.7	0	0.6
0	12	0.33	15.3	252.5	5.9	0.12	236.0	0.05	0.14	0.03	0.39	0.12	11.8	7.7	0	0.4
0	16	0.26	8.5	90.7	13.9	0.16	4.7	0.03	0.09	0.02	0.63	0.08	6.2	2.7	0	0.2
0	24	0.32	21.6	172.8	0.9	0.35	9.9	0.04	0.13	0.04	0.46	0.06	18.0	4.9	0	0.2
0	23	0.45	13.3	247.0	1.0	0.23	267.5	0.02	0.12	0.07	0.51	0.17	20.5	9.8	0	0.4
0	16	0.33	17.1	237.6	2.7	0.30	50.4	0.03	0.11	0.02	0.46	0.07	25.2	11.6	0	0.2
0	19	0.54	14.5	216.3	2.2	0.22	10.0	0.05	0.13	0.04	0.43	0.05	8.9	4.1	0	0.2
0	38	0.69	27.0	475.0	36.0	0.32	961.0	0.11	0.71	0.11	1.49	0.29	6.0	19.6	0	0.2
0	44	1.18	29.5	377.2	44.3	0.33	1290.7	0.09	1.54	0.08	0.88	0.27	9.8	21.0	0	0.3
8	24	0.72	11.9	162.6	108.7	0.15	318.7	0.02	0.82	0.04	0.37	0.05	5.5	8.2	0	0.7
0	22	0.89	22.0	312.0	53.0	0.18	399.0	0.04	1.00	0.06	0.74	0.19	17.0	26.4	0	0.7
0	31	0.47	18.4	330.1	7.0	0.26	913.3	0.06	0.67	0.05	0.49	0.16	19.3	8.0	0	0.5
0	10	0.35	12.2	455.6	5.4	0.14	4.1	0.06	0.23	0.02	0.38	0.16	10.9	8.2	0	0.5
0	10	0.29	5.6	240.8	1.4	0.38	2.1	0.03	—	0.04	0.57	0.08	2.1	13.2	0	0.7
0	5	0.42	13.6	182.2	0.7	0.15	4.1	0.03	0.26	0.02	1.26	0.04	4.8	8.0	0	0.3
0	5	0.41	13.2	176.9	0.7	0.15	4.0	0.03	0.25	0.02	1.22	0.04	4.6	7.7	0	0.3
0	9	0.23	9.4	201.6	4.3	0.15	35.7	0.03	0.46	0.02	0.51	0.07	12.8	11.7	0	0
0	12	0.33	13.5	291.5	6.2	0.21	51.7	0.05	0.66	0.02	0.73	0.10	18.5	16.9	0	0
0	12	0.52	13.4	278.2	326.8	0.18	27.9	0.06	0.39	0.04	0.82	0.13	24.3	22.2	0	0.4
0	12	0.52	13.4	278.2	12.1	0.18	27.9	0.06	0.39	0.04	0.82	0.13	24.3	22.2	0	0.4
0	12	0.98	13.8	332.6	259.1	0.21	24.9	0.02	1.41	0.05	1.01	0.07	3.9	7.2	0	1.7
0	12	0.95	13.4	324.5	18.9	0.20	24.3	0.02	1.38	0.05	0.98	0.07	3.8	7.0	0	1.7
0	11	1.11	14.4	274.4	249.4	0.22	16.3	0.02	1.23	0.05	0.92	0.08	6.9	6.6	0	0.4
0	11	1.11	14.4	274.4	17.5	0.22	16.3	0.02	1.23	0.05	0.92	0.08	6.9	6.6	0	0.4
0	13	0.82	10.8	261.6	13.2	0.17	28.8	0.04	0.67	0.03	0.64	0.09	15.6	27.4	0	0.6

DA+ Code	Food Description	Quantity	Measure	Wt (g)	H₂O (g)	Ener (kcal)	Prot (g)	Carb (g)	Fiber (g)	Fat (g)	Fat Breakdown (g)		
											Sat	Mono	Poly
Vegetables, Legumes—*continued*													
3952	Red, diced	½	cup(s)	90	85.1	16	0.8	3.5	1.1	0.2	0	0	0.1
1120	Red, stewed, canned	½	cup(s)	128	116.7	33	1.2	7.9	1.3	0.2	0	0	0.1
1125	Sauce, canned	¼	cup(s)	61	55.6	15	0.8	3.3	0.9	0.1	0	0	0
8778	Sun dried	½	cup(s)	27	3.9	70	3.8	15.1	3.3	0.8	0.1	0.1	0.3
8783	Sun dried in oil, drained	¼	cup(s)	28	14.8	59	1.4	6.4	1.6	3.9	0.5	2.4	0.6
Turnips													
678	Turnip greens, chopped, boiled, drained	½	cup(s)	72	67.1	14	0.8	3.1	2.5	0.2	0	0	0.1
679	Turnip greens, frozen, chopped, boiled, drained	½	cup(s)	82	74.1	24	2.7	4.1	2.8	0.3	0.1	0	0.1
677	Turnips, cubed, boiled, drained	½	cup(s)	78	73.0	17	0.6	3.9	1.6	0.1	0	0	0
Vegetables, mixed													
1132	Mixed vegetables, canned, drained	½	cup(s)	82	70.9	40	2.1	7.5	2.4	0.2	0	0	0.1
680	Mixed vegetables, frozen, boiled, drained	½	cup(s)	91	75.7	59	2.6	11.9	4.0	0.1	0	0	0.1
7489	V8 100% vegetable juice	½	cup(s)	122	114.1	25	1.0	5.0	1.0	0	0	0	0
7490	V8 low-sodium vegetable juice	½	cup(s)	122	114.1	25	1.0	5.0	1.0	0	0	0	0
7491	V8 Spicy Hot vegetable juice	½	cup(s)	122	114.2	25	1.0	5.0	1.0	0	0	0	0
Water chestnuts													
31073	Sliced, drained	½	cup(s)	75	—	20	0	5.0	1.0	0	0	0	0
31087	Whole	½	cup(s)	75	—	20	0	5.0	1.0	0	0	0	0
1135	Watercress	1	cup(s)	34	32.3	4	0.8	0.4	0.2	0	0	0	0
Nuts, Seeds, and Products													
Almonds													
32940	Almond butter with salt added	1	tablespoon(s)	16	0.3	98	3.4	3.0	1.6	8.9	1.0	5.2	2.2
1137	Almond butter, no salt added	1	tablespoon(s)	16	0.3	98	3.4	3.0	1.6	8.9	0.7	5.2	2.2
32886	Blanched	¼	cup(s)	36	1.6	214	7.8	6.8	3.6	19.0	1.4	12.1	4.5
32887	Dry roasted, no salt added	¼	cup(s)	35	0.9	205	7.3	7.3	3.8	18.0	1.4	11.2	4.5
29724	Dry roasted, salted	¼	cup(s)	35	0.9	205	7.3	7.3	3.8	18.0	1.4	11.2	4.5
29725	Oil roasted, salted	¼	cup(s)	39	1.1	238	8.3	6.9	4.1	21.7	1.7	13.7	5.3
508	Slivered	¼	cup(s)	27	1.3	155	5.7	5.9	3.3	13.3	1.0	8.3	3.3
1138	**Beechnuts, dried**	¼	cup(s)	57	3.8	328	3.5	19.1	5.3	28.5	3.3	12.5	11.5
517	**Brazil nuts, dried, unblanched**	¼	cup(s)	33	1.2	218	4.8	4.1	2.5	22.1	5.0	8.2	6.8
1166	**Breadfruit seeds, roasted**	¼	cup(s)	57	28.3	118	3.5	22.9	3.4	1.5	0.4	0.2	0.8
1139	**Butternuts, dried**	¼	cup(s)	30	1.0	184	7.5	3.6	1.4	17.1	0.4	3.1	12.8
Cashews													
32931	Cashew butter with salt added	1	tablespoon(s)	16	0.5	94	2.8	4.4	0.3	7.9	1.6	4.7	1.3
32889	Cashew butter, no salt added	1	tablespoon(s)	16	0.5	94	2.8	4.4	0.3	7.9	1.6	4.7	1.3
1140	Dry roasted	¼	cup(s)	34	0.6	197	5.2	11.2	1.0	15.9	3.1	9.4	2.7
518	Oil roasted	¼	cup(s)	32	1.1	187	5.4	9.6	1.1	15.4	2.7	8.4	2.8
Coconut													
32896	Dried, not sweetened	¼	cup(s)	23	0.7	150	1.6	5.4	3.7	14.6	13.0	0.6	0.2
1153	Dried, shredded, sweetened	¼	cup(s)	23	2.9	116	0.7	11.1	1.0	8.3	7.3	0.4	0.1
520	Shredded	¼	cup(s)	20	9.4	71	0.7	3.0	1.8	6.7	5.9	0.3	0.1
Chestnuts													
1152	Chinese, roasted	¼	cup(s)	71	28.7	171	3.2	37.4	—	0.9	0.1	0.4	0.2
32895	European, boiled and steamed	¼	cup(s)	45	30.9	59	0.9	12.6		0.6	0.1	0.2	0.2
32911	European, roasted	¼	cup(s)	36	14.5	88	1.1	18.9	1.8	0.8	0.1	0.3	0.3
32922	Japanese, boiled and steamed	¼	cup(s)	35	30.5	20	0.3	4.5	—	0.1	0	0	0
32923	Japanese, roasted	¼	cup(s)	36	17.8	72	1.1	16.1	—	0.3	0	0.1	0.1
4958	**Flax seeds or linseeds**	¼	cup(s)	28	2.2	150	5.5	8.1	7.8	11.6	1.0	2.1	8.0
32904	**Ginkgo nuts, dried**	¼	cup(s)	39	4.8	134	4.0	27.9	—	0.8	0.1	0.3	0.3
Hazelnuts or filberts													
32901	Blanched	¼	cup(s)	45	2.6	285	6.2	7.7	5.0	27.7	2.1	21.9	2.5
32902	Dry roasted, no salt added	¼	cup(s)	29	0.7	185	4.3	5.0	2.7	17.8	1.3	13.3	2.4
1156	**Hickorynuts, dried**	¼	cup(s)	30	0.8	197	3.8	5.5	1.9	19.3	2.1	9.8	6.6
Macadamias													
32905	Dry roasted, no salt added	¼	cup(s)	34	0.5	241	2.6	4.5	2.7	25.5	4.0	19.9	0.5
32932	Dry roasted, with salt added	¼	cup(s)	34	0.5	240	2.6	4.3	2.7	25.5	4.0	19.9	0.5
1157	Raw	¼	cup(s)	34	0.5	241	2.6	4.6	2.9	25.4	4.0	19.7	0.5
Mixed nuts													
1159	With peanuts, dry roasted	¼	cup(s)	34	0.6	203	5.9	8.7	3.1	17.6	2.4	10.8	3.7
32933	With peanuts, dry roasted, with salt added	¼	cup(s)	34	0.6	203	5.9	8.7	3.1	17.6	2.2	10.8	3.7

APPENDIX H

Chol (mg)	Calc (mg)	Iron (mg)	Magn (mg)	Pota (mg)	Sodi (mg)	Zinc (mg)	Vit A (µg)	Thia (mg)	Vit E (mg α)	Ribo (mg)	Niac (mg)	Vit B$_6$ (mg)	Fola (µg)	Vit C (mg)	Vit B$_{12}$ (µg)	Sele (µg)
0	9	0.24	9.9	213.3	4.5	0.15	37.8	0.03	0.49	0.02	0.53	0.07	13.5	12.3	0	0
0	43	1.70	15.3	263.9	281.8	0.22	11.5	0.06	1.06	0.04	0.91	0.02	6.4	10.1	0	0.8
0	8	0.62	9.8	201.9	319.6	0.12	13.4	0.01	0.87	0.04	0.59	0.06	6.7	4.3	0	0.1
0	30	2.45	52.4	925.3	66.7	0.54	11.9	0.14	0	0.13	2.44	0.09	18.4	10.6	0	1.5
0	13	0.74	22.3	430.4	73.2	0.21	17.6	0.05	—	0.11	1.00	0.09	6.3	28.0	0	0.8
0	99	0.58	15.8	146.2	20.9	0.10	274.3	0.03	1.35	0.05	0.30	0.13	85.0	19.7	0	0.6
0	125	1.59	21.3	183.7	12.3	0.34	441.2	0.04	2.18	0.06	0.38	0.05	32.0	17.9	0	1.0
0	26	0.14	7.0	138.1	12.5	0.09	0	0.02	0.02	0.02	0.23	0.05	7.0	9.0	0	0.2
0	22	0.86	13.0	237.2	121.4	0.33	475.1	0.04	0.24	0.04	0.47	0.06	19.6	4.1	0	0.2
0	23	0.75	20.0	153.8	31.9	0.45	194.7	0.06	0.35	0.11	0.77	0.07	17.3	2.9	0	0.3
0	20	0.36	—	235.0	210.0	—	50.0	0	0	0	0	0	—	36.0	0	—
0	10	0.18	—	450.0	70.0	—	50.0	0	0	0	0	0	—	36.0	0	—
0	20	0.54	—	325.0	240.0	—	31.3	—	—	—	—	—	—	36.0	0	—
0	0	0	—	—	5.0	—	0	—	—	—	—	—	—	2.4	—	—
0	7	0	—	—	5.0	—	0	—	—	—	—	—	—	2.0	—	—
0	41	0.07	7.1	112.2	13.9	0.04	54.4	0.03	0.34	0.04	0.07	0.04	3.1	14.6	0	0.3
0	56	0.56	44.6	119.7	36.3	0.53	0	0.01	3.87	0.15	0.50	0.02	8.5	0	0	0.4
0	56	0.56	44.6	119.7	1.1	0.53	0	0.01	3.87	0.15	0.50	0.02	8.5	0	0	0.4
0	86	1.19	97.2	238.9	6.9	1.08	0	0.07	8.61	0.26	1.27	0.04	17.8	0	0	1.2
0	92	1.32	96.9	245.6	1.0	1.14	0	0.03	8.21	0.33	1.23	0.04	18.3	0	0	0.8
0	92	1.32	96.9	245.6	226.3	1.14	0	0.03	8.21	0.33	1.23	0.04	18.3	0	0	0.8
0	114	1.44	107.5	274.4	133.1	1.20	0	0.04	10.19	0.31	1.44	0.05	10.6	0	0	1.1
0	71	1.00	72.4	190.4	0.3	0.83	0	0.06	7.08	0.27	0.91	0.04	13.5	0	0	0.7
0	1	1.40	0	579.7	21.7	0.21	0	0.17	—	0.21	0.50	0.39	64.4	8.8	0	4.0
0	53	0.81	125.0	219.1	1.0	1.35	0	0.21	1.91	0.01	0.10	0.03	7.3	0.2	0	637.4
0	49	0.51	35.3	616.7	16.0	0.59	8.6	0.23	—	0.14	4.22	0.24	33.6	4.3	0	8.0
0	16	1.21	71.1	126.3	0.3	0.94	1.8	0.11	—	0.04	0.31	0.17	19.8	1.0	0	5.2
0	7	0.80	41.3	87.4	65.0	0.83	0	0.05	0.15	0.03	0.26	0.04	10.9	0	0	1.8
0	7	0.80	41.3	87.4	2.4	0.83	0	0.05	—	0.03	0.26	0.04	10.9	0	0	1.8
0	15	2.06	89.1	193.5	5.5	1.92	0	0.07	0.32	0.07	0.48	0.09	23.6	0	0	4.0
0	14	1.95	88.0	203.8	4.2	1.73	0	0.12	0.30	0.07	0.56	0.10	8.1	0.1	0	6.5
0	6	0.75	20.4	123.2	8.4	0.46	0	0.01	0.10	0.02	0.14	0.07	2.0	0.3	0	4.2
0	3	0.45	11.6	78.4	60.9	0.42	0	0.01	0.09	0	0.11	0.06	1.9	0.2	0	3.9
0	3	0.49	6.4	71.2	4.0	0.22	0	0.01	0.05	0	0.11	0.01	5.2	0.7	0	2.0
0	14	1.07	64.3	341.0	2.9	0.66	0	0.11	—	0.06	1.07	0.31	51.5	27.5	0	5.0
0	21	0.78	24.5	324.3	12.2	0.11	0.5	0.07	—	0.05	0.33	0.11	17.2	12.1	0	—
0	10	0.33	11.8	211.6	0.7	0.20	0.4	0.09	0.18	0.06	0.48	0.18	25.0	9.3	0	0.4
0	4	0.19	6.4	42.2	1.8	0.14	0.4	0.04	—	0.02	0.19	0.04	6.0	3.4	0	—
0	13	0.75	22.9	152.7	6.8	0.51	1.4	0.16	—	0	0.25	0.15	21.1	10.0	0	—
0	94	1.40	102.6	232.7	7.8	1.20	0	0.46	0.09	0.05	0.86	0.13	24.4	0.2	0	7.1
0	8	0.62	20.4	384.7	5.0	0.26	21.2	0.17	—	0.07	4.52	0.25	40.9	11.3	0	—
0	68	1.50	72.6	298.4	0	1.00	0.9	0.22	7.94	0.05	0.70	0.27	35.4	0.9	0	1.9
0	35	1.25	49.4	215.8	0	0.71	0.9	0.10	4.37	0.04	0.59	0.18	25.2	1.1	0	1.2
0	18	0.64	51.9	130.8	0.3	1.29	2.1	0.26	—	0.04	0.27	0.06	12.0	0.6	0	2.4
0	23	0.89	39.5	121.6	1.3	0.43	0	0.24	0.19	0.03	0.76	0.12	3.3	0.2	0	3.9
0	23	0.89	39.5	121.6	118.3	0.43	0	0.24	0.19	0.03	0.76	0.12	3.3	0.2	0	3.9
0	28	1.24	43.5	123.3	1.7	0.44	0	0.40	0.18	0.05	0.83	0.09	3.7	0.4	0	1.2
0	24	1.27	77.1	204.5	4.1	1.30	0.3	0.07	—	0.07	1.61	0.10	17.1	0.1	0	1.0
0	24	1.27	77.1	237.4	118.2	1.30	0	0.07	3.75	0.07	1.61	0.10	17.1	0.1	0	2.6

DA+ Code	Food Description	Quantity	Measure	Wt (g)	H₂O (g)	Ener (kcal)	Prot (g)	Carb (g)	Fiber (g)	Fat (g)	Fat Breakdown (g)		
											Sat	Mono	Poly
Nuts, Seeds, and Products—*continued*													
32906	Without peanuts, oil roasted, no salt added	¼	cup(s)	36	1.1	221	5.6	8.0	2.0	20.2	3.3	11.9	4.1
	Peanuts												
2807	Dry roasted	¼	cup(s)	37	0.6	214	8.6	7.9	2.9	18.1	2.5	9.0	5.7
2806	Dry roasted, salted	¼	cup(s)	37	0.6	214	8.6	7.9	2.9	18.1	2.5	9.0	5.7
1763	Oil roasted, salted	¼	cup(s)	36	0.5	216	10.1	5.5	3.4	18.9	3.1	9.4	5.5
1884	Peanut butter, chunky	1	tablespoon(s)	16	0.2	94	3.8	3.5	1.3	8.0	1.2	3.7	2.2
30303	Peanut butter, low sodium	1	tablespoon(s)	16	0.3	94	4.0	3.1	1.0	8.1	1.6	3.8	2.2
30421	Peanut butter, natural	1	tablespoon(s)	16	—	100	3.5	3.0	1.0	8.0	1.3	4.0	2.3
30422	Peanut butter, natural, no salt added	1	tablespoon(s)	16	—	105	3.5	3.0	1.0	8.0	1.3	—	—
30305	Peanut butter, reduced fat	1	tablespoon(s)	18	0.2	94	4.7	6.4	0.9	6.1	1.0	2.9	1.7
524	Peanut butter, smooth	1	tablespoon(s)	16	0.3	94	4.0	3.1	1.0	8.1	1.7	3.9	2.3
2804	Raw	¼	cup(s)	37	2.4	207	9.4	5.9	3.1	18.0	2.5	8.9	5.7
	Pecans												
32907	Dry roasted, no salt added	¼	cup(s)	28	0.3	195	2.6	3.7	2.6	20.4	1.7	12.1	5.7
32936	Dry roasted, with salt added	¼	cup(s)	28	0.3	195	2.6	3.7	2.6	20.4	1.7	12.1	5.7
1162	Oil roasted	¼	cup(s)	28	0.3	197	2.5	3.6	2.6	20.7	2.0	11.3	6.5
526	Raw	¼	cup(s)	27	1.0	188	2.5	3.8	2.6	19.6	1.7	11.1	5.9
12973	**Pine nuts or pignolia, dried**	1	tablespoon(s)	9	0.2	58	1.2	1.1	0.3	5.9	0.4	1.6	2.9
	Pistachios												
1164	Dry roasted	¼	cup(s)	31	0.6	174	6.4	9.0	3.0	13.8	1.7	7.3	4.1
32938	Dry roasted, with salt added	¼	cup(s)	32	0.6	180	6.7	9.2	3.2	14.3	1.7	7.6	4.3
1167	**Pumpkin or squash seeds, roasted**	¼	cup(s)	57	1.2	326	16.9	8.3	3.7	27.8	4.8	8.9	11.3
	Sesame												
32912	Sesame butter paste	1	tablespoon(s)	16	0.3	94	2.9	3.8	0.9	8.1	1.1	3.1	3.6
32941	Tahini or sesame butter	1	tablespoon(s)	15	0.4	89	2.6	3.2	0.7	8.0	1.1	3.0	3.5
1169	Whole, roasted, toasted	3	tablespoon(s)	27	0.9	153	4.6	6.9	3.8	13.0	1.8	4.9	5.7
	Soy nuts												
34173	Deep sea salted	¼	cup(s)	28	—	130	11.0	9.0	6.0	6.0	1.0	1.0	3.5
34174	Unsalted	¼	cup(s)	28	—	130	11.0	9.0	6.0	6.0	1.0	1.0	3.5
	Sunflower seeds												
528	Kernels, dried	1	tablespoon(s)	9	0.4	53	1.9	1.8	0.8	4.6	0.4	1.7	2.1
29721	Kernels, dry roasted, salted	1	tablespoon(s)	8	0.1	47	1.5	1.9	0.7	4.0	0.4	0.8	2.6
29723	Kernels, toasted, salted	1	tablespoon(s)	8	0.1	52	1.4	1.7	1.0	4.8	0.5	0.9	3.1
32928	Sunflower seed butter with salt added	1	tablespoon(s)	16	0.1	99	2.8	3.7	0.9	8.8	0.7	6.2	1.6
	Trail mix												
4646	Trail mix	¼	cup(s)	38	3.5	173	5.2	16.8	2.0	11.0	2.1	4.7	3.6
4647	Trail mix with chocolate chips	¼	cup(s)	38	2.5	182	5.3	16.8	—	12.0	2.3	5.1	4.2
4648	Tropical trail mix	¼	cup(s)	35	3.2	155	2.2	23.0	—	6.0	3.0	0.9	1.8
	Walnuts												
529	Dried black, chopped	¼	cup(s)	31	1.4	193	7.5	3.1	2.1	18.4	1.1	4.7	11.0
531	English or Persian	¼	cup(s)	29	1.2	191	4.5	4.0	2.0	19.1	1.8	2.6	13.8
Vegetarian Foods													
	Prepared												
34222	Brown rice & tofu stir-fry (vegan)	8	ounce(s)	227	183.7	227	12.3	12.6	2.8	16.3	1.3	4.0	9.5
34368	Cheese enchilada casserole (lacto)	8	ounce(s)	227	88.3	400	17.6	37.3	3.8	19.5	10.3	6.6	1.5
34247	Five bean casserole (vegan)	8	ounce(s)	227	164.0	224	4.8	40.3	4.6	5.0	0.9	2.3	1.6
34261	Lentil stew (vegan)	8	ounce(s)	227	150.6	127	7.3	24.2	7.3	0.5	0.1	0.1	0.2
34397	Macaroni and cheese (lacto)	8	ounce(s)	227	163.3	182	8.2	17.3	0.6	8.8	4.5	3.0	0.9
34238	Steamed rice and vegetables (vegan)	8	ounce(s)	227	99.7	263	4.9	39.2	2.5	10.3	1.8	4.2	3.9
34308	Tofu rice burgers (ovo-lacto)	1	piece(s)	218	78.2	446	22.5	67.0	6.0	10.4	2.2	2.5	4.7
34276	Vegan spinach enchiladas (vegan)	1	piece(s)	80	59.1	86	4.8	12.7	2.1	2.3	0.4	0.6	1.0
34243	Vegetable chow mein (vegan)	8	ounce(s)	227	163.5	164	6.3	22.2	2.0	6.2	0.7	2.8	2.4
34454	Vegetable lasagna (lacto)	8	ounce(s)	227	154.8	178	11.6	25.4	2.3	3.6	2.0	0.9	0.3
34339	Vegetable marinara (vegan)	8	ounce(s)	227	180.6	93	2.8	14.9	1.4	2.8	0.4	1.3	0.9
34356	Vegetable rice casserole (lacto)	8	ounce(s)	227	172.6	222	9.0	23.5	3.8	11.3	4.4	3.4	2.0
34311	Vegetable strudel (ovo-lacto)	8	ounce(s)	227	99.9	754	18.8	50.9	3.9	53.5	18.2	26.4	6.1
34371	Vegetable taco (lacto)	1	item(s)	85	54.6	140	4.9	16.7	3.0	6.3	2.9	1.8	1.3
34282	Vegetarian chili (vegan)	8	ounce(s)	227	195.6	115	5.6	21.3	6.8	1.5	0.3	0.3	0.7

Chol (mg)	Calc (mg)	Iron (mg)	Magn (mg)	Pota (mg)	Sodi (mg)	Zinc (mg)	Vit A (µg)	Thia (mg)	Vit E (mg α)	Ribo (mg)	Niac (mg)	Vit B$_6$ (mg)	Fola (µg)	Vit C (mg)	Vit B$_{12}$ (µg)	Sele (µg)
0	38	0.93	90.4	195.8	4.0	1.68	0.4	0.18	—	0.17	0.71	0.06	20.2	0.2	0	—
0	20	0.82	64.2	240.2	2.2	1.21	0	0.16	2.53	0.04	4.94	0.09	52.9	0	0	2.7
0	20	0.82	64.2	240.2	247.8	1.21	0	0.16	2.85	0.04	4.94	0.09	52.9	0	0	2.7
0	22	0.55	63.4	261.4	115.2	1.18	0	0.03	2.50	0.03	4.98	0.17	43.2	0.3	0	1.2
0	7	0.30	25.6	119.2	77.8	0.45	0	0.02	1.01	0.02	2.19	0.07	14.7	0	0	1.3
0	7	0.30	24.6	103.8	2.7	0.47	0	0.01	1.44	0.02	2.14	0.09	11.8	0	0	0.9
0	0	0.18	—	—	52.5	—	0	—	—	—	—	—	—	0	—	—
0	0	0.18	—	—	0	—	0	—	—	—	—	—	—	0	—	—
0	6	0.34	30.6	120.4	97.2	0.50	0	0.05	1.63	0.01	2.63	0.06	10.8	0	0	1.4
0	7	0.30	24.6	103.8	73.4	0.47	0	0.01	1.44	0.02	2.14	0.09	11.8	0	0	0.9
0	34	1.67	61.3	257.3	6.6	1.19	0	0.23	3.04	0.05	4.40	0.13	87.6	0	0	2.6
0	20	0.77	36.3	116.6	0.3	1.39	1.9	0.12	0.36	0.03	0.32	0.05	4.4	0.2	0	1.1
0	20	0.77	36.3	116.6	105.4	1.39	1.9	0.12	0.36	0.03	0.32	0.05	4.4	0.2	0	1.1
0	18	0.68	33.3	107.8	0.3	1.23	1.4	0.13	0.70	0.03	0.33	0.05	4.1	0.2	0	1.6
0	19	0.69	33.0	111.7	0	1.23	0.8	0.18	0.38	0.04	0.32	0.06	6.0	0.3	0	1.0
0	1	0.48	21.6	51.3	0.2	0.55	0.1	0.03	0.80	0.02	0.38	0.01	2.9	0.1	0	0.1
0	33	1.24	33.5	309.7	1.8	0.72	4.0	0.21	0.74	0.07	0.42	0.35	15.7	0.9	0	3.1
0	34	1.29	34.9	322.2	137.0	0.75	4.2	0.22	0.77	0.07	0.44	0.36	16.3	1.0	0	3.2
0	30	4.58	312.1	447.2	10.2	4.34	0	0.04	0.32	0.09	2.51	0.06	32.3	1.0	0	5.3
0	154	3.07	57.9	93.1	1.9	1.17	0.5	0.04	—	0.03	1.07	0.13	16.0	0	0	5.7
0	21	0.66	14.3	68.8	5.3	0.69	0.4	0.24	—	0.02	0.85	0.02	14.7	0.6	0	5.2
0	267	3.99	96.1	128.3	3.0	1.93	0	0.22	—	0.07	1.24	0.22	26.5	0	0	9.3
0	40	1.80	—	—	95.0	—	0	—	—	—	—	—	—	1.2	—	—
0	40	1.80	—	—	2.0	—	0	—	—	—	—	—	—	1.2	—	—
0	7	0.47	29.3	58.0	0.8	0.45	0.3	0.13	3.17	0.03	0.75	0.12	20.4	0.1	0	4.8
0	6	0.30	10.3	68.0	29.4	0.42	0	0.01	2.09	0.02	0.56	0.06	19.0	0.1	0	6.3
0	5	0.57	10.8	41.1	51.3	0.44	0	0.03	—	0.02	0.35	0.07	19.9	0.1	0	5.2
0	10	0.66	49.8	92.2	53.0	0.78	0.5	0.01	3.66	0.03	1.08	0.09	37.9	0.4	0	16.7
0	29	1.14	59.3	256.9	85.9	1.21	0.4	0.17	—	0.07	1.77	0.11	26.6	0.5	0	—
2	41	1.27	60.4	243.0	45.4	1.18	0.8	0.15	—	0.08	1.65	0.10	24.4	0.5	0	—
0	20	0.92	33.6	248.1	33.3	0.41	0.7	0.16	—	0.04	0.52	0.11	14.7	2.7	0	—
0	19	0.98	62.8	163.4	0.6	1.05	0.6	0.02	0.56	0.04	0.15	0.18	9.7	0.5	0	5.3
0	29	0.85	46.2	129.0	0.6	0.90	0.3	0.10	0.20	0.04	0.33	0.16	28.7	0.4	0	1.4
0	268	4.67	88.0	379.7	111.4	1.53	0	0.13	0.68	0.11	1.07	0.28	39.0	15.5	0	10.9
43	478	2.52	34.7	215.7	1350.3	1.87	0	0.33	0.48	0.36	2.28	0.12	128.6	22.2	0.4	20.7
0	45	1.25	37.2	376.6	776.2	0.73	0	0.08	1.01	0.08	0.67	0.10	64.5	4.6	0	4.0
0	25	2.09	33.9	388.5	288.8	1.08	0	0.18	1.05	0.09	1.47	0.16	134.2	12.3	0	8.9
22	193	0.68	20.4	126.3	772.5	1.14	0	0.14	0.29	0.24	0.98	0.05	75.4	0.1	0.5	15.6
0	42	1.40	67.5	355.8	1397.4	0.92	175.8	0.16	1.66	0.11	2.75	0.29	31.3	13.1	0	8.3
51	528	5.83	96.8	426.9	1800.8	2.46	0	0.85	0.64	0.45	5.83	0.31	167.7	3.2	0.2	35.1
0	106	1.12	37.2	161.3	104.4	0.68	0	0.06	0.21	0.05	0.54	0.11	19.8	1.4	0	5.3
0	187	3.47	28.5	305.1	367.0	0.75	0	0.13	0.85	0.12	1.37	0.14	76.8	8.1	0	6.4
10	146	1.82	34.5	396.7	647.7	1.12	0	0.19	1.03	0.26	2.02	0.22	107.4	15.2	0.4	19.1
0	17	0.77	17.1	182.0	376.5	0.38	18.2	0.12	0.86	0.08	1.20	0.11	71.6	17.7	0	10.0
16	172	2.01	29.4	393.3	579.0	1.28	0	0.17	1.70	0.28	2.33	0.20	148.5	53.2	0.2	11.6
46	317	3.36	39.0	298.5	815.8	1.98	0	0.45	1.47	0.50	4.51	0.16	175.8	27.1	0.3	31.1
8	79	0.85	26.5	205.7	331.4	0.68	0	0.09	0.26	0.07	0.60	0.08	32.7	4.2	0.1	2.3
0	68	2.42	40.9	531.0	358.9	1.23	0	0.14	1.43	0.13	1.25	0.22	48.6	16.6	0	3.6

APPENDIX H

APPENDIX H

DA+ Code	Food Description	Quantity	Measure	Wt (g)	H₂O (g)	Ener (kcal)	Prot (g)	Carb (g)	Fiber (g)	Fat (g)	Sat	Mono	Poly
											Fat Breakdown (g)		

Vegetarian Foods—*continued*

DA+ Code	Food Description	Quantity	Measure	Wt (g)	H₂O (g)	Ener (kcal)	Prot (g)	Carb (g)	Fiber (g)	Fat (g)	Sat	Mono	Poly
34367	Vegetarian vegetable soup (vegan)	8	ounce(s)	227	204.6	92	2.5	13.6	2.5	4.0	0.7	1.8	1.2
	Boca												
68284	All American Classic patty w/ organic soy	1	item(s)	71	—	140	15.0	9.0	4.0	5.0	1.5	—	—
32072	Breakfast links	2	item(s)	45	—	70	8.0	5.0	2.0	3.0	1.0	—	—
68285	Bruschetta tomato basil parmesan patty	1	item(s)	71	—	70	10.0	9.0	3.5	1.5	0.5	0.5	—
35780	Cheeseburger meatless burger patty	1	item(s)	71	—	100	13.0	6.0	4.0	4.5	1.5	—	—
68290	Chik'n nuggets	1	item(s)	22	—	45	3.5	4.3	0.8	1.8	0.1	1.3	—
68289	Chik'n patty, original	1	item(s)	71	—	160	11.0	15.0	2.0	6.0	0	—	—
68291	Chik'n patty, spicy	1	item(s)	71	—	160	11.0	15.0	2.0	6.0	0	—	—
35781	Grilled vegetable patty	1	item(s)	71	—	80	12.0	7.0	4.0	1.0	0	—	—
68287	Meatless ground crumbles	½	cup(s)	57	—	60	13.0	6.0	3.0	0.5	0	—	—
68282	Original vegan patty	1	item(s)	71	—	70	13.0	6.0	4.0	0.5	0	—	—
	Gardenburger												
39661	Black bean chipotle burger	1	item(s)	71	45.8	100	5.0	16.0	5.0	3.0	0	1.6	0.8
29913	Original	1	item(s)	71	43.7	100	5.0	18.0	5.0	3.0	1.0	1.2	0.7
29920	Portabella veggie burger	1	item(s)	71	44.9	100	4.0	17.0	5.0	2.5	1.0	0.7	0.5
39662	Sun-dried tomato basil burger	1	item(s)	71	46.3	100	4.0	17.0	4.0	2.5	0.5	0.9	0.6
29915	Veggie medley	1	item(s)	71	46.5	100	3.0	17.0	5.0	2.5	0	1.0	1.0
	Loma Linda												
9311	Big franks, canned	1	item(s)	51	30.1	110	11.0	3.0	2.0	6.0	1.0	1.5	3.5
9323	Fried Chik'n with gravy	2	piece(s)	80	—	150	12.0	5.0	2.0	10.0	1.5	2.5	5.0
9326	Linketts, canned	1	item(s)	35	21.4	70	7.0	1.0	1.0	4.0	0.5	1.0	2.5
9336	Redi-Burger patties, canned	1	slice(s)	85	55.1	120	18.0	7.0	4.0	2.5	0.5	0.5	1.5
9354	Tender Rounds meatball substitute, canned in gravy	6	piece(s)	80	55.3	120	13.0	6.0	1.0	4.5	0.5	1.5	2.5
9356	Vege-Burger, canned	¼	cup(s)	55	39.6	60	12.0	2.0	2.0	0.5	0	0	0.5
	Morningstar Farms												
9371	Breakfast bacon strips	2	item(s)	16	6.7	60	2.0	2.0	1.0	4.5	0.5	1.0	3.0
9368	Breakfast sausage links	2	item(s)	45	28.3	80	9.0	3.0	2.0	3.0	0.5	1.0	1.5
33705	Chik'n nuggets	4	piece(s)	86	43.5	190	12.0	19.0	4.0	9.0	1.5	2.5	4.5
11587	Chik patties	1	item(s)	71	38.3	140	8.0	16.0	2.0	5.0	0.5	1.0	3.5
2531	Garden veggie patties	1	item(s)	67	40.1	110	10.0	9.0	3.0	3.5	0.5	0.5	1.5
9376	Grillers patties, original	1	item(s)	64	35.6	130	15.0	5.0	2.0	6.0	1.0	2.0	3.0
50552	Meal Starters Chik'n strips	12	piece(s)	85	50.6	140	23.0	6.0	1.0	3.5	0.5	1.5	1.5
33715	Meal Starters Grillers recipe crumbles	½	cup(s)	42	26.9	61	7.6	3.8	2.3	1.9	0	0.4	1.1
33702	Spicy black bean veggie burger	1	item(s)	67	37.7	120	11.0	13.0	4.0	4.0	0.5	1.0	2.0
	Worthington												
9418	Chic-ketts, frozen	2	slice(s)	55	31.7	110	14.0	3.0	2.0	5.0	1.0	1.0	3.0
9424	Chili, canned	1	cup(s)	230	167.1	280	24.0	25.0	8.0	10.0	1.5	1.5	7.0
9436	Diced Chik, canned	¼	cup(s)	55	43.0	50	9.0	2.0	1.0	0	0	0	0.0
9440	Dinner roast, frozen	1	slice(s)	85	52.4	180	14.0	6.0	3.0	11.0	1.5	4.5	5.0
9446	Fripats patties, frozen	1	item(s)	64	36.4	130	15.0	5.0	3.0	6.0	1.0	1.5	3.5
9452	Leanies links, frozen	1	item(s)	40	21.6	100	8.0	2.0	1.0	7.0	1.0	1.5	4.5
36702	Meatless chicken style roll, frozen	1	slice(s)	55	—	90	9.0	2.0	1.0	4.5	0.5	1.0	2.5
9428	Meatless corned beef, sliced, frozen	3	slice(s)	57	30.8	145	10.4	5.2	0	8.3	1.0	2.1	5.2
9462	Prosage links	2	item(s)	45	28.9	80	9.0	3.0	2.0	3.0	0.5	0.5	2.0
9484	Stakelets patty beef steak substitute, frozen	1	piece(s)	71	40.8	150	14.0	7.0	2.0	7.0	1.0	2.5	3.5
9486	Stripples bacon substitute	2	item(s)	16	6.7	60	2.0	2.0	0.5	4.5	0.5	1.0	3.0
9496	Vegetable Skallops meat substitute, canned	½	cup(s)	85	61.8	90	17.0	4.0	3.0	1.0	0	0	0.5

Dairy

DA+ Code	Food Description	Quantity	Measure	Wt (g)	H₂O (g)	Ener (kcal)	Prot (g)	Carb (g)	Fiber (g)	Fat (g)	Sat	Mono	Poly
	Cheese												
1433	Blue, crumbled	1	ounce(s)	28	12.0	100	6.1	0.7	0	8.1	5.3	2.2	0.2
884	Brick	1	ounce(s)	28	11.7	105	6.6	0.8	0	8.4	5.3	2.4	0.2
885	Brie	1	ounce(s)	28	13.7	95	5.9	0.1	0	7.8	4.9	2.3	0.2
34821	Camembert	1	ounce(s)	28	14.7	85	5.6	0.1	0	6.9	4.3	2.0	0.2
888	Cheddar or colby	1	ounce(s)	28	10.8	112	6.7	0.7	0	9.1	5.7	2.6	0.3
32096	Cheddar or colby, low fat	1	ounce(s)	28	17.9	49	6.9	0.5	0	2.0	1.2	0.6	0.1
32121	Cheddar or colby, low sodium	1	ounce(s)	28	11.1	113	6.9	0.5	0	9.2	5.9	2.6	0.3

Chol (mg)	Calc (mg)	Iron (mg)	Magn (mg)	Pota (mg)	Sodi (mg)	Zinc (mg)	Vit A (µg)	Thia (mg)	Vit E (mg α)	Ribo (mg)	Niac (mg)	Vit B$_6$ (mg)	Fola (µg)	Vit C (mg)	Vit B$_{12}$ (µg)	Sele (µg)
0	38	1.32	27.9	444.6	504.1	0.45	0	0.11	0.73	0.08	1.55	0.21	39.7	23.8	0	1.0
5	150	1.44	—	—	500.0	—	0	—	—	—	—	—	—	0	—	—
0	20	1.44	—	—	330.0	—	0	—	—	—	—	—	—	0	—	—
5	150	2.70	—	—	290.0	—	0	—	—	—	—	—	—	0	—	—
10	100	1.44	—	—	320.0	—	0	—	—	—	—	—	—	0	—	—
0	10	0.36	—	—	125.0	—	—	—	—	—	—	—	—	0	—	—
0	40	1.80	—	—	430.0	—	—	—	—	—	—	—	—	0	—	—
0	40	1.80	—	—	560.0	—	—	—	—	—	—	—	—	0	—	—
0	60	1.80	—	—	300.0	—	—	—	—	—	—	—	—	0	—	—
0	60	1.80	—	—	270.0	—	0	—	—	—	—	—	—	0	—	—
0	60	1.80	—	—	280.0	—	—	—	—	—	—	—	—	0	—	—
0	20	1.44	18.5	140.0	390.0	0.35	10.0	—	—	—	—	—	—	6.0	—	—
10	40	1.08	2.8	110.2	400.7	0.78	—	—	—	—	—	—	—	3.6	—	—
3	40	0.72	6.4	105.2	490.9	0.07	—	—	—	—	—	—	—	1.2	—	—
3	20	0.72	2.8	55.0	270.0	0.07	—	—	—	—	—	—	—	2.4	—	—
0	20	0.72	12.8	125.2	380.7	0.21	62.6	—	—	—	—	—	—	12.0	0	—
0	0	0.77	—	50.0	220.0	1.17	0	0.22	—	0.10	2.00	0.70	—	0	2.4	—
0	20	1.80	—	70.0	430.0	—	0	1.05	—	0.34	4.00	0.30	—	0	2.4	—
0	0	0.36	—	20.0	160.0	0.52	0	0.12	—	0.20	0.80	0.16	—	0	0.9	—
0	0	1.06	—	140.1	450.3	1.45	0	0.15	—	0.26	4.00	0.40	—	0	1.2	—
0	20	1.08	—	80.0	340.0	0.64	0	0.75	—	0.17	2.00	0.16	—	0	1.2	—
0	0	0.36	—	40.0	130.0	0.71	0	0.12	—	0.10	0.78	0.30	—	0	2.4	—
0	0	0.36	—	15.0	230.0	0.10	0	0.75	—	0.04	1.20	0.07	—	0	0.2	—
0	0	1.80	—	50.0	300.0	—	0	0.38	—	0.17	7.00	0.50	—	0	3.0	—
0	60	1.80	3.4	320.0	600.0	0.69	0	0.38	—	0.10	4.00	0.20	—	0	1.8	—
0	20	1.80	—	280.0	590.0	0.35	0	0.22	—	0.10	3.00	0.16	—	0	1.2	—
0	40	0.72	—	180.0	350.0	0.67	—	7.50	—	0.20	0.94	0.60	—	0	—	—
0	43	1.80	—	130.0	260.0	0.77	0	1.80	—	0.17	4.00	0.40	—	0	2.7	—
0	40	5.40	24.0	109.9	499.8	4.50	0	0.45	—	0.25	7.00	0.40	—	0	1.8	—
0	0	1.36	—	75.8	174.2	0.46	0	0.23	—	0.08	3.03	0.23	0	0	2.3	—
0	60	1.44	34.2	250.0	350.0	0.67	0	9.11	—	0.20	1.21	0.11	32.8	0	0	—
0	0	1.80	—	85.0	390.0	0.88	0	0.22	—	0.14	2.00	0.11	3.8	0	1.2	—
0	40	3.60	—	330.0	1130.0	1.84	0	0.30	—	0.14	2.00	0.70	—	0	1.5	—
0	0	1.08	—	100.0	220.0	0.38	0	0.06	—	0.10	4.00	0.08	—	0	0.2	—
0	20	1.80	—	120.1	580.3	0.77	0	1.80	—	0.26	6.00	0.60	—	0	1.5	—
0	60	1.80	—	120.0	320.0	0.77	0	1.80	—	0.17	3.00	0.60	—	0	1.2	—
0	20	0.72	—	40.0	430.0	0.32	0	0.22	—	0.14	0.80	0.20	—	0	0.9	—
0	100	1.08	—	240.0	240.0	—	0	0.38	—	0.14	4.00	0.30	—	0	1.8	—
0	0	1.87	—	134.7	466.4	0.46	0	1.87	—	0.18	5.18	0.31	—	0	1.9	—
0	0	1.44	—	50.0	320.0	0.45	0	1.80	—	0.17	2.00	0.30	—	0	3.0	—
0	40	1.08	—	130.0	480.0	0.85	0	1.20	—	0.14	3.00	0.30	—	0	1.5	—
0	0	0.36	—	15.0	220.0	0.10	0	0.75	—	0.03	0.40	0.08	—	0	0.2	—
0	0	0.36	—	10.0	390.2	0.85	0	—	—	—	—	—	—	0	0	—
21	150	0.09	6.5	72.6	395.5	0.75	56.1	0.01	0.07	0.11	0.29	0.05	10.2	0	0.3	4.1
27	191	0.12	6.8	38.6	158.8	0.74	82.8	0	0.07	0.10	0.03	0.02	5.7	0	0.4	4.1
28	52	0.14	5.7	43.1	178.3	0.67	49.3	0.02	0.07	0.15	0.11	0.07	18.4	0	0.5	4.1
20	110	0.09	5.7	53.0	238.7	0.67	68.3	0.01	0.06	0.14	0.18	0.06	17.6	0	0.4	4.1
27	194	0.22	7.4	36.0	171.2	0.87	74.8	0	0.08	0.11	0.03	0.02	5.1	0	0.2	4.1
6	118	0.12	4.5	18.7	173.5	0.52	17.0	0	0.02	0.06	0.01	0.01	3.1	0	0.1	4.1
28	199	0.20	7.7	31.8	6.0	0.88	74.8	0.01	0.08	0.11	0.02	0.02	5.1	0	0.2	4.1

APPENDIX H

(Computer code is for Cengage Diet Analysis program)   (For purposes of calculations, use "0" for t, <1, <0.1, 0.01)

DA+ Code	Food Description	Quantity	Measure	Wt (g)	H₂O (g)	Ener (kcal)	Prot (g)	Carb (g)	Fiber (g)	Fat (g)	Fat Breakdown (g)		
											Sat	Mono	Poly
Dairy—*continued*													
5	Cheddar, shredded	¼	cup(s)	28	10.4	114	7.0	0.4	0	9.4	6.0	2.7	0.3
889	Edam	1	ounce(s)	28	11.8	101	7.1	0.4	0	7.9	5.0	2.3	0.2
890	Feta	1	ounce(s)	28	15.7	75	4.0	1.2	0	6.0	4.2	1.3	0.2
891	Fontina	1	ounce(s)	28	10.8	110	7.3	0.4	0	8.8	5.4	2.5	0.5
8527	Goat cheese, soft	1	ounce(s)	28	17.2	75	5.3	0	0	6.0	4.1	1.4	0.1
893	Gouda	1	ounce(s)	28	11.8	101	7.1	0.6	0	7.8	5.0	2.2	0.2
894	Gruyere	1	ounce(s)	28	9.4	117	8.5	0.1	0	9.2	5.4	2.8	0.5
895	Limburger	1	ounce(s)	28	13.7	93	5.7	0.1	0	7.7	4.7	2.4	0.1
896	Monterey jack	1	ounce(s)	28	11.6	106	6.9	0.2	0	8.6	5.4	2.5	0.3
42324	Mozarella, low sodium	1	ounce(s)	28	14.1	79	7.8	0.9	0	4.8	3.1	1.4	0.1
13	Mozzarella, part skim milk	1	ounce(s)	28	15.2	72	6.9	0.8	0	4.5	2.9	1.3	0.1
12	Mozzarella, whole milk	1	ounce(s)	28	14.2	85	6.3	0.6	0	6.3	3.7	1.9	0.2
897	Muenster	1	ounce(s)	28	11.8	104	6.6	0.3	0	8.5	5.4	2.5	0.2
898	Neufchatel	1	ounce(s)	28	17.9	72	2.6	1.0	0	6.5	3.6	1.6	0.3
14	Parmesan, grated	1	tablespoon(s)	5	1.0	22	1.9	0.2	0	1.4	0.9	0.4	0.1
17	Provolone	1	ounce(s)	28	11.6	100	7.3	0.6	0	7.5	4.8	2.1	0.2
19	Ricotta, part skim milk	¼	cup(s)	62	45.8	85	7.0	3.2	0	4.9	3.0	1.4	0.2
18	Ricotta, whole milk	¼	cup(s)	62	44.1	107	6.9	1.9	0	8.0	5.1	2.2	0.2
20	Romano	1	tablespoon(s)	5	1.5	19	1.6	0.2	0	1.3	0.9	0.4	0
900	Roquefort	1	ounce(s)	28	11.2	105	6.1	0.6	0	8.7	5.5	2.4	0.4
21	Swiss	1	ounce(s)	28	10.5	108	7.6	1.5	0	7.9	5.0	2.1	0.3
53952	Swiss, low sodium	1	ounce(s)	28	10.7	106	8.1	1.0	0	7.8	5.0	2.1	0.3
Imitation cheese													
42245	Imitation American cheddar	1	ounce(s)	28	15.1	68	4.7	3.3	0	4.0	2.5	1.2	0.1
53914	Imitation cheddar	1	ounce(s)	28	15.1	68	4.7	3.3	0	4.0	2.5	1.2	0.1
Cottage cheese													
9	Low fat, 1% fat	½	cup(s)	113	93.2	81	14.0	3.1	0	1.2	0.7	0.3	0
42265	Low fat, 1% fat, no sodium added	½	cup(s)	113	94.4	81	14.0	3.1	0	1.1	0.7	0.3	0
8	Low fat, 2% fat	½	cup(s)	113	91.2	97	13.4	4.1	0	2.8	1.1	0.5	0.1
Cream cheese													
11	Cream cheese	2	tablespoon(s)	29	15.8	99	1.7	1.2	0	9.9	5.6	2.5	0.4
17366	Cream cheese, fat free	2	tablespoon(s)	30	21.6	32	4.7	2.3	0	0.3	0.2	0.1	0
10438	Tofutti Better than Cream Cheese	2	tablespoon(s)	30	—	85	1.0	9.0	0	5.0	2.0	—	—
Processed cheese													
24	American cheese food, processed	1	ounce(s)	28	12.5	94	4.8	2.4	0	7.3	4.3	1.9	0.3
25	American cheese spread, processed	1	ounce(s)	28	13.5	82	4.7	2.5	0	6.0	3.8	1.8	<0.2
22	American cheese, processed	1	ounce(s)	28	11.2	104	5.1	1.4	0	8.7	5.1	2.3	0.4
42285	Cheddar or American, pasteurized process, low sodium	1	ounce(s)	28	12.1	107	6.3	0.5	0	8.8	5.6	2.5	0.3
9110	Kraft Deluxe Singles pasteurized process American cheese	1	ounce(s)	28	—	108	5.4	0	0	9.5	5.4	—	—
23	Swiss cheese, processed	1	ounce(s)	28	12.0	95	7.0	0.6	0	7.1	4.5	2.0	0.2
Soy cheese													
65625	Galaxy Foods vegan American cheese slices alternative	1	slice(s)	19	—	40	1.0	5.0	0	2.0	0	0.5	1.5
10437	Galaxy Foods vegan grated parmesan cheese alternative	1	tablespoon(s)	8	—	23	3.0	1.5	0	0	0	0	0
Cream													
26	Half and half cream	1	tablespoon(s)	15	12.2	20	0.4	0.7	0	1.7	1.1	0.5	0.1
32	Heavy whipping cream, liquid	1	tablespoon(s)	15	8.7	52	0.3	0.4	0	5.6	3.5	1.6	0.2
28	Light coffee or table cream, liquid	1	tablespoon(s)	15	11.1	29	0.4	0.5	0	2.9	1.8	0.8	0.1
30	Light whipping cream, liquid	1	tablespoon(s)	15	9.5	44	0.3	0.4	0	4.6	2.9	1.4	0.1
34	Whipped cream topping, pressurized	1	tablespoon(s)	3	1.8	8	0.1	0.4	0	0.7	0.4	0.2	0
Sour cream													
36	Sour cream	2	tablespoon(s)	24	17.9	46	0.5	0.7	0	4.7	2.8	1.2	0.2
30556	Sour cream, fat free	2	tablespoon(s)	32	25.8	24	1.0	5.0	0	0	0	0	0
Imitation cream													
3659	Coffeemate nondairy creamer, liquid	1	tablespoon(s)	15	—	20	0	2.0	0	1.0	0	0	0
40	Cream substitute, powder	1	teaspoon(s)	2	0	11	0.1	1.1	0	0.7	0.7	0	0
904	Imitation sour cream	2	tablespoon(s)	29	20.5	60	0.7	1.9	0	5.6	5.1	0.2	0

Chol (mg)	Calc (mg)	Iron (mg)	Magn (mg)	Pota (mg)	Sodi (mg)	Zinc (mg)	Vit A (µg)	Thia (mg)	Vit E (mg α)	Ribo (mg)	Niac (mg)	Vit B$_6$ (mg)	Fola (µg)	Vit C (mg)	Vit B$_{12}$ (µg)	Sele (µg)
30	204	0.19	7.9	27.7	175.4	0.88	74.9	0.01	0.08	0.11	0.02	0.02	5.1	0	0.2	3.9
25	207	0.12	8.5	53.3	273.6	1.06	68.9	0.01	0.07	0.11	0.02	0.02	4.5	0	0.4	4.1
25	140	0.18	5.4	17.6	316.4	0.82	35.4	0.04	0.05	0.24	0.28	0.12	9.1	0	0.5	4.3
33	156	0.07	4.0	18.1	226.8	0.99	74.0	0.01	0.08	0.06	0.04	0.02	1.7	0	0.5	4.1
13	40	0.54	4.5	7.4	104.3	0.26	81.6	0.02	0.05	0.11	0.12	0.07	3.4	0	0.1	0.8
32	198	0.07	8.2	34.3	232.2	1.11	46.8	0.01	0.07	0.09	0.02	0.02	6.0	0	0.4	4.1
31	287	0.05	10.2	23.0	95.3	1.11	76.8	0.02	0.08	0.08	0.03	0.02	2.8	0	0.5	4.1
26	141	0.04	6.0	36.3	226.8	0.60	96.4	0.02	0.07	0.14	0.04	0.02	16.4	0	0.3	4.1
25	211	0.20	7.7	23.0	152.0	0.85	56.1	0	0.07	0.11	0.03	0.02	5.1	0	0.2	4.1
15	207	0.07	7.4	26.9	4.5	0.89	38.8	0.01	0.04	0.10	0.03	0.02	2.6	0	0.3	4.5
18	222	0.06	6.5	23.8	175.5	0.78	36.0	0.01	0.04	0.09	0.03	0.02	2.6	0	0.2	4.1
22	143	0.12	5.7	21.5	177.8	0.83	50.7	0.01	0.05	0.08	0.03	0.01	2.0	0	0.6	4.8
27	203	0.12	7.7	38.0	178.0	0.80	84.5	0	0.07	0.09	0.03	0.02	3.4	0	0.4	4.1
21	33	0.04	2.8	43.1	94.7	0.23	68.3	0.01	0.11	0.04	0.06	0.01	4.0	0	0.1	0.9
4	55	0.05	1.9	6.3	76.4	0.19	11.4	0	0.01	0.02	0.01	0	0.5	0	0.1	0.9
20	214	0.15	7.9	39.1	248.3	0.92	66.9	0.01	0.07	0.09	0.04	0.02	2.8	0	0.4	4.1
19	167	0.27	9.2	76.9	76.9	0.82	65.8	0.01	0.04	0.11	0.05	0.01	8.0	0	0.2	10.3
31	127	0.23	6.8	64.6	51.7	0.71	73.8	0.01	0.07	0.12	0.06	0.03	7.4	0	0.2	8.9
5	53	0.04	2.0	4.3	60.0	0.13	4.8	0	0.01	0.02	0	0	0.3	0	0.1	0.7
26	188	0.16	8.5	25.8	512.9	0.59	83.3	0.01	—	0.17	0.21	0.04	13.9	0	0.2	4.1
26	224	0.06	10.8	21.8	54.4	1.24	62.4	0.02	0.11	0.08	0.03	0.02	1.7	0	0.9	5.2
26	272	0.05	10.2	31.5	4.0	1.11	61.2	0.01	0.10	0.10	0.03	0.02	1.7	0	0.5	3.5
10	159	0.09	8.2	68.6	381.3	0.73	32.3	0.01	0.08	0.12	0.04	0.03	2.0	0	0.1	4.3
10	159	0.09	8.2	68.6	381.3	0.73	32.3	0.01	0.08	0.12	0.04	0.03	2.0	0	0.1	4.3
5	69	0.16	5.7	97.2	458.8	0.43	12.4	0.02	0.01	0.19	0.14	0.08	13.6	0	0.7	10.2
5	69	0.16	5.7	97.2	14.7	0.43	12.4	0.02	0.01	0.18	0.15	0.08	13.6	0	0.7	9.5
11	103	0.17	7.9	94.9	372.9	0.46	22.6	0.05	0.05	0.22	0.12	0.02	11.3	0	0.5	11.2
32	28	0.11	2.6	40.0	93.1	0.15	106.1	0.01	0.08	0.04	0.04	0.01	3.2	0	0.1	0.7
4	105	0.06	6.6	83.4	210.6	0.45	3.3	0.01	0.01	0.08	0.07	0.01	10.5	0	0.3	1.5
0	0	0	—	—	160.0	—	—	—	—	—	—	—	—	0	—	—
28	193	0.07	7.7	72.3	364.0	0.65	57.0	0.01	0.18	0.10	0.04	0.03	2.0	0	0.4	5.6
16	159	0.09	8.2	68.6	460.7	0.73	49.0	0.01	0.05	0.12	0.04	0.03	2.0	0	0.1	3.2
28	296	0.18	7.4	37.4	473.7	0.71	89.9	0	0.23	0.07	0.02	0.01	2.3	0	0.4	5.7
27	175	0.11	6.2	45.9	2.0	0.85	72.0	0.01	0.08	0.10	0.02	0.02	2.3	0	0.2	3.6
27	338	0	—	33.8	459.0	—	—	—	—	—	—	—	—	0	—	—
24	219	0.17	8.2	61.2	388.4	1.02	56.1	0	0.10	0.08	0.01	0.01	1.7	0	0.3	4.5
0	200	0	—	—	120.0	—	0	—	—	—	—	—	—	0	—	—
0	60	0	—	75.0	97.5	—	15.0	—	—	—	—	—	—	0	—	—
6	16	0.01	1.5	19.7	6.2	0.08	14.7	0.01	0.05	0.02	0.01	0.01	0.5	0.1	0	0.3
21	10	0	1.0	11.3	5.7	0.03	61.7	0	0.16	0.02	0.01	0	0.6	0.1	0	0.1
10	14	0.01	1.4	18.3	6.0	0.04	27.1	0	0.08	0.02	0.01	0	0.3	0.1	0	0.1
17	10	0	1.0	14.6	5.1	0.04	41.8	0	0.13	0.02	0.01	0	0.6	0.1	0	0.1
2	3	0	0.3	4.4	0.2	0.01	5.6	0	0.02	0	0	0	0.1	0	0	0
12	26	0.04	2.4	33.8	19.2	0.09	42.2	0.01	0.11	0.04	0.03	0.01	1.7	0.2	0.1	0.6
3	40	0	3.2	41.3	45.1	0.16	23.4	0.01	0	0.05	0.02	0.01	3.5	0	0.1	1.7
0	0	0	—	—	0	—	0	—	—	—	—	—	—	0	—	—
0	0	0.02	0.1	16.2	3.6	0.01	0	0	0.01	0	0	0	0	0	0	0
0	1	0.11	1.7	46.3	29.3	0.34	0	0	0.21	0	0	0	0	0	0	0.7

DA+ Code	Food Description	Quantity	Measure	Wt (g)	H₂O (g)	Ener (kcal)	Prot (g)	Carb (g)	Fiber (g)	Fat (g)	Fat Breakdown (g)		
											Sat	Mono	Poly
Dairy—*continued*													
35972	Nondairy coffee whitener, liquid, frozen	1	tablespoon(s)	15	11.5	20	0.1	1.7	0	1.5	0.3	1.1	0
35976	Nondairy dessert topping, frozen	1	tablespoon(s)	5	2.4	15	0.1	1.1	0	1.2	1.0	0.1	0
35975	Nondairy dessert topping, pressurized	1	tablespoon(s)	4	2.6	12	0	0.7	0	1.0	0.8	0.1	0
	Fluid milk												
60	Buttermilk, low fat	1	cup(s)	245	220.8	98	8.1	11.7	0	2.2	1.3	0.6	0.1
54	Low fat, 1%	1	cup(s)	244	219.4	102	8.2	12.2	0	2.4	1.5	0.7	0.1
55	Low fat, 1%, with nonfat milk solids	1	cup(s)	245	220.0	105	8.5	12.2	0	2.4	1.5	0.7	0.1
57	Nonfat, skim or fat free	1	cup(s)	245	222.6	83	8.3	12.2	0	0.2	0.1	0.1	0
58	Nonfat, skim or fat free with nonfat milk solids	1	cup(s)	245	221.4	91	8.7	12.3	0	0.6	0.4	0.2	0
51	Reduced fat, 2%	1	cup(s)	244	217.7	122	8.1	11.7	0	4.8	3.1	1.4	0.2
52	Reduced fat, 2%, with nonfat milk solids	1	cup(s)	245	217.7	125	8.5	12.2	0	4.7	2.9	1.4	0.2
50	Whole, 3.3%	1	cup(s)	244	215.0	149	7.7	11.7	0	7.9	4.6	2.0	0.5
	Canned milk												
62	Nonfat or skim evaporated	2	tablespoon(s)	32	25.3	25	2.4	3.6	0	0.1	0	0	0
63	Sweetened condensed	2	tablespoon(s)	38	10.4	123	3.0	20.8	0	3.3	2.1	0.9	0.1
61	Whole evaporated	2	tablespoon(s)	32	23.3	42	2.1	3.2	0	2.4	1.4	0.7	0.1
	Dried milk												
64	Buttermilk	¼	cup(s)	30	0.9	117	10.4	14.9	0	1.8	1.1	0.5	0.1
65	Instant nonfat with added vitamin A	¼	cup(s)	17	0.7	61	6.0	8.9	0	0.1	0.1	0	0
5234	Skim milk powder	¼	cup(s)	17	0.7	61	6.0	8.9	0	0.1	0.1	0	0
907	Whole	¼	cup(s)	32	0.8	159	8.4	12.3	0	8.5	5.4	2.5	0.2
909	**Goat milk**	1	cup(s)	244	212.4	168	8.7	10.9	0	10.1	6.5	2.7	0.4
	Chocolate milk												
33155	Chocolate syrup, prepared with milk	1	cup(s)	282	227.0	254	8.7	36.0	0.8	8.3	4.7	2.1	0.5
33184	Cocoa mix with aspartame, added sodium and vitamin A, no added calcium or phosphorus, prepared with water	1	cup(s)	192	177.4	56	2.3	10.8	1.2	0.4	0.3	0.1	0
908	Hot cocoa, prepared with milk	1	cup(s)	250	206.1	193	8.8	26.9	2.5	5.8	3.6	1.7	0.2
69	Low fat	1	cup(s)	250	205.9	178	8.1	31.5	1.3	2.5	1.5	0.8	0.1
68	Reduced fat	1	cup(s)	250	205.4	190	7.5	30.3	1.8	4.8	2.9	1.1	0.2
67	Whole	1	cup(s)	250	205.8	208	7.9	25.9	2.0	8.5	5.3	2.5	0.3
70	**Eggnog**	1	cup(s)	254	209.7	224	11.6	20.4	0	10.6	6.6	3.3	<0.5
	Breakfast drinks												
10089	Carnation Breakfast Essentials classic chocolate malt, prepared with skim milk	1	cup(s)	273	—	220	12.5	39.0	0.8	1.0	0.8	—	—
10092	Carnation Breakfast Essentials classic French vanilla, prepared with skim milk, no sugar added	1	cup(s)	257	—	150	12.9	24.0	3.0	0.6	0.2	—	—
57630	Carnation Breakfast Essentials milk chocolate, ready to drink, no sugar added	1	cup(s)	237	—	109	9.5	11.6	1.5	3.6	1.1	—	—
10091	Carnation Breakfast Essentials Strawberry Sensation, prepared with skim milk	1	cup(s)	273	—	220	12.5	39.0	0	0.6	0.4	—	—
1417	Ovaltine rich chocolate flavor, prepared with skim milk	1	cup(s)	258	—	170	8.5	31.0	0	0	0	0	0
8539	**Malted milk, chocolate mix, fortified, prepared with milk**	1	cup(s)	265	215.3	231	8.7	29.7	1.1	8.6	5.0	2.2	0.5
	Milkshakes												
73	Chocolate	1	cup(s)	227	164.0	270	6.9	48.1	0.7	6.1	3.8	1.8	0.2
3163	Strawberry	1	cup(s)	226	167.8	256	7.7	42.8	0.9	6.3	3.9	—	—
74	Vanilla	1	cup(s)	227	169.2	254	8.8	40.3	0	6.9	4.3	2.0	0.3
	Ice cream												
4776	Chocolate	½	cup(s)	66	36.8	143	2.5	18.6	0.8	7.3	4.5	2.1	0.3
12137	Chocolate fudge, no sugar added	½	cup(s)	71	—	100	3.0	16.0	2.0	3.0	1.5	—	—
16514	Chocolate, soft serve	½	cup(s)	87	51.7	192	3.5	19.2	0.6	11.2	6.5	3.0	0.4
16523	Sherbet, all flavors	½	cup(s)	97	63.8	139	1.1	29.3	1.3	1.9	1.1	0.5	0.1
4778	Strawberry	½	cup(s)	66	39.6	127	2.1	18.2	0.6	5.5	3.4	—	—

Chol (mg)	Calc (mg)	Iron (mg)	Magn (mg)	Pota (mg)	Sodi (mg)	Zinc (mg)	Vit A (µg)	Thia (mg)	Vit E (mg α)	Ribo (mg)	Niac (mg)	Vit B$_6$ (mg)	Fola (µg)	Vit C (mg)	Vit B$_{12}$ (µg)	Sele (µg)
0	1	0	0	28.5	11.8	0.0	0.1	0	0.12	0	0	0	0	0	0	0.2
0	0	0.01	0.1	0.8	1.2	0.0	0.3	0	0.05	0	0	0	0	0	0	0.1
0	0	0	0	0.8	2.7	0.0	0.2	0	0.04	0	0	0	0	0	0	0.1
10	284	0.12	27.0	370.0	257.3	1.03	34.3	0.08	0.12	0.38	0.14	0.08	12.3	2.5	0.5	4.9
12	305	0.07	26.8	366.0	107.4	1.02	141.5	0.05	0.02	0.45	0.23	0.09	12.2	0	1.1	8.1
10	314	0.12	34.3	396.9	127.4	0.98	144.6	0.10		0.42	0.22	0.11	12.3	2.5	0.9	5.6
5	299	0.07	27.0	382.2	102.9	1.03	149.4	0.11	0.02	0.45	0.23	0.09	12.3	0	1.2	7.6
5	316	0.12	36.8	419.0	129.9	1.00	156.8	0.10	0.02	0.43	0.22	0.11	12.3	2.5	1.0	5.4
20	293	0.05	26.8	341.6	114.7	1.17	134.2	0.10	0.07	0.45	0.22	0.09	12.2	0.5	1.3	6.1
20	314	0.12	34.3	396.9	127.4	0.98	137.2	0.10	—	0.42	0.22	0.11	12.3	2.5	0.9	5.6
24	276	0.07	24.4	322.1	104.9	0.90	112.2	0.11	0.17	0.41	0.22	0.09	12.2	0	1.1	9.0
1	93	0.09	8.6	105.9	36.7	0.29	37.6	0.01	0	0.10	0.06	0.02	2.9	0.4	0.1	0.8
13	109	0.07	9.9	141.9	48.6	0.36	28.3	0.03	0.06	0.16	0.08	0.02	4.2	1.0	0.2	5.7
9	82	0.06	7.6	95.4	33.4	0.24	20.5	0.01	0.04	0.10	0.06	0.02	2.5	0.6	0.1	0.7
21	359	0.09	33.3	482.5	156.7	1.22	14.9	0.12	0.03	0.48	0.27	0.10	14.2	1.7	1.2	6.2
3	209	0.05	19.9	289.9	93.3	0.75	120.5	0.07	0	0.30	0.15	0.06	8.5	1.0	0.7	4.6
3	211	0.05	20.0	291.9	94.0	0.75	114.5	0.07	0	0.30	0.15	0.06	8.6	1.0	0.7	4.7
31	292	0.15	27.2	425.6	118.7	1.07	82.6	0.09	0.19	0.39	0.21	0.10	11.8	2.8	1.0	5.2
27	327	0.12	34.2	497.8	122.0	0.73	139.1	0.12	0.17	0.34	0.68	0.11	2.4	3.2	0.2	3.4
25	251	0.90	50.8	408.9	132.5	1.21	70.5	0.11	0.14	0.47	0.39	0.09	14.1	0	1.1	9.6
0	92	0.75	32.6	405.1	138.2	0.52	0	0.04	0	0.21	0.16	0.05	1.9	0	0.2	2.5
20	285	1.05	57.5	492.5	110.0	1.58	127.5	0.10	0.08	0.46	0.33	0.10	12.5	0.5	1.2	6.8
8	290	0.68	32.5	425.0	152.5	1.02	145.0	0.09	0.05	0.41	0.32	0.10	12.5	2.3	0.8	4.8
20	273	0.60	35.0	422.5	165.0	0.98	160.0	0.11	0.10	0.46	0.41	0.06	5.0	0	0.8	8.5
30	280	0.60	32.5	417.5	150.0	1.02	67.5	0.09	0.17	0.41	0.31	0.10	12.5	2.3	0.8	4.8
150	330	0.51	48.3	419.1	137.2	1.17	149.9	0.09	0.53	0.48	0.27	0.13	2.5	3.8	1.1	10.7
6	500	4.47	100.0	665.0	240.0	3.75	675.7	0.38	3.38	0.51	5.07	0.50	—	30.0	1.5	—
9	500	4.50	100.0	595.0	168.0	3.75	675.7	0.38	3.38	0.51	5.00	0.50	—	30.0	1.5	—
7	364	3.28	87.4	393.1	174.7	2.73	491.9	0.27	2.46	0.31	3.64	0.36	—	21.8	1.1	—
9	500	4.47	100.0	665.0	288.0	3.75	675.7	0.38	3.38	0.51	5.07	0.50	—	30.0	1.5	8.8
5	350	3.60	100.0	—	240.0	3.75	—	0.38	—	0.43	4.00	0.40	—	12.0	1.2	—
27	368	3.76	45.0	575.0	230.6	1.14	946.0	0.76	0.19	1.28	11.03	1.01	13.3	31.8	1.1	9.5
25	300	0.70	36.4	508.9	252.2	1.09	40.9	0.11	0.11	0.50	0.28	0.06	11.4	0	0.7	4.3
25	256	0.25	29.4	412.0	187.9	0.82	58.9	0.10	—	0.44	0.40	0.10	6.8	1.8	0.7	4.8
27	332	0.23	27.3	415.8	215.8	0.89	56.8	0.07	0.11	0.44	0.33	0.10	15.9	0	1.2	5.2
22	72	0.61	19.1	164.3	50.2	0.38	77.9	0.03	0.20	0.13	0.15	0.04	10.6	0.5	0.2	1.6
10	100	0.36	—	—	65.0	—	—	—	—	—	—	—	—	0	—	—
79	113	0.18	10.4	153.1	52.8	0.45	140.1	0.04	0.53	0.16	0.08	0.04	7.8	0.7	0.4	2.6
1	52	0.14	7.7	92.6	44.4	0.46	11.6	0.03	0.01	0.09	0.06	0.02	3.9	2.2	0.1	1.4
19	79	0.14	9.2	124.1	39.6	0.22	63.4	0.03	—	0.17	0.11	0.03	7.9	5.1	0.2	1.3

DA+ Code	Food Description	Quantity	Measure	Wt (g)	H₂O (g)	Ener (kcal)	Prot (g)	Carb (g)	Fiber (g)	Fat (g)	Fat Breakdown (g)		
											Sat	Mono	Poly
Dairy—*continued*													
76	Vanilla	½	cup(s)	72	43.9	149	2.5	17.0	0.5	7.9	4.9	2.1	0.3
12146	Vanilla chocolate swirl, fat free, no sugar added	½	cup(s)	71	—	100	3.0	14.0	2.0	3.0	2.0	—	—
82	Vanilla, light	½	cup(s)	76	45.5	137	3.6	22.4	0.2	3.7	2.2	1.0	0.2
78	Vanilla, light, soft serve	½	cup(s)	88	61.2	111	4.3	19.2	0	2.3	1.4	0.7	0.1
Soy desserts													
15726	Tofutti premium Better Pecan nondairy frozen dessert	½	cup(s)	70	—	210	1.0	21.0	0	13.0	2.0	—	—
15721	Tofutti premium Chocolate Supreme nondairy frozen dessert	½	cup(s)	70	—	180	3.0	18.0	0	11.0	2.0	—	—
15720	Tofutti premium vanilla non-dairy frozen dessert	½	cup(s)	70	—	210	2.0	21.0	0	13.0	2.0	—	—
Ice milk													
16517	Chocolate	½	cup(s)	74	45.5	138	3.7	19.0	0.6	5.3	3.2	1.5	0.2
16516	Flavored, not chocolate	½	cup(s)	74	44.3	133	3.5	21.8	0.2	3.6	2.2	0.9	0.2
Pudding													
25032	Chocolate	½	cup(s)	144	109.8	153	5.2	22.7	0.8	5.3	3.1	2.2	0.3
1923	Chocolate, sugar free, pre-pared with 2% milk	½	cup(s)	133	—	100	5.0	14.0	0.3	3.0	1.5	—	—
1722	Rice	½	cup(s)	113	75.5	151	4.1	30.0	0.4	1.9	1.1	0.5	0.1
4747	Tapioca, ready to eat	1	item(s)	142	102.0	185	2.8	30.8	0	5.5	1.4	3.6	0.1
25031	Vanilla	½	cup(s)	136	109.7	116	4.6	17.6	0	2.9	1.7	1.5	0.3
1924	Vanilla, sugar free, prepared with 2% milk	½	cup(s)	133	—	90	4.0	12.0	0.2	2.0	1.5	—	—
Frozen yogurt													
4785	Chocolate, soft serve	½	cup(s)	72	45.9	115	2.9	17.9	1.6	4.3	2.6	1.3	0.2
29649	Flavors other than chocolate, nonfat	½	cup(s)	80	52.1	103	3.5	22.2	5.9	0	0	0	0
4786	Vanilla, soft serve	½	cup(s)	72	47.0	114	2.9	17.4	0	4.0	2.5	1.1	0.2
Milk substitutes													
Lactose free													
16081	Fat free, calcium fortified milk	1	cup(s)	240	—	90	8.0	13.0	0	0	0	0	0
36486	Low fat milk	1	cup(s)	240	—	110	8.0	13.0	0	2.5	1.5	—	—
36487	Reduced fat milk	1	cup(s)	240	—	130	8.0	13.0	0	5.0	3.0	—	—
36488	Whole milk	1	cup(s)	240	—	160	8.0	13.0	0	8.0	5.0	—	—
Rice													
10083	Rice Dream carob rice beverage	1	cup(s)	240	—	150	1.0	30.0	0.5	2.5	0	1.5	0.5
17089	Rice Dream original rice bev-erage, enriched	1	cup(s)	240	—	120	1.0	23.0	0	2.5	0	1.5	0.5
10087	Rice Dream vanilla enriched rice beverage	1	cup(s)	240	—	130	1.0	26.0	0	2.5	0	1.5	0.5
Soy													
66705	Silk chocolate soymilk	1	cup(s)	243	209.9	140	5.0	23.0	2.0	3.0	0.5	1.0	1.5
66710	Silk DHA omega-3 enriched soymilk	1	cup(s)	243	221.4	110	7.0	8.0	1.0	5.0	0.5	1.0	3.0
66703	Silk soymilk, unsweetened	1	cup(s)	243	226.4	80	7.0	4.0	1.0	4.0	0.5	1.0	2.5
66704	Silk vanilla soymilk	1	cup(s)	243	221.9	100	6.0	10.0	1.0	3.5	0.5	1.0	2.0
34750	Soy Dream chocolate enriched soy beverage	1	cup(s)	240	205.5	150	7.0	21.0	3.0	4.0	0.5	1.0	2.5
34749	Soy Dream vanilla enriched soy beverage	1	cup(s)	240	213.5	120	7.0	14.0	2.0	4.0	0.5	1.0	2.5
Yogurt													
3615	Custard style, fruit flavors	6	ounce(s)	170	—	180	7.0	31.0	0	2.5	1.5	—	—
3617	Custard style, vanilla	6	ounce(s)	170	—	180	7.0	31.0	0	2.5	1.5	—	—
32101	Fruit, low fat	1	cup(s)	245	184.5	243	9.8	45.7	0	2.8	1.8	0.8	0.1
29638	Fruit, non fat, sweetened with low calorie sweetener	1	cup(s)	241	210.3	108	9.0	19.0	1.0	0.4	0.2	0.1	0
68613	Greek style, vanilla	1	cup(s)	245	—	173	23.0	18.7	0	0	0	0	0
93	Plain, low fat	1	cup(s)	245	208.4	154	12.9	17.2	0	3.8	2.5	1.0	0.1
94	Plain, nonfat	1	cup(s)	245	208.8	137	14.0	18.8	0	0.4	0.3	0.1	0
32100	Vanilla, low fat	1	cup(s)	245	193.6	208	12.1	33.8	0	3.1	2.0	0.8	0.1
5242	Yogurt beverage	1	cup(s)	245	199.8	172	5.9	28.7	0	3.7	2.5	0.6	0.1
Soy yogurt													
29767	Silk blueberry cultured soy yogurt	5.3	ounce(s)	150	117.5	132	3.5	25.6	0.9	1.8	0	0.4	0.9

Chol (mg)	Calc (mg)	Iron (mg)	Magn (mg)	Pota (mg)	Sodi (mg)	Zinc (mg)	Vit A (µg)	Thia (mg)	Vit E (mg α)	Ribo (mg)	Niac (mg)	Vit B$_6$ (mg)	Fola (µg)	Vit C (mg)	Vit B$_{12}$ (µg)	Sele (µg)
32	92	0.06	10.1	143.3	57.6	0.50	85.0	0.03	0.22	0.17	0.08	0.03	3.6	0.4	0.3	1.3
10	100	0	—	—	65.0	—	—	—	—	—	—	—	—	0	—	—
21	122	0.14	10.6	158.1	56.2	0.55	97.3	0.04	0.09	0.19	0.10	0.03	4.6	0.9	0.4	1.5
11	138	0.05	12.3	194.5	61.6	0.47	25.5	0.05	0.05	0.17	0.10	0.04	4.4	0.8	0.4	3.2
0	0	0	—	22.0	220.0	—	—	—	—	—	—	—	—	0	—	—
0	0	0	—	—	180.0	—	—	—	—	—	—	—	—	0	—	—
0	0	0	—	—	130.0	—	—	—	—	—	—	—	—	0	—	—
21	118	0.51	14.8	125.8	52.5	0.32	52.5	0.02	0.15	0.09	0.09	0.02	3.0	0.9	0.1	1.6
20	119	0.14	10.4	153.9	54.8	0.54	94.7	0.04	0.09	0.19	0.10	0.03	4.4	0.9	0.3	1.5
35	149	0.95	28.6	211.0	134.8	1.07	0	0.06	0.11	0.25	0.18	0.05	9.9	0.2	0.7	5.4
10	150	0.72	—	330.0	310.0	—	—	0.06	—	0.26	—	—	—	0	—	—
7	115	0.27	15.8	194.6	72.0	0.58	24.8	0.04	0.05	0.18	0.34	0.06	4.5	0.2	0.3	4.3
1	101	0.16	8.5	130.6	205.9	0.31	0	0.03	0.21	0.14	0.09	0.03	4.3	0.4	0.3	0
35	144	0.16	13.8	173.2	133.7	0.63	0	0.05	0.09	0.24	0.11	0.05	8.7	0.2	0.7	5.0
10	150	0	—	190.0	380.0	—	—	0.03	—	0.17	—	—	—	0	—	—
4	106	0.90	19.4	187.9	70.6	0.35	31.7	0.03	—	0.15	0.22	0.05	7.9	0.2	0.2	1.7
0	117	0	7.2	155.8	87.4	0.25	105.7	0.02	0	0.10	0.06	0.02	3.2	0	0.4	1.5
1	103	0.22	10.1	151.9	62.6	0.30	42.5	0.03	0.08	0.16	0.21	0.06	4.3	0.6	0.2	2.4
3	500	0	—	410.0	125.0	—	150.1	—	—	0.43	—	—	—	0	0.9	—
15	300	0	—	410.0	125.0	—	150.1	—	—	0.43	—	—	—	0	0.9	—
20	300	0	—	410.0	125.0	—	150.1	—	—	0.43	—	—	—	0	0.9	—
35	300	0	—	410.0	125.0	—	90.1	—	—	0.43	—	—	—	0	0.9	—
0	0	0	—	0	80.0	—	0	—	0.80	—	—	—	—	0	—	—
0	300	0.72	—	0	100.0	—	25.0	—	—	—	—	—	—	0	1.5	—
0	300	0.36	—	0	105.0	—	25.0	—	—	—	—	—	—	0	1.5	—
0	300	1.44	40.0	350.0	100.0	0.60	25.0	—	—	0.51	—	—	—	0	3.0	4.2
0	350	1.08	40.0	350.0	120.0	1.50	50.0	—	—	0.51	—	0.60	—	21.0	3.0	5.6
0	300	1.08	40.0	300.0	85.0	0.60	25.0	—	—	0.51	—	—	—	0	3.0	5.6
0	300	1.08	40.0	300.0	95.0	0.60	25.0	—	—	0.51	—	—	—	0	3.0	5.6
0	400	1.80	60.0	290.0	125.0	0.75	25.0	0.15	5.03	0.63	1.20	0.20	—	0	3.0	11.2
0	350	8.10	60.0	250.0	135.0	0.62	25.0	0.15	5.03	0.44	1.02	0.20	21.6	0	3.0	5.5
15	300	0	—	290.0	110.0	—	—	—	—	—	—	—	—	0	—	—
15	300	0	—	290.0	110.0	—	—	—	—	—	—	—	—	0	—	—
12	338	0.15	31.9	433.6	129.9	1.64	27.0	0.08	0.05	0.40	0.21	0.09	22.0	1.5	1.1	6.9
5	323	0.58	33.7	470.0	127.7	1.57	4.8	0.09	0.14	0.38	0.42	0.10	26.5	21.9	1.0	6.7
0	288	0	—	—	108.0	—	—	—	—	—	—	—	—	0	—	—
15	448	0.20	41.7	573.3	171.5	2.18	34.3	0.11	0.07	0.52	0.28	0.12	27.0	2.0	1.4	8.1
5	488	0.22	46.5	624.8	188.6	2.38	4.9	0.12	0	0.57	0.30	0.13	29.4	2.2	1.5	8.8
12	419	0.17	39.2	536.5	161.7	2.03	29.4	0.10	0.05	0.49	0.26	0.11	27.0	2.0	1.3	12.0
15	225	0.22	27.0	303.8	95.6	1.10	14.7	0.11	0	0.27	0.30	0.15	29.4	2.1	0.7	—
0	265	0.95	—	—	22.1	—	—	—	—	—	—	—	—	26.5	—	—

APPENDIX H

DA+ Code	Food Description	Quantity	Measure	Wt (g)	H₂O (g)	Ener (kcal)	Prot (g)	Carb (g)	Fiber (g)	Fat (g)	Fat Breakdown (g) Sat	Mono	Poly
Dairy—*continued*													
34617	Stonyfield Farm O'Soy strawberry-peach pack organic cultured soy yogurt	1	item(s)	113	—	100	5.0	15.0	1.0	2.0	0	—	—
34616	Stonyfield Farm O'Soy vanilla organic cultured soy yogurt	1	item(s)	170	—	150	7.0	24.0	1.0	3.0	0	—	—
Eggs													
	Eggs												
99	Fried	1	item(s)	46	32.0	90	6.3	0.4	0	6.8	2.0	2.8	1.5
100	Hard boiled	1	item(s)	50	37.3	78	6.3	0.6	0	5.3	1.6	2.0	0.7
101	Poached	1	item(s)	50	37.9	72	6.3	0.4	0	4.7	1.6	1.8	1.0
97	Raw, white	1	item(s)	33	28.9	17	3.6	0.2	0	0.1	0	0	0
96	Raw, whole	1	item(s)	50	38.1	72	6.3	0.4	0	4.8	1.6	1.8	1.0
98	Raw, yolk	1	item(s)	17	8.9	55	2.7	0.6	0	4.5	1.6	2.0	0.7
102	Scrambled, prepared with milk and butter	2	item(s)	122	93.2	182	12.2	2.0	0	13.4	4.1	5.4	3.0
	Egg substitutes												
4028	Egg Beaters	¼	cup(s)	61	—	30	6.0	1.0	0	0	0	0	0
920	Frozen	¼	cup(s)	60	43.9	96	6.8	1.9	0	6.7	1.2	1.5	3.7
918	Liquid	¼	cup(s)	63	51.9	53	7.5	0.4	0	2.1	0.4	0.6	1.0
Seafood													
	Cod												
1573	Atlantic cod, cooked, dry heat	3	ounce(s)	85	64.6	89	19.4	0	0	0.7	0.1	0.1	0.2
6040	Atlantic cod or scrod, baked or broiled	3	ounce(s)	85	64.6	89	19.4	0	0	0.7	0.1	0.1	0.2
2905	**Eel, raw**	3	ounce(s)	85	58.0	156	15.7	0	0	9.9	2.0	6.1	0.8
	Fish fillets												
25079	Baked	3	ounce(s)	85	79.9	99	21.7	0	0	0.7	0.1	0.1	0.3
8615	Batter coated or breaded, fried	3	ounce(s)	85	45.6	197	12.5	14.4	0.4	10.5	2.4	2.2	5.3
25082	Broiled fish steaks	3	ounce(s)	85	68.1	128	24.2	0	0	2.6	0.4	0.9	0.8
25083	Poached fish steaks	3	ounce(s)	85	67.1	111	21.1	0	0	2.3	0.3	0.8	0.7
25084	Steamed	3	ounce(s)	85	72.2	79	17.2	0	0	0.6	0.1	0.1	0.2
25089	**Flounder, baked**	3	ounce(s)	85	64.4	113	14.8	0.4	0.1	5.5	1.1	2.2	1.4
1825	**Grouper, cooked, dry heat**	3	ounce(s)	85	62.4	100	21.1	0	0	1.1	0.3	0.2	0.3
	Haddock												
6049	Baked or broiled	3	ounce(s)	85	63.1	95	20.6	0	0	0.8	0.1	0.1	0.3
1578	Cooked, dry heat	3	ounce(s)	85	67.7	77	17.0	0	0	0.5	0.1	0.1	0.2
1886	**Halibut, Atlantic and Pacific, cooked, dry heat**	3	ounce(s)	85	64.7	94	19.2	0	0	1.4	0.3	0.5	0.3
1582	**Herring, Atlantic, pickled**	4	piece(s)	60	33.1	157	8.5	5.8	0	10.8	1.4	7.2	1.0
1587	**Jack mackerel, solids, canned, drained**	2	ounce(s)	57	39.2	88	13.1	0	0	3.6	1.1	1.3	0.9
8580	**Octopus, common, cooked, moist heat**	3	ounce(s)	85	51.4	139	25.4	3.7	0	1.8	0.4	0.3	0.4
1831	**Perch, mixed species, cooked, dry heat**	3	ounce(s)	85	62.3	99	21.1	0	0	1.0	0.2	0.2	0.4
1592	**Pacific rockfish, cooked, dry heat**	3	ounce(s)	85	63.5	93	18.9	0	0	1.4	0.4	0.4	0.4
	Salmon												
1594	Broiled or baked with butter	3	ounce(s)	85	53.9	155	22.9	0	0	6.3	1.2	2.3	2.3
2938	Coho, farmed, raw	3	ounce(s)	85	59.9	136	18.1	0	0	6.5	1.5	2.8	1.6
75533	Pink, canned, boneless, drained	3	ounce(s)	85	60.1	116	20.9	0	0	3.6	0.6	0.8	1.1
50124	Pink, canned, with bones, drained	3	ounce(s)	85	60.1	117	19.6	0.0	0	4.3	0.8	1.0	1.3
75532	Red (sockeye), canned, boneless, drained	3	ounce(s)	85	57.6	134	22.4	0	0	5.0	1.0	1.6	1.3
33210	Red (sockeye), canned, with bones, no salt, drained	3	ounce(s)	85	58.4	130	17.4	0	0	6.2	1.4	2.4	1.9
29727	Smoked chinook (lox)	2	ounce(s)	57	40.8	66	10.4	0	0.0	2.4	0.5	1.1	0.6
154	**Sardines, Atlantic with bones, canned in oil**	3	ounce(s)	85	50.7	177	20.9	0	0	9.7	1.3	3.3	4.4
	Scallops												
155	Mixed species, breaded, fried	3	item(s)	47	27.2	100	8.4	4.7	—	5.1	1.2	2.1	1.3
1599	Steamed	3	ounce(s)	85	69.7	59	10.2	2.7	0	0.4	0.1	0	0.1
1839	**Snapper, mixed species, cooked, dry heat**	3	ounce(s)	85	59.8	109	22.4	0	0	1.5	0.3	0.3	0.5
	Squid												
1868	Mixed species, fried	3	ounce(s)	85	54.9	149	15.3	6.6	0	6.4	1.6	2.3	1.8
16617	Steamed or boiled	3	ounce(s)	85	63.3	89	15.2	3.0	0	1.3	0.3	0.1	0.5

Chol (mg)	Calc (mg)	Iron (mg)	Magn (mg)	Pota (mg)	Sodi (mg)	Zinc (mg)	Vit A (µg)	Thia (mg)	Vit E (mg α)	Ribo (mg)	Niac (mg)	Vit B6 (mg)	Fola (µg)	Vit C (mg)	Vit B12 (µg)	Sele (µg)
0	100	1.08	32.0	220.0	25.0	—	—	0.22	—	0.10	—	0.08	—	0	0	—
0	150	1.44	40.0	310.0	40.0	—	—	0.38	—	0.17	—	0.08	—	0	0	—
184	29	0.87	6.0	69.9	95.2	0.64	100.7	0.02	0.60	0.23	0.04	0.08	23.5	0	0.4	15.2
187	25	0.60	5.0	63.0	62.0	0.52	74.5	0.03	0.51	0.26	0.03	0.06	22.0	0	0.6	15.4
185	28	0.88	6.0	69.0	148.5	0.64	80.0	0.02	0.52	0.19	0.03	0.07	17.5	0	0.4	15.3
0	2	0.03	3.6	53.8	54.8	0.01	0.0	0	0	0.14	0.03	0	1.3	0	0	6.6
186	28	0.88	6.0	69.0	71.0	0.64	80.0	0.02	0.52	0.23	0.04	0.09	23.5	0	0.4	15.4
184	22	0.46	0.9	18.5	8.2	0.39	64.8	0.03	0.44	0.09	0.00	0.06	24.8	0	0.3	9.5
338	81	1.60	13.4	161.0	176.9	1.27	196.4	0.05	1.40	0.46	0.09	0.16	43.9	0	0.9	28.7
0	20	1.08	—	95.0	115.0	0.60	225.0	0.15	—	0.85	—	0.08	—	0	1.2	
1	44	1.19	9.0	127.8	119.4	0.59	6.6	0.07	0.95	0.23	0.08	0.08	9.6	0.3	0.2	24.8
1	33	1.32	5.6	207.1	111.1	0.82	11.3	0.07	0.17	0.19	0.07	0	9.4	0	0.2	15.6
47	12	0.42	35.7	207.5	66.3	0.49	11.9	0.07	0.69	0.07	2.14	0.24	6.8	0.9	0.9	32.0
47	12	0.42	35.7	207.5	66.3	0.49	11.9	0.07	0.69	0.07	2.14	0.24	6.8	0.9	0.9	32.0
107	17	0.43	17.0	231.3	43.4	1.38	886.9	0.13	3.40	0.03	2.98	0.06	12.8	1.5	2.6	5.5
44	8	0.32	29.1	489.0	86.1	0.49	0	0.03	0.78	0.05	2.48	0.46	8.5	3.0	1.0	44.3
29	15	1.79	20.4	272.1	452.4	0.37	9.4	0.09	—	0.09	1.79	0.09	17.0	0	0.9	7.7
37	55	0.98	96.7	524.3	62.9	0.49	0	0.06	0.99	0.08	6.81	0.36	14.0	0	1.2	42.5
32	48	0.85	84.0	455.6	54.7	0.43	0	0.06	0.86	0.08	5.92	0.33	12.1	0	1.1	37.0
41	12	0.29	24.7	319.3	41.8	0.35	0	0.07	0.62	0.06	1.89	0.21	6.8	0.8	0.8	32.0
44	19	0.35	47.3	224.7	280.2	0.21	0	0.06	0.42	0.08	2.03	0.19	7.8	2.8	1.6	33.5
40	18	0.97	31.5	403.9	45.1	0.43	42.5	0.07	—	0.01	0.32	0.30	8.5	0	0.6	39.8
63	36	1.15	42.5	339.3	74.0	0.41	16.2	0.03	0.43	0.04	3.94	0.29	6.8	0	1.2	34.4
56	12	0.18	22.1	298.5	222.0	0.34	17.9	0.02	0.47	0.06	3.50	0.28	11.1	0	1.8	27.0
51	8	0.17	23.8	449.0	69.7	0.37	20.4	0.05	0.63	0.03	6.73	0.54	11.9	0	1.1	47.1
8	46	0.73	4.8	41.4	522.0	0.32	154.8	0.08	1.03	0.08	1.98	0.10	1.2	0	2.6	35.1
45	137	1.16	21.0	110.0	214.9	0.58	73.7	0.02	0.58	0.12	3.50	0.12	2.8	0.5	3.9	21.4
82	90	8.11	51.0	535.8	391.2	2.86	76.5	0.05	1.02	0.06	3.21	0.55	20.4	6.8	30.6	76.2
98	87	0.99	32.3	292.5	67.2	1.22	8.5	0.07	—	0.10	1.62	0.12	5.1	1.4	1.9	13.7
52	14	0.31	28.1	397.1	75.7	0.37	4.3	0.02	0.37	0.19	2.46	0.20	8.5	0	1.4	64.8
40	15	1.02	26.9	376.5	98.6	0.56		0.14	1.15	0.05	8.33	0.19	—	1.8	2.3	41.0
43	10	0.29	26.4	382.7	40.0	0.37	47.6	0.08	—	0.09	5.79	0.56	11.1	0.9	2.3	10.7
71	51	0.48	20.4	277.3	321.5	0.55	17.0	0.02	1.09	0.17	6.32	0.09	3.4	0	4.2	33.7
71	241	0.65	27.2	283.2	324.0	0.82	17.0	0.02	1.09	0.17	6.30	0.09	3.4	0	4.2	33.6
56	31	0.41	20.4	265.4	328.3	0.49	47.6	0.03	1.78	0.18	6.54	0.10	3.4	0	4.7	29.4
37	203	0.90	24.7	320.6	63.8	0.87	45.1	0.01	—	0.16	4.66	0.26	8.5	0	0.3	30.1
13	6	0.48	10.2	99.2	1133.8	0.18	14.7	0.01	—	0.06	2.68	0.16	1.1	0	1.8	21.6
121	325	2.48	33.2	337.6	429.5	1.11	27.2	0.07	1.74	0.19	4.46	0.14	8.5	0	7.6	44.8
25	20	0.38	27.4	154.8	215.8	0.49	10.2	0.02	—	0.05	0.70	0.07	23.3	1.1	0.6	12.5
20	5	0.29	18.7	155.6	547.7	0.76	0.8	0.01	0	0.01	0.53	0.06	10.2	0	1.1	10.8
40	34	0.20	31.5	443.9	48.5	0.37	29.8	0.05	—	0	0.29	0.39	5.1	1.4	3.0	41.7
221	33	0.86	32.3	237.3	260.2	1.48	9.4	0.05	—	0.39	2.21	0.05	17.0	3.6	1.0	44.1
227	31	0.63	28.9	192.2	356.3	1.50	8.5	0.01	1.17	0.32	1.70	0.04	3.4	3.2	1.0	43.7

APPENDIX H

(Computer code is for Cengage Diet Analysis program)   (For purposes of calculations, use "0" for t, <1, <0.1, 0.01)

DA+ Code	Food Description	Quantity	Measure	Wt (g)	H₂O (g)	Ener (kcal)	Prot (g)	Carb (g)	Fiber (g)	Fat (g)	Fat Breakdown (g)		
											Sat	Mono	Poly
Seafood—*continued*													
1570	**Striped bass, cooked, dry heat**	3	ounce(s)	85	62.4	105	19.3	0	0	2.5	0.6	0.7	0.9
1601	**Sturgeon, steamed**	3	ounce(s)	85	59.6	111	17.0	0	0	4.3	1.0	2.0	0.7
1840	**Surimi, formed**	3	ounce(s)	85	64.9	84	12.9	5.8	0	0.8	0.2	0.1	0.4
1842	**Swordfish, cooked, dry heat**	3	ounce(s)	85	58.0	146	19.9	0	0	6.7	1.6	3.0	1.2
1846	**Tuna, yellowfin or ahi, raw**	3	ounce(s)	85	63.0	93	20.7	0	0	0.4	0.1	0.1	0.1
	Tuna, canned												
159	Light, canned in oil, drained	2	ounce(s)	57	33.9	112	16.5	0	0	4.7	0.9	1.7	1.6
355	Light, canned in water, drained	2	ounce(s)	57	44.3	49	11.0	0	0	0.5	0.1	0.1	0.2
33211	Light, no salt, canned in oil, drained	2	ounce(s)	57	33.9	112	16.5	0	0	4.7	0.9	1.7	1.6
33212	Light, no salt, canned in water, drained	2	ounce(s)	57	42.6	66	14.5	0	0	0.5	0.1	0.1	0.2
2961	White, canned in oil, drained	2	ounce(s)	57	36.3	105	15.0	0	0	4.6	0.7	1.8	1.7
351	White, canned in water, drained	2	ounce(s)	57	41.5	73	13.4	0	0	1.7	0.4	0.4	0.6
33213	White, no salt, canned in oil, drained	2	ounce(s)	57	36.3	105	15.0	0	0	4.6	0.9	1.4	1.9
33214	White, no salt, canned in water, drained	2	ounce(s)	57	42.0	73	13.4	0	0	1.7	0.4	0.4	0.6
	Yellowtail												
8548	Mixed species, cooked, dry heat	3	ounce(s)	85	57.3	159	25.2	0	0	5.7	1.4	2.2	1.5
2970	Mixed species, raw	2	ounce(s)	57	42.2	83	13.1	0	0	3.0	0.7	1.1	0.8
	Shellfish, meat only												
1857	Abalone, mixed species, fried	3	ounce(s)	85	51.1	161	16.7	9.4	0	5.8	1.4	2.3	1.4
16618	Abalone, steamed or poached	3	ounce(s)	85	40.7	177	28.9	10.1	0	1.3	0.3	0.2	0.2
	Crab												
1851	Blue crab, canned	2	ounce(s)	57	45.2	47	10.1	0	0	0.4	0.1	0.1	0.1
1852	Blue crab, cooked, moist heat	3	ounce(s)	85	67.8	71	15.2	0	0	0.6	0.2	0.1	0.2
8562	Dungeness crab, cooked, moist heat	3	ounce(s)	85	62.3	94	19.0	0.8	0	1.1	0.1	0.2	0.3
1860	**Clams, cooked, moist heat**	3	ounce(s)	85	54.1	126	21.7	4.4	0	1.7	0.2	0.1	0.5
1853	**Crayfish, farmed, cooked, moist heat**	3	ounce(s)	85	68.7	74	14.9	0	0	1.1	0.2	0.2	0.4
	Oysters												
8720	Baked or broiled	3	ounce(s)	85	71.6	77	4.5	2.2	0	5.4	1.1	2.1	1.6
152	Eastern, farmed, raw	3	ounce(s)	85	73.3	50	4.4	4.7	0	1.3	0.4	0.1	0.5
8715	Eastern, wild, cooked, moist heat	3	ounce(s)	85	66.5	87	9.7	4.6	0	2.9	0.8	0.4	0.9
8584	Pacific, cooked, moist heat	3	ounce(s)	85	54.5	139	16.1	8.4	0	3.9	0.9	0.7	1.5
1865	Pacific, raw	3	ounce(s)	85	69.8	69	8.0	4.2	0	2.0	0.4	0.3	0.8
1854	**Lobster, northern, cooked, moist heat**	3	ounce(s)	85	66.4	76	16.2	0	0	0.7	0.2	0.2	0.3
1862	**Mussels, blue, cooked, moist heat**	3	ounce(s)	85	52.0	146	20.2	6.3	0	3.8	0.7	0.9	1.0
	Shrimp												
158	Mixed species, breaded, fried	3	ounce(s)	85	45.0	206	18.2	9.8	0.3	10.4	1.8	3.2	4.3
1855	Mixed species, cooked, moist heat	3	ounce(s)	85	60.9	101	19.4	1.3	0	1.4	0.4	0.3	0.5
Beef, Lamb, Pork													
	Beef												
4450	Breakfast strips, cooked	2	slice(s)	23	5.9	101	7.1	0.3	0	7.8	3.2	3.8	0.4
174	Corned beef, canned	3	ounce(s)	85	49.1	213	23.0	0	0	12.7	5.3	5.1	0.5
33147	Cured, thin siced	2	ounce(s)	57	32.9	100	15.9	3.2	0	2.2	0.9	1.0	0.1
4581	Jerky	1	ounce(s)	28	6.6	116	9.4	3.1	0.5	7.3	3.1	3.2	0.3
	Ground beef												
5898	Lean, broiled, medium	3	ounce(s)	85	50.4	202	21.6	0	0	12.2	4.8	5.3	0.5
5899	Lean, broiled, well done	3	ounce(s)	85	48.4	214	23.8	0	0	12.5	5.0	5.7	0.3
5914	Regular, broiled, medium	3	ounce(s)	85	46.1	246	20.5	0	0	17.6	6.9	7.7	0.7
5915	Regular, broiled, well done	3	ounce(s)	85	43.8	259	21.6	0	0	18.4	7.5	8.5	0.5
	Beef rib												
4241	Rib, small end, separable lean, 0" fat, broiled	3	ounce(s)	85	53.2	164	25.0	0.0	0	6.4	2.4	2.6	0.2
4183	Rib, whole, lean and fat, 1/4" fat, roasted	3	ounce(s)	85	39.0	320	18.9	0.0	0	26.6	10.7	11.4	0.9
	Beef roast												
16981	Bottom round, choice, separable lean and fat, 1/8" fat, braised	3	ounce(s)	85	46.2	216	27.9	0	0	10.7	4.1	4.6	0.4

Chol (mg)	Calc (mg)	Iron (mg)	Magn (mg)	Pota (mg)	Sodi (mg)	Zinc (mg)	Vit A (µg)	Thia (mg)	Vit E (mg α)	Ribo (mg)	Niac (mg)	Vit B$_6$ (mg)	Fola (µg)	Vit C (mg)	Vit B$_{12}$ (µg)	Sele (µg)
88	16	0.92	43.4	278.9	74.8	0.43	26.4	0.10	—	0.03	2.18	0.29	8.5	0	3.8	39.8
64	11	0.59	29.8	239.8	316.3	0.36	199.8	0.07	0.53	0.07	8.33	0.19	14.5	0	2.2	13.3
26	8	0.22	36.6	95.2	121.6	0.28	17.0	0.02	0.54	0.02	0.19	0.03	1.7	0	1.4	23.9
66	5	0.38	29.8	424.3	82.5	0.66	36.6	0.08	2.05	0.05	7.87	0.52	1.7	0	1.4	58.3
33	3	0.65	29.8	375.0	38.3	0.31	15.3	0.10	0.20	0.10	15.71	0.79	1.7	0	1.8	77.0
10	7	0.79	17.6	117.3	200.6	0.51	13.0	0.02	0.49	0.07	7.03	0.06	2.8	0	1.2	43.1
20	10	0.92	13.0	101.4	140.0	0.39	9.6	0.02	0.19	0.05	5.74	0.18	2.3	0	1.4	40.0
10	7	0.79	17.6	117.3	28.3	0.51	—	0.02	—	0.07	7.03	0.06	2.8	0	1.2	43.1
17	6	0.87	15.3	134.4	28.3	0.44	—	0.02		0.04	7.53	0.20	2.3	0	1.7	45.6
18	2	0.37	19.3	188.8	224.5	0.27	2.8	0.01	1.30	0.04	6.63	0.24	2.8	0	1.2	34.1
24	8	0.55	18.7	134.3	213.6	0.27	3.4	0	0.48	0.02	3.29	0.12	1.1	0	0.7	37.2
18	2	0.37	19.3	188.8	28.3	0.27	—	0.01	—	0.04	6.63	0.24	2.8	0	1.2	34.1
24	8	0.55	18.7	134.4	28.3	0.27	3.4	0		0.02	3.29	0.12	1.1	0	0.7	37.2
60	25	0.54	32.3	457.5	42.5	0.57	26.4	0.15		0.04	7.41	0.16	3.4	2.5	1.1	39.8
31	13	0.28	17.0	238.1	22.1	0.29	16.4	0.08		0.02	3.85	0.09	2.3	1.6	0.7	20.7
80	31	3.23	47.6	241.5	502.6	0.81	1.7	0.19	—	0.11	1.62	0.13	17.0	1.5	0.6	44.1
144	50	4.85	68.9	295.1	980.6	1.39	3.4	0.29	6.74	0.13	1.90	0.22	6.0	2.6	0.7	75.6
55	52	0.28	20.4	146.8	223.9	2.16	0.6	0.01	1.04	0.05	1.56	0.09	28.9	1.9	1.9	24.3
82	77	0.43	30.6	220.3	335.9	3.24	0.9	0.02	1.56	0.08	2.34	0.13	43.4	2.8	2.8	36.5
65	50	0.37	49.3	347.0	321.5	4.65	26.4	0.05	—	0.17	3.08	0.15	35.7	3.1	8.8	40.5
57	78	23.78	15.3	534.1	1022.2	2.32	145.4	0.13	—	0.36	2.85	0.09	24.7	18.8	84.1	54.4
117	43	0.94	28.1	202.4	82.5	1.26	12.8	0.04	—	0.07	1.42	0.11	9.4	0.4	2.6	29.1
31	48	3.67	14.5	125.0	308.7	31.26	49.3	0.01	1.12	0.06	0.70	0.02	5.1	0	6.6	15.6
21	37	4.92	28.1	105.5	151.4	32.25	6.8	0.09	—	0.06	1.08	0.05	15.3	4.0	13.8	54.2
67	99	7.83	29.8	118.2	141.2	66.84	22.1	0.03	1.45	0.15	1.57	0.05	11.9	0	14.9	33.6
85	14	7.82	37.4	256.8	180.3	28.27	124.2	0.11	0.72	0.38	3.08	0.08	12.8	10.9	24.5	131.0
43	7	4.35	18.7	142.9	90.1	14.13	68.9	0.06	—	0.20	1.71	0.04	8.5	6.8	13.6	65.5
124	82	0.25	36.6	195.6	413.3	3.44	0.8	0.02	0.85	0.01	1.56	0.10	9.4	0	1.2	62.2
48	28	5.71	31.5	227.9	313.8	2.27	77.4	0.26		0.36	2.55	0.09	64.6	11.6	20.4	76.2
117	57	1.07	34.0	191.4	292.6	1.17	47.6	0.11	1.11	0.12	2.61	0.08	33.2	1.3	1.6	35.5
179	77	0.27	31.5	144.6	805.3	1.39	76.5	0.03	1.87	0.02	2.28	0.21	20.4	0	1.4	42.1
27	2	0.71	6.1	93.1	509.2	1.44	0	0.02	0.07	0.06	1.46	0.07	1.8	0	0.8	6.1
73	10	1.77	11.9	115.7	762.8	3.04	0	0.02	0.13	0.13	2.07	0.11	7.7	0	1.4	36.5
23	6	1.53	10.8	243.2	815.8	2.26	0	0.05	0	0.11	2.99	0.19	6.2	0	1.5	16.0
14	6	1.54	14.5	169.2	590.0	2.30	0	0.04	0.14	0.04	0.49	0.05	38.0	0	0.3	3.0
58	6	2.00	17.9	266.2	59.5	4.64	0	0.05	—	0.23	4.22	0.23	7.7	0	1.8	16.0
69	12	2.21	18.5	250.0	62.4	5.87	0	0.09	—	0.24	5.10	0.16	9.4	0	1.7	19.1
62	9	2.08	17.0	248.3	70.6	4.41	0	0.03	—	0.16	4.91	0.23	7.7	0	2.5	16.2
71	12	2.30	18.5	242.4	72.4	5.19	0	0.09	—	0.23	4.93	0.18	8.5	0	1.6	18.0
77	16	1.59	21.3	319.8	51.9	4.64	0	0.07	0.34	0.13	7.16	0.54	8.5	0	1.4	29.2
72	9	1.96	16.2	251.7	53.6	4.46	0	0.06	—	0.14	2.86	0.20	6.0	0	2.1	18.7
86	6	2.30	17.9	223.7	35.7	4.59	0	0.06	0.42	0.16	5.05	0.37	8.5	0	1.7	29.3

DA+ Code	Food Description	Quantity	Measure	Wt (g)	H₂O (g)	Ener (kcal)	Prot (g)	Carb (g)	Fiber (g)	Fat (g)	Fat Breakdown (g)		
											Sat	Mono	Poly
Beef, Lamb, Pork—*continued*													
16979	Bottom round, separable lean and fat, 1/8" fat, roasted	3	ounce(s)	85	52.4	185	22.5	0	0	9.9	3.8	4.2	0.4
16924	Chuck, arm pot roast, separable lean and fat, 1/8" fat, braised	3	ounce(s)	85	42.9	257	25.6	0	0	16.3	6.5	7.0	0.6
16930	Chuck, blade roast, separable lean and fat, 1/8" fat, braised	3	ounce(s)	85	40.5	290	22.8	0	0	21.4	8.5	9.2	0.8
5853	Chuck, blade roast, separable lean, 0" trim, pot roasted	3	ounce(s)	85	47.4	202	26.4	0	0	10.0	3.9	4.3	0.3
4296	Eye of round, choice, separable lean, 0" fat, roasted	3	ounce(s)	85	56.5	138	24.4	0.0	0	3.7	1.3	1.5	0.1
16989	Eye of round, separable lean and fat, 1/8" fat, roasted	3	ounce(s)	85	52.2	180	24.2	0	0	8.5	3.2	3.6	0.3
	Beef steak												
4348	Short loin, t-bone steak, lean and fat, 1/4" fat, broiled	3	ounce(s)	85	43.2	274	19.4	0.0	0	21.2	8.3	9.6	0.8
4349	Short loin, t-bone steak, lean, 1/4" fat, broiled	3	ounce(s)	85	52.3	174	22.8	0.0	0	8.5	3.1	4.2	0.3
4360	Top loin, prime, lean and fat, 1/4" fat, broiled	3	ounce(s)	85	42.7	275	21.6	0.0	0	20.3	8.2	8.6	0.7
	Beef variety												
188	Liver, pan fried	3	ounce(s)	85	52.7	149	22.6	4.4	0	4.0	1.3	0.6	0.5
4447	Tongue, simmered	3	ounce(s)	85	49.2	242	16.4	0.0	0	19.0	6.9	8.6	0.6
	Lamb chop												
3275	Loin, domestic, lean and fat, 1/4" fat, broiled	3	ounce(s)	85	43.9	269	21.4	0	0	19.6	8.4	8.2	1.4
	Lamb leg												
3264	Domestic, lean and fat, 1/4" fat, cooked	3	ounce(s)	85	45.7	250	20.9	0	0	17.8	7.5	7.5	1.3
	Lamb rib												
182	Domestic, lean and fat, 1/4" fat, broiled	3	ounce(s)	85	40.0	307	18.8	0	0	25.2	10.8	10.3	2.0
183	Domestic, lean, 1/4" fat, broiled	3	ounce(s)	85	50.0	200	23.6	0	0	11.0	4.0	4.4	1.0
	Lamb shoulder												
186	Shoulder, arm and blade, domestic, choice, lean and fat, 1/4" fat, roasted	3	ounce(s)	85	47.8	235	19.1	0	0	17.0	7.2	6.9	1.4
187	Shoulder, arm and blade, domestic, choice, lean, 1/4" fat, roasted	3	ounce(s)	85	53.9	174	21.2	0	0	9.2	3.5	3.7	0.8
3287	Shoulder, arm, domestic, lean and fat, 1/4" fat, braised	3	ounce(s)	85	37.6	294	25.8	0	0	20.4	8.4	8.7	1.5
3290	Shoulder, arm, domestic, lean, 1/4" fat, braised	3	ounce(s)	85	41.9	237	30.2	0	0	12.0	4.3	5.2	0.8
	Lamb variety												
3375	Brain, pan fried	3	ounce(s)	85	51.6	232	14.4	0	0	18.9	4.8	3.4	1.9
3406	Tongue, braised	3	ounce(s)	85	49.2	234	18.3	0	0	17.2	6.7	8.5	1.1
	Pork, cured												
29229	Bacon, Canadian style, cured	2	ounce(s)	57	37.9	89	11.7	1.0	0	4.0	1.3	1.8	0.4
161	Bacon, cured, broiled, pan fried or roasted	2	slice(s)	16	2.0	87	5.9	0.2	0	6.7	2.2	3.0	0.7
42274	Bacon, cured, broiled, pan fried or roasted, reduced sodium	2	slice(s)	16	2.0	87	5.9	0.2	0	6.7	2.2	3.0	0.7
35422	Breakfast strips, cured, cooked	3	slice(s)	34	9.2	156	9.8	0.4	0	12.5	4.3	5.6	1.9
189	Ham, cured, boneless, 11% fat, roasted	3	ounce(s)	85	54.9	151	19.2	0	0	7.7	2.7	3.8	1.2
29215	Ham, cured, extra lean, 4% fat, canned	2	ounce(s)	57	41.7	68	10.5	0	0	2.6	0.9	1.3	0.2
1316	Ham, cured, extra lean, 5% fat, roasted	3	ounce(s)	85	57.5	123	17.8	1.3	0	4.7	1.5	2.2	0.5
16561	Ham, smoked or cured, lean, cooked	1	slice(s)	42	28.8	58	9.1	0.4	0	2.1	0.7	1.0	0.3
42257	Ham, smoked or cured, lean, cooked, low sodium	1	slice(s)	42	28.4	61	8.8	0.6	0	2.3	0.8	1.1	0.2
	Pork chop												
32671	Loin, blade, chops, lean and fat, pan fried	3	ounce(s)	85	49.3	218	21.3	0	0	14.1	3.8	4.3	1.7
32672	Loin, center cut, chops, lean and fat, pan fried	3	ounce(s)	85	49.7	202	23.5	0	0	11.3	4.1	4.7	1.8

Chol (mg)	Calc (mg)	Iron (mg)	Magn (mg)	Pota (mg)	Sodi (mg)	Zinc (mg)	Vit A (µg)	Thia (mg)	Vit E (mg α)	Ribo (mg)	Niac (mg)	Vit B6 (mg)	Fola (µg)	Vit C (mg)	Vit B12 (µg)	Sele (µg)
72	5	1.84	14.5	182.0	29.8	3.77	0	0.05	0.35	0.12	3.93	0.30	6.8	0	1.3	23.0
102	14	2.15	17.0	205.8	42.5	5.94	0	0.05	0.45	0.15	3.63	0.25	7.7	0	1.9	24.1
88	11	2.66	16.2	198.1	55.3	7.15	0	0.06	0.17	0.20	2.07	0.22	4.3	0	1.9	20.9
73	11	3.13	19.6	223.7	60.4	8.73	0	0.07	—	0.24	2.27	0.25	5.1	0	2.1	22.7
63	5	2.17	16.2	200.7	32.3	4.29	0	0.05	0.31	0.15	4.70	0.34	8.5	0	1.4	28.0
73	5	1.98	15.3	193.0	31.5	3.95	0	0.05	0.35	0.14	4.38	0.32	7.7	0	1.5	25.2
58	7	2.56	17.9	233.9	57.8	3.56	0	0.08	0.19	0.18	3.29	0.28	6.0	0	1.8	10.0
50	5	3.11	22.1	278.1	65.5	4.35	0	0.09	0.12	0.21	3.94	0.33	6.8	0	1.9	8.5
67	8	1.89	19.6	294.2	53.6	3.85	0	0.07	—	0.15	3.96	0.31	6.0	0	1.6	19.5
324	5	5.25	18.7	298.5	65.5	4.45	6586.3	0.15	0.39	2.91	14.86	0.87	221.1	0.6	70.7	27.9
112	4	2.22	12.8	156.5	55.3	3.48	0	0.02	0.26	0.25	2.97	0.13	6.0	1.1	2.7	11.2
85	17	1.54	20.4	278.1	65.5	2.96	0	0.09	0.11	0.21	6.04	0.11	15.3	0	2.1	23.3
82	14	1.60	19.6	263.6	61.2	3.79	0	0.09	0.12	0.21	5.66	0.11	15.3	0	2.2	22.5
84	16	1.60	19.6	229.6	64.6	3.40	0	0.08	0.10	0.19	5.95	0.09	11.9	0	2.2	20.3
77	14	1.88	24.7	266.2	72.3	4.48	0	0.09	0.15	0.21	5.57	0.13	17.9	0	2.2	26.4
78	17	1.68	19.6	213.5	56.1	4.45	0	0.08	0.12	0.20	5.23	0.11	17.9	0	2.2	22.3
74	16	1.81	21.3	225.4	57.8	5.14	0	0.08	0.15	0.22	4.90	0.13	21.3	0	2.3	24.2
102	21	2.03	22.1	260.2	61.2	5.17	0	0.06	0.13	0.21	5.66	0.09	15.3	0	2.2	31.6
103	22	2.30	24.7	287.4	64.6	6.21	0	0.06	0.15	0.23	5.38	0.11	18.7	0	2.3	32.1
2129	18	1.73	18.7	304.4	133.5	1.70	0	0.14	—	0.31	3.87	0.20	6.0	19.6	20.5	10.2
161	9	2.24	13.6	134.4	57.0	2.54	0	0.07	—	0.36	3.14	0.14	2.6	6.0	5.4	23.8
28	5	0.39	9.6	195.0	512.5	0.79	0	0.43	0.12	0.10	3.53	0.22	2.3	0	0.4	14.2
18	2	0.23	5.3	90.4	274.7	0.56	1.8	0.06	0.05	0.04	1.78	0.06	0.3	0	0.2	9.9
18	2	0.23	5.3	90.4	164.8	0.56	1.8	0.06	0.05	0.04	1.78	0.06	0.3	0	0.2	9.9
36	5	0.67	8.8	158.4	713.7	1.25	0	0.25	0.09	0.13	2.58	0.12	1.4	0	0.6	8.4
50	7	1.14	18.7	347.9	1275.8	2.10	0	0.62	0.26	0.28	5.23	0.26	2.6	0	0.6	16.8
22	3	0.53	9.6	206.4	711.5	1.09	0	0.47	0.10	0.13	3.01	0.26	3.4	0	0.5	8.2
45	7	1.26	11.9	244.1	1023.0	2.45	0	0.64	0.21	0.17	3.42	0.34	2.6	0	0.6	16.6
22	3	0.37	8.4	131.5	540.1	0.94	2.5	0.23	0.10	0.09	2.18	0.18	1.3	0	0.2	11.9
22	3	0.62	5.9	120.5	407.0	1.21	0	0.31	0.11	0.08	1.69	0.17	1.3	0	0.3	8.2
70	41	0.71	17.0	270.4	72.3	2.45	4.3	0.41	0.17	0.28	7.07	0.40	0	0	0.6	31.2
67	45	0.79	19.6	300.2	79.9	2.72	4.3	0.46	0.17	0.31	7.81	0.45	0	0	0.7	34.7

Table of Food Composition H-45

DA+ Code	Food Description	Quantity	Measure	Wt (g)	H₂O (g)	Ener (kcal)	Prot (g)	Carb (g)	Fiber (g)	Fat (g)	Fat Breakdown (g)		
											Sat	Mono	Poly
Beef, Lamb, Pork—*continued*													
32682	Loin, center rib, chops, boneless, lean and fat, braised	3	ounce(s)	85	49.5	217	22.4	0	0	13.4	5.2	6.1	1.1
32603	Loin, center rib, chops, lean, broiled	3	ounce(s)	85	55.4	158	21.9	0	0	7.1	2.4	3.0	0.8
32478	Loin, whole, lean and fat, braised	3	ounce(s)	85	49.6	203	23.2	0	0	11.6	4.3	5.2	1.0
32481	Loin, whole, lean, braised	3	ounce(s)	85	52.2	173	24.3	0	0	7.8	2.9	3.5	0.6
	Pork leg or ham												
32471	Rump portion, lean and fat, roasted	3	ounce(s)	85	52.7	178	23.0	0	0	8.8	2.9	3.7	1.9
32468	Whole, lean and fat, roasted	3	ounce(s)	85	46.8	232	22.8	0	0	15.0	5.5	6.7	1.4
	Pork ribs												
32693	Loin, country style, lean and fat, roasted	3	ounce(s)	85	41.1	305	18.5	0	0	25.1	9.1	10.9	2.5
32696	Loin, country style ribs, lean, roasted	3	ounce(s)	85	49.9	193	24.8	0	0	9.7	3.2	3.7	1.2
	Pork shoulder												
32626	Shoulder, arm picnic, lean and fat, roasted	3	ounce(s)	85	44.3	270	20.0	0	0	20.4	7.5	9.1	2.0
32629	Shoulder, arm picnic, lean, roasted	3	ounce(s)	85	51.3	194	22.7	0	0	10.7	3.7	5.1	1.0
	Rabbit												
3366	Domesticated, roasted	3	ounce(s)	85	51.5	168	24.7	0	0	6.8	2.0	1.8	1.3
3367	Domesticated, stewed	3	ounce(s)	85	50.0	175	25.8	0	0	7.2	2.1	1.9	1.4
	Veal												
3391	Liver, braised	3	ounce(s)	85	50.9	163	24.2	3.2	0	5.3	1.7	1.0	0.9
3319	Rib, lean only, roasted	3	ounce(s)	85	55.0	151	21.9	0	0	6.3	1.8	2.3	0.6
1732	Deer or venison, roasted	3	ounce(s)	85	55.5	134	25.7	0	0	2.7	1.1	0.7	0.5
Poultry													
	Chicken												
29562	Flaked, canned	2	ounce(s)	57	39.3	97	10.3	0.1	0	5.8	1.6	2.3	1.3
	Chicken, fried												
73336	Breast, meat and skin, breaded, fried, fast food	3	ounce(s)	85	45.1	214	18.6	6.6	0.4	12.6	3.3	4.7	3.4
36413	Broiler breast, meat and skin, flour coated, fried	3	ounce(s)	85	48.1	189	27.1	1.4	0.1	7.5	2.1	3.0	1.7
35327	Broiler breast, meat only, fried	3	ounce(s)	85	51.2	159	28.4	0.4	0	4.0	1.1	1.5	0.9
36414	Broiler drumstick, meat and skin, flour coated, fried	3	ounce(s)	85	48.2	208	22.9	1.4	0.1	11.7	3.1	4.6	2.7
35389	Broiler drumstick, meat only, fried	3	ounce(s)	85	52.9	166	24.3	0	0	6.9	1.8	2.5	1.7
35406	Broiler leg, meat only, fried	3	ounce(s)	85	51.5	177	24.1	0.6	0	7.9	2.1	2.9	1.9
35484	Broiler wing, meat only, fried	3	ounce(s)	85	50.9	179	25.6	0	0	7.8	2.1	2.6	1.8
29580	Patty, fillet or tenders, breaded, cooked	3	ounce(s)	85	39.1	246	15.9	13.4	0.7	14.2	2.3	5.4	5.8
	Chicken, roasted, meat only												
35409	Broiler leg, meat only, roasted	3	ounce(s)	85	55.0	162	23.0	0	0	7.2	1.9	2.6	1.7
35486	Broiler wing, meat only, roasted	3	ounce(s)	85	53.4	173	25.9	0	0	6.9	1.9	2.2	1.5
35138	Roasting chicken, dark meat, meat only, roasted	3	ounce(s)	85	57.0	151	19.8	0	0	7.4	2.1	2.8	1.7
35136	Roasting chicken, light meat, meat only, roasted	3	ounce(s)	85	57.7	130	23.1	0	0	3.5	0.9	1.3	0.8
35132	Roasting chicken, meat only, roasted	3	ounce(s)	85	57.3	142	21.3	0	0	5.6	1.5	2.1	1.3
	Chicken, stewed												
1268	Gizzard, simmered	3	ounce(s)	85	57.8	131	25.8	0.0	0	2.3	0.6	0.4	0.3
1270	Liver, simmered	3	ounce(s)	85	56.8	142	20.8	0.7	0	5.5	1.8	1.2	1.7
3174	Meat only, stewed	3	ounce(s)	85	56.8	151	23.2	0	0	5.7	1.6	2.0	1.3
	Duck												
1286	Domesticated, meat and skin, roasted	3	ounce(s)	85	44.1	287	16.1	0	0	24.1	8.2	11.0	3.1
1287	Domesticated, meat only, roasted	3	ounce(s)	85	54.6	171	20.0	0	0	9.5	3.4	3.3	1.3
	Goose												
35507	Domesticated, meat and skin, roasted	3	ounce(s)	85	44.2	259	21.4	0	0	18.6	5.8	8.7	2.1
35524	Domesticated, meat only, roasted	3	ounce(s)	85	48.7	202	24.6	0	0	10.8	3.9	3.7	1.3
1297	Liver pate, smoked, canned	4	tablespoon(s)	52	19.3	240	5.9	2.4	0	22.8	7.5	13.3	0.4

Chol (mg)	Calc (mg)	Iron (mg)	Magn (mg)	Pota (mg)	Sodi (mg)	Zinc (mg)	Vit A (µg)	Thia (mg)	Vit E (mg α)	Ribo (mg)	Niac (mg)	Vit B$_6$ (mg)	Fola (µg)	Vit C (mg)	Vit B$_{12}$ (µg)	Sele (µg)
62	4	0.78	14.5	329.1	34.0	1.76	1.7	0.45	—	0.21	3.67	0.26	3.4	0.3	0.4	28.4
56	22	0.58	21.3	291.7	48.5	1.91	0	0.48	0.09	0.19	6.68	0.57	0	0	0.4	38.6
68	18	0.91	16.2	318.0	40.8	2.02	1.7	0.54	0.20	0.22	3.76	0.31	2.6	0.5	0.5	38.5
67	15	0.96	17.0	329.1	42.5	2.11	1.7	0.56	0.18	0.23	3.90	0.33	3.4	0.5	0.5	41.0
72	14	0.77	21.3	353.8	65.5	1.96	2.6	0.42	0.23	0.32	6.33	0.43	0	0	0.6	21.3
80	12	0.86	18.7	299.3	51.0	2.52	2.6	0.54	0.19	0.27	3.89	0.34	8.5	0.3	0.6	38.5
77	21	0.85	18.7	273.8	44.2	1.87	1.7	0.70	0	0.28	3.57	0.36	3.4	0	0.6	29.3
84	26	0.83	18.7	331.7	77.4	3.35	0.9	0.49	0.23	0.39	6.65	0.44	0	0	0.8	40.5
80	16	1.00	14.5	276.4	59.5	2.93	1.7	0.44	—	0.26	3.33	0.30	3.4	0.2	0.6	28.6
81	8	1.21	17.0	298.5	68.0	3.46	0	0.49	—	0.30	3.67	0.35	4.3	0.3	0.7	32.7
70	16	1.93	17.9	325.7	40.0	1.93	0	0.08	—	0.18	7.17	0.40	9.4	0	7.1	32.7
73	17	2.02	17.0	255.1	31.5	2.02	0	0.05	0.37	0.14	6.09	0.29	7.7	0	5.5	32.7
435	5	4.35	17.0	279.8	66.3	9.55	17981.7	0.15	0.58	2.43	11.18	0.78	281.5	0.9	71.9	16.4
98	10	0.82	20.4	264.5	82.5	3.82	0	0.05	0.31	0.25	6.38	0.23	11.9	0	1.3	9.4
95	6	3.80	20.4	284.9	45.9	2.34	0	0.15		0.51	5.71		—	0		11.0
35	8	0.90	6.8	147.4	408.2	0.80	19.3	0.01	—	0.07	3.59	0.20	2.3	0	0.2	—
75	23	0.56	21.3	232.2	468.6	0.66	16.2	0.06	0.22	0.12	7.55	0.31	25.3	0	0.3	21.9
76	14	1.01	25.5	220.3	64.6	0.94	12.8	0.07	0.39	0.11	11.69	0.49	6.3	0	0.3	20.3
77	14	0.97	26.4	234.7	67.2	0.92	6.0	0.07	0.36	0.11	12.57	0.54	3.4	0	0.3	22.3
77	10	1.14	19.6	194.7	75.7	2.46	21.3	0.07	0.66	0.19	5.13	0.30	9.7	0	0.3	15.6
80	10	1.12	20.4	211.7	81.6	2.74	15.3	0.07	—	0.20	5.23	0.33	7.7	0	0.3	16.7
84	11	1.19	21.3	216.0	81.6	2.53	17.0	0.07	0.38	0.21	5.69	0.33	7.7	0	0.3	16.0
71	13	0.97	17.9	176.9	77.4	1.80	15.4	0.04	0.41	0.11	6.16	0.50	3.4	0	0.3	21.6
38	20	0.55	23.8	265.3	728.8	0.51	4.3	0.06	0.76	0.06	7.22	0.42	37.4	0	0.3	15.4
80	10	1.11	20.4	205.8	77.4	2.43	16.2	0.06	0.23	0.20	5.37	0.31	6.8	0	0.3	18.8
72	14	0.99	17.9	178.6	78.2	1.82	15.3	0.04	0.23	0.11	6.22	0.50	3.4	0	0.3	21.0
64	9	1.13	17.0	190.5	80.8	1.81	13.6	0.05	—	0.16	4.88	0.26	6.0	0	0.2	16.7
64	11	0.92	19.6	200.7	43.4	0.66	6.8	0.05	0.23	0.08	8.90	0.46	2.6	0	0.3	21.9
64	10	1.03	17.9	194.7	63.8	1.29	10.2	0.05	—	0.13	6.70	0.35	4.3	0	0.2	20.9
315	14	2.71	2.6	152.2	47.6	3.76	0	0.02	0.17	0.18	2.65	0.06	4.3	0	0.9	35.0
479	9	9.89	21.3	223.7	64.6	3.38	3385.4	0.25	0.70	1.69	9.39	0.64	491.5	23.7	14.3	70.1
71	12	0.99	17.9	153.1	59.5	1.69	12.8	0.04	0.23	0.14	5.20	0.22	5.1	0	0.2	17.8
71	9	2.30	13.6	173.5	50.2	1.58	53.6	0.15	0.60	0.23	4.10	0.15	5.1	0	0.3	17.0
76	10	2.30	17.0	214.3	55.3	2.21	19.6	0.22	0.60	0.40	4.34	0.21	8.5	0	0.3	19.0
77	11	2.41	18.7	279.8	59.5	2.23	17.9	0.07	1.48	0.27	3.54	0.31	1.7	0	0.3	18.5
82	12	2.44	21.3	330.0	64.6	2.70	10.2	0.08	—	0.33	3.47	0.40	10.2	0	0.4	21.7
78	36	2.86	6.8	71.8	362.4	0.48	520.5	0.05	—	0.16	1.31	0.03	31.2	0	4.9	22.9

APPENDIX H

DA+ Code	Food Description	Quantity	Measure	Wt (g)	H₂O (g)	Ener (kcal)	Prot (g)	Carb (g)	Fiber (g)	Fat (g)	Fat Breakdown (g)		
											Sat	Mono	Poly
Poultry—*continued*													
	Turkey												
3256	Ground, cooked	3	ounce(s)	85	52.8	173	23.3	0	0	8.8	2.3	2.9	2.5
3263	Patty, batter coated, breaded, fried	1	item(s)	94	46.7	266	13.2	14.8	0.5	16.9	4.4	7.0	4.4
219	Roasted, dark meat, meat only	3	ounce(s)	85	55.5	147	23.6	0	0	5.1	1.5	1.8	1.4
222	Roasted, fryer roaster breast, meat only	3	ounce(s)	85	58.2	115	25.6	0	0	0.6	0.2	0.1	0.2
220	Roasted, light meat, meat only	3	ounce(s)	85	57.7	125	25.6	0	0	1.8	0.5	0.5	0.4
1303	Turkey roll, light and dark meat	2	slice(s)	57	39.8	84	10.3	1.2	0	4.0	1.2	1.3	1.0
1302	Turkey roll, light meat	2	slice(s)	57	42.9	56	9.2	1.4	0	1.2	0.3	0.3	0.3
Processed Meats													
	Beef												
1331	Corned beef loaf, jellied, sliced	2	slice(s)	57	39.2	87	13.0	0	0	3.5	1.5	1.5	0.2
	Bologna												
13459	Beef	1	slice(s)	28	15.4	90	3.0	1.0	0	8.0	3.5	2.4	0.2
13461	Light, made with pork and chicken	1	slice(s)	28	—	60	3.0	2.0	0	4.0	1.0	2.0	0.4
13458	Made with chicken and pork	1	slice(s)	28	15.0	90	3.0	1.0	0	8.0	3.0	2.6	0.7
3260	Turkey	1	slice(s)	57	36.6	119	6.5	2.7	0.3	9.1	2.5	3.9	2.2
	Chicken												
56941	Deli-sliced, oven roasted	1	slice(s)	9	6.3	13	1.2	0.3	0	0.8	0.3	0.2	0.1
	Ham												
7127	Deli-sliced, honey	1	slice(s)	10	—	10	1.5	0.5	0	0.3	0	—	—
7126	Deli-sliced, smoked	1	slice(s)	9	—	10	1.5	0.2	0	0.2	0.1	—	—
8614	**Mortadella, beef & pork, sliced**	2	slice(s)	46	24.1	143	7.5	1.4	0	11.7	4.4	5.2	1.4
1323	**Pork olive loaf**	2	slice(s)	57	33.1	133	6.7	5.2	0	9.4	3.3	4.5	1.1
1324	**Pork pickle & pimento loaf**	2	slice(s)	57	34.2	128	6.4	4.8	0.9	9.1	3.0	4.0	1.6
	Sausages & frankfurters												
37296	Beerwurst beef, beer salami (bierwurst)	1	slice(s)	29	16.6	74	4.1	1.2	0	5.7	2.5	2.7	0.2
37257	Beerwurst pork beer salami	1	slice(s)	21	12.9	50	3.0	0.4	0	4.0	1.3	1.9	0.5
35338	Berliner, pork & beef	1	ounce(s)	28	17.3	65	4.3	0.7	0	4.9	1.7	2.3	0.4
37298	Bratwurst pork, cooked	1	piece(s)	74	42.3	181	10.4	1.9	0	14.3	5.1	6.7	1.5
37299	Braunschweiger pork liver sausage	1	slice(s)	15	7.7	50	2.2	0.5	0	4.3	1.4	1.9	0.5
1329	Cheesefurter or cheese smokie, beef & pork	1	item(s)	43	22.6	141	6.1	0.6	0	12.5	4.5	5.9	1.3
1330	Chorizo, beef & pork	2	ounce(s)	57	18.1	258	13.7	1.1	0	21.7	8.2	10.4	2.0
8600	Frankfurter, beef	1	item(s)	45	24.3	141	5.0	1.7	0	12.5	4.9	5.4	0.3
202	Frankfurter, beef & pork	1	item(s)	45	25.2	137	5.2	0.8	0	12.4	4.8	6.2	1.2
1293	Frankfurter, chicken	1	item(s)	45	28.1	100	7.0	1.2	0.2	7.3	1.7	2.7	1.7
42303	Frankfurter, low sodium	1	item(s)	57	32.3	178	6.8	1.0	0	16.3	6.9	7.8	0.8
3261	Frankfurter, turkey	1	item(s)	45	28.3	100	5.5	1.7	0	7.8	1.8	2.6	1.8
37275	Italian sausage, pork, cooked	1	item(s)	68	32.0	235	13.0	2.9	0.1	18.6	6.6	8.7	2.4
37307	Kielbasa, kolbassa, pork & beef	1	slice(s)	30	16.6	94	3.7	0.9	0	8.3	2.8	3.6	1.2
1333	Knockwurst or knackwurst, beef & pork	2	ounce(s)	57	31.3	174	6.3	1.8	0	15.7	5.8	7.3	1.7
37285	Pepperoni, beef & pork	1	slice(s)	11	3.4	54	2.5	0	0	4.8	1.6	1.9	0.4
37313	Polish sausage, pork	1	slice(s)	21	11.1	68	2.9	0.3	0	6.0	2.2	2.8	0.6
206	Salami, beef, cooked, sliced	2	slice(s)	52	31.2	136	6.6	1.0	0	11.5	5.1	5.5	0.5
54045	Salami, pork & beef, dry, sliced, 50% less sodium	1	slice(s)	6	2.3	20	1.2	0.4	0	1.5	0.5	0.7	0.1
37272	Salami, pork, dry or hard	1	slice(s)	13	4.6	52	2.9	0.2	0	4.3	1.5	2.0	0.5
40987	Sausage, turkey, cooked	2	ounce(s)	57	36.9	111	13.5	0.0	0	5.9	1.3	1.7	1.5
8620	Smoked sausage, beef & pork	2	ounce(s)	57	30.6	181	6.8	1.4	0	16.3	5.5	6.9	2.2
8619	Smoked sausage, pork	2	ounce(s)	57	32.0	175	6.8	0.5	0	16.0	5.3	6.4	2.1
37273	Smoked sausage, pork link	1	piece(s)	76	42.8	233	9.1	0.1	0	21.4	7.1	8.5	2.7
1336	Summer sausage, thuringer, or cervelat, beef & pork	2	ounce(s)	57	25.6	205	9.9	1.9	0	17.3	6.5	7.4	0.7
37294	Vienna sausage, cocktail, beef & pork, canned	1	piece(s)	16	10.4	37	1.7	0.4	0	3.1	1.1	1.5	0.2
	Spreads												
1318	Ham salad spread	¼	cup(s)	60	37.6	130	5.2	6.4	0	9.3	3.0	4.3	1.6
32419	Pork and beef sandwich spread	4	tablespoon(s)	60	36.2	141	4.6	7.2	0.1	10.4	3.6	4.6	1.5

Chol (mg)	Calc (mg)	Iron (mg)	Magn (mg)	Pota (mg)	Sodi (mg)	Zinc (mg)	Vit A (µg)	Thia (mg)	Vit E (mg α)	Ribo (mg)	Niac (mg)	Vit B$_6$ (mg)	Fola (µg)	Vit C (mg)	Vit B$_{12}$ (µg)	Sele (µg)
79	24	1.29	25.5	250.0	66.3	2.64	20.4	0.07	0.09	0.18	7.42	0.54	6.0	0	1.1	26.4
71	13	2.07	14.1	258.5	533.0	1.35	9.4	0.09	0.87	0.18	2.16	0.19	57.3	0	0.2	20.8
109	14	1.22	23.0	193.0	88.4	2.98	4.3	0.05	0.06	0.32	5.69	0.37	7.7	0	1.4	26.7
71	10	1.30	24.7	248.3	44.2	1.48	0	0.04	0.08	0.11	6.37	0.48	5.1	0	0.3	27.3
68	8	0.60	27.2	211.7	84.2	1.46	2.6	0.03	0.05	0.17	9.99	0.69	8.5	0	0.3	25.7
31	18	0.77	10.2	153.1	270.5	1.13	0	0.05	0.19	0.16	2.72	0.15	2.8	0	0.1	16.6
28	5	0.20	11.3	281.8	526.2	0.48	0	0.01	0.07	0.03	4.23	0.11	2.3	0	0.3	10.8
27	6	1.16	6.2	57.3	540.4	2.32	0	0	—	0.06	1.00	0.07	4.5	0	0.7	9.8
20	0	0.36	4.0	47.6	310.0	0.58	0	0.01	—	0.03	0.69	0.05	3.7	0	0.4	—
20	40	0.36	—	45.6	300.0	—	0	—	—	—	—	—	—	0	—	—
30	20	0.36	6.0	43.1	300.0	0.40	—	—	—	—	—	—	—	0	—	—
43	70	1.70	9.1	76.5	607.3	0.74	5.1	0.03	0.26	0.05	1.48	0.14	5.1	7.5	0.1	8.7
5	0	0.06	2.2	27.2	108.3	0.10	0.8	—	—	—	—	—	—	0	—	—
5	0	0.06	—	—	125.0	—	0	—	—	—	—	—	—	0	—	—
5	0	0.06	—	—	111.6	—	0	—	—	—	—	—	—	0	—	—
26	8	0.64	5.1	75.0	573.2	0.97	0	0.05	0.10	0.07	1.23	0.06	1.4	0	0.7	10.4
22	62	0.31	10.8	168.7	547.6	0.78	34.1	0.17	0.14	0.15	1.04	0.13	1.1	0	0.7	9.3
33	62	0.76	19.3	210.7	590.7	0.95	44.3	0.22	0.23	0.07	1.41	0.24	21.0	4.4	0.3	4.5
18	3	0.44	3.5	66.5	264.9	0.71	0	0.02	0.06	0.04	0.99	0.05	0.9	0	0.6	4.7
12	2	0.16	2.7	53.3	261.0	0.36	0	0.12	0.04	0.04	0.68	0.07	0.6	0	0.2	4.4
13	3	0.33	4.3	80.2	367.7	0.70	0	0.11	—	0.06	0.88	0.06	1.4	0	0.8	4.0
44	33	0.95	11.1	156.9	412.2	1.70	0	0.37	0.01	0.14	2.37	0.16	1.5	0.7	0.7	15.7
27	1	1.70	1.7	30.2	176.2	0.43	641.0	0.04	0.05	0.23	1.27	0.05	6.7	0	3.1	8.8
29	25	0.46	5.6	88.6	465.3	0.97	2.6	0.11	0.10	0.07	1.25	0.06	1.3	0	0.7	6.8
50	5	0.90	10.2	225.6	700.1	1.93	0	0.36	0.12	0.17	2.91	0.30	1.1	0	1.1	12.0
25	6	0.56	4.9	130.5	466.6	0.93	0	0.01	0.09	0.02	0.94	0.08	4.5	0	0.6	5.0
23	5	0.52	4.5	75.2	368.5	0.83	8.1	0.09	0.11	0.05	1.18	0.06	1.8	0	0.6	6.2
43	33	0.53	9.0	90.9	340.2	0.50	0	0.03	0.10	0.12	2.11	0.15	4.9	0	0.2	10.4
35	11	0.82	1.7	94.6	177.3	1.24	0	0.03	0.10	0.06	1.38	0.07	2.3	0	0.9	6.8
35	67	0.66	6.3	176.4	410.0	0.83	0	0.02	0.28	0.08	1.66	0.06	4.5	0	0.4	6.8
39	14	0.97	12.2	206.7	820.8	1.63	6.8	0.42	0.17	0.16	2.83	0.22	3.4	0.1	0.9	15.0
20	5	0.25	4.0	91.4	275.4	0.48	0	0.06	0.07	0.06	1.00	0.06	0.3	0	0.2	5.9
34	6	0.37	6.2	112.8	527.2	0.94	0	0.19	0.32	0.08	1.55	0.10	1.1	0	0.7	7.7
12	2	0.18	2.3	30.7	181.8	0.28	0	0.04	0	0.03	0.51	0.04	0.6	0.1	0.2	3.8
15	2	0.30	2.9	49.4	182.5	0.40	0	0.10	0.05	0.03	0.72	0.04	0.4	0.2	0.2	3.7
37	3	1.14	6.8	97.8	592.8	0.92	0	0.05	0.10	0.10	1.68	0.09	1.0	0	1.6	7.6
5	0	0.08	1.0	21.2	52.4	0.18	0	0.03	0.02	0.02	0.27	0.03	0.1	0	0.4	1.5
10	2	0.17	2.8	48.4	289.3	0.54	0	0.12	0.03	0.04	0.72	0.07	0.3	0	0.4	3.3
52	12	0.84	11.9	169.0	377.1	2.20	7.4	0.05	0.10	0.14	3.24	0.18	3.4	0.4	0.7	0
33	7	0.43	7.4	101.5	516.4	0.71	7.4	0.11	0.07	0.06	1.67	0.09	1.1	0	0.3	0
35	6	0.33	6.2	273.8	468.8	0.74	0	0.12	0.14	0.10	1.59	0.10	0.6	0	0.4	10.4
46	8	0.45	8.3	366.0	626.6	0.99	0	0.16	0.19	0.14	2.13	0.14	0.8	0	0.5	13.9
42	5	1.16	7.9	147.4	737.0	1.45	0	0.09	0.12	0.19	2.44	0.15	1.1	9.4	3.1	11.5
14	2	0.14	1.1	16.2	155.0	0.26	0	0.01	0.04	0.02	0.26	0.02	0.6	0	0.2	2.7
22	5	0.35	6.0	90.0	645.0	0.66	0	0.26	1.04	0.07	1.26	0.09	0.6	0	0.5	10.7
23	7	0.47	4.8	66.0	607.8	0.61	15.6	0.10	1.04	0.08	1.04	0.07	1.2	0	0.7	5.8

APPENDIX H

DA+ Code	Food Description	Quantity	Measure	Wt (g)	H₂O (g)	Ener (kcal)	Prot (g)	Carb (g)	Fiber (g)	Fat (g)	Fat Breakdown (g)		
											Sat	Mono	Poly
Processed Meats—*continued*													
	Turkey												
71632	Breast, honey roasted & smoked, presliced	1	slice(s)	28	—	30	5.0	2.0	0	0.5	0	—	—
71612	Breast, oven roasted deli slices, lower sodium	1	slice(s)	28	—	25	5.0	0.5	0	0.5		—	—
71631	Breast, oven roasted, presliced	1	slice(s)	28	—	30	5.0	1.0	0	0.5	0	—	—
7124	Breast, oven-roasted	1	slice(s)	9	—	8	1.5	0.3	0	0.1	0	—	—
71633	Breast, smoked, presliced	1	slice(s)	28	—	30	5.0	1.0	0	0.5	0	—	—
71629	Ham	1	slice(s)	28	—	40	5.0	0.5	0	2.0	0.5	—	—
37270	Pastrami	1	slice(s)	28	20.3	38	4.6	0.5	0	1.8	0.5	0.6	0.5
3262	Salami	2	slice(s)	57	39.1	98	10.9	0.9	0.1	5.2	1.6	1.8	1.4
37318	Salami, cooked	1	slice(s)	28	19.5	49	5.5	0.4	0	2.6	0.8	0.9	0.7
Beverages													
	Beer												
866	Ale, mild	12	fluid ounce(s)	360	331.1	155	1.7	12.8	0	0	0	0.0	0
686	Beer	12	fluid ounce(s)	356	327.7	153	1.6	12.7	0	0	0	0	0
16886	Beer, non alcoholic	12	fluid ounce(s)	360	328.1	133	0.8	29.0	0	0.4	0.1	0.1	0.2
31609	Bud Light beer	12	fluid ounce(s)	355	337.3	110	0.9	6.6	0	0	0	0	0
31608	Budweiser beer	12	fluid ounce(s)	355	329.3	145	1.3	10.6	0	0	0	0	0
869	Light beer	12	fluid ounce(s)	354	335.9	103	0.8	5.8	0	0	0	0.0	0
31613	Michelob Beer	12	fluid ounce(s)	355	326.5	164	1.7	15.0	0	0	0	0	0
31614	Michelob Light beer	12	fluid ounce(s)	355	336.8	123	1.4	8.8	0	0	0	0	0
	Gin, rum, vodka, whiskey												
687	Distilled alcohol, 80 proof	1	fluid ounce(s)	28	18.5	64	0	0	0	0	0	0	0
688	Distilled alcohol, 86 proof	1	fluid ounce(s)	28	17.8	70	0	0	0	0	0	0	0
689	Distilled alcohol, 90 proof	1	fluid ounce(s)	28	17.3	73	0	0	0	0	0	0	0
856	Distilled alcohol, 94 proof	1	fluid ounce(s)	28	16.8	76	0	0	0	0	0	0.0	0
857	Distilled alcohol, 100 proof	1	fluid ounce(s)	28	16.0	82	0	0	0	0	0	0.0	0
	Liqueurs												
33187	Coffee liqueur, 53 proof	1	fluid ounce(s)	35	10.8	113	0	16.3	0	0.1	0	0	0
3142	Coffee liqueur, 63 proof	1	fluid ounce(s)	35	14.4	107	0	11.2	0	0.1	0	0	0
736	Cordials, 54 proof	1	fluid ounce(s)	30	8.9	104	0	13.3	0	0.1	0	0	0
	Wine												
861	California red wine	5	fluid ounce(s)	150	133.4	125	0.3	3.7	0	0	0	0.0	0
858	Domestic champagne	5	fluid ounce(s)	150	—	105	0.3	3.8	0	0	0	0.0	0
690	Sweet dessert wine	5	fluid ounce(s)	147	103.7	235	0.3	20.1	0	0	0	0	0
1481	White wine	5	fluid ounce(s)	148	128.1	121	0.1	3.8	0	0	0.0	0	0
1811	Wine cooler	10	fluid ounce(s)	300	266.8	165	0.2	19.6	0	0.1	0	0	0
	Carbonated												
31898	7 Up	12	fluid ounce(s)	360	320.4	150	0	39.0	0	0	0	0	0
692	Club soda	12	fluid ounce(s)	355	354.8	0	0	0	0	0	0	0	0
12010	Coca-Cola Classic cola soda	12	fluid ounce(s)	360	325.1	210	0	58.5	0	0	0	0	0
693	Cola	12	fluid ounce(s)	368	332.7	136	0.3	35.2	0	0.1	0	0	0
2391	Cola or pepper-type soda, low calorie with saccharin	12	fluid ounce(s)	355	354.5	0	0	0.4	0	0	0	0	0
9522	Cola soda, decaffeinated	12	fluid ounce(s)	372	333.4	153	0.0	39.4	0	0			
9524	Cola, decaffeinated, low calorie with aspartame	12	fluid ounce(s)	355	354.3	4	0.4	0.5	0	0	0	0	0
1415	Cola, low calorie with aspartame	12	fluid ounce(s)	355	353.6	7	0.4	1.0	0	0.1	0	0	0
1412	Cream soda	12	fluid ounce(s)	371	321.5	189	0	49.3	0	0	0	0	0
31899	Diet 7 Up	12	fluid ounce(s)	360	—	0	0	0	0	0	0	0	0
12031	Diet Coke cola soda	12	fluid ounce(s)	360	358.3	2	0	0.2	0	0	0	0	0
29392	Diet Mountain Dew soda	12	fluid ounce(s)	354	—	0	0	0	0	0	0	0	0
29389	Diet Pepsi cola soda	12	fluid ounce(s)	354	352.8	0	0	0	0	0	0	0	0
12034	Diet Sprite soda	12	fluid ounce(s)	360	—	4	0	0	0	0	0	0	0
695	Ginger ale	12	fluid ounce(s)	366	333.9	124	0	32.1	0	0	0	0	0
694	Grape soda	12	fluid ounce(s)	372	330.3	160	0	41.7	0	0	0	0	0
1876	Lemon lime soda	12	fluid ounce(s)	368	330.7	147	0.2	37.4	0	0.1	0	0	0
29391	Mountain Dew soda	12	fluid ounce(s)	354	—	170	0	46.0	0	0	0	0	0
3145	Orange soda	12	fluid ounce(s)	372	325.9	179	0	45.8	0	0	0	0	0
1414	Pepper-type soda	12	fluid ounce(s)	368	329.3	151	0	38.3	0	0.4	0.3	0	0
29388	Pepsi regular cola soda	12	fluid ounce(s)	354	313.3	150	0.8	41.0	0	0	0	0	0
696	Root beer	12	fluid ounce(s)	370	330.1	152	0	39.2	0	0	0	0	0
12044	Sprite soda	12	fluid ounce(s)	360	323.2	144	0	39.0	0	0	0	0	0

Chol (mg)	Calc (mg)	Iron (mg)	Magn (mg)	Pota (mg)	Sodi (mg)	Zinc (mg)	Vit A (μg)	Thia (mg)	Vit E (mg α)	Ribo (mg)	Niac (mg)	Vit B$_6$ (mg)	Fola (μg)	Vit C (mg)	Vit B$_{12}$ (μg)	Sele (μg)
10	0	0	—	—	240.0	—	0	—	—	—	—	—	—	0	—	—
15	0	0.18	—	—	175.0	—	0	—	—	—	—	—	—	0	—	—
10	0	0	—	—	240.0	—	0	—	—	—	—	—	—	0	—	—
3	0	0.06	—	—	103.3	—	0	—	—	—	—	—	—	0	—	—
10	0	0	—	—	240.0	—	0	—	—	—	—	—	—	0	—	—
18	0	0.72	—	—	315.0	—	0	—	—	—	—	—	—	0	—	—
19	3	1.19	4.0	97.8	278.1	0.61	1.1	0.02	0.06	0.07	1.00	0.08	1.4	2.3	0.1	4.6
43	23	0.71	12.5	122.5	627.7	1.32	1.1	0.24	0.14	0.17	2.26	0.24	5.7	0	0.6	15.0
22	11	0.35	6.2	61.2	284.6	0.66	0.6	0.12	0.07	0.09	1.13	0.12	2.8	0	0.3	7.5
0	14	0.07	21.6	97.2	14.4	0.04	0	0.02	0	0.09	1.85	0.17	21.6	0	0.1	2.2
0	14	0.07	21.4	96.2	14.3	0.04	0	0.02	0	0.09	1.83	0.16	21.4	0	0.1	2.1
0	25	0.22	25.2	28.8	46.8	0.07	0	0.06	0	0.17	4.01	0.10	50.4	1.8	0.1	4.3
0	14	0.11	24.9	92.3	10.6	0	0	0.02	0	0.05	1.39	0.12	21.3	0	0.1	1.4
0	14	0	24.9	117.2	10.6	0	0	0.02	0	0.09	1.82	0.16	21.3	0	0.1	2.1
0	14	0.11	17.7	74.3	14.2	0.04	0	0.02	0	0.05	1.38	0.12	21.2	0	0.1	1.4
0	14	0.07	21.3	95.8	9.0	0.04	0	0.02	0	0.09	1.82	0.16	21.3	0	0.1	2.1
0	14	0.11	17.8	74.6	9.0	0.04	0	0.02	0	0.05	1.39	0.12	21.3	0	0.1	1.4
0	0	0.01	0	0.6	0.3	0.01	0	0	0	0	0	0	0	0	0	0
0	0	0.01	0	0.6	0.3	0.01	0	0	0	0	0	0	0	0	0	0
0	0	0.01	0	0.6	0.3	0.01	0	0	0	0	0	0	0	0	0	0
0	0	0.01	0	0.6	0.3	0.01	0	0	—	0	0	0	0	0	0	0
0	0	0.01	0	0.6	0.3	0.01	0	0	—	0	0	0	0	0	0	0
0	0	0.02	1.0	10.4	2.8	0.01	0	0	0	0	0.05	0	0	0	0	0.1
0	0	0.02	1.0	10.4	2.8	0.01	0	0	—	0	0.05	0	0	0	0	0.1
0	0	0.02	0.6	4.5	1.8	0.01	0	0	0	0	0.02	0	0	0	0	0.1
0	12	1.43	16.2	170.6	15.0	0.15	0	0.02	0	0.04	0.12	0.05	—	0	0	—
0	—	—	—	—	—	—	—	—	—	—	—	—	—	0	—	—
0	12	0.35	13.2	135.4	13.2	0.10	0	0.03	0	0.03	0.31	0	0	0	0	0.7
0	13	0.40	14.8	104.7	7.4	0.18	0	0.01	0	0.02	0.16	0.07	1.5	0	0	0.1
0	18	0.78	18.0	147.0	21.0	0.24	0	0.02	0.03	0.03	0.25	0.08	3.0	5.4	0	0.3
0	7	0.07	3.6	3.6	37.5	0.04	0	0	0	0	0.05	0	0	0	0	0
0	18	0.04	3.6	7.1	74.6	0.36	0	0	0	0	0	0	0	0	0	0
0	7	0.00	0	0	67.5	0.07	0	0	0	0	0	0	0	0	0	0.4
0	7	0.41	0	7.4	14.7	0.07	0	0	0	0	0	0	0	0	0	0.4
0	14	0.07	3.6	14.2	56.8	0.11	0	0	0	0	0	0	0	0	0	0.4
0	7	0.07	0	11.2	14.9	0.04	0	0	0	0	0	0	0	0	0	0.4
0	11	0.07	0	24.9	14.2	0.04	0	0.02	0	0.08	0	0	0	0	0	0.4
0	11	0.39	3.6	28.4	28.4	0.04	0	0.02	0	0.08	0	0	0	0	0	0
0	19	0.19	3.7	3.7	44.5	0.26	0	0	0	0	0	0	0	0	0	0
0	—	—	—	—	45.0	—	—	—	—	—	—	—	—	—	—	—
0	11	0.40	3.6	18.0	42.0	0.04	0	0.02	0	0.08	0	0	0	0	0	0
0	0	0	—	80.0	50.0	—	0	—	—	—	—	—	—	0	—	—
0	0	0	3.6	30.0	35.0	0.04	0	0.02	0	0.08	0	0	0	0	0	0
0	0	0.00	—	109.5	36.0	—	0	—	—	—	—	—	—	0	—	—
0	11	0.66	3.7	3.7	25.6	0.18	0	0	0	0	0	0	0	0	0	0.4
0	11	0.30	3.7	3.7	55.8	0.26	0	0	—	0	0	0	0	0	0	0
0	7	0.41	3.7	3.7	33.2	0.15	0	0	0	0	0.06	0	0	0	0	0
0	0	0	—	5.0	65.0	—	0	—	—	—	—	—	—	0	—	—
0	19	0.22	3.7	7.4	44.6	0.37	0	0	—	0	0	0	0	0	0	0.4
0	11	0.15	0	3.7	36.8	0.15	0	0	—	0	0	0	0	0	0	0.4
0	0	0	0.0	10.0	30.0	0.07	0	0	0	0	0	0	0	0	0	0.4
0	18	0.18	3.7	3.7	48.0	0.26	0	0	0	0	0	0	0	0	0	0.4
0	7	0.40	3.6	0	70.5	0.14	0	0	0	0	0.05	0	0	0	0	0

DA+ Code	Food Description	Quantity	Measure	Wt (g)	H₂0 (g)	Ener (kcal)	Prot (g)	Carb (g)	Fiber (g)	Fat (g)	Fat Breakdown (g)		
											Sat	Mono	Poly
	Beverages—*continued*												
	Coffee												
731	Brewed	8	fluid ounce(s)	237	235.6	2	0.3	0	0	0	0	0	0
9520	Brewed, decaffeinated	8	fluid ounce(s)	237	234.3	5	0.3	1.0	0	0	0	0	0
16882	Cappuccino	8	fluid ounce(s)	240	224.5	77	4.0	6.2	0.2	4.0	2.3	1.0	0.2
16883	Cappuccino, decaffeinated	8	fluid ounce(s)	240	224.4	74	4.0	6.2	0.2	4.0	2.3	1.0	0.2
16880	Espresso	8	fluid ounce(s)	237	231.8	5	0.3	0	0	0.4	0.2	0	0.2
16881	Espresso, decaffeinated	8	fluid ounce(s)	237	231.8	0	0.2	0	0	0.4	0.2	0	0.2
732	Instant, prepared	8	fluid ounce(s)	239	236.5	5	0.2	0.8	0	0	0	0	0
67526	Starbucks Doubleshot Energy, coffee, canned	15	fluid ounce(s)	443	—	210	12.0	36.0	0	2.0	2.5	—	—
67525	Starbucks Doubleshot, canned	6.5	fluid ounce(s)	192	—	138	4.1	17.9	—	5.7	3.7	—	—
32561	Starbucks Frappuccino, bottled, coffee	9.5	fluid ounce(s)	281	—	200	6.0	37.0	0	3.0	2.0	—	—
32562	Starbucks Frappuccino, bottled, mocha	9.5	fluid ounce(s)	281	—	180	7.0	33.0	0	3.0	2.0	—	—
	Fruit drinks												
29357	Crystal Light sugar-free lemonade drink	8	fluid ounce(s)	237	—	592	0	0	0	0	0	0	0
6012	Fruit punch drink with added vitamin C, canned	8	fluid ounce(s)	248	218.2	119	0	29.7	0	0	0	0	0
31143	Gatorade Thirst Quencher, all flavors	8	fluid ounce(s)	240	—	50	0	14.0	0	0	0	0	0
260	Grape drink, canned	8	fluid ounce(s)	250	210.5	153	0	39.4	0	0	0	0	0
17372	Kool-Aid (lemonade/punch/fruit drink)	8	fluid ounce(s)	248	220.0	108	0.1	27.8	0.2	0	0	0	0
17225	Kool-Aid sugar-free, low-calorie tropical punch drink mix, prepared	8	fluid ounce(s)	240	—	5	0	0	0	0	0	0	0
266	Lemonade, prepared from frozen concentrate	8	fluid ounce(s)	248	221.6	99	0.2	25.8	0	0.1	0	0	0
268	Limeade, prepared from frozen concentrate	8	fluid ounce(s)	247	212.6	128	0	34.1	0	0	0	0	0
57054	Odwalla Pomegranate Limeade	8	fluid ounce(s)	240	—	122	0	29.7	0	0	0	0	0
14266	Odwalla Strawberry C Monster smoothie blend	8	fluid ounce(s)	240	—	240	0	56.0	1.0	0	0	0	0
10099	Snapple fruit punch fruit drink	8	fluid ounce(s)	240	—	110	0	27.0	0	0	0	0	0
10096	Snapple kiwi strawberry fruit drink	8	fluid ounce(s)	240	211.2	110	0	27.0	0	0	0	0	0
14818	SunnyD Tangy Original	6 ¾	fluid ounce(s)	203	—	76	0	18.6	0	0	0	0	0
	Slim Fast ready-to-drink shake												
16054	French Vanilla	10	fluid ounce(s)	325	—	220	10.0	40.0	5.0	2.5	0.5	1.5	0.5
10095	Rich Chocolate Royale	10	fluid ounce(s)	325	—	220	10.0	40.0	5.0	3.0	1.0	1.5	0.5
16055	Strawberries 'n' Cream	10	fluid ounce(s)	325	—	220	10.0	40.0	5.0	2.5	0.5	1.5	0.5
	Tea												
33179	Decaffeinated, prepared	8	fluid ounce(s)	237	236.3	2	0	0.7	0	0	0	0	0
1877	Herbal, prepared	8	fluid ounce(s)	237	236.1	2	0	0.5	0	0	0	0	0
735	Instant tea mix, lemon flavored with sugar, prepared	8	fluid ounce(s)	259	236.2	91	0	22.3	0.3	0.2	0	0	0
734	Instant tea mix, unsweetened, prepared	8	fluid ounce(s)	238	237.1	2	0.1	0.4	0	0	0	0	0
16097	Nestea with lemon, canned	12	fluid ounce(s)	360	328.2	125	0	34.5	0	0	0	0	0
733	Prepared	8	fluid ounce(s)	237	236.3	2	0	0.7	0	0	0	0	0
	Water												
1413	Mineral water, carbonated	8	fluid ounce(s)	237	236.8	0	0	0	0	0	0	0	0
33183	Poland spring water, bottled	8	fluid ounce(s)	237	237.0	0	0	0	0	0	0	0	0
1821	Tap water	8	fluid ounce(s)	237	236.8	0	0	0	0	0	0	0	0
1879	Tonic water	8	fluid ounce(s)	244	222.3	83	0	21.5	0	0	0	0	0
	Fats and Oils												
	Butter												
104	Butter	1	tablespoon(s)	14	2.3	102	0.1	0	0	11.5	7.3	3.0	0.4
2522	Butter Buds, dry butter substitute	1	teaspoon(s)	2	0.1	7	0	1.6	0	0.1	0.1	—	—
921	Unsalted	1	tablespoon(s)	14	2.5	102	0.1	0	0	11.5	7.3	3.0	0.4
107	Whipped	1	tablespoon(s)	9	1.5	67	0.1	0	0	7.6	4.7	2.2	0.3
944	Whipped, unsalted	1	tablespoon(s)	9	1.7	67	0.1	0	0	7.6	4.8	2.0	0.3

Chol (mg)	Calc (mg)	Iron (mg)	Magn (mg)	Pota (mg)	Sodi (mg)	Zinc (mg)	Vit A (µg)	Thia (mg)	Vit E (mg α)	Ribo (mg)	Niac (mg)	Vit B$_6$ (mg)	Fola (µg)	Vit C (mg)	Vit B$_{12}$ (µg)	Sele (µg)
0	5	0.02	7.1	116.1	4.7	0.05	0	0.03	0.02	0.18	0.45	0	4.7	0	0	0
0	9	0.09	9.5	85.2	9.5	0.02	0	0	0	0.03	0.67	0	0	0	0	0.5
12	144	0.10	16.8	220.8	55.2	0.48	55.2	0.07	0.10	0.29	0.34	0.04	7.2	0	0.4	4.6
12	144	0.14	19.2	225.6	55.2	0.48	55.2	0.05	0.10	0.21	0.38	0.04	4.8	0	0.4	4.6
0	5	0.31	189.6	272.5	33.2	0.12	0	0	0.02	0.42	12.34	0	2.4	0.5	0	0
0	5	0.31	189.6	272.5	33.2	0.12	0	0	0	0.42	12.34	0	2.4	0.5	0	0
0	10	0.10	9.5	71.6	9.5	0.02	0	0	0	0	0.56	0	0	0	0	0.2
15	4	0	—	—	170.0	—	0	—	—	1.70	20.00	2.00	—	0.3	3.0	—
20	162	0	—	20.3	69.1	—	0	—	—	—	—	—	—	0	—	—
15	3	0	—	—	100.0	—	0	—	—	—	—	—	—	0	—	—
15	2	0	—	—	95.0	—	0	—	—	—	—	—	—	0	—	—
0	0	0	—	—	4140.5	—	—	—	—	—	—	—	—	0	—	—
0	20	0.22	7.4	62.0	24.8	0.03	1.7	0.05	0.05	0.06	0.05	0.03	9.9	89.3	0	0.5
0	2	0	—	30.0	110.0	—	0	—	—	—	—	—	—	0	—	—
0	130	0.17	2.5	30.0	40.0	0.30	0	0	0	0.01	0.03	0	0	78.5	0	0.3
0	14	0.46	5.0	49.6	31.0	0.20	—	0.04	—	0.05	0.05	0.01		41.6	0	1.0
0	0	0	—	10.1	10.1	—	0	—	—	—	—	—	—	6.0	—	—
0	10	0.40	5.0	37.2	9.9	0.05	0	0.01	0.02	0.05	0.04	0.01	2.5	9.7	0	0.2
0	5	0	4.9	24.7	7.4	0.02	0	0	0	0.01	0.02	0.01	2.5	7.7	0	0.2
0	0	0	—	—	10.1	—	0	—	—	—	—	—	—	0	—	—
0	40	1.08	—	600.0	30.0	—	0	—	—	—	—	—	—	900.0	—	—
0	0	0	—	20.0	5.0	—	0	—	—	—	—	—	—	0	—	—
0	0	0	—	40.0	10.0	—	0	—	—	—	—	—	—	0	—	—
0	0	0	—	—	143.4	—	0	0.19	—	—	—	—	—	50.6	—	—
5	400	2.70	140.0	600.0	220.0	2.25	525.1	0.52	—	0.60	7.00	0.70	—	60.0	2.1	17.5
5	400	2.70	140.0	600.0	220.0	2.25	525.1	0.53	—	0.60	7.00	0.70	—	60.0	2.1	17.5
5	400	2.70	140.0	600.0	220.0	2.25	525.1	0.52	—	0.60	7.00	0.70	—	60.0	2.1	17.5
0	0	0.05	7.1	87.7	7.1	0.05	0	0	0	0.03	0	0	11.9	0	0	0
0	5	0.19	2.4	21.3	2.4	0.09	0	0.02	0	0.01	0	0	2.4	0	0	0
0	5	0.05	2.6	38.8	5.2	0.03	0	0	0	0	0.03	0	0	0	0	0.3
0	7	0.02	4.8	42.8	9.5	0.02	0	0	0	0.01	0.08	0	0	0	0	0
0	11	0	3.6	68.4	46.5	0.22	0	—	—	—	—	—	—	—	—	—
0	0	0.05	7.1	87.7	7.1	0.05	0	0	0	0.03	0	0	11.9	0	0	0
0	33	0	0	0.0	2.4	0	0	0	—	0	0	0	0	0	0	0
0	2	0.02	2.4	0	2.4	0	0	0	—	0	0	0	0	0	0	0
0	7	0	2.4	2.4	7.1	0	0	0	0	0	0	0	0	0	0	0
0	2	0.02	0	0	29.3	0.24	0	0	0	0	0	0	0	0	0	0
31	3	0.00	0.3	3.4	101.4	0.01	97.1	0	0.33	0	0.01	0	0.4	0	0	0.1
0	8	0	0	—	80.2	0	0	0	0	0	0	0	—	0	0	0
31	3	0	0.3	3.4	1.6	0.01	97.1	0	0.33	0	0.01	0	0.4	0	0	0.1
21	2	0.02	0.2	2.4	77.7	0	64.3	0	0.22	0	0	0	0.3	0	0	0.1
20	2	0	0.2	2.3	1.0	0.01	64.3	0	0.22	0	0	0	0.3	0	0	0.1

APPENDIX H

DA+ Code	Food Description	Quantity	Measure	Wt (g)	H₂O (g)	Ener (kcal)	Prot (g)	Carb (g)	Fiber (g)	Fat (g)	Fat Breakdown (g)		
											Sat	Mono	Poly
Fats and Oils—*continued*													
	Fats, cooking												
2671	Beef tallow, semisolid	1	tablespoon(s)	13	0	115	0	0	0	12.8	6.4	5.4	0.5
922	Chicken fat	1	tablespoon(s)	13	0	115	0	0	0	12.8	3.8	5.7	2.7
5454	Household shortening with vegetable oil	1	tablespoon(s)	13	0	115	0	0	0	13.0	3.4	5.5	3.5
111	Lard	1	tablespoon(s)	13	0	115	0	0	0	12.8	5.0	5.8	1.4
	Margarine												
114	Margarine	1	tablespoon(s)	14	2.3	101	0	0.1	0	11.4	2.1	5.5	3.4
5439	Soft	1	tablespoon(s)	14	2.3	103	0.1	0.1	0	11.6	1.7	4.4	2.1
928	Unsalted	1	tablespoon(s)	14	2.6	101	0.1	0.1	0	11.3	2.1	5.2	3.5
119	Whipped	1	tablespoon(s)	9	1.5	67	0	0.1	0	7.5	1.3	3.3	2.4
	Spreads												
54657	I Can't Believe It's Not Butter!, tub, soya oil (non-hydrogenated)	1	tablespoon(s)	14	2.3	102	0.1	0.1	0	11.4	2.8	2.0	5.1
2708	Mayonnaise with soybean and safflower oils	1	tablespoon(s)	14	2.1	99	0.2	0.4	0	11.0	1.2	1.8	7.6
57582	Promise buttery spread	1	tablespoon(s)	14	—	80	0	0	0	8.0	1.5	2.5	4.0
	Oils												
2681	Canola	1	tablespoon(s)	14	0	120	0	0	0	13.6	1.0	8.6	3.8
120	Corn	1	tablespoon(s)	14	0	122	0	0	0	13.6	1.8	3.8	7.4
122	Olive	1	tablespoon(s)	14	0	119	0	0	0	13.5	1.9	9.8	1.4
124	Peanut	1	tablespoon(s)	14	0	119	0	0	0	13.5	2.3	6.2	4.3
2693	Safflower	1	tablespoon(s)	14	0	120	0	0	0	13.6	1.0	10.2	1.7
923	Sesame	1	tablespoon(s)	14	0	120	0	0	0	13.6	1.9	5.4	5.7
128	Soybean, hydrogenated	1	tablespoon(s)	14	0	120	0	0	0	13.6	2.0	5.8	5.1
130	Soybean, with soybean and cottonseed oil	1	tablespoon(s)	14	0	120	0	0	0	13.6	2.4	4.0	6.5
2700	Sunflower	1	tablespoon(s)	14	0	120	0	0	0	13.6	1.8	6.3	5.0
357	**Pam original no stick cooking spray**	1	serving(s)	0	—	0	0	0	0	0	0	0	0
	Salad dressing												
132	Blue cheese	2	tablespoon(s)	30	11.9	143	0.4	1.4	0.1	15.3	2.5	4.0	8.3
133	Blue cheese, low calorie	2	tablespoon(s)	31	25.4	32	1.6	0.9	0	2.3	0.8	0.6	0.8
1764	Caesar	2	tablespoon(s)	30	10.3	163	0.7	1.0	0.2	17.4	2.6	4.1	9.9
29654	Creamy, reduced calorie, fat free, cholesterol free, sour cream and/or buttermilk and oil	2	tablespoon(s)	33	24.7	35	0.5	6.6	0	0.9	0.2	0.2	0.5
29617	Creamy, reduced calorie, sour cream and/or buttermilk and oil	2	tablespoon(s)	30	22.2	48	0.4	2.1	0	4.2	0.6	1.0	2.4
134	French	2	tablespoon(s)	32	11.7	146	0.2	5.0	0	14.3	1.8	2.7	6.7
135	French, low fat	2	tablespoon(s)	32	17.4	71	0.2	10.0	0.5	3.7	0.3	1.4	1.2
136	Italian	2	tablespoon(s)	29	18.6	71	0.1	3.6	0	6.2	0.9	1.7	3.2
137	Italian, diet	2	tablespoon(s)	30	24.0	31	0.1	3.0	0	2.0	0.2	0.5	1.0
139	Mayonnaise-type	2	tablespoon(s)	29	11.7	115	0.3	7.0	0	9.8	1.4	2.6	5.3
942	Oil and vinegar	2	tablespoon(s)	32	15.2	144	0	0.8	0	16.0	2.9	4.7	7.7
1765	Ranch	2	tablespoon(s)	30	11.5	145	0.3	2.0	0.2	15.4	2.4	3.4	8.5
3666	Ranch, reduced calorie	2	tablespoon(s)	30	20.5	63	0.1	2.0	0	6.1	1.6	3.6	0.6
940	Russian	2	tablespoon(s)	30	11.6	107	0.2	9.6	0.2	7.9	0.7	1.8	4.4
939	Russion, low calorie	2	tablespoon(s)	32	20.8	45	0.2	8.8	0.1	1.3	0.2	0.3	0.7
941	Sesame seed	2	tablespoon(s)	30	11.8	133	0.9	2.6	0.3	13.6	1.9	3.6	7.5
142	Thousand Island	2	tablespoon(s)	32	14.9	118	0.3	4.7	0.3	11.2	1.6	2.5	5.8
143	Thousand Island, low calorie	2	tablespoon(s)	30	18.2	59	0.2	7.2	0.4	3.4	0.2	1.9	0.8
	Sandwich spreads												
140	Mayonnaise, low calorie	1	tablespoon(s)	16	8.9	52	0.1	1.3	0	5.3	0.8	1.3	2.9
138	Mayonnaise with soybean oil	1	tablespoon(s)	14	3.0	94	0.1	0.1	0	10.3	1.6	2.3	6.2
141	Tartar sauce	2	tablespoon(s)	29	18.8	59	0.3	3.7	0.1	4.7	0.9	1.0	2.5
Sweets													
4799	**Butterscotch or caramel topping**	2	tablespoon(s)	41	13.1	103	0.6	27.0	0.4	0	0	0	0
	Candy												
1786	Almond Joy candy bar	1	item(s)	45	3.7	220	2.0	26.0	2.0	13.0	8.0	2.4	0.5
1785	Bit-O-Honey candy	6	item(s)	40	—	150	1.0	32.0	0	3.0	2.0	—	—
33375	Butterscotch candy	2	piece(s)	12	0.6	47	0	10.8	0	0.4	0.2	0.1	0
1701	Chewing gum, stick	1	item(s)	3	0.1	11	0	2.9	0.1	0	0	0	0
33378	Chocolate fudge with nuts, prepared	2	piece(s)	38	2.9	175	1.7	25.8	0.9	7.2	2.5	1.5	2.9

APPENDIX H

Chol (mg)	Calc (mg)	Iron (mg)	Magn (mg)	Pota (mg)	Sodi (mg)	Zinc (mg)	Vit A (µg)	Thia (mg)	Vit E (mg α)	Ribo (mg)	Niac (mg)	Vit B$_6$ (mg)	Fola (µg)	Vit C (mg)	Vit B$_{12}$ (µg)	Sele (µg)
14	0	0	0	0	0	0	0	0	0.35	0	0	0	0	0	0	0
11	0	0	0	0	0	0	0	0	0.35	0	0	0	0	0	0	0
0	0	0	0	0	0	0	0	0	0	0	0	0	0	0	0	0
12	0	0	0	0	0	0.01	0	0	0.08	0	0	0	0	0	0	0
0	4	0.01	0.4	5.9	133.0	0	115.5	0	1.27	0.01	0	0	0.1	0	0	0
0	4	0	0.3	5.5	155.4	0	142.7	0	1.01	0	0	0	0.1	0	0	0
0	2	0	0.3	3.5	0.3	0	115.5	0	1.80	0	0	0	0.1	0	0	0
0	0	0	0.1	1.5	59.1	0	73.7	0	1.39	0	0	0	0.1	0	0	0
0	4	0	0.3	5.4	153.1	0	140.6	0	0.71	0	0	0	0.1	0	0	0
8	2	0.07	0.1	4.7	78.4	0.02	11.6	0	3.04	0	0	0.08	1.1	0	0	0.2
0	0	0	—	—	85.0	—	—	—	2.01	—	—	0.70	—	0	1.2	—
0	0	0	0	0	0	0	0	0	2.37	0	0	0	0	0	0	0
0	0	0	0	0	0	0	0	0	1.94	0	0	0	0	0	0	0
0	0	0.08	0	0.1	0.3	0	0	0	1.94	0	0	0	0	0	0	0
0	0	0	0	0	0	0	0	0	2.12	0	0	0	0	0	0	0
0	0	0	0	0	0	0	0	0	4.64	0	0	0	0	0	0	0
0	0	0	0	0	0	0	0	0	0.19	0	0	0	0	0	0	0
0	0	0	0	0	0	0	0	0	1.10	0	0	0	0	0	0	0
0	0	0	0	0	0	0	0	0	1.65	0	0	0	0	0	0	0
0	0	0	0	0	0	0	0	0	5.59	0	0	0	0	0	0	0
0	0	0	—	—	0	—	—	—	—	—	—	—	—	0	—	—
9	11	0.03	2.4	26.4	312.3	0.06	5.7	0	1.28	0.03	0.03	0.01	2.4	0.2	0	0.3
0	28	0.16	2.2	1.6	300.5	0.08	0	0.01	0.08	0.03	0.02	0.01	1.0	0.1	0.1	0.5
12	14	0.32	0.6	8.7	362.7	0.03	2.7	0	1.42	0	0.01	0.01	0.6	0.1	0	0.5
0	12	0.08	1.6	43.9	296.0	0.06	0.3	0	0.22	0.02	0.01	0.01	2.0	0	0	0.5
0	2	0.04	0.6	10.8	336.0	0.01	0	0	0.72	0	0.01	0.01	0	0.1	0	0.5
0	8	0.26	1.6	21.4	267.5	0.09	7.4	0.01	1.60	0.02	0.06	0	0	1.2	0	0
0	4	0.23	2.6	34.2	268.2	0.06	8.6	0.01	0.32	0.02	0.15	0.02	0.6	1.5	0	0.5
0	4	0.08	1.5	24.7	291.9	0.02	0.6	0	0.64	0	0.04	0.02	0	0.1	0	0.6
0	5	0.08	1.2	27.0	300.9	0.02	0.3	0	1.28	0	0.03	0.02	0.9	0	0	0.5
8	4	0.06	0.6	2.6	209.0	0.05	6.2	0	0.61	0.01	0	0	1.8	0	0.1	0.5
0	0	0	0	2.6	0.3	0	0	0	1.48	0	0	0	0	0	0	0.5
10	9	0.19	1.5	18.6	328.2	0.12	3.0	0.03	1.27	0.02	0	0.01	1.2	1.0	0.1	0.6
0	5	0.01	0.9	8.1	414.0	0.02	0.6	0	0.73	0.01	0	0	0.3	0.1	0	0.1
0	4	0.18	3.0	51.9	339.9	0.07	8.7	0.01	1.00	0.01	0.18	0.03	1.5	1.8	0	0.5
2	6	0.19	0	50.2	277.8	0.03	0.6	0	0.13	0	0	0	1.0	1.9	0	0.5
0	6	0.18	0	47.1	300.0	0.03	0.6	0	1.50	0	0	0	0	0	0	0.5
8	5	0.38	2.6	34.2	276.2	0.08	4.5	0.46	1.28	0.02	0.13	0	0	0	0	0.5
3	8	0.27	2.1	60.6	286.5	0.06	4.8	0.01	0.30	0.01	0.13	0	0	0.4	0	0
6	1	0.05	0.3	6.4	107.7	0.03	3.4	0	0.49	0	0	0	0.6	0	0	0.4
6	1	0.03	0.1	2.8	87.6	0.02	2.2	0	0.45	0	0	0	0.7	0	0	0.3
2	7	0.07	1.7	19.0	186.8	0.03	3.1	0	0.47	0.01	0.03	0.01	1.4	0.6	0	0.3
0	22	0.08	2.9	34.4	143.1	0.08	11.1	0	—	0.04	0.02	0.01	0.8	0.1	0	0
0	18	0.33	—	114.3	50.0	—	0	—	—	—	—	—	—	0	—	—
0	20	0	—	—	120.0	—	0	—	—	—	—	—	—	0	—	—
1	0	0	0	0.4	46.9	0.01	3.4	0	0.01	0	0	0	0	0	0	0.1
0	0	0	0	0.1	0	0	0	0	0	0	0	0	0	0	0	0
5	22	0.75	20.9	69.5	14.8	0.54	14.4	0.03	0.10	0.04	0.12	0.03	6.1	0.1	0	1.1

APPENDIX H

DA+ Code	Food Description	Quantity	Measure	Wt (g)	H₂O (g)	Ener (kcal)	Prot (g)	Carb (g)	Fiber (g)	Fat (g)	Fat Breakdown (g) Sat	Mono	Poly
	Sweets—*continued*												
1787	Jelly beans	15	item(s)	43	2.7	159	0	39.8	0.1	0	0	0	0
1784	Kit Kat wafer bar	1	item(s)	42	0.7	210	3.0	27.0	1.0	11.0	7.0	2.5	0.4
4674	Krackel candy bar	1	item(s)	41	0.6	210	2.0	28.0	0.5	10.0	6.0	3.9	0.4
4934	Licorice	4	piece(s)	44	7.3	154	1.1	35.1	0	1.0	0	0.1	0
1780	Life Savers candy	1	item(s)	2	—	6	0	1.4	—	0	0	0	0
1790	Lollipop	1	item(s)	28	—	108	0	28.0	0	0	0	0	0
4679	M & Ms peanut chocolate candy, small bag	1	item(s)	49	1.2	250	5.0	30.0	2.0	13.0	5.0	4.0	1.7
1781	M & Ms plain chocolate candy, small bag	1	item(s)	48	0.8	240	2.0	34.0	1.0	10.0	6.0	2.5	0.4
4673	Milk chocolate bar, Symphony	1	item(s)	91	0.9	483	7.7	52.8	1.5	27.8	16.7	7.2	0.6
1783	Milky Way bar	1	item(s)	58	3.7	260	2.0	41.0	1.0	10.0	7.0	1.3	0.2
1788	Peanut brittle	1 ½	ounce(s)	43	0.3	207	3.2	30.3	1.1	8.1	1.8	3.4	1.9
1789	Reese's peanut butter cups	2	piece(s)	42	0.6	210	5.0	24.0	1.0	13.0	4.5	5.5	2.3
4689	Reese's pieces candy, small bag	1	item(s)	43	0.4	200	5.0	25.0	2.0	9.0	8.0	1.9	0.8
33399	Semisweet chocolate candy, made with butter	½	ounce(s)	14	0.1	68	0.6	9.0	0.8	4.2	2.5	1.4	0.1
1782	Snickers bar	1	item(s)	59	3.3	280	4.0	35.0	1.0	14.0	5.0	4.6	1.8
4694	Special Dark chocolate bar	1	item(s)	41	—	190	2.0	25.0	3.0	12.0	8.0	—	—
4695	Starburst fruit chews, original fruits	1	package(s)	40	3.5	160	0	33.0	0	3.5	3.0	0	0
4698	Taffy	3	piece(s)	45	2.2	179	0	41.2	0	1.5	0.9	0.4	0.1
4699	Three Musketeers bar	1	item(s)	60	3.5	260	2.0	46.0	1.0	8.0	5.0	1.4	0.2
11764	Turtles	1	piece(s)	17	—	86	1.0	10.1	0.5	5.0	2.0	—	—
4702	Twix caramel cookie bars	2	item(s)	51	2.4	250	2.0	33.0	1.0	12.0	9.0	1.7	0.4
4705	York Peppermint Pattie	1	item(s)	39	3.5	140	1.0	31.0	1.0	2.5	1.5	0.2	0
	Frosting, icing												
4760	Chocolate frosting, ready to eat	2	tablespoon(s)	31	5.2	122	0.3	19.4	0.3	5.4	1.7	2.8	0.7
4771	Creamy vanilla frosting, ready to eat	2	tablespoon(s)	35	5.3	146	0	23.8	0	5.7	1.0	1.7	2.8
536	White icing	2	tablespoon(s)	40	6.0	167	0	27.2	0	6.5	1.2	1.9	3.2
	Gelatin												
13697	Gelatin snack, all flavors	1	item(s)	96	—	70	1.0	17.0	0	0	0	0	0
2616	Sugar-free, low-calorie mixed fruit gelatin mix, prepared	½	cup(s)	121	—	10	1.0	0	0	0	0	0	0
548	**Honey**	1	tablespoon(s)	21	3.6	64	0.1	17.3	0	0	0	0	0
	Jams, jellies												
550	Jam or preserves	1	tablespoon(s)	20	6.1	56	0.1	13.8	0.2	0	0	0	0
42199	Jams, preserves, dietetic, all flavors, w/sodium saccarin	1	tablespoon(s)	14	6.4	18	0	7.5	0.3	0	0	0	0
552	Jelly	1	tablespoon(s)	21	6.3	56	0	14.7	0.2	0	0	0	0
545	**Marshmallows**	4	item(s)	29	4.7	92	0.5	23.4	0	0.1	0	0	0
4800	**Marshmallow cream topping**	2	tablespoon(s)	40	7.9	129	0.3	31.6	0	0.1	0	0	0
555	**Molasses**	1	tablespoon(s)	20	4.4	58	0	14.9	0	0	0	0	0
4780	**Popsicle or ice pop**	1	item(s)	59	47.5	47	0.0	11.3	0	0.1	0	0	0
	Sugar												
559	Brown sugar, packed	1	teaspoon(s)	5	0.1	17	0	4.5	0	0	0	0	0
563	Powdered sugar, sifted	⅓	cup(s)	33	0.1	130	0	33.2	0	0	0	0	0
561	White granulated sugar	1	teaspoon(s)	4	0	16	0	4.2	0	0	0	0	0
	Sugar substitute												
1760	Equal sweetener, packet size	1	item(s)	1	—	0	0	0.5	0	0	0	0	0
13029	Splenda granular no-calorie sweetener	1	teaspoon(s)	1	—	0	0	0.5	0	0	0	0	0
1759	Sweet N Low sugar substitute, packet	1	item(s)	1	0.1	4	0	0.5	0	0	0	0	0
	Syrup												
3148	Chocolate syrup	2	tablespoon(s)	39	12.1	109	0.8	25.4	1.0	0.4	0.2	0.1	0
29676	Maple syrup	¼	cup(s)	79	25.5	205	0	52.8	0	0	0	0	0
4795	Pancake syrup	¼	cup(s)	80	30.4	187	0	49.2	0	0	0	0	0
	Spices, Condiments, Sauces												
	Spices												
807	Allspice, ground	1	teaspoon(s)	2	0.2	5	0.1	1.4	0.4	0.2	0	0	0
1171	Anise seeds	1	teaspoon(s)	2	0.2	7	0.4	1.1	0.3	0.3	0	0.2	0.1
729	Bakers' yeast, active	1	teaspoon(s)	4	0.2	13	1.6	1.6	1.1	0.3	0	0.2	0
683	Baking powder, double acting with phosphate	1	teaspoon(s)	5	0.2	2	0	1.1	0	0	0	0	0

APPENDIX H

Chol (mg)	Calc (mg)	Iron (mg)	Magn (mg)	Pota (mg)	Sodi (mg)	Zinc (mg)	Vit A (µg)	Thia (mg)	Vit E (mg α)	Ribo (mg)	Niac (mg)	Vit B$_6$ (mg)	Fola (µg)	Vit C (mg)	Vit B$_{12}$ (µg)	Sele (µg)
0	1	0.06	0.9	15.7	21.3	0.02	0	0	0	0	0	0	0	0	0	0.5
5	60	0.36	15.5	97.0	30.0	0.04	0	0.05	0.14	0.09	0.21	0.01	7.6	0	0.2	2.1
3	40	0.37	—	168.8	50.0	—	0	—	—	—	—	—	—	0	—	—
0	0	0.22	2.6	28.2	126.3	0.07	0	0.01	0.08	0.02	0.04	0	—	0	0	—
0	0	0	—	—	0	—	0	—	—	—	—	—	—	0	—	—
0	0	0	—	—	10.8	—	0	0	—	0	0	—	—	0	—	1.0
5	40	0.36	34.0	171.1	25.0	0.87	—	0.03	1.40	0.07	1.60	0.04	27.1	0.6	0.2	2.0
5	40	0.36	21.1	125.0	30.0	0.77	13.3	0.04	0.17	0.07	0.13	0.01	3.8	0.6	0.3	1.6
22	228	0.83	61.0	398.6	91.9	1.00	—	0.06	—	0.25	0.15	0.10	—	2.0	0.4	—
5	60	0.18	11.6	72.1	95.0	0.40	14.0	0.03	0.52	0.07	0.09	0.01	2.3	0.6	0.1	1.2
5	11	0.52	17.9	71.4	189.2	0.37	16.6	0.06	1.09	0.02	1.13	0.03	19.6	0	0	1.1
5	20	0.72	26.0	144.1	150.0	0.54	0	0.07	0.06	0.05	1.89	0.04	21.0	0	0.1	0.6
0	0	0.72	37.8	154.4	55.0	0.50	0	0.08	0.43	0.09	2.61	0.05	23.6	0	0	0.3
3	5	0.44	16.3	51.7	1.6	0.23	0.4	0.01	—	0.01	0.06	0.01	0.4	0	0	0.5
5	40	0.36	42.3	189.6	140.0	1.47		0.03	0.88	0.08	2.11	0.05	18.2	0	0.1	4.6
5	0	1.80	—	—	15.0	—	0	—	—	—	—	—	—	0	—	—
0	0	0	0.6	1.2	0	0	0	0	0.19	0	0	0	0.6	23.5	0	0.5
4	4	0	0	1.4	23.4	0.09	12.1	0.01	0.04	0.01	0	0	0	0	0	0.3
5	20	0.36	17.5	80.3	110.0	0.33	14.9	0.02	0.59	0.03	0.14	0.01	2.4	0.6	0.1	1.1
0	20	0.36	—	—	15.1	—	0	—	—	—	—	—	—	0	—	—
5	39	0.35	15.2	104.5	100.0	0.60	—	0.08	0.42	0.11	0.57	0.01	19.9	0.6	0.2	2.4
0	0	0.33	—	43.3	10.0	—	0	—	—	—	—	—	—	0	—	—
0	2	0.43	6.4	60.0	56.1	0.09	0	0	0.48	0.01	0.04	0	0.3	0	0	0.2
0	1	0.06	0.3	11.9	64.4	0.02	0	0	0.54	0.11	0.08	0	2.8	0	0	0
0	1	0.06	0.4	13.6	73.6	0.03	0	0	0.61	0.12	0.09	0	3.2	0	0	0
0	0	0	—	0	40.0	—	0	—	—	—	—	—	—	0	—	—
0	0	0	0	45.0	—	0	0	0	0	0	0	0	—	0	—	—
0	1	0.09	0.4	10.9	0.8	0.05	0	0	0	0.01	0.03	0.01	0.4	0.1	0	0.2
0	4	0.10	0.8	15.4	6.4	0.01	0	0	0.02	0.02	0.01	0	2.2	1.8	0	0.4
0	1	0.06	0.7	9.7	0	0.01	0	0	0.01	0	0	0	1.3	0	0	0.2
0	1	0.04	1.3	11.3	6.3	0.01	0	0	0	0.01	0.01	0	0.4	0.2	0	0.1
0	1	0.07	0.6	1.4	23.0	0.01	0	0	0	0	0.02	0	0.3	0	0	0.5
0	1	0.09	0.8	2.0	32.0	0.02	0	0	0	0	0.03	0	0.4	0	0	0.8
0	41	0.94	48.4	292.8	7.4	0.06	0	0.01	0	0	0.19	0.13	0	0	0	3.6
0	0	0.32	0.6	8.9	4.1	0.09	0	0	0	0	0	0	0	0.4	0	0.1
0	4	0.03	0.4	6.1	1.3	0	0	0	0	0	0.01	0	0	0	0	0.1
0	0	0.02	0.0	0.7	0.7	0	0	0	0	0.01	0	0	0	0	0	0.2
0	0	0	0	0.1	0	0	0	0	0	0	0	0	0	0	0	0
0	0	0	—	0	0	—	—	—	—	—	—	—	—	—	0	—
0	0	0	—	—	0	—	0	—	—	—	—	—	—	—	0	—
0	0	0	—	—	0	—	—	—	—	—	—	—	—	—	0	—
0	5	0.82	25.4	87.4	28.1	0.28	0	0	0.01	0.02	0.13	0	0.8	0.1	0	0.5
0	80	0.09	16.5	166.9	9.4	1.16	0	0.05	0	1.00	0.06	0	0	0	0	0.5
0	2	0.02	1.6	12.0	65.6	0.06	0	0.02	0	0.01	0	0	0	0	0	0
0	13	0.13	2.6	19.8	1.5	0.02	0.5	0	—	0	0.05	0	0.7	0.7	0	0.1
0	14	0.78	3.6	30.3	0.3	0.11	0.3	0.01	—	0.01	0.06	0.01	0.2	0.4	0	0.1
0	1	0.09	2.2	38.2	2.0	0.32	0	0.44	0	0.16	1.61	0.06	93.6	0	0	0.3
0	339	0.52	1.8	0.2	363.1	0	0	0	0	0	0	0	0	0	0	0

APPENDIX H

DA+ Code	Food Description	Quantity	Measure	Wt (g)	H₂O (g)	Ener (kcal)	Prot (g)	Carb (g)	Fiber (g)	Fat (g)	Fat Breakdown (g)		
											Sat	Mono	Poly
Spices, Condiments, Sauces—*continued*													
1611	Baking soda	1	teaspoon(s)	5	0	0	0	0	0	0	0	0	0
8552	Basil	1	teaspoon(s)	1	0.8	0	0	0	0	0	0	0	0
34959	Basil, fresh	1	piece(s)	1	0.5	0	0	0	0	0	0	0	0
808	Basil, ground	1	teaspoon(s)	1	0.1	3	0.3	0.7	0.5	0.1	0	0	0
809	Bay leaf	1	teaspoon(s)	1	0	2	0	0.4	0.2	0.1	0	0	0
11720	Betel leaves	1	ounce(s)	28	—	17	1.8	2.4	0	0	—	—	—
11710	Capers	1	teaspoon(s)	5	—	0	0	0	0	0	0	0	0
1172	Caraway seeds	1	teaspoon(s)	2	0.2	7	0.4	1.0	0.8	0.3	0	0.1	0.1
1173	Celery seeds	1	teaspoon(s)	2	0.1	8	0.4	0.8	0.2	0.5	0	0.3	0.1
1174	Chervil, dried	1	teaspoon(s)	1	0	1	0.1	0.3	0.1	0	0	0	0
810	Chili powder	1	teaspoon(s)	3	0.3	7	0.3	1.3	0.9	0.4	0.1	0.1	0.2
8553	Chives, chopped	1	teaspoon(s)	1	0.9	0	0	0	0	0	0	0	0
51420	Cilantro (coriander)	1	teaspoon(s)	0	0.3	0	0	0	0	0	0	0	0
811	Cinnamon, ground	1	teaspoon(s)	2	0.2	6	0.1	1.9	1.2	0	0	0	0
812	Cloves, ground	1	teaspoon(s)	2	0.2	6	0.1	1.4	0.7	0.3	0.1	0	0.2
1175	Coriander leaf, dried	1	teaspoon(s)	1	0	2	0.1	0.3	0.1	0	0	0	0
1176	Coriander seeds	1	teaspoon(s)	2	0.2	5	0.2	1.0	0.8	0.3	0	0.2	0
1706	Cornstarch	1	tablespoon(s)	8	0.7	30	0	7.3	0.1	0	0	0	0
1177	Cumin seeds	1	teaspoon(s)	2	0.2	8	0.4	0.9	0.2	0.5	0	0.3	0.1
11729	Cumin, ground	1	teaspoon(s)	5	—	11	0.4	0.8	0.8	0.4	—	—	—
1178	Curry powder	1	teaspoon(s)	2	0.2	7	0.3	1.2	0.7	0.3	0	0.1	0.1
1179	Dill seeds	1	teaspoon(s)	2	0.2	6	0.3	1.2	0.4	0.3	0	0.2	0
1180	Dill weed, dried	1	teaspoon(s)	1	0.1	3	0.2	0.6	0.1	0	0	0	0
34949	Dill weed, fresh	5	piece(s)	1	0.9	0	0	0.1	0	0	0	0	0
1181	Fennel seeds	1	teaspoon(s)	2	0.2	7	0.3	1.0	0.8	0.3	0	0.2	0
1182	Fenugreek seeds	1	teaspoon(s)	4	0.3	12	0.9	2.2	0.9	0.2	0.1	—	—
11733	Garam masala, powder	1	ounce(s)	28	—	107	4.4	12.8	0	4.3	—	—	—
1067	Garlic clove	1	item(s)	3	1.8	4	0.2	1.0	0.1	0	0	0	0
813	Garlic powder	1	teaspoon(s)	3	0.2	9	0.5	2.0	0.3	0	0	0	0
1068	Ginger root	2	teaspoon(s)	4	3.2	3	0.1	0.7	0.1	0	0	0	0
1183	Ginger, ground	1	teaspoon(s)	2	0.2	6	0.2	1.3	0.3	0.1	0	0	0
35497	Leeks, bulb and lower-leaf, freeze-dried	¼	cup(s)	1	0	3	0.1	0.6	0.1	0	0	0	0
1184	Mace, ground	1	teaspoon(s)	2	0.1	8	0.1	0.9	0.3	0.6	0.2	0.2	0.1
1185	Marjoram, dried	1	teaspoon(s)	1	0	2	0.1	0.4	0.2	0.0	0	0	0
1186	Mustard seeds, yellow	1	teaspoon(s)	3	0.2	17	0.9	0.9	0.4	1.2	0.1	0.7	0.3
814	Nutmeg, ground	1	teaspoon(s)	2	0.1	12	0.1	1.1	0.5	0.8	0.6	0.1	0
2747	Onion flakes, dehydrated	1	teaspoon(s)	2	0.1	6	0.1	1.4	0.2	0	0	0	0
1187	Onion powder	1	teaspoon(s)	2	0.1	7	0.2	1.7	0.3	0	0	0	0
815	Oregano, ground	1	teaspoon(s)	2	0.2	5	0.2	1.2	0.8	0.1	0	0	0
816	Paprika	1	teaspoon(s)	2	0.2	6	0.3	1.1	0.7	0.3	0	0	0.2
817	Parsley, dried	1	teaspoon(s)	0	0	1	0.1	0.2	0.1	0	0	0	0
818	Pepper, black	1	teaspoon(s)	2	0.3	5	0.2	1.3	0.5	0.1	0	0	0
819	Pepper, cayenne	1	teaspoon(s)	2	0.1	6	0.2	1.0	0.5	0.3	0.1	0	0.2
1188	Pepper, white	1	teaspoon(s)	2	0.3	7	0.2	1.6	0.6	0.1	0	0	0
1189	Poppy seeds	1	teaspoon(s)	3	0.2	15	0.5	0.8	0.5	1.2	0.1	0.2	0.8
1190	Poultry seasoning	1	teaspoon(s)	2	0.1	5	0.1	1.0	0.2	0.1	0	0	0
1191	Pumpkin pie spice, powder	1	teaspoon(s)	2	0.1	6	0.1	1.2	0.3	0.2	0.1	0	0
1192	Rosemary, dried	1	teaspoon(s)	1	0.1	4	0.1	0.8	0.5	0.2	0.1	0	0
11723	Rosemary, fresh	1	teaspoon(s)	1	0.5	1	0	0.1	0.1	0	0	0	0
2722	Saffron powder	1	teaspoon(s)	1	0.1	2	0.1	0.5	0	0	0	0	0
11724	Sage	1	teaspoon(s)	1	—	1	0	0.1	0	0	—	—	—
1193	Sage, ground	1	teaspoon(s)	1	0.1	2	0.1	0.4	0.3	0.1	0	0	0
30189	Salt substitute	¼	teaspoon(s)	2	—	0	0	0	0	0	0	0	0
30195	Salt, kosher	¼	teaspoon(s)	2	—	0	0	0	0	0	0	0	0
822	Salt, table	¼	teaspoon(s)	2	0	0	0	0	0	0	0	0	0
1194	Savory, ground	1	teaspoon(s)	1	0.1	4	0.1	1.0	0.6	0.1	0	—	—
820	Sesame seed kernels, toasted	1	teaspoon(s)	3	0.1	15	0.5	0.7	0.5	1.3	0.2	0.5	0.6
11725	Sorrel	1	teaspoon(s)	3	—	1	0.1	0.1	0	0	0	—	—
11721	Spearmint	1	teaspoon(s)	2	1.6	1	0.1	0.2	0.1	0	0	0	0
35498	Sweet green peppers, freeze-dried	¼	cup(s)	2	0	5	0.3	1.1	0.3	0	0	0	0
11726	Tamarind leaves	1	ounce(s)	28	—	33	1.6	5.2	0	0.6	—	—	—
11727	Tarragon	1	ounce(s)	28	—	14	1.0	1.8	0	0.3	—	—	—
1195	Tarragon, ground	1	teaspoon(s)	2	0.1	5	0.4	0.8	0.1	0.1	0	0	0.1
11728	Thyme, fresh	1	teaspoon(s)	1	0.5	1	0	0.2	0.1	0	0	0	0

Chol (mg)	Calc (mg)	Iron (mg)	Magn (mg)	Pota (mg)	Sodi (mg)	Zinc (mg)	Vit A (µg)	Thia (mg)	Vit E (mg α)	Ribo (mg)	Niac (mg)	Vit B6 (mg)	Fola (µg)	Vit C (mg)	Vit B12 (µg)	Sele (µg)
0	0	0	0	0	1258.6	0	0	0	0	0	0	0	0	0	0	0
0	2	0.03	0.6	2.6	0	0.01	2.3	0	0.01	0	0.01	0	0.6	0.2	0	0
0	1	0.02	0.3	1.5	0	0	1.3	0	—	0	0	0	0.3	0.1	0	0
0	31	1.26	10.0	36.8	1.1	0.10	0.5	0	0.15	0.02	0.07	0.02	4.3	0	0	0
0	5	0.26	0.7	3.2	0.1	0.02	1.9	0	—	0	0.01	0.01	1.1	0.3	0	0
0	110	2.29	—	155.9	2.0	—	—	0.04	—	0.07	0.20	—	—	0.9	0	—
0	—	—	—	—	105.0	—	—	—	—	—	—	—	—	—	0	—
0	14	0.34	5.4	28.4	0.4	0.12	0.4	0.01	0.05	0.01	0.08	0.01	0.2	0.4	0	0.3
0	35	0.90	8.8	28.0	3.2	0.14	0.1	0.01	0.02	0.01	0.06	0.02	0.2	0.3	0	0.2
0	8	0.19	0.8	28.4	0.5	0.05	1.8	0	—	0	0.03	0.01	1.6	0.3	0	0.2
0	9	0.45	3.9	50.7	42.6	0.11	38.6	0.01	0.99	0.02	0.30	0.05	0.7	0	0	0.5
0	1	0.02	0.4	3.0	0	0.01	2.2	0	0	0	0.01	0	1.0	0.6	0	0
0	0	0.01	0.1	1.7	0.2	0	1.1	0	0.01	0	0	0	0.2	0.1	0	0
0	23	0.19	1.4	9.9	0.2	0.04	0.3	0	0.05	0	0.03	0	0.1	0.1	0	0.1
0	13	0.25	5.4	21.4	5.8	0.05	0.2	0	0.19	0	0.03	0.01	0.5	0	0	0.2
0	7	0.25	4.2	26.8	1.3	0.03	1.8	0.01	0.01	0.01	0.06	0	1.6	3.4	0	0.2
0	13	0.29	5.9	22.8	0.6	0.08	0	0	—	0.01	0.04	—	0	0.4	0	0.5
0	0	0.04	0.2	0.2	0.7	0	0	0	0	0	0	0	0	0	0	0.2
0	20	1.39	7.7	37.5	3.5	0.10	1.3	0.01	0.07	0.01	0.10	0.01	0.2	0.2	0	0.1
0	20	—	—	43.6	4.8	—	—	—	—	—	—	—	—	—	0	—
0	10	0.59	5.1	30.9	1.0	0.08	1.0	0.01	0.44	0.01	0.07	0.02	3.1	0.2	0	0.3
0	32	0.34	5.4	24.9	0.4	0.11	0.1	0.01	—	0.01	0.06	0.01	0.2	0.4	0	0.3
0	18	0.49	4.5	33.1	2.1	0.03	2.9	0	—	0	0.03	0.02	—	0.5	0	—
0	2	0.07	0.6	7.4	0.6	0.01	3.9	0	0.02	0	0.02	0	1.5	0.9	0	—
0	24	0.37	7.7	33.9	1.8	0.07	0.1	0.01	—	0.01	0.12	0.01	—	0.4	0	—
0	7	1.24	7.1	28.5	2.5	0.09	0.1	0.01	—	0.01	0.06	0.02	2.1	0.1	0	0.2
0	215	9.24	93.6	411.1	27.5	1.07	—	0.10	—	0.09	0.71		—	0	0	—
0	5	0.05	0.8	12.0	0.5	0.03	0	0.01	0	0	0.02	0.04	0.1	0.9	0	0.4
0	2	0.16	2.2	33.4	1.7	0.08	0	0.01	0.02	0	0.02	0.05	1.3	0	0	0.7
0	1	0.02	1.7	16.6	0.5	0.01	0	0	0.01	0	0.03	0.01	0.4	0.2	0	0
0	2	0.36	3.9	23.8	0.5	0.07	0	0	0	0	0.17	0.01	0.2	0	0	1.0
0	3	0.06	1.3	19.2	0.3	0.01	0.1	0.01	—	0	0.03	0.01	2.9	0.9	0	0
0	4	0.24	2.8	7.9	1.4	0.04	0.7	0.01	—	0.01	0.02	0	1.3	0.4	0	0
0	12	0.50	2.1	9.1	0.5	0.02	2.4	0	0.01	0	0.02	0.01	1.6	0.3	0	0
0	9	0.30	12.2	24.4	0.4	0.20	0.1	0.03	0.17	0.01	0.16	0.01	5.3	0.2	0	6.9
0	4	0.07	4.0	7.7	0.4	0.05	0.1	0.01	—	0	0.03	0	1.7	0.1	0	0
0	4	0.03	1.5	27.1	0.4	0.03	0	0.01	0	0	0.02	0.03	2.8	1.3	0	0.1
0	8	0.08	2.4	20.7	1.5	0.09	0	0.01	0.01	0	0.01	0.02	1.3	0.5	0	0.3
0	29	0.66	4.9	22.7	0.5	0.05	1.5	0	0.33	0.01	0.08	0.02	4.3	0	0	0.1
0	5	0.44	3.7	47.9	1.4	0.09	51.7	0.01	0.61	0.03	0.21	0.04	1.0	0	0	0.1
0	3	0.07	1.2	8.0	1.4	0.02	0.3	0	0.03	0.01	0.03	0	0.5	0.4	0	0
0	9	0.20	3.6	27.9	0.4	0.02	0.6	0	0.02	0	0.01	0.01	0.4	0	0	0.1
0	3	0.14	2.7	36.3	0.5	0.04	37.5	0.01	0.54	0.02	0.16	0.04	1.9	1.4	0	0.2
0	6	0.34	2.2	1.8	0.1	0.03	0	0	—	0	0.01	0	0.2	0.5	0	0.1
0	40	0.27	9.7	20.1	0.7	0.22	0	0.02	0.05	0	0.03	0.01	2.3	0	0	0.4
0	15	0.53	3.4	10.3	0.4	0.05	2.0	0	0.02	0	0.04	0.02	2.1	0.2	0	0.1
0	12	0.34	2.3	11.3	0.9	0.04	0.2	0	0.03	0	0.04	0.01	0.4	0.4	0	0.2
0	15	0.35	2.6	11.5	0.6	0.04	1.9	0.01	—	0.01	0.01	0.02	3.7	0.7	0	0.1
0	2	0.05	0.6	4.7	0.2	0.01	1.0	0	—	0	0.01	0	0.8	0.2	0	—
0	1	0.08	1.8	12.1	1.0	0.01	0.2	0	—	0	0.01	0.01	0.7	0.6	0	0
0	4	—	1.1	2.7	0	0.01	—	0	—	—	—	—	—	—	0	—
0	12	0.20	3.0	7.5	0.1	0.03	2.1	0.01	0.05	0	0.04	0.02	1.9	0.2	0	—
0	8	0	0	754.5	0.1	—	0	—	—	—	0	—	—	0	—	—
0	0	0	—	—	600.0	—	0	—	—	—	—	—	—	0	—	—
0	0	0	0	0.1	581.4	0	0	0	0	0	0	0	0	0	0	0
0	30	0.53	5.3	14.7	0.3	0.06	3.6	0.01	—	—	0.06	0.03	—	0.7	0	0.1
0	3	0.21	9.2	10.8	1.0	0.27	0.1	0.03	0.01	0.01	0.15	0	2.6	0	0	0.9
0	—	—	—	—	0.1	—	—	—	—	—	—	—	—	—	0	—
0	4	0.23	1.2	8.7	0.6	0.02	3.9	0	—	0	0.02	0	2.0	0.3	0	—
0	2	0.17	3.0	50.7	3.1	0.04	4.5	0.02	0.06	0.02	0.12	0.04	3.7	30.4	0	0.1
0	85	1.48	20.2	—	—	—	—	0.07	—	0.03	1.16	—	—	0.9	0	—
0	48	—	14.5	128.1	2.6	0.17	—	0.04	—	—	—	—	—	0.6	0	—
0	18	0.52	5.6	48.3	1.0	0.06	3.4	0	—	0.02	0.14	0.04	4.4	0.8	0	0.1
0	3	0.14	1.3	4.9	0.1	0.01	1.9	0	—	0	0.01	0	0.4	1.3	0	—

APPENDIX H

DA+ Code	Food Description	Quantity	Measure	Wt (g)	H₂O (g)	Ener (kcal)	Prot (g)	Carb (g)	Fiber (g)	Fat (g)	Fat Breakdown (g) Sat	Mono	Poly
Spices, Condiments, Sauces—*continued*													
821	Thyme, ground	1	teaspoon(s)	1	0.1	4	0.1	0.9	0.5	0.1	0	0	0
1196	Turmeric, ground	1	teaspoon(s)	2	0.2	8	0.2	1.4	0.5	0.2	0.1	0	0
11995	Wasabi	1	tablespoon(s)	14	10.7	10	0.7	2.3	0.2	0	—	—	—
	Condiments												
674	Catsup or ketchup	1	tablespoon(s)	15	10.4	17	0.2	3.9	0	0	0	0	0
5812	Catsup or ketchup, low sodium	1	tablespoon(s)	15	10.4	15	0.3	3.8	0.3	0	0	0	0
703	Dill pickle	1	ounce(s)	28	26.7	3	0.2	0.7	0.3	0	0	0	0
32128	Guacamole	2	tablespoon(s)	32	25.1	42	0.7	2.1	1.1	3.8	0.6	2.4	0.4
1814	Hummus	½	cup(s)	123	79.8	218	6.0	24.7	4.9	10.6	1.4	6.0	2.6
138	Mayonnaise with soybean oil	1	tablespoon(s)	14	3.0	94	0.1	0.1	0	10.3	1.6	2.3	6.2
140	Mayonnaise, low calorie	1	tablespoon(s)	16	8.9	52	0.1	1.3	0	5.3	0.8	1.3	2.9
1682	Mustard, brown	1	teaspoon(s)	5	4.1	5	0.3	0.3	0	0.3	—	—	—
700	Mustard, yellow	1	teaspoon(s)	5	4.1	3	0.2	0.3	0.2	0.2	0	0.1	0
706	Sweet pickle relish	1	tablespoon(s)	15	9.3	20	0.1	5.3	0.2	0.1	0	0	0
141	Tartar sauce	2	tablespoon(s)	29	18.8	59	0.3	3.7	0.1	4.7	0.9	1.0	2.5
	Sauces												
685	Barbecue sauce	2	tablespoon(s)	31	17.1	54	0.3	12.7	0.3	0.2	0	0	0
40849	Barbecue sauce, low sodium	2	tablespoon(s)	31	17.1	54	0.3	12.7	0.3	0.2	0	0	0
834	Cheese sauce	¼	cup(s)	63	44.4	110	4.2	4.3	0.3	8.4	3.8	2.4	1.6
32123	Chili enchilada sauce, green	2	tablespoon(s)	57	53.0	15	0.6	3.1	0.7	0.3	0	0	0.1
32122	Chili enchilada sauce, red	2	tablespoon(s)	32	24.5	27	1.1	5.0	2.1	0.8	0.1	0	0.4
29688	Hoisin sauce	1	tablespoon(s)	16	7.1	35	0.5	7.1	0.4	0.5	0.1	0.2	0.3
1641	Horseradish sauce, prepared	1	teaspoon(s)	5	3.5	9	0.1	0.1	0	0.9	0.5	0.2	0
16670	Mole poblano sauce	½	cup(s)	133	102.8	156	5.4	10.9	2.7	11.6	2.7	5.2	3.0
29689	Oyster sauce	1	tablespoon(s)	16	12.8	8	0.2	1.7	0	0	0	0	0
1655	Pepper sauce or Tabasco	1	teaspoon(s)	5	4.5	1	0.1	0	0	0	0	0	0
347	Salsa	2	tablespoon(s)	32	28.5	9	0.5	2.2	0.6	0.1	0	0	0
2835	Soy sauce, low sodium	1	tablespoon(s)	18	12.8	10	0.9	1.5	0.1	0	0	0	0
52206	Soy sauce, tamari	1	tablespoon(s)	18	11.8	11	1.9	1.0	0.1	0	0	0	0
25292	Sweet and sour sauce	2	tablespoon(s)	35	29.0	22	0.5	5.0	0.1	0.1	0	0	0
1613	Teriyaki sauce	1	tablespoon(s)	18	12.2	16	1.1	2.8	0	0	0	0	0
25294	Tomato sauce	½	cup(s)	150	133.4	61	2.4	11.2	2.7	1.6	0.2	0.4	0.8
42233	Tomato sauce, no salt added	½	cup(s)	122	111.2	35	1.6	8.2	1.8	0.2	0	0	0.1
5187	White sauce, medium-thick	¼	cup(s)	59	44.3	87	2.3	5.4	0.1	6.3	1.7	2.6	1.7
1654	Worcestershire sauce	1	teaspoon(s)	6	4.5	4	0	1.1	0	0	0	0	0
	Vinegar												
30853	Balsamic	1	tablespoon(s)	15	—	10	0	2.0	0	0	0	0	0
727	Cider	1	tablespoon(s)	15	14.0	3	0	0.1	0	0	0	0	0
5176	Distilled	1	tablespoon(s)	15	14.0	3	0	0	0	0	0	0	0
12948	Tarragon	1	tablespoon(s)	15	14.3	3	0	0.1	0	0	0	0	0
Mixed Foods, Sandwiches, and Soups													
	Mixed dishes												
16652	Almond chicken	1	cup(s)	242	187.5	283	20.4	15.9	3.4	15.6	2.1	7.1	5.2
25224	Barbecued chicken	1	serving(s)	177	100.1	325	27.2	15.1	0.1	17.1	4.8	6.8	3.8
25227	Bean burrito	1	item(s)	147	81.7	318	16.2	30.9	5.7	14.9	8.3	4.7	1.0
9516	Beef and vegetable fajita	1	item(s)	223	145.3	377	23.3	34.9	2.7	15.7	5.1	6.9	2.5
16796	Beef or pork egg roll	2	item(s)	128	59.6	355	12.0	35.6	2.6	18.3	4.2	8.6	4.1
73314	Beef stew w/potatoes & vegetables	1	cup(s)	252	214.4	161	18.0	12.9	1.7	4.0	1.6	1.9	0.3
30233	Beef stroganoff with noodles	1	cup(s)	256	192.9	330	19.7	22.1	1.5	18.0	6.6	5.5	3.2
16651	Cashew chicken	1	cup(s)	242	187.5	283	20.4	15.9	3.4	15.6	2.1	7.1	5.2
30274	Cheese pizza with vegetables, thin crust	2	slice(s)	224	101.7	603	25.4	63.0	5.4	27.8	12.2	7.2	6.0
30330	Cheese quesadilla	1	item(s)	142	51.9	484	20.0	37.3	1.8	28.2	13.8	9.7	3.5
25240	Chicken and noodles	1	cup(s)	227	163.7	176	18.3	15.2	0.8	4.2	1.0	1.7	1.0
30239	Chicken and vegetables with broccoli, onion, bamboo shoots in soy based sauce	1	cup(s)	217	164.8	289	22.9	10.6	2.8	17.2	3.0	6.6	6.4
25093	Chicken cacciatore	1	cup(s)	244	176.2	283	29.8	5.4	1.3	15.3	4.3	6.2	3.3
28020	Chicken fried turkey steak	3	ounce(s)	85	62.7	115	12.7	11.3	0.7	1.8	0.5	0.3	0.7
218	Chicken pot pie	1	cup(s)	252	147.9	539	15.3	58.0	3.0	27.3	9.1	9.3	3.5
30240	Chicken teriyaki	1	cup(s)	244	153.2	451	66.5	4.7	0	16.6	4.6	5.9	3.8
25119	Chicken waldorf salad	½	cup(s)	100	67.5	179	14.0	6.5	1.0	10.9	1.8	3.2	5.0
25099	Chili con carne	¾	cup(s)	215	174.7	196	13.7	20.9	6.6	6.9	2.6	3.0	0.5
1062	Coleslaw	¾	cup(s)	90	73.4	70	1.2	11.2	1.3	2.3	0.3	0.6	1.2
1574	Crab cakes, from blue crab	1	item(s)	60	42.6	93	12.1	0.3	0	4.5	0.9	1.7	1.4

Chol (mg)	Calc (mg)	Iron (mg)	Magn (mg)	Pota (mg)	Sodi (mg)	Zinc (mg)	Vit A (µg)	Thia (mg)	Vit E (mg α)	Ribo (mg)	Niac (mg)	Vit B6 (mg)	Fola (µg)	Vit C (mg)	Vit B12 (µg)	Sele (µg)
0	26	1.73	3.1	11.4	0.8	0.09	2.7	0.01	0.10	0.01	0.07	0.01	3.8	0.7	0	0.1
0	4	0.91	4.2	55.6	0.8	0.10	0	0	0.07	0.01	0.11	0.04	0.9	0.6	0	0.1
0	13	0.11	—	—	—	—	—	0.02	—	0.01	0.07	—	—	11.2	0	—
0	2	0.06	2.3	47.3	136.1	0.04	3.9	0	0.22	0.03	0.22	0.02	1.4	0.6	0	0
0	3	0.08	2.8	57.3	3.0	0.04	7.0	0	0.22	0.02	0.21	0.02	1.5	2.3	0	0
0	12	0.10	2.0	26.1	248.1	0.03	2.6	0.01	0.03	0.01	0.03	0.01	0.3	0.2	0	0
1	12	0.27	10.2	152.1	79.1	0.13	—	0.03	0.28	0.04	0.43	0.07	—	2.5	0	0
0	60	1.92	35.7	212.8	297.7	1.34	0	0.11	0.92	0.06	0.49	0.49	72.6	9.7	0	3.0
6	1	0.03	0.1	2.8	87.6	0.02	2.2	0	0.45	0	0	0	0.7	0	0	0.3
6	1	0.05	0.3	6.4	107.7	0.03	3.4	0	0.49	0	0	0	0.6	0	0	0.4
0	6	0.09	0.9	6.8	68.1	0.02	0	0	0.09	0	0.01	0	—	0.1	0	—
0	3	0.08	2.5	6.9	56.8	0.03	0.2	0.02	0.02	0	0.03	0	0.3	0.1	0	1.6
0	0	0.13	0.8	3.8	121.7	0.02	9.1	0	0.09	0	0.03	0	0.2	0.2	0	0
2	7	0.07	1.7	19.0	186.8	0.03	3.1	0	0.47	0.01	0.03	0.01	1.4	0.6	0	0.3
0	10	0.20	4.1	72.5	320.9	0.05	3.4	0.01	0.25	0.02	0.19	0.02	0.6	0.2	0	0.4
0	10	0.20	4.1	72.5	41.6	0.05	3.4	0.01	0.25	0.02	0.19	0.02	2.5	0.2	0	0.4
18	116	0.13	5.7	18.9	521.6	0.62	50.4	0	—	0.07	0.02	0.01	2.5	0.3	0.1	2.0
0	5	0.36	9.5	125.7	61.9	0.11	—	0.03	0	0.02	0.63	0.06	—	43.9	0	0
0	7	1.05	11.1	231.3	113.8	0.14	—	0.02	0	0.22	0.61	0.34	—	0.3	0	0.3
0	5	0.16	3.8	19.0	258.4	0.05	0	0	0.04	0.03	0.19	0.01	3.7	0.1	0	0.3
2	5	0.01	0.5	6.6	15.8	0.02	8.0	0	0.02	0.01	0	0	0.3	0.1	0	0.1
1	36	1.56	58.3	278.3	294.1	1.06	13.3	0.07	1.76	0.09	1.95	0.09	14.6	3.6	0.1	2.1
0	5	0.03	0.6	8.6	437.3	0.01	0	0	0	0.02	0.24	0	2.4	0	0.1	0.7
0	1	0.05	0.6	6.0	29.8	0.01	3.9	0	0	0	0.01	0.01	0.1	0.2	0	0
0	10	0.13	4.8	91.2	225.6	0.06	7.7	0.01	0.39	0.01	0.36	0.06	1.3	0.6	0	0.3
0	3	0.36	6.1	32.4	599.9	0.07	0	0.01	0	0.02	0.60	0.03	2.9	0	0	0.1
0	4	0.43	7.2	37.9	999.3	0.08	0	0.01	0	0.03	0.71	0.04	3.2	0	0	0.1
0	5	0.20	3.1	51.2	183.6	0.05	0	0.01	0.09	0.01	0.33	0.02	2.3	1.4	0	1.4
0	5	0.31	11.0	40.5	689.9	0.02	0	0.01	0	0.01	0.23	0.02	1.4	0	0	0.2
0	28	1.36	26.2	573.3	265.2	0.40	0	0.06	2.11	0.07	1.58	0.16	20.4	19.6	0	1.9
0	16	1.24	19.5	403.8	13.4	0.24	26.8	0.03	1.73	0.08	1.19	0.12	13.4	8.5	0	0.2
4	70	0.20	8.3	92.3	209.4	0.24	61.5	0.04	0.17	0.11	0.24	0.02	6.0	0.5	0.2	2.4
0	6	0.30	0.7	45.4	55.6	0.01	0.3	0	0	0.01	0.04	0	0.5	0.7	0	0
—	0	0	—	—	0	—	—	—	—	—	—	—	—	0	—	—
0	1	0.03	0.7	10.9	0.7	0.01	0	0	0	0	0	0	0	0	0	0
0	1	0	0.1	0.3	0.3	0	0	0	0	0	0	0	0	0.0	0	0.1
0	0	0.08	—	2.4	0.8	—	—	0.08	—	0.08	0.08	—	—	0.3	0	—
44	63	1.55	55.7	617.1	532.4	1.33	33.9	0.07	3.97	0.24	9.03	0.53	29.0	5.8	0.2	23.2
120	26	1.56	30.0	372.8	476.7	2.68	0	0.07	1.12	0.26	6.89	0.39	25.8	5.0	0.3	19.3
38	331	2.91	47.3	451.1	548.1	1.85	0	0.27	1.42	0.27	1.80	0.20	132.3	4.4	0.3	15.3
58	98	3.46	35.7	468.3	876.4	3.66	17.8	0.35	1.07	0.17	7.01	0.44	133.8	23.0	0.8	33.9
17	41	1.87	24.3	256.0	633.6	1.09	471.0	0.29	0.41	0.13	2.72	0.21	124.2	35.3	0.2	27.5
38	30	2.20	32.8	484.2	589.5	4.31	178.4	0.13	0.24	0.23	5.02	0.24	25.1	8.7	1.5	15.6
74	77	2.82	35.8	378.9	829.4	3.58	66.6	0.24	1.36	0.25	6.39	0.28	87.0	1.0	0.7	33.5
44	63	1.55	55.7	617.1	532.4	1.33	33.9	0.07	3.97	0.24	9.03	0.53	29.0	5.8	0.2	23.2
54	569	3.65	49.3	436.8	1480.6	3.18	143.4	0.41	2.64	0.33	5.32	0.17	266.6	12.5	1.1	39.6
60	487	2.77	31.2	215.8	1036.6	2.17	154.8	0.28	0.89	0.30	2.42	0.08	102.2	6.2	0.3	24.9
49	45	1.34	25.3	275.7	399.5	0.96	58.2	0.18	0.41	0.18	8.21	0.31	71.4	0.7	0.4	25.4
61	54	1.59	34.7	429.7	557.7	1.87	342.9	0.10	2.52	0.21	5.54	0.41	54.3	26.9	0.2	16.9
109	49	2.35	39.4	471.4	488.9	2.14	0	0.11	1.39	0.22	9.75	0.59	25.3	8.6	0.3	22.2
25	64	1.21	17.7	205.8	129.7	1.03	4.0	0.14	0.06	0.17	3.28	0.20	29.1	0.4	0.3	14.8
60	73	2.77	35.3	312.5	957.6	1.41	138.6	0.58	0.40	0.28	5.62	0.30	126.0	0.8	0.8	17.9
200	41	3.22	75.6	612.4	1361.5	4.73	36.6	0.16	0.61	0.42	20.90	1.03	14.6	0	0.7	49.5
42	20	0.78	23.7	196.1	245.7	1.13	0	0.04	0.58	0.10	4.05	0.25	13.1	1.7	0.2	10.6
27	44	3.17	50.0	645.1	788.7	3.44	0	0.13	1.38	0.24	3.02	0.24	31.0	10.2	0.6	6.6
7	41	0.53	9.0	162.9	20.7	0.18	47.7	0.06	—	0.06	0.24	0.11	24.3	29.4	0	0.6
90	63	0.65	19.8	194.4	198.0	2.45	34.2	0.05	—	0.05	1.74	0.10	36.6	1.7	3.6	24.4

DA+ Code	Food Description	Quantity	Measure	Wt (g)	H₂O (g)	Ener (kcal)	Prot (g)	Carb (g)	Fiber (g)	Fat (g)	Fat Breakdown (g)		
											Sat	Mono	Poly
Mixed Foods, Sandwiches, and Soups—*continued*													
32144	Enchiladas with green chili sauce (enchiladas verdes)	1	item(s)	144	103.8	207	9.3	17.6	2.6	11.7	6.4	3.6	1.0
2793	Falafel patty	3	item(s)	51	17.7	170	6.8	16.2	—	9.1	1.2	5.2	2.1
28546	Fettuccine Alfredo	1	cup(s)	240	87.2	267	12.3	45.8	1.8	3.5	1.8	0.8	0.4
32146	Flautas	3	item(s)	162	78.0	438	24.9	36.3	4.1	21.6	8.2	8.8	2.3
29629	Fried rice with meat or poultry	1	cup(s)	198	122.0	323	17.1	52.2	1.8	4.9	1.1	1.4	1.8
16649	General Tso chicken	1	cup(s)	146	65.9	431	18.8	35.0	1.3	23.9	4.0	5.7	11.0
1826	Green salad	¾	cup(s)	104	98.9	17	1.3	3.3	2.2	0.1	0	0	0
16650	Kung Pao chicken	1	cup(s)	162	87.1	436	28.9	11.4	2.3	30.7	4.5	13.0	11.8
16622	Lamb curry	1	cup(s)	236	157.6	474	25.6	11.1	2.6	37.1	6.7	14.9	13.3
25253	Lasagna with ground beef	1	cup(s)	238	157.2	288	17.6	22.2	2.5	14.6	7.6	4.8	0.8
442	Macaroni and cheese, prepared	1	cup(s)	243	144.7	484	18.5	54.2	2.7	21.0	8.4	7.2	3.7
29637	Meat filled ravioli with tomato or meat sauce, canned	1	cup(s)	251	196.5	248	8.6	34.5	4.0	8.6	3.6	4.0	0.6
25105	Meat loaf	1	slice(s)	115	84.6	244	17.0	6.6	0.4	15.9	6.1	6.9	0.8
16646	Moo shi pork	1	cup(s)	151	77.6	504	18.9	5.2	0.6	45.4	6.8	18.2	18.6
16788	Nachos with beef, beans, cheese, tomatoes and onions	1	serving(s)	195	102.8	505	23.0	35.4	4.9	30.8	12.2	10.2	5.6
6116	Pepperoni pizza	2	slice(s)	142	66.1	362	20.2	39.7	2.9	13.9	4.5	6.3	2.3
29601	Pizza with meat and vegetables, thin crust	2	slice(s)	232	97.7	698	30.0	60.5	4.9	37.2	15.5	11.3	6.7
655	Potato salad	½	cup(s)	125	95.0	179	3.3	14.0	1.6	10.3	1.8	3.1	4.7
25109	Salisbury steaks with mushroom sauce	1	serving(s)	135	101.8	250	17.0	9.5	0.5	15.4	6.0	6.7	0.8
16637	Shrimp creole with rice	1	cup(s)	243	178.2	296	24.7	27.6	1.7	8.9	1.6	3.6	3.0
41866	Spaghetti and meatballs, canned	1	cup(s)	248	196.7	241	11.1	27.0	—	9.9	3.6	4.0	1.2
28585	Spicy thai noodles (pad thai)	8	ounce(s)	227	72.7	218	8.8	35.1	3.0	6.4	0.8	3.4	1.7
33073	Stir fried pork and vegetables with rice	1	cup(s)	236	173.9	351	15.4	33.7	1.8	16.3	5.6	7.0	2.6
28588	Stuffed shells	2 ½	item(s)	249	157.6	242	15.1	27.7	2.7	8.2	3.1	3.0	1.3
16821	Sushi with egg in seaweed	6	piece(s)	156	116.7	192	8.9	20.0	0.3	7.8	2.1	3.2	1.9
16819	Sushi with vegetables and fish	6	piece(s)	156	97.6	236	6.9	48.1	0.6	0.8	0.2	0.2	0.2
16820	Sushi with vegetables in seaweed	6	piece(s)	156	109.4	187	3.5	41.0	0.8	0.4	0.1	0.1	0.1
25266	Sweet and sour pork	¾	cup(s)	249	206.0	265	29.3	16.9	1.0	8.1	2.7	3.6	1.4
16824	Tabouli, tabbouleh or tabuli	1	cup(s)	160	124.1	198	2.6	15.5	3.7	15.0	2.1	10.7	1.7
25276	Three bean salad	½	cup(s)	99	82.3	96	1.9	9.6	2.6	6.0	0.8	1.6	3.2
160	Tuna salad	½	cup(s)	103	64.7	192	16.4	9.6	0	9.5	1.6	3.0	4.2
25241	Turkey and noodles	1	cup(s)	227	162.4	193	17.1	15.2	0.8	6.6	1.7	2.5	1.6
16794	Vegetable egg roll	2	item(s)	128	63.3	323	8.0	38.6	3.3	15.3	3.2	7.3	3.8
16818	Vegetable sushi, no fish	6	piece(s)	156	92.1	256	4.2	56.7	0.9	0.4	0.1	0.1	0.1
	Sandwiches												
1744	Bacon, lettuce and tomato with mayonnaise	1	item(s)	164	95.8	346	11.7	35.7	2.5	17.4	3.7	5.4	6.6
30287	Bologna and cheese with margarine	1	item(s)	111	45.7	339	13.6	29.9	1.4	18.0	7.0	5.7	3.3
30286	Bologna with margarine	1	item(s)	83	33.4	251	8.8	27.2	1.4	11.6	3.3	3.9	2.9
16546	Cheese	1	item(s)	83	31.2	254	9.4	27.9	1.4	11.5	4.5	2.9	3.1
8789	Cheeseburger, large, plain	1	item(s)	166	70.8	506	28.7	34.5	2.3	28.3	11.2	9.2	0.9
8624	Cheeseburger, large, with bacon, vegetables and condiments	1	item(s)	195	91.4	550	30.8	36.8	2.5	30.9	11.9	10.6	1.3
1745	Club with bacon, chicken, tomato, lettuce and mayonnaise	1	item(s)	246	137.8	546	40.4	40.1	2.7	24.0	5.6	7.6	8.5
1908	Cold cut submarine with cheese and vegetables	1	item(s)	228	131.8	456	21.8	51.0	2.0	18.6	6.8	8.2	2.3
30247	Corned beef	1	item(s)	130	74.1	268	18.8	25.6	1.7	9.4	3.6	3.4	1.2
25283	Egg salad	1	item(s)	126	72.0	278	10.4	28.5	1.5	13.4	3.0	4.0	4.8
16686	Fried egg	1	item(s)	96	49.2	225	10.7	26.2	1.4	8.1	2.2	3.0	2.3
16547	Grilled cheese	1	item(s)	83	27.8	282	9.4	28.0	1.4	14.6	5.2	5.2	3.2
16659	Gyro with onion and tomato	1	item(s)	390	255.6	593	44.2	73.7	3.9	12.2	4.4	4.3	1.6
1906	Ham and cheese	1	item(s)	146	74.2	352	20.7	33.3	2.0	15.5	6.4	6.7	1.4
31890	Ham with mayonnaise	1	item(s)	112	55.8	272	13.7	28.0	2.1	11.4	2.7	3.9	4.0
756	Hamburger, double patty, large, with condiments and vegetables	1	item(s)	226	121.5	540	34.3	40.3	—	26.6	10.5	10.3	2.8
8793	Hamburger, large, plain	1	item(s)	137	57.7	426	22.6	31.7	1.5	22.9	8.4	9.9	2.1

Chol (mg)	Calc (mg)	Iron (mg)	Magn (mg)	Pota (mg)	Sodi (mg)	Zinc (mg)	Vit A (µg)	Thia (mg)	Vit E (mg α)	Ribo (mg)	Niac (mg)	Vit B$_6$ (mg)	Fola (µg)	Vit C (mg)	Vit B$_{12}$ (µg)	Sele (µg)
27	266	1.08	38.5	251.4	276.3	1.27	—	0.07	0.03	0.16	1.28	0.18	—	59.3	0.2	6.0
0	28	1.74	41.8	298.4	149.9	0.76	0.5	0.07	—	0.08	0.53	0.06	53.0	0.8	0	0.5
10	181	1.55	36.3	167.2	412.1	1.36	0	0.34	0.09	0.34	2.69	0.10	221.3	0.8	0.5	38.4
73	146	2.66	61.3	222.9	885.7	3.44	0	0.10	0.10	0.17	3.00	0.27	—	0	1.2	36.7
63	28	1.52	25.7	223.7	689.0	1.58	17.8	0.05	0.34	0.08	5.14	0.23	11.9	0	0.1	22.2
77	18	1.69	26.3	293.5	635.1	1.90	16.1	0.04	1.77	0.17	4.16	0.29	21.9	2.3	0.3	21.0
0	13	0.65	11.4	178.0	26.9	0.22	59.0	0.03	—	0.05	0.57	0.08	38.3	24.0	0	0.4
65	50	1.94	63.2	432.5	895.9	1.52	38.9	0.15	3.73	0.15	13.05	0.58	42.1	7.5	0.3	23.0
78	45	3.68	54.3	835.4	1064.4	6.04	30.7	0.10	5.90	0.31	7.52	0.28	30.7	14.6	2.5	28.1
66	226	2.22	39.6	438.3	495.6	2.87	0	0.20	1.14	0.29	3.07	0.20	91.1	9.6	1.1	26.1
39	671	2.55	48.6	301.3	928.3	1.99	226.0	0.40	1.36	0.52	2.73	0.12	126.4	0	0.6	44.2
13	33	2.63	32.6	424.2	888.5	1.13	37.7	0.14	1.38	0.15	3.23	0.16	70.3	0	0.4	24.1
85	63	1.95	21.8	295.6	422.2	3.42	0	0.08	0.09	0.29	3.77	0.14	29.0	0.4	1.6	17.2
157	32	1.40	27.2	350.3	1051.0	1.78	54.4	0.51	5.54	0.41	4.04	0.44	18.1	7.9	0.5	31.0
74	310	2.89	93.6	481.6	520.7	4.60	115.1	0.09	2.42	0.26	2.92	0.37	50.7	5.7	1.6	17.5
28	129	1.87	17.0	305.3	533.9	1.04	105.1	0.27	—	0.47	6.09	0.11	92.7	3.3	0.4	26.1
79	554	3.83	51.0	471.0	1719.1	3.85	136.9	0.47	2.55	0.38	6.20	0.20	257.5	9.0	1.4	45.0
85	24	0.81	18.8	317.5	661.3	0.39	40.0	0.10	—	0.08	1.11	0.18	8.8	12.5	0	5.1
60	77	2.00	23.7	308.4	369.1	3.47	0	0.11	0.03	0.30	4.00	0.14	33.3	0.3	1.7	16.9
262	192	4.59	51.0	291.6	984.2	2.48	75.3	0.24	2.45	0.09	3.19	0.24	136.1	14.8	0.8	58.3
17	89	3.03	34.7	498.5	781.2	1.36	47.1	0.24	1.17	0.21	5.54	0.18	—	5.5	0.6	23.6
37	32	1.55	49.2	183.9	587.1	1.06	0	0.17	1.29	0.13	1.85	0.17	55.7	16.0	0.1	3.2
46	40	2.66	32.8	400.0	569.4	2.08	0	0.51	0.60	0.20	5.00	0.30	163.1	18.3	0.4	22.5
30	194	2.50	54.3	382.8	466.6	1.48	0	0.24	1.87	0.30	3.77	0.28	161.0	9.8	0.2	29.9
187	45	1.79	18.7	137.3	455.5	1.06	115.4	0.12	0.89	0.28	1.31	0.15	82.7	1.9	0.5	20.0
12	23	2.12	18.7	115.4	424.3	0.65	7.8	0.23	0.23	0.04	2.37	0.11	135.7	0.9	0.3	14.8
0	17	1.54	17.2	85.8	154.4	0.69	10.9	0.20	0.08	0.04	1.89	0.13	120.1	2.2	0	10.1
74	41	1.72	34.0	612.1	623.9	2.52	0	0.80	0.73	0.37	6.69	0.65	14.6	9.8	0.7	49.7
0	32	1.20	35.2	246.4	798.4	0.48	54.4	0.07	2.43	0.04	1.12	0.11	32.0	25.3	0	0.5
0	27	0.96	15.5	145.0	207.4	0.53	0	0.04	0.94	0.06	0.26	0.06	25.0	9.2	0	2.5
13	17	1.02	19.5	182.4	412.0	0.57	24.6	0.03	—	0.07	6.87	0.08	8.2	2.3	1.2	42.2
53	50	1.74	23.3	283.1	396.6	1.85	84.2	0.17	0.30	0.22	4.60	0.22	78.0	0.8	0.8	27.1
0	55	2.36	32.0	302.1	677.1	0.67	199.7	0.30	0.73	0.18	2.40	0.17	130.6	16.5	0.1	15.6
0	20	2.43	17.2	90.5	425.9	0.70	0	0.27	0.06	0.04	2.47	0.10	156.0	0.2	0	9.7
21	82	2.43	29.5	344.4	944.6	1.08	42.6	0.34	1.16	0.26	4.42	0.21	108.2	10.2	0.2	27.9
39	518	2.47	26.6	213.1	894.7	1.64	73.3	0.35	0.57	0.29	3.25	0.18	96.6	0.2	0.9	22.8
17	159	2.21	17.4	143.6	534.5	1.05	10.0	0.34	0.34	0.18	3.17	0.15	91.3	0.2	0.5	18.1
22	486	2.11	21.6	126.2	627.5	1.00	66.4	0.29	0.58	0.23	2.53	0.10	93.8	0	0.4	15.9
93	277	4.02	39.8	360.2	884.8	5.16	—	0.32	—	0.69	7.42	0.44	129.5	0.0	2.5	34.9
98	267	4.04	44.8	464.1	1314.3	5.21	—	0.34	—	0.68	8.25	0.47	122.8	1.4	2.4	6.6
96	221	4.08	54.1	511.7	949.6	2.24	36.9	0.57	1.35	0.35	17.36	0.74	147.6	4.7	0.6	51.7
36	189	2.51	68.4	394.4	1650.7	2.58	70.7	1.00	—	0.80	5.49	0.14	109.4	12.3	1.1	30.8
44	139	3.02	20.8	135.2	1099.8	2.28	2.6	0.28	0.21	0.20	3.69	0.11	92.3	0.3	0.8	33.4
217	106	2.58	18.6	145.0	494.1	0.94	0	0.27	1.10	0.44	2.27	0.16	112.1	0.7	0.6	24.5
174	164	2.71	19.2	126.7	348.5	1.05	95.0	0.30	0.68	0.34	2.56	0.13	112.3	0	0.4	25.9
21	491	2.13	21.6	128.6	669.0	1.00	131.1	0.22	1.09	0.22	2.30	0.06	62.3	0	0.2	15.9
98	179	5.66	78.0	799.5	873.6	7.99	35.1	0.85	1.09	0.69	14.11	0.62	234.0	12.5	2.5	70.2
58	130	3.24	16.1	290.5	770.9	1.37	96.4	0.31	0.29	0.48	2.69	0.20	78.8	2.8	0.5	23.1
32	150	2.44	24.6	213.9	986.7	1.16	5.6	0.61	0.49	0.22	4.05	0.26	94.1	2.1	0.2	22.6
122	102	5.85	49.7	569.5	791.0	5.67	—	0.36	—	0.38	7.57	0.54	110.7	1.1	4.1	25.5
71	74	3.58	27.4	267.1	474.0	4.11	0	0.29	—	0.29	6.25	0.23	80.8	0	2.1	27.1

APPENDIX H

DA+ Code	Food Description	Quantity	Measure	Wt (g)	H₂O (g)	Ener (kcal)	Prot (g)	Carb (g)	Fiber (g)	Fat (g)	Fat Breakdown (g) Sat	Mono	Poly
Mixed Foods, Sandwiches, and Soups—*continued*													
8795	Hamburger, large, with vegetables and condiments	1	item(s)	218	121.4	512	25.8	40.0	3.1	27.4	10.4	11.4	2.2
25134	Hot chicken salad	1	item(s)	98	48.8	239	15.6	23.4	1.1	8.8	2.8	2.6	2.8
25133	Hot turkey salad	1	item(s)	98	50.4	221	16.0	23.4	1.1	6.6	2.2	1.8	2.3
1411	Hotdog with bun, plain	1	item(s)	98	52.9	242	10.4	18.0	1.6	14.5	5.1	6.9	1.7
30249	Pastrami	1	item(s)	134	82.6	221	16.5	26.6	1.7	4.8	1.8	1.4	0.9
16701	Peanut butter	1	item(s)	93	24.2	337	13.0	37.4	3.4	16.1	3.4	7.1	5.0
30306	Peanut butter and jelly	1	item(s)	93	24.7	323	11.0	41.6	3.1	13.7	2.9	6.1	4.2
1909	Roast beef submarine with mayonnaise and vegetables	1	item(s)	216	127.4	410	28.6	44.3	—	13.0	7.1	1.8	2.6
1910	Roast beef, plain	1	item(s)	139	67.6	346	21.5	33.4	1.2	13.8	3.6	6.8	1.7
1907	Steak with mayonnaise and vegetables	1	item(s)	204	104.2	459	30.3	52.0	2.3	14.1	3.8	5.3	3.3
25288	Tuna salad	1	item(s)	179	102.2	415	24.1	28.9	1.6	22.2	3.6	5.5	11.4
30283	Turkey submarine with cheese, lettuce, tomato and mayonnaise	1	item(s)	277	167.2	529	29.3	50.9	3.0	22.9	6.5	5.8	8.3
31891	Turkey with mayonnaise	1	item(s)	143	73.9	330	29.1	26.7	1.4	11.1	2.4	2.5	4.9
	Soups												
25296	Bean	1	cup(s)	243	204.8	153	10.8	23.2	8.2	1.9	0.5	0.7	0.6
711	Bean with pork, condensed, prepared with water	1	cup(s)	266	227.0	168	7.7	22.1	7.7	5.7	1.5	2.1	1.8
713	Beef noodle, condensed, prepared with water	1	cup(s)	244	224.9	83	4.7	8.7	0.7	3.0	1.1	1.2	0.5
825	Cheese, condensed, prepared with milk	1	cup(s)	251	206.9	231	9.5	16.2	1.0	14.6	9.1	4.1	0.5
826	Chicken broth, condensed, prepared with water	1	cup(s)	244	234.1	39	4.9	0.9	0	1.4	0.4	0.6	0.3
25297	Chicken noodle soup	1	cup(s)	241	218.2	99	9.1	9.2	0.8	2.5	0.7	0.9	0.6
827	Chicken noodle, condensed, prepared with water	1	cup(s)	248	232.7	62	3.1	7.3	0.5	2.4	0.6	1.0	0.7
724	Chicken noodle, dehydrated, prepared with water	1	cup(s)	245	230.7	56	2.1	9.0	0.2	1.3	0.3	0.5	0.4
1769	Chicken noodle, low sodium, condensed, prepared with water	1	cup(s)	248	232.7	62	3.1	7.3	0.5	2.4	0.6	1.0	0.7
823	Cream of asparagus, condensed, prepared with milk	1	cup(s)	248	213.3	161	6.3	16.4	0.7	8.2	3.3	2.1	2.2
824	Cream of celery, condensed, prepared with milk	1	cup(s)	248	214.4	164	5.7	14.5	0.7	9.7	3.9	2.5	2.7
708	Cream of chicken, condensed, prepared with milk	1	cup(s)	248	210.4	191	7.5	15.0	0.2	11.5	4.6	4.5	1.6
715	Cream of chicken, condensed, prepared with water	1	cup(s)	244	221.1	117	3.4	9.3	0.2	7.4	2.1	3.3	1.5
709	Cream of mushroom, condensed, prepared with milk	1	cup(s)	252	219.3	161	5.9	14.5	0.8	9.0	2.8	2.2	3.6
716	Cream of mushroom, condensed, prepared with water	1	cup(s)	248	229.3	97	1.6	8.3	0.7	6.4	1.2	1.5	3.5
9558	Cream of mushroom, low sodium, canned, ready to serve	1	cup(s)	244	218.1	139	2.5	15.6	0	7.4	2.0	1.6	3.3
25298	Cream of vegetable	1	cup(s)	260	229.2	151	6.4	13.8	1.7	7.9	1.4	4.2	1.8
16689	Egg drop	1	cup(s)	244	226.6	66	2.8	10.5	1.0	1.5	0.4	0.5	0.3
25138	Golden squash	1	cup(s)	258	224.0	144	7.5	20.6	1.5	3.9	0.7	2.1	0.9
16663	Hot and sour	1	cup(s)	244	221.2	95	6.3	10.6	1.2	3.0	0.6	0.7	0.8
28054	Lentil chowder	1	cup(s)	248	204.1	168	10.8	30.9	13.2	0.6	0.1	0.1	0.3
28560	Macaroni and bean	1	cup(s)	246	138.8	147	6.6	23.3	5.0	3.5	0.5	2.2	0.5
714	Manhattan clam chowder, condensed, prepared with water	1	cup(s)	249	229.7	75	2.1	11.9	1.5	2.1	0.4	0.4	1.3
28561	Minestrone	1	cup(s)	240	184.6	105	5.2	17.3	4.8	2.2	0.3	1.4	0.3
717	Minestrone, condensed, prepared with water	1	cup(s)	241	220.1	82	4.3	11.2	1.0	2.5	0.6	0.7	1.1
53690	Minestrone, reduced sodium, canned, ready to serve	1	cup(s)	241	210.4	120	4.8	21.7	5.8	1.9	0.3	0.7	0.6
28038	Mushroom & wild rice	1	cup(s)	241	196.6	88	4.7	13.3	1.6	0.3	0.1	0	0.1
828	New England clam chowder, condensed, prepared with milk	1	cup(s)	252	215.4	154	8.2	18.8	0.8	5.1	2.8	0.7	1.3
75714	New England clam chowder, reduced sodium, canned, ready to serve	1	cup(s)	256	221.5	179	6.0	14.5	2.0	10.8	1.8	2.3	5.1
28036	New England style clam chowder	1	cup(s)	254	235.0	67	3.0	11.5	1.2	0.1	0	0	0
28566	Old country pasta	1	cup(s)	245	175.9	151	6.5	21.0	3.4	3.4	1.6	2.1	0.7

Chol (mg)	Calc (mg)	Iron (mg)	Magn (mg)	Pota (mg)	Sodi (mg)	Zinc (mg)	Vit A (µg)	Thia (mg)	Vit E (mg α)	Ribo (mg)	Niac (mg)	Vit B$_6$ (mg)	Fola (µg)	Vit C (mg)	Vit B$_{12}$ (µg)	Sele (µg)
87	96	4.93	43.6	479.6	824.0	4.88	—	0.41	—	0.37	7.28	0.33	115.5	2.6	2.4	33.6
39	114	1.93	20.4	150.3	469.7	1.22	0	0.20	0.31	0.23	4.94	0.20	79.5	0.2	0.3	16.9
37	113	2.04	21.8	166.9	459.2	1.09	0	0.19	0.25	0.21	4.36	0.23	79.6	0.2	0.3	20.3
44	24	2.31	12.7	143.1	670.3	1.98	0	0.24	—	0.27	3.65	0.05	60.8	0.1	0.5	26.0
36	142	3.19	22.8	178.2	1069.3	3.10	4.0	0.31	0.20	0.22	4.78	0.17	93.8	0.4	1.0	21.0
0	181	2.86	59.5	255.8	447.3	1.36	0	0.37	2.65	0.19	6.85	0.21	132.1	0	0	15.9
0	154	2.45	51.2	225.1	384.1	1.16	0	0.31	2.25	0.16	5.82	0.18	112.5	0.1	0	13.6
73	41	2.81	67.0	330.5	844.6	4.38	30.2	0.41	—	0.41	5.96	0.32	88.6	5.6	1.8	25.7
51	54	4.23	30.6	315.5	792.3	3.39	11.1	0.38	—	0.31	5.87	0.26	68.1	2.1	1.2	29.2
73	92	5.16	49.0	524.3	797.6	4.53	—	0.41	—	0.37	7.30	0.37	128.5	5.5	1.6	42.0
53	97	3.23	35.6	296.5	794.9	1.09	0	0.27	1.76	0.27	12.29	0.48	99.6	1.1	2.4	71.5
66	521	4.68	52.6	567.8	1795.0	2.63	105.3	0.70	1.36	0.60	4.38	0.33	166.2	11.6	0.7	44.9
64	162	3.42	34.3	307.5	471.9	3.00	5.7	0.34	0.74	0.28	7.05	0.46	98.7	0	0.3	42.2
4	63	2.35	45.2	491.2	539.4	1.12	28.0	0.24	0.23	0.13	3.00	0.16	116.2	2.1	0.2	4.3
3	82	2.00	45.2	391.0	928.3	1.01	45.2	0.09	1.14	0.03	0.55	0.04	31.9	1.6	0	8.2
5	20	1.07	7.3	97.6	793.0	1.51	12.2	0.07	1.22	0.06	1.04	0.04	29.3	0.5	0.2	7.3
48	289	0.80	20.1	341.4	1019.1	0.68	358.9	0.06	—	0.33	0.50	0.08	10.0	1.3	0.4	7.0
0	10	0.51	2.4	209.8	746.6	0.24	0	0.01	0.05	0.07	3.35	0.02	4.9	0	0.2	0
19	22	1.03	13.6	283.4	631.4	0.68	54.2	0.13	0.18	0.13	4.74	0.12	46.8	1.1	0.3	10.3
12	15	1.64	9.9	54.6	865.5	0.40	27.3	0.14	0.07	0.11	1.34	0.05	29.8	0	0	11.9
10	5	0.49	7.3	31.9	561.0	0.20	2.5	0.20	0.12	0.07	1.06	0.02	27.0	0	0	9.3
12	15	1.64	9.9	54.6	429.0	0.40	27.3	0.14	0.07	0.11	1.34	0.05	29.8	0	0	11.9
22	174	0.87	19.8	359.6	1041.6	0.92	62.0	0.10	—	0.28	0.88	0.06	29.8	4.0	0.5	8.0
32	186	0.69	22.3	310.0	674.6	0.20	114.1	0.07	—	0.25	0.44	0.06	7.4	1.5	0.5	4.7
27	181	0.67	17.4	272.8	897.8	0.67	178.6	0.07	—	0.26	0.92	0.07	7.4	1.2	0.5	8.0
10	34	0.61	2.4	87.8	846.7	0.63	163.5	0.03	—	0.06	0.82	0.02	2.4	0.2	0.1	7.0
10	171	0.25	17.6	259.6	899.6	0.78	75.6	0.07	0.65	0.26	0.54	0.07	7.6	0.3	0.7	6.8
0	17	0.22	5.0	76.9	843.2	0.15	2.5	0.01	0.60	0.02	0.42	0.02	2.5	0	0	3.5
8	33	0	—	155.6	61.4	—	0	—	—	—	—	—	—	0	—	—
1	71	1.10	16.3	311.2	694.3	0.55	202.3	0.10	1.35	0.19	2.99	0.10	49.2	9.2	0.3	4.5
56	17	0.63	4.9	53.7	902.8	0.22	48.8	0.05	0.32	0.05	0.38	0.05	17.1	15.9	0.1	1.0
4	261	0.76	42.4	522.3	496.7	0.92	0	0.18	1.39	0.38	1.15	0.16	42.9	10.5	0.9	7.9
51	46	1.56	22.0	134.2	917.4	0.54	22.0	0.06	0.95	0.08	1.24	0.16	19.5	0	0.2	1.0
0	54	3.86	64.9	660.5	28.3	2.00	0	0.38	0.98	0.11	1.82	0.33	195.9	13.1	0	3.7
0	70	1.92	35.6	270.1	519.7	0.89	0	0.14	1.33	0.14	1.42	0.13	92.3	4.9	0	9.3
2	27	1.59	10.0	184.3	562.7	0.90	49.8	0.03	1.25	0.04	0.79	0.10	10.0	4.0	3.9	9.2
0	71	1.80	31.8	264.9	434.9	0.74	0	0.10	1.12	0.10	0.69	0.10	61.8	10.9	0	4.0
2	34	0.92	7.2	313.3	612.1	0.75	118.1	0.05	—	0.04	0.94	0.10	50.6	1.2	0	8.0
0	48	1.73	31.3	448.2	518.1	0.77	79.5	0.14	1.69	0.10	1.54	0.14	37.8	13.7	0	5.1
0	29	1.11	27.3	347.4	281.1	0.93	0	0.06	0.16	0.22	3.03	0.15	26.9	3.8	0.1	4.9
18	176	3.05	30.2	451.1	688.0	1.08	93.2	0.20	0.55	0.44	2.00	0.18	22.7	5.3	12.2	11.1
8	36	1.43	33.3	862.7	496.6	0.69	20.5	0.04	1.02	0.04	1.08	0.33	15.4	8.4	2.4	11.0
4	74	1.06	29.6	465.3	349.9	0.45	0	0.07	0.41	0.11	0.93	0.17	16.1	10.0	6.1	6.2
7	55	2.44	50.7	472.7	346.8	0.76	0	0.20	0.94	0.16	2.55	0.26	102.3	18.8	0.1	9.2

APPENDIX H

DA+ Code	Food Description	Quantity	Measure	Wt (g)	H₂O (g)	Ener (kcal)	Prot (g)	Carb (g)	Fiber (g)	Fat (g)	Fat Breakdown (g) Sat	Mono	Poly
Mixed Foods, Sandwiches, and Soups—*continued*													
725	Onion, dehydrated, prepared with water	1	cup(s)	230	220.3	28	0.7	6.4	0.7	0	0	0	0
16667	Shrimp gumbo	1	cup(s)	244	207.2	163	8.9	18.3	2.9	6.8	3.9	1.6	0.6
28037	Southwestern corn chowder	1	cup(s)	240	211.2	109	4.8	19.7	2.2	0.5	0.1	0.1	0.2
30282	Soybean (miso)	1	cup(s)	240	218.8	84	6.0	7.8	1.9	3.4	0.6	1.2	1.3
25140	Split pea	1	cup(s)	165	116.6	83	4.5	18.5	1.5	0.4	0.1	0	0.2
718	Split pea with ham, condensed, prepared with water	1	cup(s)	253	206.9	190	10.3	28.0	2.3	4.4	1.8	1.8	0.6
32411	Split pea with ham, reduced sodium & fat, ready to serve	1	cup(s)	243	196.8	185	12.6	27.5	—	2.7	0.7	1.0	0.5
726	Tomato vegetable, dehydrated, prepared with water	1	cup(s)	245	230.9	54	1.9	9.9	0.7	0.8	0.4	0.3	0.1
710	Tomato, condensed, prepared with milk	1	cup(s)	252	216.7	139	6.3	22.6	1.5	3.3	1.8	0.8	0.3
719	Tomato, condensed, prepared with water	1	cup(s)	248	226.6	74	2.0	16.3	1.5	0.7	0.2	0.2	0.2
40840	Tomato, low sodium, prepared wtih water	1	cup(s)	244	223.0	73	1.9	16.0	1.5	0.7	0.2	0.2	0.2
28595	Turkey noodle	1	cup(s)	244	217.0	113	8.1	14.9	1.9	2.4	0.3	1.2	0.7
28051	Turkey vegetable	1	cup(s)	241	215.4	106	12.1	9.0	1.9	1.1	0.3	0.2	0.3
25141	Vegetable	1	cup(s)	252	225.3	92	4.5	19.6	4.3	0.4	0.1	0	0.2
720	Vegetable beef, condensed, prepared with water	1	cup(s)	244	224.0	76	5.4	9.9	2.0	1.9	0.8	0.8	0.1
28598	Vegetable gumbo	1	cup(s)	244	179.4	162	4.2	27.5	3.4	4.6	0.7	3.1	0.6
40841	Vegetable, low sodium, prepared with water	1	cup(s)	253	231.1	83	2.8	15.3	2.8	1.1	0.2	0.3	0.5
721	Vegetarian vegetable, condensed, prepared with water	1	cup(s)	241	222.7	67	2.1	11.8	0.7	1.9	0.3	0.8	0.7
Fast Food													
Arby's													
751	Beef 'n cheddar sandwich	1	item(s)	195	—	440	23.0	47.0	2.0	18.0	5.0	—	—
36131	Chocolate shake, regular	1	serving(s)	468	—	570	14.0	99.0	1.0	15.0	10.0	—	—
80611	Cravin' chicken sandwich, crispy	1	item(s)	221	—	510	26.0	51.0	4.0	22.0	4.0	—	—
36045	Curly fries, large size	1	serving(s)	210	—	658	7.3	77.3	8.4	36.6	5.2	—	—
36044	Curly fries, medium size	1	serving(s)	170	—	540	6.0	62.0	7.0	29.0	4.0	—	—
57171	French dip & Swiss sandwich w/ au jus sauce	1	item(s)	286	—	430	26.0	52.0	2.0	14.0	6.0	—	—
34778	Jalapeno bites, 5-piece	1	serving(s)	110	—	280	5.0	31.0	2.0	16.0	6.0	—	—
9296	Jamocha shake, regular	1	serving(s)	468	—	560	14.0	98.0	1.0	15.0	10.0	—	—
9249	Junior roast beef sandwich	1	item(s)	87	—	210	12.0	25.0	1.0	6.0	2.0	—	—
9251	Large roast beef sandwich	1	item(s)	281	—	580	45.0	49.0	3.0	22.0	9.0	—	—
56537	Market Fresh crispy chicken farmhouse chopped salad	1	serving(s)	337	—	430	29.0	26.0	4.0	24.0	9.0	—	—
34769	Market Fresh roast turkey & Swiss sandwich	1	serving(s)	326	—	700	39.0	77.0	5.0	27.0	7.0	—	—
80607	Market Fresh roast turkey & Swiss wrap	1	item(s)	272	—	490	34.0	39.0	7.0	25.0	6.0	—	—
80608	Market Fresh roast turkey farmhouse chopped salad	1	serving(s)	300	—	240	22.0	10.0	3.0	13.0	7.0	—	—
9267	Market Fresh roast turkey ranch & bacon sandwich	1	serving(s)	344	—	800	45.0	78.0	5.0	36.0	9.0	—	—
34780	Mozzarella sticks, 4-piece	1	serving(s)	137	—	420	21.0	35.0	2.0	21.0	9.0	—	—
9275	Potato cakes	1	serving(s)	100	—	230	2.0	25.0	3.0	14.0	2.0	—	—
750	Roast beef sandwich, regular	1	item(s)	154	—	350	23.0	39.0	2.0	12.0	4.0	—	—
2009	Super roast beef sandwich	1	item(s)	229	—	430	23.0	45.0	3.0	17.0	5.0	—	—
36130	Vanilla shake, regular	1	serving(s)	425	—	470	14.0	75.0	0	15.0	10.0	—	—
Auntie Anne's													
35351	Almond soft pretzel	1	item(s)	127	—	350	8.0	74.0	2.0	2.0	1.0	—	—
35371	Cheese dipping sauce	1	serving(s)	31	—	78	2.3	1.6	0	6.2	2.3	—	—
35353	Cinnamon sugar soft pretzel	1	item(s)	136	—	380	8.0	84.0	2.0	1.0	0	—	—
35354	Cinnamon sugar soft pretzel with butter	1	item(s)	147	—	470	8.0	84.0	2.0	12.0	7.0	—	—
56651	Garlic soft pretzel	1	item(s)	119	—	310	8.0	65.0	2.0	1.0	0	—	—
35372	Marinara dipping sauce	1	serving(s)	37	—	45	2.0	6.0	1.0	2.0	0	—	—
35357	Original soft pretzel	1	serving(s)	119	—	310	8.0	65.0	2.0	1.0	0	—	—
35358	Original soft pretzel with butter	1	item(s)	123	—	340	8.0	65.0	2.0	5.0	3.0	—	—
56653	Original soft preztel without salt	1	item(s)	117	—	310	8.0	65.0	2.0	1.0	0	—	—

Chol (mg)	Calc (mg)	Iron (mg)	Magn (mg)	Pota (mg)	Sodi (mg)	Zinc (mg)	Vit A (µg)	Thia (mg)	Vit E (mg α)	Ribo (mg)	Niac (mg)	Vit B$_6$ (mg)	Fola (µg)	Vit C (mg)	Vit B$_{12}$ (µg)	Sele (µg)
0	21	0.12	9.2	71.3	795.8	0.12	0	0.03	0.02	0.03	0.14	0.06	0	0.2	0	0.5
90	132	2.90	46.4	429.4	646.6	1.12	65.9	0.17	1.63	0.13	2.21	0.24	104.9	18.5	0.2	17.6
1	83	0.94	26.1	417.2	208.9	0.58	0	0.09	0.28	0.15	1.75	0.24	32.9	35.9	0.3	2.1
0	62	1.82	38.4	367.2	976.8	0.77	230.4	0.06	0.96	0.16	2.62	0.16	55.2	4.6	0.2	1.4
0	30	1.25	32.7	352.3	607.9	0.57	0	0.12	0.14	0.09	1.67	0.21	72.0	8.8	0	0.4
8	23	2.28	48.1	399.7	1006.9	1.32	22.8	0.15	—	0.08	1.47	0.07	2.5	1.5	0.3	8.0
15	—	2.24	—	—	833.5	—	318.3	—	—	—	—	—	—	10.2	—	—
0	20	0.59	9.8	164.1	323.4	0.20	9.8	0.06	0.42	0.09	1.22	0.06	12.3	2.9	0	2.0
10	174	1.39	30.2	461.2	529.2	0.93	95.8	0.10	0.45	0.32	1.37	0.15	5.0	15.9	0.7	9.3
0	20	1.34	17.4	277.8	471.2	0.30	24.8	0.05	0.42	0.08	1.25	0.10	0	15.6	0	6.2
0	20	1.32	17.1	273.3	80.5	0.29	24.4	0.05	0.41	0.08	1.23	0.10	0	15.4	0	6.1
25	31	1.44	23.7	217.1	398.4	0.76	0	0.22	0.63	0.11	2.91	0.16	65.1	5.9	0.1	16.1
21	38	1.37	23.5	405.4	347.8	0.96	0	0.09	0.21	0.09	3.55	0.30	26.1	10.5	0.2	10.0
0	41	2.44	39.4	684.8	720.6	0.67	0	0.12	1.82	0.13	2.37	0.27	38.0	23.5	0.1	1.8
5	20	1.10	7.3	168.4	851.6	1.51	190.3	0.04	0.59	0.05	1.01	0.07	9.8	2.4	0.3	2.7
0	56	2.05	37.3	334.5	501.9	0.60	0	0.18	1.13	0.08	1.72	0.16	104.1	18.8	0	3.9
0	30	0.83	32.9	549.0	490.8	0.51	108.8	0.14	1.82	0.12	1.95	0.21	15.2	1.0	0	5.1
0	24	1.06	7.2	207.3	814.6	0.46	171.1	0.05	1.40	0.05	0.90	0.06	9.6	1.4	0	4.3
45	150	4.50	—	—	1290.0	—	—	—	—	—	—	—	—	1.2	—	—
50	500	1.08	—	—	450.0	—	—	—	—	—	—	—	—	4.8	—	—
50	100	2.70	—	—	1110.0	—	—	—	—	—	—	—	—	9.0	—	—
0	42	2.82	—	—	1483.6	—	0	—	—	—	—	—	—	0	—	—
0	20	1.80	—	—	1200.0	—	0	—	—	—	—	—	—	0	—	—
55	150	4.50	—	—	2120.0	—	—	—	—	—	—	—	—	4.8	—	—
25	40	0.72	—	—	600.0	—	—	—	—	—	—	—	—	0	—	—
50	500	0.72	—	—	440.0	—	—	—	—	—	—	—	—	4.8	—	—
25	40	2.70	—	—	520.0	—	0	—	—	—	—	—	—	0	—	—
110	80	8.10	—	—	1870.0	—	0	—	—	—	—	—	—	0	—	—
65	250	1.44	—	—	1000.0	—	—	—	—	—	—	—	—	12.0	—	—
80	450	5.40	—	—	1770.0	—	—	—	—	—	—	—	—	6.0	—	—
80	350	3.60	—	—	1550.0	—	—	—	—	—	—	—	—	6.0	—	—
60	250	1.44	—	—	760.0	—	—	—	—	—	—	—	—	12.0	—	—
105	450	5.40	—	—	2200.0	—	—	—	—	—	—	—	—	6.0	—	—
50	600	0.72	—	—	1690.0	—	—	—	—	—	—	—	—	0	—	—
0	20	0.36	—	—	460.0	—	0	—	—	—	—	—	—	0	—	—
45	60	4.50	—	—	950.0	—	0	—	—	—	—	—	—	0	—	—
45	80	4.50	—	—	1060.0	—	—	—	—	—	—	—	—	6.0	—	—
50	500	0.36	—	—	390.0	—	—	—	—	—	—	—	—	4.8	—	—
0	20	0.72	—	—	400.0	—	0	—	—	—	—	—	—	0	—	—
12	78	0	—	—	364.3	—	23.3	—	—	—	—	—	—	0	—	—
0	20	0.72	—	—	400.0	—	0	—	—	—	—	—	—	0	—	—
25	20	0.72	—	—	400.0	—	—	—	—	—	—	—	—	0	—	—
0	20	0.72	—	—	990.0	—	0	—	—	—	—	—	—	0	—	—
0	0	1.08	—	—	140.0	—	20.0	—	—	—	—	—	—	36.0	—	—
0	20	0.72	—	—	990.0	—	0	—	—	—	—	—	—	0	—	—
10	20	0.72	—	—	990.0	—	—	—	—	—	—	—	—	0	—	—
0	20	0.72	—	—	400.0	—	0	—	—	—	—	—	—	0	—	—

APPENDIX H

DA+ Code	Food Description	Quantity	Measure	Wt (g)	H₂O (g)	Ener (kcal)	Prot (g)	Carb (g)	Fiber (g)	Fat (g)	Sat	Mono	Poly
												Fat Breakdown (g)	
Fast Food—*continued*													
35361	Sesame soft pretzel	1	item(s)	128	—	360	10.0	67.0	3.0	6.0	1.0	—	—
35362	Sesame soft pretzel with butter	1	item(s)	132	—	400	10.0	67.0	3.0	10.0	3.5	—	—
35364	Sour cream & onion soft pretzel	1	item(s)	123	—	330	9.0	68.0	2.0	1.5	0	—	—
35366	Sour cream & onion soft pretzel with butter	1	item(s)	127	—	360	9.0	68.0	2.0	5.0	3.0	—	—
35373	Sweet mustard dipping sauce	1	serving(s)	35	—	60	2.0	10.0	0	2.0	1.0	—	—
	Boston Market												
51194	Apple pie	1	slice(s)	163	—	580	43.0	74.0	3.0	30.0	13.0	—	—
35058	Brownie, chocolate	1	item(s)	85	—	320	5.0	49.0	3.0	13.0	3.0	—	—
35013	Chicken Carver sandwich with cheese and sauce	1	item(s)	321	—	750	57.0	64.0	3.0	29.0	8.0	—	—
34979	Chicken gravy	4	ounce(s)	113	—	50	0	7.0	0	2.0	0.5	—	—
35053	Chicken noodle soup	¾	cup(s)	171	—	103	9.1	9.5	0.8	3.3	1.0	—	—
34973	Chicken pot pie	1	item(s)	425	—	800	32.0	59.0	4.0	48.0	18.0	—	—
35054	Chicken tortilla soup with toppings	¾	cup(s)	171	—	193	8.0	14.1	0.9	12.2	3.3	—	—
35064	Chocolate cake	1	slice(s)	145	—	580	5.0	67.0	3.0	34.0	11.0	—	—
35057	Cornbread	1	item(s)	57	—	180	2.0	31.0	0	5.0	1.5	—	—
35008	Cranberry walnut relish	¾	cup(s)	210	—	350	3.0	75.0	3.0	4.5	0	—	—
34980	Creamed spinach	¾	cup(s)	191	—	280	9.0	12.0	4.0	23.0	15.0	—	—
34998	Fresh vegetable stuffing	1	cup(s)	136	—	190	3.0	25.0	2.0	8.0	1.0	—	—
34991	Garlic dill new potatoes	¾	cup(s)	156	—	140	3.0	24.0	3.0	3.0	1.0	—	—
34982	Green beans	¾	cup(s)	91	—	60	2.0	7.0	3.0	3.5	1.5	—	—
34984	Homestyle mashed potatoes	¾	cup(s)	221	—	270	5.0	36.0	4.0	11.0	5.0	—	—
34985	Homestyle mashed potatoes and gravy	1	cup(s)	227	—	153	3.4	22.4	2.0	6.4	4.1	—	—
34988	Hot cinnamon apples	¾	cup(s)	145	—	210	0	47.0	3.0	3.0	0	—	—
34989	Macaroni and cheese	¾	cup(s)	221	—	300	11.0	35.0	2.0	11.0	7.0	—	—
34970	Meatloaf	1	serving(s)	218	—	520	29.0	21.0	0	36.0	16.0	—	—
39383	Nestle Toll House chocolate chip cookie	1	item(s)	78	—	370	4.0	49.0	2.0	19.0	9.0	—	—
34963	Quarter chicken, white meat, no skin or wing	1	item(s)	173	—	250	41.0	4.0	0	8.0	2.5	—	—
34964	Quarter chicken, white meat, with skin and wing	1	item(s)	110	—	330	50.0	3.0	0	12.0	4.0	—	—
34968	Roasted turkey breast	5	ounce(s)	141	—	188	38.8	0	0	3.1	1.3	—	—
35011	Seasonal fresh fruit salad	1	serving(s)	142	—	60	1.0	15.0	1.0	0	0	0	0
51192	Spinach with garlic butter sauce	1	serving(s)	170	—	130	5.0	9.0	5.0	9.0	6.0	—	—
34999	Squash casserole	¾	cup(s)	187	—	330	7.0	20.0	3.0	24.0	13.0	—	—
35003	Steamed vegetables	1	cup(s)	136	—	60	2.0	8.0	3.0	2.0	0	—	—
35005	Sweet corn	¾	cup(s)	176	—	170	6.0	37.0	2.0	4.0	1.0	—	—
35004	Sweet potato casserole	¾	cup(s)	198	—	460	4.0	77.0	3.0	16.0	4.5	—	—
	Burger King												
29731	Biscuit with sausage, egg & cheese	1	item(s)	191	—	550	20.0	34.0	1.0	37.0	19.0	—	—
14249	Cheeseburger	1	item(s)	121	—	310	16.0	28.0	1.0	15.0	7.0	—	—
14251	Chicken sandwich	1	item(s)	218	—	630	24.0	46.0	3.0	39.0	7.0	—	—
14259	Chocolate shake, 12 ounces	1	item(s)	340	—	340	7.0	60.0	1.0	9.0	7.0	—	—
29732	Croissanwich with sausage & cheese	1	item(s)	106	37.2	380	14.0	26.0	0	24.0	10.0	12.7	3.3
14261	Croissanwich with sausage, egg & cheese	1	item(s)	159	71.4	460	19.0	27.0	0	31.0	11.0	15.8	6.1
3809	Double cheeseburger	1	item(s)	171	—	460	27.0	28.0	1.0	27.0	13.0	—	—
14244	Double Whopper sandwich	1	item(s)	373	—	920	48.0	51.0	3.0	58.0	19.0	—	—
14245	Double Whopper with cheese sandwich	1	item(s)	398	—	1010	53.0	53.0	3.0	65.0	24.0	—	—
14250	Fish sandwich	1	item(s)	248	—	640	23.0	67.0	3.0	31.0	5.0	—	—
14255	French fries, medium, salted	1	serving(s)	148	—	440	5.0	56.0	5.0	22.0	4.5	—	—
14262	French toast sticks (5)	1	serving(s)	109	37.6	380	5.0	49.0	2.0	18.0	3.0	10.6	2.9
14248	Hamburger	1	item(s)	110	—	260	14.0	27.0	1.0	11.0	4.0	—	—
14263	Hash brown rounds, small	1	serving(s)	78	27.1	240	2.0	23.0	3.0	15.0	3.5	—	—
14256	Onion rings, medium	1	serving(s)	117	—	400	6.0	47.0	4.0	21.0	3.5	—	—
39000	Tendercrisp chicken sandwich	1	item(s)	284	—	800	32.0	68.0	3.0	46.0	8.0	—	—
37514	TenderGrill chicken sandwich	1	item(s)	259	—	410	38.0	49.0	2.0	7.0	1.5	—	—
14258	Vanilla shake, 12 ounces	1	item(s)	340	—	290	7.0	46.0	0	9.0	7.0	—	—
1736	Whopper sandwich	1	item(s)	290	—	670	29.0	51.0	3.0	40.0	11.0	—	—
14243	Whopper with cheese sandwich	1	item(s)	315	—	770	33.0	52.0	3.0	48.0	16.0	—	—

Chol (mg)	Calc (mg)	Iron (mg)	Magn (mg)	Pota (mg)	Sodi (mg)	Zinc (mg)	Vit A (µg)	Thia (mg)	Vit E (mg α)	Ribo (mg)	Niac (mg)	Vit B6 (mg)	Fola (µg)	Vit C (mg)	Vit B12 (µg)	Sele (µg)
0	100	1.80	—	—	990.0	—	0	—	—	—	—	—	—	0	—	—
10	100	1.80	—	—	990.0	—	—	—	—	—	—	—	—	0	—	—
0	60	0.72	—	—	1180.0	—	0	—	—	—	—	—	—	0	—	—
10	60	0.72	—	—	1180.0	—	—	—	—	—	—	—	—	0	—	—
35	0	0	—	—	0	—	0	—	—	—	—	—	—	0	—	—
0	—	—	—	—	690.0	—	—	—	—	—	—	—	—	—	—	—
50	—	—	—	—	220.1	—	—	—	—	—	—	—	—	—	—	—
160	211	2.85	—	—	1960.0	—	—	—	—	—	—	—	—	15.8	—	—
0	0	0	—	—	690.0	—	—	—	—	—	—	—	—	0	—	—
39	—	—	—	—	584.8	—	—	—	—	—	—	—	—	—	—	—
140	40	4.50	—	—	1090.0	—	—	—	—	—	—	—	—	1.2	—	—
33	—	—	—	—	986.4	—	—	—	—	—	—	—	—	—	—	—
45	36	1.64	—	—	360.0	—	—	—	—	—	—	—	—	0	—	—
10	0	0.71	—	—	320.0	—	—	—	—	—	—	—	—	0	—	—
0	0	5.40	—	—	0	—	0	—	—	—	—	—	—	0	—	—
70	264	2.85	—	—	580.0	—	—	—	—	—	—	—	—	9.5	—	—
0	—	—	—	—	580.0	—	—	—	—	—	—	—	—	—	—	—
0	0	0.86	—	—	120.0	—	—	—	—	—	—	—	—	14.3	—	—
0	43	0.38	—	—	180.0	—	—	—	—	—	—	—	—	5.1	—	—
25	51	0.46	—	—	820.0	—	—	—	—	—	—	—	—	19.2	—	—
17	68	0.41	—	—	834.1	—	—	—	—	—	—	—	—	16.9	—	—
0	16	0.29	—	—	15.0	—	—	—	—	—	—	—	—	0	—	—
30	345	1.66	—	—	1100.0	—	—	—	—	—	—	—	—	0	—	—
145	140	3.78	—	—	1030.0	—	—	—	—	—	—	—	—	1.8	—	—
20	0	1.32	—	—	340.0	—	—	—	—	—	—	—	—	0	—	—
125	0	0.89	—	—	480.0	—	0	—	—	—	—	—	—	0	—	—
165	0	0.78	—	—	960.0	—	0	—	—	—	—	—	—	0	—	—
69	25	2.25	—	—	625.0	—	—	—	—	—	—	—	—	0	—	—
0	16	0.29	—	—	20.0	—	—	—	—	—	—	—	—	29.5	—	—
20	—	—	—	—	200.0	—	—	—	—	—	—	—	—	—	—	—
70	200	0.72	—	—	1110.0	—	—	—	—	—	—	—	—	4.8	—	—
0	53	0.47	—	—	40.0	—	—	—	—	—	—	—	—	24.0	—	—
0	0	0.43	—	—	95.0	—	—	—	—	—	—	—	—	5.8	—	—
5	44	1.18	—	—	270.0	—	—	—	—	—	—	—	—	9.8	—	—
210	250	2.70	—	—	1520.0	—	—	—	—	—	—	—	—	0	—	—
40	150	2.70	—	—	740.0	—	—	0.25	—	0.32	4.18	—	—	1.2	—	—
65	64	2.90	—	—	1390.0	—	—	0.50	—	0.32	10.29	—	—	0	—	—
30	333	0.80	—	—	270.0	—	—	0.12	—	0.62	0.26	—	—	2.7	0	—
50	99	1.78	20.1	—	780.0	1.52	—	0.35	1.04	0.34	4.34	—	—	0	0.6	22.2
215	146	2.63	28.6	—	1000.0	2.08	—	0.39	1.67	0.52	4.72	0.29	—	0	1.1	38.0
80	250	4.50	—	—	990.0	—	—	0.26	—	0.45	6.37	—	—	1.2	—	—
140	150	8.08	—	—	1090.0	—	—	0.40	—	0.60	11.05	—	—	9.0	—	—
160	299	8.08	—	—	1530.0	—	—	0.40	—	0.67	11.04	—	—	9.0	—	—
45	101	3.63	—	—	1560.0	—	—	—	—	—	—	—	—	3.6	—	—
0	20	0.71	—	—	670.0	—	0	0.16	—	0.48	2.30	—	—	8.9	—	—
0	60	1.80	21.3	—	430.0	0.57	0	0.32	0.99	0.20	2.88	0.06	—	0	0	13.7
30	80	2.70	—	—	520.0	—	—	0.25	—	0.28	4.26	—	—	1.2	—	—
0	0	0.36	—	—	380.0	—	0	0.12	0.83	0.06	1.36	0.17	—	1.2	—	—
0	100	0	—	—	630.0	—	0	0.14	—	0.09	2.33	—	—	0	—	—
70	79	4.44	—	—	1640.0	—	—	—	—	—	—	—	—	8.9	—	—
75	57	6.83	—	—	1310.0	—	—	—	—	—	—	—	—	5.7	—	—
30	348	—	—	—	230.0	—	—	0.11	—	0.63	0.21	—	—	2.4	0	—
75	100	5.38	—	—	1020.0	—	—	0.39	—	0.44	7.30	—	—	9.0	—	—
100	249	5.38	—	—	1450.0	—	—	0.39	—	0.51	7.29	—	—	9.0	—	—

APPENDIX H

DA+ Code	Food Description	Quantity	Measure	Wt (g)	H₂O (g)	Ener (kcal)	Prot (g)	Carb (g)	Fiber (g)	Fat (g)	Fat Breakdown (g) Sat	Mono	Poly
Fast Food—*continued*													
Carl's Jr.													
33962	Carl's bacon Swiss crispy chicken sandwich	1	item(s)	330	—	800	39.0	67.0	5.0	41.0	10.0	—	—
10801	Carl's Catch Fish sandwich	1	item(s)	298	—	730	20.0	77.0	5.0	37.0	6.0	—	—
10862	Carl's Famous Star hamburger	1	item(s)	269	—	620	25.0	56.0	3.0	34.0	10.0	—	—
10785	Charbroiled chicken club sandwich	1	item(s)	269	—	570	42.0	45.0	3.0	26.0	7.0	—	—
10866	Charbroiled chicken salad	1	item(s)	417	—	260	32.0	15.0	5.0	8.0	3.0	—	—
10855	Charbroiled Santa Fe Chicken sandwich	1	item(s)	273	—	640	39.0	45.0	3.0	33.0	8.0	—	—
10790	Chicken stars (6 pieces)	6	item(s)	84	—	320	12.0	14.0	2.0	24.0	6.0	—	—
34864	Chocolate shake, small	1	serving(s)	397	260.7	710	14.0	86.0	1.0	33.0	23.0	4.3	0.6
10797	Crisscut fries	1	serving(s)	139	—	450	5.0	42.0	4.0	29.0	5.0	—	—
10799	Double Western Bacon cheeseburger	1	item(s)	326	—	980	52.0	73.0	3.0	53.0	23.0	—	—
14238	French fries, small	1	serving(s)	119	—	320	4.0	42.0	4.0	15.0	3.0	—	—
10798	French toast dips without syrup, 5 pieces	1	serving(s)	110	—	430	8.0	51.0	2.0	22.0	4.0	—	—
10802	Onion rings	1	serving(s)	128	—	530	8.0	61.0	3.0	28.0	4.5	—	—
34858	Spicy chicken sandwich	1	item(s)	206	—	550	17.0	61.0	4.0	28.0	5.0	—	—
34867	Strawberry shake, small	1	serving(s)	397	—	700	14.0	85.0	0	33.0	23.0	—	—
10865	Super Star hamburger	1	item(s)	358	—	820	41.0	58.0	3.0	49.0	17.0	—	—
38925	The Six Dollar burger	1	item(s)	406	—	910	46.0	63.0	3.0	54.0	20.0	—	—
10818	Vanilla shake, small	1	item(s)	397	260.9	710	14.0	86.0	0	33.0	23.0	7.1	1.3
10770	Western Bacon cheeseburger	1	item(s)	225	—	667	30.6	67.6	2.8	30.6	12.0	—	—
10770	Western Bacon cheeseburger	1	item(s)	243	—	720	33.0	73.0	3.0	33.0	13.0	—	—
Chick-Fil-A													
38746	Biscuit with bacon, egg & cheese	1	item(s)	159	—	500	21.0	43.0	2.0	27.0	12.0		
38747	Biscuit with egg	1	item(s)	127	—	420	14.0	42.0	2.0	21.0	8.5		
38748	Biscuit with egg & cheese	1	item(s)	154	—	470	17.0	43.0	2.0	25.0	11.0		
38752	Biscuit with sausage, egg & cheese	1	item(s)	204	—	760	28.0	43.0	2.0	52.0	20.0		
72643	Chargrilled chicken and fruit salad	1	item(s)	347	—	230	22.0	23.0	4.0	6.0	3.5		
38760	Chargrilled chicken club sandwich	1	item(s)	258	—	410	37.0	39.0	3.0	12.0	5.0		
38761	Chargrilled chicken Cool Wrap	1	item(s)	291	—	410	33.0	50.0	9.0	12.0	4.0	—	—
38758	Chargrilled chicken sandwich	1	item(s)	228	—	300	29.0	38.0	3.0	3.5	1.0	—	—
38763	Chick-n-Strips	4	item(s)	218	—	500	47.0	24.0	1.0	24.0	4.5	—	—
38742	Chicken biscuit	1	item(s)	143	—	440	17.0	47.0	3.0	20.0	8.0	—	—
38743	Chicken biscuit with cheese	1	item(s)	157	—	490	20.0	48.0	3.0	24.0	10.5	—	—
38764	Chicken salad sandwich on wheat bun	1	item(s)	233	—	500	29.0	52.0	4.0	20.0	3.5	—	—
38756	Chicken sandwich	1	item(s)	179	—	430	31.0	39.0	3.0	17.0	3.5	—	—
38757	Chicken sandwich, deluxe	1	item(s)	257	—	490	35.0	43.0	3.0	21.0	6.0	—	—
38770	Cole slaw	1	item(s)	298	—	580	3.0	31.0	5.0	50.0	8.0	—	—
38776	Diet lemonade, small	1	cup(s)	226	—	11	0	4.2	0	0.0	0	0	0
50867	Fruit cup, large	1	item(s)	194	—	100	1.0	27.0	3.0	0.0	0	0	0
38755	Hashbrowns	1	serving(s)	77	—	280	3.0	25.0	2.0	19.0	4.0	—	—
38765	Hearty breast of chicken soup	1	cup(s)	226	—	114	5.7	15.4	1.6	3.3	0.8	—	—
38741	Hot buttered biscuit	1	item(s)	86	—	310	5.0	41.0	2.0	13.0	6.0	—	—
38778	IceDream, small cone	1	item(s)	135	—	170	5.0	31.0	0	4.0	2.0	—	—
38774	IceDream, small cup	1	serving(s)	227	—	290	8.0	50.0	0	7.0	4.5	—	—
38775	Lemonade, small	1	cup(s)	226	—	120	0	32.5	0	0.0	0	0	0
38777	Nuggets	8	item(s)	113	—	270	28.0	12.0	1.0	12.0	2.5	—	—
38769	Side salad	1	item(s)	113	—	70	5.0	5.0	2.0	4.5	3.0	—	—
72642	Spicy chicken sandwich	1	item(s)	196	—	490	31.0	46.0	4.0	20.0	4.0	—	—
72626	Spicy chicken sandwich, deluxe	1	item(s)	281	—	580	36.0	48.0	4.0	27.0	8.0	—	—
38772	Waffle potato fries, small, salted	1	serving(s)	85	—	290	3.0	34.0	3.0	16.0	3.0	—	—
72633	Yogurt parfait w/ granola	1	item(s)	227	—	290	7.0	54.0	2.0	5.0	2.0	—	—
Cinnabon													
39571	Cinnabon Bites	1	serving(s)	149	—	510	8.0	77.0	2.0	19.0	5.0	—	—
39570	Cinnabon Stix	5	item(s)	85	—	379	6.0	41.0	1.0	21.0	6.0	—	—
39567	Classic roll	1	item(s)	221	—	813	15.0	117.0	4.0	32.0	8.0	—	—
39568	Minibon	1	item(s)	92	—	339	6.0	49.0	2.0	13.0	3.0	—	—

Chol (mg)	Calc (mg)	Iron (mg)	Magn (mg)	Pota (mg)	Sodi (mg)	Zinc (mg)	Vit A (µg)	Thia (mg)	Vit E (mg α)	Ribo (mg)	Niac (mg)	Vit B_6 (mg)	Fola (µg)	Vit C (mg)	Vit B_{12} (µg)	Sele (µg)
75	200	4.50	—	—	2130.0	—	—	—	—	—	—	—	—	4.8	—	—
40	100	3.60	—	—	1320.0	—	—	—	—	—	—	—	—	4.8	—	—
65	100	4.50	—	—	1060.0	—	—	—	—	—	—	—	—	6.0	—	—
90	150	3.60	—	—	1310.0	—	—	—	—	—	—	—	—	6.0	—	—
70	200	2.70	—	—	620.0	—	—	—	—	—	—	—	—	27.0	—	—
95	200	3.60	—	—	1440.0	—	—	—	—	—	—	—	—	6.0	—	—
35	19	0.72	—	—	460.0	—	0	—	—	—	—	—	—	0	—	—
100	500	0.72	67.5	794.0	300.0	1.63	419.4	0.23	0.44	0.97	0.64	0.20	19.9	0	1.4	6.7
0	20	1.08	—	—	900.0	—	—	—	—	—	—	—	—	0	—	—
140	250	5.40	—	—	1790.0	—	—	—	—	—	—	—	—	0	—	—
0	20	1.08	—	—	830.0	—	0	—	—	—	—	—	—	1.2	—	—
0	80	1.44	—	—	480.0	—	0	—	—	—	—	—	—	0	—	—
0	20	1.08	—	—	590.0	—	—	—	—	—	—	—	—	1.2	—	—
25	100	3.60	—	—	1220.0	—	—	—	—	—	—	—	—	1.2	—	—
100	500	0	—	—	250.0	—	—	—	—	—	—	—	—	9.0	—	—
115	100	5.40	—	—	1120.0	—	—	—	—	—	—	—	—	9.0	—	—
130	250	5.40	—	—	2080.0	—	—	—	—	—	—	—	—	9.0	—	—
100	500	0.00	51.6	659.2	240.1	2.26	449.1	0.10	0.99	2.62	0.85	0.24	0	0	0.9	12.7
69	185	4.17	—	—	1342.6	—	—	—	—	—	—	—	—	0	—	—
75	200	4.50	—	—	1450.0	—	—	—	—	—	—	—	—	0	—	—
230	150	3.60	—	—	1370.0	—	—	—	—	—	—	—	—	0	—	—
260	—	—	—	—	830.0	—	—	—	—	—	—	—	—	—	—	—
275	—	—	—	—	1070.0	—	—	—	—	—	—	—	—	—	—	—
325	—	—	—	—	1620.0	—	—	—	—	—	—	—	—	—	—	—
55	150	1.80	—	—	650.0	—	—	—	—	—	—	—	—	96.0	—	—
80	250	3.60	—	—	1460.0	—	—	—	—	—	—	—	—	12.0	—	—
55	200	4.50	—	—	1300.0	—	—	—	—	—	—	—	—	24.0	—	—
55	100	3.60	—	—	1120.0	—	—	—	—	—	—	—	—	12.0	—	—
95	60	3.60	—	—	1630.0	—	—	—	—	—	—	—	—	0	—	—
25	80	1.80	—	—	1240.0	—	—	—	—	—	—	—	—	0	—	—
40	—	—	—	—	1480.0	—	—	—	—	—	—	—	—	—	—	—
80	200	4.50	—	—	1240.0	—	—	—	—	—	—	—	—	4.8	—	—
65	150	2.70	—	—	1370.0	—	0	—	—	—	—	—	—	0	—	—
75	250	2.70	—	—	1620.0	—	—	—	—	—	—	—	—	9.0	—	—
35	80	1.08	—	—	450.0	—	—	—	—	—	—	—	—	72.0	—	—
0	—	—	—	—	7.1	—	—	—	—	—	—	—	—	—	—	—
0	20	0.36	—	—	0	—	0	—	—	—	—	—	—	186.0	—	—
0	20	0.72	—	—	410.0	—	0	—	—	—	—	—	—	6.0	—	—
20	—	—	—	—	902.4	—	—	—	—	—	—	—	—	—	—	—
0	60	1.08	—	—	700.0	—	—	—	—	—	—	—	—	0	—	—
15	150	0.36	—	—	115.0	—	—	—	—	—	—	—	—	1.2	—	—
25	—	—	—	—	200.0	—	—	—	—	—	—	—	—	—	—	—
0	—	—	—	—	3.5	—	—	—	—	—	—	—	—	—	—	—
70	—	—	—	—	990.0	—	—	—	—	—	—	—	—	—	—	—
15	150	0.72	—	—	110.0	—	—	—	—	—	—	—	—	24.0	—	—
60	150	4.50	—	—	1730.0	—	—	—	—	—	—	—	—	0	—	—
80	300	4.50	—	—	1880.0	—	—	—	—	—	—	—	—	9.0	—	—
0	—	—	—	—	140.1	—	—	—	—	—	—	—	—	—	—	—
10	200	0.72	—	—	70.0	—	—	—	—	—	—	—	—	30.0	—	—
35	—	—	—	—	530.0	—	—	—	—	—	—	—	—	—	—	—
16	—	—	—	—	413.2	—	—	—	—	—	—	—	—	—	—	—
67	—	—	—	—	801.0	—	—	—	—	—	—	—	—	—	—	—
27	—	—	—	—	337.0	—	—	—	—	—	—	—	—	—	—	—

APPENDIX H

DA+ Code	Food Description	Quantity	Measure	Wt (g)	H₂O (g)	Ener (kcal)	Prot (g)	Carb (g)	Fiber (g)	Fat (g)	Fat Breakdown (g) Sat	Mono	Poly
Fast Food—*continued*													
39573	Mochalatta Chill with whipped cream	16	fluid ounce(s)	473	—	425	10.0	63.0	2.0	17.0	11.0	—	—
39569	Pecanbon	1	item(s)	272	—	1100	16.0	141.0	8.0	56.0	10.0	—	—
	Dairy Queen												
1466	Banana split	1	item(s)	374	—	520	9.0	94.0	3.0	13.0	10.0	—	—
38561	Chocolate chip cookie dough Blizzard, small	1	item(s)	319	—	710	13.0	103.0	1.0	27.0	14.0	—	—
1464	Chocolate malt, small	1	item(s)	427	—	650	15.0	110.0	0	16.0	10.0	—	—
38541	Chocolate shake, small	1	item(s)	406	—	570	13.0	92.0	0	15.0	10.0	—	—
1463	Chocolate sundae, small	1	item(s)	163	—	280	5.0	48.0	0	7.0	4.5	—	—
1462	Dipped cone, small	1	item(s)	156	—	330	6.0	36.0	0	15.0	6.0	—	—
58376	Mint Oreo Blizzard, small	1	item(s)	297	—	580	12.0	89.0	1.0	20.0	10.0	—	—
38552	Oreo Brownie Earthquake	1	serving(s)	304	—	760	11.0	117.0	2.0	27.0	16.0	—	—
38555	Oreo cookies Blizzard, small	1	item(s)	283	—	550	12.0	81.0	1.0	20.0	10.0	—	—
38547	Royal Treats Peanut Buster Parfait	1	item(s)	304	—	700	16.0	94.0	2.0	30.0	16.0	—	—
17256	Vanilla soft serve	½	cup(s)	94	—	145	4.0	21.5	0	4.5	3.0	—	—
	Domino's												
31604	Breadsticks	1	item(s)	30	—	110	2.0	11.0	0	6.0	1.5	—	—
31605	Cheesy bread	1	item(s)	36	—	120	4.0	11.0	0	6.0	2.0	—	—
37548	CinnaStix	1	item(s)	33	—	120	2.0	14.0	1.0	6.0	1.0	—	—
31606	Wings w/ barbecue sauce	1	item(s)	30	—	63	4.3	3.5	0.3	3.3	0.9	—	—
31607	Wings w/ hot sauce	1	item(s)	30	—	50	4.3	1.3	0.3	3.3	0.9	—	—
	Domino's Artisan pizza												
81942	Chicken & Bacon Carbonara	1	slice(s)	76	—	150	8.0	18.0	1.0	6.0	2.5	—	—
81939	Italian Sausage & Pepper Trio	1	slice(s)	74	—	160	7.0	18.0	1.0	7.0	2.5	—	—
81940	Spinach & Feta	1	slice(s)	62	—	150	6.0	17.0	1.0	7.0	3.0	—	—
81941	Tuscan Salami & Roasted Veggie	1	slice(s)	64	—	150	5.0	17.0	1.0	6.0	2.0	—	—
	Domino's hand-tossed pizza												
31573	America's favorite feast, 12"	1	slice(s)	109	—	250	10.0	27.0	2.0	12.0	5.0	—	—
31574	America's favorite feast, 14"	1	slice(s)	151	—	350	14.0	36.0	2.0	17.0	7.0	—	—
37543	Bacon Cheeseburger Feast, 12"	1	slice(s)	106	—	270	12.0	26.0	2.0	13.0	5.5	—	—
37545	Bacon Cheeseburger Feast, 14"	1	slice(s)	146	—	380	17.0	36.0	2.0	19.0	9.0	—	—
81919	Cali Chicken Bacon Ranch, 12"	1	slice(s)	114	—	320	14.0	25.0	1.0	18.0	6.5	—	—
81920	Cali Chicken Bacon Ranch, 14"	1	slice(s)	154	—	430	19.0	33.0	1.0	25.0	9.0	—	—
31569	Cheese, 12"	1	slice(s)	93	—	210	8.0	25.0	1.0	8.0	3.5	—	—
31570	Cheese, 14"	1	slice(s)	128	57.4	290	12.0	35.0	2.0	11.0	5.5	2.8	2.3
31685	Deluxe Feast, 12"	1	slice(s)	106	—	226	9.0	26.0	2.0	9.5	3.5	—	—
31694	Deluxe Feast, 14"	1	slice(s)	145	—	320	13.0	36.0	2.0	14.0	6.0	—	—
31686	ExtravaganZZa Feast, 12"	1	slice(s)	132	—	290	13.0	28.0	2.0	14.0	6.5	—	—
31695	ExtravaganZZa Feast, 14"	1	slice(s)	176	89.4	390	17.0	37.0	2.0	19.0	8.0	6.4	2.5
81931	Fiery Hawaiian, 12"	1	slice(s)	118	—	250	12.0	27.0	2.0	11.0	5.0	—	—
81932	Fiery Hawaiian, 14"	1	slice(s)	160	—	350	15.0	36.0	2.0	16.0	6.5	—	—
39033	Ham and pineapple, 12"	1	slice(s)	100	—	200	8.0	26.0	1.0	6.5	2.5	—	—
39034	Ham and pineapple, 14"	1	slice(s)	128	—	290	12.0	35.0	2.0	11.0	5.5	—	—
31687	MeatZZa Feast, 12"	1	slice(s)	117	—	280	12.0	27.0	2.0	14.0	5.5	—	—
31696	MeatZZa Feast, 14"	1	slice(s)	160	—	380	18.0	36.0	2.0	19.0	8.0	—	—
81923	Memphis BBQ Chicken, 12"	1	slice(s)	103	—	260	13.0	28.0	1.0	12.0	5.5	—	—
81924	Memphis BBQ Chicken, 14"	1	slice(s)	140	—	360	16.0	37.0	1.0	15.0	7.5	—	—
31571	Pepperoni Feast, 12"	1	slice(s)	99	—	260	11.0	25.0	1.0	13.0	5.5	—	—
31572	Pepperoni Feast, 14"	1	slice(s)	136	—	360	15.0	34.0	2.0	18.0	8.0	—	—
81927	Philly Cheese Steak, 12"	1	slice(s)	101	—	260	11.0	24.0	1.0	12.0	6.0	—	—
81928	Philly Cheese Steak, 14"	1	slice(s)	135	—	330	15.0	32.0	1.0	16.0	8.0	—	—
81935	Wisconsin 6 Cheese, 12"	1	slice(s)	101	—	250	12.0	25.0	1.0	12.0	6.0	—	—
81936	Wisconsin 6 Cheese, 14"	1	slice(s)	137	—	340	15.0	34.0	2.0	16.0	7.5	—	—
	Domino's thin crust pizza												
31583	America's Favorite Feast, 12"	1	slice(s)	81	—	210	8.0	15.0	1.0	13.0	5.0	—	—
31584	America's Favorite Feast, 14"	1	slice(s)	112	—	280	11.0	21.0	2.0	17.5	6.5	—	—
31579	Cheese, 12"	1	slice(s)	64	—	165	7.0	15.0	1.5	9.0	3.5	—	—
31580	Cheese, 14"	1	slice(s)	88	36.6	230	9.0	20.0	1.0	11.5	5.0	3.3	3.3
31688	Deluxe Feast, 12"	1	slice(s)	79	—	190	7.0	15.5	1.0	11.0	4.0	—	—
31697	Deluxe Feast, 14"	1	slice(s)	106	—	250	10.0	21.0	2.0	14.5	5.5	—	—
31689	ExtravaganZZa Feast, 12"	1	slice(s)	103	—	240	10.5	16.0	1.5	15.0	6.0	—	—
31698	ExtravaganZZa Feast, 14"	1	slice(s)	137	—	320	14.0	22.0	2.0	19.5	7.5	—	—
39037	Ham and pineapple, 12"	1	slice(s)	71	—	155	6.5	16.0	1.0	7.5	2.8	—	—
39038	Ham and pineapple, 14"	1	slice(s)	97	—	220	8.0	21.0	1.0	10.0	3.5	—	—

Chol (mg)	Calc (mg)	Iron (mg)	Magn (mg)	Pota (mg)	Sodi (mg)	Zinc (mg)	Vit A (µg)	Thia (mg)	Vit E (mg α)	Ribo (mg)	Niac (mg)	Vit B6 (mg)	Fola (µg)	Vit C (mg)	Vit B12 (µg)	Sele (µg)
63	—	—	—	—	287.0	—	—	—	—	—	—	—	—	—	—	—
63	—	—	—	—	600.0	—	—	—	—	—	—	—	—	—	—	—
30	250	2.70	—	—	160.0	—	—	—	—	—	—	—	—	18.0	—	—
55	350	2.70	—	—	350.0	—	—	—	—	—	—	—	—	0	—	—
55	500	3.60	—	—	310.0	—	—	—	—	—	—	—	—	0	—	—
50	500	1.80	—	—	250.0	—	—	—	—	—	—	—	—	0	—	—
25	200	1.80	—	—	115.0	—	—	—	—	—	—	—	—	0	—	—
25	200	1.44	—	—	105.0	—	—	—	—	—	—	—	—	0	—	—
40	350	3.60	—	—	410.0	—	—	—	—	—	—	—	—	0	—	—
60	250	3.60	—	—	400.0	—	—	—	—	—	—	—	—	0	—	—
40	350	2.70	—	—	410.0	—	—	—	—	—	—	—	—	0	—	—
35	400	3.60	—	—	360.0	—	—	—	—	—	—	—	—	0	—	—
15	125	0.90	—	—	65.0	—	112.6	—	—	—	—	—	—	0	—	—
0	0	0.72	—	—	100.0	—	0	—	—	—	—	—	—	1.2	—	—
5	40	0.72	—	—	140.0	—	—	—	—	—	—	—	—	1.2	—	—
0	0	0.72	—	—	85.0	—	0	—	—	—	—	—	—	1.2	—	—
21	5	0.36	—	—	162.5	—	—	—	—	—	—	—	—	0.6	—	—
21	5	0.27	—	—	340.0	—	—	—	—	—	—	—	—	0.9	—	—
20	60	1.44	—	—	360.0	—	—	—	—	—	—	—	—	3.6	—	—
15	60	1.44	—	—	330.0	—	—	—	—	—	—	—	—	9.0	—	—
10	80	1.08	—	—	250.0	—	—	—	—	—	—	—	—	2.4	—	—
10	60	1.08	—	—	280.0	—	—	—	—	—	—	—	—	6.0	—	—
25	100	2.16	—	—	630.0	—	—	—	—	—	—	—	—	3.6	—	—
35	150	2.52	—	—	870.0	—	—	—	—	—	—	—	—	6.0	—	—
35	150	1.80	—	—	590.0	—	—	—	—	—	—	—	—	6.0	—	—
45	203	2.56	—	—	830.0	—	—	—	—	—	—	—	—	8.5	—	—
40	190	1.80	—	—	660.0	—	—	—	—	—	—	—	—	4.8	—	—
55	260	2.52	—	—	900.0	—	—	—	—	—	—	—	—	7.2	—	—
20	150	1.44	—	—	460.0	—	—	—	—	—	—	—	—	3.6	—	—
25	220	1.80	30.7	206.1	640.0	1.59	78.1	0.41	1.27	0.22	4.29	0.10	—	6.0	0.6	29.4
15	80	1.80	—	—	505.0	—	—	—	—	—	—	—	—	6.0	—	—
30	150	2.52	—	—	730.0	—	—	—	—	—	—	—	—	11.4	—	—
35	150	2.52	—	—	770.0	—	—	—	—	—	—	—	—	7.2	—	—
45	200	2.88	38.7	306.2	1020.0	2.23	89.8	0.24	1.41	0.39	5.72	—	—	11.4	1.0	42.2
25	170	1.80	—	—	760.0	—	—	—	—	—	—	—	—	10.2	—	—
35	240	2.52	—	—	1030.0	—	—	—	—	—	—	—	—	14.4	—	—
20	100	1.47	—	—	490.0	—	—	—	—	—	—	—	—	4.9	—	—
25	220	1.80	—	—	640.0	—	—	—	—	—	—	—	—	6.0	—	—
35	150	2.16	—	—	760.0	—	—	—	—	—	—	—	—	3.6	—	—
50	200	2.52	—	—	1030.0	—	—	—	—	—	—	—	—	6.0	—	—
30	210	1.80	—	—	500.0	—	—	—	—	—	—	—	—	2.4	—	—
40	260	2.16	—	—	680.0	—	—	—	—	—	—	—	—	4.8	—	—
30	150	1.80	—	—	650.0	—	—	—	—	—	—	—	—	3.6	—	—
40	200	2.52	—	—	880.0	—	—	—	—	—	—	—	—	6.0	—	—
30	190	1.80	—	—	540.0	—	—	—	—	—	—	—	—	4.8	—	—
40	260	2.16	—	—	690.0	—	—	—	—	—	—	—	—	6.0	—	—
25	220	1.80	—	—	500.0	—	—	—	—	—	—	—	—	3.6	—	—
35	290	2.16	—	—	690.0	—	—	—	—	—	—	—	—	6.0	—	—
25	120	0.72	—	—	500.0	—	—	—	—	—	—	—	—	3.0	—	—
35	170	1.08	—	—	690.0	—	—	—	—	—	—	—	—	3.6	—	—
20	170	0.54	—	—	330.0	—	—	—	—	—	—	—	—	2.4	—	—
25	240	0.72	21.1	177.8	460.0	1.36	—	0.05	1.31	0.07	0.89	0.06	—	3.6	0.6	17.4
20	120	0.72	—	—	410.0	—	—	—	—	—	—	—	—	6.0	—	—
30	170	1.08	—	—	550.0	—	—	—	—	—	—	—	—	9.0	—	—
35	170	1.08	—	—	640.0	—	—	—	—	—	—	—	—	6.0	—	—
45	220	1.44	—	—	840.0	—	—	—	—	—	—	—	—	9.0	—	—
15	120	0.36	—	—	350.0	—	—	—	—	—	—	—	—	3.0	—	—
25	190	0.72	—	—	485.0	—	—	—	—	—	—	—	—	4.8	—	—

APPENDIX H

DA+ Code	Food Description	Quantity	Measure	Wt (g)	H₂O (g)	Ener (kcal)	Prot (g)	Carb (g)	Fiber (g)	Fat (g)	Fat Breakdown (g)		
											Sat	Mono	Poly
Fast Food—*continued*													
31690	MeatZZa Feast, 12"	1	slice(s)	89	—	235	10.5	15.5	1.0	14.5	6.0	—	—
31699	MeatZZa Feast, 14"	1	slice(s)	121	—	310	15.0	21.0	2.0	19.5	7.5	—	—
31581	Pepperoni Feast, 12"	1	slice(s)	71	—	215	9.0	14.0	1.0	13.5	5.5	—	—
31582	Pepperoni Feast, 14"	1	slice(s)	97	—	290	12.0	19.0	2.0	18.5	7.5	—	—
	In-n-Out Burger												
34391	Cheeseburger with mustard & ketchup	1	serving(s)	268	—	400	22.0	41.0	3.0	18.0	9.0	—	—
34374	Cheeseburger	1	serving(s)	268	—	480	22.0	39.0	3.0	27.0	10.0	—	—
34390	Cheeseburger, lettuce leaves instead of buns	1	serving(s)	300	—	330	18.0	11.0	3.0	25.0	9.0	—	—
34377	Chocolate shake	1	serving(s)	425	—	690	9.0	83.0	0	36.0	24.0	—	—
34375	Double-Double cheeseburger	1	serving(s)	330	—	670	37.0	39.0	3.0	41.0	18.0	—	—
34393	Double-Double cheeseburger with mustard & ketchup	1	serving(s)	330	—	590	37.0	41.0	3.0	32.0	17.0	—	—
34392	Double-Double cheeseburger, lettuce leaves instead of buns	1	serving(s)	362	—	520	33.0	11.0	3.0	39.0	17.0	—	—
34376	French fries	1	serving(s)	125	—	400	7.0	54.0	2.0	18.0	5.0	—	—
34373	Hamburger	1	item(s)	243	—	390	16.0	39.0	3.0	19.0	5.0	—	—
34389	Hamburger with mustard & ketchup	1	serving(s)	243	—	310	16.0	41.0	3.0	10.0	4.0	—	—
34388	Hamburger, lettuce leaves instead of buns	1	serving(s)	275	—	240	13.0	11.0	3.0	17.0	4.0	—	—
34379	Strawberry shake	1	serving(s)	425	—	690	9.0	91.0	0	33.0	22.0	—	—
34378	Vanilla shake	1	serving(s)	425	—	680	9.0	78.0	0	37.0	25.0	—	—
	Jack in the Box												
30392	Bacon ultimate cheeseburger	1	item(s)	315	—	980	43.0	52.0	2.0	67.0	27.0	—	—
1740	Breakfast Jack	1	item(s)	113	—	300	16.0	29.0	1.0	14.0	5.0	—	—
14074	Cheeseburger	1	item(s)	118	—	320	16.0	30.0	1.0	15.0	7.0	—	—
14106	Chicken breast strips, 4 piece	4	piece(s)	201	—	500	35.0	36.0	3.0	25.0	6.0	—	—
37241	Chicken club salad, plain, without salad dressing	1	serving(s)	415	—	480	33.0	28.0	6.0	27.0	10.0	—	—
14064	Chicken sandwich	1	item(s)	145	—	400	15.0	38.0	2.0	21.0	4.5	—	—
14111	Chocolate ice cream shake, small	1	serving(s)	351	—	750	12.0	95.0	1.0	36.0	24.0	—	—
14073	Hamburger	1	item(s)	106	—	280	14.0	29.0	1.0	12.0	4.5	—	—
14090	Hash browns	1	serving(s)	73	—	230	2.0	20.0	2.0	16.0	4.0	—	—
14072	Jack's Spicy Chicken sandwich	1	item(s)	251	—	550	24.0	59.0	4.0	24.0	5.0	—	—
1468	Jumbo Jack hamburger	1	item(s)	249	—	580	20.0	51.0	2.0	33.0	11.0	—	—
1469	Jumbo Jack hamburger with cheese	1	item(s)	274	—	670	24.0	53.0	2.0	40.0	15.0	—	—
14099	Natural cut french fries, large	1	serving(s)	229	—	620	9.0	75.0	8.0	32.0	7.0	—	—
14098	Natural cut french fries, medium	1	serving(s)	169	—	460	6.0	55.0	6.0	24.0	6.0	—	—
1470	Onion rings	1	serving(s)	119	—	500	6.0	51.0	3.0	30.0	6.0	—	—
33141	Sausage, egg & cheese biscuit	1	item(s)	184	—	590	20.0	38.0	2.0	40.0	16.0	—	—
14095	Seasoned curly fries, medium	1	serving(s)	130	—	420	6.0	46.0	5.0	24.0	5.0	—	—
14077	Sourdough Jack	1	item(s)	228	—	680	26.0	41.0	2.0	46.0	17.0	—	—
37249	Southwest grilled chicken salad, plain, w/o salad dressing	1	serving(s)	442	—	310	31.0	28.0	7.0	12.0	5.0	—	—
14112	Strawberry ice cream shake, small	1	serving(s)	349	—	730	11.0	90.0	0	35.0	24.0	—	—
14078	Ultimate cheeseburger	1	item(s)	304	—	920	38.0	52.0	2.0	63.0	26.0	—	—
14110	Vanilla ice cream shake, small	1	serving(s)	314	—	650	11.0	70.0	0	35.0	24.0	—	—
	Jamba Juice												
55682	Acai Super-Antioxidant smoothie	24	fluid ounce(s)	680	—	415	5.5	85.1	5.5	6.5	2.2	—	—
31645	Aloha Pineapple smoothie	24	fluid ounce(s)	680	—	447	6.5	105.8	4.4	1.6	0.5	—	—
31646	Banana Berry smoothie	24	fluid ounce(s)	680	—	436	4.4	102.5	4.4	1.6	0.5	—	—
31647	Carribean Passion smoothie	24	fluid ounce(s)	680	—	393	3.3	89.5	3.3	1.6	0.5	—	—
38422	Carrot juice	16	fluid ounce(s)	473	—	130	4.0	30.0	0	0.5	0	—	—
31648	Chocolate Moo'd smoothie	24	fluid ounce(s)	680	—	622	16.4	126.5	3.3	5.5	2.7	—	—
31657	Mango-a-go-go smoothie	24	fluid ounce(s)	680	—	436	3.3	102.5	3.3	1.6	0.5	—	—
55684	Matcha Green Tea Blast smoothie	24	fluid ounce(s)	680	—	458	10.9	99.3	2.2	0	0	0	0
55673	Mega Mango smoothie	24	fluid ounce(s)	680	—	371	4.4	92.7	5.5	0.5	0	—	—
36945	Orange Dream Machine smoothie	24	fluid ounce(s)	680	—	513	10.9	112.4	1.1	1.6	1.1	—	—
38424	Orange juice, freshly squeezed	16	fluid ounce(s)	473	—	200	3.0	52.0	1.0	1.0	0	—	—
31660	Orange-a-peel smoothie	24	fluid ounce(s)	680	—	404	7.6	92.7	3.3	0	0	—	—

Chol (mg)	Calc (mg)	Iron (mg)	Magn (mg)	Pota (mg)	Sodi (mg)	Zinc (mg)	Vit A (µg)	Thia (mg)	Vit E (mg α)	Ribo (mg)	Niac (mg)	Vit B6 (mg)	Fola (µg)	Vit C (mg)	Vit B12 (µg)	Sele (µg)
35	170	0.90	—	—	630.0	—	—	—	—	—	—	—	—	3.0	—	—
50	220	1.08	—	—	850.0	—	—	—	—	—	—	—	—	3.6	—	—
30	145	0.72	—	—	510.0	—	—	—	—	—	—	—	—	3.0	—	—
40	220	1.08	—	—	700.0	—	—	—	—	—	—	—	—	3.6	—	—
60	200	3.60	—	—	1080.0	—	—	—	—	—	—	—	—	12.0	—	—
60	200	3.60	—	—	1000.0	—	—	—	—	—	—	—	—	9.0	—	—
60	200	2.70	—	—	720.0	—	—	—	—	—	—	—	—	12.0	—	—
95	300	0.72	—	—	350.0	—	—	—	—	—	—	—	—	0	—	—
120	350	5.40	—	—	1440.0	—	—	—	—	—	—	—	—	9.0	—	—
115	350	5.40	—	—	1520.0	—	—	—	—	—	—	—	—	12.0	—	—
120	350	4.50	—	—	1160.0	—	—	—	—	—	—	—	—	12.0	—	—
0	20	1.80	—	—	245.0	—	0	—	—	—	—	—	—	0	—	—
40	40	3.60	—	—	650.0	—	—	—	—	—	—	—	—	9.0	—	—
35	40	3.60	—	—	730.0	—	—	—	—	—	—	—	—	12.0	—	—
40	40	2.70	—	—	370.0	—	—	—	—	—	—	—	—	12.0	—	—
85	300	0	—	—	280.0	—	—	—	—	—	—	—	—	0	—	—
90	300	0	—	—	390.0	—	—	—	—	—	—	—	—	0	—	—
135	308	7.39	—	490.0	1880.0	—	—	—	—	—	—	—	—	0.6	—	—
215	145	3.49	—	180.0	730.0	—	—	—	—	—	—	—	—	3.5	—	—
45	151	3.61	—	230.0	730.0	—	—	—	—	—	—	—	—	0	—	—
80	18	1.60	—	530.0	1260.0	—	—	—	—	—	—	—	—	1.1	—	—
75	280	3.36	—	790.0	1050.0	—	—	—	—	—	—	—	—	50.4	—	—
35	100	2.70	—	240.0	740.0	—	—	—	—	—	—	—	—	4.8	—	—
115	460	0.47	—	740.0	280.0	—	—	—	—	—	—	—	—	0	—	—
30	100	3.60	—	210.0	540.0	—	0	—	—	—	—	—	—	0	—	—
0	10	0.18	—	160.0	330.0	—	0	—	—	—	—	—	—	0	—	—
50	150	1.80	—	420.0	1050.0	—	—	—	—	—	—	—	—	9.0	—	—
50	164	4.92	—	350.0	920.0	—	—	—	—	—	—	—	—	9.8	—	—
75	234	4.21	—	380.0	1290.0	—	—	—	—	—	—	—	—	8.4	—	—
0	20	1.42	—	1580.0	1150.0	—	0	—	—	—	—	—	—	8.9	—	—
0	19	1.01	—	1160.0	850.0	—	0	—	—	—	—	—	—	5.6	—	—
0	40	2.70	—	140.0	420.0	—	9.8	—	—	—	—	—	—	18.0	—	—
245	88	2.37	—	260.0	1140.0	—	—	—	—	—	—	—	—	0	—	—
0	40	1.80	—	610.0	920.0	—	—	—	—	—	—	—	—	0	—	—
75	200	4.50	—	390.0	1200.0	—	—	—	—	—	—	—	—	9.0	—	—
90	274	4.11	—	930.0	820.0	—	—	—	—	—	—	—	—	43.8	—	—
115	466	0	—	630.0	240.0	—	—	—	—	—	—	—	—	0	—	—
120	308	7.39	—	440.0	1530.0	—	—	—	—	—	—	—	—	0.6	—	—
115	532	0	—	630.0	230.0	—	—	—	—	—	—	—	—	0	—	—
5	109	1.96	—	—	60.0	—	—	—	—	—	—	—	—		—	—
5	164	1.96	—	—	60.0	—	—	—	—	—	—	—	—	98.2	—	—
5	164	1.18	—	—	98.2	—	—	—	—	—	—	—	—	16.2	—	—
5	106	1.57	—	—	54.5	—	—	—	—	—	—	—	—	72.0	—	—
0	100	2.72	—	—	230.0	—	2325.0	—	—	—	—	—	—	18.1	—	—
16	545	—	—	—	414.5	—	—	—	—	—	—	—	—	1.3	—	—
5	112	0.79	—	—	49.1	—	—	—	—	—	—	—	—	72.0	—	—
0	164	1.57	—	—	229.1	—	—	—	—	—	—	—	—	9.8	—	—
0	44	1.96	—	—	10.9	—	354.5	—	—	—	—	—	—	183.3	—	—
11	218	0.79	—	—	218.2	—	—	—	—	—	—	—	—	52.4	—	—
0	57	1.03	—	—	0	—	49.8	—	—	—	—	—	—	246.0	—	—
0	218	1.57	—	—	136.4	—	—	—	—	—	—	—	—	150.5	—	—

DA+ Code	Food Description	Quantity	Measure	Wt (g)	H₂O (g)	Ener (kcal)	Prot (g)	Carb (g)	Fiber (g)	Fat (g)	Fat Breakdown (g)		
											Sat	Mono	Poly
Fast Food—*continued*													
55675	Peach Perfection smoothie	24	fluid ounce(s)	680	—	327	2.2	81.8	5.5	0.5	0	—	—
31662	Peach Pleasure smoothie	24	fluid ounce(s)	680	—	404	3.3	96.0	4.4	1.6	1.1	—	—
55676	Pomegranate Paradise smoothie	24	fluid ounce(s)	680	—	371	2.2	92.7	5.5	0.5	0	—	—
55665	Protein Berry Workout smoothie w/ soy protein	24	fluid ounce(s)	680	—	412	18.2	84.3	4.8	1.4	0	—	—
31668	Razzmatazz smoothie	24	fluid ounce(s)	680	—	425	3.3	99.3	4.4	1.6	1.1	—	—
31669	Strawberries Wild smoothie	24	fluid ounce(s)	680	—	404	5.5	94.9	3.3	0	0	—	—
38421	Strawberry Surf Rider smoothie	24	fluid ounce(s)	680	—	469	3.3	112.4	4.4	1.6	0.5	—	—
55674	Strawberry Whirl smoothie	24	fluid ounce(s)	680	—	327	2.2	81.8	6.5	0.5	0	—	—
38428	Wheatgrass juice, freshly squeezed	1	ounce(s)	28	—	5	1.0	1.0	0	0	0	0	0
Kentucky Fried Chicken (KFC)													
31850	BBQ baked beans	1	serving(s)	130	—	200	8.0	39.0	9.0	1.5	0	—	—
31853	Biscuit	1	item(s)	57	—	220	4.0	24.0	1.0	11.0	2.5	—	—
31851	Cole slaw	1	serving(s)	130	—	180	1.0	22.0	3.0	10.0	1.5	—	—
31849	Corn on the cob	1	item(s)	146	—	140	5.0	33.0	4.0	1.0	0	—	—
31842	Crispy strips	3	item(s)	151	—	370	28.0	17.0	1.0	20.0	4.0	—	—
74456	Doublicious chicken sandwich, Original Recipe	1	item(s)	188	—	520	32.0	40.0	2.0	25.0	7.0	—	—
3761	Extra Crispy chicken, breast	1	item(s)	176	—	510	39.0	16.0	0	33.0	7.0	—	—
3762	Extra Crispy chicken, drumstick	1	item(s)	60	—	150	12.0	4.0	0	10.0	2.5	—	—
3763	Extra Crispy chicken, thigh	1	item(s)	114	—	290	17.0	16.0	1.0	18.0	4.0	—	—
3764	Extra Crispy chicken, whole wing	1	item(s)	52	—	150	11.0	11.0	1.0	7.0	1.5	—	—
51218	Famous Bowls mashed potatoes with gravy	1	serving(s)	525	—	680	26.0	74.0	6.0	31.0	8.0	—	—
39387	Green beans	1	serving(s)	86	—	20	1.0	3.0	1.0	0	0	—	—
31841	Honey BBQ chicken sandwich	1	item(s)	162	—	310	23.0	42.0	1.0	4.0	1.0	—	—
10859	Hot wings	6	piece(s)	132	—	420	24.0	24.0	0	24.0	3.0	—	—
51223	Hot wings, fiery buffalo	6	item(s)	204	—	480	30.0	36.0	6.0	21.0	3.0	—	—
39386	Hot wings, honey BBQ	6	item(s)	198	—	480	30.0	42.0	6.0	21.0	3.0	—	—
31848	Macaroni & cheese	1	serving(s)	135	—	160	5.0	19.0	1.0	7.0	2.5	—	—
31847	Mashed potatoes with gravy	1	serving(s)	145	—	120	2.0	19.0	1.0	4.0	1.0	—	—
74460	Original Recipe chicken filet	1	item(s)	100	—	200	22.0	8.0	1.0	9.0	1.5	—	—
10825	Original Recipe chicken, breast	1	item(s)	161	—	340	38.0	9.0	2.0	17.0	4.0	—	—
10826	Original Recipe chicken, drumstick	1	item(s)	59	—	140	13.0	3.0	0	8.0	2.0	—	—
10827	Original Recipe chicken, thigh	1	item(s)	126	—	350	19.0	7.0	1.0	27.0	7.0	—	—
10828	Original Recipe chicken, whole wing	1	item(s)	47	—	140	10.0	4.0	0	9.0	2.0	—	—
31844	Popcorn chicken, small or individual	1	item(s)	114	—	370	19.0	21.0	2.0	24.0	4.5	—	—
31852	Potato salad	1	serving(s)	128	—	200	2.0	24.0	3.0	10.0	2.0	—	—
10845	Potato wedges, small	1	serving(s)	102	—	250	4.0	32.0	3.0	12.0	2.0	—	—
Long John Silver's													
3777	Batter dipped fish sandwich	1	item(s)	176	—	470	18.0	49.0	3.0	23.0	5.0	—	—
37568	Battered fish	1	item(s)	92	—	260	12.0	17.0	0	16.0	4.0	—	—
37569	Breaded clams	1	serving(s)	85	—	320	9.0	29.0	2.0	19.0	4.5	—	—
39398	Cocktail sauce	1	ounce(s)	28	—	25	0	6.0	0	0	0	0	0
3770	Coleslaw	1	serving(s)	113	—	200	1.0	15.0	3.0	15.0	2.5	—	—
39401	Crab cake, lobster stuffed	1	item(s)	62	—	170	6.0	16.0	1.0	9.0	2.0	—	—
78975	Fish taco	1	item(s)	117	—	360	9.0	30.0	3.0	23.0	4.5	—	—
39400	French fries, large	1	item(s)	113	—	310	3.0	45.0	4.0	14.0	3.5	—	—
3774	Fries, regular	1	serving(s)	85	—	230	3.0	34.0	3.0	10.0	2.5	—	—
3779	Hushpuppy	1	piece(s)	23	—	60	1.0	9.0	1.0	2.5	0.5	—	—
56963	Shrimp scampi	1	serving(s)	130	—	200	17.0	3.0	0	13.0	2.5	—	—
3781	Shrimp, batter-dipped, 1 piece	1	piece(s)	14	—	43	1.7	2.7	0	3.0	0.8	—	—
39399	Tartar sauce	1	ounce(s)	28	—	100	0	4.0	0	9.0	1.5	—	—
39395	Ultimate Fish sandwich	1	item(s)	206	—	530	21.0	50.0	3.0	27.0	8.0	—	—
McDonald's													
2262	Baked apple pie	1	item(s)	77	28.5	250	2.0	32.0	4.0	13.0	7.0	7.1	0.8
2247	Barbecue sauce	1	item(s)	28	16.1	50	0	12.0	0	0	0	0	0
737	Big Mac hamburger	1	item(s)	215	112.3	550	25.0	46.0	3.0	29.0	10.0	7.5	0.7
738	Cheeseburger	1	item(s)	114	45.0	300	15.0	33.0	2.0	12.0	6.0	4.1	0.4
1873	Chicken McNuggets 6 piece	6	item(s)	97	44.8	280	13.0	18.0	1.0	18.0	3.0	7.7	5.3
3792	Chicken McNuggets, 4 piece	4	item(s)	65	30.2	190	9.0	12.0	1.0	12.0	2.0	5.2	3.6
2264	Chocolate shake, large	1	item(s)	356	223.2	580	13.0	102.0	1.0	14.0	8.0	4.0	0.8

Chol (mg)	Calc (mg)	Iron (mg)	Magn (mg)	Pota (mg)	Sodi (mg)	Zinc (mg)	Vit A (µg)	Thia (mg)	Vit E (mg α)	Ribo (mg)	Niac (mg)	Vit B$_6$ (mg)	Fola (µg)	Vit C (mg)	Vit B$_{12}$ (µg)	Sele (µg)
0	65	1.18	—	—	32.7	—	245.5	—	—	—	—	—	—	65.5	—	—
5	87	0.77	—	—	54.5	—	—	—	—	—	—	—	—	16.4	—	—
0	109	1.96	—	—	38.2	—	245.5	—	—	—	—	—	—	85.1	—	—
0	1054	3.45	76.7	690.0	220.4	0.86	—	0.11	0.39	0.16	1.15	0.67	76.7	51.7	0	5.4
5	109	1.57	—	—	60.0	—	—	—	—	—	—	—	—	58.9	—	—
5	218	1.92	—	—	152.7	—	—	—	—	—	—	—	—	52.4	—	—
5	87	1.13	—	—	10.9	—	—	—	—	—	—	—	—	85.1	—	—
0	87	1.96	—	—	27.3	—	5.5	—	—	—	—	—	—	78.5	—	—
0	—	1.73	—	—	0	—	—	—	—	—	—	—	—	3.7	—	—
0	—	—	—	—	680.0	—	—	—	—	—	—	—	—	—	—	—
0	40	1.80	—	—	640.0	—	—	—	—	—	—	—	—	0	—	—
5	40	0.72	—	—	270.0	—	—	—	—	—	—	—	—	12.0	—	—
0	—	—	—	—	5.0	—	—	—	—	—	—	—	—	—	—	—
65	40	1.44	—	—	1220.0	—	0	—	—	—	—	—	—	1.2	—	—
85	—	—	—	—	1300.0	—	—	—	—	—	—	—	—	—	—	—
110	—	—	—	—	1010.0	—	—	—	—	—	—	—	—	—	—	—
55	0	1.44	—	—	300.0	—	0	—	—	—	—	—	—	0	—	—
95	20	2.70	—	—	700.0	—	—	—	—	—	—	—	—	—	—	—
45	20	1.08	—	—	340.0	—	—	—	—	—	—	—	—	0	—	—
45	—	—	—	—	2130.0	—	—	—	—	—	—	—	—	—	—	—
0	—	—	—	—	290.0	—	—	—	—	—	—	—	—	—	—	—
70	—	—	—	—	810.0	—	—	—	—	—	—	—	—	—	—	—
120	—	—	—	—	840.0	—	—	—	—	—	—	—	—	—	—	—
60	—	—	—	—	2340.0	—	—	—	—	—	—	—	—	—	—	—
60	—	—	—	—	2040.0	—	—	—	—	—	—	—	—	—	—	—
5	—	—	—	—	720.0	—	—	—	—	—	—	—	—	—	—	—
0	—	—	—	—	530.0	—	—	—	—	—	—	—	—	—	—	—
55	—	—	—	—	670.0	—	—	—	—	—	—	—	—	—	—	—
135	20	2.70	—	—	960.0	—	—	—	—	—	—	—	—	6.0	—	—
70	20	1.08	—	—	340.0	—	—	—	—	—	—	—	—	0	—	—
110	20	2.70	—	—	870.0	—	—	—	—	—	—	—	—	1.2	—	—
50	20	1.44	—	—	350.0	—	0	—	—	—	—	—	—	1.2	—	—
25	40	1.80	—	—	1110.0	—	0	—	—	—	—	—	—	0	—	—
5	—	—	—	—	540.0	—	—	—	—	—	—	—	—	—	—	—
0	20	1.08	—	—	700.0	—	0	—	—	—	—	—	—	0	—	—
40	60	1.80	—	—	1180.0	—	—	—	—	—	—	—	—	1.2	—	—
35	0	0	—	—	790.0	—	0	—	—	—	—	—	—	0	—	—
35	20	1.44	—	—	1190.7	—	0	—	—	—	—	—	—	0	—	—
0	0	0	—	—	250.0	—	—	—	—	—	—	—	—	0	—	—
20	40	0.36	—	—	340.0	—	—	—	—	—	—	—	—	18.0	—	—
30	60	0.72	—	—	390.0	—	0	—	—	—	—	—	—	0	—	—
25	60	1.44	—	—	810.0	—	0	—	—	—	—	—	—	0	—	—
0	0	0	—	—	460.0	—	0	—	—	—	—	—	—	18.0	—	—
0	0	0	—	370.2	350.2	—	0	—	—	—	—	—	—	15.0	—	—
0	20	0.36	—	—	200.0	—	0	—	—	—	—	—	—	0	—	—
135	60	0.72	—	—	650.0	—	—	—	—	—	—	—	—	0	—	—
15	0	0	—	—	160.0	—	0	—	—	—	—	—	—	0	—	—
15	0	0	—	—	250.0	—	0	—	—	—	—	—	—	0	—	—
55	100	1.80	—	—	1500.0	—	—	—	—	—	—	—	—	1.2	—	—
0	20	1.08	5.4	48.5	170.0	0.18	—	0.23	1.49	0.16	2.03	0.04	—	15.0	—	—
0	0	0	3.6	55.2	260.0	0.05	3.4	0.01	0.30	0.01	0.19	0.02	—	0	—	—
75	250	4.50	43.0	396.0	970.0	4.11	—	0.38	—	0.45	7.28	—	—	1.2	1.9	—
40	202	2.72	22.8	200.0	680.0	2.18	—	0.25	—	0.30	4.59	—	—	1.2	1.0	—
40	20	0.72	23.0	239.0	540.0	0.56	0	0.15	—	0.10	7.03	0.38	—	1.2	0.3	—
25	0	0.36	15.0	161.0	360.0	0.38	0	0.10	—	0.07	4.74	0.25	—	1.2	0.2	—
50	450	1.80	57.0	804.6	250.0	1.78	324.0	0.14	0	0.76	0.47	0.18	—	0	1.9	—

APPENDIX H

DA+ Code	Food Description	Quantity	Measure	Wt (g)	H₂O (g)	Ener (kcal)	Prot (g)	Carb (g)	Fiber (g)	Fat (g)	Fat Breakdown (g)		
											Sat	Mono	Poly
Fast Food—*continued*													
29774	Crispy chicken sandwich	1	item(s)	213	110.7	510	24.0	55.0	3.0	22.0	3.5	6.8	7.9
743	Egg McMuffin	1	item(s)	138	72.6	300	18.0	30.0	2.0	12.0	5.0	3.8	2.4
742	Filet-O-Fish sandwich	1	item(s)	142	65.0	390	15.0	39.0	2.0	19.0	4.0	5.7	8.5
2257	French fries, large	1	serving(s)	154	58.0	500	6.0	63.0	6.0	25.0	3.5	12.0	7.2
1872	French fries, small	1	serving(s)	71	27.1	230	3.0	29.0	3.0	11.0	1.5	5.5	3.3
33822	Fruit 'n Yogurt Parfait	1	item(s)	149	110.9	150	4.0	30.0	1.0	2.0	1.0	0.2	0.1
739	Hamburger	1	item(s)	100	44.5	250	12.0	31.0	2.0	9.0	3.5	3.8	1.4
2003	Hash browns	1	item(s)	56	29.5	150	1.0	15.0	2.0	9.0	1.5	4.8	2.8
2260	Hot fudge sundae	1	item(s)	179	105.3	330	8.0	53.0	1.0	9.0	7.0	1.9	0.4
1874	Plain Hotcakes with syrup and margarine	3	item(s)	221	89.7	570	8.0	105.0	3.0	13.0	3.5	1.9	4.6
38388	Premium Bacon Ranch salad with grilled chicken, no dressing	1	serving(s)	306	249.1	230	30.0	10.0	4.0	9.0	4.0	3.1	1.3
38391	Premium Caesar salad with grilled chicken, no dressing	1	serving(s)	296	245.5	190	27.0	10.0	4.0	5.0	3.0	1.5	0.6
38393	Premium Caesar salad, no dressing	1	serving(s)	213	191.1	90	7.0	9.0	3.0	4.0	2.5	1.0	0.3
29775	Premium Grilled Chicken Classic sandwich	1	item(s)	200	114.6	350	28.0	42.0	3.0	9.0	2.0	2.3	4.1
57683	Premium Southwest salad w/grilled chicken	1	item(s)	335	—	290	27.0	28.0	7.0	8.0	2.5	—	—
740	Quarter Pounder hamburger	1	item(s)	173	87.1	420	25.0	38.0	2.0	19.0	7.0	7.3	0.5
741	Quarter Pounder hamburger with cheese	1	item(s)	202	98.8	520	30.0	41.0	3.0	26.0	12.0	9.3	0.9
2005	Sausage McMuffin with egg	1	item(s)	164	81.9	450	21.0	30.0	2.0	27.0	10.0	10.8	4.5
50831	Side salad	1	item(s)	87	81.7	20	1.0	4.0	1.0	0.0	0	0	0
	Pizza Hut												
39009	Hot chicken wings	2	item(s)	44	—	100	10.0	1.0	0	6.0	2.0	—	—
14025	Meat Lovers hand-tossed pizza	1	slice(s)	105	—	300	14.0	26.0	1.0	16.0	7.0	—	—
14026	Meat Lovers pan pizza	1	slice(s)	113	—	330	14.0	27.0	1.0	18.0	7.0	—	—
31009	Meat Lovers stuffed crust pizza	1	slice(s)	165	—	480	22.0	39.0	2.0	26.0	11.0	—	—
14024	Meat Lovers thin 'n crispy pizza	1	slice(s)	85	—	280	13.0	22.0	1.0	16.0	6.0	—	—
14031	Pepperoni Lovers hand-tossed pizza	1	slice(s)	95	—	270	13.0	26.0	1.0	13.0	6.0	—	—
14032	Pepperoni Lovers pan pizza	1	slice(s)	101	—	290	13.0	27.0	1.0	14.0	6.0	—	—
31011	Pepperoni Lovers stuffed crust pizza	1	slice(s)	149	—	430	20.0	40.0	2.0	21.0	10.0	—	—
14030	Pepperoni Lovers thin 'n crispy pizza	1	slice(s)	75	—	250	12.0	22.0	1.0	13.0	6.0	—	—
10834	Personal Pan pepperoni pizza	1	slice(s)	50	—	153	6.5	16.8	0.8	6.5	2.5	—	—
10842	Personal Pan supreme pizza	1	slice(s)	64	—	180	7.5	17.3	1.0	9.0	3.5	—	—
39013	Personal Pan Veggie Lovers pizza	1	slice(s)	58	—	138	5.5	17.5	1.0	5.0	2.0	—	—
14028	Veggie Lovers hand-tossed pizza	1	slice(s)	102	—	200	9.0	27.0	2.0	6.0	3.0	—	—
14029	Veggie Lovers pan pizza	1	slice(s)	107	—	230	9.0	28.0	2.0	9.0	3.5	—	—
31010	Veggie Lovers stuffed crust pizza	1	slice(s)	155	—	330	15.0	41.0	2.0	12.0	6.0	—	—
14027	Veggie Lovers thin 'n crispy pizza	1	slice(s)	86	—	180	8.0	23.0	1.0	6.0	3.0	—	—
39012	Wing blue cheese dipping sauce	1	item(s)	43	—	230	1.0	2.0	0	24.0	4.5	—	—
39011	Wing ranch dipping sauce	1	item(s)	43	—	220	0	2.0	0	23.0	3.5	—	—
	Starbucks												
33107	Caffè mocha, tall nonfat, w/o whipped cream	12	fluid ounce(s)	354	—	170	10.0	32.0	0.5	2.0	1.0	—	—
33108	Caffè mocha, tall whole milk, w/whipped cream	12	fluid ounce(s)	354	—	290	10.0	34.0	0.5	16.0	9.0	—	—
38052	Cappuccino, tall	12	fluid ounce(s)	354	—	110	6.0	9.0	0	6.0	3.0	—	—
38053	Cappuccino, tall nonfat	12	fluid ounce(s)	354	—	60	6.0	9.0	0	0	0	0	0
38054	Cappuccino, tall soymilk	12	fluid ounce(s)	354	—	100	5.0	12.0	0.5	3.0	0	—	—
68043	Caramel Apple Spice, tall, w/o whipped cream	12	fluid ounce(s)	354	—	210	0	53.0	0	0	0	0	0
68256	Espresso, single shot	1	fluid ounce(s)	30	—	5	0	0.5	0	0	0	0	0
38088	Flavored syrup, 1 pump	1	serving(s)	10	—	20	0	5.0	0	0	0	0	0
38067	Frappuccino, tall caramel w/o whipped cream	12	fluid ounce(s)	354	—	200	3.0	42.0	0	2.5	1.5	—	—
39891	Frappuccino, tall caramel, light	12	fluid ounce(s)	354	—	130	4.0	25.0	2.0	1.0	0	—	—
67330	Frappuccino, tall Cinnamon Dolce, light	12	fluid ounce(s)	354	—	105	3.7	21.7	2.2	0.4	0.0	—	—

KEY: H-4 = Breads/Baked Goods H-8 = Cereal/Rice/Pasta H-14 = Fruit H-18 = Vegetables/Legumes H-28 = Nuts/Seeds H-30 = Vegetarian H-32 = Dairy H-40 = Eggs H-40 = Seafood H-42 = Meats H-46 = Poultry
H-8 = Processed Meats H-50 = Beverages H-52 = Fats/Oils H-54 = Sweets H-56 = Spices/Condiments/Sauces H-60 = Mixed Foods/Soups/Sandwiches H-66 = Fast Food H-84 = Convenience H-86 = Baby Foods

Chol (mg)	Calc (mg)	Iron (mg)	Magn (mg)	Pota (mg)	Sodi (mg)	Zinc (mg)	Vit A (µg)	Thia (mg)	Vit E (mg α)	Ribo (mg)	Niac (mg)	Vit B$_6$ (mg)	Fola (µg)	Vit C (mg)	Vit B$_{12}$ (µg)	Sele (µg)
45	150	3.60	57.5	483.5	990.0	1.41	38.3	0.42	—	0.36	11.80	—	—	3.6	—	—
260	300	2.70	27.6	238.7	780.0	1.78	—	0.36	0.81	0.51	4.28	0.20	—	0	—	—
40	150	1.80	38.3	312.4	590.0	0.82	—	0.30	1.66	0.19	3.15	—	—	0	1.5	36.2
0	20	1.44	49.3	862.4	350.0	0.68	0	0.50	—	0.05	4.25	0.80	—	12.0	—	—
0	20	0.72	22.7	397.6	160.0	0.31	0	0.23	—	0.03	1.96	0.37	—	4.8	—	—
5	100	0.67	20.9	248.8	70.0	0.54	—	0.07	—	0.17	0.35	—	—	9.0	0.3	—
25	100	2.70	21.0	192.0	480.0	1.95	—	0.25	—	0.24	4.54	—	—	1.2	0.8	26.2
0	0	0.36	11.8	219.0	310.0	0.19	0	0.06	—	0.01	1.26	0.14	—	1.2	—	—
25	249	1.44	34.0	440.3	170.0	1.00	145.0	0.08	0.34	0.41	0.27	0.09	—	0	1.0	—
20	150	2.70	28.7	276.3	660.0	0.64	—	0.45	—	0.40	3.23	0.12	—	0	0	—
85	150	1.80	—	—	700.0	—	—	0.17	—	0.25	11.33	—	—	21.0	—	—
70	200	1.78	—	754.8	580.0	—	—	0.17	—	0.20	11.31	—	—	21.0	—	—
10	200	1.44	19.2	460.1	180.0	—	—	0.09	—	0.08	0.45	—	—	18.0	0	0.4
65	150	3.60	56.0	456.0	820.0	1.38	—	0.39	1.06	0.38	13.04	0.61	—	4.8	0.3	—
70	150	2.70	—	—	650.0	—	—	—	—	—	—	—	—	21.0	—	—
65	150	4.50	38.1	392.7	700.0	4.65	—	0.32	—	0.60	7.70	—	—	1.2	2.2	—
95	300	4.50	44.4	442.4	1100.0	5.31	—	0.33	—	0.71	7.78	—	—	1.2	2.5	—
285	300	3.60	29.5	280.4	890.0	2.00	—	0.43	0.82	0.56	4.80	0.24	—	1.2	1.1	—
0	20	0.72	—	191.4	10.0	—	—	0.04	—	0.03	0.18	—	—	15.0	0	—
55	—	—	—	—	430.0	—	—	—	—	—	—	—	—	—	—	—
40	—	—	—	—	860.0	—	—	—	—	—	—	—	—	—	—	—
40	—	—	—	—	830.0	—	—	—	—	—	—	—	—	—	—	—
70	—	—	—	—	1380.0	—	—	—	—	—	—	—	—	—	—	—
40	—	—	—	—	860.5	—	—	—	—	—	—	—	—	—	—	—
35	—	—	—	—	770.0	—	—	—	—	—	—	—	—	—	—	—
35	—	—	—	—	730.0	—	—	—	—	—	—	—	—	—	—	—
60	—	—	—	—	1230.0	—	—	—	—	—	—	—	—	—	—	—
35	—	—	—	—	760.0	—	—	—	—	—	—	—	—	—	—	—
14	—	—	—	—	352.5	—	—	—	—	—	—	—	—	—	—	—
20	—	—	—	—	420.0	—	—	—	—	—	—	—	—	—	—	—
9	—	—	—	—	297.5	—	—	—	—	—	—	—	—	—	—	—
15	—	—	—	—	530.0	—	—	—	—	—	—	—	—	—	—	—
15	—	—	—	—	500.0	—	—	—	—	—	—	—	—	—	—	—
35	—	—	—	—	880.0	—	—	—	—	—	—	—	—	—	—	—
15	—	—	—	—	530.0	—	—	—	—	—	—	—	—	—	—	—
20	—	—	—	—	420.0	—	—	—	—	—	—	—	—	—	—	—
10	—	—	—	—	420.0	—	—	—	—	—	—	—	—	—	—	—
2	300	2.70	—	—	0	—	—	—	—	—	—	—	—	0	—	—
45	300	2.70	—	—	0	—	—	—	—	—	—	—	—	0	—	—
15	200	0	—	—	0	—	—	—	—	—	—	—	—	0	—	—
2	200	0	—	—	0	—	—	—	—	—	—	—	—	0	—	—
0	200	1.08	—	—	0	—	—	—	—	—	—	—	—	0	—	—
0	0	0	—	—	0	0	—	—	—	—	—	—	—	0	—	—
0	0	0	—	—	0	0	—	—	—	—	—	—	—	0	—	—
0	0	0	—	—	0	0	—	—	—	—	—	—	—	0	0	—
10	100	0	—	—	0	—	—	—	—	—	—	—	—	0	—	—
2	100	0	—	—	180.0	—	0	—	—	—	—	—	—	0	—	—
0	112	0	—	—	0	—	0	—	—	—	—	—	—	0	—	—

APPENDIX H

APPENDIX H

DA+ Code	Food Description	Quantity	Measure	Wt (g)	H₂O (g)	Ener (kcal)	Prot (g)	Carb (g)	Fiber (g)	Fat (g)	Fat Breakdown (g)		
											Sat	Mono	Poly
Fast Food—*continued*													
67329	Frappuccino, tall Cinnamon Dolce, w/whipped cream	12	fluid ounce(s)	354	—	270	3.0	42.0	0	11.0	6.0	—	—
38070	Frappuccino, tall coffee	12	fluid ounce(s)	354	—	180	3.0	36.0	0	2.5	1.5	—	—
39894	Frappuccino, tall coffee, light	12	fluid ounce(s)	354	—	90	3.0	18.0	0	0	0	—	—
67339	Frappuccino, tall Double Chocolaty Chip, w/whipped cream	12	fluid ounce(s)	354	—	300	5.0	42.0	0.5	14.0	9.0	—	—
38071	Frappuccino, tall espresso	12	fluid ounce(s)	354	—	150	2.0	34.0	0	1.0	0	—	—
39883	Frappuccino, tall Java Chip, w/o whipped cream	12	fluid ounce(s)	354	—	240	4.0	47.0	0.5	5.0	3.5	—	—
38073	Frappuccino, tall mocha w/o whipped cream	12	fluid ounce(s)	354	—	200	3.0	42.0	0.5	3.0	2.0	—	—
39897	Frappuccino, tall mocha, light	12	fluid ounce(s)	354	—	110	3.0	23.0	0.5	0.5	0	—	—
39887	Frappuccino, tall Strawberries & Creme, w/o whipped cream	12	fluid ounce(s)	354	—	190	3.0	38.0	0	3.0	1.5	—	—
38063	Frappuccino, tall Tazo chai creme w/o whipped cream	12	fluid ounce(s)	354	—	160	3.0	32.0	0	2.0	1.0	—	—
38080	Frappuccino, tall vanilla w/o whipped cream	12	fluid ounce(s)	354	—	200	3.0	39.0	0	3.5	2.0	—	—
39898	Frappuccino, tall white chocolate mocha, light blend	12	fluid ounce(s)	354	—	210	4.0	37.0	0	5.0	3.0	—	—
38074	Frappuccino, tall white chocolate w/o whipped cream	12	fluid ounce(s)	354	—	240	4.0	48.0	0	4.0	2.5	—	—
68822	Hot chocolate, tall soy, w/o whipped cream	12	fluid ounce(s)	354	—	230	9.0	40.0	2.0	6.0	1.5	—	—
67421	Iced mocha, tall white chocolate, nonfat, w/whipped cream	12	fluid ounce(s)	354	—	230	8.0	42.0	0	4.5	3.5	—	—
33111	Latte, tall w/nonfat milk	12	fluid ounce(s)	354	—	100	10.0	15.0	0	0	0	0	0
33112	Latte, tall w/whole milk	12	fluid ounce(s)	354	—	180	10.0	14.0	0	9.0	5.0	—	—
33109	Macchiato, tall caramel w/nonfat milk	12	fluid ounce(s)	354	—	140	7.0	25.0	0	1.0	1.0	—	—
33110	Macchiato, tall caramel w/whole milk	12	fluid ounce(s)	354	—	200	8.0	25.0	0	8.0	5.0	—	—
38089	Mocha syrup	1	serving(s)	17	—	25	1.0	6.0	0	0.5	0	—	—
67546	Tazo black shaken iced tea lemonade, tall	12	fluid ounce(s)	354	—	100	0	25.0	0	0	0	0	0
38076	Tazo black shaken iced tea, tall	12	fluid ounce(s)	354	—	60	0	15.0	0	0	0	0	0
67551	Tazo iced chai tea latte, tall w/nonfat milk	12	fluid ounce(s)	354	—	150	6.0	33.0	0	0	0	0	0
67417	Tazo iced green tea latte, tall w/nonfat milk	12	fluid ounce(s)	354	—	180	7.0	37.0	0.5	0	0	0	0
38090	Whipped cream	1	serving(s)	30	—	90	1.0	3.0	0	9.0	5.0	—	—
38062	White chocolate mocha, tall nonfat w/o whipped cream	12	fluid ounce(s)	354	—	270	12.0	47.0	0	4.5	3.5	—	—
38061	White chocolate mocha, tall w/ whipped cream	12	fluid ounce(s)	354	—	390	11.0	48.0	0	18.0	11.0	—	—
38048	White hot chocolate, tall nonfat w/o whipped cream	12	fluid ounce(s)	354	—	270	12.0	47.0	0	4.5	3.5	—	—
38050	White hot chocolate, tall soymilk w/whipped cream	12	fluid ounce(s)	354	—	380	11.0	52.0	0.5	14.0	8.0	—	—
38047	White hot chocolate, tall w/whipped cream	12	fluid ounce(s)	354	—	410	12.0	48.0	0	19.0	12.0	—	—
	Subway												
80576	Breakfast BMT Melt	1	item(s)	142	—	240	16.0	25.0	6.0	10.0	4.0	—	—
80594	Buffalo chicken sandwich, 6", wheat bread	1	item(s)	268	—	420	25.0	46.0	5.0	15.0	3.0	—	—
15842	Cheese steak sandwich, 6", wheat bread	1	item(s)	297	—	500	38.0	51.0	6.0	17.0	9.0	—	—
40478	Chicken & bacon ranch sandwich, 6", white or wheat bread	1	serving(s)	292	—	570	35.0	47.0	5.0	28.0	10.0	—	—
32045	Chocolate chip cookie	1	item(s)	45	—	220	2.0	30.0	1.0	10.0	5.0	—	—
32048	Chocolate chip M&M cookie	1	item(s)	45	—	210	2.0	32.0	0.5	10.0	5.0	—	—
32049	Chocolate chunk cookie	1	item(s)	45	—	220	2.0	30.0	0.5	10.0	5.0	—	—
4024	Classic Italian B.M.T. sandwich, 6", white bread	1	item(s)	221	—	400	19.0	44.0	2.0	16.0	6.0	—	—
15838	Classic tuna sandwich, 6", wheat bread	1	item(s)	233	—	470	20.0	44.0	5.0	24.0	4.0	—	—
15837	Classic tuna sandwich, 6", white bread	1	item(s)	228	—	460	19.0	42.0	2.0	24.0	4.0	—	—

Chol (mg)	Calc (mg)	Iron (mg)	Magn (mg)	Pota (mg)	Sodi (mg)	Zinc (mg)	Vit A (µg)	Thia (mg)	Vit E (mg α)	Ribo (mg)	Niac (mg)	Vit B$_6$ (mg)	Fola (µg)	Vit C (mg)	Vit B$_{12}$ (µg)	Sele (µg)
35	100	0	—	—	0	—	—	—	—	—	—	—	—	0	—	—
10	100	0	—	—	0	—	—	—	—	—	—	—	—	0	—	—
0	100	0	—	—	0	—	—	—	—	—	—	—	—	0	—	—
40	150	2.70	—	—	0	—	—	—	—	—	—	—	—	0	—	—
2	60	0	—	—	0	—	—	—	—	—	—	—	—	0	—	—
10	100	2.70	—	—	0	—	—	—	—	—	—	—	—	0	—	—
10	100	0.72	—	—	0	—	—	—	—	—	—	—	—	0	—	—
0	100	0.72	—	—	0	—	—	—	—	—	—	—	—	0	—	—
10	100	0.36	—	—	0	—	—	—	—	—	—	—	—	3.6	—	—
10	100	0	—	—	0	—	—	—	—	—	—	—	—	0	—	—
10	100	0	—	—	0	—	—	—	—	—	—	—	—	0	—	—
10	150	0	—	—	0	—	—	—	—	—	—	—	—	0	—	—
10	2	0	—	—	0	—	—	—	—	—	—	—	—	0	—	—
0	300	3.60	—	—	0	—	—	—	—	—	—	—	—	0	—	—
2	300	0	—	—	0	—	—	—	—	—	—	—	—	1.2	—	—
2	350	0	—	—	0	—	—	—	—	—	—	—	—	0	—	—
30	300	0	—	—	0	—	—	—	—	—	—	—	—	0	—	—
2	250	0	—	—	0	—	—	—	—	—	—	—	—	0	—	—
25	250	0	—	—	0	—	—	—	—	—	—	—	—	0	—	—
0	0	0.72	—	—	0	—	0	—	—	—	—	—	—	0	—	—
0	0	0	—	—	10.0	—	0	—	—	—	—	—	—	6.0	—	—
0	0	0	—	—	0	—	0	—	—	—	—	—	—	0	—	—
2	200	0.36	—	—	0	—	—	—	—	—	—	—	—	0	—	—
2	250	0.36	—	—	0	—	—	—	—	—	—	—	—	6.0	—	—
30	20	0.36	—	—	10.0	—	90.1	—	—	—	—	—	—	0	—	—
2	400	0	—	—	0	—	—	—	—	—	—	—	—	1.2	0	—
50	400	0	—	—	0	—	—	—	—	—	—	—	—	1.2	—	—
2	450	0	—	—	0	—	—	—	—	—	—	—	—	1.2	—	—
25	400	1.44	—	—	0	—	—	—	—	—	—	—	—	1.2	—	—
50	400	0	—	—	0	—	—	—	—	—	—	—	—	1.2	—	—
130	200	1.44	—	—	830.0	—	—	—	—	—	—	—	—	1.2	—	—
55	300	3.60	—	—	1190.0	—	—	—	—	—	—	—	—	15.0	—	—
85	500	4.50	—	—	1310.0	—	—	—	—	—	—	—	—	12.0	—	—
95	500	3.60	—	—	1080.0	—	—	—	—	—	—	—	—	15.0	—	—
15	0	1.08	—	—	130.0	—	—	—	—	—	—	—	—	0	—	—
15	20	1.00	—	—	100.0	—	0	—	—	—	—	—	—	0	—	—
10	0	1.00	—	—	100.0	—	0	—	—	—	—	—	—	0	—	—
45	300	3.60	—	—	1280.0	—	—	—	—	—	—	—	—	12.0	—	—
35	300	3.60	—	—	620.0	—	—	—	—	—	—	—	—	12.0	—	—
35	300	3.60	—	—	600.0	—	—	—	—	—	—	—	—	12.0	—	—

APPENDIX H

DA+ Code	Food Description	Quantity	Measure	Wt (g)	H₂O (g)	Ener (kcal)	Prot (g)	Carb (g)	Fiber (g)	Fat (g)	Sat	Mono	Poly
											Fat Breakdown (g)		

Fast Food—*continued*

DA+ Code	Food Description	Quantity	Measure	Wt (g)	H₂O (g)	Ener (kcal)	Prot (g)	Carb (g)	Fiber (g)	Fat (g)	Sat	Mono	Poly
32043	Club sandwich, 6", wheat bread	1	item(s)	347	—	420	39.0	50.0	4.0	8.0	3.5	—	—
4030	Cold cut combo sandwich, 6", white bread	1	item(s)	228	—	360	17.0	44.0	2.0	13.0	4.0	—	—
385	Ham sandwich, 6", wheat bread	1	item(s)	219	—	290	18.0	46.0	5.0	4.5	1.0	—	—
3888	Meatball marinara sandwich, 6", wheat bread	1	item(s)	301	—	480	21.0	59.0	8.0	18.0	7.0	—	—
4651	Meatball sandwich, 6", white bread	1	item(s)	296	—	470	20.0	57.0	5.0	18.0	7.0	—	—
80596	Melt sandwich, 6", wheat bread	1	item(s)	240	—	370	23.0	47.0	5.0	11.0	5.0	—	—
32046	Oatmeal raisin cookie	1	item(s)	45	—	200	3.0	30.0	1.0	8.0	4.0	—	—
16379	Oven-roasted chicken breast sandwich, 6", wheat bread	1	item(s)	233	—	320	23.0	47.0	5.0	5.0	1.5	—	—
32047	Peanut butter cookie	1	item(s)	45	—	220	4.0	26.0	1.0	12.0	5.0	—	—
4655	Roast beef sandwich, 6", wheat bread	1	item(s)	233	—	320	24.0	45.0	5.0	5.0	1.5	—	—
3957	Roast beef sandwich, 6", white bread	1	item(s)	228	—	310	23.0	43.0	2.0	5.0	1.5	—	—
16378	Roasted chicken breast sandwich, 6", white bread	1	item(s)	228	—	310	22.0	45.0	2.0	5.0	1.5	—	—
4032	Spicy Italian sandwich, 6", white bread	1	item(s)	216	—	470	19.0	44.0	2.0	24.0	9.0	—	—
4031	Steak & cheese sandwich, 6", white bread	1	item(s)	239	—	370	25.0	46.0	2.0	10.0	4.5	—	—
32050	Sugar cookie	1	item(s)	45	—	220	2.0	28.0	0.5	12.0	6.0	—	—
80579	Sunrise Melt with egg	1	item(s)	149	—	230	18.0	26.0	6.0	8.0	3.0	—	—
40477	Sweet onion chicken teriyaki sandwich, 6", white or wheat bread	1	serving(s)	276	—	380	26.0	59.0	5.0	4.5	1.0	—	—
15834	Turkey breast & ham sandwich, 6", white bread	1	item(s)	214	—	270	17.0	44.0	2.0	4.0	1.0	—	—
16376	Turkey breast sandwich, 6", white bread	1	item(s)	214	—	270	17.0	44.0	2.0	3.5	1.0	—	—
15843	Veggie Delite salad	1	item(s)	271	—	50	3.0	9.0	4.0	1.0	0	—	—
15841	Veggie Delite sandwich, 6", wheat bread	1	item(s)	162	—	230	8.0	44.0	5.0	2.5	0.5	—	—
16375	Veggie Delite sandwich, 6", white bread	1	item(s)	157	—	220	7.0	42.0	2.0	2.5	0	—	—
32051	White chip macadamia nut cookie	1	item(s)	45	—	220	2.0	29.0	0.5	11.0	5.0	—	—
	Taco Bell												
29906	7-Layer burrito	1	item(s)	283	—	510	18.0	68.0	12.0	18.0	6.0	—	—
744	Bean burrito	1	item(s)	198	105.3	370	13.0	56.0	10.0	10.0	3.5	2.1	3.7
749	Beef burrito supreme	1	item(s)	248	—	420	17.0	52.0	9.0	15.0	7.0	—	—
33417	Beef Chalupa Supreme	1	item(s)	153	—	370	14.0	31.0	3.0	21.0	6.0	—	—
29910	Beef Gordita Supreme	1	item(s)	153	—	300	13.0	31.0	4.0	13.0	5.0	—	—
2014	Beef soft taco	1	item(s)	99	—	210	10.0	21.0	3.0	9.0	4.0	—	—
10860	Beef soft taco supreme	1	item(s)	135	—	240	11.0	24.0	3.0	11.0	5.0	—	—
66167	Cheesy Fiesta potatoes	1	serving(s)	135	—	270	4.0	28.0	3.0	16.0	2.5	—	—
34472	Chicken burrito supreme	1	item(s)	248	—	390	21.0	51.0	7.0	12.0	5.0	—	—
33418	Chicken Chalupa Supreme	1	item(s)	153	—	350	17.0	30.0	2.0	18.0	4.0	—	—
29909	Chicken quesadilla	1	item(s)	184	—	520	28.0	41.0	4.0	28.0	12.0	—	—
50171	Chicken soft taco	1	item(s)	99	57.8	187	13.2	19.5	1.2	6.3	2.5	1.6	1.6
29907	Chili cheese burrito	1	item(s)	156	—	370	16.0	40.0	4.0	16.0	8.0	—	—
10794	Cinnamon twists	1	serving(s)	35	—	170	1.0	26.0	1.0	7.0	0.0	—	—
66033	Fresco bean burrito	1	item(s)	213	—	340	12.0	56.0	11.0	8.0	2.5	—	—
66025	Fresco chicken burrito supreme	1	item(s)	241	—	340	18.0	50.0	8.0	8.0	2.5	—	—
66020	Fresco crunchy taco	1	item(s)	92	—	150	7.0	13.0	3.0	7.0	2.5	—	—
66026	Fresco steak burrito supreme	1	item(s)	241	—	330	16.0	49.0	8.0	8.0	3.0	—	—
29911	Grilled chicken Gordita Supreme	1	item(s)	153	—	270	17.0	29.0	2.0	10.0	3.5	—	—
29912	Grilled Steak Gordita Supreme	1	item(s)	153	—	270	14.0	29.0	2.0	11.0	4.0	—	—
29904	Grilled steak soft taco	1	item(s)	128	—	250	11.0	20.0	2.0	14.0	4.0	—	—
2021	Mexican pizza	1	serving(s)	213	—	540	21.0	47.0	8.0	30.0	8.0	—	—
10772	Meximelt	1	serving(s)	128	—	280	15.0	23.0	4.0	14.0	7.0	—	—
2011	Nachos	1	serving(s)	99	33.9	330	4.0	31.0	2.0	20.0	2.0	12.6	2.6
2012	Nachos BellGrande	1	serving(s)	305	—	770	20.0	78.0	15.0	42.0	7.0	—	—
2023	Pintos 'n cheese	1	serving(s)	128	—	180	10.0	19.0	9.0	7.0	3.0	—	—

Chol (mg)	Calc (mg)	Iron (mg)	Magn (mg)	Pota (mg)	Sodi (mg)	Zinc (mg)	Vit A (µg)	Thia (mg)	Vit E (mg α)	Ribo (mg)	Niac (mg)	Vit B6 (mg)	Fola (µg)	Vit C (mg)	Vit B12 (µg)	Sele (µg)
70	80	7.20	—	—	2100.0	—	—	—	—	—	—	—	—	18.0	—	—
50	350	4.50	—	—	1120.0	—	—	—	—	—	—	—	—	12.0	—	—
25	300	2.70	—	—	830.0	—	—	—	—	—	—	—	—	12.0	—	—
30	350	4.50	—	—	950.0	—	—	—	—	—	—	—	—	21.0	—	—
30	350	4.50	—	—	930.0	—	—	—	—	—	—	—	—	21.0	—	—
45	400	2.70	—	—	1210.0	—	—	—	—	—	—	—	—	12.0	—	—
15	20	1.08	—	—	130.0	—	0	—	—	—	—	—	—	0	—	—
45	300	2.70	—	—	640.0	—	—	—	—	—	—	—	—	18.0	—	—
10	20	1.08	—	—	130.0	—	—	—	—	—	—	—	—	0	—	—
45	300	4.50	—	—	700.0	—	—	—	—	—	—	—	—	12.0	—	—
45	300	4.50	—	—	680.0	—	—	—	—	—	—	—	—	12.0	—	—
45	300	2.70	—	—	620.0	—	—	—	—	—	—	—	—	18.0	—	—
50	300	3.60	—	—	1500.0	—	—	—	—	—	—	—	—	12.0	—	—
50	400	4.50	—	—	1040.0	—	—	—	—	—	—	—	—	12.0	—	—
15	0	0.72	—	—	130.0	—	0	—	—	—	—	—	—	0	—	—
130	200	1.44	—	—	810.0	—	—	—	—	—	—	—	—	1.2	—	—
50	350	3.60	—	—	900.0	—	—	—	—	—	—	—	—	18.0	—	—
20	300	3.60	—	—	800.0	—	—	—	—	—	—	—	—	12.0	—	—
20	300	3.60	—	—	790.0	—	—	—	—	—	—	—	—	12.0	—	—
0	40	1.08	—	—	65.0	—	—	—	—	—	—	—	—	27.0	—	—
0	300	2.70	—	—	310.0	—	—	—	—	—	—	—	—	12.0	—	—
0	300	2.70	—	—	290.0	—	—	—	—	—	—	—	—	12.0	—	—
15	20	0.72	—	—	130.0	—	—	—	—	—	—	—	—	0	—	—
20	—	—	—	—	1410.0	—	—	—	—	—	—	—	—	—	—	—
5	246	4.69	67.3	516.8	980.0	1.70	5.9	0.38	0.99	0.20	4.00	0.20	134.6	0.8	0.3	20.6
35	—	—	—	—	1380.0	—	—	—	—	—	—	—	—	—	—	—
35	—	—	—	—	610.0	—	—	—	—	—	—	—	—	—	—	—
35	—	—	—	—	590.0	—	—	—	—	—	—	—	—	—	—	—
30	—	—	—	—	620.0	—	—	—	—	—	—	—	—	—	—	—
35	—	—	—	—	650.0	—	—	—	—	—	—	—	—	—	—	—
5	—	—	—	—	840.0	—	—	—	—	—	—	—	—	—	—	—
40	—	—	—	—	1420.0	—	—	—	—	—	—	—	—	—	—	—
40	—	—	—	—	650.0	—	—	—	—	—	—	—	—	—	—	—
75	—	—	—	—	1420.0	—	—	—	—	—	—	—	—	—	—	—
29	121	1.57	23.8	214.8	606.9	0.75	1.0	0.16	0.35	0.11	5.12	0.11	68.3	0.2	0.1	12.6
40	—	—	—	—	1080.0	—	—	—	—	—	—	—	—	—	—	—
0	—	—	—	—	200.0	—	—	—	—	—	—	—	—	—	—	—
0	—	—	—	—	1290.0	—	—	—	—	—	—	—	—	—	—	—
25	—	—	—	—	1410.0	—	—	—	—	—	—	—	—	—	—	—
20	—	—	—	—	350.0	—	—	—	—	—	—	—	—	—	—	—
15	—	—	—	—	1340.0	—	—	—	—	—	—	—	—	—	—	—
35	—	—	—	—	620.0	—	—	—	—	—	—	—	—	—	—	—
30	—	—	—	—	550.0	—	—	—	—	—	—	—	—	—	—	—
30	—	—	—	—	710.0	—	—	—	—	—	—	—	—	—	—	—
45	—	—	—	—	1020.0	—	—	—	—	—	—	—	—	—	—	—
45	—	—	—	—	870.0	—	—	—	—	—	—	—	—	—	—	—
5	88	1.11	51.5	180.2	370.0	1.14	1.0	0.12	1.04	0.13	0.90	0.18	12.9	—	—	3.4
30	—	—	—	—	1300.0	—	—	—	—	—	—	—	—	—	—	—
15	—	—	—	—	720.0	—	—	—	—	—	—	—	—	—	—	—

DA+ Code	Food Description	Quantity	Measure	Wt (g)	H₂0 (g)	Ener (kcal)	Prot (g)	Carb (g)	Fiber (g)	Fat (g)	Fat Breakdown (g) Sat	Mono	Poly
Fast Food—*continued*													
34473	Steak burrito supreme	1	item(s)	248	—	380	18.0	51.0	7.0	12.0	5.0	—	—
33419	Steak Chalupa Supreme	1	item(s)	153	—	340	15.0	29.0	2.0	18.0	4.0	—	—
747	Taco	1	item(s)	78	—	170	8.0	12.0	3.0	10.0	3.5	—	—
2015	Taco salad with salsa, with shell	1	serving(s)	463	—	770	27.0	75.0	12.0	41.0	10.0	—	—
748	Tostada	1	item(s)	170	—	250	11.0	29.0	10.0	10.0	3.5	—	—
Convenience Meals													
Banquet													
14788	Boneless white fried chicken meal	1	item(s)	286	—	350	12.0	35.0	5.0	17.0	4.0		
14773	Chicken pasta marinara meal	1	item(s)	184	—	290	12.0	29.0	3.0	14.0	3.0	—	—
29960	Fish sticks meal	1	item(s)	207	—	310	11.0	44.0	4.0	10.0	2.5	—	—
29957	Lasagna with meat sauce meal	1	item(s)	227	—	250	12.0	34.0	4.0	7.0	2.5	—	—
14777	Macaroni and cheese meal	1	item(s)	227	—	260	10.0	39.0	3.0	6.0	3.0	—	—
1741	Meatloaf meal	1	item(s)	269	—	280	12.0	28.0	4.0	13.0	5.0	—	—
39418	Pepperoni pizza meal	1	item(s)	163	—	340	11.0	47.0	4.0	12.0	3.5	—	—
1743	Salisbury steak meal	1	item(s)	269	—	290	11.0	25.0	4.0	16.0	7.0	—	—
14772	Turkey meal	1	item(s)	262	—	250	13.0	32.0	5.0	7.0	2.0	—	—
Healthy Choice													
9337	Beef pot roast complete meal	1	item(s)	312	—	290	17.0	45.0	6.0	4.5	1.5	2.0	1.0
57075	Café Steamers Cajun style chicken & shrimp	1	item(s)	295	—	250	17.0	33.0	5.0	5.0	1.0	—	—
57078	Café Steamers chicken linguini with red pepper Alfredo	1	item(s)	292	—	280	22.0	35.0	5.0	6.0	2.5	—	—
57074	Café Steamers sweet sesame chicken	1	item(s)	292	—	330	17.0	50.0	6.0	6.0	1.0	2.5	2.0
9316	Lemon pepper fish meal	1	item(s)	303	—	310	14.0	50.0	5.0	5.0	1.0	—	—
9322	Traditional salisbury steak meal	1	item(s)	354	—	310	18.0	46.0	9.0	6.0	2.5	2.0	1.0
9359	Traditional turkey breast meal	1	item(s)	298	—	290	18.0	44.0	8.0	4.5	1.0	2.0	1.5
Lean Cuisine													
11043	Culinary Collection baked chicken	1	item(s)	245	—	240	15.0	34.0	3.0	4.5	1.0	1.0	2.0
360	Simple Favorites chicken chow mein	1	item(s)	255	—	260	14.0	41.0	3.0	4.0	1.0	1.5	1.5
11054	Simple Favorites chicken enchilada suiza	1	serving(s)	255	190.9	270	12.0	47.0	3.0	4.0	1.5	1.0	1.0
9467	Simple Favorites fettuccini Alfredo	1	item(s)	262	—	280	14.0	42.0	1.0	6.0	3.0	1.5	1.0
9479	Simple Favorites French bread deluxe pizza	1	item(s)	174	—	340	16.0	46.0	4.0	10.0	3.5	2.5	1.5
11055	Simple Favorites lasagna with meat sauce	1	item(s)	298	—	320	20.0	43.0	4.0	7.0	3.5	2.0	1.0
58267	Spa Collection ginger garlic stir fry with chicken	1	item(s)	280	—	290	17.0	46.0	4.0	4.0	1.0	1.0	1.5
Michelina's													
55332	Cheese manicotti with marinara sauce entrée	1	item(s)	227	—	270	11.0	34.0	3.0	11.0	5.0	—	—
1915	Fettuccini Alfredo with chicken & broccoli entrée	1	item(s)	241	—	310	15.0	38.0	2.0	10.0	6.0	—	—
55320	Lean Gourmet five cheese lasagna entrée	1	item(s)	227	—	290	13.0	50.0	8.0	5.0	2.0	—	—
55329	Lean Gourmet Meatloaf entrée	1	item(s)	227	—	180	11.0	21.0	2.0	6.0	3.0	—	—
55325	Lean Gourmet Santa Fe style rice & beans entrée	1	item(s)	255	—	320	10.0	52.0	3.0	8.0	3.5	—	—
Stouffer's													
36564	Beef pot roast entrée	1	item(s)	255	—	210	14.0	30.0	2.0	4.0	1.5	—	—
2313	Cheese French bread pizza	1	serving(s)	294	—	370	14.0	44.0	4.0	15.0	6.0	—	—
2366	Chicken pot pie entrée	1	item(s)	454	—	590	19.0	54.0	1.0	34.0	13.0	—	—
11116	Homestyle baked chicken breast with mashed potatoes and gravy entrée	1	item(s)	252	—	250	20.0	20.0	1.0	10.0	3.0	—	—
11152	Homestyle roast turkey breast with stuffing and mashed potatoes entrée	1	item(s)	273	—	290	16.0	30.0	2.0	12.0	3.5	—	—
41781	Lasagna with meat & sauce entrée	1	item(s)	595	441.9	690	47.8	74.8	6.5	22.4	11.3	7.6	1.4
Weight Watchers													
11164	Smart Ones chicken enchiladas suiza entrée	1	item(s)	255	—	290	11.0	49.0	3.0	5.0	2.0	1.5	1.0
39763	Smart Ones chicken oriental entrée	1	item(s)	255	—	230	14.0	41.0	2.0	1.5	0	0	0

APPENDIX H

Chol (mg)	Calc (mg)	Iron (mg)	Magn (mg)	Pota (mg)	Sodi (mg)	Zinc (mg)	Vit A (µg)	Thia (mg)	Vit E (mg α)	Ribo (mg)	Niac (mg)	Vit B$_6$ (mg)	Fola (µg)	Vit C (mg)	Vit B$_{12}$ (µg)	Sele (µg)
30	—	—	—	—	1340.0	—	—	—	—	—	—	—	—	—	—	—
30	—	—	—	—	580.0	—	—	—	—	—	—	—	—	—	—	—
30	—	—	—	—	330.0	—	—	—	—	—	—	—	—	—	—	—
60	—	—	—	—	1650.0	—	—	—	—	—	—	—	—	—	—	—
15	—	—	—	—	730.0	—	—	—	—	—	—	—	—	—	—	—
35	100	1.08	40.0	450.0	930.0	1.50	0	0.09	—	0.14	1.20	0.12	—	4.8	0.5	—
15	60	1.80	40.0	380.0	550.0	0.90	—	—	—	0.10	1.60	0.08	—	2.4	1.8	—
25	100	1.80	40.0	290.0	540.0	0.60	—	0.12	—	0.17	0.80	—	—	0	3.0	—
10	60	2.70	40.0	420.0	510.0	1.20	—	0.30	—	0.25	3.00	—	—	1.2	2.7	—
15	100	1.44	32.0	220.0	760.0	0.90	—	0.30	—	0.51	0.80	—	—	0	0.4	—
40	20	1.08	32.0	420.0	1000.0	1.50	0	—	—	—	—	0.16	—	0	0.6	—
10	150	1.80	40.0	310.0	730.0	1.50	—	0.12	—	0.34	0.80	0.12	—	0	—	—
30	20	1.44	32.0	380.0	1100.0	1.50	0	0.06	—	0.14	1.60		—	0	0.9	—
25	40	1.80	32.0	360.0	1060.0	1.20	—	0.15	—	0.17	3.00	0.12	—	6.0	0.9	—
40	40	1.80	40.0	720.0	500.0	3.00	—	0.03	—	0.17	3.00	0.20	—	30.0	1.5	21.0
45	40	1.08	40.0	480.0	590.0	1.20	—	0.22	—	0.25	6.00	0.20	—	6.0	0.5	0
35	100	1.80	40.0	440.0	570.0	1.50	—	0.22	—	0.34	5.00	0.20	—	30.0	0.6	63.0
30	40	0.72	40.0	440.0	400.0	0.90	—	0.09	—	0.14	4.00	0.30	—	1.2	0.6	0
25	60	1.08	40.0	480.0	450.0	0.60	—	0.22	—	0.07	2.00	0.40	—	24.0	1.2	14.0
35	100	2.70	60.0	880.0	590.0	3.00	—	0.15	—	0.17	3.00	0.16	—	24.0	1.2	21.0
25	60	1.80	40.0	680.0	450.0	1.20	—	0.22	—	0.25	5.00	0.20	—	4.8	0.5	31.5
25	40	1.17	—	500.0	650.0	—	—	—	—	—	—	—	—	3.6	—	—
25	60	1.08	—	380.0	550.0	—	—	—	—	—	—	—	—	3.6	—	—
20	150	1.08	—	380.0	550.0	—	—	0.20	—	0.15	1.68	—	—	1.2	—	—
15	200	0.72	—	270.0	690.0	—	0	—	—	—	—	—	—	0	—	—
20	150	2.70	—	330.0	760.0	—	—	—	—	—	—	—	—	12.0	—	—
30	200	2.70	—	710.0	590.0	—	—	—	—	—	—	—	—	3.6	—	—
30	60	1.44	—	550.0	640.0	—	—	—	—	—	—	—	—	15.0	—	—
30	200	1.08	—	—	980.0	—	—	—	—	—	—	—	—	12.0	—	—
45	150	1.80	—	—	700.0	—	—	—	—	—	—	—	—	9.0	—	—
10	150	3.60	—	—	560.0	—	—	—	—	—	—	—	—	12.0	—	—
35	40	1.44	—	—	860.0	—	—	—	—	—	—	—	—	6.0	—	—
20	150	1.44	—	—	670.0	—	—	—	—	—	—	—	—	6.0	—	—
25	20	1.80	—	—	710.0	—	—	—	—	—	—	—	—	4.8	—	—
20	250	1.80	—	—	600.0	—	—	—	—	—	—	—	—	2.4	—	—
50	100	3.60	—	—	930.0	—	—	—	—	—	—	—	—	1.2	—	—
60	40	0.36	—	—	730.0	—	0	—	—	—	—	—	—	0	—	—
45	40	1.08	—	490.0	970.0	—	—	—	—	—	—	—	—	3.6	—	—
77	411	3.63	—	940.1	1856.4	—	—	0.36	—	0.30	3.81	—	—	3.6	—	—
25	100	0.72	—	—	640.0	—	—	—	—	—	—	—	—	6.0	—	—
25	20	0.36	—	—	700.0	—	—	—	—	—	—	—	—	2.4	—	—

DA+ Code	Food Description	Quantity	Measure	Wt (g)	H₂O (g)	Ener (kcal)	Prot (g)	Carb (g)	Fiber (g)	Fat (g)	Fat Breakdown (g)		
											Sat	Mono	Poly
Convenience Meals—*continued*													
11187	Smart Ones pepperoni pizza	1	item(s)	173	—	410	19.0	63.0	4.0	9.0	3.0	2.5	1.5
58556	Smart Ones spaghetti with meat sauce entrée	1	item(s)	326	—	290	14.0	44.0	5.0	6.0	1.5	2.0	1.0
31512	Smart Ones spicy szechuan style vegetables & chicken	1	item(s)	255	—	240	11.0	38.0	4.0	5.0	1.0	2.0	2.5
Baby Foods													
787	Apple juice	4	fluid ounce(s)	127	111.6	60	0	14.8	0.1	0.1	0	0	0
778	Applesauce, strained	4	tablespoon(s)	64	56.7	26	0.1	6.9	1.1	0.1	0	0	0
779	Bananas with tapioca, strained	4	tablespoon(s)	60	50.4	34	0.2	9.2	1.0	0.1	0	0	0
604	Carrots, strained	4	tablespoon(s)	56	51.7	15	0.4	3.4	1.0	0.1	0	0	0
770	Chicken noodle dinner, strained	4	tablespoon(s)	64	54.8	42	1.7	5.8	1.3	1.3	0.4	0.5	0.3
801	Green beans, strained	4	tablespoon(s)	60	55.1	16	0.7	3.8	1.3	0.1	0	0	0.1
910	Human milk, mature	2	fluid ounce(s)	62	53.9	43	0.6	4.2	0	2.7	1.2	1.0	0.3
760	Mixed cereal, prepared with whole milk	4	ounce(s)	113	89.8	109	4.7	13.9	0.9	3.8	2.0	1.0	0.4
772	Mixed vegetable dinner, strained	2	ounce(s)	57	50.3	23	0.7	5.4	0.9	0.1	—	—	0.1
762	Rice cereal, prepared with whole milk	4	ounce(s)	113	92.9	96	3.9	11.7	0	3.8	2.0	1.0	0.4
758	Teething biscuits	1	item(s)	11	0.7	43	1.2	8.4	0.2	0.5	0.1	0.2	0.1

APPENDIX H

GE KEY: H-4 = Breads/Baked Goods H-8 = Cereal/Rice/Pasta H-14 = Fruit H-18 = Vegetables/Legumes H-28 = Nuts/Seeds H-30 = Vegetarian H-32 = Dairy H-40 = Eggs H-40 = Seafood H-42 = Meats H-46 = Poultry
18 = Processed Meats H-50 = Beverages H-52 = Fats/Oils H-54 = Sweets H-56 = Spices/Condiments/Sauces H-60 = Mixed Foods/Soups/Sandwiches H-66 = Fast Food H-84 = Convenience H-86 = Baby Foods

Chol (mg)	Calc (mg)	Iron (mg)	Magn (mg)	Pota (mg)	Sodi (mg)	Zinc (mg)	Vit A (µg)	Thia (mg)	Vit E (mg α)	Ribo (mg)	Niac (mg)	Vit B$_6$ (mg)	Fola (µg)	Vit C (mg)	Vit B$_{12}$ (µg)	Sele (µg)
30	100	1.44	—	—	730.0	—	—	—	—	—	—	—	—	0	—	—
10	100	2.70	—	—	520.0	—	—	—	—	—	—	—	—	12.0	—	—
10	60	2.70	—	—	710.0	—	—	—	—	—	—	—	—	0	—	—
0	5	0.72	3.8	115.4	10.1	0.04	1.3	0.01	0.76	0.02	0.11	0.04	0	73.4	0	0.1
0	3	0.14	1.9	45.4	0	0.01	0.6	0.01	0.38	0.02	0.04	0.02	1.3	24.5	0	0.2
0	3	0.12	6.0	52.8	0	0.04	1.2	0.01	0.36	0.02	0.11	0.07	3.6	10.0	0	0.4
0	12	0.21	5.0	109.8	38.6	0.08	320.9	0.01	0.29	0.02	0.26	0.04	8.4	3.2	0	0.1
10	17	0.41	9.0	89.0	24.3	0.35	70.4	0.03	0.13	0.04	0.46	0.04	8.3	0.1	0	2.4
0	23	0.40	12.0	87.6	4.2	0.13	10.8	0.02	0.04	0.05	0.22	0.02	14.4	0.2	0	0.1
9	20	0.02	1.8	31.4	10.5	0.10	37.6	0.01	0.05	0.02	0.11	0.01	3.1	3.1	0	1.1
10	249	11.83	22.7	188.2	47.6	0.67	46.5	0.35	0.53	0.51	4.40	0.06	10.2	0.3	0.5	6.9
—	12	0.19	6.2	68.6	21.5	0.09	77.1	0.01	—	0.02	0.28	0.04	4.5	1.6		0.4
11	191	4.12	28.4	171.2	47.6	0.56	47.6	0.28	0.50	0.37	2.78	0.08	6.8	0.2	0.5	4.9
1	29	0.39	3.8	35.5	25.0	0.10	3.1	0.03	0.12	0.06	0.48	0.01	7.6	1.0	0	2.6

APPENDIX H

Appendix I   WHO Nutrition Recommendations

The World Health Organization (WHO) has assessed the relationships between diet and the development of chronic diseases. This appendix presents its nutrition recommendations for adults:

- Energy: sufficient to support growth, physical activity, and a healthy body weight (BMI between 18.5 and 24.9) and to avoid weight gain greater than 11 pounds (5 kilograms) during adult life
- Total fat: 15 to 30 percent of total energy
- Saturated fatty acids: <10 percent of total energy
- Polyunsaturated fatty acids: 6 to 10 percent of total energy
- Omega-6 polyunsaturated fatty acids: 5 to 8 percent of total energy
- Omega-3 polyunsaturated fatty acids: 1 to 2 percent of total energy
- *Trans*-fatty acids: <1 percent of total energy
- Total carbohydrate: 55 to 75 percent of total energy
- Sugars: <10 percent of total energy, preferably <5 percent of total energy
- Protein: 10 to 15 percent of total energy
- Cholesterol: <300 mg per day
- Salt (sodium): <5 g salt per day (<2 g sodium per day), appropriately iodized
- Potassium: ≥3510 mg per day
- Fruits and vegetables: ≥400 g per day (about 1 pound)
- Total dietary fiber: >25 g per day from foods
- Physical activity: one hour of moderate-intensity activity, such as walking, on most days of the week

Table 1-4 (p. 25) lists the objectives from the Nutrition and Weight Status section of the Healthy People 2020 initiative. Table J-1 presents additional nutrition-related objectives from other topic areas.

TABLE J-1 Nutrition-Related Objectives from Other Topic Areas

Access to Health Services

- Increase the proportion of persons who receive appropriate evidence-based clinical preventive services.

Adolescent Health

- Increase the proportion of schools with a school breakfast program.

Arthritis, Osteoporosis, and Chronic Back Conditions

- Reduce hip fractures among older adults.
- Reduce the proportion of adults with osteoporosis.

Cancer

- Reduce the cancer death rate.
- Increase the mental and physical health-related quality of life of cancer survivors.

Diabetes

- Reduce the annual number of new cases of diagnosed diabetes in the population.
- Reduce the death rate among the population with diabetes.
- Reduce the diabetes death rate.
- Improve glycemic control among the population with diagnosed diabetes.
- Improve lipid control among persons with diagnosed diabetes.
- Increase the proportion of the population with diagnosed diabetes whose blood pressure is under control.
- Increase the proportion of persons with diagnosed diabetes who receive formal diabetes education.
- Increase prevention behaviors in persons at high risk for diabetes with prediabetes.

Early and Middle Childhood

- Increase the proportion of elementary, middle, and senior high schools that require school health education.

Educational and Community-Based Programs

- Increase the proportion of preschool Early Head Start and Head Start programs that provide health education to prevent health problems in the following areas: unintentional injury; violence; tobacco use and addiction; alcohol and drug use, unhealthy dietary patterns; and inadequate physical activity, dental health, and safety.
- Increase the proportion of elementary, middle, and senior high schools that provide comprehensive school health education to prevent health problems in the following areas: unintentional injury; violence; suicide; tobacco use and addiction; alcohol or other drug use; unintended pregnancy, human immunodeficiency virus/acquired immune deficiency syndrome (HIV/AIDS), and STD (sexually transmitted diseases) infection; unhealthy dietary patterns; and inadequate physical activity.
- Increase the proportion of college and university students who receive information from their institution on each of the priority health risk behavior areas (all priority areas; unintentional injury; violence; suicide; tobacco use and addiction; alcohol and other drug use; unintended pregnancy, human immunodeficiency virus/acquired immune deficiency syndrome (HIV/AIDS), and STD (sexually transmitted diseases) infection,; unhealthy dietary patterns; and inadequate physical activity).
- Increase the proportion of worksites that offer an employee health promotion program to their employees.
- Increase the number of community-based organizations (including local health departments, tribal health services, nongovernmental organizations, and state agencies) providing population-based primary prevention services.

Environmental Health

- Reduce blood lead levels in children.
- Reduce the number of US homes that are found to have lead-based paint or related hazards.

(continued)

Food Safety

- Reduce infections caused by key pathogens transmitted commonly through food.
- Reduce the number of outbreak-associated infections due to Shiga toxin-producing *Escherichia coli* O157:H7, or *Campylobacter, Listeria,* or *Salmonella* species associated with food commodity groups.
- Reduce severe allergic reactions to food among adults with a food allergy diagnosis.
- Increase the proportion of consumers who follow key food safety practices.
- Improve food safety practices associated with foodborne illness in foodservice and retail establishments.

Heart Disease and Stroke

- Increase overall cardiovascular health in the US population.
- Reduce coronary heart disease deaths.
- Reduce stroke deaths.
- Reduce the proportion of persons in the population with hypertension.
- Reduce the proportion of adults with high total blood cholesterol levels.
- Reduce the mean total blood cholesterol levels among adults.
- Increase the proportion of adults with prehypertension who meet the recommended guidelines.
- Increase the proportion of adults with hypertension who meet the recommended guidelines.
- Increase the proportion of adults with elevated LDL cholesterol who have been advised by a health-care provider regarding cholesterol-lowering management including lifestyle changes and, if indicated, medication.
- Increase the proportion of adults with elevated LDL-cholesterol who adhere to the prescribed LDL cholesterol–lowering management lifestyle changes and, if indicated, medication.

Maternal, Infant, and Child Health

- Reduce low birth weight (LBW) and very low birth weight (VLBW).
- Reduce preterm births.
- Increase the proportion of pregnant women who receive early and adequate prenatal care.
- Increase the proportion of mothers who achieve a recommended weight gain during their pregnancies.
- Increase the proportion of women of childbearing potential with intake of at least 400 micrograms of folic acid from fortified foods or dietary supplements.
- Reduce the proportion of women of childbearing potential who have low red blood cell folate concentrations.
- Increase the proportion of women delivering a live birth who received preconception care services and practiced key recommended preconception health behaviors.
- Increase the proportion of infants who are breastfed.
- Increase the proportion of employers that have worksite lactation support programs.
- Reduce the proportion of breastfed newborns who receive formula supplementation within the first 2 days of life.
- Reduce the occurrence of fetal alcohol syndrome (FAS).
- Reduce occurrence of neural tube defects.

Mental Health and Mental Disorders

- Reduce the proportion of adolescents who engage in disordered eating behaviors in an attempt to control their weight.

Older Adults

- Increase the proportion of the health-care workforce (including dietitians) with geriatric certification.

Oral Health

- Increase the proportion of the US population served by community water systems with optimally fluoridated water.

Physical Activity

- Reduce the proportion of adults who engage in no leisure-time physical activity.
- Increase the proportion of adolescents and adults who meet current federal physical activity guidelines for aerobic physical activity and for muscle-strengthening activity.
- Increase the proportion of the nation's public and private schools that require daily physical education for all students.
- Increase the proportion of adolescents who participate in daily school physical education.
- Increase regularly scheduled elementary school recess in the United States.
- Increase the proportion of children and adolescents who do not exceed recommended limits for screen time.

SOURCE: Adapted from Healthy people 2020: www.healthypeople.gov.

Appendix K   Aids to Calculation

Many mathematical problems have been worked out in the "How To" features of the text. This appendix offers additional help and examples.

Conversions

A conversion factor is a fraction that converts a measurement expressed in one unit to another unit—for example, from pounds to kilograms or from feet to meters. To create a conversion factor, an equality (such as 1 kilogram = 2.2 pounds) is expressed as a fraction:

$$\frac{1 \text{ kg}}{2.2 \text{ lb}} \text{ and } \frac{2.2 \text{ lb}}{1 \text{ kg}}$$

To convert the units of a measurement, use the fraction with the desired unit in the numerator.

Example 1 Convert a weight of 130 pounds to kilograms. Multiply 130 pounds by the conversion factor that includes both pounds and kilograms, with the desired unit (kilograms) in the numerator:

$$130 \text{ lb} \times \frac{1 \text{ kg}}{2.2 \text{ lb}} = \frac{130 \text{ kg}}{2.2} = 59 \text{ kg}$$

Alternatively, to convert a measurement from one unit of measure to another, multiply the given measurement by the appropriate equivalent found on the next page of weights and measures.

Example 2 Convert 64 fluid ounces to liters.
Locate the equivalent measure from the volume section on the next page (1 ounce = 0.03 liter) and multiply the number of ounces by 0.03:

$$64 \text{ oz} \times 0.03 \text{ oz/L} = 1.9 \text{ L}$$

Percentages

A percentage is a fraction whose denominator is 100. For example:

$$50\% = \frac{50}{100}$$

Like other fractions, percentages are used to express a portion of a quantity. Fractions whose denominators are numbers other than 100 can be converted to percentages by first dividing the numerator by the denominator and then multiplying the result by 100.

Example 3 Express 5/8 as a percent.

$$\frac{5}{8} = 5 \div 8 = 0.625$$

$$0.625 \times 100 = 62.5\%$$

The following examples show how to calculate specific percentages.

Example 4 Suppose your energy intake for the day is 2000 kcalories (kcal) and your recommended energy intake is 2400 kcalories. What percent of the recommended energy intake did you consume?

> Divide your intake by the recommended intake.
>   2000 kcal (intake) ÷ 2400 kcal (recommended) = 0.83
> Multiply by 100 to express the decimal as a percent.
>   0.83 × 100 = 83%

Example 5 Suppose a man's intake of vitamin C is 120 milligrams and his RDA is 90 milligrams. What percent of the RDA for vitamin C did he consume?

> Divide the intake by the recommended intake.
>   120 mg (intake) ÷ 90 mg (RDA) = 1.33
> Multiply by 100 to express the decimal as a percent.
>   1.33 × 100 = 133%

Example 6 Dietary recommendations suggest that carbohydrates provide 45 to 65 percent of the day's energy intake. If your energy intake is 2000 kcalories, how much carbohydrate should you eat?

Because this question has a range of acceptable answers, work the problem twice. First, use 45% to find the least amount you should eat.

> Divide 45 by 100 to convert to a decimal.
>   45 ÷ 100 = 0.45
> Multiply kcalories by 0.45.
>   2000 kcal × 0.45 = 900 kcal
> Divide kcalories by 4 to convert carbohydrate kcal to grams.
>   900 kcal ÷ 4 kcal/g = 225 g

Now repeat the process using 65% to find the maximum number of grams of carbohydrates you should eat.

Divide 65 by 100 to convert it to a decimal.

$65 \div 100 = 0.65$

Multiply kcalories by 0.65.

$2000 \text{ kcal} \times 0.65 = 1300 \text{ kcal}$

Divide kcalories by 4 to convert carbohydrate kcal to grams.

$1300 \text{ kcal} \div 4 \text{ kcal/g} = 325 \text{ g}$

If you plan for between 45% and 65% of your 2000-kcalorie intake to be from carbohydrates, you should eat between 225 grams and 325 grams of carbohydrates.

Weights and Measures

Length

1 centimeter (cm) = 0.39 inches (in)
1 foot (ft) = 30 centimeters (cm)
1 inch (in) = 2.54 centimeters (cm)
1 meter (m) = 39.37 inches (in)

Weight

1 gram (g) = 0.001 kilograms (kg)
        = 1000 milligrams (mg)
        = 0.035 ounces (oz)
1 kilogram (kg) = 1000 grams (g)
           = 2.2 pounds (lb)
1 microgram (μg) = 0.001 milligrams (mg)
1 milligram (mg) = 0.001 grams (g)
           = 1000 micrograms (μg)
1 ounce (oz) = 28 grams (g)
        = 0.03 kilograms (kg)
        = 1/16 or 0.0625 pound (lb)
1 pound (lb) = 454 grams (g)
       = 0.45 kilograms (kg)
       = 16 ounces (oz)

Volume

1 cup = 16 tablespoons (tbs or T)
     = 0.25 liters (L)
     = 236 milliliters (mL, commonly rounded to 250 mL)
     = 8 ounces (oz)
1 liter (L) = 33.8 fluid ounces (fl oz)
      = 0.26 gallons (gal)
      = 2.1 pints (pt)
      = 1.06 quarts (qt)
      = 1000 milliliters (mL)
1 milliliter (mL) = 0.001 liters (L)
         = 0.03 fluid ounces (fl oz)
         = 1/5 teaspoon (tsp)

1 ounce (oz) = 0.03 liters (L)
        = 30 milliliters (mL)
        = 2 tablespoons (tbs)
1 pint (pt) = 2 cups (c)
      = 0.47 liters (L)
      = 16 ounces (oz)
      = 0.5 quarts (qt)
1 quart (qt) = 4 cups (c)
       = 0.95 liters (L)
       = 32 ounces (oz)
       = 1/4 or 0.25 gallon (gal)
       = 2 pints (pt)
1 tablespoon (tbs or T) = 3 teaspoons (tsp)
                 = 15 milliliters (mL)
1 teaspoon (tsp) = 5 milliliters (mL)
1 gallon (gal) = 16 cups (c)
        = 3.8 liters (L)
        = 128 ounces (oz)
        = 8 pints (pt)
        = 4 quarts (qt)
1 cup (c) = 8 ounces (oz)
      = 16 tablespoons (tbs)
      = 250 milliliters (mL)

Energy

1 megajoule (MJ) = 240 kcalories (kcal)
1 kilojoule (kJ) = 0.24 kcalories (kcal)
1 kcalorie (kcal) = 4.2 kilojoule (kJ)
1 g alcohol = 7 kcal = 29 kJ
1 g carbohydrate = 4 kcal = 17 kJ
1 g fat = 9 kcal = 37 kJ
1 g protein = 4 kcal = 17 kJ

Temperature

To change from Fahrenheit (°F) to Celsius (°C), subtract 32 from the Fahrenheit measure and then multiply that result by 0.56.

To change from Celsius (°C) to Fahrenheit (°F), multiply the Celsius measure by 1.8 and add 32 to that result.

A comparison of some useful temperatures is given below.

	Celsius	Fahrenheit
Boiling point	100°C	212°F
Body temperature	37°C	98.6°F
Freezing point	0°C	32°F

Glossary

Many medical terms have their origins in Latin or Greek. By learning a few common derivations, you can glean the meaning of words you have never heard of before. For example, once you know that "hyper" means above normal, "glyc" means glucose, and "emia" means blood, you can easily determine that "hyperglycemia" means high blood glucose. The derivations below will help you to learn many terms presented in this glossary.

General

a- or **an-** = not, without
ana- = (build) up
ant- or **anti-** = against
ante- or **pre-** = before
bi- or **di-** = two, twice
bio- or **-biotic** = life
bovine = of cattle
calor = heat
cata- or **kata-** = (break) down
chele = claw
chroma = color
co- = with, together
dys- or **mal-** = bad, difficult, painful
endo- = inner, within, inside
epi- = upon (over)
erythro- = red
exo- = outside of, without
extra- = outside of, beyond, in addition
gen or **genesis** = producing, arising, making
homeo- = like, similar, the same, constant unchanging state
hyper- = over, above, excessive
hypo- = below, under, beneath, too little
in- = not
inter- = between, in the midst
intra- = within
lac- or **lacto-** = milk
-lysis = breaking, breakdown
macro- = large, great
malacia = softening
metallo- = metal
micro- = small
mono- = one, single
neo- = new, recent
oligo- = few, small
-osis or **-asis** = condition
para- = near
peri- = around, about
phag- or **phago-** = eat
-philia, -phil, -philic = love
-phobia = fear
phyto- = plant
poly- = many, much
pro- = for, in front of
re- = back, again
semi- = half
-stat or **-stasis-** = stationary, staying
sub- = beneath
tri- = three
xero- = dry

Body

angi- or **vaso-** = vessel
arterio- = artery
cardi-, cardiac, cardio-, or **cardial** = heart
cerebro = brain
cyst = closed sac
-cyte or **-cytic** = cell
encephalic = brain
entero- or **enteric** = intestine
fibro- = fibrous tissue
gastro- = stomach
globin = globular protein
hema-, hemo-, or **-emia** = blood
hepatic = liver
myo- = muscle
nephr- or **renal** = kidney
neuro- = nerve
osteo- = bone
pulmo- = lung
sarco- = flesh
soma = body
ure- or **-uria** = urine
vascular = blood vessels
vena = vein

Chemistry

-al = aldehyde
amino- or **amine** = containing nitrogen
-ase = enzyme
-ate = salt
carbo- = carbon (C)
glyc-, glyco-, gluc-, or **gluco-** = sweet (glucose)
glyceride = of glycerol
hydro- or **hydrate** = water
lipo- = lipid
-ol = alcohol
-ose = carbohydrate
peptide = referring to amino acids
saccha- or **sucro-** = sugar

Disease

athero- = porridge, fatty plaque
carcin- = cancer
-itis = infection, inflammation
-osis or **-asis** = condition
thrombo = clot

24-hour dietary recall: a record of foods eaten by a person for one 24-hour period.

A

A1C: a test that measures the percentage of hemoglobin that has glucose attached, which helps to diagnose diabetes and evaluate long-term glycemic control.

absorption: the uptake of nutrients by the cells of the small intestine for transport into either the blood or the lymph.

Academy of Nutrition and Dietetics: the professional organization of dietitians in the United States; formerly the American Dietetic Association.

Acceptable Daily Intake (ADI): the estimated amount of a sweetener that individuals can safely consume each day over the course of a lifetime without adverse effect.

Acceptable Macronutrient Distribution Ranges (AMDR): ranges of intakes for the energy nutrients that provide adequate energy and nutrients and reduce the risk of chronic diseases.

accredited: approved; in the case of medical centers or universities, certified by an agency recognized by the US Department of Education.

acetaldehyde (ass-et-AL-duh-hide): an intermediate in alcohol metabolism.

acetyl CoA (ASS-eh-teel or ah-SEET-il, coh-AY): a 2-carbon compound (acetate, or acetic acid) to which a molecule of CoA is attached.

acid controllers: medications used to prevent or relieve indigestion by suppressing production of acid in the stomach; also called *H2 blockers*.

acid-base balance: the equilibrium in the body between acid and base concentrations.

acidosis (assi-DOE-sis): higher-than-normal acidity in the blood and body fluids.

acids: compounds that release hydrogen ions in a solution.

acne: a chronic inflammation of the skin's follicles and oil-producing glands, which leads to an accumulation of oils inside the ducts that surround hairs; usually associated with the maturation of young adults.

acupuncture (AK-you-PUNK-cher): a technique that involves piercing the skin with long thin needles at specific anatomical points to relieve pain or illness. Acupuncture sometimes uses heat, pressure, friction, suction, or electromagnetic energy to stimulate the points.

acute malnutrition: malnutrition caused by recent severe food restriction; characterized in children by underweight for height (*wasting*).

adaptive thermogenesis: adjustments in energy expenditure related to changes in environment such as extreme cold and to physiological events such as overfeeding, trauma, and changes in hormone status.

added sugars: sugars and other kcaloric sweeteners that are added to foods during processing, preparation, or at the table. Added sugars do not include the naturally occurring sugars found in fruits and milk products.

additives: substances not normally consumed as foods but added to food either intentionally or by accident.

adequacy (dietary): providing all the essential nutrients, fiber, and energy in amounts sufficient to maintain health.

Adequate Intake (AI): the average daily amount of a nutrient that appears sufficient to maintain a specified criterion; a value used as a guide for nutrient intake when an RDA cannot be determined.

adipokines (ADD-ih-poe-kines): proteins synthesized and secreted by adipose cells.

adiponectin: a protein produced by adipose cells that inhibits inflammation and protects against insulin resistance, type 2 diabetes, and cardiovascular disease.

adipose (ADD-ih-poce) **tissue:** the body's fat tissue; consists of masses of triglyceride-storing cells.

adolescence: the period from the beginning of puberty until maturity.

adrenal glands: glands adjacent to, and just above, each kidney.

adverse reactions: unusual responses to food (including intolerances and allergies).

aerobic (air-ROE-bic): requiring oxygen.

aerobic physical activity: activity in which the body's large muscles move in a rhythmic manner for a sustained period of time. Aerobic activity, also called *endurance activity*, improves cardiorespiratory fitness. Brisk walking, running, swimming, and bicycling are examples.

AIDS (acquired immune deficiency syndrome): the late stage of HIV infection, in which severe complications of opportunistic infections and cancers develop.

alcohol: a class of organic compounds containing hydroxyl (OH) groups.

alcohol abuse: a pattern of drinking that includes failure to fulfill work, school, or home responsibilities; drinking in situations that are physically dangerous (as in driving while intoxicated); recurring alcohol-related legal problems (as in aggravated assault charges); or continued drinking despite ongoing social problems that are caused by or worsened by alcohol.

alcohol dehydrogenase (dee-high-DROJ-eh-nayz): an enzyme active in the stomach and the liver that converts ethanol to acetaldehyde.

alcoholism: a pattern of drinking that includes a strong craving for alcohol, a loss of control and an inability to stop drinking once begun, withdrawal symptoms (nausea, sweating, shakiness, and anxiety) after heavy drinking, and the need for increasing amounts of alcohol to feel "high."

alcohol-related birth defects (ARBD): malformations in the skeletal and organ systems (heart, kidneys, eyes, ears) associated with prenatal alcohol exposure.

alcohol-related neurodevelopmental disorder (ARND): abnormalities in the central nervous system and cognitive development associated with prenatal alcohol exposure.

aldosterone (al-DOS-ter-own): a hormone secreted by the adrenal glands that regulates blood pressure by increasing the reabsorption of sodium by the kidneys. Aldosterone also regulates chloride and potassium concentrations.

alkalosis (alka-LOE-sis): higher-than-normal alkalinity (base) in the blood and body fluids.

alpha cells: cells of the pancreas that secrete glucagon in response to low blood glucose concentration.

alpha-lactalbumin (lact-AL-byoo-min): a major protein in human breast milk, as opposed to *casein* (CAY-seen), a major protein in cow's milk.

alpha-tocopherol: the active vitamin E compound.

Alzheimer's (AHLZ-high-merz) **disease:** a degenerative disease of the brain involving memory loss and major structural changes in neuron networks; also known as *senile dementia of the Alzheimer's type (SDAT), primary degenerative dementia of senile onset,* or *chronic brain syndrome.*

amenorrhea (ay-MEN-oh-REE-ah): the absence of or cessation of menstruation. *Primary amenorrhea* is menarche delayed beyond 16 years of age. *Secondary amenorrhea* is the absence of three to six consecutive menstrual cycles.

amino (a-MEEN-oh) **acids:** building blocks of proteins. Each contains an amino group, an acid group, a hydrogen atom, and a distinctive side group, all attached to a central carbon atom.

amino acid pool: the supply of amino acids derived from either food proteins or body proteins that collect in the cells and circulating blood and stand ready to be incorporated in proteins and other compounds or used for energy.

amino acid score: a measure of protein quality assessed by comparing a protein's amino acid pattern with that of a reference protein; also called the *chemical score.*

ammonia: a compound with the chemical formula NH_3, produced during the deamination of amino acids.

amniotic (am-nee-OTT-ic) sac: the "bag of waters" in the uterus, in which the fetus floats.

amylase (AM-ih-lace): an enzyme that hydrolyzes amylose (a form of starch). Amylase is a *carbohydrase*, an enzyme that breaks down carbohydrates.

anabolic steroids: drugs related to the male sex hormone, testosterone, that stimulate the development of lean body mass.

anabolism (an-AB-o-lism): reactions in which small molecules are put together to build larger ones. Anabolic reactions require energy.

anaerobic (AN-air-ROE-bic): not requiring oxygen.

anaphylactic (ana-fill-LAC-tic) shock: a life-threatening, whole-body allergic reaction to an offending substance.

androstenedione: hormones made in the adrenal glands that serve as precursors to the male sex hormone, testosterone; falsely promoted as burning fat, building muscle, and slowing aging.

anecdote: a personal account of an experience or event; not reliable scientific information.

anemia (ah-NEE-me-ah): literally, "too little blood." Anemia is any condition in which too few red blood cells are present, or the red blood cells are immature (and therefore large) or too small or contain too little hemoglobin to carry the normal amount of oxygen to the tissues. Anemia is not a disease itself but can be a symptom of many different disease conditions, including many nutrient deficiencies, bleeding, excessive red blood cell destruction, and defective red blood cell formation.

anencephaly (AN-en-SEF-a-lee): an uncommon and always fatal type of neural tube defect, characterized by the absence of a brain.

aneurysm (AN-you-rizm): an abnormal enlargement or bulging of a blood vessel (usually an artery) caused by damage to or weakness in the blood vessel wall.

angina (an-JYE-nah or AN-ji-nah): a painful feeling of tightness or pressure in and around the heart, often radiating to the back, neck, and arms; caused by a lack of oxygen to an area of heart muscle.

angiotensin I (AN-gee-oh-TEN-sin): an inactive precursor that is converted by an enzyme to yield active angiotensin II.

angiotensin II: a hormone involved in blood pressure regulation.

angiotensinogen: a precursor protein that is hydrolyzed to angiotensin I by renin.

anions (AN-eye-uns): negatively charged ions.

anorexia (an-oh-RECK-see-ah) nervosa: an eating disorder characterized by a refusal to maintain a minimally normal body weight and a distortion in perception of body shape and weight.

antacids: medications used to relieve indigestion by neutralizing acid in the stomach.

antagonist: a competing factor that counteracts the action of another factor. When a drug displaces a vitamin from its site of action, the drug renders the vitamin ineffective and thus acts as a vitamin antagonist.

anthropometric (AN-throw-poe-MET-rick): relating to measurement of the physical characteristics of the body, such as height and weight.

antibodies: large proteins of the blood and body fluids, produced by the immune system in response to the invasion of the body by foreign molecules (usually proteins called *antigens*). Antibodies combine with and inactivate the foreign invaders, thus protecting the body.

antidiuretic hormone (ADH): a hormone produced by the pituitary gland in response to dehydration (or a high sodium concentration in the blood) that stimulates the kidneys to reabsorb more water and therefore to excrete less. In addition to its antidiuretic effect, ADH elevates blood pressure and so is also called *vasopressin* (VAS-oh-PRES-in).

antigens: substances that elicit the formation of antibodies or an inflammation reaction from the immune system. A bacterium, a virus, a toxin, and a protein in food that causes allergy are all examples of antigens.

antioxidants: in the body, substances that significantly decrease the adverse effects of free radicals on normal physiological functions.

antioxidants: as a food additive, preservatives that delay or prevent rancidity of fats in foods and other damage to food caused by oxygen.

antipromoters: factors that oppose the development of cancer.

antiscorbutic (AN-tee-skor-BUE-tik) factor: the original name for vitamin C.

anus (AY-nus): the terminal outlet of the GI tract.

aorta (ay-OR-tuh): the large, primary artery that conducts blood from the heart to the body's smaller arteries.

apoptosis: cell death.

appendix: a narrow blind sac extending from the beginning of the colon that contains bacteria and lymph cells.

appetite: the integrated response to the sight, smell, thought, or taste of food that initiates or delays eating.

aquaculture: the practice of fish farming.

arachidonic (a-RACK-ih-DON-ic) acid: an omega-6 polyunsaturated fatty acid with 20 carbons and four double bonds; present in small amounts in meat and other animal products and synthesized in the body from linoleic acid.

ariboflavinosis (ay-RYE-boh-FLAY-vin-oh-sis): riboflavin deficiency.

aroma therapy: a technique that uses oil extracts from plants and flowers (usually applied by massage or baths) to enhance physical, psychological, and spiritual health.

arteries: vessels that carry blood from the heart to the tissues.

artesian water: water drawn from a well that taps a confined aquifer in which the water is under pressure.

arthritis: inflammation of a joint, usually accompanied by pain, swelling, and structural changes.

artificial fats: zero-energy fat replacers that are chemically synthesized to mimic the sensory and cooking qualities of naturally occurring fats but are totally or partially resistant to digestion.

artificial sweeteners: sugar substitutes that provide negligible, if any, energy; sometimes called *nonnutritive sweeteners*.

ascorbic acid: one of the two active forms of vitamin C. Many people refer to vitamin C by this name.

-ase (ACE): a suffix denoting an enzyme. The root of the word often identifies the compounds the enzyme works on. Examples include: *carbohydrase* (KAR-boe-HIGH-drase), an enzyme that hydrolyzes carbohydrates; *lipase* (LYE-pase), an enzyme that hydrolyzes lipids (fats); and *protease* (PRO-tee-ase), an enzyme that hydrolyzes proteins.

atherogenic: able to initiate or promote atherosclerosis.

atherosclerosis (ATH-er-oh-scler-OH-sis): a type of artery disease characterized by plaques (accumulations of lipid-containing material) on the inner walls of the arteries.

atoms: the smallest components of an element that have all of the properties of the element.

ATP, or adenosine (ah-DEN-oh-seen) triphosphate (try-FOS-fate): a common high-energy compound composed of a purine (adenine), a sugar (ribose), and three phosphate groups. ATP = A–P~P~P, with each ~ denoting a "high-energy" bond.

atrophic (a-TRO-fik) gastritis (gas-TRY-tis): chronic inflammation of the stomach accompanied by a diminished size and functioning of the mucous membranes and glands. This condition is also characterized by inadequate hydrochloric acid and intrinsic factor—two substances needed for vitamin B_{12} absorption.

atrophy (AT-ro-fee): becoming smaller; with regard to muscles, a decrease in size (and strength) because of disuse, undernutrition, or wasting diseases.

autoimmune disorder: a condition in which the body develops antibodies to its own proteins and then proceeds to destroy cells containing these proteins. In type 1 diabetes, the body develops antibodies to its insulin and destroys the pancreatic cells that produce the insulin, creating an insulin deficiency.

autonomic nervous system: the division of the nervous system that controls the body's automatic responses. Its two branches are the *sympathetic* branch, which helps the body respond to stressors from the outside environment, and the *parasympathetic* branch, which regulates normal body activities between stressful times.

avidin (AV-eh-din): the protein in egg whites that binds biotin.

ayurveda (AH-your-VAY-dah): a traditional Hindu system of improving health by using herbs, diet, meditation, massage, and yoga to stimulate the body, mind, and spirit to prevent and treat disease.

B

bacteriophages (bak-TIR-ee-oh-fayjz): viruses that infect bacteria.

balance (dietary): providing foods in proportion to one another and in proportion to the body's needs.

bariatric: pertaining to the field of medicine that specializes in treating obesity.

basal metabolic rate (BMR): the rate of energy use for metabolism under specified conditions: after a 12-hour fast and restful sleep, without any physical activity or emotional excitement, and in a comfortable setting. It is usually expressed as kcalories per kilogram body weight per hour.

basal metabolism: the energy needed to maintain life when a body is at complete digestive, physical, and emotional rest.

bases: compounds that accept hydrogen ions in a solution.

B-cells: lymphocytes that produce antibodies. *B* stands for *bone marrow*, where the B-cells develop and mature.

beer: an alcoholic beverage traditionally brewed by fermenting malted barley and adding hops for flavor.

behavior modification: the changing of behavior by the manipulation of antecedents (cues or environmental factors that trigger behavior), the behavior itself, and consequences (the penalties or rewards attached to the behavior).

belching: the release of air or gas from the stomach through the mouth.

benign: an abnormal mass of cells that is noncancerous.

beriberi: the thiamin-deficiency disease characterized by muscle weakness, edema, or both.

beta-carotene (BAY-tah KARE-oh-teen): one of the carotenoids; an orange pigment and vitamin A precursor found in plants.

beta cells: cells in the pancreas that secrete insulin in response to elevated blood glucose concentration.

beta-alanine: a nonessential amino acid that is the rate-limiting precursor for the synthesis of the dipeptide carnosine. Carnosine acts primarily as a buffer in skeletal muscle. Beta-alanine supplements raise carnosine concentrations, which enhance the muscles' buffering capacity.

beta-hydroxymethylbutryate (HMB): a metabolite of the amino acid leucine promoted to increase muscle mass and strength.

BHA and BHT: preservatives commonly used to slow the development of off-flavors, odors, and color changes caused by oxidation; BHA is butylated hydroxyanisole and BHT is butylated hydroxytoluene.

bicarbonate: an alkaline compound with the formula HCO_3 that is secreted from the pancreas as part of the pancreatic juice. Bicarbonate is also produced in all cell fluids from the dissociation of carbonic acid to help maintain the body's acid-base balance.

bile: an emulsifier that prepares fats and oils for digestion; an exocrine secretion made by the liver, stored in the gallbladder, and released into the small intestine when needed.

binders: chemical compounds in foods that combine with nutrients (especially minerals) to form complexes the body cannot absorb. Examples include *phytates* (FYE-tates) and *oxalates* (OCK-sa-lates).

binge drinking: pattern of drinking that raises blood alcohol concentration to 0.08 percent or higher; usually corresponds to four or more drinks for women and five or more drinks for men on a single occasion, generally within a couple of hours.

binge-eating disorder: an eating disorder characterized by recurring episodes of eating a significant amount of food in a short period of time with marked feelings of lack of control.

bioaccumulation: the accumulation of contaminants in the flesh of animals high on the food chain.

bioavailability: the rate at and the extent to which a nutrient is absorbed and used.

bioelectromagnetic medical applications: the use of electrical energy, magnetic energy, or both to stimulate bone repair, wound healing, and tissue regeneration.

biofeedback: the use of special devices to convey information about heart rate, blood pressure, skin temperature, muscle relaxation, and the like to enable a person to learn how to consciously control these medically important functions.

biofield therapeutics: a manual healing method that directs a healing force from an outside source (commonly God or another supernatural being) through the practitioner and into the client's body; commonly known as "laying on of hands."

biological value (BV): a measure of protein quality assessed by measuring the amount of protein nitrogen that is retained from a given amount of protein nitrogen absorbed.

biotechnology: the use of biological systems or organisms to create or modify products. Examples include the use of bacteria to make yogurt, yeast to make beer, and cross-breeding to enhance crop production.

biotin (BY-oh-tin): a B vitamin that functions as a coenzyme in metabolism.

blastocyst (BLASS-toe-sist): the developmental stage of the zygote when it is about 5 days old and ready for implantation.

blind experiment: an experiment in which the subjects do not know whether they are members of the experimental group or the control group.

bloating: uncomfortable abdominal fullness or distention.

blood lipid profile: results of blood tests that reveal a person's total cholesterol, triglycerides, and various lipoproteins.

body composition: the proportions of muscle, bone, fat, and other tissue that make up a person's total body weight.

body mass index (BMI): a measure of a person's weight relative to height; determined by dividing the weight (in kilograms) by the square of the height (in meters).

bolus (BOH-lus): a portion; with respect to food, the amount swallowed at one time.

bomb calorimeter (KAL-oh-RIM-eh-ter): an instrument that measures the heat energy released when foods are burned, thus providing an estimate of the potential energy (kcalories) of the foods.

bone density: a measure of bone strength. When minerals fill the bone matrix (making it dense), they give it strength.

bone meal: crushed or ground bone preparations intended to supply calcium to the diet. Calcium from bone is not well absorbed and is often contaminated with toxic minerals such as arsenic, mercury, lead, and cadmium.

bottled water: drinking water sold in bottles.

botulism (BOT-chew-lism): an often fatal foodborne illness caused by the ingestion of foods containing a toxin produced by bacteria that grow without oxygen.

bovine growth hormone (BGH): a hormone produced naturally in the pituitary gland of a cow that promotes growth and milk production; now produced for agricultural use by bacteria.

bovine spongiform encephalopathy (BOH-vine SPON-jih-form in-SEF-eh-LOP-eh-thee) or **BSE:** an often fatal illness of cattle and wild game that affects the nervous system and is transmitted to people by eating infected meats; commonly called *mad cow disease*.

bran: the protective coating around the kernel of grain, rich in nutrients and fiber.

branched-chain amino acids: the essential amino acids leucine, isoleucine, and valine, which are present in large amounts in skeletal muscle tissue; falsely promoted as fuel for exercising muscles.

breast milk bank: a service that collects, screens, processes, and distributes donated human milk.

brite adipocytes: white fat cells with brown fat cell characteristics; also called *beige adipocytes*.

brown adipose tissue: masses of specialized fat cells packed with pigmented mitochondria that produce heat instead of ATP.

brown sugar: refined white sugar crystals to which manufacturers have added molasses syrup with natural flavor and color; 91 to 96 percent pure sucrose.

buffers: compounds that keep a solution's pH constant when acids or bases are added.

bulimia (byoo-LEEM-ee-ah) **nervosa:** an eating disorder characterized by repeated episodes of binge eating usually followed by self-induced vomiting, misuse of laxatives or diuretics, fasting, or excessive exercise.

C

caffeine: a natural stimulant found in many common foods and beverages, including coffee, tea, and chocolate; may enhance endurance by stimulating fatty acid release. High doses cause headaches, trembling, rapid heart rate, and other undesirable side effects.

calbindin: a calcium-binding transport protein that requires vitamin D for its synthesis.

calcidiol: vitamin D found in the blood that is made from the hydroxylation of calciol in the liver; also called *25-hydroxyvitamin D*.

calciferol (kal-SIF-er-ol): vitamin D.

calciol: vitamin D derived from animals in the diet or made in the skin from 7-dehydrocholesterol, a precursor of cholesterol, with the help of sunlight; also called *cholecalciferol* or *vitamin D₃*.

calcitonin (KAL-seh-TOE-nin): a hormone secreted by the thyroid gland that regulates blood calcium by lowering it when levels rise too high.

calcitriol: vitamin D that is made from the hydroxylation of calcidiol in the kidneys; the biologically active hormone; also called *1,25-dihydroxyvitamin D* or *active vitamin D*.

calcium: the most abundant mineral in the body; found primarily in the body's bones and teeth.

calcium-binding protein: a protein in the intestinal cells, made with the help of vitamin D, that facilitates calcium absorption.

calcium rigor: hardness or stiffness of the muscles caused by high blood calcium concentrations.

calcium tetany (TET-ah-nee): intermittent spasm of the extremities due to nervous and muscular excitability caused by low blood calcium concentrations.

calmodulin (cal-MOD-you-lin): a calcium-binding protein that regulates such cell activities as muscle contractions.

calories: a measure of *heat* energy. Energy provided by foods and beverages is measured in *kilocalories* (1000 calories equal 1 kilocalorie), abbreviated *kcalories* or *kcal*. One kcalorie is the amount of heat necessary to raise the temperature of 1 kilogram (kg) of water 1°C. The scientific use of the term *kcalorie* is the same as the popular use of the term *calorie*.

cancers: malignant growths or tumors that result from abnormal and uncontrolled cell division.

capillaries (CAP-ill-aries): small vessels that branch from an artery. Capillaries connect arteries to veins. Exchange of oxygen, nutrients, and waste materials takes place across capillary walls.

carbohydrase (KAR-boe-HIGH-drase): an enzyme that hydrolyzes carbohydrates.

carbohydrate loading: a regimen of moderate exercise followed by the consumption of a high-carbohydrate diet that enables muscles to store glycogen beyond their normal capacities; also called *glycogen loading* or *glycogen super compensation*.

carbohydrates: compounds composed of carbon, oxygen, and hydrogen arranged as monosaccharides or multiples of monosaccharides. Most, but not all, carbohydrates have a ratio of one carbon molecule to one water molecule: $(CH_2O)_n$.

carbonated water: water that contains carbon dioxide gas, either naturally occurring or added, that causes bubbles to form in it; also called *bubbling* or *sparkling water*. The FDA defines seltzer, soda, and tonic waters as soft drinks; they are not regulated as water.

carbonic acid: a compound with the formula H_2CO_3 that results from the combination of carbon dioxide (CO_2) and water (H_2O); of particular importance in maintaining the body's acid-base balance.

carcinogenesis (CAR-sin-oh-JEN-eh-sis): the process of cancer development.

carcinogen (CAR-sin-oh-jen or car-SIN-oh-jen): a substance that can cause cancer; the adjective is *carcinogenic*.

cardiac output: the volume of blood discharged by the heart each minute; determined by multiplying the stroke volume by the heart rate. The stroke volume is the amount of oxygenated blood the heart ejects toward the tissues at each beat. Cardiac output (volume/minute) = stroke volume (volume/beat) × heart rate (beats/minute).

cardiorespiratory conditioning: improvements in heart and lung function and increased blood volume, brought about by aerobic training.

cardiorespiratory endurance: the ability to perform large-muscle, dynamic exercise of moderate to high intensity for prolonged periods.

cardiovascular disease (CVD): diseases of the heart and blood vessels throughout the body. Atherosclerosis is the main cause of CVD. When the arteries that carry blood to the heart muscle become blocked, the heart suffers damage known as *coronary heart disease (CHD)*.

carnitine (CAR-neh-teen): a nonessential, nonprotein amino acid made in the body from lysine that helps transport fatty acids across the mitochondrial membrane. As a supplement, carnitine supposedly "burns" fat and spares glycogen during endurance events, but in reality it does neither.

carotenoids (kah-ROT-eh-noyds): pigments commonly found in plants and animals, some of which have vitamin A activity. The carotenoid with the greatest vitamin A activity is beta-carotene.

cartilage therapy: the use of cleaned and powdered connective tissue, such as collagen, to improve health.

catabolism (ca-TAB-o-lism): reactions in which large molecules are broken down to smaller ones. Catabolic reactions release energy.

catalyst (CAT-uh-list): a compound that facilitates chemical reactions without itself being changed in the process.

cataracts (KAT-ah-rakts): clouding of the eye lenses that impairs vision and can lead to blindness.

cathartic (ka-THAR-tik): a strong laxative.

cations (CAT-eye-uns): positively charged ions.

CDC (Centers for Disease Control and Prevention): a branch of the Department of Health and Human Services that is responsible for, among other things, monitoring foodborne diseases.

celiac disease: an intestinal disorder in which the inability to absorb the protein portion of gluten results in an immune response that damages intestinal cells; also called *celiac sprue* or *gluten-sensitive enteropathy*.

cell: the basic structural unit of all living things.

cell differentiation (DIF-er-EN-she-AY-shun): the process by which immature cells develop specific functions different from those of the original that are characteristic of their mature cell type.

cell membrane: the thin layer of tissue that surrounds the cell and encloses its contents, made primarily of lipid and protein.

central nervous system: the central part of the nervous system; the brain and spinal cord.

central obesity: excess fat around the trunk of the body; also called *abdominal fat* or *upper-body fat*.

cerebral thrombosis: a clot that blocks blood flow through an artery that feeds the brain.

certified lactation consultants: health-care providers who specialize in helping new mothers establish a healthy breastfeeding relationship with their newborn. These consultants are often registered nurses with specialized training in breast and infant anatomy and physiology.

certified nutritionist or **certified nutritional consultant** or **certified nutrition therapist:** a person who has been granted a document declaring his or her authority as a nutrition professional.

cesarean (si-ZAIR-ee-un) **delivery:** a surgically assisted birth involving removal of the fetus by an incision into the uterus, usually by way of the abdominal wall.

chaff: the outer inedible part of a grain; also called the *husk*.

CHD risk equivalents: disorders that raise the risk of heart attacks, strokes, and other complications associated with cardiovascular disease to the same degree as existing CHD. These disorders include symptomatic carotid artery disease, peripheral arterial disease, abdominal aortic aneurysm, and diabetes mellitus.

cheilosis (kye-LOH-sis or kee-LOH-sis): a condition of reddened lips with cracks at the corners of the mouth.

chelate (KEY-late): a substance that can grasp the positive ions of a mineral.

chelation (kee-LAY-shun) **therapy:** the use of ethylene diamine tetraacetic acid (EDTA) to bind with metallic ions, thus healing the body by removing toxic metals.

chiropractic (KYE-roh-PRAK-tik): a manual healing method of manipulating the spine to restore health.

chloride (KLO-ride): the major anion in the extracellular fluids of the body. Chloride is the ionic form of chlorine, Cl⁻.

chlorophyll (KLO-row-fil): the green pigment of plants, which absorbs light and transfers the energy to other molecules, thereby initiating photosynthesis.

cholecalciferol (KO-lee-kal-SIF-er-ol): vitamin D derived from animals in the diet and made in the skin from 7-dehydrocholesterol, a precursor of cholesterol, with the help of sunlight; also called *vitamin D₃*.

cholecystokinin (COAL-ee-SIS-toe-KINE-in), or **CCK:** a hormone produced by cells of the intestinal wall. Target organ: the gallbladder. Response: release of bile and slowing of GI motility.

cholesterol (koh-LESS-ter-ol): one of the sterols containing a four-ring carbon structure with a carbon side chain.

cholesterol-free: less than 2 milligrams of cholesterol per serving and 2 grams or less of saturated fat and *trans* fat combined per serving.

choline (KOH-leen): a nitrogen-containing compound found in foods and made in the body from the amino acid methionine. Choline is part of the phospholipid lecithin and the neurotransmitter acetylcholine.

chromium: an essential trace mineral that enhances the activity of insulin.

chromium picolinate (CROW-mee-um pick-oh-LYN-ate): a trace mineral supplement; falsely promoted as building muscle, enhancing energy, and burning fat. *Picolinate* is a derivative of the amino acid tryptophan, which seems to enhance chromium absorption.

chromosomes: structures within the nucleus of a cell made of DNA and associated proteins. Human beings have 46 chromosomes in 23 pairs. Each chromosome has many genes.

chronic diseases: diseases characterized by slow progression and long duration. Examples include heart disease, diabetes, and some cancers.

chronic malnutrition: malnutrition caused by long-term food deprivation; characterized in children by short height for age (*stunting*).

chronological age: a person's age in years from his or her date of birth.

chylomicrons (kye-lo-MY-cronz): the class of lipoproteins that transport lipids from the intestinal cells to the rest of the body.

chyme (KIME): the semiliquid mass of partly digested food expelled by the stomach into the duodenum.

cirrhosis (seer-OH-sis): advanced liver disease in which liver cells turn orange, die, and harden, permanently losing their function; often associated with alcoholism.

cis: on the near side of; refers to a chemical configuration in which the hydrogen atoms are located on the same side of a double bond.

citric acid cycle: a series of metabolic reactions that break down molecules of acetyl CoA to carbon dioxide and hydrogen atoms; also called the *TCA cycle*, *tricarboxylic acid cycle*, or the *Krebs cycle*.

clinically severe obesity: a BMI of 40 or greater or a BMI of 35 or greater with additional medical problems. A less preferred term used to describe the same condition is *morbid obesity*.

clone: a genetic copy of an organism created without sexual reproduction; similar to identical twins, but born at different times.

CoA (coh-AY): coenzyme A; the coenzyme derived from the B vitamin pantothenic acid and central to energy metabolism.

coenzymes: complex organic molecules that work with enzymes to facilitate the enzymes' activity. Many coenzymes have B vitamins as part of their structures.

cofactor: a small, inorganic or organic substance that facilitates the action of an enzyme.

colitis (ko-LYE-tis): inflammation of the colon.

collagen (KOL-ah-jen): the structural protein from which connective tissues such as scars, tendons, ligaments, and the foundations of bones and teeth are made.

colonic irrigation: the popular, but potentially harmful practice of "washing" the large intestine with a powerful enema machine; also called *colonic hydrotherapy*.

colostrum (ko-LAHS-trum): a milklike secretion from the breast, present during the first few days after delivery before milk appears; rich in protective factors.

complementary and alternative medicine (CAM): diverse medical and health-care systems, practices, and products that are not currently considered part of conventional medicine; also called *adjunctive, unconventional,* or *unorthodox therapies*.

complementary proteins: two or more dietary proteins whose amino acid assortments complement each other in such a way that the essential amino acids limited in one are supplied by the other.

complex carbohydrates: polysaccharides (starches and fibers).

compound: a substance composed of two or more different atoms—for example, water (H_2O).

conception: the union of the male sperm and the female ovum; fertilization.

condensation: a chemical reaction in which water is released as two molecules combine to form one larger product.

conditionally essential amino acid: an amino acid that is normally nonessential, but must be supplied by the diet in special circumstances when the need for it exceeds the body's ability to make it.

conditionally essential nutrient: a nutrient that is normally nonessential, but must be supplied by the diet in special circumstances when the need for it exceeds the body's ability to make it.

confectioners' sugar: finely powdered sucrose, 99.9 percent pure.

congregate meals: nutrition programs that provide food for the elderly in conveniently located settings such as community centers.

conjugated linoleic acids: several fatty acids that have the same chemical formula as linoleic acid (18 carbons, two double bonds) but with different configurations (the double bonds occur on adjacent carbons).

constipation: the condition of having infrequent or difficult bowel movements.

contaminants: substances that make a food impure and unsuitable for ingestion.

contamination iron: iron found in foods as the result of contamination by inorganic iron salts from iron cookware, iron-containing soils, and the like.

control group: a group of individuals similar in all possible respects to the experimental group except for the treatment. Ideally, the control group receives a placebo while the experimental group receives a real treatment.

convenient dietary supplements: liquid mean replacers, energy drinks, energy bars, and energy gels that athletes and active people use to replenish energy and nutrients when time is limited.

conventional medicine: diagnosis and treatment of diseases as practiced by medical doctors (M.D.), doctors of osteopathy (D.O.), and allied health professionals such as physical therapists and registered nurses; also called *allopathy; Western, mainstream, orthodox,* or *regular medicine;* and *biomedicine.*

cool-down: 5 to 10 minutes of light activity, such as walking or stretching, following a vigorous workout to gradually return the body's core to near-normal temperature.

copper: an essential trace mineral that is part of many enzymes.

Cori cycle: the pathway in which glucose is metabolized to lactate (by anaerobic glycolysis) in the muscle, lactate is converted back to glucose in the liver, then glucose is returned to the muscle; named after the scientist who elucidated this pathway.

corn sweeteners: corn syrup and sugars derived from corn.

corn syrup: a syrup made from cornstarch that has been treated with acid, high temperatures, and enzymes to produce glucose, maltose, and dextrins. It may be dried and used as *corn syrup solids.* See also *high-fructose corn syrup (HFCS).*

cornea (KOR-nee-uh): the transparent membrane covering the outside of the eye.

coronary arteries: blood vessels that supply blood to the heart.

coronary heart disease (CHD): the damage that occurs when the blood vessels carrying blood to the heart (the *coronary arteries*) become narrow and occluded.

coronary thrombosis: a clot that blocks blood flow through an artery that feeds the heart muscle.

correlation (CORE-ee-LAY-shun): the simultaneous increase, decrease, or change in two variables. If A increases as B increases, or if A decreases

as B decreases, the correlation is *positive.* (This does not mean that A causes B or vice versa.) If A increases as B decreases, or if A decreases as B increases, the correlation is *negative.* (This does not mean that A prevents B or vice versa.) Some third factor may account for both A and B.

cortical bone: the very dense bone tissue that forms the outer shell surrounding trabecular bone and comprises the shaft of a long bone.

coupled reactions: pairs of chemical reactions in which some of the energy released from the breakdown of one compound is used to create a bond in the formation of another compound.

covalent bonds: strong chemical bonds formed between atoms by sharing electrons.

covert (KOH-vert): hidden, as if under covers.

C-reactive protein (CRP): a protein released during the acute phase of infection or inflammation that enhances immunity by promoting phagocytosis and activating platelets. Its presence may be used to assess a person's risk of an impending heart attack or stroke.

creatine (KREE-ah-tin): a nitrogen-containing compound that combines with phosphate to form the high-energy compound creatine phosphate (or phosphocreatine) in muscles.

creatine phosphate (CP): a high-energy compound in muscle cells that acts as a reservoir of energy that can maintain a steady supply of ATP. CP provides the energy for short bursts of activity; also called *phosphocreatine.*

cretinism (CREE-tin-ism): a congenital disease characterized by mental and physical retardation and commonly caused by maternal iodine deficiency during pregnancy.

critical periods: finite periods during development in which certain events occur that will have irreversible effects on later developmental stages; usually a period of rapid cell division.

cross-contamination: the contamination of food by bacteria that occurs when the food comes into contact with surfaces previously touched by raw meat, poultry, or seafood.

cruciferous vegetables: vegetables of the cabbage family, including cauliflower, broccoli, and brussels sprouts.

crypts (KRIPTS): tubular glands that lie between the intestinal villi and secrete intestinal juices into the small intestine.

cultural competence: having an awareness and acceptance of cultures and the ability to interact effectively with people of diverse cultures.

cytokines (SIGH-toe-kines): special proteins that direct immune and inflammatory responses.

cytoplasm (SIGH-toh-plazm): the cell contents, except for the nucleus.

cytosol: the fluid of cytoplasm that contains water, ions, nutrients, and enzymes.

D

Daily Values (DV): reference values developed by the FDA specifically for use on food labels.

dead zones: oxygen-depleted areas of bodies of water in which marine life cannot survive; often caused by algae overgrowth that occurs when agricultural fertilizers and wastes enter the waterways.

deamination (dee-AM-ih-NAY-shun): removal of the amino (NH_2) group from a compound such as an amino acid.

defecate (DEF-uh-cate): to move the bowels and eliminate waste.

deficient: inadequate; a nutrient amount that fails to meet the body's needs and eventually results in deficiency symptoms.

dehydration: the condition in which body water output exceeds water input. Symptoms include thirst, dry skin and mucous membranes, rapid heartbeat, low blood pressure, and weakness.

Delaney Clause: a 1958 amendment to the Food, Drugs, and Cosmetic Act of 1938, named after Congressman James Delaney of New York that states that no substance that is known to cause cancer in animals or human beings at any dose level shall be added to foods.

de minimis **rule:** a guideline that defines risk as a cancer rate of less than one cancer per million people exposed to a contaminant over a 70-year lifetime.

denaturation (dee-NAY-chur-AY-shun): the change in a protein's shape and consequent loss of its function brought about by heat, agitation, acid, base, alcohol, heavy metals, or other agents.

dental caries: decay of teeth.

dental plaque: a gummy mass of bacteria that grows on teeth and can lead to dental caries and gum disease.

dextrose: the name food manufacturers use for the sugar that is chemically the same as glucose; *anhydrous dextrose* is similar, differing primarily in the temperature of crystallization.

DHEA (dehydroepiandrosterone) and **androstenedione:** hormones made in the adrenal glands that serve as precursors to the male sex hormone, testosterone; falsely promoted as burning fat, building muscle, and slowing aging.

DHF (dihydrofolate): a coenzyme form of folate.

DIAAS (digestible indispensable amino acid score): a measure of protein quality similar to PDCAAS, except it determines protein digestibility at the end of the small intestine, which more accurately reflects the extent of amino acid absorption.

diabetes (DYE-uh-BEE-teez): metabolic disorders characterized by elevated blood glucose resulting from insufficient insulin, ineffective insulin, or both; the complete medical term is *diabetes mellitus* (meh-LIE-tus).

diarrhea: the frequent passage of watery bowel movements.

diet: the foods and beverages a person eats and drinks.

diet history: a record of eating behaviors and the foods a person eats.

dietary fibers: in plant foods, the *nonstarch polysaccharides* that are not digested by human digestive enzymes, although some are digested by GI tract bacteria.

dietary folate equivalents (DFE): the amount of folate available to the body from naturally occurring sources, fortified foods, and supplements, accounting for differences in the bioavailability from each source. DFE = μg food folate + (1.7 × μg synthetic folate).

Dietary Reference Intakes (DRI): a set of nutrient intake values for healthy people in the United States and Canada. These values are used for planning and assessing diets and include: Estimated Average Requirements (EAR), Recommended Dietary Allowances (RDA), Adequate Intakes (AI), and Tolerable Upper Intake Levels (UL).

dietary supplement: any pill, capsule, tablet, liquid, or powder that contains vitamins, minerals, herbs, or amino acids intended to increase dietary intake of these substances.

dietetic technician: a person who has completed a minimum of an associate's degree from an accredited university or college and an approved dietetic technician program that includes a supervised practice experience. See also *dietetic technician, registered (DTR)*.

dietetic technician, registered (DTR): a dietetic technician who has passed a national examination and maintains registration through continuing professional education.

dietitian: a person trained in nutrition, food science, and diet planning. See also *registered dietitian nutritionist*.

digestion: the process by which food is broken down into absorbable units.

digestive enzymes: proteins found in digestive juices that act on food substances, causing them to break down into simpler compounds.

digestive system: all the organs and glands associated with the ingestion and digestion of food.

dioxins (dye-OCK-sins): a class of chemical pollutants created as byproducts of chemical manufacturing, incineration, chlorine bleaching of paper pulp, and other industrial processes. Dioxins persist in the environment and accumulate in the food chain.

dipeptide (dye-PEP-tide): two amino acids bonded together.

diploma mills: entities without valid accreditation that provide worthless degrees.

direct calorimetry: a means of estimating energy expenditure by measuring the amount of heat released.

disaccharides (dye-SACK-uh-rides): pairs of monosaccharides linked together.

discretionary kcalories: the kcalories remaining in a person's energy allowance after consuming enough nutrient-dense foods to meet all nutrient needs for a day.

disordered eating: eating behaviors that are neither normal nor healthy, including restrained eating, fasting, binge eating, and purging.

dispensable amino acids: nonessential amino acids.

dissociates (dis-SO-see-aites): physically separates.

distilled water: water that has been vaporized and recondensed, leaving it free of dissolved minerals.

diverticula (dye-ver-TIC-you-la): sacs or pouches that develop in the weakened areas of the intestinal wall (like bulges in an inner tube where the tire wall is weak).

diverticulitis (DYE-ver-tic-you-LYE-tis): infected or inflamed diverticula.

diverticulosis (DYE-ver-tic-you-LOH-sis): the condition of having diverticula.

DNA (deoxyribonucleic acid): the double helix molecules of which genes are made.

docosahexaenoic (DOE-cossa-HEXA-ee-NO-ick) **acid (DHA):** an omega-3 polyunsaturated fatty acid with 22 carbons and six double bonds; present in fatty fish and synthesized in limited amounts in the body from linolenic acid.

dolomite: a compound of minerals (calcium magnesium carbonate) found in limestone and marble. Dolomite is powdered and is sold as a calcium-magnesium supplement. However, it may be contaminated with toxic minerals, is not well absorbed, and interferes with absorption of other essential minerals.

double-blind experiment: an experiment in which neither the subjects nor the researchers know which subjects are members of the experimental group and which are serving as control subjects, until after the experiment is over.

Down syndrome: a genetic abnormality that causes mental retardation, short stature, and flattened facial features.

drink: a dose of any alcoholic beverage that delivers ½ ounce of pure ethanol: 5 ounces of wine, 10 ounces of wine cooler, 12 ounces of beer, or 1½ ounces of liquor (80 proof whiskey, scotch, rum, or vodka).

drug: a substance that can modify one or more of the body's functions.

drug history: a record of all the drugs, over-the-counter and prescribed, that a person takes routinely.

DTR: see *dietetic technician, registered*.

duodenum (doo-oh-DEEN-um or doo-ODD-num): the top portion of the small intestine (about "12 fingers' breadth" long in ancient terminology).

dysphagia (dis-FAY-jah): difficulty swallowing.

E

eating disorders: disturbances in eating behavior that jeopardize a person's physical or psychological health.

eating pattern: customary intake of foods and beverages over time.

eclampsia (eh-KLAMP-see-ah): a condition characterized by extremely high blood pressure, elevated protein in the urine, seizures, and possibly coma.

edema (eh-DEEM-uh): the swelling of body tissue caused by excessive amounts of fluid in the interstitial spaces; seen in protein deficiency (among other conditions).

edentulous (ee-DENT-you-lus): lack of teeth.

eicosanoids (eye-COSS-uh-noyds): derivatives of 20-carbon fatty acids; biologically active compounds that help to regulate blood pressure, blood clotting, and other body functions. They include *prostaglandins* (PROS-tah-GLAND-ins), *thromboxanes* (throm-BOX-ains), and *leukotrienes* (LOO-ko-TRY-eens).

eicosapentaenoic (EYE-cossa-PENTA-ee-NO-ick) **acid** (EPA): an omega-3 polyunsaturated fatty acid with 20 carbons and five double bonds; present in fatty fish and synthesized in limited amounts in the body from linolenic acid.

electrolyte solutions: solutions that can conduct electricity.

electrolytes: salts that dissolve in water and dissociate into charged particles called ions.

electron transport chain: the final pathway in energy metabolism that transports electrons from hydrogen to oxygen and captures the energy released in the bonds of ATP; also called the *respiratory chain.*

element: a substance composed of atoms that are alike—for example, iron (Fe).

embolism (EM-boh-lizm): the obstruction of a blood vessel by an *embolus* (EM-boh-luss), or traveling clot, causing sudden tissue death.

embryo (EM-bree-oh): the developing infant from 2 to 8 weeks after conception.

emetic (em-ETT-ic): an agent that causes vomiting.

empty-kcalorie foods: a popular term used to denote foods that contribute energy but lack protein, vitamins, and minerals.

emulsifier (ee-MUL-sih-fire): a substance with both water-soluble and fat-soluble portions that promotes the mixing of oils and fats in a watery solution.

endogenous (en-DODGE-eh-nus): from within the body.

endoplasmic reticulum (en-doh-PLAZ-mic reh-TIC-you-lum): a complex network of intracellular membranes. The *rough endoplasmic reticulum* is dotted with ribosomes, where protein synthesis takes place. The *smooth endoplasmic reticulum* bears no ribosomes.

endosperm: the inner edible part of a kernel of grain, rich in starch and proteins.

enema: solution inserted into the rectum and colon to stimulate a bowel movement and empty the lower large intestine.

energy: the capacity to do work. The energy in food is chemical energy. The body can convert this chemical energy to mechanical, electrical, or heat energy.

energy balance: the energy (kcalories) consumed from foods and beverages compared with the energy expended through metabolic processes and physical activities.

energy density: a measure of the energy a food provides relative to the weight of the food (kcalories per gram).

energy-yielding nutrients: the nutrients that break down to yield energy the body can use (carbohydrate, fat, and protein).

enhanced water: water that is fortified with ingredients such as vitamins, minerals, protein, oxygen, or herbs. Enhanced water is marketed as *vitamin water, sports water, oxygenated water,* and *protein water.*

enriched: the addition to a food of specific nutrients to replace losses that occur during processing so that the food will meet a specified standard.

enteropancreatic (EN-ter-oh-PAN-kree-AT-ik) **circulation:** the circulatory route from the pancreas to the intestine and back to the pancreas.

enzymes: proteins that facilitate chemical reactions without being changed in the process; protein catalysts.

EPA (Environmental Protection Agency): a federal agency that is responsible for, among other things, regulating pesticides and establishing water quality standards.

epidemic (ep-ih-DEM-ick): the appearance of a disease (usually infectious) or condition that attacks many people at the same time in the same region.

epigenetics: the study of heritable changes in gene function that occur without a change in the DNA sequence.

epiglottis (epp-ih-GLOTT-iss): cartilage in the throat that guards the entrance to the trachea and prevents fluid or food from entering it when a person swallows.

epinephrine (EP-ih-NEFF-rin): a hormone of the adrenal gland that modulates the stress response; formerly called *adrenaline*. When administered by injection, epinephrine counteracts anaphylactic shock by opening the airways and maintaining heartbeat and blood pressure.

epithelial (ep-i-THEE-lee-ul) **cells:** cells on the surface of the skin and mucous membranes.

epithelial tissue: the layer of the body that serves as a selective barrier between the body's interior and the environment. Examples are the cornea of the eyes, the skin, the respiratory lining of the lungs, and the lining of the digestive tract.

ergocalciferol (ER-go-kal-SIF-er-ol): vitamin D derived from plants in the diet; also called *vitamin D_2*.

ergogenic (ER-go-JEN-ick) **aids:** substances or techniques used in an attempt to enhance physical performance.

erythrocyte (eh-RITH-ro-cite) **hemolysis** (he-MOLL-uh-sis): the breaking open of red blood cells (erythrocytes); a symptom of vitamin E–deficiency disease in human beings.

erythrocyte protoporphyrin (PRO-toe-PORE-fe-rin): a precursor to hemoglobin.

erythropoietin (eh-RITH-ro-POY-eh-tin): a hormone that stimulates red blood cell production.

esophageal (ee-SOFF-ah-GEE-al) **sphincter:** a sphincter muscle at the upper or lower end of the esophagus. The *lower esophageal sphincter* is also called the *cardiac sphincter* because of its proximity to the heart.

esophagus (ee-SOFF-ah-gus): the food pipe; the conduit from the mouth to the stomach.

essential amino acids: amino acids that the body requires but cannot make, and so must be obtained from the diet; also called *indispensable amino acids.*

essential fatty acids: fatty acids that the body requires but cannot make, and so must be obtained from the diet; both linoleic acid and linolenic acid are essential fatty acids.

essential nutrients: nutrients a person must obtain from food because the body cannot make them for itself in sufficient quantity to meet physiological needs; also called *indispensable nutrients*. About 40 nutrients are currently known to be essential for human beings.

Estimated Average Requirement (EAR): the average daily amount of a nutrient that will maintain a specific biochemical or physiological function in half the healthy people of a given age and gender group.

Estimated Energy Requirement (EER): the average dietary energy intake that maintains energy balance and good health in a person of a given age, gender, weight, height, and level of physical activity.

estrogens: hormones responsible for the menstrual cycle and other female characteristics.

ethanol: a particular type of alcohol found in beer, wine, and liquor; also called *ethyl alcohol.*

ethnic foods: foods associated with particular cultural groups.

excessive drinking: heavy drinking, binge drinking, or both.

exchange lists: diet-planning tools that organize foods by their proportions of carbohydrate, fat, and protein. Foods on any single list can be used interchangeably.

exercise: planned, structured, and repetitive body movements that promote or maintain physical fitness.

exogenous (eks-ODGE-eh-nus): from outside the body.

experimental group: a group of individuals similar in all possible respects to the control group except for the treatment. The experimental group receives the real treatment.

extra lean: less than 5 grams of fat, 2 grams of saturated fat and *trans* fat combined, and 95 milligrams of cholesterol per serving and per 100 grams of meat, poultry, and seafood.

extracellular fluid: fluid outside the cells. Extracellular fluid includes two main components—the interstitial fluid between cells and the intravascular fluid inside blood vessels. Extracellular fluid accounts for approximately one-third of the body's water.

F

FAD (flavin adenine dinucleotide): a coenzyme form of riboflavin.

fad diets: popular eating plans that promise quick weight loss. Most fad diets severely limit certain foods or overemphasize others (for example, never eat potatoes or pasta, or eat cabbage soup daily).

faith healing: healing by invoking divine intervention without the use of medical, surgical, or other traditional therapy.

false negative: a test result indicating that a condition is not present (negative) when in fact it is present (therefore false).

false positive: a test result indicating that a condition is present (positive) when in fact it is not (therefore false).

famine: widespread and extreme scarcity of food in an area that causes starvation and death in a large portion of the population.

FAO (Food and Agriculture Organization): an international agency (part of the United Nations) that has adopted standards to regulate pesticide use among other responsibilities.

fasting plasma glucose: a test that measures plasma glucose after a person has fasted (no food or kcaloric beverages) for at least 8 hours.

fat replacers: ingredients that replace some or all of the functions of fat and may or may not provide energy.

fat-free: less than 0.5 gram of fat per serving (and no added fat or oil); synonyms include *zero-fat, no fat,* and *nonfat.*

fats: lipids that are solid at room temperature (77°F, or 25°C).

fatty acid oxidation: the metabolic breakdown of fatty acids to acetyl CoA; also called *beta oxidation.*

fatty acids: organic compounds composed of a carbon chain with hydrogens attached and an acid group (COOH) at one end and a methyl group (CH_3) at the other end.

fatty liver: an early stage of liver deterioration seen in several diseases, including obesity and alcoholic liver disease. Fatty liver is characterized by an accumulation of fat in the liver cells.

fatty streaks: accumulations of cholesterol and other lipids along the walls of the arteries.

FDA (Food and Drug Administration): the federal agency responsible for ensuring the safety and wholesomeness of all dietary supplements and foods processed and sold in interstate commerce except meat, poultry, and eggs (which are under the jurisdiction of the USDA); inspecting food plants and imported foods; and setting standards for food composition and product labeling.

female athlete triad: a potentially fatal combination of three medical problems—disordered eating, amenorrhea, and osteoporosis.

fermentable: the extent to which bacteria in the GI tract can break down fibers to fragments that the body can use.

ferritin (FAIR-ih-tin): the iron storage protein.

fertility: the capacity of a woman to produce a normal ovum periodically and of a man to produce normal sperm; the ability to reproduce.

fetal alcohol spectrum disorder: a range of physical, behavioral, and cognitive abnormalities caused by prenatal alcohol exposure.

fetal alcohol syndrome (FAS): a cluster of physical, behavioral, and cognitive abnormalities associated with prenatal alcohol exposure, including facial malformations, growth retardation, and central nervous disorders.

fetal programming: the influence of substances during fetal growth on the development of diseases in later life.

fetus (FEET-us): the developing infant from 8 weeks after conception until term.

fibrocystic (FYE-bro-SIS-tik) **breast disease:** a harmless condition in which the breasts develop lumps, sometimes associated with caffeine consumption. In some, it responds to abstinence from caffeine; in others, it can be treated with vitamin E.

fibrosis (fye-BROH-sis): an intermediate stage of liver deterioration seen in several diseases, including viral hepatitis and alcoholic liver disease. In fibrosis, the liver cells lose their function and assume the characteristics of connective tissue cells (fibers).

filtered water: water treated by filtration, usually through *activated carbon filters* that reduce the lead in tap water, or by *reverse osmosis* units that force pressurized water across a membrane removing lead, arsenic, and some microorganisms from tap water.

fitness: the characteristics that enable the body to perform physical activity; more broadly, the ability to meet routine physical demands with enough reserve energy to rise to a physical challenge; or the body's ability to withstand stress of all kinds.

flatulence: passage of excessive amounts of intestinal gas.

flavonoids (FLAY-von-oyds): yellow pigments in foods; phytochemicals that may exert physiological effects on the body.

flaxseeds: the small brown seeds of the flax plant; valued in nutrition as a source of fiber, lignans, and omega-3 fatty acids.

flexibility: the capacity of the joints to move through a full range of motion; the ability to bend and recover without injury.

fluid balance: maintenance of the proper types and amounts of fluid in each compartment of the body fluids.

fluorapatite (floor-APP-uh-tite): the stabilized form of tooth crystal, in which fluoride has replaced the hydroxyl groups of hydroxyapatite.

fluoride: an essential trace mineral that makes teeth stronger and more resistant to decay.

fluorosis (floor-OH-sis): discoloration and pitting of tooth enamel caused by excess fluoride during tooth development.

FMN (flavin mononucleotide): a coenzyme form of riboflavin.

folate (FOLE-ate): a B vitamin; also known as folic acid, folacin, or pteroylglutamic (tare-o-EEL-glue-TAM-ick) acid (PGA). The coenzyme forms are *DHF (dihydrofolate)* and *THF (tetrahydrofolate).*

follicle-stimulating hormone (FSH): a hormone that stimulates maturation of the ovarian follicles in females and the production of sperm in males. (The ovarian follicles are part of the female reproductive system where the eggs are produced.) The release of FSH is mediated by *follicle-stimulating hormone releasing hormone (FSH–RH).*

food allergy: an adverse reaction to food that involves an immune response; also called *food-hypersensitivity reaction.*

food and nutrition history: a record of eating behaviors and the foods a person eats.

food aversions: strong desires to avoid particular foods.

food banks: facilities that collect and distribute food donations to authorized organizations feeding the hungry.

food chain: the sequence in which living things depend on other living things for food.

food cravings: strong desires to eat particular foods.

food crisis: a sharp rise in the rates of hunger and malnutrition, usually set off by a shock to either the supply of, or demand for, food and a sudden spike in food prices.

food deserts: neighborhoods and communities characterized by limited access to nutritious and affordable foods.

food frequency questionnaire: a checklist of foods on which a person can record the frequency with which he or she eats each food.

food group plans: diet-planning tools that sort foods into groups based on nutrient content and then specify that people should eat certain amounts of foods from each group.

food insecurity: limited or uncertain access to foods of sufficient quality or quantity to sustain a healthy and active life.

food insufficiency: an inadequate amount of food due to a lack of resources.

food intolerances: adverse reactions to foods that do not involve the immune system.

food poverty: hunger resulting from inadequate access to available food for various reasons, including inadequate resources, political obstacles, social disruptions, poor weather conditions, and lack of transportation.

food record: an extensive, accurate log of all foods eaten over a period of several days or weeks. A food record that includes associated information such as when, where, and with whom each food is eaten is sometimes called a *food diary*.

food recovery: collecting wholesome food for distribution to low-income people who are hungry.

food security: access to enough food to sustain a healthy and active life.

food substitutes: foods that are designed to replace other foods.

foodborne illness: an illness transmitted to human beings through food and water, caused by either an infectious agent (foodborne infection) or a poisonous substance (food intoxication); commonly known as *food poisoning*.

foods: products derived from plants or animals that can be taken into the body to yield energy and nutrients for the maintenance of life and the growth and repair of tissues.

fortified: the addition to a food of nutrients that were either not originally present or present in insignificant amounts. Fortification can be used to correct or prevent a widespread nutrient deficiency or to balance the total nutrient profile of a food.

fossil fuels: coal, oil, and natural gas.

fraudulent: the promotion, for financial gain, of devices, treatments, services, plans, or products (including diets and supplements) that alter or claim to alter a human condition without proof of safety or effectiveness.

free: "nutritionally trivial" and unlikely to have a physiological consequence; synonyms include *without*, *no*, and *zero*. A food that does not contain a nutrient naturally may make such a claim, but only as it applies to all similar foods (for example, "applesauce, a fat-free food").

free radical: an unstable molecule with one or more unpaired electrons.

fructose (FRUK-tose or FROOK-tose): a monosaccharide; sometimes known as *fruit sugar* or *levulose*. Fructose is found abundantly in fruits, honey, and saps.

fuel: compounds that cells can use for energy. The major fuels include glucose, fatty acids, and amino acids; other fuels include ketone bodies, lactate, glycerol, and alcohol.

full term: births occurring at 39 through 40 weeks of gestation.

functional foods: foods that have a potentially beneficial effect on health when consumed as part of a varied diet on a regular basis at effective levels.

G

g: grams; a unit of weight equivalent to about 0.03 ounces.

galactose (ga-LAK-tose): a monosaccharide; part of the disaccharide lactose.

gallbladder: the organ that stores and concentrates bile. When it receives the signal that fat is present in the duodenum, the gallbladder contracts and squirts bile through the bile duct into the duodenum.

gastric glands: exocrine glands in the stomach wall that secrete gastric juice into the stomach.

gastric juice: the digestive secretion of the gastric glands of the stomach.

gastrin: a hormone secreted by cells in the stomach wall. Target organ: the glands of the stomach. Response: secretion of gastric acid.

gastroesophageal reflux: the backflow of stomach acid into the esophagus, causing damage to the cells of the esophagus and the sensation of heartburn; commonly known as *heartburn* or *acid indigestion*.

gastroesophageal reflux disease (GERD): a condition characterized by symptoms of reflux occurring two or more times a week.

gastrointestinal (GI) tract: the digestive tract. The principal organs are the stomach and intestines.

gatekeepers: with respect to nutrition, key people who control other people's access to foods and thereby exert profound impacts on their nutrition. Examples are the spouse who buys and cooks the food, the parent who feeds the children, and the caregiver in a day-care center.

gene expression: the process by which a cell converts the genetic code into RNA and protein.

gene pool: all the genetic information of a population at a given time.

generally recognized as safe (GRAS): food additives that have long been in use and are believed to be safe. First established by the FDA in 1958, the GRAS list is subject to revision as new facts become known.

genes: sections of chromosomes that contain the instructions needed to make one or more proteins.

genetic engineering: the use of biotechnology to modify the genetic material of living cells so that they will produce new substances or perform new functions. Foods produced via this technology are called *genetically modified (GM)* or *genetically engineered (GE) foods*.

genetics: the study of genes and inheritance.

genome (GEE-nome): the complete set of genetic material (DNA) in an organism or a cell. The study of genomes is called *genomics*.

genomics: the study of all the genes in an organism and their interactions with environmental factors.

genotoxicant: a substance that mutates or damages genetic material.

geophagia: the specific craving for nonfood items such as clay, baby powder, chalk, ash, ceramics, paper, paint chips, charcoal, or dirt.

germ: the seed that grows into a mature plant, especially rich in vitamins and minerals.

gestation (jes-TAY-shun): the period from conception to birth. For human beings, the average length of a healthy gestation is 40 weeks. Pregnancy is often divided into 3-month periods, called *trimesters*.

gestational diabetes: glucose intolerance with onset or first recognition during pregnancy.

gestational hypertension: high blood pressure that develops in the second half of pregnancy and resolves after childbirth, usually without affecting the outcome of the pregnancy.

ghrelin (GRELL-in): a protein produced by the stomach cells that enhances appetite and decreases energy expenditure.

glands: cells or groups of cells that secrete materials for special uses in the body. Glands may be *exocrine* (EKS-oh-crin) *glands*, secreting their materials "out" (into the digestive tract or onto the surface of the skin), or *endocrine* (EN-doe-crin) *glands*, secreting their materials "in" (into the blood).

glossitis (gloss-EYE-tis): an inflammation of the tongue.

glucagon (GLOO-ka-gon): a hormone secreted by special cells in the pancreas in response to low blood glucose concentration. Glucagon elicits release of glucose from liver glycogen stores.

glucocorticoids: hormones from the adrenal cortex that affect the body's management of glucose.

glucogenic amino acids: amino acids that can make glucose via either pyruvate or TCA cycle intermediates.

gluconeogenesis (gloo-ko-nee-oh-JEN-ih-sis): the making of glucose from a noncarbohydrate source such as amino acids or glycerol.

glucose (GLOO-kose): a monosaccharide; sometimes known as *blood sugar* in the body or *dextrose* in foods.

glucose polymers: compounds that supply glucose, not as single molecules, but linked in chains somewhat like starch. The objective is to attract less water from the body into the digestive tract (osmotic attraction depends on the number, not the size, of particles).

glucose tolerance factors (GTF): small organic compounds that enhance insulin's actions.

gluten-free: a food that contains less that 20 parts per million of gluten from any source; synonyms include *no gluten, free of gluten,* or *without gluten.*

glycemic (gly-SEEM-ic) **index:** a method of classifying foods according to their potential for raising blood glucose.

glycemic response: the extent to which a food raises the blood glucose concentration and elicits an insulin response.

glycerol (GLISS-er-ol): an alcohol composed of a three-carbon chain, which can serve as the backbone for a triglyceride.

glycobiology: the study of sugars and their derivatives.

glycogen (GLY-ko-jen): an animal polysaccharide composed of glucose; a storage form of glucose manufactured and stored in the liver and muscles. Glycogen is not a significant food source of carbohydrate and is not counted as a dietary carbohydrate in foods.

glycolipids: sugars attached to lipids in a cell's membrane.

glycolysis (gly-COLL-ih-sis): the metabolic breakdown of glucose to pyruvate. Glycolysis does not require oxygen (anaerobic).

glycoproteins: sugars attached to a protein.

goblet cells: cells of the GI tract (and lungs) that secrete mucus.

goiter (GOY-ter): an enlargement of the thyroid gland due to an iodine deficiency, malfunction of the gland, or overconsumption of a goitrogen. Goiter caused by iodine deficiency is sometimes called *simple goiter.*

goitrogen (GOY-troh-jen): a substance that enlarges the thyroid gland and causes *toxic goiter.* Goitrogens occur naturally in such foods as cabbage, kale, brussels sprouts, cauliflower, broccoli, and kohlrabi.

Golgi (GOAL-gee) **apparatus:** a set of membranes within the cell where secretory materials are packaged for export.

good source of: the product provides between 10 and 19 percent of the Daily Value for a given nutrient per serving.

gout (GOWT): a common form of arthritis characterized by deposits of uric acid crystals in the joints.

growth hormone (GH): a hormone secreted by the pituitary that regulates the cell division and protein synthesis needed for normal growth; also called *somatotropin.* The release of GH is mediated by *GH-releasing hormone (GHRH)* and *GH-inhibiting hormone.*

H

hard water: water with a high calcium and magnesium content.

hazard: a source of danger; used to refer to circumstances in which harm is possible under normal conditions of use.

Hazard Analysis Critical Control Points (HACCP): a systematic plan to identify and correct potential microbial hazards in the manufacturing, distribution, and commercial use of food products; commonly referred to as "HASS-ip."

HDL (high-density lipoprotein): the type of lipoprotein that transports cholesterol back to the liver from the cells; composed primarily of protein.

health claims: statements that characterize the relationship between a nutrient or other substance in a food and a disease or health-related condition.

health history: an account of a client's current and past health status and disease risks.

healthy: on food labels, a food that is low in fat, saturated fat, cholesterol, and sodium and that contains at least 10 percent of the Daily Values for vitamin D, potassium, iron, calcium, protein, or fiber.

Healthy Eating Index: a measure that assesses how well a diet meets the recommendations of the *Dietary Guidelines for Americans.*

Healthy People: a national public health initiative under the jurisdiction of the US Department of Health and Human Services (DHHS) that identifies the most significant preventable threats to health and focuses efforts toward eliminating them.

heart attack: sudden tissue death caused by blockages of vessels that feed the heart muscle; also called *myocardial* (my-oh-KAR-dee-al) *infarction* (in-FARK-shun) or *cardiac arrest.*

heartburn: a burning sensation in the chest area caused by backflow of stomach acid into the esophagus; medically known as *gastroesophageal reflux.*

heat stroke: a dangerous accumulation of body heat with accompanying loss of body fluid.

heavy drinking: more than three drinks on any day for women and more than four drinks on any day for men.

heavy metals: mineral ions such as mercury and lead, so called because of their relatively high atomic weight; many heavy metals are poisonous.

Heimlich (HIME-lick) **maneuver (abdominal thrusts):** a technique for dislodging an object from the trachea of a choking person; named for the physician who developed it.

hematocrit (hee-MAT-oh-krit): the percentage of total blood volume that consists of red blood cells.

heme (HEEM) **iron:** the iron in foods that is bound to the hemoglobin and myoglobin proteins; found only in meat, fish, and poultry.

hemochromatosis (HE-moh-KRO-ma-toe-sis): a genetically determined failure to prevent absorption of unneeded dietary iron that is characterized by iron overload and tissue damage.

hemoglobin (HE-moh-GLO-bin): the globular protein of the red blood cells that transports oxygen from the lungs to tissues throughout the body; hemoglobin accounts for 80 percent of the body's iron.

hemolytic (HE-moh-LIT-ick) **anemia:** the condition of having too few red blood cells as a result of erythrocyte hemolysis.

hemophilia (HE-moh-FEEL-ee-ah): a hereditary disease in which the blood is unable to clot because it lacks the ability to synthesize certain clotting factors.

hemorrhagic (hem-oh-RAJ-ik) **disease:** a disease characterized by excessive bleeding.

hemorrhoids (HEM-oh-royds): painful swelling of the veins surrounding the rectum.

hemosiderin (heem-oh-SID-er-in): an iron-storage protein primarily made in times of iron overload.

hepatic portal vein: the vein that collects blood from the GI tract and conducts it to the liver.

hepatic vein: the vein that collects blood from the liver and returns it to the heart.

hepcidin: a hormone produced by the liver that regulates iron balance.

herbal (ERB-al) **medicine:** the use of plants to treat disease or improve health; also known as *botanical medicine* or *phytotherapy*.

hGH (human growth hormone): a hormone produced by the brain's pituitary gland that regulates normal growth and development; also called *somatotropin*.

high: on food labels, 20 percent or more of the Daily Value for a given nutrient per serving; synonyms include *rich in* or *excellent source*.

high fiber: 5 grams or more of fiber per serving. A high-fiber claim made on a food that contains more than 3 grams of fat per serving and per 100 grams of food must also declare total fat.

high food security: no indications of food-access problems or limitations.

high potency: 100% or more of the Daily Value for the nutrient in a single supplement and for at least two-thirds of the nutrients in a multinutrient supplement.

high-fructose corn syrup (HFCS): a syrup made from cornstarch that has been treated with an enzyme that converts some of the glucose to the sweeter fructose; made especially for use in processed foods and beverages, where it is the predominant sweetener. With a chemical structure similar to sucrose, HFCS has a fructose content of 42 or 55 percent, with glucose making up the remainder.

high-quality proteins: dietary proteins containing all the essential amino acids in relatively the same amounts that human beings require. They may also contain nonessential amino acids.

high-risk pregnancy: a pregnancy characterized by indicators that make it likely the birth will be surrounded by problems such as premature delivery, difficult birth, restricted growth, birth defects, and early infant death.

histamine (HISS-tah-mean or HISS-tah-men): a substance produced by cells of the immune system as part of a local immune reaction to an antigen.

HIV (human immunodeficiency virus): the virus that destroys lymphocytes and impairs immunity, eventually causing AIDS.

HMB (beta-hydroxybetamethylbutyrate): a metabolite of the branched-chain amino acid leucine promoted to increase muscle mass and strength.

homeopathy (hoh-me-OP-ah-thee): a practice based on the theory that "like cures like," that is, that substances that cause symptoms in healthy people can cure those symptoms when given in very dilute amounts.

homeostasis (HOME-ee-oh-STAY-sis): the maintenance of constant internal conditions (such as blood chemistry, temperature, and blood pressure) by the body's control systems. A homeostatic system is constantly reacting to external forces to maintain limits set by the body's needs.

honey: sugar (mostly sucrose) formed from nectar gathered by bees. Composition and flavor vary, but honey always contains a mixture of sucrose, fructose, and glucose.

hormones: chemical messengers. Hormones are secreted by a variety of glands in response to altered conditions in the body. Each hormone travels to one or more specific target tissues or organs, where it elicits a specific response to maintain homeostasis. The study of hormones and their actions is called *endocrinology*.

hormone-sensitive lipase: an enzyme inside adipose cells that responds to the body's need for fuel by hydrolyzing triglycerides so that their parts (glycerol and fatty acids) enter the general circulation and thus become available to other cells for fuel. The signals to which this enzyme responds include epinephrine and glucagon, which oppose insulin.

hourly sweat rate: the amount of weight lost plus fluid consumed during exercise per hour. One pound equals roughly 2 cups (500 milliliters) of fluid.

human carrying capacity: the maximum number of people the earth can support over time.

human genome (GEE-nome): the complete set of genetic material (DNA) in a human being.

human microbiome: the collection of microbes found in or on the human body.

hunger: the painful sensation caused by a lack of food that initiates food-seeking behavior; a consequence of food insecurity that, because of prolonged, involuntary lack of food, results in discomfort, illness, weakness, or pain that goes beyond the usual uneasy sensation.

husk: the outer inedible part of a grain; also called the *chaff*.

hydrochloric acid: an acid composed of hydrogen and chloride atoms (HCl) that is normally produced by the gastric glands.

hydrogenation (HIGH-dro-jen-AY-shun or high-DROJ-eh-NAY-shun): a chemical process by which hydrogens are added to monounsaturated or polyunsaturated fatty acids to reduce the number of double bonds, making the fats more saturated (solid) and more resistant to oxidation (protecting against rancidity). Hydrogenation produces *trans*-fatty acids.

hydrolysis (high-DROL-ih-sis): a chemical reaction in which one molecule is split into two molecules, with hydrogen (H) added to one and a hydroxyl group (OH) added to the other (from water, H_2O). The noun is *hydrolysis*; the verb is *hydrolyze*.

hydrophilic (high-dro-FIL-ick): a term referring to water-loving, or water-soluble, substances.

hydrophobic (high-dro-FOE-bick): a term referring to water-fearing, or non-water-soluble, substances; also known as *lipophilic* (fat loving).

hydrotherapy: the use of water (in whirlpools, as douches, or packed as ice, for example) to promote relaxation and healing.

hydroxyapatite (high-drox-ee-APP-ah-tite): crystals made of calcium and phosphorus.

hyperactivity: inattentive and impulsive behavior that is more frequent and severe than is typical of others a similar age; professionally called *attention-deficit/hyperactivity disorder (ADHD)*.

hypercalcemia: high blood calcium that may develop from a variety of disorders, including vitamin D toxicity. It does *not* develop from a high calcium intake.

hyperglycemia: elevated blood glucose concentrations.

hyperplastic obesity: obesity due to an increase in the *number* of fat cells.

hypertension: consistently higher-than-normal blood pressure. Hypertension that develops without an identifiable cause is known as *essential* or *primary hypertension*; hypertension that is caused by a specific disorder such as kidney disease is known as *secondary hypertension*.

hyperthermia: an above-normal body temperature.

hypertrophic obesity: obesity due to an increase in the *size* of fat cells.

hypertrophy (high-PER-tro-fee): growing larger; with regard to muscles, an increase in size (and strength) in response to use.

hypnotherapy: a technique that uses hypnosis and the power of suggestion to improve health behaviors, relieve pain, and heal.

hypoglycemia (HIGH-po-gly-SEE-me-ah): an abnormally low blood glucose concentration.

hyponatremia (HIGH-poe-na-TREE-mee-ah): a decreased concentration of sodium in the blood.

hypothalamus (high-po-THAL-ah-mus): a brain center that controls activities such as maintenance of water balance, regulation of body temperature, and control of appetite.

hypothermia: a below-normal body temperature.

hypothesis (hi-POTH-eh-sis): an unproven statement that tentatively explains the relationships between two or more variables.

hypothyroidism: underactivity of the thyroid gland that may be caused by iodine deficiency or any number of other causes.

I

ileocecal (ill-ee-oh-SEEK-ul) valve: the sphincter separating the small and large intestines.

ileum (ILL-ee-um): the last segment of the small intestine.

imagery: a technique that guides clients to achieve a desired physical, emotional, or spiritual state by visualizing themselves in that state.

imitation foods: foods that substitute for and resemble another food, but are nutritionally inferior to it with respect to vitamin, mineral, or protein content. If the substitute is not inferior to the food it resembles and if its name provides an accurate description of the product, it need not be labeled "imitation."

immune response: the body's reaction to foreign antigens, which neutralizes or eliminates them, thus preventing damage.

immune system: the body's natural defense against foreign materials that have penetrated the skin or mucous membranes.

immunity: the body's ability to defend itself against diseases.

immunoglobulins (IM-you-noh-GLOB-you-linz): proteins capable of acting as antibodies.

implantation (IM-plan-TAY-shun): the embedding of the blastocyst in the inner lining of the uterus.

indigestion: incomplete or uncomfortable digestion, usually accompanied by pain, nausea, vomiting, heartburn, intestinal gas, or belching.

indirect calorimetry: a means of estimating energy expenditure by measuring the amount of oxygen consumed.

indirect or incidental additives: substances that can get into food as a result of contact during growing, processing, packaging, storing, cooking, or some other stage before the foods are consumed; sometimes called *accidental additives*.

indispensable amino acids: essential amino acids.

indispensable nutrients: nutrients a person must obtain from food because the body cannot synthesize them in amounts sufficient to meet physiological needs; also called *essential nutrients*.

infectious diseases: diseases caused by bacteria, viruses, parasites, or other microorganisms that can be transmitted from one person to another through air, water, or food; by contact; or through vector organisms such as mosquitoes.

inflammation: an immunological response to cellular injury characterized by an increase in white blood cells.

initiators: factors that cause mutations that give rise to cancer, such as radiation and carcinogens.

inorganic: not containing carbon or pertaining to living organisms. The two classes of nutrients that are inorganic are minerals and water.

inositol (in-OSS-ih-tall): a nonessential nutrient that can be made in the body from glucose. Inositol is a part of cell membrane structures.

insoluble fibers: nonstarch polysaccharides that do not dissolve in water. Examples include the tough, fibrous structures found in the strings of celery and the skins of corn kernels.

insulin (IN-suh-lin): a hormone secreted by special cells in the pancreas in response to (among other things) elevated blood glucose concentration. Insulin controls the transport of glucose from the bloodstream into the muscle and fat cells.

insulin resistance: the condition in which a normal amount of insulin produces a subnormal effect in muscle, adipose, and liver cells, resulting in an elevated fasting glucose; a metabolic consequence of obesity that precedes type 2 diabetes.

integrative medicine: care that combines conventional and complementary therapies for which there is some high-quality scientific evidence of safety and effectiveness. Integrative medicine emphasizes the importance of the relationship between the practitioner and the patient and focuses on wellness, healing, and the whole person.

intentional food additives: additives intentionally added to foods, such as nutrients, colors, and preservatives.

intermittent claudication (klaw-dih-KAY-shun): severe calf pain caused by inadequate blood supply. It occurs when walking and subsides during rest.

Internet (the Net): a worldwide network of millions of computers linked together to share information.

interstitial (IN-ter-STISH-al) fluid: fluid between the cells (intercellular), usually high in sodium and chloride. Interstitial fluid is a large component of extracellular fluid.

intestinal ischemia (is-KEY-me-ah): a diminished blood flow to the intestines that is characterized by abdominal pain, forceful bowel movements, and blood in the stool.

intra-abdominal fat: fat stored within the abdominal cavity in association with the internal abdominal organs, as opposed to the fat stored directly under the skin (subcutaneous fat); also called *visceral fat*.

intracellular fluid: fluid within the cells, usually high in potassium and phosphate. Intracellular fluid accounts for approximately two-thirds of the body's water.

intravascular fluid: fluid within blood vessels.

intrinsic factor: a glycoprotein (a protein with short polysaccharide chains attached) secreted by the stomach cells that binds with vitamin B_{12} in the small intestine to aid in the absorption of vitamin B_{12}.

invert sugar: a mixture of glucose and fructose formed by the hydrolysis of sucrose in a chemical process; sold only in liquid form and sweeter than sucrose. Invert sugar is used as a food additive to help preserve freshness and prevent shrinkage.

iodide: the ion form of iodine.

iodine: an essential trace mineral that is needed for the synthesis of thyroid hormones.

ions (EYE-uns): atoms or molecules that have gained or lost electrons and therefore have electrical charges. Examples include the positively charged sodium ion (Na^+) and the negatively charged chloride ion (Cl^-).

iridology: the study of changes in the iris of the eye and their relationships to disease.

iron: an essential trace mineral that is needed for the transport of oxygen and the metabolism of energy nutrients.

iron deficiency: the state of having depleted iron stores.

iron-deficiency anemia: severe depletion of iron stores that results in low hemoglobin and small, pale red blood cells. Iron-deficiency anemia is a *microcytic* (my-cro-SIT-ic) *hypochromic* (high-po-KROME-ic) *anemia*.

iron overload: toxicity from excess iron.

irradiation: sterilizing a food by exposure to energy waves, similar to ultraviolet light and microwaves; sometimes called *ionizing radiation*.

irritable bowel syndrome: an intestinal disorder of unknown cause. Symptoms include abdominal discomfort and cramping, diarrhea, constipation, or alternating diarrhea and constipation.

IU: international units; an old measure of vitamin activity determined by biological methods (as opposed to new measures that are determined by direct chemical analyses). Many fortified foods and supplements use IU on their labels.

J

jejunum (je-JOON-um): the first two-fifths of the small intestine beyond the duodenum.

joule: a measure of *work* energy; the amount of energy expended when 1 kilogram is moved 1 meter by a force of 1 newton.

K

kcal: abbreviation of kcalories; a unit by which energy is measured.

kcalorie: a unit by which energy is measured. One kcalorie is the amount of heat necessary to raise the temperature of 1 kilogram (kg) of water 1°C. The scientific use of the term *kcalorie* is the same as the popular use of the term *calorie*.

kcalorie (energy) control: management of food energy intake.

kcalorie-free: fewer than 5 kcalories per serving.

kefir (keh-FUR): a fermented milk created by adding *Lactobacillus acidophilus* and other bacteria that break down lactose to glucose and galactose, producing a sweet, lactose-free product.

keratin (KARE-uh-tin): a water-insoluble protein; the normal protein of hair and nails.

keratinization: accumulation of keratin in a tissue; a sign of vitamin A deficiency.

keratomalacia (KARE-ah-toe-ma-LAY-shuh): softening of the cornea that leads to irreversible blindness; a sign of severe vitamin A deficiency.

Keshan (KESH-an or ka-SHAWN) **disease:** the heart disease associated with selenium deficiency; named for one of the provinces of China where it was first studied. Keshan disease is characterized by heart enlargement and insufficiency; fibrous tissue replaces the muscle tissue that normally composes the middle layer of the walls of the heart.

keto (KEY-toe) **acid:** an organic acid that contains a carbonyl group (C=O).

ketogenic amino acids: amino acids that are degraded to acetyl CoA.

ketone (KEE-tone) **bodies:** acidic compounds produced by the liver during the incomplete breakdown of fat when carbohydrate is not available.

ketosis (kee-TOE-sis): an undesirably high concentration of ketone bodies in the blood and urine.

Krebs cycle: named after the scientist who elucidated this biochemistry, a series of metabolic reactions that break down molecules of acetyl CoA to carbon dioxide and hydrogen atoms; also called the *citric acid cycle* or the *TCA cycle*.

kwashiorkor (kwash-ee-OR-core or kwash-ee-or-CORE): severe malnutrition characterized by failure to grow and develop, edema, changes in the pigmentation of hair and skin, fatty liver, anemia, and apathy.

L

lactase: an enzyme that hydrolyzes lactose.

lactase deficiency: a lack of the enzyme required to digest the disaccharide lactose into its component monosaccharides (glucose and galactose).

lactate: a 3-carbon compound produced from pyruvate during anaerobic metabolism.

lactation: production and secretion of breast milk for the purpose of nourishing an infant.

lacteals (LACK-tee-als): the lymphatic vessels of the intestine that take up nutrients and pass them to the lymph circulation.

lacto-ovo-vegetarian diet: an eating pattern that includes milk, milk products, and eggs, but excludes meat, poultry, and seafood from the diet.

lactose (LAK-tose): a disaccharide composed of glucose and galactose; commonly known as *milk sugar*.

lactose intolerance: a condition that results from the inability to digest the milk sugar lactose; characterized by bloating, gas, abdominal discomfort, and diarrhea. Lactose intolerance differs from milk allergy, which is caused by an immune reaction to the protein in milk.

lactovegetarian diet: an eating pattern that includes milk and milk products, but excludes meat, poultry, seafood, and eggs from the diet.

large intestine or colon (COAL-un): the lower portion of intestine that completes the digestive process. Its segments are the *ascending colon*, the *transverse colon*, the *descending colon*, and the *sigmoid colon*.

larynx (LAIR-inks): the entryway to the trachea that contains the vocal cords; also called the *voice box*.

laxatives: substances that loosen the bowels and thereby prevent or treat constipation.

LDL (low-density lipoprotein): the type of lipoprotein derived from very-low-density lipoproteins (VLDL) as triglycerides are removed and broken down; composed primarily of cholesterol.

lean: less than 10 grams of fat, 4.5 grams of saturated fat and *trans* fat combined, and 95 milligrams of cholesterol per serving and per 100 grams of meat, poultry, and seafood. For mixed dishes such as burritos and sandwiches, less than 8 grams of fat, 3.5 grams of saturated fat, and 80 milligrams of cholesterol per reference amount customarily consumed.

lean body mass: the body minus its fat.

lecithin (LESS-uh-thin): one of the phospholipids. Both nature and the food industry use lecithin as an emulsifier to combine water-soluble and fat-soluble ingredients that do not ordinarily mix, such as water and oil.

legumes (lay-GYOOMS or LEG-yooms): plants of the bean and pea family, with seeds that are rich in protein compared with other plant-derived foods.

leptin: a protein produced by fat cells under direction of the *ob* gene that decreases appetite and increases energy expenditure.

less: on food labels, at least 25 percent less of a given nutrient or kcalories than the comparison food; synonyms include *fewer* and *reduced*.

less cholesterol: 25 percent or less cholesterol than the comparison food (reflecting a reduction of at least 20 milligrams per serving), and 2 grams or less of saturated fat and *trans* fat combined per serving.

less fat: 25 percent or less fat than the comparison food.

less saturated fat: 25 percent or less saturated fat and *trans* fat combined than the comparison food.

let-down reflex: the reflex that forces milk to the front of the breast when the infant begins to nurse.

levulose: an older name for fructose.

license to practice: permission under state or federal law, granted on meeting specified criteria, to use a certain title (such as dietitian) and offer certain services. *Licensed dietitians* may use the initials *LD* after their names.

life expectancy: the average number of years lived by people in a given society.

life span: the maximum number of years of life attainable by a member of a species.

light or lite: one-third fewer kcalories than the comparison food; 50 percent or less of the fat or sodium than the comparison food; any use of the term other than as defined must specify what it is referring to (for example, "light in color" or "light in texture").

lignans: phytochemicals present in flaxseed that are converted to phytosterols by intestinal bacteria and are under study as possible anticancer agents.

limiting amino acid: the essential amino acid found in the shortest supply relative to the amounts needed for protein synthesis in the body. Four amino acids are most likely to be limiting: lysine, methionine, threonine, and tryptophan.

lingual: pertaining to the tongue.

linoleic (lin-oh-LAY-ick) **acid:** an essential fatty acid with 18 carbons and two double bonds.

linolenic (lin-oh-LEN-ick) **acid:** an essential fatty acid with 18 carbons and three double bonds.

lipases (LYE-pasez): enzymes that hydrolyze lipids (fats). *Lingual lipase* is a fat-digesting enzyme secreted from the salivary gland at the base of the tongue; *gastric lipase* is a fat-digesting enzyme secreted from the cells of the stomach.

lipids: a family of compounds that includes triglycerides, phospholipids, and sterols. Lipids are characterized by their insolubility in water. (Lipids also include the fat-soluble vitamins.)

lipoprotein-associated phospholipase A(2) or **Lp-PLA(2):** a lipoprotein-bound enzyme that generates potent proinflammatory and proatherogenic products such as oxidized free fatty acids and lysophosphatidylcholine. Lp-PLA(2) is a specific marker of plaque inflammation.

lipoprotein lipase (LPL): an enzyme that hydrolyzes triglycerides passing by in the bloodstream and directs their parts into the cells, where they can be metabolized for energy or reassembled for storage.

lipoproteins (LIP-oh-PRO-teenz): clusters of lipids associated with proteins that serve as transport vehicles for lipids in the lymph and blood.

lipotoxicity: the adverse effects of fat in nonadipose tissues.

liquor or **distilled spirits:** an alcoholic beverage traditionally made by fermenting and distilling a carbohydrate source such as molasses, potatoes, rye, beets, barley, or corn.

Listeriosis (lis-TEAR-ee-OH-sis): an infection caused by eating food contaminated with the bacterium *Listeria monocytogenes*, which can be killed by pasteurization and cooking but can survive at refrigerated temperatures; certain ready-to-eat foods, such as hot dogs and deli meats, may become contaminated after cooking or processing, but before packaging.

liver: the organ that manufactures bile, among many other functions.

longevity: long duration of life.

low: on food labels, an amount that would allow frequent consumption of a food without exceeding the Daily Value for the nutrient. A food that is naturally low in a nutrient may make such a claim, but only as it applies to all similar foods (for example, "fresh cauliflower, a low-sodium food"); synonyms include *little, few,* and *low source of.*

low birthweight (LBW): a birthweight of 5½ pounds (2500 grams) or less; indicates probable poor health in the newborn and poor nutrition status in the mother during pregnancy, before pregnancy, or both. Optimal birthweight for a full-term baby is 6½ to 8 pounds.

low cholesterol: 20 milligrams or less of cholesterol per serving and 2 grams or less of saturated fat and *trans* fat combined per serving.

low fat: 3 grams or less of fat per serving.

low food security: reduced quality of life with little or no indication of reduced food intake; formerly known as *food insecurity without hunger.*

low kcalorie: 40 kcalories or less per serving.

low saturated fat: 1 gram or less of saturated fat and less than 0.5 gram of *trans* fat per serving.

low sodium: 140 milligrams or less per serving.

low-risk pregnancy: a pregnancy characterized by factors that make it likely the birth will be normal and the infant healthy.

lumen (LOO-men): the space within a vessel such as the intestine.

lutein (LOO-teen): a plant pigment of yellow hue; a phytochemical believed to play roles in eye functioning and health.

luteinizing (LOO-tee-in-EYE-zing) **hormone (LH):** a hormone that stimulates ovulation and the development of the corpus luteum (the small tissue that develops from a ruptured ovarian follicle and secretes hormones); so called because the follicle turns yellow as it matures. In men, LH stimulates testosterone secretion. The release of LH is mediated by *luteinizing hormone–releasing hormone (LH–RH).*

lycopene (LYE-koh-peen): a pigment responsible for the red color of tomatoes and other red-hued vegetables; a phytochemical that may act as an antioxidant in the body.

lymph (LIMF): a clear yellowish fluid that is similar to blood except that it contains no red blood cells or platelets. Lymph from the GI tract transports fat and fat-soluble vitamins to the bloodstream via lymphatic vessels.

lymphatic (lim-FAT-ic) **system:** a loosely organized system of vessels and ducts that convey fluids toward the heart. The GI part of the lymphatic system carries the products of fat digestion into the bloodstream.

lymphocytes (LIM-foh-sites): white blood cells that participate in acquired immunity; B cells and T cells.

lysosomes (LYE-so-zomes): cellular organelles; membrane-enclosed sacs of degradative enzymes.

M

macrobiotic diet: a philosophical eating pattern based on mostly plant foods such as whole grains, legumes, and vegetables, with small amounts of fish, fruits, nuts, and seeds.

macrocytic: abnormally large blood cells.

macronutrients: carbohydrate, fat, and protein; the nutrients the body requires in relatively large amounts (many grams daily).

macrophages (MAK-roe-fay-jez): large phagocytic cells that serve as scavengers of the blood, clearing it of old or abnormal cells, cellular debris, and antigens.

macrosomia (mak-roh-SO-me-ah): abnormally large body size. In the case of infants, a birthweight at the 90th percentile or higher for gestational age (roughly 9 lb—or 4000 g—or more); macrosomia results from prepregnancy obesity, excessive weight gain during pregnancy, or uncontrolled gestational diabetes.

macular (MACK-you-lar) **degeneration:** deterioration of the macular area of the eye that can lead to loss of central vision and eventual blindness. The *macula* is a small, oval, yellowish region in the center of the retina that provides the sharp, straight-ahead vision so critical to reading and driving.

magnesium: a cation within the body's cells, active in many enzyme systems.

major minerals: essential mineral nutrients the human body requires in relatively large amounts (greater than 100 milligrams per day); sometimes called *macrominerals.*

malignant (ma-LIG-nant): describes a cancerous cell or tumor, which can injure healthy tissue and spread cancer to other regions of the body.

malnutrition: any condition caused by excess or deficient food energy or nutrient intake or by an imbalance of nutrients.

malt syrup: a sweetener made from sprouted barley and containing mostly maltose.

maltase: an enzyme that hydrolyzes maltose.

maltose (MAWL-tose): a disaccharide composed of two glucose units; sometimes known as *malt sugar.*

mammary glands: glands of the female breast that secrete milk.

manganese: an essential trace mineral that acts as a cofactor for many enzymes.

maple sugar: a sugar (mostly sucrose) purified from the concentrated sap of the sugar maple tree.

marasmus (ma-RAZ-mus): severe malnutrition characterized by poor growth, dramatic weight loss, loss of body fat and muscle, and apathy.

margin of safety: when speaking of food additives, a zone between the concentration normally used and that at which a hazard exists. For common table salt, for example, the margin of safety is 1/5 (five times the amount normally used would be hazardous).

marginal food security: one or two indications of food-access problems but with little or no change in food intake.

massage therapy: a healing method in which the therapist manually kneads muscles to reduce tension, increase blood circulation, improve joint mobility, and promote healing of injuries.

mastication: the process of chewing.

matrix (MAY-tricks): the basic substance that gives form to a developing structure; in the body, the formative cells from which teeth and bones grow.

matter: anything that takes up space and has mass.

Meals on Wheels: a nutrition program that delivers food for the elderly to their homes.

meat replacements: products formulated to look and taste like meat, fish, or poultry; usually made of textured vegetable protein.

medical history: an account of a patient's current and past health status and disease risks.

medication and supplement history: a record of all the drugs, over-the-counter and prescribed, as well as dietary and herbal supplements that a person takes routinely.

meditation: a self-directed technique of relaxing the body and calming the mind.

megaloblastic: abnormally large blood cells.

menadione (men-uh-DYE-own): the synthetic form of vitamin K.

menaquinone (men-ah-KWYN-own): the bacteria-produced form of vitamin K; also called *vitamin K_2*.

Menkes disease: a genetic disorder of copper transport that creates a copper deficiency and results in mental retardation, poor muscle tone, seizures, brittle kinky hair, and failure to thrive.

MEOS or microsomal (my-krow-SO-mal) ethanol oxidizing system: a system of enzymes in the liver that oxidize not only alcohol but also several classes of drugs.

metabolic syndrome: a combination of risk factors—elevated fasting blood glucose, hypertension, abnormal blood lipids, and abdominal obesity—that greatly increase a person's risk of developing coronary heart disease; also called *Syndrome X, insulin resistance syndrome,* or *dysmetabolic syndrome.*

metabolic water: water generated during metabolism.

metabolism: the sum total of all the chemical reactions that go on in living cells. *Energy metabolism* includes all the reactions by which the body obtains and expends the energy from food.

metalloenzymes (meh-TAL-oh-EN-zimes): enzymes that contain one or more minerals as part of their structures.

metallothionein (meh-TAL-oh-THIGH-oh-neen): a sulfur-rich protein that avidly binds with and transports metals such as zinc.

metastasize (me-TAS-tah-size): the spread of cancer from one part of the body to another.

methylation: the addition of a methyl group (CH_3).

MFP factor: a peptide released during the digestion of meat, fish, and poultry that enhances nonheme iron absorption.

mg: milligrams; one-thousandth of a gram.

mg NE: milligrams of niacin equivalents; a measure of niacin activity.

micelles (MY-cells): tiny spherical complexes of emulsified fat that arise during digestion; most contain bile salts and the products of lipid digestion, including fatty acids, monoglycerides, and cholesterol.

microangiopathies: disorders of the small blood vessels.

microarray technology: research tools that analyze the expression of thousands of genes simultaneously and search for particular gene changes associated with a disease. DNA microarrays are also called *DNA chips.*

microbes (MY-krobes): microscopically small organisms including bacteria, viruses, fungi, and protozoa; also called *microorganisms.*

microcytic (my-cro-SIT-ic) hypochromic (high-po-KROME-ic) anemia: small, pale red blood cells that develop in iron-deficiency anemia.

microgram (μg): one millionth of a gram.

microgram DFE (μg DFE): micrograms dietary folate equivalents; a measure of folate activity.

microgram RAE (μg RAE): micrograms retinol activity equivalents; a measure of vitamin A activity.

micronutrients: vitamins and minerals; the nutrients the body requires in relatively small amounts (milligrams or micrograms daily).

microvilli (MY-cro-VILL-ee or MY-cro-VILL-eye): tiny, hairlike projections on each cell of every villus that can trap nutrient particles and transport them into the cells; singular *microvillus.*

milk anemia: iron-deficiency anemia that develops when an excessive milk intake displaces iron-rich foods from the diet.

milliequivalents per liter (mEq/L): the concentration of electrolytes in a volume of solution. Milliequivalents reveal characteristics about the solution that are not evident when the concentration is expressed in terms of weight.

mineral oil: a purified liquid derived from petroleum and used to treat constipation.

mineral water: water from a spring or well that naturally contains at least 250 parts per million (ppm) of minerals. Minerals give water a distinctive flavor. Many mineral waters are high in sodium.

mineralization: the process in which calcium, phosphorus, and other minerals crystallize on the collagen matrix of a growing bone, hardening the bone.

minerals: inorganic elements. Some minerals are essential nutrients required in small amounts by the body for health.

misinformation: false or misleading information.

mitochondria (my-toh-KON-dree-uh): the cellular organelles responsible for producing ATP aerobically; made of membranes with enzymes mounted on them. (The singular is *mitochondrion.*)

mmol: millimoles; one thousandth of a mole, the molecular weight of a substance. To convert mmol to mg, multiply by the atomic weight of the substance.

moderate-intensity physical activity: physical activity that requires some increase in breathing and/or heart rate and expends 3.5 to 7 kcalories per minute. Walking at a speed of 3 to 4.5 miles per hour (about 15 to 20 minutes to walk 1 mile) is an example.

moderation (alcohol): up to one drink per day for women and up to two drinks per day for men.

moderation (dietary): providing enough but not too much of a substance.

molasses: the thick brown syrup produced during sugar refining. Molasses retains residual sugar and other by-products and a few minerals; blackstrap molasses contains significant amounts of calcium and iron.

molecule: two or more atoms of the same or different elements joined by chemical bonds. Examples are molecules of the element oxygen, composed of two oxygen atoms (O_2), and molecules of the compound water, composed of two hydrogen atoms and one oxygen atom (H_2O).

molybdenum (mo-LIB-duh-num): an essential trace mineral that acts as a cofactor for many enzymes.

monoglycerides: molecules of glycerol with one fatty acid attached. A molecule of glycerol with two fatty acids attached is a *diglyceride*.

monosaccharides (mon-oh-SACK-uh-rides): carbohydrates of the general formula $C_nH_{2n}O_n$ that typically form a single ring. The monosaccharides important in nutrition are *hexoses*, sugars with six atoms of carbon and the formula $C_6H_{12}O_6$.

monosodium glutamate (MSG): a sodium salt of the amino acid glutamic acid commonly used as a flavor enhancer. The FDA classifies MSG as a "generally recognized as safe" ingredient.

monounsaturated fatty acid: a fatty acid that lacks two hydrogen atoms and has one double bond between carbons; abbreviated *MUFA*. Examples include palmitoleic acid and oleic acid. A *monounsaturated fat* is composed of triglycerides in which most of the fatty acids are monounsaturated.

more: on food labels, at least 10 percent more of the Daily Value for a given nutrient than the comparison food; synonyms include *added* and *extra*.

motility (moh-TIL-ih-tee): the spontaneous movement of the GI tract muscles.

mouth: the oral cavity containing the tongue and teeth.

MSG symptom complex: an acute, temporary intolerance reaction that may occur after the ingestion of the additive MSG (monosodium glutamate). Symptoms include burning sensations, chest and facial flushing and pain, and throbbing headaches.

mucous (MYOO-kus) **membranes:** the membranes, composed of mucus-secreting cells, that line the surfaces of body tissues.

mucus (MYOO-kus): a slippery substance secreted by cells of the GI lining (and other body linings) that protects the cells from exposure to digestive juices (and other destructive agents). The lining of the GI tract with its coat of mucus is a *mucous membrane*. (The noun is *mucus*; the adjective is *mucous*.)

muscle dysmorphia (dis-MORE-fee-ah): a psychiatric disorder characterized by a preoccupation with building body mass.

muscle endurance: the ability of a muscle to contract repeatedly without becoming exhausted.

muscle power: the product of force generation (strength) and movement velocity (speed); the speed at which a given amount of exertion is completed.

muscle strength: the ability of muscles to work against resistance.

mutations: permanent changes in the DNA that can be inherited.

myoglobin: the oxygen-holding protein of the muscle cells.

N

NAD (nicotinamide adenine dinucleotide): the main coenzyme form of the vitamin niacin. Its reduced form is NADH.

NADP (the phosphate form of NAD): a coenzyme form of niacin.

nanoceuticals: substances with extremely small particles that have been manufactured by nanotechnology.

nanotechnology: a manufacturing technology that manipulates atoms to change the structure of matter.

narcotic (nar-KOT-ic): a drug that dulls the senses, induces sleep, and becomes addictive with prolonged use.

natural water: water obtained from a spring or well that is certified to be safe and sanitary. The mineral content may not be changed, but the water may be treated in other ways such as with ozone or by filtration.

naturopathic (nay-chur-oh-PATH-ick) **medicine:** a system that taps the natural healing forces within the body by integrating several practices, including traditional medicine, herbal medicine, clinical nutrition, homeopathy, acupuncture, East Asian medicine, hydrotherapy, and manipulative therapy.

nectar: a sugary fluid secreted by plants to encourage pollination by insects.

neotame (NEE-oh-tame): an artificial sweetener composed of two amino acids (phenylalanine and aspartic acid); approved for use in the United States.

net protein utilization (NPU): a measure of protein quality assessed by measuring the amount of protein nitrogen that is retained from a given amount of protein nitrogen eaten.

neural tube: the embryonic tissue that forms the brain and spinal cord.

neural tube defects: malformations of the brain, spinal cord, or both during embryonic development that often result in lifelong disability or death. The two main types of neural tube defects are *spina bifida* (literally "split spine") and *anencephaly* ("no brain").

neurofibrillary tangles: snarls of the threadlike strands that extend from the nerve cells, commonly found in the brains of people with Alzheimer's dementia.

neurons: nerve cells; the structural and functional units of the nervous system. Neurons initiate and conduct nerve impulse transmissions.

neuropeptide Y: a chemical produced in the brain that stimulates appetite, diminishes energy expenditure, and increases fat storage.

neurotransmitters: chemicals that are released at the end of a nerve cell when a nerve impulse arrives there. They diffuse across the gap to the next cell and alter the membrane of that second cell to either inhibit or excite it.

neutrophils (NEW-tro-fills): the most common of white blood cell. Neutrophils destroy antigens by phagocytosis.

niacin (NIGH-a-sin): a B vitamin. The coenzyme forms are *NAD (nicotinamide adenine dinucleotide)* and *NADP (the phosphate form of NAD)*. Niacin can be eaten preformed or made in the body from its precursor, tryptophan, an essential amino acid.

niacin equivalents (NE): the amount of niacin present in food, including the niacin that can theoretically be made from its precursor, tryptophan, present in the food. 1 NE = 1 mg niacin or 60 mg tryptophan.

niacin flush: a temporary burning, tingling, and itching sensation that occurs when a person takes a large dose of nicotinic acid; often accompanied by a headache and reddened face, arms, and chest.

night blindness: slow recovery of vision after flashes of bright light at night or an inability to see in dim light; an early symptom of vitamin A deficiency.

nitrites (NYE-trites): salts added to food to prevent botulism. One example is sodium nitrite, which is used to preserve meats.

nitrogen balance: the amount of nitrogen consumed (N in) as compared with the amount of nitrogen excreted (N out) in a given period of time.

nitrosamines (nye-TROHS-uh-meens): derivatives of nitrites that may be formed in the stomach when nitrites combine with amines. Nitrosamines are carcinogenic in animals.

nonessential amino acids: amino acids that the body can make; also called *dispensable amino acids*.

nonexercise activity thermogenesis (NEAT): energy expended in everyday spontaneous activities.

nonheme iron: the iron in foods that is not bound to proteins; found in both plant-derived and animal-derived foods.

nonnutritive sweeteners: sweeteners that yield no energy (or insignificant energy in the case of aspartame).

nucleotide bases: the nitrogen-containing building blocks of DNA and RNA—cytosine (C), thymine (T), uracil (U), guanine (G), and adenine (A). In DNA, the base pairs are A–T and C–G and in RNA, the base pairs are A–U and C–G.

nucleotides: the subunits of DNA and RNA molecules, composed of a phosphate group, a 5-carbon sugar (deoxyribose for DNA and ribose for RNA), and a nitrogen-containing base.

nucleus: a major membrane-enclosed body within cells, which contains the cell's genetic material (DNA), embedded in chromosomes.

nursing bottle tooth decay: extensive tooth decay due to prolonged tooth contact with formula, milk, fruit juice, or other carbohydrate-rich liquid offered to an infant in a bottle.

nutrient claims: statements that characterize the quantity of a nutrient in a food.

nutrient density: a measure of the nutrients a food provides relative to the energy it provides. The more nutrients and the fewer kcalories, the higher the nutrient density.

nutrient profiling: ranking foods based on their nutrient composition.

nutrients: chemical substances obtained from food and used in the body to provide energy, structural materials, and regulating agents to support growth, maintenance, and repair of the body's tissues. Nutrients may also reduce the risks of some diseases.

nutrigenetics: the science of how genes affect the activities of nutrients.

nutrigenomics: the science of how nutrients affect the activities of genes.

nutrition: the science of the nutrients in foods and of their actions within the body. A broader definition includes the study of human behaviors related to food and eating.

nutrition assessment: a comprehensive analysis of a person's nutrition status that uses health, socioeconomic, drug, and diet histories; anthropometric measurements; physical examinations; and laboratory tests.

nutrition screening: the use of preliminary nutrition assessment techniques to identify people who are malnourished or are at risk for malnutrition.

nutritional genomics: the science of how nutrients affect the activities of genes (*nutrigenomics*) and how genes affect the activities of nutrients (*nutrigenetics*).

nutritionist: a person who specializes in the study of nutrition. Note that this definition does not specify qualifications and may apply not only to registered dietitian nutritionists but also to self-described experts whose training is questionable. Most states have licensing laws that define the scope of practice for those calling themselves nutritionists.

nutritive sweeteners: sweeteners that yield energy, including both sugars and sugar alcohols.

O

obese: too much body fat with adverse health effects; BMI 30 or more.

obesogenic (oh-BES-oh-JEN-ick) **environment:** all the factors surrounding a person that promote weight gain, such as increased food intake, especially of unhealthy choices, and decreased physical activity.

obligatory (ah-BLIG-ah-TORE-ee) **water excretion:** the minimum amount of water the body has to excrete each day to dispose of its wastes—about 500 milliliters (about 2 cups, or 1 pint).

oils: lipids that are liquid at room temperature (77°F, or 25°C).

olestra: a synthetic fat made from sucrose and fatty acids that provides 0 kcalories per gram; also known as *sucrose polyester*.

oligopeptide (OL-ee-go-PEP-tide): string of four to nine amino acids.

omega: the last letter of the Greek alphabet (ω), used by chemists to refer to the position of the closest double bond to the methyl (CH_3) end of a fatty acid.

omega-3 fatty acid: a polyunsaturated fatty acid in which the closest double bond to the methyl (CH_3) end of the carbon chain is three carbons away.

omega-6 fatty acid: a polyunsaturated fatty acid in which the closest double bond to the methyl (CH_3) end of the carbon chain is six carbons away.

omnivorous: an eating pattern that includes foods derived from both animals and plants.

opsin (OP-sin): the protein portion of visual pigment molecules.

oral rehydration therapy (ORT): the administration of a simple solution of sugar, salt, and water, taken by mouth, to treat dehydration caused by diarrhea.

organelles: subcellular structures such as ribosomes, mitochondria, and lysosomes.

organic: in agriculture, crops grown and processed according to USDA regulations defining the use of fertilizers, herbicides, insecticides, fungicides, preservatives, and other chemical ingredients.

organic: in chemistry, substances or molecules containing carbon-carbon bonds or carbon-hydrogen bonds that are characteristic of living organisms. The four classes of nutrients that are organic are carbohydrates, lipids (fats), proteins, and vitamins.

organic: on food labels, that at least 95 percent of the product's ingredients have been grown and processed according to USDA regulations defining the use of fertilizers, herbicides, insecticides, fungicides, preservatives, and other chemical ingredients.

organic halogens: an organic compound containing one or more atoms of a halogen—fluorine, chlorine, iodine, or bromine.

orthomolecular medicine: the use of large doses of vitamins to treat chronic disease.

osmosis: the movement of water across a membrane *toward* the side where the solutes are more concentrated.

osmotic pressure: the amount of pressure needed to prevent the movement of water across a membrane.

osteoarthritis: a painful, degenerative disease of the joints that occurs when the cartilage in a joint deteriorates; joint structure is damaged, with loss of function; also called *degenerative arthritis*.

osteoblasts: cells that build bone during growth.

osteocalcin (os-teo-KAL-sen): a calcium-binding protein in bones, essential for normal mineralization.

osteoclasts: cells that destroy bone during growth.

osteomalacia (OS-tee-oh-ma-LAY-shuh): a bone disease characterized by softening of the bones. Symptoms include bending of the spine and bowing of the legs. The disease occurs most often in adult women.

osteoporosis (OS-tee-oh-pore-OH-sis): a disease in which the bones become porous and fragile due to a loss of minerals; also called *adult bone loss*.

outbreaks: two or more cases of a similar illness resulting from the ingestion of a common food.

overload: an extra physical demand placed on the body; an increase in the frequency, duration, or intensity of an activity. A principle of training that states for a body system to improve, it must be worked at frequencies, durations, or intensities that increase by increments.

overnutrition: excess energy or nutrients.

overt (oh-VERT): out in the open and easy to observe.

overweight: body weight greater than the weight range that is considered healthy; BMI 25 to 29.9.

ovum (OH-vum): the female reproductive cell, capable of developing into a new organism upon fertilization; commonly referred to as an egg.

oxaloacetate (OKS-ah-low-AS-eh-tate): a carbohydrate intermediate of the TCA cycle.

oxidants (OKS-ih-dants): compounds (such as oxygen itself) that oxidize other compounds. Compounds that prevent oxidation are called *antioxidants,* whereas those that promote it are called *prooxidants.*

oxidation (OKS-ee-DAY-shun): the process of a substance combining with oxygen; oxidation reactions involve the loss of electrons.

oxidative stress: a condition in which the production of oxidants and free radicals exceeds the body's ability to handle them and prevent damage.

oxytocin (OCK-see-TOH-sin): a hormone that stimulates the mammary glands to eject milk during lactation and the uterus to contract during childbirth.

oyster shell: a product made from the powdered shells of oysters that is sold as a calcium supplement, but it is not well absorbed by the digestive system.

ozone therapy: the use of ozone gas to enhance the body's immune system.

P

pancreas: a gland that secretes digestive enzymes and juices into the duodenum. (The pancreas also secretes hormones into the blood that help to maintain glucose homeostasis.)

pancreatic (pank-ree-AT-ic) **juice:** the exocrine secretion of the pancreas that contains enzymes for the digestion of carbohydrate, fat, and protein as well as bicarbonate, a neutralizing agent. The juice flows from the pancreas into the small intestine through the pancreatic duct. (The pancreas also has an endocrine function, the secretion of insulin and other hormones.)

pantothenic (PAN-toe-THEN-ick) **acid:** a B vitamin. The principal active form is part of coenzyme A, called "CoA" throughout Chapter 7.

parasympathetic nervous system: the part of the autonomic nervous system that dominates during nonstressful conditions and includes such effects as normal heart rate, pupil dilation, and peristalsis.

parathyroid hormone: a hormone from the parathyroid glands that regulates blood calcium by raising it when levels fall too low; also known as *parathormone* (PAIR-ah-THOR-moan).

pasteurization: heat processing of food that inactivates some, but not all, microorganisms in the food; not a sterilization process. Bacteria that cause spoilage are still present.

pathogens (PATH-oh-jenz): microorganisms capable of producing disease.

PBB (polybrominated biphenyl) and **PCB (polychlorinated biphenyl):** toxic organic halogens used in pesticides, paints, and flame retardants.

PDCAAS (protein digestibility–corrected amino acid score): a measure of protein quality assessed by comparing the amino acid score of a food protein with the amino acid requirements of preschool-age children and then correcting for the true digestibility of the protein.

peak bone mass: the highest attainable bone density for an individual, developed during the first three decades of life.

peer review: a process in which a panel of scientists rigorously evaluates a research study to ensure that the scientific method was followed.

pellagra (pell-AY-gra): the niacin-deficiency disease, characterized by diarrhea, dermatitis, dementia, and eventually death.

pepsin: a gastric enzyme that hydrolyzes protein. Pepsin is secreted in an inactive form, *pepsinogen,* which is activated by hydrochloric acid in the stomach.

pepsinogen: an inactive compound that is activated by hydrochloric acid in the stomach to form pepsin.

peptic ulcer: a lesion in the mucous membrane of either the stomach (a *gastric ulcer*) or the duodenum (a *duodenal ulcer*).

peptidase: a digestive enzyme that hydrolyzes peptide bonds. *Tripeptidases* cleave tripeptides; *dipeptidases* cleave dipeptides.

peptide bond: a bond that connects the acid end of one amino acid with the amino end of another, forming a link in a protein chain.

percent Daily Value (%DV): the percentage of a Daily Value recommendation found in a specified serving of food for key nutrients based on a 2000-kcalorie diet.

percent fat-free: may be used only if the product meets the definition of *low fat* or *fat-free* and must reflect the amount of fat in 100 grams (for example, a food that contains 2.5 grams of fat per 50 grams can claim to be "95 percent fat-free").

perinatal: referring to the time between the twenty-eighth week of gestation and 1 month after birth.

peripheral (puh-RIFF-er-ul) **nervous system:** the peripheral (outermost) part of the nervous system; the vast complex of wiring that extends from the central nervous system to the body's outermost areas. It contains both *somatic* and *autonomic* components.

peripheral resistance: the resistance to pumped blood in the small arterial branches (arterioles) that carry blood to the tissues.

peristalsis (per-ih-STALL-sis): wavelike muscular contractions of the GI tract that push its contents along.

pernicious (per-NISH-us) **anemia:** a blood disorder that reflects a vitamin B_{12} deficiency caused by lack of intrinsic factor and characterized by abnormally large and immature red blood cells. Other symptoms include muscle weakness and irreversible neurological damage.

persistence: stubborn or enduring continuance; with respect to food contaminants, the quality of persisting, rather than breaking down, in the bodies of animals and human beings.

personal and social history: a record of a person's economic and social background, including such factors as education, income, and ethnic identity.

pesticides: chemicals used to control insects, weeds, fungi, and other pests on plants, vegetables, fruits, and animals. Used broadly, the term includes herbicides (to kill weeds), insecticides (to kill insects), and fungicides (to kill fungi).

pH: the unit of measure expressing a substance's acidity or alkalinity. The lower the pH, the higher the H^+ ion concentration and the stronger the acid. A pH above 7 is alkaline, or base (a solution in which OH^- ions predominate).

phagocytes (FAG-oh-sites): white blood cells (neutrophils and macrophages) that have the ability to ingest and destroy foreign substances.

phagocytosis (FAG-oh-sigh-TOH-sis): the process by which phagocytes engulf and destroy foreign materials.

pharmacological effect: the body's response to a large dose of a nutrient (levels commonly available only from supplements) that overwhelms some body system and acts like a drug.

pharynx (FAIR-inks): the passageway leading from the nose and mouth to the larynx and esophagus, respectively.

phenylketonuria (FEN-il-KEY-toe-NEW-ree-ah) or **PKU:** an inherited disorder characterized by failure to metabolize the amino acid phenylalanine to tyrosine.

phlebotomy: the withdrawal of blood from the body.

phosphocreatine (PC): a high-energy compound in muscle cells that acts as a reservoir of energy that can maintain a steady supply of ATP and provides the energy for short bursts of activity; also called *creatine phosphate (CP).*

phospholipid (FOS-foe-LIP-id): a compound similar to a triglyceride but having a phosphate group (a phosphorus-containing salt) and choline (or another nitrogen-containing compound) in place of one of the fatty acids.

phosphorus: a major mineral found mostly in the body's bones and teeth.

photosynthesis: the process by which green plants use the sun's energy to make carbohydrates from carbon dioxide and water.

phylloquinone (FILL-oh-KWYN-own): the plant form of vitamin K; also called *vitamin K₁*.

physical activity: bodily movement produced by muscle contractions that substantially increase energy expenditure.

physiological age: a person's age as estimated from her or his body's health and probable life expectancy.

physiological effect: the body's response to a normal dose of a nutrient (levels commonly found in foods) that provides a normal blood concentration.

physiological fuel value: the number of kcalories that the body derives from a food, in contrast to the number of kcalories determined by calorimetry.

phytic (FYE-tick) **acid:** a nonnutrient component of plant seeds; also called *phytate* (FYE-tate). Phytic acid occurs in the husks of grains, legumes, and seeds and is capable of binding minerals such as zinc, iron, calcium, magnesium, and copper in insoluble complexes in the intestine, which the body excretes unused.

phytochemicals (FIE-toe-KEM-ih-cals): nonnutrient compounds found in plants. Some phytochemicals have biological activity in the body.

phytoestrogens: phytochemicals structurally similar to human estrogen that weakly mimic or modulate estrogen's action in the body. Phytoestrogens include the isoflavones *genistein, daidzein,* and *glycitein.*

pica (PIE-ka): a craving for and consumption of nonfood substances. Pica is known as *geophagia* (gee-oh-FAY-gee-uh) when referring to eating clay, baby powder, chalk, ash, ceramics, paper, paint chips, or charcoal; *pagophagia* (pag-oh-FAY-gee-uh) when referring to eating large quantities of ice; and *amylophagia* (AM-ee-low-FAY-gee-ah) when referring to eating uncooked starch (flour, laundry starch, or raw rice).

pigment: a molecule capable of absorbing certain wavelengths of light so that it reflects only those that we perceive as a certain color.

placebo (pla-SEE-bo): an inert, harmless medication given to provide comfort and hope; a sham treatment used in controlled research studies.

placebo effect: a change that occurs in response to expectations about the effectiveness of a treatment that actually has no pharmaceutical effects.

placenta (plah-SEN-tuh): the organ that develops inside the uterus early in pregnancy, through which the fetus receives nutrients and oxygen and returns carbon dioxide and other waste products to be excreted.

plant-based diets: an eating pattern that derives most of its protein from plant products (although some animal products may be included).

plant-pesticides: pesticides made by the plants themselves.

plant sterols: phytochemicals that have structural similarities to cholesterol and lower blood cholesterol by interfering with cholesterol absorption. Plant sterols include *sterol esters* and *stanol esters.*

plaque (PLACK): an accumulation of fatty deposits, smooth muscle cells, and fibrous connective tissue that develops in the artery walls in atherosclerosis. Plaque associated with atherosclerosis is known as *atheromatous* (ATH-er-OH-ma-tus) *plaque.*

platelets: tiny, disc-shaped bodies in the blood, important in blood clot formation.

PLP (pyridoxal phosphate): the primary active coenzyme form of vitamin B₆.

point of unsaturation: the double bond of a fatty acid, where hydrogen atoms can easily be added to the structure.

polar: characteristic of a neutral molecule, such as water, that has opposite charges spatially separated within the molecule.

polypeptide: many (10 or more) amino acids bonded together.

polysaccharides: compounds composed of many monosaccharides linked together. An intermediate string of 3 to 10 monosaccharides is an *oligosaccharide.*

polyunsaturated fatty acid: a fatty acid that lacks four or more hydrogen atoms and has two or more double bonds between carbons; abbreviated *PUFA.* Examples include linoleic acid (two double bonds) and linolenic acid (three double bonds). A *polyunsaturated fat* is composed of triglycerides in which most of the fatty acids are polyunsaturated.

portion sizes: the quantity of a food served or eaten at one meal or snack; *not* a standard amount.

postpartum amenorrhea (ay-MEN-oh-REE-ah): the normal temporary absence of menstrual periods immediately following childbirth.

potable (POT-ah-bul): suitable for drinking.

potassium: the principal cation within the body's cells; critical to the maintenance of fluid balance, nerve impulse transmissions, and muscle contractions.

powdered bone: crushed or ground bone preparations intended to supply calcium to the diet. Calcium from bone is not well absorbed and is often contaminated with toxic minerals such as arsenic, mercury, lead, and cadmium.

prebiotics: food components (such as fibers) that are not digested by the human body but are used as food by the GI bacteria to promote their growth and activity.

precursors: substances that precede others; with regard to vitamins, compounds that can be converted into active vitamins; also known as *provitamins.*

prediabetes: condition in which blood glucose levels are higher than normal, but below the diagnosis of diabetes; formerly called *impaired glucose tolerance.* Prediabetes is considered a major risk factor for future diabetes and cardiovascular diseases.

preeclampsia (PRE-ee-KLAMP-see-ah): a condition characterized by high blood pressure and some protein in the urine.

preformed vitamin A: dietary vitamin A in its active form.

prehypertension: slightly higher-than-normal blood pressure, but not as high as hypertension.

prenatal alcohol exposure: subjecting a fetus to a pattern of excessive alcohol intake characterized by substantial regular use or heavy episodic drinking.

preservatives: antimicrobial agents, antioxidants, and other additives that retard spoilage or maintain desired qualities, such as softness in baked goods.

pressure ulcers: damage to the skin and underlying tissues as a result of compression and poor circulation; commonly seen in people who are bedridden or chair-bound.

preterm (premature): births occurring before 37 weeks of gestation; births occurring at 37 to 38 weeks of gestation are designated *early term.*

primary deficiency: a nutrient deficiency caused by inadequate dietary intake of a nutrient.

probiotics: living microorganisms found in foods and dietary supplements that, when consumed in sufficient quantities, are beneficial to health.

processed foods: foods that have been treated to change their physical, chemical, microbiological, or sensory properties.

proenzyme: the inactive form of an enzyme; also called a *zymogen.*

progesterone: the hormone of gestation (pregnancy).

prolactin (pro-LAK-tin): a hormone secreted from the anterior pituitary gland that acts on the mammary glands to promote the production of milk. The release of prolactin is mediated by *prolactin-inhibiting hormone (PIH).*

promoters: factors that favor the development of cancers once they have begun.

proof: a way of stating the percentage of alcohol in distilled liquor. Liquor that is 100 proof is 50 percent alcohol; 90 proof is 45 percent, and so forth.

prooxidants: substances that significantly induce oxidative stress.

proteases (PRO-tee-aces): enzymes that hydrolyze proteins.

protein digestibility: a measure of the amount of amino acids absorbed from a given protein intake.

protein efficiency ratio (PER): a measure of protein quality assessed by determining how well a given protein supports weight gain in growing rats; used to establish the protein quality for infant formulas and baby foods.

protein turnover: the degradation and synthesis of protein.

protein-sparing action: the action of carbohydrate (and fat) in providing energy that allows protein to be used for other purposes.

proteins: compounds composed of carbon, hydrogen, oxygen, and nitrogen atoms, arranged into amino acids linked in a chain. Some amino acids also contain sulfur atoms.

proteome: all proteins in a cell. The study of all proteins produced by a species is called *proteomics*.

puberty: the period in life in which a person becomes physically capable of reproduction.

public health dietitians: dietitians who specialize in providing nutrition services through organized community efforts.

public water: water from a municipal or county water system that has been treated and disinfected.

purified water: water that has been treated by distillation or other physical or chemical processes that remove dissolved solids. Because purified water contains no minerals or contaminants, it is useful for medical and research purposes.

purines: compounds of nitrogen-containing bases such as adenine, guanine, and caffeine. Purines that originate from the body are *endogenous* and those that derive from foods are *exogenous*.

pyloric (pie-LORE-ic) **sphincter:** the circular muscle that separates the stomach from the small intestine and regulates the flow of partially digested food into the small intestine; also called *pylorus* or *pyloric valve*.

pyruvate (PIE-roo-vate): a 3-carbon compound that plays a key role in energy metabolism.

Q

qi gong (chée GUNG): a Chinese system that combines movement, meditation, and breathing techniques to enhance the flow of *qi* (vital energy) in the body.

quality of life: a person's perceived physical and mental well-being.

R

rachitic (ra-KIT-ik) **rosary:** the poorly formed rib attachments that may develop in a vitamin D deficiency; literally, "the rosary of rickets."

radura: an international symbol used to identify retail foods that have been irradiated.

randomization (RAN-dom-ih-ZAY-shun): a process of choosing the members of the experimental and control groups without bias.

raw sugar: the first crop of crystals harvested during sugar processing. Raw sugar cannot be sold in the United States because it contains too much filth (dirt, insect fragments, and the like). Sugar sold as "raw sugar" domestically has actually gone through more than half of the refining steps.

RDN: see *registered dietitian nutritionist*.

Recommended Dietary Allowance (RDA): the average daily amount of a nutrient considered adequate to meet the known nutrient needs of practically all healthy people; a goal for dietary intake by individuals.

rectum: the muscular terminal part of the intestine, extending from the sigmoid colon to the anus.

reduced kcalorie: at least 25 percent fewer kcalories per serving than the comparison food.

reference protein: a standard against which to measure the quality of other proteins.

refined: the process by which the coarse parts of a food are removed. When wheat is refined into flour, the bran, germ, and husk are removed, leaving only the endosperm.

refined flour: finely ground endosperm that is usually enriched with nutrients and bleached for whiteness; sometimes called *white flour*.

reflux: a backward flow.

registered dietitian (RD): an alternative term for an RDN.

registered dietitian nutritionist (RDN): a person who has completed a minimum of a bachelor's degree from an accredited university or college, has completed approved course work and a supervised practice program, has passed a national examination, and maintains registration through continuing professional education; also called *registered dietitian (RD)*.

registration: listing; with respect to health professionals, listing with a professional organization that requires specific course work, experience, and passing of an examination.

relaxin: the hormone of late pregnancy.

remodeling: the dismantling and re-formation of a structure.

renin (REN-in): an enzyme from the kidneys that hydrolyzes the protein angiotensinogen to angiotensin I, which results in the kidneys reabsorbing sodium.

rennin: an enzyme that coagulates milk; found in the gastric juice of cows, but not human beings.

replication (REP-lih-KAY-shun): repeating an experiment and getting the same results.

requirement: the lowest continuing intake of a nutrient that will maintain a specified criterion of adequacy.

residues: whatever remains. In the case of pesticides, those amounts that remain on or in foods when people buy and use them.

resistance training: the use of free weights or weight machines to provide resistance for developing muscle strength, power, and endurance; also called *weight training*. A person's own body weight may also be used to provide resistance such as when a person does push-ups, pull-ups, or abdominal crunches.

resistant starches: starches that escape digestion and absorption in the small intestine of healthy people.

resistin (ree-ZIS-tin): a protein produced by adipose cells that promotes inflammation and causes insulin resistance.

respiratory chain: the final pathway in energy metabolism that transports electrons from hydrogen to oxygen and captures the energy released in the bonds of ATP; also called the *electron transport chain*.

resting metabolic rate (RMR): similar to the basal metabolic rate (BMR), a measure of the energy use of a person at rest in a comfortable setting, but with less stringent criteria for recent food intake and physical activity. Consequently, the RMR is slightly higher than the BMR.

retina (RET-in-uh): the innermost membrane of the eye, composed of several layers, including one that contains the rods and cones.

retinal (RET-ih-nal): the aldehyde form of vitamin A.

retinoic (RET-ih-NO-ick) **acid:** the acid form of vitamin A.

retinoids (RET-ih-noyds): chemically related compounds with biological activity similar to that of retinol; metabolites of retinol.

retinol (RET-ih-nol): the alcohol form of vitamin A.

retinol activity equivalents (RAE): a measure of vitamin A activity; the amount of retinol that the body will derive from a food containing preformed retinol or its precursor, beta-carotene.

retinol-binding protein (RBP): the specific protein responsible for transporting retinol.

rheumatoid (ROO-ma-toyd) **arthritis:** a disease of the immune system involving painful inflammation of the joints and related structures.

rhodopsin (ro-DOP-sin): a light-sensitive pigment of the retina that contains the retinal form of vitamin A and the protein opsin.

riboflavin (RYE-boh-flay-vin): a B vitamin. The coenzyme forms are *FMN (flavin mononucleotide)* and *FAD (flavin adenine dinucleotide)*.

ribose: a naturally occurring 5-carbon sugar needed for the synthesis of ATP.

ribosomes (RYE-boh-zomes): protein-making organelles in cells that are composed of RNA and protein.

rickets: the vitamin D–deficiency disease in children characterized by inadequate mineralization of bone (manifested in bowed legs or knock-knees, outward-bowed chest, and "beads" on ribs). A rare type of rickets, not caused by vitamin D deficiency, is known as *vitamin D–refractory rickets.*

risk: a measure of the probability and severity of harm.

risk factor: a condition or behavior associated with an elevated frequency of a disease but not proved to be causal. Leading risk factors for chronic diseases include obesity, cigarette smoking, high blood pressure, high blood cholesterol, physical inactivity, and a diet high in saturated fats and low in vegetables, fruits, and whole grains.

RNA (ribonucleic acid): a compound similar to DNA, but RNA is a single strand with a ribose sugar instead of a deoxyribose sugar and uracil instead of thymine as one of its bases.

S

safety: the condition of being free from harm or danger.

saliva: the secretion of the salivary glands. Its principal enzyme begins carbohydrate digestion.

salivary glands: exocrine glands that secrete saliva into the mouth.

salt: a compound composed of a positive ion other than H^+ and a negative ion other than OH^-. An example is sodium chloride (Na^+Cl^-).

sarcopenia (SAR-koh-PEE-nee-ah): loss of skeletal muscle mass, strength, and quality.

satiating: having the power to suppress hunger and inhibit eating.

satiation (say-she-AY-shun): the feeling of satisfaction and fullness that occurs during a meal and halts eating. Satiation determines how much food is consumed during a meal.

satiety (sah-TIE-eh-tee): the feeling of fullness and satisfaction that occurs after a meal and inhibits eating until the next meal. Satiety determines how much time passes between meals.

saturated fat-free: less than 0.5 gram of saturated fat and 0.5 gram of *trans* fat per serving.

saturated fatty acid: a fatty acid carrying the maximum possible number of hydrogen atoms—for example, stearic acid. A *saturated fat* is composed of triglycerides in which most of the fatty acids are saturated.

scurvy: the vitamin C–deficiency disease.

secondary deficiency: a nutrient deficiency caused by something other than an inadequate intake such as a disease condition or drug interaction that reduces absorption, accelerates use, hastens excretion, or destroys the nutrient.

secretin (see-CREET-in): a hormone produced by cells in the duodenum wall. Target organ: the pancreas. Response: secretion of bicarbonate-rich pancreatic juice.

sedentary: physically inactive (literally, "sitting down a lot").

segmentation (SEG-men-TAY-shun): a periodic squeezing or partitioning of the intestine at intervals along its length by its circular muscles.

selenium (se-LEEN-ee-um): an essential trace mineral that is part of an antioxidant enzyme.

senile dementia: the loss of brain function beyond the normal loss of physical adeptness and memory that occurs with aging.

senile plaques: clumps of the protein fragment beta-amyloid on the nerve cells, commonly found in the brains of people with Alzheimer's dementia.

serotonin (SER-oh-TONE-in): a neurotransmitter important in sleep regulation, appetite control, and sensory perception, among other roles. Serotonin is synthesized in the body from the amino acid tryptophan with the help of vitamin B_6.

serving sizes: the standardized quantity of a food; such information allows comparisons when reading food labels and consistency when following the *Dietary Guidelines.*

set point: the point at which controls are set (for example, on a thermostat). The set-point theory that relates to body weight proposes that the body tends to maintain a certain weight by means of its own internal controls.

sickle-cell anemia: a hereditary form of anemia characterized by abnormal sickle- or crescent-shaped red blood cells. Sickled cells interfere with oxygen transport and blood flow. Symptoms are precipitated by dehydration and insufficient oxygen (as may occur at high altitudes) and include hemolytic anemia (red blood cells burst), fever, and severe pain in the joints and abdomen.

simple carbohydrates: monosaccharides and disaccharides (the sugars).

small for gestational age (SGA): term describing an infant whose birthweight is low compared with the number of weeks in utero, often reflecting growth failure.

small intestine: a 10-foot length of small-diameter intestine that is the major site of digestion of food and absorption of nutrients. Its segments are the *duodenum, jejunum,* and *ileum.*

sodium: the principal cation in the extracellular fluids of the body; critical to the maintenance of fluid balance, nerve impulse transmissions, and muscle contractions.

sodium bicarbonate (baking soda): a white crystalline powder that is used to buffer acid that accumulates in the muscles and blood during high-intensity exercise.

sodium-free and **salt-free:** less than 5 milligrams of sodium per serving.

soft water: water with a high sodium or potassium content.

solanine (SOH-lah-neen): a poisonous narcotic-like substance present in potato skins and sprouts.

solid fats: fats that are not usually liquid at room temperature; commonly found in most foods derived from animals and vegetable oils that have been hydrogenated. Solid fats typically contain more saturated and *trans* fats than most oils.

soluble fibers: nonstarch polysaccharides that dissolve in water to form a gel. An example is pectin from fruit, which is used to thicken jellies.

solutes (SOLL-yutes): the substances that are dissolved in a solution. The number of molecules in a given volume of fluid is the *solute concentration.*

somatic (so-MAT-ick) **nervous system:** the division of the nervous system that controls the voluntary muscles, as distinguished from the autonomic nervous system, which controls involuntary functions.

somatostatin (GHIH): a hormone that inhibits the release of growth hormone; the opposite of *somatotropin (GH)*.

sperm: the male reproductive cell, capable of fertilizing an ovum.

sphincter (SFINK-ter): a circular muscle surrounding, and able to close, a body opening. Sphincters are found at specific points along the GI tract and regulate the flow of food particles.

spina (SPY-nah) bifida (BIFF-ih-dah): one of the most common types of neural tube defects, characterized by the incomplete closure of the spinal cord and its bony encasement.

sports anemia: a transient condition of low hemoglobin in the blood, associated with the early stages of sports training or other strenuous activity.

spring water: water originating from an underground spring or well. It may be bubbly (carbonated), or "flat" or "still," meaning not carbonated. Brand names such as "Spring Pure" do not necessarily mean that the water comes from a spring.

starches: plant polysaccharides composed of many glucose molecules.

sterile: free of microorganisms, such as bacteria.

sterols (STARE-ols or STEER-ols): compounds containing a four-ring carbon structure with any of a variety of side chains attached.

stomach: a muscular, elastic, saclike portion of the digestive tract that grinds and churns swallowed food, mixing it with acid and enzymes to form chyme.

stools: waste matter discharged from the colon; also called *feces* (FEE-seez).

stress: any threat to a person's well-being; a demand placed on the body to adapt.

stress fractures: bone damage or breaks caused by stress on bone surfaces during exercise.

stress response: the body's response to stress, mediated by both nerves and hormones.

stressors: environmental elements, physical or psychological, that cause stress.

stroke: an event in which the blood flow to a part of the brain is cut off; also called *cerebrovascular accident (CVA)*.

structure-function claims: statements that characterize the relationship between a nutrient or other substance in a food and its role in the body.

subclavian (sub-KLAY-vee-an) vein: the vein that provides passageway from the lymphatic system to the vascular system.

subclinical deficiency: a deficiency in the early stages, before the outward signs have appeared.

subcutaneous fat: fat stored directly under the skin.

subjects: the people or animals participating in a research project.

successful weight-loss maintenance: achieving a weight loss of at least 5 to 10 percent of initial body weight and maintaining the loss for at least 1 year.

sucrase: an enzyme that hydrolyzes sucrose.

sucrose (SUE-krose): a disaccharide composed of glucose and fructose; commonly known as *table sugar*, *beet sugar*, or *cane sugar*. Sucrose also occurs in many fruits and some vegetables and grains.

sudden infant death syndrome (SIDS): the unexpected and unexplained death of an apparently well infant; the most common cause of death of infants between the second week and the end of the first year of life; also called *crib death*.

sugar alcohols: sugarlike compounds that can be derived from fruits or commercially produced from dextrose; also called *polyols*. Examples include *erythritol, isomalt, lactitol, maltitol, mannitol, sorbitol,* and *xylitol*.

sugar-free: less than 0.5 gram of sugar per serving.

sugars: simple carbohydrates composed of monosaccharides, disaccharides, or both.

sulfate: a salt produced from the oxidation of sulfur.

sulfites: salts containing sulfur that are added to foods to prevent spoilage.

sulfur: a mineral present in the body as part of some proteins.

sushi: vinegar-flavored rice and seafood, typically wrapped in seaweed and stuffed with colorful vegetables. Some sushi is stuffed with raw fish; other varieties contain cooked seafood.

sustainable: able to continue indefinitely; using resources at such a rate that the earth can keep on replacing them and producing pollutants at a rate with which the environment and human cleanup efforts can keep pace, so that no net accumulation of pollution occurs.

sustainable agriculture: ability to produce food indefinitely, with little or no harm to the environment.

sympathetic nervous system: the part of the autonomic nervous system that dominates during stressful conditions and includes such effects as increased heart rate, dilated pupils, slowed peristalsis, and secretion of epinephrine and norepinephrine.

symptomatic allergy: an immune response that produces antibodies and symptoms.

synergistic (SIN-er-JIS-tick): multiple factors operating together in such a way that their combined effects are greater than the sum of their individual effects.

T

tagatose (TAG-ah-tose): poorly absorbed monosaccharide similar in structure to fructose; naturally occurring or derived from lactose.

TCA cycle or tricarboxylic (try-car-box-ILL-ick) acid cycle: a series of metabolic reactions that break down molecules of acetyl CoA to carbon dioxide and hydrogen atoms; also called the *citric acid cycle* or the *Krebs cycle* after the biochemist who elucidated its reactions.

T-cells: lymphocytes that attack antigens. *T* stands for the *thymus gland*, where the T-cells mature.

tempeh (TEM-pay): a fermented soybean food, rich in protein and fiber.

teratogen (ter-AT-oh-jen): a substance that causes abnormal fetal development and birth defects.

teratogenic (ter-AT-oh-jen-ik): causing abnormal fetal development and birth defects.

testosterone: a steroid hormone from the testicles, or testes. The steroids are chemically related to, and some are derived from, the lipid cholesterol.

textured vegetable protein: processed soybean protein used in vegetarian products such as soy burgers.

theory: a tentative explanation that integrates many and diverse findings to further the understanding of a defined topic.

thermic effect of food (TEF): an estimation of the energy required to process food (digest, absorb, transport, metabolize, and store ingested nutrients); also called the *specific dynamic effect (SDE)* of food or the *specific dynamic activity (SDA)* of food. The sum of the TEF and any increase in the metabolic rate due to overeating is known as *diet-induced thermogenesis (DIT)*.

thermogenesis: the generation of heat; used in physiology and nutrition studies as an index of how much energy the body is expending.

THF (tetrahydrofolate): a coenzyme form of folate.

thiamin (THIGH-ah-min): a B vitamin. The coenzyme form is *TPP (thiamin pyrophosphate)*.

thirst: a conscious desire to drink.

thoracic (thor-ASS-ic) **duct:** the main lymphatic vessel that collects lymph and drains into the left subclavian vein.

thrombosis (throm-BOH-sis): the formation of a *thrombus* (THROM-bus), or a blood clot, that may obstruct a blood vessel, causing gradual tissue death.

thyroid-stimulating hormone (TSH): a hormone secreted by the pituitary that stimulates the thyroid gland to secrete its hormones—thyroxine and triiodothyronine. The release of TSH is mediated by *TSH-releasing hormone (TRH).*

thyrotropin: another name for thyroid-stimulating hormone (TSH).

tocopherols (tuh-KOFF-uh-rawls): members of the vitamin E family having the chemical structure of a complex ring structure with a long saturated side chain.

tocotrienols (TOE-koh-try-EE-nawls): members of the vitamin E family having the chemical structure of a complex ring structure with a long unsaturated side chain.

tofu (TOE-foo): a curd made from soybeans, rich in protein and often fortified with calcium; used in many Asian and vegetarian dishes in place of meat.

Tolerable Upper Intake Level (UL): the maximum daily amount of a nutrient that appears safe for most healthy people and beyond which there is an increased risk of adverse health effects.

tolerance level: the maximum amount of residue permitted in a food when a pesticide is used according to the label directions.

toxicity: the ability of a substance to harm living organisms. All substances are toxic if high enough concentrations are used.

TPP (thiamin pyrophosphate): the coenzyme form of thiamin.

trabecular (tra-BECK-you-lar) **bone:** the lacy inner structure of calcium crystals that supports the bone's structure and provides a calcium storage bank.

trace minerals: essential mineral nutrients the human body requires in relatively small amounts (less than 100 milligrams per day); sometimes called *microminerals.*

trachea (TRAKE-ee-uh): the air passageway from the larynx to the lungs; also called the *windpipe.*

trans: on the other side of; refers to a chemical configuration in which the hydrogen atoms are located on opposite sides of a double bond.

trans fat-free: less than 0.5 gram of *trans* fat and less than 0.5 gram of saturated fat per serving.

transamination (TRANS-am-ih-NAY-shun): the transfer of an amino group from one amino acid to a keto acid, producing a new nonessential amino acid and a new keto acid.

transcription: the process of messenger RNA being made from a template of DNA.

transcription factors: proteins that bind to specific sites in DNA and alter gene expression.

trans-fatty acids: fatty acids with hydrogens on opposite sides of the double bond.

transferrin (trans-FAIR-in): the iron transport protein.

transient ischemic (is-KEY-mik) **attack (TIA):** a temporary reduction in blood flow to the brain, which causes temporary symptoms that vary depending on the part of the brain affected. Common symptoms include light-headedness, visual disturbances, paralysis, staggering, numbness, and inability to swallow.

translation: the process of messenger RNA directing the sequence of amino acids and synthesis of proteins.

travelers' diarrhea: nausea, vomiting, and diarrhea caused by consuming food or water contaminated by any of several organisms, most commonly, *E. coli, Shigella, Campylobacter jejuni,* and *Salmonella.*

triglycerides (try-GLISS-er-rides): the chief form of fat in the diet and the major storage form of fat in the body; composed of a molecule of glycerol with three fatty acids attached; also called *triacylglycerols* (try-ay-seel-GLISS-er-ols).

tripeptide: three amino acids bonded together.

tumor: an abnormal tissue mass with no physiological function; also called a *neoplasm* (NEE-oh-plazm).

turbinado (ter-bih-NOD-oh) **sugar:** sugar produced using the same refining process as white sugar, but without the bleaching and anticaking treatment. Traces of molasses give turbinado its sandy color.

type 1 diabetes: the less common type of diabetes in which the pancreas produces little to no insulin. Type 1 diabetes usually results from autoimmune destruction of pancreatic beta cells.

type 2 diabetes: the more common type of diabetes in which the cells fail to respond to insulin. Type 2 diabetes usually accompanies obesity and results from insulin resistance coupled with insufficient insulin secretion.

U

ulcer: a lesion of the skin or mucous membranes characterized by inflammation and damaged tissues. See also *peptic ulcer.*

ultrahigh temperature (UHT) treatment: sterilizing a food by brief exposure to temperatures above those normally used.

umbilical (um-BILL-ih-cul) **cord:** the ropelike structure through which the fetus's veins and arteries reach the placenta; the route of nourishment and oxygen to the fetus and the route of waste disposal from the fetus. The scar in the middle of the abdomen that marks the former attachment of the umbilical cord is the *umbilicus* (um-BILL-ih-cus), commonly known as the "belly button."

uncoupled reactions: chemical reactions in which energy is released as heat.

undernutrition: deficient energy or nutrients.

underweight: body weight lower than the weight range that is considered healthy; BMI less than 18.5.

unsaturated fatty acid: a fatty acid that lacks hydrogen atoms and has at least one double bond between carbons (includes monounsaturated and polyunsaturated fatty acids). An *unsaturated fat* is composed of triglycerides in which most of the fatty acids are unsaturated.

urea (you-REE-uh): the principal nitrogen-excretion product of protein metabolism. Two ammonia fragments are combined with carbon dioxide to form urea.

USDA (US Department of Agriculture): the federal agency responsible for enforcing standards for the wholesomeness and quality of meat, poultry, and eggs produced in the United States; conducting nutrition research; and educating the public about nutrition.

uterus (YOU-ter-us): the muscular organ within which the infant develops before birth.

V

validity (va-LID-ih-tee): having the quality of being founded on fact or evidence.

variables: factors that change. A variable may depend on another variable (for example, a child's height depends on his age), or it may be independent (for example, a child's height does not depend on the color of her eyes). Sometimes both variables correlate with a third variable (a child's height and eye color both depend on genetics).

variety (dietary): eating a wide selection of foods within and among the major food groups.

vasoconstrictor (VAS-oh-kon-STRIK-tor): a substance that constricts or narrows the blood vessels.

vasopressin (VAS-oh-PRES-in): another name for antidiuretic hormone, so called because it elevates blood pressure.

vegan (VEE-gan) **diet:** an eating pattern that excludes all animal-derived foods (including meat, poultry, fish, eggs, and dairy products) from the diet; also called *pure vegetarian, strict vegetarian,* or *total vegetarian.*

vegetarian diet: a general term used to describe an eating pattern that excludes meat, poultry, fish, or other animal-derived foods from the diet.

veins (VANES): vessels that carry blood to the heart.

very low food security: multiple indications of disrupted eating patterns and reduced food intake; formerly known as *food insecurity with hunger.*

very low sodium: 35 milligrams or less per serving.

vigorous-intensity physical activity: physical activity that requires a large increase in breathing and/or heart rate and expends more than 7 kcalories per minute. Walking at a very brisk pace (>4.5 miles per hour) or running at a pace of at least 5 miles per hour are examples.

villi (VILL-ee or VILL-eye): fingerlike projections from the folds of the small intestine; singular *villus.*

visceral fat: fat stored within the abdominal cavity in association with the internal abdominal organs; also called *intra-abdominal fat.*

viscous: a gel-like consistency.

vitamin A: all naturally occurring compounds with the biological activity of *retinol,* the alcohol form of vitamin A.

vitamin A activity: a term referring to both the active forms of vitamin A and the precursor forms in foods without distinguishing between them.

vitamin B$_6$: a family of compounds—pyridoxal, pyridoxine, and pyridoxamine. The primary active coenzyme form is *PLP* (*pyridoxal phosphate*).

vitamin B$_{12}$: a B vitamin characterized by the presence of cobalt. The active forms of coenzyme B$_{12}$ are *methylcobalamin* and *deoxyadenosylcobalamin.*

vitamin D$_2$: vitamin D derived from plants in the diet; also called *ergocalciferol* (ER-go-kal-SIF-er-ol).

vitamin D$_3$: vitamin D derived from animals in the diet or made in the skin from 7-dehydrocholesterol, a precursor of cholesterol, with the help of sunlight; also called *cholecalciferol* (KO-lee-kal-SIF-er-ol) or *calciol.* After hydroxylation in the liver, calciol becomes *calcidiol* and after hydroxylation in the kidneys, calcidiol becomes *calcitriol.*

vitamins: organic, essential nutrients required in small amounts by the body for health. Vitamins regulate body processes that support growth and maintain life.

VLDL (very-low-density lipoprotein): the type of lipoprotein made primarily by liver cells to transport lipids to various tissues in the body; composed primarily of triglycerides.

VO$_{2max}$: the maximum rate of oxygen consumption by an individual at sea level.

vomiting: expulsion of the contents of the stomach up through the esophagus to the mouth.

vulnerable plaque: plaque that is susceptible to rupture because it has only a thin fibrous barrier between its lipid-rich core and the artery lining.

W

waist circumference: an anthropometric measurement used to assess a person's abdominal fat.

warm-up: 5 to 10 minutes of light activity, such as easy jogging or cycling, prior to a workout to prepare the body for more vigorous activity.

water balance: the balance between water intake and output (losses).

water intoxication: the rare condition in which body water contents are too high in all body fluid compartments.

water stress: intense demands on water resources by human activities such as municipal water supplies, industries, power plants, and agriculture.

wean: to gradually replace breast milk with infant formula or other foods appropriate to an infant's diet.

websites: Internet resources composed of text and graphic files, each with a unique URL (Uniform Resource Locator) that names the site (for example, www.usda.gov).

weight management: maintaining body weight in a healthy range by preventing gradual weight gains over time and losing weight if overweight, and by preventing weight losses and gaining weight if underweight.

well water: water drawn from ground water by tapping into an aquifer.

Wernicke-Korsakoff (VER-nee-key KORE-sah-kof) **syndrome:** a neurological disorder typically associated with chronic alcoholism and caused by a deficiency of the B vitamin thiamin; also called *alcohol-related dementia.*

wheat flour: any flour made from the endosperm of the wheat kernel.

whey protein: a by-product of cheese production; falsely promoted as increasing muscle mass. Whey is the watery part of milk that separates from the curds.

white sugar: granulated sucrose or "table sugar," produced by dissolving, concentrating, and recrystallizing raw sugar.

WHO (World Health Organization): an international agency concerned with promoting health and eradicating disease.

whole grain: a grain that maintains the same relative proportions of starchy endosperm, germ, and bran as the original (all but the husk); not refined.

whole-wheat flour: any flour made from the entire wheat kernel.

Wilson's disease: a genetic disorder of copper metabolism that creates a copper toxicity and results in neurologic symptoms such as tremors, impaired speech, inappropriate behaviors, and personality changes.

wine: an alcoholic beverage traditionally made by fermenting a sugar source such as grape juice.

without appreciable risk: practical certainty that injury will not result even after a lifetime of exposure.

X

xanthophylls (ZAN-tho-fills): pigments found in plants responsible for the color changes seen in autumn leaves.

xerophthalmia (zer-off-THAL-mee-uh): progressive blindness caused by inadequate mucus production due to severe vitamin A deficiency.

xerosis (zee-ROW-sis): abnormal drying of the skin and mucous membranes; a sign of vitamin A deficiency.

Y

yields: production per acre.

yogurt: milk product that results from the fermentation of lactic acid in milk by *Lactobacillus bulgaricus* and *Streptococcus thermophilus.*

Z

zinc: an essential trace mineral that is part of many enzymes and a constituent of insulin.

zygote (ZY-goat): the initial product of the union of ovum and sperm; a fertilized ovum.

zymogen (ZYE-mo-jen): the inactive precursor of an enzyme; sometimes called a *proenzyme.*

Index

STUDY IT

1 An Overview of Nutrition

Review the key points of this chapter below, learn the definitions of bold terms, and then take the practice quiz on the back of this card.

1.1 Food Choices

- People select **foods** based on such factors as taste and convenience, but **diets** based on **nutrition** knowledge may better support good health.
- Individual foods are neither "good" nor "bad"; daily food choices made over a lifetime may improve health or contribute to **chronic diseases.**

1.2 The Nutrients

- Foods provide **nutrients**—substances that provide **energy,** structural materials, and regulating agents to support the growth, maintenance, and repair of the body's tissues. **Essential nutrients** *must* be obtained from foods.
- The six classes of nutrients include carbohydrates, lipids (fats), proteins, **vitamins, minerals,** and water. Carbohydrates, lipids, proteins, and vitamins are **organic,** meaning they contain carbon; minerals and water are **inorganic,** meaning they do not contain carbon.
- Energy is measured in **kcalories**—a measure of heat energy. One kcalorie is the amount of heat necessary to raise the temperature of 1 kg water 1°C.
- The **energy-yielding nutrients** are carbohydrate (4 kcal/g), fat (9 kcal/g), and protein (4 kcal/g). Alcohol provides 7 kcal/g, but it is not considered a nutrient.
- Vitamins, minerals, and water do not provide energy; instead, they facilitate a variety of activities in the body.

1.3 The Science of Nutrition

- The science of nutrition is the study of nutrients and other substances in foods and the body's handling of them.
- Researchers follow the scientific method (review Figure 1-3, p. 13). They randomly assign **control** and **experimental groups,** use large sample sizes, provide **placebos,** and are **blind** to treatments. Their findings are reviewed and **replicated** by other scientists before being accepted as **valid.**
- **Correlations** indicate an association between **variables,** not a cause.

1.4 Dietary Reference Intakes

- **Dietary Reference Intakes (DRI)** are a set of nutrient intake values used to plan and evaluate diets for healthy people.
- **Estimated Average Requirement (EAR)** defines the amount of a nutrient that supports a specific function in the body for half of the population. **Recommended Dietary Allowance (RDA)** is based on the EAR and establishes a goal for dietary intake that will meet the needs of almost

all healthy people. **Adequate Intake (AI)** serves a similar purpose when an RDA cannot be determined.
- **Estimated Energy Requirement (EER)** defines the average amount of energy intake needed to maintain energy balance, and **Acceptable Macronutrient Distribution Ranges (AMDR)** define the proportions contributed by carbohydrate, fat, and protein to a healthy diet. Carbohydrates should consist of 45 to 65 percent of kcalories, fat 20 to 35 percent, and protein 10 to 35 percent.
- **Tolerable Upper Intake Level (UL)** establishes the highest amount that appears safe for regular consumption.

1.5 Nutrition Assessment

- **Malnutrition** develops when people get too little **(undernutrition),** too much **(overnutrition),** or an imbalance of energy or nutrients.
- Four **nutrition assessment** methods include historical information on diet and health, **anthropometric** measurements, physical examinations, and laboratory tests. Together, these methods reveal the stages of a nutrient deficiency (review Figure 1-7, p. 24).
- A **primary deficiency** is caused by an inadequate intake of a nutrient; a **secondary deficiency** is caused by a condition that reduces absorption, accelerates use, increases excretion, or destroys the nutrient.

1.6 Diet and Health

- **Risk factors** such as obesity and cigarette smoking increase the likelihood of disease development (review Table 1-6).
- Some risk factors, such as genetics, are important but cannot be changed. Recommendations focus on changeable, personal life choices such as diet and activity habits.
- Diet has no influence on some diseases but is linked closely to others (review Table 1-5).

TABLE 1-5 Leading Causes of Death in the United States

	Percentage of Total Deaths
1. **Heart disease**	23.7
2. **Cancers**	22.9
3. Chronic lung diseases	5.7
4. **Strokes**	5.1
5. Accidents	4.9
6. Alzheimer's disease	3.4
7. **Diabetes mellitus**	2.9
8. Pneumonia and influenza	2.1
9. Kidney disease	1.8
10. Suicide	1.5

NOTE: The diseases highlighted in bold have relationships with diet.
SOURCE: Deaths: Preliminary data for 2011, *National Vital Statistics Reports*, October 10, 2012, Centers for Disease Control and Prevention, www.cdc.gov/nchs.

TABLE 1-6 Factors Contributing to Deaths in the United States

Factors	Percentage of Deaths
Tobacco	18
Poor diet/inactivity	15
Alcohol	4
Microbial agents	3
Toxic agents	2
Motor vehicles	2
Firearms	1
Sexual behavior	<1
Illicit drugs	<1

SOURCE: A. H. Mokdad and coauthors, Actual causes of death in the United States, 2000, *Journal of the American Medical Association* 291 (2004): 1238–1245, with corrections from *Journal of the American Medical Association* 293 (2005): 298.

© Cengage Learning

Take the quiz below to test your mastery of the key chapter concepts.

1.1 Food Choices (pp. 4–6)

LEARN IT Describe how various factors influence personal food choices.

1. When people eat the foods typical of their families or geographic region, their choices are influenced by:
   a. habit.
   b. nutrition.
   c. personal preference.
   d. heritage or tradition.

1.2 The Nutrients (pp. 6–12)

LEARN IT Name the six major classes of nutrients and identify which are organic and which yield energy.

2. What is the difference between organic and inorganic?

3. How much energy do carbohydrates, fats, and proteins yield per gram? How is energy measured?

4. Describe how alcohol resembles nutrients. Why is alcohol not considered a nutrient?

5. The nutrient found most abundantly in both the human body and most foods is:
   a. fat.
   b. water.
   c. minerals.
   d. proteins.

6. The inorganic nutrients are:
   a. proteins and fats.
   b. vitamins and minerals.
   c. minerals and water.
   d. vitamins and proteins.

7. The energy-yielding nutrients are:
   a. fats, minerals, and water.
   b. minerals, proteins, and vitamins.
   c. carbohydrates, fats, and vitamins.
   d. carbohydrates, fats, and proteins.

1.3 The Science of Nutrition (pp. 12–18)

LEARN IT Explain the scientific method and how scientists use various types of research studies and methods to acquire nutrition information.

8. What is the science of nutrition?

9. Explain how variables might be correlational but not causal.

10. Studies of populations that reveal correlations between dietary habits and disease incidence are:
    a. clinical trials.
    b. laboratory studies.
    c. case-control studies.
    d. epidemiological studies.

11. An experiment in which neither the researchers nor the subjects know who is receiving the treatment is known as:
    a. double blind.
    b. double control.
    c. blind variable.
    d. placebo control.

1.4 Dietary Reference Intakes (pp. 18–22)

LEARN IT Define the four categories of the DRI and explain their purposes.

12. What judgment factors are involved in setting the energy and nutrient recommendations?

13. An RDA represents the:
    a. highest amount of a nutrient that appears safe for most healthy people.
    b. lowest amount of a nutrient that will maintain a specified criterion of adequacy.
    c. average amount of a nutrient considered adequate to meet the known nutrient needs of practically all healthy people.
    d. average amount of a nutrient that will maintain a specific biochemical or physiological function in half the people.

1.5 Nutrition Assessment (pp. 22–26)

LEARN IT Explain how the four assessment methods are used to detect energy and nutrient deficiencies and excesses.

14. What methods are used in nutrition surveys? What kinds of information can these surveys provide?

15. Historical information, physical examinations, laboratory tests, and anthropometric measurements are:
    a. techniques used in diet planning.
    b. steps used in the scientific method.
    c. approaches used in disease prevention.
    d. methods used in a nutrition assessment.

16. A deficiency caused by an inadequate dietary intake is called a(n):
    a. overt deficiency.
    b. covert deficiency.
    c. primary deficiency.
    d. secondary deficiency.

1.6 Diet and Health (pp. 26–28)

LEARN IT Identify several risk factors and explain their relationship to chronic diseases.

17. Behaviors such as smoking, dietary habits, physical activity, and alcohol consumption that influence the development of disease are known as:
    a. risk factors.
    b. chronic causes.
    c. preventive agents.
    d. disease descriptors.

HIGHLIGHT 1 Nutrition Information and Misinformation (pp. 30–35)

LEARN IT Recognize misinformation and describe how to identify reliable nutrition information.

18. Nutrition misinformation is often based on:
    a. clinical trials.
    b. anecdotal evidence.
    c. double blind studies.
    d. epidemiological research.

STUDY IT

2 Planning a Healthy Diet

Review the key points of this chapter below, learn the definitions of bold terms, and then take the practice quiz on the back of this card.

2.1 Principles and Guidelines

- A well-planned **eating pattern** delivers *adequate* nutrients, a *balanced* array of nutrients, and an appropriate amount of *energy* (*kcalories*). It is based on **nutrient-dense** foods, **moderate** in substances that can be detrimental to health, and *varied* in its selections.

- The *Dietary Guidelines for Americans* offer practical advice on how to eat for good health (review Table 2-1, p. 41).

2.2 Diet-Planning Guides

- The USDA Food Patterns help consumers select the types and amounts of foods to provide **adequacy, balance,** and **variety** in the diet. It makes it easier to plan a diet that includes a balance of grains, vegetables, fruits, protein foods, and milk products. In making any food choice, remember to view the food in the context of the total diet.

- The combination of many different foods provides the array of nutrients that are essential to a healthy diet (review Figure 2-2, pp. 44–45).

- MyPlate reminds consumers to make healthy choices from the five **food groups** (review Figure 2-4, p. 48).

2.3 Food Labels

- Food labels list ingredients in descending order of predominance by weight, nutrition facts based on standard **serving sizes,** and **Daily Values** based on a 2000-kcalorie diet (review Figure 2-9).

- **Nutrient claims** reflect the quantity of a nutrient (high or low), **health claims** reflect relationships between a nutrient and a disease (potassium reduces risk of hypertension), and **structure-function claims** reflect relationships between a nutrient and its function in the body (calcium builds bones).

> FIGURE 2-9   **Example of a Food Label**

© 2016 Cengage Learning

TEST IT

Take the quiz below to test your mastery of the key chapter concepts.

2.1 Principles and Guidelines (pp. 38–42)

LEARN IT Explain how each of the diet-planning principles can be used to plan a healthy diet.

1. What recommendations appear in the *Dietary Guidelines for Americans*?

2. The diet-planning principle that provides all the essential nutrients in sufficient amounts to support health is:
   a. balance.
   b. variety.
   c. adequacy.
   d. moderation.

3. A person who chooses a chicken leg that provides 0.5 milligram of iron and 95 kcalories instead of 2 tablespoons of peanut butter that also provides 0.5 milligram of iron but 188 kcalories is using the principle of nutrient:
   a. control.
   b. density.
   c. adequacy.
   d. moderation.

4. Which of the following is consistent with the *Dietary Guidelines for Americans*?
   a. Choose a diet restricted in fat and cholesterol.
   b. Balance the food you eat with physical activity.
   c. Choose a diet with plenty of milk products and meats.
   d. Eat an abundance of foods to ensure nutrient adequacy.

2.2 Diet-Planning Guides (pp. 42–56)

LEARN IT Use the USDA Food Patterns to develop a meal plan within a specified energy allowance.

5. Review the *Dietary Guidelines*. What types of grocery selections would you make to achieve those recommendations?

6. According to the USDA Food Patterns, cheese is grouped as a:
   a. meat.
   b. protein food.
   c. milk product.
   d. miscellaneous fat.

7. Foods within a given food group of the USDA Food Patterns are similar in their contents of:
   a. energy.
   b. proteins and fibers.
   c. vitamins and minerals.
   d. carbohydrates and fats.

8. In the exchange system, each portion of food on any given list provides about the same amount of:
   a. energy.
   b. satiety.
   c. vitamins.
   d. minerals.

9. Enriched grain products are fortified with:
   a. fiber, folate, iron, niacin, and zinc.
   b. thiamin, iron, calcium, zinc, and sodium.
   c. iron, thiamin, riboflavin, niacin, and folate.
   d. folate, magnesium, vitamin B_6, zinc, and fiber.

2.3 Food Labels (pp. 56–63)

LEARN IT Compare the information on food labels to make selections that meet specific dietary and health goals.

10. What information can you expect to find on a food label? How can this information help you choose between two similar products?

11. What are the Daily Values? How can they help you meet health recommendations?

12. Describe the differences between nutrient claims, health claims, and structure-function claims.

13. Food labels list ingredients in:
    a. alphabetical order.
    b. ascending order of predominance by weight.
    c. descending order of predominance by weight.
    d. manufacturer's order of preference.

14. "Milk builds strong bones" is an example of a:
    a. health claim.
    b. nutrition fact.
    c. nutrient content claim.
    d. structure-function claim.

15. Daily Values on food labels are based on a:
    a. 1500-kcalorie diet.
    b. 2000-kcalorie diet.
    c. 2500-kcalorie diet.
    d. 3000-kcalorie diet.

HIGHLIGHT 2 Vegetarian Diets (pp. 64–69)

LEARN IT Develop a well-balanced vegetarian meal plan.

16. Which of the following nutrients is most likely to be lacking in a vegan diet?
    a. fiber
    b. folate
    c. vitamin C
    d. vitamin B_{12}

STUDY IT

3 Digestion, Absorption, and Transport

Review the key points of this chapter below, learn the definitions of the bold terms, and then take the practice quiz on the back of this card.

3.1 Digestion

- **Digestion** breaks down foods into nutrients. **Absorption** brings the nutrients into the cells of the **small intestine** for transport to the body's cells.

- Food enters the **mouth** and travels along the **gastrointestinal tract**—down the **esophagus** and through the upper and lower esophageal **sphincters** to the **stomach,** then through the **pyloric sphincter** to the small intestine, on through the **ileocecal valve** to the **large intestine,** past the **appendix** to the **rectum,** ending at the **anus** (review Figure 3-1).

- The wavelike contractions of **peristalsis** and the periodic squeezing of **segmentation** keep things moving at a reasonable pace. Along the way, secretions from the **salivary glands,** stomach, **pancreas, liver** (via the **gallbladder**), and small intestine deliver fluids and **digestive enzymes** (review Table 3-1).

3.2 Absorption

- The many folds and **villi** of the small intestine increase its surface area, making nutrient absorption efficient.

- Nutrients pass through the cells of the intestinal villi and enter either the blood (if they are water soluble or small fat fragments) or the **lymph** (if they are fat soluble).

3.3 The Circulatory Systems

- Nutrients leaving the **digestive system** via the blood are routed directly to the liver before being transported to the body's cells.

- Nutrients leaving via the lymphatic system bypass the liver at first, but eventually enter the vascular system via the **thoracic duct,** which opens into the **subclavian vein.**

3.4 The Health and Regulation of the GI Tract

- A diverse and abundant population of **microbes** supports GI health.

- The regulation of GI processes depends on the coordinated efforts of the hormonal system and the nervous system.

- Together, digestion and absorption break down foods into nutrients for the body's use.

- To function optimally, a healthy GI tract needs a balanced diet, adequate rest, and regular physical activity.

> FIGURE 3-1   **The Gastrointestinal Tract**

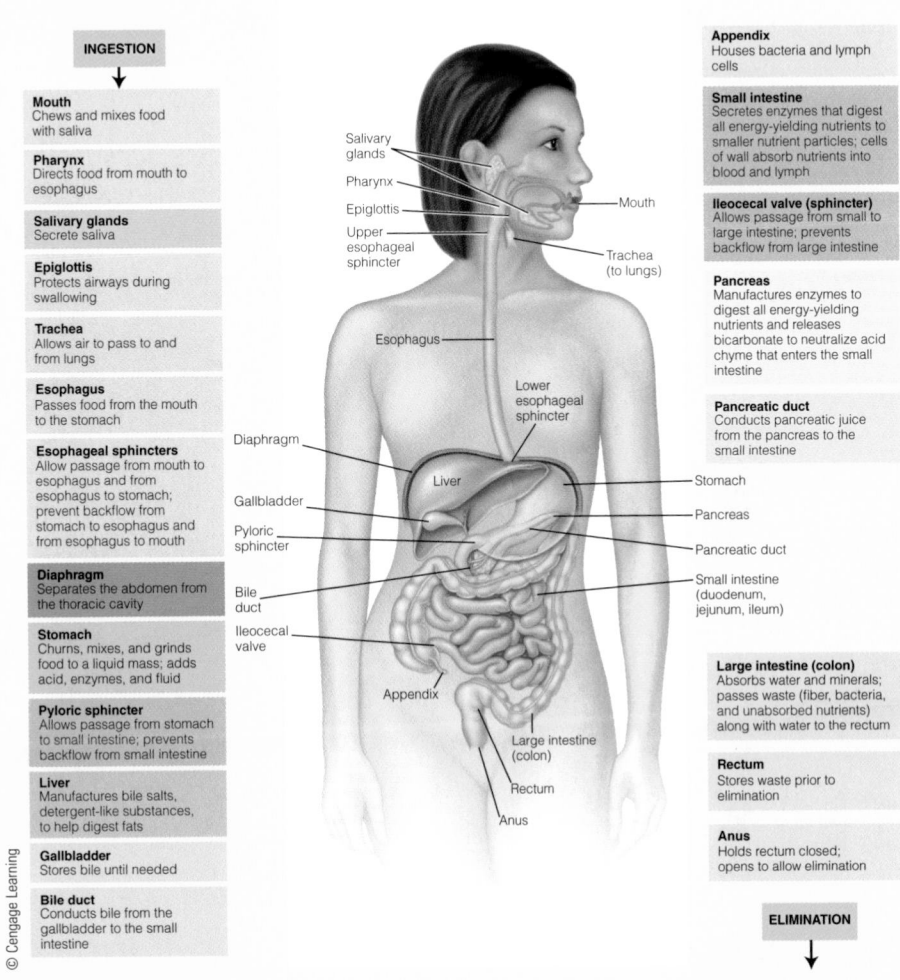

INGESTION

Mouth
Chews and mixes food with saliva

Pharynx
Directs food from mouth to esophagus

Salivary glands
Secrete saliva

Epiglottis
Protects airways during swallowing

Trachea
Allows air to pass to and from lungs

Esophagus
Passes food from the mouth to the stomach

Esophageal sphincters
Allow passage from mouth to esophagus and from esophagus to stomach; prevent backflow from stomach to esophagus and from esophagus to mouth

Diaphragm
Separates the abdomen from the thoracic cavity

Stomach
Churns, mixes, and grinds food to a liquid mass; adds acid, enzymes, and fluid

Pyloric sphincter
Allows passage from stomach to small intestine; prevents backflow from small intestine

Liver
Manufactures bile salts, detergent-like substances, to help digest fats

Gallbladder
Stores bile until needed

Bile duct
Conducts bile from the gallbladder to the small intestine

Appendix
Houses bacteria and lymph cells

Small intestine
Secretes enzymes that digest all energy-yielding nutrients to smaller nutrient particles; cells of wall absorb nutrients into blood and lymph

Ileocecal valve (sphincter)
Allows passage from small to large intestine; prevents backflow from large intestine

Pancreas
Manufactures enzymes to digest all energy-yielding nutrients and releases bicarbonate to neutralize acid chyme that enters the small intestine

Pancreatic duct
Conducts pancreatic juice from the pancreas to the small intestine

Large intestine (colon)
Absorbs water and minerals; passes waste (fiber, bacteria, and unabsorbed nutrients) along with water to the rectum

Rectum
Stores waste prior to elimination

Anus
Holds rectum closed; opens to allow elimination

Labels on figure: Salivary glands, Pharynx, Epiglottis, Upper esophageal sphincter, Mouth, Trachea (to lungs), Esophagus, Lower esophageal sphincter, Diaphragm, Liver, Stomach, Gallbladder, Pancreas, Pyloric sphincter, Pancreatic duct, Bile duct, Small intestine (duodenum, jejunum, ileum), Ileocecal valve, Large intestine (colon), Appendix, Rectum, Anus

ELIMINATION

© Cengage Learning

TABLE 3-1   Summary of Digestive Secretions and Their Major Actions

Organ or Gland	Target Organ	Secretion	Action
Salivary glands	Mouth	Saliva	Fluid eases swallowing; salivary enzyme breaks down some **carbohydrate.***
Gastric glands	Stomach	Gastric juice	Fluid mixes with bolus; hydrochloric acid uncoils **proteins;** enzymes break down proteins; mucus protects stomach cells.*
Pancreas	Small intestine	Pancreatic juice	Bicarbonate neutralizes acidic gastric juices; pancreatic enzymes break down **carbohydrates, fats,** and **proteins.**
Liver	Gallbladder	Bile	Bile is stored until needed.
Gallbladder	Small intestine	Bile	Bile emulsifies **fat** so that enzymes can have access to break it down.
Intestinal glands	Small intestine	Intestinal juice	Intestinal enzymes break down **carbohydrate, fat,** and **protein** fragments; mucus protects the intestinal wall.

*Saliva and gastric juice also contain lipases, but most fat breakdown occurs in the small intestine.

© Cengage Learning

TEST IT

Take the quiz below to test your mastery of the key chapter concepts.

3.1 Digestion (pp. 72–78)

LEARN IT Explain how foods move through the digestive system, describing the actions of the organs, muscles, and digestive secretions along the way.

1. Describe the challenges associated with digesting food and the solutions offered by the human body.

2. Name five organs that secrete digestive juices. How do the juices and enzymes facilitate digestion?

3. The semiliquid, partially digested food that travels through the intestinal tract is called:
   a. bile.
   b. lymph.
   c. chyme.
   d. secretin.

4. The muscular contractions that move food through the GI tract are called:
   a. hydrolysis.
   b. sphincters.
   c. peristalsis.
   d. bowel movements.

5. The main function of bile is to:
   a. emulsify fats.
   b. catalyze hydrolysis.
   c. slow protein digestion.
   d. neutralize stomach acidity.

6. The pancreas neutralizes stomach acid in the small intestine by secreting:
   a. bile.
   b. mucus.
   c. enzymes.
   d. bicarbonate.

7. Which nutrient passes through the GI tract mostly undigested and unabsorbed?
   a. fat
   b. fiber
   c. protein
   d. carbohydrate

3.2 Absorption (pp. 78–82)

LEARN IT Describe the anatomical details of the intestinal cells that facilitate nutrient absorption.

8. The fingerlike projections on the small intestine that dramatically increase its surface area are called:
   a. villi.
   b. crypts.
   c. goblet cells.
   d. chylomicrons.

9. Absorption occurs primarily in the:
   a. mouth.
   b. stomach.
   c. small intestine.
   d. large intestine.

3.3 The Circulatory Systems (pp. 82–84)

LEARN IT Explain how nutrients are routed in the circulatory systems from the GI tract into the body and identify which nutrients enter the blood directly and which must first enter the lymph.

10. All blood leaving the GI tract travels first to the:
    a. heart.
    b. liver.
    c. kidneys.
    d. pancreas.

11. Which nutrients leave the GI tract by way of the lymphatic system?
    a. water and minerals
    b. proteins and minerals
    c. all vitamins and minerals
    d. fats and fat-soluble vitamins

3.4 The Health and Regulation of the GI Tract (pp. 85–89)

LEARN IT Describe how bacteria, hormones, and nerves influence the health and activities of the GI tract.

12. How does the composition of the diet influence the functioning of the GI tract?

13. What steps can you take to help your GI tract function at its best?

14. Digestion and absorption are coordinated by the:
    a. pancreas and kidneys.
    b. liver and gallbladder.
    c. hormonal system and the nervous system.
    d. vascular system and the lymphatic system.

15. Gastrin, secretin, and cholecystokinin are examples of:
    a. crypts.
    b. enzymes.
    c. hormones.
    d. goblet cells.

HIGHLIGHT 3 Common Digestive Problems (pp. 90–96)

LEARN IT Outline strategies to prevent or alleviate common GI problems.

16. Which of the following is most likely to aggravate heartburn?
    a. eat small meals
    b. drink liquids between meals
    c. eat slowly and chew thoroughly
    d. lie down to rest immediately after eating

3.c 4.c 5.a 6.d 7.b 8.a 9.c 10.b 11.d 14.c 15.c 16.d
Multiple Choice Answers

STUDY IT

4 The Carbohydrates: Sugars, Starches, and Fibers

Review the key points of this chapter below, learn the definitions of bold terms, and then take the practice quiz on the back of this card.

4.1 The Chemist's View of Carbohydrates

- **Carbohydrates** include **monosaccharides, disaccharides,** and **polysaccharides** (review Table 4-1).

- Carbohydrates are made of carbon (C), oxygen (O), and hydrogen (H); each atom forms a specified number of chemical bonds: carbon forms four, oxygen forms two, and hydrogen forms one (review Figure 4-1, p. 100).

- Monosaccharides (**glucose, fructose,** and **galactose**) all have the same chemical formula ($C_6H_{12}O_6$), but their structures differ. Disaccharides (**maltose, sucrose,** and **lactose**) each contain a glucose paired with one of the three monosaccharides.

- A **condensation** reaction can bond two monosaccharides together to form a disaccharide and water (review Figure 4-4, p. 102). A **hydrolysis** reaction can use water to split a disaccharide into its two monosaccharides (review Figure 4-5, p. 102).

- Chains of monosaccharides are called **polysaccharides** and include **glycogen, starches,** and **dietary fibers.** Both glycogen and starch are storage forms of glucose—glycogen in the body, and starch in plants—and both yield energy.

- Dietary fibers contain glucose (and other monosaccharides), but their bonds cannot be broken by human digestive enzymes; they yield little, if any, energy.

- **Soluble fibers** dissolve in water to form gels and are easily digested by bacteria in the colon. **Insoluble fibers** do not dissolve in water or form gels and are less readily fermented.

TABLE 4-1 The Carbohydrate Family

Monosaccharides	
Glucose	
Fructose	
Galactose	
Disaccharides	
Maltose (glucose + glucose)	
Sucrose (glucose + fructose)	
Lactose (glucose + galactose)	
Polysaccharides	
Glycogen[a]	
Starches (amylose and amylopectin)	
Fibers (soluble and insoluble)	

© Cengage Learning

[a]Glycogen is a polysaccharide, but not a common dietary source of carbohydrate.

4.2 Digestion and Absorption of Carbohydrates

- The body digests starches into the disaccharide maltose. Maltose and the other disaccharides (lactose and sucrose) from foods are broken down into monosaccharides, which are absorbed (review Figure 4-9, p. 107).

- Fibers help to regulate the passage of food through the GI tract and slow the absorption of glucose.

- **Lactose intolerance** occurs when there is insufficient lactase to digest the disaccharide lactose found in milk and milk products. Symptoms include GI distress.

4.3 Glucose in the Body

- Dietary carbohydrates provide glucose that can be used by the cells for energy, stored by the liver and muscles as glycogen, or converted into fat if intakes exceed needs.

- All of the body's cells depend on glucose; those of the central nervous system are especially dependent on it.

- Without glucose, the body is forced to break down its protein tissues to make glucose (**gluconeogenesis**) and to alter energy metabolism to make **ketone bodies** from fats.

- Blood glucose regulation depends on two pancreatic hormones: **insulin** to move glucose from the blood into the cells when levels are high and **glucagon** to free glucose from glycogen stores and release it into the blood when levels are low (review Figure 4-10, p. 111).

4.4 Health Effects and Recommended Intakes of Sugars

- Excessive intakes of **sugars** may increase the risk of dental caries, displace needed nutrients and fiber, and contribute to obesity when energy intake exceeds needs.

- Concentrated sweets are relatively low in nutrients, high in kcalories, and may need to be limited; sugars that occur naturally in fruits, vegetables, and milk are acceptable.

- To control weight gain, blood glucose, and dental caries, consumers may use alternative sweeteners (**artificial sweeteners,** herbal products, and **sugar alcohols**) to limit kcalories and minimize sugar intake (review Table 4-8, p. 119).

4.5 Health Effects and Recommended Intakes of Starch and Fibers

- Adequate intake of fiber fosters weight management, lowers blood cholesterol, and may help prevent colon cancer, diabetes, hemorrhoids, appendicitis, and diverticulosis.

- Excessive intake of fiber displaces energy- and nutrient-dense foods, causes intestinal discomfort and distention, and may interfere with mineral absorption.

- Because starches and fibers help control body weight and prevent heart disease, cancer, diabetes, and GI disorders, the *Dietary Guidelines* suggest plenty of whole grains, vegetables, legumes, and fruits—enough to provide 45 to 65 percent of the daily energy intake from carbohydrate.

TEST IT

Take the quiz below to test your mastery of the key chapter concepts.

4.1 The Chemist's View of Carbohydrates

(pp. 100–105)

LEARN IT Identify the monosaccharides, disaccharides, and polysaccharides common in nutrition by their chemical structures and major food sources.

1. What happens in a condensation reaction? In a hydrolysis reaction?

2. How are starch and glycogen similar, and how do they differ? How do the fibers differ from the other polysaccharides?

3. Disaccharides include:
   a. starch, glycogen, and fiber.
   b. amylose, pectin, and dextrose.
   c. sucrose, maltose, and lactose.
   d. glucose, galactose, and fructose.

4. The making of a disaccharide from two monosaccharides is an example of:
   a. digestion.
   b. hydrolysis.
   c. condensation.
   d. gluconeogenesis.

5. The significant difference between starch and cellulose is that:
   a. starch is a polysaccharide, but cellulose is not.
   b. animals can store glucose as starch, but not as cellulose.
   c. hormones can make glucose from cellulose, but not from starch.
   d. digestive enzymes can break the bonds in starch, but not in cellulose.

4.2 Digestion and Absorption of Carbohydrates (pp. 105–108)

LEARN IT Summarize carbohydrate digestion and absorption.

6. What role does fiber play in digestion and absorption?

7. Describe lactose intolerance and its symptoms.

8. The ultimate goal of carbohydrate digestion and absorption is to yield:
   a. fibers.
   b. glucose.
   c. enzymes.
   d. amylase.

9. The enzyme that breaks a disaccharide into glucose and galactose is:
   a. amylase.
   b. maltase.
   c. sucrase.
   d. lactase.

4.3 Glucose in the Body (pp. 108–113)

LEARN IT Explain how the body maintains its blood glucose concentration and what happens when blood glucose rises too high or falls too low.

10. What are the possible fates of glucose in the body? What is the protein-sparing action of carbohydrate?

11. The storage form of glucose in the body is:
    a. insulin.
    b. maltose.
    c. glucagon.
    d. glycogen.

12. With insufficient glucose in metabolism, fat fragments combine to form:
    a. dextrins.
    b. mucilages.
    c. phytic acids.
    d. ketone bodies.

13. What does the pancreas secrete when blood glucose rises? When blood glucose falls?
    a. insulin; glucagon
    b. glucagon; insulin
    c. insulin; glycogen
    d. glycogen; epinephrine

4.4 Health Effects and Recommended Intakes of Sugars (pp. 113–120)

LEARN IT Describe how added sugars can contribute to health problems.

14. What are the dietary recommendations regarding concentrated sugar intakes?

15. Describe the risks and benefits of using alternative sweeteners.

4.5 Health Effects and Recommended Intakes of Starch and Fibers (pp. 120–127)

LEARN IT Identify the health benefits of, and recommendations for, starches and fibers.

16. What foods provide starches and fibers?

17. Carbohydrates are found in virtually all foods except:
    a. milks.
    b. meats.
    c. breads.
    d. fruits.

18. What percentage of the daily energy intake should come from carbohydrates?
    a. 15 to 20
    b. 25 to 30
    c. 45 to 50
    d. 45 to 65

HIGHLIGHT 4 Carbs, kCalories, and Controversies (pp. 128–131)

LEARN IT Summarize the key scientific evidence behind some of the current controversies surrounding carbohydrates and their kcalories.

19. Which of the following is true?
    a. High-fructose corn syrup is chemically similar to sucrose.
    b. Diets with a high glycemic index are most effective in fostering weight loss.
    c. Energy intake from carbohydrates has remained stable over the past 40 years.
    d. Some people have a physiological addiction to sugar that makes it impossible for them to moderate use.

Multiple Choice Answers
3.c 4.c 5.d 8.b 9.d 11.d 12.d 13.a 17.b 18.d 19.a

STUDY IT

5 The Lipids: Triglycerides, Phospholipids, and Sterols

Review the key points of this chapter below, learn the definitions of the bold terms, and then take the practice quiz on the back of this card.

5.1 The Chemist's View of Fatty Acids and Triglycerides

- The predominant **lipids** both in foods and in the body are **triglycerides: glycerol** with three **fatty acids** attached by way of condensation reactions (review Figure 5-3, p. 137). Other lipids include phospholipids and sterols (review Table 5-2).

- Fatty acids vary in the length of their carbon chains, their degrees of unsaturation, and the location of their double bond(s). **Saturated fatty acids** are fully loaded with hydrogens; **unsaturated** (**monounsaturated** or **polyunsaturated**) fatty acids are missing hydrogens and have double bonds.

- **Hydrogenation** protects against oxidation (thereby promoting shelf-life) and alters the texture of foods by making liquid vegetable **oils** more solid. This process makes polyunsaturated **fats** more saturated and creates *trans*-**fatty acids.**

5.2 The Chemist's View of Phospholipids and Sterols

- The chemical structure of **phospholipids,** including **lecithin,** allows them to be soluble in both water and fat. In the body, phospholipids are part of cell membranes; in foods, phospholipids act as emulsifiers to mix fats with water.

- **Sterols** have a multiple-ring structure that differs from the structure of other lipids. In the body, sterols include **cholesterol,** bile, vitamin D, and some hormones. Animal-derived foods contain cholesterol.

5.3 Digestion, Absorption, and Transport of Lipids

- **Bile emulsifies** fats, making them accessible to the **lipases** that dismantle triglycerides to **monoglycerides** and fatty acids for absorption (review Figure 5-12, p. 143).

- Four types of **lipoproteins** carry triglycerides, phospholipids, and cholesterol throughout the body: **chylomicrons** are the

largest and contain mostly dietary triglycerides, **VLDL** are smaller and are about half triglycerides, **LDL** are smaller still and contain mostly cholesterol, and **HDL** are the densest and are rich in protein (review Figure 5-16, p. 146).

5.4 Lipids in the Body

- In the body, triglycerides provide energy, insulate against temperature extremes, protect against shock, and help the body use carbohydrate and protein efficiently.

- **Linoleic acid** (18 carbons, omega-6) and **linolenic acid** (18 carbons, omega-3) are **essential fatty acids,** serving as structural parts of cell membranes and as precursors to the longer fatty acids that can make **eicosanoids.**

- The body stores fat if given excesses, and uses body fat for energy when needed. (The liver can also convert excess carbohydrate and protein into fat.) Fat breakdown requires carbohydrate for maximum efficiency; without carbohydrate, fatty acids break down to **ketone bodies.**

5.5 Health Effects and Recommended Intakes of Saturated Fats, *Trans* Fats, and Cholesterol

- Some fat in the diet is necessary, but too much fat provides energy (kcalories) without nutrients, which leads to obesity and nutrient inadequacies.

- Too much saturated fat, *trans* fat, and cholesterol increases the risk of heart disease and possibly cancer.

- Recommendations advise that a diet be moderate in total fat and low in saturated fat, *trans* fat, and cholesterol.

5.6 Health Effects and Recommended Intakes of Monounsaturated and Polyunsaturated Fats

- Some fat in the diet has health benefits, especially the monounsaturated and polyunsaturated fats that protect against heart disease and possibly cancer.

- The *Dietary Guidelines* recommend replacing saturated fats with monounsaturated and polyunsaturated fats, particularly **omega-3** fatty acids from foods such as fatty fish, not from supplements.

- Many selection and preparation strategies can help bring these goals within reach, and food labels help to identify foods consistent with these guidelines.

TABLE 5-2 The Lipid Family

Triglycerides

- 1 Glycerol (per triglyceride) and
- 3 Fatty acids (per triglyceride); depending on the number of double bonds, fatty acids may be:
  - *Saturated* (no double bonds)
  - *Monounsaturated* (one double bond)
  - *Polyunsaturated* (more than one double bond); depending on the location of the double bonds, polyunsaturated fatty acids may be:
    - *Omega-3* (double bond closest to methyl end is 3 carbons away)
    - *Omega-6* (double bond closest to methyl end is 6 carbons away)

Phospholipids (such as lecithin)

Sterols (such as cholesterol)

© Cengage Learning

Take the quiz below to test your mastery of the key chapter concepts.

5.1 The Chemist's View of Fatty Acids and Triglycerides (pp. 134–140)

LEARN IT Recognize the chemistry of fatty acids and triglycerides and differences between saturated and unsaturated fats.

1. Name three classes of lipids found in the body and in foods. What are some of their functions in the body? What features do fats bring to foods?

2. What features distinguish fatty acids from each other?

3. What does the term *omega* mean with respect to fatty acids? Describe the roles of the omega fatty acids in disease prevention.

4. Describe the structure of a triglyceride.

5. What does hydrogenation do to fats? What are *trans*-fatty acids, and how do they influence heart disease?

6. Saturated fatty acids:
   a. are always 18 carbons long.
   b. have at least one double bond.
   c. are fully loaded with hydrogens.
   d. are always liquid at room temperature.

7. A triglyceride consists of:
   a. three glycerols attached to a lipid.
   b. three fatty acids attached to a glucose.
   c. three fatty acids attached to a glycerol.
   d. three phospholipids attached to a cholesterol.

8. The difference between *cis*- and *trans*-fatty acids is:
   a. the number of double bonds.
   b. the length of their carbon chains.
   c. the location of the first double bond.
   d. the configuration around the double bond.

5.2 The Chemist's View of Phospholipids and Sterols (pp. 140–141)

LEARN IT Describe the chemistry, food sources, and roles of phospholipids and sterols.

9. Which of the following is *not* true? Lecithin is:
   a. an emulsifier.
   b. a phospholipid.
   c. an essential nutrient.
   d. a constituent of cell membranes.

5.3 Digestion, Absorption, and Transport of Lipids (pp. 142–147)

LEARN IT Summarize fat digestion, absorption, and transport.

10. What do lipoproteins do? What are the differences among the chylomicrons, VLDL, LDL, and HDL?

11. Chylomicrons are produced in the:
    a. liver.   c. gallbladder.
    b. pancreas.   d. small intestine.

12. Transport vehicles for lipids are called:
    a. micelles.   c. blood vessels.
    b. lipoproteins.   d. monoglycerides.

5.4 Lipids in the Body (pp. 148–150)

LEARN IT Outline the major roles of fats in the body, including a discussion of essential fatty acids and the omega fatty acids.

13. Which of the following is *not* true? Fats:
    a. contain glucose.
    b. provide energy.
    c. protect against organ shock.
    d. carry vitamins A, D, E, and K.

14. The essential fatty acids include:
    a. stearic acid and oleic acid.
    b. oleic acid and linoleic acid.
    c. palmitic acid and linolenic acid.
    d. linoleic acid and linolenic acid.

5.5 Health Effects and Recommended Intakes of Saturated Fats, *Trans* Fats, and Cholesterol (pp. 150–152)

LEARN IT Explain the relationships among saturated fats, *trans* fat, and cholesterol and chronic diseases, noting recommendations.

15. How does excessive fat intake influence health? What factors influence LDL, HDL, and total blood cholesterol?

16. What are the dietary recommendations regarding saturated fat and cholesterol intake? List ways to reduce intake.

17. The lipoprotein most associated with a high risk of heart disease is:
    a. CHD.   c. LDL.
    b. HDL.   d. LPL.

5.6 Health Effects and Recommended Intakes of Monounsaturated and Polyunsaturated Fats (pp. 153–162)

LEARN IT Explain the relationships between monounsaturated and polyunsaturated fats and health, noting recommendations.

18. What is the Daily Value for fat (for a 2000-kcalorie diet)?

19. A person consuming 2200 kcalories a day who wants to meet health recommendations should limit daily fat intake to:
    a. 20 to 35 grams.   c. 75 to 100 grams.
    b. 50 to 85 grams.   d. 90 to 130 grams.

HIGHLIGHT 5 High-Fat Foods–Friend or Foe? (pp. 164–169)

LEARN IT Identify which fats support health and which impair it.

20. List foods that are high in saturated fats and foods that are high in unsaturated fats that could easily be used to replace them.

Review the key points of this chapter below, learn the definitions of the bold terms, and then take the practice quiz on the back of this card.

6.1 The Chemist's View of Proteins

- **Proteins** are more chemically complex than carbohydrates or lipids; they are made of 20 different **amino acids,** 9 of which the body cannot make. These 9 are the **essential amino acids**—histidine, isoleucine, leucine, lysine, methionine, phenylalanine, threonine, tryptophan, and valine.

- Each amino acid contains an amino group, an acid group, a hydrogen atom, and a distinctive side group, all attached to a central carbon atom.

- Cells link amino acids together by a **peptide bond** in a series of condensation reactions to create proteins (review Figure 6-3, p. 173). The distinctive sequence of amino acids in each protein determines its unique shape and function.

6.2 Digestion and Absorption of Proteins

- The stomach's hydrochloric acid first denatures dietary proteins, then **enzymes** cleave them into smaller polypeptides and some amino acids.

- Pancreatic and intestinal enzymes split **polypeptides** further, to **oligo-, tri-,** and **dipeptides,** and then split most of these to single amino acids that can be absorbed into the intestinal cells (review Figure 6-6, p. 176).

6.3 Proteins in the Body

- Cells synthesize proteins according to the **genetic** information provided by the **DNA** in the nucleus of each cell (review Figure 6-7, p. 178). This information dictates the sequence in which amino acids are linked together to form a given protein. Sequencing errors occasionally occur, sometimes with significant consequences.

- A sampling of protein functions are summarized in Table 6-3.

- Proteins are constantly being synthesized and broken down as needed, a process known as **protein turnover.**

- The body's assimilation of amino acids into proteins and its release of amino acids via protein degradation and excretion can be tracked by measuring **nitrogen balance,** which should be positive during growth and steady in adulthood. An energy deficit or an inadequate protein intake may force the body to break down lean body tissue and use amino acids as fuel, creating a negative nitrogen balance.

- Protein eaten in excess of need is degraded and stored as body fat.

6.4 Protein in Foods

- **High-quality proteins** deliver all of the essential amino acids in adequate amounts, which ensures protein synthesis. Mixtures of foods containing **complementary proteins** can each supply the amino acids missing in the other.

- In addition to its amino acid content, the quality of a protein is measured by its **digestibility** and its ability to support growth.

6.5 Health Effects and Recommended Intakes of Protein

- Protein deficiency impairs the body's ability to grow and function optimally.

- Excess protein offers no advantage and may incur health problems as well.

- The optimal diet is adequate in energy from carbohydrate and fat and delivers 0.8 grams of protein per kilogram of healthy body weight each day.

- Healthy people do not need protein or amino acid supplements.

TABLE 6-3 **Protein Functions in the Body**

Structural materials	Proteins form integral parts of most body tissues and provide strength and shape to skin, tendons, membranes, muscles, organs, and bones.
Enzymes	Proteins facilitate chemical reactions.
Hormones	Proteins regulate body processes. (Some, but not all, hormones are proteins.)
Fluid balance	Proteins help to maintain the volume and composition of body fluids.
Acid-base balance	Proteins help to maintain the acid-base balance of body fluids by acting as **buffers.**
Transportation	Proteins transport substances, such as lipids, vitamins, minerals, and oxygen, around the body.
Antibodies	Proteins inactivate foreign invaders, thus protecting the body against diseases.
Energy and glucose	Proteins provide some fuel, and glucose if needed, for the body's energy needs.
Other	The protein fibrin creates blood clots; the protein **collagen** forms scars; the protein opsin participates in vision.

© Cengage Learning

Take the quiz below to test your mastery of the key chapter concepts.

6.1 The Chemist's View of Proteins
(pp. 172–175)

LEARN IT Recognize the chemical structures of amino acids and proteins.

1. Explain how the sequence of amino acids affects protein shape and function.

2. What are essential amino acids?

3. Which part of its chemical structure differentiates one amino acid from another?
   a. its side group
   b. its acid group
   c. its amino group
   d. its double bonds

4. Isoleucine, leucine, and lysine are:
   a. proteases.
   b. polypeptides.
   c. essential amino acids.
   d. complementary proteins.

6.2 Digestion and Absorption of Proteins (pp. 175–176)

LEARN IT Summarize protein digestion and absorption.

5. In the stomach, hydrochloric acid:
   a. denatures proteins and activates pepsin.
   b. hydrolyzes proteins and denatures pepsin.
   c. emulsifies proteins and releases peptidase.
   d. condenses proteins and facilitates digestion.

6.3 Proteins in the Body (pp. 177–184)

LEARN IT Describe how the body makes proteins and uses them to perform various roles.

6. What are enzymes? What roles do they play in chemical reactions? Describe the differences between enzymes and hormones.

7. How does the body use amino acids? What is deamination? Define *nitrogen balance*. What conditions are associated with zero, positive, and negative balance?

8. Proteins that maintain the acid-base balance of the blood and body fluids by accepting and releasing hydrogen ions are:
   a. buffers.
   b. enzymes.
   c. hormones.
   d. antigens.

9. If an essential amino acid that is needed to make a protein is unavailable, the cells must:
   a. deaminate another amino acid.
   b. substitute a similar amino acid.
   c. break down proteins to obtain it.
   d. synthesize the amino acid from glucose and nitrogen.

10. Protein turnover describes the amount of protein:
    a. found in foods and the body.
    b. absorbed from the diet.
    c. synthesized and degraded.
    d. used to make glucose.

6.4 Protein in Foods (pp. 185–186)

LEARN IT Explain the differences between high-quality and low-quality proteins, including notable food sources of each.

11. How can vegetarians meet their protein needs without eating meat?

12. Which of the following foods provides the highest quality protein?
    a. egg   c. gelatin
    b. corn   d. whole grains

6.5 Health Effects and Recommended Intakes of Protein (pp. 186–191)

LEARN IT Identify the health benefits of, and recommendations for, protein.

13. How might protein excess, or the type of protein eaten, influence health?

14. What factors are considered in establishing recommended protein intakes?

15. What are the benefits and risks of taking protein and amino acid supplements?

16. The protein RDA for a healthy adult who weighs 180 pounds is:
    a. 50 milligrams/day.
    b. 65 grams/day.
    c. 180 grams/day.
    d. 2000 milligrams/day.

17. Which of these foods has the least protein per ½ cup?
    a. rice   c. pinto beans
    b. broccoli   d. orange juice

HIGHLIGHT 6 Nutritional Genomics (pp. 194–199)

LEARN IT Explain how nutrients influence gene activity (nutrigenomics) and how genes influence the activities of nutrients (nutrigenetics).

18. The area of study that examines how environmental factors influence gene expression without changing the DNA is known as:
    a. epigenetics.
    b. chromosomal biology.
    c. methylation translation.
    d. microarray technologies.

Multiple Choice Answers
3.a 4.c 5.a 8.a 9.c 10.c 12.a 16.b 17.d 18.a

STUDY IT

7 Energy Metabolism

Review the key points of this chapter below, learn the definitions of the bold terms, and then take the practice quiz on the back of this card.

7.1 Chemical Reactions in the Body

- During digestion, the energy-yielding nutrients—carbohydrates, lipids, and proteins—are broken down to glucose (and other monosaccharides), glycerol, fatty acids, and amino acids. These compounds may enter metabolic pathways to yield energy (review Figure 7-5).

- **Enzymes** with their **coenzymes** help cells use nutrients to build compounds **(anabolism)** or break them down to release energy **(catabolism)**—review Figure 7-2, p. 203.

- **ATP**—a high-energy compound—captures the energy released during catabolism (review Figure 7-4, p. 204).

7.2 Breaking Down Nutrients for Energy

- Glucose breakdown begins with **glycolysis,** a pathway that produces **pyruvate** (review Figure 7-6, p. 208).

- Pyruvate may be converted to **lactate anaerobically** (without oxygen) or to **acetyl CoA aerobically** (with oxygen).

- Pyruvate can make glucose; acetyl CoA cannot make glucose (review Figure 7-9, p. 210).

- The glycerol part of a triglyceride can make either pyruvate (and then glucose) or acetyl CoA. The fatty acids of a triglyceride *cannot* make glucose; they can provide abundant acetyl CoA (review Figure 7-11, p. 211).

- Some amino acids can be used to make glucose; others can be used either to provide energy or to make fat. Before an amino acid enters these metabolic pathways, its nitrogen-containing amino group must be removed through deamination.

- The digestion of carbohydrate yields glucose (and other monosaccharides); some glucose is stored as glycogen, and some is broken down to pyruvate and acetyl CoA.

- The digestion of fat yields glycerol and fatty acids; some are reassembled and stored as body fat, and others are broken down to acetyl CoA.

- The digestion of protein yields amino acids; most amino acids are used to build body protein or other nitrogen-containing compounds, some are broken down to acetyl CoA, and others enter the **TCA cycle** directly.

- Acetyl CoA may enter the TCA cycle to release energy (review Figure 7-15, p. 214) or combine with other molecules of acetyl CoA to make body fat.

7.3 Feasting and Fasting

- If energy intake exceeds energy needs, the result will be weight gain—regardless of whether the excess is from protein, carbohydrate, or fat (review Table 7-2). The body is most efficient at storing excess energy from dietary fat.

- When fasting, the body adapts to conserve energy and minimize losses by increasing fat

> FIGURE 7-5 **Simplified Overview of Energy-Yielding Pathways**

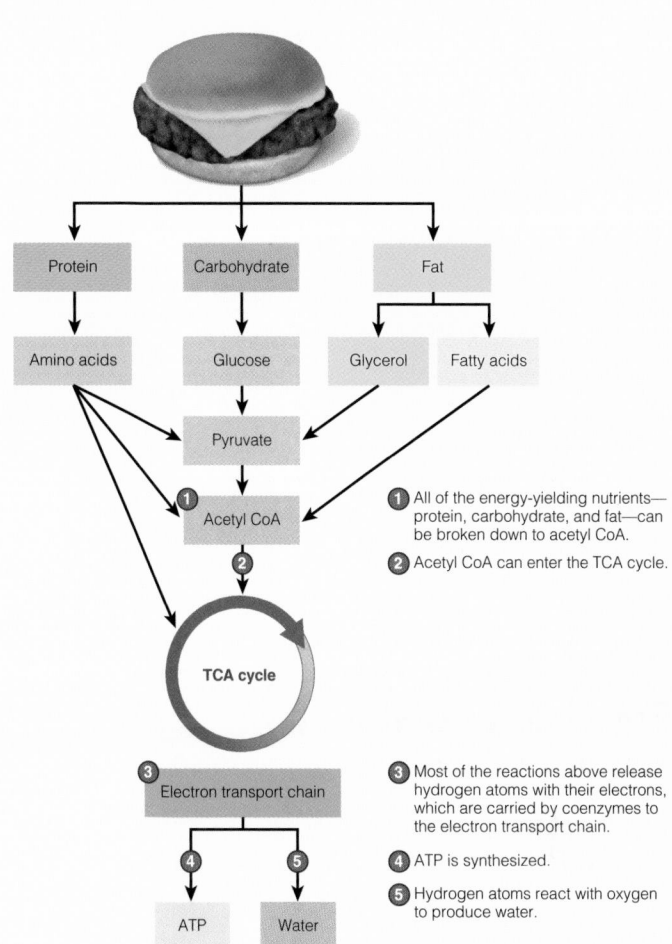

1. All of the energy-yielding nutrients—protein, carbohydrate, and fat—can be broken down to acetyl CoA.
2. Acetyl CoA can enter the TCA cycle.
3. Most of the reactions above release hydrogen atoms with their electrons, which are carried by coenzymes to the electron transport chain.
4. ATP is synthesized.
5. Hydrogen atoms react with oxygen to produce water.

© Cengage Learning

breakdown to **fuel** most cells, using glycerol and amino acids to make glucose for the brain and red blood cells, producing **ketone bodies** for the brain, suppressing appetite, and slowing **metabolism.**

TABLE 7-2 Review of Energy-Yielding Nutrient End Points

Nutrient	Yields energy?	Yields glucose?	Yields amino acids and body proteins?	Yields fat stores?
Carbohydrates (glucose)	Yes	Yes	Yes—when nitrogen is available, can yield *nonessential* amino acids	Yes
Lipids (fatty acids)	Yes	No	No	Yes
Lipids (glycerol)	Yes	Yes—when carbohydrate is unavailable	Yes—when nitrogen is available, can yield *nonessential* amino acids	Yes
Proteins (amino acids)	Yes	Yes—when carbohydrate is unavailable	Yes	Yes

© Cengage Learning

Take the quiz below to test your mastery of the key chapter concepts.

7.1 Chemical Reactions in the Body

(pp. 202–205)

LEARN IT Identify the nutrients involved in energy metabolism and the high-energy compound that captures the energy released during their breakdown.

1. Define *metabolism*, *anabolism*, and *catabolism*; give an example of each.

2. What are coenzymes, and what service do they provide in metabolism?

3. Hydrolysis is an example of a(n):
   a. coupled reaction.
   b. anabolic reaction.
   c. catabolic reaction.
   d. synthesis reaction.

4. During metabolism, released energy is captured and transferred by:
   a. enzymes.
   b. pyruvate.
   c. acetyl CoA.
   d. adenosine triphosphate.

7.2 Breaking Down Nutrients for Energy

(pp. 205–215)

LEARN IT Summarize the main steps in the energy metabolism of glucose, glycerol, fatty acids, and amino acids.

5. Name the four basic units, derived from foods, that are used by the body in energy metabolism. How many carbons are in the "backbones" of each?

6. Define *aerobic* and *anaerobic*. How does insufficient oxygen influence metabolism?

7. How does the body dispose of excess nitrogen?

8. The body derives most of its energy from:
   a. proteins and fats.
   b. vitamins and minerals.
   c. glucose and fatty acids.
   d. glycerol and amino acids.

9. Glycolysis:
   a. requires oxygen.
   b. generates abundant energy.
   c. converts glucose to pyruvate.
   d. produces ammonia as a by-product.

10. The pathway from pyruvate to acetyl CoA:
    a. produces lactate.
    b. is known as gluconeogenesis.
    c. is metabolically irreversible.
    d. requires more energy than it produces.

11. For complete oxidation, acetyl CoA enters:
    a. glycolysis.
    b. the TCA cycle.
    c. the Cori cycle.
    d. the electron transport chain.

12. Deamination of an amino acid produces:
    a. vitamin B_6 and energy.
    b. pyruvate and acetyl CoA.
    c. ammonia and a keto acid.
    d. carbon dioxide and water.

13. Before entering the TCA cycle, each of the energy-yielding nutrients is broken down to:
    a. ammonia.
    b. pyruvate.
    c. electrons.
    d. acetyl CoA.

7.3 Feasting and Fasting (pp. 216–221)

LEARN IT Explain how an excess of any of the three energy-yielding nutrients contributes to body fat and how an inadequate intake of any of them shifts metabolism.

14. What adaptations does the body make during a fast? What are ketone bodies? Define *ketosis*.

15. Distinguish between a loss of *fat* and a loss of *weight*, and describe how each might happen.

16. The body stores energy for future use in:
    a. proteins.
    b. acetyl CoA.
    c. triglycerides.
    d. ketone bodies.

17. During a fast, when glycogen stores have been depleted, the body begins to synthesize glucose from:
    a. acetyl CoA.
    b. amino acids.
    c. fatty acids.
    d. ketone bodies.

18. During a fast, the body produces ketone bodies by:
    a. hydrolyzing glycogen.
    b. condensing acetyl CoA.
    c. transaminating keto acids.
    d. converting ammonia to urea.

HIGHLIGHT 7 Alcohol in the Body (pp. 222–233)

LEARN IT Describe how alcohol disrupts metabolism and impairs health.

19. Alcohol metabolism occurs primarily in the:
    a. liver.
    b. brain.
    c. kidneys.
    d. muscles.

Multiple Choice Answers
3.c 4.d 8.c 9.c 10.c 11.b 12.c 13.d 16.c 17.b 18.b 19.a

Review the key points of this chapter below, learn the definitions of the bold terms, and then take the practice quiz on the back of this card.

8.1 Energy Balance

- When energy consumed equals energy expended, a person is in **energy balance** and body weight is stable.
- If more energy is taken in than is expended, a person gains weight. If more energy is expended than is taken in, a person loses weight.

8.2 Energy In: The kCalories Foods Provide

- Scientists use a **bomb calorimeter** to estimate the potential energy of foods by measuring the heat energy released when foods are burned.
- **Hunger** and **appetite** initiate eating, whereas **satiation** and **satiety** stop and delay eating, respectively (review Figure 8-2, p. 238). Each responds to messages from the nervous and hormonal systems. Superimposed on these signals are complex factors involving emotions, habits, and other aspects of human behavior.

8.3 Energy Out: The kCalories the Body Expends

- A person in energy balance takes in energy from food and expends much of it on **basal metabolism,** some of it on physical activities, and a little on the **thermic effect of food** (review Figure 8-4).
- Energy requirements vary from person to person based on such factors as gender, age, weight, and height as well as the intensity and duration of physical activity.

8.4 Body Weight and Body Composition

- Body weight standards are based on a person's weight in relation to height, called the **body mass index (BMI),** and reflect disease risks, not **body composition.** BMI does not identify body fat or its distribution, and it may misclassify muscular people as overweight.

- $\text{BMI} = \dfrac{\text{weight (kg)}}{\text{height (m)}^2}$ or $\dfrac{\text{weight (lb)}}{\text{height (in)}^2} \times 703$.

> FIGURE 8-4   **Components of Energy Expenditure**

The amount of energy expended in voluntary physical activities has the greatest variability, depending on a person's activity patterns. For a sedentary person, physical activities may account for less than half as much energy as basal metabolism, whereas an extremely active person may expend as much on activity as for basal metabolism.

30–50% Physical activities

10% Thermic effect of food

50–65% Basal metabolism

© Cengage Learning

The amount of energy expended in a day differs for each individual, but in general, basal metabolism is the largest component of energy expenditure and thermic effect of food is the smallest.

> FIGURE 8-7   **Distribution of Body Weights in US Adults**

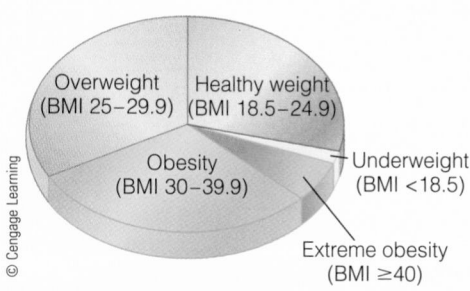

Overweight (BMI 25–29.9)

Healthy weight (BMI 18.5–24.9)

Obesity (BMI 30–39.9)

Underweight (BMI <18.5)

Extreme obesity (BMI ≥40)

© Cengage Learning

Underweight: BMI <18.5
Healthy: BMI 18.5 to 24.9
Overweight: BMI 25 to 29.9
Obese: BMI ≥30

- Two-thirds of US adults have a BMI greater than 25 (review Figure 8-7).
- The ideal amount of body fat varies from person to person, but excess body fat poses health risks.
- **Central obesity**—excess abdominal fat distributed around the trunk of the body— is more common in men and presents greater health risks than excess fat distributed on the lower body, which is more common in women (review Figure 8-8 and Figure 8-9, p. 250).
- Women with a **waist circumference** greater than 35 inches and men with a waist circumference greater than 40 inches have a high risk of central obesity–related health problems.

8.5 Health Risks Associated with Body Weight and Body Fat

- The healthiest weight for an individual depends on personal factors such as body fat distribution, family health history, and current health status. At the extremes, both overweight and underweight impose health risks (review Figure 8-11).

> FIGURE 8-11   **BMI and Mortality**

This J-shaped curve describes the relationship between body mass index (BMI) and mortality and shows that both underweight and overweight present risks of a premature death.

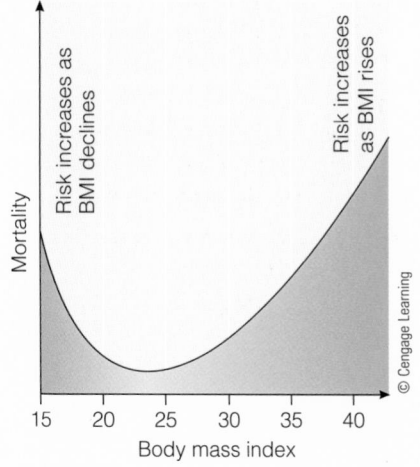

Mortality

Risk increases as BMI declines

Risk increases as BMI rises

15   20   25   30   35   40

Body mass index

© Cengage Learning

TEST IT

Take the quiz below to test your mastery of the key chapter concepts.

8.1 Energy Balance (p. 236)

LEARN IT Describe energy balance and the consequences of not being in balance.

1. A person who consistently consumes 1700 kcalories a day and expends 2200 kcalories a day for a month would be expected to:
   a. lose ½ to 1 pound.
   b. gain ½ to 1 pound.
   c. lose 4 to 5 pounds.
   d. gain 4 to 5 pounds.

8.2 Energy In: The kCalories Foods Provide

(pp. 236–240)

LEARN IT Discuss some of the physical, emotional, and environmental influences on food intake.

2. A bomb calorimeter measures:
   a. physiological fuel.
   b. energy available from foods.
   c. kcalories a person derives from foods.
   d. heat a person releases in basal metabolism.

3. The psychological desire to eat that accompanies the sight, smell, or thought of food is known as:
   a. hunger.
   b. satiety.
   c. appetite.
   d. palatability.

4. A person watching television after dinner reaches for a snack during a commercial in response to:
   a. external cues.
   b. hunger signals.
   c. stress arousal.
   d. satiety factors.

8.3 Energy Out: The kCalories the Body Expends (pp. 240–244)

LEARN IT List the components of energy expenditure and factors that might influence each.

5. The largest component of energy expenditure is:
   a. basal metabolism.
   b. physical activity.
   c. indirect calorimetry.
   d. thermic effect of food.

6. A major factor influencing BMR is:
   a. hunger.
   b. food intake.
   c. body composition.
   d. physical activity.

7. The thermic effect of an 800-kcalorie meal is about:
   a. 8 kcalories.
   b. 80 kcalories.
   c. 160 kcalories.
   d. 200 kcalories.

8.4 Body Weight and Body Composition

(pp. 245–251)

LEARN IT Distinguish between body weight and body composition, including methods to assess each.

8. What problems are involved in defining "ideal" body weight?

9. What is central obesity, and what is its relationship to disease?

10. For health's sake, a person with a BMI of 21 might want to:
    a. lose weight.
    b. maintain weight.
    c. gain weight.

11. Which of the following reflects height and weight?
    a. body mass index
    b. central obesity
    c. waist circumference
    d. body composition

8.5 Health Risks Associated with Body Weight and Body Fat (pp. 251–254)

LEARN IT Identify relationships between body weight and chronic diseases.

12. Which of the following increases disease risks?
    a. BMI 19–21
    b. BMI 22–25
    c. lower-body fat
    d. central obesity

HIGHLIGHT 8 Eating Disorders (pp. 256–263)

LEARN IT Compare the diagnoses, characteristics, and treatments of the different eating disorders.

13. Which of the following is *not* commonly seen in RED-S?
    a. amenorrhea
    b. osteoporosis
    c. disordered eating
    d. muscle dysmorphia

STUDY IT

9 Weight Management: Overweight, Obesity, and Underweight

Review the key points of this chapter below, learn the definitions of the bold terms, and then take the practice quiz on the back of this card.

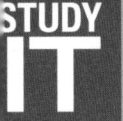

9.1 Overweight and Obesity

- Fat cells develop by increasing in number and size (review Figure 9-2, p. 267).

- Preventing weight gain depends on limiting the number of fat cells; weight loss depends on decreasing the size of fat cells.

- **Lipoprotein lipase (LPL)** removes triglycerides from the blood for storage in both adipose tissue and muscle cells. LPL activity is influenced by weight and gender.

- With weight gains or losses, the body adjusts in an attempt to return to its previous weight (**set point** theory).

9.2 Causes of Overweight and Obesity

- Obesity has multiple causes and different combinations of causes in different people.

- Some environmental causes, such as overeating and physical inactivity, may be within a person's control, and some, such as genetics, may be beyond it.

- Proteins such as **ghrelin** and **leptin** regulate food intake and energy homeostasis. Ghrelin enhances appetite and decreases energy expenditure, whereas leptin suppresses appetite and increases energy expenditure.

9.3 Problems of Overweight and Obesity

- Whether a person should lose weight depends on factors such as the extent of overweight, age, health risks, and genetics.

- Not all obesity will cause disease or shorten life expectancy. Just as there are unhealthy, normal-weight people, there are healthy, overweight people.

- Some people risk more in the process of losing weight than in remaining overweight.

- **Fad diets** and weight-loss supplements can be as physically and psychologically damaging as excess body weight.

9.4 Aggressive Treatments for Obesity

- Obese people with high risks of medical problems may need aggressive treatment, including drugs or surgery.

- Others may benefit most from improving eating and physical activity habits.

9.5 Weight-Loss Strategies

- A person who adopts a lifelong "eating plan for good health" rather than a "diet for weight loss" will be more likely to keep the lost weight off.

- Table 9-3 (p. 281) provides several tips for successful **weight management.**

> FIGURE 9-7 **Influence of Physical Activity on Discretionary kCalories**

- Physical activity can increase energy expenditure (review Figure 9-7), improve body composition, help control appetite, reduce stress and stress eating, and enhance physical and psychological well-being.

- A surefire remedy for obesity has yet to be found; a combination of approaches is most effective.

- Weight loss depends on adjusting diet and physical activity so that more energy is expended than is taken in.

- For weight loss, energy intake should be reduced by 500 to 750 kcalories per day, depending on starting body weight and usual food intake.

- Safe rate for weight loss is ½ to 2 pounds per week or 5 to 10 percent body weight within 6 months.

- **Behavior modification** and cognitive restructuring retrain habits to support a healthy eating and activity plan.

- Treatment requires time, individualization, and sometimes the assistance of a registered dietitian nutritionist or support group.

- Preventing weight gains and **successful weight-loss maintenance** require vigilant attention to diet and physical activity; taking care of oneself is a lifelong responsibility.

9.6 Underweight

- Both the incidence of **underweight** and the health problems associated with it are less prevalent than overweight and its associated problems.

- To gain weight, a person must train physically and increase energy intake by selecting energy-dense foods, eating regular meals, taking larger portions, and consuming extra snacks and beverages.

- Table 9-5 (p. 291) includes a summary of weight-gain strategies.

Take the quiz below to test your mastery of the key chapter concepts.

9.1 Overweight and Obesity (pp. 266–268)

LEARN IT Describe how body fat develops and why it can be difficult to maintain weight gains and losses.

1. With weight loss, fat cells:
   a. decrease in size only.
   b. decrease in number only.
   c. decrease in both number and size.
   d. decrease in number, but increase in size.
2. Describe the role of lipoprotein lipase (LPL).

9.2 Causes of Overweight and Obesity
(pp. 268–272)

LEARN IT Review some of the causes of obesity.

3. Obesity is caused by:
   a. overeating.
   b. inactivity.
   c. defective genes.
   d. multiple factors.
4. The protein produced by the fat cells under the direction of the *ob* gene is called:
   a. leptin.
   b. serotonin.
   c. sibutramine.
   d. phentermine.

9.3 Problems of Overweight and Obesity
(pp. 272–274)

LEARN IT Discuss the physical, social, and psychological consequences of overweight and obesity.

5. Which of the following is *not* used to evaluate the risks to health from obesity?
   a. body mass index
   b. blood leptin levels
   c. waist circumference
   d. disease risk profiles

9.4 Aggressive Treatments for Obesity
(pp. 274–276)

LEARN IT Explain the risks and benefits, if any, of aggressive ways to treat obesity.

6. Gastric bypass surgery:
   a. is the best noninvasive treatment for obesity.
   b. limits food intake by reducing the capacity of the stomach.
   c. allows a person to eat unlimited amounts of food without weight gain.
   d. suppresses hunger by increasing production of gastrointestinal hormones.

9.5 Weight-Loss Strategies (pp. 276–289)

LEARN IT Outline reasonable strategies for achieving and maintaining a healthy body weight.

7. What are the benefits of increased physical activity in a weight-loss program?
8. Describe the behavioral strategies for changing an individual's dietary habits. What role does personal attitude play?
9. A realistic goal for weight loss is to reduce body weight:
   a. down to the weight a person was at age 25.
   b. down to the ideal weight in the weight-for-height tables.
   c. by 10 percent over 6 months.
   d. by 15 percent over 3 months.
10. A nutritionally sound weight-loss diet might restrict daily energy intake to create a:
    a. 1000-kcalorie-per-month deficit.
    b. 500-kcalorie-per-month deficit.
    c. 500-kcalorie-per-day deficit.
    d. 3500-kcalorie-per-day deficit.
11. Successful weight loss depends on:
    a. avoiding fats and limiting water.
    b. taking supplements and drinking water.
    c. increasing proteins and restricting carbohydrates.
    d. reducing energy intake and increasing physical activity.
12. Physical activity does *not* help a person to:
    a. lose weight.
    b. retain muscle.
    c. maintain weight loss.
    d. lose fat in trouble spots.
13. Which strategy would *not* help an overweight person to lose weight?
    a. Exercise.
    b. Eat slowly.
    c. Limit high-fat foods.
    d. Eat energy-dense foods regularly.

9.6 Underweight (pp. 289–291)

LEARN IT Summarize strategies for gaining weight.

14. Which strategy would *not* help an underweight person to gain weight?
    a. Exercise.
    b. Drink plenty of water.
    c. Eat snacks between meals.
    d. Eat large portions of foods.

HIGHLIGHT 9 The Latest and Greatest Weight-Loss Diet–Again (pp. 296–299)

LEARN IT Contrast the differences between popular fad diets and weight-loss diets based on sound nutrition.

15. A real benefit typical of many fad diets is:
    a. ketosis.
    b. a diet plan.
    c. changes in metabolism.
    d. quick and permanent weight loss.

1.a 3.d 4.a 5.b 6.b 9.c 10.c 11.d 12.d 13.d 14.b 15.b
Multiple Choice Answers

STUDY IT

10 The Water-Soluble Vitamins: B Vitamins and Vitamin C

Review the key points of this chapter below, learn the definitions of the bold terms, and then take the practice quiz on the back of this card.

10.1 The Vitamins—An Overview

- The **vitamins** are organic, essential nutrients needed in tiny amounts in the diet both to prevent deficiency diseases and to support optimal health. The body handles the vitamins differently depending on whether they are water- or fat-soluble (review Table 10-2).

- The water-soluble vitamins are the B vitamins and vitamin C; the fat-soluble vitamins are vitamins A, D, E, and K. The B vitamins include thiamin, niacin, riboflavin, vitamin B_6, folate, vitamin B_{12}, pantothenic acid, and biotin.

10.2 The B Vitamins

- The B vitamins serve as **coenzymes**—small organic molecules closely associated with enzymes that facilitate the work of cells (review Figure 10-2).

- **Thiamin** is part of the coenzyme TPP, which assists in energy metabolism. Deficiency can result in **beriberi.** Thiamin occurs in small quantities in many nutritious foods; pork is an exceptionally good source.

- **Riboflavin** is part of the coenzymes FMN and FAD that accept and donate hydrogens during energy metabolism. Milk and milk products are good sources.

- **Niacin** is part of the coenzymes NAD and NADP that participate in many metabolic reactions. The deficiency disease, **pellagra,** causes diarrhea, dermatitis, dementia, and eventually death ("the 4 Ds"). Toxicity produces **"niacin flush"**—a tingling, painful sensation. The amino acid tryptophan can be converted to niacin in the body: 60 mg tryptophan = 1 **NE (niacin equivalent)**. Good sources of niacin are protein-rich foods.

- **Biotin** plays a critical role in energy metabolism, replenishing oxaloacetate in the TCA cycle. Biotin is widespread in foods; deficiencies and toxicities are rare.

- **Pantothenic acid** is part of coenzyme A that forms acetyl CoA in many metabolic pathways. Pantothenic acid is widespread in foods; deficiencies and toxicities are rare.

- **Vitamin B_6** occurs as pyridoxal, pyridoxine, and pyridoxamine; all can become part of the coenzyme PLP, which is active in amino acid metabolism. Deficiency causes convulsions; toxicity causes nerve damage.

- **Folate** is part of the coenzyme THF that activates vitamin B_{12}, synthesizes DNA, and regenerates the amino acid methionine from homocysteine. Folate helps prevent **neural tube defects.** Excessive folate can mask the **anemia** of a vitamin B_{12} deficiency, but it will not prevent the associated nerve damage. Folate is abundant in legumes, fruits, and vegetables.

TABLE 10-2 Water-Soluble and Fat-Soluble Vitamins Compared

	Water-Soluble Vitamins: B Vitamins and Vitamin C	Fat-Soluble Vitamins: Vitamins A, D, E, and K
Absorption	Directly into the blood	First into the lymph, then the blood
Transport	Travel freely	Many require transport proteins
Storage	Circulate freely in water-filled parts of the body	Stored in the cells associated with fat
Excretion	Kidneys detect and remove excess in urine	Less readily excreted; tend to remain in fat-storage sites
Toxicity	Possible to reach toxic levels when consumed from supplements	Likely to reach toxic levels when consumed from supplements
Requirements	Needed in frequent doses (perhaps 1 to 3 days)	Needed in periodic doses (perhaps weeks or even months)

© Cengage Learning

NOTE: Exceptions occur, but these differences between the water-soluble and fat-solule vitamins are valid generalizations.

- **Vitamin B_{12}** activates folate, synthesizes DNA, regenerates methionine from homocysteine, and maintains the sheath that protects nerve fibers. Deficiencies typically occur when either hydrochloric acid or **intrinsic factor** is lacking. Vitamin B_{12} is found primarily in foods derived from animals.

- Many substances (including **inositol and carnitine**) that people claim as B vitamins are not. Fortunately, a variety of foods from each food group provides an adequate supply of all B vitamins.

10.3 Vitamin C

- Vitamin C acts as an **antioxidant**—a substance that decreases the adverse effects of **free radicals** in the body.

- Vitamin C works as a **cofactor** in the synthesis of **collagen,** neurotransmitters (serotonin and norepinephrine), hormones (thyroxine), and other compounds.

- Vitamin C deficiency causes **scurvy.**

> FIGURE 10-2 **Coenzyme Action**

Some vitamins form part of the coenzymes that enable enzymes either to synthesize compounds (as illustrated by the lower enzymes in this figure) or to dismantle compounds (as illustrated by the upper enzymes).

Without coenzymes, compounds A, B, and CD don't respond to their enzymes.

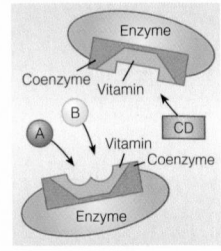

With the coenzymes in place, compounds are attracted to their sites on the enzymes . . .

. . . and the reactions proceed instantaneously. The coenzymes often donate or accept electrons, atoms, or groups of atoms.

The reactions are completed with either the formation of a new product, AB, or the breaking apart of a compound into two new products, C and D, and the release of energy.

© Cengage Learning

Take the quiz below to test your mastery of the key chapter concepts.

10.1 The Vitamins—An Overview
(pp. 302–304)

LEARN IT Describe how vitamins differ from the energy nutrients and how fat-soluble vitamins differ from water-soluble vitamins.

1. Vitamins:
   a. are inorganic compounds.
   b. yield energy when broken down.
   c. are soluble in either water or fat.
   d. perform best when linked in long chains.

2. The rate at and the extent to which a vitamin is absorbed and used in the body is known as its:
   a. bioavailability.
   b. intrinsic factor.
   c. physiological effect.
   d. pharmacological effect.

10.2 The B Vitamins (pp. 304–327)

LEARN IT Identify the main roles, deficiency symptoms, and food sources for each of the B vitamins (thiamin, riboflavin, niacin, biotin, pantothenic acid, vitamin B_6, folate, and vitamin B_{12}).

3. Which B vitamins are involved in energy metabolism? Protein metabolism? Cell division?

4. What is the relationship of tryptophan to niacin?

5. Describe the relationship between folate and vitamin B_{12}.

6. What risks are associated with high doses of niacin? Vitamin B_6? Vitamin C?

7. Many of the B vitamins serve as:
   a. coenzymes.
   b. antagonists.
   c. antioxidants.
   d. serotonin precursors.

8. With respect to thiamin, which of the following is the most nutrient dense?
   a. 1 slice whole-wheat bread (69 kcalories and 0.1 milligram thiamin)
   b. 1 cup yogurt (144 kcalories and 0.1 milligram thiamin)
   c. 1 cup snow peas (69 kcalories and 0.22 milligram thiamin)
   d. 1 chicken breast (141 kcalories and 0.06 milligram thiamin)

9. The body can make niacin from:
   a. tyrosine.
   b. serotonin.
   c. carnitine.
   d. tryptophan.

10. The vitamin that protects against neural tube defects is:
    a. niacin.
    b. folate.
    c. riboflavin.
    d. vitamin B_{12}.

11. A lack of intrinsic factor may lead to:
    a. beriberi.
    b. pellagra.
    c. pernicious anemia.
    d. atrophic gastritis.

12. Which of the following is a B vitamin?
    a. inositol
    b. carnitine
    c. vitamin B_{15}
    d. pantothenic acid

10.3 Vitamin C (pp. 327–333)

LEARN IT Identify the main roles, deficiency symptoms, and food sources for vitamin C.

13. Vitamin C serves as a(n):
    a. coenzyme.
    b. antagonist.
    c. antioxidant.
    d. intrinsic factor.

14. The requirement for vitamin C is highest for:
    a. smokers.
    b. athletes.
    c. alcoholics.
    d. the elderly.

See p. 332–333 for a summary of the water-soluble vitamins.

HIGHLIGHT Vitamin and Mineral Supplements (pp. 335–341)

LEARN IT Present arguments for and against the use of dietary supplements.

15. The best vitamin and mineral supplements:
    a. are high-potency.
    b. are nanoceuticals.
    c. are made from organic vegetables.
    d. provide 50 to 150 percent of the Daily Value.

Multiple Choice Answers
1.c 2.a 7.a 8.c 9.d 10.b 11.c 12.d 13.c 14.a 15.d

STUDY IT

11 The Fat-Soluble Vitamins: A, D, E, and K

Review the key points of this chapter below, learn the definitions of the bold terms, and then take the practice quiz on the back of this card.

11.1 Vitamin A and Beta-Carotene

- **Vitamin A** is found in the body in three forms: **retinol, retinal,** and **retinoic acid.** Together, they are essential to vision, healthy epithelial tissues, and growth.

- Vitamin A deficiency is a major health problem worldwide, leading to infections, blindness, and **keratinization.**

- Toxicity can also cause problems and is most often associated with supplement abuse.

- Animal-derived foods such as liver and whole or fortified milk provide **retinoids,** whereas brightly colored plant-derived foods such as spinach, carrots, and pumpkins provide **beta-carotene** and other **carotenoids.**

- In addition to serving as a precursor for vitamin A, beta-carotene may act as an antioxidant in the body.

11.2 Vitamin D

- **Vitamin D** can be synthesized in the body with the help of sunlight (see Figure 11-8) or obtained from fortified milk.

- Vitamin D sends signals to three primary target sites: the GI tract to absorb more calcium and phosphorus, the bones to release more, and the kidneys to retain more. These actions maintain blood calcium concentrations and support bone formation.

- A vitamin D deficiency causes **rickets** in childhood and **osteomalacia** in later life.

11.3 Vitamin E

- Vitamin E **(alpha-tocopherol)** acts as an antioxidant, defending lipids and other components of the cells against oxidative damage.

- Deficiencies are rare, but they do occur in premature infants, the primary symptom being **erythrocyte hemolysis** (red blood cell breakage).

- Vitamin E is found predominantly in vegetable oils and appears to be one of the least toxic of the fat-soluble vitamins.

11.4 Vitamin K

- Vitamin K helps with blood clotting (review Figure 11-11, p. 359), and its deficiency causes **hemorrhagic disease** (uncontrolled bleeding).

- Bacteria in the GI tract can make vitamin K; people typically receive about half of their requirements from bacterial synthesis and half from foods such as green vegetables and vegetable oils.

- Because people depend on bacterial synthesis for vitamin K, deficiency is most likely in newborn infants and in people taking antibiotics.

> FIGURE 11-8 **Vitamin D Synthesis and Activation**

The final activation step in the kidneys is tightly regulated by hormones.

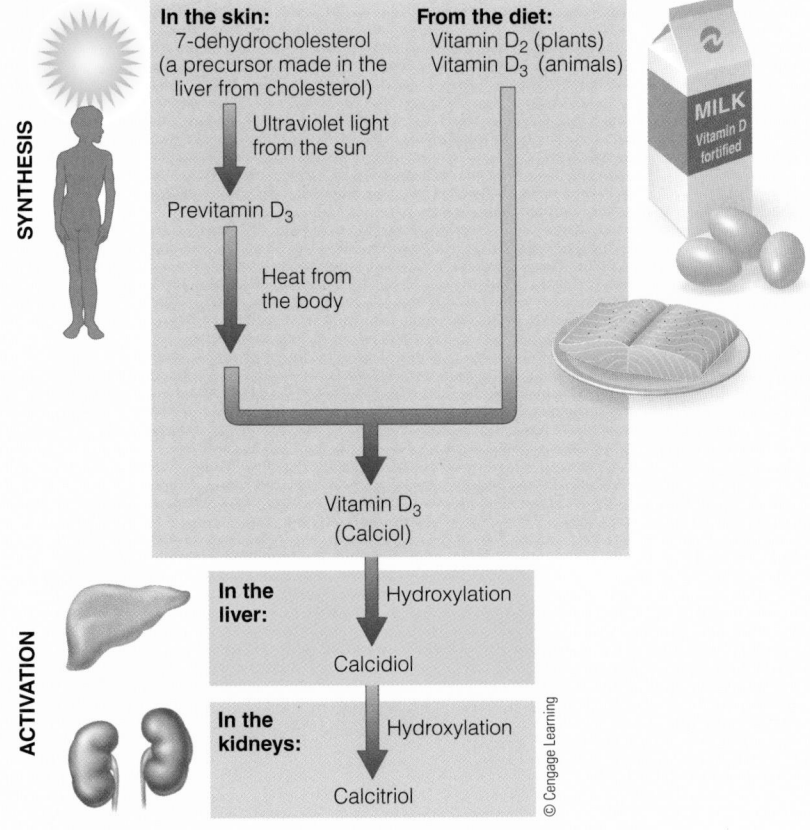

In the skin:
7-dehydrocholesterol (a precursor made in the liver from cholesterol)

From the diet:
Vitamin D_2 (plants)
Vitamin D_3 (animals)

SYNTHESIS

Ultraviolet light from the sun

Previtamin D_3

Heat from the body

Vitamin D_3 (Calciol)

ACTIVATION

In the liver:
Hydroxylation
Calcidiol

In the kidneys:
Hydroxylation
Calcitriol

© Cengage Learning

TEST IT

Take the quiz below to test your mastery of the key chapter concepts.

11.1 Vitamin A and Beta-Carotene

(pp. 344–351)

LEARN IT Identify the main roles, deficiency symptoms, and food sources for vitamin A.

1. What are vitamin precursors? Name the precursors of vitamin A, and tell in what classes of foods they are found. Give examples of foods with high vitamin A activity.

2. The form of vitamin A active in vision is:
   a. retinal.
   b. retinol.
   c. rhodopsin.
   d. retinoic acid.

3. Vitamin A–deficiency symptoms include:
   a. rickets and osteomalacia.
   b. hemorrhaging and jaundice.
   c. night blindness and keratomalacia.
   d. fibrocystic breast disease and erythrocyte hemolysis.

4. Good sources of vitamin A include:
   a. oatmeal, pinto beans, and ham.
   b. apricots, turnip greens, and liver.
   c. whole-wheat bread, green peas, and tuna.
   d. corn, grapefruit juice, and sunflower seeds.

11.2 Vitamin D (pp. 351–356)

LEARN IT Identify the main roles, deficiency symptoms, and sources for vitamin D.

5. How is vitamin D unique among the vitamins?

6. To keep minerals available in the blood, vitamin D targets:
   a. the skin, the muscles, and the bones.
   b. the kidneys, the liver, and the bones.
   c. the intestines, the kidneys, and the bones.
   d. the intestines, the pancreas, and the liver.

7. Vitamin D can be synthesized from a precursor that the body makes from:
   a. bilirubin.
   b. tocopherol.
   c. cholesterol.
   d. beta-carotene.

11.3 Vitamin E (pp. 357–358)

LEARN IT Identify the main roles, deficiency symptoms, and food sources for vitamin E.

8. Vitamin E's most notable role is to:
   a. protect lipids against oxidation.
   b. activate blood-clotting proteins.
   c. support protein and DNA synthesis.
   d. enhance calcium deposits in the bones.

9. The classic sign of vitamin E deficiency is:
   a. rickets.
   b. xeropthalmia.
   c. muscular dystrophy.
   d. erythrocyte hemolysis.

11.4 Vitamin K (pp. 358–361)

LEARN IT Identify the main roles, deficiency symptoms, and sources for vitamin K.

10. What conditions may lead to vitamin K deficiency?

11. Without vitamin K:
    a. muscles atrophy.
    b. bones become soft.
    c. skin rashes develop.
    d. blood fails to clot.

12. A significant amount of vitamin K comes from:
    a. vegetable oils.
    b. sunlight exposure.
    c. bacterial synthesis.
    d. fortified grain products.

See p. 361 for a summary of the fat-soluble vitamins.

HIGHLIGHT 11 Antioxidant Nutrients in Disease Prevention (pp. 364–368)

LEARN IT Describe how antioxidants defend against free radicals that contribute to diseases.

13. Free radicals:
    a. are highly reactive.
    b. defend against cancer.
    c. act as antioxidants in the body.
    d. are found abundantly in fruits and vegetables.

Review the key points of this chapter below, learn the definitions of the bold terms, and then take the practice quiz on the back of this card.

12.1 Water and the Body Fluids

- Water makes up about 60 percent of an adult's body weight.
- Water assists with the transport of nutrients and waste products throughout the body, participates in chemical reactions, acts as a solvent, serves as a shock absorber, and regulates body temperature.
- To maintain **water balance,** intake from liquids, foods, and metabolism must equal losses from the kidneys, skin, lungs, and GI tract (review Table 12-3, p. 373).
- Whenever the body experiences low blood volume, low blood pressure, or highly concentrated body fluids, the actions of **ADH, renin, angiotensin,** and **aldosterone** restore homeostasis (review Figure 12-3).
- **Electrolytes** (charged minerals) in the fluids help distribute the fluids inside and outside the cells, thus ensuring the appropriate water balance and acid-base balance to support all life processes.
- Excessive losses of fluids and electrolytes upset these balances, and the kidneys play a key role in restoring homeostasis.

> **FIGURE 12-3 How the Body Regulates Blood Volume and Blood Pressure**

The renin-angiotensin-aldosterone system helps regulate blood volume and therefore blood pressure.

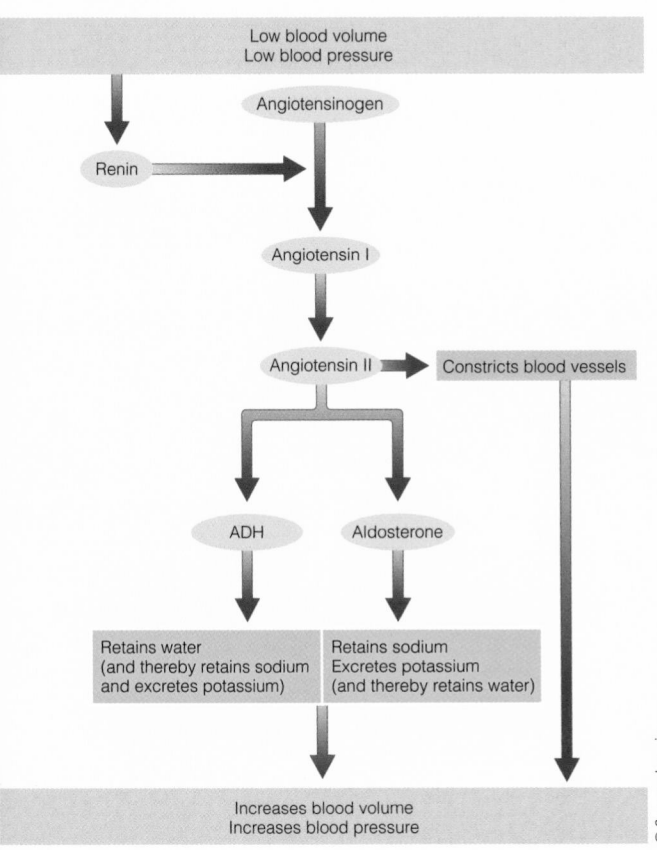

Low blood volume
Low blood pressure

Angiotensinogen

Renin

Angiotensin I

Angiotensin II → Constricts blood vessels

ADH   Aldosterone

Retains water
(and thereby retains sodium
and excretes potassium)

Retains sodium
Excretes potassium
(and thereby retains water)

Increases blood volume
Increases blood pressure

© Cengage Learning

12.2 The Minerals—An Overview

- The **major minerals** are needed in the diet and found in the body in larger quantities than the **trace minerals** (review Figure 12-9, p. 381).
- Minerals are inorganic elements that retain their chemical identities; receive special handling and regulation in the body; and may bind with other substances or interact with other minerals, thus limiting their absorption.
- The major minerals, especially sodium, chloride, and potassium, influence the body's fluid balance; whenever an **anion** moves, a **cation** moves—always maintaining homeostasis.
- Sodium, chloride, potassium, calcium, and magnesium are key members of the team of nutrients that direct nerve impulse transmission and muscle contraction. They are also the primary nutrients involved in regulating blood pressure.
- Phosphorus and magnesium participate in many reactions involving glucose, fatty acids, amino acids, and the vitamins. Calcium, phosphorus, and magnesium combine to form the structure of the bones and teeth. Each major mineral also plays other specific roles in the body.

12.3 The Major Minerals

- **Sodium** is the main cation outside cells and one of the primary electrolytes responsible for maintaining fluid balance. Dietary deficiency is rare; excesses seem to aggravate hypertension, and so health professionals advise a diet moderate in **salt** and sodium.
- **Chloride** is the major anion outside cells, and it associates closely with sodium. In addition to its role in fluid balance, chloride is part of the stomach's hydrochloric acid.
- **Potassium** is the primary cation inside cells and plays an important role in maintaining fluid balance. Fresh foods, notably fruits and vegetables, are its best sources.
- **Calcium** is found primarily in the bones, where it provides a rigid structure and a reservoir of calcium for the blood. Blood calcium participates in muscle contraction, blood clotting, and nerve impulses, and it is closely regulated by a system of hormones and vitamin D (review Figure 12-12, p. 389). Milk and milk products are good sources of calcium, but certain vegetables and tofu also provide calcium. Even when calcium intake is inadequate, blood calcium remains normal, but at the expense of bone loss, which can lead to **osteoporosis.**
- **Phosphorus** accompanies calcium both in the crystals of bone and in many foods such as milk. Phosphorus is also important in energy metabolism, as part of phospholipids, and as part of the genetic materials DNA and RNA.
- **Magnesium** supports bone **mineralization** and participates in numerous enzyme systems and in heart function. It is found abundantly in legumes and dark green, leafy vegetables and, in some areas, in water.
- **Sulfate** is found in all protein-containing foods. Its primary role in amino acids is to stabilize proteins by forming disulfide bridges.

Take the quiz below to test your mastery of the key chapter concepts.

12.1 Water and the Body Fluids (pp. 372–381)

LEARN IT Explain how the body regulates fluid balance.

1. List the roles of water in the body.

2. List the sources of water intake and routes of water excretion.

3. What is ADH? Where does it exert its action? What is aldosterone? How does it work?

4. How does the body use electrolytes to regulate fluid balance?

5. The body generates water during the:
   a. buffering of acids.
   b. dismantling of bone.
   c. metabolism of minerals.
   d. oxidation of energy nutrients through the electron transport chain.

6. Regulation of fluid and electrolyte balance and acid-base balance depends primarily on the:
   a. kidneys.
   b. intestines.
   c. sweat glands.
   d. specialized tear ducts.

12.2 The Minerals—An Overview (pp. 381–382)

LEARN IT List some of the ways minerals differ from vitamins and other nutrients.

7. What do the terms major and trace mean when describing the minerals in the body?

8. Describe some characteristics of minerals that distinguish them from vitamins.

9. The distinction between the major and trace minerals reflects the:
   a. ability of their ions to form salts.
   b. amounts of their contents in the body.
   c. importance of their functions in the body.
   d. capacity to retain their identity after absorption.

12.3 The Major Minerals (pp. 382–398)

LEARN IT Identify the main roles, deficiency symptoms, and food sources for each of the major minerals (sodium, chloride, potassium, calcium, phosphorus, magnesium, and sulfate).

10. What is the major function of sodium in the body? Describe how the kidneys regulate blood sodium. Is a dietary deficiency of sodium likely? Why or why not?

11. List calcium's roles in the body. How does the body keep blood calcium constant regardless of intake?

12. Name significant food sources of calcium. What are the consequences of inadequate intakes?

13. List the roles of phosphorus in the body. Discuss the relationships between calcium and phosphorus. Is a dietary deficiency of phosphorus likely? Why or why not?

14. State the major functions of chloride, potassium, magnesium, and sulfur in the body. Are deficiencies of these nutrients likely to occur in your own diet? Why or why not?

15. The principal cation in extracellular fluids is:
    a. sodium.
    b. chloride.
    c. potassium.
    d. phosphorus.

16. The role of chloride in the stomach is to help:
    a. support nerve impulses.
    b. convey hormonal messages.
    c. maintain a strong acidity.
    d. assist in muscular contractions.

17. Which would provide the most potassium?
    a. bologna
    b. potatoes
    c. pickles
    d. whole-wheat bread

18. Calcium homeostasis depends on:
    a. vitamin K, aldosterone, and renin.
    b. vitamin K, parathyroid hormone, and renin.
    c. vitamin D, aldosterone, and calcitonin.
    d. vitamin D, calcitonin, and parathyroid hormone.

19. Calcium absorption is hindered by:
    a. lactose.
    b. oxalates.
    c. vitamin D.
    d. stomach acid.

20. Phosphorus assists in many activities in the body, but not:
    a. energy metabolism.
    b. the clotting of blood.
    c. the transport of lipids.
    d. bone and teeth formation.

21. Most of the body's magnesium can be found in the:
    a. bones.
    b. nerves.
    c. muscles.
    d. extracellular fluids.

See p. 397 for a summary of the major minerals.

HIGHLIGHT 12 Osteoporosis and Calcium (pp. 400–405)

LEARN IT Describe factors that contribute to the development of osteoporosis and strategies to prevent it.

22. Which of the following is not a risk factor for osteoporosis?
    a. overweight
    b. cigarette smoking
    c. sedentary lifestyle
    d. inadequate dietary calcium and vitamin D

5. d 6. a 9. b 15. a 16. c 17. b 18. d 19. b 20. c 21. a 22. a

Multiple Choice Answers

STUDY IT

13 The Trace Minerals

Review the key points of this chapter below, learn the definitions of the bold terms, and then take the practice quiz on the back of this card.

13.1 The Trace Minerals—An Overview

- The body needs tiny amounts of the **trace minerals.**
- The trace minerals can be toxic at levels not far above estimated requirements—a consideration for supplement users (review Figure 13-1, p. 408).
- Like the other nutrients, the trace minerals are best obtained by eating a variety of foods.

13.2 The Trace Minerals

- **Iron** is found in **hemoglobin** and **myoglobin** where it carries oxygen for energy metabolism; iron also acts as a **cofactor** for some enzymes. **Iron deficiency** is most common among infants and young children, teenagers, women of childbearing age, and pregnant women. Symptoms include fatigue and anemia. **Iron overload** is most common in men. **Heme iron,** which is found only in meat, fish, and poultry, is better absorbed than **nonheme iron,** which occurs in most foods (review Figure 13-4). Nonheme iron absorption is improved by eating iron-containing foods with foods containing the **MFP factor** and vitamin C; absorption is limited by phytates and oxalates.

- Zinc-requiring enzymes participate in reactions affecting growth, vitamin A activity, and pancreatic digestive enzyme synthesis. Both dietary **zinc** and zinc-rich pancreatic secretions (via **enteropancreatic circulation**) are available for absorption. Absorption is monitored by a special binding protein (**metallothionein**) in the small intestine. Protein-rich foods derived from animals are the best sources of bioavailable zinc. Fiber and phytates in cereals bind zinc, limiting absorption. Symptoms of deficiency include growth retardation and sexual immaturity.

- Iodide, the **iodine** ion, is an essential component of the thyroid hormone. A deficiency can lead to **goiter** (enlargement of the thyroid gland) and can impair fetal development, causing **cretinism.** Iodization of salt has largely eliminated iodine deficiency in the United States.

- **Selenium** is an antioxidant nutrient that works closely with the glutathione peroxidase enzyme and vitamin E. Selenium is found with protein in foods. Deficiencies are associated with a predisposition to a type of heart abnormality known as **Keshan disease.**

- **Copper** is a component of several enzymes, all of which are involved with oxygen or oxidation. Some act as antioxidants; others are essential to iron metabolism. Legumes, whole grains, and shellfish are good sources of copper.

- Manganese-dependent enzymes are involved in bone formation and various metabolic processes. Because **manganese** is widespread in plant foods, deficiencies are rare, although regular use of calcium and iron supplements may limit manganese absorption.

- **Fluoride** makes teeth more resistant to decay. Fluoridation of public water reduces the incidence of dental caries; excess fluoride during tooth development can discolor and pit tooth enamel (**fluorosis).**

- **Chromium** enhances insulin's action. Deficiency can result in a diabetes-like condition. Chromium is widely available in unrefined foods including brewer's yeast, whole grains, and liver.

- **Molybdenum** is a part of many **metalloenzymes.** Deficiencies are unknown and toxicity is rare. It is found in a variety of foods.

13.3 Contaminant Minerals

- Contaminant minerals include the **heavy metals** lead, mercury, and cadmium that enter the food supply by way of soil, water, and air pollution.

- Lead typifies the ways all heavy metals behave in the body: they interfere with nutrients that are trying to do their jobs. The "good guy" nutrients are shoved aside by the "bad guy" contaminants. Then, when the contaminants cannot perform the role of the nutrients, health diminishes.

- To safeguard our health, we must defend ourselves against contamination by eating nutrient-rich foods and preserving a clean environment.

> FIGURE 13-4   **Heme and Nonheme Iron in Foods**

About 40% of the iron in meat, fish, and poultry is bound into heme; the other 60% is nonheme iron.

Key:
- ● Heme
- ■ Nonheme

All of the iron in foods derived from plants is nonheme iron.

Heme accounts for about 10% of the average daily iron intake, but it is well absorbed (about 25%).

Nonheme iron accounts for the remaining 90%, but it is less well absorbed (about 17%).

© Cengage Learning

Take the quiz below to test your mastery of the key chapter concepts.

13.1 The Trace Minerals—An Overview
(pp. 408–409)

LEARN IT Summarize key factors unique to the trace minerals.

1. Discuss the importance of balanced and varied diets in obtaining the essential minerals and avoiding toxicities.

2. Describe some of the ways trace minerals interact with one another and with other nutrients.

13.2 The Trace Minerals (pp. 410–428)

LEARN IT Identify the main roles, deficiency symptoms, and food sources for each of the essential trace minerals (iron, zinc, iodine, selenium, copper, manganese, fluoride, chromium, and molybdenum).

3. Distinguish between heme and nonheme iron. Discuss the factors that enhance iron absorption.

4. Iron absorption is impaired by:
   a. heme.
   b. phytates.
   c. vitamin C.
   d. MFP factor.

5. Which of these people is *least* likely to develop an iron deficiency?
   a. 3-year-old boy
   b. 52-year-old man
   c. 17-year-old girl
   d. 24-year-old woman

6. Which of the following would *not* describe the blood cells of a severe iron deficiency?
   a. anemic
   b. microcytic
   c. pernicious
   d. hypochromic

7. Which provides the most absorbable iron?
   a. 1 apple
   b. 1 cup milk
   c. 3 ounces steak
   d. ½ cup spinach

8. What causes iron overload? What are its symptoms?

9. Describe the similarities and differences in the absorption and regulation of iron and zinc.

10. Discuss possible reasons for a low intake of zinc. What factors affect the bioavailability of zinc?

11. The intestinal protein that helps to regulate zinc absorption is:
    a. albumin.
    b. ferritin.
    c. hemosiderin.
    d. metallothionein.

12. A classic sign of zinc deficiency is:
    a. anemia.
    b. goiter.
    c. mottled teeth.
    d. growth retardation.

13. What public health measure has been used in preventing simple goiter?

14. Cretinism is caused by a deficiency of:
    a. iron.
    b. zinc.
    c. iodine.
    d. selenium.

15. The mineral best known for its role as an antioxidant is:
    a. copper.
    b. selenium.
    c. manganese.
    d. molybdenum.

16. What measure has been recommended for protection against tooth decay?

17. Fluorosis occurs when fluoride:
    a. is excessive.
    b. is inadequate.
    c. binds with phosphorus.
    d. interacts with calcium.

18. Which mineral enhances insulin activity?
    a. zinc
    b. iodine
    c. chromium
    d. manganese

See p. 430 for a summary of the trace minerals.

13.3 Contaminant Minerals (p. 429)

LEARN IT Describe how contaminant minerals disrupt body processes and impair nutrition status.

19. Which of the following does lead *not* compete with?
    a. iron
    b. zinc
    c. fluoride
    d. calcium

HIGHLIGHT 13 Phytochemicals and Functional Foods (pp. 433–439)

LEARN IT Define *phytochemicals* and explain how they might defend against chronic diseases.

20. Examples of phytochemicals include:
    a. LDL and DNA.
    b. lycopene and lutein.
    c. soybeans and flaxseeds.
    d. carcinogens and goitrogens.

Multiple Choice Answers
4.b 5.b 6.c 7.c 11.d 12.d 14.c 15.b 17.a 18.c 19.c 20.b

STUDY IT

14 Fitness: Physical Activity, Nutrients, and Body Adaptations

Review the key points of this chapter below, learn the definitions of the bold terms, and then take the practice quiz on the back of this card.

14.1 Fitness

- **Physical activity** promotes good health and reduces the risk of developing a number of diseases.
- The components of **fitness** are **flexibility, muscle strength, muscle endurance,** and **cardiorespiratory endurance.**
- Participating in cardiorespiratory, strength, and flexibility activities improves fitness and benefits health (review Table 14-2, p. 444).

14.2 Energy Systems and Fuels to Support Activity

- The high-energy compound ATP provides the energy that powers all the activities of living cells.
- Physical activity uses ATP.
- Three major energy systems enable muscle cells to replenish ATP during activity: the phosphagen system, the lactic acid system, and the aerobic system.
- The phosphagen system relies on the breakdown of another high-energy compound **CP,** to replenish ATP. CP can split anaerobically to release phosphate and restore ATP.

    During rest: ATP + creatine → CP

    During activity: CP → ATP + creatine

- The lactic acid system involves the incomplete oxidation of glucose anaerobically to lactate.
- The aerobic system generates ATP from the oxidation of carbohydrate, fat, and some amino acids.
- Fuel mixture during physical activity depends on diet, the intensity and duration of activity, and training (review Table 14-4).
- During intense activity, the fuel mix is mostly glucose; during less intense activity, fat makes a greater contribution.

- With endurance training, muscle cells adapt to store more glycogen and to rely less on glucose and more on fat for energy.
- Active athletes may need more protein than **sedentary** people do; they typically eat more food and therefore obtain enough protein without supplements (review Table 14-5, p. 455).

14.3 Vitamins and Minerals to Support Activity

- With the possible exception of iron for women, well-nourished active people and athletes do not need dietary supplements.

14.4 Fluids and Electrolytes to Support Activity

- **Heat stroke** and **hypothermia** can be fatal (review Table 14-6, p. 458).
- Active people need to drink plenty of water (review Table 14-7, p. 458).
- Endurance athletes need to drink both water and carbohydrate-containing beverages, especially during training and competition.
- During events lasting longer than 3 hours, athletes may need to replace sodium losses to prevent **hyponatremia.**

14.5 Diets for Physically Active People

- To enhance athletic performance and overall health, a diet should provide ample fluid and a variety of nutrient-dense foods in quantities to meet energy needs.
- Carbohydrate-rich foods that are light and easy to digest are best for pregame and recovery meals.

TABLE 14-4 Primary Fuels Used for Activities of Different Intensities and Durations

Activity Intensity	Activity Duration	Energy System	Preferred Fuel Source	Oxygen Needed?	Activity Example
Extreme	5 to 10 sec	Phosphagen system	ATP-CP (immediate availability)	No	100-meter sprint, shot put, golf or baseball bat swing, tennis or volleyball serve
Very high	20 sec to 2 min	Lactic acid system	ATP from carbohydrate (anaerobic glycolysis)	No	400-meter run, 100-meter swim, gymnastic routines
High	2 min to 20 min	Aerobic system	ATP from carbohydrate (glycolysis and TCA cycle)	Yes	Cycling, swimming, running
Moderate	> 20 min	Aerobic system	ATP from fat (fatty acid-oxidation and TCA cycle)	Yes	Hiking

© Cengage Learning

NOTE: All energy systems function at all times, but depending on the intensity of the activity and the conditioning of the athlete, one system will predominate at any given time.

TEST
IT

Take the quiz below to test your mastery of the key chapter concepts.

14.1 Fitness (pp. 442–448)

LEARN IT Describe the health benefits of being physically fit and explain how to develop the components of fitness.

1. Explain physical activity overload.

2. Define *cardiorespiratory conditioning* and list some of its benefits.

3. Physical inactivity is linked to all of the following diseases except:
   - a. cancer.
   - b. diabetes.
   - c. emphysema.
   - d. hypertension.

4. Physical activity overload can be applied by performing:
   - a. an activity less often.
   - b. an activity with more intensity.
   - c. an activity in a different setting.
   - d. a different activity each day of the week.

14.2 Energy Systems and Fuels to Support Activity (pp. 448–455)

LEARN IT Identify the factors that influence fuel use during physical activity and the types of activities that depend more on glucose or fat, respectively.

5. What types of activity are anaerobic? Which are aerobic? Describe the relationships among energy expenditure, type of activity, and oxygen use.

6. The process that regenerates glucose from lactate is known as the:
   - a. Cori cycle.
   - b. ATP-CP cycle.
   - c. adaptation cycle.
   - d. cardiac output cycle.

7. "Hitting the wall" is a term runners sometimes use to describe:
   - a. dehydration.
   - b. competition.
   - c. indigestion.
   - d. glucose depletion.

8. The technique endurance athletes use to maximize glycogen stores is called:
   - a. aerobic training.
   - b. muscle conditioning.
   - c. carbohydrate loading.
   - d. progressive overloading.

9. Conditioned muscles rely less on _____ and more on _____ for energy.
   - a. protein; fat
   - b. fat; protein
   - c. glycogen; fat
   - d. fat; glycogen

14.3 Vitamins and Minerals to Support Activity (pp. 455–457)

LEARN IT List which vitamins and mineral supplements, if any, athletes may need and why.

10. Why are some athletes likely to develop iron-deficiency anemia? Compare iron-deficiency anemia and sports anemia, explaining the differences.

11. Vitamin or mineral supplements taken just before an event are useless for improving performance because the:
    - a. athlete sweats the nutrients out during the event.
    - b. stomach can't digest supplements during physical activity.
    - c. nutrients are diluted by all the fluids the athlete drinks.
    - d. body needs hours or days for the nutrients to do their work.

12. Physically active young women, especially those who are endurance athletes, are prone to:
    - a. energy excess.
    - b. iron deficiency.
    - c. protein overload.
    - d. vitamin A toxicity.

14.4 Fluids and Electrolytes to Support Activity (pp. 457–462)

LEARN IT Identify the factors that influence an athlete's fluid needs and describe the differences between water and sports drinks.

13. Discuss the importance of hydration during training, and list recommendations to maintain fluid balance.

14. The body's need for _____ far surpasses its need for any other nutrient.
    - a. water
    - b. protein
    - c. vitamins
    - d. carbohydrate

14.5 Diets for Physically Active People (pp. 462–464)

LEARN IT Discuss an appropriate daily eating pattern for athletes and list one example of a recommended pregame and recovery meal.

15. A recommended pregame meal includes plenty of fluids and provides between:
    - a. 300 and 800 kcalories, mostly from fat-rich foods.
    - b. 50 and 100 kcalories, mostly from fiber-rich foods.
    - c. 1000 and 2000 kcalories, mostly from protein-rich foods.
    - d. 300 and 800 kcalories, mostly from carbohydrate-rich foods.

HIGHLIGHT 14 Supplements as Ergogenic Aids (pp. 467–471)

LEARN IT Present arguments for and against the use of ergogenic aids.

16. Discuss two ergogenic aids that seem to perform as claimed and two that are dangerous, banned, or illegal.

3.c 4.b 6.a 7.d 8.c 9.c 11.d 12.b 14.a 15.d

Multiple Choice Answers

STUDY IT

15 Life Cycle Nutrition: Pregnancy and Lactation

Review the key points of this chapter below, learn the definitions of the bold terms, and then take the practice quiz on the back of this card.

15.1 Nutrition prior to Pregnancy

- Prior to pregnancy, the health and behaviors of both men and women can influence **fertility** and fetal development. In preparation, they can achieve and maintain a healthy body weight, choose an adequate and balanced diet, be physically active, receive regular medical care, manage chronic conditions, and avoid harmful influences.

15.2 Growth and Development during Pregnancy

- The infant develops through three stages—the **zygote, embryo,** and **fetus** (review Figure 15-2, p. 476). Each organ and tissue grows on its own schedule (review Figure 15-4, p. 477).
- Times of intense development are **critical periods** that depend on nutrients (review Figure 15-3, p. 477).
- Without folate, the neural tube fails to develop completely during the first month of pregnancy; all women of childbearing age should take folate daily. Common **neural tube defects** include **anencephaly** and **spina bifida.**

15.3 Maternal Weight

- A healthy pregnancy depends on a sufficient weight gain.
- Women who begin their pregnancies at a healthy weight need to gain about 30 pounds to cover the growth and development of the **placenta, uterus,** blood, breasts, and infant (review Figure 15-7, p. 482).
- Physical activity throughout pregnancy can help a woman develop the strength she needs to carry the extra weight and maintain habits that will help her lose it after the birth.

15.4 Nutrition during Pregnancy

- Energy and nutrient needs are high during pregnancy; the diet should include an extra serving from each of the five food groups.
- Supplements of iron and folate are recommended.

- Nausea, constipation, and heartburn commonly accompany pregnancy and can usually be alleviated with a few simple strategies (review Table 15-2, p. 487). **Food cravings** do not typically reflect physiological needs.

15.5 High-Risk Pregnancies

- High-risk pregnancies (review Table 15-3) threaten the life and health of both mother and infant and are likely to produce an infant with a low birth weight.
- Proper nutrition and abstinence from smoking, alcohol, and other drugs improve the outcome as can prenatal care to monitor for **gestational diabetes** and **preeclampsia.**

15.6 Nutrition during Lactation

- During **lactation,** women need extra fluid and enough energy and nutrients to produce about 25 ounces of milk a day.
- Breastfeeding is contraindicated for those with HIV/AIDS.
- Alcohol, other drugs, smoking, and contaminants may reduce milk production or enter breast milk and impair infant development.

TABLE 15-3 High-Risk Pregnancy Factors

Factor	Condition that Raises Risk
Maternal weight	
• Prior to pregnancy	Prepregnancy BMI either <18.5 or ≥25
• During pregnancy	Insufficient or excessive pregnancy weight gain (see Table 15-1, p. 481)
Maternal nutrition	Nutrient deficiencies or toxicities; eating disorders
Socioeconomic status	Poverty, lack of family support, low level of education, limited food available
Lifestyle habits	Smoking, alcohol or other drug use
Age	Teens, especially 15 years or younger; women 35 years or older
Previous pregnancies	
• Number	Many previous pregnancies (3 or more to mothers younger than age 20; 4 or more to mothers age 20 or older)
• Interval	Short or long intervals between pregnancies (<18 months or >59 months)
• Outcomes	Previous history of problems
• Multiple births	Twins or triplets
• Birthweight	Low- or high-birthweight infants
Maternal health	
• High blood pressure	Development of gestational hypertension
• Diabetes	Development of gestational diabetes
• Chronic diseases	Diabetes; heart, respiratory, and kidney disease; certain genetic disorders; special diets and medications

TEST IT

Take the quiz below to test your mastery of the key chapter concepts.

15.1 Nutrition prior to Pregnancy (p. 474)

LEARN IT List the ways men and women can prepare for a healthy pregnancy.

1. The health and behaviors of men and women prior to pregnancy can influence all of the following except:
   a. fertility.
   b. infant's gender.
   c. placenta development.
   d. infant's mental development.

15.2 Growth and Development during Pregnancy (pp. 475–479)

LEARN IT Describe fetal development from conception to birth and explain how maternal malnutrition can affect critical periods.

2. Describe the placenta and its function.

3. Explain why women of childbearing age need folate in their diets. How much is recommended, and how can women ensure that these needs are met?

4. The spongy structure that delivers nutrients to the fetus and returns waste products to the mother is called the:
   a. embryo.
   b. uterus.
   c. placenta.
   d. amniotic sac.

15.3 Maternal Weight (pp. 480–483)

LEARN IT Explain how both underweight and overweight can interfere with a healthy pregnancy and how weight gain and physical activity can support maternal health and infant growth.

5. Which of these strategies is *not* a healthy option for an overweight woman?
   a. Limit weight gain during pregnancy.
   b. Postpone weight loss until after pregnancy.
   c. Follow a weight-loss diet during pregnancy.
   d. Try to achieve a healthy weight before becoming pregnant.

6. A reasonable weight gain during pregnancy for a normal-weight woman is about:
   a. 10 pounds.
   b. 20 pounds.
   c. 30 pounds.
   d. 40 pounds.

15.4 Nutrition during Pregnancy (pp. 483–488)

LEARN IT Summarize the nutrient needs of women during pregnancy.

7. Which nutrients are needed in the greatest amounts during pregnancy? Why are they so important? Describe wise food choices for the pregnant woman.

15.5 High-Risk Pregnancies (pp. 488–496)

LEARN IT Identify factors predicting low-risk and high-risk pregnancies and describe ways to manage them.

8. What is the significance of infant birthweight in terms of the child's future health?

9. Describe some of the special problems of the pregnant adolescent. Which nutrients are needed in increased amounts?

10. What practices should be avoided during pregnancy? Why?

11. To help prevent neural tube defects, grain products are now fortified with:
    a. iron.
    b. folate.
    c. protein.
    d. vitamin C.

12. Pregnant women should *not* take supplements of:
    a. iron.
    b. folate.
    c. vitamin A.
    d. vitamin C.

13. The combination of high blood pressure, protein in the urine, and edema signals:
    a. jaundice.
    b. preeclampsia.
    c. gestational diabetes.
    d. gestational hypertension.

15.6 Nutrition during Lactation (pp. 496–501)

LEARN IT Summarize the nutrient needs of women during lactation.

14. To facilitate lactation, a mother needs:
    a. about 5000 kcalories a day.
    b. adequate nutrition and rest.
    c. vitamin and mineral supplements.
    d. a glass of wine or beer before each feeding.

15. A breastfeeding woman should drink plenty of water to:
    a. produce more milk.
    b. suppress lactation.
    c. prevent dehydration.
    d. dilute nutrient concentration.

16. A woman may need iron supplements during lactation:
    a. to enhance the iron in her breast milk.
    b. to provide iron for the infant's growth.
    c. to replace the iron in her body's stores.
    d. to support the increase in her blood volume.

HIGHLIGHT 15 Fetal Alcohol Syndrome (pp. 505–507)

LEARN IT Explain how drinking alcohol endangers the fetus and how women can prevent fetal alcohol syndrome.

17. Public health recommendations for alcohol consumption during pregnancy advise:
    a. abstinence from alcohol.
    b. beer or wine, but not liquor.
    c. alcohol in moderation (one drink a day).
    d. abstinence in the first trimester, moderation in the second trimester, and no restrictions in the third trimester.

STUDY IT

16 Life Cycle Nutrition: Infancy, Childhood, and Adolescence

Review the key points of this chapter below, learn the definitions of the bold terms, and then take the practice quiz on the back of this card.

16.1 Nutrition during Infancy

- The primary food for infants during the first 12 months is either breast milk or iron-fortified formula. Cow's milk is not appropriate until after age 1.
- Breast milk offers both nutrients and immunological protection.
- Breast milk provides many factors that protect infants from illnesses and support better nutrition (review Table 16-3, p. 514).
- At 4 to 6 months of age, infants should gradually begin eating solid foods, including iron-fortified cereals and vitamin-C rich fruits and vegetables.
- By 1 year, infants are drinking from a cup and eating a variety of foods.
- Infants may benefit from supplements containing vitamin D, iron, and fluoride (review Table 16-2, p. 513).

16.2 Nutrition during Childhood

- Children's appetites and nutrient needs reflect their stage of growth.
- Children who are chronically hungry and malnourished suffer growth retardation; temporary hunger and mild nutrient deficiencies produce more subtle problems—such as poor academic performance.
- Iron deficiency is widespread and has many physical and behavioral consequences.
- **Hyperactivity** is not caused by poor nutrition; misbehavior may be due to lack of sleep, too little physical activity, or too much television, among other factors.
- Some children have **food allergies**—**adverse reactions** that involve an immune response—most often caused by peanuts, tree nuts, milk, eggs, wheat, soybeans, fish, or shellfish.

- Childhood obesity is a major health problem (review Figure 16-9, p. 529).
- Children need to eat nutrient-dense foods and learn how to make healthful diet and activity choices.

16.3 Nutrition during Adolescence

- The need for iron increases during **adolescence** for both males and females; blood losses incurred through menstruation increase iron needs for females further (review Table 16-15, p. 540).
- Sufficient calcium intake during adolescence supports optimal bone growth and density.
- The adolescent growth spurt increases the need for energy and nutrients.
- Adolescents who drink soft drinks regularly have a higher energy intake and a lower calcium intake; they are also more likely to be overweight.

TABLE 16-9 **Recommended Eating and Physical Activity Behaviors to Prevent Obesity**

The Expert Committee of the American Medical Association recommends the following healthy habits for children 2 to 18 years of age to help prevent childhood obesity:

- Limit consumption of sugar-sweetened beverages, such as soft drinks and fruit-flavored punches.
- Eat the recommended amounts of fruits and vegetables every day (2 to 4.5 cups per day based on age).
- Learn to eat age-appropriate portions of foods.
- Eat foods low in energy density such as those high in fiber and/or water and modest in fat.
- Eat a nutritious breakfast every day.
- Eat a diet rich in calcium.
- Eat a diet balanced in recommended proportions for carbohydrate, fat, and protein.
- Eat a diet high in fiber.
- Eat together as a family as often as possible.
- Limit the frequency of restaurant meals.
- Limit television watching or other screen time to no more than 2 hours per day and do not have televisions or computers in bedrooms.
- Engage in at least 60 minutes of moderate to vigorous physical activity every day.

SOURCE: S. E. Barlow, Expert Committee recommendations regarding the prevention, assessment, and treatment of child and adolescent overweight and obesity: Summary report, *Pediatrics* 120 (2007): S164–S192.

© Cengage Learning

TEST
IT

Take the quiz below to test your mastery of the key chapter concepts.

16.1 Nutrition during Infancy (pp. 510–520)

LEARN IT List some of the components of breast milk and describe the appropriate foods for infants during the first year of life.

1. What are the appropriate uses of formula feeding? What criteria would you use in selecting an infant formula?

2. Why are solid foods not recommended for an infant during the first few months of life? When is an infant ready to start eating solid food?

3. Identify foods that are inappropriate for infants and explain why they are inappropriate.

4. A reasonable weight for a healthy 5-month-old infant who weighed 8 pounds at birth might be:
   a. 12 pounds.
   b. 16 pounds.
   c. 20 pounds.
   d. 24 pounds.

5. Dehydration can develop quickly in infants because:
   a. much of their body water is extracellular.
   b. they lose a lot of water through urination and tears.
   c. only a small percentage of their body weight is water.
   d. they drink lots of breast milk or formula, but little water.

6. An infant should begin eating solid foods between:
   a. 2 and 4 weeks.
   b. 1 and 3 months.
   c. 4 and 6 months.
   d. 8 and 10 months.

16.2 Nutrition during Childhood (pp. 520–539)

LEARN IT Explain how children's appetites and nutrient needs reflect their stage of growth and why iron deficiency and obesity are often concerns during childhood.

7. What nutrition problems are most common in children? What strategies can help prevent these problems?

8. Describe the relationships between nutrition and behavior. How does television influence nutrition?

9. Describe a true food allergy. Which foods most often cause allergic reactions? How do food allergies influence nutrition status?

10. Describe the problems associated with childhood obesity and the strategies for prevention and treatment.

11. List strategies for introducing nutritious foods to children.

12. What impact do school meal programs have on the nutrition status of children?

13. Among US children, the most prevalent nutrient deficiency is of:
    a. iron.
    b. folate.
    c. protein.
    d. vitamin D.

14. A true food allergy always:
    a. elicits an immune response.
    b. causes an immediate reaction.

c. creates an aversion to the offending food.
d. involves symptoms such as headaches or hives.

15. Which of the following strategies is *not* effective?
    a. Play first, eat later.
    b. Provide small portions.
    c. Encourage children to help prepare meals.
    d. Use dessert as a reward for eating vegetables.

16.3 Nutrition during Adolescence (pp. 539–543)

LEARN IT Discuss some of the challenges in meeting the nutrient needs of adolescents.

16. Describe the changes in nutrient needs from childhood to adolescence. Why is an adolescent girl more likely to develop an iron deficiency than an adolescent boy?

17. How do adolescents' eating habits influence their nutrient intakes?

18. To help teenagers consume a balanced diet, parents can:
    a. monitor the teens' food intake.
    b. give up—parents can't influence teenagers.
    c. keep the pantry and refrigerator well stocked.
    d. forbid snacking and insist on regular, well-balanced meals.

19. During adolescence, energy and nutrient needs:
    a. reach a peak.
    b. fall dramatically.
    c. rise, but do not peak until adulthood.
    d. fluctuate so much that generalizations can't be made.

20. The nutrients most likely to fall short in the adolescent diet are:
    a. sodium and fat.
    b. folate and zinc.
    c. iron and calcium.
    d. protein and vitamin A.

21. To balance the day's intake, an adolescent who eats a hamburger, fries, and cola at lunch might benefit most from a dinner of:
    a. fried chicken, rice, and banana.
    b. ribeye steak, baked potato, and iced tea.
    c. pork chop, mashed potatoes, and apple juice.
    d. spaghetti with meat sauce, broccoli, and milk.

HIGHLIGHT 16 Childhood Obesity and the Early Development of Chronic Diseases (pp. 548–553)

LEARN IT Describe the lifestyle factors that can help prevent childhood obesity and the development of type 2 diabetes and heart disease.

22. Describe the dietary strategies to help prevent the early development of chronic diseases.

Multiple Choice Answers
4.b 5.a 6.c 13.a 14.a 15.d 18.c 19.c 20.a 21.d

STUDY IT

17 Life Cycle Nutrition: Adulthood and the Later Years

Review the key points of this chapter below, learn the definitions of the bold terms, and then take the practice quiz on the back of this card.

17.1 Nutrition and Longevity

- **Life expectancy** in the United States increased dramatically in the 20th century.
- Factors that enhance **longevity** include limited or no alcohol use, regular balanced meals, weight control, abstinence from smoking, regular physical activity, and adequate sleep.
- Energy restriction in animals seems to lengthen their lives; whether such dietary intervention in human beings is beneficial remains unknown.
- Nutrition—especially when combined with regular physical activity—can influence aging and longevity by supporting good health and preventing disease.

17.2 The Aging Process

- Changes that accompany aging can impair nutrition status. Among physiological changes, hormone activity alters body composition, immune system changes raise the risk of infections, **atrophic gastritis** interferes with digestion and absorption, and tooth loss limits food choices.
- Psychological changes such as depression, economic changes such as loss of income, and social changes such as loneliness contribute to poor food intake.

17.3 Energy and Nutrient Needs of Older Adults

- Some nutrients need special attention in the diet, but supplements are not routinely recommended.
- Table 17-5 helps determine the risk of malnutrition in older adults.

17.4 Nutrition-Related Concerns of Older Adults

- **Senile dementia** and cognitive losses afflict many older adults; some face loss of vision due to **cataracts** or **macular degeneration** or cope with the pain of **arthritis.**

TABLE 17-5 Risk Factors for Malnutrition in Older Adults

	These questions help *determine* the risk of malnutrition in older adults:
Disease	• Do you have an illness or condition that changes the types or amounts of foods you eat?
Eating poorly	• Do you eat fewer than two meals a day? Do you eat fruits, vegetables, and milk products daily?
Tooth loss or mouth pain	• Is it difficult or painful to eat?
Economic hardship	• Do you have enough money to buy the food you need?
Reduced social contact	• Do you eat alone most of the time?
Multiple medications	• Do you take three or more different prescribed or over-the-counter medications daily?
Involuntary weight loss or gain	• Have you lost or gained 10 pounds or more in the last 6 months?
Needs assistance	• Are you physically able to shop, cook, and feed yourself?
Elderly person	• Are you older than 80?

© Cengage Learning

- Table 17-2 provides a summary of the nutrient concerns of aging.
- Some problems may be inevitable, but others are preventable and good nutrition may play a key role.

17.5 Food Choices and Eating Habits of Older Adults

- **Congregate meals** provide older adults with both nutrients and social interactions; those who are homebound receive delivered meals **(Meals on Wheels).**
- Adults living alone can prepare nutritious, inexpensive meals with a little creativity and careful shopping.

TABLE 17-2 Nutrient Concerns of Aging

Nutrient	Effect of Aging	Comments
Water	Lack of thirst and decreased total body water make dehydration likely.	Mild dehydration is a common cause of confusion. Difficulty obtaining water or getting to the bathroom may compound the problem.
Energy	Need decreases as muscle mass decreases **(sarcopenia).**	Physical activity moderates the decline.
Fiber	Likelihood of constipation increases with low intakes and changes in the GI tract.	Inadequate water intakes and lack of physical activity, along with some medications, compound the problem.
Protein	Needs may stay the same or increase slightly.	Low-fat, high-fiber legumes and grains meet both protein and other nutrient needs.
Vitamin B$_{12}$	Atrophic gastritis is common.	Deficiency causes neurological damages; supplements may be needed.
Vitamin D	Increased likelihood of inadequate intake; skin synthesis declines.	Daily sunlight exposure in moderation or supplements may be beneficial.
Calcium	Intakes may be low; osteoporosis is common.	Stomach discomfort commonly limits milk intake; calcium substitutes or supplements may be needed.
Iron	In women, status improves after menopause; deficiencies are linked to chronic blood losses and low stomach acid output.	Adequate stomach acid is required for absorption; antacid or other medicine use may aggravate iron deficiency; vitamin C and meat increase absorption.
Zinc	Intakes are often inadequate and absorption may be poor, but needs may also increase.	Medications interfere with absorption; deficiency may depress appetite and sense of taste.

© Cengage Learning

Take the quiz below to test your mastery of the key chapter concepts.

17.1 Nutrition and Longevity (pp. 556–560)

LEARN IT Describe the role nutrition plays in longevity.

1. Life expectancy in the United States is about:
   a. 48 to 60 years.
   b. 58 to 70 years.
   c. 68 to 80 years.
   d. 78 to 90 years.

2. The human life span is about:
   a. 85 years.
   b. 100 years.
   c. 115 years.
   d. 130 years.

3. A 72-year-old person whose physical health is similar to that of people 10 years younger has a(n):
   a. chronological age of 62.
   b. physiological age of 72.
   c. physiological age of 62.
   d. absolute age of minus 10.

4. Rats live longest when given diets that:
   a. eliminate all fat.
   b. provide lots of protein.
   c. allow them to eat freely.
   d. restrict their energy intakes.

17.2 The Aging Process (pp. 560–564)

LEARN IT Summarize how nutrition interacts with the physical, psychological, economic, and social changes involved in aging.

5. Which characteristic is *not* commonly associated with atrophic gastritis?
   a. inflamed stomach
   b. vitamin B_{12} toxicity
   c. bacterial overgrowth
   d. lack of intrinsic factor

17.3 Energy and Nutrient Needs of Older Adults (pp. 564–567)

LEARN IT Explain why the needs for some nutrients increase or decrease during aging.

6. Why does the risk of dehydration increase as people age?

7. Why do energy needs usually decline with advancing age?

8. Identify some factors that complicate the task of setting nutrient standards for older adults.

9. On average, adult energy needs:
   a. decline 5 percent per year.
   b. decline 5 percent per decade.
   c. remain stable throughout life.
   d. rise gradually throughout life.

17.4 Nutrition-Related Concerns of Older Adults (pp. 567–572)

LEARN IT Identify how nutrition might contribute to, or prevent, the development of age-related problems associated with vision, arthritis, the brain, and alcohol use.

10. Which nutrients seem to protect against cataract development?
    a. minerals
    b. lecithins
    c. antioxidants
    d. amino acids

11. The best dietary advice for a person with osteoarthritis might be to:
    a. avoid milk products.
    b. take fish oil supplements.
    c. take vitamin E supplements.
    d. lose weight, if overweight.

17.5 Food Choices and Eating Habits of Older Adults (pp. 572–576)

LEARN IT Instruct an adult on how to shop for groceries and prepare healthy meals for one person on a tight budget.

12. What characteristics contribute to malnutrition in older people?

13. Congregate meal programs are preferable to Meals on Wheels because they provide:
    a. nutritious meals.
    b. referral services.
    c. social interactions.
    d. financial assistance.

14. The OAA Nutrition Program is available to:
    a. all people 65 years and older.
    b. all people 60 years and older.
    c. homebound people only, 60 years and older.
    d. low-income people only, 60 years and older.

HIGHLIGHT 17 Nutrient-Drug Interactions (pp. 579–583)

LEARN IT Explain why certain nutrients and medications interact.

15. Why are elderly people with chronic diseases most vulnerable to interactions that can create nutrient imbalances or interfere with drug effectiveness?

Multiple Choice Answers

1.c 2.d 3.c 4.d 5.b 9.b 10.c 11.d 13.c 14.b

STUDY IT

18 Diet and Health

Review the key points of this chapter below, learn the definitions of the bold terms, and then take the practice quiz on the back of this card.

18.1 Nutrition and Infectious Diseases

- Public health measures such as purification of water and safe handling of food help prevent the spread of infection in populations; immunizations and antibiotics protect individuals.
- Nutrition cannot prevent or cure **infectious diseases,** but adequate intakes of all the nutrients can help support the immune system as the body defends against disease-causing agents.
- If the immune system is impaired because of malnutrition or diseases such as **AIDS,** a person becomes vulnerable to infectious disease.
- Inflammation underlies many chronic diseases.

18.2 Nutrition and Chronic Diseases

- Heart disease, cancers, and strokes are the three leading causes of death in the United States, and diabetes also ranks among the top 10.
- These four chronic diseases have significant links with nutrition and with one another (review Figure 18-2); other lifestyle risk factors and genetics also play a role.

18.3 Cardiovascular Disease

- **Atherosclerosis** is characterized by a build-up of plaque in an artery wall.
- Rupture of **plaque** or abnormal blood clotting can cause heart attacks and strokes.
- Dietary recommendations to lower the risks of **cardiovascular disease** are summarized in Table 18-5 (p. 596); quitting smoking and engaging in regular physical activity also improve heart health.

18.4 Hypertension

- The most effective dietary strategy for preventing **hypertension** is weight control.
- The **DASH eating pattern** is rich in fruits, vegetables, nuts, and low-fat milk products and low in fat, saturated fat, and sodium (review Table 18-7, p. 600).

> FIGURE 18-2   **Interrelationships among Chronic Diseases**

Notice that many chronic diseases are themselves risk factors for other chronic diseases and that all of them are linked to obesity. The risk factors highlighted in blue define the **metabolic syndrome.**

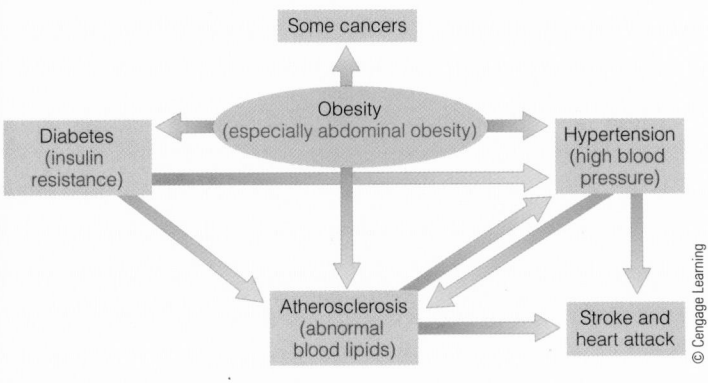

© Cengage Learning

TABLE 18-12 Dietary Guidelines and Recommendations for Chronic Diseases Compared

Dietary Guidelines	Heart Disease	Hypertension	Diabetes	Cancer
Prevent and/or reduce overweight and obesity.	✓	✓	✓	✓
Increase physical activity and reduce time spent in sedentary behaviors.	✓	✓	✓	✓
Keep total fat 20 to 35 percent of kcalories.	✓			
Consume less than 10 percent of kcalories from saturated fatty acids and keep *trans*-fatty acid consumption as low as possible.	✓		✓	✓
Increase vegetable and fruit intake. Consume at least half of all grains as whole grains.	✓	✓	✓	✓
Reduce the intake of kcalories from added sugars.	✓		✓	
Reduce daily sodium intake to less than 2300 milligrams.[a]	✓	✓	✓	✓
If alcohol is consumed, it should be consumed in moderation.	✓	✓	✓	✓

© Cengage Learning

[a]Further reduce daily sodium intake to 1500 milligrams among persons who are 51 and older and those of any age who are African American or have hypertension, diabetes, or chronic kidney disease.

18.5 Diabetes Mellitus

- **Diabetes** is characterized by high blood glucose and either insufficient insulin, ineffective insulin, or a combination of the two.
- People with **type 1 diabetes** coordinate diet, insulin, and physical activity to help control their blood glucose.
- People with **type 2 diabetes** benefit most from a diet and physical activity program that controls glucose fluctuations and promotes weight loss.

18.6 Cancer

- Some dietary factors, such as alcohol and heavily smoked foods, may **initiate cancer** development; others, such as animal fats, may **promote** cancer once it develops; and still others, such as fiber, antioxidant nutrients, and phytochemicals, may serve as **antipromoters** that protect against cancer development.
- To obtain the best possible nutrition at the lowest possible cancer risk, eat many fruits, vegetables, legumes, and whole grains and little saturated fat; minimize weight gain through regular physical activity and a healthy diet (review Table 18-11, p. 609).

18.7 Recommendations for Chronic Diseases

- Optimal nutrition keeps people healthy and reduces the risk of chronic diseases.
- Dietary recommendations are aimed at populations and focus on controlling weight; limiting saturated and *trans* fat; increasing fiber-rich fruits, vegetables, and whole grains; and balancing food intake with physical activity (review Table 18-12).

Take the quiz below to test your mastery of the key chapter concepts.

18.1 Nutrition and Infectious Diseases

(pp. 586–588)

LEARN IT Identify factors that protect people from the spread of infectious diseases and describe the role of nutrition in immunity.

1. The immune cells of the body do *not* include:
   - a. B-cells.
   - b. T-cells.
   - c. antigens.
   - d. phagocytes.

2. Which of the following produce antibodies?
   - a. phagocytes
   - b. T-cells
   - c. antigens
   - d. B-cells

18.2 Nutrition and Chronic Diseases

(pp. 589–590)

LEARN IT List the leading nutrition-related causes of death in the United States.

3. The leading cause of death in the United States is:
   - a. AIDS.
   - b. cancer.
   - c. diabetes.
   - d. heart disease.

18.3 Cardiovascular Disease (pp. 590–597)

LEARN IT Describe how atherosclerosis develops and strategies to lower blood cholesterol levels.

4. Plaques in the arteries contribute to the development of:
   - a. cancer.
   - b. diabetes.
   - c. atherosclerosis.
   - d. infectious diseases.

5. Which blood lipid correlates directly with heart disease?
   - a. HDL
   - b. LDL
   - c. VLDL
   - d. triglycerides

6. Weight loss and physical activity may protect against heart disease by:
   - a. raising LDL and insulin levels.
   - b. raising HDL and lowering blood pressure.
   - c. lowering LDL and increasing clot formation.
   - d. improving insulin sensitivity and raising blood pressure.

18.4 Hypertension (pp. 598–600)

LEARN IT Present strategies to lower blood pressure.

7. What is the most effective strategy for most people to lower their blood pressure?
   - a. lose weight
   - b. restrict salt
   - c. monitor glucose
   - d. supplement protein

18.5 Diabetes Mellitus (pp. 600–605)

LEARN IT Compare the dietary strategies to manage type 1 diabetes with those to prevent and treat type 2 diabetes.

8. Describe the differences between type 1 diabetes and type 2 diabetes.

9. All of the following factors increase the risk of type 2 diabetes *except:*
   - a. aging.
   - b. inactivity.
   - c. obesity.
   - d. smoking.

10. The most important dietary strategy in diabetes is to:
    - a. provide for a consistent carbohydrate intake.
    - b. restrict fat to 30 percent of daily kcalories.
    - c. limit carbohydrate intake to 300 milligrams a day.
    - d. take multiple vitamin and mineral supplements daily.

18.6 Cancer (pp. 605–609)

LEARN IT Differentiate among cancer initiators, promoters, and antipromoters and describe how nutrients or foods might play a role in each category.

11. Describe the characteristics of a diet that might offer the best protection against the onset of cancer.

12. Which of the following help(s) to protect against cancer?
    - a. alcohol
    - b. pickled foods
    - c. phytochemicals
    - d. omega-6 fatty acids

18.7 Recommendations for Chronic Diseases (pp. 610–612)

LEARN IT Summarize dietary recommendations to prevent chronic diseases.

13. Describe the dietary choices that best protect against the development of most chronic diseases.

HIGHLIGHT 18 Complementary and Alternative Medicine (pp. 617–625)

LEARN IT Present arguments for and against the use of complementary and alternative medicine.

14. Which of the following is generally considered safe?
    - a. chaparral to cleanse blood
    - b. aloe to promote wound healing
    - c. ephedra to promote weight loss
    - d. yohimbe to enhance male sexual performance

Multiple Choice Answers
1.c 2.d 3.d 4.c 5.b 6.b 7.a 9.d 10.a 12.c 14.b

Daily Values (DV) for Food Labels

The Daily Values (DV) are standards developed by the Food and Drug Administration (FDA) for use on food labels. The values are based on 2000 kcalories a day for adults and children aged 4 years and older. Chapter 2 provides more details.

Nutrient	Amount
Vitamins	
Biotin	30 µg
Choline	550 mg
Folate	400 µg DFE
Niacin	16 mg NE
Pantothenic acid	5 mg
Riboflavin	1.3 mg
Thiamin	1.2 mg
Vitamin A	900 µg RAE
Vitamin B_6	1.7 mg
Vitamin B_{12}	2.4 µg
Vitamin C	90 mg
Vitamin D	20 µg
Vitamin E (α-tocopherol)	15 mg
Vitamin K	120 µg
Minerals	
Calcium	1300 mg
Chloride	2300 mg
Chromium	35 µg
Copper	0.9 mg
Iodine	150 µg
Iron	18 mg
Magnesium	420 mg
Manganese	2.3 mg
Molybdenum	45 µg
Phosphorus	1250 mg
Potassium	4700 mg
Selenium	55 µg
Sodium	2300 mg
Zinc	11 mg

Food Component	Amount	Calculation Factors
Fat	65 g	30% of kcalories
Saturated fat	20 g	10% of kcalories
Cholesterol	300 mg	Same regardless of kcalories
Carbohydrate (total)	300 g	60% of kcalories
Fiber	28 g	14 g per 1000 kcalories
Protein	50 g	10% of kcalories

GLOSSARY
OF NUTRIENT MEASURES

kcal: kcalories; a unit by which energy is measured (Chapter 1 provides more details).

g: grams; a unit of weight equivalent to about 0.03 ounces.

mg: milligrams; one-thousandth of a gram.

µg: micrograms; one-millionth of a gram.

IU: international units; an old measure of vitamin activity determined by biological methods (as opposed to new measures that are determined by direct chemical analyses).For those still using IU, the following factors can be used for conversions.

- For vitamin A, 1 IU = 0.3 µg retinol

- For vitamin D, 1 IU = 0.02 µg cholecalciferol

- For vitamin E, 1 IU = 0.67 mg α-tocopherol

mg NE: milligrams niacin equivalents; a measure of niacin activity (Chapter 10 provides more details).

- 1 NE = 1 mg niacin
  = 60 mg tryptophan (an amino acid)

µg DFE: micrograms dietary folate equivalents; a measure of folate activity (Chapter 10 provides more details).

- 1 µg DFE = 1 µg food folate
  = 0.6 µg folic acid from fortified food or as a supplement taken with food

µg RAE: micrograms retinol activity equivalents; a measure of vitamin A activity (Chapter 11 provides more details).

- 1 µg RAE = 1 µg retinol
  = 12 µg β-carotene
  = 24 µg other vitamin A carotenoids

mmol: millimoles; one-thousanth of a mole, the molecular weight of a substance. To convert mmol to mg, multiply by the atomic weight of the substance.

- For sodium, mmol × 23 = mg Na
- For chloride, mmol × 35.5 = mg Cl
- For sodium chloride, mmol × 58.5 = mg NaCl